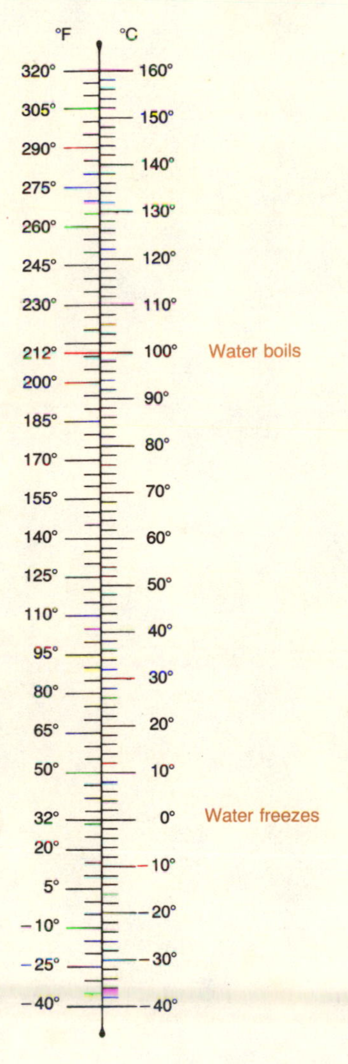

To convert Fahrenheit to centrigrade, use this formula:

$$°C = \frac{5}{9}(°F - 32)$$

To convert centigrade to Fahrenheit, use this formula:

$$°F = \frac{9}{5}(°C + 32)$$

MICROBIOLOGY

PREFACE

Microbiology is written for the college freshman or sophomore student. The text is aimed at the student who is interested in a broad coverage of microbiology and who has little or no background in chemistry or biology. The book can be used for a one-semester course or for a one-year course in general microbiology. The topics discussed in the text can be organized so that they form a core that is suitable to students in allied health sciences, biological sciences, agriculture, natural resources management, animal and veterinary science, food science, dietetics, home economics, and health education.

Throughout the text, we have strived to convey basic biological principles and have emphasized, wherever we could, the enormous impact that microorganisms have on every facet of human life. *Microbiology* provides the student with a balanced discussion of fundamental principles and applications of microbiology. Through this book, first-time students of microbiology should come to appreciate and be awed by the large variety of microorganisms, their activities, and their impact on everyday life.

In order to provide students from different majors an appropriate core of subjects, we have divided the textbook into four parts: "The Basic Concepts," "A Survey of Microorganisms," "Medical Microbiology," and "Applied and Environmental Microbiology."

PART I: THE BASIC CONCEPTS constitutes a core of topics that should be common to all curricula. The chapters in Part I cover important, modern concepts of microbial biology and emphasize their importance to microbiology, to the environment, and to humans. We have included separate chapters on microbial genetics, metabolic regulation, and genetic engineering in order to give broader coverage to these very current and pertinent areas of microbiology. In these chapters, the student is introduced to the basic concepts of microbial genetics, including the traditional experiments and the modern discoveries, as well as the applications of these concepts in the rapidly-developing science of genetic engineering.

PART II: A SURVEY OF MICROORGANISMS discusses the various groups of microorganisms, their biology, and importance to humans. A chapter discussing the origin and classification of microorganisms is included in order to emphasize the relationship among the various groups of microorganisms and to unify the chapters within Part II.

PART III: MEDICAL MICROBIOLOGY emphasizes not only the clinical and epidemiological features of important diseases, but also the intricate interactions that take place between a parasite and its host and the consequences of these interactions. A chapter covering the diagnosis of infectious diseases is included in this part to emphasize the importance of properly taking samples so that patients are accurately diagnosed and treated. This chapter also illustrates that cultural and immunological procedures are indispensible in the diagnosis of infectious diseases.

PART IV: APPLIED AND ENVIRONMENTAL MICROBIOLOGY discusses many of the activities that microorganisms carry out in nature that influence human life. This part is concerned with the remarkable variety of microbial activities that occur in nature and the use of microorganisms to transform chemical substances into valuable products. In this part, we emphasize how and why microorganisms carry out various metabolic activities and how these activities affect the economy, environment, and public health. In addition, the chapters in this part discuss the evergrowing knowledge of how microbial metabolism can be employed by humans to improve their standard of living.

PEDAGOGICAL AIDS

We have incorporated many features that should help the student to study and learn this complex and broad subject.

CHAPTER PREVIEWS introduce the student to practical applications of the subject matter discussed in the chapter or significant events in microbiology. For example, some of the chapter previews discuss famous microbiologists who made important contributions to the knowledge of microorganisms and their biology. Chapter previews are aimed at illustrating that microbiology is not a sterile laboratory science of interest only to scientists, but a dynamic field of endeavor with practical applications, which constantly affect all of our lives.

LEARNING OBJECTIVES outline key concepts in the chapter and provide the student with study goals so that he or she can establish a systematic approach to studying the chapter.

CROSS REFERENCES provide the student with rapid access to additional information on key topics discussed in more than one chapter. Key topics are indicated by a small box outline. In the inner margin, on the same line that the box appears, there is a larger box with the page number to where the student is referred, if additional information is desired. The page numbers may indicate where a given section begins, or where a table or figure with the pertinent information appears. Since subjects are often discussed in more than one chapter, the student is exposed to different aspects or perspectives of a subject. This gives the student a broader base of knowledge and a better understanding of the multifaceted nature of microbiology.

FOCUS BOXES introduce practical, interesting, or additional information to expand upon topics discussed in the chapter. Some of the focus boxes are historical vignettes that serve to remind the student that the information covered in class and in the text is based upon data provided by scientists and teachers like their instructors.

KEY TERMS in **bold face type** will greatly help the student recognize the important terms in the chapter. Some terms are printed in bold in more than one chapter so that students will come across important terms regardless of the course emphasis.

TABLES in most chapters compile pertinent data and aid the student in quickly grasping large quantities of information. In addition, the tables serve as easily accessible reference material.

LINE DRAWINGS and **PHOTOGRAPHS** complement the text. All chapters in *Microbiology* have line drawings rendered in two colors to aid the student in understanding the subject matter. In addition, we have made extensive use of photographs and electron micrographs to illustrate important structures, diseases, or concepts.

The **SUMMARIES** are extensive and written in outline form. Each summary recaps the major sections in the chapter and provides the student with a condensation of most of the important concepts discussed in the chapter.

STUDY QUESTIONS test the student's grasp of the information and help to reinforce the stated goals through short answer and essay questions. The essay questions frequently require the student to analyze and gather information from various parts of the chapter before answering the question.

SUPPLEMENTAL READINGS provide the student with a source of information with which to expand upon some of the key concepts discussed in the chapter.

APPENDICES include a temperature conversion chart, a scheme for the classification of bacteria, and a guide to infectious diseases. These appendices provide the student with easy access to information not readily available in the text.

An extensive **GLOSSARY/PRONUNCIATION GUIDE** has been prepared to provide the student with easy access to the definitions of many of the important terms used throughout the text and the accepted way of pronouncing them. Pronunciations have also been provided for most of the microorganisms commonly discussed in the textbook.

Microbiology is accompanied by a student study guide, *Study Guide to Accompany Microbiology*, written by Dr. Geraldine Ross at Bellevue Community College and a laboratory manual, *Laboratory Exercises in Microbiology*, written by Drs. J. S. Colomé, A. M. Kubinski, R. J. Cano, and D. V. Grady at California Polytechnic State University. These study aids should further enhance the student's understanding and appreciation of microbial life. An instructor's manual, written by Drs. J. S. Colomé and R. J. Cano is also available.

We would like to express our appreciation to all those who helped us complete this text; without their help, our tasks would have been extremely difficult or impossible. We wish to thank our students for helping us test our ideas and approaches to teaching. We also would like to express our deep gratitude to the following reviewers of our text for their insightful, although sometimes painful, comments, suggestions, and opinions.

Alfred E. Brown, Auburn University;
James Craig, San Jose State University;
Donald Deters, University of Texas, Austin;
David Filmer, Purdue University;
Diane O. Fleming, Wright State University;
David V. Grady, California Polytechnic State, San Luis Obispo;
Joan Handley, University of Kansas;
Dennis K. Huff, California State University, Sacramento;
Robert J. Janssen, University of Arizona;
John Larkin, Louisiana State University;

John Lewis, Loma Linda University;
Gene Martin, University of Nebraska;
Karen Messley, Rock Valley College, Illinois;
Charlene Mims, San Antonio College;
Elinor M. O'Brien, Boston College;
R. T. O'Brien, New Mexico State University;
William O'Dell, University of Nebraska;
Geraldine Ross, Bellevue Community College;
Richard D. Sagers, Brigham Young University;
Robert H. Satterfield, College of DuPage;
Lou Shainberg, Mt. San Antonio College;
Robert E. Sjogren, University of Vermont;
Kathy Talaro, Pasadena City College;
James E. Urban, Kansas State University;
Patricia S. Vary, Northern Illinois University;
Robie Vestal, University of Cincinnati;
Robert T. Vinopal, University of Connecticut;
Carl E. Warnes, Ball State University; and
Brian J. Wilkinson, University of Illinois, Normal.

We would like to give a special thanks to Dr. Alfred E. Brown at Auburn University, Dr. William O'Dell at the University of Nebraska, and Dr. Geraldine Ross at Bellevue Community College for all their valuable comments and suggestions, which have made this text much better than it would otherwise have been.

PART I THE BASIC CONCEPTS

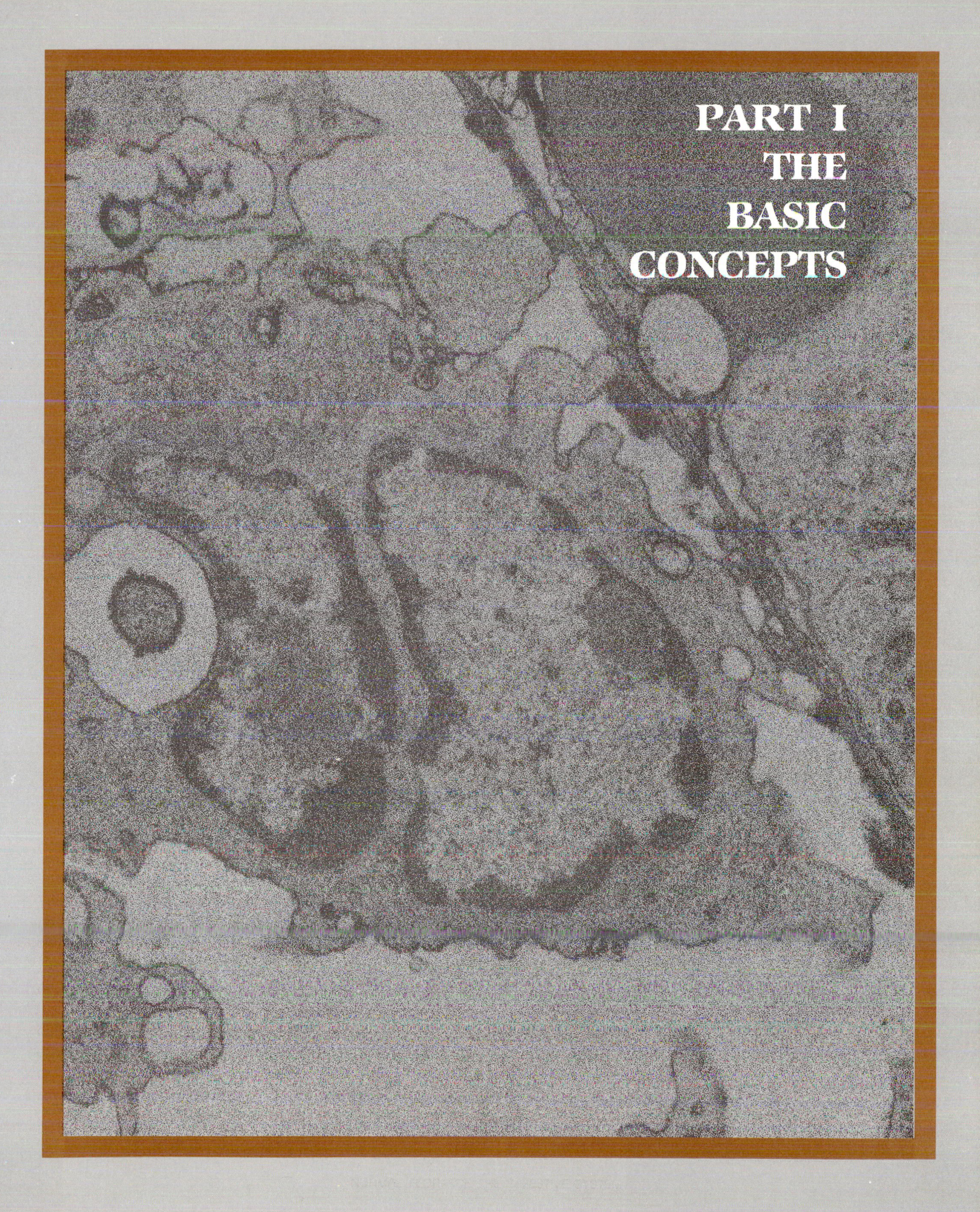

CHAPTER 1
AN INTRODUCTION TO MICROBIOLOGY

CHAPTER PREVIEW

MICROORGANISMS AND HUMAN DISEASE

A change in behavior or living conditions may cause the incidence of common infectious diseases to increase or decrease or new diseases to appear. There are many examples throughout history. Whenever rural people congregated in cities, the incidence of diseases such as typhoid fever, hepatitis, typhus, plague, and influenza often increased because of poor sanitation, filth, malnutrition, and overcrowding. Conversely, when people improved their living conditions, the incidence of many infectious diseases decreased significantly.

When humans traveled and made contact with isolated peoples, new diseases were often acquired by the travelers or given to the isolated peoples. These diseases have often destroyed whole cultures. For example, in the early 1500s, when Cortez was conquering Mexico, smallpox spread from the invading army to the Indian population. It is estimated that smallpox killed 13 million Indians, of a population of 25 million, in the first year of the epidemic. Before 1490, syphilis did not exist in Europe. It was introduced by explorers or soldiers sometime in the 1490s. It is not clear exactly where the disease originated, but it is hypothesized that it may have come from Africa or from America as a mutated form of a disease that existed on these continents. Syphilis, when first introduced, was a deadly disease that ravaged Europe for many years. It rapidly spread from Europe to India, China, and Japan, and then to Africa and the American continent.

In the early 1800s, when Americans settled in the Hawaiian Islands, they introduced measles. This disease almost completely eliminated the native Hawaiians. A consequence was the immigration of Chinese and Japanese to the Hawaiian Islands.

Recently, leprosy has been on the increase in the U.S. because of the recent wave of immigration from Southeast Asia. There are now about 4,000 lepers in the United States who have to take drugs for the rest of their lives in order to control the disease. Leprosy is a major problem in certain parts of the world, despite the development of antibiotics. There are approximately 12 million lepers worldwide and 5 million of them are not being treated. The untreated lepers are crippled by the disease and generally abandoned by their families and friends.

The viruses that cause influenza are responsible for many millions of cases of the disease and over a million deaths per year worldwide. During the 1918 flu epidemic, 20 million died worldwide while 500,000 died in the U.S. alone. More recently, the 1981–82 influenza epidemic in the U.S. was responsible for over 60,000 deaths.

Until quite recently, humans did not understand what was responsible for infectious diseases. Many people believed that angry gods or immoral acts were responsible for these afflictions. It is now known that various types of microorganisms are responsible for infectious diseases of plants, animals, and humans. Microorganisms are also involved in many processes that benefit plants, animals, and humans, and are now being manipulated by humans to their benefit. In this chapter you will be introduced to some of the discoveries that led us to our understanding of the roles microorganisms play in nature.

LEARNING OBJECTIVES

A STUDY OF THIS CHAPTER SHOULD ENABLE YOU TO:

DISCUSS HOW HUMAN BEINGS HAVE EXPLAINED DISEASE AND PUTREFACTION (DECAY)

DISCUSS THE TWO NONSUPERNATURAL EXPLANATIONS FOR INFECTIOUS DISEASES

EXPLAIN HOW HUMANS HAVE TRIED TO CONTROL DISEASE AND PUTREFACTION

LIST SOME OF THE MAJOR HISTORIC MOMENTS IN THE UNDERSTANDING AND CONTROL OF PATHOGENIC MICROORGANISMS

EXPLAIN THE IDEA OF SPONTANEOUS GENERATION

EXPLAIN THE GERM THEORY OF DISEASE AND PUTREFACTION

DISCUSS PASTEUR'S AND KOCH'S CONTRIBUTIONS TO THE ADVANCEMENT OF MICROBIOLOGY

LIST AND EXPLAIN KOCH'S POSTULATES

LIST PAUL EHRLICH'S CONTRIBUTIONS TO MICROBIOLOGY

DESCRIBE WINOGRADSKY'S AND BEIJERINCK'S CONTRIBUTIONS TO THE STUDY OF ECOLOGY

INDICATE WHAT DEVELOPMENTS HAVE RECENTLY OCCURRED IN MICROBIOLOGY

As humans have become more aware of their world and of themselves, they have discovered a world of microscopic life forms called **microorganisms** (fig. 1-1). Microorganisms have helped to shape the world we take for granted, and they have also influenced the evolution of all plants and animals. In fact, life is completely dependent upon the metabolic activities of many different microorganisms. Although microorganisms carry out numerous chemical transformations that are essential to human existence, many people not acquainted with microorganisms believe that all are harmful and undesirable. This view of microorganisms is due to the fact that some of them spoil food and cause serious diseases. The notion that microorganisms are for the most part harmful is dispelled once their roles in nature are understood and appreciated.

MICROBIOLOGY: A MULTIFACETED SCIENCE

Microbiology is much more than just a taxonomic study of microorganisms. It is also concerned with the roles microorganisms play in causing disease, the changes they make in the environment, and the products they generate. Microbiology encompasses such a vast area of knowledge that it is broken down into a number of specialty areas. Scientists often divide microorganisms into groups based upon shape and physiology, and restrict their study to one or more groups (fig. 1-2a). This approach to studying microorganisms, sometimes called the taxonomic approach to microbiology, divides microorganisms into viruses, bacteria, protozoa, algae, fungi, and microscopic animals. **Virology** is the study of the noncellular organisms called viruses. Virology is concerned with the structure and reproduction of viruses, how they cause disease, and how they can be controlled. **Bacteriology** is the study of the non-nucleated cellular organisms known as bacteria. Bacteriology deals with the uses of bacteria in industry; the role of bacteria in the ecology of the world; bacterial genetics, structure, and multiplication; the mechanisms by which bacteria cause disease; and the means by which bacteria can be controlled. **Protozoology, phycology** (algology), and **mycology** are the studies of protozoa, algae, and fungi, respectively. Although some of the organisms in these groups are not microscopic, the vast majority of them are; thus, their study is considered part of microbiology. Protozoans, parasitic animals, and anthropods of medical importance are the subject matter of **parasitology.**

Scientists often study microorganisms from a functional rather than a taxonomic point of view (fig. 1-2b). For example, the science of **microbial ecology** is concerned with the interactions that take place between microorganisms and their environment. **Industrial microbiology** is a study of microbial activities that are valuable to humans. The production of antibiotics, alcoholic beverages, and cheeses is carried out by industrial microbiologists. **Medical microbiology** deals with the microbial activities that lead to health and disease. The study of how organisms cause disease and the diagnosis and prevention of infectious diseases are major aspects of medical microbiology. Studies of microorganisms involved in soil fertility, livestock health, and plant disease are in the realm of **agricultural microbiology.** Since **immunology** is concerned with aspects of the immune systems that protect animals from microorganisms, it is generally considered to be part of microbiology.

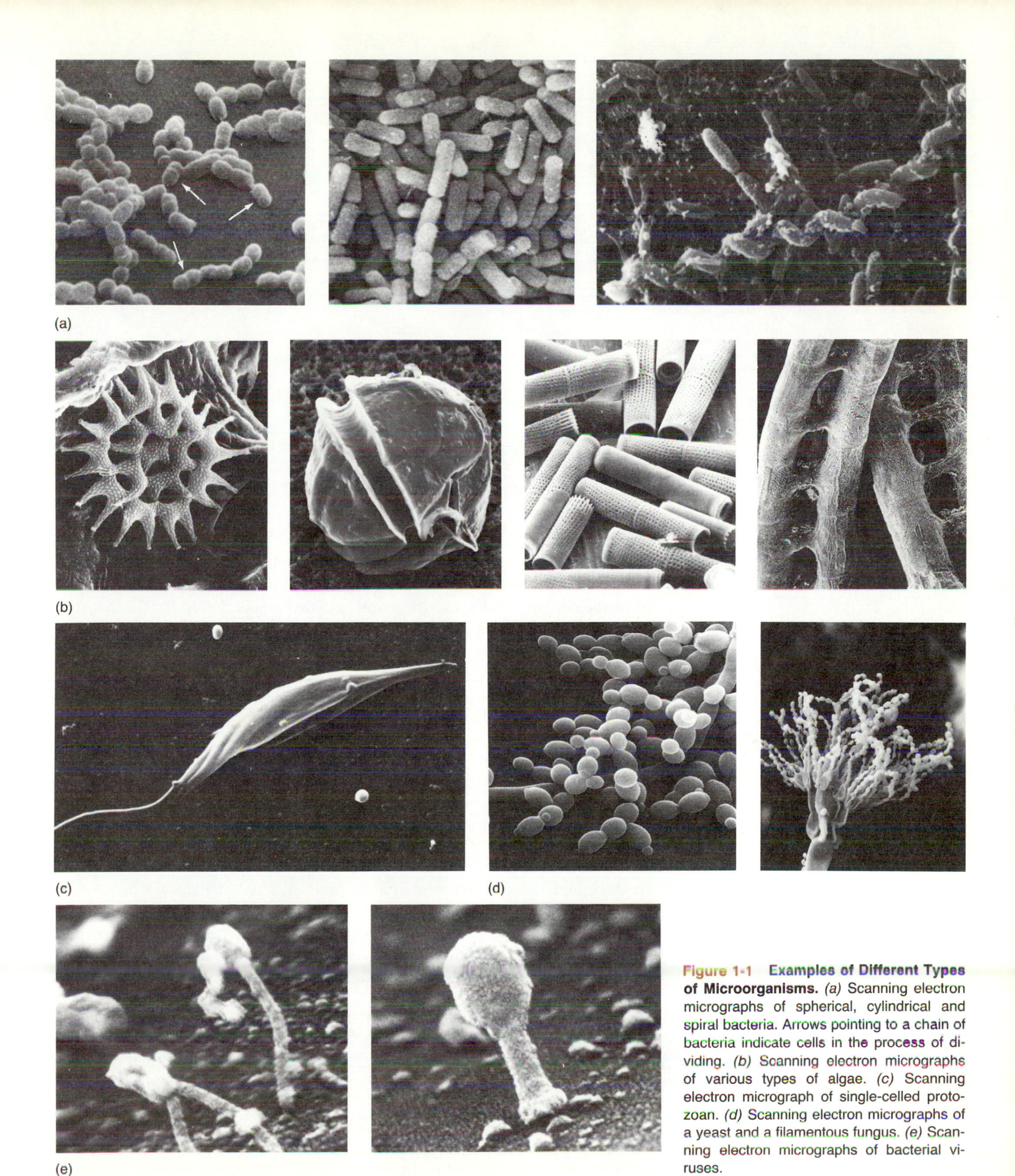

Figure 1-1 Examples of Different Types of Microorganisms. *(a)* Scanning electron micrographs of spherical, cylindrical and spiral bacteria. Arrows pointing to a chain of bacteria indicate cells in the process of dividing. *(b)* Scanning electron micrographs of various types of algae. *(c)* Scanning electron micrograph of single-celled protozoan. *(d)* Scanning electron micrographs of a yeast and a filamentous fungus. *(e)* Scanning electron micrographs of bacterial viruses.

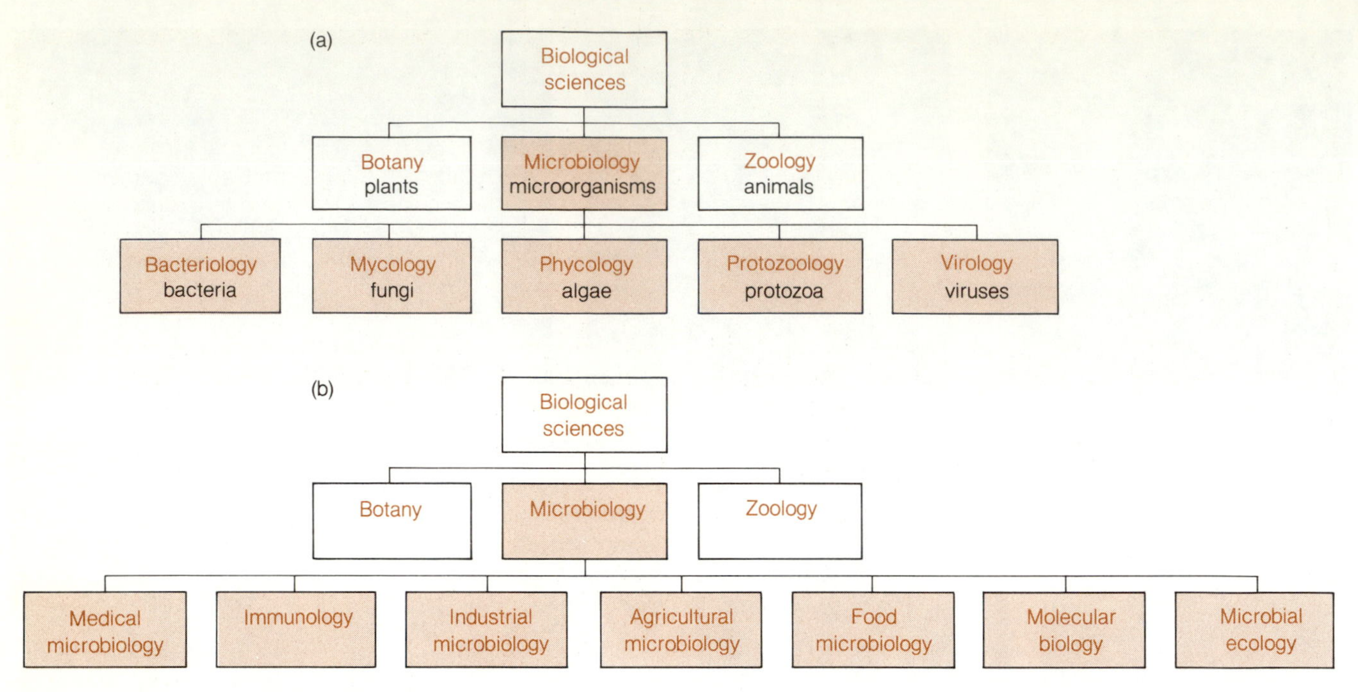

Figure 1-2 Disciplines within the Field of Microbiology. *(a)* Disciplines within the field of microbiology based on a taxonomic approach. *(b)* Microbiology may also be divided up into fields of study based on a functional approach.

Microbiology is an integral part of many fields of study, such as immunology, genetics, molecular biology, physiology, and ecology. Its central role in medicine, agriculture, industry, and genetic engineering indicates that microbiology is a major field of study and that an understanding of this field is essential to our welfare.

ANCIENT IDEAS OF DISEASE AND PUTREFACTION

Microorganisms include a myriad of free-living and parasitic forms. They are ubiquitous (everywhere): in the environment, on our bodies, in the air we breathe, in the food we eat, and in the water we drink. Throughout the ages, humans have been aware of the effects of microbial activities even though they were not aware of the microorganisms as living entities. Most peoples have made and consumed fermented foods and beverages, observed the spoilage and decomposition of matter, and suffered the misery of various infectious diseases.

The fact that putrefaction and many diseases are caused by microorganisms was unknown until relatively recently. Nonsupernatural explanations of these phenomena generally postulated that "poisoned air" or "seeds of decomposition," rather than unseen life forms, were responsible. Diseases were also known to be contagious. For instance, in the *Ordinances of Manu*, an Indian doctrine dated at about 1300 B.C., bridegrooms were warned about marrying into a family that was prone to tuberculosis because the bridegroom might acquire the disease. Clearly, the disease was suspected of being contagious at this early date. An Indian medical document dated at about 500 B.C., called the *Bhagavata Purana*, describes the first signs of bubonic plague. It warns that dying rats, falling from the roofs, are the first sign of this plague that can spread to humans. Here we have an example of a disease

FOCUS

A PRIMITIVE HEALTH CODE

Many of the ideas and laws of Moses found in the Book of Numbers (1000 B.C.) and in the Book of Deuteronomy (650 B.C.) represent an early public health code. The Israelites clearly realized that many diseases could be contracted from the peoples they subjugated. For instance, venereal diseases could be acquired from infected men and women. Moses' solution was to kill all the men and women old enough to have had sexual intercourse and possibly be infectious. It must also have been realized that diseases could be transmitted on animal skins (anthrax) and on clothing (louse-infected materials can transmit typhus fever), since the codes recommended that these items be purified by washing.

The laws attributed to Moses in Deuteronomy indicate that the Israelites were aware of the connection between pork and disease. Swine are known to be infected by parasites (in particular, nematodes that burrow into the lymph and muscle) that cause very serious diseases (trichinosis). Eating undercooked pork spreads the parasites to humans. In addition, it may have been noticed that the consumption of shellfish often resulted in diseases (typhoid fever, cholera, hepatitis, polio, etc.). The easiest way to deal with the problem was to ban the food. Today we know that shellfish in sewage-polluted waters generally accumulate bacteria and viruses that can cause disease.

that spreads from animals to humans. Tiberius Claudius Nero Caesar, a Roman emperor (A.D. 14–37), is reported to have banned kissing in Rome to avoid the spread of oral herpes (cold sores). This ban indicated a realization that herpes was contracted by touching sores. In addition, sailors throughout the ages were aware of "plague ships" and "cholera ships" and were careful to avoid contact with individuals from these ships.

One of the oldest manuscripts on diseases was written by an Italian scientist, Girolamo Fracastoro (1478–1553). In his 1546 manuscript *About Contagion and Contagious Diseases*, Fracastoro proposed the idea that diseases were due to a *contagion* that could be transferred from one person to another. Contagions, according to Fracastoro, were transferred by direct contact; by touching contaminated inanimate objects (fomites), such as clothing, cups, and eating utensils; or through the air. He believed (incorrectly) that contagions were not responsible for the putrefaction of meat or the spoilage of milk, but he understood that fruit putrefaction (decay), as well as plant and animal infections, were due to contagions. Fracastoro's writings indicate that he considered contagions to be destructive particles whose heat, moisture, or other characteristics caused the destructive changes. He believed that foot and mouth disease was spread through the air (incorrect); that tuberculosis was transmitted through the air or on contaminated clothing (correct); that rabies was transmitted by the bite of a rabid dog (correct); and that syphilis was transmitted to children in the milk of infected mothers (possible, but it is usually transmitted through the placental barrier before the child is born) and through sexual intercourse (correct). In a manuscript published in 1530 *(Syphilus Suffering from the French Disease)*, Fracastoro described the symptoms of venereal disease in a shepherd named Syphilus. Eventually, this disease—which was known as the French Disease in Italy, and as the English Disease in France—became known as syphilis. Fracastoro also pointed out that many contagions attacked only certain crops or trees, while others attacked only specific animals or humans. He observed that diseases often attacked specific organs of the body. For example, tuberculosis generally affected the lungs, while trachoma damaged the eyes.

Even though most scientists were ignorant as to the actual cause of disease until the 1880s, they discovered how to deal with one of the most feared contagious diseases; smallpox. This disease killed and disfigured great numbers of individuals each year worldwide. Indian Vedic writings of about 1100 B.C. explained how to protect against smallpox. It was suggested that pus from the pox blisters should be injected into individuals wishing to be protected. A fever and a mild disease would result, but the person would become resistant to smallpox. This advice clearly indicated that smallpox was due to something in the pox blisters which could be transferred to other persons. By the 1700s, Europeans who had learned about the ancient Eastern practice of smallpox "vaccination" (known as **variolation**) were attempting to introduce this procedure into Europe. There was resistance to this practice, however, because it often resulted in outbreaks of smallpox. In 1798 Edward Jenner, an English physician, reported that material from cowpox lesions could be used to vaccinate humans with no danger of smallpox and that vaccinated individuals became resistant to smallpox (fig. 1-3). Even though Jenner's observations were resisted by the medical profession of the time, vaccination soon caught on, not only in England but also in Europe and the United States.

In the 1800s puerperal sepsis, better known as childbed fever, often killed 15% of the women giving birth in some hospital wards. Childbed fever is due to a blood infection that can occur during childbirth, when many blood vessels are ruptured and microorganisms can enter the mother's bloodstream. The frequency of infection increases if improperly sterilized instruments are brought into contact with the damaged tissue or if midwives, nurses, or doctors introduce microorganisms into the wound.

An Austrian physician, Ignaz Semmelweis, noticed that one of the wards in the hospital where he worked had four times as many puerperal sepsis deaths as did other wards, and that expectant mothers tried to avoid this ward. He was also aware of the fact that doctors attending the ward often examined expectant mothers or aided in childbirth without washing their hands after doing autopsies. Semmelweis concluded that the doctors were picking up "putrefying organic matter" on their hands from corpses and that they were transmitting this matter to the mothers, thus causing puerperal

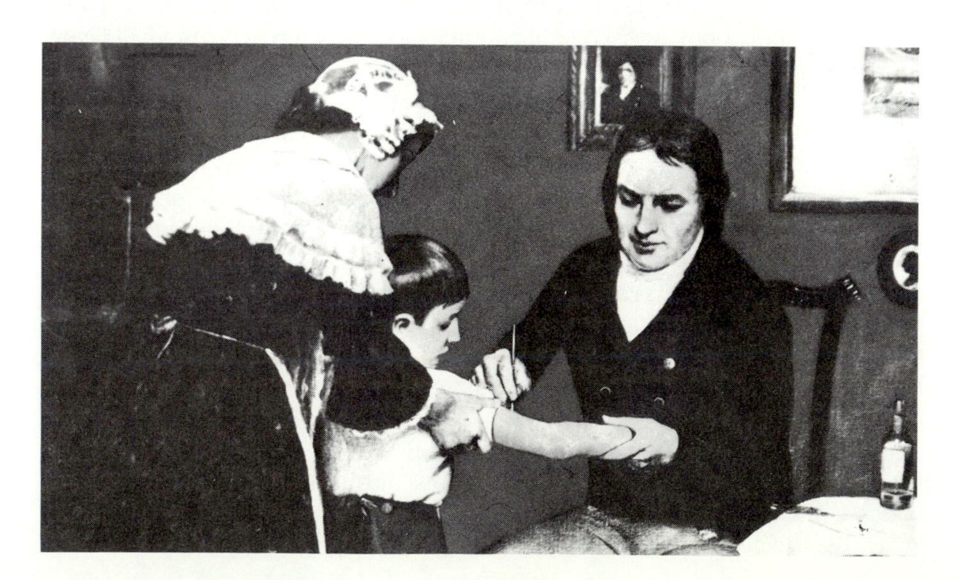

Figure 1-3 Edward Jenner. A painting illustrating Edward Jenner (1749–1823) administering the cowpox vaccine.

sepsis. In order to alleviate this situation, Dr. Semmelweis required that all hospital aides wash their hands carefully in a solution of lime chloride in order to destroy the "putrefying organic matter" before they examined expectant mothers or women who had just given birth. This preventive measure decreased the death rate from 8.3% to 2.3%, the rate in private practice and in other hospital wards. In an 1850 manuscript, Semmelweis concluded that many cases of puerperal fever were due to something on instruments or on the hands of student obstetricians and physicians, and that the medical staff was responsible for infecting the patients. Most physicians refused to accept this idea and did not bother to test its validity. Instead, the physicians reviled Semmelweis and were instrumental in having him dismissed from his position at the hospital where he worked. It was not for another thirty years or so that physicians realized that Dr. Semmelweis was indeed correct in his assessment of the source of infection in childbed fever.

THE DISCOVERY OF MICROORGANISMS

Our present understanding of the microbial world and the causes of infectious diseases has depended upon the development of the microscope. This tool allows scientists to view many microorganisms that cannot be seen by the unaided eye. It would have been very difficult to go beyond Girolamo Fracastoro's observations on contagions if microorganisms could not have been visualized in some manner. Although a number of scientists were using primitive microscopes in the 1600s, microorganisms were first described by Antony van Leeuwenhoek (1632–1723), a Dutch dry goods merchant and minor city official who built simple microscopes as a hobby (fig. 1-4a). His

(a)

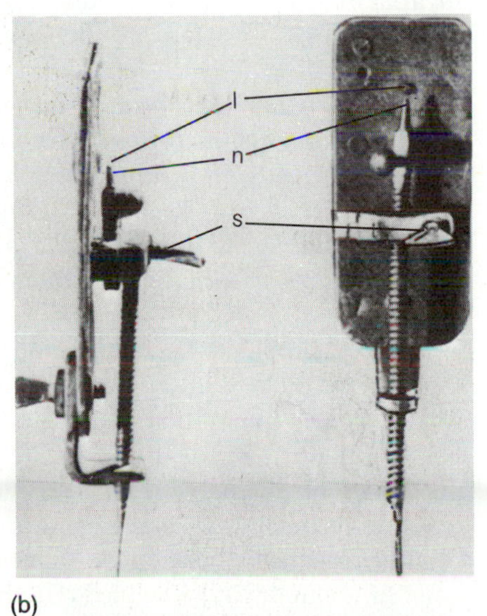

(b)

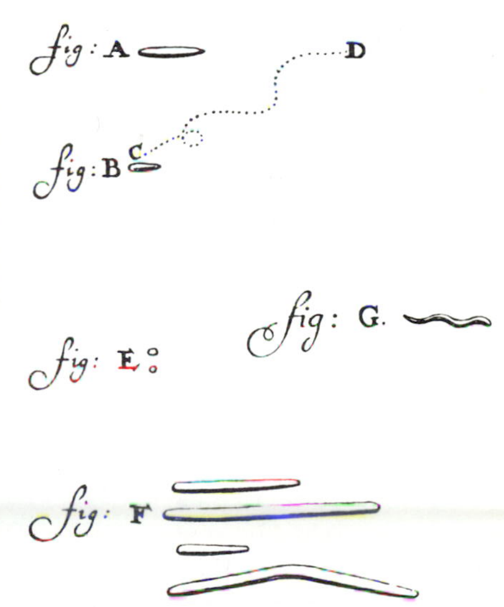

(c)

Figure 1-4 Antony van Leeuwenhoek and His Microscope. *(a)* A portrait of Antony van Leeuwenhoek (1632–1723). *(b)* Front and side views of a hand-held microscope similar to that used by van Leeuwenhoek. The sample was placed at the tip of the needle (n) and brought into focus by turning the screw (s). The sample was viewed through the lens (l). The light from a candle or from a bright window was used to illuminate the specimen. *(c)* Antony van Leeuwenhoek's drawings of "animalcules" from the human mouth. The "animalcules" in these drawings are large bacteria. In B, the movement of a motile organism is illustrated.

microscopes resembled magnifying glasses (fig. 1-4b). They consisted of a single, tiny lens sandwiched between two thin metal plates. The object of interest was placed on one side of the lens, on the point of a needle, or inside a fine glass tube. The object was brought into focus by turning a screw that moved the object farther from or closer to the lens. The eye was positioned on the other side of the lens. Van Leeuwenhoek's simple microscope □ was capable of magnifying objects approximately 275 times, enough to see protozoans, algae, fungi, and many of the larger bacteria clearly. For many years he reported his microscopic observations of water, pepper infusions, dental plaque, blood, and semen in papers to the *Philosophical Transactions of the Royal Society of London.* In these papers he meticulously described many different types of bacteria, fungi, protozoans, algae, blood cells, and sperm (fig. 1-4c). Van Leeuwenhoek called the microorganisms he discovered **"animalcules,"** because many of them were motile and darted around rapidly like tiny animals in the drops of liquid where they were found. Van Leeuwenhoek provided the groundwork for future microbiologists, for he introduced the scientific community to a previously unseen world and showed that animalcules were found nearly everywhere they were sought. The idea that microbes are intimately involved in our everyday lives, and that they are the cause of putrefaction and disease, was not seriously considered for another 200 years.

58

THE SPONTANEOUS GENERATION CONTROVERSY

The idea that life forms could arise from nonliving materials was held by many scientists and philosophers until the late 1860s. Many people believed that horse hairs in stagnant water could give rise to snakes and that rotting flesh would breed maggots. One of the most outlandish ideas was a recipe for making mice from soiled undergarments and wheat, written by J. B. van Helmont (1577–1644), a prominent chemist and philosopher:

> *If a dirty undergarment is squeezed into the mouth of a vessel containing wheat, within a few days (say 21) a ferment drained from the garments and transformed by the smell of the grain, encrusts the wheat itself with its own skin and turns it into mice. And what is more remarkable, the mice from the grain and undergarments are neither weanlings or sucklings nor premature but they jump out fully formed.*

Van Helmont's observations on how to make mice is an indication of the state of biology in the 1600s. Much opinion and pure fabrication substituted for experimentation and basic facts. During the next 250 years, a number of scientists showed convincingly that the spontaneous generation hypothesis was untenable. Francesco Redi (1626–1697), a physician and poet, was one of the first to challenge seriously the idea of spontaneous generation of large organisms by showing that rotting meat and fish did not breed maggots. In about 1665, he carried out an experiment that demonstrated the fact that maggots develop from fly eggs laid on meat and do not arise spontaneously from the meat (fig. 1-5). In his experiment, he placed meat inside each of three containers. One of the containers was covered with a glass top, another was covered with gauze, and the third was left uncovered. Redi observed that flies landed on the uncovered meat and on the gauze. His observations showed that the meat in the container left uncovered eventually be-

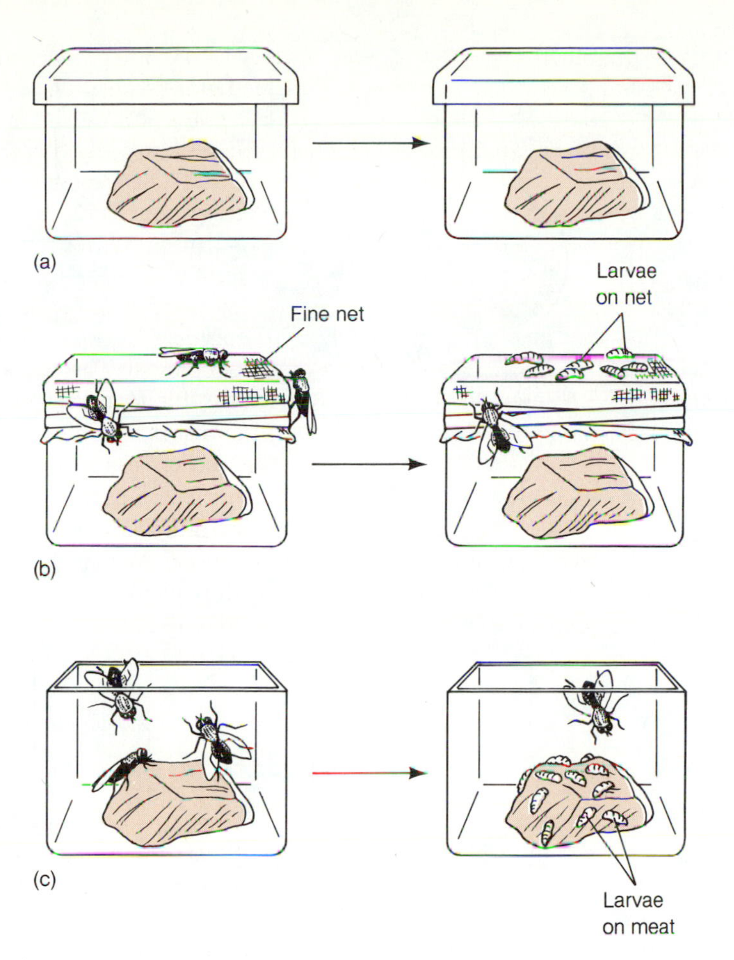

Figure 1-5 Francesco Redi's Experiment Disproving Spontaneous Generation of Maggots. Francesco Redi (1626–1697), in an experiment similar to the one illustrated, demonstrated that maggots (fly larvae) do not arise spontaneously from meat or fish but from eggs laid by flies. The meat that is open to the flies develops maggots in 2–3 days *(c)*, but the meat that is covered with cheesecloth *(b)* or glass *(a)* does not develop maggots. Maggots appear on the cheesecloth covering the meat because the flies lay their eggs in response to the smell of the meat.

came infested with maggots, while the meat in the glass- and gauze-covered containers remained free of maggots, although maggots were found on the gauze. From these observations, Redi correctly reasoned that the smell of rotting meat attracts flies, which then lay their eggs on the meat or as close to it as possible. The eggs hatch within a few days, giving rise to maggots that consume the meat or crawl around on the gauze.

The idea that microorganisms could arise spontaneously from nonliving matter was challenged by a number of scientists, including Lazzaro Spallanzani (1729–1799) and Theodor Schwann (1810–1882). In their experiments they demonstrated that vegetable and meat juices, called infusions, did not give rise to microorganisms if the infusions were heated to kill all organisms and were not subsequently contaminated by organisms in the air. Spallanzani prevented contamination by sealing the infusions in flasks, while Schwann heated a bent or coiled tube that opened the flasks to the air (fig. 1-6). Their experiments showed that air was laden with microbes and capable of contaminating infusions. If microorganisms present in the air could

Figure 1-6 Experiments by Spallanzani and Schwann, Aimed at Disproving that Microorganisms Arise by Spontaneous Generation. *(a)* Spallanzani (1729–1799) heated flasks containing nutrient infusions and sealed the hot flasks. These flasks showed no evidence of growth after an extended incubation period. Once the seal was broken and air entered into the flasks, microbial growth was evident within a day. Spallanzani concluded that microorganisms did not arise spontaneously but came from contaminating air. Proponents of the spontaneous generation theory argued that the spontaneous generation of microorganisms required fresh air. *(b)* Like Spallanzani, Schwann (1810–1882) heated flasks contraining nutrient infusions. Schwann, however, allowed heated air to reenter the flasks. The unsealed flasks showed no evidence of growth no matter how long he allowed heated air to enter the flask. Schwann concluded that microorganisms in the air were killed by the heating and this was why there was no growth in the infusion. Proponents of the spontaneous generation theory argued that Schwann was destroying a "vital force" when he heated the air entering the flask.

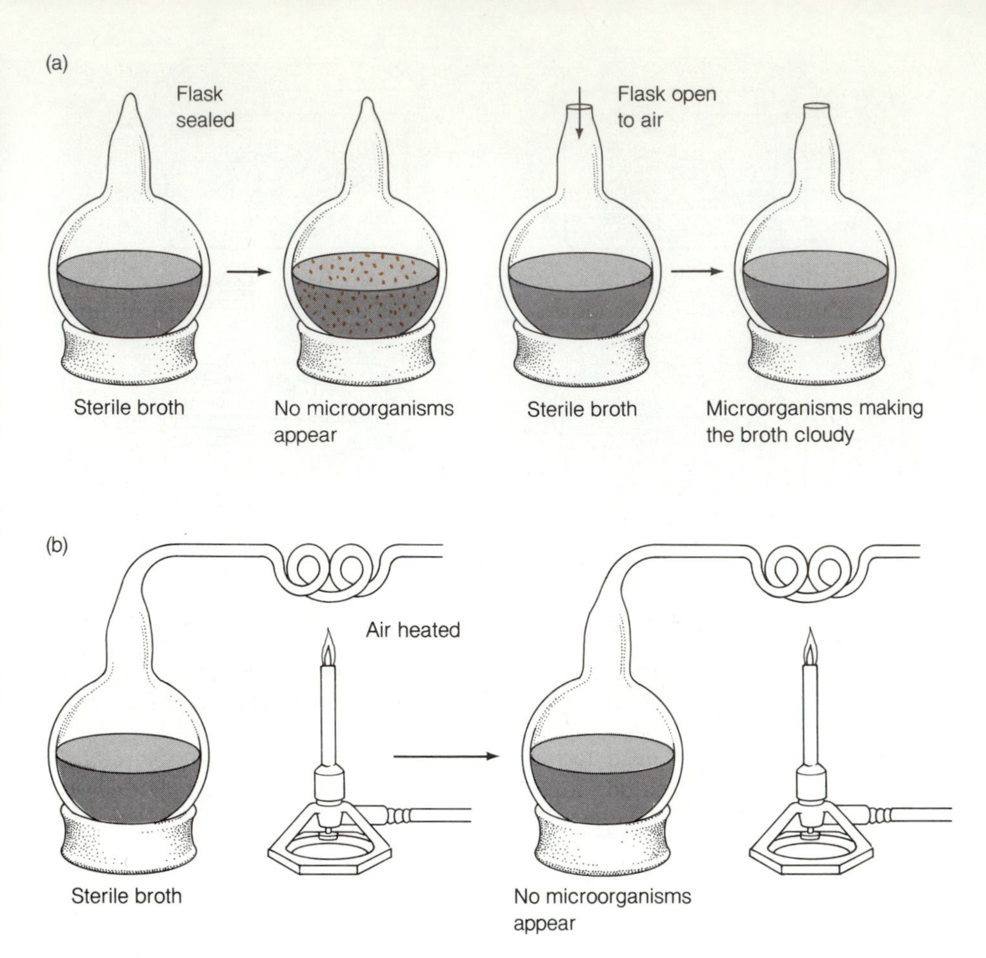

not come into contact with the infusions, or were killed by heating the air before it entered the infusions, the infusions remained free of life. Although these experiments provided evidence that even microorganisms do not arise spontaneously, proponents of the theory of spontaneous generation (of whom John Needham was one of the most active) still found fault with them. When spontaneous generation did not occur in heated infusions, proponents argued that the heating process necessary to sterilize the air also destroyed a "vegetative force" essential for the spontaneous development of microorganisms. Support for the spontaneous generation of microorganisms came from many eminent scientists in the early 1800s, because they often found microorganisms within infusions when they repeated Spallanzani's and Schwann's experiments. In retrospect, it is possible to understand why heated and sealed infusions sometimes showed growth: some microorganisms are extremely resistant to heat and are not always killed by boiling for short periods of time. The idea that microorganisms could develop by spontaneous generation died slowly and was accepted on faith by many scientists, possibly because they wanted a nonsupernatural explanation for the origin of life.

Louis Pasteur (1822–1895) showed convincingly that certain groups of microorganisms are responsible for specific types of metabolism (fermentations) and that microorganisms do not arise spontaneously within infusions (fig. 1-7). Pasteur hypothesized that the appearance of microbes within infusions was due to contaminating organisms from the air. To test his hypothesis, Pasteur filtered large volumes of air through "clean" guncotton to

Figure 1-7 Louis Pasteur. Portrait of Louis Pasteur (1822–1895) in his laboratory, checking infusions for evidence of microbial growth.

catch any microorganisms that might be in the air. He then dissolved the guncotton in a mixture of alcohol and ether and collected the freed microorganisms that precipitated in the solvent. He showed that microorganisms were in the air by comparing the number of organisms on "clean" guncotton to that on "used" guncotton under a microscope.

The most convincing evidence against the idea of spontaneous generation was provided by Pasteur in 1861, when he showed that heated infusions open to the air did not become contaminated. Pasteur placed infusions such as sugared yeast water, sugar beet juice, or pepper water into round-bottomed flasks. He heated the openings of the flasks and then drew them out so that the flasks had thin, curved necks (fig. 1-8). Pasteur then boiled the infusions to kill all the microorganisms already present. The flasks were then allowed to stand at room temperature for many days, and although the infusions were exposed directly to unheated air—which, according to many who believed in spontaneous generation, contained the "vegetative force"—no growth occurred in the flasks. Pasteur reasoned that the flasks remained sterile because bacteria in the air settled out along the curved necks of the flasks before they could reach the infusions. In order to see if microorganisms were really present in the necks, he tipped several of the flasks so that the infusions flowed into their necks and then back into the flasks. When he did this, the infusions became turbid within a few days. Pasteur's experiments left little doubt that the spontaneous generation hypothesis was untenable and that living organisms that appeared in infusions originated from contaminating organisms in the infusion or in the air.

By 1857, Pasteur had established that alcoholic fermentations were carried out by yeasts and that lactic acid fermentations were due to much smaller, spherical organisms. In 1861, one of Pasteur's manuscripts indicated that butyric acid fermentations were due to small rod-shaped organisms that required an oxygen-free environment. Pasteur's experiments disproving the

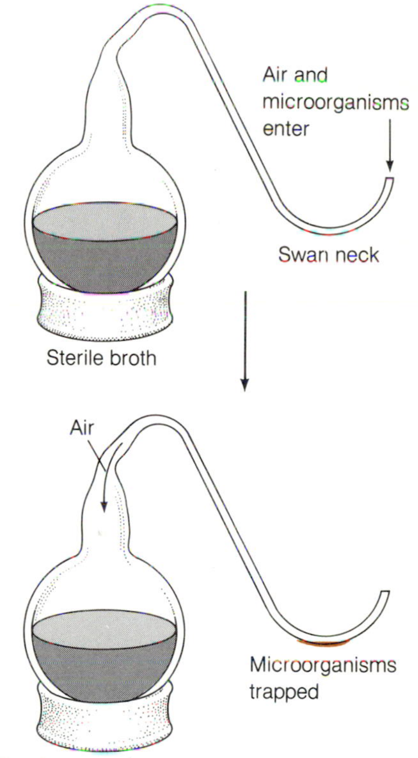

Figure 1-8 Louis Pasteur's "Swan-Necked" Flasks. Pasteur heated flasks containing nutrient infusions and then drew out the necks of the flasks so that microorganisms would have difficulty reaching the infusions. Many of the flasks were not sealed, so as to allow fresh, cool air to enter the flasks. The open flasks showed no evidence of growth unless the infusion was tipped into the neck area, where microorganisms were trapped. These experiments showed that microorganisms, not spontaneous generation, were responsible for growth in the flasks.

spontaneous generation hypothesis, and those showing that particular microorganisms were responsible for specific fermentations, finally convinced many scientists that microorganisms did not arise spontaneously and that they could be the cause of fermentations, putrefaction, and disease.

Nevertheless, Pasteur was challenged by a number of scientists who could not repeat his experiments. The reason their experiments failed was that they often used infusions containing hay, which in turn contained extremely heat-resistant bacteria. In 1877, the British physicist John Tyndall demonstrated that media contaminated with heat-resistant bacteria could not be sterilized by boiling. At about the same time, the German botanist Ferdinand Cohn showed that the heat-resistant bacteria were those that formed endospores □ and that the endospores were a heat-resistant form of the bacteria. Tyndall developed a simple method whereby everyone could sterilize media containing heat-resistant bacteria: after boiling the medium for a minute or two, he allowed it to cool for a few hours so that the endospores might germinate into growing cells that were sensitive to subsequent boiling. Tyndall found that boiling and cooling several times would generally sterilize a medium. Tyndall's method for sterilizing media became known as **Tyndallization.** The successful sterilization of all types of media quieted those who had been unable to repeat Pasteur's experiments. Today, heat sterilization at 121°C for 15–25 minutes is used to sterilize most media. Tyndallization is occasionally used to sterilize certain heat-sensitive media that are altered at 121°C.

94

THE GERM THEORY OF DISEASE

Until about 1880 there were many ideas regarding the cause of infectious diseases. Most people believed that infectious diseases were not caused by organisms, but were a punishment sent by the gods. Even today, there are many people who believe that sexually transmitted diseases are God's punishment for sexual transgressions. Those who had nonsupernatural explanations to account for infectious diseases invariably blamed nonliving entities, such as "contagion" (Fracastoro, 1546), "morbid matter" (Jenner, 1798), or "putrefying organic materials" (Semmelweis, 1850). The first pathogenic microorganisms discovered were fungi. In 1807, Bénédict Prévost observed the germination of wheat-bunt spores and proposed that the fungus caused the plant disease. Agostino Bassi in 1835 showed that a fungus was causing a disease (muscardine) of silkworms, and in 1839 J. Schoenlein and David Gruby demonstrated that fungi were responsible for a chronic ringworm (such as athlete's foot). At about the same time, B. Lagenbeck showed that yeasts were the cause of thrush, an infection of the mouth. In 1845 M. J. Berkeley characterized the fungus that causes late blight of potato. Anton De Bary's 1853 manuscripts described fungi as the cause of plant smuts and rusts. In addition, De Bary described the life cycle of the fungus that causes late blight of potato. Despite this evidence, however, it was not until Louis Pasteur and Robert Koch published their observations that the role of microorganisms in causing disease became generally accepted. These two men could be called the fathers of the **germ theory of disease,** which postulates that infectious diseases are caused by microorganisms.

Louis Pasteur contributed considerable evidence to support the idea that microorganisms were responsible for disease. For instance, he developed vaccines that protected humans and domestic animals from diseases such as swine erysipelas, chicken cholera, anthrax, and rabies. He made the vaccines by making the organisms responsible for the diseases nonvirulent (unable to cause disease). The nonvirulent organisms, when injected into an animal or human, did not cause the disease but made the animal or human immune or resistant to it. The successful development of these vaccines was very strong circumstantial evidence that these microorganisms caused the diseases.

Robert Koch (1843–1910) provided irrefutable evidence linking specific microorganisms with infectious diseases (fig. 1-9). His first series of experiments demonstrated that anthrax was caused by a bacterium. Anthrax is a devastating disease that killed large numbers of livestock and inflicted severe financial losses on farmers in Europe during Koch's time. In his initial studies (1876), Koch noticed the presence of rod-shaped structures in the spleen and blood of animals killed by anthrax but not in the spleen or blood of normal animals. Consequently, he guessed that these rods might be the cause of anthrax. He transferred some of the rods to a drop of sterile beef serum or aqueous humor from the eye of a cow and noticed that they grew and multiplied in the media. He observed the rod-shaped organisms with his microscope and watched them grow into long filaments and produce spores. In addition, he found that the spores germinated into rod-shaped bacteria similar to the ones that gave rise to the spores. This observation demonstrated that the rod-shaped structures were living bacteria and that they could be grown outside the living animal for many generations. By growing the microorganisms in fresh media a number of times, he was diluting out any "contagion" or "putrefying organic materials" that might be the cause of the disease. Koch then injected the bacteria into mice, which subsequently died of anthrax. When the diseased animals were examined, the bacteria that had been injected were found in large numbers in the animals' blood. These studies provided the first experimental evidence that a specific bacterium caused an infectious disease.

Figure 1-9 Robert Koch (1843–1910)

Koch is also credited with proving, in 1882–1884, that tuberculosis is caused by a bacterium. He removed infected mouse spleen or tubercles from the lungs of guinea pigs and spread the material over the surface of jelled cow or sheep blood in a tube. It took approximately 10 days at 37°C for very small colonies of bacteria to appear on the denatured blood. Koch was able to show by a special staining procedure that the colonies were made up of one type of bacteria. Thus, he was dealing with a pure culture of bacteria. When the pure culture was injected into mice, the animals developed tuberculosis, and the bacteria could be isolated once more in pure culture.

In order to prove that an organism was the cause or **etiological agent** of a disease, Koch applied a specific series of objective criteria which eventually became known as **Koch's postulates** (fig. 1-10). This series of criteria generally must be satisfied if an etiological agent is to be accepted as the cause of an infectious disease.

Koch's Postulates

1. The specific microorganism thought to be the cause of a disease must be consistently isolated from individuals suffering from the disease, but not from healthy individuals.

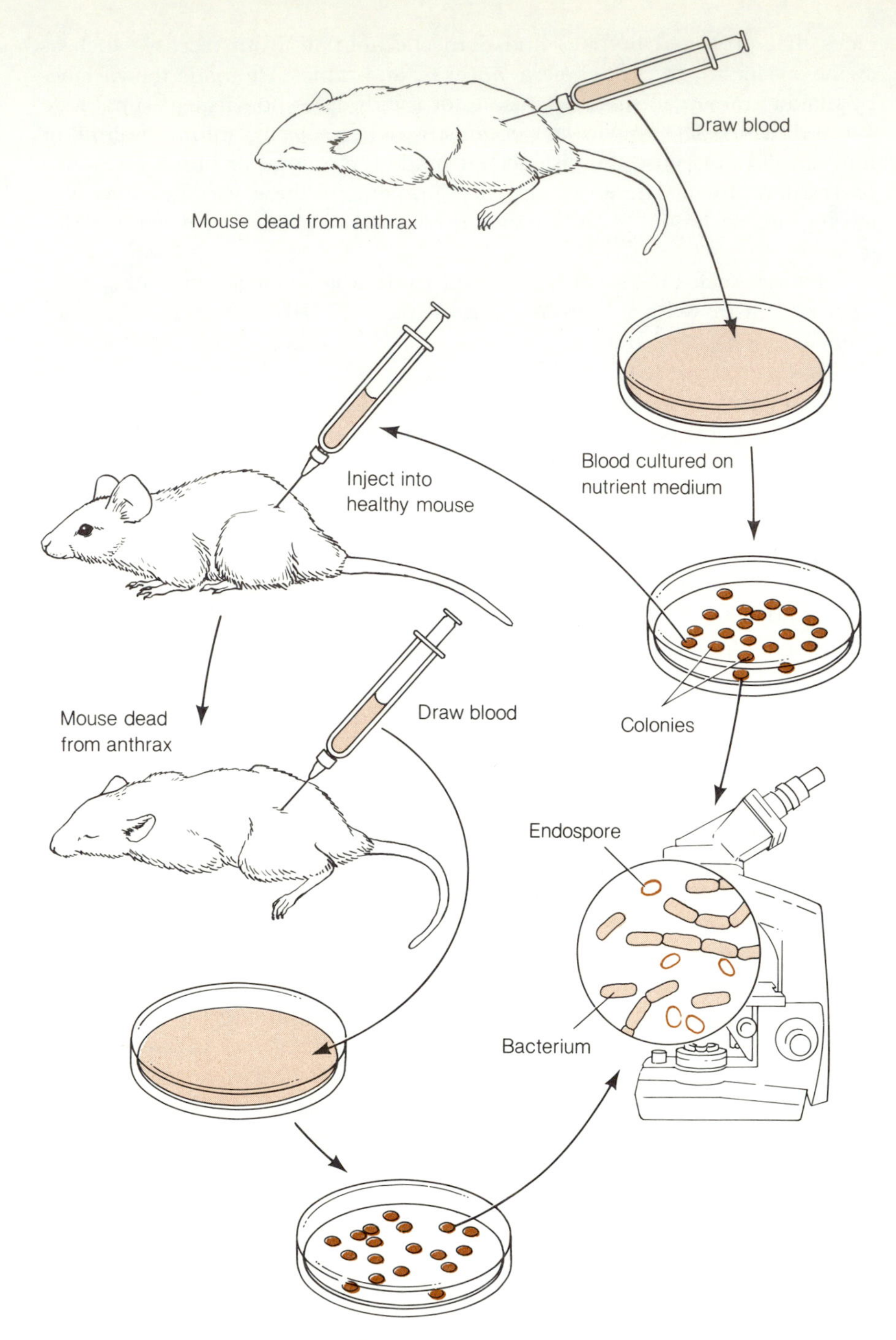

Figure 1-10 Illustration of Koch's Postulates. Koch's postulates are a set of rules that are followed in order to prove that an organism is responsible for a disease. The organism must be isolated and observed in each case of the disease. In this illustration, a bacterium is isolated on a solid nutrient from the blood of a dead animal. The organism can be observed with the aid of the microscope and characterized. A pure culture of the organism is injected into healthy animals and then isolated and observed in each case of the disease. The organism isolated initially should be identical to the organisms isolated from the inoculated animals.

2. The suspected etiologic agent of the disease must be cultivated in pure form outside the host *in vitro.*

3. Pure cultures of the suspected pathogen, when introduced into a suitable and susceptible host, must produce the signs and symptoms characteristic of the disease.

4. The same organisms must be consistently isolated in pure culture from the afflicted experimental host and be cultivated again *in vitro.*

Koch's postulates cannot always be followed in establishing the cause of a disease, because in some cases it is impossible to find the etiological agent in the diseased organism. For example, some viruses that cause cancer do not behave as typical viruses, but simply exist as a piece of genetic material (DNA) integrated into the host's genetic information. Microorganisms that cause diseases such as food poisoning and tetanus because of a toxin □ 431 they produce often are not present during the disease. In many cases it is not possible to cultivate the organism outside the host. For instance, the bacteria that cause syphilis and leprosy and certain viruses that cause disease cannot be grown *in vitro* and so pure cultures are difficult to prove. Finally, many diseases are caused by organisms that disappear from the host, sometimes long before the onset of the disease; for example, the viruses that cause an animal's immune system to turn upon itself.

SIGNIFICANT DISCOVERIES IN MICROBIOLOGY

Louis Pasteur and Robert Koch were instrumental in establishing some of the roles microorganisms play in nature, but more importantly, they stimulated a systematic study of microbes by many other scientists. By showing that vaccines could be made which afforded protection against a number of diseases, and that some infectious diseases were caused by specific microorganisms, Pasteur and Koch promoted the discovery of new vaccines and of other pathogens. Table 1-1 illustrates the rapid discovery of over 25 pathogens by numerous scientists over a period of about 40 years (1875–1915), sometimes called the **golden age of microbiology.**

Technology

Robert Koch, his students, and his colleagues, in particular, are credited with the development of techniques that made the study of microbiology relatively simple. For example, stained smears of bacteria on glass slides not only made bacteria easier to see using the microscope but also helped in the identification of bacteria, such as the tubercle bacillus. Paul Ehrlich (1854–1915), one of Koch's colleagues, is credited with the development of a staining procedure similar to the one used today to identify the tubercle bacillus. Ehrlich's staining technique, developed in 1882, was used to test for the tubercle bacillus in the sputum of suspected tuberculosis patients at medical centers all over Europe, including the hospital where Koch worked. Walther Hesse (1846–1911), a student of Koch, initiated the use of agar to solidify or jell nutrients □ 120 upon which microorganisms would grow. Unlike gelatin, which was used by Koch, agar remains solid at temperatures at which many microorganisms grow best (35°C–40°C) and it is not degraded by most microorganisms. Solidified growth media are extremely important in microbiology

TABLE 1-1
INFECTIOUS AGENTS DISCOVERED IN THE GOLDEN AGE OF MICROBIOLOGY

DISEASE	INFECTIOUS AGENT*	YEAR	DISCOVERER(S)
Pear fire blight	*Erwinia amylovora*	1877	Burrill
Anthrax	*Bacillus anthracis*	1877	Koch
Gonorrhea	*Neisseria gonorrhoeae*	1879	Neisser
Malaria	*Plasmodium malariae*	1880	Laverans
Wound infections	*Staphylococcus aureus*	1881	Ogston
Tuberculosis	*Mycobacterium tuberculosis*	1882	Koch
Erysipelas	*Streptococcus pyogenes*	1882	Fehleisen
Diphtheria	*Corynebacterium diphtheriae*	1883	Klebs & Loeffler
Cholera	*Vibrio cholera*	1883	Koch
Typhoid fever	*Salmonella typhi*	1884	Eberth & Gaffky
Bladder infections	*Escherichia coli*	1885	Escherich
Smallpox	Smallpox virus	1887	Buist
Food poisoning	*Salmonella enteritidis*	1888	Gaertner
Tetanus	*Clostridium tetani*	1889	Kitasato
Gas gangrene	*Clostridium perfringens*	1892	Welch & Nuttall
Plague	*Yersinia pestis*	1894	Yersin & Kitasato
Botulism	*Clostridium botulinum*	1897	Van Ermengem
Bacillary dysentery	*Shigella dysenteriae*	1898	Shiga
Foot and mouth disease	Foot & Mouth Disease Virus	1898	Loeffler & Frosch
Tobacco mosaic disease	Tobacco Mosaic Virus	1898	Beijerinck
Yellow fever	Yellow Fever Virus	1900	Reed
Syphilis	*Treponema pallidium*	1905	Schaudinn & Hoffman
Whooping cough	*Bordetella pertussis*	1906	Bordet & Gengou
Rocky mountain spotted fever	*Rickettsia rickettsii*	1909	Ricketts
Typhus	*Rickettsia prowazekii*	1910	Ricketts & Wilder
		1916	Lima
Tularemia	*Franciscella tularensis*	1912	McCoy & Chapin

*The names for the infectious agents are those in current use rather than those given at the time of their discovery.

because microorganisms can be separated and purified on these media. Richard Petri (1852–1921), another of Koch's students, inspired the use of shallow flat-bottomed dishes with flat, overhanging covers to grow microorganisms (1887). The nutrient-agar dishes provided a large area over which microorganisms could be spread, and the lids helped to reduce contamination. These dishes are currently used in all microbiology laboratories and are known as **petri dishes** or petri plates. The Danish scientist Christian Gram (1853–1935) developed a staining method that demonstrated bacteria in infected animal tissues (1884) and which was used years later to distinguish between two basic types of bacteria. The staining procedure became known as the **Gram stain** and is still used universally today.

From about 1890 to 1910, Raymond Sabouraud intensively studied the fungi that cause ringworm, and in 1910 he published a classic manuscript called *Les Teignes* (the skin diseases). He also developed a culture medium that is still used to grow fungi.

Pasteur's studies of microorganisms demonstrated that they are responsible for many of the chemical changes that take place in foods and beverages. For instance, in the early 1860s Pasteur noticed that the presence or absence of oxygen had an effect on the growth rate and yield of yeasts in a juice. The presence of oxygen provoked a rapid growth rate and a high yield of yeast, but no alcohol production. The absence of oxygen resulted in a slow growth rate and a low yield, but ethanol was produced. Pasteur noticed that organisms grown in the presence of oxygen used glucose less rapidly than organisms grown in the absence of oxygen. The slower rate of glucose utilization and the lack of alcohol or acid production when an organism begins to grow aerobically (with O_2) is known as the **Pasteur Effect** 190 □. Pasteur also demonstrated that many microorganisms produce specific waste materials (or combinations of waste materials), such as ethanol, butyric acid, and lactic acid. These discoveries have led to the manipulation of environmental conditions and to the selection of specific microorganisms in order to produce valuable products. For instance, certain microorganisms are used to make ethanol, vinegar, lysine, monosodium glutamate, and antibiotics. The use of microorganisms to make various products has resulted in the development of the very profitable branch of microbiology called **industrial microbiology.**

Pasteur reasoned that he could prevent spoilage of wines and beers by inhibiting or killing the microorganisms that chemically alter them. He found that the simplest way of destroying microorganisms in wines and beers was to heat the alcoholic beverages, after they were bottled, to temperatures just below their boiling point. This treatment generally prevented spoilage because it killed the spoilage organisms. Heating to kill spoilage or disease-causing microorganisms is now referred to as **pasteurization.** It is used not only on beer, wine, juices, and nonalcoholic beverages to prevent spoilage but also on milk to kill pathogenic organisms, such as those that cause tuberculosis, undulant fever (brucellosis), and Q fever.

Microbial Ecology

The discoveries made by Pasteur and Koch stimulated many scientists not only to investigate medically important organisms but also to study microorganisms in nature. Two of the most famous scientists of the golden age of microbiology were the Russian Sergei Winogradsky and the Dutch Martinus Beijerinck (fig. 1-11), who studied soil microorganisms. These individuals are considered to be the founders of microbial ecology. They developed a medium for growing unusual bacteria that used CO_2 to make carbon compounds and ammonium to generate energy. In addition, they found a medium for growing bacteria that could use atmospheric nitrogen (N_2) as a nitrogen source, rather than ammonium (NH_4^+) and nitrate (NO_3^-).

The photosynthetic purple and green sulfur bacteria, as well as the nonphotosynthetic sulfur bacteria, were first studied by Winogradsky in the 1880s. Winogradsky's studies of the nonphotosynthetic sulfur bacteria provided the first evidence that organisms could use an inorganic molecule such as H_2S to obtain energy. In addition, Winogradsky's studies of another group of bacteria, called the **nitrifying bacteria,** demonstrated that they could use inorganic chemicals (NH_3 and NO_2^-) as their sole source of energy and that they could use CO_2 as their source of carbon in the absence of light and chlorophyll. This was a revolutionary idea, because at that time it was be-

(a)

(b)

Figure 1-11 Martinus Beijerinck and Sergei Winogradsky. *(a)* Martinus Beijerinck (1851–1931) discovered the virus that causes tobacco mosaic disease of tobacco plants, and bacteria that fix atmospheric nitrogen into organic compounds. *(b)* Sergei Winogradsky (1856–1953) discovered bacteria that could use hydrogen sulfide as a source of energy and fix atmospheric carbon dioxide into organic compounds. In addition, he discovered bacteria that fix atmospheric nitrogen into organic compounds.

lieved that only green plants could synthesize organic matter from carbon dioxide. Winogradsky's studies were the first to show the importance of microorganisms in cycling nutrients such as sulfur, nitrogen, and carbon, and have stimulated many subsequent studies in microbial ecology.

Bacteria that could use atmospheric nitrogen (N_2) as their source of nitrogen were isolated by Winogradsky and Beijerinck at about the beginning of the twentieth century. In 1888, Beijerinck reported the presence of bacteria within root nodules of legumes, such as beans, peas, and alfalfa. He postulated that some soil bacteria were able to invade the roots of these plants and that they were responsible for the root nodules.

The pioneering work carried out by Winogradsky and Beijerinck has stimulated many scientists to study the roles microorganisms play in nature. Because of these scientists' contributions over the last century, we are now beginning to understand how microorganisms shape the environment and affect the evolution of all plants, animals, and humans.

Immunology

The development of solidified media, staining techniques, and pure culture methods was of great importance to the study of microbiology. These developments promoted the widespread study of microorganisms and the discovery and characterization of pathogens. In addition, Pasteur's development of vaccines that afforded protection against anthrax, rabies, chicken cholera, and swine erysipelas stimulated scientists to attempt to find vaccines against

(a)

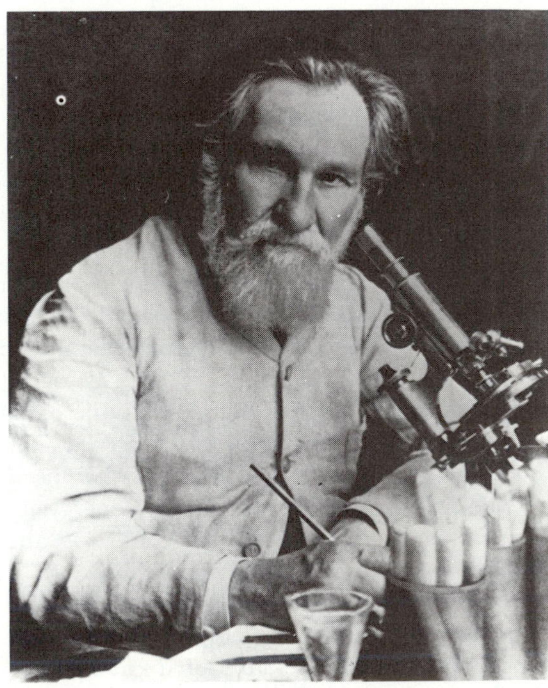

(b)

Figure 1-12 Paul Ehrlich and Elie Metchnikoff. *(a)* Paul Ehrilch (1854–1915) proposed that the immune response consists of circulating molecules, now known as antibodies, that destroy foreign molecules. His idea is known as the humoral theory. In addition, Ehrlich worked with Emile von Behring to develop a diphtheria antitoxin. *(b)* Elie Metchnikoff (1845–1916) proposed that the immune response consists of cells, now called phagocytes, that ingest and destroy foreign materials. His idea is known as the cellular theory of immunity.

other diseases. Such discoveries were possible for an increasing number of pathogens once pure culture techniques had been developed.

Paul Ehrlich and Emil von Behring worked together on the development of a diphtheria antitoxin from blood serum during the 1890s (fig. 1-12). Von Behring's studies showing that diphtheria toxins could be neutralized by antitoxin won him the first Nobel Prize in Medicine or Physiology in 1901, and the diphtheria antitoxin made von Behring a rich man. Ehrlich, on the other hand, signed away his interest in the diphtheria antitoxin in order to obtain a government post.

In the early 1890s, Ehrlich observed that immunity to toxins was transferred from immune mouse mothers to nursing baby mice. In addition, he noticed that immunity could be transferred from one mouse to another by injection of blood serum from immune mice. Based upon these observations, Ehrlich proposed one of the first good explanations for how vaccines protect an animal. He hypothesized that a foreign material or toxin entering the body would bind to receptors on cells and stimulate the multiplication of the receptors, which were then released into the blood, where they neutralized the toxins. On the other hand, Elie Metchnikoff, a student of Koch's, thought that immunity was mediated by cells that phagocytize (ingest) foreign materials and microorganisms. Today, we know that the immune system is much more complicated than envisioned by Ehrlich (**humoral hypothesis**) and Metchnikoff (**cellular hypothesis**) □. There are specific cells that produce antibodies, others that synthesize toxic inhibitory substances, some that regulate immune responses, and still others that are involved in phagocytosis. Ehrlich shared the 1908 Nobel Prize with Elie Metchnikoff for their experiments and ideas about the immune system (fig. 1-12). The contributions made by Pasteur, Ehrlich, Metchnikoff, von Behring, and other scientists of that period provided the foundation for the dynamic branch of microbiology known as immunology.

452

Industrial Microbiology

Paul Ehrlich was also intimately involved in the search for chemicals that might inhibit or kill pathogenic microorganisms. In 1910 he reported the discovery of an arsenic compound he called **Salvarsan,** which inhibited and killed the bacteria that cause yaws, syphilis, and relapsing fever. Although the drug was very toxic to humans, it proved to be most effective against syphilis. Not until 1935 was another drug discovered that would cure diseases. Gerhard Domagk, in Germany, reported that a **sulfonamide** called **Prontosil** was effective in controlling a number of bacteria.

Drugs that could be used to cure diseases were discovered and produced in the 1940s with increasing frequency. In 1941, Howard Florey and Ernst Chain were able to isolate from the fungus *Penicillium* the antibiotic **penicillin.** Antibiotics are organic chemicals produced by various organisms which inhibit or kill other organisms. Alexander Fleming had reported the discovery of penicillin in 1928 but had been unable to purify it. By the mid 1940s, Selman Waksman and his colleagues had discovered a number of antibiotics produced by various species of the bacterial genus *Streptomyces.* The most important of the antibiotics were streptomycin and tetracycline, which were effective against a large spectrum of microorganisms. Since the 1940s, thousands of antibiotics have been isolated from microorganisms and used to treat infectious diseases. Numerous multi-billion-dollar pharmaceutical industries exist in many of the industrialized countries and capitalize on the antibiotic-producing capabilities of microorganisms. New antibiotics and drugs continue to be developed and tested in order to find chemical agents that are more effective against specific pathogens and cancers and that have less severe side effects on the host.

Virology

During the golden age of microbiology, numerous scientists discovered that certain diseases were due not to typical bacteria but to something that was much smaller. In 1892, Dmitri Iwanowski reported that mosaic disease of tobacco could be transmitted among plants by something that passed through filters fine enough to stop all known bacteria. Beijerinck in 1899 reported that this infectious agent could be dried and heated to temperatures as high as 90°C and still remain infectious. These filterable agents (viruses) were much smaller than endospores and therefore invisible in the light microscope; also, they could not be grown on artificial media. In 1898, Friedrich Loeffler and Paul Frosch showed that foot and mouth disease was also caused by a filterable agent that could not be observed with the microscope. William Twort in 1915, and Felix-Hubert d'Herelle in 1917, reported a filterable agent that destroyed bacterial lawns (confluent growth on solid media) or cleared turbid broth cultures. The filterable agent was believed to be a new microscopic life form, and was named **bacteriophagium** (bacterial eater). Eventually, all the filterable agents were regarded as one type of organism and were given the name virus.

With the development of the electron microscope during the late 1930s and early 1940s, it soon became possible to "see" that viruses and bacteria were very different from each other. The bacteria and other microorganisms were found to be essentially typical cells, while the viruses had no cellular organization and consisted of a protein coat enclosing the genetic material. Because its resolution □ is much greater than that of the light microscope (1 nm vs. 200 nm), the electron microscope has been extremely useful, not only in characterizing viruses but also in establishing the fundamental differences between bacteria and cells of higher organisms.

59

Medicine

During the golden age of microbiology (1875–1915), the scientific community became convinced that infectious diseases were caused by microorganisms. In addition, the microorganisms responsible for many diseases were being isolated and characterized. Scientists, for the first time, had specific targets to attack and they did so by developing vaccines (Pasteur and von Behring) and chemotherapeutic agents (Ehrlich). The knowledge that microorganisms cause infectious diseases prompted Joseph Lister (1827–1912) to employ carbolic acid as a disinfectant during surgery (fig. 1-13). Lister based his technique on the assumption that airborne microbes falling on the wound were the cause of postoperative infections. Lister was successful in reducing the number of postoperative infections by using dressings soaked in carbolic acid. Thus, we see how Pasteur's study of microorganisms led to an understanding of how pathogens might spread through the air, and how Lister used this information and his knowledge about carbolic acid to stop the spread of pathogens in wounds. The understanding that microorganisms were responsible for spoilage and disease led to pasteurization of foods, the

(a)

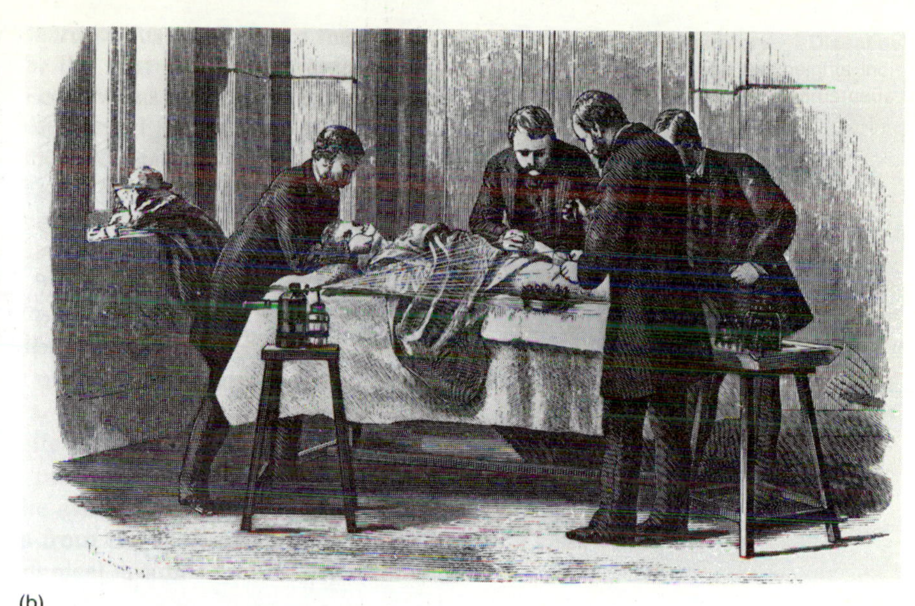

(b)

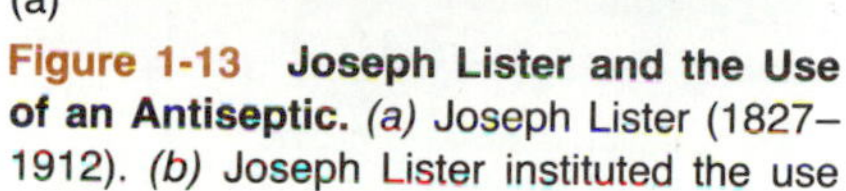

Figure 1-13 Joseph Lister and the Use of an Antiseptic. *(a)* Joseph Lister (1827–1912). *(b)* Joseph Lister instituted the use of the antiseptic carbolic acid (phenol) to kill microorganisms and prevent infection during surgery. The carbolic acid was sprayed over the area of the incision and used on dressings. Lister's use of carbolic acid during and after surgery reduced the death rate caused by infection from 45% to about 15%.

development of sewage and garbage systems that improved sanitation, and the filtering and eventually the chlorination of drinking water supplies. During the golden age of microbiology there were major new discoveries almost every year (table 1-2).

Today, medical microbiology remains just as fertile. Research continues to explore the fundamental nature of pathogenic microorganisms, how they cause disease, and ways of combating them. As human populations increase in density and their patterns of behavior change, unusual diseases and pathogens become important while others almost disappear. For example, outbreaks of Legionnaires' disease are now occurring at numerous locations where individuals congregate; genital herpes has become epidemic in industrialized nations; and an immunodeficiency disease called AIDS (acquired immune deficiency syndrome) is becoming prevalent among homosexuals.

The identification of pathogens is still an important part of microbiology. Since each pathogen is sensitive to particular treatments, it must be identified promptly in order to combat it effectively. For this reason, constant research is directed toward the development of faster and more reliable methods of isolating and identifying bacteria.

Molecular Biology

During the last 30 to 35 years, the genetics of microorganisms has provided fundamental insights into all areas of genetics and physiology. Two important scientific papers initiated the present era of microbiology, which is dominated by molecular biology: the 1944 report by Oswald Avery, Colin MacLeod, and Maclyn McCarty, describing bacterial transformation by DNA, and the 1953 paper by James Watson and Francis Crick which described the structure and possible functioning of DNA. The work of Francois Jacob and Jacques Monod, published in 1960, initiated numerous studies that clarified

TABLE 1-2
A PARTIAL LIST OF PIONEERS IN MICROBIOLOGY

INVESTIGATOR	YEAR	CONTRIBUTION
Antony Van Leeuwenhoek	1677	Reported the discovery of microorganisms in a letter to the Royal Society of London. Built simple microscopes that could magnify more than 100x.
Lazzaro Spallanzani	1767	Conducted the first experiments that challenged the idea that microorganisms arise by spontaneous generation.
Edward Jenner	1796	Carried out the first successful vaccination with cowpox against smallpox.
Benedict Prevost	1807	Proposed that a fungus was responsible for wheat-bunt disease, and promoted the germ theory of disease.
Theodor Schwann	1837	Reported that yeasts are required for an alcoholic fermentation to occur.
Anton De Bary	1853	Reported that fungi were responsible for plant smuts and rusts. Challenged the theory of spontaneous generation and promoted the germ theory of disease.
Louis Pasteur	1857	Each type of fermentation is dependent upon a specific type of organism. Yeast carry out an alcoholic ferment, specific bacteria carry out a lactic acid ferment.
	1861	Oxygen inhibits butyric acid ferments and kills the bacteria responsible. The theory that microorganisms can evolve from nutrients in a sterile medium is disproved. "Spontaneous generation" is shown to be due to contaminating microorganisms.
	1864	Developed a technique, now called pasteurization, to destroy spoilage organisms in wine.
	1880	Developed vaccines against chicken cholera, anthrax, swine erysipelas, and rabies.
Joseph Lister	1867	Published a study on the use of antiseptics during surgery.
Robert Koch	1876	Observed the growth of *Bacillus anthracis* and spore formation. Showed that this bacterium caused anthrax in animals.
	1881	Obtained the first pure culture of a bacterium on solid medium. Developed techniques for staining microorganisms. Developed various complex media to isolate in different bacteria in pure culture. Koch's Postulates were made public.
Ferdinand Cohn	1877	Discovered bacterial endospores and their extreme resistance to heat.
Joseph Lister	1878	Obtained pure cultures of a lactic acid bacterium by a dilution method.
Elie Metchnikoff	1884	Proposed a cellular theory of immunity.
Hans Christian Gram	1884	Developed a differential staining technique, now called the gram stain, for bacteria.
Richard J. Petri	1887	Introduced a covered dish for growing microorganisms on a solidified medium.
Dmitri Iwanowski	1892	Discovered a filterable organism that caused tobacco mosaic disease.
Martinus Beijerinck	1898	Further characterized the organism that caused tobacco mosaic disease.
Erwin F. Smith	1890s	Showed that bacteria were the cause of a number of plant diseases.
Emil Von Behring	1890s	Developed a diphtheria antitoxin.
Paul Ehrlich	1890s	Proposed a humoral theory of immunity which proposed that antibodies were responsible for the immunity.
	1910	Announced the discovery of an arsenic compound that could be used to treat syphilis.

how the information contained in DNA directed the synthesis of proteins and how the synthesis of proteins could be regulated. Their work, and the research of many scientists all over the world, showed how various sugars and amino acids could turn specific gene systems on or off and drastically alter the physiology of a cell. The nature of the DNA's genetic code was worked out in the early 1960s by Marshall Nirenberg, Heinrich Matthaei, and Har Gobind Khorana. Also, during the late 1960s, a large number of researchers were determining how the genetic information (DNA) directed the synthesis of proteins □ and how the antibiotics blocked cellular processes.

209

During the 1960s and 1970s, our idea of how cells obtain their energy was revolutionized almost singlehandedly by Peter Mitchell. He proposed that some bacteria generate electrical potentials across their cytoplasmic membranes and then use the electrical potential to concentrate nutrients, 183 propel the cell, or synthesize energy □.

In the 1970s and early 1980s, microbiologists learned how to cut DNA with specific enzymes and integrate and exchange pieces of DNA in the test tube. Microbiologists are now able to put almost any gene into a bacterial cell and produce billions of copies of it. This new science, known as genetic engineering, has already made possible the synthesis of lifesaving products such as insulin, interferon, a foot and mouth disease vaccine, and a hepatitis 262 vaccine. In the future, new products made by genetic engineering □ will save countless human lives.

The tremendous amount of knowledge gained by studying microbial genetics and physiology and the powerful tool of genetic engineering are providing a solid foundation for an understanding of the genetics, physiology, and evolution of higher organisms. Prominent scientists who have contrib-

TABLE 1-3
NOBEL LAUREATES WHO CONTRIBUTED TO MICROBIOLOGY

NOBEL LAUREATES	YEAR	CONTRIBUTION
Emil von Behring	1901	Developed a diphtheria antitoxin.
Ronald Ross	1902	Discovered that the malaria parasite entered the human host from the bite of certain mosquitos.
Robert Koch	1905	Proved that a bacterium was the cause of tuberculosis and developed techniques for diagnosing the disease.
Charles Laveran	1907	Discovered that protozoans can cause human disease.
Paul Ehrlich and Elie Metchnikoff	1908	Hypotheses and studies on the immune response.
Charles Richet	1913	Described one type of allergic response, called anaphylaxis.
Jules Bordet	1919	Discovered and studied the complement system.
Charles Nicolle	1928	Studied the cause of typhus.
Karl Landsteiner	1930	Described the ABO scheme for grouping human blood.
Gerhard Domagk	1939	Discovered the antimicrobial sulfa drug protonsil.
Alexander Fleming, Ernst Chain and Howard Flory	1945	Discovered penicillin and how to isolate it.
James Sumner, John Northrup, and Wendell Stanley	1945	Crystallized enzymes and viruses.
Max Theiler	1951	Developed a yellow fever vaccine.
Selman Waksman	1952	Discovered and isolated the antibiotic streptomycin.
Frits Zernicke	1953	Developed the phase contrast microscope.
John Enders, Thomas Weller, and Frederick Robbins	1954	Developed culture techniques for the polio virus.
Edward Tatum, George Beadle and Joshua Lederberg	1958	Studied the relationship between genes and biochemical defects in microorganisms.
Severo Ochoa and Arthur Kornberg	1959	Isolated and characterized enzymes that synthesized nucleic acids.
Macfarlane Burnet and Peter Medawar	1960	Studied immunological tolerance.
Francis Crick, James Watson, and Maurice Wilkins	1962	Proposed a structure for DNA.
Francois Jacob, Andre Lwoff, and Jacques Monod	1965	Proposed a model for the control of enzyme synthesis.

TABLE 1-3 CONTINUED

NOBEL LAUREATES	YEAR	CONTRIBUTION
Peyton Rous and Charles Huggins	1966	Discovered a cancer-producing virus and discovered that hormones could be used to treat some cancers.
Robert Holley, Har Gobind Khorana and Marshall Nirenberg	1968	Determined how tRNA was involved in protein synthesis and what the genetic code was.
Max Delbruck, Alfred Hershey and Salvador Luria	1969	Studied the genetics of bacterial viruses.
Earl Sutherland	1971	Discovered the effects of cAMP in bacteria and in eukaryotes.
Gerald Edelman and Rodney Porter	1972	Studied the structure of antibodies.
Albert Claude, George Palade and Christian De Duve	1974	Elucidated the structure and function of eukaryotic organelles.
Renato Dulbecco	1975	Studied DNA tumor viruses.
Howard Temin and David Baltimore	1975	Studied RNA tumor viruses and discovered reverse transcriptase.
Carlton Gajdusek and Baruch Blumberg	1976	Studied the epidimiology of slow virus diseases. Discovered the relationship between Australian antigen and hepatitis B virus.
Werner Arber, Hamilton Smith, and Daniel Nathans	1978	Discovered restriction enzymes and used them to map genes on DNA molecules.
Peter Mitchell	1978	Proposed the chemiosmotic hypothesis to explain oxidative phosphorylation.
Baruj Benacerraf, George Snell, and Jean Dausset	1980	Studied the gentic regulation of the immune response.
Paul Berg, Walter Gilbert, and Fredrick Sanger	1980	Developed various techniques to sequence nucleic acids.
Barbara McClintock	1983	Discovered mobile genetic elements.
Niels Jerne, George Koehler, and Cesar Milstein	1984	Studied the development and control of the immune system. Developed techniques for production of monoclonal antibodies.

uted significantly to our knowledge of life are sometimes awarded the Nobel Prize. Nobel laureates who have been honored for their work in the fields of chemistry and medicine or physiology are listed in table 1-3, along with a brief summary of their accomplishments.

SUMMARY

MICROBIOLOGY: A MULTIFACETED SCIENCE

1. Microbiology is often subdivided into fields of study based upon various types of microorganisms. This subdivision is sometimes known as the taxonomic approach to microbiology. Virology, bacteriology, mycology, phycology, and protozoology are studies that are concerned with viruses, bacteria, fungi, algae, and protozoa, respectively.
2. Microbiology is often subdivided into fields of study based upon microbial activity. This subdivision may be referred to as the functional approach to microbiology. Microbial ecology, industrial microbiology, agricultural microbiology, medical microbiology, physiology, genetics, and biochemistry are subdivisions of microbiology.

ANCIENT IDEAS OF DISEASE AND PUTREFACTION

1. For most of recorded human history, disease has generally been regarded as a punishment from gods or demons. Humans have also known, since very ancient times, that some diseases were spread from person to person, on solid materials (fomites), and through the air.
2. Some of the oldest public health codes are found in the Bible in the Book of Numbers and in the Book of Deuteronomy. These laws, attributed to Moses, attempted to establish how subjugated peoples should be dealt with after a war so as not to bring disease into the Israeli nation. The laws also spelled out which land animals and water creatures could be eaten and which could not. It is believed that

these restrictions were established in order to reduce the spread of parasites and disease.

3. Girolamo Fracastoro proposed in a manuscript (1546) that disease and some types of putrefaction were due to things he called "contagions," which were transferred by direct contact, by touching inanimate objects (fomites), or through the air.
4. Edward Jenner (1798) developed a smallpox vaccine from material taken from cowpox lesions. Soon after his discovery, smallpox vaccination was commonly practiced in England, Europe, and the United States.
5. Ignaz Semmelweis (1850) proposed that childbed fever was due to "putrefying organic matter." He believed that doctors were picking up this material from corpses they were studying and transmitting it to new or expectant mothers.

THE DISCOVERY OF MICROORGANISMS

1. The person credited with first observing bacteria and other microorganisms through a microscope was Antony van Leeuwenhoek. In the late 1600s and early 1700s van Leeuwenhoek published a number of papers in which he described bacteria, fungi, protozoans, algae, blood cells, and sperm.
2. Van Leeuwenhoek used a simple microscope that resembled a magnifying lens to observe his "animalcules."

THE SPONTANEOUS GENERATION CONTROVERSY

1. Francesco Redi showed with a simple experiment that maggots do not arise spontaneously from rotting meat, but instead arise from eggs laid by flies.
2. Spallanzani and Schwann demonstrated that boiled infusions remain sterile until contaminated with air.
3. Louis Pasteur showed that boiled infusions remain sterile as long as airborne microorganisms do not contaminate the infusions. His experiments demonstrated that microorganisms do not arise by spontaneous generation in sterile infusions.
4. John Tyndall was able to sterilize hay infusions containing heat-resistant bacteria by alternately heating and cooling the infusions. The sterilization process became known as tyndallization. Ferdinand Cohn discovered that the heat-resistant bacteria were those that produced endospores.

THE GERM THEORY OF DISEASE

1. In the 1800s, scientists believed that putrefaction and infectious diseases were due either to (a) decomposing substances that imparted their characteristics to living or nonliving things, or (b) microorganisms carrying out normal metabolic processes on living or nonliving things.
2. Not until the 1880s was it generally accepted that microorganisms were responsible for infectious diseases and putrefaction. The French microbiologist Louis Pasteur and the German physician Robert Koch were largely responsible for convincing the scientists of the 1880s that microorganisms have a central role in fermentations, putrefactions, and infectious diseases.
3. Louis Pasteur demonstrated that specific organisms were responsible for a particular type of fermentation, and that microorganisms which frequently appeared in infusions were not arising spontaneously but were contaminants from the air. Pasteur's experiments showed that it was possible to study specific microorganisms and to determine their effects on the environment and on living organisms. He succeeded despite the fact that he did not use pure cultures, but only enrichment cultures in which one microorganism predominated. Pasteur's experiments and conclusions stimulated scientists all over the world to realize the importance of microorganisms in fermentations, putrefactions, and infectious diseases.
4. Robert Koch, together with some of his students and colleagues, developed many of the basic techniques, media, and equipment necessary for handling and studying microorganisms. His laboratory is credited with the development of agar-solidified media (Hesse), covered culture dishes (Petri), and staining techniques (Koch and Enrlich). In addition, Koch demonstrated convincingly that the bacteria *Bacillus anthracis* and *Mycobacterium tuberculosis* were responsible for anthrax and tuberculosis, respectively. The technical developments credited to Koch's laboratory allowed many other scientists to purify and begin studying microorganisms in a sensible way.
5. The rules for establishing whether or not an organism is responsible for a disease, now known as Koch's postulates, helped develop microbiology into a systematic and believable science.

SIGNIFICANT DISCOVERIES IN MICROBIOLOGY

1. Louis Pasteur is responsible for the development of vaccines against rabies, anthrax, chicken cholera, and swine erysipelas.
2. Paul Ehrlich is credited with one of the first good models to explain immunity. He is also responsible

for the development of an arsenic compound, called Salvarsan, that was useful in treating syphilis.

3. During a study of water fleas, Elie Metchnikoff discovered that phagocytosis was an important part of the immune system.
4. Emil von Behring developed an antitoxin vaccine against the diphtheria toxin.
5. Sergei Winogradsky studied the sulfur, nitrifying, and nitrogen-fixing bacteria. He clarified their role in chemical conversions of sulfur and nitrogen in nature. He postulated that the sulfur and nitrifying bacteria obtained energy by converting H_2S to SO_4^{-2} and NH_3 to NO_3^-, respectively and that they used CO_2 as their sole source of carbon.
6. Martinus Beijerinck investigated various nitrogen-fixing bacteria. He showed that the bacteria-like structures found within root nodules on legumes were bacteria of a single genus.
7. William Twort and Felix-Hubert d'Herelle are credited with the discovery of bacterial viruses, which d'Herelle called bacteriophagium (now called bacteriophage).
8. Joseph Lister obtained the first pure culture of a bacterium in 1878, but is most famous for his use of carbolic acid as an antiseptic during surgery.
9. In 1944 Oswald Avery, Colin MacLeod, and Maclyn McCarty demonstrated that the transforming material was DNA.
10. James Watson and Francis Crick in 1953 published a model for the structure of DNA and how its replication might occur.
11. In 1960 Francois Jacob and Jacques Monod presented a model for how genes might be regulated in bacteria and in viruses.
12. Marshall Nirenberg, Heinrich Matthaei, and Har Gobind Khorana are credited with working out the genetic code during the early 1960s.
13. During the 1960s and 1970s, Peter Mitchell demonstrated that some bacteria generate electrical potentials across the cytoplasmic membrane. Bacteria use this potential to concentrate nutrients, rotate flagella, and synthesize ATP.

STUDY QUESTIONS

1. Discuss two nonsupernatural explanations for infectious diseases. Who were the early supporters of these ideas?
2. What is the current scientific explanation for infectious diseases?
3. Outline major historical moments in the understanding and control of infectious diseases. Be sure to consider Fracastoro, Jenner, Semmelweis, De Bary, Pasteur, Koch, Lister, Ehrlich, Domagk, Fleming, and Waksman.
4. Explain the idea of spontaneous generation and why so many people believed in it. Consider the role that Redi and Pasteur played in disproving spontaneous generation.
5. Did Pasteur prove that life could not have developed naturally on earth? Explain.
6. Explain the germ theory of disease and putrefaction. Did Fracastoro, Jenner, or Semmelweis believe that microorganisms were responsible for infectious diseases and putrefaction?
7. Describe Pasteur's and Koch's contributions to the advancement of microbiology.
8. List and explain Koch's postulates. How do you prove that an organism is responsible for a disease if you cannot fulfill one or more of Koch's postulates?
9. List Ehrlich's and Metchnikoff's contributions to the science of immunology.
10. Describe Winogradsky's and Beijerinck's contributions to microbiology and the science of ecology.
11. What discoveries in microbiology are attributed to the following persons? Van Leeuwenhoek; von Behring; Iwanowski; Beijerinck; Twort; d'Herelle; Avery, MacLeod, and McCarty; Watson and Crick; Jacob and Monod; Nirenberg and Khorana; Mitchell.

SUPPLEMENTAL READINGS

Bendiner, E. 1980. Ehrlich: Immunologist, chemotherapist, prophet. *Hospital Practice 15(11):* 129–157.

Brock, T. D. 1961. *Milestones in microbiology.* Englewood Cliffs, N. J.: Prentice-Hall Biological Sciences Series.

Bulloch, W. 1938. *The history of bacteriology.* London: Oxford University Press.

Cartwright, F. and Biddiss, M. 1972. *Disease and history.* New York: Thomas Y. Crowell Company.

De Kruif, P. 1926. *Microbe hunters.* New York: Harcourt, Brace & World.

Dobell, C. 1932. *Antony van Leeuwenhoek and his little animals.* London: Constable and Co., Ltd.

Groschel, D. 1982. The etiology of tuberculosis: A tribute to Robert Koch on the occasion of the centenary of his discovery of the tubercle bacillus. *American Society for Microbiology News* 48(6):248–250.

Hoyle, F. and Wickramasinghe, N. C. 1979. *Diseases from space.* New York: Harper & Row.

Lechevalier, H. and Solotorovsky, M. 1965. *Three centuries of microbiology.* New York: McGraw-Hill.

Porter, J. R. 1972. Louis Pasteur sesquicentennial (1822–1972). *Science* 178:1249–1254.

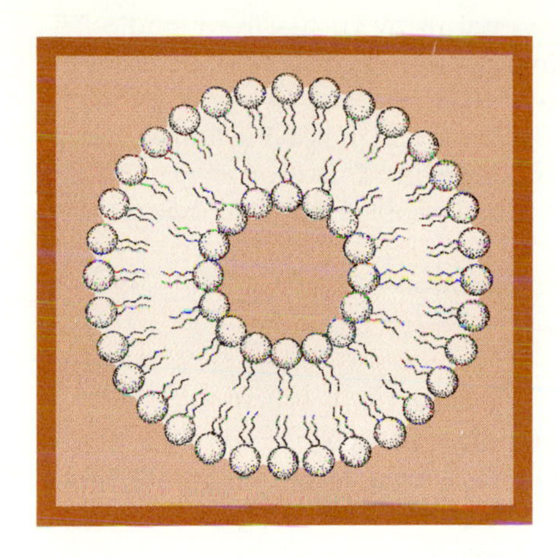

CHAPTER 2
SYNOPSIS OF BIOLOGICAL CHEMISTRY

CHAPTER PREVIEW

CHEMICAL REACTIONS BY MICROORGANISMS AND THE LEAVENING OF BREAD

Chemical reactions are a part of all life processes. Some chemical reactions are involved in transporting nutrients into the cell, while others produce the energy necessary for the cell to function. As a consequence of these chemical reactions, microorganisms break down and synthesize numerous chemicals. Generally, waste products are discharged into the environment. Some of these chemical reactions drastically change the environment.

The leavening of bread occurs because a waste product, carbon dioxide, is released by a microorganism, such as a yeast or bacterium. The rising of bread occurs because the leavening agent produces gas that makes the dough light and fluffy.

Bakers' yeast is used to leaven most breads (one notable exception is sourdough bread), coffee cakes, Danish pastries, and a variety of dinner rolls. This yeast produces the gas carbon dioxide and grain alcohol (ethanol), using the sugar in the dough as its source of energy and carbon. The ethanol may contribute to the aroma and flavor of the baked bread.

Sourdough breads are made in much the same way as other leavened breads, except that the leavening agents also include **lactic acid bacteria.** These bacteria produce lactic acid and carbon dioxide, instead of ethanol, from their energy-producing chemical activities. The carbon dioxide causes the dough to rise, while the lactic acid gives the sourdough bread its characteristic sour, tart flavor.

Because chemical reactions constitute the bases for all life processes, it is important to be familiar with basic chemical principles in order to appreciate microbiology and the impact that this science has on human affairs. In this chapter we will review some basic chemical principles and describe chemical substances that are frequently associated with life processes.

LEARNING OBJECTIVES

A STUDY OF THIS CHAPTER SHOULD ENABLE YOU TO:

DRAW THE STRUCTURE OF ATOMS AND EXPLAIN HOW THEY COME TOGETHER TO FORM MOLECULES

DESCRIBE THE DIFFERENCES AMONG COVALENT, IONIC, AND HYDROGEN BONDS

EXPLAIN WHAT ORGANIC AND INORGANIC COMPOUNDS ARE

DESCRIBE THE VARIOUS MACROMOLECULES IN THE CELL, THEIR COMPOSITION AND FUNCTION

EXPLAIN WHAT A CHEMICAL REACTION IS

DESCRIBE THE STRUCTURE AND FUNCTION OF ENZYMES

Much of our present knowledge of microorganisms comes from studies involving their chemistry. An understanding of basic chemical principles helps us appreciate the remarkably ordered life processes that take place in the cell and the influence of microorganisms on living and nonliving things everywhere. This chapter introduces some fundamental chemical concepts that will help broaden your understanding of microorganisms and the role(s) they play in our world.

THE STRUCTURE OF MATTER

Matter is anything that has a mass and that occupies space. It is found throughout the universe and exists either in a solid, a liquid, or a gaseous state. In our environment, some substances can exist in all three states depending upon environmental factors. Water, for example, which has the chemical formula H_2O, can exist in any one of the three states depending upon temperature and pressure. At temperatures below 0°C, water generally freezes into a solid. When the temperature is between 0°C and 100°C, water is usually a liquid, while above 100°C it is a gas. Most of the water inside cells is in a liquid state. The state of water in or around the cell determines whether the cell can carry out life processes. Since almost all biological processes take place in an aqueous environment, any change in the state of the water usually stops life processes. For example, if the water around the cell becomes frozen, cell multiplication ceases because there is no available water for the cell to carry out its necessary activities.

If two different substances (in the same or different states) are mixed, the resulting **mixture** may exist as a) a **suspension,** b) a **solution,** or c) a **colloid** (fig. 2-1). **Suspensions** are heterogeneous mixtures in which particles are scattered throughout a solid, a liquid, or a gas for a relatively short period of time, after which the particles will settle out of the mixture. In these types of mixtures, the suspended solid does not alter the chemical properties of the suspending medium. Microorganisms often store their reserve nutrients as suspensions for long periods of time, without affecting the functioning of the cells in any adverse way.

Solutions are homogeneous mixtures between two or more substances in which one substance, the **solvent,** separates the other substances, the **solutes,** which disperse themselves evenly throughout the solvent. For example, if a tablespoon of salt (NaCl) is mixed in a glass of water, the components of the salt (Na^+ and Cl^-) become evenly distributed throughout the water and change the chemistry of the liquid. The salt is called the solute and the water the solvent. The resulting solution has properties that neither the salt nor the water has on its own. Most of the chemical activities that take place in the cell occur in solutions where water is the solvent.

Colloids are homogeneous mixtures of very small particles that, because of their size, remain evenly dispersed throughout a medium. Examples of colloids are smoke, gelatin, homogenized milk, and fog. Raw milk, if allowed to stand undisturbed, will separate into fats and liquid. The fat in raw milk is suspended in the milk liquid as it is produced by the mammal. When the milk fat is broken up into tiny globules during homogenization, it forms colloidal particles that remain dispersed in the liquid. The cell cytoplasm (the matrix that makes up the cell) is largely colloidal, with large molecules (macromolecules) dispersed in the cell water.

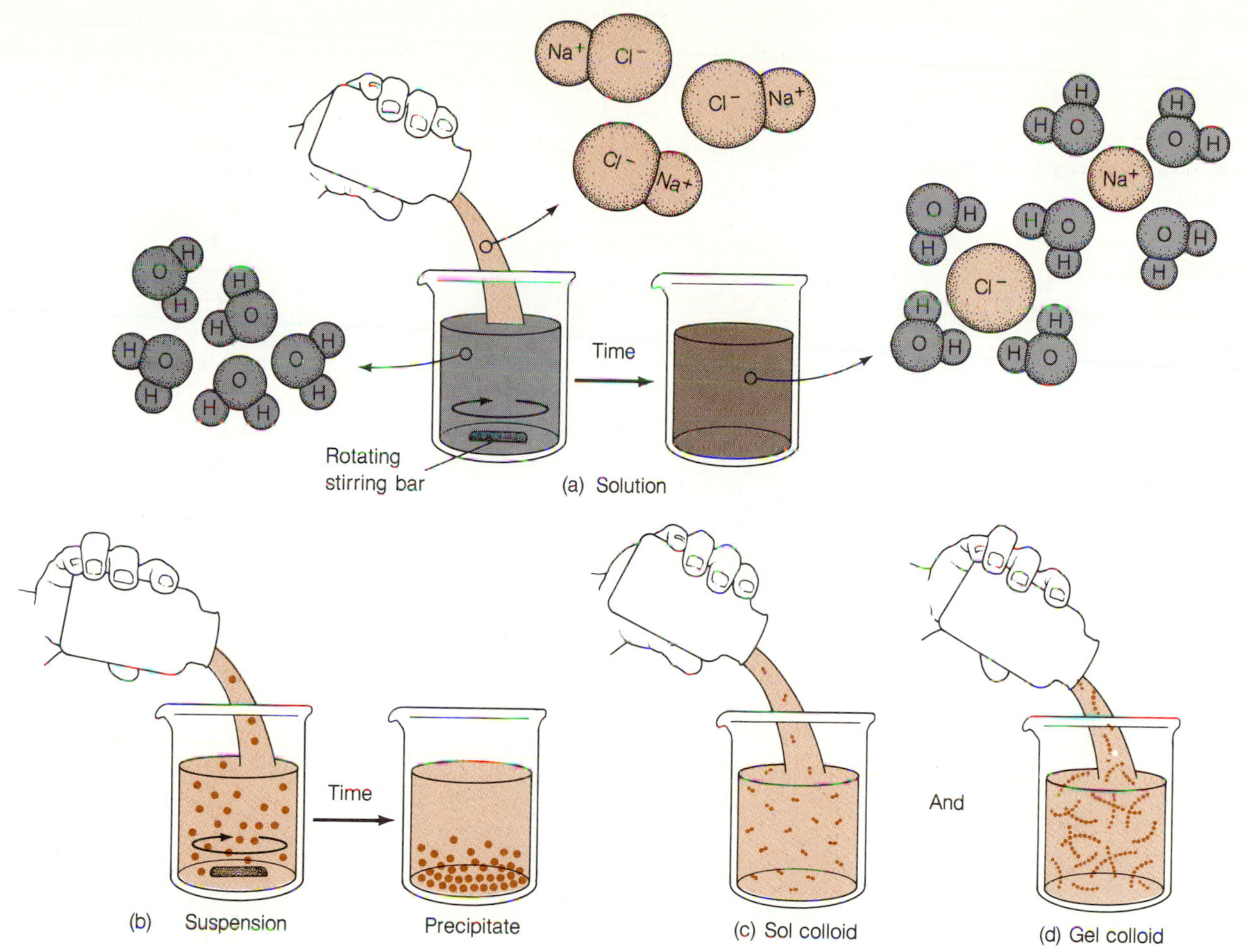

Figure 2-1 The Three Different Types of Mixtures. *(a) A solution.* In solutions, the solute particles are separated by the solvent and become evenly distributed. The solute particles do not settle out. *(b) A suspension.* In suspensions, large particles become suspended while they are agitated. After agitation stops, the particles settle out. *(c)* and *(d) Colloids.* In colloidal mixtures, particles are homogeneously dispersed in a liquid medium for extended periods.

Atoms and Ions

Atoms (fig. 2-2) are uncharged units of matter constructed of three types of particles: **protons, neutrons** and **electrons.** The single exception to this definition is the hydrogen (H) atom, which consists of a single proton and a single electron. **Protons** are particles of atomic matter with a positive charge and a mass of 1.673×10^{-24} grams (g), while **neutrons** have no electrical charge and a mass of 1.675×10^{-24}. Every atom consists of a dense core, called the **nucleus,** and orbiting electrons. The atomic nucleus, because it contains protons, is positively charged. **Electrons** have a negative charge and a mass of 9.110×10^{-28}. Since the mass of an electron is 1/1837 that of a proton or neutron, it contributes little to the total mass of the atom. The number of electrons orbiting around the nucleus equals the number of protons in the nucleus. These two opposing electrical charges offset each other, making the atom neutral in charge. Figure 2-2 illustrates a number of simplified atoms and the relationships among the various subatomic particles. From this figure it can be seen that each atom has its own unique number of electrons and protons. Table 2-1 illustrates that each type of atom has an

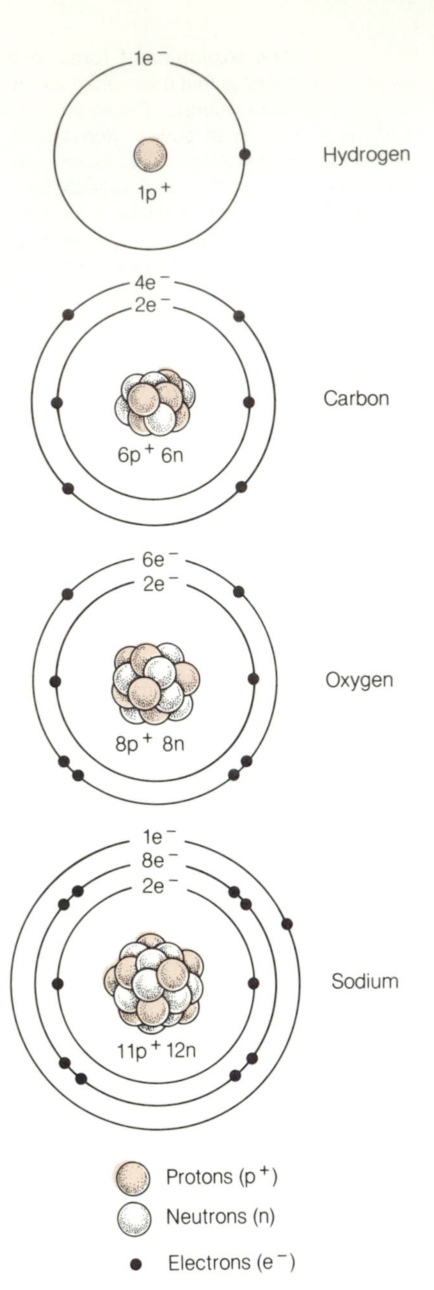

Figure 2-2 The Structure of Atoms. Atoms of hydrogen (H), carbon (C), oxygen (O), and sodium (Na) are illustrated. The atom consists of a nucleus surrounded by an electron cloud. The nucleus contains protons and neutrons and is positively charged. The electron cloud has a negative charge equal in strength to the positive charge of the nucleus. Hence, the atom is neutral in charge because the positive nucleus and the negative electron cloud offset each other. The electron cloud is arranged in orbitals around the nucleus. The outermost orbit may contain no more than eight electrons.

TABLE 2-1
COMMON ELEMENTS FOUND IN CELLS

ELEMENT	SYMBOL	ATOMIC NUMBER	APPROXIMATE ATOMIC WEIGHT
Hydrogen	H	1	1
Carbon	C	6	12
Nitrogen	N	7	14
Oxygen	O	8	16
Fluorine	F	9	19
Sodium	Na	11	23
Magnesium	Mg	12	24
Silicon	Si	14	28
Phosphorus	P	15	31
Sulfur	S	16	32
Chlorine	Cl	17	35
Potassium	K	19	39
Calcium	Ca	20	40
Chromium	Cr	24	52
Manganese	Mn	25	55
Iron	Fe	26	56
Cobalt	Co	27	59
Copper	Cu	29	64
Zinc	Zn	30	65
Selenium	Se	34	79
Iodine	I	53	127

assigned number. For example, hydrogen (H) is number 1, calcium (Ca) is number 20, and oxygen (O) is number 8. Such a number is called the **atomic number,** and it represents the number of protons in the nucleus of an atom. The atomic number also indicates the number of electrons orbiting around the nucleus, since this number equals the number of protons. Electrons exist around the nucleus in orbitals (fig. 2-2). The electrons in the outermost orbitals are extremely important, because they determine what chemical properties an atom has and with which other atoms it can combine.

When atoms gain or lose electrons, they become charged and are called **ions** (fig. 2-3). The difference between an atom and an ion can be illustrated by considering the sodium and chlorine atoms. The sodium (Na) atom (fig. 2-3) has 11 electrons: 2 in the first orbital, 8 in the second orbital, and 1 in the outermost (third) orbital. When the sodium atom donates the electron in its outer orbit to another atom, the sodium atom becomes a positively charged ion or **cation,** (Na^+) because the Na^+ still has 11 protons but only 10 electrons. When chlorine (Cl), which has 17 protons and 17 electrons, accepts an electron from another atom (e.g., Na), it becomes a negatively charged ion or **anion** (Cl^-) with 17 protons and 18 electrons. An understanding of ions is important because they are extremely common in living organisms and play essential roles in the proper functioning of the cell, including the passage of materials in and out of the cell and many energy-producing chemical reactions.

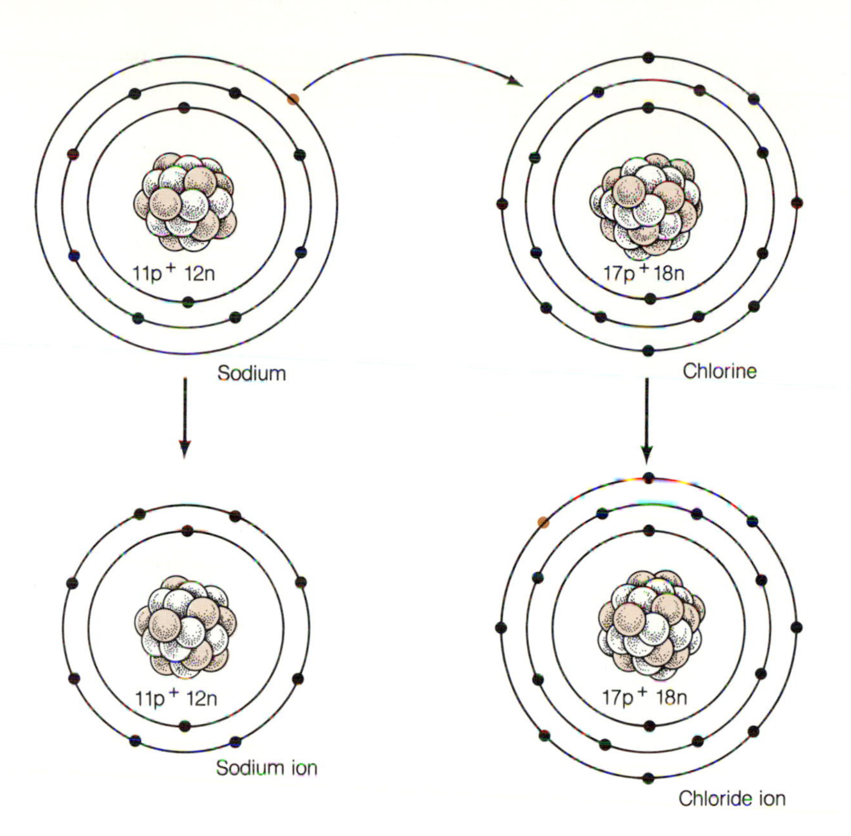

Figure 2-3 The Structure of Ions. Ions are charged particles that arise when atoms donate or accept electrons. The sodium ion (Na^+) is formed when sodium donates an electron to another atom, such as chlorine (Cl). The sodium ion is positively-charged because it has one more proton than electrons. The chloride ion (Cl^-) is negatively-charged because it has an extra electron in its outermost shell.

Elements

An **element** is generally thought of as a pure substance that consists of atoms, all of which have the same atomic number. Presently, there are 109 different elements that cannot be chemically purified into simpler substances. Ninety-two of the elements are naturally occurring substances and 17 are relatively unstable, synthetic ones. Of the many elements that can be found in living cells (table 2-2), hydrogen (H), carbon (C), oxygen (O), nitrogen (N), phosphorus (P) and sulfur (S) constitute more than 99% of all atoms found in living matter.

TABLE 2-2
APPROXIMATE ABUNDANCE OF CHEMICAL ELEMENTS FOUND IN BACTERIAL CELLS

ELEMENT	SYMBOL*	ATOMS (%)**	WEIGHT (%)
Carbon	C	9.5	20
Nitrogen	N	1.4	3
Oxygen	O	25.5	62
Hydrogen	H	63.0	9
Phosphorus	P	0.1	1.1
Sulfur	S	0.05	0.1
Sodium	Na	0.005	0.1
Potassium	K	0.005	0.1
Magnesium	Mg		0.07
Calcium	Ca		0.06

*Each element is assigned a 1–2 letter symbol.
**Percent of dry weight of cell mass of *Escherichia coli.*

Elements can combine with other elements to form **compounds.** The properties of the individual elements are lost and the resulting compound has new properties. For example, H_2 and O_2 are two elements that are present in the earth's atmosphere as gases. When these two elements combine to form water (H_2O), however, the resulting compound has new properties and can be either a liquid, solid or a gas depending upon the temperature.

Isotopes and Radioisotopes

The **mass number** of an atom is the sum of the number of protons and neutrons in the nucleus. Although all atoms that make up an element have the same number of protons, the atoms may have different numbers of neutrons (or different mass numbers). Various atoms of an element that have different numbers of neutrons are called **isotopes.** For example, all the atoms in the element hydrogen (H) have 1 proton (atomic number of 1) and 1 electron, but the atoms may have 0, 1, or 2 neutrons. The isotope of hydrogen with zero neutrons is called **protium.** This is the most common isotope of H in the universe. The isotope of H called **deuterium** has 1 neutron, 1 proton, and 1 electron, while the isotope called **tritium** has 2 neutrons, 1 proton, and 1 electron. Tritium is a **radioactive** isotope, because its nucleus gives off particles. Radioactivity consists of energy and/or subatomic particles released by unstable atoms due to transformations occurring within the nucleus (table 2-3). Radioactivity converts unstable nuclei to stable nuclei. For example, when the radioactive isotope tritium (H-3) releases an electron (a beta particle) from its nucleus, it becomes the stable helium isotope (He-3).

Chemical Bonds and Molecules

When two or more atoms or ions are bonded together in a stable fashion, they form a **molecule.** For example, the joining of H and O atoms in the ratio of 2:1 results in the formation of a water molecule, while the association of Na^+ and Cl^- ions results in table salt (NaCl). Very complicated chemical molecules consisting of many thousands of atoms are found in living organisms. We will consider some of these molecules in the following sections of this chapter.

There are basically two types of bonds (table 2-4) that atoms and ions make with each other. These are **ionic bonds** and **covalent bonds.** (There are other types of weak attractive forces between atoms, such as **hydrogen bonds,** which are discussed later.) **Ionic bonds** result from strong electrical

TABLE 2-3
PROPERTIES OF RADIOACTIVE PARTICLES

TYPE OF DECAY	CHARGE	MASS*	DESCRIPTION
Alpha	+2	4	Helium nucleus
Beta	−1	1/1800	Electron
	+1	1/1800	Positron
Gamma	no charge	0	High-energy radiation

*Mass is expressed in atomic mass units (AMU). See page glossary for definition of AMU.

TABLE 2-4

TYPES OF CHEMICAL BONDS

TYPE OF BOND	DESCRIPTION	EXAMPLE
Covalent bond	One or more pairs of electrons shared between nuclei of bonded atoms	CH_4 (H—C—H with H above and below)
Ionic bond	Ions with opposite charges attract each other	Na^+Cl^-
Hydrogen bond	Electrostatic attractions between 2 partially-charged molecules	δ^-O—Hδ^+ ---- δ^-O (water molecules: Hδ^+, Oδ^-, H-bond H----O)
Van der Waals forces	Weak forces developing from atom-atom interactions (repelling and attractive)	

attractions between two or more ions (fig. 2-4). For example, NaCl results from an ionic bond between the Na^+ and the Cl^- (ions). A specific distribution of ions both inside and outside the cell is absolutely essential in order for the cell to function properly.

Covalent bonds result when two atoms share the electrons in their outer orbitals (fig. 2-5). Covalent bonds are strong and relatively stable. Because of this property, most biological molecules that provide structural integrity, or require stability for their activity, are made up of covalently-bonded atoms (table 2-5).

Molecules constructed of covalently bonded carbon atoms are known as **organic molecules** and are very common in living organisms. Large organic

FOCUS

RADIOISOTOPES AND THEIR USE IN MEDICINE

Radioactive atoms (radioisotopes) give off energy and/or particles that often interact with other forms of matter. Sometimes the radiation given off by radioisotopes can interact with other atoms and cause their ionization. Such ionization may be detected by using appropriate electronic detectors (counters), special films, or techniques designed to measure or detect the energy given off by these isotopes. Medical science has taken advantage of such techniques, and has capitalized on the fact that certain chemical elements have an affinity for certain organs of the body.

A radioactive isotope of iodine (I-131) is used to treat overactive thyroid glands. Since the thyroid takes up iodine selectively, the isotope accumulates in this gland, where it gives off radioactive particles that kill the reproducing cells. This method of treatment (often called the "atomic cocktail") obviates the need for surgical removal of much of the thyroid.

Many tumors and cancerous growths are treated by irradiating the affected area with radium-226, cobalt-60, or phosphorus-32. Sometimes "atomic cocktails" are given not to destroy diseased tissue, but simply to detect the presence of diseased tissue in various organs of the body. For example, copper-64 is used to detect Wilson's disease (a copper storage disease); xenon-133 is used to perform lung scans; and selenium-75 is used to carry out pancreas scans.

Although the term "radioactivity" is generally taken to mean the potential for severe tissue destruction and/or death, medical science is learning to harness radioactivity to improve the quality of human life.

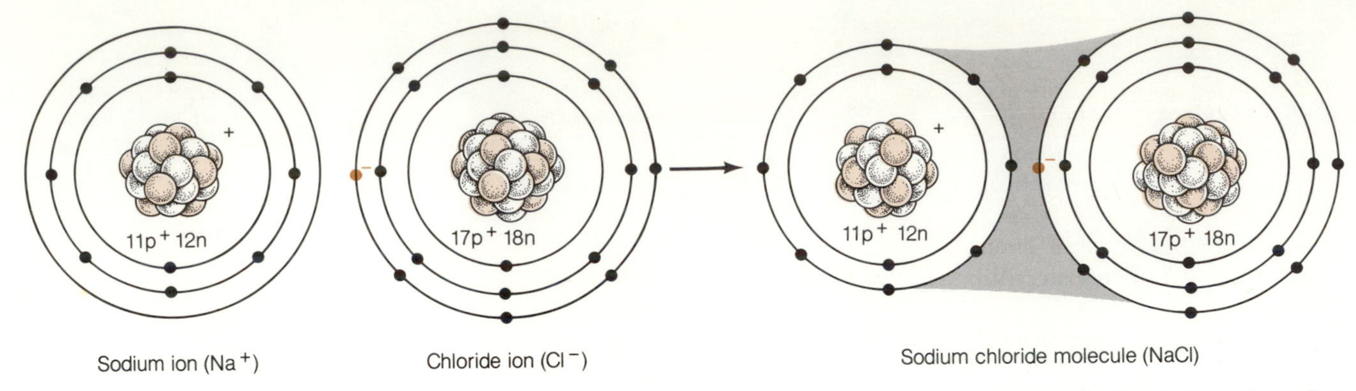

Figure 2-4 **The Formation of an Ionic Bond.** Ionic bonds form when ions of opposite charge attract each other. Table salt (NaCl) forms when sodium and chloride ions attract each other to form the sodium chloride molecule.

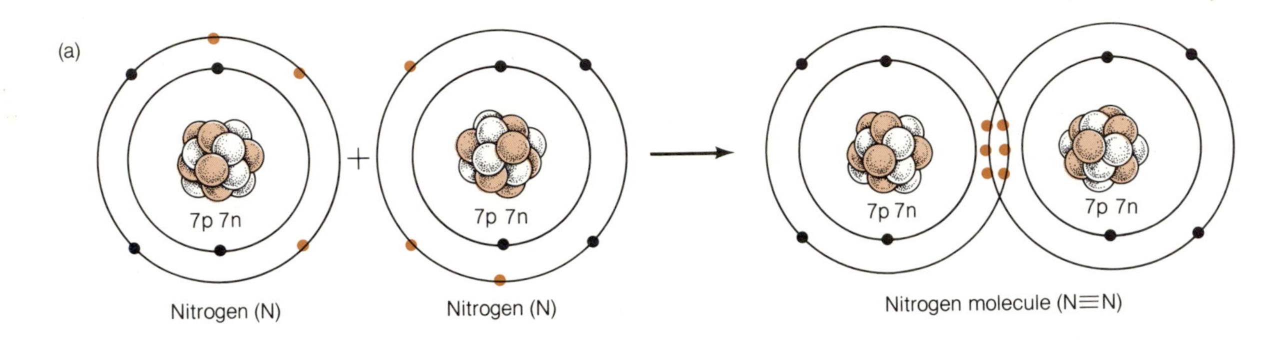

Figure 2-5 **The Formation of Covalent Bonds.** Covalent bonds form when two atoms share their electrons. Sharing electrons insures that each atom has a full outer orbital, hence enhancing its stability. Covalent bonds usually form between nonmetallic atoms. *(a)* Nitrogen gas forms when two atoms of nitrogen are covalently bonded. *(b)* Methane (natural gas) forms when one carbon atom and four hydrogen atoms all share their electrons.

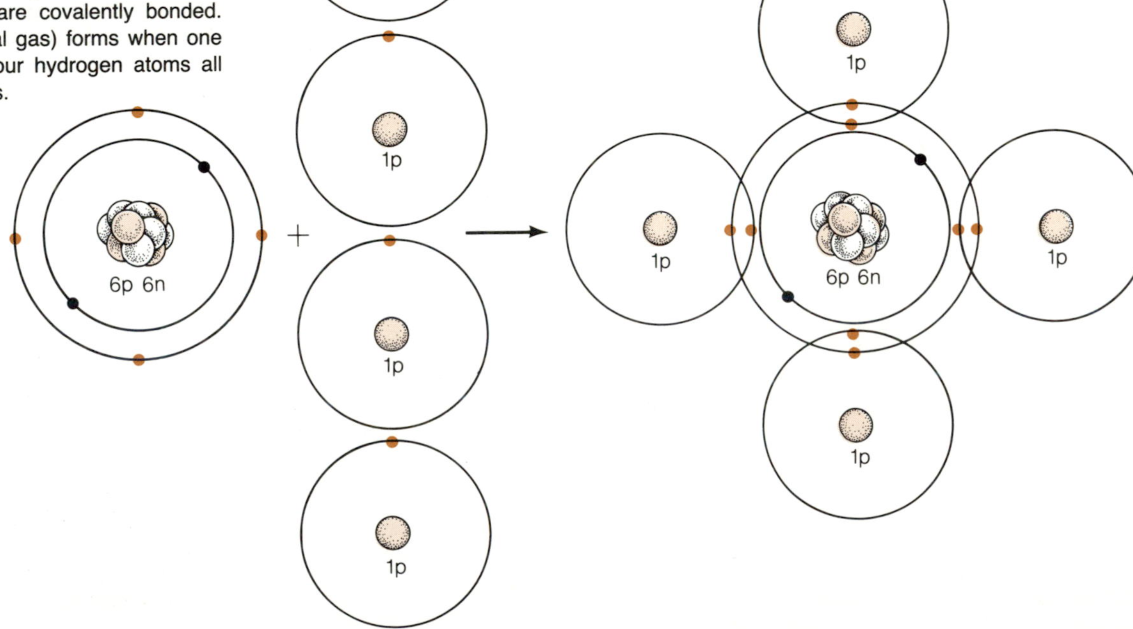

TABLE 2-5
SOME COMMON COVALENT BONDS SEEN IN MACROMOLECULES

MACROMOLECULE	MONOMERIC UNIT	TYPE BOND	ILLUSTRATION
Protein	Amino acid	Peptide	—C(=O)—NH—C—
Nucleic acid	Nucleotide	Ester	—O—P(=O)(O—H)—O—CH₂—H
Polysaccharide	Monosaccharide	Glycosidic	C—C—O—C—O—C—C
Triglyceride (fat)	Fatty acids and glycerol	Ester	H—C—O—CO—C . . . H—C—O—CO—C . . . H—C—O—CO—C . . .

molecules in nature occur almost exclusively because of biological activities. The branch of chemistry that studies these carbon compounds is called **organic chemistry.** Because carbon atoms share their outer electrons with other atoms in a number of ways, a variety of linear, branched, or ring-shaped organic molecules can be made (fig. 2-6). This variety of carbon compounds is the basis for the diversity of molecules we see in living systems.

When some atoms form covalent bonds, they do not share their bond electrons evenly. For example, the oxygen atom in a water molecule (fig. 2-7) is said to be electronegative because it has a high affinity for electrons; therefore, electrons tend to spend more time orbiting around the electronegative (oxygen) nucleus than around the electropositive (hydrogen) nuclei. As a consequence, the portion of the molecule with the electronegative nucleus has a slightly negative charge, while the hydrogen atoms have a slightly positive charge. This situation can result in a type of interaction commonly called **hydrogen bonding.** For example, each water molecule forms a num-

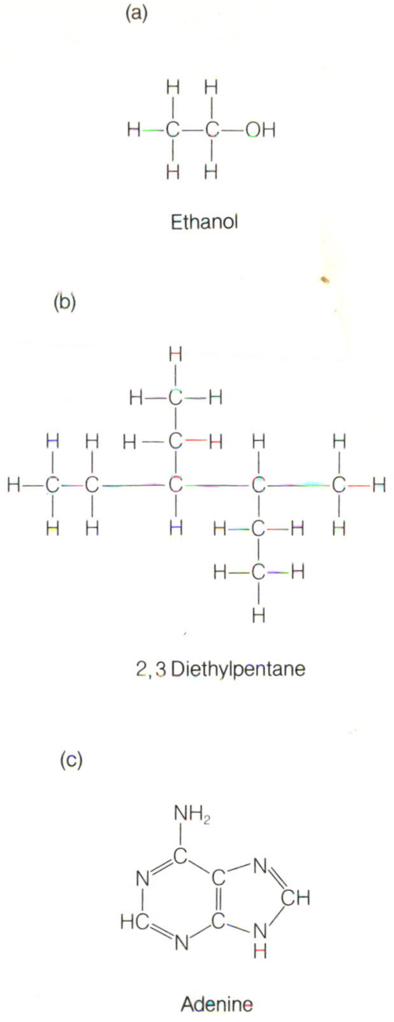

(a)

Ethanol

(b)

2,3 Diethylpentane

(c)

Adenine

Figure 2-6 Examples of Molecules Found in Living Cells. *(a)* Linear molecule. *(b)* Branched molecule. *(c)* Ringed molecule.

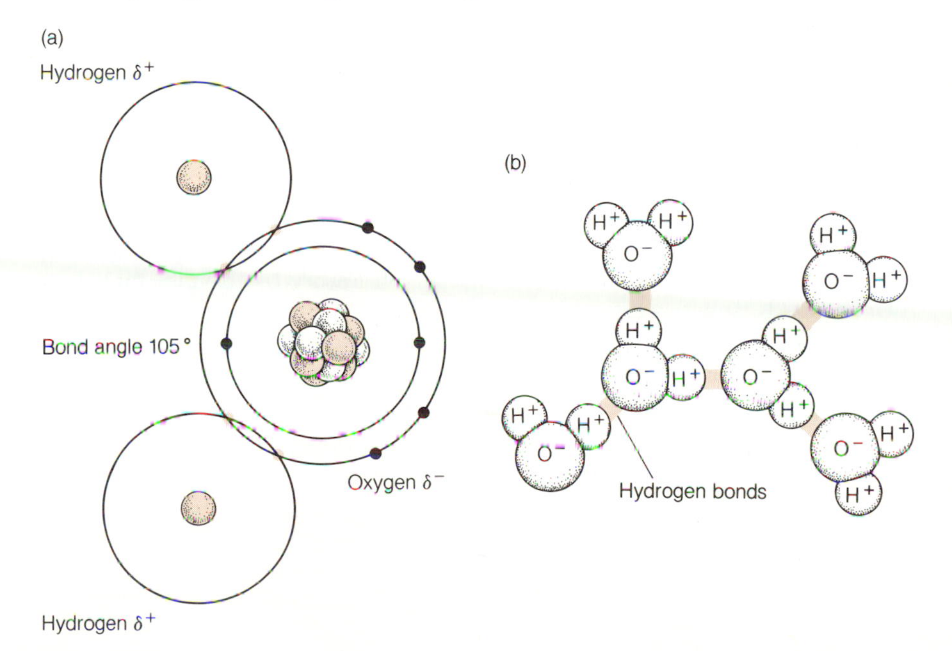

Figure 2-7 Hydrogen Bonding in Water Molecules. *(a)* A water molecule is formed when one atom of oxygen is covalently bonded to two hydrogen atoms. The atoms are bonded to each other forming a molecule with a 105° angle. The oxygen nucleus attracts the shared electrons. Because the electrons are attracted to the oxygen nucleus, it has a partial negative charge, leaving the hydrogen atoms with partial positive charges. *(b)* Water molecules orient themselves so that positively-charged regions of the molecule are in close proximity to the negatively-charged regions. This interaction results in weak attractive forces called hydrogen bonds.

ber of hydrogen bonds in water (fig. 2-7). Since oxygen is highly electronegative, it attracts electropositive H atoms on nearby water molecules. Thus, water molecules attract each other and create a lattice network of molecules that are joined together by hydrogen bonds (fig. 2-7).

In general, hydrogen bonds are weak attractive forces that occur between chemical groups, in which H atoms are covalently bonded to electronegative atoms, usually N or O. Hydrogen bonding is extremely important in living organisms because it determines many of the interactions molecules can have with each other. Hydrogen bonding also determines the shape of some molecules (fig. 2-8).

Atomic Weight and Molecular Weight

The term **atomic weight** is sometimes used by chemists to express the mass of a given element. It is defined as the average mass of all isotopes of the element as they occur in nature. Every element has an atomic weight (table 2-1). For example, the atomic weight for the element H is 1.0079, and that for oxygen is 15.9994. The mass of the element is usually expressed in **atomic mass units** (AMU), where 1 AMU = 1.660×10^{-24} g. This figure is 1⁄12 the mass of an atom of the carbon isotope with a mass number of 12. **Molecular weight** is a measurement of the mass of a molecule. It is calculated by adding the atomic weights of all the atoms in a molecule. For example, the molecular weight of water (H_2O) is 18. This is determined by adding the atomic weight of each of two hydrogen atoms ($2 \times 1 = 2$) to that of one oxygen atom (16).

ACIDS, BASES, AND SALTS

Acids, bases, and **salts** are substances that are used as nutrients or discharged as waste products by some microorganisms. Often, it is essential to know what these molecules are and how they are made in order to understand microbial processes.

Acids are chemicals that can release H^+ (protons) in aqueous (water) solutions. For example, hydrochloric acid, HCl, dissociates to produce H^+ and Cl^- (ions) when mixed in water. Acids commonly encountered are usually dissolved in water, and the hydrogen ions are bound to water molecules, forming hydronium ions (H_3O^-). These ions impart a sour or tart flavor to the solution. Hydronium ions account for the tart flavor of common household acids, such as vinegar (acetic acid) and lemon juice (ascorbic and citric acids). Acids may be identified easily in the laboratory by the fact that they turn **litmus** (a substance found in certain plants) red.

Bases are molecules or ions that can bind protons or that produce OH^- (hydroxyl ions) when dissolved in water. Sodium hydroxide (NaOH) is a base because it dissociates into Na^+ and OH^- (ions) in aqueous solutions. Hydroxyl ions give the water a bitter taste and a slippery, soapy feeling. Bases can be recognized easily because they turn litmus blue.

Salts are ionic compounds containing one or more positively-charged ions (not H^+) and one or more negatively-charged ions (not OH^-). Salts result when an acid and a base in a solution neutralize each other. Consider the reaction between HCl (an acid) and NaOH (a base), which produces

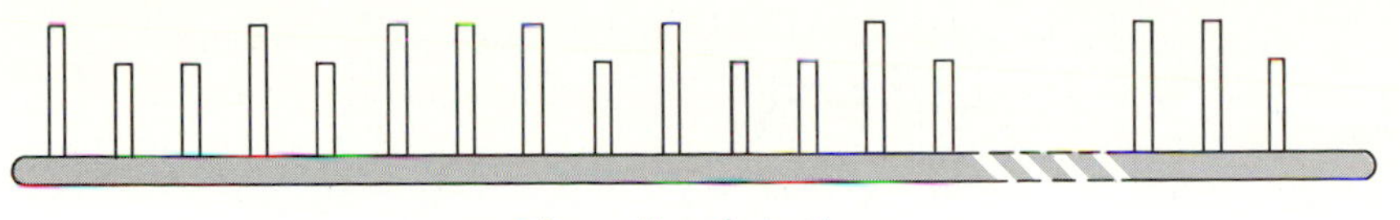

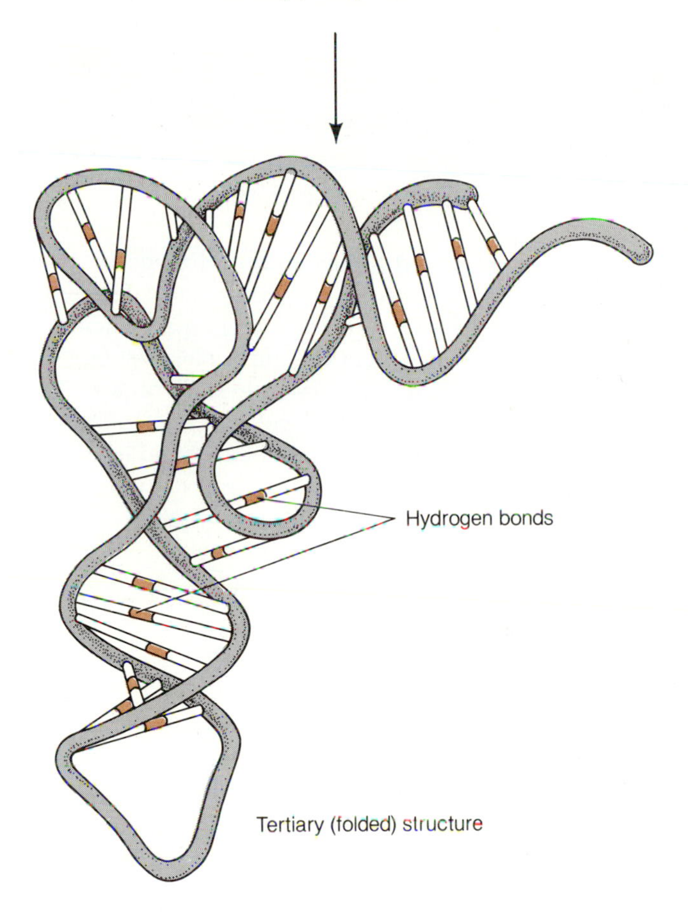

Figure 2-8 Folding of Macromolecules Due to Hydrogen Bonding. The 3-dimensional shape of the transfer RNA molecule is due in part to the spatial arrangement of the atoms in the molecule and to attractive forces such as hydrogen bonds between atoms in the macromolecule.

NaCl and H_2O. The NaCl is a salt in which Na^+ is the cation and Cl^- is the anion:

$$\text{HCl (acid)} + \text{NaOH (base)} \rightarrow \text{NaCl (salt)} + \text{H}_2\text{O (water)}$$

The concentration of hydrogen ions in a solution is generally measured as the **pH.** The pH of a solution is equal to the negative logarithm of the H^+ concentration (in moles/liter):

$$\text{pH} = -\log_{10} (\text{H}^+)$$

The calculation of the negative $\log_{10}$ of the concentration of H^+ is not as difficult as it may seem at first. Let us calculate the pH of pure water. The hydrogen ion concentration of pure water is 1.0×10^{-7} moles/liter. The number of moles is given as the ratio of the weight of the substance in grams to the molecular weight of 6×10^{23} units of the substance. From the formula for determining the pH given above:

$$pH = -\log_{10} (H^+)$$
$$pH = -\log_{10} (1.0 \times 10^{-7})$$
$$pH = -(-7)$$
$$pH = 7$$

Thus pH of pure water is 7. Figure 2-9 lists the pH and the H^+ concentration of some common chemicals.

BIOLOGICALLY IMPORTANT MOLECULES

Molecules are generally classified as **organic** or **inorganic.** Organic molecules as previously noted, are defined as those molecules containing covalently-bonded carbon atoms. Sucrose (table sugar), acetic acid (vinegar), and methane (swamp gas) are examples of organic molecules. Organic molecules generally result from biological activities, although organic chemists have been able to synthesize many organic molecules in the laboratory. **Inorganic molecules** lack covalently bonded carbon atoms and usually result from nonbiological processes. Ammonium sulfate, hydrochloric acid, and sodium chloride are examples of inorganic molecules.

Although organisms are more than 70% water, most of the remaining cellular material consists of various organic molecules. Many of the organic

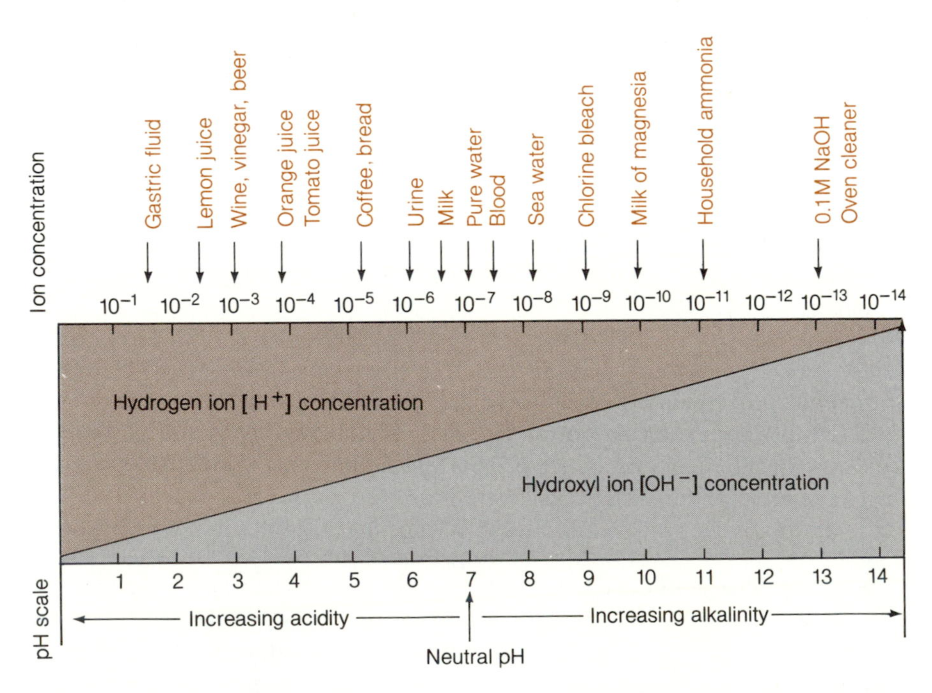

Figure 2-9 The pH Scale and the pH of Some Common Substances. The pH scale is a measure of the hydrogen ion concentration of aqueous substances. The pH is the negative log of the hydrogen ion concentration. Acidic substances have pH values lower than 7, basic or alkaline substances have pH values over 7, and neutral substances have pH values of 7.

molecules (proteins, starches, nucleic acids, etc.) are called **polymers,** because they are constructed by joining a large number of similar molecules called **monomers** (table 2-5). Polymers are also referred to as **macromolecules** because they are so large.

Proteins

Proteins are fundamental components of all living cells. In a bacterium such as *Escherichia coli*, about 15–20% of the cell's weight is protein. In living organisms, proteins function as structural components, catalysts, hormones, toxins, reserve materials, and physiological regulators. Table 2-6 summarizes some of the basic functions proteins play in living organisms.

Proteins are large polymers of **amino acids** (Table 2-7), joined together by a type of covalent bond called a peptide bond (fig. 2-10). Because these molecules are made up of many amino acids joined together by peptide bonds, they are called **polypeptides.** Amino acids are organic acids that contain an amino (NH_2) group, an R group (made up of various atoms), and a carboxyl group (COOH) (fig. 2-10). The 20 common amino acids used by cells to make proteins (table 2-7) all have similarly arranged carboxyl and amino groups. They differ, however, in the structure of the R group, which is attached to the same carbon as the amino group. R can be anything from an H atom (in the amino acid glycine) to a complex molecule such as indole (in the amino acid tryptophan).

The **primary structure** of a protein consists of the linear arrangement of amino acids making up the protein (fig. 2-10). This linear arrangement ultimately determines the shape and function of the protein. The **secondary structure** of a protein is determined by disulfide and hydrogen bonding, which gives rise to bends, helices, or pleated sheets (fig. 2-10). The **tertiary structure** of a protein refers to the intricate bending and folding of the protein (fig. 2-10). The folding is achieved by disulfide bonds or by hydrogen

TABLE 2-6
SOME BIOLOGICAL FUNCTIONS OF PROTEINS

PROTEIN FUNCTION	EXAMPLE
Catalysis	DNA polymerase—replicates and repairs DNA
Transport	Hemoglobin—transports O_2 in blood
Storage	Casein—a milk protein
Motion	Flagellin—movement of flagella
Structure	Collagen—fibrous connective tissue
Protection	Antibodies—form complexes with foreign proteins
Chemical messengers (hormones)	Insulin—regulates glucose metabolism
Ion transport	Bacteriorhodopsin—energy transformation
Toxins	Diphtheria toxin—causes diphtheria Snake venoms—Enzymes that hydrolyze phosphoglycerides

Adapted by permission from *Introduction to General, Organic, and Biological Chemistry* by Martha J. Gilleland. p. 458.

TABLE 2-7

CHEMICAL STRUCTURES OF COMMON AMINO ACIDS

Amino acid	Structure
AMINO ACIDS WITH NEUTRAL (UNCHARGED) POLAR SIDE CHAINS	
Cysteine Cys C Mol wt 121	$HS{-}CH_2{-}CH(NH_3^+){-}COO^-$
Tyrosine Tyr Y Mol wt 181	$HO{-}C_6H_4{-}CH_2{-}CH(NH_3^+){-}COO^-$
Asparagine Asn N Mol wt 132	$H_2N{-}C(=O){-}CH_2{-}CH(NH_3^+){-}COO^-$
Glutamine Gln Q Mol wt 146	$H_2N{-}C(=O){-}CH_2{-}CH_2{-}CH(NH_3^+){-}COO^-$
AMINO ACIDS WITH ACIDIC (NEGATIVELY CHARGED) SIDE CHAINS	
Aspartic Acid Asp D Mol wt 133	$^-O{-}C(=O){-}CH_2{-}CH(NH_3^+){-}COO^-$
Glutamic acid Glu E Mol wt 147	$^-O{-}C(=O){-}CH_2{-}CH_2{-}CH(NH_3^+){-}COO^-$
AMINO ACIDS WITH NEUTRAL (UNCHARGED) POLAR SIDE CHAINS	
Glycine Gly G Mol wt 75	$H{-}CH(NH_3^+){-}COO^-$
Serine Ser S Mol wt 105	$HO{-}CH_2{-}CH(NH_3^+){-}COO^-$
Threonine Thr T Mol wt 119	$CH_3{-}CH(OH){-}CH(NH_3^+){-}COO^-$

TABLE 2-7 (CONTINUED)

AMINO ACIDS WITH BASIC (POSITIVELY CHARGED) SIDE CHAINS	
Lysine Lys K Mol wt 146	$H_3\overset{+}{N}-CH_2-CH_2-CH_2-CH_2-C(H)(\overset{+}{N}H_3)-COO^-$
Arginine Arg R Mol wt 174	$H_2N-C(=\overset{+}{N}H_2)-NH-CH_2-CH_2-CH_2-C(H)(\overset{+}{N}H_3)-COO^-$
Histidine (at pH 6.0) His H Mol wt 155	$HC{=}C-CH_2-C(H)(\overset{+}{N}H_3)-COO^-$ (ring: HC=C, $\overset{+}{H}N$, NH, C–H)
AMINO ACIDS WITH NONPOLAR (HYDROPHOBIC) SIDE CHAINS	
Alanine Ala A Mol wt 89	$CH_3-C(H)(\overset{+}{N}H_3)-COO^-$
Valine Val V Mol wt 117	$(CH_3)_2CH-C(H)(\overset{+}{N}H_3)-COO^-$
Leucine Leu L Mol wt 131	$(CH_3)_2CH-CH_2-C(H)(\overset{+}{N}H_3)-COO^-$
Isoleucine Ile I Mol wt 131	$CH_3-CH_2-CH(CH_3)-C(H)(\overset{+}{N}H_3)-COO^-$
Proline Pro P Mol wt 115	ring: H_2C, H_2C, $\overset{H_2}{C}$, N–H, C(H)–COO^-
Phenylalanine Phe P Mol wt 165	$C_6H_5-CH_2-C(H)(\overset{+}{N}H_3)-COO^-$

TABLE 2-7 (CONTINUED)

AMINO ACIDS WITH NONPOLAR (HYDROPHOBIC) SIDE CHAINS	
Tryptophan Trp W Mol wt 204	(indole ring)—CH_2—C(H)(NH_3^+)—COO^-
Methionine Met M Mol wt 149	CH_3—S—CH_2—CH_2—C(H)(NH_3^+)—COO^-

Adapted by permission from *Introduction to General, Organic, and Biological Chemistry* by Martha J. Gilleland. p. 434–435.

bonds. The tertiary structure determines the proteins' functions. Sometimes the functional protein consists of more than one polypeptide. In this case, the protein is said to have a **quaternary structure** (fig. 2-10). An example of a protein with quaternary structure is the immunoglobulin molecule, found in serum, which is made up of four separate polypeptides joined together by disulfide bonds (fig. 2-10). The concept of primary, secondary, tertiary, and quaternary structures applies not only to proteins but to all macromolecules.

Proteins are macromolecules whose shape (tertiary and/or quaternary structures) largely determines their activity. Proteins are sensitive to environmental conditions, such as pH and temperature. Most proteins are active between 0°C and 40°C, although proteins isolated from some heat-loving microorganisms can be active at much higher temperatures (about 75°C). The range of pH tolerance of proteins is usually between 6 and 8. Some microbial proteins, however, can tolerate extreme pH values.

Environmental changes beyond the normal limits cause the tertiary structure and/or quaternary structure of proteins to unfold or change shape. When this happens, the protein is said to have **denatured.** Denatured proteins are not functional and may even precipitate out of solution. When proteins that participate in vital cell functions become denatured, the organism ceases to reproduce.

Polysaccharides

Polysaccharides are polymers of simple sugars, joined together by covalent bonds called glycosidic bonds. **Carbohydrate** is a more general term that is sometimes used to describe these polymers. Strictly speaking, however, carbohydrates are organic compounds with one or more hydroxyl groups (OH) and either an aldehyde or a ketone group (table 2-8). The term carbohydrate indicates that these molecules are made up of carbon, hydrogen, and oxygen. The ratio of H:O is 2:1, the same as in water. This ratio is seen in the general formula for carbohydrates: $(CH_2O)_x$ where x is the number of carbon atoms in the molecule. For example, $(CH_2O)_6$ is the chemical formula for a six-carbon sugar such as glucose and fructose. The suffix **ose** is sometime added

(a) Generalized formula for an amino acid

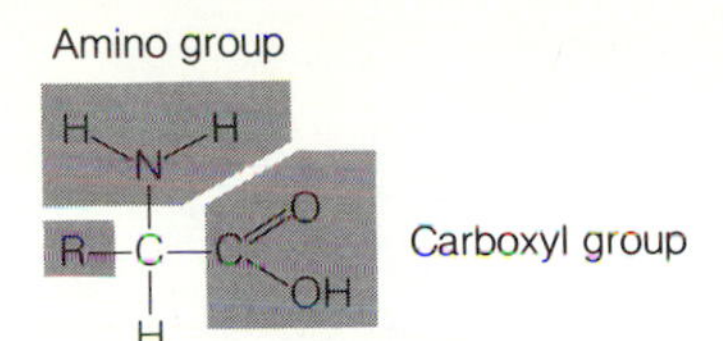

(b) Formation of a peptide bond

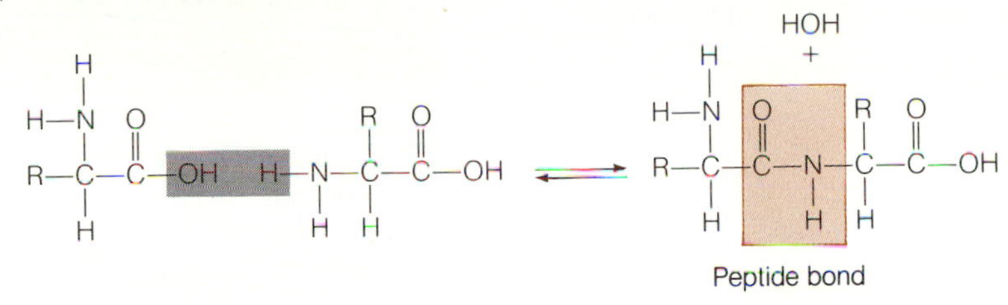

(c) Primary structure

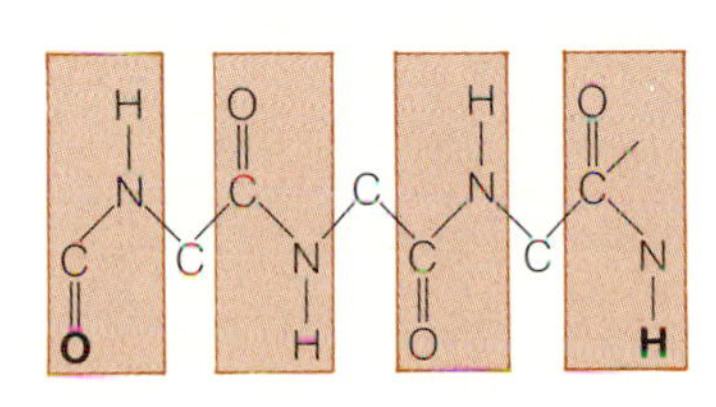

(d) Secondary structure

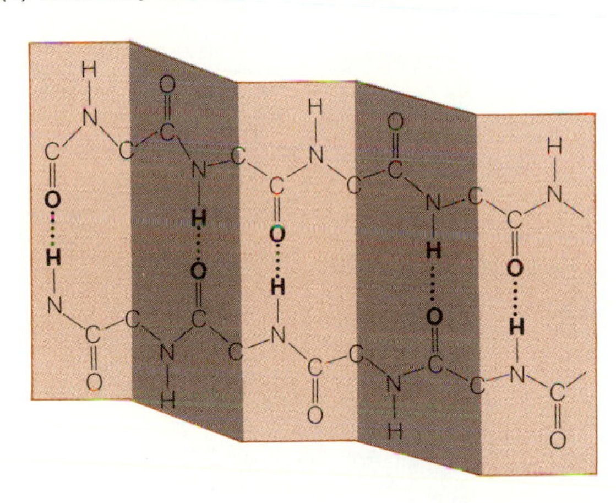

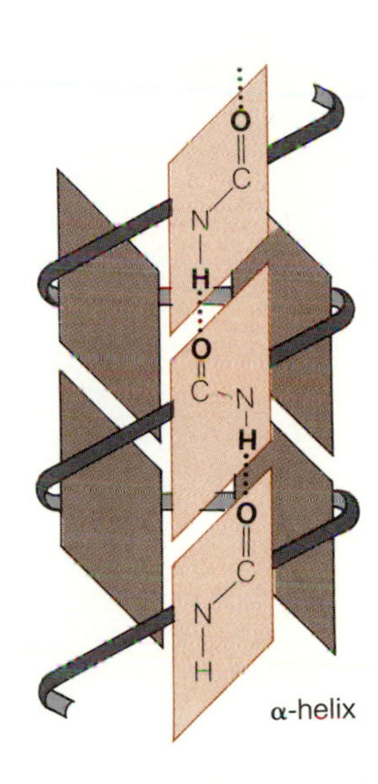

(e) Tertiary structure

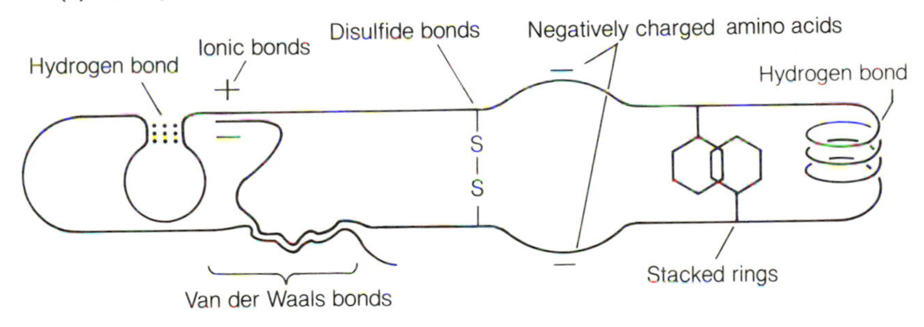

(f) Quaternary structure

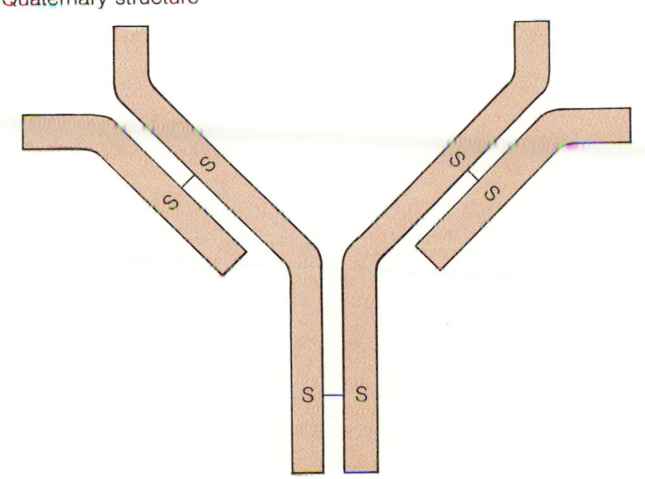

Figure 2-10 Structure and Composition of Proteins. *(a)* Generalized formula of an amino acid. Amino acids consist of a carboxyl group, an amino group, and an R group. The R group may be an atom or a group of atoms (see table 2-7). *(b)* A peptide bond. Amino acids are joined together by covalent bonds, called peptide bonds, to form polypeptides. Peptide bonds form between the carboxyl group of one amino acid and the amino group of the other. *(c)* The primary structure of proteins consists of the linear arrangement of amino acids. *(d)* The secondary structure of proteins is determined by the spatial orientation of amino acids in relation to each other. They usually form helices or pleated sheets. *(e)* The tertiary structure of proteins consists of the folding and looping of the secondary structure to impart a 3-dimensional structure to the protein. The tertiary structure is determined by hydrogen bonding, disulfide bonding, and other attractive interactions between atoms. *(f)* The quaternary structure of a protein. Antibody molecules are proteins made up of four different polypeptides joined together to form a functional protein. The 3-dimensional structure of the assembled protein determines its activity.

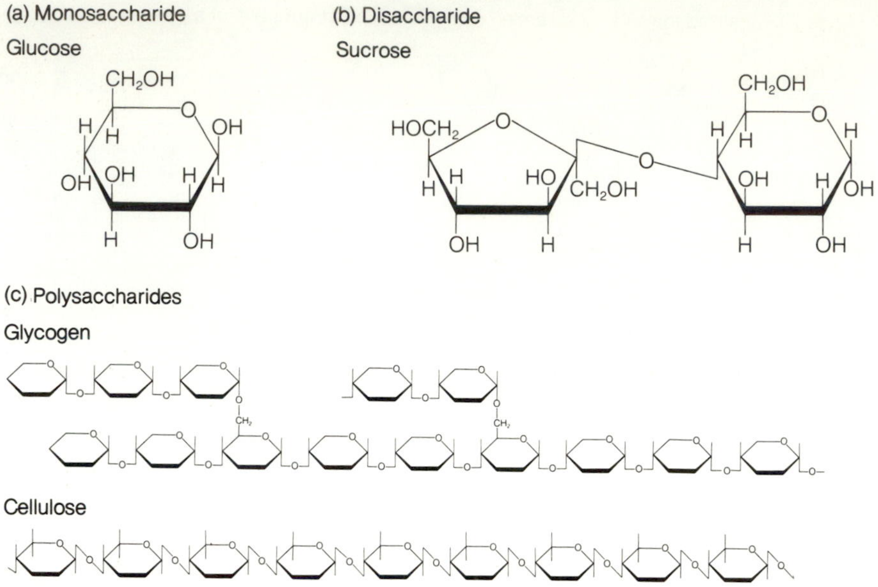

Figure 2-11 Some Common Saccharides Seen in Biological Systems

(a) Monosaccharide

Glucose

(b) Disaccharide

Sucrose

(c) Polysaccharides

Glycogen

Cellulose

TABLE 2-8 FUNCTIONAL GROUPS SEEN IN ORGANIC MOLECULES			
FUNCTIONAL GROUP	NAME OF FUNCTIONAL GROUP	NAME OF CLASS OF COMPOUNDS	EXAMPLE
>C=C<	Double bond	Alkene	$H_2C{=}CH_2$ Ethene
—C≡C—	Triple bond	Alkyne	H—C≡C—H Ethyne
—OH	Hydroxyl	Alcohol	CH_3OH Methyl alcohol
—C—O—C—	Ether	Ether	$H_3C—O—CH_3$ Dimethyl ether
>C=O	Carbonyl	Aldehyde	$H_3C—C(H){=}O$ Acetaldehyde
		Ketone	$H_3C—C({=}O)—CH_3$ Acetone
—C(=O)OH	Carboxyl	Carboxylic acid	$H_3C—C({=}O)OH$ Acetic acid

TABLE 2-8 (CONTINUED)

FUNCTIONAL GROUP	NAME OF FUNCTIONAL GROUP	NAME OF CLASS OF COMPOUNDS	EXAMPLE
$-C(=O)-O-\overset{\mid}{\underset{\mid}{C}}-$	Ester	Ester	$H_3C-C(=O)-OCH_2CH_3$ Ethyl acetate
$-\underset{\mid}{N}-$	Amino	Amine	$CH_3-CH(NH_2)-COOH$ Alanine
$-C(=O)-NH_2$	Amide	Amide	$H_3C-C(=O)-NH_2$ Acetamide
$-SH$	Thiol	Thiol	H_3C-SH Methanethiol
$-S-S-$	Disulfide	Disulfide	$H_3C-S-S-CH_3$ Dimethyl disulfide

Adapted by permission from *Introduction to General, Organic, and Biological Chemistry* by Martha J. Gilleland. pp. 282–283.

to the name of the molecule in order to identify it as a sugar. A **triose** is a 3-carbon sugar, a **pentose** is a 5-carbon sugar, and a **hexose** is a 6-carbon sugar. The prefix indicates the number of carbon atoms in the sugar molecule. Examples of carbohydrates are potato starch (amylose) and cane sugar (sucrose). Sugars, or **saccharides,** are sweet-tasting, water-soluble carbohydrates. A carbohydrate consisting of only one sugar is called a monosaccharide, a two-sugar carbohydrate is called a disaccharide, and a many-sugar carbohydrate is called a polysaccharide. Figure 2-11 illustrates some common polysaccharides.

Carbohydrates are frequently used as carbon and energy sources by living organisms. For example, the monosaccharide glucose is used as an energy source by a large number of different microorganisms. Microorganisms also use polysaccharides to construct cell material; cell walls, for instance, are often made of polysaccharides such as peptidoglycan, cellulose, or chitin. When these fibrous polysaccharides are knitted together they form strong cell walls. In addition to being integral parts of many cell walls, various polysaccharides may exist in the cell matrix (cytoplasm) as reserve materials (e.g., glycogen and starch).

Lipids

Lipids constitute a heterogeneous group of substances that are insoluble in water, but soluble in nonpolar solvents such as benzene or acetone. Many lipids feel greasy or oily. Lipids are used by living organisms in a variety of ways. For example, steroids and prostaglandins (fig. 2-12) are used by higher animals as hormones to send messages from the brain to other parts of the body. Many lipids (e.g., fats) are stored as reserve nutrients or are used as insulation by many animals. In addition, some lipids, such as the **phospholipids** (fig. 2-13) are structural components of the plasma membrane in all living organisms. The plasma membrane carries out many vital functions. For instance, it regulates the passage of nutrients and wastes into and out of the cell, recognizes environmental conditions, and functions as the site for many biochemical activities of the cell.

Complex lipids are composed of an alcohol covalently bonded to one or more **fatty acids** (fig. 2-12). For example, the triglycerides, which are very common lipids, are made up of three long-chain fatty acids covalently

(a) Simple lipids

Steroid

Progesterone

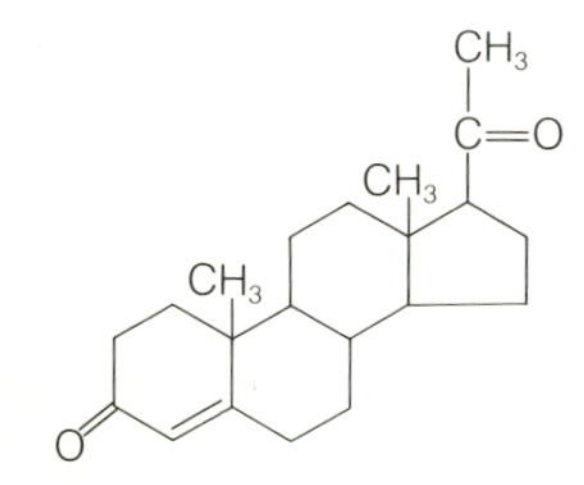

(b) Complex lipids

Triglyceride

Glycerol group

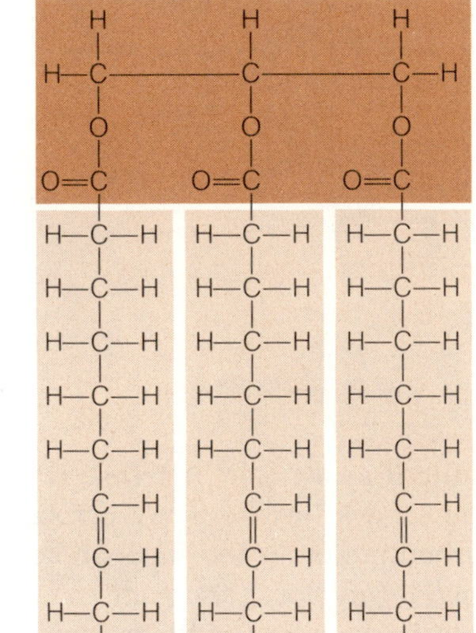

Fatty acid

Phospholipid

Polar group (hydrophilic)

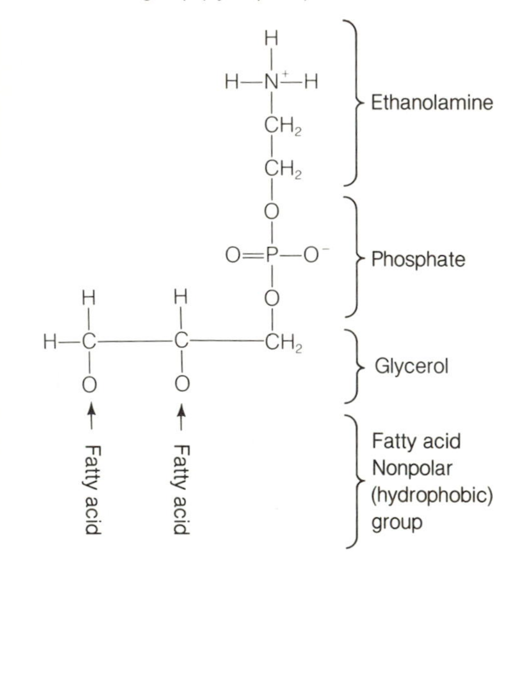

Figure 2-12 Chemical Structures of Representative Lipids

bonded to an alcohol called glycerol. The fatty acids generally have 10–20 carbon atoms.

Saturated fats are those lipids in which the carbon atoms in the fatty acids form single bonds with each other. Two adjacent C atoms share only one pair of electrons and the rest of the electrons are shared with H atoms:

$$. . . H_2C - CH_2 - COOH$$

Unsaturated fats, on the other hand, have fatty acids with carbons joined together by double bonds (adjacent C atoms share two pair of electrons).

$$. . . HC = CH - COOH$$

Triglycerides often serve as reserve nutrients in plant and animal cells.

Complex lipids, such as lipopolysaccharides (glycolipids) and lipoproteins (proteolipids), are those in which fatty acids are covalently bonded to carbohydrates or to proteins, respectively.

Phospholipids are derivatives of triglycerides. In phospholipids, one of the fatty acids generally is replaced by a phosphate-containing group (fig. 2-12). Phospholipids are essential components of plasma membranes □. 80

If several phospholipid molecules are spread over a water surface, they form a layer one molecule thick, and orient themselves so that the phosphate group points in one direction and the fatty acids in the opposite direction. The reason for this phenomenon is that part of the molecule, the phosphate group, has an affinity for water and is said to be **hydrophilic,** while the fatty acid portion has an aversion for water **hydrophobic**. Phospholipids can spontaneously form a two-layer structure in water, as shown in figure 2-13, with the hydrophobic tails toward the center and the hydrophilic heads toward the aqueous portion of the environment. This self-assembly process is believed to have contributed to the origin of the first cells about 3.8 billion years ago.

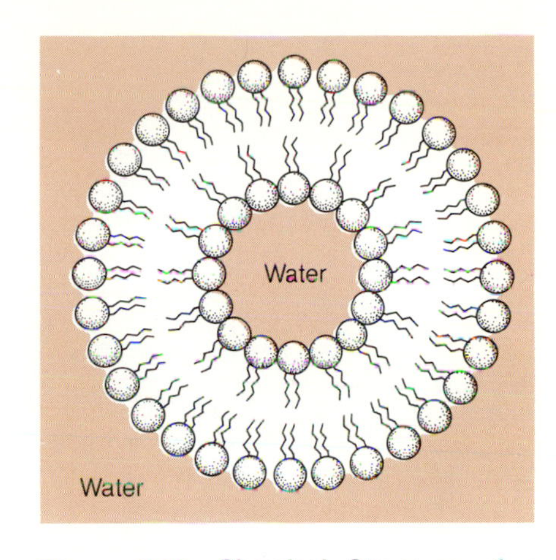

Figure 2-13 Chemical Structure of a Membrane. Phospholipids are the principal component of membranes. These molecules are composed of a polar head (hydrophilic) and a nonpolar tail (hydrophobic). In water, the polar heads are oriented toward the water while the nonpolar tails are aimed away from the water.

Nucleic Acids

Nucleic acids are large, acidic biological polymers that are found in all living cells. Nucleic acids get their name from the fact that they are concentrated in the nucleus of the cell. We are all aware that certain characteristics are passed on from parents to offspring. The information for creating these characteristics is contained in one type of nucleic acid, generally referred to as **genetic** or **hereditary material** □. 89 The genetic material in all cells is **deoxyribonucleic acid,** abbreviated **DNA.** Within the cell, the genetic information is used to direct the production of proteins, which are responsible for the functioning of the cell and all of its characteristics. These proteins are made through the intervention of another type of nucleic acid, called **ribonucleic acid** or **RNA**.

Nucleic acids are polymers of small molecules called **nucleotides.** Nucleotides consist of one or more phosphates, a 5-carbon sugar (pentose), and a heterocyclic nitrogen base (fig. 2-14). The heterocyclic nitrogen base can be of at least five different types (fig. 2-14): the bases adenine, cytosine, guanine, and thymine are found associated with DNA, while thymine is replaced by uracil in RNA. The pentoses are ribose in the RNA molecule and deoxyribose in the DNA molecule (fig. 2-14).

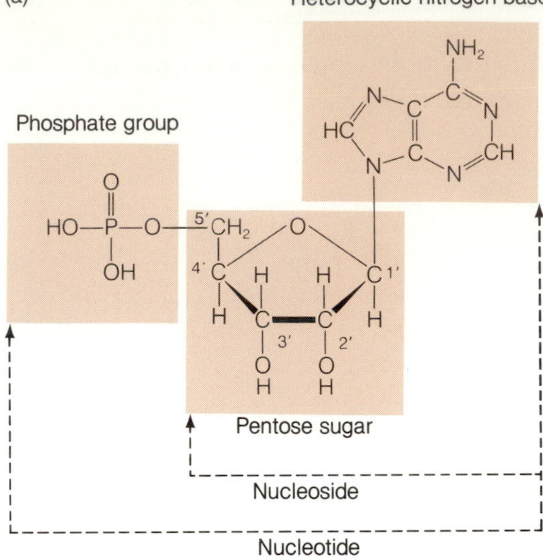

Figure 2-14 Components of Nucleotides. Nucleotides are the building units of nucleic acids. Each is composed of a pentose, a heterocyclic nitrogen base, and a phosphate group. *(a)* A heterocyclic nitrogen base covalently bonded to the pentose forms a nucleoside. A phosphate group covalently bonded to a nucleoside forms a nucleotide. *(b)* Various components of nucleotides.

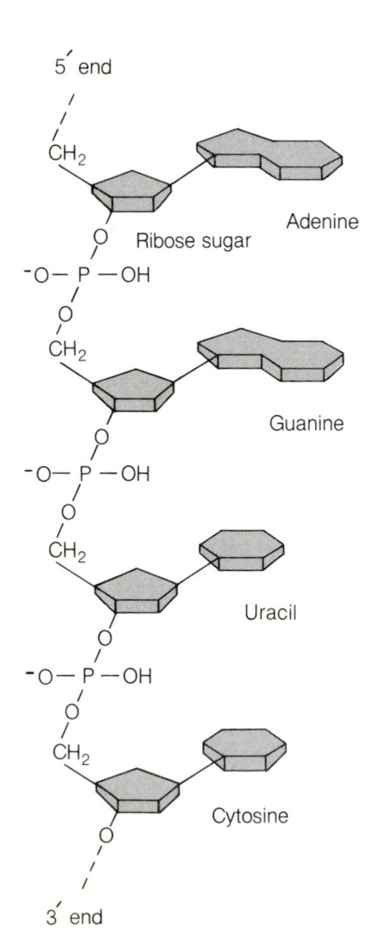

Figure 2-15 Primary Structure of a Ribonucleic Acid (RNA) Molecule

The "backbone" of a nucleic acid consists of a long string of alternating pentoses and phosphates. The bases are joined to the pentoses by covalent bonds. The sugar ribose resembles a pentagon, made up of an O atom and four C atoms (fig. 2-14). A fifth carbon atom is attached to an apex of the pentagon. By convention, the carbon atoms are counted from the carbon atom that bonds with only one other carbon atom and are named 1′, 2′, 3′, 4′ and 5′. In this way, we can make specific references to the individual atoms in the sugar molecule. The sugar deoxyribose is basically the same as the sugar ribose, except that deoxyribose is missing a hydroxyl group on the 2′ carbon.

The pentose molecule (ribose or deoxyribose) with a nitrogenous base attached to the 1′ C atom is called a **nucleoside** (fig. 2-14). If the nucleoside has one or more phosphate groups attached to the 5′ C, then it is called a **nucleotide** (fig. 2-14). The nucleosides in a nucleic acid are joined together by phosphate groups, which serve as linkages between the 5′ C of one nucleoside and the 3′ C of the other. RNA in cells generally takes on a secondary or even a tertiary structure and does not exist as a single, long strand (fig. 2-15).

DNA is assembled in the same manner as RNA. The molecule of DNA is double-stranded however, (fig. 2-16), each strand wrapped around the other in a helical fashion, forming what is called a **double helix.** The double helix is maintained by hydrogen bonding between heterocyclic nitrogen bases on the two strands. This hydrogen bonding is not random, but instead is quite ordered along the entire length of the DNA molecule: adenine pairs with thymine and cytosine pairs with guanine. It is this specificity of hydrogen bonding that insures a true replication of the DNA molecule when the cell makes new copies of the molecule during cell division.

(b)

Nucleic acid	Heterocyclic nitrogen base	Pentose sugar	Phosphate
Deoxyribonucleic acid (DNA)	Purines: 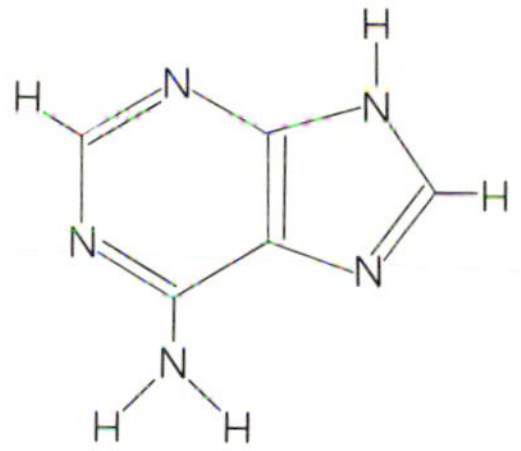 Adenine, 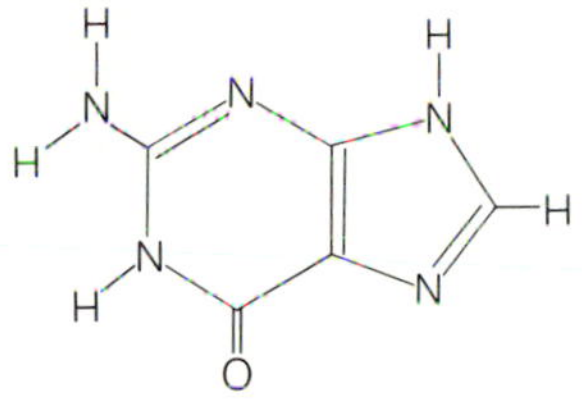 Guanine Pyrimidines: 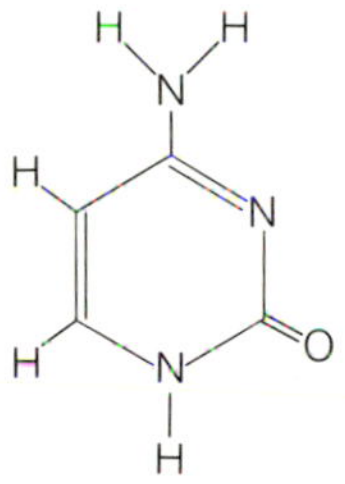 Cytosine, 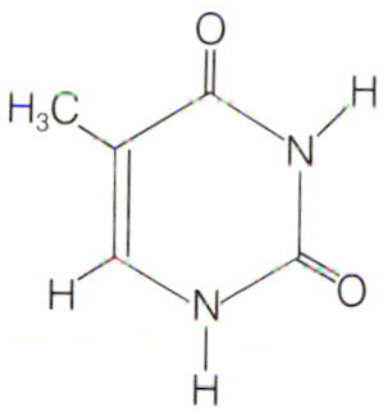 Thymine	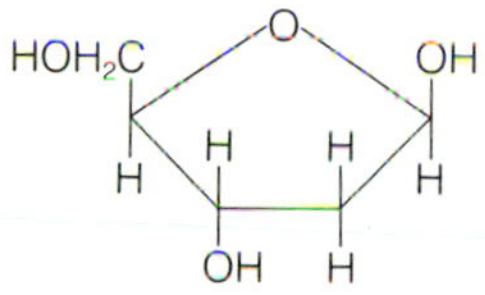 Deoxyribose	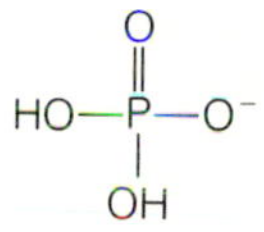 HO—P—O⁻ (with =O and OH)
Ribonucleic acid (RNA)	Purines: 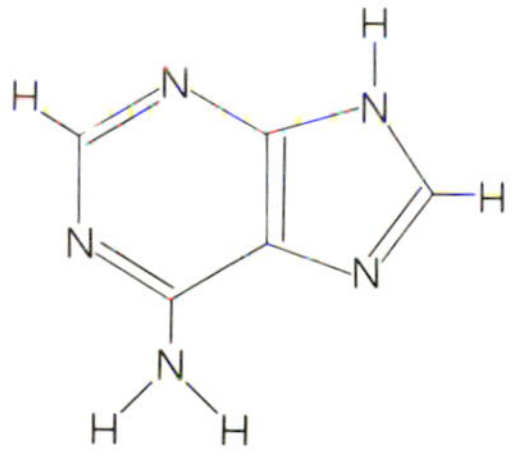 Adenine, 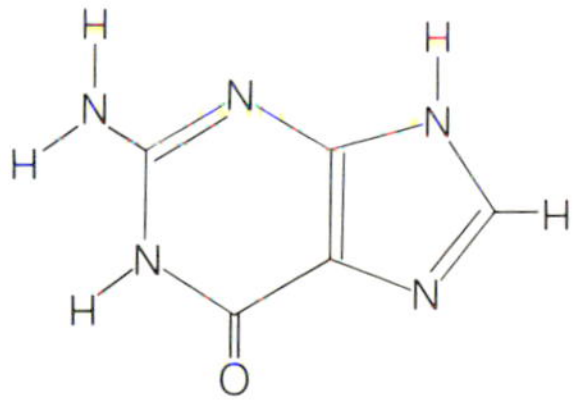 Guanine Pyrimidines: 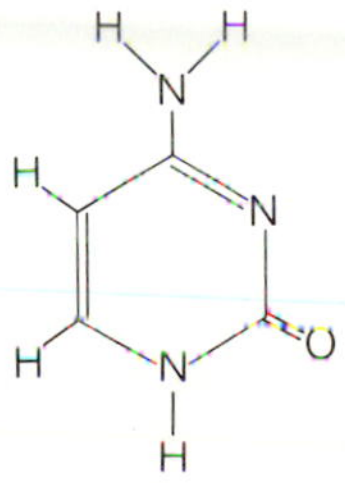 Cytosine, 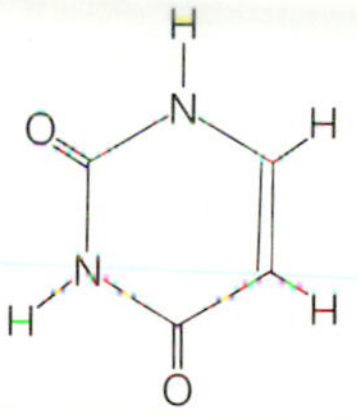 Uracil	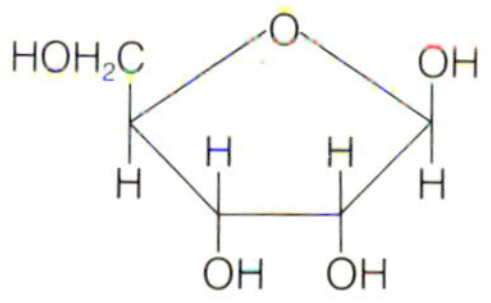 Ribose	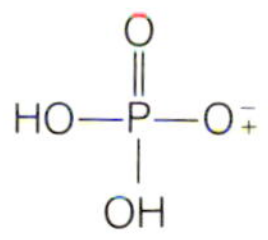 HO—P—O⁻ (with =O and OH)

Figure 2-14 (continued)

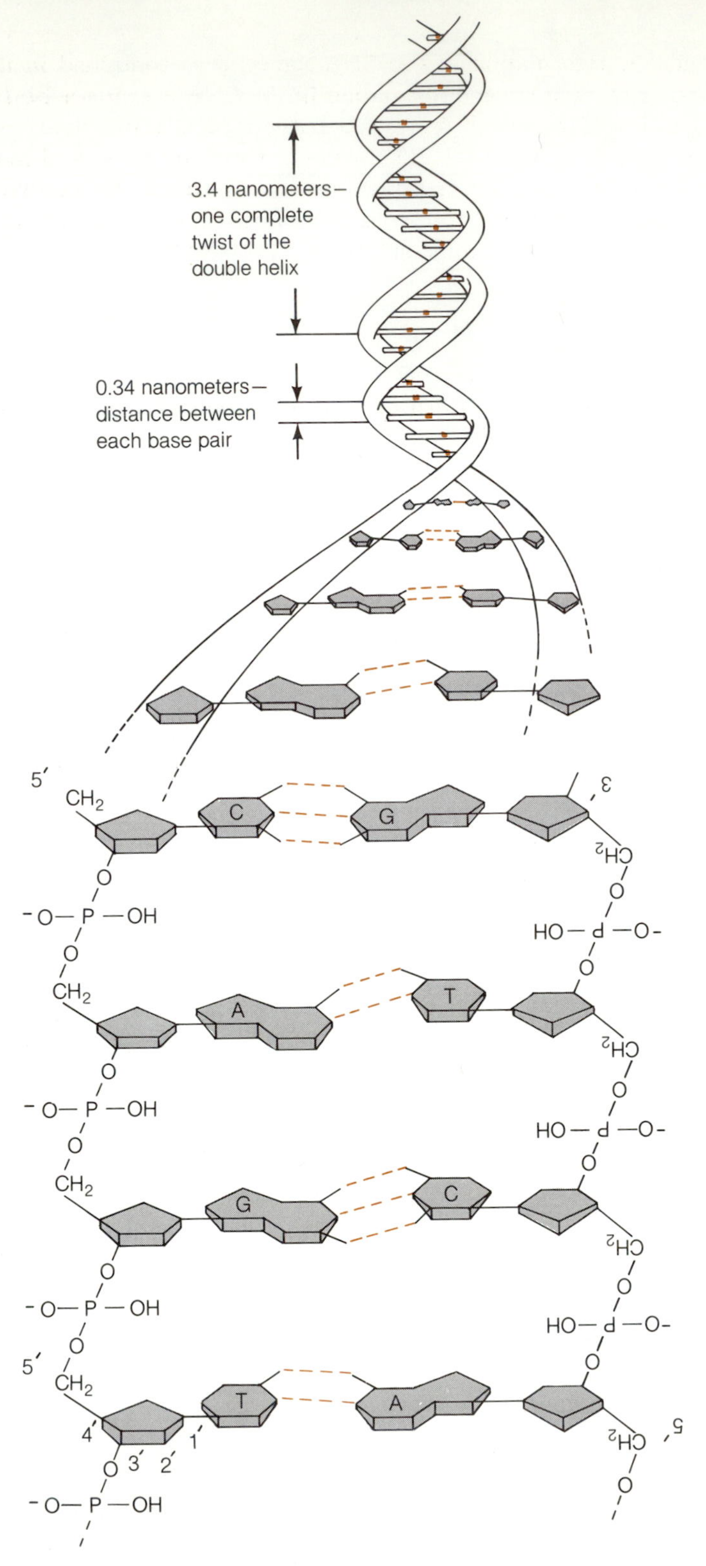

Figure 2-16 **Structure of a Deoxyribonucleic Acid (DNA) Molecule.** Diagram of the DNA double helix. The two helical strands are joined together by hydrogen bonds between nitrogenous bases in opposite strands. Adenine (A) hydrogen-bonds with thymine (T) and cytosine (C) with guanine (G).

The genetic information in the DNA molecule is contained in the nucleotide sequence. The linear information in the DNA is **transcribed** into a linear molecule of RNA which, in turn, is **translated** □ into a linear protein. The idea that DNA contains the information for making RNA, and that RNA contains the information for making proteins, is called the **central dogma** □. In later chapters, we will learn more about the mechanism of information storage and the synthesis of RNA and proteins by the cell.

209
205

CHEMICAL REACTIONS

Living organisms carry out numerous activities, such as moving about, obtaining energy, taking up nutrients, and reproducing. All of these activities involve chemical reactions in which some bonds are broken and others are formed. In fact, each cell can be thought of as a highly sophisticated chemical factory. The sum total of all the chemical reactions in the cell is referred to as **metabolism** □.

178

In order for some chemical reactions to take place, energy must be introduced. Chemical reactions that absorb energy are called **endergonic** reactions. The synthesis (buildup) of cell constituents, such as proteins or nucleic acids, involves endergonic reactions. On the other hand, chemical reactions in which energy is released are called **exergonic** reactions. When molecules are broken down into simpler compounds they generally yield energy. Consequently, these reactions are exergonic. Exergonic reactions are used by cells to power endergonic reactions.

Some chemical reactions in the cell involve the transfer of electrons from one molecule to another. These are called **oxidation-reduction** or **redox reactions,** in which one of the reactants is oxidized while the other is reduced. These reactions occur in pairs, and for every oxidation there is a concurrent and equivalent reduction. A molecule or an atom is said to be oxidized when it loses electrons, while a molecule is reduced when it gains electrons. An example of an oxidation-reduction reaction is the reduction of pyruvate to lactate, accompanied by the oxidation of nicotinamide adenine dinucleotide (NAD), in the following reaction:

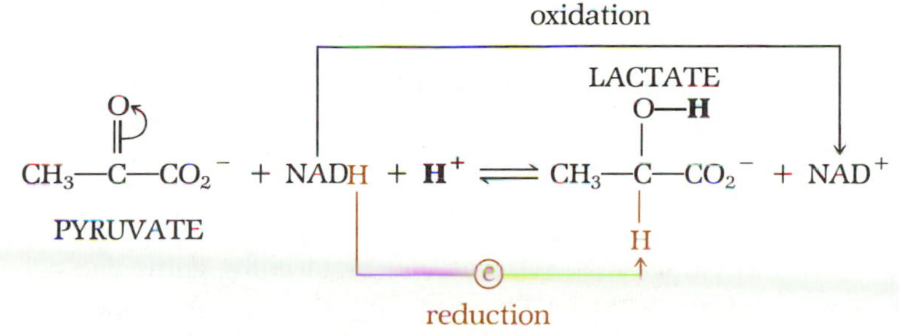

In order for most chemical reactions to get started, some energy (from exergonic reactions) must initially be made available. This energy is called the **activation energy** (fig. 2-17). For example, if oxygen gas is mixed with hydrogen gas in a container, usually nothing will happen. If an electrical spark is then introduced into the container, however, there will be a violent explosion and water will form (fig. 2-17). The electrical spark provided the needed activation energy, which triggered the reaction:

$$2H_2 + O_2 \longrightarrow 2H_2O$$

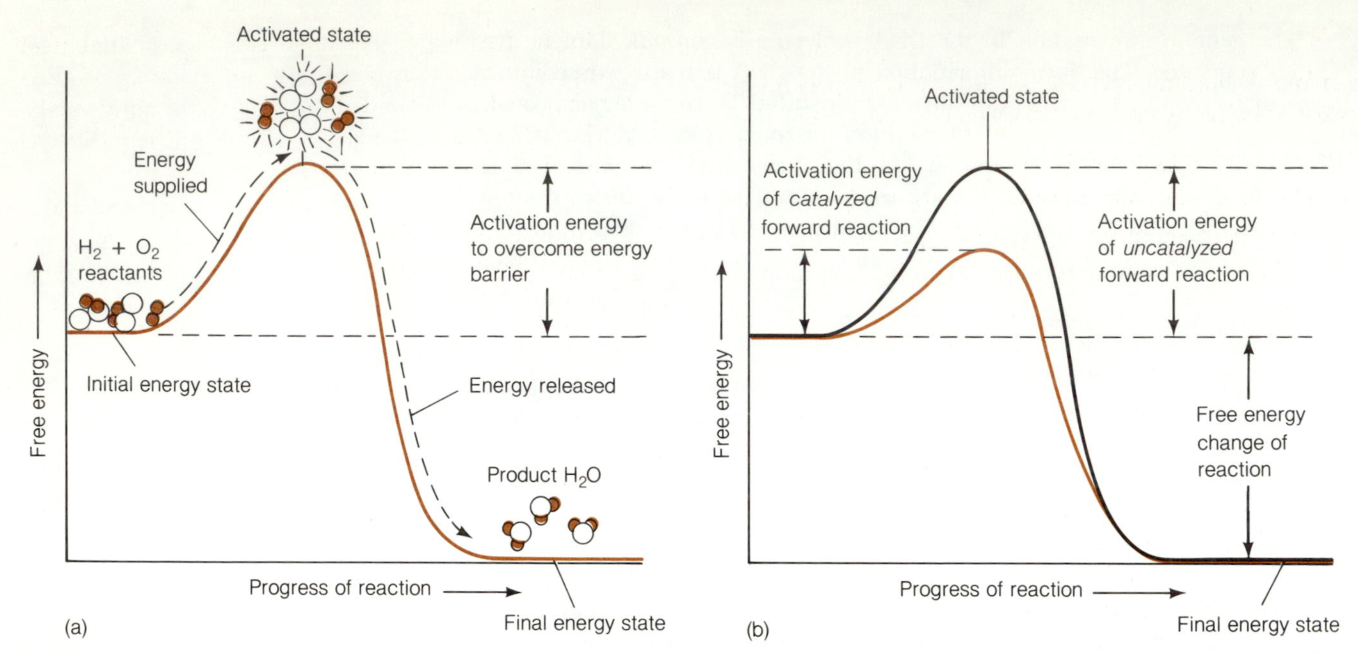

Figure 2-17 Energy of Activation. *(a)* Diagram illustrating the amount of energy required to initiate a chemical reaction using water formation as an example. The reactants must acquire additional energy so that they can react with each other and form water. This extra energy required is the energy of activation and is used in the reaction to break some bonds and form others. *(b)* Enzymes reduce the energy of activation required for a chemical reaction.

A **catalyst** is a substance that lowers the activation energy or reduces the amount of energy required to initiate a chemical reaction. A catalyst speeds up the rate of a reaction. Protein catalysts in cells are not permanently altered in reactions (fig. 2-17) and consequently can speed up one reaction right after another.

Most chemical reactions that take place in the cell do not take place in the absence of catalysts. Catalysts of biological reactions are proteins called **enzymes** □. Enzymes are very specific, and are active only when their specific reactants are present.

179

Some enzymes require a nonprotein molecule, or **coenzyme,** in order to function. Coenzymes are generally involved in transferring chemical groups from one molecule to another. Since coenzymes are derived from **vitamins,** any vitamin missing in the diet can lead to serious problems. For example, individuals with vitamin C-deficient diets can develop a disease called **scurvy.** The disease is characterized by bleeding gums and subcutaneous tissues, accompanied by the loosening of teeth and abnormal bone development. Sailors on British ships were required to eat limes (thus the nickname "limey") or drink lime juice, which is rich in vitamin C, in order to avoid this disease. Vitamin C, or ascorbic acid, is a factor necessary for the activity of enzymes involved in the synthesis of connective tissue and the cementing material that holds many tissues together. Another disease, **beriberi,** results from a vitamin B_1 deficiency, while **pellagra** is due to a lack of niacin in the diet. The reason that vitamin deficiencies cause disease is that, in the absence of their coenzymes, some enzymes cannot function properly.

SUMMARY

An understanding of basic chemistry is important because it helps us understand how cells function.

THE STRUCTURE OF MATTER

1. Matter is any substance with mass occupying space.
2. Matter exists in solid, liquid, or gaseous states.
3. Mixtures of matter result in suspensions, solutions, or colloids.
4. Matter is made up of elements which, combined with other elements, form compounds.
5. An atom is the smallest unit of an element and is made up of protons, neutrons, and electrons.
6. Protons and neutrons are closely associated with each other, forming a dense mass called the atomic nucleus. Electrons spin around the nucleus along paths called orbitals.
7. Isotopes are atoms of the same element having a different mass number. The difference results from variations in the nucleus.
8. Chemical bonds result when two or more atoms are joined together by attractive forces.
9. Chemical bonds can be covalent, ionic, or hydrogen bonds, or other weak bonds.

ACIDS, BASES, AND SALTS

1. Molecules are groups of atoms joined together by chemical bonds.
2. Acids are molecules that can release protons (H^+), while bases are molecules that accept protons.
3. Salts result when an acid and a base neutralize each other.
4. The H^+ concentration in a solution can be expressed as: $pH = -\log_{10} (H^+)$.

BIOLOGICALLY IMPORTANT MOLECULES

1. Cellular material is composed of water, ions, and large organic molecules called macromolecules.
2. Proteins, polysaccharides, and nucleic acids are examples of common macromolecules found in all cells.
3. Proteins are macromolecules made up of amino acids joined together by peptide bonds. They make up 15% to 20% of the cell's weight and are important because they participate in most of the functions of the cell.
4. Polysaccharides are made up of many sugars joined together by glycosidic bonds. They can function as reserve materials or as structural components of the cell wall.
5. Lipids are water-insoluble substances that feel greasy or oily. Complex lipids are made up of alcohols covalently bonded to long-chain fatty acids. They are used by the cell as reserve nutrients and as the basis of cell membranes.
6. Nucleic acids (DNA and RNA) are macromolecules composed of nucleosides joined together by phosphates. They function in the storage and processing of genetic information.

CHEMICAL REACTIONS

1. Biological activities involve chemical reactions during which some bonds are broken and others are made.
2. Some chemical reactions in the cell involve transferring electrons from one molecule to another. These are called redox reactions. The molecule that donates the electrons is said to be oxidized, while the molecule that accepts the electrons is said to be reduced.
3. For most chemical reactions to get started, a certain amount of energy must be introduced into the system. This is called activation energy.
4. Enzymes are biological catalysts that serve to lower the activation energy. Without enzymes, cellular activities would not take place in appreciable amounts.
5. Some enzymes require the presence of a coenzyme to be active. Coenzymes are nonprotein components of some enzymes and are derived from vitamins.

STUDY QUESTIONS

1. Discuss briefly the three states of matter.
2. Describe the structure of an atom.
3. How are isotopes, atoms, and elements different from one another?
4. Describe the structures of four types of macromolecules commonly found in cells.
5. How does DNA differ from RNA?
6. What are oxidation-reduction reactions?
7. Compare ionic and covalent bonds.
8. What are hydrogen bonds? How are they formed?
9. What is a chemical reaction?
10. What is activation energy?
11. How do enzymes participate in the chemical activities of the cell?
12. What are coenzymes?

13. Calculate the molecular weight of the following compounds:
    a. $C_6H_{12}O_6$
    b. H_2SO_4
    c. CH_3COOH
    d. CH_3CH_2OH
    e. NaCl

SUPPLEMENTAL READINGS

Baker, J. J. W. and Allen, G. E. 1975. *Matter, energy and life.* 3d ed. Reading, Mass: Addison-Wesley.

Dickerson, R. and Geis, I. 1973. *The structure and action of proteins.* Menlo Park, Calif.: Benjamin-Cummings.

Frieden, E. 1972. The chemical elements of life. *Scientific American* 227(1): 52–64.

Gilleland, M. J. 1982. *Introduction to general, organic and biological chemistry.* St. Paul, Minn.: West.

Kamp, G. 1984. *Cell Biology.* 2d ed. New York: McGraw-Hill.

King, E. L. 1963. *How chemical reactions occur.* New York: Benjamin-Cummings.

Lehninger, A. 1975. *Biochemistry.* 2d ed. New York: Worth.

Watson, J. 1976. *Molecular biology of the gene.* 3d ed. Menlo Park, Calif.: Benjamin-Cummings.

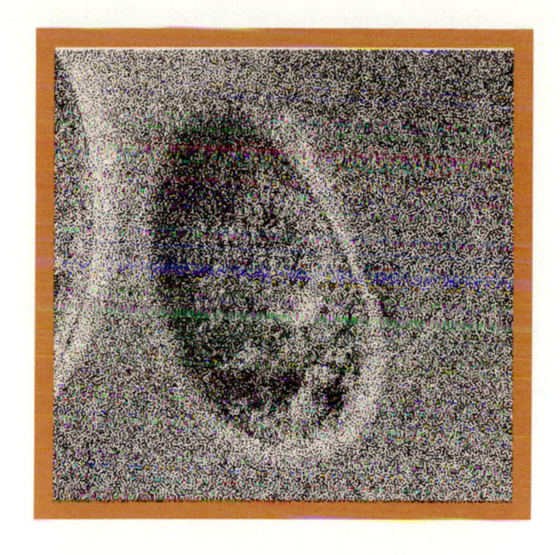

CHAPTER 3
MICROSCOPY AND THE STUDY OF MICROORGANISMS

CHAPTER PREVIEW
THE MOST POWERFUL MICROSCOPES IN THE WORLD

In their quest to see more and more detail, scientists have built high-voltage electron microscopes that can be used to study cellular structures on the molecular level. Currently, there are only three of these powerful microscopes being used for biological research in the United States. The high-voltage electron microscope at the University of Colorado is 32 feet high and weighs 22 tons (fig. 3-1). It is large because of the need to accelerate electrons to high energies, and because of the need to have large electromagnets to focus the electrons. Lead shielding protects the operator from X-rays that are produced when the electrons strike the specimen. The electrons in the high-voltage electron microscope are accelerated to energies between 500,000 and 1,000,000 volts so that the electrons have short wavelengths, which enable the viewer to see objects with dimensions as small as 0.5 to 1.0 nanometer (nm). In contrast, the electrons in conventional transmission electron microscopes reach energies of "only" 50,000 to 100,000 volts. Consequently, scientists cannot observe objects with dimensions smaller than 2.0 nm using these conventional instruments.

One of the problems with the high-voltage electron microscope is the chemical damage that occurs when the specimen is irradiated. This damage can introduce artifacts and destroy the normal structure of the specimen. In order to alleviate this problem, new microscopes are being built that can cool specimens to nearly −273°C. At this temperature the specimens are more resistant to the radiation that destroys and distorts specimens at normal temperatures.

The new electron microscopes and techniques being developed to prepare specimens will allow biologists to locate chemical elements, small diffusible ions, and specific molecules in various regions of the cell. This procedure will help in determining where chemical reactions occur in the cell and how the cell works on a molecular level. Already, the movement of Ca^{2+} and its sites of storage have been visualized in muscle cells by means of new techniques. In addition, the composition of magnets within some bacteria has been determined. The elemental composition is determined by observing the X-rays that are emitted when the specimen is struck by electron beams. Each element emits a characteristic spectrum of energies that gives away its identity.

In the future, the high-voltage electron microscope and newly developed techniques for preparing specimens will allow scientists to analyze the cellular structure down to the atomic level. In this chapter we will discuss some of the microscopes used by biologists and common techniques used to study the structure of microorganisms.

LEARNING OBJECTIVES

A STUDY OF THIS CHAPTER SHOULD ENABLE YOU TO:

EXPLAIN HOW THE VARIOUS TYPES OF MICROSCOPES WORK, AND THEIR IMPORTANCE TO THE FIELD OF MICROBIOLOGY

OUTLINE THE METHODS USED TO PREPARE SPECIMENS FOR THE LIGHT MICROSCOPE AND FOR THE ELECTRON MICROSCOPE

Antony van Leeuwenhoek's development of powerful hand lenses (simple light microscopes) that could magnify organisms 100 to 250× allowed him to see single-celled microorganisms. At about the same time, another scientist, Robert Hooke used a compound microscope to show that the unit of life in higher organisms was the cell. It was not until the middle 1800s, however, that scientists began to develop a precise **cell theory.** The study of cells with the light microscope during the 1800s and early 1900s revealed many of the differences and similarities between microorganisms and cells from higher organisms. The study of cellular structure was and is important because it allows us not only to see how organisms differ structurally, but also to understand what kind of structures are involved in various cellular processes. For example, the light microscope made it possible to observe flagella and cilia propelling certain microorganisms. Observations on chromosomes demonstrated that they were involved in the heredity of the cell.

Much of our knowledge about the structure of microorganisms has been linked to technological advances in the field of microscopy. Before the advent of the electron microscope, our understanding of the structure of microorganisms was restricted to those structures that were visible using a light microscope. As a consequence, a great many unique features of microbial anatomy were unknown. With the development of the electron microscope in the 1930s, much has been learned about the anatomy of microorganisms. Even today, the electron microscope is being used to learn new things about the anatomy of microorganisms and cells from higher organisms.

Figure 3-1 High Voltage Electron Microscope

LIGHT MICROSCOPES

In order for observations with microscopes to have any meaning, the observer must have a standard means of measurement and a way to determine the size of the specimens under investigation. Many of the early investigators used no standard system of measurement, but instead compared their specimens to common objects. For example, in his letters to the Royal Academy, Van Leeuwenhoek described his "animalcules" and compared them to the size of a grain of sand or to a louse's hair. Since not all sand grains nor all louse hairs are of the same size, these types of standards are not very useful when precise measurements are required. A system of measurements called the **metric system** has been adopted by the scientific community in order to establish a uniform standard of measurement. The metric system is now used by most scientists for all quantitative determinations (table 3-1).

Light microscopes that use two sets of lenses in series to view an object are called **compound microscopes.** The **objective** lenses are close to the object and produce a magnified image of it. The **ocular** lenses are near the eye and magnify the image. Van Leeuwenhoek's microscopes were **simple** microscopes because only one lens was used between the object and the eye. Many of the modern compound microscopes have two eyepieces in order to reduce eyestrain. Because of the two eyepieces, they are called **binocular microscopes.** Those microscopes that have only one eyepiece are called **monocular microscopes.**

The Bright Field Microscope

Bright field microscopes are used to provide the viewer an enlarged image of the object being examined. The enlargement of the image is called a **mag-**

TABLE 3-1
THE METRIC SYSTEM

UNIT	MEASURE		SYMBOL	ENGLISH EQUIVALENT
Linear measure				
1 kilometer	= 1000 meters	10^3 m	km	0.62137 mile
1 meter		10^0 m	m	39.37 inches
1 decimeter	= 1/10 meter	10^{-1} m	dm	3.937 inches
1 centimeter	= 1/100 meter	10^{-2} m	cm	0.3937 inch
1 millimeter	= 1/1000 meter	10^{-3} m	mm	
1 micrometer (or micron)	= 1/1,000,000,000 meter	10^{-6} m	μm (or μ)	English equivalents
1 nanometer	= 1/1,000,000,000	10^{-9} m	nm	infrequently used
1 angstrom*	= 1/10,000,000,000	10^{-10}m	Å	
Measures of capacity (for fluids and gases)				
1 liter			L	1.0567 U.S. liquid quarts
1 milliliter	= 1/1000 liter		ml	
1 milliliter	= volume of 1 g of water at standard temperature and pressure (stp)			
Measures of volume				
1 cubic meter			m^3	
1 cubic decimeter	= 1/1000 cubic meter = 1 liter (L)		dm^3	
1 cubic centimeter	= 1/1,000,000 cubic meter = 1 milliliter (mL)		cm^3 = ml	
1 cubic millimeter	= 1/100,000,000 cubic meter		mm^3	
Measures of mass				
1 kilogram	= 1000 grams		kg	2.2046 pounds
1 gram			g	15.432 grains
1 milligram	= 1/1000 gram		mg	.01 grain (about)
1 microgram	= 1/1,000,000 gram		μg (or mcg)	

*The angstrom is not part of the metric system but is so frequently encountered in the literature that it is included.

nification and is the end result of the combined effects of the objective and ocular lens systems (fig. 3-2). The objective lens provides an enlarged image of the object, usually 4–100 times, and the ocular lens magnifies the image projected by the objective lens. Ocular lenses generally magnify the image no more than 10–15 times. The **total magnification** of the lens system is the product of the objective's magnification and the ocular's magnification. For example, if the object is viewed with a 100× objective lens, the image in the microscope tube will be 100 times larger than the object. If this image is then viewed with a 10× ocular lens, it will be magnified 10 times. The total magnification of the object will be 100 × 10 = 1000 times.

The **resolution** of a bright field microscope is defined as the ability of a lens system to produce an image that allows the viewer to distinguish between two structures near each other in a specimen. The resolving power of a microscope is determined by the wavelength of light (λ) used to illuminate the object; half the angle the light has entering the lens (θ); and the **index of refraction** (n) of the air or oil between the condenser and the objective lens (fig. 3-3). The product of n and sin θ is generally given as the **numerical aperture** (NA). The **limit of resolution** of a microscope is determined mathematically using the following formula:

$$R = \frac{0.61\ \lambda}{n\ (\sin\ \theta)} = \frac{0.61\ \lambda}{NA}$$

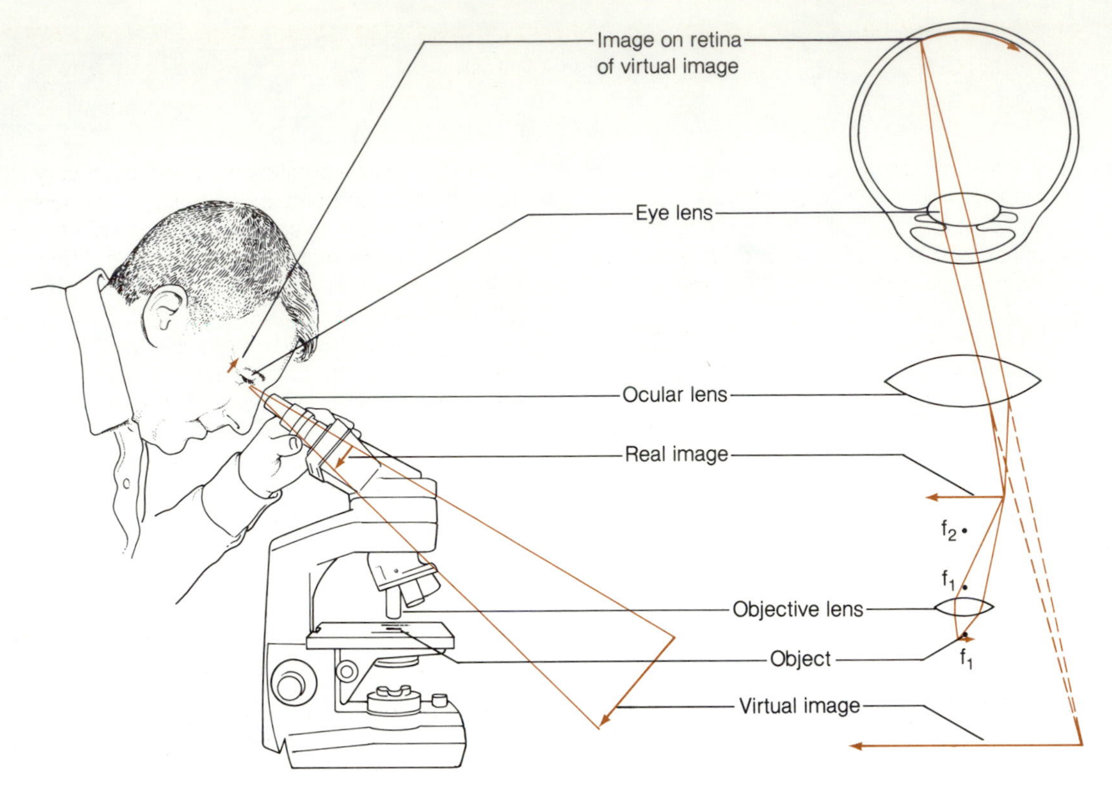

Figure 3-2 How the Bright Field Light Microscope Works. Light from the specimen is bent by the objective lens so that an inverted, magnified real image is created between the objective lens and the ocular lens. The image is magnified further by the ocular lens, so that an inverted, magnified virtual image is created. The lens in the eye focuses the inverted magnified virtual image on the retina so that the specimen can be seen.

where R is the limit of resolution of the lens system; λ is the wavelength of light employed to illuminate the object; n is the index of refraction ($n_{air} = 1$ or $n_{oil} = 1.5$) of the material between the specimen and the objective lens; and θ is half the angle of the cone of light from the specimen entering the objective lens. Typical values for λ, n, and sin θ are $\lambda = 550$ nm, $n = 1.5$, and $\sin\theta = 0.8$ (NA = 1.2). The wavelength 550 nm was chosen arbitrarily

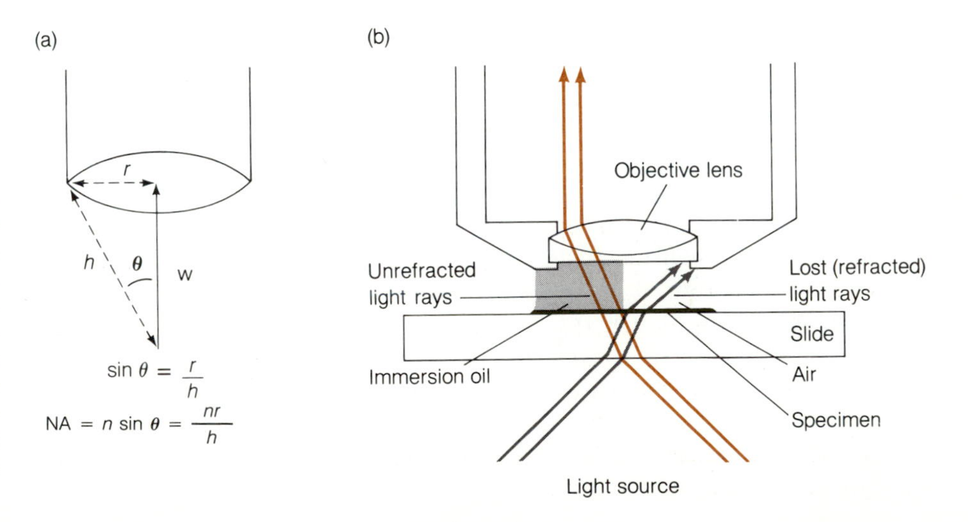

Figure 3-3 How the Limit of Resolution Is Determined. *(a)* The limit of resolution of the bright field microscope is proportional to the wavelength of light used (in nanometers) and inversely proportional to the numerical aperture of the lens. *(b)* The resolving power of the oil immersion lens can be improved by placing oil between the specimen and the objective lens. The oil eliminates refraction of light so that more light enters the lens to give definition to the specimen.

FOCUS
RESOLUTION AND THE WAVE NATURE OF LIGHT

Because of its wave nature, light behaves in an unexpected manner when it passes through small apertures. Instead of producing a sharp image, light under these conditions results in a large, diffuse image. The smaller the hole, the greater is the diffraction or spreading out of the light.

Diffraction is a problem in microscopes because high-magnification lenses are quite small. Lenses that magnify 50× have diameters of about 5 mm, while those that magnify 100× have diameters of about 2–3 mm. If you check your microscope, you will discover that the diameter of the lens decreases as the magnification increases. The smaller the lens, the more the light spreads out (diffracts) and the more the image is diffused and blurred (fig. 3-4).

A lens that magnifies 50× results in very little visible diffraction because of its large diameter, but a 100× objective lens spreads out the light and makes the images more diffuse than expected. The images may blur into each other, yet still be distinguished. A 150× lens severely diffracts the light, however, so that the images may overlap completely and two objects would appear to be one. Thus, an increase in the magnification, which would be expected to make the images larger and clearer, actually results in a decrease in the resolution of enlarged objects.

Since the diffraction is due to the waves of light interacting with the edge of the lens, it can be minimized by using a large lens and light with a short wavelength.

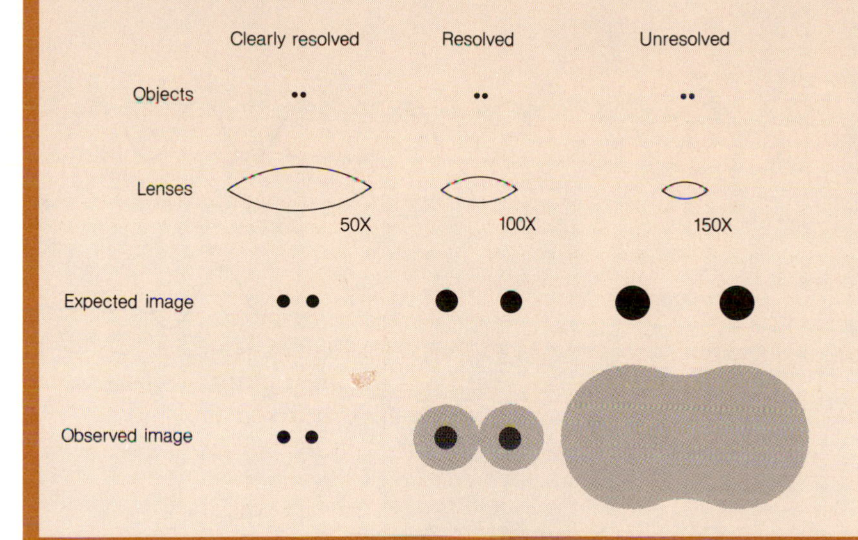

Figure 3-4 Resolution and the Wave Nature of Light. *(a)* Very little diffraction of the light is produced by the 50x objective lens. *(b)* The 100x objective lens causes some diffraction because of its size. Consequently, the image is more diffuse and blurred than expected. Nevertheless, the objects can be resolved. *(c)* A 150x objective lens results in a severe diffraction of the light, so that the objects cannot be resolved. The light from the objects is so diffuse that it overlaps, producing a single, blurred image.

to represent white light because it is in the center of the visible spectrum. These values give a limit of resolution of 0.28 μm. A limit of resolution of 0.28 μm means that structures smaller than 0.28 μm can not be resolved with the microscope. If air rather than oil is used between the condenser and the objective lens, the limit of resolution is "increased" to about 0.42 μm, which is not as good as 0.28 μm.

Although it might appear that magnifications much greater than 1000× could be obtained by using objective and ocular lenses above 100×, such magnification generally is not possible. If the total magnification of microorganisms exceeds about 1500×, the resolution and clarity of the image decrease because the light becomes **diffracted** (spread out) as it passes through very small lenses. The diffraction becomes worse as the magnification increases because the lens becomes smaller. Consequently, the image becomes more and more blurry.

The Phase Contrast Microscope

Most microorganisms are large enough to be resolved using a compound microscope, but they are often difficult to see in unstained preparations because there is little contrast between them and their surroundings. Furthermore, it is frequently impossible to see structures within cells because the structures and the cytoplasm are transparent and have similar refractive indices. Light passes through the aqueous environment, the cytoplasm, and transparent structures in much the same way: it is absorbed, reflected (scattered), and refracted (bent) about the same amount. Consequently, it reaches the eye with approximately the same intensity. Only those objects that absorb, reflect, and refract the light sufficiently are darker than the background and consequently can be seen (fig. 3-5).

Transparent objects that do not significantly reflect or refract light can be seen with the **phase contrast microscope.** This type of microscope amplifies small differences in refractive indices, such as those between the object and its aqueous environment. A difference in refractive index causes light to diffract (spread out). The phase contrast microscope has a special optical design that takes advantage of the diffracted light from translucent objects

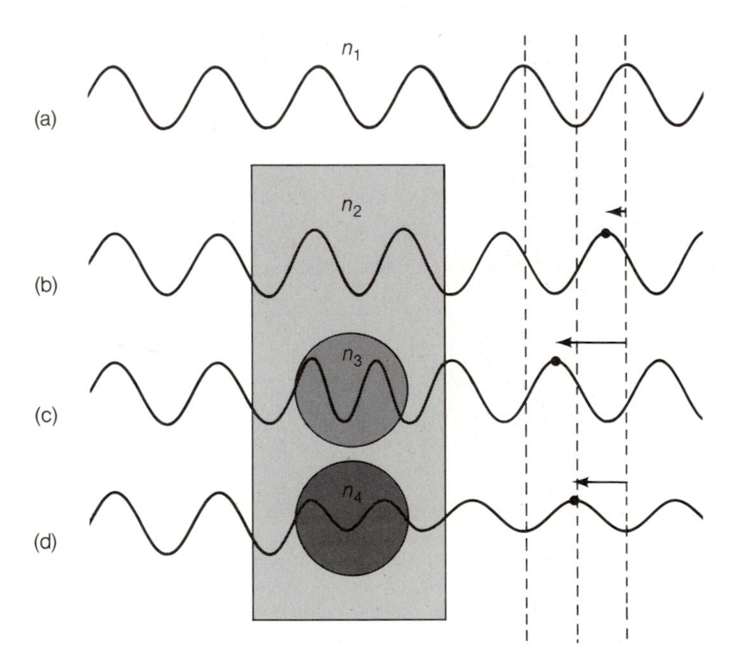

Figure 3-5 Phase Changes. As light from an aqueous environment penetrates transparent objects, the phase of the light is retarded. The higher the refractive index (n) of the object, the greater is the retardation or phase change. Phase changes in the light cannot normally be detected. Consequently, objects that only cause phase changes cannot be distinguished from their surroundings. Only if the light intensity (height of the electric and magnetic fields) is increased or decreased can the objects be seen. *(a)* Light passing through the aqueous environment shows little or no decrease in intensity. *(b)* Light passing through a transparent cell has a phase shift but shows little or no decrease in intensity. Consequently the cell cannot be seen. *(c)* Light passing through another transparent object has a very large phase shift, but no decrease in intensity. Therefore, the object cannot normally be seen. *(d)* Light passing through a transparent object in the cell has a phase shift and shows a decrease in intensity. Because there is a decrease in intensity, the object can be seen. The notation $n_3 > n_4 > n_2 > n_1$ indicates the relative magnitudes of the indices of refraction.

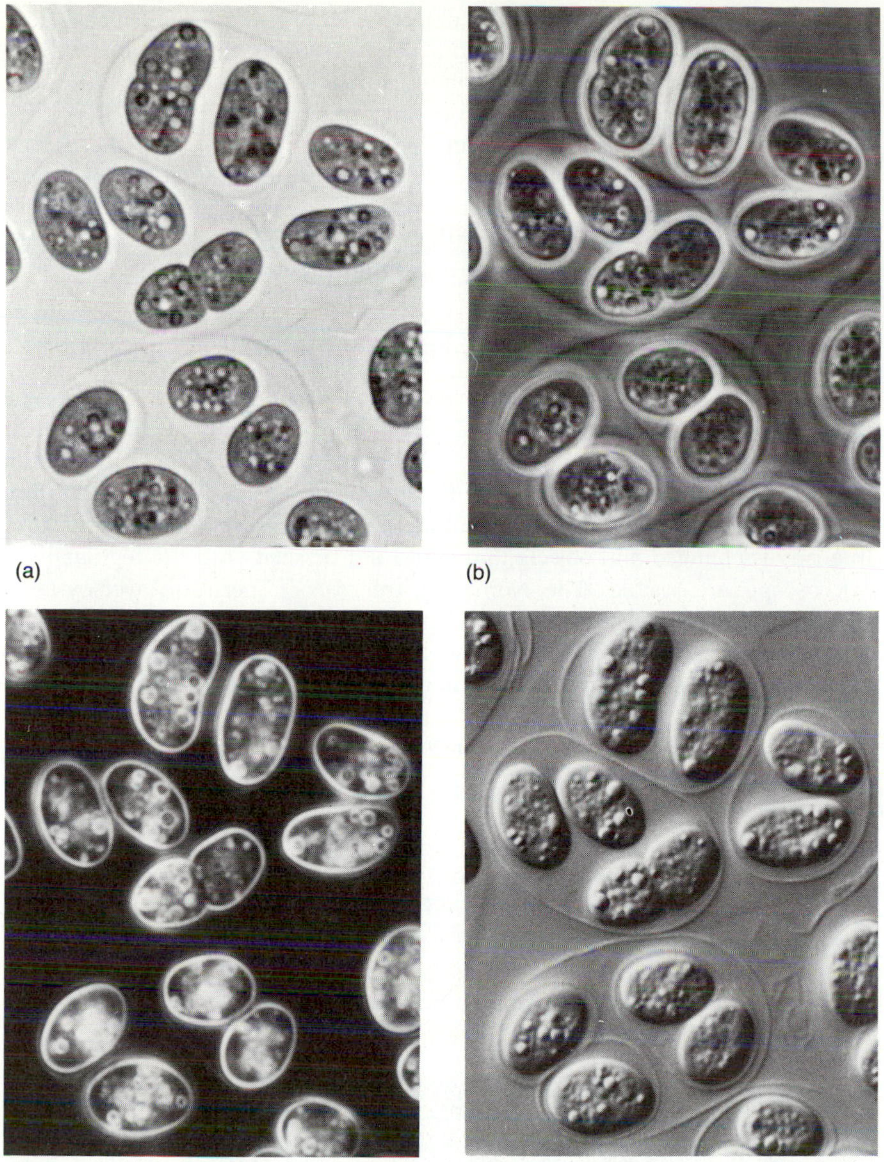

(a)

(b)

(c)

(d)

Figure 3-6 Images Produced with Bright Field, Phase Contrast, Dark Field, and Interference Microscopes. Shown are the same cells as seen using different microscope techniques. *(a)* Bright field. *(b)* Phase contrast. *(c)* Dark field. *(d)* Interference.

to increase the contrast between them and the background. Figure 3-6 compares the images of an unstained bacterium examined with a phase contrast microscope and a similar bacterium examined with an ordinary light microscope. The phase contrast microscope has become an extremely important tool in the microbiology laboratory because it intensifies the contrast between translucent objects in living, unstained specimens.

The Dark Field Microscope

The **dark field microscope** is essentially a light microscope with a dark field stop (fig. 3-7) placed between the light source and the condenser. Only light at very oblique angles hits the specimen, since most of the light rays are blocked by the dark field stop. When a microorganism or any other object

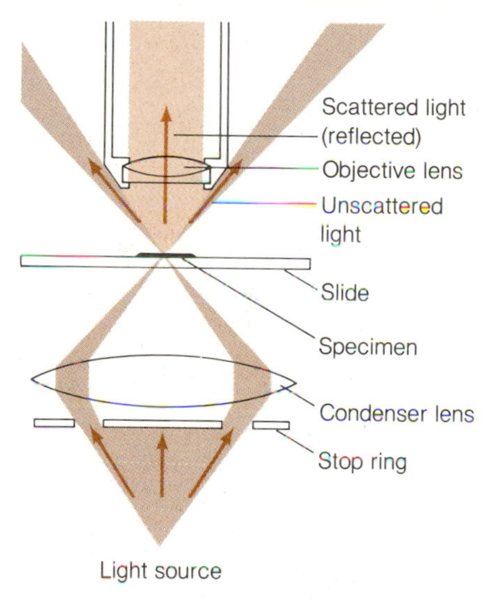

Figure 3-7 How the Dark Field Microscope Works. The dark field microscope is very similar to the bright field microscope, except that an opaque disk eliminates all of the light in the center of the beam. The only light that reaches the specimen comes in at an angle. Consequently, no direct light reaches the objective lens. The only light that reaches the objective lens is light that is scattered from the specimen. Thus, the specimen and/or objects within it appear bright, while the background looks dark.

reflects (scatters) the light into the objective lens, the subject appears as a bright spot against a dark background (fig. 3-6). Many microorganisms that are not visible under bright field or phase contrast microscopes, because they are too small, can be detected because they reflect (scatter) light in a dark field and so appear much larger than they really are. Dark field microscopy is frequently used to verify the presence of *Treponema pallidum*, the spirochete that causes syphilis, in scrapings from chancres (syphilitic sores). These organisms are generally invisible with bright field microscopes and phase contrast microscopes because they are too thin (fig. 3-8).

The Nomarski Differential Interference Contrast Microscope

The **Nomarski interference contrast microscope** is a light microscope that uses two beams of linearly polarized light to detect slight differences in indices of refraction in the specimen. Differences in the indices of refraction alter the phase relationship between the beams of light so that they interfere with each other when they are recombined. Consequently, the resultant light intensity is changed and objects that could not be seen with the regular light microscope are visible. The Nomarski microscope permits layer-by-layer scanning of specimens because the depth of focus is very shallow. In addition, the microscope gives a sharply defined, relief-like image with excellent contrast (fig. 3-9). The Nomarski microscope works better than the phase contrast microscope because it does not produce halos and because its resolution is almost twice that of the phase contrast microscope (fig. 3-6).

The Fluorescence Microscope

The **fluorescence microscope** is similar to a dark field microscope, except that "invisible" ultraviolet light is used to illuminate the object. The object becomes visible if it absorbs the ultraviolet light and then gives off light at longer wavelengths in the visible spectrum. Certain substances, when exposed to ultraviolet radiation, absorb it and almost immediately release it as

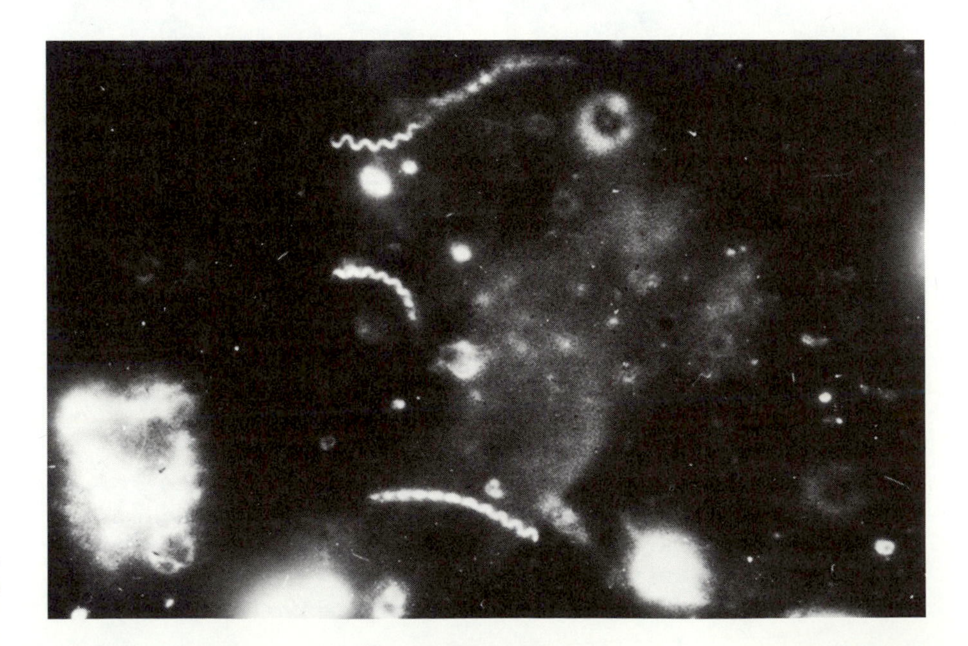

Figure 3-8 Dark Field View of *Treponema Pallidum*. The spiral shaped bacteria can be seen amidst blood cells.

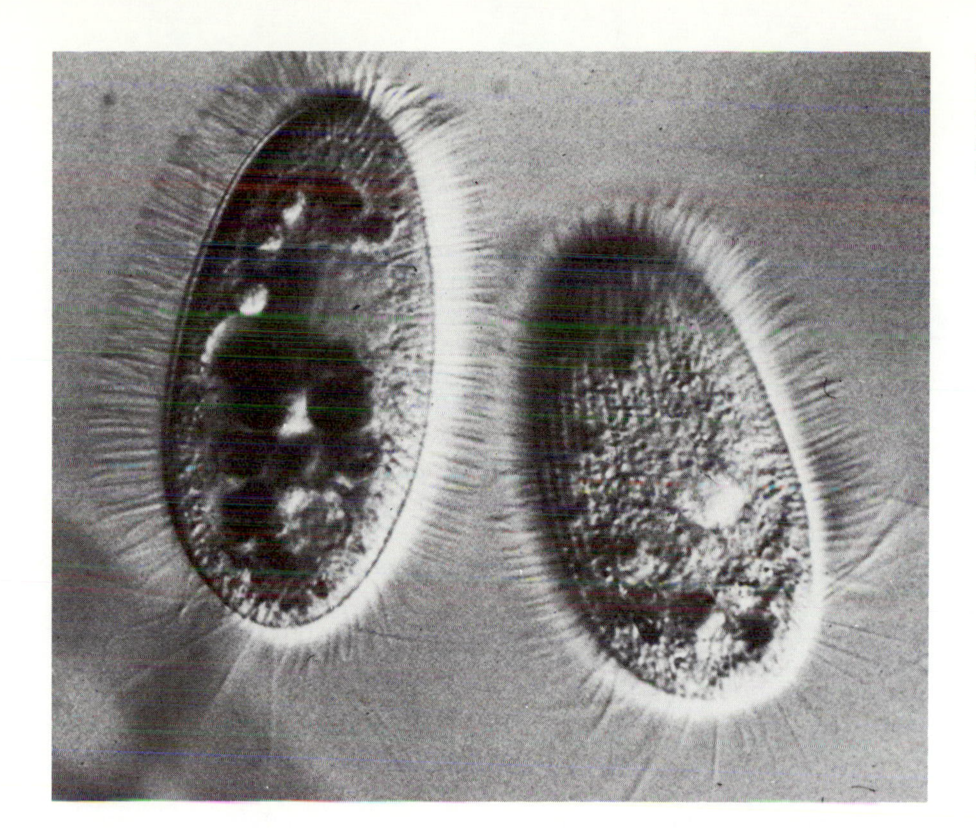

Figure 3-9 A View of A Ciliate Using Nomarski Interference. This technique permits viewing of unstained specimens and provides a three-dimensional image of the specimen.

visible light of a longer wavelength. Materials that give off light in this way are said to be **luminescent.** If the emissions by a luminescent substance occur only during the time of exposure to the ultraviolet light, the emissions are known as **fluorescence** and the material fluoresces. If the emissions persist at an appreciable level of intensity after the excitation ends, the radiation is called **phosphorescence** and the substance is said to phosphoresce.

ELECTRON MICROSCOPES

The **electron microscope** has enabled microbiologists to "see" and study structures that are too small to be resolved with the light microscope. The electron microscope uses electrons rather than light to produce images. This is possible because electrons have wavelike properties similar to those of light. Since electron beams have extremely short wavelengths (0.0055 nm for an accelerating voltage of 50,000 volts), the theoretical limit of resolution is much lower than for the light microscope. For example, a 50,000 volt electron beam with a wavelength of 0.0055 nm gives a limit of resolution of 2.7 nm, nearly 100 times better than the light microscope (approximately 280 nm):

$$\lambda = \frac{1.23\ \text{nm}}{\sqrt{V}} = \frac{1.23\ \text{nm}}{\sqrt{5 \times 10^4}} = 0.0055\ \text{nm}$$

$$R = \frac{\lambda}{2(\text{NA})} = \frac{0.0055\ \text{nm}}{2(0.001)} = 2.7\ \text{nm}$$

The equation for the limit of resolution of the electron microscope is different than the equation used for the light microscope because the physical characteristics of the two microscopes are different.

There are two types of electron microscopes: the **transmission electron microscope** (TEM) and the **scanning electron microscope** (SEM) (fig. 3-10). Both the TEM and SEM are used extensively by microbiologists. The TEM is used when information regarding subcellular details is desired, while the SEM is used for visualization of a surface structure (fig. 3-11). In practice, the transmission electron microscope can resolve objects as close as 2.5 nm. Objects are generally magnified between 10,000 and 100,000×. In contrast, the scanning electron microscope can resolve objects as close as 20 nm. With this microscope, objects are usually magnified between 1000 and 10,000×.

Electrons from an **electron gun** are accelerated to high velocities in both the TEM and SEM. They are focused on the specimen by magnetic lenses, rather than glass lenses as in the light microscope (fig. 3-10). In the TEM, an electromagnetic **condenser lens** focuses the electron beam on a small area of the specimen and thus performs roughly the same function as the condenser in a light microscope. The beam of electrons passes through the object and then through the electromagnetic **objective,** resulting in a magnified image. Finally, the electrons are focused onto a fluorescent screen or photographic plate by the electromagnetic **projector lens.** The final image appears as an assortment of bright and dark areas, depending upon how many electrons have been absorbed by the specimen. For example, bright areas represent portions of the specimen which absorbed few electrons, while dark areas represent electron-dense portions of the specimen.

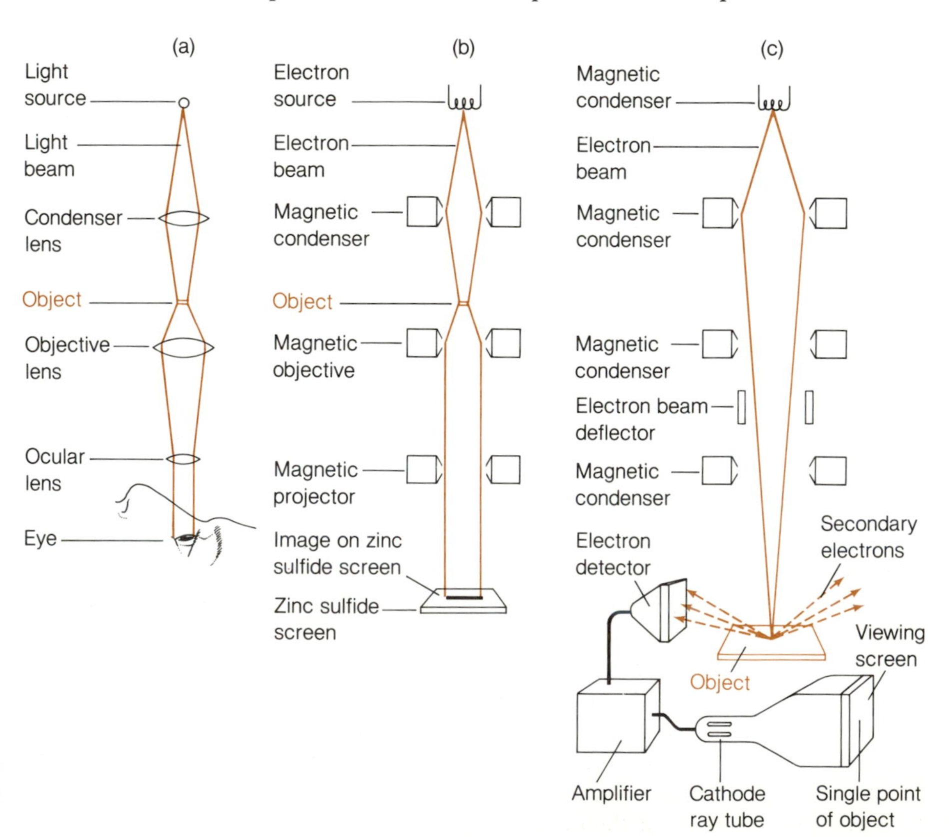

Figure 3-10 Comparison of *(a)* Light Microscope, *(b)* Transmission Electron Microscope, and *(c)* Scanning Electron Microscope

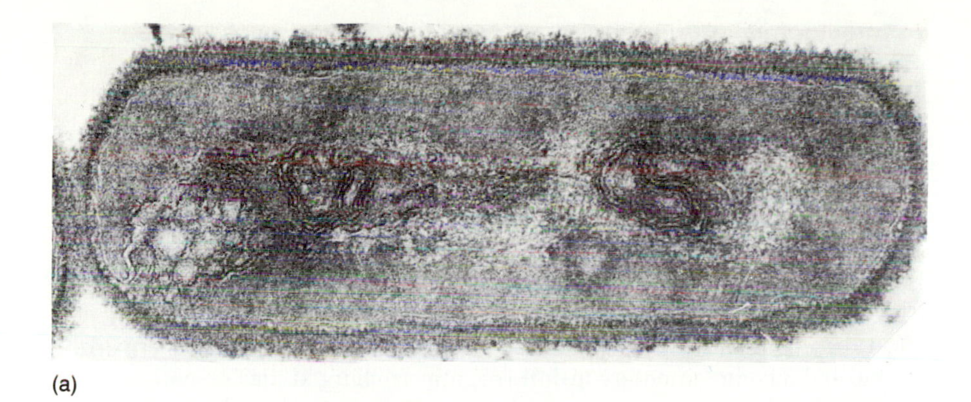

(a)

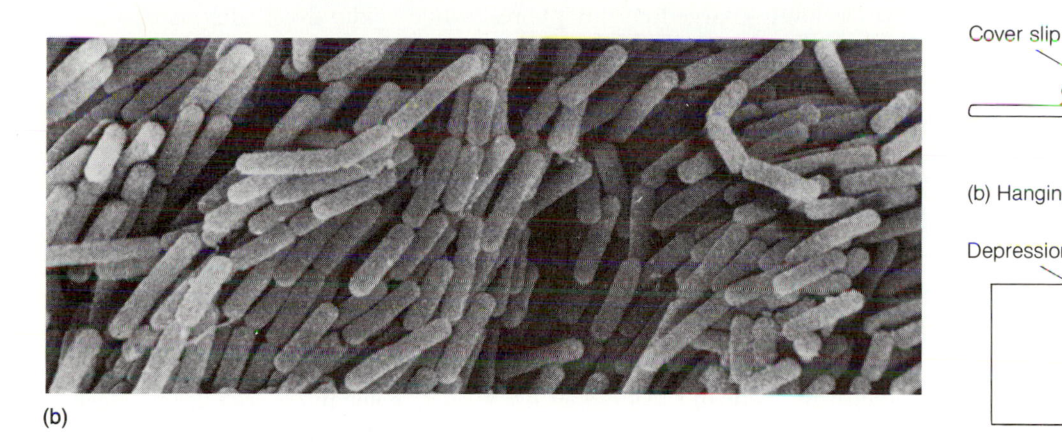

(b)

Figure 3-11 Images Produced by Transmission Electron Microscope and Scanning Electron Microscope. *(a)* Transmission and *(b)* scanning electron micrographs of *Bacillus*. Notice that internal details are best seen using the TEM while surface features are readily viewed with the SEM.

In contrast to the TEM, the electron beam in the SEM scans the entire surface of the specimen and induces the release of secondary electrons from the specimen. This secondary radiation controls the formation of an image on a cathode ray tube in much the same way an image is formed on a television screen (fig. 3-10).

LIGHT MICROSCOPE TECHNIQUES

The Examination of Living Microorganisms

Living microorganisms are studied in order to determine their size and shape, whether or not they are motile, how they divide, and sometimes whether they produce spores and granules. These observations are made using a light microscope with unstained and stained preparations. **Wet mount** (fig. 3-12) preparations are generally used to study living microorganisms. Wet mounts are made by placing an aqueous suspension of microorganisms on a glass slide and covering it with a cover glass (fig. 3-12a). The slide can then be viewed with a light microscope. A variation of the wet mount is the **hanging drop** preparation (fig. 3-12b), which requires the use of a **depression slide,** a cover glass, and some petroleum jelly. The advantage of this technique is that it allows for extended observations of the microbial population in a more or less "natural" environment. The petroleum jelly is used to prevent evaporation of the drop.

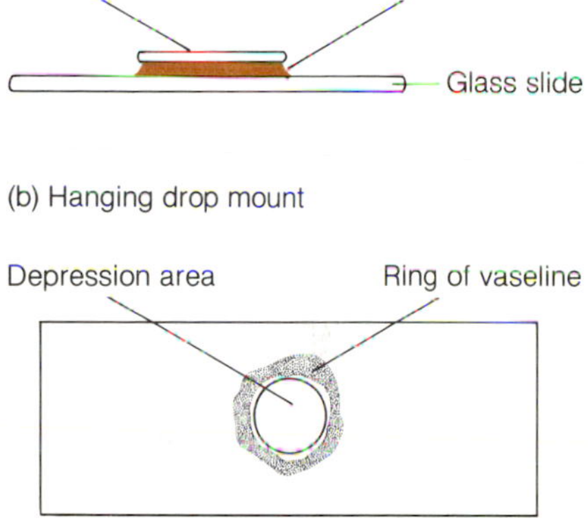

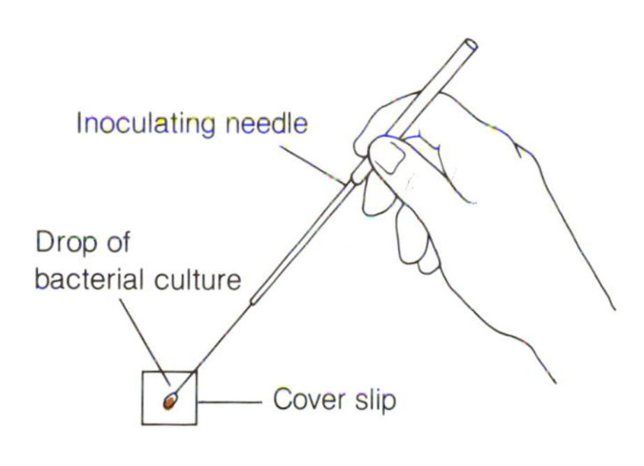

Cover slip
Drop of culture
Vaseline
Hanging drop preparation

Figure 3-12 Wet Mount and Hanging Drop Mount. *(a)* The wet mount is prepared by placing a small drop of culture onto a clean glass slide and then covering the drop with a coverslip. *(b)* The hanging drop mount is prepared by placing a small drop of culture onto a coverslip and then placing the inverted coverslip onto a ring of vaseline that surrounds the depression in a depression slide.

Examination of Stained Preparations

Although much can be learned from observing unstained microorganisms, the addition of colored substances can reveal structures that are invisible in unstained preparations. Colored organic compounds called **dyes** are often used to stain microbiological specimens on slides in order to enhance the contrast between the microorganisms and their background. Staining techniques are used routinely to differentiate among major groups of microorganisms, to detect chemical and structural differences in bacterial cell walls (Gram stain and acid fast stain), and to observe specific cellular components.

Dyes are made up of a charged colored portion, called the **chromophore,** and a complementary **anion** (negatively-charged ion) or **cation** (positively-charged ion). Basic dyes, such as methylene blue and crystal violet, have positively-charged chromophores, while acidic dyes, such as eosin and picric acid, have negatively-charged chromophores. Since cells generally have many more negatively-charged groups than positively-charged groups, they will be stained readily by basic dyes because the opposite charges attract each other. Acidic dyes are almost always used to stain the background.

Before microorganisms can be stained, usually they must be attached or fixed to a glass slide. To accomplish this fixation, a thin **smear** of the microorganism is prepared by spreading a small portion of an aqueous suspension of microorganisms on a clean glass slide. The smear is allowed to air-dry and is then heated for about 3 seconds over an open flame. The heating process is called **heat fixation.** It serves to attach the microorganisms to the slide so that they will not wash away during the staining procedures.

Simple staining techniques are used to color microorganisms or cellular inclusions so that they can be seen readily. Basic dyes are generally used in simple staining procedures, because they attach efficiently to the many negatively-charged groups in the cytoplasm, cell membrane, and cell wall. Sometimes, dyes have a higher affinity for some components within the cell than others and an unevenly stained cell results.

Differential staining techniques are commonly used in microbiology because they provide a great deal of information about the structure and chemical composition of the organisms being stained. There are two differential staining procedures in common use: the **Gram stain** and the **acid-fast stain.**

The gram stain, developed by Hans Christian Gram in 1884, was originally used to reveal bacteria in diseased animal tissue. Soon after its development, however, microbiologists realized how valuable the Gram stain could be in distinguishing among bacteria. The Gram stain is undoubtedly the most widely utilized differential staining procedure in the bacteriology laboratory, because it allows microbiologists to divide bacteria into two types: the **gram positive** bacteria and the **gram negative** bacteria (table 3-2). The separation of bacteria into these two types is one of the first steps in the identification of a bacterium. After being Gram stained, bacteria are colored either by **crystal violet** (blue-violet) or by **safranin** (light red). The blue-violet bacteria are called gram positive, while the light red bacteria are gram negative (table 3-2). It is believed that differences in the thickness, charge concentration, and chemical composition of the cell wall □ are responsible for the differential staining of bacteria. The gram positive bacteria retain the purple dye better than the gram negative bacteria because the gram positive bacteria have thicker walls and more positive charges.

81

TABLE 3-2
THE GRAM STAIN PROCEDURE

REAGENTS	TIME APPLIED	REACTIONS	APPEARANCE
Unstained smear	—	—	Cells are colorless and difficult to see.
Crystal violet	1 minute, then rinse with water.	Basic dye attaches to negatively charged groups in the cell wall, membrane, & cytoplasm.	Both gram negative and gram positive cells are dark blue-violet.
Gram's iodine (mordant)	1 minute, then rinse with water.	Iodine strengthens the attachment of crystal violet to the negatively charged groups.	Both gram negative and gram positive cells remain dark blue-violet.
Ethanol or acetone-ethanol mix (decolorizer or leaching agent)	10–15 seconds, then rinse with water.	Decolorizer leaches the crystal violet & iodine from the cells. The color diffuses out of gram positive cells more slowly than out of gram negative cells because of the chemical composition and thickness of the cell walls.	Gram positive cells remain dark blue-violet, but the gram negative cells become colorless and difficult to see.
Safranin (counterstain)	1 minute, then rinse thoroughly, blot dry, observe under oil immersion.	Basic dye attaches to negatively charged groups in both cell types. Few negative groups are free in gram positive cell, while most negative groups are free of crystal violet in gram negative bacteria. Consequently, gram positive bacteria remain dark blue-violet, while gram negative bacteria become light red.	Gram positive cells remain dark blue-violet, while gram negative cells are stained a light red.

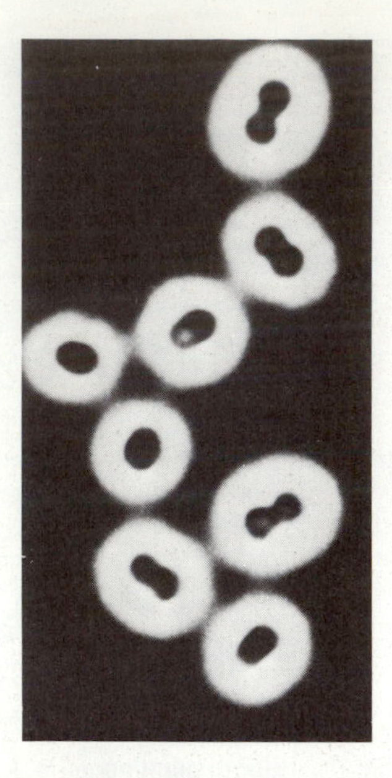

Figure 3-13 Negative Staining. A smear of the bacterium *Streptococcus pneumoniae* is negatively stained. The capsule does not stain and shows up as a bright area around the cells.

The acid-fast stain measures the resistance of a bacterial cell to decolorization by acid. It was originally developed by Paul Ehrlich and later modified by Franz Ziehl and Friedrich Neelsen in order to demonstrate the presence of *Mycobacterium tuberculosis* in patients afflicted with tuberculosis. The acid-fast characteristic is a distinctive property of the genus *Mycobacterium* and some species of *Nocardia.* Pathogenic bacteria in these genera include the agents of tuberculosis, leprosy, and lumpy jaw (nocardiosis). Acid-fast organisms, stained with a hot phenolic solution of carbol fuchsin, retain the red dye even after treatment with dilute sulfuric or hydrochloric acid-alcohol solutions. On the other hand, non-acid fast organisms are decolorized when treated with acid-alcohol solutions. Non-acid-fast organisms appear blue after the acid-fast staining procedure because they are counterstained with methylene blue. The acid-fast characteristic is generally associated with the very high content of complex waxes in the cell wall.

Negative staining techniques are used to reveal the overall shape and arrangement of microorganisms and particular structures that are not easy to stain directly (positively). In negative stains, the background is colored while the structure of interest frequently remains colorless. Figure 3-13 is a photograph of bacteria that have thick coverings called capsules □, made of polysaccharide or polypeptide. The capsule is unstained and is seen as a clear area between the stained cell and the stained background. Acid dyes such as congo red or nigrosin, which have little affinity for the negatively-charged cellular structures, are allowed to dry on the slide and are responsible for the colored background. Basic dyes are used to stain the cell.

Fluorescent Staining

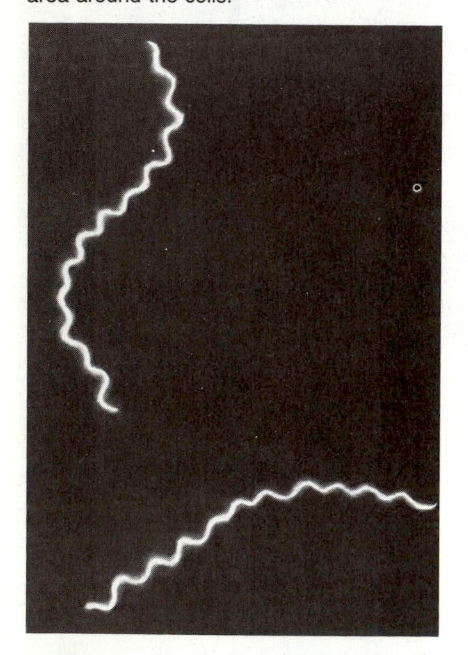

Figure 3-14 Fluorescent Staining. This photograph illustrates the fluorescent staining of *Treponema pallidum,* the cause of syphilis. When cells "stained" with appropriate fluorescent dyes or reagents are subjected to ultraviolet light, they give off visible light and therefore appear bright. The background is dark because the ultraviolet light is invisible and filtered out so that it cannot reach the viewer.

A few microorganisms, when irradiated with ultraviolet light (UV), exhibit a pale blue fluorescence called **autofluorescence.** This autofluorescence is due to special pigments the organisms produce. Most biological specimens do not exhibit any autofluorescence, but can be examined with a fluorescence microscope by "staining" them with special fluorescing dyes called **fluorochromes** or simply **fluors.** One fluorochrome used extensively in microbiology is **fluorescein isothiocyanate** (FITC). FITC is often used as a fluorochrome in a technique called **fluorescent antibody staining.** In this technique, fluorochromes attached to serum proteins called **antibodies** are used to "stain" microorganisms. Antibodies bind specifically to certain chemical groups called **antigens.** Fluorescent antibody tests take advantage of the specificity of the antibodies to stain specific portions of specimens. This technique is widely used in clinical and public health laboratories to diagnose certain infectious diseases. Figure 3-14 illustrates two bacteria stained by this technique.

ELECTRON MICROSCOPE TECHNIQUES

Preparation of Specimens

Specimens for the transmission electron microscope are usually **fixed** with chemicals such as glutaraldehyde so that they will not decompose when dried and sliced. After specimens are chemically fixed in glutaraldehyde and dehydrated in acetone and alcohol, they may then be embedded in a plastic matrix and cut into very thin sections with an **ultramicrotome.** Since biological specimens have a limited capacity to scatter electrons, they generally are "stained" with electron dense materials, such as salts of tungsten, manganese, uranium, or osmium, before they are examined.

Metal shadowing or **shadow casting** is a TEM technique used to give objects a three-dimensional appearance. To achieve this effect, the specimen is coated with a heavy metal, such as gold or palladium. The metal is generally deposited at an angle to the specimen so that a shadow of the specimen is created.

Osmium tetraoxide is commonly used to "stain" and fix tissue. Sometimes the specimen is negatively stained with phosphotungstic acid. In this case, those portions of the specimen that exclude or lack phosphotungstic acid transmit electrons and consequently look lighter than the areas where electrons are scattered. Most viruses are viewed in this fashion (fig. 3-15). Viruses are often coated with metal rather than negatively stained, because this procedure creates more contrast between the viruses and the background.

Freeze fracture techniques have been used extensively to study the structure of cellular organelles and membranes. In this technique the specimen is rapidly frozen in Freon 12, a refrigerant, to $-100°C$ and then sectioned in a vacuum. Under these conditions, the cell and its organelles are fractured. The fractured specimens are then shadowed with platinum and carbon to create a metal replica of the fractured material. The organic material is dissolved with acid and the remaining platinum and carbon shell is examined in the TEM. The resulting image appears three-dimensional (fig. 3-16). This technique has been useful in studying the structure of chloroplast, mitochondrial, and cytoplasmic membranes.

Specimens for SEM examination are prepared differently from those to be studied in the TEM. The specimens are usually mounted on stubs and then dried so that the water evaporates at approximately $-70°C$. Drying under these conditions prevents the specimen from becoming distorted. Once the specimens are properly **freeze dried,** they are generally **sputter coated** with a layer of a heavy metal, such as gold or palladium. This coating provides the electrons that are used to create the image.

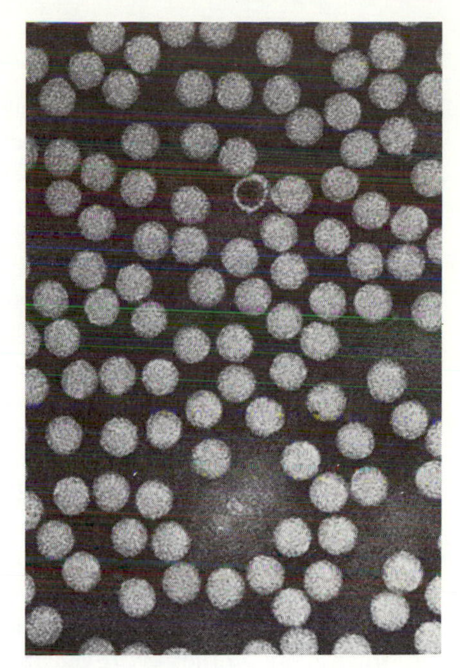

(a)

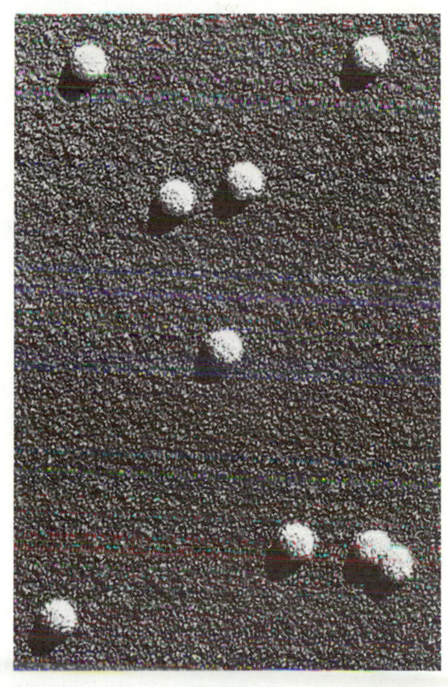

(b)

Figure 3-15 Phosphotungstic Acid Staining and Shadow Casting. *(a)* Virus stained with phosphotungstic acid. *(b)* Shadow casting of a virus.

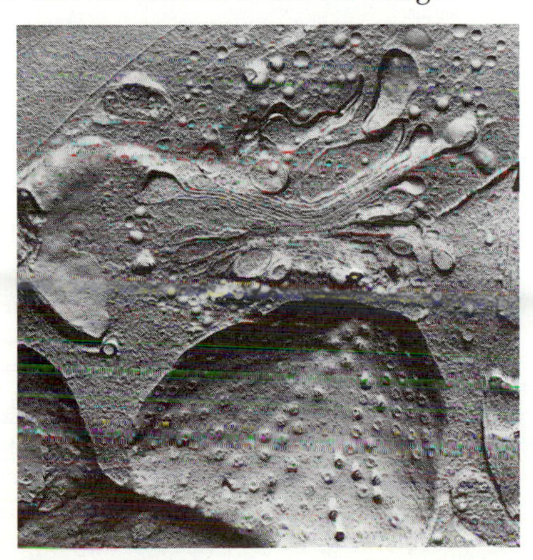

Figure 3-16 Freeze Fracture Technique. Freeze fracture replica of a cell. Notice the three-dimensional appearance of organelles within the cell. The nucleus and pores of the nuclear membrane as well as the Golgi complex are seen.

SUMMARY

LIGHT MICROSCOPES

1. A compound microscope is a light microscope that has two lens systems in series: the objective lens and the ocular lens. There are several different types of light microscopes: bright field, dark field, phase contrast, differential interference contrast, and fluorescence microscopes.
2. The limit of resolution of all light microscopes is a function of the wavelength of light (λ), the index of refraction of the medium between the specimen and the objective lens (n), and half the angle of the cone of light entering the lens (θ). The product $n(\sin\theta)$ is called the numerical aperture. Objects closer together than the resolvable distance cannot be distinguished as separate entities. In addition, objects that have a dimension much smaller than the limit of resolution generally cannot be seen.
3. The bright field microscope is the most widely used microscope in biology. It generally is used to determine the shape, size, characteristic arrangement, and motility of bacteria and other microorganisms in their natural environments.
4. The phase contrast microscope is used to increase the contrast between cells and their background, as well as between structures within cells and the cytoplasm. The phase contrast microscope takes advantage of the fact that very small translucent objects and objects with different refractive indices diffract (spread) light. A phase shifting element in the objective lens system alters the light so that there is a difference in phase between the diffracted and undiffracted light. When the diffracted and undiffracted light are brought together again, there is a decrease in light intensity because of the phase difference. The smaller the translucent object, the greater the diffraction and the darker the object will appear.
5. The dark field microscope is used to view bacteria that are too small to be resolved with the bright field microscope. The dark field microscope is frequently used to view the very thin spirochetes, such as *Treponema* (syphilis) and *Leptospira* (leptospirosis), which cannot be resolved clearly by bright field microscopes. A light stop is placed between the light source and the object and the condenser focuses the light so that it hits the specimen from the side. The only light that reaches the objective lens is light that is reflected (scattered) from objects.
6. The Nomarski differential interference contrast microscope uses light that is split into two beams in order to detect objects that have very similar refractive indices. When the light beams penetrate different materials, they undergo different phase shifts. Consequently, when the light beams are recombined, they interfere constructively or destructively and change the intensity of the light reaching the observer's eyes. The interference microscope produces an image that is three-dimensional.
7. The fluorescence microscope is used to detect certain microorganisms, such as those that are too small to be seen with other types of microscopes or that are difficult to find because they are few in number or in a mixed culture. The fluorescence microscope uses "invisible" ultraviolet light. Organisms are stained with a dye that fluoresces visible light when hit by ultraviolet. The staining can be made very specific if the dye is attached to antibodies that bind to only one type of organism.

ELECTRON MICROSCOPES

1. Electron microscopes have provided detailed knowledge about the structure of microorganisms. Electron microscopes employ electrons instead of light to form an image. Magnetic lenses are used to direct electrons against the object and to focus the resulting electrons onto a screen or photographic plate. Because electrons have a much shorter wavelength than visible light or ultraviolet radiation, the theoretical limit of resolution for an electron microscope is about $100\times$ less than the limit of resolution for a light microscope.
2. There are two types of electron microscopes: the transmission electron microscope (TEM) and the scanning electron microscope (SEM). The TEM is generally used when subcellular structures and their relationships to each other are to be studied. In the TEM, the electrons that penetrate the specimen are used to make an image of the object. Staining or coating with heavy metals increases the contrast between different parts of the object by reflecting (scattering) electrons. The SEM is most often employed when information about the surface structure of a microorganism is desired. In the SEM, the electrons that hit the specimen cause the release of secondary electrons from the specimen. Secondary electrons and reflected (scattered) electrons are used to form an image of the object.
3. The electron microscope has been used to visualize viruses that are too small to be seen with any light microscope.

LIGHT MICROSCOPE TECHNIQUES

1. Living microorganisms are generally observed by using wet and hanging drop mounts.
2. In order to see microorganisms more clearly and study their anatomy, various staining techniques are used: a) simple staining is the staining of a specimen with a basic dye so that a good contrast between the specimen and the background is achieved; b) negative staining is the staining of the background with an acidic dye so that the specimen can be distinguished from the background. An example of negative staining is the capsule stain. Negative staining and simple staining are often combined to improve contrast; c) differential staining is the staining of a specimen with two or more basic dyes so as to differentiate cellular structures. The gram stain, and the acid fast stain are examples of differential staining; d) fluorescent staining is the staining of the specimen with fluorescing dyes.

ELECTRON MICROSCOPE TECHNIQUES

1. Specimens for the transmission microscope are usually fixed to prevent decomposition, embedded in plastic to maintain the specimens' shape, and cut into thin sections with an ultramicrotome.
2. Metal shadowing is used to give specimens a three-dimensional appearance. It is generally used with intact viruses and bacteria whenever the shape of the organisms is to be studied.
3. Freeze fracture techniques allow researchers to view the surfaces of organelles and membranes. Fractured specimens are coated with platinum and carbon to create metal replicas of the specimens. The replicas appear three-dimensional.
4. A specimen for the SEM is freeze-dried and then sputter coated with a layer of heavy metal, such as gold or palladium, which provides the secondary electrons for creating an image.

STUDY QUESTIONS

1. What is a compound microscope? What is a monocular microscope?
2. Determine the limit of resolution in μm of a light microscope when the wavelength is 500 nm and the numerical aperture is 1. Determine the limit of resolution in μm when the wavelength is 400 nm and the numerical aperture is 1.20.
3. Why is oil often put between stained specimens and the objective lens?
4. What is the limit of resolution of (a) the human eye; (b) a good bright field microscope; (c) an electron microscope? Which lens system should be used to see (a) a flat human cheek cell 100 μm × 100 μm; (b) a rod-shaped bacterium 5 μm × 1 μm; (c) a virus 0.1 μm in diameter?
5. Of what use is a dark-field microscope? Why is the field dark and the object light?
6. Of what use is a phase-contrast microscope? Explain why small objects are dark compared to the bright field.
7. How does the Nomarski differential interference microscope differ from the phase-contrast microscope? How do the images differ?
8. Of what use is a fluorescence microscope? What kind of light is used to illuminate the specimen? Explain why the field is dark and the object colored.
9. Discuss two different ways in which a specimen might be prepared for the bright field microscope.
10. Of what use is a transmission electron microscope? How is contrast achieved in the TEM?
11. Of what use is a scanning electron microscope? How is contrast achieved in the SEM?
12. Discuss three ways in which a specimen might be prepared for the TEM. Discuss how a specimen is prepared for the SEM.

SUPPLEMENTAL READINGS

Beer, M., Carpenter, R., Eyring, L., Lyman, C., and Thomas, J., 1981. Chemistry viewed through the eyes of high-resolution microscopy. *Chemical and Engineering News* Aug. 17, 40–61.

Jensen, W., and Park, R. 1967. *Cell Ultrastructure.* California: Wadsworth Publ. Co.

Porter, K., and Tucker, J., 1981. The ground substance of the living cell. *Scientific American* 244(3): 56–67.

Spencer, M., 1982. *Fundamentals of Light Microscopy.* Cambridge University Press.

Valkenburg, J. A. C., Woldringh, C. L., Brakenhoff, G. J., van der Voort, H. T. M., and Nanninga, N. 1985. Confocal scanning light microscopy of the *Escherichia coli* nucleoid: Comparison with phase-contrast and electron microscope images. *Journal of Bacteriology* 161(2):478–483.

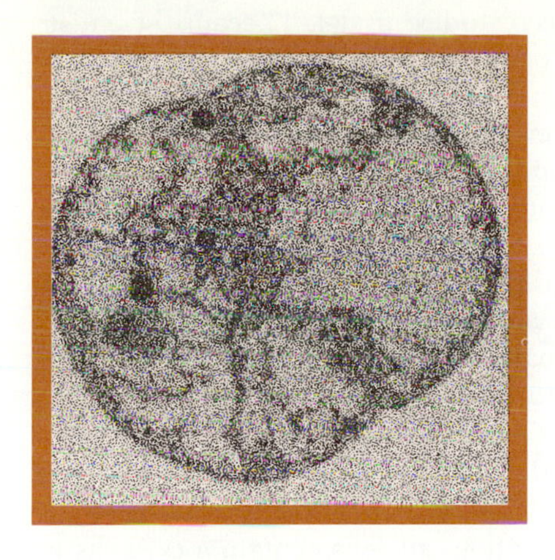

CHAPTER 4
STRUCTURE OF PROKARYOTIC AND EUKARYOTIC CELLS

CHAPTER PREVIEW

THE BACTERIAL CELL WALL IS IMPORTANT TO HUMANS TOO

Most bacteria have a tough, rigid outer layer called the cell wall, which gives the cell its shape and protects it from environmental damage. In addition, the bacterial cell wall may control the passage of chemicals into and out of the cell. Clearly, the bacterial cell wall is an asset to bacteria because it improves their chances for survival in hostile environments.

Study of bacterial cell walls has led to the discovery that most bacteria have one of two basic types of cell walls, which can readily and easily be differentiated by performing the **Gram stain.** This stain is used routinely in microbiology laboratories and constitutes a very important criterion for the identification of bacteria.

The type of cell wall a bacterium has is extremely important to physicians who are interested in how to treat a patient suffering from a bacterial infection. For example, the use of the antibiotic penicillin is recommended primarily for the control of infections caused by gram positive bacteria, since it is these bacteria, not the gram negative ones, that are more susceptible to the antibiotic. Similarly, there are numerous antibiotics that are used against gram negative bacteria because they usually work better than penicillin.

The cell wall of gram negative bacteria is important to humans in another way. It is now known that gram negative cell walls have a component called **lipopolysaccharide,** or simply "LPS," that has toxic properties. LPS is also commonly known as "endotoxin." Persons whose blood becomes infected by gram negative bacteria may suffer severe ill effects, including high fever, extremely low blood pressure, and sometimes death. It is important to insure that medicines and pharmaceuticals administered by injection are free of gram negative bacteria or lipopolysaccharide from their cell walls, because injection of bacteria or LPS can have grave consequences.

Scientists have learned that LPS also has a significant effect on the functioning of the immune system, so it is now used to study the immune system.

As can be seen from these examples, the bacterial cell wall has characteristics that are important to humans. The study of biological structures such as bacterial cell walls can lead to our understanding of how microorganisms affect us. In this chapter we will discuss the major structural characteristics of microorganisms, emphasizing their biological importance and impact on humans.

LEARNING OBJECTIVES

A STUDY OF THIS CHAPTER SHOULD ENABLE YOU TO:

LIST THE DIFFERENCES AND SIMILARITIES BETWEEN A PROKARYOTIC CELL AND A EUKARYOTIC CELL

DIAGRAM A BACTERIAL CELL AND INDICATE ITS SALIENT STRUCTURES

EXPLAIN WHERE RESPIRATION AND PHOTOSYNTHESIS OCCUR IN PROKARYOTES

DIAGRAM A TYPICAL EUKARYOTIC CELL

EXPLAIN WHERE RESPIRATION AND PHOTOSYNTHESIS OCCUR IN EUKARYOTES

The structure of microorganisms has been studied in detail because scientists are interested in understanding how organisms are constructed and how their various parts function. In addition, a knowledge of what microorganisms are and the various forms that they can have is often helpful in identifying and controlling them. Microorganisms may be noncellular, like the viruses, or cellular, like bacteria and protozoans. In this chapter, we will be concerned only with cellular organisms.

The term "cell" was first introduced by Robert Hooke in 1665 to describe the chambers in thin sections of cork and wood. Although we now know that these chambers represent the remains of dead cells, Hooke's observations were the first of many that led to the **cell theory** of life, which was formulated in the 1800s by a number of scientists such as Theodor Schwann and Rudolf Virchow. Schwann proposed that all organisms are made up of cells, while Virchow suggested that all cells come from preexisting cells. Although cells vary greatly in form and function, all cells contain a cytoplasm, which is enclosed by a plasma membrane, and a DNA genome, which contains the hereditary material of the cell. The cytoplasm is an aqueous solution as well as a colloid of many different small and large molecules. In addition, it is the site of many of the cell's chemical reactions. The cytoplasm may contain large, membranous structures called **organelles** that are associated with many of the cell's vital functions. For example, some cells have one or more **nuclei** (singular, **nucleus)**. A nucleus is an organelle that consists of the cell's genome and two surrounding membranes.

PROKARYOTIC AND EUKARYOTIC CELLS: A DEFINITION

Examination of a variety of cells with the light and electron microscopes has shown that there are two different types of cells: **prokaryotic** cells and **eukaryotic** cells. Prokaryotic cells (prokaryotes) get their name from the fact that they do not contain a nucleus: *pro* means "primitive," while *karyote* (from the Greek *karyon*) refers to the nucleus. Thus, prokaryote denotes a cell with a primitive or nonexistent nucleus. Eukaryotic cells (eukaryotes) get their name because they have a nucleus (*eu* means "normal" or "true"). Prokaryotic cells are generally much simpler than eukaryotic cells (fig. 4-1). Electron photomicrographs of typical prokaryotes show a cell wall, a cytoplasmic membrane, a rather homogeneous granular cytoplasm, and a compact mass of chromatin (DNA genome) in the cytoplasm. In some prokaryotic cells, membranes such as mesosomes, chromatophores, and thylakoids are visible within the cytoplasm. Granules of starch, glycogen, polyphosphate, sulfur, and poly-β-hydroxybutyric acid, as well as gas vesicles, are also found in some prokaryotic cells. All the different types of bacteria (**eubacteria, archaebacteria, cyanobacteria, chlamydiae, rickettsiae,** and **mycoplasmas**) are prokaryotic cells. In fact, the terms prokaryotic and bacterial may be used interchangeably. Eukaryotic cells are generally much larger than prokaryotic cells, and numerous different organelles can be observed in the eukaryote's cytoplasm (fig. 4-1). Almost all eukaryotic cells have one or more nuclei that contain most of the cell's hereditary information (DNA genomes). Another organelle, called the **mitochondrion** (plural, **mitochondria**) is generally found in eukaryotic cells in addition to the nucleus. Organelles such as chloroplasts, endoplasmic reticulum, Golgi membranes, food

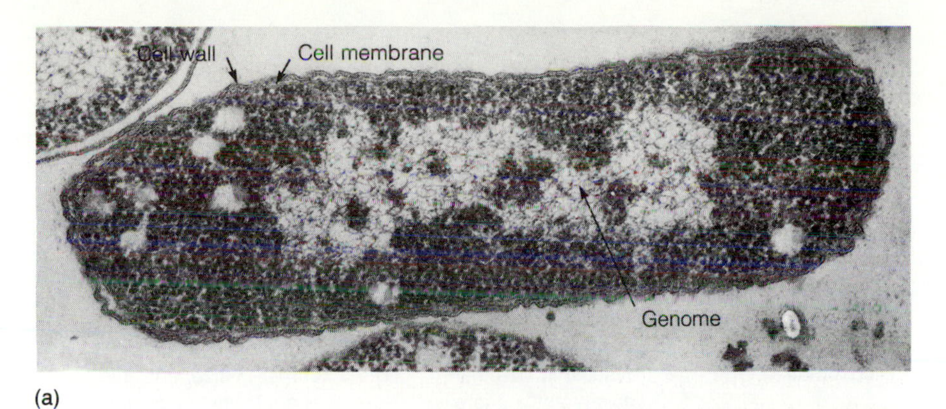

(a)

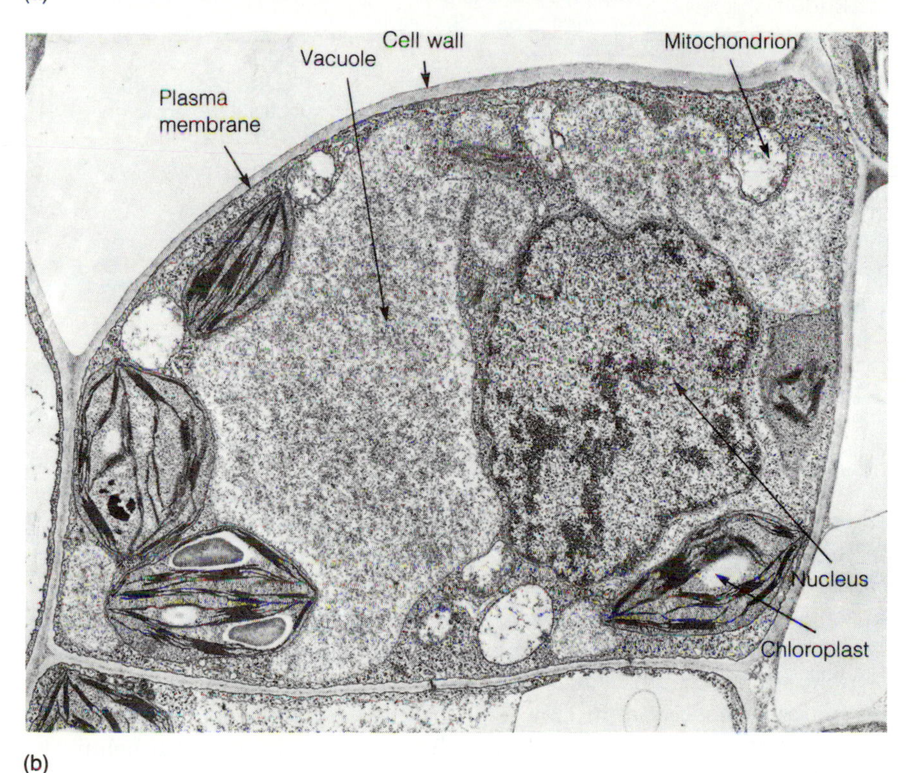

(b)

Figure 4-1 Prokaryotic and Eukaryotic Cells. *(a)* A transmission electron micrograph shows a cross section along the length of a prokaryotic (bacterial) cell, *Bacillus cereus.* The cell wall, cell membrane, and genome are regularly seen in prokaryotes. *(b)* A transmission electron micrograph shows a cross section through a eukaryotic (plant) cell. The plant cell is from the bean plant, *Phaseolus vulgaris.* The cell wall, plasma membrane, nucleus with chromatin, chloroplasts, mitochondria, vacuoles, Golgi membranes, endoplasmic reticulum, and ribosomes are generally seen in plant cells.

vacuoles, water vacuoles, contractile vacuoles, lysosomes, and centrioles are found in various eukaryotic cells. Some of the differences between prokaryotic cells and eukaryotic cells are outlined in table 4-1.

GROSS CELLULAR MORPHOLOGY OF BACTERIA

Bacteria, as a group, exhibit a wide range of sizes. Some, like the agents of typhus fever and Rocky Mountain spotted fever (rickettsia), are quite small and are barely resolved with the light microscope. Others are very large and can be seen with a magnifying lens or simple microscope. Most commonly encountered bacteria, however, have diameters between 0.2 μm and 3.0 μm. Rod-shaped bacteria usually have lengths between 0.5 μm and 15 μm. The average rod-shaped bacterium is about the size of an average mitochondrion

TABLE 4-1
SUMMARY OF CHARACTERISTICS OF PROKARYOTIC AND EUKARYOTIC CELLS.

CHARACTERISTICS	PROKARYOTIC CELL	EUKARYOTIC CELL
Genetic material	Circular DNA molecule. No discrete nucleus.	Arranged in chromosomes. Nucleus present in cell.
Membrane-enclosed organelles	Absent.	Present. These include mitochondria, chloroplasts, Golgi complexes, lysosomes and endoplasmic reticulum.
Ribosomes	Smaller in size than eukaryotes, generally referred to as 70S. Free in cytoplasm.	Larger in size. 80S. On endoplasmic reticulum.
Locomotion	Flagella that rotates, of simple composition. Some glide.	Flagella and cilia that undulate. Ameboid movements.
Plasma membrane	Sterols usually absent.	Sterols usually present.
Size of cell	Usually smaller.	Usually larger.
Cell wall	Contains peptidoglycan.	No peptidoglycan present.
Mitosis and meiosis	Absent.	Present.
Site of respiration	Plasma membrane.	Mitochondria.
Site of photosynthesis	Internal membranes.	Chloroplasts.

in a eukaryotic cell. Eukaryotic cells generally have diameters between 7 and 100 μm. A mouse cell infected with an intracellular bacterium, *Streptococcus pyogenes*, illustrates the difference in size between a prokaryote and a eukaryote (fig. 4-2). Because bacteria are so small, they have a large surface-to-volume ratio. Thus, they are able to concentrate nutrients rapidly and the nutrients are able to diffuse quickly to all sites inside the cell. Their small size is one of the factors that enables some bacteria to multiply rapidly. For example, the bacterium *Escherichia coli* can duplicate itself in about 20 minutes in a rich medium at 37°C.

The shape of a bacterial cell is a characteristic that is often helpful in identifying it. Most of the commonly encountered bacteria are spheres, rods, curved rods, spiral rods, or helical rods (fig. 4-3). The spherical bacteria are commonly called **cocci** (singular, **coccus**); the rods are known as **bacilli** (singular, **bacillus**); the curved rods are designated **vibrios** (singular, **vibrio**); the spiral rods are referred to as **spirilla** (singular, **spirillum**); and the helical rods are identified as **spirochetes** (singular, **spirochete**). The shape of a bacterium is determined by the shape of a rigid layer called the **cell wall.** If bacteria are treated with chemicals that disrupt or remove their cell walls and are placed in an isotonic environment, they assume a spherical shape, regardless of their original shape. A cell wall determines the shape of some eukaryotic cells, but many eukaryotic cells maintain a rigid shape without one. In these cells, the cytoplasmic membrane is strengthened by a layer of **stress fibers,** called the **cortex,** along the inside surface of the cytoplasmic membrane. The stress fibers consist of proteins, such as microtubules, actin, and myosin.

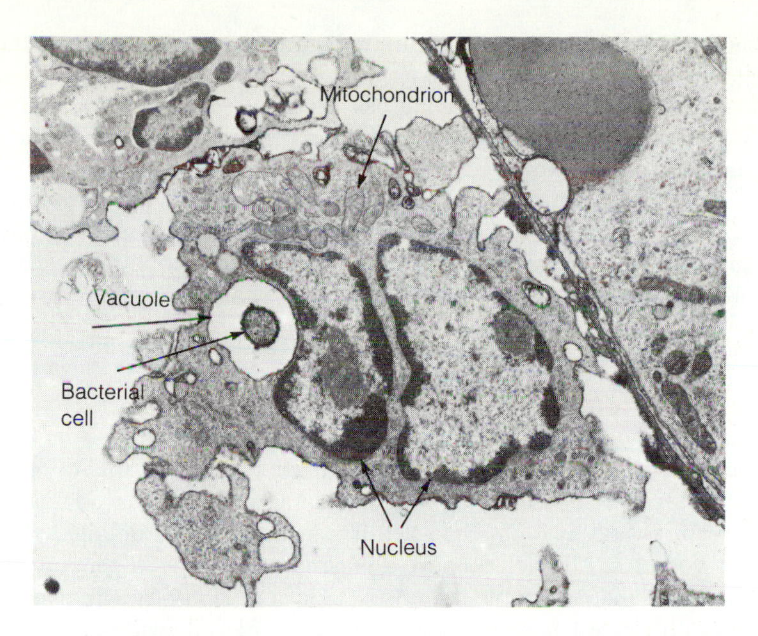

Figure 4-2 Comparison between a Eukaryotic Cell and a Prokaryotic Cell. A lung mononuclear phagocyte (a eukaryotic cell), which has ingested a cell of *Streptococcus pyogenes* (a prokaryotic cell) is shown here. The bacterial cell is contained within a vacuole. The phagocyte's nucleus and mitochondrion are also indicated.

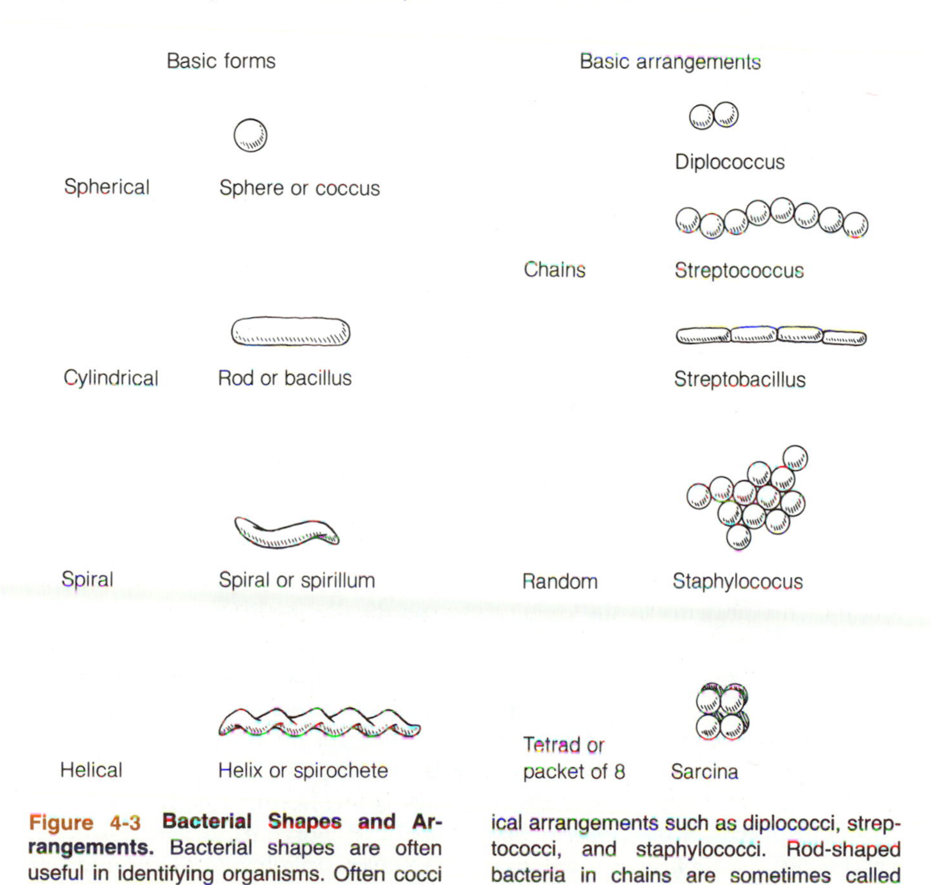

Figure 4-3 Bacterial Shapes and Arrangements. Bacterial shapes are often useful in identifying organisms. Often cocci remain attached to each other and form typical arrangements such as diplococci, streptococci, and staphylococci. Rod-shaped bacteria in chains are sometimes called streptobacilli.

Bacteria, for the most part, multiply by a process called **binary fission.** During this process, a mature bacterial cell divides in half to produce two identical "daughter" cells. Following cell division, each daughter cell grows until it is capable of dividing again. Among certain groups of bacteria, the daughter cells remain attached after cell division and characteristic arrangements of the cells occur. For example, bacteria in the genus *Streptococcus* often form long chains, while those in the genus *Staphylococcus* frequently form grapelike clusters. Some of the basic cell arrangements that bacteria may assume are illustrated in figure 4-3.

THE FINE STRUCTURE OF BACTERIAL CELLS

Although the light microscope is very useful for identifying microorganisms, the electron microscope has been an invaluable tool in the study of subcellular structures and has allowed scientists to uncover the detailed structure of the prokaryotes. All bacteria have a plasma membrane, a genome (hereditary material) that is double-stranded DNA, and ribosomes that carry out protein synthesis. All bacteria have some kind of cell wall, except for the mycoplasmas. There are many structures such as endospores, capsules, thylakoids, gas vesicles, reserve materials, etc., that are associated only with certain bacteria. In the sections that follow, some of the major bacterial structures will be discussed.

Cell Envelopes

Plasma Membrane

The **plasma membrane** (or cytoplasmic membrane) is the thin, pliable, lipid and protein envelope that defines a cell and controls the movement of small molecules in and out of the cell. Plasma membranes are approximately 7 nm in thickness and are therefore below the limit of resolution of the light microscope (approximately 280 nm), but they can be seen with the transmission electron microscope (TEM) when properly stained (fig. 4-1). The membrane consists of two layers of phospholipids (lipid bilayer) in which proteins and glycoproteins are embedded □ (fig. 4-4). According to the **fluid mosaic model,** the proteins in the lipid bilayer are free to move like "icebergs" in the phospholipid "sea." This model suggests that the membrane is a dynamic structure, capable of rapidly changing its protein and lipid composition. The phospholipid bilayer is the barrier that limits the movement of material back and forth between the cytoplasm and the outside environment. Many of the proteins in the membrane specifically facilitate the movement of materials through the membrane. For example, a number of membrane proteins form **pores** that allow water and a few other materials to rapidly move back and forth across the membrane. In addition, membrane proteins called **permeases** transport nutrients into the cell. In some bacteria, respiratory enzymes are associated with the bacterial plasma membrane. Thus, the plasma membrane in some prokaryotes is involved in **respiration.** Respiration is a type of oxidative metabolism whereby organisms generate ATP. The bacterial cytoplasmic membrane also plays a role in the distribution of the genetic information to daughter cells. The genetic material or genome in bacteria is bound to the plasma membrane. After the genome replicates,

49

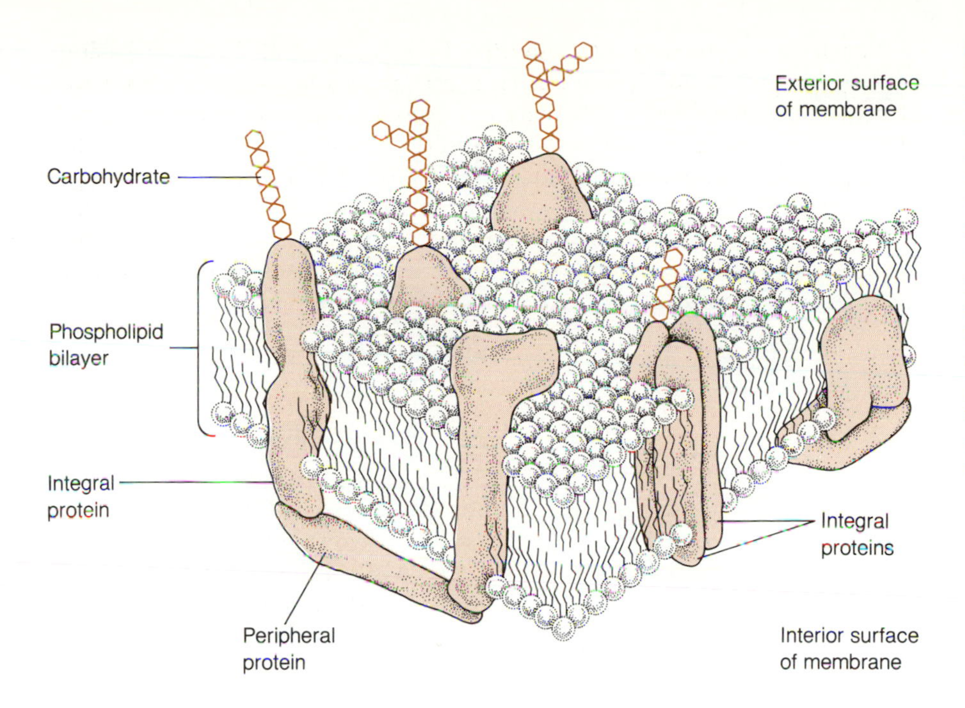

Figure 4-4 Bacterial Plasma Membrane. The bacterial plasma membrane consists of a lipid bilayer and various proteins that are associated with the lipid bilayer. Proteins that penetrate into the interior of the lipid bilayer are called integral proteins. Some integral proteins form channels, function in the transport of materials, and strengthen the membrane. Proteins that lie on the surface and do not penetrate the lipid bilayer are called peripheral or extrinsic proteins. Extrinsic proteins generally strengthen the membrane, bind nutrients, and carry out chemical reactions. Polysaccharides are frequently attached to membrane proteins and membrane lipids on the exterior side of the membrane.

both copies are attached to the plasma membrane. As the membrane grows, a copy of the bacterial genome ends up near each end of the cell; therefore, daughter cells have a complete copy of the genetic material after cell division. Bacterial flagella are anchored in the plasma membrane and are powered by the energy derived from energy-generating systems found in the membrane.

Cell Wall

Almost all bacteria have a rigid layer of material that surrounds the plasma membrane. This rigid layer is called the **cell wall** and is analogous to the cell walls found in plants and fungi. The cell wall protects the membrane and maintains the shape of the cell (fig. 4-1). All bacteria (except the **mycoplasmas,** which lack a cell wall, and the **archaebacteria,** which have unusual polysaccharide or proteinaceous walls) have a wall that is constructed of a polysaccharide made up of two alternating sugars, N-acetylglucosamine and N-acetylmuramic acid, joined together by glycosidic bonds. The polysaccharides are joined together by short peptides into a three-dimensional framework called **peptidoglycan** (fig. 4-5). **Tetrapeptide side chains,** which are attached to each N-acetylmuramic acid, and **peptide cross bridges,** consisting of as many as 5 amino acids, cross-link the tetrapeptides and hold the polysaccharides together.

The cell walls of gram positive bacteria are 30 to 60 nm thick and consist almost exclusively of the peptidoglycan layer and attached **teichoic acid** polymers (fig. 4-5). In some cases the teichoic acid polymers, alternating phosphate and glycerol (or ribitol), account for 50% of the dry weight of the cell wall. The cell walls of gram negative bacteria contain a thin peptidoglycan layer that is about 2–3 nm thick (fig. 4-5). This layer contains no teichoic acids. An **outer membrane** 7 nm thick surrounds the peptidoglycan layer in the gram negative bacterium and is considered to be part of the cell wall. The outer membrane is similar to the plasma membrane in that it consists

Figure 4-5 Cell Wall of Gram Positive and Gram Negative Bacteria. *(a)* The cell wall and cell membrane of a gram positive bacterium are illustrated. The cell wall consists of peptidoglycan, wall teichoic acids, and membrane teichoic acids. The wall teichoic acids are covalently linked to the muramic acid residues of the peptidoglycan layer, while the membrane teichoic acids are linked to glycolipids. The molecules found in the membrane include various proteins, phospholipids, glycolipids, and phosphatidyl glycolipids. *(b)* The cell wall and cell membrane of a gram negative bacterium are illustrated. The cell wall consists of a layer of peptidoglycan and a lipid bilayer called the outer envelope. The outer envelope contains lipopolysaccharides on the outer surface, which are called endotoxins because they cause allergic reactions in animals. The cell membrane of gram negative bacteria is similar to the cell membrane in gram positive bacteria. *(c)* Most bacteria have a peptidoglycan layer in their cell wall. The peptidoglycan consists of a polysaccharide made up of alternating N-acetylglucosamine and N-acetylmuramic acid and short peptides attached to the N-acetylmuramic acid.

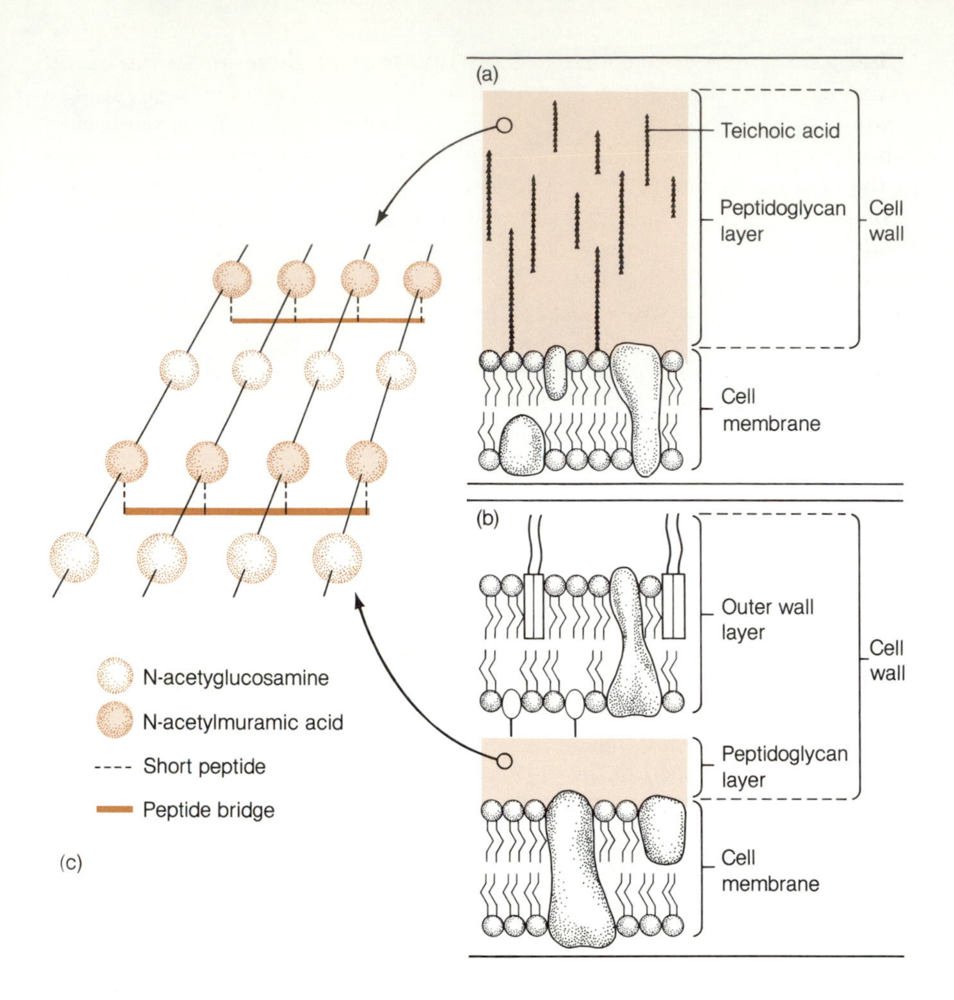

of a lipid bilayer. Many of the lipids making up the external side of the outer membrane have polysaccharide attached to them. Consequently, these lipids are referred to as **lipopolysaccharides.** The outer membrane is a selective permeability barrier that excludes compounds with molecular weights above 800. Many small molecules pass through the outer membrane, however, because it contains various types of pores. The peptidoglycan layer and the outer membrane form a wall that is approximately 10 nm thick. When the lipopolysaccharide fraction of the outer membrane is toxic to mammals, it is known as an **endotoxin.**

The concentration of solutes (sugars, amino acids, and salts) in an environment often have an effect on the flow of water into or out of cells. A high concentration of small molecules attracts water and is a region with a high **osmotic pressure.** With regard to cells, an **isotonic** environment is one in which the osmotic pressure is the same inside and outside the cell. A **hypotonic** environment is one that has a lower concentration of solutes than the cell's cytoplasm, while a **hypertonic** environment is one that has a higher concentration of solutes than the cell's cytoplasm. Prokaryotes without walls assume a spherical shape in an isotonic environment. When cells lacking walls are placed in a hypotonic environment, however, they fill with water and eventually burst, while cells introduced into a hypertonic medium lose water and shrink. A **protoplast** is said to form when the cell wall is completely removed from gram positive bacteria in an isotonic environment.

Spheroplasts look very much like protoplasts, but these structures result from the partial destruction of the cell wall in gram negative bacteria. The treatment of these bacteria with penicillin in isotonic media for a number of generations results in daughter cells that are missing the peptidoglycan layer of the cell wall. Bacteria in their natural habitat often grow in aqueous environments where the concentration of salts and sugars is much lower than inside the cell. In such environments, water tends to diffuse into the cell. Withough a cell wall, the influx of water would cause the cell to swell and ultimately burst. The cell wall prevents this from happening by offering resistance to the excessive influx of water and the internal water pressure (osmotic pressure). Water flows into the cell until the osmotic pressure is balanced by the resistance of the cell wall.

Capsules

Capsules, which vary in thickness, are slimy or gummy layers of polysaccharide or polypeptide that surround the cell wall (fig. 4-6). Most capsules are composed of polysaccharide. The production of capsules is sometimes influenced by the cultural conditions under which bacteria grow. For example, some organisms produce polysaccharide capsules only if sucrose is present. Capsules appear to have two major functions: to protect bacteria from predators, and to promote their attachment to various objects and to each other.

The bacterium that frequently causes pneumonia in humans, *Streptococcus pneumoniae*, generally must possess a capsule in order to cause disease. Strains of *S. pneumoniae* unable to synthesize a capsule are readily ingested by ameboid cells, such as macrophages and neutrophils, which protect the host from microorganisms □. On the other hand, those strains of *S. pneumoniae* that are encapsulated resist ingestion by phagocytic cells and

440

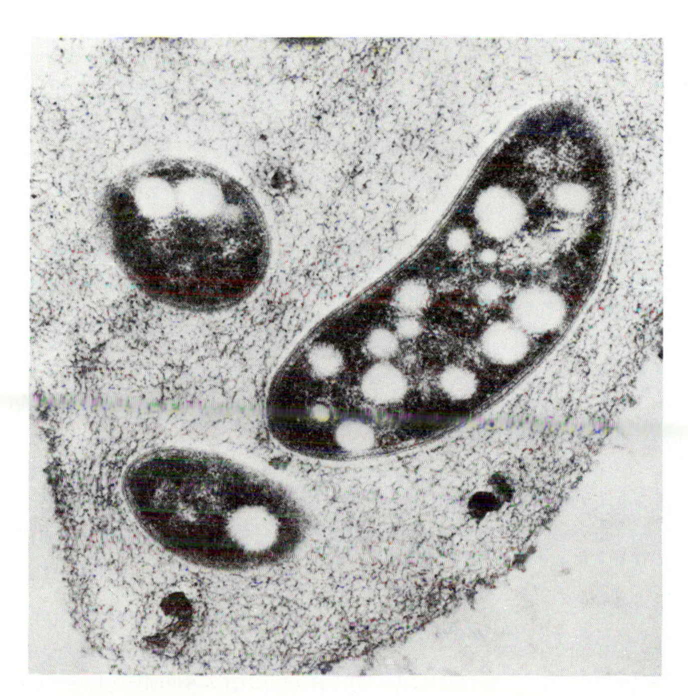

Figure 4-6 Bacterial Capsule. A transmission electron micrograph of a thin section of *Rhizobium trifolii* shows the capsule around these bacteria.

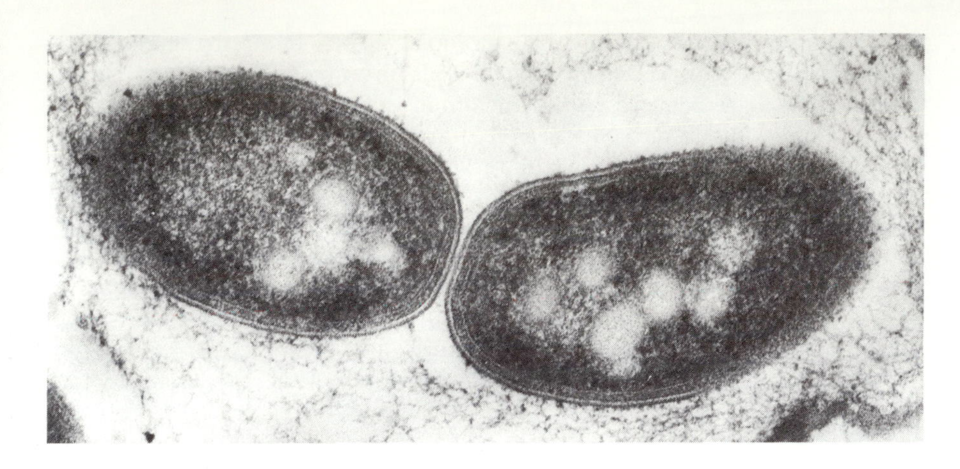

Figure 4-7 Bacterial Glycocalyx. Bacterial glycocalyx is a network of polysaccharide that extends from the surface of many bacteria. A thick, dense glycocalyx is equivalent to a capsule. In this transmission electron micrograph, bacteria are attached by their glycocalyx to each other.

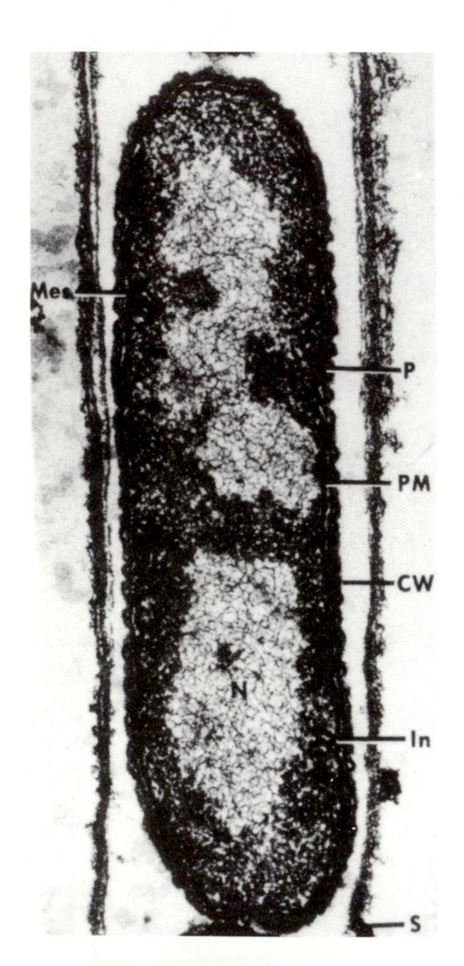

Figure 4-8 Bacterial Sheath. A transmission electron micrograph of a thin section of bacteria covered by a sheath. Mesosome (M), periplasmic space (P), plasma membrane (PM), cell wall (CW), inclusions (In), sheath (S), nuclear material (N).

are capable of causing disease. It is believed that similar functions can be ascribed to capsules of microorganisms that normally inhabit extrahuman habitats. The capsule may prevent predatory microorganisms from ingesting the encapsulated bacterium or interfere with its destruction once ingested. The presence of a capsule does not mean that an organism is pathogenic. There are many capsule-forming organisms, such as *Azotobacter* and *Leuconostoc*, which do not cause disease.

The **glycocalyx** consists of polysaccharide fibers that extend from the bacterial surface (fig. 4-7) and that allow bacteria to attach to inanimate objects, other bacteria, plant cells, or animal cells. The glycocalyx may become so extensive that it forms a capsule. Adhesion to inanimate objects can be advantageous to microorganisms in a rapidly moving stream, while attachment to plant or animal tissue provides bacteria with nutrients and a physical environment that is conducive to growth. Many diseases are caused by bacteria partly because they are able to synthesize a glycocalyx □ or a sticky capsule. For example, the bacteria that cause dental caries are able to do so because they can attach to tooth surfaces and reproduce there. The network of glycocalyx and bacteria on tooth surfaces, called **dental plaque,** often grows quite thick. This means that bacterial enzymes, which degrade the dental enamel, are highly concentrated at some locations and consequently cause cavities. In addition, it is believed that the bacteria that cause cholera and gonorrhea are able to do so partly because they are able to attach to specific tissues, because of either their glycocalyx or the glycocalyx of the host tissue.

429

A number of other reasons for capsules have been proposed. For instance, it has been hypothesized that capsules may be a way of storing large amounts of organic materials, protecting against desiccation, and protecting against virus infections. Capsules are best seen with negative stains, although simple stains have been developed to demonstrate their presence in biological materials.

Sheaths

Sheaths are stiff polysaccharide coverings that some bacteria secrete. Polysaccharides form a stiff sheath around a chain or group of bacteria (fig. 4-8). The sheath provides an extra layer of material that protects the bacteria.

Appendages and Structures Used for Locomotion

Flagella

Bacterial **flagella** are hairlike appendages 10–20 μm long and about 0.02 μm in diameter that originate in the plasma membrane (fig. 4-9). Because they are so thin, bacterial flagella cannot be seen with the light microscope and special staining techniques must be used to demonstrate their presence. These flagellar stains employ compounds called **mordants** that coat the fla-

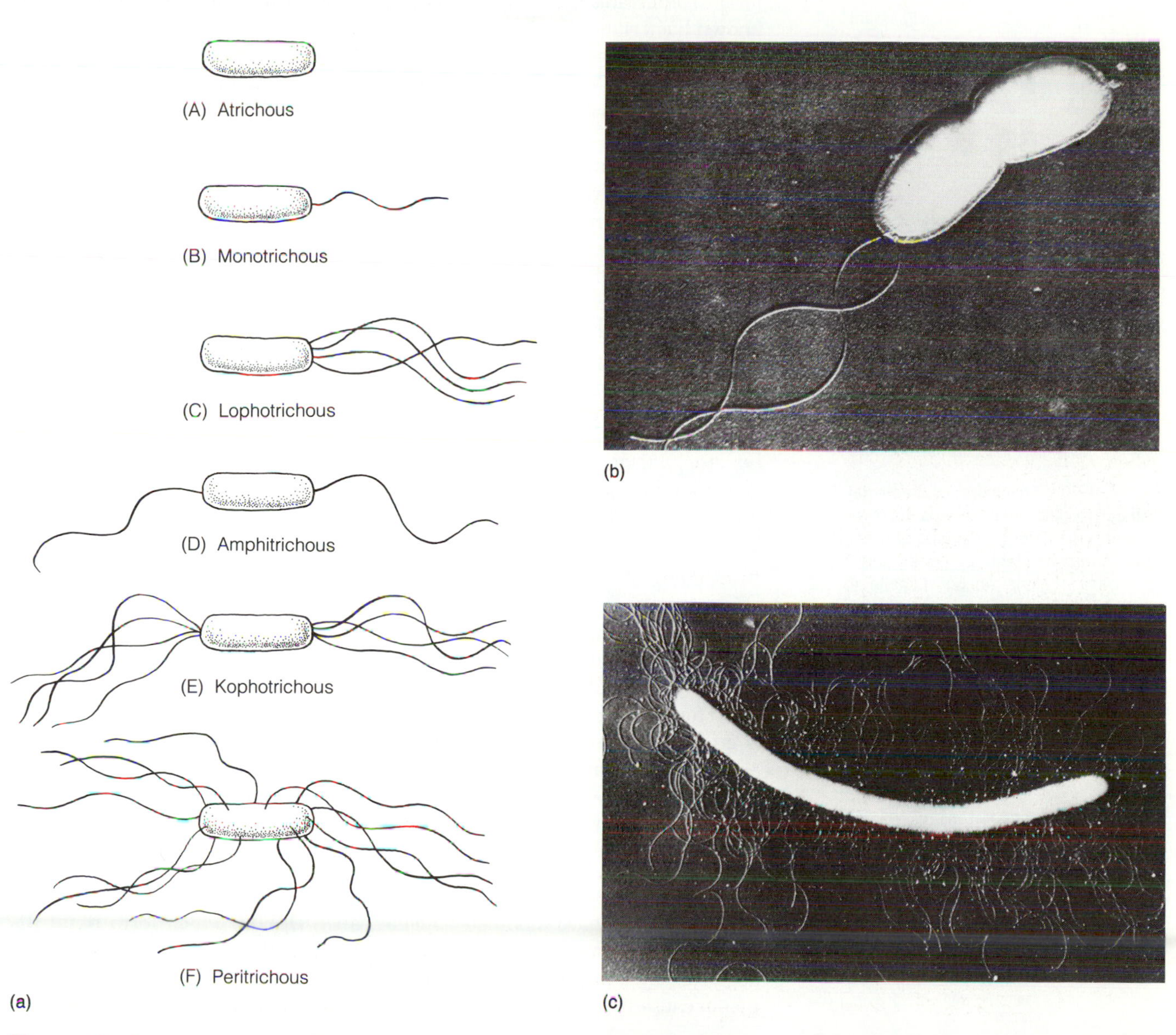

Figure 4-9 Arrangement of Bacterial Flagella. *(a)* This drawing illustrates the arrangement of bacterial flagella. Bacteria that have flagella only at their ends are polarly flagellated, while those that have flagella over much of their surface are peritrichously flagellated. The following nomenclature is sometimes used to describe the types of flagellation: A) atrichous; B) monotrichous; C) lophotrichous; D) amphitrichous; E) kophotrichous; F) peritrichous. *(b)* This transmission electron micrograph of *Pseudomonas fluorescens,* shadowed with a heavy metal, shows that it has two flagella at one pole. *(c)* This transmission electron micrograph of *Proteus vulgaris*, shadowed with a heavy metal, shows the flagella over the entire surface.

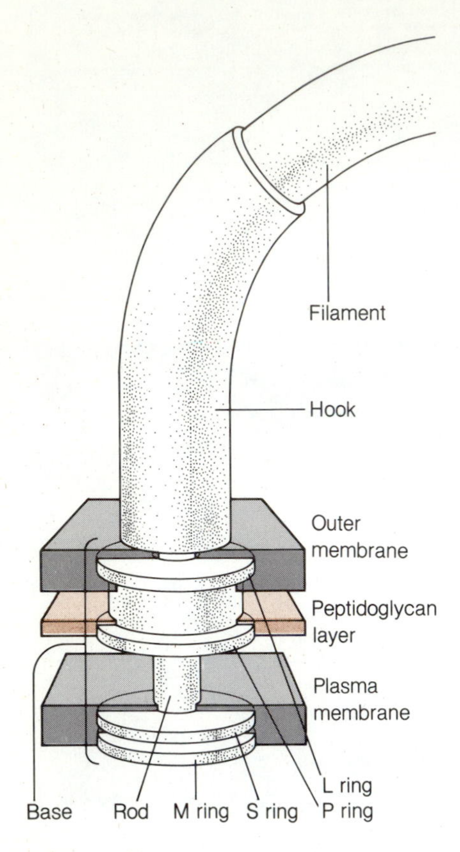

Figure 4-10 Structure of Bacterial Flagella. This diagram illustrates the three major components of the flagella of gram negative bacteria: filament, hook, and base. The base regions anchor flagella to the bacterial plasma membrane and cell wall. The hook and the filament, when rotating, impart motility to the cell.

gella and make them thicker. The mordant is stained with a dye, thus coloring it and making the flagella easier to see. Bacteria that have flagella distributed around the cell are said to be **peritrichously** flagellated, while those that have flagella only at the poles of the cell are said to be **polarly** flagellated. Organisms with polar flagella may be further subdivided into those that have a single flagellum and those that have flagellar tufts at their poles. Bacteria with a single polar flagellum are referred to as **monotrichous** bacteria, while those with tufts are called **lophotrichous** bacteria. The arrangement of flagella is sometimes used in identifying an unknown bacterium.

A bacterial flagellum consists of a filament, a hook, and a basal structure (fig. 4-10). The filament is the hollow, whiplike part of the flagellum that drives the cell. It is constructed from a polypeptide called **flagellin.** The hook structure is located at the base of the filament and is slightly thicker than the filament. The basal structure attaches the flagellum to the cell and presumably is involved in driving the rotation of the filament and hook. The rotating flagellum bends and takes on a sinusoidal shape because of its structure and length. As the flagellum rotates, it pushes against the water and consequently propels the bacterium. The basal structure of gram negative bacteria has four ringlike elements. The outer two rings (L and P) are attached to the lipopolysaccharide layer and the peptidoglycan layer of the wall, respectively, while the inner two rings (S and M) are bound to the plasma membrane. Gram positive flagellated bacteria have only two rings; the outer ring is associated with the peptidoglycan of the wall and the inner ring with the plasma membrane.

When a flagellated bacterium moves in a liquid, it generally moves in one direction for a while and then **tumbles** before moving off in another direction (fig. 4-11). Light and chemicals affect the movement of bacteria. Some bacteria respond to chemicals or to light by moving toward or away from them. Movement in response to chemicals is called **chemotaxis,** while movement in response to light is known as **phototaxis.** In a concentration gradient of a nutrient, bacteria moving up the gradient tumble less frequently than usual and therefore have long runs. When they are moving away from a nutrient, however, they tumble more frequently than usual. Therefore, bacteria eventually end up where the concentration of the nutrient is greatest. If the bacterium is subjected to a repellent, it tumbles frequently as it moves toward it and the runs are short. On the other hand, when the bacterium is moving away from the repellent, tumbling is inhibited. Eventually the bacterium moves away from the repellent.

Axial Filaments

The spirochetes have modified flagella, called **fibrils,** which form **axial filaments.** Axial filaments coil around the body of the cell between the peptidoglycan layer and the outer membrane (fig. 4-12). The rotating axial filaments cause the spirochete to corkscrew through an aqueous environment.

Gliding Microorganisms

The **gliding bacteria** have neither flagella nor axial filaments, yet they are capable of a slow, steady movement over surfaces. There are a number of explanations for **gliding motility.** One explanation proposes that the gliding bacteria release a polysaccharide slime at the forward end of the cell which

Figure 4-11 Flagellar Motility. This drawing illustrates how a bacterium such as *Escherichia coli* might move in an aqueous environment. It moves in a straight line (run) when its flagella rotate in synchrony. When the flagella rotate asynchronously, the cell begins to tumble. When the flagella rotate synchronously again, the cell changes direction.

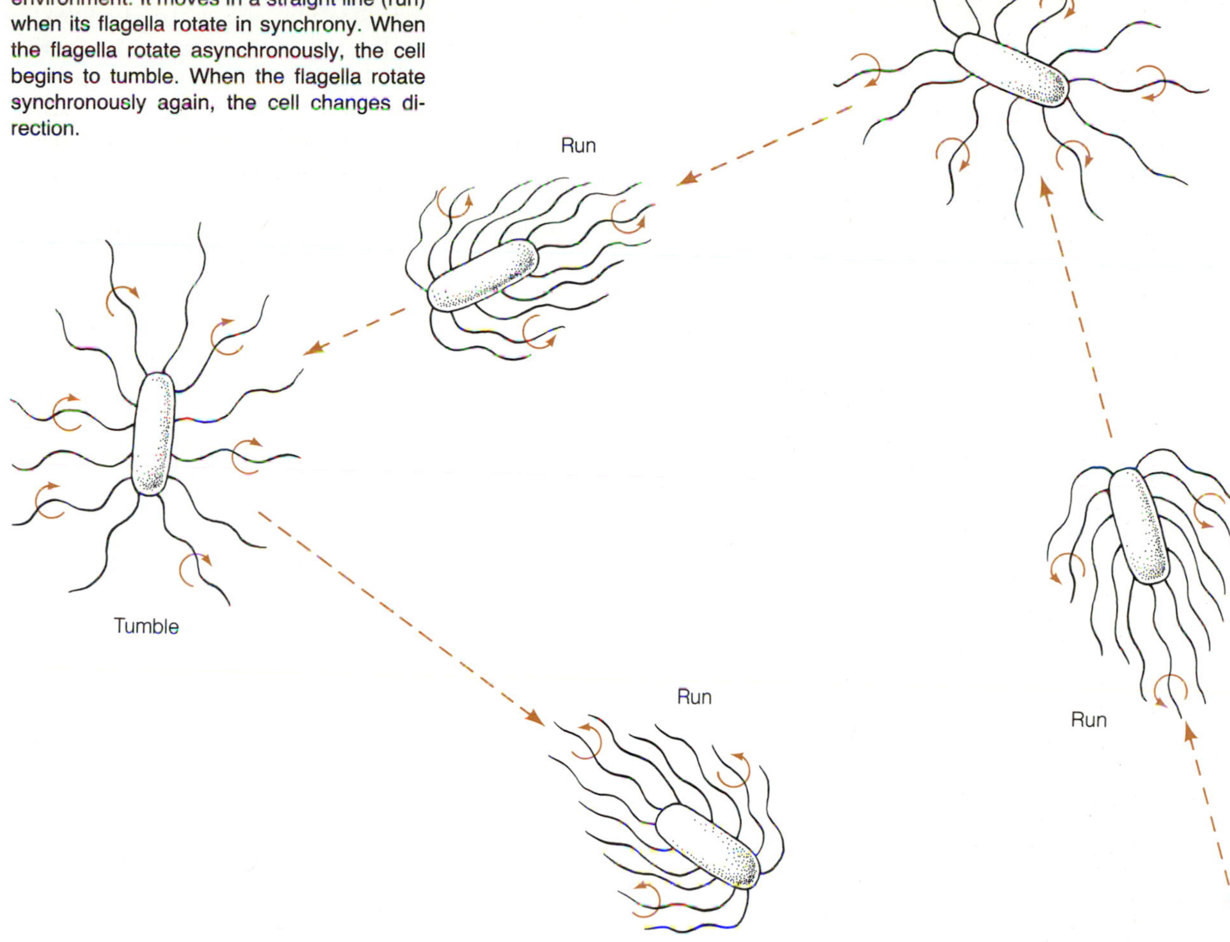

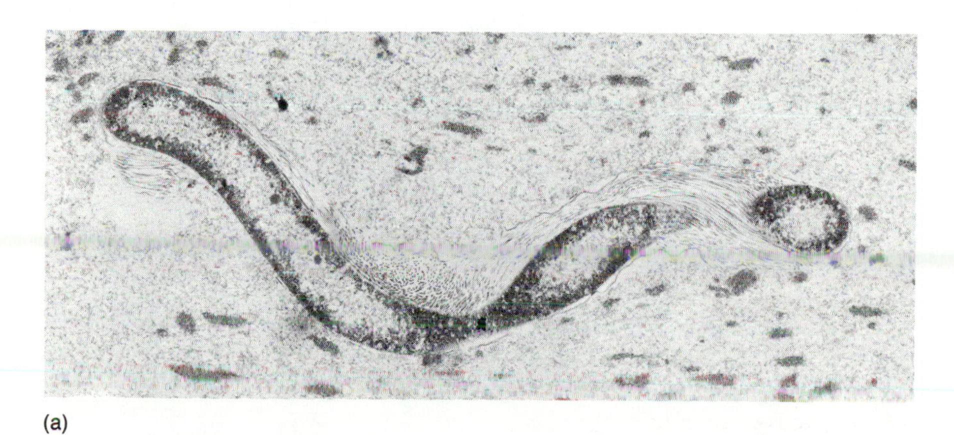

(a)

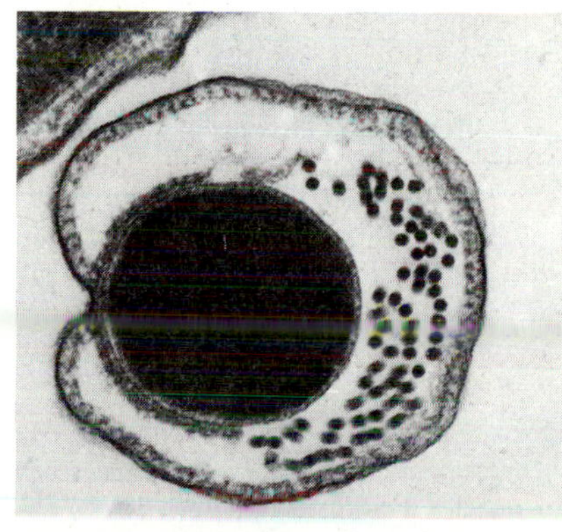

(b)

Figure 4-12 Axial Filaments. These are transmission electron micrographs of a spirochete from the human mouth. *(a)* Numerous fibrils that make up the axial filament can be seen wound along the length of this helical cell. Fibrils originate near each end of the cell and wind back along more than half a cell length, where they overlap. *(b)* A cross section of a spirochete shows many fibrils between the peptidoglycan layer and the outer envelope (or membrane) of the cell wall.

contracts between the cell and the surface. The contraction of the polysaccharide supposedly propels the cell forward. Another hypothesis proposes that the gliding bacteria excrete hydrogen ions and that these induce a negative charge in the surface. When the hydrogen ions are reabsorbed by the cell, the negatively-charged surface repels the negatively-charged bacterium and forces it forward. A third idea is that the gliding bacteria have flagellar stubs that act as paddle wheels to push the bacteria forward.

Pili

Pili (singular, **pilus**), or **fimbriae** (singular, **fimbria**), are long, hollow tubes, which protrude peritrichously from some bacteria, and are made of a protein called **pilin** (fig. 4-13). Pili are about 0.02 μm in diameter and 0.2–20 μm in length. There are at least six different groups of pili, classified on the basis of their adhesive and morphological properties. Pili are for the most part found in gram negative bacteria, such as the **enterics** and the **pseudomonads,** although at least one gram positive bacterium, *Corynebacterium renale*, has them.

One group of pili **(sex pili)** are involved in the transfer of DNA from donor or "male" bacteria to recipient or "female" bacteria. The sex pili bind a male and female bacterium together and facilitate the transfer of DNA from the donor to the recipient. The pilus □ may provide a passageway through which the DNA can be transferred, or it may simply draw the cells together so that a more intimate connection forms between the cells.

219

Pili allow bacteria to stick to one another, to other organisms, and to inanimate objects. Thus, like the glycocalyx, pili contribute to the adhesiveness of bacteria. There are many examples indicating that the adhesiveness of piliated cells is one factor that determines the **virulence** (disease-causing ability) of some bacteria. For example, *Neisseria gonorrhoeae,* which causes gonorrhea, and *Escherichia coli,* a common cause of urinary tract infections,

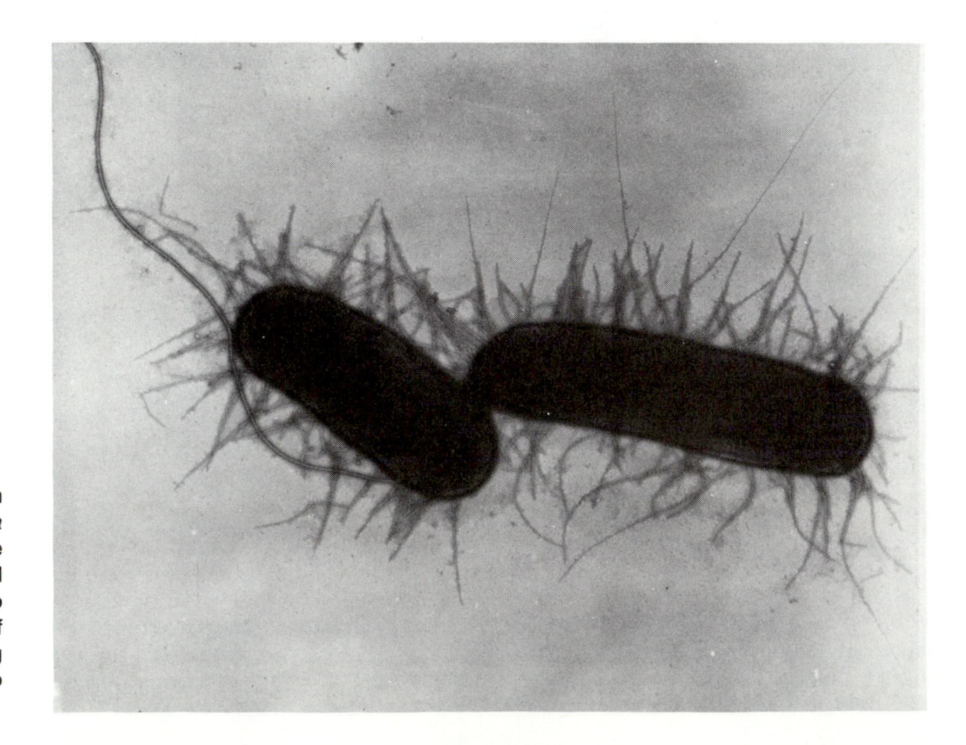

Figure 4-13 Bacterial Pili. Transmission electron photomicrograph of *Escherichia coli* exhibiting pili and flagella. Pili are the short appendages that protrude all around the cell and may help bacteria attach to each other or to solid surfaces. One type of pilus, the F pilus is involved in transfering genetic information from one bacterium to another.

both must be piliated in order to cause disease. This requirement is the exception rather than the rule, however, since pili are not required by most pathogenic bacteria.

Pili and the glycocalyx allow bacteria on the surface of a liquid medium to form thin films, or **pellicles.** The formation of pellicles that do not sediment is advantageous because the bacteria float on the surface, where the oxygen concentration is the greatest.

Prosthecae and Stalks

A number of bacteria have cellular extensions called **prosthecae** (singular, **prostheca**). The prosthecae in some bacteria increase their surface area so that they can absorb nutrients more efficiently (fig. 4-14). In other bacteria, the prosthecae allow attachment to solid surfaces.

328

A few bacteria form **stalks** □ from polysaccharide they secrete. Sometimes the stalks become encrusted with inorganic materials, such as $Fe(OH)_3$. The stalks allow these bacteria to attach themselves to objects.

Intracellular Structures

The Genome

The genetic material in bacteria is a double-stranded DNA molecule called the **genome** (fig. 4-15). It contains all the information for controlling the development and metabolic activities characteristic of the cell □. The typical bacterial genome consists of a single, circular molecule of DNA with a contour length (circumference) of about 1,100 μm and containing about 4,000 genes. The bacterial genome is folded into a tight mass, often less than 0.2 μm in diameter, and is complexed with small amounts of protein and RNA (fig. 4-15). Thin sections of bacteria examined with a transmission electron microscope show a tightly bundled genome occupying 15–25% of the cell's cytoplasm (fig. 4-1). Some bacterial genomes, such as those found in many of the mycoplasmas, are very small and are estimated to contain fewer than 1,000 genes, while others, such as those found in some cyanobacteria, are larger than average and may have more than 5,000 genes.

204

Many bacteria possess multiple copies of small self-duplicating pieces of circular DNA called **plasmids** (sometimes called episomes), in addition to their genome (fig. 4-16). Small plasmids contain only a few genes, while the larger plasmids may consist of hundreds of genes. In general, the plasmid's

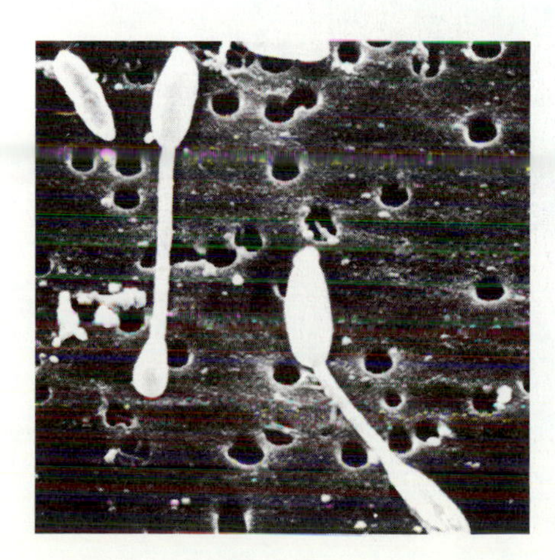

Figure 4-14 The Appendage of *Hyphomicrobium Vulgare.* This organism is viewed with a scanning electron microscope. The photograph illustrates the prostheca of *Hyphomicrobium* with a bud at its tip.

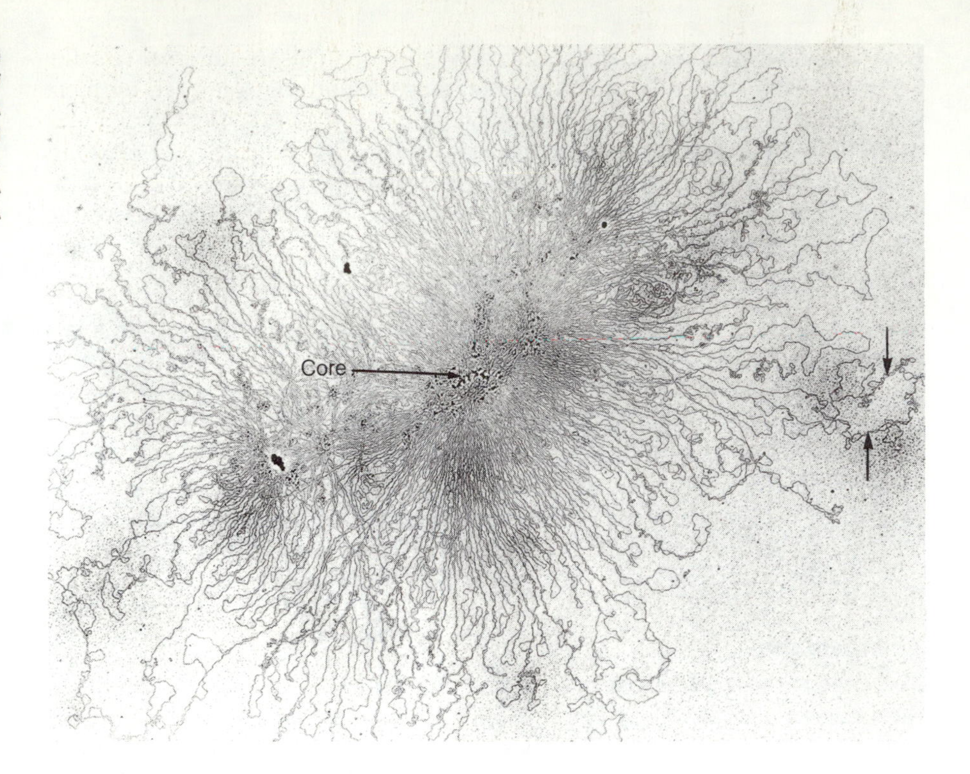

Figure 4-15 A Bacterial Genome. The bacterial genome from *Escherichia coli* consists of a circular, double-stranded DNA molecule. The genome is supercoiled and folded (arrows) so that it can fit into the cell. The protein-RNA core that keeps the DNA folded and loops of the supercoiled DNA are indicated.

genetic information is not necessary for the survival of the cell. Plasmids often carry genetic information that makes the bacterium resistant to certain antibiotics and heavy metals or able to synthesize or break down unusual compounds. Plasmids can be transferred from one bacterium to another, and some can mobilize the transfer of the main genome by a process called **conjugation** □.

219

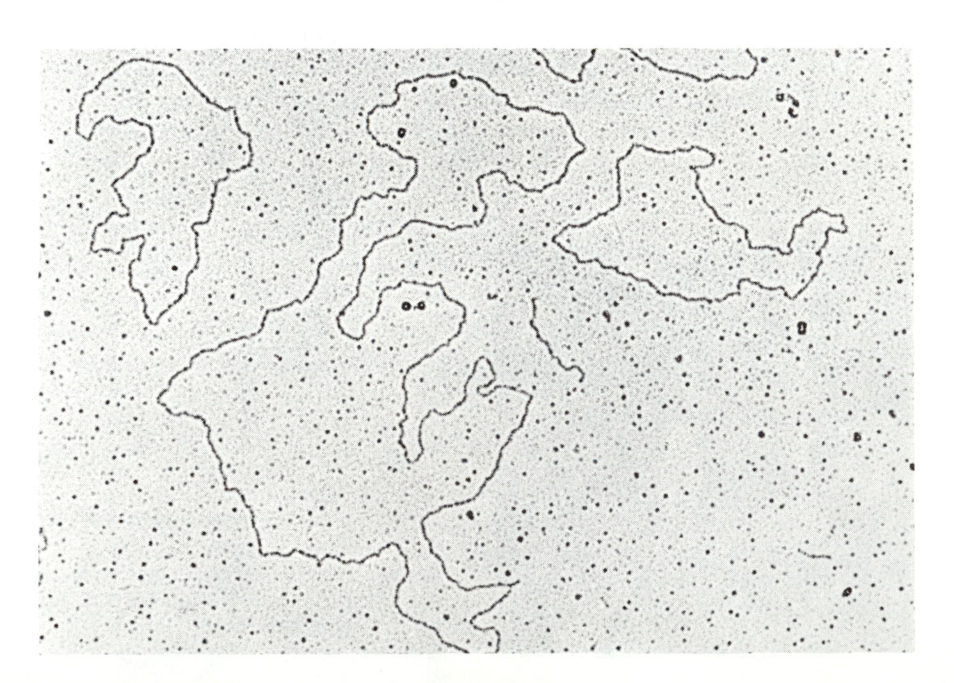

Figure 4-16 Bacterial Plasmids. The larger molecule (plasmid) is pBF4 isolated from *Bacteroides fragilis* and codes resistance to clindamycin and erythromycin. The smaller plasmids are pSC101 isolated from *Escherichia coli* and code resistance to tetracycline.

Mesosomes

In a few bacteria the plasma membrane invaginates into the cytoplasm to form what are called **mesosomes** (fig. 4-17). In these bacteria, the mesosomes are often associated with the invaginating plasma membrane during cell division and with the bacterial genome. This association has led to the hypothesis that mesosomes are involved in genome separation and cell division in these organisms. Since most cells do not have mesosomes, yet are able to segregate their chromosomes and divide, mesosomes clearly are not required for segregation and division in most bacteria. It has been hypothesized that mesosomes may serve to increase the surface area of the plasma membrane where respiratory enzymes are located, and hence to increase respiration, but researchers have had difficulty demonstrating the presence of any respiratory enzymes in these membranes. Some researchers claim that mesosomes are involved in penicillinase excretion and that in some strains of *Bacillus* much of the cellular penicillinase is associated with mesosomes.

A few scientists believe that mesosomes are membrane artifacts caused by the methods used to fix the cells. Freeze fractured cells that are fixed show mesosomes, but freeze fractured cells that are not fixed lack mesosomes. The controversy over mesosomes will no doubt continue for some time. Presently, most scientists believe that mesosomes are real and not due to methods of fixation.

Thylakoids, Photsynthetic Vesicles, and Chlorobium Vesicles

Thylakoids and **photosynthetic vesicles** develop from the cytoplasmic membrane into the cytoplasm and are continuous with the cytoplasmic membrane (fig. 4-18). These membranes contain the enzymes, chlorophyll, and accessory pigments that allow some bacteria to carry out photosynthesis □. Although the thylakoids and photosynthetic vesicles are continuous with the cytoplasmic membrane, they may become detached from the plasma membrane during preparation of the bacteria. Some scientists have referred to the photosynthetic vesicles in the purple bacteria as **chromatophores.**

194

The green photosynthetic bacteria contain **chlorobium vesicles** that lie just under the cytoplasmic membrane (fig. 4-18). The chlorobium vesicles are not continuous with the cytoplasmic membrane.

Some nonphotosynthetic bacteria, such as the nitrogen-fixers *(Azotobacter)* and the nitrifying bacteria *(Nitrobacter)*, have invaginating mem-

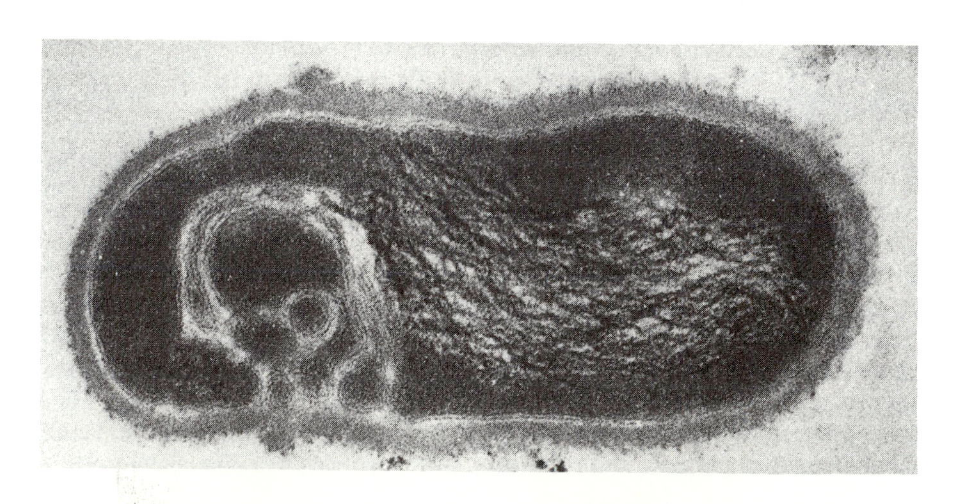

Figure 4-17 Mesosomes. Mesosomes are vesicular membranes that arise from the cytoplasmic membrane and often appear to be associated with the genome.

Figure 4-18 Photosynthetic and Internal Membranes. *(a)* The photosynthetic membranes (chlorobium vesicles or chlorosomes) of the green sulfur bacterium *Pelodictyon* are cigar-shaped vesicles and lie just under the plasma membrane. They are involved in the synthesis of ATP, which is required in large quantities in order to fix CO_2. The chlorobium vesicles are not continuous with the plasma membrane. *(b)* The photosynthetic membranes (lamellar vesicles) of the purple sulfur bacterium *Ectothiorhodospira* resemble stacked, folded sheets. The membranes are continuous with the plasma membrane. *(c)* Vesicular membranes (chromatophores) of *Rhodospirillum rubrum*. The photosynthetic membranes are continuous with the plasma membrane.

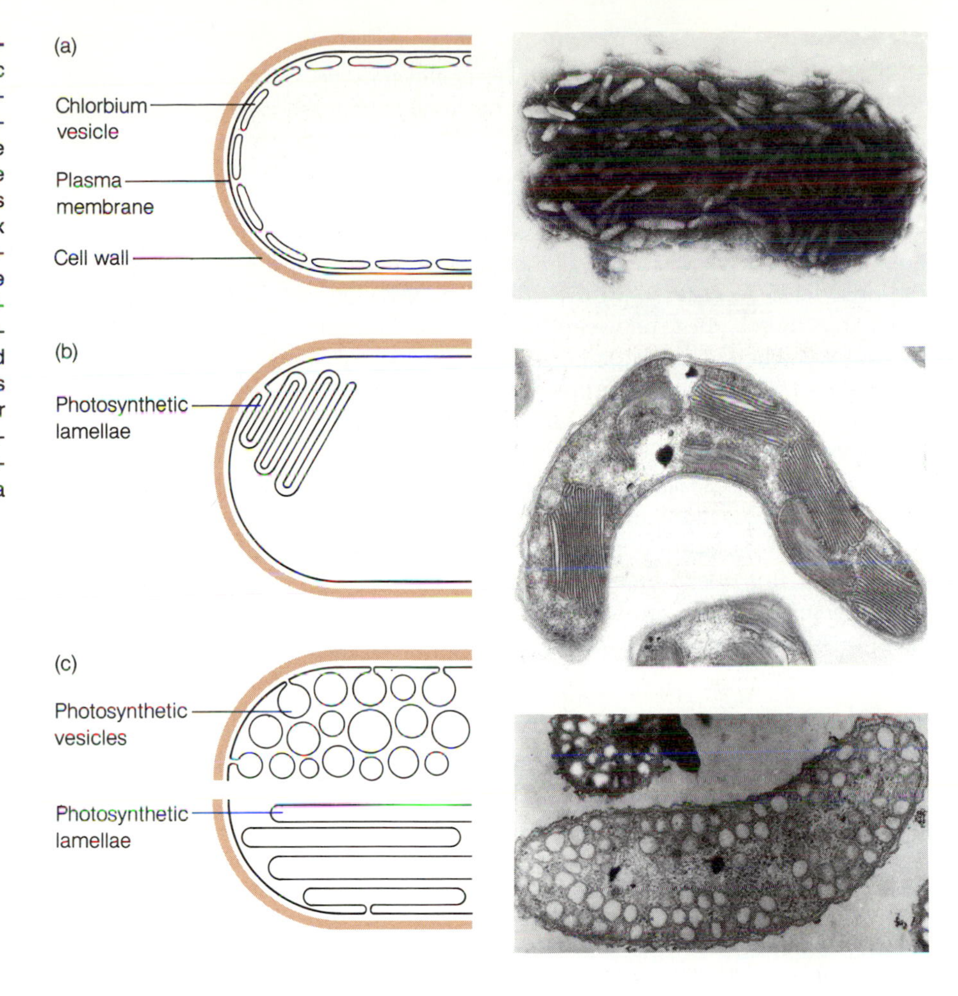

branes that are not involved in photosynthesis. The function of these invaginating membranes is not known. It is believed that they may contain the enzymes necessary for the fixation of CO_2 into organic molecules or for the generation of extra ATP used in fixing CO_2.

Gas Vesicles

Gas vesicles are hollow protein cylinders measuring 75 nm in diameter and 200–1000 nm in length which are found in a few of the cyanobacteria, a group of photosynthetic bacteria (fig. 4-19). The gas vesicles are filled with a gas that helps the bacteria float near the surface of water so that they can absorb sufficient light.

Storage Granules

Granules or **inclusions** are observed in many different types of bacteria (fig. 4-20). Inclusions are frequently reserve materials, such as polymetaphosphate (called metachromatic granules or volutin), cyanophycin (a polymer of arginine and aspartic acid), polyhydroxybutyrate (PHB), protein, sulfur, starch, and glycogen. The reserves are usually large polymers that are osmotically insignificant and do not upset the osmotic balance of the cell.

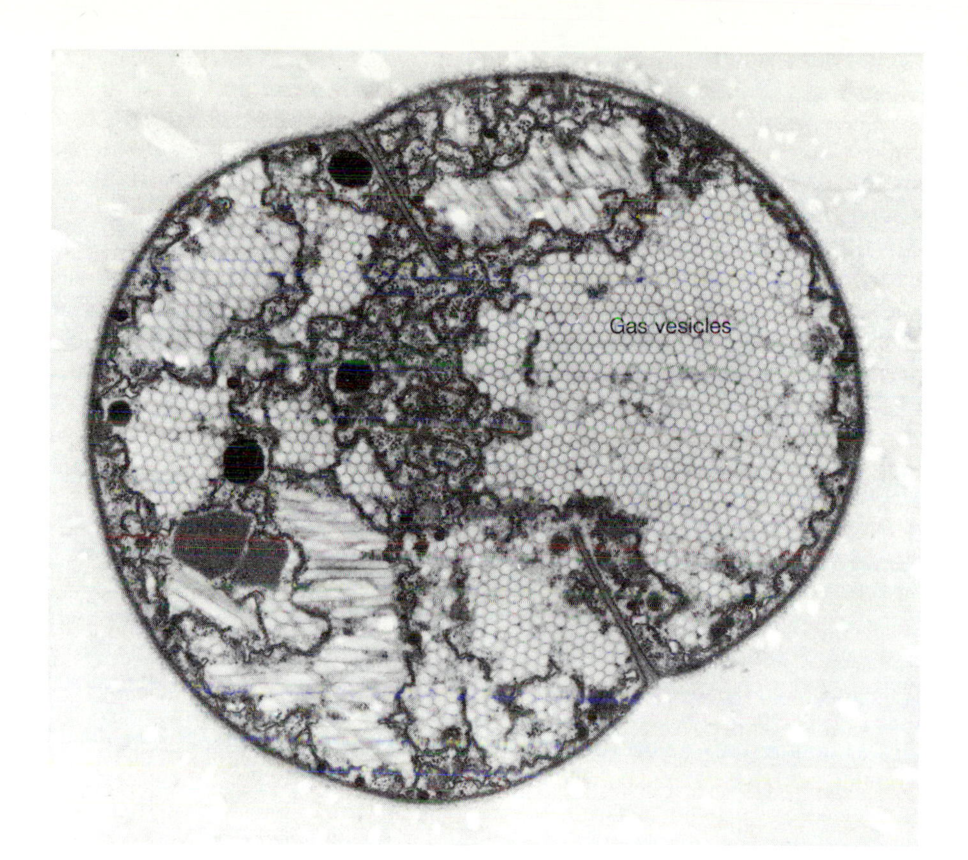

Figure 4-19 Gas Vesicles. Gas vesicles in the cyanobacterium *Microcystis* are seen in this transmission electron micrograph.

Some species of bacteria characteristically produce one type of granule. For example, *Corynebacterium diphtheriae* produces a type of granule called **volutin.** Volutin, a polyphosphate that may function as a source of phosphate for the synthesis of DNA and RNA, can be demonstrated using dyes such as methylene blue. By far the most common organic reserve material among the prokaryotes is polyhydroxybutyrate. Actually, polyhydroxybutyrate granules consist of a collection of short-chained hydroxy fatty acids. Granules of polyhydroxybutyrate can be detected by staining with nonionic fat-soluble dyes, such as sudan black. These granules are used by the cell as reserves for building new cell material. A number of bacteria store sulfur as a reserve material that they use as a source of energy. The sulfur can be made readily visible in a wet mount by using a phase contrast microscope.

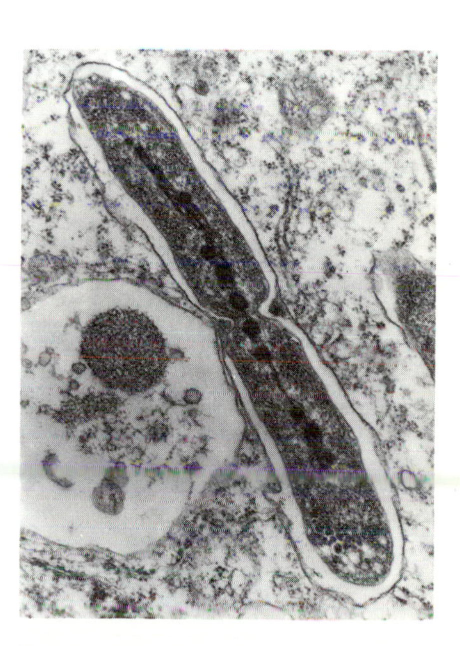

Figure 4-20 Granules. The photosynthetic purple non-sulfur bacterium *Rhodospirillum* contains several storage granules. The granules are 0.25 to 0.3 μm in diameter.

Ribosomes

Ribosomes are cellular structures composed of proteins and RNA. They are involved in protein synthesis, an essential process for all cells. Protein synthesis in bacteria is carried out by **70s ribosomes** (fig. 4-21). 70s refers to the sedimentation velocity, in Svedberg units, of the ribosome in gradients. Since Svedberg units reflect the size and density of objects, a 50s structure is larger and denser than a 30s structure. When bacterial ribosomes are not engaged in protein synthesis, they split into two parts known as the 50s and 30s ribosomal subunits. The 50s and 30s subunits are scattered throughout the cytoplasm and join together to form a 70s ribosome □ when protein synthesis begins.

211

Figure 4-21 70s Ribosomes. The bacterial 70s ribosome is approximately 25 nm in diameter and is made up of rRNA and proteins. It consists of two subunits, the 30s subunit and the 50s subunit.

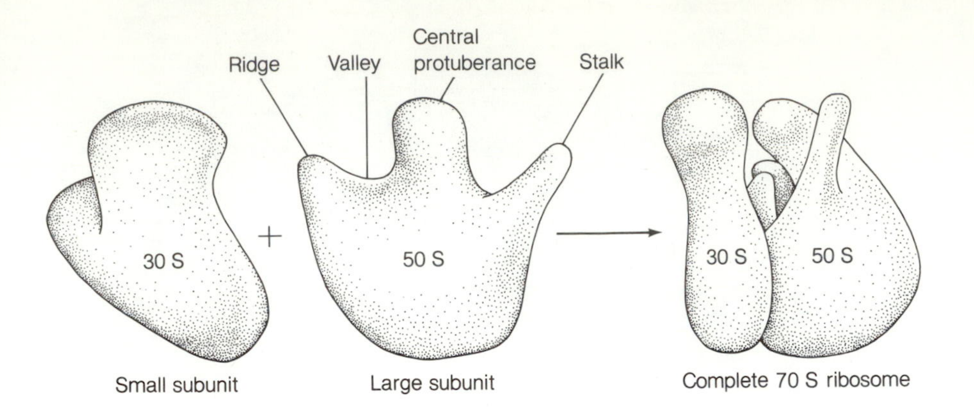

Endospores

A small number of soil bacteria *(Bacillus, Clostridium, Desulfotomaculum, Sporosarcina, Sporolactobacillus, Thermoactinomyces*, and possibly *Metabacterium)* produce a special type of spore that forms within the bacterial cell, called an **endospore** (fig. 4-22). Endospores are extremely resistant to high temperatures and desiccation. For example, *Thermoactinomyces* produces heat-resistant endospores that can survive boiling at 100°C for as long as an hour. *Thermoactinomyces*, in its cellular form, is killed at temperature above 75°C. Endospores are also very resistant to chemicals such as pesticides, antibiotics, and dyes. Endospores contain only about 15% water, in comparison to vegetative cells, which are more than 75% water. Because of their low water content, no metabolism occurs within endospores. As a consequence, they do not assimilate chemicals. Since endospores are resistant to all but extreme conditions, they protect the bacterial genome.

Under conditions of limiting carbon, nitrogen, or phosphorus, the endospore-forming bacteria undergo a developmental process called **sporulation,** which results in the formation of an endospore (fig. 4-23). Each cell

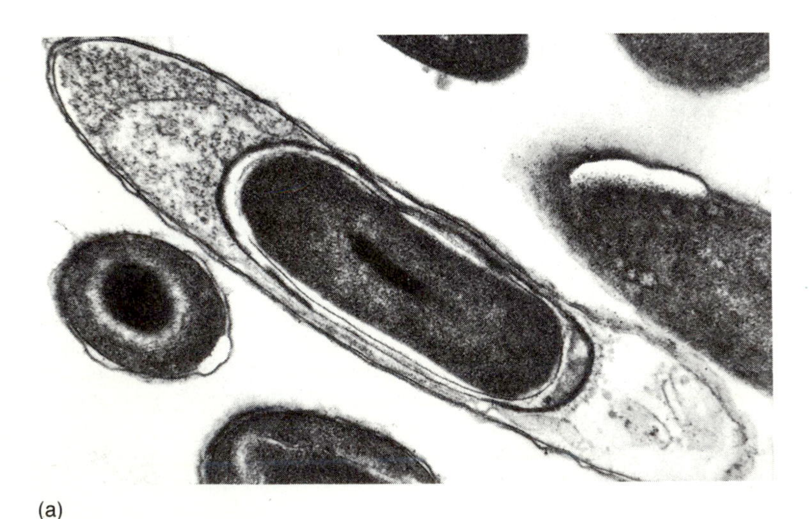

(a)

(b)

Figure 4-22 Endospores. Endospores may be spherical or oval. Each cell produces only one endospore. *(a)* An endospore produced by *Bacillus fastidiosus* is visible within the cell. *(b)* The endospore consists of a dehydrated core where the genome (G) is found, an inner membrane (IM) that surrounds the dehydrated core, a surrounding cortex (C), an outer membrane (OM), a spore coat (SC), and an exosporium (E).

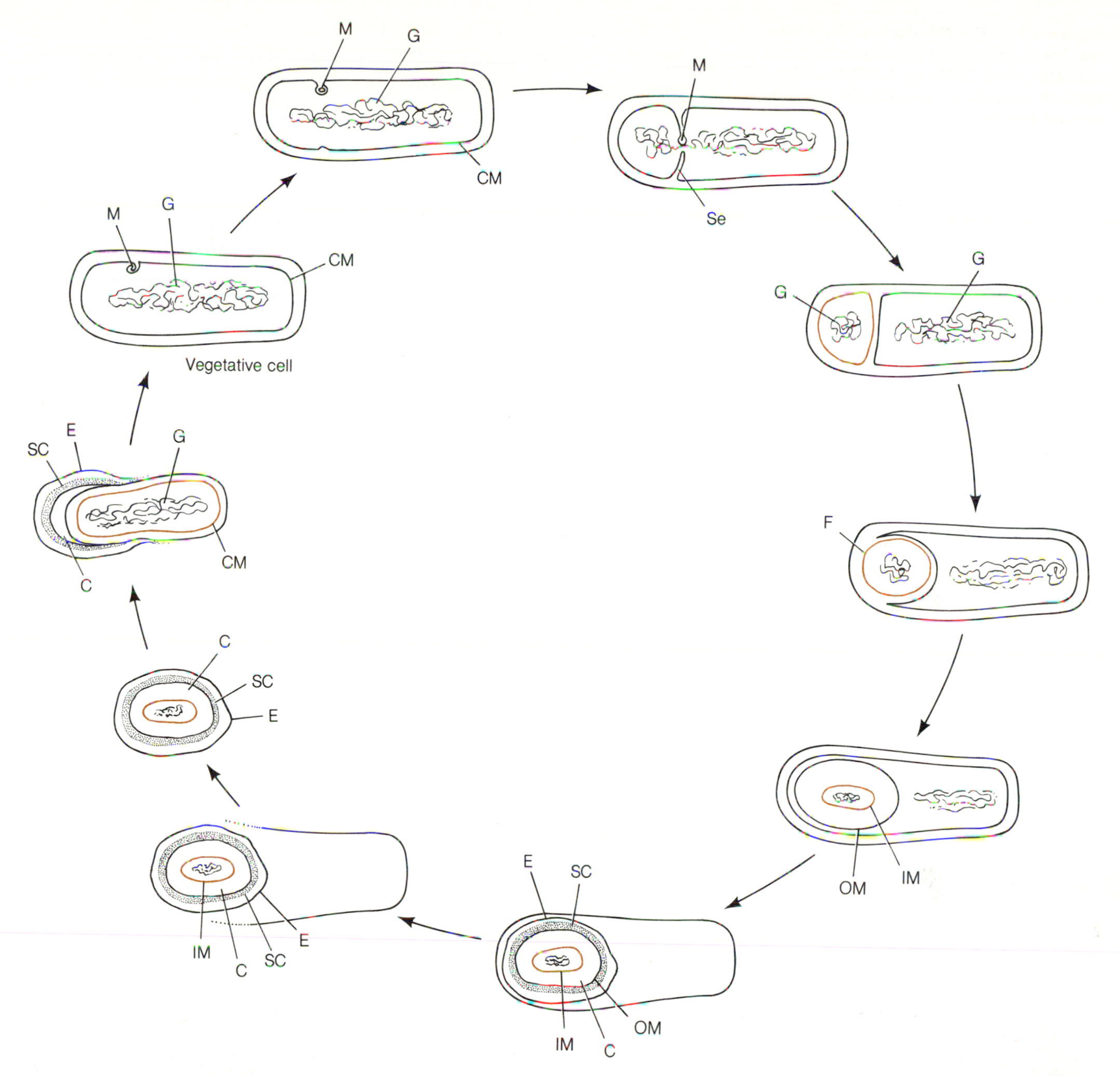

Figure 4-23 Life Cycle of Typical Spore-forming Bacterium. Sporulation. Sporulation begins when a committed cell divides into a mini-cell and a large cell. The sporulation process involves the engulfment of the mini-cell by the large cell, the loss of water by the mini-cell, and the development of various layers of material within and around the developing endospore. At 37°C, the sporulation process takes approximately 15 hours. **Germination.** The germination process involves the reconstitution of the endospore, the breakdown of the cortex and the rupture of the endospore coat. From the endospore emerges a vegetative cell. Germination takes approximately 1 hour. Mesosome (M), septum (Se), cell membrane (CM), forespore (F); for other labels see figure 4-22.

produces only one endospore, whose location within the cell and size vary depending upon the species. Thus, sporulation is not a form of reproduction. Eventually the vegetative cell degenerates and the endospore is released. Since endospores often survive temperatures as high as 100°C and the lack of nutrients and water, they persist for long periods of time in foods and soils.

Even after boiling for as long as 5 minutes, most endospores commonly encountered survive and are capable of germinating when favorable environmental conditions return. **Germination** involves the emergence of a single vegetative cell from the endospore (fig. 4-23).

The failure to sterilize canned vegetables or meats that are contaminated with *Clostridium botulinum* endospores can result in a deadly situation if these canned goods are subsequently consumed without cooking. Under the anaerobic conditions within the jar, the endospores germinate and the

FOCUS

THE MAGNETIC BACTERIA

Various aquatic bacteria have tiny deposits of magnetic iron in their cytoplasm that orient them along the lines of force of the earth's magnetic field (fig. 4-24). Consequently, when these bacteria swim, they move along magnetic field lines toward the bottom of the water. Bacteria that respond to magnetic fields are called **magnetotactic bacteria.** The magnetic field lines of the earth are vertical at the poles and tangential at the equator, but between the poles and the equator the field lines are inclined at angles that increase with latitude. In the Northern Hemisphere the field is inclined downward, while in the Southern Hemisphere it is inclined upward. The magnetic bacteria in the Northern Hemisphere swim downward and toward the north pole unless a strong magnet is brought near them. These bacteria also swim downward in the Southern Hemisphere, because their magnets are reversed. The north and south poles of a strong magnet determine the direction in which the magnetotactic bacteria swim. If the direction of the magnet changes, the bacteria change their direction.

The electron microscope has helped scientists see the magnets within the magnetotactic bacteria. The magnets consist of small crystals of magnetite (Fe_3O_4) which are usually arranged in a thin line along the length of the bacterial cells (fig. 4-24). The particles of magnetite, called magnetosomes, function like a compass, orienting the bacteria along the lines of force of the earth's magnetic field. The magnetosomes appear to be individually enclosed in membranes.

The bacteria synthesize magnetosomes from iron in the environment. If magnetotactic bacteria are grown in an iron-poor medium, their descendents lack magnetosomes and are not magnetotactic. How the bacteria synthesize a magnetic crystal rather than a nonmagnetic crystal of iron is not understood. The bacteria in the northern and southern hemispheres must orient their magnets in opposite directions with respect to their flagella, so that they swim northward and southward, respectively. This orientation insures that they always swim down toward the sediment. The earth's magnetic field appears to be partially responsible for the orientation of the bacterial magnets.

Magnetotactic bacteria use their magnetosomes to navigate toward the bottom of aqueous environments, where anaerobic conditions prevail. Since most of the magnetotactic bacteria are anaerobic (survive only in the absence of oxygen) or microaerophilic (grow best in low concentrations of oxygen), it is to their advantage to be forced consistently toward the bottom, where conditions are anaerobic and where sedimented nutrients might be plentiful. Although magnetosomes are not absolutely essential for the survival of magnetotactic bacteria, those populations with magnetosomes are more successful.

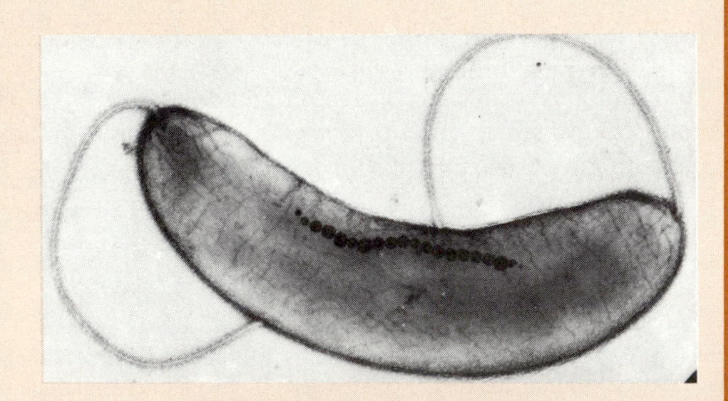

Figure 4-24 Magnetite Crystals within Magnetic Bacteria. A transmission electron micrograph of a magnetotactic bacterium shows a chain of magnetite crystals within the bacterium aligned along the length of the cell. The magnetite crystals, called magnetosomes, act like a magnetic compass needle and align the bacterium along the earth's magnetic field in the same direction as the lines of force. This particular bacterium has flagella at both ends so that it is able to swim in either direction.

resulting vegetative cells produce one of the most potent poisons known, the botulism toxin. This toxin, when ingested even in very small amounts, causes severe illness or death.

Alternate Cell Structures

Conidia

335

Many soil bacteria produce spores that arise by a sexual means called **conidia** (singular, **conidium**) □. The main function of conidia is to spread the organism into other environments. Many conidia appear to be very similar to vegetative cells, except that they generally have thickened cell walls that protect them as they are blown, tumbled, and scraped by wind and water. Many conidia are about as sensitive to heat as a vegetative cell, and they cannot exist for long periods of time without water or nutrients. The genus *Streptomyces* comprises a large group of soil bacteria that show substantial branching. Many of the aerial branches fragment into small cells that develop into conidia.

Cysts

A few bacteria develop another type of differentiated cell, called a **cyst.** Blue-green bacteria produce **heterocysts,** long, thick-walled cells, which are the sites where N_2 fixation occurs □. The function of this heterocyst may be to tide the bacterium over periods of drought. *Azotobacter* also produces a cyst, but it appears to be a "resting" form of the bacterium (fig. 4-25). It is more resistant to desiccation than the vegetative cell, but not much more resistant to heat.

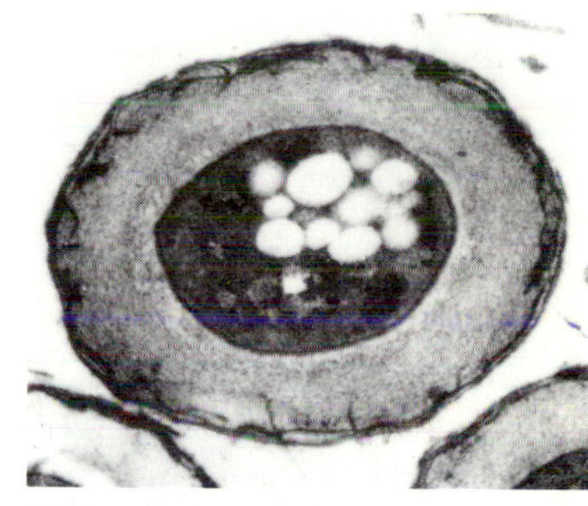

(a)

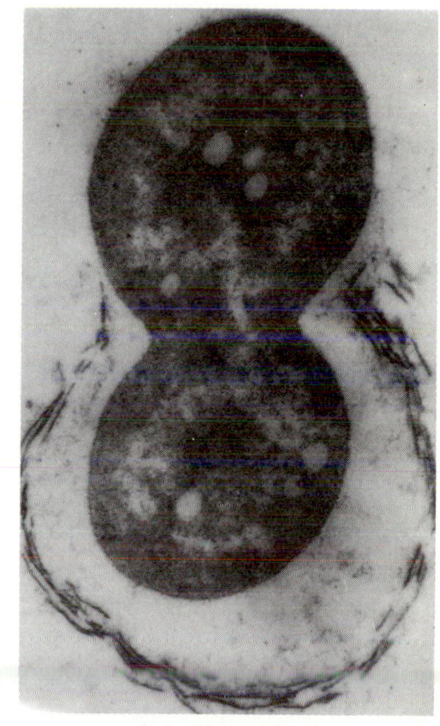

(b)

Figure 4-25 ***Azotobacter* Cysts.** In *Azotobacter* a single cell develops into a cyst *(a).* The cysts have extremely thick cell walls and may contain one or more inner cysts. The cysts contain numerous granules of poly-β-hydroxybutyric acid. When the cyst germinates, *(b)* one or more vegetative cells are released.

THE FINE STRUCTURE OF EUKARYOTIC CELLS

Cell envelopes

Plasma Membrane

The plasma membrane of eukaryotic cells consists of a lipid bilayer and associated proteins, and so resembles that of the prokaryotes. The eukaryotic plasma membrane however, generally contains high concentrations of lipids such as cholesterol, while the prokaryotic membrane generally does not. One notable exception is the plasma membranes of some of the mycoplasmas, which contain cholesterol. There are numerous proteins in the eukaryotic plasma membrane which are involved in controlling the movement of materials back and forth, but there are no respiratory enzymes located in the eukaryotic plasma membrane. All respiratory enzymes in the eukaryotic cell are located in the inner membranes of **mitochondria.**

Cell Wall

Most of the photosynthetic microorganisms, the fungi, and plant cells have cell walls that are chemically and structurally very different from one another and from the bacterial cell walls. For example, the cell walls in fungi may be constructed from the polysaccharides cellulose and chitin. The cell walls of the photosynthetic microorganisms can consist of cellulose, silicon, or calcium carbonate, while those of plant cells are generally constructed from cellulose and other polysaccharides. A great many eukaryotes do not have cell walls. For example, most of the protozoans (amebas and paramecia) lack walls. Their plasma membranes are reinforced by **stress fibers** that consist of proteins such as microtubules, actin, and myosin.

Glycocalyx

Most eukaryotic cells appear to have a glycocalyx, a tangled mat of polysaccharide fibers that covers the cell. The glycocalyx arises from glycoproteins in the cytoplasmic membrane, and its chemical composition varies from tissue to tissue in animals. The glycocalyx on eukaryotic cells allows them to adhere to each other and form specific tissues. In addition, the glycocalyx □ may bind nutrients, be involved in cellular communication, and protect cells from abrasion and microorganisms.

429

Appendages and Structures Used for Locomotion

Eukaryotic flagella (singular, flagellum) and prokaryotic flagella are very different from each other structurally and are powered by completely different mechanisms. The eukaryotic flagellum develops from a **basal body** in the cytoplasm, sometimes called a **centriole** (fig. 4-26), in contrast to the prokaryotic flagellum, which originates from the cytoplasmic membrane. Microtubules and other proteins that make up the skeleton of the eukaryotic flagellum mold the membrane into a thin, whiplike extension of the cell. In contrast, the proteins that make up the prokaryotic flagellum are entirely

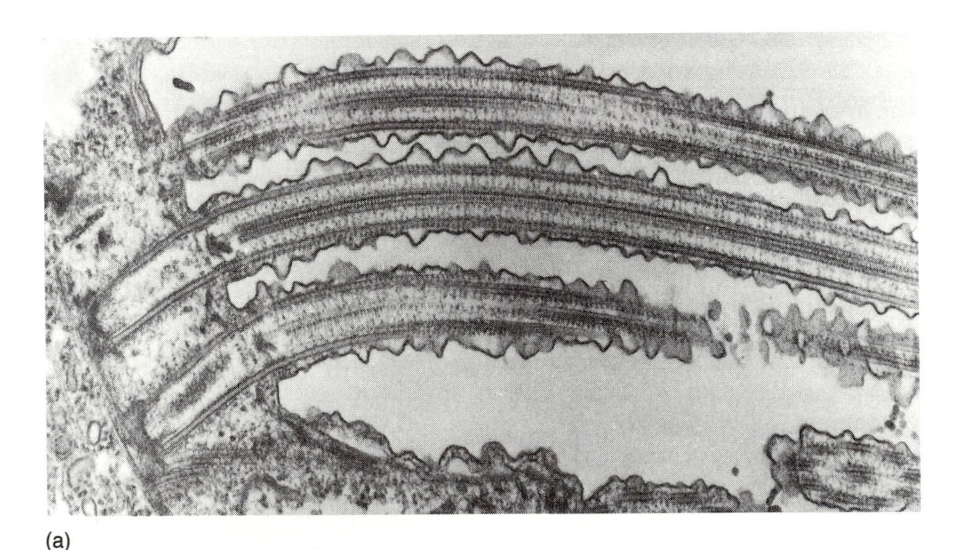

(a)

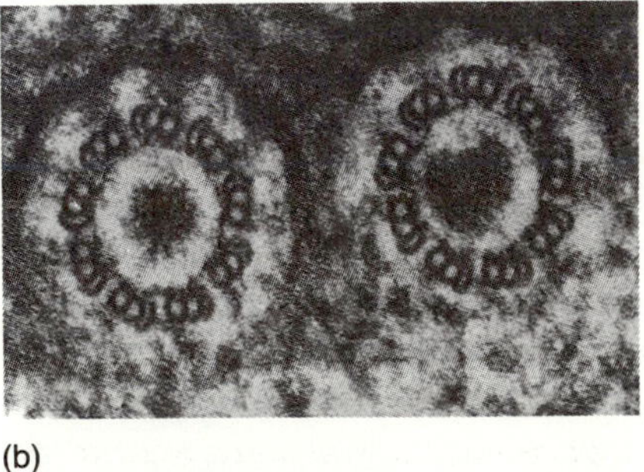

(b)

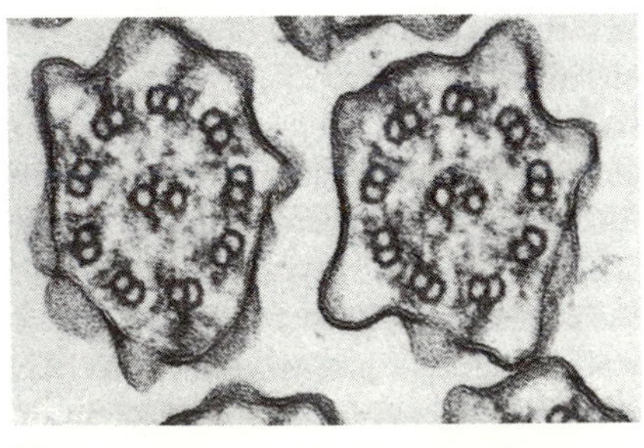

(c)

Figure 4-26 The Structure of Cilia. Cilia *(a)* arise from basal bodies located in the cytoplasm of the cell. A cross-section through the base of the basal body *(b)* shows microtubules arranged in 9 groups of 3 around a central core. A cross section through the shaft of the cilium *(c)* shows that the microtubules are arranged in 9 groups of 2 around a central doublet.

outside the cell. The eukaryotic flagellum propels the cell by bending and twisting against its watery environment. The bending and twisting of the flagella are due to microtubules in the flagellar skeleton that are forced to slide past each other. Eukaryotic flagella are powered by the hydrolysis of ATP. In contrast, prokaryotic flagella rotate like propeller shafts and are powered by the movement of ions across the cytoplasmic membrane.

Many eukaryotic cells have **cilia** rather than flagella. Cilia are very similar to flagella, both structurally and functionally, but they are much shorter.

Intracellular Structures

Nucleus

The **nucleus** is a double-membraned organelle that contains the eukaryotic cell's genetic information (fig. 4-1). The nuclear membranes contain numerous large pores through which proteins and RNA can move. The outer nuclear membrane often gives rise to the **endoplasmic reticulum,** a network of cytoplasmic membranes where proteins are sometimes synthesized and modified. The fluid and dissolved materials within the nucleus constitute the **nucleoplasm.** The nucleoplasm may contain one or more **nucleoli** (singular, **nucleolus**), where the synthesis of eukaryotic ribosomal subunits begins.

Genome

The genetic material in eukaryotes is found in the nucleus and looks like a mass of threads when examined with the electron microscope. The threads consist of DNA and protein and are called **chromatin.** During cell division, the chromatin condenses into chromosomes that are visible under the light microscope (fig. 4-27). The eukaryotic DNA is coiled around basic proteins called **histones.** The histones and coiled DNA structures are known as **nucleosomes.** The chromatin in some cells is attached to the inner nuclear membrane; the significance of this attachment is unknown.

The plasma membrane in eukaryotic cells does not play a role in distributing the genetic information to the poles of the cell during cell multiplication. Instead, proteins called **microtubules** control the movement of the 233 chromosomes to the poles in eukaryotic cells □.

Mitochondria

Mitochondria (singular, **mitochondrion**) are cytoplasmic organelles in- 183 volved in the production of chemical energy in the form of ATP □. Because of this function, the mitochondria are often called the "powerhouses" of the eukaryotic cell (fig. 4-28). Quite simply, a mitochondrion consists of a convoluted inner membrane and an outer membrane. Invaginations formed by the inner membrane are called **cristae** (singular, **crista**) and contain the enzymes for creating proton gradients that are used to synthesize adenosine triphosphate (ATP). All respiratory enzymes in the eukaryotic cell are located in the inner membranes of mitochondria.

Many mitochondria appear to multiply by binary fission, but the division is coordinated with the multiplication of the cell. Mitochondria have circular DNA genomes similar to those found in bacteria but about 100 times smaller. Because the mitochondrial genome codes for only a few proteins, most of the mitochondrial proteins and enzymes are coded for by genes on the chromosomes (nuclear genetic information). Mitochondria have the enzymes (coded for by nuclear genes) for duplicating their genome and for the

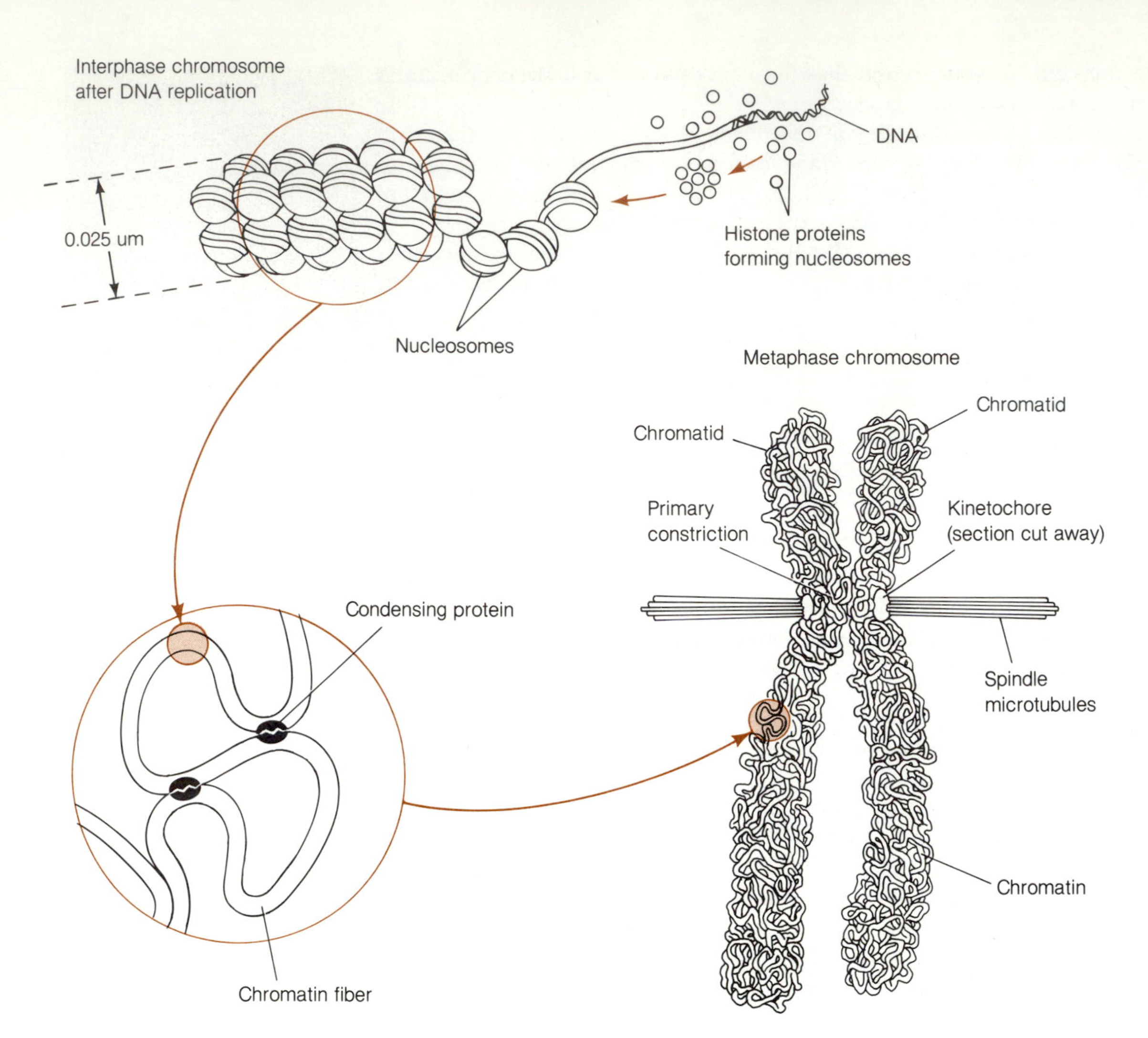

Figure 4-27 Structure of the Eukaryotic Genome. The interphase chromosomes are not visible using the light microscope because they are only 0.025 μm in diameter. The interphase chromosomes consist of DNA wound around nucleosomes. The nucleosomes are complexed so that a 0.025 μm strand is formed. The metaphase chromosomes are visible using the light microscope because they reach diameters of 0.5 μm. The metaphase chromosomes form when the interphase chromosomes condense. The metaphase chromosome illustrated actually consists of two daughter chromosomes that have not separated after DNA replication. The daughter chromosomes are called chromatids.

synthesis of RNA. In addition, mitochondria possess ribosomes that carry out protein synthesis. The ribosomes found in mitochondria are not the usual eukaryotic 80s ribosomes found in the cytoplasm, but instead are 55s to 70s ribosomes that are very similar to bacterial 70s ribosomes. Many scientists have suggested that mitochondria may have originated from prokaryotic cells that established endosymbiotic relationships with larger prokaryotes □.

285

Chloroplasts

In eukaryotic cells, double-membraned organelles called **chloroplasts** are involved in photosynthesis □ (fig. 4-29). Within the **stroma** of the chloroplast are found the thylakoids that contain the enzymes, chlorophyll, and accessory pigments necessary for photosynthesis. In some chloroplasts, pan-

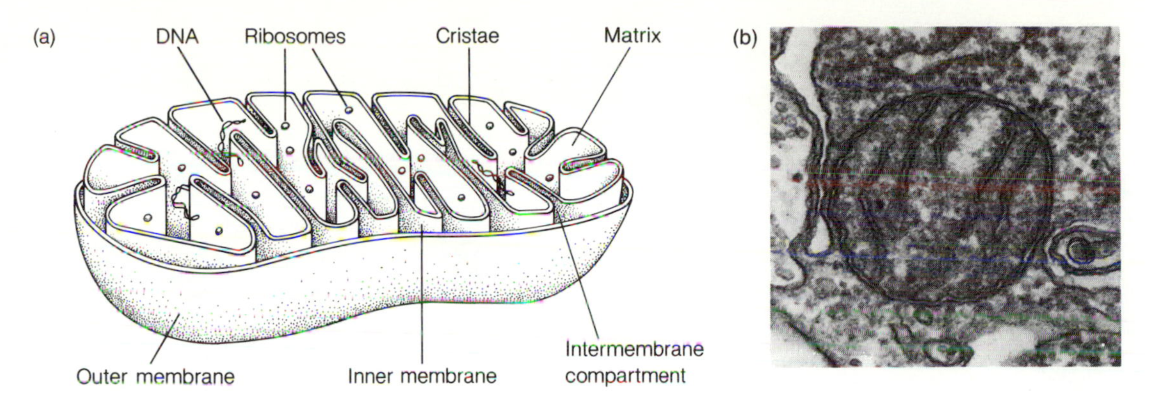

Figure 4-28 Mitochondria. *(a)* Mitochondria are double-membraned organelles that are involved in respiration. The inner membrane invaginates to form cristae. Most of the enzymes involved in respiration are located in the cristae and in the matrix. *(b)* Transmission electron photomicrograph of a mitochondrion.

cake-like thylakoids are stacked like coins. The stacked thylakoids are called **grana.** In contrast, thylakoids in some algal chloroplasts are layered on top of one another like blankets on a bed. Chloroplasts have small, circular DNA genomes and 70s ribosomes that resemble those found in bacteria. Chloroplasts reproduce by binary fission, but do not have the capacity to reproduce outside the cell because of their small genome.

Ribosomes

Eukaryotic **ribosomes** are 80s ribosomes and are found in the cytoplasm. Some are attached to the endoplasmic reticulum, while others are attached to the cytoskeleton and appear to be "free" in the cytoplasm (fig. 4-30). The 80s ribosomes found in eukaryotes are larger than the 70s ribosomes found in bacteria. In addition, the two types of ribosomes are sensitive to different sets of antibiotics and drugs. Nevertheless, both types of ribosomes carry out protein synthesis in much the same way.

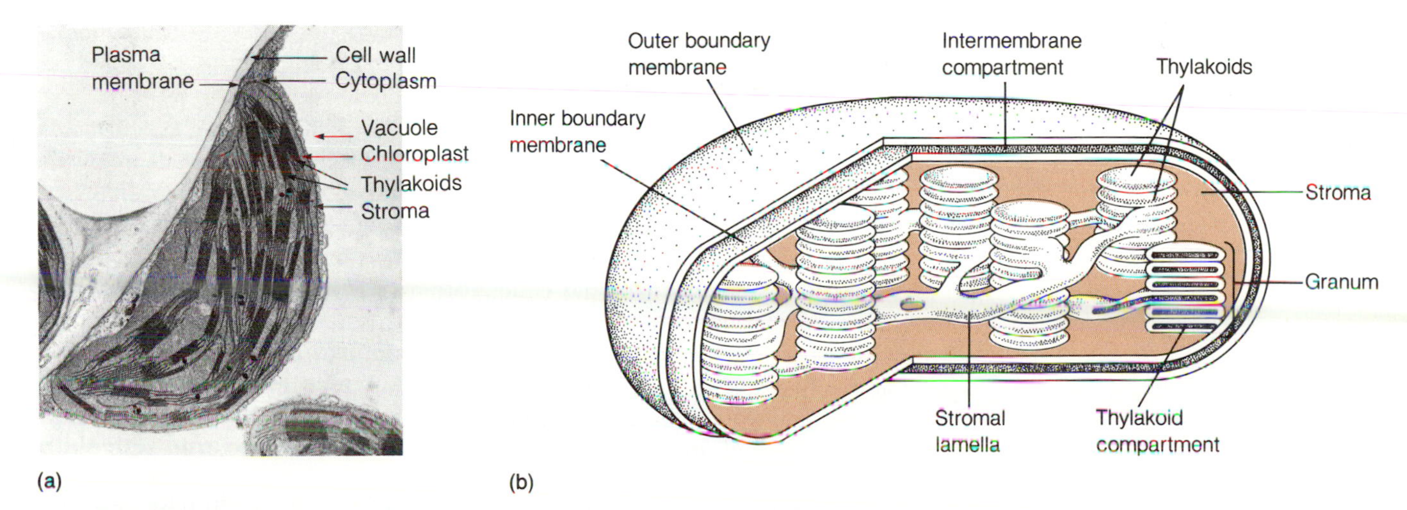

Figure 4-29 Structure of Chloroplasts. Chloroplasts are organelles involved in photosynthesis. They consist of two outer membranes and numerous stacked membranes within the stroma. Most of the enzymes involved in photosynthesis are located in the stacked membranes and in the stroma. *(a)* Transmission electron micrograph of a cross section of a corn chloroplast. *(b)* Diagram illustrating the various components of chloroplasts.

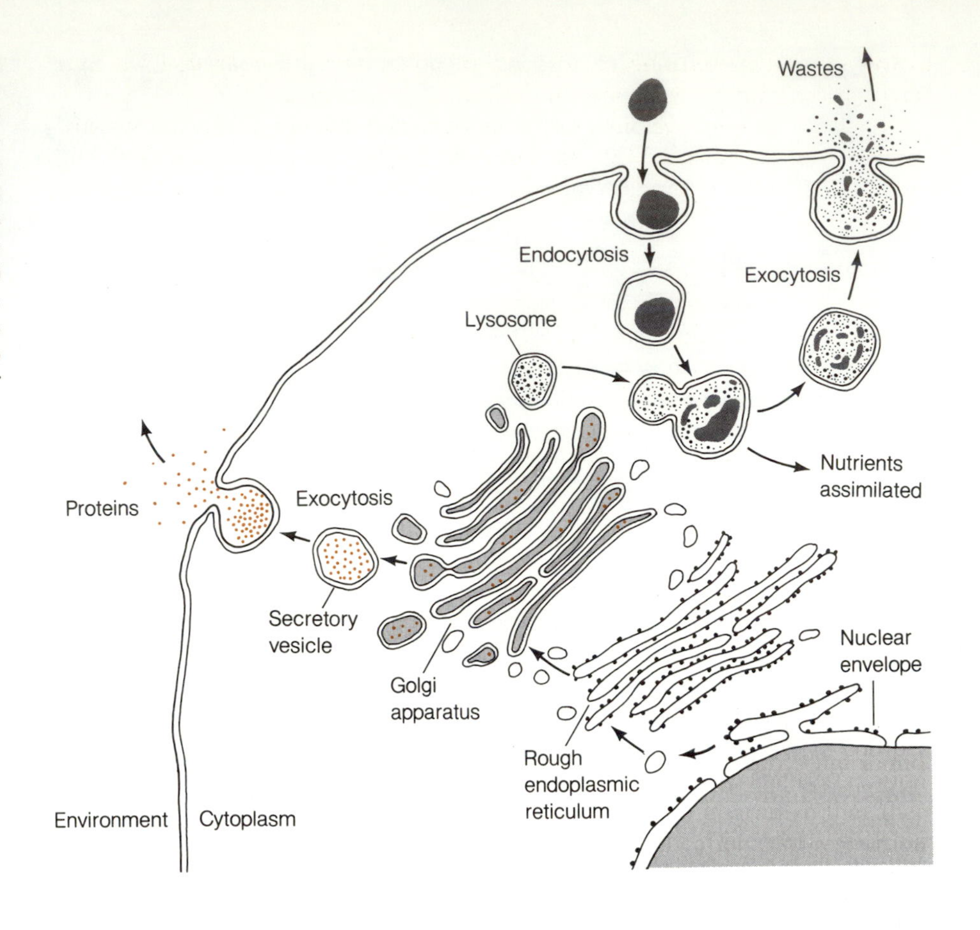

Figure 4-30 Endoplasmic Reticulum and Golgi Membranes. Extensive endoplasmic reticulum and numerous Golgi bodies are often visible in the cytoplasm of eukaryotic cells. The endoplasmic reticulum is a site where protein synthesis occurs and where the proteins are modified. Vesicles that pinch off from the endoplasmic reticulum carry the modified proteins to the Golgi body, where they may be modified further. Membrane material in the nuclear envelope, the endoplasmic reticulum, the Golgi body, and the plasma membrane are cycled continuously, as illustrated.

The 80s ribosomes attached to the endoplasmic reticulum synthesize proteins that enter the lumen (interior) of the endoplasmic reticulum. These proteins are often modified in the endoplasmic reticulum and then transported within vesicles to other parts of the cell, or to the cytoplasmic membrane, where they may be excreted. The 80s ribosomes "free" in the cytoplasm synthesize proteins that function as enzymes within the cytoplasm or as structural components of the cell.

Endoplasmic Reticulum and Golgi Membranes

The **endoplasmic reticulum** is an intracellular membrane system that creates channels within the cytoplasm of eukaryotic cells (fig. 4-30). Often it appears to develop from the outer nuclear membrane. In cells that are secreting proteins, much of the endoplasmic reticulum is studded with 80s ribosomes that are catalyzing the synthesis of proteins. As these proteins are synthesized, they enter the lumen of the endoplasmic reticulum and are eventually secreted. The endoplasmic reticulum with attached ribosomes is usually referred to as **rough endoplasmic reticulum.** Eukaryotic cells involved in steroid and lipid synthesis have **smooth endoplasmic reticulum** because it is devoid of ribosomes.

Golgi membranes (in nonphotosynthetic cells) or **dictyosomes** (in photosynthetic cells) are stacked membranes that can be distinguished from the continuous endoplasmic reticulum (fig. 4-30). In the Golgi membranes, proteins are modified and packaged into membranous vesicles for transport to the plasma membrane or to food vacuoles. Membranous vesicles that carry hydrolyzing enzymes are referred to as **lysosomes.**

Bacteria do not have any type of endoplasmic reticulum or Golgi system. Proteins that are released into the environment by prokaryotes are generally synthesized on the plasma membrane in such a way that the protein appears outside the cell.

Vacuoles

Various membranous **vacuoles** or vesicles may be found in eukaryotic cells. For example, the single-celled amoeba often contains **food vacuoles.** These vacuoles arise when the amoeba engulfs bacteria or other single-celled organisms. Within the food vacuoles, the engulfed organisms are decomposed by enzymes released into the vacuoles when **lysosomes** fuse with them. The single-celled *paramecium* contains two **contractile vacuoles** that pump water out of the cell and so protect the paramecium from swelling with water and bursting like an overfilled balloon. Many plant cells have **water vacuoles** that absorb and store water (fig. 4-1). Water vacuoles expand plant cells and push them against their walls, thus making the plant tissue turgid. No membranous vacuoles are present in bacterial cells. The gas vesicles (also called gas vacuoles) in bacteria are not membranous.

Granules

Eukaryotic cells generally store fats, starches, and glycogen. In photosynthetic cells, granules of starch are often found within the chloroplast (fig. 4-1) and fats are found in the cytoplasm. In nonphotosynthetic cells, granules of fat and glycogen are usually found within the cytoplasm. Starchy material can be seen by staining a wet mount with Gram's iodine. Granules of starch, glycogen, and fat are used by the cell as a source of carbon and energy for making new cell material and for powering the cell.

Cytoskeleton

The cytoskeleton is the network of proteins that fills the cytoplasm in eukaryotic cells (fig. 4-31). The proteins that make up the cytoskeleton include ac-

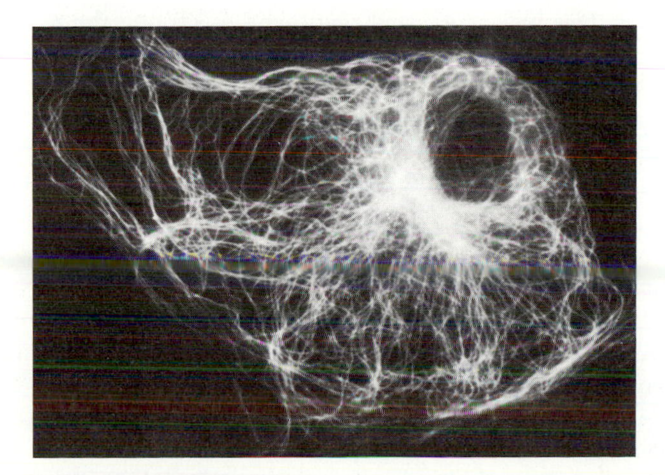

Figure 4-31 The Cytoskeleton of Eukaryotic Cells. The cytoskeleton in this photograph is revealed using immunofluorescence microscopy. The cytoskeleton strengthens the cytoplasmic membrane, confers shape to the cell and is involved in cellular movements.

tin, myosin, tropomyosin, tonofilaments, and microtubules. The cytoskeleton determines the shape of the eukaryotic cell and in some cases is involved in cell motility. The network of proteins called the **cortex,** which lies along the inside surface of the plasma membrane and strengthens the cytoplasmic membrane, is also part of the cytoskeleton. Ameboid motion displayed by the protozoan *Amoeba proteus* is believed to be due to the depolymerization and repolymerization of the cytoskeleton as well as the movement of plasma membrane from the rear of the cell to the front of the cell. Some eukaryotes, such as the diatoms, show a gliding motility, but this is believed to be very different from prokaryotic gliding motility. It is thought that gliding motility in eukaryotes is due to the transport of plasma membrane from the back of the cell to the front of the cell.

No bacteria, except for some spirochetes, have anything that approximates a cytoskeleton. Some spirochetes have microtubule-like proteins that run the length of the cell and appear to be responsible for the bending and flexing motion seen in spirochetes.

Alternate Cell Structures

Spores and Conidia

Spores and **conidia** of various types are produced by many different eukaryotes. The fungi produce many different types of spores, such as sporangiospores, conidiospores, ascospores, and basidiospores.

Cysts

Cysts □ are produced by many protozoans during part of their life cycle. 367 Often, the protozoan will round up and produce a thick wall around itself. This form of the organism is resistant to conditions that the vegetative form of the protozoan could not survive.

Prokaryotic and eukaryotic cells are structurally distinct, even though they both often carry out very similar activities. For example, both cell types frequently have flagella, but these are physically and mechanically very different. In addition, both cell types may have respiratory or photosynthetic abilities, but they are associated with very different structures. The prokaryotes carry out their respiration and photosynthesis along the plasma membrane and internal membranes, respectively, while the eukaryotes carry out these processes within mitochondria and chloroplasts.

A study of prokaryotic and eukaryotic structure and function has provided numerous insights into cellular physiology and into ways that microorganisms can be controlled and put to use for the benefit of humans.

SUMMARY

PROKARYOTIC AND EUKARYOTIC CELLS: A DEFINITION

1. The electron microscope has been used to demonstrate that there are two fundamental types of cells: the prokaryotic cell and the eukaryotic cell. All bacteria (eubacteria, cyanobacteria, archaebacteria, rickettsiae, mycoplasmas) have a prokaryotic cell structure. All other types of cells have a nucleus and organelles, such as mitochondria and chloroplasts, and consequently are considered eukaryotic cells.
2. In general, prokaryotic cells are much smaller than eukaryotic cells and their internal structure is much simpler. Even though prokaryotes lack the organelles commonly found in the eukaryotes (nuclei, mitochondria, chloroplasts, endoplasmic reticulum, Golgi membranes, vacuoles, etc.), prokaryotes carry out many of the same functions associated with eukaryotes. For example, many bacteria are capable of respiration and photosynthesis.

GROSS CELLULAR MORPHOLOGY OF BACTERIA

1. Most bacteria studied in the laboratory can be classified as spheres (cocci), rods (bacilli), spirals (spirilla), or helices (spirochetes). There are also many oddly shaped bacteria, such as the bacteria with prosthecae.
2. Some bacteria remain attached to each other after cell division and form long chains or clusters. Rod-shaped bacteria in chains are called streptobacilli. On the other hand, cocci in chains are referred to as streptococci, while cocci in random clumps are called staphylococci. Cocci that are found mostly in pairs are called diplococci.
3. Most bacteria studied in the laboratory divide by binary fission.

THE FINE STRUCTURE OF BACTERIAL CELLS

1. Bacteria may have envelopes that surround their cell membrane. Most bacteria have a cell wall that protects them from osmotic lysis. Many bacteria have a capsule that protects them from predators and mechanical damage. A few bacteria have sheaths that protect them from predators and mechanical damage. Many, if not most, bacteria are believed to have a glycocalyx that helps them attach to each other and to solid substrates.
2. Bacteria have appendages that are unique to the prokaryotes. Some bacteria have a number of different pili. Some of these pili are believed to help these bacteria stick to each other and to solid substrates, while others are known to be involved in the transfer of genetic information. A few bacteria have elongated cellular regions called prosthecae which may increase the surface area of the cells so that their uptake of nutrients is enhanced. Many bacteria are motile because they have flagella. One group of bacteria has modified flagella, called axial filaments, which are used to propel them through aqueous environments.
3. Bacteria have many intracellular structures. All bacteria are believed to have only one main genome, although it may be present in multiple copies when they reproduce rapidly. Some bacteria have small pieces of circular DNA called plasmids. A few bacteria have membranes that invaginate into the cell's cytoplasm. These membranes are called mesosomes. Their function is in doubt, and some researchers claim that they are simply artifacts. Photosynthetic bacteria have thylakoids and photosynthetic membranes that are involved in photosynthesis. A number of aquatic bacteria contain gas vesicles that help them float near the surface of the water, where they can obtain sufficient light. Many bacteria store reserve material in their cytoplasm. Often, electron micrographs will show numerous very small granules in the cytoplasm. These very small granules are ribosomes, where protein synthesis occurs.
4. Bacteria have a number of alternate cell structures. An endospore can be considered an alternate cell structure when it is released from the bacterial cell that produced it. Endospores are extremely resistant to heat, desiccation, chemicals, and the lack of nutrients.

THE FINE STRUCTURE OF EUKARYOTIC CELLS

1. Eukaryotes may have envelopes that surround their plasma membrane. For example, most of the photosynthetic microorganisms and plant cells have cell walls that protect them from osmotic lysis and mechanical damage. Eukaryotic cells may be covered by a glycocalyx, which allows them to bind together.
2. The most obvious appendages of eukaryotic cells are their flagella and cilia.
3. Eukaryotes generally have a great many internal structures. Almost all eukaryotic cells have one or more nuclei. The nucleus contains the cell's genetic information, which is packaged in a number of genomes. Almost all eukaryotic cell's have one or more mitochondria, which are involved with respiration. All photosynthetic eukaryotes have one or more chloroplasts, which carry out photosynthesis. Most eukaryotic cells have some endoplasmic reticulum. Some of the cell's protein synthesis occurs on the endoplasmic reticulum. In addition, proteins may be modified in the endoplasmic reticulum and packaged in membranous vesicles. These vesicles may form lysosomes or fuse with food vacuoles, Golgi membranes, or the plasma membrane. Golgi membranes can also be found in most eukaryotic cells; they are another site where proteins are modified and packaged into membanous vesicles. Eukaryotic cells contain a protein skeleton called the cytoskeleton. Actin, myosin, and tubulin are examples of some of the proteins found in the cytoskeleton. The cytoskeleton strengthens the plasma membrane, gives the cell its shape, and is involved in amoeboid and gliding motion.
4. Eukaryotes may develop alternate cellular forms such as spores, conidia, and cysts.

STUDY QUESTIONS

1. Compare a prokaryotic cell with a eukaryotic cell.
2. What is a bacterial capsule? What is the importance of capsules to a bacterial population?
3. Discuss the structure and chemical composition of bacterial cell walls. Compare the wall of a gram positive bacterium with that of a gram negative bacterium. What is the importance of a cell wall to a bacterium?
4. Discuss the structure and chemical composition of a cytoplasmic membrane. What are some of the functions associated with bacterial cell membranes?
5. Discuss the structure and function of pili.
6. Compare the structures of prokaryotic and eukaryotic flagella.
7. Explain the significance of the following structures in prokaryotic cells: mesosomes, thylakoids, endospores, genomes (chromosomes), plasmids, gas vesicles, and granules.
8. Explain the significance of the following structures in eukaryotic cells: endoplasmic reticulum, Golgi apparatus, lysosomes, food vacuoles, contractile vacuoles, water vacuoles, centrioles, basal bodies, cytoskeleton, mitochondria, and chloroplasts.
9. Compare prokaryotic ribosomes with eukaryotic ribosomes.

SUPPLEMENTAL READINGS

Albersheim, P. 1975. The wall of growing plant cells. *Scientific American* 232(4):80–90.

Burchard, R. 1980. Gliding motility of bacteria. *BioScience* 30(3):157–162.

Costerton, J., Geesey, G., and Gheng, K. 1978. How bacteria stick. *Scientific American* 238:86–95.

Farquhar, M. and Palade, G. 1981. The Golgi apparatus (complex). *The Journal of Cell Biology* 91(1):77s–103s.

Govindjee and Govindjee, R. 1974. The primary events of photosynthesis. *Scientific American* 231(6):68–82.

Hinkle, P. and McCarty, R. 1978. How cells make ATP. *Scientific American* 238(3):104–123.

Ingraham, J., Maaløe, O., and Neidhardt, F. 1983. *Growth of the Bacterial Cell.* Sinauer Associates Inc.

Lake, J. 1981. The ribosome. *Scientific American* 245(2):84–97.

Lazarides, E. and Revel, J. 1979. The molecular basis of cell movement. *Scientific American* 240(5):100–113.

Miller, K. 1979. The photosynthetic membrane. *Scientific American* 241(4):102–113.

Nikaido, H. and Vaara, M. 1985. Molecular basis of bacterial outer membrane permeability. *Microbiological Reviews* 49:1–32.

Olins, D. and Olins, A. 1978. Nucleosomes: the structural quantum in chromosomes. *American Scientist* 66:704–711.

Ordal, G. 1980. Bacterial chemotaxis: a primitive sensory system. *BioScience* 30(6):408–410.

Porter, K. and Tucker, J. 1981. The ground substance of the living cell. *Scientific American* 244(3):56–67.

Rothman, J. 1981. The Golgi apparatus: two organelles in tandem. *Science* 213:1212–1219.

Satir, P. 1974. How cilia move. *Scientific American* 231(4): 44–52.

Shaw, P. J., Hills, G. J., Henwood, J. A., Harris, J. E., and Archer, D. B. 1985. Three-dimensional architecture of the cell sheath and septa of *Methanospirillum hungatei. Journal of Bacteriology* 161(2):750–757.

Staehelin, A. and Hull, B. 1978. Junctions between living cells. *Scientific American* 238(5):140–152.

Vreeland, R. H., Anderson, R., and Murray, R. G. E. 1984. Cell wall and phospholipid composition and their contribution to the salt tolerance of *Halomonas elongata. Journal of Bacteriology* 160(3):879–883.

CHAPTER 5
NUTRITION AND CULTIVATION OF MICROORGANISMS

CHAPTER PREVIEW

MAKING PROTEIN FROM AIR POLLUTANTS USING BACTERIA

Single-cell protein is a term frequently used to describe protein-rich microorganisms that can be used as human or domestic animal food. The ever-increasing world population and ever-decreasing natural resources have stimulated research in the area of single-cell protein. The idea is to grow large quantities of protein-rich microorganisms using cheap, readily available materials or industrial waste products. Microorganisms such as yeasts, bacteria, fungi, and algae have been grown on a variety of inexpensive substrates, including whey, potato starch, paper pulp waste, coffee, and molasses. Studies on the growth habits of these microorganisms and their nutritional requirements have proved fruitful, and much single-cell protein is being produced using modern technology.

These studies have also revealed that a group of bacteria called the **carboxydobacteria** can use carbon monoxide (CO) resulting from automobile exhausts as their sole source of carbon atoms, as well as their source of energy for life processes.

Carbon monoxide is a principal pollutant from the internal combustion engine, and in sufficiently high quantities can cause death by poisoning. The discovery of the carboxydobacteria served as an important first step in the bioconversion of the poisonous carbon monoxide into a life-giving product.

The bacterium *Pseudomonas carboxydovorans*, one of several CO-consuming bacteria, has been shown to reproduce well in an environment containing automobile exhaust and air in a 50:50 mixture. From this mixture, *P. carboxydovorans* produces a population containing approximately 65% crude protein. Using simple mathematical computations it has been determined that the estimated 350,000,000 cars in the world can be used to make about 500,000 tons of dry cell protein/year, or about 2.5 g dry weight protein/car/year, a significant contribution to the nutrition of countless humans who are presently living under near-starvation conditions.

The discovery of the carboxydobacteria represents one of the many discoveries of useful microorganisms by microbiologists. These discoveries were made only after these scientists had a clear understanding of the nutritional requirements of the desired microorganisms and how to grow them in the laboratory. This chapter will introduce you to some of the basic concepts in microbial cultivation and nutrition and how these can be used to grow the desired organisms.

LEARNING OBJECTIVES

A STUDY OF THIS CHAPTER SHOULD ENABLE YOU TO:

- DISCUSS WHY NUTRIENTS ARE REQUIRED BY MICROORGANISMS
- OUTLINE THE BASIC NUTRITIONAL NEEDS OF MICROORGANISMS
- CITE THE DIFFERENCE BETWEEN PHOTOTROPHS AND CHEMOTROPHS
- CITE THE DIFFERENCE BETWEEN AUTOTROPHS AND HETEROTROPHS
- DEFINE THE TERMS AEROBE, ANAEROBE, AND FACULTATIVE ANAEROBE
- DEFINE WHAT A CULTURE MEDIUM IS AND EXPLAIN HOW IT IS PREPARED
- CITE THE VARIOUS TYPES OF CULTURE MEDIA AND DESCRIBE SOME OF THEIR USES IN THE LABORATORY
- OUTLINE THE VARIOUS TECHNIQUES USED IN CULTURING MICROORGANISMS

All living organisms require **nutrients** for their growth and reproduction. Nutrients are the raw materials that are used to build new cellular components and to provide the energy required to carry out the cell's life processes (fig. 5-1).

The ability to grow and multiply rapidly is essential to the survival of microorganisms and is linked to the efficiency with which they gather nutrients. In nature, nutrients are often limiting and competition for them is great. In order to overcome these difficulties, many microorganisms have developed great flexibility in their utilization of various substrates available in their environment and have evolved forms that aid them in assimilating nutrients. For example, a soil bacterium, *Pseudomonas cepacia*, can utilize as many as 105 different organic compounds as sources of carbon and energy. *Caulobacter*, a bacterium that is commonly found in aquatic environments, has an appendage (prostheca) that allows the bacterium to absorb nutrients more efficiently from highly diluted (oligotrophic) environments.

The study of microorganisms in the laboratory is an important aspect of understanding microbial life. In order to understand microorganisms, we must be able to cultivate and maintain them in pure form. The techniques for growing microorganisms have numerous applications. For example, in the hospital laboratory these techniques are used routinely for isolating and identifying microorganisms suspected of causing disease. Also, microorganisms are isolated from nature and studied in the laboratory so that we can better understand their roles in natural environments. Many industries, such as the petroleum, agriculture, and food industries, capitalize on the knowledge obtained from laboratory studies of microorganisms to improve their products or to acquire materials from nature.

In this chapter we will discuss some of the basic concepts of microbial nutrition and illustrate the various ways in which microorganisms can satisfy their need for nutrients. This chapter also discusses the basic techniques em-

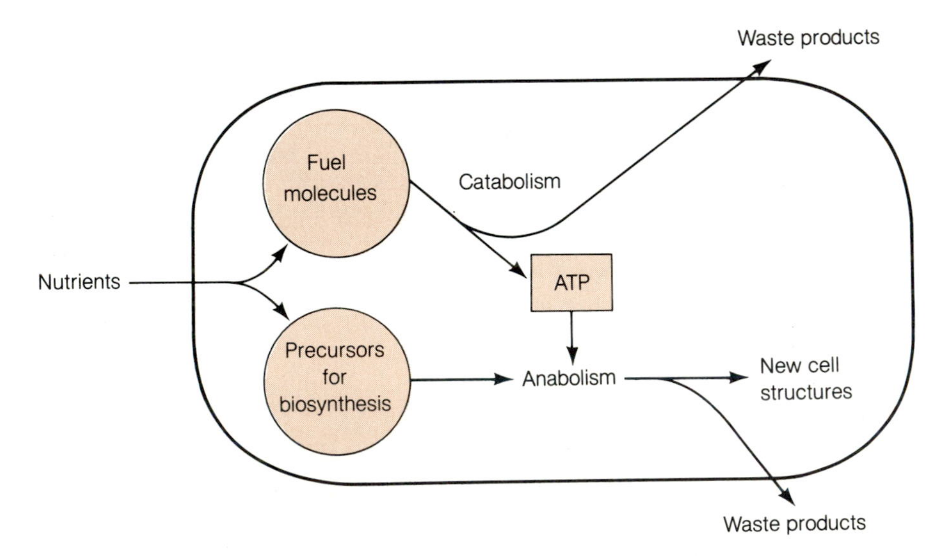

Figure 5-1 Relationship between Nutrient Intake and Cellular Activities. Nutrients in the environment are taken up by the cell. These nutrients are then oxidized to produce energy-rich compounds such as adenosine triphosphate (ATP). Nutrients are also used by the cell to synthesize cellular structures. The synthesis of cell structures requires that ATP be present as the source of energy.

ployed for cultivating microorganisms in the laboratory and the various types of culture media used to grow them.

PRINCIPLES OF MICROBIAL NUTRITION

Microorganisms obtain nutrients from their environment. Although they have the same basic nutritional requirements, they differ in the ways they take up these nutrients. The bacteria and fungi are called **osmotrophs** because they obtain their nutrients as solutions that pass through the plasma membranes. Osmotrophic microorganisms are unable to transport large molecules (such as proteins or polysaccharides) through their membranes, and consequently must digest them outside the cell. Large molecules are broken down by **exoenzymes** that are secreted by the cell to the external environment (fig. 5-2). The small breakdown products are then transported into the cell. Other microorganisms, such as the protozoa, ingest food particles by a process called **phagocytosis** and are called **phagotrophs** (fig. 5-3). During the process of phagocytosis, the organism surrounds the food particle with its plasma membrane and then brings the particle inside the cell surrounded by a bit of plasma membrane (vacuole). Inside the vacuole, enzymes digest the food particle into simpler molecules, such as amino acids and sugars. This digestive process resembles the digestion of molecules by exoenzymes. The resulting small molecules are used by the cell as nutrients.

Once the nutrients are inside the cell, they are chemically modified to meet the specific nutritional needs of the cell. For example, many bacteria can utilize proteins to satisfy some of their nutritional requirements. Exoenzymes called **proteases** degrade these proteins into amino acids, which are then brought into the cell and used to build bacterial proteins. Sometimes

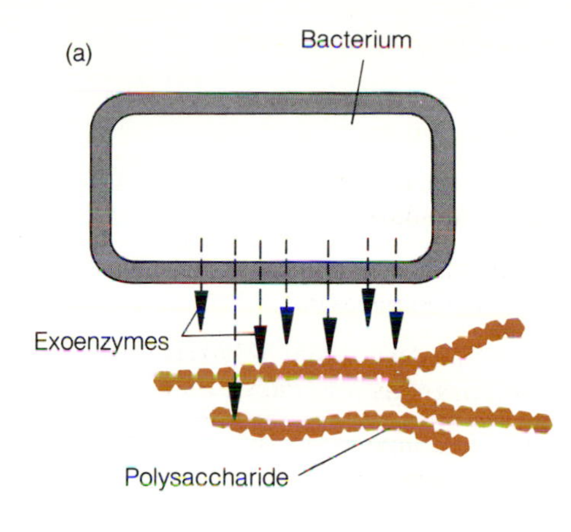

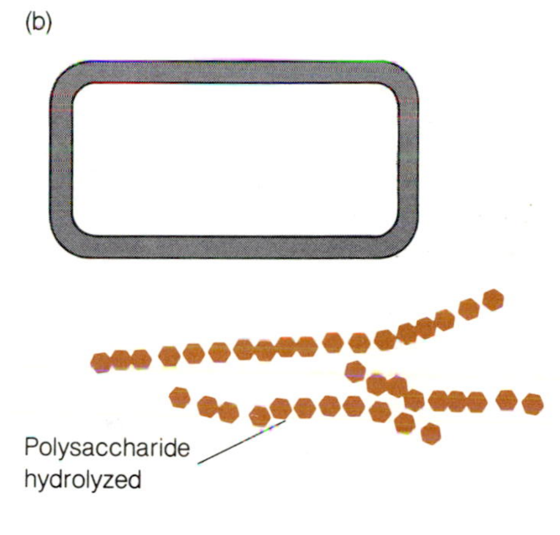

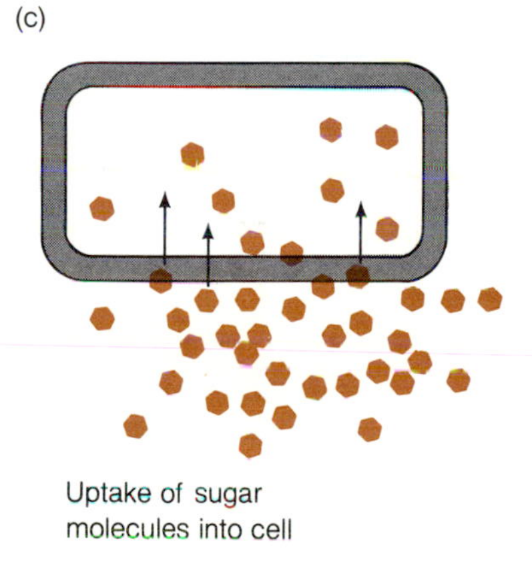

Figure 5-2 Function of Exoenzymes in Nutrient Digestion. Exoenzymes are secreted by the cell to act on large, insoluble food molecules (carbohydrates, proteins, fats, etc.) present in the immediate environment. The exoenzymes break these large molecules into constituent molecules (simple sugars, amino acids, etc.), which are then transported into the cell to be used as nutrients.

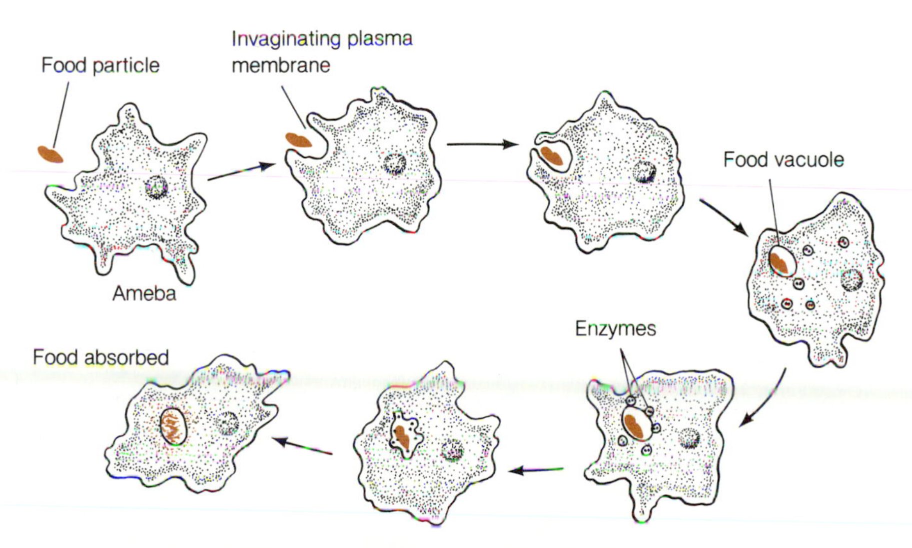

Figure 5-3 Phagocytosis. Phagotrophic organisms obtain their nutrients by engulfing food particles with their plasma membrane. The particle is surrounded by an invaginating bit of plasma membrane, which then forms a food vacuole. The food particles within the food vacuole are acted upon by digestive enzymes, which then break down the food into small molecules (sugars, amino acids, nucleosides, etc.). These simple molecules are then used by the cell as nutrients.

the amino acids are transformed into other compounds that are needed to build cellular components, or they are used as energy sources. For example, amino acids are used to build cell wall material as well as flagella or pili.

NUTRITIONAL REQUIREMENTS OF MICROORGANISMS

Microorganisms, like all living cells, require nutrients for their energy and biosynthesis. These nutrients are incorporated as components of the cell (table 5-1) and they make up cellular structures such as the genome, the plasma membrane, and the cell wall. The nutrient requirements that all organisms have are often satisfied in very different ways, depending upon the species of microorganism and the availability of specific nutrients. For example, although all microorganisms require the element carbon, some can take up the carbon only in an organic form, while others may assimilate it as carbon dioxide instead. Similarly, nitrogen is universally required by microorganisms, but some can use protein-nitrogen, while others can use ammonium- or nitrate-nitrogen. By knowing the nutritional requirements of microorganisms, microbiologists can carry out detailed studies of these organisms and determine their role(s) in nature and their usefulness in industrial processes. In the following paragraphs, we will discuss some of the basic nutritional requirements of microorganisms and indicate how microorganisms use these nutrients.

Energy Requirements

All cellular processes, such as the synthesis of new cell material, concentration of nutrients within the cell, and cell motility, require energy. The energy is obtained from sunlight or from the oxidation of chemical compounds. Whether the initial source of energy is the sun or chemical compounds, microorganisms convert this energy into energy-rich compounds such as adenosine triphosphate (ATP) □, which can be used as "energy currency" for cellular activities. Those organisms that obtain their energy from sunlight are called **phototrophs,** while those that derive their energy from the oxidation of chemicals are called **chemotrophs.**

183

TABLE 5-1
CHEMICAL COMPOSITION OF A TYPICAL BACTERIAL CELLS*

CHEMICAL COMPONENT	% OF CELL WEIGHT	NUMBER OF MOLECULES PER CELL ($\times 10^6$)	TYPES OF EACH MOLECULE
Water	70	4,000	1
Inorganic ions	0.3	200	20
Carbohydrates and precursors	0.1	210	180
Amino acids and precursors	0.4	200	100
Nucleotides and precursors	0.2	100	100
Lipids and precursors	3.0	245	70
Macromolecules			
DNA	1.0	0.000002	1
RNA	6.0	0.3	460
Proteins	16.5	2	1100
Polysaccharides	2.5	1	10

*Values are approximate values of wet weight cells *(Escherichia coli)* growing aerobically at 35°C.

Phototrophs obtain their energy by a process called **photosynthesis** (fig. 5-4). In this process, the radiant energy from the sun is "trapped" by light-sensitive pigments such as chlorophyll. The ultimate result is the production of chemical energy in the form of molecules of adenosine triphosphate, or ATP.

There are two major groups of phototrophic bacteria: the cyanobacteria and the photosynthetic bacteria. One important difference between these two groups is that the cyanobacteria produce O_2 as a byproduct of photosynthesis, while the photosynthetic bacteria produce other compounds. Photosynthesis will be discussed in greater detail in Chapter 8 □.

194

Chemotrophic microorganisms obtain their energy from the oxidation of chemical compounds. They can accomplish this by oxidizing molecules such as glucose or ammonium, which release sufficient energy to power energy-requiring functions of the cell. These biological oxidations involve a sequence of steps during which fuel molecules are slowly oxidized and their energy stored in energy-rich compounds such as ATP. These processes include fermentation and respiration (fig. 5-5) which are discussed in detail in Chapter 8.

Carbon Requirements

Carbon (C) atoms are present in all cellular components. Macromolecules such as proteins, lipids, polysaccharides, and nucleic acids, which collectively

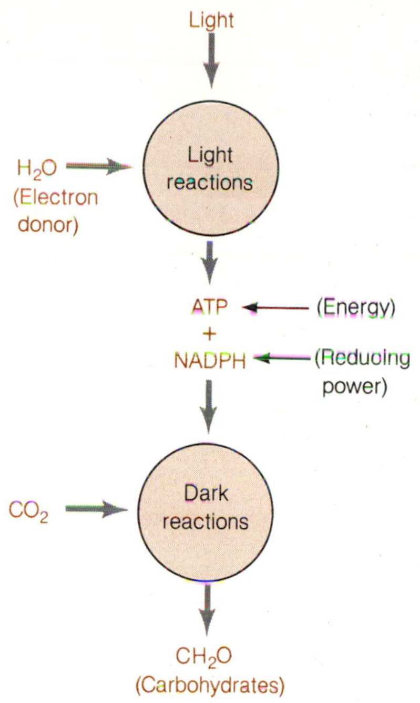

Light
H_2O (Electron donor)
Light reactions
ATP + NADPH
(Energy)
(Reducing power)
CO_2
Dark reactions
CH_2O (Carbohydrates)

Figure 5-4 Simplified Illustration of Phototrophic Metabolism. Phototrophic organisms obtain their energy from the sun. Radiant energy in light stimulates light-sensitive pigments, initiating a series of light-dependent chemical reactions which result in the production of usable energy in the form of ATP and reducing electrons in the form of NADPH. Part of the energy is used to build cellular compounds and part of it is used to convert carbon dioxide into sugars.

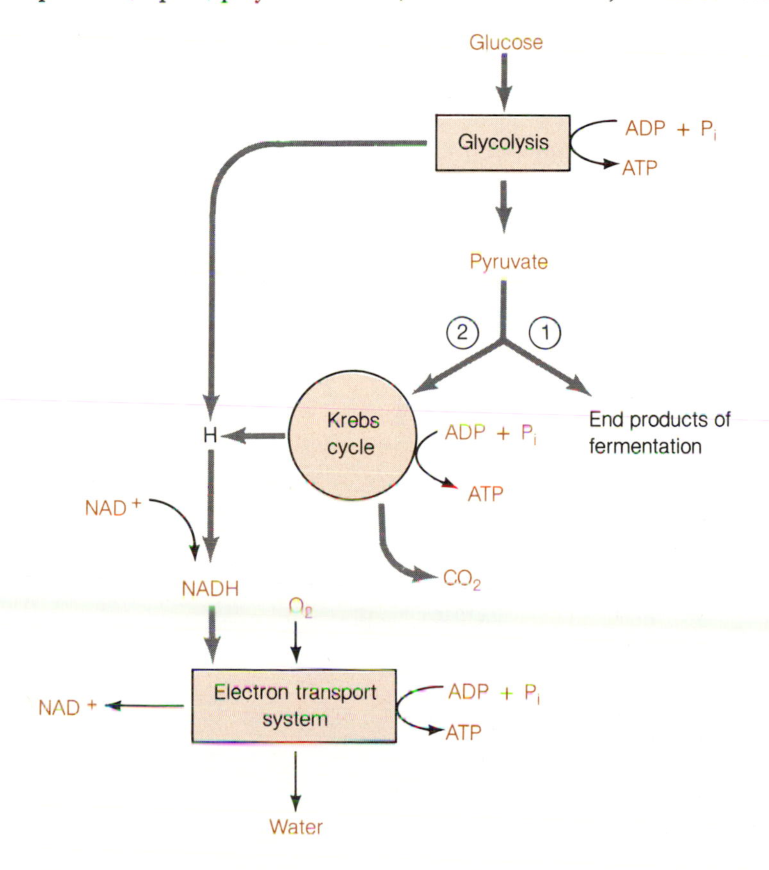

Figure 5-5 Simplified Illustration of Chemotrophic Metabolism. Chemotrophic organisms oxidize fuel molecules, such as glucose, in a series of enzyme-catalyzed reactions such as glycolysis, the Krebs cycle, and the electron transport system. The end result of these oxidations is the extraction of energy from fuel molecules plus its conservation in ATP molecules. If the oxidation of fuel molecules includes the Krebs cycle and the electron transport system, the process is called respiration. If the oxidation of fuel molecules is only to pyruvate, the process is called a fermentation.

make up the bulk of the cell's organic material, are constructed with many C atoms. Carbon-containing molecules participate in all phases of cellular metabolism. They participate in **catabolic** (biodegrading) activities leading to the production of energy-rich molecules such as ATP, as well as in **anabolic** (biosynthetic) activities (fig. 5-1), which lead to the synthesis of cellular components.

Some microorganisms can satisfy all of their carbon requirements from carbon dioxide in the atmosphere. These organisms are called **autotrophs** (self-feeders). Other microorganisms must have organic molecules as carbon sources and are called **heterotrophs** (table 5-2). Microorganisms that obtain their energy from the sun, and their carbon in the form of carbon dioxide, are called **photoautotrophs.** Those phototrophic organisms that use organic molecules to satisfy their carbon needs are called **photoheterotrophs.** Similarly, microorganisms that oxidize chemical compounds for their energy and use carbon dioxide as their carbon source are called **chemoautotrophs,** and those that oxidize chemical compounds for their energy and require organic forms of carbon are called **chemoheterotrophs.**

Nitrogen Requirements

All living organisms require nitrogen to make amino acids, nucleotides, and vitamins. Microorganisms exhibit remarkable versatility in their ability to use various nitrogenous compounds to satisfy their nitrogen requirements. Some bacteria use proteins or polypeptides to obtain their nitrogen. These large molecules are digested to amino acids, which in turn are chemically altered to make other nitrogen-containing molecules or are incorporated into cellular proteins. Many microorganisms, however, can obtain all of their needed nitrogen from inorganic salts of nitrate or ammonium. Other inorganic sources of nitrogen that sometimes are used by microorganisms include molecular nitrogen and cyanide.

A unique characteristic of some microorganisms is their ability to **fix** molecular nitrogen. The process of nitrogen fixation involves the reduction of molecular nitrogen to ammonia and its subsequent incorporation into organic molecules to form amino acids. Some microorganisms, such as the bacteria in the genus *Rhizobium*, can fix nitrogen only when associated with

TABLE 5-2

SUMMARY OF THE NUTRITIONAL TYPES OF MICROORGANISMS BASED ON THEIR ENERGY SOURCE AND ON THEIR CARBON SOURCE

NUTRITIONAL TYPE	ENERGY SOURCE	REDUCING POWER	CARBON SOURCE
*Photo*trophs			
a. Photo*lithotrophs*	Light	Inorganic molecules	—
b. Photo*organotrophs*	Light	Organic molecules	—
c. Photo*autotrophs*	Light	—	Carbon dioxide
d. Photo*heterotrophs*	Light	—	Organic carbon
*Chemo*trophs			
a. Chemo*lithotrophs*	Chemicals	Inorganic molecules	—
b. Chemo*organotrophs*	Chemicals	Organic molecules	—
c. Chemo*autotrophs*	Chemicals	—	Carbon dioxide
d. Chemo*heterotrophs*	Chemicals	—	Organic carbon
*Myxo*trophs			Carbon dioxide or Organic carbon

a leguminous plant such as sweet pea or clover. These microorganisms are said to be **symbiotic nitrogen fixers.** Others, such as the bacterium *Azotobacter*, can fix nitrogen independently of other organisms and are referred to as **nonsymbiotic nitrogen fixers.** The process of nitrogen fixation will be discused in greater detail in Chapter 31.

Oxygen Requirements

Oxygen is a common atom found in many biological molecules. It is an integral part of amino acids, nucleotides, glycerides, and other molecules. It is generally taken into the cell as part of nutrients, such as proteins and lipids. In addition, this element in the form of molecular oxygen (O_2) is required by most eukaryotes and many prokaryotes in order to generate energy by respiration.

Organisms requiring O_2 for cellular respiration are called **aerobes.** The aerobes constitute a very large group of organisms which includes most of the algae, the fungi, and many of the protozoa and bacteria. There are some microorganisms, mostly bacteria and protozoa, that have oxygen-sensitive enzymes and cannot function in the presence of molecular oxygen. As a consequence, these organisms are unable to multiply in any environment that contains O_2 and are called **anaerobes.** Anaerobes obtain their energy from oxygen-independent metabolism (fig. 5-5). **Microaerophilic** organisms are aerobes that require low O_2 tensions. Other microorganisms, called **facultative anaerobes,** can multiply either in the presence or in the absence of O_2.

Sulfur and Phosphorus Requirements

Most of the sulfur in cells is found in the sulfur-containing amino acids cysteine and methionine. Sulfur is also found in some polysaccharides, such as agar, and in certain coenzymes. Microorganisms can obtain their sulfur as inorganic salts of sulfate, hydrogen sulfide, sulfur granules, thiosulfate, or as organic compounds (cysteine or methionine). When microorganisms use sulfate as their source of sulfur, they reduce it to hydrogen sulfide, which is then incorporated into existing organic molecules. If the microorganism cannot reduce sulfate, then the sulfur must be obtained in a reduced form, such as that found in amino acids or in sulfides.

Phosphorus (P) is found in the cell primarily in nucleic acids, phospholipids, and coenzymes. This element can be obtained by microorganisms in either an organic or an inorganic form. Phosphate salts of sodium or potassium are the most common sources of this element for microorganisms, although some phosphorus is made available when they take up organic molecules that contain phosphorus (*e.g.*, nucleotides).

Mineral Requirements

Numerous minerals such as cobalt (Co), potassium (K), molybdenum (Mo), magnesium (Mg), manganese (Mn), calcium (Ca), and iron (Fe) are required by all cells. Minerals are required for the activity of a number of enzymes and other organic molecules. For example, magnesium is an integral component of the light-sensitive pigment chlorophyll, and cobalt is required for the enzyme **nitrogenase** to catalyze the fixation of nitrogen. Iron is an essential component of cytochromes (pigments involved in cellular respiration) and of hemoglobin.

Minerals are needed only in very small amounts, so they are often referred to as **trace elements.** Although the trace elements are sometimes supplied to microorganisms in the form of mineral salts, usually it is not necessary to do so. Trace elements are required in such low concentrations by the cell that sufficient quantities usually are found as contaminants in the other nutrients supplied to the cell or dissolved in water.

Growth Factor Requirements

All cells require approximately 20 different amino acids and several vitamins and cofactors. All of these must be either synthesized by the cell or obtained in the cell's diet. Those substances that cannot be synthesized by cells and that must be obtained from the environment are called **growth factors.** Growth factors are compounds such as vitamins, amino acids, and nucleosides that an organism is unable to synthesize from inorganic salts and the carbon and energy source (table 5-3).

Some microorganisms can grow in an environment consisting of water, inorganic salts, and a single organic compound such as glucose (table 5-4). For example, *Escherichia coli* can synthesize all of its required cellular components (*e.g.*, amino acids, vitamins, and nucleotides) from inorganic chemicals and glucose in its environment. Many microorganisms, however, require additional preformed substances in order to grow and multiply. Lactic acid bacteria such as *Leuconostoc mesenteroides*, for example, lack the ability to synthesize many of their required amino acids and vitamins and can grow only in environments containing these nutrients.

Water Requirements

Nearly all metabolic activities of a cell are carried out in aqueous environments. A dry environment generally results in the loss of the cell's water to its surroundings. This disrupts cellular activities.

The available water in an organism's surroundings is sometimes expressed as **water activity** (a_w). The water activity of a substance is determined by measuring the relative humidity (RH) of the air space in the immediate environment of the substance with an instrument called an **isoteniscope.** The water activity of nutrient broth, which has a relative humidity of about 98.5%, can be determined using the formula:

$$a_w = RH/100 = 98.5/100 = 0.985$$

TABLE 5-3
GROWTH FACTOR REQUIREMENTS OF SOME BACTERIA

SPECIES OF BACTERIA	GROWTH FACTOR REQUIRED
Bacillus anthracis	Thiamine (B_1)
Bacteroides melaninogenicus	Vitamin K
Brucella abortus	Niacin
Clostridium tetani	Riboflavin
Lactobacillus sp.	Pyridoxine (B_6)/Cobalamin (B_{12})
Leuconostoc dextranicum	Folic acid
Leuconostoc mesenteroides	Biotin
Proteus morganii	Pantothenic acid
Purple nonsulfur bacteria	Thiamine (B_1)
Purple sulfur bacteria	Cobalamin (B_{12})

TABLE 5-4
CHEMICALLY-DEFINED MEDIUM AND COMPLEX MEDIUM

A. MEDIUM FOR CULTIVATION OF *E. coli*—MINERAL SALTS—GLUCOSE—	
$NH_4H_2PO_4$	1g
Glucose (Energy & carbon source)	5 g
NaCl	5 g
$MgSO_4 \cdot 7H_2O$	0.2 g
K_2HPO_4	1 g
H_2O	1,000 ml

B. MEDIUM FOR CULTIVATION OF LACTOBACILLI*	
Casein hydrolyzate	5 g
Glucose	10 g
Solution A	10 ml
Solution B	5 ml
L-Asparagine*	250 ml
L-Tryptophan*	50 mg
L-Cystine*	100 mg
DL-Methionine*	100 mg
Cysteine*	100 mg
Ammonium citrate	2 g
Sodium acetate (anhydrous)	6 g
Adenine; guanine; xanthine; uracil; each*	10 mg
Riboflavin; thiamin; pantothenate; niacin; each*	500 μg
Pyridoxamine*	200 μg
Pyridoxal*	100 μg
Pyridoxin*	200 μg
Inositol and choline, each*	10 mg
p-Aminobenzoic acid*	200 μg
Biotin*	5 μg
Folic acid (synthetic)*	3 μg

Make up to 1 l with distilled water.
Solution A. K_2HPO_4 and KH_2PO_4, each 25 g, into distilled water to a volume of 250 ml.
Solution B. $FeSO_4 \cdot 7H_2O$, 0.5 g; $MnSO_4 \cdot 2H_2O$,2.0 g; NaCl, 0.5 g; and $MgSO_4 \cdot 7H_2O$, 10 g. Dissolve in distilled water to a volume of 250 ml.

*Growth factors.
From *Microbiology,* 4th ed., by Pelczar, Reed, and Chan. Copyright © 1977 by McGraw-Hill Book Company. Reprinted by permission. Data from M. Rogosa, et al., *J Bacteriol, 54*:13, 1947.

Microorganisms commonly require water activities above 0.90 in order to grow and multiply, but optimal metabolism occurs at water activities above 0.95. Some microorganisms, especially certain fungi, can grow in environments with water activities as low as 0.60.

The water activities of foods can be used to predict how rapidly food spoilage can occur. Foods with high water activities spoil more rapidly than those with low water activities. For example, meat and fish, foods with a high water activity, spoil much more rapidly than beans and grains, which have low water activities. In some cases, the water activity of foods can be artificially lowered in order to prevent food spoilage. For example, freeze-dried foods, like those used by hikers, are resistant to spoilage by microorganisms because these foods have been desicated in vacuum and hence have a very low water activity. There are many other methods of reducing the water activities of foods. These are discussed in the sections dealing with food preservation.

CULTURAL CONDITIONS

The growth characteristics of pure cultures of microorganisms can provide valuable information about their biology. The cultivation of a microorganism requires that suitable environmental conditions as well as the needed nutrients be provided. Microorganisms are able to grow in many different environments, although no single species is able to grow in all possible environments. In order to grow a given species of microorganism in the laboratory, it is necessary to know what specific cultural and environmental factors it requires for reproduction. The four principal environmental parameters that must be carefully controlled for the successful cultivation of microorganisms are a) temperature; b) atmospheric conditions; c) pH; and d) osmotic pressure.

Temperature

Microorganisms are usually cultivated in cabinets called **incubators** that are designed to maintain a constant temperature (fig. 5-6). Incubators are equipped with a heating and/or cooling unit to hold the temperature above or below that of the outside. In the clinical laboratory, most incubators and water baths are set to maintain temperatures between 30°C and 40°C, because most microorganisms that cause human or animal disease reproduce best at these temperatures. Microbes that grow in the soil, that damage crop and ornamental plants, or that spoil foods are often adapted to grow best at temperatures between 15°C and 30°C. Those organisms that grow best at temperatures between 20°C and 45°C are called **mesophiles.** Sometimes, the incubation temperature is elevated above 40°C to cultivate heat-loving microorganisms called **thermophiles.** Thermophiles reproduce rapidly at temper-

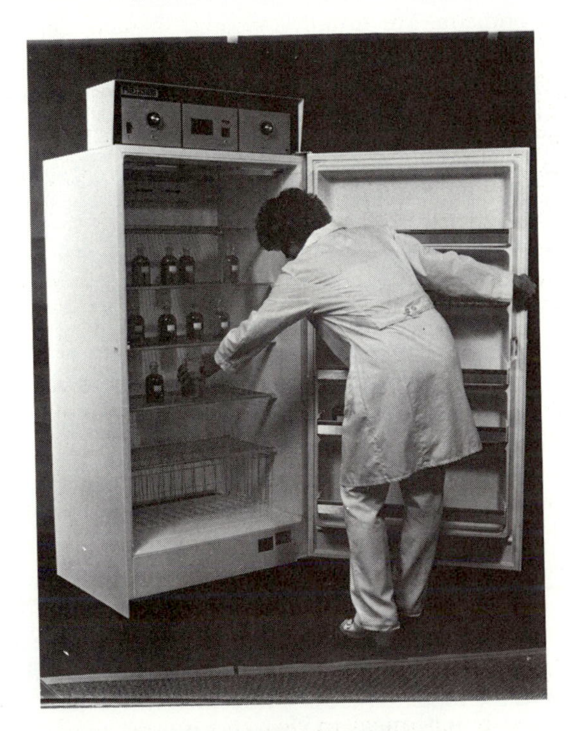

Figure 5-6 An Incubator. Air-heated incubator. This is the most common type of incubator in the microbiology laboratory.

atures between 45°C and 80°C, although there are certain extreme thermophiles that can reproduce at even higher temperatures. Incubators equipped with refrigerating units are used to grow microorganisms known as **psychrophiles,** which prefer low temperatures (between 0°C and 20°C) for multiplication. Many of the bacteria that inhabit glacier-fed lakes and arctic (and antarctic) regions are psychrophiles; the temperature in these regions seldom reaches a few degrees above freezing.

Sometimes culture vessels (flasks or tubes) are partially immersed in water baths maintained at a desired temperature. Water baths are used when it is necessary to maintain the environmental temperature at a constant level with a minimum of fluctuation. One method for the cultivation of a group of bacteria called **fecal coliforms** □ involves using a water bath set a 44.5°C ± 0.2°C so that they can be differentiated from other coliforms. Incubation temperatures that flucutate within 0.2°C are readily attained with water baths, but not with air-heated incubators.

315

The Atmospheric Environment

148

Many microorganisms require free molecular oxygen (O_2) in order to reproduce □. Many anaerobic microorganisms, however, can be inhibited or even killed as a result of brief exposures (as little as 30 minutes) to free oxygen. This occurs because some essential enzymes are denatured or inhibited by oxygen.

There are groups of microorganisms that require an atmosphere composed of O_2 and elevated levels of carbon dioxide. For example, *Neisseria gonorrhoeae*, the bacterium that causes gonorrhea, grows best in an atmosphere enriched with 5–10% carbon dioxide. Microorganisms requiring elevated CO_2 environments are usually cultured in incubators equipped with devices that permit the regulation of the gases inside the chamber. These environments can also be achieved with a device called a **candle jar** which consists of a large jar (a one-gallon mayonnaise jar will do) with a candle in it. After the cultures are placed inside the jar, the candle is lit and then the jar is sealed. The candle inside the jar burns until there is not enough oxygen left inside the jar to maintain combustion. The jar then contains a reduced concentration of free oxygen and about 3.5% CO_2.

Organisms that require oxygen-free environments must be cultured in special incubators or chambers that exclude molecular oxygen. **Anaerobic incubators** are specially designed to culture anaerobes. These incubators are equipped with ports and valves so that the air can be extracted from the growth chamber and replaced with inert gases such as nitrogen, helium, or carbon dioxide. Hydrogen gas is seldom used because of its explosive nature. Sometimes the incubator chamber is filled with a mixture of inert gases, such as nitrogen and carbon dioxide in a ratio of 95:5. The carbon dioxide is important to the reproduction of certain anaerobes and hence must be made available in the growth environment.

In many clinical and industrial laboratories, the **anaerobic jar** (fig. 5-7) is routinely used as an economical alternative to the anaerobic incubator. There are several versions of the anaerobic jar in current use in the microbiology laboratory. All these versions work in essentially the same way. Oxygen is removed by chemical means and is replaced with an inert gas. Figure 5-7 illustrates how the anaerobic jar, is used to create an oxygen-free environment for the cultivation of anaerobes.

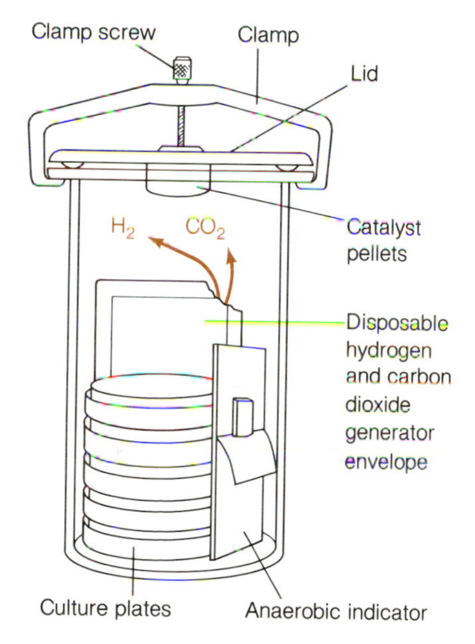

Figure 5-7 An Anaerobic System. The anaerobic system illustrated here is commonly used in the microbiology laboratory to cultivate anaerobic microorganisms. It consists of a jar with a screw-on lid, a hydrogen-carbon dioxide generator envelope, and a palladium catalyst. Water is added to the hydrogen-carbon dioxide generating envelope to activate it, and the envelope is placed inside the jar along with the cultures. An anaerobic indicator strip may also be included. The strip consists of a pad saturated with methylene blue solution, and it changes from blue to colorless when an anaerobic atmosphere inside the jar is developed. The hydrogen gas generated by the envelope reacts with any oxygen that may be present in the jar to form water. The reaction is catalyzed by the palladium catalyst, which permits the formation of water from hydrogen and oxygen to take place at room temperature. This reaction removes any free oxygen from inside the jar, creating an anaerobic environment. The CO_2 is needed to stimulate the growth of certain anaerobes.

The Hydrogen Ion Concentration (pH)

The concentration of hydrogen ions in the growth environment can drastically influence the reproduction of microorganisms, because it can affect the activity of many enzymes. Many bacteria reproduce only at pH levels between 6 and 8. Very few proliferate at pH levels much below 4. Many fungi, however, grow well at low pH values (5–6) at which most bacteria do not grow.

When microorganisms reproduce, they release waste products that may change the pH of their environment. If the pH change is extensive, their environment becomes inhospitable. Therefore, if extensive microbial growth is desirable, changes in the pH of the environment must be avoided. Buffers (chemicals that resist changes in pH) can be added to the growth environment in order to eliminate wide changes in pH.

Foods can be preserved by lowering their pH □. For example, sauerkraut and pickles are relatively free of microbial spoilage because their low pH inhibit microbial growth. 38

The Osmotic Pressure

Osmotic pressure is defined as the minimum amount of pressure that must be applied to a solution in order to prevent the flow of water across a membrane within the solution. For example, if a dialysis bag (a membrane-like material that allows the free passage of water molecules) is filled with 10% sucrose (table sugar) inside a beaker filled with water, the water in the beaker will tend to flow into the bag to dilute the sucrose solution. The amount of pressure that must be exerted from inside the bag to counteract the flow of water inward is the osmotic pressure.

The osmotic pressure of the environment influences whether or not a particular microorganism can reproduce. Microorganisms usually inhabit slightly **hypotonic** environments, or those in which the concentration of solutes (dissolved nutrients) is lower than that of the cytoplasm (fig. 5–8). In these environments, water tends to move into the cell and make it turgid.

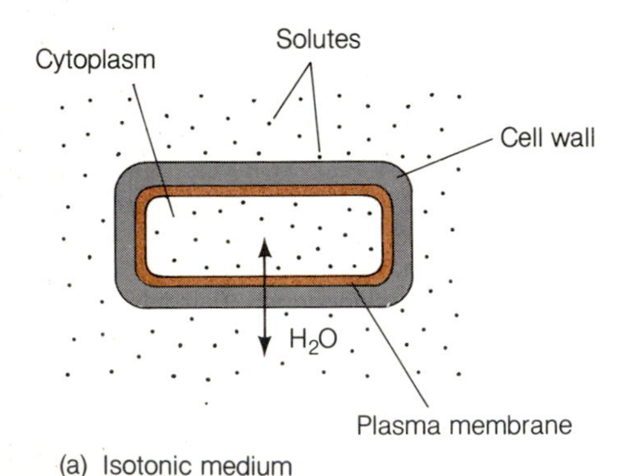

(a) Isotonic medium

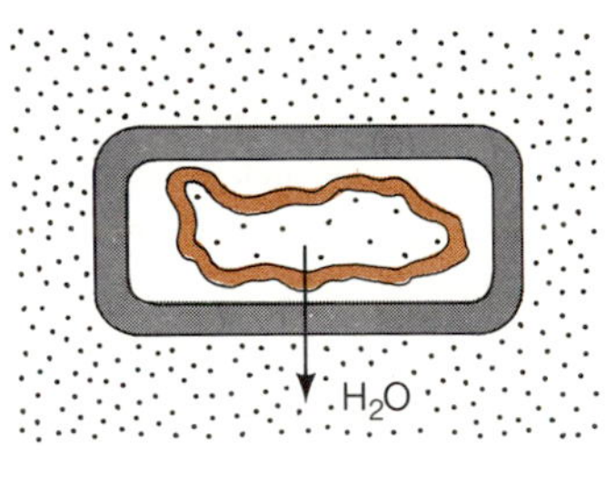

(b) Hypertonic medium

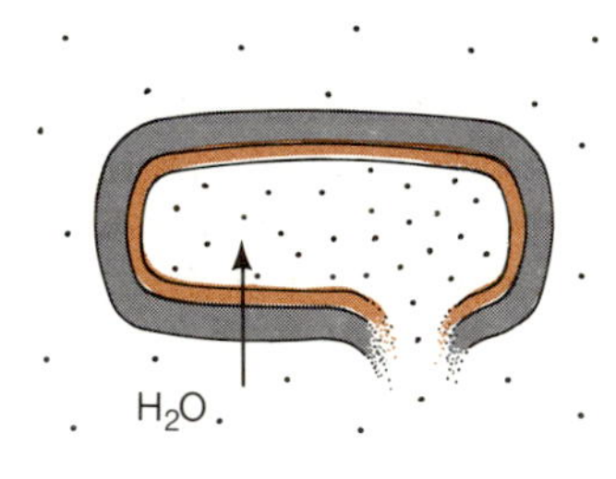

(c) Hypotonic medium

Figure 5-8 Flow of Water in Isotonic, Hypertonic, and Hypotonic Environments. *(a)* Cell in an isotonic environment. In isotonic environments, the concentration of environmental solutes is equal to that of the cell. In this environment, cellular water shows no net movement in or out of the cell. *(b)* Cell in a hypertonic environment. In hypertonic environments the concentration of solutes is greater than that of the cell. Hence, there will be a tendency for cellular water to flow out into the environment. This results in the dehydration and shrinking of the cell. *(c)* Cell in hypotonic environment. In hypotonic environments the concentration of solutes is lower than that of the cell. Hence, environmental water will tend to flow into the cell. In the process, the cell swells from an excess of water. If the flow of water is too great, the cell may burst. Although cells with cell walls may simply swell and not burst, cells with weak cell walls (gram negative bacteria) may burst due to excessive water intake.

The rigid cell wall of bacteria, fungi, algae, and some protozoa eventually limits the amount of water coming into the cell and also prevents the cell from swelling and bursting. On the other hand, if the solute concentration of the environment is greater than that of the cell (**hypertonic** environment), the cell becomes dehydrated and ceases to function because cellular water tends to flow out. The osmotic pressure of the culture environment is particularly important when culturing bacteria such as the mycoplasmas, which lack a rigid wall and hence are susceptible to osmotic lysis. The cultivation of pure cultures of mycoplasmas and bacteria with weak cell walls requires that the proper concentration of nutrients be used, so that the osmotic pressure of the environment does not cause the lysis of cells.

Foods such as fruit preserves and salted fish are resistant to spoilage by microorganisms because of their high osmotic pressure. Microorgaisms present in these foods are unable to grow because the hypertonic environment draws out the cell's water, thus inhibiting microbial metabolism.

CULTURE MEDIA AND THEIR USE IN THE LABORATORY

A **culture medium** is an aqueous solution of the various nutrients required by a microorganism (table 5-4). It generally contains a source of energy, carbon, fixed nitrogen, sulfur, phosphorus, hydrogen and oxygen, a buffer, trace elements, and water. Various growth factors and other ingredients may be added to (or removed from) this basic medium in order to grow the desired microorganism.

The culture medium may be in a liquid or a gel form. Liquid culture media are usually referred to as **broths,** while gel or semisolid media are often called **agars** because they are solidified with a red algae polysaccharide called **agar-agar.** This polysaccharide is particularly useful as a jelling agent for several reasons: a) agar is rarely used as a nutrient by microorganisms and therefore is not readily degraded when microbes grow on culture media solidified with agar; b) unlike gelatin, which originally was used as a solidifying agent, agar remains solid over a wide range of incubating temperatures (0°C–80°C); c) agar liquefies at about the boiling temperature of water, but does not jell until it cools to approximately 42°C. This property allows one to make suspensions of most microorganisms in liquefied agar at about 45°C without killing them. As the agar jells, the individual microorganisms suspended in the agar are trapped in the gel matrix and their movement is restricted. This process results in the formation of discrete clumps of cells called **colonies** (clones). Distributing the microorganisms throughout the agar provides more growing room and reduces crowding. Microorganisms growing in this fashion can then be counted. This procedure represents the 132 principle involved in plate counts □.

In order to prepare solid media, a broth or liquid culture medium is mixed with 1.5–2.0% agar and brought to a boil to dissolve the agar in the water. The culture medium is not yet ready for use, however, because many microorganisms may have contaminated it. The contamination is due to microorganisms and endospores associated with the nutrients, the water, the glassware, the media-maker, and the air. Even though the medium may have been boiled, these contaminating endospores may germinate in the newly-prepared culture medium and spoil it. Thus, it is imperative to **sterilize** the culture medium before it is used. The process of sterilization kills or removes

FOCUS

THE FIRST USE OF AGAR AS A SOLIDIFYING AGENT

The first use of agar as a solidifying agent is credited to Walther Hesse, who worked in Dr. Robert Koch's laboratory. Koch experienced difficulties isolating disease-causing bacteria on culture media solidified with gelatin. At room temperature (18–22°C) the gelatin remained solid, but when the bacteria grew, many of them digested the gelatin and the medium became liquid. This problem was further compounded when disease-causing bacteria were cultured at human body temperature (37°C). At this temperature the gelatin liquefied, regardless of whether the bacteria digested the gelatin. With the gelatin turning liquid, it became impossible for Koch to culture bacteria in pure form.

This problem was solved by Hesse's resourceful wife. Apparently, Koch had been discussing with Hesse the difficulties of preparing a stable solid culture medium, and Hesse mentioned this discussion to his wife. Frau Hesse suggested—to the benefit of countless microbiologists—that Koch try an ingredient her family in the West Indies used to thicken sauces, jams, and the like. This ingredient was called **agar-agar.** It is reported that Hesse told Dr. Koch about his wife's suggestion, Koch tried it, and it worked! Agar is now used by microbiologists throughout the world as *the* hardening agent for nearly all culture media.

all living microorganisms from the culture medium. After sterilization, the medium can be poured into special dishes called **petri dishes.** As the sterile agar solution cools, it hardens and forms a gel upon which microorganisms can reproduce. The petri dishes provide a large surface area for the proliferation of microorganisms.

Categories of Culture Media

The formulation of culture media largely determines which types of microorganisms can grow in these media. The judicious preparation and use of culture media can afford a microbiologist a powerful tool with which to study the biology of microorganisms and isolate them in pure form. There are three different categories of culture media: a) chemically-defined or simple; b) chemically-undefined or complex; and c) living.

Chemically-defined media are formulated so that the concentration of each ingredient is known. The ingredients used to build chemically-defined media are usually inorganic salts and simple organic compounds, such as glucose or purified amino acids. By knowing the exact composition of the culture medium, it is possible to maintain a high degree of consistency from batch to batch. One of the drawbacks of a defined medium is that it can be very expensive to prepare if purified nutrients such as amino acids, vitamins, and nucleotides have to be added. Since a particular defined medium may support the growth of only a few species of bacteria, it cannot be used to grow a large variety of microorganisms. Consequently, complex media are used when it is necessary to grow a large variety of organisms in the same medium.

Complex media are prepared using natural products, such as meat extracts or vegetable infusions. Although natural products contain many of the essential nutrients necessary for microbial reproduction, the exact concentration of each nutrient is unknown. Complex culture media are used extensively in the microbiology laboratory because they are easy to prepare, inexpensive, and able to support the growth of a wide variety of microorganisms. The major shortcoming of complex media is the variability of their composition from batch to batch. Variability in the chemical composition of the medium may alter the growth characteristics of the microorganisms. Thus, precise physiological studies may be difficult to conduct using complex media. For applications such as these, chemically-defined media are highly desirable.

Some microorganisms will proliferate only when growing in a living host, which serves as the culture medium for the microorganisms. For example, viruses require living cells in order to reproduce. They are grown in the laboratory in living organisms (e.g. a chick embryo or a mouse) or in **tissue cultures.** Tissue cultures consist of a nutrient solution that can support the growth of the host cells. Viruses, chlamydias, rickettsias, and some spirochetes are cultured in living organisms.

Selective and Differential Media

Culture media can be formulated to support the growth of a wide variety of microorganisms. For example, nutrient agar is a culture medium consisting of beef extract and peptones (short polypeptides derived from the partial digestion of proteins) that can support the growth of a wide variety of chemoheterotrophic bacteria, including many human pathogens. Although it is impossible to prepare a single culture medium that can support the growth of all bacteria, many complex media can support a wide array of microbial forms.

Culture media can also be formulated to select for a particular microorganism or group of microorganisms, such as *Staphylococcus* or coliforms. These types of culture media are called **selective media** (table 5-5). Other media can be used to differentiate between (or among) groups or species of microorganisms and are referred to as **differential media.**

Selective media are culture media containing at least one ingredient that inhibits the reproduction of unwanted organisms, but permits the reproduction of specific microorganisms. The ingredients are usually called **selective agents.** For example, the isolation of *Staphylococcus* from the human skin can be accomplished using a simple selective medium. It is possible to grow

TABLE 5-5
SELECTED CULTURE MEDIA FOR THE ISOLATION AND DIFFERENTIATION OF REPRESENTATIVE GROUPS OF MICROORGANISMS

MICROORGANISM OR GROUP DESIRED	SELECTIVE MEDIA (FOR ISOLATION)	DIFFERENTIAL MEDIA (FOR DIFFERENTIATION)
Streptococccus	Azide-blood agar base	Mitis-Salivarius agar
Staphylococci	*Staphylococcus* 110 agar Chapman-Stone medium	Mannitol-Salts agar Chapman-Stone medium
Neisseria	Thayer-Martin medium	Phenol red-carbohydrate media
Mycobacterium	Lowenstein-Jensen medium 7H10 and 7H11	7H10 and 7H11
Coliforms	Violet Red Bile agar Levine EMB MacConkey agar	Violet Red Bile agar Levine EMB MacConkey agar
Salmonella and *Shigella*	SS agar XLD agar Hektoen-Enteric agar	SS agar XLD agar Hektoen-Enteric agar
Fungi	Sabouraud Glucose agar	Corn Meal agar Chlamydospore agar

this bacterium on a culture medium such as nutrient agar supplemented with 7.5% NaCl. Sodium chloride constitutes the selective agent in the culture medium, because it will inhibit the reproduction of skin microorganisms other than *Staphylococcus.* Selective agents often employed in culture media include dyes, such as crystal violet and malachite green; antibiotics, such as penicillin and streptomycin; and chemicals, such as sodium azide or sodium chloride.

Enrichment cultures are used to increase the relative concentration of a desired microorganism with respect to others that may also be present in a sample. An enrichment culture provides an environment that is conducive to the reproduction of the desired microbe and that is antagonistic (or inhibitory) to other microorganisms. Enrichment cultures may also take advantage of selective agents, or may simply capitalize on certain cultural requirements, such as light or elevated temperatures, to promote the reproduction of the desired microorganism. A simple way to enrich for endospore-formers is to heat a soil suspension (which contains endospores as well as many other soil microorganisms) at 80°C for 15 minutes and then culture the heated suspension in a nutrient medium. The heat treatment kills most of the soil microorganisms, except those that form endospores. Hence, this simple procedure enriches the suspension for endospore formers because they are the only microorganisms that can survive the treatment. The use of enrichment cultures is usually the first step in isolating a desired microorganisms from a contaminated specimen.

Differential media are special formulations designed to differentiate among microorganisms or groups of microorganisms. Differential media usually contain a chemical that is utilized or altered by some microorganisms but not by others. When different microorganisms grow on this medium, they change its appearance. These changes can be used to differentiate between different microorganisms. For example, the selective medium we described previously for the isolation of *Staphylococcus* can be modified so that it is possible to differentiate among various species in the genus. The modification involves the addition of 0.5% mannitol and a pH indicator, such as phenol red. On this medium, both *S. aureus* and *S. epidermidis* will grow because they can multiply in the presence of 7.5% NaCl. *Staphylococcus aureus* will ferment the mannitol, however, releasing acidic byproducts, whereas *S. epidermidis* will not. The acidic products released by *S. aureus* will cause the pH indicator surrounding its colonies to change from red to yellow. Since *S. epidermidis* does not ferment mannitol, the area around its colonies remains red. By using this differential medium, it is possible to differentiate between these species of staphylococci. Differential media are widely used in microbiology because they aid in the isolation and identification of microorganisms isolated from nature, foods, or diseased tissues. Some of the most common differential media in current use are listed in table 5-5.

METHODS FOR CULTIVATING MICROORGANISMS

In order to study and characterize a microorganism, it is first necessary to isolate the microbe from all others and to maintain it in pure form throughout the study. Cultures consisting of only one type of microorganism are called **pure** or **axenic cultures.** The isolation of a microorganism and maintenance of its purity are achieved by procedures known as **aseptic techniques.** Aseptic techniques include methods for the isolation of microorganisms from contaminated sources and for the transfer and maintenance of pure cultures (fig. 5-9).

One technique for isolating pure cultures from contaminated sources involves a progressive dilution of the sample, so that individual cells can proliferate in uncrowded areas of an agar plate and form colonies. Another common method for isolating pure cultures of microorganisms is the **streak technique** (fig. 5-9), which is performed by spreading a small amount of culture over the surface of an agar-solidified medium, using an **inoculating loop.** As the microorganisms are spread over the surface of the agar, their concentration on the inoculating loop decreases. Eventually, a single microbial cell lands on an isolated area of the agar-solidified medium. As the isolated microorganism multiplies, a population develops, forming a clone of cells called a **colony.** If the original specimen possessed several different types of microorganisms, each type may form a distinct colony that can generally be distinguished from the others (fig. 5-10). The different organisms can then be purified by picking a colony with an inoculating loop and streaking it onto fresh medium.

The isolation of pure cultures of microorganisms can be accomplished by a variety of different techniques. All of these techniques involve some form of dilution, so that cells can be separated and grown individually (fig. 5-11).

CULTURAL CHARACTERISTICS OF BACTERIA

Bacteria growing in laboratory culture media display growth patterns that are often unique. An organism's growth characteristics can often be used to differentiate one group from another. The various growth patterns are a reflection of an organism's morphology, mechanism of cell division, and unique metabolic activities. Thus, the cultural characteristics of bacteria are studied in order to gain a better understanding of their biology. For example, in liquid culture media, bacteria can form pellicles or sediments, or cause the medium to become cloudy (turbid). The formation of **pellicles** (surface films) by bacteria growing on broth cultures is generally due to pili or glycocalyx, which hold the cells together, forming the film. Some pellicle-forming bacteria, such as *Mycobacterium phlei*, have hydrophobic cell walls that cause them to cluster. Their cell walls contain a high concentration of lipids that tend to adhere to each other and repel water. The cluster of mycobacteria at the surface of the broth gives rise to pellicles. **Sediments,** on the other hand, are precipitates formed by clustering bacteria that are too dense to remain suspended in the broth. For instance, *Staphylococcus aureus*, a nonmotile bacterium that frequently forms clusters of spherical cells, forms sediments because its clusters are not buoyant and therefore sink to the bottom of the cultural tube. **Turbidity** refers to the ability of certain bacteria to turn the medium cloudy because they are evenly distributed throughout the broth. Many bacteria forming turbid cultures are motile; as they swim, they remain dispersed throughout the medium rather than sinking to the bottom.

Agar-solidified media can also provide important clues as to the culture's identity and biology. Typically, agar slants and agar plates are exam-

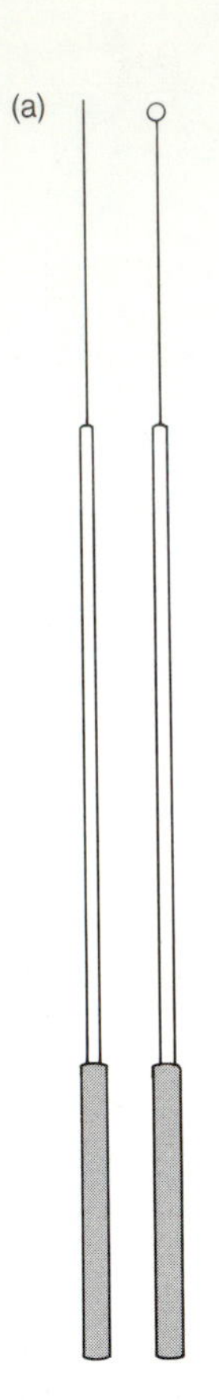

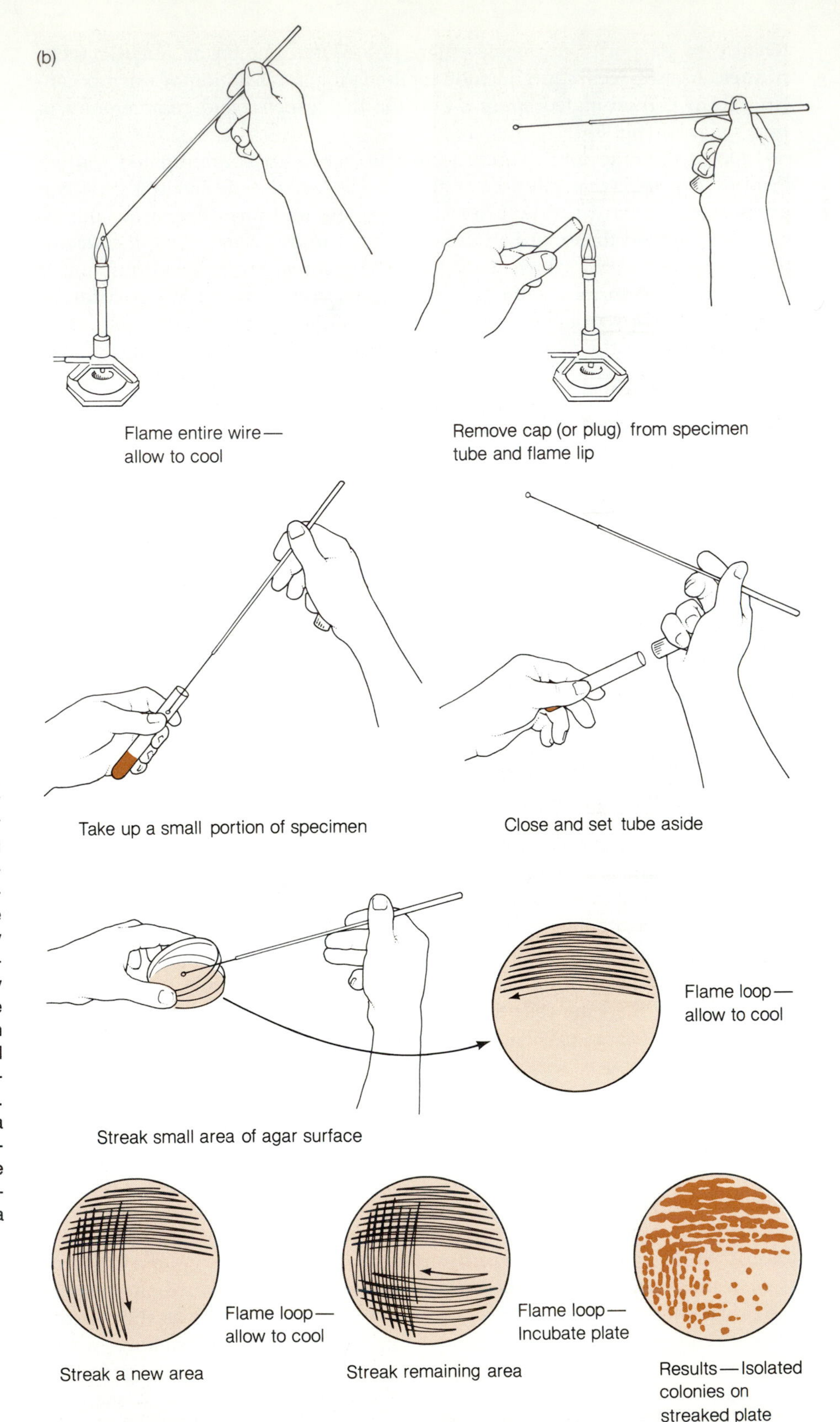

Figure 5-9 The Streaking Technique. *(a)* Inoculating loop and needle. These are very important tools used in the microbiology laboratory. The loops are used to spread microorganisms over agar surfaces. Needles are used to pick out small bits of colonies for subculture. *(b)* Inoculating the streak plate. The streak plate is commonly used to obtain pure cultures of microorganisms. The streak plate is performed by spreading a portion of a bacterial sample over the entire surface of the plate with an inoculating loop. In this manner, bacterial cells are separated from each other, forming distinct colonies when they multiply. Each colony, theoretically, originates from a single cell and hence represents a population of genetically-identical bacteria. The colonies can then be picked with an inoculating needle and subcultured to obtain a pure culture.

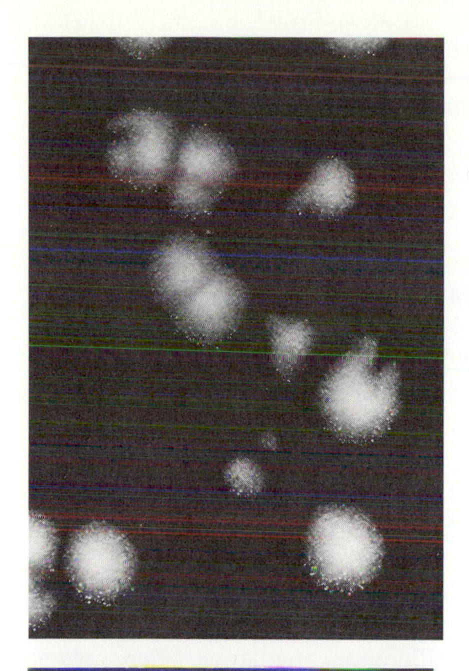

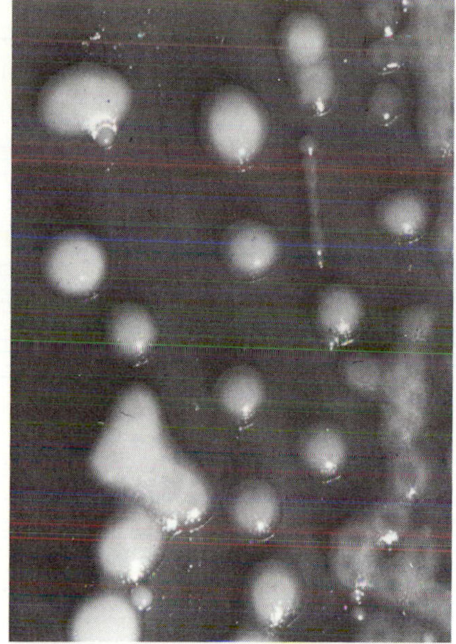

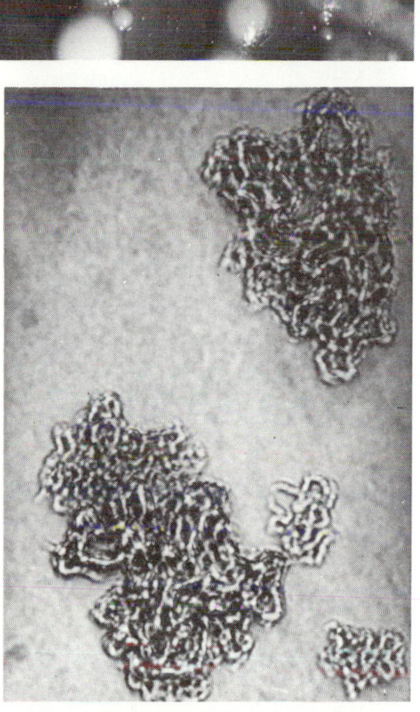

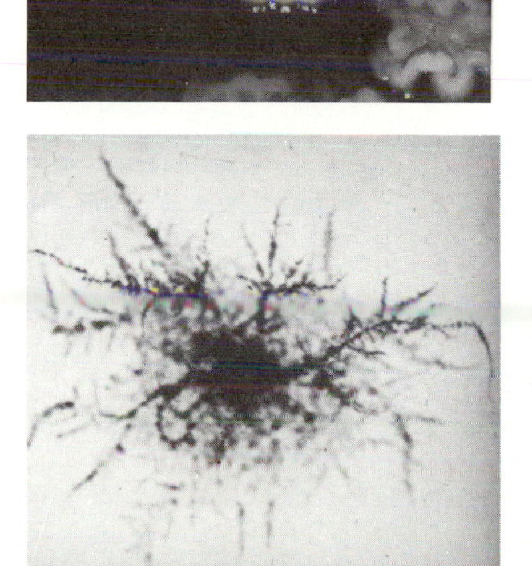

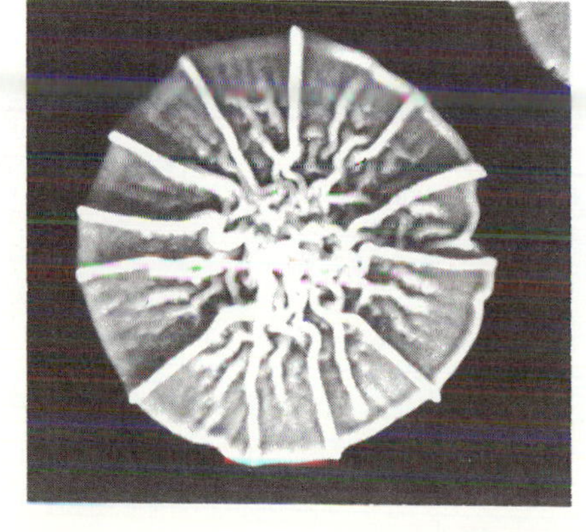

Figure 5-10 Various Types of Bacterial Colonies

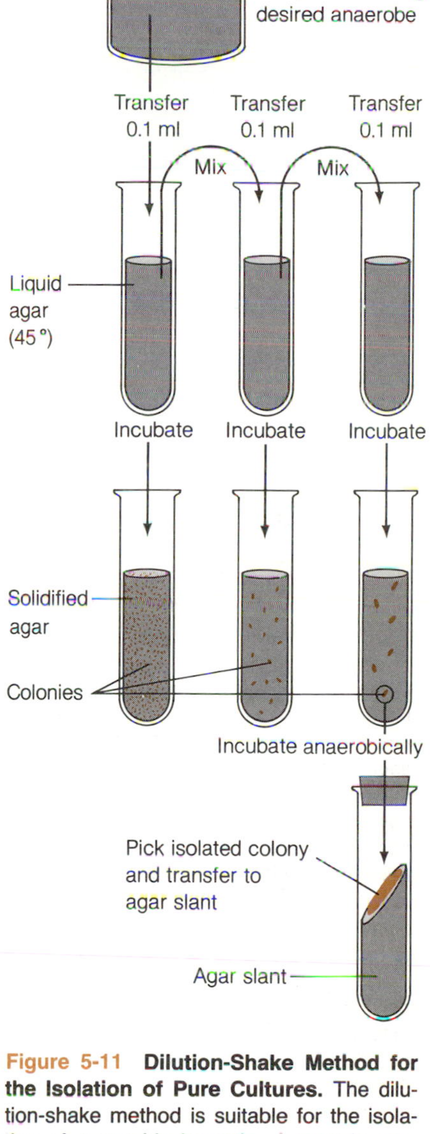

Figure 5-11 Dilution-Shake Method for the Isolation of Pure Cultures. The dilution-shake method is suitable for the isolation of anaerobic bacteria. A sample containing the desired bacteria is diluted several times and small quantities (0.1–0.5 ml) of the various dilutions are thoroughly mixed with melted, cooled agar. When the agar hardens, the bacteria will be trapped within the agar matrix, in an oxygen-free environment where they will form colonies. The colonies can then be picked and subcultured in an anaerobic environment.

ined to determine how the bacteria grow on their surfaces. They are examined for cultural characteristics, such as amount of growth, pigment production, shape of the colonies, and any other characteristic that may be helpful in differentiating one type of colony from another.

SUMMARY

All forms of life require nutrients for reproduction. These nutrients provide the raw materials for building all cellular components and the energy necessary to power these activities.

PRINCIPLES OF MICROBIAL NUTRITION

1. Microorganisms obtain their nutrients from their immediate environment. Often, these nutrients must be chemically digested before they can be brought into the cell. To accomplish this digestion, many microorganisms secrete exoenzymes that serve to digest large molecules into simpler ones which can then be transported into the cell.
2. Osmotrophic microorganisms obtain their nutrients in solutions, while phagotrophic microorganisms obtain their nutrients by engulfing food particles and digesting them.

NUTRITIONAL REQUIREMENTS OF MICROORGANISMS

1. Microorganisms require nutrients for their energy and biosynthesis.
2. All microorganisms require an energy source and a source of carbon, nitrogen, oxygen, sulfur, phosphorus, trace elements, and water. In addition, many microorganisms require growth factors, such as vitamins and amino acids.

CULTURAL CONDITIONS

1. Environmental factors such as temperature, atmosphere composition, pH, and solute concentration all play important roles in determining whether or not microorganisms can reproduce in the environment.
2. The culturing of microorganisms in the laboratory is carried out in chambers called incubators. These can be equipped to provide the culture with the proper environmental temperature and atmospheric gases.

CULTURE MEDIA AND THEIR USE IN THE LABORATORY

1. There are three categories of culture media used to propagate microorganisms in the laboratory. These include complex, chemically-defined, and "living" culture media.
2. Complex media consist of natural products and water. These support the growth of many microorganisms, but the exact chemical composition of the medium is unknown.
3. Chemically-defined media are formulated so that the exact composition of the medium is known. The chemicals employed in constructing chemically-defined media are usually inorganic salts and simple organic molecules.
4. Living culture media are used primarily to cultivate obligate parasites such as the viruses and rickettsias, which cannot reproduce outside a living host.
5. Selective and differential media are often used in the laboratory, since they can be used to isolate desired groups of microorganisms by capitalizing on their cultural and/or biochemical characteristics.
6. Enrichment cultures are used to increase the chances of isolating a microorganism from a heavily contaminated specimen. These cultures are often employed in the hospital laboratory. For example, the isolation of *Shigella* from stool specimens begins with an enrichment procedure to increase the chances of finding this disease-causing bacterium in human feces.

METHODS FOR CULTIVATING MICROORGANISMS

1. Pure culture techniques are essential to all phases of microbiology.
2. Pure (axenic) cultures of microorganisms are usually obtained by streaking a small amount of sample over the surface of a petri dish containing a suitable agar medium. As the individual bacteria are deposited over uncrowded portions of the medium, they reproduce and form colonies. Each colony is the result of the multiplication of a single bacterium over many generations and consists of genetically very similar cells.
3. Bacteria, when growing in laboratory culture media, often display unique cultural characteristics that can be useful in their recognition and identification.

CULTURAL CHARACTERISTICS OF BACTERIA

1. Bacteria cultivated in the laboratory often display growth characteristics that are unique, and can be

used to identify the bacteria. The growth characteristics reflect cellular morphology, mode of cell division, or unique metabolic activities.

2. In broth cultures, bacteria may form pellicles, sediments, or turn the medium turbid (cloudy).
3. Agar-solidified media also provide information about the growth characteristics of the bacteria including production of pigments, shape of the colonies, texture of the colonies and amount of growth.

STUDY QUESTIONS

1. List the basic nutritional requirements of a chemoautotroph.
2. Construct a culture medium that would be suitable for the cultivation of chemoheterotrophic microorganisms.
3. Describe two different types of microorganisms, based on their requirements for free oxygen.
4. You have a soil sample containing a bacterium that is highly resistant to a chemical called sodium azide. Outline a procedure for the isolation of an axenic culture of this organism from the soil sample. (Hint: The soil sample contains many other microorganisms, and the desired bacterium is in relatively low concentration in the sample.)
5. What is the difference between a selective medium and a differential medium? Can you make a medium that is both selective and differential?
6. Discuss the various ways microorganisms meet their requirements for the various elements in their cellular structures and activities: a. carbon; b. nitrogen; c. oxygen; d. sulfur; e. phosphorus; f. hydrogen.
7. Explain the use of the streak plate for isolation of microorganisms.
8. Outline a basic cultural procedure to isolate: a. chemoheterotrophic microorganisms; b. chemoautotrophic microorganisms; c. photoheterotrophic microorganisms; d. photoautotrophic microorganisms.
9. From what you know about the requirements of microorganisms for suitable environmental conditions, how might you manipulate the environmental conditions in order to preserve foods?
10. List three important properties of agar.

SUPPLEMENTAL READINGS

American Society for Microbiology. 1981. *Methods in general microbiology*. Washington: American Society for Microbiology.

Difco Laboratories. 1984. *Difco manual of dehydrated culture media and reagents for microbiological and clinical laboratory procedures*. 10th ed. Detroit: Difco Laboratories.

Ingraham, J., Maaløe, O., and Neidhardt, F. 1983. *Growth of the Bacterial Cell*, Sinauer Associates Inc.

Lenette, E. H., Balows, A., Hausler, W. J., Jr., and Shadomy, J. P. 1985. *Manual of clinical microbiology*. 4th ed. Washington: American Society for Microbiology.

Payne, W. J. and Wiebe, W. J. 1978. Growth yield and efficiency in chemosynthetic microorganisms. *Annual Review of Microbiology* 32:155–183.

Randall, L. L. and Hardy, S. 1984. Export of protein in bacteria. *Microbiological Reviews* 48(4):290–298.

Shapiro, J. A., 1985. Photographing bacterial colonies. *ASM News* 51(2):62–69.

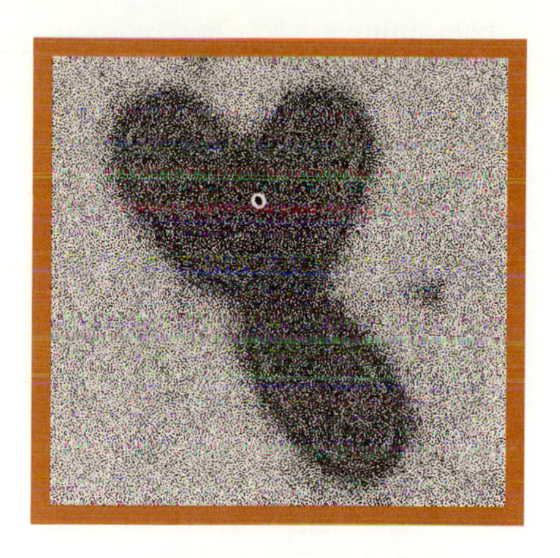

CHAPTER 6
MICROBIAL GROWTH

CHAPTER PREVIEW

ANTIBIOTICS: A CONSEQUENCE OF MICROBIAL GROWTH

When certain spore-forming microorganisms reproduce, they release chemicals that inhibit the reproduction of other microorganisms or even kill them. These microbial products that kill or inhibit microorganisms are called **antibiotics**.

Microbiology made its debut in the pharmaceutical industry in the 1940s with the isolation of two antibiotics: penicillin by Ernst Chain and Howard Florey, and streptomycin by Selman Waksman. These discoveries revolutionized the pharmaceutical industry. Until then, physicians were virtually helpless in combating infectious diseases. Today, antibiotics save countless lives and the pharmaceutical industry nets billions of dollars annually from the exploitation of antibiotics produced by microorganisms.

Much research has been directed toward the study of the growth of antibiotic-producing organisms, so that optimal conditions for antibiotic production can be achieved. This information is important because, even though a certain microorganism can produce an antibiotic, it may not do so under all growth conditions. Furthermore, a microorganism may not produce the antibiotic at a constant rate throughout its growth cycle. Some may produce antibiotics when they are dividing actively, while others may produce them only when their rate of reproduction is slowing down. For example, many of the bacteria in the genus *Bacillus*—which is characterized by the production of endospores—synthesize antibiotics at about the time they are forming their endospores. Since the production of endospores is influenced by the physiologic state of the cells, antibiotic production can be optimized by establishing the growth conditions that are conducive to endospore formation. Hence, knowing the growth conditions that each microorganism must have in order to produce antibiotics is essential for the optimal production of these life-saving substances.

The pharmaceutical industry relies heavily on available knowledge of microbial growth and reproduction in order to maximize the production of antibiotics. This chapter will discuss some of the characteristics of microbial growth and the factors that affect it.

LEARNING OBJECTIVES

A STUDY OF THIS CHAPTER SHOULD ENABLE YOU TO:

- **OUTLINE AND APPLY THE VARIOUS TECHNIQUES USED TO MEASURE POPULATIONS, AND TO KNOW THE LIMITATIONS OF THESE TECHNIQUES**
- **EXPLAIN HOW TO CALCULATE THE GENERATION TIMES FOR BACTERIAL POPULATIONS**
- **DRAW A BACTERIAL GROWTH CURVE, IDENTIFYING THE VARIOUS PHASES OF GROWTH AND DESCRIBING THE FACTORS THAT INFLUENCE THEM**
- **EXPLAIN HOW VARIOUS ENVIRONMENTAL FACTORS AFFECT GROWTH**
- **DISCUSS BRIEFLY HOW MICROBIAL POPULATIONS GROW IN NATURE**

All living organisms are characterized by their ability to grow and reproduce in their environment. Microorganisms grow by increasing their size (volume), a vital process that normally culminates in cell division and, thus, in an increase in the number of cells. To understand how microorganisms carry out decomposition, cause disease, and participate in useful industrial processes, it is necessary that we know how they grow, how they multiply, and what effect the environment has on their rate of reproduction. This chapter will discuss how microbial populations grow and how the environment affects their growth.

CELLULAR GROWTH

Cellular growth can be defined as the sum of all metabolic activities of a cell leading to the orderly increase of all its constituents. Cellular growth involves the synthesis of the vital cell components, such as the ribosomes, plasma membrane, and genetic material. This cellular growth results in an increase in the size and mass of the cell. An increase in cell mass, however, does not always reflect cellular growth. It is possible for microorganisms to synthesize reserve materials, such as starch or volutin, yet not increase their size.

Although it is relatively simple to study the growth of plants and animals, microbial cells are much too small to be studied individually. Instead, microbiologists study microbial populations to gain an understanding of the biology of the microorganism. **Population growth** refers to the increase in the number of cells in a population, whereas cellular growth refers to the increase in the size of the cell. The rate at which populations grow is determined largely by the environment they inhabit and by factors such as moisture, heat, and nutrient availability. The following sections will deal largely with these concepts.

MEASURING MICROBIAL POPULATIONS

If a flask of nutrient broth is seeded with a bacterium such as *Escherichia coli* and is allowed to stand in a warm place for a few days, it will become turbid. The increase in turbidity of the nutrient broth roughly reflects the growth of the bacterial population inside the flask and can be used as an indication of how fast the population is growing. The most obvious and direct way of measuring population growth is to count the number of bacteria present in a broth culture at various time intervals to see if the number of bacteria increases with time. Population growth can also be studied by measuring the increase in mass (weight) of the cells making up the population, or by measuring a metabolic activity usually associated with multiplying cells. For example, measuring the amount of oxygen consumed by a population at various time intervals can serve as an indicator of population growth. The amount of oxygen consumed by a population is proportional to the number of cells present in the population; hence, a growing population would consume oxygen at a rate proportional to its size.

In general, the dynamics of a growing microbial population can be studied by measuring a) the number of cells, b) cell mass, and c) metabolic activities.

Measuring the Number of Cells

There are two basic procedures commonly employed in the laboratory to measure the increase in the number of individual cells in a microbial population. These are the direct microscopic count and the viable count.

Direct Microscopic Count

The **direct microscopic count** consists of counting the number of cells in the population with the aid of a counting chamber. This procedure can be carried out simply by spreading a premeasured volume of culture (usually 0.01 ml) over a known area of a microscope slide (usually 1 cm^2). The resulting smear can then be stained and examined with the aid of a compound microscope. If the area of the microscope field is known, the following formula can then be used to estimate the number of cells in the culture:

$$\text{cells/ml} = A_s/A_f \times 1/V \times N = NA_s/A_fV$$

where A_s is the area of the smear; A_f is the area of the microscopic field; V is the volume of sample in the smear; and N is the number of cells counted per field. This method of direct enumeration of microorganisms, sometimes called the **Breed slide method,** can provide satisfactory results when properly performed. The quantitative measurement of microorganisms in milk and other dairy products is sometimes conducted using the Breed slide method.

Counting chambers, such as the **Petroff-Hausser chamber** and the **hemocytometer,** are constructed especially for the direct enumeration of cells in suspensions. These counting chambers are special slides with chambers of known depth which have been ruled into squares of known area (fig. 6-1) so that they can be used to determine the volume of suspension being counted. The specimens may be stained before they are counted with these chambers in order to increase the ease with which the microorganisms are viewed.

Direct microscopic counts using the Breed slide method or the Petroff-Hausser chamber, although they are fast and easy to perform, are somewhat inaccurate because all microorganisms present, both living and dead, are counted. Since microscopic examination of the population is necessary, artifacts may be counted along with the cells. In addition, since extremely small volumes are used, each microorganism seen in the slide represents a very large number of microorganisms per ml of suspension; hence large numbers of microorganisms must be present in the population before they can be counted accurately. For example, the Petroff-Hausser chamber holds .00002 ml, so each bacterial cells seen represents approximately 50,000 bacteria per ml. For counts to be accurate, the population must therefore contain more than 500,000 cells per ml. The sensitivity of this technique is quite low.

Electronic counters, such as the **Coulter counter** (fig. 6-2), are routinely used in the clinical laboratory to count red and white blood cells. They are less frequently used to count microbial populations, although the results obtained are accurate and reproducible. Electronic counting is based on changes of electrical conductivity that occur in a liquid medium when cells are present. These changes in conductivity are recorded as individual cells pass through a narrow slit. Each change is registered in the electronic counter, and after the entire volume of suspension has passed through the counter, the total number of cells (or the number of cells/ml of suspension)

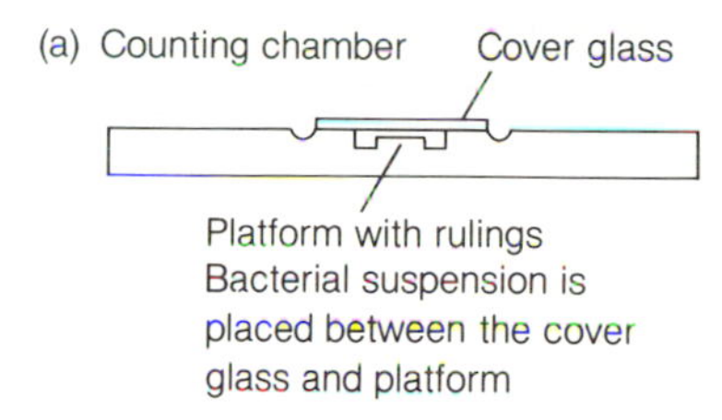

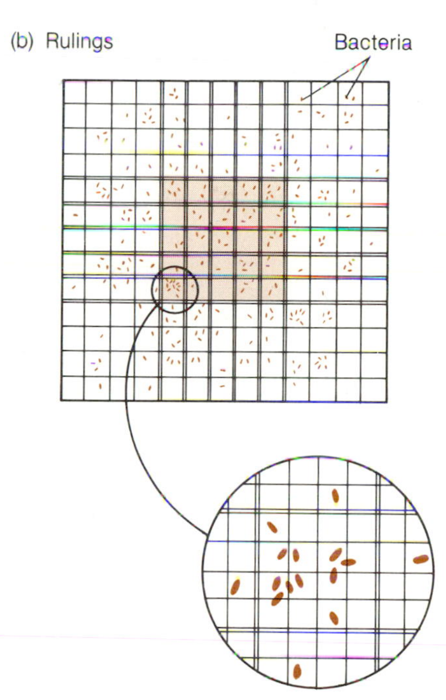

Figure 6-1 Direct Enumeration of Bacterial Populations Using the Petroff-Hausser Counting Chamber. The Petroff-Hausser bacterial counter consists of a platform with rulings. The bacterial suspension is placed between the cover glass and the ruled chambers. The suspension enters the chamber by capillary action and the bacteria are distributed throughout the ruled chamber. The bacteria are then counted using a compound microscope. The central chamber of the Petroff-Hausser bacterial counter consists of 25 large squares. The bacteria in all the squares are counted and the number is multiplied by 50,000 to give the value of bacteria/ml of suspension.

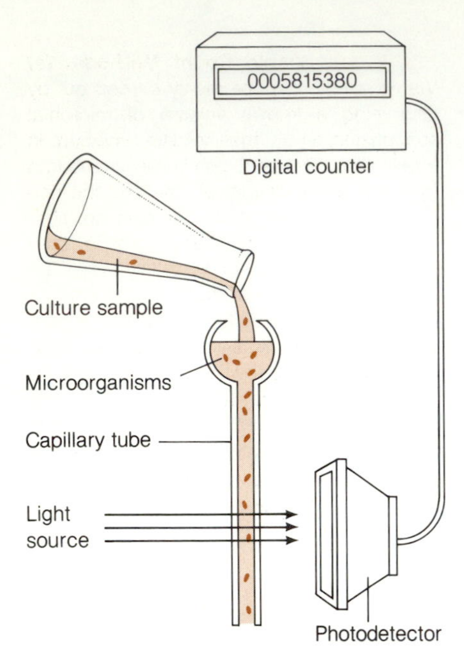

Figure 6-2 Counting Bacteria Using the Coulter Counter. Bacteria are counted with a Coulter counter by placing a known volume of suspension inside a capillary tube that is part of the apparatus. As each bacterial cell passes through a slit in the Coulter counter, it causes changes in conductivity. These changes are recorded and tallied automatically. When the entire suspension has been counted, the results are displayed. The total count represents all of the conductivity changes recorded by the Coulter counter.

can be obtained. Electronic counters are expensive pieces of equipment that require a certain amount of skill to use properly. Also, because small slits are required to count bacteria, special culture media are required in order to avoid the formation of crystals that would plug up the slits. For these reasons, electronic counters are infrequently used to carry out routine counts of microbial populations. An intrinsic shortcoming of this procedure is that the cell counting is based on changes in conductivity created by particles passing through a beam of light. Hence, *any* particle passing through the beam will be recorded as a cell even though it may not be a microbial cell.

Viable Counts

The methods most commonly used to determine **viable counts** are based on the assumption that each cell in a solid medium can multiply repeatedly and eventually form a distinct colony. The spread plate and the pour plate are used extensively to obtain these counts.

In the **spread plate** method, a bent glass rod is used to spread 0.1–0.5 ml of a suspension of microorganisms evenly over the surface of a suitable solid culture medium in a petri dish (fig. 6-3a). The glass rod (shaped like a hockey stick) is used to spread the microbial cells over the agar surface so that they are not crowded and can multiply freely, forming colonies. The colonies that develop after an incubation period can be counted to obtain the number of microorganisms originally present in the suspension.

To perform a viable count by the **pour plate** method (fig. 6-3b), a sample (0.1–1.0 ml) of a liquid culture or suspension is mixed with a melted agar medium in a sterile petri dish and incubated for a suitable length of time. After incubation, the colonies in (or on) the medium are counted, usually with the aid of a **colony counter** (fig. 6-4). This device may be a magnifying glass or a sophisticated electronic colony counter. The **standard plate count** (SPC) is a standardized pour plate count for determining the number of aerobic and facultative anaerobic chemoheterotrophs □ in a suspension. The SPC method is used routinely to count the number of microorganisms in water, milk, and foods, because it affords microbiologists a means of comparing results with those of other laboratories.

112

Some specimens (or cultures) may contain so many microorganisms that, after plating, the developing colonies grow together and cannot be counted. When **confluent** (grown together) growth is expected, the specimens must be diluted before they are plated so that most colonies appearing on the plate will be isolated. The degree of dilution of the specimen before plating is extremely important. Excessive dilution will usually result in too few colonies, while insufficient dilution will result in crowded and confluent growth. Either situation will lead to inaccurate counts. It is best to dilute the sample so that it will yield between 30 and 300 colonies per plate. Insufficient dilution results in microorganisms landing very near one another in the agar, where their populations merge and give rise to a single colony. When this happens, the number of microorganisms estimated to be in the sample is lower than the actual number. Excessive dilutions will also lead to statistically inaccurate counts. Plates with many more than 300 colonies, or with fewer than 30 colonies, give inaccurate counts and should not be used.

Viable counts can also be determined by the **most probable number** (MPN) method. This is a rapid and reproducible method of estimating the number of microorganisms in a suspension. In essence, the MPN test is

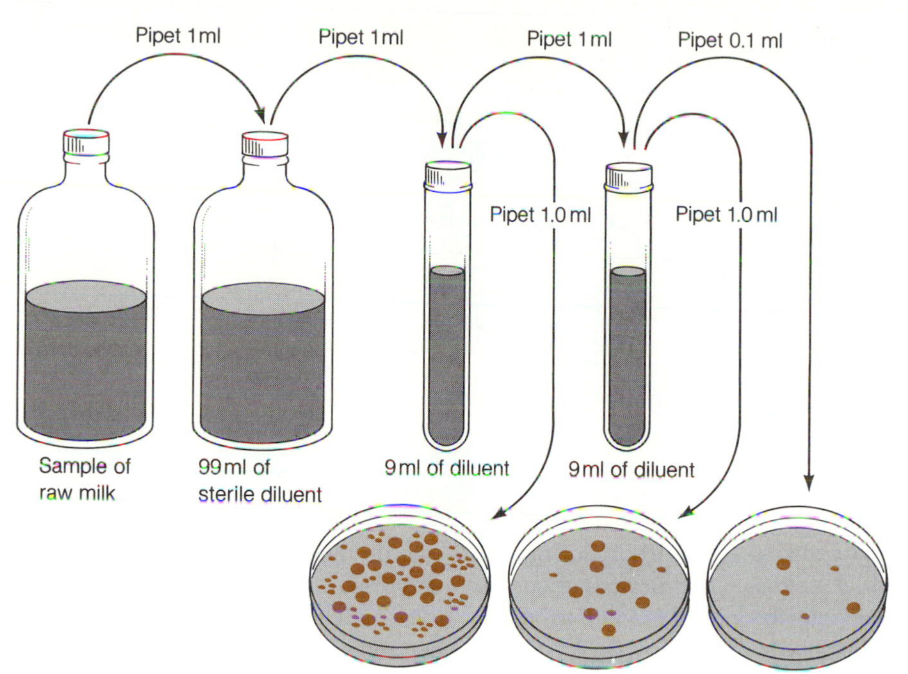

Figure 6-3 Viable Count Methods. *(a)* Viable counts are commonly carried out by dispersing a known volume of microbial suspension on an agar-solidified medium in a petri dish. The petri dish is then incubated for a specified amount of time and the colonies are counted. If the samples are suspected to contain too many bacteria, they must be diluted before plating. *(b)* The **spread-plate method** involves spreading a known volume (usually 0.1 ml) of suspension over the surface of an agar-solidified medium. *(c)* The **pour-plate method** involves mixing a known volume (0.1-1.0 ml) of suspension in 15–20 ml of liquefied, cooled agar medium. After an incubation period, colonies that appear throughout the medium are counted.

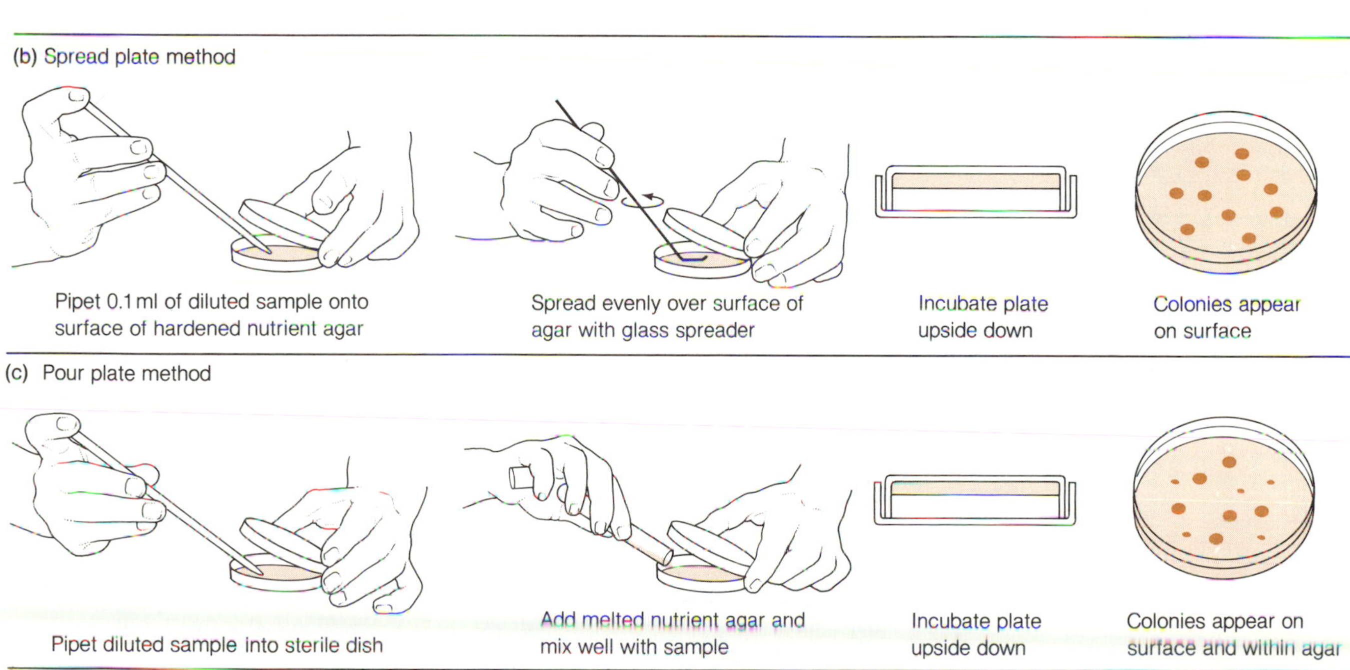

conducted by inoculating several tubes of liquid culture medium with a measured volume of suspension (fig. 6-5). The broth cultures are then incubated under the desired conditions and subsequently examined for evidence of growth. The MPN test is based on the assumption that the more microorganisms present in a sample, the greater the chance for growth in the tubes, and the more tubes in which growth is evident. For example, in the test

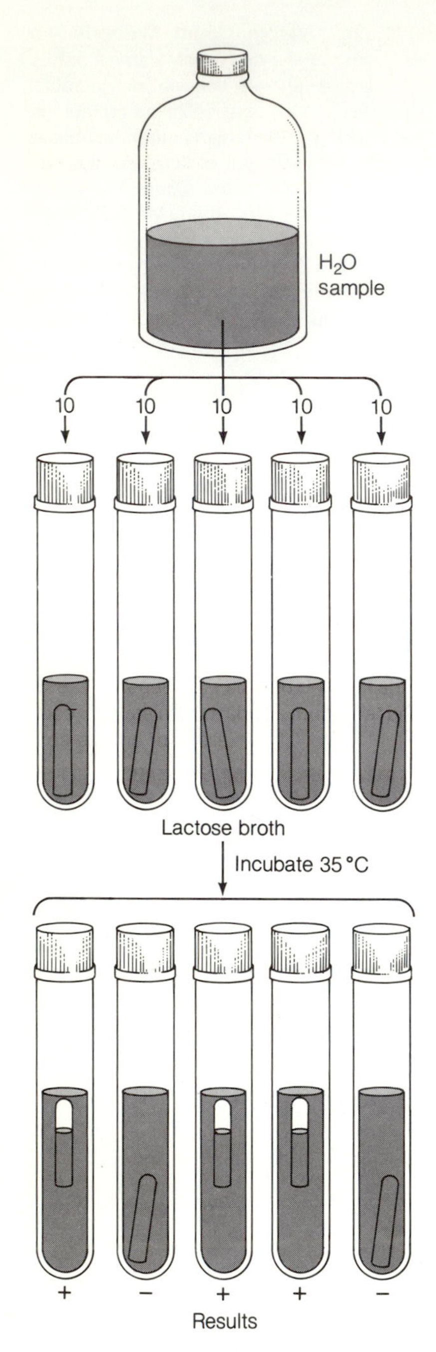

Figure 6-5 The Most Probable Number Method for Counting Bacteria in Suspensions. Five 10-ml samples are inoculated into 5 tubes of lactose broth and incubated for 48 hours at 35°C. After the incubation period the positive tubes are counted. In this example, 3 out of 5 tubes were positive. These results can then be used to estimate the number of bacteria/100 ml of sample (MPN/100 ml) by using table 6-1. From the table it can be estimated that the MPN value is 9.2/100 ml.

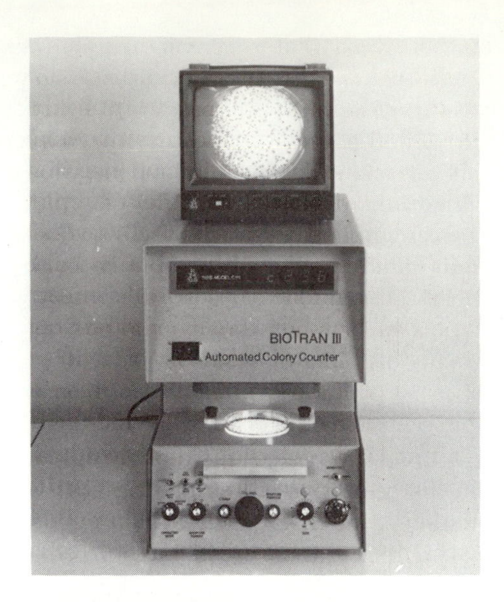

Figure 6-4 Electronic Colony Counter. The plate to be counted is inserted in the counting chamber. The colonies appearing on the plate are counted electronically and the total count is displayed automatically.

outlined in figure 6-5 there are 3 positive tubes out of a total of 5 tubes inoculated. From that information it has been determined statistically that the MPN is 9.2 bacteria/100 ml of sample (table 6-1). If 4 tubes had been positive instead of 3, the average count would have been 16. In general, only samples suspected of containing relatively few bacteria are tested using this procedure, although samples with numerous microorganisms can also be tested after dilution.

A wide variety of different organisms can be tested using the most probable number method. An MPN test that is routinely used to determine the number of coliform bacteria present in water and shellfish meat is the **multiple tube fermentation** test for coliforms. This procedure will be described in detail in Chapter 31.

Viable count methods, unlike direct microscopic methods, are extremely sensitive. In theory, even a single microorganism present in a sample can be detected by viable counts. Like all other methods for enumerating populations of microorganisms, however, viable counts have shortcomings. Cultural conditions can affect the viable counts, and the procedure may therefore underestimate the actual number of microbes in the sample. It is obviously impossible to satisfy the growth requirements of all microorganisms with one set of cultural conditions; therefore, viable count methods detect only those microorganisms capable of reproducing under the cultural conditions established. Such methods indicate the *minimum* number of microorganisms present in the sample, because some microorganisms present may not have been able to reproduce due to inappropriate cultural conditions. For instance, a standard plate count of a dairy product that has been incubated aerobically for 48 hours will detect only the number of aerobic and facultative anaerobic microorganisms present. Any anaerobic bacteria present in the dairy product will not form a colony, because they are inhibited by oxygen. In addition, certain groups of microorganisms have a tendency to form aggregates of two or more cells that will give rise to a single colony. This process will also result in viable counts that are lower than the actual number of cells present in the sample. In spite of these shortcomings, viable counts are widely used in microbiology and they yield accurate and acceptable results.

TABLE 6-1
TABLE OF MOST PROBABLE NUMBERS (MPN) FOR DRINKING WATER

NUMBER OF POSITIVE TUBES	MPN VALUE/100 ML
0	< 2.2
1	2.2
2	5.1
3	9.2
4	16.0
5	>16.0

Measuring Cell Mass

Many procedures have been developed to measure cell mass. These procedures include dry weight, packed cell volume, total nitrogen, total protein, and turbidity measurements. Each of these methods has specific uses and is suitable under certain conditions.

Dry weight determinations are reliable, quantitative methods for measuring the size of microbial populations. These procedures are based on the assumption that the dry weight of a culture is directly proportional to the mass of cells in the culture. One of several commonly used procedures is to separate the microbial cells from the culture medium, either by filtration or by centrifugation. The collected cells are then washed several times with water to remove all traces of extraneous materials. The washed cells are either oven-dried or lyophilized (freeze-dried) and subsequently weighed. The dry weight determination is frequently used in research because it yields a precise measure of microbial mass; however, the mass of the cells must be great enough to be weighed accurately.

The **total nitrogen** in a culture is directly proportional to the mass of cells present and hence can be used to measure the size of populations. First the cells are harvested and washed thoroughly to remove all traces of culture media and other extraneous nitrogen-containing materials. The washed cells are then digested chemically and the amount of nitrogen present is determined by chemical analysis. This procedure is more sensitive than dry weight determinations, but it is more difficult to perform and requires specialized equipment that may not be available in the average microbiology laboratory.

Estimation of population size by measurement of population mass or total nitrogen can yield inaccurate results if the cultural conditions are not well controlled or if the cells are actively synthesizing reserve materials. If a bacterial population is storing large amounts of reserve materials, the population mass is greater than it would be otherwise, so mass determinations overestimate the population size.

Turbidimetric Measurements

Turbidimetric measurements are based on the principle that the number of tiny particles (e.g., bacteria) in a suspension is directly proportional to the amount of light scattered (and inversely proportional to the amount of light transmitted) when a beam of light is directed through the suspension. Turbidity is measured in **optical density** (O.D.) units that are proportional to the percent transmittance (%T). Optical density is calculated by the following formula:

$$\text{O.D.} = \log_{10} 100 - \log_{10} \%\text{T}$$

Turbidimetric measurements of bacterial population size are usually made with a colorimeter (fig. 6-6). This is an apparatus that measures the

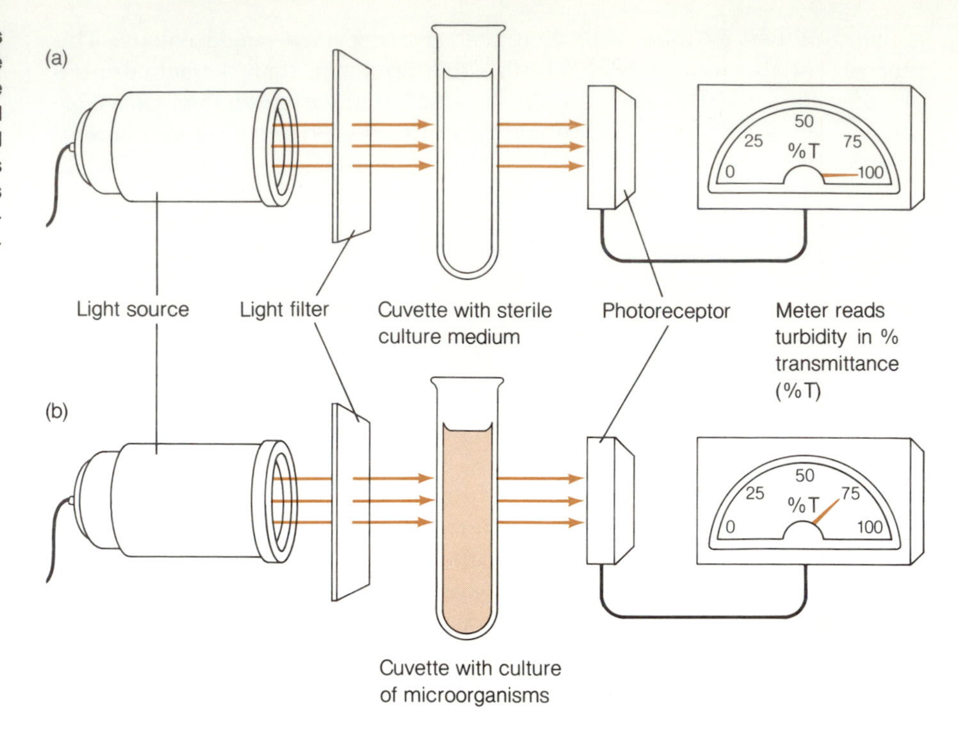

Figure 6-6 Turbidimetric Measurements of Bacterial Populations. Measuring the turbidity of a bacterial population. *(a)* The turbidity of a cell-free medium is measured and the colorimeter adjusted to read 100%T. *(b)* The turbidity of the culture is measured and the %T recorded. The %T is inversely proportional to the number of bacteria in the sample and can be used to estimate the population size.

amount of light absorbed or scattered by a suspension. The colorimeter consists of a light source, a chamber in which the specimen is placed, and a photocell to measure the amount of transmitted light.

Turbidimetric methods provide rapid and reproducible results when properly performed, but they only approximate population size, because they measure the amount of light scattered by the population rather than counting the cells themselves. For this reason, the technique may yield inaccurate results when the concentration of cells is high or the suspension contains particles such as dust or crystals that scatter light. Furthermore, the age of the culture and the physiologic state of the cells can greatly influence O.D. values. Accurate results are obtained only when cultural and nutritional conditions are properly monitored and maintained throughout the study.

Measurements of O.D. do not directly yield the number of cells in the culture; the O.D. values are converted to cell numbers or cell mass by using a calibration curve (fig. 6-7). A calibration curve is constructed by plotting the optical density against some parameter that reflects population size, such as viable count or dry weight. Since the nutritional and environmental conditions of the culture affect the mass and size of the population, they also affect the O.D. For this reason, it is essential that well-defined conditions be employed when estimating population size by optical density methods.

Measuring Metabolic Activities

The quantity of a metabolite consumed per unit time by a population of microorganisms is proportional to the number of cells in the culture, and the rate of metabolite consumption also reflects the rate at which the popu-

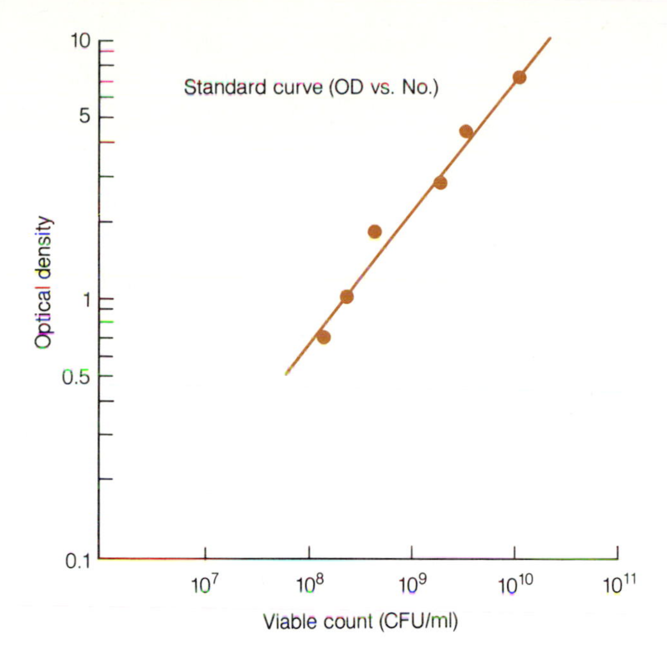

Figure 6-7 Optical Density versus Viable Count Calibration Curve. The calibration curve is used to estimate the size of a population using optical density measurements. To construct the curve, the optical density and viable counts of various dilutions of a population are made. The results are plotted on graph paper as illustrated. A straight line can be obtained if the O.D. values are plotted on a logarithmic scale and the viable counts on an arithmetic scale. CFU = colony forming units.

lation is growing. For example, the amount of glucose in a medium decreases as the population size increases. The rate of glucose uptake should remain constant for each species of microorganism, provided that other cultural conditions remain unchanged. Also, the rate of accumulation of metabolic products (such as organic acids) in a culture is proportional to the size of the population. Measurements of metabolic activities □ to determine population size are of limited use and are not widely employed in routine studies of growth. Metabolic activities are measured, however, when microbial assays are performed to measure the amount of a vitamin or amino acid present in a solution.

178

GROWTH OF BACTERIAL POPULATIONS

Transverse binary fission is the method by which most bacteria multiply (fig. 6-8). During **vegetative growth,** a newly-formed cell undergoes a gradual increase in volume, reflecting an increase in cellular constituents in preparation for cell division. After the cell reaches a certain size, it begins to delimit (form) a septum that ultimately divides the enlarged cell into two identical daughter cells. During this time, vital cellular components □ are divided equally between the two developing cells. Each of the daughter cells then begins to increase in volume in preparation for the next division cycle (fig. 6-8). Thus, each time the cells divide, the population doubles in size. If the starting population consists of 1 cell, the next generation will contain 2

33

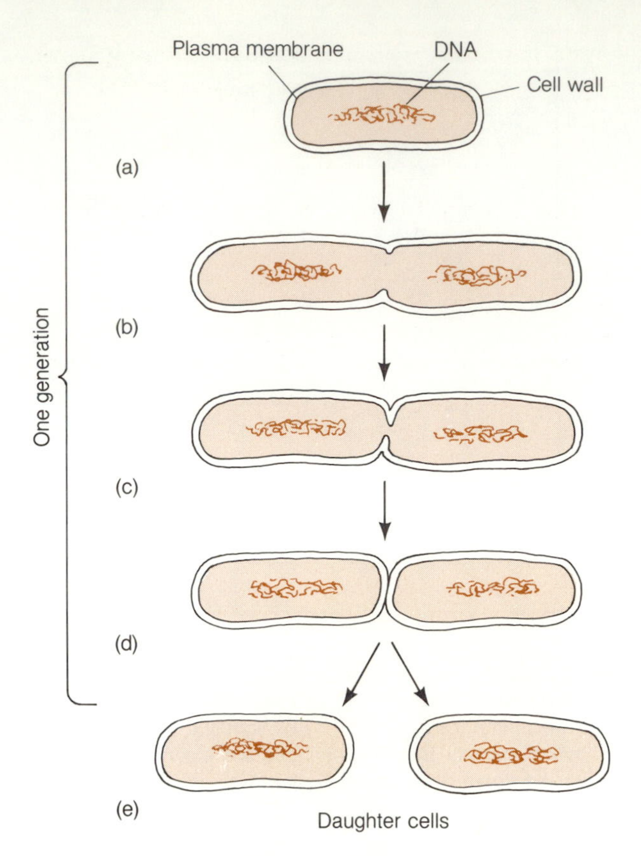

Figure 6-8 Binary Fission by Bacteria. The process of bacterial cell division is an orderly sequence of steps that results in the formation of two bacteria from a parental cell with an identical genetic makeup. *(a)* Newly formed parental cell. *(b)* Cell elongation and septum formation. *(c)* Invagination of cell wall and distribution of genetic material. *(d)* Separation of two new cells. Each new cell then begins a new cell division cycle.

cells, the next 4 cells, then 8, 16, 32, 64, etc. The size of the population can be determined by the following formula:

$$N_f = (N_i)2^n$$

where N_f is the final size of the population; N_i is the initial size of the population; and n is the number of generations (doublings). The speed at which a population increases in size is usually expressed as its **generation time,** defined as the time it takes for a population to double in size. The generation time of microbial populations varies from about 20 minutes to several days, depending on the species of organism and the cultural conditions in which it grows.

The generation time can be determined with the aid of a compound microscope, simply by watching a single cell in the population go through cell division. The time it takes for a cell to grow, divide, and become two cells is the generation time for that particular cell. If this observation is made for several cells, the average time between cell divisions can be considered to be an approximation of the generation time for the population as a whole. This method is tedious, time consuming, and seldom used in the laboratory.

Instead, generation times for microbial populations are generally calculated by employing the simple algebraic expression given below:

$$T_g = t/n = t/(\log_2 N_f - \log_2 N_i)$$

where t is the time interval between N_i and N_f and n is the number of generations in time t.

In order to calculate the generation time, the initial size of the population (N_i), the size of the population at a later time (N_f), and the time interval (t) between N_f and N_i are required. The values for N_i and N_f can be obtained by carrying out viable counts or direct microscopic counts. With these data it is possible to calculate the generation time. Because it is much easier to convert the size of populations using $\log_{10}$ than $\log_2$, this converted formula is used:

$$T_g = 0.301t/(\log_{10} N_f - \log_{10} N_i)$$

Consider a bacterial population that increases from 100,000 bacteria/ml to 1,000,000 bacteria/ml in 2 hours. To calculate the generation time, first we must convert the population size into logarithmic values. By inspection of a table of logarithms to the base 10 (or by using an electronic calculator), it can be determined that log 100,000 is 5 and log 1,000,000 is 6. Hence, the value of log N_i is 5 and that of log N_f is 6. The time of growth (t) is 2 hours. The generation time can be calculated simply by applying the formula given above:

$$T_g = 0.301t/(\log_{10} N_f - \log_{10} N_i) = 2(0.301)/6 - 5 \cong 0.6 \text{ hours}$$

The Bacterial Growth Curve

Bacterial populations, like all other populations, increase or decrease in numbers in response to environmental changes. When a pure culture of bacteria multiplies in a liquid culture medium, the population undergoes a predictable sequence of changes in its growth rate. When these changes are plotted on graph paper, they give rise to a curve known as the **growth curve** (fig. 6-9).

The size of the population is expressed on a logarithmic scale rather than on an arithmetic scale. Since the bacteria are dividing by binary fission and increasing exponentially, the resulting growth curve would rise sharply if the number of cells were plotted on an arithmetic (linear) scale (fig.

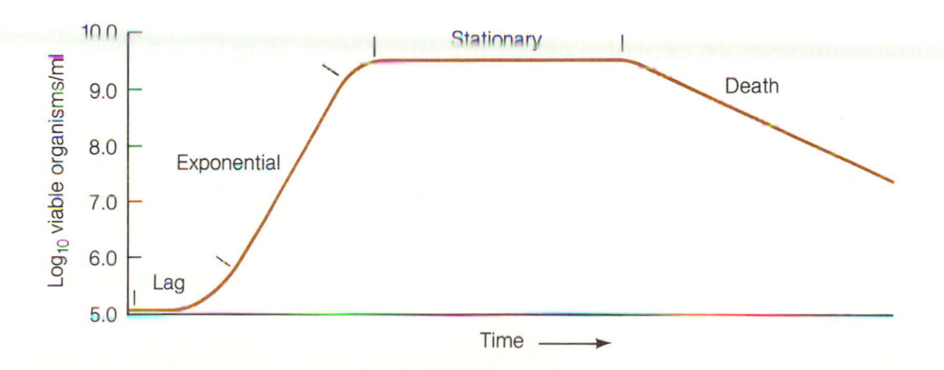

Figure 6-9 The Bacterial Growth Curve. A population of bacteria growing in a broth medium will undergo a series of changes in its rate of growth that reflect changes in its environment and physiology. These phases are the lag phase, the exponential growth phase, the maximum stationary phase, and the death phase.

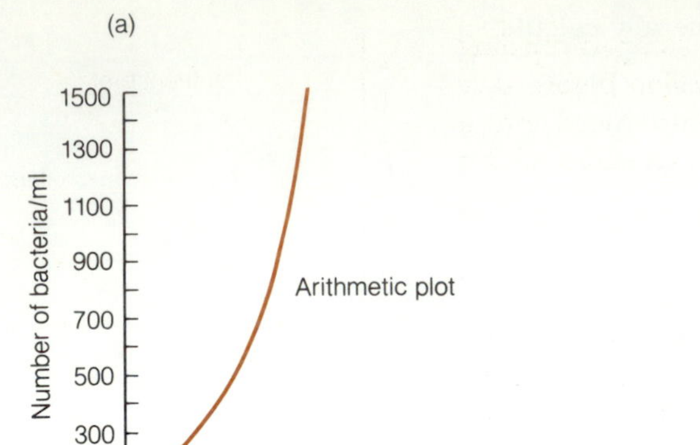

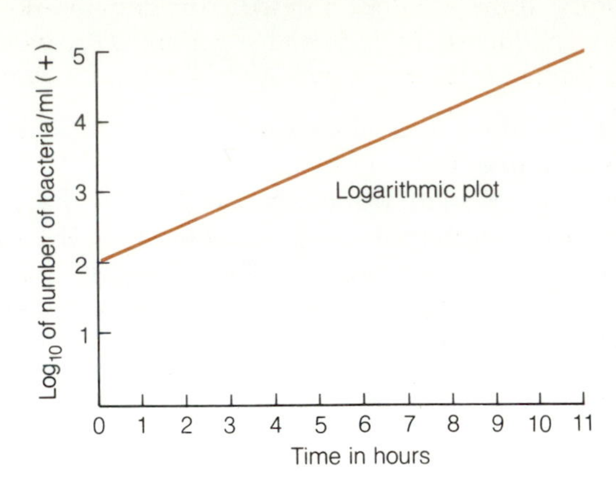

Figure 6-10 Growth Rates of Bacterial Populations Plotted on Arithmetic and Logarithmic Scales. The same population plotted using an arithmetic *(a)* and a logarithmic *(b)* scale. Notice the sharp rise of the population growth curve when the population size is plotted arithmetically. At this rate of climb, the curve will shoot off the graph paper in a few generations. The logarithmic plot gives a gradually rising straight line. The slope of the line can be used to visualize the rate of population growth: the steeper the slope, the faster the rate of growth.

6-10). An arithmetic plot would make difficult long-term observations of the population, because an enormous sheet of graph paper would be required to plot the growth over a two-day period. In addition, an arithmetic scale would not clearly delineate the characteristic phases of growth displayed by populations. If the number of cells is plotted on a logarithmic scale *versus* time (fig. 6-10), the growth curve clearly demonstrates the various phases of growth (fig. 6-9). In addition, the slope of the straight line can be used to obtain the growth rate of the population because it graphically reflects the rate of growth. Hence, the steeper the slope, the higher the growth rate and *vice versa.*

The different **growth phases** of the culture (fig. 6-9) reflect the physiological states of the cells and the condition of the culture medium. From figure 6-9 it can be seen that there are at least four recognizable phases of growth: a) the lag phase; b) the exponential growth phase; c) the maximum stationary growth phase; and d) the death phase.

The Lag Phase

When a population of bacteria is transfered into a new culture medium □ it's growth may be delayed until it adjusts to the new growth conditions. This period of adjustment is called the **lag phase.** During much of the lag phase, the cells are increasing in size (mass) in preparation for cell division and actively synthesizing the enzymes necessary to reproduce. During the latter part of the lag phase, the cells begin to multiply more quickly until the maximum growth rate is reached.

119

The duration of the lag phase depends on factors such as the age of the culture and the nature of the medium. If an old culture is inoculated into fresh medium, there is a lag phase. If a young, actively-dividing culture is inoculated into a similar medium, however, the lag phase is short or even absent, because young cultures are already dividing and can resume their reproduction in the new medium. Older cultures that have stopped dividing must synthesize the enzymes necessary for growth and division before they can start reproducing. Even actively-dividing cells, when introduced into a new environment (e.g., a different type of culture medium), typically exhibit a lag phase.

The Exponential Growth Phase

During the **exponential growth phase** all cells are dividing at a maximum rate. This rate is characteristic of the microorganism species and is also influenced by cultural conditions. During the exponential phase, the number of growing, dividing cells far exceeds the number of dying cells. The rate of cell reproduction is controlled by a variety of factors, such as temperature, hydrogen ion concentration, and nutrients present in the culture medium. For example, when a bacterium such as *Escherichia coli* is growing in a complex medium, many of the essential amino acids, nucleosides, and vitamins are already present and need not be synthesized by the cell. As a consequence, the bacterial generation time is short (fig. 6-11). When the same bacterium is growing in a culture medium consisting of mineral salts and glucose, however, it must synthesize all the amino acids, nucleosides, and vitamins required for growth. This synthesis takes time and is reflected in a longer generation time.

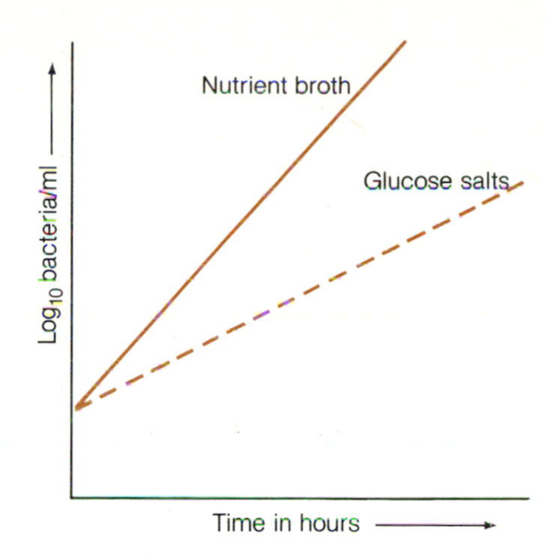

Figure 6-11 Generation Time of *Escherichia coli* Growing in a Minimal Medium (Glucose-Salts) and a Rich Culture Medium (Nutrient Broth). Notice that the generation time, as measured by the slope of the line, is shorter (steeper slope) in rich media than in poor media. This is a consequence of the time required to build all the biosynthetic precursors from the minerals provided. The rich medium already has these precursors.

The Maximum Stationary Phase

The exponential growth phase lasts as long as ample nutrients are available in the environment, or until the population pollutes its environment and makes it inhospitable. During exponential growth, large amounts of organic acids and other toxic substances may accumulate in the culture medium and inhibit the growth of the culture. The accumulation of toxic materials and the depletion of essential metabolites are largely responsible for the onset of the **maximum stationary phase** (fig. 6-9). The maximum stationary phase is characterized by a static population growth rate, in which the number of growing, dividing bacteria is offset by an equivalent number of dying cells.

A population of bacteria growing exponentially will enter the maximum stationary phase of growth before all of the available nutrients are depleted. For example, if a sample of broth culture that just entered the maximum stationary phase is filtered to remove all bacteria and is then inoculated with a fresh culture, further growth takes place in the culture medium. This growth indicates that the accumulation of toxic materials, and/or the depletion of essential nutrients, cannot be exclusively responsible for the cessation of growth. It appears that each cell must have its own "biological space" and be surrounded by a minimum concentration of nutrients before it can reproduce. The total cell yield per ml of broth culture usually remains constant for a given species of bacteria growing under a given set of cultural conditions. This yield has been named the **M concentration,** and it is determined by crowding as well as by nutrient depletion and accumulation of toxic factors in the culture medium.

The Death Phase

The **death phase** usually follows the maximum stationary phase and it is characterized by a progressive decline in the number of viable cells in the population. During the death phase, the dying cells far outnumber those still capable of reproducing. The rate of population decline during the death phase is a characteristic of the species, and is also influenced by cultural conditions and by the presence of toxic materials. It is important to note that, although the population as a whole is declining, many individual cells may still be viable. For this reason, when a dying population is placed in fresh culture medium, it usually resumes growing.

SYNCHRONOUS GROWTH OF BACTERIAL POPULATIONS

When a sample of a bacterial population in exponential growth is examined with a microscope, it is apparent that the bacteria are not dividing in synchrony (at the same time). Some are elongating; others are undergoing cell division; still others have just finished dividing (fig. 6-12). In this usual type of population growth, called **asynchronous growth,** each cell in the population is dividing according to its own biological clock. Although asynchronously growing populations can be used for many microbiological studies, certain aspects of population dynamics, cell division, and metabolic regulation can be studied only if all the cells are dividing at the same rate and at the same time. That is, these studies require **synchronous growth** of the population.

Several techniques are used to achieve synchronous growth. Most often, a dividing population is passed through a filter that retains all but the smallest cells. Presumably, these cells represent the daughter cells resulting from newly completed binary fission. When these young cells are introduced into a fresh culture medium, they elongate and divide at the same time (in synchrony) for a few generations until they get out of phase.

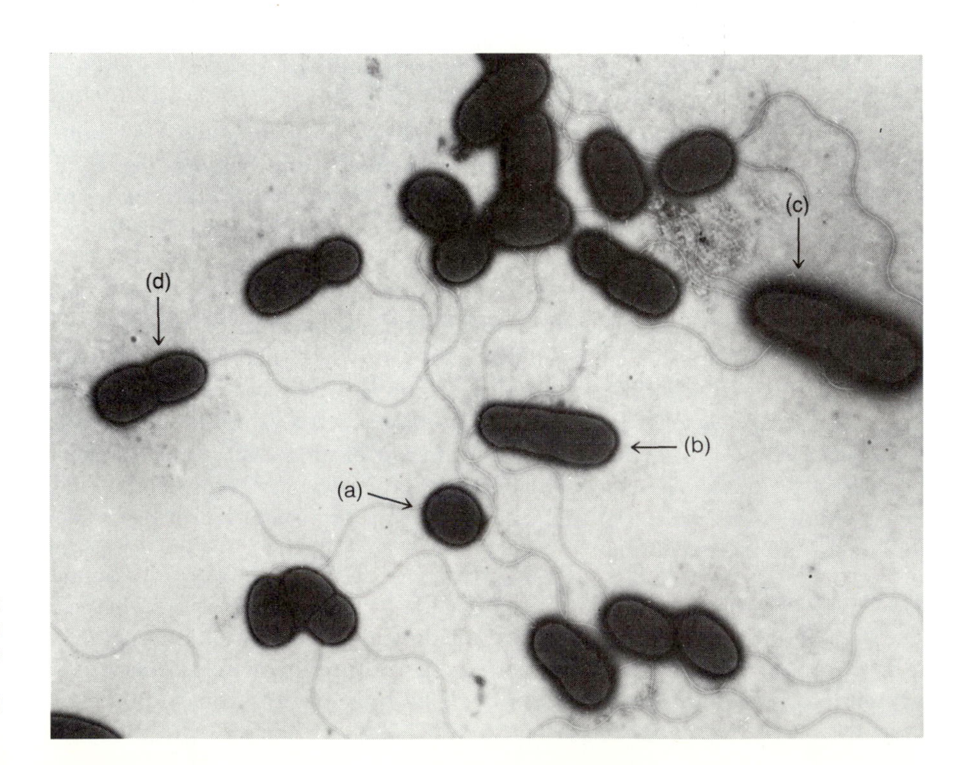

Figure 6-12 Asynchronously-Dividing Bacteria. This population is reproducing asynchronously. Notice that some cells are newly formed *(a)*, others are elongating *(b)*, and still others are in various stages of cell division (*c* and *d*).

CONTINUOUS CULTURES OF MICROORGANISMS

Up to this point we have discussed the growth of microorganisms in closed systems. That is, the microorganisms have been inoculated into a flask of sterile broth and left to reproduce. These types of cultures are called **batch cultures,** in which such factors as toxic waste accumulation, crowding, and nutrient depletion eventually halt microbial growth. If these inhibitory factors are alleviated, however, it is possible to maintain a microbial population growing exponentially for an extended period of time. These types of cultures are called **continuous cultures.**

Continuous cultures may be maintained in an apparatus called a **chemostat.** The chemostat consists of a growth chamber, a reservoir of sterile broth, and an outlet to discard the old culture medium (fig. 6-13). The culture medium usually consists of an abundant supply of all necessary nutrients, except for one that is in growth-limiting concentration. The new medium is introduced into the growth chamber at a constant rate and the spent medium is removed from the chamber at the same rate. The rate at which the culture medium is introduced regulates the growth rate of the microorganisms in the growth chamber. When first inoculated into the fresh chemostat, the microorganisms undergo a period of rapid reproduction until the limiting nutrient is depleted. As fresh medium is introduced into the chamber, the microorganisms grow just fast enough to replace those removed along with the old medium through the chemostat outlet. The rate of nutrient input determines how fast the population grows. The chemostat sometimes is used to mimic natural environments and is used to study microbial populations in nature.

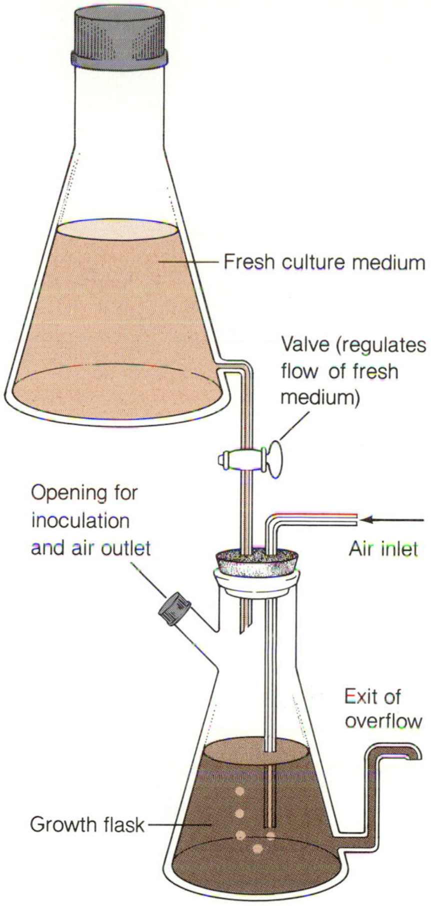

Figure 6-13 The Chemostat. This device is used to maintain microbial populations growing exponentially for extended periods. The chemostat consists of: a reservoir of sterile culture medium with a valve to control the flow rate of fresh medium, a growth chamber, and a siphon, used to discard the used culture medium.

MICROBIAL GROWTH ON SOLID CULTURE MEDIA

119 Microorganisms growing on solid culture media generally remain together and consequently form colonies □. A colony often arises from a single cell (fig. 6-14) and consists of a population of microorganisms at various stages

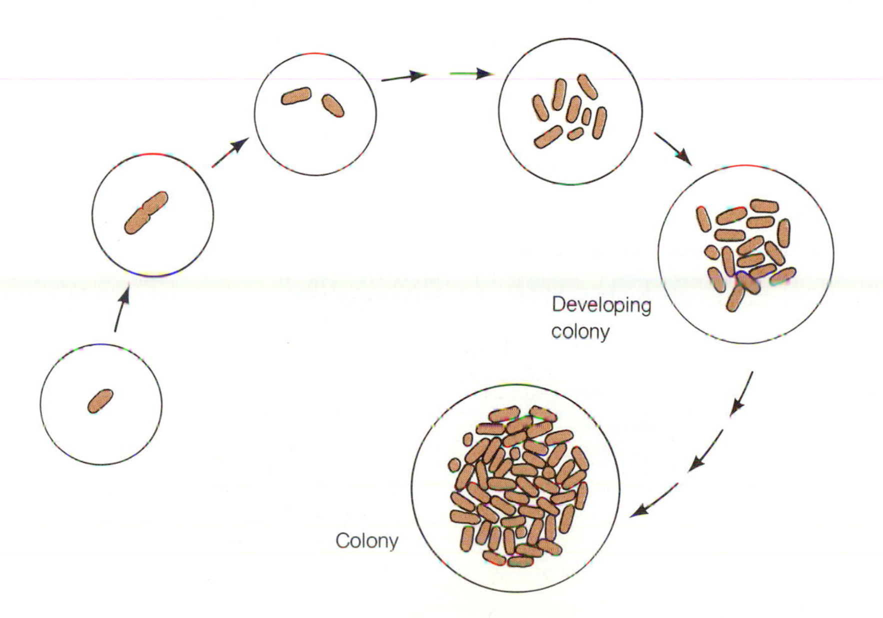

Figure 6-14 Bacterial Colony Formation on an Agar Medium. A colony results from sequential cell divisions of a single bacterium. Since they are growing in an agar-solidified medium, the bacterial cells remain together, forming a discrete mass (colony).

in their growth cycle. Some of the microorganisms are entering the exponential growth phase, while others are already in their maximum stationary phase or even in their death phase.

Since some microorganisms in a colony are not in contact with the culture medium, or have used up the nutrients near them, the rate of nutrient uptake varies from cell to cell. The response of individual miroorganisms to nutrient availability can affect colony size. In addition, crowding influences the size of the colony. This is a result of two adjacent colonies competing for nutrients in their immediate area. Metabolic wastes released by the growing populations may also reduce their growth rate. Both these factors result in a reduction in the sizes of the two adjacent colonies.

FACTORS AFFECTING MICROBIAL GROWTH

The chemical and physical environment can alter the rate at which a population grows. The concentration of oxygen, nutrients, heavy metals, antibiotics, and water, as well as the temperature and electromagnetic radiation, can all dramatically affect the growth of microbial populations.

Nutrient Concentration

The concentration of nutrients in a culture medium can affect either the microbial population growth rate or the total cell yield of the culture. At low nutrient concentrations, the rate at which populations grow is directly proportional to the concentration of nutrients in the culture medium. This phenomenon may be due to the efficiency with which the cell can transport nutrients □. Cells accumulate nutrients by means of proteins called **permeases,** which are part of the plasma membrane. At low nutrient concentrations these permeases are not fully utilized, and so cells do not have sufficient nutrients to reproduce at their maximum rate. As the nutrient concentration increases, more and more permeases are utilized, until a point is reached at which all the permeases are transporting nutrients at their maximum rate into the cell. Beyond this point, nutrient concentration no longer affects growth rate and the growth rate reaches a maximum and remains constant.

182

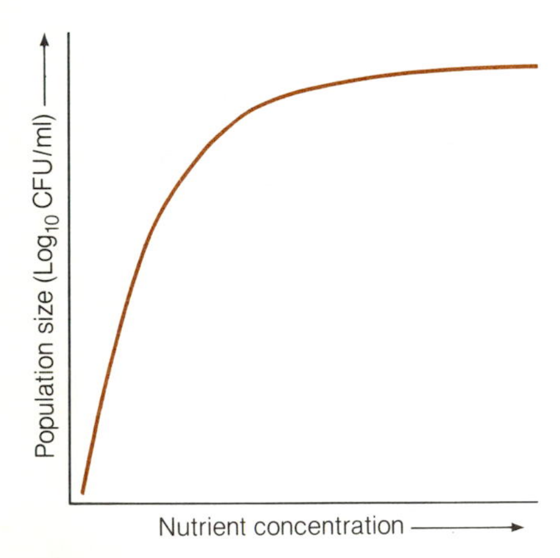

Figure 6-15 Microbial Assay Curve for the Amino Acid Alanine. The curve expresses the relationship between the size of a *Lactobacillus* population and the amount of alanine supplied in the medium. The size of the *Lactobacillus* population in the presence of an unknown amount of alanine can be determined by using the microbial assay curve.

Biological assays are frequently used in the laboratory to determine the concentration of a particular amino acid or vitamin. Biological assays employ microorganisms that are known to require the chemical to be assayed, and they are based on the principle that, at low nutrient concentrations, the yield of microbial mass is proportional to the concentration of the substance (fig. 6-15). The lactic acid bacteria are often used as indicator microorganisms in biological assays because they respond dramatically to slight changes in nutrient concentrations. Biological assays are carried out by first determining the cell yield (e.g., dry weight) of a microorganism in a culture medium containing a known quantity of the substance to be assayed. This cell yield is compared with those in culture media containing various known concentrations of the substance (fig. 6-15). By comparing the yield in the assayed sample with that in the standard curve, one can ascertain the concentration of the assayed substance (e.g., an amino acid or a vitamin) in the unknown sample.

As previously noted, the concentration of nutrients affects the total cell yield, or **maximum crop,** of a culture. As the concentration of nutrients in a culture medium increases, more of these nutrients can be converted to cellular material, and the result is a net increase in the number of cells in the culture. Studies of maximum crop yield are important when it is necessary to obtain large quantities of cells for physiological studies, or for the preparation of single-cell protein or other industrially valuable products.

Temperature

Microorganisms have evolved enzymes and cell membranes that function best within the temperature range they normally encounter. This temperature range may be as wide as 45°C, such as that exhibited by *Bacillus subtilis* (8° to 53°C), or as narrow as 10°C, as that of *Neisseria gonorrhoeae* (30° to 40°C).

The range of temperatures within which various microorganisms reproduce is largely determined by whether or not cellular enzymes can function. If the temperature is too high, the enzymes denature permanently and the membrane becomes excessively fluid, resulting in autolysis. On the other hand, if the temperature is too low, chemical reactions become exceedingly slow or stop altogether. The lowest temperature at which a microbial population can grow is referred to as the **minimum growth temperature** (fig. 6-16). As the temperature increases from the minimum growth temperature, the population multiplies at an increasing rate until it reaches a point

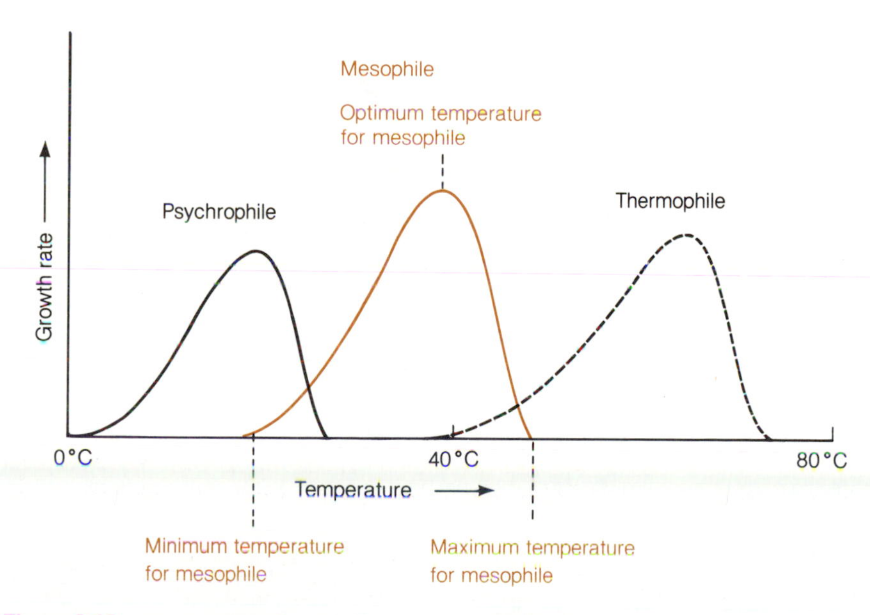

Figure 6-16 Temperature Growth Range of Microorganisms. *Thermophilic* microorganisms grow best at temperatures between 45°C and 80°C with optimum temperatures above 60°C. *Mesophilic* microorganisms grow best between 20°C and about 45°C, with optimum growth temperatures between 30°C and 40°C. *Psychrophilic* microorganisms grow best in cooler environments (−5°C to 20°C).

at which it multiplies at a maximum rate. The temperature at which a population grows at its maximum rate is called the **optimum growth temperature** (fig. 6-16). Above the optimum growth temperature, the rate of growth slows down, because higher temperatures cause some of the cellular enzymes to denature. At some temperature above the optimum, one or more essential enzymes become nonfunctional and the population ceases to reproduce. The elevated temperature at which growth ceases is called the **maximum growth temperature.** This response to increasing temperature gives rise to the skewed curve illustrated in figure 6-16.

Microorganisms that grow most rapidly in cold environments, at temperatures between about −5°C (still in liquid state) and +20°C, are called **psychrophilic** (cold-loving) organisms (fig. 6-16). Psychrophiles grow well in refrigerators at temperatures near 4°C and are responsible for the spoilage of refrigerated foods. In nature, these organisms are common in glacier-fed lakes, cold ocean bottoms, and arctic and subarctic environments. Those microorganisms that grow best at temperatures between 20°C and 45°C are called **mesophilic** (middle-loving) organisms (fig. 6-16). Organisms growing in the human body, and in most topsoils in the United States, are mesophiles. **Thermophilic** (heat-loving) organisms grow best at temperatures between 45° and 80°C (fig. 6-16). These are found growing predominantly in hot springs and in compost piles where the temperature is above 50°C.

Hydrogen Ion Concentration (pH)

The enzymes, electron transport systems, and nutrient transport systems found in the cell membrane are sensitive to the concentration of hydrogen ions (H^+) because this concentration affects the three-dimensional structures of most proteins, including enzymes necessary for growth. Every organism grows most rapidly at its optimum level of H^+. Each organism also has a range of H^+ concentrations outside which it fails to reproduce (fig. 6-17). Microorganisms have evolved enzymes and transport systems that function within the pH ranges of the environments they normally inhabit. Bacteria

FOCUS
BACTERIA THAT GROW DEEP IN THE OCEAN

In the past few years, bacteria from deep oceanic trenches have been discovered that can grow only at pressures considerably above atmospheric pressure. These bacteria inhabit environments 2,500 meters below sea level, where the pressure is more than 250 bars (250 times atmospheric pressure). Pressure-dependent bacteria, called **barophiles,** have been discovered that can grow at pressures between 500 and 1,000 bars but die at atmospheric pressure. Their death has been attributed to the fact that their gas vesicles expand on decompression, pushing the cytoplasm against the membrane and cell wall with such force that the cell ruptures.

Equally noteworthy are bacteria that reproduce near hot oceanic vents. These vents fill the water with hydrogen sulfide and other minerals and heat the water to temperatures in excess of 350°C. Experiments have indicated that some bacteria may be able to grow at temperatures as high as 120°C and at pressures of more than 250 atmospheres. Even under these environmental conditions, the population doubles every 40 minutes or so.

The discovery of barophiles and thermophiles that live thousands of meters below sea level indicates that bacteria may also exist within the earth's crust, where sufficient nutrients exist.

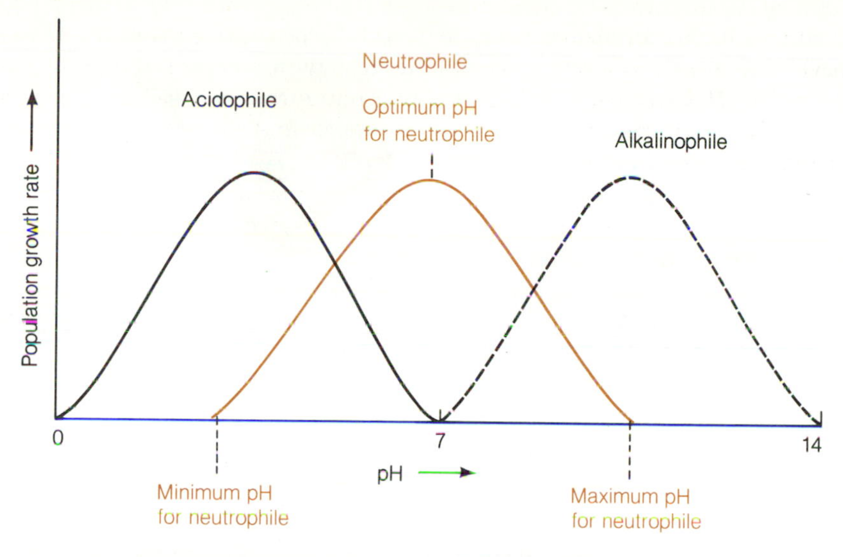

Figure 6-17 Relationship between pH and Population Growth Rate. Each microbial population responds to changes in environmental pH with changes in its growth rate. The response of the growth rate to changes in pH is expressed as a bell-shaped curve with a maximum, a minimum, and an optimum pH.

have been found to grow at pH values between 1 and 10 (table 6-2), although most bacteria multiply best between pH 6.5 and 7.5.

Solute Concentration

Microorganisms generally live in aqueous environments in which the nutrients are dissolved. The concentration of materials dissolved in the aqueous environment can influence the passage of water and nutrients into the cell, and can therefore determine whether or not the cell can reproduce. Cells in hypertonic environments have a tendency to lose water to their environment, because the concentration of solutes outside the cell is greater than that inside the cell. Under these conditions, most bacteria cannot reproduce because they lack sufficient cellular water. Thus, most microorganisms live in hypotonic or isotonic environments with an ample supply of available water.

Bacteria that normally inhabit salty environments are called **halophiles** (salt-lovers). Some halophiles require more than 15% salt to reproduce and are called **extreme halophiles.** These organisms require the salt to stabilize their membranes. Other microorganisms can grow over a wide range of salinity (salt concentration) and are called **moderate halophiles** or facultative halophiles. Most of these organisms are marine bacteria. Halophiles are able

TABLE 6-2
GROWTH RANGES OF SELECTED BACTERIA

ORGANISM	pH RANGE	pH OPTIMUM
Escherichia coli	4.5– 9.0	6.0–7.0
Erwinia caratovora	5.6– 9.3	7.1
Nitrobacter sp.	6.6–10.0	7.6–8.6
Proteus vulgaris	4.4– 8.4	6.0–7.0
Thiobacillus thiooxidans	1.0– 6.0	2.0–2.8

to survive high salt environments and avoid becoming dehydrated because they concentrate solutes that do not interfere with enzyme activities. Halophiles like *Halococcus* or *Halobacterium* concentrate K^+ so that the cytoplasmic concentration of solutes becomes approximately equal to that of the environment.

Molecular Oxygen

Free oxygen is usually thought to be absolutely necessary for life, but this is not always the case. Many microorganisms (anaerobes and facultative anaerobes) can reproduce in the absence of oxygen □. In fact, oxygen inhibits the growth of anaerobic microorganisms and in most cases kills them. Such microorganisms lack enzymes that remove superoxide ions and hydrogen peroxide, common byproducts of an aerobic existence which are extremely reactive and toxic to living organisms. Their toxicity is attributed to their ability to react with enzymes and other cellular components. Aerobes must have enzymes such as superoxide dismutase, peroxidase, and catalase, so that they can remove superoxide ions and hydrogen peroxide as these compounds are formed.

117

REPRODUCTION OF MICROORGANISMS IN NATURE

This chapter has dealt primarily with pure populations of microorganisms reproducing under laboratory conditions, in which nutrients are plentiful and competition for them is relatively insignificant. In nature, microorganisms seldom exist as pure populations; they must interact with other populations on a constant basis. The interactions may be positive or **synergistic,** in which both populations reproduce better in each other's presence than alone. One example is the phenomenon of **cross-feeding** in which one population releases a metabolic waste that is an essential nutrient for another population. The second population may reciprocate by producing a metabolite that is essential for the first population. Two populations may also interact in a negative or **antagonistic** way. They may compete for nutrients, so that each will reproduce better alone than in the presence of the competing population. Another negative interaction occurs when one (or both) of the populations produces toxic substances, such as antibiotics, that inhibit the growth of the other population.

In nature, where nutrients are usually limiting, microbial growth rates are generally much lower than in the laboratory. The effect of limited nutrient concentration on populations can be studied by using a chemostat in which the nutrient concentration can be controlled. Competing microbial populations must be able to reproduce as rapidly as they can, in order to have access to their share of the nutrients, or else face extinction. Microbial populations must also be capable of adapting rapidly to changing environmental conditions. Unlike those growing in the laboratory, microbial populations in nature are faced with the problem of reproducing under conditions of changing temperature, moisture, and nutrient concentration. All of these changes affect the rate at which populations grow. The effects of these changes on microbial populations is the realm of microbial ecology.

SUMMARY

All living microorganisms are characterized by their ability to grow and reproduce.

CELLULAR GROWTH

1. The growth of a microbial cell is defined as the orderly increase of all the components of the cell.
2. The growth of a microbial population is defined as the increase in the numbers (or mass) of the population.

MEASURING MICROBIAL POPULATIONS

1. The growth of a population can be measured by determining the rate: a) of increase in the number of cells; b) of increase in the mass of the culture; or c) at which a metabolic activity takes place.
2. Direct microscopic counts can be carried out by counting the number of microorganisms in a premeasured volume of sample with the aid of a microscope. Counting chambers such as the Petroff-Hausser chamber and the hemocytometer are used for this purpose.
3. Viable counts are based on the fact that any evidence of microbial growth results from microorganisms originally present in the culture tested. Evidence of growth may be the formation of a colony or the turbidity of a broth.
4. Viable counts can be carried out by plate counts or by most probable number counts.
5. Measurement of cell mass can be done by determining the dry weight of a population, the amount of a chemical in the culture, or the turbidity of a culture.
6. Turbidity measurements are commonly used to determine the sizes of microbial populations, and are based on the fact that the turbidity of a liquid culture is directly proportional to the number of microorganisms in the culture.
7. Metabolic activities of populations can also be used to determine population size.

GROWTH OF BACTERIAL POPULATIONS

1. A population of bacteria growing in a liquid culture medium undergoes a predictable sequence of changes in its growth rate. These changes are called phases of growth.
2. The lag phase is a period of adjustment in which cells prepare for cell division. Initially, the cells may increase in size. As the cells begin to divide, they multiply at an increasing rate.
3. The exponential growth phase is characterized by a rapid increase in the size of the population. During this phase, the cells are dividing at a maximum rate and the number of births far exceeds the number of deaths.
4. The maximum stationary phase is characterized by a population that does not increase in number. During this phase, the population has ceased to grow and the number of deaths equals the number of newly formed cells.
5. The death phase is characterized by a progressive decline in the size of the population. The number of deaths is far greater than the number of dividing cells.

SYNCHRONOUS GROWTH OF BACTERIAL POPULATIONS

1. Bacterial cells dividing under laboratory conditions are out of phase with respect to other dividing cells in the population. This type of population growth is said to be asynchronous.
2. Synchronous growth occurs when all the cells in the population are dividing at the same time. This result can be accomplished by filtration or by selectively inhibiting cell functions, such as DNA replication.

CONTINUOUS CULTURES OF MICROORGANISMS

1. Continuous cultures of microorganisms are those in which exponential growth is maintained for an extended period of time.
2. Continuous cultures are maintained in chemostats, where the rate of fresh medium input determines the rate of poulation growth.

MICROBIAL GROWTH ON SOLID CULTURE MEDIA

1. Microorganisms growing on solid culture media form aggregates called colonies.
2. A colony represents a population whose cells are in various phases of growth.

FACTORS AFFECTING MICROBIAL GROWTH

1. Nutrient concentration, temperature, osmotic pressure, pH, and oxygen concentration are environmental factors that influence microbial growth.
2. In general, the above factors influence cellular growth by affecting the functioning of enzymes or membranes.

REPRODUCTION OF MICROORGANISMS IN NATURE

1. In nature, microorganisms seldom exist as pure populations. Instead, they usually interact either synergistically or antagonistically with other populations.
2. Microorganisms in nature must contend with limited nutrient availability and a changing environment, both of which factors can affect their growth. Microorganisms must be able to adapt quickly to these changing conditions or become extinct.

STUDY QUESTIONS

1. Compare the results obtained using a direct microscopic count with those obtained using a plate count.
2. Why are viable counts considered to be minimal counts?
3. What is an important shortcoming of the turbidimetric technique for determining population size?
4. A bacterial culture is diluted 100 times by mixing 1 ml of the culture with 99 ml of sterile water. A 1-ml aliquot of this dilution is plated using the pour plate method, and after 48 hours there are 80 colonies in the plate. What is the viable count of bacteria per ml of the culture?
5. Determine the generation time of an exponentially growing population that increases from 500,000 to 1,500,000 colony-forming units/ml in 1.5 hours.
6. Draw a bacterial population growth curve, label all the phases of growth, and indicate the factors that determine the slope and the duration of the phases.
7. Discuss how the following factors affect population growth:
   a. temperature;
   b. nutrient concentration;
   c. pH;
   d. salt concentration.
8. How does low nutrient concentration influence growth rates? How does the chemostat illustrate this fact?
9. What is a biological assay and how is it conducted?
10. How are microbial populations growing in nature different from those growing in the laboratory?

SUPPLEMENTAL READINGS

American Public Health Association. 1985. *Standard methods for the analysis of water and wastewater.* 16th ed. Washington: American Public Health Association.

Koch, A. L. 1984. Turbidity measurements in microbiology. *ASM News* 50(10):473–477.

Kushner, D. J. 1978. *Microbial life in extreme environments.* New York: Academic Press.

Laskin, A. I. and Lechevalier, H. eds. 1974. *Microbial ecology.* Cleveland: CRC Press.

Meynell, G. G. and Meynell, E. 1965. *Theory and practice in experimental bacteriology.* New York: Cambridge University Press.

Monod, J. 1949. The growth of bacterial cultures. *Annual Review of Microbiology* 3:371–394.

Norris, J. R. and Ribbons, D. W. eds. 1969. *Methods in microbiology.* Volume 1. New York: Academic Press.

Payne, W. J. and Wiebe, W. J. 1978. Growth yield and efficiency in chemosynthetic microorganisms. *Annual Review of Microbiology* 32:155–183.

CHAPTER 7 CONTROL OF MICROORGANISMS

CHAPTER PREVIEW

CONTROL OF INFECTIONS ACQUIRED IN A HOSPITAL

Between 5% and 15% of hospital patients develop infections during the course of their hospitalization. Sometimes as many as 5% of these infected patients die as a direct result of the infections. Since these statistics amount to more than two million infections and as many as a hundred thousand deaths each year, hospitals are constantly studying outbreaks of infections to discover their source and to develop procedures to control the microorganisms responsible.

In one outbreak of hospital infections, laboratory technicians observed an increase in isolates of the bacterium *Acinetobacter calcoaceticus,* which can cause lung and blood infections. In order to determine which patients were at risk and why, the hospital bacteriology laboratory technicians checked personnel, patients, air, floors, walls, sinks, hospital equipment, and furniture for microorganisms. It was discovered that the ventilators used to help patients breathe were contaminated. Seventy-five percent of the spirometers (used to measure air volume) on the ventilators were contaminated with *Acinetobacter.* Even after the ventilator parts were sterilized, approximately 30% of the spirometers were found to be contaminated with *Acinetobacter.* A review of records indicated that *Acinetobacter* had been cultured from the respiratory tracts of 45 patients during a six-month period. All the patients had been mechanically ventilated in intensive care units. In addition, *Acinetobacter* was found on the hands of approximately 10% of the respiratory therapists and nurses working in the intensive care wards.

After it was discovered that condensate was being wiped from the inside of the spirometers with paper towels six to nine times a day, investigators suspected that respiratory technologists and nurses were contaminating the spirometers. The spirometers served as a growth chamber and reservoir for *Acinetobacter* because of the condensate. The investigators also assumed that the hospital personnel were responsible for transmitting the bacteria among the patients and the ventilators, but they did not rule out the possibility that some patient infections were caused by contaminated ventilators.

In order to control the outbreak of *Acinetobacter* infections in the intensive care units, all spirometers were removed from the ventilators, hands were washed religiously, and personnel used sterile gloves. Approximately six weeks after the control measures were instituted, the infection rate for *Acinetobacter* in the intensive care units decreased steadily from 9% to about 0.5%. Forty-five patients were found to have been infected by *Acinetobacter* during the study. Several infected patients died, but the cause of death was not definitely linked to the *Acinetobacter* infections.

Acinetobacter infections in hospitals can be serious. Studies indicate that as many as 35% of patients with *Acinetobacter* pneumonia die as a result of the pneumonia. In one outbreak where 53 patients were infected with *Acinetobacter,* 25 had pneumonia. Of these 25 patients, 11 died. Nine of the deaths were caused directly by the pneumonia.

Hospital programs that monitor the incidence of infections are extremely important because they allow hospital personnel to discover and eliminate outbreaks of infection. Microorganisms are controlled in a number of ways that depend upon the situation. For example, the control of *Acinetobacter* that was growing in ventilators required that the equipment be modified and that personnel wash their hands and wear sterile gloves. Because control is such an important part of microbiology, the methods and principles of control are considered in the following chapter.

LEARNING OBJECTIVES

A STUDY OF THIS CHAPTER SHOULD ENABLE YOU TO:

- DIFFERENTIATE AMONG ANTISEPTICS, DISINFECTANTS, AND STERILANTS
- DISCUSS THE PRINCIPLES OF MICROBIAL CONTROL
- LIST THE FACTORS THAT AFFECT THE RATE AT WHICH ANTIMICROBIALS KILL MICROORGANISMS
- OUTLINE THE METHODS USED TO STERILIZE MATERIALS
- DISCUSS THE IMPORTANCE OF A QUALITY ASSURANCE PROGRAM
- EXPLAIN THE IMPORTANCE OF ASEPTIC TECHNIQUES

Microorganisms are responsible for the death and suffering of millions of people each year. These organisms also cause disease in farm animals and crops and spoil food and other manufactured goods, thus inflicting serious financial losses on farmers and on many industries. In order to reduce the death and suffering caused by microorganisms and to alleviate financial losses, much research in microbiology and in industry is devoted to discovering new and more efficient ways of controlling microorganisms. In the ensuing sections of this chapter we will discuss some of the materials and methods commonly used to control populations of microorganisms.

MICROBIOSTATIC AND MICROBIOCIDAL AGENTS

Since chemicals and physical agents (such as heat and radiation) play such an important role in the control of microorganisms, their identities and modes of action are considered in detail in this chapter. Chemical agents and physical agents can be classified on the basis of how they affect microbes. If the control agent inhibits or completely stops the growth of a population, it is called a **microbiostatic** agent (fig. 7-1). If, on the other hand, the agent actually kills and therefore reduces the number of viable organisms in the population, it is called a **microbiocidal** agent.

There are various related terms commonly used in place of microbiostatic and microbiocidal. **Bacteriocidal** agents kill bacteria, while **bacteriostatic** agents inhibit the growth of bacteria. Agents that kill viruses, fungi, or algae are called **virucidal, fungicidal,** or **algicidal,** respectively. Antimicrobial agents that kill a wide variety of different microorganisms are called **germicides.**

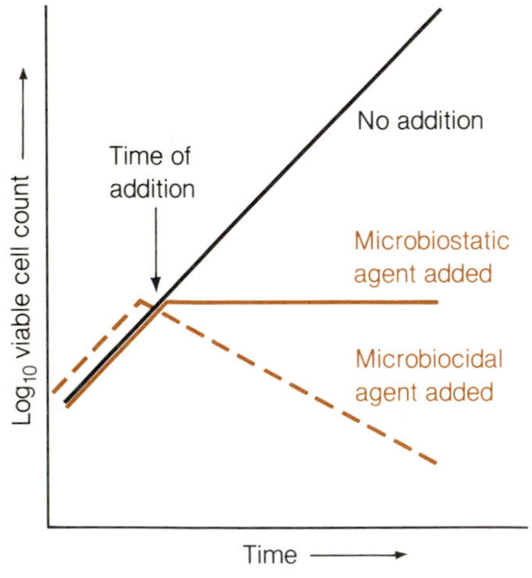

Figure 7-1 Microbiostatic and Microbiocidal Agents. Agents that kill (microbiocidal) stop the growth of a population and decrease the number of viable cells, while those that simply inhibit (microbiostatic) stop growth without an immediate decrease in the number of viable cells.

ANTISEPTICS AND DISINFECTANTS

The diseased state of living tissue due to the colonization and growth of pathogenic microorganisms is called **sepsis.** This is a condition that must be prevented or eliminated in order to insure the health and well-being of the individual. Sepsis is often prevented by using **antiseptics** and **disinfectants.**

Antiseptics are chemical compounds that reduce the number of microorganisms on body surfaces. Antiseptics are relatively mild chemicals, because they are used primarily on human tissues. Tincture of iodine and isopropyl alcohol are antiseptics that are often used to treat superficial wounds.

Disinfectants are chemicals employed to reduce or kill microorganisms present on inanimate objects. They are harsher chemicals than antiseptics and consequently are not used to decontaminate or disinfect plants or animals. Occasionally, a highly diluted disinfectant can be used as an antiseptic. **Phenolic compounds** are disinfectants commonly used in homes, industrial plants, laboratories, and hospitals to destroy microroganisms on tables and equipment. Phenolic compounds are very effective against a wide variety of microorganisms, but must be applied with caution because they can cause skin irritation and nervous system degeneration.

DISINFECTION, SANITIZATION, STERILIZATION, AND ASEPTIC TECHNIQUES

Disinfection is the treatment of materials with disinfectants in order to eliminate or reduce the number of disease-causing microorganisms. This definition implies that there may be residual living microorganisms present in or on the material after disinfection. Disinfectants are usually employed to reduce the possibility of infection when handling contaminated materials. Often the term **decontamination** is used in place of disinfection, but decontamination is also used to describe the reduction or elimination of harmful chemicals or radioactive substances.

Sanitization is a procedure that reduces the number of microorganisms to a low level so that they will not be a problem. The reduction in microorganisms is achieved through the use of chemical or physical agents.

Sterilization is a process that removes or *kills all* living microorganisms. A substance is said to be sterile if it is free of all living microorganisms. Sterilization can involve the use of heat, electromagnetic radiation, filters, and germicidal antimicrobials. Each method has advantages and disadvantages that should be considered before one attempts to sterilize any substance. Sterilization procedures are routinely employed in the laboratory in order to remove contaminating microbes from culture media and the other materials employed in the isolation and cultivation of microorganisms. Sterilization procedures are also used in hospitals to remove all microorganisms from surgical supplies and equipment in order to prevent post-surgical sepsis.

123 **Aseptic techniques** □ are procedures used to prevent contamination of previously uncontaminated materials and to obtain and perpetuate pure cultures of microorganisms. A surgical team performing even the most insignificant of procedures employs aseptic techniques in order to prevent the contamination of the surgical incision or internal organs with unwanted microorganisms. Rudimentary aseptic techniques include personnel washing their hands with antiseptics and mild disinfectants (fig. 7-2); treating bench tops and tables with disinfectants; sterilization of all materials and equipment used in an operation, such as instruments, gloves, surgical drapes, masks and gowns; and maintenance of the sterility of materials by appro-

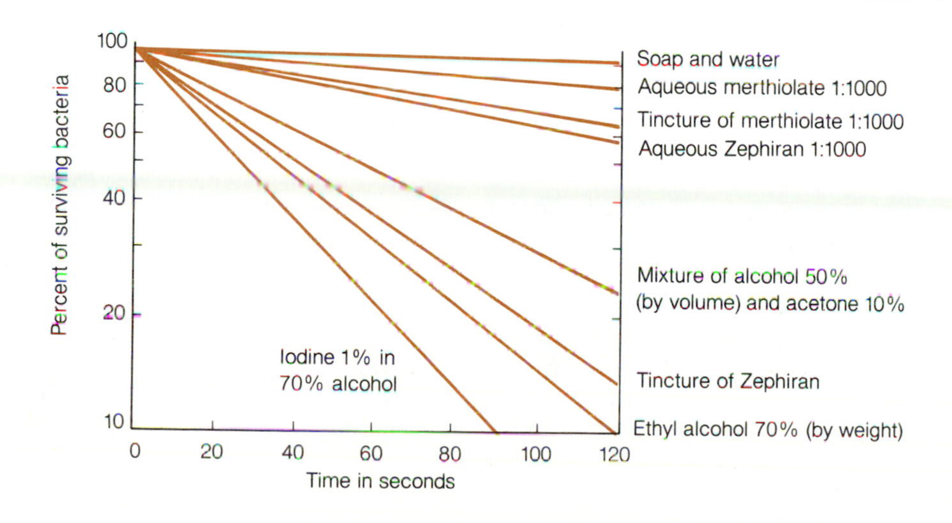

Figure 7-2 Effectiveness of Antiseptics. The effectiveness of various antiseptic solutions on skin bacteria indicates that alcoholic solutions are more effective antiseptics than aqueous solutions. (Based on P. B. Price, 1957, skin antisepsis in *Lectures on Sterilization,* J. H. Brewer, ed., Duke University Press, Durham, N.C.)

priate wrapping and handling. This control of microorganisms is essential to prevent infection or sepsis. **Asepsis** is a non-diseased state of living tissue that is free of proliferating pathogenic microorganisms.

PRINCIPLES OF MICROBIAL KILLING

In order to determine the most efficient way of dealing with contaminating microbes, scientists have studied the behavior of microbial populations exposed to various killing agents. Some chemical and physical control agents are effective over a wide range of microorganisms, while others are more restricted in their scope. Regardless of the nature of the control agent, its effectiveness is influenced by the conditions under which it is applied. The following discussion covers some of the most important factors that influence the effectiveness of control agents.

Population Size and Volume

When a population of microorganisms is exposed to a killing agent, only a fraction of the microbes present dies during a given time interval (fig. 7-3). The longer the population is exposed to a killing agent, the more individuals that are killed. Generally, the decline in viable count of a population occurs exponentially □, and the time it takes to achieve sterility is directly proportional to the number of organisms initially present. For example, if the initial population contains 100,000 viable cells/ml of culture medium, and it declines by 90% every 60 seconds, the death curve (or survivor curve) would look like curve C in figure 7-3. It can be seen that the initial size of the population (at time 0) was 100,000 viable cell/ml and that it took 5 minutes to reduce the population to 1 cell/ml. Curve B has the same slope—that is, the population is dying at the same rate—but it takes 1 minute longer to reduce the population to 1 cell/ml, since the initial poulation was 1,000,000 viable cells/ml instead of 100,000. It takes even longer to reduce a population of 10^7 cells/ml (curve A).

141

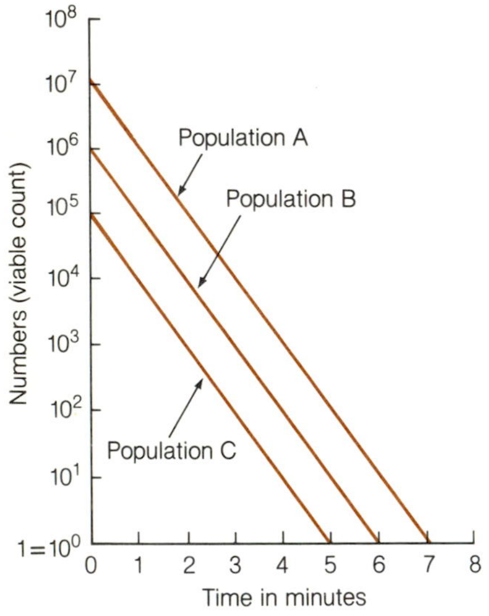

In general, the larger the size of the population subjected to a sterilizing agent, the longer it takes to achieve sterility. It has also been found that the larger the volume of material to be sterilized, the longer it takes to achieve sterility.

Figure 7-3 Death Curves. The time required to kill a population of microorganisms is directly proportional to the size of the population. The graph shows three hypothetical populations of the same organism exposed to the same killing agent (*e.g.*, heat). Notice that populations A, B, and C die at the same rate (all three lines are parallel) although population C dies sooner than B and B dies sooner than A because the initial sizes (microorganisms/ml) are smaller.

Population Susceptibility

Another characteristic of microbial populations to keep in mind when developing an appropriate control method is that different microorganisms die at different rates when exposed to a given sterilizing agent. In other words, not all microorganisms have the same susceptibility to a given sterilizing agent.

Microbiologists in various industries employ a value known as the **death rate constant** (K) to compare the susceptibilities of different microorganisms to a given control agent. The more rapidly a population can be sterilized, the larger the death rate constant. The death rate constant is the same and independent of the initial population size. It would only be different if you were comparing different organisms. The death rate constant can be calculated using the following formula:

$$K = [\log_{10}(N_o/N_t)]/t$$

Where time, t, is in minutes of exposure to a killing agent: N_o is the initial number of microorganisms; and N_t is the final number of microbes after treatment. If at time 0 the size of the population is 10,000,000, and after 3 minutes the population is reduced to 10,000, the K value is found to be 1.00 (fig. 7-3, curve A). Now calculate the K value for the population of bacteria illustrated in curve C in figure 7-3.

Exposure Time

Figure 7-3 shows how long it takes different populations to be reduced to one organism. By treating the populations for another minute, the chances of having one viable organism is reduced 1⁄10. If the population were treated for longer periods of time, the chances of having one viable organism in the population is reduced even further. The question is: how much longer should the microbial population be treated with a sterilizing agent before it can be considered sterile? Many microbiologists consider a culture medium sterile if there is only one chance in a million of having a living microorganism left in the culture. This would require 6 extra minutes of sterilization (fig. 7-3). The actual number of microorganisms in a contaminated solution is generally not known. Consequently, the length of treatment of the solution with a sterilizing agent must be determined empirically (by experimentation).

Generally, sterilization times are determined using microorganisms that are highly resistant to the sterilizing agent. For example, the endospores of *Bacillus stearothermophilus* are highly resistant to heat and are therefore used to establish routine times for heat sterilization. Routine laboratory sterilizations using steam under pressure are carried out at 121°C for 15 minutes. During this time and at this temperature, there is only an infinitesimal chance that even one endospore of *B. stearothermophilus* remains viable. Standard times for achieving sterilization of various materials with different killing agents are determined using highly resistant organisms in order to insure the sterility of the material, even when it is heavily contaminated with other types of microorganisms.

Age of the Microorganisms

The rate at which a given population of microorganisms dies is also a function of its age. Generally, young, actively growing cultures are more susceptible to sterilizing agents than are older ones in stationary phase. This tendency appears to be true of populations exposed to heat and to disinfectants. It may be explained, at least in part, by the fact that older cells are biochemically less active than younger ones. Because they are biochemically less active, they also have a diminished capacity to take up chemical control agents that damage them. Since many antimicrobials must penetrate the cell in order to be active, they are not as effective on stationary phase cells as they are on actively metabolizing cells.

Hydration

The presence of water in an environment affects the rate at which a particular sterilizing agent kills a microbial population. For example, a high water content expedites the coagulation or jelling of cellular proteins when heat is the killing agent. A 50% solution of egg albumin in water is coagulated when the solution is heated to 55°C, whereas a 75% solution is not coagulated until it reaches 75°C. Hot, moist air kills more efficiently than hot, dry air, because the moist air has a greater heat content than does dry air. In addition, many chemicals must be in solution in order to be effective as control agents. Water affects the activity of most chemical control agents by promoting chemical reactions and the ionization of the agents.

Heat

Excessive moist heat promotes the denaturation of proteins, including essential enzymes. This denaturation inhibits or kills microorganisms. Mild heat generally enhances the activity of disinfectants because it promotes chemical reactions, metabolic activities, and possibly a more rapid uptake of disinfectants by the cell. As the temperature of an environment increases toward the optimum for a microorganism, it stimulates its rate of growth and thus increases the rate at which the germicide penetrates cells. As the concentration of disinfectant increases inside the cell, so does the rate of death. For example, a 1.1% aqueous solution of phenol, maintained at 10°C, sterilizes a standard bacterial culture in approximately 160 minutes. The same solution of phenol at 20°C sterilizes a similar culture in less than 60 minutes. Conversely, low temperatures slow down chemical reactions and metabolic activities so that disinfectants are either less active or totally inactive.

Antimicrobial Concentration

Generally, the rate of killing is directly proportional to the concentration of the antimicrobial. That is, the higher the concentration of the antimicrobial applied, the faster it kills microorganisms. Optimal concentrations for each routinely-used antimicrobial should be determined empirically. A compromise should be reached between rate of killing and economy. A highly concentrated solution of antimicrobial undoubtedly is more effective than a more dilute one, but it is also more expensive to use. Thus, the use of optimal concentrations insures that the antimicrobial will be effective for its desired purpose but not overly expensive.

Hydrogen Ion Concentration (pH)

The concentration of H^+ in a solution can influence the effectiveness of a killing agent, as well as affect the metabolic activities of microorganisms. The pH of a disinfectant can influence its effectiveness in several ways. Some of these effects are antagonistic, while others contribute to the killing efficiency of the chemical control agent. At extreme pH values, such as 2 or 14, the drastically altered hydrogen ion concentration in the cell's environment inactivates many surface components. This inactivation results in altered membrane potentials and transport and the leakage of cell materials. Less dramatic departures of the pH from neutrality (pH 7) also inhibit (or at least slow down) metabolic activities, since many membrane-bound enzymes function less efficiently. An inhibition of transport proteins can result in a reduction of the activity of the germicide because the germicide is less able to penetrate the cell. Changes from neutral pH may also reduce the activity of the germicide by promoting its ionization and therefore making it less effective as a killing agent. Germicides in nonionized form are usually more effective as killing agents.

Organic Matter

The presence of extraneous organic matter may reduce a disinfectant's activity. This reduction in activity usually comes about in two ways: a) the organic material may bind the disinfectant and consequently take it out of solution, or b) the organic material may bind the disinfectant and inactivate it. Chlorine, a chemical used to disinfect water supplies and swimming pools, is bound by extraneous organic matter; thus, the effective concentration of the chlorine is reduced. Similarly, organic matter found in pus or serum often reacts with a disinfectant, producing a less active or totally inactive molecule.

PHYSICAL AGENTS USED TO CONTROL MICROORGANISMS

The removal or killing of contaminating microoorganisms from surfaces or materials can be accomplished using a variety of different control agents (table 7-1). These control agents can be either physical agents, such as heat and electromagnetic radiation, or chemical agents, such as disinfectants and antibiotics. In this section we will discuss some of the most common physical control agents, their applications, and their modes of action.

TABLE 7-1
TYPES OF ANTIMICROBIAL AGENTS

TYPE OF AGENT	AGENTS	APPLICATIONS
Physical	Dry heat (160–180°C)	Sterilization
	Moist heat (115–150°C)	Sterilization
	Moist heat (65–100°C)	Disinfection
	Ionizing radiation (gamma, electrons)	Sterilization
	Ultraviolet radiation	Disinfection
Chemical (vapors)	Ethylene oxide	Sterilization
	Formaldehyde	Sterilization or disinfection
	β-propiolactone	Sterilization
Chemical (low selectivity)	Alcohols, aldehydes, halogens, phenols, Quarternary ammonium compounds	Disinfection or preservation
Chemical (moderate selectivity)	Antibiotics (bacitracin, polymyxins)	Topical chemotherapy
	Dyes (acridines, triphenylmethanes)	Antisepsis
	Metal chelate complexes	Antisepsis
	Organic arsenic compounds	Chemotherapy
	Organic mercury compounds	Preservation or antisepsis
Chemical (high selectivity)	Synthetic (*p*-aminosalicylic acid, isonicotinic acid hydrazide, sulfonamides, trimethoprim)	Chemotherapy
	Antibiotics (aminoglycosides, amphotericin, cepholosporins, chloramphenicol, erythromycin, griseofulvin, lincomycin, nystatin, penicillins, rifamycins, tetracyclines)	Chemotherapy

Gardner, J. F.: Principles of Antimicrobial Activity. In *Disinfection, Sterilization, and Preservation,* 2nd ed. Edited by S. S. Block. Lea & Febiger, Philadelphia, 1977.

Heat

Of all the control agents, high heat is the most efficient and cost-effective sterilant. If a material is not likely to be damaged at high temperatures, its sterilization with heat is recommended. Generally, when psychrophilic and mesophilic □ microorganisms are exposed to temperatures above 55°C, their proteins denature irreversibly and they die. As the temperature is increased, they die at a faster rate (fig. 7-4). The loss of viability in a population exposed to high heat correlates well with the denaturation of proteins essential for microbial activities, particularly when the heat is applied in moist environments (moist heat). It also appears, however, that membranes become much more fluid and extremely permeable at high temperatures. The increased permeability results in the leakage of intracellular materials. High temperatures also cause the loss of membrane potentials, thus blocking the transport of nutrients into the cell and, in some cases, the generation of ATP. Thus, denaturation of cellular proteins and disruption of membrane integrity, whether operating alone or together, contribute to efficient sterilization.

145

Heat affects microorganisms in dry environments differently than those in moist environments. Dry heat tends to dehydrate the cell, causing reversible protein denaturation. Since desiccation tends to "preserve" some cells in an inactive state, however, it is generally agreed that a greater amount of heat is required to kill cells in dry environments than in moist ones. The physiological state of desiccated cells is such that their cellular contents become much more resistant to irreversible damage. In general, it takes considerably more time to kill endospores using dry heat at 180°C than it does using moist heat at 121°C. Bacteria and other microorganisms eventually are killed in dry heat environments when their proteins denature and they become charred (oxidized).

Several terms are used to express the susceptibility of microorganisms to high heat. These terms are: a) **Thermal Death Point** (TDP); b) **Thermal Death Time** (TDT); and c) **Decimal Reduction Time** (D). The thermal death point for a particular population of microorganisms in an aqueous suspension is defined as the lowest temperature that sterilizes the culture within 10 minutes. The thermal death time is the shortest period of time necessary to achieve sterility at a given temperature under standardized conditions. Decimal reduction time, or **D value,** is another way of expressing the susceptibility of microbes to the killing effect of heat. The D value is defined as the time required for a pure culture of microorganisms to decrease by 90% when exposed to a specific temperature (fig. 7-4). The D value expresses the rate at which a population dies when exposed to heat: the smaller the D value, the more rapidly the organisms die.

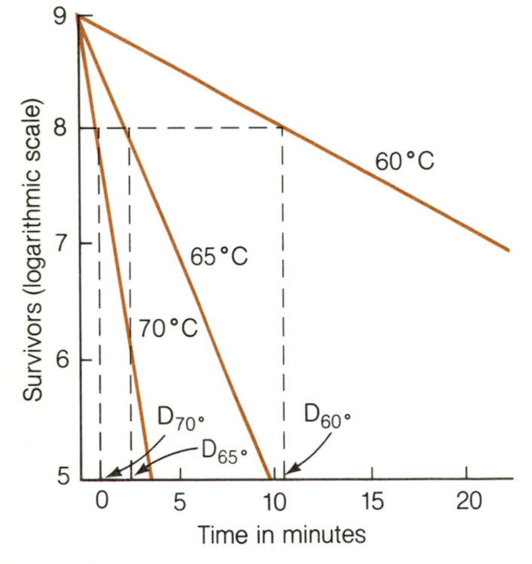

Figure 7-4 Decimal Reduction Times. The decimal reduction time is the time required to reduce a population by 90% at a particular temperature. The decimal reduction time decreases as the temperature increases. For example, in the graph above, the D_{60} is about 10.5 minutes, the D_{65} is 3 minutes and the D_{70} is 1 minute. The decimal reduction time is calculated by finding the points on the death curve between which the population changed by one log unit (e.g., from 10^4 to 10^3), and then determining the time it took for this change.

Steam Sterilization

Steam is an ideal sterilizing agent because, when under pressure, it can be used to heat materials to temperatures above 100°C, and because it heats porous substances evenly and quickly. In order to increase the efficiency of sterilization using steam, the materials are usually sterilized in a chamber in which the steam pressure is above that of the atmosphere. At elevated pressures, steam can reach temperatures above 100°C. The increase in steam temperature is proportional to the pressure (table 7-2). Routine laboratory sterilizations are carried out in **autoclaves** at 121°C for 15 to 30 minutes,

TABLE 7-2
RELATIONSHIPS BETWEEN TIME OF AUTOCLAVING, PRESSURE AND TEMPERATURE

STEAM PRESSURE (PSI[1])	TEMPERATURE (°C)	(°F)	TIME OF EXPOSURE (min[2])
0	100	212	>60
5	110	230	>60
10	115	239	60
15	121	250	15
20	126	259	10
25	134	273	5

1. PSI refers to pounds per square inch above atmospheric pressure.
2. Approximate time required to kill a standard suspension of endospores.

depending upon the volume of material to be sterilized and the amount of contamination expected (fig. 7-5). It is usually recommended that large volumes of materials, such as culture media or surgical dressings, be dispensed into small aliquots or portions to reduce the time required for sterilization and, thus, decrease the chance of overcooking or damaging the material (table 7-3). Although autoclaving is a highly recommended sterilization procedure, its major drawback is that it may cause damage to heat-sensitive materials. Therefore, steam sterilization cannot be used to sterilize all materials.

Boiling

Boiling is an inexpensive and relatively effective means of disinfecting substances. Many contaminating microorganisms and pathogens are readily destroyed by boiling for a few minutes. Boiling is often used to disinfect and sterilize materials at home and in places where autoclaves are not readily available. This method of moist heat sterilization has one major shortcoming, however: it does not destroy many of the bacterial endospores and some of the viruses. As a consequence, boiled materials are often not sterile. Nevertheless, boiling has been used extensively for centuries and has been found

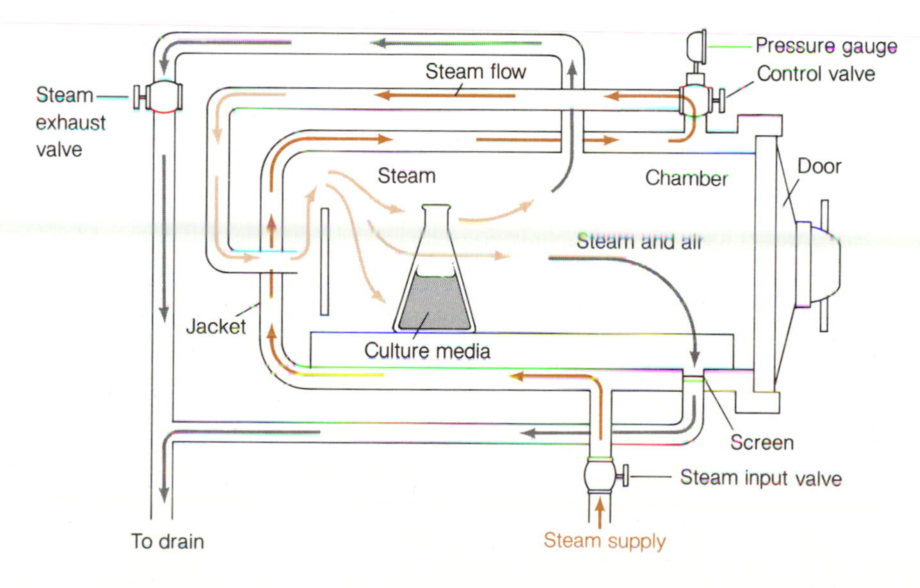

Figure 7-5 Autoclave. Autoclaves are used to sterilize heat-stable media and equipment. Steam initially enters the steam jacket around the autoclave chamber and heats up the autoclave. An object is placed in the chamber, the door is shut securely, and steam from the jacket is allowed to enter the chamber. Air is forced out of the chamber by the incoming steam until only pure steam is being forced out. Then the air exit is closed by the high temperature of the steam, and the steam pressure builds to 15 pounds per square inch. Eventually, the temperature rises to 121°C. When the sterilization is completed, the steam exhaust valve opens and the steam flows out of the chamber. When liquid media are being sterilized, the steam must be cooled and released slowly in order to avoid the evaporation and boiling over of the liquid media.

TABLE 7-3
STERILIZATION TIMES AND METHODS FOR SELECTED INSTRUMENTS, SUPPLIES AND CULTURE MEDIA

ITEMS TO BE STERILIZED	TIME REQUIRED (min)	METHOD OF STERILIZATION		
		AUTOCLAVING (121°C)	DRY HEAT (160°C)	FILTRATION
Gloves and gauze pads	30	*****		
Surgical drapes and dressings	30	*****		
Surgical instruments	30	*****		
Glassware and metalware	30	*****		
Culture media in tubes	15	*****		
Culture media in flasks	15	*****		
Culture media in large volumes (more than 10L)	70	*****		
Glassware	60		*****	
Surgical instruments	120		*****	
Oils, vaseline, gauze	120		*****	
Sharp instruments	270		*****	
Heat-sensitive solutions	—			*****
Serum-containing media	—			*****
Culture media	—			*****

quite satisfactory for disinfecting and sterilizing certain materials, such as baby bottles and rubber nipples.

Pasteurization

Pasteurization is a process whereby disease or spoilage-causing microorganisms in liquids are destroyed by heating the liquids to temperatures below 100°C. For example, some liquids can be **batch pasteurized** in large tanks by subjecting them to temperatures of 63°C for 30 minutes, or **flash pasteurized** by heating them to 72°C for 15 seconds as they flow through a heater. Pasteurization was developed by Louis Pasteur in the mid-1800s to kill microorganisms that were spoiling French wines. Unlike boiling, the pasteurization process did not alter the flavor or quality of the wines. The procedure developed by Pasteur is now used to kill pathogenic organisms in a number of beverages and foods. For instance, most of the milk products on supermarket shelves are pasteurized before packaging to kill all the disease-causing microorganisms that may be present. Of particular importance are the bacteria that cause tuberculosis *(Mycobacterium)*, undulant fever *(Brucella)*, and Q fever *(Coxiella)*. Since pasteurization kills all of the heat-sensitive microorganisms, it prolongs the products' shelf life. Pasteurized milk is not sterile, however, because many **thermodurics** (heat-resistant organisms) survive. Foods and beverages usually are not boiled or sterilized, because these procedures alter their flavor, texture, and quality.

Incineration

Microorganisms exposed to open flames burn and consequently are removed from materials that can stand the heat. This form of sterilization is used to sterilize inoculating loops and some other equipment used in the cultivation

122 of microorganisms □. In addition, many contaminated materials and carcasses of infected animals are **incinerated** in order to kill potentially dangerous organisms.

Hot Air Ovens

Dry heat is routinely used in laboratories and in hospitals to kill microorganisms. Glassware such as pipets, glass petri dishes, and other heat-resistant materials are often sterilized in hot air ovens, usually at 160°C to 180°C for 1 to 2 hours. These ovens are equipped with a heating unit to raise the temperature inside the chamber, and a fan to circulate the hot air so that the materials are heated evenly. Successful sterilizations can be carried out using a gas or electric stove.

Dry heat sterilization has at least one advantage over steam sterilization in that it does not require water, which may damage many materials or alter their properties. In addition, certain materials, such as glass pipets and glass petri dishes, need to be completely dry before they can be used. With moist heat, some of the water vapor condenses on the glass.

Refrigeration

Refrigeration is employed to control microbial growth in order to preserve easily spoiled materials. In general, the lack of heat (low temperature) does not kill microorganisms; rather, it inhibits their growth by slowing enzymatic reactions. It is only at temperatures much below 0°C that cellular water begins to freeze and to create an environment that is not conducive to metabolic activities. A certain amount of cell damage occurs during freezing because of the formation of ice crystals within the cell. Leakage of cell materials may occur as a result of damage to the plasma membrane. In addition, ice forming around the exterior of the cell causes water to be lost from within the cell, thus causing some proteins to be denatured. Although normal metabolism and cell repair may cease, some enzymes may remain active and degrade parts of the cell. The damage these enzymes cause may in some cases be so extensive that the cells are killed. If cells are frozen rapidly at a low temperature of −80°C in an appropriate medium, however, they can be preserved indefinitely in a viable form. The low temperature and the frozen state of the cytoplasm protect these cells from damage.

Freezing is commonly used to preserve foods and other rapidly spoiled materials. At temperatures of −20°C (the approximate temperature inside the freezer compartment of most refrigerators), there is no spoilage due to microbial growth.

Radiation

Ultraviolet light, X-rays, and gamma rays are high-energy electromagnetic radiations that are microbiocidal (fig. 7-6). They cause cell death by inducing extensive changes in the cell's DNA or by ionizing cellular components. X-rays, gamma rays, high-energy alpha and beta particles, and neutrons are known as **ionizing radiation.** When a cell is in the path of any of these types of radiation, considerable ionization occurs. Cellular constitutents hit by the radiation lose electrons and protons. This process results in the formation of ionized compounds and free radicals, chemicals that are highly

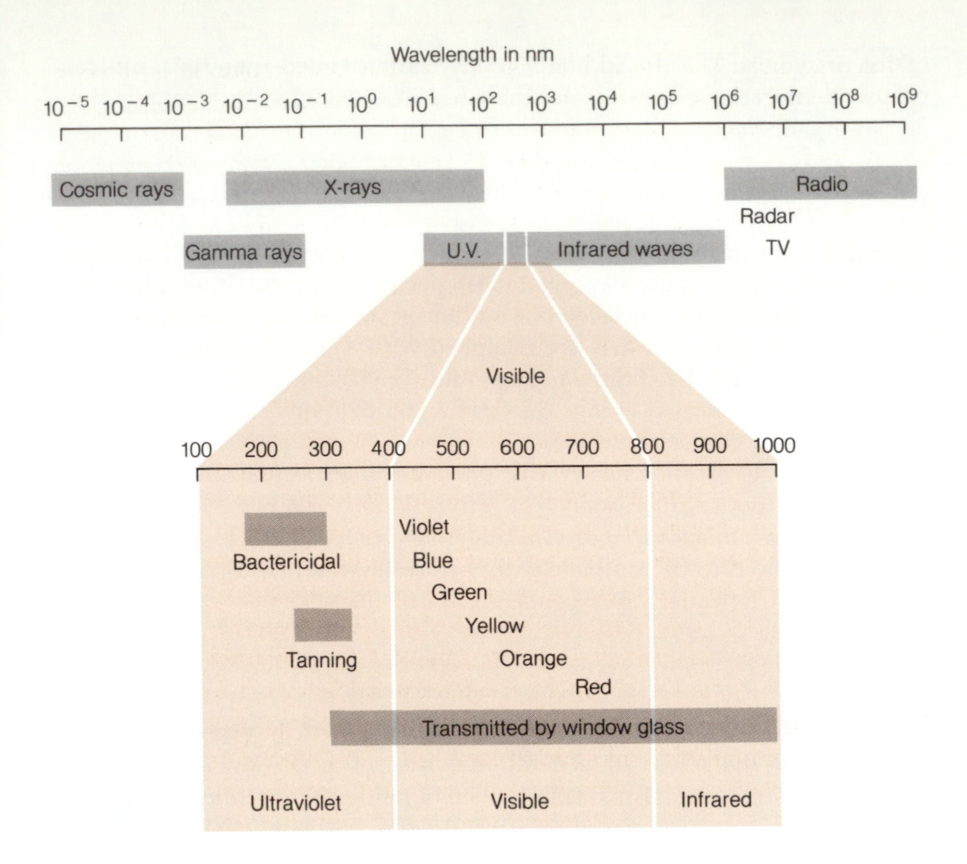

Figure 7-6 Electromagnetic Radiation. Ultraviolet light, X-rays, gamma rays, and cosmic rays all destroy cells and viruses. Ultraviolet light is often used to sanitize the air and the tops of work benches in laboratories and manufacturing plants where microbial contamination must be minimal. Gamma rays from radioactive elements are often used to sterilize packaged materials, such as plastic petri dishes and filtering equipment. X-rays and cosmic rays are not used to sterilize.

reactive and can cause severe damage to the cell by reacting with proteins and nucleic acids. One of the more permanent changes believed to occur as a result of damage caused by ionizing radiation is the formation of large numbers of single- and double-strand breaks in the DNA molecule. These breaks can alter the DNA so that it can no longer function as a template for its own replication or for the synthesis of RNA.

Microorganisms differ in their sensitivity to ionizing radiation. One explanation for these differences is that some microorganisms have multiple copies of their genome. In such a microorganism, if one gene or cluster of genes is destroyed by the ionizing radiation, there are other copies of the gene(s) present to direct the necessary cellular function(s). Another explanation for the variation in susceptibility is that some microorganisms have more efficient DNA repair ☐ systems than others. Those with efficient DNA repair systems can neutralize more of the damage than can other microorganisms, and consequently can tolerate greater amounts of ionizing radiation.

217

Ultraviolet Light

Electromagnetic radiation with wavelengths between 100 and 400 nm is called **ultraviolet (UV) light** (fig. 7-6). These wavelengths, especially those near 260 nm, are particularly important from a biological point of view because they are absorbed by nucleic acids. This energy absorption induces chemical changes that alter the structure of the DNA. During the repair process, the sequence of nucleic acids may be altered, thus leading to the death of the organism. The amount of damage to the DNA molecule is directly proportional to the amount of radiation absorbed by the cell.

Most ultraviolet radiation does not readily penetrate glass, opaque solids, or liquids, so it is generally used to sterilize surfaces in work areas or the air in rooms and surgical suites. Because many microbes possess a DNA repair system that is activated by visible light, UV light is most effective when applied in dark or dimly lit rooms. Darkness prevents the activation of light repair mechanisms, thus maximizing the damage to the DNA molecule. Some microorganisms, however, have DNA repair systems that function in the absence of light, and hence, do not need to inhabit lit environments to repair DNA that has been damaged by ultraviolet light.

Radioactivity

34 Ionizing radiation □ from radioactive elements is occasionally used to sterilize heat-sensitive medical supplies, such as plastic syringes, drugs, and single-service surgical equipment. Some consumer goods (for example, bacon) have also been "sterilized" with ionizing radiation in order to increase their shelf life and kill any pathogens that may be present. Cobalt-60 is one of the most common sources of gamma rays employed in industry. It is useful because it produces high energy radiation capable of penetrating materials so that the products can be prepackaged before sterilization. Aside from medical supplies, ionizing radiation is seldom employed as a routine sterilizer because it is expensive and dangerous to use.

Filtration

Filtration is an effective and reasonably economical way to "sterilize" liquids and gases (fig. 7-7). It is recommended for the elimination of microorganisms in **thermolabile** (heat-sensitive) fluids and air. Sugar solutions used for the cultivation of microorganisms tend to caramelize during autoclaving, so they are often filter-sterilized. Beer and wine, which some breweries and wineries have traditionally pasteurized after the fermentation process is completed, are now being passed through special filters to remove spoilage microorganisms. An advantage of this process is that it sterilizes and clarifies the beverage in one step. Many pharmaceuticals and biological research fluids are sterilized by filtration. Currently, filters are manufactured which can remove viruses.

Materials such as asbestos, sintered glass, porcelain, and diatomaceous earth have been used successfully to filter liquids and gases. Because these materials usually exclude only bacteria and large microorganisms, and because they are expensive, they have been replaced by economical **membrane filters** made of cellulose acetate and similar materials. These filters are manufactured with pores of different sizes. The diameter of the pore and the electrical charge on the filter determine the kinds of microorganisms that will be filtered out. For example, filters with pore size of 0.1–0.2 μm are routinely used to filter out contaminating bacteria. Many membrane filters have the advantage of being semitransparent when moistened, so that direct microscopic examinations can be made of organisms trapped on the filter. In addition, nutrients can pass virtually unimpeded through these filters; thus, microorganisms can be grown on the surface of a filter by placing the filter on an agar medium. This procedure allows microbiologists to count the number of microorganisms captured on the filter and consequently to estimate the number of organisms per ml of fluid or gas.

All filters work in much the same way, whether the material to be filtered is a fluid or a gas. When a substance is filtered, the fluid or gas passes through the pores, but microorganisms and inert particles that are too large to fit into the pores remain trapped (fig. 7-7). In addition, many particles and microorganisms that are small enough to pass through the pores are trapped by the filter due to electrostatic charges. Because suction devices are generally used on filtering apparatuses to speed up the process, certain bacteria, such as the **mycoplasmas** □ (which lack cell walls and are therefore 321

(a)

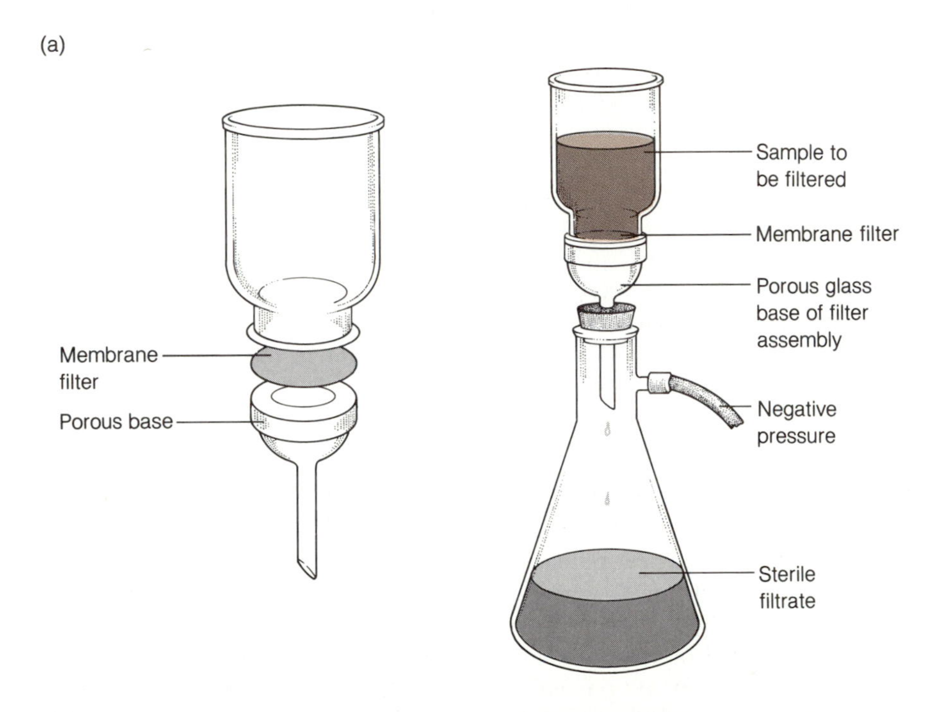

(b)

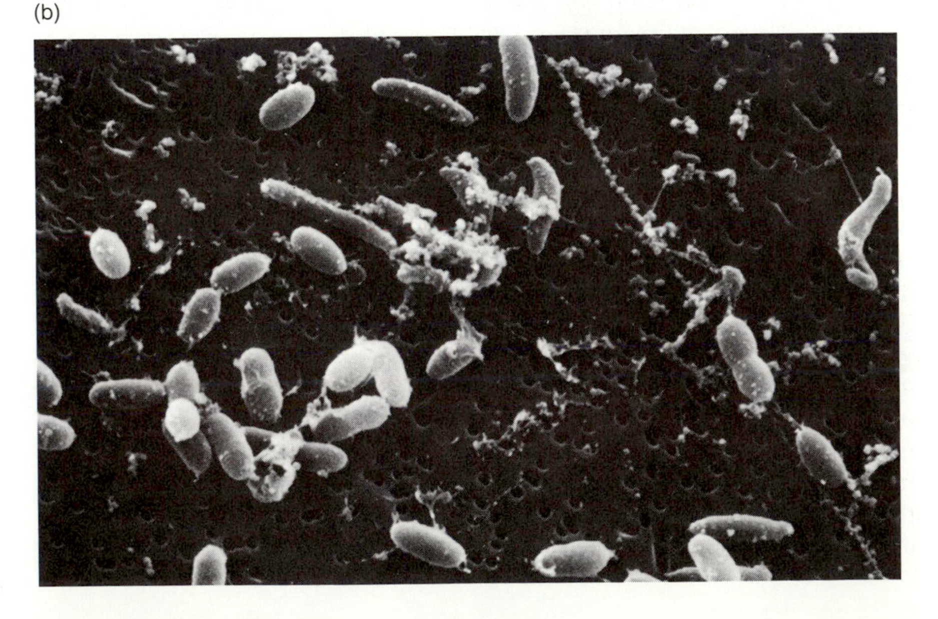

Figure 7-7 Filtration. *(a)* Heat-labile (sensitive) media are generally "sterilized" by filtration through membrane filters. In fact, the medium is not really sterilized because viruses generally pass through the filters. Water samples are often checked for microorganisms by filtering. The microorganisms stuck on the filters can be detected by placing the filter on a selective and differential agar medium. The resulting colonies can then be counted and identified. *(b)* Bacteria are shown caught on a millipore membrane filter, in which the pores are too small to allow the organisms to pass through.

pliable), can squeeze through pores as small as 0.2 μm. It must also be recognized that materials filtered through pores as small as 0.1 μm in diameter may not really be sterile, because many viruses can pass through these pores.

CHEMICAL AGENTS USED TO CONTROL MICROORGANISMS

Many chemicals disrupt cellular metabolism or vital cellular structures and consequently inhibit growth or kill microbes. Some of these substances are used in laboratories and in hospitals to decontaminate or disinfect work areas, such as surgical suites and media preparation rooms. Other chemicals are used as antiseptics to prevent infections.

Any one germicide may not be effective under all conditions or against all microorganisms. Consequently, in order to choose an effective sterilant or disinfectant it is important to understand which compounds are best to use in specific situations.

The type of microbial contamination determines which chemical control agent to use. Chemical control agents are generally effective against most vegetative bacteria, viruses, and fungi, but few are effective against endospores 94 □. Glutaraldehyde, formaldehyde, ethylene dibromide, and ethylene oxide are four of the most effective sporicides in common use. They are alkylating agents that are capable of inducing mutations, not only in microorganisms but also in humans. Consequently, these mutagens and carcinogens (cancer-causing agents) should be used carefully. The control of spore-forming bacteria is important because they are common airborne organisms that often contaminate pharmaceuticals and culture media. For these reasons, it is important to select a chemical control agent with sporicidal properties when attempting to decontaminate areas that are likely to contain these microorganisms. The tubercle bacillus, *Mycobacterium tuberculosis*, is another bacterium that is highly resistant to the action of chemical control agents and thus poses a considerable health hazard. Viruses exhibit varying resistances to chemical control agents. The hepatitis virus, the rhinoviruses (which cause the common cold), and the enteroviruses (which include the poliovirus) are among the more resistant ones.

The strength of a chemical control agent is also a factor in determining which should be used. Chemical germicides are classified on the basis of how efficiently they kill (table 7-4). All germicides classified in the **high level** category are effective against all forms of life, including bacterial endospores. High level germicides include ethylene oxide and 2% buffered glutaraldehyde. Even though these agents have a high level of activity, they may require 10 hours before they kill an entire population of endospore-forming bacteria. **Intermediate level** control agents are defined as **tuberculocidal;** that is, they are capable of killing *M. tuberculosis*. In addition, these chemical control agents are usually effective against the more resistant viruses, such as the hepatitis viruses and the rhinoviruses. Intermediate level control agents are not very effective against endospores. **Low level** chemical control agents are not effective against *M. tuberculosis*, bacterial endospores, many fungal spores, or viruses that lack a membrane. They are effective against most vegetative bacteria and fungi, however, and are used extensively in routine decontaminations because they are economical and not excessively toxic to humans. Many of these chemicals are available in supermarkets and are sold

TABLE 7-4

ANTIMICROBIAL ACTIVITY OF COMMONLY USED COLD "STERILANTS"*

MICROBIOCIDAL AGENT	RANGE OF EFFECTIVENESS					
	BACTERIA	MYCOBACTERIUM TUBERCULOSIS	ENDOSPORES	FUNGAL SPORES	VIRUSES	
Ethylene Oxide (450–800 mg/L)	+ + +	+ + +	+ + +	+ + +	+ + +	High
β-Propiolactone (1.6 mg/L)	+ +	+ +	+ +	+ +	+ +	High
Glutaraldehyde (buffered, 2%)	+	+	+ +	+	+	High
Formalin (37% aqueous formaldehyde)	+	+	+	+	+	High
Alcohol-formaldehyde (70%–8%)	+ +	+	+	+	+	Intermediate
Alcohol-iodine (70%–2%)	+ +	+	±	+	+	Intermediate
Iodine (2% to 5% aqueous)	+ +	+	±	+	+	Intermediate
Iodophors (1%)	+	+	±	±	+	Intermediate
Alcohol—ethyl (70% to 90%)	+	+	0	+	±	Intermediate
Alcohol—isopropyl (70% to 90%)	+ +	+	0	+	±	Intermediate
Phenolic derivatives (0-syl 1% to 3%)	+	+	0	+	±	Intermediate
Quaternary Ammonium Compounds	+	0	0	+	±	Low
Hexachlorophene (phenol)	+	0	0	+	±	Low
Benzalkonium chloride (1:750)	+ +	0	0	+	0	Low
Merthiolate	±	0	0	+	±	Low

+ + + superior, + +, Very good; +, good; ±, fair (greater concentration or more time needed); 0, no activity.

*Based on data from DiPalma, J. R., editor: Drill's pharmacology in medicine, ed. 4, New York, 1971, McGraw-Hill Book Co. Reprinted by permission.

as mouthwashes, disinfectants, and room deodorizers. Detergents such as quaternary ammonium compounds, Lysol® (a phenol derivative), and mercurials (such as Merthiolate) fall in this category.

Evaluating the Germicidal Activity of Chemical Control Agents

The need to control infectious organisms in hospitals and in other institutions has been an incentive for chemical industries to develop new and more effective germicides. In order to test the effectiveness of a new germicide, it is necessary to compare its activity to known, proven standards, such as the disinfectant phenol (table 7-5). There are several methods for comparing the activity of germicides, but the best known is the **Phenol Coefficient Test.** The **phenol coefficient** is a measure of the effectiveness of a germicide, and it is expressed as the ratio of the effectiveness of the test germicide to that of phenol against a test organism. For example, if a 1:250 dilution of a test germicide kills a standard population of *Staphylococcus aureaus,* while the highest dilution of phenol showing the same results is 1:60, the phenol coefficient of the test germicide is 250/60 or 4.2. This phenol coefficient means that the test germicide is 4.2 times more effective than phenol in killing *S. aureus* at least *in vitro.*

Chemical control agents can usually be placed in one of the following categories: alcohols, aldehydes, detergents, halogens, heavy metals, phenols, or alkylating gases. Some of these chemicals are extremely toxic to humans and should be used with care and in appropriate facilities. As an aid in choosing the best possible chemical control agent and understanding how they work and why they may be dangerous, some of the more important chemicals are considered.

TABLE 7-5
CHEMICAL STRUCTURES AND PHENOL COEFFICIENTS OF SOME COMMON PHENOLICS

FORMULA	NAME	PHENOL COEFFICIENT*
C_6H_5OH	Phenol	1.0
$C_6H_4(OH)CH_3$ (OH with ortho CH_3)	o-cresol (Lysol®)	2.3
$C_6H_4(OH)CH_3$ (OH with meta CH_3); $C_6H_4(OH)CH_3$ (OH with para CH_3)	m-cresol p-cresol	2.3 2.3
$C_6H_3(OH)(CH_3)_2$	2, 4 dimethylphenol	5.0
HO–C_6H_4–$C(CH_3)_3$	Butylphenol	43.7
$C_6H_3(OH)_2$–CH_2–$(CH_2)_4$–CH_3	Hexylresorcinol	313.2
(C_6HCl_3OH)–CH_2–(C_6HCl_3OH)	Hexachlorophene	125.0

*Phenol coefficients were determined at 37°C against *Staphylococcus aureus*.

Phenolic Compounds

Phenols are aromatic compounds consisting of a benzene ring with a hydroxyl group attached (table 7-5). Phenol, or carbolic acid, has long been recognized for its germicidal properties. In the late 1800s, Joseph Lister used phenol extensively as an antiseptic to treat surgical incisions and as a disinfectant to decontaminate surgical areas. Using this germicide, Lister was successful in reducing the number of post-surgical sepsis cases in the hospital where he worked. Many products currently available are derivatives of phenol (table 7-5). Most of these phenol derivatives are more germicidal than phenol, with phenol coefficients ranging from 2.3 to about 300 when *Staphylococcus aureus* is used as the test organism. Most of the phenol derivatives are used as disinfectants rather than as antiseptics. A noteworthy exception is **hexachlorophene,** which is mixed with some soaps. These soaps were once used in hospitals to wash newborn babies in order to prevent staphylococcal infections. It is now known, however, that prolonged use of 3% hexachlorophene results in brain damage in rats, so it is no longer used routinely to wash babies. Nevertheless, hexachlorophene is very effective against staphylococci and its judicious use is still recommended to control staphylococcal infections in nurseries.

Phenolics can be either germistatic or germicidal, depending upon the concentration used. If the concentration is high, phenolics are bactericidal, tuberculocidal, fungicidal, and virucidal, but they are not effective against endospores. The phenolics kill cells by disrupting the plasma membrane, resulting in the cessation of vital membrane-associated functions of the cell, and by inactivating intracellular components such as proteins and nucleic acids. **Amphyl,** a preparation used in many laboratories to disinfect work areas, contains phenolic compounds, alcohol, and soap.

Alcohols

Ethanol (CH_3CH_2OH) and isopropanol [$(CH_3)_2CH_2OH$] are very effective germicides. At concentrations of 70% to 80%, these alcohols kill fungi, most viruses, and the vegetative forms of most bacteria. Ethanol and isopropanol are widely used, primarily as antiseptics for cleansing wounds and for disinfecting the skin before injections. They are also used to reduce the microbial flora on thermometers.

Alcohols are believed to kill cells by denaturing vital proteins and by solubilizing and disrupting the integrity of the plasma membrane. Since 100% alcohol is a dehydrating agent, it can extract intracellular water and thus enhance the survival of treated cells. This fact may explain why 70% ethanol is more effective than 100% ethanol in killing cells (table 7-6).

TABLE 7-6

EFFECTIVENESS OF ALCOHOL (ETHANOL) AS A BACTERIOCIDAL AGENT[1]

CONCENTRATION OF ETHANOL[2]	TIME REQUIRED TO STERILIZE CULTURE[3] (seconds)
100%	>60
95%	20
90%	<10
80%	<10
70%	<10
60%	<10
50%	30
40%	>60
30%	>60

1. Source of data: Morton, E. H. 1950. *Annals of the New York Academy of Sciences* 51:191–196
2. Percentages refer to the volume of 100% ethanol mixed in distilled water.
3. Time of sterilization is an approximation of the time required to sterilize (no colonies formed) a standard suspension of *Streptococcus pyogenes.*

Detergents

Detergents are organic molecules that consist of hydrophobic (water-repelling) and hydrophilic (water-loving) chemical groups on the same molecule (fig. 7-8). Two examples of common detergents are **sodium lauryl sulfate** (SLS) and **Zephiran** (benzalkonium chloride). In water, sodium lauryl sulfate dissociates into a sodium ion (Na^+) and a lauryl sulfate ion (LS^-), while Zephrian dissociates into a chloride ion (Cl^-) and a benzalkonium ion (BA^+). The organic portions of these molecules have detergent properties because they bind tightly to organic materials, such as greases and oils. Detergents such as SLS, with negative charges on the organic portions of the molecule, are called **anionic detergents.** Detergents such as Zephiran with positive charges on the organic portion of the molecule, are called **cationic detergents.** Some detergents do not ionize and are therefore called **nonionic detergents.** The nonionic detergents do not possess germicidal activity.

Detergents are used primarily to cleanse surfaces by removing miroorganisms and organic matter. Detergents are microbiocidal because they emulsify and disrupt lipids that make up the plasma membrane □, causing leakage of intracellular materials and ultimately cell death. In addition to disrupting the plasma membrane, some detergents (quaternary ammonium

80

compounds) have been reported to inactivate enzymes and other cellular proteins.

Quaternary ammonium compounds (fig. 7-8) are cationic (positively charged) detergents that have considerable bactericidal and fungicidal and some virucidal activity. *Zephiran* and *Roccal* are two brand names for preparations that contain quanternary ammonium compounds and are used in many laboratories. These detergents are inactive, however, against endospores and the tubercle bacillus. Another, very important exception to the bactericidal activity of quaternary ammonium compounds is their inability to kill *Psuedomonas aeruginosa.* This bacterium is characteristically resistant to the action of quaternary ammonium compounds. Failure to recognize this fact has led to serious outbreaks of pseudomonal infections in hospitals and nurseries treated only with such disinfectants.

Sodium lauryl sulfate (anionic detergent)

$$CH_3{-}(CH_2)_{11}{-}O{-}S(=O)_2{-}O^{\ominus} \quad Na^{\oplus}$$

Zephiran (cationic detergent)

$$C_6H_5{-}CH_2{-}N^{\oplus}(CH_3)_2{-}C_{18}H_{37} \quad Cl^{\ominus}$$

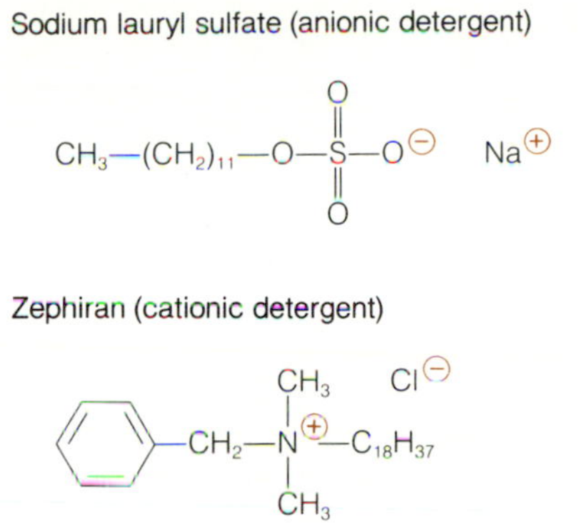

Figure 7-8 Detergents. Anionic detergents are those that have a negative charge, while cationic detergents have a positive charge.

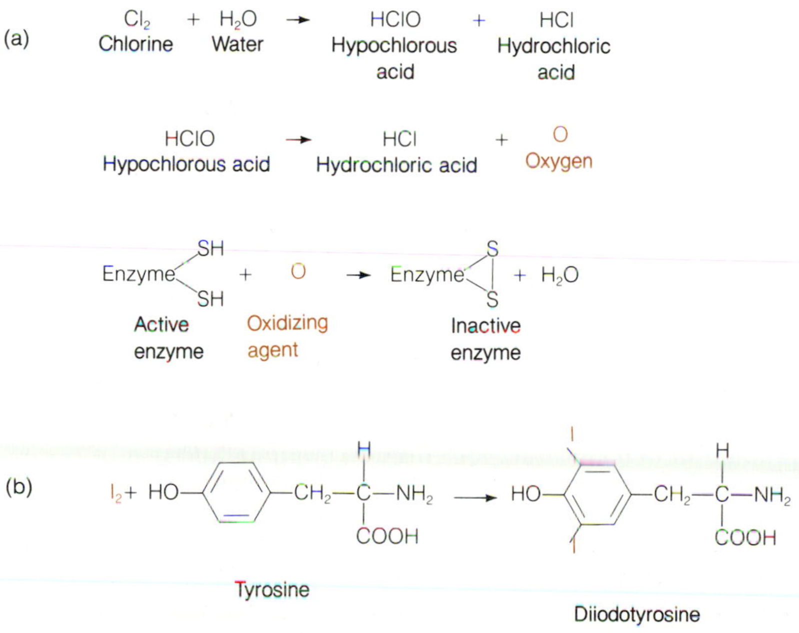

Halogens

The halogens, fluorine (F), bromine (Br), chlorine (Cl), and iodine (I), constitute a family of chemical elements with a high affinity for electrons. Because of this affinity, the halogens are extremely reactive with and toxic to most forms of life. The halogens often react with cellular components or produce powerful oxidizing agents, such as monoatomic oxygen (O), which react with various parts of the cell. For instance, halogens may oxidize the

48 double bonds in unsaturated fatty acids □ or inactivate enzymes (fig. 7-9). These effects disrupt vital functions within the organism.

Chlorine compounds are not used as antiseptics, but are routinely used as disinfectants in water supplies, hot tubs, and swimming pools. When chlorine gas (Cl_2) reacts with water it forms hypochlorous acid (HClO). The

(a)

$$\underset{\text{Chlorine}}{Cl_2} + \underset{\text{Water}}{H_2O} \rightarrow \underset{\text{Hypochlorous acid}}{HClO} + \underset{\text{Hydrochloric acid}}{HCl}$$

$$\underset{\text{Hypochlorous acid}}{HClO} \rightarrow \underset{\text{Hydrochloric acid}}{HCl} + \underset{\text{Oxygen}}{O}$$

$$\underset{\text{Active enzyme}}{\text{Enzyme}(SH)_2} + \underset{\text{Oxidizing agent}}{O} \rightarrow \underset{\text{Inactive enzyme}}{\text{Enzyme}(S{-}S)} + H_2O$$

(b)

$$I_2 + \underset{\text{Tyrosine}}{HO{-}C_6H_4{-}CH_2{-}CH(COOH){-}NH_2} \rightarrow \underset{\text{Diiodotyrosine}}{HO{-}C_6H_2I_2{-}CH_2{-}CH(COOH){-}NH_2}$$

Figure 7-9 Halogens. Chlorine (Cl_2) and iodine (I_2) react directly with various molecules, altering them chemically and interfering with vital processes. *(a)* Chlorine reacts with water to produce hypochlorous acid, which yields "reactive" oxygen (O), a very powerful oxidizing agent that readily inactivates proteins. *(b)* Iodine reacts with tyrosine residues in proteins to form diiodotyrosine. This reaction inactivates the proteins.

HClO is believed to dissociate into a reactive form of oxygen (O) that oxidizes various molecules. Hypochlorous acid may also give rise to chlorine gas (Cl_2), which reacts with cellular components. Household bleach is a 5% aqueous solution of sodium hypochlorite (NaClO). Bleach or chlorine should not be used with compounds that contain ammonium or with acids because explosive and poisonous gases (nitrogen trichloride and chlorine, respectively) are produced.

Iodine has been used for over a century as an antiseptic for treating superficial wounds. Dissolved in 70% ethanol, it is known as **tincture of iodine** and is one of the most effective antiseptics in use today. **Betadine** is also an effective iodine-containing antiseptic used to treat superficial wounds and for preparing areas of skin for surgery. Betadine consists of an iodine-containing compound mixed with a detergent. Since betadine is not painful and does not irritate the skin, it is favored over tincture of iodine as a disinfectant, especially for children. Iodine interacts with proteins by iodinating the amino acid tyrosine and forming **diiodotyrosine** (fig. 7-9). Iodination of tyrosine residues irreversibly inactivates many cellular proteins by changing their secondary and tertiary structures □.

41

Heavy Metal Ions

Heavy metal ions of arsenic, copper, mercury, and silver are toxic to most forms of life because they combine with cellular proteins and denature them. Heavy metal ions add to sulfhydryl groups in proteins and disrupt the protein structure. The mode of action of mercury (Hg) is illustrated below:

$$\text{Active protein}\begin{matrix}\diagup \text{SH} \\ \diagdown \text{SH}\end{matrix} + HgCl_2 \longrightarrow \text{Inactive protein}\begin{matrix}\diagup \text{S} \diagdown \\ \diagdown \text{S} \diagup\end{matrix}\text{Hg} + 2HCl$$

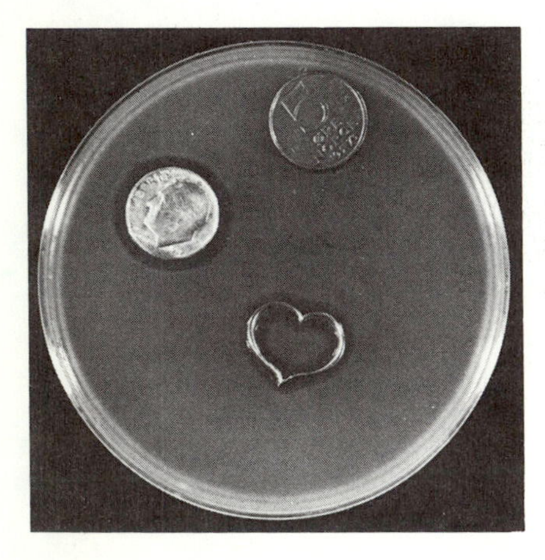

Figure 7-10 **Heavy Metals.** Many heavy metal compounds containing mercury, lead, copper, silver, and arsenic are extremely toxic to living organisms, because heavy metals react with proteins and inactivate them. The inhibition and killing of microorganisms by small amounts of a heavy metal is referred to as the oligodynamic action of the metal. The dime that contains silver and the heart with gold inhibit the growth of the bacterial lawn near them.

Some of the heavy metals (e.g., mercury and silver) are toxic even in minute quantities. This toxic property is sometimes called the oligodynamic action of metals (fig. 7-10).

Mercury, in the form of inorganic salts such as mercuric chloride, is germicidal and has been used as the active ingredient in antiseptic ointments. Its activity as a germicide is reduced by extraneous organic material; therefore, skin surfaces should be scrupulously clean before application. Merthiolate and Mercurochrome are organic mercurials that are often used as antiseptics in the home to treat minor skin abrasions and wounds.

Silver nitrate ($AgNO_3$) is an antiseptic that is sometimes used on newborn babies. Eye drops containing 1% $AgNO_3$ are used to prevent a disease called **opthalmia neonatorum,** which results from an infection of the eye by *Neisseria gonorrhoeae.* The bacterium is acquired during delivery as the baby passes through an infected birth canal. Presently, $AgNO_3$ is infrequently used on newborn babies because it is irritating and better agents are available such as the antibiotic erythromycin.

Copper sulfate ($CuSO_4$) is an effective algicide that is used to control algal blooms in lakes and reservoirs; the cupric ion (Cu^{2+}) is the active agent. Algae in drinking water is undesirable because the algae impart obnoxious flavors and odors to the water. **Bordeaux** mixtures, which also contain $CuSO_4$, are used to control fungal growth on plants. For example, **peach**

leaf curl disease, a fungal infection that causes the leaves of peach trees to curl, is prevented by several applications of copper sulfate to the peach trees before the buds open.

Alkylating Agents

Alkylating gases are chemicals that generally attach methyl or ethyl groups to cellular molecules. The alkylating gases generally cause the death of microorganisms by alkylating proteins and DNA. The degree of alkylation can be so great that proteins and DNA become totally nonfunctional. The two most commonly used alkylating gases are **ethylene oxide** and **β-propiolactone.** These chemosterilants can be used to sterilize heat-sensitive and packaged materials because they are generally capable of penetrating such materials.

Increases in labor costs for cleaning and sterilizing glassware and medical supplies have increased the demand for single-service, disposable items and have therefore increased the use of alkylating gases. Gaseous sterilizations are carried out at temperatures near 60°C for 1–10 hours. Ethylene oxide is extremely flammable, however, and β-propiolactone causes blisters when it comes in contact with the skin; thus, special precautions must be taken in order to reduce the hazards they present. Because these gases are so toxic to humans, materials that have been sterilized with ethylene oxide or β-propiolactone must be set aside in detoxification chambers for days to allow the gases to dissipate. Nevertheless, ethylene oxide–and ethylene dibromide–sterilized materials still retain residues of the gases. Ethylene oxide–sterilized petri dishes, for example, retain enough ethylene oxide to cause mutations in bacteria. Gaseous sterilizations are usually carried out in special autoclave-like chambers designed specifically for such a purpose (fig. 7-11).

Figure 7-11 Ethylene Oxide Sterilization Chamber. Hospital equipment and surgical supplies can be decontaminated in an ethylene oxide sterilization chamber.

FOCUS

DANGERS OF GASEOUS STERILANTS

In 1981 consumer groups presented evidence that ethylene oxide caused an increase in certain types of cancer and genetic damage. One consumer group sued the Occupational Safety and Health Administration (OSHA) in order to force the government agency to establish a limit of one part per million (1 ppm) per exposure and a limit of five parts per million (5 ppm) for repeated exposures in the work place. OSHA refused to issue these standards and stalled in setting any standards.

In 1982 a report in the *British Medical Journal* indicated that thousands of female hospital workers in the United States, at exposure levels of only 0.01 ppm, had a miscarriage rate three times that of the general population. This finding indicated that a limit of 1 ppm may not be sufficient. About 144,000 American workers are consistently exposed to ethylene oxide. 100,000 are affected when they use it as a sterilant in the workplace and as a fumigant in libraries, in museums, and on crops. The Public Citizen Health Research Group associated with Ralph Nader has urged the House Science and Technology Subcommittee to look into the matter and to force OSHA to set a limit for ethylene oxide exposure in the workplace.

In 1984, a number of muffin mixes were withdrawn from California stores after state tests showed that they had concentrations of ethylene dibromide ranging from 0.372 to 5.4 ppm. These muffin mixes were not withdrawn from stores in any other state. Ethylene dibromide is a known cause of cancer in laboratory animals and is a suspected cause of cancer in humans. It gets into wheat and corn products when the grains are treated with ethylene dibromide to kill insects and insect larvae. The California State Department of Health Services reported finding traces of ethylene dibromide, between 0.001 and 0.189 ppm, in 30 other baking products, but these were not removed from stores. Safe levels have not been established; nevertheless, government agencies have established that 0.030 ppm will be accepted in foods.

Alkylating solutions such as a 37% aqueous solution of formaldehyde gas (formalin) and a 2% aqueous solution of glutaraldehyde (buffered at pH 7.5–8.5) are sporicidal, tuberculocidal, virucidal, bactericidal, and fungicidal. Aldehydes inactivate cells primarily by alkylating proteins and cross-linking them, but nucleic acids and some membrane lipids apparently are also affected. Since aldehydes are such effective chemosterilants, they can be used to sterilize heat-sensitive materials such as anesthesia tubing and surgical instruments. Glutaraldehyde is more effective and less irritating to the skin than formaldehyde. In addition, glutaraldehyde rinses off easily with running water, so sterilized materials can be washed with sterile water to eliminate toxic residues. Low concentrations of glutaraldehyde are used as antiseptics, but long-term usage damages the skin.

CHEMOTHERAPEUTIC AGENTS

Chemotherapeutic agents are chemical substances that possess a high degree of antimicrobial activity and that can safely be used internally. Many of the chemotherapeutic agents used to inhibit or kill microorganisms are called **antibiotics.** These are produced by microorganisms. Most of the useful antibiotics are produced by bacteria such as *Streptomyces* and *Bacillus* and by fungi such as *Penicillium.* The antibiotics produced by bacteria and fungi are usually produced during late exponential or early stationary phase, when these organisms begin to sporulate. The chemotherapeutic agents that are synthesized by scientists are called **synthetic drugs.** Antibiotics and synthetic drugs affect the growth of microorganisms by interfering with specific cellular functions.

The best chemotherapeutic agents show **selective toxicity**, that is, they are more effective against a microbe than they are against the patient. The selective toxicity of many drugs relies on morphological or chemical differences between the host and the infecting microbe. For example, the antibiotic called **tetracycline** inhibits prokaryotic protein synthesis but not eukaryotic protein synthesis, because it binds only to bacterial 70S ribosomes. 209 Many antibiotics that specifically block protein synthesis □ in bacteria can have adverse effects on animals if taken in large doses, since they can block mitochondrial protein synthesis as well. Penicillin also exhibits high selective 81 toxicity, because it acts by inhibiting the synthesis of peptidoglycan □ and consequently affects only multiplying bacteria. Animal cells do not have cell walls, let alone any peptidoglycan, so they are not adversely affected by penicillin.

Some drugs show little selectivity in their target. This is particularly true of those drugs developed to combat eukaryotic pathogens. For example, **Amphotericin B,** a very effective antibiotic against a number of pathogenic fungi, often causes serious damage to human cells. In fact, this drug is so dangerous that it must be administered to individuals only while in the hospital. Amphotericin B acts by disrupting the plasma membrane. Its selective toxicity is based on subtle differences in chemical composition between human and fungal plasma membranes. In chapter 29 we will consider some of the chemotherapeutic drugs in more detail.

QUALITY ASSURANCE

The use of contaminated medical supplies, drugs, vaccines, or food can result in infections and death. Consequently, it is important to insure that products are free of dangerous microorganisms by having a **quality assurance** check as part of the production procedure. Quality assurance is simply a set of controls to insure that the product does what it is supposed to do and that it contains few microorganisms or is sterile. If a drug used for the treatment of a particular disease is administered by injection, not only should it be effective against the disease being treated; it should also be sterile, so that it does not cause an infection when injected. In any quality assurance program, a representative portion of the drug should be tested for sterility before it is marketed. In addition to testing the final product for sterility, it is customary to test the effectiveness of the sterilizing agents and the sterilization protocol on a periodic basis. For example, routine laboratory sterilizations using steam are carried out at 121°C for 15 to 30 minutes. Under these conditions even the most resistant of life forms should be killed. To insure that this assumption is indeed valid, a standard suspension of *Bacillus stearothermophilus* (usually 1,000,000 endospores/ml) is subjected to a routine sterilizing procedure, together with other materials such as culture media and dressings. If the standard suspension of test bacteria is sterile, it can be assumed that the sterilization procedure is working correctly. If the suspension is not sterile, then it is necessary to reevaluate the sterilization procedure. Quality assurance should be as routine in microbiology laboratories as lighting Bunsen burners or inoculating culture media.

SUMMARY

MICROBIOSTATIC AND MICROBIOCIDAL AGENTS

1. Controlling the growth of microbial populations is important because it allows us to prevent the transmission of disease, spoilage of foods, and contamination of materials.
2. Microbiostatic drugs *inhibit* the growth of microorganisms, while microbiocidal drugs *kill* microorganisms.

ANTISEPTICS AND DISINFECTANTS

1. Antiseptics are chemical compounds that can be used on the surfaces of animals and plants to kill or reduce the number of microorganisms.
2. Disinfectants are chemical compounds that generally cannot be used on the surface of animals and plants, because they are too toxic. They are usually used to kill microorganisms on the surface of inanimate objects.

DISINFECTION, SANITIZATION, STERILIZATION, AND ASEPTIC TECHNIQUES

1. Disinfection is the treatment of materials with disinfectants to eliminate or reduce the number of disease-causing microorganisms.
2. Sanitization is a procedure that reduces the number of microorganisms to a low level so that they will not be a problem.
3. A material is said to be sterile if it is free of all living microorganisms.
4. Aseptic techniques are procedures designed to prevent the contamination of sterile materials.

PRINCIPLES OF MICROBIAL KILLING

1. Populations of microorganisms, when exposed to the influence of a killing agent, exhibit a progressive decline in numbers until the culture becomes sterile.
2. Many factors affect the killing activity of control agents. Some of these factors are exposure time, age of the culture, pH, amount of organic matter present, concentration of microorganisms, temperature, and hydration.

PHYSICAL AGENTS USED TO CONTROL MICROORGANISMS

1. Microorganisms can be controlled by steaming under pressure (autoclaving), boiling, pasteurizing, incinerating, heating in ovens, refrigerating, irradiating, and filtering.
2. Steaming under pressure is the most common method of sterilizing heat-stable materials, such as glassware and most media.
3. Boiling generally is not used to sterilize materials, because some endospores are quite resistant to boiling.
4. Pasteurization is used to destroy spoilage- and disease-causing microorganisms in beverages, because generally it does not alter the taste of the beverage.
5. Incineration is used to destroy materials that are no longer wanted, but which are contaminated with dangerous microorganisms.
6. Heating in ovens is used not only to kill microorganisms but also to dry heat-stable equipment.
7. Refrigeration is used to inhibit the growth of microorganisms, but even at very low temperatures, generally it does not kill all microorganisms.
8. Radiation is often used to sterilize bench tops and the air in rooms.
9. Filtration is used to sterilize heat-labile liquid media.

CHEMICAL AGENTS USED TO CONTROL MICROORGANISMS

1. Microorganisms can be controlled by exposing them to phenols, alcohols, detergents, halogens, heavy metals, alkylating gases, and aldehydes.
2. Alcohols, such as ethanol and isopropanol, are good antiseptics and disinfectants.
3. Aldehydes, such as formaldehyde and glutaraldehyde, are infrequently used as disinfectants.
4. Detergents, such as Roccal and Zephyran, are frequently used as disinfectants to clean work spaces.
5. The halogen chlorine is used as a disinfectant in swimming pools, while the halogen-containing compound known as bleach is used as a disinfectant on floors and on work benches. Iodine-containing compounds, such as tincture of iodine and betadine, are used as antiseptics.
6. Compounds containing heavy metals, such as Mercurochrome, silver nitrate, and copper sulfate, are used as antiseptics.
7. Phenols are generally used as disinfectants, but some phenols, such as carbolic acid and hexachlorophene, can also be used as antiseptics.
8. Alkylating gases, such as ethylene oxide and propiolactone, are used to sterilize heat-labile materials.

CHEMOTHERAPEUTIC AGENTS

1. Chemotherapeutic agents are drugs that show selective toxicity for the pathogen. These drugs are used to control infections of microorganisms and can be injected into the human body.
2. Antibiotics are microbial products that are active against other microorganisms.

QUALITY ASSURANCE

1. Quality assurance programs are essential in industrial and other processes involving the control of microorganisms to insure that the products are safe to use.

STUDY QUESTIONS

1. Define the following terms: a) sepsis; b) asepsis; c) aseptic technique; d) antiseptic; e) disinfectant; f) sterilant; g) antimicrobial.
2. Give an example of both an antiseptic and a disinfectant.
3. What is the difference between a microbiostatic agent and a microbiocidal agent? Give an example of each.
4. What evidence demonstrates that the time it takes to achieve sterility is directly proportional to the number of organisms present?
5. What is the death rate constant (K)? If two populations of microorganisms have death rate constants of 1 and 5 respectively, which is killed more rapidly?
6. Initially, a population consists of 10^5 organisms. After exposure to a killing agent or 5 minutes, the population is reduced to 1 organism. What is the death rate constant? How much longer should the population be sterilized so that you are sure there is only a one in a million chance that an organism remains alive?
7. Explain how the age of a microbial population affects the rate of sterilization.
8. Explain how the hydration of the environment affects the rate of sterilization of microbial populations.
9. Explain how the concentration of an antimicrobial affects the rate of killing.
10. Explain how the pH affects the rate of killing by antimicrobials.
11. Explain how organic debris affects the rate of killing by antimicrobials.
12. Explain how temperature affects the rate of killing by antimicrobials.
13. Explain why moist heat kills microorganisms more efficiently than dry heat.
14. Why does freezing at low temperatures preserve many microorganisms?
15. How does refrigeration control microbial growth?
16. Equal portions of a microbial population are heated to 100°C, 121°C, 150°C, and 200°C. These portions become sterile after 60 minutes, 15 minutes, 10 minutes, and 5 minutes, respectively. What is the thermal death point for this population?
17. Define thermal death time and decimal reduction time.
18. List the methods used to sterilize materials, and discuss any problems associated with the following methods: a) steam under pressure; b) boiling; c) incineration; d) hot ovens; e) radiation; f) ethylene oxide.
19. What is pasteurization, and how does it differ from sterilization?
20. List the methods used to control the growth and multiplication of microorganisms in foods, and discuss problems associated with the following methods: a) pasteurization; b) refrigeration; c) desiccation (drying); d) immersion in hypertonic environments.
21. Explain the difference between high level, intermediate level, and low level chemical control agents.
22. Explain the phenol coefficient test.
23. Under which categories (antiseptic, disinfectant, sterilant, chemotherapeutic agent) would the following be placed? a) Isopropanol and ethanol. b) Formaldehyde and glutaraldehyde. c) sodium lauryl sulfate. d) Roccal and Zephiran. e) Tincture of iodine. f) Sodium hypochlorite. g) Arsenic. h) Mercuric chloride. i) Silver nitrate. j) Copper sulfate. k) Hexachlorophene. l) Ethylene oxide.
24. Discuss the importance of selective toxicity.
25. Describe a typical quality assurance check.

SUPPLEMENTAL READINGS

Baldry, P. 1977. *The battle against bacteria.* Cambridge, England: Cambridge University Press.

Ball, A. P., Gray, J. A., and Murdoch, J. M. 1978. *Antibacterial drugs today.* Baltimore: University Park Press.

Block, S. S. ed. 1977. *Disinfection, sterilization, and preservation.* 2d ed. Philadelphia: Lea & Febriger.

Castle, M. 1980. *Hospital infection control.* New York: John Wiley.

Laskin, A. I. and Lechevalier, H. 1984. *CRC Handbook of Microbiology.* 2d ed. Vol. VI. *Growth and Metabolism.* Florida: CRC Press Inc.

McInnes, B. 1977. *Controlling the spread of infection.* 2d ed. St. Louis: Mosby.

Spaulding, E. H. and Gröschel, D. H. M. 1985. Hospital disinfectants and antiseptics. *Manual of Clinical Microbiology,* 4th ed. Lennette et al., Eds, American Society for Microbiology, Washington D.C.

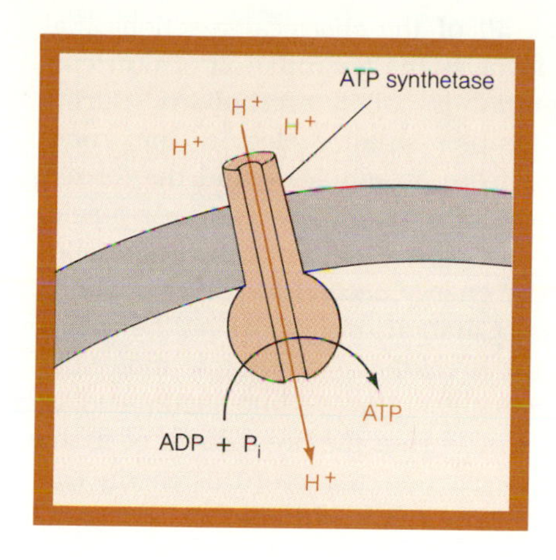

CHAPTER 8
MICROBIAL METABOLISM

CHAPTER PREVIEW

PRODUCTION OF NATURAL GAS BY BACTERIA

Methane, also known as natural gas, is an important energy source for human populations—especially now, in view of dwindling petroleum reserves and the high cost of oil. What makes methane such an attractive alternative to petroleum is that methane is a resource that can be made from cheap starting materials.

Methane production occurs naturally in swamps, black muds, lake sediments, and sewage treatment plants. The starting products include methanol (wood alcohol), various organic acids, and carbon dioxide, all of which are readily available waste products of microbial activities. Plant and animal wastes as well as vegetation have been found to be suitable starting materials for the large-scale production of methane.

Sewage treatment plants produce ample supplies of methane, which is used to heat the plant or to operate machinery that would otherwise use gasoline, diesel fuel, or electricity. The methane that is formed in sewage treatment plants develops in large tanks called **anaerobic sludge digestors,** which are filled with solid wastes collected during the treatment of sewage. The methane produced in the sludge digestors is due to the metabolic activities of many different types of anerobic and facultative anaerobic microorganisms that break down the organic matter, converting it to methane.

The organisms that actually produce the methane are the **methanogenic bacteria.** These bacteria, one of several groups of **archaebacteria,** represent the descendants of life that dominated the early earth, where the environment was anaerobic and simple organic molecules were plentiful.

The methanogenic bacteria obtain energy for growth by oxidizing simple organic molecules or H_2 in the presence of CO_2. The waste product of their metabolism is methane. This gas is insoluble in water, so it escapes to the atmosphere, making it easy to collect.

The power industry is taking advantage of methanogenic bacteria to produce a cheap energy source, and in the near future, a significant proportion of the energy consumed in the United States will come from the action of methanogenic bacteria on organic waste products. In order to optimize the yield of methane it is essential that the metabolism of methanogenic bacteria, as well as that of the other microorganisms that participate in methanogenesis, be fully understood. This chapter will discuss some important concepts of microbial metabolism and how microogranisms use energy to carry out life processes.

LEARNING OBJECTIVES

A STUDY OF THIS CHAPTER SHOULD ENABLE YOU TO:

EXPLAIN WHAT ENZYMES ARE AND HOW THEY FUNCTION

DESCRIBE HOW NUTRIENTS ARE TRANSPORTED INTO THE CELL

EXPLAIN THE VARIOUS WAYS IN WHICH MICROORGANISMS CAN MAKE ATP

DISCUSS HOW ATP AND NAD ARE USED BY THE CELL IN ENERGY METABOLISM

SUMMARIZE THE FUNCTIONS OF THE EMBDEN-MEYERHOF PATHWAY, THE KREBS CYCLE, AND THE ELECTRON TRANSPORT SYSTEM IN CATABOLISM

EXPLAIN THE DIFFERENCES BETWEEN FERMENTATION AND RESPIRATION

SUMMARIZE THE DIFFERENCES BETWEEN CHEMOORGANOTROPHIC AND CHEMOLITHOTROPHIC METABOLISM

DISCUSS THE BASIC DIFFERENCES BETWEEN THE PHOTOSYNTHETIC PROCESSES CARRIED OUT BY THE PHOTOGROPHIC BACTERIA AND THE CYANOBACTERIA

Metabolism is a term used to describe all of the chemical reactions that take place in the cell. Metabolism that involves the assimilation of nutrients and the construction of new cell material is called **anabolism.** During anabolism, or biosynthesis, the cell assembles small molecules into more complex ones. The formation of proteins from amino acids and the formation of DNA from nucleotides are examples of anabolic processes. As a rule, the assemblage of small molecules into biological polymers during anabolism requires that energy be provided. The energy needed to power anabolic reactions is generated during **catabolism.** Catabolism includes all the chemical reactions of the cell that result in the breakdown of organic molecules into simpler ones. The breakdown of the sugar glucose into lactic acid or carbon dioxide is an example of catabolism. During the breakdown of many organic molecules (*e.g.*, glucose), energy is released. Some of this energy can be conserved in nucleotides, such as **adenosine triphosphate** (ATP) or **guanosine triphosphate** (GTP), and can subsequently be used to power anabolic reactions. In this chapter we will study the various ways in which microorganisms conserve energy in ATP molecules, and how these molecules are used to power biosynthetic reactions and ultimately cell reproduction.

METABOLISM, AN OVERVIEW

Microorganisms can extract energy from foods and conserve it in molecules of ATP. This extraction of energy requires the modification of the food (fig. 8-1). The first step toward obtaining energy is sometimes called **digestion,** which involves the breakdown of large molecules into simpler ones. For example, the digestion of proteins yields amino acids, while the digestion of starch yields simple sugars. Bacteria and many eukaryotic microogranisms cannot bring large molecules into the cell and hence must digest them extracellularly. This digestion is accomplished by the secretion of exoenzymes □ 109 that catalyze the breakdown of materials outside the cell. Examples of exoenzymes are **proteases** (protein-digesting enzymes) **amylases** (starch-digesting enzymes), and **lipases** (lipid-digesting enzymes). Sometimes the exoenzymes released by microorganisms in the human body digest host tissues and cause severe damage and disease. The second step toward obtaining energy consists of the transport of small molecules into the cell. Most molecules require transport mechanisms that use cellular energy. The third step in the extraction of energy from foods involves the degradation of small organic molecules into simpler molecules. For example, the degradation of glucose gives rise to molecules of pyruvate or acetate. During this stage, organism molecules are oxidized and the electrons are donated to electron carriers (coenzymes), such as nicotinamide adenine dinucleotide (NAD) or flavin adenine dinucleotide (FAD). Additionally, during this stage some of the energy in the organic molecules is conserved in molecules of ATP.

The fourth stage in the catabolism of nutrients involves the complete oxidation of pyruvate or acetate to carbon dioxide. The complete oxidation of these molecules results in the reduction of additional molecules of NAD and FAD, which are then used in an electron transport system (ETS) that is coupled to the synthesis of ATP. Thus, much of the energy in organic molecules is conserved in molecules of ATP.

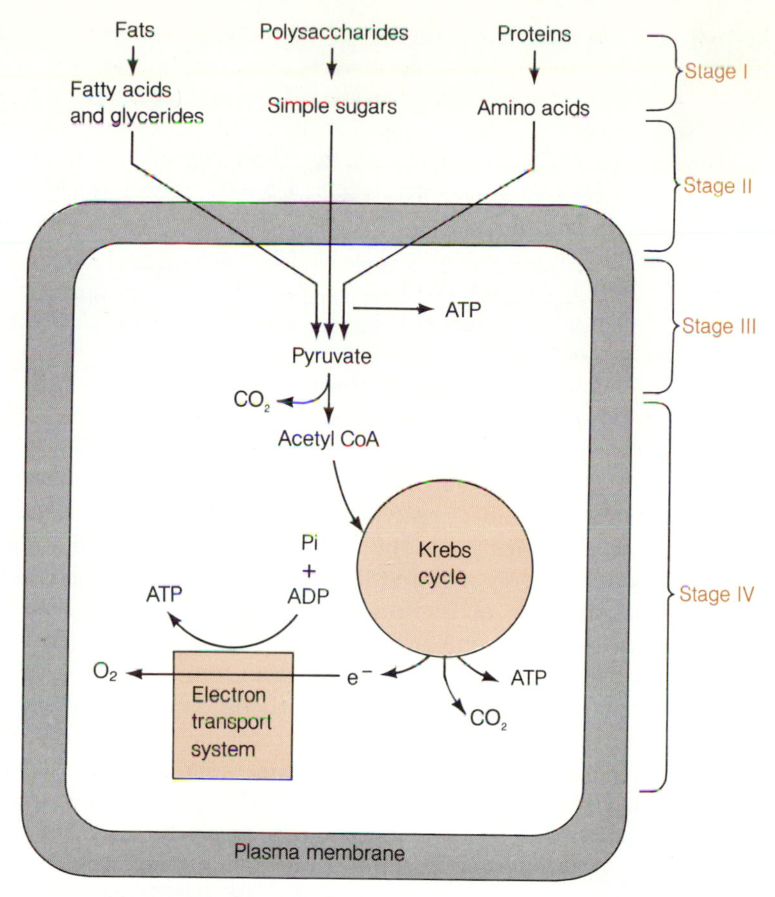

Figure 8-1 Stages in the Metabolism of Foods. Stage I involves the breakdown of large molecules into small ones. Stage II involves the transport of small molecules into the cell. Stage III involves the partial oxidation of small molecules, with the production of chemical energy (ATP). Some microorganisms cannot catabolize foods further. Stage IV involves the complete oxidation of food molecules and the subsequent production of large quantities of ATP.

ENZYMES AND HOW THEY WORK

54

Enzymes are biological catalysts □ that participate in most cellular reactions. In the absence of enzymes, cellular reactions would be very slow and uncoordinated. Enzymes speed up chemical reactions without becoming altered during the process. Because they are not altered, enzyme molecules can be reused by the cell, sometimes as often as 100,000 times each second. Enzymes are given names that are descriptive of their activity. Each enzyme has a **recommended** name and a **systematic** name. The recommended name is a convenient name that is constructed by adding the suffix *ase* to the name of the substrate on which the enzyme acts. For example, the recommended name of an enzyme that catalyses the breakdown of a lipid is *lipase* and that of a protein-hydrolyzing enzyme is *protease*. The systematic name of an enzyme describes the enzyme unambiguously according to the reaction it catalyzes. On the basis of systematic names, there are six classes of enzymes. The six classes and the types of reactions they catalyze are outlined in table 8-1.

TABLE 8-1 SYSTEMATIC CLASSIFICATION OF ENZYMES	
SYSTEMATIC NAME	ACTIVITY
Oxido-reductases	Catalyze oxidation-reduction reactions
Transferases	Transfer group of atoms from one compound to another
Hydrolases	Catalyze hydrolysis of ester, glycosidic, and peptide bonds
Lyases	Catalyze addition of water, ammonia, and carbon dioxide to compounds; also the removal of these substances to form double bonds
Isomerases	Catalyze isomerization reactions
Ligases	Catalyze the formation of covalent bonds

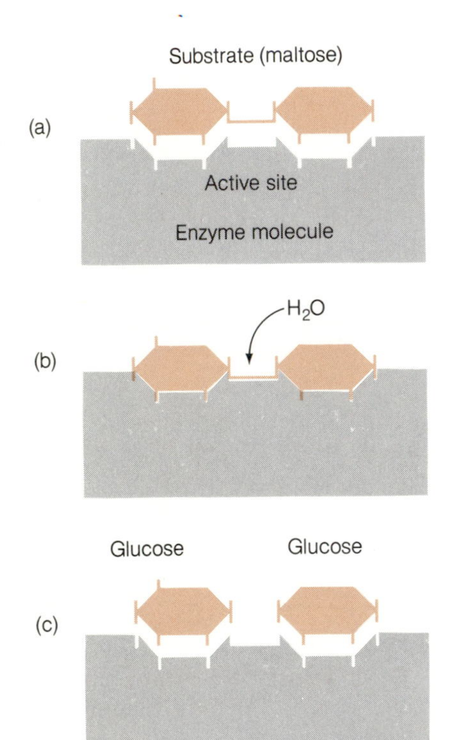

Figure 8-2 Lock and Key Explanation of Enzyme Activity. *(a)* A molecule of maltose (two glucose molecules covalently bonded) coming into the active site of the enzyme. Notice that the active site has a shape that is complementary to the shape of the substrate. *(b)* Enzyme-substrate complex. *(c)* Enzyme catalyzes reaction and the products (two glucose molecules) are formed and released from the active site. Enzyme is now available to catalyze the hydrolysis of another molecule of maltose.

In order to have a chemical reaction, an intermediate state, called the **activated state,** must be achieved. The activated state requires the input of energy. The amount of energy required for a substrate molecule to reach the activated state is called the **activation energy** □. Activation energy is expressed in calories* per mole and is the amount of energy required to bring all the molecules of one mole of a substrate (at a given temperature) to the activated state. Enzymes speed up cellular reactions because they lower the activation energy requirement. These catalysts combine with their substrate to form an activated state with less energy than the activated state in the absence of the enzyme.

53

The function of an enzyme is determined largely by its three-dimensional shape (fig. 8-2). Substrate molecules bind to specific sites, called **active sites.** It has been suggested that the active site may have a shape that is complementary to the shape of its substrate. A popular way of expressing this complementarity is by using a "lock and key" model of enzyme activity, where the substrate (the key) fits into the active site of the enzyme (the lock). This concept is illustrated in figure 8-2. Since it is the shape of the active site that determines the specificity of an enzyme, any changes that occur in the three-dimensional shape of the active site will affect the catalytic ability of the enzyme.

How Enzymes Work

Many activities that take place in the cell involve the breaking of some bonds and the formation of others. In the absence of enzymes, these reactions would be very sluggish because the activation energy required is far greater than that available at normal cell temperatures. By lowering the necessary activation energy, enzymes make it possible for biological reactions to take place rapidly at cell temperatures. Exactly how enzymes reduce the activation energy is not well understood, although a simplified explanation has been advanced. When an enzyme binds its substrate, it may produce stresses on the substrate's bonds. These stresses are a form of activation energy that facilitate the breakage of bonds. Once the bonds are broken the resulting compound rearranges itself to form the end product. The formation of mol-

*A calorie (cal) is the quantity of energy that will raise the temperature of 1 gram of water from 14.5°C to 15.5°C.

ecules from substrates requires that the component atoms (or groups of atoms) collide to form new bonds. When these atoms are held fast by enzymes and brought into close proximity, the likelihood of collision and formation of a new bond is increased. In this way, enzymes not only help in reducing the energy required to break the reactants' bonds but also enhance the likelihood that the reactants will come together to form a new bond. Together, these factors are responsible for the catalytic property of enzymes.

Factors That Influence Enzyme Activity

144

In Chapter 6 we discussed some of the factors that influence microbial growth □ and saw how these factors could alter the rate of population growth. It is easier to understand these influences if we realize that cellular reproduction is the outcome of many different enzymes working more or less synergistically within the cell. Hence, any factor that influences enzyme activity will ultimately influence cellular reproduction.

Temperature

Temperature is a measurement of the amount of heat in an environment. All enzymes require a certain amount of heat in order to be active. As the amount of available heat increases, the kinetic energy of the substrate also increases, thus speeding up the rate at which the enzyme catalyzes the conversion of reactants into products. Increased heat also increases molecule motion and therefore the number of molecules entering the activated state. As a general rule, for every 10°C increase above a minimum temperature, the enzyme activity doubles. The enzyme activity increases at this rate until the temperature reaches an **optimum** point. Above this optimum temperature, the increased amount of kinetic energy in the form of heat causes weak bonds (*e.g.*, hydrogen bonds) within the enzyme to break, causing structural changes in the enzyme. These changes usually reduce the efficiency at which the enzyme catalyzes the reaction. As the temperature is raised from an optimum temperature, the enzyme progressively diminishes in activity until a **maximum** temperature is reached, above which the enzyme becomes **denatured** and nonfunctional.

Substrate Concentration

The speed at which enzymes catalyze reactions is dependent on the concentration of substrate present, and can easily be determined by measuring the rate of product formation. The speed depends upon how fast the substrate is bound by enzymes; hence, the availability of substrate determines the velocity of the reaction, given a fixed amount of enzyme. If the initial velocity of an enzyme-catalyzed reaction is measured at various concentrations of substrate, a curve similar to that illustrated in figure 8-3 will likely be obtained. At low substrate concentrations, the enzyme has not attained its maximum rate of conversion because it is relatively difficult for the enzyme to find a substrate molecule with which to react. As the concentration of substrate increases, the enzyme velocity also increases, up to a point at which a maximum rate of conversion is attained. At this point (saturation point), all the active sites on the enzyme are constantly bound to substrate molecules. Any further increase in the concentration of substrate in the environment will no longer affect the velocity of the reaction, because the enzymes are already working as fast as they can. Microbial populations respond in much the same way to increased nutrient concentrations □, because the pro-

144

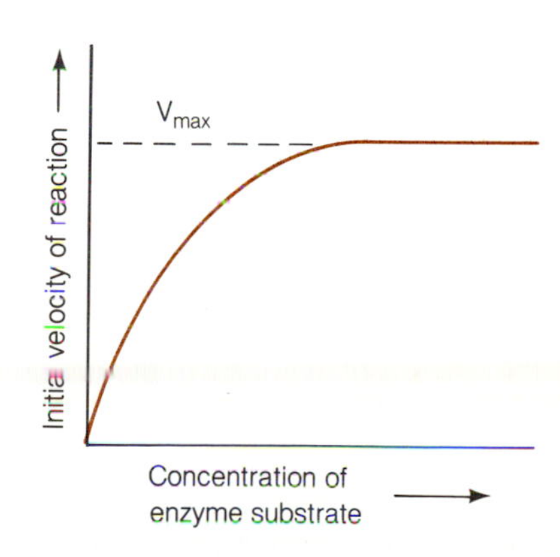

Figure 8-3 Effect of Substrate Concentration on Enzyme Activity. The enzyme activity increases gradually as the substrate concentration increases, up to a point at which all the active sites are constantly filled (V_{max}). Beyond this point, the increase of substrate concentration does not affect enzyme activity.

teins involved in transporting nutrients inside the cell work essentially like the enzymes considered here.

pH

The environmental pH also affects the rate at which enzymes catalyze reactions, because the hydrogen ion concentration affects the three-dimensional structure of the enzyme □ and therefore its activity. Every enzyme has an optimum pH at which its three-dimensional structure is most conducive to binding of the substrate. If the concentration of hydrogen ions is changed from this optimal concentration, the enzyme's activity progressively diminishes until the enzyme becomes nonfunctional. Cells regulate their internal pH within the range that is optimum for activity of their enzymes.

54

NUTRIENT TRANSPORT

Nutrients and energy sources found in the environment are used by microorganisms for growth and other biological activities. Before these nutrients can do any good, however, they must be brought into the cell. A few of these nutrients (e.g., water and glycerol) pass readily into the cell by diffusion, but the plasma membrane is impermeable to most ions and molecules. This is a desirable property, because otherwise many of the cell's essential components would leak out.

Living organisms accumulate nutrients and discharge cellular wastes in a number of ways. Figure 8-4 summarizes the most important processes involved in moving chemicals across membranes: **passive diffusion, facilitated diffusion,** and **active transport.** In passive and facilitated diffusion (figs. 8-4a and 8-4b), molecules pass through the cell membrane from areas of high concentration to areas of low concentration. Neither of these mechanisms can account for the transport of materials against a concentration gradient. Cells do not expend energy in transporting molecules into and out of the cell by passive or by facilitated diffusion; however, these mechanisms

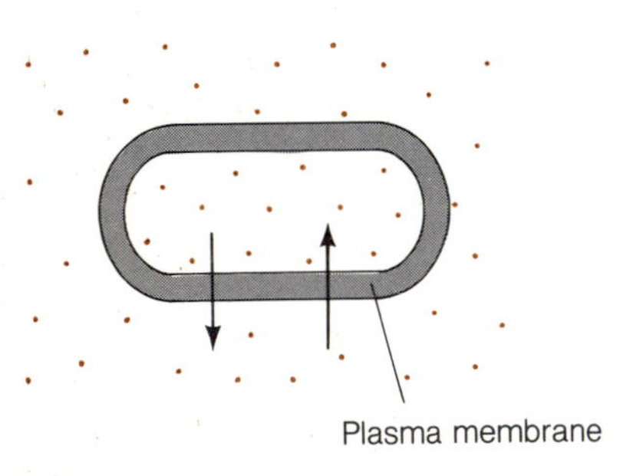

(a) Simple diffusion

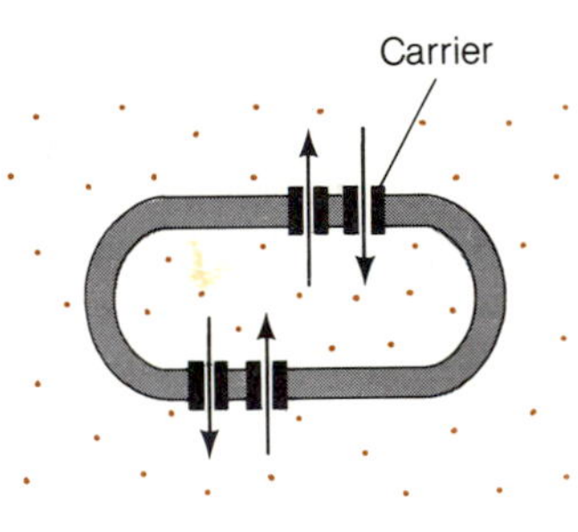

(b) Facilitated diffusion

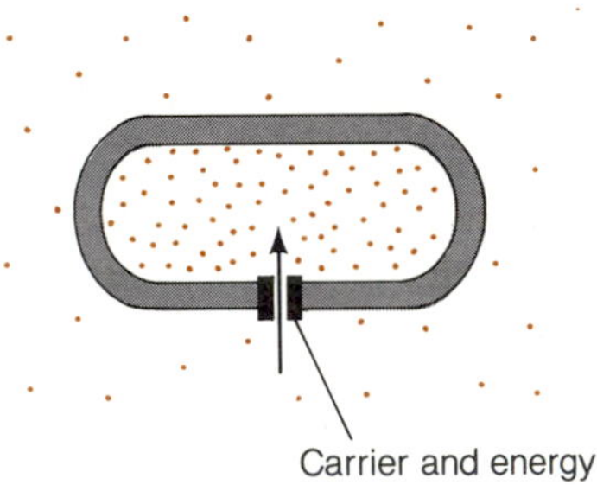

(c) Active transport

Figure 8-4 Transport of Nutrients Across Biological Membranes. *(a)* Simple diffusion. No energy expenditure by the cell is required. The rate at which nutrients pass into the cell is directly proportional to the difference in concentration between the environment and the interior of the cell. The concentration of nutrients inside the cell never exceeds that of the environment. *(b)* Facilitated diffusion. No energy expenditure by the cell is required. Carrier proteins transfer nutrient molecules from the environment into the cell and increase the rate of transport into the cell. The concentration of nutrients inside the cell never exceeds that of the environment. *(c)* Active transport. Requires that the cell use up energy. Proteins called permeases transfer nutrients from the environment into the cell. The rate of transfer is increased (as in facilitated diffusion). Because of the expenditure of energy (ATP $\rightarrow$ ADP + Pi), it is possible for the cell to concentrate nutrients, even from areas of lower solute concentration to areas of higher concentration.

do not allow the concentration of nutrients inside the cell to exceed that outside the cell. Thus, the cell makes only limited use of diffusion as a means of transporting nutrients into the cell.

Most of the nutrients that enter the cell are concentrated by active transport (fig. 8-4c). During active transport, the cell expends energy to bring nutrients into the cell. The energy derived from the hydrolysis of ATP, for example, can be used to bring nutrients into the cell against a concentration gradient. In this way, the cell can accumulate nutrients inside the cell at a concentration far exceeding that outside the cell.

BIOENERGETICS

Not all chemical reactions result in products with less energy than the reactants (i.e., exergonic reactions). Some chemical reactions catalyzed by enzymes absorb or retain energy so that some of the products have more energy than initially (i.e., energonic reactions). For example, the synthesis of cellular components such as peptidoglycan, proteins, and polysaccharides are all energy-requiring reactions, even though they are catalyzed by enzymes. When reactions like these take place in the cell, they often occur concurrently with an exergonic reaction such as the hydrolysis of ATP. The endergonic biosynthetic reaction is said to be **coupled** (fig. 8-5) to the energy-yielding reaction. Some of the energy released by the exergonic reaction is used to allow the endergonic reaction to take place.

Instead of having a variety of coupled reactions, the cell usually employs a few energy-yielding reactions to couple with endergonic reactions. Phosphorylated nucleotides, such as ATP and GTP, are generally employed as the source of energy. ATP is most commonly used by the cell as the "energy currency" for reactions that require energy. Thus, ATP serves to couple energy-yielding reactions with energy-requiring, biosynthetic reactions.

Bioenergetics is a field of study that is concerned with the production and use of energy by living organisms. The energy available to these organisms may be in the form of electromagnetic radiation or stored in chemical

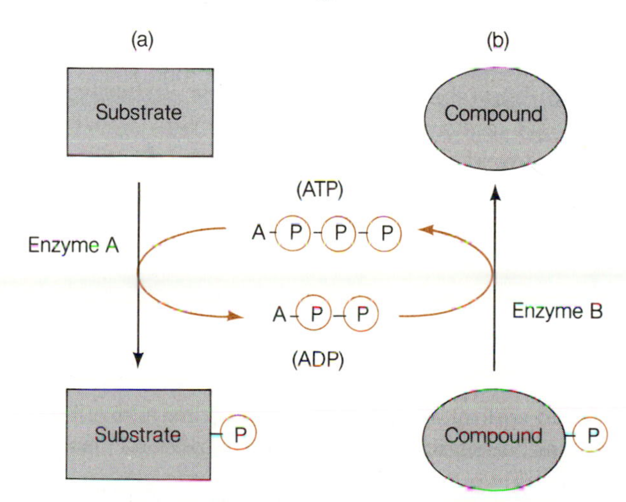

Figure 8-5 An Example of a Coupled Reaction. Reaction *(a)* shows how a substrate may be phosphorylated (a phosphate group added) by ATP. Reaction *(b)* shows how a phosphorylated compound can be used by the cell to make ATP. These types of reactions are said to be coupled by ATP.

bonds of nutrient molecules. Part of this energy is conserved in phosphorylated nucleotides so that it can be used at a later time to power energy-requiring cellular activities.

Chemotrophic organisms obtain their energy from the oxidation of molecules. Most microorganisms use organic molecules as the electron donor in energy-yielding metabolism and are classified as **chemoorganotrophs** □. 112 Certain bacteria, however, use inorganic molecules such as ammonia, molecular hydrogen, and hydrogen sulfide as their electron donors. These bacteria are called **chemolithotrophs.** Regardless of the source of electrons, the vast majority of chemotrophic organisms obtain and store their energy by similar chemical processes.

The principal energy storage molecule used by the cell is ATP. This molecule consists of the sugar ribose, the heterocyclic nitrogen base adenine, and three phosphate groups linked together by **anhydride linkages** (fig. 8-6). The molecule of ATP is charged because the double bonded oxygen atoms attract electrons, thus becoming partially electronegative, while the phosophorus atoms become partially electropositive. The negative electrical charges in the oxygens make the ATP molecule rather unstable, and it is particularly susceptible to hydrolysis at the anhydride linkages. When the molecule breaks at these locations, it releases **free energy** in the reaction:

$$\text{ATP} + H_2O \longrightarrow \text{ADP} + \text{Pi} + \text{energy}$$

This reaction releases about 7.3 kilocalories (kcal) of energy per mole and can be used to power other reactions. This property makes ATP admirably well suited to be the energy currency for the cell.

If the hydrolysis of ATP by the cell results in the net output of free energy, then the synthesis of ATP must require that energy be furnished. The energy required by the cell comes from metabolism and the oxidation of chemicals. The resulting intermediate (**phosphorylated substrates**) and **electrons** provide the energy necessary to synthesize ATP. Among microorganisms there are three basic mechanisms of ATP synthesis: a) substrate-level phosphorylation; b) oxidative phosphorylation; and c) photophosphorylation.

Substrate-level phosphorylation is the name given to the synthesis of ATP when a phosphorylated substrate donates its phosphate group to ADP (fig. 8-7). For example, during the biochemical reaction in which 1,3 diphos-

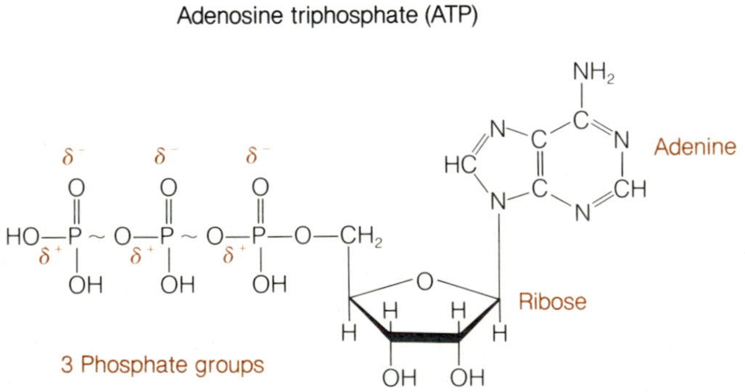

Adenosine triphosphate (ATP)

3 Phosphate groups — Ribose — Adenine

Figure 8-6 Chemical Structure of ATP. A ~ indicates a high-energy phosphate bond, a δ shows partial charges of P and O atoms.

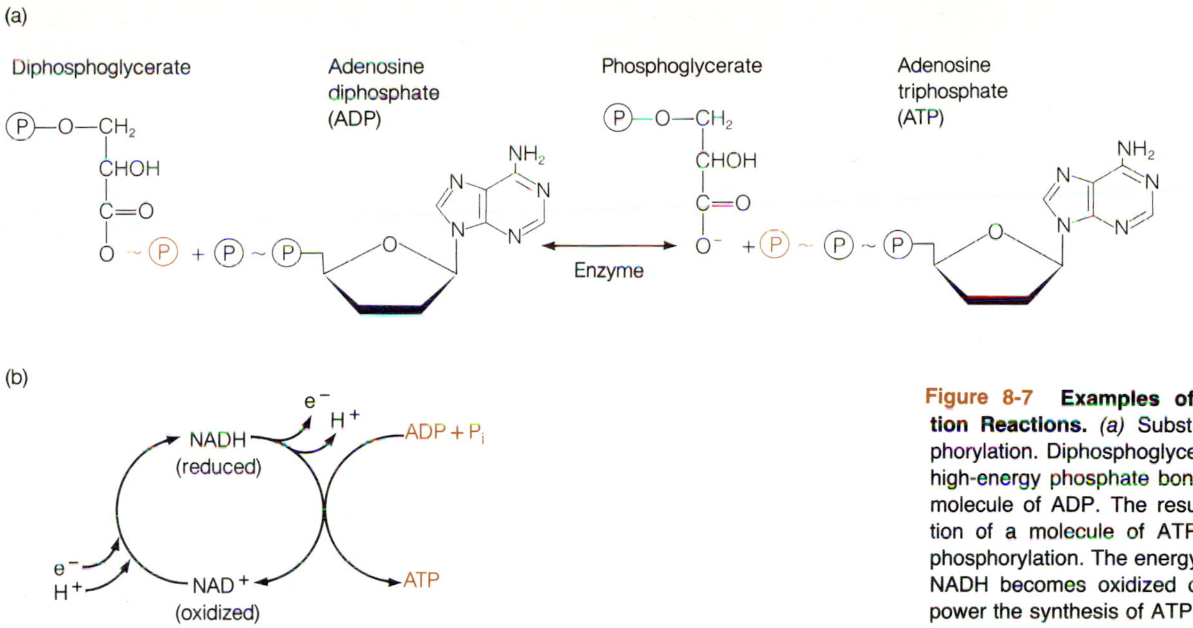

Figure 8-7 Examples of Phosphorylation Reactions. *(a)* Substrate-level phosphorylation. Diphosphoglycerate transfers a high-energy phosphate bond (colored) to a molecule of ADP. The result is the formation of a molecule of ATP. *(b)* Oxidative phosphorylation. The energy released when NADH becomes oxidized can be used to power the synthesis of ATP.

phoglyceric acid (1,3-diPGA) is converted into 3-phosphoglyceric acid, in the presence of ADP, the anhydride linkage in the phosophate group is preserved and transferred to ADP to form ATP (fig. 8-7).

Oxidative phosphorylation is the name given to the synthesis of ATP when electrons from an oxidized substrate are used. During oxidative phosphorylation, electrons are passed from one electron carrier (e.g., cytochromes) to another in a membrane-associated **electron transport system** 192 □. The free energy liberated during the transport of electrons from one carrier to another is sufficient to power the synthesis of several molecules of ATP (fig. 8-7).

Photophosphorylation is the name given to the synthesis of ATP when electrons from a chlorophyll are used. The synthesis of ATP takes place in essentially the same way as oxidative phosphorylation, except that solar energy is used to oxidize light-sensitive pigments such as chlorophylls 91 □. The electrons from the light-sensitive pigments are passed through an electron transport system in much the same way as in oxidative phosphorylation. The passage of electrons from one carrier to another also results in the release of sufficient free energy to power the synthesis of several ATP molecules.

In Chapter 2 we learned that for every oxidation there is a concurrent, equivalent reduction. We know that during ATP synthesis some chemicals become oxidized while others become reduced. The chemicals that are oxidized can be light-sensitive pigments or food molecules. The chemicals that are reduced can be other organic molecules or, as is frequently the case, **coenzymes** such as **nicotinamide adenine dinucleotide** (NAD) or **flavin adenine dinucleotide** (FAD).

Many coenzymes used by the cell become reduced and hence are electron acceptors (fig. 8-8). Coenzymes such as NAD, NADP, FAD, and **flavin mononucleotide** (FMN) are commonly used as electron carriers in metabolic reactions. For example, prior to oxidative phosphorylation the cell oxidizes chemical substrates, reducing electron carriers such as NAD during the

Figure 8-8 Nicotinamide Adenine Dinucleotide (NAD). The coenzyme NAD is used by the cell in oxidation-reduction reactions during catabolism and is a major free energy carrier in the cell (reducing power). It exists in the cell in two different forms. Its oxidized form, NAD^+, has a positive charge on the nitrogen atom. When NAD is reduced to NADH, it gains an H atom and an electron so it loses its positive charge.

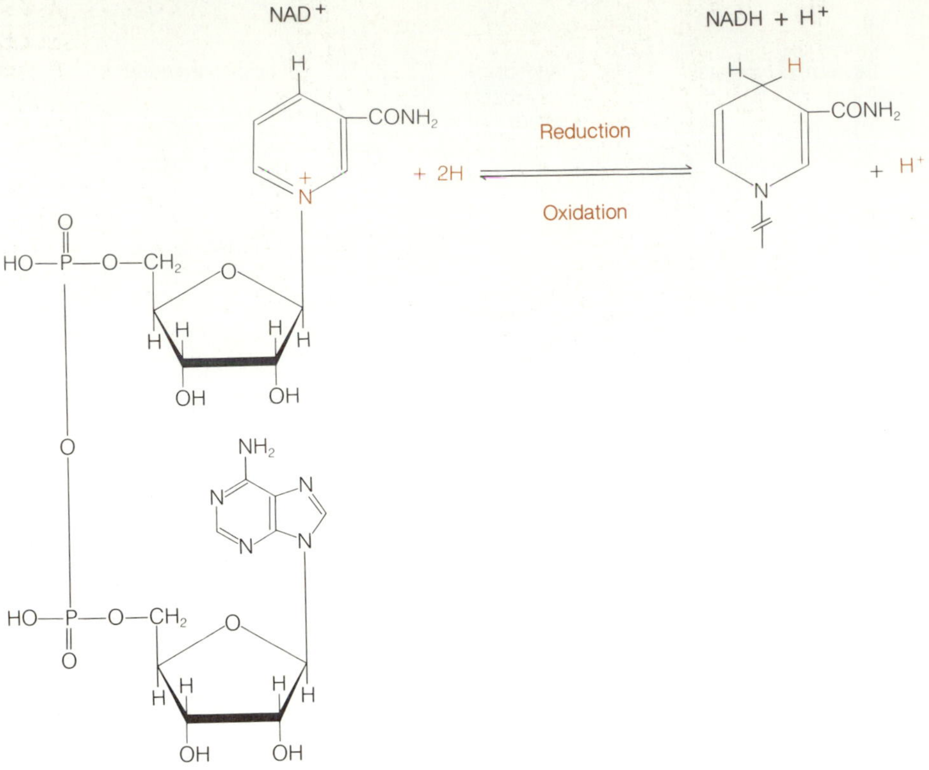

process. These, in turn, carry the electrons to an electron transport system where ATP is synthesized. Hence, the cell extracts electrons from molecules and uses them as the energy source to power the synthesis of ATP, which is the source of energy to power all other cellular reactions.

CHEMOORGANOTROPHIC METABOLISM

All eukaryotic cells, and a large number of bacteria, carry out chemoorganotrophic metabolism in which organic molecules are oxidized. The free energy released from this oxidation is partly conserved in molecules of ATP synthesized by substrate level and/or oxidative phosphorylation. Although many organic molecules can be used as the electron source for chemoorganotrophic metabolism, carbohydrates are especially widely used for this purpose. One reason is that carbohydrates, which are readily available in nature, have a large amount of potential energy in their bonds that can be used to power the synthesis of ATP. For example, the oxidation of 1 mole of glucose to carbon dioxide and water releases approximately 688 kcal (1 kcal = 1,000 calories). Some of this energy is conserved and can yield as many as 38 moles of ATP.

The oxidation of glucose in chemoorganotrophic metabolism can be broken down into three stages. The first stage involves the oxidation of glucose to 2 pyruvates, resulting in the net production of 2 ATP and 2 NADH + $2H^+$. This first stage is also called the **Embden-Meyerhof pathway** (glycolysis). The second stage involves the oxidation of the pyruvates to carbon dioxide and the production of 8 NADH + $8H^+$ + 2 $FADH_2$ + 2 GTP. This stage is called the **Krebs cycle**, tricarboxylic acid cycle, or citric acid cycle. The third stage involves the oxidation of NADH + 4 + and $FADH_2$ in an **electron transport system and** the subsequent production of 34 ATP. All three stages together are called **respiration.** Some microorganisms can carry out only the oxidation of glucose to pyruvate and then a subsequent reduction of pyruvic acid. This process, called fermentation, results in the production of 2 ATP/glucose.

The Embden-Meyerhof Pathway

The Embden-Meyerhof pathway (fig. 8-9), also called **glycolysis,** has been studied in considerable detail. By 1940, through the efforts of Gustav Embden, Otto Meyerhof, Carl Neuberg, Jacob Parnus, and others, all the major

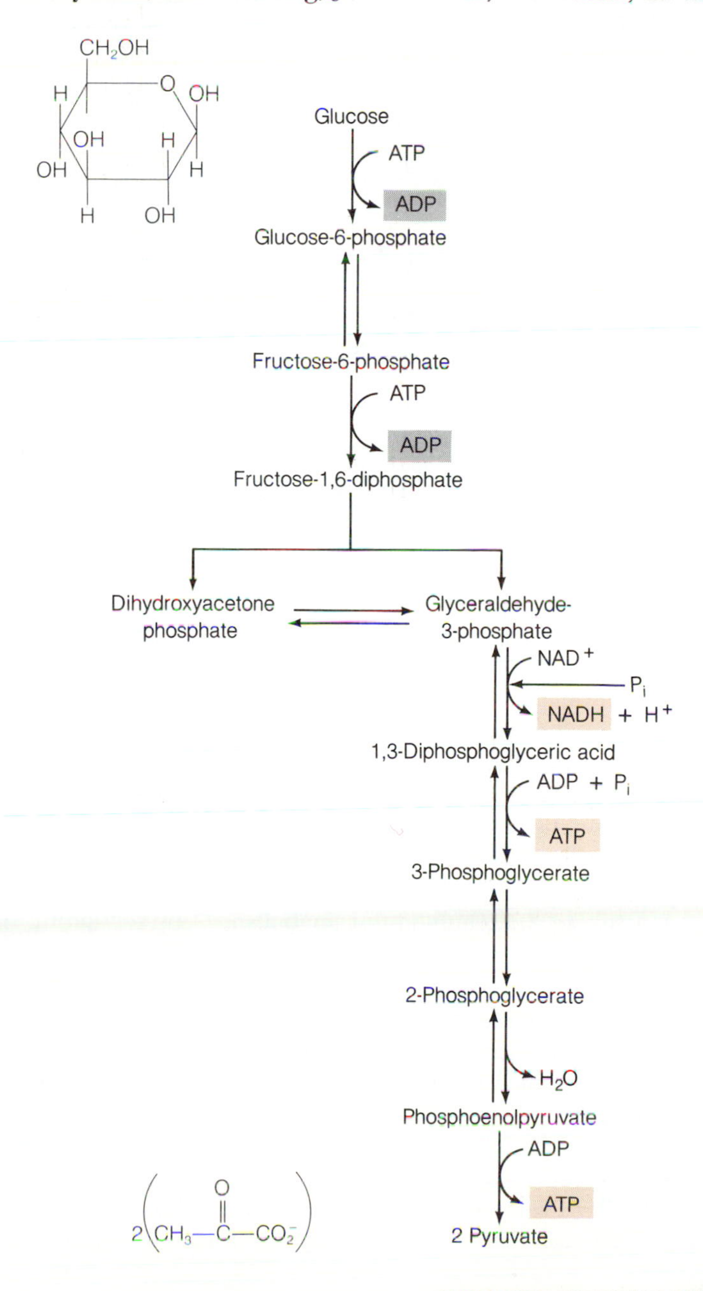

Figure 8-9 The Embden-Meyerhof Pathway of Glycolysis

steps in the pathway were known. The Embden-Meyerhof pathway consists of ten reactions, each of which is catalyzed by a different enzyme, that oxidize glucose to pyruvate. These enzymes are found in the cytoplasm of both prokaryotes and eukaryotes. The first three reactions in the Embden-Meyerhof pathway do not involve any oxidation. They represent preparatory steps for a subsequent oxidation. During these three steps, 2 ATP molecules are used. The subsequent reactions of the Embden-Meyerhof pathway include one oxidation-reduction reaction leading to the synthesis of ATP by substrate level phosphorylation and of reducing power in molecules of NADH. In the sixth step, 2 molecules of reduced NAD ($NADH + H^+$) are produced for every molecule of glucose oxidized to 1,3-diPGA. During the metabolism of 1,3-diPGA to pyruvic acid, 4 molecules of ATP are synthesized by substrate-level phosphorylation. Hence, after the completion of the Embden-Meyerhof pathway, the cell has a net total of 2 ATP molecules and reducing power in the form of $2\ NADH + 2\ H^+$.

Adding together all of the reactions of the Embden-Meyerhof pathway leading to the formation of pyruvate, we can get an overall view of glycolysis:

$$\text{Glucose} + 2\text{ATP} + 2\text{NAD}^+ + 2\text{ADP} + 2\text{Pi} \longrightarrow$$

$$\longrightarrow 2\text{Pyruvate} + 2\text{NADH} + 2\text{H}^+ + 4\text{ATP} + 2\text{H}_2\text{O}$$

If the 2ATP molecules required to initiate the Embden-Meyerhof reactions are subtracted from the total yield, the equation can be summarized as:

$$\text{Glucose} + 2\text{ADP} + 2\text{Pi} + 2\text{NAD}^+ \longrightarrow$$

$$\longrightarrow 2\text{Pyruvate} + 2\text{NADH} + 2\text{H}^+\ 2\text{ATP} + 2\text{H}_2\text{O}$$

In addition to the Embden-Meyerhof pathway, microorganisms may oxidize glucose via the **pentose phosphate** pathway or the **Entner-Doudoroff** pathway. The pentose phosphate pathway may be carried out along with the Embden-Meyerhof pathway by certain bacteria. If microorganisms possess the enzymes for both the Embden-Meyerhof and the pentose phosphate pathways (not all microorganisms do), they use the pentose phosphate pathway mainly to produce biosynthetic reducing power. Through this pathway the cell carries out the breakdown of most 5-carbon sugars (pentoses), using some of the intermediate products as precursors for nucleic acid, glucose, and amino acid synthesis. Additionally, the reducing power generated by the pentose phosphate pathway is reduced NADP (not NAD, as in catabolic reactions). This characteristic of the pentose phosphate pathway suggests that it may play a role in the regulation of anabolism □ by regulating the levels of NADPH. This phosphorylated nucleotide is involved in oxidation-reduction reactions in anabolism. NAD, by contrast, is involved in catabolic oxidation-reduction reactions. By altering the levels of NADPH in the cell it is possible to regulate the extent of anabolism. The Entner-Doudoroff pathway provides yet another way in which hexoses (6-carbon sugars) can be oxidized to pyruvate. The reducing power generated during this pathway includes NADPH and NADH, so that pathway may be involved in anabolism as well as in catabolism.

108

The fate of pyruvate depends upon a number of factors. In the absence of an external electron acceptor (e.g., oxygen or nitrate ions), or if the cell is

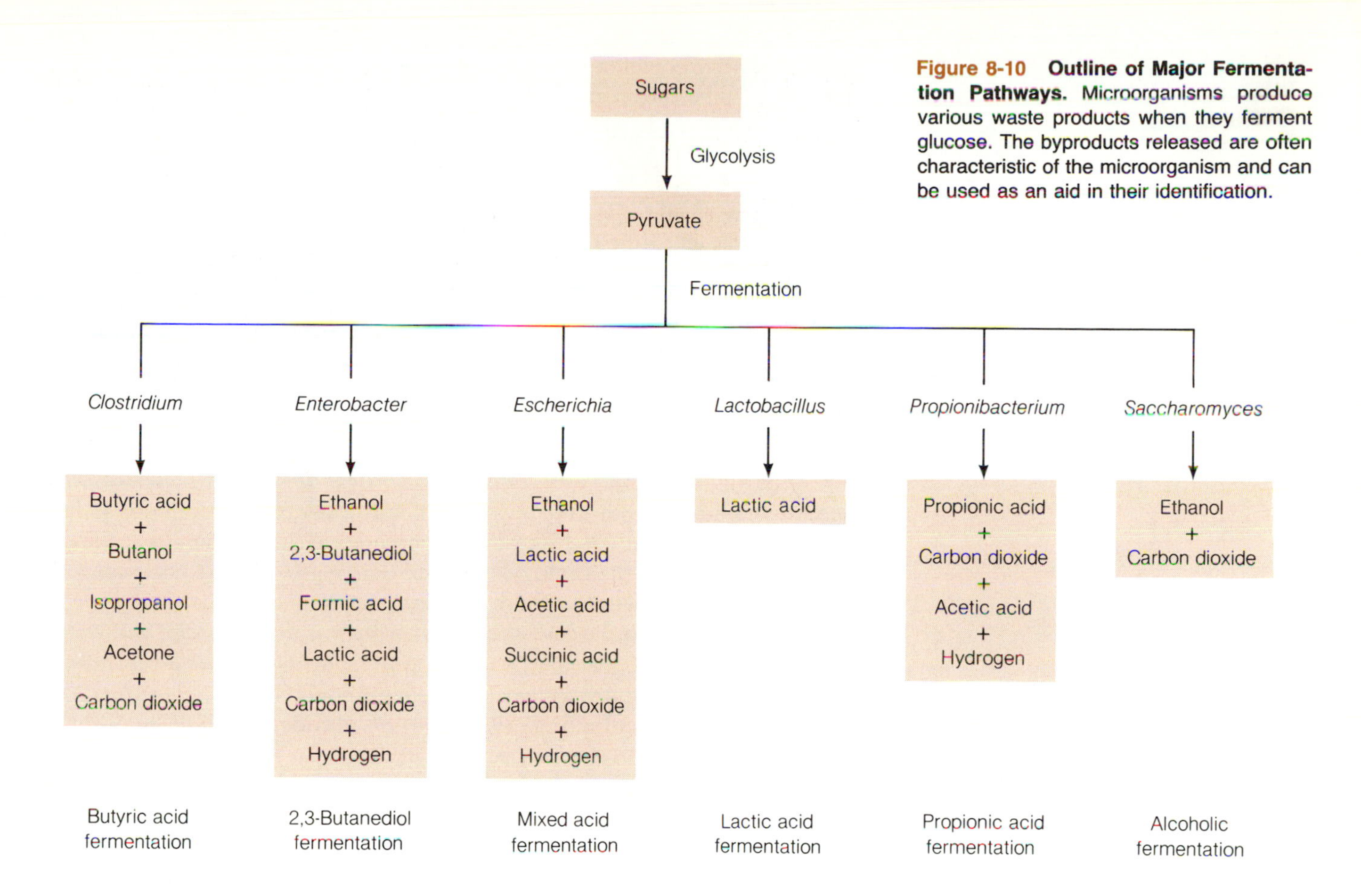

Figure 8-10 Outline of Major Fermentation Pathways. Microorganisms produce various waste products when they ferment glucose. The byproducts released are often characteristic of the microorganism and can be used as an aid in their identification.

deficient in some enzymes of the Krebs cycle, the energy metabolism is restricted to fermentation. If the Embden-Meyerhof pathway is to continue functioning, there must be a readily available supply of oxidized NAD (NAD^+) to accept electrons from glyceraldehyde (fig. 8-9). Since there is a limited supply of NAD^+ in a cell, there must be a mechanism to regenerate it from reduced NAD ($NADH + H^+$). One such mechanism involves giving up the electrons in NADH to organic molecules. Figure 8-10 shows how various chemoorganotrophs dispose of their electrons. When an organism synthesizes its ATP by substrate-level phosphorylation and disposes of its electrons by giving them to organic molecules, the organism has carried out a **fermentation.**

No single organism has all the enzymes required for the fermentation pathways shown in figure 8-10. In fact, many microorganisms use only one pathway preferentially, even though they might be able to use a number of pathways. Thus, organisms can often be characterized by the kinds of waste products they release into the environment when they ferment.

Those chemoorganotrophs that can ferment and respire (depending on conditions) are known as **facultative anaerobes.** The term facultative anaerobe indicates that the organism can do without oxygen and can generate its energy either by fermention or by anaerobic respiration. These organisms will respire rather than ferment when the oxygen concentration is sufficiently high. When the oxygen concentration is high, oxygen rather than an organic molecule is the final electron and proton acceptor. A facultative anaerobe consumes a carbohydrate such as glucose more rapidly when it ferments than when it respires. In addition, when an organism ferments, it releases wastes, such as acids, alcohols, and gases. The decrease in the rate of glucose consumption and the inhibition of acid and/or alcohol production when respiration occurs is known as the **Pasteur Effect.** Even though cells consume glucose more slowly when they respire than when they ferment, their growth rate is greater when they respire. Also the cell yield of a population is greater when it is respiring than when it is fermenting. Respiration utilizes the carbohydrate more efficiently to produce ATP, leaving much of the carbohydrate available for the synthesis of cell material. In fermentation, most of the carbohydrate is consumed to generate ATP (energy) rather than new cell material.

Respiration

The oxidation of glucose to pyruvate results in a net release of about 45 kcal of energy, but only about 15 kcal are conserved as ATP. Therefore, the potential energy in the molecule of pyruvate (688 − 45 = 643 kcal/mole) remains largely untapped. During respiration, nearly half of this energy is conserved in ATP through the oxidation of pyruvate to carbon dioxide and water.

Organisms that are capable of respiration oxidize pyruvate to acetate (fig. 8-11). During this conversion, acetate combines with coenzyme A (CoA), forming acetyl-CoA and a molecule of carbon dioxide and reducing a molecule of NAD (to NADH), which is released. This is a necessary first step for the complete oxidation of pyruvate. The acetyl group enters the Krebs cycle (fig. 8-11), where it is further oxidized. Notice that the Krebs cycle begins when acetyl-CoA reacts with oxaloacetic acid to form citric acid, a 6-carbon compound, and CoA is released. The release of CoA provides the free energy necessary for the synthesis of citric acid. In the Krebs cycle, a series of oxidation-reduction reactions takes place in which the molecule of acetate entering the cycle becomes oxidized (fig. 8-11). Notice also that several molecules of NADH are produced as a consequence of acetate oxidation. The electrons used to reduce NAD^+ come from the Krebs cycle intermediates. The oxidation of isocitric acid to α-ketoglutaric acid yields one molecule of NADH; the oxidation of α-ketoglutaric acid to succinyl-CoA yields another; and the oxidation of malic acid to oxaloacetic acid yields yet another. Hence, each turn of the Krebs cycle produces three molecules of reduced NAD. There is one further reduction of a coenzyme that takes place during the oxidation of succinic to fumaric acid; however, the electron carrier in this reaction is the coenzyme FAD. In addition to all of the electron carriers reduced in the Krebs cycle, a molecule of GTP is synthesized by substrate-level phosphorylation during the conversion of succinyl-CoA to succinic acid (fig. 8-11). GTP, which yields approximately the same amount of free energy upon hydrolysis as ATP, is used by the cell to power protein synthesis (or produce ATP). In prokaryotic cells, the enzymes involved in the Krebs cycle are located in the cytoplasm of the cell. In eukaryotic cells, these enzymes are located in the matrix of the mitochondrion □.

99

The electrons released during the oxidation of glucose in the Embden-Meyerhof pathway and the Krebs cycle are carried by electron carriers (e.g., NAD or FAD) to certain sites on the plasma membrane of prokaryotes □, or to the inner mitochondrial membrane of eukaryotes, where an electron

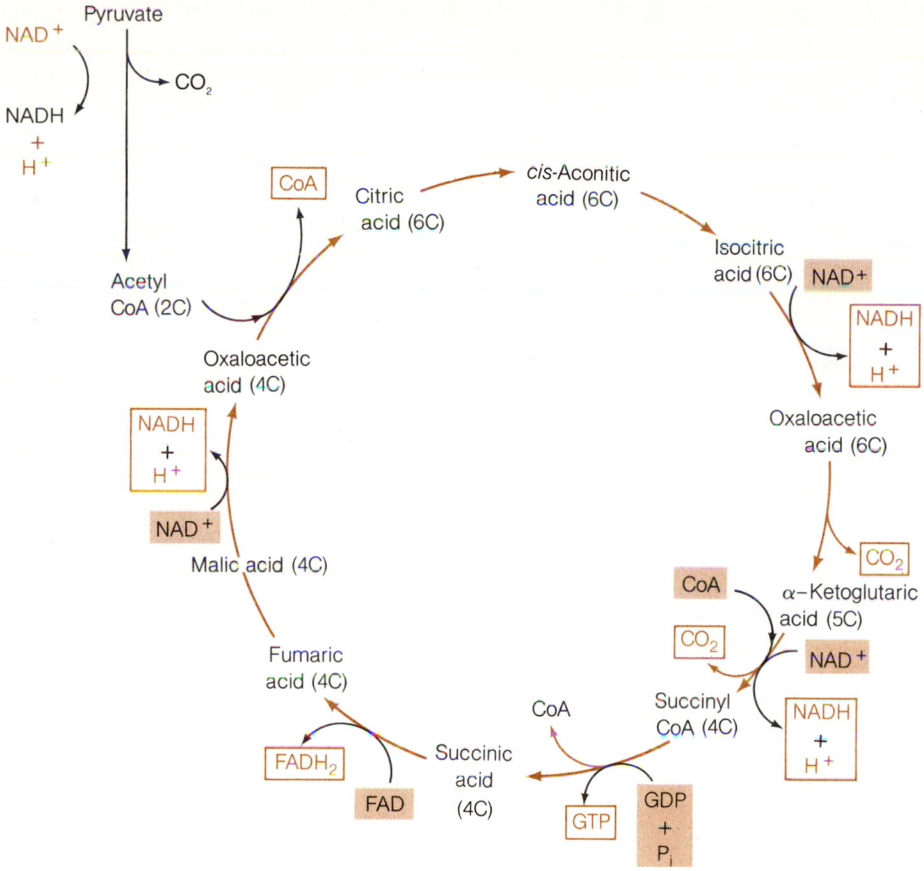

Figure 8-11 **The Krebs Cycle.** Materials entering the Krebs cycle are in shaded boxes. Important products of the Krebs cycle are in outlined boxes. Reduced NAD and FAD donate electrons to an electron transport system.

transport system is located. The electron transport system (fig. 8-12) consists of a sequence of various types of molecules that are capable of becoming reduced and then reoxidized. The transport of electrons and protons across the plasma membrane (in prokaryotes) or the inner mitochondrial membrane (in eukaryotes) can release sufficient free energy to power the synthesis of ATP. The electrons, once they have passed through the electron transport system, are used to reduce molecules such as oxygen or inorganic ions called external electron acceptors, which are found in the environment. If oxygen is the final electron acceptor, the process is called **aerobic respiration.** If other inorganic molecules serve as the final electron acceptor then the process is called **anaerobic respiration.**

The overall chemical reaction for an aerobic respiration of glucose can be written as follows:

$$\text{GLUCOSE } (C_6H_{12}O_6) + 38\text{ADP} + 38\text{Pi} + 6O_2 \longrightarrow 38\text{ATP} + 6CO_2 + 6H_2O$$

Of the 688 kcal per mole of glucose oxidized which are available for ATP synthesis, 277 kcal are conserved in 38 moles of ATP synthesized by

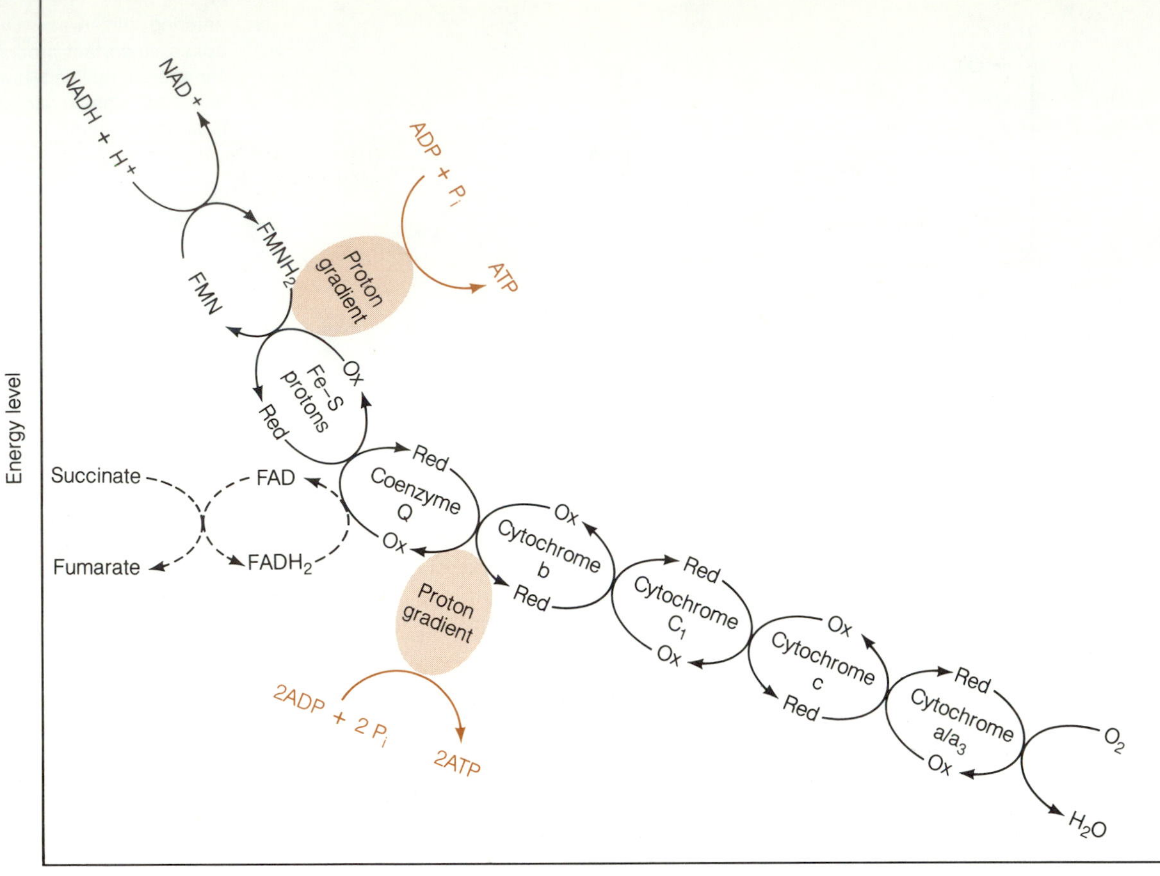

Figure 8-12 The Electron Transport System. The Krebs cycle is not directly involved in the synthesis of ATP. All electrons produced during the oxidation of glucose enter the electron transport system via NADH and FADH. The electrons are transported through a series of membrane-associated cytochromes (electron acceptors) and ATP is produced. Three ATP molecules are produced for every NADH donating to the electron transport system, but only 2 ATP molecules when $FADH_2$ is the electron carrier.

microorganisms (fig. 8-13). This figure represents a conservation of about 40% of the potential energy of glucose. In higher organisms, the remaining 60% is not completely lost, because some of those calories have been used to provide the heat necessary to maintain the cellular temperature at levels conducive to biochemical reactions. Table 8-2 compares the efficiency of energy conservation in respiration with that of fermentation.

The Chemiosmotic Hypothesis of Energy Conservation

The **chemiosmotic hypothesis** explains how electrons and hydrogen ions (protons) may be used to generate chemical energy (ATP) during their transport through an electron transport system. According to the chemiosmotic hypothesis, H^+ and electrons are separated from each other across the bacterial cytoplasmic membrane because of their divergent pathways along an electron transport system (fig. 8-14). This creates a potential across the membrane that can be used to power cellular processes or to synthesize ATP.

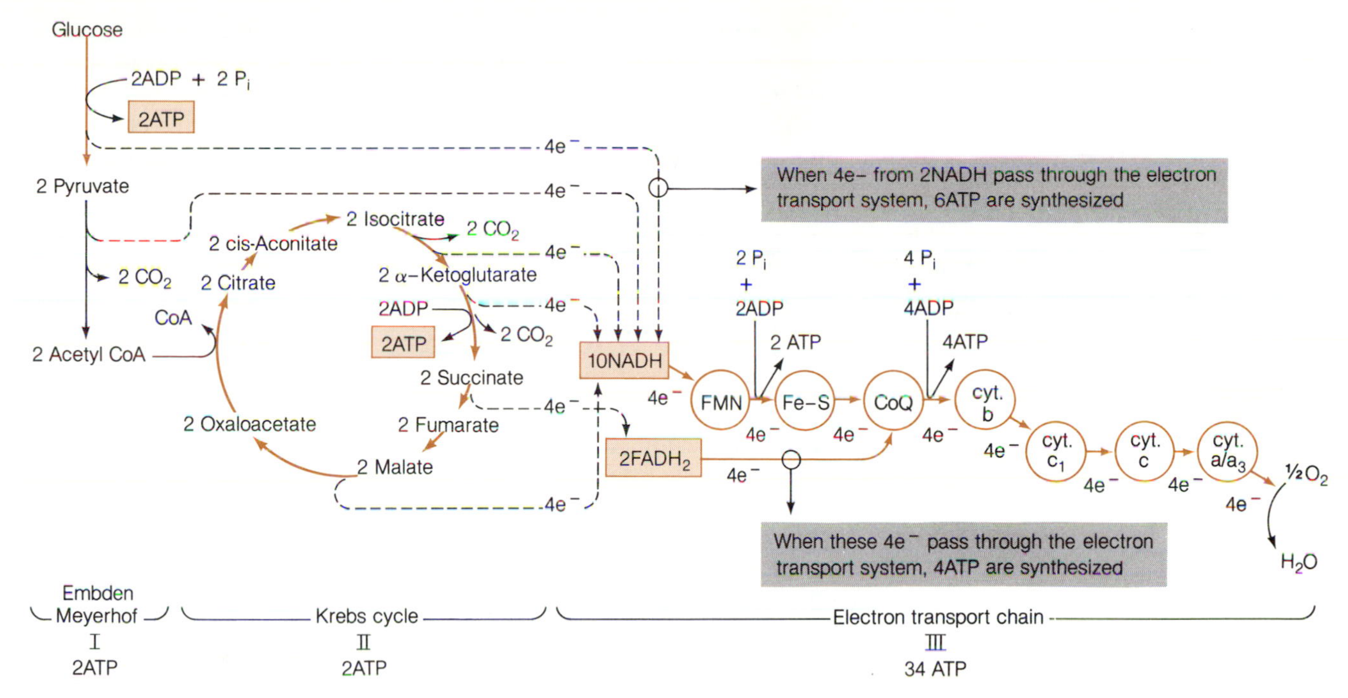

Figure 8-13 Summary of Chemoorganotrophic Metabolism in Bacteria. Glucose is oxidized to pyruvate with the net production of 2 ATP molecules/glucose oxidized. NADH is also produced. Pyruvate is oxidized to acetyl-coenzyme A, producing an NADH/pyruvate. The acetyl group is oxidized to CO_2 in the Krebs cycle, producing additional NADH as well as reduced FAD. The $FADH_2$ and NADH transfer their electrons to cytochromes of an electron transport system. Electrons passing through the electron transport system release sufficient energy to catalyze the synthesis of several molecules of ATP. In bacteria, the net output of ATP is 38 molecules per molecule of glucose oxidized to CO_2 and H_2O.

The electrons and H^+ from a reduced coenzyme (NADH + H^+) are given to a membrane-associated protein that has the coenzyme flavin mononucleotide (FMN) attached to it. The H^+ travel along the protein to the outside of the cell. The protons are released while the electrons travel back across the membrane. Additional electrons and H^+ are picked up by coenzyme Q and transported across the membrane. Again, H^+ ions are released to the outside of the cell, while the electrons travel back across the membrane along cytochromes b, c and a.

TABLE 8-2
BALANCE SHEET OF BACTERIAL CHEMOORGANOTROPHIC METABOLISM

REACTION	ATP/GLUCOSE	ENERGY (KCAL) CONSERVED IN ATP	EFFICIENCY OF ENERGY CONSERVATION
Fermentation			
Glucose → Pyruvate → Lactate	2	15	2.2%
Respiration			
Glucose + $6O_2$ → $6CO_2$ + $4H_2O$			
Glucose → 2Pyruvate	2		
→ 2 NADH	6		
2Pyruvate → 2AcCoA + 2NADH	6		
2AcCoA → $4CO_2$ + $4H_2O$	24		
TOTAL	38	277	40.3%

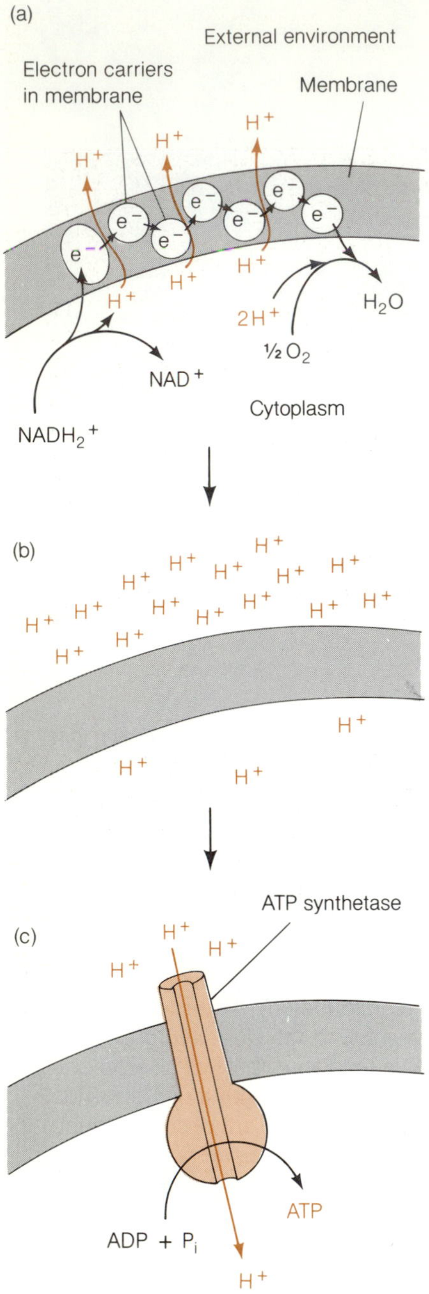

Figure 8-14 The Chemiosmotic Hypothesis. *(a)* Electron carriers in the electron transport system in the bacterial plasma membrane pump protons (H^+) to the outside of the cell. Electrons remain inside the cell. *(b)* The pumping of protons and electrons on either side of the membrane creates a charge imbalance across the membrane, resulting in a potential. *(c)* Protons return from the outside of the cell, passing through a membrane-associated ATP synthetase (ATPase). The energy released when the potential is discharged is coupled to the synthesis of ATP.

The increased concentration of H^+ outside the cell results in a pH difference between the two sides of the membrane. The concentration of H^+ outside the cell also creates a potential difference (voltage) across the membrane. The energized state of the membrane conserved by this charge separation is called the **proton motive force** (PMF). The PMF provides the energy necessary for enzymes called **ATP synthetases** (ATPases) to catalyze the synthesis of ATP from ADP and Pi. Thus, the membrane serves as a tiny battery that can power the synthesis of ATP.

CHEMOLITHOTROPHIC METABOLISM

Chemolithotrophic microorganisms carry out their energy metabolism in much the same way chemoorganotrophs do, except that chemolithotrophs obtain their energy from the oxidation of reduced inorganic compounds (table 8-3) and do not carry out fermentations. For example, *Nitrosomonas* is an obligately aerobic chemolithotroph that obtains its energy from the oxidation of ammonia. During this oxidation, ammonia is oxidized to nitrite ions and the electrons are passed through an electron transport system that uses flavin adenine dinucleotide (FAD), instead of NAD, as its first electron carrier. In this metabolism, ATP is generated when the electrons are passed through the electron transport system, creating a proton motive force that powers the synthesis of ATP. The final electron acceptor is oxygen, which is reduced to water. Hence, *Nitrosomonas* conserves part of the energy in the ammonia molecule in the form of ATP. Other bacteria (table 8-3) utilize other inorganic substances, such as hydrogen sulfide, molecular hydrogen, ferrous ions, nitrite ions, and sulfur, as the source of electrons for their chemolithotrophic metabolism.

PHOTOTROPHIC METABOLISM

Phototrophs are organisms that absorb light and use it to power the synthesis of ATP. These organisms utilize light energy to oxidize chlorophyll. The electrons obtained during this oxidation are passed along various electron carriers in an electron transport system to produce ATP and/or NADPH (reducing power). Both of these molecules are employed by phototrophic organisms to power anabolic reactions, such as the fixation of carbon dioxide (fig. 8-16).

Cyclic Photophosphorylation

Let us begin our discussion of phototrophic metabolism with the anaerobic photosynthetic bacteria □, because they carry out a simple form of photosynthesis. The phototrophic bacteria possess extensive **intracytoplasmic membranes** which are extensions of the plasma membrane except in the case of chlorobium vesicles, which are not continuous with the plasma membrane. The intracytoplasmic membranes contain photosynthetic pigments called **bacteriochlorophylls** (Bchl). There are several types of bacteriochlorophylls, called Bchl-a, Bchl-c, Bchl-d, and Bchl-e. In addition, the cytoplasmic membranes contain light-sensitive pigments, such as carotenoids, as well as cytochromes and other components of electron transport system (fig. 8-17).

326

FOCUS

MICROORGANISMS THAT GIVE OFF LIGHT

Bioluminescence is the production and release of light by living organisms. This characteristic is widely distributed in nature, so it may be a beneficial trait. Bioluminescent organisms include certain bacteria, fungi, algae, sponges, worms, clams, snails, squids, and insects. The light emission of fireflies, which is used to communicate between sexes, is an example of bioluminescence.

Even though some organisms (e.g., fireflies and mushrooms) produce their own bioluminescence, many of the marine animals rely on bacteria to produce light. Marine fishes that live deep in the ocean often have "light organs" that are colonized by bioluminescent bacteria. These fishes "cultivate" bioluminescent bacteria and use them to lure other animals, attract sexual partners, or warn invaders.

Bacterial bioluminescence results when electrons flowing through the electron transport system are diverted to a flavoprotein called **luciferin.** An enzyme called **luciferase** catalyzes the oxidation of luciferin in the presence of O_2, resulting in the emission of visible light.

The conversion of chemical energy to radiant energy is an interesting example of energy transduction and points out the remarkable versatility of microorganisms in adapting to their environment (fig. 8-15).

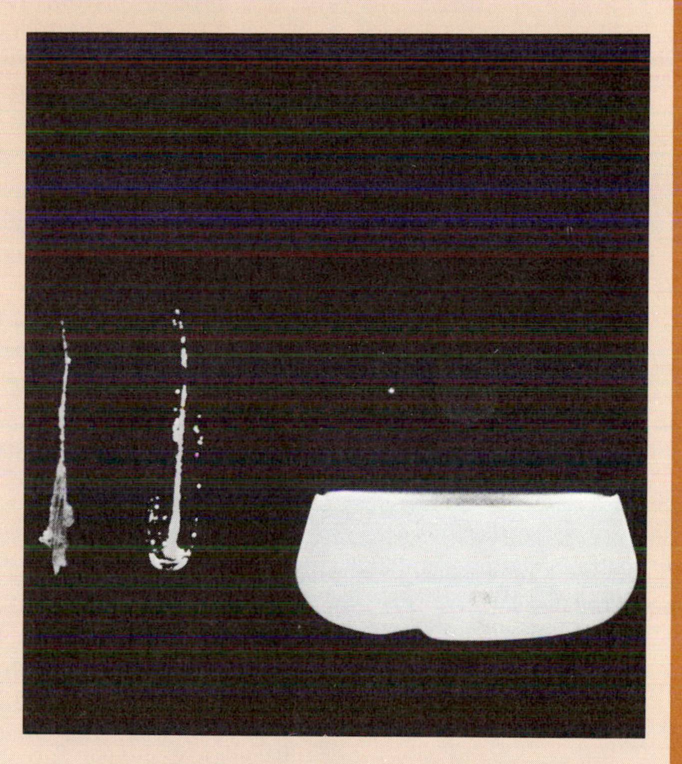

Figure 8-15 Bioluminescent Bacteria Photographed Using Their Own Light

The end product of bacterial photosynthesis is glucose (or fixed carbon). The energy and reducing electrons required for the synthesis of glucose comes from the ATP and NADPH generated during light-dependent reactions. ATP is generated when electrons and hydrogen ions flow through an electron transport system. The coenzyme $NADP^+$ can be reduced with electrons if the cell has sufficient ATP. Thus, the cell can make either ATP or NADPH, depending upon its needs, by diverting the flow of electrons from the electron transport system to $NADP^+$ or vice versa (fig. 8-17).

The process of ATP synthesis in the anaerobic photosynthetic bacteria is known as **cyclic photophosphorylation**. Light is required to drive electrons from photosystem-I into an electron transport system where they are used to generate an electrical potential, using hydrogen ions, across the intracytoplasmic membranes. The potential is used to drive the phosphorylation of ADP. Since, phosphorylations are dependent upon light, they are referred to as photophosphorylations. Because the electrons return to chlorophyll, they can be cycled through the system and used to drive phosphorylations over and over again. Thus, the entire process is called cyclic photophosphorylation.

The chemiosmotic hypothesis can explain how energy is conserved in photophosphorylation. Light is absorbed by **photosystem-I** (PS-I), which

TABLE 8-3

SOURCES OF ELECTRONS FOR SELECTED CHEMOLITHOTROPHS

ORGANISM (OR GROUP)	SOURCE OF ELECTRONS
Sulfur bacteria	H_2S, S
Iron bacteria	Fe^{++}
Nitrosomonas	NH_3
Nitrobacter	NO_2^-
Hydrogen bacteria	H_2

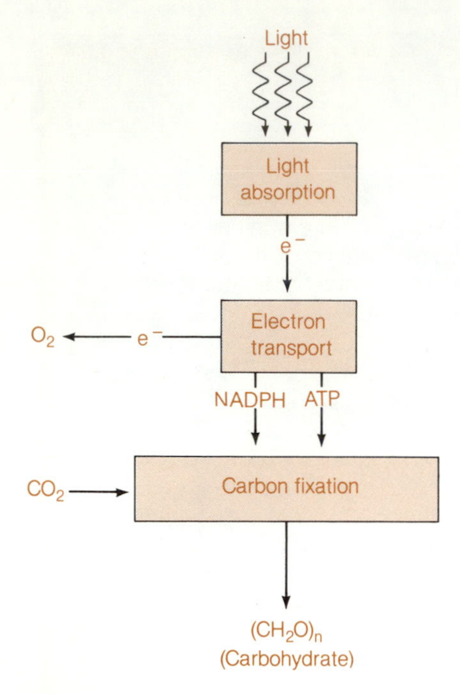

Figure 8-16 An Overview of Phototrophic Metabolism

consists of carotenoids and bacteriochlorophyll. Energetic electrons accumulate in a special bacteriochlorophyll, which eventually ejects them to be picked up by electron acceptors called **quinones** (Q′). Hydrogen ions are picked up from the cytoplasm of the cell. The electrons and H^+ are transported across the membrane, the H^+ are released within the photosynthetic vesicles □, and the electrons move to a cytochrome. In this way a charge separation is achieved, creating a proton motive force. The proton motive force is used to provide the free energy necessary for the formation of ATP from ADP and Pi. The reaction is catalyzed by ATPases in the photosynthetic membranes. The electrons return to oxidized chlorophylls in photosystem-I (remember that light oxidized these molecules, so now they must be reduced if the process is to continue).

92

The electrons ejected from PS-I may not always return to an oxidized photosystem-I, but may instead be used to reduce $NADP^+$. Reduced $NADP^+$ (NADPH) is requred for the fixation of carbon dioxide into organic molecules (fig. 8-17). Thus, electrons and H^+ are permanently lost from PS-I and the electron transport system. In order to replace these electrons and H^+, molecules such as H_2 and H_2S are consumed. Figure 8-17 shows how hydrogen sulfide supplies electrons and H^+ to the electron transport system.

Noncyclic Photophosphorylation

The cyanobacteria □ and eukaryotic phototrophs carry out a more complex type of photophosphorylation, which involves two pigment systems instead of one (fig. 8-18). In addition, during this type of photophosphorylation electrons are extracted from water molecules to re-reduce oxidized chlorophylls; therefore, molecular oxygen is released as a major byproduct of this type of phototrophic metabolism. The cyanobacteria contain the enzymes and pigments required for their phototrophic metabolism in intracytoplasmic membranes called **thylakoids** □.

308

91

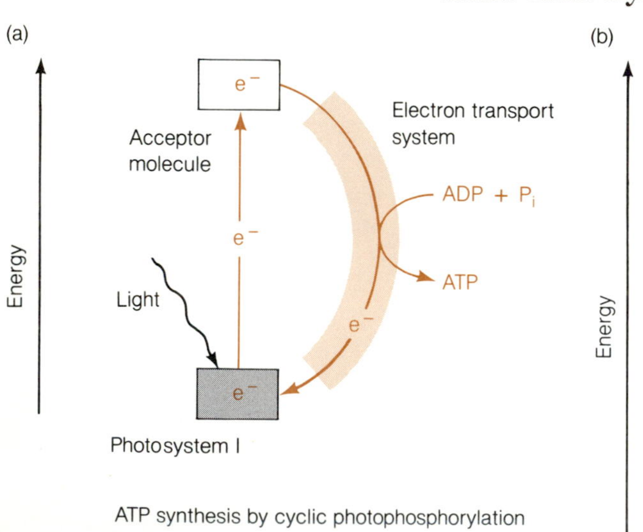

Figure 8-17 Cyclic Photophosphorylation. *(a)* Bacteriochlorophyll in photosystem-I (PS-1) becomes excited when it absorbs light, releasing electrons. The electrons are passed through an electron transport system, producing ATP. *(b)* An alternative path of electron flow that produces NADPH. The electrons required to reduce the oxidized PS-I come from the oxidation of H_2S.

FOCUS

HALOPHILIC BACTERIA USE BACTERIORHODOPSIN INSTEAD OF CHLOROPHYLL TO CONVERT LIGHT ENERGY INTO CHEMICAL ENERGY

Sometimes dried, salted fish, or other consumer goods preserved by the addition of salt, display reddish areas that result from bacterial growth. The bacteria that cause this red discoloration are not ordinary bacteria. They are **extreme halophiles** that require very high concentrations of NaCl and magnesium in their environment. The extreme halophiles are found in marine environments such as the Dead Sea, where the salt concentration is very high. The red color is due to the fact that these bacteria have in their plasma membrane red and purple pigments. The formation of the purple pigment occurs only under anaerobic conditions.

The purple pigment is of interest because it is involved in an unusual form of energy generation. The purple pigment is **bacteriorhodopsin.** This photosensitive pigment bears a remarkable resemblance to the retinal pigment **rhodopsin,** found in the vertebrate eye. Although both pigments are involved in light-dependent reactions, rhodopsin is intimately involved in vision, while bacteriorhodopsin is involved in energy-yielding metabolism.

The mechanism of chlorophyll-independent photophosphorylation is not entirely known, although studies have revealed at least part of the mechanism of energy generation by extreme halophiles. Bacteriorhodopsin absorbs light, which drives protons (H^+) out of the cell, creating a proton motive force similar to that created during chlorophyll and light driven photophosphorylation or oxidative phosphorylation. The proton motive force is then used to provide the energy for the synthesis of ATP.

The discovery of these photosynthetic bacteria in 1971 provided yet another example of the remarkable versatility that exists in the world of microorganisms.

Cyanobacterial (and green plant) photophosphorylation begins when a molecule of chlorophyll-a (Chl-a) in **photosystem-II** (PS-II) is oxidized by the light absorbed by the pigments. This Chl-a is extremely reactive when oxidized and can readily remove electrons from water (fig. 8-18). The electrons from chlorophyll-a and H^+ are picked up by a quinone and transported

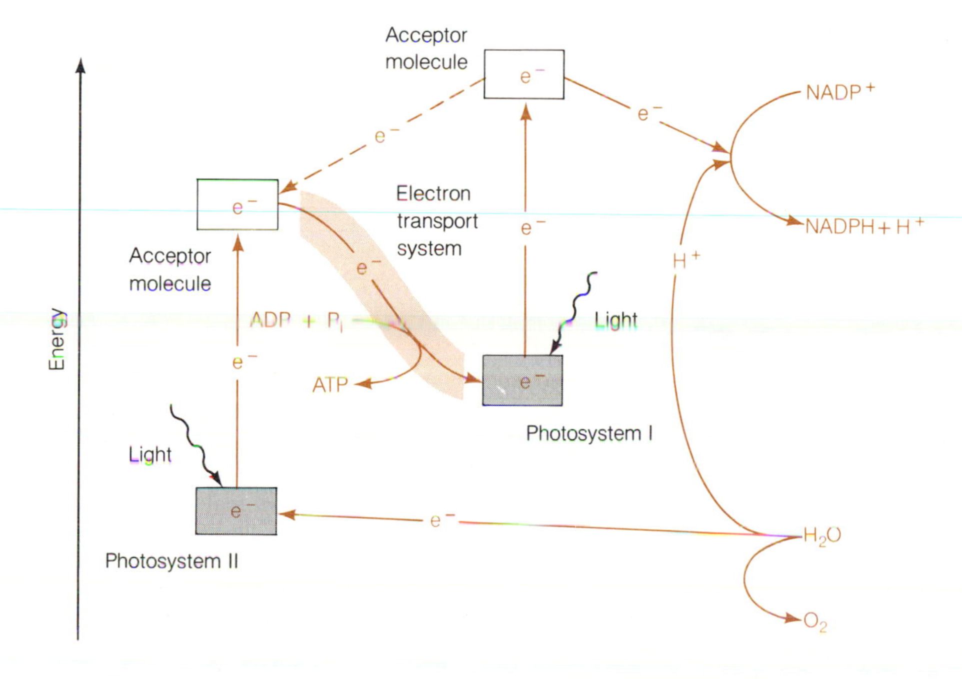

Figure 8-18 Noncyclic Photophosphorylation. These series of reactions occur in cyanobacterial photosynthesis and in eukaryotic cell photosynthesis. Noncyclic photophosphorylation does not occur in the phototrophic bacteria.

across the thylakoids, creating a proton motive force of sufficient intensity to power the synthesis of ATP by an ATPase. The electrons and H^+ flowing through an electron transport system eventually are absorbed by chlorophyll-a in photosystem-I, which supplies the electrons for the reduction of $NADP^+$. Compare figure 8-17 with figure 8-18 and notice that in phototrophic bacteria (which have only PS-I), PS-I supplies electrons for both ATP synthesis and the reduction of $NADP^+$. In the cyanobacteria, however (which have both PS-I and PS-II), PS-I and PS-II, working in series, are involved in the reduction of $NADP^+$. Photosystem-I also functions in photophosphorylation, when electrons from PS-I are run repeatedly through an electron transport system and back to PS-I.

BIOSYNTHESIS

Microorganisms conserve energy in the form of ATP so that they can use it to power mechanical movements such as flagellar motion, the transport of nutrients into the cell, and biosynthesis. Biosynthesis encompasses all cellular activities in which biological polymers and their precursors are made. As a rule, all these reactions require a net input of energy, much of which comes from the hydrolysis nucleotides.

Carbon Assimilation

Carbon assimilation is an important aspect of microbial metabolism, because it is required for the synthesis of all organic molecules that make up the cell □.

33

Autotrophic microorganisms can build their organic compounds by fixing carbon dioxide in a process called the Calvin cycle (fig. 8-19). In this process, carbon dioxide reacts with ribulose 1,5-diphosphate to form two

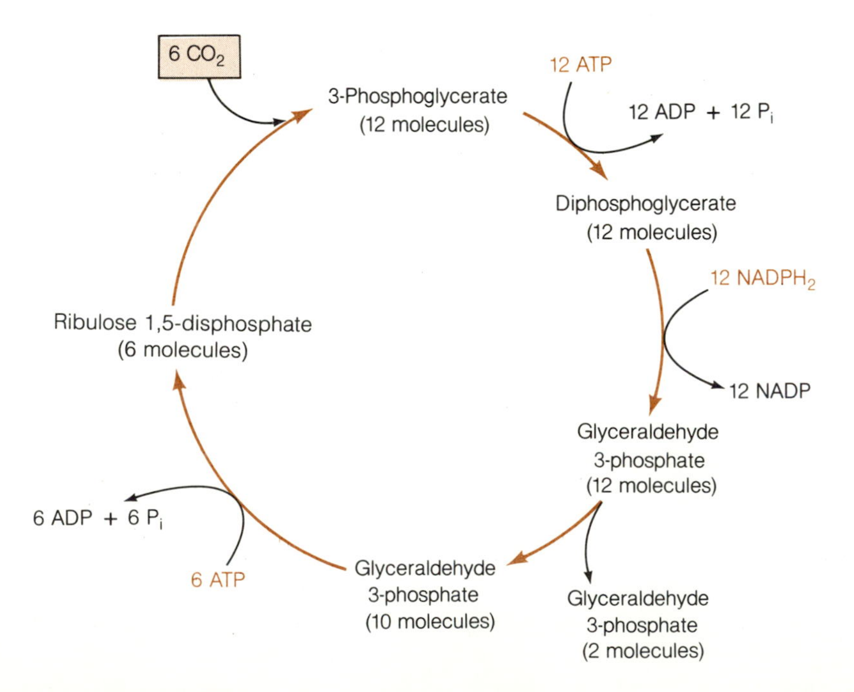

Figure 8-19 A Summary of the Calvin Cycle. The Calvin cycle, also known as the dark reaction of photosynthesis, is involved in the fixation of carbon dioxide to produce organic molecules. Carbon dioxide molecules enter the cycle one at a time. Six turns of the cycle, summarized in the figure, produce two glyceraldehyde-3-phosphate molecules. The energy required to drive the Calvin cycle is derived from ATP and NADPH, which frequently come from cyclic and noncyclic photophosphorylation.

molecules of 3-phosphoglyceric acid (3-PGA). This reaction is catalyzed by an enzyme called **ribulose diphosphate carboxylase.** The fixation of carbon dioxide to the ribulose 1,5-diphosphate molecule, with the resultant formation of two molecules of 3-PGA, requires that the 3-PGA be reduced. This reduction requires both energy and electrons. The energy comes from the hydrolysis of ATP and the electrons come from the oxidation of NADPH. The reactions of the Calvin cycle produce a number of different sugars that may be used to build cellular components, such as nucleotides and polysaccharides.

Heterotrophic microorganisms, on the other hand, must have a ready source of organic molecules for their biosynthesis. These molecules can be used by the cell not only to supply its carbon but also as a source of energy. The same pathways that serve to generate electrons (e.g., the Krebs cycle) can also be used to provide essential metabolic intermediates for biosynthesis (fig. 8-20). For this reason, the Krebs cycle is said to be an **amphibolic** pathway, used by the cell in both anabolism and catabolism. The intermediate product of the Krebs cycle called α-ketoglutaric acid is used by the cell

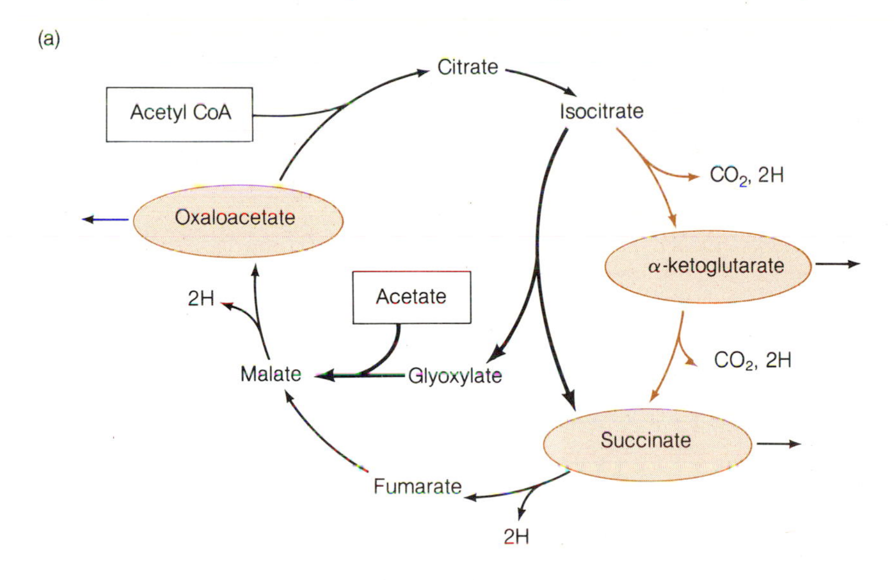

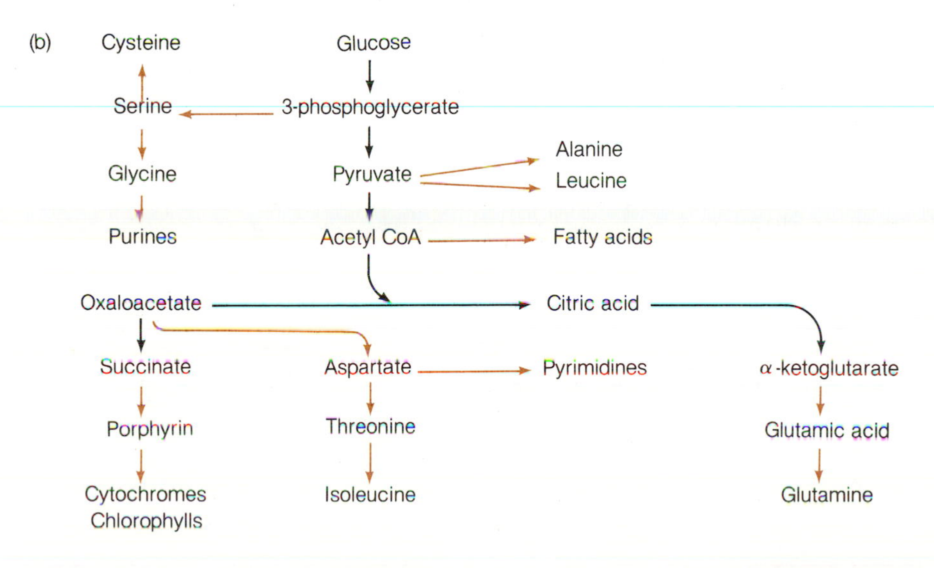

Figure 8-20 An Amphibolic Pathway. *(a)* Amphibolic pathways are involved in both anabolism and catabolism. The Krebs cycle (colored lines) and the glyoxylate bypass (black lines) are involved in the production of important anabolic intermediates for the synthesis of various amino acids. *(b)* Major biosynthetic pathways. Names in black represent precursors involved in the synthesis of molecules (in color). These precursors are also part of glycolysis or the citric acid cycle.

as a precursor in the synthesis of the amino acid called glutamic acid, which in turn is used by the cell to build proteins. Similarly, oxaloacetic acid is a precursor of the amino acid aspartic acid, which is a precursor for pyrimidine biosynthesis and therefore participates in nucleic acid biosynthesis. Even some anaerobes that are unable to respire have most of the Krebs cycle enzymes, mainly for the generation of biosynthetic intermediates.

Regulation of Metabolism

The cell "fine tunes" its use and synthesis of materials so as to utilize its resources efficiently. For example, if all the cell did was oxidize glucose to produce ATP, eventually there would be a surplus of ATP and no organic compounds to make new cell materials. This would be a waste of glucose. Instead, when the cell "senses" that there is sufficient ATP, it shifts its metabolism toward biosynthesis. In this fashion, some of the intermediates in energy metabolism can be used to fuel biosynthesis (fig. 8-20).

Biosynthetic reactions take place in small steps in which the product of one reaction serves as the substrate for another. Also, the fact that the energy release is small aids the cell in optimizing its energy usage. The sequence of reactions leading to the formation of a product is called a **pathway.** The loss of any one step in the pathway inhibits the synthesis of the final product. The cell can take advantage of this fact to regulate its metabolism □.

242

SUMMARY

METABOLISM, AN OVERVIEW

1. Metabolism is the term used to describe all the chemical reactions that take place in a cell. Anabolism refers to those reactions that result in the building of cell material, while catabolism refers to processes that break down cell material and nutrients.
2. Microorganisms can extract energy from foods and conserve it in molecules of phosphorylated nucleotides.
3. The steps involved in extracting energy from foods may include digestion, transport, oxidation, and phosphorylation.

ENZYMES AND HOW THEY WORK

1. Enzymes are catalysts that participate in nearly all cellular activities. They function by lowering the amount of energy required for a reactant to reach an activated state. This energy is called activation energy.
2. The activity of an enzyme is largely determined by its 3-dimensional shape. An enzyme's active site has a shape that is complementary to its substrate.
3. Enzymes reduce the activation energy of a reaction by binding to their substrate. This stresses the substrates' bonds and makes it easier to break them. Enzymes also bring reacting atoms into close proximity, thus enhancing the likelihood of a reaction between them.
4. Temperature, pH, and substrate concentration all influence the speed at which enzymes catalyze reactions.

NUTRIENT TRANSPORT

1. Microorganisms bring nutrients into the cell and get rid of wastes by transport mechanisms.
2. Transport mechanisms can be classified as passive or active mechanisms. Passive transport mechanisms do not require the expenditure of energy, but these mechanisms cannot transport nutrients against a concentration gradient. Active transport mechanisms require the expenditure of energy, but they can be used to bring nutrients into the cell even against a concentration gradient.

BIOENERGETICS

1. Bioenergetics is the study of energy production and utilization by the cell.
2. Cells couple exergonic reactions with endergonic reactions so that the combined reaction is exergonic.

3. Reactions that hydrolyze ATP or other nucleotides are exergonic reactions that are commonly coupled to biosynthetic reactions in the cell.
4. Energy from food molecules and from electromagnetic radiation is conserved by microorganisms in molecules of ATP, either through substrate-level phosphorylation, oxidative phosphorylation, or photophosphorylation.
5. Substrate-level phosphorylation involves the synthesis of ATP, utilizing phosphorylated molecules.
6. Oxidative phosphorylation is a process in which molecules are oxidized and energy is extracted from the electrons. The electrons are passed through an electron transport system and the resulting free energy is used to power the synthesis of ATP.
7. Photophosphorylation involves the synthesis of ATP with energy initially derived from electromagnetic waves.
8. Coenzymes such as FAD and NAD serve as electron carriers in energy metabolism.

CHEMOORGANOTROPHIC METABOLISM

1. Chemoorganotrophs obtain their energy by oxidizing organic molecules. They may ferment or respire the molecules.
2. In a fermentation, ATP is synthesized by substrate-level phosphorylation and organic molecules act as the final electron acceptors. The net yield of ATP during a fermentation is generally 2 molecules of ATP/glucose molecule fermented.
3. Respiring microorganisms oxidize substrates and use the resulting electrons and H^+ to synthesize ATP by oxidative phosphorylation. The electrons and H^+ come from the Embden-Meyerhof pathway and the Krebs cycle. The total amount of energy conserved during respiration may be as high as 38 ATP/glucose respired.
4. In an aerobic respiration, oxygen serves as the final electron acceptor in the electron transport system. In an anaerobic respiration, other inorganic molecules are the acceptors.
5. The chemiosmotic hypothesis suggests that nutrient and radiant energy are conserved by the cell when electrons and H^+ are separated by membrane-associated electron transport systems. The concentration of hydrogen ions on one side of a membrane creates a proton motive force (PMF) that can be used to drive the synthesis of ATP by ATPases.

CHEMOLITHOTROPHIC METABOLISM

Chemolithotrophs obtain their energy from inorganic compounds. They oxidize inorganic molecules and pass the resulting electrons through an electron transport system, where oxidative phosphorylation takes place.

PHOTOTROPHIC METABOLISM

1. Phototrophs absorb light energy and conserve it in ATP. The electrons are obtained when light oxidizes chlorophyll. The electrons are then passed through an electron transport system, where ATP is synthesized.
2. Phototrophic bacteria (other than cyanobacteria) carry on anoxygenic (no oxygen evolved) photosynthesis, involving only cyclic photophosphorylation.
3. The cyanobacteria and eukaryotic phototrophs carry oxygenic (oxygen evolved) photosynthesis, involving both cyclic and noncyclic photophosphorylation.

BIOSYNTHESIS

1. Biosynthesis includes all cellular reactions in which biological polymers and their precursor molecules are made.
2. Autotrophs can satisfy all of their carbon needs by fixing carbon dioxide in a pathway called the Calvin cycle.
3. Heterotrophs use organic molecules as their source of carbon. They use the Embden-Meyerhof pathway and the Krebs cycle intermediates to provide the carbon skeletons for certain steps in biosynthesis.
4. The cell coordinates its metabolism so as to optimize the utilization of nutrients available in the environment.

STUDY QUESTIONS

1. Compare anabolism with catabolism.
2. Discuss how a molecule of starch found in the environment might be used by:
   a) chemoorganotrophs; b) chemolithotrophs.
3. Discuss three ways in which ATP can be synthesized by bacteria.
4. How are ATP and NADH used by the cell in metabolism?
5. Discuss how the following pairs of cellular processes differ from each other:
   a. fermentation and respiration;
   b. aerobic and anaerobic respiration;
   c. chemoorganotrophic and chemolithotrophic metabolism;
   d. cyclic and noncyclic photophosphorylation;
   e. cyanobacterial and noncyanobacterial photosynthesis;
   f. active and passive transport.
6. Discuss the nature and function of the Calvin cycle. Why is it also called the dark reaction of photosynthesis?

7. Discuss the function(s) of the Krebs cycle in: a) catabolism; b) anabolism.
8. How does the chemiosmotic hypothesis explain the synthesis of ATP?
9. What are amphibolic pathways?
10. In which ways are all three types of phosphorylation similar?

SUPPLEMENTAL READINGS

Chory, J., Hoger, J., Kiley, P., Yen, G., and Kaplan, S. 1984. Structure, function, and synthesis of the photosynthetic membranes of *Rhodopseudomonas sphaeroides. ASM News* 50(4):144–149.

Dickerson, R. E. 1980. Cytochrome C and the evolution of energy metabolism. *Scientific American* 242(3):136–153.

Drews, G. 1985. Structure and functional organization of light-harvesting complexes and photochemical reaction centers in membranes of phototrophic bacteria. *Microbiological Reviews* 49:59–70.

Gottschalk, G. 1978. *Bacterial metabolism.* New York: Springer-Verlag.

Hinckle, P. C. and R. E. McCarthy. 1978. How cells make ATP. *Scientific American* 238(3):104–123.

Ingledew, W. J. and Poole, R. K. 1984. The respiratory chains of *Escherichia coli. Microbiological Reviews* 48(3):222–271.

Lascelles, J. 1973. *Microbial photosynthesis.* Stroudsberg, Penn.: Dowden, Hutchinson and Ross.

Lehninger, A. L. 1971. *Bioenergetics: The molecular basis of biological energy transformations.* 2d ed. Menlo Park, Calif.: Benjamin/Cummings.

Michels, M. and Bakker, E. P. 1985. Generation of a large, protonophore-sensitive proton motive force and pH difference in the acidophilic bacteria *Thermoplasma acidophilum* and *Bacillus acidocaldarius. Journal of Bacteriology* 161(1):231–237.

Phillips, D. C. 1966. The three-dimensional structure of an enzyme molecule. *Scientific American* 217(5):78–90.

Stryer, L. 1980. *Biochemistry.* 2d ed. San Francisco: W. H. Freeman.

CHAPTER 9 MICROBIAL GENETICS

CHAPTER PREVIEW

THE DISCOVERY AND CHARACTERIZATION OF THE HEREDITARY MATERIAL

In 1944 Oswald Avery and his colleagues demonstrated that a cell's hereditary information is contained in a nucleic acid called deoxyribonucleic acid (DNA) and not in protein. They were able to change the characteristics of bacteria with purified DNA. For a number of years after this discovery, however, most scientists preferred to believe that protein contained the genetic information.

By 1953, James Watson and Francis Crick had created a model for the structure of DNA and showed that there were two strands of DNA wound helically around each other, but running in opposite directions. X-ray data indicated that the nucleotides that made up each DNA strand were stacked on each other like a pile of plates approximately 3.4 angstroms apart. The DNA model created by Watson and Crick was important because it aided scientists in their understanding of how DNA could control its own replication. Watson and Crick proposed that the bases adenine and thymine as well as guanine and cytosine, hydrogen-bonded to each other. The specificity of the base pairing could explain what controlled the replication of complementary strands of DNA and the order of bases in RNA.

The problem to be solved in the 1960s was how the sequence of nucleotides in the DNA could specify the order of amino acids in a protein. Once more, the DNA model suggested to scientists how DNA might control the synthesis of protein. The physicist George Gamow had proposed in 1954 that trios of the 4 nucleotides would provide 64 different combinations (4^3), more than enough to code for the 20 amino acids usually found in proteins. In the early 1960s Marshall Nirenberg, Heinrich Matthaei, Philip Leder, and Har Gobind Khorana uncovered the genetic code and showed how the order of nucleotides in messenger RNA directed the synthesis of a specific protein.

Microbiologists in the 1970s and 1980s have been concerned with the mechanisms of DNA exchange between cells and with genetic engineering.

The knowledge of genetics gained through the efforts of many scientists have provided us with a basic understanding of how genes function and have allowed us to predict the occurence of certain diseases and devise methods to prevent them. This chapter will demonstrate how DNA functions as the hereditary material through a discussion of DNA replication, RNA synthesis, and protein synthesis. In addition, the transfer of genetic information between organisms will be discussed.

LEARNING OBJECTIVES

A STUDY OF THIS CHAPTER SHOULD ENABLE YOU TO:

EXPLAIN HOW GENES CONTROL THE CHARACTERISTICS OF AN ORGANISM

DISCUSS THE CENTRAL DOGMA

COMPARE AND CONTRAST DNA REPLICATION, TRANSCRIPTION, AND TRANSLATION

EXPLAIN HOW SOME TYPES OF MUTATIONS OCCUR

DISCUSS HOW PROKARYOTES REPAIR MUTATIONS

DIFFERENTIATE AMONG CONJUGATION, TRANSDUCTION, AND TRANSFORMATION

EXPLAIN HOW GENES ARE MAPPED IN PROKARYOTES

Figure 9-1 James Watson and Francis Crick with Their Model of the DNA Molecule

Genetics is the study of **deoxyribonucleic acid** (DNA), how the information contained within the DNA molecule is expressed, and how it accounts for the heredity of an organism. The DNA found in cells consists of two complementary strands of DNA coiled helically around each other; consequently, it is sometimes called a **double helix** (fig. 9–1).

Each DNA strand is constructed from four nucleotides: deoxyadenosine monophosphate (dAMP), deoxythymidine monophosphate (dTMP), deoxycytidine monophosphate (dCMP), and deoxyguanosine monophosphate (dGMP). For simplicity, the nucleotides (or the nitrogenous bases of each nucleotide) are symbolized by A, T, C and G. If the nucleotide sequence of one DNA strand of the double helix is known, then the sequence of nucleotides in the complementary strand can be determined, because the nucleotides in the two strands are complementary. That is, A in one strand always pairs with T in the complementary strand and G always pairs with C.

Modern genetics has been able to explain, at least in some cases, how DNA determines the characteristics of an organism and maintains and controls the vital processes of all cells. The first step in the expression of DNA is the copying of various stretches of nucleotides along the DNA into **ribonucleic acids** (RNA) (fig. 9-2). RNA is very similar to DNA except that it consists of the four nucleotides: adenosine monophosphate (AMP), uridine monophosphate (UMP), cytidine monophosphate (CMP), and guanosine monophosphate (GMP). RNA found in cells is divided into three classes: **transfer RNA** (tRNA), **ribosomal RNA** (rRNA), and **messenger RNA** (mRNA). Messenger RNA contains the information for synthesizing proteins that function as structural components or as enzymes in all cells. Ribosomal RNAs (which are structural components of ribosomes) and transfer RNAs, which carry amino acids from the cytoplasm to ribosomes, participate in protein synthesis.

The RNAs and proteins produced by a cell are important because they determine the cell's characteristics and control its vital processes. We shall see in the following sections that the structure of each RNA or protein is determined by a specific sequence of nucleotides on one strand of the DNA. The sequence of nucleotides that codes for an RNA, or ultimately for a protein, is known as a **gene.**

The theory that genes guide the synthesis of RNAs and that some RNAs determine which proteins are synthesized is known as the **central dogma** (fig. 9-2). The central dogma also postulates that a DNA molecule can give rise to new, identical DNA molecules. This replication is necessary when a cell is reproducing, since each daughter cell must inherit a complete set of all the essential genes from its progenitor cell if it is to be viable. There have been a number of modifications of the central dogma since Francis Crick proposed it in the early 1950s. It is now known, for example, that RNA can direct the synthesis of DNA in cells infected by certain viruses. Also, in cells infected with RNA viruses, the viral RNA can direct the synthesis of daughter RNA molecules.

DNA REPLICATION AND ITS SEGREGATION IN THE PROKARYOTE

The DNA molecule in bacteria is generally referred to as the **genome** □. It is a circular, double-stranded molecule that is attached to the plasma mem-

89

FOCUS

ONE-GENE-ONE-ENZYME HYPOTHESIS

Between 1937 and 1946, George Beadle and Edward Tatum isolated a large number of mutants of the fungus *Neurospora* that were unable to synthesize various amino acids and vitamins. Their genetic analyses showed that genes controlled the expression of enzymes involved in the synthesis of amino acids and vitamins. In addition, their work provided evidence for the idea that one gene controlled the synthesis of one enzyme.

George Beadle and Edward Tatum shared the Nobel Prize in Medicine or Physiology in 1958 for their research demonstrating that genes controlled the presence or absence of enzymes and for their discoveries in microbial genetics.

brane. Some mycoplasmas have the smallest bacterial genomes, with fewer than 10^6 nucleotide base pairs. These genomes would be large enough to code for approximately 1,000 average-sized proteins. In contrast, the common intestinal bacterium *Escherichia coli* has a genome with about 4.5×10^6 nucleotide base pairs. This genome is approximately 4.5 times larger than that of the mycoplasma genome and might code for 4,500 average-sized RNAs and proteins.

When a bacterial genome replicates, each of its strands is duplicated. Each strand functions as a **template** that specifies what the complementary strand will be, because of the specific hydrogen bonding between the bases (A-T and G-C). Figure 9-3 illustrates how a complementary strand is synthesized. During DNA replication, enzymes called DNA polymerases pick up nucleotides from the cytoplasm that are complementary to the template DNA and fit them into place. Then the polymerases chemically attach the nucleotides to a piece of RNA or DNA called the **primer.** Once DNA replication is underway, the RNA primer is removed and replaced by DNA. The strand of DNA that is read to make the complementary DNA is frequently referred to as the template. DNA replication in bacteria begins at a specific site on the DNA, called the **origin** (fig. 9-3). The genome is replicated in both directions, so DNA replication is found to terminate halfway around the circular genome.

When DNA replication is completed, each daughter genome consists of an old DNA strand and a new DNA strand. This type of DNA replication is known as **semiconservative replication.** The first evidence indicating that newly replicated DNA consists of one new strand and one old strand was presented by Matthew Meselson and Franklin Stahl in 1959. They grew bacteria using nutrients containing the rare heavy nitrogen (^{15}N) for a large number of generations in order to make the bacterial DNA (which usually contains light nitrogen) heavier than normal. Then the bacteria were transferred to a medium containing the normal light isotope of nitrogen (^{14}N) for one, two, or more generations. The DNA was isolated from the bacteria at various times and then centrifuged in a cesium chloride gradient. Because of the heavy and light nitrogen, the DNAs sedimented to different points in the gradient. The heavest DNA had ^{15}N in both strands, while the lightest DNA had only ^{14}N in both strands. Meselson and Stahl found that, after one generation in light nitrogen, the DNA was intermediate in weight. After two generations, half of the DNA was intermediate and the other half was light. Eventually most of the DNA became light.

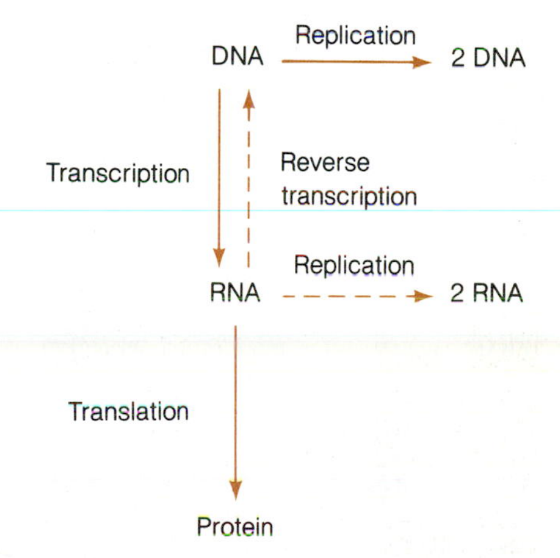

Figure 9-2 Central Dogma. The central dogma illustrates how DNA controls its own replication and the synthesis of RNA. In addition, the relationship between RNA and the synthesis of protein is indicated.

Figure 9-3 The Role of DNA Polymerase in DNA Replication. *(a)* The main genome in bacteria is a circular DNA molecule that is replicated bidirectionally. The arrows indicate the replication forks. *(b)* A detailed view of the replication fork illustrates how DNA polymerases synthesize new strands in the 5′ to 3′ direction. Because the templates are oriented in opposite directions, the DNA polymerases are moving in opposite directions in the replication fork. *(c)* DNA polymerase promotes hydrogen bonding of the correct nucleotide to the template. In addition, the DNA polymerase chemically joins each nucleotide to the primer DNA.

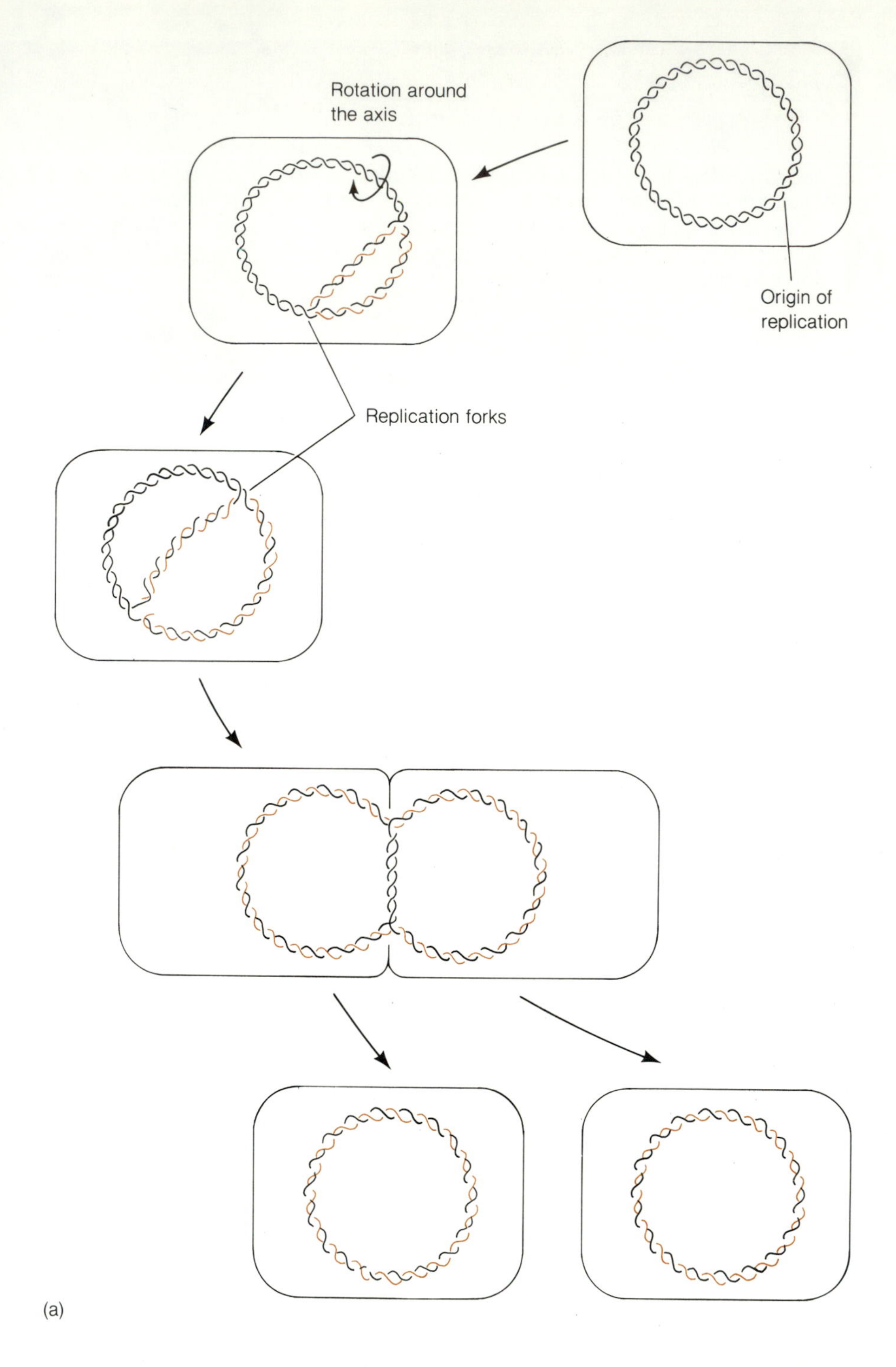

Figure 9-3 illustrates how daughter genomes are believed to be segregated from each other in bacteria. The two original strands of the genome apparently are attached to different sites on the plasma membrane. As the membrane grows between the attachment points, the daughter genomes are moved to the poles of the cell. When the cell membrane invaginates between the daughter genomes, two new cells are formed, containing identical genomes.

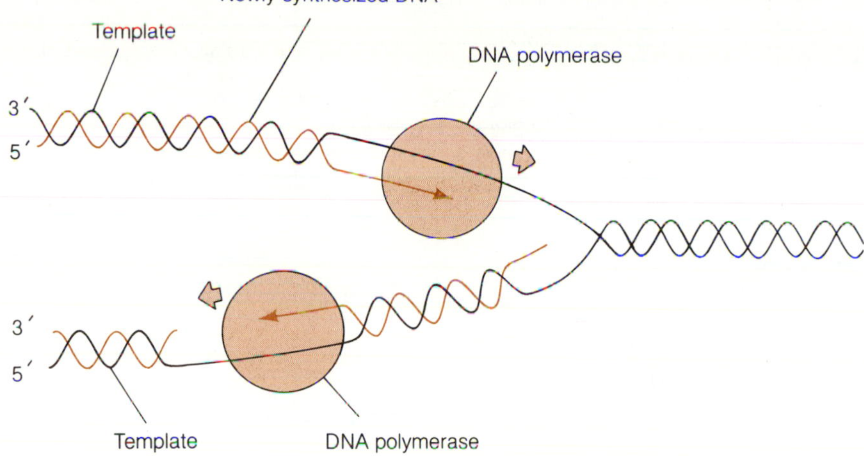

(b) Replication fork

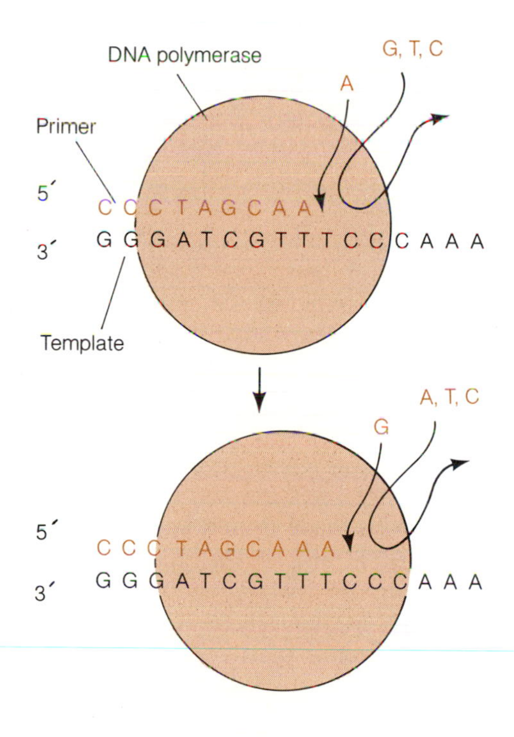

(c)

Figure 9-3 (continued)

RNA SYNTHESIS IN PROKARYOTES

Although the DNA template determines the sequence of nucleotides in RNA, usually only one of the complementary strands of the DNA is copied into RNA. The copying of DNA into RNA, known as **transcription,** is carried out by an **RNA polymerase**. RNA polymerase initiates RNA synthesis at a **promoter site** on the DNA and transcribes the DNA until a **termination site** is encountered (fig. 9-4). Notice that the RNA is displaced from the DNA template, even before it is completed, because hydrogen bonding between two strands of DNA is more stable than hydrogen bonding between DNA and RNA.

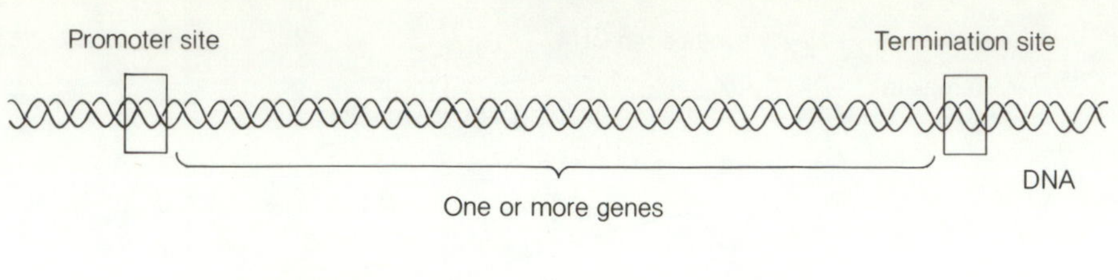

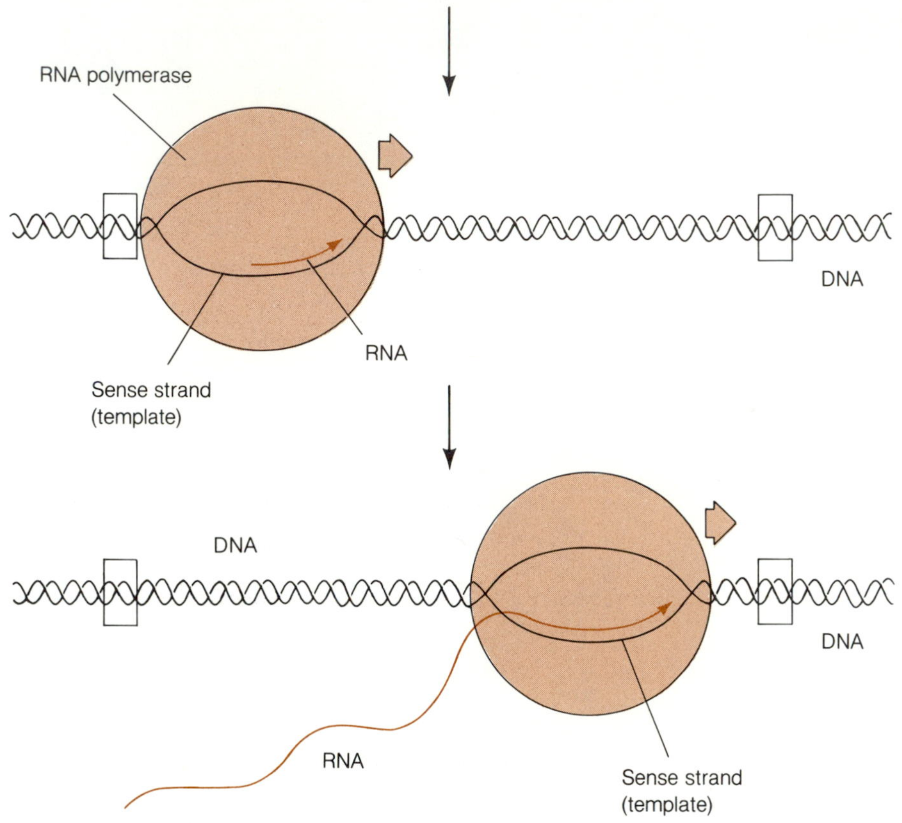

Figure 9-4 Transcription. Transcription begins at promoter sites and ends at termination sites on the DNA. An RNA polymerase transcribes one strand of the DNA, sometimes called the sense strand, into RNA.

Ribosomal RNA (rRNA) is made from a large precursor molecule that is enzymatically cleaved to tRNA and to 16s, 23s, and 5s rRNA. The rRNAs initially serve as focal points for the aggregation of ribosomal proteins □. 93 The rRNAs and the proteins are the structural components of the ribosomal subunits. The 30s subunit contains the 16s rRNA and 21 proteins, while the 50s subunit contains the 23s and 5s rRNAs and approximately 34 proteins.

Transfer RNA (tRNA) is made from a number of different precursor RNA molecules. The precursor RNAs may contain many tRNAs that are separated from each other by enzymatic cleavage. Once the tRNAs are produced, many of the nucleotides are converted into unusual nucleotides (fig. 9-5). Notice that one of the loops, called the TψC-loop, contains thymine (T) rather than uracil (U) and an unusual nucleoside called pseudouridine (ψ). Thymine is made from uracil by adding a methyl group (CH_3) to the fifth position in the ring (fig. 9-5). Pseudouridine is an alternate form of the nucleoside uridine. Another loop is called the **D-loop** because it contains dihydrouridine. The **anticodon loop** normally has unusual nucleotides on either side of the **anticodon.** An anticodon is a triplet of nucleotides on the tRNA which binds to a triplet of nucleotides on the mRNA, called the **codon,** dur-

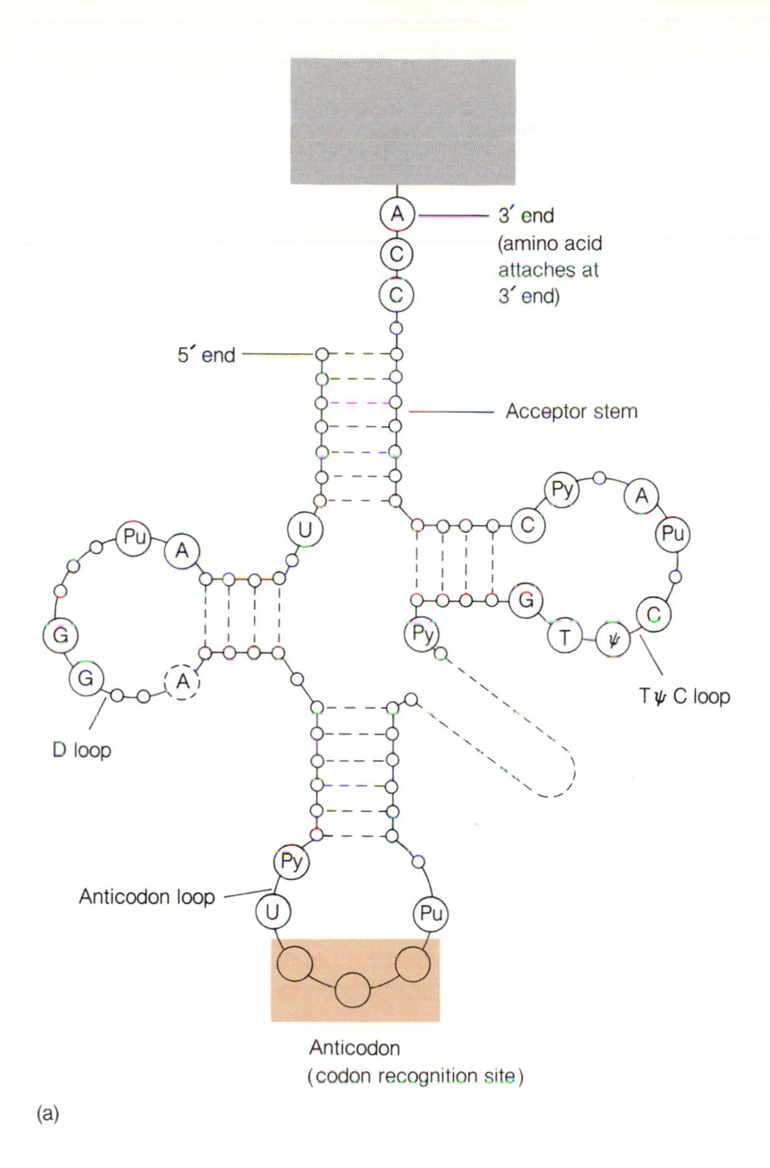

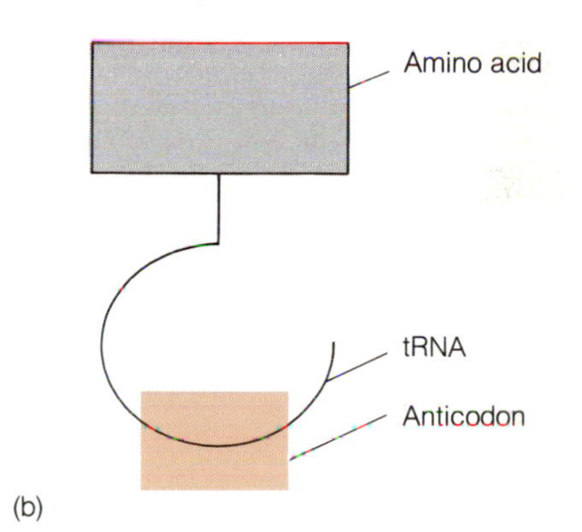

Figure 9-5 The Structure of Transfer RNA. *(a)* Transfer RNA (tRNA) is processed from various large RNA transcripts. The final tRNAs are all about the same size and have a sedimentation value of approximately 4s. Many of the nucleotides in tRNA are modified chemically to produce nucleotides such as thymidine (T), methylguanosine (mG), pseudouridine (ψ), dihydrouridine (D), and 7-methylguanosine (7mG). *(b)* Simplified notation used to illustrate a tRNA molecule.

ing protein synthesis. The tRNAs are differentiated on the basis of their anticodons and other physical and chemical characteristics. Each type of tRNA is covalently bonded to one of the 20 amino acids used to make protein. When an amino acid is attached to a tRNA, the tRNA is said to be **charged.** On the other hand, when no amino acid is attached to a tRNA, the tRNA is **uncharged.**

In bacteria, mRNA is usually not cleaved or modified before it is used. In many eukaryotes, however, the mRNA is extensively processed before it can be translated.

PROTEIN SYNTHESIS IN PROKARYOTES

41 The amino acid sequence in proteins □ is determined by the nucleotide sequence in mRNA. The sequence is "read" in groups of three nucleotides by tRNAs with complementary anticodons (fig. 9-6). Ribosomes promote the

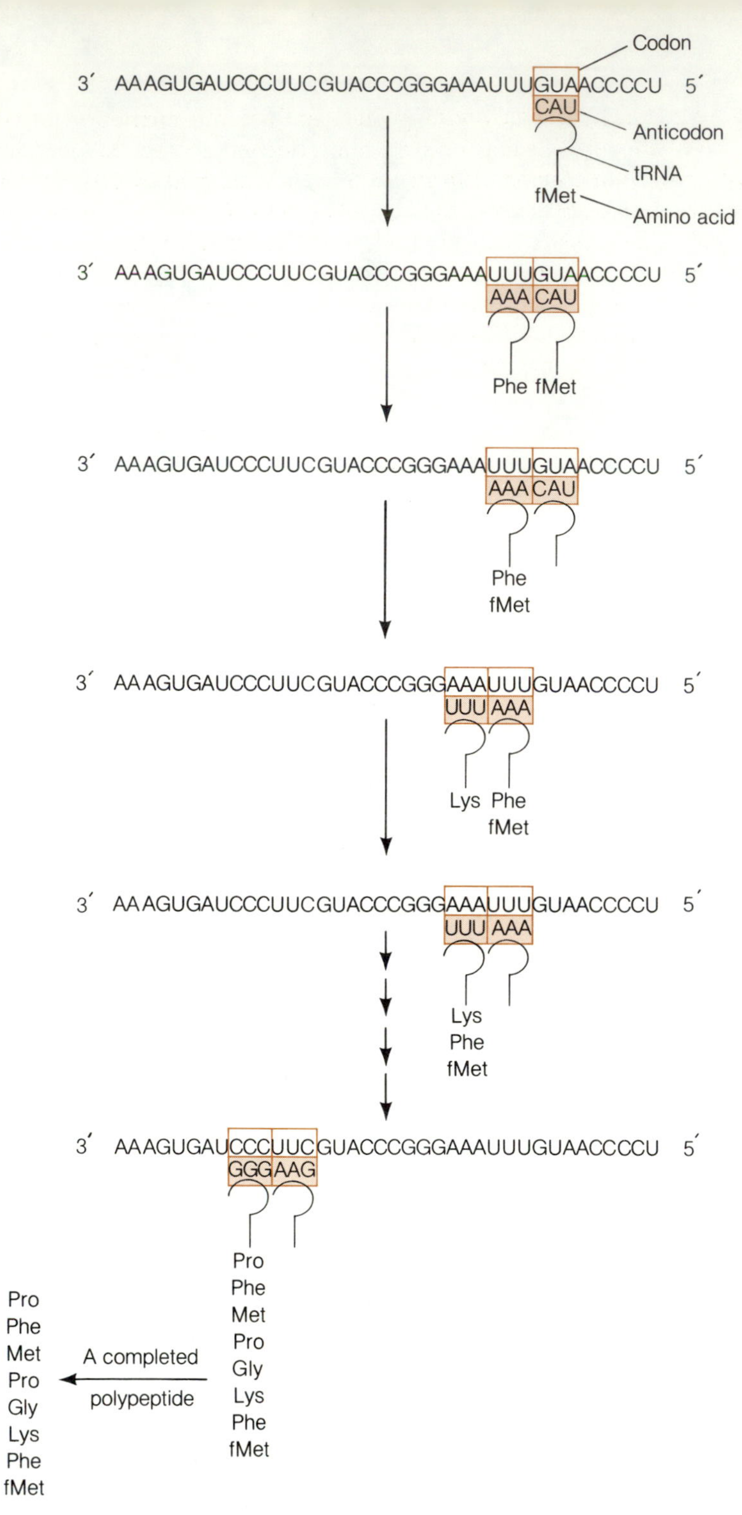

Figure 9-6 Relationship Between Messenger-RNA and Protein. The nucleotide sequence in a mRNA determines the amino acid sequence in a protein. Protein synthesis in bacteria begins near the 5′ end of a mRNA with the first start codon 5′AUG 3′, and ends at the first nonsense codon (termination codon): 5′UAA 3′, 5′UAG 3′, or 5′UGA 3′. The anticodon of the appropriate tRNA hydrogen bonds to triplet codons starting with 5′AUG 3′. The ribosome promotes the hydrogen bonding between the codons and anticodons and catalyzes the formation of peptide bonds between adjacent amino acids attached to the tRNAs.

codon-anticodon hydrogen bonding and catalyze the formation of covalent bonds between neighboring amino acids attached to the tRNAs. The polymerization of amino acids to form proteins is catalyzed by 70s ribosomes in prokaryotes, but by 80s ribosomes in eukaryotes. Protein synthesis begins at the 5′ end of the mRNA with the first **start codon** (5′ AUG 3′), and ends at the first **nonsense codon** (termination codon) (5′UAG 3′, 5′UAA 3′, or 5′UGA 3′). The **aminoacyl-tRNA** determined by each possible triplet codon is indicated in figure 9-7. The first ribonucleotide of the triplet codon is the one on the 5′ end. Notice that some amino acids (leucine and arginine) are coded for by as many as six codons while methionine and tryptophan are each coded for by a single codon. For the purpose of discussion, the synthesis of protein is divided into three stages: **initiation, elongation,** and **termination.**

Initiation of Translation

The initiation of translation in bacteria begins when the 30s subunit of the ribosome binds to a start codon (AUG) in messenger RNA, and to a **start region** complementary to 5–6 bases of the 16s rRNA in the 30s subunit (fig. 9-8a). A molecule of GTP bound to the 30s subunit and initiation factors (IF) promote the attachment of the first **aminoacyl-tRNA,** which in bacteria is always **N-formylmethionyl-tRNA.** Aminoacyl-tRNAs are formed in the cytoplasm when amino acids are attached to their tRNAs by enzymes called **aminoacyl-tRNA synthetases.** N-formylmethionyl-tRNA binds to the

Second ribonucleoside

First ribonucleoside	U	C	A	G	Third ribonucleoside
U	UUU Phe	UCU Ser	UAU Tyr	UGU Cys	U
	UUC Phe	UCC Ser	UAC Tyr	UGC Cys	C
	UUA Leu	UCA Ser	UAA non	UGA non	A
	UUG Leu	UCG Ser	UAG non	UGG Trp	G
C	CUU Leu	CCU Pro	CAU His	CGU Arg	U
	CUC Leu	CCC Pro	CAC His	CGC Arg	C
	CUA Leu	CCA Pro	CAA Gln	CGA Arg	A
	CUG Leu	CCG Pro	CAG Gln	CGG Arg	G
A	AUU Ile	ACU Thr	AAU Asn	AGU Ser	U
	AUC Ile	ACC Thr	AAC Asn	AGC Ser	C
	AUA Ile	ACA Thr	AAA Lys	AGA Arg	A
	AUG Met	ACG Thr	AAG Lys	AGG Arg	G
G	GUU Val	GCU Ala	GAU Asp	GGU Gly	U
	GUC Val	GCC Ala	GAC Asp	GGC Gly	C
	GUA Val	GCA Ala	GAA Glu	GGA Gly	A
	GUG Val	GCG Ala	GAG Glu	GGG Gly	G

non—Nonsense or polypeptide chain-terminating codons
AUG when functioning as a start codon codes for N-formylmethionine

Figure 9-7 The Genetic Code. The genetic code is a list of all the possible triplet nucleotides (codons) in mRNA and the amino acids these triplet nucleotides specify. Three of the codons do not specify any amino acid because there are no tRNAs with matching anticodons. The three codons that do not specify any amino acid are called nonsense codons or termination codons.

Figure 9-8 Protein Synthesis. Protein synthesis, or translation, is divided into three stages: initiation, elongation, and termination. *(a)* During the initiation of translation, the 30s and 50s subunits of the 70s ribosome bind to the mRNA at a start codon. In addition, the first aminoacyl-tRNA (N-formylmethionyl-tRNA) is attached to the start codon and to the peptidyl-tRNA site (P-site) on the 50s subunit. *(b)* During the elongation process, aminoacyl-tRNAs bind one by one to the mRNA and the aminoacyl-tRNA site (A-site) on the 50s subunit. The ribosome catalyzes the formation of a peptide bond between the amino acid in the A-site and the growing peptide in the P-site. *(c)* Eventually, nonsense codons are reached and the termination process is initiated. A release factor binds to the ribosome and causes the release of the growing peptide from the tRNA and a dissociation factor causes the dissociation of the 70s ribosome.

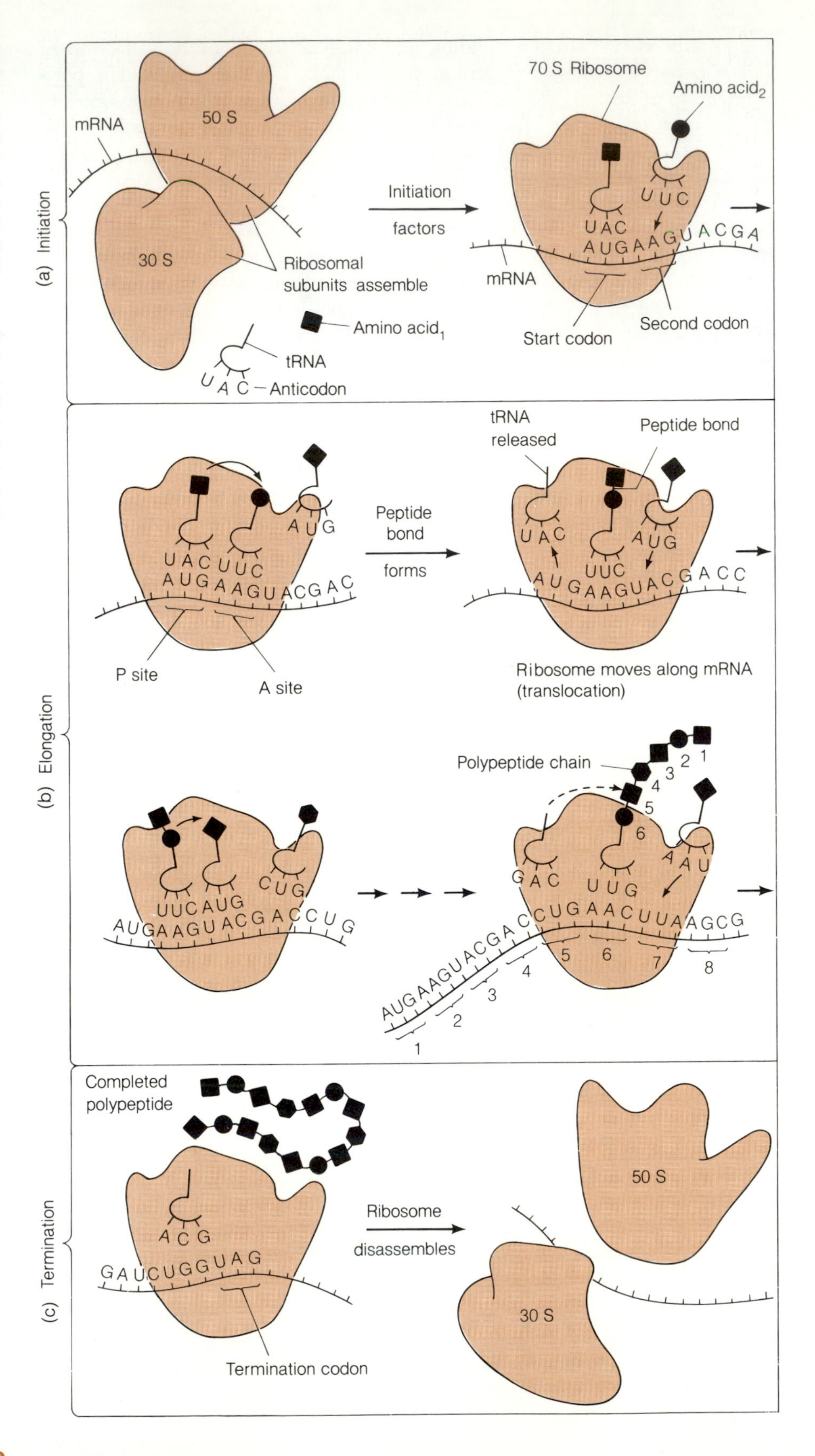

mRNA because the anticodon on the tRNA (UAC) hydrogen-bonds to the start codon on the mRNA (AUG). The hydrolysis of GTP stimulates the binding of the 50s ribosomal subunit to both the 30s subunit and the N-formylmethionyl-tRNA. Although the anticodon of the N-formylmethionyl-tRNA is attached to the start codon on the mRNA, the rest of the tRNA and the amino acid are attached to a site on the 50s subunit called the **peptidyl-tRNA site** (P-site).

Elongation in Translation

Once the initiation of protein synthesis has occurred, subsequent aminoacyl-tRNAs associate with the mRNA and the 70s ribosome. The codon following the start codon determines which of the aminoacyl-tRNAs is to be next. The appropriate aminoacyl-tRNA, an **elongation factor** (Tu), and a molecule of GTP form a complex that binds to the ribosome (fig. 9-8b). The hydrolysis of GTP causes the aminoacyl-tRNA to bind strongly to the **aminoacyl-tRNA site** (A-site) on the 50s subunit of the ribosome. An enzyme called **peptidyl transferase,** located in the 50s subunit of the ribosome, catalyzes the formation of a peptide bond between the carboxyl group of N-formylmethionine and the amino group of the newly arrived amino acid. The formation of the peptide bond cleaves the N-formylmethionine from its tRNA. The two amino acids are now joined together and attached to the tRNA in the aminoacyl-tRNA binding site (fig. 9-8). A complex of the enzyme called **translocase** and GTP bind to the ribosome at this time. The hydrolysis of GTP by the translocase provides the stimulus for ejecting the uncharged tRNA in the P-site and moving the **peptidyl-tRNA** (the tRNA with the growing protein) from the A-site to the P-site (fig. 9-8). Since the peptidyl-tRNA is still attached to the codon just after the start codon, the mRNA is pulled past the ribosome. Amino acids are polymerized into macromolecules by repeating these elongation steps.

Termination of Translation

Protein synthesis ends when the ribosome comes to a **nonsense codon** or **termination codon** (UAA, UAG, UGA) (fig. 9-8c). The termination codons are commonly called nonsense codons because they provide no sense or information for making a protein. There are no normal tRNAs with anticodons that can recognize the termination codons.

Proteins called **release factors** bind to the ribosome when it is stalled in front of a nonsense codon and stimulate the peptidyl transferase to cleave the newly formed peptide from the tRNA. When this happens, a protein called **dissociation factor** binds to the 30s subunit and causes the ribosome to dissociate into its subunits. The uncharged tRNA dissociates from the mRNA.

The **genetic code** was elucidated in the 1960s by a number of scientists. In 1961, Marshall Nirenberg and Heinrich Matthaei reported that chemically synthesized RNAs composed of only one type of ribonucleotide would stimulate the polymerization of just one of the 20 amino acids into a polypeptide. By 1964, Marshall Nirenberg and Philip Leder discovered that specific aminoacyl-tRNAs would bind to ribosomes only when certain triplet codons were present. This finding allowed them to show which of the codons corresponded to which of the aminoacyl-tRNAs. In addition, they found that three codons did not stimulate the binding of any known aminoacyl-tRNA.

Har Gobind Khorana used another approach to determine the genetic code during the early 1960s: he chemically synthesized RNAs with different proportions of the four nucleotides and then determined what types of proteins were synthesized. This allowed him to determine which triplet codons corresponded to which aminoacyl-tRNAs. He also discovered that three of the triplet codons did not promote the incorporation of amino acids, but instead terminated protein synthesis. Marshall Nirenberg and Har Gobind Khorana shared the 1968 Nobel Prize in Medicine or Physiology for their clarification of the genetic code and how it specifies the synthesis of proteins (fig. 9-7).

THE MOLECULAR BASIS FOR VARIATION

The sequence of nucleotides in DNA determines the structure of rRNA, tRNA, mRNA, and the multitude of proteins required for the maintenance and multiplication of all living cells. The genetic information must be accurately replicated and passed on intact to daughter cells if a species is to survive. If the DNA is not faithfully replicated, or if only part of it is replicated, one or both of the daughter cells may not survive.

Even though the integrity of an organism's DNA is of paramount importance, DNA is not a static, unchanging molecule. In fact, for any given gene, one out of every 10^7 to 10^9 bacteria will have an altered form of the gene. Proteins of related bacteria are almost identical, but proteins from unrelated bacteria can be quite different. Similar enzymes from prokaryotes and from eukaryotes may be even more diverse in structure. Not all proteins diverge at the same rate. Some proteins that carry out very specific processes in conjunction with other macromolecules, such as the histones and cytochrome c □, do not differ as much as those proteins which work alone or have broad functions, such as the hemoglobins and the fibrinopeptides. Cytochrome c diverges at a rate of 1 amino acid change/100 every 20 million years, while fibrinopeptides diverge at a rate of 1 amino acid change/100 every 1.1 million years. Even in rapidly evolving proteins, there are regions that show little or no variation. These regions are said to be conserved and are usually intimately involved with the protein's function.

192

A change in an organism's genetic information is referred to as a **mutation.** A mutation occurs when a base pair is substituted for another base pair (**transversion** or **transition**), or when a base pair is added or deleted (**addition** or **deletion**) (fig. 9-9). A DNA molecule can be altered not only by changes in its bases but also by exchanging and rearranging regions of DNA (**recombination**).

A great many mutations are due to the instability of the nucleotide bases. Very infrequently, the bases absorb sufficient energy from bombarding water molecules to undergo a chemical change called a **tautomeric shift** (fig. 9-10). A tautomeric shift changes the distribution of electrons and protons in the nucleotide bases so that they no longer pair normally. Figure 9-10 illustrates how the bases pair with one another when they are in their unstable high-energy states (*). Instead of the normal G≡C and A=T pairing, the pairing is between G*≡T, A*=C, T*≡G, and C*=A.

Spontaneous tautomeric shifts can result in permanent mutations. For example, suppose that as a DNA molecule is being replicated, a base (A) undergoes a tautomeric shift just as the DNA polymerase reaches it (fig. 9-

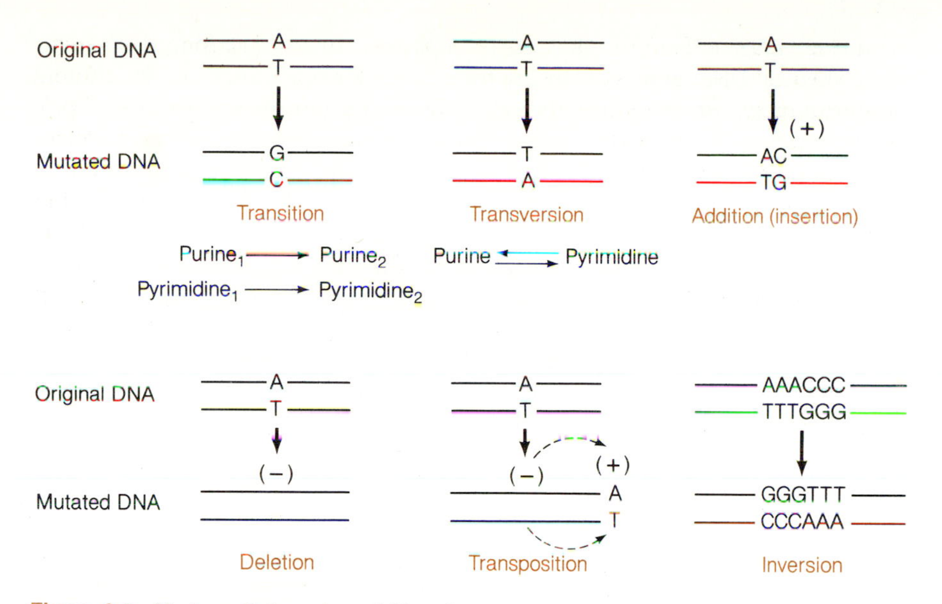

Figure 9-9 Various Categories of Mutations

11). The base that is complementary to A* is cytosine rather than thymine, so cytosine becomes part of one of the new strands. Notice that one of the daughter chromosomes is perfectly normal while the other has a mispaired region. When the chromosome with the mispaired region is replicated, however, a perfectly matched mutant chromosome (different from the grandparent) is formed (fig. 9-11). Spontaneous mutations of this sort can occur any

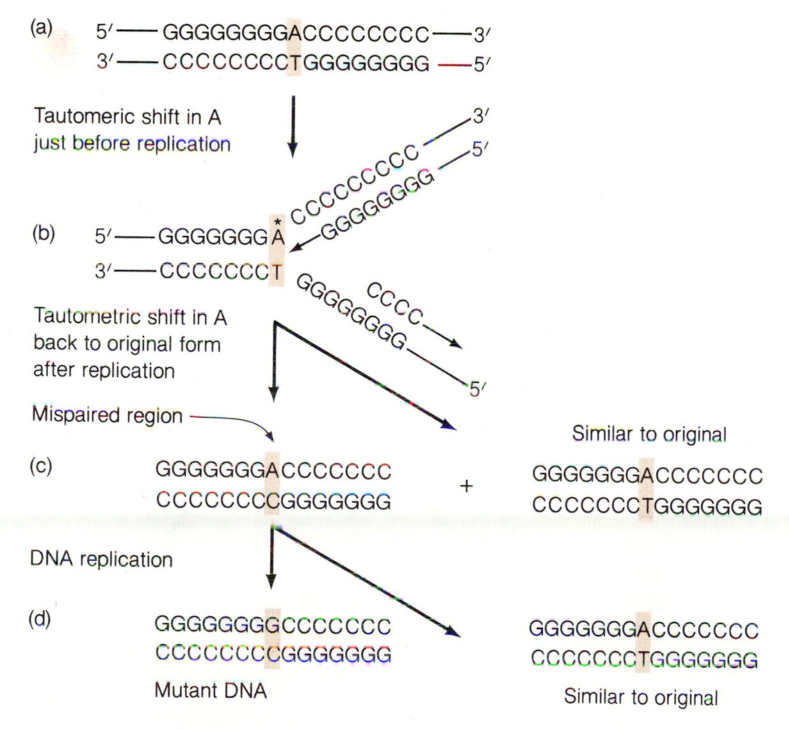

Figure 9-11 Consequence of Abnormal Base Pairing During DNA Synthesis. Tautomeric shifts during DNA synthesis cause the wrong nucleotides to be incorporated into the new strands of DNA. When the DNA is replicated again, one of the daughter DNAs contains a mutation.

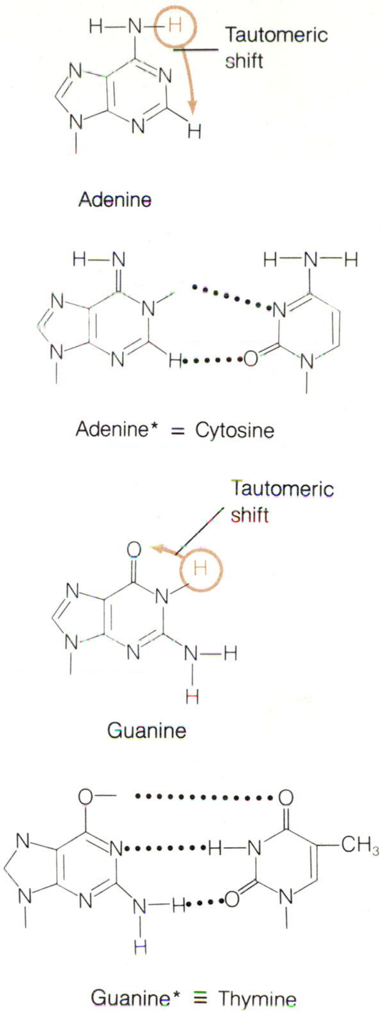

Tautomeric shift

Adenine

Adenine* = Cytosine

Tautomeric shift

Guanine

Guanine* ≡ Thymine

Figure 9-10 Tautomeric Shifts and Abnormal Base Pairing. A tautomeric shift occurs when molecules redistribute their electrons and protons. Tautomeric shifts can cause the bases in nucleotides to become unstable and pair abnormally with each other. The unstable adenine bonds to cytosine (A*═C); the unstable guanine pairs with thymine (G*═T); the unstable thymine bonds to guanine (T*═G); and the unstable cytosine joins with adenine (C*═A).

time the DNA is being replicated or repaired. Since the amount of energy absorbed by DNA is related to the temperature of its aqueous environment, the mutation rate should be directly proportional to the temperature of the environment.

Mutations are caused by many chemical and physical agents that are referred to as **mutagens** (table 9-1). Ultraviolet light □ is a powerful mutagen because it is able to penetrate cells and because it is absorbed by thymine and cytosine. The absorption of ultraviolet light causes a redistribution of electrons and protons in T and C (pyrimidine bases) that makes them reactive. If there are two pyrimidines stacked together, they may react with each other, forming **pyrimidine dimers** (fig. 9-12a). The formation of thymine dimers appears to occur more frequently than the formation of other dimers. The formation of dimers distorts the DNA and thus interferes with DNA replication and transcription. If not corrected, this distortion leads to cell death; however, cells have a number of repair systems that correct the distortion. One repair system consists of an **endonuclease** (an enzyme that cuts within a DNA strand) that recognizes the distortion on the DNA and nicks the DNA near the dimer. A **DNA polymerase** excises the distortions and replaces the deleted nucleotides. Eventually, a **DNA ligase** joins the nicks left by the DNA polymerase (fig. 9-12b).

162

Most of the time, DNA repair restores the DNA to its original state, but the DNA polymerase is not perfect. It makes mistakes at a rate of about 1 in 10^7 to 10^9 repairs. Therefore, if there are many dimers to repair in each cell and the number of cells considered are in the millions, the number of mutations permanently introduced into the population will be high. If genes are drastically altered so that essential RNAs or proteins are not produced, cells do not survive.

Chemicals are frequently used to induce mutation. For example, the mutagen HNO_2 (nitrous acid) manages to get into cells and deaminate the bases (fig. 9-13a). This **deamination** alters the way in which the bases pair and consequently introduces mutations into daughter DNA molecules when the DNA is replicated. Nitrous acid has an effect on three of the bases in DNA: adenine (A) is changed to hypoxanthine (HX), cytosine (C) is changed to uracil (U), guanine is changed to xanthine (X), and thymine is not affected. Even though guanine is changed to xanthine, there is no alteration in the base pairing. Pairing by the unusual bases is HX = C and X = C.

Base analogues are molecules that resemble the normal nucleotides so closely that they are incorporated into DNA (or RNA). Base analogues such

TABLE 9-1
MUTAGENS AND THEIR MODE OF ACTION

MUTAGEN	MODE OF ACTION
Heat	Causes tautomeric shifts in the bases.
Chemicals	Interact directly with the DNA and cause disruptive changes.
Ultraviolet light	Chemically alters the DNA and stimulates excessive repair.
X-rays, Gamma rays (electromagnetic waves)	Chemically alter the DNA by ionization and cause disruptive changes and excessive repair.
Radioactive atoms beta particles alpha particles (subatomic particles)	Chemically alter the DNA by ionization and cause disruptive changes and excessive repair.

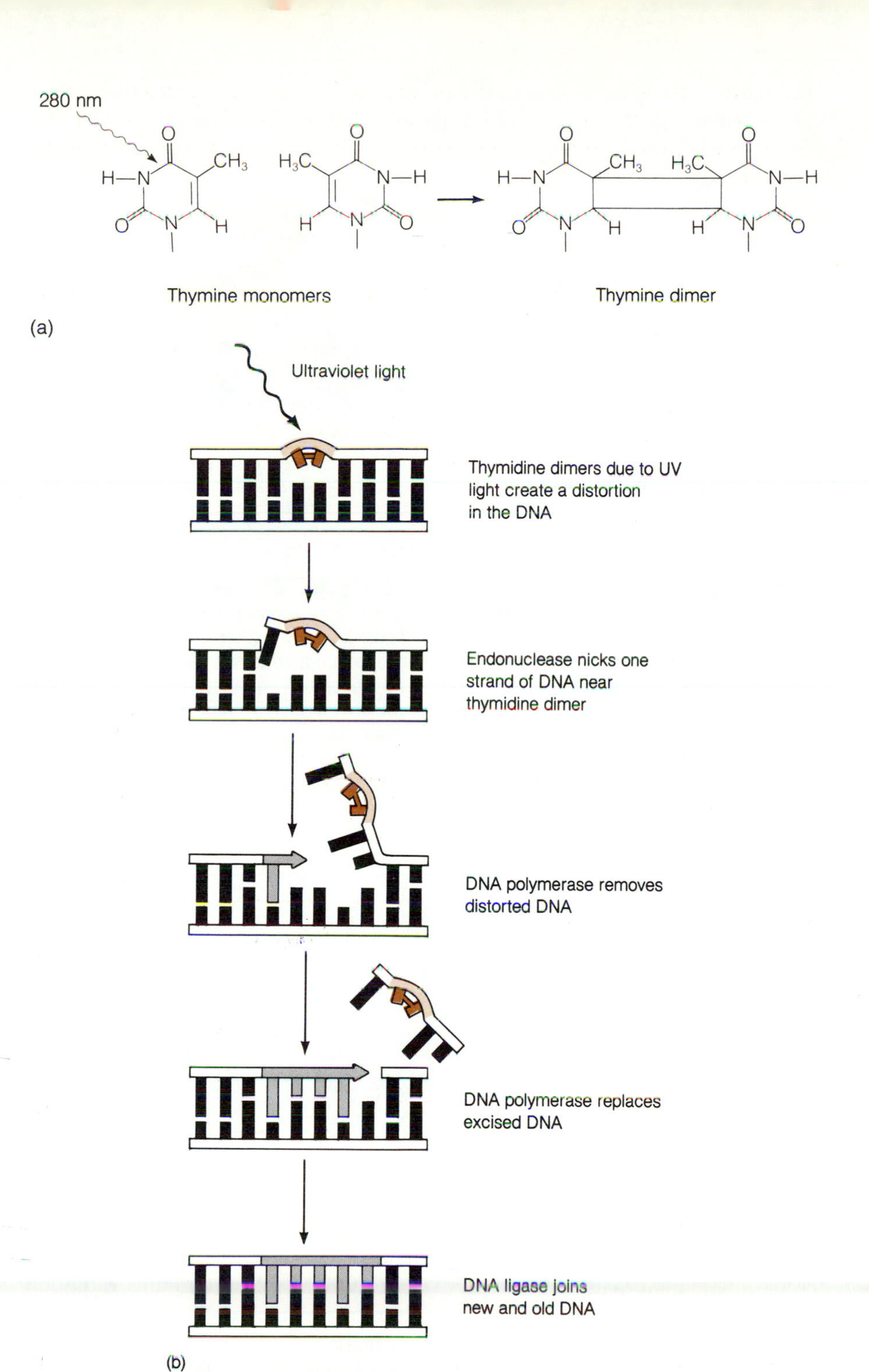

Figure 9-12 The Formation of Thymidine Dimers. *(a)* When two thymidine nucleotides lie next to each other in the same DNA strand, they can react together if they are destabilized by the absorption of ultraviolet light with a wavelength of 280 nm. The thymidines form thymidine dimers. *(b)* Thymidine dimers distort the DNA molecule and prevent DNA replication and transcription. Most bacteria have a repair system that removes thymidine dimers and replaces them with the appropriate nucleotides. Some of the steps involved in the repair of DNA distorted by thymidine dimers are outlined.

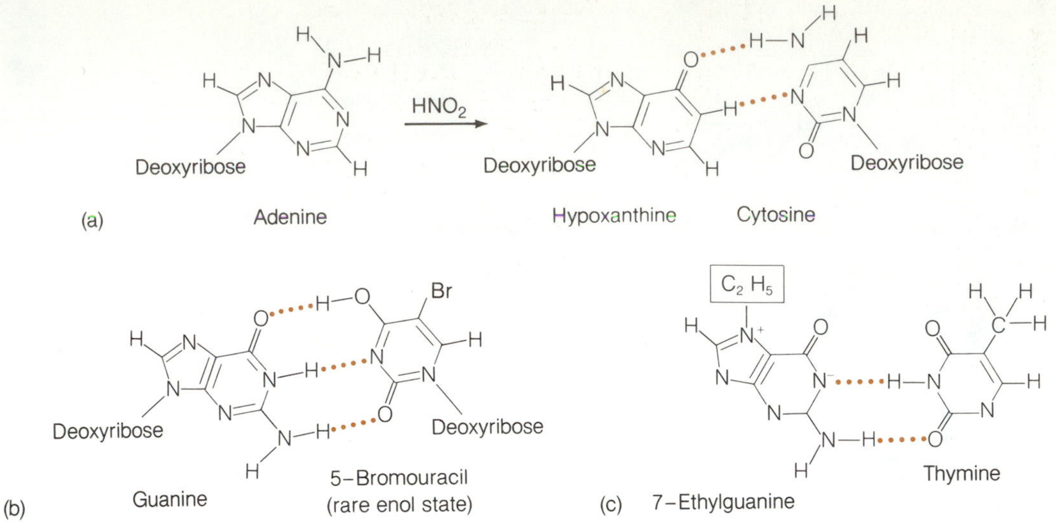

Figure 9-13 The Effect of Mutagens on Nucleotides. *(a) Nitrous acid* is a mutagen that deaminates DNA bases, disrupting base pairing. Consequently, when DNA is repaired or replicated, incorrect nucleotides are introduced into the DNA. Adenine is deaminated to hypoxanthine, which hydrogen-bonds to cytosine; cytosine is deaminated to uracil, which pairs with adenine; and guanine is deaminated to xanthine, which associates with cytosine. The deamination of guanine to xanthine is not mutagenic, because xanthine still pairs with cytosine *(b) 5-bromouracil* (5BU) is a base analogue that is incorporated into DNA during DNA repair or replication. 5BU frequently undergoes a tautomeric shift and then pairs with guanine. Consequently, when DNA is repaired or replicated, incorrect nucleotides are introduced into the DNA. *(c) Alkylating agents,* such as ethylmethane sulfonate, methyl bromide, or ethylene oxide, add methyl or ethyl groups to guanine or adenine. When guanine is alkylated by ethylmethane sulfonate, an ethyl group is added to the 7-carbon, creating 7-ethylguanine. The latter pairs with thymine rather than with cytosine. Consequently, when DNA is repaired or replicated, thymine is incorporated into the DNA instead of cytosine.

as 5-bromouracil (5-BU) that frequently undergo tautomeric shifts can induce mutations (fig. 9-13b). Since many of the base analogues hydrogen-bond differently than the bases they replace, they are often used to induce mutations in DNA.

Many of the most potent mutagens are **alkylating agents,** such as nitrosamines, methyl bromide, and ethylene oxide. These mutagens add methyl or ethyl groups (alkylate) to G and A (fig. 9-13c). Alkylation of the bases causes them to hydrogen-bond incorrectly, and this effect in turn introduces mutations. In addition, alkylation often causes the DNA to break.

X-rays, gamma rays, and high-energy particles □ (such as neutrons, beta particles, and alpha particles) from radioactive atoms are very potent mutagens because they are able to penetrate cells and disrupt many molecules along their path. This type of radiation is called ionizing radiation because many ions and **radicals** (highly reactive uncharged molecules) are formed within the cell. These ions and radicals react with the nucleotides in the DNA to cause the release of bases and breaks in the DNA strands. The loss of bases can result in deletions, and the DNA breaks can lead to genome breaks and recombination.

161

Although most mutations are harmful because they drastically alter the function of essential components, some mutations are beneficial to the bearer and to the population because mutations introduce variability. Diversity promotes the survival of a population by enabling it to survive a broad spectrum of conditions. The accumulation of mutations has led to the evolution of many bacteria that differ ever so slightly from each other and live in many different environments.

Most bacteria proliferate rapidly (some generation times are as short as 20 minutes) in comparison to eukaryotes and frequently reach cell densities greater than 1 billion cells per ml (10^9/ml). If the mutation rate is 10^{-7} for a given gene, there should be 100 organisms per ml which contain mutations in a given gene (10^9 cells/ml $\times$ 10^{-7} mutations/cells $=$ 100 mutants/ml). These mutants might do better than **wild-type** (normal) organisms under some environmental conditions and more poorly under other environmental conditions. Variation of this sort in all the genes of a population often insures enough diversity so that some organisms can survive changes in the environment or live in several environments.

CONJUGATION, TRANSDUCTION, AND TRANSFORMATION

Many bacteria have evolved mechanisms to create more variability than would be possible from spontaneous mutations alone. These mechanisms for creating variation allow a beneficial characteristic to spread throughout a population within hours. The spread of a spontaneous mutation throughout a population by natural selection would require the replacement of most of the population and the extensive proliferation of the mutant carrying the beneficial genes. This process would be very wasteful in terms of time and nutrients and the lag in time might endanger the population. **Conjugation, transduction,** and **transformation** are processes that promote variability in a bacterial population.

Conjugation

89

Numerous bacteria have, in addition to their main genome, small circular DNA molecules called **plasmids** □. Plasmids usually have 5 to 100 genes and are able to replicate themselves. Gram negative bacteria may have sex plasmids, or **fertility factors** (F factors) as they are sometimes called. These fertility factors code for a protein known as pilin that self-assembles to form sex pili □. The hollow sex pili bind bacteria together and establish a connection between the bacterial cytoplasms (fig. 9-14). It is unclear whether or not DNA can pass through the pilus from the donor cell to the recipient cell. Some scientists believe that the pilus simply draws the donor and recipient cells together so that the outer membranes fuse and a passageway is created for the transfer of the DNA. In either case, one strand of the plasmid is able to pass from a male bacterium (one that has a plasmid) to a female bacterium (one that lacks a plasmid) through the pilus or pore that is created between the bacteria.

88

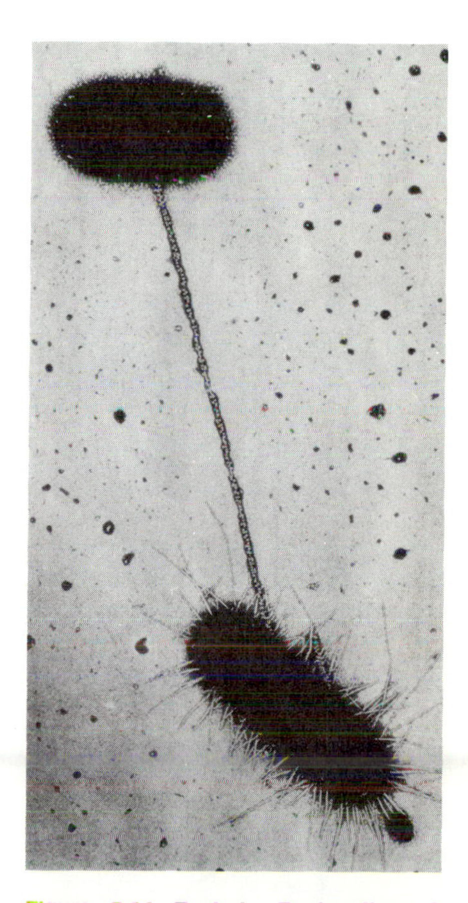

Figure 9-14 Bacteria Conjugating. A male bacterium (bottom) has produced a long (about 4 μm) sex pilus that has attached to a female bacterium (top). Other types of pili are visible on the male bacterium, but the female bacterium has no pili on its surface.

One model for the transfer of DNA proposes that the plasmid DNA is cut at a specific site and that the 5′ end of one of the DNA strands enters a growing pilus (fig. 9-15). As the pilus grows, the single strand of DNA is pulled into the pilus. When the pilus binds to a female bacterium, a channel forms that allows the plasmid to enter the female. Possibly, the female bacterium breaks down the pilus, thus enabling the plasmid to enter.

Even though the plasmid in the male cell is transferring one of its strands to the female, it remains a double-stranded plasmid because a DNA

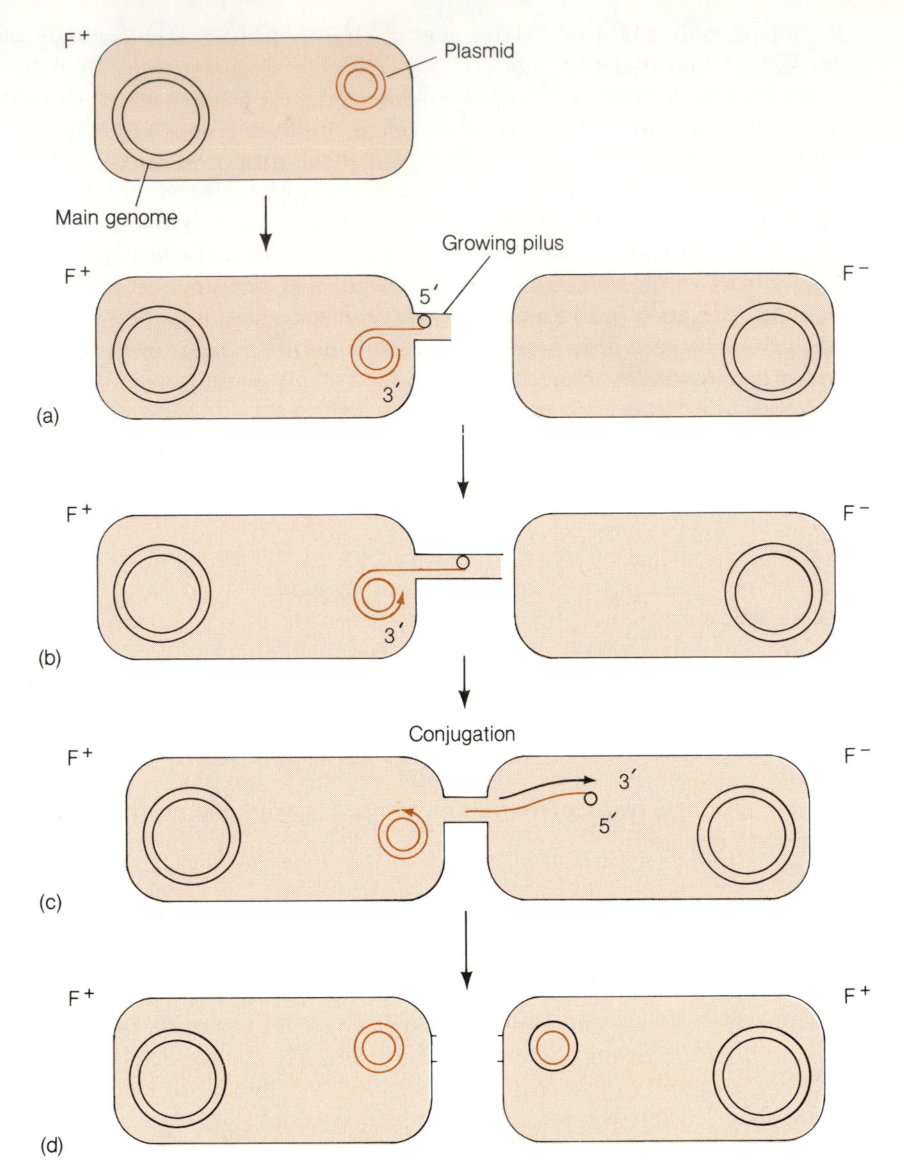

Figure 9-15 Transfer of DNA During Conjugation in Bacteria. *(a)* One strand of the plasmid in a F$^+$ bacterium is cut and the 5′ end of the strand attaches to the inside of a growing pilus. *(b)* As the pilus grows, the DNA strand unwinds from the episome and is pulled into the pilus. The 3′ end of the strand is elongated by a DNA polymerase so that the plasmid continues to be double-stranded. *(c)* Eventually, the pilus makes contact with a F$^-$ bacterium that absorbs the pilus. When the pilus is absorbed by the female bacterium, the strand of DNA enters the female, in which a DNA polymerase synthesizes a complementary strand. *(d)* If a complete plasmid enters the female, it forms a circular plasmid and the female becomes a male.

polymerase is replacing the strand as fast as it leaves the cell. The single-stranded plasmid DNA that is entering the female cell serves as a template for the synthesis of a complementary strand.

When the female bacterium obtains a sex plasmid, it is converted into a male and can synthesize pili. Not all the cells in a population become males, because as males reproduce they frequently do not segregate plasmids. Thus, a good number of the daughter cells are females.

172

The plasmid usually transmits genes that are not found on the main genome. Genes that make the bacterium resistant to a variety of antibiotics □ (sulfanilamide, streptomycin, chloramphenicol, and tetracycline) are common and protect these bacteria from many antibiotics produced by other microorganisms in their environment. Plasmids may also carry alternate forms (**alleles**) of genes found on the main genome. Alleles can benefit a bacterium by making it partially **diploid** (two copies of the hereditary material), hence, providing additional genetic information, which may be of use should environmental changes occur that affect the reproduction of the cell. Some plasmids, called **episomes,** are able to integrate into the main genome (fig. 9-16). Episomes are defined as plasmids that can replicate either as an integrated part of the main genome or as an autonomous piece of DNA unlinked to the main genome. When a plasmid is part of the main genome, the genome can be transferred to female cells in much the same way that the plasmid is transferred. Since the main genome (3,000–5,000 genes) is so much larger than the plasmid (5–100 genes), the entire genome usually is not transferred. The fragment of DNA that enters the female cell does not form a plasmid, but instead **recombines** with the female's main genome. Any part of the DNA that does not recombine is promptly destroyed. Apparently, there is a region on the plasmid DNA that is required for circularization. This region enters the female cell last. If this region does not make it into the female cell (it usually does not), the DNA can not circularize and it is broken down.

There are three types of male bacteria defined by the type of plasmid or fertility factor they have (fig. 9-16). **F$^+$** bacteria have fertility factors (plasmids) with a limited number of genes. These genes are not duplicates or allelic forms of those on the main genome. **F$'$** bacteria have fertility factors with a limited number of genes, many of which are alleles or duplicates of those on the main genome. **Hfr** bacteria have a fertility factor, but it is integrated into the main genome. Hfr is an acronym for "high frequency of

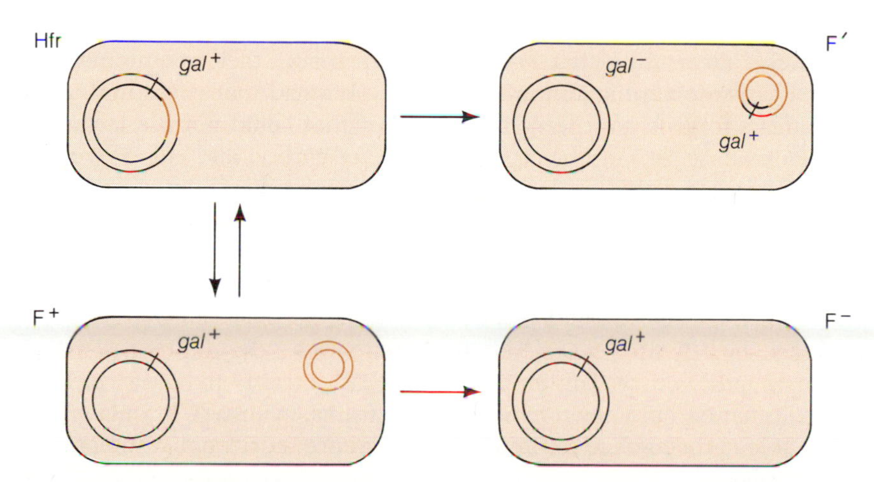

Figure 9-16 The Relationships Among Different Mating Types. The F$^+$ bacterium is haploid and contains one or more plasmids. The F$'$ bacterium contains one or more plasmids with genes from the main genome. The Hfr bacterium is haploid and its plasmid is integrated into the main genome. The F$^-$ bacterium is haploid and lacks a plasmid or an integrated plasmid. The bacteria may change from one form to another, as illustrated.

recombination" and refers to the fact that these bacteria transfer chromosomal markers to females at a much higher frequency than do F^+ strains. F^+ bacteria transfer chromosomal markers infrequently, because they must first become Hfrs and this occurs very infrequently. In an $F^+ \times F^-$ mating, genes on the main genome are transferred to the female only when an F^+ undergoes a transition to an Hfr. Thus, only a very few females will show recombination of the genome genes. On the other hand, in an Hfr $\times$ F^- mating, all the male cells can transfer genes on the genome to the female without waiting for an infrequent transition, so a large proportion of the females will show recombination of the main genome genes.

Conjugation in the gram positive bacteria does not involve pili. All that is required is the intimate contact between the donor and recipient cells. In one case, *Streptococcus faecalis*, the recipient releases a chemical attractant that initiates the "adhesion" of the donor cell. Once the donor and the recipient cells are joined together, the transfer of DNA begins. Little is known about the "pore" that connects the male and female cells.

Gene Mapping in the Prokaryote

Different Hfr strains allowed the early workers in bacterial genetics to determine where in the genome □ certain genes were located. The ordering of the genes on a genome is known as **mapping** (fig. 9-17). The mapping of genes is an important step toward the understanding of microbial genetics and physiology. If it is known where genes are in relation to one another, it is possible to devise experiments to study how the physiology of a cell is determined by genes and how the genes themselves function.

89

In the middle 1940s and early 1950s, geneticists isolated numerous **auxotrophic** mutants of *Escherichia coli*. **Auxotrophs** are mutant cells that have lost the ability to make an essential cellular component, such as an amino acid, a vitamin, or a nucleoside. An Hfr auxotroph that cannot synthesize the amino acids threonine, proline, histidine, or metionine would be symbolized as follows: Hfr $Thr^-Pro^-His^-Met^-$. If an organism is able to synthesize these amino acids, it would be represented by Hfr Thr^+Pro^+ His^+Met^+. Organisms that are able to synthesize their requirements are known as **prototrophs.** Mutants were also obtained that could not use various sugars to generate energy □. A mutant that could not use lactose, galactose, maltose, or L-arabinose as a source of carbon and energy would be indicated by Hfr $Lac^-Gal^-Mal^-Ara^-$. On the other hand, an organism that could use the sugars would be symbolized by Hfr $Lac^+Gal^+Mal^+Ara^+$. Still other mutants were resistant to various antibiotics and bacterial viruses (bacteriophages). Strains resistant to the antibiotics streptomycin, tetracycline, chloramphenicol, and penicillin, and to the bacteriophage □ T6, are indicated by Hfr $Str^rTet^rChl^rPen^rT6^r$. Sensitivity to the antibiotics and the phage is indicated by Hfr $Str^sTet^sChl^sPen^sT6^s$. Strains that are commonly found in nature are known as **wild-type** strains. Wild-type strains are generally able to metabolize common sugars and are sensitive to antibiotics and bacteriophages.

184

Various Hfr strains and F^- mutants were used to map the genes on the *E. coli* genome. In a typical mapping experiment, 10^7/ml wild-type Hfr were mixed with 10^8/ml $F^-Thr^-Pro^-His^-Met^-Gal^-Str^rT6^r$ (fig. 9-17a). At various times after mixing, samples were taken and agitated so as to break apart mating pairs. The samples were then spread onto chemically defined

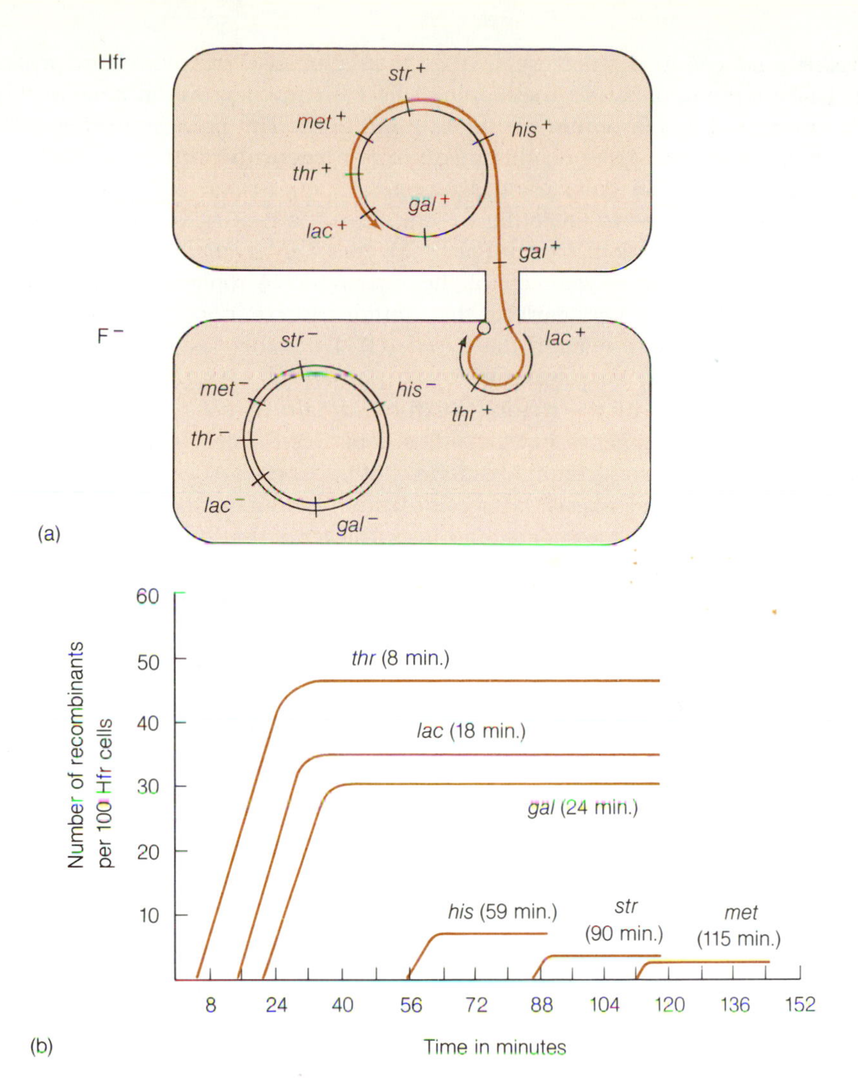

Figure 9-17 The Use of Conjugating Bacteria to Map Genes. *(a)* A prototrophic Hfr and an auxotrophic F^- are mixed together and allowed to mate for various periods of time. Samples of the bacteria are agitated to discontinue conjugation and are then plated on selective media that do not allow the parents to grow. *(b)* The order of the genes entering the female bacterium is Thr, Lac, Gal, His, Met, Str, and Met. Consequently, if the mating is disrupted soon after it is initiated, only the first gene system (Thr) has time to enter the female. Thus, bacteria grow initially only on the plates that have all the nutrients except threonine. The time at which recombinants appear on each type of plate indicates the time the gene systems enter the female bacterium. The order of the gene systems can be determined from this data.

media that contained streptomycin. The media lacked one of the amino acids, or contained only one of the sugars as the sole source of energy. On a mineral-glucose-streptomycin medium that lacks threonine, neither the F^- nor the Hfr will grow because of the lack of threonine and the presence of streptomycin. The only organism that can grow on a medium of this type is an F^- that has picked up the threonine gene from the Hfr. After conjugation, the F^- might have the following genotype: $F^-Thr^+Pro^-His^-Met^-Gal^-Str^rT6^r$. It would be able to grow because it can synthesize its own threonine (Thr^+), and because it is still resistant to streptomycin (Str^r). In a second plate that contains galactose (rather than glucose), streptomycin, proline, histidine, threonine, and methionine, the F^- cannot grow on galactose because it is Gal^-. Streptomycin again inhibits the Hfr because it is Str^s. The only cell that can grow on this medium is a **recombinant** F^- that has received the genes for galactose utilization.

The first genes to enter the F^- cell during a mating are those involved in plasmid replication. Not all plasmid genes enter the F^- cell, however, because the **origin of transfer** of the entering DNA molecule is in the middle of the F-factor. Very early in the mating process, none of the genes that can be deleted have entered the female (F^-) possibly because they are still in the passageway or pilus (fig. 9-17). Somewhat later (10 minutes after mating), samples of the mating mixture contain a few F^- cells that have received threonine genes but no other genes. Still later, samples taken 50 minutes after mating, contain bacteria that have received genes for threonine synthesis, proline synthesis, lactose utilization, and galactose utilization, but not the His, Str, or Met genes. Limiting the time that the Hfr and F^- can mate reveals the time of entry of the genes. This timing indicates the order of the genes on the bacterial genome. For this particular Hfr, threonine genes enter first (8 minutes after mating), while methionine genes enter last (115 minutes after initiation of mating). The remaining part of the fertility factor carrying genes for the transfer of the plasmid follows the methionine genes. The frequency of recombinants drops drastically for the last genes, because more and more of the mating pairs (Hfr $\times$ F^-) break apart during the mating process. Brownian motion and agitation of the mixture separate the bacteria.

Data from many mapping experiments have been used to generate a map of the *E. coli* genome. Average times have been used to construct the map. The theoretical time required for all genes to be transferred is 100 minutes. Deviations from this theoretical time are due to peculiarities of the organisms used in the experiments.

Joshua Lederberg is credited with the discovery of conjugation in bacteria in 1947, when he showed that contact between an Hfr auxotroph (Met^-Bio^-) and an F^- auxotroph (Thr^-Pro^-) converted the female cell into a prototroph. In addition, Lederberg was the first to map bacterial genes and show that they were arranged in a linear fashion. In 1958, Joshua Lederberg received the Nobel Prize in Medicine or Physiology for his discovery of conjugation in bacteria.

Transduction

The changing of a bacterial genotype as a result of the transfer of bacterial genes from one bacterium to another by bacterial viruses (bacteriophages) □ 404 is referred to as a **transduction.** While conjugation experiments can be used to get a rough idea of the spatial relationship among genes, transduction experiments can be used to map genes much more accurately. Transducing bacteriophages (or phages, for short) are categorized either as **generalized transducing** phages or as **specialized transducing** phages. Both types can be isolated and used to transfer genes between bacteria. **Generalized transducing phages,** such as PLT22 or P1, reproduce by injecting their DNA into *E. coli.* Virus progeny form in the cytoplasm and eventually rupture the host cell (fig. 9-18). During the reproduction of these viruses, the bacterial genome breaks into fragments, and some of the newly forming viruses pick up

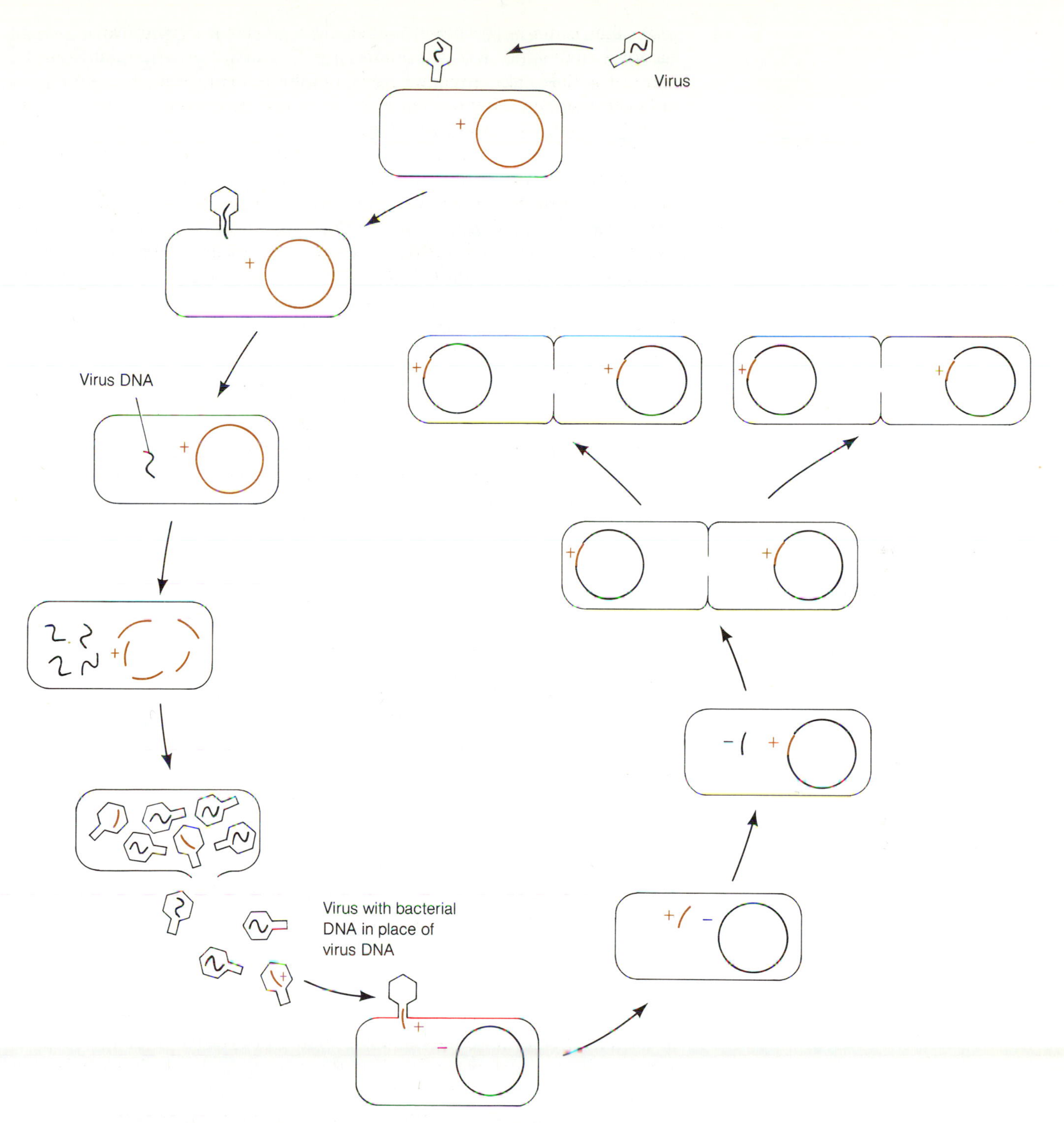

Figure 9-18 Generalized Transduction. Generalized transducing bacterial viruses invade bacteria by injecting their hereditary material into the bacteria. The virus DNA directs the synthesis of new virus DNA and proteins. As the viruses develop within the bacterial cytoplasm, some of the fragmented bacterial DNA is incorporated within virus coats instead of virus DNA. The "pseudoviruses" with the bacterial DNA are able to inject the bacterial DNA into other bacteria. The bacterial DNA injected by the pseudoviruses may recombine with the bacterial genome giving the bacterium a new trait.

(package) the bacterial DNA fragments instead of virus DNA. Thus, a small fraction of the virus progeny are not viruses but simply carriers of bacterial DNA. These virus-like structures are able to bind to other bacteria and inject the bacterial DNA, just as viruses do. Because the virus-like structures can pick up any bacterial fragment, they can be used to transduce any bacterial gene.

In contrast, a **specialized transducing phage,** such as lambda □, frequently integrates its DNA into a bacterial genome at one specific site (fig. 9-19). When the **provirus** (virus DNA) excises itself from the bacterial genome, it sometimes carries with it a region of the bacterial DNA near its site of integration. All the altered viruses carry only genes from around their integration site. Since these viruses can transduce only a restricted group of genes, they are called specialized transducing phages. 404

Generalized transducing bacteriophages have been used to map and order genes that are closely linked. Data from transduction experiments can be used to map genes and regions within genes that are fractions of a minute apart. This level of accuracy contrasts with data from conjugation experiments, which cannot readily resolve the order of closely-spaced genes.

Joshua Lederberg and Norton Zinder are credited with the discovery, in 1961, that viruses can carry bacterial DNA from one bacterium to another. They showed that prototrophs and auxotrophs, separated by a filter that prevented conjugation, still exchanged genetic information (fig. 9-20).

Transformation

Small pieces of naked DNA (DNA isolated from cells or viruses) can be taken up by bacteria and can recombine with the main genome. This phenomenon may occur in nature frequently and may be an important means of creating variability in the prokaryotic kingdom. The ability to take up DNA in this manner is called **competency,** and it occurs for a short period of time during exponential growth.

One well-known example of transformation in the laboratory occurs when heat-killed, encapsulated *Streptococcus pneumoniae* provide the ability to produce capsules □ to pneumococci that lack this ability. The nonencapsulated cells and the heat-killed encapsulated pneumococci, when injected individually into mice, do not kill them (fig. 9-21). When injected together, however, they kill the mice, and all the pneumococci found in the dead animals have capsules. The heat-killed, encapsulated pneumococci are unable to reproduce because their proteins have been denatured irreversibly by heating, but not their DNA. The DNA released from the dead cells as they lyse can be taken up by the unencapsulated, viable cells, and acquire the ability to form a capsule. Since the nonencapsulated pneumococci are destroyed by the animals' immune system □, only those bacteria that pick up the ability to make capsules, and subsequently synthesize capsules will survive. 83 434

It was Frederick Griffith who first reported, in 1928, that mice died when inoculated with both a nonpathogenic streptococcus and a heat-killed, pathogenic streptococcus. Griffith correctly concluded that the nonpathogenic streptococcus had been transformed by something that was not destroyed by heating in the dead pathogenic bacteria. It was not until 1944 that Oswald Avery, Colin MacLeod, and Maclyn McCarty finally isolated the substance responsible for the transformation and demonstrated that it was DNA.

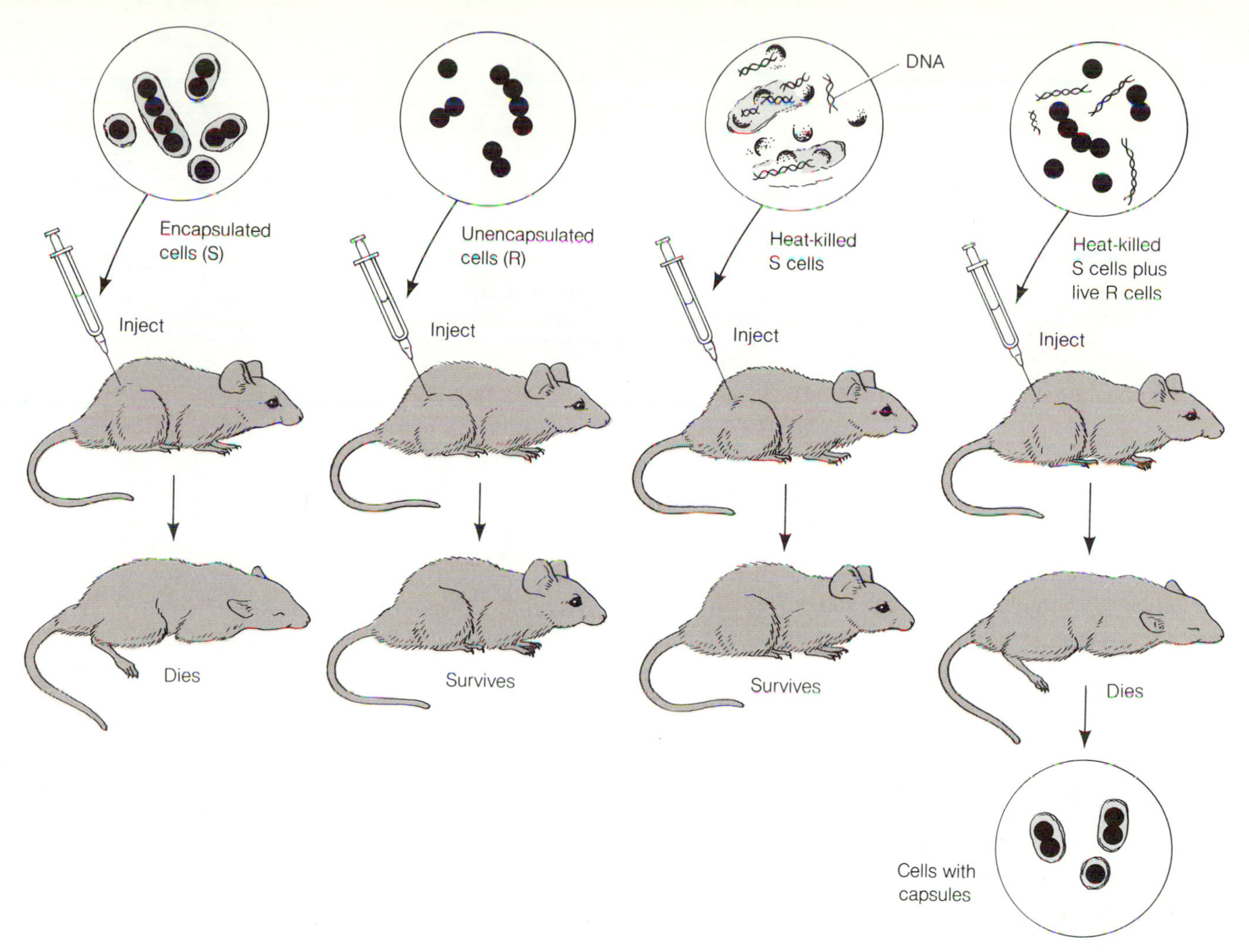

Figure 9-21 Bacterial Transformation. Encapsulated (smooth or S) pneumococci cause a fatal infection in mice, while nonencapsulated (rough or R) pneumococci do not. Heat-killed smooth pneumococci do not kill mice; however, if a mixture of heat-killed smooth strains and living rough strains are injected into mice, the mice are killed. DNA from the heat-killed bacteria enters the rough bacteria and transforms them into smooth bacteria, which can kill the mice.

RECOMBINATION

There are a number of mechanisms by which DNA molecules recombine with each other. One of the most important types of recombination is that which occurs between homologous regions of DNA. The mechanism of this type of recombination is based upon a model proposed by R. Holliday in 1964 (fig. 9-22).

Recombination between homologous regions of DNA begins with the pairing of the DNAs. An endonuclease nicks a single strand of each DNA, and DNA-binding proteins and unwinding proteins promote the creation of single-stranded DNA. Certain other proteins, such as the product of the *E. coli recA* gene, catalyze the pairing of the single-stranded DNAs to the complementary strands on the homologous DNA molecules. DNA polymerase □ may extend the exchanged strands, after which a DNA ligase eliminates the gaps between the strands. The DNA at this stage has been isolated from *E. coli* and is referred to as the **chi** (X) form or **Holliday intermediate** (fig. 9-

206

Figure 9-22 Genetic Recombination. The figure illustrates two proposed mechanisms of genetic recombination. (See text for explanation.)

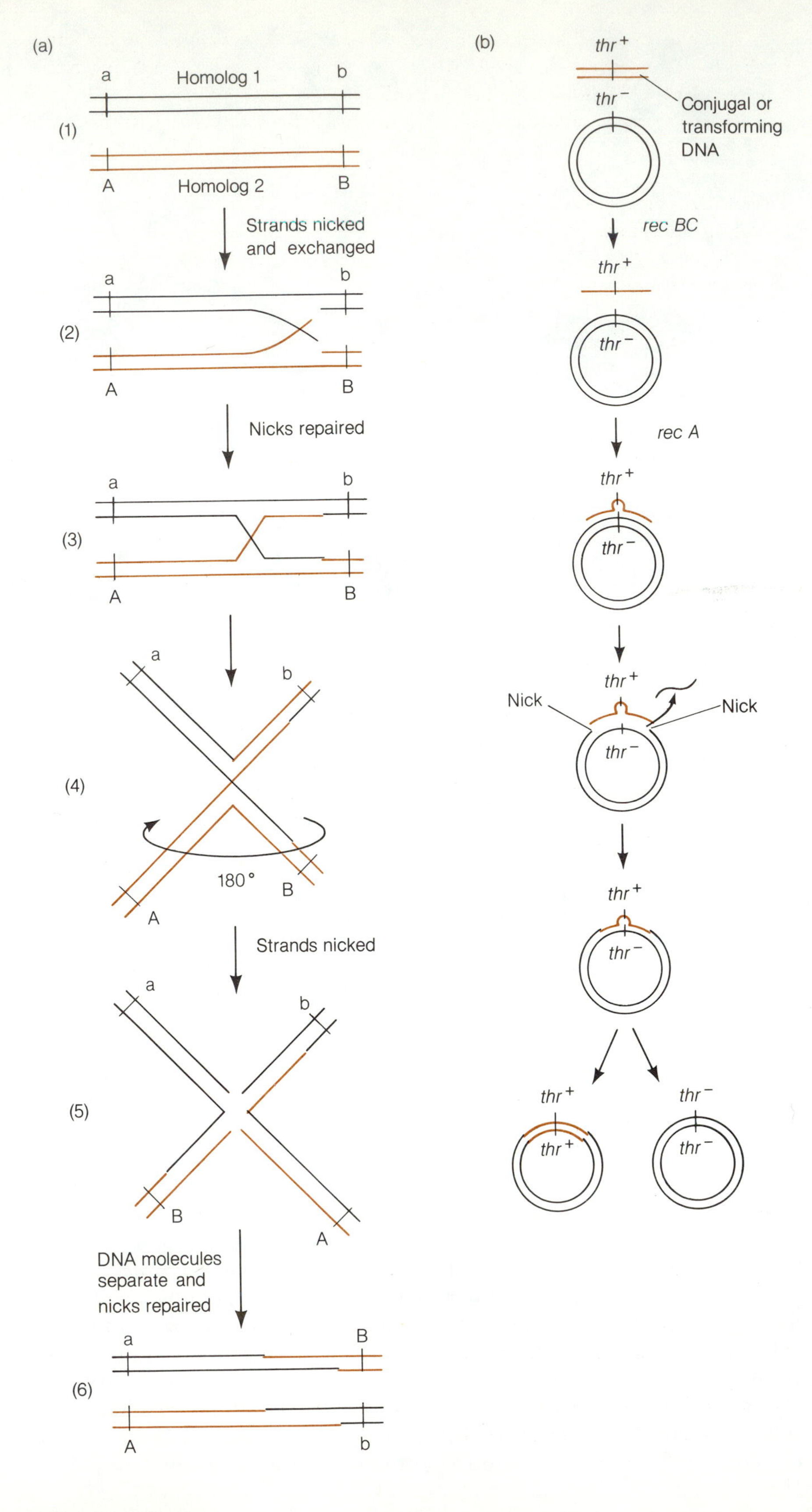

22). The *chi* form is converted into two linear DNA molecules by an endonuclease that cleaves the previously uncut strands of each DNA molecule. The resulting DNA molecules are recombinants of each other.

Some plasmids and virus genomes normally are unable to integrate into the bacterial genome, because there are no homologous regions in the two genomes where recombination can take place. How do some plasmids (episomes) and temperate phage such as lambda recombine into the bacterial genome? Integration of plasmids and temperate phages is known to be dependent upon homologous **insertion sequences** (or **IS elements**) in the bacterial genome and in the plasmid or virus DNA. The insertion sequences, which are usually between 700 and 5,000 base pairs long, mediate recombination between nonhomologous regions of DNA.

The F-plasmid of *E. coli* K-12 contains a number of insertion sequences, as does the *E. coli* K-12 genome. The genome contains eight copies of IS1, five of IS2, and one or more of IS3 and IS4, while the F-plasmid contains one copy of IS2 and two copies of IS3. Since recombination can occur between homologous insertion sequences, the F-factor can integrate at a number of locations within the bacterial genome. The integration of plasmids (episomes) and virus DNA into bacterial genomes occurs through a crossover event between homologous IS elements. The modified Holliday model for crossing over nicely explains the integration of plasmids (episomes) and virus DNA into bacterial genomes.

255

Transposons □ are pieces of DNA 2,000–20,000 base pairs long, which are able to direct the synthesis of copies of themselves and which jump from place to place. In addition, transposons are able to switch directions in bacterial genomes. They often carry genes that confer antibiotic resistance to the bacterium. For instance, the transposon Tn-4 carries the genes that make some bacteria resistant to ampicillin, streptomycin, and sulfonamide. Other transposons may carry the genes for resistance to kanamycin, trimethoprim, chloramphenicol, tetracycline, fusidic acid, erythromycin, gentamicin, and tobramycin.

Plasmids that carry genes for drug resistance are known as resistant factors or **R-plasmids.** *E. coli* R-plasmids sometimes are transmitted to other genera, such as *Shigella*, *Salmonella*, *Proteus*, *Haemophilus*, and *Pasteurella.* This spread of R-plasmids among genera increases the chances of bacteria picking up a wide variety of genes conferring drug resistance. R-plasmids recombine with transposons in the different bacteria when they have homologous insertion sequences. It is likely that transposons and R-plasmids carrying genes controlling drug resistance have been and are very important in each other's evolution. Recombination between these DNAs has created molecules with numerous drug resistant genes.

The mechanism by which transforming DNA and bacterial genomes recombine is an example of another type of recombination (fig. 9-22). The insertion of exogenous DNA takes place in several steps, each mediated by different enzymes. Initially, an exonuclease chews away one strand of the DNA fragment and the homologous region of the bacterial genome is melted. Next, the strand of transforming DNA hydrogen-bonds the homologous strand of the bacterial genome and an endonuclease cuts out the unpaired strand of DNA. Replication of the genome results in the strands being completely complementary. When one strand is thr^- and the other is thr^+, the DNA is not perfectly complementary.

EUKARYOTIC DNA

The DNA found in eukaryotic nuclei is double-stranded and appears chemically identical to the genetic information found in bacteria □, but while the bacterial genomes are circular double helices, the eukaryotic chromosomes are linear molecules. It is believed that the 5′ and 3′ ends of complementary strands in the eukaryotic chromosomes are joined together by covalent bonds.

49

Although bacteria generally have a single genome, diploid eukaryotes have from 6 (fruitflies *Drosophila willistoni* and *D. prosaltans*) to 124 (cordgrass, *Spartina anglica*) chromosomes (table 9-2). Table 9-2 also shows that eukaryotes have much more DNA than bacteria and viruses, and that the

TABLE 9-2
BASE PAIR AND CHROMOSOME NUMBER OF SELECTED ORGANISMS AND VIRUSES

COMMON NAME	SPECIES	DIPLOID NO.	HAPLOID NO.	BASE PAIRS
Viruses				
$\phi \times 174$	—	—	1	4.5×10^3
Polyoma	—	—	1	0.5×10^4
λ	—	—	1	0.5×10^5
T2	—	—	1	1.9×10^5
Bacteria				
	Aerobacter	—	1	1.9×10^6
	Escherichia	—	1	4.0×10^6
	Bacillus	—	1	3.0×10^7
Fungi and algae				
Mold (fungus)	*Aspergillus nidulans*	—	4 or 8	4×10^7
Pink bread mold	*Neurospora crassa*	—	7	
Penicillin mold	*Penicillium* species	—	2	
Yeast	*Saccharomyces cerevisiae*	4 or 8	2 or 4	7×10^7
Green algae	*Chlamydomonas reinhardi*	—	8	
Green algae	*Acetabularia mediterranea*	ca. 20		
Insects				
Silkworm	*Bombyx mori*	56	28	
Red ant	*Formica sanguinea*	48	24	
House fly	*Musca domestica*	12	6	
Fruit fly	*Drosophila melanogaster*	8	4	0.8×10^8
Mosquito	*Culex pipiens*	6	3	
Plants				
Wheat	*Triticum aestivum*	42		
Corn	*Zea mays*	20		
Garden pea	*Pisum sativum*	14		
Potato	*Solanum tuberosum*	48		
White clover	*Trifolium repens*	32		
Broad bean	*Vicia faba*	12		
Animals				
Chicken	*Gallus domesticus*	ca. 78	39	1.1×10^9
Carp	*Cyprinus carpio*	104	52	1.7×10^9
Dog	*Canis familiaris*	78	39	2.5×10^9
Human	*Homo sapiens*	46	23	2.8×10^9
Cat	*Felis catus*	38	19	—
Horse	*Equus calibus*	64	32	2.8×10^9
Alligator	*Alligator mississipiensis*	32	16	7.0×10^9
Toad	*Bufo americanus*	22	11	7.0×10^9
Frog	*Rana pipiens*	26	13	23.0×10^9

"higher" eukaryotes (worms to humans generally have much more DNA than the "lower" eukaryotes (fungi, sponges, etc.). Since the chromosomes in eukaryotes come in all sizes, there is no direct relationship between the number of chromosomes and the amount of DNA. For example, a carp, a human, and a frog have 104, 46, and 26 chromosomes, respectively, but they have 1.7×10^9, 2.8×10^9, and 23×10^9 base pairs per haploid nucleus. Also, much of the DNA (perhaps 99%) in the higher eukaryotes does not appear to code for rRNAs, tRNAs, or proteins, but apparently functions in the regulation of gene expression and recombination. Much of the primary transcripts (mRNA, and RNA for rRNA and tRNA) in the higher eukaryotes consists of "spacer" material, called **introns,** that is excised from the mRNA before it is translated. Therefore, even though some of the higher eukaryotes contain enough DNA for as many as 20 million genes (assuming the average gene consists of 1,000 base pairs), they are believed to have between 50,000 and 100,000 genes per diploid nucleus. These numbers are obtained by assuming that more than 99% of the DNA in organisms with large amounts of DNA does not code for functional products.

The Structure of Eukaryotic DNA

Unlike the bacterial genome, which is condensed by being supercoiled and then folded, the eukaryotic DNA is condensed by being supercoiled and wound around clumps of histone proteins. The spherical clumps of histone proteins and DNA are about 10 nanometers (nm) in diameter and are called **nucleosomes.** The nucleosomes usually bind together to form a structure that is 25 nm in diameter. Since this structure is below the limit of resolution of the light microscope (0.2 μm = 200 nm), it can be seen only with the electron microscope.

The DNA in eukaryotes is duplicated during part of **interphase,** a period when chromosomes are not visible under the light microscope. The daughter DNAs, called **chromatids,** remain together until they are partitioned into daughter cells. Just before the daughter DNAs are partitioned, they become extremely compact and visible with the aid of the light microscope. Chromosomes become visible during **mitosis** and **meiosis.**

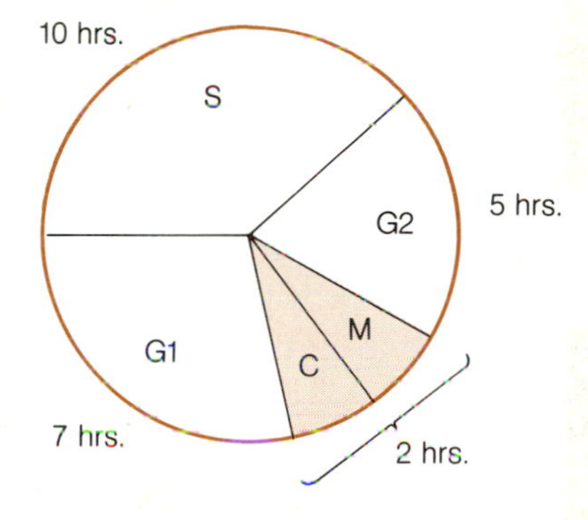

S = DNA synthesis (replication)

G2 = Gap in chromosome activity

M = Mitosis

C = Cytokinesis (cell division)

G1 = Gap in chromosome activity

G1 + S + G2 = Interphase

Figure 9-23 Eukaryotic Cell Cycle. A typical eukaryotic cell cycle consists of a gap in chromosome activity (G1), DNA synthesis (S), a second gap in chromosome activity (G2), mitosis (M), and cytokinesis or cell division (C). G1 + S + G2 is often referred to as *interphase.* During interphase, the chromosomes are not visible with the light microscope. Chromosomes are visible with the light microscope only during mitosis and cytokinesis, when they condense sufficiently.

The Eukaryotic Cell Cycle

The eukaryotic cell cycle is often divided into periods as shown in figure 9-23. DNA synthesis (or duplication) takes place during the **S-period. G2** represents a gap in activity with respect to the chromosomes. During the **M-period,** known as mitosis, the chromosomes generally condense so that they become visible in the light microscope. Then they are pulled to the poles of the cell. **Cytokinesis,** or cell division, occurs during the **C-period. G1** represents a gap in the activity of the chromosomes just before DNA replication. The cell cycle in eukaryotes varies in duration from a few hours to a number of days. For example, bakers' yeast has a cell cycle that is typically 2 to 3 hours long under common laboratory conditions, while certain animal cell lines may have cell cycles that take 24 hours.

Mitosis

Mitosis is the term used to describe the separation of daughter chromosomes in eukaryotes into individual nuclei. Mitosis does not necessarily involve cell division (cytokinesis), but often the two processes overlap. Eukaryotic chromatids are separated from each other by a number of mechanisms. In some of the simple eukaryotes, the chromosomes do not become visible under the light microscope during mitosis. In addition, the chromosomes are attached to the inner nuclear membrane. In these organisms, the chromatids are separated from each other as the nuclear membrane expands and divides. Cytokinesis or cytoplasmic division may also take place during or after mitosis. In such organisms, no microtubules are involved in the separation of the chromosomes.

The mitosis that is most familiar to students of biology is that found in the higher eukaryotes. In these eukaryotes, the chromosomes condense at the

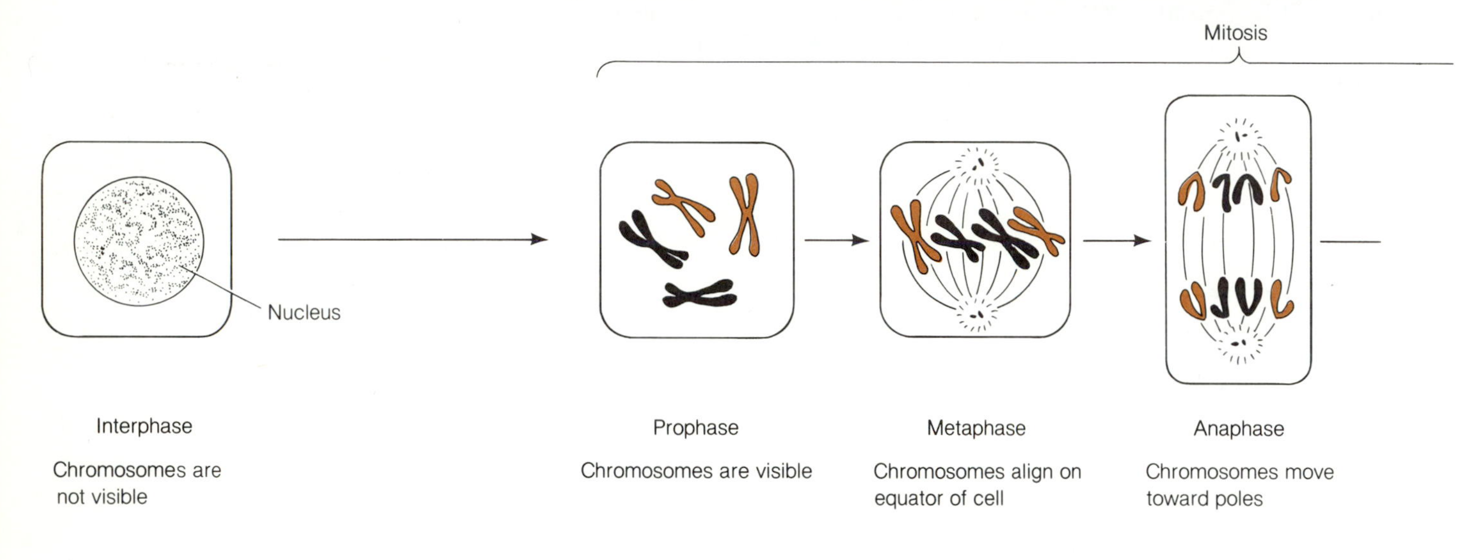

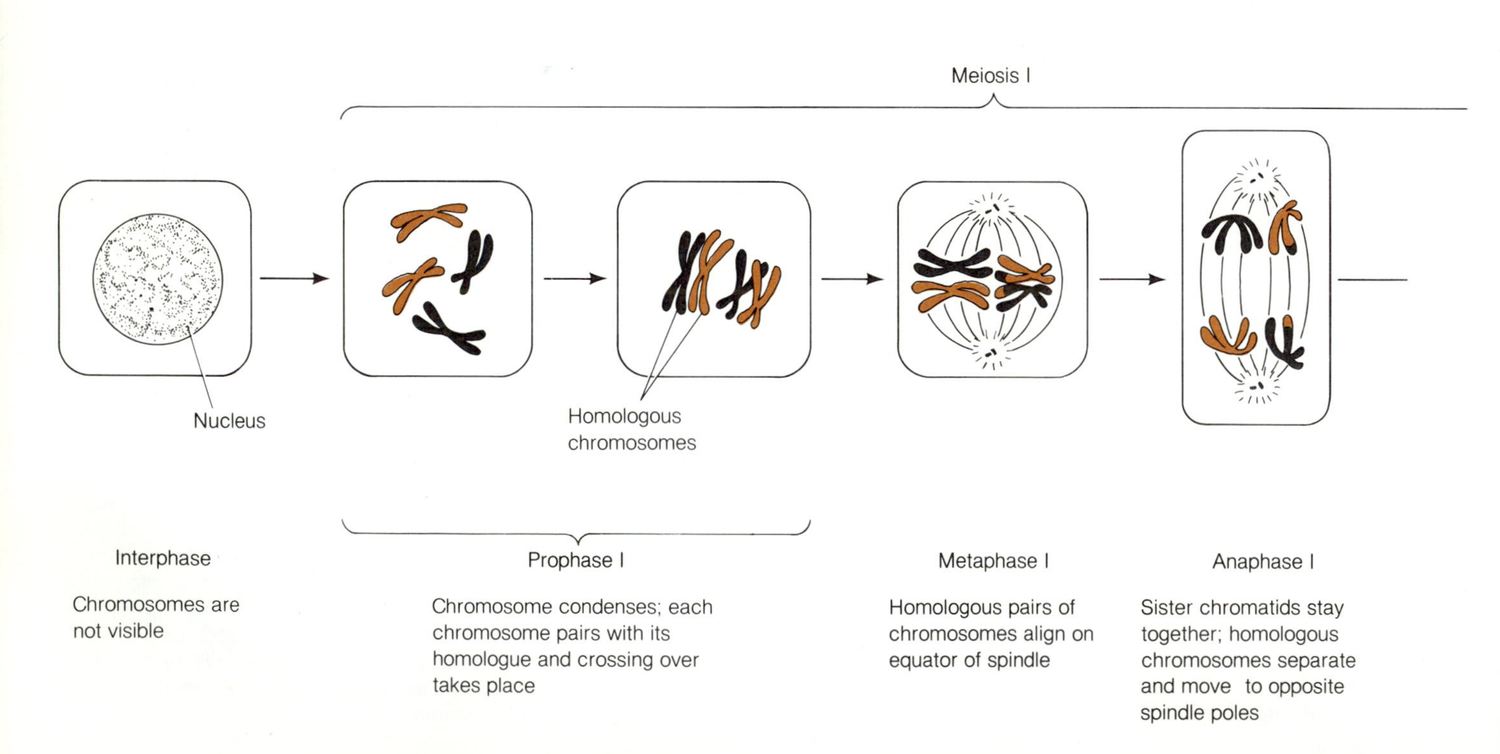

beginning of mitosis and thus become visible in the light microscope (fig. 9-24). They are not attached to the nuclear membrane; however, they soon become attached to **microtubules** that pull the chromatids apart and to the opposite poles of the cell. The nuclear membranes degenerate, so that the microtubules can attach to the chromatids, and then reappear around the chromosomes when they have reached the poles.

Mitosis in the higher eukaryotes has been divided into four stages: **prophase, metaphase, anaphase,** and **telophase.** In early prophase the chromosomes condense so that they become visible in the light microscope. The daughter chromosomes, or chromatids, are still attached to each other. The two chromatids are referred to as chromosomes when they separate from

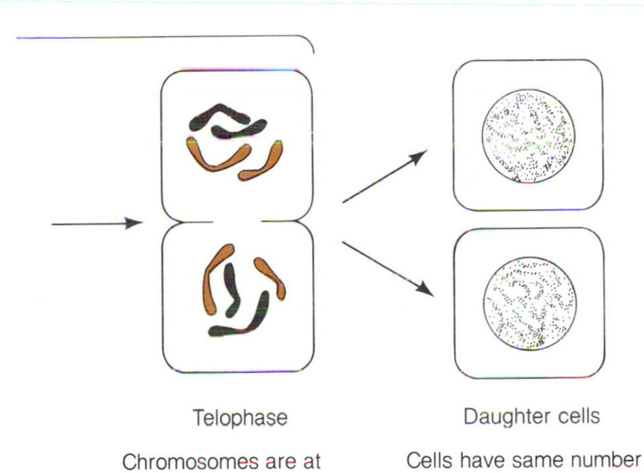

Figure 9-24 Mitosis and Meiosis. Illustrations compare mitosis and meiosis. Notice that in mitosis the number of chromosomes in each of the daughter cells is the same as that of the parental cell. In meiosis the number of chromosomes is reduced by half. Hence, during mitosis, diploid cells remain diploid while in meiosis they become haploid.

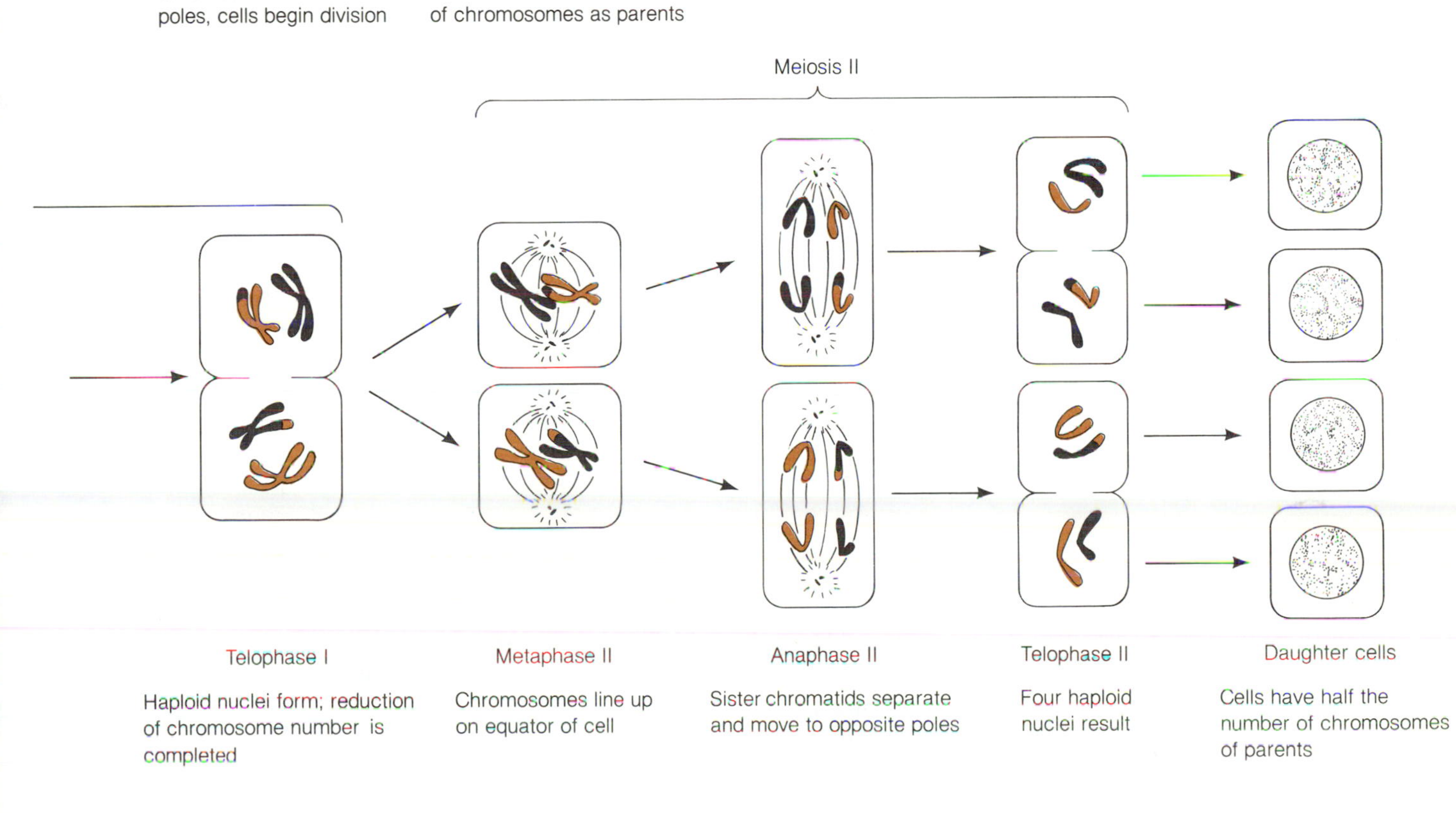

each other. By late prophase, the chromosomes have attached to microtubules and the nuclear membranes have dissipated. The microtubules begin to line up the chromosomes, which by metaphase have been aligned along the equatorial plane of the cell. Anaphase begins when the chromatids are pulled away from each other. In telophase, the chromosomes have reached their greatest separation and nuclear membranes have begun to form around the chromosome at the poles of the cell.

CONJUGATION AND MEIOSIS IN THE EUKARYOTE

Since eukaryotic populations (especially the higher eukaryotes) are generally much smaller than prokaryotic populations, it is very important for the survival of eukaryotes that most individuals in a species be genetically diverse. In order to create genetic variability, reproduction in the higher eukaryotes generally follows recombination. Variability in the eukaryotes is also achieved by conjugation.

Conjugation in eukaryotic microorganisms differs from that of the prokaryotes. In the eukaryotes, two mating types fuse with each other, partially or completely, and recombine their entire genetic material (fig. 9-25). For example, the green alga *Chlamydomonas* fuses completely with its mating type. During the process, the two haploid nuclei fuse to form a diploid nucleus.

Since nuclei are fusing with each other every generation, there must be some mechanism to keep the chromosome number the same from generation to generation. An increase in the number of chromosomes often disrupts or drastically alters the physiology of the cell. In many cases this is a lethal condition. Eukaryotes have evolved a mechanism that eliminates the problem of increasing chromosome number: before fusion, they reduce the chromosome number by one-half. An eukaryote that is diploid (has two of each chromosome) eliminates one of each of the **homologous chromosomes** so that a haploid nucleus or cell is produced (with one of each chromosome). The haploid cell produced by meiosis is sometimes called a **gamete.** An eukaryote that is tetraploid (has four of each chromosome) eliminates two of each of the homologous chromosomes, so that a diploid nucleus or cell is produced. A diploid cell produced from meiosis is also called a gamete. The mechanism by which a nucleus or a cell becomes reduced in chromosome number is called **meiosis** (fig. 9-24).

Many studies have shown that eukaryotes can be transduced and transformed in the laboratory, but the importance of these mechanisms in creating variability in nature has not been estimated.

TRANSCRIPTION AND TRANSLATION IN THE EUKARYOTE

The transcription and translation process in eukaryotes is more involved than in prokaryotes. RNA is transcribed by RNA polymerases □ that are quite similar in function to the RNA polymerase in bacteria. In eukaryotes, however, protein synthesis does not begin before transcription is completed. Before translation can occur, the mRNA must be modified and transported into the cytoplasm. No protein synthesis occurs in the nucleus.

207

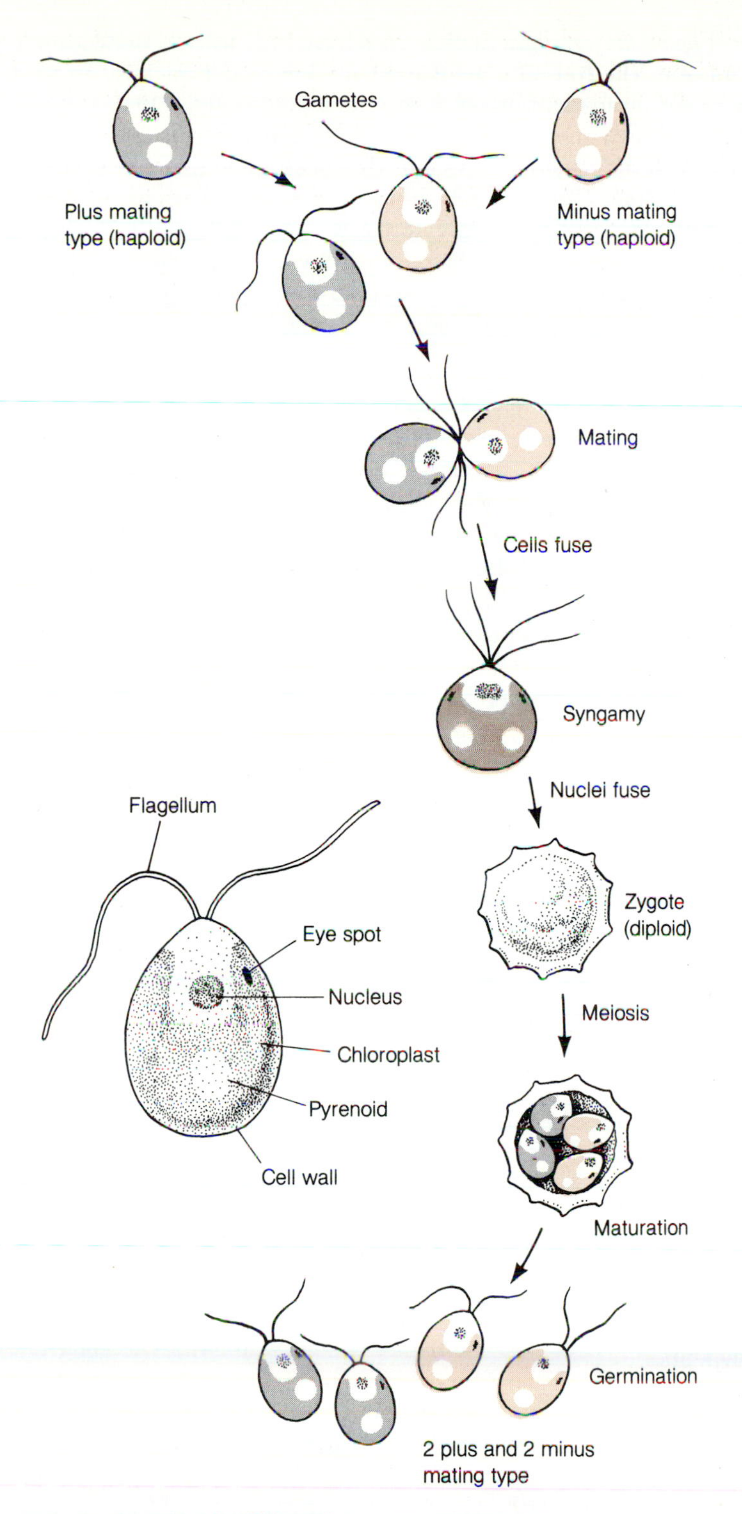

Figure 9-25 Conjugation in *Chlamydomonas.* After conjugation in *Chlamydomonas* begins, the haploid cells fuse completely. Subsequently, the two haploid nuclei fuse to form a diploid nucleus. When the zygote undergoes meiosis, four haploid cells are produced. The haploid cells reproduce by mitosis.

Most genes in the higher eukaryotes have 50% to 75% noncoding regions called introns. The coding regions are called **exons.** Typically, an RNA polymerase begins transcription at a promoter site and produces a transcript complementary to one of the DNA strands. After the 5′ and 3′ ends have been modified, the noncoding regions (introns) are excised, bringing the coding regions (exons) next to each other. The modified mRNA is transported through nuclear pores into the cytoplasm. In the cytoplasm, the mRNA is translated into protein by 80s ribosomes. Some researchers believe that the introns provide space between the exons so that recombination is more likely between the coding regions of the gene.

SUMMARY

1. Oswald Avery and his colleagues demonstrated in 1944 that DNA and not protein was the hereditary material.
2. James Watson and Francis Crick proposed in 1953 that cellular DNA consisted of two strands of DNA hydrogen-bonded and wound around each other in the form of a double helix. The central dogma, proposed by Francis Crick, states that the information contained in DNA is transcribed into RNA and that the information in RNA is translated into protein.
3. A copy of the information in a DNA molecule is made by DNA replication.
4. Since the central dogma was first presented, it has been learned that some viruses produce enzymes that can transcribe RNA into DNA and that some viruses replicate their RNA genomes. The synthesis of DNA using RNA as a template is called reverse transcription.
5. Transcription, the synthesis of RNA using a DNA template, is catalyzed by an RNA polymerase; reverse transcription, the synthesis of DNA using an RNA template, by a reverse transcriptase; translation by a ribosome; and DNA replication by a "DNA polymerase complex."
6. George Beadle and Edward Tatum are credited with obtaining supporting data for the hypothesis that one gene controls the activity of one polypeptide (one-gene-one-enzyme hypothesis) in the lower eukaryotes.

DNA REPLICATION AND SEGREGATION IN THE PROKARYOTE

1. Bacteria generally contain a single circular DNA genome attached to the cytoplasmic membrane.
2. The bacterial genome is replicated bidirectionally from a single start site. Some genomes can be replicated in less than one half hour. Numerous enzymes and proteins are involved in DNA replication. The most noteworthy are an RNA polymerase (which makes the RNA primer), a DNA ligase, and DNA polymerases.
3. Matthew Meselson and Franklin Stahl provided the first evidence that DNA replication is semiconservative.
4. In bacteria, daughter genomes are separated from each other by the growth of the cytoplasmic membrane between their points of attachment. Daughter cells form when the plasma membrane and the cell wall invaginate and create a membrane partition between the chromosomes.

RNA SYNTHESIS IN PROKARYOTES

1. Prokaryotes (as well as eukaryotes) synthesize three types of ribonucleic acid (RNA): ribosomal RNA (rRNA), transfer RNA (tRNA) and messenger RNA (mRNA).
2. Ribosomal RNAs are structural components of the ribosomes. In bacteria, the 5s and 23s rRNAs are found in the 50s ribosomal subunit, while the 16s rRNA is found in the 30s ribosomal subunit.
3. There are 1 to 6 different transfer RNAs for each of the 20 amino acids found in protein. Because of their anticodons, tRNA molecules hydrogen-bond to specific triplet codons in the mRNA. The mRNA determines the order in which charged tRNAs (tRNA molecules with covalently attached amino acids) bind to an mRNA-ribosome complex.
4. Ribosomes catalyze the formation of peptide bonds between adjacent amino acids.

PROTEIN SYNTHESIS IN PROKARYOTES

1. Protein synthesis consists of three stages: initiation, elongation, and termination. During initiation (in bacteria), a 70s ribosome forms from 30s and 50s

subunits at a start codon (AUG) on the mRNA. During elongation, amino acids are covalently linked in a sequential manner as the mRNA moves along the ribosome. Termination of protein synthesis occurs when the ribosome comes to a nonsense codon (UAA, UGA, or UAG). The peptide is cleaved from the tRNA holding it to the last codon of the mRNA and the ribosome dissociates into its subunits.

2. The genetic code is a triplet code because it is based upon groups of three nucleotides, called codons. The codon consists of any three of the four nucleotides (A, U, G, C) in any order. This means that there are 64 possible codons ($4^3 = 64$). All except the three nonsense codons are recognized by tRNAs. Since more than one codon may specify the same amino acid, the code is said to be degenerate.
3. Marshall Nirenberg and Har Gobind Khorana are credited with working out the genetic code. The genetic code is the same in all organisms, from bacteria to humans.

THE MOLECULAR BASIS FOR VARIATION

1. In bacteria, spontaneous mutations in a particular gene occur at a rate of about 10^{-7} to 10^{-9} (i.e., once in about every 10^7 to 10^9 bacteria). In the "higher" eukaryotes, such mutations occur at much higher rates, 10^{-4} to 10^{-7}. Most spontaneous mutations in bacteria are caused by heat and ultraviolet light.
2. Scientists commonly induce mutations with ultraviolet light, alkylating chemical, nitrous acid, and base analogues.

CONJUGATION, TRANSDUCTION, AND TRANSFORMATION

1. There are three types of male bacteria: F^+, Hfr, and F′. Male bacteria are capable of transmitting their genetic information to females (F^-) by conjugation.
2. Gram negative male bacteria generally attach to female bacteria by long, hollow sex pili through which they may transmit their genetic information. Gram positive male bacteria do not produce visible sex pili. Mating gram positive cells appear to bind together and then generate a connection between the cytoplasms.
3. Conjugation between prototrophic Hfr strains and auxotrophic F^- strains has been used to map genes. Joshua Lederberg is credited with showing that bacteria are capable of conjugation.
4. Transduction occurs when a virus introduces DNA into a cell and recombination occurs. Transductions using generalized and specialized transducing bacteriophages have been used to map genes with respect to each other and to map regions within genes. Joshua Lederberg and Norton Zinder are credited with the discovery of transduction in bacteria.
5. Transformation occurs when naked DNA enters a cell and recombines with the cell's genetic information. Frederick Griffith first observed transformation, but Oswald Avery, Colin MacLeod, and Maclyn McCarty are credited with first showing that transformations were due to naked DNA.

RECOMBINATION

Recombination between DNA molecules appears to proceed by at least two different mechanisms. Transposons, episomes and viruses appear to integrate into genomes in the same way, while fragments of DNA are integrated into genomes by a second mechanism.

EUKARYOTIC DNA

1. The hereditary material in eukaryotes is double-stranded DNA. In many of the higher eukaryotes, more than 99% of the DNA does not code for rRNA, tRNA, or protein.
2. The eukaryotic cell cycle is divided into periods: first gap (G1), DNA synthesis (S), second gap (G2), mitosis (M), and cytokinesis (C).
3. During the gap periods, there is no visible or cytological activity of the hereditary material. During DNA synthesis, the hereditary material is replicated. In mitosis and cytokinesis the chromosomes become highly condensed and move rapidly within the cytoplasm.

CONJUGATION AND MEIOSIS IN THE EUKARYOTE

1. Conjugation in eukaryotes involves partial or complete fusion between two mating types.
2. Eukaryotes avoid the problem of increasing chromosome number when they conjugate by reducing the number of chromosomes to one-half the original number before they mate. The haploid cell is produced by meiosis and is often referred to as a gamete.

TRANSCRIPTION AND TRANSLATION IN THE EUKARYOTE

1. Fifty to seventy-five percent of genes in the higher eukaryotes do not code for a product. The noncoding regions are called introns, while the coding regions are called exons.
2. Translation in the eukaryote occurs in the cytoplasm and is carried out by 80s ribosomes.

STUDY QUESTIONS

1. Who proposed the central dogma, and what is it?
2. Explain the following: a) DNA replication; b) reverse transcription; c) RNA replication; d) transcription; e) translation.
3. Describe how DNA is replicated.
4. Define semiconservative DNA replication.
5. Explain what a gene is, and what the relationship is between a gene's structure and an organism's phenotype.
6. Describe how each of the different species of RNA (rRNA, tRNA, mRNA) functions in a cell.
7. Outline the details of protein synthesis in the prokaryote.
8. Compare and contrast the following: a) codons and anticodons; b) template and primer; c) start codon and nonsense codon; d) promoter site and termination site.
9. Explain what the genetic code is. Who is credited with working out the genetic code?
10. Explain how spontaneous mutations occur.
11. Explain how ultraviolet light, alkylating agents, nitrous acid, and base analogues induce mutations.
12. Who is credited with showing that the one-gene-one-enzyme hypothesis was applicable to microorganisms?
13. Explain the differences among the following bacteria: F^-, F^+, F', and Hfr.
14. Explain how conjugation can be used to map genes in bacteria.
15. Contrast conjugation, transduction, and transformation.
16. What is the difference between generalized and specialized transducing bacteriophages?
17. Who is credited with the discovery of transduction in bacteria?
18. Who is credited with the discovery of transformation in bacteria?
19. Compare and contrast the two types of recombination.
20. Who is credited with discovering the structure of DNA?

SUPPLEMENTAL READINGS

Barany, F. and Kahn, M. 1985. Comparison of transformation mechanisms of *Haemophilus parainfluenzae* and *Haemophilus influenzae*. *Journal of Bacteriology* 161(1):72–79.

Chambon, P. 1981. Split genes. *Scientific American 244(5):* 60–71.

Croce, C. M. 1985. Chromosomal translocations, oncogenes, and B-cell tumors. *Hospital Practice* 20:41–48.

Dickerson, R. 1972. The structure and history of an ancient protein. *Scientific American 226*:58–72.

Dickerson, R. 1980. Cytochrome C and the evolution of energy metabolism. *Scientific American 242(3)*:137–149.

Howard-Flanders, P. 1981. Inducible repair of DNA. *Scientific American 245(5)*:72–80.

Judson, H. F. 1979. *The eighth day of creation.* New York: Toughstone Books, Simon and Schuster, 686 pp.

Kazazian, H., H., Jr. 1985. The nature of mutation. *Hospital Practice* 20:55–69.

Kelly, R., Atkinson, M., Huberman, J., and Kornberg, A. 1969. Excision of thymine dimers and other mismatched sequences by DNA polymerase of *Escherichia coli. Nature 224*:495–501.

Kornberg, A. 1969. Active center of DNA polymerase. *Science 163*:1410–1418.

Kronberg, R. and Klug, A. 1981. The nucleosome. *Scientific American 244(2):* 52–64.

Lake, J. 1981. The ribosome. *Scientific American 245(2)*:84–97.

Nomura, M. 1984. The control of ribosome synthesis. *Scientific American* 250(1):102–114.

Walker, G. 1984. Mutagenesis and inducible responses to deoxyribonucleic acid damage in *Escherichia coli. Microbiological Reviews* 48(1):60–93.

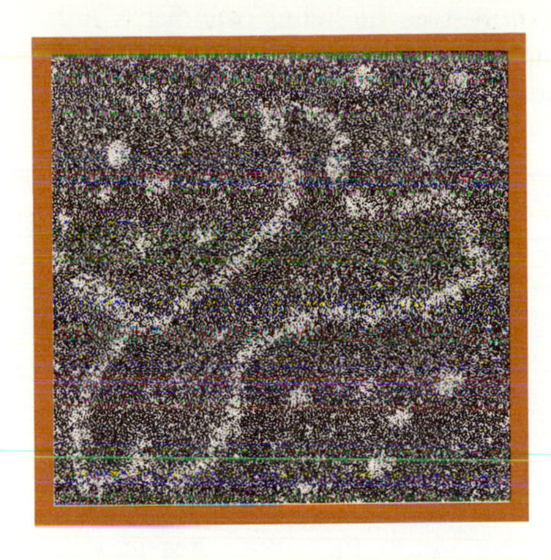

CHAPTER 10 REGULATION OF METABOLISM

CHAPTER PREVIEW

"JUMPING GENES" AND AFRICAN SLEEPING SICKNESS

African sleeping sickness is an often fatal disease caused when a protozoan called *Trypanosoma gambiense* infects humans. The trypanosome is transferred among humans by the bite of the tsetse fly *(Glossina palpalis)*. Millions of human beings in western and central Africa die each year from this disease. The patients first experience fever, chills, headache, and loss of appetite. The protozoa multiply within the human host and invade the spleen, liver, and lymph nodes. The invasion leads to involvement of the nervous system. The destruction of nervous tissue leads to the characteristic symptoms of the disease: sleepiness, coma, emaciation, and death.

Physicians at the turn of the century were aware that patients with African sleeping sickness had a cyclic appearance of large numbers of trypanosomes in the blood every week or so. This cyclic appearance and disappearance of trypanosomes in the blood lasted for many weeks to months, until the patient died.

Modern research has elucidated part of what happens inside the host which promotes this cyclic appearance of the parasites. Shortly after the patient becomes infected, he or she develops a resistance that is specific to the invading parasite and fights off the infection vigorously. The resistance is directed at destroying the membranous surface of the parasite. Apparently the attack is aimed at a specific protein that constitutes more than 10% of the parasite's surface. When the degree of resistance of the host is sufficiently high, the vigorous attack eliminates many of the trypanosomes—but not all. Some of the trypanosomes spared from the host attack develop a new membrane protein that the host defense mechanisms do not recognize. Hence, a change in surface protein composition protects the surviving parasites from host attack. The parasites take advantage of this lull in the host attack to resume multiplying. The new variants, when subjected to the host attack, will also be destroyed, but they will resume multiplying as soon as they change their surface protein. The cyclic appearances of trypanosomes in the blood occur after the parasites have changed their surface proteins.

How can the trypanosomes change so frequently their surface composition? Apparently, trypanosomes can change the structure of their membrane protein by changing the gene that codes for the protein. "Jumping genes," or more properly **insertion elements,** in a segment of DNA (gene) coding for the surface protein account for some of the hypervariability: bits of DNA within the surface protein gene "jump" from one portion of the gene to another, creating a section of DNA sufficiently different to code for protein variants. Each of these variants is different enough from the others that the host is unable to recognize it and cannot react promptly. This delay allows the trypanosomes to multiply unmolested for a period of time, until the host either develops resistance to the new variant or dies.

The trypanosomes adapt to hostile environments by regulating the chemical structure of their cells. They also regulate their metabolism in order to insure their survival in a hostile environment. This is the case with all organisms. Regulation plays a central role in the survival of all organisms and involves all facets of life processes. In this chapter we will learn some of the ways in which microorganisms regulate their metabolism and what survival advantages this regulation affords them.

LEARNING OBJECTIVES

A STUDY OF THIS CHAPTER SHOULD ENABLE YOU TO:

- OUTLINE SOME OF THE REASONS WHY CELLS NEED TO REGULATE THEIR METABOLISM
- DEFINE WHAT OPERONS ARE AND DESCRIBE HOW THEY FUNCTION TO REGULATE METABOLISM
- OUTLINE THE MAJOR STEPS IN THE REGULATION OF LACTOSE CATABOLISM BY THE LACTOSE OPERON
- DISCUSS HOW CATABOLITE REPRESSION WORKS
- OUTLINE THE MAJOR STEPS IN THE REGULATION OF TRYPTOPHAN BIOSYNTHESIS BY THE TRYPTOPHAN OPERON
- DEFINE WHAT ALLOSTERIC ENZYMES ARE, AND EXPLAIN HOW THEY FUNCTION
- DEFINE TRANSPOSONS AND INSERTION SEQUENCES

The chemical reactions and mechanical processes of living organisms are well coordinated to achieve orderly cellular growth and division and to optimize the utilization of nutrient and energy resources. If metabolism were not carefully regulated and integrated, death would ensue.

Regulation of metabolism can be achieved by transcriptional control. Microorganisms achieve transcriptional control of metabolism by repressing or activating the transcription of genes that code for enzymes. Often, genetically linked genes are regulated together and constitute part of what is called an **operon** (fig. 10-1). Operons have been described in a number of different organisms; two operons will be discussed in detail in this chapter.

Another way of regulating metabolism is by translational control, or control of the process of protein synthesis itself. This mechanism of control is not common in prokaryotes, but is prevalent among eukaryotes.

Controlling the rate at which enzymes catalyze reactions is another common mechanism of metabolic regulation, both in prokaryotes and in eukaryotes. The activity of some enzymes can be regulated by altering the shape of the enzyme so that it is less (or more) efficient in catalyzing biochemical reactions (fig. 10-2). Enzymes whose activity can be altered are called **allosteric** enzymes and are commonly found in anabolic and catabolic pathways.

In this chapter we will study the lactose and tryptophan operons of *Escherichia coli* to exemplify how microorganisms regulate their metabolism. We will also consider briefly how metabolism is regulated by allosteric enzymes.

REGULATION OF LACTOSE METABOLISM

In 1961, Francois Jacob and Jacques Monod published an explanation of how the information in 2'deoxyribonucleic acid □ (DNA) might be used by the cell to direct the synthesis of proteins. 49 They also proposed that some gene systems, which they called operons, were controlled by the binding of repressor proteins to specific regions on the DNA molecule. These regions on the DNA molecule are called **operator** sites. Since their original publications, much has been learned regarding the genetic control of mRNA and protein synthesis. We now know how the information contained in the DNA mole-

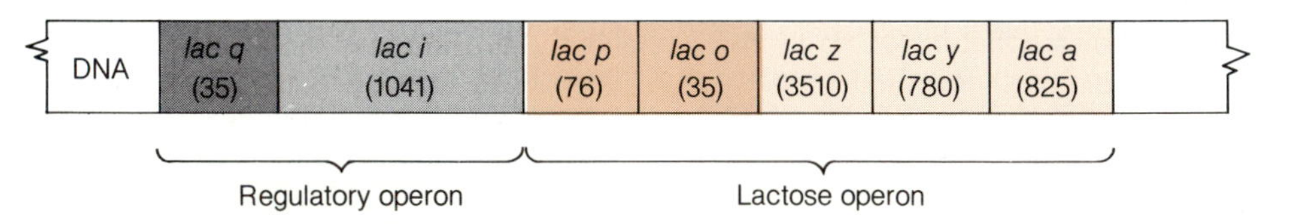

Figure 10-1 The Lactose Operon of *Escherichia Coli.* The lactose operon in *Escherichia coli* consists of three structural genes (*lac z, lac y* and *lac a*), a promoter gene *(lac p)*, an operator gene *(lac o)* and a regulatory gene *(lac i)*. The structural genes code for enzymes involved in the transport and catabolism of lactose. Transcription begins near the promoter site, where RNA polymerase attaches. The *lac i* gene codes for a regulatory protein that binds to the operator gene *(lac o)*. The *lac q* gene is a promoter site for the *lac i* gene. The number in parenthesis indicate the estimated number of nucleotide base pairs that make up each gene.

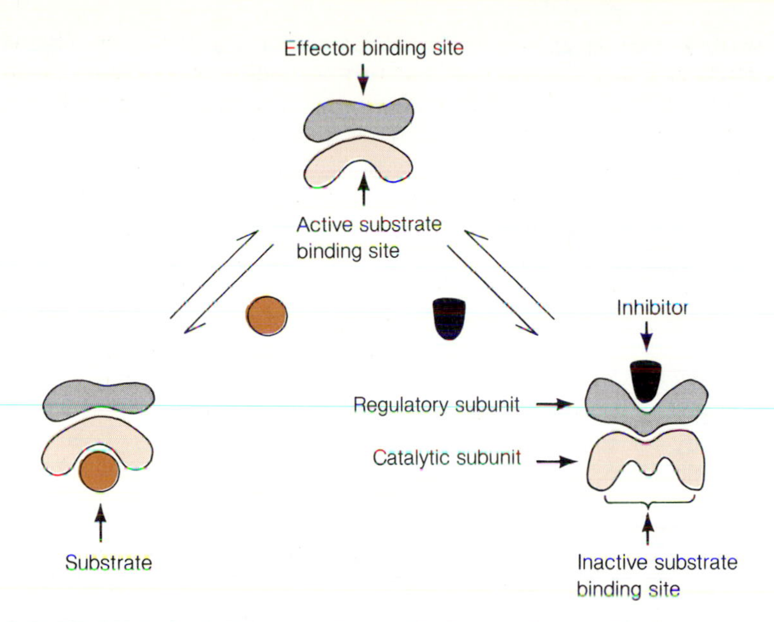

Figure 10-2 Diagram Illustrating Allosteric Regulation of an Enzyme. The allosteric enzyme illustrated has two distinct subunits. One of the subunits plays a regulatory role, while the other plays a catalytic role. The active form of the enzyme (left) has an active site that can bind with the substrate and hence catalyzes the reaction. If an inhibitor binds to the allosteric site (right), the three-dimensional structure of the active site is altered so that the substrate can no longer bind to it. This prevents the enzyme from catalyzing the reaction.

cule directs the synthesis of proteins and how the appearance and disappearance of certain proteins regulate metabolism.

The catabolism of lactose by bacteria such as *Escherichia coli* is dependent upon a cluster of genes known as the **lactose operon.** Lactose metabolism is regulated by controlling the transcription of the genes that code for the enzymes involved in lactose transport and breakdown. In lactose-free environments, these enzymes will be present in very low levels in the cell, while in the presence of lactose, their concentration will rise sharply.

The Lactose Operon

The genes that contain the information for making the enzymes involved in the catabolism of lactose are called **structural genes** (fig. 10-1). The lactose operon in *Escherichia coli* consists of three structural genes and at least two controlling sites that regulate the expression of the structural genes. The structural genes are *lac z*, *lac y*, and *lac a*, while the controlling sites are *lac p* and *lac o* (fig. 10-1).

The Structural Genes

50

The structural gene *lac z* is approximately 3,510 nucleotides □ long and codes for the subunit of the enzyme **β-galactosidase.** β-galactosidase catalyzes the cleavage of the disaccharide lactose into galactose and glucose (fig. 10-3). The structural gene *lac y* is approximately 780 nucleotides long and codes for the enzyme called **lactose permease.** This enzyme is found in the plasma membrane and is involved in the transport □ of lactose into the cell. The structural gene *lac a* is 825 nucleotides long and codes for the enzyme

182

Figure 10-3 Reaction Catalyzed by β-galactosidase. β-galactosidase catalyzes the hydrolysis of lactose (a disaccharide) into glucose and galactose (monosaccharides).

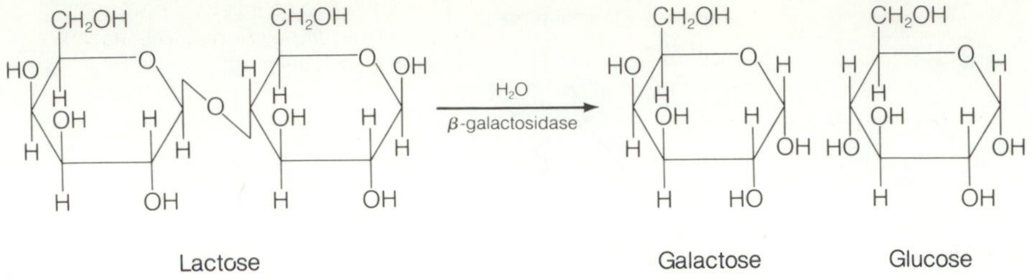

called β-galactoside transacetylase. This enzyme catalyzes the attachment of an acetyl group from acetyl-CoA to lactose. β-galactoside transacetylase is not necessary for the metabolism of lactose, although it has been found to detoxify toxic biproducts of lactose transport.

The Controlling Sites

The controlling site *lac o*, known as the **operator site** (fig. 10-1), consists of only 35 nucleotides. It does not code for a polypeptide, but instead serves as an attachment site for a regulatory protein called the **lactose repressor** (fig. 10-4). The controlling site *lac p*, called the **promoter site** (fig. 10-4), does not code for a polypeptide either; it serves as the attachment site for the enzyme RNA polymerase.

Contiguous with the lactose operon, and adjacent to the *lac p* site, is a regulatory gene that controls the expression of the lactose operon (fig. 10-4). The **regulatory gene** consists of a single structural gene, *lac i*, and a single controlling site. The structural gene *lac i* is approximately 1,041 nucleotides long and codes for the subunit of the lactose **repressor protein.** The *lac q* site is adjacent to *lac i* and serves as the attachment site for the RNA polymerase that transcribes the *lac i* gene. The *lac i* gene product is continually synthesized by the cell at a low rate; thus it is said to be a **constitutive** protein. It is estimated that there are 15 molecules of the lactose repressor in each *E. coli* cell.

How the Lactose Operon Works

A population of cells growing in an environment lacking lactose synthesizes very few molecules of β-galactosidase, lactose permease, and transacetylase. This repressed state of the cells results because the lactose repressor, coded for by the *lac i* gene, binds to the *lac o* region and blocks the transcription of the structural genes in the lactose operon (fig. 10-4). In the absence of lactose, RNA polymerase cannot bind to the *lac p* site because the lactose repressor, which is bound to the *lac o* site, blocks the RNA polymerase from binding to *lac p* (fig. 10-4). As a consequence, transcription of the lactose operon does not occur. Since little lactose mRNA is synthesized, very few of the proteins involved in the catabolism of lactose are made by the cell. We see, then, that the lactose operon is turned off by the lactose repressor protein, which is constitutively (constantly) synthesized in small amounts. In the absence of lactose, the cell does not waste energy making large amounts of mRNA or lactose-catabolizing enzymes for which it has no use. Very low levels of the lactose-catabolyzing enzymes are present in the cell, however, because the lactose repressor infrequently detaches itself from the *lac o* site (every 50 minutes or so), permitting the attachment of mRNA to *lac p*. This

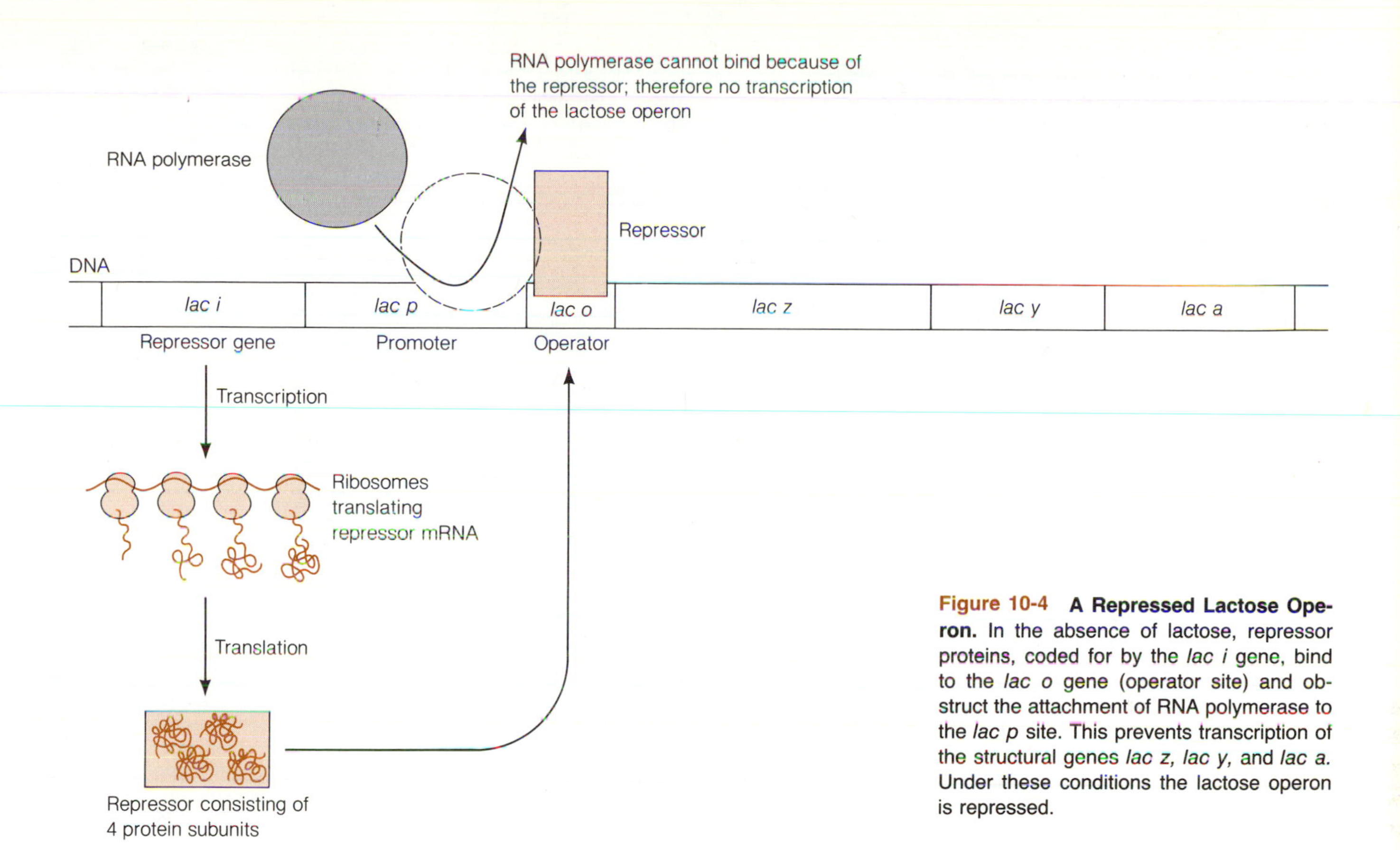

Figure 10-4 A Repressed Lactose Operon. In the absence of lactose, repressor proteins, coded for by the *lac i* gene, bind to the *lac o* gene (operator site) and obstruct the attachment of RNA polymerase to the *lac p* site. This prevents transcription of the structural genes *lac z, lac y,* and *lac a.* Under these conditions the lactose operon is repressed.

"sneak" synthesis of lactose mRNA is believed to be responsible for the **basal levels** of lactose enzymes found in cells growing in the absence of lactose.

If lactose is added to a population of *E. coli* cells, initially it is transported into the bacterial cells very slowly because of the low levels of lactose permease in their plasma membranes. The small amount of lactose that does get into the cell is acted upon by β-galactosidase. It is believed that minute quantities of lactose are rearranged so as to form a molecule called **allolactose.** This molecule (and possibly others) acts as an **inducer** of the lactose operon, turning it on (fig. 10-5). The inducer binds to the lactose repressor, causing a change in the repressor's three-dimensional shape. The altered repressor molecule has very little affinity for the operator site *(lac o)* and does not bind to it as readily as does the normal repressor. As the concentration of effector (lactose or its derivatives) increases, the operator is free of repressor more frequently, and more RNA polymerase can initiate transcription of the structural genes beginning at *lac p.* Multiple transcriptions of the lactose operon result in an increase in the concentration of the lactose enzymes. Permease systems increase in number in the plasma membrane, resulting in an accelerated rate of lactose transport into the cell. The operon eventually is expressed at its maximum rate, which is influenced by the affinity of the RNA polymerase for the *lac p* site.

The lactose operon is often referred to as an **inducible operon** because it is generally off until turned on (induced) by the presence of lactose (the inducer) in the medium. It is noteworthy that the regulation of the lactose

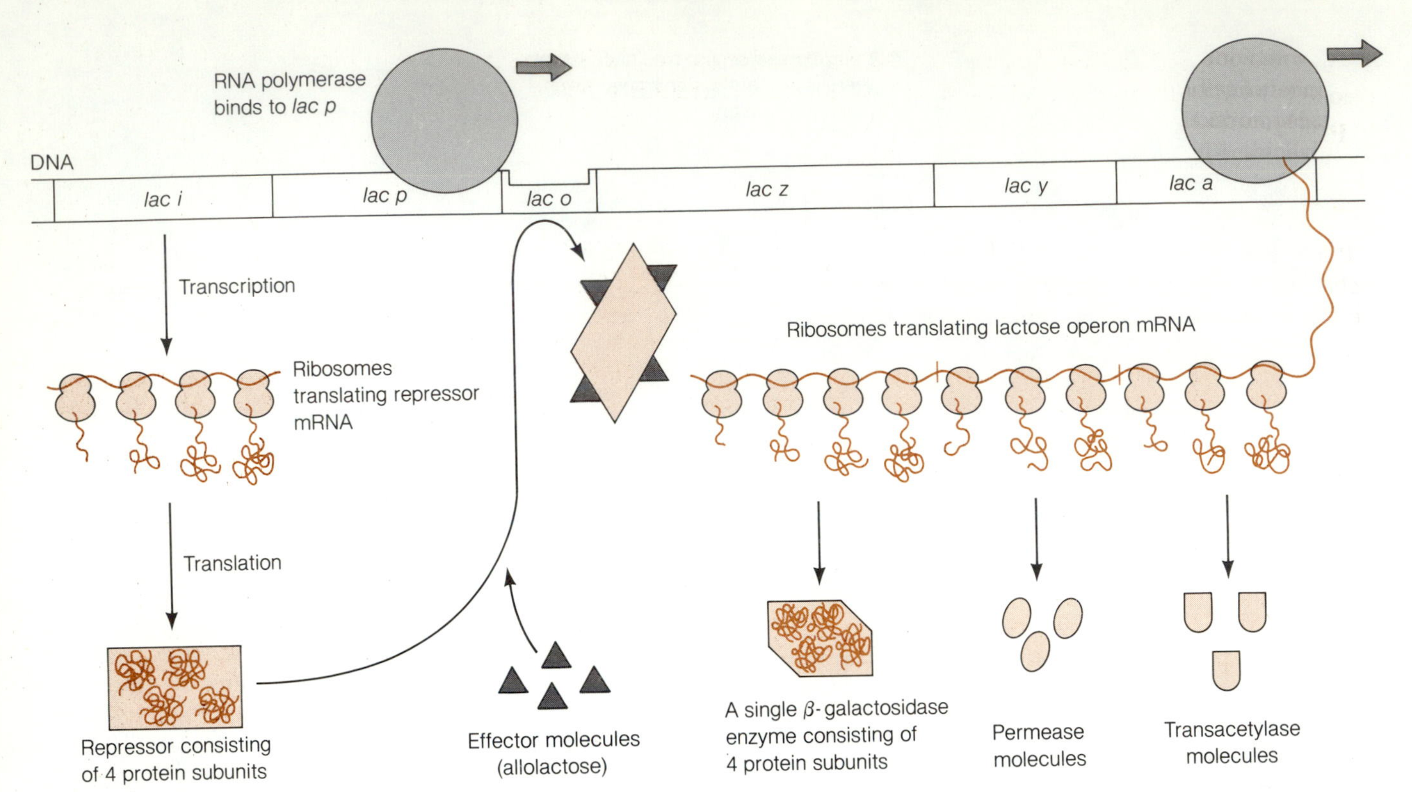

Figure 10-5 An Induced Lactose Operon. When lactose is present, the lactose operon is induced and the structural genes are transcribed. The effector (allolactose) interacts with the repressor and inactivates it so that it can no longer bind to the operator site *(lac o)*. This allows RNA polymerase to attach to the promoter site *lac p* and initiate transcription of the structural genes. The resulting mRNA (polycistronic mRNA) contains the information for the synthesis of all three enzymes necessary for the catabolism of lactose.

operon involves the inhibition of transcription by a repressor, which inhibits the binding of an RNA polymerase to the DNA. This type of regulation is common to most operons that are involved with catabolism □. Regulation of this sort affords maximum energy conservation, because the cell does not waste any energy making mRNA or proteins that it cannot use. The only energy expenditure is involved in the synthesis of a repressor molecule, which is made only in very small amounts.

CATABOLITE REPRESSION

If the lactose operon functioned at the maximum rate possible, it would produce excessive amounts of the lactose mRNA and lactose enzymes. This would be wasteful, not only because it takes a lot of energy to synthesize the mRNA and enzymes, but also because the excess glucose and other catabolites produced (e.g. pyruvate, α-ketoglutarate, succinate, etc.) would leak out of the cell. Also, if lactose and another carbohydrate such as glucose were present in the environment at the same time, there would be excessive metabolism of both carbohydrates and this would be wasteful. In order to avoid the excessive production of catabolites, lactose mRNA, and lactose enzymes, a system of control that limits the synthesis of these molecules exists. This control system is superimposed upon that exerted by the lactose repressor (*lac i* gene product) and is called **catabolite repression.** Catabolite repression involves a small effector molecule called **cyclic adenosine 3′,5′-monophosphate** (cAMP) and a cAMP-binding protein called **catabolite activator protein** (CAP).

Cyclic AMP

ATP is converted into cAMP (fig. 10-6) by an enzyme called **adenylcyclase.** Cyclic AMP is required for the attachment of RNA polymerase to the *lac p* site of the lactose operon. It also serves as an indicator of the concentration of catabolites within the cell and the availability of nutrients to the cell. If the concentration of catabolites is high, or if a lot of glucose is being transported into the cell, cAMP rapidly exits the cell. A large decline in the concentration of cAMP within the cell can occur within a minute or so. On the other hand, if the concentration of catabolites is low and/or if few nutrients are being transported into the cell, cAMP accumulates inside the cell. Hence, high levels of cAMP inside the cell reflect the need for additional carbon and energy sources, while low levels of cAMP indicate the need to limit the cell's catabolism and production of energy □. By this mechanism, the cell can monitor its energy status and keep itself from producing excessive amounts of enzymes, catabolites, and energy.

183

Catabolite Repression

If the RNA polymerase is to bind and initiate transcription at the *lac p* site, a CAP-cAMP complex must also be associated with the *lac p* site (fig. 10-7). When cells begin to catabolize a sugar such as lactose, the concentration of catabolites is low. This fact is reflected by the high levels of cAMP within the cell. The higher the level of cAMP in the cell, the more CAP becomes complexed with cAMP and hence the more efficient its association with the *lac p* site. This increased efficiency promotes the binding of the RNA polymerase to the promoter site and the subsequent transcription of the operon (fig. 10-7). The maximum rate of expression of the lactose operon is determined not only by the efficiency of the RNA polymerase binding to the *lac p* site but also by the presence of CAP-cAMP complexes bound to *lac p*.

As the concentrations of lactose mRNA and enzymes increase, the level of catabolites also increases, resulting in a decrease in the level of cAMP within the cell. At low concentrations of cAMP, much of the CAP is free of

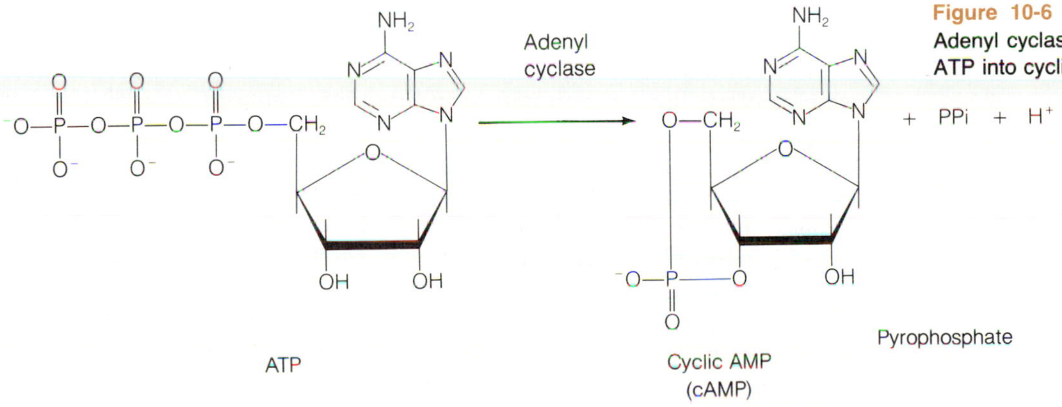

Figure 10-6 Synthesis of Cyclic AMP. Adenyl cyclase catalyzes the conversion of ATP into cyclic AMP.

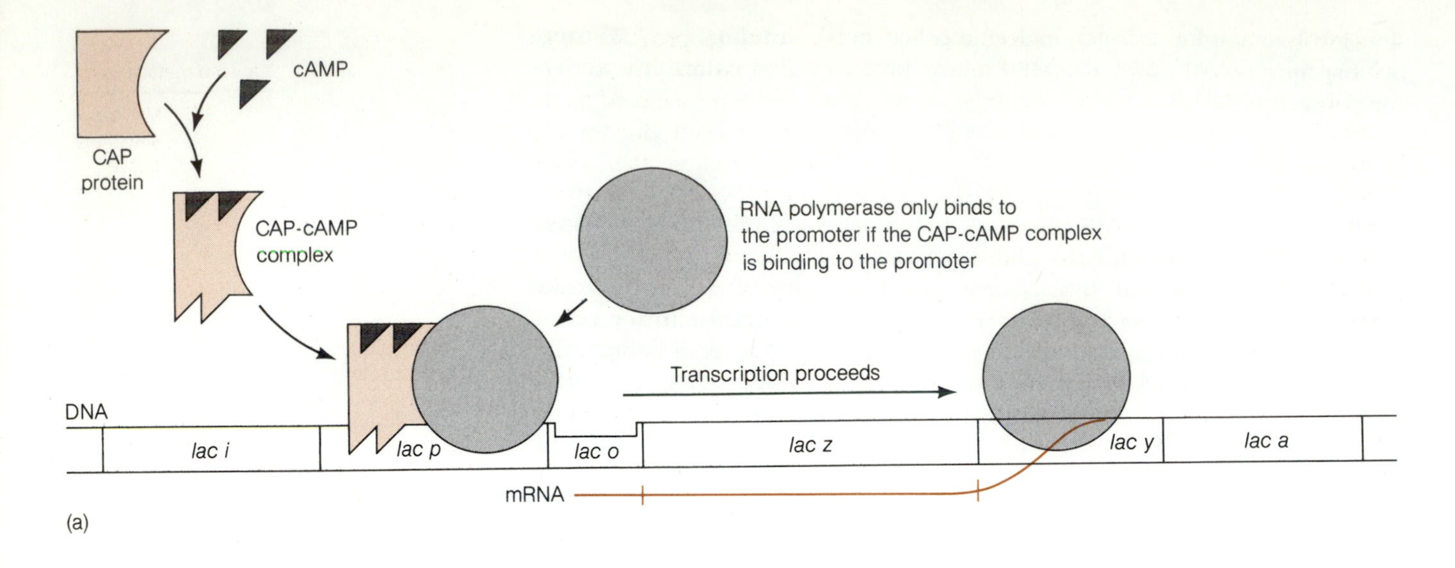

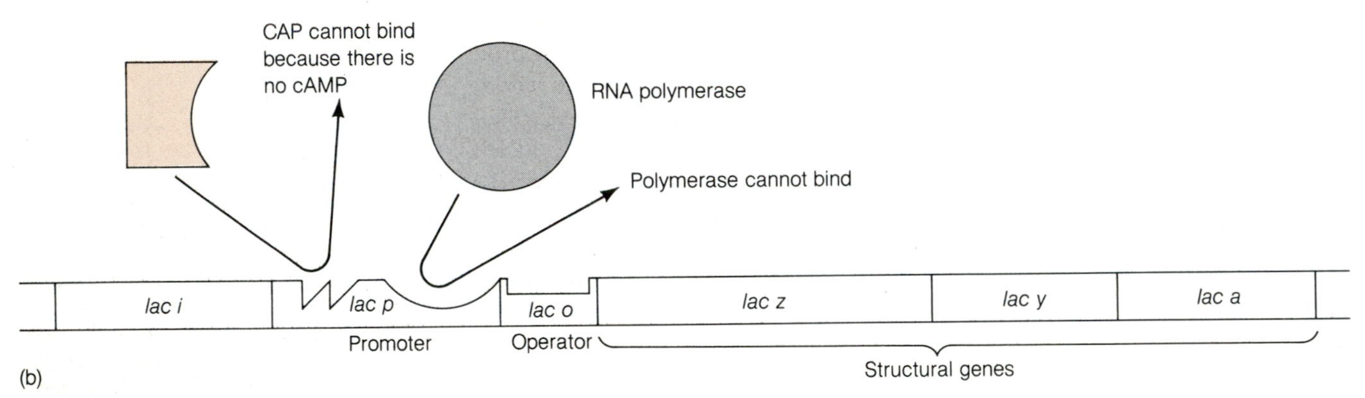

Figure 10-7 Catabolite Repression. Catabolites, molecules resulting from the degradation of larger molecules, can regulate the rate at which structural genes in an operon are transcribed. In order for transcription to begin, a complex consisting of CAP and cAMP must bind to a specific region on the promoter site. The rate of transcription can be affected by the levels of cAMP in the cell. *(a)* When the levels of cAMP are high the cell is in need of energy, and transcription of the operon proceeds at a high rate. *(b)* When the levels of cAMP are low, very little CAP can bind to the promoter site and transcription does not take place.

cAMP and does not bind to the *lac p* site (fig. 10-7). Thus, the RNA polymerase cannot bind to the promoter site and subsequently initiate transcriptions either. As a result, no new lactose mRNA is synthesized, and the enzymes involved in lactose catabolism do not increase in concentration. As the cells reproduce, the level of lactose enzymes decreases to a point at which the concentration of catabolites from lactose no longer increases, but instead declines. When the concentration of catabolites decreases because they are being utilized more rapidly than they are being synthesized, the concentration of cAMP rises once again, leading to new rounds of transcription. Eventually, a steady state equilibrium is reached among the concentration of catabolites, the concentration of cAMP, and the initiation of transcription.

The accumulation of substrates and catabolites which accompanies rapid metabolism turns off the lactose operon, a phenomenon called **catabolite repression.** Catabolite repression is found to operate in many operons besides the lactose operon. Those carbohydrates that are most rapidly transported into the cell, and most rapidly metabolized to produce large pools of catabolites, cause the most severe catabolite repression.

Glucose generally induces the most severe catabolite repression. If a culture of *E. coli* is grown in a culture medium that contains both glucose and lactose, the microorganisms will first utilize the glucose, because the lactose operon is shut off by catabolite repression and it will not begin to transcribe the lactose structural genes until the glucose is almost gone (fig. 10-8). When little glucose remains in the medium, the microorganisms begin using the lactose. At this point, there is little or no catabolite repression and the lactose operon is transcribed. During the period of induction of the lactose enzymes, there is a period of no apparent growth by the population, representing a lag phase in the middle of the population's exponential growth phase 141 □ (fig. 10-8). This growth pattern, in which the population exhibits two lag phases, one for each of the two carbohydrates utilized, is called **diauxic** growth. Although glucose induces the most severe catabolite repression, other molecules such as fructose, ribose, acetate, succinate, glycerol, and lactate also promote catabolite repression to varying degrees.

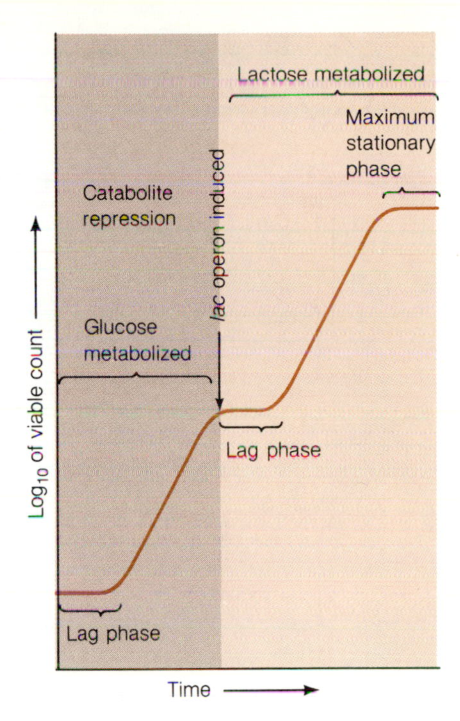

Figure 10-8 Diauxic Growth. In an environment containing both glucose and lactose, the growth curve of a population of *Escherichia coli* will look like the one illustrated. Glucose will be metabolized first due to catabolite repression of the lactose operon as well as to the declining levels of cAMP. When glucose becomes exhausted, the lactose operon will be induced. The plateau seen in the middle of the growth curve is the lag phase for the population before it begins reproducing in the presence of lactose (i.e., the time required to synthesize the lactose-catabolizing enzymes).

REGULATION OF TRYPTOPHAN SYNTHESIS

Each of the 20 different amino acids used by the cell to make proteins must be readily available to the cell in order for protein synthesis to occur. If these amino acids are not readily available in the environment, the cell must synthesize them. When an amino acid is supplied exogenously (from the outside) to a bacterial population, that amino acid is no longer made by the cell, because the amino acid shuts off transcription of the genes coding for the enzymes involved in its synthesis (fig. 10-9). If the exogenously supplied amino acid is removed, the genes involved in the synthesis of the amino acid are again turned on. The following section will discuss how the synthesis of the amino acid tryptophan is regulated, and it will serve as a model to illustrate the basic strategy by which bacteria regulate anabolic pathways.

The Tryptophan Operon

The tryptophan operon in *Escherichia coli* (fig. 10-10) consists of five structural genes *(trpE, trpD, trpC, trpB,* and *trpA)* and at least three controlling sites *(trpP, trpO,* and *trpL).* It is negatively controlled (turned off) by a complex of the amino acid tryptophan, and the tryptophan regulatory protein coded for by the *trpR* gene. As shown in figure 10-10, the enzymes coded 207 □ for by the structural genes of the tryptophan operon convert chorismic acid to L-tryptophan.

When tryptophan is limiting or absent, the regulatory protein is unable to form a complex with tryptophan and consequently is unable to bind to the operator site *(trpO).* Thus, an RNA polymerase is able to bind to the promoter *trpP* and initiate transcription 207 □. The RNA polymerase synthesizes a mRNA molecule that is translated into the enzymes involved in the synthesis of tryptophan from chorismic acid.

When tryptophan is abundant in the cell, it acts as a **corepressor** of the operon: it binds to the inactive *trpR* gene product and induces a change in the shape of the repressor protein (called the **aporepressor**) so that it can

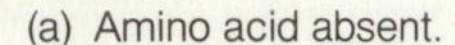

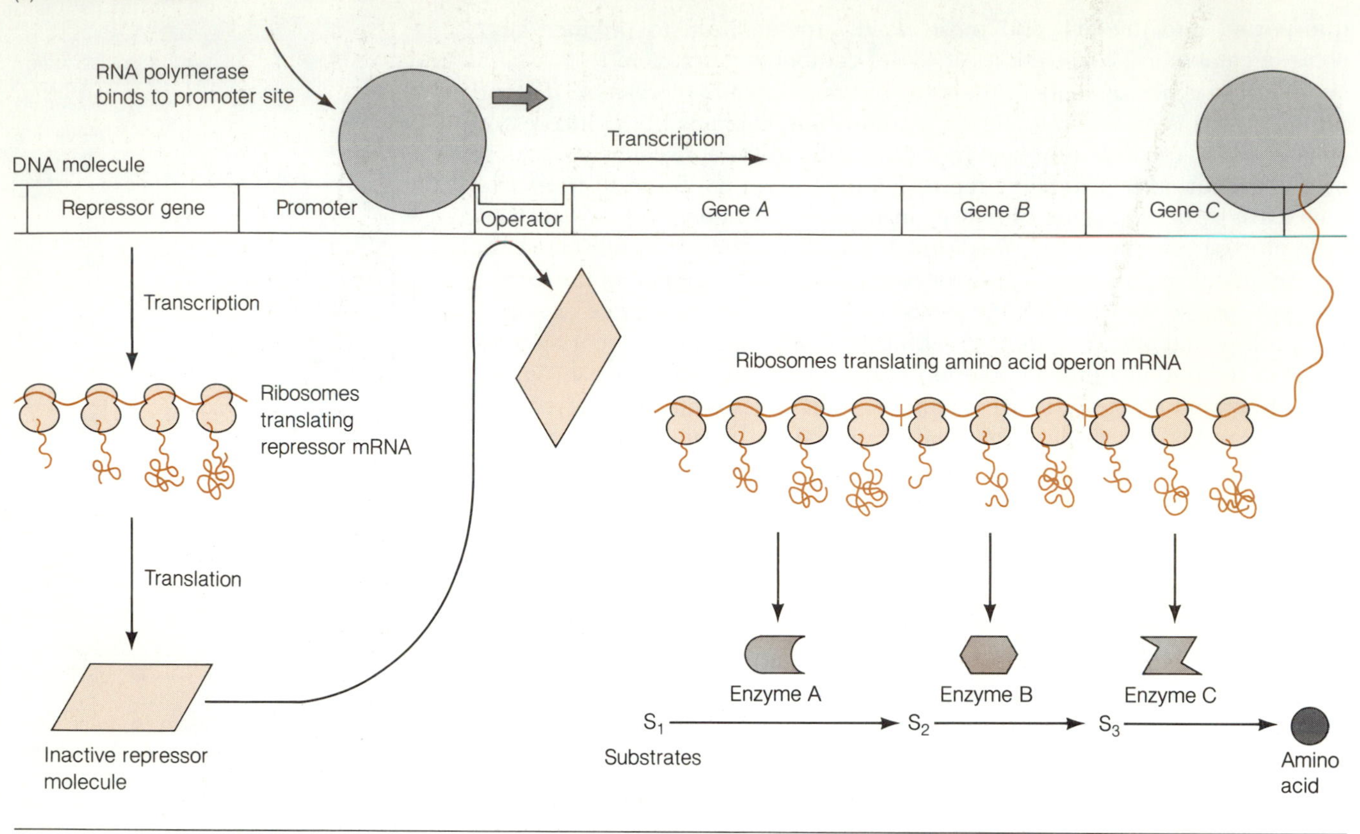

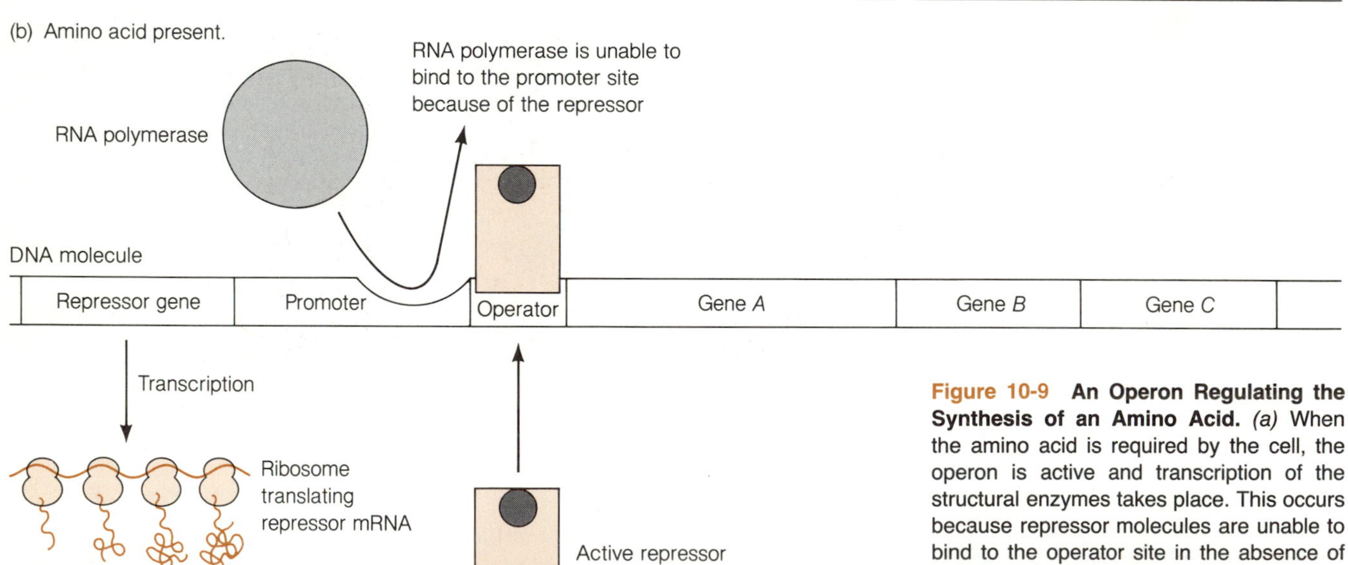

Figure 10-9 An Operon Regulating the Synthesis of an Amino Acid. *(a)* When the amino acid is required by the cell, the operon is active and transcription of the structural enzymes takes place. This occurs because repressor molecules are unable to bind to the operator site in the absence of corepressor. *(b)* When the amino acid end product (corepressor) is present in enough quantity in the cell, the operon is turned off and transcription of the structural genes does not take place. When there is a plentiful supply the amino acid in the cell, some of it binds to the repressor molecules, which activates them. The active repressor can then bind to the operator site and block transcription.

bind to the *trpO* site. Binding of the active repressor (aporepressor + corepressor) at the *trpO* site blocks transcription and consequently shuts off the operon (fig. 10-9). In this way, the cell can regulate the synthesis of tryptophan by turning on or off the transcription of the structural genes. This operon, like most other anabolic operons, is said to be a **repressible** operon because it is turned off by a repressor when sufficient tryptophan has been made. Repression reduces transcription approximately 70-fold.

Unlike the lactose operon, where the promoter and the operator sites can be viewed as more or less separate sites, the *trpO* site is completely contained within the *trpP* site (fig. 10-10). When the RNA polymerase binds to the promoter site *(trpP)*, it begins to transcribe from that point, making a single mRNA coding for all five polypeptides (a **polycistronic message**). 204 The first part of the polycistronic mRNA, called the **leader region** □ (fig. 10-10), extends to the beginning of the *trpE* gene. This first portion of the mRNA (about 165 nucleotides long) codes for a very short polypeptide that, along with the aporepressor, is involved in the regulation of the tryptophan operon.

Attenuation

The expression of the tryptophan operon is dependent upon whether or not 209 the leader region is translated □. If the leader region is completely translated, the tryptophan operon is shut off. If the leader peptide is not synthesized, however, because of the lack of tyrptophan, then the tryptophan operon is expressed. This type of negative regulation is called **attenuation** (fig. 10-10), and it reduces transcription about tenfold. The total reduction of transcription due to repression and attenuation is nearly 700-fold (70 × 10).

In general, the attenuation of the tryptophan operon works in the following fashion: a) if tryptophan is abundant, the bacterium shuts off the synthesis of tryptophan enzymes, both by attenuation and by repression of transcription; b) if tryptophan is the only amino acid in short supply, the cell promotes transcription of the tryptophan operon structural genes, and c) if a bacterial cell is starved for many amino acids, it turns off the transcription of the tryptophan operon by attenuation.

When tryptophan is available to the cell, the leader region is translated because there is an ample supply of charged tRNAs (tRNAs with their corresponding amino acids attached), including tryptophanyl-tRNA. Translation allows the remaining leader RNA to form a secondary structure that causes transcription to terminate (fig. 10-10). The mechanism by which the leader polypeptide shuts off the tryptophan is not entirely clear, but some scientists believe that the leader RNA binds to the RNA polymerase and alters its structure.

When tryptophan is in short supply, translation of the leader mRNA stalls at the first codon in the leader mRNA molecule coding for tryptophan because there is little or no tryptophanyl-tRNA in the cell. Since the leader polypeptide is not made, the leader RNA forms a secondary structure that does not inhibit the RNA polymerase. Thus, transcription of the entire polycistronic mRNA *(trpE* through *trpA)* can take place. This transcription eventually results in the production of the tryptophan enzymes and the synthesis of tryptophan (fig. 10-10).

Apparently, starvation for amino acids other than tryptophan attenuates or inhibits the transcription of the tryptophan structural genes. For example,

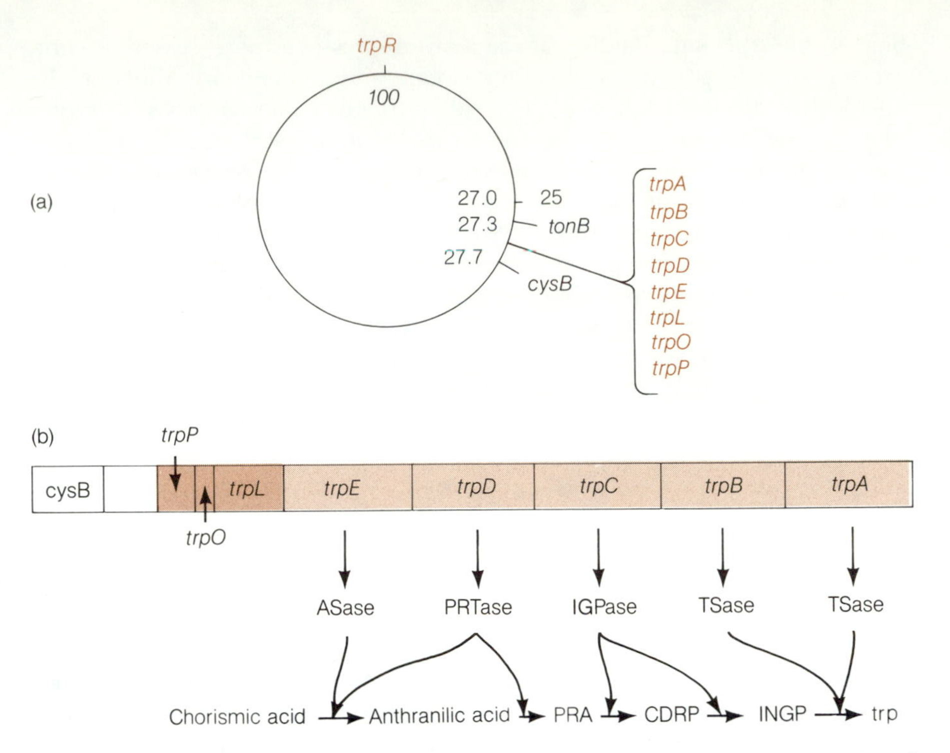

Figure 10-10 **The Tryptophan *(Trp)* Operon in *Escherichia coli.*** *(a)* Location of the *Trp* operon in the bacterial chromosome. (The numbers are map units). *(b)* Genes making up the *Trp* operon consist of five structural genes *(E, D, C, B* and *A),* a regulatory gene *(R),* an operator site *(O),* and a promoter *(P). (c)* Attenuation of the Tryptophan Operon. *1.* When tryptophan is available to the cell, the leader region is translated. Synthesis of the polypeptide blocks the transcription of the tryptophan structural genes because it allows hairpin loop-3 of the leader RNA to form. Any time that loop-3 forms, transcription is terminated. *2.* When tryptophan is in short supply, the leader region will not be entirely translated (because tryptophan is one of the amino acids that make up the polypeptide). Under these circumstances, transcription of the tryptophan genes will occur and tryptophan will be synthesized. When the leader peptide is partially synthesized, hairpin loop-3 does not form. *3.* When various amino acids are in short supply, the leader region will not be translated at all. The leader RNA will assume a tertiary structure (forming loops 1 and 3), which will cause the RNA polymerase to terminate transcription. ASase = Anthranilate synthetase. PRTase = Anthranilate phosphoribosyl transferase. IGPase = Indole glycerol phosphate synthetase. TSase = Tryptophan synthetase. PRA = Phosphoribosyl anthranilic acid. CDRP = Carboxylphenylamino-deoxyribosyl phosphate. INGP = Indole glycscol phosphate.

if *E. coli* is starved for alanine as well as for tryptophan, one would expect the cell to allow the transcription of the tryptophan operon to alleviate the lack of tryptophan. This is not the case, however. Instead, transcription of the tryptophan operon is turned off by attenuation. Starvation for amino acids such as alanine, isoleucine, phenylalanine, valine, methionine, leucine, and serine causes attenuation of the tryptophan operon even when tryptophan is limiting. In the absence of any of these amino acids, the leader mRNA is not translated (or only a very short segment of it is translated) and the leader RNA forms the structure that inhibits the RNA polymerase. Thus, transcription terminates when tryptophan and other amino acids are limiting (or absent from the cell). Starvation for more than one amino acid inhibits its transcription of the tryptophan operon by not allowing the RNA polymer-

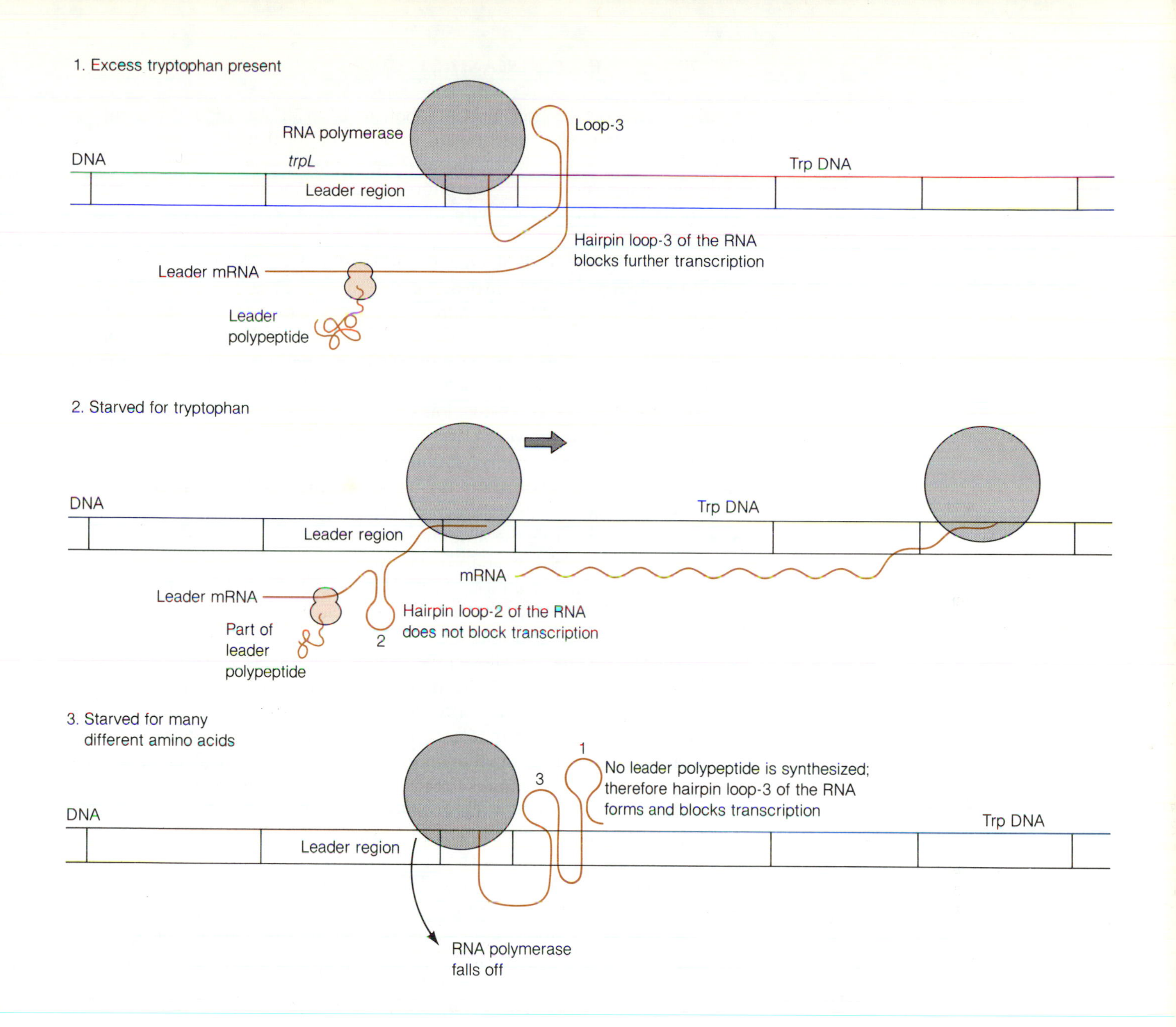

ase to transcribe the tryptophan genes beyond the leader region (fig. 10-10). Attenuation is advantageous to the cell, because in the absence of many amino acids, very few of the cell's proteins can be made. Therefore, by turning off the tryptophan operon, the cell economizes its energy and nutrient resources by not making unnecessary enzymes.

In summary, the repression of the tryptophan operon responds to the concentration of the specific amino acid, while the attenuation system monitors both the concentration of the specific amino acid and the concentrations of other amino acids in the cell. This dual control allows a much finer expression of the operon than would otherwise be possible, thus providing the appropriate concentration of nutrients to satisfy the cell's needs.

Another method of controlling metabolism is to inhibit or activate the enzymes that catalyze □ critical steps in a biochemical pathway. The cell accomplishes this control by using **allosteric enzymes** at strategic points in metabolic pathways. Their enzymatic activity is modulated by small regulatory molecules that interact with the enzyme at specific regulatory or **effector sites.**

54

The activity of allosteric enzymes follows a sigmoidal (S-shaped) curve, as shown in figure 10-11, instead of the hyperbolic curve of nonallosteric enzymes (fig. 10-11). When an allosteric enzyme binds to inhibitors or stimulators, there is a change in activity, as measured by the velocity of conversion of substrate to product. At low substrate concentrations, allosteric enzymes that are stimulated by their substrate catalyze reactions at a rate that is proportional to the concentration of the substrate. As the substrate concentration increases, however, the activity of allosteric enzymes increases much faster than the activity of nonallosteric proteins. The stimulation of the enzyme by the substrate is known as **cooperativity.** This mechanism provides a fine level of control over enzymes in catabolic and anabolic pathways and hence controls the amount of end product synthesized.

Feedback inhibition is a process in which the activity of an allosteric enzyme is diminished or inhibited completely as the concentration of a product increases. The velocity at which an allosteric enzyme catalyzes a reaction is dependent upon the concentration of the end product of the reaction. What is the advantage of having allosteric enzymes in a cell? Consider the synthesis of the amino acid isoleucine from threonine (fig. 10-12). The synthesis of isoleucine from threonine is controlled by derepression and attenuation of the isoleucine-valine operon. This operon contains the information for the synthesis of the enzymes that function in the synthesis of isoleucine and valine. Because the enzymes that catalyze the synthesis of isoleucine are relatively stable, they would continue to catalyze the production of large amounts of isoleucine even when this amino acid was no longer required. This production would waste threonine and isoleucine, because these molecules would have to be synthesized from material that could be used for other purposes. The overproduction of threonine and isoleucine can be avoided, however, if the first enzyme in the isoleucine pathway (fig. 10-12) is responsive to the concentration of threonine and isoleucine. In fact, the first enzyme in the pathway is an allosteric enzyme called threonine deaminase. It becomes inactive when threonine is limiting or when the concentration of isoleucine becomes excessive. In other words, the enzyme requires threonine in order to be active, and it is inhibited by high concentrations of isoleucine (fig. 10-12).

Allosteric enzymes such as threonine deaminase can respond to a variety of different nutritional states of the cell. For example, this enzyme is not only feedback-inhibited by isoleucine and stimulated by the binding of the substrate threonine, but it is also stimulated by valine (fig. 10-12). This response is advantageous to the cell. When the isoleucine pool gets larger than it should be for the proper functioning of the cell, its synthesis is slowed down by inhibition of the first enzyme in its biosynthetic pathway. If, however, the supply of valine or other amino acids is at the appropriate level and isoleu-

(a)

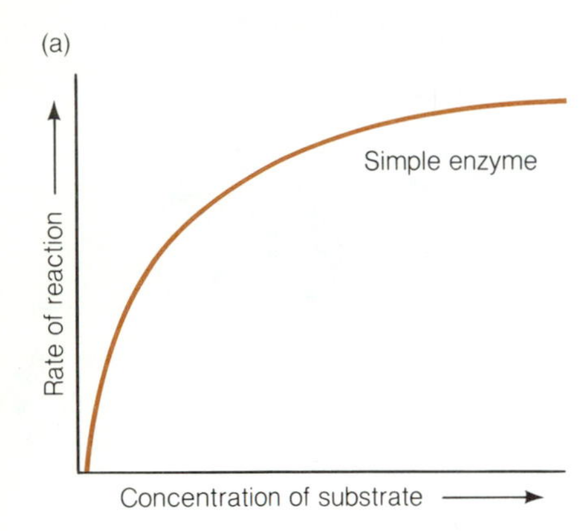

(b)

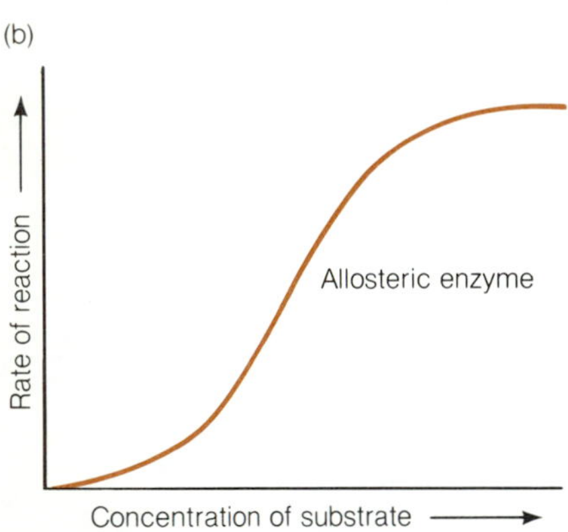

Figure 10-11 Activity of Allosteric and Nonallosteric Enzymes. *(a)* Nonallosteric enzyme. The rate at which nonallosteric enzymes catalyze reactions increases as the concentration of substrate increases until a maximum substrate concentration is reached, where it saturates the enzyme. *(b)* Allosteric enzymes respond to varying substrate concentrations with changes in catalytic activity. At low substrate concentration, the catalytic activity of allosteric enzymes is low. As the substrate concentration increases, so does the catalytic activity of the enzyme. This results in the characteristic sigmoid (S-shape) of allosteric enzymes.

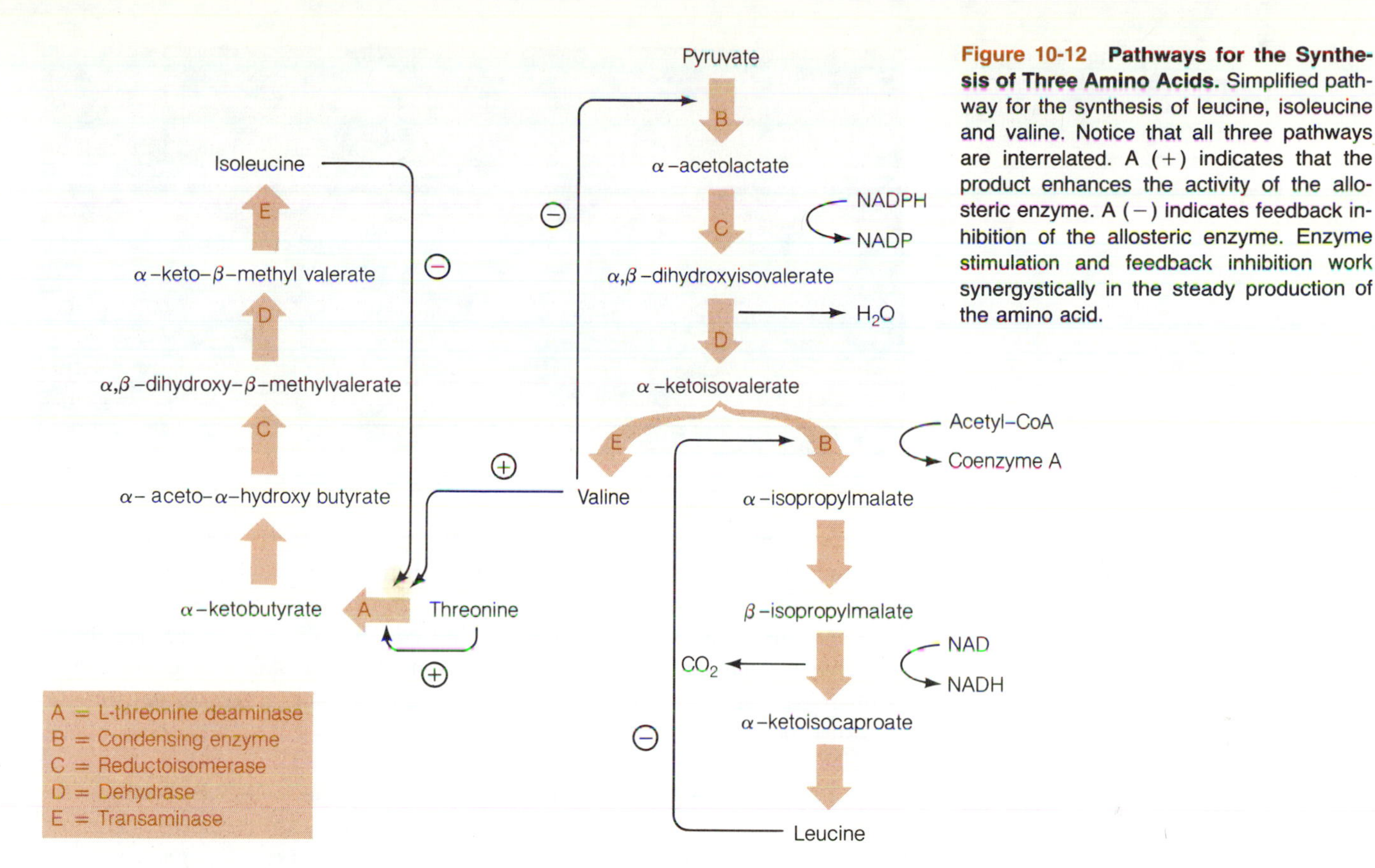

Figure 10-12 Pathways for the Synthesis of Three Amino Acids. Simplified pathway for the synthesis of leucine, isoleucine and valine. Notice that all three pathways are interrelated. A (+) indicates that the product enhances the activity of the allosteric enzyme. A (−) indicates feedback inhibition of the allosteric enzyme. Enzyme stimulation and feedback inhibition work synergystically in the steady production of the amino acid.

cine is lagging behind, it is advantageous for the cell to be able to stimulate isoleucine synthesis by stimulating the enzyme with valine. Allosteric proteins may have multiple subunits. They are able to respond simultaneously to several **effector molecules** and to different concentrations of the effector molecules. Thus, the enzymes are inhibited or stimulated to varying degrees.

Jacques Monod, Jean-Pierre Changeux, and Francois Jacob in the early 1960s proposed that allosteric proteins had binding sites for effectors which were different from the binding sites for the substrates. It is now known that some enzymes have substrate binding sites on what are called **catalytic** subunits and effector binding sites on **regulatory** subunits. Allosteric enzymes abound in the various biosynthetic and catabolic pathways of the cell. They are usually located strategically at the beginning of pathways or at forks, so that the synthesis of various products can be finely tuned.

TRANSPOSABLE ELEMENTS

DNA mapping, using recent advances in genetic engineering, has shown that there are pieces of DNA measuring 700 to 20,000 base pairs long which invert themselves or "jump" from one region to another. The "jumping" genes (fig. 10-13) are known as **insertion sequences** (IS) and **transposons** (Tn). Insertion sequences and transposons have been found in bacteria, fungi, protozoa, fruitflies, and corn, and there is some evidence suggesting

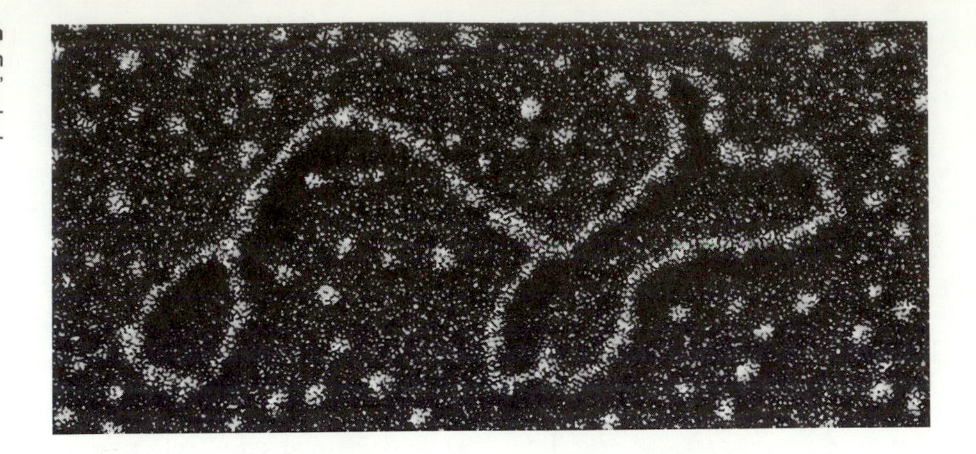

Figure 10-13 Electron Photomicrograph of a Transposable Element. Transposon exhibiting the characteristic "stem and loop" structure resulting from inverted repeat nucleotide sequences at each end of the section of DNA.

that insertion sequences and transposable elements may be found in all types of cells. Insertion sequences range in size from 700 to 5,000 base pairs. Bacterial transposons are generally larger than insertion sequences (table 10-1). Transposons frequently contain genes that confer antibiotic resistance. These transposons have been found mainly in bacteria. Regulatory genes, genes conferring resistance to heavy metals, and genes for enterotoxins are also found in bacterial transposons.

By removing whole segments of DNA from one section of the genome □ 89 and by inserting segments of DNA into the genome, cells can regulate the expression of certain phenotypes. For example, the gram negative bacterium *Salmonella* can synthesize two chemically distinct types of flagella □ 85. The genes coding for the flagellins (proteins from which flagella are made) are H1 and H2, and they are located in separate operons. Normally, the two genes are not expressed at the same time. The orientation of an insertion

TABLE 10-1
CHARACTERISTICS OF TRANSPOSABLE ELEMENTS

TRANSPOSABLE ELEMENT	LENGTH (base pairs)	CHARACTERISTICS
Insertion sequences	700–5,000	Carry regulatory genes and controlling sites. Promote inversions. Nonsense codons in all reading frames.
Transposons	2,000–20,000	Carry inverted or direct repeats of insertion sequences.
Episomes	(?)–100,000+	Carry insertion sequences and numerous genes. Promote deletions and additions. Self-replicating.
Temperature viruses	(?)–100,000+	Carry insertion sequences and numerous genes. Self-replicating. Require living host for replication.

sequence next to the H2 gene determines whether the H2 or H1 gene is expressed (fig. 10-14). Sometimes, an insertion sequence is oriented in such a way that the H2 promoter allows the initiation of transcription to the right (fig. 10-14). The operon has two structural genes, one of which codes for the H2-flagellin and the other for a repressor of the H1 operon. When the insertion sequence is oriented so that transcription is to the right, both of these genes are expressed, resulting in the synthesis of H2-type flagella and the

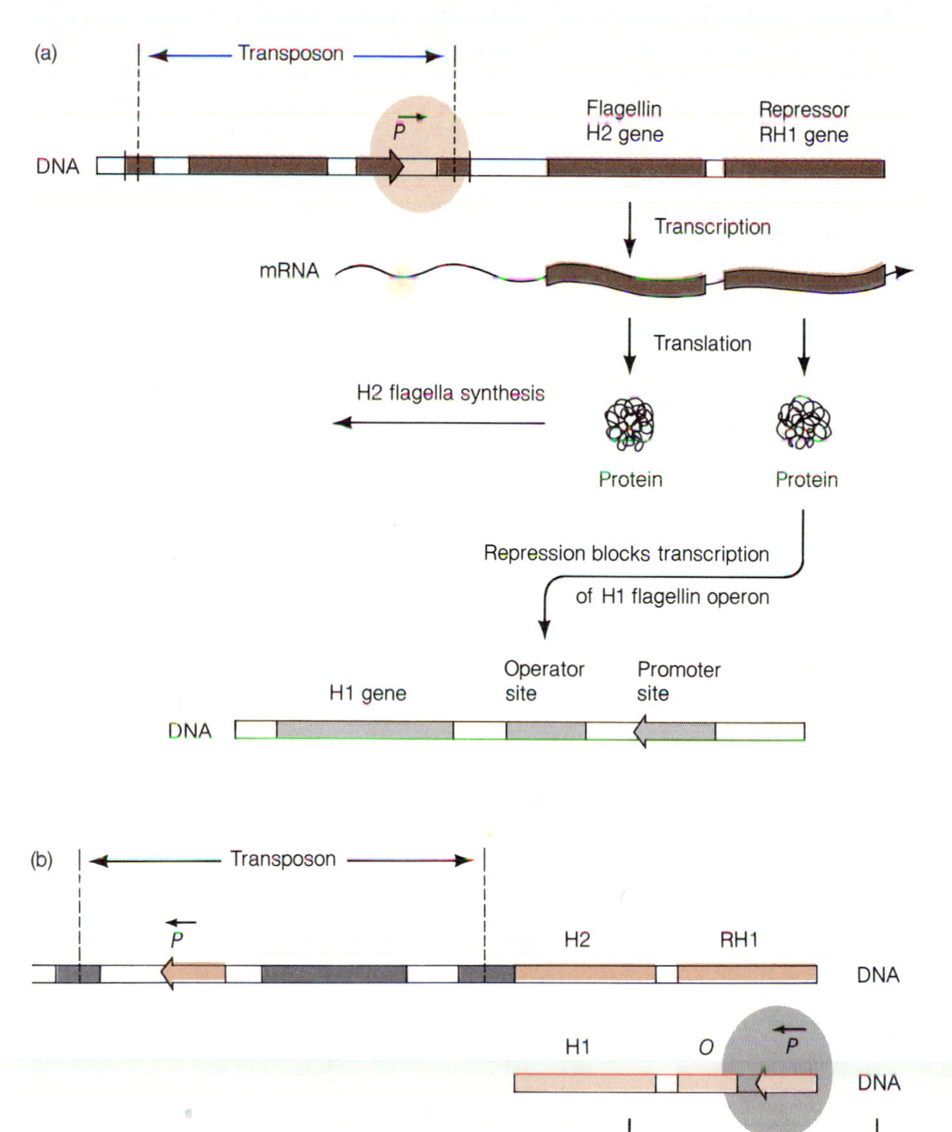

Figure 10-14 Regulation of Flagellar Synthesis in *Salmonella* by an Insertion Segment. *(a)* Insertion sequence is oriented so that the H2-flagellin gene is transcribed as well as the H1 flagellin repressor gene. This allows for synthesis of a flagellum made of H2 flagellin and a repression of the H1 flagellin gene. *(b)* Insertion sequence is inverted so that the H1 gene and the H2 repressor protein are transcribed. This results in the synthesis of a flagelum out of H1 flagellin and the repression of the H2 gene.

repression of the H1 operon. If the insertion sequence changes its orientation, the H2 operon is not transcribed, because there is no place for the RNA polymerase to bind (no promoter site). Hence, no H1 repressor is formed and the H1 operon is turned on. Under these circumstances, the H2 operon is not expressed. Thus, an insertion sequence, strategically located in the bacterial chromosome, can regulate the expression of two different gene clusters by changing its orientation ("jumping") in the chromosome.

Insertion sequences and transposons are found in bacterial genomes, in plasmids, and in viruses (table 10-1). This widespread occurrence is not surprising, because transposons are able to jump from plasmid to plasmid, from genome to plasmid or to virus and vice versa. Some plasmids (episomes) and some viruses (temperate viruses) can integrate themselves into bacterial genomes. It is known that the integration of these plasmids and viruses is made possible by the presence of insertion sequences that are part of the plasmid and the virus. Those plasmids and viruses that normally do not become integrated in bacterial genomes lack IS regions that are homologous to IS regions in the bacterial genomes.

FOCUS

METABOLIC REGULATION IN EUKARYOTES

Prokaryotic organisms regulate their metabolism mainly by controlling transcription and allosteric enzymes. Eukaryotes, on the other hand, regulate their cellular activities by controlling transcription, translation and their allosteric proteins.

Eukaryotic regulation is extremely complex at each level. For example, transcription of DNA can be controlled by regulatory proteins that bind to the DNA and alter its three-dimensional structure. DNA can be converted from the normal **B-form**, right-handed double helix, to the unusual **Z-form**, left-handed double helix that has zigzagging sugar-phosphate backbones. Transcription of DNA can also be controlled by the methylation of guanine and cytosine bases.

Eukaryotes regulate the expression of genes by altering the mRNA. The initial mRNA transcript, called **heterogenous nuclear RNA** (hnRNA), is drastically reduced in size by the excision of introns and protein coding sequences (exons) are joined together. Also, the 5′ and 3′ ends of the mRNA have nucleotides added to them. **Post transcriptional**, as the alterations of hnRNA are called, control the production of proteins.

Eukaryotes also regulate their metabolism by controlling protein sythesis. For example, initiation factors required for the initiation of protein synthesis may be modified, thus influencing the rate of translation.

Regulation of metabolism by eukaryotes results in a finely-tuned cell that efficiently uses its resources and minimizes waste. The figure below outlines the various levels of eukaryotic metabolic regulation (fig. 10-15).

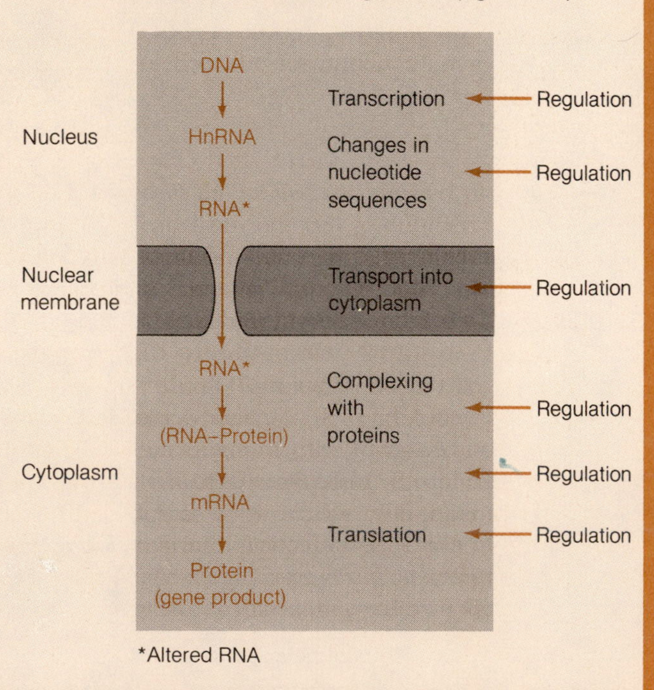

*Altered RNA

SUMMARY

1. Metabolic reactions in the cell are well integrated so as to optimize the cell's use of energy and resources.
2. The cell controls its metabolism by regulating the transcription of genes, the translation of mRNAs, and the activity of enzymes.

REGULATION OF LACTOSE METABOLISM

1. The *E. coli* lactose operon consists of three structural genes, which code for the enzymes involved in the transport and catabolism of lactose, and two controlling sites, which influence the expression of the structural genes.
2. Adjacent to the lactose operon is a gene *(lac i)* whose sole function is to regulate the expression of the lactose operon by producing a protein called the lactose repressor.
3. In the absence of lactose, the lactose repressor binds to the operator site *(lac o)* and prevents the transcription of the lactose structural genes by RNA polymerase.
4. In the presence of lactose, an effector, such as allolactose, binds to the lactose repressor and prevents binding to *lac o*. This event permits the RNA polymerase to attach the *lac p* site and transcribe the lactose structural genes.

CATABOLITE REPRESSION

1. Catabolite repression is a means of turning off certain catabolic operons either when glucose becomes available or when sufficient amounts of energy and catabolites have been produced.
2. The energy and nutritional status of the cell can be monitored by measuring the concentration of cAMP in the cell. When the cell is starved, the levels of cAMP inside the cell are high, but when the cell has an ample supply of energy nutrients, and they are being rapidly transported into the cell and metabolized, the levels of cAMP inside the cell are low.
3. In order for an RNA polymerase to bind to *lac p*, a complex of cAMP and CAP must bind to *lac p*. This condition requires that high concentrations of cAMP be present.
4. Catabolite repression is most severe when the levels of cAMP in the cell are low. Catabolite repression occurs because a CAP-cAMP complex does not bind to *lac p* and stimulate the attachment of an RNA polymerase to the *lac p* site. Without the initiation of transcription, the lactose operon structural genes are not expressed.

REGULATION OF TRYPTOPHAN SYNTHESIS

1. The tryptophan operon is a repressible operon that functions so long as the levels of tryptophan inside the cell are low. High levels of tryptophan turn off the tryptophan operon.
2. The tryptophan operon consists of five structural genes and at least three controlling sites.
3. When trypotphan is absent (or in very low concentration), the tryptophan repressor molecule (*trpR* gene product) is unable to bind to the operator site, and transcription of the tryptophan structural genes results.
4. When tryptophan is abundant in the cell, tryptophan binds to the inactive repressor, forming an active repressor that binds to the operator site and prevents transcription.
5. Attenuation is another mechanism that reduces transcription of the tryptophan operon. This mechanism of control monitors not only tryptophan levels in the cell but also the levels of other amino acids.

FEEDBACK INHIBITION AND STIMULATION OF ENZYMES

1. Allosteric enzymes can be either stimulated or inhibited by their effector molecules.
2. Allosteric enzymes that are feedback-inhibited become inactive as the concentration of the pathway and product increases.
3. Allosteric enzymes that are stimulated become increasingly active as concentrations of their substrate increases.
4. Allosteric enzymes are found at all key points (at the beginning of pathways and at the forks) in metabolic pathways so as to increase the efficiency at which microorganisms metabolize.

TRANSPOSABLE ELEMENTS

1. Transposable elements are pieces of DNA that invert themselves or jump from one region of DNA to another.
2. Some genes, like the flagellar phase variation in *Salmonella*, are under the control of transposable elements.
3. It is believed that various cellular processes, in both eukaryotes and prokaryotes, are under the regulation of transposable elements.

STUDY QUESTIONS

1. Explain how the lactose operon is affected by:
   a. high levels of lactose;
   b. low levels of lactose;
   c. high levels of glucose and low levels of lactose;
   d. high levels of glucose and lactose.

2. Explain how the tryptophan operon will be affected by:
   a. high levels of tryptophan only;
   b. high levels of tryptophan and valine;
   c. low levels of tryptophan only;
   d. low levels of tryptophan and valine.

3. Explain how the allosteric enzymes W, X, and Y should respond to changes in the levels of various metabolites in the hypothetical biosynthetic pathway given below:

```
          F → G → H
   ⟨X⟩ ↗
A → B  → C → D → E
         ⟨Y⟩
```

   a. concentration of B is low;
   b. concentration of both E and H are high;
   c. concentration of E is low and K is high;
   d. concentration of H is very low;
   e. concentration of H is high.

4. Outline the process of translation (protein synthesis) and determine in which steps a cell might exert translational regulation. Explain how this might be achieved.

5. Explain how attenuation takes place in the tryptophan operon. Of what use is attenuation to the cell?

6. Why do some carbohydrates induce more severe catabolite repressions than others?

7. Why is translational control important in eukaryotic cells?

8. Explain what would happen to a cell that lacked catabolite repression.

9. Explain what would happen to a cell that did not have allosteric enzymes.

10. Distinquish between feedback (or end product) inhibition of enzymes and feedback (or end product) repression of operons.

11. Explain the difference between an inducible operon and a repressible operon.

12. How are catabolic pathways that involve inducible enzymes regulated?

13. How are catabolic pathways that involve constifulive enzymes regulated?

14. Explain what diauxic growth is and why it occurs.

15. Explain how some bacteria change the proteins they use to make their flagella.

SUPPLEMENTAL READINGS

Beckwith, J., Davies, J., and Gallant, J. A. 1983. *Gene Function in Prokaryotes.* New York: Cold Spring Harbor Laboratory.

Buck, G. A. and Eisen, H. 1985. Regulation of the genes encoding variable surface antigens in African trypanosomes. *ASM News* 51:118–122.

Crose, C. and Klein, G. 1985. Chromosome translocations and human cancer. *Scientific American* 252:54–60.

Donelson, J. E. and Turner, M. J. 1985. How the trypanosome changes its coat. *Scientific American* 252:44–52.

Koshland, D. E. 1973. Protein shape and biological control. *Scientific American* 229:52–64.

Maniatis, T., and Ptashne, M. 1976. A DNA operator-repressor system. *Scientific American* 234:64–69.

Nierlich, D. P. 1978. Regulation of bacterial growth, RNA and protein synthesis. *Annual Review of Microbiology* 32:393–432.

Pastan, I. 1972. Cyclic AMP. *Scientific American* 227:97–105.

Watson, J. D. 1976. *Molecular biology of the gene.* 3d ed. Menlo Park, Calif.: Banjamin-Cummings Publishing Co.

CHAPTER 11 GENETIC ENGINEERING

CHAPTER PREVIEW

GENETIC ENGINEERING IN NATURE

Many isolates of the soil bacterium *Agrobacterium tumefaciens* contain a plasmid known as the Ti-plasmid (**t**umor-**i**nducing plasmid). When this bacterium infects wounds in plants such as tobacco, tomatoes, and begonias, it stimulates a "cancerous growth" or tumor at the site of infection. The plant tumors generally develop on the stems, and the disease is known as Crown-gall. It becomes noticeable three to four weeks after infection.

Agrobacterium attaches to plant cells within a lesion, and Ti-plasmids from the bacterium gain entrance to the plant cells. Part of the Ti-plasmid, called T-DNA, recombines with the plant's genome and transforms the plant cell. The transforming bacterial DNA is believed to cause an imbalance in the production of plant hormones. This imbalance results in the uncontrolled growth of plant cells and the formation of a crown-gall.

The effects of hormones on plant cells in tissue culture give us some insight as to how hormones act in whole plants. Most normal plant cells will not grow in tissue culture without plant hormones. Crown-gall tissue free of *Agrobacterium* can be cultivated on simple media without the addition of hormones, evidence that some of the plant cells from the crown-gall have been transformed.

In addition to causing the uncontrolled growth of plant cells, T-DNA directs plant cells to produce unusual nitrogenous compounds called **opines,** which are released into the soil around the crown-gall. Agrobacteria in the soil which carry the Ti-plasmid use these opines as a source of energy, nitrogen, and carbon. The opines induce *Agrobacterium* to conjugate and spread the Ti-plasmid throughout the population. It has been reported that *Agrobacterium* conjugate very infrequently in the absence of opines.

The Ti-plasmids from *Agrobacterium* are of interest because they are expressed in plant cells. Genetic engineers are attempting to splice various bacterial and plant genes to the controlling sites in the T-DNA, introduce these genes into plant cells and have them expressed. The genetically modified Ti-plasmids may, in the future, allow biologists to introduce a number of genes that would improve the plant. Genetic engineers are looking into the possibility of introducing genes that would allow plants to fix their own nitrogen, increase their nutritive value, and make them more resistant to plant pathogens and desiccation.

In this chapter you will be introduced to some procedures used by genetic engineers to create recombinant DNAs, to make genes, to clone genes, and to make new types of cells.

LEARNING OBJECTIVES

A STUDY OF THIS CHAPTER SHOULD ENABLE YOU TO:

EXPLAIN THE IMPORTANCE OF RESTRICTION ENDONUCLEASES, VECTORS, AND HOST CELLS IN GENETIC ENGINEERING

LIST THE STEPS INVOLVED IN CLONING A GENE

EXPLAIN HOW USEFUL PROTEINS CAN BE PRODUCED IN GENETICALLY ENGINEERED CELLS

EXPLAIN WHAT HYBRIDOMAS ARE AND HOW THEY ARE ISOLATED

Figure 11-1 Genetic Engineering in Nature. A common soil bacterium, *Agrobacterium tumefaciens,* is able to infect certain plants like tomatoes, tobaccos, and begonias. During some of these infections, bacterial plasmids enter the plant cells and become part of the plant chromosomes. The bacterial plasmids disrupt the plant cells so that they reproduce and grow in an uncontrolled manner. Tumor-like growths on the stem are the result. The hybrid DNA (bacterial and plant) represents the product of a natural form of genetic engineering.

Recombination □ between DNA molecules from different types of organisms is a common phenomenon in nature. Viruses such as λ insert their genomes into the *Escherichia coli* chromosome, and in the process, may change the genetic make up of the cell. Sometimes a bacterium can become pathogenic (disease-causing) when infected by a virus. For example, when *Corynebacterium diphtheriae* carries in its genome the prophage for the B virus, it produces a toxin that is responsible for the disease called diphteria. In the absence of the B prophage, *C. diphtheriae* cannot produce the toxin. In certain instances, inserted genetic material alters drastically the appearance of the host organism (fig. 11-1). 229

The study of bacterial viruses and plasmids have enabled scientists to develop techniques for inserting genes into bacterial and eukaryotic □ cells. 76 By using genetic recombination techniques, genetic engineers have been able to change the genetic make up of microorganisms like bacteria and yeasts and use them to produce valuable proteins and genes economically and in large quantities (table 11-1).

Human hormones such as insulin, growth hormone, and thymosin are being produced in microorganisms and are being used to treat genetic deficiencies. Proteins from various viruses (foot and mouth disease, herpes, hepatitis, etc.) are being tested to determine whether or not they can be used as effective vaccines. In addition, the human interferons □ are being studied to 437 determine if they can be used to inhibit some cancers and virus infections. In the future, many other proteins may be produced in genetically engineered microorganisms. For example, bacterial proteases, which break down proteins, are of value as cleaning aids. The enzymes α-amylase, glucamylase, and glucose isomerase are used to convert starch into fructose, a sugar that is used to sweeten a great number of foods and soft drinks. Rennin is an enzyme that is used to denature milk protein and is employed in making cheeses. Since more than $200 million worth of these five enzymes are sold each year, it may turn out to be profitable to make these enzymes in genetically engineered bacteria.

Attempts are being made to introduce genes into various animals and plants to cure genetic disorders and to produce more valuable livestock and crops. It is hoped that the genes for β-globin (a subunit of hemoglobin, the protein that carries O_2 in red blood cells) can be introduced into people suffering from the degenerative and deadly genetic diseases sickle-cell anemia and β-thalassemia. It is also hoped that genes that allow certain bacteria to convert atmospheric nitrogen (N_2) into ammonia (NH_3) can be introduced into cereal crops such as wheat, rice, barley, and corn, so that these crops will not require nitrogen fertilizers (table 11-1).

Various techniques developed for genetic engineering have made it possible to sequence structural genes and their controlling sites and so learn much more about the genetics of both prokaryotes and eukaryotes.

THE DEVELOPMENT OF GENETIC ENGINEERING

In 1962, Werner Arber at the University of Geneva reported that bacteriophage lambda (λ) grown on one strain of *E. coli* did not efficiently infect a second strain. He found that 100% of the λ virus grown on *E. coli* strain K proliferated on *E. coli* K, but that only 0.1% grew on *E. coli* strain B (fig.

TABLE 11-1

GENETIC ENGINEERING: TODAY AND FUTURE

	ECONOMIC PRODUCTION OF USEFUL PROTEINS		
	PRODUCT	FIELD	SPECIFIC USE
Today	Human insulin	Medical	Treatment for diabetes
	Human interferons (8 types)	Medical	Treatment for virus infections, cancers
	Human growth hormone	Medical	Prevention of growth deficiencies
	Human Thymosin-alpha 1	Medical	Treatment for cancers
	Foot and mouth proteins	Medical	Vaccine against foot and mouth disease
	Hepatitis-B proteins	Medical	Vaccine against Hepatitis-B
	Human somatostatin	Medical	Inhibits release of insulin and growth hormone
	Colibacillus proteins	Medical	Vaccine against calf and piglet disease
	Urokinase	Medical	Enzyme that dissolves blood clots
Future	Blood coagulation proteins	Medical	Treatment for hemophilia
	Complement proteins	Medical	Treatment for immune defects
	Various enzymes	Chemistry, industry, medical	Used in industry, science
	Immunoglobulins	Medical	Vaccines against toxins
	Surface antigens	Medical	Vaccines against microorganisms
	Toxins	Military	Toxins used on people and animals
	Tissue plasminogen activator	Medical	Protein that stimulates the body to dissolve blood clots

	PRODUCTION OF USEFUL GENES ISOLATED FROM ANY ORGANISM		
	PRODUCT	FIELD	SPECIFIC USE
Today	Human beta-globin genes	Medical	Treatment for sickle-cell anemia, beta thalassemia
Future	Human immunoglobulin genes	Medical	Treatment for immune deficiencies
	HLA (human lymphocyte antigens) genes	Medical	Treatment for immune deficiencies
	Bacterial nitrogen fixation genes	Agriculture	Make plants independent of nitrogen fertilizers
	Plant genes for altering protein, carbohydrate, lipid, vitamin, mineral content	Agriculture	Makes nutritious plants
	Genes for hydrocarbon production	Chemistry	Fuel production

	MAPPING GENES AND DETERMINING CHROMOSOME STRUCTURE		
	TYPE OF DNA	FIELD	SPECIFIC USE
Today	Introns and exons	Genetics and evolution	
	Insersion sequences and transposons	Genetics and evolution	
	Centromere structure	Biology	
	Regulatory genes in developing organisms	Biology	

	HYBRIDOMAS		
	PRODUCT	FIELD	SPECIFIC USE
Today	Monoclonal antibodies	Medicine, chemistry, industry	For tissue typing, purification of proteins, diagnosis of infectious diseases.

11-2a). Similarly, if λ grown on *E. coli* B were used to infect *E. coli* B and *E. coli* K, 100% reproduced on *E. coli* B but only 0.4% produced progeny on *E. coli* K (fig. 11-2b). In order to explain this phenomenon, Arber and co-workers proposed that most virus genomes (99.9%) are destroyed by bacterial enzymes that they called **restriction endonucleases.** Today it is known that restriction endonucleases bind DNA and cut it in one or more locations. Each species or strain of bacterium has a specific restriction endonuclease that breaks down foreign DNA. Some of the virus genomes (0.1%) avoid destruc-

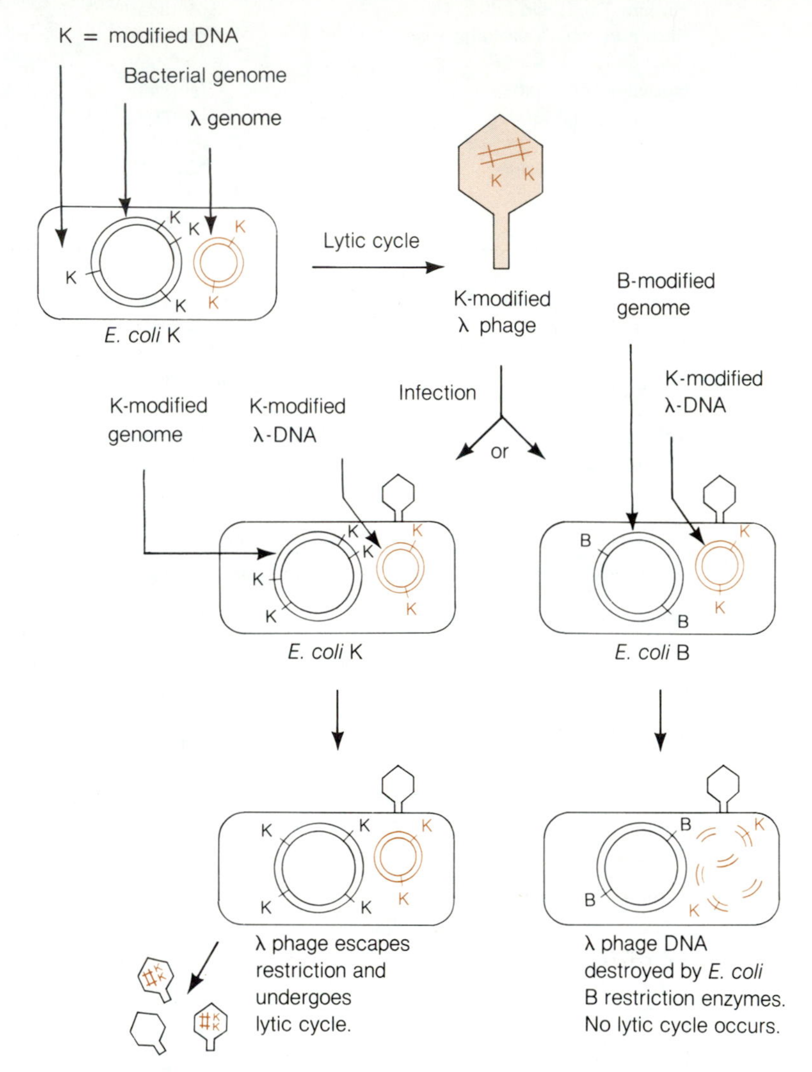

Figure 11-2 Restriction-Modification Systems in *E. coli*. Many strains of *E. coli* have their own DNA modification enzymes and restriction endonucleases. Modification enzymes protect an organism's DNA from its restriction endonucleases generally by adding methyl groups to the DNA at specific sites. Generally, DNA modified in one species to protect against a restriction endonuclease will not be protected in another type of organism. When virus DNA synthesized and modified in *E. coli* strain-K enters *E. coli* strain-B, most of it is destroyed by the restriction enzymes because modification in strain-K does not protect against strain-B restriction endonucleases. When lambda bacteriophage DNA synthesized and modified in strain-B enters strain-K, most of the lambda DNA is destroyed by the restriction enzymes because modification in strain-B does not protect against strain-K restriction endonucleases.

tion when they become methylated by bacterial enzymes called **modifying enzymes.** This methylation of the DNA at sites where restriction endonucleases bind or cleave the DNA generally protects the genome. Once a viral DNA is protected against a particular type of degrading enzyme, it can reproduce efficiently in the same type of bacterium in which it was modified. Modification of a viral DNA in one bacterium, however, does not necessarily protect that DNA in a second strain of the same bacterium, because restriction endonucleases from different strains generally are not identical. Consequently, viral DNA modified in one strain of bacterium is frequently destroyed in another strain because different DNA sequences are recognized. Bacterial cells produce modifying enzymes to protect their own DNA against their restriction endonucleases; thus, bacteria generally have methylases (or other modifying enzymes) which recognize the same sequences that their endonucleases recognize.

In 1970, Hamilton Smith and his co-workers at the John Hopkins University Department of Microbiology reported the purification of a restriction endonuclease which cleaved DNA at specific sites. This enzyme was isolated from *Haemophilus influenzae* Rd and became known as **Hind II** restriction endonuclease. It cuts DNA so that **blunt ends** are produced (fig. 11-3). Soon after the isolation of Hind II, Herbert Boyer and others at the University of California at San Francisco characterized several *E. coli* R-factor-specified restriction endonucleases. In particular, they isolated the enzyme called **Eco R1,** which produces **staggered ends** (fig. 11-3). Since then, restriction endonucleases have been isolated from many different bacteria. Table 11-2 lists some of the endonucleases which have been isolated and characterized.

In 1971, Paul Berg and his colleagues at Stanford University School of Medicine used Herbert Boyer's **Eco R1** to construct the first recombinant DNA molecule between a monkey virus called simian virus-40 (SV40) and bacteriophage lambda (λ) in a test tube *(in vitro).* Berg found that Eco Rl cut SV40 DNA at one site, while it severed lambda DNA at five sites. In spite of the way Eco Rl cut λ virus DNA, Berg managed to join a complete linear λ, genome with a linear SV40 genome. Berg had intended to grow the λ-SV40 hybrid in *E. coli* to obtain large amounts of SV40 DNA for study, but he was persuaded not to try the experiment because of the chance that λ-SV40-infected *E. coli* might infect researchers. It was feared that SV40 might escape from the bacterium and subsequently infect the researcher. SV40 is a dangerous virus because it can **transform** numerous mammalian cell types into cancerous cells that can form tumors.

In the mid 1970s, numerous researchers became concerned about the possible dangers involved in cloning mammalina DNA and animal viruses. It was feared that the indiscriminate proliferation of mammalian DNA might create or activate cryptic proviruses (hidden virus DNAs) that could cause cancer or degenerative diseases. As more was learned about recombinant DNA and hosts in which the DNA replicates, the controversy over the possible dangers of recombinant DNA research subsided. Eventually, Berg spliced together SV40 and *E. coli* genes and introduced the resulting recombination DNA into cultured mammalian cells. He then obtained mammalian cells that expressed the *E. coli* genes. Later, Berg spliced animal globin, animal histone, and bacterial dihydrofolate genes to the SV40–*E. coli* recombinant DNA and introduced the hybrid into cultured mammalian cells. The successful introduction of functional genes into higher organisms was a significant

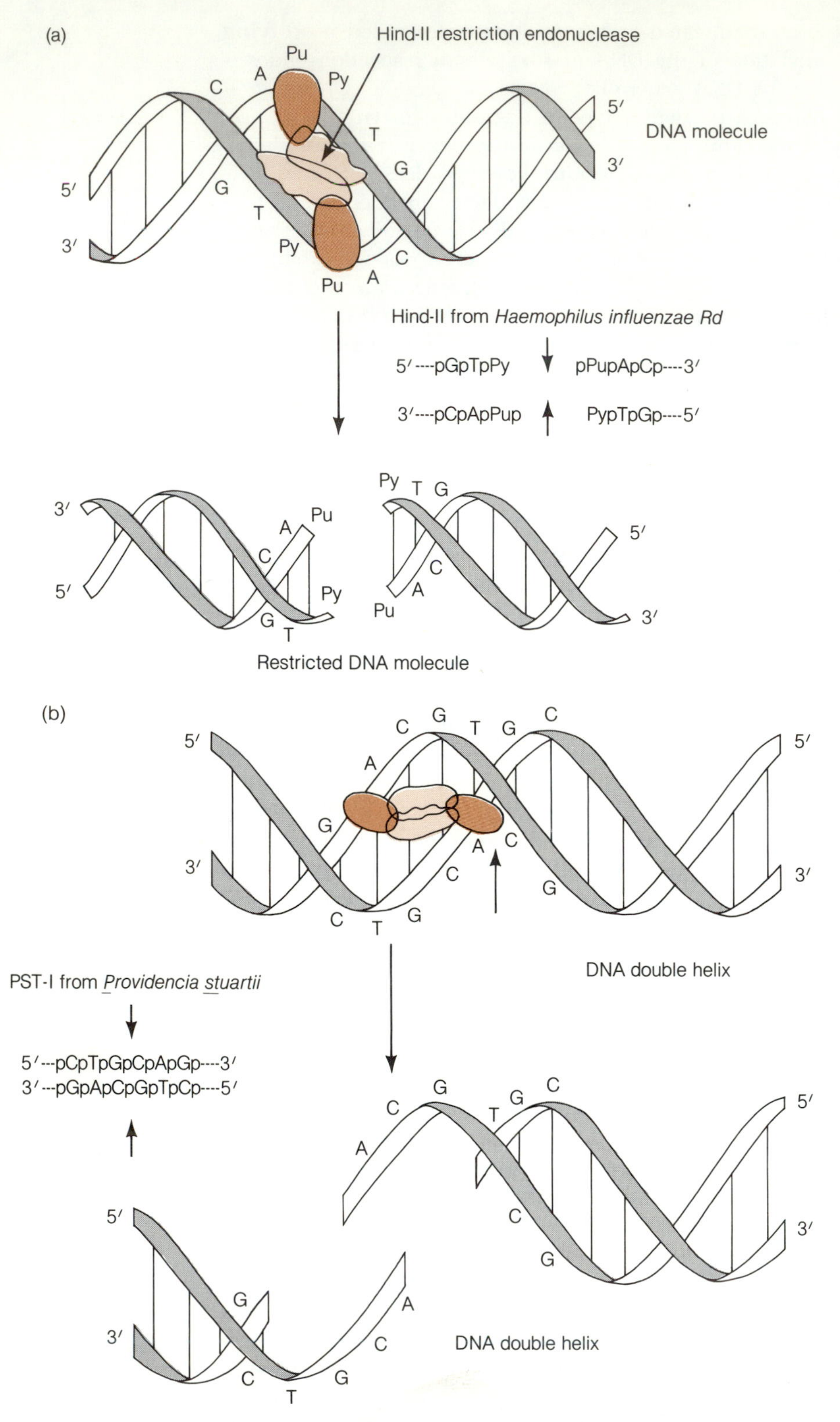

Figure 11-3 DNA Cuts Made by Restriction Nucleases. Each restriction endonuclease illustrated cuts DNA at a precise location. *(a)* Hind II cuts the DNA so that it has blunt ends while Pst-I *(b)* cuts the DNA so that it has staggered ends. Py = pyrimidine (T or C), Pu = purine (A or G). T = thymine (thymidine), C = cytosine (cytidine), A = adenine (adenosine), G = guanine (guanosine).

TABLE 11-2
RESTRICTION ENDONUCLEASES

SOURCE	SYMBOL	*RECOGNITION SITE ON DNA MOLECULE	NUMBER OF CLEAVAGE SITES: PHAGE λ	ADENOVIRUS AD2	SV40
Anabaena variabilis	*Ava* I	CGA↓CCG	8	?	?
Bacillus amyloliquefaciens H	*Bam* HI	G↓GATCC	5	3	1
Bacillus globigii	*Bgl* II	A↓GATCT	5	10	0
Escherichia coli RY13	*Eco* RI	G↓AATTC	5	5	1
Escherichia coli R245	*Eco* RII	A ↓CC(T)GG	>35	>35	16
Haemophilus aegyptius	*Hae* III	GG↓CC	>50	>50	18
Haemophilus gallinarum	*Hga* I	GACGC	>30	>30	0
Haemophilus haemolyticus	*Hha* I	GCG↓C	>50	>50	2
Haemophilus influenzae Rd	*Hind* II	C A GT(T)↓(G)AC	34	20	7
	Hind III	A↓AGCTT	6	11	6
Haemophilus parainfluenzae	*Hpa* I	GTT↓AAC	11	6	4
	Hpa II	C↓CGG	>50	>50	1
Klebsiella pneumoniae	*Kpn* I	GGTAC↓C			
Moraxella bovis	*Mbo* I	↓GATC			
Providencia stuartii	*Pst* I	CTGCA↓G	18	25	2
Serratia marcescens	*Sma* I	CCC↓GGG	3	12	0
Streptomyces stanford	*Sst* I	GAGCT↓C			
Xanthomonas malvacearum	*Xma* I	C↓CCGGG	3	12	0

*Cleavage site is indicated by an arrow

contribution to the science of genetic engineering, because it made it theoretically possible to alter plants and animals. Berg's work demonstrated that viruses could serve as **vectors** to carry genes and that both *E. coli* and cultured mammalian cells could be used to proliferate or clone the genes.

In 1978, Werner Arber of the Biozentrum in Basel, Switzerland, and Hamilton Smith and Daniel of Nathans of Johns Hopkins University shared the Nobel Prize in Physiology or Medicine for their part in the discovery and characterization of restriction endonucleases. Paul Berg of Stanford University shared the 1980 Nobel Prize in Chemistry for his work in constructing SV40-lambda hybrids using recombinant DNA techniques.

Genetic engineering requires not only the use of restriction endonucleases for precisely cutting DNA molecules, but also DNA ligase to join the DNA segments an appropriate DNA molecule (vector) to promote the reproduction of spliced genes, and a cell in which the recombinant DNA can be proliferated and selected **(cloned).** Because viruses such as λ or SV40 replicate in *E. coli* and in cultured monkey cells, respectively, they are often used as vectors to carry spliced genes. In the early 1970s, however, Stanley Cohen and others at Stanford University and Herbert Boyer and co-workers at UCSF developed bacterial plasmids (R-factors) that could be used to proliferate foreign DNA in *E. coli.* The R-factors □ generally used as vectors have antibiotic-resistant genes that make it possible to detect the presence of the R-factors and to select for them.

88

A METHOD FOR CLONING GENES

One way of cloning a gene is to cut it out of an organism's genome with a restriction endonuclease, introduce it into a vector cut with the same endonuclease, and then insert the recombinant vector into cells where it can proliferate. Figure 11-4a illustrates how purified plasmids carrying genes that confer tetracycline resistance are cut with a restriction endonuclease (Pst I), and how foreign genes with ends that match the cut plasmid are inserted into the middle of the plasmid. Since Pst I creates staggered or overlapping ends, foreign DNAs with staggered ends that are complementary to the staggered ends in the plasmid bind to the plasmid DNA because of the hydrogen bonding between the single-stranded ends. When the recombinant plasmids are mixed with *E. coli*, some of the bacteria become transformed to tetracycline resistance (fig. 11-4b). Many bacteria do not take up a plasmid and consequently remain tetracycline-sensitive. If the bacteria are grown on a medium containing tetracycline, only those with the plasmid having the tetracycline-resistant gene will grow (fig. 11-4c). The cloning procedure allows one to take a single piece of foreign DNA and obtain billions of copies within a day. It is very easy to grow *E. coli* to concentrations greater than 10^9 bacteria/ml. The bacteria are concentrated by centrifugation and gently lysed so as not to disrupt the plasmids. The plasmids and *E. coli* debris are separated from each other by centrifugation. The plasmids and bacterial genomes remain in the **supernatant fluid** because they are very light in comparison to the debris (fig. 11-4d). The plasmids and bacterial genomes can be separated from each other by centrifugation in a cesium chloride gradient, because they have different densities. The plasmids are concentrated in a band that is easily drawn off with a syringe (fig. 11-4d). The plasmids may then be cleaved with a restriction endonuclease so that the foreign DNA is released. The foreign DNA and the plasmid DNA can be separated from each other by **electrophoresis** using thin **agarose gels.** Purified genes are valuable because they can be used to determine nucleotide sequences. Some genes may also be used in the future to transform plants, animals, and humans in various ways.

A METHOD FOR PRODUCING USEFUL PROTEINS

A number of useful proteins, such as human insulin, human interferon, and human growth hormone, are already being produced in bacteria. **Human insulin** obtained from genetically engineered cells will eventually replace bovine (cow) and swine (pig) insulin for treating diabetes. Production of human insulin will be advantageous because it does not have the side effects that the bovine and swine insulins do. **Human growth hormone** for treating children with growth hormone defficiencies, and various **human interferons** for inhibiting some cancers and virus infections, have already been produced in bacteria and tested clinically. No doubt, these and many other human products obtained from genetic engineering will soon be available to the public at a reasonable price. The following section will discuss some of the techniques used to produce valuable proteins and genes in bacteria.

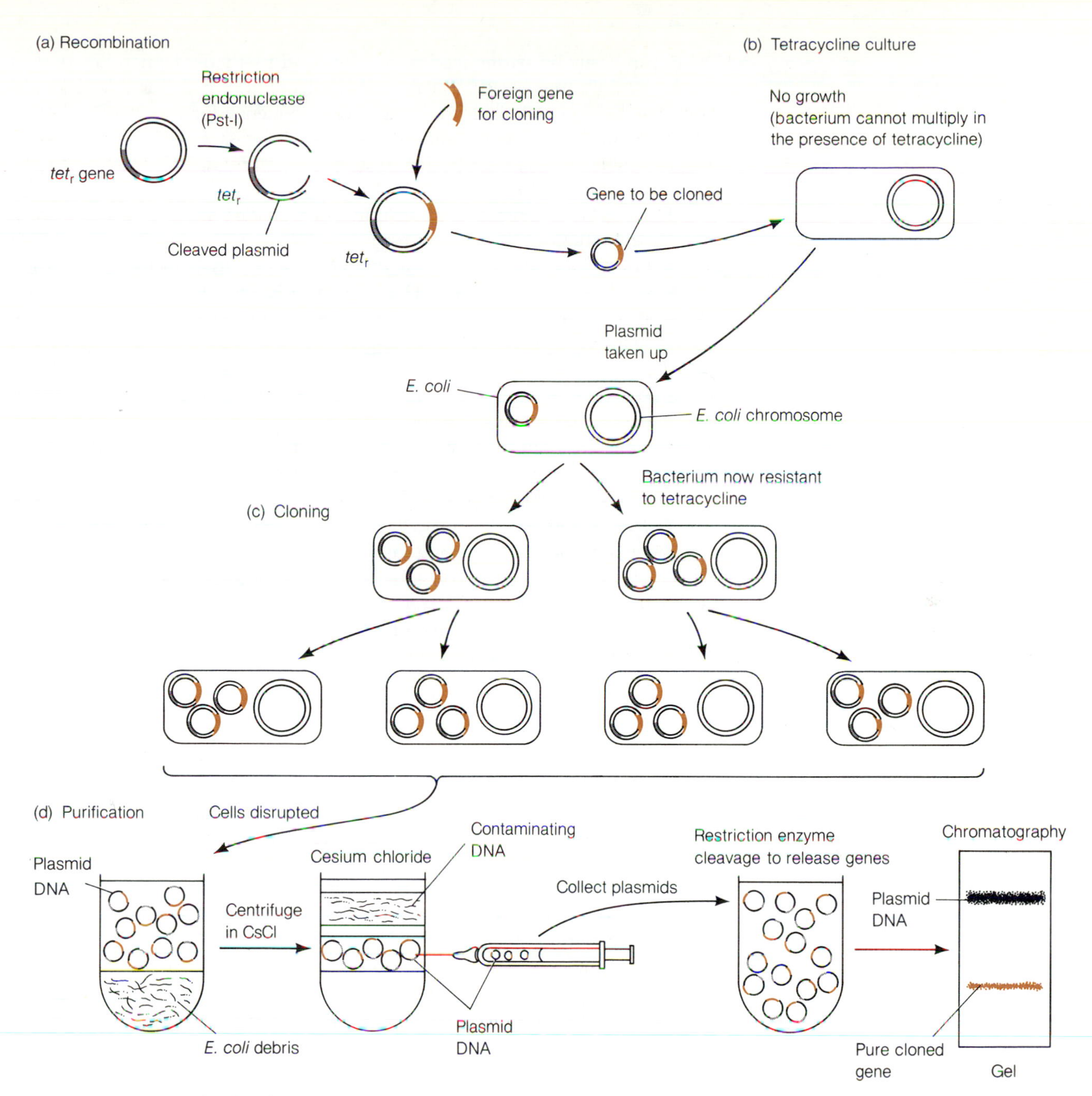

Figure 11-4 A Method for Cloning DNA. *(a)* Bacterial plasmids that can confer resistance to tetracycline are cut with a restriction endonuclease and foreign genes are spliced into the plasmids. *(b)* The bacteria are plated on a tetracycline containing medium. Only bacteria that contain plasmids and are resistant to tetracycline multiply and form colonies. *(c)* Consequently, all the bacteria in the final population contain the plasmid with the foreign gene. *(d)* The bacterial plasmids can be isolated from the bacteria and purified. The foreign genes can be extracted from the plasmids by using restrictions endonucleases.

Genetically Engineered Human Insulin

Active human insulin consists of two short polypeptide chains (A and B) hooked together by two disulfide bonds. The A-polypeptide is 21 amino acids long, while the B-polypeptide is 30 amino acids long. Since the amino acid sequences are known for both chains and the proteins are short, synthetic

DNAs can easily be synthesized using chemical methods. Remember that, if you know the amino acid sequence of a polypeptide, it is possible to determine from the genetic code □ a nucleotide sequence for the RNA and DNA that code for the protein. Since the genetic code is degenerate, however, (there is more than one codon for most of the amino acids), the DNA made may not be identical to the natural gene.

211

The synthetic genes for the insulin polypeptides are introduced into *E. coli* plasmids that are genetically engineered so as to contain the controlling sites from the lactose operon □ *(lac p* and *lac o)* and most of the gene coding for β-galactosidase *(lac z)* (fig. 11-5a-b). In addition, the plasmids contain the gene for resistance to the antibiotic ampicillin. The artificial A and B genes are inserted at the end of the β-galactosidase gene in separate plasmids in order to produce stable proteins. If short proteins such as the insulin A and B polypeptides were to be synthesized in *E. coli*, they would be degraded. When they are part of a larger polypeptide, however, such as the *lac z* gene product, they are not destroyed. The reason for splicing the A and B genes to the beginning of the lactose operon is to provide the insulin genes with a promoter site. Without a promoter site in the proper reading frame, the A and B genes could not be transcribed. The gene for ampicillin resistance is not affected by the gene splicing and hence can be used to select for ampicillin resistant clones of *E. coli* after the hybrid plasmids are picked up by the bacteria (fig. 11-5c).

243

Transcription □ of the plasmid DNA and translation of the mRNA result in the accumulation of a β-galactosidase-insulin hybrid protein. It has been found that transcription and translation are very efficient. Approximately 20% of the total cellular protein consists of the β-galactosidase-insulin hybrid. This protein is isolated by concentrating the cells and lysing them gently under pressure in what is called a **French press.** The hybrid protein (also called a **fusion protein**) is somewhat insoluble and consequently is present in the cell pellet after low-speed centrifugation and precipitation (fig. 11-5d). The treatment of the fusion proteins with cyanogen bromide cleaves the A and B peptide chains away from the β-galactosidase (fig. 11-5e). Cyanogen bromide cleaves after a methionine that begins the insulin peptides. Since the β-galactosidase fragments are much larger than the insulin peptides, the two are easily separated (fig. 11-5e). The sulfhydryl groups on the A and B peptides are chemically altered so that they will react with each other when the peptides are mixed and form disulfide bonds (fig. 11-5f). When the A and B polypeptides form disulfide bonds with each other, active human insulin results. Active human insulin has been produced in large amounts from the polypeptides produced in *E. coli*. It is now cheaper to produce insulin using *E. coli* than to extract the hormone from the pancreas of animals.

207

HYBRIDOMAS

One of the most recent developments in genetic engineering is the production of **hybridomas.** Hybridomas are cells that result from the fusion of normal eukaryotic cells with cancerous eukaryotic cells. The most valuable hybridoma cells produced so far have been from fusions between plasma cells □ (differentiated B-lymphocytes that produce antibodies) and cancerous plasma cells called **myeloma** cells. The hybrid cells, or hybridomas, are able

457

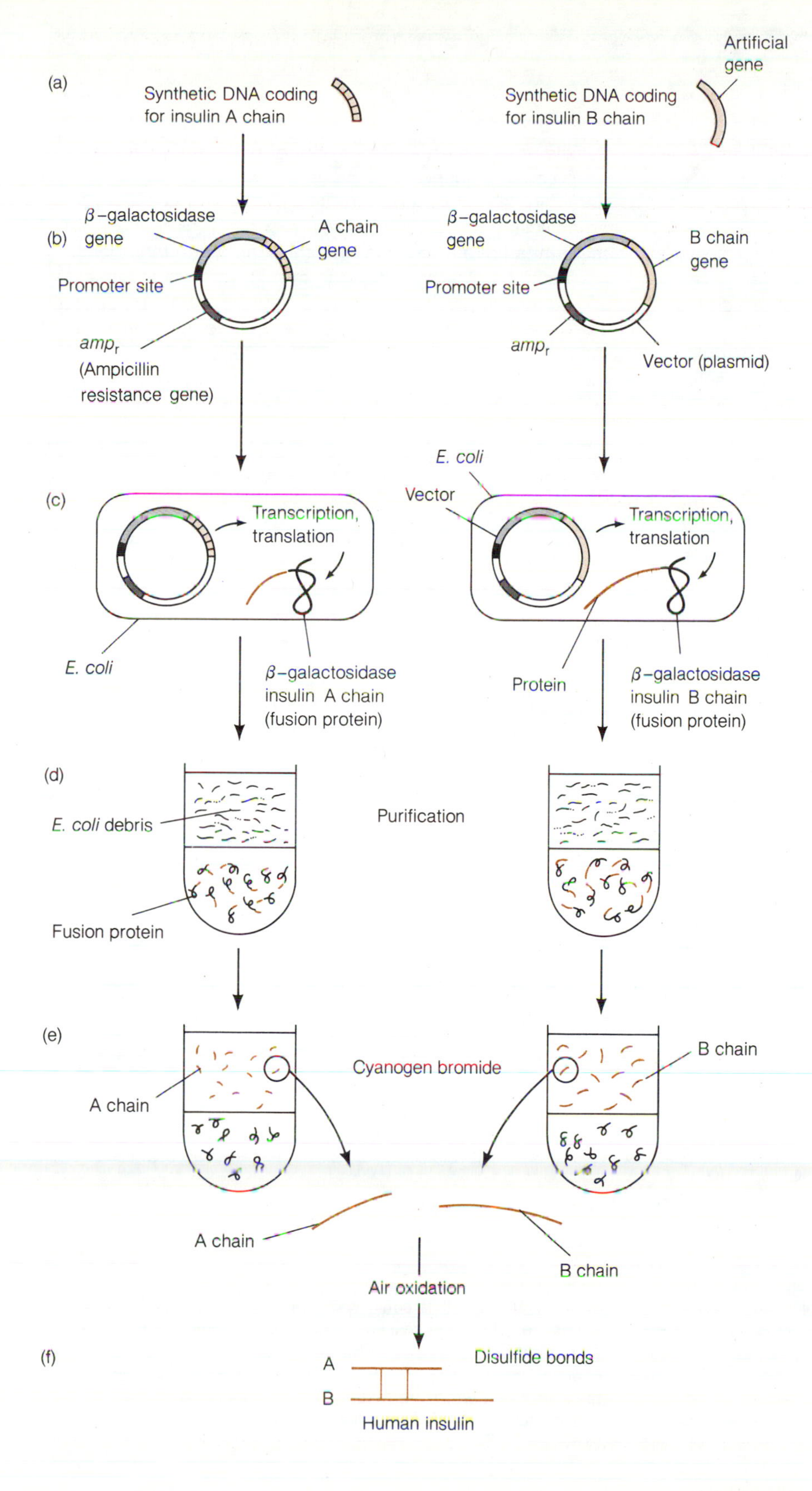

Figure 11-5 A Method for Cloning Insulin Genes and Producing Insulin. *(a)* DNA coding for the insulin chains is synthesized chemically. *(b)* The insulin DNA is spliced into plasmids in such a way that it will be expressed. The insulin DNA is spliced near the end of the gene for β-galactosidase. This gene can easily be regulated. If it is expressed, the gene for the insulin chains will also be expressed. *(c)* Each of the plasmids with the two different insulin genes are mixed with separate populations of bacteria. Some transform the bacteria to ampicillin resistance. The bacteria are grown on an ampicillin containing medium to select for bacteria that contain the plasmids. After the ampicillin resistant populations have been selected, the β-galactosidase gene is turned on. The resultant products are fusion proteins that consists of β-galactosidase and the insulin proteins. *(d)* Next, the fusion proteins are isolated from their respective host bacteria. Then, the fusion proteins are split into β-galactosidase and the insulin proteins. Subsequently, the insulin proteins are purified. *(f)* Finally, the two different insulin proteins are used to make insulin.

FOCUS

AUTOMATED GENETIC ENGINEERING

Many valuable proteins are extremely difficult or expensive to isolate in more than minute quantities. One of the goals of genetic engineering is to produce proteins in large amounts and economically for use in medicine or industry. One way to produce a protein is to synthesize the gene for the protein and then introduce the gene into a microorganism in which the protein can be synthesized.

Recently, a number of machines called **protein sequences** have been developed that allow scientists to determine the amino acid sequence of proteins with as little as a few μg (10^{-6} grams) of protein. With the amino acid sequence of a protein, it is possible to determine a genetic code for the protein and construct an artificial gene. After much development, **gene machines** have also been built that can piece together nucleotides in the right order to create gene fragments 20–30 nucleotides long in 10–15 hours (fig. 11-6). These gene fragments can then be spliced together in the right order to create a gene (sometimes 1,000 or more nucleotides long), or they can be radioactively labeled and used as a probe to detect cellular mRNA

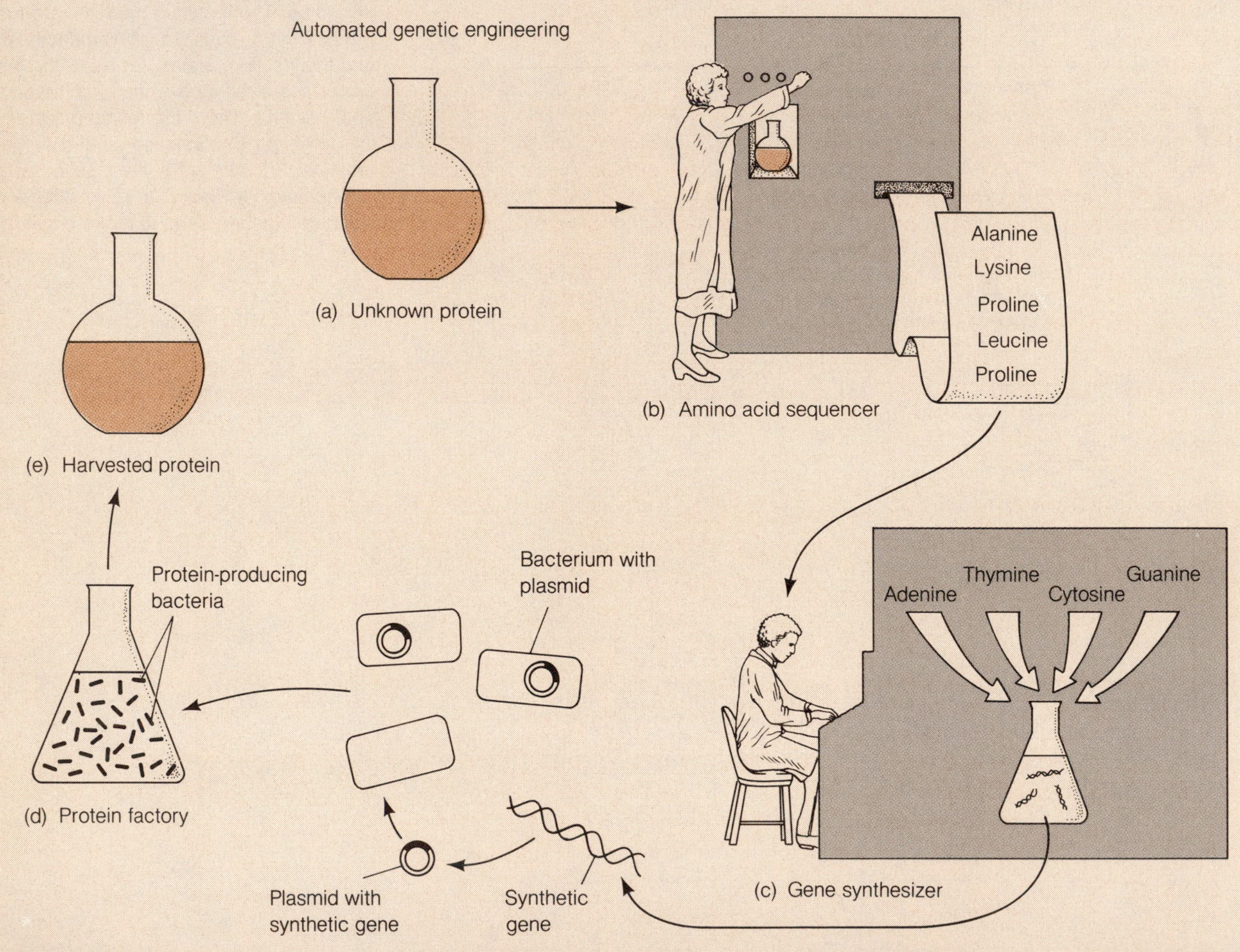

Figure 11-6 Protein Sequencer and Gene Machine. The protein sequencer and gene machine will allow scientists to automatically make a gene from minute samples of an unknown protein. This in turn will allow them to make large amounts of the protein for study. The making of a gene and then protein begins when minute samples of an unknown protein are put into the amino acid sequencer. This machine gives the sequence of amino acids in the protein. In order to make a gene that will code for the protein, a scientist types a genetic code into the gene synthesizer that could code for the protein's amino acid sequence. The gene machine would then chemically hook together the correct nucleotides to make the gene. By other chemical means the DNA would be integrated into a bacterial plasmid which would be taken up by a population of bacteria. The bacteria would function like a factory and produce the protein.

from which a complementary DNA can be synthesized. Once the gene has been synthesized, it can be spliced into a vector and cloned in a host cell, where a large number of copies can be synthesized.

The protein sequencer and the gene machine should facilitate the production of hundreds of medically and industrially important proteins, such as interferons and catalysts. Possibly, medically important genes will be synthesized using the gene machine. For instance, the genes for **urokinase** and **tissue plasminogen activator** might be produced and used to make these proteins. Urokinase is medically important because it dissolves blood clots, and tissue plasminogen activator is valuable because it stimulates the body's blood system to dissolve clots. These proteins will become more and more important in preventing heart attacks due to blood clots.

The protein sequencer and gene machine will facilitate the mapping of animal and plant genes and the determination of their nucleotide sequence. Because it is impossible (or extremely difficult) to select for mutations in numerous genes, these genes cannot be mapped, nor can their boundaries be determined. By making labeled DNA probes from the proteins that these genes code for, however, it is possible to locate the genes in the chromosomes.

After genes have been located and sequenced, accurate copies can be synthesized to establish a "genetic library" of animal and plant genes. A genetic library has many uses. For example, a genetic library for humans would allow the screening of fetuses for the presence of abnormal genes. A sample of a fetus's genes obtained from fetal cells in the amniotic fluid could be compared with the genetic library of normal DNAs. Prospective parents could then be counseled in cases where it was discovered that there were deviations in the fetus's genes that would produce birth defects.

to proliferate indefinitely, as do myelomas. They are also able to produce antibodies with one specificity, as do plasma cells. Each hybridoma clone is a factory for the synthesis of a single type of antibody, which can be produced in tissue culture or within an animal.

Because antibodies derived from a single clone **(monoclonal antibodies)** have a single specificity, they are extremely useful in microbiology. Monoclonal antibodies can be used to destroy disease-causing organisms and to diagnose the diseases that they cause. Monoclonal antibodies are becoming more and more important as a tool in the identification of blood types (ABO, Rh, MN, etc.), transplantation antigens, and cancer cells. Researchers are presently employing monoclonal antibodies to identify, quantify, and purify membrane proteins involved in regulating the physiology and development of eukaryotic cells. Valuable proteins such as interferons, neurotransmitters, and hormones, which are produced naturally in minute quantities, can be purified using single-specificity antibodies. Passive immunization against toxins (*e.g.* snake venom, bee and wasp toxins, tetanus and botulism toxins,) and disease-causing organisms can be greatly improved by using monoclonal antibodies. Monoclonal antibodies made against specific tumor antigens can be used to treat many types of cancers. The antibodies may aid the patient's immune system to destroy the cancerous tissue naturally, or they may be used to direct toxic drugs at cancerous tissue and selectively destroy the abnormality.

Figure 11-7a summarizes the immune response in an animal and how polyclonal (mixed) sera are obtained, while figure 11-7b outlines how hybridomas are made and how monoclonal antibodies are produced. A cell (or a large molecule) with a number of different chemical groups is called an **antigen,** while the different chemical groups are known as **antigenic determinants** (fig. 11-7a). When an antigen is injected into an animal, numerous different B-lymphocytes, which produce immunoglobulin against the antigenic determinants of the antigen, are stimulated to develop into plasma

Figure 11-7 Production of Antibodies. *(a)* The production of polyclonal antibodies. Complex molecules such as polysaccharides and proteins, when injected into animals, are capable of inducing an immune response. The molecules, which are able to induce an immune response, are called antigens. The immune response is usually against a number of sites on the antigen known as antigenic determinants. In this illustration, a bacterium is shown with three different antigenic determinants. In a typical immune response, clones of B-lymphocytes multiply and release antibodies which bind to the antigenic determinants on the antigen which stimulated their proliferation and production of antibodies. Each clone of B-lymphocytes produces only one type of antibody, which generally only binds to one type of antigenic determinant. A mixture of antibodies that bind to an antigen can be isolated from the blood serum of a stimulated animal. The antibodies are known as gamma globulins or immunoglobulins. Their range of diversity is illustrated. *(b)* The production of monoclonal antibodies. B-lymphocytes isolated from the spleens of antigen stimulated rats and cell cultured cancerous B-lymphocytes (myeloma line) are mixed together in polyethylene glycol to stimulate cell fusion. The fused B-lymphocytes known as hybridomas and the parent cells are placed in a special medium (HAT medium) which selects for the hybridomas. B-lymphocytes and myeloma cells die in the HAT medium. Each of the hybridomas (many thousands) is grown in its own well so that a clone develops which produces only one type of antibody. The hybridoma clones are assayed to determine whether they produce specific antibodies against the antigen used to stimulate the rat. The clones may be injected into rats where they induce myelomas which secrete only one type of antibody. Alternatedly, the clones may be grown in mass culture where they secrete only one type of antibody.

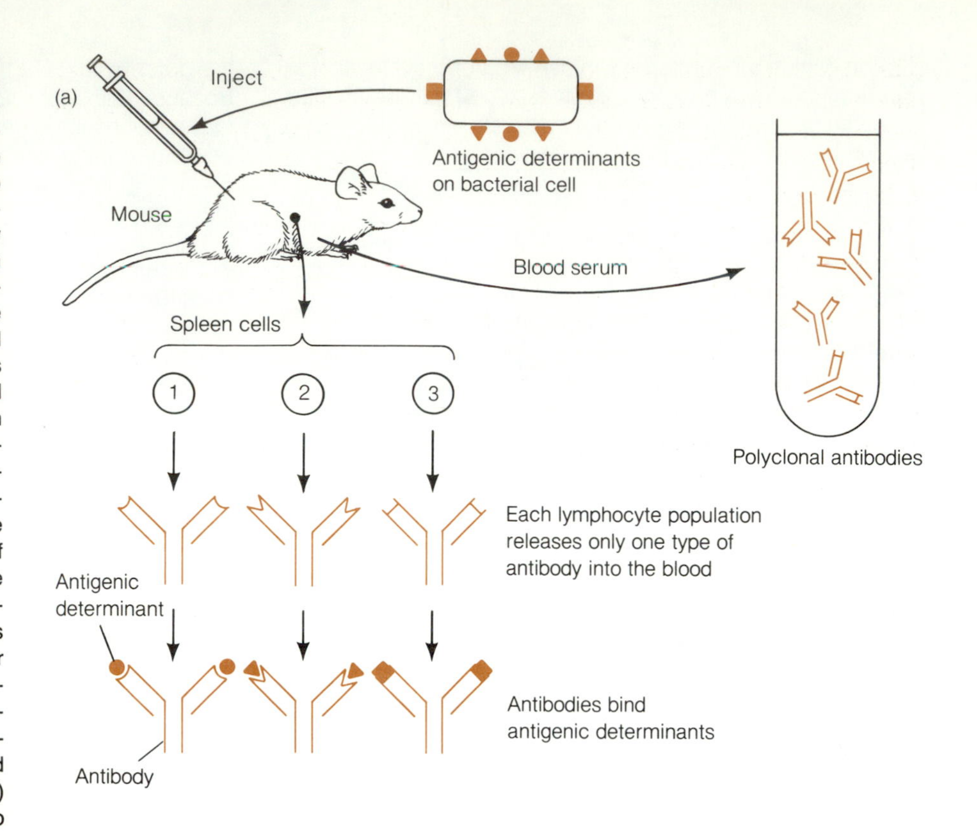

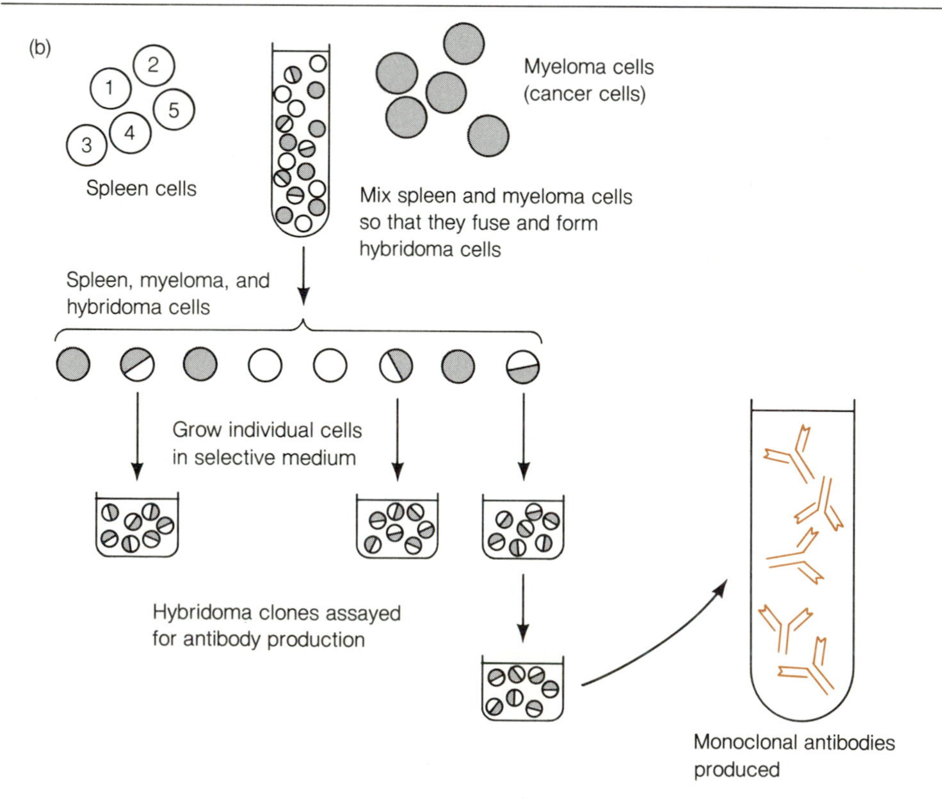

457 cells. Each different population of plasma cells (which arose from a B-lymphocyte □ stimulated by one of the antigenic determinants) produces only one immunoglobulin, and this immunoglobulin binds specifically to the antigenic determinant that induced its production. These lymphocytes are said to be **committed.** The committed populations of lymphocytes concentrate in the animal's spleen and lymph nodes, where they produce large amounts of immunoglobulins. The mixture of immunoglobulins moves from the lymphatic system into the animal's blood, where the immunoglobulins bind to foreign antigenic determinants that induced their production. Serum that contains immunoglobulins that bind to cells or molecules is often referred to as **antiserum**. Antiserum can be drawn from an animal, and the proteins in the antiserum can be electrophoresed on a polyacrylamide gel. If the amount of protein is plotted against the distance the proteins in the antiserum move along the gel, a heterogeneous spread of proteins is observed. The variety of immunoglobulins or gamma globulins found in a normal animal's serum when it has been subjected to an antigen with numerous antigenic determinants is illustrated (fig. 11-7a).

Hybridomas that produce only one of the many antibodies made against an antigen can be produced by mixing mouse spleen cells with myeloma cells from a plasma cell tumor. A mixed monolayer of B-lymphocytes and myeloma cells is treated for one minute with a solution consisting of polyethylene glycol (PEG) and balanced salt solution (BSS) to stimulate cell fusion (fig. 11-7b). After 24 hours, the cells are transferred to a medium that selects for hybridomas. The various hybridoma clones are assayed for immunoglobulin production. Those clones that are of value are recloned and then stored for future projects by freezing in liquid nitrogen. Some of the cells can be grown in mass culture or in mice, and the immunoglobulins produced can be harvested.

SUMMARY

THE DEVELOPMENT OF GENETIC ENGINEERING

1. The discovery, isolation, and characterization of restriction endonucleases that cleave DNA molecules at specific sites has allowed researchers to recombine DNA molecules from different organisms *in vitro*.
2. DNA molecules of interest are generally spliced into vectors, such as bacterial plasmids, bacteriophages, and hybrid eukaryotic and prokaryotic viruses.

A METHOD FOR CLONING GENES

1. Genetically engineered bacterial plasmids (vectors) carrying human, animal, and chemically synthesized genes are taken up by *E. coli*. The bacteria are proliferated in a medium that selects for the vectors. The selection and proliferation of a particular genetically-engineered vector in a host cell is known as cloning.
2. Presently, genes for human insulin, interferons, growth hormone, urokinase, and tissue plasminogen activator have been spliced into genetically engineered bacterial plasmids.

A METHOD FOR PRODUCING USEFUL PROTEINS

Genes can be expressed in the host cell if they are placed next to a promoter site that the host cell's RNA polymerase can recognize. Genes that are not under the control of a promoter will not be expressed.

HYBRIDOMAS

1. Hybridomas are eukaryotic cells formed from the fusion of a normal plasma cell and a cancerous plasma cell.

2. Hybridomas are of use because they produce only one type of antibody (monoclonal antibody) against a specific antigenic determinant.
3. Monoclonal antibodies can be used a) as a diagnostic tool; b) to identify, quantify, and purify valuable proteins; c) as passive vaccines against toxins and disease-causing organisms; and d) to treat many types of cancers.
4. In the years ahead, genetic engineering will play an important role in the development of useful genes and proteins.

STUDY QUESTIONS

1. Explain what restriction endonucleases and DNA methylases are.
2. Explain why lambda bacteriophage grown on *E. coli* B does not proliferate on *E. coli* K.
3. Write down the sequence of nucleotides in the single-stranded ends of a piece of DNA cut with **Pst I.** Do the same for a piece of DNA cut with **Eco R1.**
4. Show with a drawing why two pieces of DNA that have been cut with **Eco R1** will join together, while two pieces of DNA, one cut with **Pst I** and the other cut with **Eco R1,** will not join together.
5. Explain what cloning means.
6. What does the term vector mean to the molecular biologist? What does this term mean to the epidemiologist?
7. Why are plasmids, bacterial viruses, and animal viruses used as vectors rather than any piece of DNA?
8. Explain how vectors and the spliced DNA are selected for in bacteria.
9. Outline the procedure used to produce the protein subunits of human insulin in *E. coli.*
10. Explain why the lactose-controlling sites and most of the β-galactosidase gene must be genetically engineered into the vector used to make the protein subunits of human insulin.
11. Explain what a hybridoma is.
12. Define a monoclonal antibody.
13. List the uses for monoclonal antibodies.

SUPPLEMENTAL READINGS

Abelson, J. 1983. Biotechnology: An overview. *Science* 219: 611–613.

Baxter, J. 1980. Recombinant DNA and medical progress. *Hospital Practice* 15:57–67.

Brill, W. J. 1985. Safety concerns and genetic engineering in agriculture. *Science* 227:381–384.

Chambon, P. 1981. Split genes. *Scientific American* 244:60–71.

Chilton, M. 1983. A vector for introducing new genes into plants. *Scientific American* 248:50–59.

Collier, J. R. and Kaplan, D. A. 1984. Immunotoxins. *Scientific American* 251(1):56–64.

Gilbert, W. and Villa-Komarooff, L. 1980. Useful proteins from recombinant bacteria. *Scientific American* 242:74–94.

Hirsch, A. M., Drake, D. Jacobs, T. W., and Long, S. R. 1985. Nodules are induced on alfalfa roots by *Agrobacterium tumefaciens* and *Rhizobium trifolli* containing small segments of *Rhizobium meliloti* nodulation region. *Journal of Bacteriology* 161(1):223–230.

Itakura, K. and Riggs, A. 1980. Chemical DNA synthesis and recombinant DNA studies. *Science* 209:1401–1405.

Macario, E. and Macario, A. 1983. Monoclonal antibodies for bacterial identification and taxonomy. *ASM News* 49:1–7.

Maxam, A. and Gilbert, W. 1977. A new method for sequencing DNA. *Proc. Natl. Acad. Sci. USA* 74:560–564.

Milstein, C. 1980. Monoclonal antibodies. *Scientific American* 243:66–74.

Novick, R. 1980. Plasmids. *Scientific American* 243:102–127.

Nowinski, R., Tam, M., Goldstein, L., Stong, L., Kuo, C., Caorey, L., Stamm, W., Handsfield, H., Knapp, J., and Holmes, K. 1983. Monoclonal antibodies for diagnosis of infectious diseases in humans. *Science* 219:637–644.

Old, R. and Primrose, S. 1980. *Principles of gene manipulation*, Berkeley, Los Angeles, London: University of California Press.

PART II A SURVEY OF MICROORGANISMS

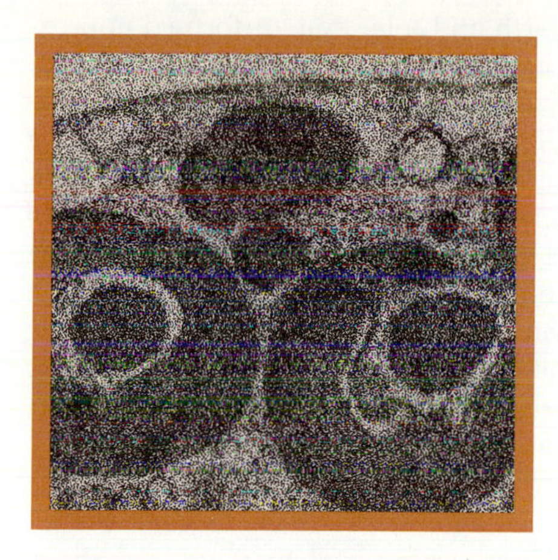

CHAPTER 12
THE ORIGIN AND CLASSIFICATION OF MICROORGANISMS

CHAPTER PREVIEW

CLUES TO THE EVOLUTION OF EUKARYOTIC ORGANELLES

The study of microorganisms and eukaryotic organelles has provided strong circumstantial evidence that evolution does not always occur through the gradual accumulation of minute changes. For instance, one theory proposes that eukaryotic organelles may have developed from small prokaryotes that were engulfed by large ones or primitive eukaryotes. Sometimes the engulfed bacteria were able to live within the host cells and a beneficial arrangement or symbiotic relationship developed. The engulfed bacteria, called **endosymbionts** because of their internal existence, sometimes proceeded to evolve into organelles through the exchange of hereditary material. This sequence of events explaining the origin of mitochondria, chloroplasts, and nuclei is founded on the observation that many eukaryotes engulf other prokaryotes, which then develop into stable endosymbionts that sometimes begin to evolve into organelles.

An example of a bacterial symbiont that has developed into an organelle is found in the eukaryote *Cyanophora paradoxa*. This protozoan contains a number of photosynthetic organelles that at one time were cyanobacteria (blue-green algae). The organelles still have a thin petidoglycan cell wall, unstacked thylakoids identical to those found in cyanobacteria, and circular DNA genomes, but have retained only a little of their hereditary material. The organelles still contain the genes for the large and small subunits of ribulose 1, 5-bisphosphate carboxylase, an enzyme required for CO_2 fixation. In most photosynthetic eukaryotes, the small subunit is coded by the nuclear genome while the large subunit is coded by the chloroplast genome. These data indicate that the photosynthetic organelles in *Cyanophora* have not yet lost the hereditary material for the small subunit to the nucleus and that the organelles are still evolving.

Other evidence supporting the theory that mitochondria and chloroplasts have evolved from bacterial endosymbionts is that similar DNA sequences have been found in chloroplasts, mitochondria, and in nuclei. This information indicates that genes may be readily exchanged between these organelles.

The foregoing discussion indicates that present-day microorganisms carry a number of clues as to the origin of the eukaryotic cell. In this chapter you will learn about other characteristics of chloroplasts, mitochondria, and nuclei which demonstrate their similarity to prokaryotes and which, elucidate the origin of these organelles and the eukaryotic cell. In addition, you will learn how taxonomic techniques such as nucleic acid hybridization are used to discover DNA sequences common to chloroplasts, mitochondria, and nuclei.

LEARNING OBJECTIVES

A STUDY OF THIS CHAPTER SHOULD ENABLE YOU TO:

DISCUSS THE MAJOR EVENTS IN CHEMICAL EVOLUTION AND ITS IMPORTANCE TO THE ORIGIN OF LIFE

SUMMARIZE THE EVIDENCE INDICATING THAT BACTERIA WERE THE FIRST ORGANISMS TO EVOLVE ON EARTH

LIST THE CHARACTERISTICS THAT MAKE THE BACTERIA DIFFERENT FROM OTHER ORGANISMS

OUTLINE THE EVOLUTION OF EUKARYOTIC CELLS FROM PROKARYOTIC CELLS

DIAGRAM THE EVOLUTIONARY RELATIONSHIPS AMONG WHITTAKER'S KINGDOMS

PRESENT A HYPOTHESIS FOR THE ORIGIN OF VIRUSES

SUMMARIZE THE VARIOUS WAYS IN WHICH MICROORGANISMS CAN BE CLASSIFIED AND IDENTIFIED

Geological evidence indicates that the earth and solar system formed approximately 4.5 billion years ago. In the early oceans and lakes that formed, chemical evolution was responsible for the development of a myriad of complex organic molecules that eventually assembled into cells resembling the prokaryotes of today. **Microfossils** (microscopic fossils) that resemble prokaryotes have been found in ancient rocks called **stromatolites**, estimated to be about 3.5 billion years old (fig. 12-1). Microfossils resembling blue-green bacteria (cyanobacteria) have been found in stromatolites dated at 3.2 billion years, a fact indicating that highly advanced microorganisms existed at that time (fig. 12-2). There is no evidence, however, for the presence of eukaryotic cells at this early date. Microfossils resembling eukaryotes (at least in size) do not appear until 1.5 billion years ago.

Biological evolution has led to a great diversity of microorganisms since their origin more than 3,000 million years ago. Many of the original microorganisms have disappeared, while others have evolved to share and shape the environment. The multicellular animals and plants that are so familiar to us are very recent products of biological evolution. For example, the first fishes appeared approximately 550 million years ago, the first primitive reptiles about 300 million years ago, and the first modern mammals only 60 million years ago. Humans are among the most recent products of biological evolution. Animals resembling humans appeared approximately 3 million years ago, and modern humans appeared only 75,000 years ago (fig. 12-3).

This chapter will be concerned not only with chemical evolution, the origin of life, and biological evolution, but also with the present methods of grouping organisms and identifying them.

CHEMICAL EVOLUTION

The early earth's oceans are believed to have formed from the vast amounts of water that welled up from the earth's interior. Its atmosphere contained large amounts of N_2, H_2, and water vapor (H_2O), and lesser amounts of CO_2,

FOCUS

The "UNIVERSAL" GENETIC CODE HAS EVOLVED IN SOME ORGANELLES

Organelles have evolved not only by exchanging and losing hereditary material to other organelles, but in some cases, they have developed their own genetic code. Scientists have discovered that the "universal" genetic code is not followed in some mitochondria. For instance, the codon UGA, which functions as a termination codon in the "universal" genetic code, specifies tryptophan in all mammalian and fungal mitochondrial genetic systems studied. Also, the codons AGA and AGG, which code for arginine in the "universal" code, are instead termination codons in mammalian mitochondrial systems, but arginine codons in fungal mitochondria.

In the fly *Drosophila melanogaster*, AGA codes for serine instead of specifying arginine. Similarly, in the "universal" code, AUA specifies isoleucine, but in mammalian and yeast mitochondria it codes for methionine. Finally, the initiation of protein synthesis in mitochondria does not always begin at UAG start codons. In some cases, the start codons in mitochondria can be AUA (triplet) and AUAA (quadruplet).

(a)

(b)

Figure 12-1 Stromatolites. *(a)* Stromatolites are rocky outcrops that have formed from mats of blue-green algae and other debris. The remains of microorganisms often become fossilized within stromatolites. *(b)* A cross section of a fossilized stromatolite from the 3.5-billion-year-old Warrawoona Group in the North Pole Dome region of northwestern Australia. The cross section shows the layers of bacteria and debris that made up the stromatolite.

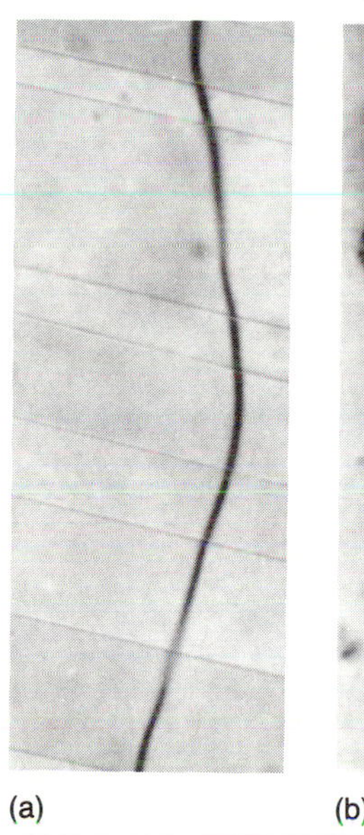

(a)

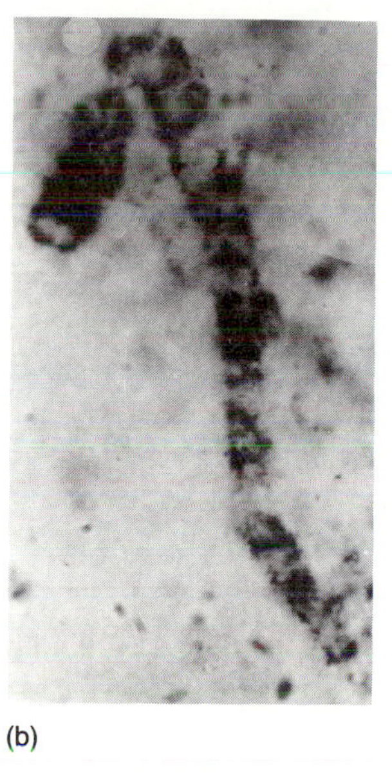

(b)

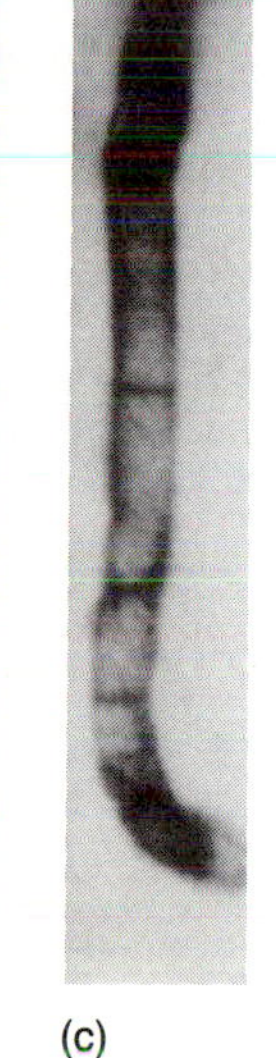

(c)

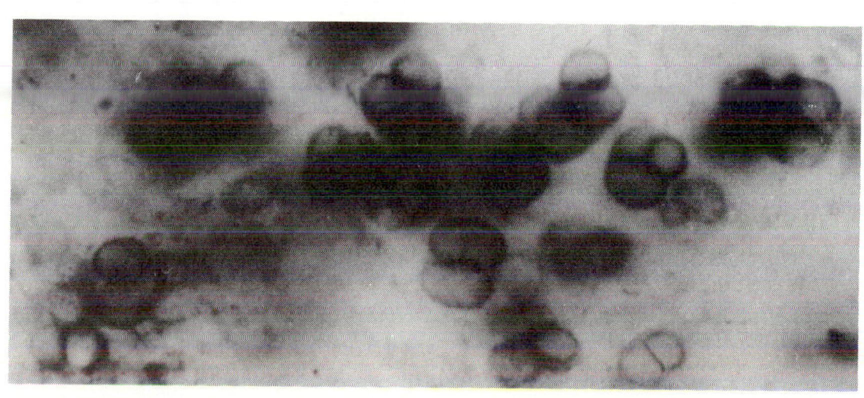

(d)

Figure 12-2 Microfossils. *(a)* Microfossil resembling a filamentous bacterium from a thin section of a stromatolite from the 3.5-billion-year-old Warrawoona Group. *(b)* Microfossil resembling a septated, filamentous prokaryote from a thin section of a stromatolite from the 3.5-billion-year-old Warrawoona Group. *(c)* Microfossil resembling a septated, filamentous bacterium from a thin section of a stromatolite from the 2-billion-year-old Duck Creek Dolomite. *(d)* Microfossils resembling packets of spherical bacteria from a thin section of a stromatolite from the 1.6-billion-year-old Bungle Bungle Dolomite.

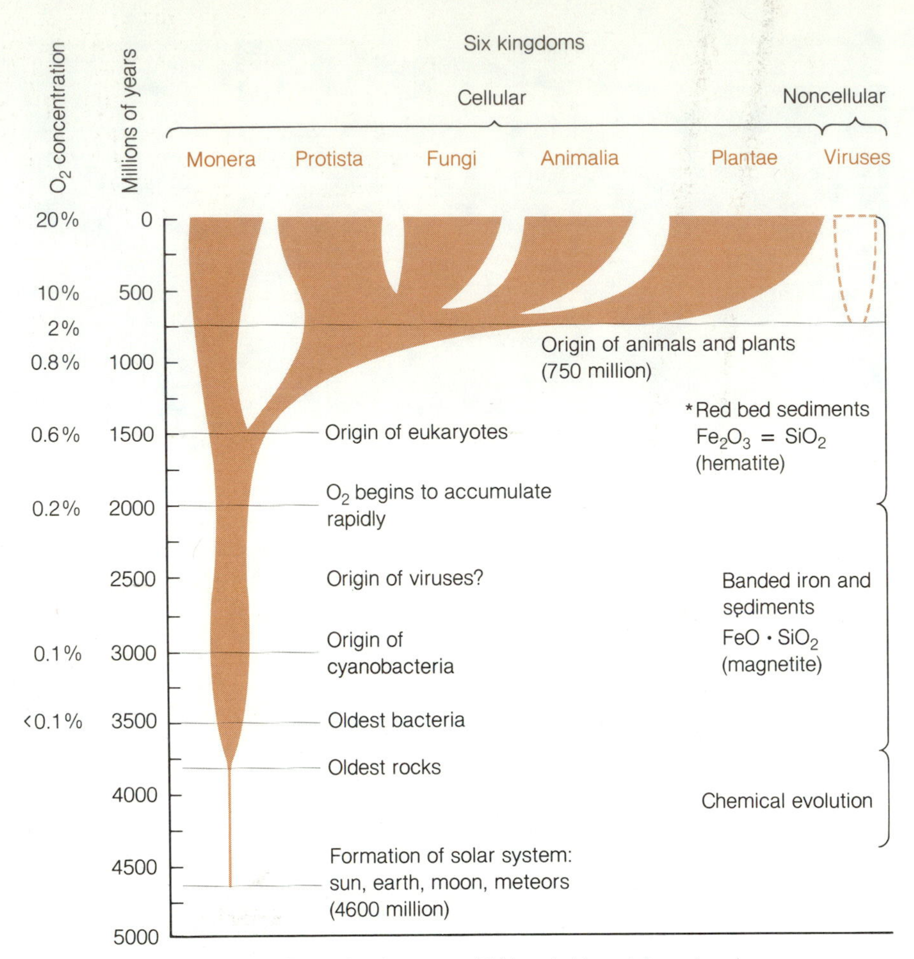

Figure 12-3 Evolutionary Tree. Bacteria-like organisms evolved 3.5 billion years ago on an earth that contained little molecular oxygen (O_2) in its atmosphere. Organisms resembling the blue-green algae released molecular oxygen for over 2 billion years, increasing the oxygen concentration to about 0.6%. Approximately 1.5 billion years ago, prokaryotic organisms evolved into eukaryotic organisms by forming endosymbiotic and exosymbiotic relationships with one another. The appearance of eukaryotic algae and plants caused the concentration of oxygen in the atmosphere to increase even more rapidly. Multicellular animals and plants began to evolve approximately 0.75 billion years ago. The viruses are noncellular organisms and are thought to have evolved from the genetic information in cellular organisms, and in intimate association with their hosts. For this reason, viral diversity occurred in parallel with organismal evolution.

CO, CH_4, NH_3, and H_2S. There was little or no molecular oxygen (O_2) present in the early atmosphere.

It is believed that the inorganic molecules in the atmosphere and in the oceans interacted with one another to produce simple organic molecules. These molecules then reacted among themselves to form a variety of amino acids, fatty acids, sugars, and numerous other types of complex organic compounds. The formation of organic molecules from simple inorganic compounds is referred to as **chemical evolution** and has been reproduced in the laboratory by numerous scientists.

In the 1950s, Stanley Miller and Harold Urey, in one of the first experiments on chemical evolution, discharged electric sparks into a heated mixture of H_2O, CH_4, NH_3, and H_2 in a closed container for a number of days (fig. 12-4). They found that these simple molecules reacted with one another to form some of the simple amino acids and a number of other organic compounds. Since these early experiments, scientists have used electric sparks, ultraviolet light, X-rays, high energy electrons, and high temperatures to produce a variety of organic molecules from various mixtures of com-

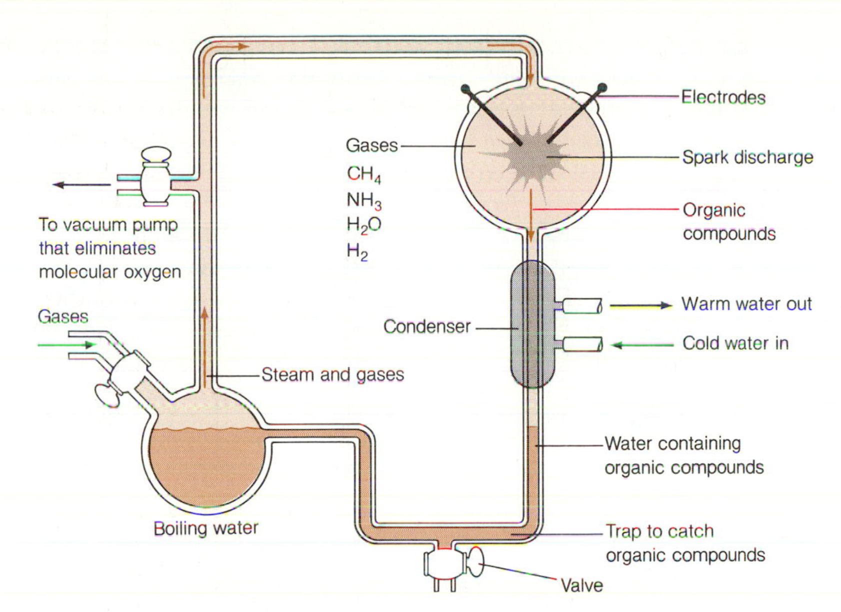

Figure 12-4 The Miller and Urey Experiment. Miller and Urey in 1953 used an apparatus similar to the one illustrated to demonstrate that many organic compounds can form from a mixture of simple molecules such as H_2O, H_2, NH_3, and CH_4. Most of the amino acids that are found in proteins have been synthesized from simple molecules.

pounds believed to have been present in the earth's atmosphere. In these experiments, most of the 20 amino acids commonly found in proteins have been produced.

Researchers have shown that heated amino acids will assemble to form branched proteins, often referred to as **proteinoids**. Sidney Fox has synthesized proteinoids containing 40–100 amino acids simply by heating a mixture of dried amino acids. Proteinoids have been formed at temperatures as low as 65°C, a temperature often reached on the surface of the earth. When boiling water is added to the proteinoids, they form what Fox calls **microspheres**. Some of these microspheres show a striking resemblance to spherical bacteria (fig. 12-5). Microspheres are of interest because they represent self-ordering structures that grow by the accretion of new proteinoids and can multiply by budding or by fission. The self-ordering of molecules into various structures, and their multiplication by budding or fission, are characteristics of present-day bacteria. In addition, enzymatic and other activities commonly associated with living cells have been detected in proteinoid mixtures.

It is hypothesized that microspheres formed in the oceans and lakes of the primitive earth and that some contained proteinoids that acted as catalysts in the formation of new proteins. The proteins synthesized were added to the microspheres, causing them to grow and eventually divide. It is proposed that some of the protein catalysts eventually acted as templates, guiding their own synthesis as well as the synthesis of other proteins. The basis for these ideas is that present-day bacteria still synthesize some short polypeptides on **protein templates** (fig. 12-6). For example, the antibiotics gramicidin, tyrocidine, and bacitracin are polypeptides synthesized on protein templates.

(a)

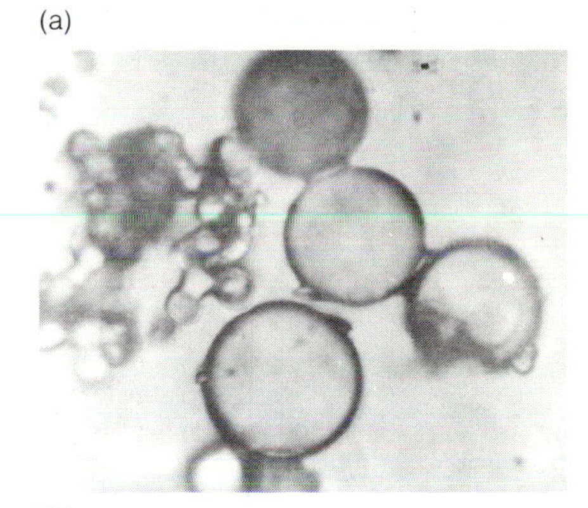

(b)

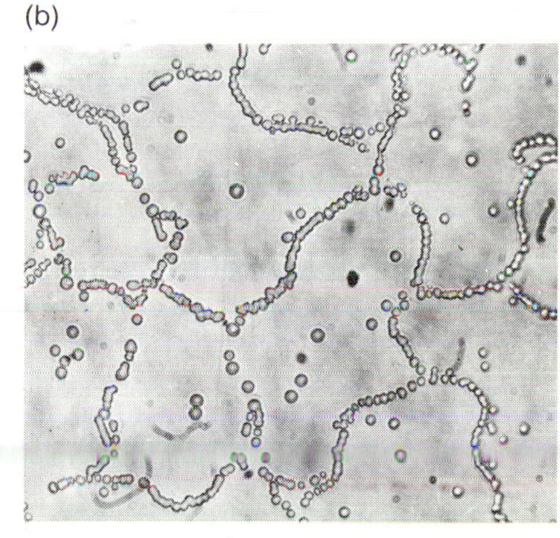

Figure 12-5 Microspheres. A mixture of amino acids added to hot water polymerizes into short proteins. The proteins aggregate into spherical structures called microspheres *(a)*, which resemble spherical bacteria *(b)*. Some of the proteins associated with the microspheres catalyze various chemical reactions.

It is also hypothesized that fatty acids were synthesized by protein catalysts, and that these fatty acids mixed with the microsphere proteins to form primitive membranes resembling those found in present-day cells. These cell-like microspheres have been called **progenotes**, because they did not contain DNA or RNA as their genetic information (fig. 12-7). It is proposed that the first cells arose from progenotes when protein templates were replaced by **RNA templates** □. These nucleic acid templates directed the formation of proteins and these in turn catalyzed the formation of all cellular components. Hypothetical cells with RNA as their genetic information are referred to as **eugenotes**. The 3.5 billion–year-old microfossils may be the remains of eugenotes or early prokaryotes with DNA as their hereditary information. Approximately 3.2 billion years ago, cells that resemble contemporary prokaryotes evolved (fig. 12-7). Because these fossilized organisms re-

207

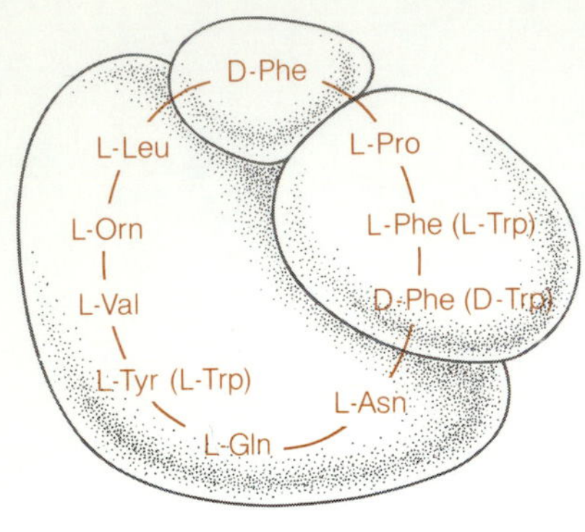

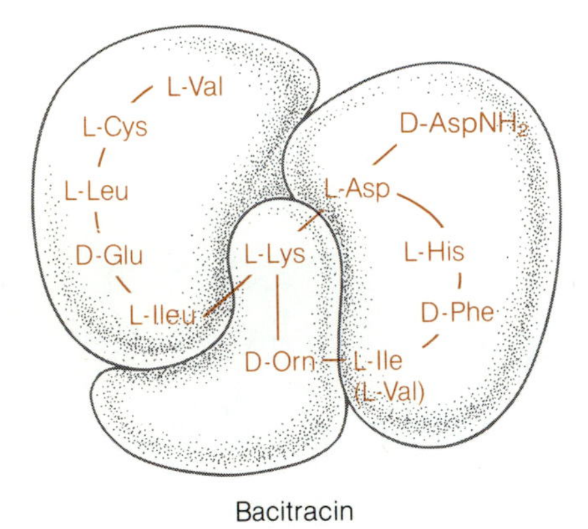

Figure 12-6 Protein Templates. Proteins in some bacteria serve as templates for the synthesis of short peptides. Tyrocidine and bacitracin are synthesized on a protein template in *Bacillus.*

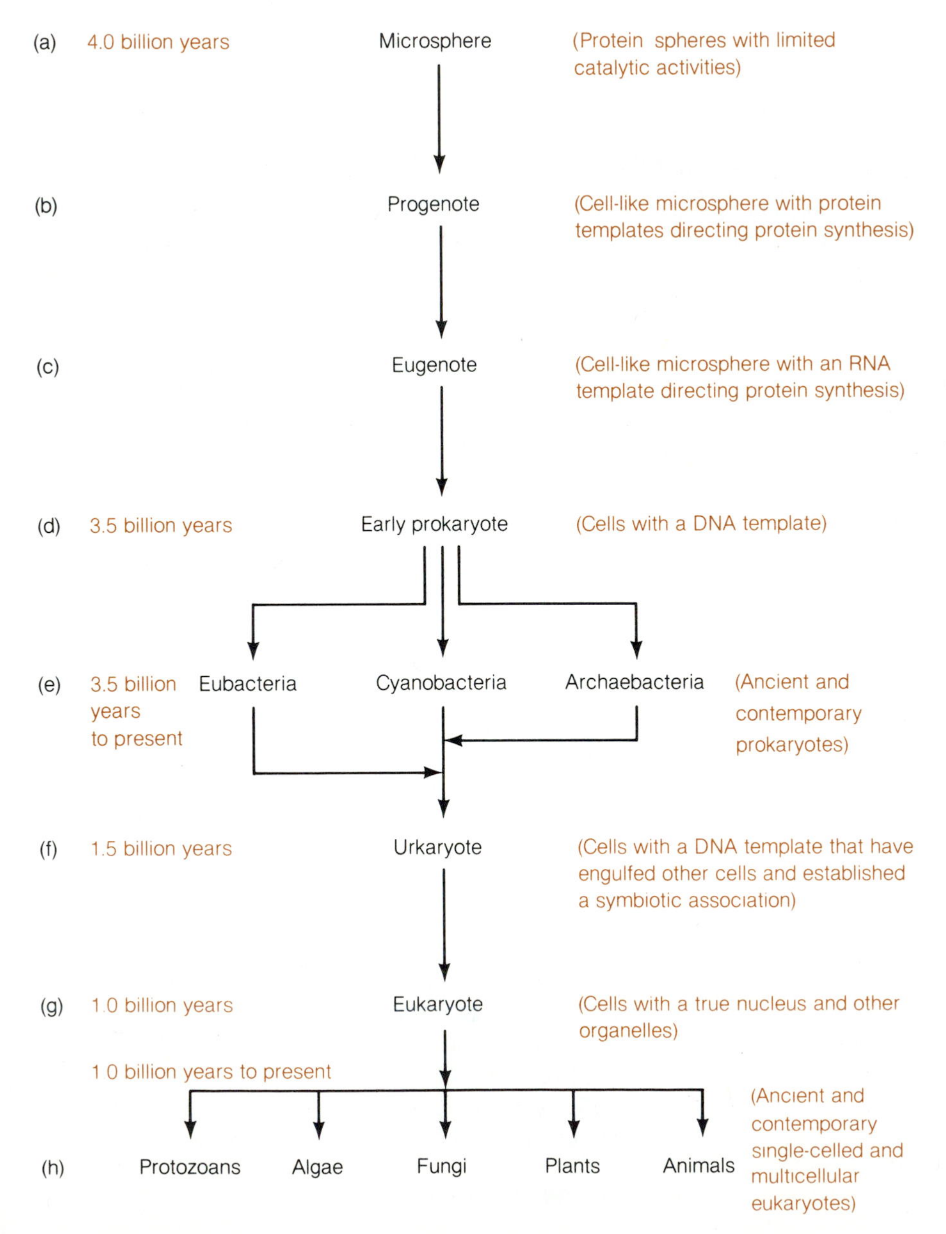

Figure 12-7 Hypothetical Stages in the Evolution of Prokaryotes and Eukaryotes. *(a)* Microspheres formed spontaneously from proteinoids (branched and unbranched polypeptides). The microspheres had limited catalytic activities. *(b)* Progenotes directed protein synthesis by using protein templates. *(c)* Eugenotes were organisms that used an RNA template to direct protein synthesis. *(d)* Early prokaryotes were organisms that used a DNA template to direct RNA and protein synthesis. *(e)* The early prokaryotes evolved into at least 3 major groups which still exist today. *(f)* Prokaryotes forming endosymbiotic and exosymbiotic relationships with each other evolved into the urkaryotes. *(g)* The eukaryotes evolved when the symbionts formed stable nuclei. *(h)* The eukaryotes evolved into the many ancient and contemporary single and multicellular organisms.

semble modern bacteria, some scientists have speculated that the ancient prokaryotes probably carried out many of the functions found in modern prokaryotes and that they included eubacteria, cyanobacteria, and archaebacteria.

BACTERIAL EVOLUTION

Microfossils resembling a variety of bacteria have been found in 3.5 billion–year-old stromatolites from Western Australia. These microfossils indicate that the first living organisms colonizing the earth were similar to prokaryotic cells or bacteria. Once prokaryotes appeared on the earth, apparently they evolved rapidly. They used a variety of available nutrients and dramatically changed their environment. For example, the photosynthetic bacteria converted H_2S and CO_2 to SO_4^{-2} and organic matter, while the blue-green algae (cyanobacteria) converted H_2O and CO_2 to atmospheric oxygen and organic matter. Eventually, the O_2 accumulated in the atmosphere and significantly influenced the evolution of microorganisms by stimulating the development of aerobic respiration 190 □. Molecular oxygen in the atmosphere is believed to be an important factor in the development of the eukaryotic cell, approximately 1.5 billion years ago.

Fossilized bacteria represent the ancestors of all living organisms. For more than 2 billion years, the bacteria ruled the earth as the only living organisms. Eventually, eukaryotic cells appeared. These cells were dramatically different from the bacteria. The following discussion is an attempt to explain a plausible mechanism by which the eukaryotes developed from the prokaryotes.

There are various hypotheses that attempt to explain the origin of organelles in the eukaryotic cell. These hypotheses can be divided into two major types: 1) the **symbiotic hypothesis** and 2) the **membrane proliferation hypothesis** (fig. 12-8). Proponents of the symbiotic hypothesis suggest that some of the eukaryote organelles, such as nuclei, mitochondria, chloroplasts, and flagella, arose from bacteria forming symbiotic relationships, in which one symbiont lived within the other. The internal bacteria **(endosymbionts)** are believed to have evolved into the organelles, eventually losing all their autonomy. Supporters of the membrane proliferation hypothesis usually argue that nuclei, mitochondria, and chloroplasts arose from the invagination of the cytoplasmic membrane and not from endosymbiotic associations.

Membrane systems such as the endoplasmic reticulum, the Golgi membranes, and the various vacuoles are thought to have originated from the nuclear membrane and cytoplasmic membrane.

THE ORIGIN OF THE EUKARYOTIC CELL

The Evolution of Mitochondria and Chloroplasts

Certain bacteria, such as the cyanobacteria, are larger than the typical prokaryote. The increased volume may be advantageous because it allows for more internal membranes where photosynthesis can occur. In addition, larger cytoplasmic membranes and more internal membranes may be advantageous to cells because they contain more respiratory enzymes (electron transport systems) 192 □ and so can produce more ATP.

Figure 12-8 Hypotheses for the Origin of Organelles. *(a)* and *(b)* The symbiotic hypothesis postulates that organelles evolved from endo- and exosymbiotic relationships among prokaryotes. *(c)* and *(d)* The membrane proliferation hypothesis proposes that organelles evolved from the growth of the cytoplasmic membrane.

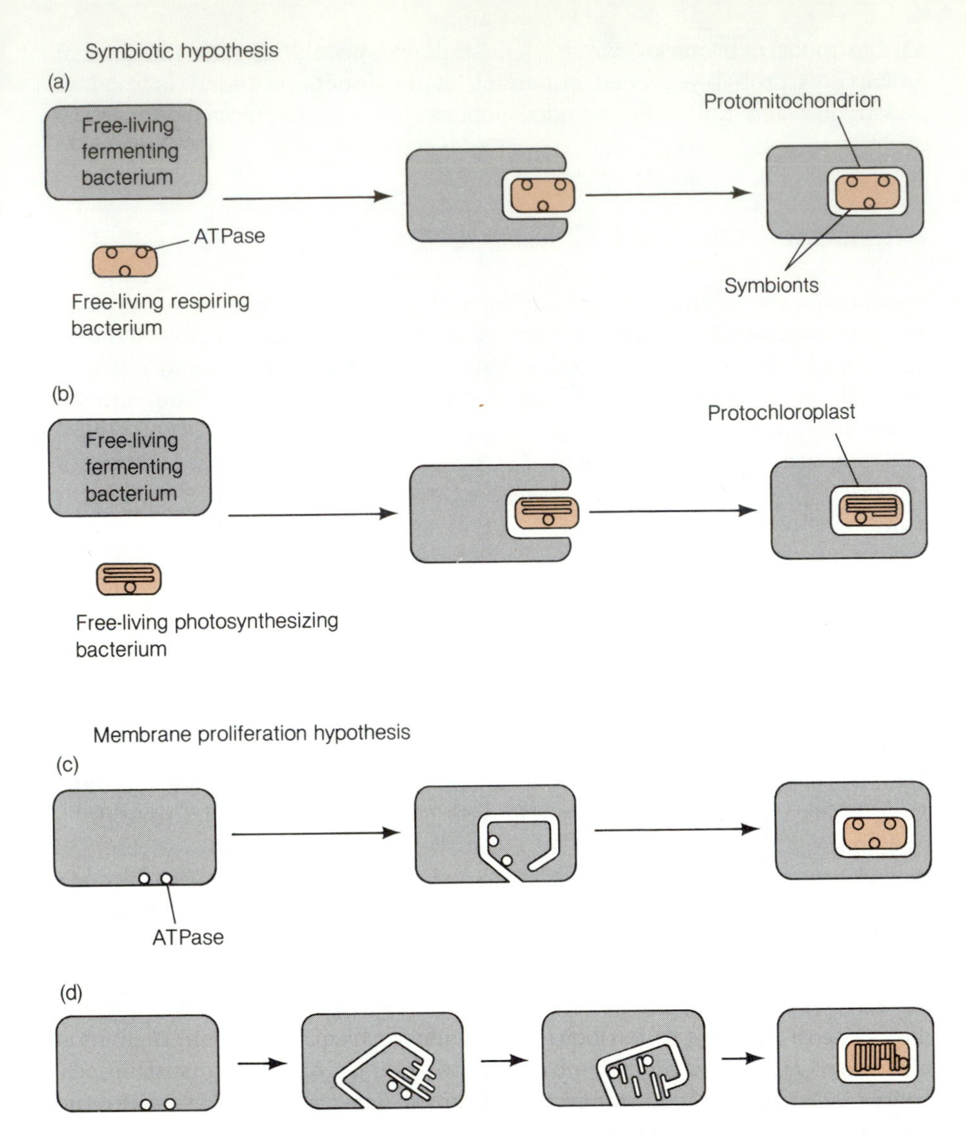

An increase in cell size beyond that reached by most contemporary bacteria is apparent in the fossil record approximately 1.5 billion years ago. At about the same time, the concentration of O_2 was increasing in the atmosphere. It is hypothesized that the production of molecular oxygen (O_2) by the cyanobacteria and its accumulation in the atmosphere may have stimulated the development of mitochondria and chloroplasts, which in turn led to the increase in cell size. Cells could increase their rate of energy metabolism either by engulfing smaller respiring bacteria or by proliferating their cytoplasmic membranes. Similarly, cells could protect their photosynthesis from the effects of molecular oxygen either by establishing endosymbiotic relationships with larger cells or by surrounding their internal membranes with other membranes. The endosymbiotic bacteria (or proliferated membranes) eventually evolved into mitochondria and chloroplasts (fig. 12-9).

There is considerable support for the idea that mitochondria and chloroplasts originally were derived from endosymbiotic bacteria. For example, there are protozoans that lack mitochondria but have endosymbiotic bacteria

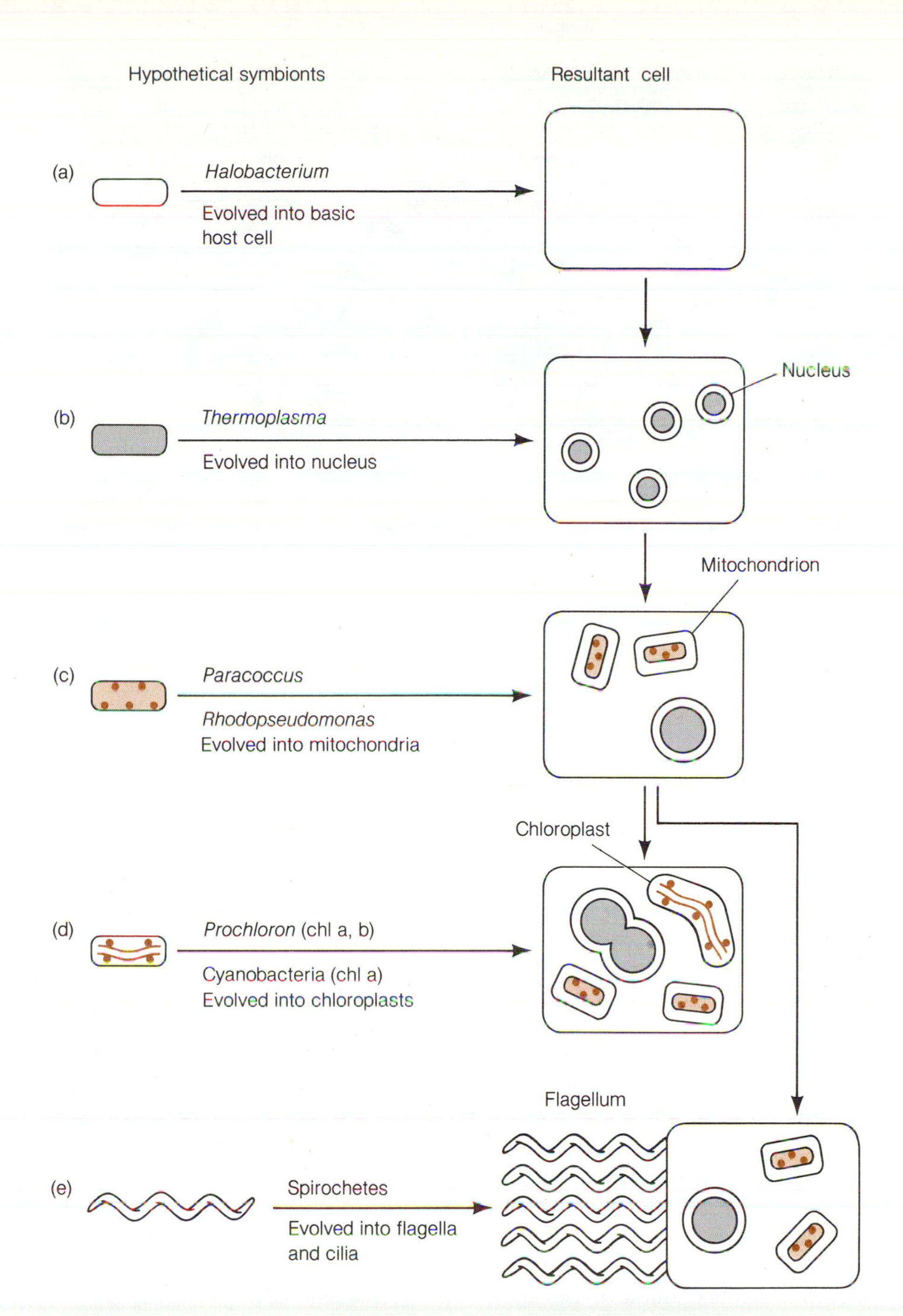

Figure 12-9 The Symbiotic Hypothesis for the Origin of Organelles. *(a)* Prokaryotic cells, like those of the present day *Halobacterium,* may have served as the basic cell for the evolution of eukaryotic cells. Bacteria similar to the ones illustrated may have given rise to the various organelles. *(b)* Bacteria that lack cell walls, such as *Thermoplasma,* may have given rise to nuclei and readily allowed genetic recombination. *(c)* Bacteria similar to *Paracoccus* and *Rhodopseudomonas*, which have electron transport systems that closely resemble those found in eukaryotes, may have given rise to mitochondria. *(d)* Similarly, *Prochloron* and various cyanobacteria, which have photosynthetic systems similar to those found in eukaryotes, may have led to the evolution of chloroplasts. *(e)* Spirochetes are suggested as symbionts to explain the origin of flagella and cilia.

that apparently function as mitochondria (fig. 12-10). In addition, there are protozoans that are packed with symbiotic cyanobacteria that provide the protozoans with nutrients. Mitochondria and chloroplasts still retain some characteristics that hint at their origin. They contain circular genomes that resemble those found in bacteria. The genomes are much too small, however, to code for more than a fraction of the proteins and enzymes required for the synthesis of the organelles or their metabolism. This fact explains

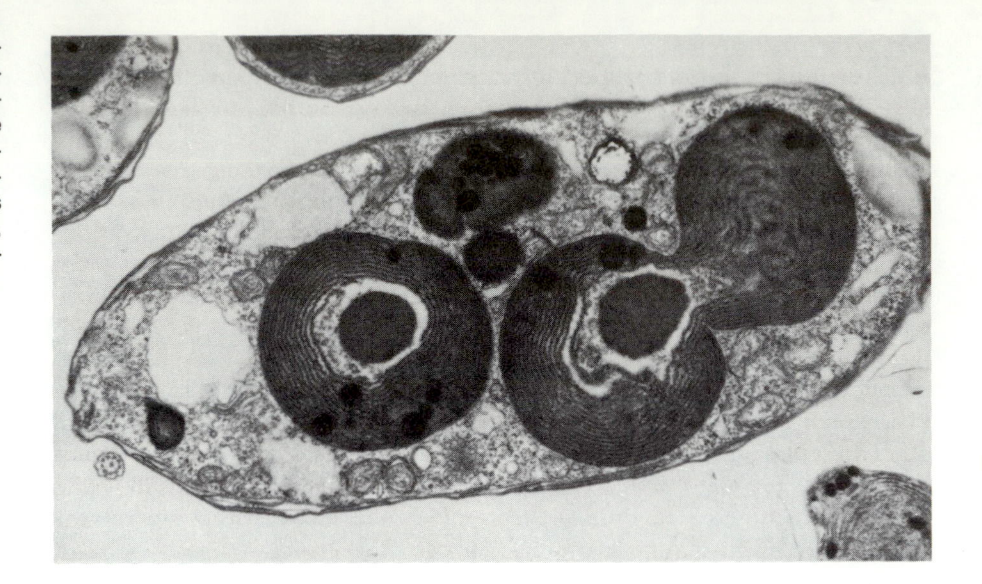

Figure 12-10 ***Cyanophora paradoxa.*** *Cyanophora paradoxa* is a protozoan that harbors organelles that evolved from blue-green endosymbionts. The organelles have a residual peptidoglycan cell wall, a prokaryotic genome, and the typical stacked thylakoids of blue-green algae. The organelles photosynthesize and nourish the protozoan, but are incapable of an independent existence because they have lost many genes.

why mitochondria and chloroplasts cannot be grown on artificial media □, as most bacteria can. Mitochondria and chloroplasts have DNA and RNA polymerases for DNA replication and transcription. Most significant of all, these organelles contain ribosomes that resemble bacterial ribosomes not eukaryotic ribosomes □.

119

93

The Evolution of Nuclei

The origin of nuclei may have occurred as a consequence of an increase in cell size. Since large cells require more nutrients and energy for synthesis and maintenance than small cells, and large cells take longer to multiply, large organisms generally form smaller populations than do small microorganisms. Small populations are always in danger of extinction, however, because there is not enough genetic variability to meet changing environmental conditions. Thus, large cells that develop mechanisms to create extensive variability are more likely to survive. One way of creating variability may have been to engulf and/or fuse with other bacteria, thus increasing the amount of genetic information. This genetic information could be kept in separate nuclei or within a single nucleus. Copies of the same genome, when in the same nucleus, may recombine to create even more variability. Once cells maintained stable nuclei they would be called **eukaryotes** (fig. 12-7). The endosymbiotic hypothesis and the membrane proliferation hypothesis have also been used to explain the origin of eukaryotic nuclei.

In order to insure extensive variability, nucleated cells evolved mechanisms that forced them to fuse with each other and to recombine their DNA before they multiplied. The fusion of cells and nuclei, and the subsequent recombination of DNA, is part of the process referred to as **sexual reproduction**. Some bacteria are capable of recombination and a one-way transfer of DNA. Consequently, some bacteria are capable of a primitive form of sexual reproduction.

The Evolution of Flagella and Cilia

Flagella in bacteria are anchored to the cytoplasmic membrane and are powered by the respiratory system in the cytoplasmic membrane. In eukaryotes the respiratory system is located in mitochondria, consequently, the rotation of bacterial-like flagella could not be driven directly by the flow of electrons and protons through an electron transport system. Thus, a new type of flagellum had to evolve.

Proponents of the symbiotic origin of organelles believe that symbiotic bacteria on the surface of eukaryotes may have evolved into flagella. The basis for this idea is that many protozoans are covered by spirochetes that propel the eukaryotes by their undulations. The attachment of these bacteria to the surface of the eukaryote appears to stimulate the synthesis of structures resembling basal bodies in the eukaryotes' cytoplasm. Presumably, these structures support the membrane where the bacteria are intimately attached. These present-day examples of symbiotic bacteria acting like flagella may represent intermediate steps in the development of the true flagella seen in eukaryotes (fig. 12-9).

A second hypothesis for the origin of eukaryotic flagella and cilia is based upon the appearance of a cytoskeleton within cells. A **cytoskeleton** is a framework of protein in the cytoplasm that gives eukaryotic cells their shape and helps hold them together so that they do not fragment. It is hypothesized that parts of the cytoskeleton □ could polymerize and push the cytoplasmic membrane out, thus forming flagella, cilia, or microvilli. Flagella and cilia would form from microtubules polymerizing from basal bodies in the cytoplasm.

103

THE EVOLUTION OF MULTICELLULAR ORGANISMS

The single-celled eukaryotes and prokaryotes are believed to have shared the earth for more than 800 million years before any multicellular eukaryotes evolved. The photosynthetic activity of the eukaryotic algae and the cyanobacteria continued to alter the environment, selecting for oxygen-utilizing organisms (aerobes) and creating new opportunities for diversification. The origin of multicellular organisms occurred when some of the algae and protozoans evolved recognition systems in their membranes which allowed not only cell fusion for genetic recombination, but also endocytosis and phagocytosis of nutrients. Other protists evolved recognition systems in their membranes which caused cells to remain together and specialize to their mutual advantage. Eventually, organized masses of cells working together evolved unique shapes that could be recognized easily in the fossil record. The earliest multicellular fossil organisms, approximately 600 million years old, include sponges, worms, jellyfish, sea pens, filamentous algae, nonvascular multicellular algae, and fungus-like organisms.

THE EVOLUTION OF VIRUSES

It has not been possible to determine from the fossil record when the viruses appeared, because most viruses are cylindrical or spherical and so small that they cannot easily be distinguished from artifacts. As we shall see, however, it is reasonable to believe that viruses appeared soon after eugenotes.

Viruses □ are obligate intracellular parasites with genomes of DNA or RNA surrounded by a coat of protein. The viruses are not cellular in nature and are unable to carry out metabolism unless they get their genetic information into a suitable host cell. The only way in which viruses can reproduce is by invading a cell and using its energy, nutrients, and biosynthetic machinery.

390

It is hypothesized that the first viruses originated from small pieces of RNA or DNA, similar to the DNA plasmids □ found in bacteria today. The early plasmids may have occurred as RNA or DNA genomes. Only DNA plasmids are found in contemporary bacteria, but RNA and DNA viruses are found in all organisms from bacteria to humans. Since the viruses are obligate parasites, they probably evolved with the eugenotes when they evolved into the bacteria. These viruses would have evolved with the bacteria into the numerous and specific **bacteriophages** (bacterial viruses) observed today. Since the eukaryotes evolved from bacteria, viruses would also be expected in the eukaryotes. Today, all animals and plants can be infected by viruses. Some of these viruses may have evolved from ancestral viruses, while others may have originated *de novo* (from new genetic material) quite recently.

89

Some scientists have postulated that naked nucleic acids (coatless viruses) were intermediates between nonliving macromolecules and cells. The nucleic acids would have been part of a soup of organic and inorganic molecules on the primitive earth. The soup, according to this hypothesis, functioned as a gigantic cytoplasm without a membrane. This hypothesis suffers from one major problem: under these circumstances, any random genetic information that might be present would have no way of being expressed as proteins that could be involved in the synthesis of new nucleic acids or other molecules. Thus, free nucleic acids in a soup of organic and inorganic molecules would be useless in directing the formation of new nucleic acids, much less the parts of a cell.

TAXONOMY

Taxonomy is the science that identifies and names organisms and arranges them into categories called **taxa** (singular, **taxon**). Consequently, taxonomy consists of **identification** (distinguishing), **nomenclature** (naming), and **classification** (ordering).

In the mid-1700s, Carl von Linné (Carolus Linnaeus) began a system of classification of living organisms which is still used today. He divided all living organisms into two obvious kingdoms, the **Plantae** and the **Animalia** (table 12-1). In Linnaeus's system of classification, the plants and animals are grouped to indicate their degree of similarity. A frequently used system of classification places organisms into groups on the basis of similarities in their morphology and physiology. Organisms are first assigned to a **kingdom**, then to a division or **phylum**, a **class**, an **order**, a **family**, a **genus**, and finally a **species**. Organisms in a kingdom may be very different from one another, but they share some major characteristics. Organisms within a family, genus, or species are more similar to one another and share major

FOCUS

DID VIRUSES AND BACTERIA ORIGINATE IN OUTER SPACE?

Some astronomers have hypothesized that viruses and bacteria evolved in outer space and became part of the earth during its formation or soon after it had cooled. In addition, some astronomers have proposed that microorganisms from outer space are constantly raining down upon the earth and are responsible for the epidemics that occur from time to time. Few biologists believe that there is any evidence to support the hypothesis that microorganisms first populated the earth from space **(pangenesis)**, or the idea that epidemics are due to alien microorganisms presently hitting the earth.

The idea that viruses could have evolved in outer space, or in the icy head of comets, does not fit in with what is known about viruses and chemical evolution. The concentration of organic molecules and the temperatures in space are so low that chemical reactions would not lead to the formation of macromolecules, such as proteins, nucleic acids, polysaccharides, or lipids. In fact, astronomers have found no evidence for these macromolecules anywhere in space or on numerous types of rock samples from space (meteorites and moon rocks). If viruses or cellular organisms have been raining down upon the earth and moon from outer space, their remains should have been found in the moon rocks, which have been studied extensively. On the other hand, a myriad of small molecules such as amino acids and macromolecules such as polypeptides have been easily produced in the Miller-Urey experiments, evidence that the earth is a much better environment than outer space for the production of such molecules.

Chemical reactions in living organisms have evolved in an aqueous environment and would not occur in the nonaqueous environment of outer space or in the icy heads of comets. Only when comets are near a sun might the temperature be favorable for chemical reactions to build organic molecules. Energetic radiations (such as ultraviolet light) found near a sun would be expected to ionize and rapidly destroy any guiding templates, such as nucleic acids or proteins, that might form in space. All forms of life on earth, from the viruses and bacteria to animals and plants, are extremely sensitive to radiation and are rapidly inactivated by it. Even if the concentration of molecules, the temperature, the nonaqueous environment, and the radiation in space were not sufficient constraints for macromolecules to evolve by chemical evolution, it would not be expected that viruses could evolve in space in the absence of specific types of cells. Virus replication is a cell-dependent process and would not occur in space or on the prebiotic earth.

and minor characteristics. Sometimes species are subdivided into **races**, **subspecies**, or **strains**. Linnaeus is also credited with establishing a **binomial system of nomenclature** that gives each organism a generic (genus) name and a specific (species) name that, in many cases, indicate something about the organism. For example, *Streptococcus pyogenes* indicates that it is a spherical (coccus) bacterium arranged in chains (strepto) that causes pus-containing (pyogenic) lesions. Very closely related organisms (species) that do not normally interbreed may be in the same genus (plural, genera). Many genera of bacteria are not placed in categories higher than families. Instead they are placed in parts □.

306

After Linnaeus established a two-kingdom system of classification, it became obvious that the kingdoms Plantae and Animalia were not appropriate for microorganisms such as the fungi, protozoans, and bacteria. Ernst Haeckel, one of Charles Darwin's students, proposed in 1866 that microorganisms be placed in a third kingdom called the Protista (primitive or first organisms) (table 12-1). Because bacteria were found to be fundamentally different from other microorganisms, scientists later proposed that they be placed in their own kingdom. In 1969 R. H. Whittaker proposed that all cellular organisms be placed in one of five kingdoms: Animalia, Plantae, Fungi, Protista, and Monera. The prokaryotic bacteria and cyanobacteria are

TABLE 12-1
THE CLASSIFICATION OF MICROORGANISMS INTO KINGDOMS

NUMBER OF KINGDOMS	SCIENTIST CREDITED	KINGDOMS AND ORGANISMS
2	Linnaeus (1753)	Plantae Plants, Algae, Fungi, Bacteria Animalia Animals, Protozoans
3	Haeckel (1865)	Plantae Plants, Multicellular Algae Animalia Animals Protista Protozoans, Single-celled Algae, Fungi, Bacteria, and Sponges (which are animals).
5	Whittaker (1969)	Plantae Plants, Multicellular Algae Animalia Animals Protista Protozoans, Single-celled Algae Fungi Molds and Yeasts Monera Cyanobacteria (Blue-green Algae), Eubacteria (Photosynthetic and Nonphotosynthetic Bacteria), Archaebacteria.
7	—	Plantae Plants, Multicellular Algae Animalia Animals Protista Protozoans, Single-celled Algae Fungi Molds and Yeasts Monera (Prokaryotae) Cyanobacteria (Blue-green Algae), Eubacteria (Photosynthetic and Nonphotosynthetic Bacteria), Archaebacteria. Viruses Viruses, Viroids Prions Prions

single-celled organisms, so different from the eukaryotic, multicellular, multitissued, and multiorganed animals and plants that the kingdom **Monera** (or **Prokaryotae**) was established for them. In addition, because the single-celled algae and protozoans have very complex cell structures and long evolutionary histories that make them very different from bacteria, plants, and animals, a kingdom called the **Protista** was established for them. The fungi □, which are quite different physiologically and morphologically from bacteria, protista, plants, and animals, are placed in a kingdom of their own, called the **Fungi**.

322

The Whittaker system of classification is concerned only with cellular microorganisms, since it ignores the noncellular microorganisms: viruses, viroids, and prions (table 12-1). Some scientists have introduced a sixth kingdom, **Viruses**, which included all the viruses and viroids. Some textbook writers do not consider the viruses, viroids, and prions to be organisms, because they are noncellular in nature. Since these particles are "organized" structures (some of the viruses are extremely complex) that engage in many processes associated with life, we will continue to call them organisms.

Methods Used to Identify and Classify Microorganisms

The DNA of a cell determines its RNAs and proteins and consequently its structure and behavior. Thus, there are several aspects of a cell that can be used to identify it and to determine its relationships to other organisms. For example, cells can be classified and identified on the basis of their genomes, proteins, cell components, or morphology (structure) (table 12-2). Each of

TABLE 12-2
CRITERIA FOR CLASSIFYING BACTERIA

CHARACTERISTICS OF BACTERIA			
CELLULAR	STAINING	COLONY	GROWTH IN BROTH
Shape Size Arrangement Flagella and their arrangement Endospores Cysts Heterocysts Capsules Sheaths Intracellular granules: Sulfur Volutin	Gram Acid fast	Color Texture Shape Size Diffusible pigment	Surface (pellicle formation) Sediment formation Diffusible pigment

CHARACTERISTICS OF BACTERIA			
NUTRITIONAL AND ENVIRONMENTAL REQUIREMENTS	PHYSIOLOGICAL	GENETIC	BIOCHEMICAL
Carbon Energy Nitrogen Amino acid Vitamin Complex Atmospheric: Aerobic Microaerophilic Anaerobic	Range of carbohydrates that can serve as carbon and energy sources Optimum temperature Range of temperature Optimum atmospheric conditions (aerobic, microaerophilic, anaerobic) Optimum osmotic pressure and range Optimum pH Range of pH Optimum salt concentration and range Sensitivity to antibiotics and drugs Products of respirations and fermentations Photoautotroph Photoheterotroph Chemoautotroph Chemoheterotroph	Tm and density of DNA indicates % G + C Amount of DNA hybridizations indicates DNA similarities Ribosomal RNA	Types of storage granules Antigenic determinants Chemical structure of cell wall

these characteristics is now being investigated and used in microbial taxonomy. Traditionally, morphology and physiology have been used almost exclusively to identify and classify eukaryotic organisms such as fungi and protozoa. This generalization is also partly true of viruses, which are differentiated on the basis of their morphology, host range, and type of nucleic acid. In this section we will discuss how a cell's genome, proteins, cell components, and morphology can be used to identify and classify organisms.

Conjugation

With reference to eukaryotes, a generally accepted definition of a species is a group of interbreeding organisms. Only members of the same species can interbreed normally; thus, the exchange or transfer of genetic information can be used to determine whether or not two individuals are of the same species. For example, fungi called dermatophytes, which cause skin infections, are sometimes classified by mating them with known strains. If the two fungi mate successfully, the unknown is considered to be of the same species as the sister strain.

Defining a bacterial species as a population that shares genes, or has a **common gene pool**, is not very useful because bacterial plasmids and pieces of DNA are often transferred (by conjugation or transformation) among different species and genera of bacteria. On the other hand, it is sometimes difficult to determine the conditions that make bacteria competent, and hence, able to transfer hereditary material. Thus, the observation that bacteria do not exchange hereditary material is no guarantee that they belong to different species. In addition, specific modification and restriction enzymes □ that alter and destroy DNA can block the transfer of hereditary material between very closely related organisms. Organisms that have different restriction and modification enzymes may be identified as strains of a species and not as different species or genera. For example, *Escherichia coli* K and *E. coli* B are two strains of bacteria that are unable to exchange genetic information although they are in the same species. In general, the transfer of hereditary material and genetic recombination cannot be used to determine whether or not two bacteria are of the same species.

264

Percent G + C

The evolutionary relationships among cellular microorganisms can often be deduced by comparing the **base composition** □ of their DNA. Organisms that are closely related have DNAs with very similar sequences of nucleotides. Consequently, closely related organisms have almost identical *guanosine* (G) + *cytosine* (C) percentages (%G + C). Organisms that are not closely related generally have different %G + C. There is much experimental evidence indicating that the %G + C among members of a single species does not vary by more than 5%, and that the %G + C from members of a genera does not vary more than 10%. Thus, the %G + C is one indication of how closely related organisms are and is used extensively in taxonomic studies of microorganisms (table 12-3). It should be realized that very distantly related organisms may have evolved similar G + C percentages even though the nucleotide sequences of their DNA are quite different. For example, humans (38.4–40.5) and *Streptococcus pneumoniae* (38.5) have very similar %G + C values, yet the nucleotide sequences of their DNAs are very different and they are very dis-

49

TABLE 12-3
%G+C VALUES

ORGANISMS	%G+C	RANGE
Clostridium spp.	22–28	6
Sarcina spp.	29–31	2
Staphylococcus spp.	30–40	10
Streptococcus spp.	33–44	9
Lactobacillus spp.	35–51	16
Desulfotomaculum spp.	42–46	4
Sporosarcina spp.	40–43	3
Bacillus spp.	33–51	18
Thermoactinomyces spp.	44–55	11
Leucothrix spp.	46–51	5
Cytophaga spp.	33–42	9
Neisseria spp.	46–52	6
Branhamella spp.	40–45	5
Moraxella spp.	40–46	6
Proteus spp.	37–51	14
Yersinia spp.	46–47	1
Escherichia spp.	50–53	3
Erwinia spp.	50–58	8
Serratia spp.	56–59	3
Nitrosomonas spp.	50–51	1
Nitrobacter spp.	60–62	2
Azotobacter spp.	63–66	3
Rhizobium spp.	59–66	7
Pseudomonas spp.	58–69	11
Corynebacterium spp.	57–60	3
Propionibacterium spp.	59–66	7
Actinomyces spp.	60–63	3
Mycobacterium spp.	62–70	8
Streptomyces spp.	69–73	4
Halobacterium spp.	66–69	3
Micrococcus spp.	66–75	9

tantly related. Thus, organisms with similar %G+C are not necessarily closely related. The study of %G+C values of closely related organisms sometimes elucidates the relationships of nearly identical organisms.

The base composition of a DNA sample may be determined by measuring its absorbancy at 260 nm at various temperatures (fig. 12-11). The midpoint of the absorbancy increase is the temperature (T_m) at which approximately half of the hydrogen bonds holding the two strands together have been broken. The higher the T_m value, the greater is the %G+C of the DNA. The more guanine and cytosine in DNA, the more strongly the strands are held together, because guanine and cytosine form three hydrogen bonds with each other while adenine and thymine form only two hydrogen bonds with each other. Thus, DNA with a high %G+C requires more heat to melt (break hydrogen bonds), than DNA with a high %A+T (low %G+C).

The mole %G+C is given by the following formula:

$$\text{mole } \%G+C = (T_m - 16.6 \log M - 81.5)/0.41$$

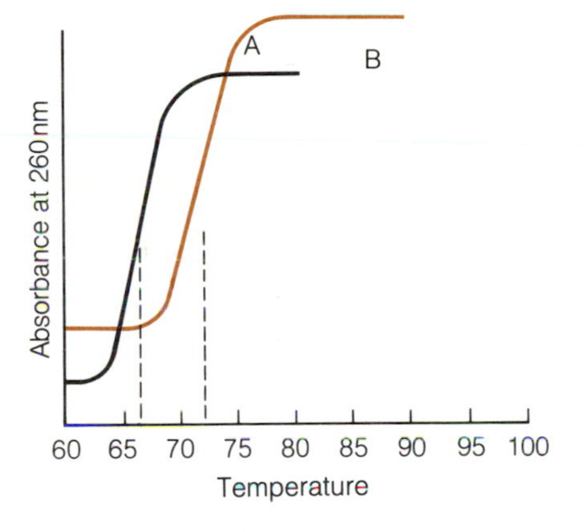

Figure 12-11 DNA Melting Curves. The separation or melting of DNA strands occurs when DNA is heated to temperatures that break the hydrogen bonds. Single-stranded DNA absorbs more light at 260 nm than double-stranded DNA. The melting curve is a reflection of the amount of strand separation. The temperature (T_m) at which the absorbancy has reached half the maximum is proportional to the GC content of the DNA. Since G and C form three hydrogen bonds while A and T form two hydrogen bonds, the higher the T_m the greater the GC content of the DNA. Curve A, T_m = 66°C. Curve B, T_m = 72°C.

where T_m is obtained from the melting curve for a particular DNA and M is the concentration of monovalent positive ions, usually between 0.0001 and 0.2 molar. The concentration of monovalent positive ions affects the melting temperature because they stabilize the hydrogen bonding.

The %G+C can also be determined by centrifuging DNA in a CsCl (cesium chloride) gradient and determining where the DNA bands in the gradient, with respect to known DNAs. The position of the DNA reflects its buoyant density, which is directly related to the G+C/A+T ratio. Since GC base pairs are more dense than AT base pairs, a DNA with a high G+C content will band in zones of high density, while a DNA with a low G+C content will band in zones of low density.

The mole %G+C is given by the following formula:

$$\text{mole }\%G+C = (p - 1.66)/0.98 \times 10^{-3}$$

where p is the buoyant density of the DNA, as determined from where the DNA resides in the CsCl gradient.

The %G+C values for numerous bacteria that have been grouped together because of morphological and physiological similarities often indicate that the bacteria are quite unrelated. The %G+C has been useful in demonstrating that some bacteria grouped together as "slime-bacteria" □, because of their unusual differentiation cycle, were in fact unrelated. For example, the "slime-bacteria" *Sporocytophaga* and *Myxococcus* have G+C percentages of 38 and 68 respectively. In addition, *Staphylococcus* and *Micrococcus* have been shown to have quite different DNAs. *Staphylococcus* has a %G+C of 35, while *Micrococcus* has a %G+C of 70. The bacterium *Listeria* was thought to be related to *Lactobacillus* and to *Corynebacterium*. The %G+C indicates, however, that *Listeria* (38%G+C) is probably not related to *Lactobacillus* (47–51%G+C) nor to *Corynebacterium* (50–60%G+C). Finally, the %G+C has indicated that there are problems with the traditional classification of organisms in the genus *Bacillus*, since its %G+C values range from 32 to 55. This wide range in the DNA content indicates that organisms in *Bacillus* should be divided into at least two genera.

327

TABLE 12-4
HYBRIDIZATIONS AND DEGREE OF DNA SIMILARITY

SOURCE OF LABELED DNA	RELATIVE HYBRIDIZATION TO *ESCHERICHIA* DNA (%)
Escherichia	100
Shigella	89
Salmonella	45
Enterobacter	35
Serratia	20
Proteus	10
Pseudomonas	1
Bacillus	1
Brucella	<0.1

SOURCE: From Jocklik, Willett, and Amos. *Zinsser Microbiology,* 18th ed., p. 17, 1984. Reprinted with permission from Appleton-Century Crofts. Modified from Brenner: *Int J Syst Bacteriol* 23:298, 1973; Brenner et al.: *J Bacteriol* 94:486, 1967; Kingsbury: *J Bacteriol* 94:870, 1967; McCarthy and Bolton: *Proc Natl Acad Sci USA* 50:156, 1963

DNA Hybridizations

The evolutionary relationships between organisms can often be determined by the degree of hybridization between their DNAs or between one organism's DNA and another's RNA (table 12-4). Organisms that are closely related have nearly identical proteins. This means that the DNA and RNA that code for these proteins are very similar. Consequently, if the complementary DNA strands □ from closely related organisms are allowed to hydrogen-bond with each other, there should be an almost perfect match and the two strands will hybridize with each other. The nucleotide sequences from the nucleic acids of two unrelated organisms are quite different. If the complementary strands from different DNAs are allowed to hydrogen-bond, they will hybridize poorly or not at all. The more unrelated organisms are, the more poorly their DNAs hydrogen-bond with each other. If the strands hydrogen-bond poorly, they form DNA hybrids poorly also. Thus, the degree of relatedness can be determined by the extent of hybridization.

49

The extent of hybridization between two strands of DNA, or between a strand of DNA and a strand of RNA, is determined by radioactively labeling one strand and then measuring the amount of radioactivity associated with the hybrid. The nucleic acids are allowed to hybridize for 8–12 hours at high temperatures by adding labeled, single-stranded DNA in solution to unlabeled DNA immobilized on a nitrocellulose filter. The filter is treated with a DNase that attacks single-stranded DNA and is then washed free of unhybridized labeled DNA. The amount of hybridization is determined by measuring the radioactivity that remains attached to the filter. DNA that does not anneal or hybridize efficiently is degraded by the enzyme that attacks ss-DNA (fig. 12-12).

RNA hybridizations to DNA are also useful in determining the extent to which microorganisms are related. For example, ribosomal RNA from any of the bacteria in the family Enterobacteriaceae anneals well with *Escherichia coli* DNA. This fact indicates that their ribosomal RNAs have not changed significantly during evolution.

Phage Typing

394
Bacteriophages □ (bacterial viruses) generally have a very narrow host range, that is, they infect only very closely related bacteria. Thus, the ability of specific bacteriophages to grow on a bacterial population can be used to identify or **type** bacteria.

544
Phage typing □ is very simple and has been used successfully in typing a variety of bacteria. A lawn of the unknown bacterium is spotted with various bacteriophages. The identity of the phages that lyse the bacterial lawn

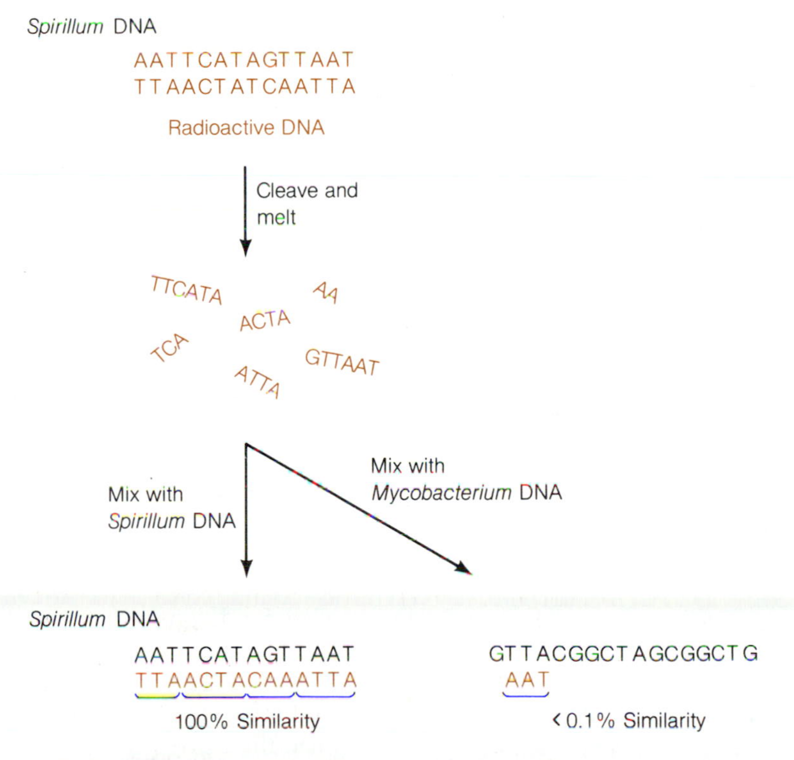

Figure 12-12 Hybridization of DNA. Strands of DNA from very closely related organisms are, for the most part, complementary and consequently will hydrogen-bond to each other and form hybrid DNA. On the other hand, strands of DNA from unrelated organisms are usually not very complementary and so there will be little or no hybrid DNA formed. The degree of hybridization between two DNAs is directly proportional to how similar the DNAs are.

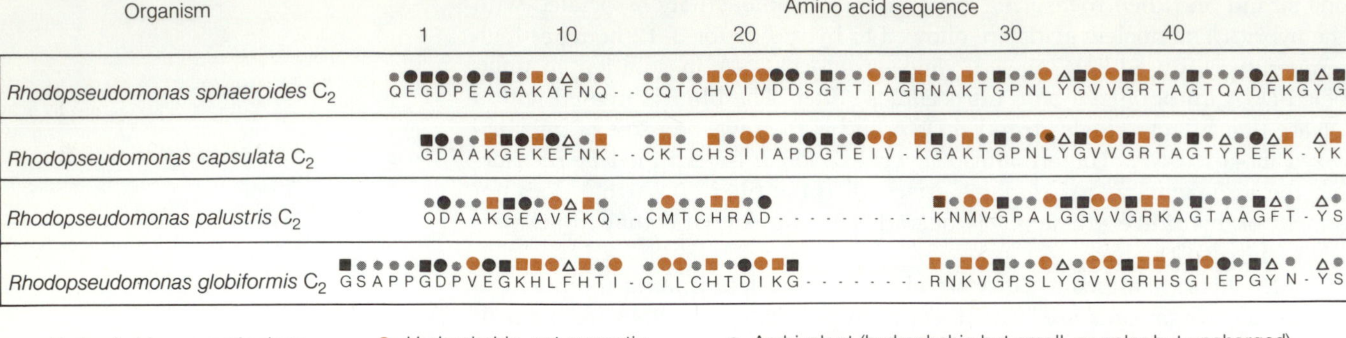

△ Hydrophobic, aromatic rings
F Phenylalanine
W Tryptophan
Y Tyrosine

● Hydrophobic, not aromatic
I Isoleucine
L Leucine
M Methionine
V Valine

• Ambivalent (hydrophobic but small, or polar but uncharged)
A Alanine
B Asparagine or aspartic acid
C Cysteine
N Asparagine
P Proline
Q Glutamine
S Serine
T Threonine
Z Glutamine or glutamic acid

Figure 12-13 Amino Acid Sequences of Cytochrome C in *Rhodopseudomonas.* The amino acid sequences of cytochrome c indicate the extent of evolution in this protein within a genus. Notice that even within a genus there is variation in the amino acid sequence of C_2. The numbers appearing on top of the chart indicate the position of the amino acid in the protein.

indicates what the unknown organism is. Phage typing is often used in epidemiological studies of bacteria such as *Staphylococcus aureus.*

Organisms that are closely related to each other have proteins with similar amino-acid sequences and similar shapes. Often, evolutionary relationships between organisms can be deduced from the differences between equivalent proteins. For example, cytochrome c, an enzyme found in respiring organisms, has been used to determine how closely related organisms are by calculating how many amino acid changes would be required to derive the protein in one organism from that in another (fig. 12-13).

Serology

Some bacteria can be distinguished from others by the use of antibodies □ that are known to bind specifically to proteins and polysaccharides on the bacterial cell. If antibodies added to a drop of bacteria bind specifically to some cell component, the bacteria clump together (agglutinate) □ . If, however, the antibodies do not bind to the bacteria, the bacteria will not be clumped together. The slide agglutination test is so sensitive that very slight differences in surface proteins and polysaccharides can be distinguished and different strains within a species can be defined.

463

Pyrolysis

Microbial cells grown under carefully controlled conditions, when vaporized in an inert gas atmosphere, release products that can be separated and analyzed by gas-liquid chromatography or mass spectrometry. The analysis produces a number of peaks that vary in height and represent vaporized cell components. Different microorganisms produce the same peaks, but different quantities of each product. A comparison of the peak heights can be used to establish relationships among organisms. For example, if three peaks are used, each organism can be described in a three-dimensional space (fig. 12-14). If 50 or 100 peaks are used, then the organisms are described using 50 or 100 characteristics.

50 60 70 80 90 100

EGMKEAGA - - KGLAWDEEHFVQYVQDPTKFLKEYTGDAK - - - - - - AKGKM - - - - TFK - - LKK - EADAHNIWAYLQQVAVRP

DSIVALGA - - SGFAWTEEDIATYVKDPGAFLKEKLDDKK - - - - - - AKTGM - - - - AFK - - LAK - - - GGEDVAAYALASVVK

PLNHNSGE - - AGLVWTQENIIAYLPDPNAYLKKFLTDKGQADKATGSTKM - - - - TFK - - LAN - - DQQRKDVAAYLATLK

EANIK - - - - - SGIVWTPDVLFKYIEHPQKIVP - - - - - - - - - - - - - G - TKM - - - - GYPG - QPD - PQKRADIIAYLETLK

■ Ambivalent (no side chain)
G Glycine

■ Hydrophilic, basic
H Histidine
J Monomethyl lysine
K Lysine
R Arginine

● Hydrophilic, acidic
D Aspartic acid
E Glutamic acid

Morphology and Physiology

Organisms are generally classified on the basis of their morphology, staining characteristics, and physiology, but may at times be classified instead on the basis of growth, biochemical, and genetic characteristics (table 12-2). For instance, in the ninth edition of *Bergey's Manual of Systematic Bacteriology*, bacteria are grouped into 30 sections on the basis of shape, gram reaction, metabolism motility, and characteristic structures. In some cases, the highest classification in a section is the genus and the bacteria within a Section may not very closely related. For example, the budding and appendaged bacteria represent a very diverse group of organisms that are grouped together simply because they form appendages. Even though bacteria resemble one another, many can be identified from a few characteristics such as their shape, staining properties metabolism, and obvious structures, such as sheaths, flagella, endospores, and appendages (fig. 12-15). Fungi, algae, and protozoans are also classified on their morphology and physiology.

The shape of the cell is one of the first characteristics that is used in the identification of bacteria. Most bacteria can be characterized as spheres, rods, spirals, coils, or branches. These cells may exist singly or in chains or clumps. The shape and size of cells can also be used to classify algae, fungi, and protozoans.

68
The Gram stain □ divides bacteria into two groups based on the type of cell wall they have. The acid-fast stain can indicate whether or not a bacterium is acid-fast. If it is, this information narrows the possibilities to *Mycobacterium* or *Nocardia*.

189
The ability or inability to ferment □ carbohydrates is another important characteristic that helps to identify microorganisms. Many closely related organisms can be distinguished by their abilities to ferment various carbohydrates and metabolize assorted compounds.

The presence or absence of endospores can aid in the identification of an unknown, because very few genera of bacteria produce endospores: *Bacillus*, *Clostridium*, *Desulfotomaculum*, *Sporolactobacillus*, *Thermoactinomyces*, and *Sporosarcina*.

121
Growth on selective media □ often can help in the identification of an organism. For example, soil organisms that grow on a mineral medium that

(a) Three-dimensional space

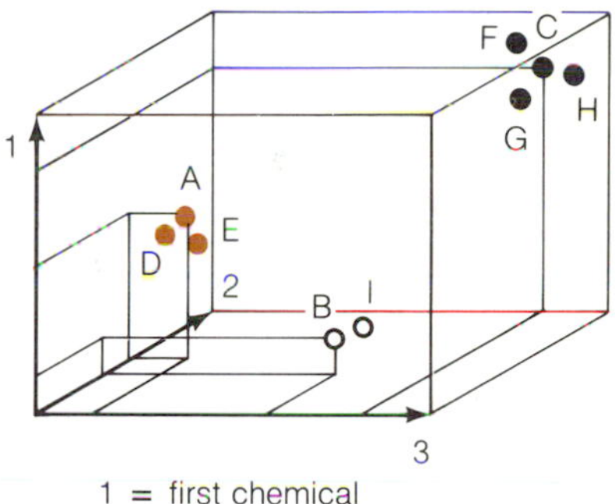

1 = first chemical
2 = second chemical
3 = third chemical

(b) Multidimensional space represented in two dimensions

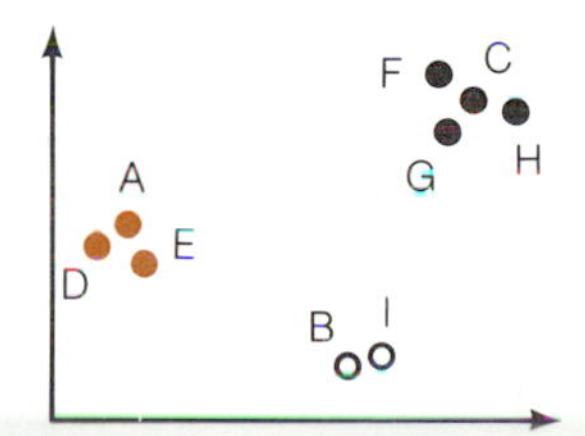

Figure 12-14 Taxonomy. Three characteristics can be used to differentiate among microorganisms in a three-dimensional space. Those organisms that are most similar will cluster together in the three-dimensional space, while those that are unlike each other will not. If n characteristics are used to differentiate among microorganisms, their relationships can be determined in n-dimensional space, which can be represented in two dimensions.

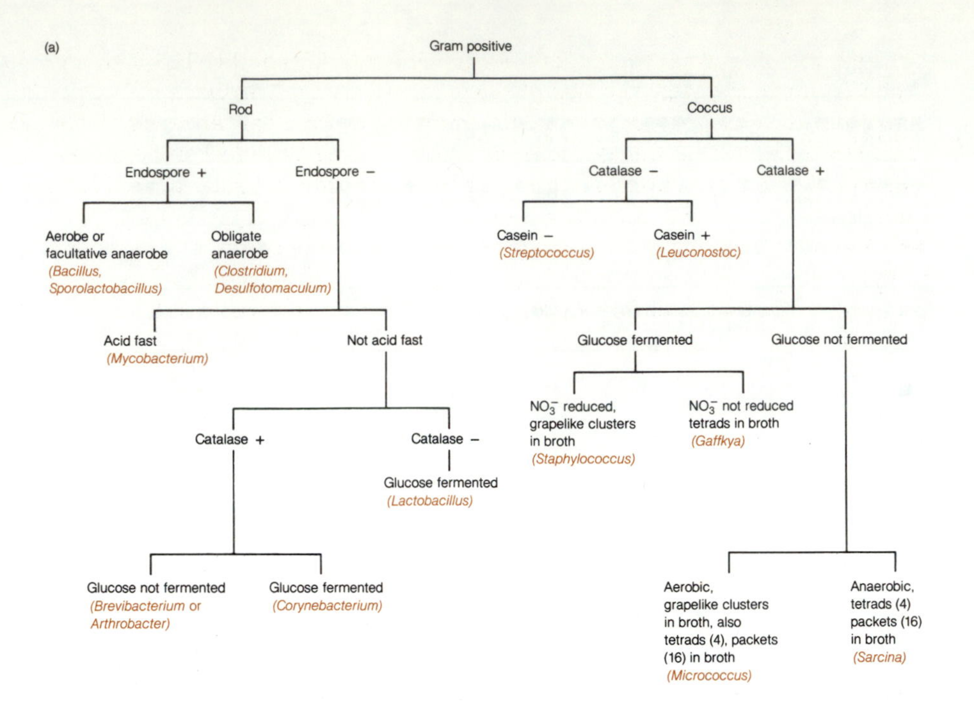

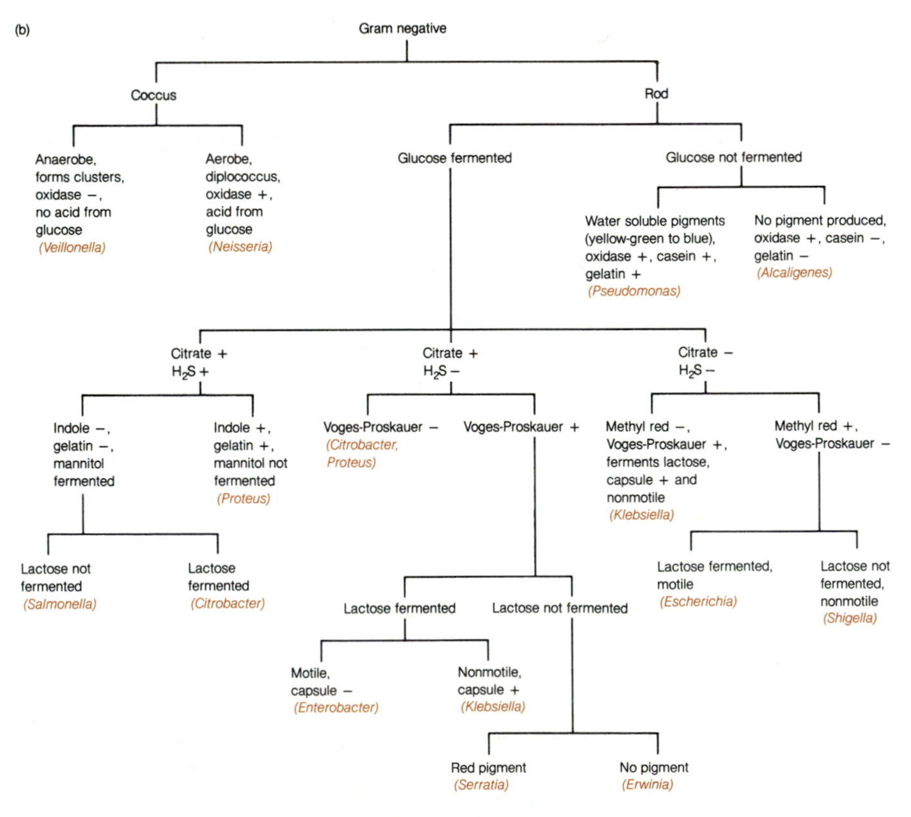

Figure 12-15 A Key for the Identification of Common Bacteria. *(a)* Key to selected gram positive bacteria. *(b)* Key to selected gram negative bacteria.

lacks an organic carbon source, but is supplied with CO_2 and $(NH_4)_2SO_4$, are probably nitrifying bacteria, such as *Nitrosomonas* or *Nitrobacter*. Organisms from the skin that grow on media containing 7.5% salt are most likely *Staphylococcus*. Within this genus, *S. aureus* and *S. epidermidis* can be distinguished by using the differential medium called mannitol salts agar. Mannitol salts agar is both differential and selective because it contains phenol red, mannitol, salt. The 7.5% salt is inhibitory to most bacteria, but not those in the genus *Staphylococcus*. The phenol red, a pH indicator, detects the acid produced from mannitol fermentation, which is a characteristic of *S. aureus*.

LINNAEAN, ARTIFICIAL, PHYLOGENIC, AND NUMERICAL TAXONOMY

The **Linnaean system** of taxonomy groups organisms on the basis of major anatomical and physiological patterns. In Linnaeus's *Systema Naturae*, published in 1735, plants were arranged according to the type of sexual system they possessed. Following Linnaeus's example, animal taxonomists have grouped together animals with similar physical characteristics. For instance, animals that are built like fish are grouped together, while animals that have characteristics similar to those of mammals are placed in another category. This type of classification is very useful because the taxonomic position of an organism, when determined in this manner, indicates what it is like. The fact that a plant, for instance, belongs to the conifers (pine trees, firs, etc.) indicates that it has a number of characteristics that distinguish it from ferns, cycads, ginkgos, and the myriad of flowering plants. **Artifical** or **utilitarian** schemes of classification are used routinely in microbiology. In this system, organisms are often unrelated, although they share some morphological or physiological characteristics that are readily determined and useful in their classification. **Phylogenetic taxonomy**, by contrast, attempts to group organisms according to their evolutionary relatedness as well as their structural and physiological similarities. In many cases, the phylogenic system groups organisms in the same way that the Linnaean system does. **Numerical taxonomy** groups organisms by quantifying the similarities and differences among them. Each characteristic is weighed, and then the "taxonomic distance" between them is determined by comparing the number of characteristics they share in terms of the total number of characteristics.

A very simple form of numerical taxonomy would weight a number of traits equally and then compare organisms with one another. Very complicated numerical methods, requiring the use of computers, are now being used to group and identify organisms.

An example of a numerical method is that used to analyze pyrolyzed (vaporized) bacteria. Pyrolyzed bacteria release all their chemical components in gaseous form. Since the organic chemistry of most microorganisms varies, the concentrations of organic components can be used for taxonomic purposes (fig. 12-14).

Computer Assisted Identification

Computers are being used to identify organisms in half the time this process once took (fig. 12-16). The computer can easily analyze 30 biochemical tests

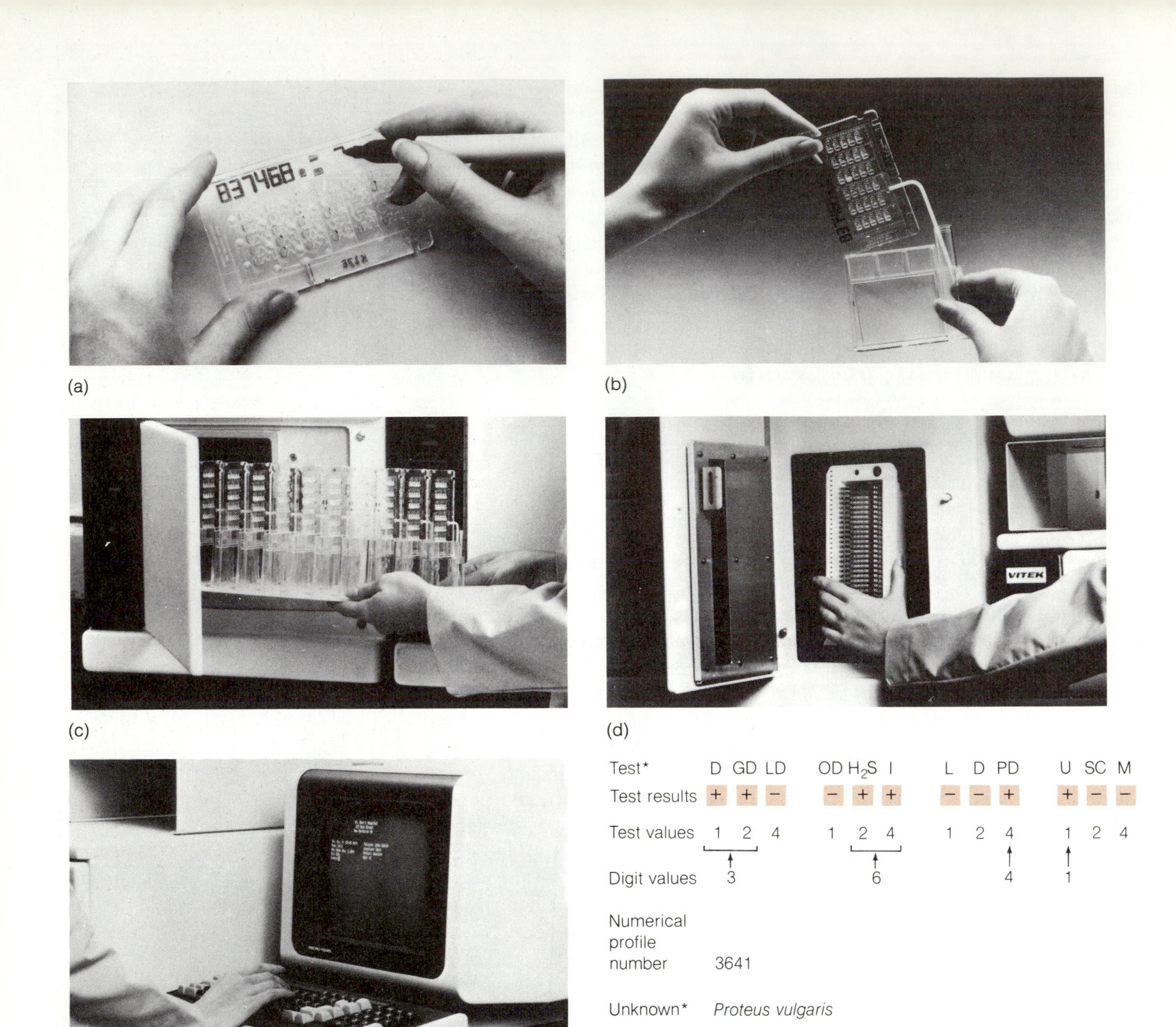

Figure 12-16 A Computer Used to Identify and Determine Antimicrobial Susceptibilities. *(a)* A test card contains as many as 30 wells for determining biochemical characteristics or antimicrobial susceptibilities of a microorganism. The test card is marked with an instrument-readable sample number. *(b)* A test card is placed in a holder, with its capillary straw in an aqueous sample of the microorganism to be tested. *(c)* The wells of the test card are inoculated by the filler/sealer module of the computer. A vacuum is created which draws the sample into the test card. Subsequently, the test card is sealed. *(d)* The inoculated cards are loaded into the reader/incubator module of the computer system. *(e)* Status reports are available through prompting at the keyboard (data terminal) by entering the sample number. *(f)* Determination of a numerical profile number. Tests are divided up into groups of three. In each group, the tests are given the values 1, 2, and 4, respectively. When all the tests are negative, the digit is zero (0+0+0). If the first and second tests are positive and the third test is negative, the digit is 3 (1 + 2 + 0 = 3). If the first test is negative and the second and third tests are positive, the digit is 6 (0 + 2 + 4 = 6). Depending upon the pattern of positive and negative tests in a group, the digit may range from 0 to 7. If 30 tests are carried out on a microorganism, the numerical profile number will have 10 digits, each of which may range from 0 to 7. From table 12-5, it can be determined that the unknown is *Proteus vulgaris*.

TABLE 12-5

REACTION CHART FOR THE IDENTIFICATION OF SOME GRAM NEGATIVE, ENTERIC BACTERIA*

BIOCHEMICAL TESTS	NAME OF MICROORGANISM																
				Enterobacter						Proteus							
	Arizona sp.	*Citrobacter* sp.	*Edwardsiella* sp.	*E. aerogenes*	*E. cloacae*	*E. hafniae*	*E. liquefaciens*	*Escherichia coli*	*Klebsiella* sp.	*P. mirabilis*	*P. morganii*	*P. rettgeri*	*P. vulgaris*	*Providencia* sp.	*Salmonella* sp.	*Serratia* sp.	*Shigella* sp.
Glucose fermentation	+	+	+	+	+	+	+	+	+	+	+	+	+	+	+	+	+
Gas production from glucose	+	+	+	+	+	+	+	+	+	+	d	d	+	d	+	+	−
Lysine decarboxylation	+	−	+	+	−	+	d	d	+	−	−	−	−	−	+	+	−
Ornithine decarboxylation	+	d	+	+	+	+	+	d	−	+	+	−	−	−	+	+	d
H_2S production	+	d	+	−	−	−	−	−	−	+	−	−	+	−	+	−	−
Indole production	−	−	+	−	−	−	−	+	−	−	+	+	+	+	−	−	d
Lactose fermentation	d	d	−	+	+	d	d	+	+	−	−	−	−	−	−	d	−
Dulcitol fermentation	−	d	−	−	−	−	−	d	d	−	−	−	−	−	+	−	d
Phenylalanine deamination	−	−	−	−	−	−	−	−	−	+	+	+	+	+	−	−	−
Urea hydrolysis	−	d	−	−	d	−	d	−	+	+	+	+	+	−	−	d	−
Growth on Simmon's citrate agar	+	+	−	+	+	d	+	−	+	d	−	+	d	+	d	+	−
Mannitol fermentation	+	+	−	+	+	+	+	+	d	−	−	+	−	−	+	+	+
Numerical profile number (assumed d = +)	7316	3337	7700	7716	3117	7116	7117	7534	7037	3343	3241	3447	3643 or 3641	3442	5326	7117	1524

+ = >90% of strains are positive for test

− = >90% of strains are negative for test

d = results vary depending on strain

*Based on data from Ewing, W. H.: Differentiation of *Enterobacteriaceae* by Biochemical Reactions, Center for Disease Control, Atlanta, 1968. (Revised and amended 1970.)

in 10 to 12 hours and then determine which organisms are most likely to fit the results. The computer program compares the test results with a large list of isolates that have previously been characterized. In addition, the computer can indicate which antibiotics are most effective.

How does the computer determine which is the most probable organism? Several systems can be used. One system compares the **numerical profile number** determined for an unknown organism to the numerical profile numbers of a large list of organisms. In order to determine the numerical profile number, tests A, B, C, etc., may be taken in groups of three (fig. 12-16). The tests might be assigned the numbers 1, 2, and 4, respectively. If the first test in a group is positive and the remaining tests are negative, the number would add to 1. If all tests in a group are positive, however, the number would add to 7 (1+2+4). Depending upon which of the three tests in a group is positive, it is possible to derive any number from 1 to 7. Table 12-5 contains the reactions and numerical profile numbers for a number of organisms in the family Enterobacteriaceae. If the computer determines that the unknown has a numerical profile number of 3,641, for example, the best choice is *Proteus vulgaris*.

Computers are now being used to determine, in one trial, the susceptibility of a microorganism to as many as 13 antimicrobials in multiple concentration. Inoculum for the trial is made from three or four colonies emulsified in 1.8 ml of sterile saline solution. After 4 hours of incubation, the computer checks the control sample to determine whether there has been sufficient growth. Susceptibilities for fast-growing organisms, such as the gram negative rods, generally can be determined after 4 hours of incubation. If the growth is not adequate, however, the computer automatically expands the incubation period and records results when there is sufficient growth of the control. The computer printout indicates whether the organism is sensitive, intermediate in sensitivity, or resistant to the antimicrobial. The computer system eliminates the time-consuming setup of the Kirby-Bauer or microdilution minimum inhibitory concentration (MIC) procedures, and removes subjective error in reading disc zones and end points. Best of all, the computer system can give susceptibilities within 4 to 10 hours, while traditional tests require 24 hours.

SUMMARY

CHEMICAL EVOLUTION

1. The atmosphere of the primitive earth contained N_2, CO_2, H_2, H_2O, NH_3, and H_2S, but little or no O_2.
2. Miller-Urey experiments have demonstrated that the simple compounds found in the primitive earth's atmosphere could have evolved into complicated organic compounds.
3. Experiments by Fox indicate that amino acids can easily polymerize into proteinoids and that proteinoids can form cell-like structures called microspheres. Many of the proteinoids formed in experiments have enzymatic activities.
4. It is hypothesized that protein templates guided the formation of early proteins in microspheres and that this development in turn led to the growth and multiplication of microspheres. Microspheres with protein templates are referred to as progenotes. Nucleic acids eventually replaced proteins as the template for protein synthesis. Microspheres or cells with nucleic acid templates are known as eugenotes.
5. The oldest microfossils are 3.5 billion years old and resemble many of the simple bacteria that exist today.

BACTERIAL EVOLUTION

1. Stromatolites 3.2 billion years old contain microfossils that resemble today's cyanobacteria. It is thought that they may have produced O_2 if they photosynthesized, as do present-day cyanobacteria.
2. The production of O_2 by cyanobacteria selected for aerobic bacteria.

THE ORIGIN OF THE EUKARYOTIC CELL

1. There is no evidence for eukaryotic cells in the fossil record until about 1.5 billion years ago. At this time, various types of cells that were larger than the typical prokaryote appear in the fossil record. It is hypothesized that the large cells were primitive eukaryotes.
2. Eukaryotes evolved from prokaryotes (bacteria).
3. The endosymbiotic hypothesis for the origin of mitochondria, chloroplasts, and nuclei proposes that bacteria formed symbiotic relationships with one another and that one of the symbionts existed within the other. The membrane proliferation hypothesis for the origin of organelles proposes that the cytoplasmic membrane invaginated so as to form mitochondria, chloroplasts, and nuclei.
4. The symbiotic hypothesis for the origin of flagella and cilia proposes that spirochetes attached to the surface of primitive eukaryotes and eventually evolved into flagella or cilia. The cytoskeleton polymerization hypothesis for the development of these organelles proposes that the cytoskeleton polymerized in such a way that flagella and cilia are formed.

EVOLUTION OF MULTICELLULAR ORGANISMS

1. The oldest multicellular organisms are approximately 800 million years old.
2. Animals and plants evolved within the last 600–800 million years. Modern humans are a recent arrival on earth, being only about 75 thousand years old.

THE EVOLUTION OF VIRUSES

The viruses are believed to have evolved from nucleic acids that resembled present-day plasmids.

TAXONOMY

1. Taxonomy is the science that arranges organisms into categories called taxa, that names organisms, and that identifies them. Thus, classification, nomenclature, and identification are important aspects of taxonomy.
2. Organisms are generally classified (grouped) on the basis of similarities in morphology and physiology, and are placed in kingdoms, divisions, classes, orders, families, genera, and species.
3. Every cellular organism is given a genus name and a species name. The system of naming is called the binomial system of nomenclature and was developed by the Swedish botanist Carolus Linnaeus in the mid-1700s.
4. R. H. Whittaker, in 1969, proposed that all cellular organisms be placed in one of five kingdoms: Animalia, Plantae, Fungi, Protista, and Monera. The eubacteria, archaebacteria, and cyanobacteria are in the kingdom Monera. All bacteria are placed in a genus and given a species name. Depending upon the classification system used, bacteria may also be placed in a family, in an order, and in a class.
5. Since an organism's DNA specifies its RNAs and proteins and since these in turn determine the molecules that make up the cellular structure, an organism can be studied on a number of levels. For instance, cells can be identified and classified on the basis of their nucleic acids, proteins, cell components, or morphology.
6. The %G+C in an organism's DNA can be used to indicate the degree of difference between organisms.
7. The degree to which DNAs from different organisms hybridize indicates their similarity.
8. Bacteriophages generally have very narrow host ranges. Consequently, the pattern of growth of various phages on an unknown bacterium can be used to identify the bacterium. This procedure is called phage typing.
9. Amino-acid sequences of certain proteins can reveal how closely related organisms are.
10. Because each type of organism produces specific cell components, subtle similarities or differences between organisms can often be determined by using antibodies found in serum. The use of antibodies in identifying microorganisms or chemical compounds is part of the science of serology.
11. Again, since organisms produce specific cell components in varying amounts, organisms can often be identified by investigating a large spectrum of compounds from a cell. One way to do this is to vaporize cells and investigate the gases released. The vaporization of cells is known as pyrolysis.

12. Usually organisms are identified and classified on the basis of their gross morphology and physiology. Many bacteria can be identified and classified by a few characteristics, such as shape (rod, sphere, coil), staining characteristics (gram positive or negative, etc.), type of metabolism (aerobic, anaerobic, facultative), special structures (endospores, appendages), and type of physiology (growth on selective and differential media).
13. There are several ways to classify organisms. The Linnaean system groups on the basis of major anatomical and physiological patterns. The artificial or utilitarian system classifies in any manner that is useful, and consequently relies heavily on morphology and physiology. Phylogenic taxonomy groups organisms according to their evolutionary relatedness. Numerical taxonomy groups organisms by quantifying the similarities and differences among organisms.

STUDY QUESTIONS

1. What evidence is there for chemical evolution?
2. Explain what proteinoids are, and their relationship to microspheres.
3. Explain how proteinoids may have served as templates for the synthesis of proteins, and how this function could lead to the multiplication of microspheres.
4. Diagram a possible relationship among microspheres, progenotes, and eugenotes. Explain what microspheres, progenotes, and eugenotes are.
5. Diagram the evolutionary relationships among the organisms in the five kingdoms (Monera, Protista, Fungi, Plantae, and Animalia) by drawing an evolutionary "tree." Along the ordinate (Y axis) of your drawing, clearly indicate the times at which the organisms in each kingdom diverged.
6. Compare the endosymbiotic and membrane proliferation hypotheses for the origin of chloroplasts by drawing the steps necessary to evolve this organelle from a photosynthetic prokaryote.
7. Diagram possible relationships among prokaryotes (eubacteria, archaebacteria, cyanobacteria), urkaryotes, and eukaryotes.
8. Explain how viruses might have originated from cellular nucleic acids.
9. Explain why it is unlikely that viruses originated from cellular nucleic acids.
10. Explain why it is unlikely that viruses are intermediates between organic molecules and cellular organisms.
11. What are the three areas of taxonomy?
12. Define classification, identification, and nomenclature.
13. Discuss how %G+C is used to identify and classify bacteria.
14. Explain how DNA-DNA and DNA-RNA hybridizations are used to identify and classify bacteria.
15. Explain how amino acid sequences of proteins can be used to classify bacteria.
16. Discuss how antibodies can be used to identify and classify bacteria.
17. Discuss how phage typing is used to identify and classify bacteria.
18. How can the pyrolysis of bacteria be used to identify and classify bacteria?
19. Compare and contrast Linnaean, utilitarian, phylogenic, and numerical taxonomy.

SUPPLEMENTAL READINGS

Dickerson, R. 1978. Chemical evolution and the origin of life. *Scientific American* 239:70–86.

Dickerson, R. 1980. Cytochrome c and the evolution of energy metabolism. *Scientific American* 242(3):137–149.

Fox, S. 1980. Metabolic microspheres. *Naturwissenschaften* 67:378–383.

Fox, S. and Dose, K. 1977. *Molecular evolution and the origin of life*; Marce Dekker Inc.

Gray, M. and Doolittle, F. 1982. Has the endosymbiont hypothesis been proven? *Microbiological Reviews* 46(1):1–42.

Hoyle, F. and Wickramasinghe, C. 1979. *Diseases from space.* New York: Harper & Row.

Karachewski, N. O., Busch, E. L., and Wells, C. L. 1985. Comparison of PRAS II, RapID ANA, and API 20A systems for identification of anaerobic bacteria. *Journal of Clinical Microbiology* 21:122–126.

Katz, E. and Demain, A. 1977. The peptide antibiotics of *Bacillus*: chemistry, biogenesis, and possible functions. *Bacteriological Reviews* 41(2): 449–474.

Margulis, L. 1970. *Origin of eukaryotic cells*. New Haven, Conn.: Yale University Press.

Margulis, L. 1981. *Symbiosis in cell evolution*. San Francisco: W. H. Freeman.

Pace, N. R., Stahl, D. A., Lane, D. J., and Olsen, G. J. 1985. Analyzing natural microbial populations by RNA sequences. *ASM News* 51(1):4–12.

Vidal, G. 1984. The oldest eukaryotic cells. *Scientific American* 250(2):48–57.

CHAPTER 13
THE BACTERIA

CHAPTER PREVIEW

THE ARCHAEBACTERIA: A SEPARATE KINGDOM?

Microbiologists are constantly searching for—and sometimes discovering—new microorganisms inhabiting our biosphere. The search is often directed toward the discovery of useful microorganisms, such as antibiotic producers, or of new pathogens. Much of this search is financially supported by industrial concerns or governmental agencies. Sometimes microorganisms are discovered which are extremely important in the understanding of biology and evolutionary processes. Such is the case with the archaebacteria. Some scientists claim that these bacteria represent one of the oldest forms of living organisms on the earth, and that they are different enough from the rest of the bacteria to warrant their own kingdom.

The archaebacteria are currently found dispersed throughout the various groups of bacteria described in Bergey's Manual and include the methanogens, the extreme halophiles, and the thermoacidophiles. These bacteria have in common a cell wall composed of proteinaceous materials that form mosaic patterns, rather than a cell wall of peptidoglycan fibers, like that found in all other walled bacteria. In addition, these bacteria have RNAs, DNAs, and proteins that differ from those of other bacteria. All this evidence has suggested to some scientists that the archaebacteria were once widely distributed in the primitive earth and that, as the environment became aerobic, they were restricted to those anaerobic environments conducive to their growth.

The archaebacteria are considered by some scientists to be a third form of life, sharing the biosphere with the other bacteria and the eukaryotes. The archaebacteria, cyanobacteria, and the eubacteria are believed to have originated from a now extinct, common prokaryotic ancestor called the **progenote**.

The study of bacteria often results in the discovery of new and exciting life forms that perform important functions in nature (as do the archaebacteria). Some of these functions are essential for the survival of the human race, while others are harmful. In this chapter we will study some of the prominent microbial groups and emphasize the role that they play in nature or in human affairs.

LEARNING OBJECTIVES

A STUDY OF THIS CHAPTER SHOULD ENABLE YOU TO:

DESCRIBE THE WAY IN WHICH BACTERIA ARE GROUPED IN BERGEY'S MANUAL

OUTLINE THE SALIENT FEATURES OF THE TWO DIVISIONS IN THE KINGDOM PROKARYOTAE

OUTLINE CHARACTERISTICS OF SOME OF THE MOST IMPORTANT GROUPS OF BACTERIA AND HOW THEY INFLUENCE HUMAN AFFAIRS

The bacteria are an extremely heterogeneous group of prokaryotic microorganisms. They inhabit very diverse environments in which they perform numerous functions, many of which are essential for life on earth. For example, the bacteria recycle nutrients in the biosphere, making them available to all living organisms. Bacteria also perform activities that are economically and medically valuable to humans such as the production of antibiotics and the fermentation of foods and beverages. Not all the bacteria carry out useful activities in nature. There are various groups of bacteria that are best known by their ability to cause disease in plants and animals.

In order to appreciate the impact that these microorganisms have on human affairs, it is necessary that we know a little about their biology. This chapter summarizes the characteristics of the major groups of bacteria, emphasizing those of medical or industrial importance.

BERGEY'S MANUAL OF SYSTEMATIC BACTERIOLOGY

Bergey's manual of systematic bacteriology, also called **Bergey's Manual** for short, is a comprehensive treatise on the classification of bacteria. It is used extensively by bacteriologists as a reference source because it classifies bacteria into more or less logical groups. The first edition of Bergey's Manual, published in 1923, classified the bacteria into conventional taxonomic hierarchies (kingdom, phylum, class, order, family, etc.). This approach was continues in the succeeding six editions. In the eighth edition, however, the conventional taxonomic approach was abandoned and a scheme of classification based on a few readily-determined characteristics was introduced. The classification of bacteria in the eighth edition of Bergey's Manual is summarized in table 13-1. The bacteria were classified into 19 parts, based on features such as gram reaction, cell shape and arrangement, metabolic and nutritional capabilities, and oxygen requirements. The cyanobacteria (blue-green algae) are largely ignored in the eighth edition of Bergey's Manual. The ninth edition retains the utilitarian classification approach and subdivides the bacteria into four groups, each discussed in a separate volume: (1) the gram-negatives of general, medical, or industrial importance; (2) the gram positives and other than actinomycetes; (3) the archaeobacteria, cyanobacteria, and remaining gram-negatives; and (4) the actinomycetes.

The utilitarian classification scheme used in the eighth and ninth editions of Bergey's Manual reflects the difficulty microbiologists have encountered in placing bacteria in genetically-related groups, especially in the higher hierarchical taxa such as orders, classes, or phyla. As presently prepared, Bergey's Manual is a descriptive compilation of bacterial species, taxonomic keys to the various genera in families, and practical tables to identify species. All the bacteria are classified in one of four divisions within the kingdom Procaryotae, based on the chemical nature and the presence of the cell wall (see Appendix B).

THE CYANOBACTERIA (BLUE-GREEN ALGAE)

The **cyanobacteria** is a group of photoautotrophs □ that share the ability to release molecular oxygen during photosynthesis. These microorganisms 194

TABLE 13-1

SUMMARY OF THE VARIOUS GROUPS OF BACTERIA ACCORDING TO THE 8TH EDITION OF BERGEY'S MANUAL OF DETERMINATIVE BACTERIOLOGY

PART	NAME OF GROUP	BRIEF DESCRIPTION OF GROUP	SAMPLE GENERA
1	Phototrophic bacteria	Gram negative spheres or rods. Carry out anoxygenic photosynthesis. Colonies pigmented red, orange, green, purple.	*Chromatium, Chlorobium, Rhodospirillum*
2	Gliding bacteria	Gram negative rods surrounded by a slime coat. May aggregate to form fruiting bodies. Move on solid surfaces by gliding.	*Beggiatoa, Cytophaga, Stigmatella*
3	Sheathed bacteria	Gram negative rods arranged in chains enclosed within a sheath.	*Sphaerotilus, Leptothrix*
4	Budding and/or appendaged bacteria	Gram negative rods, cocci or vibrios that divide by budding or binary fission. Division products are dissimilar.	*Caulobacter, Hyphomicrobium Gallionella*
5	Spirochetes	Gram negative, slender, coiled cells with flexible walls. Divide by transverse fission.	*Treponema, Borrelia, Leptospira*
6	Spiral and curved bacteria	Gram negative, helically-curved rods with rigid cell walls.	*Bdellovibrio, Campylobacter*
7	Gram negative, aerobic rods and cocci	Generally motile organisms that carry out aerobic respiration. Many species are pathogens.	*Pseudomonas, Brucella, Azotobacter*
8	Gram negative facultatively anaerobic rods	Straight and curve rods and vibrios that carry out various types of fermentation. Many species are motile by flagella.	*Escherichia, Vibrio, Salmonella, Shigella*
9	Gram negative anaerobic rods	Obligate anaerobes that tend to be pleomorphic. No endospores formed.	*Bacteroides, Fusobacterium, Desulfovibrio*
10	Gram negative cocci and coccobacilli	Spherical organisms that tend to cluster into parts and chains. Facultative anaerobes or aerobes.	*Neisseria, Acinetobacter*
11	Gram negative anaerobic cocci	Spherical cells with anaerobic type of metabolism.	*Veillonella*
12	Chemolithotrophic bacteria	Gram negative rods and cocci. Use inorganic substances as energy sources.	*Nitrobacter, Nitrosomonas, Thiobacillus*
13	Methanogenic bacteria	Gram positive and gram negative anaerobes that produce methane during metabolism.	*Methanosarcina, Methanococcus*
14	Gram positive cocci	Spherical bacteria arranged in clusters and chains. Aerobes, anaerobes, or facultative anaerobes.	*Staphylococcus, Streptococcus, Micrococcus*
15	Endospore forming rods and cocci	Gram positive bacteria that form endospores. Aerobes, anaerobes, and facultative anaerobes.	*Bacillus, Clostridium, Sporosarcina*
16	Gram positive asporogenous rods	Aerobes, anaerobes, and facultative anaerobes.	*Listeria, Lactobacillus*
17	Actinomycetes and related organisms	Gram positive rods that tend to form filaments. Some species are acid-fast.	*Mycobacterium, Nocardia, Corynebactierum*
18	Rickettsias	Gram negative bacteria with typical cell walls. Most species are obligate intracellular parasites. Chlamydiae lack peptidoglycan in their cell walls.	*Chlamydia, Rickettsia, Coxiella*
19	Mycoplasmas	Bacteria without cell walls. Highly pleomorphic. Aerobes and anaerobes.	*Mycoplasma, Spiroplasma*

SOURCE: From *Bergey's Manual of Determinative Bacteriology,* 8th ed. Copyright © 1983 by Williams & Wilkins Co., Baltimore. Reprinted by permission.

are a highly diversified group, consisting of various morphological types (fig. 13-1). The cyanobacteria represent the oldest living ancestors of the eukaryotic algae. Their appearance on the early earth about 3 billion years ago, and their subsequent success, altered the earth significantly. These cyanobacteria produced most of the oxygen found in the developing atmosphere. The increase in atmospheric oxygen to about 0.6%, approximately 1.5 billion years ago, served as a stimulus for the evolution of aerobic organisms.

All cyanobacteria have intracytoplasmic networks of membranes called **thylakoids** (fig. 13-2). These membranes are distributed throughout the cytoplasm and contain many of the enzyme systems involved in photosynthesis. The cyanobacterial thylakoids also contain chlorophyll *a*. Associated with the thylakoids of the cyanobacteria are structures called **phycobilisomes**, which function as light receptor molecules during photosynthesis.

Cells of some species of cyanobacteria remain attached to each other and thus form filaments called **trichomes** (fig. 13-1). Individual cyanobacterial cells are held together by a mucilaginous sheath.

Filamentous cyanobacteria often fragment into shorter gliding filaments called **hormongonia** (fig. 13-1). Hormogonia develop within the trichome and are released when the trichome fragments. The hormogonia show a gliding motion that allows them to disperse to and colonize other habitats. Cyanobacteria also form **akinetes**. These are resting cells that differentiate

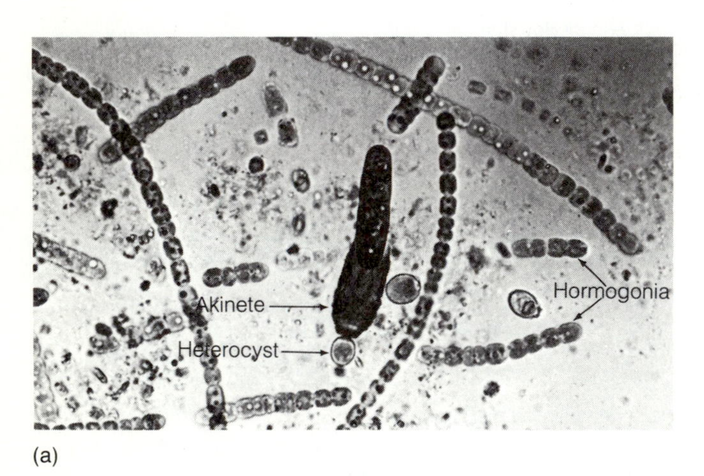

(a)

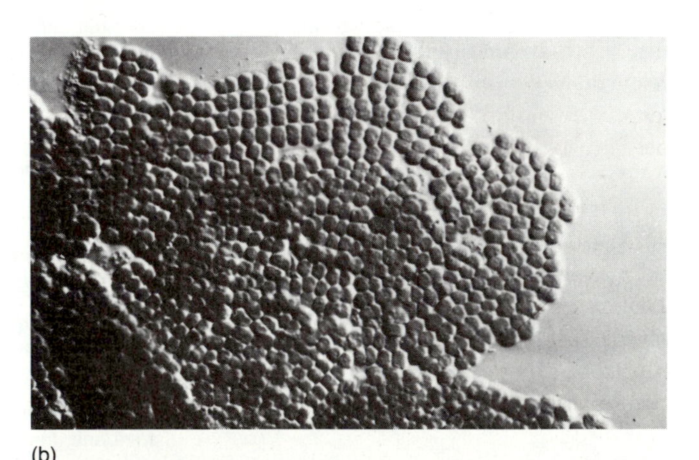

(b)

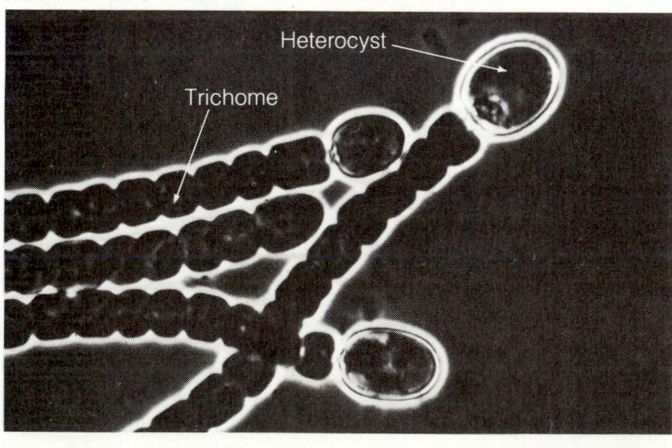

(c)

Figure 13-1 Various Species of Cyanobacteria. *(a) Calothrix* trichomes (filaments), exhibiting hormogonia, a heterocyst, and an akinete. *(b) Gloeocapsa.* The spherical cells are enclosed within a gelatinous sheath. *(c) Anabaena,* a filamentous cyanobacterium that forms heterocysts.

from vegetative cells by a developmental process that thickens the cell wall and fills the cytoplasm with reserve materials. The akinete protects the organism from stressful environmental changes, such as desiccation and freezing. **Heterocysts** (fig. 13-1) are also commonly seen in filamentous cyanobacteria. These structures are round cells with thick, transparent cell walls. Heterocysts arise from vegetative cells and are found at the tips of filaments or distributed along the trichomes. The heterocysts are thought to be a site where nitrogen fixation takes place.

The cyanobacteria inhabit a wide range of aquatic and terrestrial environments. Many species of cyanobacteria are tolerant to extreme environments, such as those found in alkaline hot springs (where the temperature reaches 74°C) and desert soils.

The cyanobacteria exert a significant influence on human affairs. Their growth in drinking water can impart "earthy" odors and undesirable flavors to the water. Some species of cyanobacteria, such as *Anabaena flos-aquae* and *Lyngbya majuscula*, produce metabolic products that are toxic to humans and animals. For example, the toxin produced by *L. majuscula* can cause skin irritations and dermatitis in humans and other primates.

The cyanobacteria are often the primary colonizers of bare soils, and in this way they contribute to soil fertility. Sometimes they form large mats on certain soils, releasing large amounts of organic nutrients into the soil and consequently increasing the soil's fertility. The cyanobacteria have also been used to fertilize rice paddies. Species like *Anabaena*, which can fix nitrogen from the atmosphere, are allowed to grow in rice paddies, where they fix nitrogen and hence increase the nitrogen content of the paddy. These cyanobacteria represent an inexpensive and natural source of fertilizer that significantly enhances rice yield. Some cyanobacteria, particularly *Spirulina*, have been grown on a large scale and are sold as a food supplement.

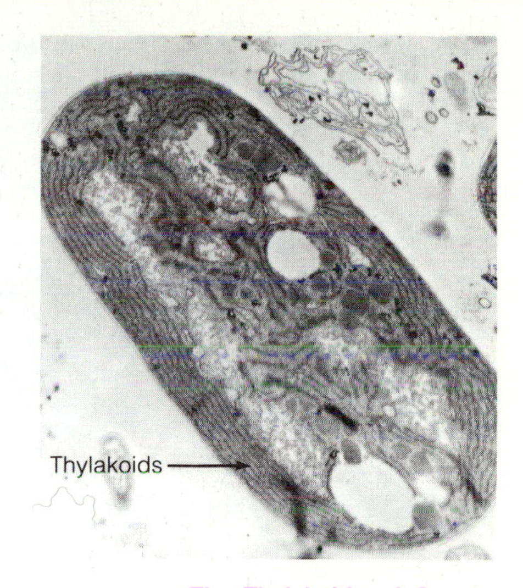

Figure 13-2 The Thylakoids of Cyanobacteria. Thylakoids are intracytoplasmic membranes where photosynthesis takes place. Electron photomicrograph of *Anabaena.* The membranous structures layered around the cell are the thylakoids.

THE BACTERIA

The bacteria are found in nearly every conceivable habitat in the biosphere, because they can tolerate a wide variety of environmental conditions and use a wide variety of substrates as carbon and energy sources. Many of these carbon and energy sources are synthetic (manmade) and toxic.

Bacterial activities affect humans in many ways. For example, pathogenic (disease-causing) bacteria are of importance because they can destroy crops or kill domestic animals, as well as cause disease and death in human populations. In addition, the metabolic activities of bacteria sometimes release substances that are economically useful. This is the case with antibiotic-producing bacteria. The sale of antibiotics brings billions of dollars in revenues to pharmaceutical companies, while their use prevents much disease and death due to infections. Bacterial metabolic activities are also important because they affect most aspects of the earth's ecology 740 □. The following sections will familiarize you with the various groups of bacteria so that you will have a basic knowledge of the bacteria discussed in future chapters.

Gram Negative Bacteria of Medical or Applied Importance

This is a broad group that includes primarily human pathogens (disease-causing bacteria). This group contains the spirochetes, helical and curved

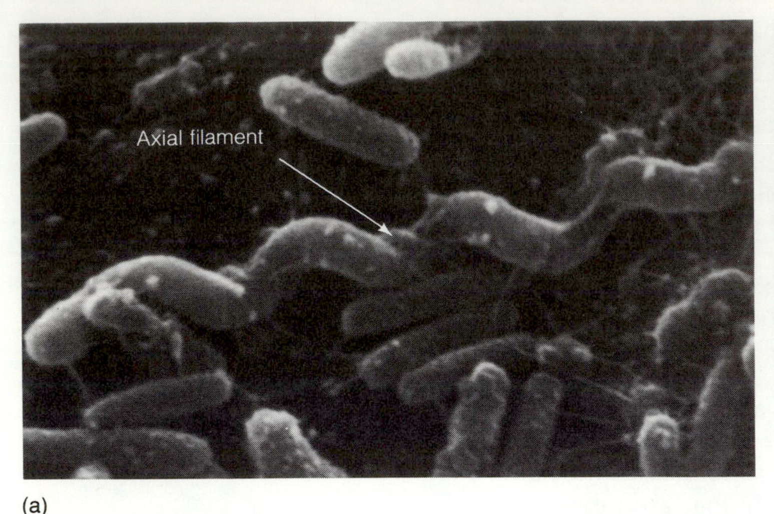

(a)

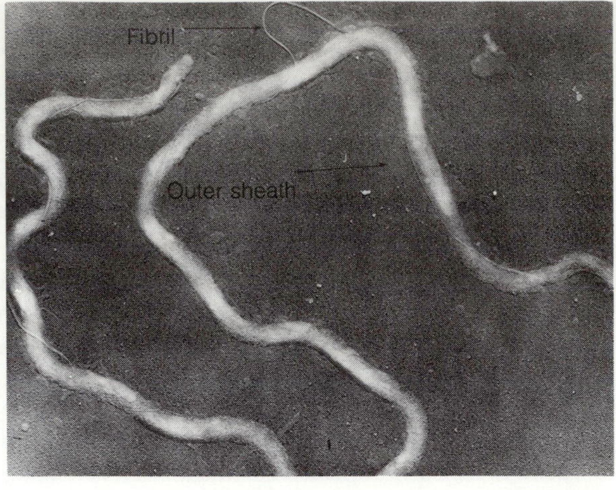

(b)

Figure 13-3 Structural Features of the Spirochetes. *(a)* Scanning electron micrograph of a spirochete showing the axial filament wound around the body of the cell. *(b)* Transmission electron micrograph of a palladium-platinum shadowed spirochete exhibiting a fibril and the outer sheath. *(c)* Transmission electron micrograph of *Cristispira.* Note the numerous axial fibrils running lengthwise along the cell.

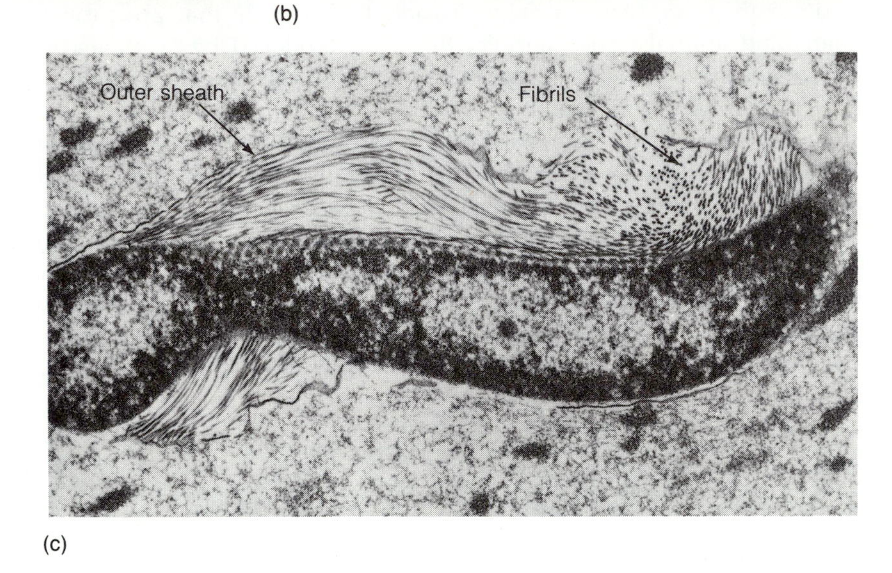

(c)

bacteria, aerobes, anaerobes and facultative anaerobes, rickettsieas, and mycoplasmas.

Spirochetes

The **spirochetes** (fig. 13-3.) are gram negative, slender, flexible bacteria that are helically coiled. Different genera of spirochetes show considerable variability in their lenght, width, and amount of coiling. The human pathogenic species *Treponema pallidum* and *Leptospira interrogans* measure approximately 0.1 μm in diameter and 5–18 μm in length (table 13-2). These organisms are generally invisible using routine light microscopic techniques □ and special procedures are required in order to view them. In dark field preparations, the spirochetes can be seen in tight coils, flexing and bending in a characteristic fashion.

Modified flagella called **fibrils** allow these bacteria to be motile. The fibrils may be arranged in a bundle known as the **axial filament** (fig. 13-3). The fibrils originate near the two ends of the cell and fold back along the cell, between the peptidoglycan layer and an **outer sheath** (fig. 13-3). The

TABLE 13-2
SUMMARY OF CHARACTERISTICS OF MEDICALLY-IMPORTANT SPIROCHETES

CHARACTERISTIC	*BORRELIA*	*LEPTOSPIRA*	*TREPONEMA*
Length of cell (μm)	5–30	5–18	6–15
Width of cell (μm)	0.4	0.1	0.2
Coils			
Number	<10	30–40	5–20
Arrangement	Irregular	Tightly wound	Regular
Number of flagella	15–20	1	3
O_2 requirement	Anaerobic	Aerobic	Anaerobic
Energy metabolism	Fermentative	Respiratory	Fermentative
Cultivated *in vitro*	Yes	Yes	No
Diseases caused	Vincent's angina, Lyme's disease	Leptospirosis	Syphilis, Yaws

fibrils are not free to whirl about like flagella of other motile bacteria, because they are contained within the outer sheath and wound around the cell. As the fibrils try to rotate, they make the cell spin like a corkscrew, imparting a characteristic form of motility to the cell.

Some species of spirochetes can cause disease in humans. For example, *Treponema pallidum* causes the disease syphilis; *Borrelia recurrentis* causes relapsing fever; and *Leptospira interrogans* causes leptospirosis or Weil's disease. These diseases are of great public health importance and will be discussed in Part III.

Spiral and Curved Bacteria

The **spiral and curved bacteria** are rigid, helically-curved or S-shaped rods (fig. 13-4). Many species of these gram negative bacteria are motile by means

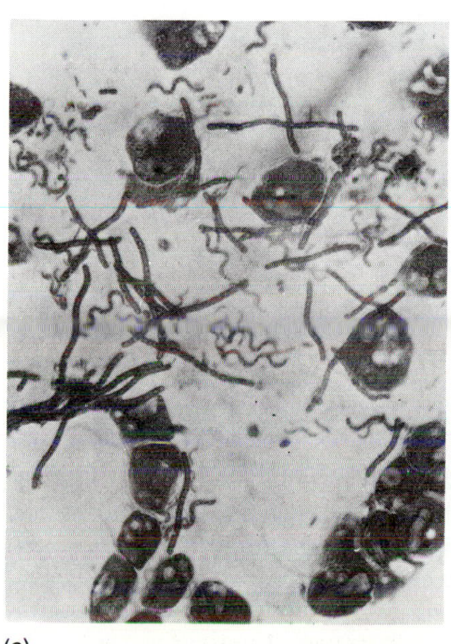

(a)

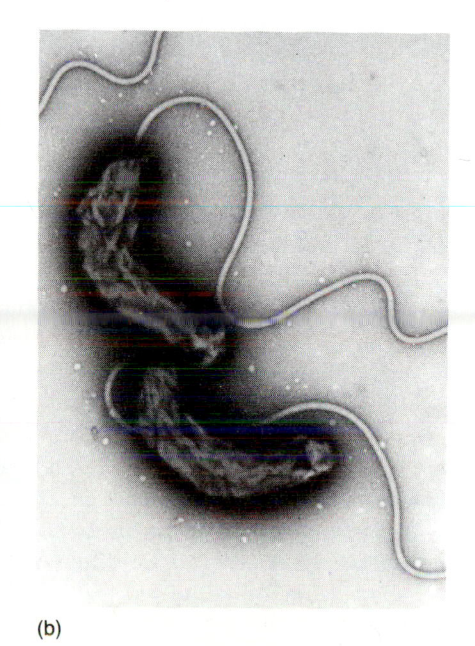

(b)

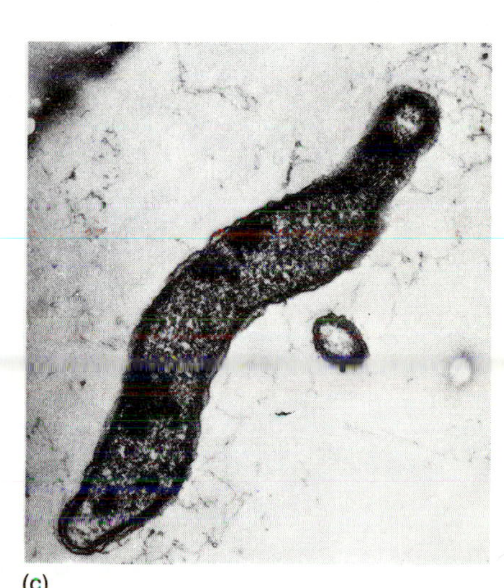

(c)

Figure 13-4 Various Species of Spiral and Curved Bacteria. *(a) Spirillum. (b) Bdellovibrio bacteriovorus. (c) Campylobacter.*

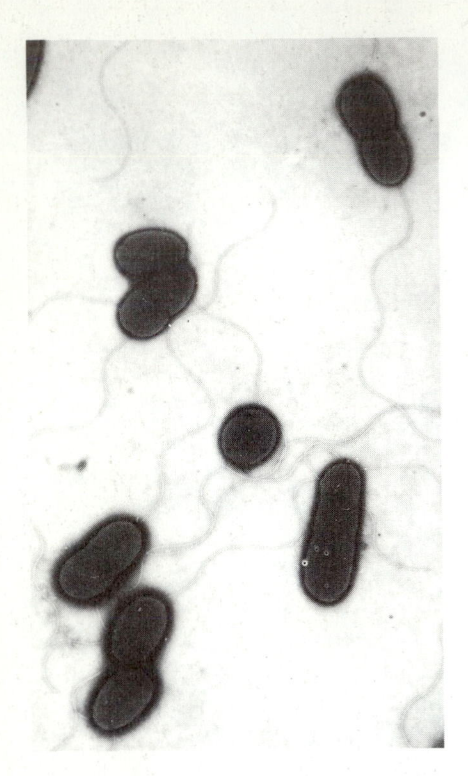

Figure 13-5 ***Pseudomonas,* a Typical Gram Negative, Aerobic Bacterium.** A TEM of *Pseudomonas aeruginosa* stained with phosphotungstic acid.

of polar flagella. Some members of this broad group parasitize mammals and even other bacteria. They are readily distinguished from the spirochetes because they are rigid and their flagella are not enclosed by a sheath □. 84

Campylobacter has recently been recognized to include species pathogenic to humans. Some other members of the genus, however, like *C. sputorum,* are found in the human oral cavity in the absence of disease. *Campylobacter jejuni* has been found to infect humans and to cause a variety of diseases, such as enteritis (inflammation of the intestines), subacute bacterial endocarditis (inflammation of the heart), meningitis (inflammation of the membranes surrounding the brain and spinal cord), and septicemia (invasion of the blood or tissues). Recent studies have revealed that *C. jejuni* is one of the most common causes of gastroenteritis in human populations. Its role in the disease became apparent when microbiologists started culturing fecal specimens on a routine basis at 40–42°C in an elevated CO_2 atmosphere. These cultural conditions are ideal for the cultivation of *Campylobacter,* and until this fact was broadly recognized, the role of this bacterium in human disease was largely unknown.

Gram Negative Aerobic Rods

The **gram negative aerobic rods** are all chemoheterotrophs. These bacteria are remarkably versatile in their ability to assimilate organic materials and to perform biochemical activities.

Pseudomonas aeruginosa is an important member of this group (fig. 13-5). Because it causes disease only when hosts are weakened or compromised, *P. aeruginosa* is considered to be an **opportunistic** □ pathogen. 427 This bacterium is commonly found in moist environments and in many hospitals. *Pseudomonas aeruginosa* is of major concern in hospitals because it is capable of aggravating surgical wounds or initiating pulmonary (lung) infections in debilitated patients. This organism grows well on water faucet aerators, so washing may contaminate materials such as oxygen tent air hoses, ventilators, nebulizers, and nursery incubators. If these pieces of equipment are not thoroughly dried after washing, contaminating bacteria may proliferate and increase the likelihood of an infection. Other pathogenic pseudomonads include *Pseudomonas cepacia,* which causes **onion bulb rot;** *P. mallei,* which causes the respiratory tract disease **glanders,** which affects horses and humans; and *P. pseudomallei,* which causes **melioidosis,** another disease of the respiratory tract.

Other medically-important bacteria in this group include *Bordetella, Franciscella, Brucella,* and *Legionella. Bordetella pertussis* is an important pathogen of humans. This short, gram negative rod causes whooping cough, a disease afflicting primarily children. Until relatively recently, whooping cough was prevalent in human populations. With the advent of a vaccine, however, the incidence of this disease in children has been significantly reduced.

Legionella pneumophila is an aerobic, gram negative bacterium that causes a pneumonia (lung inflammation with fluid accumulation) better known as Legionnaire's disease. The bacterium responsible for the 1976 outbreak of Legionnaire's disease in Philadelphia has recently been recognized as a common cause of pneumonia in human populations. *Legionella pneumophila* can exist in many different acquatic environments. The source of the bacterium that infected the legionnaires in Philadelphia was contaminated water used to cool the air in the hotel's air conditioning system.

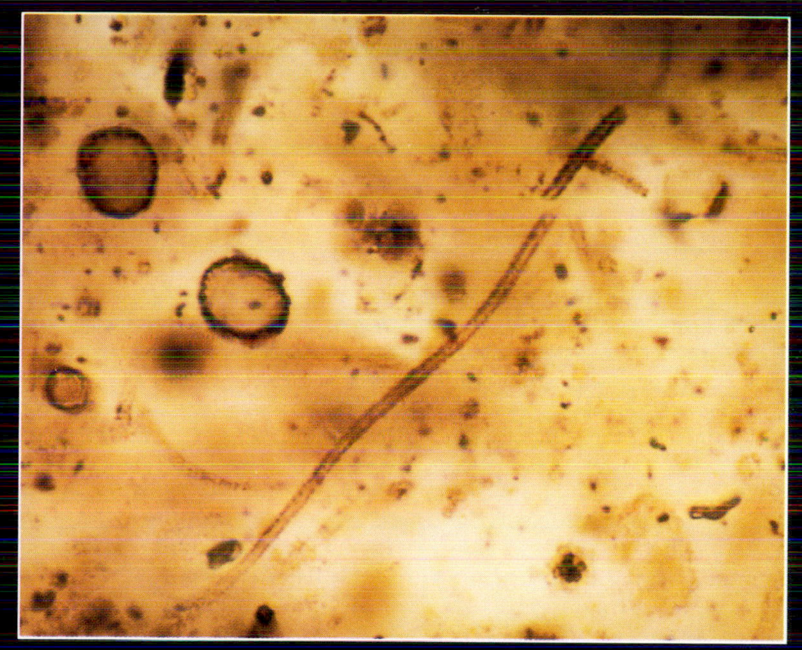
Fossil prokaryotes in gunflint chert.

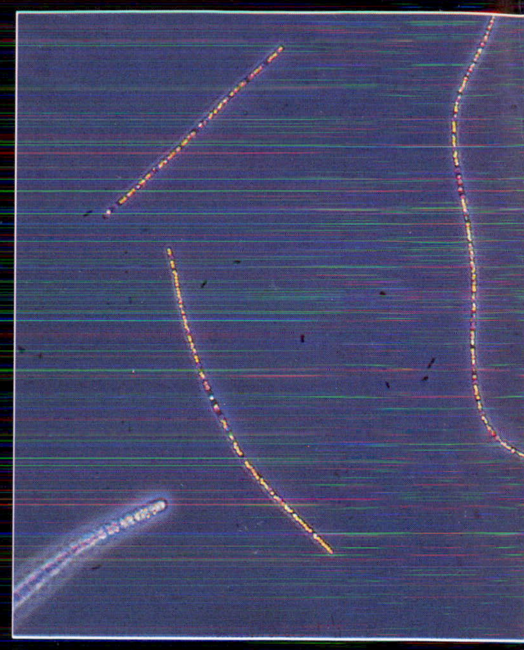
The bacterium *Thiothrix* obtained from marine muds.

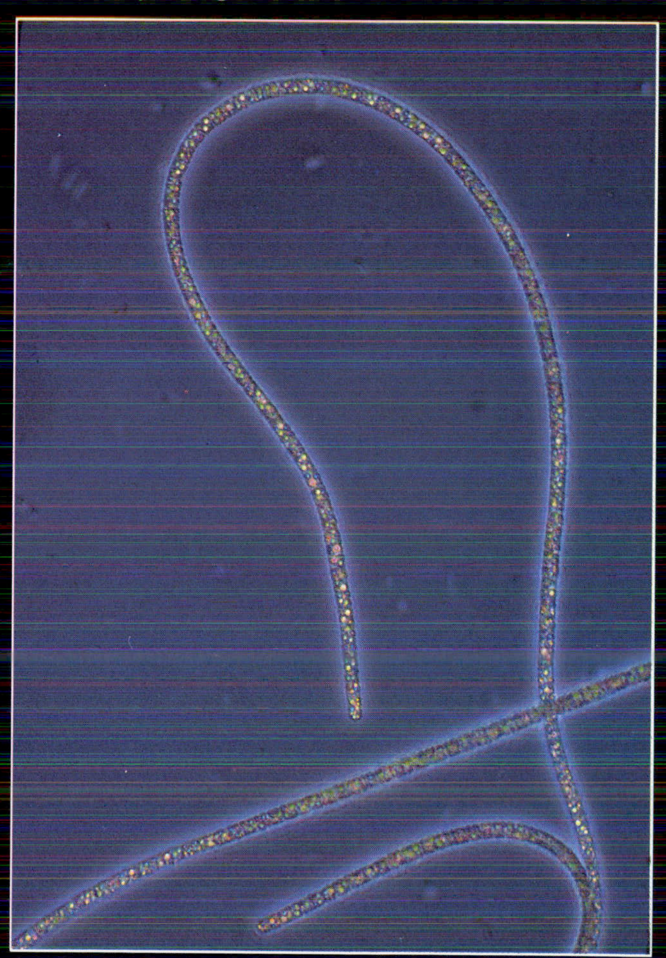
The bacterium *Beggiatoa* with sulfur crystals throughout its cytoplasm.

Phase contrast of a colony of *Beggiatoa*.

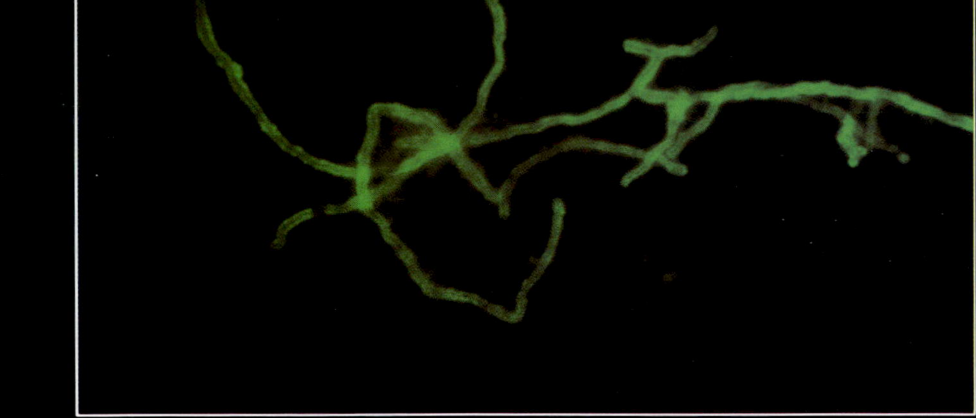

Gram positive cocci (*Staphylococcus aureus*) and gram negative rods (*Escherichia coli*).

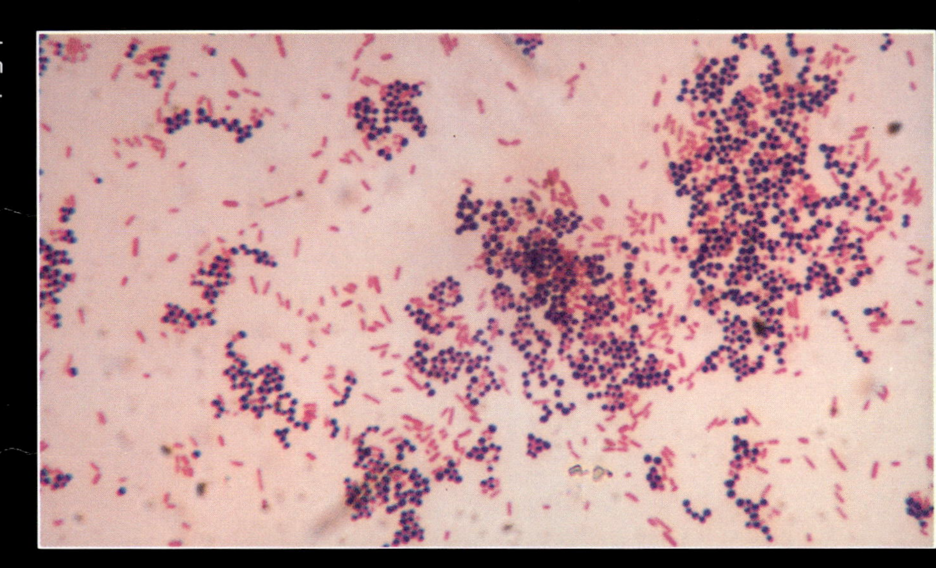

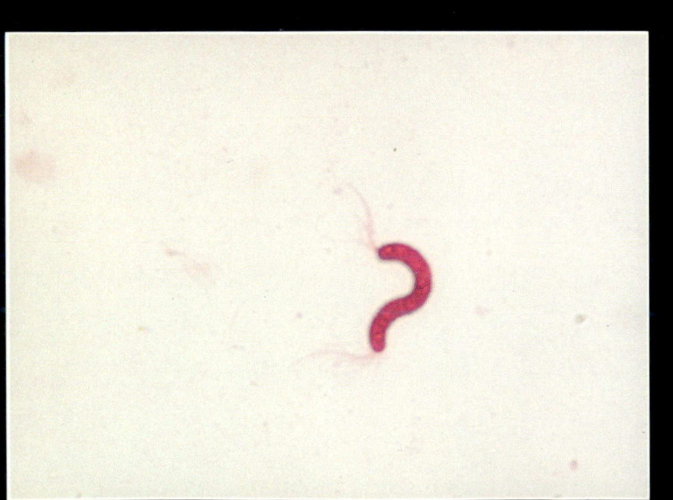

Flagellar stain of *Spirillum* showing spiral-shaped

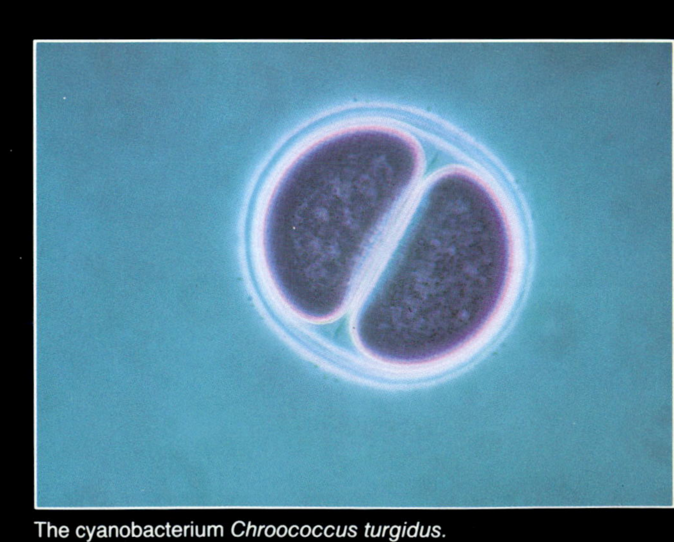

The cyanobacterium *Chroococcus turgidus*.

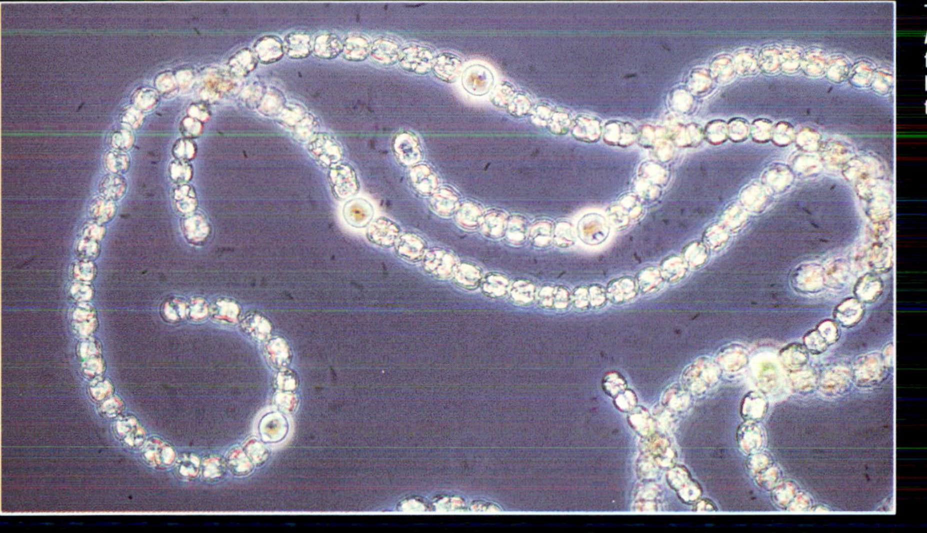

The cyanobacterium *Anabaena flos-aquae*. The filament contains several heterocysts (round, thick-walled structures).

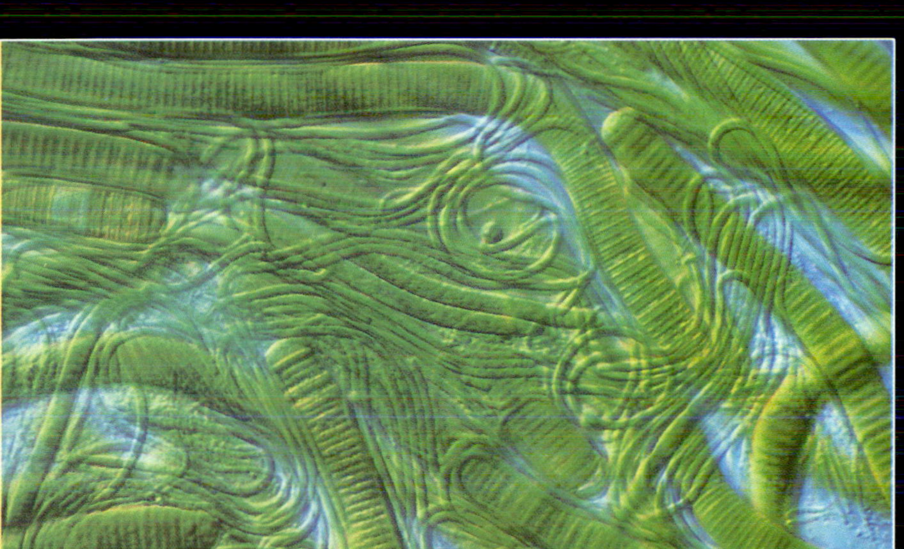

Trichomes of the cyanobacterium *Lyngbia* obtained from a bloom on the coast of Texas.

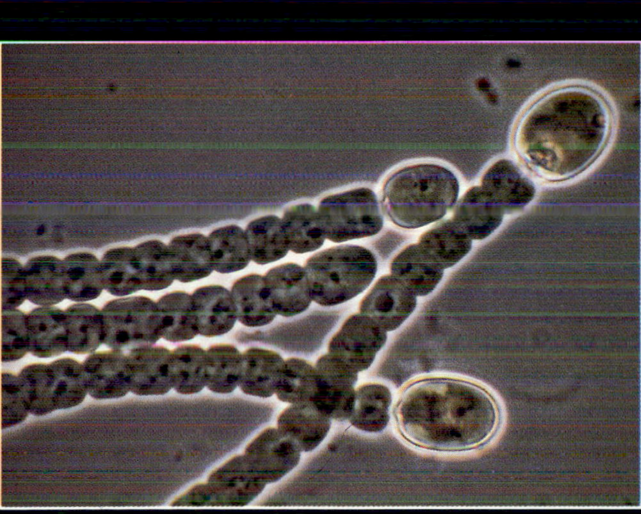

A cyanobacterium with terminal heterocysts.

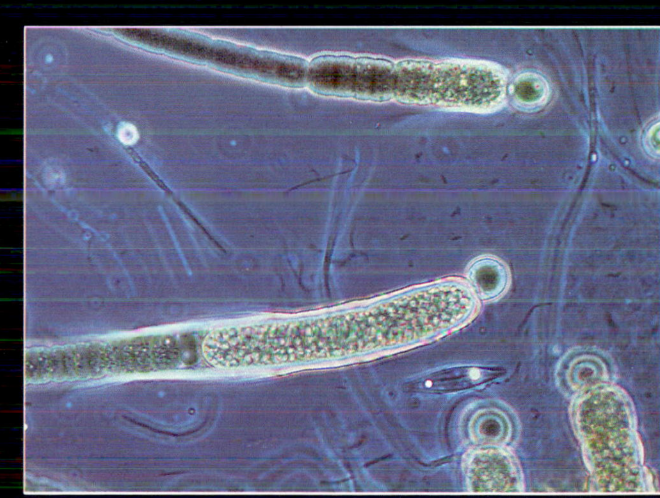

The cyanobacterium *Gloeotrichia* with subterminal akinetes and terminal heterocysts.

A crustose lichen.

A crustose lichen.

The British soldier lichen.

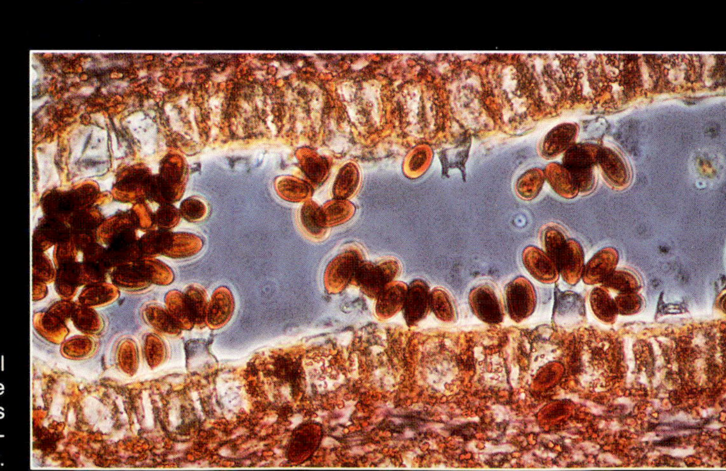

Stained section of a gill of the basidiomycete *Coprinus*. Numerous basidiospores and the club-shaped basidia are evident.

Basidiocarps of *Coprinus micaceous* (the inky cap).

The ascomycete *Morchella esculenta*, a common morel.

Basidiocarp of the bracket mushroom *Polyporus*.

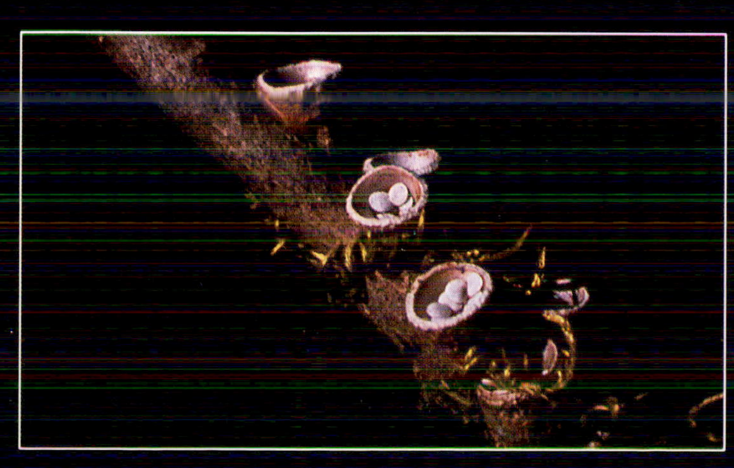

Bird's nest fungi.

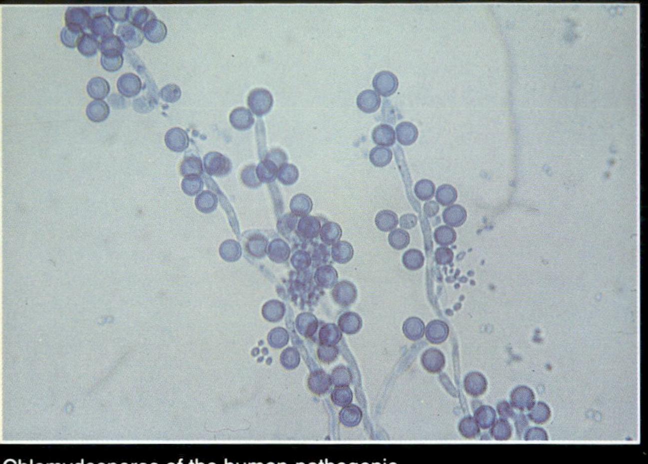

Chlamydospores of the human-pathogenic yeast *Candida albicans.*

Mycelial stage of the human-pathogenic fungus *Coccidioides immitis.* The spores, when inhaled, can cause a severe lung infection.

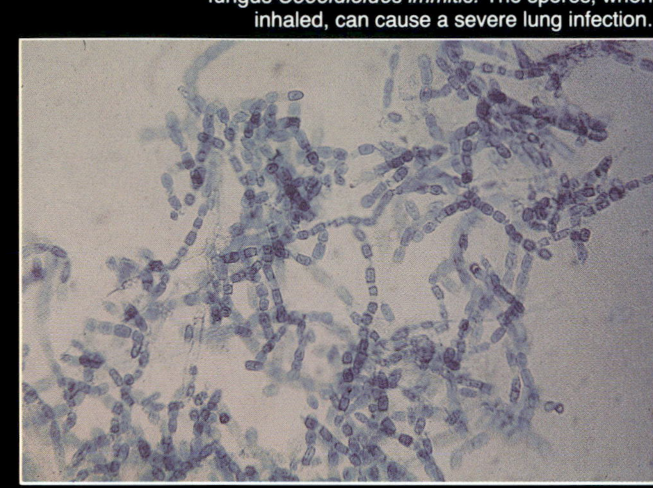

Macroconidia of *Microsporum gypseum.* This fungus can cause very severe cases of ringworm of the scalp in humans.

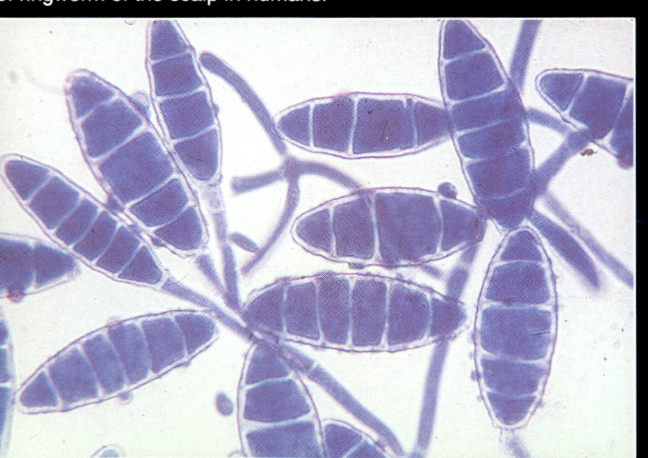

Parasitic phase of *Coccidioides immitis.* The endosporulating spherules are filled with endospores, which perpetuate the infection within the host.

Conidial head of the Deuteromycete *Aspergillus fumigatus*. This fungus can cause pulmonary diseases in humans.

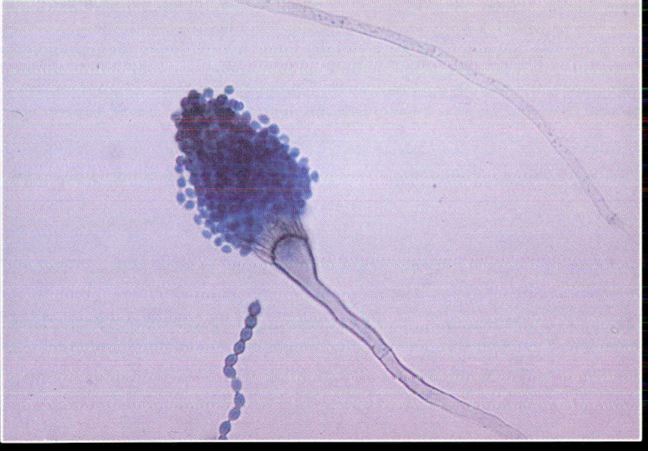

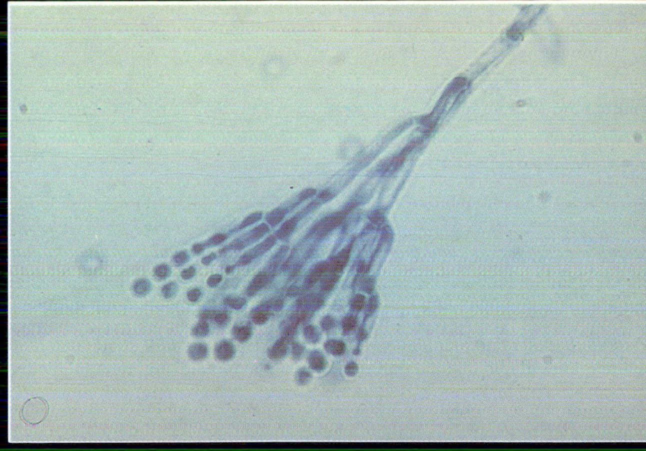

Typical conidial head of *Penicillium*.

Fruiting bodies of the slime mode *Hemitrichia*.

Plasmodium of the Myxomycete *Physarum*.

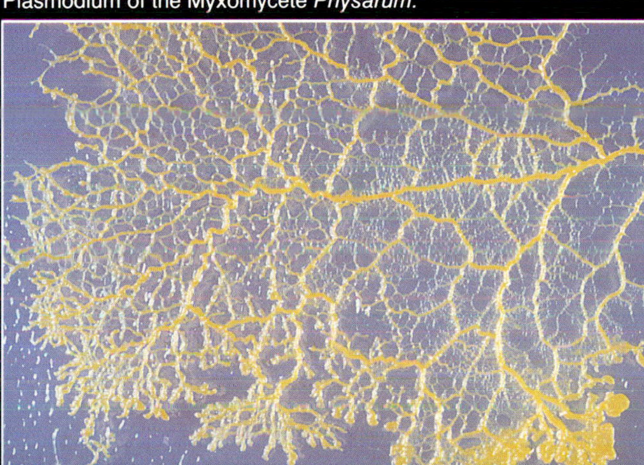

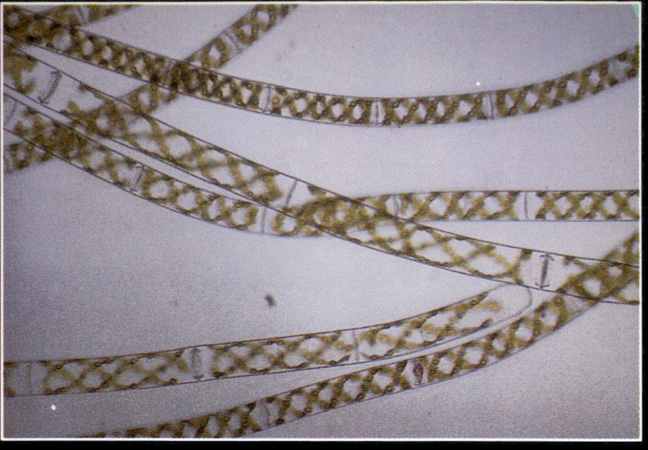
The green alga *Spirogyra*.

The green alga *Scenedesmus*.

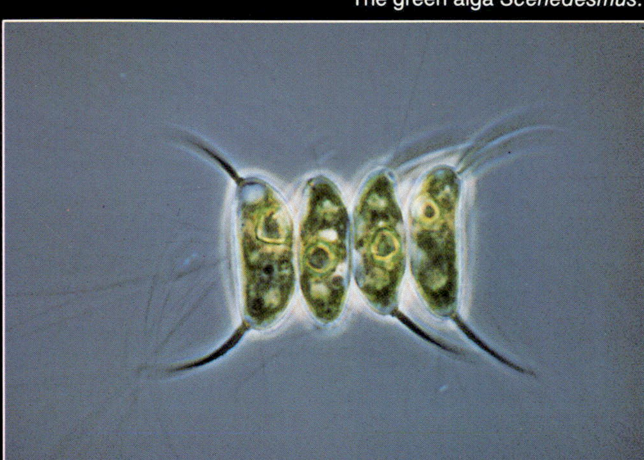

Large, motile colony of the alga *Volvox aureus*.

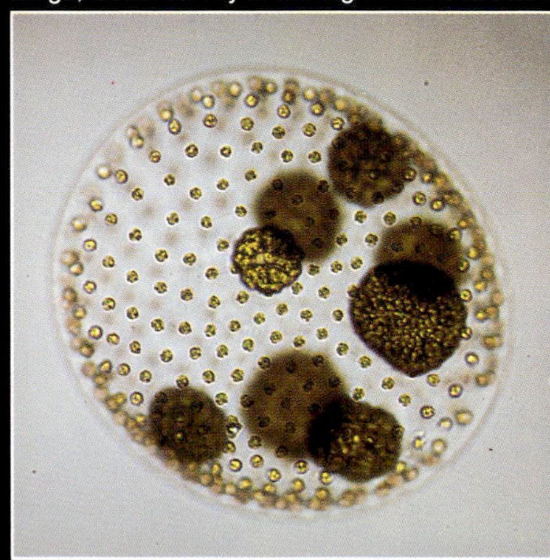

The diatom *Astrolampra affinis*.

Nomarski interference of marine ciliates showing numerous internal structures and cilia covering the surface.

FOCUS

PSEUDOMONAS HELPS PREVENT PLANT DISEASE

Pseudomonas is a gram negative rod that is widely distributed in nature. These bacteria have a remarkable ability to assimilate a wide variety of organic substrates, and hence can reproduce in many different environments. Certain species of *Pseudomonas* can extensively colonize plant root surfaces. It has been noted that plants that have these bacteria colonizing their roots are less likely to acquire certain diseases than plants devoid of *Pseudomonas*. Plant pathogens like the fungi *Fusarium*, *Pythium*, and *Phytophthora* are inhibited, or prevented, from causing disease in these plants by the growth of *Pseudomonas*.

The reason that *Pseudomonas* suppresses the growth of plant pathogens is that it produces iron-binding compounds known as **siderophores.** The siderophores bind iron and sequester it from the area around the root. In the absence of available iron, plant pathogens cannot reproduce.

The discovery of *Pseudomonas* sp. that protect plants from disease is one of the many examples supporting the fact that bacteria perform useful functions in nature.

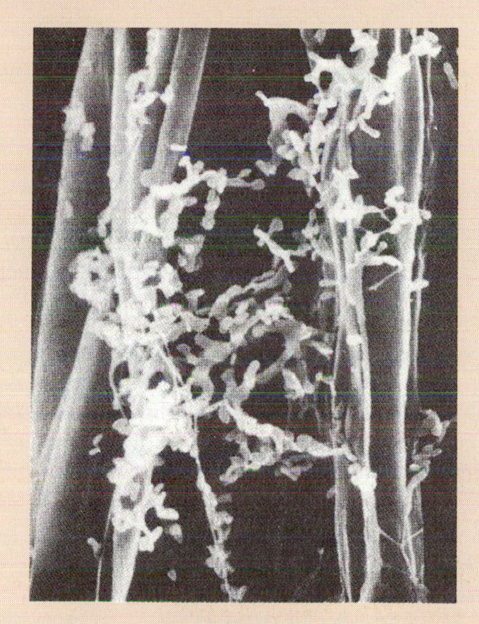

Figure 13-6 ***Pseudomonas*** **Attached to the Roots of a Plant Host**

Bacteria in the genera *Azotobacter* and *Rhizobium* are of applied importance because they carry out a process called nitrogen fixation. These gram negative, aerobic rods are soil inhabitants that contribute to soil fertility by converting molecular nitrogen, a form of nitrogen largely unusable to living organisms, into organic nitrogen, which can readily be used by most living organisms. These microorganisms will be discussed in Part IV.

Gram Negative, Facultatively Anaerobic Rods

The **gram negative, facultatively anaerobic rods** are a large group of chemoheterotrophic 109 □ bacteria that are widely distributed in nature (table 13-3) and includes many important human pathogens. These bacteria are currently grouped into two families: the Enterobacteriaceae and the Vibrionaceae. The two families are differentiated largely on the basis of biochemical and morphological features.

The **enterics,** or the Enterobacteriaceae family, are straight rods that characteristically inhabit the intestines of humans and other animals. Some genera, however, can also be isolated from soils or from acquatic environments. The enterics include some very important pathogens such as the agents of typhoid *(Salmonella typhi)*, bacillary dysentery *(Shigella dysenteriae)*, and gastroenteritis *(Salmonella* and *Shigella)*. Although the enterics cause gastrointestinal disease predominantly, they are also capable of causing urinary tract infections, wound infections, pneumonia, meningitis, and septicemia. *Escherichia coli* has been implicated in many of these diseases.

TABLE 13-3
CHARACTERISTICS OF GRAM NEGATIVE, FACULTATIVELY ANAEROBIC BACTERIA

ORGANISM	SALIENT CHARACTERISTICS
Family: Enterobacteriacieae (enterics)	Usually straight rods.
Escherichia coli	Inhabit the gastrointestinal tract of mammals. May cause enteric (intestinal) diseases. Is used as an indicator of fecal contamination.
Salmonella *S. typhi* *S. enteritidis* *S. arizona* *S. paratyphi*	Pathogenic bacteria. Cause typhoid fever *(S. typhi)* and gastroenteritis. Common agents of food poisoning.
Shigella *S. dysenteriae* *S. sonnei* *S. flexneri* *S. boydii*	Agents of shigellosis or bacillary dysentery. Some species produce exotoxins. Inhabit the gastrointestinal tract of mammals. Transmitted in food and water.
Citrobacter freundii	Saprophyte in water, foods, and body wastes. May be associated with a variety of infections.
Klebsiella pneumoniae	Widely distributed in nature. Commensal in the intestinal tract of mammals. Can cause a variety of diseases including gastroenteritis, pneumonia, and urinary tract infections.
Enterobacter *E. aerogenes* *E. cloacae*	Inhabit the gastrointestinal tract of mammals. Can cause enteric and urinary tract infections.
Serratia marcescens	Opportunistic pathogen. Form red-colored colonies. Widely distributed in nature.
Proteus *P. vulgaris* *P. mirabilis* *P. inconstans*	Some species cause urinary tract infections, others cause diarrheas. All found in intestines of mammals.
Yersinia	
Y. pestis	*Y. pestis* is the agent of bubonic plague. It is found in nature infecting small feral rodents. It is transmitted to humans by fleas.
Y. enterocolitica	*Y. enterocolytica* causes an enteric disease called enterocolitis.
Family: Vibrionaceae	Usually curved rods.
Vibrio	
V. cholerae	Causes cholera. Organisms produce choleragen, a toxin that causes the disease. Found in water.
V. parahaemolyticus	Causes food poisoning. Shellfish is a common source of infection. Found in coastal marine environments.
Aeromonas hydrophila	Associated with wound infections, meningitis and septicemia. Normally found in aquatic environments and marine life.
Plesiomonas shigelloides	May cause gastroenteritis. Found in aquatic environments.

FOCUS

PSEUDOMONAS HELPS PREVENT PLANT DISEASE

Pseudomonas is a gram negative rod that is widely distributed in nature. These bacteria have a remarkable ability to assimilate a wide variety of organic substrates, and hence can reproduce in many different environments. Certain species of *Pseudomonas* can extensively colonize plant root surfaces. It has been noted that plants that have these bacteria colonizing their roots are less likely to acquire certain diseases than plants devoid of *Pseudomonas*. Plant pathogens like the fungi *Fusarium, Pythium,* and *Phytophthora* are inhibited, or prevented, from causing disease in these plants by the growth of *Pseudomonas*.

The reason that *Pseudomonas* suppresses the growth of plant pathogens is that it produces iron-binding compounds known as **siderophores.** The siderophores bind iron and sequester it from the area around the root. In the absence of available iron, plant pathogens cannot reproduce.

The discovery of *Pseudomonas* sp. that protect plants from disease is one of the many examples supporting the fact that bacteria perform useful functions in nature.

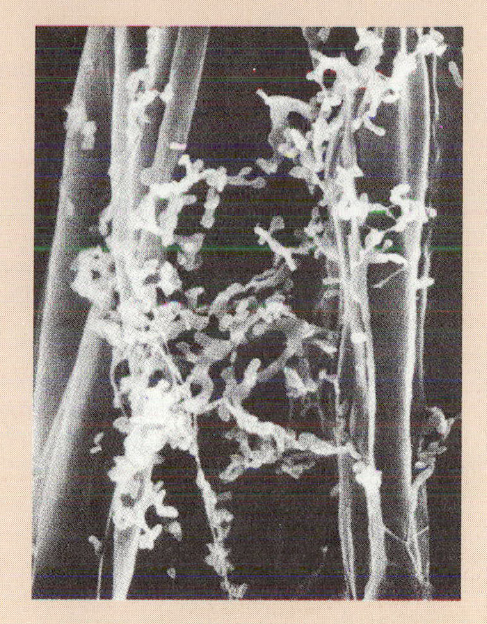

Figure 13-6 *Pseudomonas* Attached to the Roots of a Plant Host

Bacteria in the genera *Azotobacter* and *Rhizobium* are of applied importance because they carry out a process called nitrogen fixation. These gram negative, aerobic rods are soil inhabitants that contribute to soil fertility by converting molecular nitrogen, a form of nitrogen largely unusable to living organisms, into organic nitrogen, which can readily be used by most living organisms. These microorganisms will be discussed in Part IV.

Gram Negative, Facultatively Anaerobic Rods

The **gram negative, facultatively anaerobic rods** are a large group of chemoheterotrophic 109 □ bacteria that are widely distributed in nature (table 13-3) and includes many important human pathogens. These bacteria are currently grouped into two families: the Enterobacteriaceae and the Vibrionaceae. The two families are differentiated largely on the basis of biochemical and morphological features.

The **enterics,** or the Enterobacteriaceae family, are straight rods that characteristically inhabit the intestines of humans and other animals. Some genera, however, can also be isolated from soils or from acquatic environments. The enterics include some very important pathogens such as the agents of typhoid *(Salmonella typhi)*, bacillary dysentery *(Shigella dysenteriae)*, and gastroenteritis *(Salmonella* and *Shigella)*. Although the enterics cause gastrointestinal disease predominantly, they are also capable of causing urinary tract infections, wound infections, pneumonia, meningitis, and septicemia. *Escherichia coli* has been implicated in many of these diseases.

TABLE 13-3
CHARACTERISTICS OF GRAM NEGATIVE, FACULTATIVELY ANAEROBIC BACTERIA

ORGANISM	SALIENT CHARACTERISTICS
Family: Enterobacteriacieae (enterics)	Usually straight rods.
Escherichia coli	Inhabit the gastrointestinal tract of mammals. May cause enteric (intestinal) diseases. Is used as an indicator of fecal contamination.
Salmonella *S. typhi* *S. enteritidis* *S. arizona* *S. paratyphi*	Pathogenic bacteria. Cause typhoid fever *(S. typhi)* and gastroenteritis. Common agents of food poisoning.
Shigella *S. dysenteriae* *S. sonnei* *S. flexneri* *S. boydii*	Agents of shigellosis or bacillary dysentery. Some species produce exotoxins. Inhabit the gastrointestinal tract of mammals. Transmitted in food and water.
Citrobacter freundii	Saprophyte in water, foods, and body wastes. May be associated with a variety of infections.
Klebsiella pneumoniae	Widely distributed in nature. Commensal in the intestinal tract of mammals. Can cause a variety of diseases including gastroenteritis, pneumonia, and urinary tract infections.
Enterobacter *E. aerogenes* *E. cloacae*	Inhabit the gastrointestinal tract of mammals. Can cause enteric and urinary tract infections.
Serratia marcescens	Opportunistic pathogen. Form red-colored colonies. Widely distributed in nature.
Proteus *P. vulgaris* *P. mirabilis* *P. inconstans*	Some species cause urinary tract infections, others cause diarrheas. All found in intestines of mammals.
Yersinia	
Y. pestis	*Y. pestis* is the agent of bubonic plague. It is found in nature infecting small feral rodents. It is transmitted to humans by fleas.
Y. enterocolitica	*Y. enterocolytica* causes an enteric disease called enterocolitis.
Family: Vibrionaceae	Usually curved rods.
Vibrio	
V. cholerae	Causes cholera. Organisms produce choleragen, a toxin that causes the disease. Found in water.
V. parahaemolyticus	Causes food poisoning. Shellfish is a common source of infection. Found in coastal marine environments.
Aeromonas hydrophila	Associated with wound infections, meningitis and septicemia. Normally found in aquatic environments and marine life.
Plesiomonas shigelloides	May cause gastroenteritis. Found in aquatic environments.

Rickettsias

The **rickettsias** (fig. 13-8) constitute a group of pleomorphic rods or coccobacilli that generally give a gram negative reaction and have cell walls that are similar in composition to those of other gram negative bacteria. They are intracellular parasites □ of eukaryotes, with life cycles that usually include an arthropod (joint-legged invertebrate) vector □. One notable exception is *Coxiella burnetii*, which is not transmitted by vectors but rather carried in air currents. The rickettsias live in a state of mutual tolerance with their arthropod hosts and are transmitted to warm-blooded animals by the arthropod's bite. The rickettsias have their own metabolic enzymes and possess a complete Krebs cycle and electron transport system, but these bacteria apparently cannot synthesize NAD or coenzyme A and must obtain these compounds from their host. One important feature of the rickettsias is that their membranes are "leaky," that is, nutrients and metabolic products pass into and out of the cell through the plasma membrane much more freely than in other bacteria. The increased permeability of their membranes appears to be a necessary adaptation to intracellular parasitism, since they must obtain their growth factors from the host. The rickettsias have an affinity for the cells that constitute the endothelial lining (the inner lining of blood vessels) of capillaries. It is because of this affinity that all diseases caused by rickettsias (table 13-6) are characterized by a rash. The rash results when the rickettsial infection obstructs the small blood vessels of the skin causing lesions (the rash). The rickettsias can cause severe epidemics in human populations, and have influenced the outcome of wars by causing widespread disease and debilitating entire armies. Epidemic typhus caused by *Rickettsia prowazekii*, and Rocky Mountain Spotted Fever caused by *R. rickettsii*, have often accounted for the deaths of many thousands of humans.

429
536

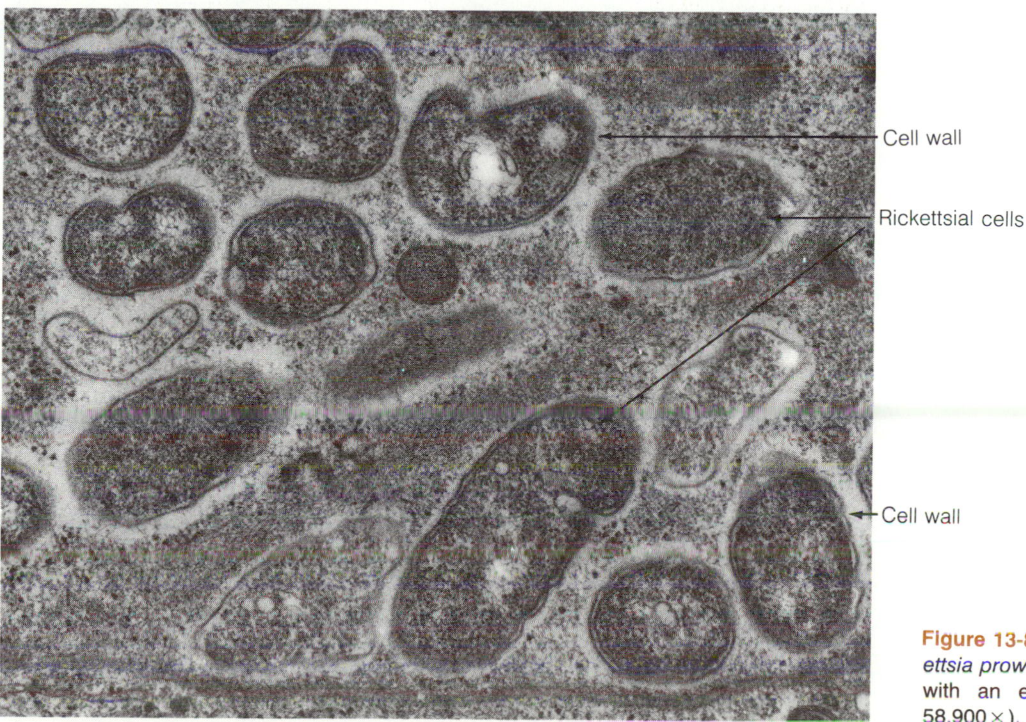

Figure 13-8 ***Rickettsia prowazekii.*** *Rickettsia prowazekii* in a chick embryo as seen with an electron microscope (magnified 58,900×).

TABLE 13-6
RICKETTSIAL DISEASES AND THEIR CAUSATIVE AGENTS

ORGANISM	DISEASE IT CAUSES	VEHICLE OF TRANSMISSION OR VECTOR*
Rickettsia rickettsii	Rocky Mountain Spotted Fever	Hard (ixodid) ticks
Rickettsia prowazekii	Epidemic typhus	Lice
Rickettsia typhi	Endemic typhus	Fleas
Rickettsia tsutsugamushi	Scrub typhus	Mites (trombiculid)
Rochalimaea quintana	Trench fever	Body louse
Coxiella burnetii	Q-fever	Arthropod? aerosol, food consumption

*Vector is an arthropod that is involved in the transmission of an infectious agent from one individual to another.

Chlamydias

The **chlamydias** (fig. 13-9) are very small gram negative bacteria that resemble the rickettsias in that they are obligate parasites. These microorganisms are much smaller than the rickettsias, however, and have a life cycle that involves two different morphological types: the infectious stage or **elementary body** and the vegetative stage or **reticulate body** (fig. 13-9). The chlamydia are unable to synthesize their own ATP and hence are sometimes

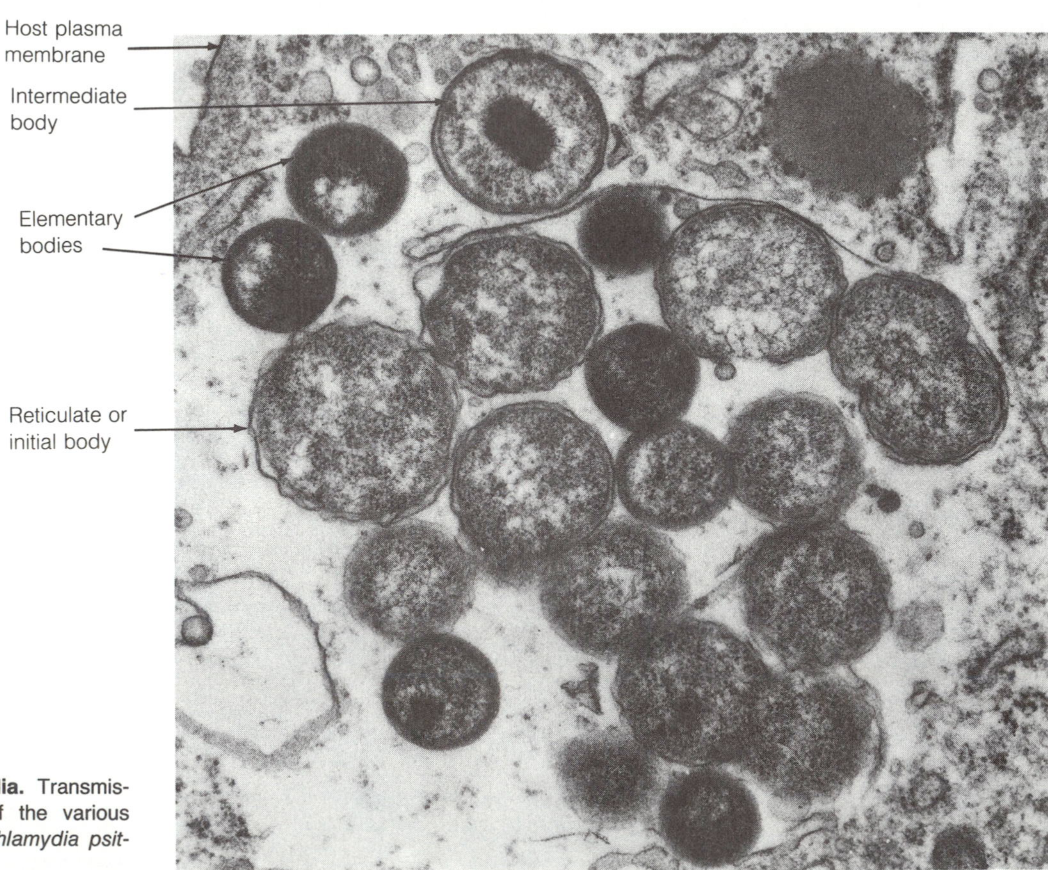

Figure 13-9 The Chlamydia. Transmission electron micrograph of the various stages in the life cycle of *Chlamydia psittaci.*

TABLE 13-7
COMMON CHLAMYDIAL DISEASES

ORGANISM	DISEASE IT CAUSES	MODE OF INFECTION
Chlamydia trachomatis	Trachoma (an eye infection)	Aerosols, personal contact
	Lymphogranuloma venereum	Sexual intercourse
	Nongonococcal urethritis	Sexual intercourse
Chlamydia psittaci	Psittacosis (a disease of the lungs)	Aerosols, contact with infected birds (psittaccine birds)

referred to as "energy parasites." Diseases such as trachoma, psittacosis, and lymphogranuloma venereum are caused by chlamydia (table 13-7). *Chlamydia trachomatis* has been known to cause various diseases, notably trachoma, a highly infectious disease of the eye that often leads to blindness. More recently, it has been noted that this chlamydia is a common agent of a sexually-transmitted disease, which is called nongonococcal urethritis because the symptoms may resemble those caused by the gonococcus.

Mycoplasmas

The **mycoplasmas** are a highly variable group of bacteria lacking a cell wall (fig. 13-10). These bacteria are quite small: their average size is just below the limit of resolution 58 □ of the light microscope (0.1–0.2μm). The mycoplasmas are the simplest prokaryotic microorganisms that are still capable

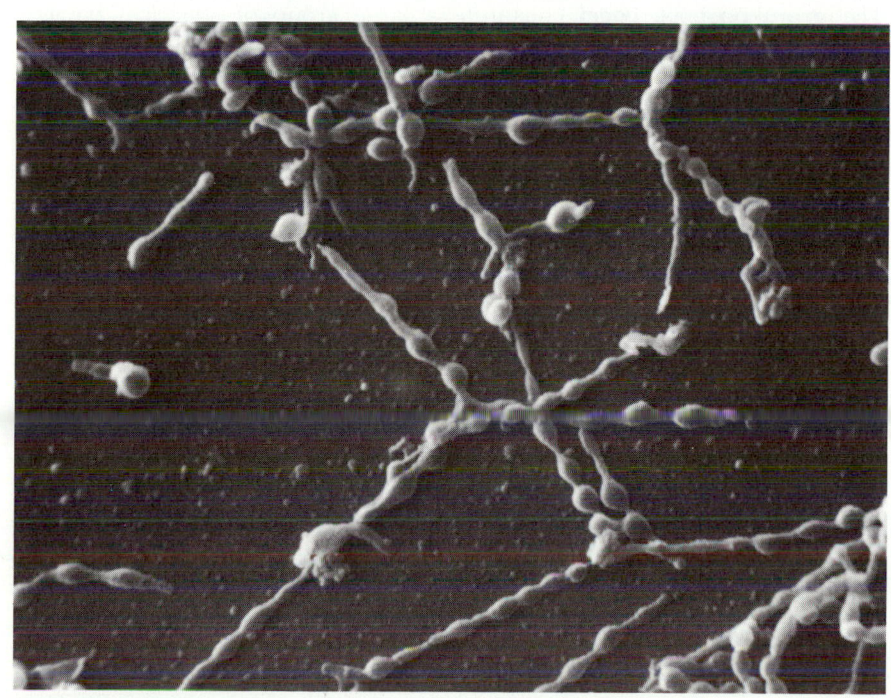

(a)

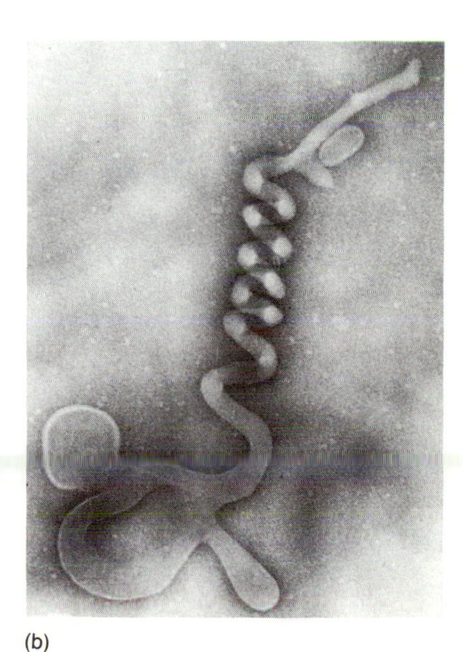

(b)

Figure 13-10 The Mycoplasmas. *(a)* Scanning electron photomicrograph of *Mycoplasma pneumoniae.* *(b)* Electron photomicrograph of a spiroplasma.

of a free-living existence. In general, each mycoplasma is a highly **pleomorphic** (variable in shape), forming spheres, filaments, and irregularly-shaped structures (fig. 13-10). This characteristic may be due to the absence of the rigid cell wall that maintains the shape of other bacteria. Most mycoplasmas require cholesterol in their diets. Apparently, this lipid is needed to strengthen their plasma membrane in order to prevent osmotic lysis of the cell.

Some species of *Mycoplasma* and *Ureaplasma* are pathogenic for humans and have been described as **membrane parasites,** because they are always found in close association with the host's plasma membrane. *Mycoplasma pneumoniae* is the cause of an inflammation of the lung called primary atypical pneumonia. *Ureaplasma urealyticum* is transmitted from one individual to another by sexual intercourse, causing urethritis (inflammation of the urethra). Spiroplasmas have recently been discovered to be the cause of more than 10 different plant diseases previously thought to be caused by viruses. These helically-shaped, motile mycoplasmas have been found in the phloem of infected plants.

Gram Positive Bacteria of Medical or Applied Importance

The gram positive bacteria of medical or applied importance are predominantly pathogens of humans and mammals. These include the agents of tuberculosis, diptheria, and streptococcal sore throat. All of these diseases can have serious consequences if not treated promptly. Also included in this group are bacteria that are important in the preparation of fermented foods, or in the synthesis of antibiotics and certain insecticides.

Gram Positive Cocci

The **gram positive cocci** constitute a large group of spherical bacteria (fig. 13-11) that are extremely important from both a medical and an industrial point of view. Although all the gram positive cocci are chemoheterotrophic, some obtain their energy exclusively by aerobic respiration while others obtain it by fermentation (table 13-8). The individual cells of these gram positive bacteria are often arranged in characteristic patterns. They may exist as individual cells, pairs (diplococci), packets of eight, chains, or grapelike clusters. These arrangements are often characteristic of a particular genus and can be used in naming □ the organisms. For example, *Streptococcus* typically forms chains of cocci, while *Staphylococcus* forms an irregular cluster of spherical cells roughly resembling a bunch of grapes. 290

Some gram positive cocci are part of the resident microbiota of many animals (including humans), while others are normal inhabitants of soils. Some species in this group cause severe diseases of humans and livestock, most of them characterized by the formation of lesions containing pus. For this reason, these pathogenic bacteria are sometimes called **pyogenic** (pus-forming) □ cocci. The disease commonly called **strep throat** because its most obvious manifestation is a sore throat) is usually caused by *Streptococcus pyogenes.* Many wound infections and **folliculites** (inflamation of hair follicles), such as carbuncles, furuncles, and pimples, are caused by another pyogenic bacterium called *Staphylococcus aureus.* More recently, *S. aureaus* has been identified as the most likely cause of the disease called **toxic shock syndrome,** which affects a small but significant proportion of menstruating 563

(a)

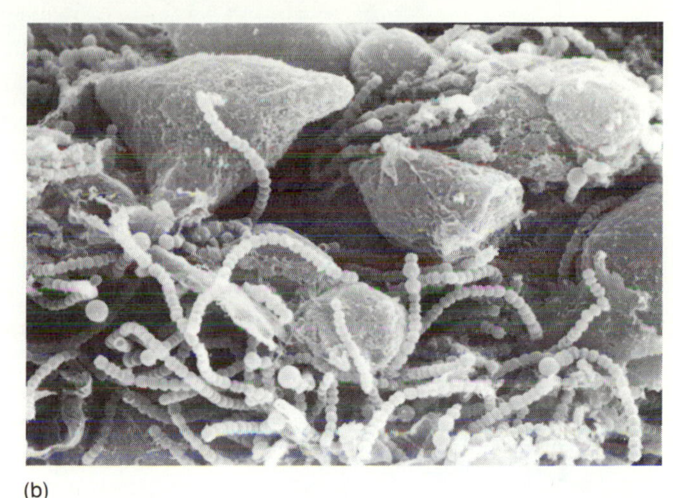

(b)

Figure 13-11 Scanning Electron Photomicrograph of Representative Gram Positive Cocci. *(a)* Transmission electron micrograph of *Streptococcus salivarius* showing various stages of cell division. Notice that the chain of cells is fragmenting. *(b)* Scanning electron micrograph of *Streptococcus pneumoniae* adhering to human conjunctival epithelial cells. Notice the long chains of cells.

TABLE 13-8
CHARACTERISTICS OF SOME GRAM POSITIVE COCCI

GENUS	O_2 REQUIREMENTS	ARRANGEMENT OF CELLS	CATALASE ACTIVITY*
Staphylococcus	Facultative anaerobe	Clusters (grape-like)	+
Micrococcus	Aerobes	Single, pairs, packets	+
Streptococcus	Aerotolerant anaerobe	Chains, pairs	−
Aerococcus	Microaerophilic	Single, pairs, tetrads	−/+

*Catalase activity refers to the ability to synthesize the enzyme catalase. Activity is detected by the production of gas bubbles (O_2) from hydrogen peroxide (H_2O_2).

females along with other individuals suffering from certain staphylococcal infections. Mastitis (inflammation of the udder) in cattle, a very difficult disease to control, is caused by *Streptococcus agalactiae* and *Staphylococcus aureus*.

Other species, collectively called the **lactic acid bacteria,** participate in the production of fermented foods such as yogurt and buttermilk. The lactic acid bacteria, which include those in the genera *Streptococcus* and *Leuconostoc*, are catalase negative and produce lactic acid as their major fermentation end product. These bacteria are very important economically because they are used to prepare yogurt, sauerkraut, kefir, cheese, buttermilk, and other fermented foods.

Endospore-forming Rods

This group of prokaryotes includes several genera of gram positive heterotrophic bacteria that undergo a process of cellular differentiation that results in the formation of an endospore (fig. 13-12). Some are strict anaerobes (e.g., *Clostridium*), while others are aerobes o facultative anaerobes (e.g., *Bacillus*). The endospore-forming rods are widely distributed in soils and occasionally infect animals (including humans), causing various diseases.

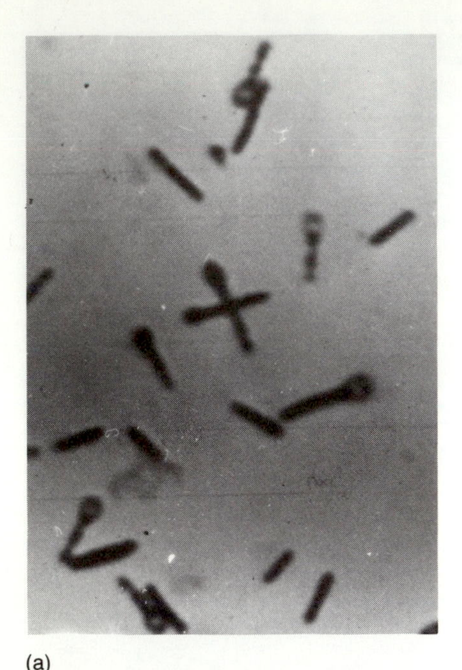

(a)

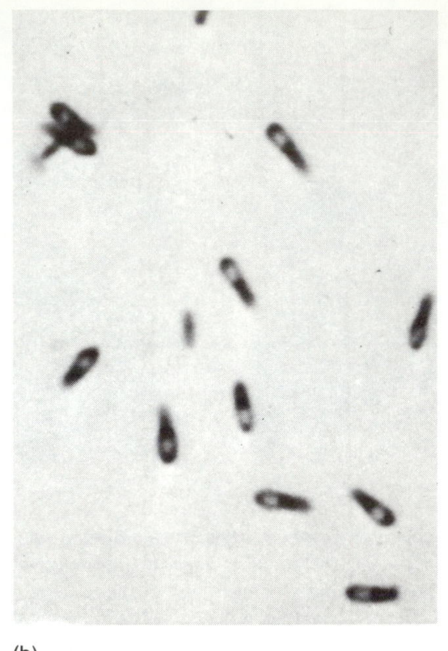

(b)

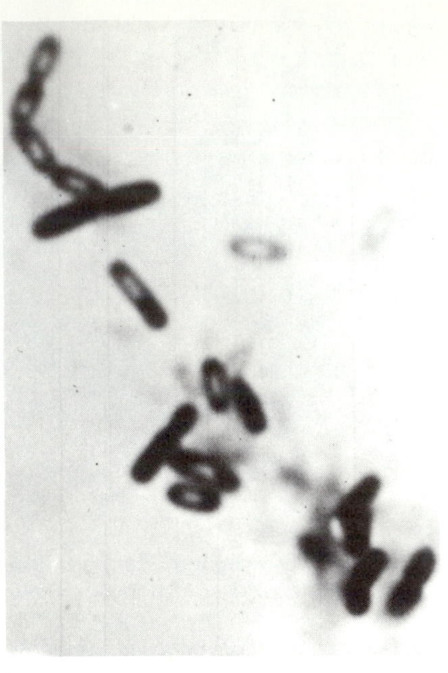

(c)

Figure 13-12 Endospore-Forming Bacteria. *(a)* Photomicrograph of *Clostridium tetani* showing the round, terminal spores. *(b)* Photomicrograph of *Clostridium subterminale* showing oval, subterminal endospores. *(c)* Photomicrograph of *Clostridium bifermentans* showing oval, central to subterminal endospores.

Many of the bacteria in this group are capable of producing antibiotics. For this reason, some of them, particularly those in the genus *Bacillus*, are studied by the pharmaceutical industry. Some species of this genus, notably *B. thuringiensis*, produce a crystal of protein called the **parasporal body**. This protein, which can be seen adjacent to the endospore, kills the larvae of some insects. Thus the bacterium is grown in large quantities by industrial concerns and sold as an insecticide under various trade names (e.g., Dipel®). Some species of spore-forming rods are capable of causing severe diseases, such as anthrax, botulism, gas gangrene, and tetanus.

Coryneform Bacteria

The **coryneform bacteria** constitute an aggregation of ubiquitous, gram positive, chemoheterotrophic bacteria. These bacteria are generally pleomorphic rods with a tendency to form angular arrangements (fig. 13-13). They do not form endospores and frequently have polyphosphate (volutin) granules.

Some bacteria in this group can cause disease in humans. For example, *Corynebacterium diphtheriae* causes the disease **diphtheria,** which afflicts predominantly children. Vaccines against diptheria have reduced the incidence of the disease to a few cases each year in the U.S.

The **propionic acid bacteria** are coryneform bacteria of economic importance. They are found in dairy products, and some species participate in the ripening of Swiss cheese. The characteristic flavor and appearance of this cheese is partially the result of the reproduction of the propionic acid bacteria on the cheese curd.

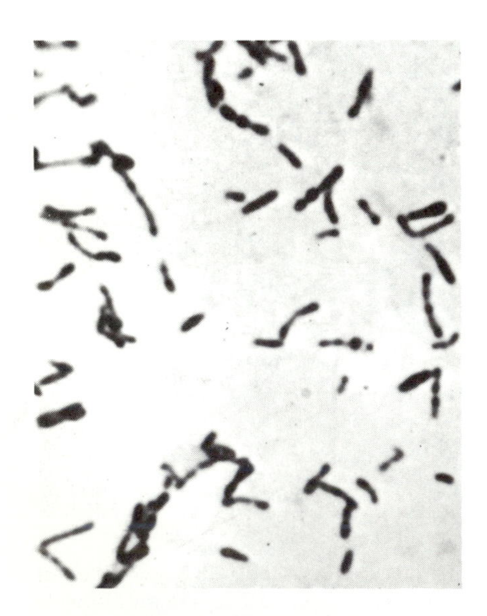

Figure 13-13 *Corynebacterium diphtheriae.* Photomicrograph of Gram stained smear of *Corynebacterium diphtheriae.* Notice the strongly stained, swollen areas that are typical of this organism.

Mycobacteria

The **mycobacteria** are gram positive, acid-fast rods with a tendency to form filaments. All of these bacteria are aerobic and generally grow slowly. The mycobacteria have a high concentration of lipids associated with their cell wall (as much as 60% of their dry weight). The lipids are responsible for the waxy nature of the mycobacteria, their **serpentine** growth pattern (fig. 13-14) and their acid-fast characteristics. A cell wall component called **cord factor** is toxic to eukaryotic cells. This toxicity is partly responsible for the disease-causing capabilities of some species of *Mycobacterium.* There are two very important pathogens in this group *Mycobacterium tuberculosis* and *M. leprae,* which cause tuberculosis and leprosy, respectively. Both of these diseases are of considerable public health importance and afflict millions of people throughout the world.

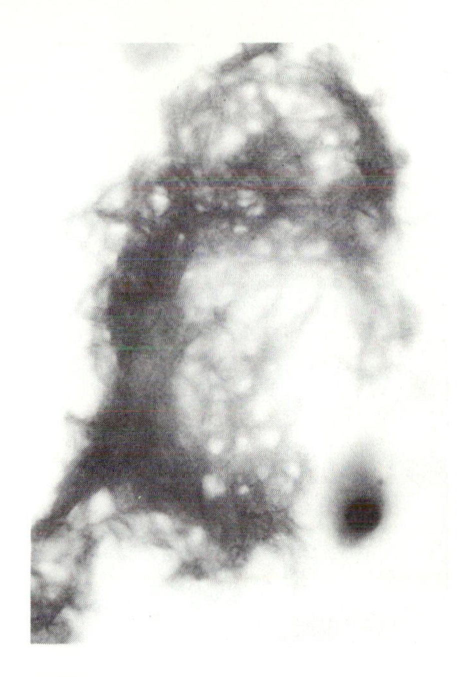

Figure 13-14 Cultural Characteristics of *Mycobacterium tuberculosis.* Serpentine growth is a characteristic of the tubercle bacillus growing on an agar plate.

Nocardias

The **nocardias** are aerobic, gram positive rods resembling the mycobacteria. Their colonies are often orange- or red-pigmented. Some species, such as *Nocardia asteroides,* are acid-fast. The genus *Nocardia* includes very important human pathogens that cause mycetoma (subcutaneous) and pulmonary (lung) infections.

Streptomycetes and Related Organisms

The **streptomycetes** are all gram positive bacteria that form well-developed, branching filaments (mycelium), and some species form very elaborate arrays of aerial spores or **conidia** (fig. 13-15). The streptomycetes are primarily soil inhabitants; in fact, the "earthy" odor of moist soils and compost heaps is due to the formation of **geosmin,** a chemical synthesized by various species of streptomycetes.

Some species of streptomycetes are important human pathogens, but the most notable characteristic of these bacteria is that they produce a large array of antibiotics. Many of the antibiotics in common use today are synthesized by these bacteria. **Steptomycin** is one such antibiotic that is syn-

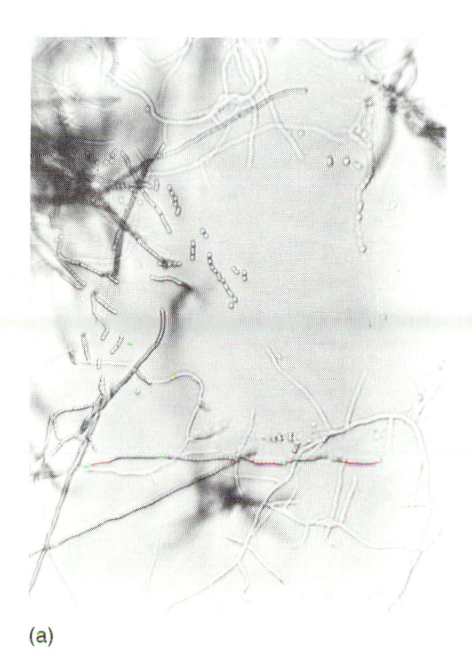

(a)

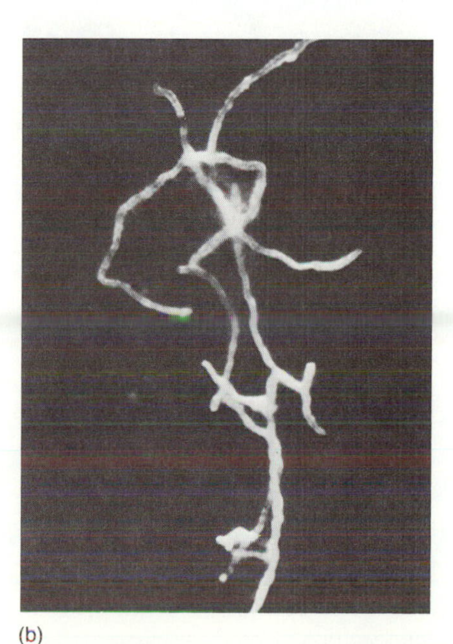

(b)

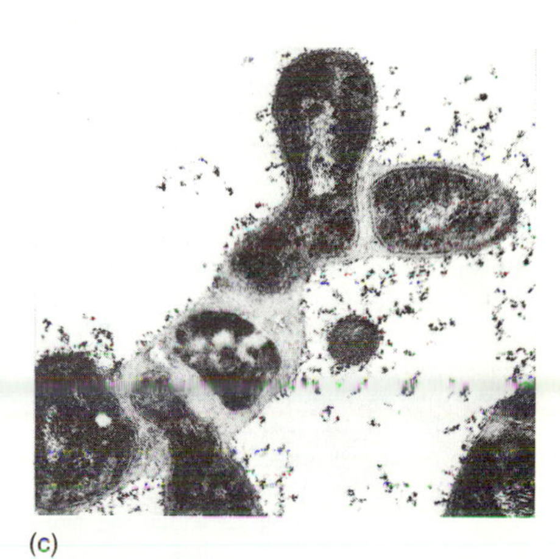

(c)

Figure 13-15 Growth Characteristics of some Actinomycetes. *(a)* Hyphal filaments of *Streptomyces. (b) Actinomyces israelii* stained with fluorescent antibodies. *(c)* Transmission electron micrograph of *Actinomyces* showing central cell producing branching filaments.

thesized by *Streptomyces griseus.* Discovered by Selman Waksman in 1947, this was the first bacterial antibiotic produced in large quantities. Since the discovery of streptomycin, many other antibiotics have been isolated from streptomycetes, and some of these are presently in common use to treat infectious diseases.

Other Gram Negative Bacteria

There are a few groups of bacteria that neither cause diseases nor perform industrially-useful functions. Nevertheless, these bacteria are an integral part of our environment, and without their presence, life as we know it could not exist. Bacteria play a central role in recycling of nutrients □ in the biosphere. 745 Some species convert organic materials into inorganic substances, while others transform these inorganic substances into other inorganic or organic materials. By the concerted effort of all these organisms, nutrients are constantly released from dead organisms as waste products that can be used as nutrients by other organisms.

Phototrophic Bacteria

The **phototrophic bacteria** include all the bacteria (other than the cyanobacteria) that obtain their energy from the sun by photosynthesis (fig. 13-

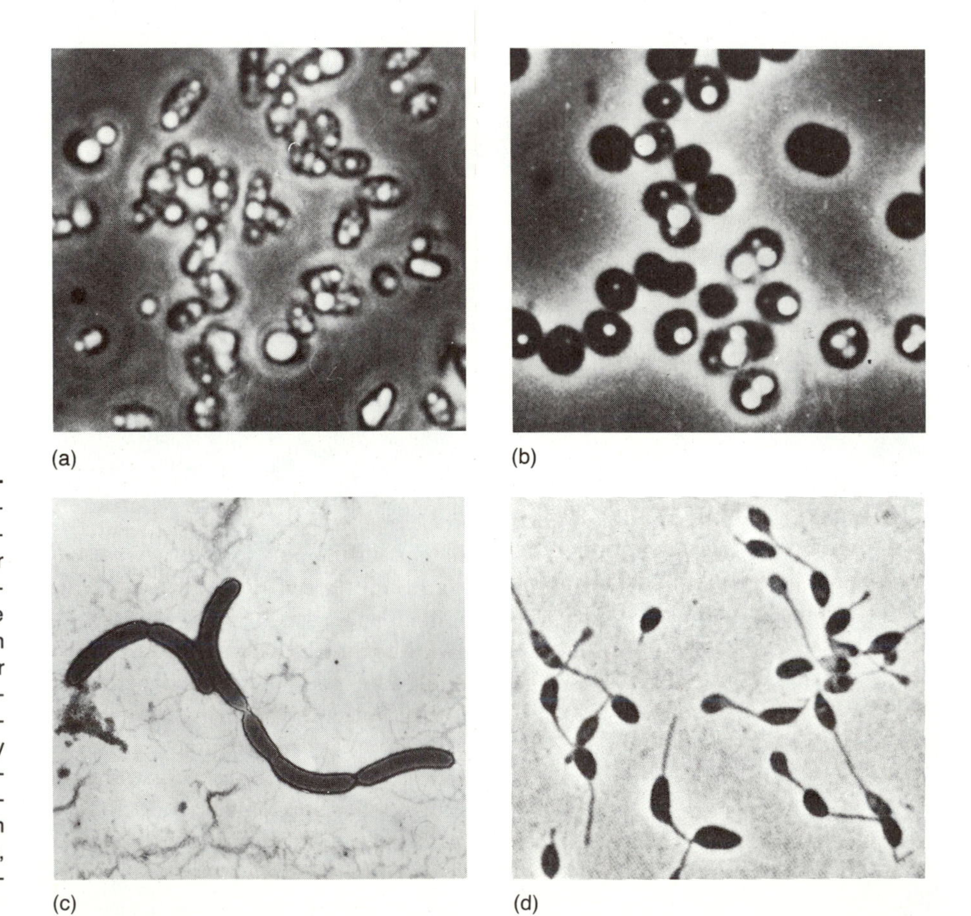

Figure 13-16 Representative Phototrophic Bacteria. *(a)* Phase contrast photomicrograph of *Chromatium* sp., an anaerobic, marine photosynthetic, purple-sulfur bacterium. The bright spots within the bacteria represent sulfur granules. *(b)* Phase contrast photomicrograph of *Thiocapsa,* an anaerobic, photosynthetic, purple-sulfur bacterium. The bright spots within the bacteria represent sulfur granules. *(c)* Transmission electron micrograph of negatively stained *Rhodospirillum rubrum,* an anaerobic photosynthetic, purple-non-sulfur bacterium. *(d)* Phase contrast photomicrograph of *Rhodomicrobium vannielii,* an anaerobic, photosynthetic, appendaged, purple-non-sulfur bacterium.

196

16). Unlike the cyanobacteria, the phototrophic bacteria carry out **anoxygenic** □ photosynthesis (no oxygen released), because they use only cyclic photophosphorylation to generate their ATP (see Chapter 8). The phototrophic bacteria play a very important role in the recycling of nutrients in nature. They are anaerobic organisms that live predominantly in aquatic habitats that are rich in sulfides or organic matter. Some phototrophic bacteria obtain their carbon from the atmosphere as carbon dioxide, but others obtain it from organic acids. Those that use reduced inorganic compounds such as hydrogen sulfide, sulfur, or molecular hydrogen to reduce carbon dioxide remove these toxic sustances from the water. Thus, some phototrophic bacteria convert carbon dioxide and inorganic sulfur compounds into chemical compounds that other organisms can use.

The phototrophic bacteria can be subdivided into the purple and green bacteria. These bacteria are so named because of the colors of their predominant photosynthetic pigments. In both groups, the intensity of the color is inversely proportional to light intensity: the most vivid colors are found in bacteria inhabiting areas of dim light. The intensity of color is due to the amount of **photosynthetic membrane** □ in the purple bacteria and the number of **chlorobium vesicles** in the green bacteria.

91

Gliding Bacteria

The **gliding bacteria** constitute a complex group of microorganisms exhibiting a variety of morphological and functional types (fig. 13-17). They have in common, however, the ability to move by the process of gliding when associated with solid surfaces.

The fruiting myxobacteria (fig. 13-17) are gram negative, chemoheterotrophic rods that move by gliding. Most of the myxobacteria are found in terrestrial habitats, where they are associated with decaying organic matter in the form of animal feces or vegetation. These myxobacteria are unique among the prokaryotes in that they aggregate and subsequently form **fruiting bodies.** The fruiting bodies are filled with **myxospores,** which are capable of germinating into vegetative cells.

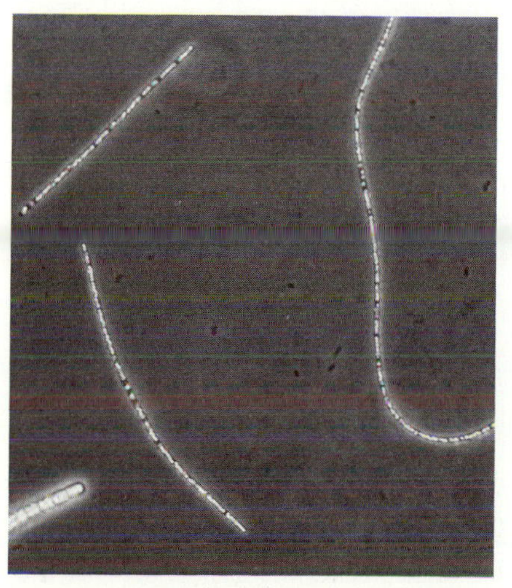

(a)

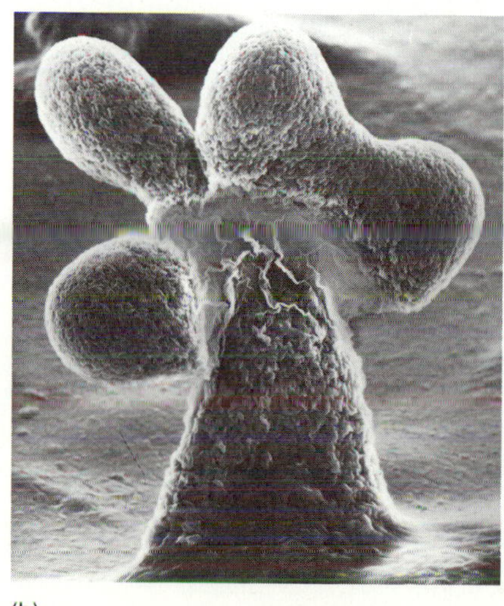

(b)

Figure 13-17 Gliding Bacteria. *(a)* Phase contrast micrograph of *Thiothrix* sp., a marine gliding bacterium that uses H_2S as an electron source. Notice the numerous sulfur granules within the cytoplasm. *(b) Stigmatella,* a fruiting myxobacterium. The fruiting body is made up of numerous individual cells aggregated in an organized fashion.

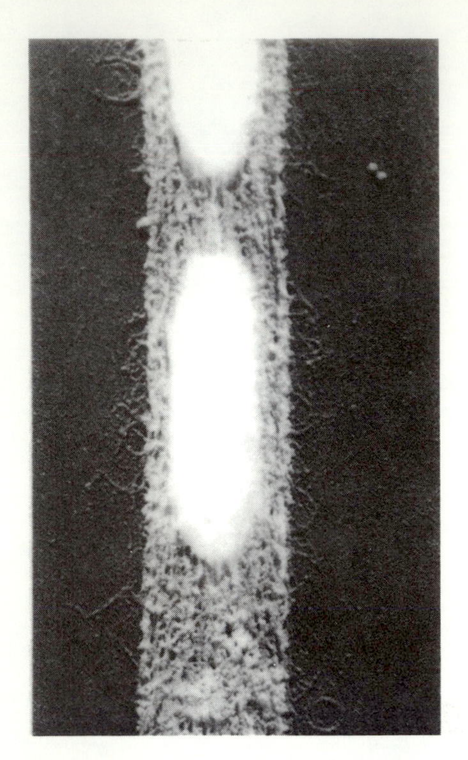

Figure 13-18 **The Sheathed Bacterium *Leptothrix***

The fruiting myxobacteria form red, yellow, or orange colonies and are found speckling speckling pieces of rotting wood in shady, damp areas of forests. These bacteria are capable of digesting a variety of large, insoluble molecules, such as chitin, cellulose, and peptidoglycan, breaking them down in to smaller units that can be used as nutrients by other organisms.

Sheathed Bacteria

The **sheathed bacteria** are filamentous, gram negative bacteria that divide by binary fission within a mucilaginous sheath (fig. 13-18). The sheath is composed of proteins, polysaccharides, and lipids. The sheathed bacteria are common in aquatic habitats, where they sometimes form **blooms** (large masses of aquatic microorganisms) in habitats that are rich in nutrients. For example, *Sphaerotilus* can become a nuisance in sewage treatment plants because, in the presence of large quantities of organic matter commonly found there, these bacteria proliferate rapidly and form blooms that foul up filters.

Budding and/or Appendaged Bacteria

The **budding and/or appendaged bacteria** develop various forms of appendages other than flagella (fig. 13-19). Generally, these microorganisms undergo a form of cell division in which the products are asymmetrical. Some members of this group divide by binary fission, while others divide by budding. The appendages may be either cellular extensions of the cytoplasm or acellular materials deposited outside the cell.

The **Caulobacters** are appendaged bacteria that have a prominent prostheca (fig. 13-19), with a **holdfast** that serves to attach the cell to other cells or to solid materials. The prostheca is an extension of the cytoplasmic ma-

(a)

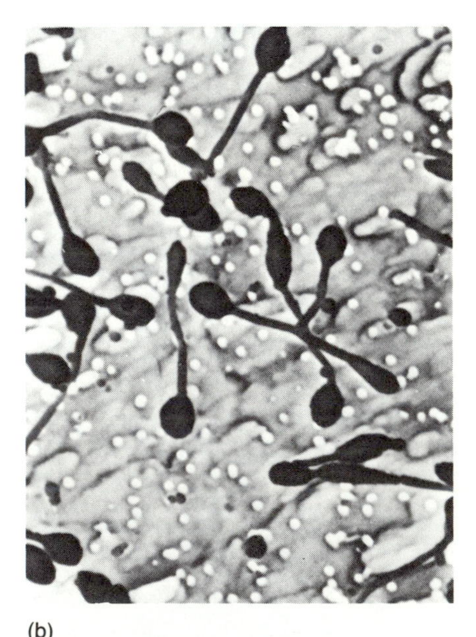

(b)

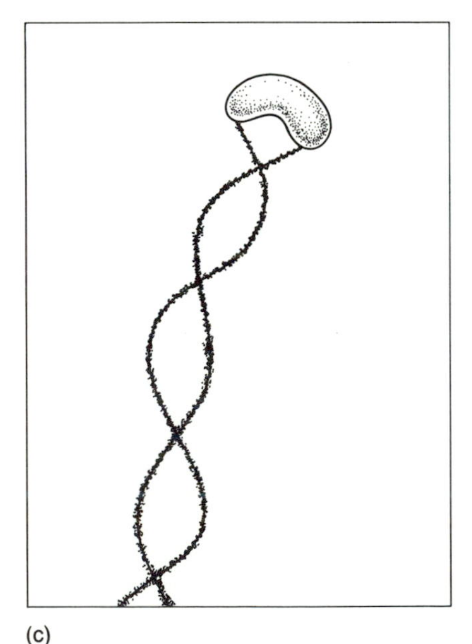

(c)

Figure 13-19 **Appendaged Bacteria.** *(a) Caulobacter.* The appendage is an extension of the cytoplasm. *(b) Hyphomicrobium,* a budding bacterium. *(c) Gallionella.* Note the twisted stalk. The stalk is acellular and consist of iron deposits.

terial and provides additional membrane area for absorption of nutrients, thus allowing the bacteria to concentrate nutrients more efficiently in nutrient-poor environments.

Hyphomicrobium, another important inhabitant of aquatic environments, reproduces by budding. It synthesizes a stalk called a **hypha** (fig. 13-19), which has a primarily reproductive function, because newly-formed cells bud out from the tips of these hyphae.

The appendaged budding bacteria play an important role in the recycling of nutrients. They prefer one-carbon (C-1) organic compounds, such as formate, methanol, and formaldehyde, as carbon sources. Many of the C-1 compounds are toxic to aquatic organisms; hence, these bacteria detoxify their environment as they reproduce by using these C-1 compounds as energy and carbon sources.

Chemolithotrophic Bacteria

The **chemolithotrophic bacteria** are gram negative microorganisms that oxidize a variety of reduced inorganic ions for their energy. These bacteria, which may be rods or cocci, play a major role in nutrient recycling in both aquatic and terrestrial environments. For example, the **nitrifying bacteria** □ 194 obtain their energy from the oxidation of ammonia or nitrite. *Nitrosomonas* oxidizes ammonia to nitrite, while *Nitrobacter* oxidizes nitrite to nitrate. Both of these bacteria are commonly found in soils, where they work to convert the ammonia resulting from the decomposition of orgnic matter into nitrate. This nitrate can be used by plants and other organisms (such as bacteria, algae, and fungi) as their source of nitrogen to build cellular components. The **sulfur bacteria** are the counterparts of the nitrifying bacteria, except that they oxidize inorganic sulfur compounds instead of nitrogen.

Methane-producing Bacteria

The **methane-producing** (methanogenic) bacteria are a heterogeneous group consisting of gram variable, anaerobic bacteria, all of which can convert carbon dioxide into methane. These bacteria have proteinaceous cell walls that lack peptidoglycan. This characteristic, coupled with other genetic differences, set these bacteria apart from most other prokaryotes.

The methanogenic bacteria are widely distributed in anaerobic environments that are rich in decayed organic matter. These environments include the intestinal tracts of humans and other animals, the rumen of livestock, stagnant water, swamps (bogs), garbage dumps, and sewage treatment plants. Marsh gas (methane) is produced in profusion by these bacteria in areas that are anaerobic and contain organic materials. The methane produced by these bacteria is sometimes used by sewage treatment plant operators to produce heat or to power some of the mechanical operations of the plant.

SUMMARY

1. The bacteria constitute an extremely heterogeneous group of microorganisms that share a prokaryotic cell type.
2. As a result of their prolific metabolic activities, the bacteria perform numerous beneficial functions for humans and for their environment.

3. An important role of bacteria in nature is that of decomposing organic matter so that it can be used by other life forms.

BERGEY'S MANUAL OF DETERMINATIVE BACTERIOLOGY

1. Bergey's Manual is a comprehensive treatise on the classification of bacteria. It includes descriptions of all the bacteria, keys to the identification of the various genera, and tables containing useful information on the identification of the various species.
2. All the bacteria are classified in the kingdom Monera (Prokaryotae). This kingdom is subdivided into two divisions; the cyanobacteria and the bacteria.

THE CYANOBACTERIA (BLUE-GREEN ALGAE)

1. The cyanobacteria are all those prokaryotic microorganisms that carry out a process of photosynthesis similar to that of the green plants, evolving oxygen from the process.
2. Many cyanobacteria have the ability to convert atmospheric nitrogen into organic nitrogen. Nitrogen-fixing cyanobacteria have been used as "fertilizers" of rice paddies, where they increase the nitrogen content of the paddy when they fix atmospheric nitrogen.
3. The cyanobacteria are widely distributed in aquatic and terrestrial habitats. Some of them cause undesirable odor and flavor changes in drinking water when they grow in profusion in reservoirs and lakes. Other species of cyanobacteria produce toxins that cause diseases in humans and domestic animals.

THE BACTERIA

1. There are many gram negative bacteria that are human or animal pathogens. These have various morphological and physiological characteristics. Some of these pathogens include *Treponema pallidum*, *Salmonella*, *Shigella*, *Vibrio cholerae*, *Neisseria gonorrhoeae*, *Pseudomonas aeruginosa*, and *Escherichia coli*. All of these bacteria cause serious diseases and hence are of considerable public health significance.
2. There are many gram positive bacteria, both rods and cocci, that are capable of causing human disease. These include *Mycobacterium tuberculosis*, *Corynebacterium diphtheriae*, *Streptococcus pyogenes*, and *Staphylococcus aureus*.
3. The genus *Bacillus* and several species of streptomycetes are gram positive rods that produce a large variety of antibiotics, many of which are in common use today.
4. There are other groups of gram negative bacteria that do not cause disease or produce industrially important substances. Nevertheless, these bacteria, which include the phototrophic, gliding, appendaged, and chemolithotrophic bacteria, play important roles in the cycles of matter in nature and in this way make life on earth possible for all other organisms.

STUDY QUESTIONS

1. Describe four characteristics that are common to all bacteria.
2. Explain how the cyanobacteria are different from the phototrophic bacteria.
3. Cite the criteria that we used to differentiate the vibrios from the enterics.
4. What difficulty is encountered by anyone who attempts to classify the bacteria along phylogenetic lines?
5. Construct a table outlining the salient characteristics of the following groups of bacteria. Also include an important function or characteristic) they perform.
   a. Gram positive bacteria
   b. Gram negative bacteria
   c. Prosthecate or appendaged bacteria
   d. Chemolithotrophic bacteria
6. How do cyanobacteria differ from other bacteria?

SUPPLEMENTAL READINGS

Atkinson, W. H. and Winkler, H. H. 1985. Transport of ATP by *Rickettsia prowazekii*. *Journal of Bacteriology* 161(1):32–38.

Hackstadt, T., Todd, W., and Caldwell, H. 1985. Disulfide-mediated interactions of the chlamydial major outer membrane protein: role in the differentiation of chlamydiae? *Journal of Bacteriology* 161(1):25–31.

Krieg, N. R. Chief editor and Holt, J. G., editor. 1984. *Bergey's manual of systematic bacteriology*. vol. 1 9th ed. Baltimore: Williams & Wilkins.

Krogman, D. W. 1981. Cyanobacteria (blue-green algae)—Their evolution and relation to other photosynthetic organisms. *BioScience* 31:121–124.

Lenette, E. H., Balows, A., Hausler, W. J., and Truant, J. P. eds. 1980. *Manual of clinical microbiology*. Washington: American Society for Microbiology.

Schachter, J. and Caldwell, H. D. 1980. Chlamydiae. *Annual Review of Microbiology* 34:285–311.

Woese, C. R. 1981. The archaebacteria. *Scientific American* 244:98–126.

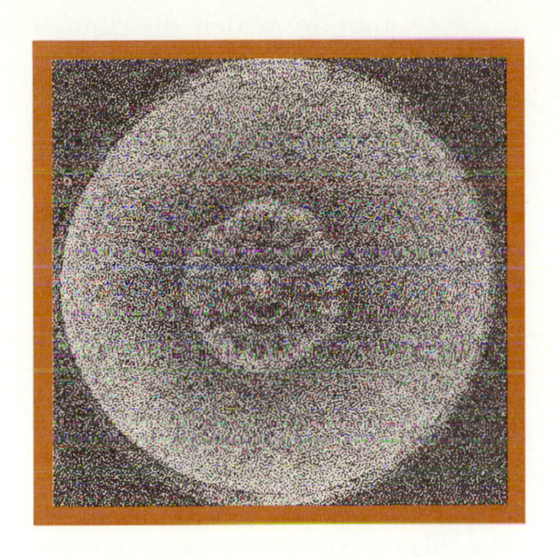

CHAPTER 14
THE FUNGI

CHAPTER PREVIEW
TURKEY-X DISEASE

In 1960 more than 100,000 turkeys died in England due to a previously unknown disease. The disease struck suddenly and without warning. Because the cause of the disease was unknown, the disease was dubbed the "Turkey-X Disease." The disease was not contagious and it was not caused by any known microorganism.

Poultry feed was suspected of being the cause of the disease. Studies revealed that the groundnut meal in the feed was contaminated with a toxic substance. The toxin could be extracted with methanol, and when injected into ducklings it mimicked the signs and symptoms of Turkey-X Disease.

Eventually, it was discovered that the peanut meal fed to the turkeys was contaminated with a fungus called *Aspergillus flavus*. This fungus, when grown under the appropriate cultural conditions, secreted a toxin that caused Turkey-X Disease. The toxin, called **aflatoxin** (**A***spergillus* **fla***vus* **toxin**), causes liver damage in poultry and liver cancer in laboratory animals.

The reports of Turkey-X Disease heralded the development of modern **mycotoxicology** (the study of fungal toxins and their effects). It is now known that many fungi produce toxins and that these toxins cause a variety of human and animal diseases.

Toxin-producing fungi have a significant impact on humans by causing disease and killing livestock. Fungi also participate in many natural processes that are essential to the survival of many organisms. This chapter will introduce you to the characteristics of the major groups of fungi, their biology, and their impact on humans.

LEARNING OBJECTIVES

A STUDY OF THIS CHAPTER SHOULD ENABLE YOU TO:

OUTLINE THE MAJOR CHARACTERISTICS OF THE FUNGI

DESCRIBE THE MOST COMMON GROSS AND MICROSCOPIC FEATURES OF THE FILAMENTOUS FUNGI

DESCRIBE SOME OF THE ACTIVITIES THAT FUNGI CARRY OUT IN NATURE, AND WHERE THESE ORGANISMS CAN BE FOUND

CITE THE CHARACTERISTICS OF THE MAJOR GROUPS OF FUNGI

The fungi are a group of heterotrophic eukaryotes that are widely distributed in nature. They are considered to be **saprophytes** because they obtain their nutrients from the decomposition of dead organic matter. Although early investigators thought that the fungi were related to plants, the fungi are quite different from the green plants and their cells lack chlorophyll. Presently most biologists recognize this difference and have classified the fungi in a kingdom of their own, the kingdom **Fungi.**

BIOLOGY OF THE FUNGI

The fungi characteristically form filaments called **hyphae** (fig. 14-1). A hypha consists of either a single, elongated cell or a chain of cells whose cytoplasms are generally shared. The hyphae lack chlorophyll and serve to absorb nutrients from the environment. Hyphae also play a role in the reproduction of the fungus, because they can give rise to other hyphae or to specialized reproductive structures called **spores** (fig. 14-2). The hyphae have cell walls consisting of **chitin** or **cellulose,** along with other minor polysaccharides. The fungal colony, or **thallus,** consists of a mass of hyphae called **mycelium** (fig. 14-3). The fungal thallus may originate from a single fungal spore or from a piece of hypha. As simple as the above description of a fungus may be, it is not without exceptions. This is because many microorganisms that are considered to be fungi do not form hyphae. For example, the **yeasts** are one-celled fungi that characteristically form colonies resembling those of bacteria and do not produce demonstrable hyphae. Other fungi produce structures during one stage of their life cycle that are definitely fungus-like, while at another stage they have a protozoa-like existence. These organisms that have both fungal and protozoal □ characteristics are called the **slime molds.** 363

Materials and nutrients move throughout the hyphae by cytoplasmic streaming. This is possible because hyphae are composed of interconnecting

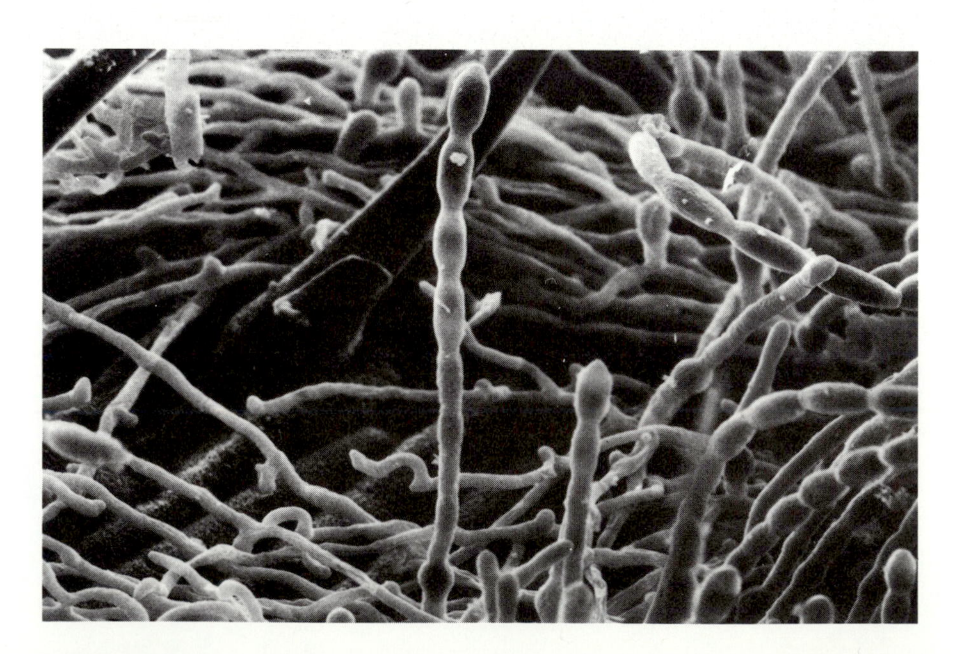

Figure 14-1 Fungal Hyphae. Fungal hyphae of *Erysiphe graminis* on a grass leaf as seen with a scanning electron microscope.

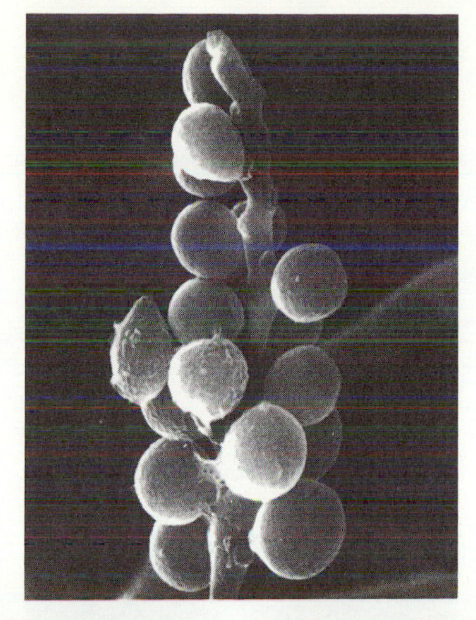

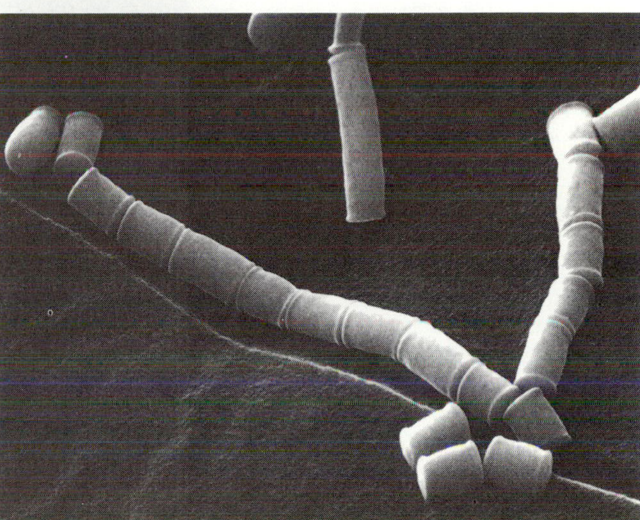

Figure 14-2 Representative Types of Fungal Spores. Fungi produce spores of many different shapes. Spores may develop from asexual or sexual processes and are frequently used to identify organisms. Spores efficiently spread fungi throughout the biosphere.

cells. Some hyphae are completely devoid of cross walls and are called **coenocytic hyphae,** while others have distinct cross walls and are called **septate hyphae.** Even septate hyphae can permit the passage of nutrients and cellular structures, because the cross walls are perforated at the centers.

The growth of fungi occurs by extension of the hyphal tip (fig. 14-4). As the tips of the hyphae in the thallus grow, they extend outward radially, forming a more or less rounded colony. Studies have revealed a possible mechanism for hyphal tip growth (fig. 14-5). A typical cross section of an actively growing hyphal tip generally exhibits a large number of **vesicles** (globules) concentrated at the tip. These vesicles contain the enzymes that soften the chitinous wall and subsequently permit the extension of the tip. The hypothesis that attempts to explain how fungal hyphae grow is called the "vesicular hypothesis of hyphal tip growth." According to this hypothesis, the enzymes required for the hyphal tip elongation are synthesized on the

Figure 14-3 Fungal Colonies. Colonies of four different fungi as seen growing on agar-solidified media. The colony, also known as a thallus, consists of a mass of fungal hyphae.

Figure 14-4 Development of a Fungal Colony. A fungal spore *(a)* germinates producing the beginning of a *(b)* hypha (a germ tube). The hypha elongates *(c)*, branches *(d)*, and eventually forms a thallus *(e)*.

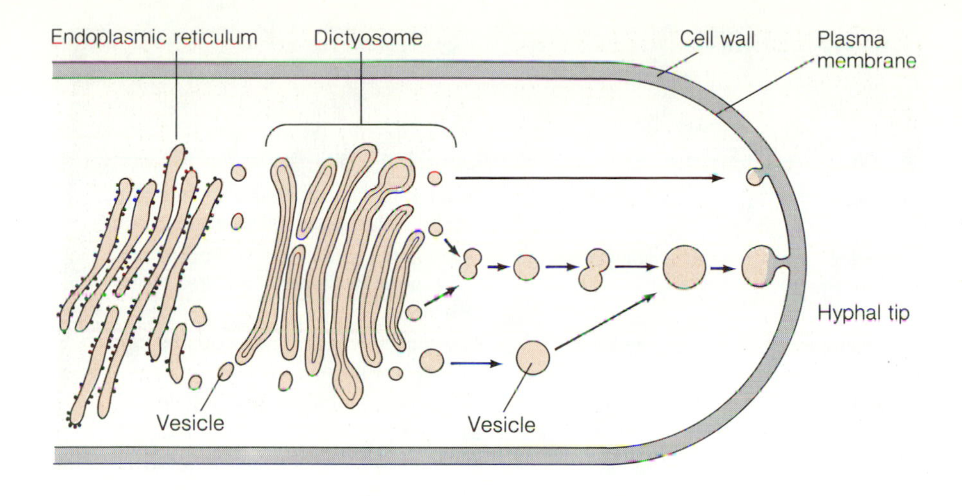

Figure 14-5 The Process of Hyphal Tip Elongation. Diagram illustrating a section of hyphal tip. Notice the movement of vesicles from the dictyosome to the hyphal tip. The vesicles contain enzymes that catalyze the breakdown of fungal cell wall and the synthesis of new wall material.

102 endoplasmic reticulum □. The enzymes eventually end up in vesicles that bud from the Golgi membranes and diffuse toward the hyphal tip. When the vesicles fuse with the plasma membrane at the hyphal tip, enzymes for the breakdown of the old cell wall or synthesis of the new wall are liberated.

The cells that make up the fungal thallus generally possess only one **haploid** nucleus, that is, each nucleus has only one copy of each of the chromosomes. At certain stages of their life cycle, some fungi may be **diploid** (having a nucleus with two copies of each chromosome). In most fungi, however, diploidy is a brief condition and is usually evident only during certain sexual reproductive stages. The exceptions are the Oomycetes, which are diploid except when forming gametes. Some fungi possess more than one haploid nucleus per cell and are said to be **dikaryotic** (di = two, karyon = nucleus). As in most other fungi, haploidy is the prevalent nuclear state in the dikaryotic fungi.

Fungal Reproduction

The formation of a new thallus (colony) from a preexisting one can take place by either a **sexual** or an **asexual** process. In asexual reproduction, new 233 cells are formed by mitosis □ and then spread to new sites to form new thalli. This form of reproduction represents the most common form of fungal multiplication during periods of rapid growth and does not require that two compatible fungal elements fuse to form a zygote.

In many fungi, mitotic divisions result in asexual spores called **conidia.** These structures are often produced in profusion and are carried away by air currents that spread the fungus to other habitats. Many fungi, especially those in the group called the **Fungi Imperfecti,** produce a large array of different types of conidia, many of which can often be used to differentiate the various species of these fungi (table 14-1).

During sexual reproduction, two compatible nuclei fuse together to form a **diploid zygote.** The zygote subsequently undergoes a meiotic division that results in the production of four or more haploid fungal cells. These cells are generally called **sexual spores.** The sexual spores separate from the fungus and eventually germinate to form new haploid thalli.

Certain fungi can carry out sexual reproduction by fusing two nuclei within the same thallus; that is, they are **self-fertile.** Self-fertile fungi are

TABLE 14-1
VARIOUS TYPES OF FUNGAL SPORES

SPORE TYPE	GROUP OF FUNGI*	CHARACTERISTICS
Ascospores	Ascomycetes	Develop by sexual means. Spores are formed inside a sac called ascus.
Basidiospores	Basidiomycetes	Develop by sexual means. Spores are borne on club-shaped structures called basidia.
Zygospores	Zygomycetes	Develop by sexual means. Spores form from the fusion of 2 hyphae.
Conidia	Various groups	Produced by asexual means. Easily airborne.
Chlamydospores	Various groups	Thick-walled spores, develop asexually. Resting spore.
Sporangiospores	Zygomycetes	Asexual spores formed inside a sac called sporangium. Easily airborne.

*See table 14-3 for a more complete classification of the Fungi.

usually called **homothallic** fungi. Fungi that require a compatible nucleus from another individual in order to undergo sexual reproduction. These fungi are **self-sterile** and are said to be **heterothallic.** Fungi are unique among microorganisms in that their hyphae can carry two or more genetically different nuclei. These nuclei can arise from the fusion of two genetically different hyphal cells, or from a mutational event (change in the genetic material) within the same hypha. Since the hyphae of most fungi have septal pores, nuclei can pass from one cell to another virtually unimpeded (fig. 14-6). As a consequence of nuclear migration, a given hyphal cell may have more than one nucleus and the nuclei may be genetically different. These cells are **heterokryotic** (fig. 14-7). The dissimilar nuclei may multiply independently by mitosis, so that a thallus is created with genetic information derived from two or more nuclei. **Heterokaryosis** apparently is a desirable condition that enhances genetic variability among fungi in the absence of sexual reproduction.

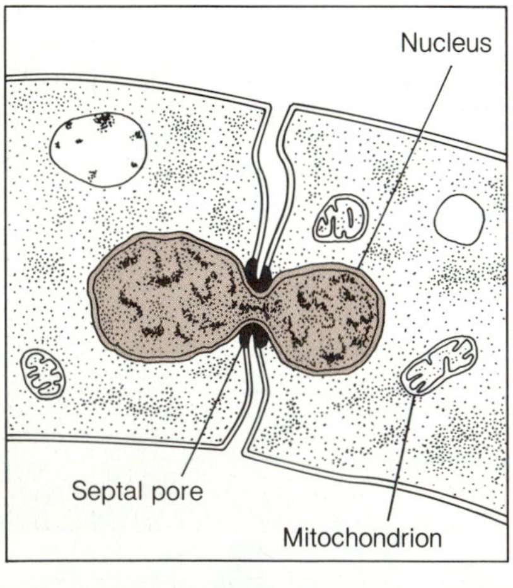

Figure 14-6 Nucleus Passing through the Septal Pore of a Hypha of the Ascomycete *Neurospora crassa*

Dimorphism

Certain fungi exhibit two different morphologies. For example, they may grow as filamentous forms in the soil, whereas in a suitable animal host they may reproduce like a yeast (fig. 14-8). This phenomenon of having two forms is called **dimorphism.** Some dimorphic fungi can be made to display their dimorphism simply by changing the environmental and/or nutritional conditions in which they are grown. For example, the dimorphic fungus *Histoplasma capsulatum*, when grown at room temperature on a culture medium such as Sabouraud Glucose Agar, grows in a filamentous, **mycelial** form, producing typical large conidia with knobs (tuberculations) on thin hyphae (fig. 14-8). If however, this fungus is grown on blood agar at 37°C, it will form a single-celled, rounded yeast phase (fig. 14-8). The same dimorphic change occurs when the mycelial form of the fungus is introduced into a

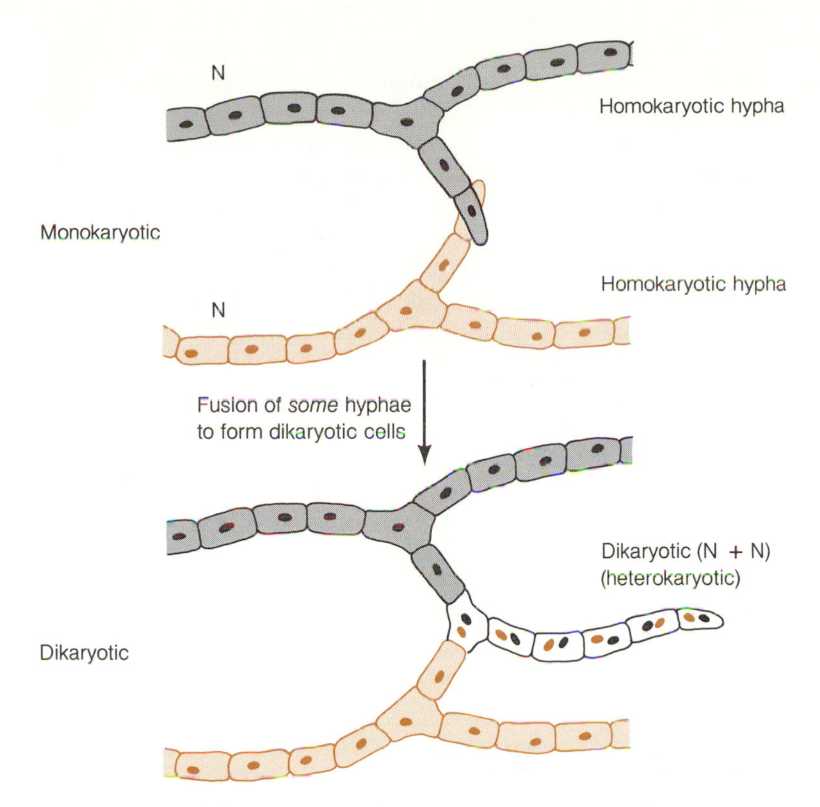

Figure 14-7 The Formation of a Heterokaryotic Hypha from the Fusion of Two Homokaryotic Hyphae

(a)

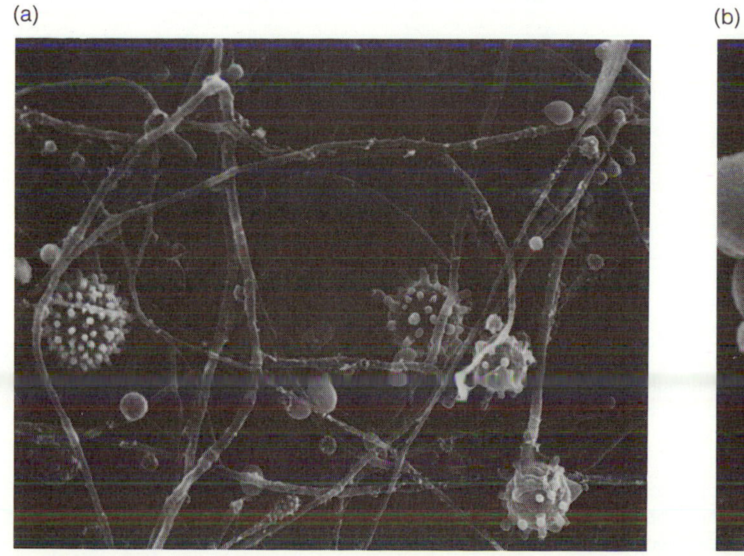

(b)

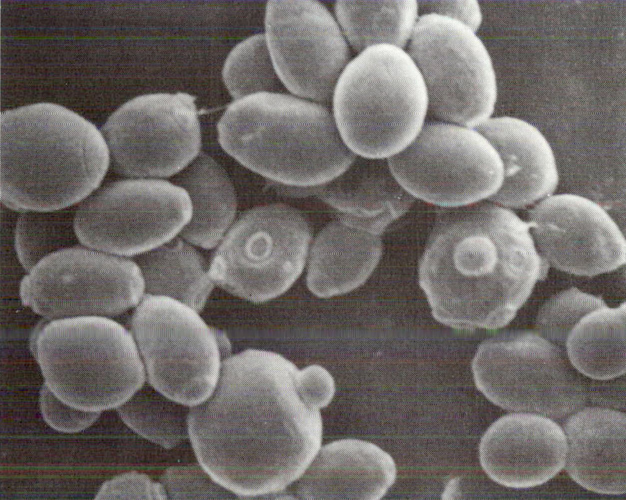

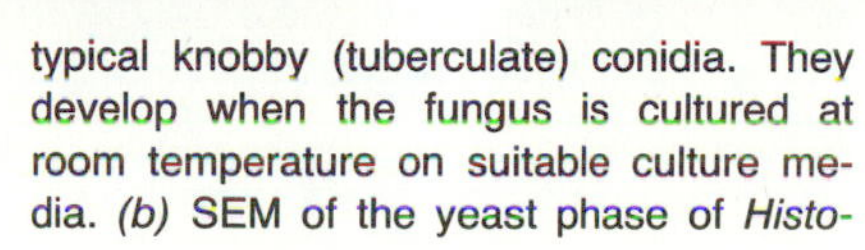

Figure 14-8 Dimorphism of *Histoplasma capsulatum*. *(a)* Scanning electron photomicrograph showing the mycelial stage of *Histoplasma capsulatum.* Note the typical knobby (tuberculate) conidia. They develop when the fungus is cultured at room temperature on suitable culture media. *(b)* SEM of the yeast phase of *Histoplasma capsulatum.* This phase develops when the fungus reproduces inside a suitable host or when cultured at body temperature on blood-containing agar.

susceptible host, such as a human or a laboratory mouse. This shift is generally called the M to Y shift, because the fungus changes from **m**ycelial form to **y**east form. Another fungus, *Candida albicans*, exists as a yeast at room temperature, but upon inoculation into a susceptible host, or when cultivated under low oxidation-reduction potentials, the fungus forms filaments referred to as **pseudomycelium.** Table 14-2 indicate some important human pathogens that are dimorphic.

Dimorphism is advantageous to fungi because it allows them to reproduce more rapidly in adverse conditions or to fend off immune host responses. Yeast cells generally reproduce faster than filamentous fungi. This rapid growth may also afford the dimorphic fungus with a means of obtaining nutrients more rapidly from its environment, thus giving the fungus a competitive edge over other organisms that may be sharing a food resource. The Y to M shift of *C. albicans* allows this fungus to resist host defense mechanisms □ more effectively than its yeast form. Hence, it is possible that dimorphic changes may also be beneficial to pathogenic fungi because they aid in preventing the host defense mechanisms from destroying the invading fungus.

434

DISTRIBUTION AND ACTIVITIES OF FUNGI

Fungi have a tremendous impact on our environment. They convert complex organic matter (plant and animal remains) into simple chemical compounds, which can then be used as nutrients by other organisms. Because fungi can act on a large variety of substrates, both living and nonliving, they have been influencing human populations for centuries. Certain fungi are used in making wines, beers, cheeses, breads, and many other foods. Others serve as sources of food (the truffles, morelles, and mushrooms), and still others are the source of life-saving drugs such as antibiotics. The human-

TABLE 14-2
IMPORTANT HUMAN PATHOGENS

NAME OF FUNGUS	NAME OF DISEASE	TAXONOMIC GROUP
Malassezia furfur	Tinea Versicolor	Deuteromycetes
Microsporum	Ringworms	Deuteromycetes
Trichophyton	Ringworms	Deuteromycetes
Epidermophyton	Ringworms	Deuteromycetes
**Sporothrix schenckii*	Sporotrichosis	Deuteromycetes
Fonsecaea	Chromoblastomycosis	Deuteromycetes
**Candida albicans*	Candidiasis	Deuteromycetes
Pseudoallescheria	Mycetoma	Ascomycetes
**Ajellomyces capsulatus*	Histoplasmosis	Ascomycetes
**Coccidioides immitis*	Coccidioidomycosis	Deuteromycetes
**Ajellomyces dermatitidis*	Blastomycosis	Ascomycetes
Aspergillus fumigatus	Aspergillosis	Deuteromycetes
Filobasidiella neoformans	Cryptococcosis	Basidiomycetes
Mucor sp.	Mucormycosis	Zygomycetes
Rhizopus sp.	Zygomycosis	Zygomycetes

*Names of organisms preceded by an asterisk indicate that the fungus is dimorphic (see section on dimorphism).

fungus association is not exclusively a mutually beneficial one, however. For example, fungi constitute the single most important cause of diseases in plants, resulting in billions of dollars in damage to crops annually throughout the world. Fungi also cause spoilage of foods and destruction of wood and textiles. In addition, certain of these microorganisms can cause diseases in humans and in domestic animals. Fungal diseases of humans result in the loss of millions of dollars due to the lack of productivity. The fungi generally cause disease when they grow within living tissue, when they release exoenzymes that destroy tissue, or when they produce toxins that disrupt the proper functioning of living cells.

Mycorrhiza

749 **Mycorrhiza** □ is the fungal growth that appears on the surface and in the cortex of the roots of certain green plants. This association is mutually beneficial and is widespread among terrestrial plants. **Conifers,** such as pine trees or spruces, are often found forming mycorrhizal associations with **basidiomycetes** such as *Boletus*. Mycorrhizal associations often permit economically-important trees to grow on poor soils that normally would not support the tree's growth. The mycorrhizal association is believed to increase the plant's surface area for absorption of nutrients in nutrient-deficient environments and supply the fungus with plant carbohydrates. Many of the mycorrhizal fungi are nutritionally dependent on the plant host. We will discuss mycorrhizae in greater detail in Chapter 30.

FOCUS
NEMATODE-TRAPPING FUNGI

Fungi inhabit many different environments. They are commonly found sharing soils with a myriad of other organisms. Many of the soil fungi obtain their nutrients by decomposing organic matter, such as animal carcasses and plant debris. Some fungi, however, have evolved unique mechanisms to ensnare live organisms and subsequently feed on them. Such is the case with the predatory fungi known as nematode (roundworm) trappers (fig. 14-9).

The nematode-trapping fungi produce a network of hyphae with loops interspersed throughout the thallus. The hyphal loops serve as snares for the nematodes. When a nematode slithers into one of these loops, the loop constricts, trapping the nematode. The hyphal loops also secrete adhesive materials so that the trapped nematode cannot escape. The fungi then produce specialized hyphae that penetrate the cuticle (outer skin) of the worm. Once the hyphae penetrate, they proliferate rapidly and the slow digestion process begins. Eventually, all that remains of the nematode is the cuticle around a mass of hyphae.

The nematode trappers represent yet another adaptation of microorganisms to environments where nutrients are low and competition for them is great.

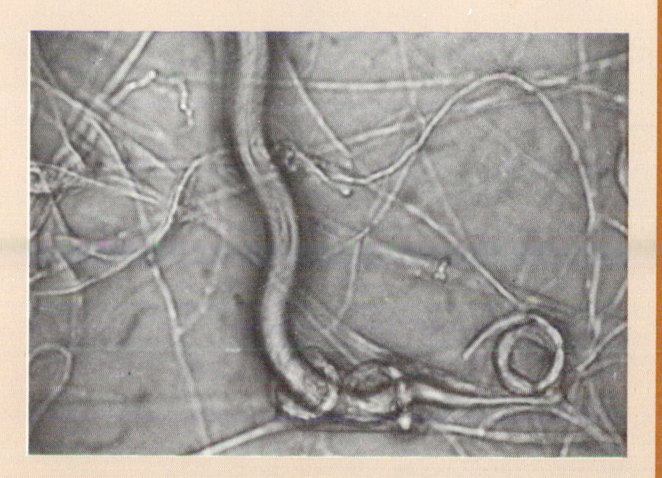

Figure 14-9 Nematode-Trapping Fungus

THE MAJOR GROUPS OF FUNGI

The fungi are subdivided into various groups (table 14-3) based upon their structural and reproductive characteristics. Many of these subdivisions are named after the type of sexual reproductive structure the fungus produces. For example, the basidiomycetes characteristically form **basidia** during their sexual reproduction, while the ascomycetes produce **asci** instead. In the following paragraphs some of the major groups of fungi will be discussed, with emphasis on their biology and impact on humans.

Zygomycetes

The **zygomycetes** are a group of predominantly terrestrial (land-dwelling) fungi that are widely distributed in nature. The hyphae of these fungi vary

TABLE 14-3
SUMMARY OF THE MAJOR GROUPS OF FUNGI

CLASSIFICATION	CHARACTERISTICS
Kingdom: Fungi	
Division: Gymnomycota	Naked (wall-less) cells.
Subdivision: Acrasogymnomycotina Class: Acrasiomycetes	Myxamoeba → pseudoplasmodium (made up of individual amoeba) → sporocarp. Example: *Dictyostelium*
Subdivision: Plasmodiogymnomycotina Class: Protosteliomycetes Class: Myxomycetes	Myxamoeba → plasmodium (a single multinucleated macroscopic cell) → sporocarp. Examples: *Physarum, Fuligo*
Division: Mastigomycota	Flagellated cells.
Subdivision: Haploidmastigomycotina Class: Chytridiomycetes Class: Hyphochytridiomycetes Class: Plasmodiophoromycetes.	Haploid hyphae, flagellated zoospores, hyphae coenocytie, many aquatic forms. Examples: *Synchytrium, Allomyces; Rhizidiomyces;* Plasmo-*diophora*.
Subdivision: Diplomastigomycotina Class: Oomycetes	Diploid hyphae, flagellated zoospores, Hyphae coenocytic. Many aquatic forms. Examples: *Phytophthora* and *Plasmopara*
Division: Amastigomycota	Nonflagellated cells.
Subdivision: Zygomycotina Class: Zygomycetes Class: Trichomycetes	Haploid, coenocytic hyphae. Sexual spore is a zygospore. Asexual spore is the sporangiospore borne inside sporangia. Examples: *Rhizopus* and *Mucor*.
Subdivision: Ascomycotina Class: Ascomycetes	Haploid, septated hyphae with septal pore. Sexual spore is the ascospore borne inside ascus. Asci frequently develop inside ascocarp. Asexual spore is usually the conidium. Examples: *Neurospora, Morchella,* and *Saccharomyces* (yeast).
Subdivision: Basidiomycotina Class: Basidiomycetes	Hyphae with dolipore septum. Clamp connections seen in certain hyphae. Sexual spores are basidiospores borne on basidia. Examples: *Puccinia* and *Amanita*
Subdivision: Deuteromycotina Class: Deuteromycetes (Fungi Imperfecti)	Yeast-like or filaments. Hyphae resemble those of the Ascomycetes. Sexual stage unknown. Parasexual cycle in some species. Conidia of various types produced. Examples: *Trichosporon* and *Candida*.

in diameter and are characteristically wide, measuring 3–5 μm in diameter. In addition, their hyphae usually lack septa (cell boundaries) and are said to be **coenocytic** (or aseptate). When septa are present, they are often spaced at irregular intervals along the length of the hyphae. The zygomycetes are important decomposers ☐ in terrestrial habitats because they break down organic matter and recycle nutrients. Some species are human pathogens, while others produce industrially-useful chemicals. Their life cycle includes both sexual and asexual reproduction. Figure 14-10 illustrates the typical life cycle of the zygomycete *Rhizopus*.

745

The zygomycetes are characterized by the formation of sexual spores called **zygospores.** These spores form after two compatible fungal hyphae fuse: the site where the hyphae fuse becomes enlarged and diffentiates into a zygospore. When the zygospore germinates, it gives rise to a **sporangium,** a balloon-shaped structure filled with spores called **sporangiospores** (fig. 14-10). Each sporangiospore can give rise to a new fungal colony. These structures can also develop from hyphae in the absence of sexual reproduction, and constitute a very common means of asexual reproduction in the zygomycetes. One single sporangium can release numerous sporangiospores,

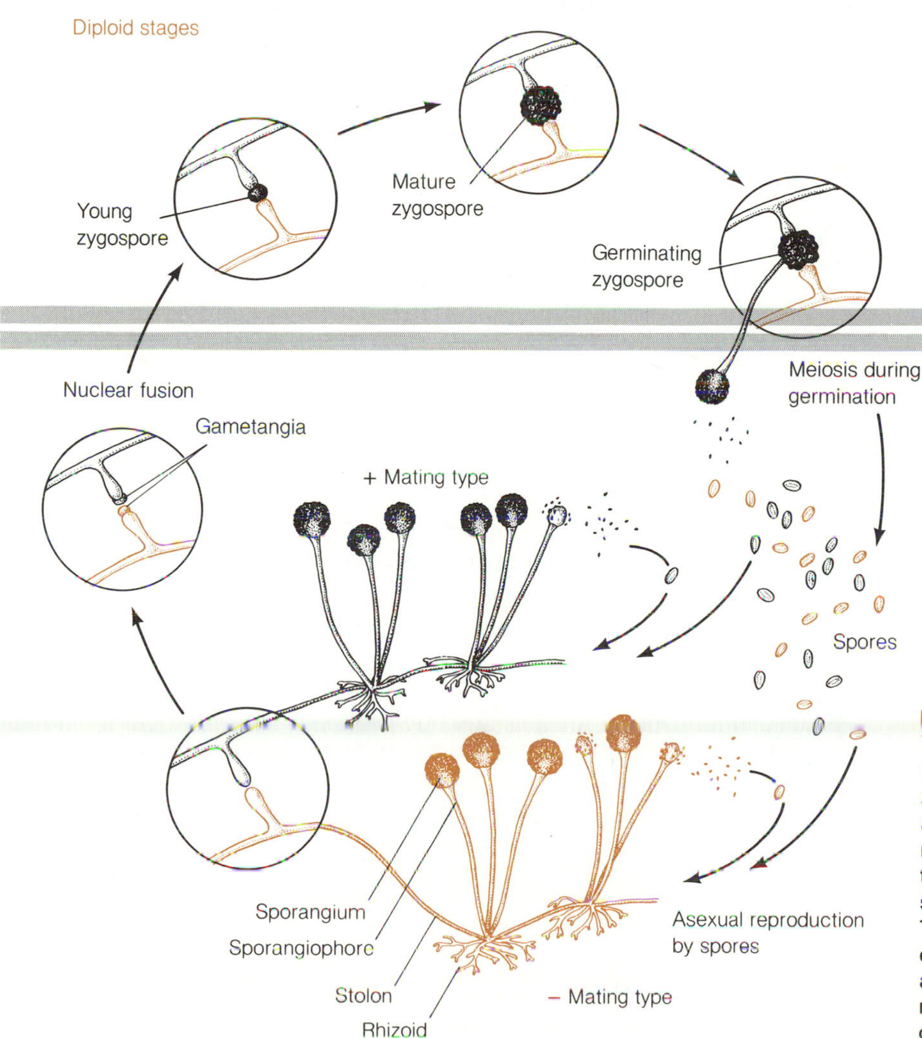

Figure 14-10 Life Cycle of *Rhizopus*, a Typical Zygomycete. Compatible hyphae (+ and −) fuse to form a zygote called the zygospore. The mature zygospore develops a thick coat and can remain dormant for many months. After the dormancy period, the zygospore cracks open, producing a sporangium filled with sporangiospores. The spores can germinate (fig. 14-4), producing new thalli. Vegetative hyphae can also give rise to sporangia by asexual means. When compatible hyphae fuse, the cycle begins again.

which can then be transported for many miles by air and water currents. The sporangiospore represents the major means of dispersal for these fungi. The arrangement of the sporangia and their location along the hyphae sometimes serve to differentiate among the various zygomycetes.

The zygomycetes, such as *Mucor* and *Rhizopus*, are capable of very rapid and extensive growth that can cause rapid food spoilage. Because of the rapid rate at which these fungi multiply, they occasionally invade human tissue in individuals who have been debilitated by diseases such as cancer, immune deficiencies, or diabetes. Their growth and invasion of organs often results in the death of the individual within 2–5 days. Fungi in the genera *Rhizopus*, *Mucor*, and *Absidia* have often been implicated in human disease and death.

The zygomycetes can exhibit a fermentative metabolism □, so they are used to make fermented food items. Soybean products such as **tempeh** and **sufu** are made using zygomycetes such as *Rhizopus*. Their fermentation of carbohydrates causes the soybean protein to curdle and therefore contributes to the formation of the soybean cakes that characterize these food items. Certain of these fungi are employed in industry for the synthesis of economically important products such as enzymes, organic acids, and alcohols. For example, the enzyme rennin, which is used in the cheese industry to curdle milk as a first step in cheesemaking, is produced by the zygomycete *Mucor pusillus*. Cortisone is another product that is synthesized using zygomycetes, such as *Rhizopus*.

189

Oomycetes

Until recently, the oomycetes (fig. 14-11) were classified with the zygomycetes in a group that was called the **phycomycetes.** Most mycologists believe, however, that the oomycetes are quite different from other fungi and should be classified separately. This argument is based on the facts that the Oomycetes are diploid rather than haploid, that they form motile spores, and that their cell walls have cellulose instead of chitin as the major structural polysaccharide. The motile spores develop from mitotic divisions (asexual) and are called **zoospores.** The zoospores have two types of flagella (fig. 14-11): a **whiplash** and a **tinsel.**

The oomycetes as a group include some very important plant pathogens such as *Phytophthora infestans* and *Plasmopara viticola*. *Phytophthora infestans* causes a disease of potatoes called **late blight.** It was this fungus that destroyed the potato crops in Ireland in the nineteenth century. The destruction of the potato crop led to starvation, disease, and death in Ireland. Because of crop failures, millions of Irish citizens emigrated to various countries, including the United States. *Plasmopara viticola* causes **downy mildew** of grapes, a devastating disease that can result in significant financial losses to viticulturists. Other oomycetes can cause diseases of fish; **ich,** for example, a disease of aquaria fish, can be caused by the oomycete *Saprolegnia*.

Ascomycetes

The ascomycetes produce sexual spores (derived from the union of two compatible fungal elements) called **ascospores.** Ascospores develop inside a sac-

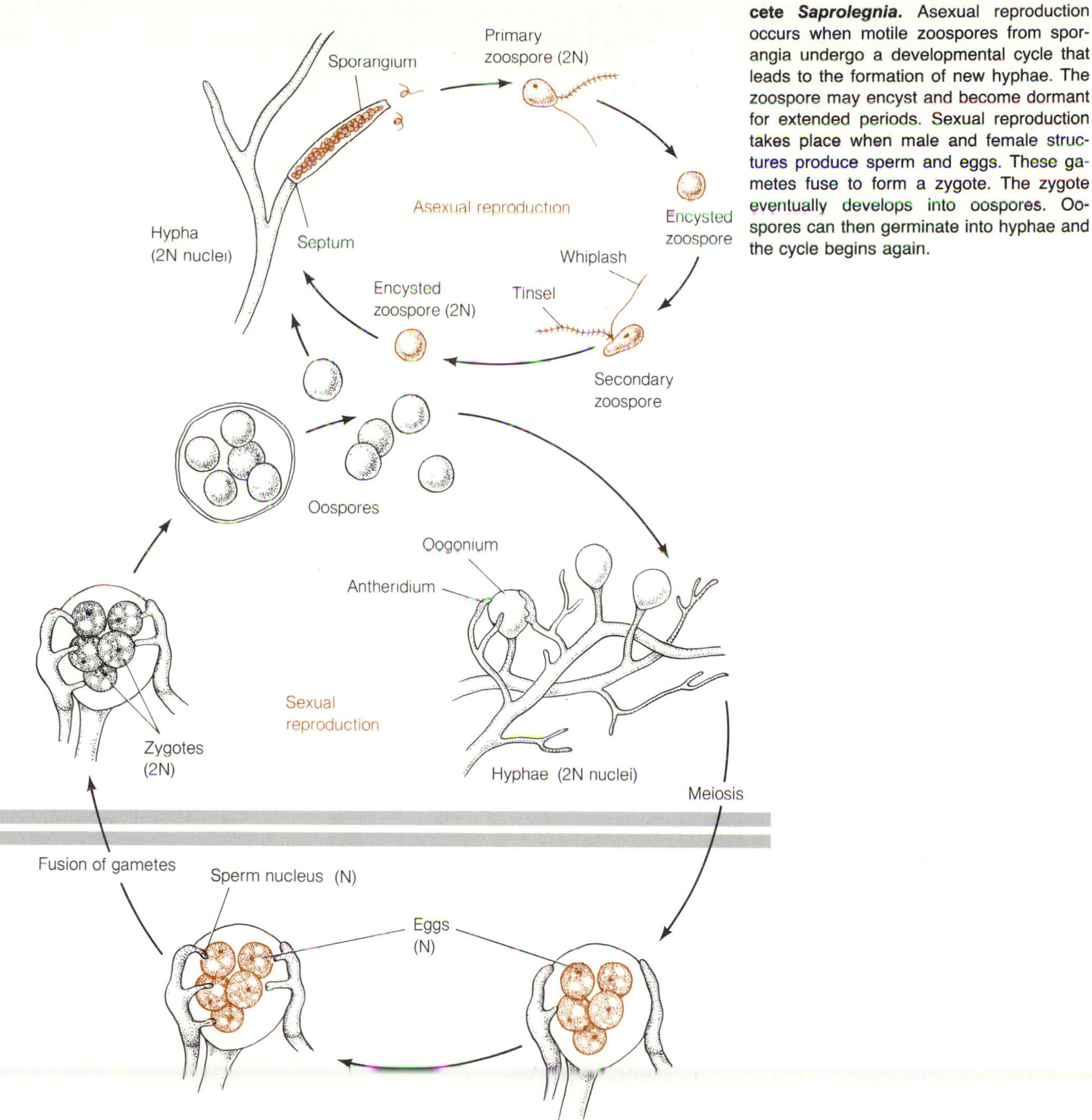

Figure 14-11 Life Cycle of the Oomycete *Saprolegnia*. Asexual reproduction occurs when motile zoospores from sporangia undergo a developmental cycle that leads to the formation of new hyphae. The zoospore may encyst and become dormant for extended periods. Sexual reproduction takes place when male and female structures produce sperm and eggs. These gametes fuse to form a zygote. The zygote eventually develops into oospores. Oospores can then germinate into hyphae and the cycle begins again.

like structure called the **ascus.** All the fungi that produce ascospores within asci are called **sac fungi** and are placed in the Class **Ascomycetes.**

The ascomycetes are widely distributed in nature. Some of the ascomycetes parasitize crops, causing tremendous financial losses (billions of dollars annually) and starvation throughout the world. Others cause diseases in humans and other animals. Diseases such as coccidioidomycosis and sporotrichosis are caused by fungi in this group. The various fungal diseases of humans will be discussed in Section III of this text.

Ascomycetes can cause severe damage to textiles because of the production of enzymes such as **cellulases** and **proteases,** which break down the fibers of cotton, wool, and silk, particularly in humid, warm climates. Some fungi also produce toxic substances that can cause the death of animals and humans. For example, *Claviceps purpurea,* the ascomycete that causes **ergot of rye,** produces toxic alkaloids that cause a disease called **ergotism.** Ergotism can cause the death of anyone who consumes bread made with contaminated rye grain. On the other hand, ergot alkaloids have been used in medicine to stop bleeding, induce smooth muscle contraction, or induce uterine contractions in pregnant women to expedite their labor.

The hyphae of ascomycetes, unlike those of the zygomycetes, are thin and delicate with septations at regular intervals along their length. The septa are often closely associated with membrane-bounded, electron-dense structures and granules called **Woronin bodies.** It is hypothesized that the Woronin bodies serve as plugs of the septal pore in older cells so as to isolate them from the rest of the thallus. The hyphal wall may be pigmented, consists primarily of chitin, and generally contains a single nucleus.

The ascomycetes reproduce asexually by producing a variety of spores, predominantly **conidia.** Sexual reproduction in the ascomycetes (fig. 14-12) results in the formation of an **ascus,** which generally occurs within an **ascocarp.** The ascocarp is a complex structure, consisting of hyphae knitted together to form a bed upon which the asci develop (fig. 14-12). The ascomycetes may show male and female differentiation. The male structure can be a conidium, a hyphal fragment, or a nucleus. The female structure is called the **ascogonium** and it has a tubelike structure called the **trichogyne.** When a male structure comes in contact with a trichogyne, it fuses with the trichogyne and the nucleus travels down it to the ascogonium. There it initiates the development of specialized hyphae (also known as **ascogenous hyphae**) that eventually develop into asci. The asci develop from the tips of the ascogenous hyphae when two nuclei in the ascogenous hyphae fuse. Nuclear fusion is followed by meiosis after a period of ascus development. The ascus matures within the ascocarp and contains usually four or eight haploid ascospores, which contain genetic characteristics from both of the parent nuclei.

The Yeasts

The yeasts are a group of fungi that are typically unicellular and reproduce asexually by budding or by fission (fig. 14-13). In addition to reproducing asexually, many yeasts also exhibit sexual reproduction. The majority of the sexual yeasts are ascomycetes.

Yeasts have a tremendous impact on human activities. For example, yeast infections are commonplace in medical practice. These infections are commonly caused by *Candida albicans*, *Torulopsis glabrata*, and *Cryptococcus neoformans.* Such yeast infections sometimes are fatal. Yeasts are also important to humans because of the products they make, such as wine, beer, and bread. The alcohol in wine and the distinct flavors of various wines are due, at least in part, to the action of yeasts on the grape juice. Yeasts are also used as a source of protein for human consumption.

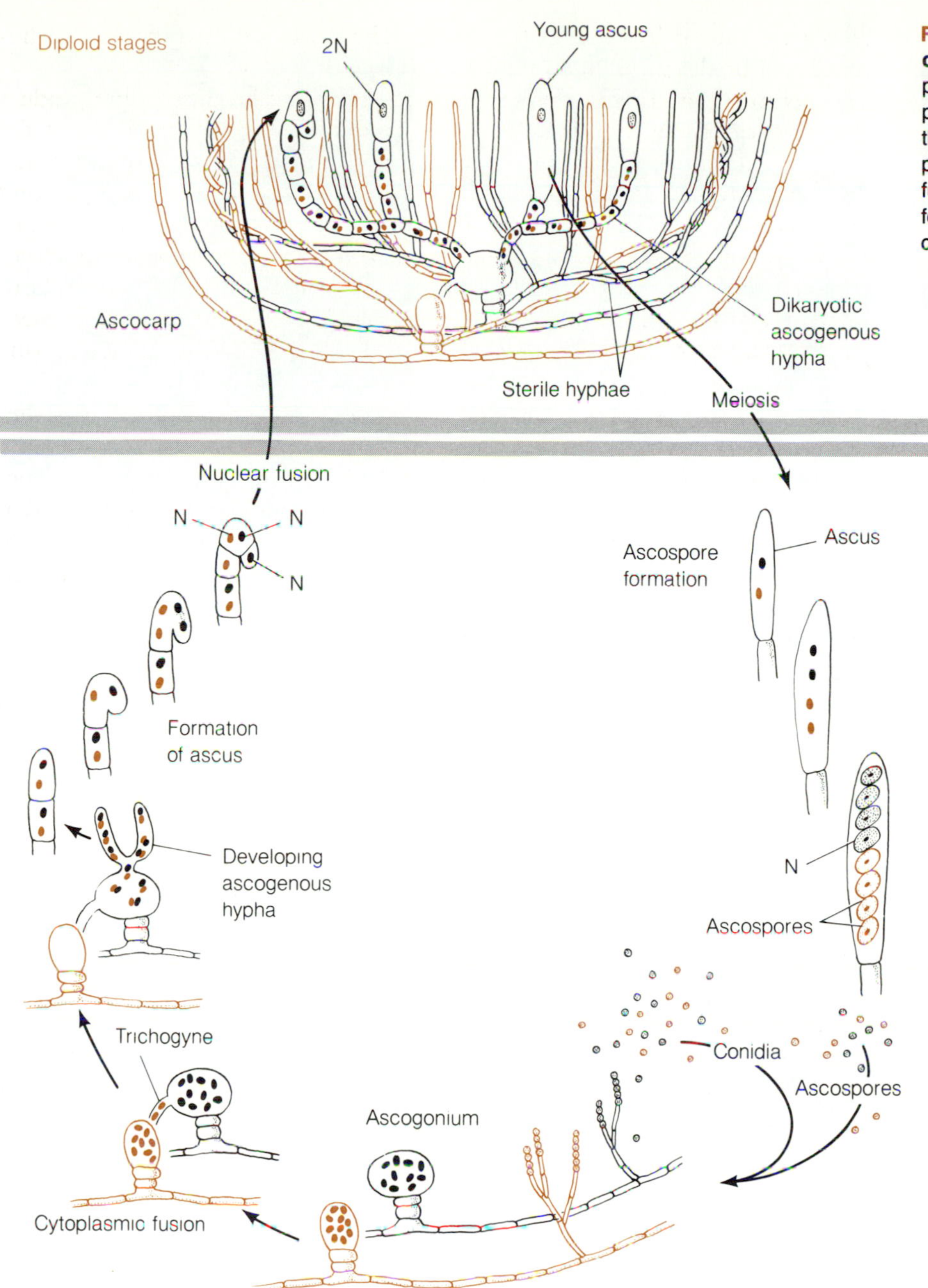

Figure 14-12 Life Cycle of a Typical Ascomycete. Asexual reproduction takes place when conidia, which develop from hyphae, germinate to produce new fungal thalli. During sexual reproduction, two compatible fungal elements come together. The fusion of the fungal elements result in the formation of asci containing four to eight ascospores.

Basidiomycetes

The basidiomycetes constitute a very large group of fungi that are widely distributed in terrestrial habitats. Some of the basidiomycetes have a macroscopic stage and are visible with the unaided eye (fig. 14-14). More than 20,000 species of basidiomycetes have been described in the literature. These include **mushrooms, toadstools, rusts, smuts, stink horns, bracket fungi, puffballs, coral fungi,** and **bird's nest fungi.** In this group are some very important plant pathogens, such as *Puccinia graminis,* which causes

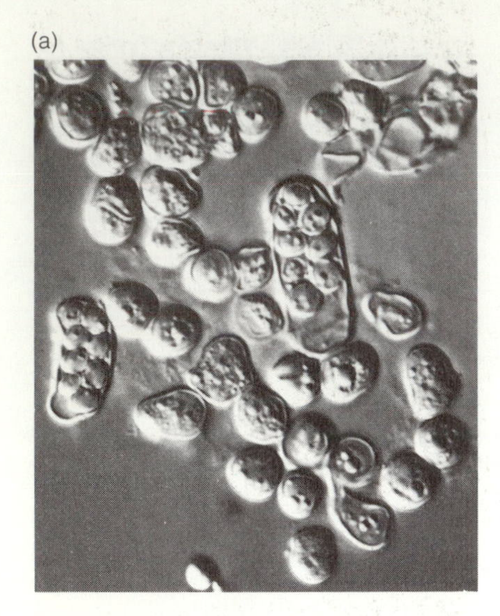

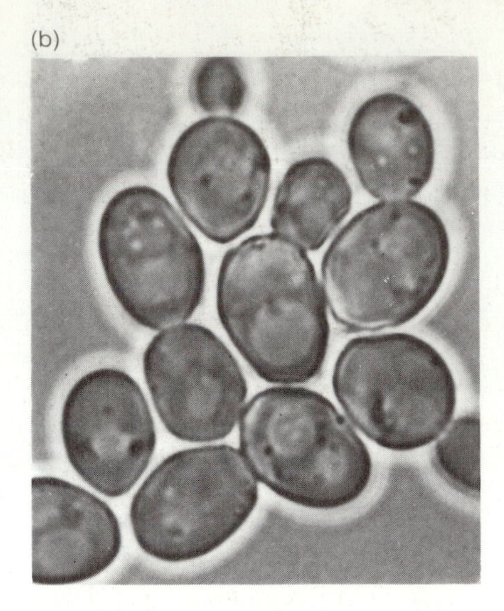

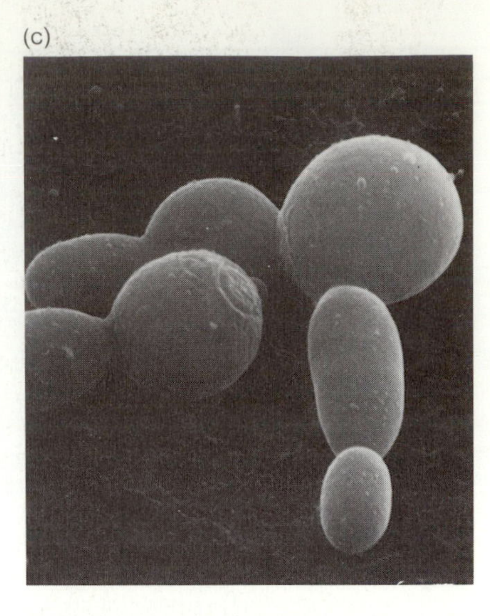

Figure 14-13 Representative Yeasts. *(a)* Nomarski interference photomicrograph of the yeast *Schizosaccharomyces* showing vegetative cells and asci containing ascospores. *(b)* Phase contrast photomicrograph of the yeast *Saccharomyces cerevisiae* showing budding yeast cells. *(c)* Scanning electron micrograph of *Candida albicans* showing budding yeast cells.

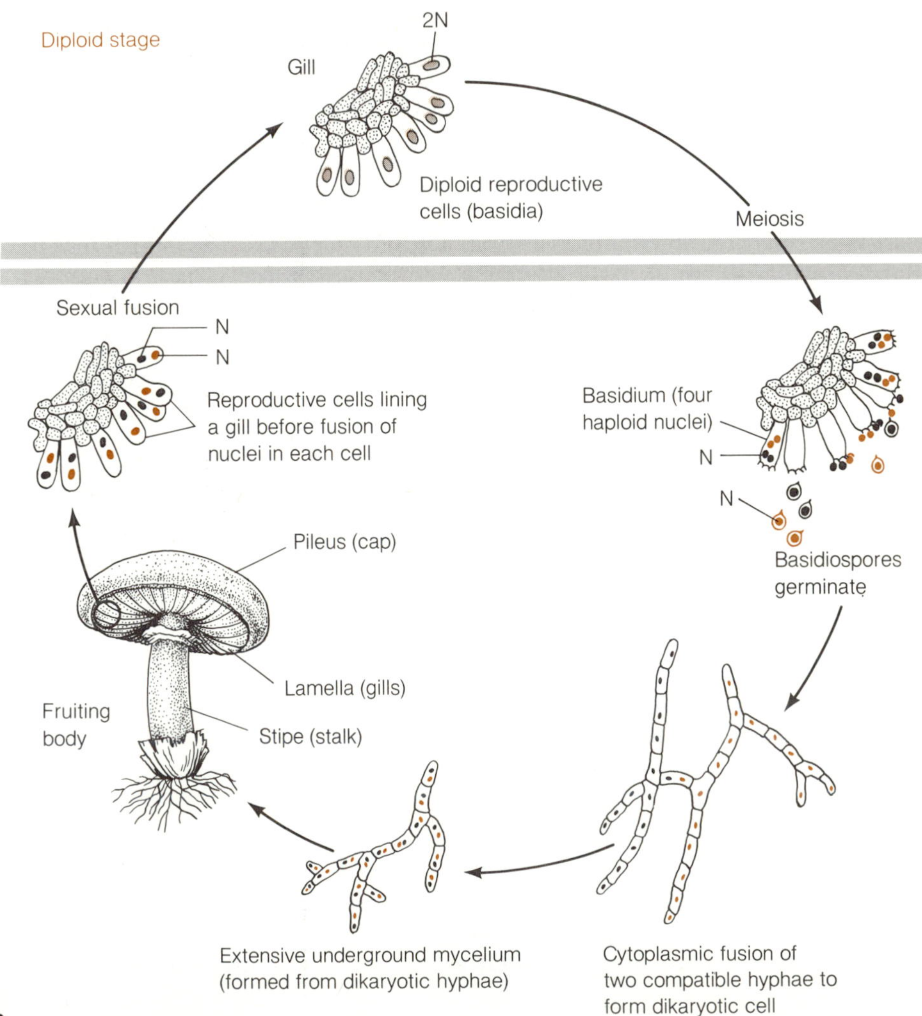

Figure 14-14 Structure and Life Cycle of a Basidiomycete

wheat rust, and *Ustilago zeae*, the cause of **corn smut.** Together, these two plant pathogens cause the loss of billions of dollars each year worldwide. Other basidiomycetes are important human pathogens. For example, *Filobasidiella neoformans*, the basidiomycete state of *Cryptococcus neoformans*, infects human nervous tissue, causing a disease called cryptococcosis.

The hyphae of basidiomycetes are tubular structures that consist of typical eukaryotic cells separated by septa. Except for certain rusts and smuts, which have ascomycete-like hyphae, the basidiomycete hypha have a septum called the **dolipore septum** (fig. 14-15). The dolipore septum consists of a central pore with swellings or doughnut-shaped structures surrounding the opening. The dolipore system is "capped' on either side of the septal wall with a membranous structure that has been called the **septal pore cap.** The function of the septal pore cap is still unclear, although it may function in controlling the passage of cellular materials from one cell to another.

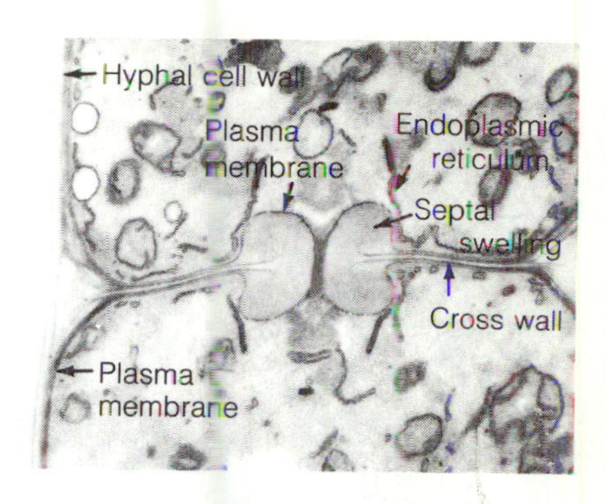

Figure 14-15 The Dolipore Septum of the Secondary Hyphae in Basidiomycetes. This electron photomicrograph illustrates two hyphal cells of *Rhizoctonia solani* separated by a dolipore septum. The various components of the septum, including the cross walls and septal swellings are indicated.

All basidiomycetes form sexual spores called **basidiospores** (fig. 14-16). The basidiospores develop from supporting structures called **basidia.** The mushroom, or **basidiocarp,** is a fruiting body formed by many basidiomycetes (fig. 14-14). The basidiocarp, a specialized structure that results from the fusion of two compatible hyphae, is composed of three major parts: a) the **pileus** or cap; b) the **stipe** or stem; and c) the **lamellae** or gills. The basidiocarps are found above ground and are generally large enough to be seen easily with the unaided eye. Beneath the ground, the basidiocarp is supported and nourished by vegetative hyphae. On the underside of the pileus one can find the lamellae or gills, which consist of a thin layer of hyphae that give rise to the basidia and their basidiospores (fig. 14-14). The function of the cap or pileus is to bear the basidia and their basidiospores and to participate in spore dispersal.

When hyphae of two compatible mating strains (for example + and −) fuse, the resulting hypha, containing nuclei from both fungi, is called a **secondary hypha**. Some of the secondary hyphae in the basidiocarp differentiate into basidia (fig. 14-14). The basidia are dikaryotic at first, but as the basidium matures, the nuclei fuse to form diploid nuclei that undergo meiosis almost immediately. The four haploid nuclei that develop from meiosis migrate toward the tip of the basidium and form the basidiospores. Each of these basidiospores is then capable of germinating and giving rise to primary hyphae.

Some of the basidiomycetes are edible. In certain countries, mushroom eating or **mycophagy** is widely practiced. There is always the possibility however, of ingesting poisonous or hallucinogenic mushrooms instead. This points out the importance of carefully determining the identity of the fungus before eating it, because fungi similar to edible ones can be extremely toxic or even lethal. The hallucinogenic mushrooms have been used by many ancient cultures. The Maya and Inca civilizations, for example, used hallucinogenic mushrooms in their religious ceremonies. Some of these religious ceremonies involved the consumption of a toxic mushroom called *Amanita muscaria*, which can be hallucinogenic if eaten in minute quantities but lethal if eaten in even slightly greater quantities.

Figure 14-16 Basidium with Basidiospores. Interpretive drawing of a scanning electron micrograph of basidia from the gill of a basidiomycete. The basidia are shown with and without attached basidiospores.

Fungi Imperfecti or Deuteromycetes

The **Fungi Imperfecti,** or **Deuteromycetes**, constitutes a very large group of fungi with unknown genetic affinities. As you may have noticed, all major groups of fungi discussed thus far are characterized by their unique sexual reproductive structures. The fungi imperfecti cannot be classified using sexual criteria, because they have no demonstrable sexual reproductive cycle. Some of the fungi in this group may have the ability to reproduce sexually, but their sexual cycle has not yet been observed; others may have lost this ability altogether. In either case, all fungi lacking a demonstrable sexual cycle are placed in the fungi imperfecti. As **mycologists** (students of fungi) learn more about the fungi and develop new methods for their cultivation and for promoting the sexual reproductive cycles, more of the fungi imperfecti will probably exhibit their sexual cycles. Great strides have already been made in this respect and mycologists have uncovered the sexual cycles of many fungi, most of which have turned out to be ascomycetes.

TABLE 14-4
TYPES OF CONIDIA BASED ON THEIR DEVELOPMENT

DEVELOPMENTAL PROCESS	COMMON NAME	APPEARANCE	EXAMPLE
Thallic	Arthroconidium	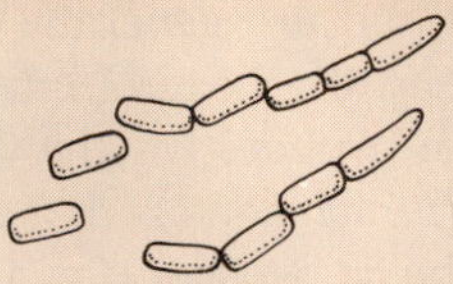	*Geotrichum candidum* *Coccidioides immitis*
Blastic	Blastoconidium	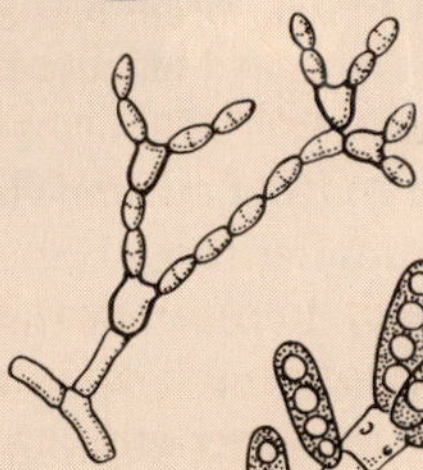	Yeasts *Cladosporium*
	Poroconidium	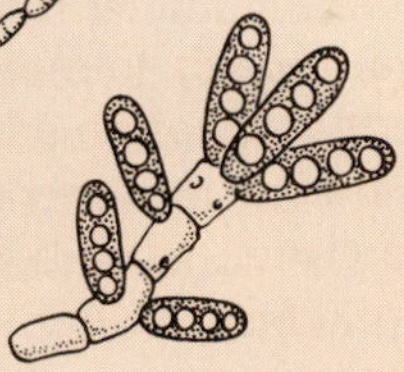	*Alternaria* sp. *Drechslera* sp.
	Sympoduloconidium		*Sporothrix schenckii*
	Annelloconidium	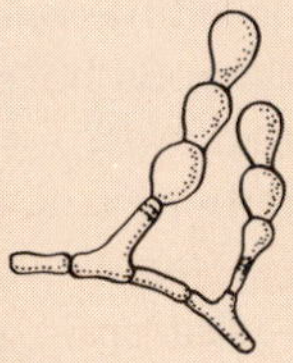	*Scopulariopsis*
	Phialoconidium	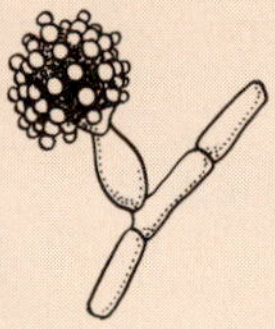	*Phialophora* *Aspergillus* *Penicillium*

The fungi imperfecti produce a large variety of conidia (from the Greek, meaning "fine dust") that can be used as a means of classifying these organisms. Development of the various types of conidia is controlled by a variety of different genes and is characteristic of species. Mycologists have developed a scheme for the classification and identification of the fungi imperfecti based on how their conidia develop. Table 14-4 summarizes the various types of conidia and their mechanisms of development.

Basically, there are two types of conidial development: **blastic** and **thallic** (fig. 14-17). During blastic development, the hyphal tip that will give rise to a new conidium swells and differentiates before the septum separating the hyphae and the new conidium is formed. In thallic development, the septum is formed prior to the differentiation of the conidium.

Many of the fungi imperfecti are plant and animal pathogens, which can cause disease either by destroying cells during their reproduction or by producing toxins. **Aflatoxin**, which is produced by *Aspergillus flavus*, causes severe liver damage and sometimes death in animals. Other fungi imperfecti are of economic importance because they make useful products. For example, *Pencillium notatum* produces the antibiotic penicillin, which is widely used to treat a large variety of infectious diseases caused by bacteria.

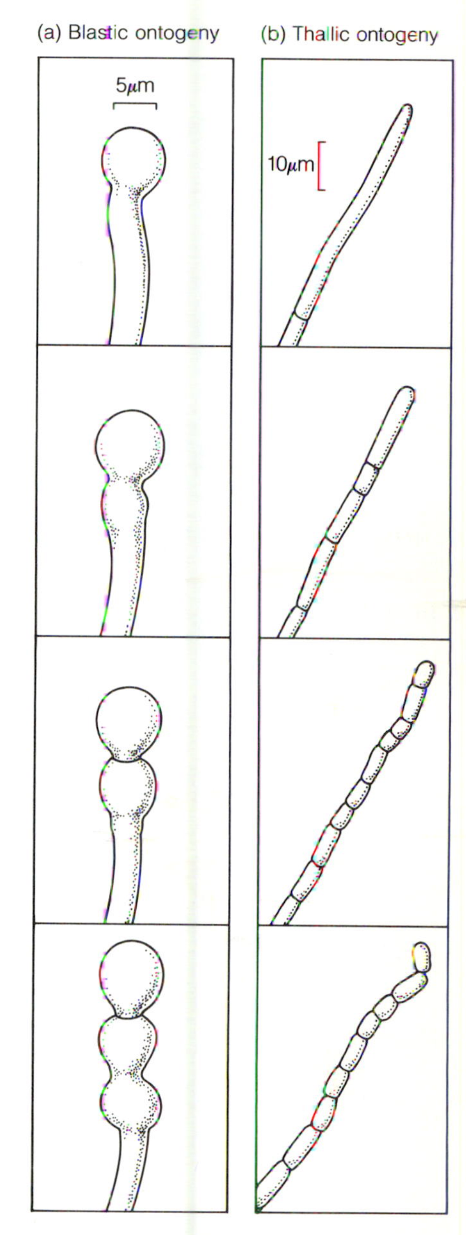

Figure 14-17 Conidial Classification Based on Ontogeny (Mechanism of Development). *(a)* Blastic ontogeny. *(b)* Thallic ontogeny.

Slime Molds

The **slime molds** (table 14-3) are a group of eukaryotic organisms that share characteristics with both the fungi and the protozoa. The slime molds are fungi-like during certain stages of their life cycle and protozoa-like during others.

Myxomycetes (a class of slime molds) form sporangia with spores that can germinate and give rise to amoebae and, therefore, a new generation of slime molds. The amebas, which represent the vegetative stage of these organisms, are similar to protozoa. When the amebas coalesce, they form a structure called a **plasmodium**, which lacks a cell wall. The plasmodium can obtain nutrients by ingestion of food particles. A mature plasmodium can feed upon bacteria, yeast cells, and fungal spores, as well as absorb dissolved materials.

The slime molds live in cold, moist, shady places, such as the environment under the canopy of forests. They feed upon decaying vegetable matter such as leaves, dead logs, or animal dung. The slime molds are a very important group of decomposers of leaf litter in forests. During the spring months, while it is still cold, one can visit the forest and be awed by the colorful display of fruiting bodies of the organisms on fallen logs and decaying vegetable matter. Some species of slime molds form enormous pseudoplasmodia (see below), measuring several feet in width.

A group of slime molds called the **cellular slime molds** (table 14-3) which belong to the class **Acrasiomycetes,** has been studied extensively beause these organisms present a good model for the study of cellular differentiation in eukaryotes. Certain cellular slime molds can be grown easily in the laboratory and therefore serve as a useful organism for research.

When the spores of a cellular slime mold, such as *Dictyostelium discoideum*, are placed in a nutrient medium containing *Escherichia coli* cells, the spores germinate and give rise to motile cells known as **myxamebas** (fig. 14-18). The myxamebas creep over the surface of the agar and feed upon the

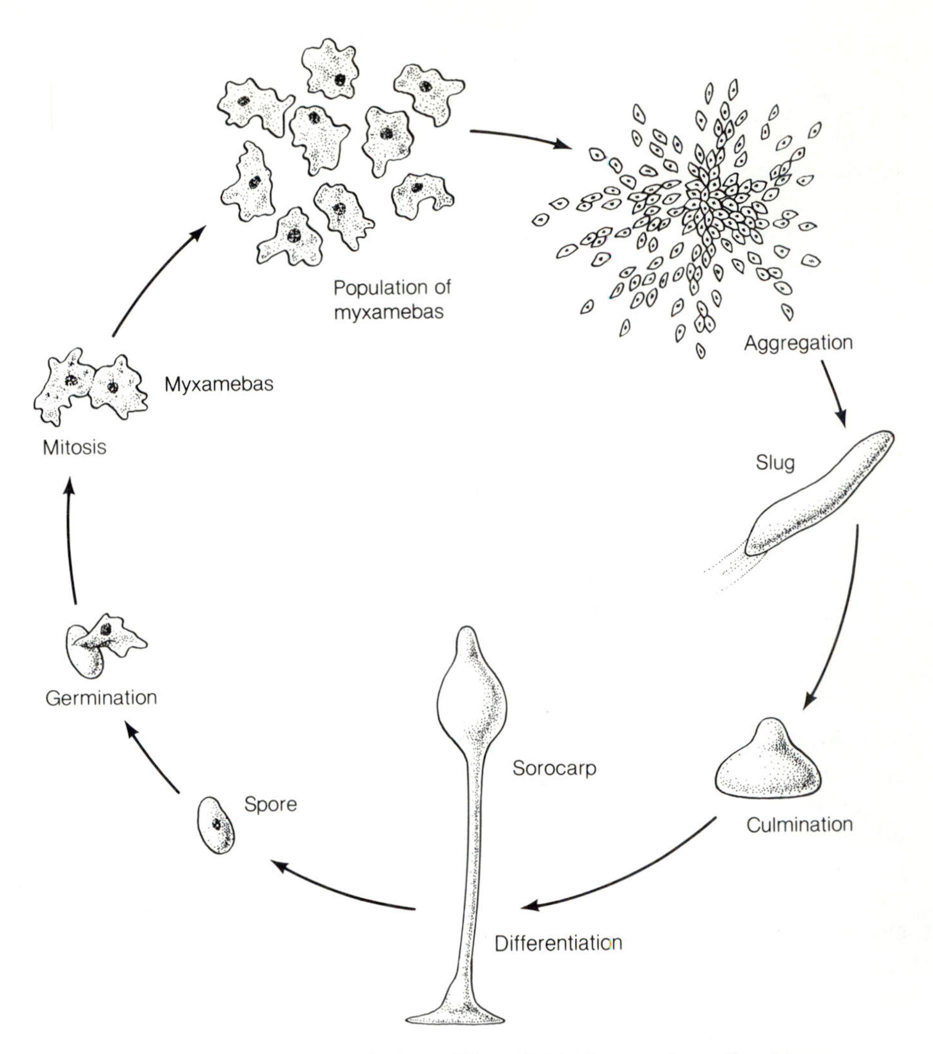

Figure 14-18 Life Cycle of the Cellular Slime Mold *Dictyostelium discoideum*

bacterial cells. The myxamebas multiply on the surface of the agar medium by mitosis. When the supply of nutrients is nearly exhausted, the myxamebas aggregate. They stop feeding and set up **centers of aggregation**, orienting themselves toward these centers. Eventually, they move toward the centers of aggregation, fuse with each other, and develop into what is called a **pseudoplasmodium**.

What is the nature of the stimulus that is responsible for the aggregation of the myxoamoebae and the formation of a pseudoplasmodium? In the case of *D. discoideum*, the best studied species in this group, the starving myxamebas secrete a substance called **acrasin**. Acrasin is now known to be cyclic adenosine monophosphate (cAMP). This substance serves as a chemoattractant that guides the organisms toward the centers of aggregation. The pseudoplasmodium further differentiates into a mature **sorocarp** with walled spores. At this point, the sorocarp resembles the sporangia of the zygomycetes.

SUMMARY

Fungi are heterotrophic, eukaryotic microorganisms that lack chlorophyll.

BIOLOGY OF THE FUNGI

1. The fungi characteristically form **hyphae** that consist of elongated cells or chains of cells. The hyphae absorb nutrients from the environment.
2. The hyphae of fungi grow by extension of their tips.
3. The yeasts are unicellular fungi and rarely form hyphae.
4. Most fungi can produce sexual and asexual spores. These reproductive structures may be used as an aid in differentiating among the fungi.
5. Dimorphism is a phenomenon in which certain fungi can have two different morphologies, depending upon the environmental conditions under which they reproduce. Some human pathogens exhibit dimorphism.

DISTRIBUTION AND ACTIVITIES OF FUNGI

1. The fungi play a major ecological role as decomposers. They are largely responsible for the decaying of organic matter and the recycling of nutrients from organic materials in many environments.
2. As decomposers, fungi may cause a variety of diseases, both in animals (and humans); and in plants.
3. Many metabolic waste products produced by fungi are of industrial importance. For example, the production of antibiotics and alcohol are both industrial processes carried out primarily using fungi.

THE MAJOR GROUPS OF FUNGI

1. Fungi are classified on the basis of their reproductive structures. Table 14-3 summarizes the different groups of fungi and their salient characteristics.
2. Fungi reproduce sexually by fusing two compatible haploid cells to form a diploid zygote. The zygote subsequently undergoes **meiosis** to form haploid sexual spores.
3. Fungi reproduce asexually by forming various types of spores by **mitotic** divisions.

STUDY QUESTIONS

1. Describe and diagram a typical fungus.
2. How do yeasts differ from other fungi?
3. Outline the process of hyphal growth. Why are fungal colonies generally circular?
4. Outline the major characteristics of the various groups of fungi.
5. Cite three industrial processes that employ fungi.
6. Describe briefly three fungal activities that are undesirable to humans.
7. What is dimorphism? How can it benefit the fungus? How can microbiologists take advantage of this characteristic to diagnose a fungal disease?
8. What is a conidium? Of what use is it in identification of a fungus? Of what use is it to the fungus? Do sporangiospores serve the same function?
9. Why is sexual reproduction important to the fungi?
10. How do fungi (including yeasts) differ from bacteria?
11. How do the fungi imperfecti promote genetic variation? Accompany your explanation with a drawing.

SUPPLEMENTAL READINGS

Alexopoulos, C. J. and Mims, C. W. 1979. *Introductory mycology*. 3d ed. New York: John Wiley & Sons.

American Society for Microbiology, Board of Education and Training. 1978. *Identification of saprophytic fungi commonly encountered* in the clinical laboratory. Washington-American Society for Microbiology.

Christensen, C. M. 1951. *The molds and man: An introduction to the fungi*. Minneapolis: University of Minnesota Press.

Kendrick, B., ed. 1979. *The whole fungus*. Ottawa: National Museums of Canada.

Lennette, E. H., Balows, A., Hausler Jr., W. J., and Shadomy, H. J. 1985. *Manual of Clinical Microbiology*. 4th ed. Washington D.C. American Society for Microbiology.

McGinnis, M. R., D'Amato, R. F., and Land, G. A. 1982. *Pictorial* handbook of medically important fungi and aerobic actinomycetes. New York: Praeger.

CHAPTER 15
THE ALGAE AND THE PROTOZOA

CHAPTER PREVIEW

NAEGLERIA FOWLERI: AN AGENT OF FATAL MENINGITIS

Natural bodies of water are inhabited by a myriad of microorganisms, many of which are important to the aquatic community. These microorganisms, which include bacteria, fungi, algae, and protozoa, contribute to the overall "health" of the lake and are prominent members of lakes and streams. Because of them, many lakes and streams are suitable for human recreation (e.g., swimming and fishing) and as reservoirs for drinking water.

Amebas, single-celled, nonphotosynthetic eukaryotes, abound in these aquatic environments. Like other members of the aquatic community, they contribute positively to the lake and stream ecology by decomposing organic matter and consuming excess microorganisms that may foul the water.

Some of these amebas, however, may cause severe and often fatal diseases in humans. For example, *Naegleria fowleri*, an ameba that measures 10 to 20 μm in diameter, reproduces in stagnant waters and can cause severe (and fatal) swelling of the brain and spinal chord.

In 1965, three fatal cases of meningitis were reported in Australia. All three cases were caused by the amebas multiplying in the membranes that envelop the nervous system (meninges). In 1966 and 1968, similar cases were reported in Florida. All the cases had one thing in common: the victims were young people who had been swimming in lakes and ponds about a week before their affliction. By 1973, more than 70 cases of amebic meningitis caused by *Naegleria fowleri* had been reported from throughout the world.

In 1978 a young English girl was reported dead of amebic meningitis. The girl was known to frequent the Roman Baths in England. Since the Roman Baths were routinely cleaned and chlorinated, the source of the infection was a puzzle. On close examination, however, the amebas were found to reproduce in pockets behind the small cracks in the pool's wall, where they were protected from the harmful effects of chlorine.

A large number of cases of amebic meningitis have been reported from all over the world and in some cases bathing places have been closed or quarantined because they represented public health hazards.

How does *Naegleria* enter the body and infect the central nervous system? Apparently, the amebas enter through the nose and penetrate by breaching the mucous membranes. From there they enter the brain and multiply at its base. The amebas may then invade the spinal fluid, from which they can be isolated and cultured. The disease begins with a frontal headache, a fever, and a blocked nose. An altered sense of smell and/or taste, along with other central nervous system disorders, are evident as the disease progresses, until the patient is successfully treated (with a drug called amphotericin B) or dies.

The discovery that free-living amebas can cause a fatal disease in humans emphasizes the fact that microorganisms influence human lives significantly. Some of their activities are essential to our well-being, while others are detrimental. *Naegleria fowleri* plays both roles. As a free-living organism, this ameba helps maintain the health of the aquatic environment. As a pathogen, it causes severe disease. In this chapter we will discuss the major groups of protozoa and algae and emphasize their importance to humans.

LEARNING OBJECTIVES

A STUDY OF THIS CHAPTER SHOULD ENABLE YOU TO:

DESCRIBE WHAT PROTISTS ARE AND HOW THEY ARE CLASSIFIED

OUTLINE THE MAJOR CHARACTERISTICS OF THE ALGAE AND THE FUNCTIONS THEY PERFORM IN NATURE

DESCRIBE HOW THE VARIOUS ALGAE ARE GROUPED AND THE CHARACTERISTICS OF EACH GROUP

EXPLAIN WHAT PROTOZOA ARE AND HOW THEY DIFFER FROM OTHER EUKARYOTES

CITE SOME OF THE WAYS IN WHICH PROTOZOA AFFECT HUMANS

OUTLINE THE CHARACTERISTICS OF THE PRINCIPAL GROUPS OF PROTOZOA

The **protists** traditionally have included all those organisms that are predominantly single-celled eukaryotes: many of the algae and all the protozoa. Because there are numerous species in each of these two groups, many of which share characteristics, biologists have had difficulty in agreeing how best to classify the protists. This difficulty has been aggravated by the fact that the fossil record contains limited evidence on the existence of these organisms. This chapter will discuss the major groups of algae and protozoa, emphasizing their structure, means of reproduction, and medical and ecological importance.

INTRODUCTION TO THE EUKARYOTIC ALGAE

The vast expanses of oceans and other bodies of water are the most common habitat for a group of photosynthetic eukaryotes called the **algae**. Although the algae are primarily aquatic organisms, many live in terrestrial habitats, and several species of algae may be found growing on soils, wood, and "bare" rock.

The algae are found in two different kingdoms: the **Protista,** and the **Plantae.** The large, multicellular algae, such as the marine **kelp**, are placed in the kingdom Plantae, while the microscopic, single-celled eukaryotic algae are in the kingdom Protista.

The algae, as a group, play an essential part in the biosphere. For example, the algae produce much of the oxygen in the atmosphere and are a very important source of food for aquatic communities. In the open sea, the microscopic algae float near the surface of the water and constitute a part of the **plankton**. Plankton, a term derived from the Greek *planktos* or "wandering," describes microscopic animals and plants that move with the wind and currents in fresh and salt waters. **Phytoplankton** is a name given to the algal component of plankton, while **zooplankton** is the name given to the animal constituents of plankton. Together, the zooplankton and the phytoplankton constitute a very important source of food for all other water-dwelling organisms. The phytoplankton contribute to our well-being, not only because they provide oxygen but also because they produce much of the food for higher organisms □ (fig. 15-1). Besides their role as producers of biomass, the algae are important in many other respects. For instance, kelp such as *Laminaria* and *Nereocystis*, commonly found beyond the breaker zone, provide shelter as well as food for a vast community of marine animals. Certain algae, the **dinoflagellates** in particular, can produce substances that are toxic to humans.

744

The Biology of the Algae

The algae are quite varied in their morphology (fig. 15-2). Some forms are unicellular, while others are colonial (aggregates of single cells), cylindrical, or foliose (leaflike). Some single-celled algae are as small as bacteria, while others, like the giant kelp, are as large as trees. Regardless of their morphological variability, however, algae are photosynthetic, eukaryotic organisms. Most of the multicellular algae have simple sex organs, such as antheridia (male reproductive structures) and oogonia (female reproductive structures). Nearly all algae display some form of sexual reproduction. In this process,

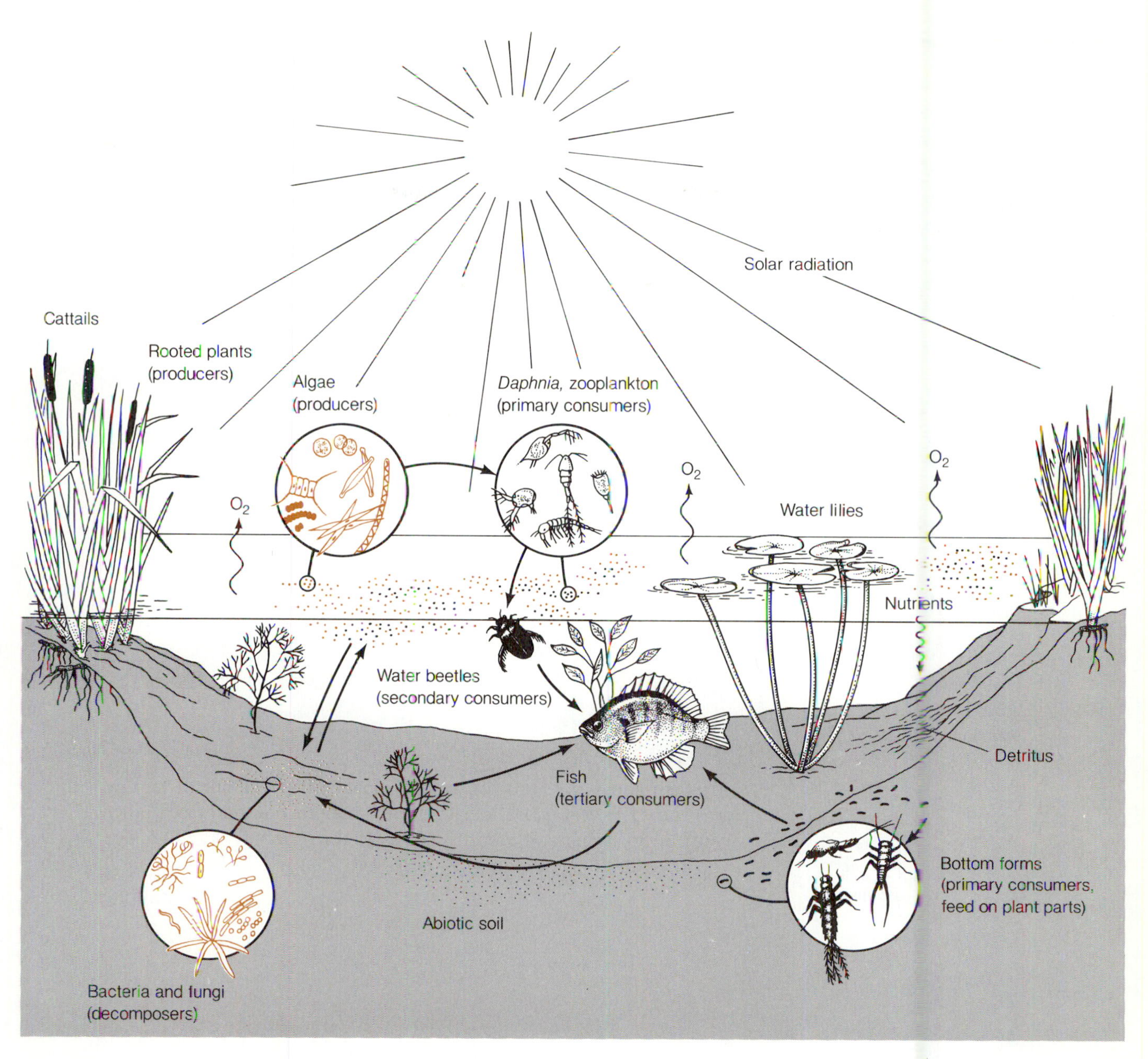

Figure 15-1 A Simplified Food Chain. Phytoplankton algae comprises the primary food and energy source for aquatic ecosystems. They utilize the energy from the sun to generate stored chemical energy and then convert atmospheric carbon dioxide into organic materials. These organisms are the primary source of food for other aquatic organisms, which in turn serve as food for others. Together, all the organisms comprise what is called a food chain.

229 there is an exchange of genetic material □ between compatible organisms, giving rise to genetically similar (but not identical) progeny. Some of the algae have specialized cells that function as gametes, while in others, such as *Chlamydomonas*, vegetative cells can function as gametes. Although all 194 the algae are potentially photoautotrophs □, they vary extensively in their morphology and ecology. Table 15-1 summarizes a common classification scheme for the eukaryotic algae.

Figure 15-2 **Illustrations of Representative Algae.**

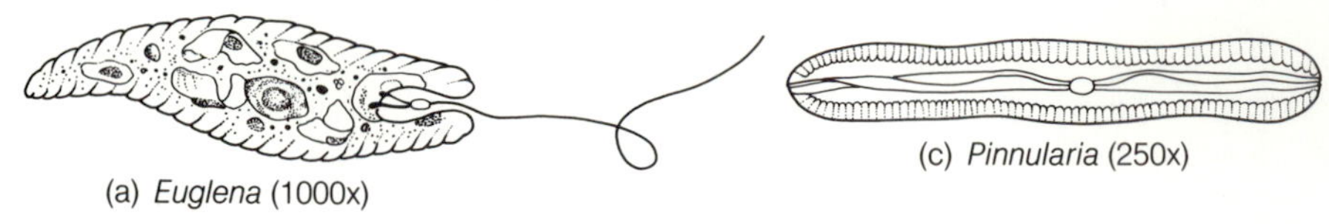

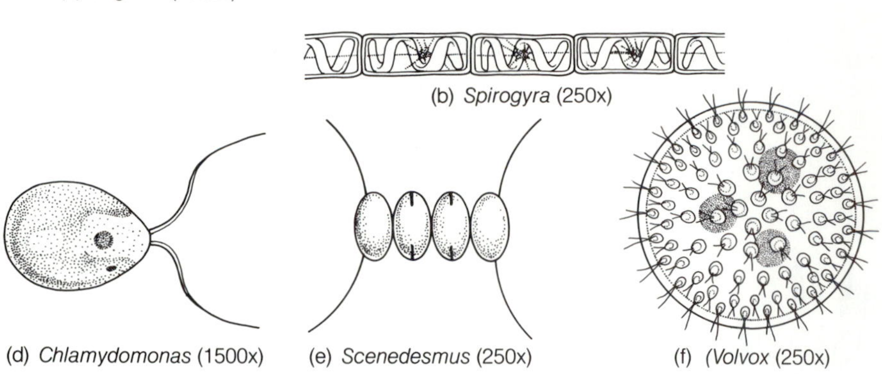

TABLE 15-1

SUMMARY OF CHARACTERISTICS OF MAJOR GROUPS OF EUKARYOTIC ALGAE

DIVISION	PIGMENTS FOR PHOTOSYNTHESIS	NUMBER OF THYLAKOIDS PER STACK	FLAGELLAR CHARACTERISTICS	CELL WALL COMPOSITION	NUTRIENT RESERVE
Chlorophyta (Green Algae)	Chl* *a*, Chl *b* α, β, γ,–carotenes xanthophylls	2–5	1 to many apical, equal in length	Cellulose	Starch
Phaeophyta (Brown algae)	Chl *a*, Chl c α, β -carotenes xanthophylls fucoxanthine	2–6	2, lateral, unequal in length	Cellulose, Alginic acid	Laminarin, Mannitol
Rhodophyta (Red algae)	Chl *a* α, β-carotenes phycocyanin allophycocyanin phycoerythrin xanthophylls	1	Not present	Cellulose, Xylans	Floridean, Starch
Chrysophyta (Diatoms)	Chl *a*, Chl *c* β, ϵ -carotenes xanthines fucoxanthines	3	2, apical	Cellulose, Silica, $CaCO_3$	Chrysolaminarin, Oils, and fats
Pyrrophyta (Dinoflagellates)	Chl *a*, Chl *c* β-carotene xanthophylls	3	2, one apical and 1 lateral	Cellulose, mucilagenous substances	Starch
Euglenophyta (Euglenoids)	Chl *a*, Chl *b* β-carotene xanthophylls	2–6	1 to 3 (to 7) apical or subapical	None	Paramylon

*Chl = chlorophyll

Classification of the Algae

The eukaryotic algae can be classified into six divisions (table 15-1) based on characteristics such as motility, types of photosynthetic pigments, nature of the reserve materials, mode of reproduction, and chemical composition of the cell wall. These characteristics, when analyzed together, constitute a chemico-morphological profile characteristic of each group. We would like to note, however, that we have chosen a six-Division classification scheme because of its simplicity. Some authorities classify the eukaryotic algae in eight Divisions and include a ninth one, the **Cyanochloronta**, that includes all of the blue-green algae (cyanobacteria).

Many of the algae have a combination of photosynthetic pigments that give them their characteristic colors (table 15-1). The algae generally all have chlorophyll *a*, but some groups also possess chlorophylls *b* and/or *c* as well as a variety of carotenoids. Besides the variation in photosynthetic pigments, algae differ in their principal food reserve. Most of these reserve materials are carbohydrates, such as **starch**, **paramylon**, **mannitol**, or **laminarin** (table 15-1).

Many algae, or their gametes, are motile by means of flagella; one group, the **diatoms**, may be motile by gliding. Algal flagella are typically eukaryotic and resemble those found in fungi and protozoans. Motile algae can be differentiated on the basis of a) the number of flagella; b) their attachment site on the cell) equal or unequal); and d) whether the flagella are **whiplash** or **tinsel**.

All algae except the euglenoids have a rigid cell wall. The cell wall gives the cell its shape and rigidity, provides protection from physical stress, and may serve as a passive osmoregulator by resisting excessive swelling of the cell in hypotonic environments □. The various groups of algae can also be differentiated by the chemical composition of their cell walls, which typically consist of cellulose and pectinate materials. Molecules such as silicon, alginilic acid, and calcium carbonate also contribute to the structure of cell walls.

118

THE VARIOUS GROUPS OF ALGAE

Green Algae (Chlorophyta)

The **green algae** constitute a large and extremely important group of photoautotrophic microorganisms. They exhibit a wide range of morphological types, including the unicellular alga *Chlamydomonas*, the filamentous alga *Spirogyra*, (fig. 15-3) and leafy forms such as *Ulva* (sea lettuce).

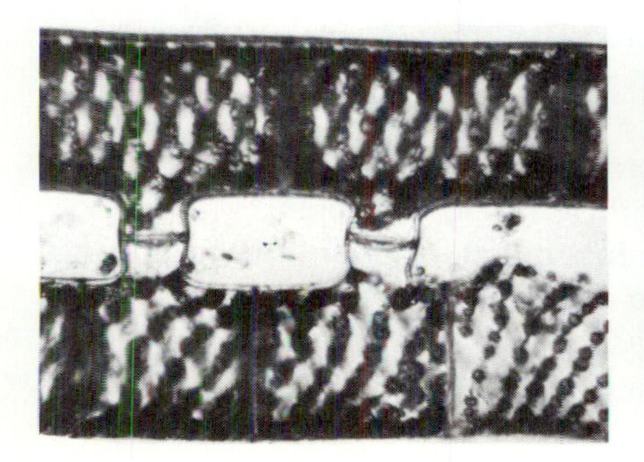
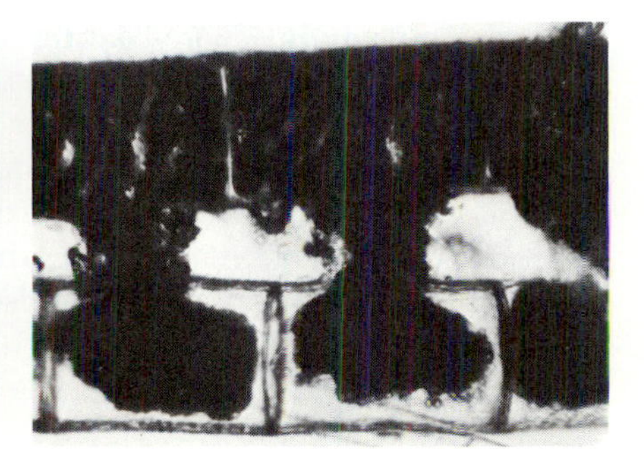

Figure 15-3 Sexual Reproduction in *Spirogyra*. Sexual reproduction takes place when conjugation tubes form between adjacent filaments. The cell's contents pass through these tubes. The nuclei fuse to form a thick-walled zygospore. When the zygospore germinates, vegetative filaments of *Spirogyra* form.

Green algae reproduce both sexually and asexually. Asexual forms of reproduction involve filament fragmentation, binary fission, or the formation of motile cells called **zoospores**. The green algae also reproduce by sexual means. One type of sexual reproduction, characteristic of *Chlamydomonas*, involves the fusion of haploid (n) cells to form diploid (2n) zygotes. The zygotes subsequently give rise to haploid daughter cells by the process of meiosis □. Another type, which takes place in *Spirogyra*, involes the transfer of hereditary material from one cell to another (fig. 15-3).

234

Most species of green algae are found in freshwater habitats, but several species occur on tree trunks, in soils, and in marine environments. Some species of free-living algae can also be found in symbiotic associations □ with fungi to form **lichens**. The wide distribution of these algae may be the result of their ability to tolerate a variety of different conditions such as salinity, nutrient concentrations, and temperature.

426

The green algae provide much of the oxygen necessary for aerobic respiration. In addition, they serve as a source of nourishment for many aquatic animals found in lakes, ponds, and other bodies of water.

The green algae are capable of growing in **oligotrophic** (low-nutrient) environments because of their ability to fix CO_2 and to utilize inorganic nutrients for their metabolism and reproduction. When they multiply in these environments, the algal population increases in size and creates a source of nutrients for other organisms. In sewage treatment plants, the filamentous green algae often form masses of filaments in drainpipes and on filters, thus clogging up these water passages. Some species of green algae are also capable of releasing toxic materials that can make animals ill if they drink waters contaminated with toxigenic (toxin-producing) algae.

Brown Algae (Phaeophyta)

The **brown algae**, also known as **kelp**, are mostly marine forms that vary greatly in size. Some brown algae, such as *Macrocystis*, may be 100–250 feet long. These marine brown algae generally grow in cool waters along the coast beyond the surf and can form extensive kelp beds that provide food and shelter to a variety of marine animals. *Sargassum* is a complex brown alga that forms enormous masses of kelp floating on torrid waters. The Sargasso Sea, an expanse of Atlantic Ocean northeast of the Caribbean Islands, gets its name because these brown algae grow there profusely. *Sargassum* has a very complex anatomy that resembles that of the green plants.

Many species of kelp have structures called **holdfasts** that attach the **sporophyte** (fig. 15-4) or spore-producing form of the alga to the ocean bottom. Many species have gass-filled bladders called **floats** that maintain the kelp floating in an upright position. The **blades** are leaflike structures that bear **sporangia** (spore-bearing structures), which are attached to the main trunk of the sporophyte by means of **stipes**. The brown algae obtain their brown to olive drab color from a xanthophyll (a photosynthetic pigment) called **fucoxanthine** that predominates in these algae. The kelp also have various other carotenoids, chlorophyll *a*, and chlorophyll *c*. The most common food storage material is **laminarin** rather than starch, although mannitol and lipids are sometimes also accumulated.

The brown algae have been used as food by oriental cultures for centuries. The importance of these algae as a potential source of food has recently

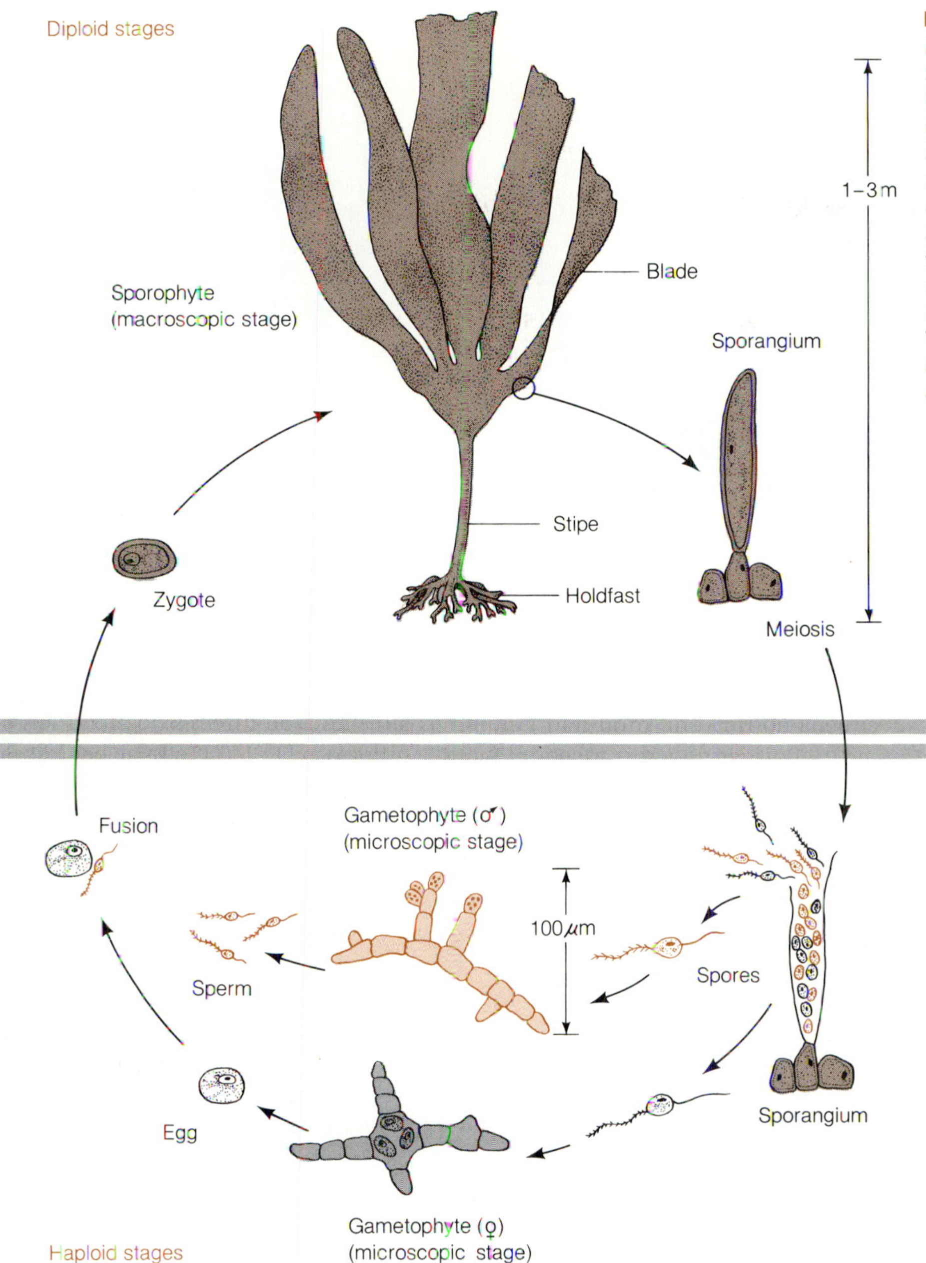

Figure 15-4 Life Cycle of the Kelp *Laminaria*. *Laminaria* produces motile, haploid zoospores. These give rise to gamete-bearing microscopic organisms called gametophytes, which, in turn, produce eggs or sperm. The sperm are motile. When an egg and sperm fuse, they form a diploid zygote, and this develops into a sporophyte. The sporophyte is a very conspicuous structure that is seen floating on the ocean or strewn along the shoreline. The sporophyte contain many sporangia with numerous zoospores. The zoospores serve to disperse the brown alga to other areas and to form new generations of kelp.

been realized by western civilizations, and several companies have begun harvesting them. The brown algae are used in pet foods because they are highly nutritious and rich in vitamins. One algal product, **algin**, deserves special mention in view of its many uses: it has been used as a constituent of adhesives and plastics, and is added to many dairy products in order to improve their texture. Algal products have also been used as lubricants, jellies, and ointments.

Red Algae (Rhodophyta)

The red algae (fig. 15-5) constitute a large group of marine and freshwater algae, which includes more than 4,000 species. These algae have a variety of

(a)

(b)

Figure 15-5 The Red Algae *(a)* Gelidium and *(b)* Gracilaria

photosynthetic pigments, including chlorophyll *a* and phycoerythrin; the latter gives the red algae their characteristic color.

The red algae are generally found in warm tropical seas, although a few species inhabit freshwater habitats. In contrast to other algae, the red algae are able to colonize deep waters by virtue of their photosynthetic pigments. The phycobilins permit these algae to absorb light in the shorter wavelengths (blue and violet), which penetrate water to a greater depth than the red and orange wavelengths. Many of the red algae grow and become attached to the surface of other algal types and are called **epiphytic** algae.

Several species of red algae, such as *Gracilaria* and *Gelidium* (fig. 15-5), are sources of the polysaccharide **agar** □ that is used routinely in microbiology to harden culture media. Agar is used not only for the preparation of solid culture media but also for pill capsules and for the preparation of jellies. Another red algal product, **carrageenan**, is a colloid used to stabilize emulsions such as paints, cosmetics, and dairy products.

119

Diatoms (Chrysophyta)

The diatoms are unicellular, phototrophic microorganisms that constitute the bulk of the phytoplankton. These microorganisms are enclosed by sculptured shells of various geometrical shapes, and are made of silica (fig. 15-6). The cell walls or **frustules** of diatoms consist of two shells, or **valves**, that fit together like the lid and bottom of a petri dish. The valves contain pores that allow nutrients and wastes to be exchanged with the environment.

The diatoms reproduce, for the most part, by asexual means. The most familiar stage in their life cycle is the asexual stage or the valved structure. Sexual reproduction involves the fusion of motile sperm with eggs, giving rise to diploid zygotes or **auxospores**. It is the mature auxospores that develop into the valved structure. Diatom multiplication occurs when the valves separate at intervals and each synthesizes another, complementary valve. This asexual reproduction is diagrammed in figure 15-7. Occasionally, some of the diatoms undergo meiosis, giving rise to haploid gametes.

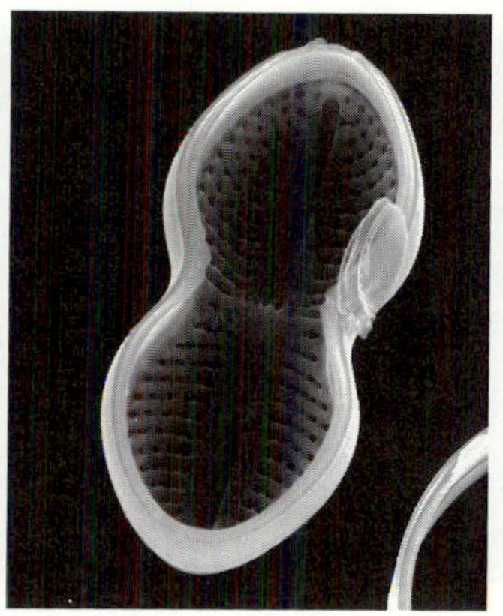
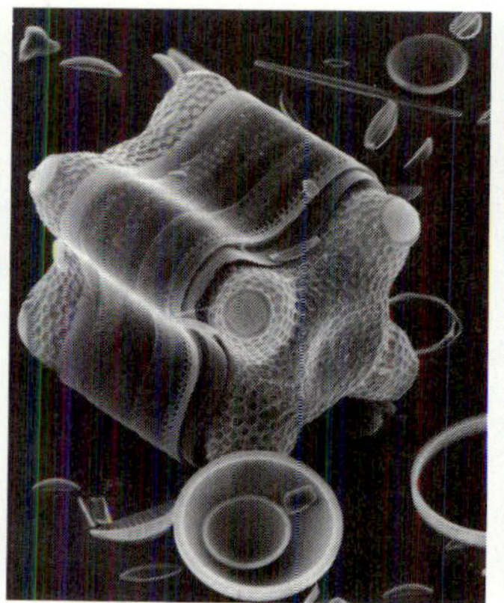
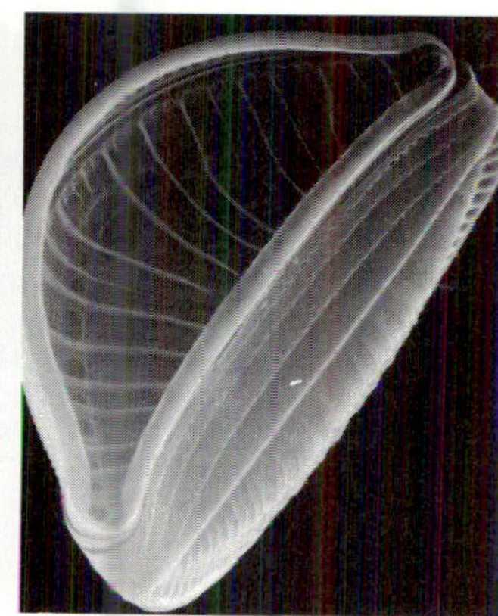

Figure 15-6 Diatoms

110

The diatoms, like other algae, are photoautotrophs □. Some species, however, have lost the ability to photosynthesize and must obtain their energy from chemotrophic metabolism and their carbon from organic molecules.

Many species of diatoms exhibiting bilateral symmetry are capable of gliding motility. Their motility apparently is related to the secretion of a slime through the groove in their shell called the **raphe** (figure 15-6). It is hypothesized that the slime released from the diatom connects the diatom's plasma membrane to the surface on which it is moving. As the membrane moves along the raphe, the diatom also moves.

FOCUS

ALGAE ARE USED IN MANY MODERN INDUSTRIES

Brown and red algae, or seaweeds, have been used by many Eastern cultures as food sources. For example, the red alga *Porphyra*, known also as **nori**, has been cultivated by many generations of Chinese and Japanese people. This red alga is presently very popular in the Far East as a food. In Japan alone, the exploitation of nori involves more than 40,000 workers. Other cultures eat their native seaweeds instead of *Porphyra*.

Although seaweed, composed mainly of cellulose, is not a particularly nutritive source of amino acids, it contains many ions, vitamins, and minerals that are required in a balanced diet. Seaweed, considered by some as a "health food," has increased in popularity as a source of food as well as economically valuable materials.

Kelp such as *Macrocystis* are brown algae commonly found in temperate waters. They are grown commercially and harvested for fertilizers and as sources of iodine, sodium, and potassium salts. *Macrocystis* also contains certain substances, collectively called **alginates**, which are used commercially as thickening and/or stabilizing agents in such diverse industries as the cosmetic, food, paper, pharmaceutical, and textile industries.

Certain red algae contain agar in their cell walls. In addition to its use as a solidifying agent for culture media, agar has been used in bakery goods to prevent drying, in pills, and as a preservative. Much of the agar used in the world is produced in Japan. Carrageenan is another red algal product of economic importance. It is used primarily to stabilize emulsions, such as paints, cosmetics, and dairy products.

The algal industry is still in its nascent stage. In the ensuing years we will probably see an increase in the number of industries that exploit algae as foods or for economically important products. The renewed interest in the algae will no doubt reveal new uses for it.

Figure 15-7 Life Cycle of a Diatom. Diatoms reproduce primarily by asexual means, producing bivalved structures. During this process, valves separate, each forming a complimentary valve. Occasionally, diatoms undergo meiosis, forming gametes that eventually fuse to form zygotes called auxospores. The auxospores synthesize valves, which reproduce asexually.

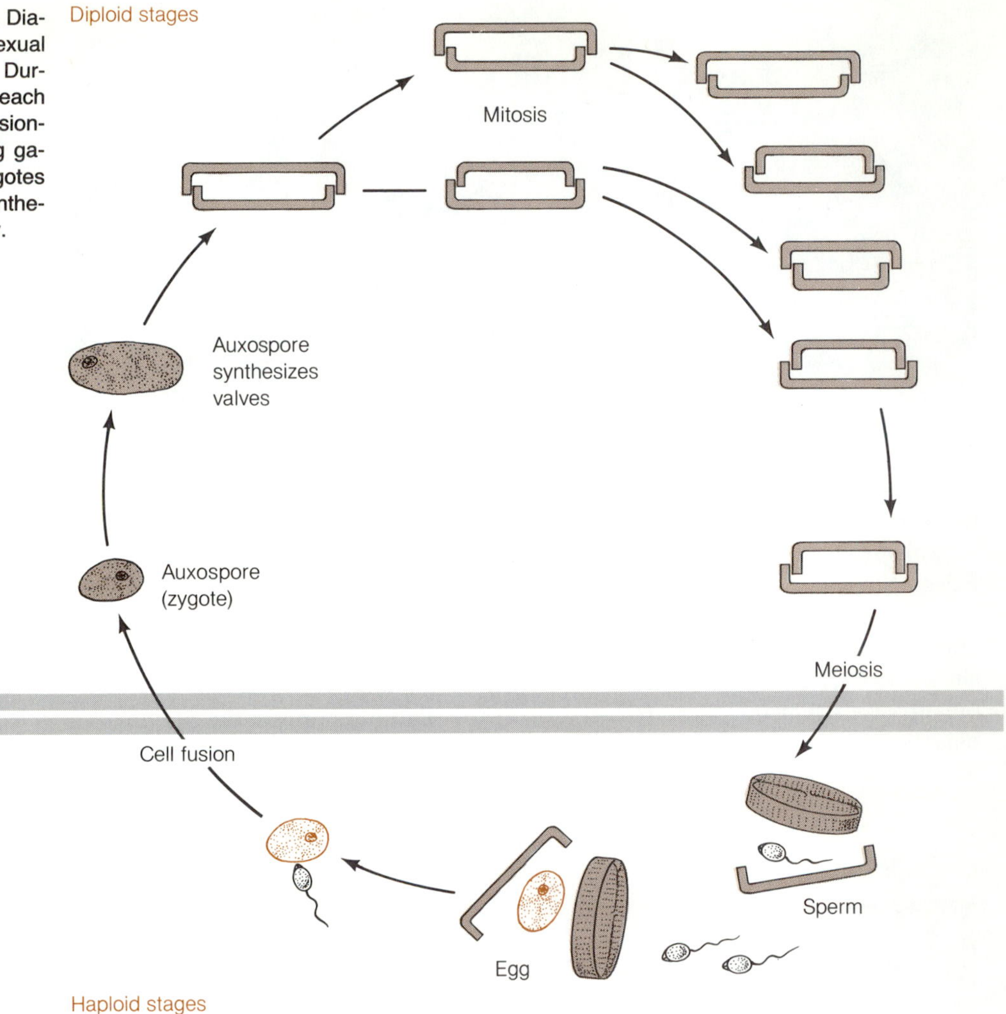

Since diatoms constitute a major portion of the phytoplankton, they contribute significantly to the biomass (mass of living organisms) that produces nutrients for higher organisms. Over many millions of years, the remains of diatom shells have formed deep sediments all over the world, called **diatomaceous earth**. Diatomaceous earth is used in many cleansers and polishing agents because of its abrasive nature. In addition, diatomaceous earth is used to filter drinking water because the rough texture and pores in the diatom shells catch organic debris and microorganisms.

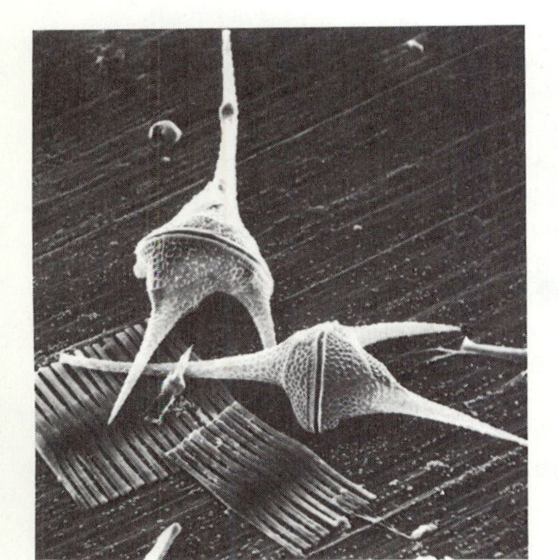

Figure 15-8 The Dinoflagellate *Ceratium*

Dinoflagellates (Pyrrophyta)

The dinoflagellates are single-celled microorganisms with two lateral flagella (fig. 15-8) that make the organisms whirl. While the majority of the algae obtain their nutrients by absorbing dissolved materials (osmotrophic), certain of the dinoflagellates can ingest food particles as well. Some species of dinoflagellates are protected by a thick cell wall called the **theca**, which is generally made of cellulose. The photosynthetic pigments of the dinoflagellates include chlorophyll *a* and chlorophyll *c*, as well as a variety of carotenoid pigments that contribute to the overall color of these organisms. Like some of the diatoms, a few dinoflagellates are **achlorophyllous** (without chlorophyll) and therefore heterotrophic □. In fact, some wall-less dinoflagellates resemble protozoa.

110

The dinoflagellates do not have very complex life cycles. They multiply primarily by mitosis, but they may also proliferate by fragmentation or by spore formation. The cell division of dinoflagellates differs from that of many eukaryotes in that the chromosomes of dinoflagellates are attached to the nuclear membrane during cell division. The attachment of chromosomes to a membrane is a characteristic of prokaryotes.

The dinoflagellates, like the diatoms, are very important sources of food for a variety of aquatic organisms. Dinoflagellates have achieved notoriety because some of them produce toxins that are lethal to fish. During periods of extensive algal growth, fish often die when they feed upon the toxigenic dinoflagellates. In addition, dinoflagellates such as *Gonyaulax excavata* or *G. catanella* are eaten by shellfish. The shellfish are not affected by the dinoflagellate, even though it produces toxins inside them. The most potent of these is the **saxitoxin** produced by *G. catanella.* Saxitoxin affects the central nervous system of many animals and can be lethal to humans. When extensive dinoflagellate growth occurs (a condition also called **red tides**), public health agencies often quarantine the shellfish to prevent their consumption by human populations. The disease caused by this toxin is called **paralytic shellfish poisoning** and is contracted by ingestion of shellfish that have been feeding upon these dinoflagellates.

Euglenoids (Euglenophyta)

The euglenoids are a small group of algae that bear a remarkable resemblance to the protozoa. The euglenoids lack a cell wall, although they have a reinforced cell membrane called the **pellicle** (fig. 15-9). The pellicle consists of a plasma membrane that is corrugated and strengthened with underlying microtubules and other proteins. Much of the energy for cellular functions is derived from photosynthetic reactions. In the dark, however, euglenoids can obtain energy by chemotrophic □ metabolism.

186

Some species of euglenoids move by ameboid (creeping) motion, while other species are motile by means of flagella. Since the euglenoids lack a cell wall, they are sensitive to changes in osmotic pressure □. To avoid cell lysis, many euglenoids form **contractile vacuoles** (fig. 15-9) that serve to rid the cell of excess water.

118

The euglenoids have chlorophyll *a* and chlorophyll *b* as well as several carotenoids in their chloroplasts. Some species of euglenoids have an eyespot, or **stigma,** that serves as a photosensitive organ to orient the cell toward the source of light. These organisms store their excess photosynthetic products in the form of a starchlike polysaccharide called **paramylon.**

The euglenoids are freshwater life forms that serve as a source of food for other freshwater organisms. The euglenoids contribute to the productivity of lakes and ponds by providing some of the necessary organic matter for many aquatic organisms.

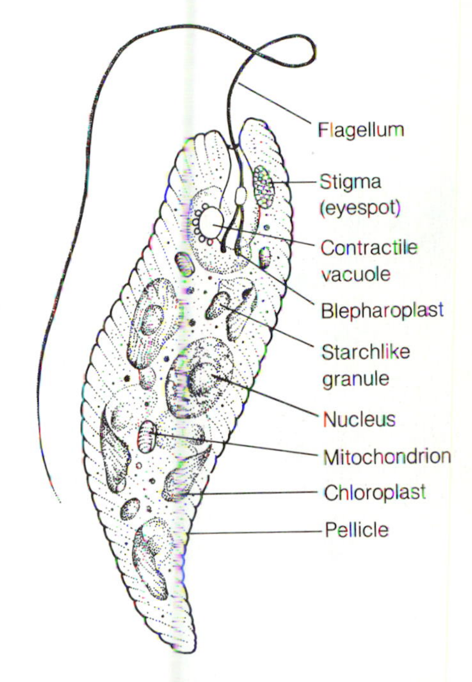

Figure 15-9 Diagram of *Euglena gracilis*

Lichens

The lichens are plantlike growths that consist of algae and fungi (fig. 15-10). The algal symbiont is capable of free-living existence as a phototroph, but

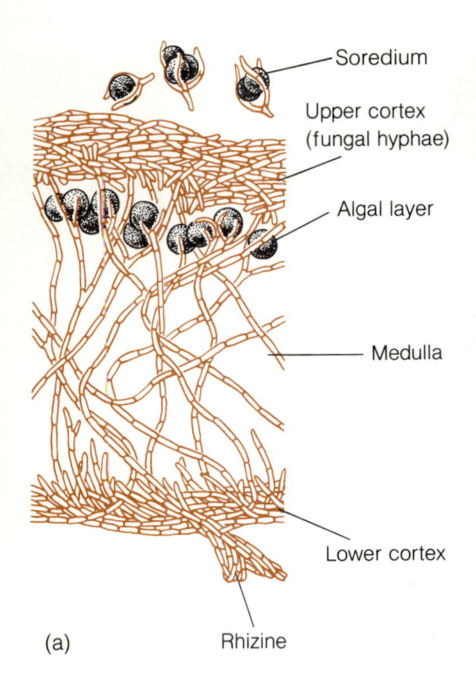

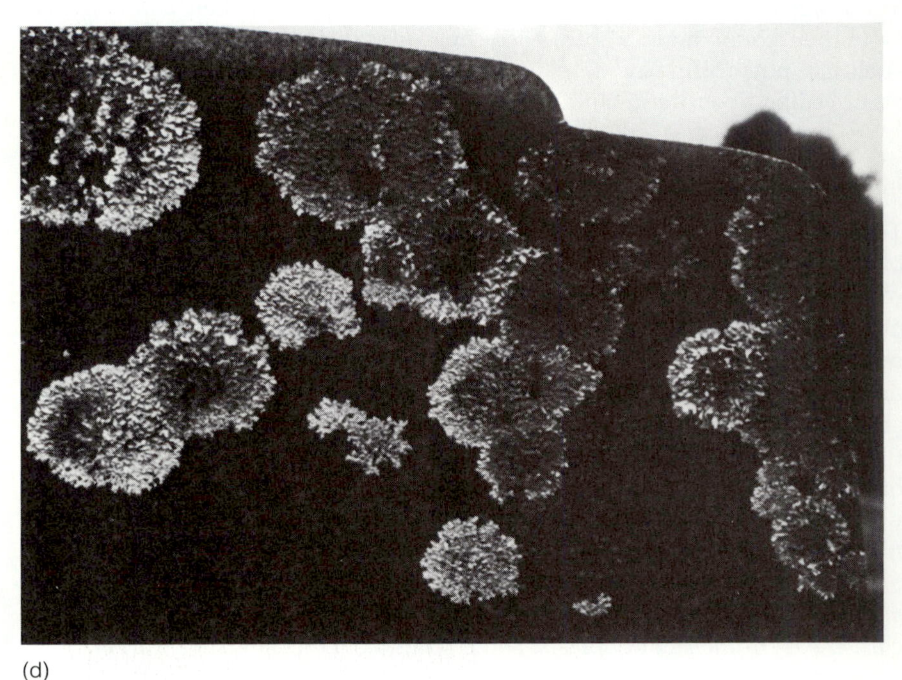

Figure 15-10 The Structure of a Lichen. *(a)* Diagram of the cross section of a lichen and the arrangement of the algal and fungal symbiont. *(b)* A foliose lichen on a tree trunk. *(c)* A fruticose lichen on a tree trunk. *(d)* A crustose lichen on a gravestone.

the fungal symbiont seldom exists as a free living organism. The lichen-forming fungi are nearly all ascomycetes. Species of green algae such as *Trebouxia* and *Trentepholia*, together with the cyanobacterium *Nostoc*, make up the bulk of the lichen-forming algae.

The lichens are classified on the basis of their morphology and include crustose, foliose, and fruticose forms. **Crustose** lichens are crustlike growths found on the surfaces of tree trunks and rocks. Next time you are on a hike, examine closely the large rocks along your path. No doubt you will find an abundance of gray, brown, and green crustose lichens. **Foliose** lichens are leaflike. The predominant color of these lichens is gray-green. These lichens are usually attached to rocks and trees. **Fruticose** lichens are shrublike, highly branched structures attached to tree trunks, branches, etc.

The lichens may reproduce by the formation of **soredia,** small masses of algal and fungal cells knitted together, which serve to disperse these organisms to other areas (fig. 15-10).

INTRODUCTION TO THE PROTOZOA

The protozoa constitute a diverse group of chemoheterotrophic eukaryotic microorganisms widely distributed in nature (table 15-2). They all lack a cell wall and their cytoplasm contains prominent nuclei, mitochondria, Golgi bodies, lysosomes, and vacuoles (fig. 15-11). In other words, they are typical eukaryotic microorganisms. Protozoa gather nutrients by absorbing them through the plasma membrane or by engulfing particles and microorganisms by phagocytosis.

Reproduction

Protozoa reproduce by asexual and sexual means. Asexual reproduction may consist of binary fission (fig. 15-12) or multiple fission referred to as **schizogony** (fig. 15-12). Sexual reproduction may involve the transfer of hereditary material between two cells or fusion of two cells or gametes to give rise to diploid zygotes. Both sexual and asexual reproduction may be integrated into a single, complex life cycle.

TABLE 15-2
CHARACTERISTICS OF THE MAJOR GROUPS OF PROTOZOA

TAXONOMIC GROUP	LOCOMOTION	ASEXUAL REPRODUCTION	SEXUAL REPRODUCTION	GENERA OF IMPORTANCE
Phylum: Ciliophora Subphylum: Ciliata (ciliates)	Ciliary movement	Transverse fission	Conjugation	*Balantidium* *Paramecium* *Vorticella*
Phylum: Sarcomastigophora* Subphylum: Sarcodina (amebas)	Pseudopodia	Binary fission	Gamete fusion	*Entamoeba* *Amoeba* *Naegleria*
Subphylum: Mastigophora (flagellates)	Flagellar movement	Binary fission	Not seen	*Giardia* *Trypanosoma* *Trichomonas*
Phylum: Apicomplexa Class: Sporozoa (sporozoans)	Usually not motile	Schizogony (multiple fission)	Gamete fusion	*Plasmodium* *Toxoplasma* *Eimeria* *Cryptosporidium*

*fleshy, flagellated organisms.

Figure 15-11 Representative Protozoa

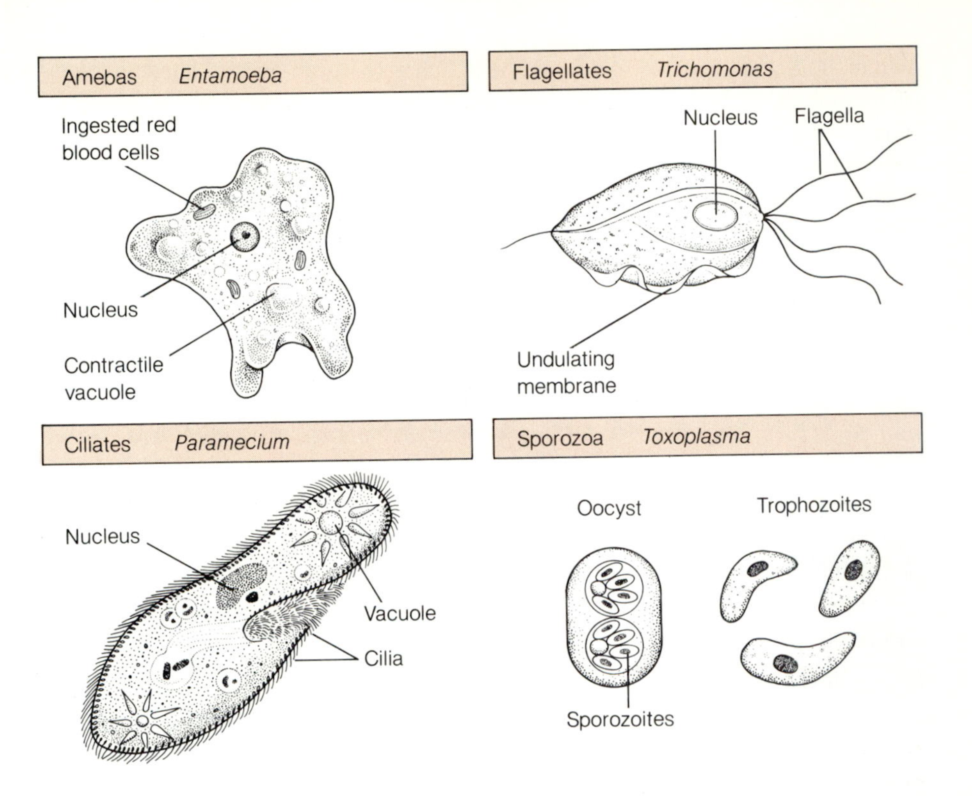

Importance

The protozoa are extremely important to humans because of their ability to cause infectious diseases. Malaria, African sleeping sickness, and kala-azar are among the many protozoal diseases that afflict millions of humans each year. In addition, many parasites, especially in a group called the **coccidians**, are capable of infecting and causing disease in nearly all domestic and wild animals. Protozoa also participate in beneficial associations with humans or other animals. For example, a ciliated protozoan forms a long-lasting association with termites □, in which the protozoan digests much of the cellulose consumed by the termites, thus allowing the insect to use wood as a source of food. Ruminants such as cattle and sheep depend on protozoa (among other microorganisms) to digest some of the foods in the rumen. Without these microorganisms, cattle and sheep would not use plant materials efficiently as food sources.

751

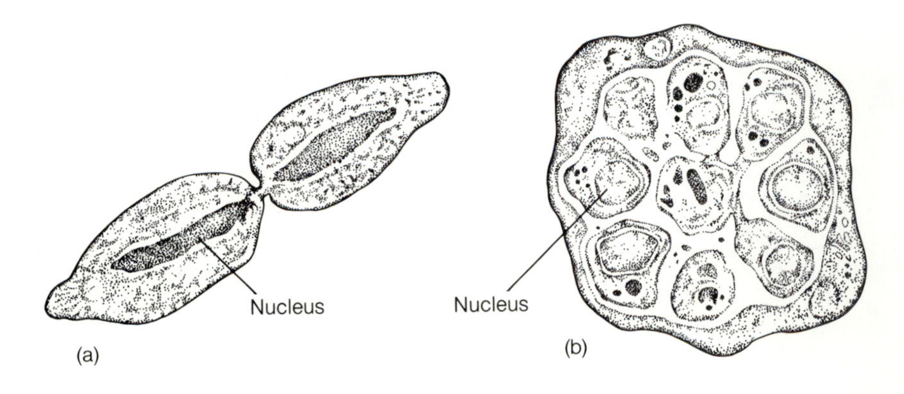

Figure 15-12 Asexual Reproduction of Protozoa. *(a)* Binary fission by *Paramecium.* *(b)* Multiple fission by *Plasmodium.*

THE VARIOUS GROUPS OF PROTOZOA

There are four major group of protozoa of importance to humans: the amebas, the flagellates, the ciliates, and the sporozoans (table 15-2). Their characteristic means of locomotion and morphological characteristics are important critieria used for classification (fig. 15-11).

Amebas (Subphylum Sarcodina)

In essence, **amebas** are single-celled microscopic organisms that float or creep in aquatic environments. They resemble eukaryotic animal cells in that they are chemoheterotrophic and lack a cell wall. The amebas are motile by pseudopodia that serve not only as a means of locomotion but also as a means of gathering food. The amebas obtain nutrients by phagocytosis 109 □ (fig. 15-13). During phagocytosis, their pseudopodia extend around food particles and envelop them. The phagocytized food particles are surrounded by a membrane within the ameba and are then digested by lysosomal enzymes. Table 15-3 summarizes the salient characteristics of some amebas of public health importance.

The life cycle of the amebas is relatively simple. One notable feature of their life cycle is the occurance of a resting, highly resistant structure called the **cyst** (fig. 15-14). Cysts develop from **trophozoites,** which are the active, feeding, reproducing form of the amebas. When nutrients are depleted or other adverse changes occur in their environment, the trophozoite differentiates into a cyst. This process is usually called **encystment.** When the cyst is in an environment conducive to growth, it **excysts,** producing a new trophozoite. This life cycle (fig. 15-14) takes place in certain free-living protozoa, as well as in parasitic forms within their hosts.

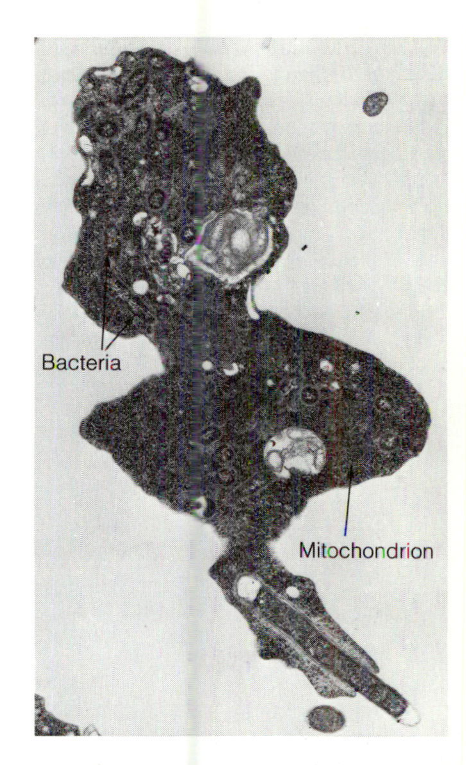

Figure 15-13 Phagocytosis Carried Out by an Ameba. Transmission electron micrograph of *Vahlkampfa,* (a small soil ameba) engulfing a large bacterium. Notice the large number of bacterial cells (arrows) in phagosomes.

TABLE 15-3
CHARACTERISTICS OF REPRESENTATIVE PARASITIC PROTOZOA AND THE DISEASES THEY CAUSE

ORGANISM	GROUP*	DISEASE(S) CAUSED	MODE OF TRANSMISSION	GEOGRAPHY
Entamoeba histolytica	A	Amebic dysentery	Fecal-oral route	Worldwide, tropics
Balantidium coli	C	Balantidiasis	Fecal-oral route	Worldwide
Giardia intestinalis	F	Gastroenteritis	Fecal-oral route	Worldwide
Babesia sp.	S	Babesiasis (cattle)	Tick bites	Worldwide
Eimeria tenella	S	Coccidiosis in chicken	Fecal-oral route	Worldwide
Trypanosoma gambiense	F	Sleeping sickness nagana in cattle	Bite of tsetse fly	Tropical Africa
Trypanosoma cruzi	F	Chagas' Disease	Bite of reduviid bug	Tropical S. America
Leishmania sp.	F	Kala-azar, oriental sore.	Bite of sandfly	Tropics and subtropics
Trichomonas vaginalis	F	Vaginitis	Sexually-transmitted	Worldwide
Plasmodium sp.	S	Malaria	Bite of mosquito	Tropic and subtropics
Naegleria fowleri	A	Meningoencephalitis	Invasion of nasal mucosa	Worldwide
Toxoplasma gondii	S	Toxoplasmosis	Ingestion of cysts	Worldwide

*Group: A = ameba; C = ciliate; F = flagellate; S = sporozoan

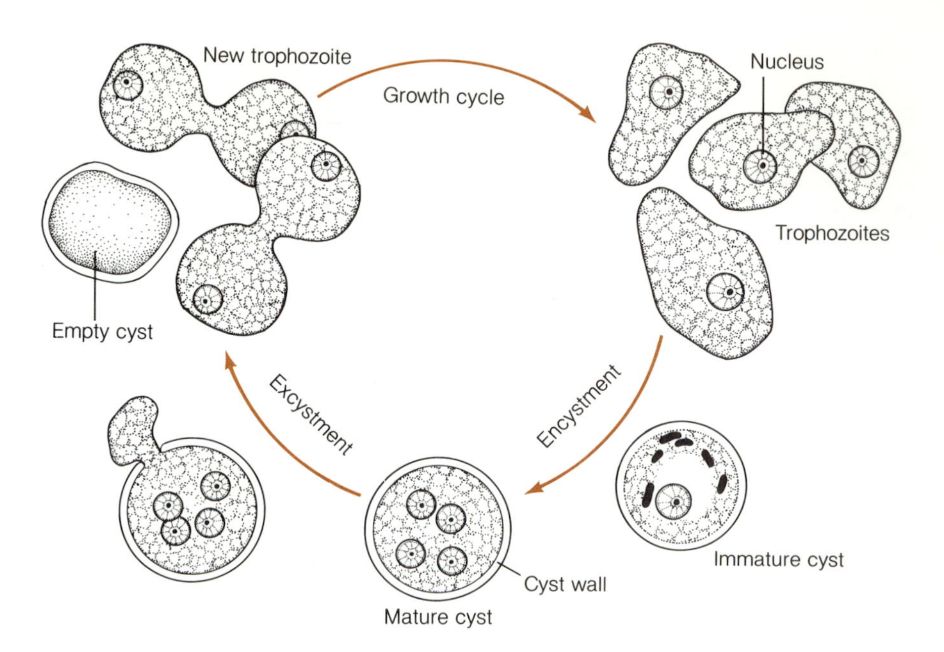

Figure 15-14 Life Cycle of *Entamoeba histolytica*. *Entamoeba histolytica* exists in nature predominantly in the *cyst* form. When humans or other animals consume food contaminated with cysts, they become infected. The cyst germinates (excysts) in the intestines and yields a trophozoite. The trophozoites reproduce by binary fission. Later during infection, the trophozoites may encyst and some of the cysts are passed in the feces. The cysts in the feces may then contaminate foods, which are then infective to other humans or animals.

Amebas are generally free-living protozoa and many of them (e.g., radiolarians and foraminiferans) inhabit marine environments. The amebas participate in the recycling of nutrients by digesting organic matter and by serving as a source of food for other organisms.

Some species of amebas are capable of colonizing humans and causing disease. For example, *Entamoeba histolytica* is an ameba that can be ingested along with contaminated food or berverages and can cause a disease called **amebic dysentery**. This disease is characterized by diarrhea, painful intestinal contractions, and blood and mucus in the stool (feces). Occasionally, this organism can invade other organs such as the liver and cause pus-filled lesions called **abscesses.** When this invasion occurs, the outcome of the infection can be very grave.

There are several species of amebas that parasitize the oral cavity and intestines of humans. For example, *Entamoeba gingivalis* can be found colonizing the human oral cavity, and *Entamoeba coli*, *E. histolytica*, *Endolimax nana*, and *Iodamoeba butschlii* can sometimes be found reproducing in the intestines. Some free-living amebas, such as *Naegleria fowleri*, may cause serious disease of the central nervous system in humans ☐.

353

Flagellates (Subphylum Mastigophora)

The flagellates (table 15-2) are single-celled protozoa that move by means of flagella. They are chemoheterotrophic microorganisms that store starch as a reserve material and reproduce by means of longitudinal binary fission (fig. 15-15). The flagellates are thought to be closely related to certain algae, such as the dinoflagellates.

The flagellates are found in aquatic and terrestrial environments as free-living organisms or forming various types of symbiotic associations ☐ with other organisms. Flagellates such as *Giardia intestinalis*, *Trichomonas vaginalis*, *Leishmania donovani*, *Trypanosoma cruzi*, and *T. gambiense* can cause severe diseases in humans.

746

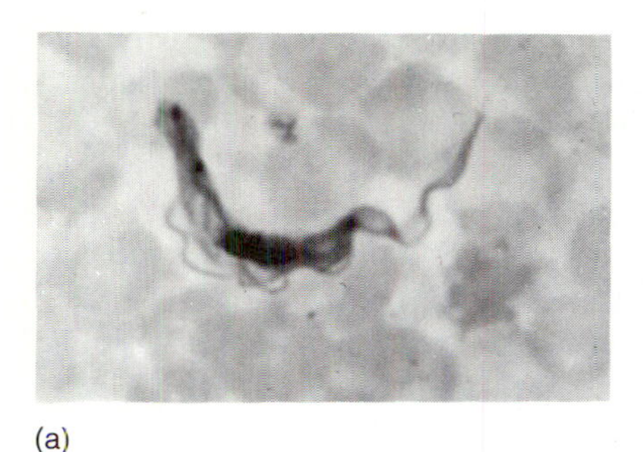

(a)

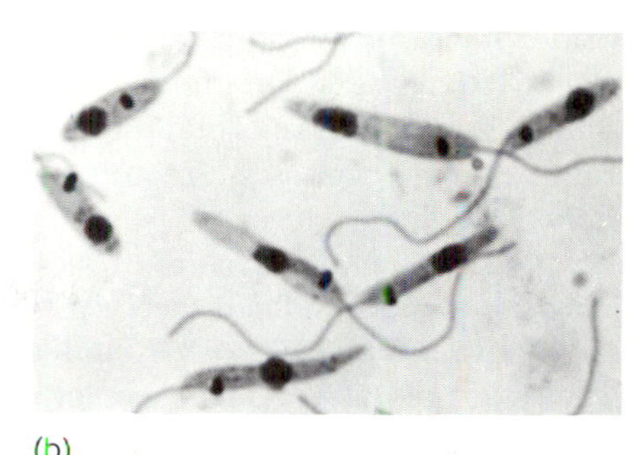

(b)

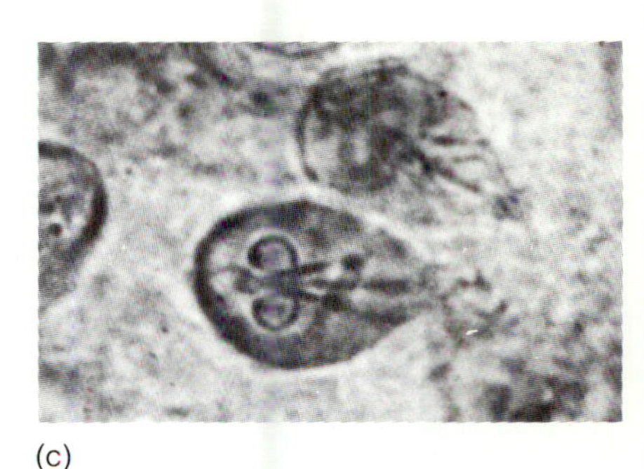

(c)

Figure 15-15 Representative Flagellates. *(a) Trypanosoma.* This flagellate causes serious diseases in humans, including African sleeping sickness and Chaga's disease. *(b) Leishmania.* This organism is the cause of kala-azar, a severe infectious disease of the internal organs. *(c) Giardia intestinalis.* This is a common cause of traveler's diarrhea.

Giardia intestinalis (fig. 15-15) is a common human parasite that inhabits the small intestine and may cause severe infections. Sometimes, *Giardia* infections result in gastrointestinal disorders such as diarrhea, cramps, bleeding, and anorexia. *Giardia intestinalis* exists in two forms: the trophozoite stage and the cyst stage. The pear-shaped trophozoite exhibits bilateral symmetry and has a pair of nuclei. On the ventral side of the trophozoite there is a sucking disc that serves to attach the parasite to the intestinal wall. The trophozoite is motile by means of four pairs of flagella: lateral (side), anterior (front), posterior (rear), and ventral. The beating flagella cause *G. intestinalis* trophozoites to have a typical "falling leaf" motility. The nuclei, when stained, resemble a pair of eyes, giving the impression that this tiny microorganism is looking back at you through the miscroscope. The cyst stage is oval with four prominent nuclei (fig. 15-15). This structure is quite resistant to adverse environmental conditions, and it represents the infectious form of this parasite. The cysts of *Giardia* represent a public health hazard when present in drinking water supplies. They are not killed by the chlorine used to decontaminate the water, and can be infectious if ingested along with the water.

Trichomonas vaginalis (fig. 15-11) is another important human parasite. This organism is transmitted by sexual intercourse. In heavily infected females, *T. vaginalis* causes an inflamation of the vagina that is characterized by a profuse, malodorous discharge. The organism differs from *G. intestinalis* in that it lacks bilateral symmetry. *Trichomonas* is characterized by a tuft of anterior flagella and undulating membrane associated with a posterior flagellum.

The **trypanosomes** are very important human pathogens. This family includes a rather diverse group of flagellates that can exist in the latex of plants, the gut of insects, and the blood of most vertebrates. These organisms cause very serious diseases that afflict millions of humans throughout the world, including **kala-azar, Chagas' disease,** and **African sleeping sickness.**

Leishmania donovani (fig. 15-15) causes the often fatal disease kala-azar. Inside the human, it exists in a stage called the **amastigote**, oval cells that lack a visible flagellum. The amastigotes of *L. donovani* exist primarily inside cells of the **reticulo-endothelial** □ system of humans. The vector for this protozoan is the sandfly *Phlebotomos*. When it bites an infected host, the

461

sandfly ingests some of the amastigotes in the blood. In the sandfly's midgut, the amastigotes develop into **promastigotes**, elongated forms of the parasite with a short flagellum. The promastigotes divide in the midgut of the fly and migrate to its proboscis (mouth parts). When the fly bites an uninfected individual, it introduces into the new host numerous promastigotes, which can then initiate an infection in the host. The promastigotes are ingested by phagocytes, within which they again differentiate into amastigotes.

Trypanosoma (fig.15-15) causes Chagas' disease and African sleeping sickness (table 15-3). These are primarily parasites that invade the blood, although sometimes they can be found multiplying in brain or heart tissue. The agent of African sleeping sickness, *Trypanosoma gambiense*, is transmitted to humans by the bite of the tsetse fly (*Glossina*). The agent of Chagas' disease, *T. cruzi*, is transmitted to humans by reduviid bugs.

Ciliates (Subphylum Ciliophora)

The ciliated protozoa are prodominantly single-celled organisms characterized by the presence of cilia (fig. 15-16). The cilia of these microorganisms not only function in locomotion but also participate in the feeding process. Beating cilia create currents that draw food particles into the oral (mouth) opening of the protozoa. Morphologically, the cilia of these organisms are similar to the cilia of other eukaryotes □, consisting of nine pairs of peripheral tubular fibrils and a central pair of fibrils in a 9 + 2 arrangement, sometimes called **axonemes**. Like all cilia, the axoneme is surrounded by an outer envelope that is indistinguishable from the plasma membrane. The cilia grow out from **basal bodies** underlying the plasma membrane.

98

Some species of ciliates, particularly those that inhabit the intestines of mammals (e.g., *Balantidium coli*), have ciliated trophozoites and nonciliated cysts. Many species of ciliates are free-living organisms that act as decomposers □ of organic matter. In sewage treatment plants, ciliates help in the purification of water by breaking down the organic matter. Some ciliates, like *B. coli*, are capable of causing human disease. The ciliates are seldom parasitic, however, although they are often symbionts. They are found in the rumen (a stomach) of cattle, where they help the ruminants digest their food, and in the gut of termites, where they help in the digestion of cellulose.

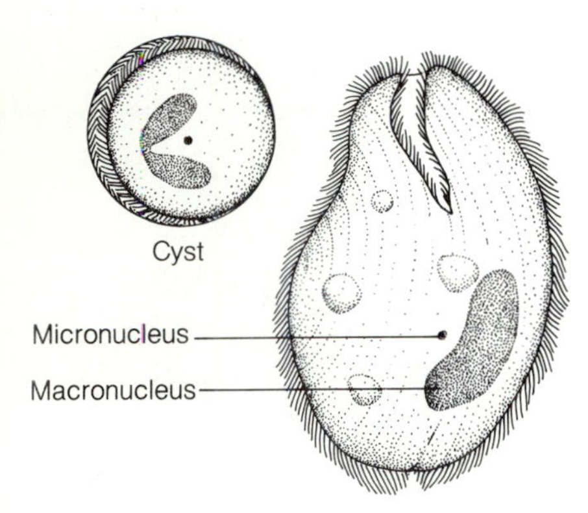

Figure 15-16 ***Balantidium coli.*** This organism infects humans when they ingest foods or beverages containing cysts. In the intestines, the parasite excysts, releasing a trophozoite. The trophozoite multiplies by transverse fission. Some of the trophozoites may encyst and be passed out in the feces. Encysted organisms may remain dormant for many days until they are consumed by another individual.

The Sporozoans (Phylum Apicomplexa)

The sporozoa are protozoa that are characterized by the production of spores at some stage in their life cycle. Without exception, the sporozoa are parasites □ of animals. Their life cycles are often quite complex and involve alternating sexual and asexual reproductions in different hosts. Usually, the **final host** harbors the sexual forms while the **intermediate host** harbors the asexual forms. In the case of malaria, a disease that afflicts humans, the final host is the mosquito while the intermediate host is the human. There are several important species of sporozoa. Several sporozoans, collectively called **coccidians**, affect livestock. Coccidians reproduce in the digestive tract of many animals, causing severe diarrheal diseases. One such disease is coccidiosis, which can spread very rapidly among a population of animals. For instance, *Eimeria tenella* can destroy an entrie flock of chickens in a few days. These sporozoans have a tremendous biotic potential: *one* infectious

426

coccidian (sporozoite) ultimately results in the formation of over one million infectious cells.

Toxoplasma gondii is another sporozoan (a coccidian) that infects humans. The final host for *T. gondii* is the cat and other felines. In the cat, male and female *T. gondii* gametes fuse to form a zygote that develops into an **oocyst**. The mature oocyst, containing **sporocysts** with **sporozoites**, is shed in the feces. Mature oocysts can last for more than a year in soils, sandboxes, and cat litterboxes. The most common route of infection for humans is by ingestion of the oocyst. House flies or cockroaches can serve as a vehicle of transmission for the disease agent. Inside the host, the oocyst releases sporocysts that contain sporozoites. The sporozoites migrate to the intestinal epithelial cells and infect them. The sporozoites then transform and undergo a process of multiple fission called **schizogony** within the infected cells. The cells derived from schizogony are called **merozoites** (or trophozoites). The merozoites can then invade other intestinal epithelial cells and divide by schizogony, thereby inititating the cycle once more. This is the most common course of the human disease. The merozoites can exist inside cysts that contain slowly-dividing merozoites. The merozoites can migrate to the brain and heart and develop into cysts in these organs. Merozoites of *T. gondii* can be transferred across the placenta to the fetus, so this organism is of great public health importance. Infected fetuses can be born with congenital malformation as a result of toxoplasmal infection.

In order to diagnose and treat diseases caused by sporozoa, it is necessary to become familiar with several characteristic stages in their life cycle, which may include many different hosts.

The sporozoan *Plasmodium* (fig. 15-17) causes malaria in humans. The sporozoite, the infective stage for humans, develops from the oocyst in the stomach epithelium of the mosquito. The oocyst forms from the fusion of a male gamete (microgamete) and a female gamete (macrogamete). The sporozoites liberated from the oocyst in the mosquito stomach migrate to the salivary glands and are then injected into the susceptible human host by the bite of the mosquito. The sporozoites leave the blood and immediately enter

Figure 15-17 Various Stages of *Plasmodium* in the Blood of Humans. *(a)* Trophozoites of *Plasmodium vivax*. *(b)* Schizont of *Plasmodium vivax*. *(c)* Gametocyte of *Plasmodium falciparum*.

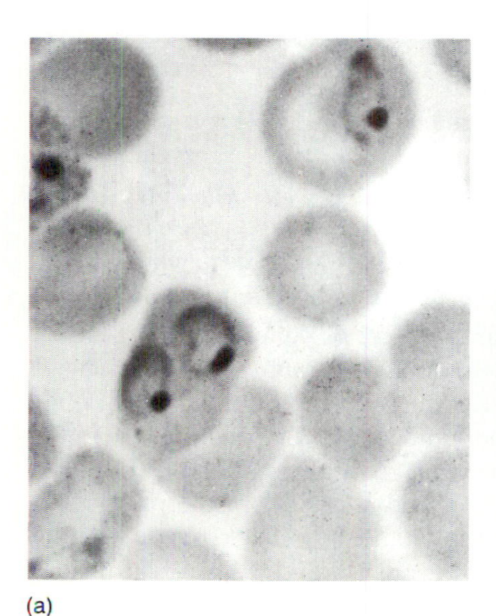

(a)

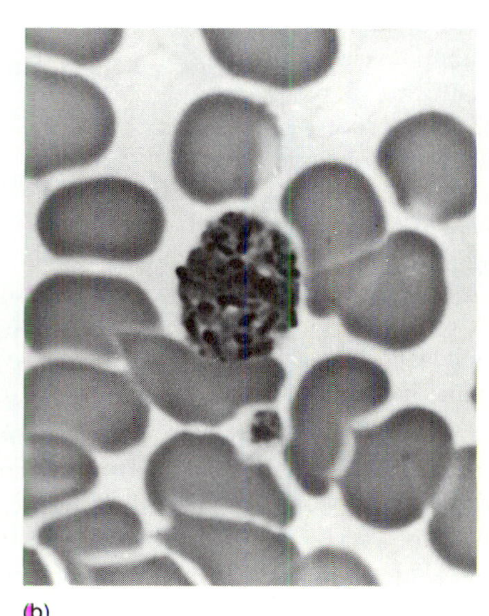

(b)

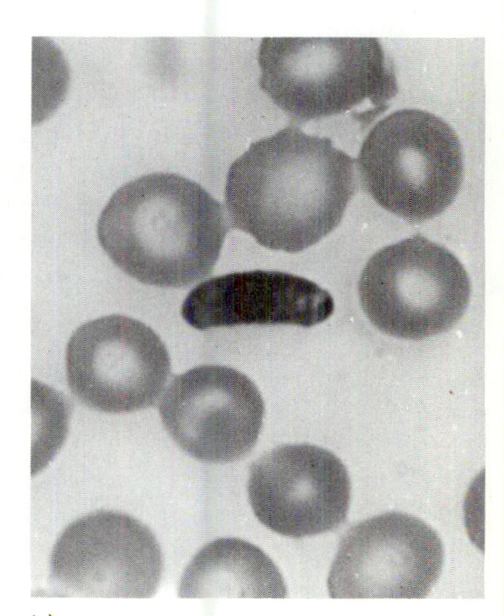

(c)

the liver. There the parasite undergoes asexual reproduction, producing numerous merozoites. Eventually (7–10 days post-infection) the merozoites leave the liver and enter the bloodstream, infecting the red blood cells. In the red blood cells, the malarial parasites continue to reproduce rapidly. When the merozoites complete each of their reproductive cycles in the blood, they burst the red blood cells, releasing into the circulation toxic substances that cause chills and fever, the characteristic symptoms of malaria □. The infecting merozoites may also develop into gametocytes, which are then released from the red blood cells and may be picked up by another mosquito upon biting an infected host. In the mosquito, the gametocytes differentiate into gametes that mate and give rise to oocysts.

699

The protists constitute an integral part of our environment. Many of the functions they perform are essential for the survival of organisms in the biosphere. Certain species, however, are capable of producing metabolic products or colonizing human and animal organs, causing disease, death, and/or financial loss. Regardless of these detrimental influences that certain protists have on human populations, these organisms represent a very important part of our environment and must be studied in order to understand the effects that our actions may have on other organisms sharing this planet.

SUMMARY

1. The algae and the protozoa are widely distributed in nature and have a significant impact on nature, because they are responsible for the accumulation of biomass and for organic decomposition.
2. The protozoans are important to humans because they are responsible for a variety of serious human and animal diseases, many of which affect millions of human throughout the world.

INTRODUCTION TO THE EUKARYOTIC ALGAE

1. The eukaryotic algae are classified into six divisions based on morphological and structural criteria.
2. The algae perform two essential functions in our environment: they release large quantities of oxygen, which is necessary for the respiratory processes of other life forms, and they constitute a significant portion of the biomass of terrestrial and aquatic habitats. It is this biomass that provides the nutrients required by many other organisms.

THE VARIOUS GROUPS OF ALGAE

1. Certain algae, such as the dinoflagellates, produce toxins that occasionally cause human and animal deaths if ingested.
2. Certain red algae produce a polysaccharide called agar that is routinely used in microbiology laboratories as a hardening agent for culture media.
3. Most algae reproduce both asexually and sexually. Asexual reproduction involves mitosis, while sexual reproduction involves the fusion of two gametes to form a diploid zygote, which undergoes meiosis.

INTRODUCTION TO THE PROTOZOA

1. The protozoa are a group of single-celled, chemoheterotrophic eukaryotes that are widely distributed in nature. They generally contain nuclei, mitochondria, endoplasmic reticula, and Golgi bodies.
2. In contrast to the algae, which are mostly free-living forms, the protozoa include many important human and animal parasites.

THE VARIOUS GROUPS OF PROTOZOA

1. The protozoa are classified into three major groups, based on their mode of locomotion, mode of reproduction, and chemical and morphological characteristics.
2. Many of the protozoa have very complicated life cycles that may involve more than one host.
3. The primary importance of the protozoa to humans resides in their ability to cause disease. Many of these diseases, such as malaria and giardiasis, are very common infections. In addition, diseases caused by coccidians can have a significant impact on domestic and wild animal populations. It is believed that coccidian infections have played a significant role in the evolution and distribution of wild animals.

STUDY QUESTIONS

1. Define a protist. How do the Pha phyta and Rhodophyta fit in your definition?
2. Differentiate between algae protozoa.
3. What are two important ns that algae carry out in nature?
4. Discuss briefly three ways in which protozoa affect humans.
5. Discuss the m f algae and explain how they differ from e
6. Describe s of protozoa and explain how they dif her.
7. Outl f:
   a.
   site

e of ten important human protozoal ating what disease they cause and how cquired.

9. Are protozoans important to farmers and ranchers? Explain.
10. How do algae differ from photosynthetic bacteria and from cyanobacteria?

SUPPLEMENTAL READINGS

Bold, H. C. and Wynne, M. J. 1978. *Introduction to the algae: Structure and reproduction*. Englewood Cliffs, N. J.: Prentice-Hall.

Boney, A. D. 1966. *A biology of marine algae*. London: Hutchison & Co., Ltd.

Ciferri, O. 1983. *Spirulina*, the edible microorganism. *Microbiological Review* 47(4): 551–578.

Farmer, J. N. 1980. *The protozoa: Introduction to protozoology*. St. Louis: C. V. Mosby.

Lennette, E. H., Balows, A., Hausler, W. J., and Shadomy, H. J. 1985. *Manual of Clinical Microbiology*. 4th ed. Washington, D.C.: American Society for Microbiology.

Markell, E. K. and Voge, M. 1981. *Medical parasitology*. 5th ed. Philadelphia: W. B. Saunders Co.

Whittaker, R. H. and Margulis, L. 1978. Protist classification and the kingdom of organisms. *BioSystems* 10:3–18.

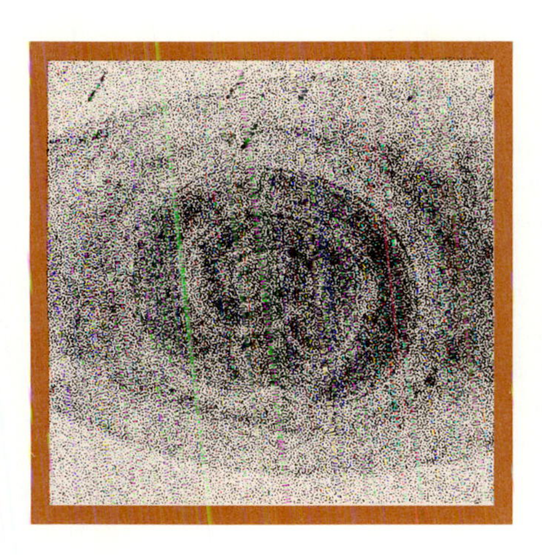

CHAPTER 16
THE MULTICELLULAR PARASITES

CHAPTER PREVIEW

RIVER BLINDNESS, A CRUEL PARASITIC DISEASE

Parasitic diseases ravage more than 2 billion humans throughout the world. Most of these people reside in Third World countries, although more than 60 million Americans are infected with some kind of worm. It is remarkable, however, that in spite of the magnitude of worm-caused diseases in the world, less than 4% of the funds for research is earmarked for the study and control of these diseases. Nevertheless, vigorous public health programs are being instituted in order to eradicate parasitic diseases. For example, the World Health Organization (WHO) has spent more than $125 million trying to control **river blindness** alone.

River blindness, more properly called **onchocerciasis** (ahn-ko-sir-KI-ah-sis), is a disease caused by the roundworm *Onchocerca volvulus.* This worm infects more than 2 million people in the equatorial regions of Central and South America and Africa, with a potential for infecting an additional 10 million people. The disease ravages entire villages, causing blindness in more than 30% of the inhabitants.

Onchocerca volvulus is transmitted by the bite of bloodsucking female blackflies. The blackfly, upon biting an infected individual, sucks up numerous larval (juvenile worms) into its gut. There the juvenile worms develop into infectious forms, which can then infect another human when the blackfly bites again. The larvae travel throughout the host and eventually develop into adults, which subsequently mate and produce millions of offspring called **microfilariae.** The microfilariae are tiny worms that migrate throughout the body, notably to the skin and eyes. The numerous microfilariae penetrating the retina cause scarring of the tissue and subsequently blindness.

To control river blindness, it is necessary that biologists know intimately the life cycle of the causative organism. For example, it is theoretically possible to control river blindness by controlling the population of blackflies. In practice, such control has proven more difficult than predicted. It is also essential that the anatomy and physiology of the causative organism be known in order to develop effective drugs to cure and/or prevent the disease.

Since it is necessary to learn about parasites before control measures can be applied, this chapter includes some of the more important human parasitic worms as well as their life histories. This chapter also considers a group of organisms called **arthropods** (joint-legged invertebrate animals), not only because they are capable of causing disease, but also because they act as vehicles for the transmission of many infectious diseases.

LEARNING OBJECTIVES

A STUDY OF THIS CHAPTER SHOULD ENABLE YOU TO:

OUTLINE THE MAJOR CHARACTERISTICS OF MEDICALLY-IMPORTANT MULTICELLULAR PARASITES, AND EXPLAIN HOW THEY INFLUENCE HUMAN AFFAIRS

DIAGRAM THE LIFE CYCLES OF A TREMATODE, A CESTODE, AND A NEMATODE

OUTLINE THE MAJOR GROUPS OF PARASITIC ARTHROPODS AND VECTORS

Multicellular parasites are multitissued and multiorganed (metazoan) animals that can live at the expense of other animals or plants (their hosts). The multicellular parasites include animals such as the tapeworms, the flukes, and the roundworms. In addition, a group of arthropods (joint-appendaged invertebrate animals), such as the spiders and the insects, are considered to be parasites because some can colonize body surfaces and cause much discomfort and disease. Although many of the multicellular parasites (like the insects) are not microorganisms, they are frequently studied by microbiologists because many act as carriers of microorganisms that cause human diseases. This chapter will discuss some of the most common multicellular parasites that attack humans, and some of the arthropods that can transmit infectious agents among human populations or that can themselves cause annoying infestations of the skin and hair.

THE FLATWORMS (PHYLUM PLATYHELMINTHES)

The **flatworms** are multicellular, dorsoventrally flattened animals in the phylum Platyhelminthes (fig. 16-1). These animals have primitive digestive system and generally have both sexes contained in the same worm. One notable exception is the **schistosomes,** which have separate sexes. Many of the flatworms have very complex life cycles involving more than one host. These parasites are widely distributed in nature and may infect millions of humans annually (table 16-1). Among the flatworms there are four classes, two of which include numerous important human pathogens. These are the **trematodes,** or flukes, and the **cestodes,** or tapeworms.

The Trematodes (Flukes)

The trematodes are parasitic worms that can live either inside their host or attached to external surfaces of the host. Generally, the life cycles of these

Figure 16-1 A Typical Human-Parasitic Flatworm. A photomicrograph of a human liver fluke *clonorchis sinensis.* The flattened appearance of the parasite is characteristic of all the flukes. The opening at the top is a sucker, used for attachment and feeding.

TABLE 16-1
REPRESENTATIVE PARASITIC DISEASES

NAME OF DISEASE	NAME OF PARASITE	PEOPLE AFFECTED (in millions)
	Trematodes:	320+
Schistosomiasis	*Schistosoma,* spp.	250
Paragonomiasis	*Paragonimus westermanni*	4
Opistorchiasis	*Opistorchis sinensis*	20
Fasciolopsiasis	*Fasciolopsis* sp.	11
	Cestodes:	100+
Tapeworm disease	*Taenia, Taeniarrhyncus, Diphyllobothrium*	69
	Nematodes:	1800+
Filariasis	*Loa loa, Onchocerca, Dracunculus, Wuchereria*	280
Ascariasis	*Ascaris lumbricoides*	650
Hookworm disease	*Necator, Ancylostoma*	460
Whipworm disease	*Trichuris trichura*	360
Strongyloidasis	*Strogyloides stercoralis*	36

soni, and *S. haematobium*, cause **schistosomiasis** in humans. This disease is prevalent in more than 70 countries throughout the world and affects more than 200 million people. In countries where schistosomiasis is prevalent, such as Africa and Asia, the disease not only causes much human misery but also inflicts severe financial losses due to lack of productivity as a result of illness and disability. Infections by the schistosomes usually result in enlarged and inflamed livers, spleens, kidneys and bladder, depending upon the species of schistosome infecting the host.

The life cycle of all schistosomes (fig. 16-3) is essentially the same, regardless of the species, and involves a freshwater snail as an intermediate host. The human-infectious stage, called the **cercaria,** develops from a **sporocyst** and matures inside the snail. The cercaria is motile and possesses a forked tail. After emerging from the snail, the cercaria moves incessantly until it either finds a susceptible human or dies. The most common sites of human infection are the feet and legs, because these extremities are usually exposed when the host wades in shallow, cercariae-infested waters. The cercaria penetrates the skin and invades the bloodstream, where it develops into an adult worm. The developing worms travel throughout the human body, and upon reaching sexual maturity they mate in the small blood vessels that feed the intestines. The female worm deposits the fertilized eggs in these blood vessels. Many of the eggs erode the blood vessels and work their way into the intestinal lumen or the urinary bladder, where they are shed into the environment in the feces or urine. It is the erosion of host tissue caused by the eggs that is responsible for the signs and symptoms of the disease. If the eggs reach the water, they hatch, each releasing a **miracidium** that swims actively until it finds and infects a suitable snail host. Within the snail, the miracidium develops into a **sporocyst.** The sporocycst, in turn, eventually gives rise to several cercariae and the infectious cycle begins once again.

The schistosomes are not the only important human parasites. Many other flukes, including *Paragonimus westermani* (fig. 16-4) and *Opistorchis sinensis* (fig. 16-2), afflict more than 100 million humans throughout the world. Some of these parasites and the diseases they cause are listed in table 16-1.

The Cestodes (Tapeworms)

The cestodes or tapeworms have a segmented body and a head with a holdfast called the **scolex** (fig. 16-5). Each of the body segments, called **proglottids,** has a complete male and female reproductive system. Some of the tapeworms that infect humans may have more than 3,000 proglottids attached to the scolex and measure several meters in length. The scolex, which is usually attached to the host, aids in the absorption of nutrients for the rest of the worm. The proglottids, when gravid (after mating), are nothing more than bags filled with fertilized eggs. The eggs and/or the proglottids are periodically shed from the host in the feces. The eggs can then be ingested by susceptible hosts and hence initiate a new infectious cycle.

Tapeworms often affect the health of infected humans. In some countries, it is not uncommon for skinny individuals who consume large quantities of food to be infected with tapeworms. Infected individuals are thin possibly because the tapeworms derive their nutrition from the host they parasitize. In the early 1900s, some enterprising individuals in the United

parasites are very complex and may involve two or more different hosts. Most of the trematodes are parasites of fish, although a few (table 16-1) are important human pathogens. The adult flatworms have powerful suckers with which they adhere to host surfaces and feed upon them (fig. 16-2). Some of the immature stages of trematode parasites have penetrating organs that can pierce human skin and invade the body.

One of the best known of all the flukes are those called the schistosomes (fig. 16-3). Three species of these organisms, *Schistosoma japonicum*, *S. man-*

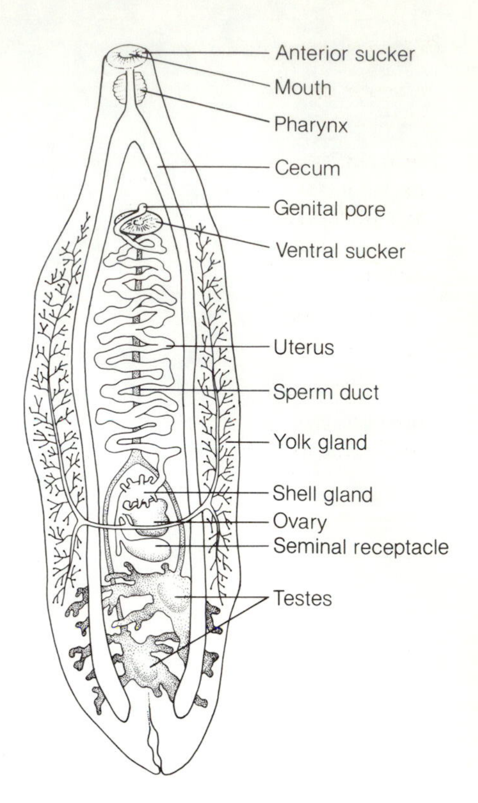

Figure 16-2 Salient Anatomical Features of the Flukes. Flukes are animals with well-developed sexual reproductive systems. Both male and female organs are present in each animal. This drawing is of the Chinese liver fluke *Opistorchis sinensis*.

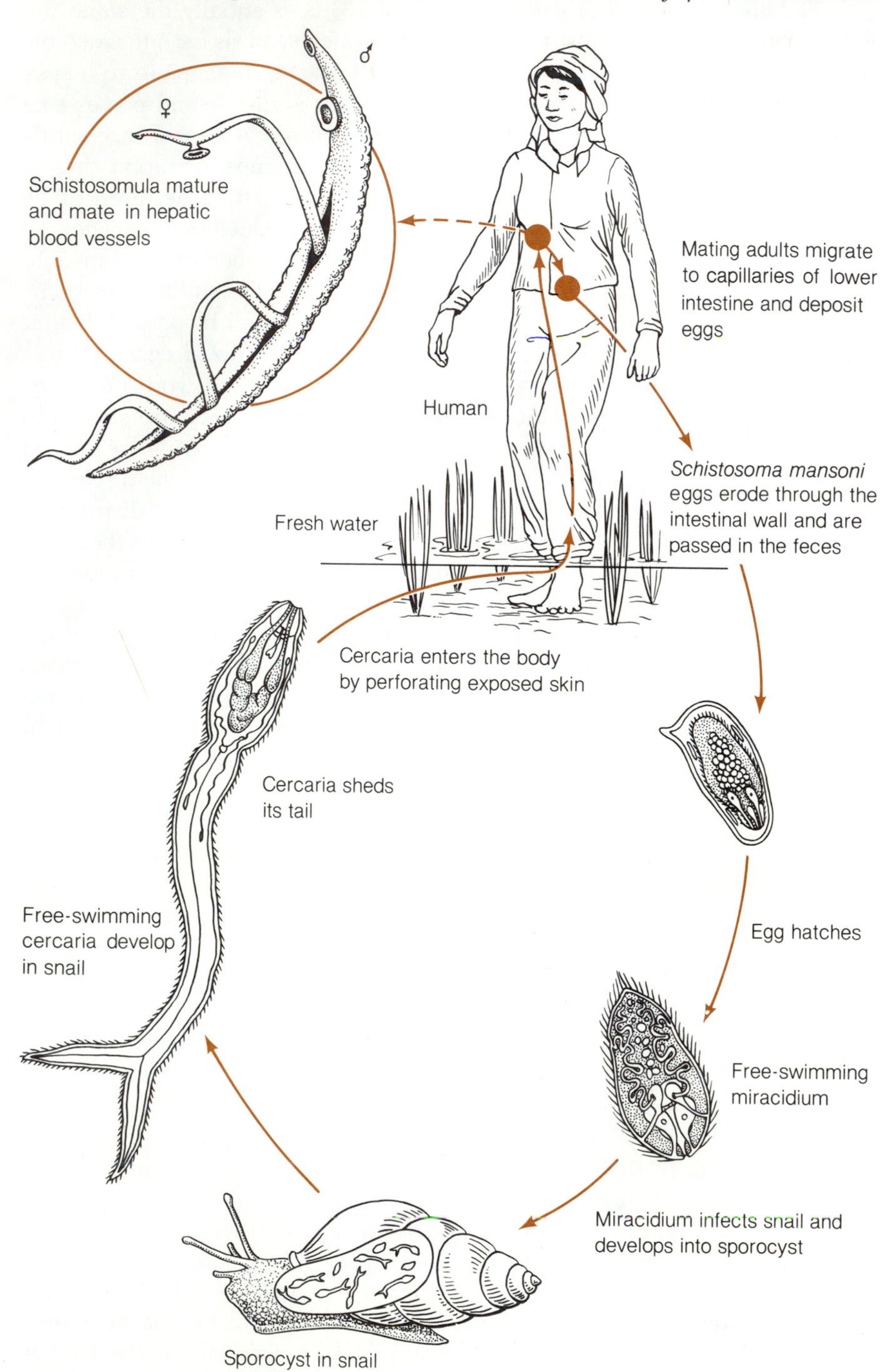

Figure 16-3 Life Cycle of *Schistosoma mansoni*. The adult schistosomes mate and deposit eggs in the capillaries of the intestinal walls. The eggs erode their way into the lumen (cavity) of the intestines and are shed into the environment in the feces. If the feces contaminate freshwater environments, the egg hatches to produce a miracidium, which can then infect a snail host. The miracidium continues to develop within the snail and eventually cercariae emerge from the snail host. The cercariae swim actively until they encounter a suitable human host. The cercariae penetrate the skin of individuals wading in the water and enter the circulation system. The cercaria develop into schistosomula, which eventually mature to adulthood within the human host. Upon reaching the capillaries of the intestines, the adults mate and deposit eggs.

STUDY QUESTIONS

1. Define a protist. How do the Phaeophyta and Rhodophyta fit in your definition?
2. Differentiate between algae and protozoa.
3. What are two important functions that algae carry out in nature?
4. Discuss briefly three different ways in which protozoa affect humans.
5. Discuss the major groups of algae and explain how they differ from each other.
6. Describe the major groups of protozoa and explain how they differ from each other.
7. Outline the life cycles of:
   a. a brown alga
   b. a green alga
   c. an ameba
   d. a trypanosome
   e. a malarial parasite
8. Construct a table of ten important human protozoal pathogens, indicating what disease they cause and how the disease is acquired.
9. Are protozoans important to farmers and ranchers? Explain.
10. How do algae differ from photosynthetic bacteria and from cyanobacteria?

SUPPLEMENTAL READINGS

Bold, H. C. and Wynne, M. J. 1978. *Introduction to the algae: Structure and reproduction*. Englewood Cliffs, N. J.: Prentice-Hall.

Boney, A. D. 1966. *A biology of marine algae*. London: Hutchison & Co., Ltd.

Ciferri, O. 1983. *Spirulina*, the edible microorganism. *Microbiological Review* 47(4): 551–578.

Farmer, J. N. 1980. *The protozoa: Introduction to protozoology*. St. Louis: C. V. Mosby.

Lennette, E. H., Balows, A., Hausler, W. J., and Shadomy, H. J. 1985. *Manual of Clinical Microbiology*. 4th ed. Washington, D.C.: American Society for Microbiology.

Markell, E. K. and Voge, M. 1981. *Medical parasitology*. 5th ed. Philadelphia: W. B. Saunders Co.

Whittaker, R. H. and Margulis, L. 1978. Protist classification and the kingdom of organisms. *BioSystems* 10:3–18.

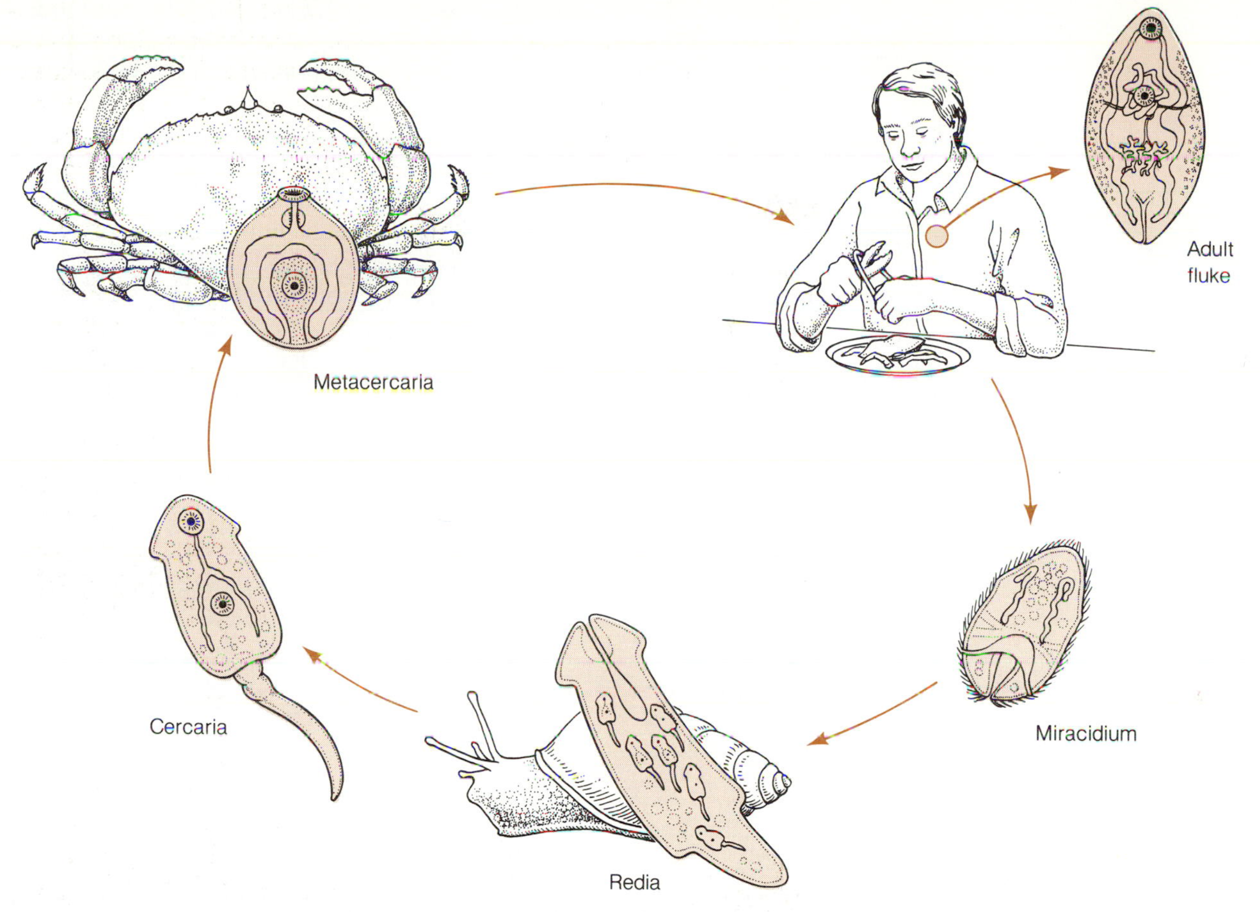

Figure 16-4 Life Cycle of the Lung Fluke *Paragonimus Westermanni.* Humans become infected with *Paragonimus westermani* when they consume uncooked crab containing encysted metacercaria. The metacercaria make their way to the lungs from the intestines and there they develop into adults. The adults deposit eggs in the lungs and the eggs are shed into freshwater habitats in sputum or feces (after the eggs are swallowed). In freshwater environments the eggs hatch into miracidia, which then infect snails. In the snail, the miracidia develop into cercariae, which then infect freshwater crabs. In the crab meat, the cercariae develop into metacercariae and encyst. The life cycle is completed when humans consume metacercaria-infected crabs.

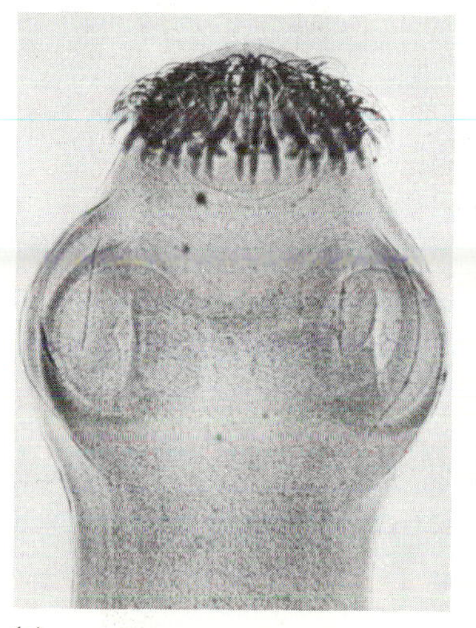

(a)

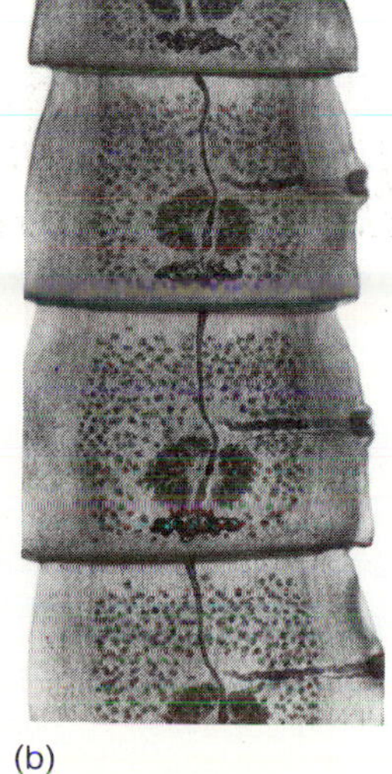

(b)

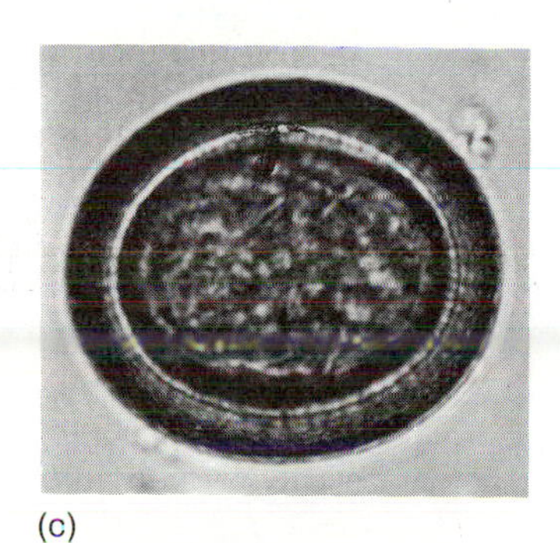

(c)

Figure 16-5 Salient Anatomical Features of Tapeworms. *(a)* The scolex of the beef tapeworm *Taeniarhynchus saginatus.* *(b)* Mature proglottids. *(c)* Egg.

States sold tapeworm eggs as "weight reduction pills" (would this have worked?). In spite of common belief, many of the tapeworms that infect humans can exist as parasites for extended periods of time without eliciting signs or symptoms.

Three tapeworms that commonly infect humans are the fish tapeworm *(Diphyllobothrium latum)*, the pork tapeworm *(Taenia solium)*, and the beef tapeworm *(Taeniarhynchus saginatus)*. Each of these parasites can develop long-lasting parasitic associations with humans without causing notable disease. The parasites' life cycles involve at least two hosts. The host in which the parasite reproduces sexually is known as the **definitive** host.

One of the most common tapeworm parasites of humans is *Taeniarhynchus saginatus* (the beef tapeworm). This parasite averages 8 meters in length (at least one has been reported to measure 25 m). The stage of the parasite that is infective for humans is the **cysticercus** (fig. 16-6). This is an embryonic stage of the parasite that develops in beef tissues and consists of a fluid-filled sac containing an invaginated (inside-out) scolex. Humans be-

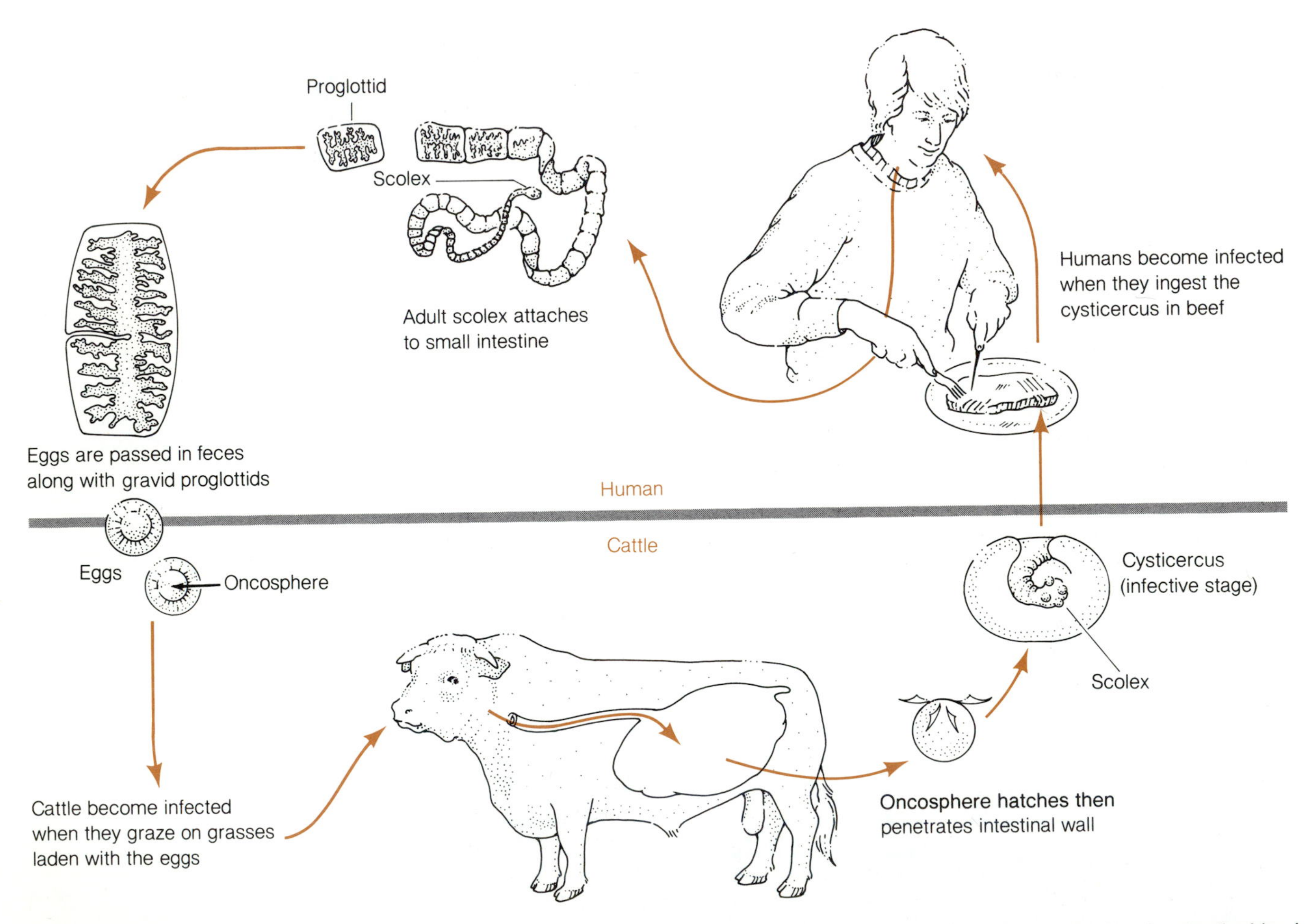

Figure 16-6 Life Cycle of *Taeniarhyncus saginatus.* Humans become infected with the beef tapeworm *Taeniarhynchus saginatus* by ingesting rare beef contaminated with tapeworm embryos called cysticerci. The cysticercus yields a scolex, which attaches to the intestines and begins to develop proglottids. Eggs develop in the proglottids and are shed in the feces. The eggs may be ingested by grazing cattle and hatch into oncospheres. The oncospheres penetrate the intestinal wall, enter the blood circulation and migrate to muscles, where they develop into cirsticerci. The cycle is completed when measly (contaminated) beef is ingested by humans.

come infected when they consume uncooked, "measly" (infected) beef containing cysticerci. Once in the intestines, the scolex emerges from the cysticercus and attaches itself to intestinal epithelium. There it begins to feed upon its host and to develop proglottids. As each proglottid reaches maturity, eggs in the proglottid are fertilized. The eggs are shed in the feces, and if ingested by a suitable intermediate host (cattle or buffalo), the eggs hatch, releasing embryos that migrate throughout the animal and develop into cysticerci. The cycle begins again when a human consumes the measly meat.

The signs and symptoms of tapeworm disease, which are often totally absent, are related to the presence of the worms in the intestines. Intestinal disorders and weight loss are the most common manifestations of tapeworm disease. Occasionally, some of the eggs may be regurgitated by the patient and then swallowed. These eggs may develop into cysticerci, which will affect the cerebrum, cerebellum, meninges, skeletal muscle, or heart, causing a disease called **cysticercosis.** This is a relatively rare occurrence with the beef tapeworm, but occurs more often with *Taenia solium* (the pork tapeworm).

Diseases caused by tapeworms (table 16-1) afflict more than 60 million humans throughout the world and are caused primarily by *Taeniarhynchus saginatus*, *Taenia solium*, *Diphyllobothrium latum*, *Hymenolepis nana*, and *H. dimunuta* (table 16-1). The single most important preventive measure for tapeworm disease is to cook all meats were before eating. This precaution will reduce the incidence of tapeworm disease by almost 99%.

THE ROUNDWORMS (CLASS NEMATODA)

The **roundworms** (nematodes) constitute a very large—possibly the largest—group of animals that are widely distributed in nature. Some are found living in water and soil, while many others are parasites of plants or animals. These organisms cause extensive damage to crop plants, inflicting billions of dollars in financial losses to farmers. Similarly, infectious diseases caused by the roundworms aflict more than a billion humans throughout the world.

The roundworms are generally small (usually smaller than the flatworms), round, slender worms with bodies that taper at both ends (fig. 16-7). The mouth is near the anterior portion of the worm and the anus near the posterior. The nematodes have separate sexes, the female generally larger than the male, and when they mate they produce fertilized eggs. Some species of nematodes, however, like *Trichinella spiralis*, give birth to live young, because the eggs are kept in the uterus of the female until they hatch.

Many nematodes are important human parasites (table 16-1) that cause a variety of diseases, with symptoms ranging in severity from an itch to brain damage and death. The life cycles of the nematodes are generally simpler than those of the cestodes or trematodes, although they do have one or two different larval stages. The simplest life cycle is that of the **human whipworm** *Trichuris trichura* (fig. 16-8). *Trichuris* has a slender anterior end and a broader posterior (fig. 16-8) and measures 30 to 50 mm in length. The males are slightly smaller than the females. Fertilized eggs represent the human-infectious stage of the parasite and are found in moist, shady soils. When these eggs are swallowed, they hatch, releasing larvae that make their way into the intestines, attach to the intestinal wall, and develop into a more mature worm. After a period of development, the whipworms make their

Figure 16-7 Salient Anatomical Features of Roundworms. Internal anatomy of the nematodes. Note that the sexes are separate in these organisms and that they have well-developed reproductive and digestive systems.

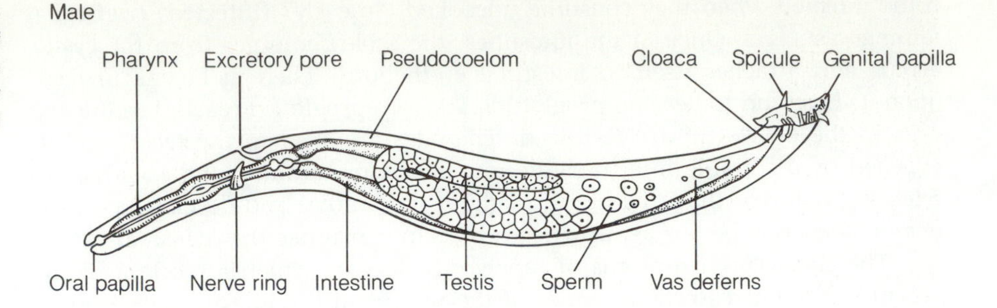

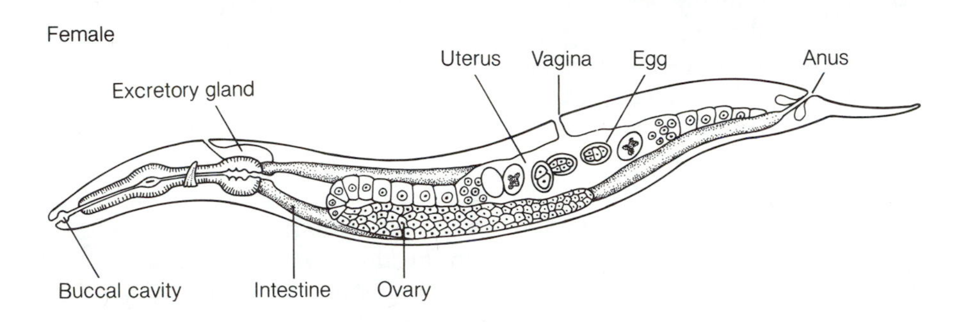

way into the lower part of the large intestine and complete their maturation, becoming adult worms. They mate in the intestine and release as many as 7,000 eggs each day. The eggs are shed in the feces and complete their development into infectious larvae in the soil. Heavy infections may lead to intestinal bleeding and anemia, because the adult worms burrow into the intestinal mucosa, where they feed on blood. Bacteria may also infect the site, causing further damage. Inflammation of the colon (colitis) and inflammation of the rectum (proctitis) are common manifestations of heavy para-

(a)

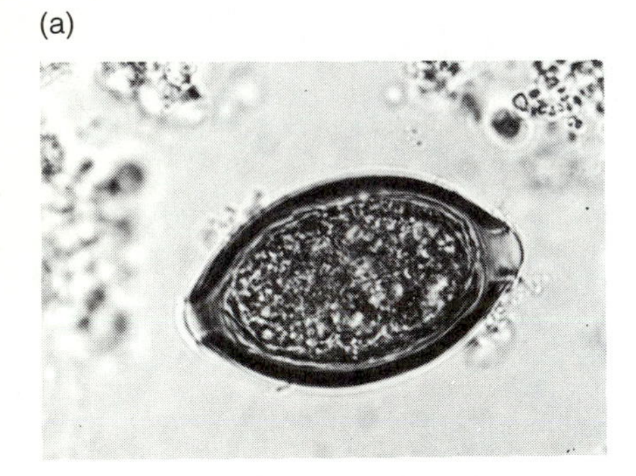

(b)

Figure 16-8 ***Trichuris trichura.*** *(a)* When infective eggs are ingested, they hatch into larvae in the intestines, the larvae penetrate and develop in the mucosa. Adults in the cecum mucosa mate and produce eggs. The eggs are passed in the feces and mature into the infective stage. The life cycle is completed when infective eggs are ingested. *(b) Trichuris trichura* is called the whipworm because it resembles a whip with a broad posterior and a thin anterior.

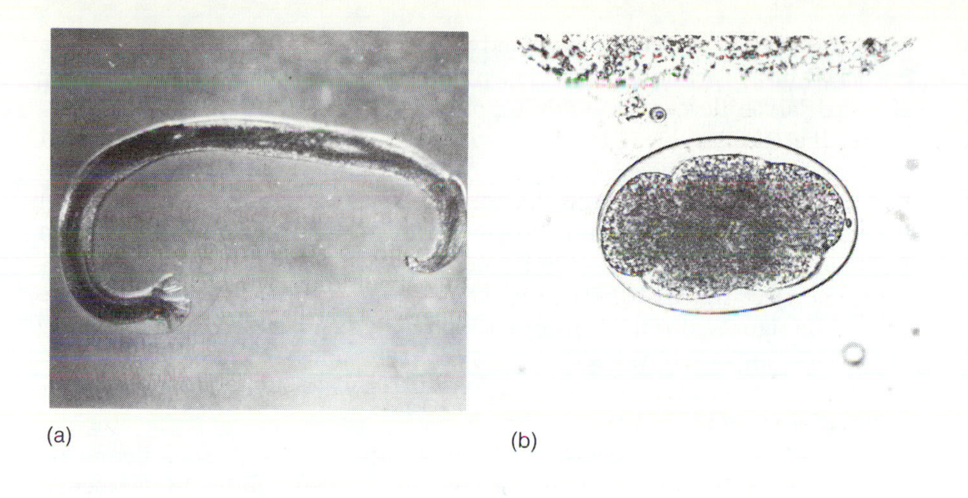

(a) (b)

Figure 16-9 Hookworms. *(a)* Photomicrograph of a hoookworm. *(b)* The anterior portion of the worm is used to attach to the intestines. Eggs are passed in the feces and develop into larvae. The infective larvae penetrate the skin of the susceptible hosts and migrate to the intestines where they attach, develop into adults, and mate.

site burdens. Other symptoms of whipworm disease include insomnia, vomiting, rash, constipation, loss of appetite, and diarrhea.

The **hookworms** *Ancylostoma duodenale* and *Necator americanus* (fig. 16-9) are very important human pathogens. The World Health Organization (WHO) estimates that more than 450 million people are infected with these parasites. It is thought that approximately 2 million people in the United States carry hookworms in their bowels. Because the hookworms attach themselves to the intestinal lining with very sharp "teeth," they lacerate intestinal tissue and cause bleeding. Throughout the course of a hookworm infection, a patient may lose 150–200 ml of blood due to intestinal bleeding. Anemia, abdominal pain, and loss of appetite are common manifestations of hookworm disease. Occasionally, the patient may experience a desire to eat soil (geophagy). In heavy infestations, the patient may exhibit hair loss, mental dullness, protein deficiency; in extreme cases, death may result. The life cycle of this parasite involves larval forms that penetrate the skin and travel throughout the body before the adult worms colonize the bowels.

Trichinella spiralis (fig. 16-10) is a very small nematode that causes one of the most important and widespread nematode diseases in the world. The

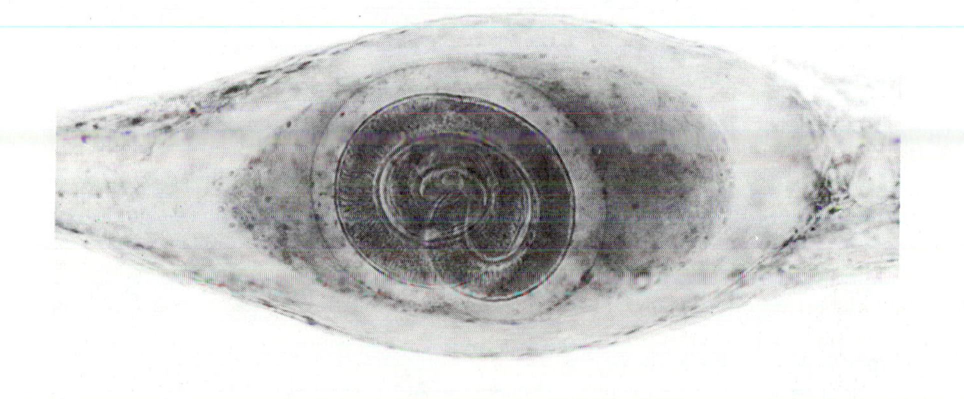

Figure 16-10 Encysted Larva of *Trichinella spiralis.* Humans become infected when they ingest uncooked pork meat laden with encysted larvae of *Trichinella spiralis.* The larvae excyst and develop into adults in the intestines. Adults mate producing larvae. The larvae penetrate the intestinal tissues and migrate in the blood to skeletal muscle. There they penetrate the muscle fibers and become encysted.

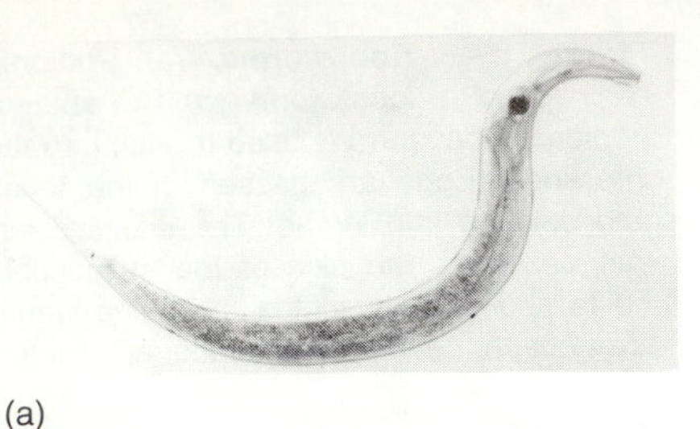

(a)

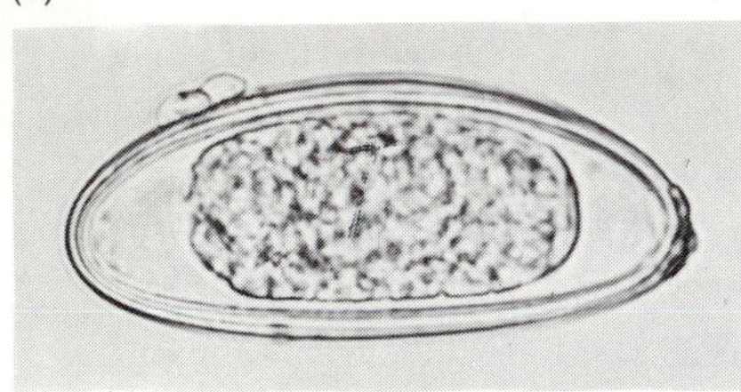

(b)

Figure 16-11 The Pinworm, *Enterobius vermicularis.* The pinworm *Enterobius vermicularis* (*a*) mates in the lumen of the cecum. The gravid females migrate to the anus to deposit the eggs (*b*). The eggs contaminate clothing or bedding or are picked up on the fingers. The eggs are ingested and develop into larvae in the intestines. The larvae migrate to the cecum and mature into adults.

disease is known as **trichinosis,** and it is acquired when a human ingests uncooked meat (primarily pork) laden with encysted larvae (fig. 16-10). The encysted larvae develop into adult worms that live in the host's intestines. There, the adults mate. The male worms die shortly after copulation, but the females live and give birth to live offspring. The young worms penetrate host tissue and enter the venous blood, where they may migrate anywhere in the body. Striated muscle is primarily invaded, although larvae have been found in the stomach, brain, liver, and lungs. In the muscle, the larvae become encysted (fig. 16-10). The migration and encystment of the larvae give rise to the signs and symptoms of trichinosis, which vary because the larvae can invade any organ. Consequently, *Trichinella spiralis* is the "great imitator" because the disease produces signs and symptoms that are very similar to those produced by other organisms.

The **pinworm** *Enterobius vermicularis* (fig. 16-11) is a small nematode that infects more than 300 million people, predominantly in temperate zones. It is estimated that 40 to 50 million humans in the United States, mainly children, are infected. The infection is acquired by ingesting eggs, and the usual route of infection is the fecal-oral route. The ingested eggs hatch in the intestines, where the resulting adults mate. The female carries the eggs to the anus and deposits them there, usually at night, causing the itching and restlessness that are the most prominent manifestations of the disease.

FOCUS

THE SERPENT ON A STAFF

The serpent on a staff was said to be carried by Aesculapius, the Roman god of medicine. This symbol was adopted by the American Medical Association (AMA) and the Army Medical Corps (fig. 16-12). What was the origin of such an emblem? Was it really a snake on a stick, or was it *Dracunculus medinensis* on a stick?

The female *Dracunculus medinensis* is a rather large, slender nematode worm that migrates through human tissue to deposit its eggs near the skin. Sometimes, the skin ulcerates and the worm emerges from the ruptured skin and dangles from the sore. Modern medicine treats these "guinea worms" with drugs or by surgically removing the worm, but many individuals inhabiting Third World countries treat the disease differently: they simply wrap the worm around a stick (it could be a matchstick or something much fancier than that) and wrap the worm a few turns every day until it comes free of the tissue. Undoubtedly, this technique has been passed from generation to generation since ancient times. Woodcuts dating back to the 1600s illustrate the technique for removing guinea worms by winding the worm on a stick. The disease is prevalent in desert areas of the Middle East and Africa.

Whether or not *Dracunculus medinensis* is the "snake" wrapped around the staff in the AMA's emblem will probably never be known for certain, but it is certainly a plausible explanation.

Figure 16-12 Emblem of Army Medical Corps

There are many other nematodes that cause human diseases, many of which are deforming or fatal. Some of these organisms and the diseases they cause are listed in table 16-1.

THE ARTHROPODS

The **arthropods** (class Arthropoda) constitute a very large group of invertebrate animals that have jointed appendages and a segmented body. Most arthropods have life cycles that include several immature or embryonic stages.

The arthropods are very important economically because they destroy billions of dollars worth of crops each year and can serve as vehicles for the transmission of infectious diseases. Many species of arthropods, primarily the **insects** and **arachnids** (a group that includes the spiders and ticks), can cause human and animal disease by parasitizing the skin and other body surfaces. The majority of arthropods of public health importance participate in the transmission of disease.

Parasitic Arthropods

A few arthropods are parasitic on humans (table 16-2). Among these are the itch mite, *Sarcoptes scabiei;* the head or human body louse, *Pediculus humanus;* and the crab louse, *Phthirus pubis* (fig. 16-13). The itch mite is an 427 ectoparasite □ (parasitizes outer body surfaces) of humans and domestic animals. The adult mites penetrate the host's skin and burrow channels or trenches in the skin. In their burrows, the female itch mites deposit eggs that hatch within five days. The larvae dig new burrows as they mature into adult mites in three to five days. The mite infestation causes severe itching, which in turn results in excessive scratching and sores that eventually become covered with scabs. Sometimes the trenches become infected with bacteria that further aggravate the condition. The disease in humans is called **scabies,** but in domestic animals it is called **mange.**

The head or body louse (fig. 16-13) is common in populations where personal hygiene is poor. Some races of lice invade primarily the head, while other races infest the human body. The head louse deposits its eggs (nits) on the hair shaft. These eggs hatch into nymphs that resemble the adult louse. The infestation can be transmitted from person to person by direct contact, in contaminated clothing, or on personal articles. The body louse colonizes

(a)

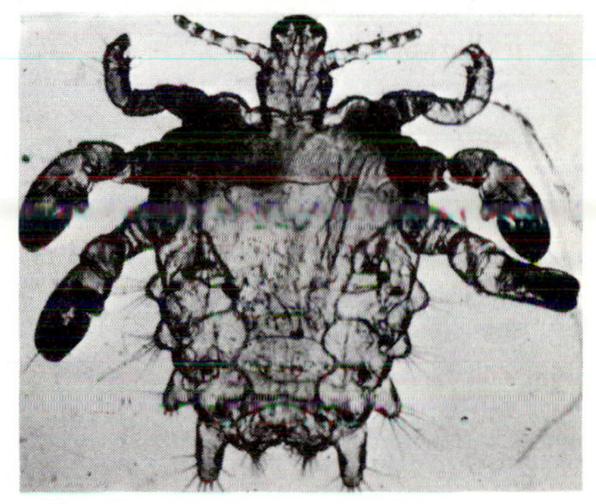

(b)

Figure 16-13 Representative Parasitic Arthropods *(a) Pediculus humanus. (b) Phthirus pubis.*

TABLE 16-2
REPRESENTATIVE DISEASES CAUSED BY PARASITIC ARTHROPODS

NAME OF DISEASE	NAME OF ORGANISM(S)
Scabies	*Sarcoptes scabiei* (itch mite)
Pediculosis	*Pediculus humanus* (body louse)
Crabs	*Phthirus pubis* (crab louse)
Myasis	*Dermatobia hominis, Callitroga; Sarcophaga, Phaenicia, Phormia* (all flies)
Chigoe flea lesions	*Tunga penetrans* (flea)

the body instead of the hair. People suspected of being infested with lice are diagnosed by making a careful examination of the head, body, and clothing for the presence of nits, nymphs, or adults. The prompt detection of *Pediculus humanus* is important because this ectoparasite spreads rapidly among humans and is capable of transmitting various infectious diseases (table 16-3).

The crab louse, *Phthirus pubis* (fig. 16-13) colonizes primarily the genital areas. The female louse deposits her eggs on pubic hairs. The louse may cause severe itching, and some sensitive hosts develop a rash. The crab louse can also transmit diseases such as trench fever, epidemic typhus, and relapsing fever.

Arthropods as Agents of Disease Transmission

Arthropods, particularly insects and arachnids, are often agents of disease transmission. Some of these arthropods act as **vectors** □ 536 (table 16-3): animals that harbor a parasitic microorganism and transmit it from one individual to another. For example, the *Anopheles* mosquito (fig. 16-14) is a vector for the transmission of malaria □ 699. A portion of the life cycle of the malarial parasites must be completed within the mosquito, which then transmits the parasite by biting a suitable host. Notice that in malaria, as in all vector-borne diseases, the parasite has an intimate (and often necessary) association with the vector. Ticks are also important vectors (table 16-3). For example, the tick *Dermacentor andersonii* (fig. 16-14) is the principal vector for the often fatal disease Rocky Mountain spotted fever.

TABLE 16-3
REPRESENTATIVE INFECTIOUS DISEASES THAT ARE TRANSMITTED BY ARTHROPODS

INFECTIOUS DISEASE	GEOGRAPHICAL DISTRIBUTION	ARTHROPOD VECTOR	GROUP
Diphyllobothriasis	Finland, United States, Russia	*Cyclops sp.*	Copepod
Dracunculosis	Africa, India, Middle East	*Cyclops sp.*	Copepod
Paragonomiasis	Africa, Asia, Phillipines, S. America	Crabs	Crustacean
Scrub typhus	Far East, Phillipines, S. Pacific	*Trombicula sp.*	Mites
Tularemia	Europe, Japan, N. America	*Dermacentor sp.*	Ticks
Rocky Mountain Spotted Fever	N. and S. America	*Dermacentor, Ixodes*	Ticks
Relapsing fever	Worldiwde	*Ornithodorus*	Ticks
Epidemic typhus	Worldwide	*Pediculus humanus*	Lice
Trench fever	Europe	*Pediculus humanus*	Lice
Plague	Worldwide	*Xenopsylla cheopis*	Fleas
Murine typhus	Tropics and subtropics	*Xenopsylla cheopis*	Fleas
Chagas' disease	N. and S. America	*Triatoma, Panstrongylus*	Bugs
Hymenolepiasis	Worldwide	Flour beetles	Beetles
African sleeping sickness (nagana in cattle)	Africa	*Glossina sp.* (tsetse)	Flies
Onchocerciasis	Africa, Central and S. America,	*Simulium sp.*	Flies
Leishmaniasis	Asia, E. Africa, Mediterranean, C. and S. America	*Phlebotomus sp.*	Sandfly
Malaria	Worldwide	*Anopheles sp.*	Mosquitoes
Yellow fever	Africa, C. and S. America	*Aedes aegypti*	Mosquitoes
Dengue fever	Tropics and subtropics	*Aedes sp.*	Mosquitoes
Equine encephalites	N. and S. America	*Culex*	Mosquitoes
Filariasis (bancroftian)	Africa, Asia, Australia, S. Pacific	*Culex, Anopheles, Aedes*	Mosquitoes

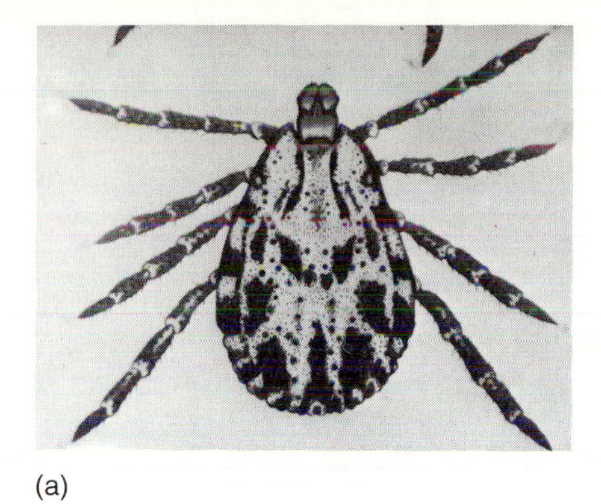

(a)

(b)

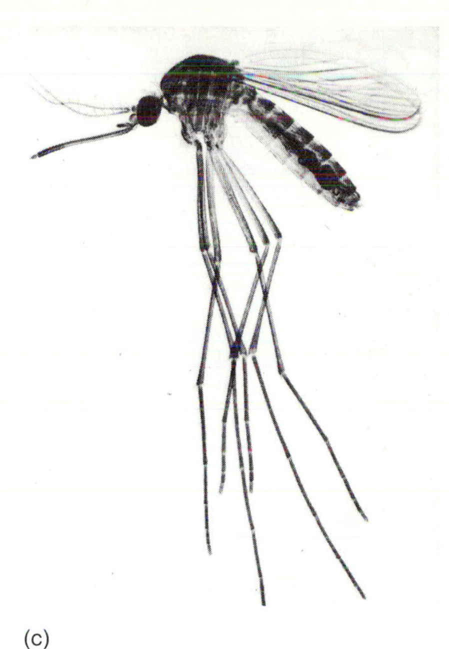

(c)

Figure 16-14 Representative Vectors of Human Disease. *(a) Dermacentor. (b) Xenopsylla cheopis. (c)* Culex.

Some arthropods act as **phoronts;** that is, they act as a mechanical means of transmission. For example, a domestic fly may land in a patch of soil contaminated with cat feces containing oocysts of *Toxoplasma gondii* □. 702 The fly may pick up some of the oocysts on its feet and legs and transfer them to foods. A human may then consume the food and become infected. The fly did not act as a true vector, because the event was an accident. Table 16-3 lists some of the arthropod vectors and the diseases they transmit.

SUMMARY

1. Multicellular parasites such as trematodes, cestodes, and nematodes, are metazoan animals with the ability to live at the expense of their host.
2. There are four major groups of multicellular parasites that are important to humans. These are the flukes, the tapeworms, the roundworms, and the arthropods.

THE FLATWORMS

1. The flatworms (Platyhelminthes) are parasitic worms that are flattened dorsoventrally.
2. There are two large groups of flatworms: the flukes and the tapeworms.
3. The flukes (trematodes) are flatworms lacking a segmented body and with one or more suckers, which they use to adhere to hosts' surfaces. The flukes cause human diseases such as schistosomiasis, paragonomiasis, and opistorchiasis, which cause millions of deaths each year.
4. The tapeworms (cestodes) are segmented flatworms with a scolex (head) and one or more segments called proglottids. Tapeworm disease in humans may cause severe symptoms and death.

THE ROUNDWORMS

1. Roundworms are slender, cylindrical worms with tapering bodies. Free-living and parasitic roundworms are widely distributed in nature.
2. These organisms are economically important because they cause extensive damage to plants and kill millions of domestic animals and humans each year.

THE ARTHROPODS

1. The arthropods are invertebrate animals with jointed appendages and segmented bodies.
2. These animals are widely distributed in nature and can cause severe damage to crop plants.

3. Some arthropods, like the itch mite, the head louse, and the crab louse, can cause annoying infestations in animals.
4. Many arthropods, such as mosquitoes, lice, mites, fleas, and ticks, transfer diseases from one person to another. These arthropods are also referred to as vectors.

STUDY QUESTIONS

1. Construct a table outlining the differences and similarities among flukes, tapeworms, and roundworms.
2. Using the life cycle of a schistosome illustrated in figure 16-3, discuss two ways that may be used to control the spread of schistosomiasis.
3. Using figure 16-6 as an aid, discuss two ways of controlling the spread of tapeworm disease.
4. In which way(s) do tapeworms differ from flukes? from roundworms? How are they similar?
5. Explain how you might control the spread of pinworms.
6. In which way(s) are arthropods different from the parasitic worms? How are they similar?
7. Can a mosquito transmit parasitic worms from one human to another? Explain.
8. Using figure 16-14 as an aid, construct a simple table differentiating among mosquitoes, fleas, ticks, and lice.
9. Define the meaning of the term vector, and explain how it differs from the term phoront.

SUPPLEMENTAL READINGS

Askew, R. R. 1971. *Parasitic insects.* New York: Elsevier.

Kolata, G. 1985. Avoiding the schistosome's tricks. *Science.* 227:285–287.

Markell, E. K. and Vogue, M. 1981. *Medical parasitology.* 5th ed. Philadelphia: W. B. Saunders Co.

Noble, E. R. and Noble G. A. 1982. *Parasitology: The biology of animal parasites.* 5th ed. Philadelphia: Lea & Febiger.

Rothschild, M. and Clay, T. 1952. *Fleas, flukes and cuckoos.* London: Wm. Collins and Sons.

Schmidt, G. D. and Roberts, L. S. 1981. *Foundations of parasitology.* 2nd ed. St. Louis: C. V. Mosby.

Snow, K. R. 1975. *Insects and disease.* New York: Halsted Press.

Weller, P. F. 1984. Helminthic infections. *Scientific American Medicine* Section 7–XXXV. New York: Freeman.

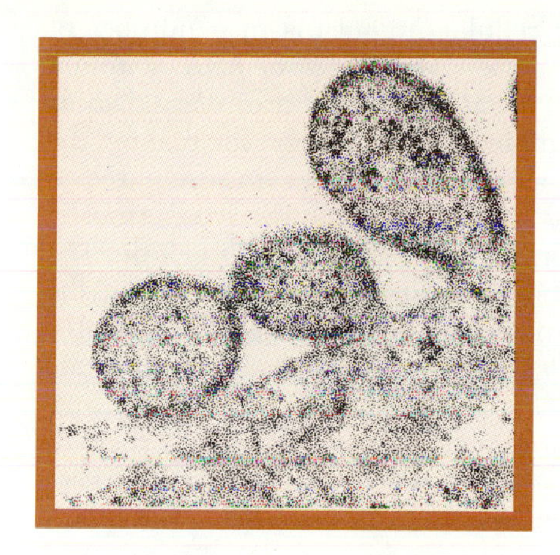

CHAPTER 17 THE BIOLOGY OF VIRUSES

CHAPTER PREVIEW

SOME VIRUSES SUPPRESS THE IMMUNE SYSTEM

Several viruses have been discovered that suppress the immune system and predispose a host to numerous infections. For example, feline leukemia virus, which causes cat leukemia, frequently causes death by making the host animal susceptible to severe infections by other viruses, bacteria, fungi, and protozoans that are not usually pathogens. The feline leukemia virus is believed to suppress the immune system by killing lymphocytes called T-helper cells that are required for the development of immunity.

Human T-cell leukemia virus (HTLV-III), a virus related to feline leukemia virus, is beliveved by many scientists to be the cause of acquired immune deficiency syndorme (AIDS) in humans. In five years (1980–1985), AIDS was diagnosed in more than 10,000 persons in the United States. The groups most susceptible to AIDS were homosexual and bisexual men (72%), intravenous drug users (17%), and Haitians (4%). Sex partners and children associated with the groups at high risk for AIDS accounted for another 5% of the AIDS cases. AIDS victims also include hemophiliacs and persons receiving blood transfusions (1%). AIDS is a very severe disease, as indicated by the fact that 40% of the victims of the disease have died, only 14% have survived more than three years, and none of the AIDS patients has recovered.

AIDS frequently leads to herpes infections (herpes simplex, cytomegalovirus, Epstein-Barr), pneumonia (*Pneumocystis carinii*), hepatitis (hepatitis A and B), toxoplasmosis, cryptosporidiosis, and the cancer Kaposi's sarcoma.

The evidence that HTLV-III may cause the suppression of the immune system comes from a number of studies. In one study, 24% of the AIDS patients studied had antibodies against surface antigens of the human T-cell leukemia virus, while only 0.6% of control patients had the antibodies. In another study, 6% of the AIDS patients had the virus DNA integrated into T-cell DNA. Human T-cell leukemia virus has also been cultured from one AIDS patient (figure 17-1). Finally, AIDS patients have a greatly reduced number of T-helper cells. Because of the discovery that HTLV-III may be the cause of AIDS scientists can now begin to work on a treatment for the disease and a vaccine to prevent it.

Since viruses infect all cellular organisms and often cause serious diseases, the fundamental characteristics of various types of viruses are considered in this chapter. A number of very different viruses, such as the retroviruses that cause cancer and possibly AIDS, are discussed in detail so that the student can appreciate their similarities and differences.

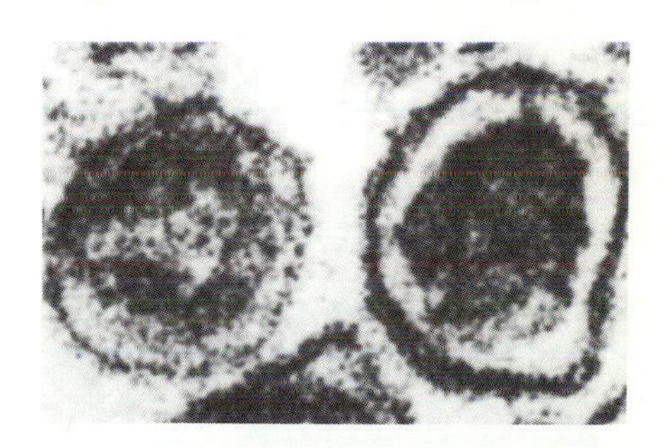

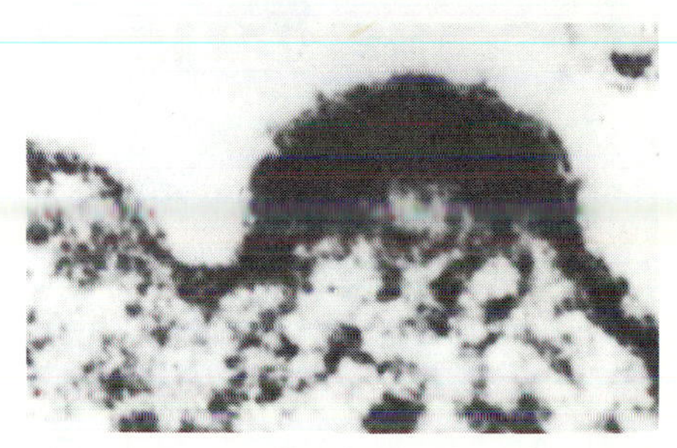

Figure 17-1 Human T-Cell Leukemia Viruses

LEARNING OBJECTIVES

A STUDY OF THIS CHAPTER SHOULD ENABLE YOU TO:

- EXPLAIN WHAT VIRUSES ARE AND HOW THEY MULTIPLY
- COMPARE AND CONTRAST BACTERIAL, PLANT, AND ANIMAL VIRUSES
- DISCUSS THE EFFECTS VIRUSES HAVE ON THEIR HOSTS
- EXPLAIN HOW VIRUSES ARE CHARACTERIZED AND CLASSIFIED
- COMPARE AND CONTRAST VIRUSES WITH VIROIDS
- DESCRIBE THE LYTIC CYCLE OF A THE T-4 BACTERIOPHAGE
- OUTLINE THE LYSOGENIC CYCLE OF A TEMPERATE BACTERIOPHAGE
- EXPLAIN HOW A TYPICAL ENVELOPED VIRUS REPRODUCES
- EXPLAIN HOW BACTERIOPHAGES, PLANT VIRUSES, AND ANIMAL VIRUSES ARE CULTIVATED

Viruses differ from bacteria and other cellular organisms in a number of ways. Viruses consist of a single type of nucleic acid (DNA or RNA) within a protein coat or **capsid.** The nucleic acid and the capsid are referred to as the **nucleocapsid**. The nucleic acids contain the information for making the proteins found in the viral coat as well as many of the enzymes required to invade the host and to replicate the virus's nucleic acid. Some viruses (mostly animal viruses) are covered by a membrane, often called the **envelope**. The term **virion** is used to describe the complete viral particle, including the nucleocapsid and any envelope that it may have. None of the viruses is able to replicate in the absence of host cells. Thus, all the viruses are obligate parasites of cellular organisms.

Many viruses produce gross morphological changes in the cells they infect because they transform or destroy them when they multiply. These changes are sometimes referred to as **cytopathic effects** (CPE). For example, when a confluent layer of bacteria is infected by bacterial viruses called **bacteriophages**, morphological changes can be seen (fig. 17-2a). The confluent layer of bacteria, known as a **bacterial lawn**, makes the solidified nutrient in the plate cloudy. When a bacteriophage reproduces in a bacterium, it destroys the cell and the viral progeny spread to adjacent bacteria. Within 10–20 hours, a large area of the bacterial lawn around the original infection becomes lysed, producing a clear area in the lawn. Actually, each clear area represents a virus "colony" that has arisen from a single virus. These virus colonies are known as **plaques**.

Figure 17-2b shows a leaf (which can be thought of as a lawn of plant cells) that has been invaded by **tobacco mosaic virus**. The destruction of large areas of tissue by the virus results in visible lesions. Each lesion probably originated from a single virus.

Figure 17-2c illustrates a human infected by smallpox virus. The smallpox virus is proliferating in the skin cells and causing extensive destruction. The lesions represent virus colonies, and each lesion may have arisen from a single virus. In general, viral infections are detected when their reproduction causes noticeable changes in a host's cells.

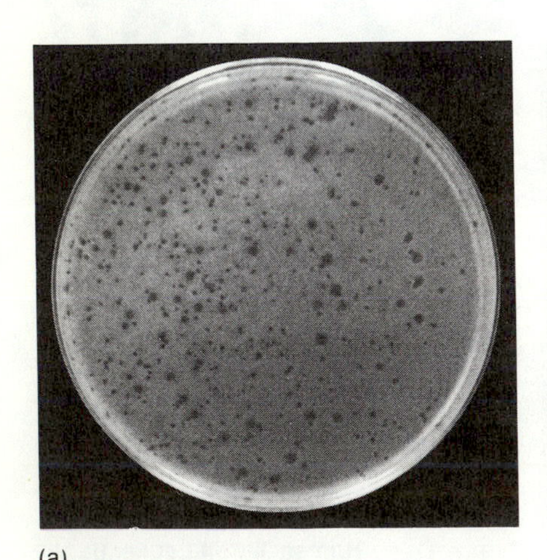
(a)

(b)

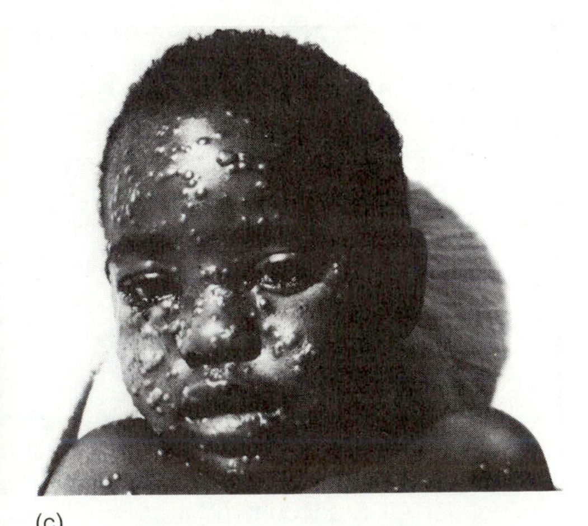
(c)

Figure 17-2 Viral Lesions. *(a)* A lawn of bacteria contains clear areas that represent bacteriophage plaques or virus colonies that have developed from a single bacteriophage. *(b)* A leaf with light-colored zones (arrows) that represent tobacco mosaic virus lesions. The leaf can be thought of as a lawn of plant cells. *(c)* A person exhibiting numerous smallpox lesions caused by the smallpox virus.

CLASSIFICATION OF VIRUSES

Bacteriophages

Bacteriophage, or **"phage"** for short, is the term used for the viruses that infect bacteria. The bacteriophages occur in a wide assortment of shapes and sizes, and they may contain double-stranded DNA (ds-DNA), single-stranded DNA (ss-DNA), double-stranded RNA (ds-RNA), or single-stranded RNA (ss-RNA) (table 17-1). Bacteriophages are generally classified on the basis of their nucleic acid and their structure, which may be **complex**, **icosahedral** (20-sided), **cylindrical**, or **enveloped**. Most of the bacteriophages are complex or icosahedral.

The ds-DNA bacteriophages are very complicated viruses. Their genomes code for 50–200, proteins depending upon the phage. On the other hand, the ss-DNA and ss-RNA bacteriophages are very simple and code for only 3–5 proteins. The ds-RNA bacteriophage $\phi 6$ is one of the largest RNA viruses and codes for approximately 20 proteins (table 17-1).

Plant Viruses

In contrast to the bacteriophages, almost all the plant viruses are ss-RNA viruses (table 17-2). The exceptions are the **caulimoviruses** (ds-DNA), the **geminiviruses** (ss-DNA), and the **reoviruses** (ds-RNA). The protein coat or capsid of plant viruses is generally in the shape of an icosahedron or a cylinder. The plant viruses are generally transmitted from plant to plant by 381 insects, nematodes □ (roundworms), fungi, or contaminated machinery. Some viruses are also transmitted from parent to offspring in pollen or in seeds.

Many of the plant viruses are unusual in that their genetic information is segmented and packaged into separate capsids. For example, **ilarviruses** packages four different genomes into separate capsids, and all four are required for virus multiplication. The genomes of the RNA plant viruses are about the same size as those found in the RNA bacteriophages. The RNA genomes in plants code for 3–15 proteins.

Animal Viruses

Almost all of the animal viruses are either icosahedral (20-sided polyhedron) viruses or enveloped viruses and have ss-RNA or ds-DNA genomes. Some, however, have ds-RNA or ds-DNA genomes (table 17-3). The DNA animal viruses fall into two size ranges: those that code for 5–10 proteins and those that code for 30–300 proteins. One of the pox viruses has the largest genome and is believed to code for more than 300 proteins.

The large percentage of animal viruses that are enveloped (approximately 50%) may reflect the fact that their hosts do not have cell walls and entrance to the host's cytoplasm may be expedited by having an envelope that can fuse with the plasma membrane or a pinocytic vesicular membrane. Alternatively, enveloped animal viruses may reflect the need of a foreign particle to hide from the host's immune system. At least two enveloped animal viruses (arenaviruses and bunyaviruses) appear to lack a capsid around their RNAs. Nevertheless, their RNAs are associated with protein. The RNA-protein complex is referred to as the **ribonucleoprotein** core.

TABLE 17-1

CHARACTERISTICS OF COMMON BACTERIOPHAGES

CLASSIFICATION	REPRESENTATIVE VIRUSES	SHAPE (MORPHOLOGY)	TYPE OF NUCLEIC ACID	NUMBER OF GENES	CHARACTERISTICS
Class I					
Myoviruses	T2, T4, T6, Mu P2, P4, SP8 SP50, PBS1		Linear ds-DNA	160 (T4)	Prolate icosahedral head 80 × 110 nm, complex tail 110 nm long. Adsorb to wall except PBS1 which attaches to flagella.
Styloviruses	Lambda, Chi, T1, T5		Linear ds-DNA	55 (λ)	Regular icosahedral head 54 nm diameter, simple tail 140 nm long. λ adsorbs to wall, Χ adsorbs to flagella.
Pedoviruses	T3, T7 P22		Linear ds-DNA	40 (P22)	Regular icosahedral head 60 nm diameter, short tail 20 nm long. Adsorb to wall.
Corticoviruses	PM2		Circular ds-DNA	10 (PM2)	Icosahedral capsid 60 nm diameter, enclosed in a membrane & covered with protein. No tail.
Plasmaviruses	MVL2		Circular ds-DNA	12 (MVL2)	Helical capsid? enclosed in a membrane. May be spherical or linear.
Class II					
Microviruses	ϕX174, G4, G6 G13, G14, S13 M12		Circular ss-DNA	4 (ϕX174)	Icosahedral capsid 30 nm diameter. No tail.
Inoviruses	Fd, M13, f1 MVL1		Circular ss-DNA	4 (fd)	Filamentous (helical?) capsid 8 × 800 nm. Adsorb to end of pili. Host cell not lysed.
Mycoplasmaviruses	L51		Circular ss-DNA	5 (L51)	Helical capsid? enclosed in a membrane. Bullet shaped 14 × 80 nm
Class III					
Cystoviruses	ϕ6		Linear ds-RNA 3 genomes	20 (ϕ6)	Icosahedral capsid 70 nm diameter enclosed in a membrane.
Class IV					
Leviviruses	R17, MS2, f2 fr, Qβ, C1		Linear ss-RNA	3 (R17)	Icosahedral capsid 25 nm diameter. No tail. Adsorb along the length of pili

TABLE 17-2

CHARACTERISTICS OF PLANT VIRUSES

CLASSIFICATION	SHAPE (MORPHOLOGY)	NUMBER OF GENONES	SIZE OF GENONE (IN KILOBASES)	SIZE OF VIRION (IN NM)	MODE OF TRANSMISSION
RNA VIRUSES					
Naked, isometric or icosahedral single-stranded RNA genomes					
Bromovirus (3) (Brome mosaic)		4, linear	3.7 + ?	23	Beetle
Comovirus (2) (Cowpea mosaic)	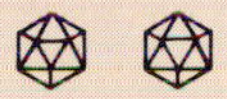	2, linear	7.7 + 4.7	30	Beetle
Cucumovirus (3) (Cucumber mosaic)		4, linear	3.7 + 3.3 + 2.3 + 1.0	30	Aphid
Nepovirus (2) (Tobacco ringspot)	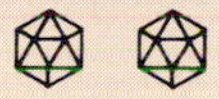	2, linear	8.0 + ?	30	Nematode
Ilarvirus (4) (Tobacco streak)		4, linear	4.3 + 3.7 + 2.7 + 1.3	26, 30, 35	Seed and pollen
Penamovirus (2) Pea enation mosaic virus group		2, linear	5.6 + 4.7	30	Aphid
Luteovirus		1, linear	6.4	25	Aphid
Tombusvirus (Tomato bushy stunt)		1, linear	5.0	30	?
Tymovirus (Turnip yellow mosaic)		1, linear	6.4	30	Beetle
Tobanecrovirus (Tobacco necrosis virus group)		1, linear	5.0	28	Fungus
Naked, isometric or icosahedral, double-stranded RNA genomes					
Reovirus		10, linear or 12, linear	6.6 1.5	70–80	Leaf hoppers & planthoppers
Naked, cylindrical single-stranded RNA genomes					
Tabamovirus (Tobacco mosaic)		1, linear	6.6	300 × 18	
Potexvirus (Potato X)		1, linear	7.0	480–580 × 13	
Potyvirus (Potato Y)	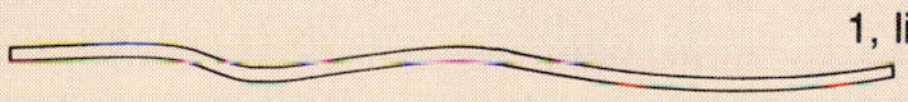	1, linear	10.0	680–900 × 12	Aphid
Carlavirus	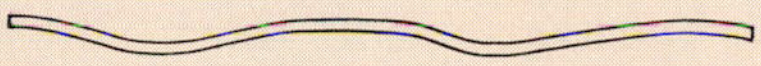	1, linear	?	690 × 12	Aphid
Closterovirus		1, linear	14.0	600–2000 × 12	Aphid
Tobravirus (2) (Tobacco rattle)		2, linear	8.0 + 4.3	300 × 18	Nematode
Hordeivrus (Barley stripe mosaic)		3, linear	5.0 + 3.3 + ?	110–160 × 23	?
Enveloped, bullet-shaped, single-stranged RNA genomes					
Rhabdovirus	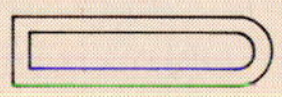 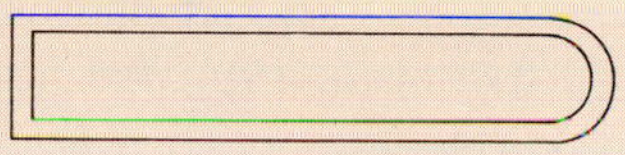	1, linear	?	130–150 × 45 430–500 × 110	Insects

* genome/capsid. Number of viruses illustrated indicates the number of different genomes required for an infection.

**11 to 12 genomes/capsid. Only one virus required for an infection.

***1 genome/capsid. Only largest particle is infectious by itself.

TABLE 17-2 (CONTINUED)

CLASSIFICATION	SHAPE (MORPHOLOGY)	NUMBER OF GENONES	SIZE OF GENONE (IN KILOBASES)	SIZE OF VIRION (IN NM)	MODE OF TRANSMISSION
Naked, bacilli-shaped, single-stranded RNA genomes					
Almovirus (Alfalfa mosaic)		4, linear	3.0 + 2.3 + 2.0 + 0.8	58 & 48 & 36 & 18 × 18	Aphid
Enveloped (spherical) single-stranded RNA genome					
Tospovirus		1, linear (?)	24.7	75	Thrips
DNA VIRUSES					
Naked, icosahedral Double-stranded DNA genomes					
Caulimovirus (Cauliflower mosaic)		1, circular	6.0	50	Aphid
Naked, fused capsid, single-stranded DNA genomes					
Geminivirus (Maize streak)		1, circular	3.0	2(18)	White fly & leafhopper

* genome/capsid. Number of viruses illustrated indicates the number of different genomes required for an infection.

**11 to 12 genomes/capsid. Only one virus required for an infection.

***1 genome/capsid. Only largest particle is infectious by itself.

BACTERIOPHAGES

Bacteriophage or phage is the name given to bacterial viruses by Felix d'Herelle in 1917 because they seemed to "eat" (Greek, *phagos*) bacterial lawns. Bacteriophages cannot be seen with the light microscope, and they pass through filters that would trap bacteria. Consequently, when phages were first discovered they were thought to be related to other filterable infectious agents. With the development of the electron microscope □ in the late 1930s, it was possible to see that viruses were very different from bacteria and other cellular organisms. The bacteriophages came in assorted sizes and shapes, but each type maintained its major characteristics as it reproduced. Thus, viruses were true breeding infectious agents much like cellular microorganisms are.

65

Bacteriophages are generally classified on the basis of their structure and the type of nucleic acid they have (table 17-1). They may be extremely complex. For example, the T-even phages have complicated tails associated with their capsids or head regions. **T4** has an elongated (prolate) icosahedral capsid with an attached cylindrical core and sheath. It is a ds-DNA phage with approximately 200 genes. On the other hand, viruses such as **φX174** and **Qβ** have small icosahedral capsids and their genome may contain as few as 3 genes. φX174 has a closed circular ss-DNA, while Qβ has a linear ss-RNA genome. In contrast, some single-stranded DNA viruses, such as **fd** and **M13**, are filamentous (cylindrical). Unusual phages are **MVL2** and **φ6**, which are covered by membranous envelopes derived from the host bacterium.

Viruses are unable to propagate outside a cell because they lack the necessary enzymes and structures to carry on the metabolism required to reproduce themselves. Viruses can reproduce only if they get their genetic

TABLE 17-3
CHARACTERISTICS OF ANIMAL VIRUSES

CLASSIFICATION	SHAPE (MORPHOLOGY)	TYPE OF NUCLEIC ACID	SIZE OF GENOME (IN KILOBASES)	SIZE OF VIRION (IN NM)	REPRESENTATIVE DISEASES
RNA VIRUSES					
Naked, icosahedral capsid, single stranded RNA genome					
Picornaviruses		1, linear + ssRNA	7.7	22–30	Poliomyelitis, hepatitis A, foot and mouth disease
Naked, icosahedrad capsid, double stranded, RNA genomes					
Reoviruses		10–11, linear dsRNA	31 + 37 + 46	60–80	Respiratory diseases, polyhedrosis disease of insects
Enveloped, icosahedral capsid, single stranded, RNA genome					
Togarviruses		1, linear + ssRNA	13	40–75	Encephalitis, yellow fever
Enveloped, cylindrical capsid, single stranded, RNA genomes					
Orthomyxoviruses		8, linear − ssRNA	18	80–120	Influenza
Bunyaviruses		3, linear − ssRNA	12 + 6 + 2.4	100	Encephalitis
Coronaviruses		1, linear + ssRNA	18	60–220	Common cold
Arenaviruses		2, linear − ssRNA	11 + 4.8	50–300	Hemorrhagic fevers, lassa fever, lymphocyte choriomeningitis
Paramyxoviruses		1, linear − ssRNA	20	150–300	Mumps, measles, newcastle disease of chickens, distemper in dogs
Retroviruses		2, linear + ssRNA	5	100	Cancers, hepatitis non A non B, AIDS
Enveloped, bullet-shaped, cylindrical capsid, single stranded RNA genome					
Rhabdovirus		1, linear − ssRNA	11.7 –15.3	70–80 × 130–240	Rabies

TABLE 17-3 (CONTINUED)

CLASSIFICATION	SHAPE (MORPHOLOGY)	TYPE OF NUCLEIC ACID	SIZE OF GENOME (IN KILOBASES)	SIZE OF VIRION (IN NM)	REPRESENTATIVE DISEASES
DNA VIRUSES					
Naked, icosahedral capsid, single stranded DNA genome					
Parvovirus		1, linear ssDNA	6.0	50	Diseases in rodents
Naked, icosahedral capsid, double stranded DNA genome					
Papoviruses		1, circular dsDNA	5 + 9	45–55	Warts and tumors
Adenoviruses		1, linear dsDNA	30 + 55	70–90	Cancers
Naked, icosahedral double capsid, double stranded DNA genome					
Hepatitis B		1, circular dsDNA	3.2	42	Hepatitis B
Enveloped, icosahedral capsid, double stranded DNA genome					
Herpes viruses		1, linear dsDNA	154 + 231	150–200	Chicken pox, genital herpes, shingles, mononucleosis, cancer, fever blisters
Iridoviruses		1, linear dsDNA	167–417	130–300	African swine fever
Complex coats, complex cylindrical capsid, double stranded DNA genome					
Pox viruses		1, linear dsDNA	231 + 307	200–260X	Smallpox, cowpox

Virus Attachment

The bacteriophages begin their multiplication by attaching to a specific host cell (figure 17-3). The tailed bacteriophages, such as **T4** and **lambda (λ)**, attach to and puncture the outer envelope of their host's cell wall, while **Chi** attaches along the length of the flagella. The icosahedral bacteriophage **MS2** attaches along the length of F pili, while the cylindrical bacteriophage **M13** binds to the ends of F pili. The enveloped bacteriophages that attack the wall-less mycoplasmas attach to the host cell's plasma membrane.

Genome Penetration

After the virus attaches to its host, it introduces its genetic material into the cell. Bacteriophage genomes enter the bacterial hosts in various ways. **T4** "drills" a hole through the bacterial cell wall and then pushes the core of its

(a)

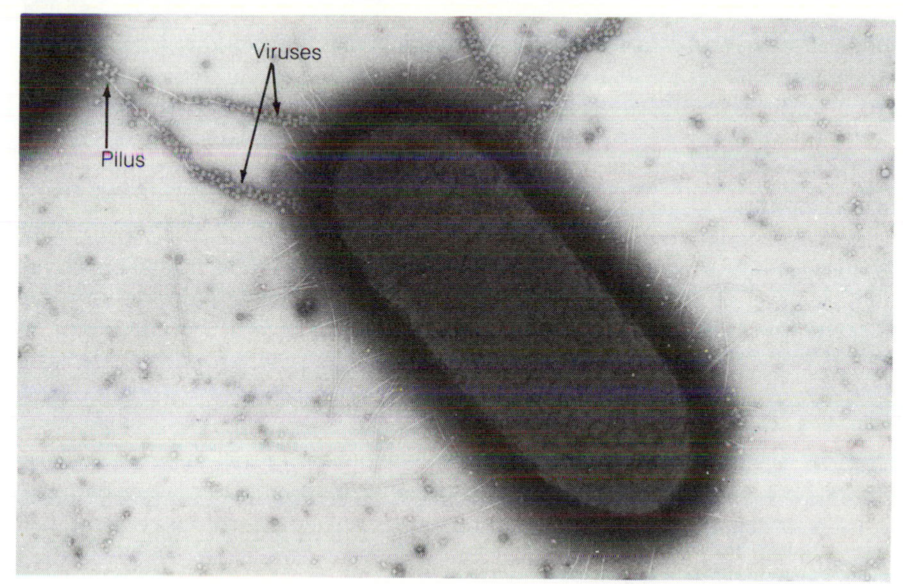

(b)

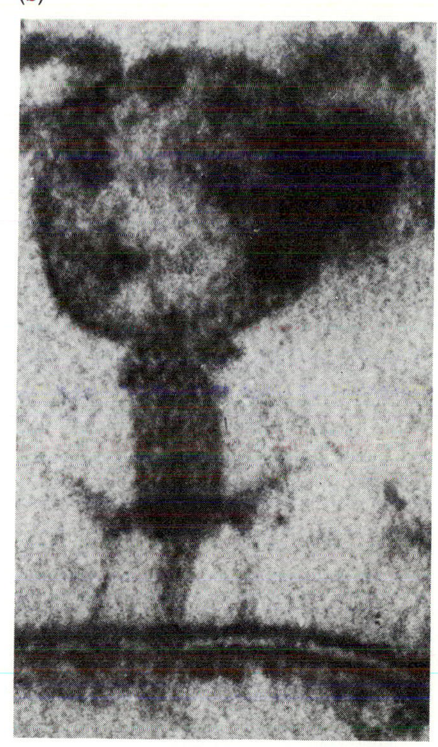

(c)

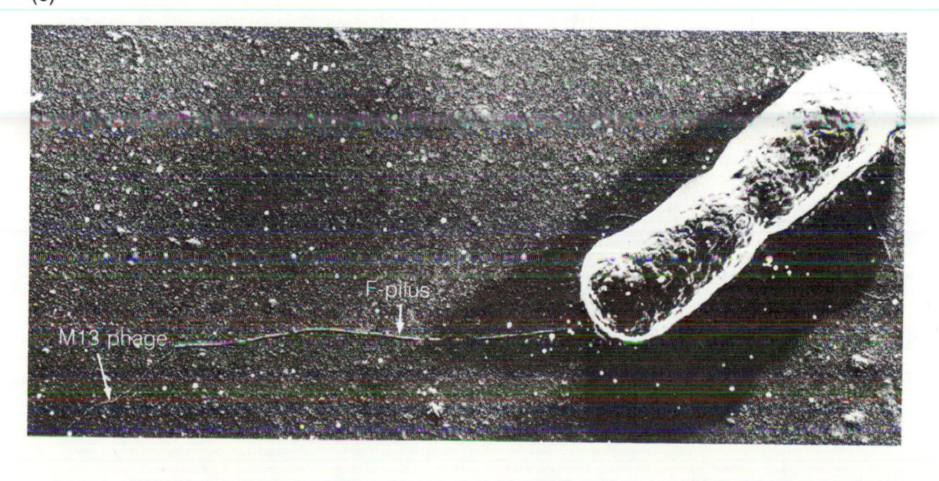

Figure 17-3 Bacteriophage Attachment. *(a)* The icosahedral bacteriophage MS2 attached along the length of F-pili. *(b)* The complex bacteriophage T4 attached to the bacterial cell wall. *(c)* The cylindrical bacteriophage M13 attached to the end of an F-pilus.

Figure 17-4 **Bacteriophage Infection.** A number of T4 are attached to the cell wall of a bacterium. When the phage sheath contracts, the core is forced through the wall and plasma membrane, and the bacteriophage DNA passes from the phage head into the cytoplasm. Strands of phage DNA can be visualized entering the cell's cytoplasm (arrows).

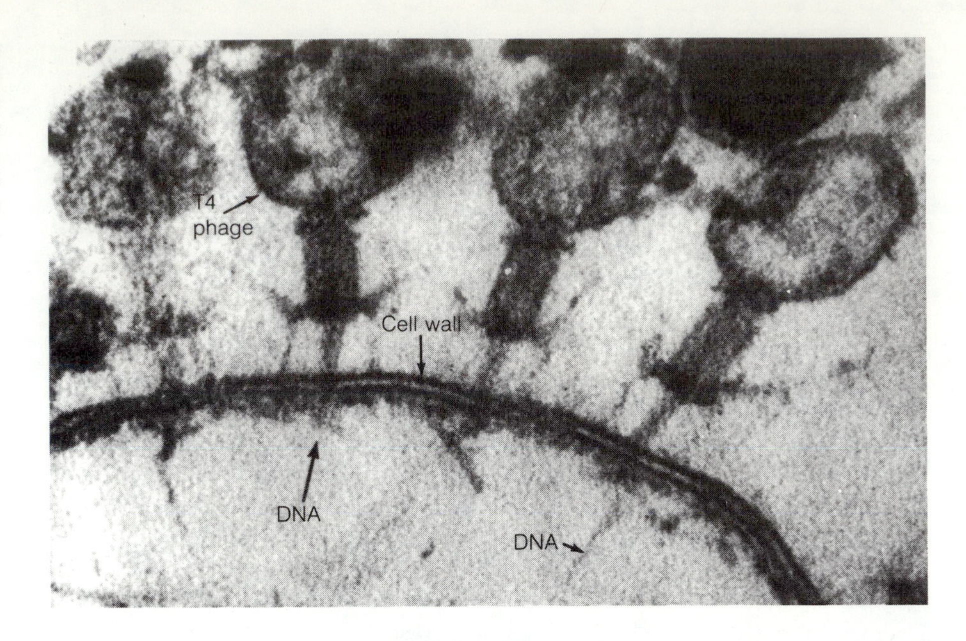

tail into the host cell (fig. 17-4). The DNA genome is then injected into the cell. **Chi** may inject its hereditary material into the hollow flagella, which must somehow migrate from the flagella to the cytoplasm. MS2 and M13 fuse their capsids with bacterial sex pili. How the virus genomes reach the cytoplasm is unknown; perhaps the virus genomes enter the cytoplasm when the pili are retracted by the cell. The enveloped bacteriophage may enter the wall-less mycoplasmas by pinocytosis or by fusing their envelopes with the host's plasma membrane.

Virus RNA Synthesis, Protein Synthesis, and Genome Replication

Once the bacteriophage genome enters the cytoplasm, it directs the synthesis of virus mRNA and proteins (fig. 17-5). The virus proteins catalyze the replication of the virus genome and function as structural components of the virus particle.

Virus Assembly

As the virus proteins are synthesized, they self-assemble into virus components such as the head, tail, and tail fibers (fig. 17-6). The assembly of many virus components is catalyzed by virus proteins.

Release

In some infections, virus proteins cause the lysis of the cell so that the assembled particles accumulating in the cytoplasm escape (fig. 17-7). In a few cases (Fd, Ml3, Fl), the viruses appear to be excreted from the cell without lysing it. When bacteriophages repeatedly cause the death and lysis of host cells, the infectious cycle is known as the **lytic cycle** (fig. 17-6). Some viruses, such as the DNA bacteriophage λ and the animal RNA retroviruses, are able to delay or avoid virus assembly and release by integrating a DNA genome

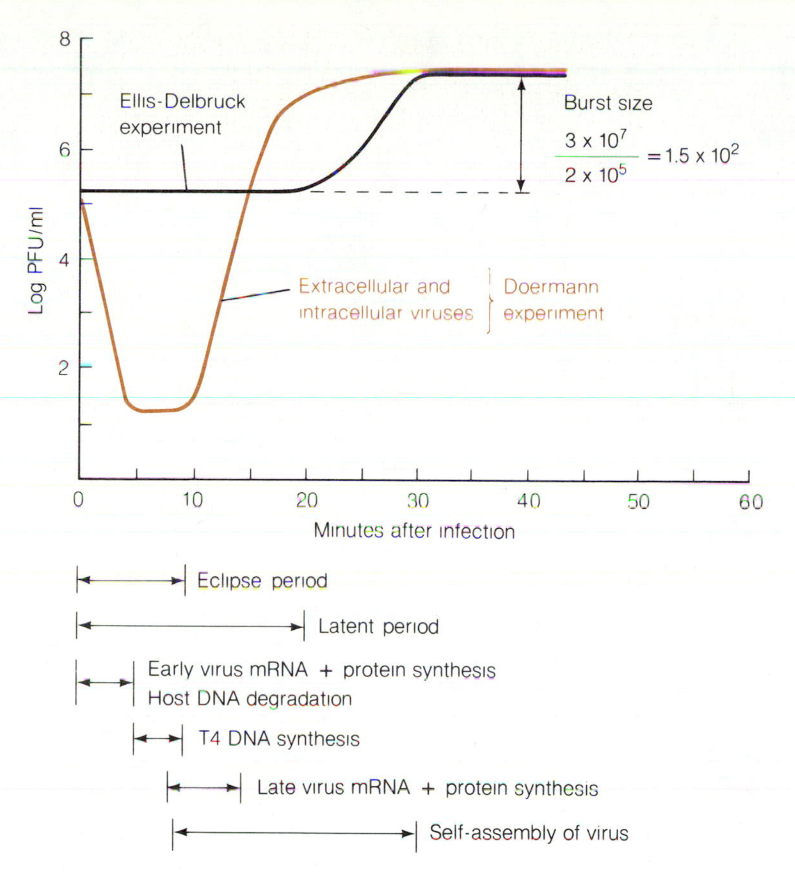

Figure 17-5 One-Step Growth Curve. Within minutes after T4 DNA enters *E. coli,* early virus mRNA and proteins are synthesized. In addition, the host's DNA is degraded. Approximately 5 minutes after infection, T4 DNA begins to be replicated. Viral proteins self-assemble into mature viruses 10 to 15 minutes after infection. During this same period, late viral mRNA and proteins begin to be synthesized. Some of these late proteins will promote the lysis of the host cell and the escape of the viruses. Cells begin to break open approximately 20 minutes after infection and extracellular phage begin to appear. **Ellis-Delbruck experiment.** Bacteriophage (2×10^5/ml) are mixed with bacteria (5×10^8/ml) so that only one phage adsorbs to a bacterium. The diluted culture of infected bacteria is subsequently sampled every 5 minutes. For the first 20–25 minutes after T4 infects *E. coli,* the number of plaques remains the same. The time during which no "new" phage appear is called the **latent period.** After the latent period, the number of phage in the culture suddenly increases approximately 70-fold and then levels off. The rise in phage number represents the release of virus progeny. **Doermann experiment.** If infected bacteria are broken open at various times after infection, the number of plaques decreases more than 100-fold during the first 10 minutes of the infection. Phage begin to form within the cell approximately 10 minutes after infection. The time during which no phage progeny can be detected within the cells is called the **eclipse period.**

into the host's genome (fig. 17-8). Generally, almost all transcription and translation of virus genes is inhibited when the virus genome integrates into the cell's genetic information. The integrated bacteriophage DNA is called a **prophage**. Since the prophage is replicated along with the cell's genome, each daughter cell carries a prophage. A bacterium that carries a prophage is called a **lysogen**, and this unusual infection is called **lysogeny**.

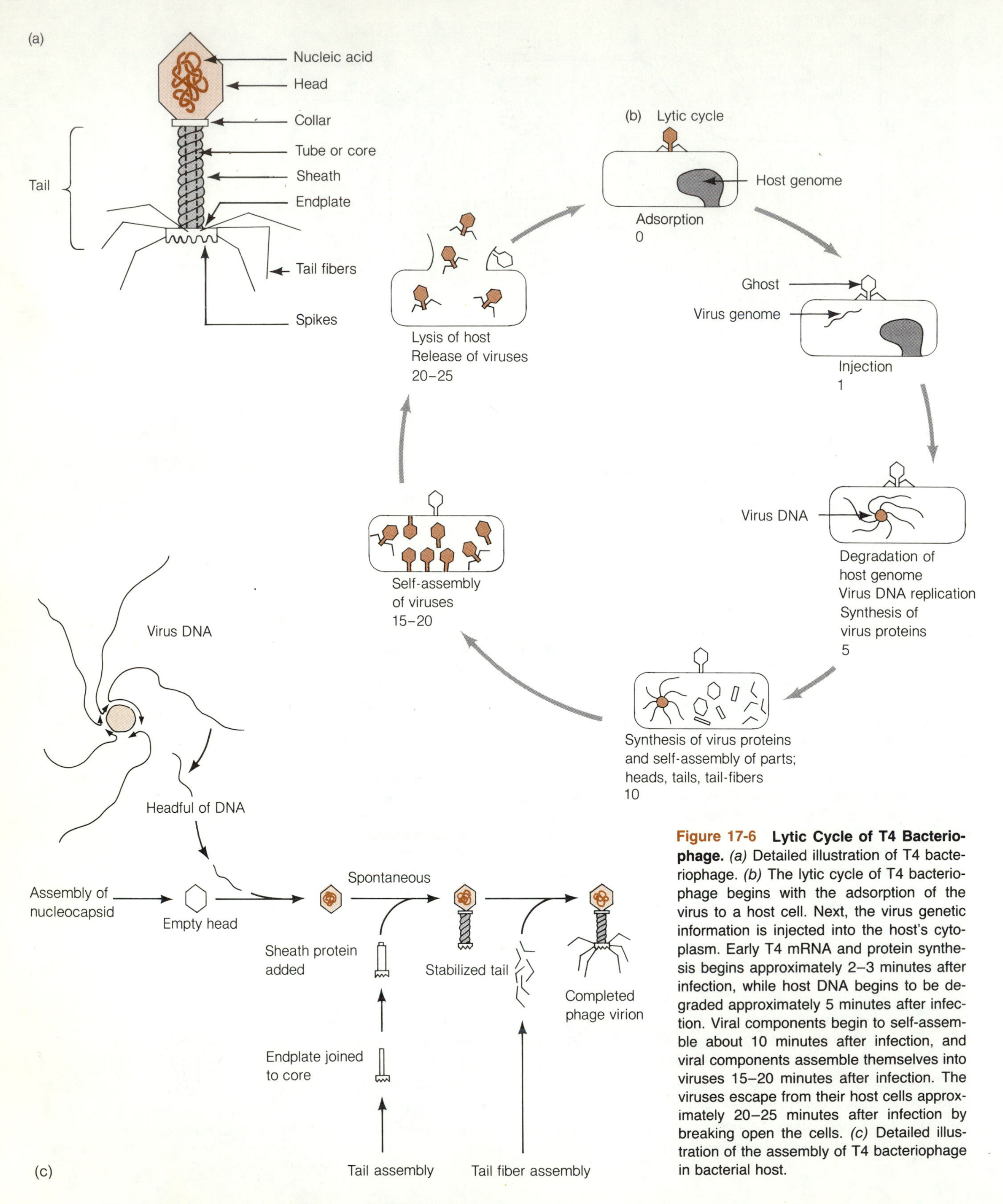

Figure 17-6 Lytic Cycle of T4 Bacteriophage. *(a)* Detailed illustration of T4 bacteriophage. *(b)* The lytic cycle of T4 bacteriophage begins with the adsorption of the virus to a host cell. Next, the virus genetic information is injected into the host's cytoplasm. Early T4 mRNA and protein synthesis begins approximately 2–3 minutes after infection, while host DNA begins to be degraded approximately 5 minutes after infection. Viral components begin to self-assemble about 10 minutes after infection, and viral components assemble themselves into viruses 15–20 minutes after infection. The viruses escape from their host cells approximately 20–25 minutes after infection by breaking open the cells. *(c)* Detailed illustration of the assembly of T4 bacteriophage in bacterial host.

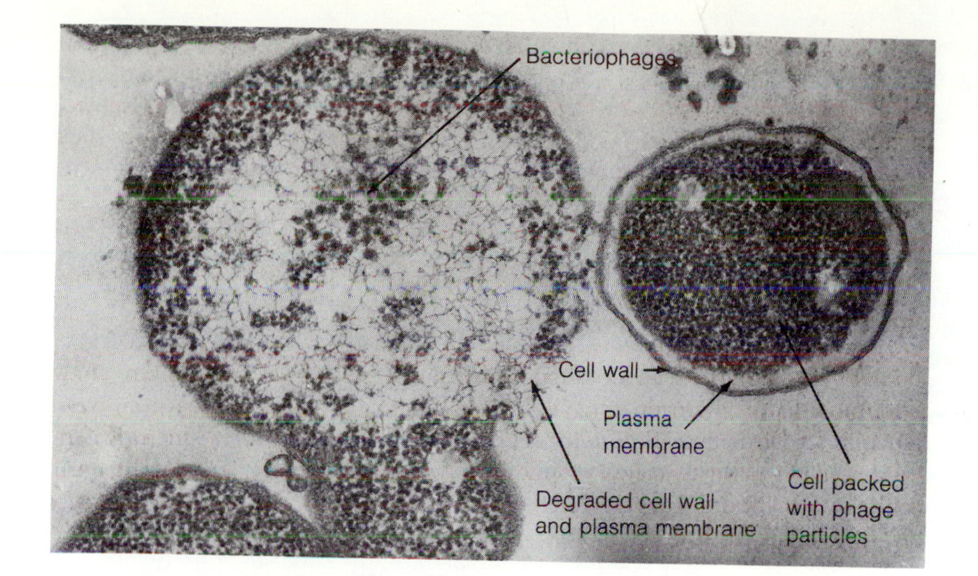

Figure 17-7 Release of Bacteriophage 0X174. Numerous bacteriophages can be seen in the *E. coli* cytoplasm. The viruses are escaping through the partially degraded cell wall.

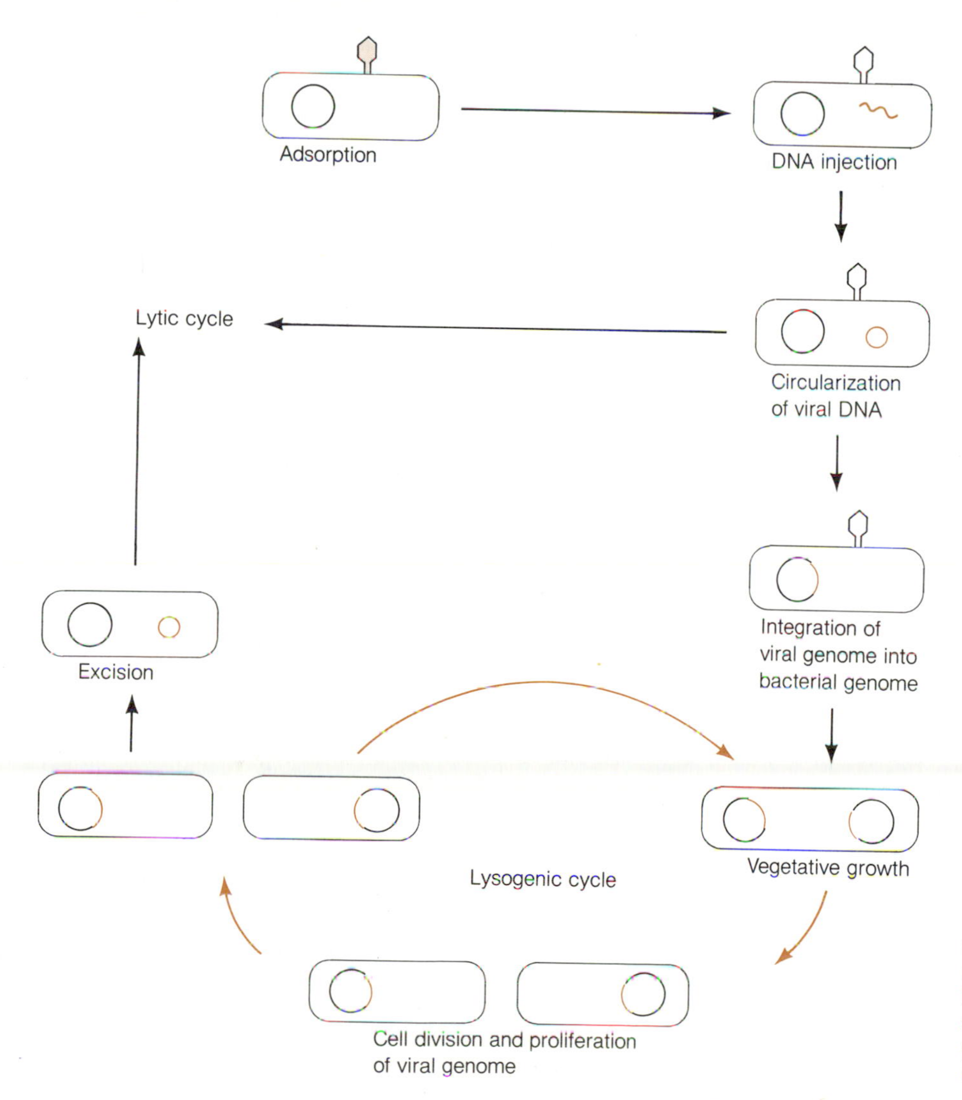

Figure 17-8 Lysogenic Cycle of Lambda Bacteriophage. Lambda bacteriophage frequently goes through a lysogenic cycle rather than a lytic cycle. After adsorption, injection, and genome circularization, the virus genome recombines with the host's genome and becomes part of it. A bacterium that contains an integrated virus genome is called a lysogen. The virus genome is replicated as the lysogen proliferates, so each daughter cell contains a virus genome. The replication of the virus genome as the lysogen proliferates represents the lysogenic cycle. Generally, when the lysogen's DNA is damaged, lambda prophage excises itself from the lysogen's DNA and enters a lytic cycle.

FOCUS
THE DISCOVERY OF VIRUSES

During the "golden age of microbiology" (1875–1915), numerous scientists discovered that some plant and animal diseases were caused by a group of organisms that were much smaller than bacteria and that could not be grown on artificial media. Martinus Beijerinck in the late 1890s hypothesized that plant and animal diseases caused by agents that passed through filters that retained bacteria were not bacteria but something quite different. Eventually these **filterable agents** of disease became known as **viruses**. Because of his studies and insights on these infectious agents, Beijerinck is considered the "father" of virology.

It was not until 1915, however, that virus infections of bacteria were recognized. Frederic Twort observed that bacterial colonies were destroyed by a filterable agent that he hypothesized was either a bacterial virus or a toxin. In 1917, Felix d'Herelle studied the destruction of bacteria by a filterable agent. The virus produced clear holes called **plaques** in bacterial lawns and cleared turbid broth cultures within a few hours, killing most of the bacteria. He reasoned that the infectious agent was reproducing at the expense of the bacteria and consequently causing their destruction. D'Herelle called these bacterial viruses **bacteriophages** because they appeared to consume or eat the bacteria.

In 1935 Wendell Stanley, using techniques for purifying and crystallizing proteins, crystallized tobacco mosaic virus (TMV) and concluded that it was mostly protein. Near the end of 1939, Stanley and Frederic Bawden independently isolated a nucleic acid from the crystallized TMV. This nucleic acid was later shown to be necessary for virus infectivity. Stanley received the 1946 Nobel Prize in chemistry for his crystallization of TMV. Tobacco mosaic virus was the first virus to be observed in the electron microscope in 1939. TMV turned out to be a long, thin, tubelike structure, very much smaller than the smallest bacteria and very different in structure.

In 1939, Emory Ellis and Max Delbruck showed that T4 bacteriophage multiplied within bacteria and that daughter bacteriophages were released when the bacteria lysed. Thomas Anderson and Salvador Luria obtained electron micrographs of T4 bacteriophage in 1942 that clearly indicated it had a "head" and a tubular "tail." In 1952, Martha Chase and Alfred Hershey performed what is now known as the Hershey-Chase experiment, which demonstrated that T4 DNA entered *E. coli* cells but little or no phage protein penetrated. Their experiment showed that the viral genetic information was a nucleic acid rather than protein. Max Delbruck, Salvador Luria, and Alfred Hershey shared the 1969 Nobel Prize for Physiology or Medicine because of their contributions to the understanding of bacteriophage structure, function, and genetics.

The Ellis and Delbruck Experiment

In 1939, Emory Ellis and Max Delbruck discovered that T4 bacteriophages did not proliferate exponentially like cells but increased in spurts or steps (fig. 17-5). For approximately the first 20 minutes after T4 infects *E. coli*, no new phages are evident in the culture. The time during which no new phages appear is called the **latent period.** During this period, viruses are replicating within host cells. After the latent period, the number of phages in the culture suddenly increases approximately 70-fold and then levels off. The rise in phage number represents the release of virus progeny from the infected cells. The release occurs within a short period of time because all the infected cells lyse at about the same time. The **one-step growth curve** is characteristic of many viral infections that lyse the host cells. In those cases in which the host is not lysed and the viruses are shed continuously, however, there is no one-step growth curve. The Ellis-Delbruck experiment demonstrates that one viral particle gives rise to numerous progeny in a short period of time. The number of progeny released per infected cell is called the **burst size.**

The Doermann Experiment

A. H. Doermann reported in 1952 that phage development during the latent period occurred inside the infected bacteria, and that it was not until the bacteria lysed, 20–30 minutes after infection, that progeny phage were released. If infected bacteria are broken open at various times after infection, the number of bacteriophages detected (plaque-forming units/ml) for the first 10 minutes is more that 100 times less than expected (fig. 17-5). This phenomenon occurs because adsorbed phages that have not reproduced are inactivated when the bacteria are lysed. Not until phages begin to form, 10–20 minutes after infection, are the parental phages replaced by the progeny. The time during which no progeny can be detected in the host cells is called the **eclipse period.**

The Doermann experiment supported the idea that virus formation takes place like the building of automobiles on an assembly line. The different parts of the virus are synthesized, and then they rapidly self-assemble into a complete virus.

Bacteriophage T4

The T4 bacteriophage is an obligate parasite of *Escherichia coli* and has a very complicated structure (fig. 17-6). T4 consists of an elongated icosahedral (20-sided) head and a cylindrical tail. The tail is a complex structure that has a contractile sheath, base plate, tail fibers, and spikes (pins). The protein head or capsid protects the double-stranded DNA genome □, while the hollow tail functions as a passageway for the movement of the genome from the head to a host cell's cytoplasm.

89

Lytic Infection

A lytic infection by T4 results in the lysis and death of the host cell. The tips of the T4 tail fibers initially bind the phage to the outer envelope of the *E. coli* cell wall. The binding of T4 to a host cell apparently induces the sheath to contract and plunge the tail core through the cell wall and plasma membrane. The tip of the tail contains lysozyme, an enzyme that degrades the peptidoglycan in the cell wall. ATP associated with the sheath rings is believed to power the contraction of the sheath. The contraction of the sheath apparently induces a conformational change in the head that forces the ds-DNA out of the head, into the tube, and then into the host (fig. 17-6).

207

Soon after T4 DNA is injected into *E. coli*, it is transcribed □ into mRNA. The mRNA codes for the T4 early proteins, which are synthesized 2–5 minutes after infection and are involved in: a) controlling which T4 mRNA are to be synthesized and translated; b) the synthesis of nucleotides; and c) the synthesis of new T4 DNA. The intermediate proteins, synthesized 7–10 minutes after infection, appear to be concerned with processing and recombining the new T4 DNA. The late proteins act as structural components of the phage (head proteins, tail proteins, etc.) and catalyze the development of virus parts. In addition, late proteins such as lysozyme promote the escape of the virus from the host cell.

Morphogenesis of T4

T4 proteins self-assemble into a large number of viruses within the cytoplasm (fig. 17-6). Eventually, lysozyme and other enzymes degrade the membrane and cell wall of the host cell so that numerous T4 phages are released. A single virus can give rise to 150 new particles in about 25 minutes at 35°C.

T4 mutants that are unable to synthesize tail fibers, tails, or heads have facilitated the study of T4 morphogenesis. The inability of a virus to synthesize one component does not inhibit it from synthesising other components. This could be visualized with the aid of the transmission electron microscope. Many mutated genes have been found that cause the accumulation of normal and abnormal structures, but nonfunctional viruses. From the pattern of structures produced, the functions of the genes can be deduced.

Bacteriophage Lambda

Lambda is a complex bacteriophage with a linear ds-DNA genome that codes for approximately 55 proteins (fig. 17-8). This bacterial virus is of interest because it can undergo either a lytic or it can integrate itself into the host's genome an remain in that state for many generations. This phenomenon is called **lysogeny**.

Lysogeny

The lambda genome may direct the synthesis of new viruses, or it may integrate into the host's genome and become quiescent. This viral state will not cause the lysis and death of the host cell. It has been observed that lysogeny occurs most frequently when a large number lambda viruses infect a cell. When there is an excess of lambda viruses and many viruses infect each cell, there is said to be a **high multiplicity of infection**. In addition, the metabolic state of the host is important in determining whether or not the phage genome will become a **prophage** (integrated into the host genome).

Once the lambda genome has integrated into the bacterial genome, it is quite stable and is replicated along with the host's genetic information. Thus, each daughter cell that is produced will contain a prophage. Lysogeny is maintained by a **repressor** protein coded for by the lambda prophage. The lambda repressor blocks the expression of all the other lamdba genes. A cell that contains a prophage is known as a **lysogen.**

Excision

Certain physiological states of the host cell determine whether or not the prophage will remain committed to the lysogenic state or begin a lytic infection. For example, if a lysogen's DNA is damaged (by enzymes, chemicals, ultraviolet radiation, etc.) so that single-stranded regions of DNA are formed, lysogeny is no longer favored and a host protein (a protease) cleaves the lambda repressor that is maintaining lysogeny and inhibiting the expression of the the lambda prophage. When the lambda repressor is destroyed by the protease, RNA polymerase initiates transcription, and lambda proteins required for prophage excision are synthesized. Lambda proteins as well as bacterial proteins are involved in the excision process, which culminates in a lytic life cycle.

Specialized Transduction

Specialized transducing phages □ are bacterial viruses that carry a particular region of the host's hereditary material. Consequently, these phages

224

are capable of introducing (transducing) special genes into their hosts. Specialized transducing phages were previously discussed in Chapter 9 (Microbial Genetics). Lambda is a specialized transducing phage because it integrates into the host's genome at only one site and packages only host DNA near the integration site. Figure 17-9 illustrates how the linear lambda DNA circularizes, integrates into the host's DNA between the galactose (*gal*) operon and the biotin (*bio*) operon, and then disengages from the host DNA in an abnormal fashion so that it carries host genes. When the abnormal phage injects its DNA into a host, a recombination between the host's *gal* region and the phage's *gal* region can take place, or the phage can simply become a prophage. If the host was originally gal^- and the phage was carrying a gal^+ gene, the host would be transformed into one that could metabolize the sugar galactose.

PLANT AND ANIMAL VIRUSES

The plant and animal viruses rely on a number of mechanisms for penetrating cells, replicating themselves, and escaping from their hosts. For ease of discussion, virus multiplication can be divided into a number of steps: a) penetration of cell walls (plant viruses); b) adsorption to cytoplasmic membranes; c) penetration into the cytoplasm of the cell; d) uncoating; e) RNA synthesis, protein synthesis, and genome replication; f) assembly; and g) release.

Virus Penetration of Cell Walls

The plant viruses have not evolved any mechanism for drilling through cell walls to reach the plasma membrane. Consequently, they are dependent upon insects or other mechanisms that damage or break the cell wall. Once plant viruses have entered a plant cell, however, they can spread from cell to cell through the **plasmodesmata** (singular, **plasmodesma**), the openings that connect the cytoplasms of adjacent plant cells. Tobacco mosaic virus (TMV) is generally transmitted from plant to plant when contaminated leaves are damaged. On the other hand, plant rhabdoviruses (bullet-shaped viruses) are transmitted from leaf to leaf by mites, aphids, and leafhoppers. Many of the plant rhabdoviruses are known to multiply in the insects (**vectors**) that spread them as well as in the plant. Thus, these viruses have an extremely broad **host range.**

Virus Adsorption to the Cytoplasmic Membrane

Once plant viruses reach the plasma membrane, they adsorb to the surface and enter the host cell by pinocytosis. In some cases, insects that puncture the plasma membrane may deposite viruses within the cells. Animal viruses are not troubled by cell walls, because these structures are not found in animal cells. Animal viruses adsorb specifically to the membranes of host cells and gain entrance either through pinocytosis or, in some cases, by fusing their envelopes with the plasma membrane. The host range of some animal viruses can be very broad. For example, the togaviruses, which cause encephalitis (an inflammation of the brain), commonly infect domestic and wild birds, mosquitoes, horses, and humans.

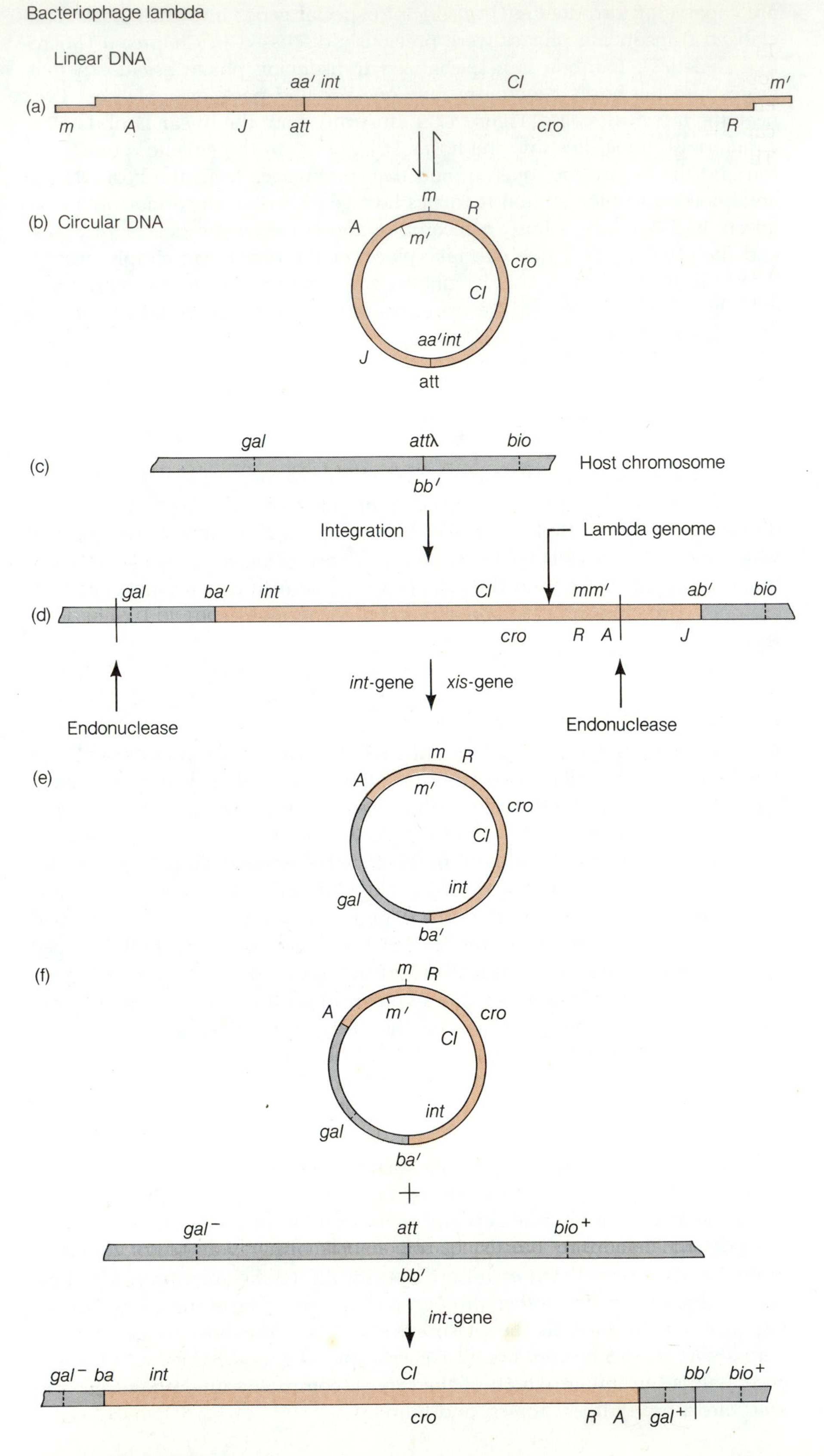

Figure 17-9 Lambda Genome and Prophage. *(a)* The lambda genome exists as a linear molecule within the virus capsid. *(b)* After infection, the lambda genome circularizes. *(c)* The lambda genome *(att)* integrates between the genes for galactose utilization *(gal)* and the genes for biotin synthesis *(bio)* on the host chromosome. *(d)* When the lambda genome integrates into the host's chromosome, it is linear once more, but the genes have a different order than in the original linear molecule. *(e)* Occasionally, the provirus does not excise itself properly but instead may leave some viral genes behind and take with it some host genes. When this abnormal provirus forms the hereditary material of a virus, the virus may function as a specialized transducing virus. *(f)* An abnormal provirus carrying the *gal* genes may integrate into the chromosome of a host cell that is unable to use galactose and convert the host cell into one that can use the sugar.

Virus Penetration into the Cell's Cytoplasm

In most cases, the entire plant or animal virus (genome, capsid, and envelope) enters the host. In most bacteriophages, by contrast, only the genome enters the cell. In the case of some enveloped viruses, only the genome and capsid enter the cell because the envelope fuses with the plasma membrane. The entire virus is brought into the cell by **pinocytosis.**

Virus Uncoating

Virus uncoating is believed to take place within pinocytotic vesicles or within the cytoplasm after the virus enters the cell. Envelopes are removed when they fuse with pinocytotic vesicles. In the case of viruses lacking a membrane, the capsid depolymerizes and the nucleic acid is released into the vesicle. The factors stimulating the depolymerization of the capsid are not known.

Virus RNA Synthesis, Protein Synthesis, and Genome Replication

There is a great deal of variety in the way that plant and animal viruses are expressed and replicated. Part of the reason for this variety is the fact that virus genomes can be ss-RNA, ds-RNA, ss-DNA, or ds-DNA. In addition, the ss-RNA genomes may be either sense strands ("+RNA") that can function as mRNA or nonsense strands ("−RNA") that are unable to code for proteins.

The plant and animal rhabdoviruses and the animal orthomyxoviruses are "−RNA" viruses (tables 17-2 and 17-3). These viruses carry an **RNA-dependent RNA polymerase** called a **transcriptase** for synthesizing mRNA ("+RNA") from the virus genome (fig. 17-10). The mRNA is translated into capsid proteins and other proteins that are required for the development of the viruses. The "+RNA" also serves as a template for the synthesis of new virus genomes ("−RNA"). The virus genomes are synthesized by an RNA-dependent RNA polymerase called **RNA replicase**. In some cases, the transcriptase that makes mRNA and the replicase that makes new virus genomes can be the same enzyme.

The picornaviruses and togaviruses are "+RNA" viruses and consequently can be translated immediately (tables 17-2 and 17-3). An RNA replicase catalyzes the synthesis of "−RNA" and then new "+RNA" genomes (fig. 17-10). The new "+RNA" genomes often function as mRNA.

The plant and animal reoviruses are ds-RNA viruses. Those that infect plants contain 10 or 12 separate genomes, while those that infect animals have 10 or 11 (tables 17-2 and 17-3). Messenger RNA is synthesized by a transcriptase carried within the virus capsid (fig. 17-10). A second enzyme, the RNA replicase, is believed to make "−RNA" complementary to the "+RNA."

The double-stranded DNA viruses (herpes, pox, hepatitis-B, adenovirus, and papovavirus) are expressed when a host **DNA-dependent RNA polymerase** synthesizes mRNA. The replication of the DNA viruses is catalyzed by various DNA polymerases.

An unusual group of viruses are the retroviruses (table 17-3). They are enveloped "+RNA" viruses that carry an enzyme called **reverse transcriptase.** Once the RNA is free of the capsid, reverse transcriptase synthesizes a complementary "−DNA" strand (fig. 17-10). Then a DNA-dependent DNA

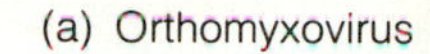

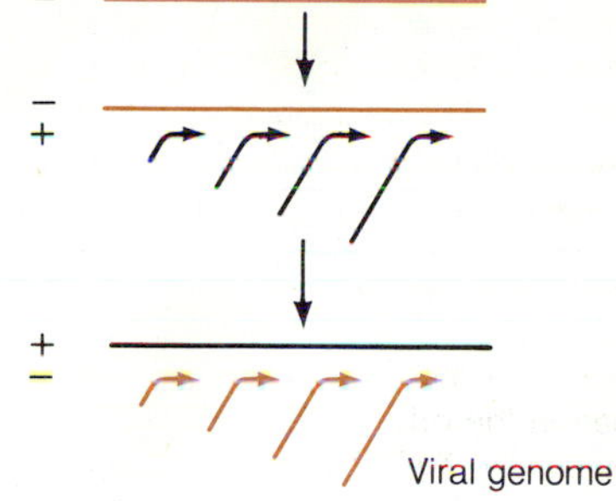

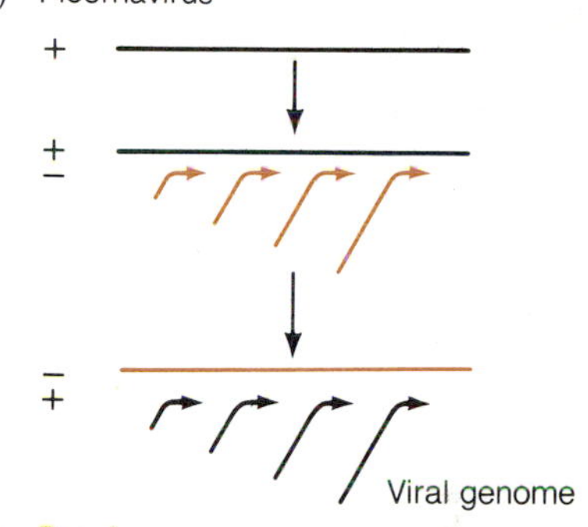

(c) Reovirus

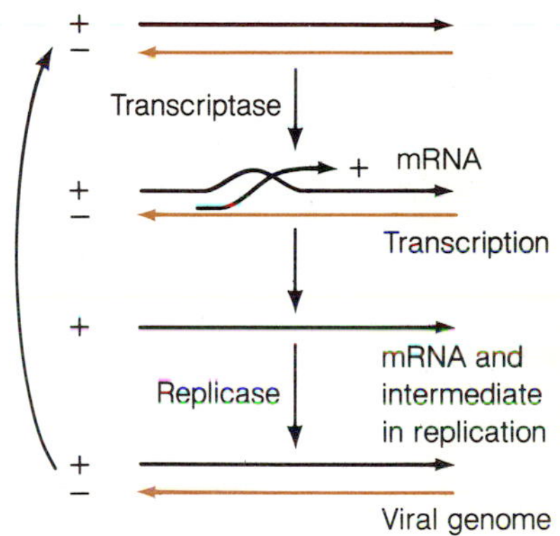

(d) Retrovirus

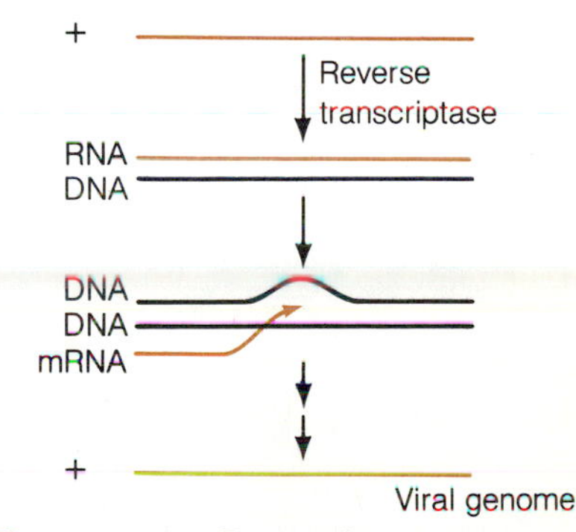

Figure 17-10 Expression of RNA Virus Genomes. *(a)* Expression of "−RNA" virus genome. *(b)* Expression of "+RNA" virus genome. *(c)* Expression of "ds-RNA" virus genome. *(d)* Expression of "+RNA" retrovirus genome.

polymerase synthesizes a "+DNA" strand complementary to the first DNA strand. The double-stranded DNA integrates into the host cell's genetic information. Messenger RNA and progeny RNA are made by transcription of the proviruses.

Virus Assembly and Release

In most cases, the packaging of plant and animal virus genomes begins when the nucleic acid associates with capsid proteins or with an incomplete capsid. In the case of tobacco mosaic virus (TMV), double-layered protein discs bind a virus genome, thus stimulating the polymerization of discs into a cylindrical capsid. TMV is released as a nonenveloped virus when the plant cell lyses. Very few plant viruses are enveloped (table 17-2). Those that are enveloped appear to be both insect and plant pathogens, such as the rhabdoviruses.

Many of the animal viruses (herpes, pox, orthomyxovirus, paramyxovirus, rhabodovirus, togavirus, hepatitis B, retrovirus) become enveloped either as they develop within the cell or when they escape from the cell. The herpes virus nucleocapsid becomes enveloped as it passes through the nuclear membrane into the cytoplasm, while the poxvirus obtains its membranes from the cytoplasm of the host cell. Most of the other enveloped viruses obtain their membrane as they pass through the cytoplasmic membrane. Generally, the nonenveloped animal and plant viruses escape from the host cell when it lyses, while the enveloped viruses escape from the host cell when they pinch off from the membrane. Enveloped viruses cause the lysis of the host cell if they disturb the cell's metabolism sufficiently.

Representative Plant and Animal Viruses

There is no simple way of discussing a "typical" virus, since each virus has its peculiarities and affects its host in different ways. Consequently, a few noteworthy viruses will be discussed in order to understand how diverse the viruses are.

Tobacco Mosaic Virus

Tobacco mosaic virus is typical of many plant viruses in the way that it invades plant cells and multiplies. It is an important virus because it is responsible for damage to numerous crops. TMV is a ss-RNA helical plant virus (fig. 17-11). The coat is made from 2,130 identical subunits called **capsomeres.** This virus measures about 300 nm by 18 nm and is generally transmitted from plant to plant when contaminated leaves are damaged. Insects are not involved in the spread of this virus. The damaged cell wall allows the virus to come into contact with the plasma membrane of the host cell. One end of the TMV rod is adsorbed by the plasma membrane, and then the membrane invaginates, forming a pinocytotic vesicle. This process takes approximately 10 minutes. What happens next is not clear. There is a 4-hour "eclipse period" during which no virus particles can be detected. Some 50 hours after infection, much of the cytoplasm has been converted into a crystalline array of viruses.

The ss-RNA contains enough genetic information for about six proteins, but it is not known exactly how many proteins are actually involved in TMV

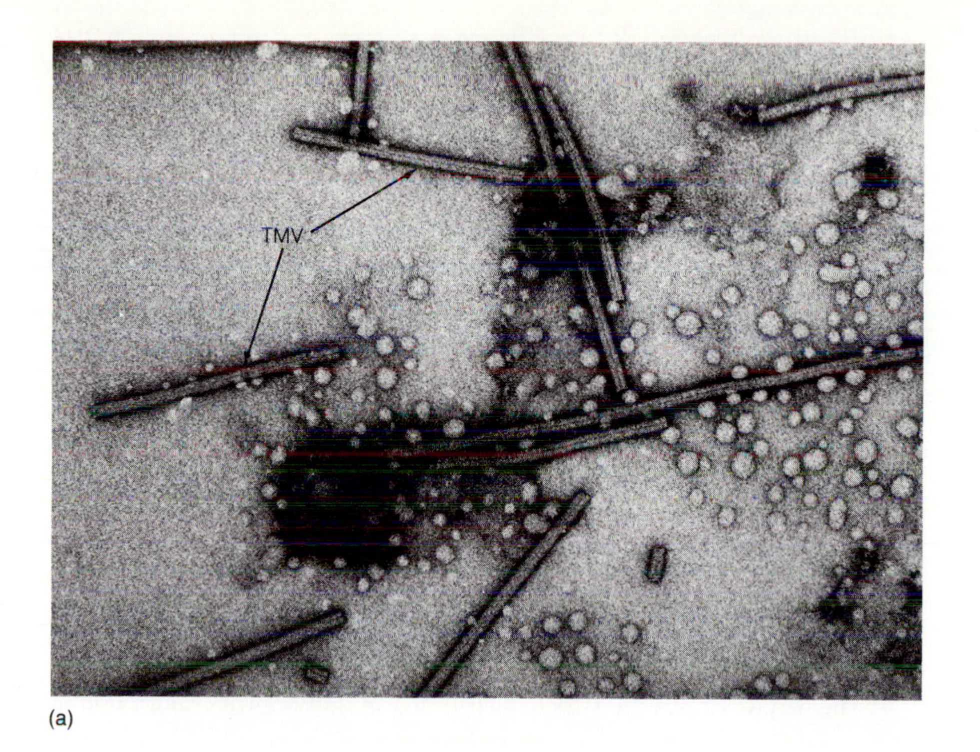

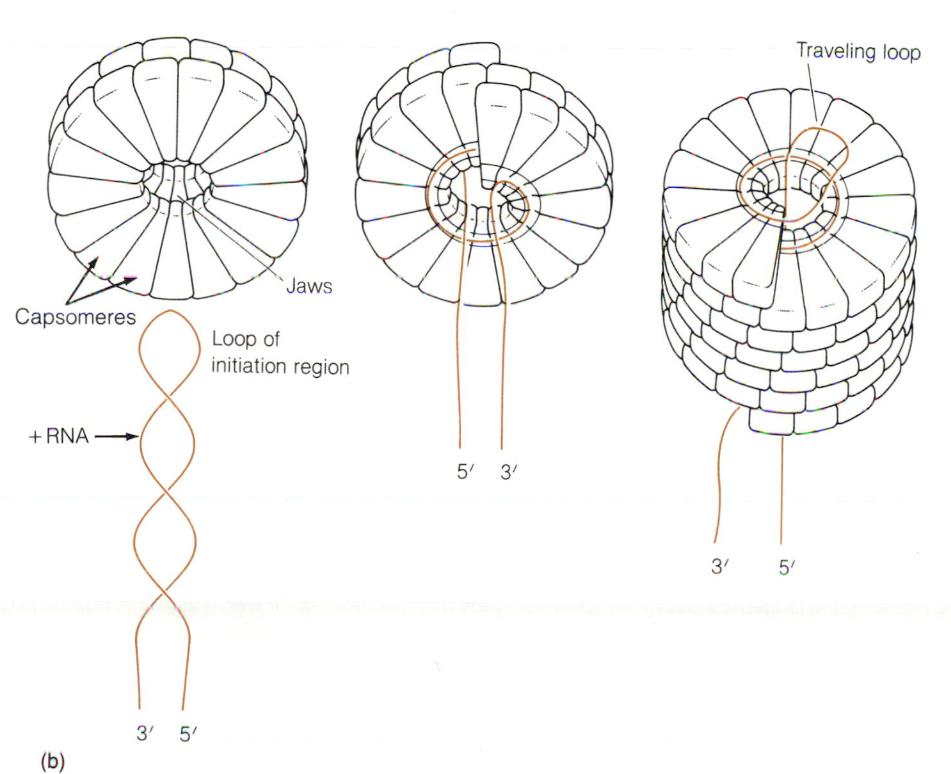

Figure 17-11 Tobacco Mosaic Virus Infection. *(a)* Tobacco mosaic virus (TMV) is a cylindrical "+RNA" plant virus that enters cells by pinocytosis after adsorbing to the plasma membrane. *(b)* After an eclipse period, TMV proteins and RNA self-assemble into viruses.

development. Once RNAs and coat proteins have been synthesized, they self-assemble into the infectious virus particles. The A-protein polymerizes under physiological conditions into double-layered discs (fig. 17-11). Some double-layered discs associate with the virus RNA near its 3' end. The RNA arranges itself between the two layers of the disc, thus apparently inducing a conformational change in the discs so that the RNA becomes locked between the layers. The conformational change converts the double disc into a double-helical disc. As further double discs add to the double-helical discs, the RNA associates with them, and conformational changes occur to lock in the RNA and convert the double disc into a double-helical disc.

Herpes Viruses

The herpes viruses are double-stranded DNA (ds-DNA) viruses with an icosahedral capsid and a membrane envelope. They are 125–200 nm in diameter and replicate in the nuclei of animal cells (fig. 17-12). The herpes viruses include **herpes simplex** I and II, **herpes zoster** (varicella-zoster), the **cytomegalovirus,** and the **Epstein-Barr virus.** The virus genome exists as a linear ds-DNA molecule and codes for more than 150 proteins. These viruses have a wide host range including mice, guinea pigs, hamsters, rabbits, chickens, monkeys, humans, and many types of cultured animal cells.

The herpes viruses are widely distributed in nature. They cause human diseases that are of public health importance, such as cold sores, genital herpes, chicken pox, shingles, and mononucleosis. One of these diseases, genital herpes, is a sexually-transmitted disease of epidemic proportions in the

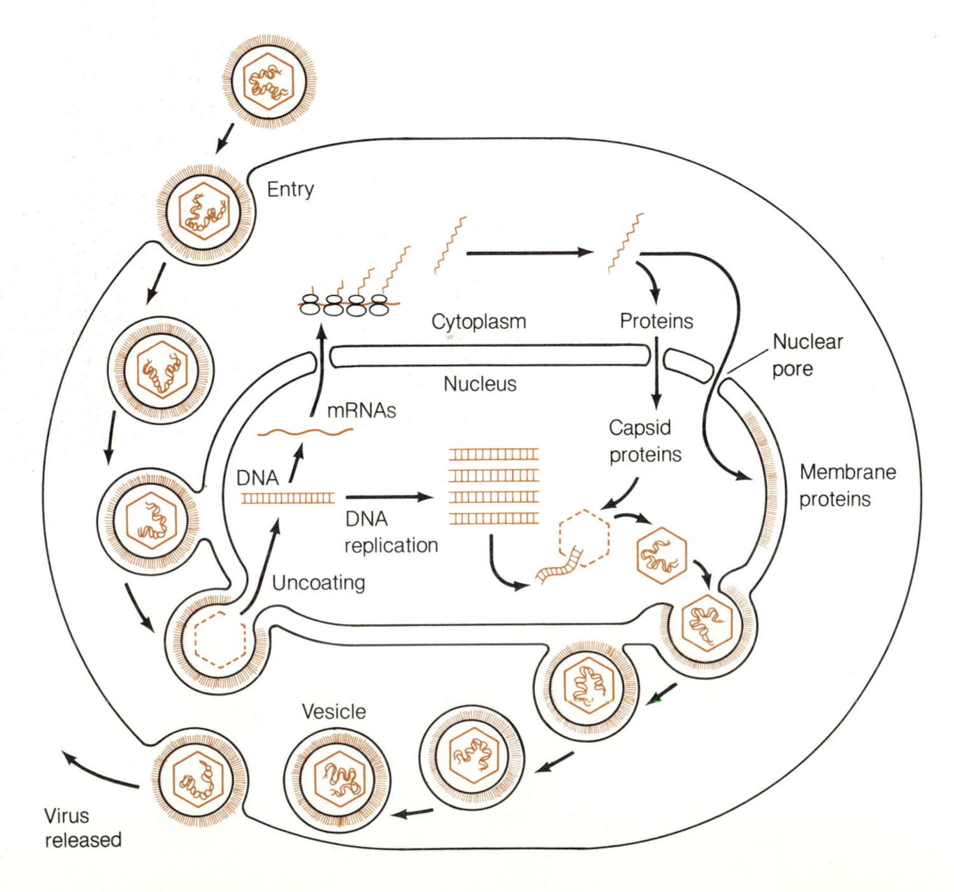

Figure 17-12 Herpes Infection. Herpes viruses are enveloped "ds DNA" animal viruses that enter cells by pinocytosis. After the virus DNA is uncoated in the cytoplasm, it enters the nucleus. DNA replication and transcription occur in the nucleus, but mRNA translation takes place in the cytoplasm. Virus proteins synthesized in the cytoplasm enter the nucleus, where self-assembly of the virus capsid occurs. The virus obtains its envelope when the capsid associates with the inner nuclear membrane. When the enveloped viruses pass through the outer nuclear membrane, they end up in vesicles. The virus escapes from the cell when the vesicles fuse with the plasma membrane.

United States. Cytomegalovirus is 125 nm in diameter (slightly smaller than herpes simplex) and replicates in the salivary glands, brain, kidney, liver, and lungs. It is responsible for causing a large percentage of the mental retardation in newborn children each year. The Epstein-Barr virus is about 125 nm in diameter and is responsible for infectious mononucleosis, a disease that generally causes extreme exhaustion and swollen tonsils. The virus replicates in lymphocytes and lymph tissue and can be isolated from pharyngeal secretions of patients with infectious mononucleosis.

Infections caused by herpes viruses usually begin when the virion (nucleocapsid plus envelope) attaches to the cell's plasma membrane through proteins found in the viral envelope (fig. 17-12). The virion enters the cell's cytoplasm by pinocytosis and then moves toward the nucleus within the pinocytotic vesicle. Upon reaching the nucleus, the pinocytotic vesicle fuses with the outer nuclear membrane. The viral envelope then fuses with the inner nuclear membrane (fig. 17-12), and the nucleocapsid is released into the nucleus. At this point, it is believed that the genetic information (core) is uncoated and released (fig. 17-12). DNA replication begins about 3 hours after infection and is completed 12 hours later. Virus particles are evident approximately 18 hours after infection and are released from the cell shortly thereafter (fig. 17-12).

An important characteristic of most herpes viruses is that they can remain latent within cells for many years and reappear periodically. A good example of this is herpes zoster, which is responsible for both chickenpox (varicella) and shingles. Chickenpox generally subsides within a few days, as a result of the host immune response, and the person never has it again. In some cases, however, the virus that causes chickenpox remains latent for years in neurons. From time to time this latent virus (provirus) is activated to produce virus particles. These viruses travel from the infected nerves to the skin, where they initiate a lytic infection called shingles. The development and/or movement of the virus in the nerves causes severe pain in the infected area. The vesicular eruptions and pain caused by herpes zoster may last as long as a month.

255

The herpes genome resembles two transposons □ in series. Because of its structure, it has the potential to integrate into host chromosomes and jump from one site to another. The ability of the herpes viruses to integrate into host chromosomes and to excise themselves allows the herpes viruses to remain latent for long periods of time, occasionally giving rise to lytic infections. The similarity between transposons (which cause mutations and chromosome breaks) and the herpes genome may shed light on how herpes viruses can transform cells and cause some cancers.

Orthomyxoviruses (Influenza viruses)

The orthomyxoviruses include all of the influenza viruses that infect humans and animals. These viruses are important because of the widespread illness and death they cause each year in the United States. The orthomyxoviruses are single-stranded segmented RNA viruses that consist of eight nucleocapsids within a membrane or envelope (fig. 17-13). At times the virus may have a filamentous structure.

The eight single-stranded RNAs code for the 10 virus proteins. The **neuraminidase spikes** on the surface of the virus allow it to cut through glycoproteins in respiratory secretions and bind to host cells rather than to the proteins in the mucus. The **hemagglutinin spikes** on the surface of the

Figure 17-13 Orthomyxovirus Infection. Orthomyxoviruses (influenza viruses) are enveloped "−RNA" animal viruses that enter the cell by pinocytosis. The virus genome is segmented into eight pieces of RNA. After the virus RNAs are uncoated in the cytoplasm, they enter the nucleus. Transcription and RNA replication occur in the nucleus, but mRNA translation takes place in the cytoplasm. A cylindrical capsid forms around each of the eight pieces of the genome. The pieces associate with the cytoplasmic membrane, which undergoes exocytosis. M-protein is the matrix protein, NA represents the neuraminidase spike, and HA indicates the hemaglutinin spike.

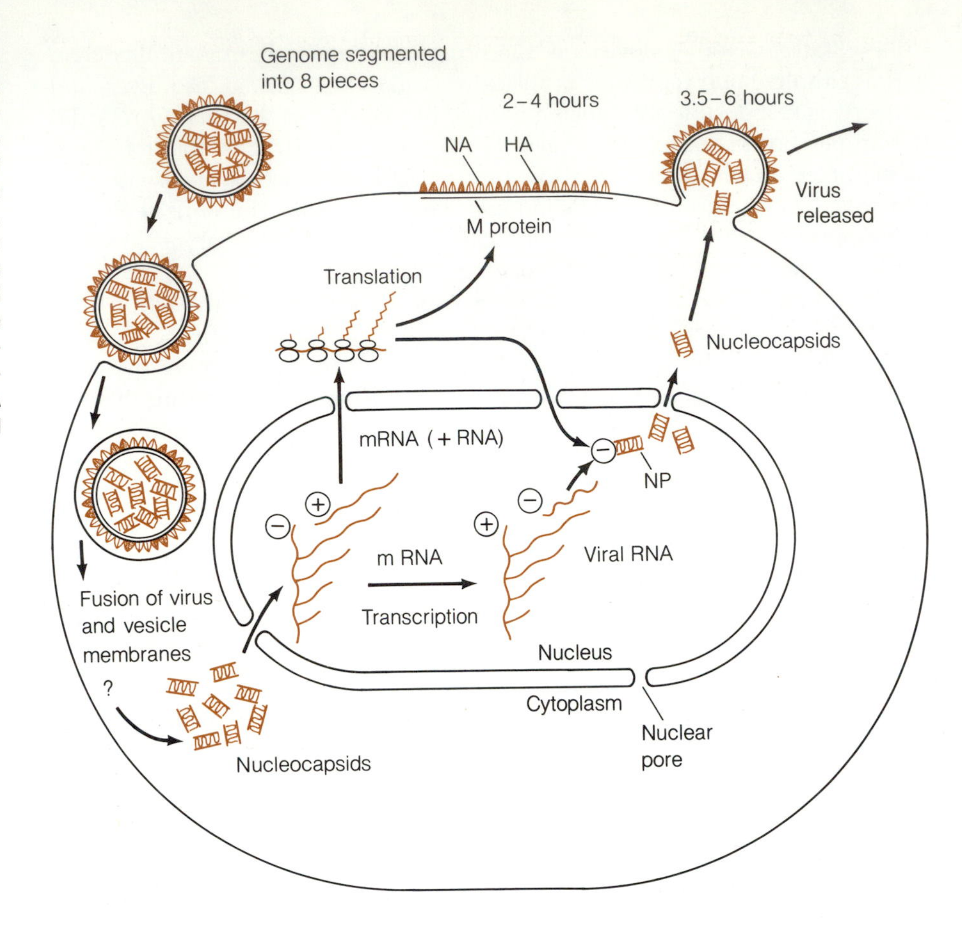

virus bind it to specific glycoproteins on the host's plasma membrane. After the virus attaches to the host cell, the host plasma membrane invaginates and the virus enters a pinocytotic vesicle. The virus envelope then fuses with the vesicle membrane and the nucleocapsids are released into the cytoplasm (fig. 17-13). A virus RNA-dependent RNA polymerase (RNA transcriptase) and a virus replicase are required for multiplication. The RNA transcriptase synthesizes "+RNA" (mRNA) from the complementary "−RNA" genomes (fig. 17-13).

Figure 17-13 indicates how the nucleocapsids leave the nucleus and associate with the mass of virus proteins on the plasma membrane. The plasma membrane then buds out, taking with it the eight nucleocapsids. The virus envelope consists of host membrane that has been altered by virus-coded proteins.

Studies of the virus proteins have indicated that there are three distinct types of influenza viruses: influenza A, B, and C. The hemagglutinin (H) and neuraminidase (N) spikes in influenza A vary slightly through time. This minor variation in virus proteins is called **antigenic drift,** and it is due to spontaneous mutations in the RNAs coding for these proteins. Major changes in the hemagglutinin and neuraminidase occur infrequently and apparently depend upon the recombination of genomes from different virus strains found in widely separated populations or in different hosts. The appearance of mutant viruses in a host population that is partially or totally unfamiliar with the new surface proteins has been responsible for the influenza epidemics that occur every 10 years or so.

Figure 17-14 indicates the major changes that have occurred in influenza A surface antigens through the years. It is believed that the 1889–1890 epidemic that hit Asia, Europe, and America, and caused the death of more than 20 million persons worldwide, was due to an uncommon virus with a hemagglutinin called H2 and a neuraminidase called N2. In 1900, a less severe epidemic occurred due to an H3N2 influenza virus. This virus was responsible for most of the influenza until 1918, when another epidemic occurred due to a virus that was HswNl (antigenically similar to swine flu). This virus was in turn responsible for most of the influenza until 1933, when a new type known as H0Nl became predominant.

599

The influenza A viruses that have appeared in the last 100 years have five significantly different hemagglutinins □ (H0, H1, H2, H3, and Hsw) and two significantly different neuraminidases (N1 and N2). Viruses with different combinations of these surface antigens are believed to reside in animal and human populations by causing mild disease in immunized individuals and virulent infections in susceptible individuals. Previously uninfected individuals may have little or no immunity to infectious agents that are foreign to them, and so the symptoms caused by the influenza virus are severe. Similarly, sick or old individuals frequently have a poorly functioning immune system □ and consequently are not protected against an influenza virus. The infection in sick or old individuals may be severe enough to cause their death. When the antibodies to surface antigens decrease in a population, the virus with these surface antigens is able to infect, multiply, and spread from person to person. An epidemic occurs when large numbers of individuals are no longer immune to a particular virus.

455

Influenza B viruses undergo changes in their envelope proteins also, but these changes are not as extreme or as frequent as for influenza A. Since influenza B surface antigens show a constant, slow antigenic drift, they do not cause major epidemics.

Morphologically, influenza C differs from A and B by showing a hexagonal surface pattern and by having a large core. Since influenza C surface antigens do not change significantly, it does not cause epidemics. It is responsible for mild respiratory infections.

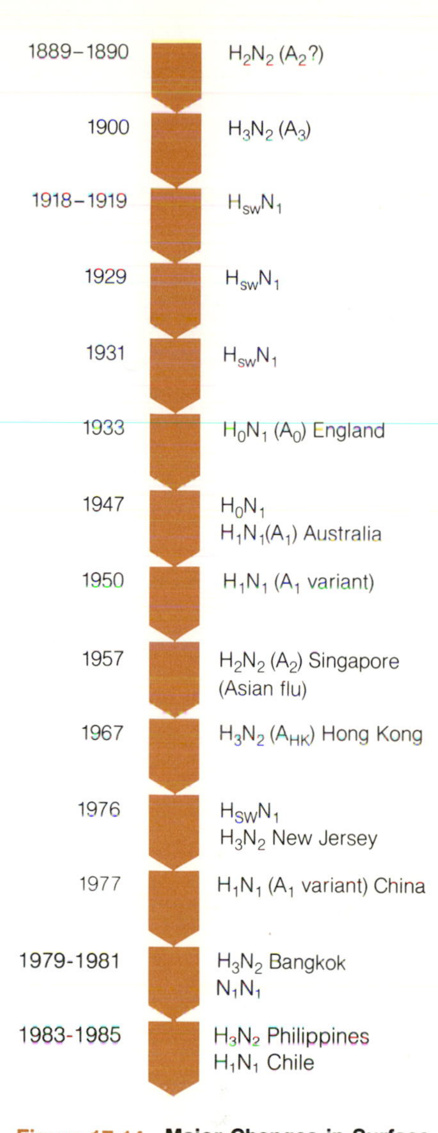

Figure 17-14 Major Changes in Surface Antigens of Orthomyxoviruses. In the influenza A virus, hemaglutinin spikes (H) since the 1890s have undergone five major changes (H2, H3, Hsw, H0, H1), while neuraminidase spikes (N) have undergone only two major changes (N2 and N1). The major changes in surface antigens are responsible for the epidemics caused by influenza A virus approximately every 10 to 20 years. When an animal or human population is not immune to altered virus antigens, the virus is able to infect most of the population. Epidemics depend upon major changes in virus antigens and the lack of immunity to the mutated antigens. Influenza B virus generally does not cause major epidemics, because its surface antigens do not undergo major changes. The surface antigens of influenza B change slowly.

ONCOGENIC VIRUSES

Oncogenic viruses are those viruses that cause cancers and tumors in animals (table 17-4). A tumor is a large mass of cells that reproduce abnormally. If the mass of abnormal cells is encapsulated by a layer of cells or by a basement membrane, the tumor is said to be benign. If, however, cells from the tumor spread throughout the body, the tumor is malignant. Cancer is a synonym for a malignant tumor.

A number of DNA viruses can cause cancers. For example, there are several herpes viruses that cause carcinomas and sarcomas in animals. **Carcinomas** are derived from the innermost (endoderm) and outermost (ectoderm) embryonic layers of cells, while **sarcomas** develop from the middle (mesoderm) layer of embryonic cells (table 17-4). The papovaviruses are responsible for sarcomas and carcinomas in animals and in humans, while the

TABLE 17-4
VARIOUS TYPES OF CANCERS

TYPE OF CANCER	CELLS AFFECTED
Sarcoma (Cells from the mesoderm that spread through the blood):	
Osteosarcoma	Bone cells
Liposarcoma	Fat cells
Fibrosarcoma	Connective cells
Leukemia (leukosarcoma)	White blood cells
Lymphoma (lymphosarcoma)	Lymph cells
Reticulosarcoma	Blood forming cells
Retinoblastosarcoma (retinoblastoma)	Eye cells
Carcinoma (Cells from the endoderm and ectoderm that spread through the lymph):	
Hepatocarcinoma (hepatoma)	Liver cells
Melanocarcinoma (melanoma)	Pigment-forming skin cells
Lung cancer	Lung cells
Skin cancer	Skin cells

adenoviruses and poxviruses generally cause cancers only in animals. Retroviruses are the only RNA viruses that are able to cause cancers and tumors in animals and humans.

Retroviruses

The retroviruses are single-stranded RNA viruses. The genome is unusual in that it is a dimer of two identical RNAs. The RNA is long enough to code for 10 average-sized proteins. It is not known exactly how the ss-RNA is coiled within the virus particle, but it is found in association with a **nucleoprotein** and a **reverse transcriptase** (RNA-dependent DNA polymerase). The genome is coated by a capsid protein that is reported to form an icosahedron. The nucleocapsid (RNA, RNA-associated proteins, and capsid) is covered by an outer protein coat, which in turn is surrounded by an envelope derived from the host cell (fig. 17-15).

The retroviruses adsorb to specific host cell receptors by their glycoprotein knobs. The host cell membrane and the virus envelope subsequently fuse. Once the nucleocapsid enters the cell, the RNA genome is uncoated and the reverse transcriptase begins to polymerize a complementary DNA. Figure 17-15 outlines the reproductive cycle of a retrovirus.

After the double-stranded DNA genome is synthesized, it circularizes and moves into the nucleus. Twenty-four hours after infection, some of the DNA molecules may have integrated into host genomes, becoming proviruses. Messenger RNA and progeny RNA are made by transcription of the integrated proviruses.

Retroviruses with a transforming ability □ can induce tumors such as fibrosarcomas, carcinomas, lymphomas, and erythroleukemias. The genes that promote transformation are designated *onc* for **oncogenic genes**. A **protein kinase** has been identified in mouse sarcoma and in avian sarcoma viruses, which transformed chicken and hamster cells. It is thought that the phosphorylation of Na^{+}/K^{+}–ATPase pumps by the protein kinase disrupts the ionic balance and wastes ATP. The waste of ATP is thought to result in

699

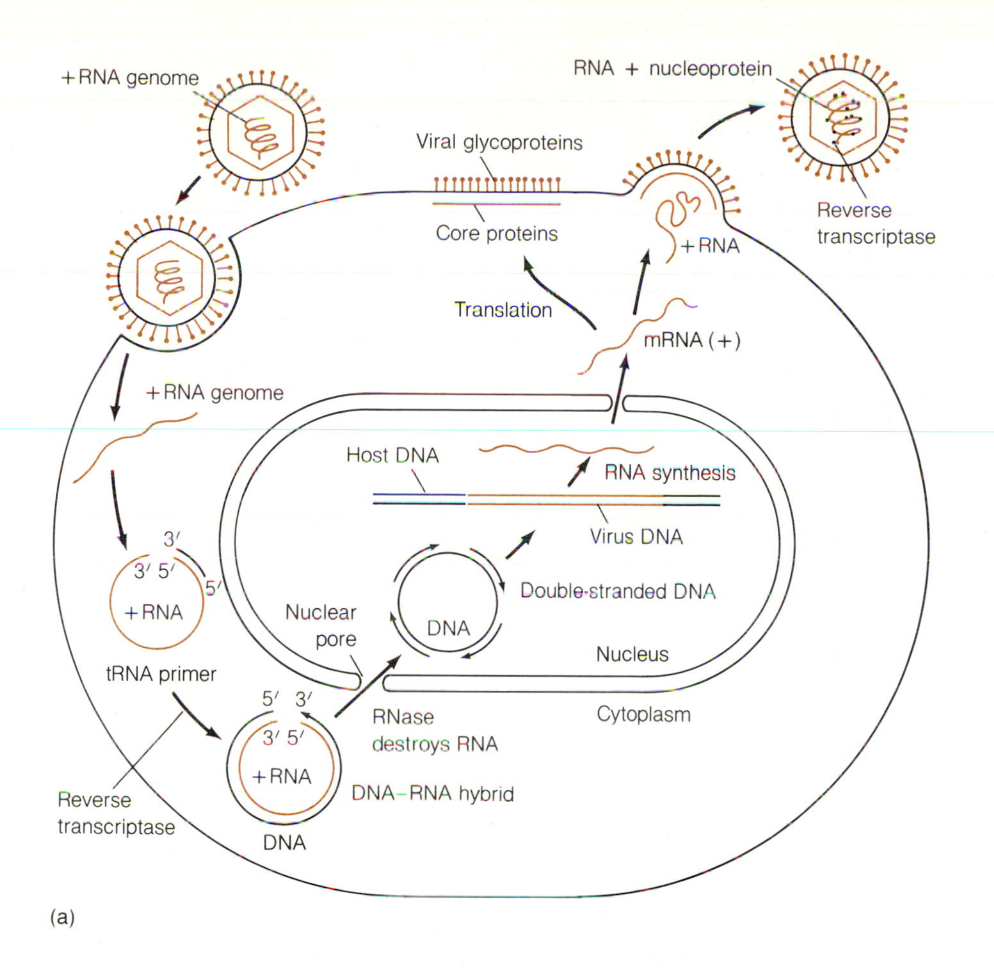

(a)

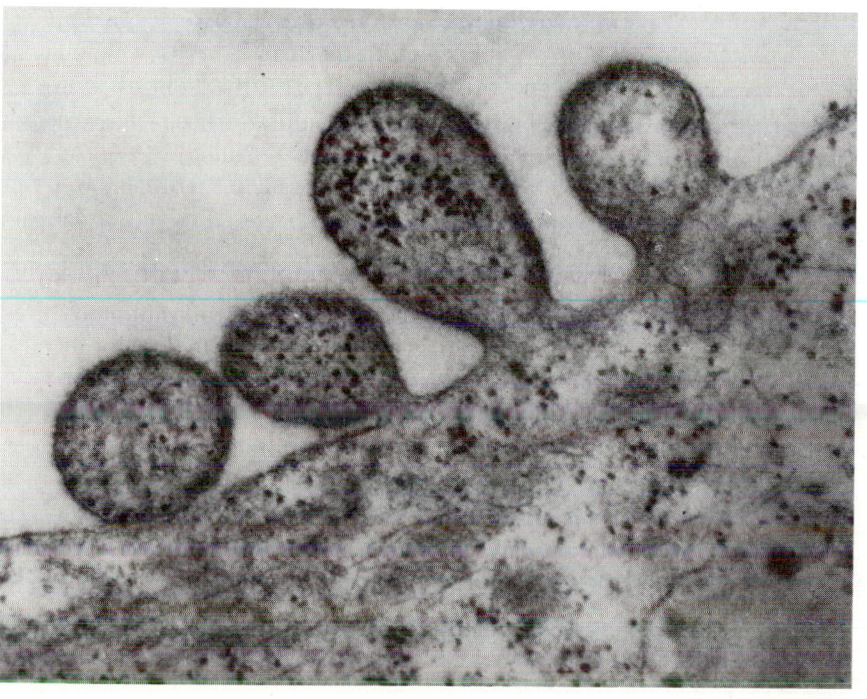

(b)

Figure 17-15 **Retrovirus Infection.** *(a)* Retroviruses are enveloped RNA viruses that infect animals. The retroviruses are of interest because they frequently make a DNA copy of their RNA genome and integrate into host chromosomes. In addition, a number of retroviruses carry genes that can transform their host cells into cancerous cells. Retroviruses enter host cells by pinocytosis. The envelope is removed when it fuses with the pinocytotic vesicle. After the RNA genome is uncoated, a virus reverse transcriptase makes a DNA copy of the virus RNA. If a lytic cycle occurs, virus capsid protein and RNA associate with the plasma membrane. The virus forms as the plasma membrane undergoes exocytosis. *(b)* C-type retroviruses, exiting the host's cell.

a more active glycolysis. □ In addition, phosphorylation of membrane proteins that affect cytoskeleton attachment to the plasma membrane can eliminate **contact inhibition** and alter the cell cycle. Contact inhibition is the inhibition of cell division and movement that occurs when normal animal cells come in contact with each other. Cells that do not show contact inhibition can spread through an animal and may represent precancerous or cancerous cells. The *onc* gene of simian sarcoma virus is closely related to the gene that codes for **platelet-derived growth factor** (PDGF), which is a protein that stimulates mitosis and cytokinesis. The abnormal functioning of this *onc* gene stimulates cell division, which represents one of the steps in the development of a malignancy.

Rous sarcoma virus, a retrovirus, transforms not only cells in the chicken but also chick embryo cells in tissue culture. The virus causes the cells to round up and stain abnormally. Transformed cells often show one or more of the following characteristics: 1) loss of contact inhibition; 2) random orientation in culture; 3) change in chromosome number; 4) infinite number of multiplications; 5) capacity to produce cancer in animals; 6) shorter generation time.

Normal cells are capable of a limited number of generations and are contact-inhibited. Some transformed cells called **cell lines** are able to proliferate indefinitely, but are not cancerous because their multiplication ceases when they come into contact with other cells. Such cells are said to be contact-inhibited. On the other hand, a transformed cell line that reproduces indefinitely and is not contact-inhibited is usually a **cancerous cell.** When the tumor or cancer cells spread through an animal's body, the cancer is said to have **metastasized.** Sarcomas usually spread via the blood system, while carcinomas generally spread through the lymphatic system. A metastasized cancer usually kills the animal.

GROWING VIRUSES IN THE LABORATORY

In order to study viruses, large quantities of the infectious particles are necessary. This generally means that they must be cultivated in the laboratory. With high concentration of viruses, it is possible to make detailed studies of their chemistry and biology. The animal and plant viruses, for the most part, are the most difficult organisms to cultivate and purify because they require a live host for propagation and often only a few viruses are formed. Of all the viruses, the bacteriophages are the easiest to propagate and purify, because they multiply in bacteria that are generally easy to grow and because large numbers of viral prugeny are formed.

Bacteriophages

A phage such as P1 can be grown on lawns of *E. coli* proliferating in a layer of soft agar (fig. 17-2). Phage P1 goes through a number of lytic cycles, destroying most of the lawn in about 10 hours. The soft agar containing the phage and surviving bacteria is scraped into a centrifuge tube and shaken with a few drops of chloroform to kill the bacteria. Then the tube is centrifuged to sediment the bacterial debris. The phage remain in the supernatant fluid. The phage concentration, or **titer,** can be determined by mixing dilutions of the phage sample with fresh *E. coli* and soft agar and plating the mixture onto nutrient agar plates. The plaques (virus colonies) that appear in the bacterial lawn 6 to 8 hours later are counted. The number of phages in the undiluted lysate is determined by multiplying the plaque number by the inverse of the dilution. The phage titers generally range from 10^8 to 10^{11} phage/ml of phage solution. The phage solution can be stored for many months in the refrigerator without a significant loss of infectivity.

Plant Viruses

Plant viruses are generally grown on living host plants rather than in cell culture. For example, tobacco mosaic virus (TMV) is grown to high titers by rubbing a solution of these viruses into the surface of tobacco leaves attached to healthy plants (fig. 17-2). The rubbing breaks some of the cell walls and allows the viruses to infect. Within two weeks, the plant leaves develop visible lesions that resemble plaques. The TMV can be harvested by grinding the infected leaves with a small amount of water. The grinding releases the TMV into the water.

Animal Viruses

506 The isolation and identification of viruses in clinical specimens □ is often accomplished by growing the viruses in an appropriate host. In addition, the production of large titers of animal viruses is of value in making virus vaccines. Animal viruses are generally studied and manipulated by growing them in animal hosts, in eggs, or in cell (tissue) cultures.

Bird embryos (fertilized eggs) have been very useful for cultivating a number of animal viruses (fig. 17-16). Five- to 10-day-old duck or chicken embryos developing within their eggs are generally used for growing viruses.

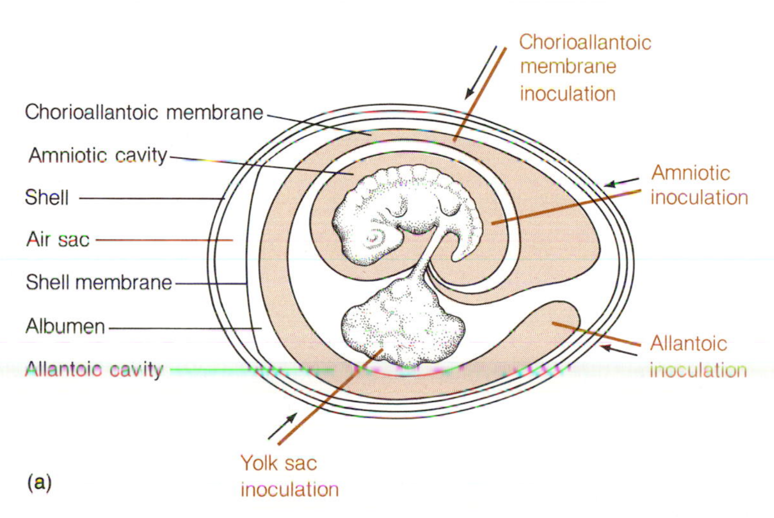

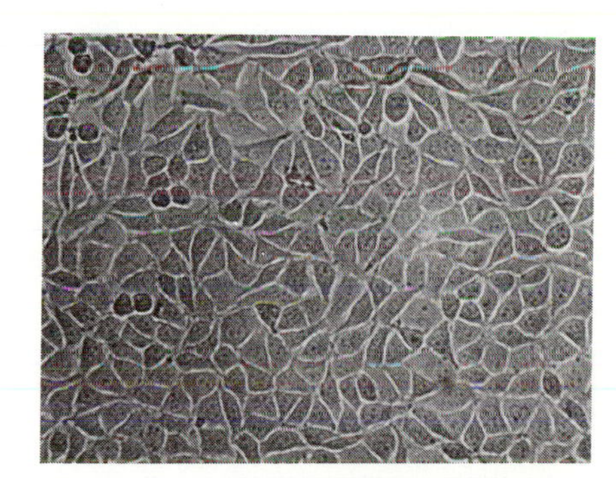

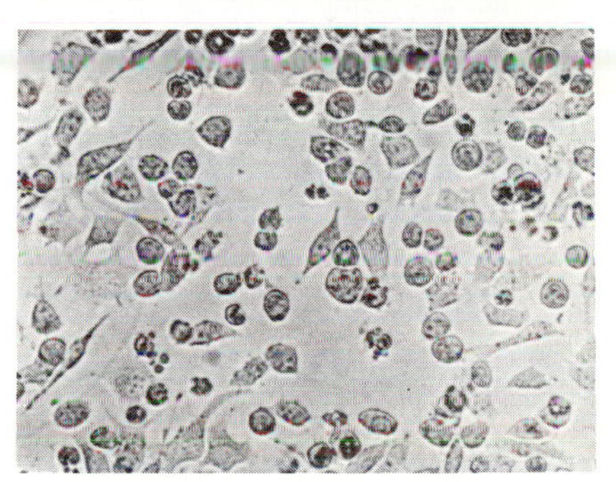

Figure 17-16 Growing Animal Viruses *(a)* Some animal viruses taken from infected tissue are grown on various tissues in chicken eggs. *(b)* Most animal viruses are now proliferated on cells growing in cultures. Mouse L-cells are shown growing in a monolayer. The cells are flattened against the wall of the growth chamber and have a spindle shape. 24 hours after the mouse L-cells have been infected with vesicular stomatitis virus, most of the cells have been transformed. The transformed cells are no longer attached to the wall of the growth chamber and they have become spherical.

Viruses are introduced with a hypodermic needle into the allantoic cavity, chorioallantoic membrane, yolk sac, amniotic cavity, or embryo. The egg is incubated for as long as it takes the viruses to reproduce. Inoculation of bird embryos has been used to prepare vaccines □ against the viruses that cause rabies, poliomyelitis, influenza, smallpox, and yellow fever. Presently, rabies and polio viruses for vaccines are grown in cell cultures rather than in eggs.

477

Cell cultures are being used with increasing frequency to grow animal viruses (fig. 17-16). Cell cultures are often started from fresh animal tissue that is separated into individual cells by mechanical and enzymatic treatments. The cells are placed into the bottom of petri dishes, covered with an appropriate growth medium, and cultured in a high CO_2 environment. The cells attach themselves to the plastic petri dishes and grow into a monolayer. Viruses are added to the multiplying animal cells. The destruction or transformation of the cells is referred to as the cytopathic effect (CPE). Not all animal viruses demonstrate cytopathic changes. For example, influenza viruses produce little or no visible change in the cultured cells. Influenza viruses can be detected, however, because red blood cells added to infected cells bind to the surface of the cells.

Unfortunately, cell cultures started from fresh animal tissue do not last very long, because the cells multiply only a few times before they die. This means that fresh tissue must be prepared frequently in order to grow viruses. The longest lasting **primary cell cultures** are derived from human embryo tissue, in which the cells are able to multiply between 50 and 100 times. Occasionally, primary cell cultures become transformed and are able to multiply indefinitely. Transformed cell cultures that do not have a limited number of generations are called **continuous cell lines.** A number of cell lines that are used to grow viruses have been derived from cancerous tissue. The most notable cell line derived from a human tumor is the **HeLa** cell line.

VIROIDS AND PRIONS

Viroids

Viroids are pieces of RNA found in the nucleus of infected plant cells which are able to replicate and to spread from one cell to another within the host plant. These pieces of RNA are capable of extensive folding (secondary structure), and consequently form a structure with multiple hairpin loops. The viroids exist as naked ribonucleic acids (fig. 17-17) and are not enclosed in a protein coat. For example, **potato spindle-tuber viroids** are single-stranded circular RNAs, 359 bases long, with extensive secondary structure. In electron micrographs, they have the appearance of rods approximately 50 nm long. Viroids differ in the nucleotide sequences of their RNAs. For example, potato spindle-tuber viroid and **citrus exocortis viroid** have very different nucleotide sequences, while various strains of potato sprindle-tuber viroid differ at only a few sites. Analyses of the viroid nucleotide sequences indicate that the nucleotides do not code for proteins. Consequently, it is believed that the viroid RNA does not code for proteins that catalyze its replication. Instead, it is thought that the host provides the necessary RNA polymerase. So far, viroids have not been discovered in animals, the lower eukaryotes, or bacteria, but there is no reason to believe that they will not be found in all these organisms.

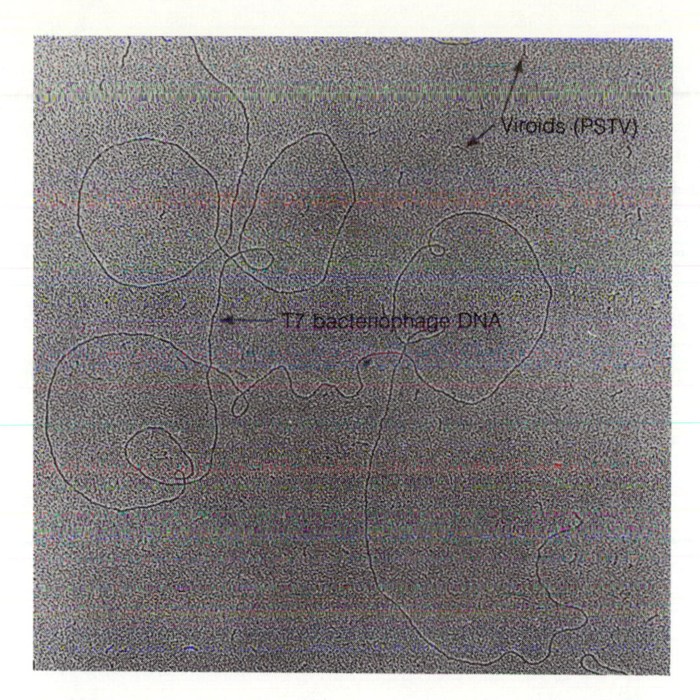

Figure 17-17 Viroid. The potato spindle tuber viroid (arrows) consists of 359 ribonucleotides. The viroid does not appear to code for a protein, since there are no start or termination codons in the viroid RNA. T7 bacteriophage DNA is included in the TEM to show relative sizes.

Viroids have attracted attention because they cause a number of plant diseases and because they are one of the smallest and simplest biological entities known to proliferate within cells and to spread from cell to cell. Viroids are responsible for a number of plant diseases (table 17–5). The most notable of the diseases is **cadang-cadang** of coconuts, which has killed 12 million coconut trees in the Phillipines; and **potato spindle tuber disease** that destroys more than $3.5 million worth of potatoes in the U.S. each year. In the 1950s, a viroid nearly wiped out the chrysanthemum industry in the U.S., and another disrupted efforts to graft orange and lemon trees to a virsus-resistant root stock.

TABLE 17-5
THE VIROIDS

DISEASE	VIROID
Potato spindle tuber disease	Potato spindle tuber viroid (PSTV)
Tomato bunchy top disease	Tomato bunchy top viroid (TBYV)
Citrus exocortis disease	Citrus exocortis viroid (CEV)
Chrysanthemum stunt disease	Chrysanthemum stunt viroid (CSV)
Chrysanthemum chlorotic mottle disease	Chrysantemum chlorotic mottle viroid (ChCMV)
Coconut cadang-cadang disease	Coconut cadang-cadang viroid (CCCV)
Cucumber pale fruit disease	Cucumber pale fruit viroid (CPFV)
Hop stunt disease	Hop stunt viroid (HSV)

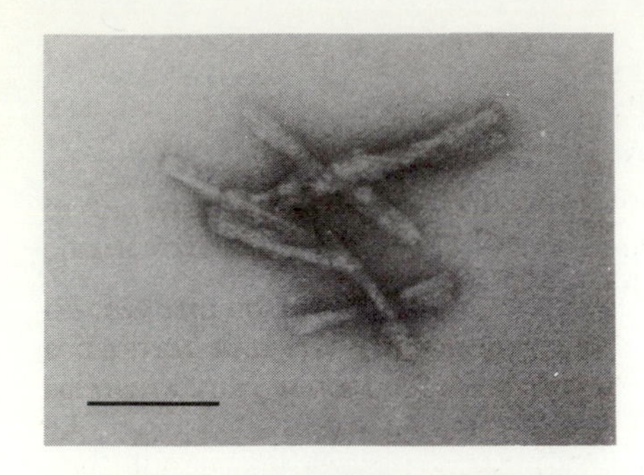

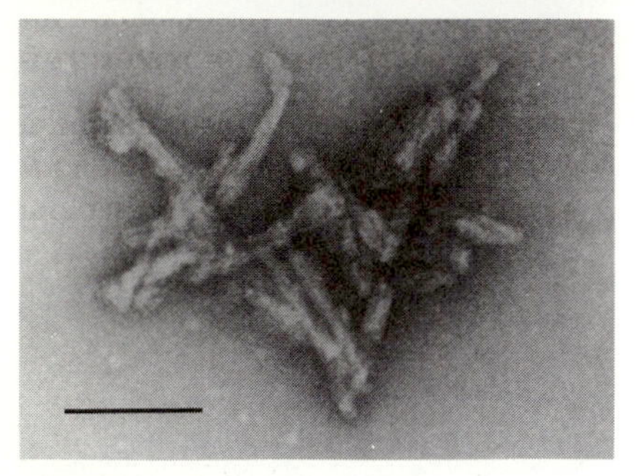

Figure 17-18 **Prion.** Electron micrographs of extensively purified fractions of prions. Bars are 100 nm. Prions are negatively stained with uranyl formate.

Prions

At one time, scientists believed that "DNA viroids" caused several slowly developing brain diseases, such as **kuru, Creutzfeldt-Jakob disease,** and **Gerstmann-Sträussler syndrom** in humans; **scrapie** in sheep and goats; **chronic wasting disease** in mule, deer, and elk; and **transmissible encephalopathy** in mink. This belief in a "DNA viroid" was based on experiments that indicated the scrapie infectious agent was DNA. It supposedly was inactivated by deoxyribonucleases (there are contradictory experiments in the literature), but not by ribonucleases. In addition, the infectious agent was unusually resistant to ionizing and ultraviolet radiation, as are RNA viroids. This evidence indicated that the infectious agent resembled single-stranded viroids. Recently, however, the scrapie infectious agent has been isolated and characterized. It appears to be pure protein and free of any type of nucleic acid.

Research conducted at the University of California at San Francisco indicates that the infectious agent causing Creutzfeldt-Jakob disease and scrapie in sheep is not a slow virus, a DNA plasmid, or an RNA viroid. Protein appears to be an integral part of the infectious agent, since proteases destroy the agent's ability to cause disease while DNases and RNases do not. It is suggested that the infectious agent, called a **prion,** is a self-replicating protein (fig. 17-18).

SUMMARY

CLASSIFICATION OF VIRUSES

1. The bacteriophages are classified on the basis of their structure or shape and type of nucleic acid. The majority of the bacteriophages have ds-DNA, ss-DNA, or ss-RNA; however, a few have ds-RNA (ϕ6). Almost all of the bacteriophages can be characterized as complex, icosahedral, or cylindrical. A few bacteriophages are enveloped (MVL2, ϕ6). In general, the complex viruses have large ds-DNA genomes that code for 50–200 proteins, depending upon the phage. The icosahedral and cylindrical phages are very simple and usually have small ss-RNA, ss-DNA, or ds-RNA genomes that code for 3–5 proteins.

2. The plant and animal viruses are generally classified on the basis of their structure and nucleic acid. The majority of the plant viruses contain ss-RNA, but there are a few with ds-RNA, ss-DNA, and ds-DNA. Most of the plant viruses are icosahedral and cylindrical. A small number are enveloped or fused polyhedrons. The majority of the animal viruses contain ss-RNA or ds-DNA and are icosahedral or enveloped.

BACTERIOPHAGES

1. The phage genetic information is known as the genome while the protein coat that directly covers the genome is referred to as the capsid. The individual proteins that make up the capsid are called capsomeres.
2. The infection of a bacterium by a bacteriophage is divided into steps: virus attachment (adsorption); virus genome penetration; virus RNA synthesis, protein synthesis, and genome replication; virus assembly; and virus release.
3. The Ellis and Delbruck experiments demonstrated that bacteriophages multiply in an unusual manner. The one-step growth curve for viruses indicates that a single virus gives rise to a large number of viruses in a short period of time. The period of time between adsorption and release is called the latent period.
4. The Doermann Experiment indicated that virus parts are synthesized soon after genome penetration and that the parts self-assemble into large numbers of complete viruses. The period of time between adsorption and the formation of the first viruses in the host cell is called the eclipse period.

5. The tailed bacteriophages have tails to enable them to penetrate cell walls or flagella. The icosahedral and cylindrical bacteriophages bind to pili and may leak their genomes into these hollow tubes that connect with the cell cytoplasm. The enveloped bacteriophages that attack the wall-less mycoplasmas probably enter the host cell by fusing their envelope with the host's cytoplasmic membrane.
6. Bacteriophage T4 is a complex ds-DNA virus that has a narrow host range in that it infects only *Escherichia coli*. Bacteriophage T4 undergoes a lytic cycle of infection when it invades *E. coli*.
7. The synthesis and self-assembly of T4 proteins and parts occurs independently of each other, as shown by the fact that complete capsids (heads), tails, and tail fibers can form even when other parts do not self-assemble.
8. Extracts from virus-infected cells that are missing one virus part can, when mixed together *in vitro* (in a test tube), complement each other and result in the self-assembly of complete and infectious virus particles.
9. Lambda is a complex ds-DNA virus that has a narrow host range, since it infects only *E. coli*. Bacteriophage lambda is capable of both a lytic infection and a lysogenic infection.
10. During a lytic infection, lambda multiplies at the expense of the host cell, which it lyses. When the lambda genome becomes a prophage, it remains quiescent and replicates along with the host's genome. When the host contains a prophage, the host cell is called a lysogen.
11. Lambda integrates at only one site on the *E. coli* genome. Occasionally, it excises itself incorrectly from the *E. coli* genome so that some of the bacterial genes are carried with the virus genome. The defective virus genome and the attached bacterial genes are packaged within infective viruses. These viruses are capable of specialized transduction.

PLANT AND ANIMAL VIRUSES

1. Most of the plant viruses are icosahedral or cylindrical and contain a ss-RNA genome. Those proteins that make up the cylindrical viruses are usually arranged helically; thus, the cylindrical plant capsids are often called helices. When the nucleic acid is associated with noncapsid protein, the complex is referred to as ribonucleoprotein. When the virus is covered by an envelope, the nucleic acid and capsid are referred to as a nucleocapsid.
2. The plant viruses are generally transmitted from plant to plant by contaminated machinery and by insect, nematode, and fungal vectors. Some viruses are also transmitted in pollen and in seeds.
3. Many of the plant viruses are unusual in that their genetic information is segmented and packaged into separate capsids. For example, ilarvirus packages four different genomes into separate capsids, and all four are required for virus multiplication. The genomes of the RNA plant viruses are about the same size as those found in the RNA bacteriophages. The RNA genomes in plants code for 3–15 proteins.
4. Almost all of the animal viruses are either icosahedal viruses or enveloped viruses. There are no cylindrical animal viruses. The DNA animal viruses fall into two size classes: those that code for 5–10 proteins and those that code for 30–300 proteins. One of the pox viruses has the largest genome.
5. At least two enveloped animal viruses (arenaviruses and bunyaviruses) appear to consist of simply a ribonucleoprotein core. A capsid apparently is not formed.
6. The plant and animal viruses differ from the typical bacteriophages in that the entire virus generally enters the host cell.

GROWING VIRUSES IN THE LABORATORY

1. Bacteriophages are generally cultured on a lawn of bacteria growing in soft agar. The bacteriophages can be isolated by centrifuging the lysed bacteria and soft agar. The bacteriophages are found in the supernatant fluid.
2. Plant viruses are usually cultured on plants that support the growth of the viruses. The viruses are isolated by grinding up the affected plant tissue and centrifuging the plant debris. The plant viruses are found in the supernatant fluid.
3. Animal viruses may be grown in experimental animals, egg embryos, and cultured cells.

VIROIDS AND PRIONS

1. Viroids are pieces of RNA that cause a number of plant diseases. These pieces of RNA are able to replicate and spread from one cell to another within the host plant. In addition, they can spread from plant to plant through insect vectors or contaminated equipment.

2. Viroids are approximately 360 nucleotides long (about one-third the size of an average gene) and their nucleotide sequence does not appear to code for a protein, since there are no start and stop codons in phase with each other.
3. Viroids do not develop capsids, nor do they become enveloped. Thus, viroids appear to be fundamentally different from viruses.
4. Prions are a new type of infectious agent that is extremely resistant to physical and chemical treatments that inactivate bacteria, viruses, and viroids. Prions are small hydrophobic particles that appear to consist only of protein.

STUDY QUESTIONS

1. What are viruses and how are they classified?
2. What kinds of nucleic acids do the bacterial, plant, and animal viruses have?
3. How do viruses differ from cellular organisms?
4. List the important stages in a bacteriophage infection.
5. Describe the Ellis and Delbruck experiment, and explain what it showed about bacteriophages.
6. Describe the Doermann experiment, and explain what it showed about bacteriophages.
7. Draw bacteriophage T4 and label its parts.
8. Discuss the lytic cycle of T4.
9. Discuss the lytic cycle and the lysogenic cycle of lambda.
10. Describe the mechanism that decides whether lambda will undergo a lytic cycle or a lysogenic cycle.
11. What is the lambda genome called when it integrates into the host's genome?
12. What is the host cell called when it contains an integrated virus genome?
13. Compare and contrast a bacteriophage infection with a plant or animal virus infection.
14. Discuss a TMV infection.
15. Describe a herpes virus infection.
16. Describe an orthomyxovirus infection.
17. Explain what an oncovirus is.
18. Compare and contrast the growth of bacteriophages, plant viruses, and animal viruses in the laboratory.
19. Compare and contrast a viroid with a virus.
20. What is a prion?

SUPPLEMENTAL READINGS

Bolton, D., McKinley, M., and Prusiner, S. 1982. Identification of a protein that purifies with the scrapie prion. *Science* 218:1309–1311.

Butler, P. and Klug, A. 1978. The assembly of a virus. *Scientific American* 239:62–68.

Cohen, S. and Shapiro. 1980. Transposable genetic elements. *Scientific American* 242(2):40–49.

Diener, T. D. 1980. Viroids. *Scientific American* 244(1):66–73.

Diener, T. O. 1982. Viroids: Minimal biological systems," *BioScience* 32(1):38–44.

Eigen, M., Gardiner, W., Schuster, P., and Winkler-Oswatitsch, R. 1981. The origin of genetic information. *Scientific American* 224(4):88–118.

Gajdusek, C. 1977. Unconventional viruses and the origin and disappearance of kuru. *Science* 197(4307):943–959.

Gonda, M. A., Wong-Staal, F., Gallo, R. C., Clements, J. E., Narayan, O., and Gilden, R. V. 1985. Sequence homology and morphologic similarity of HTLV-III and Visna Virus, a pathogenic Lentivirus. *Science* 227:173–177.

Hendrix, R. W., Roberts, J. W., Stahl, F. W., and Weisberg, R. A. 1983. *Lambda II* New York: Cold Spring Harbor.

Henle, W., Henle, G., and Lennette, E. 1979. The Epstein-Barr virus. *Scientific American* 241(1):48–59.

Karpas, A. 1982. Viruses and leukemia. *American Scientist* 70:277–28.

Kornberg, A. 1979. The enzymatic replication of DNA. *CRC Critical Reviews in Biochemistry* 7(1):23–43.

Mathews, C. K., Kotler, E. H., Mosig, G., and Berget, P. B. 1983. *Bacteriophage T4* Washington D.C.: American Society for Microbiology.

Merz, P., Sommerville, R., Wisniewski, H., Manuelidis, L., and Manuelidis, E. 1983. Scrapie-associated fibrils in Creutzfeldt-Jakob disease. *Nature* 306:473–476.

Muesing, M. A., Smith, D. H., Cabradilla, C. D., Benton, C. V., Lasky, L. A., and Capon, D. J. 1985. Nucleic acid structure and expression of the human AIDS/Lymphoadenopathy retrovirus. *Nature* 313:450–457.

Prusiner, S. 1984. Prions. *Scientific American 251*:50–59.

Prusiner, S., McKinley, M., Bowman, K., Bolton, D., Bendheim, P., Groth, D., and Glenner, G. 1983. Scrapie prions aggregate to form amyloid-like birefringent rods. *Cell* 35:349–358.

Rapp, F. 1978. Herpesviruses, venereal disease, and cancer. *American Scientist* 66:670–674.

Stuart-Harris 1981. The epidemiology and prevention of influenza. *American Scientist* 69:166–172.

Tiollais, P., Charnay, P., and Vyas, G., 1981. Biology of hepatitis B virus. *Science* 213:406–411.

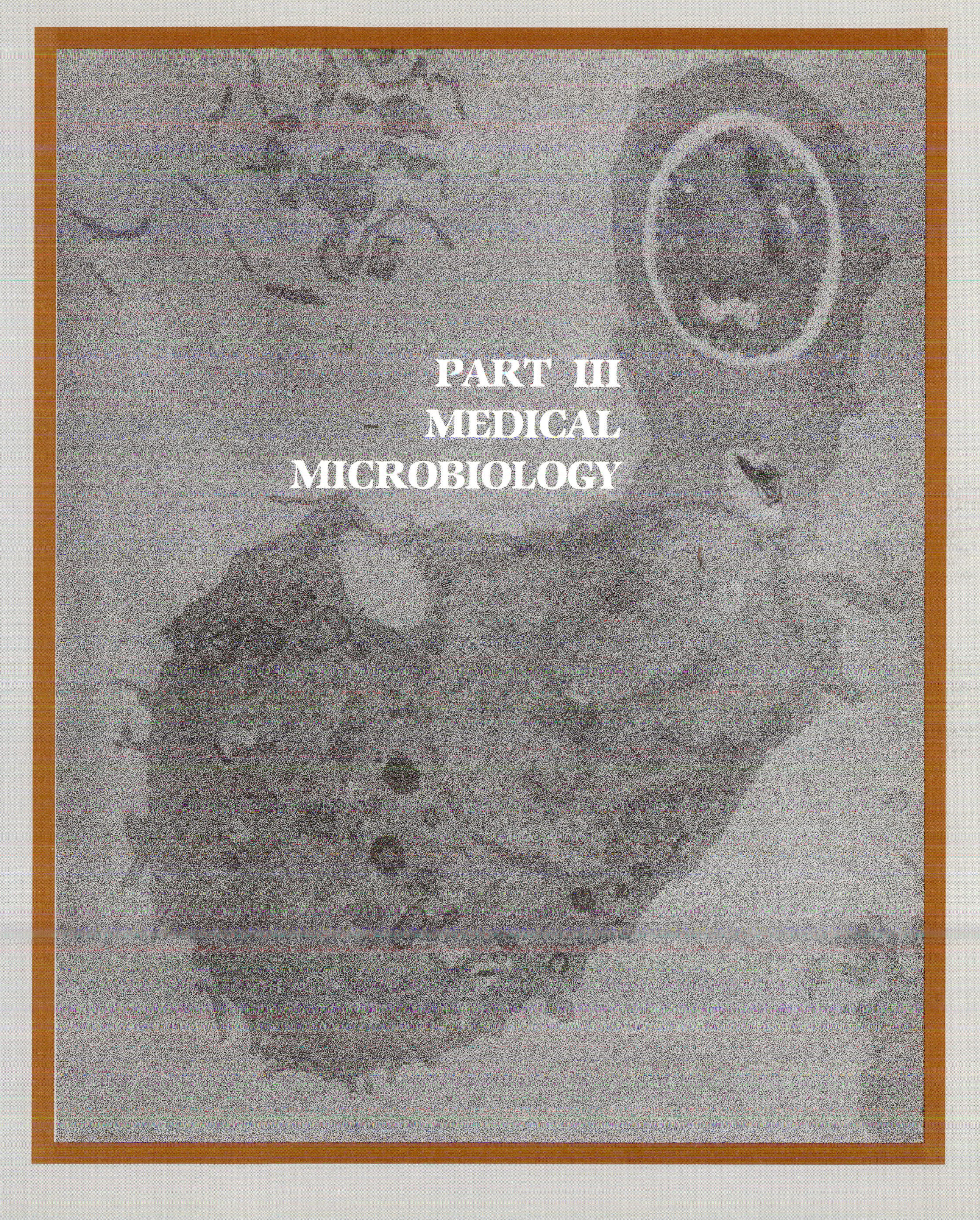

PART III
MEDICAL MICROBIOLOGY

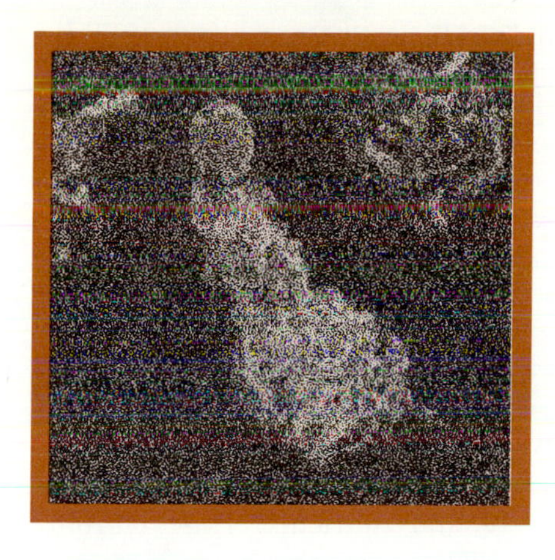

CHAPTER 18 DETERMINANTS OF HEALTH AND DISEASE

CHAPTER PREVIEW

LEPROSY OR HANSEN'S DISEASE

Leprosy is an infectious disease that disfigures and cripples many millions of people throughout the world. In ancient times, lepers were banished from their villages and towns and forced to roam the countryside begging until they became crippled or died of other diseases. Leprosy evoked such terror in people that many lepers were forced to wear bells so that everyone would know that a leper was near. Even in this century, before the development of antibiotics and drugs for treating leprosy, the disease was so feared that those unfortunate enough to be afflicted by it were ostracized from society and placed in "leper colonies" or locked away for life in insititutions or on remote islands. As recently as the 1930s in the United States, leprosy patients at the U.S. Public Health Service hospital in Carville, Louisiana, were restricted to the institution and not allowed to marry or even vote. In many parts of the world, a person with leprosy is still excluded from village life and rejected by family members because of fear of the disease.

There are between 11 and 15 million lepers in the world, about 4,000 of whom are in the United States. Although 4,000 is a rather small fraction of the leper population, it represents a 500% increase in the incidence of leprosy in the United States since 1960.

In recent years, scientists have learned much about this misunderstood disease, and prospects are good that more will soon be known. Perhaps the shroud of "unclean" will be removed from those who suffer from this insidious disease.

Leprosy is caused by the bacterium *Mycobacterium leprae,* a close relative of the organism that causes tuberculosis. Leprosy is contracted by intimate contact with infected individuals or their possessions. Sometimes it may take 15–20 years for the disease to manifest itself.

Leprosy can occur as a mild disease called **tuberculoid** leprosy, characterized by a few inconspicuous lesions, or as a severe disease, called **lepromatous** leprosy, which is manifested by the appearance of numerous skin lesions on the ears, nose, and fingers. The nerves may also be affected resulting in a feeling of numbness around the lesions. As the lepromatous leprosy develops, nerves are destroyed and the individual becomes crippled. One notable difference between the two forms of leprosy is that lesions of tuberculoid leprosy contain very few bacteria, while those of lepromatous leprosy contain many.

Scientists have discovered that a human host who reacts promptly and efficiently with an immune response will develop the less severe form of the disease. Those who have a sluggish or slow immune response develop the more severe form of the disease. The intensity of the immune response is believed to depend upon factors such as the genetic makeup and overall health of the host, as well as the number of infecting microorganisms and their ability to invade host cells.

Much is yet to be learned about leprosy. We presently know, however, that the severity of the disease and the ultimate recovery of the patient depend upon the invasive properties of the parasite and the defense mechanisms that can be mustered to combat the invading microbe. This chapter discusses some of the most important characteristics of microorganisms which make them able to cause disease and the host factors that prevent infections and disease.

LEARNING OBJECTIVES

A STUDY OF THIS CHAPTER SHOULD ENABLE YOU TO:

DISCUSS THE MAJOR TYPES OF SYMBIOTIC RELATIONS THAT MAY EXIST BETWEEN MICROORGANISMS AND HUMANS

DEFINE PATHOGENICITY, VIRULENCE, INFECTION, AND DISEASE

EXPLAIN HOW MICROORGANISMS MAY CAUSE DISEASE

OUTLINE THE VARIOUS MECHANISMS THAT THE HOST HAS TO AVERT INFECTIOUS DISEASES

DISCUSS THE INTRICACY OF THE INTERACTIONS BETWEEN HOST AND PATHOGENS IN HEALTH AND IN DISEASE STATES

Shortly after birth, we are colonized by a variety of microorganisms from the environment. These microbes colonize all of our body surfaces and cavities, where they live, metabolize, and reproduce. Some populations of colonizing microorganisms become an integral part of our bodies and remain there for our entire lives. These microorganisms and their host become attuned to each other and live together in harmony. Besides these permanent residents, we are colonized by transient microorganisms. Some are harmless, while others have the potential to cause disease. These microorganisms may come from various sources, such as the air, water, food, and soil. Whether or not colonization by these microorganisms results in an infectious disease is a consequence of the genetic ability of the colonizer to cause detrimental changes in the host and the host's defense mechanisms.

In previous chapters we have pointed out the remarkable versatility of microorganisms in obtaining their energy and nutrients. This versatility sometimes plays a role in the disease process, because many microorganisms obtain their nutrients from the animals and plants they colonize. Their growth and metabolic activities may cause detrimental changes in the host. On the other hand, activities of the invading microbe may not elicit much of a response from its host, and both may live in mutual tolerance. Such is the case with most of the permanent residents. Sometimes the colonizing microbes, however, elicit such severe responses that they are either expelled or quickly kill the host. The intensity of the response to the invading microbe is a function of both the host and its colonizer. The realm of host-parasite interactions is extremely complicated and involves many factors. It is the sum total of these interactions which determines the outcome of microbial colonization (fig. 18-1). This chapter discusses some of the ways in which hosts and microorganisms interact in health and in disease.

SOME DEFINITIONS

Before we delve into the realm of host-parasite interactions, it is imperative that we master a few pertinent terms that will serve as a foundation for the ensuing discussion.

Symbiotic Relations

Many organisms have evolved so that they spend their lives together. These close associations are called **symbiotic relations.** The term **symbiosis** means "living together" and generally describes any interaction, more or less permanent, between two or more organisms of different species. For convenience purposes, biologists have subdivided symbiotic associations into three different types, based upon the impact that one of the members, or **symbionts,** has on the other symbiont. Symbiotic associations that result in a mutually advantageous association are called **mutualistic** associations. The lichens □ represent a good example of a mutualistic association: both the algal and fungal symbionts benefit from the association. The associations that exist between many microorganisms and their human hosts are also mutualistic. The microorganisms are nourished by the host's secretions and surplus food, while the microorganisms protect the host from invading microbes by competing for space and food. **Commensalistic** associations are

363

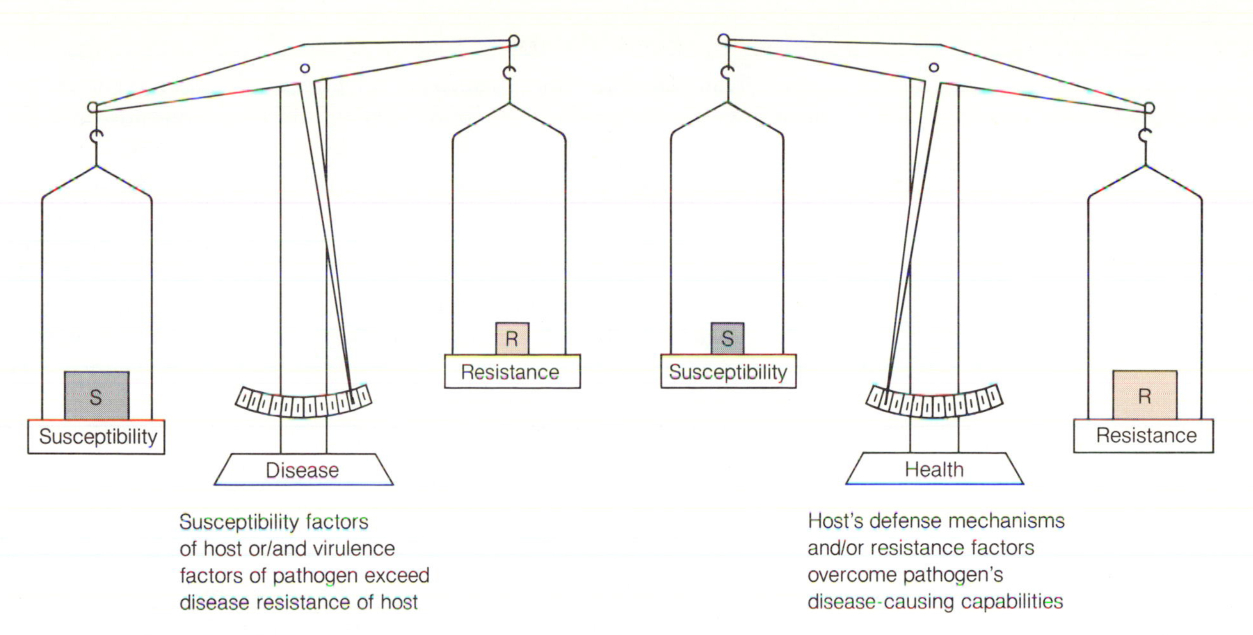

Figure 18-1 Determinants of Health and Disease. The host's overall health is dependent upon various factors, including its genetic makeup and that of the parasite, the effectiveness of the host's defense mechanisms, and the pathogen's virulence.

symbioses in which one symbiont derives a benefit from the association while the other is unaffected. The word commensalism is derived from a Latin word meaning "eating at the same table." Many microorganisms that are transient colonizers are called commensals because they feed upon their temporary host but generally do not cause any harm. When the host becomes debilitated by an illness or the resident flora has been killed, some commensals can cause disease and are referred to as **opportunists. Parasitic** associations are symbiotic relations in which one of the symbionts, the **parasite,** lives at the expense of the other symbiont, the **host**. The parasite is usually smaller than its host. The term parasite can be modified to define the location of the parasitic association. **Ectoparasites** parasitize the host's external surfaces. For example, the ectoparasitic mite *Sarcoptes* parasitizes the human skin, causing irritation and itching. The disease known as scabies is caused by this mite. **Endoparasites** parasitize internal body areas, such as organs and tissues, usually the latter. **Obligate parasites** are organisms that *must* lead a parasitic existence, while **facultative parasites** may obtain their nutrients either from a parasitic association or as a free-living form.

The symbiotic association between a parasite and its host may not always result in the ultimate destruction of the host. In fact, a "good parasite" does not kill its host, because in the process it would deprive itself of its only source of nutrients. In general, parasites tend to evolve mechanisms that ensure their survival within the host. Such mechanisms may involve reduction of irritation so that the host reacts less violently in the presence of the parasite. For the parasite to be able to obtain sufficient nutrients for growth and multiplication and yet not kill its host, both host and parasite must evolve together toward a state of mutual tolerance. Some of these parasitic conditions eventually evolve into mutualistic ones.

Infection Versus Disease

The terms **infection** and **disease** often are used interchangeably. If these terms were synonymous, we would constantly be ill, because we are constantly infected. Infection is a term that denotes the establishment and proliferation of a microorganism within a host. For infection to occur, access to the host is necessary. Access can occur through the skin and mucous membranes. Frequently, microorganisms enter animals from the respiratory tract or from the alimentary tract. Once inside the host, the invading microbes begin to multiply. This multiplication constitutes an infection; through common usage, however, this term is generally restricted to invasion of body organs and tissues by pathogenic (disease-causing) microorganisms. Disease is usually defined as a harmful alteration of the host's tissues or metabolic activities. Such alteration is not always the result of an infection, for there are many noninfectious diseases, such as strokes and heart attacks.

Pathogenicity Versus Virulence

Pathogenic microorganisms are those microorganisms that have the genetic capacity to cause disease; that is, they produce metabolic products or cause tissue changes that are harmful to the host. Many parasitic microorganisms can cause disease. In addition, certain commensals, given an alteration of the balance between the host and the microorganism, can become pathogenic. These are often called opportunists because they cause diseases only when the host has been compromised somehow.

The fact that pathogenic microorganisms such as *Mycobacterium tuberculosis* or *Streptococcus pneumoniae* infect a host does not necessarily mean that the host will become ill. Not all infections by pathogens ultimately result in disease. Even when disease does occur, it may not be of the same severity in all individuals affected. For this reason, the term **virulence** has been coined to describe the degree of pathogenicity or the pathogenic potential of a given microorganism. **Avirulent** strains of *S. pneumoniae* do not cause disease in humans, while **virulent** strains do. Nearly all pathogenic microorganisms exhibit wide variability in their virulence. To aid in measuring the degree of virulence for a certain microorganism, expressions such as **LD_{50}**,

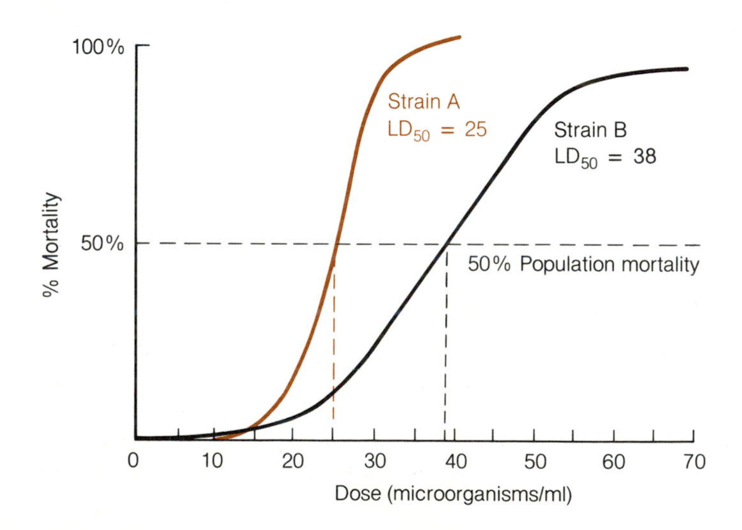

Figure 18-2 Determination of the LD_{50} of a Pathogen. The LD_{50} of a population is determined by injecting various doses of the pathogen into members of the population. Deaths are recorded and a graph is constructed. The graph represents the susceptibility of a population to two different strains of the same species of pathogen: strain A is more virulent than strain B. The dose (microorganisms/ml) of pathogen A that killed 50% of the population is 25, while that of pathogen B is 38. Hence the LD_{50} of strain A is 25 and that of strain B is 40.

LD_{100}, and **ID_{50}** have been introduced. The expression LD_{50} denotes the *lethal dose* of microorganisms which will kill 50% of the tested population. For example, an LD_{50} of 150 for a bacterium such as *Streptococcus pneumoniae* might mean that a dose of 150 microorganisms per mouse will kill 50% of the mice in a population. A strain of *S. pneumoniae* with an LD_{50} of 25 is much more virulent than another strain with an LD_{50} of 1,500. Similarly, an LD_{100} is the dose of microorganisms that will kill the entire population. Figure 18-2 illustrates how the LD_{50} of a hypothetical microorganism and its host is calculated. Along the same lines, the ID_{50} indicates the dose of a given microorganism which will *infect* 50% of a population. These values are extremely important in assessing the disease-causing potentials of pathogens and the impact that they may have on the well-being of a susceptible animal population.

THE INFECTIOUS PROCESS

In order for parasitic microorganisms to initiate an infection in a host, they must first reach those host tissues where microbial growth is favored. Many microorganisms find favorable environments in the respiratory tract and frequently penetrate the host from these tissues. Other microorganisms enter the host through the alimentary tract, and still others penetrate through the skin.

Once a parasite reaches a suitable site for growth inside a host, it may then become established and begin to reproduce. Many invasions, however, do not result in infections. Consider, for example, that every time you inhale a breath of air or eat a morsel of food, hundreds of thousands of microorganisms find their way into your lungs or intestines. But this is as far as many of the infections go because host defenses prevent the pathogens from becoming established. Similarly, many microorganisms find their way into wounds, but the probability of their initiating a serious infection generally is slight.

In order for a microorganism to cause an infection, it must have access to nutrients. Microorganisms may obtain nutrients by attaching to tissues. Researchers have established that the attachment often is tissue-specific and is mediated by surface characteristics of the microorganism. One well-studied example is the specific adherence of *Neisseria gonorrhoeae* to the genital tract epithelium. This adherence is mediated by pili □, so only pilated strains of *N. gonorrhoeae* can adhere to the epithelium and hence initiate an infection.

88

Adherence may also be due to a **glycocalyx** □ produced by certain bacteria, fungi, and other microorganisms. The glycocalyx is a fibrous matrix of polysaccharide which is found on the surface of many cells. Recent investigations have revealed that, in nature, bacteria grow in glycocalyx-enclosed microcolonies (fig. 18-3). Apparently this covering affords the bacteria a means of binding and channeling nutrients into their cells, as well as a means of protection from phagocytosis by predatory organisms and from infectious virus particles. Certain eukaryotic parasites, such as hookworms, tapeworms, and flukes (fig. 18-4), have evolved structures for attachment and for obtaining nutrients from the host. Once a parasite has attached to a suitable nutritive surface and become established there, it can begin to multiply.

84

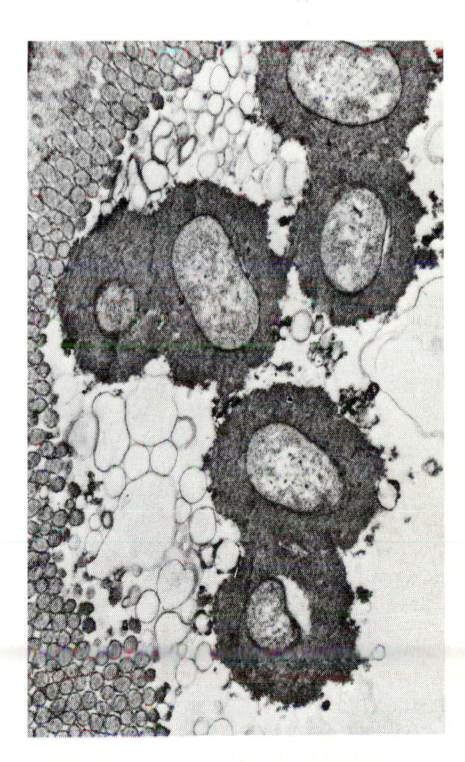

Figure 18-3 Microcolony of *Escherichia coli* Enclosed in a Glycocalyx. Transmission electron micrograph showing *E. coli* with a thick glycocalyx attached to the microvilli of a calf's ileum. The glycocalyx serves to attach the bacteria to solid surfaces and to shield them from host defense mechanisms.

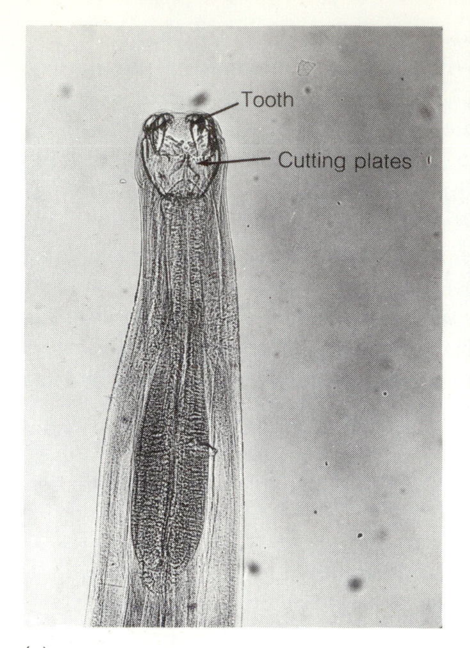

(a)

(b)

Figure 18-4 Attachment Organs of Some Human-Parasitic Worms. *(a)* Photomicrograph (SEM) of the head of a hookworm. The mouth has some prominent teeth that are used to attach to intestinal epithelium, where the worms feed and reproduce. *(b)* Photomicrograph of a tapeworm. The scolex (head portion) has four suckers, which are used to adhere to the intestinal epithelium.

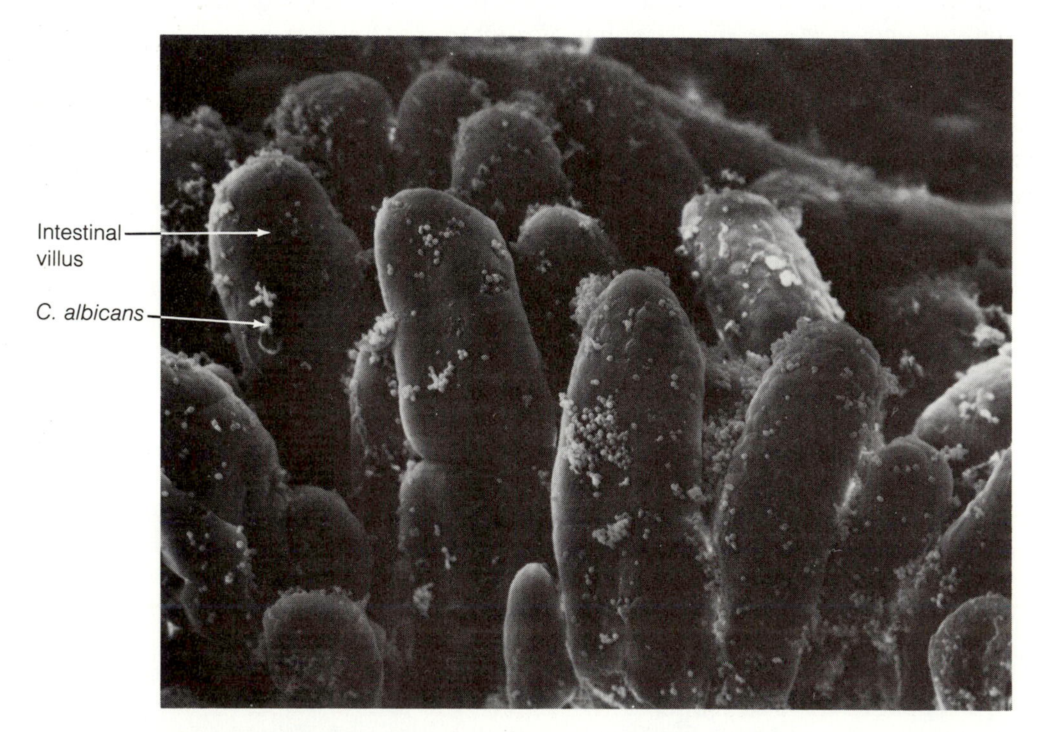

Figure 18-5 *Candida albicans* Colonizing Intestinal Villi. Experimental infection of a mouse, in which *Candida albicans* (arrow) is reproducing on the surface of intestinal villi.

Some microorganisms are well suited for growth on the surface of tissues (fig. 18-5). For example, *Shigella dysenteriae* adheres to the intestinal epithelium and multiplies by using host materials as nutrients. As they grow, the bacteria can extend to adjacent sites. Some microorganisms invade cells and multiply inside them (fig. 18-6). These microorganisms, called **intracellular parasites,** can exhibit remarkable adaptations to life inside cells. Extracellular parasites generally cause acute, short-lasting diseases such as sore throats, wound infections, and pneumonia, because they can be attacked and eliminated successfully by the host's immune system. Intracellular parasites, on the other hand, can grow virtually unaffected by the host's defenses. Often, the intracellular parasite and the host achieve a balance that results in a chronic or long-lasting infection. Certain groups of intracellular parasites have evolved nutrient-gathering mechanisms that make them totally dependent upon their host for survival. These parasites are called **obligate intracellular parasites.** The rickettsiae, the chlamydias, the viruses, and some of the protozoa are obligate intracellular parasites.

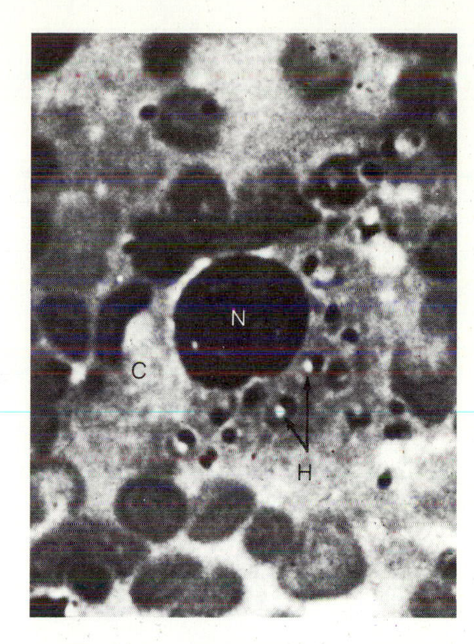

Figure 18-6 ***Histoplasma capsulatum*, an Intracellular Parasite.** The fungus *Histoplasma capsulatum* (H) invades macrophages and reproduces within them. The photograph illustrates a macrophage infected with reproducing yeast cells of this fungus. Macrophage nucleus (N) and cytoplasm (C) are indicated.

PATHOGENIC PROPERTIES OF MICROORGANISMS

An infectious disease is the result of a parasite's activities in a host and the host's response to the invading parasite. A microorganism alone, no matter how virulent, cannot cause disease in the absence of a suitable host. In this section we will examine the factors that determine the disease-causing potentials of parasites.

Diseases caused by microorganisms can be classified into two distinct categories, based on the level of participation of the pathogen itself in the disease process. These categories are **intoxications** and **infections.** Intoxications are diseases that result from the entrance into the body of a toxin that can cause disease in the absence of the toxin-producing microbe. For example, the disease commonly known as **botulism** is an intoxication caused by the ingestion of a toxin present in the food. This toxin, called **botulin** or **botulinal toxin,** can cause the disease in the absence of *Clostridium botulinum.* Another disease, **straphylococcal food poisoning**, results from the ingestions of **enterotoxin** produced by *Staphylococcus aureus.* Tetanus, cholera, and diptheria are also diseases that result from the production of toxins by microorganisms. **Infectious diseases** result from the physico-chemical alterations of a host as a result of the parasite's growth and extension to adjacent tissues. For example, *S. aureus* commonly causes **pyogenic** □ or pus-producing infections, such as wound infections, impetigo, and furuncles. All these diseases result from the colonization of subepidermal tissues, where *S. aureus* multiplies and produces a variety of tissue-destroying enzymes that allow the microorganism to invade adjacent tissues. Table 18-1 summarizes various infectious diseases resulting from intoxications or invasions.

563

Toxigenicity

Many microorganisms, during their growth and reproduction, release an array of metabolic products into their environment. Others, as part of their anatomical makeup, have chemical components that are toxic. Some of these products can be extremely harmful to humans and can cause severe damage in minute quantities (table 18-1). Such noxious substances are called **toxins.**

TABLE 18-1
INFECTIOUS DISEASES CAUSED BY BACTERIA OR FUNGI

DISEASE	TYPE* OF DISEASE	MICROORGANISM RESPONSIBLE FOR DISEASE	FACTOR(S) INVOLVED IN DISEASE	MODE OF ACTION
Anthrax	I/T	*Bacillus anthracis*	Invasive properties, toxin	Affects CNS
Botulism	T	*Clostridium botulinum*	Neurotoxins	Flaccid paralysis
Candidiasis	I	*Candida albicans*	Neuraminidase?, lipase?	
Cholera	T	*Vibrio cholerae*	Choleragen	Alters intestinal permeability
Coccidioidomycosis	I	*Coccidioides immitis*	Invasive properties	
Diphtheria	T	*Corynebacterium diphtheriae*	Diphtheria toxin	Halts protein synthesis
Dysentery	I	*Shigella* sp.	Invasive properties	
Food poisoning	T	*Staphylococcus aureus*	Enterotoxins	Diarrhea/vomiting
Gas gangrene	I	*Clostridium perfringens*	Various enzymes	
Histoplasmosis	I	*Histoplasma capsulatum*	Invasive properties	
Pharyngotonsillitis	I	*Streptococcus pyogenes*	Various enzymes	
Plague	I/T	*Yersinia pestis*	Invasive enzymes plus plague toxin	Cytotoxic
Pyoderma	I	*Staphylococcus aureus*	Coagulase, other enzymes?	
Salmonellosis	I	*Salmonella* sp.	Invasive properties	
Scarlet fever	I	*Streptococcus pyogenes*	Erythrogenic toxin	Cytotoxic
Tetanus	T	*Clostridium tetani*	Neurotoxin	Spastic paralysis
Toxic shock syndrome	T	*Staphylococcus aureus*	Pyrogenic exotoxin	Cytotoxic/shock
Tuberculosis	I	*Mycobacterium tuberculosis*	Invasive properties	
Typhoid	I	*Salmonella typhi*	Invasive properties	
Whooping cough	T/I	*Bordetella pertussis*	Various toxins	Cytotoxic

*I = disease caused primarily by invasion of tissue by multiplying bacteria.
T = disease caused by ingestion or production of toxins at site of infection.

TABLE 18-2
CHARACTERISTICS OF EXOTOXINS AND ENDOTOXINS

CHARACTERISTIC	EXOTOXIN	ENDOTOXIN
Source of toxin	Synthesized by gram^{+} and gram^{-} as a result of metabolic activity	Gram negative cells walls
Chemical composition	Proteinaceous materials	Lipopolysaccharides
Resistance to heat (100°C)	Heat sensitive	Heat stable
Toxicity	Extremely toxic in minute doses	Toxic at high doses only
Mode of action	Unique for each toxin	Generally the same regardless of source
Example of action	Spastic paralysis, flaccid paralysis, citotoxicity, alter protein synthesis, plasma membrane permeability	Fever, shock, and weakness
Immunogenicity (induce immune response)	Very immunogenic	Weakly immunogeneic
Conversion to toxoid (detoxified with formalin)	Readily converted	Not converted

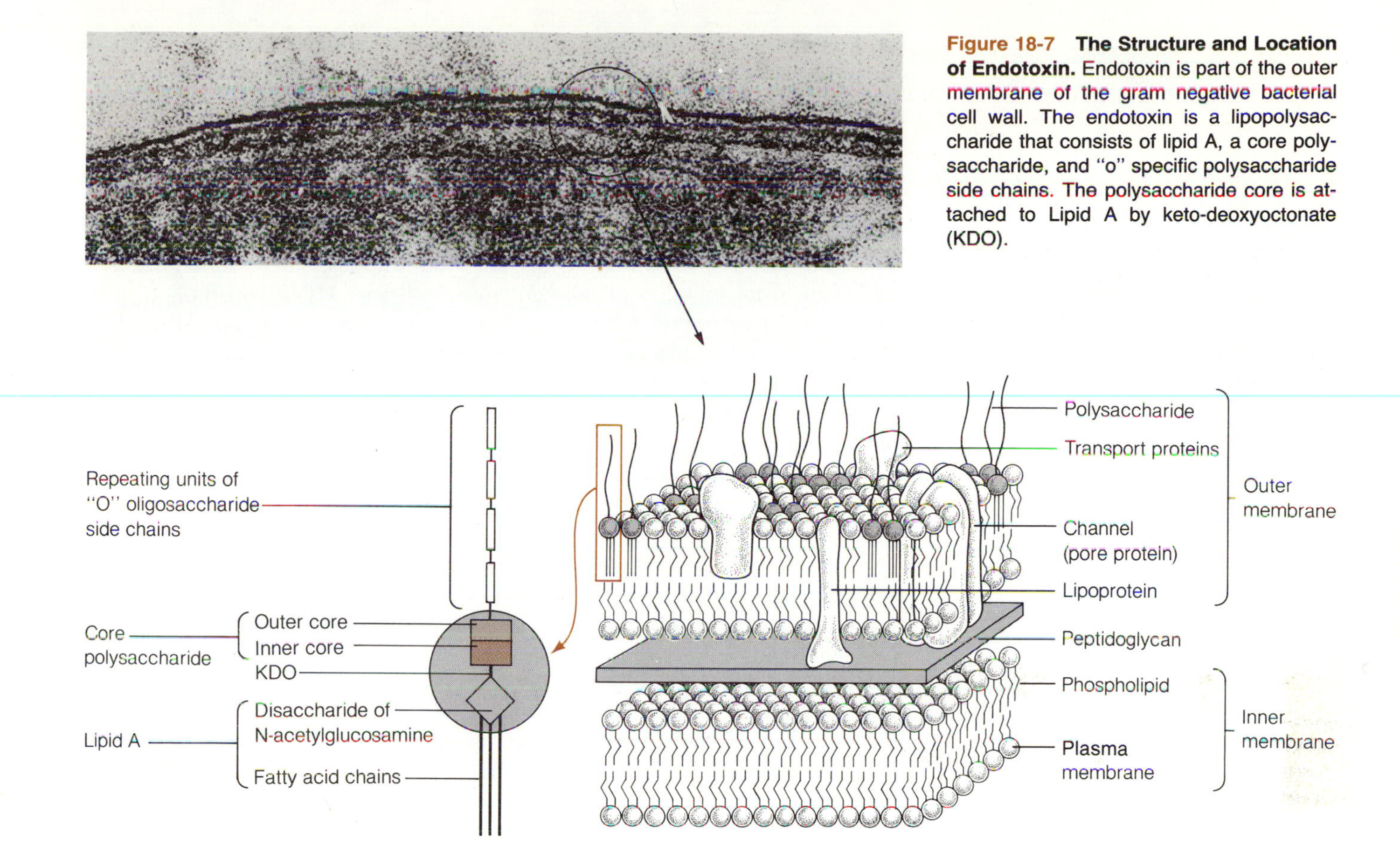

Figure 18-7 The Structure and Location of Endotoxin. Endotoxin is part of the outer membrane of the gram negative bacterial cell wall. The endotoxin is a lipopolysaccharide that consists of lipid A, a core polysaccharide, and "o" specific polysaccharide side chains. The polysaccharide core is attached to Lipid A by keto-deoxyoctonate (KDO).

209 Some toxins interfere with specific physiologic processes, such as protein synthesis □ or nerve impulse conduction, while others interfere with body defenses. Microbial toxins are classified into two categories: **exotoxins** and **endotoxins.** In general, exotoxins are heat-labile (heat-sensitive) proteins produced by a specific microorganism and are not structural components of the cell. Exotoxins have specific and unique modes of action that differ from one species to another. Botulinal toxin, for example, typically causes flaccid paralysis, because it inhibits the release of **acetylcholine** at the synaptic junction of motor nerve fibers, thus preventing the transmission of a nerve impulse to muscle cells. By contrast, the **diphtheria toxin** inhibits protein synthesis and therefore causes cell death. Other characteristics of exotoxins are summarized in table 18-2.

Endotoxins are toxic components of the cell wall of gram negative bacteria (fig. 18-7). These toxins are not proteins but **lipopolysaccharides**. All gram negative bacteria possess endotoxins which unlike exotoxins, have similar modes of action (table 18-2). It is the lipid portion, **Lipid A** (fig. 18-7), of the lipopolysaccharide which is responsible for the toxicity of the endotoxins. The characteristic signs and symptoms that are caused by endotoxin include fever, shock (1–2 hours after introduction of endotoxin), severe diarrhea, and altered immunity states.

Infectivity

Certain pathogens cause disease as a result of their growth and multiplication within the host. As these microorganisms grow, they produce enzymes and metabolic products that act on host tissues. For example, *Clostridium perfringens*, a causative agent of gas gangrene, causes extensive tissue damage because it produces many exoenzymes when it reproduces within the host. One such enzyme, **collagenase,** breaks down the collagen that provides structural integrity to muscle tissue. The resulting tissue destruction aids the pathogen in its invasion of adjacent muscle tissue. **Hyaluronidase** is produced by *C. perfringens* and by a variety of other microorganisms, including streptococci and staphylococci. This enzyme has been called **spreading factor** because it breaks down hyaluronic acid of connective tissue, which helps to hold cells together. As we can see from table 18-1, many microorganisms produce enzymes that can play a role in infection. These enzymes act on various tissues and components of the body and therefore can cause the signs and symptoms characteristic of various diseases.

Sometimes a pathogen not only invades adjacent tissues but also erodes its way into blood or lymphatic vessels. Once in the circulation, the pathogen gains access to other sites in the host. This grave condition is called **septicemia** □. While in the blood, the parasite may release toxins or metabolic products and cause severe diseases. For example, the parasite that causes **malaria** invades the red blood cells and reproduces inside them. At the end of its reproductive cycle, it disrupts the red blood cells, releasing progeny parasites as well as metabolic waste products and cellular debris. This release of toxic wastes results in the characteristic signs and symptoms of malaria, which include chills, fever, and malaise. Of course, some microorganisms owe their virulence to both invasive properties and toxigenic properties. Examples include the agents of anthrax, and plague.

684

HOST DEFENSE MECHANISMS

When invaded by a pathogenic microorganism, the host is not passive, but rather it confronts the pathogen with a variety of barriers aimed at aborting the infectious process. Vertebrate hosts have evolved remarkably effective mechanisms to deal with invading microbes. Some mechanisms are nonspecific and are effective against a wide variety of microbial invaders, while others are quite specific □ (table 18.3). Because of these mechanisms, many infections are stopped before the parasite has had a chance to cause detrimental changes in the host. On the other hand, the host responses can be intense enough to contribute to the signs and symptoms of the disease. In the ensuing discussions we will examine various host responses and the impact they have on the overall well-being of the host.

455

Nonspecific Physical and Mechanical Barriers

Skin and Mucous Membranes

The unbroken skin very effectively protects animals against invading microbes, since most microorganisms cannot penetrate this barrier. A few microorganisms, such as the blood flukes and hookworms, can penetrate the intact skin by secreting hydrolytic enzymes. For the most part, however, mi-

TABLE 18-3
SUMMARY OF DEFENSE MECHANISMS

DEFENSE MECHANISM	FUNCTION
I. Nonspecific Defense Mechanisms	
A. Physical Barriers	
1. Flushing mechanisms	
a. Coughing and sneezing	Explusion of invaders from respiratory track
b. Urination	Flushes the urinary tract
c. Lacrimation	Washes out invaders from eyes
d. Ciliary action	Moves microbes out of body
e. Peristalsis	Flushes microbes from intestines
2. Skin and mucous membranes	Prevent entry of pathogens into body
B. Chemical Barriers	
1. pH of body fluids	Inhibit growth of many pathogens
2. Lysozyme	Breaks down cell walls of bacteria
3. Complement	Cell lysis and enhances phagocytosis
4. Interferon	Inhibits viral multiplication
C. Biological Barriers	
1. Natural resistance	Not affected by certain infectious agents
2. Normal flora	Competes and antagonizes invaders
3. Inflammation	Localize pathogens & repair tissue damage
4. Phagocytosis	Engulfs and destroy invaders
II. Specific Defense Mechanisms	
A. Humoral Immunity	Antibodies bind to foreign particles, neutralize toxins and viruses. Also enhance phagocytosis.
B. Cell Mediated Immunity	Cells of the immune response attack invading microorganisms and cancer cells. Produce cytotoxins and recruit cells to augment the specific immune response.

croorganisms penetrate through breaks in the skin, such as cuts, puncture wounds, or insect bites.

The lining of the respiratory tract is coated with mucus, which traps most microorganisms that enter the trachea and bronchi. The ciliary action of the respiratory epithelium brings the microbe-laden mucus into the mouth, where it is swallowed. Many invading microorganisms are removed from the respiratory passages in this manner. Also, the external nares are coated with hairs that can trap some of the microbes in the incoming air before they can enter the host further.

Flushing Mechanisms

In addition to physical barriers, the vertebrate host has other mechanisms antagonistic to microbial invasion. Flushing mechanisms, such as coughing, sneezing, ciliary action (above), urination, peristalsis, lacrimation, and salivation, play central roles in ridding the various organs of infectious microorganisms. Impaired function of any of these mechanisms generally leads to infection. For example, individuals with a partially blocked urethra are more likely to suffer from urinary tract infection than individuals who are capable of normal urination.

Coughing and sneezing are reflexes that serve to clear the respiratory tract of noxious particles. These mechanisms cause the forceful exit of air from the lungs through the mouth and nose, respectively, thus serving to maintain the respiratory tract relatively free of infectious agents and irritating particles.

In the intestines, potentially pathogenic microorganisms are moved into the colon by **peristalsis,** or contractions of the intestines. These periodic movements serve to flush microorganisms from the epithelium of the intestines and out of the body in the feces.

Urination is another flushing mechanism that plays an important role in the overall health of the vertebrate host. This process keeps the urethra relatively free of microorganisms. As the urine flows through the urethra from the bladder, it washes out the urethral lumen and walls and carries with it many of the microorganisms that have entered the urethra. Similarly, tears and saliva serve to flush the eyes and the mouth, respectively, of potential pathogens.

Nonspecific Chemical Barriers

Vertebrate hosts also have a chemical arsenal with which to combat infectious agents. Body cavities often are coated with acidic secretions that are germicidal or germistatic. For example, the stomach has gastric juices, primarily HCl, with a pH of less than 2. Most microorganisms are killed at this pH and thus have very little chance of surviving in the stomach. Millions of ingested salmonellae are required in order to infect a normal human host because most of the bacteria are killed in their passage through the stomach. If the patient is hypochlorohydric (little HCl is produced by the stomach), however, far fewer salmonellae are required to initiate an intestinal infection.

The skin generally is free of harmful microbes, not only because of the dry conditions that exist there but also because of the secretion of fatty acids by sebaceous glands. Some fatty acids, in particular **oleic acid,** are bactericidal for some potentially pathogenic transients.

Saliva and tears function not only as a means of flushing out invading microbes but also as germicidal agents. Both lacrimal and salivary secretions contain **lysozyme,** an enzyme that weakens or disrupts bacterial cell walls □.

81

Tissue extracts and body fluids possess a host of antimicrobial agents. Polypeptides with high quantities of lysine or arginine are bactericidal (at least *in vitro*) against the anthrax bacillus, staphylococci, streptococci, and the tubercle bacillus. Blood serum contains **β-lysins,** cationic proteins that are active against a variety of bacterial cells. The β-lysins apparently act by disrupting the plasma membrane. It is believed that the combined action of proteins such as lysozyme and the β-lysins plays an important role in the host's defense against microorganisms.

Complement

Complement □ is the name given to a group of proteins found in serum. These proteins interact with one another to produce active enzymes and proteins that have several biological functions. Among the most important functions complement proteins carry out are: a) lysis of microorganisms; b) neutralization of viruses; c) enhancement of phagocytosis; d) anaphylatoxin

472

activity; e) recruitment of phagocytes; and f) damage to plasma membranes. Complement is generally considered to be part of the immune response and will be discussed in greater depth in Chapter 19.

Interferon

Isaacs and Lindenmann in the late 1950s described a substance called **interferon** that had antiviral properties. Their search for interferon was initiated by the observation that once an animal was infected with a virus, it was more resistant to an unrelated viral infection than an animal not infected by a virus. In other words, one viral infection "interfered" with another. It is now known that interferon is really a group of **glycoproteins** of related molecular structure with antiviral properties.

Uninfected animals generally do not contain detectable levels of interferon. Upon infection by a virus, however, the host begins to produce interferon, which "interferes" with viral replication. Because of the action of interferon, many severe viral infections are avoided.

Interferon produced in virus-infected cells initiates the production of molecules in neighboring, uninfected cells which inhibit transcription or translation of viral mRNA (fig. 18-8).

Nonspecific Biological Barriers

Natural Resistance

Every infectious microorganism has a **host range;** that is, it can infect only certain hosts. The host range is a function of both the parasite's characteristics and the host's traits. For example, the first step in the infection of a 405 cell by a virus is its attachment □, which is possible only if the host cells

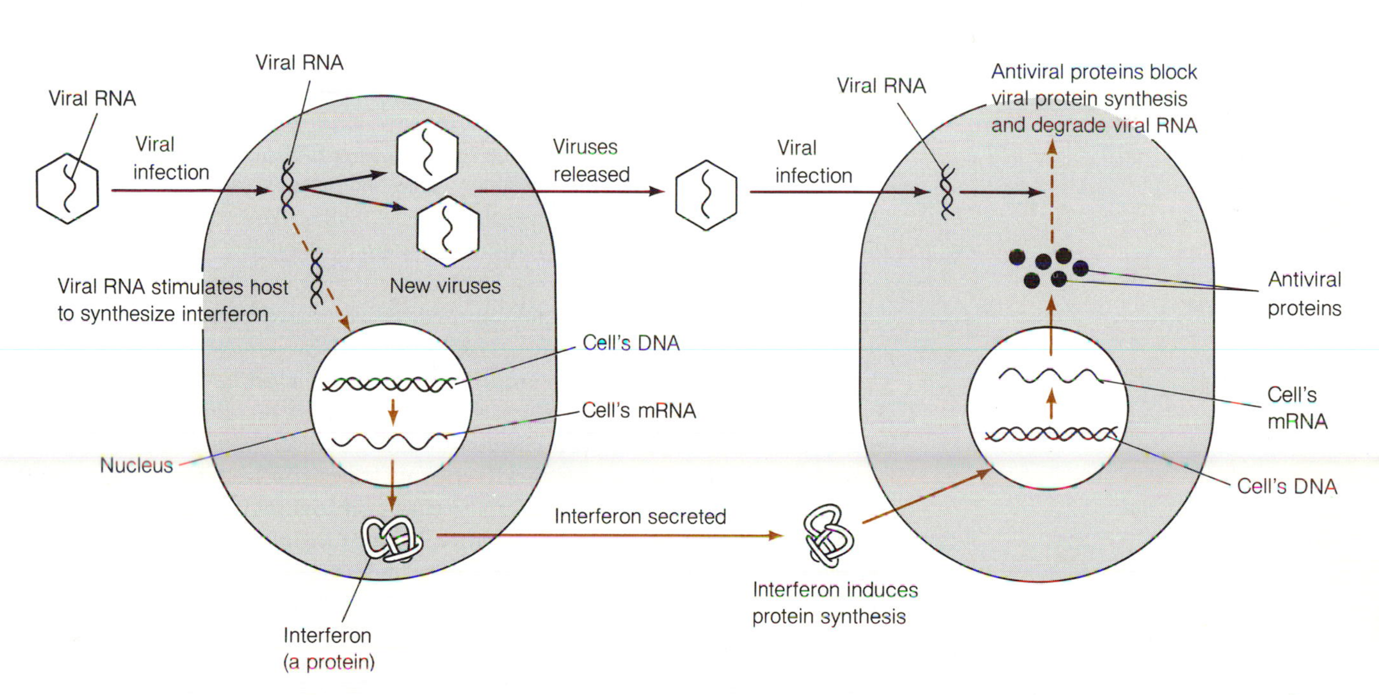

Figure 18-8 Mode of Action of Interferon. Interferon is a protein synthesized by virus-infected cells which ultimately leads to the inhibition of viral replication. When certain RNA viruses infect cells, they induce the infected cells to synthesize interferon. This protein is secreted by the cell and stimulates other cells to produce antiviral proteins.

have surface structures that can serve as attachment sites for the virus. A simple example is a bacteriophage that attaches to bacterial pili: species of bacteria that lack pili would be **naturally resistant** to this virus. Also, if a given human cell lacks receptor sites that can be recognized by a virus, the cell is naturally resistant to the virus. This fact may explain why humans are resistant to diseases such as distemper and mousepox. It is likely that human cells lack surface receptors for the pathogens that cause these diseases.

Even if a given parasite can attach to a host cell, the attachment may not culminate in an infection. The intracellular conditions of the host cell must be conducive to the parasite's multiplication, or the infection will be aborted. Any number of physiologic or anatomic characteristics of a particular host may make it naturally resistant to a given infection. An example is body temperature. Humans and other mammals are quite susceptible to a disease called anthrax, caused by the invasion of host tissue by *Bacillus anthracis.* Fowl (*e.g.*, chickens), however, are naturally resistant to the disease. Human body temperature is around 36.7°C, while chicken body temperature is closer to 40°C. At 40°C, the bacterium cannot carry out essential metabolic activities and is therefore unable to initiate an infection. At the lower temperature, the disease occurs as a result of the bacterial multiplication. If the chicken's body temperature is lowered to that of humans, it too will succumb to anthrax.

The Normal Flora

Immediately after birth, newborns are colonized by many different bacteria, viruses, fungi, and algae. Some of these microorganisms remain associated with the newborn host for only a short time, while others remain throughout the host's lifetime. The microorganisms that colonize the surfaces of cavities of humans and other animals are called the **normal flora.**

Each body surface is colonized by its own unique normal flora (table 18-4). The skin and hair have about 10^4 to 10^5 bacteria/cm^2. The most common colonizers of the skin are staphylococci, diphtheroids, and yeasts. The intestines normally are colonized by enterics, yeasts, and anaerobic bacteria.

The normal flora affords its host a certain amount of protection against invading pathogens. For example, *E. coli*, *Fusobacterium*, and *Bacteroides*, prominent inhabitants of the human intestine, make it difficult for enteric pathogens such as *Salmonella* and *Shigella* to colonize the human intestine and cause disease. Evidence that the normal flora protects against invasion by other microorganisms comes from studies using experimental animals. **Gnotobiotic** (or germ-free) animals are easily infected with *Salmonella* or *Shigella.* These infections usually result in the death of the animal. If these gnotobiotic animals are fed a meal containing *E. coli*, however (which subsequently colonizes the intestines), and are then infected with the enteric pathogens, the animals are much more resistant to infection. There are other lines of evidence supporting the idea that the normal flora protects its host partly by competing with invaders for space and nutrients. Studies have shown that patients treated with antibiotics such as streptomycin become more susceptible to certain intestinal pathogens, apparently because of the destruction of the normal intestinal flora by the antibiotic. Similarly, vaginal infections by certain strains of *Neisseria gonorrhoeae* are inhibited by the presence of *Candida albicans* □, which is a common commensal of the human vagina. It is very difficult to ascertain the type of relationship between

572

TABLE 18-4

REPRESENTATIVE ORGANISMS FOUND AS RESIDENT FLORA OF HUMANS

NAME OF ORGANISM	GROUP	ANATOMICAL SITE WHICH IT COLONIZES
Actinomyces sp.	Bacterium	Mouth, oral cavity
Bacteroides fragilis	Bacterium	Large intestine
Bacteroides sp.	Bacterium	Mouth and tooth surfaces
Candida albicans	Yeast	Oral cavity, small intestine
Clostridium sp.	Bacterium	Vagina, cervix
Diphtheroids	Bacterium	Skin, oropharynx, nasopharynx, intestines, vagina, cervix
Entamoeba coli	Protozoan	Large intestine
Entamoeba gingivalis	Protozoan	Oral cavity
Enteric bacteria	Bacterium	Vagina
Enterobacter sp.	Bacterium	Large intestine
Enterococci	Bacterium	Small intestine
Escherichia coli	Bacterium	Large intestine
Fusobacteria	Bacterium	Oral cavity, large intestine
Group D streptococci	Bacterium	Vagina
Haemophilus influenzae	Bacterium	Oral cavity, oropharynx
Lactobacilli	Bacterium	Small intestine, vagina
Neisseria meningitidis	Bacterium	Oropharynx
Peptostreptococci	Bacterium	Saliva and teeth, large intestine
Pityrosporum	Yeast	Skin
Propionibacterium acnes	Bacterium	Skin
Proteus sp.	Bacterium	Large intestine
Staphylococcus aureus	Bacterium	Skin, nasopharynx
Staphylococcus epidermidis	Bacterium	Skin, nose, mouth, vagina
Streptococcus sp.	Bacterium	Mouth, oropharynx
Treponema sp.	Bacterium	Oral cavity
Trichomonas vaginalis	Protozoan	Vagina

a host and its normal flora, but some of the colonizers benefit their host, while others may cause disease.

The protective action of normal colonizers such as *E. coli* has been explained in several ways. It is thought that the normal flora consumes all the available nutrients, thus preventing the invaders from obtaining nutrients. It is also suggested that the normal flora inhibits infection by preventing the invaders from attaching to host surfaces. If a given surface area is heavily colonized by the normal flora, the invaders may be unable to bind to specific receptors and may therefore be flushed from the body surface. The production of inhibitory metabolites and antibiotics by the normal flora has also been cited as playing a role in protecting the host.

Competition for available nutrients is a powerful mechanism whereby the normal flora inhibits the growth of pathogens and opportunists. The yeast *Candida albicans,* for example, is normally found in small numbers in the human vagina because of competition with the normal flora for available nutrients. If a patient is subjected to prolonged antibiotic treatment, however, very often vaginitis caused by *C. albicans* occurs. The antibiotic treatment kills the gram positive vaginal flora and thus eliminates the competition, so *C. albicans* proliferates rapidly, causing vaginitis. Patients with vaginal yeast infections are sometimes treated with douches containing fermented milk 812 □ or sweet acidophilus milk, which have high concentrations of lactic acid bacteria that can recolonize the vagina and reestablish a competition for nutrients. Very often, this treatment is all that is required for the successful control of yeast infections.

Nonspecific Defense Mechanisms Associated with Blood and Lymph

Blood and Its Components

Blood serves to deliver nutrients, oxygen, and informational molecules (hormones) from one part of the body to another. It also contains numerous molecules that afford the host an additional dimension in anti-infection defenses. Blood consists of cells (and cell fragments known as **platelets**) suspended in a liquid called **plasma** (table 18-5). If you draw a volume of blood

TABLE 18-5
BLOOD CELLS, THEIR FUNCTION AND DISTRIBUTION

TYPE OF CELL	MORPHOLOGY	FUNCTION OF CELL
I. ERYTHROCYTES (red blood cells)		Carry oxygen to tissues and exchange CO_2, comprise 35–45% of blood volume
II. LEUKOCYTES (white blood cells)		
A. Granulocytes		Cells with granules in cytoplasm and lobed nucleus
1. Neutrophils		Phagocytosis and digestion of particles and microorganisms, 60% of all white cells
2. Basophils		Mediate inflammation, release histamine, 1% of all white cells
3. Eosinophils		Immunity to some parasites, modulate inflammation, 3% of all white cells
B. Lymphocytes		Participate in immune reactions 30% of all white cells
C. Monocytes		Phagocytosis and digestion of particles, 5% of all white cells

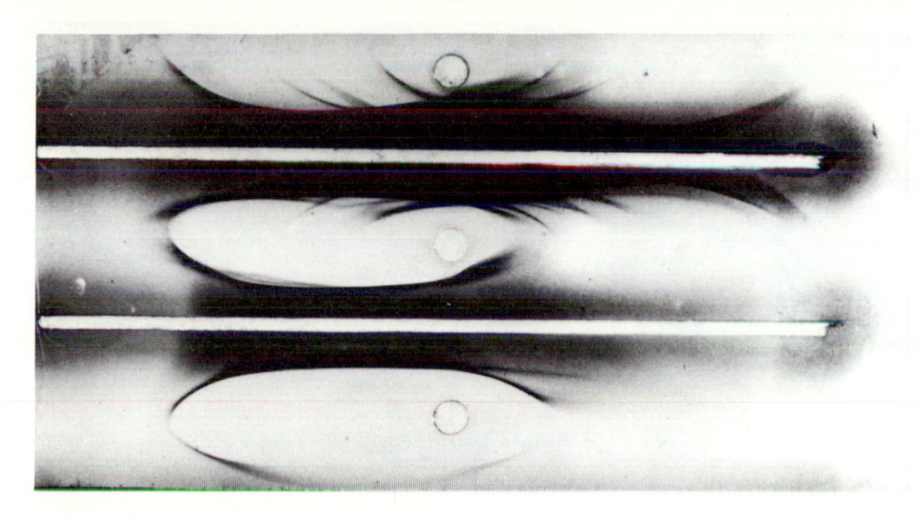

Figure 18-9 The Various Components of Serum. The photograph shows the results of immunoelectrophoresis (see page 525) of normal serum. The technique allows us to separate and visualize the various components of serum. Each curved line represents a different component of serum.

from a patient, place it in a test tube, and allow it to stand at room temperature for a few minutes, a **clot** begins to form. The clot eventually will develop into a red mass in the bottom of the tube. The straw-colored liquid above the clot, called **serum,** contains many different proteins such as globulins, complement, albumen, and interferon (fig. 18-9). The clot consists of blood cells knitted together by fibers of **fibrin.**

Blood includes the red blood cells (**erythrocytes**), white blood cells (**leukocytes**), and platelets. Each of these **formed elements** has a specific function and is central to the overall health of the individual. The platelets' primary function is that of participating in the clotting mechanism, so they are essential in the healing process. The leukocytes have various functions, the most important of which is to defend the host against infection.

Erythrocytes. The erythrocytes' function is to carry oxygen from the lungs to the rest of the body. Human erythrocytes measure about 7.5 μm in diameter and are biconcave in shape. They develop in the bone marrow and lose their nuclei when they mature. Erythrocytes constitute the largest portion of formed elements in the blood, making up nearly 45% of the blood volume. When the concentration of erythrocytes falls much below this figure, a condition known as **anemia** develops. Associated with the plasma membrane of erythrocytes is **hemoglobin,** a molecule that binds oxygen in the lungs and releases it to cells in other parts of the body. New erythrocytes are constantly being made in the bone marrow to replace the dead erythrocytes. The life span of an erythrocyte is about 100 days.

Leukocytes. The blood also contains white cells called leukocytes (fig. 18-10). These cells are nucleated and participate actively in protecting the host against infection. If a blood smear stained using the Wright's procedure is examined with a microscope, various types of white blood cells can be seen.

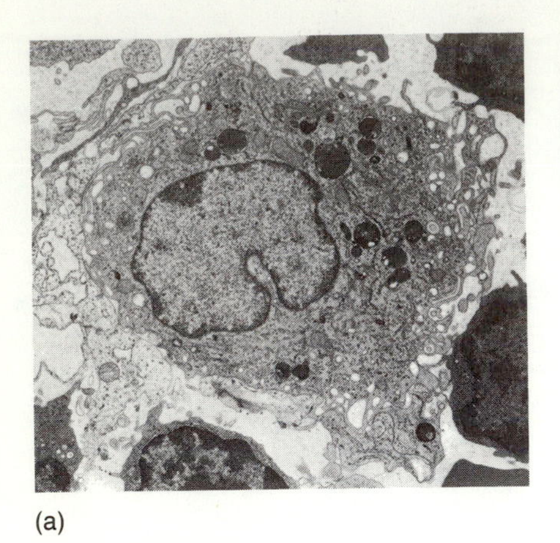

(a)

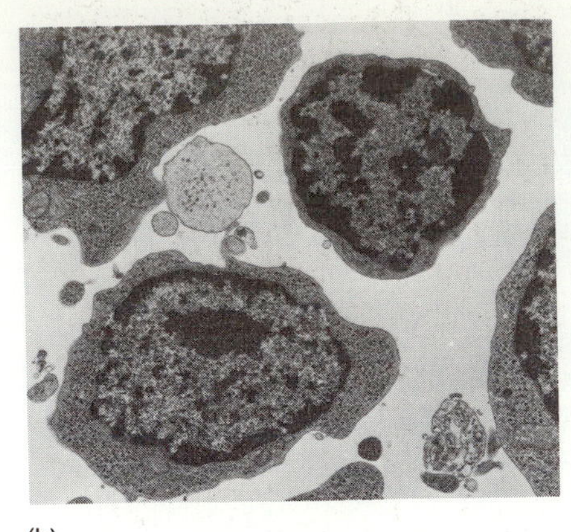

(b)

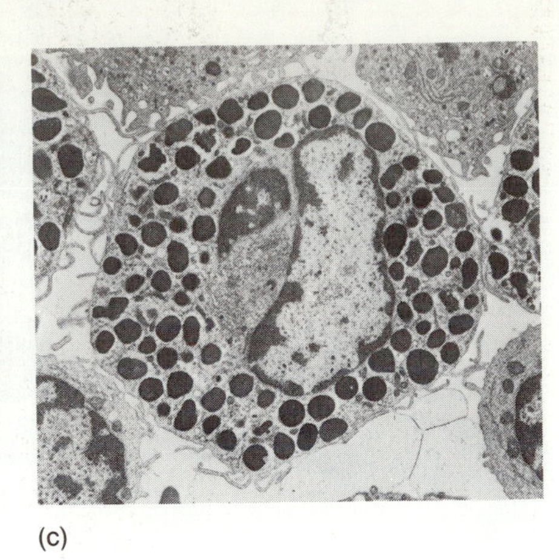

(c)

Figure 18-10 Electron Photomicrographs of Cells Involved in Host Defense Mechanisms. *(a)* Macrophage. *(b)* Lymphocyte. *(c)* Granulocyte (eosinophil).

Those called **granulocytes** constitute 60–65% of the leukocytes in the blood and contain prominent granules in their cytoplasm. They have a multilobed nucleus, so they are sometimes called **polymorphonuclear** leukocytes or **PMN** leukocytes, although this term is usually reserved to describe only neutrophils.

Granulocytes can be differentiated from one another by their staining properties. Some have granules that react with the acidic dye **eosin** in Wright's stain and stain red, so they are called **eosinophils.** Other granulocytes have granules that take up the basic dye **hematoxylin** and stain blue; these are called **basophils.** There are also granulocytes whose granules stain with neutral dyes as well as with hemotoxylin and eosin. These granulocytes, which stain both red and blue, are called **neutrophils.** Granulocytes are very important in protecting the host against infection. They participate in **inflammation** and in **phagocytosis,** and their numbers increase or decrease in response to various types of infections.

The nongranulated leukocytes constitute the remaining 35–40% of the leukocytes in the blood and include **monocytes** and **lymphocytes.** Monocytes are large leukocytes, 12–15 μm in diameter, with a centrally-located, kidney-shaped nucleus. These cells also participate in phagocytosis. The lymphocytes are rounded cells, 10–15 μm in diameter, with a very large ovoid nucleus that takes up most of the cytoplasm. As we shall see in Chapter 19, the lymphocytes play a vital role in specific immunity to infectious agents.

Differential blood counts, which determine the proportions of the various leukocytes in the blood, are important because they serve as an indicator of the state of health of an individual. For example, some diseases characteristically alter the white blood cell count by increasing or decreasing the total number of leukocytes (leukocytosis or leukopenia, respectively). In addition, certain diseases cause an increase in some types of leukocytes but not in others. This information may eventually be of use in determining the cause, treatment, and prognosis of diseases.

The Lymphatic System

Lymph is a pale fluid with a composition resembling that of blood plasma. This fluid bathes tissues of the body and circulates via the lymphatic vessels.

The lymph contains lymphocytes and a type of phagocytic cell called the **macrophage** (fig. 18-10), but no erythrocytes or PMN leukocytes. There is a connection between the circulatory and lymphatic systems via the **thoracic duct.** The lymphatic system consists of **lymphatic vessels, lymph nodes, lymph,** and **lymphocytes.** Located throughout the lymphatic system are lymph nodes, organs containing numerous lymphocytes and macrophages (table 18-5). Most of the specific immune mechanisms are initiated in these tissues.

Inflammation

Inflammation is a very effective host defense mechanism that develops in response to tissue injury. (fig. 18-11). The injury may result from mechanical damage such as a puncture wound or scratch, a chemical substance, sunburn, or an infectious agent. The destruction of cells by any of these mechanisms sets in motion a sequence of events that is aimed at the repair of the injury and the removal of the injurious agent. Although this is primarily a protective mechanism, the degree of inflammation can be so severe that at times it is responsible for most of the disease. For example, fever, headaches, and other aches and pains characteristic of many diseases are directly or indirectly due to inflammation.

Regardless of the cause of inflammation, the sequence of events is essentially the same. Immediately upon damage to tissue, the destroyed cells release cellular materials that initiate the inflammatory response. Initially there is a marked dilatation (widening) of blood vessels near the damaged site which results in an increased blood flow to the area. This event is paralleled by an increase in vascular permeability, which results in the leakage of plasma and a few blood cells to the damaged area. The increased amount of fluid in the damaged area causes a swelling called **edema.** Following the leakage of fluid, leukocytes pass through the blood vessels to the damaged site (fig. 18-11). Initially, the leukocytes are predominantly granulocytes. The

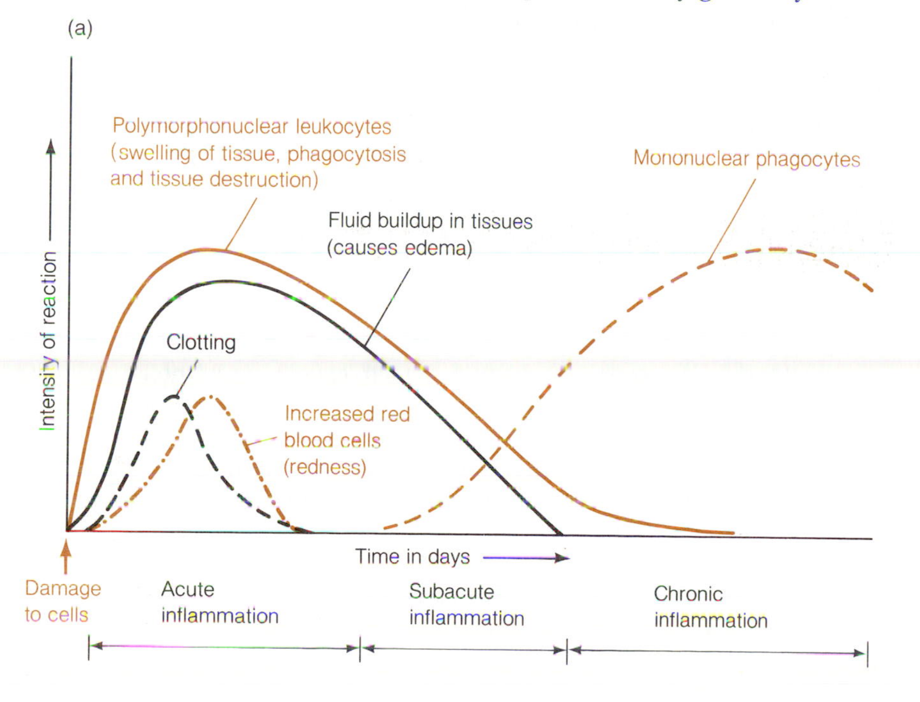

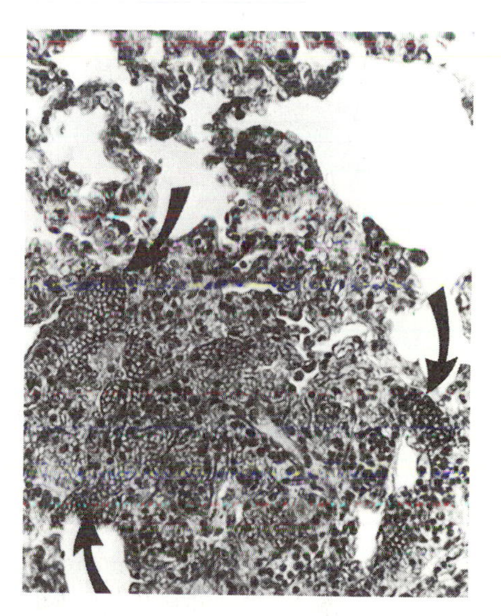

Figure 18-11 Kinetics of Inflammation. *(a)* Inflammation is characterized by redness, swelling, and tenderness of the affected area, often accompanied by an increase in body temperature. These signs reflect the cellular events that take place when tissue is damaged. The redness is due to an accumulation of red blood cells in the area. The swelling is due to fluid (plasma) and leukocyte (PMN) accumulation. During acute inflammation, the predominant cell is the PMN, while during long-lasting (chronic) inflammation macrophages predominate. The pain is due partly to the repair of damaged tissue by leukocytes during the healing process. *(b)* Photomicrograph illustrating acute inflammation of lung. Arrows indicate engorged capillaries with many red blood cells.

neutrophils act primarily as phagocytes, engulfing and digesting microorganisms and damaged host tissue. During this process, the basophils release granules containing **histamine** and **serotonin.** This process, sometimes called **degranulation,** serves to enhance the degree of inflammation. The eosinophils, on the other hand, are thought to play a modulator role in the inflammatory process by releasing antihistaminic substances which counteract, at least in part, the effect of histamine at the site of inflammation.

The presence of endotoxin or other microbial products, or substances produced by the PMN leukocytes themselves, induces fever. These substances, called **pyrogens,** can stimulate the central nervous system (CNS) to elevate the body temperature.

Inflammation is characterized by redness, swelling, heat, and pain. The redness **(rubor)** is due to the increased blood flow to the area of injury. The swelling **(tumor)** is the combined effect of increased extravascular fluid and PMN infiltration in the damaged area. The heat **(calor)** is due to the increased blood flow to the area and to the action of pyrogens, while pain **(dolor)** results from the local tissue destruction and irritation of sensory nerve receptors.

The first cells to arrive at the site of inflammation are usually the PMN leukocytes. This kind of inflammation, which occurs within 15 minutes or so after the damage, is called **acute inflammation** (fig. 18-11). As the inflammation progresses as a result of the presence of infectious microorganisms, the PMN leukocytes are gradually replaced by monocytes and lymphocytes, resulting in **chronic inflammation** (fig. 18-11). This type of inflammation is of longer duration and is accompanied by the development of specific immune responses.

Phagocytosis

Phagocytosis is a very important host defense mechanism that is carried out by phagocytes such as polymorphonuclear leukocytes, monocytes, and macrophages. It rids the body of harmful microorganisms and dead or dying cells (fig. 18-12). Particulate matter, whether dead or alive, that enters host tissues, is taken up by cells (phagocytized) and then destroyed intracellularly.

Phagocytosis is a continuous process that culminates in the ingestion of a particle and its subsequent intracellular digestion. For convenience, the process has been subdivided into four stages: chemotaxis, attachment (adherence), ingestion, and digestion. **Chemotaxis,** the first step in phagocytosis, involves the migration of the phagocyte toward its prey. A random collision between the phagocyte and the prey could occur, but usually the phagocytes are attracted to the prey by chemotactic substances, which may be microbial or host-derived products. As an example, consider the dimorphic fungus *Coccidioides immitis*, which causes a pulmonary disease called valley fever □. Inside a host, the fungus develops **endosporulating spherules.** The mature spherules are not particularly chemotactic, but when they rupture they release endospores which, along with other fungus-derived substances inside the spherule, attract PMN leukocytes. The process of chemotaxis is also promoted by certain host components. For example, certain complement components, such as C3a, C5a, and C567, attract phagocytes. Damaged host cells also release chemotactic substances. The influx of PMN leukocytes from the peripheral circulation into the site of injury during inflammation is due to the chemotactic substances released by damaged cells.

600

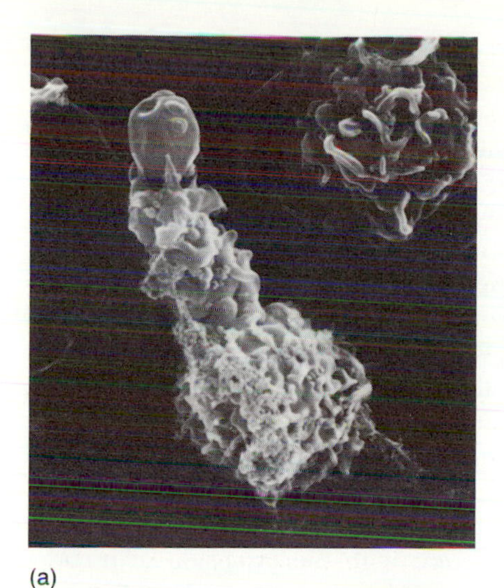

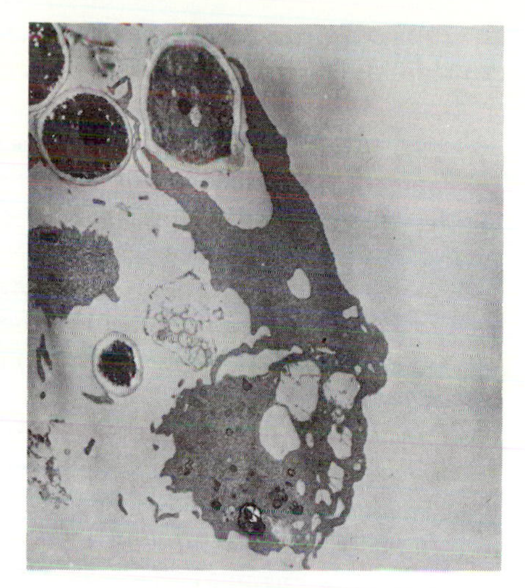

(a)

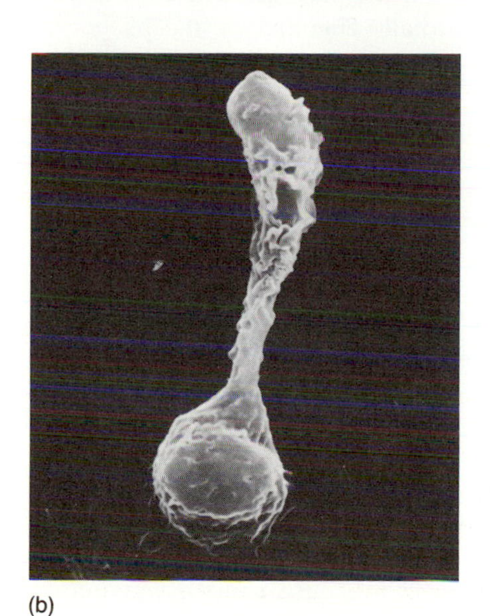

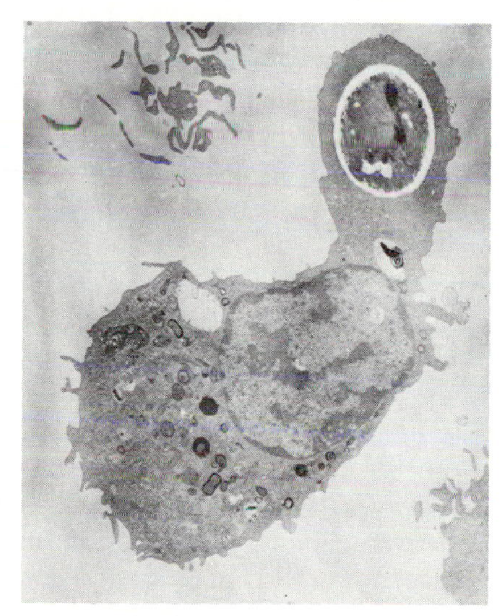

(b)

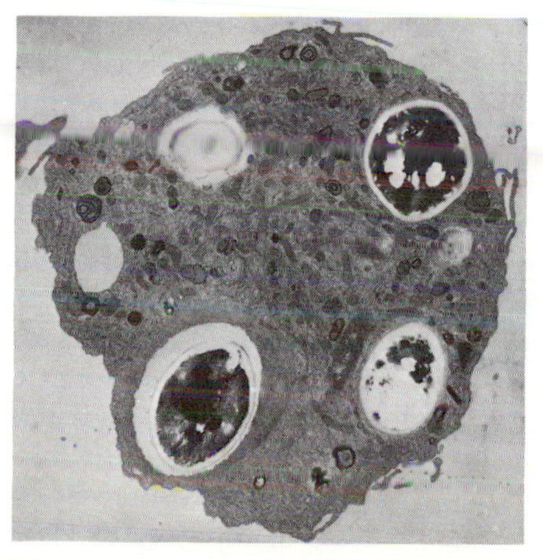

(c)

Figure 18-12 The Process of Phagocytosis. *(a)* Transmission and scanning electron micrographs of alveolar (lung) macrophages attaching to a yeast cell by extended pseudopods. *(b)* Transmission and scanning electron micrographs of alveolar (lung) macrophages of various stages of phagocytosis of the yeast cells. Note the lack of subcellular organelles in the pseudopod and that the pseudopod is being withdrawn into the body of the macrophage. *(c)* Transmission electron micrograph of alveolar (lung) macrophage containing several yeast cells. The yeasts have now been completely drawn into the cell and are digested within phagosomes. Note that the yeasts inside the phagosomes are in various stages of digestion.

Attachment is the necessary next step in phagocytosis: there must be an intimate contact between the phagocyte and the microorganism (or particle) in order for phagocytosis to take place. In fact, microorganisms that inhibit the attachment of the phagocytes to their cell surfaces generally are resistant to phagocytosis. For example, encapsulated strains of *Streptococcus pneumoniae* are not readily ingested by phagocytes because of their capsules, which do not permit the attachment of the phagocyte to the bacterial surface. *Streptococcus pyogenes* **M protein** inhibits the attachment of phagocytes to their surfaces. To improve the attachment of phagocytes to microbial surfaces, the host produces **opsonins:** proteins, either antibodies □ or components of complement, that promote the attachment of the phagocyte to its prey. Occasionally, phagocytes trap foreign particles or microorganisms against the surface of a vessel or organ, thus enhancing the cell-to-cell contact.

463

Engulfment of particles involves the extension of pseudopodia by the phagocyte to surround the prey. Concurrent with the extension of pseudopodia around the prey is the invagination of the plasma membrane around the particle. Eventually, the particle is totally engulfed by the phagocyte. Together, the particle and the membrane that surrounds it are called the **phagosome** (fig. 18-12).

Killing and digestion constitute the last and most important step in phagocytosis. The best-known mechanism of killing and digestion occurs in the neutrophil, which contains two different types of granules: **azurophilic** granules and **secondary** granules. The azurophilic granules are **lysosomes,** which contain acid hydrolases, lysozyme, cationic proteins, collagenase, and myeloperoxidase (MPO). The secondary granules contain lysozyme, lactoferrin, alkaline phosphatase, collagenase, and vitamin B_{12} binding protein. When the components of these two types of granules are introduced into the phagosome, they are very effective microbiocidal agents.

Phagolysosome Formation

When the lysosomes fuse with the phagosomal membrane, the resulting structure is called a **phagolysosome** (fig. 18-12). The lysosomal products act on the ingested microbe or particle. Enzymes such as lysozyme and cationic proteins kill microorganisms; another protein, lactoferrin, impairs microbial growth within the phagolysosome by sequestering iron, which is needed for microbial growth. The enzyme myeloperoxidase, in conjunction with hydrogen peroxide (H_2O_2) and halide ions (e.g., I^- or ClO^-), is very effective in killing microbes. During phagocytosis by PMNs, O_2 uptake increases, a process known as the **respiratory burst.** An enzyme called NADPH oxidase reduces the O_2 taken up during the respiratory burst to produce hydrogen peroxide. Other microbiocidal mechanisms involve superoxide ions and hydroxyl radicals as the killing agents. The digestion of killed cells within the phagolysosome is carried out by hydrolases, such as nucleases, lipases, elastases, collagenases, and proteases found in lysosomes.

The macrophage is the primary phagocytic cell of the **reticuloendothelial system (RES).** The RES comprises a group of cells and a network of loose connective tissue (recticulum) that serves to filter out and destroy foreign particles and damaged host material that may be present in the blood or body tissues. The macrophages are either fixed to organ tissues or wandering throughout the lymphatics and blood. The fixed macrophages are found in organs such as the spleen, liver, tonsils, lymph nodes, bone marrow, and brain. The wandering macrophages are seen throughout the lymphatics, blood, peritoneal cavity, and lungs. The macrophages have special names, depending upon the site in which they are found. For example, liver macrophages are also called **Kupffer cells;** lung macrophages are called **dust cells;** and brain macrophages are also called **microglial cells.**

The mechanism of phagocytosis by the macrophage is similar to that by the PMN. Certain microbes are known to inhibit macrophage degranulation and therefore can live within these phagocytic cells. The microbiocidal action of the macrophage can be enhanced significantly in the **activated macrophage.** Apparently, an interaction between the macrophage and specific host defense mechanisms renders the macrophage an extremely effective microbiocidal agent.

Many parasitic infections are characterized by a condition called **eosinophilia,** an increase in the number of eosinophils in the blood. It has been suggested that the eosinophils may play a role in extracellular digestion. Eosinophils may adhere to the surface of a parasite (such as a roundworm or a fluke) and degranulate. This process, which has been called **exocytosis,** involves the release of lysosomal granules so that they act on the surface layers of the parasite and damage or kill the worm.

Specific Immunity

In addition to the nonspecific host defense mechanisms previously described, the vertebrate host has the ability to attack specifically the invading mi-

FOCUS

METCHNIKOFF ON PHAGOCYTOSIS

During the golden age of microbiology, the concept that infectious diseases were caused by specific microorganisms grew considerably in popularity in a few short years. Scientists throughout the world searched for, and often found, the microbes that were responsible for many of the maladies that afflicted the human race. No sooner was the connection made between infectious diseases and microorganisms than scientists began asking how the survivors of a disease fight off microbes.

One theory, the **phagocytic theory,** was advanced by Elie Metchnikoff in 1884. The Russian-born Metchnikoff theorized that, when harmful intruders invade animals, certain host cells move toward the invaders and engulf them. According to Metchnikoff, these cells, called **phagocytes,** are constantly on the lookout for intruders and represent the main line of defense against invading microbes.

Metchnikoff's phagocytic theory grew from his observation that motile cells of starfish larvae move toward and accumulate around rose thorns, which he had used to impale the larvae. The development of Metchnikoff's theory was based on numerous observations in various areas of biology and occupied 25 years of his life.

The most convincing evidence for the phagocytic theory came from studies of a disease that afflicts the water flea *Daphnia.* The disease Metchnikoff studied is caused by a fungus he called *Monospora bicuspidata.* Metchnikoff showed that spores of *Monospora* were quickly surrounded, and often killed, by the water flea's white blood cells (phagocytes). Metchnikoff's study revealed that in some cases, when phagocytes did not respond to the invaders, the fungi would proliferate and kill *Daphnia.* These studies emphasized the importance of phagocytes in protecting the host against disease.

Metchnikoff's studies on phagocytosis created a foundation for the science of *cellular immunology* (immunity conferred by cells). Cellular immunology represents only one part of the complex immune system that protects animals from invading microorganisms.

crobes. This type of immunity, called specific immunity, is mediated by serum proteins called **immunoglobulins** (antibodies) lymphocytes, and plasma cells. The macrophage also plays an important role in this process. Specific immunity mediated by immunoglobulins is called **humoral immunity** □, while that mediated by cells is called **cell-mediated immunity** □. These types of immunity are the subject matter of Chapter 19.

466
468

EVASIVE STRATEGIES OF PARASITES

The development of a parasitic existence requires that the parasite adapt to the host's environment. Not only must the parasite adhere to and penetrate the host, but it must also establish a long-standing relationship with the host. This task is not particularly easy, in view of the vertebrate host's powerful arsenal of protective responses. A parasite must either evolve mechanisms to evade the host response or perish. Table 18-6 summarizes some of the evasive strategies used by parasites. The ensuing discussion will attempt to describe some, but not all, of the strategies that successful parasites may use to evade host defenses.

Encystation or the formation of an impervious structure around the cell, serves as a mechanism to avoid host antagonistic responses. For example, the cysts of certain parasitic protozoa, such as *Entamoeba histolytica* and *Giardia intestinalis*, have a tough outer layer that makes them resistant to gastric juices and other host-protective responses. Thus, encystment allows them to pass unaffected into the intestines, where they can excyst and initiate an infection.

The **development of a complex life cycle** can afford the infectious agent several avenues for escaping the host response. For example, specific immunity directed toward one stage of the parasite life cycle will likely not be effective against another stage. This is the case with many protozoans, such as the malaria parasite, and parasitic worms, such as the flukes and roundworms.

TABLE 18-6
SOME STRATEGIES PARASITES USE TO EVADE HOST RESPONSES

EVASIVE STRATEGY	FUNCTION	EXAMPLE(S)
Encystation	Forms a resistant outer shell to avoid host attack	*Giardia intestinalis* *Entamoeba histolytica*
Multiple forms/stages	Presents the host with a wide array of antigens to which it has to react; allows pathogen to get a foothold on host tissue	*Malarial parasites (Plasmodium), Trypanosoma* *Leishmania* sp. dimorphic fungi
Anti-phagocytic substances	Inhibits phagocytosis	*Staphylococcus aureus* *Histoplasma capsulatum*
Intracellular parasitism	Protects from host defenses	*Mycobacterium tuberculosis* *Rickettsia rickettsii*
Leukocydin production	Kills phagocytes	*Staphylococcus aureus*
Inhibition of immune response	Prevents development of specific immunity	Malarial parasites *(Plasmodium)*
Changes in antigenic composition	Prevents host from developing an effective immune response	*Trypanosoma brucei gambiense*

Antiphagocytic defenses are also evident among parasites. The polysaccharide capsule of the pneumococcus, the A protein of staphlococci, and the M protein of streptococci are three examples of antiphagocytic strategies. Certain intracellular parasites are readily phagocytised, but they prevent the formation of phagolysosomes and therefore can live and multiply within the phagosome. In this way, the intracellular parasite can live within host cells and be protected from other host antagonistic actions.

Many parasites secrete substances that can **inhibit or depress the immune response.** The malaria parasite apparently secretes substances that depress the immune response, particularly the humoral immune response. This may be an adaptation to its mode of reproduction: since its life cycle involves the infection of blood cells, the invading parasite must make its way into the circulation and be exposed to antimalarial immunoglobulins. By reducing the amount of immunoglobulins produced by an infected host, the parasite increases its chance of reaching and infecting an erythrocyte before being destroyed.

An outstanding example of evasive strategy, in which there are **periodic changes in antigenic composition** can be illustrated with the flagellated protozoan *Trypanosoma gambiense*, the organism that causes African sleeping sickness. One characteristic feature of this disease is the periodic appearance of the parasite in the blood and an abnormally elevated serum immunoglobulin concentration (IgM). The appearance of the parasite in the blood is paralleled by fever, and its disappearance relieves the fever. Apparently when the parasite appears in the blood it induces a specific immune response, which then drives the parasite out of the blood and into the lymph nodes. There, the parasite changes the composition of its surface layers and reappears in the blood. The immunity against the "previous" parasite is not effective. The parasite then induces an additional humoral immune response. Thus, the appearance and disappearance of the parasite in the blood is accompanied by a change in the chemical composition of its surface layers. In this way, the parasite can withstand long periods of parasitic existence within a host. The shift in antigenic composition is thought to result from transposable elements (jumping genes, Chapter 10) in the trypanosome chromosomes.

SUMMARY

This chapter summarizes the interactions between hosts and parasites and the consequences of such interactions.

SOME DEFINITIONS

1. Symbiosis, or "living together," signifies close association between two or more organisms whose biological activities are closely linked.
2. Mutualistic relations are generally permanent symbiotic associations in which both symbionts benefit from the association.
3. Commensalistic associations are symbioses in which one of the symbionts benefits while the other is unaffected.
4. Parasitic associations are symbioses in which one of the symbionts benefits at the expense of the other.
5. Infections result from the establishment and proliferation of a parasite in or on host tissues.
6. Infectious diseases result when an infecting parasite causes detrimental changes in the host.
7. Pathogenicity is the ability of microorganisms to cause disease. This ability is due, for the most part, to inheritable traits of the microorganism.
8. Virulence is a measure of a microorganism's degree of pathogenicity. Virulent strains cause disease, while avirulent strains do not. The virulence of a strain is usually measured as the LD_{50} or the ID_{50}.

THE INFECTIOUS PROCESS

1. In order for a microorganism to infect a host, it must enter the host, attach itself to host surfaces,

and begin to multiply. Only microorganisms that can successfully carry out these steps are capable of causing an infection.

2. Entrance into a host can be accomplished through a break in the skin or through the mucous membranes of the respiratory, urogenital, or digestive system.
3. Attachment by pathogens to host surfaces can be mediated by various mechanisms, such as recognition of receptor sites on host membranes, pili, capsules, or glycocalyx.
4. Multiplication of the pathogen can take place either inside or outside host cells, depending upon the infecting microorganism.

PATHOGENIC PROPERTIES OF MICROORGANISMS

1. Microorganisms can cause disease either by invading host tissues and causing detrimental changes or by releasing toxic substances.
2. Intoxications are diseases resulting from the action of a toxin produced by a microorganism. These toxins could be proteins with specific modes of action, which are called exotoxins, or by lipopolysaccharides which are called endotoxins.
3. Infectious diseases are those that result from invasion of host tissue by a parasite. Exoenzymes such as collagenases, hyaluronidases, or proteases can be involved in the invasive process.

HOST DEFENSE MECHANISMS

1. The host has numerous mechanisms to prevent infectious diseases. These mechanisms can be specific for the invading microbe or nonspecific and effective against a variety of microorganisms. Defense mechanisms may involve physical barriers, chemical barriers, or biological barriers.
2. Physical barriers include the skin and mucous membranes, and flushing mechanisms such as coughing, sneezing, urinating, and tearing.
3. Chemical barriers are nonspecific defense mechanisms involving chemicals that are active against a variety of microorganisms. These include interferon, complement, and B-lysins.
4. Biological barriers may be nonspecific or specific. The processes of inflammation and phagocytosis are nonspecific defense mechanisms aimed at the removal and destruction of noxious microorganisms.
5. The process of phagocytosis involves the migration of phagocytes to the site where the microorganisms are causing the damage; the engulfment of the parasite by the phagocyte; and the subsequent destruction of the invader within the phagocyte.
6. The normal flora serves a useful function, in that it protects its host from invading pathogens and provides the host with some necessary nutrients such as vitamins and amino acids.
7. Specific immunity can be mediated by antibodies or by certain lymphocytes and the activated macrophage.

EVASIVE STRATEGIES OF PARASITES

Parasites evolve in conjunction with their hosts and develop mechanisms to avoid host defenses. These evasive mechanisms include : a) forming resistant structures, such as cysts; b) developing multiple life stages; c) evolving antiphagocytic defenses; d) attenuating immune responses; and e) changing antigenic composition.

STUDY QUESTIONS

1. Define symbiosis. Describe three different types of symbiotic associations and give an example of each.
2. Differentiate between:
   a. infection and disease;
   b. pathogenicity and virulence;
   c. LD_{50} and ID_{50};
   d. exotoxin and endotoxin;
   e. toxin and toxoid;
   f. viral inactivation and viral attenuation.
3. Explain how a microorganism can cause disease by invading host tissue.
4. Explain how a microorganism can cause disease by producing toxins.
5. Explain how a microorganism can cause disease by eliciting an immune response.
6. Define what is meant by "normal flora" and how it protects the host from infectious diseases.
7. What is inflammation? What role does it play in health? In disease?
8. Describe five different strategies that parasites may employ to evade host defense mechanisms.
9. Describe the process of phagocytosis.
10. What are opportunistic pathogens?

SUPPLEMENTAL READINGS

Burnet, M. and White, D. B. 1972. *Natural history of infectious diseases.* 4th ed. London: Cambridge University Press.

Boxer, G. J., Curnutte, J. T., and Boxer, L. A. 1985. Polymorphonuclear leukocyte function. *Hospital Practice* 20(3):69–90.

Dautry-Varsat, A. and Lodish, H. 1984. How receptors bring proteins and particles into cells. *Scientific American* 250(5):52–58.

Donelson, J. E. and Turner, M. J. 1985. How the trypanosome changes its coat. *Scientific American* 252(2):42–52.

Hood, L. E., Weissman, I. L., and Wood, W. B. 1978. *Immunology.* Menlo Park, Calif.: Benjamin/Cummings Publishing Co.

Horwitz, M. A. 1982. Phagocytosis of microorganisms. *Review of Infectious Diseases* 4:104–108

Iglewski, W. J. 1984. Critical roles for mono (ADP-ribosyl) transferases in cellular regulation. *ASM News* 50(5):195–198.

Lewin, R. 1984. New regulatory mechanism of parasitism. *Science* 226(4673):427.

Mechnikov, I. 1908. *The prolongation of life. Vol. 5, Lactic acid as inhibitor of intestinal putrefaction.* New York: G. P. Putnam's Sons.

Middlebrook, J. L. and Dorland, R. B. 1984. Bacterial toxins: cellular mechanisms of action. *Microbiological Reviews* 48(3):199–221.

Ranney, D. F. 1985. Targeted modulation of acute inflammation. *Science* 227:182–184.

Root, R. K. and Cohen, M. S. 1981. The microbiocidal mechanism of human neutrophils and eosinophils. *Review of Infectious Diseases* 3:565–568

Smith, H., Skehel, J. J., and Turner, M. J. 1980. *The molecular basis of microbial pathogenicity:* Deerfield Beach, Fla.: Verlag Chemie.

Taylor, P. W. 1983. Bacteriocidal and bacteriolytic activity against gram negative bacteria. *Microbiological Reviews* 47:46–65

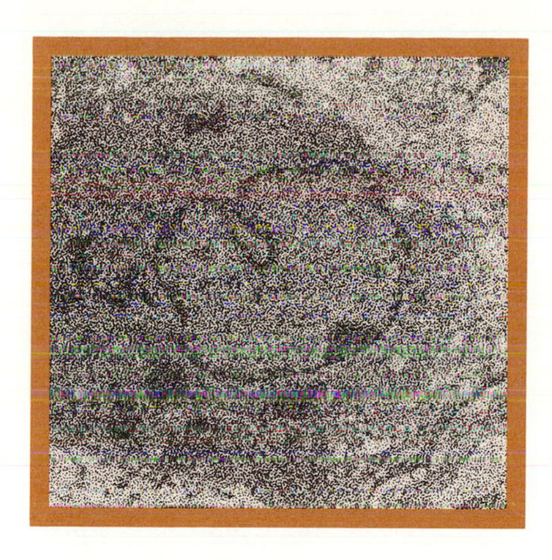

CHAPTER 19
THE IMMUNE RESPONSE

CHAPTER PREVIEW

IMMUNOLOGY, A BENEVOLENT SCIENCE

The science of immunology (the study of immune systems) developed from the experiments conducted by scientists such as Edward Jenner. Jenner's experiments in 1798 demonstrated that vaccination of individuals with the mildly virulent cowpox agent (a virus) could protect vaccinated individuals from smallpox, a disease that scarred and killed many humans until quite recently. Approximately 80 years later, Pasteur showed that the bacterium that causes anthrax (a disease that killed many farm animals as well as humans) could be tamed (attenuated) and used to induce protective immunity to the disease-causing bacterium in vaccinated animals. After this discovery, Pasteur developed vaccines against swine erysipelas, chicken cholera, and rabies.

Immunology has grown considerably in importance since its origin in the 18th century. Discoveries regarding the initiation and expression of the immune response are of fundamental importance in biology, because they have led, directly or indirectly, to the improvement of human health and the treatment of infectious diseases. The development of immunological tests has been extremely important in medicine, because these tests allow physicians to diagnose infectious diseases before debilitating or fatal illnesses have resulted.

It is as true now as it was in the eighteenth century: immunology is a pertinent science that has a direct impact on human welfare and health. In this chapter we will discuss the basic concepts of immunology and how these concepts can be put to practical use in the detection and prevention of disease.

LEARNING OBJECTIVES

A STUDY OF THIS CHAPTER SHOULD ENABLE YOU TO:

OUTLINE THE VARIOUS TYPES OF IMMUNITY AND GIVE AN EXAMPLE OF EACH

CITE THE CHARACTERISTICS OF ACQUIRED IMMUNITY

LIST THE SALIENT CHARACTERISTICS OF ANTIGENS AND ANTIBODIES AND DIAGRAM THE BASIC STRUCTURES OF EACH

EXPLAIN THE BASIS FOR ANTIBODY SPECIFICITY

OUTLINE THE VARIOUS TYPES OF CELLS THAT PARTICIPATE IN IMMUNITY, AND STATE THEIR ROLE IN THE DEVELOPMENT OF SPECIFIC IMMUNITY

UNDERSTAND THE DIFFERENCES BETWEEN ANTIBODY AND CELL-MEDIATED IMMUNITY

DESCRIBE WHAT VACCINES ARE AND HOW THEY WORK

Immunology is the study of immune systems. Immune systems consist of a number of different types of cells and their products whose primary function is to protect animals from invading microorganisms and from cancerous growths. Without part or all of the immune system, viruses, bacteria, fungi, protozoans, and cancerous cells would kill animal hosts. Normally, these invaders are inhibited or killed by the immune system. The purpose of this chapter is to acquaint you with the various cells of the immune system and the substances they produce to block the multiplication and spread of microorganisms in the body.

TYPES OF IMMUNITY

Immunity to infectious agents occurs when antimicrobial substances are produced by the body to combat a pathogen, or when various types of immune cells attack and kill infecting organisms (table 19-1). Certain antimicrobial substances, called **antibodies** or immunoglobulins, are complex proteins produced by **plasma cells** (antibody-forming cells) which bind specifically to various molecules and microorganisms. Immunity in vertebrate hosts (including humans) is expressed in two distinct but interrelated ways: **humoral immunity** and **cell-mediated immunity.** Both these types of immunity, working in unison, provide the host with a very effective means of preventing infections and aborting infectious processes. Immunity due to antibodies circulating in the lymph and blood is known as humoral immunity, while immunity due to various groups of cells is called cell-mediated immunity (table 19-1). Genetic and environmental factors may also determine whether or not an animal remains healthy.

Immunity is often referred to as either **native** or **acquired immunity**. Native immunity, the resistance that an animal has at birth, is not dependent upon antibodies or immune cells. Native immunity is often due to the chemical makeup or physiological properties of an animal which make it impossible for a microorganism to invade and reproduce in that animal. For example, humans are not subject to infections by bacteriophages or by the

TABLE 19-1
TYPES OF IMMUNITY

TYPES OF IMMUNITY	EXAMPLE
I. Native (hereditary)	
Species	Resistance of chickens to anthrax
Individual	A person's resistance to a disease
II. Acquired	
Natural	
Active	Immunity to second dose of mumps virus
Humoral	Immunity effected by antibodies
Cellular	Immunity effected by cells
Passive	Immunity to disease transmitted to babies in mother's milk
Artifical	
Active	Immunity acquired from vaccination
Humoral	Immunity effected by antibodies
Cellular	Immunity effected by cells
Passive	Treatment of tetanus by injecting antitoxin from horses

tobacco mosaic virus, because the plasma membranes of human cells lack the necessary receptor sites to which these viruses attach. Even if these viruses could enter human cells, it is unlikely that they could reproduce within them, because human enzymes for nucleic acid replication, transcription, and translation do not function with the viral nucleic acids. Another example of native immunity is the natural resistance of many birds against infections by the anthrax bacillus. The bacterium that causes anthrax *(Bacillus anthracis)* cannot reproduce at the body temperatures of birds and hence cannot infect them. Even within a species, some individuals are immune to a disease while others are not, because of slight differences in their physiologies. Factors that affect the physiology of an animal include its sex, age, nutrition, genetic background, and overall health.

Acquired or adoptive immunity is the specific immunity acquired before or after birth from humoral antibodies or from cells directed toward a specific pathogen. When the acquired immunity develops after the host comes in contact with a foreign substance in its body, the resulting immunity is called **active immunity.** Active immunity that develops when a host encounters an infectious agent is called **naturally acquired active immunity,** while active immunity that develops as a consequence of injection (vaccination) with a microorganism or its products is called **artificially acquired active immunity.**

If the acquired immunity is due to the acquisition by the host of antibodies or immune cells that were made in another host, it is called **passive immunity.** Passive immunity acquired when humoral antibodies pass from mother to fetus in the womb is called **naturally acquired passive immunity.** Babies frequently are resistant to many infectious diseases because their mothers transfer antibodies and other antimicrobial substances to their offspring in the milk. This resistance is of short duration, however, and the baby must develop its own immunity to achieve long-lasting protection. Passive immunity acquired by injection of antibodies (or immune cells) is called **artificially acquired passive immunity.** Some infectious diseases such as botulism and tetanus, which result from the activity of potent toxins on host cells, frequently are treated by administering antibodies against the toxins, because the patient is not able to form antibodies against the toxins in order to avert the disease.

CHARACTERISTICS OF ACQUIRED IMMUNITY

Acquired immunity is extremely important in protecting animals against infectious disease. It is characterized by **memory, specificity,** and **recognition of "non-self".**

Memory allows the immune system to respond swiftly and vigorously to an infectious agent that has previously been encountered. For example, the organisms that cause childhood diseases such as chickenpox and mumps seldom attack an individual more than once. An individual develops an effective, long-lasting immune response when first infected, and this first immune response is largely responsible for the clinical cure of the disease. When such an individual becomes infected with the same agent at a later date, the body "remembers" the infectious agent, because it developed long-lived "memory" cells that stimulate an immediate, powerful secondary immune response (fig. 19-1). This secondary immune response, sometimes

called the **anamnestic immune response,** is characterized by a more rapid and intense response than the primary response. The anamnestic response can be studied by following the production of antibodies when an animal is injected with a foreign substance **(antigen).** Any material that stimulates the production of immunoglobulins □ against itself is known as an antigen. 463 When an antigen such as the tetanus toxoid (tetanus vaccine) is injected into an animal, there is a lapse of several days before any antibodies against the toxoid can be detected in the blood's serum. The concentration of antibodies in the serum gradually increases, peaks out, and then begins to decline (fig. 19-1). After three to four weeks, the level of antibodies is too low for detection. If, however, the same animal is injected again five to ten weeks after the first injection (a "booster" shot), the concentration of antibodies in the serum will increase rapidly within two or three days and will reach a much higher level than after the first injection (fig. 19-1).

The dynamics of antibody production helps us understand why the primary invasion of the human body by a pathogen, such as the chickenpox or mumps viruses, can result in disease, while the secondary invasion of these pathogen has little effect on a host. Pathogens that reproduce rapidly in an unsensitized host may kill the host within a week after infection. The reason is that the immune system in these individuals does not have a chance to get going, since antibody levels generally do not reach effective serum concentrations for at least two weeks. When a previously sensitized host is invaded for the second time by the same pathogen, however, the antibodies in the blood reach effective levels within a short period of time (less than a week), aborting the infection.

The anamnestic response is the principle underlying vaccination. When children are vaccinated with antigens, such as the tetanus and diphtheria toxoids, they are stimulated to develop an anamnestic response so that they can react swiftly and effectively in the event of an infection.

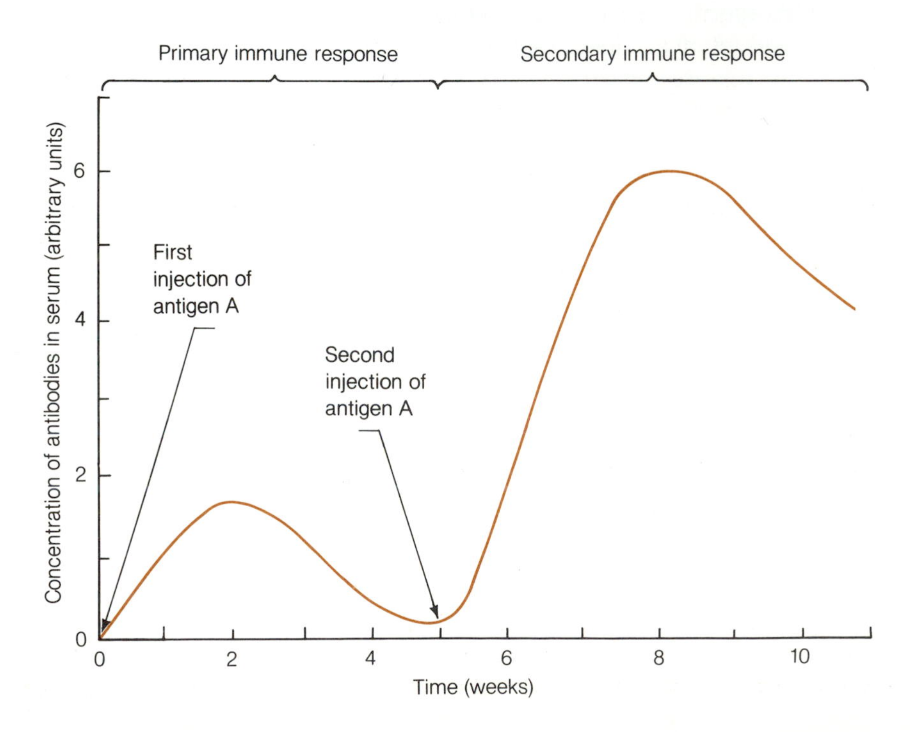

Figure 19-1 Kinetics of the Immune Response. Upon first encounter with a given antigen, the immune response is not detected immediately. The immune response develops over a period of 1 to 3 weeks. As time goes on, the level of immunity gradually declines. When the immune system encounters the same antigen for the second time, it reacts quickly and vigorously. More antibodies are produced and for a much longer period of time than after the first encounter with antigen. The second encounter with antigen is referred to as the secondary or anamnestic response.

Another characteristic of the acquired immune response is its **specificity** □. When an individual is vaccinated against a disease such as mumps, the immune response that develops is specific against the mumps virus. This is because the memory cells that arise after the first bout of mumps respond only to the mumps virus. If an individual who had mumps at an earlier date becomes infected by another virus, such as the rubella virus, the immunity against mumps will not be of use against rubella and the individual may contract rubella. Any subsequent encounter with the rubella virus, however, results in the development of a rapid, specific, and powerful immune response.

464

The normal immune response distinguishes between "self" and "nonself." Since the immune response can be induced by a variety of large molecules, it is theoretically possible that the immune system could attack the host and cause severe damage throughout the body. For this reason, the immune system becomes nonresponsive to "self" antigens. It is believed that this nonresponsiveness develops during fetal life. Sometimes, however, the mechanisms that make the immune system unresponsive to self do not function properly and the individual develops an immune response to some part of the body. This response results in what is known as an **autoimmune disease** □. Some diseases, such as systemic lupus erythematosus and rheumatic fever result when the body's defenses are directed against self antigens. The next chapter will be concerned with some of the most common autoimmune disorders that afflict humans.

493

CELLS OF THE IMMUNE SYSTEM

Several types of specialized cells are responsible for an animal's immunity (table 19-2). These cells recognize the presence of foreign substances; interact with other cells to initiate an immune response; produce substances that interact specifically with the antigen; or attack the antigen directly. The most important cells of the immune system are lymphocytes and macrophages, although certain types of granulocytes also play a role in immunity (fig. 19-2). The cells of the immune system begin to develop before the animal is born (fig. 19-3). In the embryo, cells associated with the yolk sac and blood islets in the circulatory system migrate to the fetal liver, where they develop into **stem cells.** Stem cells, in turn, migrate to other fetal tissues such as the thymus, spleen, gut, lymph nodes, and bone marrow, where they differentiate into the cells of the immune system (fig. 19-3). For example, a population of lymphocytes (fig. 19-2) called T-lymphocytes completes its maturation in the thymus. B-lymphocytes, neutrophils, basophils, eosinophils, monocytes, and macrophages develop primarily in the bone marrow, but are also found in the spleen and gut. In mammals, various stem cells that migrate into the bone marrow are stored there and give rise to fresh stem cells, which can repopulate the spleen, gut, and lymph nodes during the animal's life.

B-Lymphocytes

B-lymphocytes (table 19-3), also called bursa-derived lymphocytes (because they were first discovered to mature in an organ called the **bursa of Fabricius** in birds), make up 10–20% of the lymphocytes found in the human

TABLE 19-2
CELLS OF THE IMMUNE SYSTEM AND THEIR FUNCTIONS

CELL TYPE	PROTEINS PRODUCED	FUNCTION OF PROTEINS
Neutrophils	Lysozymes Interferon	Digest cells and degrade macromolecules, Inhibit virus replication
Monocytes	Lysozymes	Digest cells and degrade macromolecules
Macrophages	Lysozymes Complement components (C4, C2) Genetically-related macrophage factor (GRF) Lymphocyte activation factor (LAF)	Digest cells and degrade macromolecules Create holes in plasma membranes (stimulates release of histamin) Stimulates T-lymphocytes (T_H) to secrete immune response proteins Stimulates T-lymphocytes (T_H) to multiply
Plasma cells (B-Lymphocytes)	Immunoglobulins (antibodies): IgM, IgD, IgG, IgA, IgE	Bind to and agglutinate foreign cells and debris, neutralize toxins and viruses, enhance phagocytosis
T-Lymphocytes	Interferon Chemotactic factors Migration inhibitory factors Macrophage activation factors	Inhibits virus replication Attract macrophages and other cells Inhibit departure of macrophages Stimulate macrophages to fuse lysosomes with food vacuoles
T_C & T_K	Killing factor	Kills foreign cells, cancerous cells, and infected cells
T_H	Helper factor	Stimulates development of lymphocytes, initiates immune response
T_S	Suppressor factor	Inhibits development of lymphocytes and immune response
Fibroblasts	Interferon	Inhibits virus replication
Epithelial Gastrointestinal Mucosa	Complement components	Lyse plasma membranes, kill microorganisms
Liver Cells	Complement components	Lyse plasma membranes, kill microorganisms

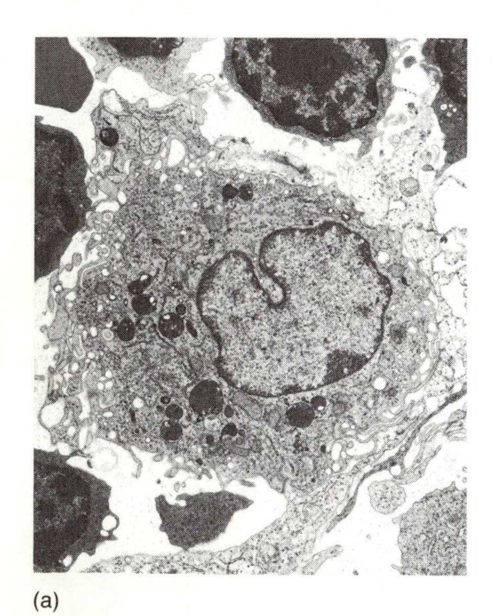

(a)

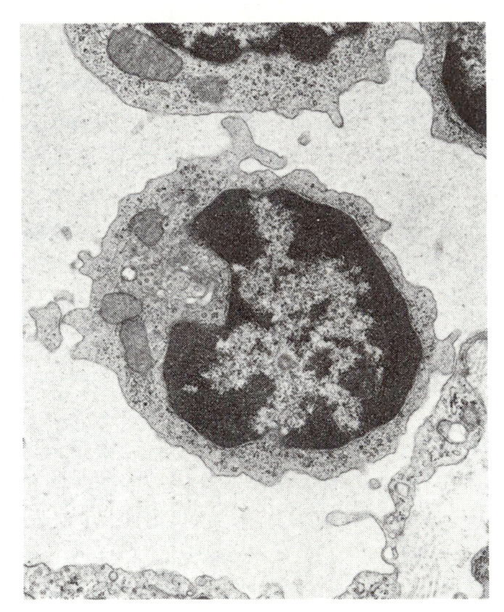

(b)

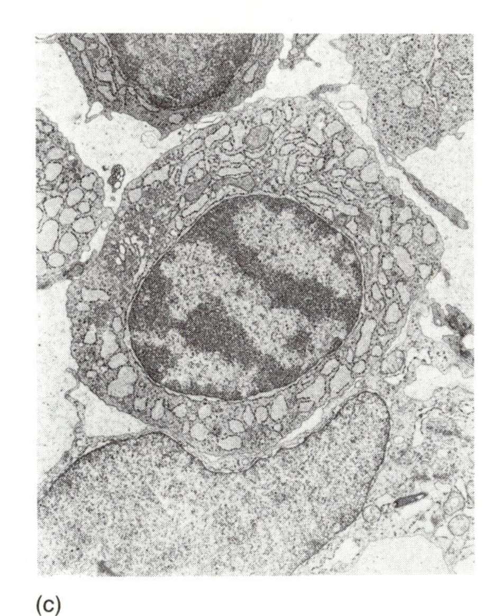

(c)

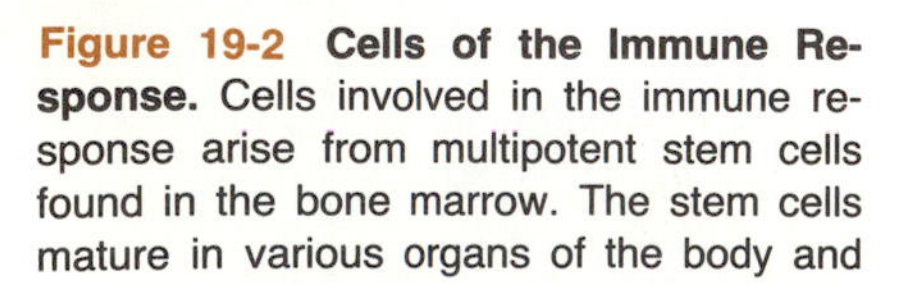

Figure 19-2 Cells of the Immune Response. Cells involved in the immune response arise from multipotent stem cells found in the bone marrow. The stem cells mature in various organs of the body and become effectors of special branches of the immune system. Thymus-processed lymphocytes (T-lymphocytes) are effectors of cell-mediated immunity; bone marrow- and gut-processed lymphocytes (B-lymphocytes) are involved in antibody-mediated immunity; plasma cells are involved in antibody production; and macrophages carry out phagocytosis. (*a*) Macrophage. (*b*) Lymphocyte. (*c*) Plasma cell.

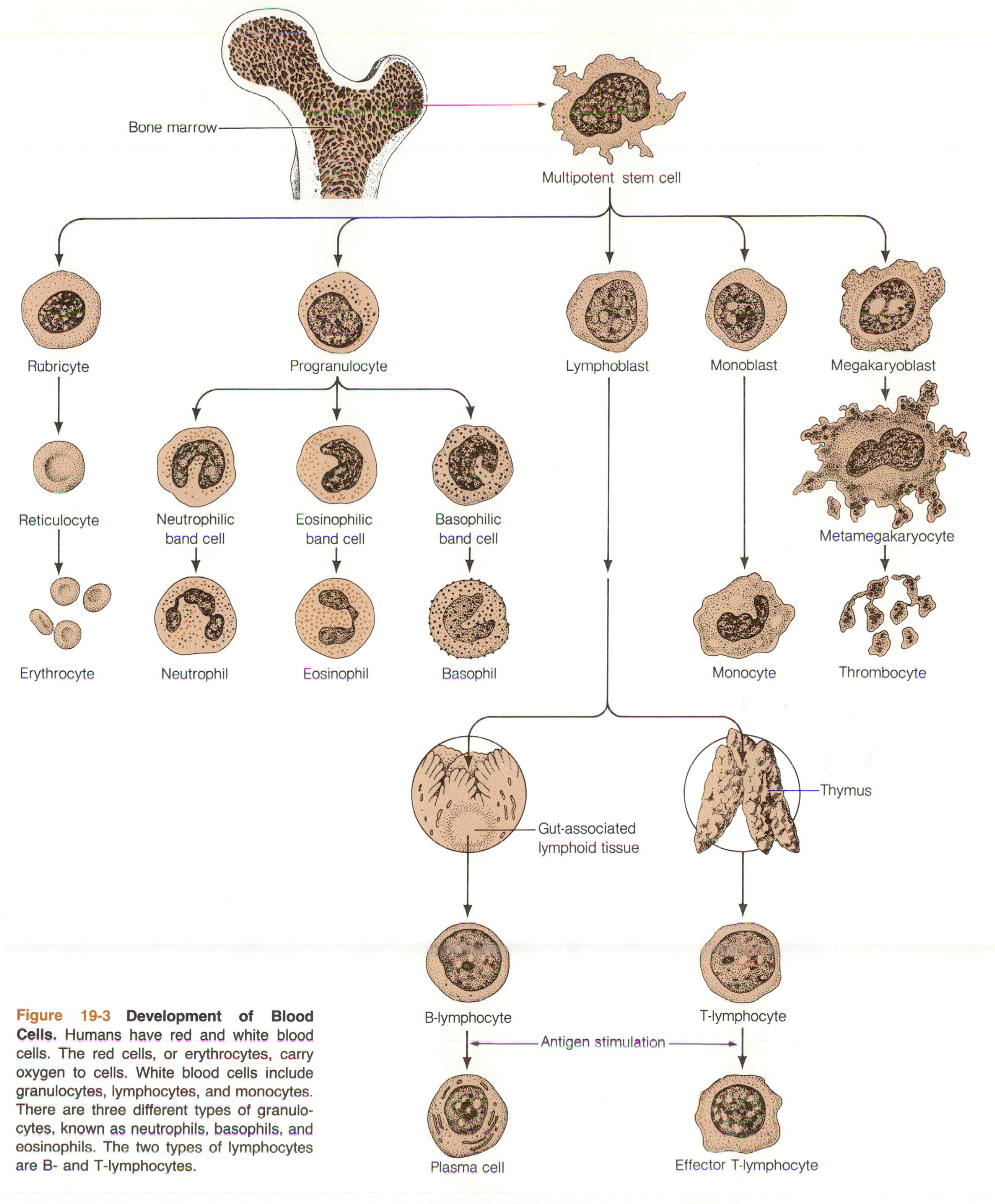

Figure 19-3 Development of Blood Cells. Humans have red and white blood cells. The red cells, or erythrocytes, carry oxygen to cells. White blood cells include granulocytes, lymphocytes, and monocytes. There are three different types of granulocytes, known as neutrophils, basophils, and eosinophils. The two types of lymphocytes are B- and T-lymphocytes.

TABLE 19-3
CHARACTERISTICS OF T- AND B-LYMPHOCYTES

CHARACTERISTIC	T-LYMPHOCYTE	B-LYMPHOCYTE
Organ where they mature	Thymus	Bone marrow
Function	Cell-mediated immunity Antigen recognition Modulate immune response	Humoral immunity
Proportion of lymphocytes	80%	20%
Membrane antigens (Markers)	Theta antigen Ly antigen	Fc receptor
Differentiate into:	Helper cells Memory cells Suppressor cells Cytotoxic cells Killer cells	Plasma cells Memory cells
Products	Lymphokines Cytotoxin	Immunoglobulins

blood. These cells are also found in the bone marrow, spleen, and gut (fig. 19-4). B-lymphocytes are involved in the development of humoral immunity; they develop into **plasma cells** (fig. 19-2) when stimulated by the presence of an antigen. The plasma cells, in turn, secrete antibodies that react specifically with these antigens. In birds, B-lymphocytes develop in a lymphoid organ called the bursa of Fabricius, which is associated with the gut. The lymphoid organs of mammals are not well characterized, although the bone marrow, the **Peyer's patches** in the intestines, and the mesenteric lymph nodes are the lymphoid organs thought to be the sites of B-lymphocyte maturation.

T-Lymphocytes

The **thymus-derived lymphocytes,** called T-lymphocytes, constitute approximately 70% of the lymphocytes in the human blood. The T-lymphocytes perform a variety of functions in the immune system (table 19-3), including antigen recognition, T and B cell cooperation in the development of an immune response □, and cytotoxicity. The T-lymphocytes include several subpopulations with different functions: cytotoxic cells (T_C), helper cells (T_H), suppressor cells (T_S), and killer cells (T_K) (table 19-2). The Tc lymphocytes secrete proteinaceous cytotoxic substances that kill or inactivate target cells, such as cancer cells or cells invaded by microorganisms. Killer cells (T_K), produce toxic proteins that destroy foreign, cancerous, and virus-infected cells. Apparently, T_C and T_K cells must be attached to their targets before they can release their toxins. The T_H lymphocytes release proteins that stimulate or "help" the B-lymphocytes to differentiate into immunoglobulin-producing plasma cells. The T_H lymphocyte also participates in cell-mediated immunity by releasing substances called **lymphokines,** which recruit other immune cells to participate in the immune response. The T_S lymphocytes release proteins that are believed to suppress the release of stimulatory pro-

469

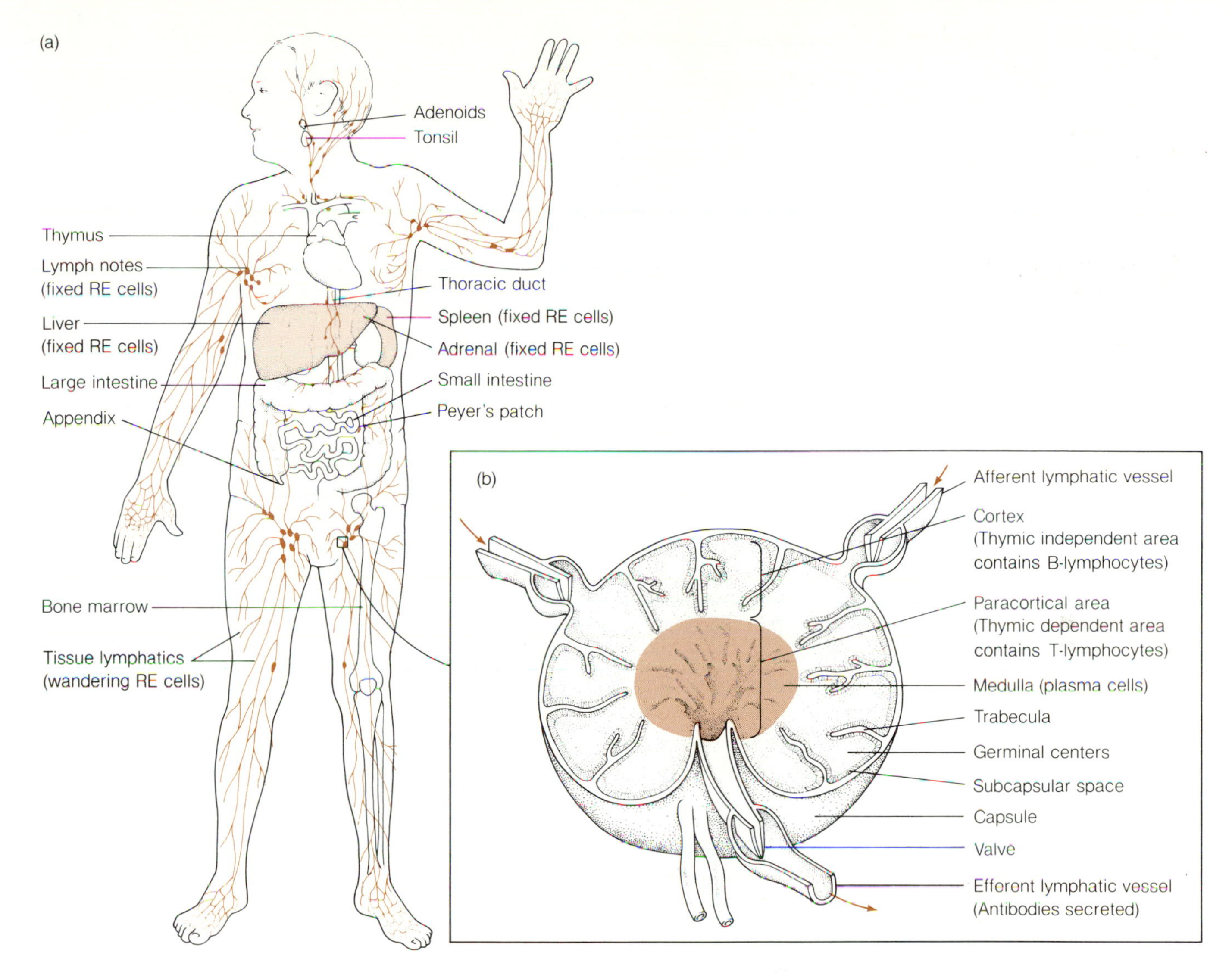

Figure 19-4 The Lymphatic and Reticuloendothelial Systems. *(a)* The lymphatic system is a network of vessels, lymph nodes (see illustration b), and organs that function in the initiation and maintenance of the immune response. The reticuloendothelial (RE) system consists of fixed and wandering phagocytic cells distributed throughout the body. They monitor and rid the body of infectious agents and cancerous cells. *(b)* Structure of the lymph node. The germinal centers of the lymph node contain primarily B-lymphocytes, while the thymic-dependent areas are composed of T-lymphocytes. The very center of the lymph node, the medulla, contains primarily antibody-forming cells or plasma cells.

teins by the T_H and thus are involved in the suppression of the immune response. Although a major function of T_S is to modulate the intensity of the immune response and to avert damage of "self," these lymphocytes also play a role in certain diseases, because invading microorganisms may stimulate T_S and hence prevent the host from mounting an adequate antimicrobial defense.

Macrophages

Macrophages and **monocytes** (fig. 19-2) are scavenger cells found in the blood, lymph, and reticuloendothelial system. Macrophages are important in the development of specific immunity because they process and present antigens to lymphocytes, an important first step in the development of immunity to antigens. Macrophages are also important effectors of cell-mediated immunity. The engulfing of antigens by macrophages seems to be enhanced by antibodies and other substances released by lymphocytes.

ANTIGENS AND ANTIBODIES

Any molecule that induces a specific immune response is called an **antigen.** Sources of antigens include microbial components such as toxins, cell walls, flagella, capsules, and proteins, as well as viral particles and cancerous cells. Antigens generally are found to be macromolecules such as polypeptides, polysaccharides, lipopolysaccharides, and glycoproteins. The chemical group in an antigen that specifically stimulates the immune response is called the **antigenic determinant** (fig. 19-5). The antigenic determinant may consist of a single sugar or a group of amino acids or other small molecules. The term **hapten** has been used to describe antigenic determinants that are incapable of stimulating an immune response by themselves. Almost any molecule that is not antigenic by itself can be made antigenic by covalently joining it to macromolecules such as polypeptides or polysaccharides, which are referred to as **carrier molecules** (fig. 19-5). Since many of the subunits in a polypeptide or polysaccharide can stimulate the development of specific antibodies, an antigen generally stimulates the production of a family of antibodies with different specificities that bind different antigenic determinants (fig. 19-5).

Sometimes two unrelated antigens have antigenic determinants that are so similar that they bind to the same antibody molecule. Antibodies that bind to different antigens are said to be **crossreacting.** Crossreacting antigens are sometimes a problem when microbiologists use antigen-antibody

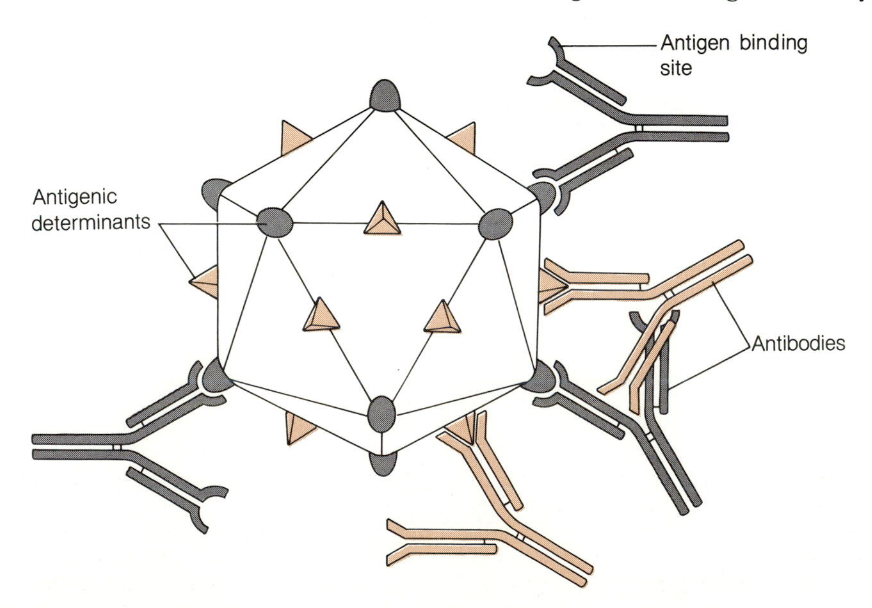

Figure 19-5 Antigens and Antigenic Determinants. The antigen represented by a virus particle has two different types of antigenic determinants, each recognized by antibodies of different specificities. The specificity of each antibody molecule is represented by the shape of the antigen-binding site of the molecule, which has a shape complementary to that of the antigenic determinants (pyramids and half spheres). When an antigen, such as a virus, is injected into an animal, it induces the animal to produce various types of antibodies against the antigen. Each type of antibody will react with a different antigenic determinant.

reactions to diagnose infectious diseases. It is always possible that an observed antigen-antibody reaction was due to crossreactivity, and this problem may result in a misdiagnosis. The topic of crossreactivity will be taken up again in Chapter 21.

Antibodies, or immunoglobulins, are a family of glycoproteins that bind specifically to the antigen that stimulated their production (fig. 19-5). Antibodies are found predominantly in the blood serum, specifically in the gamma globulin fraction of serum (fig. 19-6). Because these globulins participate in the immune response, they are called **immunoglobulins.**

The basic immunoglobulin molecule is composed of four polypeptides joined together by disulfide bonds (fig. 19-7). The four polypeptides consist of two identical heavy (H) chains, containing approximately 440 amino acids each, and two light (L) chains with approximately 220 amino acids each. Each heavy chain is covalently bonded to a light chain so that each immunoglobulin molecule is made up of two identical halves.

Both the H and L chains are composed of two regions: the variable (V) region and the constant (C) region. The constant regions of the heavy and light chains contain approximately 320 and 105 amino acids, respectively, and do not vary much among various immunoglobulin molecules. The amino terminal portions (or variable regions) of both the H and L chains show considerable variation in their amino acid sequence and are responsible for the specificity of each type of antibody.

The immunoglobulins perform a number of functions. The amino-terminal ends (the ends of the arms), called **Fab** regions, bind to antigens, while the single carboxy-terminal end, known as the **Fc** region (fig. 19-7),

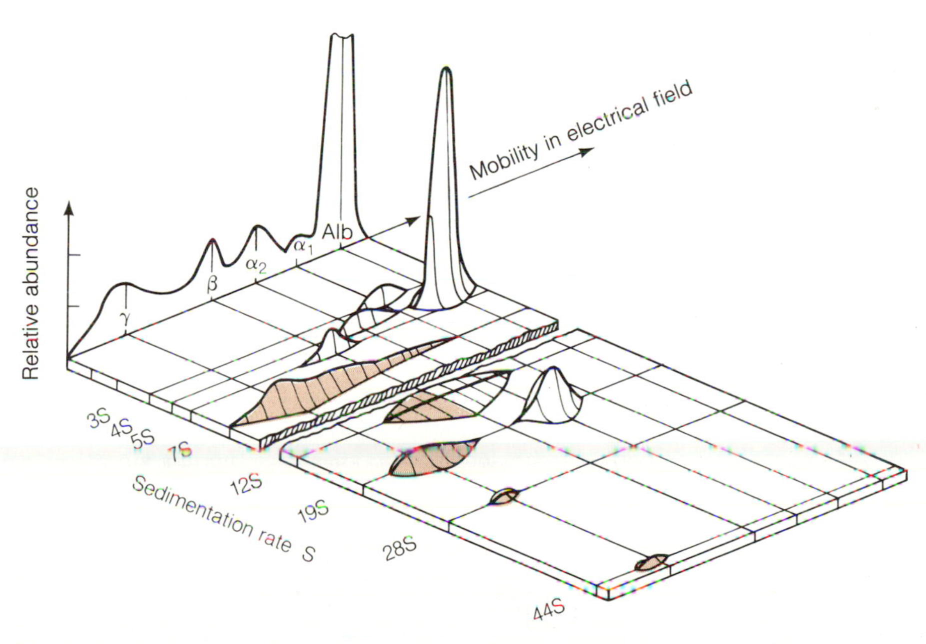

Figure 19-6 Major Proteins Found in Serum. The three-dimensional plot illustrates the relative abundance of serum proteins in human blood. Notice that serum contains five major proteins (albumins, alpha-1 and alpha-2 globulins, beta globulin, and gamma globulin). Serum albumins (Alb) are present in high concentration (highest peak) and are relatively uniform in composition (width of zone and number of distinct peaks). The gamma globulins or immunoglobulins are quite heterogeneous in composition. The three most common gamma globulins in serum are IgG, IgM, and IgA (see table 19-4).

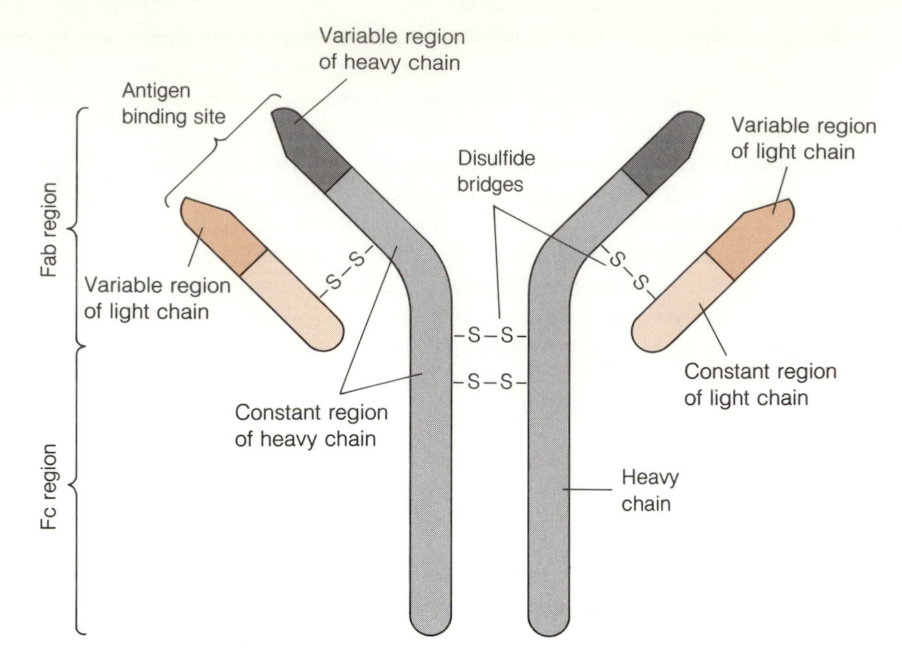

Figure 19-7 Basic Structure of Immunoglobulins. Immunoglobulins are proteins composed of four polypeptides joined by disulfide bonds (-S-S-). Two of the polypeptides are large molecules, referred to as heavy chains, and the other two polypeptides are smaller and are called light chains. Each heavy chain is bonded to a light chain. The two halves are joined to form a bivalent antibody molecule. The anterior portion of each light and heavy chain has a variable region. The variable regions of the heavy and light chains form the antigen binding site (Fab region), which confers specificity on the molecule. The three-dimensional shape of the Fab region is complementary to that of the antigen with which it reacts.

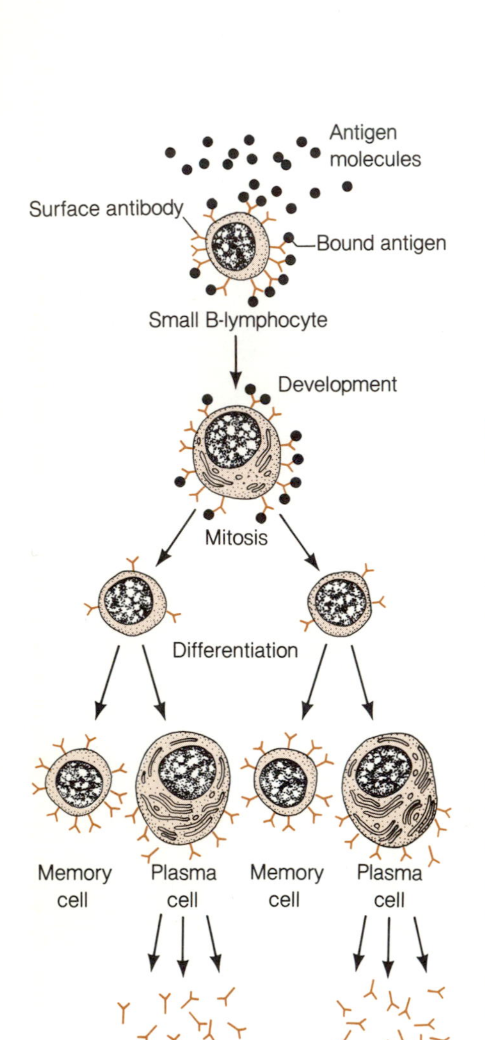

Figure 19-8 Production of Antibodies by Plasma Cells. When B-lymphocytes are stimulated by an antigen (or by a helper T-lymphocyte), they undergo a developmental change that ultimately leads to the formation of antibodies against the stimulating antigen. The B-lymphocyte first undergoes several rounds of mitosis, producing memory cells and plasma cells. The plasma cells synthesize the antibodies and secrete them into the blood and lymph.

may bind to various types of cells or be involved in activating complement. The binding between immunoglobulins and antigens is due to hydrogen bonding, electrostatic bonding, and Van der Waals attraction. The basic immunoglobulin molecule is said to be **bivalent** because its two Fab regions can attach to two antigenic determinants.

The specificity with which antibodies bind to antigens is related to the three-dimensional structure of both the antigen and the antibody. The three-dimensional structure of the Fab region is complementary to the shape of the antigenic determinant (fig. 19-5). Any antigenic determinant with a shape that fits within the Fab region of an antibody will react with the antibody. Crossreacting antibodies have binding sites that are able to fit a number of antigenic determinants.

Immunoglobulin Variability

An animal is able to produce a multitude of different immunoglobulins that can recognize and bind specifically to millions of antigens. It is believed that millions of different lymphocytes are constantly developing in animals, and that each of these lymphocytes produces one of the millions of possible immunoglobulins. If an antigen is present that can bind to some of the im-

FOCUS

IMMUNOGLOBULIN VARIABILITY CAN BE ACHIEVED BY REARRANGING GENES

When one looks at the amino acid sequences of a number of light chains, it is found that the variable regions in the light chains are made up of what are called V-regions and J-regions. Similarly, it is found that the variable region of the heavy chain has a D-region, in addition to its V and J regions (Fig. 19-9). Furthermore, the heavy chain has highly conserved regions: CH1, H, CH2 and CH3.

The hundreds of millions of different immunoglobulins that an animal needs to protect itself from the many invading microorganisms it encounters can be created from the recombination of a few hundred different V, D, and J mini-genes. If, for example, the DNA in a B-lymphocyte contains 500 V_H genes, 5 D_H genes, and 5 J_H genes, then it is possible to make 12,500 different variable regions of the heavy chain (500 × 5 × 5). Similarly, if the B-lymphocyte contains 500 V_L genes and 5 J_L genes, it is possible to make 2,500 different variable regions of the light chain. Thus, the total number of immunoglobulins possible when the heavy and light chains are combined is about 30 million (12,500 × 2,500). In fact, the number of immonoglobulins possible is 10–100 times greater than we just calculated. This is because the splicing of the gene elements introduces mutations between them so that three hypervariable regions are created in the variable region of the heavy chain. There are also three hypervariable regions in the variable region of the light chain. The hypervariable regions of both the heavy and light chains are partly responsible for the enormous number of antigen-binding sites necessary to fight off infection.

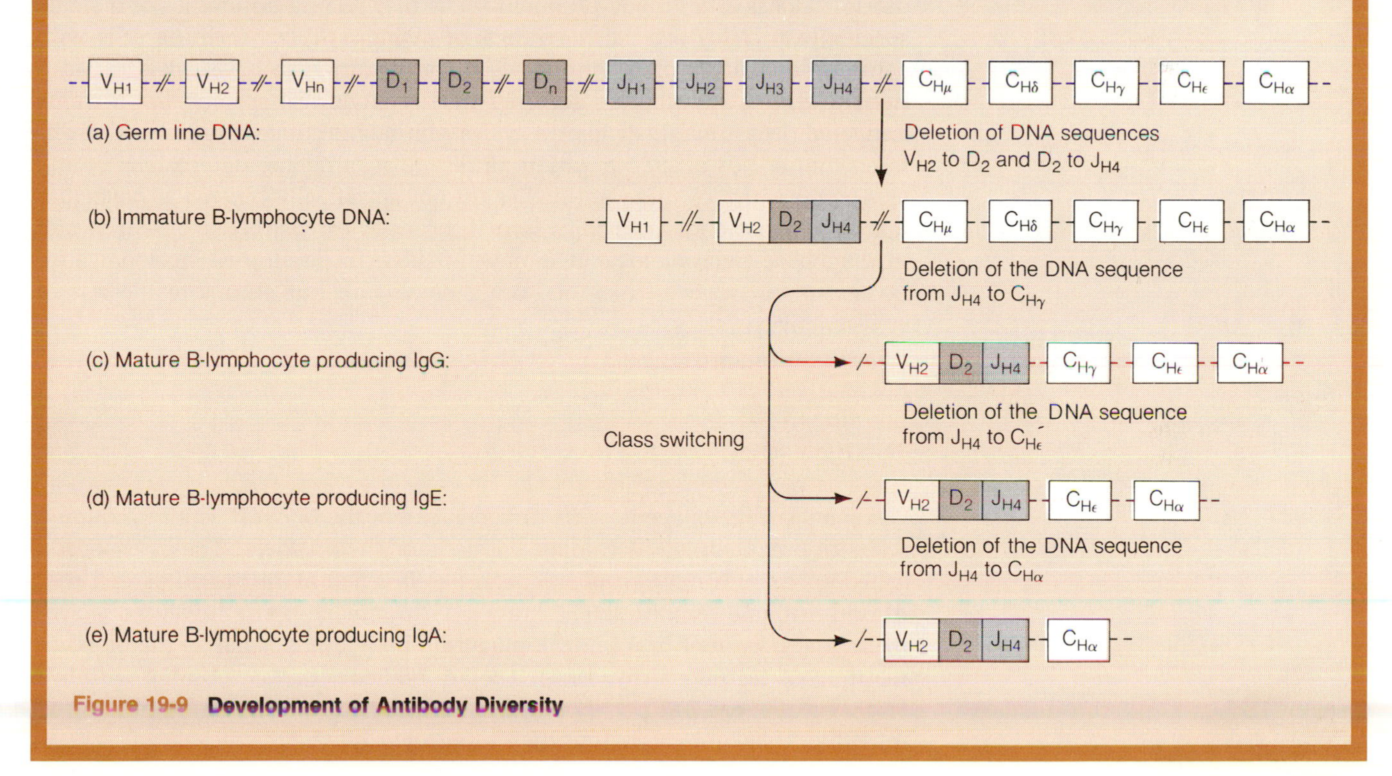

Figure 19-9 Development of Antibody Diversity

munoglobulins on the surface of B-lymphocytes, the B-lymphocytes are stimulated to proliferate and differentiate (fig. 19-8). The B-lymphocytes become plasma cells, which secrete antibodies that bind to the antigen that stimulated the B-lymphocyte to become a plasma cell.

Classes of Immunoglobulins

There are five classes of immunoglobulins (table 19-4), each of which is constructed from very similar proteins. The various types of immunoglobulins are classified according to the amino acid sequences in the constant region of the heavy chain. These immunoglobulins, which include IgG, IgM, IgA, IgE, and IgD, differ not only in their heavy chains but also in their biological and chemical properties (table 19-4). For example, IgM is a pentameric (five basic antibody molecules joined) molecule that first appears during the primary immune response. This immunoglobulin eventually is replaced by the monomeric immunoglobulin IgG. IgA molecules are most common in mucous secretions, where they interact with foreign antigens associated with the mucous membranes of the nasal passages, intestines, urethra, and vagina. The physical and biological properties of the various classes of immunoglobulins are summarized in table 19-4.

Monoclonal Antibodies

Cesar Milstein and Georges Koehler developed a technique to produce "immortal" clones of antibody-producing cells that secrete antibodies of a single specificity □ . This procedure consists of fusing antibody-forming cells with a tumorous culture of B-lymphocytes. The resulting cell hybrid is called a **hybridoma.** All antibodies secreted by the clone are specific for a single antigenic determinant. By contrast, live animals produce a spectrum of different antibodies against a variety different of antigenic determinants, and as many as 90% of these antibodies do not specifically bind to a particular antigen. Monoclonal antibodies have been used extensively in research and in diagnostic microbiology. It is now possible to develop monoclonal antibodies to very specific microbial antigens and all but eliminate the risk of crossreactivity.

270

IMMUNE RESPONSE TO ANTIGENS

When an infectious agent or its antigens enters the body, it sets in motion a series of events aimed at ridding the body of the invaders. These responses, called specific immune responses, can be classified as **humoral** or **cell mediated** immune responses. The humoral immune response involves the synthesis and release of immunoglobulins into body fluids, such as the blood or the mucous secretions (fig. 19-8). The antibodies act by combining with microbial toxins or viral particles and neutralizing □ them, or by binding to microbial cells and promoting their phagocytosis.

521

The level of antibodies (often referred to as the **titer**) produced during an immune response is dependent upon the concentration of antigen. Some materials, when introduced into a host along with an antigen, are able to stimulate the immune response to a greater degree than if the antigen were introduced by itself. Those compounds that are able to enhance the immune response to an antigen are called **adjuvants.** One of the better-known adjuvants is **complete Freund's adjuvant,** which consists of a mixture of waxes, mineral oil, and killed *Mycobacterium tuberculosis.* The antigen is usually mixed with the adjuvant to form a water-in-oil emulsion. This emulsion,

TABLE 19-4
CHARACTERISTICS OF THE FIVE MAJOR CLASSES OF HUMAN IMMUNOGLOBULINS

CLASS	DIAGRAM	SEDIMENTATION CONSTANT	RELATIVE ABUNDANCE IN SERUM (%)	SUBCLASSES	CHAIN SUBUNITS			ARRANGEMENT OF CHAINS	SALIENT CHARACTERISTICS
					LIGHT	HEAVY	OTHERS		
IgG		7S	80%	IgG_1 IgG_2 IgG_3 IgG_4	Kappa or lambda	Gamma		Monomer	Present in blood and lymph. Half-life is 23 days. Active against microbes and toxins.
IgM		19S	6%	IgM	Kappa or lambda	Mu	J	Pentamer	Present in blood and lymph. Half-life is 5 days. Appears early in infection. Effective against microorganisms.
IgA		7S 11S	13%	IgA_1 IgA_2	Kappa or lambda	Alpha	J	Monomer Dimer	Present in secretions and lymph. Half-life is 6 days. Protects body's external surfaces. Found in tears, saliva, vagina mucous membranes.
IgE		8S	0.002%	IgE	Kappa or lambda	Epsilon		Monomer	Plays role in allergic reactions. Half-life is 3 days.
IgD		7S	1%	IgD	Kappa or lambda	Delta		Monomer	Present on surface of lymphocytes. Half-life is 3 days.

when injected into an animal, remains in the tissue for long periods of time and continuously stimulates the immune system. Freund's adjuvant without the killed bacteria is called **incomplete Freund's adjuvant.**

The cell-mediated immune response is effected by "sensitized" lymphocytes and by macrophages (fig. 19-10). The lymphocytes mediate cellular immunity by releasing a variety of substances that are microbiocidal (or cytocidal), or that stimulate other cells to become microbiocidal. For example, when a macrophage becomes activated by **lymphokines**, polypeptides released by T-lymphocytes that stimulate the immune response (table 19-5), it becomes very aggressive and effective in killing microbial pathogens. The activated macrophage is an important effector of cell-mediated immunity, playing a central role in protecting mammals against intracellular parasites, such as the tubercle bacillus and viruses, by engulfing and subsequently killing them.

Cytotoxic lymphocytes (T_C) also participate in cell-mediated immunity by destroying cancerous cells and virus-transformed cells when they arise. In order to perform this function, cytotoxic lymphocytes must recognize both foreign and self antigens on the target cells. Once these T-lymphocytes recognize foreign antigens, they release killing factors **(lymphotoxins),** which associate with the target cell's plasma membrane and damage it. The killing factors create pores in the membrane or act as ionophores, allowing calcium ions to flow into the cell. The increase in calcium ions within the cell drastically disrupts its physiology and causes its death.

Another type of lymphocyte that plays a role in cell-mediated immunity is the killer T-lymphocyte (T_K). These lymphocytes attach to microorganisms, virus-transformed cells, and cancerous cells. The binding of these lym-

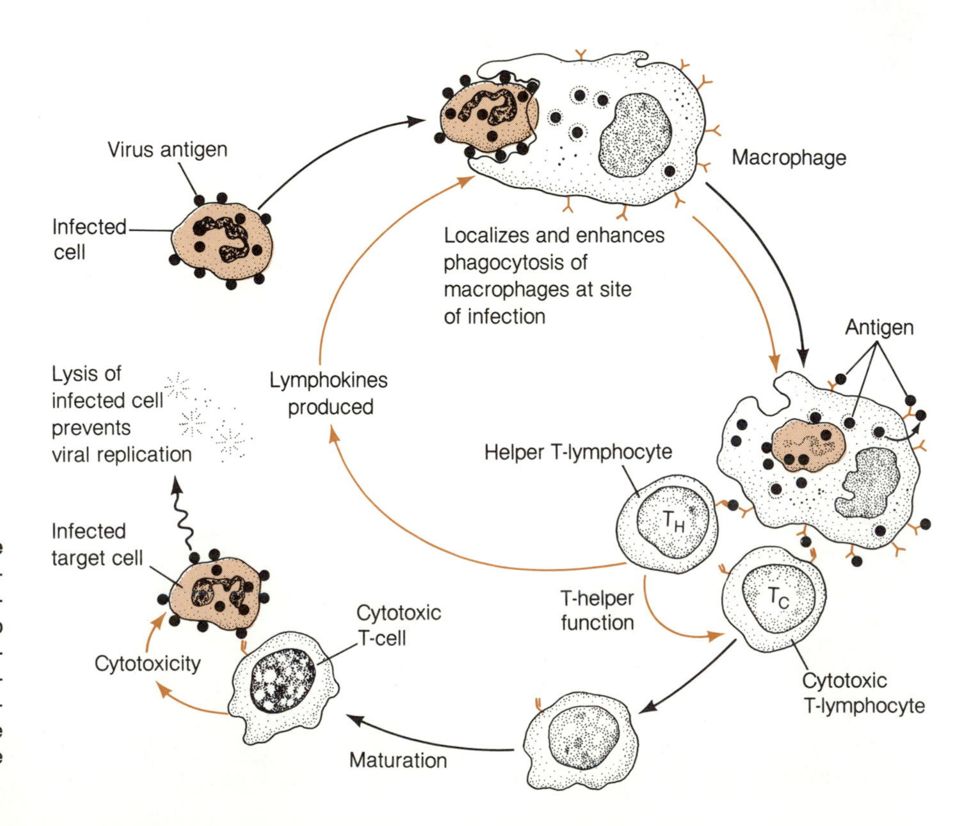

Figure 19-10 Cell-Mediated Immune Mechanisms. T-lymphocytes and macrophages are the effector cells of cell-mediated immunity. T-lymphoctes may develop into killer cells, which can then destroy virus-infected cells or tumor cells or may produce substances called lymphokines. Lymphokines have various functions (see table 19-5) aimed at mounting a strong immune response.

TABLE 19-5
PROPERTIES OF VARIOUS LYMPHOKINES PRODUCED BY T-CELLS

LYMPHOKINE	CELL AFFECTED	MODE OF ACTION
Macrophage activation factor	Macrophage	Enhances killing by macrophages
Chemotactic factor	Macrophage	Attracts macrophages to site of infection
Migration inhibition factor	Macrophages	Inhibits migration of macrophages from site of infection
Mitogenic factor	Lymphocyte	Promotes cell division
Chemotactic factor	Lymphocyte	Attracts lymphocytes to antigen
Leukocyte chemotactic factors	Eosinophils	Attracts eosinophils
	Basophils	Attracts basophils
	Neutrophils	Attracts neutrophils
Lymphotoxins	Various cells	Kills cells
Colony stimulating factor	Various cells	Promotes mitosis
Interferon	Various cells	Inhibits viral replication

phocytes to their targets stimulates them to produce killing factors that destroy their target cells. The killing factors produced by T_K lymphocytes are believed to be similar to those released by the cytotoxic T-lymphocytes.

Induction of the Immune Response

Lymphocytes and macrophages play a central role in the induction of the immune response. The intimate interaction between the plasma membrane of macrophages and lymphocytes and their products is responsible for the various manifestations of immunity. When an antigen enters the body, it is promptly attacked by macrophages and partly digested. A portion of the antigen (possibly an antigenic determinant) is incorporated into the plasma membrane of the macrophage so that it can be recognized by other cells of the immune system. These macrophages with the incorporated antigen interact with certain types of T-lymphocytes (T_H), which are then stimulated to react (fig. 19-11). This stimulation of lymphocytes by macrophages take place in lymphoid organs, such as the lymph nodes. The stimulated lymphocytes are largely helper T-lymphocytes, which induce B-lymphocytes to develop into antibody-forming cells (plasma cells) or into memory cells. The B-lymphocytes must also interact with the antigen in order to be stimulated to become plasma cells. Soluble growth factors derived largely from T-lymphocytes are also required to stimulate the B-lymphocytes to proliferate and mature into plasma cells. A plausible mechanism of the induction of the immune response is outlined in figure 19-11.

During the immune response, two major groups of cells arise: long-lived memory cells and shorter-lived effector cells. Effector cells include plasma cells and cytotoxic T-lymphocytes. These cells have a half life of about 3-5 days; in contrast, memory cells may live for several years.

How an animal's immune response gets turned on is a problem that has fascinated immunologists since the discovery of immune responses. In the past few years, partly due to the development of new techniques such as genetic engineering and monoclonal antibodies, immunologists have deter-

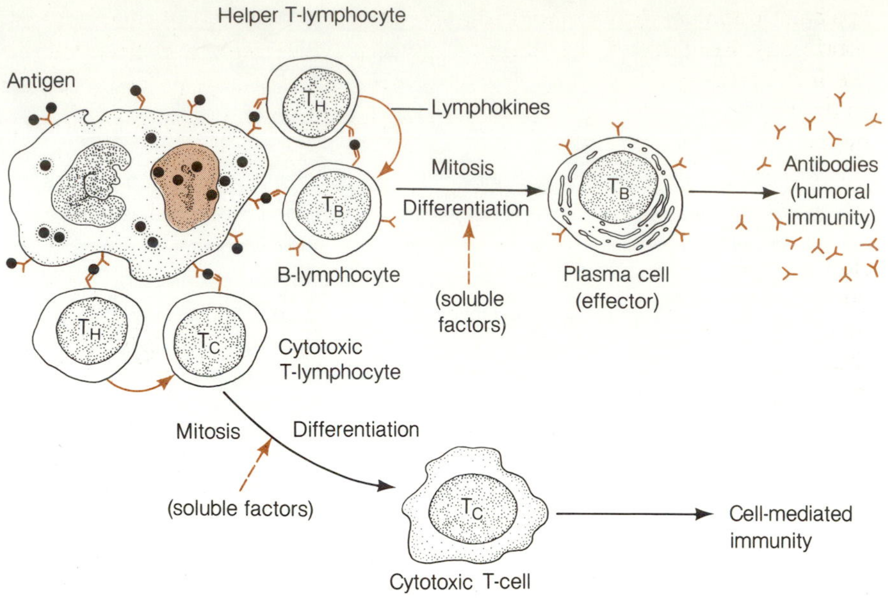

Figure 19-11 The Induction of the Immune Response. The immune response is induced by the cooperative interaction among B-lymphocytes, helper T-lymphocytes, and macrophages. The macrophages process antigen and stimulate helper T-lymphocytes. These, in turn, interact with B-lymphocytes. B-lymphocytes differentiate into plasma cells, which secrete antibodies (see figure 19-8). Soluble factors (proteins? hormones?) are also involved in the induction process. Helper T-lymphocytes may also interact with other T-lymphocytes to induce cell-mediated immunity (see figure 19-10).

mined many of the steps involved in the development of the immune response. This process is very complex, involving several types of cells and a variety of chemical substances and growth factors. At least three different steps are involved in the development of the immune response. The first step involves the **stimulation** of immune cells, almost certainly lymphocytes, by antigens. This step can be accomplished directly by an antigen or by an antigen-presenting macrophage. The second step involves **proliferation** of the immune cells, during which the stimulated cells are induced to undergo several rounds of mitosis □ aimed at increasing their number. Apparently there are a variety of factors, called **growth factors,** that act by inducing mitosis. These factors are specific for their target cells; some act by stimulating only B-lymphocytes, while others stimulate only T-lymphocytes. In the absence of these factors, the immune response will not develop and the cells eventually will die. The third step is **differentiation,** wherein some of the stimulated cells become memory cells while others become effector cells, such as plasma cells, suppressor T-lymphocytes, or cytotoxic T-lymphocytes. Again, growth factors are involved, many of which are specific for their target cells.

233

The research that is currently taking place to uncover the sequence of events leading to the induction of specific immunity is not only of academic importance but also of applied importance. Many disease states of humans and domestic animals are direct or indirect consequences of abnormal immune responses. Examples include many forms of cancer, autoimmune diseases, and chronic diseases. Understanding how the immune response develops can ultimately lead to the development of drugs to treat or alleviate these conditions. In addition, the ability to stimulate or suppress specific steps in the immune response can be extremely helpful in surgical procedures, such as organ transplants from unrelated donors, which are prone to be rejected by the recipient's immune system.

Clonal Selection Theory

How does the antigen "know" to stimulate the appropriate lymphocyte? This question has intrigued scientists since the nineteenth century. It is currently thought that the genetic information necessary to synthesize an antibody (immunocompetency) is present in each prelymphoid cell □, and that the genes for specific antibody production are expressed as antibody molecules on the plasma membranes of lymphoid cells. The antigen stimulates the multiplication of each type of lymphoid cell when it combines specifically with the antibody molecules present on the surface of the lymphoid cells. Thus the antigen-stimulated cell becomes an immunologically committed cell (fig. 19-12). This theory, called the **clonal selection theory,** was ad-

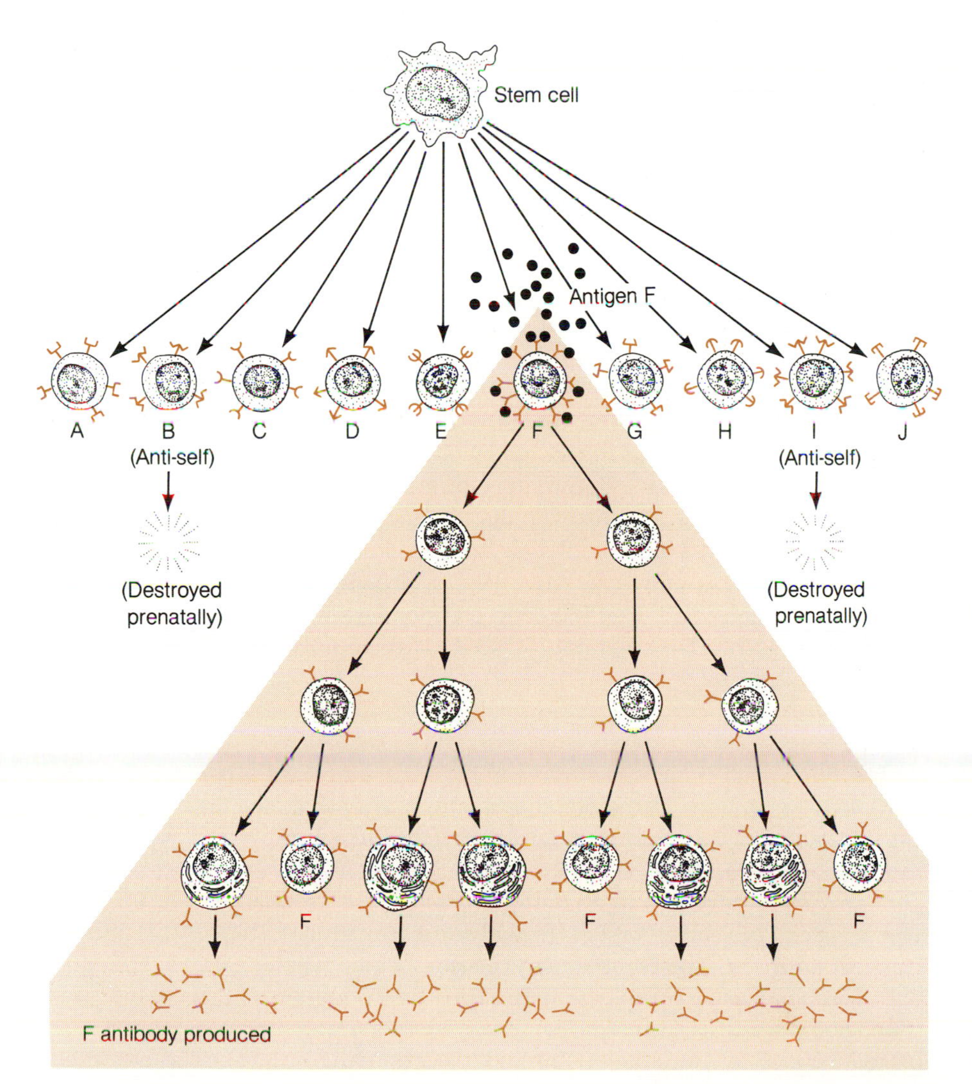

Figure 19-12 The Clonal Selection Hypothesis of Antibody Production. The clonal selection hypothesis states that the body possesses numerous small clones of immunocompetent lymphocytes, each clone is able to recognize a different antigenic determinant. This specificity of recognition is expressed on the plasma membrane in the form of antibody receptors. When an antigen (e.g., F) enters the body, it reacts (selects) clone F and induces the clone to proliferate. This response leads to the differentiation of clone F into additional memory cells and plasma cells that secrete antibodies of the same specificity. During prenatal life, certain clones arise that recognize self antigens. These clones are eliminated before birth.

vanced originally by Sir Macfarlane Burnett and has withstood considerable scientific scrutiny throughout the years.

The clonal selection theory suggests that each lymphocyte has the genetic information to synthesize a particular antibody molecule, and that these molecules can be found "advertised" on the surface (plasma membrane) of each immunocompetent lymphocyte. Each lymphocyte has a different **antibody receptor** on its surface, and when an antigen is introduced into the body, it will bind only to those surface receptors that have a complementary shape. The reaction between the antigen and the surface receptor stimulates lymphocytes to differentiate and proliferate into a **clone** of effector cells synthesizing the same type of antibody. Not all the cells in the clone will become effector cells; some will become memory cells. On first exposure to an antigen few of the cells will have surface receptors complementary to the antigen, so there will be a significant time lapse between stimulation and the presence of detectable levels of antibodies, due to the differentiation steps involved □. On second exposure to the antigen (anamnestic response), however, a clone of memory cells exists that can immediately proliferate, giving rise to numerous effector cells as well as more memory cells. For this reason, the secondary immune response appears much faster and is stronger than the primary immune response.

469

The clonal selection theory also explains how the immune response differentiates self from non-self. Apparently, during fetal development, lymphocytes carrying surface receptors that recognize self antigens are destroyed. Thus, at birth humans normally do not have lymphocytes with surface receptors that can recognize self antigens.

COMPLEMENT

Complement is a term given to a group of eleven serum proteins that are involved in disrupting plasma membranes. In addition, some complement components stimulate the release of histamine from leukocytes and so are indirectly involved in the inflammatory process. Other complement proteins stimulate chemotaxis □ of neutrophils, monocytes and eosinophils, as well as enhancing phagocytosis by phagocytes (opsonization) and platelet-stimulated neutralization of viruses (fig. 19-13). **Opsonins** are substances that stimulate phagocytosis. Complement components are thought to be produced by various cells, including those that make up the epithelium of the gastrointestinal mucosa.

444

The Classical Pathway of Complement Activation

Complement proteins in the blood and lymph are not active until they interact with antigen-bound immunoglobulins (immune complexes). Not all immunoglobulins can **fix** □ complement; only IgM and IgG can do so. A single pentameric IgM attached to an antigen can activate complement, but it takes two adjacent molecules of IgG to activate complement (fig. 19-14) by the classical pathway. The complement cascade begins when C1 (C1q, C1r, and C1s) components of complement bind to the Fc region of IgG or IgM immunoglobulins, after the antibody has undergone a conformational change brought about by the antigen-antibody reaction (fig. 19-14). The activated

525

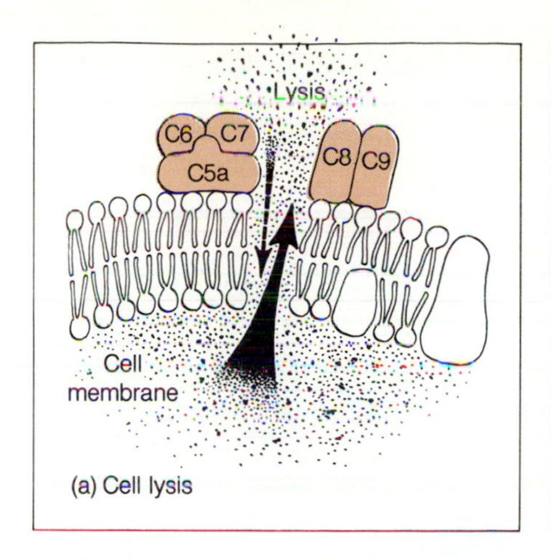

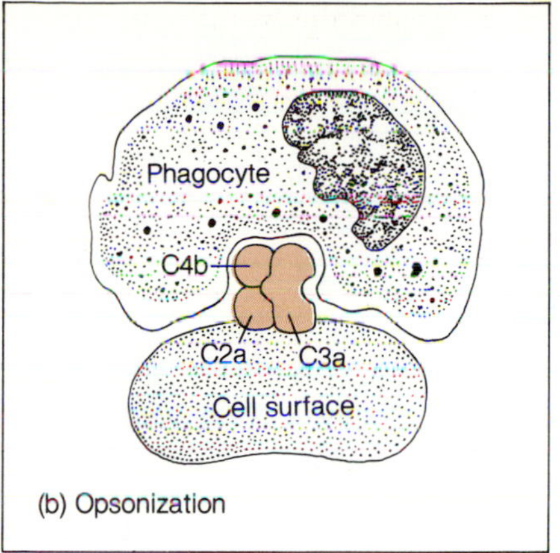

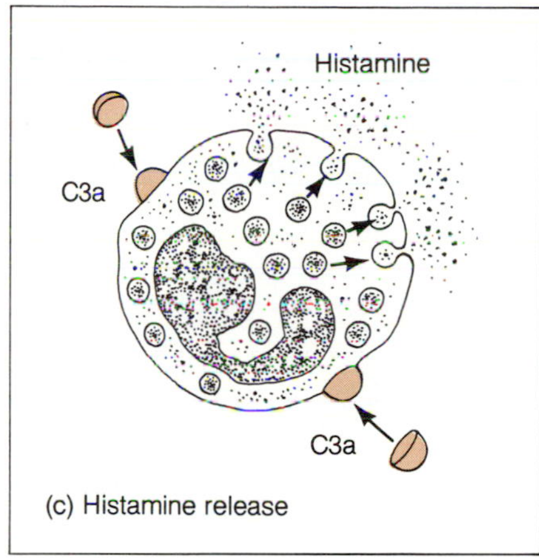

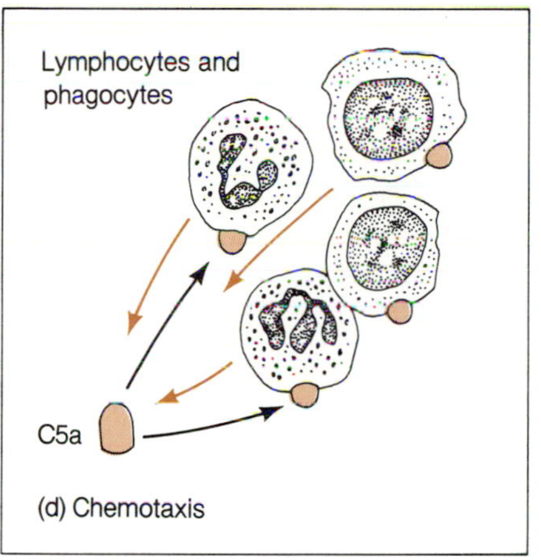

Figure 19-13 Functions of Complement. Complement has various protective functions in host responses to infection. (*a*) Its main function is to destroy foreign cells by damaging their plasma membranes. (*b*) Some components of complement can also enhance binding of phagocytes to target cells (opsonization) and hence improve the efficiency of phagocytosis. (*c*) Complement can also stimulate basophils and mast cells to release histamine (inflammatory response) and (*d*) to attract various cells of the immune system (chemotaxis).

C1 complex, in turn, activates C4 and C2. Two products of this activation (C2a and C4b) form an enzyme that binds to membranes and activates the C3 component of complement. Activated C3a component is an extremely potent peptide that causes the release of chemicals from certain leukocytes, leading to inflammation, pain, and the contraction of smooth muscle. C3a also induces chemotaxis in a variety of phagocytes. Another C3 product, C3b, associates with the membrane-bound $\overline{\text{C4b2a}}$ enzyme and promotes the activation of C5 into C5a and C5b. C5b binds to C6 and C7 forming a complex ($\overline{\text{C567}}$), which in turn binds to target cell membranes, attracting C8 and C9. The complex of $\overline{\text{C5b6789}}$ disrups the cell by creating pores in the plasma membrane. This alteration leads to cell death, because the internal physiology of the cell is irreversibly altered.

The Properdin Pathway

The **properdin pathway** of complement activation (fig. 19-15), also called the alternate pathway of complement activation, is independent of immunoglobulins. Insoluble polysaccharides and other molecules from bacteria and yeast can activate C3 in the absence of C1, C2, and C4. This antibody-independent activation may be an important initial defense against various types of infection.

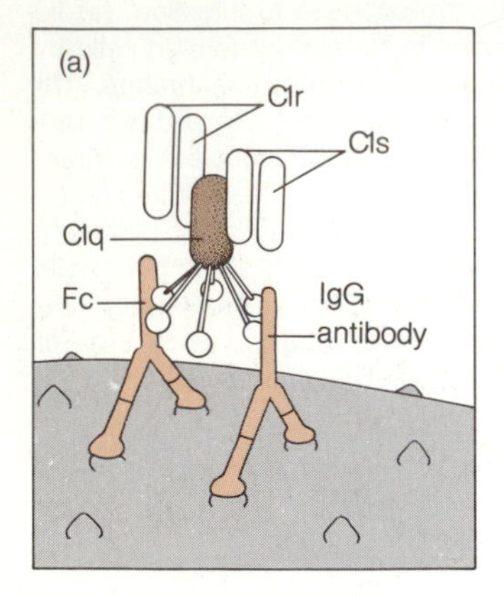

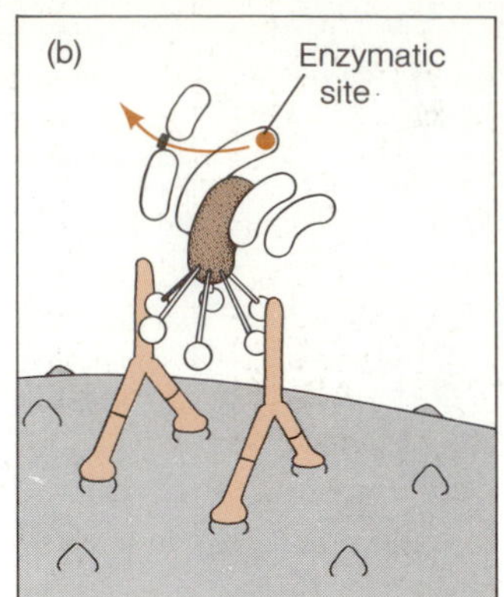

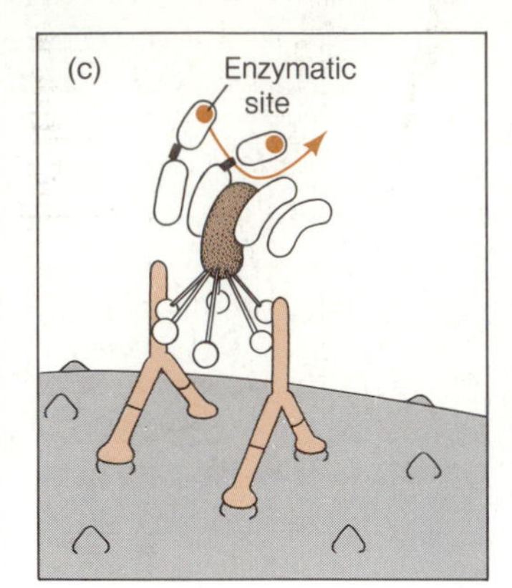

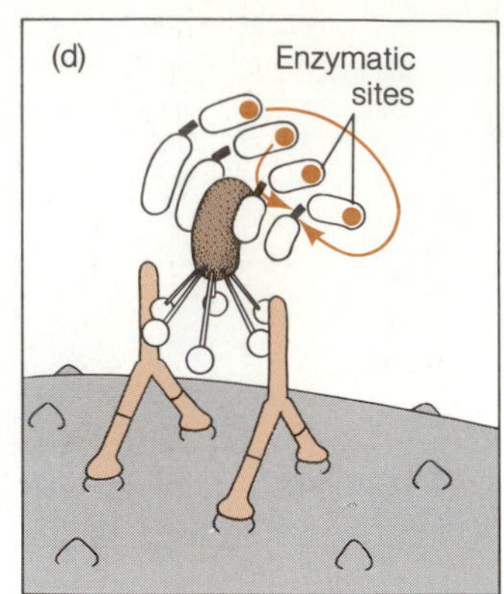

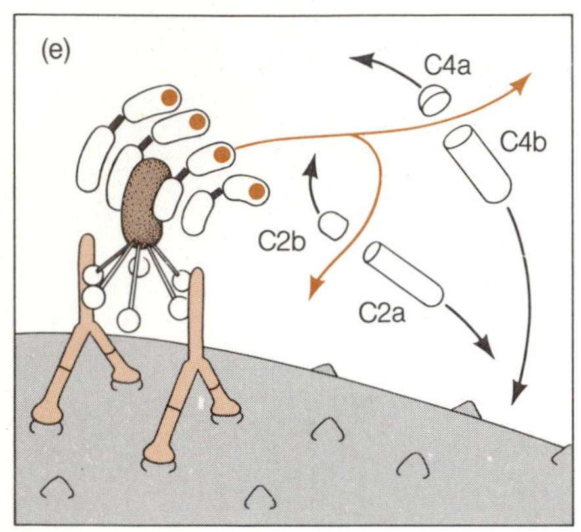

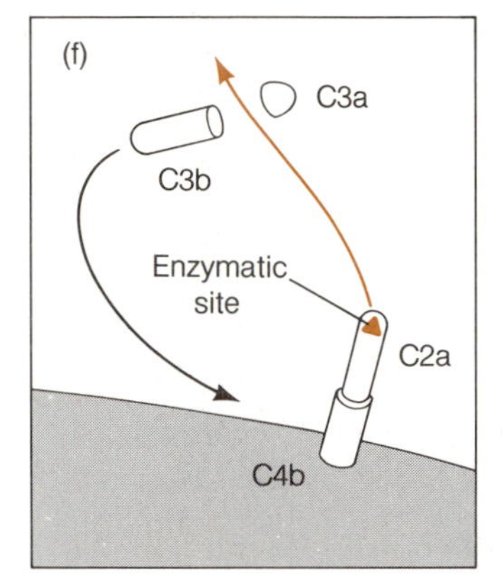

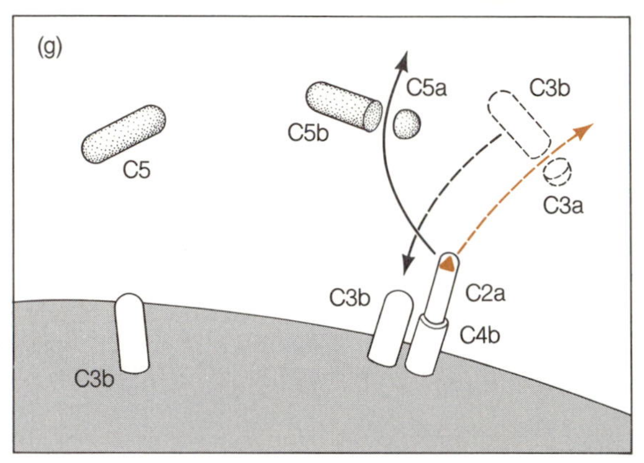

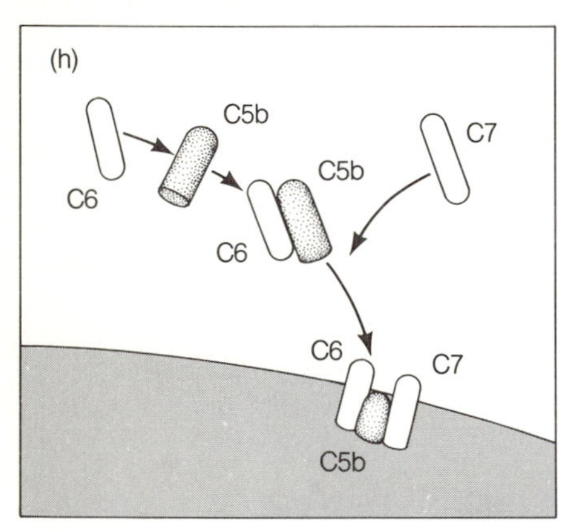

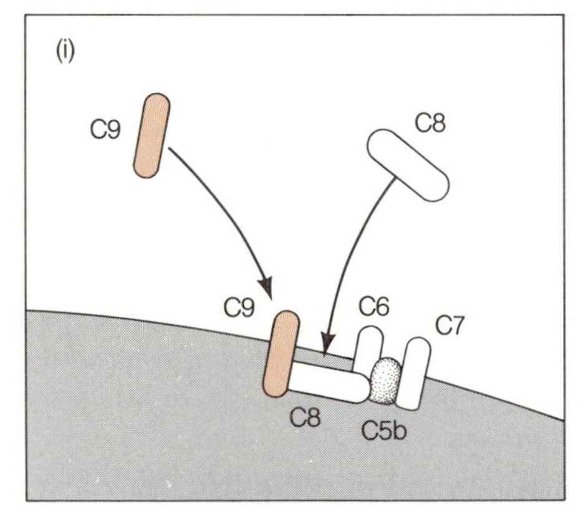

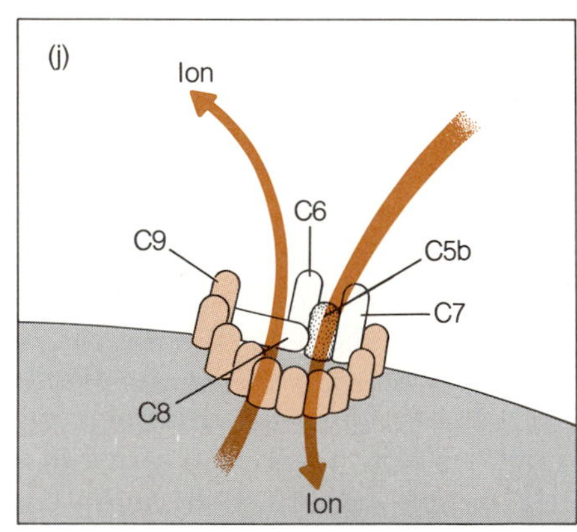

Figure 19-14 The Complement Cascade. *(a)* At the start of the classical complement activation pathway antibodies recognize and bind to the antigen (e.g., foreign cells or bacteria). The complement component C1, consisting of subunits C1q, C1r, and C1s, binds to two adjacent IgG antibody molecules. *(b)* Binding by the C1q subunit to the antibody alters its configuration, which results in conformational changes in C1r, exposing a proteolytic site on one C1r molecule. *(c)* This proteolytic site then acts on the other C1r molecule, exposing a similar enzymatic site. *(d)* These enzymatic sites act on C1s, exposing two additional enzymatic sites. *(e)* The C1s now cleaves two other complement components, C4 and C2, allowing the C4b fragment to attach to the cell membrane. *(f)* C2a then binds to C4b, producing $\overline{C4b2a}$ (C3 convertase), which then cleaves C3. C3b can attach to the cell membrane. *(g)* When C3b attaches itself close to the C3 convertase, it alters the specificity of the enzyme site on the C2a, which can now split C5, and is thus called a C5 convertase. *(h)* and *(i)* C5b binds to C6 and C7 and attaches to the cell surface. C8 and C9 now attach to the $\overline{C5b67}$. *(j)* The insertion of this late complex damages the cell membrane and permits a rapid flux of ions that ultimately leads to cell lysis.

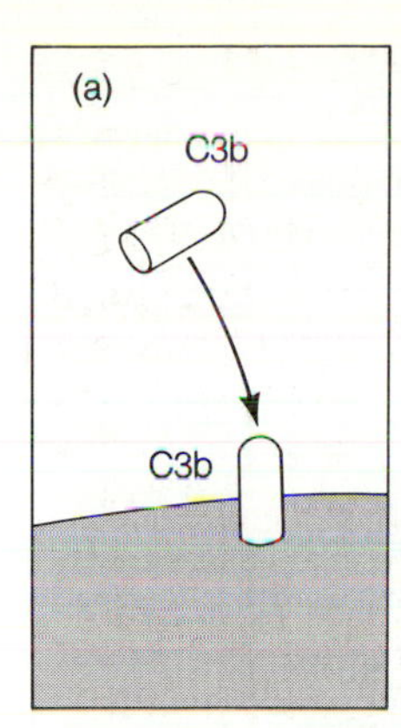

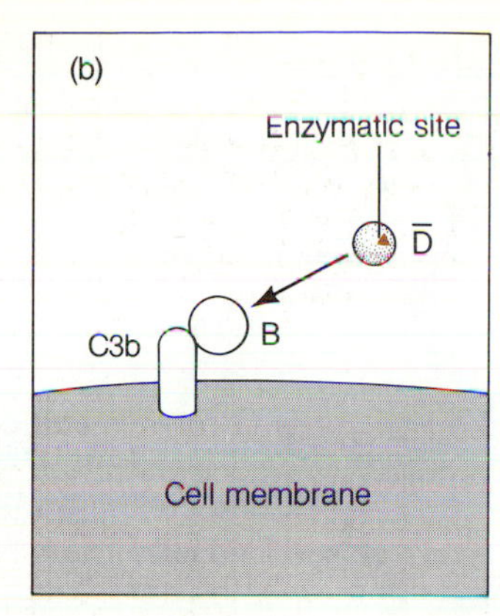

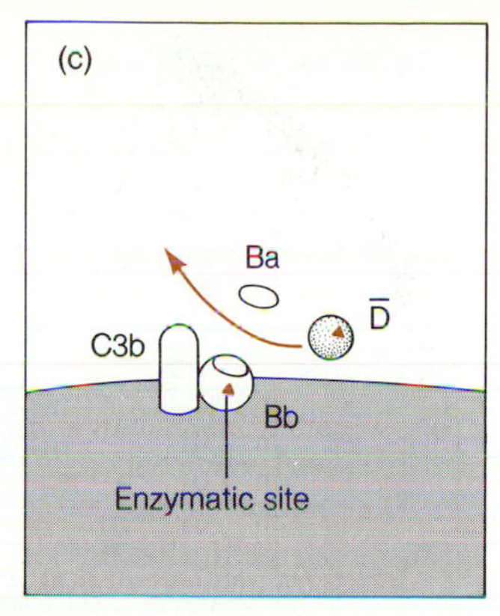

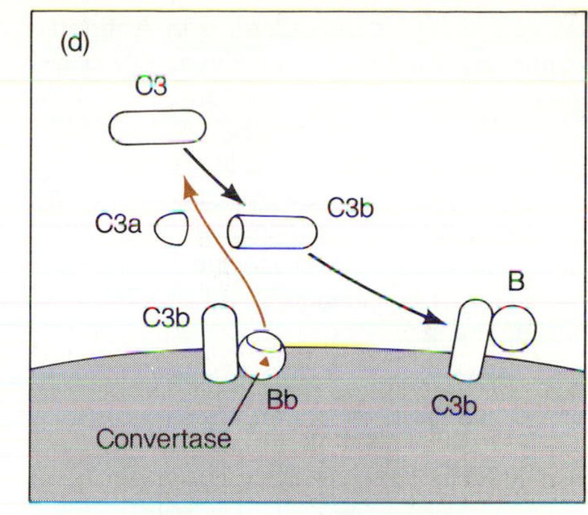

Figure 19-15 The Properdin Pathway of Complement Activation. *(a)* The alternative, or properdin, pathway is triggered when complement component C3b becomes attached to certain surfaces. *(b)* and *(c)* Component B then attaches to C3b and is activated by a serine esterase, D. *(d)* The enzymatic site on Bb now cleaves more C3b, which then attaches to the membrane. As more B attaches to the C3b, a chain reaction occurs. This results in an amplification of the complement reaction. An enzyme called convertase also activates C5. This leads to the complement cascade illustrated in figure 19-14 (steps h though j).

REACTIONS BETWEEN ANTIGENS AND ANTIBODIES

As we have come to realize, the immune response is specific for the stimulating antigen. Hence, if an individual becomes infected with a particular microbe, chances are that this individual will develop an immune response against the microbe. There are various ways of detecting the presence of immunity to infectious agents, all of which include antigen-antibody reactions. The field of immunology that deals with antigen-antibody reactions is called **serology** □. **Serological** tests are widely used in the diagnosis of infectious diseases because they are rapid and, if properly controlled, yield reliable results. We will discuss the principles and uses of some of the most important serological tests in Chapter 21.

515

The Classical Precipitin Reaction

If an antigen in solution is mixed with specific antibodies (antiserum) in the correct proportions, a precipitate will form. The precipitate results because the bivalent antibodies bind to two different antigen molecules, cross-linking them (fig. 19-16). Extensive cross-linking of antigen and antibody forms a lattice network that precipitates out of solution. This is the basis of antigen-antibody reactions.

In the **classical precipitin reaction,** a constant volume of antiserum is placed in a number of tubes and then various concentrations of antigen are added to the tubes. In the tubes where little antigen is added there will be little cross-linking because of the excess in antibodies, hence no precipitate will form (fig. 19-17). In the tubes that contain a larger proportion of antigens as compared to immunoglobulins, all the binding sites will not be covered by immunoglobulins, so again little cross-linking will occur. In the zone where antigen and antibody are in approximately the same proportion, or where there is a slight antibody excess, maximum precipitate will occur. This zone is called the **equivalence point** or the zone of optimal proportions. All

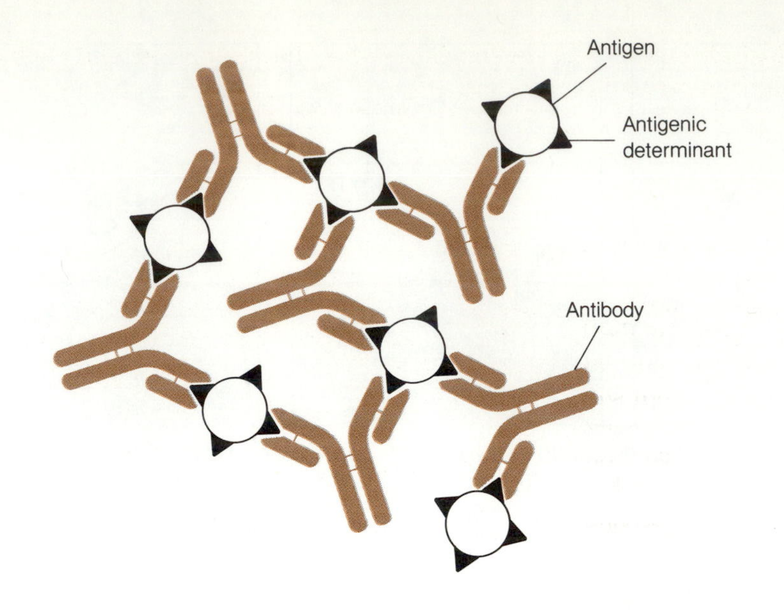

Figure 19-16 Cross-Linking in Antigen-Antibody Reactions. Antibodies are polyvalent molecules (i.e., they can bind to two or more antigen molecules). Antigen-antibody reactions can lead to the formation of lattice networks created by antibodies binding to antigens. The clumps can be so large that they settle out. Inside the host, these clumps can be phagocytosed by PMNs or macrophages.

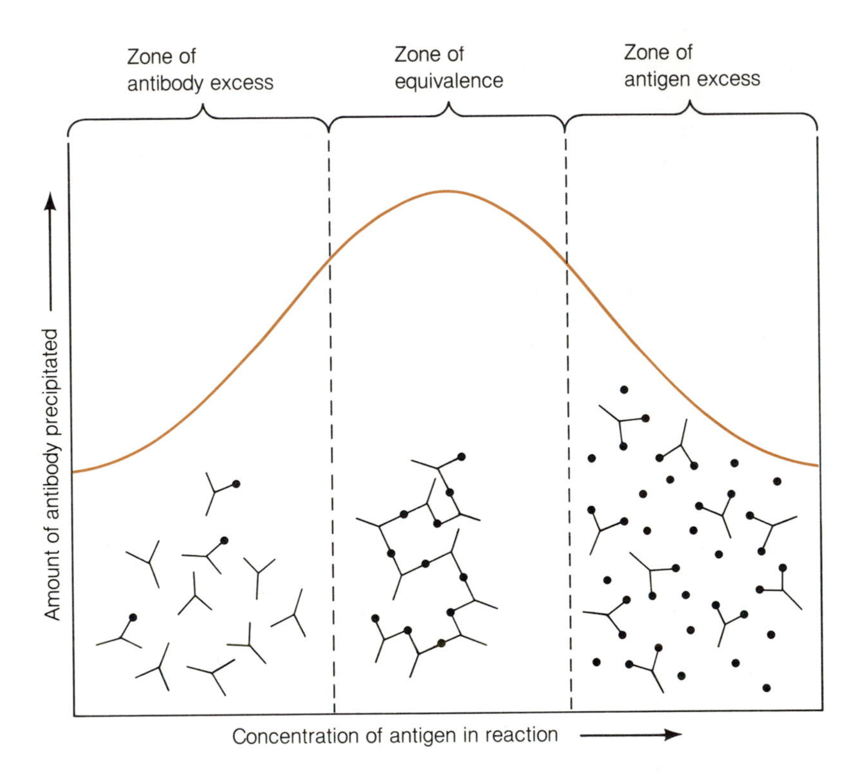

Figure 19-17 The Classical Precipitin Reaction. The degree of cross-linking in antigen-antibody reactions is determined largely by the relative concentrations of antigen and antibody present. When increasing concentrations of antigen are reacted with a standard concentration of antibody and the amount of precipitated antigen-antibody complexes measured, the results resemble those illustrated. At low antigen concentration, there is an excess of antibody, and very little cross-linking occurs. The precipitate (representing antigen-antibody complexes) is small and the supernatant fluid (containing unbound reagents) has a large concentration of antibody. At optimal concentrations of antigen, much cross-linking occurs, forming a large precipitate and little or no free antigen, and antibody will be left in the supernatant fluid. At high concentrations of antigen, again little cross-linking takes place.

serological reagents must be standardized so that the antigen and antibodies combine in optimal proportions.

VACCINATION

It has been known (or at least suspected) for the past 2,000 years that persons who have recovered from an illness such as smallpox, plague, or typhus are very resistant or completely immune to a subsequent infection by the same agent. In the late 1700s Edward Jenner carried out an experiment showing that inoculation of a person with the cowpox "germ" would protect that person from smallpox. Because of Jenner's work, the inoculation of individuals with cowpox pus to protect against smallpox became known as **vaccination.** We now use the term vaccination to describe an inoculation of any antigen (or microorganism) into an animal in order to induce a protective immune response. The material injected is often referred to as a **vaccine.**

Vaccines may consist of living or dead microorganisms or extracts of these microorganisms. Some vaccines consist of purified materials, such as protein toxins or polysaccharides. During the preparation of these vaccines, organisms generally are **attenuated** or **inactivated.** Organisms are inactivated by heating or by treating them with chemicals such as merthiolate, acetone, or formalin. Attenuated microorganisms are live organisms that are unable to cause a serious infection, but can still induce an immune reaction. For example, the Sabin polio (oral) vaccine consists of an attenuated poliomyelitis virus. This virus does not cause serious disease in humans, although it can induce a lifelong immunity in the human host. The living organisms are attenuated by various procedures, such as repeated cultivation in laboratory media until the vaccine is no longer able to cause serious disease, or by selection of avirulent mutants.

431
The exotoxins □ from several different bacteria, including those produced by *Clostridium tetani* and *Corynebacterium diphtheriae*, are used to make important vaccines. The exotoxins are denatured or inactivated by treatment with formalin, and in this form they are known as **toxoids.** Toxoids do not cause detrimental changes in the host, although they can initiate an immune response that will neutralize the active toxin. In order to make some toxoids more immunogenic, they are sometimes absorbed to adjuvants, like aluminum hydroxide or aluminum phosphate. Such a complex, when injected into an animal, releases the toxoid slowly so that the immune system becomes stimulated over a period of a few days, resulting in an enhanced immune reaction.

Viral vaccines may be made by treating the virus with formalin or with ultraviolet light, or by attentuating the virus using special culture techniques. Formalin-killed viruses are used in the Salk polio vaccine, developed by Jonas Salk and his coworkers during the early 1950s for use against the polio viruses. The vaccine consists of the three known types of polio viruses. Since inactivated viruses are eliminated rapidly from a vaccinated animal, the immune response is of short duration, lasting less than a year. The Sabin vaccine, developed by Albert Sabin in the late 1950s and early 1960s, consists of attenuated strains of the three major types of polio viruses. These viruses apparently grow to a limited extent in the epithelium of the throat and intestines, eliciting a strong and long-lasting immune response. Large numbers of memory cells can rapidly yield high levels of IgA in the respiratory tract, throat, and intestines, as well as high levels of IgG and IgM, when the vaccinated host encounters the virulent form of the poliovirus. The cell-mediated immune system is also stimulated by the live vaccine and is better prepared to meet a challenge from virulent poliomyelitis viruses.

Recombinant DNA technology □ (genetic engineering) has recently led 268 to the development of a vaccine against the hepatitis-B virus, which causes a serious liver infection affecting 200,000 to 300,000 people in the U.S. each year. The hepatitis B vaccine consists of a purified virus envelope protein that must be injected in three doses over a six-month period. This vaccination appears to induce an immune response that lasts up to five years.

A vaccine against an agent of pneumonia, *Streptococcus pneumoniae*, has also been developed and is used primarily in older persons or those with a compromised immune system. This vaccine consists of a mixture of the capsular materials from the most common strains that infect humans. The vaccine was released in 1977 and has played a central role in reducing the number of deaths due to pneumonia in older individuals. This disease ranks among the ten leading causes of death due to infection in the U.S.

Table 19-6 lists bacterial and viral diseases that can be averted by vaccination. In the United States, only a few of these vaccines are regularly used. These include the oral polio vaccine and the DPT vaccine, which includes the toxoids of the diphtheria, tetanus, and whooping cough organisms. Booster vaccinations are required with the DPT vaccination in order to build up the antibody titers against the toxins. Booster shots capitalize on the anamnestic response to increase the level and duration of antibodies in the serum.

Vaccination represents one of the most effective yet simple means of protecting populations against deadly or debilitating diseases. It represents the results of the efforts of many scientists working toward unveiling the basic mechanisms of the immune response and the application of these principles.

TABLE 19-6
SOME IMMUNIZATIONS CURRENTLY AVAILABLE FOR HUMANS

DISEASE	AGENT*	WHO SHOULD RECEIVE IT	FREQUENCY OF BOOSTER
Cholera	B	Travellers to endemic areas	6–8 months
Diphtheria	B	Children	Every 10 years
Measles	V	Children before 24 months	
Meningitis *(N. meningitidis)*	B	Individuals at risk	?
Mumps	V	Children before 24 months	As needed
Poliomyelitis	V	Children	
Pertussis	B	Children	As needed for adults
Plague	B	High-risk professions	Every 6 months
Pneumococcal pneumonia	B	Adults over 50 years	?
Rabies	V	Individuals in high-risk professions	
Rubella	V	Infants and females before pregnancy	
Smallpox	V	Laboratory personnel working with small pox virus	
Tetanus	B	All children	Every 10 years
Tuberculosis	B	High-risk individuals	
Typhus	B	Medical personnel	Every 6 months
Yellow fever	V	Travellers to endemic area	Every 10 years

*Agents: B = bacterial agent V = viral agent

SUMMARY

The immune response protects animals from invading microorganisms and cancerous outgrowths.

TYPES OF IMMUNITY

1. Humoral immunity is due to circulating antibodies, while cell-mediated immunity is due in part to lymphocytes and macrophages.
2. Native immunity is the resistance that an organism has at birth because of its distinct physiology, chemistry, or structure.
3. Acquired is that immunity obtained after exposure to foreign substances.

CHARACTERISTICS OF ACQUIRED IMMUNITY

1. Acquired immunity is characterized by memory, specificity, and recognition of self.
2. Memory allows the immune system to respond swiftly and vigorously to an infectious agent that has been previously encountered.
3. The anamnestic response, which involves the rapid and intense response to antigens that have previously been encountered, is based on the ability of the immune system to "remember" the antigen and the presence of memory cells that can respond rapidly.
4. Repeated vaccinations with the same antigen increase the anamnestic response.
5. The specificity of the immune response allows it to differentiate among antigens.
6. The immune response is capable of differentiating between self and foreign antigens, so that the immune reponse does not cause damage to the host.

CELLS OF THE IMMUNE SYSTEM

1. In mammals, the B-lymphocytes, granulocytes, macrophages, and monocytes develop primarily in the bone marrow, while the T-lymphocytes develop in the thymus.
2. When stimulated by an antigen, specific B-lymphocytes differentiate into plasma cells (antibody-secreting).
3. Specific T_H, T_S, and T_C multiply in response to antigen. T_C release a killing factor; T_H participate in helping the development and maintenance of immunity; and T_S suppress immune responses.
4. The neutrophils and macrophages are the major phagocytic cells in the immune system.

ANTIGENS AND ANTIBODIES

1. An antigen is any molecule that can stimulate an immune response. Antigens generally contain numerous antigenic determinants or chemical groups that are recognized by different antibodies.
2. A hapten is a chemical group that is not immunogenic in itself, but when attached to a large molecule, called a carrier, can induce the production of specific antibodies.
3. Antibodies, also called immunoglobulins, are glycoproteins found in blood serum which are produced by cells in the immune system in response to the presence of antigen.
4. Antibodies are bivalent molecules consisting of two light chains and two heavy chains. The arms (Fab regions) of the antibody molecule bind to antigen, while the tails (Fc region) bind to some cells of the immune system and can also fix complement.
5. There are five classes of immunoglobulins: IgG, IgM, IgD, IgA, and IgE.
6. Immunoglobulin variability is achieved by DNA recombination among numerous mini-genes and by RNA splicing.
7. Immunologic tolerance is due to the inability of T-lymphocytes and B-lymphocytes to respond to antigen.

IMMUNE RESPONSE TO ANTIGEN

1. Antibodies protect by combining with antigen and inactivating it or by enhancing its phagocytosis and destruction.
2. Adjuvants are chemicals that enhance the immune response to antigen.
3. Cell-mediated immunity is effected by various subpopulations of lymphocytes and macrophages.
4. The immune response is induced by the interaction of antigen with macrophages, B-lymphocytes, and T-lymphocytes. Antigen stimulates immune cells, and their proliferation and differentiation into effector cells.
5. The clonal selection theory explains how antigens select and stimulate the proper immunocompetent cells to become immune committed.

COMPLEMENT

1. Complement is the name for a number of serum proteins involved in immune responses.
2. Complement proteins frequently are activated when they associate with antibodies attached to antigens.

3. The complement cascade ultimately results in the disruption of plasma membranes of target cells. Certain components of complement, such as C3a and C5a, have other biological properties. Some function as chemotactic factors, allergic factors, and opsonins.

REACTIONS BETWEEN ANTIGENS AND ANTIBODIES

1. Antigens react with antibodies to form lattices of cross-linked molecules, thus causing the precipitation of these molecules. These reactions, called serological reactions, can be used as a means of diagnosing infectious disease.
2. In the classical precipitin reaction, soluble antigens react with antibodies to form a visible precipitate when they combine in optimal proportions.

VACCINATION

1. Edward Jenner is credited with the first successful vaccination against an infectious disease (smallpox).
2. Vaccines may be attenuated microorganisms, dead microorganisms, or their products. Toxoids are altered microbial toxins that are used as vaccines because they retain their immunogenicity.
3. Vaccination induces an immunological memory in the host so that it can develop an anamnestic response when the individual encounters the virulent organism or toxin.
4. Vaccinations against diphtheria, whooping cough (pertussis), and tetanus (DPT), and poliovirus, are recommended in the first months of life.

STUDY QUESTIONS

1. Comment on the following statements:
   a. An antibody molecule can bind only one type of antigen.
   b. A given antigen may bind to various antibodies.
   c. Antibody production is based on antigen-antibody reactions.
   d. Natural immunity is based on genetic traits.
2. Define the following terms:
   a. Hapten.
   b. Carrier.
   c. Antigen.
   d. Antibody.
   e. Immunogenic determinant.
3. Describe in detail the structure of:
   a. An antigen.
   b. An IgG molecule.
   c. An IgM molecule.
   d. An IgA molecule.
4. How does the clonal selection theory account for the ability of an animal to respond immunologically to thousands of different antigens?
5. What is an autoimmune disease?
6. Describe the three characteristics of the immune response.
7. Describe the various types of cells involved in immunity.
8. Describe the process of induction of antibody formation by plasma cells.
9. What would be the consequence of defects in:
   a. B-lymphocyte populations;
   b. T-lymphocyte populations;
   c. Macrophage populations.
10. What would happen if the body produced too many Ts lymphocytes in response to a pathogenic microorganism?
11. What are vaccines and how do they work?
12. Describe the kinetics of antibody formation in response to a first and second exposure to the same antigen.
13. What is the basis of antigen-antibody reactions?

SUPPLEMENTAL READINGS

Burnet, F. M. 1957. A modification of Jerne's theory of antibody production using the concepts of clonal selection. *Australian Journal of Science* 20:67–71

Edelman, G. M. 1970. The structure and functions of antibodies. *Scientific American* 223:34–41

Golub, E. S. 1982. *The cellular basis of the immune response.* Sunderland, Mass.: Sinauer Assoc.

Hood, L. E., Weissman, I. L., and Wood, W. B. 1978. *Immunology.* Menlo Park, Calif.: Benjamin/Cummings Publishing Co.

Jerne, N. K. 1973. The immune system. *Scientific American* 229:52–60

Kapp, J. A., Pierce, C. W., and Sorensen, C. M. 1984. Antigen-specific suppressor T-cell factors. *Hospital Practice* 19(8):85–99.

Landsteiner, K. 1945. *The specificity of serological reactions.* Cambridge: Harvard University Press.

Manser, T., Huang, S. Y., and Gefter, M. L. 1984. Influence of clonal selection on the expression of immunoglobulin variable region genes. *Science* 266(4680):1283–1288.

Merigan, T. C. 1985. Viral vaccines, immunoglobulins and antiviral drugs. *Scientific American Medicine* Section 7–XXXIII. New York: Freeman.

Raff, M. C. 1976. Cell surface immunology. *Scientific American* 234:30–38

Roitt, I. 1980. *Essential immunology.* 3d ed. Oxford: Blackwell Scientific Press.

Sabin, A. B. 1981. Evaluation of some currently available and prospective vaccines. *Journal of the American Medical Association* 246:236–241.

CHAPTER 20
DISORDERS ASSOCIATED WITH THE IMMUNE RESPONSE

CHAPTER PREVIEW

THE BOY IN THE PLASTIC BUBBLE

In 1971 a boy named David was born with **combined immune deficiency,** a hereditary condition in which T-lymphocytes and B-lymphocytes are missing. Without these cells, humans develop lethal viral, bacterial, and fungal infections. In order to save David from certain death, he was placed in a germ-free environment. Until his death in 1984, he was the longest-living survivor of combined immune deficiency.

David lived his entire life (except for the last two weeks) in a plastic bubble of one sort or another, in order to exclude microorganisms (fig. 20-1). His bubble consisted of three large chambers that occupied most of his family's living and dining rooms. Everything that entered the bubble had to be sterilized. Air was forced into his bubble through high-efficiency filters that trapped viruses, bacteria, and fungi. David's food was heat-sterilized and stored in jars, while his paper, books, toys, and other heat-sensitive items were sprayed with a bactericide. Materials were passed to David through a double air lock. Inside the chamber, he prepared his own food. David's garbage, dirty dishes, and soiled clothing were slipped out the air locks.

During his life in the bubble, David never touched another human or an animal. He could touch people and animals indirectly using long rubber gloves that extended outside his bubble. David's schooling was via a closed-circuit TV and telephone hookup with a local school. Teachers visited him occasionally for special tutoring, and classmates joined David at his home for class get-togethers. David was an intelligent child, in the upper 20% of all students tested. When not studying or playing he spent his time watching TV and talking to his friends on the telephone. In spite of his genetic disorder and the fact that he could never leave his bubble, David was a very healthy, intelligent, well-adjusted child. David never had an infectious disease during the first twelve years of life.

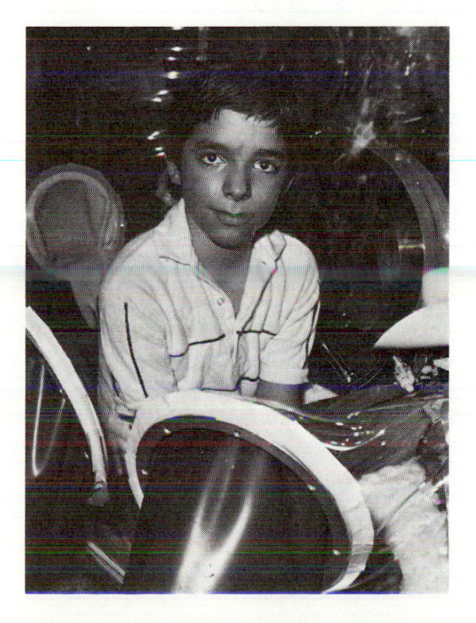

Figure 20-1 David in His Plastic Bubble

In 1983, David underwent a bone marrow transplant in order to give him an immune system. It was hoped that this operation would allow him to live a normal life in the outside world. Bone marrow cells from his sister were used. Unfortunately, the transplanted cells began to attack David approximately three months after his operation, and he began to suffer from **graft versus host disease.** David's first illness was characterized by vomiting and diarrhea. He also developed a stomach ulcer and internal bleeding. In order to be treated, he was removed from his protective bubble and placed in a "sterile" hospital room. In the hospital, David apparently contracted at least one infectious disease, as indicated by the fact that he was being treated with the drug acyclovir to counteract herpes. David died of heart failure two weeks after he emerged from his germ-free plastic bubble.

David's rare condition was due to a defect in his immune system. A great many persons suffer because their immune systems do not work to their benefit. Because of the importance of some disorders associated with the immune system, this chapter is devoted to elucidating the relationship between disease and the immune system.

LEARNING OBJECTIVES

A STUDY OF THIS CHAPTER SHOULD ENABLE YOU TO:

DESCRIBE THE CONSEQUENCES OF AN IMMUNE RESPONSE DIRECTED TOWARD THE HOST'S TISSUE

EXPLAIN THE DIFFERENCE BETWEEN IMMEDIATE TYPE, COMPLEMENT MEDIATED, AND DELAYED TYPE HYPERSENSITIVITIES

DISCUSS THE CAUSE AND CONSEQUENCES OF AUTOIMMUNE DISEASES

EXPLAIN THE PROBLEMS ASSOCIATED WITH TISSUE TRANSPLANTS

In most situations the immune system efficiently protects animals and humans from invading microorganisms and toxic substances. Sometimes, however, an animal's immune system does not do its job in a way that is beneficial and can be responsible for troublesome allergies, chronic degenerative diseases, and even an attack upon the very same animal that it is supposed to protect. Any of these conditions, in its most severe form, can result in the death of the animal. When an animal's immune system reacts strongly or inappropriately to an antigen and pathological changes result, the animal is said to be **hypersensitive** or **allergic.** The most common of the allergic states are summarized in table 20-1. The purpose of this chapter is to explain how microorganisms and noninfectious agents can lead to an immune response that is detrimental to the host. In addition, there will be a discussion on how defective genes lead to defective immune responses and to disease.

IgE-MEDIATED (IMMEDIATE TYPE) HYPERSENSITIVITY

Some antigens, when reintroduced into a sensitized animal, result in **anaphylaxis** (Latin, *ana* = against + *phylaxis* = protection), a situation that often harms rather than protects the animal. **Localized anaphylaxis** is exemplified by such conditions as hay fever, asthma, and hives. **Systemic**

TABLE 20-1
MAJOR TYPES OF HYPERSENSITIVITY REACTIONS

ANTIBODY-MEDIATED ("IMMEDIATE"-TYPE)*			CELL-MEDIATED (DELAYED-TYPE)**	
I ANAPHYLACTIC-TYPE REACTIONS	II IMMUNE-COMPLEX REACTIONS	III CYTOTOXIC REACTIONS	IV ALLERGY OF INFECTION	V TRANSPLANT REJECTION
IgE binds to mast cells or circulating basophils. Antigen combines with bound antibody to trigger the release of various vasoactive amines, such as histamine, serotonin, and slow reacting substance of anaphylaxis.	Antibody combines with a large dose of antigen to produce antigen-antibody complexes that trigger the release of histamine and other mediators. Blood vessels are damaged, resulting in inflammation and necrosis of tissue.	Antigen elicits the formation of antibody, which then combines with target cells. The attached antibody causes the cells to be destroyed by various means (phagocytosis, complement-mediated tissue lysis, and destruction by lymphoid cells).	Antigens of the infectious agent sensitize the infected individual, causing a proliferation of lymphoid cells. Sensitized lymphoid cells react to the presence of the antigens by releasing lymphokines that mediate the hypersensitivity reaction. Humoral antibody not involved.	Tissue from a donor is grafted onto a recipient. A population of sensitized lymphoid cells develops and is instrumental in the destruction of the graft by mechanisms similar to those involved in type IV allergies. Antibody-mediated hypersensitivity can also be involved in graft rejection.
Examples: Food allergies and anaphylactic shock due to bee stings or injections with penicillin.	Examples: Serum sickness, rheumatoid arthritis, systemic lupus erythematosus.	Examples: Drug reactions, agranulocytosis, hemolytic anemia, thrombocytopenic purpura, transfusion reactions, and erythroblastosis fetalis.	Examples: Tuberculin reaction and similar manifestations.	Example: Graft rejection

*Time of onset of reaction occurs within 5–12 hours after contact with antigen.
**Time of onset of reaction is delayed 24–48 hours after contact with antigen.

(throughout the body) **anaphylaxis** is characterized by dilation of blood vessels, swelling of tissue, contraction of smooth muscle, and pain. These changes in the body often lead to shock, asphyxia, and (if not reversed in time) death. The IgE-mediated **atopic** (foreign or out of place) allergies develop very rapidly and occur after frequent exposures to certain antigens.

467

The antigens that stimulate the production of IgE antibodies □ are responsible for immediate type (IgE-mediated) hypersensitivity (fig. 20-2), so called because these effects of the allergy occur shortly after exposure to an antigen. During immediate type hypersensitivity, the IgE molecules bind to **mast cells** (located in the respiratory tract lymph, and lining the blood vessels and capillaries) and to **basophils** (found in the blood) through their Fc regions □. When an antigen **(allergen)** enters a hypersensitive animal, it binds to specific IgE on mast cells and basophils. The antigen-antibody complexes stimulate these cells to release a number of chemicals, such as histamine, serotonin, eosinophil chemotactic factor of anaphylaxis (ECF-A), heparin, platelet activating factors (PAF), slow reacting substance of anaphylaxis (SRS-A), and serotonin (table 20-2). These chemicals are responsible for most of the symptoms of an immediate hypersensitive reaction. Immediate hypersensitive reactions that are not too excessive are generally beneficial, because they promote the movement of antibodies and white blood cells from the blood into aggravated or damaged tissue and also stimulate blood clotting.

463

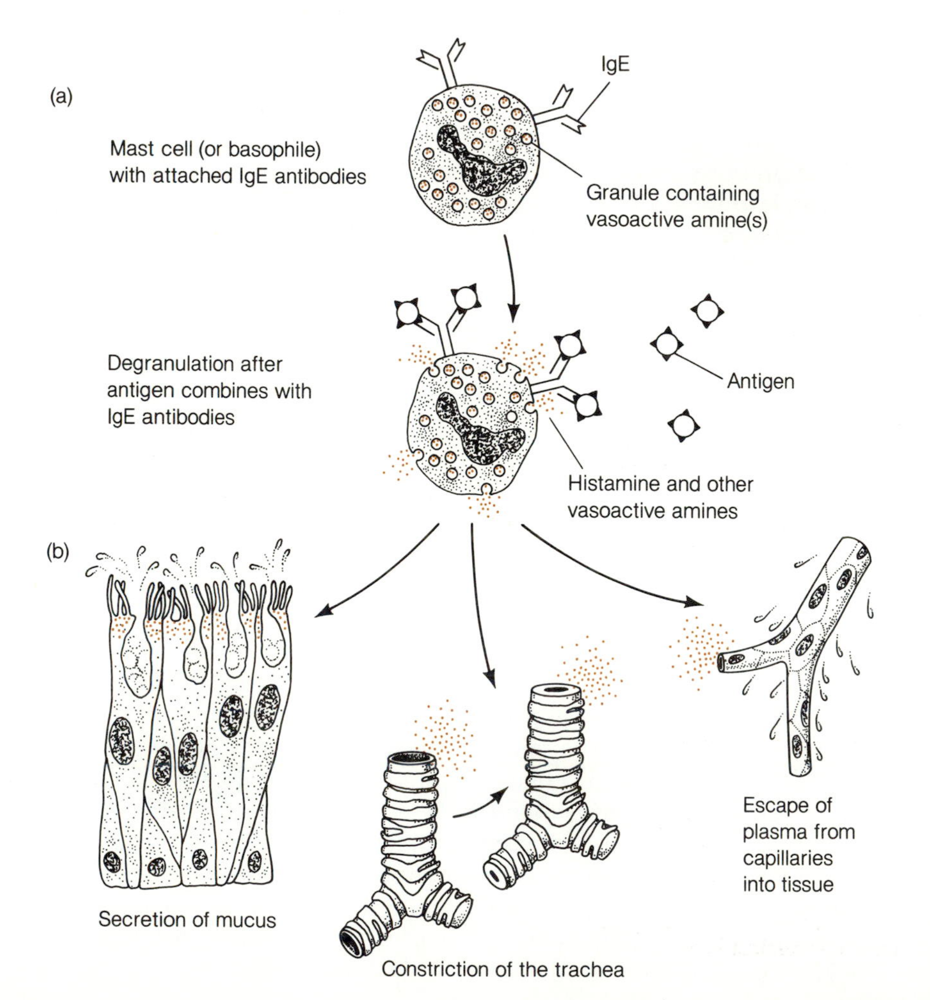

Figure 20-2 Anaphylactic Reaction. IgE are produced in response to certain antigens. These antibodies attach to mast cells in the lymph and basophils in the blood by their stems (Fc portion). When antigen binds to the arms of the cell-bound antibodies, the mast cells and basophils are stimulated to release histamine and other effector molecules. The effector molecules cause the release of plasma from the capillaries, the release of mucus from cells, and the contraction of smooth muscle in the trachea and intestines.

TABLE 20-2
MEDIATORS OF ANAPHYLAXIS

MEDIATOR	COMMENT AND FUNCTION
Histamine	Found in mast cells. Increases blood capillary permeability and smooth muscle contraction of bronchial tubes and attracts eosinophils. Increases gastric secretions of hydrochloric acid.
Eosinophil chemotactic factor of anaphylaxis (ECF-A)	Found in mast cells and basophils. Attracts eosinophils and enhances complement reactions.
Heparin	Found in mast cells and basophils. Inhibits coagulation.
Platelet-activating factors (PAFs)	Found in mast cells and basophils. Aggregates and causes the lysis of platelets, which release serotonin.
Slow-reacting substance of anaphylaxis (SRS-A)	Formed and secreted by mast cells and basophils after they are activated. Increases blood capillary permeability and smooth muscle contraction of bronchial tubes.
Serotonin	Found in blood platelets. Increases blood capillary permeability.

In addition, immediate hypersensitive reactions may have benefits such as ridding an infected gut of worms.

Histamine induces smooth muscle contraction, the release of mucus, and the dilation of blood vessels (fig. 20-2). Excessive smooth muscle contraction in the **trachea** and **bronchi** closes these air passages and causes **asphyxia** and death if not reversed. Severe uterine contraction during pregnancy can lead to abortion of the fetus. Histamine-induced dilation of blood vessels causes redness of the affected area and allows the escape of fluid from the blood into the tissues. Extensive loss of fluid into the tissues causes swelling and can lead to shock, which occurs when the volume of blood flowing through the body is decreased to the extent that tissues are not properly oxygenated or cleared of CO_2. Under these conditions, organs begin to fail and death may rapidly follow.

The effects of excessive histamine release can be partially reversed by **antihistamines,** such as diphenhydramine and chlorpheniramine maleate. Antihistamines are believed to compete with histamines by inhibiting histamine binding to tissue. Antihistamines help alleviate the symptoms of hay fever, since this allergy is IgE-mediated. **Epinephrine** (adrenaline) is a hormone that is also frequently used to reverse the effects of histamine. Epinephrine stimulates smooth muscle dilation, and this effect leads to the relaxing of the trachea and bronchi and consequently prevents asphyxiation. The action of epinephrine is very rapid, so individuals undergoing anaphylactic shock can be saved by prompt administration of this hormone.

Serotonin (5-hydroxytryptamine) functions both as a neurotransmitter in the brain and as a chemoeffector of smooth muscle. In some situations, however, serotonin has been found to be 10,000 times more effective than histamine in causing smooth muscle contraction. In addition to causing asphyxiation, serotonin increases the respiratory rate, decreases central nervous system activity, produces pain, and stimulates histamine release. A large amount of serotonin is found in blood platelets and is released when platelets are aggregated and lysed by platelet activating factors (PAFs).

Bradykinin, a small peptide released in minute amounts by mast cells and basophils, also causes smooth muscle to contract. Bradykinin enhances capillary permeability and leukocyte migration outside the blood vessels. In addition, bradykinin may stimulate peripheral nerves, since it induces pain.

Slow reacting substance of anaphylaxis (SRS-A) causes capillary dilation and smooth muscle contraction, and consequently causes many of the same symptoms as histamine, bradykinin, and serotonin.

Platelet activating factors (PAFs) cause the aggregation and lysis of platelets, and these effects, in turn, aids clot formation and the release of serotonin.

Eosinophil chemotactic factors of anaphylaxis (ECF-A) draw eosinophils to the inflamed site and are believed also to promote some complement fixation reactions □. Eosinophils are believed to release antihistaminic substances and help reverse some of the deleterious effects of anaphylaxis (fig. 20-2).

525

Heparin blocks the formation of thrombin, which converts fibrinogen into fibrin in the blood. Thus heparin decreases blood coagulation.

As you can see, histamine alone does not cause all of the annoying signs and symptoms of hay fever. This is why antihistamines do not completely alleviate hay fever, but simply reduce its severity.

Localized IgE-mediated hypersensitive reactions are very common in humans. Allergies to fungal spores, plant antigens (many of which are pollens), animal antigens (dander), insect venoms, and foods are responsible for hay fever, asthma, eczema, and hives. Figure 20-3 illustrates how an insect's venom (antigen) can stimulate the production of IgE, and how IgE bind to mast cells and basophils through their Fc regions. When antigen is reintroduced in a sensitized individual, it binds to the Fab regions of the antibody and induces the release of histamine and other cell mediators. An atopic allergy results when the cell mediators bind to goblet cells, causing the release of mucus; to capillaries, causing **edema** (swelling) and **erythema** (redness); and to smooth muscle, causing contraction of the trachea, bronchi, intestines, and uterus, as well as pain.

Hay fever (allergic rhinitis) is an IgE-mediated allergic reaction that generally affects the upper respiratory tract and includes such symptoms as nasal congestion, draining sinuses, puffiness and swelling in the nasal passages and around the eyes, itching, and sneezing. Antihistamines alleviate most hay fever symptoms, apparently because histamine is the major mediator of these allergies.

Asthma is an allergic reaction that results in the contraction of the trachea and bronchi. The inability to fill the lungs, accompanied by wheezing and the lack of response to antihistamines, is a sure sign of this type of allergic reaction. Antihistamines are not effective in reversing the smooth muscle contraction because its main mediators are SRS-A and serotonin. This condition can be reversed by epinephrine and aminophylline.

Hives (urticaria) is an IgE-mediated allergic reaction that results in wheals on the skin (fig. 20-3). Wheals are swollen white to pink blotches that appear on the skin sometimes within minutes after contact with the allergen. Allergens that do not touch the skin, such as those in foods, can also cause hives. Food allergies can cause diarrhea, vomiting, and severe pain in the intestines instead of hives. Eczema is a very severe IgE-mediated allergic reaction of the skin and is characterized by inflammation, itching, and the formation of scaly skin.

Figure 20-3 Bee Sting and the Allergic Reaction. IgE, produced in response to bee venom, attach to mast cells and basophils. If the bee venom enters a sensitized person, it binds to the IgE cell-bound antibodies and stimulates the release of histamine and other effector molecules. Local reactions (wheals) and/or systemic reactions (anaphylactic shock) may occur.

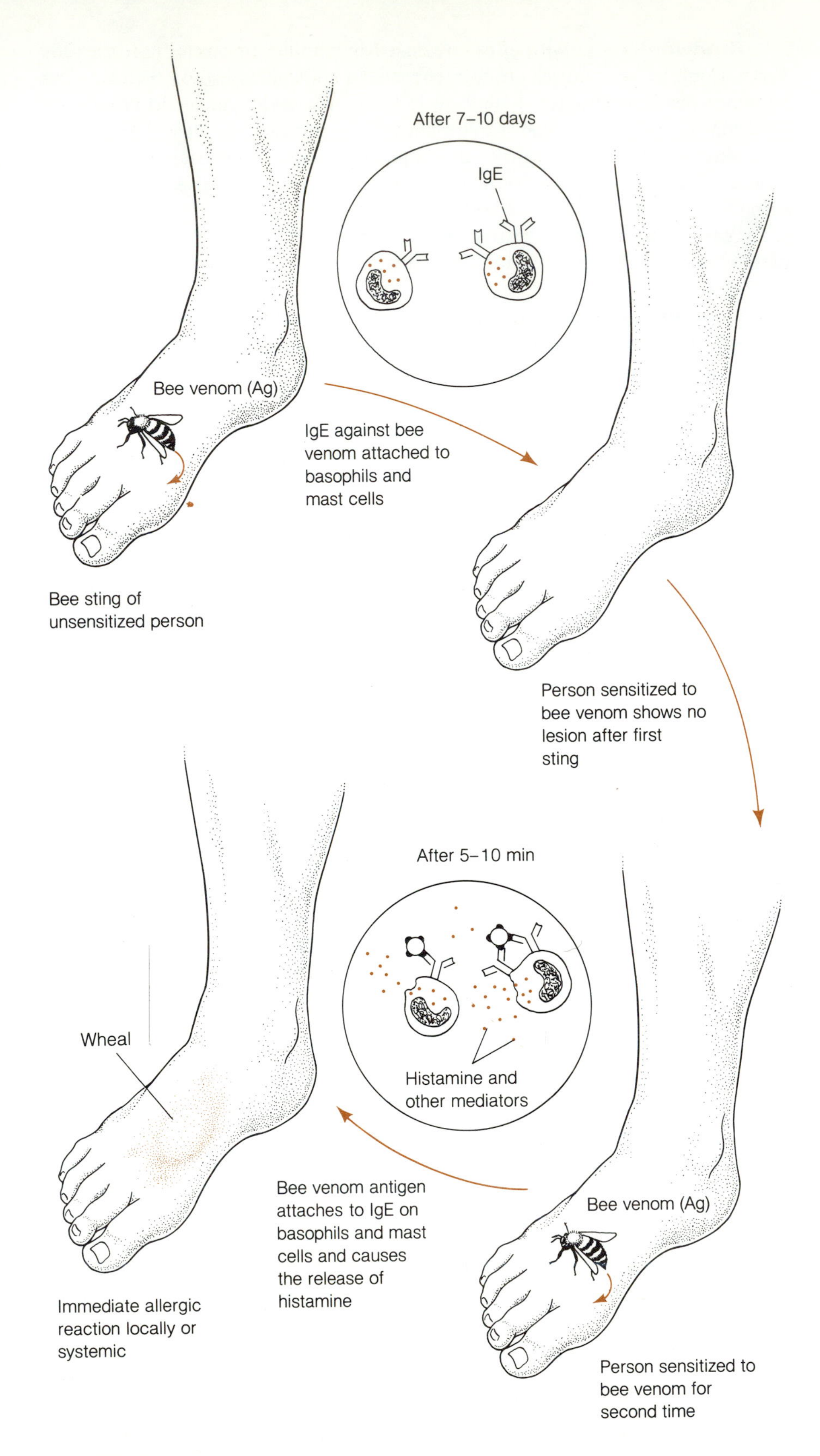

Cutaneous anaphylaxis is an IgE-mediated allergic reaction of the skin to an intradermal injection of an allergen. The allergen causes, within a few minutes, the formation of a flat, pale wheal surrounded by an inflamed red area. This type of hypersensitivity is sometimes called a **wheal and flare** reaction. **Generalized anaphylaxis** is an IgE-mediated allergic reaction throughout the body caused by an injected allergen in a sensitized animal. The allergen causes constriction of the respiratory passages, which leads to suffocation and loss of fluid from the blood. Either one of these conditions can cause death in a few minutes. Generalized anaphylaxis is sometimes observed in people and animals stung by bees or injected with penicillin. Penicillin by itself is incapable of inducing the production of specific IgE; it is thought that penicillin becomes an antigen when it combines chemically with host protein, as occurs in approximately 4% of penicillin injections. Antibodies can then be made against the hapten penicillin. As mentioned earlier, the production of antibodies against penicillin requires approximately two to three weeks. Thus, no adverse reaction occurs unless penicillin is still being injected, but the person becomes sensitized to penicillin. If penicillin is injected into a sensitized person, it binds to the Fab portions □ 463 of the IgE attached to mast cells and basophils. This antibody-hapten complex induces the release of the cell mediators (histamine, etc.) throughout the body. If high concentrations of the cell mediators are released, extensive dilation of blood vessels and excessive loss of fluid from the blood causes shock. In addition, there is a complete constriction of the trachea and bronchi, which results in suffocation. These reactions occur so rapidly that death can result within 10 to 15 minutes. Generalized anaphylaxis can be treated with an injection of epinephrine. Animals or humans who demonstrate mild allergic reactions to penicillin must not be given this antibiotic, because there is a good chance that a severe reaction will occur. We shall see later that penicillin can also cause a cell-mediated (delayed-type) hypersensitivity. Even skin contact with penicillin can result in lesions of the skin and mucous membranes.

Persons who are allergic to common plant, animal, and fungal antigens can sometimes be **desensitized** to the antigens by injecting them with small amounts of antigen over a long period of time so that high titers of IgG antibodies □ 467 against them are produced. These IgG are called **blocking antibodies** because they bind to the antigens so that they are unable to attach to IgE. Some scientists claim, however, that desensitization is due to a preferential stimulation of suppressor T-cells rather than blocking antibody. The allergen (antigen or hapten □ 462) is identified by injecting it intradermally (under the skin) into the sensitized person or animal. The development of a wheal and flare reaction within a few minutes indicates an allergy to the antigen or hapten.

IgG-COMPLEMENT-MEDIATED HYPERSENSITIVITY

When antigens are introduced repeatedly into a sensitized animal, they react with IgG and IgM antibodies and consequently stimulate complement activation □ 472, which can cause severe tissue damage. For instance, in the treatment of a disease such as tetanus by passive immunization against the toxin with immunoglobulins made in domestic animals, a condition known as **serum sickness** sometimes develops. Serum sickness is characterized by fever, rash, and swelling of the lymph nodes, ankles, and face. The immune reaction can cause serious internal problems, such as kidney dysfunction, degeneration of vascular and cardiac tissue, and lesions in the joints (arthritis). The complement fixed affects the body in a number of ways. C3a and C5a, called **anaphylatoxins,** cause the release of cell mediators such as histamine from mast cells and basophils. These mediators cause the fever, rash, and swelling characteristic of serum sickness. The lysis of host cells by $\overline{\text{C5b6789}}$ also contributes to the symptoms. The antibody-antigen complexes that are fixing complement in blood vessels and heart, in the glomeruli of the kidney, and in the joints can lead to vascular disease, glomerulonephritis, and arthritis, respectively.

A subcutaneous injection of foreign antigens into a sensitized animal can result in what is called the **Arthus reaction:** antibody-antigen complexes can also occur in any tissue where the antibody-antigen-complement complexes accumulate. Dilation of capillaries results from the anaphylatoxin activity of C3a and C5a and results in tissue swelling. C5a also contributes to Arthus reactions by attracting PMN leukocytes. IgG-complement-mediated immune reactions play an important role in such diseases as **rheumatic fever, erythroblastosis fetalis,** some **drug-induced diseases,** and some **autoimmune diseases.**

Rheumatic fever is caused by the immune system's reaction to a chronic *Streptococcus pyogenes* infection in the throat. The bacterial antigens sensitize the infected person. IgG and IgM form antigen-antibody-complement complexes with the bacterial antigens, which are ever-present in a chronic infection. When these complexes localize in the heart, they cause **endocarditis;** in the kidney glomeruli, they cause **glomerulonephritis;** and in the joints they cause **arthritis.** Some IgG and IgM made against M-protein from the bacterial cell wall crossreact with host antigens found in the heart. In this case, immunoglobulins bind to heart cells and activate complement, which result in the lysing of heart cells. When an immune system attacks its owner, we have what is known as an **autoimmune reaction** (see Autoimmune Diseases).

IgG-complement-mediated hypersensitivities frequently are divided into two categories: the immune-complex reactions and the cytotoxic reactions. In the immune complex reactions, the damage is caused by complement that has been activated on a *free* antigen-antibody complex. Immune complex reactions are found in serum sickness, Arthus reaction, rheumatoid arthritis, and systemic lupus erythematosus. The damage in cytotoxic reactions is induced by complement that has been activated on a *tissue cell*–antibody complex. In addition, there may be cellular lysis by antibody-dependent cells of the immune system. Cytotoxic reactions are found in transfusion reactions to incompatable blood; erythroblastosis fetalis; drug-induced agranulocytosis (depletion of polymorphonuclear leukocytes); infection-and drug-induced hemolytic anemia (depletion of the red blood cells); and infection-and drug-induced thrombocytopenic purpura (depletion of thrombocytes or platelets).

CELL-MEDIATED (DELAYED TYPE) HYPERSENSITIVITY

Viruses, bacteria, fungi, and noninfectious agents that continually stimulate the immune system induce T-lymphocytes and macrophages to activity □.

FOCUS

AN ALLERGIC REACTION TO A VACCINE

In the U.S., most babies and children under the age of two receive a series of three to four inoculations of the diphtheria-pertussis-tetanus (DPT) vaccine to protect them from these diseases. Before entering school, most children are required to have a recent booster shot. Thus, most people in the U.S. receive four or five inoculations of the DPT vaccine early in life. Without inoculations against diphtheria, whooping cough (pertussis), and tetanus, epidemics of these diseases would occur frequently.

There is a problem, however, with the pertussis component of the DPT vaccine. Studies on the effects of DPT indicate that some children vaccinated with DPT suffer severe reactions including crying, fever and drowsiness. These reactions occur 3–4 hours after injection and last for approximately 12 hours. More severe allergic reactions also occur: approximately one in 700 babies and children who receive DPT injections suffers convulsions and goes into shock. Brain and motor nerve damage that leads to severe retardation and paralysis occurs at a frequency of one in 50,000 babies and children vaccinated. Over the years, the pertussis component in the DPT vaccine has produced many thousands of mentally retarded and crippled children and adults in the U.S. because of allergic reactions.

Presently, there are approximately 2,000 cases of pertussis and an average of six deaths from pertussis each year in the United States. If the pertussis toxoid were left out of the DPT vaccine, the nation would risk epidemics of whooping cough in which millions would become ill and thousands would die each year. In 1934, before the DPT vaccine was used in the U.S., there were more than 250,000 cases of the disease and more than 7,500 deaths. In England, where the pertussis vaccination is no longer given to all babies and young children, whooping cough has recently become epidemic: in one year there were more than 100,000 cases and about 30 deaths.

The dangers of a whooping cough epidemic in the U.S. far outweight the dangers of allergic reactions. This knowledge is of little consolation, however, to those parents whose children become mentally retarded and crippled by the vaccine. Stimulated by public outcry and numerous lawsuits against drug firms that prepare the DPT vaccine, Congress is attempting to make the Public Health Agencies of the U.S. Government and the American Medical Association warn doctors and parents of the risks of the vaccine, so that babies who have allergic reactions to the first vaccination can be excused from further vaccinations.

This activity occasionally leads to **delayed type hypersensitivity,** so named because it appears 24 to 48 hours after the host encounters the antigen. It is characterized by **erythema** (redness), due to blood flow to the damaged area by the leukocytes, and **induration,** a consequence of the accumulation of macrophages and lymphocytes in the affected area. The activated T-lymphocytes and macrophages, in their attempt to eliminate the foreign antigens, can cause severe tissue destruction.

Much of the discomfort and damage associated with infections by organisms such as *Schistosoma japonicum* (schistosomiasis), *Mycobacterium tuberculosis* (tuberculosis), *M. leprae* (leprosy), *Treponema pallidum* (syphilis), and *Histoplasma capsulatum* (histoplasmosis) are due to long-term immune reactions against these organisms by macrophages and cytotoxic T-cells. For example, the organism that causes syphilis is an intracellular bacterium that grows very slowly. It causes little damage to the cells it infects and is not readily destroyed by the host's immune system, because it generally resides inside cells. The constant release of *Treponema* antigens from the infected tissue sensitizes the animal, however, and leukocytes, macrophages, and T-cells are drawn to the site of infection. Treponema antigens binding to the T-cell receptors cause the T-cells to release immune substances called **lymphokines.** The lymphokines include **chemotactic factors** that attract macrophages, **migration inhibitory factors** (MIF), and **macrophage activating factors** (MAF), which keep macrophages at the site of infection and cause the macrophages to actively phagocytize and destroy the free *Treponema* at the site of infection. T-cells also release **cytotoxic factors** that kill the *Treponema* and nearby host cells. If the infection is a chronic one, there is continual destruction of tissue. The **chancres** (lesions of the primary stage), **rashes** (characteristic of the secondary stage), and **gummas** □ (lesions of the tertiary stage) of syphilis result partly from tissue destruction by T-cells and macrophages at sites of infection. Since primary chancres appear three to five weeks after infection, they may be due not only to extensive *Treponema* replication but also to cell-mediated hypersensitivity. Arthus type (IgG-complement-mediated) allergic reactions and cell-mediated allergic reactions are responsible for the extensive rash and ulcers that are part of secondary syphilis and generally begin a month or so after the initial infection. IgG-complement-mediated allergic reactions and cell-mediated allergic reactions are responsible for the symptoms of tertiary syphilis, which can include brain degeneration, circulatory system disintegration, kidney and liver disfunction, arthritis, and bone deformation. Death from a chronic syphilis infection in some cases does not occur for 10–20 years after the initial infection.

649

Chronic infections of the brain by viruses, such as those that cause chickenpox and measles, are believed to cause one type of **multiple sclerosis.** These viruses are thought to damage the **glial cells** that form the myelin sheaths around nerve cells, as well as the capillaries that bring blood to the brain. Myelin protein released from the damaged cells passes from the brain into the blood because of a breakdown in the blood-brain barrier. Consequently, myelin protein, which normally does not enter the circulatory system, is recognized as a foreign antigen and stimulates the immune system against the brain myelin. Macrophages (fig. 20-4) and T-lymphocytes infiltrate the brain and further destroy the myelin sheaths around the nerve cells.

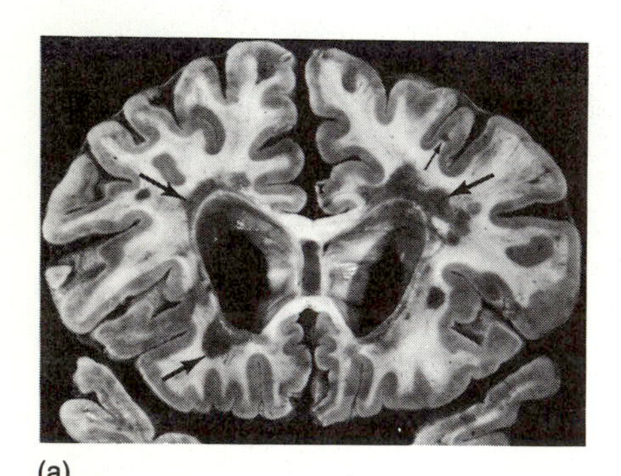

(a)

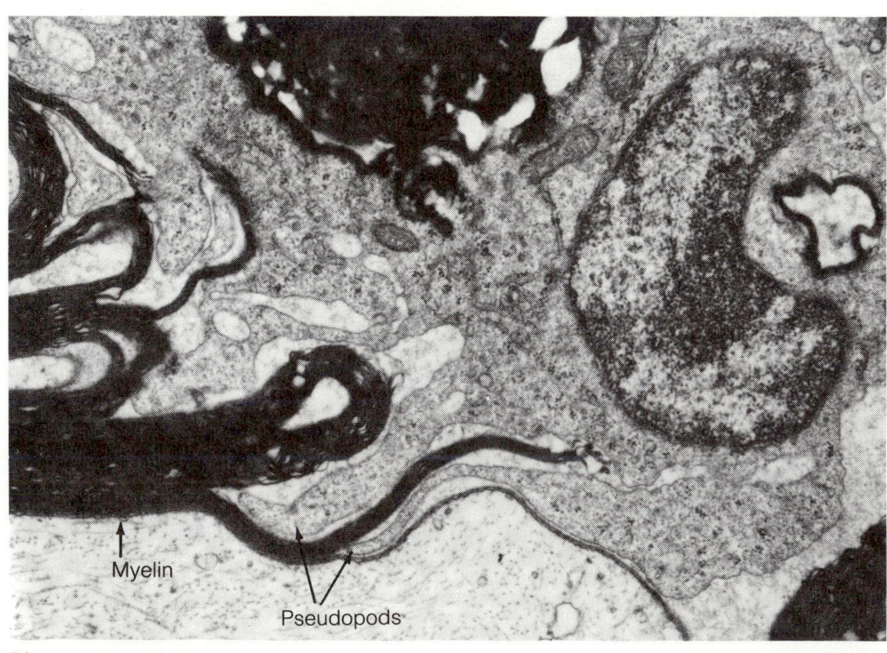

(b)

Figure 20-4 Multiple Sclerosis and Cell Mediated Immune Reaction. *(a)* The brain from a man who died of multiple sclerosis shows numerous dark areas where the nerves have lost their myelin sheaths. Arrows indicate location of lesions caused by the autoimmune response. *(b)* A macrophage is destroying the myelin sheath in the nervous system of a rabbit that was sensitized to myelin. Macrophage pseudopods are separating the myelin from an axon.

Because nerve cells in the brain do not function normally without a myelin sheath surrounding them, a person suffering from this induced autoimmune disease often loses his or her sight and muscular coordination. A person suffering from multiple sclerosis eventually becomes unable to speak and wastes away because his or her muscles are no longer stimulated. Because the host's own immune system attacks self antigens, this type of multiple sclerosis is considered an autoimmune disease (see Autoimmune Diseases).

Delayed hypersensitive reactions of the skin are often used to diagnose bacterial infections □, such as those caused by *Mycobacterium tuberculosis* (tuberculosis), *Francisella tularensis* (tularemia), and *Chlamydia psittaci* (psittacosis, lymphogranuloma venereum, trachoma), and fungal infections such as those caused by *Coccidioides immitis* (valley fever) and *Histoplasma capsulatum* (histoplasmosis).

529

One of the best known skin tests is the **tuberculin skin test,** in which a bacteria-free protein fraction (purified protein derivative—PPD) from *Mycobacterium tuberculosis* broth culture is injected under the skin. If the individual has tuberculosis, or has previously been sensitized to *Mycobacterium*, a strong cell-mediated allergic reaction occurs. The cytotoxic T-lymphocytes and macrophages cause swelling, redness, and induration in the area around the site of inoculation.

Allergic contact dermatitis is a cell-mediated allergic reaction that occurs when certain chemicals come into contact with the skin. For example, penicillin and catechols from poison ivy or poison oak leaves, rubbed on the skin of a sensitized individual, can cause severe blistering of the skin (fig. 20-5 and 20-6). Destruction of the epidermis by T-lymphocytes and macrophages is accompanied by severe itching. The small molecules that cause

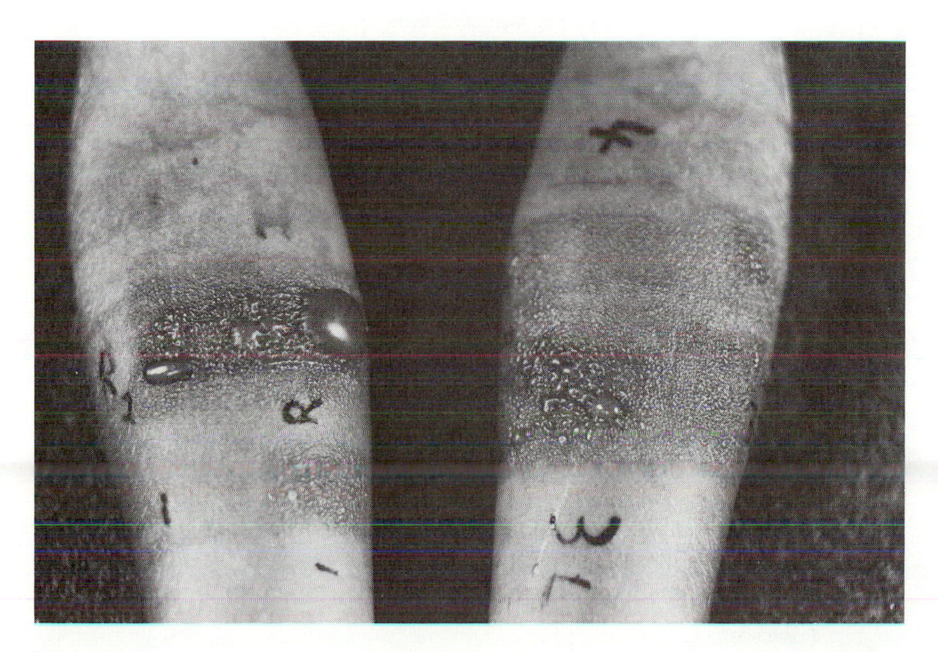

(a)

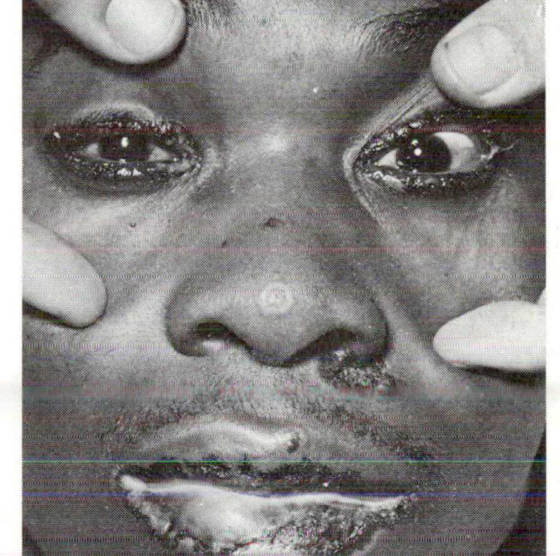

(b)

Figure 20-5 Allergic Contact Dermatitis Is a Delayed Type Hypersensitivity. *(a)* Various materials are tested on a human by taping filter paper patches treated with them. If a person is sensitized to a material, blisters appear within a day or so after contact. *(b)* This man demonstrates allergic contact dermatitis to penicillin. He has developed blisters on his eyelids, nasal mucosa, and lips.

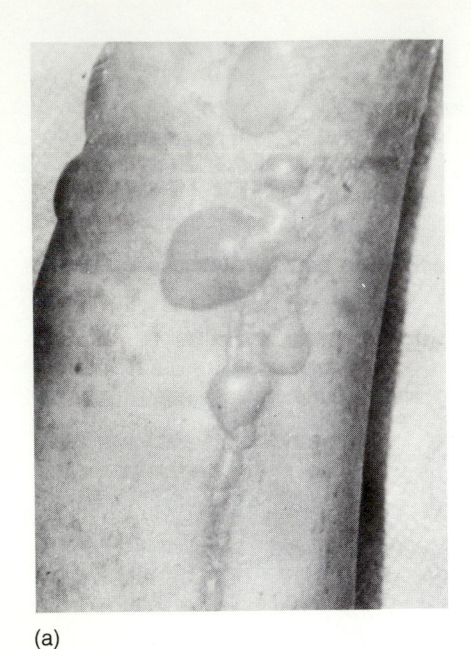

(a)

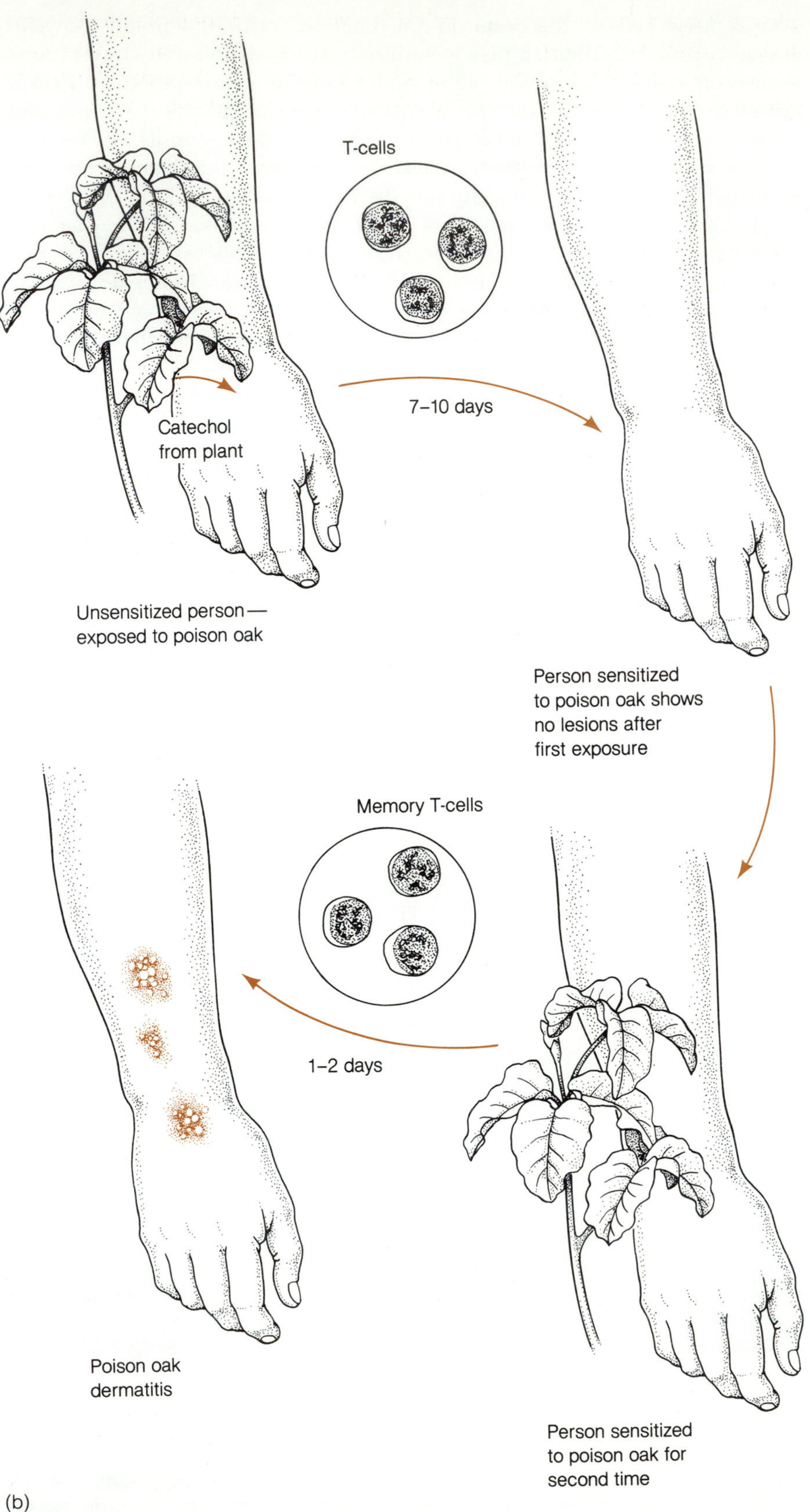

(b)

Figure 20-6 Allergy to Poison Oak is a Delayed Type Hypersensitivity. *(a)* Typical lesion of poison oak dermatitis. *(b)* Exposure to catechol molecules in poison oak (and also in poison ivy) leads to them combining with skin proteins to form an immunogenic compound (an antigen). The initial contact results in the development, in approximately one to two weeks, of T-lymphocytes that are sensitized to the catechol molecules, but no visible lesions develop. Upon subsequent contacts with the catechols, sensitized T-lymphocytes will respond immunologically to the antigen and initiate a delayed-type hypersensitivity reaction that leads to a dermatitis at the site of contact with the catechols.

allergic contact dermatitis react with skin proteins and consequently become antigens that trigger host immune reactions. The first time an animal comes into contact with the allergen, no dermatitis results, but memory T-lymphocytes can quickly respond to the chemical if exposure is repeated.

AUTOIMMUNE DISEASES

An autoimmune disease occurs when the immune system becomes sensitized to self antigens, or when it attacks new antigens on cells and then proceeds to destroy extensive amounts of tissue.

One type of arthritis, **rheumatoid arthritis,** is caused by IgM (rheumatoid factor) that bind to IgG associated with the **synovial membrane** in the joint (table 20-3). Complement fixation results in inflammation, stiffness, and deformity of the joints. It is hypothesized that IgM, possibly made against some foreign antigen, crossreacts with human IgG.

As mentioned previously, in the case of rheumatic fever, M-protein from the cell wall of *Streptococcus pyogenes* stimulates the production of antibodies that crossreact with a glycoprotein fraction from heart valve tissue. Complement fixation in the heart valve results in heart disease.

Many chemicals and drugs can react with the surfaces of cells and hence create new antigens that stimulate an immune response. One disease, called **autoimmune thrombocytopenic purpura** (lack of platelets), is caused by immunoglobulins attacking platelets that have been chemically altered. Antibodies bind to the altered platelets, complement is fixed, and the platelets

TABLE 20-3
AUTOIMMUNE DISEASES

DISEASE	PART OF IMMUNE SYSTEM INVOLVED	TISSUES AFFECTED (SPECIFIC AUTOANTIGENS)
Encephalomyelitis	T-cells and macrophages	Brain cells
Diabetes mellitus (juvenile)	T-cells	Pancreas
Multiple sclerosis	Macrophages and T-cells	Brain cells (myelin protein)
Myasthenia gravis	Antibodies	Nerve and muscle (receptor for acetylcholine)
Goodpasture glomerulonephritis	Antibodies and complement	Kidney
Hashimoto's thyroiditis	T-cells and antibodies	Thyroid gland
Hemolytic anemia	Antibodies and complement	Red blood cells
Thrombocytopenic purpura	Antibodies	Blood platelets
Pernicious anemia	Antibodies and complement	Intestines (receptor for B12 absorption)
Addison's disease	T-cells	Adrenal gland
Sympathetic ophthalmia	Antibodies	Eye
Ulcerative colitis	T-cells	Intestines
Rheumatic fever	Antibodies	Systemic; in particular heart, kidneys, joints
Rheumatoid arthritis	Antibodies	Systemic; in particular joints (mutant IgG associated with synovial membranes)
Systemic lupus erythematosus	Antibodies	Systemic (DNA from many tissues)

are lysed. In a normal person, platelets release serotonin and thromboplastin (an enzyme that converts prothrombin to thrombin). Because thromboplastin is required for blood clotting, persons with this disease have problems with internal bleeding. Multiple hemorrhages, seen as purple spots (purpura), appear in the skin and on the gums.

A number of enveloped viruses, □ such as those that cause influenza, measles, mumps, and chickenpox, insert their antigens into the plasma membranes □ of the cells they infect. This activity results in an attack by macrophages and T-lymphocytes on virus-infected cells. If the virus infection becomes latent, destruction of host cells is caused by the immune system attempting to destroy the latent viruses. It is believed that **juvenile diabetes mellitus** is caused by a virus infection of the cells found in the pancreatic islet of Langerhans. The virus is thought to modify these cells so that they are destroyed by macrophages and T-lymphocytes. Islet of Langerhans cells produce the essential hormone **insulin,** which controls the level of glocose in the blood. Without appropriate amounts of the hormone, a person slowly starves and various tissues degenerate.

395

80

The **Guillain-Barre syndrome,** a disease characterized by paralysis of many muscles, and death when the paralysis blocks the long diaphragm, generally occurs at a very low frequency after influenza virus infections, but occasionally this syndrome is also observed after infections by a few other viruses. The influenza virus is believed to infect the peripheral nervous sytem and destroy some of the Schwann cells that form myelin sheaths around peripheral nerves. If the immune system becomes sensitized to myelin proteins, a cell-mediated immune response can develop, causing extensive destruction of the myelin in the peripheral nervous system.

The lymphocytic choriomeningitis virus (LCM) infects the choroid membranes and meninges (membranes that surround the brain and spinal cord). The damage to the membranes is sufficient to allow macrophages and T-lymphocytes to infiltrate the brain. Possibly, the attack on the virus by macrophages and T-lymphocytes causes much of the disease. A cell-mediated hypersensitive reaction rather than the virus infection, may be the cause of the disease, because the virus infection is not lethal in thymectomized (thymus surgically removed) animal hosts. Since T-lymphocytes mature in the thymus, thymectomized mice have no mature T-lymphocytes with which to develop a cell-mediated immune response.

As previously mentioned, one type of multiple sclerosis may be caused by chickenpox and measles viruses. The measles virus is known to cause sclerosing panencephalitis (SSPE), a situation in which myelin is destroyed. Research has shown that the chronic infection results from the infiltration of the brain with macrophages and T-lymphocytes sensitized against myelin.

As you can see, a number of different human diseases result from immune responses directed against foreign and self antigens. The diseases we have described represent but a few examples. Table 20-3 lists some of the better-known autoimmune diseases.

IMMUNODEFICIENCIES

Phagocytic cells, B-lymphocytes, and T-lymphocytes are involved in eliminating microorganisms that manage to infect the host. The first line of cellular defense against invading microorganisms is the phagocytic cells. To be effective, the phagocytic cells must be able to recognize foreign antigens. Then they must move toward them and bind to them. After phagocytosis, the phagocytes must be able to fuse **lysosomes,** containing hydrolytic enzymes, with the phagosomes; produce toxic H_2O_2; and iodinate the cell walls of bacteria. Those organisms destroyed by this process are prevented from causing an infection, while those that escape can cause disease.

A fatal childhood disease called **Chediak-Higashi syndrome** is due to abnormal phagocytes whose lysosomes fuse slowly with the phagosomes. Because bacteria are not effectively destroyed, they grow inside the phagocytes and spread throughout the body.

Another fatal childhood disease, called **chronic granulomatous disease,** results from the inability of lysosomes to produce H_2O_2 and their inability to iodinate microorganisms. Phagocytized bacteria grow inside the phagocytes and stimulate an inflammatory reaction. The phagocytes and surrounding fibroblasts (cells found in most tissues) develop into a chronic inflammation known as a **granuloma.**

Some persons lack T-lymphocytes, or their T-lymphocytes are ineffective. Consequently, these persons are especially susceptible to viral and intracellular bacterial infections. For example, a vaccination with live attenuated viruses, such as those commonly used in the Sabin polio vaccine, can be lethal to persons who lack functional T-lymphocytes. Some of these defects are inherited, while others are not. Animals with ineffective T-lymphocytes may be producing **auto-antibodies** against the lymphocytes **(episodic lymphopenia),** or cancerous tissue may be stimulating the production of a serum factor that blocks the surface receptors on the T-lymphocytes **(Hodgkin's disease).**

Persons lacking B-lymphocytes or having nonfunctional B-cells are especially sensitive to bacterial infections of the skin and the respiratory tract. In the case of **acquired agammaglobulinemia,** the individual has normal levels of B-cells, but these are unable to develop into antibody-releasing plasma cells because of suppressor T-lymphocytes, that suppress their development.

Defective kidney or liver metabolism can cause the accumulation of toxic substances that depress the immune response. Virus infections and cancerous growths frequently cause the release into the blood of molecules that suppress the immune system.

The excessive use of cortisone and cortisol, or the abnormal release of these hormones from the adrenal cortex during periods of stress, causes a deficiency of the immune system called **Cushing's disease.** These persons are extremely susceptible to virus and intracellular bacterial infections. Adrenocortical hormones such as cortisone and cortisol are often used to alleviate chronic pain, inflammations, and allergies, and to protect transplanted tissues. If used for a prolonged period, however, they can result in a severe case of Cushing's disease. Table 20-4 summarizes some of the common immune deficiencies.

TISSUE TRANSPLANTATION AND BLOOD TRANSFUSIONS

Often it is desirable to transfer or transplant tissue or organs from one site to another or from one individual to another. For example, treatment of

TABLE 20-4
IMMUNODEFICIENCY DISEASES

DISEASE	DEFECT	CONSEQUENCE
Selective Dysgammaglobulinemia	Deficiencies of IgM, IgG, or IgA.	IgA deficiency results in gastrointestinal problems such as diarrhea and poor absorption of fats and fat-soluble vitamins. IgM and IgG deficiencies lead to bacterial infections.
Acquired Agammaglobulinemia	Suppressor T-cells inhibit development of B-lymphocytes into plasma cells	Lack of plasma cells results in serious bacterial infections of the skin and lungs.
Congenital Immune Deficiency Syndrome (CIDS)	Most B-cells, plasma cells, and T-cells are missing (inherited autosomal mutation)	Lack of plasma cells and T-cells results in virtually no protection against infection.
Acquired Immune Deficiency Syndrome (AIDS)	Helper T-cells are depressed or missing (infectious).	Lack of helper T-cells leads to infections.
DiGeorge Syndrome	Thymus does not develop correctly. Consequently, T-cells are absent (not inherited).	T-cell deficiency leads to fatal viral and intracellular bacterial infections.
Hodgkin's Disease	Cancer of lymph nodes leads to serum factors that block T-cell receptors (acquired).	T-cell deficiency results in serious viral and intracellular bacterial infections (herpes, tuberculosis).
Bruton's Disease	Total lack of B-cells (inherited, X-linked).	B-cell deficiency leads to serious bacterial infection of the skin and respiratory tract.
Cushing's Disease	Excess secretion of cortisone & cortisol leads to lysis of most T- and B-cells & decreased levels of blood monocytes (developed).	Extreme susceptibility to viral and intracellular bacterial infections.
Hereditary Angioneurotic Edema	Deficiency of C1 esterase inhibitor (inherited).	Unchecked complement activation leads to acute inflammation & possible death by suffocation.
Complement Deficiencies	Deficiency in C3 or C5 (inherited).	Susceptible to bacterial infections.
Complement Deficiencies	Deficiency in C1r, C1s, C2, or C4 (inherited).	Hypersensitive.
Chediak-Higashi Syndrome	Defective phagosome formation (inherited)	Fatal bacterial infections.
Chronic Granulomatous Disease	Defective lysosomes. They lack hydrogen peroxide & iodination enzymes (inherited).	Fatal bacterial infections & inflammation.

severe burns often necessitates that skin be transplanted from an intact region of the body to the burn site so that the wound can heal more quickly. When tissue is transplanted from one site to another in the same individual, the transplant is called an **autograft.** In order to save an individual from death in the case of kidney disease, a kidney from another person is transplanted into the diseased individual. When tissues are transplanted from one individual to another, the tissue is known as an **allograft.** If the transplant is between identical twins or highly inbred animals, the transplant is called an **isograft.** The tissue transplanted between different species, such as between a cat and a dog, is known as a **xenograft.** It is found that autografts and isografts of tissues such as skin generally are successful, because the immune system does not recognize the tissue antigens as foreign. On the other hand, allografts and xenografts generally are not successful because of cell-mediated immunity. T-lymphocytes, which recognize foreign antigens, are responsible for the rejection of grafts. The mechanism of allograft rejection also involves humoral antibodies as well as complement.

BLOOD TYPE INCOMPATABILITIES

Erythroblastosis fetalis is a disease of the fetus and newborn, caused by IgG from the mother attacking the fetus's red blood cells. During development the fetus suffers from anemia, and at birth it suffers from both anemia and jaundice. Often a woman who has Rh^- type blood becomes sensitized to Rh^+ type blood after the birth of her first Rh^+ child. At birth, some of the newborn child's Rh^+ red blood cells enter the mother and sensitize her to these foreign antigens. Subsequently, IgG that passes from her blood system through the placenta into the fetal blood system will attack the blood of any fetus that has Rh^+ blood (fig. 20-7). The IgG binds to the fetal red blood cells, complement is fixed, and the cells are lysed. The fetus responds by releasing immature red blood cells, called erythroblasts, which do not efficiently remove CO_2 or transport O_2. Thus, the fetus suffers from anemia and hypoxia. Most of the byproducts of red blood cell destruction are removed by the mother's circulatory system, so jaundice is mild in the fetus. At birth, however, the mother's circulatory system removes red blood cell debris and the cholesterol from the red blood cells. The cholesterol is turned into bile, which accumulates in the baby's blood, liver, skin, and urine. The bile turns the baby yellow.

Since jaundice develops rapidly after birth and is due the destruction of the baby's RBCs by the mother's circulating antibodies, the condition is also known as **hemolytic disease of the newborn.** If the anemia and jaundice are not alleviated, they result in the death of the infant. These conditions can be reversed by the complete removal of the baby's blood and its replacement with type O blood □. This replacement eliminates most of the mother's cir-

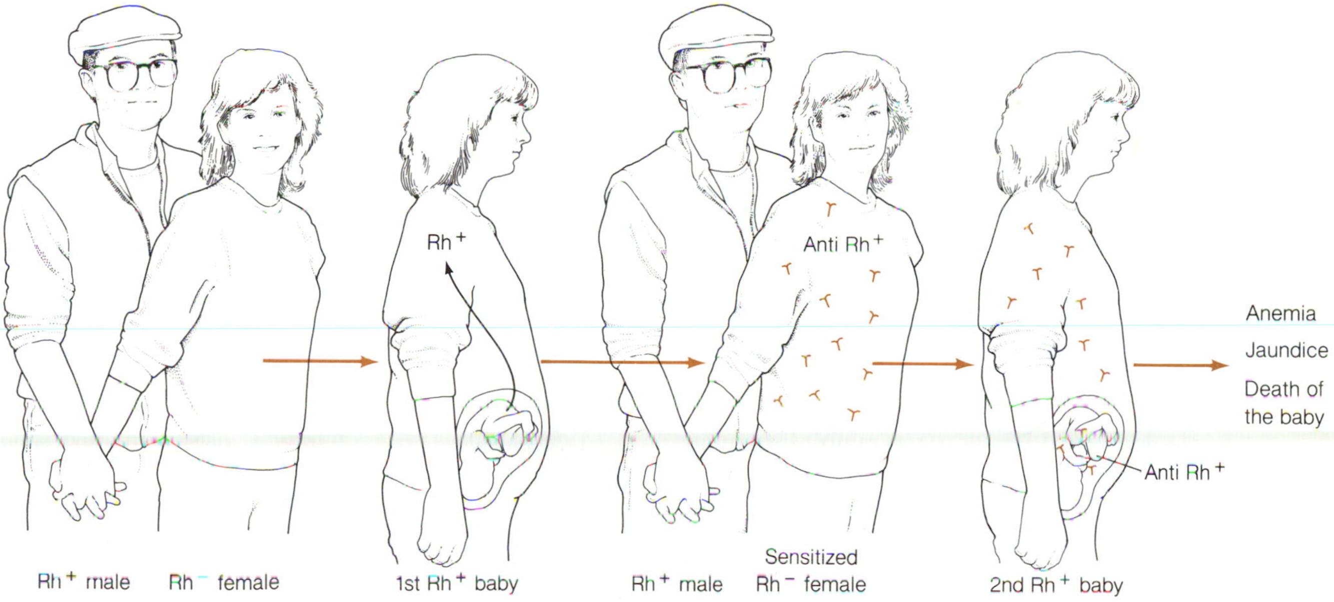

518

Figure 20-7 The Development of Blood Incompatibility. The child of an Rh^+ male and an Rh^- female may be Rh^+. When the Rh^+ baby is born, some of its blood enters the mother and sensitizes her against the Rh^+ blood. Nothing happens to the first Rh^+ baby, but subsequent Rh^+ babies are in danger of being attacked by antibodies from the sensitized Rh^- mother. The mother's antibodies against Rh^+ blood pass through the placenta, enter the baby's blood system, and attack the baby's blood cells. This attack results in anemia, jaundice, and often death of the baby. Those babies that survive the attack on their blood have the disease known as erythroblastosis fetalis.

culating antibodies and prevents further hemolysis of the baby's erythroblasts and red blood cells. Slowly, the baby's RBCs replace the transfused type O red blood cells. Type O blood is used so that the baby does not attack its new blood with antibodies against A and B antigens.

Erythroblastosis fetalis can be prevented by passive immunization of the Rh^- mother with antibodies against Rh^+ blood at the time she gives birth to her first Rh^+ baby. Any Rh^+ baby cells that might enter the mother's circulatory system at birth are rapidly destroyed by the injected immunoglobulins and so do not stimulate the mother's immune system. Thus, the next child is protected, because passive immunization blocks the development of the mother's natural immunity against the baby's foreign antigens.

SUMMARY

1. The immune system can, at times, function improperly or overreact and cause serious diseases.
2. An animal is said to be hypersensitive or allergic to an antigen when pathological changes occur.
3. A sensitized animal is one that has a memory of an antigen and is able to mount a rapid immune response to the antigen.

IgE-MEDIATED (IMMEDIATE TYPE) HYPERSENSITIVITY

1. In a sensitized animal, antigens binding to IgE attached to mast cells and basophils cause these cells to release cell mediators, such as histamine.
2. Histamine causes the contraction of smooth muscle in the trachea and bronchi; dilation of blood vessels (and consequently the outpouring of plasma from capillaries); and the release of mucous from goblet cells.
3. Excessive contraction of the bronchi and trachea and excessive loss of plasma from the blood leads to asphyxiation and shock, a condition known as anaphylaxis.
4. Anaphylactic shock can occur within 5 to 10 minutes after an antigen enters a sensitized animal, so this type of allergy is known as an immediate type (IgE mediated) hypersensitivity.
5. Antihistamines are drugs that can inhibit histamine activity, while epinephrine is a hormone that reverses some of the symptoms histamine produces.
6. Since numerous cell mediators, in addition to histamine, are released during an allergic reaction, antihistamines are unable to alleviate all the symptoms of an allergy.
7. Hay fever, asthma, eczema, and hives are all immediate type (IgE-mediated) hypersensitivities. These conditions are due to the different types of cell mediators that predominate.
8. Persons who are allergic to common antigens can sometimes be desensitized by injecting antigen over a long period of time. This treatment may result in the production of blocking antibodies (IgG or IgA), which inactivate antigen before it can bind to IgE.

IgG-COMPLEMENT-MEDIATED HYPERSENSITIVITY

1. Antigens injected into a sensitized animal generally bind to IgG and IgM, and this effect leads to complement activation.
2. Repeated injection of antigen can cause serum sickness, a condition due to activated complement components that cause anaphylaxis (C3a, C5a) and lyse host cells ($\overline{C5b789}$).
3. Subcutaneous injection of antigen can cause an Arthus reaction, a condition due to activated complement component, which causes a localized allergic reaction (C3a, C5a) and lysis of host cells ($\overline{C5b789}$)
4. Some forms of endocarditis (heart damage), nephritis (kidney damage), and arthritis (joint damage) are due to IgG-complement-mediated hypersensitivities. It is thought that antibody-antigen-complement complexes may become lodged in the heart, kidney, and joints, and so lead to immune damage.
5. IgG-complement-mediated hypersensitivities play an important part in rheumatic fever. Rheumatic fever results from a chronic infection by *Streptococcus pyogenes* which continually stimulates the formation of antigen-antibody-complement complexes.
6. Some IgG made against *Streptococcus* M-protein are known to crossreact with proteins in the heart. Thus, rheumatic fever is also partly an autoimmune disease.

CELL-MEDIATED (DELAYED TYPE) HYPERSENSITIVITY

1. Antigens not only stimulate the production of antibodies but also activate T-lymphocytes and macrophages.
2. Activated macrophages bind, phagocytize, and consume antigens that they recognize as foreign. Activated T-lymphocytes bind and attack antigens they recognize as foreign.
3. The destruction of nerve cells in the brain (one type of multiple sclerosis) is due to an attack by macrophages and T-lymphocytes on glial cells (myelin sheaths), which cover nerve cells.
4. The destruction of tissue cells which occurs in tuberculosis or leprosy is largely due to an attack by macrophages and T-lymphocytes on infected host cells. Arthus reactions may also play a role in tissue destruction.
5. Some of the destruction of tissue cells which occurs in syphilis is due to macrophages and T-lymphocytes attacking infected host cells. Arthus reactions may be important in tissue destruction.
6. Delayed skin hypersensitive reactions are used to diagnose infections by *Mycobacterium* (tuberculosis), *Francisella* (tularemia), and *Coccidioides* (valley fever).
7. Allergic contact dermatitis is a cell-mediated reaction to some chemicals that contact the skin.
8. The hapten penicillin is a very reactive molecule that often reacts chemically with animal proteins and becomes an antigen. In a sensitized animal, simple contact with penicillin can result within one or two days in blistering of the skin and tissue destruction by macrophages and T-lymphocytes.
9. Catechols are very reactive haptens that frequently react with human proteins and become antigens. In a sensitized person, simple contact with catechols results within one or two days in blistering of the skin and tissue destruction by macrophages and T-lymphocytes.

AUTOIMMUNE DISEASES

1. An autoimmune disease occurs when the immune system attacks self antigens.
2. Many autoimmune diseases are initiated by an infection, by chemicals that modify antigen on host cells, or by viruses that introduce or modify antigens on host cells.
3. Rheumatic fever is an example of an autoimmune disease caused by antibodies made against *Streptococcus* M-protein, which crossreacts with heart tissue.
4. Juvenile diabetes mellitus is an autoimmune disease caused by macrophages and T-lymphocyes attacking pancreatic cells that have been altered by a virus.
5. The Guillain-Barre syndrome is an autoimmune disease caused by macrophages and T-lymphocytes attacking Schwann cells covering peripheral nerves. A number of viruses, including influenza viruses, are believed to modify the membranes of Schwann cells.
6. One type of multiple sclerosis is an autoimmune disease caused by macrophages and T-lymphocytes attacking glial cells in the brain. Chickenpox and measles viruses have been implicated in the sclerosis of the brain.

IMMUNODEFICIENCIES

1. Immunodeficiencies are due to one or more components of the immune system not functioning correctly. An immunodeficiency can be genetic, or it can be the result of an infection.
2. Persons missing or having defective cytotoxic T-lymphocytes (or killer T-cells) are particularly sensitive to viruses, intracellular bacterial infections, fungal infections, protozoal infections, and cancer.
3. Persons missing or having defective B-lymphocytes or helper T-lymphocytes are especially sensitive to bacterial infections.
4. Those missing or having defective suppressor T-lymphocytes may develop an autoimmune disease.
5. Combined immune deficiency is a hereditary condition in which T-lymphocytes and B-lymphocytes (plasma cells) are missing.
6. A depressed immune system allows virus infections that can lead directly to cancer.
7. A depressed immune system makes virus infections, bacterial infections, fungal infections, and protozoan infections more likely.

BLOOD TYPE INCOMPATIBILITIES

Erythroblastosis fetalis is a disease of the fetus and newborn, caused by the mother's IgG attaching to the fetus's RBCs.

STUDY QUESTIONS

1. Define the term allergy.
2. What are the salient characteristics of:

a. anaphylactic reactions
b. immune complex reactions
c. cytotoxic reactions
d. delayed reactions

3. What types of conditions can result from:
   a. an immediate type hypersensitivity?
   b. a complement-mediated hypersensitivity?
   c. a delayed type hypersensitivity?
4. What is the tuberculin skin test? What is it used for?
5. What is an autoimmune disease?
6. How might an autoimmune disease develop?
7. What would be a consequence of an immunodeficiency in:
   a. phagocytic cells?
   b. B-lymphocytes?
   c. T-lymphocytes?
8. Define the following:
   a. autograft
   b. isograft
   c. allograft
   d. anaphylaxis
   e. serum sickness

SUPPLEMENTAL READINGS

Arnason, B. G. W. 1982. Multiple sclerosis: Current concepts and management. *Hospital Practice* 17:81–89.

Austen, K. F. 1984. The heterogeneity of mast cell populations and products. *Hospital Practice* 19:135–146.

Berkman, S. A. 1984. The spectrum of transfusion reactions. *Hospital Practice* 19:205–219.

Centers for Disease Control. 1985. Acquired immuno-deficiency syndrome—Europe. *Morbidity and Mortality Weekly Report* 34(11):147–156.

Herberman, R. B. 1982. Natural killer cells. *Hospital Practice* 17(4):93–103.

Hood, L. Weissman, I. W. Wood. 1978. *Immunology*. Menlo Park, Calif.: Benjamin/Cummings Publishing Company.

Koffler, D. 1980. Systemic lupus erythematosus. *Scientific American* 243(1):52–61.

Leder, P. 1982. 1979. The genetics of antibody diversity. *Scientific American* 246(5):102–115.

Muller-Eberhard, H. J. 1977. Chemistry and function of the complement system. *Hospital Practice* 12:33–43.

Muller-Eberhard, H. J. 1978. Complement abnormalities in human disease. *Hospital Practice* 13:65–76.

Rose, N. R. 1981. Autoimmune diseases. *Scientific American* 244(2):80–103.

Zinkernagel, R. M. 1978. Major transplanion antigens in host responses to infection. *Hospital Practice* 13:83–92.

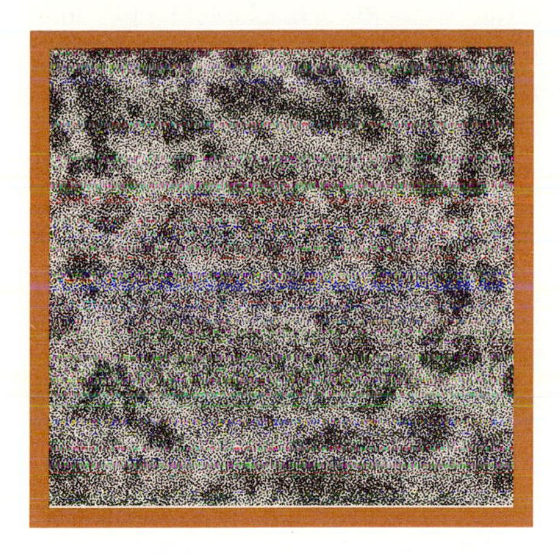

CHAPTER 21
DIAGNOSIS OF INFECTIOUS DISEASES

CHAPTER PREVIEW

TRACKING DOWN THE AGENT OF LEGIONNAIRES' DISEASE

The Pennsylvania Department of the American Legion held its 58th annual convention in 1976 at the Bellevue-Stratford Hotel in Philadelphia from July 21 to July 24. Of the more than 4,000 people attending the convention, about 221 contracted a previously unknown type of pneumonia that killed 34 individuals. This serious illness was promptly labelled by the press as "Legionnaires' disease," although not all those who became ill were Legionnaires; some were hotel or convention employees and visitors. This outbreak of a previously unknown disease, represents an important landmark in the history of infectious disease diagnosis.

The outbreak of Legionnaires' disease prompted an extensive and ultimately productive campaign to find the cause of the disease. It took about five frustrating months of research until the organism was found.

The first step in establishing the cause of Legionnaires' disease was to define the characteristics of the disease so that the population at risk could be established. The disease included a high fever, coughing, and pneumonia. Unfortunately, the clinical symptoms of the disease did not narrow down the cause sufficiently; it could have been caused by toxic organic substances, heavy metals, bacteria, fungi, or viruses, among other pathogens.

In view of the clinical findings, the decision was made to collect specimens from diseased and normal individuals in an attempt to isolate the causative agent. The samples included bits of tissue (biopsy materials), sputum, and blood. They were cultured in a variety of culture media and tissue cultures and incubated under various environmental conditions.

Finally, in January 1977, the agent of Legionnaires' disease was discovered. Joseph E. McDade, Charles C. Shepard, and other scientists at the National Centers for Disease Control (CDC) made the discovery when they examined yolk sacs of eggs that had been injected with infected material. The cause of Legionnaires' disease was a tiny gram negative bacterium, which they named *Legionella pneumophila.* This discovery represented the culmination of one of the most extensive searches for a pathogen recorded in recent history.

Once the organism of **legionellosis** was discovered, techniques were developed to culture the organism in laboratory media, and a serological test (fluorescent antibody test) was introduced as a means of diagnosing the disease.

The scientists at CDC were successful in their goal because they had previously mastered techniques employed for the isolation and cultivation of infectious disease agents. They were careful to collect the proper specimens, using aseptic techniques; transport them to the laboratory in a safe and suitable manner; and employ cultural techniques that had proved successful in isolating other pathogens. This chapter introduces you to the field of clinical microbiology. It will cover the basic principles of specimen collection and processing and discuss some of the more common techniques presently used to diagnose infectious diseases.

LEARNING OBJECTIVES

A STUDY OF THIS CHAPTER SHOULD ENABLE YOU TO:

WRITE AN ESSAY ON THE IMPORTANCE OF PROPER COLLECTION AND TRANSPORT OF CLINICAL SPECIMENS

OUTLINE THE IMPORTANT METHODS USED IN MICROSCOPIC EXAMINATION OF CLINICAL SPECIMENS

DESCRIBE THE CULTURAL PROCEDURES USED IN THE CLINICAL DIAGNOSTIC LABORATORY

OUTLINE THE CRITERIA EMPLOYED IN IDENTIFICATION OF CLINICAL ISOLATES

DISCUSS THE IMPORTANCE OF SEROLOGIC TESTS IN THE DIAGNOSIS OF INFECTIOUS DISEASES

OUTLINE THE PRINCIPLES AND USES OF SOME OF THE MOST IMPORTANT DIAGNOSTIC SEROLOGIC TESTS

Infectious diseases generally result from microbial growth, or toxin produc tion by microorganisms, in or on our body tissues. While certain infectious diseases produce characteristic signs and symptoms, which are said to be **pathognomonic** for the disease, often the signs and symptoms of a disease are not pathognomic and further steps must be taken to ascertain its cause. In these cases, it is imperative that we isolate and identify the pathogen so that the patient can be treated effectively.

In order to isolate and identify infectious agents from **clinical specimens,** it is necessary to collect and handle the specimens properly. The importance of proper specimen collection and handling cannot be overemphasized. It is at this stage that a successful pathogen isolation begins. Once the specimen is safely in the laboratory, it is then cultured for the isolation and subsequent identification of the pathogen. Identification of microorganisms generally involves a) direct microscopic observations; b) cultivation and biochemical tests; and c) serological tests. In this chapter, we will discuss techniques used in the collection of specimens and in the isolation and identification of microorganisms suspected of causing disease.

COLLECTION AND TRANSPORT OF SPECIMENS

The diagnosis of an infectious disease is based not only on its signs and symptoms but also on pathogenic microorganisms found in diseased tissues or exudates. The clinical microbiologist's task is to culture these specimens in order to isolate and identify pathogens. In addition, the microbiologist must ascertain that the results are accurate by collecting and handling the specimens properly. Often, the failure to isolate a pathogen from a clinical specimen is due not to faulty testing but to faulty collection. The following are important considerations for the successful isolation of pathogens in clinical specimens. The specimens should be:

1. collected before antimicrobials are administered;
2. collected at the time the pathogen is likely to be present;
3. collected from the active site of the infection;
4. collected in sufficient amounts;
5. collected with the proper tools and placed in suitable containers;
6. delivered promptly to the laboratory for analysis;
7. stored under appropriate conditions until processed;
8. processed as soon as possible to increase the chances of isolation.

Specimen Collection

Specimens should be collected before the administration of antimicrobials. Sometimes physicians diagnose infectious diseases based on signs and symptoms and then initiate antimicrobial therapy without the identification of the causative agent. Whether or not this therapy is successful, it may be desirable at a later date to identify the agent; however, inhibitory levels of the antimicrobials □ may be present in the clinical specimens, thus making the

718

isolation of the pathogen more difficult (or impossible). For this reason, it is desirable that attempts to isolate the pathogen be made before antimicrobial therapy has been initiated, or after treatment has been discontinued for a time, so that the concentration of antimicrobials in the specimens is no longer inhibitory.

Specimens should be collected when the parasites are likely to be present. Some parasites, including many protozoa and helminths, appear in a particular anatomical site at regular intervals and then disappear. An example is the nematode (roundworm) *Wuchereria bancrofti*. The larvae of this parasite, called **microfilariae,** can be observed in blood specimens collected during the night. Attempts to locate the microfilariae in specimens collected during the day would almost certainly fail. The protozoa that cause malaria and African sleeping sickness also make cyclic appearances in the blood of infected hosts. Many laboratories take three samples on three consecutive days to improve their chances of detecting the parasites in the specimens.

Many infectious agents are present in the clinical specimens for only a short time after the onset of signs and symptoms, or only during certain stages of the disease. The spirochete that causes syphilis is present in large numbers during early syphilis, but becomes scarce in lesions appearing during late syphilis. Specimens collected from a chancre (primary syphilis) contain numerous spirochetes that can easily be seen using a dark field microscope □. Lesions characteristic of tertiary syphilis, however, do not contain the spirochetes.

63

The clinical specimen should be sufficiently large to allow the microbiologists to conduct all the necessary tests. Frequently, a clinical specimen is split into two or more aliquots that are treated in various ways. In order for these procedures to be carried out, there must be an ample supply of specimen. For example, a sputum sample may be split into two or three 5-ml aliquots to be cultured for bacteria and fungi. Hence, at least 10 ml of sputum must be collected so that all these procedures can be carried out.

Care must be exercised to maintain the viability of the pathogens in the specimen. If the specimen is taken with a tool such as a swab or a syringe containing toxic chemicals, the chances of isolating the pathogen diminish. For example, cotton swabs may contain materials such as lipoproteins that are inhibitory for certain hard-to-grow (fastidious) microorganisms. Although cotton swabs can frequently be used to collect specimens, it is sometimes preferable to use Dacron® swabs. Swabs made with this material generally are less inhibitory than cotton and are recommended for isolating β hemolytic streptococci from throats and *Neisseria gonorrhoeae* in vaginal exudates.

The need for sterile containers for collecting clinical specimens cannot be overemphasized. If the specimen is collected in a container contaminated with other microorganisms, the contaminants may overgrow the pathogen and prevent its isolation, particularly if there are few pathogens in the specimen. Contaminating microorganisms may also release toxic substances that are inhibitory to the pathogen. Contaminated specimens also make it difficult to establish the significance of the isolated microorganism.

Specimens should be collected from the active site of the lesion. The reason is that some infectious agents are present only in certain portions of the lesions they cause, and attempts to culture the pathogen from other portions of the lesion will fail to yield the pathogen. For example, the fungi that cause ringworm spread radially from the site of infection, forming a circular lesion. The leading edge of the lesion contains the fungi, whereas the central portion of the lesion has very few fungal elements. Therefore, it is necessary to sample the leading edge of the lesion.

The proper collection of specimens requires the cooperation of the physician, the nursing staff, and the microbiologist. If the proper precautions are taken and judicious choices are made as to how and when to collect a specimen, the chances for a quick diagnosis are greatly enhanced. A rapid diagnosis will not only result in the best treatment possible for the patient, but also will reduce the cost of hospitalization.

After collection, the specimen should be delivered to the laboratory as soon as possible. In rural communities, where the closest laboratory may be miles away, the specimens often are not delivered immediately to the laboratory for processing. In such cases, it is recommended that the samples be refrigerated shortly after collection until delivery to the laboratory.

There are also problems with stored specimens. For example, toxic materials might be present in the specimen which would inhibit or kill pathogens. Specimens stored improperly for even short periods of time will allow contaminating microorganisms to multiply and outnumber the pathogen. For this reason, several transport media have been developed to preserve the viability of the pathogen while not promoting microbial growth.

MICROSCOPIC METHODS IN DIAGNOSIS

Microscopic examination of fresh clinical specimens is one quick way of determining the cause of some infectious diseases. Diagnosis of bacterial diseases generally is not made solely by microscopic examination, because bacteria, for the most part, lack distinct morphological characteristics that set them apart from all others. Fungal infections can sometimes be diagnosed by microscopic examination alone, although other diagnostic procedures usually are carried out concurrently. Infectious diseases caused by protozoa or helminths (worms), however, are almost always diagnosed using microscopic methods.

Wet mounts are a simple way to examine specimens. They are prepared by mixing a small portion of the specimen with a drop of water or saline solution on a glass slide. A coverslip is placed over the mixture, which is subsequently examined with the aid of a microscope. Wet mounts are useful in diagnosing certain infectious diseases, especially those caused by fungi, protozoa, and helminths, many of which are morphologically distinct (fig. 21-1). Vaginitis caused by *Trichomonas vaginalis* generally is diagnosed by examining a wet mount of the vaginal exudate and demonstrating the presence of flagellated protozoa (fig. 21-1). A diagnosis of syphilis can sometimes be made by examining a suspension of material obtained from the chancre, using a dark field microscope. In such preparations, the spirochetes can be seen as bright, motile, helically-shaped bacteria against a dark background.

Stained smears are commonly used for examining clinical specimens. The smears should be made with fresh specimens, preferably at bedside. Most clinical specimens are examined using a variety of staining techniques (table 21-1). For specimens suspected of containing bacterial pathogens, the Gram stain is routinely used. The gram stain allows microbiologists to view the specimens and decide which further steps are needed in order to identify

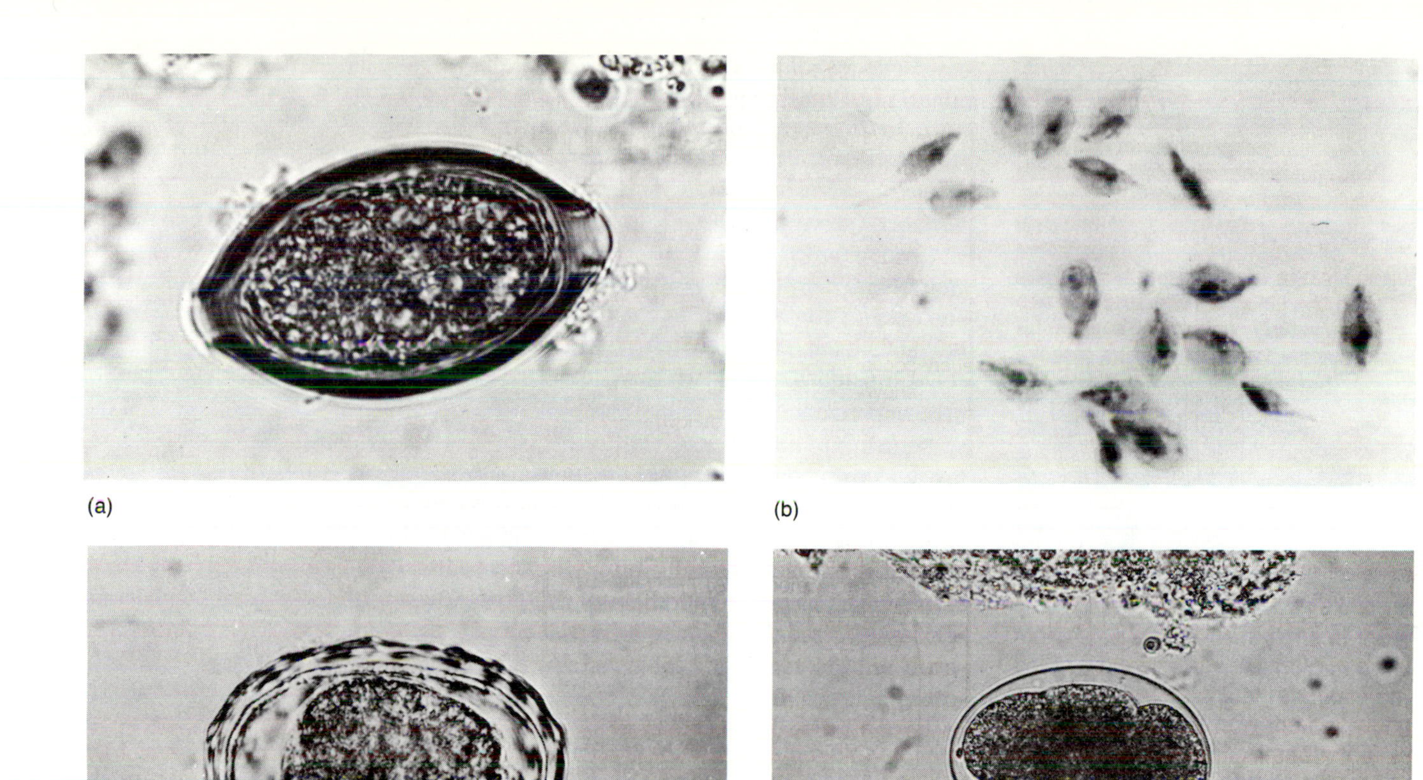

(a) (b) (c) (d)

Figure 21-1 Wet Mounts of Selected Clinical Specimens. Microscopic appearance of helminths and protozoa in wet mounts of clinical specimens. *(a) Trichuris trichura* (whipworm) egg in human feces. *(b) Trichomonas vaginalis* in vaginal exudate. *(c) Ascaris lumbricoides* egg in human feces. *(d)* Hookworm egg in human feces.

TABLE 21-1
SELECTED STAINING PROCEDURES EMPLOYED IN THE MICROBIOLOGY LABORATORY TO EXAMINE CLINICAL SPECIMENS

STAINING PROCEDURE	USE IN LABORATORY	CELLS OR ORGANISMS DETECTED
Acid-fast stain	Examine sputum, tissue, skin	*Mycobacterium, Nocardia, Cryptosporidium*
Giemsa stain	Examine blood smears	Bacteria, protozoa, fungi, blood cells
Gram stain	Examine various specimens	Bacteria
Gridley fungus stain	Examine tissue sections	Fungi
Hematoxylin and eosin stain (Wright stain)	Examine tissue sections, blood smears	Bacteria, protozoa, fungi, cancer cells
Papanicolau stain	Examine vaginal smears	Cancer cells
Periodic acid-Schiff stain	Examine sputum, tissue, skin	Fungi
Trichrome stain	Examine stool smears	Protozoa, worms

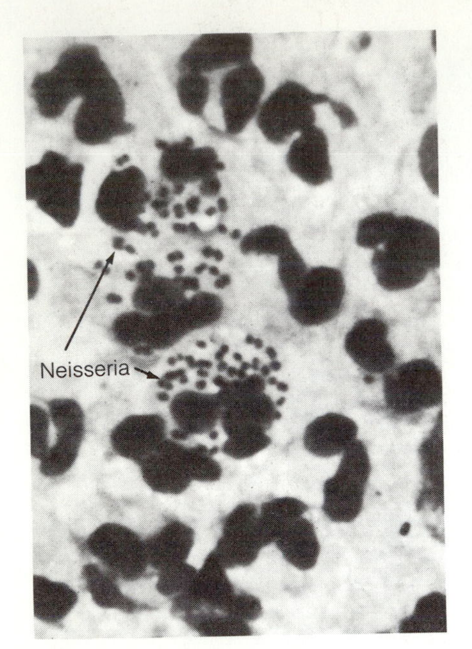

Figure 21-2 ***Neisseris gonorrhoeae* in Pus Cells.** Gram stain of pus from a patient suffering from gonorrhea. *Neisseria gonorrhoeae* is seen within white blood cells.

the pathogen. The Gram stain provides a necessary first step in □ the identification of many microorganisms.

Sometimes a preliminary diagnosis can be made based on microscopic examination of stained smears. For example, the presumptive diagnosis of gonorrhea is usually made by demonstrating the presence of *Neisseria gonorrhoeae* (gram negative cocci) inside polymorphonuclear leukocytes (PMNs) in Gram stained smears made from genital exudates (fig. 21-2). Protozoal and helminthic infections are usually diagnosed using stained smears. The presence of malarial parasites □ or trypanosomes in the blood can be made simply on the basis of Giemsa-stained thin blood smears. Fecal specimens stained using a Trichrome stain can be used to diagnose amebiasis, giardiasis, and many other protozoal and helminthic infections.

371

Stained histological sections (fig. 21-3) of diseased tissue are sometimes examined for the presence of pathogens. Most sections are examined using the Wright stain, which has the advantage of being able to demonstrate most infectious agents and at the same time give the pathologist an idea of the type of host reaction the pathogen is eliciting. Table 21-1 summarizes some common staining procedures used in the clinical laboratory to aid in the diagnosis of infectious diseases.

CULTURAL METHODS IN DIAGNOSIS

The isolation and identification of pathogens from clinical specimens in modern microbiology laboratories proceeds along very logical pathways. Depending upon the source and origin of the clinical specimen, a decision is made as to which cultural protocol to follow. This decision is made based on the anatomical site affected, all well as on the evidence obtained from the microscopic examination of stained specimen preparations.

The first step in the isolation and identification of a pathogen involves planting of the specimen on selective □ and/or differential media (table 21-

121

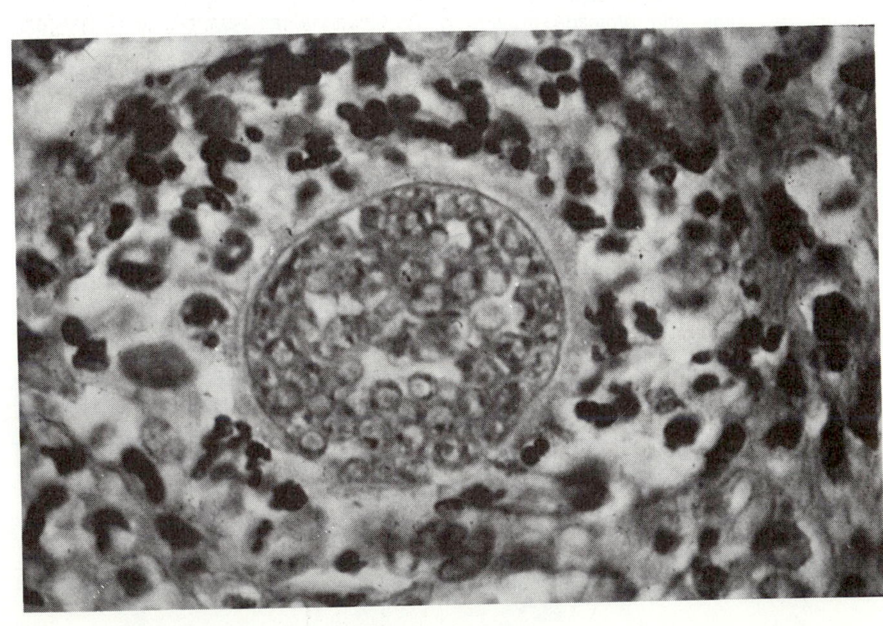

Figure 21-3 **Histological Section Showing the Spherule of *Coccidioides immitis***

TABLE 21-2
SELECTIVE MEDIA USED IN ISOLATING PATHOGENS

CULTURE MEDIUM	GROUP OF MICROORGANISM CULTURED
Baird-Parker agar	*Staphylococcus*
Brucella agar	*Brucella* sp.
EMB agar	Enteric bacteria
Hektoen-Enteric agar	Enteric bacteria
Loeffler agar	*Corynebacterium diphtheriae*
Lowenstein-Jensen agar	*Mycobacterium, Nocardia*
MacConkey's agar	Enteric bacteria
Sabouraud glucose agar	Fungi
7H10 agar	*Mycobacterium, Nocardia*
SS agar	Enteric bacteria
Staphylococcus-100 agar	*Staphylococcus aureus*
TCBS agar	*Vibrio cholerae*
Thayer-Martin agar	*Neisseria gonorrhoeae*
XLD agar	Enteric bacteria

2). The primary isolation procedure may involve treating the specimen with selective agents before planting it on the culture media to increase the chances of isolating the pathogen. A primary isolation procedure is illustrated in figure 21-4.

Let us look at one protocol used for the isolation of *Mycobacterium tuberculosis* from clinical specimens (fig. 21-4). In this procedure, the specimen, usually sputum, is treated with an alkali (NaOH or KOH) and N-acetylcysteine. The alkali does not kill mycobacteria, but is toxic to most other contaminating microorganisms and consequently selects for *M. tuberculosis*, while the mucolytic agent (N-acetylcysteine) breaks down the mucus and frees the pathogens that are trapped in the mucus. Nearly all procedures for culturing clinical specimens involve a step such as the one described above, in which the pathogen is freed from entrapping tissue or host material. Biopsy materials generally are ground in a tissue homogenizer or macerated into very small pieces before planting on the culture media. Centrifuging may be used to concentrate the mycobacteria in a specimen before planting; this procedure is particularly useful in cases where there are low numbers of pathogens in the specimen. Membrane filters are also used for concentrating pathogens in clinical specimens. The treated and/or concentrated specimens are subsequently planted onto selective media such as Lowenstein-Jensen (LJ) or 7H10. Both these media have selective agents, such as malachite green, that inhibit the growth of bacteria other than mycobacteria.

The choice of the appropriate selective or differential media is essential for the isolation of pathogens from clinical specimens, particularly in the case of specimens such as sputum or feces, which are likely to contain contaminating microorganisms in addition to the pathogen. For example, the isolation of enteric pathogens such as *Salmonella* or *Shigella* is greatly facilitated with the use of selective and differential media. Table 21-3 summarizes various specimens commonly encountered in the clinical laboratory and the possible pathogen(s) that may be isolated.

One additional procedure commonly employed in the clinical laboratory to increase the chances of isolating a pathogen is the **enrichment** procedure.

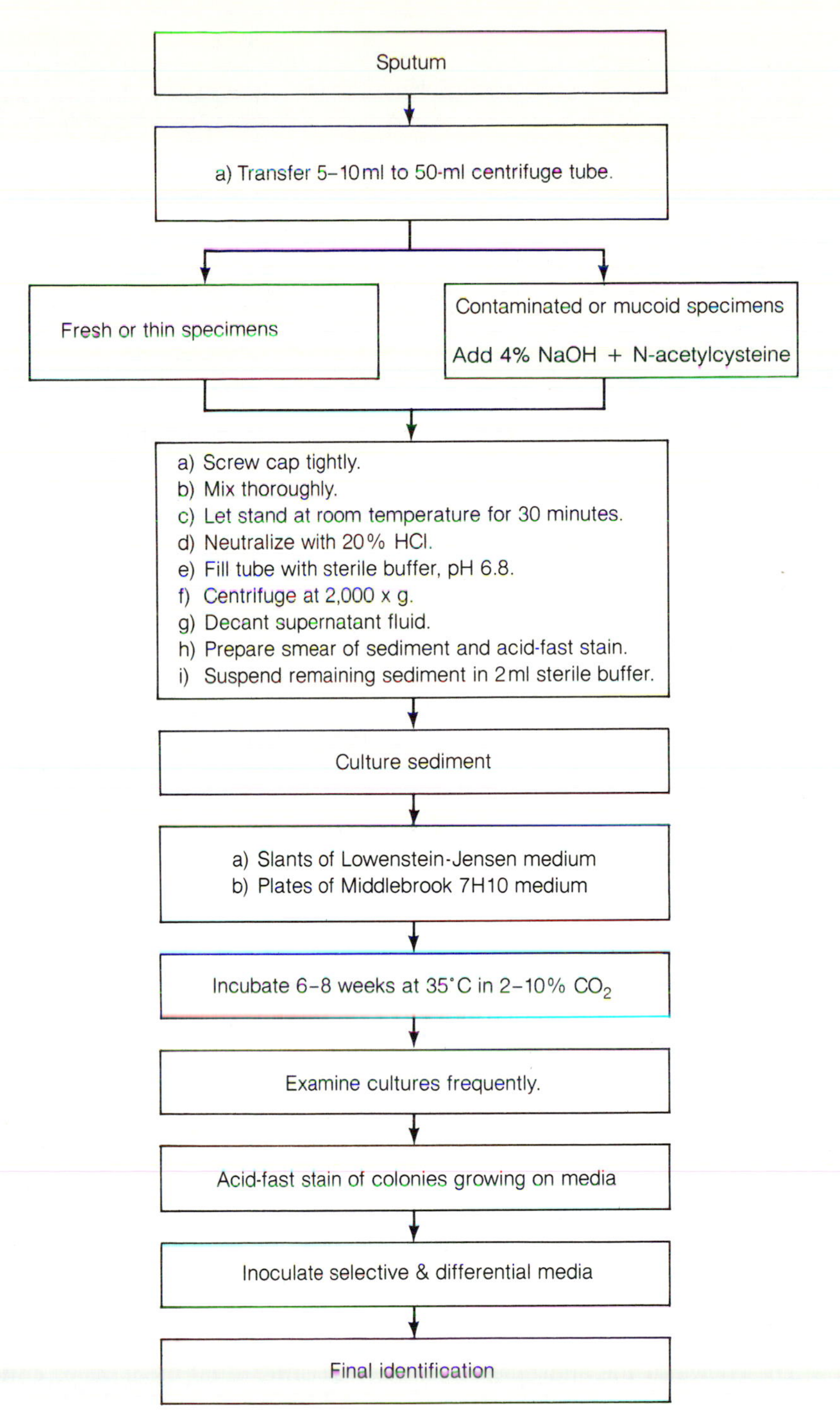

Figure 21-4 Skeleton Outline of a Common Procedure Used for the Isloation of *Mycobacterium tuberculosis* from Sputum Specimens

This technique is particularly useful when isolating fastidious microorganisms, or microorganisms that are present in low numbers in contaminated specimens. For example, **selenite F-broth** is used to enrich for *Shigella* in fecal specimens. A portion of fresh, untreated stool is inoculated into a tube containing sterile selenite F-broth. This procedure promotes the growth of

TABLE 21-3

REPRESENTATIVE PATHOGENIC MICROORGANISMS AND THE SPECIMENS FROM WHICH THEY ARE ISOLATED

CLINICAL SPECIMEN	POSSIBLE PATHOGENS ISOLATED
Blood	*Salmonella typhi, Brucella, Plasmodium, Leptospira, Streptococcus,* yeasts, *Staphylococcus, Neisseria* spp., *Streptococcus pneumoniae,* enteric bacteria, *Rickettsia* spp., *Yersinia, Franciscella,* various viruses
Feces	*Salmonella, Shigella, Escherichia, Campylobacter, Vibrio,* hepatitis virus, mumps virus, polio virus, tapeworms, flukes, nematodes
Urine	*Leptospira, Salmonella, Escherichia, Proteus, Schistosoma* (fluke), yeasts
Pus	*Neisseria gonorrhoeae, Treponema pallidum, Staphylococcus, Brucella, Streptococcus, Borrelia, Yersinia, Franciscella,* herpes virus, yeasts, smallpox viruses
Sputum, throat, and nasal discharges	*Myobacterium tuberculosis, Streptococcus pneumoniae, Klebsiella pneumoniae, Mycoplasma, Haemophilus, Corynebacterium diphtheriae, Neisseria meningitidis, Bordetella pertussis, Streptococcus, Staphylococcus, Treponema,* rhino viruses, influenza viruses, adenoviruses, pathogenic fungi, yeasts, varicella-zoster (chickenpox) virus, mumps virus. *Ascaris* (nematode), *Paragonimus* (fluke)

Shigella and inhibits the growth of other bacteria, thus increasing the relative numbers of the pathogen and the chances of isolating it from the stool specimen.

Fungi are routinely cultivated in the clinical laboratory. These organisms, which are important pathogens, generally are cultivated on selective media such as Sabouraud glucose agar. Sabouraud glucose agar selects against most bacteria because it has a pH of about 5.7, too low for the growth of many bacteria. This culture medium may be supplemented with cyclohexamide and chloroamphenicol to inhibit contaminating bacteria and many saprophytic fungi.

Viruses generally are grown in tissue cultures. These are *in vitro* cultures of human or animal cells that serve as a growth environment for the viruses. Many laboratories do not culture viruses routinely and send their specimens for virus culture to specialized laboratories. In view of the importance of viruses as human pathogens, however, and the need to diagnose viral diseases rapidly, many laboratories are implementing limited viral cultures, such as those for herpes virus and cytomegalovirus.

Identification of Microorganisms Isolated from Clinical Specimens

122

Before any attempt is made to identify a microbial isolate, the microbiologist must make sure that the culture is pure □. Microscopic examination of gram stained smears of suspect colonies can serve as a good indication as to the purity of the culture, but there is no substitute for streaking the colony on a suitable culture medium.

The choice of identification protocol is usually made after the morphology and gram staining properties of the isolate have been determined. Following the microscopic examination of a gram stain, various morphological and biochemical properties of the isolate are determined. The proper choice of tests depends upon the isolate. Some microorganisms are simple to identify, while others require extensive testing before they can be identified. Figure 21-5 illustrates a method used to identify organisms in a vaginal sample.

The first step in identifying microorganisms is to separate them into groups that can be differentiated on the basis of their staining and morphological properties. Figure 21-6 illustrates a flow chart designed for the preliminary identification of a selected group of organisms (gram positive

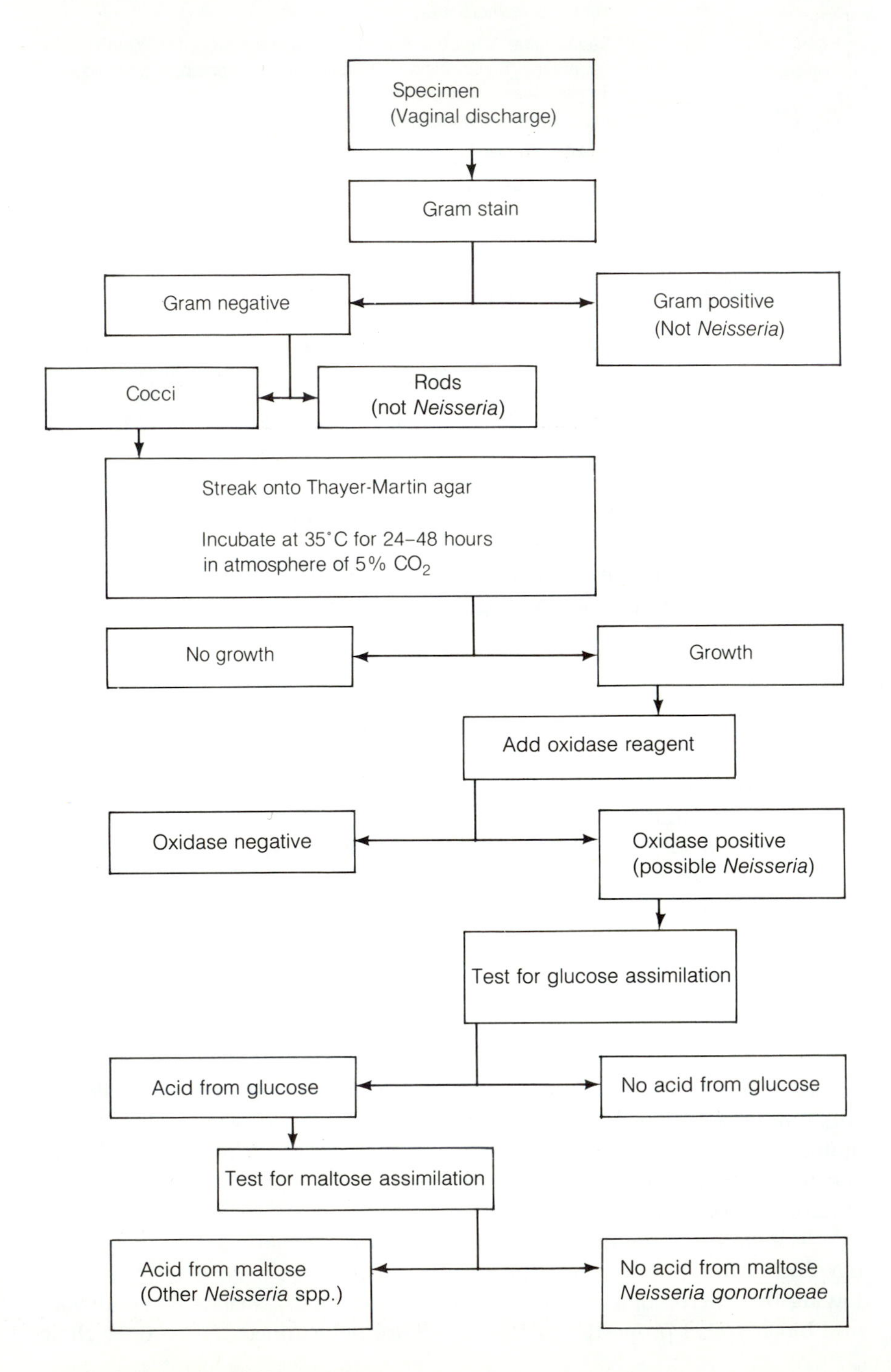

Figure 21-5 Procedure Outlining the Presumptive Identification of *Neisseria gonorrhoeae*

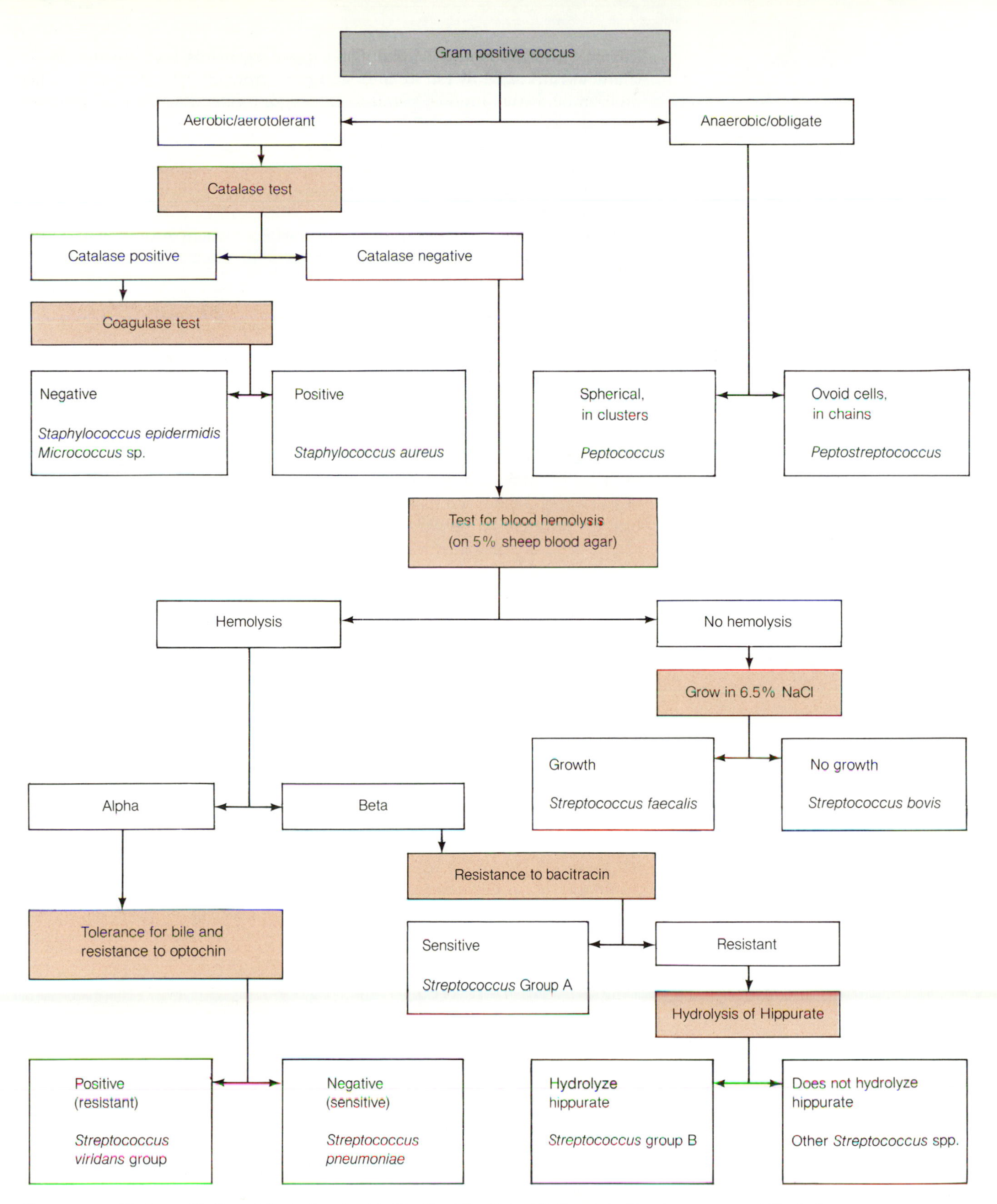

Figure 21-6 Skeleton Outline of Steps Involved in the Identification of Pathogenic *Streptococcus* and *Staphylococcus*

cocci). The second step in identifying microorganisms involves the use of characteristics such as the isolate's ability to grow on selective media, alter differential media, produce catalase, hemolyze (break down) blood, and ferment sugars.

Not all bacteria are identifiable by these simple procedures, and additional testing may be required for identification. For example, the gram negative glucose fermenting and gram negative glucose respiring □ bacteria are identified on the basis of their ability to utilize amino acids, carbohydrates, and inorganic ions, to ferment or respire various carbohydrates, or to hydrolyze various polymers. Table 21-4 shows some of the common tests used to identify one group of gram negative bacteria.

190

Rapid Methods for the Identification of Clinical Isolates

While the isolation and identification of microorganisms is essential for the control of infections, it is also very expensive and time consuming. As you can see from figure 21-6 and table 21-4 in order to identify a pathogen properly, a large number of tests may have to be carried out. Many of these tests involve the cultivation of microorganisms and therefore require 6–96 hours of incubation before results can be obtained. In addition, many types of culture media must be prepared, stored, and checked periodically for deterioration. Most of these media must be planted and read individually.

Frequently, it is desirable, either for economy or expediency (or both), to use rapid methods or "kits" to identify common pathogens. There are several of these kits on the market (fig. 21-7) that are extremely well suited for the identification of bacteria and yeasts. Most of these are made up of strips or tubes containing reagents for a large number of different tests. The kits can be stored for relatively long periods of time and occupy little laboratory space, and the cost per isolate identified usually is much lower than the cost of using conventional methods. Many of the manufacturers of these kits provide computerized identification services at little or no extra cost. These kits are extremely well suited for laboratories with relatively light work loads that would make the maintenance of a large number of conventional media prohibitive. For clinical laboratories with large work loads, there are several automated microbial identification systems □. Most of these systems are based on the ability of microorganisms to grow on media with various antibiotics or special nutrients. Microbial growth generally is measured by turbidimetric methods, and identification of the isolate is based on a computer-assisted analysis of the growth pattern of the isolate on the various culture media.

301

SEROLOGICAL METHODS

Some diseases are caused by agents that are difficult to grow or that require special media. In these cases, routine cultural techniques are either impossible or time consuming, so serological procedures are desirable. Serological procedures are also carried out when rapid results are needed. Diseases such as rubella, syphilis, and hepatitis generally are diagnosed using serological methods.

TABLE 21-4
DIFFERENTIATION OF SOME FREQUENTLY ISOLATED *PSEUDOMONAS* SPP. OF MEDICAL IMPORTANCE

CATALASE-POSITIVE	*P. AERUGINOSA*	*P. PUTIDA*	*P. FLUORESCENS*	*P. STUTZERI*	*P. PSEUDOMALLEI*	*P. MALLEI*	*P. CEPACIA*	*P. ACIDOVORANS*	*P. MALTOPHILIA*	*P. VESICULARIS*	*P. PUTREFACIENS*	*P. MENDOCINA*	*P. PICKETTII*	*P. ALCALIGENES*	*P. PSEUDOALCALIGENES*	*P. TESTOSTERONI*	*P. DIMINUTA*
Oxidase	+	+	+	+	+	V	V	+	−	+	+	+	+	+	+	+	+
Hemolysis, 5% SBA	V	γ	γ	γ	γ	NR	γ	γ	γ	NR	NR	γ	γ	γ	γ	γ	γ
Wrinkled colonies	−	−	−	+	V	NR	−	−	−	NR	−	−	NR	−	−	−	−
Sudan black granules	−	−	−	−	+	+	+	+	−	NR	−	NR	NR	−	V	+	V
Growth, at 42°C	G	NG	NG	G	G	NG	NG	NG	G	NG	G	G	V	V	V	NG	V
MacConkey agar	G	G	G	G	G	V	G	G	G	G	G	G	G	G	G	G	G
Salmonella-Shigella agar	G	G	G	V	NG	NG	NG	V	V	NG	G	G	NG	V	V	NG	NG
Cetrimide agar	G	G	G	NG	NG	NG	V	V	V	NG	NG	G	NG	V	V	NG	NG
Deoxycholate agar	G	G	G	G	V	V	G	G	G	NR	V	NR	NR	G	NR	NR	−
6.5% NaCl broth	NG	NG	NG	G	NG	NG	NG	NG	NG	NG	G	G	G	NG	NG	NG	NG
O-F glucose	O	O	O	O	O	O	O	O	O	O	O	O	O	−	−	−	−
Nitrate reduction	+	V	+	+	−	+	+	V	−	−	+	+	+	V	+	V	−
Motility, 37°C	+	+	+	+	+	−	+	+	+	+	+	+	+	+	+	+	+
Gelatin liquefaction, 22°C	V	−	+	−	+	+	V	−	+	−	+	−	−	+	−	−	−
Indole	−	−	−	−	−	−	−	−	−	−	−	−	−	−	−	−	−
Simmons citrate	+	+	+	+	+	−	+	V	−	−	V	+	+	+	V	−	−
H_2S (KIA/TSI)	−	−	−	−	−	−	−	−	−	−	+	−	−	−	−	−	−
KCN	G	NR	V	NR	G	NR	NR	NG	G	NR	NR	NR	NR	NR	NR	NR	NR
Christensen's urease	V	V	V	V	V	V	V	V	−	−	V	+	+	−	V	V	−
Methyl red	−	−	−	−	−	−	−	−	−	−	−	−	−	−	−	−	−
Voges-Proskauer	−	−	−	−	−	−	−	−	−	−	−	−	−	−	−	−	−
Arginine dihydrolase: NH_3	+	+	+	−	+	V	−	−	−	−	−	+	−	+	+	−	−
Lysine decarboxylase	−	−	−	−	−	−	+	−	+	−	−	−	−	−	−	−	−
Ornithine decarboxylase	−	−	−	−	−	−	V	−	−	−	+	−	−	−	−	−	−
Phenylalanine deaminase	−	−	−	V	−	−	−	−	−	−	−	+	V	−	V	−	−
Gluconate	+	V	V	−	+	V	V	−	−	−	−	−	−	−	−	−	−
ONPG	−	−	−	−	−	−	+	−	+	−	−	−	−	−	−	−	−
Lecithinase	V	−	+	−	+	NR	V	−	−	−	−	−	−	−	−	−	−
Starch hydrolysis	−	−	−	+	−	V	−	−	−	−	−	−	NR	−	−	−	−
Esculin hydrolysis	−	−	−	−	V	−	V	−	+	−	V	−	−	−	−	−	−
Litmus milk (peptonization)	+	−	V	−	+	−	V	−	+	−	+	−	−	−	−	−	+
DNase	V	−	−	−	−	NR	−	−	+	+	+	−	−		−	+	−
Lipase	V	−	V	V	V	NR	V	V	V	NR	NR	−	+	−	−	V	−

+ = positive results; − = negative results, V = variable results; G = growth; NG = no growth; NR = no reaction; O = oxidation; F = fermentation. From: MacFaddin, J. F. *Biochemical Tests for Identification of Medical Bacteria*. Second Edition. p. 386. Copyright © 1980 by Williams & Wilkins, Co., Balitmore, Reprinted by permission of author and publisher.

458

Serological procedures are based on the assumption that infections induce the host to produce specific antibodies or sensitized lymphocytes □. In any case, a host who is or has been infected generally will react immunologically to the infectious agent, and this response will be very specific. Serological methods are designed to detect or quantify the intensity of the immune response.

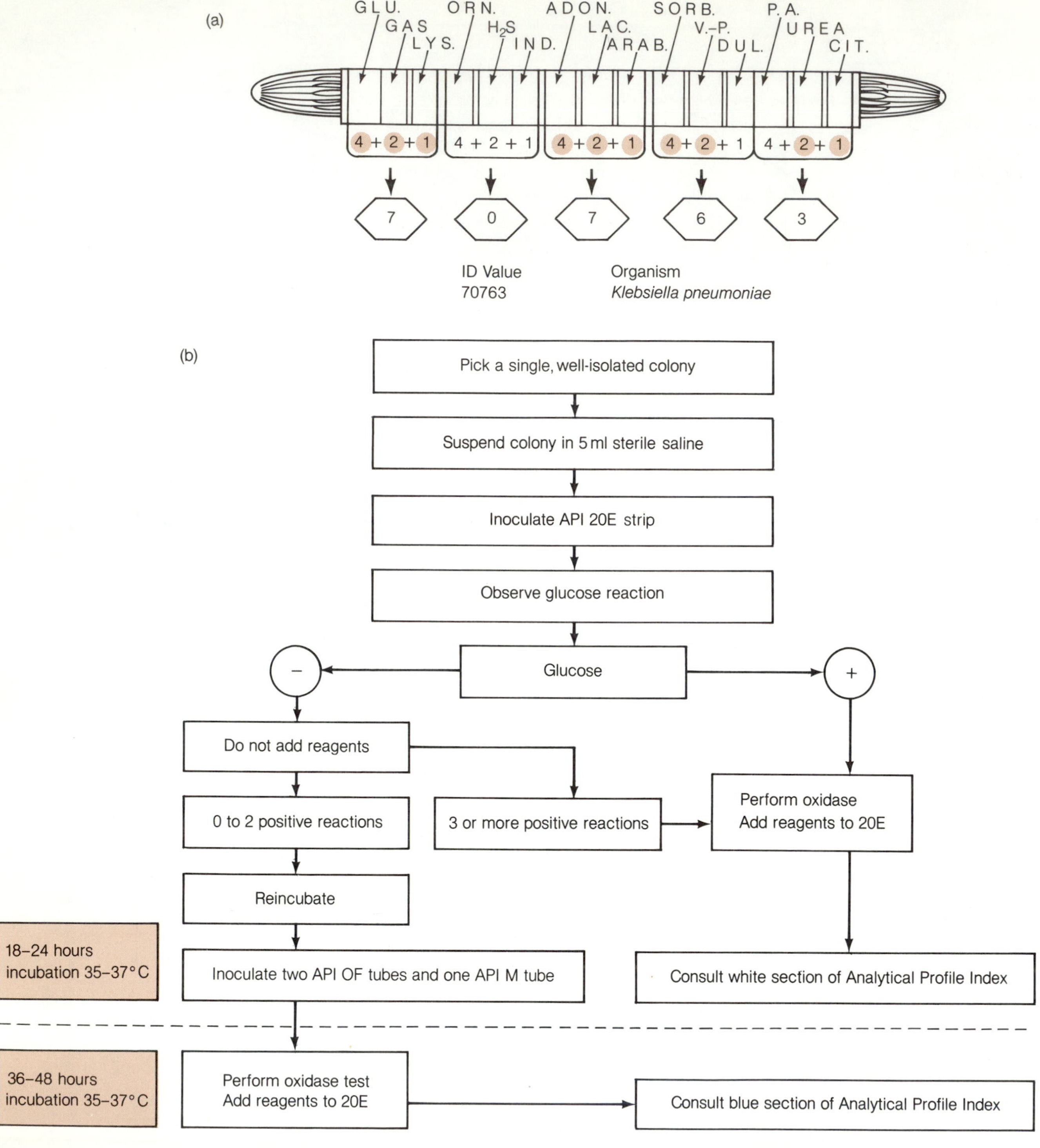

Figure 21-7 Kits Used for the Rapid Identification of Medically Important Bacteria. *(a)* The Enterotube chambers each contain a different type of culture medium, within a plastic enclosure. All chambers are inoculated at once, using the inoculating needle provided. The source of inoculum is an isolated colony. The results are assessed after 24 hours of incubation at 35°C and converted to a numerical code. The code can then be translated into a bacterial identification using a "code book." *(b)* Identification of bacteria using the API 20E strip. The strip consists of 20 wells each containing a different type of medium. A colony is suspended in 5 ml of saline and used to inoculate the 20 wells. The strip is then incubated for 18–24 hours. The results are assessed. See color insert of the API 20E strip.

Principles of Serological Reactions

Serological reactions are based on the formation of antigen-antibody complexes: serum from the patient is made to react with known antigens, and any antigen-antibody complexes that form indicate that the patient has come into contact with the antigen (usually an infectious agent). The amount of specific antibody present in a patient's serum usually reflects the intensity of the host's response, and it is sometimes important in determining the cause of the disease. Antibody concentration in the serum is usually expressed as the **titer,** which is the reciprocal of the highest dilution of antiserum that still gives a positive reaction. For example, in a series of twofold (½, ¼, ⅛, etc.) dilutions of antibody, if a 1/64 dilution gives a positive reaction but a 1/128 does not, the serum titer for this antibody is 64.

Sometimes the mere presence of antibodies in a patient's serum is clinically significant. For example, the presence of antibodies against *Salmonella typhi* in a serum sample is a good indication that the patient has typhoid fever and should be treated immediately. Humans generally do not come in contact with *Salmonella typhi* other than in a disease situation, so a specific, qualitative test is all that is needed. In certain diseases, however, it is important to demonstrate a high titer of specific antibodies in the serum because most individuals in the population are likely to have come into contact with the infectious agent and still retain some antibodies against it. In such cases, quantitative tests are essential. The titer that indicates a current infection must be determined empirically (by experimentation), by studying persons who are known to have a current infection and comparing them with persons known to have had the infection in the past. The clinical significance of a particular titer must be determined for each disease and should be based on clinical studies.

Specimens

Blood and spinal fluid are the most common specimens used for serological diagnoses, although feces, sputum, or biopsy materials may also serve as specimens. When blood is used, it is allowed to clot and the serum (which contains the desired antibodies) is drawn off. Sometimes two blood samples are taken several days apart, one during the acute stage of the illness and the other during recovery. These **paired sera** serve to determine whether antibody is present because of a current infection or a past infection. If there is a rise in the antibody titer from the **acute serum** to the **convalescent serum,** then it is likely that the antibodies detected are due to a current infection. If the titers remain about the same, however, then it is likely that the antibodies resulted from a past infection. At times, serum is collected to detect the presence of antigen, rather than antibodies, in the blood. This condition, called **antigenemia,** is common in infectious diseases that involve the spread of the pathogen or its products in the blood.

Antigens

Most diagnostic serological reactions use antigens to detect the presence of antibodies in the patient's serum. The antigens are either whole cells, cell parts (e.g., cell walls, pili, or flagella), or cell extracts. Certain diagnostic procedures, such as the fluorescent treponemal antibody (FTA) test for syphilis or the febrile agglutinin tests for typhoid fever and tularemia, involve the use of whole cells as antigens. Other tests such as the tuberculin skin test or

complement fixation □ tests, use molecules derived from cells as antigens. 525
Viral particles, either inactivated or attenuated, are also used as antigens for serological reactions.

Sensitivity and Specificity

Sensitivity and specificity are two terms with which the microbiologist should become familar. These terms describe the overall usefulness of the serological procedure and define the limits within which the procedures are useful. The **sensitivity** of an immunological procedure is indicated by the minimum quantity of detectable antigen (or antibody) that must be present in the specimen, and is usually expressed in μg of antigen (or antibody). Thus, the greater the sensitivity of a procedure, the lower the concentration of antigen (or antibody) that can be detected using the procedure.

The usefulness of a serological procedure does not depend entirely upon its sensitivity; it also depends on the **specificity** of the reaction. The specificity of a serological procedure reflects whether or not antigen-antibody reactions take place only between the desired antigens and antibodies. The phenomenon of antigen-antibody reactions involving two or more unrelated antibodies (or antigens) is called **crossreactivity.** As the sensitivity of a serological procedure increases, so does the risk of measuring crossreactivity and thereby obtaining erroneous positive reactions (false positives). When the amount of antigen-antibody complex measured is very small, antigen-antibody reactions due to crossreactivity begin to predominate. The specificity of serological reactions can be greatly enhanced by using monoclonal antibodies, because these antibodies can be made against unique antigenic determinants found only in a specific pathogen. It is essential that the microbiologist know the limitations of serological reactions in order to ascertain accurately the immune state of the patient.

Agglutination Reactions

Agglutination reactions generaly involve whole cell antigens, such as bacterial cells or red blood cells. These antigens contain many different antigenic determinants on their surfaces. Specific antibodies bind selectivity to the various antigenic determinants □ so that cells are joined together by antibodies forming a lattice or network.

Bacterial Agglutination

Bacterial agglutination tests, sometimes called **Widal** tests, are used routinely in many laboratories for diagnosing infectious diseases and identifying bacterial isolates. These procedures include tube and slide agglutination tests (fig. 21-8). To carry out such a test, the patient's serum (or a dilution of the serum) is mixed in a test tube with a standard suspension of bacterial cells (the antigen). The bacteria-antibody mixture is allowed to react at a constant temperature for a specified amount of time and is then read for agglutination.

A qualitative modification of the Widal test is the slide agglutination test, often done in the laboratory to screen bacterial isolates suspected of being in the genus *Salmonella.* This procedure is carried out by mixing a small amount of a suspected *Salmonella* colony in a drop of specific antiserum. The mixture is rocked back and forth for about 2 minutes and then

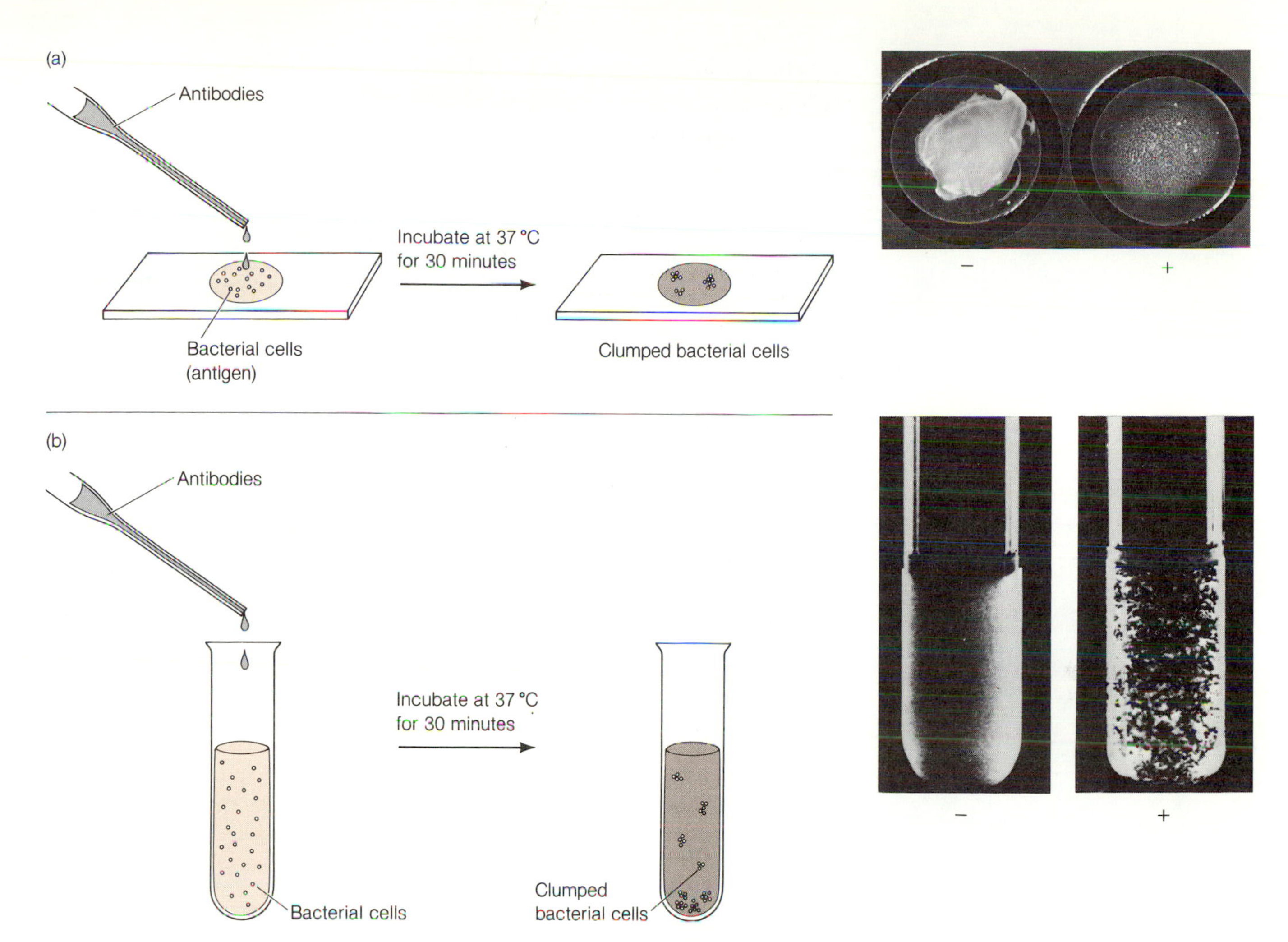

Figure 21-8 Agglutination Tests. Agglutination tests are used to diagnose certain infectious diseases or to identify unknown microorganisms. A drop of antigen (bacterial cell) is mixed with a standard volume of antiserum and mixed well. Positive results are detected by evidence of clumping of the antigen. *(a)* Slide agglutination test. *(b)* Tube agglutination test.

examined for the presence of clumping. In positive tests, the bacteria are arranged in large clumps. Negative tests are seen as a cloudy antiserum with bacterial cells evenly distributed throughout (fig. 21-8).

A modification of the slide agglutination test is the latex particle agglutination test. In this procedure, the specific antigen is adsorbed to inert particles, such as latex beads. This procedure is sometimes used to screen a patient's serum for specific antibodies against infectious agents. In the most common form of the procedure, the beads with adsorbed antigen are mixed with the patient's serum and allowed to incubate. During the incubation period, usually 2–4 minutes, specific antibodies, if present, will bind to the antigens on the surfaces of the latex beads and clump them together. Pregnancy testing is sometimes carried out by using the latex particle agglutination test to detect human chorionic gonadotropic hormone (HCGH), which appears in the urine by the 10th day of pregnancy. The patient's urine is mixed with latex beads adsorbed with anti-HCGH. Any agglutination of the latex beads indicates the presence of HCGH in the urine and therefore pregnancy. This procedure takes only a few minutes and, if properly controlled, is a reliable means of detecting pregnancy early.

Hemagglutination Reactions

Hemagglutination reactions are agglutinations involving red blood cells (fig. 21-9). For example, blood typing involves the agglutination of red blood cells and hence is a hemagglutination reaction. The human ABO antigens polysaccharides are part of glycolipids associated with the plasma membrane of erythrocytes. These antigens, A and B, determine the phenotype of the erythrocytes.

Shortly after birth, the human child has in its circulation antibodies that react against foreign blood groups. For example, an A-type human has antibodies against the B antigen and vice versa. An O-type individual, whose erythrocytes lack both A and B antigens, has antibodies against both A and B antigens. These antibodies are thought to develop in response to antigens in the environment that are extremely similar or identical to the human antigens. Bacteria such as *Escherichia coli*, *Salmonella*, and *Shigella* have antigens that are thought to stimulate the formation of antibodies against foreign blood groups.

Blood typing is carried out like a slide agglutination test: the patient's blood is mixed with antiserum directed against the A antigen and with antiserum against the B antigen. The agglutination patterns of the various blood types are illustrated in figure 21-9. Individuals with blood type A will have their blood agglutinated by anti-A antiserum, but not by anti-B antiserum. Conversely the blood of individuals with blood type B will be agglutinated by anti-B antiserum, but not by anti-A antiserum. Blood of type AB is agglutinated by both anti-A and anti-B antisera, while type O is agglutinated by neither.

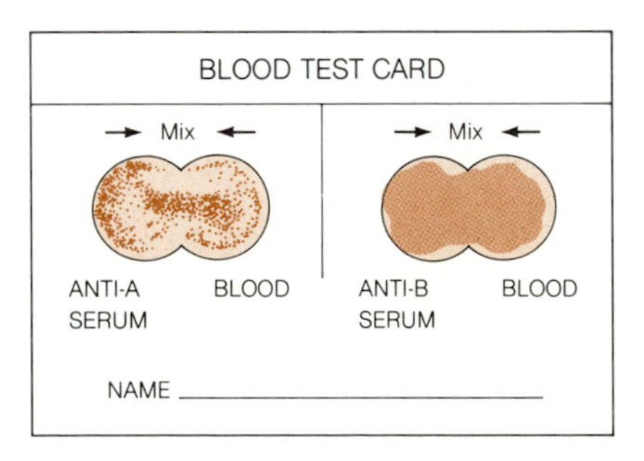

Group A

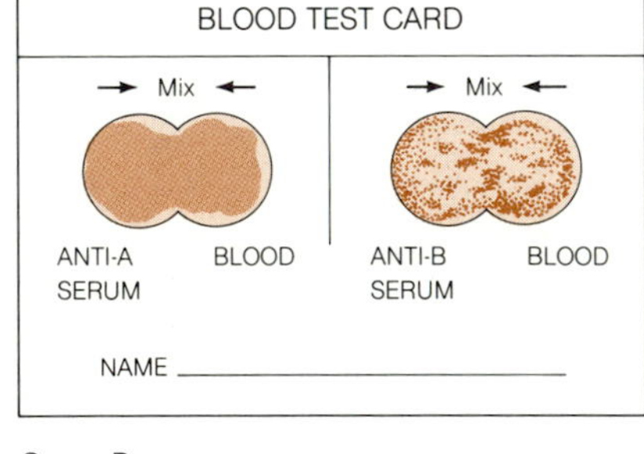

Group B

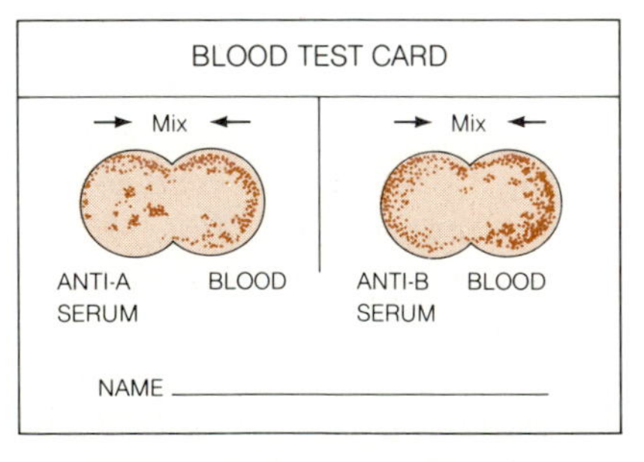

Group AB

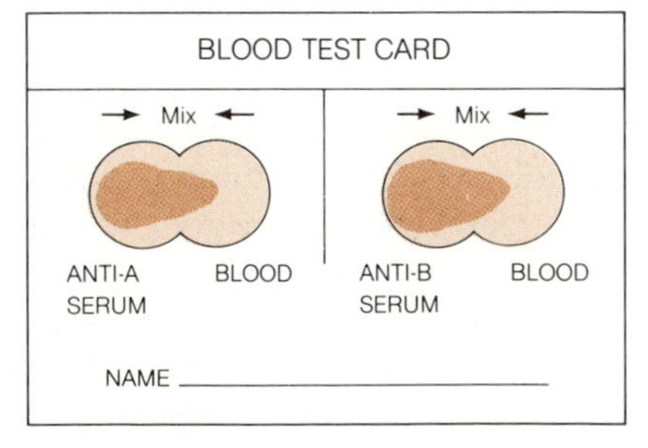

Group O

Figure 21-9 Agglutination Patterns of Human ABO Blood Groups. Blood typing is carried out by mixing known antisera against A and B antigens with blood samples. Blood that is agglutinated with anti-A antiserum only is A type blood. Blood agglutinated by anti-B antiserum only is type B blood. Blood agglutinated by both anti-A and anti-B antiserum is blood type AB. Blood that is not agglutinated by either antiserum is type O blood.

Indirect (Passive) Hemagglutination

Indirect hemagglutination techniques are widely used in clinical microbiology to diagnose viral and other infectious diseases. The procedure is based on the fact that many antigens, such as proteins and polysaccharides, can be adsorbed to red blood cells (RBCs). Thus, the RBCs serve as an easily observed indicator system for serological reactions. When RBCs suspended in a fluid such as physiological saline are left undisturbed, they settle to the bottom of the test tube, forming a distinct red button (fig. 21-10). If, however, these RBCs are agglutinated by a serological reaction between antibodies and the antigens adsorbed to the surface of the (RBCs), the button does not form. Instead, a diffuse layer of RBCs is formed in the bottom of the tube (fig. 21-10).

Microtiter plates (fig. 21-10), a relatively recent addition to the field of serology, are widely used in indirect hemagglutination tests. A microtiter plate usually contains 96 wells arranged in 8 columns of 12 wells each. The wells are generally U or V-shaped cups. In negative indirect hemagglutination tests—that is, in tests in which no antigen-antibody reactions take place—the individual RBCs roll to the bottom of the wells, forming distinct buttons. In positive tests, the cross-linked RBCs settle as a mat over the entire bottom of the well. In a typical assay, each column may represent a different patient's serum. One to three wells of a column are reserved for controls, and the remaining wells can be used for various dilutions of the serum.

Hemagglutination Inhibition

Hemagglutination inhibition tests are used to diagnose certain viral diseases such as rubella, influenza, mumps, and smallpox. Of these, the most common test is for rubella □. These viruses can agglutinate RBCs because they have receptors that recognize membrane components of RBCs (fig. 21-11). When a suspension of a hemagglutinating □ virus (influenza, mumps, reubella, or smallpox virus) is mixed with RBCs from a suitable animal species, the viruses attach to the RBCs, forming a lattice network as in agglutination reactions. Notice that the lattice network is caused by the virus itself binding red blood cells and not by antibodies. Hemagglutination inhibition tests measure the patient's serum for the presence of antiviral antibodies that will prevent the attachment of the virus to the RBCs (fig. 21-11). These antibodies bind to the receptor sites on the viruses, thus blocking their attachment to the RBCs and inhibiting hemagglutination by the virus.

579

599

A typical assay for rubella antibodies involves the use of sheep RBCs, a standard suspension of active rubella viruses, and the patient's serum. This test can also be carried out in a microtiter plate. The patient's serum is mixed with the rubella virus and allowed to react for a predetermined period of time, after which the RBCs are added. After a suitable incubation period, the plates are examined for hemagglutination inhibition. Positive results are ascertained by the lack of hemagglutination, while negative results are indicated by the presence of hemagglutination. If a patient has antiviral antibodies, some of these antibodies will bind to the surface receptors on the viruses. When the RBCs are added to the antibody-virus mixture, the viruses will not be free to bind to the RBCs, so the RBCs will roll to the bottom of the well and form a button. If the patient does not have antibodies against the virus, however, the viruses will bind to the RBCs and hemagglutination will take place.

(a)

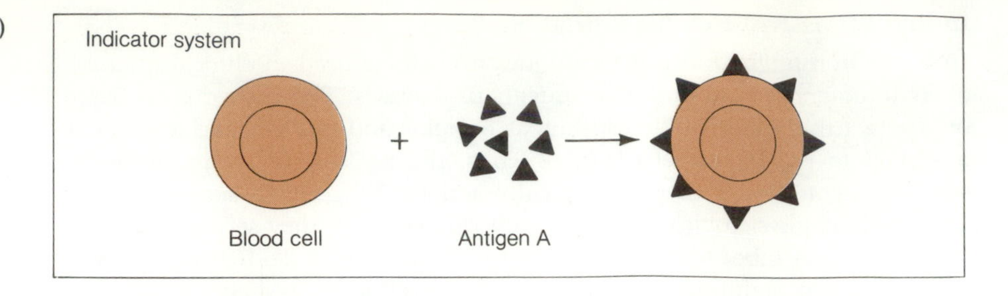

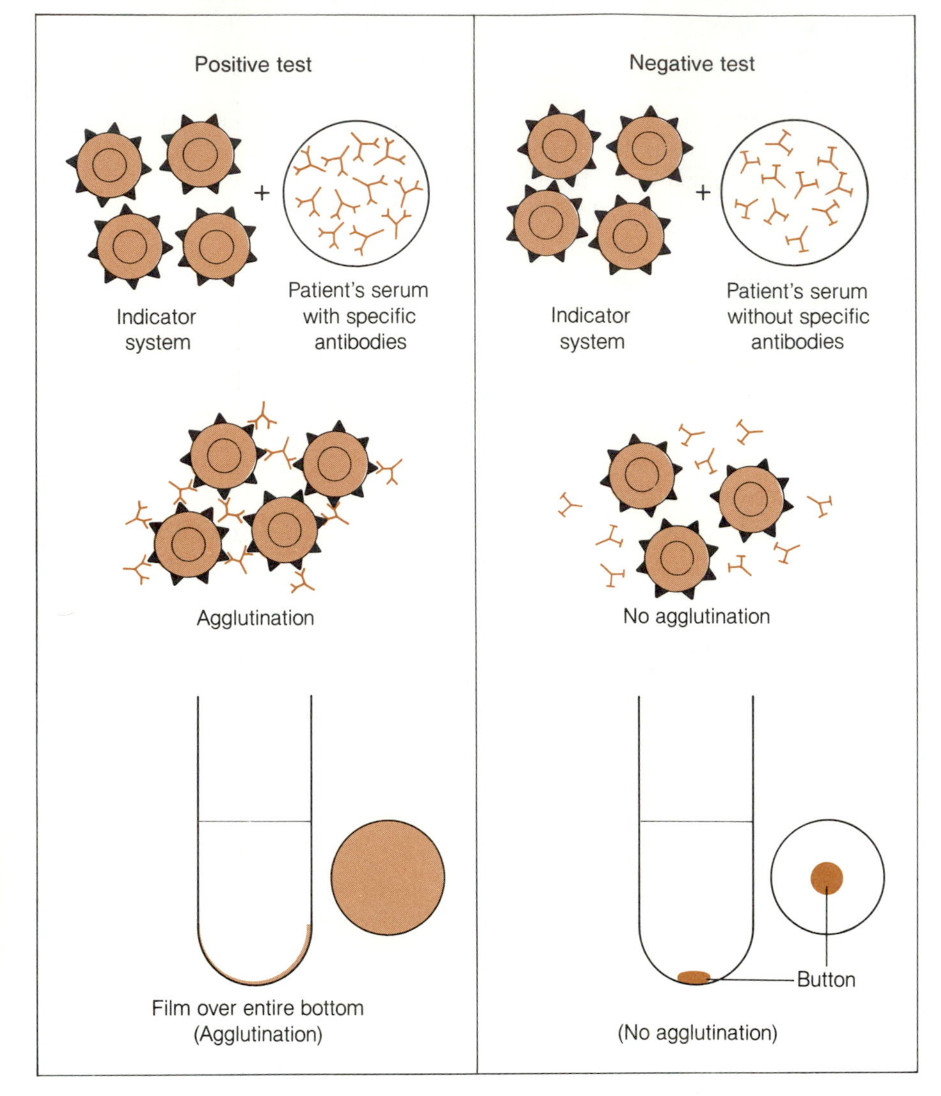

(b)

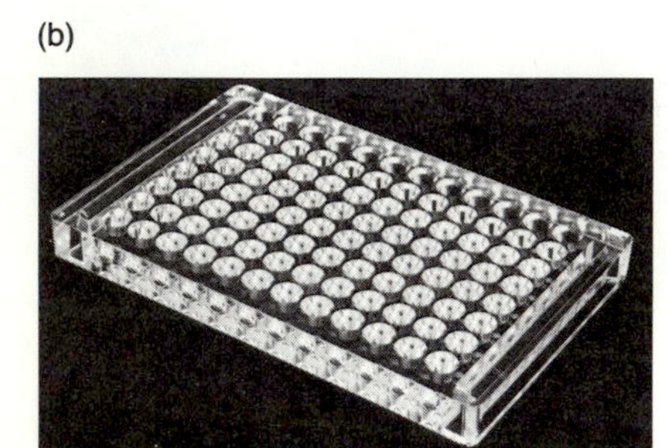

Figure 21-10 Hemagglutination Test. When red blood cells are allowed to settle in a tube, they form a small button at the bottom of the tube. When blood cells are agglutinated by antibodies (or viruses), the clumps settle and form a thin layer all over the bottom of the tube. By viewing the bottom of the tube, one can easily determine whether hemagglutination has taken place or not. *(a)* Diagrammatic representation of a passive hemagglutination test. Antigen is adsorbed to the surface of red blood cells (indicator system). If patient has antibodies against antigen A (positive test), then the patient's serum will agglutinate the red blood cells; otherwise (negative test) a button will develop as an indication of no hemagglutination. *(b)* A microtiter plate. Each well in the microtiter plate serves as a tiny tube. A large number of hemagglutination tests can be carried out and examined quickly.

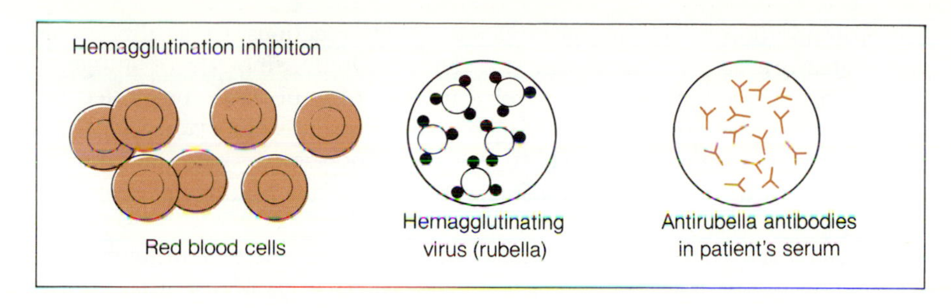

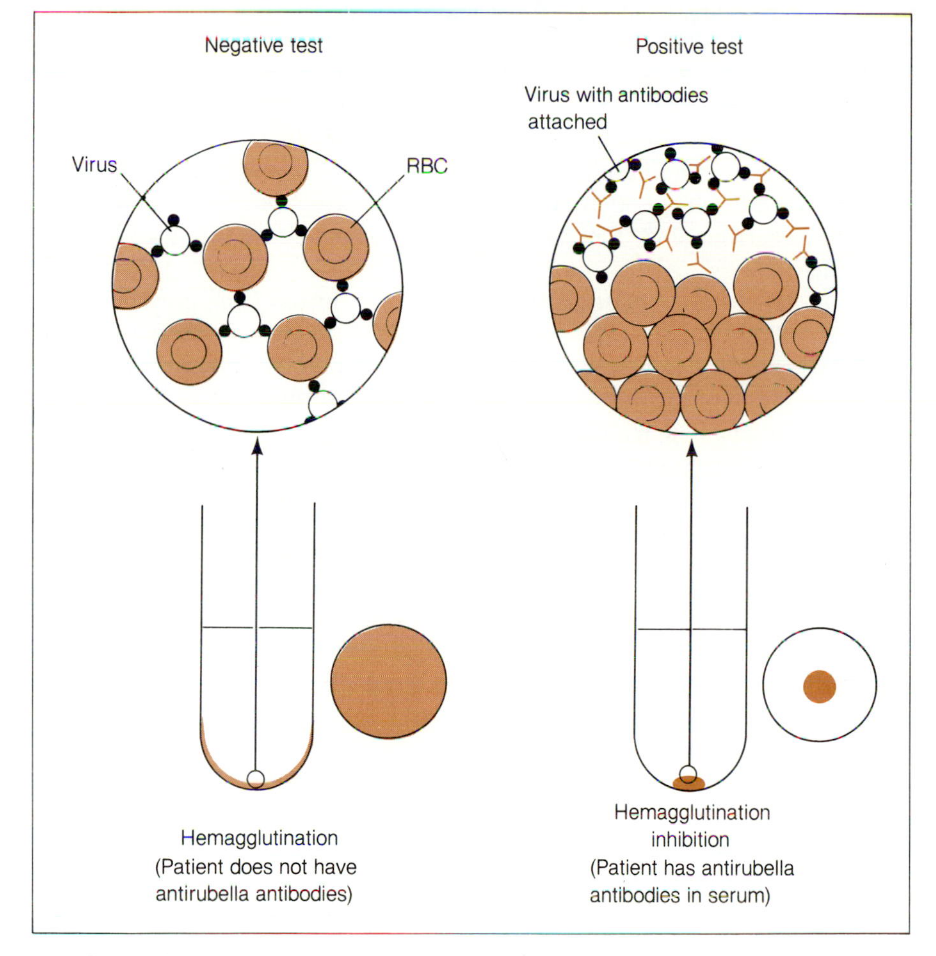

Figure 21-11 The Hemagglutination Inhibition Test. Certain viruses can bind to erythrocytes and agglutinate them. Patients with viral diseases such as rubella can be diagnosed using the patient's serum to inhibit the hemagglutination by viruses. If the patient has antibodies against the rubella virus, they will bind to the virus and prevent hemagglutination by the virus (positive hemagglutination inhibition test). If the patient has no antibodies against the rubella virus, then the virus will be free to agglutinate the erythrocytes (negative hemagglutination inhibition test).

The **viral neutralization** test works basically on the same principle as the hemagglutination inhibition test, but it is used to diagnose diseases caused by viruses that do not hemagglutinate. Virus neutralization tests are carried out by mixing the patient's serum with a standard amount of virus to determine if the patient has antibodies against the virus. If there are antiviral antibodies in the patient's serum, they will prevent the virus from infecting a tissue culture of susceptible cells and will therefore neutralize the virus. In the absence of antiviral antibodies, the viruses are able to infect the tissue culture and cause detrimental changes in the culture.

Immunofluorescence

64 **Immunofluorescence** □ techniques are very useful for detecting antigens or antibodies. They also enable the microbiologist to view the antigen in

tissue, in addition to evaluating antigen-antibody reactions. Generally, these procedures are very sensitive and can detect minute amounts of antigen. Fluorescent antibody procedures involve the use of antibodies conjugated to (or labeled with) a fluorescent dye, such as fluorescein isothiocyanate (FITC) or rhodamine isothiocyanate (RITC). These fluorescent dyes, or fluorochromes, have the characteristic of absorbing ultraviolet light and emitting a light in the visible range.

Labeled antibodies can be used to detect the presence of antigen. For example, in testing for rabies, the presence of rabies virus in brain tissue can be detected by using labeled antibodies against the virus. Another approach to this technique would be to use a known antigen to detect the presence of specific antibodies in a patient's serum. In the test for toxoplasmosis, for example, the antigen consists of killed cells (merozoites) of *Toxoplasma gondii* fixed to a microscopic slide. The patient's serum suspected of containing anti–*Toxoplasma gondii* antibodies is layered on top of the merozoite smear and incubated for about 30 minutes at 35°C. The presence of antitoxoplasmal antibodies surrounding the merozoites is detected by adding labeled antibodies.

The **direct fluorescent antibody** (DFA) test involves the use of specific labeled antibodies against the antigen (fig. 21-12). Many laboratories employ the DFA procedure to verify the identity of Group A streptococci. The reagents are purchased commercially and consist primarily of anti–Group A antibodies conjugated with fluorescein isothiocyanate. A smear of the suspected Group A streptococcal colony (or broth culture) is fixed onto a glass slide and then flooded with labeled antibodies. The bacteria and the antibodies are allowed to react for 30 minutes at 35°C and are then examined with a fluorescence microscope. Group A streptococci will be coated with FITC-conjugated antibodies and will fluoresce upon examination with a fluorescent microscope (fig. 21-12). Other bacteria will not react with the labeled antibody and therefore will not fluoresce.

Indirect fluorescent antibody (IFA) techniques are based on the same principles as DFA techniques. In an IFA test for syphilis, for example, the patient's serum is layered on top of fixed *Treponema pallidum* cells and incubated for 60 minutes. During this time, any antitreponemal antibodies that may be present in the serum will bind to the treponemes. After this incubation period, the smears are washed several times to remove unbound antibodies. The smear is then layered with labeled anti–human immunoglobulin antibodies, which will bind to the antitreponemal antibodies attached to the bacteria. These antibodies will fluoresce when examined with a fluorescence microscope (fig. 21-12). The fluorescent treponemal antibody (FTA) test is considered to be an indirect method because the fluorescent antibodies detect the presence of specific antibodies that reacted with the antigen rather than detecting the presence of the antigen itself. Compare these two techniques in figure 21-12.

Precipitin Tests

Precipitin tests involve antigen-antibody reactions in which the antigens are soluble substances. When a volume of serum known to contain antibodies against a soluble antigen is mixed with an aqueous solution of the antigen, these two components will react with each other in a fashion similar to that of agglutination. Eventually, after a period of minutes to hours, the antigen

(a) Direct Method

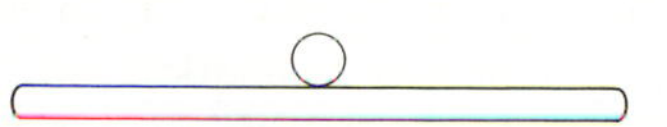

1. Organism (antigen) fixed to slide

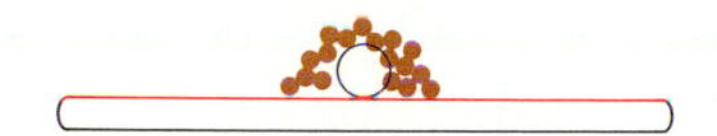

2. Cover with specific fluorescein-labeled antiserum and incubate.

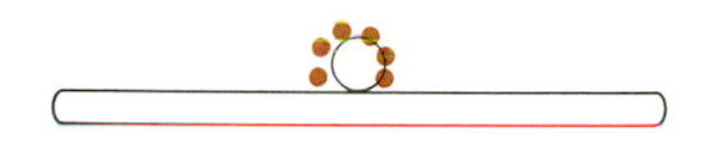

3. Wash away excess fluorescein-labeled antiserum and examine.

(b) Indirect Method

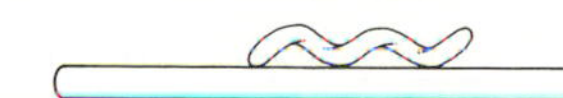

1. Spirochetes fixed to slide (antigen)

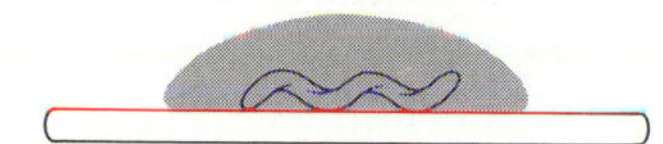

2. Cover with unlabeled serum from patient with active infection and incubate.

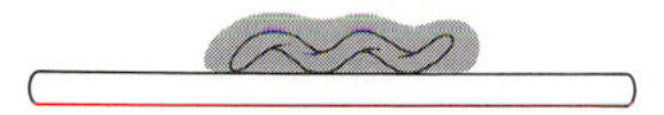

3. Wash away excess unlabeled patient's serum.

4. Cover with fluorescein-labeled antihuman gamma globulin antiserum and incubate.

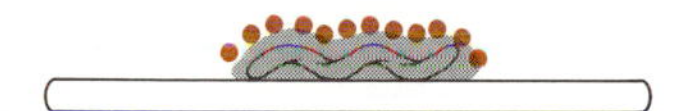

5. Wash away excess fluorescein-labeled antiserum and examine.

Figure 21-12 Fluorescent Antibody Test. The fluorescent antibody test is performed by adding antibodies labelled with a fluorescent dye to a specimen. The antibodies bind to the antigen, which can then be viewed with a fluorescence microscope. Positive results will be detected because the specimens will fluoresce. *(a)* Direct fluorescent antibody test. *(b)* Indirect fluorescent antibody test.

and the antibodies will form a cross-linked matrix that will precipitate out of solution. For example, if a volume of tetanus toxin is mixed with a suitable amount of antitoxin in a capillary tube, a line of precipitation will form where the antigen and the antibodies combine in optimal proportions. This procedure can also be carried out in small test tubes with similar results; in this case, it is called the tube precipitin test.

Gel Precipitation Techniques

All serological tests involving precipitin reactions should be closely monitored to insure that the antigen and the antibody are present in optimal proportions. Several dilutions of the reagents must be made so that optimal proportions are achieved. For this reason, various modifications of the precipitin test using gels have been developed. The purpose of the gel is to provide a medium for diffusion of antigen and antibody, thereby achieving optimal proportions in a single tube and obviating the need for multiple dilutions.

The **Ouchterlony immunodiffusion** (ID) test is the most widely used of these agar gel techniques. In a common procedure, a volume of liquefied agarose (a purified agar) is poured into a 60 mm petri dish and allowed to jell on a cool, level surface. Once the agarose has hardened, a pattern of wells (fig. 21-13) is cut into the agarose. The antigen is placed in one well (usually the center well), and the antisera are placed in peripheral wells. Both antigen and antibody diffuse radially, forming a concentration gradient from a high concentration in the well to a low concentration at the edge of the diffusion pattern. The antigen and antibody form a precipitate where they meet in optimal proportions (fig. 21-13). The ID technique can separate various antigens based on differences in diffusion rates exhibited by various components of the antigen preparation. In cases where there is more than one antigen in the well, a different line of precipitation forms for each antigen (fig. 21-13). Immunodiffusion techniques are widely used as screening tests for a variety of infectious diseases. They have the advantage of being relatively easy to set up, and the results can be obtained within 24 hours.

Radial Immunodiffusion

Radial immunodiffusion is a technique used to quantify the amount of antigen present in serum. As we discussed before, the antigen-antibody reaction at the zone of optimal proportions will give rise to a precipitin line. If an antigen is placed in a well in an agar plate in which a standard amount of antibody has been mixed, it will form a ring of precipitate as it diffuses through the agar and reacts with the antibodies in the agar. The rate of diffusion is constant for a given antigen; thus, an antigen present in high concentration will diffuse farther from the well before reaching optimal pro-

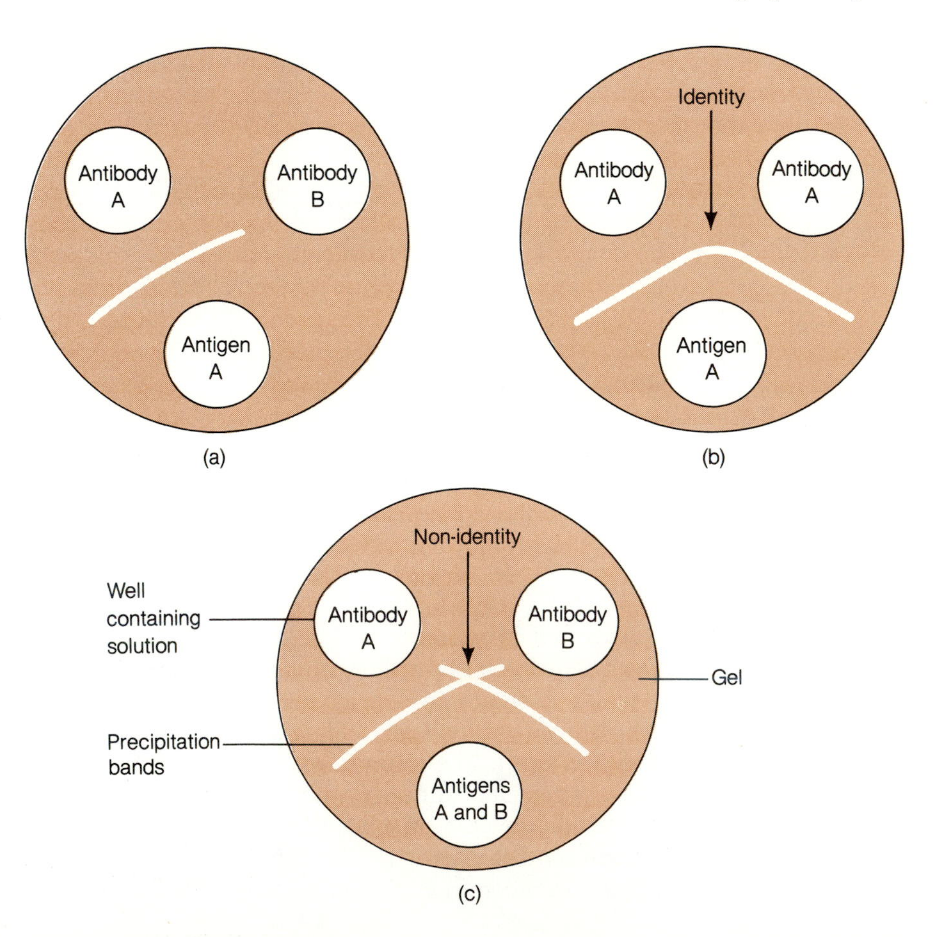

Figure 21-13 The Immunodiffusion Test. Antigen is placed in a well in an agar plate and the antiserum in another. During an incubation period, both antigen and antibody diffuse. Where the zones of diffusion meet in optimal proportions, they will form a band. Each antigen reacts with its corresponding antibody and forms a line of precipitation.

475 portion □ than the same antigen when present in lower concentration. Therefore, the diameter of the circle delimited by the ring of precipitate is directly proportional to the concentration of antigen in the well. If known standards are used, one can quantify the amount of the antigen present by comparing the diameter of the sample with unknown concentrations of antigen with those of known concentration. Radial immunodiffusion is sometimes used to determine the levels of immunoglobulins in a patient's serum. This determination is useful because certain diseases cause a significant rise in the levels of immunoglobulins. For example, African sleeping sickness is characterized by a high level of IgM antibodies in the serum. The high levels of antibodies, along with other clinical evidence, can provide a physician with sufficient information to diagnose the disease.

Immunoelectrophoresis

Immunoelectrophoresis (IEP), a variation of the ID test, is seldom used as a diagnostic test but is widely used in research. It is a desirable research tool because it permits the separation of complex antigens based on their mobility in an electrical field. Thus the investigators can get an idea of the complexity of the antigen with which they are dealing. The electrophoresis is carried out on slides that have been coated with an agar gel. The antigen is placed in a center well and allowed to migrate in an electrical field for a specified amount of time. Then a trough is cut into the agar and filled with specific antibodies. Both antigens and antibodies diffuse radially until they meet in optimal proportions, forming lines of precipitate, one for each antigen-antibody combination.

Other Types of Serological Reactions

Complement Fixation Test

The complement fixation (CF) test is widely used as a diagnostic tool in the clinical laboratory. It measures the amount of immunoglobulins (IgM and IgG) present in the patient's serum against a pathogen. In many infectious diseases, the CF test is the standard procedure by which other serological procedures are evaluated. This test is carried out primarily in large laboratories with heavy work loads, since it is a rather complicated test and requires considerable proficiency on the part of the microbiologist in order for it to yield precise and accurate results. Complement fixation tests have been used to detect immunoglobulins against very different types of infectious agents, such as viruses, bacteria, fungi, and protozoa. In the case of bacterial infections, immunoglobulins against *N. gonorrhoeae* (gonorrhea), *Corynebacterium diphtheriae* (diphtheria), *M. tuberculosis* (tuberculosis), and *Rickettsia rickettsii* (Rocky Mountain spotted fever) have been detected in serum from patients showing typical signs and symptoms of these diseases. One CF test, known as the **Wassermann test,** has been used to test for infections by *Treponema pallidum*. A positive Wassermann test implies that the patient has or has had syphilis.

The complement fixation test is aimed at detecting the presence of antibodies in the patient's serum which will be able to initiate a complement cascade □. (472) As you might recall, when antibodies bind antigens, they form immune complexes that initiate the complement cascade. If these immune complexes involve antigens on the surface of a red blood cell, lysis of the cell will result; hence the fixation of complement can be visualized, because the fluid will turn red (from hemoglobin). If complement is fixed by soluble immune complexes, however, there will be no simple way of detecting the fixation of complement. The complement fixation test is carried out using an indicator system and a test system. The indicator system consists of sheep red blood cells and antibodies (called **hemolysins**) against these cells. The indicator system provides a visible means of detecting the fixation of complement. The test system consists of the patient's serum, a known soluble antigen, and a known amount of complement (obtained from guinea pig serum).

In a complement fixation test, the patient's serum is made to react with the known antigen in the presence of complement. After a period of incubation (usually 37°C for 30 minutes), the indicator system is added (fig. 21-14). If the patient's serum has antibodies against the antigen, the immune complexes that form will fix all of the complement in the reaction tube. When the indicator system is added, the hemolysin will react with the erythrocytes, but there will be no complement left to fix. As a consequence, in cases where the patient's serum has antibodies against the test antigen, the erythrocytes will remain intact (fig. 21-14). If the patient's serum lacks antibodies, however, no immune complexes will form. Thus, when the indicator system is added, the hemolysin binding to the erythrocytes will fix complement and the red cells will be lysed.

As in the case of hemagglutination, the CF test is well suited for microtiter plates. In one microtiter plate, a microbiologist can carry out titrations of several patients' sera and all the necessary controls.

Enzyme Immunoassay

Enzyme immunoassays (EIA), also called **enzyme-linked immunosorbent assays** or ELISA tests, are widely used in the laboratory for the diagnosis of infectious diseases. The advantage of this type of procedure is that it is well adapted to automation. Most EIA procedures currently on the market are designed to detect the presence of specific antibodies against disease agents such as herpes virus, rubella virus, cytomegalovirus, and *T. gondii*, as well as a variety of bacterial and fungal pathogens.

The principle of the procedure resembles that of the indirect fluorescent antibody test. Let us examine an EIA procedure that is commonly employed for the detection of herpes simplex antibodies (fig. 21-15). All the necessary reagents can be purchased in a kit, which also provides microtiter plates with the viruses adsorbed to the bottom of each well. Upon addition of the patient's serum, any antibodies against the virus which are present will attach to the viruses adsorbed to the wells. The unbound antibody is subsequently rinsed off and a measured amount of anti-human immunoglobulin, conjugated with an enzyme such as alkaline phosophatase, is added to the wells. During this period, the enzyme-conjugated anti-immunoglobulin binds to the patient's antibodies, which are attached to the virus (fig. 21-15). After an incubation period, the contents of the cuvette are rinsed several times and the substrate for the enzyme, in this case *p*-nitrophenyl phosphate (PNPP), is added to the wells. The enzymatic breakdown of the PNPP is allowed to take place for a specified amount of time, after which the amount of substrate degraded can be seen as a color change and is measured with a spectrophotometer. The more substrate degraded, the higher the antibody concentration in the patient's serum against herpes simplex. Similar procedures are available for the detection of a variety of infectious agents.

Figure 21-14 The Complement Fixation Test

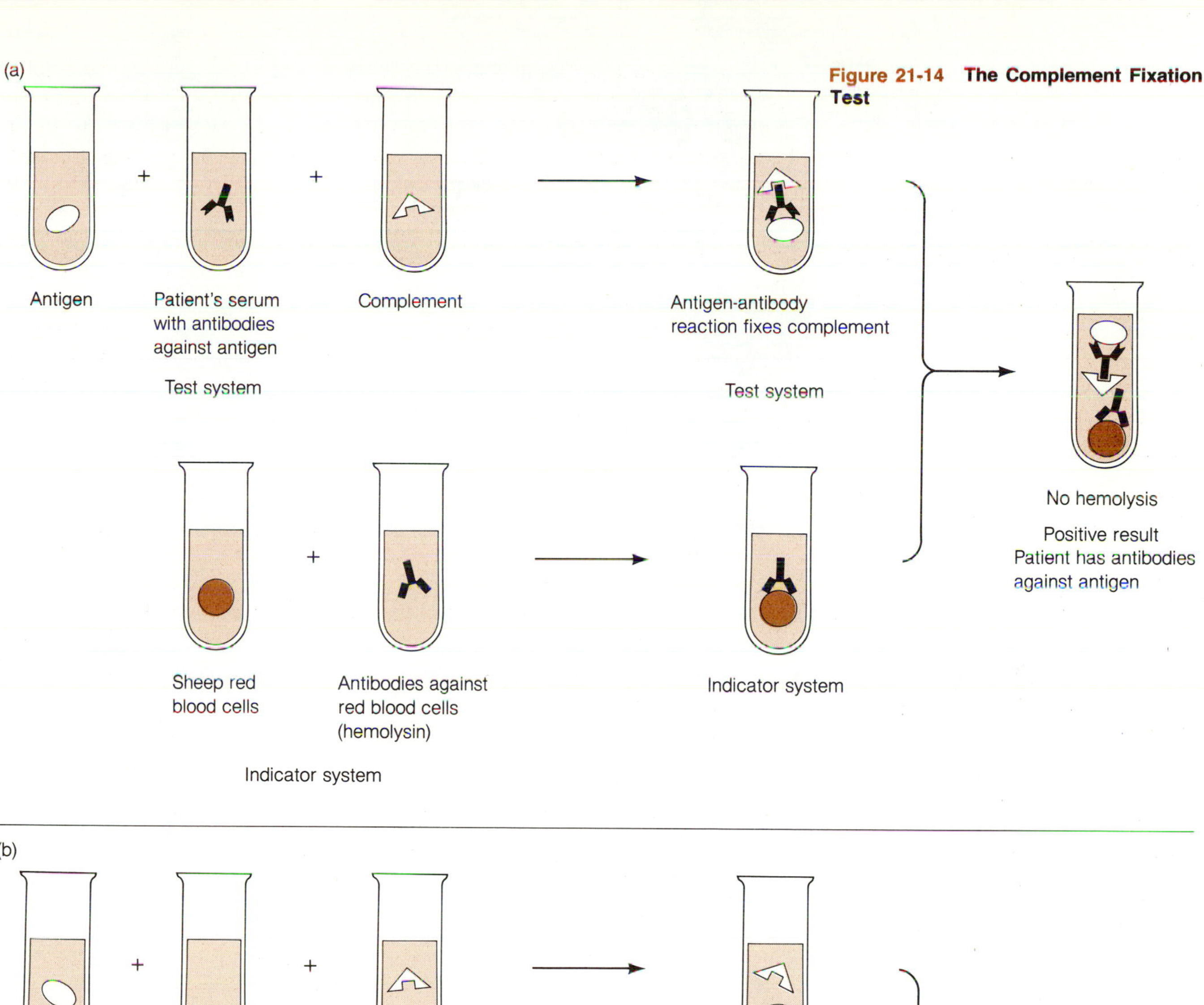

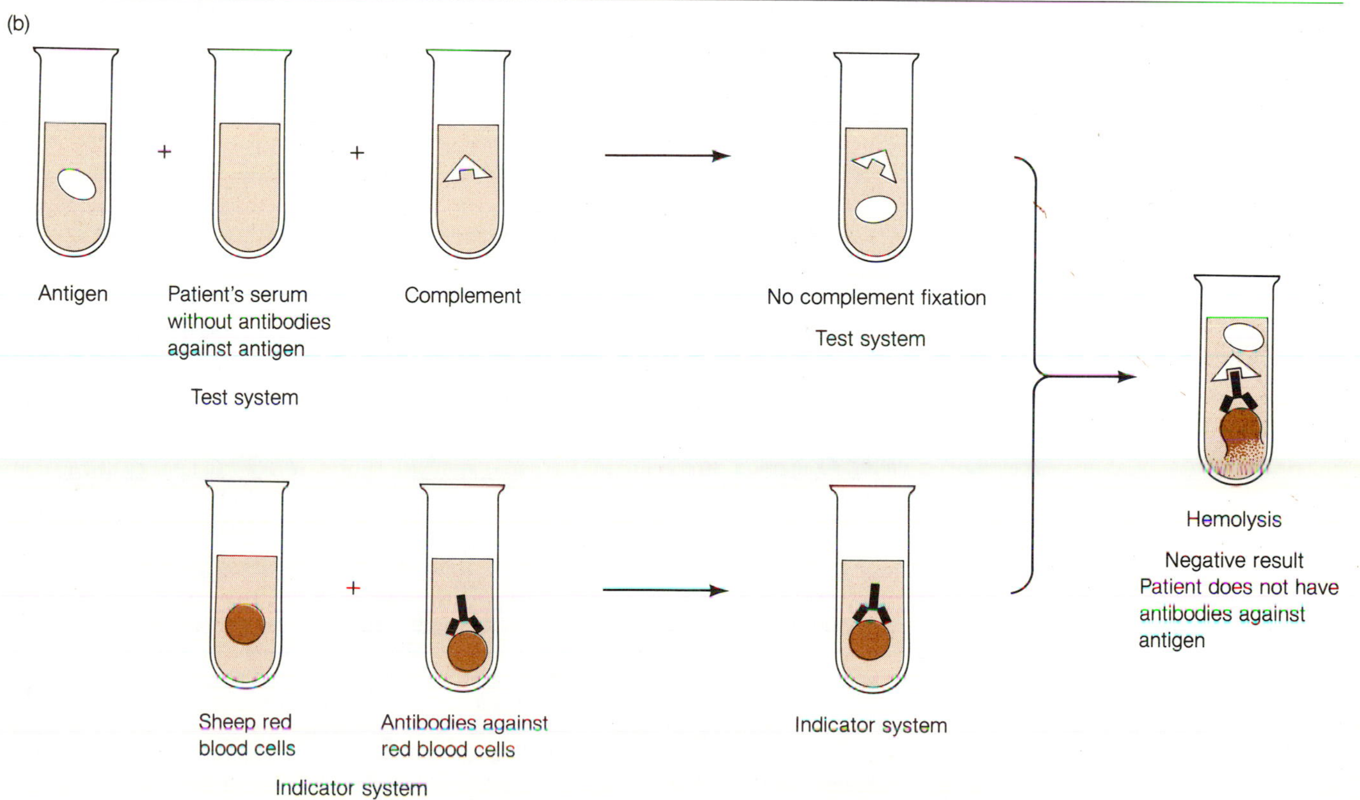

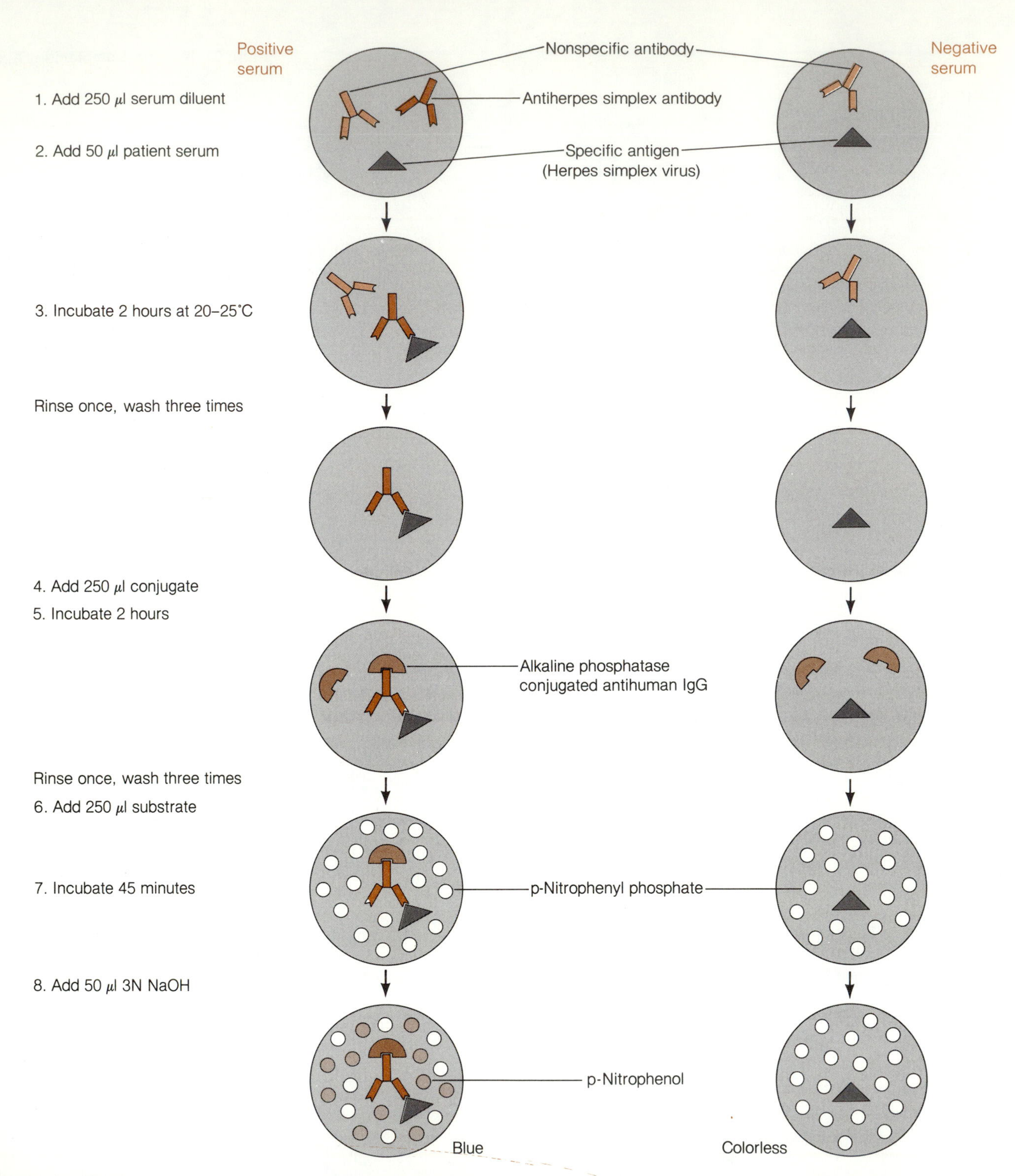

Figure 21-15 Enzyme-Linked Immunosorbent Assay (ELISA). The wells in the microtiter plate are coated with the antigen (herpes simplex viruses). The patient's serum is dispensed into the well and allowed to react with the antigen. After a period of reaction, the wells are washed, and any unbound antibody is removed from the well. At this point, the reagent, which consists of anti-human immunoglobulin antibodies conjugated with alkaline phosphatase (the enzyme), is added. The reagent reacts with the bound antibody and excess reagent is washed away by rinsing. Then the substrate for the enzyme is added (p-nitrophenyl phosphate) and the results are assessed by looking for a colored product. If the patient has antibodies against herpes simplex virus, the well will appear colored. If the patient's serum lacks those antibodies, there will be no color reaction.

FOCUS

HIGH-TECH COWS IN DIAGNOSTIC MICROBIOLOGY

Physicians have always needed rapid and reliable methods of diagnosing infectious diseases. With the development and perfection of techniques for the production of monoclonal antibodies, physicians now have the means to make fast and accurate diagnoses.

The demand for large quantities of monoclonal antibodies has increased at an enormous rate to satisfy not only the diagnostician but also the chemotherapeutic industry, veterinarians, and plant pathologists (to name a few). It is projected that the monoclonal antibody industry will gross more than $50,000,000,000 by 1988, a 1000% increase from the present volume of sales. Because of this demand, scientists have been seeking ways to produce large volumes of high-quality monoclonal antibodies at a low price. A northern California company, for example, is using cows to provide nutrients for the mass-production of substances such as interferon, and monoclonal antibodies.

Cows have a large flow of lymph, so each cow can provide nutrients to produce about 5 g of monoclonal antibodies each day. The desirable aspect of using cow lymph is that it is recycled through the cow. The lymph is tapped by inserting a catheter (tube) into the cow's lymphatic system (thoracic duct). The lymph is then mixed with a suitable growth medium and pumped into a growth chamber. In the chamber, the lymph is used to support the growth of cells, which in turn produce the monoclonal antibodies. These antibodies can then be harvested and purified. The lymph is pumped back into the cow, where it is "recycled." Hence, by using cows to produce the necessary lymph, microbiologists can achieve a continuous process of antibody production that is economical.

Skin Testing

Infections by many bacteria, viruses, protozoa, and fungi often induce a long-lasting state of cell-mediated immunity that is reflected by a delayed type hypersensitivity □ (DTH) response. The DTH state can be determined by the intradermal injection of an antigen derived from the infectious agent. Skin tests, also known as **Mantoux tests,** are often used to determine whether or not a patient has been exposed to the infectious agent. Therefore, these tests serve not only as a diagnostic tool but also as an epidemiological tool □. The procedure has been used successfully to determine the incidence and distribution of infectious diseases, such as tuberculosis and coccidioidomycosis, in various human populations.

468

543

SUMMARY

1. Isolation and identification of infectious agents essential for the proper management of infectious diseases, involves the proper collection and transport of clinical specimens as well as laboratory procedures.
2. Identification of microorganisms may involve microscopic methods, cultural methods, and/or serological methods.

COLLECTION AND TRANSPORT OF SPECIMENS

1. Collection and transport are essential to the proper isolation of pathogens.
2. The proper collection of specimens requires knowing how and when to collect the specimen. It also requires that every effort be made to prevent the introduction or removal of any source of toxic materials and to prevent contamination.
3. Specimens should be delivered promptly to the laboratory and maintained referegerated until they can be processed.

MICROSCOPIC METHODS IN DIAGNOSIS

1. Microscopic methods can help in establishing a quick diagnosis of an infectious disease, or at least

give a good indication as to the best course to follow in order to make an accurate diagnosis.

2. Wet mounts, stained smears, and stained histological sections are all useful in viewing clinical specimens for the presence of infectious agents.

CULTURAL METHODS IN DIAGNOSIS

1. Clinical specimens are planted on selective and/or differential media in order to increase the chances of isolating the pathogen from the specimen. Sometimes the clinical specimen is treated with selective agents, such as alkali or antibiotics, before planting onto culture media, in order to increase further the chances of isolating the pathogen.
2. Once a suspected pathogen has been isolated, it is then identified using logical sequences of testing which include staining procedures, cultural and biochemical tests, and serological procedures.
3. Sometimes it is desirable, for economy or expeciency, to use rapid methods or kits to identify pathogens. These kits have several advantages over the conventional methods: low cost per isolate, high reliability, and long shelf life.

SEROLOGICAL METHODS

1. Serological methods for diagnosing infectious diseases are desirable when the suspected agent is difficult or expensive to grow in the laboratory, or when the results are desired quickly.
2. Several different serological tests are commonly used in the laboratory, but all are based on the same principle: an individual infected by a microorganism will respond immunologically to the infectious agent, and the response will be detected with these serological methods.
3. Serological methods routinely used in the laboratory include agglutination, precipitation, complement fixation immunofluorescence, and skin tests.
4. Some newer methods, such as the enzyme immunoassay (EIA), are well suited for automation and therefore are very popular.

STUDY QUESTIONS

1. Discuss three important characteristics of a properly collected specimen.
2. Why are transport media used in the clinical laboratory?
3. Outline a procedure that you will follow to isolate a pathogen from a urine specimen.
4. Given that you have isolated a bacterium from a stool specimen, outline the steps that you will take in order to identify this organism.
5. Why are mucolytic agents used? Why are specimens sometimes centrifuged before plating?
6. What is the importance of selective and differential media in the identification of pathogens?
7. Why are serological techniques so popular in the clinical laboratory?
8. Of what value are monoclonal antibodies in diagnostic microbiology?
9. What is the difference between sensitivity and specificity?
10. What is the principle on which serological reactions are based?
11. How do agglutination tests differ from precipitation tests?

SUPPLEMENTAL READINGS

Centers for Disease Control. 1985. Provisional public health service inter-agency recommendations for screening donated blood and plasma for antibody to the virus causing acquired immunodeficiency syndrome. *Morbidity and Mortality Weekly Report* 34(1):5–7.

Delmee, M., Homel, M., and Wauters, G. 1985. Serogrouping of *Clostridium* difficile strains by slide agglutination. *Journal of Clinical Microbiology* 21(3):323–327.

Finegold, S. M. and Smith, W. J. 1982. *Diagnostic microbiology*. 6th ed. St. Louis: Mosby, C. V.

Forney, J. E. ed. 1968. *Collection, handling, and shipment of microbiological specimens*. Public Health Service Publication No. 976. Washington: U.S. Public Health Service.

Grimont, P. A. D., Grimont, F., Desplaces, N., and Tchen, P. 1985. DNA probe specific for *Legionella pneumophila*. *Journal of Clinical Microbiology* 21(3):431–437.

Lennette, E. H., Balows, A., Hausler, W. J. Jr., and Shadomy, H. T. eds. 1985. *Manual of clinical microbiology*. 4th ed. Washington: American Society for Microbiology.

Neu, H. C. 1978. What should the clinician expect from the microbiology laboratory? *Annals of Internal Medicine* 89:781–784.

Pál, T., Pácsa, A. S., Emödy, L., Vörös, S., and Sélley, E. 1985. Modified enzyme-linked immunosorbent assay for detecting enteroinvasive *Escherichia coli* and virulent *Shigella* strains. *Journal of Clinical Microbiology* 21(3):415–418.

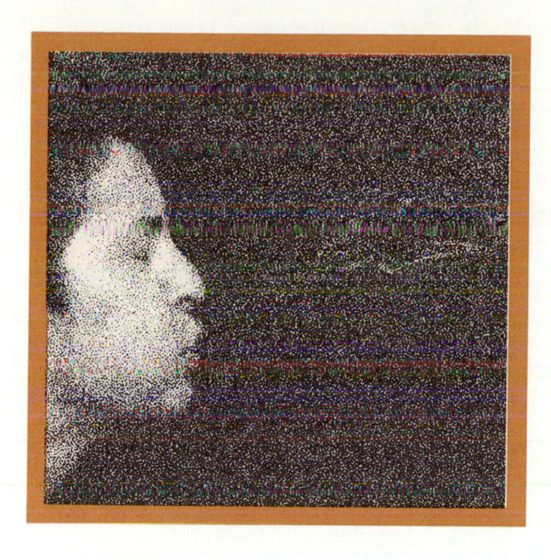

CHAPTER 22
EPIDEMIOLOGY OF INFECTIOUS DISEASES

CHAPTER PREVIEW

PERSPECTIVE: JOHN SNOW ON CHOLERA

Cholera is an infectious disease that has ravaged populations for many centuries. The agent of the disease, *Vibrio cholerae,* is spread in drinking water. The disease is characterized by a profuse diarrhea that has the color of rice water (rice water stool). The water loss due to diarrhea is so great that patients can lose 10–15% of their body weight in less than a day. Death due to cholera usually results from kidney failure or low blood volume, both consequences of fluid loss.

We now have a clear understanding of the cause of cholera, the disease process, and how it spreads in human populations. This was not the case as recently as 150 years ago. In fact, cholera was thought to be a curse of the gods as punishment for our transgressions. Much of what we know about the spread of cholera is based on studies carried out by the British physician and anesthesiologist John Snow. His studies clearly showed that cholera spreads in human populations through drinking water polluted with sewage. This research was done even before the discovery of *Vibrio cholerae* in 1883 by Robert Koch.

Snow's knowledge of cholera was acquired during a series of outbreaks in London in 1853 and 1854. At that time, the drinking water was supplied to Londoners by two different purveyors: the Southwark & Vauxhall Company (S & V) and the Lambeth Company.

Snow's studies led him to suspect that water was a plausible source of cholera. He conducted a survey of those houses in which cholera patients lived and found that the S & V company supplied the drinking water to most of those homes. He also discovered that the S & V company obtained its water from the Thames River, right in the heart of London—the very same place where Londoners discharged their sewage. In contrast, the Lambeth company obtained its water from the Thames River before it reached the city. Thus, Snow concluded that cholera was spread by drinking water contaminated with raw sewage (which contained the cholera agent). He also pointed out the need for water purification to alleviate this malady.

The studies conducted by Snow in the 1850s are considered to be classic studies in epidemiology. He employed methods that included both interviews of patients and experimentation. The results he obtained allowed him to decipher the epidemiology of cholera. In this chapter we will discuss some of the characteristics of epidemics, how epidemics are detected and how infectious diseases are spread among humans.

LEARNING OBJECTIVES

A STUDY OF THIS CHAPTER SHOULD ENABLE YOU TO:

- DISCUSS WHAT EPIDEMICS ARE AND HOW THEY TAKE PLACE
- BE FAMILIAR WITH THE WAYS IN WHICH EPIDEMICS ARE MEASURED
- DESCRIBE HOW MICROORGANISMS CAN BE TRANSMITTED
- DESCRIBE SOME SOURCES OF INFECTIOUS AGENTS
- OUTLINE THE INTERACTIONS AMONG THE ENVIRONMENT, THE MICROORGANISMS, AND THE HOSTS IN EPIDEMICS
- CITE SOME OF THE MEANS USED TO CONTROL EPIDEMICS
- DESCRIBE WHAT NOSOCOMIAL INFECTIONS ARE

Epidemiology is the science that studies the prevalence and distribution of disease in a population. This information can enable physicians to make prompt and accurate diagnoses. For example, certain infectious diseases, such as influenza, elicit signs and symptoms such as headache, malaise, fever, and achy joints, which are characteristic of a variety of diseases. Under ordinary conditions, a physician may need additional information to make a diagnosis of influenza based on these signs and symptoms. If the same physician is aware that an epidemic of influenza is taking place, however, the diagnosis is facilitated, because the signs and symptoms exhibited by the patient are consistent with those of influenza. Also, by studying groups of patients, enough information can be generated to answer such basic questions as how to prevent a given disease. For instance, epidemiological studies of patients with myocardial infarctions (heart attacks) and of normal patients have helped physicians to determine that factors such as personality, social pressures, smoking, obesity, and lack of exercise contribute to heart attacks. Therefore, heart attacks can be prevented by reducing these risk factors.

Increasing population densities and rapid worldwide travel must be considered in epidemiological studies. No longer is it possible to think of infectious diseases such as leprosy, malaria, schistosomiasis, and hemorrhagic fever as exotic diseases of distant, foreign lands. This is because a patient may contract the disease 10,000 miles from home and not show its signs and symptoms for a number of weeks after returning. For example, the number of cases of leprosy in the United States has increased consistently for the last two decades (fig. 22-1). This increase can be attributed to foreign travel as well as to increased immigration of infected individuals. Furthermore, with increasing population densities, the risk of outbreaks of infectious diseases also increases.

EPIDEMICS AND PANDEMICS

An **epidemic** or outbreak is usually a short-term increase in the occurrence of a disease in a particular population. Three cases of diphtheria in a week

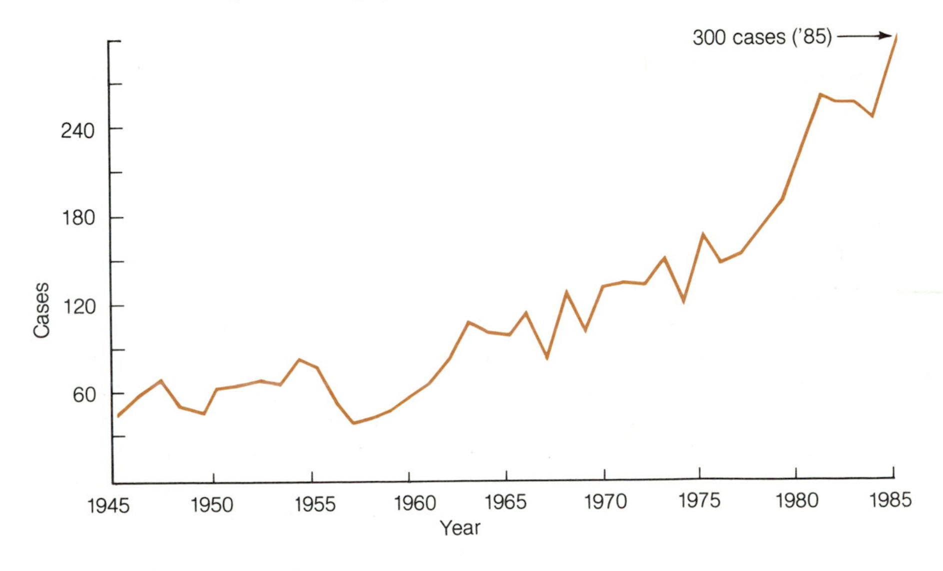

Figure 22-1 Yearly Incidence of Leprosy in the United States. The incidence of leprosy in the United States has been on the rise since 1957. This increase is due primarily to the increased immigration of individuals from areas where leprosy is prevalent.

in a small town of 1,000 inhabitants, which normally has one such case every five to ten years, is definitely an epidemic. Three cases of influenza in a day in a large metropolis of 1 million inhabitants, however, certainly is not an epidemic. Figure 22-2 illustrates a typical epidemic graph. Notice that this epidemic represents a significant increase in the number of cases above the number expected.

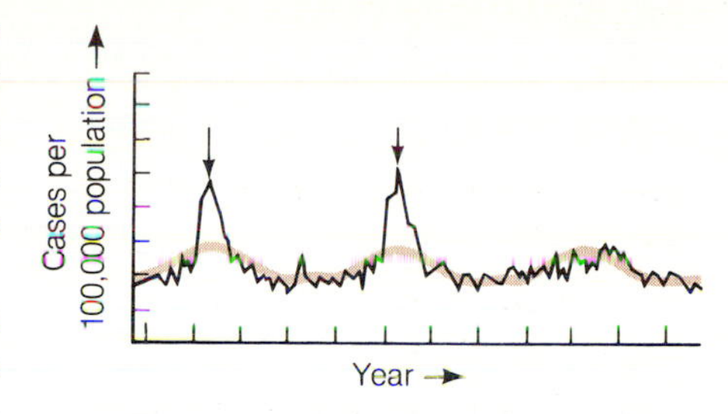

Figure 22-2 Graph Illustrating an Epidemic. The color lines indicate the expected number of cases; the black lines indicate the actual number of cases. Epidemics (arrows) are sharp increases in the number of cases of a disease above that expected.

The term **endemic** generally describes the usual number of cases of a particular disease in a population. This figure is sometimes also called the "prevalence rate." Therefore, we can define an epidemic simply as a short-term increase in the incidence of a disease over the endemic rate. An **endemic disease** is one that occurs in a population or geographic area on a constant basis. One speaks of venereal diseases being endemic in the United States, or histoplasmosis being endemic in the Mississippi River Valley. On the other hand, **sporadic** cases occur in a population without any type of periodicity. They appear in and disappear from a population at unpredictable rates.

Sometimes the term **pandemic** is used to describe a long-term increase in the incidence of a disease in a very large population. The duration is usually years, and the spread of the disease goes beyond international boundaries. For example, the most recent pandemic of cholera, originated in Indonesia in the early 1960s and spread to parts of Europe, Africa, and the Middle East within 10 years.

PARAMETERS USED BY EPIDEMIOLOGISTS TO MEASURE EPIDEMICS

In order to make sense of data obtained from studies of infectious diseases, epidemiologists have developed criteria to quantify the various characteristics of epidemics. These data include counts of the number of cases or the occurrence of a particular feature in a group of infected individuals. Simple counts are not very useful in themselves, but are used to determine proportions or rates in a population. In our examples of diphtheria and influenza, the counts were the same—3 cases—but 3 cases/1,000 (300/100,000) is much greater in significance than 3 cases/1,000,000 (0.3/100,000). Therefore, rates or proportions can provide valuable information, while simple counts generally do not.

Prevalence Rate

Prevalence rates describe the occurrence of a particular disease in a population at a specified time. The population may be a school, a hospital, a city, a state, or a country. Prevalence rates are calculated by dividing the number of individuals in the population suffering from a particular disease by the total number of individuals in the population. Consider, for example, the statement: "On August 8 there were 150 cases of influenza in a city of 13,000 inhabitants." The prevalence rate of influenza in that population is $150/13,000 \times 100 = 1.15$ percent. Prevalence rate can be measured for any time interval, such as a day, a month, or a year. If prevalence rates are determined at frequent intervals, an indication of the endemic rate of the disease in a population can be determined.

Incidence Rate

The incidence rate differs from the prevalence rate in that it describes the number of *new cases* in a population over a period of time. It is calculated by dividing the number of new cases during a given interval of time by the number of individuals in the population at risk. For example: "At one point during a measles epidemic in an elementary school of 750 students, there were 45 new cases of measles in one day." The incidence rate for this day was: $45/750 \times 100 = 6$ percent. The average number of cases per day, divided by the size of the population at risk (750), will give the incidence rate for that epidemic.

Mortality, Morbidity, and Case Fatality Rates

The **mortality rate** measures the number of individuals dying in a population, either from a specific cause (for example, subacute bacterial endocarditis) or from all causes. It is calculated by dividing the number of deaths by the size of the total population.

Morbidity rates measure the number of individuals who become ill as a result of a particular infection. It is calculated as follows: number of individuals becoming ill/population at risk $\times$ 100.

The **case fatality rate** measures the number of deaths among individuals suffering from a particular disease and it is calculated as follows: number of deaths among individuals with a particular disease/total number of patients with that particular disease.

Together, the morbidity and case fatality rates give epidemiologists an indication of the intensity of exposure and the virulence □ of the strain causing a particular disease.

SOURCES OF INFECTIOUS DISEASES

Microorganisms that cause disease can enter a host from a variety of sources. Frequently, a human host is merely an accidental host who happens to be in the wrong place at the wrong time. In cases such as this, the pathogenic microorganism leads a life outside the host and may be found in soil or water or in other hosts. These places where infectious agents are perpetuated in nature are called **reservoirs.** Microorganisms that cause disease in animals and humans have reservoirs as diverse as soil, water, rodents, and other humans. For example, the fungal disease called valley fever (coccidioidomycosis), which afflicts thousands of individuals living in endemic areas, has soil as its reservoir. The fungus reproduces in the soil, and when it becomes airborne in wind currents it can infect susceptible individuals. On the other hand, the natural reservoir for rabies consists of small feral mammals, primarily skunks. Rabies is transmitted from skunk to skunk in the wild, but occasionally an infected skunk finds its way to the suburban environment and quarrels with domestic animals, which can then become infected. These animals, as well as skunks, can transmit the disease to humans. Such diseases, which are endemic in animal populations, are called **zoonoses.** Typhoid, by contrast, which is caused by *Salmonella typhi*, is transmitted from person to person by the fecal-oral route, by ingestion of foods or water contaminated with human fecal matter from individuals carrying the typhoid

FOCUS

CONTROL OF MARY MALLON: A CARRIER OF THE TYPHOID FEVER BACILLUS

Presently, 400–500 cases of typhoid fever and 5–25 deaths from this disease are reported each year in the United States. In the early 1900s, however, there were many thousands of cases and 11,000–25,000 deaths each year. Most of these cases of typhoid fever were due to drinking water contaminated with sewage and eating foods handled or prepared by persons shedding the typhoid fever bacterium. One of the most famous carriers of the typhoid bacillus was Mary Mallon.

Mary Mallon's story began in 1906, when the New York City Department of Health hired a sanitary engineer named George Soper to investigate an outbreak of typhoid fever in a family in Oyster Bay, Long Island. Since neither the water drunk by the family nor the food in the home was contaminated, Soper suspected that the family cook, who had quit her job when the outbreaks of typhoid fever occurred, might have had the disease and spread it to the family. The family cook was Mary Mallon.

In his attempt to find Mary Mallon, Soper discovered that she had worked in seven homes from 1896 to 1906. In these homes, there had been 28 cases of typhoid fever during the time Mary had worked as a cook. In every case, Mary had left the job when persons in the homes came down with typhoid fever. In 1907, Soper caught up with Mary Mallon in Manhattan, where she was working for a family. He confronted her with his suspicions that she was a typhoid carrier and that she was responsible for the outbreaks of typhoid fever in the homes where she had worked. Soper asked for blood, urine, and fecal specimens to determine conclusively whether or not she was shedding the bacteria and promised her free medical treatment. Instead of accepting Soper's offer, Mary became belligerent and threatened Soper with a carving fork. Mary did not believe in germs nor in the possibility that she was the cause of the typhoid outbreaks where she had worked.

The New York City Health Department soon ordered Mary arrested. She was brought fighting and screaming to a city hospital for patients with infectious diseases. Examination of Mary's stool samples demonstrated that she was shedding large numbers of the typhoid bacilli, but Mary showed no external symptoms of typhoid fever. An article in the *Journal of the American Medical Association* of 1908 referred to her as Typhoid Mary, an epithet by which she is known today. Doctors believed that the bacteria were growing in her gall bladder, and recommended that it be removed in order to eliminate her carrier condition. Since there was disagreement as to exactly where the bacteria were growing in her body, she refused the operation. Mary realized that the operation might kill her because bacteria would spread into her abdominal cavity when the infected tissue was removed. Operations of this type were very dangerous because there were no antibiotics then.

In 1910, Mary was released because of public pressure on the Health Department. The public felt it was not right to imprison persons who had not committed crimes but were simply carriers of disease-causing organisms. Mary remained at large until 1915, when the Public Health Department connected a 1914 outbreak of typhoid fever at a New Jersey sanatorium and a 1915 outbreak at a New York hospital to a cook matching Mary's description. In 1915 she was returned to her island hospital prison in the middle of the New York River. She spent the next 23 years at the hospital, where she died in 1938 at the age of 70. During her life in prison, she worked in the hospital laboratory but was forced to eat alone. Mary was imprisoned for 26 years of her life because she was a danger to the community and was responsible for a number of deaths: at least 53 cases of typhoid fever and three deaths were positively connected with her.

Control of carriers is a serious moral and public health issue even today. For example, there are approximately 800,000 carriers in the United States—and more than 200 million carriers throughout the world—of the virus that causes hepatitis B. This virus is generally transmitted by sexual contact, by blood transfusions, and through needle pricks among drug addicts. What is the responsibility of Public Health Departments in controlling carriers who are prostitutes or sexually promiscuous, or who want to sell their blood to blood banks? Should these carriers be held accountable if illness and death result from their behavior? This is an ethical problem for which there may be no simple solution.

bacterium. Humans represent the only known reservoir for this disease. The dangerous aspect of typhoid resides in the fact that certain infected individuals harbor the infectious agent without themselves becoming ill. These asymptomatic individuals, called **carriers,** serve to perpetuate the disease in the human population.

MODES OF TRANSMISSION

Infectious diseases result when an individual comes in contact with an infectious agent. This contact may come about by one or more of the following routes: a) person to person contact; b) contact with infected animals; c) contact with vectors; d) contact with fomites; e) consumption of contaminated food or water; f) inhalation of airborne particles.

Person to Person Contact

Many infectious diseases are transmitted from one individual to another by direct contact. Venereal diseases □, for example, are contracted by sexual intercourse or other sexually-related activities. Touching infected lesions, such as boils and ulcers may transfer the infectious agent to a susceptible host. Other diseases can be transmitted by direct contact, such as touching hands or kissing. Infectious mononucleosis ("mono") can be transmitted by kissing, but it can also be transmitted by indirect means. The common cold can be transmitted from person to person by contact with hands contaminated with nasal secretions of infected individuals.

644

Contact with Infected Animals

Diseases such as rabies, tularemia, and ringworm can be acquired by contact with infected animals. For example, ringworm can be contracted from cats, dogs, and horses. Tularemia, a zoonosis that afflicts chiefly rabbits, is not uncommon among rabbit hunters in endemic areas. Rats, mice, and other rodents are reservoirs for a variety of infectious diseases, many of which can be transmitted to humans. Bats can transmit diseases such as rabies and histoplasmosis. It was once thought that only vampire bats attacked animals for their blood, but it is now known that other bats, if rabid, may attack and bite animals or humans and thereby transmit the rabies virus.

Vectors

Vectors are a special group of animals that can transmit infectious agents to other animals and to humans. Vectors are defined in various ways by different authors. Some liberal interpretations may include water and food as vectors. In the strict sense, however, a vector is an arthropod that has a long-lasting association with a parasite and can transmit the disease to susceptible individuals. For example, the mosquito *Anopheles* is a vector □ for malaria; that is, it can transmit the malaria parasite by biting a susceptible host while feeding. The sexual development cycle of the malarial parasite takes place in the mosquito, so the mosquito is a necessary host for the completion of the life cycle of the parasite.

386

A similar situation exists with the vector that transmits Rocky Mountain spotted fever. The tick *Dermacentor andersonii* can harbor *Rickettsia rickettsii,* a bacterium that multiples within the tick and can be transmitted from generation to generation of ticks in the eggs. The tick, upon biting a suitable host, transmits the infectious agent.

676

Arboviruses □ are a group of medically and economically important viruses that are transmitted by vectors. These viruses cause a variety of diseases, including encephalites, that afflict humans and livestock. The viruses are transmitted by the bite of arthropods, such as ticks and mosquitoes.

Now let us examine another situation in which an arthropod is not strictly a vector, but does transmit an infectious organism. Toxoplasmosis is an infectious disease caused by *Toxoplasma gondii,* a protozoan that is very common in nature. Toxoplasmosis is a common disease of cats and occasionally is transmitted to humans. The parasite is excreted in the cat's feces into soil, sandboxes, or "kitty litter" boxes. A fly, for example, can land on contaminated soil or sand and pick up the parasite with its legs. The fly may then land on food and deposit the parasite, and a human ingesting the contaminated food may become infected. The fly, although it served as a **vehicle of transmission,** cannot really be considered a vector, because the association between the parasite and the fly was accidental and not of long duration. This type of transmission is called **phoresis,** and the fly, in this example, is the **phoront.** The term **mechanical vector** has also been used to describe this mode of transmission.

Fomites

Fomites are inanimate objects such as drinking cups, combs, toilet seats, and showers that serve to transmit a disease agent but do not support its multiplication. Ringworm of the scalp in preadolescents is most often caused by a fungus called *Microsporum audouinii.* This fungus can be transmitted among children by direct contact with one another, but it can also be transmitted through fomites, such as combs or backrests of theater seats. Backrests, especially those upholstered with piled fabrics such as corduroy or velveteen, serve as thousands of tiny "inoculating needles" that can pick up spores from the heads of infected individuals and "inoculate" others. Infectious mononucleosis can be transmitted among individuals by using contaminated glasses, cups, or eating utensils.

Food and Water as Vehicles of Infection

Food and water can serve to transmit a variety of infectious diseases. Poliomyelitis, hepatitis, cholera, typhoid, and various forms of food poisoning are transmitted by contaminated food or water. Improperly cooked foods can transmit disease agents with which they are contaminated. For example, trichinosis, toxoplasmosis, and salmonellosis can be transmitted by improperly cooked foods. The **fecal-oral** route is a common mode of transmission for a variety of infectious agents. Such diseases are transmitted when feces of an infected individual enter a potable (drinkable) water supply that is subsequently consumed by other individuals. Diseases such as typhoid, cholera, poliomyelitis, dysentery, and various worm infections are transmitted by the fecal-oral route.

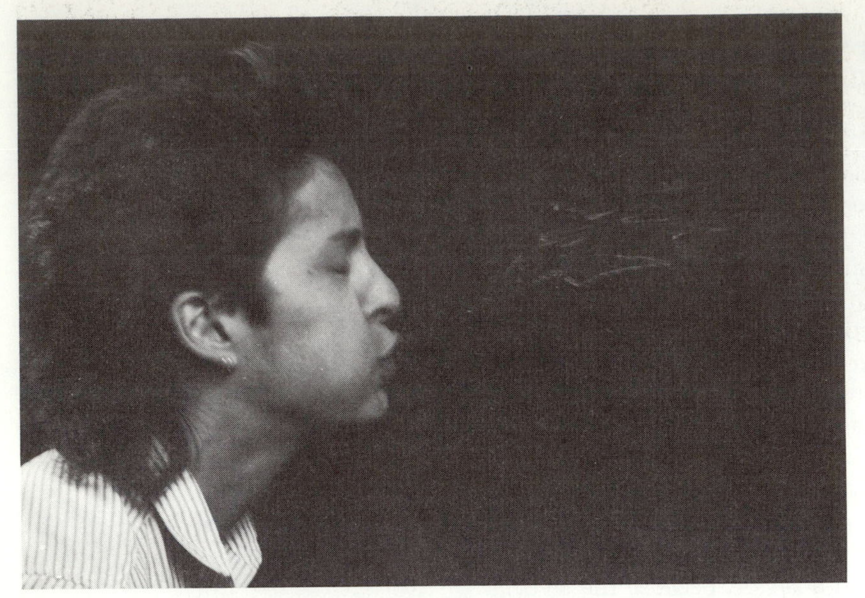

(a)

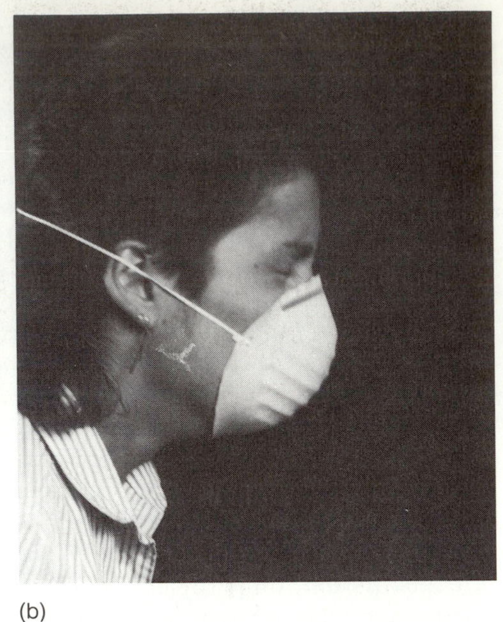

(b)

Figure 22-3 Aerosols Created by Sneezing. (*a*) An unstifled sneeze. The particles seen in the photograph are bits of saliva and mucus secretions laden with microorganisms. These particles may be infectious when inhaled by susceptible individuals. (*b*) A stifled sneeze using a surgical mask. Surgical masks such as the one illustrated reduce the formation of aerosols during sneezing to neglegible levels, thus reducing considerably the risks of infection in hospitals and surgical suites.

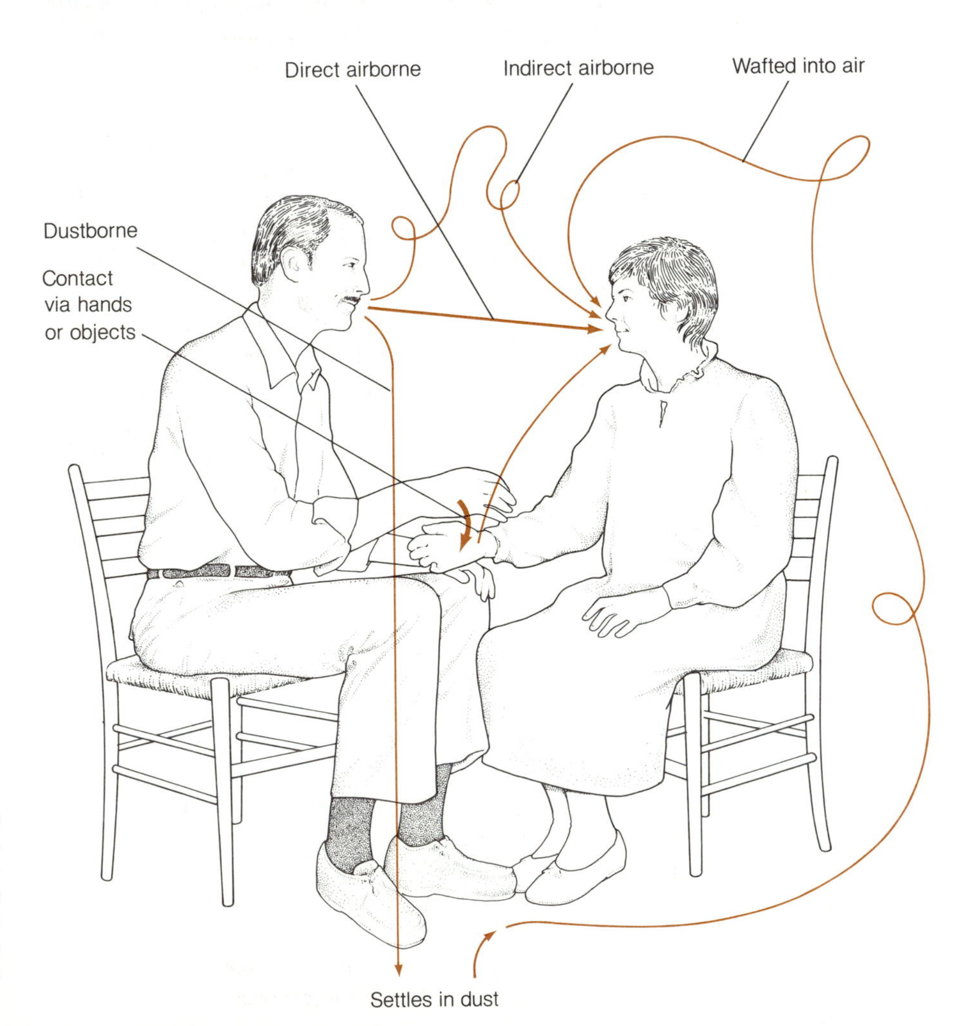

Figure 22-4 The Dispersal of Airborne Infectious Agents. Airborne microorganisms can be transmitted directly by coughing and sneezing. Indirect transmission can occur by inhaling contaminated dust or by touching contaminated objects. Insects may also transmit infectious agents when they carry microorganisms on their appendages.

Airborne Particles

The air is laden with numerous microorganisms, many of which are harmless; however, the flushing mechanisms of coughing and sneezing can serve to introduce infectious agents into the air (fig. 22-3). An unstifled sneeze can discharge into the atmosphere numerous drops of mucus and saliva, and if the individual is currently infected, numerous infectious agents are also released. Many of these infectious agents, trapped in large drops, fall to the ground and dry out and may subsequently be redistributed in the dust. The dust particles can then become airborne and infect other individuals. Tiny droplets, containing one or two infectious agents, dry up before they reach the ground. These droplets are surrounded by a dehydrated film of oral discharge and are called **droplet nuclei.** They can be airborne for long periods of time and yet retain their infectivity.

The danger of airborne infections is maximized in hospital environments and in large buildings where individuals are concentrated and droplet nuclei have a good chance of being inhaled (fig. 22-4). Air-conditioning systems are potential hazards for transmission of airborne infectious agents. For example, the organism that causes Legionnaires' disease (legionellosis) grows well in aquatic environments, including the water of air-cooling towers that are part of air-conditioning systems. The bacteria become aerosolized into the air ducts and are distributed throughout the building, resulting in a potential source of an epidemic.

FACTORS INFLUENCING EPIDEMICS

Infectious diseases result from complex interactions among the host, the parasite, and the environment (fig. 22-5). The virulence of the strain, the sus-

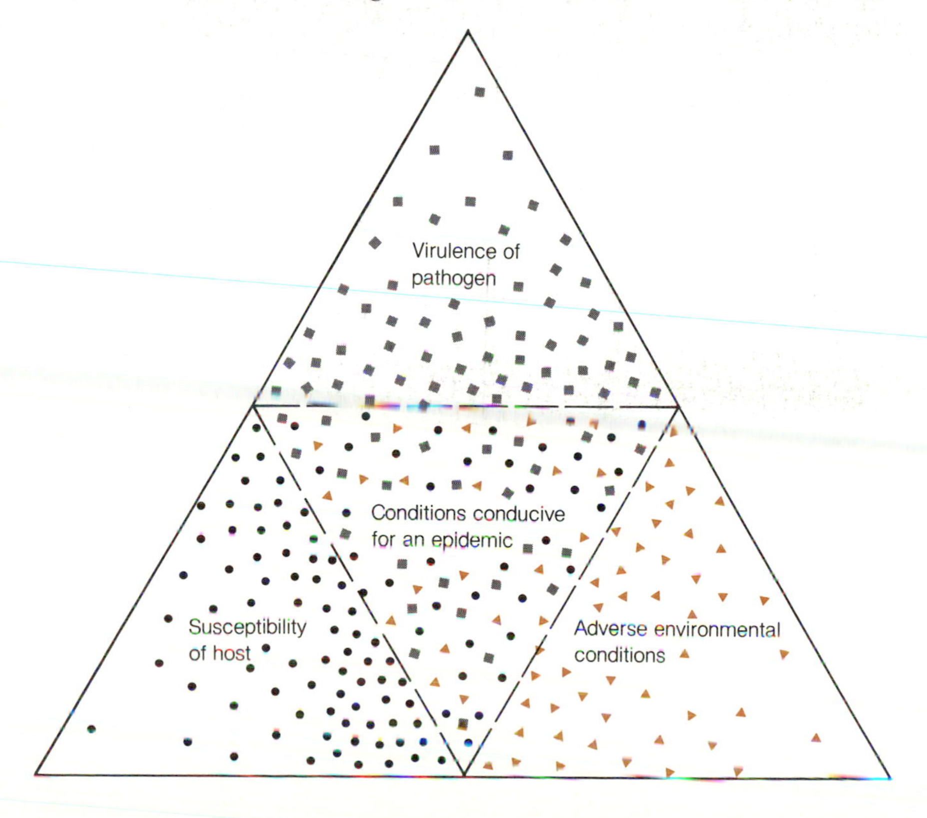

Figure 22-5 Interactions that Promote an Infectious Disease. Infectious diseases result from complex interactions among the host, the pathogen, and the environment.

ceptibility of the host, and environmental factors make each epidemic unique. Epidemics differ in character even when they are caused by the same agent, as can be ascertained from various parameters such as the incidence rate and the case fatality rate.

The environment plays an important role in the size and severity of an epidemic. For example, tuberculosis is influenced by the environment of the population at risk. Mere contact with the agent does not necessarily lead to disease; environmental factors such as crowding, stress, and malnutrition are important determinants of tuberculosis.

The physical environment also plays a role in determining the character of an epidemic. Environmental conditions such as temperature, wind, and relative humidity can drastically alter the incidence and case fatality rates of an infectious disease in the population. For example, Legionnaires' disease, a form of severe pneumonia caused by a bacterium called *Legionella pneumophila* is most common during the warm months, when air-conditioning systems are likely to be used extensively (fig. 22-6). Thus, weather conditions contribute to the high incidence of legionellosis in a population.

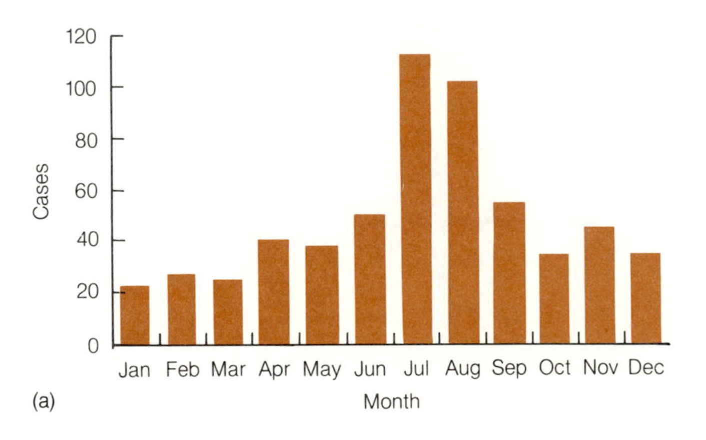

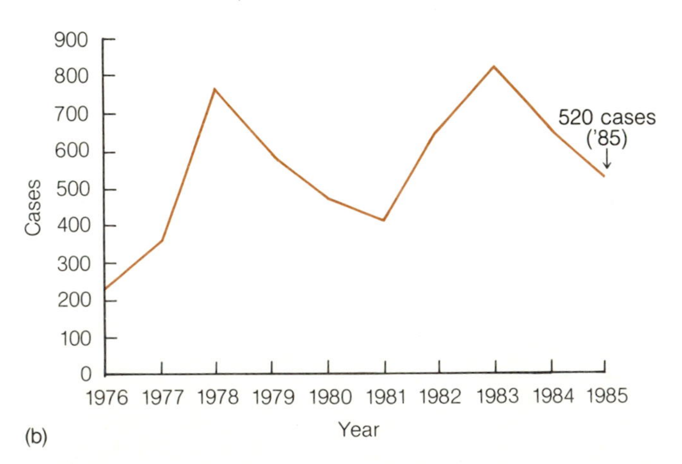

Figure 22-6 Incidence of Legionellosis in the United States. (*a*) Peak incidence of legionellosis in the United States occurs during the hot months, when air conditioning systems are more likely to be used. The plot is for the year 1979. (*b*) Yearly incidence of legionellosis in the United States.

Acquired Immunity

Since epidemics of infectious diseases occur as a result of the transmission of infectious agents from individual to individual, any factor that breaks the chain of transmission (or accelerates it) will influence the character of an epidemic.

A disease such as measles—which generally results in overt disease, and produces lifelong immunity—is an ideal disease with which to study the role 455 of acquired immunity □ in epidemics. Three epidemics of measles occurred in the Faeroes Islands, a group of islands in the North Atlantic (North of Great Britain), in the years 1781, 1846, and 1875. During the first epidemic in 1781, nearly everyone was afflicted, because the population at that time had never been exposed to measles before. During the second epidemic, in 1846, nearly all the inhabitants of the Faeroes Islands became ill, except for those few still alive who had experienced the first epidemic. The 1875 epidemic is noteworthy: of all the people in the Faeroes Islands exposed to the measles virus, only those 29 years of age or younger became ill. These individuals were not alive during the 1846 epidemic and therefore had never been exposed to measles. The rest were already immune. Hence, the immune state of the population at risk **(herd immunity)** largely influences the number of individuals that will become ill during an outbreak of a disease.

Genetic Background

If one analyzes the distribution of genetic traits in a population, it becomes apparent that certain traits are common in one subpopulation but not in others. Some of these characteristics may render a group hypersusceptible to a particular disease. To illustrate this point, let us examine two common diseases: coccidioidomycosis and malaria. Coccidioidomycosis, or valley fever, is a fungal disease that results from the inhalation of conidia of *Coccidioides immitis.* Generally speaking, infections by this fungus are either asymptomatic or mild with flulike symptoms. Occasionally, however, the pathogen spreads from the lung and causes a severe, often fatal disease. All other factors being equal, the incidence of disseminated disease (the invasive, severe form of the disease) in Blacks is much greater than in Caucasians. This fact can be seen in epidemiological studies in which the incidence rates and case fatality rates are calculated. Although the incidence rates may be similar in both populations, the case fatality rate or the incidence of severe cases will likely be greater in a predominantly Black population.

Malaria, on the other hand, occurs more often in Caucasians than in 699 some populations of Blacks. This disease, characterized by **paroxysms** □ (chills, fever, and sweats) every 24–72 hours, is caused in humans by any of four species of protozoa in the genus *Plasmodium.* The disease process is linked to and dependent upon the infection of the host erythrocytes by merozoites: malaria cannot take place in the absence of erythrocyte infection by the merozoite. It is known that individuals who are heterozygous for the allele that causes sickle-cell anemia *(s)* are more resistant to malaria than those individuals who carry only the normal allele *(S).* Individuals with sickle-cell anemia *(ss)* have defective erythrocytes, and their oxygen transport is greatly diminished. Individuals with sickle-cell trait *(Ss),* however, have a functional but fragile red blood cell. When these erythrocytes are infected with merozoites, they "sickle up," causing the death of the parasite within the erythrocyte. The sickled erythrocyte eventually is ingested by cells of the reticuloendothelial system. As a consequence, individuals with sickle-cell trait generally are more resistant to malaria than normal *(SS)* individuals. Sickle-cell trait is much more common in some Black populations than in Caucasian populations. Therefore, the incidence and case fatality rates in a malaria outbreak are much lower in the Black population than in the Caucasian one.

Cultural Factors

Religious or ethnic practices may affect the transmission of disease agents. For example, *Trichinella spiralis* is a roundworm that causes a disease called **trichinosis.** It is contracted by eating poorly-cooked meats of several domestic animals, especially pork. Individuals who, for religious or personal reasons, choose not to eat pork are unlikely to get trichinosis. In Jewish populations the incidence of trichinosis is very low (or zero) because these individuals traditionally do not eat pork □.

384

In New Guinea, members of one tribe formerly handled and ate brains of their dead ancestors. Because of this practice, a high proportion of the population suffered from **kuru,** a disease of the central nervous system, which causes a general debilitation that often leads to death. Apparently, the infectious agent responsible for this disease was infecting the brains of many people in the tribe and was transmitted when the brains were handled or consumed. High incidence rates of kuru will obviously be greater in populations that consume human brains than in populations that do not follow these cannibalistic practices.

Virulence of the Strain of Pathogen

Throughout history, various pathogens have afflicted humans and have caused widespread disease and death. In many cases, it is possible to study these epidemics and to learn about them. It has been found that epidemics caused by the same disease agent usually differ in morbidity and case fatality rates. Some epidemics have case fatality rates as high as 70%, while others have much lower ones, even in the same population. In many cases, it appears that the virulence of the infectious agent varies; i.e., it is possible that the LD-50 □ of one strain was much lower than that of the other and that this difference is reflected in the higher case fatality rate.

428

Infectious Dose

The number of individuals that become ill in a given outbreak may depend on the number of infectious doses present. For example, *Coccidioides immitis* normally has a prevalence rate of about 10 cases of valley fever per year in San Luis Obispo County, California; but a dust storm may cause 15–20 cases in a single month. The sudden increase is a direct result of the increased number of infectious particles present in the air during the dust storm.

Incubation Period

The period of time that elapses between infection and clinical manifestation of a disease is called the **incubation period.** This is the time required for microorganisms to multiply to sufficient numbers so that their damage is noticeable. Obviously, given a set generation time, large infectious doses normally will lead to shorter incubation periods than small infectious doses. Infectious diseases have characteristic incubation periods, however, that reflect the growth habits of the pathogen. Hence, epidemiologists can obtain clues as to the identity of a pathogen by determining the incubation period of the disease. Knowledge of incubation periods for various pathogens can also help the epidemiologist to establish approximately when the patient came in contact with the infectious agent and to trace the source of an epidemic.

DETERMINATION OF THE INFECTIOUS AGENT

In order to control an epidemic, its exact cause must be known. It is sometimes necessary to differentiate between the endemic cases and those that are part of an epidemic, particularly when the disease outbreak is caused by a pathogen that does not cause characteristic signs and symptoms.

Let us look at a hypothetical outbreak of diarrhea (gastroenteritis) to see the importance of determining the exact cause of an epidemic. Suppose that during a period of two weeks, 50 cases of gastroenteritis are reported to the local health agency by a hospital infection control officer. Upon examination of the stool specimens, 41 yield *Salmonella*, 1 yields *Shigella sonnei*, and 8 yield *Campylobacter jejuni*. These data indicate that *Salmonella* might be the cause of this epidemic. Further investigation of all the people (41) who yielded positive stool cultures for *Salmonella* indicated that 35 of them were at a catered affair while the other 6 were not. This information indicates that the 35 individuals at the catered affair are probably part of an outbreak, while the other 6 are not. This hypothesis can be supported further if we find that the same species of *Salmonella* is responsible for the 35 cases of gastroenteritis. Any unrelated cases may likely be caused by a different species. Current methodology for the identification of *Salmonella* is based on 516 biochemical and serological procedures (e.g., agglutination) □. Let us assume that these procedures were carried out, and that 34 isolates were identified as *S. montevideo* and 1 as *S. heidelberg*.

For all practical purposes, the study should concentrate on *S. montevideo* as the common cause of the outbreak. The importance of this identification lies in the fact that now we can trace both the immediate source of the infection and possibly the carrier. In such an outbreak, all people who attended the catered affair should be questioned as to whether they were ill or not, which food(s) they ate, and which food(s) they prepared. It is likely that all the ill patients ate the same food and that their stools would yield *S. montevideo*. Once the vehicle of infection is known (in this case a food item), the epidemiologist can begin to trace the source of the infectious agent and ultimately eliminate it. This hypothetical study shows that the precise identification of a causative agent allows the determination of the size of the epidemic, the source of infection, and any previously unrecognized carriers.

Several methods are commonly employed in the laboratory to determine the precise identity of infectious agents. The suitability of these methods depends upon the etiologic agent involved. Techniques such as serotyping, phage typing, bacteriocin typing, biotyping, and antibiogram typing are widely used in epidemiological studies of infectious diseases. All of these procedures are based on the fact that there are genetic variations even among individuals of a single species of microorganism. These differences may be related to their ability to cause disease. Below the species level, by convention, microorganisms generally are classified as "groups" or "types." The criteria are variable but generally include common biochemical, antigenic, or physicochemical surface (e.g., cell wall or plasma membrane) characteristics.

Serotyping

This form of infraspecies grouping capitalizes on the fact that antigenic differences exist on the surface layers of most microorganisms and that these differences are detectable using serological techniques. One of the best known examples is the serotyping of bacteria in the family Enterobacteriaceae. For example, the initial identification of a *Salmonella* isolate is based on variations in somatic (O) antigen, located on the lipopolysaccharide layer □ of the cell wall. The serotyping procedure generally is carried out by an agglutination test with monospecific antisera (antiserum against a single antigenic determinant or hapten). Using this scheme, species of *Salmonella* fall into 13 "O" groups (wall antigens) or **serogroups** (table 22-1). Further classification can be made using capsular (K or Vi) and flagellar (H) antigens.

433

Bacteriophage Typing

400

The fact that lytic bacteriophages □ differ in their host specificities can also be used as a grouping scheme. Bacteriophage (or phage) typing is based on the fact that various surface receptors on microorganisms bind to specific viruses. Only those viruses that can attach to surface receptors on a microorganism can infect it and cause lysis of the cell. Lytic bacteriophages cause plaques on lawns of bacteria, and these plaques represent an infection by the virus. In phage typing, one can infect the unknown bacterial isolate with each of several lytic viruses. Some of the viruses will infect the unknown bacterial isolate and cause plaques, while others will not. The pattern of infecting and noninfecting viruses can be used to determine the identity of the unknown bacterial isolate. One example of phage typing is the system used for typing strains of *Staphylococcus aureus*. The procedure involves the application of various lytic bacteriophages onto a lawn of bacteria growing

TABLE 22-1
SEROGROUPS AND ANTIGENIC COMPOSITION OF SOME *SALMONELLA SPP.*

ORGANISM	"O" GROUP*	"O" ANTIGENS	Vi ANTIGENS	H ANTIGENS
S. typhi	D	9, 12	+	d
S. paratyphi A	A	1, 2, 12	−	a
S. paratyphi B	B	1, 4, 5, 12	−	b, 1, 2
S. paratyphi C	C1	6, 7	+	c, 1, 5
S. cholera-suis	C1	6, 7	−	c, 1, 5
S. typhimurium	B	1, 4, 5, 12	−	i, 1, 2
S. enteritidis	D	1, 9, 12	−	q, m

*"O" antigens are part of the lipopolysaccharide layer of the bacterial cell wall.

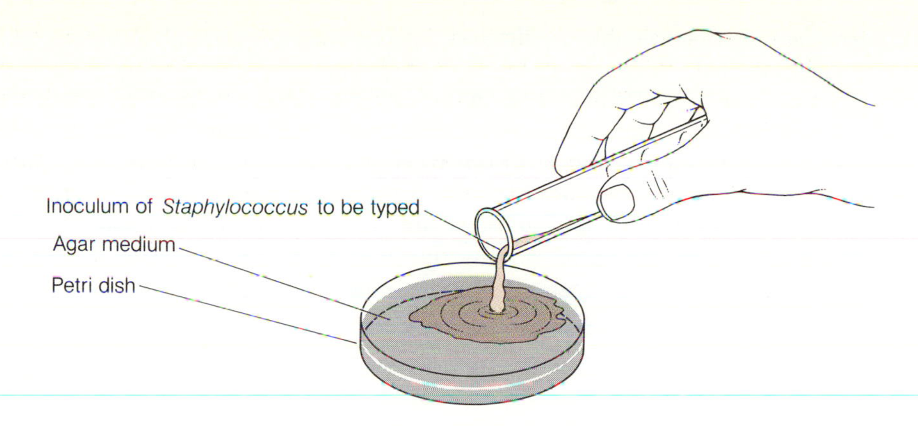

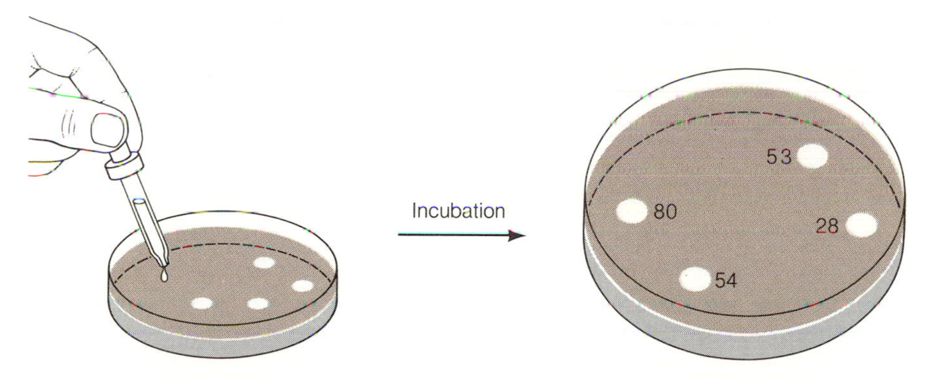

Figure 22-7 Procedure for Phage Typing of *Staphylococcus aureus.* A suspension of an unknown strain of *Staphylococcus aureus* in overlay agar is spread over the surface of a nutrient agar. Drops of 23 different suspensions of bacteriophage are applied to the plate. After incubation, the pattern of phage susceptibility and the phage group number are determined by noting which phages formed plaques on the bacterial lawn.

on an agar plate. Plaques appear on lawns of bacteria that are infected by the bacteriophage (fig. 22-7). The pattern of plaque formation by various phages can be used to determine the "phage group" to which that particular isolate of *S. aureus* belongs.

Bacteriocin Typing

Bacteriocins are antibiotics produced by one strain of microorganism which are lethal only for other strains of the same species (or closely related species). The usefulness of bacteriocins for infraspecies identification of pathogens is well documented in the case of *Shigella sonnei* and *Pseudomonas aeruginosa*.

Bacteriocins of *P. aeruginosa* are called **pyocins.** The pattern of pyocin production by a *P. aeruginosa* isolate can serve to pinpoint the source of an epidemic in hospitals, and has been used routinely for this purpose. Figure 22-8 outlines a procedure used to perform pyocin typing, as well as a typical result. The procedure is used to determine the pattern of inhibition of tester strains by the pyocins produced by the unknown isolate. Table 22-2 illustrates the patterns of pyocin inhibition by some common isolates. In this

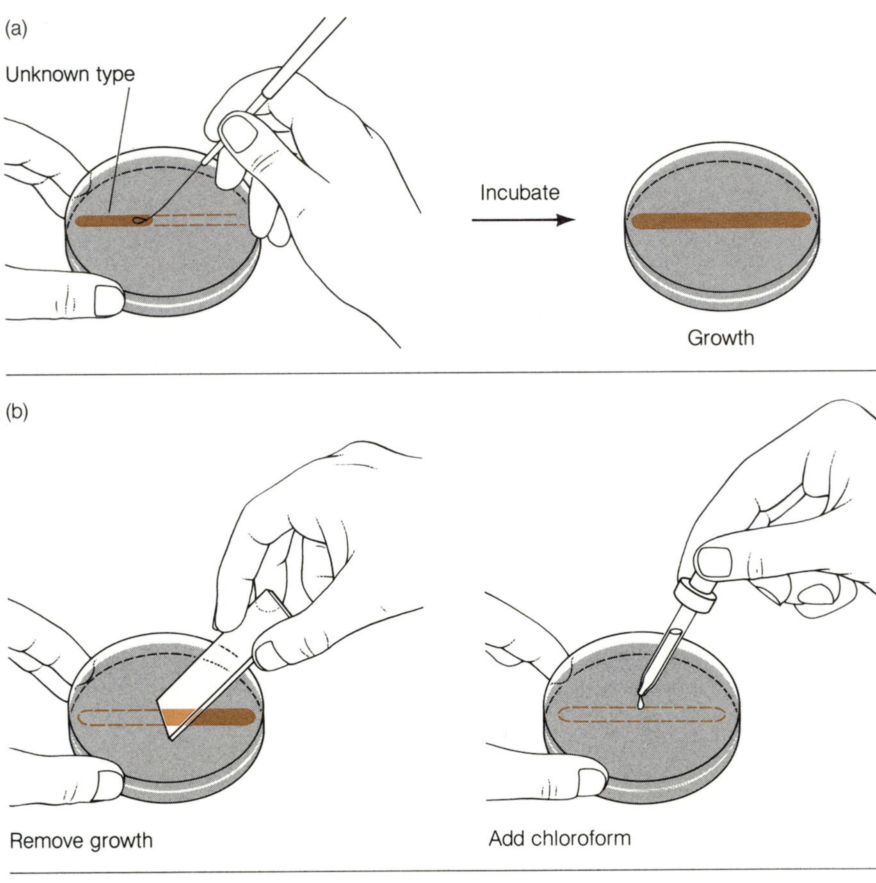

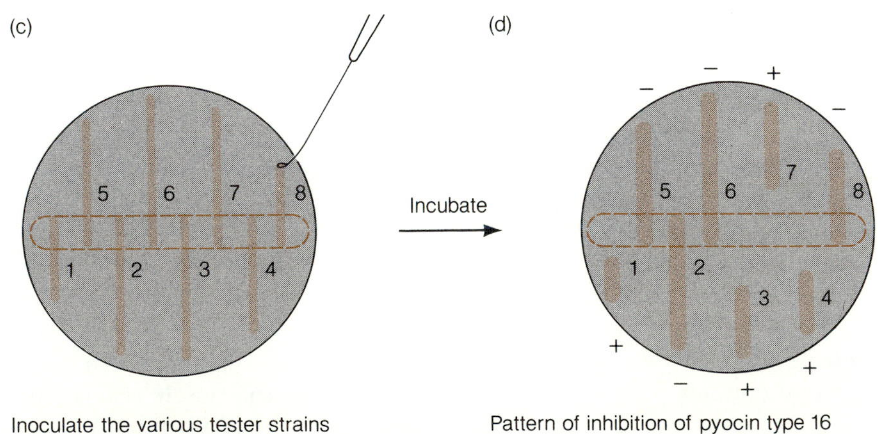

Figure 22-8 Pyocin Typing of *Pseudomonas aeruginosa.* *(a)* Unknown strain of *P. aeruginosa* is inoculated in a straight line over the surface of a nutrient agar. The culture is incubated for 24–48 hours, until growth is detectable. During this time, pyocins are produced by the bacterium and diffuse into the agar. *(b)* Growth is removed with a glass slide and *(c)* any remaining bacteria are killed with chloroform. *(d)* Known indicator strains are inoculated at right angles to the initial inoculum, and the plate is incubated once again. The pattern of growth inhibition of the tester strains is determined. The pyocin type is obtained by comparing the results with those published for various pyocin types (see Table 22-2).

TABLE 22-2
INHIBITION PATTERN OF SOME PYOCINE TYPES OF *PSEUDOMONAS AERUGINOSA*

PYOCIN TYPE	INHIBITION OF INDICATOR STRAIN #:							
	1	2	3	4	5	6	7	8
1	+	+	+	+	+	−	+	+
8	−	+	+	+	−	−	+	−
16	+	−	+	+	−	−	+	+
20	−	−	−	−	+	+	−	−
60	+	+	−	−	−	−	+	−
80	−	+	−	+	−	−	+	+
100	+	+	−	−	−	+	+	−

scheme, the unknown isolate is streaked in a straight line up and down the middle of a suitable agar plate. After a period of incubation, the growth is removed with a glass slide and any residual bacteria are killed with chloroform. During the growth period of the unknown isolate, pyocins are produced and diffuse into the medium. Once the growth has been removed and the remaining cells killed, the tester strains are streaked at right angles to the unknown strain (fig. 22-8). The plate is incubated and then examined for growth and for the pattern of inhibition.

The exact identification of *P. aeruginosa* isolates is particularly important in tracing down sources of infection and in detecting epidemics, because this bacterium is isolated from many hospital environments, even in very sanitary hospitals. Since it is next to impossible to eradicate *P. aeruginosa* from hospitals, the best approach is to determine the source of the troublesome strains and concentrate the eradication efforts on these strains.

Biotyping and Antibiograming

Other procedures, such as **biotyping** and **antibiograming,** can be used to differentiate among strains of pathogenic species. Biotyping groups strains of a species according to biochemical characteristics (table 22-3). Similarly, the patterns of antibiotic susceptibility (or resistance) for certain pathogens may

TABLE 22-3
BIOCHEMICAL PATTERNS OF TWO BIOTYPES OF *ESCHERICHIA COLI*

BIOCHEMICAL TEST	BIOTYPE I	TYPICAL	BIOTYPE II
Gram stain	−	−	−
Oxidase reaction	−	−	−
Motility	−	+	+
Gas from glucose	−	+	+
Gas from lactose	−	+	+
Indole production	+	+	+
Hydrogen sulfide production	−	−	−
Citrate utilization	−	−	−
Lysine decarboxylation	+	+	+
Urea hydrolysis	−	−	+
Potassium cyanide test	−	−	+
Mannitol fermentation	+	+	−

be distinctive enough to use in epidemiological studies. This system, more than any other, is ideal for automation. Several automatic microbial identification systems base their identification at least partly on antibiotic susceptibility patterns. The results of such tests are usually summarized in a report slip that can be used by the physician or infection control personnel to monitor infectious diseases in hospitals.

THE CONTROL OF EPIDEMICS

The successful control of epidemics requires an understanding of the factors that lead to them. Control measures vary depending upon the disease, but generally fall into several related categories. Some are short-term measures designed to control the current outbreak, while others are long-term and designed to prevent future outbreaks.

Reduction of the Source of Infection

Prompt implementation of measures that will reduce the source of infection in an ongoing epidemic is essential to its control. For example, it is essential that cases and carriers be recognized and isolated from the susceptible population. Most hospitals have an isolation ward that serves to house individuals with **contagious** diseases. Strict rules are followed in these wards in order to prevent the escape of infectious agents. Another method of separating infectious agents from the susceptible population is by **quarantine.** This is a procedure similar to isolation, except that it generally involves individuals, animals, or materials that are suspected of being contaminated with an infectious agent, but that do not show signs or symptoms of the disease. One example of a control measure is the reporting of patients with diarrheal diseases on board vessels. The master of the vessel is required to report the number of cases of diarrhea to the quarantine station portside within 24 hours of arrival, so that health authorities can control the introduction of diseases such as cholera.

International travel poses a significant obstacle to the successful control of infectious diseases. As previously mentioned, an individual might travel by jet to a distant land where an infectious disease (e.g., yellow fever) is prevalent, become infected, and introduce a "new" disease upon returning home. In cooperation with the **World Health Organization** (WHO), health authorities throughout the world enforce quarantine measures to control diseases such as plague, cholera, yellow fever, and smallpox. Along with the isolation and quarantine measures, suspected patients are treated with suitable drugs to reduce their infectivity.

In outbreaks in which the source of infectious agent is contaminated soil, soil sterilants or disinfectants are used to kill the pathogen. In agricultural areas in which valley fever (coccidioidomycosis) is endemic, farmers sometimes spray the soil with fungicides before plowing to reduce the number of infectious agents that may become airborne.

In the case of animal vectors or reservoirs, a campaign may be initiated to destroy the animals. In a recent **epizootic** (an outbreak of an infectious disease in animal populations) of rabies in California, control measures included trapping and destroying skunks in order to reduce the size of the reservoir host and therefore reduce the chances of human infection.

Disruption of the Chain of Transmission

Standards for water, milk, and food quality are established and enforced by local, state, and federal health agencies in order to prevent epidemics. Drinking water supplies, milk, and dairy products are closely monitored for their suitability for human consumption. Chlorination of water supplies, pasteurization of milk and dairy products, and inspection of restaurants are very important epidemic control measures designed to break the chain of transmission of infectious diseases.

Many diseases such as malaria, Rocky Mountain spotted fever, typhus, western equine encephalitis, and yellow fever are transmitted by vectors. Interruption of the transmission cycle is achieved by pest control. The use of insecticides is one common way of controlling vector populations in endemic areas. Schistosomiasis (which affects more than 1 million individuals annually) can be controlled by snail eradication programs, since the parasite that causes this disease requires a snail as an intermediate host to complete its life cycle.

Immunization

The immune state of the population, also known as herd immunity, plays a central role in determining the extent and course of an epidemic. Transmission of infectious agents occurs only among susceptible individuals; thus, in a population with low herd immunity, an infectious agent can spread with ease, because most individuals are able to "catch" the disease. In a population with high herd immunity, by contrast, spread of the disease is slow. A well implemented vaccination □ program can reduce the incidence of disease (fig. 22-9). Table 22-4 summarizes some of the vaccinations that are carried out routinely in the United States.

477

Travel to countries that are endemic areas for diseases such as cholera, yellow fever, and typhus could represent a health hazard. For this reason, vaccinations are recommended. Table 22-5 outlines some of the vaccinations recommended for travel to various countries. It is also common practice to

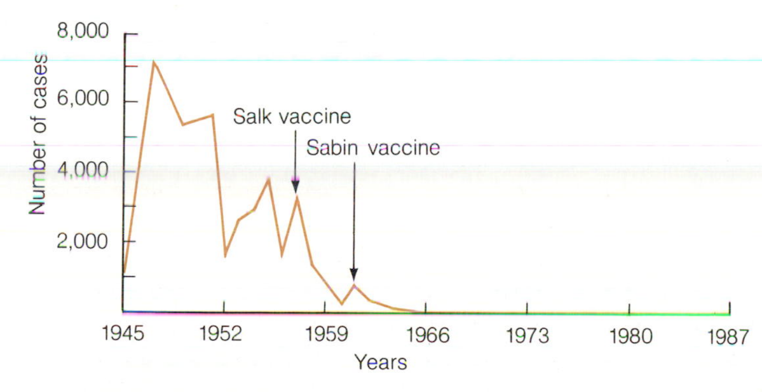

Figure 22-9 Effect of Vaccination on the Incidence of Poliomyelitis. The introduction of the Salk vaccine (inactivated poliovirus) in the late 1950s and the Sabin vaccine (attenuated poliovirus) in the early 1960s reduced the number of cases in England and Wales from 4,000–7,000 cases per year to fewer than 10 cases annually.

TABLE 22-4
VACCINES FOR CHILDHOOD IMMUNIZATION USED IN THE U.S.

VACCINE	CONTENTS OF VACCINE	WHO SHOULD RECEIVE	FREQUENCY*
Diphtheria-Pertussis-Tetanus (DPT)	Adsorbed diphtheria and tetanus toxoids, killed *Bordetella pertussis* cells	Infants: unimmunized children < 7 yrs. old	IM 2, 4 and 6 mos. Boost at 1½ and 4–6 yrs.
Tetanus-Diphtheria (Td)	Adsorbed tetanus and diphtheria toxoids	Unimmunized children 7 yrs. and older	IM 2 mos. apart, then 1 dose 6–12 mos. later. Booster every 10 yrs.
Mumps	Attenuated live virus	Children 12 mos. and older	1 dose SC
Measles	Attenuated live virus	Children 15 mos. and older	1 dose SC
Rubella	Attenuated live virus, grown in human diploid cells	Children 12 mos. and older	1 dose SC
Polio	Trivalent oral, attenuated live virus	Children 2 mos. to 18 yrs.	1 dose O at 2 and 4 mos. Booster at 1½ and 4–6 yrs.

*IM = intramuscular; SC = subcutaneous; O = oral

TABLE 22-5
VACCINATIONS OR DRUG TREATMENTS RECOMMENDED FOR FOREIGN TRAVEL

DISEASE	COUNTRIES	DISEASE PREVENTION
Cholera	Mozambique, Albania, Angola, Cape Verde, Dominican Republic, Egypt, India, Korea, Pakistan	Vaccination
Yellow fever	Benin, Cameroon, Congo, Mali, Nigeria, Mauritania, Ivory Coast, Uganda, Senegal, South America	Vaccination
Malaria	Africa, Southeast Asia, Central and South America, India	Chemoprophylaxis (Chloroquine or pyrimethamine + sulfadoxine)
Plague	Southeast Asia, Africa, South America	Vaccination
Typhoid	Tropical countries	Vaccination
Hepatitis A	Developing countries, Southeast Asia, Africa	Injection of immunoglobulins
Hepatitis B	Developing countries	Vaccination
Traveler's Diarrhea *(Escherichia, Shigella, Salmonella)*	Developing countries	Large doses of bismuth subsalicylate (e.g. Pepto-Bismol®)
Measles	Worldwide	Vaccination
Tuberculosis	Developing countries	BCG or INH prophylaxis
Typhus	Ethiopia, Mexico, Ecuador, Bolivia, Peru, Asia, Burundi	Vaccination

vaccinate a population at risk during a severe epidemic to break the chain of transmission. The impact of vaccination on a disease can be clearly seen in figure 22-9, which represents the number of reported cases of poliomyelitis in England and Wales for the years 1945–1985. The vaccination program was initiated in 1957; after that date, the annual number of polio cases dropped from about 7,000 in 1946 to fewer than 10 in 1980.

CONTROL OF HOSPITAL INFECTIONS

Hospitals are ideal invironments for the spread of infectious diseases (fig. 22-10). Many of the pathogens are opportunists and flourish in hospital environments. In addition, the high population density brings individuals into close proximity, thus favoring the transmission of infectious agents. Infections acquired in hospitals and similar institutions are called **nosocomial infections** (table 22-6). These infections can be transmitted to patients by physi-

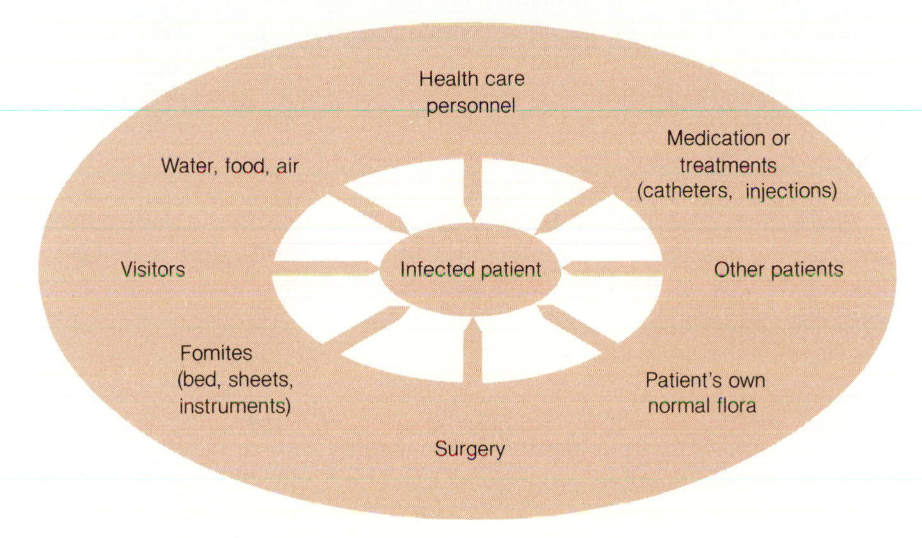

Figure 22-10 Spread of Nosocomial Infections. Institutionalized individuals, such as hospital patients, may become infected from health care personnel, other patients, or visitors. Occasionally, the patient's own resident flora may cause the nosocomial infection.

TABLE 22-6
COMMON NOSOCOMIAL INFECTIONS AND THEIR CAUSE

NOSOCOMIAL INFECTION	COMMON PATHOGENS
Gastroenteritis	*Escherichia coli, Salmonella, Shigella, Campylobacter jejuni,* viruses
Upper respiratory infections	*Haemophilus influenzae, Streptococcus pyogenes, S. pneumoniae,* viruses
Lower respiratory infections	*S. pneumoniae, Pseudomonas aeruginosa, Klebsiella pneumoniae,* influenza viruses
Septicemias	*E. coli, P. aeruginosa, Staphylococcus aureus, Candida albicans,* viruses
Burns	*P. aeruginosa, E. coli, S. aureus*
Wounds	*E. coli, P. aeruginosa, S. aureus, Streptococcus* Group A, enterococcus, *Klebsiella, Bacteroides*
Urinary tract infections	*E. coli, P. aeruginosa, Proteus, Enterobacter aerogenes,* enterococcus
Hepatitis	Hepatitis B, hepatitis A, hepatitis non A non B
Acquired immune deficiency syndrome (AIDS)	Human T-cell leukemia virus (retrovirus)

cians, nurses, or other hospital personnel during the daily routine of patient care, surgical procedures, or therapy. Illnesses caused by nosocomial infections tend to be severe, because the patient is already compromised or debilitated by a condition that required his or her hospitalization in the first place. Nosocomial infections can be caused by a variety of microorganisms and affect various tissues and organs. Table 22-6 summarizes some common nosocomial infections, as well as some of the most common pathogens involved.

Every hospital accredited by the Joint Commission for the Accreditation of Hospitals must have an infection control committee (ICC). The ICC usually includes representatives from the various departments within the hospital, such as nurses, physicians, housekeepers, laboratory technologists, engineers, and administrators. These individuals evaluate laboratory reports and the patients' charts to determine any increases in a particular disease or isolation of microorganisms that seems to be out of the ordinary.

Five to 10% of all patients entering general hospitals in the United States contract infections while at the hospital. Approximately 1.5 billion dollars is added to the cost of hospitalization as a direct result of nosocomial infections.

Control Measures

Control measures vary from one hospital to another, but certain factors must be considered. For example, all personnel involved with the care of patients must be familiar with basic infection control measures. These should in-

FOCUS

HOSPITAL INFECTIONS KILL 80,000 IN U.S.

WASHINGTON (UPI)—Accidental infections caught by hospital patients kill about 80,000 people each year and add as much as $1.5 billion to the nation's health-care costs, federal health officials told Congress Tuesday.

And the Centers for Disease Control reported progress on one front, measles—which it said may be virtually eliminated in the next two years—but problems with a second malady, herpes infections, which continue to defy treatment.

Agency witnesses appeared before a Senate appropriations subcommittee to discuss its budget for the 1983 business year. Sen. Mark Hatfield, R-Ore., chairman of the session, said the center's budget has not expanded and is less than the cost of one B-1 bomber.

The administration asked for $249 million in outlays for fiscal 1983 compared to projected outlays of $310 million in the current year.

"Perhaps we ought to transfer this (agency) to the Department of Defense," Hatfield said, adding that the real threat to national security "is the attack on the health of the people of this country" that administration budget-cutting has produced.

Center director William Foege said his agency should be able to continue doing its job, but he acknowledged that budget concerns have left it unable to work on such problems as herpes because it is untreatable where other diseases are not.

Foege said of the 2 million Americans who enter hospitals each year—5 percent of all admissions—catch infections unrelated to their original condition. Such infections add four days to their hospital stay at a cost of "over $1 billion a year, probably $1.5 billion," he said.

The accidental infections, he added, lead to 20,000 direct deaths and another 60,000 indirect deaths.

Reprinted with permission of United Press International, Inc.

clude: a) the isolation policies of the hospital; b) aseptic techniques and the proper way of handling medical equipment and supplies; c) the use of closed system urinary catheters and intravenous injections; and d) surgical wound care. Another aspect of infection control is the constant surveillance of patients. Surveillance regarding the frequency, distribution, symptomatology, and other characteristics of common nosocomial infections, as well as their etiology, can provide invaluable aid in controlling these infections. Through control measures it is possible to reduce the number of nosocomail infections and therefore patient discomfort and expense.

The Centers for Disease Control (CDC)

Many of the epidemiological studies conducted in the United States are monitored or directed by personnel affiliated with or employed by the National Centers for Disease Control (CDC). CDC, a branch of the United States Public Health Service (USPHS) located in Atlanta, Georgia, provides laboratory support to state and regional health agencies in the area of disease control and prevention. CDC also sets criteria for the safe and accurate performance of microbiological procedures, as well as training personnel and conducting updating seminars.

The National Centers for Disease Control also gathers and collates epidemiological data. Significant disease occurrences and national disease trends are published weekly in a publication entitled *Morbidity and Mortality Weekly Report (MMWR).* The extensive laboratory facilities of CDC and its highly trained personnel makes CDC one of the premier laboratories in the world.

SUMMARY

Epidemiology is the science that studies the incidence and distribution of disease in a population as well as the factors that influence them.

EPIDEMICS AND PANDEMICS

1. An epidemic is a short-term increase in the number of cases of a disease in a population.
2. The endemic rate of a disease represents the "normal" number of cases in a given population.
3. A pandemic is a very large epidemic of long duration, which may include populations in several different countries.

PARAMETERS USED BY EPIDEMIOLOGISTS TO MEASURE EPIDEMICS

1. Prevalence rates measure the usual number of cases in a population.
2. Incidence rates measure the number of *new* cases of a disease in a population.
3. Case fatality rates measure the number of fatalities among those who become ill with a particular disease.
4. Mortality rates measure the number of individuals dying in a population.

SOURCES OF INFECTIOUS DISEASES

1. Infectious diseases can be acquired from a variety of sources: a) humans; b) animals; c) water; d) food; and e) air.
2. Reservoirs are places, materials, or animals where infectious agents are perpetuated in nature.
3. Carriers are asymptomatic individuals who are capable of transmitting infectious diseases to other individuals.

MODES OF TRANSMISSION

1. Person to person transmission occurs by direct contact between two or more individuals. Sexually transmitted diseases are examples of person to person transmission.
2. Contact with infected animals may result in an infectious disease. Rabies is an example. Those diseases that are perpetuated in nature in animal reservoirs are called zoonoses.

3. Fomites (inanimate objects) serve as passive vehicles for the transmission of infectious agents. Many diseases are transmissible through fomites.
4. Food and water serve as vehicles for the transmission of infectious diseases such as gastroenteritis, cholera, and poliomyelitis. This route of transmission is sometimes called the fecal-oral route.
5. The inhalation of airborne particles containing infectious agents is a very common way of becoming infected. Many microorganisms are transmitted in this way. Many infectious diseases, particularly those that affect the respiratory system, are acquired by inhalation of airborne infectious agents.

FACTORS INFLUENCING EPIDEMICS

1. Infectious diseases result from a complex interaction among the host, the parasite, and the environment.
2. Acquired immunity determines which members of a population will be able to "catch" a disease.
3. Genetic background sometimes determines the susceptibility of an individual (or a population) to a particular infectious disease.
4. Cultural factors, including religious, ethnic, and behavioral practices, may predispose some populations to contracting an infectious disease. Such practices may protect other populations from certain diseases.
5. Virulence: the infectivity and the potential for pathogenicity may determine how devastating an epidemic is. Particularly virulent strains of microorganisms may cause many deaths in a population. The virulence of a strain may be reflected in the incidence or case fatality rates in the afflicted population.
6. The number of pathogens that infect an individual can also determine, at least in part, how severe an epidemic is.

DETERMINATION OF THE INFECTIOUS AGENT

1. In order to control an epidemic, its cause and the source of infection must be known. Identification of a pathogen can be accomplished by cultural, serological, and biochemical methods.
2. Serotyping is based on the fact that antigenic differences exist among members of the same species.
3. Bacteriophage typing is based on the fact that different strains of a single species can be infected by different lytic viruses.
4. Bacteriocin typing is based on the fact that certain strains of microorganisms produce a variety of "antibiotics" that inhibit the growth of other strains of the same species.
5. Biotyping is based on the fact that strains of microorganisms are capable of carrying out various biochemical activities, and that the patterns of these activities are unique.
6. Antibiogram typing makes use of the fact that various strains (and species) of microorganisms are susceptible to different types of antibiotics.

THE CONTROL OF EPIDEMICS

1. Controlling epidemics requires that the factors that contribute to the web of causation be reduced or eliminated. Control measures include:
2. Reduction of the source of infection, by isolation, quarantine, prophylactic chemotherapy, and reservoir eradication programs.
3. Disruption of the chain of transmission. This can be done by vector and intermediate host eradication programs, quarantine measures, and reservoir eradication.
4. Immunization.

CONTROL OF HOSPITAL INFECTIONS

1. Hospitals are ideal environments for the spread of infectious diseases.
2. Diseases that are acquired in hospitals are called nosocomial.
3. Nosocomial infections are acquired from medical personnel and are usually caused by opportunistic microorganisms.
4. Control of nosocomial infections involves knowledge of isolation policies in the hospital; the proper use of equipment; practicing aseptic techniques; and continuing education programs.

STUDY QUESTIONS

1. Define the following terms:
   a. epidemic;
   b. endemic;
   c. pandemic;
   d. morbidity;
   e. mortality.
2. Why is the science of epidemiology of central importance in public health?
3. Describe five different sources of infectious agents and give an example of each.
4. How can infectious diseases be transmitted? Are animals important in the transmission of disease? Explain.

5. Define the following terms:
   a. vector;
   b. phoront;
   c. fomite;
   d. reservoir;
   e. carrier.
6. Briefly discuss the relationship between the host, the parasite, and the environment in epidemics.
7. What factors influence the case fatality and morbidity rates of an epidemic? Explain.
8. Why is it important to identify precisely the pathogen responsible for an epidemic?
9. Cite five methods of infraspecies classification.
10. What are nosocomial infections? Of what significance are they?

SUPPLEMENTAL READINGS

Bopp, C. A., Birkness, K. A., Wachsmuth, I. K., and Barrett, T. J. 1985. In vitro antimicrobial susceptibility, plasmid analysis, and serotyping of epidemic-associated *Campylobacter jejuni*. *Journal of Clinical Microbiology* 21:4–7.

Cliff, A. and Haggett, P. 1984. Island epidemics. *Scientific American* 250(5):138–147.

Friedman, G. D. 1974. *Primer of epidemiology*. New York: McGraw-Hill Book Co.

Harris, A. A., Levin, S., and Trenholme, G. 1984. Selected aspects of nosocomial infections in the 1980s. *The American Journal of Medicine* 77(1B):3–11.

Liñares, J., Sitges-Serra, A., Garau, J., Pérez, J. L., and Martin, R. 1985. Pathogenesis of catheter sepsis: a prospective study with quantitative and semiquantitative cultures of catheter hub and segments. *Journal of Clinical Microbiology*. 21(3):357–360.

May, C. W. ed. 1980. *Microbial diseases*. Los Altos, Calif.: William Kaufmann, Inc.

Roueche, B. 1967. *Annals of epidemiology*. Boston: Little, Brown and Co.

Saxinger, W. C., Levine, P. H., Dean, A. G., Thé, G., Lange-Wantzin, G., Moghissi, J., Mei Hoh, F. L., Sarngadharan, M. G., and Gallo, R. C. 1985. Evidence for exposure of HTLV-III in Uganda before 1973. *Science* 227:1038–1040.

Zinsser, H. 1935. *Rats, lice and history*. Boston: Bantam Books.

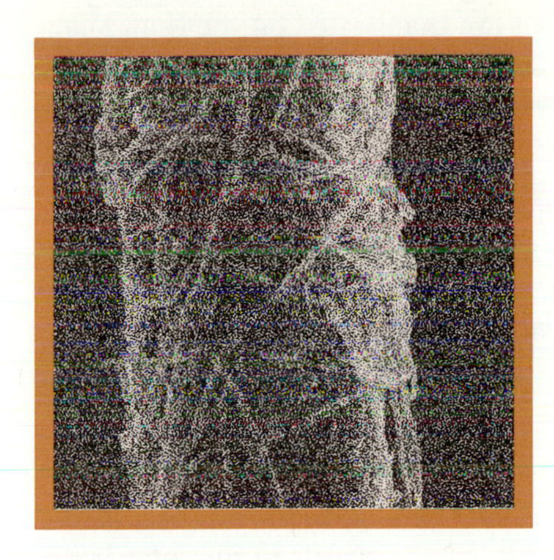

CHAPTER 23
DISEASES OF THE SKIN

CHAPTER PREVIEW

THE CONQUEST OF SMALLPOX

Smallpox, a frequently fatal disease, is characterized by the development of skin eruptions that eventually form a crust. When the crust falls off, it leaves scars called "pockmarks." In the severe form of the disease, 15–45% of those infected die. The disease, caused by a virus, spreads from person to person in aerosols, making it a particularly contagious disease. This characteristic, together with the high degree of infectivity of the virus, makes smallpox a fearsome disease.

Smallpox has ravaged human populations since antiquity. Pharaoh Ramses V is believed to have died of smallpox in 1137 B. C. During one epidemic in the early 1520s, more than 3 million Azetcs died of smallpox, and it is believed that this outbreak contributed significantly to the conquest of the Aztecs by Cortez.

The United States was not spared from smallpox. For example, during the 18th century in Boston, almost 60% of a population of about 1,000 became ill with smallpox within 10 months of the introduction of the disease. Of those who became ill, more than 50% died of the disease.

Because this disease had such an impact on the welfare of human populations, it prompted a great deal of interest that ultimately resulted in its eradication. The first significant advance in the area of smallpox control came in the area of prevention: the technique of **vaccination** was a significant first step. Later, with the growing understanding of host-parasite relationships in smallpox and with the intervention of the World Health Organization, the disease was eradicated. The last naturally occurring case of smallpox occurred in 1975, and the disease was declared "eradicated" on October 26, 1979.

Brilliant work by scientists throughout the world, capitalizing on knowledge gathered during centuries of suffering, ultimately resulted in the eradication of smallpox from the biosphere. This success was achieved because the scientists were knowledgeable about the biology of the virus, the disease process, its epidemiology, and means of prevention and treatment. In this chapter we will discuss some of the most common human diseases of the skin, their epidemiology, pathogenesis (disease process), and treatment.

LEARNING OBJECTIVES

A STUDY OF THIS CHAPTER SHOULD ENABLE YOU TO:

DESCRIBE SOME OF THE IMPORTANT STRUCTURAL FEATURES AND FUNCTIONS OF THE HUMAN SKIN

DISCUSS THE COMPOSITION AND FUNCTION OF THE HUMAN SKIN FLORA

OUTLINE THE CAUSE, PATHOGENESIS, TREATMENT, AND PREVENTION OF SOME OF THE MOST IMPORTANT BACTERIAL, VIRAL, AND FUNGAL SKIN DISEASES OF HUMANS

Many infectious agents cause diseases that are manifested on the skin. Many of these microorganisms cause disease that affect the skin directly, while others cause disease elsewhere in the body and then spread to the skin. For example, coccidioidomycosis and syphilis sometimes cause skin diseases as a result of invasion of skin tissue or allergic reactions. Diseases such as these will be discussed in the chapters dealing with the primary or most commonly recognized form of the disease. This chapter discusses some of the most important infectious diseases that affect primarily the skin, and those with signs and symptoms that are primarily cutaneous (of the skin).

STRUCTURE AND FUNCTION OF THE SKIN

The skin is an organ that covers all of the body surfaces exposed to the external environment. The human skin (fig. 23-1) consists of an outer layer, or epidermis, containing keratin, and an inner layer called the dermis. In addition, the skin has a variety of integumentary glands (sweat and sebaceous glands), blood vessels, nerves, and muscle cells and other connective tissue. In addition, hair and nails are also considered to be part of the skin.

The skin performs many important functions. It protects the individual from harsh environmental influences; integumentary pigments such as melanin protect the vertebrate animal from damage due to solar radiation. Skin also has numerous nerve endings that are sensitive to touch, pressure, heat, cold, and chemical conditions. These nerve endings collectively permit the organism to monitor the environment constantly and react accordingly.

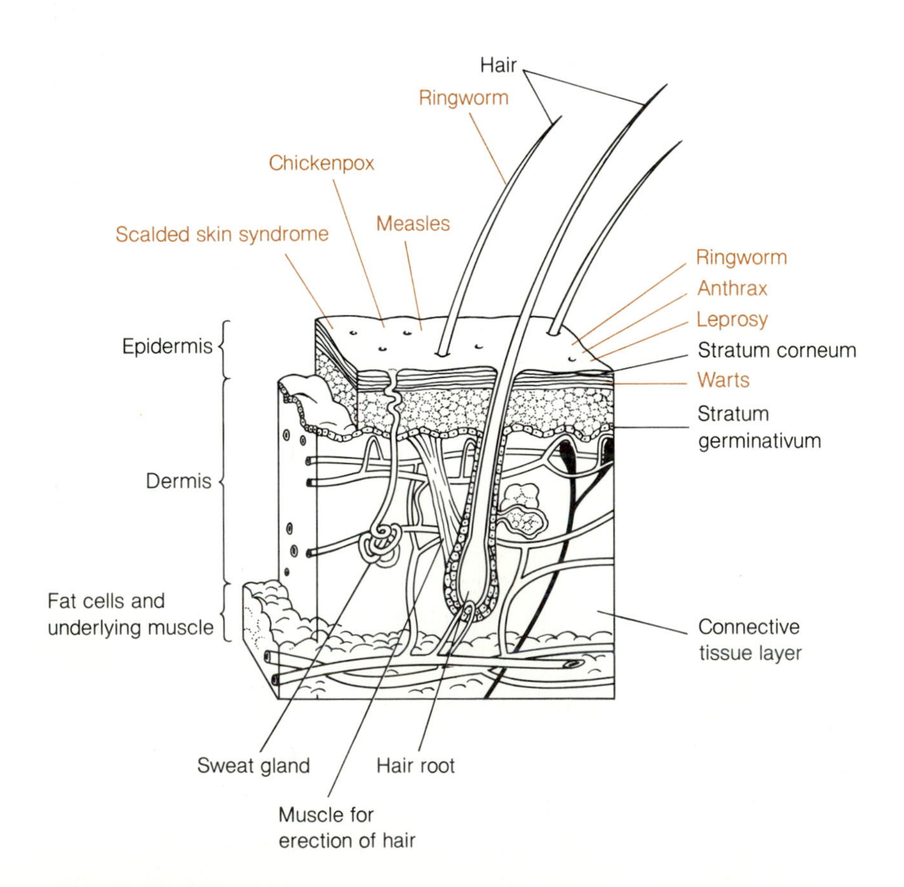

Figure 23-1 Diagram of the Various Components of the Human Skin. The diagram indicates the various components of the skin and some of the diseases that afflict the skin (in color).

The intact skin is a formidable barrier □ against infection. When its physiology or integrity is altered, however, infectious diseases can occur. The breach or normal skin by abrasions, punctures, or cuts allows microorganisms to enter the subcutaneous tissues and initiate an infection.

Conditions that alter the proper functioning of the skin also predispose the host to disease. For example, the **stratum corneum** (fig. 23-1) of the skin is continuously being shed and replaced with new stratum corneum. In this way, microorganisms present on the skin are removed along with the old stratum corneum. The rate of exfoliation (shedding of old stratum corneum) of the skin is important in the overall health of the skin. Conditions that retard the shedding of old stratum corneum can also lead to extensive colonization by microorganisms. Excessive moisture or steroid treatment, for example, can slow down the rate of exfoliation and can predispose individuals to certain diseases, such as candidiasis and tinea versicolor. Thus, the proper functioning of the skin is essential to the health of the individual.

THE NORMAL FLORA OF THE SKIN

The healthy skin is a rather harsh environment for many microorganisms because of its lack of moisture and its acidity. Although the skin is constantly exposed to a myriad of microorganisms, only a few of these are able to establish permanent residence (table 23-1). These microorganisms colonize primarily the stratum corneum and the hair follicles. The resident flora often protects the skin against infection by pathogenic microorganisms, by competing for nutrients and space and by releasing toxic substances that are inhibitory to many microorganisms. Bacteria that occupy hair follicles constitute a reservoir of microorganisms that reseed the skin surfaces as they are exfoliated or washed away.

The predominant flora of the skin consists of bacteria and fungi. The distribution and types of microorganisms on the skin vary according to the anatomical site. Skin on the forearm may harbor *Staphylococcus epidermidis*, diphteriods, and some yeasts, while the skin of the perineum (area between the anus and the genitals) may harbor these organisms in addition to enterics and coliforms. The number of microorganisms per cm^2 of skin may also vary from about 1,000 in dry skin areas to about 10 million in moist areas.

TABLE 23-1
REPRESENTATIVE MICROORGANISMS COMMONLY ISOLATED FROM THE HUMAN SKIN

MICROORGANISM	MICROBIAL GROUP	AVERAGE FREQUENCY OF ISOLATION (%)*
Candida albicans	Yeast	<10
Candida parapsilosis	Yeast	8
Diphtheroids (aerobic)	Gram + rods	60
Enteric bacteria	Gram − rods	< 5
Pityrosporum spp.	Yeast	35
Propionibacterium acnes	Gram + rods	72
Staphylococcus aureus	Gram + coccus	20
Staphylococcus epidermidis	Gram + coccus	92

*Numbers indicate the proportion of individuals in a population carrying the organisms in their skin.

Some members of the normal flora may become pathogenic if the opportunity arises. For example, the fungus *Malasezzia furfur*, under conditions of excesive moisture or slow skin exfoliation, may cause a "cosmetic" disease called tinea versicolor. Acne, a common condition in teenagers, can be caused by some bacterial members of the normal flora.

BACTERIAL DISEASES OF THE SKIN

Anthrax

Anthrax, a disease caused by *Bacillus anthracis*, is characterized by flu-like symptoms and the formation of a **malignant pustule** (fig. 23-2) at the site of infection. The lesion consists of a central black eschar (scab) and a conspicuous ring of edema (swelling due to fluid accumulation) and erythema (redness due to increased blood flow to the area). A pulmonary disease called **woolsorters' disease** is acquired by the inhalation of *B. anthracis* endospores, which germinate in the lungs and cause a severe, often fatal pulmonary infection.

Bacillus anthracis is an aerobic, gram positive, endospore-forming rod. It grows well on a variety of laboratory culture media. On blood agar containing 5% sheep blood, it causes complete hemolysis (lysis of red blood cells) and forms large, gray colonies of rough texture. Examination of the colony with a dissecting scope shows that the edges of the colonies resemble "medusa heads" (fig. 23-2). Virulent strains of *B. anthracis* synthesize a glutamyl polypeptide capsule and produce an exotoxin. The capsule protects the dividing bacterium against phagocytosis □. The exotoxin □ released by the virulent bacilli affects the central nervous system, causing respiratory failure and sometimes death.

444
432

Endospores commonly enter the skin through cuts or scratches and then germinate in the wound. Three to five days after infection, a small papule (raised lesion) develops, which later becomes filled with a dark, gelatinous

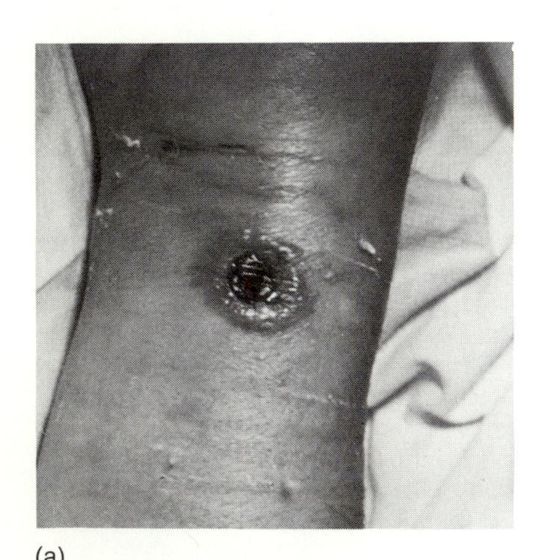

(a)

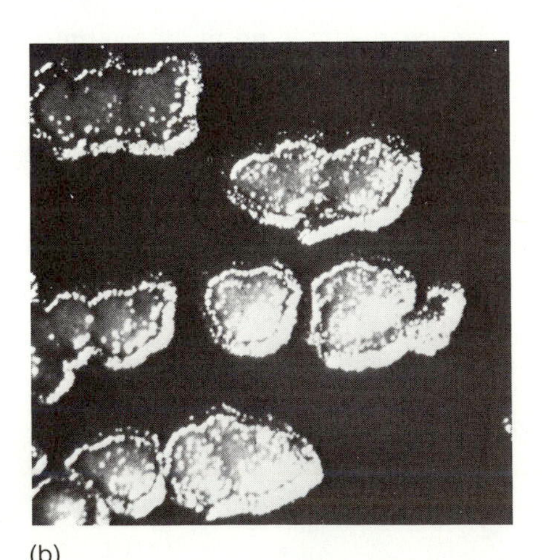

(b)

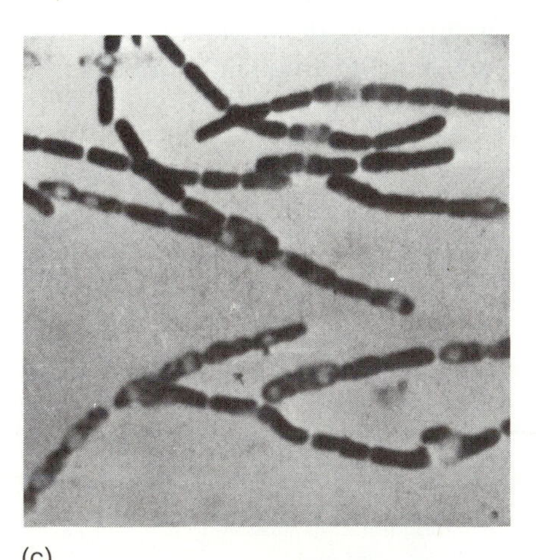

(c)

Figure 23-2 ***Bacillus anthracis.*** *(a)* The characteristic lesion caused by anthrax at the site of infection consists of a central region of dead tissue (necrosis) and a surrounding ring of swelling and redness. Anthrax bacilli can be isolated from the lesion. *(b)* Colonies of *Bacillus anthracis* on blood agar. *(c)* Gram stain of *Bacillus anthracis.* Notice the chain of bacterial cells containing endospores (clear circular areas inside cells).

fluid and surrounded by a ring of inflammation. At this stage, the lesion is called a **malignant pustule.** The bacteria multiply locally at the site of entry, producing the exotoxin, and subsequently spread via the bloodstream or lymphatics. The signs and symptoms of anthrax are similar to those of an acute respiratory infection, with cough, malaise, fever, and achy muscles. These symptoms are probably attributable to the exotoxin.

Livestock grazing on *B. anthracis*–contaminated pastures are usually infected through abrasions in the oral mucosa and mouth acquired while grazing. Infected animals generally die soon after the signs and symptoms of anthrax appear. Humans usually contract the disease from infected animals or their products. Most reported cases of anthrax in humans are acquired by individuals intimately associated with domestic animals, such as veterinarians, farmers, hunters, and slaughterhouse worker. Another group of individuals sometimes affected includes textile and felt mill workers, who get the disease by handling contaminated wool, goat's hair, or other animal-derived materials.

Vaccination with endospores from a nonencapsulated strain of *B. anthracis* prevents the disease in animals and in humans. The treatment of choice for anthrax is penicillin, which is particularly effective when given early in the course of the disease. The effectiveness of penicillin in controlling the infection diminishes as the disease progresses; this fact emphasizes the need for prompt and accurate diagnosis of anthrax. In cases in which the patient is allergic to penicillin, tetracycline can be used successfully.

Leprosy

Leprosy is characterized by the development of multiple lesions of the skin, often accompanied by loss of sensory perception of the affected areas. The disease dates back to antiquity. Ancient Hindu and Hebrew writings make reference to it. Accounts of the skin lesions and the social attitude toward this disease are commented upon in the Old Testament. Nowdays, the disease is better understood, and although the stigma of old remains, social acceptance has improved. In fact, individuals suffering from leprosy have formed a society and publish a journal called "The Star," in which they report cases and advances in the treatment and biology of the disease. The causative agent of leprosy of Hansen's disease, *Mycobacterium leprae*, was first seen by Armauer Hansen in 1878. In his report he made reference to the abundance of rod-shaped bacteria in the specimens. To date, there has been no successful cultivation of this bacterium in laboratory media.

Leprosy is a highly variable disease that ranges from benign **tuberculoid** leprosy to a severe, extensive form of the disease called **lepromatous** leprosy (fig. 23-3). In tuberculoid leprosy, a single **macule** (discolored lesion) develops, measuring about 5–20 cm in diameter. The color of the skin on the lesion is generally lighter than the uninfected skin. Many patients with tuberculoid leprosy undergo spontaneous cures. Lepromatous leprosy is a severe form of the disease which includes multiple lesions on the skin. In lepromatous leprosy, nerve involvement occurs, often with concomitant loss of sensory perception in the affected area. Intermediate forms of leprosy also occur and are referred to as **borderline** cases.

68

Mycobacterium leprae is an acid-fast □ bacillus that can be seen inside infected epithelial cells called **lepra cells**. Little known about the physiology of this organism, because it has not been grown on artificial media in the

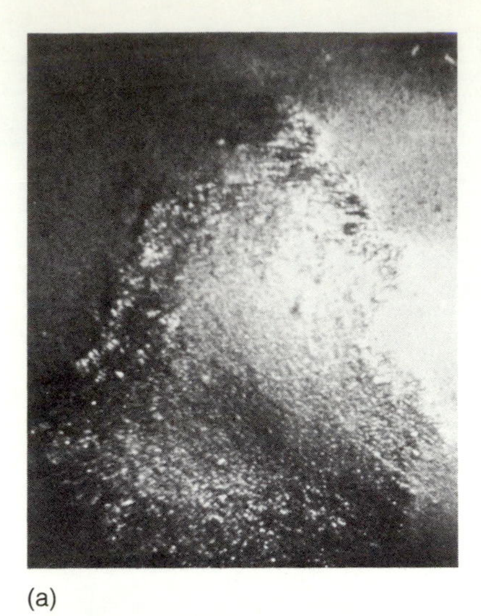

(a)

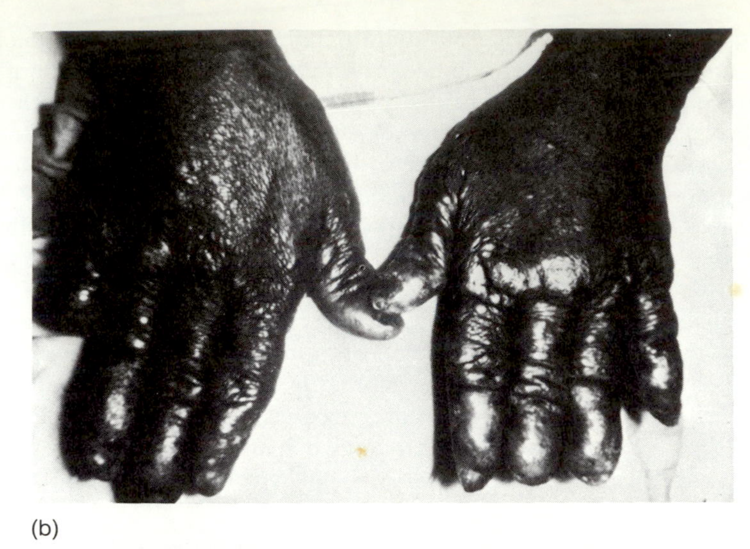

(b)

Figure 23-3 Characteristics of Leprosy. Two different clinical manifestations of leprosy. *(a)* tuberculoid leprosy (mild form) *(b)* lepromatous leprosy (severe form). The severity of leprosy is believed to be related to the speed of development and intensity of the cell-mediated immune response. In individuals whose immunity develops quickly, leprosy may not develop; when it does, it is generally of the tuberculoid type (mild). When immunity to the leprosy bacillus is slow in developing, the bacilli can reproduce without inhibition and lepromatous leprosy (severe) may then arise.

laboratory. Studies in experimental animals, principally mice, armadillos, and monkeys, suggest that the bacteria grow very slowly, with a generation time of about 12–14 days.

The bacteria generally penetrate the skin through small cuts and abrasions. They then divide slowly inside human skin and nerve cells. Humans are very resistant to infection by *M. leprae*; experimental infections using human volunteers are seldom successful. This resistance may be due to the fact that normal individuals develop prompt and strong cell-mediated immune □ (CMI) responses that inhibit the growth of the parasite. Tuberculoid, borderline, and lepromatous forms of the disease apparently are related to the time of onset of the cell-mediated immune response following infection. The later the CMI response develops, the longer the parasite can reproduce inside host cells and the more severe the infection is. In tuberculoid leprosy, evidence of a strong CMI response can be seen early in the infection as an extensive infiltration of lymphocytes and giant cells (large, multinucleated cells derived from maciophages) at the site. As a consequence, few acid-fast bacilli can be demonstrated (fewer than 100,000/cm^2). In lepromatous leprosy, little cell-mediated immunity is evident and numerous acid-fast bacilli (more than 100 million/cm^2) can be found inside macrophages.

468

Leprosy is prevalent in tropical areas of Africa, Southeast Asia, and South America. The prevalence rate for leprosy in these areas is about 2,000–4,000 per 100,000 population. Most cases of leprosy in the United States are traceable to immigrants. The incubation period □ for this disease averages more than two years, with a range of 12 weeks to 40 years.

542

Dapsone (a sulfone) has been effective in treating leprosy, but lepromatous leprosy requires prolonged treatment, sometimes for the rest of the patient's life. In cases in which sulfones cannot be used, rifampin is moderately useful. Prophylactic use of sulfones has been successful in reducing the incidence rate in endemic areas. A vaccine against *M. leprae*, consisting of live *M. bovis* and known as BCG (Bacillus Calmette-Guerin), may have some prophylactic value but has not yet been fully evaluated.

Pyoderma

Pyoderma is a diagnostic term used to define any acute inflammatory infection of the skin with **purulent** (pus containing) exudates (fig. 23-4). Pyoderma can occur as a result of primary invasion of the skin by pathogenic 427 bacteria, or as a secondary invasion of opportunistic □ bacteria following diseases such as scabies or after cuts and scratches. Some investigators believe that certain virulent strains of *Staphylococcus* or *Streptococcus* can initiate skin infections without prior breaching of the skin.

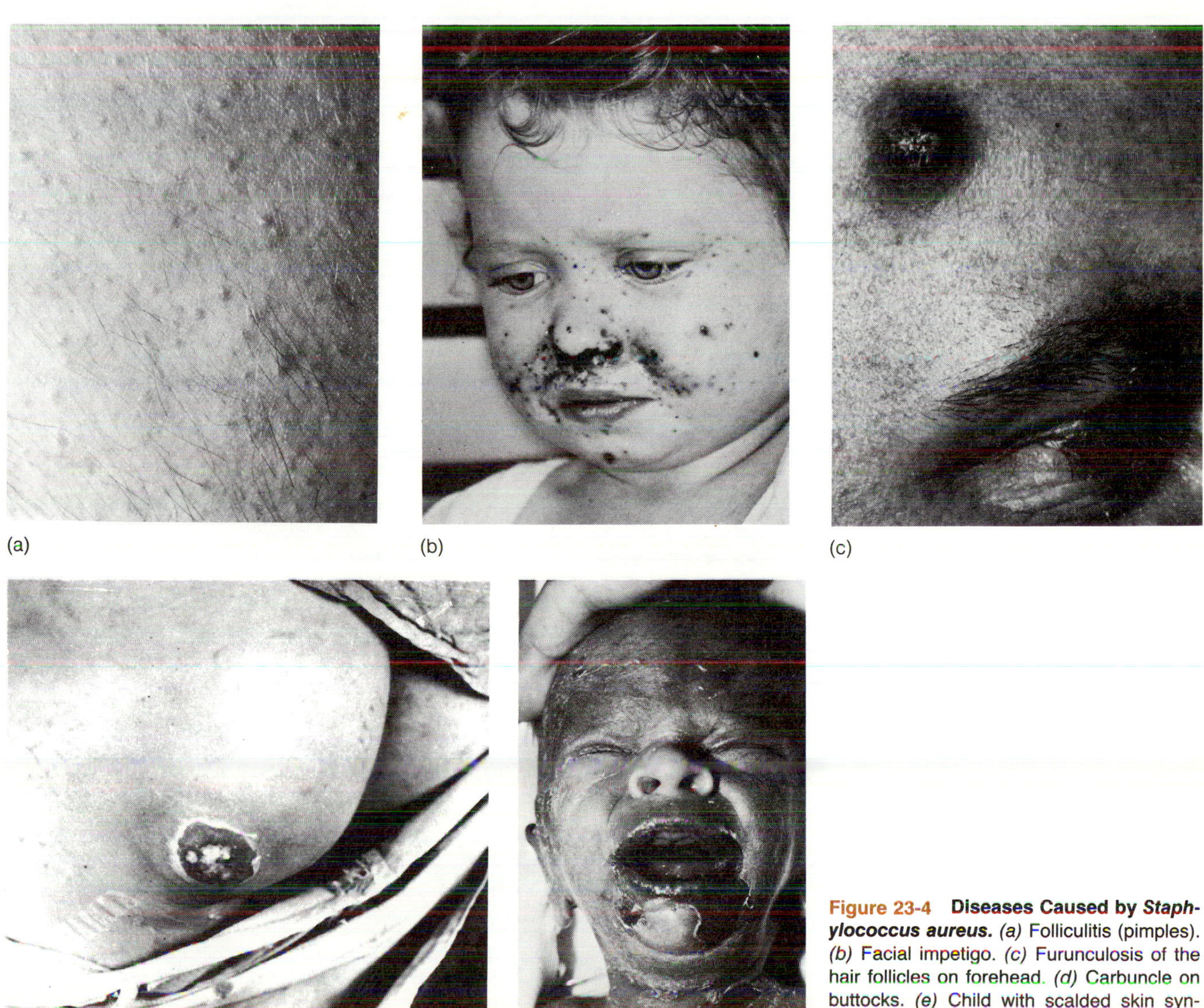

(a) (b) (c) (d) (e)

Figure 23-4 Diseases Caused by *Staphylococcus aureus.* *(a)* Folliculitis (pimples). *(b)* Facial impetigo. *(c)* Furunculosis of the hair follicles on forehead. *(d)* Carbuncle on buttocks. *(e)* Child with scalded skin syndrome (note peeling of skin around mouth, neck, and forehead).

Staphylococcal Skin Infections

The majority of the suppurative (festering, pus-discharging) skin diseases encountered in clinical practice are caused by *Staphylococcus aureus* (fig. 23-5). Staphylococcal skin lesions or abscesses are acute inflammatory lesions due to host cell destruction caused by the invading bacteria.

Staphylococci are gram positive cocci measuring approximately 1 μm in diameter. When these bacteria divide, they form characteristic grapelike clusters that can be seen in microscopic preparations of both clinical specimens and laboratory culture media. *Staphylococcus* is the only genus of facultative anaerobic cocci in the family Micrococcaceae. In the genus *Staphylococcus* there are three commonly recognized species: *S. aureus*, *S. epidermidis*, and *S. saprophyticus*. Of these three species, *S. aureus* is most frequently associated with human disease. Staphylococci grow well in routine laboratory media, forming small, opaque, white to yellow colonies.

Different strains of *S. aureus* elaborate various enzymes and toxins that have detrimental effects on the human host. These enzymes, many of which play a role in invasion of host tissues, are called **aggressins**. They aid the bacterium in spreading to adjacent tissues by degrading specific host tissue

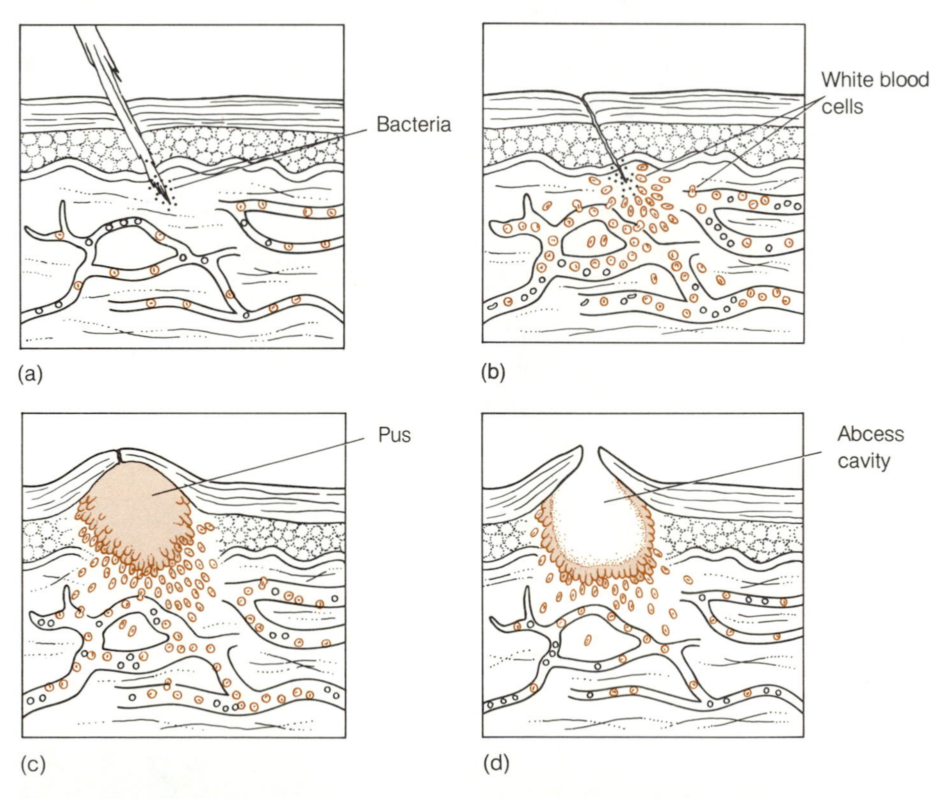

Figure 23-5 The Formation of an Abscess. When a pyogenic (pus-forming) microorganism such as *Staphylococcus aureus* is introduced into the skin, it begins to multiply at the site of infection. The developing colony produces a variety of proteins (aggressins) that cause tissue destruction and allow the colony to spread. The damage to tissue caused by the multiplying bacteria induces an inflammation, so that the area becomes swollen, reddened, and painful to the touch. Eventually, the developing abscess perforates through the skin, draining pus (dead cells, bacteria, and body fluids) to the outside. The erosion of tissue could continue if the infection is allowed to fester without treatment or removal.

components and by fighting off host defenses. Coagulases, penicillinases, and hydrolases are examples of staphylococcal aggressins.

Coagulases are enzymes that clot plasma by a mechanism resembling that of normal clotting. Coagulase activity is used by microbiologists to distinguish between pathogenic and nonpathogenic staphylococci. Although coagulase production is not required for pathogenicity, this enzyme generally is viewed as a good indicator of the pathogenic potential of *S. aureus*. It is postulated that coagulase-producing staphylococci elaborate the enzyme so as to form a fibrin lattice around themselves to avoid the host's attack.

β-lactamases, or **penicillinases**, are enzymes that break down penicillins, thus rendering these antibiotics useless in fighting off infections. The production of these enzymes is coded for by plasmids that may be transferred from cell to cell. Although β-lactamases do not participate directly in the disease process, they are important because they may protect the bacteria from antibiotics used in treating the infection.

Hydrolases are enzymes such as lipases, proteases, nucleases, and hyaluronidases, which break down (or hydrolyse) cellular materials and thereby participate in the disease process. **Staphylokinase** is a protease that breaks down fibrin. It is thought that this enzyme dissolves clots and so permits the spread of the pathogen to adjecent tissues. **Hyaluronidases** break down the hyaluronic acid that holds cells together and consequently facilitates the spread of the pathogen in the host's tissue. **Lipases** hydrolyze membrane lipids and fats, causing the lysis of host cells. The production of lipases by *S. aureus* correlates well with its ability to cause pimples and boils.

Staphylococcus aureus produces a variety of exotoxins that may be responsible for the signs and symptoms of some staphylococcal skin diseases. **Cytolytic toxins**, such as leukocydins and hemolysins, disrupt the plasma membrane of host cells. **Leukocydin**, a toxin produced by most pathogenic staphylococci, destroys PMN leukocytes and macrophages by attacking membrane phospholopids. Leukocydin appears to participate in the disease process by protecting the bacteria from phagocytes. **Hemolysins** are enzymes that destroy red blood cells. One such hemolysin is **alpha toxin** also called **alpha hemolysin.** This enzyme lyses not only erythrocytes but also skin, muscle, and renal cells. **Pyrogenic exotoxin C (PEC)** has a variety of effects on animal hosts. For example, in rabbits, PEC induces fever (pyrogenic effect); enhances the animal's susceptibility to endotoxin; and inhibits the reticuloendothelial system and the immune system. In rabbits, injection of minute amounts of PEC causes a disease that is similar to toxic shock syndrome (TSS) in humans. **Enterotoxins** are exotoxins that affect the intestines, causing vomiting and diarrhea in humans. There are various types of enterotoxins (A-F), but they all cause essentially the same signs and symptoms in humans. Enterotoxic staphylococci are the most common causes of food poisoning.

A cell wall component that may also play a role in the pathogenesis of staphylococcal infections is **Protein A.** It is believed that this protein is a major factor that contributes to the virulence of *S. aureus*. Protein A binds nonspecifically to the Fc regions of human IgGs, inactivating them; thus, it protects *S. aureus* from certain host immune responses. This protein also inhibits complement and phagocytosis, and has been found to elicit allergic reactions in mammalian hosts.

Infections of the skin by staphylococci usually induce an acute inflammatory response. As the staphylococci multiply, they produce a variety of toxins and exoenzymes that contribute to localized necrosis (cell death) and abscess formation. As the infection progresses, a fibrin network, together with fibroblasts, surrounds (or walls off) the abscess.

How the host protects itself from staphylococcal infections is poorly understood. Apparently, both cell-mediated immunity and antibody-mediated (humoral) immunity □ are important. Opsonizing antibodies (antibodies that enhance phagocytosis) have been reported to participate in protection against *S. aureus* infections. These antibodies enhance the phagocytic activities of both macrophages and PMN leukocytes. It is also believed that antibodies that neutralize the staphylococcal toxins help in the recovery of the patient.

466

The normal skin flora, which sometimes includes avirulent strains of *S. aureus* (table 23-1), inhibits the colonization of the skin by pathogenic microorganisms. For example, some medical scientists have successfully treated patients suffering from recurrent staphylococcal disease, such as furuncles and carbuncles, by seeding their skin with an avirulent strain of *S. aureus*. They hypothesized that the avirulent staphylococci would outcompete virulent strains and therefore protect the patient from disease.

Staphylococcal infections generally involve localized invasion of the skin and soft tissues. Clinical entities such as folliculitis, and impetigo fall in this category. These infections are rarely hazardous, although at times they cause a great deal of discomfort. Localized infections can become disseminated, however, and cause endocarditis, osteomyelitis, bacteremia, arthritis, pneumonia, and meningitis. When infections are caused by toxigenic (toxin-producing) strains, diseases such as toxic shock syndrome or scalded skin syndrome can result.

Staphylococcus aureus is indigenous to the human skin and external nares and is carried by a large proportion of healthy individuals. Infections presumably ensue when virulent staphylococci gain access to abraded skin.

Staphylococci present a particularly difficult problem in hospitals, where they cause a large number of nosocomial □ infections. They can be carried into hospitals by patients and by health care personnel who have small, draining lesions (such as pimples), and who can transfer this pathogen to patients in the hospital. Many of the hospital-acquired strains are resistant to penicillin, a factor that makes the treatment of nosocomial infections even more difficult.

551

Children are most susceptible to staphylococcal infections, and as they age they become colonized by their "own" strains of *S. aureus*. There is a high percentage of carriers in the population, so staphylococcal infections are difficult to control.

The drug of choice is penicillin. Many of the community-acquired strains of *S. aureus* still respond well to penicillin. Lincomycin, erythromycin, and novomycin have been used successfully to control penicillinase-producing *S. aureus*.

Impetigo. **Impetigo** is a form of pyoderma, characterized by the development of pustules that become encrusted and rupture (fig. 23-4). The external nares, face, and extremities are the most common areas involved. This condition is most prevalent among children, possible because of the high propensity they have for scratches and scrapes. In addition, person to person contact is more frequent among school-age children than among adults, so children are more likely to acquire and transmit virulent staphylococci or streptococci, either of which can cause impetigo. When toxigenic strains of *S. aureus* that produce the toxin **exfoliatin** are the cause of impetigo, the lesions are blisterlike and are known as **bullous** impetigo. The blisters eventually rupture and develop a crust.

Scalded Skin Syndrome (SSS). **Scalded skin syndrome** is the name given to a condition that results in necrosis of the epidermis with little or no dermal involvement (fig. 23-4). The name "scalded skin" is given because the afflicted skin appears to have been scalded with boiling water. Although this condition can result from drug reactions and other causes, SSS is caused by exfoliatin-producing strains of *S. aureus*. Scalded skin syndrome can be considered to be a generalized impetigo. It is most often seen in the newborn human child and may involve up to 80% of the body. Constitutional symptoms such as fever, general malaise, and prostration can accompany scalded skin syndrome.

Streptococcal Infections

Streptococcal infections are quite common in humans. The commonest are upper respiratory tract infections such as **strep throat,** but **erysipelas**, impetigo, and **scarlet fever** are also common. Like staphylococcal infections, streptococcal infections cause pyogenic (pus-forming) lesions. Generally speaking, streptococcal skin infections result from invasion of skin lesions such as scratches, insect bites, cuts, or sores. Occasionally, superficial streptococcal infections result in nonsuppurative (no pus produced) complications, such as **rheumatic fever** 585 □ or **acute glomerulonephritis.**

The genus *Streptococcus* (table 23-2) contains several species that are pathogenic for humans. All the bacteria in this genus are gram positive, fermentative cocci, typically arranged in chains and in pairs. In this genus, *S. pyogenes*, *S. agalactiae*, *S. faecalis*, and *S. pneumoniae* are the most important pathogens (table 23-2).

Streptococci are classified into groups based on their surface antigens. Rebecca Lancefield in 1933 classified the streptococci on the basis of serological reactions 512 □ involving surface polysaccharides and placed them into groups (A-H, J, and K). Of these, groups A and B streptococci contain the most important human pathogens. Their pathogenicity resides in virulence factors such as the **M-protein**, **hemolysins**, and **erythrogenic toxin**.

The **M-protein** of Group A streptococci is a structural component of the cell wall. It helps *Streptococcus pyogenes* bind to respiratory epithelial cells and inhibit phagocytosis and therefore protects the bacteria from ingestion by neutrophils.

Streptococci also produce hemolysins and are classified into three groups based on their hemolysin activity. **Alpha hemolytic streptococci** growing on blood agar break down erythrocytes and produce a zone of green discoloration around the colonies. **Beta hemolytic streptococci** produce a large, clear zone of hemolysis surrounding the colony; **gamma hemolytic streptococci** do not hemolyse blood. Two hemolysins, **streptolysin O** (SLO) and **streptolysin S** (SLS), are important. Streptolysin is an oxygen-sensitive enzyme produced by most Group A streptococci. This enzyme is active not only against RBCs but also against leukocytes and myocardial cells. Streptolysin S in contrast to SLO, is oxygen-tolerant, and contributes to the beta hemolysis of Group A streptococci.

Erythrogenic toxins (A, B, and C) are toxins that cause damage to the blood vessels underlying the skin. These toxins are produced by strains of

TABLE 23-2
SALIENT CHARACTERISTICS OF STREPTOCOCCI THAT COMMONLY AFFLICT HUMANS

SPECIES	GROUP*	HEMOLYSIS	ISOLATED FROM	DISEASES CAUSED	SALIENT FEATURES
S. pyogenes	A	Beta	Nasopharynx, skin	Strep throat, others	Major pathogen, toxigenic
S. agalactiae	B	Beta	Nose, genital tract	Endocarditis, sepsis	Also causes mastitis
S. equi	C	Beta	GU & GI tract, skin	Wound infections	Toxigenic
S. faecalis	D	Gamma (beta)	GI and GU tracts	Peritonitis, UTI	6.5% NaCl resistant
S. salivarius	K	Gamma	Oropharynx	Sinusitis, abscesses	Grow at 45°C
S. sanguis	H	Alpha	GI tract, oropharynx	Abscesses, meningitis	Resistant to bile
S. mutans	NG	Alpha	Oropharynx	Caries, endocarditis	Grows in dental plaques
S. pneumoniae	NG	Gamma (alpha)	Oropharynx	Pneumonia, meningitis	Optochin sensitive
Peptostreptococcus	NG	Gamma	Digestive system	Empyema, sinusitis	Anaerobe, causes abscesses

*Streptococci are grouped based on serological reactions involving cell surface polysaccharides.
NG = Not groupable
UTI = Urinary tract infection
GU = Genito urinary
GI = Gastrointestinal

groups A, C, and G streptococci that carry a provirus □ with the genetic material coding for erythrogenic toxins. Other enzymes, such as DNases, ATPases, neuraminidases, hyaluronidases, and streptokinases, are produced by various streptococci. **Streptokinase** breaks down fibrin and other proteins. These enzymes may play a role in pyogenic infections by streptococci.

In general, streptococci can cause disease in three different ways: a) direct invasion; b) intoxications; and c) nonsuppurative diseases resulting from a previous streptococcal infection. Postinfection sequelae, such as rheumatic fever or acute glomerulonephritis, are most commonly complications of untreated upper respiratory tract streptococcal infections. Direct invasions usually involve attachment of the streptococci to epithelial cells, with the subsequent production of toxic substances. Streptolysin O, SLS, and various exoenzymes may account for the invasive properites. Since both SLO and SLS have antileukocytic activities, pyogenic lesions may be due to the action of these enzymes on migrating leukocytes. In addition, the M-protein plays a major role in the pathogenesis of the disease, by helping the cells bind to the host and by inhibiting phagocytosis and hence augmenting the chances of survival for the streptococci. Skin diseases such as impetigo and erysipelas are consequences of direct invasion by streptococci. Intoxications due to erythrogenic toxins are responsible for the signs and symptoms characteristic of scarlet fever.

Streptococcal infections are commonly transmitted by droplet nuclei. They originate from asymptomatic carriers or from ambulatory individuals with active streptococcal infections. While "strep throat" is commonest during winter months, pyodermal infections are commonest in termperate climates during the warm months. Pyodermal diseases caused by Group A streptococci are usually transmitted by direct contact, and the bacteria colonize the breached skin. Transmission of streptococci can also involve fomites or insects. Nosocomial infections caused by streptococci are most commonly surgical wound infections; asymptomatic hospital personnel are most frequently implicated as the source of infection.

Unlike staphylococci, which have developed resistance to penicillin, streptococci are still quite sensitive to this drug. Thus penicillin remains the drug of choice. Patients allergic 491 □ to penicillin can be treated successfully with lincomycin, erythromycin, or clindamycin. Third-generation cephalosporins are also effective against streptococci.

Impetigo, erysipelas, and scarlet fever are common streptococcal diseases of the skin.

Impetigo. **Streptococcal impetigo** is similar to staphylococcal impetigo, except that the former disease is seldom of the bullous type; rather, a thick crust develops around the lesion. Beta hemolytic Group A streptococci are the most common cause.

Erysipelas. **Erysipelas** is an acute febrile disease resulting from a deep streptococcal infection, with localized inflammation and erythematous (reddish) skin. Face and head lesions are common and are often accompanied by fever, headache, nausea, and vomiting. This disease must be treated promptly, for untreated infections may result in septicemia (dissemination to the blood), multiple abscesses, nephritis, or rheumatic fever. It responds well to penicillin or erythromycin.

Scarlet Fever. **Scarlet fever** was once a common disease resulting from streptococcal infections, but since the advent of penicillin and erythromycin its incidence has declined significantly in the United States. The typical fine, red "sandpaper" rash associated with the disease results from the action of **erythrogenic toxin** on the blood vessels of the skin. This is a common consequence of streptococcal tonsillitis, but it also occurs after skin infections. The initial rash may occur on the chest, but it spreads quickly to other parts of the body. Fever, vomiting, and prostration can also accompany the rash. During the healing process, desquamation of the skin and "strawberry tongue" are characteristic.

Trachoma

Trachoma is a highly contagious conjunctivitis (an inflammation of the membrane that covers the eye, which is called the conjunctiva) caused by *Chlamydia trachomatis* (fig. 23-6). The infection often leads to the development of a cloudy layer over the eye and, in severe cases, to scar tissue formation in the infected rea of the conjunctiva. The disease is uncommon in the United States (fewer than 300 cases reported annually), but it affects more than 400 million people worldwide, primarily in Africa and Asia. The infection may lead to the formation of corneal ulcers and to blindness. It has been estimated that more than 20 million people have been blinded by this disease. It is easily transmitted from person to person by direct contact with eye secretions or fomites contaminated with the bacteria.

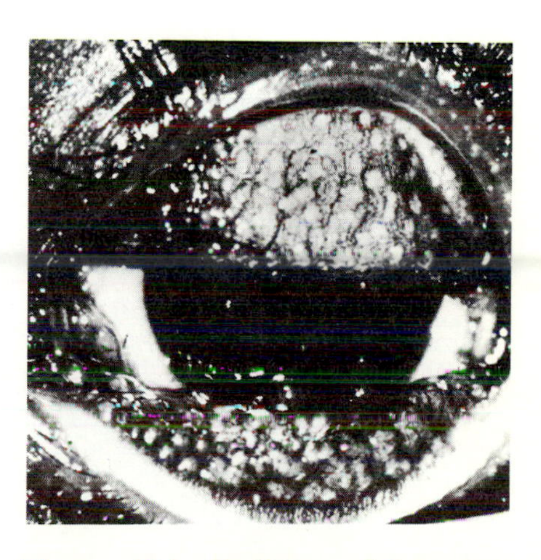

Figure 23-6 Trachoma. Inflammatory nodules of the eye characteristic of trachoma.

Chlamydia trachomatis is an obligate intracellular parasitic prokaryote. The infectious particle, called the elementary body, infects host cells and

FOCUS
LYME DISEASE

Lyme disease, first noted in 1975 in Connecticut, consists of an infectious arthritis that is characterized by a large, ringlike lesion on the skin. The lesion may measure more than 5 cm in diameter. Fever, headache, and muscle aches may also accompany the infection. Investigators have found that a spirochete is responsible for Lyme disease and have placed it in the genus *Borrelia*. The disease is effectively treated with penicillin, erythromycin, or tetracycline, especially when the medications are administered early in the course of infection.

The disease occurs primarily in the coastal regions of the northeastern United states (Connecticut, Rhode Island, New York, Maryland, Massachusetts), the western states (California, Nevada, Oregon), and the Midwest. Lyme disease is transmitted by the bite of **ixodid ticks**, and the endemic area corresponds to those areas where the ticks proliferate. Although these regions are the hotbeds for Lyme disease, it can be found in any region where ixodid ticks occur.

The incidence of Lyme disease has been increasing considerably since 1975. In 1980, the Center for Disease Control in Atlanta, Georgia reported 226 cases. In 1983 there were more than 500 cases reported, and this figure may be an understimation. The reason for the increase in Lyme disease is unknown, but it is reasonable to assume that increased awareness of the disease by physicians has contributed greatly to the increase in reported cases.

differentiates into a reticulate body □. After several rounds of binary fission, many elementary bodies arise. The elementary bodies are released following disruption of the host cell and can then infect other cells. Extensive cell disruption can occur from the infection, leading to a destructive inflammation of the affected tissue.

320

The etiology of the disease is usually determined by demonstrating the pathogen in inflammatory exudates, using staining techniques or fluorescent antibodies. Eye ointments containing tetracycline or erythromycin are effective in controlling the infection. Avoiding contact with infected persons or contaminated materials, to prevent self-inoculation, and sanitation of contaminated materials are the only methods for preventing the infection. Prompt treatment with antibiotics is essential for preventing the spread of the disease in human populations.

FUNGAL DISEASES OF THE SKIN

Fungi cause many diseases that involve the skin (fig. 23-7). Some fungi, such as *Malasezzia furfur* and *Exophiala werneckii*, cause superficial skin diseases that involve only the outermost layers of the stratum corneum (fig. 23-1). Others, such as *Sporothrix schenckii*, *Phialophora verrucosa*, and *Pseudoallescheria boydii*, are capable of invading deep tissues of the skin and penetrating to the bone. In addition, many pulmonary infections (such as cryptococcosis) caused by fungi, upon dissemination from the lung, can cause skin lesions as well. The commonest fungal infections, however, are the **ringworms** and infections caused by *Candida albicans*.

Ringworm

Ringworm, tinea, or dermatophytosis are names give to infections of the skin and scalp caused by a group of fungi called the **dermatophytes.** The

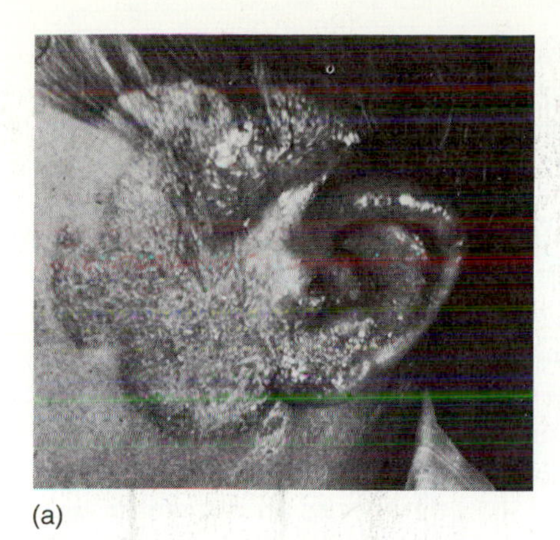

(a)

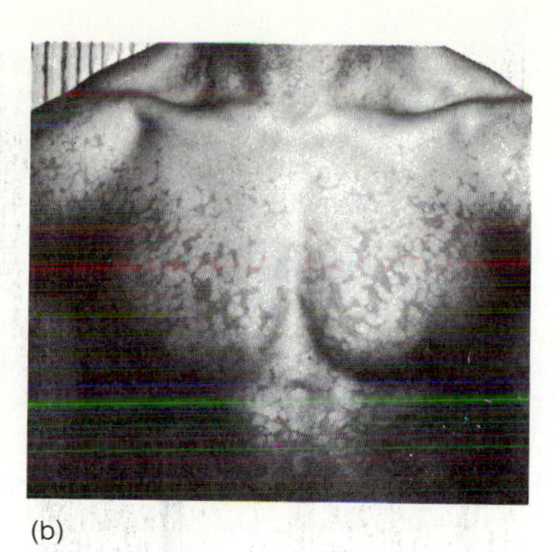

(b)

Figure 23-7 Diseases Caused by Fungi. *(a)* Child with tinea capitis (ringworm of the scalp) caused by *Microsporum audouinii.* *(b)* Child with tinea versicolor. The disease is caused by the yeast *Malasezzia furfur.*

name "ringworm" was coined because these fungi characteristically form concentric lesions on the hairless skin. The rings represent the spread of the agent outward from the site of infection, as well as an inflammatory response. Ringworm can occur in almost any part of the body, and is named based on the site of infection. For example athlete's foot is called **tinea pedis** while ringworm of the scalp is called **tinea capitis**.

The ringworms or **tineas** are caused by fungi in the genera *Epidermophyton*, *Microsporum*, and *Trichophyton*. These fungi are very closely related to one another, and all are able to digest keratin and lipids. This ability contributes to the pathogenicity of the dermatophytes.

The disease process in all the ringworms in generally the same, and different symptoms arise as a result of the virulence of the strain, the susceptibility of the host, and the site of the infection. The infection begins on the stratum corneum and spreads outward, giving rise to the typical ringworm lesion. As a consequence of this growth, itching, patchy scaling, inflammation, or eczema develop from toxic wastes or tissue damage produced by the fungi. The lesions are restricted to the keratinized layers of the epidermis. The inflammatory and allergic responses generally are due to waste products leased by the growing fungus. Rarely does a dermatophyte invade the deriis; when this does occur, it is most likely *T. rubrum* in an immune-comomised individual.

When *Microsporum audouinii* infects the human scalp, there is a period of incubation during which the fungus grows on the keratinized layers of the scalp. When the hyphae reach hair follicles, they grow into them on the hair shafts, where arthroconidia develop (fig. 23-8). After a period of spread, there is a refractory period, characterized by profuse arthrospore formation and little or no follicle invasion. After this period, infected hairs are sloughed off and other hairs return to normal.

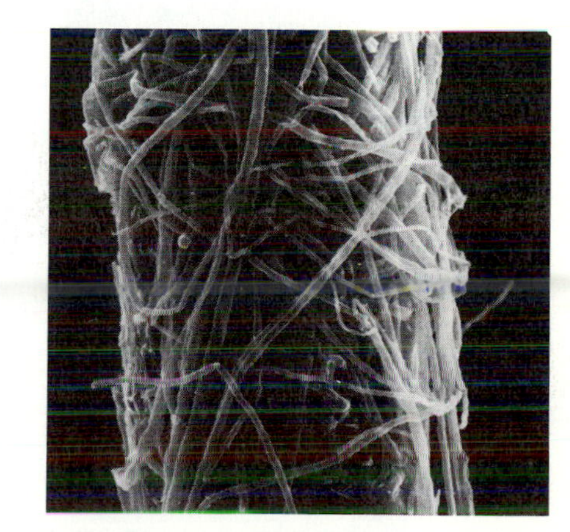

Figure 23-8 Infectious Spores of Dermatophytes in Human Hair. Longitudinal section of human hair, filled with spores of *Trichophyton violaceum*.

The dermatophytes cause a variety of clinical manifestation, depending upon the species and the antomical site of infection. Let us look briefly at tinea capitis, an infection of the scalp caused by a variety of dermatophytes including *M. audouinii*, *T. tonsurans*, and *T. schoenleinii*. Each of these dermatophytes causes a different variety of tinea capitis. *Microsporum audoui-*

nii, which generally infects children, causes a ringworm of the scalp called "grey patch" (fig. 23-8). This ringworm is due to a noninflammatory disease in which the affected hairs become lusterless and break off a few mm above the scalp, giving the appearance of a gray patch of hair. The black dot variety of tinea capitis is usually caused by *T. tonsurans*, a fungus that invades the inside of the hair shaft. The hair breaks at the follicular region, leaving a stub of hair "loaded" with spores. In dark-haired individuals, the scalp appears laden with black dots. *Trichophyton schoenleinii* causes a ringworm of the scalp called **favus** or **tinea favosa**, a chronic, noninflammatory infection. The affected portions of the scalp have **scutules**, which are flakes consisting of hyphae and scalp material knitted together.

The dermatophytes abound in nature. They are found in the soil, in a variety of animals, and in humans. More than ten species are known to cause human infections. Many of these fungi cause asymptomatic infections that are transmitted by direct contact with infected humans or animals or by fomites. Fungi that attack only human skin are said to be **anthropophilic**. Epidemics of ringworm generally are caused by anthropophilic species. Dermatophytes found associated primarily with animals are called **zoophilic**. These seldom cause epidemics; the diseases are sporadic and result from direct contact with infected animals. **Geophilic** dermatophytes grow and reproduce in the soil, occasionally causing diseases in man.

Treatment of ringworm is generally accomplished by administering the oral antibiotic griseofulvin for extended periods of time. This antibiotic accumulates in the skin and inhibits fungal growth. Ointments containing miconazole or other imidazoles are also used successfully in controlling ringworm.

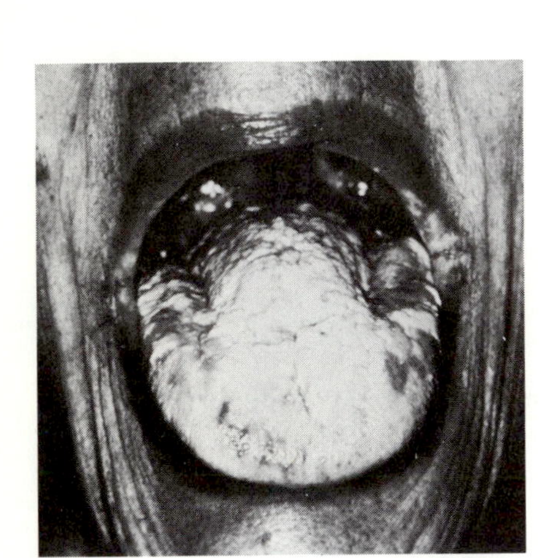

(a)

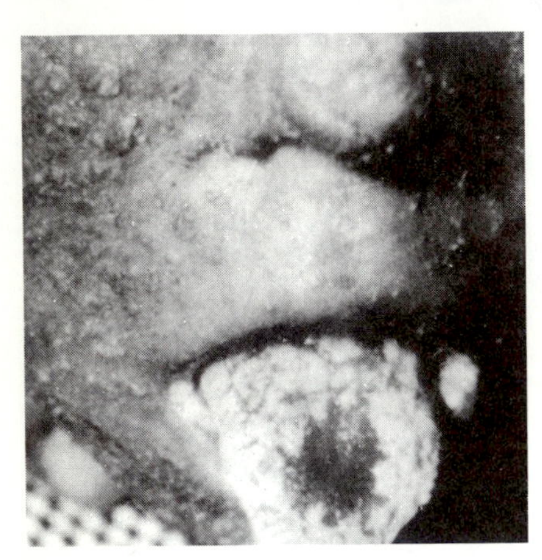

(b)

Figure 23-9 Candidiasis of Skin and Mucous Membranes. *(a)* Patient with oral candidiasis. The disease, also known as thrush, is characterized by the development of a cream-colored membranous layer over the afflicted area. The membranous layer consists of a knitted mass of affected tissue and fungal elements. *(b)* Chronic mucocutaneous candidiasis. The disease often occurs in individuals with various types of leukocyte dysfunction.

Candidiasis

Candidiasis is a name given to diseases caused by fungi in the genus *Candida*. The clinical manifestations of candidiasis vary from benign integumentary (skin and nail) diseases to severe renal (kidney) and myocardial infections. The skin diseases usually involve skin eruptions, which are papular or pustular inflammatory lesions with redness, swelling, and scaling of the skin. *Candida* infections may also involve the mucous membranes of the mouth and vagina (fig. 23-9). These forms of candidiasis are also called **thrush.** Individuals with defective immune systems □ may acquire a disease called chronic mucocutaneous candidiasis (fig. 23-9). The infection is recurrent and may involve both the integument and the mucous membranes.

494

Two species of *Candida* are most commonly associated with human disease. These are *C. albicans* and *C. tropicalis*; *C. albicans* is far more common. Both these yeasts are dimorphic □, forming yeast cells on routine culture media (e.g., Sabouraud glucose agar) and pseudohyphae in host tissue or under reduced oxygen tensions. Both these yeasts (like all other yeasts) are differentiated on the basis of their carbohydrate assimilation and fermentation patterns.

336

Candida albicans generally is unable to penetrate the intact skin, and breaking or softening of the skin is a prerequisite for infection. Virulence of *C. albicans* is believed to be related to its ability to form pseudohyphae and to produce the enzyme phospholipase. Pseudohyphae are more resistant than the yeast cell to phagocytosis by the PMN leukocytes and can penetrate host cells, possibly because of the phospholipase activity concentrated at the tips of the pseudohyphae. Phospholipase apparently breaks down plasma membrane phospholipids. The candidal infection causes abscesses that contain remnants of skin cells, leukocytes, and fungal elements.

Yeasts in the genus *Candida* are common inhabitants of the human body and are found on the skin, oropharynx, vagina, and digestive tract. Infections occur when the host's physiology allows the extensive growth of the yeast, generally as a consequence of predisposing factors such as diabetic acidosis, leukemias, excessive moisture, and obesity. An infection of the oral or vaginal mucosa often occurs after prolonged use of antibacterial antibiotics that cause a reduction of the normal flora in these areas.

The most effective treatment for candidiasis is correction of the predisponsing factors. For example, prolonged antibiotic treatment depletes the normal bacterial flora and allows *Candida albicans* to proliferate, since it is not susceptible to these antibiotics. Stopping the treatment or recolonizing the infected area with normal flora can control the infection. Ointments containing Nystatin may be helpful on skin lesions. In severe infections, the use of amphotericin B, or new drugs such as miconazole and ketoconazole, is necessary. Chronic mucocutaneous candidiasis presents a special problem because it results from an immunodeficiency □. Some patients can be treated by administering transfer factor, a substance released by sensitized T-lymphocytes which recruits other immune cells to the site of infection. Apparently, the deficiency of these patients lies in the inability of their sensitized lymphocytes to muster a strong enough cell-mediated immune response. Transfer factor obtained from normal individuals helps in amplifying the specific anticandidal response and attaining at least a temporary cure.

494

VIRAL DISEASES OF THE SKIN

Viral infections involving the skin are common (see table 23-3). Childhood diseases such as chickenpox and German measles have cutaneous manifestations. Some of the viral infections (e.g., warts) involve only the skin, while others, such as herpes, and smallpox viruses, also involve blood vessels and nervous tissue (fig. 23-10).

Dermotropic viruses, or viruses that have clinical manifestations affecting the skin, include pox, measles, varicella-zoster, herpes simplex, rubella, human wart (papilloma), some coxsackie viruses, and some echoviruses. Although these viruses can enter the host through the skin, some can also enter via the respiratory tract and others through the alimentary tract.

The life cycles of many viruses involve the disruption of the host cell as a result of viral multiplication. The newly released virus particles can, in turn, infect other cells and thereby initiate another round of infection. Eventually, the cellular destruction is so extensive that it is noticeable. The first few waves of infection cause enough cellular destruction to initiate an inflammatory response, which, in turn, contributes to tissue destruction and the subsequent formation of a viral lesion.

Specific antiviral mechanisms may also be involved in the formation of a lesion. For example, when cells are infected by a virus, the individual develops antibody and cell-mediated immunity against it. Much of this immunity is effective in aborting viral infections, but sometimes it is responsible for some of the signs and symptoms of a disease. For example, some viruses alter the host cell's surface, which acquires viral antigens; thus the host's

TABLE 23-3
SUMMARY OF INFECTIOUS DISEASES THAT AFFECT THE HUMAN SKIN AND EYE

INFECTIOUS DISEASE	CAUSATIVE AGENT	MICROBIAL GROUP
Anthrax	*Bacillus anthracis*	Gram + spore forming rod
Chickenpox	Varicella-zoster virus	Herpesviruses
Conjunctivitis	*Neisseria gonorrhoeae*	Gram − coccus
	Newcastle disease virus	Paramyxoviruses
	Streptococcus spp.	Gram + coccus
Erysipelas	*Streptococcus pyogenes*	Gram + coccus
Folliculitis	*Staphylococcus aureus*	Gram + coccus
Furunculosis	*Staphylococcus aureus*	Gram + coccus
Gas gangrene	*Clostridium perfringens*	Gram + sporogenous rod
Impetigo	*Streptococcus pyogenes*	Gram + coccus
Impetigo	*Staphylococcus aureus*	Gram + coccus
Inclusion conjunctivitis	*Chlamydia trachomatis*	Chlamydia
Leprosy	*Mycobacterium leprae*	Acid-fast rod
Lyme disease	*Borrelia* sp.	Spirochete
Measles (Rubeola)	Measles virus	Paramyxovirus
Molluscum contagiosum	Togavirus	Togaviruses
Piedra, black	*Piedraia hortae*	Fungus
Piedra, white	*Trichosporon* sp.	Yeastlike fungus
Pink eye	*Haemophilus aegypticus*	Gram − rod
Pyoderma	*Pseudomonas aeruginosa*	Gram − rod
	Streptococcus pyogenes	Gram + coccus
	Staphylococcus aureus	Gram + coccus
Ringworm (Tineas)	*Microsporum* spp.	Dermatophytic fungi
	Trichophyton spp.	Dermatophytic fungi
	Epidermophyton floccosum	Dermatophytic fungi
Rubella (German measles)	Rubella virus	Togaviruses
Scarlet fever	*Streptococcus pyogenes*	Gram + coccus
Smallpox	Smallpox virus	Poxviruses
Syphilis	*Treponema pallidum*	Spirochete
Tinea nigra	*Exophiala werneckii*	Fungus
Tinea versicolor	*Malasezzia furfur*	Yeast
Trachoma	*Chlamydia trachomatis*	Chlamydia

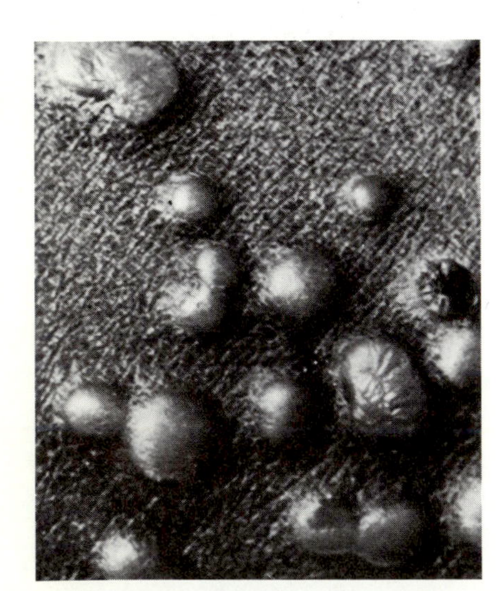

Figure 23-10 Characteristic Rash of Smallpox

immune system, in an attempt to abort the viral infection, recognizes viral antigens on the cell's surface and destroys the cell. This response might appear suicidal, but in reality it is a protective mechanism: the host destroys an infected cell before the virus has had a chance to multiply. This process, together with the inflammatory response, is partly responsible for the sores caused by some viruses.

Herpes Simplex Infections

Infections by the herpes simplex virus may involve the skin, eye, mucous membrances, and nervous system. These infections can be quite servere, even to the point of causing death. The infectious agent is usually acquired by direct contact through the broken skin or the mucous membranes. The infections eventually develop into lesions as a result of host- and viral-mediated tissue destruction. One important characteristic of herpes simplex is its tendency to remain quiescent in the ganglia of sensory neurons that innervate the infected area. Because these foci of infection persist in the host, recrudescences of the disease often occur at the same site. A tingling or burning

FOCUS

DO TOADS GIVE YOU WARTS?

Warts, or **papillomas**, have been attributed to various causes in human folklore. Probably one of the most "popular" ways of getting warts is by touching a toad.

In reality, warts are contracted by direct contact with individuals or objects carrying a type of virus called the **papillomavirus**. These are small, double-stranded DNA viruses that are highly specific for their host. The human papillomavirus, for instance, infects only humans.

In essence, human warts are benign tumors caused by the virus reproducing in host skin. Viral reproduction causes a localized, excessive proliferation of skin and an increased deposit of keratin. Human warts generally do not become malignant (cancerous) and frequently go away on their own. There are many "recipes" for curing warts, none of which work at all. In fact, simply leaving the warts alone may eventually be the best treatment of all. The more tenacious warts can be removed surgically, by freezing, or by applying ointments that dissolve keratin. One fact about warts is certain: toads won't give you warts, unless they are carrying the human papilloma virus.

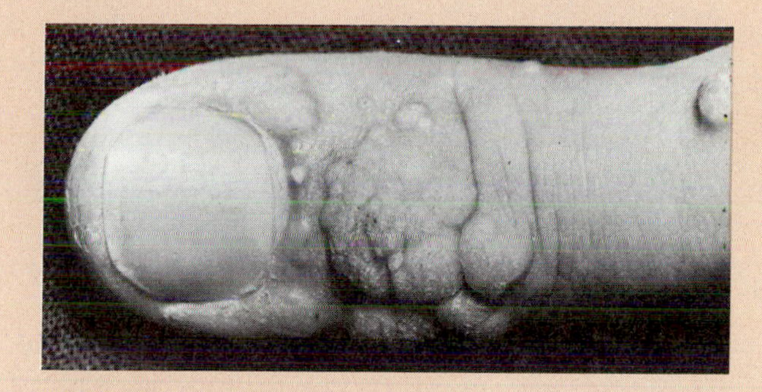

Figure 23-11 **Thumb with Warts**

sensation precedes the eruption, and many patients "feel" a cold sore coming on. Recurrent herpetic lesions often develop after a febrile disease, UV irradiation (sunbathing), or stressful situations. The latent virus, when reactivated in ganglionic cells, moves along the axon to the skin and initiates another infection. The painful lesions (fig. 23-12) initially are blisterlike, but after they rupture, they develop into a scab.

Herpes simplex virus is worldwide in distribution, and nearly everyone eventually becomes infected. There are two types of herpes simplex virus (I and II) that generally cause different diseases. Herpes simplex type I is most commonly associated with disease of the mouth, respiratory tract, skin, and central nervous system, while herpes simplex type II is an important cause of genital infection.

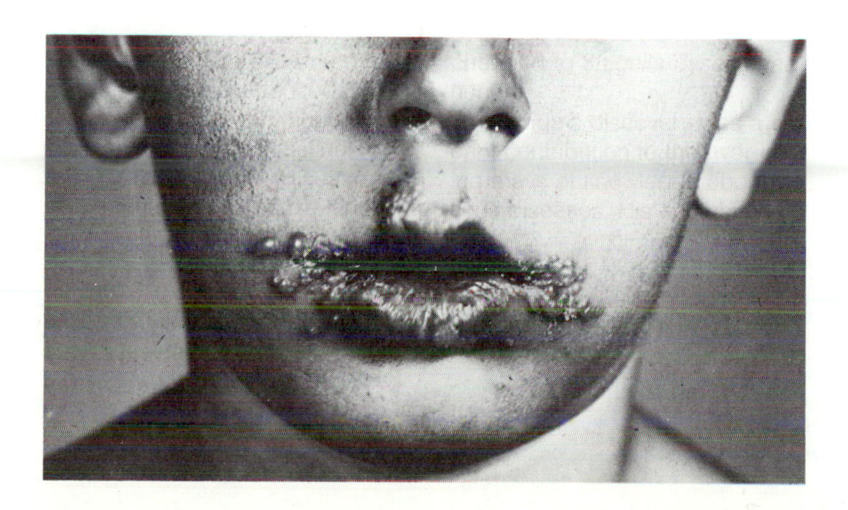

Figure 23-12 **Child Exhibiting the Typical Lesions of Herpes Simplex Infection (Cold Sores)**

Few drugs are currently available for the control of herpetic infections. Topical 5-iodo-2′-deoxyuridine (IDU) is partially effective against eye infections but not against superficial skin infections. Intravenous, intramuscular, and oral acyclovir (sold under the trade name Zovirax) reduce the severity and duration of skin and systemic herpetic infections. Zovirax in ointment form is ineffective against genital or oral herpes.

Varicella-Zoster Virus Infections

Upon first exposure to a varicella-zoster virus, the patient develops the typical rash characteristic of **chickenpox** (fig. 23-13). This infection is generally mild, lasting 5–10 days in children. Approximately 14 days after infection, a papular rash (raised spots) appears on the skin and mucous membranes. The papules eventually become vesicular and filled with a watery fluid. Fever, headache, and malaise may be evident during the **prodromal period** (initial stage of a disease). Primary varicella-zoster infections in adults generally are more severe than children, often resulting in pneumonia. Mortality is high

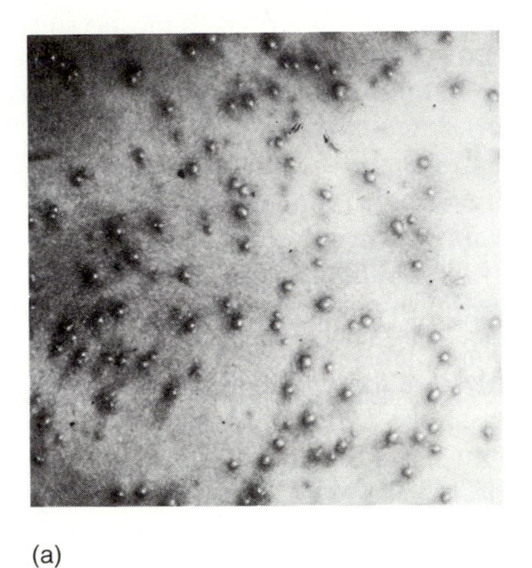

(a)

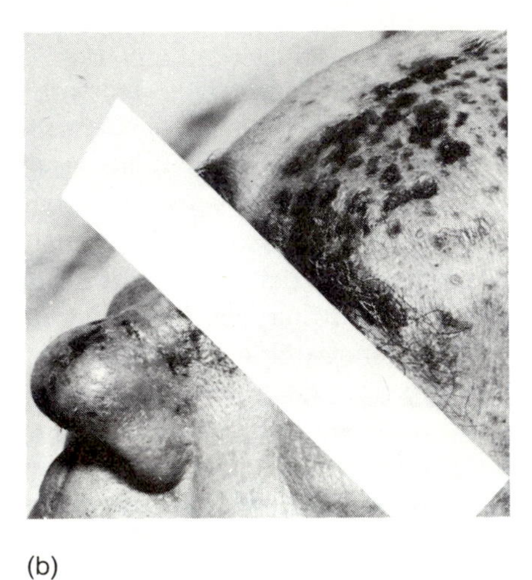

(b)

Figure 23-13 Diseases Caused by Varicella-Zoster Virus. *(a)* Patient with the typical vesicular lesions of chickenpox (varicella). *(b)* Lesions of shingles on a patient.

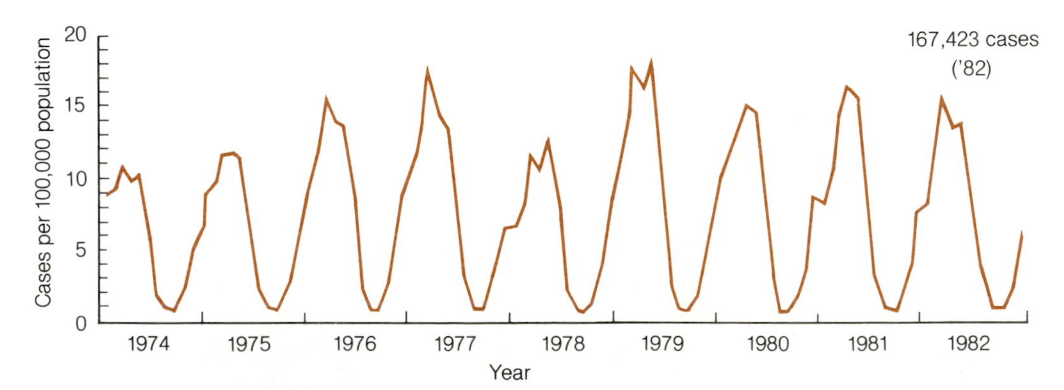

Figure 23-14 Incidence of Chickenpox in the United States. Chickenpox has appeared with predictable frequency for the past 10 years. The incidence rate is 10–20 cases/100,000 inhabitants. The majority of cases occur in children between the months of March and May.

(15%) in adults. Chickenpox seems to increase in incidence (fig. 23-14) between the months of March and May. This increase is probably due to the fact that children, the most commonly affected group, are confined together in classrooms during this period.

Adults who have had chickenpox may experience a recrudescence of varicella-zoster called **shingles** (fig. 23-13). The recrudescence involves a return of the dormant virus to the innervated skin along sensory neurons. The reactivation of the latent virus and its movement in the axons cause severe pain in the infected area. Vesicular lesions resembling those of chickenpox also occur; however, these lesions are distributed only in the skin innervated by the infected neurons. Unlike chickenpox, the vesicular lesions of shingles may last up to a month, and the pain can persist even longer.

Measles (Rubeola)

Measles (also called **rubeola**) is a highly infectious disease caused by a pseudomyxovirus. The virus infects and multiplies in respiratory tract epithelium and in draining lymph nodes. The prodromal period has signs and symptoms resembling those of the common cold, with sore throat, low-grade fever, and cough. Allergic-like symptoms (e.g., nasal congestion) are sometimes present, as is conjunctivitis. With the spread of the virus in the blood, a typical rash appears, usually heralded by the development of small red dots with blue-white centers **(Koplik dots)** located on the lateral oral mucosa. This sign is considered to be diagnostic for measles. The maculopapular rash appears first on the face and neck and later spreads throughout the body. The incubation period for the disease in 10–21 days, and the acute stage of the disease with the rash may last 4–5 days (fig. 23-15).

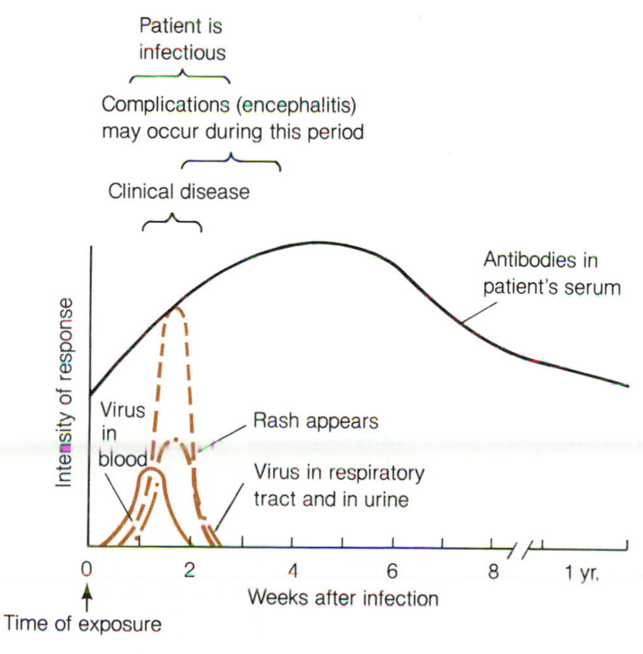

Figure 23-15 Course of Infection by the Measles (Rubeola) Virus. The characteristic rash of the disease appears 10–14 days after infection and is accompanied by a rise in serum antibodies. During the period of clinical illness, the patient is contagious because he or she is shedding viruses in saliva and urine.

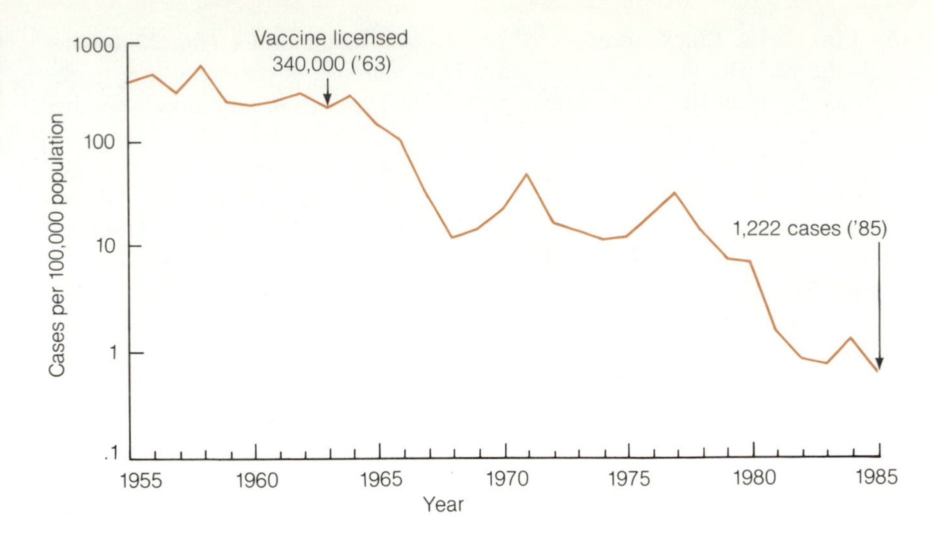

Figure 23-16 Incidence of Measles in the U.S. Before the introduction of the vaccine in 1963, there were about 250 cases/100,000 individuals. The present number of annual cases ranges between 0.6 and 1.0/100,000.

Measles is transmitted from person to person in respiratory secretions and aerosols. Because humans are the only reservoir and, once infected, are immune for long periods of time, the disease tends to disappear from small communities until it is reintroduced by travellers. It reoccurs in empidemic proportions every five to seven years, or as soon as a large number of nonimmune children have been born and the immunity has declined in a sufficient number of adults (fig. 23-16). The disease is common in children, especially during December and May. Immunity to measles may be transferred passively from mother to offspring, which explains why children under the ages of 5 or 6 months seldom are afflicted with measles. Vaccination is very effective in reducing the number of cases of measles (fig. 23-16). In 1984, a total of 2,534 cases were reported in the United States.

Rubella (German Measles)

Rubella (**German measles**) is caused by a togavirus and is characterized by a macular rash accompanied by a low-grade fever. It spreads from person to person by direct contact, through nasal secretions, and by aerosols. The virus initially infects the mucous membranes of the upper respiratory tract and then spreads to other organs of the body. Fever, nasal discharge, and lymph node enlargement are early signs and symptoms of rubella. The characteristic rash apparently is due to a hypersensitivity □ (allergic response) to circulating viral antigens. 482

Rubella is an important disease, because during the first trimester of pregnancy the virus can infect the fetus and cause major destruction of fetal tissue. Such infection can result in abortion or in developmental abnormalities, including brain damage and mental retardation, deafness, congenital cataracts and blindness, heart defects, bone lesions, enlarged spleen (splenomegaly), enlarged liver (hepatomegaly), and low birth weight.

Rubella is diagnosed almost exclusively on a serologic basis □, because this method is rapid and results can be attained the same day. Rubella testing has been implemented in many states in family planning and maternity clinics to advise potential mothers on the risk of rubella during pregnancy. Prospective mothers who have never been exposed to rubella should be vaccinated to avoid the possibility of becoming infected with the rubella virus during pregnancy. In 1984, a total of 745 cases were reported in the United States. 515

SUMMARY

Many infectious agents are capable of causing diseases of the skin. These include bacteria, viruses, and fungi

STRUCTURE AND FUNCTION OF THE SKIN

1. The skin covers the body and provides an effective barrier against infections be microorganisms.
2. Conditions that alter the proper structure of the skin predispose the hose to infectious diseases.

THE NORMAL FLORA OF THE SKIN

1. The normal flora of the skin colonizes the stratum corneum and hair follicles.
2. This normal flora protects the host against microbial pathogens, by competing for space and nutrients and by changing the skin environment so that is hostile to invading pathogens.

BACTERIAL DISEASES OF THE SKIN

1. Many bacteria may invade the human skin, causing various diseases.
2. Anthrax is an often fatal disease characterized by the formation of a malignant pustule at the site of infection. The pathogen spreads from that site and produces an exotoxin. The disease is the result of the combined action of the spreading bacterium and the toxin.
3. Leprosy is a disease of the reticuloendothelial system. *Mycobacterium leprae* invades host cells, where it multiplies. The severity of the disease is related to the time of onset of cell-mediated immunity: the longer it takes for cell-mediated immunity to develop, the more severe the disease.
4. Pyoderma includes any acute inflammatory infection of the skin which results in the formation of pus. It is most commonly caused by *Staphylococcus* and *Streptococcus*.
5. Pyoderma includes diseases such as impetigo, folliculitis, carbuncles, and furuncles.
6. The pathogenic properties of *Staphylococus aureus* reside in the ability of this bacterium to produce an array of exoenzymes as well as exotoxins.
7. Group A and B streptococci are the most important human pathogens. These microorganisms produce many exoenzymes and exotoxins that are responsibile for the disease process.
8. Trachoma is a serious eye infection caused by *Chlamydia trachomatis*. The infection causes an extensive destructive inflammation that often leads to permanent blindness.

FUNGAL DISEASES OF THE SKIN

1. Fungi cause a variety of diseases that involve the skin. The most common of these are the ringworms and candidiasis.
2. Ringworms, such as athlete's foot and tinea capitis, are caused by a group of fungi, called dermatophytes, which grow on keratinized tissues. These are superficial infections that may elicit severe inflammatory responses, leading to permanent hair loss and the formation of scar tissue.
3. *Candida albicans* causes infections of the skin and mucous membranes. These infections cause primarily inflammatory reactions that vary in severity from mild to fatal.
4. Candidal infections often result when the host's physiology is altered or when the normal bacterial flora, which normally keeps the yeasts in check, becomes depleted.

VIRAL DISEASES OF THE SKIN

1. Many viruses cause skin infections.
2. Dermotropic viruses have clinical manifestations that affect the skin and include smallpox, varicella-zoster, herpes simplex, rubella, and human wart viruses.
3. Most of the viral diseases result from host cell disruption by the reproducing virus, or from protective host reponses.
4. Herpes simplex infections are characterized by the formation of a sore at the site of entry. Repeated episodes of "cold sores" result because the virus may remain quiescent in nerve ganglia that innervate the infected area.
5. Some herpetic infections are very serious and may involve the eye and central nervous system.
6. Chickenpox and shingles are caused by the varicella-zoster virus. Chickenpox is a benign skin infection that results in the formation of vesicular lesions. The virus may remain quiescent for many years, after which it may cause a disease called shingles.
7. Measles, or rubeola, is caused by a pseudomyxovirus. The typical disease is characterized by a rash and by the formation of Koplik dots on the lateral oral mucosa. This virus can cause very serious diseases, such as encephalomyelitis and subacute sclerosing panencephalitis.
8. Rubella, or German measles, is caused by a togavirus. The disease is characterized by a rash and low-

grade fever. Rubella in pregnant women, especially in the first trimester, can lead to severe fetal deformations of fetal death.

STUDY QUESTIONS

1. Describe the salient features of the skin and the function(s) it performs. Indicate which areas of the skin are colonized by microorganisms.
2. Discuss how the normal skin flora protects its host.
3. Discuss the various mechanisms employed by microorganisms to enter the human skin and cause disease.
4. Construct a table of bacterial skin diseases, including their etiology, characteristic signs and symptoms, and mode of transmission.
5. Construct a table of fungal diseases of the skin, including their etiology, characteristic signs and symptoms, and mode of transmission.
6. Construct a table of viral diseases of the skin, including their etiology, characteristic signs and symptoms, and mode of transmission.
7. Discuss various ways of preventing skin infections.
8. Discuss the pathogenesis (disease process) of:
   a. leprosy;
   b. impetigo;
   c. scarlet fever;
   d. anthrax;
   e. shingles.
9. Why are staphylococci and streptococci considered to be pyogenic bacteria? How are exoenzymes and toxins produced by these bacteria related to their ability to cause pyogenic lesions?

SUPPLEMENTAL READINGS

Henderson, D. A. 1976. The eradication of smallpox. *Scientific American* 235:25–33.

Hirsch, M. S. 1984. *Cutaneous viral diseases. Scientific American Medicine* Section 7-XXX New York: Freeman.

Joklik, W. K., Willett, H. P., and Amos, D. B. eds. 1984. *Zinsser microbiology*. 18th ed. New York: Appleton-Century-Crofts.

Marples, M. J. 1969. Life on the human skin. *Scientific American* 220:108–115.

Rosebury, T. 1969. *Life on man*. New York: Berkeley Publishing Co.

Simon, H. B. 1984. *Gram-positive cocci. Scientific American Medicine* Section 7-I New York: Freeman.

Wolf, R. H., Gormus, B. J., Martin, L. N., Baskin, G. B., Walsh, G. P., Meyers, W. M., and Binford, C. H. 1985. Experimental leprosy in three species of monkeys. *Science* 227:529–530.

CHAPTER 24
DISEASES OF THE RESPIRATORY TRACT

CHAPTER PREVIEW

AIRBORNE INFECTIONS IN DAY-CARE CENTERS

Recent increases in the cost of living have forced both parents in many households to seek employment in order to make ends meet. As a consequence, more and more preschool children are going to day-care centers while their parents are at work. This rapid increase in the number of children attending child-care centers has created a new public health hazard: overcrowding and less than optimum hygienic facilities. This makes many children vulnerable to a vast array of infectious diseases. Of importance are diseases that are spread in airborne particles or by direct contact. This hazard is compounded by the fact that preschool children have no concept of how diseases spread or how to prevent them. Hence, unstifled sneezes, coughs, and nasal secretions create an environment in which children are continually exposed to respiratory disease agents.

Transmission of common colds, influenza cytomegaloviruses, and *Cryptosporidium* increases dramatically under these conditions. Infected children can then pass on the infectious agents to their families and neighborhood playmates.

Public health officials need to know the biology and mode of transmission of infectious agents of the respiratory tract in order to control respiratory diseases in institutions such as day-care centers. This chapter discusses some of the most common diseases of the human respiratory tract, their cause, treatment, and control.

LEARNING OBJECTIVES

A STUDY OF THIS CHAPTER SHOULD ENABLE YOU TO:

EXPLAIN THE STRUCTURE AND FUNCTION OF THE RESPIRATORY SYSTEM

DISCUSS THE ROLE OF THE NORMAL FLORA IN THE RESPIRATORY SYSTEM

DISCUSS SOME OF THE MOST COMMON INFECTIOUS DISEASES OF THE RESPIRATORY TRACT, THEIR PATHOGENESIS, TREATMENT, AND PREVENTION

An average individual, breathing quietly, inhales about 7,500 ml of air every minute along with numerous air contaminants, many of which are potentially pathogenic microorganisms. In light of this fact, it is not surprising that many infectious diseases are acquired and transmitted through the respiratory route. Since there is such a potential for lung infection, the body has evolved an array of specific and nonspecific mechanisms for protecting the lungs. Coughing, sneezing, mucus secretions, and ciliary action expel inhaled particles and microorganisms from the respiratory tract. In addition, macrophages and immunoglobulins (IgA) help destroy microorganisms that remain in the lungs.

The respiratory tract is divided into two regions: the **upper respiratory tract** (URT) and the **lower respiratory tract** (LRT). URT infections are generally benign and can be treated effectively, although some URT infections, such as diphtheria and acute epiglottitis, can be life-threatening. Conversely, LRT infections often are serious, and must be treated promptly and effectively or serious complications or death will result. In this chapter we will examine some of the most common respiratory tract infections, including their clinical and biological nature as well as their treatment.

STRUCTURE AND FUNCTION OF THE RESPIRATORY TRACT

The primary function of the respiratory tract is to bring air into contact with a respiratory surface that allows oxygen to diffuse into the blood and carbon dioxide to escape. The respiratory surface is highly vascular, not only because gases must be exchanged efficiently during breathing, but also because cells of the immune system must be delivered promptly. The respiratory system includes a system of air-transporting tubes that bring the air to the respiratory surface (fig. 24-1). The tubes in the upper respiratory tract consist of the **nasal cavity**, **eustachian tubes**, and throat, while those in the lower respiratory tract include the pharynx, larynx, trachea, and bronchi. These structures are much too thick to permit the free exchange of air with the blood, and they function solely as air passages to the respiratory **bronchioles**, **alveolar ducts**, and **alveolar sacs** (fig. 24-1).

The alveoli or air sacs are the site of exchange of gases between the blood and the environment. These thin-walled sacs, made up of a single layer of flat epithelial cells (fig. 24-1), have numerous capillaries. The alveolar sacs and capillaries exchange gases with each other, and the capillaries may also deliver immune cells to the air sacs. The alveolar surface area that functions in gas exchange in adult humans is about 50–70 m^2 ($1m^2 \approx 1$ square yard) and consists of more than 300 million alveoli. Any condition that interferes with the proper exchange of gases is harmful to the individual. In pneumonia, for example, the damaged alveoli fill with fluids from alveolar capillaries. The fluid in the alveoli reduces the efficiency of gas diffusion and blood aeration.

THE NORMAL FLORA OF THE UPPER RESPIRATORY TRACT

The microbial flora of the upper respiratory tract (table 24-1) consists mainly of bacteria and fungi. Staphylococci, streptococci, diphtheroids, actinomycetes, gram negative cocci, coccobacilli, aspergilli, yeasts, and zygomycetes may all be present in the upper respiratory tract.

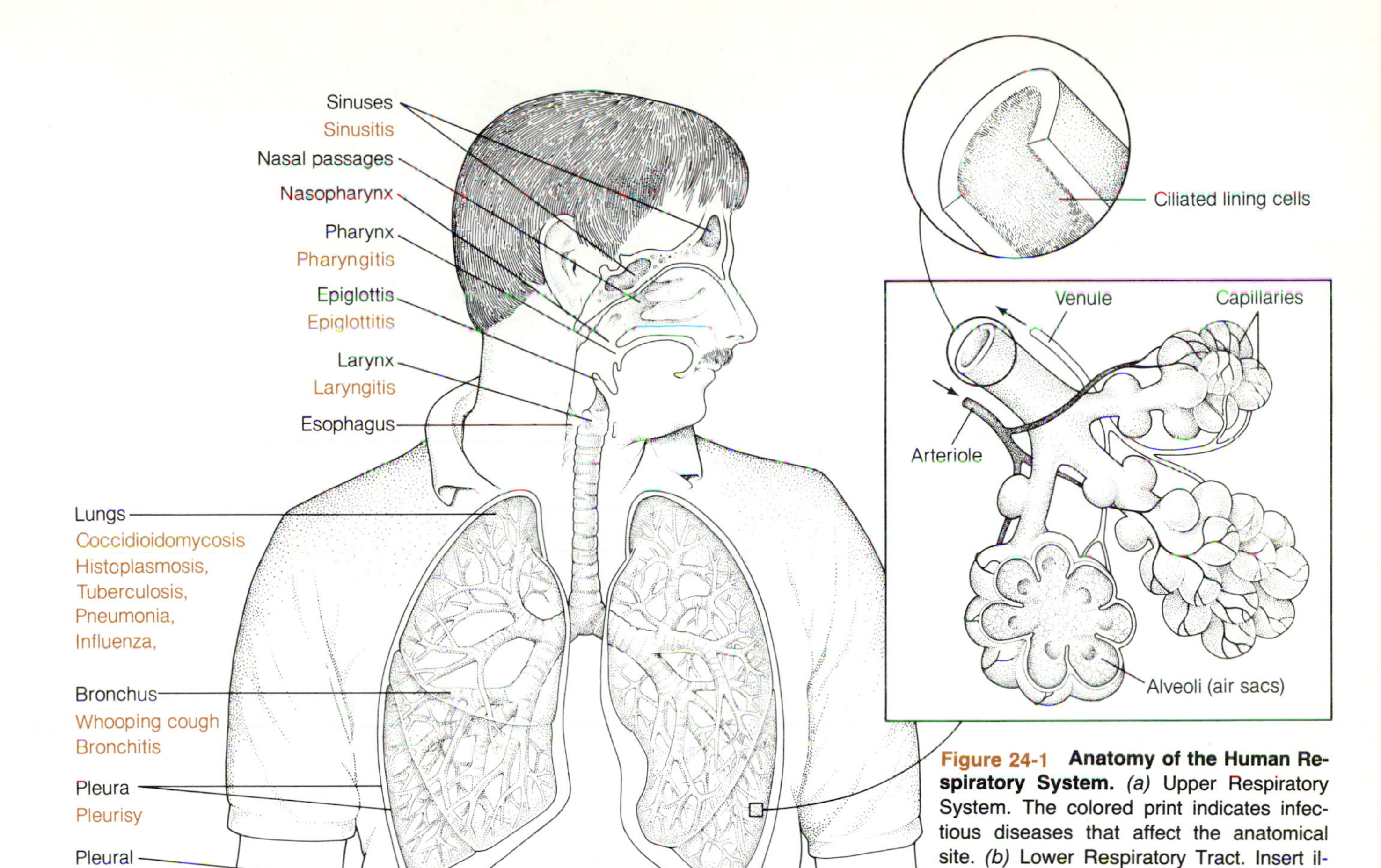

Figure 24-1 Anatomy of the Human Respiratory System. *(a)* Upper Respiratory System. The colored print indicates infectious diseases that affect the anatomical site. *(b)* Lower Respiratory Tract. Insert illustrates the structure of alveoli. Notice that respiratory tubes are lined by ciliated cells, which serve to flush out infectious agents and dust particles before they reach the alveoli. The colored print indicates infectious diseases that affect the anatomical site.

The indigenous flora of the upper respiratory tract may contain commensals as well as potential pathogens. In addition, the normal flora of the upper respiratory tract contains mutuals 426 □ that protect the host from colonizing pathogenic bacteria. For example, certain indigenous streptococci of the throat, especially those in the **viridans group**, inhibit the growth of Group A streptococci and therefore protect the host from a possible attack of "strep throat" (**pharyngotonsillitis**). Antibiotics or other inhibitory substances may be produced by the viridans streptococci, or they may deplete the available nutrients.

BACTERIAL DISEASES OF THE UPPER RESPIRATORY TRACT

Pharyngotonsillitis

Pharyngotonsillitis (table 24-2) is an inflammation of the pharynx, usually accompanied by throat pain, malaise, fever, and postnasal secretions. The throat is scarlet red, with exudative pyogenic material

Group A streptococci, such as *Streptococcus pyogenes*, are the commonest cause of bacterial pharyngotonsillitis. Occasionally, Groups C and G

TABLE 24-1
REPRESENTATIVE MICROORGANISMS FOUND IN THE UPPER RESPIRATORY TRACT

MICROORGANISM	ORAL	NASOPHARYNX	SINUSES
BACTERIA			
Bacteroides sp.	+	−	−
Borrelia sp	+	+	−
Branhamella sp.	+	+	−
Diphtheroids	+	+	−
Fusobacterium sp.	+	−	−
Haemophilus sp.	+	+	−
Mycoplasma sp	−	−	+
Neisseria sp.	+	+	−
Peptostreptococcus sp.	+	+	−
Staphylococcus aureus	+	+	−
S. epidermidis	+	+	−
Streptococcus sp.	+	+	+
Veillonella sp.	+	−	−
FUNGI			
Aspergillus sp.	+/−	+/−	−
Candida albicans	+	+	+/−
VIRUSES			
Adenoviruses	−	+	+
Coxsackieviruses	+	+	+
Epstein-Barr virus	−	+	−
Herpes simplex	+	−	−
Influenza viruses	−	+	+
Rhinoviruses	+	+	+

\+ = Organism commonly isolated from site
− = Organism infrequently or never isolated from site

TABLE 24-2
COMMON INFECTIOUS DISEASES OF THE UPPER RESPIRATORY TRACT

NAME OF DISEASE	CAUSATIVE AGENT	GROUP OF ORGANISMS
Common cold	Rhinoviruses	Virus
	Coronaviruses	Virus
Croup	Respiratory syncytial virus	Virus
	Myxoviruses	Virus
	Adenoviruses	Virus
Diphtheria	*Corynebacterium diphtheriae*	Bacteria (gran + rods)
Epiglottitis	*Haemophilus influenzae*	
	Cytomegaloviruses	Virus
Otitis Media	*Haemophilus influenzae*	Bacteria (gram − rods)
	Staphylococcus aureus	Bacteria (gram + coccus)
	Streptococcus pneumoniae	Bacteria (gram + coccus)
Pharyngotonsillitis (strep throat)	*Streptococcus pyogenes*	Bacteria (gram + coccus)
	Staphylococcus aureus	
Sinusitis	*Haemphilus influenzae*	
	Bacteroides sp.	Bacteria (gram − rod)
	Streptococcus pneumoniae	
Zygomycosis	*Rhizopus* sp.	Fungi (zygomycetes)

streptococci can also cause pharyngotonsillitis. Streptococcal pharyngotonsillitis is most common in school-age children (5–15 years) and occurs most often during the winter months. The streptococci are transmitted in aerosols 539 □ and occasionally in foods. Other bacteria that have been cited as causes of pharyngotonsillitis, albeit infrequently, are *Staphylococcus aureus*, *Haemophilus influenzae*, and *Corynebacterium diphtheriae*. In spite of a common belief that most pharyngotonsillitis is due to streptococci, most cases (70%) are in fact caused by adeno-, coxsackie-, rhino-, Epstein-Barr, influenza, parainfluenza, and respiratory syncytial viruses.

Sequelae to Pharyngotonsillitis

A small but significant percentage of individuals suffering from streptococcal pharyngotonsillitis, if they receive no treatment, may end up with nonsuppurative (no pus produced) sequelae. The most common of these are rheumatic fever and acute glomerulonephritis.

Rheumatic fever is a serious sequela that occurs in about 3% of human streptococcal infections. It is responsible for 10,000-20,000 deaths each year in the United States. The disease, also called rheumatic heart disease, is characterized by arthritis (inflammation of the joints) and carditis (inflammation of heart muscle). Rheumatic fever develops through three stages: a) acute nasopharyngitis; b) asymptomatic carriage of Group A streptococci in the throat; and c) electrocardiographic changes, accompanied by fever and by inflammation of the joints and heart muscle.

The pathogenesis of the disease is not well understood. Rheumatic fever patients characteristically have antibodies against hyaluronic acid, streptolysin O, DNase, and M-protein. It is thought that the cardiac lesions and the joint inflammation may be due to antistreptococcal antibodies that crossreact with the heart tissue. In support of this idea is the fact that streptococcal M-protein, when injected repeatedly into laboratory animals, is able to cause rheumatic fever–like diseases.

Acute glomerulonephritis is another complication of streptococcal infections which occurs in about 1–15% (average of 7%) of individuals suffering from a streptococcal infection. The disease is an acute inflammation of the glomeruli in the kidney, characterized by blood in the urine (hematuria) and hypertension (high blood pressure). Delirium and coma may accompany severe cases.

The pathogenesis of the disease appears to be related to the formation of circulating immune complexes (antigens bound by antibody). These complexes are deposited in the glomeruli, fixing complement □. 472 The complement-mediated kidney tissue destruction gives rise to the signs and symptoms of the disease. Epidemics of this disease have been documented among school-age children. The disease does not tend to recur in afflicted individuals, and deaths are rare.

Diphtheria

Diphtheria is an infectious disease characterized by fever, headache, malaise, sore throat, and eventually the formation of a pseudomembrane (fig. 24-2) on mucous membranes of the throat and nasal passages. In pharyngeal diphtheria, the airway is often obstructed due to bacterial growth and the

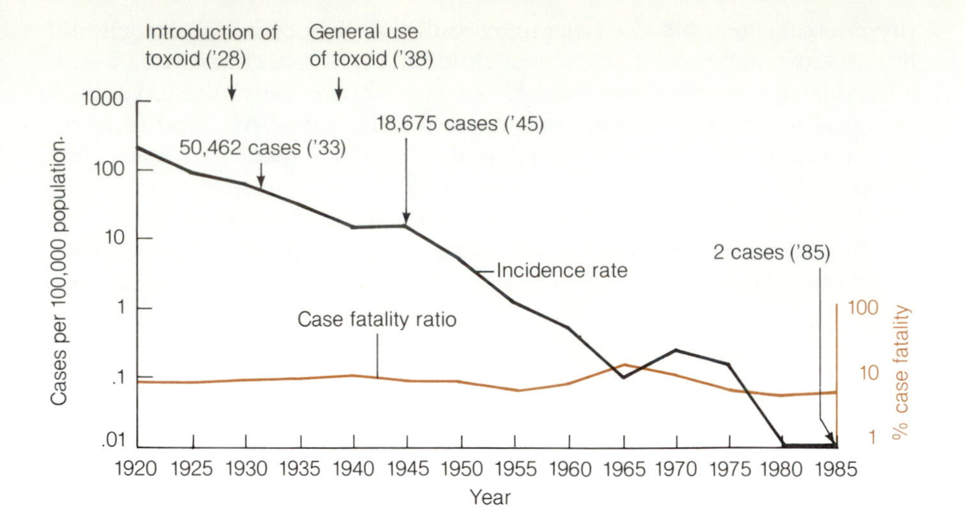

Figure 24-2 Incidence of Diphtheria. Since the introduction of the diphtheria toxoid vaccine, the incidence of diphtheria in the United States has decreased significantly. It is noteworthy, however, that the case fatality rate of the disease remains constant at about 8 percent.

pseudomembrane. Death is due to the action of the exotoxin (diphtherotoxin) on the heart and peripheral nerves. **Myocarditis** manifests itself late in the course of the disease and represents a grave turn of events.

The cause of diphtheria is *Corynebacterium diphtheriae*, a gram positive, nonmotile, asporogenous, pleomorphic rod. The individual cells are generally club-shaped with a tendency to form V's, M's, L's, and palisades. This organism produces a toxin, called **diphtherotoxin,** that is responsible for the signs and symptoms of the fatal disease. The production of toxin by *C. diphtheriae* is due to a **lysogenic conversion** □ of the bacterium by a bacteriophage. The lysogenic bacteriophage β apparently contains as part of its genome the gene coding for diphtherotoxin. In a prophage state within the host, the viral phenotype is expressed. Only lysogenised *C. diphtheriae* are capable of producing the toxin. 401

Diphtherotoxin is an exotoxin that inhibits protein synthesis □. It consists of two different polypeptides, A and B. Polypeptide A is the toxic component, but polypeptide B is required for A to be active. Apparently, B functions as a site for attachment of A to the plasma membrane of the target cell. The toxin acts by inhibiting protein synthesis at the elongation step. The A region blocks transferase activity and prevents the addition of amino acids to the growing polypeptide. The end result of this inhibition of protein syntheis is cell death. 209

The signs and symptoms of diphtheria are due not only to the diphtherotoxin but also to the growth of the bacteria and to other toxic substances that they may release. For example, the cell walls of *C. diphtheriae* contain **corynemycolic** and **corynemycolenic acids**, which inhibit oxidative phosphorylation □ and consequently are cytotoxic. Other cytotoxic factors include exoenzymes, such as the plasma membrane–degrading enzyme **neuraminidase**. Since nontoxigenic strains can cause formation of the diphtheritic pseudomembrane, but not cardiovascular and neurological changes, it is thought that the pseudomembrane is not due entirely to the exotoxin; it may also result from the inflammatory response caused by the multiplying bacteria. The bacteria growing in the throat release the diphtherotoxin, which 185 is absorbed through the mucous membranes and then distributed to distal organs (such as the heart) by the circulatory system. Myocardial lesions characteristic of the severe disease result from the destruction of the myocardial cells by the exotoxin. If the degree of myocardium destruction is sufficiently extensive, the disease is fatal. Nervous system involvement results from the destruction of neurons by the toxin. The severity of the disease is directly proportional to the amount of cell-bound diphtherotoxin in the host.

Corynebacterium diphtheriae is transmitted by direct contact with asymptomatic carriers, by inhalation of aerosols, or via fomites. With the advent of immunization, the number of annual cases in the United States has stabilized at around 2 (fig. 24-2).

Epidemics of diphtheria have been reported from economically deprived urban areas of various countries. These individuals had not been immunized—a fact that demonstrates the importance of immunization for preventing potentially devastating epidemics. The highest incidence of diphtheria occurs during the winter months in patients 15 years of age or older. This is a departure from past observations, in which individuals between the ages of 6 months and 10 years showed the highest incidence. Protection against this disease is by vaccination (DPT shots), which are compulsory in the United States. The vaccine preparation or **toxoid** consists of modified, nontoxic, but antigenic A and B regions.

The most effective treatment consists of the prompt administration of antitoxin and treatment with erythromycin. Passive immunization with equine (horse) antidiphtherotoxin antibodies is the only specific treatment for the severe form of the disease. In addition, treatment with erythromycin is very effective in eliminating the carrier state.

Otitis

Otitis is an inflammation of the ear that may or may not be of infectious nature. **Otitis externa** is a benign, inflammatory condition of the external auditory canal. The condition, very common in human populations, can be initiated by excessive moisture in the outer ear or by damage (e.g., from cotton swabs or hairpins) to the auditory canal. The most common complaints are pain and itching. Crusting and purulent discharges are also common manifestations. The etiology (cause) of otitis externa is varied and includes gram positive cocci, gram negative bacilli, and fungi. Infections of the external ear by *Pseudomonas aeruginosa* are characterized by the formation of a green pus that is discharged from the inflamed area. Central nervous system complications have been reported as a consequence of perforation of the ear by *P. aeruginosa* and require prompt treatment with carbenicillin and tetracycline.

Otitis media is a common childhood disease, but it is relatively rare in adults. This condition usually results from the invasion of the middle ear by bacteria from a pharyngeal infection. The signs and symptoms of this disease are pain, fever, and temporary hearing loss due to the acute inflammation and the resulting edema. A characteristic sign is a convex tympanic membrane, bulging outward because of the fluid buildup and pressure in the middle ear. A myriad of bacteria can cause the disease: pyogenic cocci, gram negative bacteria, *Haemophilus influenzae*, and *Branhamella catarrhalis* are often implicated.

Sinusitis

Sinusitis is an inflammation of the sinuses, especially the paranasal sinuses (fig. 24-1). Any condition that impedes the proper drainage of sinuses can lead to this disease. Nasal septum deviation, polyps, or rhinitis are common predisposing factors. The signs and symptoms of sinusitis depend upon which sinuses are involved. Paranasal sinusitis, the most common form of sinusitis, results in pain and tenderness in the forehead, with purulent drainage. It is usually caused by *Haemophilus influenzae*, *Streptococcus pneumoniae*, and Group A streptococci. Treatment usually entails draining the sinuses and administering antibiotics.

Although not strictly a sinusitis, a fungal disease called **zygomycosis** sometimes affects the sinuses. The disease is infrequent but often fatal. It is caused by opportunistic zygomycetes □ such as *Rhizopus* and *Mucor*. These opportunistic pathogens cause a rapidly developing disease in individuals with diabetic acidosis or leukemias. These fungi grow well in arterial blood and invade the vessels, causing thrombi. The lesions initiate in the paranasal sinuses, creating a black, mucoid exudate (fig. 24-3) consisting of necrotic tissue, fungal elements, and clotted blood. The fungi spread rapidly into the brain, causing severe central nervous system disorders or death. The treatment of choice is correction of the predisposing factor(s) and prompt administration of antifungal antibiotics such as amphotericin B or imidazoles.

340

Acute Epiglottitis

Acute epiglottitis is a fulminant, grave inflammation of the epiglottis (fig. 24-1). In children, the disease progresses quickly, causing fever, a painful sore throat, extreme difficulty in swallowing, and a progressive enlargement of the epiglottis. Difficulty in breathing becomes increasingly pronounced as the disease advances. If the airway is not maintained patent (open), apnea (lack of breathing) and death soon follow. The most common cause of epi-

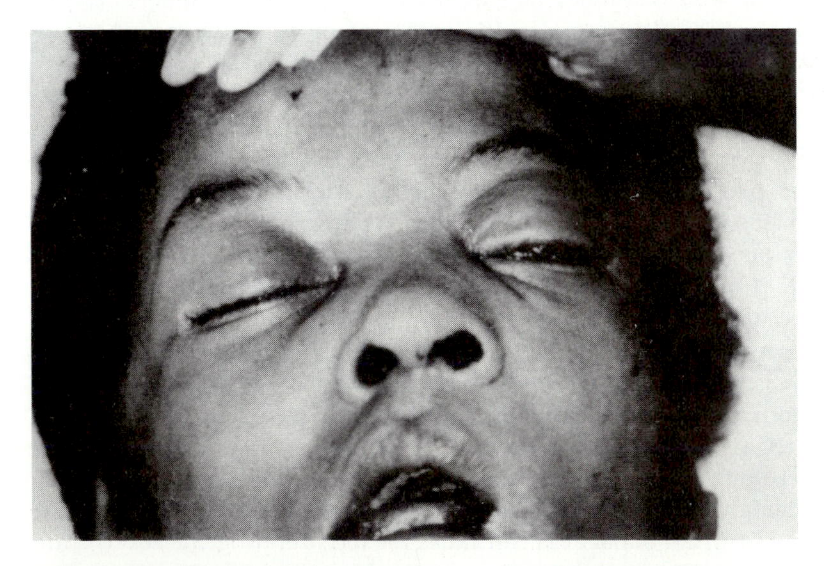

Figure 24-3 Rhinocerebral zygomycosis. The dark discharge evident in the nostrils, eyes, and mouth of the patient is characteristic of the disease and consists of fungal elements, dead tissue, and coagulated blood.

glottitis is *H. influenzae*, although *S. pneumoniae*, streptococci, and staphylococci have also been isolated from patients with the disease. The treatment involves prompt reestablishment of the airway and massive intravenous antibiotic treatment with chloroamphenicol and penicillin. Moistened, cool air also helps to alleviate the inflammation.

VIRAL DISEASES OF THE UPPER RESPIRATORY TRACT

The Common Cold

The **common cold** is a mild illness of the upper respiratory tract characterized by a "stuffy" nose, "scratchy" throat, general malaise, headache, sore throat, sneezing, and watery discharge from the nose. The nasal discharge often thickens and becomes yellowish. This is usually an afebrile disease with an incubation period of 1–4 days. The illness runs its course within 4–7 days. Sinusitis and otitis media are common complications, since drainage of the sinuses and middle ear is blocked.

The common cold results from a number of viral infections of the upper respiratory tract. Various types of viruses, including rhino-, corona-, myxo-, coxsackie-, respiratory syncytial, adeno-, and enteroviruses have been isolated from nasal secretions of patients suffering from the common cold. Clinical studies into the cause of the common cold have revealed that fewer than 60% of those individuals tested yielded positive cultures; thus, coldlike symptoms may be induced by noninfectious processes. The data do not overrule the possibility that failure to isolate the virus was due to improper cultural conditions. The rhinoviruses cause approximately 30–35% of colds. Common colds occurring in the fall and spring are usually caused by the rhinoviruses, while those in the winter months are often caused by coronaviruses.

Humans are the only natural hosts for the rhinoviruses and coronaviruses. Although chimpanzees can be infected with the virus and develop specific antibodies, the infection in these animals does not usually culminate in disease. The lack of a suitable experimental model has hampered significantly the study of the pathogenesis of the common cold. What little is known suggests that the signs and symptoms of the common cold are a 481 combination of viral-mediated and host-mediated damage □. Since the signs and symptoms of the common cold resemble those of allergic rhinitis, it is possible that at least some of the symptoms of the common cold are due to allergic reactions to viral antigens. The virus is known to adhere to and multiply in the mucous membrane of the nose and induce the formation of 417 antibodies. In tissue culture, it also causes cytopathic effects □, thus illustrating the cell-destruction potential of the virus.

At one time it was thought that coughs and sneezes created aerosols that other individuals might inhale and become infected. Recent evidence suggests, however, that person to person—or person to fomite to person—contacts are the primary modes of transmission. For example, viruses in nasal secretions contaminating the infected person's hands can spread by hand contact to other individuals, who then inoculate themselves.

Prevention of the common cold by vaccination is unlikely in the near future, because there are so many antigenically different viruses (>100 different serotypes) that can cause the common cold. Thus, it would be extremely difficult to develop a polyvalent vaccine that would contain all the common antigenic types. The problem is further compounded by the fact

FOCUS

NOSE DROPS TO CURE THE COMMON COLD?

The common cold affects nearly everyone. The multiplication of rhinoviruses (a common cause of colds) on the nasal mucosa causes the characteristic signs and symptoms of the disease. Nasal secretions are an important characteristic of the disease, because they are laden with infectious viruses and serve as the main vehicle of transmission to other individuals.

The traditional treatment of common colds, by taking aspirin, plenty of fluids, and sufficient rest, does little to reduce the nasal secretions and hence the transmissibility of the viruses. Scientists at the University of Virginia and the University of Rochester medical centers apparently have developed a mode of treatment that arrests viral shedding (in the nasal secretions) and viral multiplication. This treatment is effective in reducing the severity of the disease and at the same time eliminating its contagiousness.

The treatment involves the administration of human interferon directly into the nasal cavity. In these studies, the interferon was administered either as a nasal spray or as drops. Either way, the human interferon either prevented the disease altogether or reduced significantly its signs and symptoms. Also, the shedding of infectious viruses in nasal secretions was significantly reduced.

This study represents another example of how physicians and microbiologists can treat an infectious disease more effectively once they know the mode of action of antimicrobials and the way in which microorganisms cause disease.

that common cold viruses induce specific immunity of short duration; therefore, repeated, frequent vaccinations would be needed. Behavioral modification, such as frequent hand-washing, would reduce the chances of becoming infected. There is no useful treatment at the present time, nor is there a preventive measure. The old-fashioned remedy of drinking lots of fluid (including chicken soup), taking aspirin, and bed rest appears to be the most comforting of available treatments.

Croup (Acute Laryngotracheobronchitis)

Croup is an acute infectious disease of children (mostly under the age of 3 years) characterized by coughing, hoarseness, high-pitch sounds, and fever. Severe symptoms may lead to cyanosis (due to apnea), convulsions, and sometimes death. The disease is caused by a variety of viruses, most commonly parainfluenza viruses, respiratory syncytial viruses, and some adenoviruses.

BACTERIAL DISEASES OF THE LOWER RESPIRATORY TRACT

Pneumonia

Pneumonia (table 24-3) is an inflammation of the lungs, with accompanying fluid buildup in the alveolar sacs that can result from infectious or noninfectious processes. Bacterial pneumonia caused by pneumococci, pyogenic cocci, and bacilli appears suddenly and is characterized by high fever, chest-pain, chills, and a purulent (pus-containing) cough. If therapy is not initiated promptly, these infections have a high mortality rate. From 1980–1985 the mortality rate has stabilized at about 23 cases per 100,000 (fig. 24-4); pneumonias still rank sixth among the ten leading causes of death in the United States, higher than any other infectious disease. There are more than 1 mil-

TABLE 24-3
COMMON DISEASES OF THE LOWER RESPIRATORY TRACT

NAME OF DISEASE	CAUSATIVE AGENT(S)	GROUP OF ORGANISMS
Aspergillosis	*Aspergillus* sp.	Fungi
Blastomycosis	*Blastomyces dermatitidis*	Fungi
Coccidioidomycosis	*Coccidioides immitis*	Fungi
Cryptococcosis	*Cryptococcus neoformans*	Fungi
Histoplasmosis	*Histoplasma capsulatum*	Fungi
Influenza	Influenza viruses	Orthomyxoviruses
Hydatid cyst disease	*Echinococcus granulosus*	Cestodes
Paragonomiasis	*Paragonimus westermanni*	Trematodes
Pneumonia	*Streptococcus pneumoniae*	Bacteria (gram + cocci)
(Legionellosis)	*Legionella pneumophila*	Bacteria (gram − rods)
	Klebsiella pneumoniae	Bacteria (gram − rods)
	Streptococcus pyogenes	Bacteria (gram + cocci)
	Haemophilus influenzae	Bacteria (gram − rods)
(Atypical primary)	*Mycoplasma pneumoniae*	Bacteria (wall-less)
	Influenza viruses	Orthomyxoviruses
	Respiratory syncytial	Paramyxoviruses
	Adenovirus	Adenoviruses
	Coccidioides immitis	Fungi
	Histoplasma capsulatum	Fungi
	Blastomyces dermatitidis	Fungi
	Pneumocystis carinii	Protozoa
Pontiac fever	*Legionella* sp.	Bacteria (gram − rods)
Psittacosis	*Chlamydia psittaci*	Chlamydia (gram − rods)
Q Fever	*Coxiella burnetti*	Rickettsia (gram − rod)
Tuberculosis	*Mycobacterium tuberculosis*	Bacteria (acid fast rod)
	Mycobacterium sp.	Bacteria (acid fast rod)
Whooping cough	*Bordetella pertussis*	Bacteria (gram − rods)

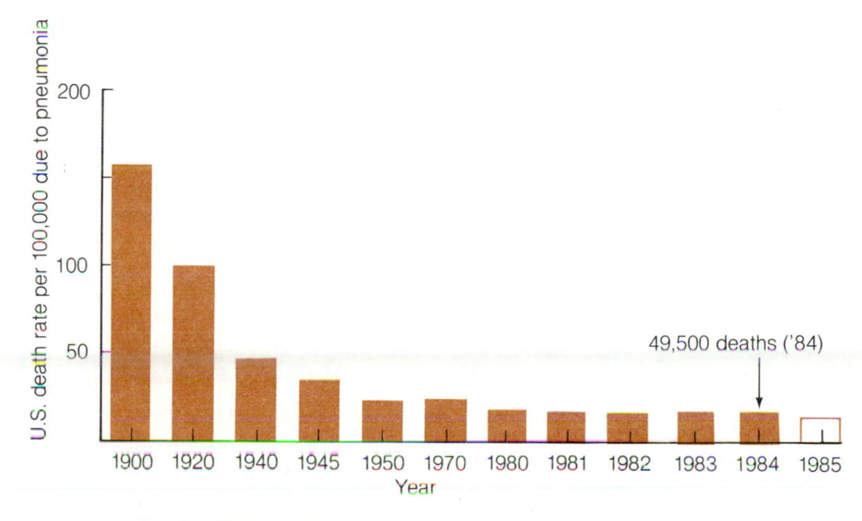

Figure 24-4 Deaths Due to Pneumonia in the United States. Although the number of deaths due to pneumonia has decreased considerably in the U.S. since the 1920s, the disease still ranks among the 10 most common causes of death in the U.S. The decrease in the number of deaths is due primarily to improved patient care and the use of antibiotics. There are about 1,000,000 cases of pneumonia each year in the U.S.

(a) Outer edema zone

Pneumococci

Alveolar wall

(b) Zone of early consolidation

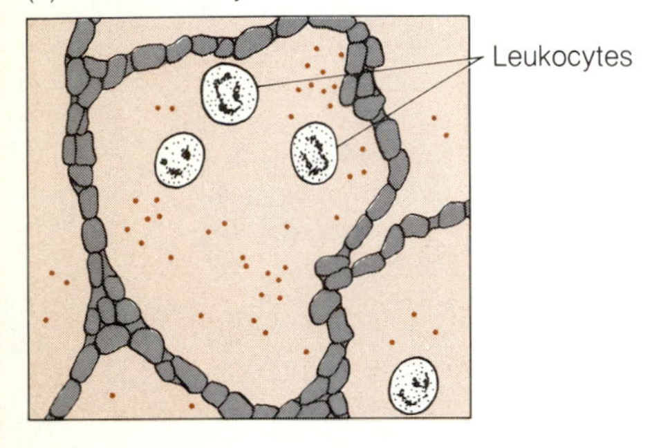

(c) Zone of advanced consolidation

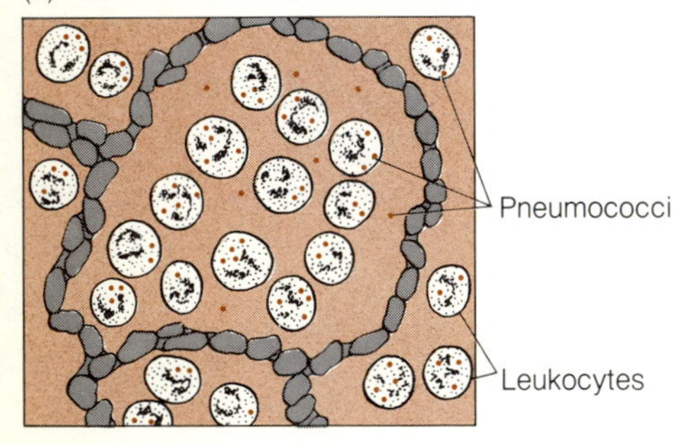

(d) Zone of resolution

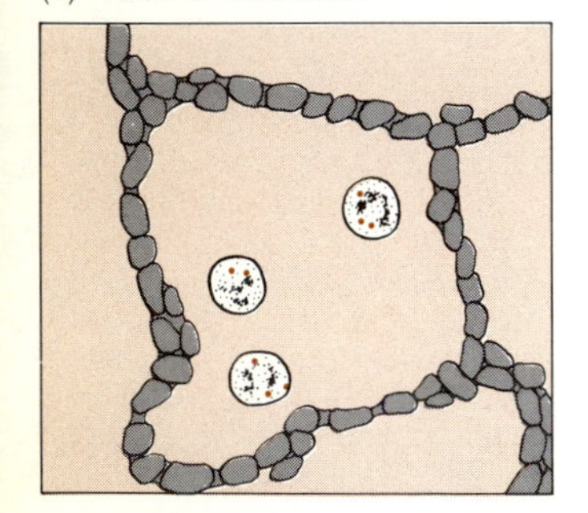

Figure 24-5 Pneumococcal Pneumonia. The pneumonia spreads from the site of infection outward until it involves the entire alveolus and adjacent alveoli. At the peak of infection, the alveoli are filled with leukocytes and fluids due to the inflammatory response that results from the damage caused by the invading pneumococci.

lion cases of pneumonia each year in the United States, and the case fatality rate is about 5–15 percent.

Pneumococcal Pneumonia

Pneumonias caused by *Streptococcus pneumoniae* account for more than 50% of all pneumonias and more than 90% of all bacterial pneumonias. The disease is characterized by an acute onset of fever, chills, dyspnea (difficulty in breathing), pleurisy (inflammation of the pleura), and productive cough. The sputum has a purulent discharge and is often spotted with blood.

Streptococcus pneumoniae is a gram positive coccus, characteristically arranged in lancet-shaped diplococci. Not infrequently, it forms chains. Virulent strains of pneumococci are encapsulated and form smooth colonies. On blood agar, the pneumococcus forms small, mucoid colonies with a central depression and a characteristic alpha hemolysis. The laboratory identification of *S. pneumoniae* is relatively simple: it is based on colony morphology, hemolysis pattern, sensitivity to surfactants such as optochin or bile, and microscopic and staining characteristics.

The normal human host is generally resistant to pneumococcal infections. It has been documented that up to 40% of normal adults carry various strains of pneumococci in their throats. Even when these bacteria gain access to the alveoli, they rarely cause disease in the normal human host, but alcoholics, patients under the influence of anesthesia, and debilitated individuals are highly susceptible. Pneumonia accompanying influenza is sometimes encountered.

Once an infection is initiated in the alveoli, the disease develops along a fairly predictable path (fig. 24-5). The pneumococci multiply extracellularly within the alveoli. Phagocytized pneumococci that are nonencapsulated are readily destroyed by PMN leukocytes, but encapsulated organisms are not. The capsule of *S. pneumoniae* inhibits phagocytosis and therefore is a virulence factor.

The reproducing pneumococci produce a variety of enzymes, including a hemolysin called **pneumolysin,** which may also play a role in the pathogenesis of pneumonia. An inflammatory reaction ensues following pneumococcal multiplication and the release of toxic substances into the alveoli. The edema fluid within the alveoli is rich in nutrients and serves to support further growth of the bacterium. Following the edema period, the area becomes infiltrated with PMN leukocytes. Eventually, the alveoli become crowded with bacteria and PMN leukocytes attempting to destroy the bacteria. Some of the bacteria, however, may escape phagocytosis □ and pass through the thin wall of the alveoli to adjacent areas. The **pleura** (membrane that lines the lungs) may be invaded, causing empyema (pus in the lungs) and pleurisy. In severe cases, the pneumococci may penetrate the walls of capillaries surrounding the alveoli, thus invading the bloodstream. From the blood, the pneumococci may invade other organs of the body, causing meningitis, encephalitis, or myocarditis. Recovery from pneumonia is associated with the development of anticapsular antibodies. These antibodies act as opsonins, enhancing phagocytosis of these bacteria by PMN leukocytes and macrophages.

444

Up to 40% of normal individuals carry *S. pneumoniae* in the nasopharynx (at least during the winter months). Most cases of pneumococcal pneumonia are due to endogenous bacteria originaying in the URT that spread from the upper respiratory tract into the lungs. The remaining cases of pneumococcal pneumonia are due to intimate contact with infected persons. Since *S. pneumoniae* readily dies when it dries out, aerosols and fomites are unimportant in the spread of this bacterium in human populations. A vaccine consisting of capsular material from the 14 most common serotypes (>75% of all the cases) has been developed and is successful in reducing the incidence rate of pneumonia, especially among high-risk individuals.

Other Bacterial Pneumonias

In addition to the pneumococcus, other bacteria, such as *Haemophilus influenzae* and *Klebsiella pneumoniae*, are capable of causing pneumonia (table 24-3). The pneumonias caused by these bacteria are similar to that caused by the pneumococcus.

Haemophilus influenzae pneumonia is rare in adults and most commonly develops in children after a bout of pharyngitis or a viral infection (e.g., influenza). In 1918, during the great influenza epidemic, many cases of pneumonia due to *H. influenzae* were noted. At that time, the viral etiology of influenza was not yet known and it was thought that this bacterium was the cause of the disease (hence, the name *H. influenzae*).

Klebsiella pneumoniae is a gram negative bacterium that can cause primary pneumonia in humans. The pneumonia caused by this opportunistic bacterium is associated with chronic alcoholics. The course of the pneumonia caused by *K. pneumoniae* is very similar to that caused by *S. pneumoniae*, although the disease generally is more destructive. For this reason, the mortality rate for untreated cases can exceed 50 percent.

Pneumonia caused by *Mycoplasma pneumoniae* is a mild disease. The onset of the disease is gradual, with fits (paroxysms) of nonproductive cough. The sputum is mucoid rather than purulent and the fever rarely exceeds 440 102° F. Monocytes and lymphocytes, instead of PMN leukocytes □, are seen in the lung lesions. These lesions are restricted to the alveoli, and the bacteria rarely invade the pleura. This type of pneumonia is sometimes called primary atypical pneumonia, and in many ways it resembles those caused by the viruses.

The isolation and identification of the organism causing pneumonia are essential for the proper management of the disease. This is because the proper choice of antibiotic depends upon the causative agent. Pneumococcal pneumonia can be treated successfully with penicillin, but pneumonias caused by gram negative bacteria cannot. Pneumonias caused by *H. influenzae* respond well to treatment with chloroamphenicol or ampicillin, while those caused by *K. pneumoniae* are best treated with gentamycin and tetracycline. Mycoplasmal pneumonias cannot be treated with penicillin, since they lack a cell wall, and are best treated with erythromycin.

Legionellosis

Legionellosis, or Legionnaires' disease, is a form of pneumonia characterized by an acute onset of high fever, bed-shaking chills, pleuritic pain, a nonproductive cough, and accelerated breathing. The disease is usually heralded by a mild headache, achy muscles, and malaise. This pneumonia is unusual in that leukocytosis (elevated white cell count) rarely exceeds 16,000 leukocytes per square mm, and an average of 10,000-12,000 is seen. Shock is a common complication in fatal cases.

The cause of legionellosis is a group of fastidious (nutritionally-demanding) gram negative bacteria in the family Legionellaceae, of which *Legionella pneumophila* is the best known member.

Epidemiological studies suggest that *L. pneumophila* is an aquatic microorganism found in natural and artificial bodies of water. The agent appears to be transmitted by aerosols, such as those created by ventilation systems that are cooled by contaminated waters in cooling towers.

Legionellosis can be diagnosed by means of an indirect fluorescent antibody test □, which detects the presence of anti-legionellae antibodies in the patient's serum. Since the implementation of serological procedures to detect legionellosis, an average of 600 cases per year in the United States has been recorded. In 1984, 651 cases were reported.

521

The shotgun approach to chemotherapy during the 1976 epidemic □ of legionellosis revealed that the highest recovery rate was seen in patients treated with erythromycin or tetracycline. To date, these two antibiotics are still the drugs of choice for the treatment of legionellosis.

532

Whooping Cough

Whooping cough is a highly infectious disease, characterized by fits of coughing with a "whooping" sound and an incubation period of about 10 days (5–21 days). During the prodromal period (before the characteristic signs and symptoms appear), which lasts about a week or two, the patient has coldlike symptoms, after which the characteristic cough appears. Each fit of coughing may last 10–20 seconds. The shortness of breath due to prolonged coughing forces the patient to inhale air forcibly, making the whooping noise as the air rushes in through the epiglottis.

The cause of whooping cough is *Bordetella pertussis*, a small, gram negative coccobacillus. This bacterium produces a capsule that is partly responsible for its virulence. *Bordetella pertussis* also produces an exotoxin called the **Pertussis toxin.** This toxin affects the adenylate cyclase system, inducing the synthesis of cAMP by the cell. Altered levels of cAMP affect adversely the coordination of cellular activities, and ultimately lead to cell death. It appears that this toxin is at least partly responsible for the signs and symptoms of whooping cough.

The infection begins when inhaled *B. pertussis* attaches to the ciliated epithelium of the bronchi (fig. 24-6). This attachment may be mediated by pili □. At the site of attachment, the bacteria multiply, causing the prodromal signs and symptoms. As the bacteria multiply locally, they elaborate their toxin(s). During the period of paroxysmal cough, there is a marked lymphocytosis (increase in the number of lymphocytes in the blood), subepithelial necrosis, and inflammation of the bronchi. These signs may be due to the actions of HSF or LPF.

Humans, the only reservoir of *B. pertussis*, transmit the disease by aerosols created during coughing paroxysms. The disease is worldwide in distribution, with no apparent seasonal variation. Since the implementation of vaccination (DPT shots) □, the incidence of this disease in the United States has dropped to an average of fewer than 3,000 cases/year and a mortality rate of <1 percent.

550

Bordetella pertussis is sensitive to erythromycin, the drug of choice. An understanding of the disease process is essential for its proper management, since the antibiotics kill the bacteria but do not inactivate the toxins already

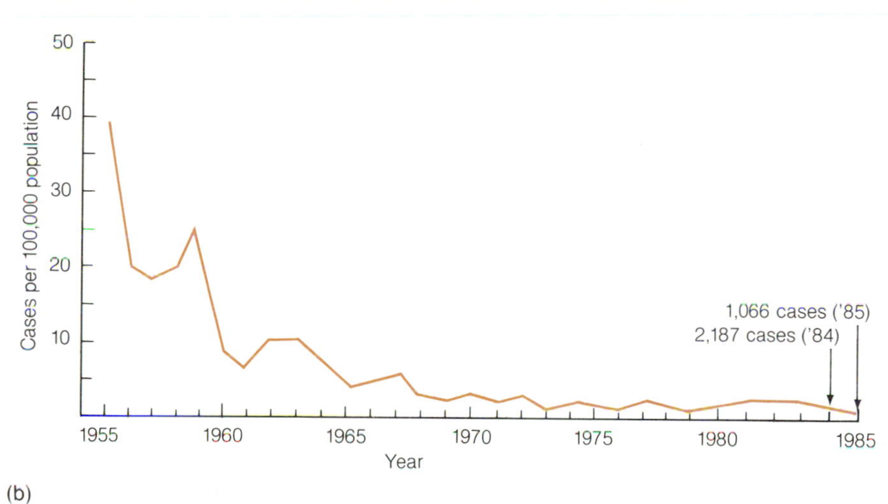

Figure 24-6 Characteristics of Whooping Cough. *(a) Bordetella pertussis* reproducing amidst ciliated epithelium. The bacteria multiply locally, releasing toxic substances that cause the characteristic disease. *(b)* Cases of pertussis in the United States. Since the 1970s the rate has been between 1–3 cases/100,000 population (2,000–6,000 cases/yr in the U.S.). The sharp reduction in cases since the early 1950s is due largely to the compulsory use of the DPT vaccine.

present. Therefore, prompt therapy reduces the severity of the disease but does not eliminate the signs and symptoms entirely, since any exotoxin present, no matter how little, has some effect on the host's tissue. Administration of **human pertussis immune serum globulin** can offset the effects of the toxins.

Tuberculosis

Tuberculosis is a chronic infectious disease of the lower respiratory tract. In its most common form, the disease is characterized by a chronic cough, low-grade fever, and malaise. The long-lasting nature of this syndrome can lead to prostration. Patients with tuberculosis are constantly shedding infectious agents in aerosols. The disease, caused by *Mycobacterium tuberculosis*, remains the #1 killer disease in the world. There are an estimated 1.5 billion infected individuals throughout the world; of those, 20 million are shedding the bacteria, causing 3–5 million new cases annually. The mortality worldwide from tuberculosis is about 600,000 deaths/year; however, countries with a high standard of living have a relatively low incidence of the disease. The morbidity in the United States is about 30,000 cases per year, with approximately 10% mortality (fig. 24-7). Humans are capable of mounting a significant and effective immune response against the highly infectious **tubercle bacillus**. Unfortunately, however, the host's immune response is responsible for much of the tissue destruction associated with some forms of tuberculosis.

Mycobacterium tuberculosis is a slender, acid-fast rod containing a high concentration of lipids in its wall. The bacterium is an obligate aerobe, and in host tissues it multiplies inside the cells (intracellular parasite). Some strains have a tendency to stick together and form colonies, showing a **serpentine growth** on artificial media. The tubercle bacillus does not produce aggressins □ or toxins, and its pathogenicity resides in its ability to induce a strong immune response and to avert these responses by growing inside mononuclear phagocytes.

434

The transmission of tuberculosis is primarily through aerosols created by persons with the active disease. During the initial stages of tuberculosis,

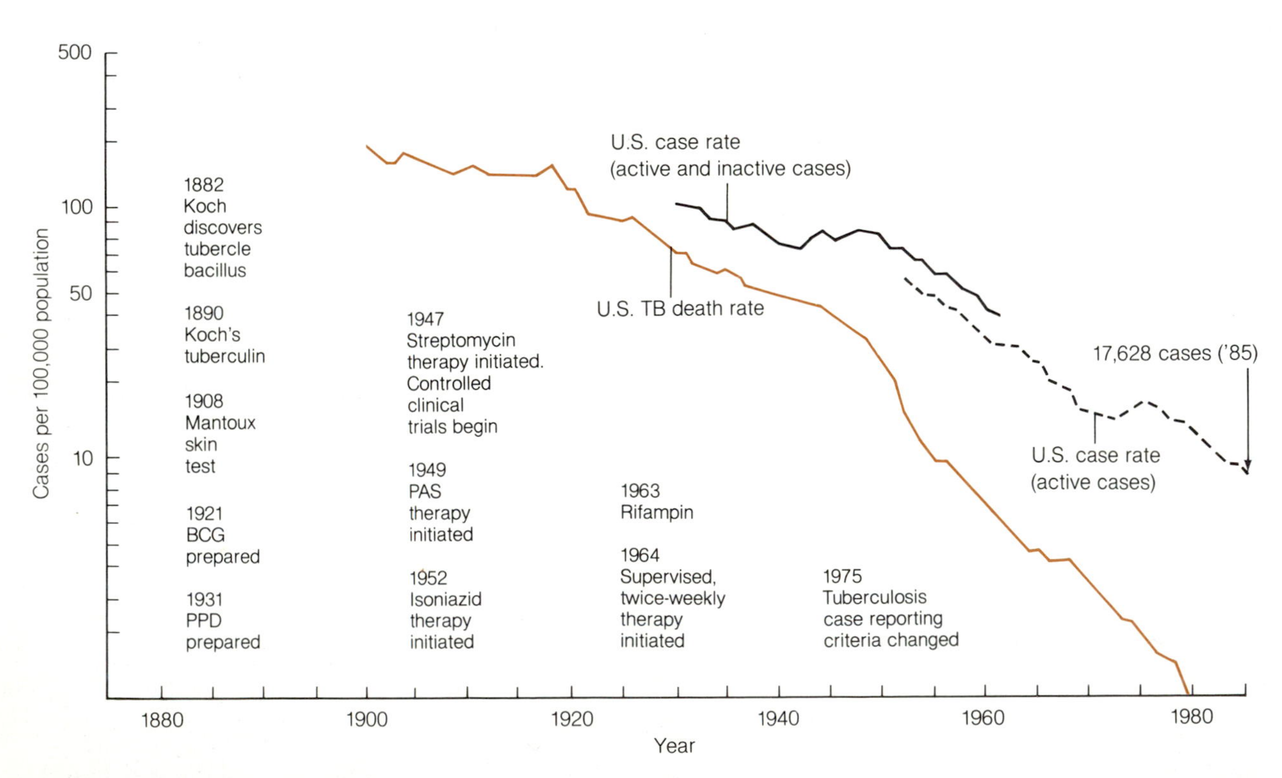

Figure 24-7 Tuberculosis in the United States. Graph illustrates that tuberculosis is still prevalent in the United States, although the number of cases/year is declining. The salient feature of the graph is that the death rate due to tuberculosis is declining at an encouraging rate. The rapid decline in the death rate began around 1947, when streptomycin therapy was initiated. Landmark events in the management of tuberculosis are indicated.

the tubercle bacillus multiplies inside alveolar macrophages in the lungs. After this initial period of multiplication, the microorganisms enter the lymphatics and are transported to the regional lymph nodes, where they continue to multiply intracellularly. After a period of multiplication, the bacteria 685 exit the lymph nodes and enter the circulation through the thoracic duct □. From there, they spread to other organs, including the lungs. At this stage, the infected host may not show any signs or symptoms. If symptoms are evident, they are due to inflammatory responses to tissue damage during bacterial multiplication. If the lung is the site of the primary infection, there may be a mild, transient pneumonia. Up to this point, the organism has been virtually unchallenged by the host. Now, the host develops a cell-mediated immunity that is tuberculocidal, and the bacilli are attacked by T-lymphocytes and activated macrophages. The cell-mediated immunity promptly and efficiently controls the spreading infection.

Soon after cell-mediated immunity is detected, the host responds to the infection by walling off the pathogen from the rest of the body within multinucleated giant cells surrounded by lymphocytes and fibroblasts. These structures, which can be seen on X-ray examinations, are called **tubercles** (fig. 24-8). The tubercle bacilli become localized in foci, primarily in the

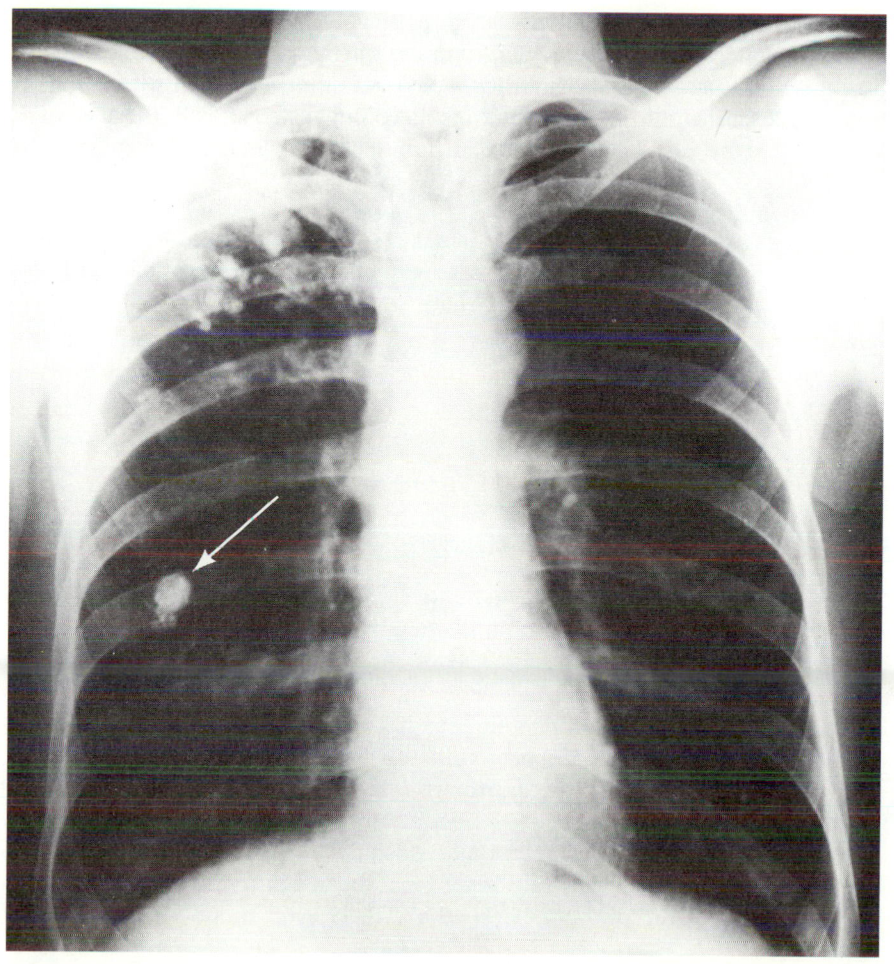

(a)

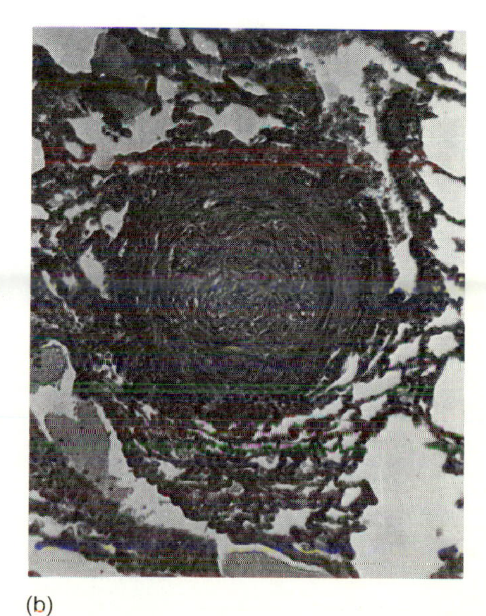

(b)

Figure 24-8 Clinical Manifestations of Tuberculosis. *(a)* X-ray showing a large tubercle (arrow) and several smaller ones in the right lung. The left lung appears normal. *(b)* Histological section of a tubercle stained with hematoxylin and eosin. The center of the tubercle contains *Mycobacterium tuberclosis* and epithelioid cells.

lungs and in other organs where the oxygen concentration is high (spleen and liver). Dormant bacteria are able to remain viable in the tubercles for many years. The tubercles may eventually undergo **caseation necrosis**, acquiring a cheesy consistency, and may ultimately become scarred or calcified. These structures are called **Ghon complexes** (fig. 24-8).

Reactivation tuberculosis is the clinical manifestation that most people associate with tuberculosis. It occurs in individuals who have cell-mediated immunity to *M. tuberculosis* antigens and who suffer a temporary lapse in their immunity or are reinfected. In either case, the bacteria in the tubercles begin to multiply, stimulating the anti–*M. tuberculosis* immune response in the host. This response causes caseation necrosis of the tubercles. This partial destruction of the tubercles provides a suitable environment for the bacteria to multiply, and this multiplication further stimulates the immune response. As the reactivation disease progresses, the lesion(s) enlarge and eventually erode through the bronchi, which then fill with the liquid caseum. The fluid in the bronchi, laden with infectious mycobacteria, induces the characteristic cough. The ruptured tubercles provide a cavity in which the bacilli continue to multiply. Tubercles that develop in the spleen, liver, or other organs can also undergo similar caseation necrosis, thus damaging the organ.

Vaccination of human volunteers with **BCG** (Bacillus Calmette-Guerin) has reduced the incidence of tuberculosis by as much as 80% in some developing countries. Presumably, BCG induces a strong cell-mediated immune response and promotes the activation of macrophages, which play a central role in the control of the disease. It has been suggested that BCG vaccination be initiated in endemic areas in order to reduce the number of cases of tuberculosis.

Patients with positive skin tests generally are treated with isoniazid (INH) daily for up to one year to prevent reactivation disease. Active infections are treated with a combination of agents such as INH, ethambutol, rifampin, streptomycin, and para-amino salicylic acid (PAS), because of the possibility that the tuberculosis has been caused by drug-resistant bacteria. In active cases, prolonged bed rest and wholesome nutrition helps expedite recovery.

VIRAL DISEASES OF THE LOWER RESPIRATORY TRACT

Viral Pneumonias

Viral pneumonia (fig. 24-9) is an acute systemic disease that can be caused by a variety of viruses, including influenza, adeno-, and respiratory syncytial viruses (table 24-3). The disease occurs most often in adults with chronic lung and heart diseases, or as a complication of other infections such as influenza or chickenpox. After influenza-like symptoms, the patient develops a fever (about 102°F), experiences difficulty in breathing, and has a cough. The sputum is usually frothy and tinged with blood. The severity of the disease varies, depending upon the individual infected and the virulence of the virus. A diagnosis of viral pneumonia is usually made after attempts to isolate and identify bacterial and fungal pathogens have failed. It has been estimated, using the above criteria, that 15–30% of all pneumonias are of viral origin.

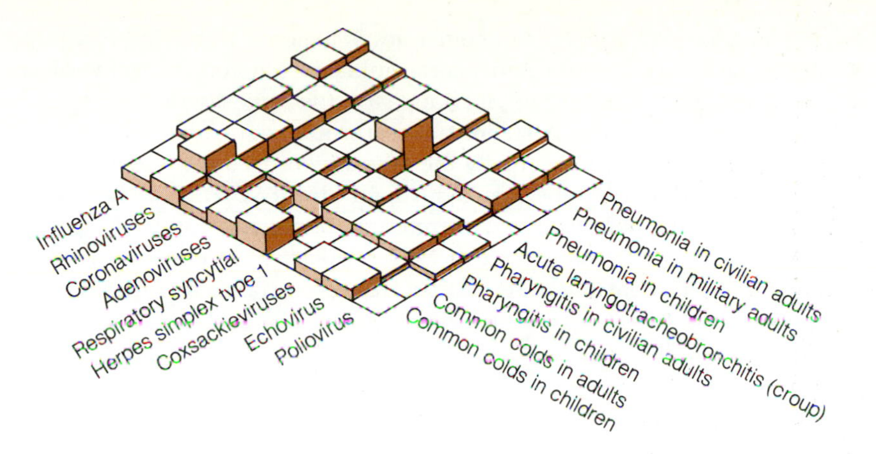

Figure 24-9 Respiratory Diseases Caused by Viruses. The bar heights indicate the relative frequency of a virus causing a specific disease.

Influenza

Influenza (grippe), or "flu," is an acute infection of the lower respiratory tract characterized by a sudden onset of fever, chills, myalgia (muscle ache), and headache. Coldlike symptoms, such as coryza (nasal inflammation with profuse discharge), sore throat, and cough, are also common. Generally speaking, the disease is benign, self-limiting, and of short duration. On occasion, "stomach flu" symptoms may appear: loss of appetite, diarrhea, and vomiting. **Reye's syndrome,** a degenerative disease of the CNS, has been implicated with certain strains of influenza and parainfluenza viruses.

Influenza is caused by a group of RNA viruses called the orthomyxoviruses. The surface of the virion contains two major components: **hemagglutinin** and **neuraminidase receptors** □. These components play very important roles in the pathogenensis and epidemiology of the disease.

413

The virus attaches to epithelial cells of the respiratory tree via the hemagglutinin receptors on its surface. Local multiplication of the virus results in the destruction of the ciliated epithelium. The virus spreads to adjacent sites and eventually causes considerable denudation of the trachea and bronchi; irritation of the airways probably induces the cough. After a period of local multiplication, the virus may spread hematogenously (via the blood). The viremia (virus in the blood) may be responsible for the fever, myalgia, and malaise associated with the illness. Viruses are shed in the cough during the early stages of the clinical disease and disappear shortly after the signs and symptoms disappear. The recovery of the patient is due to a rise in humoral immunity cell mediated immunity, and interferon activity.

Influenza is transmitted from person to person by virus-laden aerosols created by coughing and sneezing. High incidence rates occur in institutions that harbor individuals in close proximity. Such outbreaks occur in army barracks, school classrooms, and hospitals. High attack rates are more common among children than adults, perhaps because children are immunologically naive with respect to the virus. Adults may have local IgA antibodies against influenza. Influenza epidemics □ are common during the winter months, possibly because susceptible individuals spend more time indoors and in close proximity with infectious individuals.

411

The periodicity of influenza epidemics is believed to be due to antigenic shifts in the hemagglutinin and neuraminidase receptors of the viral coat. When a new type of influenza virus appears, the majority of the population is susceptible to the virus. After infection, immunity is developed against the new virus antigens, and eventually the population becomes resistant to that virus. When a new HN type appears, much of the population has no antibodies against the new type and therefore is again susceptible to the influenza. Treatment of influenza is generally supportive. In cases in which secondary bacterial infection is suspected or expected, antibiotics are administered.

FUNGAL DISEASES OF THE LOWER RESPIRATORY TRACT

Coccidioidomycosis (Valley Fever)

Coccidioidomycosis is a fungal infection of the lungs. Many of the infections (about 60%) are asymptomatic (no signs or symptoms) and are detected by indirect means such as routine X-rays, skin testing □, or serology. Approximately 35% of the people who become infected have symptoms that are quite similar to (and often mistaken for) influenza. Sometimes, during the later stages of the disease, rashes or rheumatism-like symptoms appear. In some cases, it may take many months for full recovery. The remaining 5% or so who become infected develop chronic pulmonary disease, which may disseminate to the joints, bones, central nervous system, and skin. Chronic disease is characterized by a low-grade fever, anorexia, weakness, a productive cough, and occasionally hemoptysis (spitting of blood). Since coccidioidomycosis is not a reportable disease in all states, it is difficult to estimate how many cases there are per year, but it is estimated that between 500 and 1,000 new cases occur per year in endemic areas with a mortality rate of 1–5 percent.

The cause of valley fever (coccidioidomycosis) is *Coccidioides immitis*, a dimorphic □ fungus that is commonly found in the soil in a filamentous form (fig. 24-10). The characteristic filamentous form consists of septate hyphae, which may break up to form very small, barrel-shaped arthroconidia that readily become airborne. Upon inhalation by a susceptible host, or in special culture conditions, the arthrospores develop into characteristic structures called **endosporulating spherules** (fig. 24-10). When mature, these are saclike structures filled with endospores. Each endospore, when released from the spherule, may become another spherule or, if released into the soil, may give rise to the filamentous form. The complex morphogenetic cycle of this fungus may be of significance in the pathogenesis of the disease. The constantly changing surface components may serve as an evasive mechanism to avert the host immune responses. In addition, the various stages of the life cycle of *C. immitis* produce soluble components that may participate in the disease process as well.

336

Nearly everyone who comes in contact with *Coccidioides immitis* becomes infected, yet 60% of those who become infected show no symptoms and 35% have flulike symptoms and nothing more. These facts suggest that humans are highly resistant to serious infections by this fungus. Studies have revealed that the activated macrophages play a central role in protecting against dissemination. Most of the patients who exhibit the severe form of

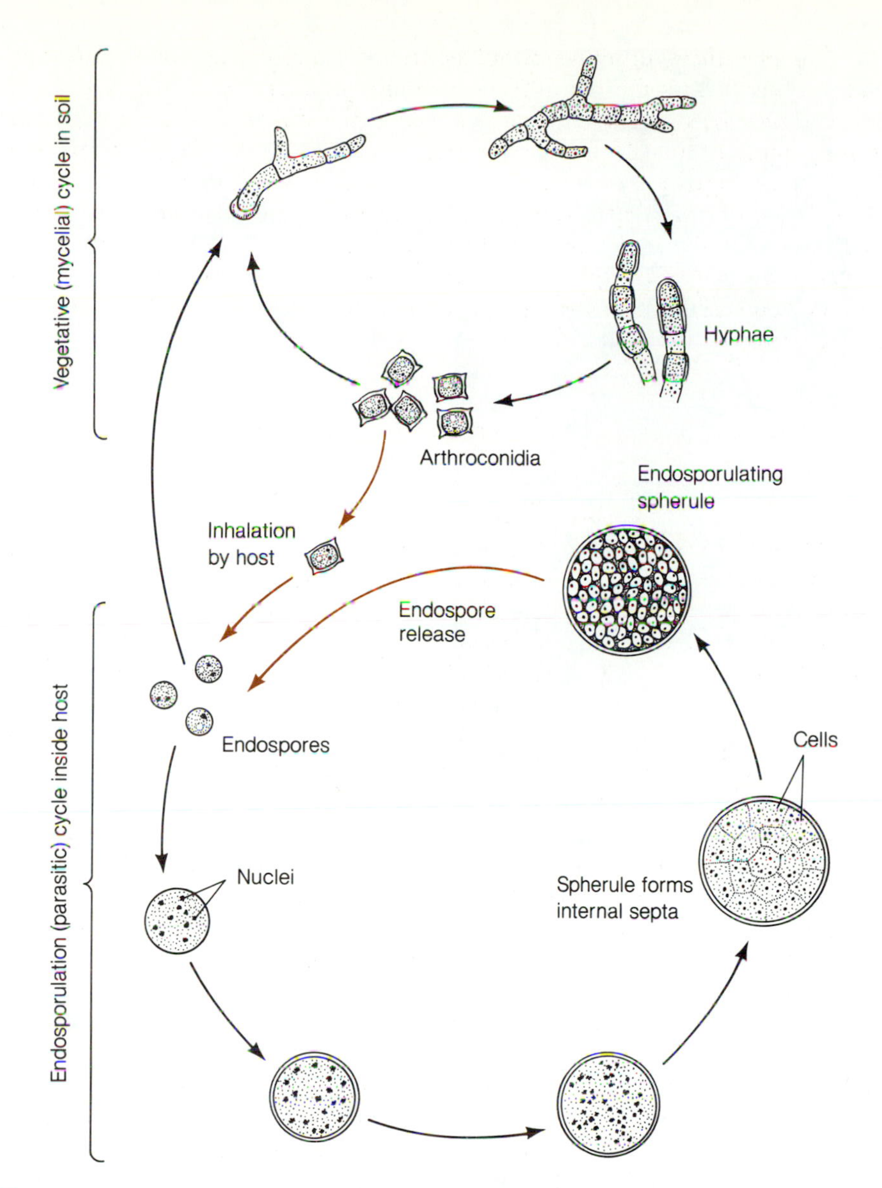

Figure 24-10 Life Cycle of *Coccidioides immitis*. *Coccidioides immitis* has two life cycles: a vegetative cycle in soils (mycelial form) and a parsitic cycle in susceptible mammalian hosts (endosporulating spherule form). Human infection occurs when arthroconidia of the fungus are inhaled. Inside the host's lung, the arthroconidia develop into endospores and eventually into endosporulating spherules. The endosporulating cycle is perpetuated inside the host. When endospores are released into the external environment, they develop into vegetative hyphae and the vegetative cycle takes place.

the disease have either a temporary or a long- lasting depression of their CMI.

Arthroconidia adhering to lung tissue initiate an inflammatory response that may be responsible for the flulike symptoms of the early stages of the disease. Although polymorphonuclear leukocytes migrate rapidly to the site of infection, they are unable to destroy all the arthroconidia. The surviving arthroconidia rapidly differentiate into endosporulating spherules, which

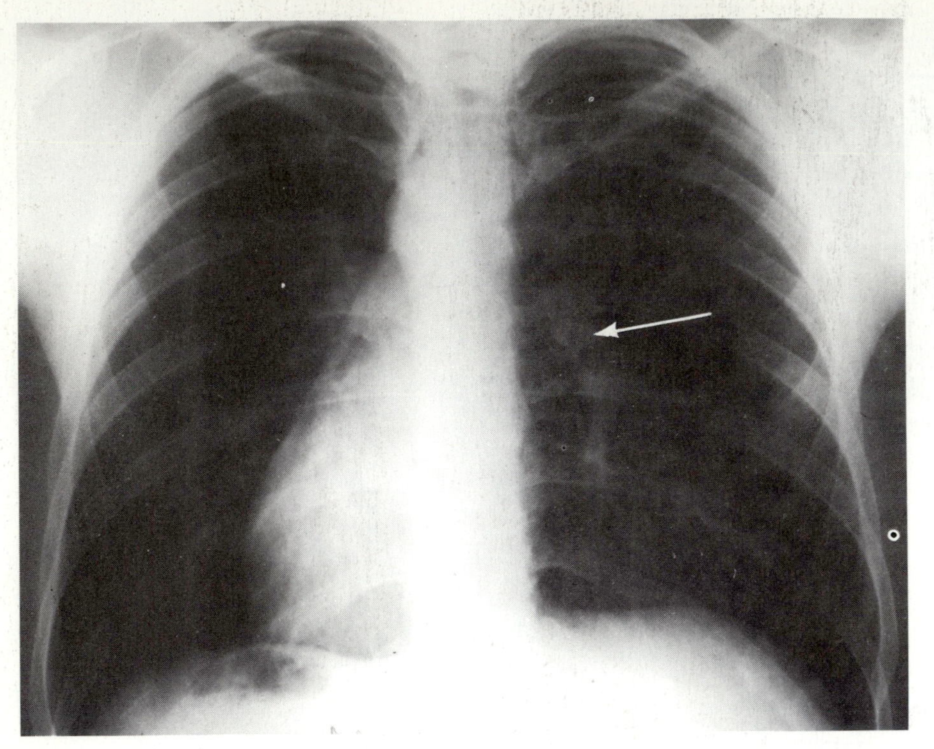

(a)

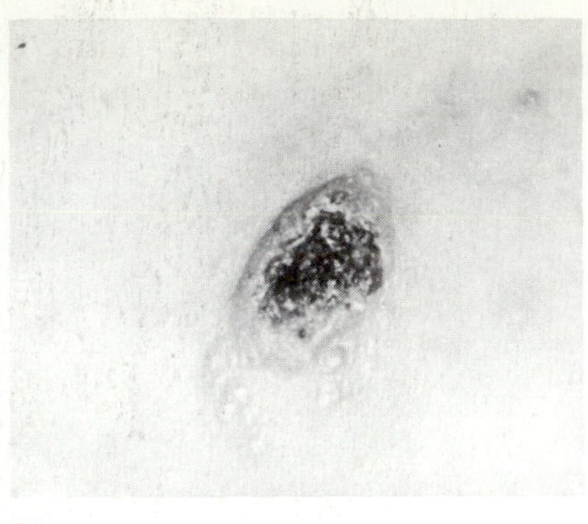

(b)

Figure 24-11 Characteristics of Coccidioidomycosis. *(a)* X-ray of patient with coccidioidomycosis, illustrating a typical "coin" lesion (arrow) in right lung. *(b)* Cutaneous manifestations of disseminated coccidioidomycosis.

eventually rupture and release numerous endospores. This event, in turn, induces the migration of yet more PMN leukocytes to the foci of infection. Some of the endospores are killed by the PMN leukocytes. Eventually, the PMN leukocytes are replaced by monocytes and macrophages. Endospores are transported to the regional lymph nodes, where they stimulate the development of cell-mediated immunity. This development leads to a chronic inflammation in the affected portions of the lung, characterized by the abundance of lymphocytes and macrophages. Once cell mediated immunity (CMI) has developed, the infection is checked and recovery usually follows. The lesions eventually heal and are reabsorbed by the body. The formation of cavities with caseation necrosis, similar to those seen in tuberculosis, are sometimes observed in valley fever. "Desert bumps," rashes, and rheumatism that sometimes accompany valley fever are believed to be due to allergic reactions to circulating fungal antigens.

Coccidioides immitis (fig. 24-11) is endemic to the southwestern United States and arid regions of Mexico and California. These areas are characterized by arid or semiarid conditions, hot summers, and mild winters, with rainfall between 10 and 20 inches annually. The infectious agent is transmitted by air currents carrying dust and arthroconidia of *C. immitis*. On windy days, large numbers of arthroconidia become airborne along with dust particles and can lead to serious consequences for animals and humans. In December 1977, a severe dust storm blowing from the southeast carried arthroconidia to northern and central coastal California. An increase of coccidioidomycosis was noted in areas not usually colonized by the fungus. Coccidioidomycosis in humans and animals (including a sea otter) was reported in California coastal communities. Presumably, burrowing mammals such as gophers, ground squirrels, and kit foxes may also serve as reservoirs for the infectious agent in nature.

Most cases of coccidioidomycosis recover without antifungal treatment; however, 60–70% of all cases of untreated, disseminated coccidioidomycosis are fatal. Hence, treatment with amphotericin B is required to reduce the mortality rate. The imidazole antifungal agents, such as miconazole and ketoconazole, have been used in clinical trials and appear to be as effective as amphotericin B but less toxic.

Histoplasmosis

Histoplasmosis is a systemic disease characterized by infection of cells of the reticuloendothelial system by the fungus *Histoplasma capsulatum*. The clinical outcome of the host-parasite relationship is quite varied. More than 75% of those who become infected show either no symptoms or those of influenza. The other 25% show various degrees of severity in their illness. The most common syndrome is a pneumonia of short duration. In these patients, a nonproductive cough, pleurisy, dyspnea (difficulty in breathing), myalgia (muscle ache), arthralgia (joint pain), fever, and night sweats are common symptoms.

The mycelial form of *H. capsulatum* grows in soils, especially those enriched with bird droppings or bat guano. On laboratory culture media such as Sabouraud glucose agar at room temperature, *H. capsulatum* forms characteristic macro- and microconidia. When the microconidia or other fungal elements are inhaled into the lungs, they develop into yeast cells, facultative intracellular parasites that multiply inside macrophages (fig. 24-12). The fungi are transported in the macrophages to the lymph nodes, where they multiply further. Many escape the lymph nodes, however, and enter the circulatory system, disseminating throughout the body. Within a week or so, the host develops a CMI that may involve fungicidal T-lymphocytes and activated macrophages. The development of CMI is paralleled by a chronic inflammation around foci of infection. The fungi are quickly destroyed, and the lesions eventually disappear or become calcified. On occasion, cavities develop, containing dormant fungi that may become reactivated at a later date. Reactivation histoplasmosis resembles tuberculosis.

More than 500,000 Americans become infected every year with *H. capsulatum*, most of them in endemic areas: moist, humid regions such as the Mississippi and Missouri river valleys. The infectious stage is found in soils that have been enriched with chicken, bird, or bat feces. In urban environments, where pigeons abound, *H. capsulatum* can be found in bird droppings. Thus, persons feeding pigeons or disturbing their roosts are in danger of contracting the disease. Construction and demolition workers who disturb contaminated soils, creating dust clouds, are also at risk. Airborne fungi can enter buildings through windows or air-conditioning systems and infect per-

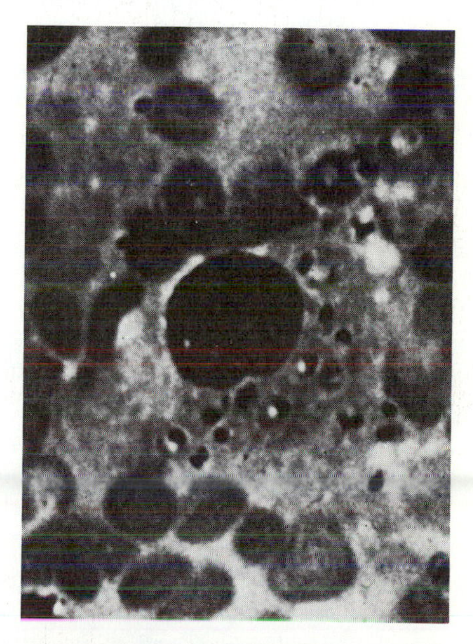

Figure 24-12 ***Histoplasma capsulatum* in histiocyte.** Hematoxylin and eosin stain of lesion, showing histiocyte filled with *Histoplasma capsulatum.*

sons inside. Severe cases of histoplasmosis are treated with amphotericin B. Miconazole and ketoconazole are being used experimentally to treat the severe infections, with encouraging results.

SUMMARY

STRUCTURE AND FUNCTION OF THE RESPIRATORY TRACT

1. The respiratory tract, consisting of the upper and lower tracts, brings air into contact with blood vessels, where an exchange of oxygen and carbon dioxide takes place.
2. Because the potential for infection via the respiratory tract is great, mammals have evolved an array of host defense mechanisms designed to abort infections.

THE NORMAL FLORA OF THE UPPER RESPIRATORY TRACT

1. The upper respiratory tract is richly colonized with microorganisms such as bacteria and fungi. Certain of these colonizers are potential pathogens that can initiate an infection if conditions are conducive to their extensive growth.
2. The normal flora protects the host from infections by inhibiting potential pathogens.
3. The lower respiratory tract is generally free of microorganisms, and microorganisms entering the lungs have a great potential for causing disease.

BACTERIAL DISEASES OF THE UPPER RESPIRATORY TRACT

1. Pharyngotonsillitis or "strep throat" is an infectious disease caused most commonly by Group A streptococci. The disease is characterized by a sore throat, malaise, and fever, which result from an acute inflammation due to the destruction caused by the reproducing bacterium.
2. Untreated streptococcal infections may give rise to autoimmune diseases, such as rheumatic fever and acute glomerulonephritis.
3. Diphtheria is a serious infectious disease characterized by fever, headache, malaise, sore throat, and the development of a pseudomembrane in the throat. It is caused by *Corynebacterium diphtheriae*, which produces a toxin called diphtherotoxin. The toxin, as well as the damage caused by the reproducing bacterium, cause the characteristic signs and symptoms of diphtheria.
4. Sinusitis is an inflammation of the sinuses as a result of hampered drainage. It is generally caused by *Haemophilus influenzae*, *Streptococcus pneumoniae*, and Group A streptococci.
5. Acute epiglottitis is a serious inflammation of the epiglottis as a consequence of microbial growth. The onset of the disease is rapid, and it can lead to death due to asphyxiation. It is most commonly caused by *Haemophilus influenzae*.

VIRAL DISEASES OF THE UPPER RESPIRATORY TRACT

1. The common cold is a mild illness characterized by headache, malaise, stuffy nose, sore throat, and nasal discharge.
2. The disease is caused by a variety of different viruses, including rhino-, corona-, myxo-, coxsackie-, respiratory syncytial, and adenoviruses.
3. The disease results from both allergic reactions to viral antigens and tissue destruction caused by the reproducing virus.
4. Transmission of the common cold is generally via person to person contact or contact with objects contaminated with nasal secretions of infected individuals.
5. Viruses such as the adeno- and respiratory syncytial viruses cause a variety of other diseases of the upper respiratory tract. One such disease is croup, which leads to difficulty in breathing and can have serious complications.

BACTERIAL DISEASES OF THE LOWER RESPIRATORY TRACT

1. Pneumonia is an inflammation of the lungs, accompanied by a buildup of fluids, and ranks in the top ten of the killer diseases in the United States.
2. *Streptococcus pneumoniae* is the most common cause of bacterial pneumonia. It causes a fulminant pneumonia characterized by a rapid onset of high fever, chills, dyspnea, and productive cough.
3. Most cases of pneumococcal pneumonia are caused

by endogenous pneumococci that spread from the upper respiratory tract.

4. Other bacteria such as *Haemophilus influenzae*, *Klebsiella pneumoniae*, and *Mycoplasma pneumoniae* cause pneumonias as well.
5. Legionellosis is a form of pneumonia characterized by a rapid onset of fever, bed-shaking chills, pleuritic pain, nonproductive cough, and accelerated breathing. It is caused by several species in the genus *Legionella*. The bacteria enter the lower respiratory tract via aerosols.
6. Whooping cough is a highly infectious disease characterized by fits of coughing. It is caused by *Bordetella pertussis*. The disease is caused by exotoxins produced by the bacteria while attached to the ciliated epithelium of the respiratory tract.
7. Tuberculosis is a chronic infectious disease caused by *Mycobacterium tuberculosis*. It is characterized by a chronic cough, low-grade fever, and malaise. The infectious bacteria are constantly being shed in the cough of patients.
8. The tubercle bacillus multiplies inside cells of the reticuloendothelial system and induces a strong immune response.
9. The immune response toward the tubercle bacillus is responsible for the control of the infection. Upon reinfection, however, it is the immune response that causes the characteristic signs and symptoms of the disease.

VIRAL DISEASES OF THE LOWER RESPIRATORY TRACT

1. Viral pneumonia is an acute infection of the lungs caused by a variety of viruses, including influenza, adeno-, and respiratory syncytial viruses.
2. Influenza (grippe) is an acute infection of the lower respiratory tract, resulting in a rapid onset of fever, chill, achy muscles, and coldlike symptoms.
3. Influenza is caused by orthomyxoviruses, primarily Influenza A and B. The virion has hemagglutinin (H) and neuraminidase (N) receptors that promote the pathogenensis of the viruses.
4. Influenza is transmitted from person to person via droplet nuclei.
5. The periodicity of influenza epidemics is believed to be due to antigenic shifts in the H and N receptors of the viral membrane envelope.

FUNGAL DISEASES OF THE LOWER RESPIRATORY TRACT

1. Coccidioidomycosis is a fungal infection of the lungs caused by *Coccidioides immitis*. The disease is generally mild with flulike symptoms, although at times the fungus may spread from the lungs, causing serious and sometimes fatal disease.
2. *Coccidioides immitis* is a dimorphic fungus that enters the lungs as arthrospores via inhalation of dust particles. The arthrospores develop into endosporulating spherules.
3. Histoplasmosis is a fungal infection caused by *Histoplasma capsulatum*, a dimorphic fungus that invades cells of the reticuloendothelial system and multiplies intracellularly as a yeast.
4. *Histoplasma capsulatum* enters the lungs via aerosols and invades phagocytic cells. It spreads throughout the body inside these cells and causes a disease similar to tuberculosis.

STUDY QUESTIONS

1. Construct a table containing the causative agent, the symptoms, and the mode of transmission of:
   a. five diseases of the upper respiratory tract;
   b. five diseases of the lower respiratory tract.
2. Most infectious diseases of the respiratory tract are characterized by a cough. Of what importance is this clinical manifestation to the host? to the pathogen?
3. Given the following signs or symptoms of respiratory tract disease, explain the host/parasite interaction that may have taken place and resulted in the clinical manifestation.
   a. cough;
   b. malaise;
   c. fever;
   d. pneumonia;
   e. scarlet-red throat;
   f. chills.
4. Briefly describe several methods that may be useful in reducing the incidence of:
   a. influenza;
   b. pneumococcal pneumonia;
   c. diphtheria;
   d. tuberculosis;
   e. common colds.
5. Why are the common cold and influenza more common during the winter months?
6. Briefly discuss the cause of rheumatic fever.
7. Discuss the value of vaccination against the common cold and influenza. Is it feasible? why?
8. Zygomycosis occurs primarily in individuals who are somehow compromised by an underlying condition. What does that fact indicate about the pathogenic potential of these organisms? Would you predict that zygomycosis is very common in human populations?

SUPPLEMENTAL READINGS

Fraser, D. W. and McDade, J. E. 1979. Legionellosis. *Scientific American* 241:82–99.

Gwaltney, J. M., Jr., Moskalski, P. B., and Hendley, J. O. 1978. Hand-to-hand transmission of rhinovirus colds. *Annals of Internal Medicine* 88:463–466.

Kaplan, M. M. and Webster, R. G. 1977. The epidemiology of influenza. *Scientific American* 237:88–106.

Merigan, T. C. 1984. Respiratory viral infections of adults. *Scientific American Medicine* Section 7-XXV. New York: Freeman.

Middlebrook, J. L. and Dorland, R. B. 1984. Bacterial toxins: cellular mechanisms of action. *Microbiological Reviews* 48(3):199–221.

Simon, H. B. 1984. Mycobacteria. *Scientific American Medicine* Section 7-VIII New York: Freeman.

Stuart-Harris, C. 1981. The epidemiology and prevention of influenza. *American Scientist* 69:166–172.

Turck, M. 1985. An AIDS patient who died too soon. *Hospital Practice* 20(1):77–80.

CHAPTER 25
DISEASES OF THE DIGESTIVE SYSTEM

CHAPTER PREVIEW

KIYOSHI SHIGA AND THE AGENT OF DYSENTERY

The discovery by Koch that microorganisms were responsible for infectious diseases sparked an intensive search for the agents of dreaded diseases such as tuberculosis, diphtheria, cholera, plague, typhoid, and dysentery. The span of time during which this search took place is referred to as the "golden age of microbiology," and it represents a bright period in the history of microbiology. During the golden age of microbiology, many scientists working in Pasteur's and Koch's laboratories achieved worldwide recognition for their contributions to the subjugation of infectious diseases. One such scientist was Kiyoshi Shiga.

Shiga became interested in dysentery—a disease characterized by a bloody diarrhea laden with mucus, intestinal pain, and painful intestinal contractions—because the disease was very common in his native land, Japan. In one of his many publications on dysentery, entitled "The agent of dysentery in Japan" (title is a translation from the original German publication), he indicated that between June and December 1897 there were 89,400 cases of dysentery in Japan, of which approximately 21,500 were fatal. His studies of the disease led him to believe that dysentery was a distinct clinical entity that was not caused by any known agent (e.g., typhoid or cholera bacteria). Based on this assumption, he studied specimens from 36 dysentery patients, using observational and cultural techniques made popular by Koch in his studies of anthrax and tuberculosis (see Chapter 1). Shiga's efforts culminated in the isolations of a bacterium that he named *Bacillus dysenteriae.* This name was later changed to *Shigella dysenteriae* in his honor.

Shiga's discovery was a significant contribution to the ever-increasing mass of knowledge about infectious diseases. His contributions did not end there; once he knew what caused dysentery, he directed his efforts toward the prevention and treatment of this disease. Again he was successful: he developed a vaccine that was useful in preventing the disease. He did this while working with Kitasato at Koch's laboratory in Germany.

Shiga's many successes with dysentery gained him prestige and earned him valuable posts in his native country, including a directorship at the Institute of Infectious Diseases in Tokyo, a post in which he served for 16 years, and a deanship of the Medical Faculty of Keijo Imperial University.

The successes of Kiyoshi Shiga came only after he spent countless hours studying the various aspects of dysentery in order to gain a thorough understanding of its clinical and epidemiological aspects. In this chapter we will discuss some of the important infectious diseases that afflict humans, the disease processes, their causes, and their control.

LEARNING OBJECTIVES

STUDY OF THIS CHAPTER SHOULD ENABLE YOU TO:

DIAGRAM THE DIGESTIVE SYSTEM AND INDICATE WHERE MICROORGANISMS CAN CAUSE INFECTIONS

DISCUSS THE ROLE OF THE NORMAL FLORA IN HEALTH AND DISEASE

STATE THE CAUSE, PATHOGENESIS, MODE OF TRANSMISSION, TREATMENT, AND PREVENTION OF SOME OF THE INFECTIOUS DISEASES OF THE DIGESTIVE SYSTEM

The body constantly requires water and nutrients for energy and metabolism. Energy is required for tasks such as locomotion, food gathering, and reproduction, while nutrients are required to replace old cells with new ones. The food and water we consume are often laden with microorganisms; many of these are harmless, but a few are capable of causing disease. This chapter discusses where these pathogens come from, how they cause disease, their impact on human populations, and their control and treatment.

STRUCTURE AND FUNCTION OF THE DIGESTIVE SYSTEM

The digestive system consists of two organ groups: the **alimentary** or **gastrointestinal tract** (GI) and the accessory organs. The GI tract is composed of the tubular organs of the system, extending from the mouth to the anus, and includes the mouth, pharynx, esophagus, stomach, small intestine, large intestine, rectum, and anal canal. The anatomical relationships among these organs can be seen in figure 25-1. Food is ground down by chewing and is partially digested by enzymes in the mouth. In the intestines, the food is further digested enzymatically, and water and small molecules are absorbed into the lymph and blood. Materials not digested or assimilated are released from the body through the anus. Accessory organs involved in digestion include the teeth, tongue, salivary glands, liver, gall bladder, and pancreas (fig. 25-1). These organs participate in digestion by macerating the food (teeth); releasing hydrolytic enzymes (salivary glands and pancreas); producing or storing chemicals that aid in absorption of nutrients (gall bladder, pancreas, and liver); and functioning as nutrient storage areas (liver).

The GI tract is composed of four distinct layers or tunics: the mucosa, the submucosa, the muscularis, and the serosa. The **mucosa,** or mucous membrane, is the outermost lining of the tract and is composed of two layers; an epithelial lining and an underlying layer of connective tissue called the **lamina propria** (fig. 25-2). The lamina propria is rich in glands that have digestive functions. The nature of the mucosa varies from section to section in the GI tract, depending upon the function. The major functions of the mucosa are to absorb nutrients and to secrete various substances needed for digestion. The epithelium consists of a single layer of cells, some of which absorb nutrients while others secrete substances such as mucus. The lamina propria is made of loose connective tissue and is rich in blood and lymph vessels, which absorb nutrients that have come through the epithelium. The blood and lymphatics of the mucosa also contain immunologically competent cells to fight off infectious agents that may have penetrated the epithelium. The **submucosa** is made of loose connective tissue and is highly vascularized to provide nourishment to adjacent tissues of the GI tract. This tissue is also innervated and contains an autonomic nerve network called the **plexus of Meisner.** The **muscularis** layer consists of two sheets of smooth muscle running perpendicular to each other. The smooth muscle is responsible for periodic contractions known as peristaltic movements. The **serosa** is a layer of connective tissue and epithelial tissue that covers the outer layers of most organs of the digestive system. The **peritoneum** is an important serosa layer, because sometimes it becomes infected and allows the spread of microorganisms into the peritoneal cavity and subsequently to internal organs. Infection and inflammation of the peritoneum is called **peritonitis.**

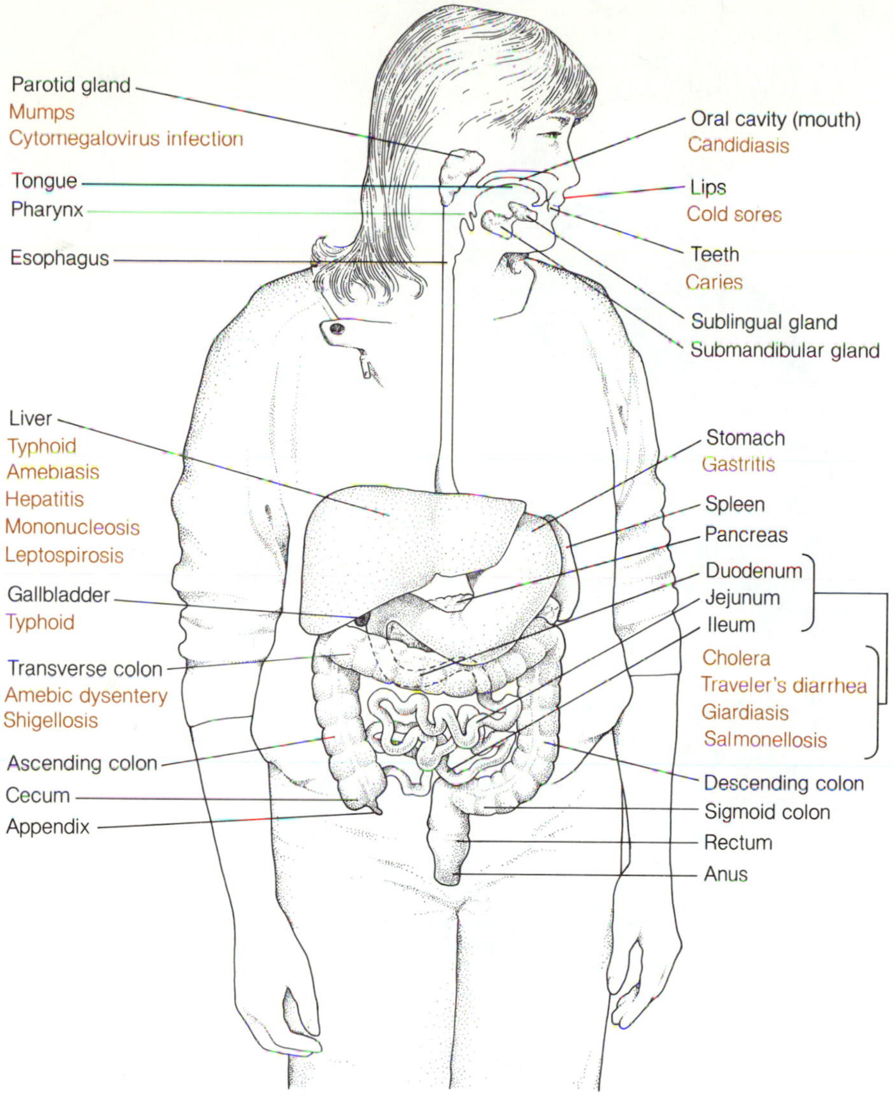

Figure 25-1 Anatomy of the Digestive System. Colored print indicates the diseases that afflict the anatomical site.

NORMAL FLORA OF THE DIGESTIVE SYSTEM

With the possible exception of the stomach, most of the alimentary canal contains numerous resident microorganisms (fig. 25-2). In contrast, most of the accessory organs are devoid of microorganisms. The microorganisms in the GI tract have very exacting environmental requirements and consequently are found only in specific parts of the tract (table 25-1).

The normal flora of the oral cavity is established shortly after birth and consists of streptococci, veillonellae, actinomycetes, spirochetes, and many other bacteria. Fungi and protozoa may also colonize the oral cavity. Microorganisms in the oral cavity derive their nutrients from food particles and soluble materials adhering to the dentition, serous exudates from the gums, the saliva, and metabolic products from other colonizing microorganisms.

Figure 25-2 The Gastrointestinal Tract. *(a)* The gastrointestinal tract consists of a series of tubes that conduct food and its by-products. The tubes consist of a number of layers called tunics, each with a function (see text for details). The mucosa is the site of many infectious diseases, although other tunics may also become involved. *(b)* Scanning electron photomicrograph of bacteria adhering to intestinal villi.

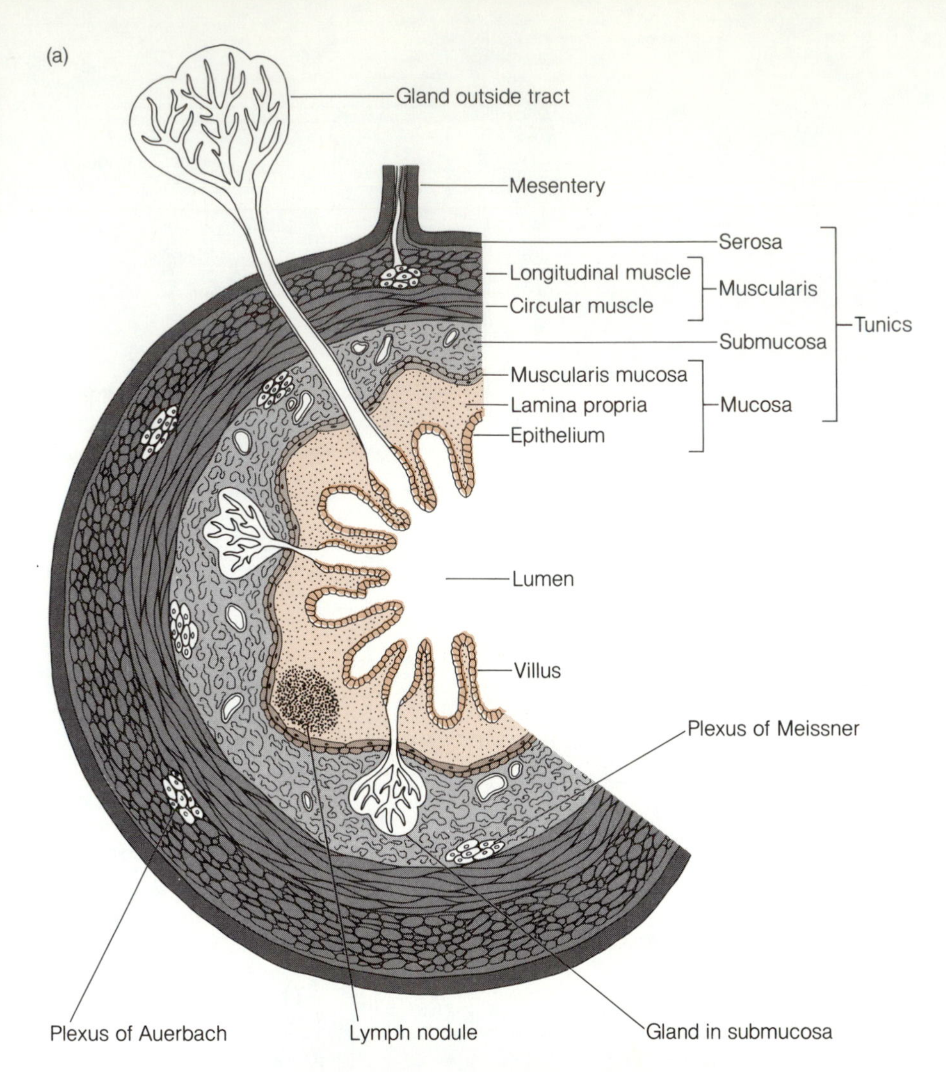

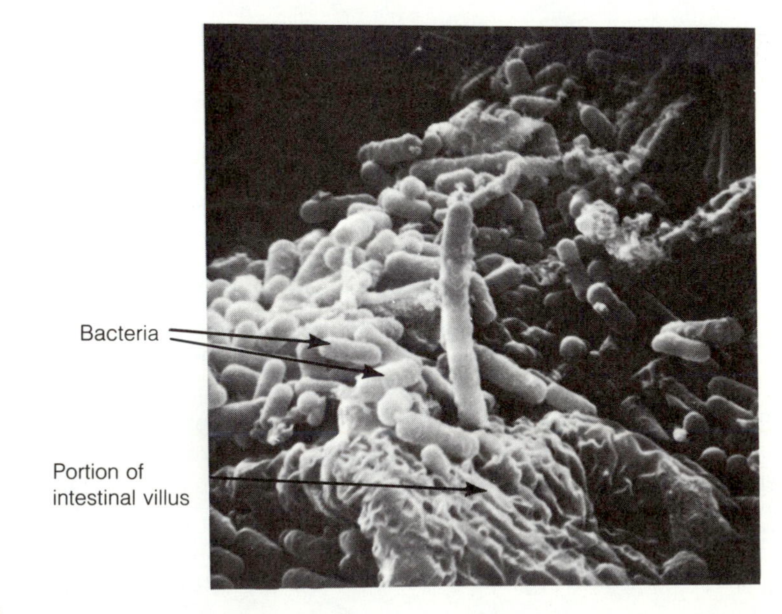

TABLE 25-1
MICROORGANISMS COMMONLY ISOLATED FROM THE HUMAN DIGESTIVE SYSTEM

GENUS OF MICROORGANISM	GROUP OF MICROORGANISMS	ORAL CAVITY	SMALL INTESTINE	LARGE INTESTINE	FECES (%)*
Actinomyces	Gram + rod	+	−	+/−	
Bacteroides	Gram − rod	+	+	+	25–30%
Borrelia	Spirochete	+	−	−	
Branhamella	Gram − coccus	+	−	−	
Candida	Yeast	+	+	+	
Clostridium	Gram + sporogenous rod	−	+	+	
Corynebacterium	Gram + rod	+	−	−	
Diphtheroids	Gram + rods	+	+	−	
Entamoeba	amebic protozoa	+	−	−	
Escherichia	Gram − rod (coliforms)	−	+	+	0.05–0.1%
Eubacterium	Gram − rod	−	+/−	+	15–18%
Fusobacterium	Gram − rod	+	−	+	5–8%
Giardia	Flagellated protozoa	−	+/−	−	
Lactobacillus	Gram + rod	+	+	+	
Neisseria	Gram − rod	+	−	−	
Peptococcus	Gram + cocci (anaerobic)	+/−	−	+	8–10%
Proteus	Gram − rod (enterics)	−	+	+	
Salmonella	Gram − rod (enterics)	−	−	+	
Shigella	Gram − rod (enterics)	−	−	+	
Staphylococcus	Gram + coccus	+	−	+	
Streptococcus	Gram + coccus	+	+	+	0.1–0.2%
Treponema	Spirochete	+	−	−	
Veillonella	Gram − coccus (anaerobic)	+	−	−	

*Percent of microbial flora that constitute the feces
+/− infrequently present
Data from Moore & Holdeman, *Appl Microbiol* 27:916, 1974.

The upper GI tract (esophagus and stomach) contains few bacteria, mostly transients. The duodenal area of the small intestine contains bacteria, mostly lactobacilli and enterococci. This flora gradually changes in the jejunum, where an increased number of coliforms and anaerobic bacteria is found. The ileal portion of the small intestine contains enteric gram negative bacteria, as well as yeasts and anaerobic bacteria. These include coliforms, *Bacteroides*, *Eubacterium*, and bifidobacteria. The large intestine contains the vast majority of microorganisms in the GI tract. These bacteria, which make up more than 20% of the mass of feces, are primarily *Bacteroides* and bifidobacteria. Other microorganisms in the large intestine include lactobacilli, coliforms, clostridia, streptococci, and yeasts. Every gram of feces contains approximately 100 billion microorganisms.

438

The normal flora protects □ the host against certain pathogens by competing for space and nutrients. The normal flora may also prevent certain infections by releasing inhibitory substances, such as bacteriocins. The contribution of the normal flora to its host may extend beyond protection; studies involving germ-free animals indicate that the normal flora is essential to the good health of the host because it stimulates the host's immune system and participates in the nutrition of the host. There is some evidence that much of the vitamin K required by animals is synthesized primarily by bacteria in the gut.

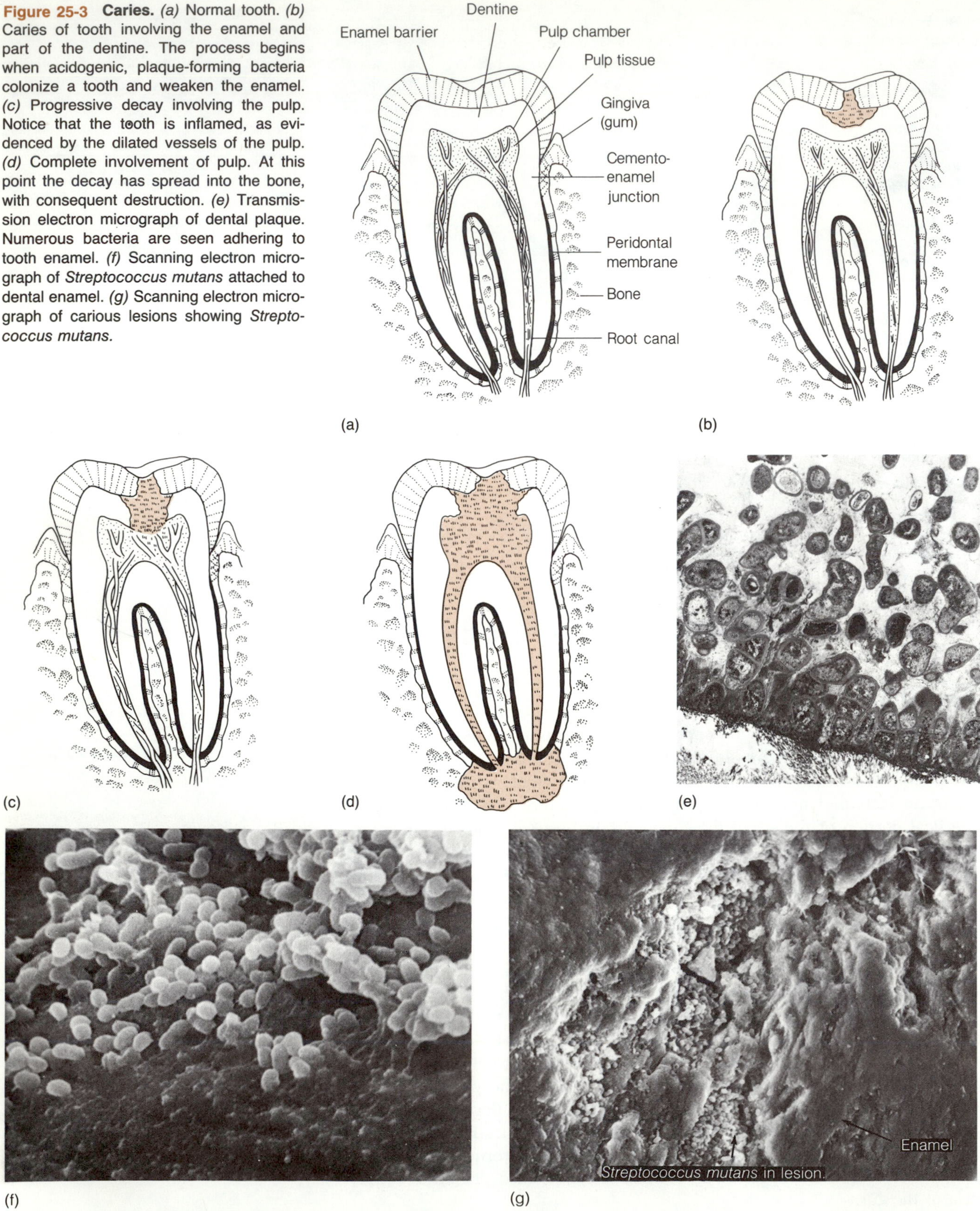

Figure 25-3 **Caries.** *(a)* Normal tooth. *(b)* Caries of tooth involving the enamel and part of the dentine. The process begins when acidogenic, plaque-forming bacteria colonize a tooth and weaken the enamel. *(c)* Progressive decay involving the pulp. Notice that the tooth is inflamed, as evidenced by the dilated vessels of the pulp. *(d)* Complete involvement of pulp. At this point the decay has spread into the bone, with consequent destruction. *(e)* Transmission electron micrograph of dental plaque. Numerous bacteria are seen adhering to tooth enamel. *(f)* Scanning electron micrograph of *Streptococcus mutans* attached to dental enamel. *(g)* Scanning electron micrograph of carious lesions showing *Streptococcus mutans.*

DISEASES OF THE ORAL CAVITY

The oral cavity is the site of entry for large numbers of microorganisms in food and water. Since there is an ample supply of nutrients from food, many microorganisms live attached to the teeth, gums, cheeks, and tongue. As a consequence, these tissues are sometimes subject to infection and disease.

Dental Caries

Dental caries, or cavities, result from the gradual decay and disintegration of teeth (fig. 25-3). Dental caries are not limited to the nonliving portions of the tooth, but may involve the pulp and the nerves within the pulp. The gradual decay and disintegration of teeth is caused by acids produced by some of the bacteria attached to the teeth. Apparently, acids produced when sugars are fermented ☐ by the tooth flora demineralize the hydroxyapatite that makes up the enamel (hard portion of the tooth). The bacterium that is largely responsible for enamel decay is *Streptococcus mutans*, which is able to attach in large numbers to the surfaces of teeth because of the dextrans (polymers of glucose) it synthesizes from the sucrose. The attached bacteria form **dental plaques** that can cover large areas of a tooth and continuously bathe these regions in acid.

186

Microorganisms growing on the teeth are the cause of dental caries, but whether or not they will cause disease in a particular instance depends on a number of factors such as dental hygiene, host immune defenses, tooth anatomy, arrangement of the dentition, and diet. Diet drastically affects the growth of cariogenic (caries-causing) bacteria on the teeth. Dental caries are most common in children and adolescents, who consume large amounts of candy and sweets that contain sucrose. Adults, in general, have fewer caries than children because of pH changes of oral secretions that occur during adulthood and reduced consumption of sucrose-containing sweets.

It is unlikely that *S. mutans* is the only microorganism involved in tooth decay. Dental caries is a progressive disease involving the destruction of the enamel, the dentine, and the pulp. The initial demineralization of the enamel is effected by acidogenic bacteria, but as soon as the dentine is reached, proteolytic bacteria (many of them anaerobes) begin to multiply. These anaerobic bacteria may be responsible for some of the damage seen in tooth decay.

Billions of dollars are spent needlessly each year by American families in the treatment of cavities. Much of the money can be saved, and the trauma of tooth repair eliminated, with proper dental hygiene. Daily brushing and flossing of the teeth and a diet low in sucrose reduce the formation of cariogenic plaques and consequently prevent dental caries. Fluoride treatment can also aid in reducing cavities by as much as 50 percent. Cavities are treated by removing the decayed portion of the tooth and replacing it with an inert filling material.

Periodontal Disease

Periodontal disease is a chronic inflammation of the anchoring and supporting tissue of the teeth (fig. 25-4). **Gingivitis** (inflammation of the gums) is a peridontal disease that develops slowly as a consequence of dental plaque formation near the gums. An inflammatory response ☐ occurs in the gums

443

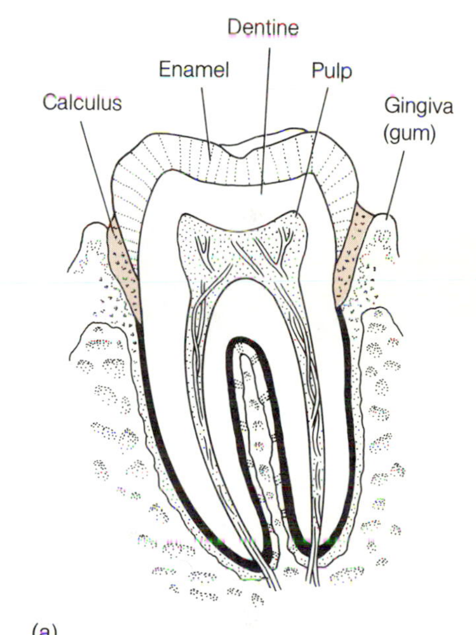

(a)

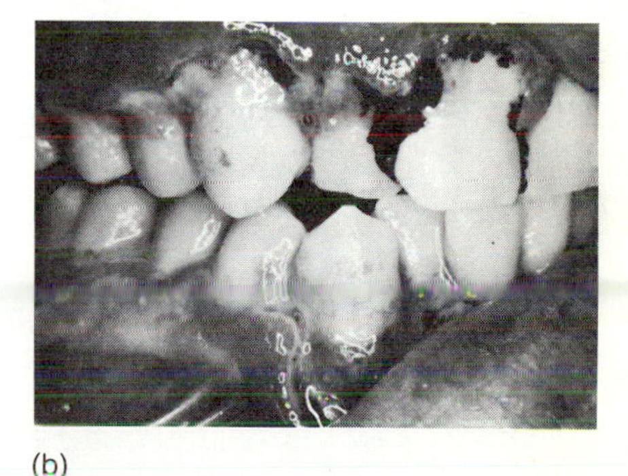

(b)

Figure 25-4 Periodontal Disease. *(a)* Diagram illustrating gingivitis. Notice the inflamed gums and heavy deposits of plaque and calculus between tooth and gum. *(b)* Photograph showing a severe case of gingivitis.

which may be due to the host's immune response to growing microorganisms or to their products (e.g., lipopolysaccharide) on the dental plaque. The inflammation causes swelling of the tissue and interdentitional crevices. Bacteria later colonize the inflamed tissue and cause more damage. The initial host response to the accumulating dental plaque is a PMN leukocyte infiltration. This acute inflammatory reaction is replaced by a chronic inflammation, containing lymphocytes and monocytes.

In gingivitis, a general increase in the number of microorganisms is seen, predominantly actinomycetes. In periodonitis, a destructive gingivitis, the microbial flora is complex and includes *Bacteroides, Capnocytophaga, Eikenella,* and *Vibrio.*

Mumps

Mumps is a disease characterized by fever and swelling of the salivary (parotid) glands. It is caused by a paramyxovirus □ known as the mumps virus. The disease occurs primarily in childhood, although some adults contract it. Complications such as orchitis (inflammation of the testes), meningitis (inflammation of the membranes lining the brain), encephalitis (inflammation of the brain), or pancreatitis (inflammation of the pancreas) occur mainly in adults. Orchitis, the commonest complication occurs in 25–30% of all adult male patients. Transmission of the mumps virus is from person to person via salivary or respiratory secretions.

395

The infection begins when a virus attaches to upper respiratory tract epithelium. There is a period of local multiplication in the epithelium and cervical lymph nodes after which the virus spreads into the bloodstream (viremia). Infection of the parotid glands results from this viremia. In the parotid glands, the virus multiplies, causing damage to the cells and an inflammatory reaction. Neutralizing antibodies and cell-mediated immunity □ play a role in blocking viral multiplication. Figure 25-5 illustrates the pathogenesis of mumps.

The highest incidence of mumps is found in individuals between 5 and 10 years old. This group of children is the most important source of infectious viruses. Adults usually get the disease from infected children. Consequently, vaccination of children eliminates (or reduces) the source of infectious virus and reduces the number of susceptible individuals in a population (fig. 25-6). Adults may be treated with hyperimmune serum containing IgG. This treatment is directed toward the prevention of orchitis (inflammation of the testes) and possible ensuing sterility in adult males.

INFECTIOUS DIARRHEAS CAUSED BY BACTERIA

Diarrhea, a symptom of many gastrointestinal disorders, is characterized by a frequent passage of watery stool. It may be due to diet, toxins, infections, drugs, or psychological factors. Diarrhea caused by infectious agents, such as viruses, bacteria, and protozoa, still acounts for a significant proportion of morbidity and mortality in developing countries.

The small and large intestines are the commonest sites of gastrointestinal disease. Diseases of the small intestine generally result in a profuse, watery discharge, resulting from an alteration of the ionic balance in the intestinal epithelium. In some infections, the mucosa and submucosa tunics

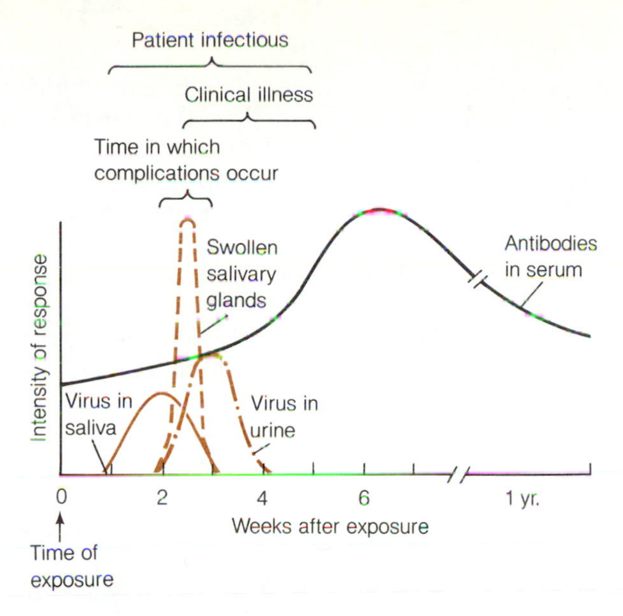

Figure 25-5 Clinical Course of Mumps. Clinical illness occurs 1–3 weeks after infection. This is the contagious period because the virus is present in the saliva. Clinical symptoms persist for 2–3 weeks. The characteristic signs of mumps, inflamed parotid glands, appear about 2 weeks after infection and last for about a week. Serious consequences, such as inflamed testicles and meninges or encephalitis, usually appear (albeit rarely) 3–5 weeks post-infection. Serum antibodies peak at about 5–6 weeks post-infection.

are not altered pathologically and no inflammatory response is noted. Generally, the invading microorganisms produce toxins that induce the profuse secretion of fluids from the epithelium. In other infections, the mucosa and submucosa are damaged directly by the invading microorganisms, causing an inflammation of the small bowel which may also cause diarrhea. Infections of the large intestine generally result in mucosal erosion accompanied by abdominal cramps, painful spasmodic contraction of the bowel (tenesmus), and inflammatory exudate (pus and PMN leukocytes) in the stool. These symptoms characterize a condition known as **dysentery.**

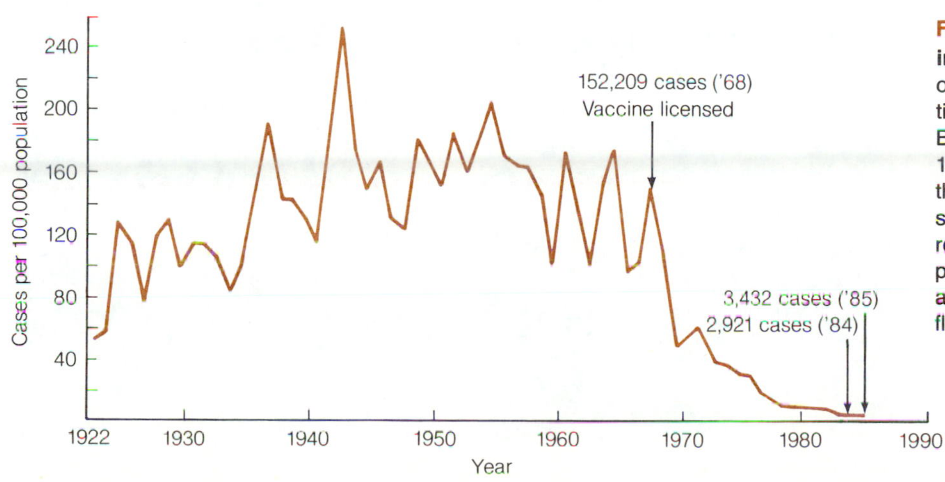

Figure 25-6 Reported Cases of Mumps in the United States. The number of cases of mumps in the U.S. waxed and waned until 1967, when the vaccine was approved. Before 1967, there was an average of about 160 cases/100,000. With the introduction of the vaccine, the number of cases dropped sharply. Presently <2 cases/100,000 are reported. About 70% of all cases are reported in children between the ages of 5 and 14. Adults (20+) are least often afflicted.

Cholera

Cholera, an acute infection of the human small intestine by *Vibrio cholerae* is characterized by a profuse, watery discharge. The diarrhea is brownish at first, but soon turns a pale whitish color. This is the characteristic "rice water" stool of cholera. Vomiting may also accompany the diarrhea.

The causative agent of cholera is *Vibrio cholerae* a gram negative motile rod exhibiting a characteristic curved or comma shape, although the curved appearance of the rods may disappear in laboratory cultures. The cholera vibrios generally enter the small intestine in contaminated water and food. They adhere to the microvilli of epithelial cells (fig. 25-7), where they multiply and release a powerful enterotoxin called **choleragen.** Like diphtherotoxin, choleragen consists of two subunits, A and B. The B fraction binds to intestinal epithelial cell membranes, and the A fraction induces the formation of cyclic AMP (cAMP) from ATP. Cyclic AMP is responsible for the efflux of electrolytes (chloride and bicarbonate) and fluids into the lumen of the small intestine (fig. 25-8). The amount of water lost is so great (4–5 gallons/day) that it cannot be reabsorbed by the large intestine. This tremendous loss of fluid causes dehydration, leading to diminished blood volume. The diminished blood volume leads in turn to **hypovolemic shock,** since organs and tissues are not oxygenated or nourished properly. Death soon follows if this condition is not reversed.

Humans are the only known reservoir □ for *V. cholerae.* Infected individuals shed the vibrios into sewage, which can contaminate water supplies. The spread of the disease is by the fecal-oral route via contaminated water. *Vibrio cholerae* is found in nearly all large bays and rivers associated with densely populated areas. In September 1978, for example, the Public Health Service reported the isolation of *V. cholerae* in 11 persons from Louisiana. Improperly cooked crab was the source of the bacterium. Since then, however, the organism has been isolated from local shrimp as well as crab. Carriers □ harbor the bacteria in their gall bladders. A vaccine has been developed which induces the development of an immunity that acts directly on

534

536

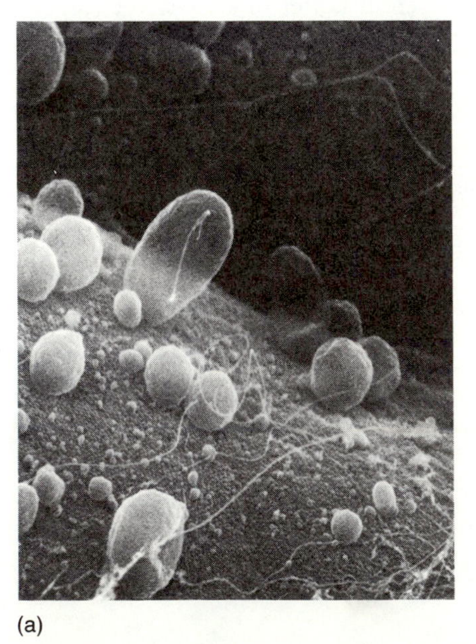
(a)

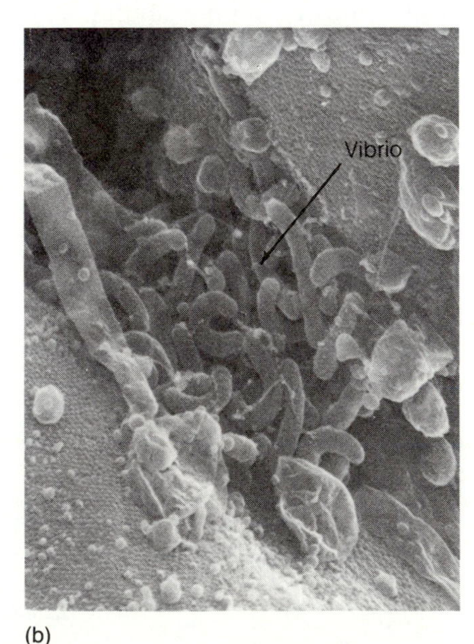

(b)

Figure 25-7 *Vibrio cholerae* in Intestines. *(a)* Scanning electron micrograph of mouse ileum showing normal flora on microvilli. *(b)* Intestine extensively colonized by *Vibrio cholerae.* Cholera is caused by the effect of the exotoxin on intestinal epithelium.

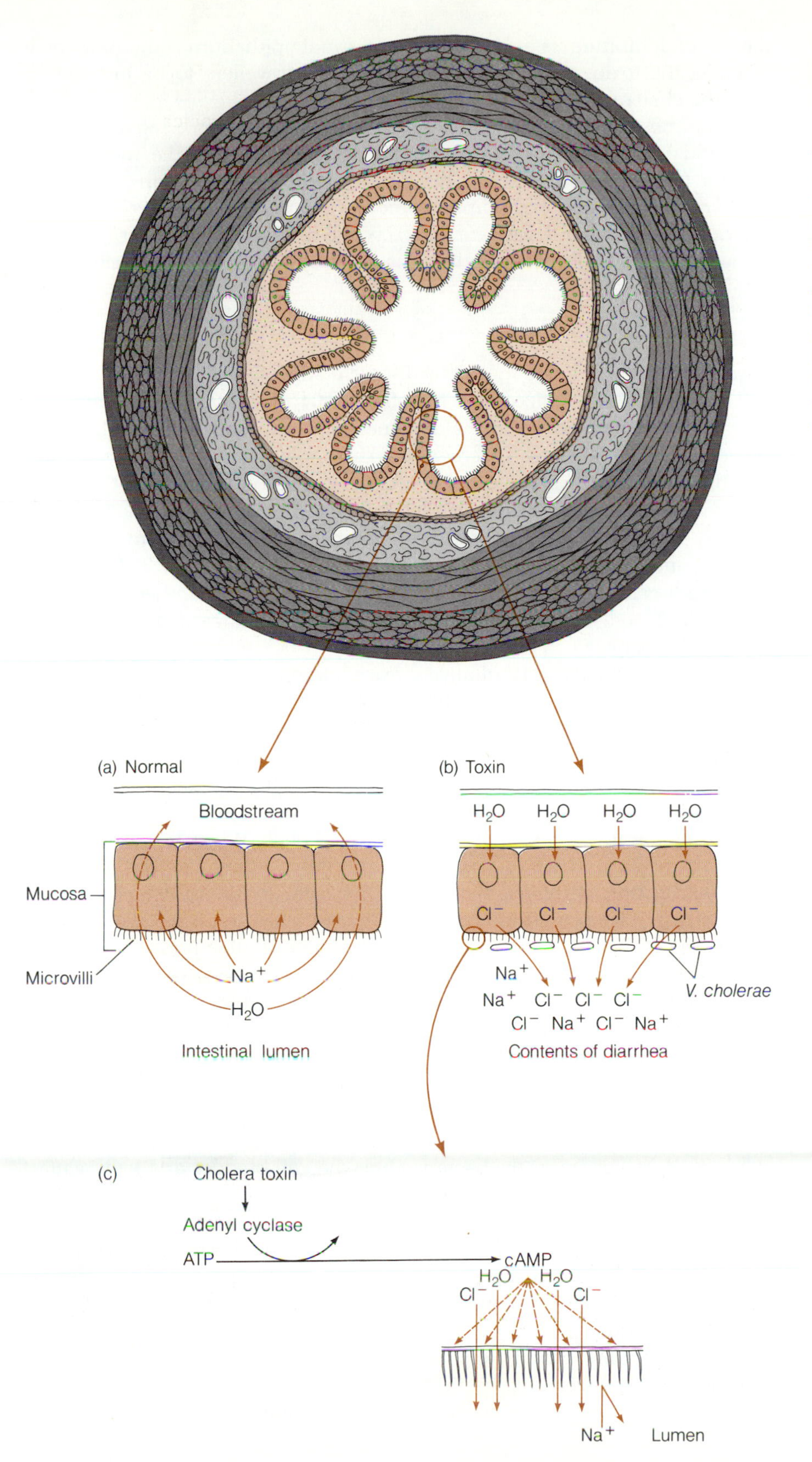

Figure 25-8 Mode of Action of the Cholera Toxin. Cholera toxin acts by causing the excessive flow of cellular fluids and electrolytes into the lumen of the intestines. *(a)* Normal direction of fluids and sodium in the intestines. Water flows from the lumen of the intestines into the bloodstream, along with sodium ions. *(b) Vibrio cholerae* colonization of intestines and production of exotoxin. *(c)* The toxin acts by stimulating adenyl cyclase to convert ATP into cAMP. cAMP blocks the flow of sodium ions into the bloodstream and causes chloride ions to be excreted from the blood. As a consequence, large volumes of water exit the blood and tissues into the lumen and are excreted, along with the electrolytes, as diarrhea.

the vibrio, inhibiting its attachment to intestinal epithelium rather than neutralizing the toxin. Some countries still require travellers to be immunized with the cholera vaccine.

The most effective treatment for cholera is the prompt replacement of lost fluids and electrolytes. This treatment alone reduces the mortality of cholera to less than 1 percent. Intravenous solutions containing Na^+, K^+, Cl^-, and HCO_3^-, and oral administration of 5% glucose in water, constitute a common method of treatment. Tetracycline reduces the number of vibrios present in the small intestine and helps in eliminating the carrier state.

Typhoid Fever

Typhoid fever, an acute infectious disease of the GI tract caused by *Salmonella typhi*, is characterized by a high fever (104°F), headache, diarrhea, and rose spots and tenderness in the abdomen. The incubation period □ for typhoid fever is 1–3 weeks (average, 2 weeks). The early symptoms include nosebleeds, general weakness, mild headache, and malaise. After these general symptoms, the characteristic tenderness and rose spots on the abdomen appear, along with a high fever, splenomegaly (an enlargement of the spleen), and diarrhea. Thin bowel movements generally occur 3–7 times per day. Shock caused by endotoxin may also occur if large numbers of *S. typhi* are in the blood. Mortality of untreated cases may be as high as 40%, but generally is between 15 and 25 percent.

542

Salmonella typhi, a facultative intracellular parasite of macrophages, is a gram negative rod in the family Enterobacteriaceae. *Salmonella typhi* has a number of antigens—capsular antigen (Vi), a somatic antigen (O), and a flagellar antigen (H)—which play a role in the pathogenesis of the disease. The microorganism enters the human small intestine in contaminated food and water. During the first 5–7 days of infection, *S. typhi* attaches to the surface of the intestinal lining, where it multiplies. When it penetrates the lamina propria and submucosa, some of the invading bacteria enter the blood and lymph, causing a bacteremia. The early symptoms (fever, malaise, and lethargy) are caused by the penetration into the intestinal wall and the bacteremia. The symptoms may be aggravated by the presence of endotoxin □ in the blood. When *S. typhi* enters the circulation, it is distributed throughout the body. The bacteria are phagocytized □, but not killed, by macrophages in the lymph nodes and other organs of the reticuloendothelial system. They multiply within the macrophages and escape into the lymph and blood in very large numbers. The high fever of typhoid is correlated with this bacteremia. Endotoxin and the host's fever-producing substances **(endogenous pyrogens)** released by the phagocytic cells also contribute to the symptoms. At this stage, the bacteria enter the gall bladder and then reinfect the intestines. Diarrhea and abdominal tenderness and rose spots parallel the reinfection of the intestines by *S. typhi*. With the development of cell-mediated immunity, activated macrophages destroy the invading bacteria. This event generally leads to the recovery of the patient; however, 1% of these individuals die and 10–15% become chronic carriers, harboring the bacteria in the gall bladder.

432
445

In the last few years, there have been fewer than 500 reported cases of typhoid fever per year in the United States and about 5 deaths annually. Approximately 45–75 new carriers enter the population each year (fig. 25-9).

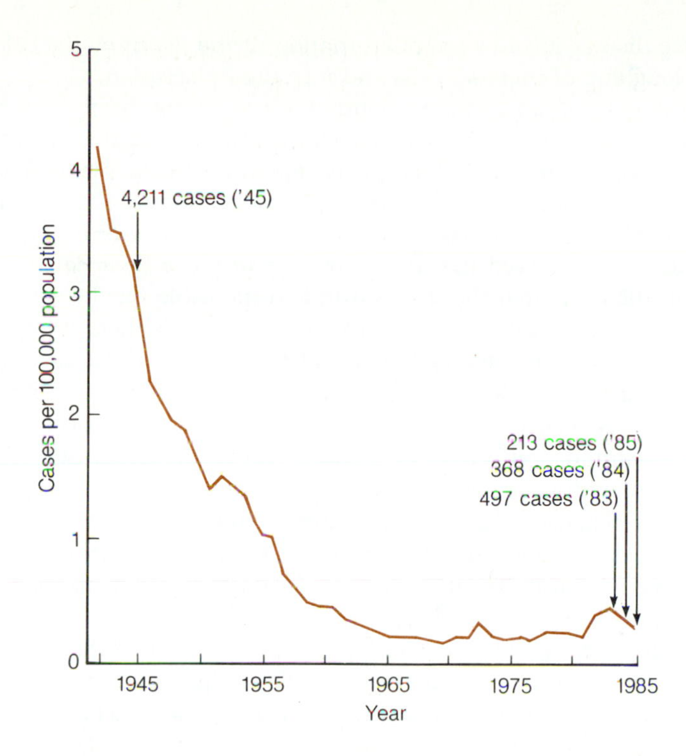

Figure 25-9 Reported Cases of Typhoid in the United States

Typhoid fever is transmitted by the fecal-oral route. Chronic carriers are the most common source of infectious agents, since their excrements contaminate food and water with virulent bacteria. The Public Health Service (PHS) maintains a very close surveillance of these carriers, because it would be dangerous for them to handle foods in such places as food processing plants, restaurants, and bakeries. Most cases of typhoid fever in the United States are traced to carriers who contaminate food or water. In most cases of typhoid fever in which water is the vehicle, the source is traced to faulty sewage systems.

535 Prevention of typhoid involves the surveillance and control of carriers □. In addition, water supplies should be properly chlorinated and reservoirs and water pipes maintained. In areas where typhoid fever is prevalent, all foods should be steaming hot when eaten, and only beverages from unopened bottles should be consumed. Foods should be cooked and/or refrigerated to prevent the growth of salmonellae.

To date, no vaccine has been developed that gives full protection against typhoid fever, although some vaccines provide partial protection. Vaccines containing the Vi antigen have given encouraging results in clinical trials. Ampicillin and chloramphenicol can be used to treat typhoid fever, but ampicillin-resistant strains of *S. typhi* have evolved. Thus, sensitivity tests should be conducted on isolates so that the best antibiotic can be used to treat the infection.

Bacillary Dysentery

Bacillary dysentery is an acute infectious disease caused by bacteria in the genus *Shigella*. It is characterized by diarrhea, tenesmus (distressing but ineffectual urge to defecate), and cramps, along with mucus and blood in the feces. The disease involves an inflammation of the ileum or the colon which causes sloughing of mucosal cells and intestinal ulceration.

Shigella is a gram negative, nonmotile rod in the family Enterobacteriaceae □. Bacillary dysentery can be caused by any of the four species of *Shigella*. Small numbers of this bacterium are able to cause dysentery in humans. Three shigellae, *S. dysenteriae*, *S. sonnei*, and *S. flexneri*, have been shown to produce an **enterotoxin** that appears to work in the same way as choleragen. It is believed that these bacteria produce their enterotoxin early in the infection and that the enterotoxin is responsible for the early diarrhea. The characteristic common to all shigellae, however, is their ability to adhere to and invade the intestinal epithelium of the ileum and colon.

315

The characteristic lesion of the intestinal mucosa is an ulcer covered by a layer of PMN leukocytes, bacteria, cellular debris, and fibrin. The lesions arise as a combined effect of the invasiveness of the bacterium, the release of endotoxin, and an acute inflammatory reaction on the part of the host. As the bacteria penetrate the intestinal mucosa, they release metabolic wastes, enterotoxin, and endotoxin, which cause cell damage and elicit an acute inflammatory response. Together, these factors induce the formation of the characteristic ulcer and diarrhea.

Dysentery caused by *S. dysenteriae* type I, also called the Shiga bacillus, produces an exotoxin called **neurotoxin.** The Shiga exotoxin causes bleeding and paralysis apparently by inhibiting protein synthesis. The role of the exotoxin in the pathogenesis of dysentery has not been clearly defined, but it appears that the neurotoxin is responsible for the severity of the dysentery caused by the Shiga bacillus as compared to the other species. Circulating antibodies appear during the infection, but they are relatively ineffective in stopping the invasion. Cell-mediated immunity □ is thought to be responsible for protective immunity and recovery of the patient.

446

Dysentery is transmitted by the fecal-oral route, and the most common source of infection is other humans. In countries with high standards of hygiene, where human wastes are properly discarded, the disease is infrequent. The majority of the cases are laboratory-associated infections and outbreaks in institutions. In institutions where coprophagy (consumption of feces) sometimes occurs, bacillary dysentery and hepatitis are common. In developing countries, where sanitary conditions and public health measures frequently are marginal, bacillary dysentery is endemic. The disease is most often transmitted in foods contaminated by infected individuals. Although the carrier stage is rare, inapparent infections do occur, and the infected person can serve as a temporary source of contamination.

Preventative measures for dysentery include the chlorination of water supplies and the maintenance of reservoirs and water pipes, as well as the treatment of raw sewage. Dairy foods have been reported as a source of infection, and therefore it is essential that milk be properly pasteurized □ before drinking and that the consumption of raw milk be avoided. There are no vaccines available with which to immunize the public. In the last few years, between 15,000 and 20,000 cases of dysentery have been reported each year in the United States. The number of deaths among these cases has been reported to be between 20 and 30.

809

Severe infections generally are treated with ampicillin. Chloroamphenicol and trimethoprim-sulfamethoxazole are used for those infections caused by ampicillin-resistant strains. The majority of cases are not treated at all, however, since the infection is self-limiting and of short duration. Supportive therapy is recommended to alleviate the discomfort and to replace the fluids and electrolytes lost in the diarrhea.

Salmonella Gastroenteritis

***Salmonella* gastroenteritis** is characterized by a sudden onset of diarrhea, headache, abdominal pain, vomiting, and fever. The disease is commonly called *Salmonella* food poisoning, but it is not really a "poisoning," because no exotoxins are clearly associated with salmonellae and the signs and symptoms of the disease are caused by bacterial invasion of the intestines. *Salmonella enteritidis*, however, has been shown to produce an enterotoxin that might be responsible for the diarrhea.

The salmonellae almost invariably enter the GI tract within contaminated foods. Unlike shigellae, of which fewer than 200 are required to initiate an infection, more than 100,000 salmonellae are needed for a successful infection. The salmonellae that survive the harsh stomach environment and reach the intestines adhere to the intestinal epithelium and initiate foci of infection in both the small and large intestines. In contrast to enteric fevers (e.g., typhoid), the salmonellae causing gastroenteritis rarely invade the blood, and their growth is limited to the GI tract. When the infecting bacteria penetrate the epithelium and enter the lamina propria, they are ingested by phagocytes. Some of the bacteria are transported to the regional lymph nodes, where they stimulate specific immunity that eventually eliminates the infectious agent from the intestines. The growth of salmonellae causes pathological changes that are thought to account for most of the symptoms of the disease. An accute inflammation is believed to be responsible for focal ulceration and microabscesses seen in biopsy materials from infected persons. Gastroenteritis caused by salmonellae is generally not treated with antibiotics, since it is a self-limiting disease.

The infection is generally acquired by ingesting food contaminated with salmonellae. Leaving contaminated food in a warm place for an extended period of time allows the multiplication of salmonellae to infectious levels in the food. In contrast to *S. typhi*, which generally infects only humans, *S. enteritidis* also infects animals; consequently, animals can serve as a source of infection. Outbreaks have been traced to chickens, pet turtles, pigs, and many other domestic animals. More than 1,400 serotypes of salmonellae have been associated with gastroenteritis. In the last few years, the number of cases of *Salmonella* gastroenteritis reported in the United States has increased drastically (fig. 25-10). Each year, the morbidity is over 30,000 cases and the mortality is between 60 and 80 cases. Since it is a rather mild disease, however, many of these cases are not reported, and it is estimated that fewer than 10% of all cases of *Salmonella* gastroenteritis are reported to health agencies.

Diarrheas Caused by *E. Coli*

Escherichia coli is a gram negative, glucose-fermenting rod that has long been known to colonize the human intestine. This organism is part of the normal flora and generally functions as a mutual □, but it can also become a pathogen. For example, more than 80% of urinary tract infections are

426

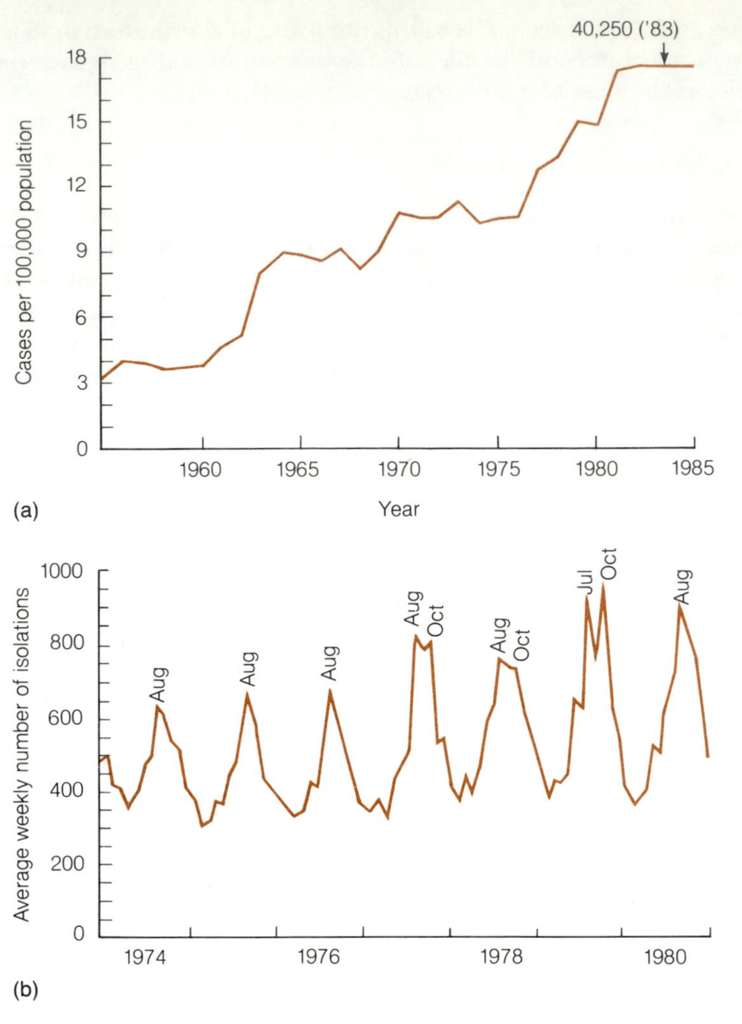

Figure 25-10 Characteristics of Salmonellosis. *(a)* Incidence of *Salmonella* gastroenteritis in the United States. *(b)* Reported isolation of *Salmonella.* It is noteworthy that most isolates occur during the summer months, when group picnics and outings are most frequent.

caused by *E. coli.* More recently, three distinct types of *E. coli* have been described which cause diarrheal diseases in humans. These are **enteropathogenic** *E. coli* (EPEC), **enteroinvasive** *E. coli* (EIEC), and **enterotoxigenic** *E. coli.* (ETEC). These three types of pathogenic *E. coli* are not normally found as part of the human intestinal flora.

Enteropathogenic strains of *E. coli* (EPEC) cause a diarrheal disease that resembles *Salmonella* gastroenteritis. This group of *E. coli* serotypes is most frequently responsible for outbreaks of diarrhea in institutions such as nurseries and children's hospitals. EPEC does not produce any kind of enterotoxin, nor is it able to invade the lamina propria of the intestines. How EPEC causes disease is not known. Stool specimens from patients under the age of 2 should be screened for these *E. coli* serotypes.

Enteroinvasive strains of *E. coli* (EIEC) cause a disease that resembles bacillary dysentery. These stereotypes of *E. coli* are capable of invading epithelial cells of the intestines. Adults and infants are equally susceptible.

Enterotoxigenic strains of *E. coli* (ETEC) cause cholera-like diseases in humans. This group of serotypes has been implicated as the cause of "travellers' diarrhea." These bacteria also cause diarrhea in piglets and calves. The disease can be very serious in children, because it appears suddenly, causing a severe dehydration. ETEC causes the disease by elaborating a powerful **heat-labile** (LT) **enterotoxin** that acts in a manner similar to the cholera toxin. Pili are important in the pathogenesis of ETEC, since they attach the bacteria to the intestinal epithelium. Attachment of ETEC to the epithelium results in the multiplication of the bacteria and the subsequent production of enterotoxin.

Pathogenic strains of *E. coli* are sensitive to a variety of drugs, including sulfonamides, tetracyclines, and ampicillin. The treatment is directed toward eliminating the bacteria and, in cases of severe diarrhea, replacing electrolytes and fluids, as for cholera patients.

The best way to control these diseases is by staying away from contaminated drinking water. Drinking bottled beverages or water can be a healthy way of quenching thirst. If domestic water suspected of contamination must be drunk, it should be boiled or treated in such a way as to kill the pathogens. Raw vegetables in salads or "munchies" have also been implicated in diarrheal diseases caused by *E. coli.* Oral sulfonamides have been recommended for travellers as a prophylactic measure for travellers' diarrhea. Bismuth subsalicylate (Pepto-Bismol®) has been shown to be effective in reducing the effect of ETEC.

Other Bacterial Diarrheas

Campylobacter jejuni has recently been discovered to be a common cause of gastroenteritis. It is a gram negative, flagellated, curved rod that is isolated from approximately 9% of all diarrheas and from 3% of formed stools. It requires a microaerophilic environment and grows best at 40–42°C. The disease is characterized by fever and malaise, followed by diarrhea and severe cramping that may mimic acute appendicitis or ulcerative colitis. The bacteria appear to penetrate the mucosal epithelium, since inflammatory cells and blood are often present in the stool. *Campylobacter jejuni* causes hemorrhagic necrotic foci in the jejunum and ileum. The disease usually lasts less than a week and is self-limiting; however, fatal cases have been reported.

Domestic animals such as cows, fowl, swine, sheep, and dogs serve as reservoirs for *C. jejuni.* The infectious agent is generally acquired by ingesting contaminated water or raw milk, or by contact with infected poultry or humans. All ages are susceptible to the infection, which has a seasonal peak during the summer. The number of annual cases of *C. jejuni*–caused diarrheas in the United States is estimated to be at least 100,000. This bacterium is now considered to be one of the most common causes of diarrhea in the United States.

Yersinia enterocolitica is a gram negative bacillus that causes a severe **enterocolitis.** It is not a frequent cause of infectious diarrhea in the United States and is more prevalent in western Europe and Scandinavian countries. The symptoms include fever, diarrhea, and abdominal pain. Animals are believed to serve as reservoirs for *Y. enterocolitica.* The bacterium has been isolated from water, food, and unpasteurized dairy products. This pathogen invades the wall of the small intestine and may spread to the regional lymph nodes. This behavior may be responsible for the abdominal pains and appendicitis-like symptoms of this enterocolitis. The enterotoxin produced by this bacterium is believed to be responsible for the diarrhea.

Clostridium difficile has recently been implicated in **antibiotic-associated pseudomembranous enterocolitis** (AAPE). The disease is characterized by diarrhea and ulcerative lesions similar to those caused by *Shigella*. The pathogenesis of the microorganism is related to the production of a cytolytic enterotoxin that is believed to cause the ulcerative colitis and the diarrhea. The symptoms are often associated with the administration of antibiotics; clindamycin and some other antibiotics at subinhibitory concentrations cause the production of enterotoxin by *C. difficile*. It is believed that the antibiotic suppresses the normal flora and hence allows the extensive multiplication of *C. difficile*. Clindamycin has also been shown to act as an inducer of the enterotoxin.

FOOD INTOXICATIONS

So far we have been discussing diseases that are caused by infectious agents multiplying in the gastrointestinal tract. The diseases they cause are attributed to their growth, metabolism, and spread. The term "food poisoning" □ 815 is commonly used to describe diseases caused by ingesting infectious agents or foods contaminated with toxic substances. In this section we will discuss diseases of the GI tract that are caused by the ingestion of toxins produced by microorganisms multiplying in foods (table 25-2). Botulism, a very serious form of food poisoning, will not be discussed in this chapter because it involves the nervous system rather than the GI tract.

Staphylococcal Gastroenteritis

Staphylococcal gastroenteritis is thought to be the most common cause of food poisoning in the United States. The disease is caused by ingesting foods containing *S. aureus* enterotoxins and is characterized by a rapid onset of nausea, vomiting, and diarrhea within an average of 4 hours after consuming the contaminated food. The severity of the intoxication varies depending on the amount and type of enterotoxin ingested. Foods such as custards, hams, hollandaise sauce, and creamy fruit salads, tuna, chicken, and macaroni salads contaminated by food handlers frequently become sources of *S. aureus* enterotoxin when the foods are not kept refrigerated. Large amounts of enterotoxin can be produced in a few hours when the food is at room temperature.

The symptoms of staphylococcal gastroenteritis are attributable to the enterotoxin(s). Since the toxins are not denatured readily by heating, disease can result even when contaminated foods are cooked. The mode of action of the enterotoxins is unknown; however, they are thought to act on emetic (vomiting) receptors in the intestine, to impede fluid absorption by the gut, and to promote the efflux of fluids into the lumen of the intestines. The toxins are also believed to be pyrogenic (fever causing), a characteristic that may account for the fever observed in some cases of food poisoning.

The source of contamination of foods is usually chronic carriers who come into contact with the food. Approximately 10% of the United States population are carriers for enterotoxigenic staphylococci. Refrigerating foods

TABLE 25-2
FOOD POISONINGS ASSOCIATED WITH THE PRODUCTION OF TOXINS

DISEASE	CAUSATIVE AGENT	SYMPTOMS OF DISEASE	FOODS IMPLICATED	TIME OF ONSET (HOURS)
Botulism	*Clostridium botulinum* (botulinal exotoxins A-E)	Flaccid paralysis, cardiac paralysis	Meat, fish, low acid foods, canned vegetables	24–48
B. cereus food poisoning	*Bacillus cereus* (toxins)	Nausea, vomiting, colic pain, diarrhea	Unrefrigerated starchy foods	8–12
Enterococcus food poisoning	*Streptococcus faecalis* (toxins)	Nausea, vomiting, diarrhea	Unrefrigerated foods	8–12
Mycetismus	*Amanita* spp. (mushroom) (Amanitin)	Nervous disorders, death	Poisonous mushrooms	24–36
Mycotoxicosis	*Aspergillus flavus* (Aflatoxin)	Liver disorders, death	Improperly stored grains	12–64
"Perfringens" food poisoning	*Clostridium perfringens* (exotoxins)	Abdominal pain, diarrhea	Unrefrigerated cooked meats	5–15
"Staph" food poisoning	*Staphylococcus aureus* (enterotoxins A-F)	Sudden nausea, vomiting, diarrhea	Potato salad, cream-filled pastries, dry skim milk, ham	4–12

until they are served and promoting safe food handling procedures insure that staphylococci do not grow and produce the enterotoxin in the food. Cases of staphylococcal food poisoning generally are not treated, because the disease is rarely fatal and the symptoms are of short duration. In severe cases, replacement of fluids and electrolytes is recommended.

Clostridium Gastroenteritis

Clostridum perfringens causes a mild form of food poisoning characterized by abdominal pain, cramps, and diarrhea. It differs from staphylococcal gastroenteritis in that nausea and vomiting are uncommon symptoms. The onset of the disease occurs 10–24 hours after ingestion of food contaminated with *C. perfringens* enterotoxins. The symptoms may last 10–20 hours. This type of food poisoning ranks second only to staphylococcal food poisoning in incidence in the United States. The cramps and diarrhea are caused by the enterotoxins produced.

The disease is most often caused by eating foods that have been cooked in bulk and then stored at temperatures that allow the bacteria to multiply. Pinto beans and other legumes are excellent substrates for the growth of these bacteria. When the foods are served without reheating to above 165°C, the toxin is not inactivated. For this reason, most outbreaks of *C. perfringens* food poisoning occur at large gatherings and in institutions, where bulk preparation of food is a necessity.

Clostridium perfringens is a gram positive, anaerobic, endospore-forming rod. It is widely distributed in soils, sewage, water, the intestinal tract of mammals, raw meats, poultry, and fish. The organism elaborates a wide variety of enzymes, many of which cause extensive tissue damage. In addition to food poisoning, *C. perfringens* causes a disease of connective tissue called **gas gangrene.**

PROTOZOAL INFECTIONS OF THE DIGESTIVE SYSTEM

Protozoal diseases such as those caused by *Entamoeba histolytica* and *Giardia intestinalis,* are quite common, especially in developing countries and in areas where poor sanitation is sometimes practiced. These organisms cause diarrheal diseases resembling those caused by bacteria. The symptoms range from mild to severe, and the infections may include accessory organs such as the liver and gall bladder.

Amebiasis

Amebiasis is a disease caused by *Entamoeba histolytica.* The classical clinical presentation of this disease is a dysentery, also called **amebic dysentery.** At the onset, frequent bowel movements and abdominal pain are experienced. Typically the stools are watery or loose and accompanied by blood and mucus. Rectal pain and tenesmus are also common manifestations of amebiasis. This symptom reflects an acute inflammatory response due to the invasion of the colonic epithelium by the amebae. The large bowel may develop an ulcer (fig. 25-11). Amebiasis may also be seen as a mild diarrhea, with no ulceration of the colon and little or no abdominal pain. The acute diarrhea is characterized by the presence of **trophozoites** □, while in chronic infections the **cyst** is most common. Liver abscesses occur when the amebae spread to the liver; pain in the right upper quadrant is the most common complaint with this condition, and hepatomegaly (an enlargement of the liver) is often felt on palpation. Peritonitis can develop from either the rupture of the abscess or the perforation of an intestinal ulcer.

367

An infection is initiated by ingestion of cysts, either in food or in water. The amebae excyst in the large intestine and then adhere to the mucosal epithelium, where they multiply and derive nutrients from cellular secretions and extracellular materials. They may remain in this state for a prolonged period of time without causing disease. It is not known why certain amebae become invasive, but it is at this stage that the dysentery syndrome appears.

Prevention is achieved by avoiding untreated water or foods contaminated with human feces. Amebicides such as metronidazole (Flagyl), diloxadine furoate (Furamide), or diiodohydroxyquin are used successfully to control amebiasis. In the last few years, the number of reported cases of amebiasis in the United States has slowly increased to over 5,000 per year. The number of fatalities has decreased to around 20 per year, however, probably due to the availablity of better amebicidal drugs.

Giardiasis

Giardiasis is an infectious disease caused by the flagellated protozoan *Giardia intestinalis* (fig. 25-12). This flagellate infects the small intestine and causes diarrhea, abdominal cramps, nausea, vomiting, flatulence, and greasy stools. The infection apparently inhibits absorption of fats and certain vitamins by the human host.

The parasite enters the host in the form of a cyst in food or water. In the duodenum, it excysts and develops into a trophozoite. The trophozoites attach to the crypts (folds of intestinal mucosa) of the duodenum and jejunum with their ventral sucking disks (fig. 25-12) and multiply there. The ability of the trophozoites to adhere to the intestinal epithelium may account

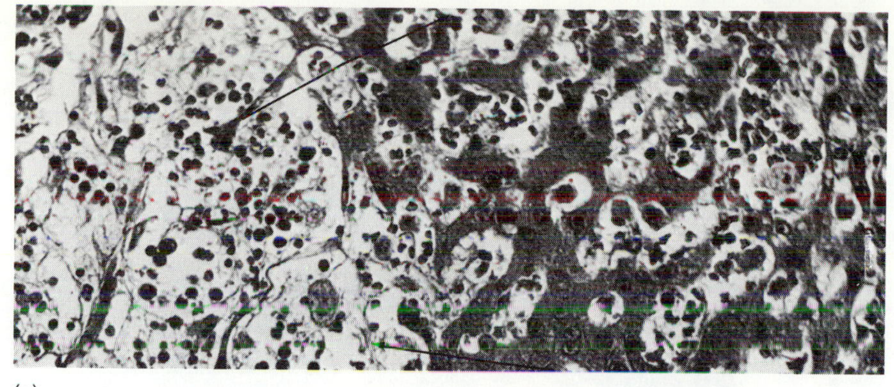

(a)

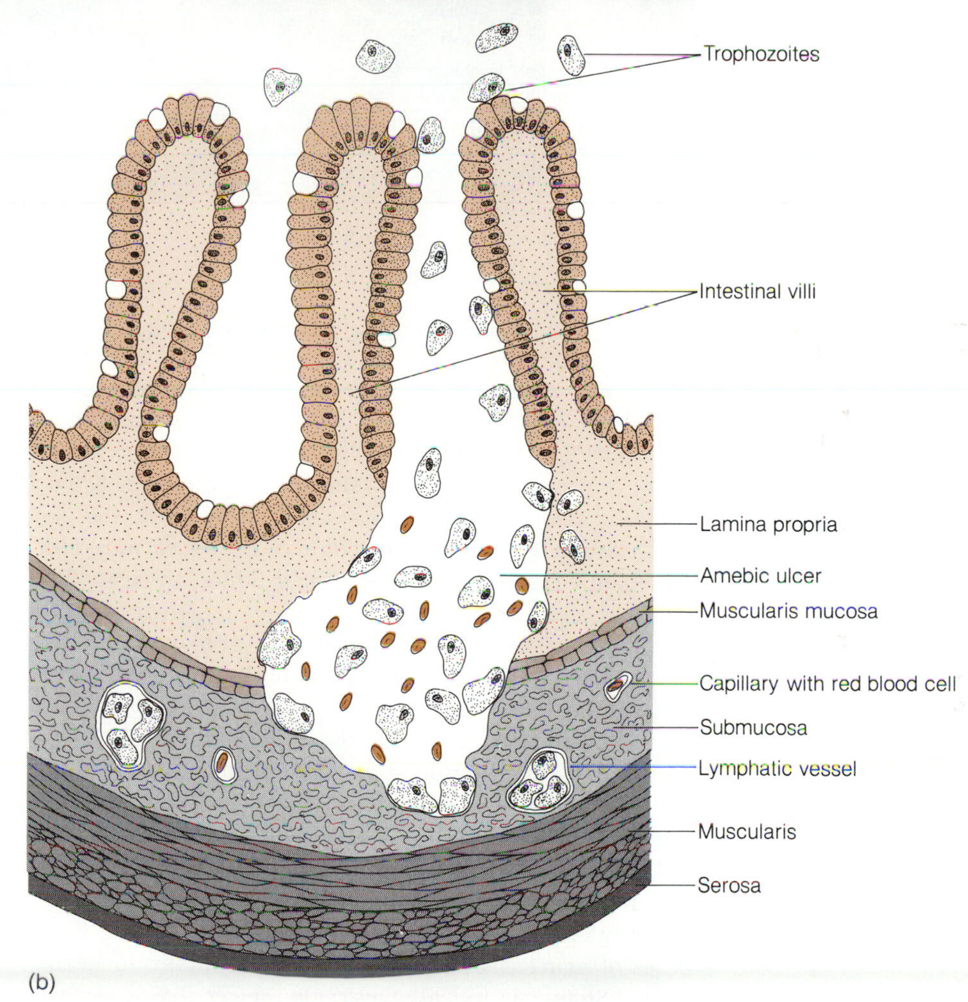

(b)

Figure 25-11 **Amebic Ulcers.** *(a)* Liver section showing amebic abscess. Arrows indicate abscess area of liver with many trophozoites of *Entamoeba histolytica*. *(b)* Diagram illustrating the formation of an intestinal abscess. The invasive trophozoites of *Entamoeba histolytica* erode the mucosa and submucosa.

for the fact that the trophozoites are rarely found in stools and can be seen only in the severest of diarrheas. The disease might be confused with amebiasis, but is characterized by the foul-smelling, greasy stool and the flatulence.

The symptoms of the disease may be explained only in part by tissue destruction caused by the sucking disks. In fact, examination of biopsy materials reveals little, if any, invasion of the mucosa. On closer examination

Figure 25-12 ***Giardia* in Mouse Intestines.** Scanning electron photomicrograph of *Giardia muris* on mouse jejunal mucosa. The parasite adheres to the mucosa, using its sucking disc. Lesions caused by the sucking discs are indicated. This adherence may be responsible for some of the signs and symptoms of giardiasis.

with an electron microscope, however, one can see that the organism may damage microvilli. This damage may result in some abnormality in the small intestine so that it does not absorb nutrients and electrolytes efficiently. The electrolyte imbalance may be the cause of the diarrhea.

Giardia intestinalis is a cosmopolitan microorganism and is a common cause of travellers' diarrhea. Most of the infections are acquired by drinking contaminated water. Like those of *E. histolytica*, the cysts of *G. intestinalis* are resistant to chlorine, so chlorinated water may still contain infectious organisms. Outbreaks in the United States are usually associated with the consumption of raw surface water or potable (drinking) water supplies contaminated with sewage. Although this protozoan is resistant to chlorination, proper sedimentation, flocculation, and filtration of potable water effectively remove the cysts.

The drug of choice for the treatment of giardiasis is quinacrine HCl (Atabrine). Because numerous side effects have been documented with this drug, however, metronidazole (Flagyl) has been used as an effective alternative. This observation is noteworthy because Flagyl is also used to control infections caused by anaerobic bacteria. Thus, the disease known as giardiasis may be a reflection of a secondary bacterial infection, as well as the primary infection by *Giardia intestinalis.* In the last few years, about 11,000 cases of giardiasis have been repoted annually to health authorities in the United States, with deaths averaging fewer than 1 per year.

VIRAL DISEASES OF THE DIGESTIVE SYSTEM

Viral Hepatitis

Hepatitis is an inflammation of the liver, manifested by jaundice, hepatomegaly, fever, and other constitutional disorders. The clinical disease has been separated into two phases: the **preicteric phase** (before jaundice) and

FOCUS

INFANTS MAY BE PROTECTED FROM *GIARDIA* BY MOTHER'S MILK

Human milk contains an array of antimicrobial agents that protect the infant from infectious diseases. These antimicrobial agents include IgA and macrophages. In addition, there are many still-undefined antimicrobial agents in milk. One of them protects infants from parasitic protozoal diseases, such as **amebiasis** and **giardiasis.** Both of these diseases are common waterborne diseases in the United States and other parts of the world.

Giardiasis is particularly common in the United States, and infected children tend not to thrive. Some infants, however, recover rapidly and without treatment. Some investigators propose that because these infants are breast-fed that the anti-*Giardia* substances in the milk protect them from infection.

To test their hypothesis, investigators incubated *Giardia intestinalis* with normal human milk. In the presence of even low concentrations of human milk, the parasite was rapidly killed. The antimicrobial substance, believed to be a **bile salt–stimulated lipase,** is very effective in killing *Giardia.*

Although this phenomenon has been demonstrated only in culture systems, it is believed that it may also work *in vivo* and that it may be responsible for the protection of breast-fed infants against giardiasis.

the **icteric phase** (fig. 25-13). The preicteric period is characterized by fever, nausea, anorexia, hepatomegaly (enlargement of the liver), and tenderness in the right upper quadrant. Liver dysfunction tests, such as alkaline phosphatase (AP) and serum glutamic-oxalacetic transaminase (SGOT) levels, may give the clinician an indication of hepatitis (fig. 25-13). By the end of

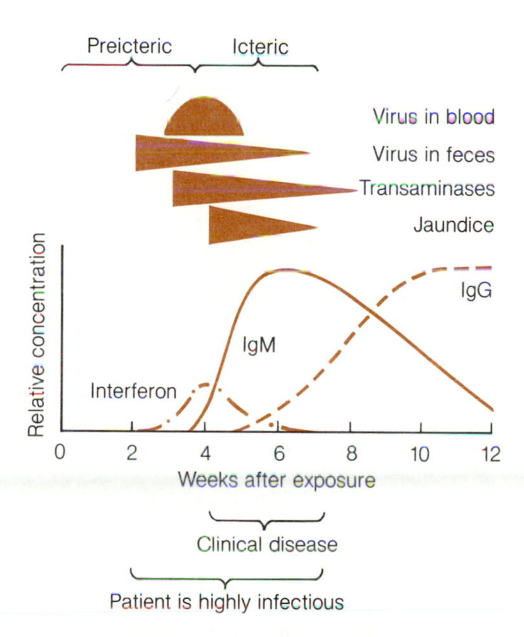

Figure 25-13 Course of Hepatitis A Virus Infection in Humans. Clinical disease caused by hepatitis A virus, which includes jaundice and liver dysfunction, is evident 3–4 weeks after infection. From week 2 to week 6, the patient is infectious and is shedding the virus in the feces. The peak of viremia (virus in the blood) is paralleled by an increase in interferon production. Immunoglobulins are not detectable until after 4 weeks post-infection.

the preicteric phase, serum bilirubin and SGOT are elevated and the patient is definitely jaundiced. These signs herald the beginning of the icteric phase, in which the patient begins to recover. Hepatitis may be caused by at least three different viruses: hepatitis A virus (HAV), hepatitis B virus (HBV), and non-A and non-B hepatitis viruses (NABV).

The hepatitis A virus traditionally was considered to be the agent of infectious hepatitis, acquired by the fecal-oral route □, while the hepatitis B virus and the non-A, non-B hepatitis viruses were considered to be the agents of serum hepatitis, acquired by parenteral injections or by puncture wounds with contaminated utensils (table 25-3). Blood transfusions using blood containing HBV and NABV and the use of contaminated syringes for injecting drugs have been cited as common ways of acquiring serum hepatitis. Recent studies have revealed, however, that HBV can be acquired by the fecal-oral route as well. HBV disease has a much longer incubation period (45–160 days) than that caused by HAV (15–50 days).

537

The host-parasite interaction in hepatitis acquired by the fecal-oral route, regardless of the agent, is essentially the same. Upon ingestion, the hepatitis virus enters the GI tract and multiplies there, probably in the cells of the mucosal epithelium. During this time, the virus may be shed in the feces. Eventually, the viruses enter the circulatory system, causing a viremia, and some of the viruses lodge in the liver parenchymal cells, where they multiply further. The virus continues to be shed in the feces. The liver damage may result from the multiplication of the pathogen or from immune reactions to viral antigens. Whether the damage results from humoral or cell-mediated immunity is still uncertain. Recovery, however, appears to coincide with the development of immunity and the inhibition of viral multiplication.

In epidemics, the source of the infection can almost always be traced to either water or food. Direct spread may also occur. Epidemics of hepatitis have been documented in which mentally handicapped children have acquired the infection by coprophagy. Shellfish and crustaceans living in con-

TABLE 25-3

COMPARISON OF TYPE A AND TYPE B HEPATITIS

CHARACTERISTIC	HEPATITIS-A VIRUS	HEPATITIS-B VIRUS	HEPATITIS-NON-A NON-B
Incubation period	Short (15–50 days)	Long (50–160 days)	Short-long (15–140 days)
Commonest route of infection	Fecal-oral	Injection, fecal-oral	Injection, fecal-oral
Nosocomial infection	Rare	Frequent	Frequent
Onset	Sudden	Insidious (gradual)	Insidious (gradual)
Serum transaminase (SGOT)	Temporary elevation	Prolonged elevation	Fluctuations
Humans most susceptible	Children	Adults	Adults
Time of year	Fall and winter months	Year-round	Year-round
High fever	Common	Rare	Rare
Serum immunoglobulin (IgM)	Elevated	Normal	?
Prevention using gamma-globulin	Successful	Doubtful	?
Viral antigens	Fecal antigen Liver antigen	Surface antigen (HB_s-Ag) Core antigen (HB_c-Ag)	?
Serum sickness	Rare (15%)	Rare (15%)	Very Rare
Cases/yr. in U.S.	>100,000	>200,000	>200,000
Carriers/yr. in U.S.	Very rare	10,000	25000
Deaths/yr. in U.S.	100	1,000	4,000

taminated water may serve as a source of infectious agent. Since there is no specific treatment available for hepatitis, the best course is prevention. A successful vaccination trial using heat-killed HV has been documented, with very encouraging results. Each year in the United States there are 55,000–60,000, reported cases of hepatitis and 500–600 deaths (fig. 25-14). The actual number of hepatitis cases is estimated to be closer to 500,000/year and the number of deaths more than 5,000/year.

Gastroenteritis of Viral Origin

Scientists estimate that fewer than 30% of the gastroenterites are of bacterial origin. What about the other 70%? Some of them are caused by protozoans, but the vast majority are caused by viruses. It is generally agreed that benign, self-limiting gastroenteritis with no definite bacterial etiology is of viral origin. The diseases are characterized by nausea, vomiting, diarrhea, fever, cramps, headache, and prostration. In the average clinical laboratory, there are no facilities for isolating these agents from the specimens; therefore, diagnosis is made on epidemiological and clinical grounds and on the absence of bacterial or protozoal etiology. Viruses suspected of causing gastroenteritis in humans include picorna-, echo-, parvo-, and rotaviruses □. Some of these viruses can cause waterborne and foodborne outbreaks. Outbreaks have also been traced to contaminated serving utensils.

A group of double-stranded DNA viruses, called the **rotaviruses,** is becoming increasingly important as an agent of viral diarrheas. These viruses are the most common cause of a widespread disease called **infantile enteritis.** The rotaviruses cause considerable vomiting and diarrhea, which lead to dehydration. The infected children, most commonly 6–25 months of age, may also exhibit fever and upper respiratory tract involvement. Infected adults generally are asymptomatic. Most cases of rotavirus diarrhea occur during the winter months and run their course within a two-week period. Apparently, the rotaviruses are widely distributed in human populations, and they are usuallly transmitted by the fecal-oral route. The treatment of rotavirus-caused diarrheas, as with most other virus-caused diarrheas, is administration of fluids and electrolytes to replace those lost during vomiting and diarrhea.

Microorganisms and viruses are responsible for a large variety of diseases of the digestive system. Some of these are summarized in table 25-4.

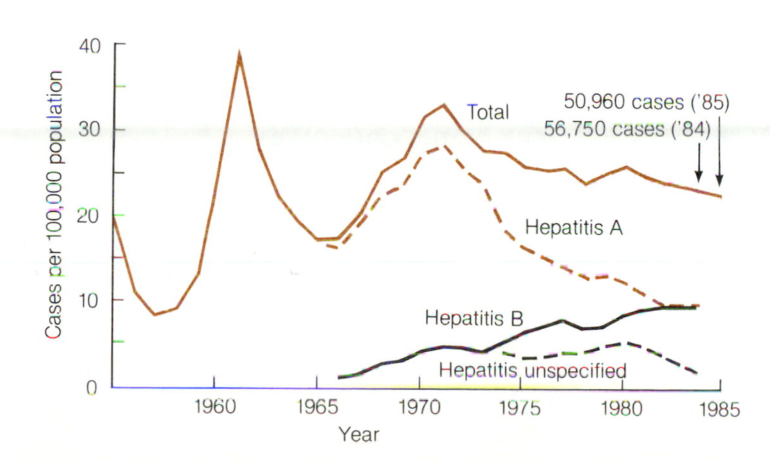

Figure 25-14 Reported Cases of Hepatitis in the United States

TABLE 25-4
SUMMARY OF INFECTIOUS DISEASES OF DIGESTIVE SYSTEM

INFECTIOUS DISEASE	CAUSE OF DISEASE	MICROBIAL GROUP
Amebiasis	*Entamoeba histolytica*	Amebic protozoa
Ascariasis	*Ascaris lumbricoides*	Nematode
Bacillary dysentery	*Shigella* spp.	Gram negative rods
Balantidiasis	*Balantidium coli*	Ciliated protozoa
Brucellosis	*Brucella melitensis*	Gram − rod
Cholera	*Vibrio cholerae*	Gram − curved rod
Diphyllobothriasis	*Diphyllobothrium latus*	Tapeworm (cestode)
Enterocolitis	*Clostridium difficile*	Gram + anaerobic sporeformer
Fasciolasis	*Fasciola hepatica*	Liver fluke (trematode)
Gastroenteritis	*Campylobacter fetus*	Gram − curved rod
Gastroenteritis	*Cryptosporidium*	Protozoa
Gastroenteritis	echoviruses	Picornaviruses
Gastroenteritis	parvoviruses	Parvoviruses
Gastroenteritis	rotaviruses	Reovirus
Giardiasis	*Giardia intestinalis*	Flagellated protozoa
Hepatitis-A	Hepatitis-A virus	Picornavirus
Hepatitis-B	Hepatitis-B virus	Comlex virus?
Hepatitis-non-A non-B	Hepatitis-non-A non-B virus	Retrovirus
Hookworm disease	*Ancylostoma* or *Necator*	Nematodes
Opistorchiasis	*Opistorchis sinensis*	Liver fluke (trematode)
Pinworms	*Enterobius vermicularis*	Nematode
Poliomyelitis	Poliovirus	Picornavirus
Salmonellosis	*Salmonella enteritidis*	Gram − rods
Schistosomiasis	*Schistosoma* spp.	Blood flukes (trematodes)
Taeniasis (tapeworm)	*Taenia solium* or *T. saginatus*	Tapeworms (cestodes)
Traveler's diarrhea	*Escherichia coli*	Gram − rod
Traveler's diarrhea	*Giardia intestinalis*	Flagellated protozoa
Trichinosis	*Trichinella spiralis*	Nematode
Typhoid fever	*Salmonella typhi*	Gram − rod
Weil's disease	*Leptospira* sp.	Spirochete
Whipworm disease	*Trichuris trichura*	Nematode
Yersiniosis	*Yersinia enterocolitica*	Gram negative rod

SUMMARY

The digestive tract serves as a common route of human infection by microorganisms.

STRUCTURE AND FUNCTION OF THE DIGESTIVE SYSTEM

1. The digestive system is made up of the gastrointestinal tract and accessory organs. The GI tract is a system of tubes through which food passes and nutrients are absorbed into the blood. The accessory organs aid in the digestion and storage of nutrients.
2. The GI tract is made up of four tunics: the mucosa, the submucosa, the muscularis, and the serosa.

NORMAL FLORA OF THE DIGESTIVE SYSTEM

1. The digestive system is colonized by a variety of microorganisms. The distribution of these microorganisms varies, depending upon the anatomical site.
2. The normal flora of the digestive system is established soon after birth. It lives on excess host nutrients or mucous exudates.
3. The normal flora protects the host from invading pathogens, and some of these organisms supply the host with essential chemicals.

DISEASES OF THE ORAL CAVITY

1. The most important diseases of the oral cavity are dental caries, periodontal disease, and mumps.
2. Dental caries result after acidogenic microorganisms in dental plaque produce acid metabolites that break down the enamel and dentine. Other microorganisms may cause further damage to the tooth. Preventing the formation of dental plaques by brushing and flossing is a successful means of avoiding caries.

3. Periodontal disease results from an inflammatory reaction caused by microorganisms colonizing dental plaques near the gums.
4. Mumps is a childhood disease, characterized by an inflammation of the salivary glands caused by an infection with the mumps virus. The disease, which can have serious consequences in adults, is prevented by vaccination.

INFECTIOUS DIARRHEAS CAUSED BY BACTERIA

1. Diarrhea, a symptom of many infections of the GI tract, is characterized by frequent, watery stools.
2. Infections of the small intestine generally result in profuse, watery discharge, while infections of the large intestine are accompanied by mucosal erosion and inflammatory exudate.
3. Cholera is caused by *Vibrio cholerae* and is characterized by profuse, whitish, watery diarrhea that can lead to dehydration, shock, and death. The disease is caused by the toxin choleragen, which is produced by the bacteria reproducing in the small intestine. The transmission of the disease is by the fecal-oral route. The disease is treated by replacement of body fluids and electrolytes. Tetracycline helps in alleviating the disease because it reduces the number of vibrios in the intestines.
4. Typhoid fever is caused by *Salmonella typhi*, and is characterized by high fever, headache, diarrhea, and rose spots. The disease results when the bacteria invade the blood and spread to the liver, multiplying inside macrophages and causing inflammatory responses. The disease is transmitted by the fecal-oral route and is treated with ampicillin or chloroamphenicol.
5. Bacillary dysentery is caused by several species of the genus *Shigella*. It is an inflammation of the large bowel, resulting in tenesmus, diarrhea, cramps, and inflammatory exudate. The disease is transmitted by the fecal-oral route. Severe infections are treated with ampicillin, chloroamphenicol, or trimethoprim-sulfamethoxazole.
6. *Salmonella* gastroenteritis is a diarrheal disease accompanied by headaches, abdominal pain, vomiting, and fever. The disease results from the multiplication of the bacteria in the small intestine and is acquired by the fecal-oral route. Treatment of the disease generally is not necessary.
7. Certain strains (or serogroups) of *Escherichia coli* are responsible for diarrheal diseases that resemble those caused by *Vibrio*, *Salmonella*, and *Shigella*. These diseases have epidemiologies similar to those of other diarrheal diseases.
8. *Campylobacter jejuni*, *Yersinia enterocolitica*, and *Clostridium difficile* are human pathogens that also cause diarrheal diseases.

FOOD INTOXICATIONS

1. Food intoxications are diseases that result from the ingestion of foods containing toxins produced by microorganisms.
2. Staphylococcal gastroenteritis is a very common form of food poisoning, characterized by a rapid onset of vomiting and diarrhea. It is caused by the action of *Staphylococcus aureus* enterotoxins on the epithelium of the digestive tract.
3. Toxins produced by *Clostridium perfringens* cause a form of gastroenteritis consisting of abdominal pain, cramps, and diarrhea. The disease is acquired by ingestion of warmed foods containing clostridial enterotoxins.

PROTOZOAL INFECTIONS OF THE DIGESTIVE SYSTEM

1. Amebiasis, a disease caused by *Entamoeba histolytica*, is characterized by dysentery. The disease results from the multiplication of the pathogen in the large bowel causing an acute inflammation. The cysts of *E. histolytica* enter the GI tract in contaminated water or foods. Treatment is by administration of Flagyl or Furamide.
2. Giardiasis is caused by *Giardia intestinalis*. It infects the small intestine, causing diarrhea, nausea, vomiting, flatulence, and greasy stools. The characteristic disease, which results from the multiplication of the flagellate on intestinal epithelium, results in malabsorption of lipids and certain vitamins. Treatment of giardiasis is achieved by administration of Atabrine or Flagyl.

VIRAL DISEASES OF THE DIGESTIVE SYSTEM

1. Hepatitis is an inflammation of the liver which is caused by hepatitis viruses. It is characterized by jaundice, hepatomegaly, and fever. The viruses, HAV, HBV, and non-A non-B virus, enter the host either by the fecal-oral route or parenterally (e.g., by injections or blood transfusions). The disease is caused by the virus replicating in GI epithelium and liver parenchyma. A delayed-type hypersensitivity, which results from prolonged viral infection, may be responsible for some of the signs and symptoms of the disease.

2. Gastroenterites caused by viruses are quite common in human populations. They are generally benign and self-limiting, with nausea, vomiting, and diarrhea. Rotaviruses and parvoviruses are very common causes of viral diarrheas.

STUDY QUESTIONS

1. What is food poisoning? How does it differ from a food intoxication?
2. Describe the epidemiology of:
   a. cholera;
   b. typhoid;
   c. infantile diarrhea;
   d. hepatitis;
   e. giardiasis;
   f. travellers' diarrhea;
   g. bacillary dysentery.
3. For each of the above-mentioned diseases, describe the causative agent and discuss the way in which it causes the disease.
4. How do infections of the small intestine differ from those of the large intestine? Why?
5. Compare and contrast the diseases caused by *Shigella*, *Vibrio cholerae*, and *Salmonella typhi*.
6. How could infectious diseases of the GI tract be diagnosed? How about food intoxications?
7. Define the following:
   a. diarrhea;
   b. tenesmus;
   c. staphylococcal food poisoning;
   d. viral gastroenteritis;
   e. enterocolitis.
8. Explain the role that the following play in the transmission of enteric pathogens:
   a. food;
   b. water;
   c. aerosols;
   d. fomites;
   e. domestic animals.
9. Describe the role(s) of the normal flora of the GI tract.

SUPPLEMENTAL READINGS

Binder, H. J. 1984. The pathophysiology of diarrhea. *Hospital Practice* 19:107–129.

Cukor, G. and Blacklow, N. R. 1984. Human viral gastroenteritis. *Microbiological Reviews* 48(2):157–179.

Elwell, L. P. and Shipley, P. L. 1980. Plasmid-mediated factors associated with virulence of bacteria to animals. *Annual Review of Microbiology* 34:465–496.

Mandel, I. D. 1979. Dental caries. *American Scientist* 67:680–687.

Middlebrook, J. L. and Dorland, R. B. 1984. Bacterial toxins: cellular mechanisms of action. *Microbiological Reviews* 48(3):199–221.

Murphy, A. M., Borden, E. C., and Crewe, E. B. 1977. Rotavirus infections of neonates. *Lancet* 2:1149–1152.

Noble, E. R. and Noble, G. A. 1982. *Parasitology: The biology of parasites.* 5th ed. Philadelphia: Lea & Febiger.

O'Brien, A. D., Newland, J., Hiller, S., Holmes, R., Smith, H., and Formal, S. 1984. Shiga-like toxin-converting phages from *Escherichia coli* strains that cause hemorrhagic colitis or infantile diarrhea. *Science* 226:694–696.

Rubin, R. H., Hopkins, C. C., and Swartz, M. N. 1985. Infections due to gram negative bacilli. *Scientific American Medicine* Section 7-II New York: Freeman.

Savage, D. C. 1977. Microbial ecology of the gastrointestinal tract. *Annual Review of Microbiology* 31:107–133.

Smibert, R. M. 1978. The genus *Campylobacter*. *Annual Review of Microbiology* 32:673–709.

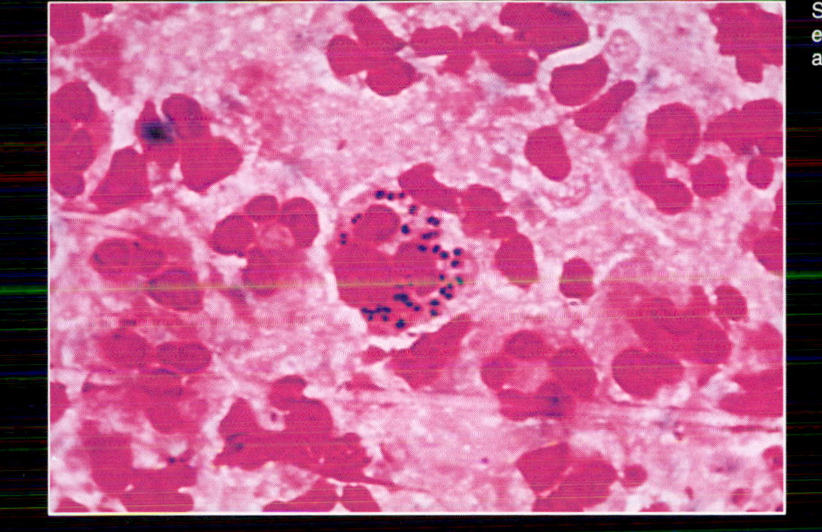

Stained smear of urethral discharge exhibiting *Neisseria gonorrhoeae* inside a neutrophil (center of photograph).

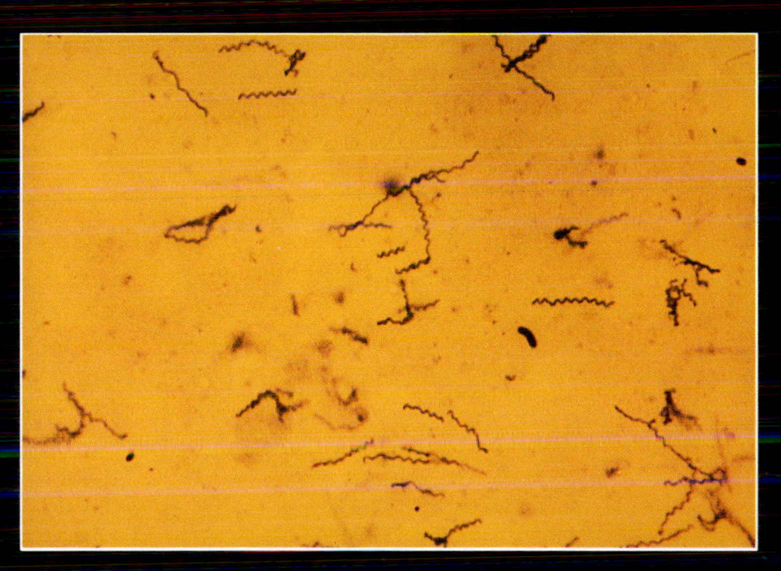

Treponema pallidum obtained from a primary chancre of a syphilitic patient.

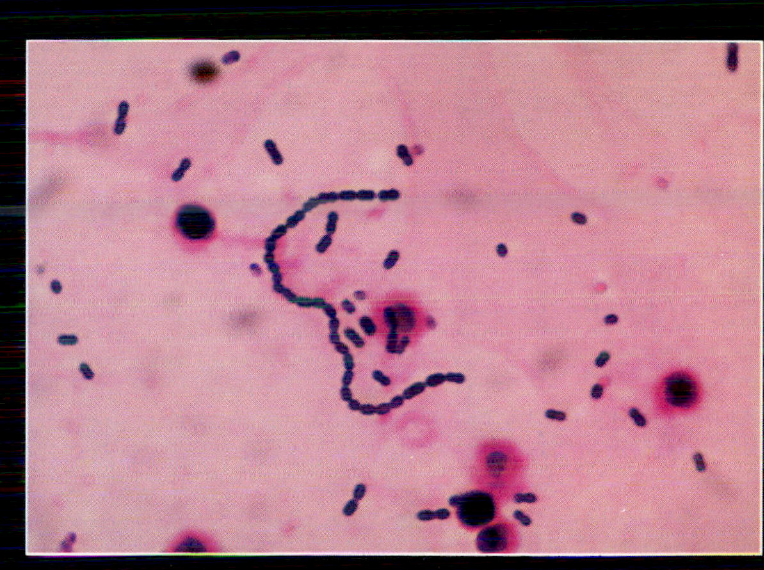

Streptococcus pneumoniae obtained from a blood broth culture.

Acid fast stain of cutaneous nerve section invaded with numerous cells of *Mycobacterium leprae*.

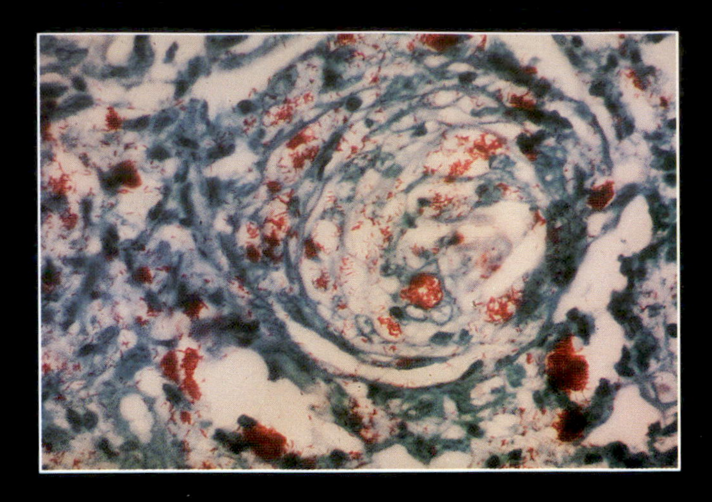

Fluorescent antibody stain of *Neisseria gonorrhoeae*.

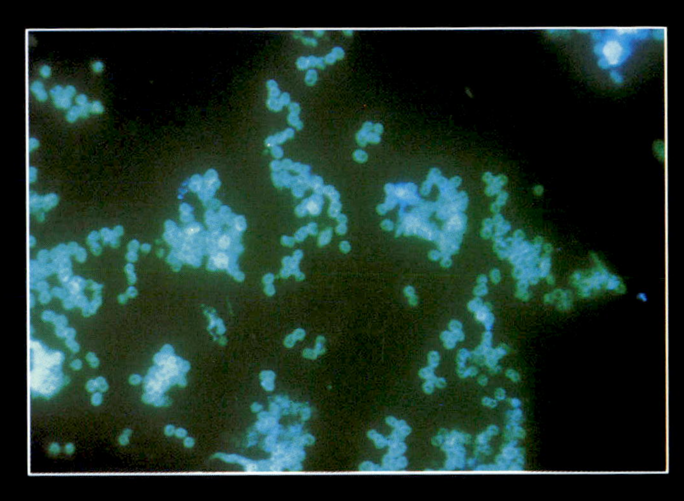

Fluorescent antibody stain of *Yersinia pestis*.

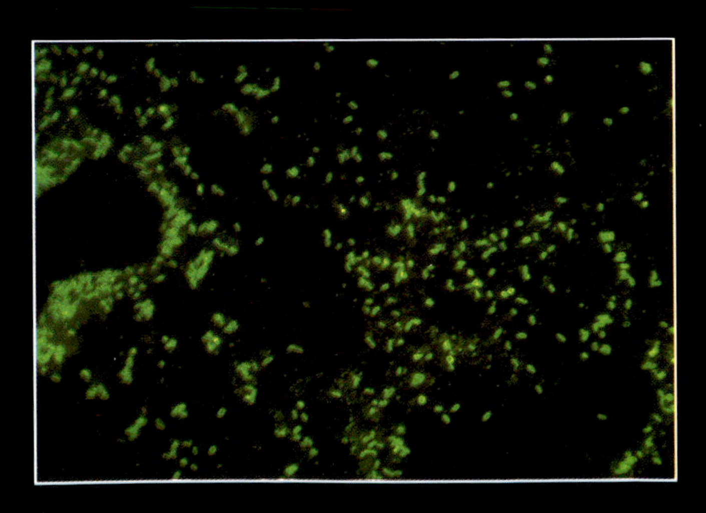

Fluorescent antibody stain of formalin-fixed lung tissue from a fatal case of Legionnaire's disease.

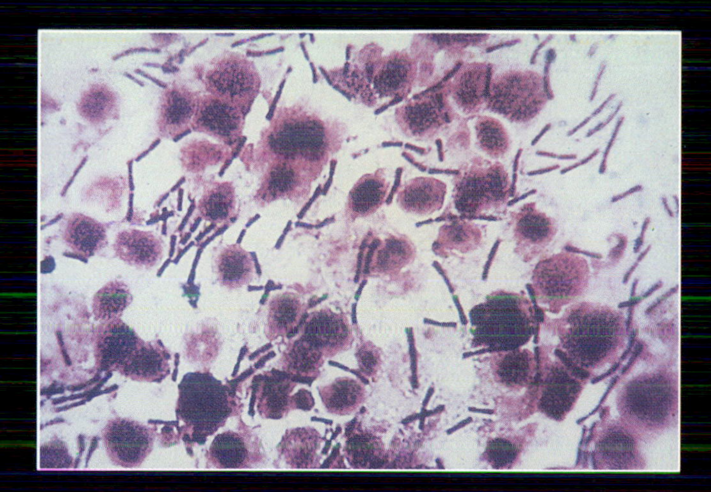

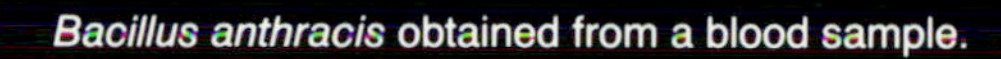

Bacillus anthracis obtained from a blood sample.

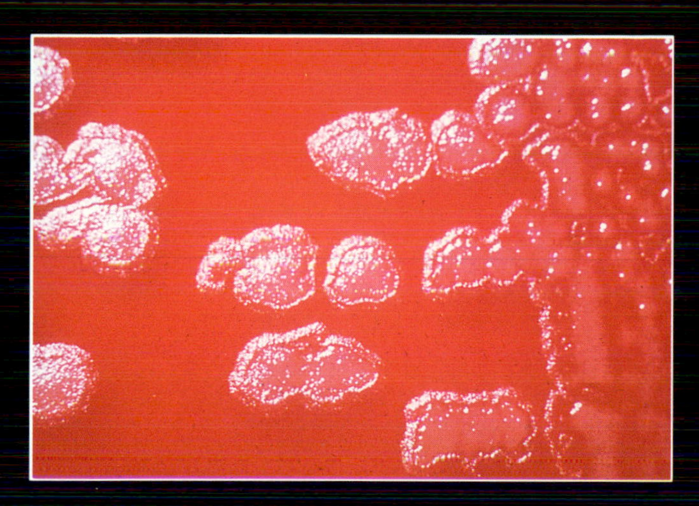

Bacillus anthracis on blood agar.

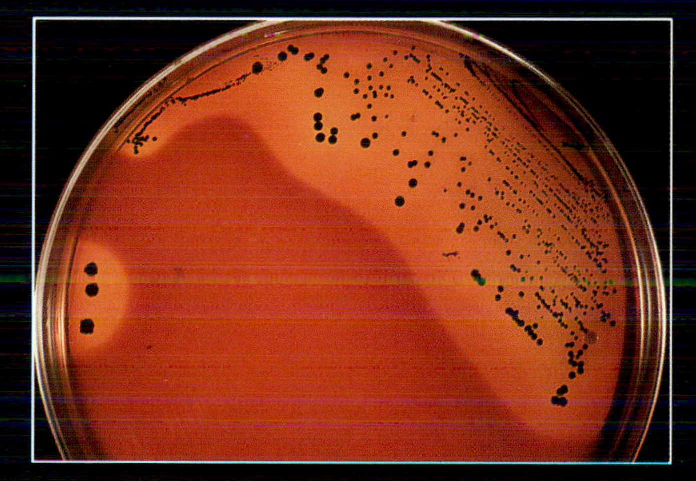

Bacteroides melaninogenicus on blood agar plate. Black colonies are surrounded by areas of β hemolysis.

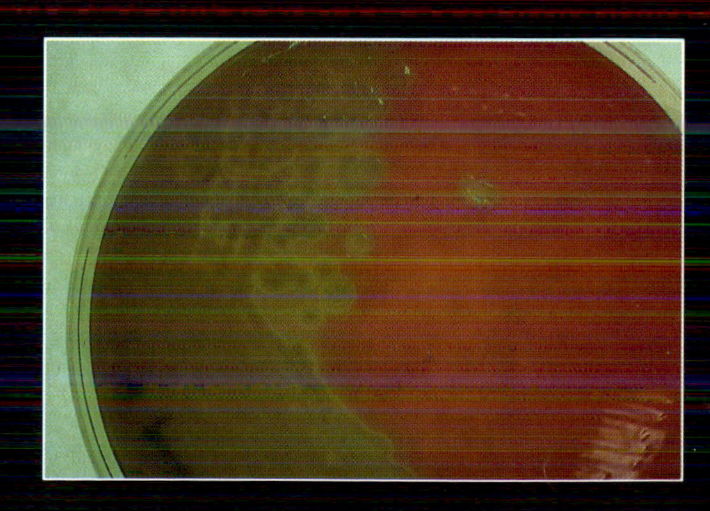

Streptococcus pneumoniae on blood agar plate. Colonies are surrounded by greenish zones of α hemolysis.

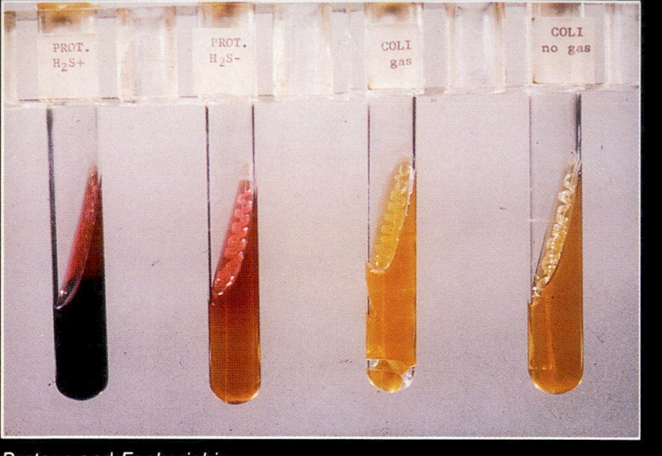

Proteus and *Escherichia coli* on triple-sugar-iron slants. The black portions indicate H_2S production. The red areas indicate peptonization of medium. The yellow areas indicate sugar fermentation. The broken agar reflects gas formation.

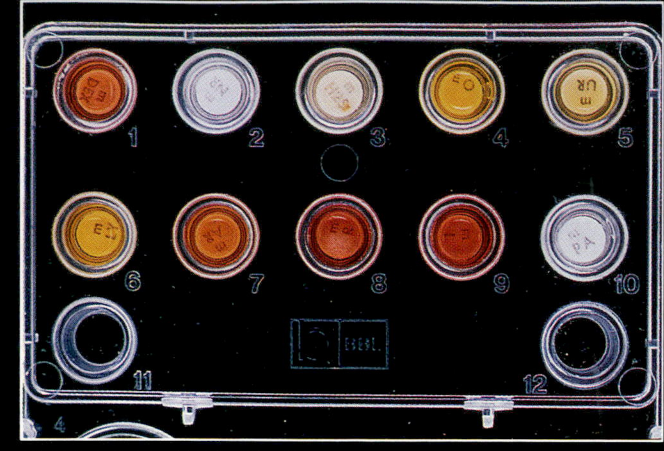

Minitek tray used in the identification of bacterial isolates.

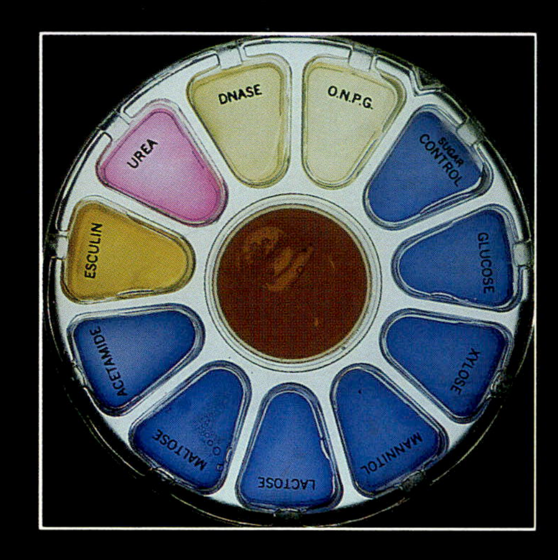

Multitest plate distributed by Flow Laboratories for the identification of gram negative, nonfermentative bacteria.

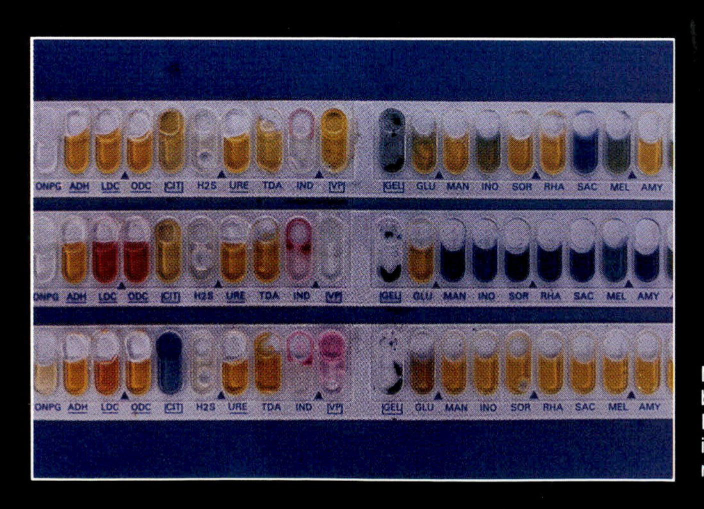

Multitest trays produced by Analytab Products Incorporated (API) for the identification of gram negative bacterial isolates.

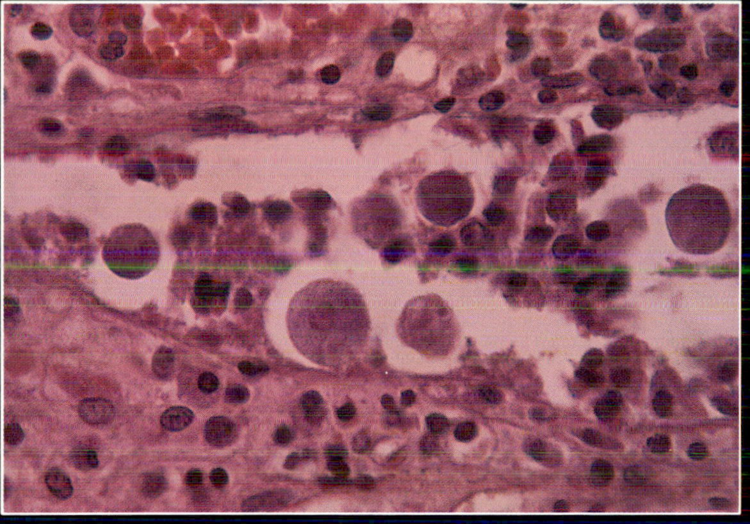

Histological section stained with hematoxylin and eosin showing *Entamoeba histolytica* in eroded region (center) of intestinal epithelium.

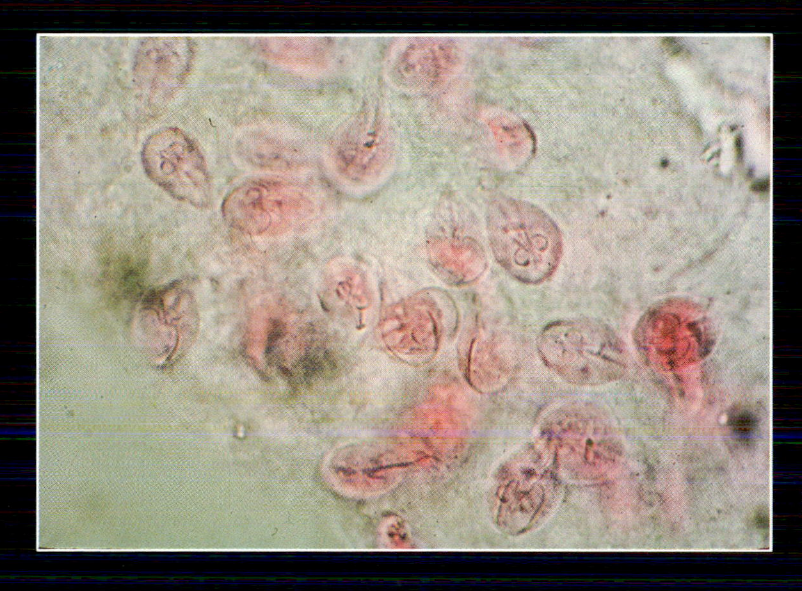

Giardia intestinalis trophozoites in human feces.

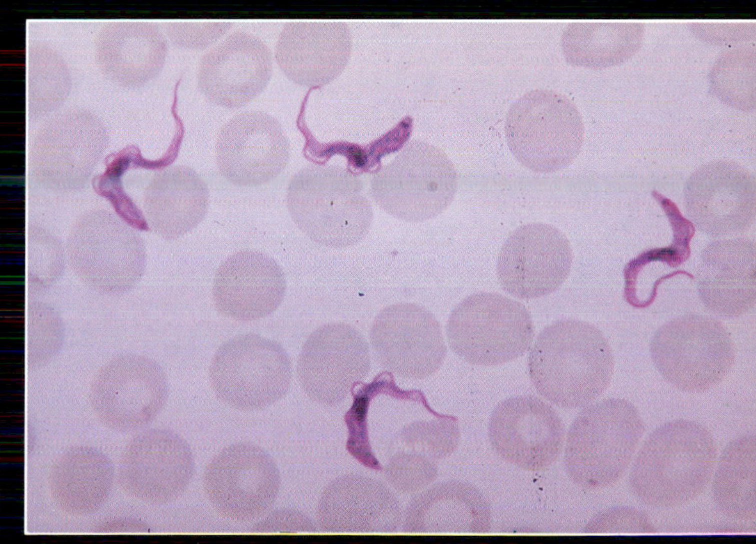

Trypomastigotes of *Trypanosoma brucei gambiense* in the blood of a 57-year-old missioinary with African sleeping sickness.

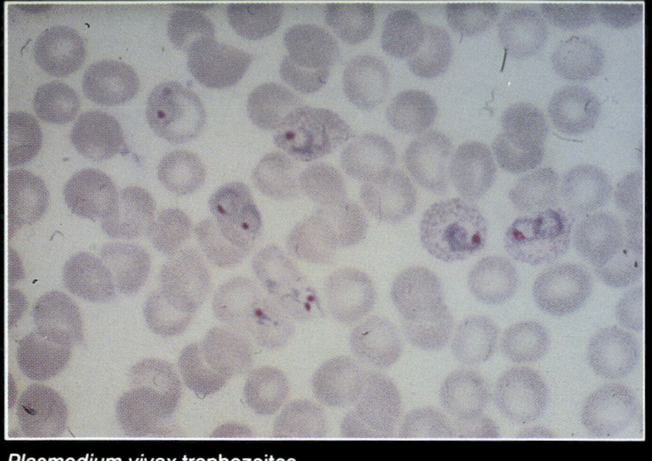

Plasmodium vivax trophozoites and rings in red blood cells.

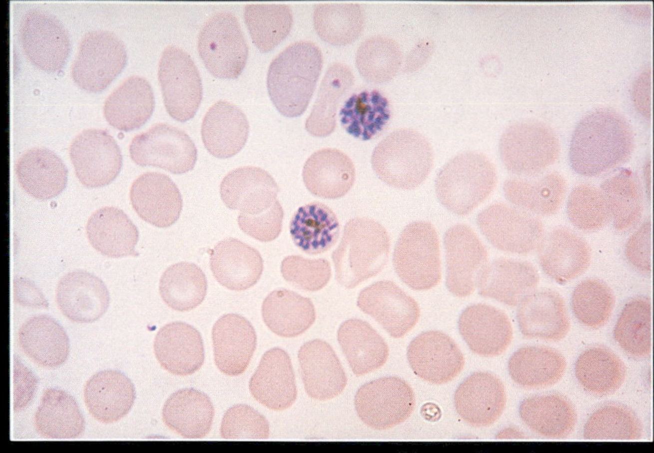

Plasmodium vivax infecting red blood cells.

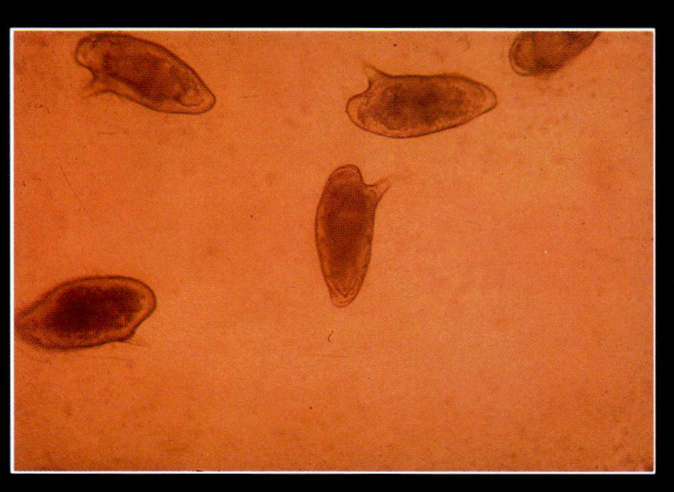

Ova of the fluke *Schistosoma mansoni* as seen in wet mounts of human stools.

Cercaria of the fluke *Schistosoma mansoni* originating from infected snails. This stage is infectious for humans.

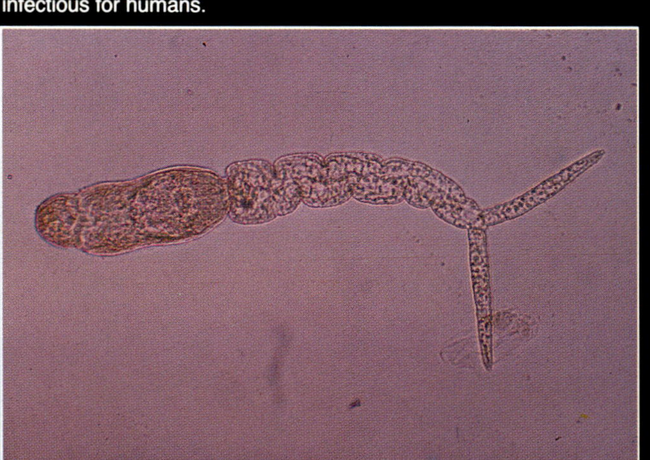

Adult male and female *Schistosoma mansoni* mating in mesenteric capillaries.

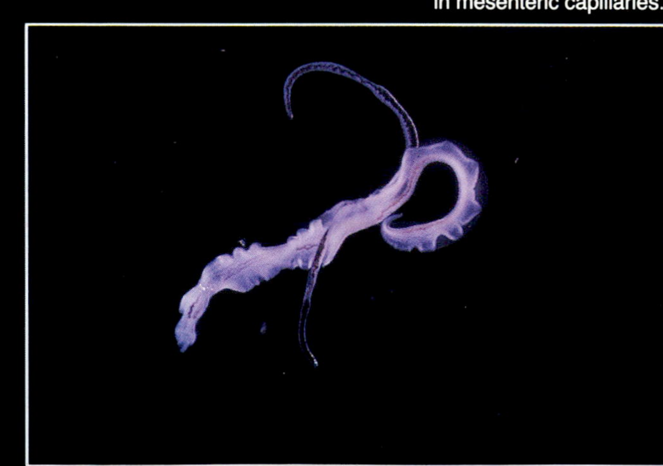

Scolex of the tapeworm *Taenia*.

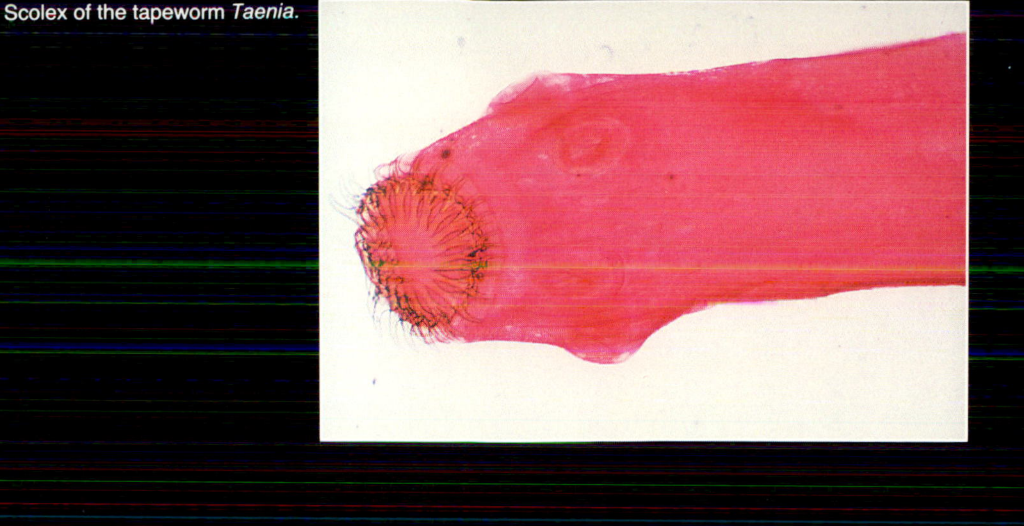

Encysted larva of the nematode *Trichinella spiralis* as seen in infected human muscle tissue.

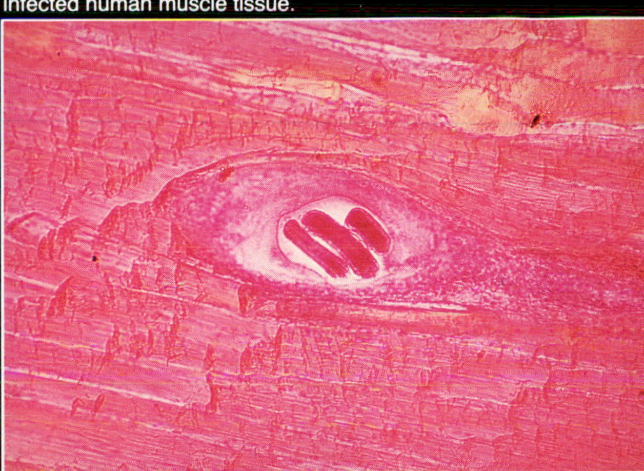

Cross section of human lung showing numerous alveoli and three broncheoli.

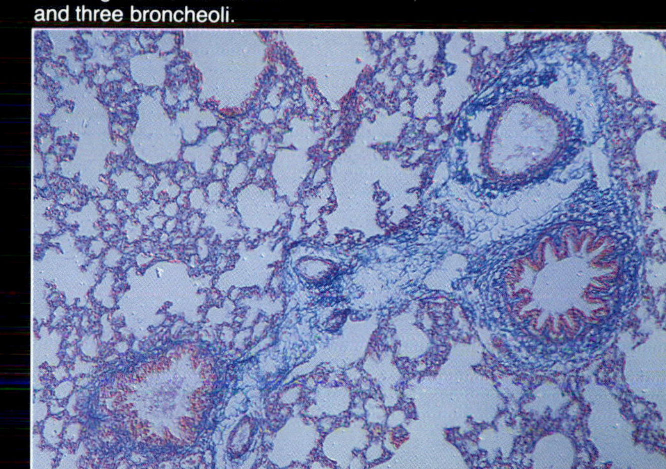

Section of mouse lung showing adiaspore of *Emmonsia crescens* completely surrounded by a multinucleated giant cell.

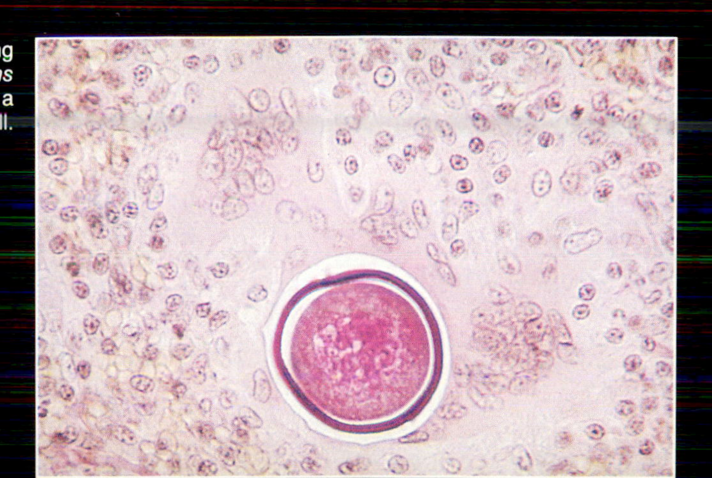

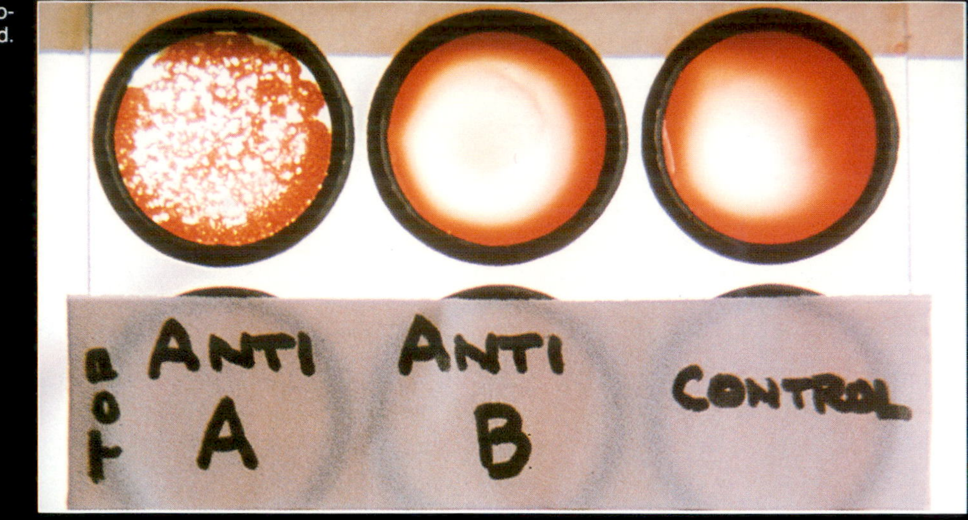

ABO blood typing. Photograph shows A-type blood.

Immunoelectrophoresis of blood serum. Each band represents a different serum component.

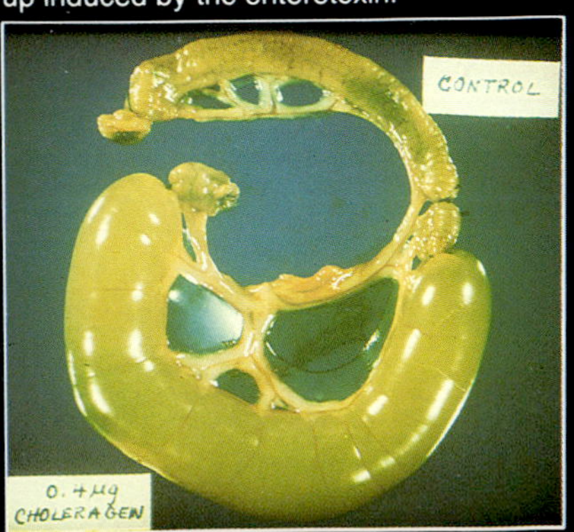

Rabbit ileal loop inoculated with cholera enterotoxin and control. The swelling is due to fluid build-up induced by the enterotoxin.

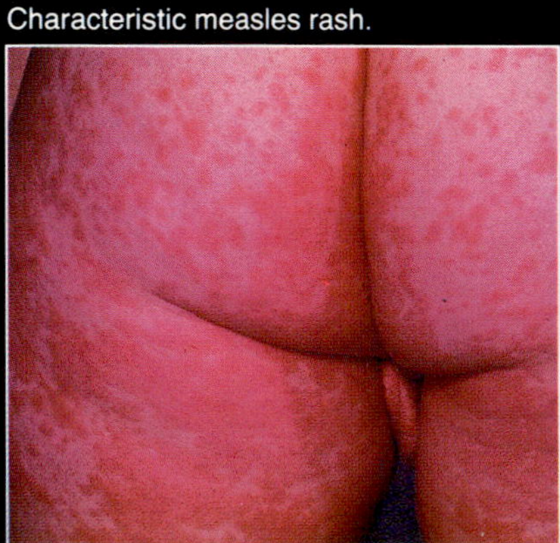
Characteristic measles rash.

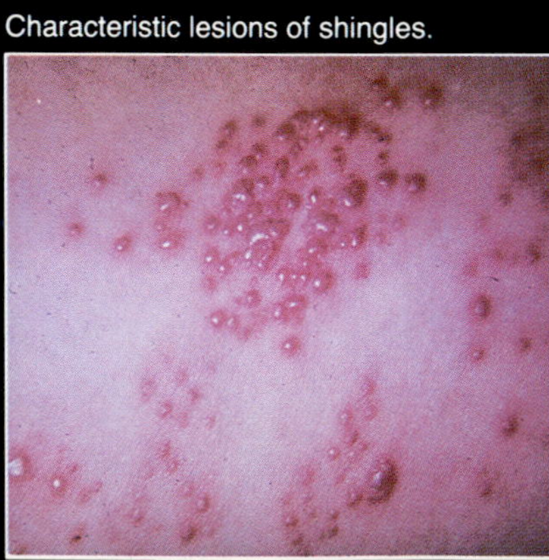
Characteristic lesions of shingles.

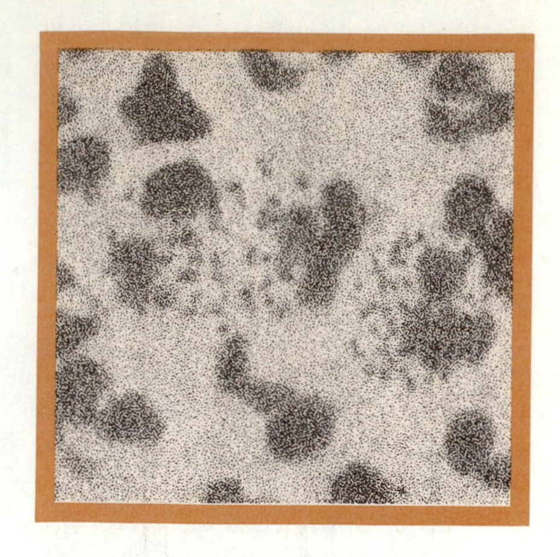

CHAPTER 26
DISEASES OF THE UROGENITAL TRACT

CHAPTER PREVIEW

TOXIC SHOCK SYNDROME AND THE USE OF TAMPONS AND CONTRACEPTIVES

During 1980 and 1981 there were numerous outbreaks of toxic shock syndrome (TSS). Most of these cases were shown to be associated with the use of superabsorbent tampons during menstruation and the practice of not changing the tampons frequently. Because many women were uncertain about the safety of tampons, they looked for other ways of absorbing their menstruation. Some women turned to sea sponges, menstrual cups, and contraceptive sponges in the hope that these would reduce the risk of toxic shock syndrome. Unfortunately, there have been cases of toxic shock syndrome associated with the use of some of these materials.

In 1983, vaginal sponges made of polyurethane and imprenated with the spermicide nonoxynol-9 were released as contraceptives. The manufacturer's label recommended that the sponges not be left in place for more than 30 hours. Late in 1983, four cases of toxic shock syndrome were associated with the use of vaginal contraceptive sponges. One of the patients had been unsuccessful in removing the sponge because it fragmented. Another patient had not removed her sponge for 32 hours, and a third patient had the sponge in place for about 5 days. By keeping sponges in place for longer than about 8 hours, women increase the risk of *Staphylococcus aureus* growth and toxic shock syndrome.

A study has shown that women who wear birth control diaphragms for 24 hours, the maximum recommended by manufactuers, promote the growth of *Staphylococcus aureus* on the cervix and in the vagina. Normally, bacteria and their toxins are shed from the body by menstruation and vaginal secretions. The use of contraceptive diaphragms and sponges inhibits the normal shedding process and creates a favorable growth environment for the bacteria. Approximately 10 women per year risk toxic shock syndrome because of their use of contraceptive diaphragms.

A minimum estimate of the danger of contracting toxic shock syndrome from contraceptive sponges is 10 cases/100,000. Since approximately 250,000 women currently use the sponges each year, this amounts to 25 cases. In comparison, 2 cases/100,000 women who use tampons are expected each year. Because there are more than 25 million women in the U.S. who use tampons, however, most of the cases of toxic shock syndrome are associated with tampon use. The death rate due to toxic shock syndrome is about 3 percent.

Toxic shock syndrome is a much better understood disease now than it was in the early 1980s, largely because of the interest that it generated due to its severity. In this chapter we will discuss some of the most common diseases of the urogenital tract, their causes, epidemiology, prevention, and treatment.

LEARNING OBJECTIVES

A STUDY OF THIS CHAPTER SHOULD ENABLE YOU TO:

- DISCUSS THE IMPACT THAT DISEASES OF THE UROGENITAL SYSTEM HAVE ON HUMAN POPULATIONS
- LIST THE VARIOUS MICROORGANISMS THAT MAKE UP THE NORMAL FLORA OF THE UROGENITAL TRACT, AND DISCUSS THEIR ROLES IN HEALTH AND DISEASE
- LIST THE MAJOR PATHOGENS OF THE UROGENITAL TRACT
- DISCUSS THE MAJOR DISEASES OF THE UROGENITAL TRACT AND SUGGEST TREATMENTS AND WAYS OF PREVENTING THE DISEASES

Diseases of the urogenital tract may be so mild as to be asymptomatic, or so severe as to cause long-lasting pain, tissue degeneration, or death. Some of these diseases, particularly sexually transmitted diseases, still carry a social stigma that makes patients reluctant to seek medical help or to reveal the identity of their contacts. This stigma makes it extremely difficult to evaluate the extent of the disease in a population or to control it. Some sexually transmitted diseases have reached pandemic proportions in human populations, and urinary tract infections (UTI) afflict nearly every human being at some time in his or her life.

Pathogens such as *Chlamydia trachomatis* and herpes simplex are now known to be the cause of some very common sexually transmitted diseases. With the increased "social acceptance" of sexual activities; the classical sexually transmitted disease agents have increased in prevalence, and microorganisms not usually transmitted sexually, have also emerged as important pathogens. Hepatitis B, Group B streptococci, *Giardia intestinalis*, and *Entamoeba histolytica* have now been associated with sexual transmission.

This chapter will examine the clinical syndromes most commonly afflicting the human urogenital tract, and the host-parasite interactions leading to the clinical disease and to recovery.

STRUCTURE AND FUNCTION OF THE UROGENITAL TRACT

The Urinary System

The principal role of the urinary system is to eliminate wastes from the blood and to maintain a steady-state equilibrium between the fluid and the solutes of the blood, by voiding (urinating) wastes and excess fluids. In addition, the urinary system controls the volume of the blood by the amount of water it allows to be passed in the urine.

The human urinary system consists of two **kidneys,** two **ureters,** a **urinary bladder,** and the **urethra** (fig. 26-1). Each kidney empties fluid that has been removed from the blood into its ureter, which in turn empties into the urinary bladder. The urethra, which originates in the bladder, serves as a conduit for transporting the urine out of the body. This flow of urine dislodges many of the microorganisms in the urethra.

On gross examination, the kidney consists of a thin surface layer, the capsule, a cortex, and a medulla. The cortex contains primary **glomeruli** (fig. 26-1), in which diffusion of fluids and solutes from the blood into the kidney tubules takes place. The electrolytes and fluids that pass across the glomeruli enter the convoluted tubules (fig. 26-1) and then drain into the papillary ducts, which merge into the calyx of the medulla. The glomerulus, convoluted tubules and papillary duct make up the **nephron**, the functional unit of the kidney.

The Reproductive System

The reproductive system generates gametes, provides a means for fertilizing eggs, and nurtures the resulting embryo. In humans, as in other mammals, male and female sexes are well differentiated. Figure 26-2 illustrates the male and female reproductive systems.

The male reproductive system includes the **testes,** which produce the male gametes, or **sperm,** and a number of ducts that serve to store the

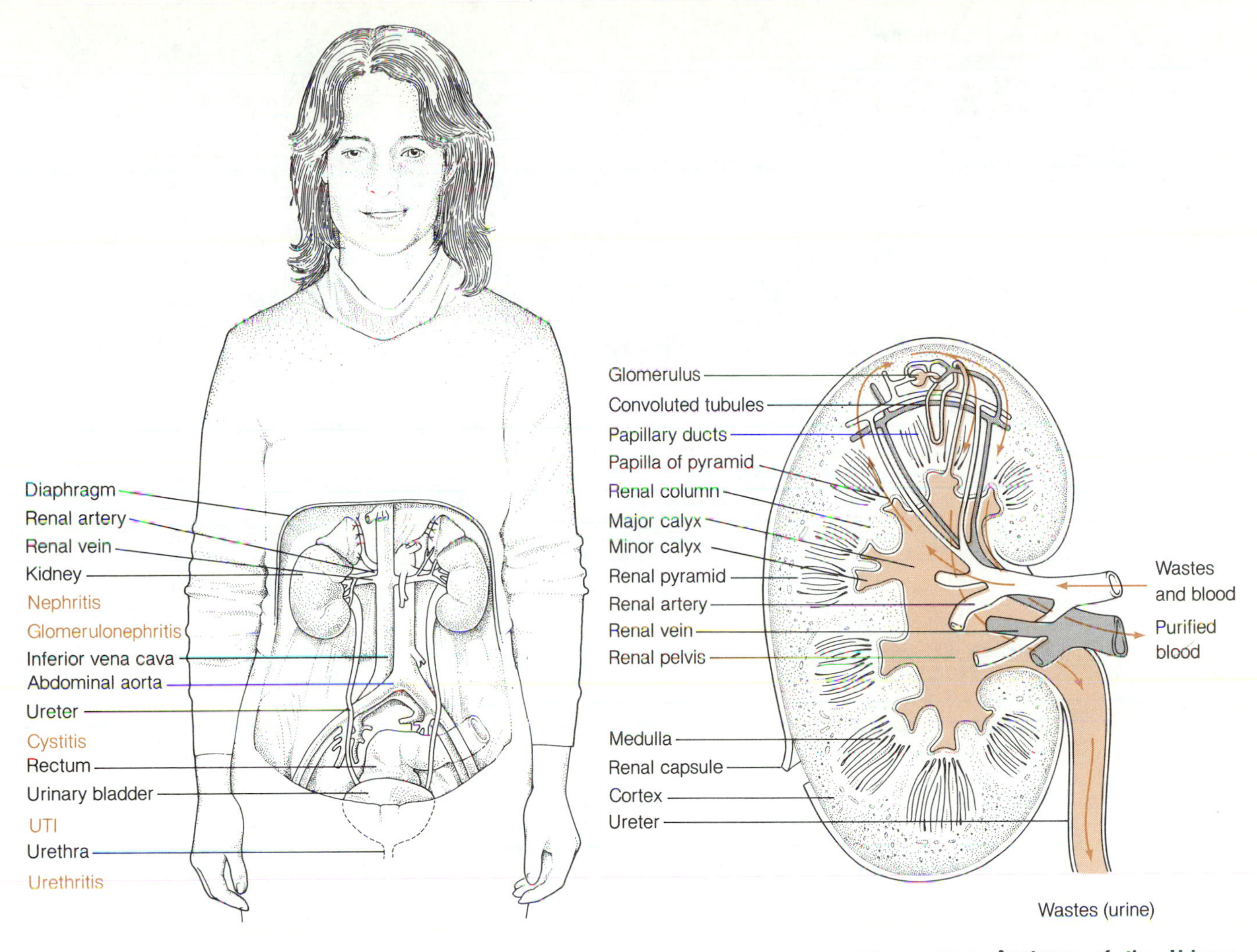

Figure 26-1 Anatomy of the Urinary System. Colored print indicates diseases that afflict the anatomical site.

sperm and transport it to the outside. In addition, there are several accessory glands that secrete substances to protect the sperm and form the semen.

The female reproductive system includes the **ovaries,** which produce the female gametes or **ova;** the **uterus;** the **fallopian tubes,** which serve as conduits for the ova from the ovaries to the **uterus;** and the **vagina.** The uterus is a muscular organ that serves as a receptacle for the fertilized ovum or zygote. Fetal development takes place in the uterus. The vagina serves as a receptacle for the penis during coitus or sexual intercourse and also as an exit for the developed fetus.

NORMAL FLORA OF THE UROGENITAL SYSTEM

Most of the urinary tract, except for the anterior urethra near the external orifice, normally is free of microorganisms because of urine flow. The anterior urethra, however, contains an array of bacteria, including staphylococci, diphtheroids, and gamma hemolytic streptococci (table 26-1). The female urethral opening may also contain lactobacilli.

Figure 26-2 The Human Reproductive System. Colored print indicates infectious diseases that afflict the anatomical site. *(a)* Female reproductive system. *(b)* Male reproductive system.

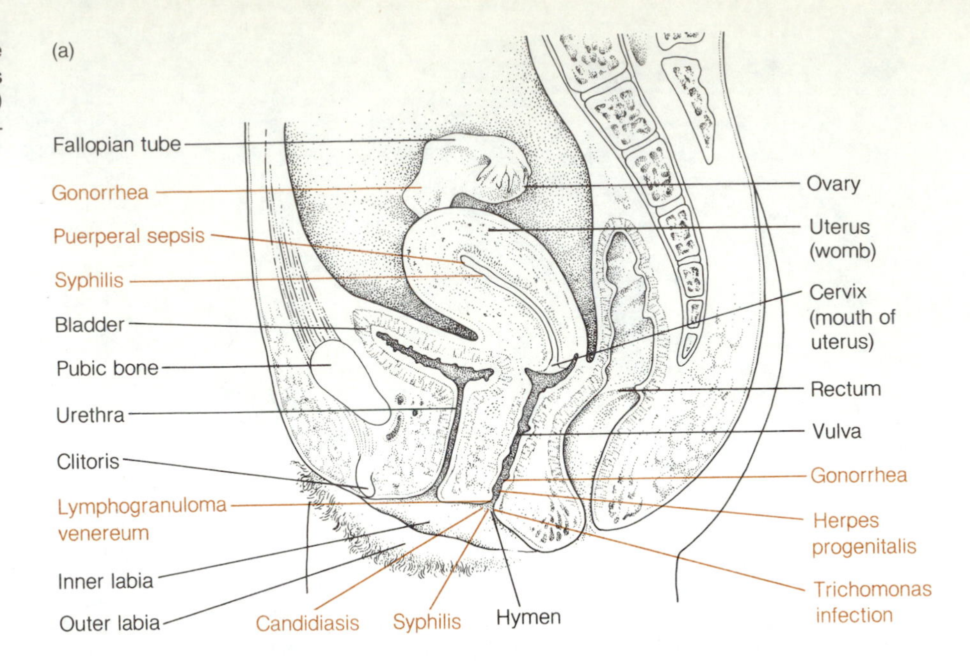

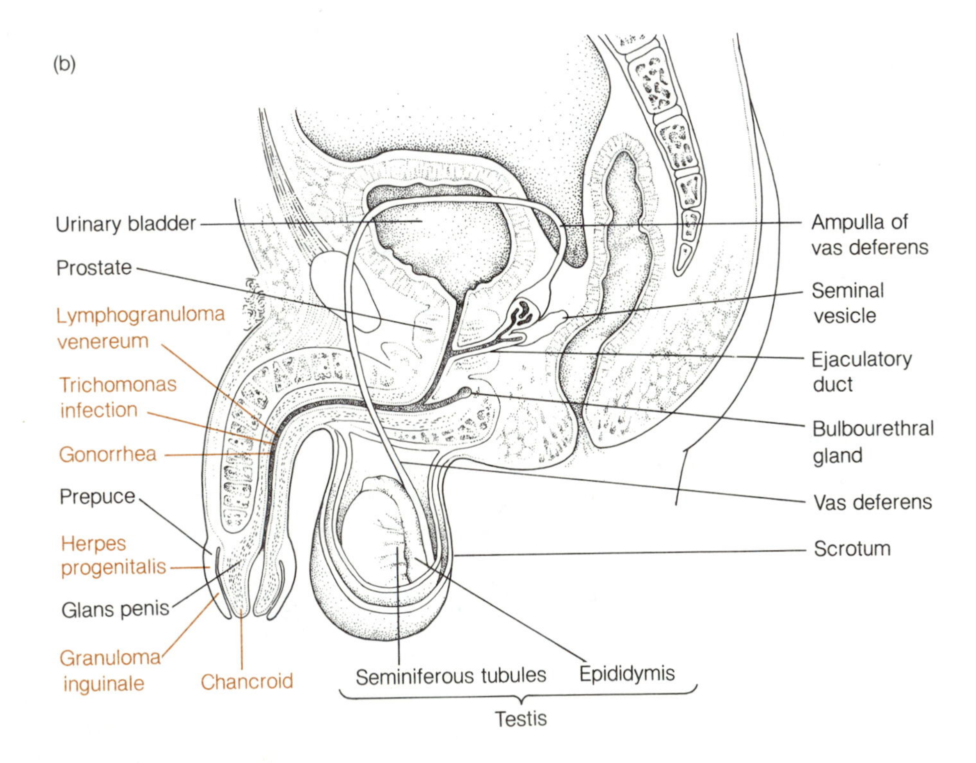

The vagina and cervix have a very complex and dynamic flora (table 26-1). Shortly after birth, the vagina is colonized predominantly by lactobacilli, because progesterone from the mother promotes the multiplication of the resident bacteria in the baby's vagina. After the maternal hormonal influences wane in the baby, the lactobacilli decline in numbers and the vagina becomes alkaline. During the time the vagina is alkaline, the normal flora

TABLE 26-1

MICROORGANISMS COMMONLY ISOLATED FROM THE UROGENITAL SYSTEM

MICROORGRANISM	MICROBIAL GROUP	AVERAGE FREQUENCY OF ISOLATION
Actinomyces sp.	Actinomycetes	25%
Bacteroides sp.	Gram − rods	70%
Bifidobacterium sp.	Gram − rod (anaerobe)	<10%
Candida albicans	Yeast	40%
Clostridium sp.	Gram + spore formers	25%
Diphtheroids	Gram + rods (aerobe)	60%
Enteric bacteria	Gram − rods	25%
Enterococci	Gram + cocci	70%
Eubacterium sp.	Gram − rod (anaerobe)	<10%
Gardnerella vaginalis	Gram − coccobacilli	30%
Lactobacillus sp.	Gram + rods	75%*
Peptostreptococcus sp.	Gram + cocci (anaerobe)	35%
Staphylococcus sp.	Gram + cocci	75%
Trichomonas vaginalis	Flagellated protozoa	17%

*Most adults carry *Lactobacillus,* but prepubertal children do not often carry these organisms.

consists of diphtheroids, enterococci, and anaerobes, such as *Bacteroides* and *Peptostreptococcus.* Hormonal influences at the onset of menstruation (10–15 years) cause the lactobacilli to reappear and become predominant, because of the new supply of glycogen. They consume most of the nutrients in the vagina and release acids, thus maintaining a low pH in the vagina which inhibits many microorganisms from proliferating and causing disease.

438

An important function of the indigenous flora □ is to keep potential pathogens in check. For example, competition for nutrients, combined with the low pH resulting from the fermentation of carbohydrates in the adult vagina by the lactobacilli, restricts the population size of the yeast *Candida albicans*. If for any reason the lactobacilli are reduced in numbers, the increased nutrients available stimulate the *C. albicans* population to increase in size and cause vaginitis. Interestingly enough, *C. albincans* is capable of inhibiting the proliferation of *Neisseria gonorrhoeae*, the cause of gonorrhea, in the normal human vagina.

DISEASES OF THE URINARY TRACT

Urinary tract infections are probably the most common infections in humans. They range from asymptomatic infections to full-blown kidney disease that can result in shock and death. Symptoms vary, depending upon the agent causing the infection and the anatomical site that is invaded. Symptoms frequently include **cystitis** (inflammation of the urinary bladder) with **dysuria** (difficulty in urination), urinary frequency, **hematuria** (blood in the urine), low back pain and fever.

The diagnosis of a urinary tract infection is generally based on the isolation of a large number of bacteria from the urine. An infection is clearly indicated when 100,000 or more bacteria of a single species are found per ml of clean voided urine. Mixed infections have been reported, but more than

90% of all urinary tract infections are caused by a single species (table 26-2). Gram negative enterics, primarily *Escherichia coli* and *Proteus mirabilis* account for most of the cases of urinary tract infections.

The source of contamination causing urinary tract infections is most commonly the normal flora encountered in the stool □. These bacteria colonize the anterior urethra and from there may reach the bladder. The kidneys may also be colonized by bacteria that infect the bladder, often causing a disease called **pyelonephritis**. The bacteria reach the kidneys most commonly via the ureters. It has been estimated that more than 95% of urinary tract infections result from an ascending infection (i.e., urethra → bladder → ureters → kidney). Another route, albeit considerably less common, is through the bloodstream. These infections result from invasion of the urinary tract from other sites via the blood. These infections are caused most often by salmonellae and staphylococci.

439

The most common cause of urinary tract infection is *E. coli* from human fecal matter. Urinary tract infections often start because of toilet habits, incontinence, or sexual habits. Since *E. coli*, *P. mirabilis*, and *Klebsiella pneumoniae* represent a minority of the total bacteria in the feces, it is noteworthy that they are responsible for 95% of all infections of the urinary tract. The pathogenicity of these organisms is believed to be due to their ability to attach to urinary tract epithelium. This ability permits the bacteria to multiply within the urinary tract without being subject to the flushing action of urine. In *E. coli*, the pili and the capsule both appear to influence the ability of the bacteria to cause disease. The pili are mainly responsible for attaching the bacteria to the epithelium. Although the capsule may also have adhesive properties, its primary role is thought to be that of protecting the bacteria from host defenses such as immunoglobulins □ or phagocytes. Similar mechanisms probably account for the virulence of *P. mirabilis*.

463

About 5–10% of pregnant women are subject to urinary tract infections with **bacteruria** (bacteria in the urine). These cases may be due to the fact that, during pregnancy, the fetus sometimes partly blocks the urethra and reduces the flow of urine. Thus, bacteria in the urethra are not flushed out efficiently, and the bladder retains residual amounts of urine, which serves as a suitable culture medium for the growth of the bacteria.

Host defenses against urinary tract infections are both local and systemic. The urinary mucosa, predominantly that of the bladder, has many antibacterial substances that inhibit microbial growth. Associated with the urinary tract epithelium is secretory IgA that may bind to pili and inhibit

TABLE 26-2
SIX MOST COMMON AGENTS OF URINARY TRACT INFECTIONS IN HUMANS

CAUSATIVE AGENT	FREQUENCY OF ISOLATION FROM PATIENT'S URINE
Escherichia coli	90%
Proteus mirabilis	3%
Klebsiella pneumoniae	2.5%
Streptococcus faecalis	2%
Enterobacter aerogenes	1%
Pseudomonas aeruginosa	0.5%

the attachment of potential pathogens. Anticapsular antibodies may also be present. Cell-mediated immunity □ is also evident in the urinary tract.

The prevalence of urinary tract infection in human populations varies depending on the age and sex of the individuals. During infancy, males are almost twice as susceptible as females to urinary tract infection. This situation quickly changes, however, and the susceptibility of females to urinary tract infection increases with age. For example, preschool girls are ten times as susceptible as boys of the same age, but adult females are almost 50 times as susceptible as adult males. The high incidence of urinary tract infection in females is partly due to the much shorter urethra and its proximity to the anus. In old age, males and females are equally susceptible, possibly due to a decreasing immune response or hormonal changes in both males and females.

Control of urinary tract infection involves toilet practices and sexual habits that prevent contamination of the urethral opening with feces. Management of urinary tract infections involves treatment with antibiotics. Large doses of amoxicillin, kanamycin, or trimethoprim-sulfamethoxathole generally are used to treat urinary tract infections.

DISEASES OF THE GENITAL TRACT

Many microorganisms are capable of colonizing the human genital tract, and some of these can cause diseases that are transmitted sexually or by intimate person to person contact. Not all infectious diseases of the genital tract are transmitted sexually, however; some are caused by commensals □ that multiply rapidly under certain conditions, causing acute inflammations. This section will discuss both sexually transmitted and opportunistic pathogens (table 26-3).

426

Candida albicans Vaginitis

Vaginitis caused by *Candida albicans*, and less frequently by *C. tropicalis* and *Torulopsis glabrata*, is characterized by a thick, cream-colored vaginal discharge. The vaginal mucosa may be spotted with patchy lesions, which are gray-white pseudomembranes consisting of host tissue and fungal material. In more serious cases, the lesions may also involve the perineum and inguinal area. Acute inflammatory reactions resulting from the infection lead to swollen, erythematous (reddened) lesions that may itch severely.

Diabetes, prolonged antibiotic treatment, and pregnancy often promote the growth of the yeast and therefore predispose to vaginitis. Conditions that promote an increase in the progesterone level, such as the use of some oral contraceptives, also promote the growth of *C. albicans*, possibly because the increase in glycogen deposition in the vagina provides additional nutrients for the yeasts. A condition similar to vaginitis in males is **balanitis,** a disease characterized by inflammatory lesions of the penis. This is an uncommon condition and is usually acquired through sexual activities with an infected individual.

The pseudomembranous lesions so characteristic of *Candida* infections consist of necrotic (dead) tissue, leukocytes, bacteria, yeast cells, and pseudohyphae. The predominant fungal element in the lesion is the pseudohyphae, which bind the individual components of the pseuodomembrane together.

TABLE 26-3

REPRESENTATIVE DISEASES OF THE UROGENITAL TRACT

INFECTIOUS DISEASE	CAUSATIVE AGENT	GROUP	COMMENTS
Actinomycosis	Actinomyces israelii	Actinomycete	Associated with use of IUDs*
Amebiasis	*Entamoeba histolytica*	Amebic protozoa	Occur in homosexual men
Chancroid	*Haemophylus ducreyi*	Gram − rods	Develop soft chancre
Condyloma acuminatum	Papilloma virus	Papilloma viruses	Warts of external genitalia
Crabs (pediculosis)	*Phthirus pubis*	Louse	Causes severe itching
Cystitis	*Escherichia coli*	Gram − rods	Commonest cause of UTI**
Genital herpes	Herpes simplex virus	Herpesviruses (HSV-2)	>1,000,000 cases/yr
Giardiasis	*Giardia intestinalis*	Flagellated protozoa	Occur in homosexual men
Glomerulonephritis	*Streptococcus pyogenes*	Gram + cocci	Nonpyogenic sequelae
Gonorrhea	*Neisseria gonorrhoeae*	Gram + cocci	More than 1 million cases
Granuloma inguinale	*Calymmatobacterium* sp.	Gram − rods	Known as Donovan bodies also
Hepatitis A	Hepatitis A virus	Picornavirus	Occurs in homosexual men
Hepatitis B	Hepatitis B virus	Complex virus	Occurs in homosexual men
Hepatitis non-A non-B	Hepatitis non-A non-B virus	Retrovirus	Occurs in homosexual men
Lymphogranuloma venereum	*Chlamydia trachomatis*	Chlamydia	Also causes trachoma
Nonspecific urethritis	*Pseudomonas*, coliforms	Gram − rods	Caused by various bacteria
Nonspecific vaginitis	*Gardnerella vaginalis*	Gram − coccobacilli	May be due to mixed flora infection
Pinta	*Treponema carateum*	Spirochetes	Prevalent in the tropics
Proctitis	*Campylobacter jejuni*	Gram − curved rods	Occurs in homosexual men
Puerperal sepsis	*Peptostreptococcus* sp.	Gram + cocci (anaerobe)	Also caused by other bacteria
Pyelonephritis	*Bacteroides*, enterics	Gram − rods	Inflamed kidney & pelvis
Reiter's syndrome	*Chlamydia* sp.	Chlamydias	Conjunctivitis + arthritis + urethritis
Shigellosis	*Shigella* sp.	Gram − rods (enterics)	Occurs in homosexual men
Syphilis	*Treponema pallidum*	Spirochete	35,000 cases annually
Thrush	*Candida albicans*	Yeasts	Also causes balanitis***
Trichomoniasis	*Trichomonas vaginalis*	Flagellated protozoa	A form of vaginitis
Urethritis, nongonococcal	*Ureaplasma urealyticum*	Mycoplasmas	> 1,000,000/yr afflicted
	Chlamydia trachomatis	Chlamydia	30–50% of NGUs

*IUD = Intrauterine device.
**UTI = Urinary Tract Infection.
***Lesions of the penis.

The primary host response to *C. albicans* infections is an acute inflammation. The PMN leukocytes effectively control most infections by this yeast; however, the lesions develop as a result not only of the infection but also of the inflammation □. Secondary colonization of the lesions by bacteria is common, and they can be seen in abundance in stained clinical specimens. These bacteria may be transient or permanent residents of the vagina.

443

Since *C. albicans* can be considered to be an opportunist, control of the infection usually involves measures to eliminate the predisposing factor. When diabetes is the predisposing factor, control of the condition with insulin will generally help resolve the infection. If the cause of vaginitis is a depleted *Lactobacillus* flora, replacement of the normal flora can often solve the problem. Douches containing buttermilk or acidophilus milk can be administered to reintroduce the lactic acid bacteria into the vagina. Ointments containing nystatin (mycostatin) may also be applied to the lesions to control the infection.

Trichomonas vaginalis Vaginitis

Vaginitis caused by *Trichomonas vaginalis* is a common condition in females and consists of burning and itching of the vulvar area, accompanied by a profuse, whitish exudate (pH 5–6). *Trichomonas vaginalis*, which may be found in profusion in the exudates, is a flagellated protozoan that colonizes the mucosal epithelium of male and female genitalia. This organism is widely distributed, and it is estimated that 20–40% of females may be infected worldwide. In males, many infections are asymptomatic and the incidence of the clinical disease is 5–10% of those infected. *Trichomonas vaginalis* in males is found in the ureters, bladder, prostate gland, epididymis, and urethra. It is thought that males serve as a source of infection for females. The infection is transmitted primarily by sexual intercourse, but infections can also be acquired by contact with fomites such as contaminated toilet articles and clothing. The growth of the parasite and the production of toxic metabolic products on the epithelium may cause an inflammation that results in the symptoms of the disease. Diagnosis of trichomoniasis is usually made by noticing the characteristic discharge and by demonstrating the parasite in wet mounts of the discharge. Oral metronidazole (Flagyl) is very effective in controlling the infection. Best results are obtained when both sexual partners are treated simultaneously.

Nonspecific Vaginitis

Nonspecific vaginitis is an inflammation of the human vagina characterized by a homogeneous, malodorous (often "fishy") vaginal discharge with a pH of 4.5–5.0. Epithelial cells (called **"clue cells"**) with adhering coccobacilli, presumably the etiologic agent, are frequently seen in wet mounts of the discharge.

The cause of nonspecific vaginitis is still in question, although a gram negative coccobacillus called *Gardnerella vaginalis* is found in large numbers in the exudates of about 50% of patients with the disease. Approximately 20% of asymptomatic women also carry this bacterium in low numbers. Anaerobic bacteria such as *Bacteroides* and *Peptostreptococcus* are also found in the human vagina and have been implicated in the disease process, along with *G. vaginalis*. At present, few laboratories routinely culture for *G. vaginalis* in clinical specimens. On rich culture media with 5% human blood at an elevated (5%) carbon dioxide atmosphere, *G. vaginalis* forms small, raised colonies with a distinct zone of diffuse beta hemolysis which is characteristic. Metronidazole (Flagyl) or broad-spectrum antibiotics are used to treat the infection successfully.

Toxic Shock Syndrome

Toxic shock syndrome is characterized by hypotension (low blood pressure), shock, fever, rash, sore throat, conjunctival infection, muscle aches, and a variety of other symptoms, such as diarrhea and vomiting. The rash resembles that seen in scarlet fever, but is usually followed by peeling (desquamation) of the hands and soles. The disease gained notoriety in 1980, when a number of cases were associated with the use of superabsorbent tampons (fig. 26-3). Even though this condition has been found primarily in menstruating individuals, it has also occurred in persons with furuncles, carbuncles, impetigo, and other forms of pyoderma 563 □. The disease has been studied extensively since 1980 and the evidence points to toxigenic strains of *S. aureus* as the cause of toxic shock syndrome. Two different toxins, **pyrogenic exotoxin C** and **enterotoxin F,** have been implicated in the disease process. Apparently, toxigenic strains colonize damaged tissue and prolifer-

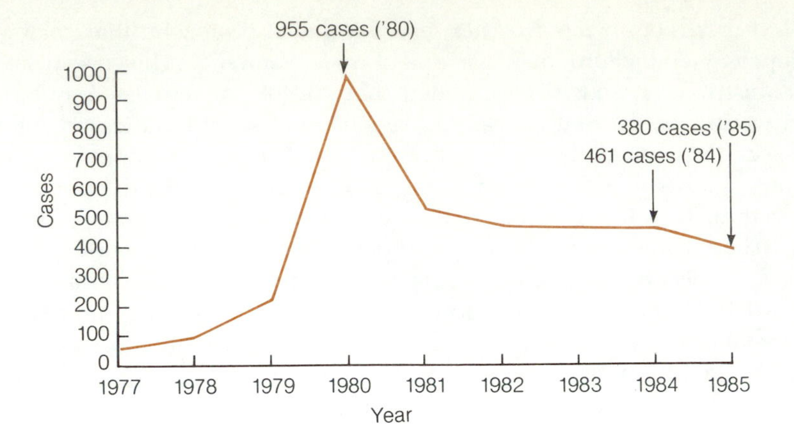

Figure 26-3 Reported Cases of Toxic Shock Syndrome in the United States. The majority of cases of toxic shock syndrome occur in menstruating females. This kind of data led investigators to deduce that certain tampons were associated with the disease.

ate locally, releasing the toxin(s), which in turn cause the characteristic signs and symptoms of toxic shock syndrome. Before the introduction of the superabsorbent tampon in about 1977, only 2 to 5 cases of toxic shock syndrome were reported in the United States each year. The number of cases jumped dramatically each year from 1978 to 1980, when over 900 cases were reported with a mortality of about 3 percent. In 1984, 461 cases were reported to CDC. Because some of the superabsorbent tampons are kept in place for extended periods of time, they may concentrate nutrients and provide a rich, moist environment for the growth of *S. aureus.* The chances of contracting the disease can be reduced by not using superabsorbent tampons, by changing tampons three times per day, and by not using them at night. Treatment of the disease requires the immediate replacement of fluids and electrolytes to reverse hypotension and shock. Abscesses, if present, must be drained, and antibiotics such as cephalosporin and β-lactamase–resistant penicillins □ must be started as soon as possible. 723

SEXUALLY TRANSMITTED DISEASES (STD)

The incidence of **sexually transmitted diseases** (table 26-3) has increased significantly in the last 20 years. The increase has included not only the classical venereal diseases, such as syphilis and gonorrhea, but also some newly recognized diseases such as nongonococcal urethritis, genital herpes, and nonspecific vaginitis. In addition, there has been a alarming increase in the number of cases of diseases that normally affect the gastrointestinal tract. For example, amebiasis, giardiasis, and hepatitis can sometimes be acquired by the sexual route. These and the more conventional diseases transmitted by sexual activities are summarized in table 26-3.

Gonorrhea

Gonorrhea is a sexually transmitted disease caused by *Neisseria gonorrhoeae* and characterized by an acute inflammation of the genital mucosal epithelium (both male and female), and a purulent discharge. The bacterium that causes the disease is a gram negative diplococcus that is normally seen in the purulent discharge inside polymorphonuclear leukocytes, where it sometimes multiplies.

When introduced into a suitable host, the gonococcus attaches to the mucosal epithelium, presumably through its pili. Only pilated strains of *N. gonorrhoeae* are thought to cause gonorrhea, but the pili alone do not make the gonococcus virulent. A cell wall component called IgA protease functions in the pathogenesis of gonorrhea by cleaving antigonococcal IgA antibodies. As you already know, IgA is associated with mucosal secretions and provides local immunity. Once *N. gonorrhoeae* is attached to the epithelium, it begins to multiply, causing localized tissue damage and eliciting an inflammation.

Gonorrhea is present in epidemic proportions in our society (fig. 26-4) and is the most commonly reported infectious disease in the United States. Gonorrhea has increased in incidence for the last 10–20 years. Let us consider the year 1979 as an example. In that year, more than 1 million new cases of gonorrhea were reported in the United States. Of these, young adults (20–24 years old) and teenagers (14–19 years old) were responsible for 65–70% of all cases, about 750,000 cases. The number of deaths due to gonorrhea has been 1–10 per year for the last few years. Although gonorrhea is a reportable disease, it has been estimated that no more than 50% (some estimate 25%) of all actual cases are reported to health authorities. If this is an accurate estimation, the true number of cases in the United States during the last few years has ranged between 2 and 4 million per year. The cost of screening for gonorrhea exceeds $1 billion per year.

Gonorrhea normally is acquired through sexual intercourse. During coitus with an infected female, the **gonococcus** is frequently transmitted to the male's urethra. The vast majority of the infected males (90%) have the typical signs and symptoms of gonorrhea, which include a purulent urethral discharge accompanied by painful urination (fig. 26-5). These individuals play a relatively minor role in the transmission of gonorrhea, because the discomfort of gonoccocal urethritis prevents sexual activity. Also, these males seek medical help to alleviate the symptoms. The remaining males who become infected (10%), however, show no evident signs or symptoms of gonorrhea. They play a very important role in the transmission of the disease, because they continue to have sexual intercourse and consequently infect their sexual partners. Assuming that 2 million cases per year is a realistic

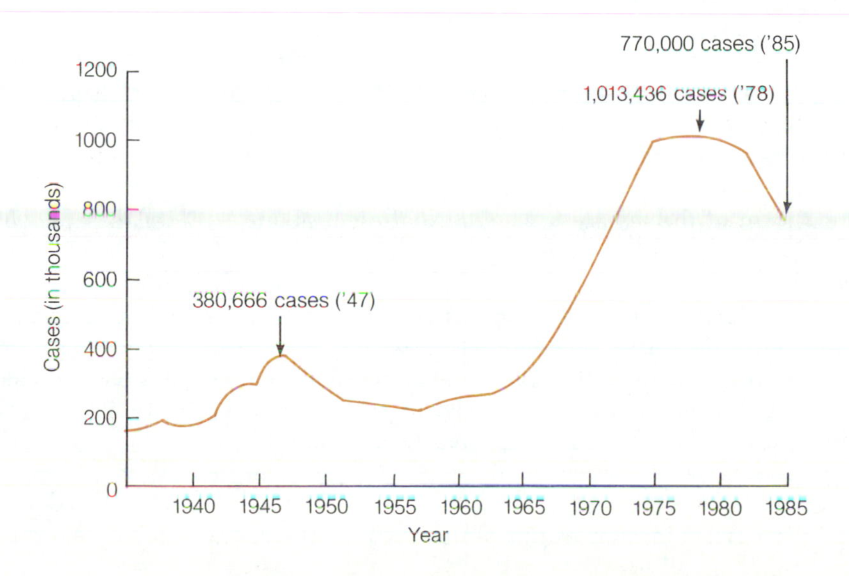

Figure 26-4 Civilian Incidence of Gonorrhea in the United States. In the past five years, the military incidence of gonorrhea has averaged about 23,000 cases per year.

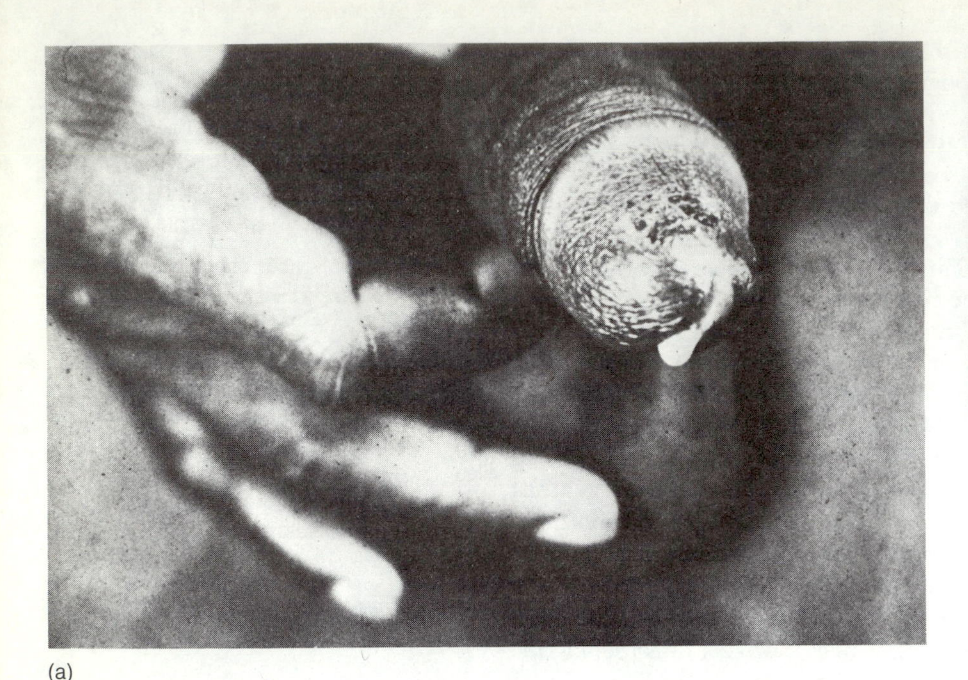

(a)

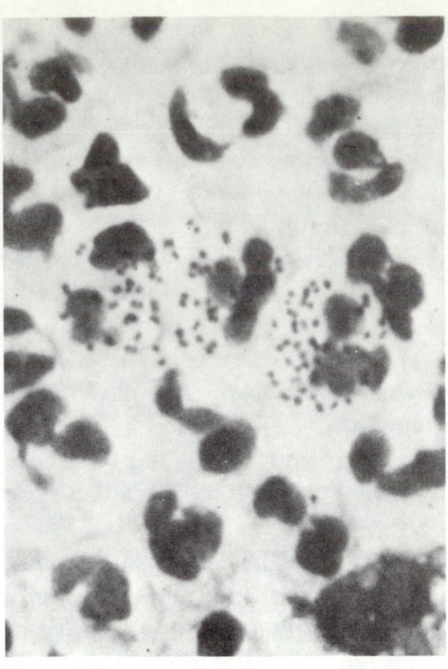

(b)

Figure 26-5 Gonorrhea. *(a)* Thick, purulent exudate characteristic of gonorrhea. *(b)* Gram stain of exudate, revealing gram negative diplococci inside leukocytes.

estimate of the actual number of new cases of gonorrhea in the United States, and that these include equal numbers of males and females, there will be approximately 100,000 new male carriers annually who can spread the disease.

Among females, the epidemiology of gonorrhea is quite different. More than 50% (perhaps closer to 75%) of the women infected are asymptomatic or have only mild, transient symptoms. These individuals rarely seek medical help on their own, so they constitute a vast reservoir □ for the infectious agent. The remaining 25–50% seek medical help, because they have a recognizable disease. Thus, there are probably over 500,000 new female carriers per year in the United States. Another picture of transmission is emerging among homosexuals: the incidence of ano-rectal and pharyngeal gonorrhea is increasing every year.

534

In heterosexual men, gonorrhea is characterized by an inflammation of the urethra, called **anterior urethritis.** The incubation period is generally 2–4 days, after which the characteristic purulent exudate and the painful urination become obvious. The exudate contains leukocytes, cellular debris, and gonococci (fig. 26-5). Even though anterior urethritis is the most common form of the disease in heterosexual men, about 2–5% of these individuals have **gonococcal pharyngitis**. Since the advent of antibiotics, formerly common complications such as urethral stricture, **epididymitis** (inflammation of epididymis) and **prostatitis** (inflammation of the prostrate glands) have all but disappeared. In homosexual men with gonococcal urethritis, involvement of the anal canal is also common. Studies have revealed that 50% of homosexual males who have gonorrhea have **ano-rectal gonorrhea.** Pharyngitis is also common. Unfortunately, about 20% of individuals carrying *N. gonorrhea* in the pharynx are asymptomatic and serve as reservoirs and sources of infection.

Gonorrhea in females most often involves the cervix. They have a puru-

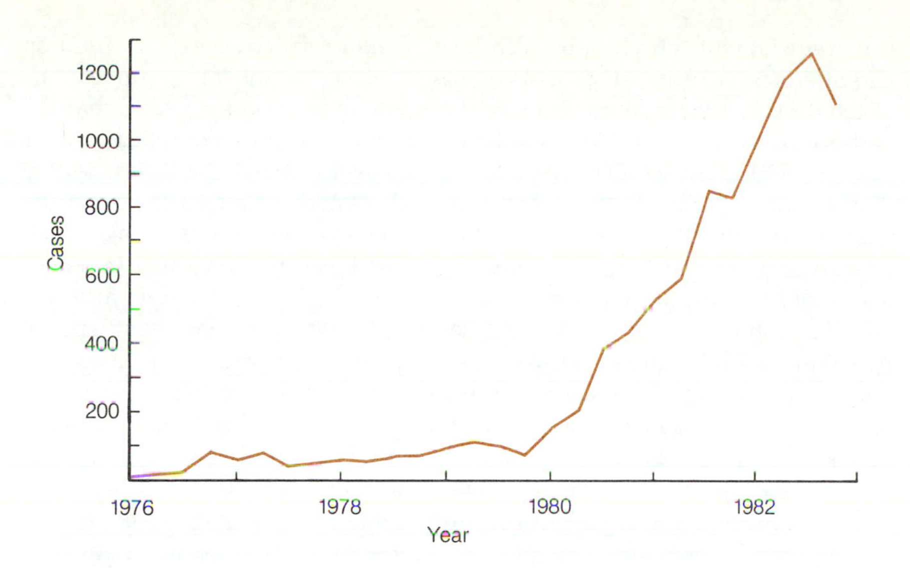

Figure 26-6 Incidence of Gonorrhea Caused by Penicillin-Resistant *Neisseria gonorrhoeae*. Note the sharp increase in penicillin resistant *N. gonorrhoeae* isolated since 1979.

lent exudate similar to that seen in males. Urethritis, anal canal involvement, and pharyngitis are also common.

Infants and children may suffer from gonorrhea as well. In infants, the disease is acquired from the infected mother as the newborn passes through the birth canal. The bacteria most often infect the eyes, causing a disease sometimes called gonococcal **ophthalmia neonatorum.** For this reason, all newborn babies are treated with eye drops containing silver nitrate, tetracycline, or erythromycin, to kill any pathogen that may be present in the eyes. Other areas, such as the pharynx, the umbilical cord, or the anus, may also be infected. Gonococcal infections of prepubertal children resemble those of adults; sexual molestation or precocious sexual activity serve to transmit the disease in this age group.

Infrequently, *Neisseria gonorrhoeae* enters the bloodstream and causes a febrile disease characterized by toxemia (probably due to the endotoxin), skin lesions **(dermatitis),** and pain in the joints **(arthralgia).** It is known that females are more prone than males to this disseminated form of gonorrhea, especially when they are menstruating or pregnant.

Penicillin remains the drug of choice for the treatment of gonorrhea, but certain strains of penicillinase-producing *Neisseria gonorrhoeae* are found to be resistant to penicillins. These strains are being isolated from patients at a rate that is alarmingly high and increasing (fig. 26-6). In these instances, tetracycline and spectinomycin have been found useful. Venereal disease, family practice, premarital, and prematernity clinics are of value because they screen the population to identify carriers and to treat those infected.

Syphilis

Syphilis is a sexually transmitted disease caused by the spirochete *Treponema pallidum* and is characterized by the formation of a firm, ulcerative lesion called a **chancre** at the site of infection. This disease has afflicted humans for many centuries. Some historians believe that this disease was introduced into Europe by the Spaniards, who, they claim, brought the disease from the Americas soon after Columbus's voyages. Other historians, however, argue that syphilis was introduced from Africa by the Portuguese.

Regardless of which hypothesis is correct, syphilis formerly was a debilitating and often fatal disease. Historical records indicate that the late complications of the disease developed much more rapidly than they do today and that the pathogens were much more virulent. Presently, serious complications or death as a result of syphilis are uncommon. For this reason, scientists believe that *T. pallidum* has been evolving toward a more balanced host-parasite relation with the human host and is becoming a "better" parasite (fig. 26-7).

Treponema pallidum is a spirochete that measures 5–20 μm in length and about 0.2 μm in diameter. Because of its small diameter, this microorganism cannot be resolved □ with the bright field microscope, and special techniques such as phase contrast or dark field microscopy must be used to see it in clinical specimens. At present, human pathogenic treponemes cannot be cultured in laboratory media, so little is known about their virulence factors. One fact is known: *T. pallidum* has a high lipid content, with cardiolipin and phosphatidyl choline. These phospholipids may be involved in eliciting immune responses that may ultimately cause damage to the host.

63

Treponema pallidum enters the host through the mucous membranes or the abraded skin. The most common sites of infection include the genitals, lips, anus, breasts, tonsils, and fingers. Once inside the host, the spirochetes multiply locally (extracellularly and intracellularly) and then spread to the regional lymph nodes.

Syphilis has traditionally been subdivided into three stages, based upon differences in signs and symptoms: primary, secondary, and tertiary. **Primary syphilis** is characterized by the appearance of a chancre at the initial site of infection (fig. 26-8 and 26-9). The chancre, an ulcer with a firm border, appears about three weeks after the infection and is laden with infectious spirochetes. The chancre is accompanied by an enlargement of the regional lymph nodes (lymphoadenopathy), reflecting the spread of the parasite and its multiplication in the lymph nodes. After 5–10 days, the chancre heals spontaneously. If the infection is not treated during the chan-

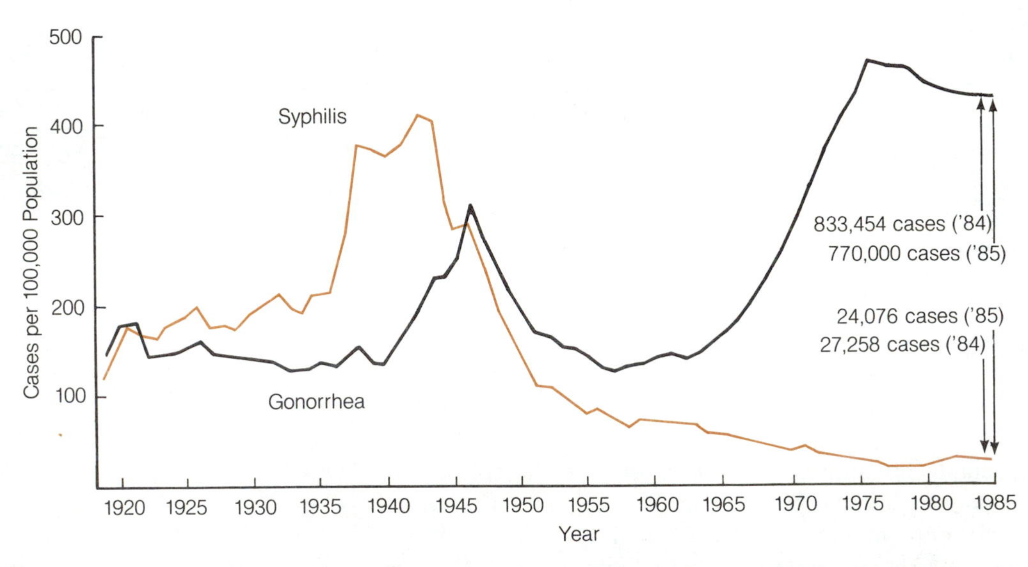

Figure 26-7 Comparison of Civilian Syphilis and Gonorrhea in the United States. Graph illustrating the incidence of primary and secondary syphilis in the United States. Of the more than 27,000 cases of syphilis in the United States in 1984, fewer than 100 were congenital syphilis. The incidence of military syphilis during the last five years has averaged at about 370 cases per year.

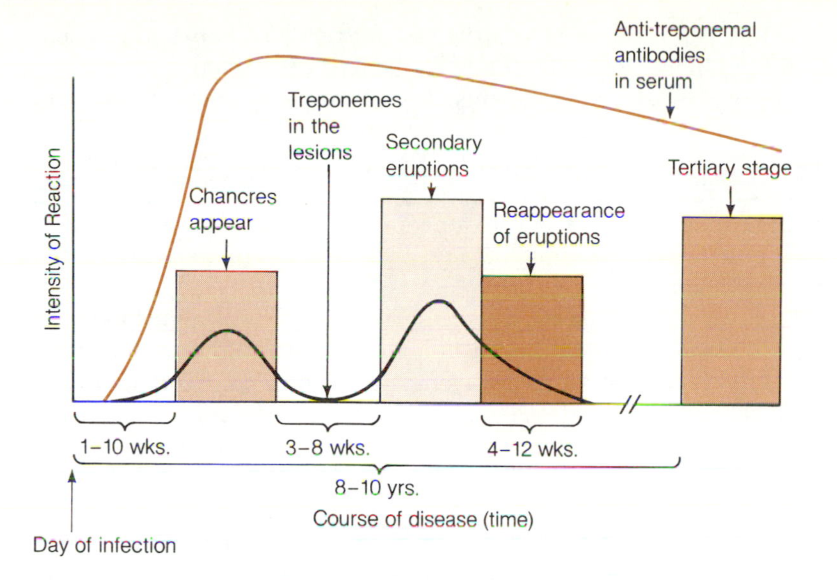

Figure 26-8 Clinical Course of the Disease in Syphilis. Primary syphilis is characterized by the development of a chancre at the site of infection. The chancre has numerous treponemes and can serve to infect sexual partners. The chancre heals after 2–3 weeks. Three to eight weeks later, a rash may appear, heralding the onset of secondary syphilis. These lesions may also contain treponemes and hence are infectious. Tertiary syphilis may appear 8–10 years after infection. Severe damage to organs may occur. The lesions are due to allergic responses and are generally free of treponemes. Antitreponemal antibodies appear early in the disease and remain high for long periods of time. These antibodies are of diagnostic value.

cre stage, **secondary syphilis** follows after a period of weeks or months. This stage is characterized by multiple skin lesions, rashes, fever, sore throat, malaise, and **lymphoadenopathy** (swelling of lymph nodes). The lesions contain numerous treponemes and therefore are highly infectious. During this stage, antitreponemal antibodies and antigen-antibody complexes can be detected. If these complexes lodge in the glomeruli, kidney disease may result from complement-mediated □ kidney damage. These same immune complexes may lodge in the joints, causing arthralgias. Because the treponemes have spread to other parts of the body, organ systems often become infected. If syphilis still remains untreated, years after the initial infection tertiary syphilis may develop. This stage is characterized by the formation of localized dermal lesions called **gummata** (singular, gumma) (fig. 26-9). These lesions contain few, if any, treponemes and arise as a result of cell-mediated

487

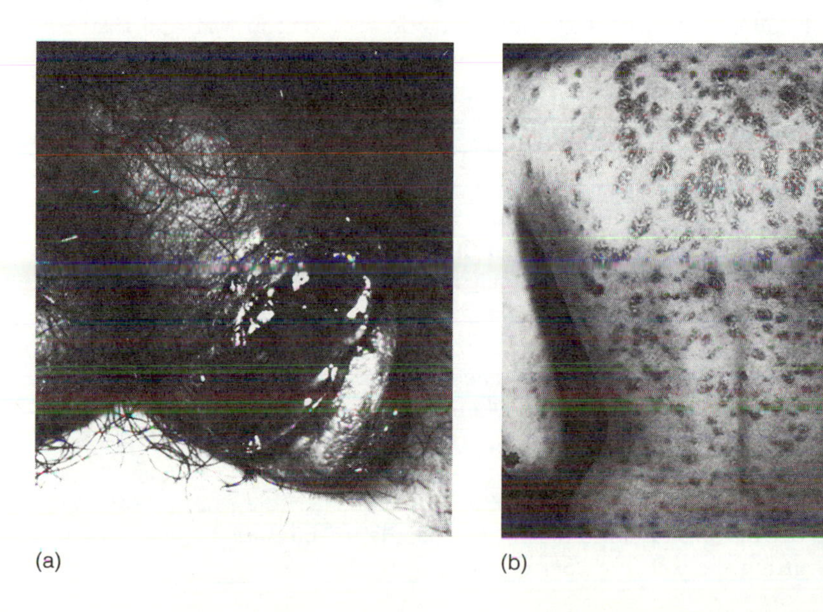

(a) (b)

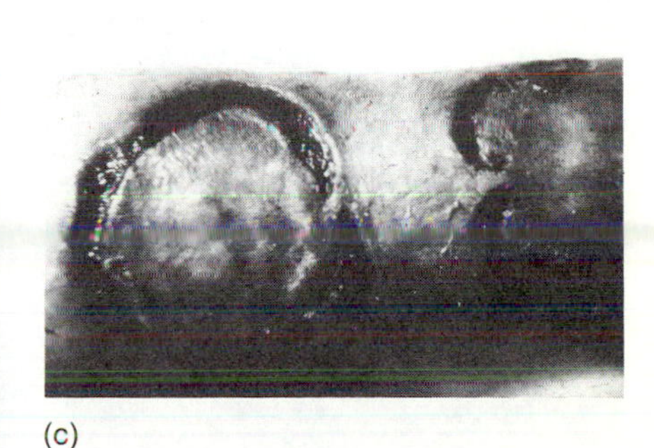

(c)

Figure 26-9 Clinical Manifestations of Syphilis. *(a)* Chancre characteristic of primary syphilis. *(b)* Rash characteristic of secondary syphilis. *(c)* Gummata characteristic of tertiary syphilis.

immunity □ to circulating *T. pallidum* antigens. Tertiary syphilis may also be reflected in cardiovascular diseases that are often fatal.

In pregnant syphilitic patients, *T. pallidum* may cross the placenta and infect the fetus. The treponemes enter the unborn child via the fetal circulation and multiply in fetal tissue. **Congenital syphilis** can cause severe damage in the fetus and lead to congenital malformation (fig. 26-10). Although routine screening and treatment of pregnant, syphilitic individuals is widely practiced in the United States, the incidence of congenital syphilis in the U.S. is about 0.4%.

Some of what is known about modern syphilis in humans was learned from a 40-year study that began in 1932 in Macon County, Alabama. This study, known as the "Tuskegee study," involved about 600 individuals, of whom 400 had some evidence of syphilis. Two hundred of these were treated while the other 200 were denied any treatment. This ethically questionable study was not terminated by the U.S. Public Health Service until 1972.

Figure 26-7 illustrates the incidence rate of primary and secondary syphilis from 1930 to 1985. As you can see, syphilis reached a peak in 1943, when over 400 cases per 100,000 population were reported. At this time, antibiotics such as penicillin were introduced, and the incidence of the disease declined rapidly to 150 cases per 100,000 in 1950. Since the 1950s, the disease has continued to decline; however, the reported cases of syphilis represent only the tip of the iceberg, since it is estimated that only 25% of cases are reported. If this estimate is correct, then there are presently over 100,000 new cases of primary and secondary syphilis in the United States each year.

The control of syphilis requires that carriers and new cases be discovered and then treated with penicillin. This procedure reduces the number of infected individuals in the population. Other control measures are similar to those employed in the control of gonorrhea. The treatment of choice is penicillin for active cases of syphilis.

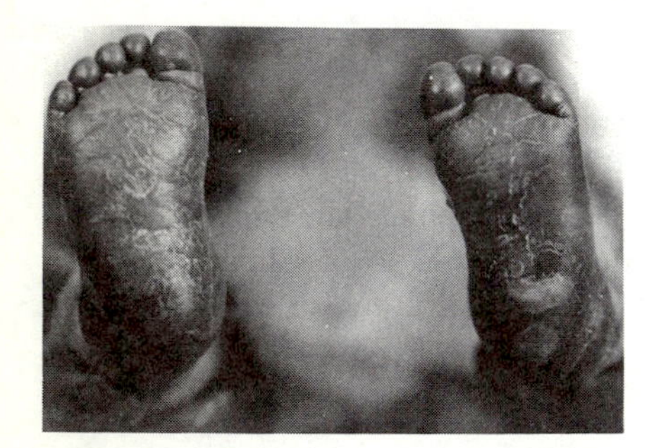

(a)

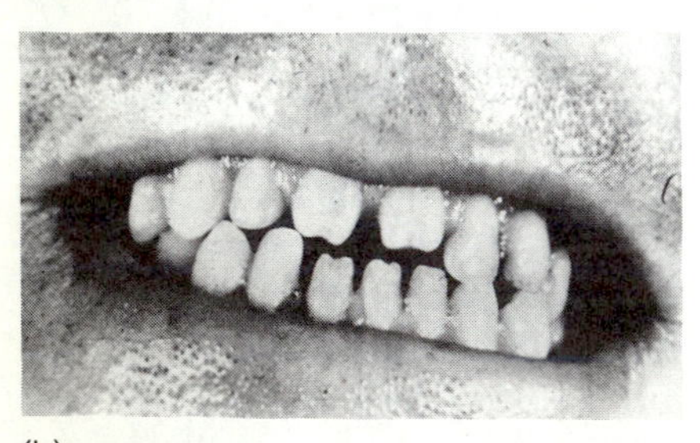

(b)

Figure 26-10 Congenital Syphilis. *(a)* Superficial skin erosion in child with congenital syphilis. *(b)* Notched incisors characteristic of congenital syphilis.

Genital Herpes

Genital herpes is a sexually transmitted disease, characterized by the development of small blisters on the skin which ulcerate after 4–5 days. The outbreak of blisters may be accompanied by fever and exhaustion. These lesions recur at frequent intervals and cause considerable discomfort and distress. The disease is caused by the herpes simplex virus, which can become latent by infecting nerve cells. There are two types of herpes simplex (I and II) that can cause genital herpes, but type II is far more common. Most infected individuals are infected for life. They appear to transmit the virus to sexual partners only when the blisters and ulcers are present (fig. 26-11).

The fetus in pregnant individuals can be infected by the circulating viruses during a genital herpes episode, thus resulting in severe damage to the fetus or in stillbirth. In addition, the virus can be transmitted to the newborn as the child passes through the birth canal if the mother is having a herpes attack. This infection can result in very severe and often fatal disease. Approximately 200 newborn babies die each year in the U.S. because of systemic herpes infections and more than 200 suffer physical or mental impairment. As if these complications were not enough, females with genital herpes have an increased incidence of genital cancer.

Genital herpes is an infectious disease of considerable public health significance because of the large number of people infected for life and capable of transmitting the infection. Some studies have revealed that genital herpes is more prevalent in human populations than gonorrhea. It reached epidemic proportions during the 1970s, and continues to be the leading cause of morbidity among sexually transmitted diseases. It is estimated that there are over 1 million new cases of genital herpes every year in the United States. The ability of the virus to become latent, and the lack of drugs that can destroy the latent virus, make it almost impossible to eradicate this disease from human populations. The disease can be prevented, however, by avoiding sexual contact during the time when the blisters and ulcers are present.

There is no effective treatment to cure genital herpes. In 1982, however, the FDA permitted the release of a new topical drug called **acyclovir** (Zovirax) that inhibits the replication of the virus. Unfortunately, the ointment does not eliminate the virus or reduce the severity of genital herpes attacks. An oral form of acyclovir has also been tested but does not appear to be effective in preventing recurrences although it reduces the severity and duration of the episodes. Preliminary tests indicate that intravenous acyclovir may eliminate the virus and recurrent outbreaks.

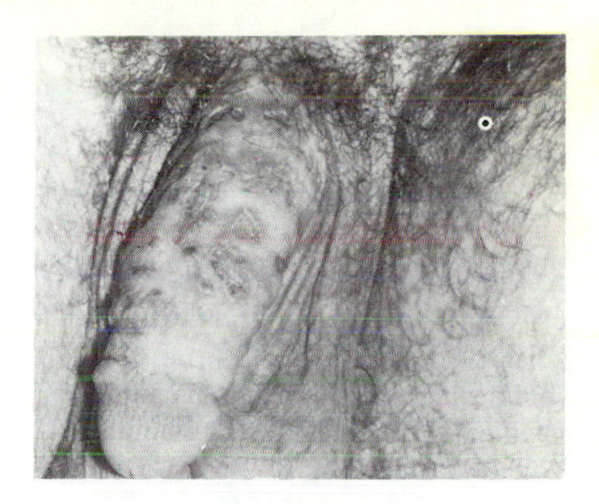

Figure 26-11 **Genital Herpes**

Nongonococcal Urethritis

Nongonococcal urethritis (NGU) is a sexually transmitted disease characterized by an inflammation of the urethra accompanied by a discharge of pus. It is estimated that, in the United States, 30–50% of the 2 million annual cases of acute urethritis (other than gonococcal) are caused by *Chlamydia trachomatis*. Other microorganisms, such as the mycoplasma *Ureaplasma urealyticum*, can also cause NGU. The diagnosis of NGU is usually made when it is discovered that the etiology is not *N. gonorrhoeae*.

FOCUS
GENITAL WARTS

One venereal disease that appears to have increased in frequency in the last few years is genital and anal warts, which are benign tumors of the skin. Genital and anal warts are caused by a number of viruses, such as the papilloma and polyoma viruses. The viruses are spread by sexual contact when infected tissue breaks down and sheds the virus.

Warts develop from infected epithelium that grows into slender cones or fingerlike projections. The warts may develop singly or in dense groups (fig. 26-12).

Individual warts on the penis or vulva or around the anus are often removed by electrodesiccation. The epidermis is subsequently treated with various chemicals that denature the viruses, so that viruses that escape from the diseased tissue do not initiate new infections. When the warts have developed in thick groups, chemicals such as podophylen in alcohol are applied, causing an often painful inflammatory reaction that sloughs off the diseased epidermis. Vaginal warts frequently are treated with triple sulfonamide-urea cream.

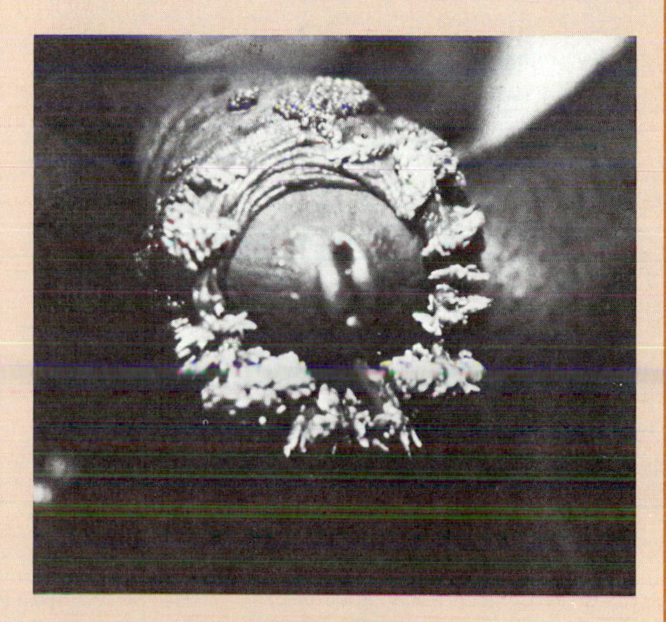

Figure 26-12 **Genital Warts of Penis**

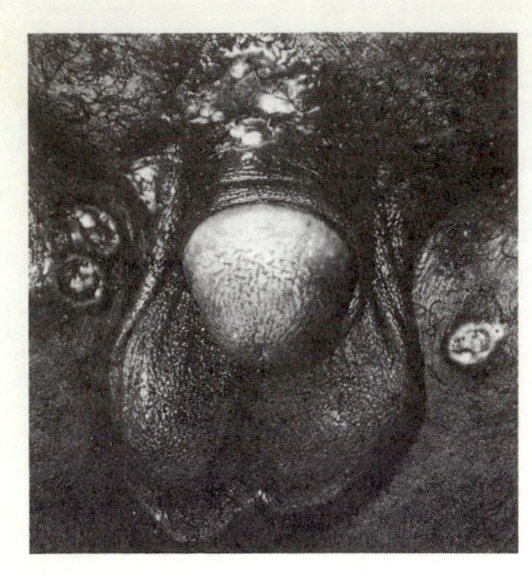

(a)

Chlamydia trachomatis (fig. 26-13) is an obligate intracellular parasite that causes, in addition to NGU, a variety of other diseases including lymphogranuloma venereum; salpingitis (an inflammation of the fallopian tubes); **Reiter's syndrome** (urethritis, arthritis, and conjunctivitis); **cervicitis** (an inflammation of the cervix); and trachoma.

Chlamydia trachomatis is transmitted by sexual activity or during childbirth. Elementary bodies of *C. trachomatis* are taken up by susceptible host cells in the urethra (or other organs) by a process resembling phagocytosis. The elementary bodies then differentiate into reticulate bodies, which, in turn, undergo a series of binary fissions inside the urethral cells. About 36–72 hours later, this process gives rise to elementary bodies. One reticulate body can give rise to as many as 1,000 elementary bodies. The parasitized host cell bursts, releasing the elementary bodies. Several such cycles can cause considerable damage to urethral epithelium and therefore incite an inflammatory reaction. The chlamydiae are thought to produce a toxin that contributes to the cell damage and subsequent inflammation.

Chlamydial infections are not routinely diagnosed, although more and more clinical laboratories are beginning to check for *C. trachomatis* in clinical specimens. Diagnostic procedures usually involve the growth of the suspected agent in tissue culture □. The infected tissue culture cells are then viewed using special staining techniques or fluorescent antibodies □. Routine screening for this pathogen is desirable in cases of urethritis, so that the appropriate treatment can be carried out and sexual contacts can be informed of their possible infection and subsequently treated. Active cases are treated effectively with tetracyclines or with erythromycin.

417
52

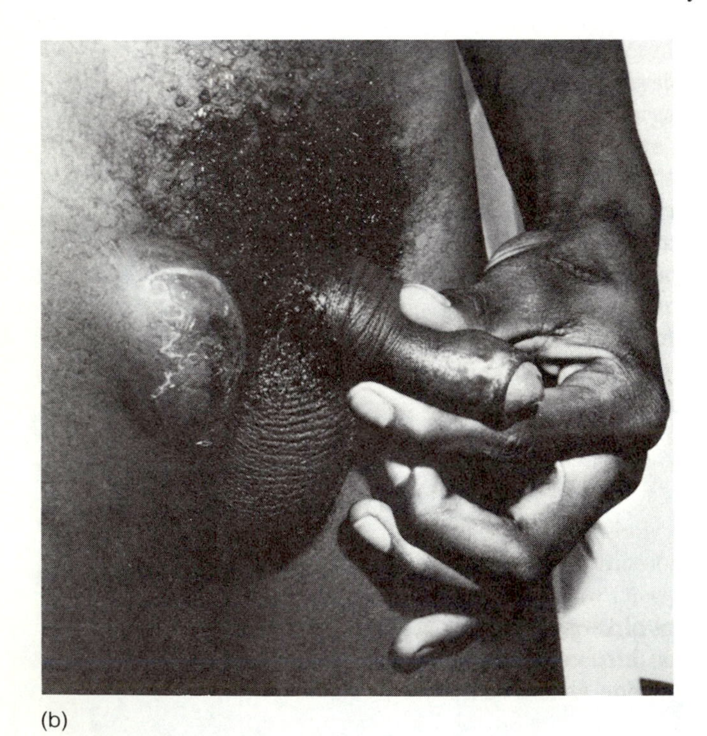

(b)

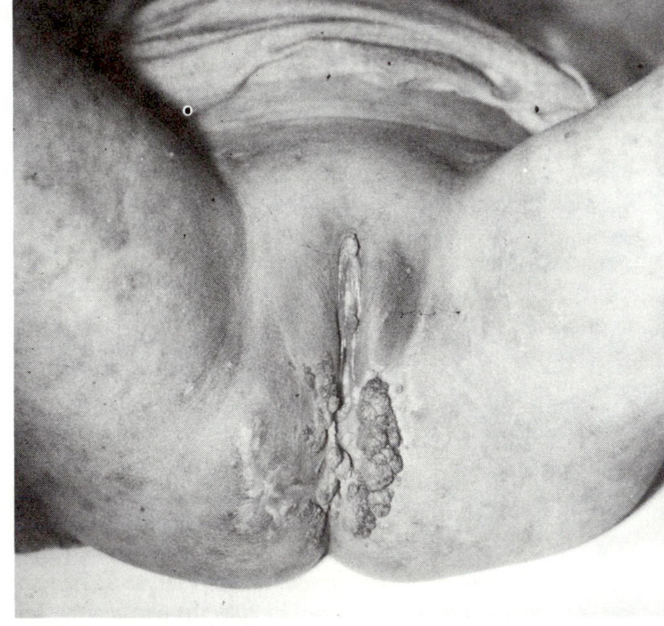

(c)

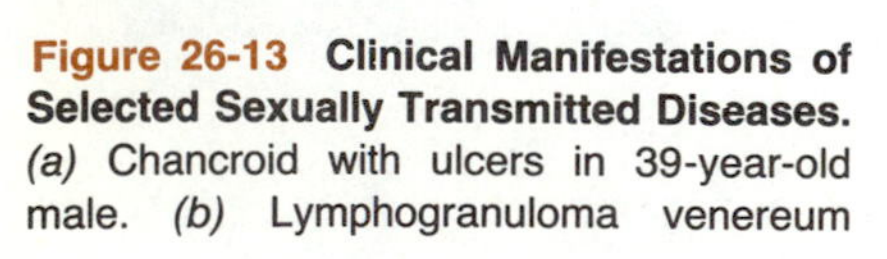

Figure 26-13 Clinical Manifestations of Selected Sexually Transmitted Diseases. *(a)* Chancroid with ulcers in 39-year-old male. *(b)* Lymphogranuloma venereum caused by *Chlamydia trachomatis*.in a male and *(c)* in a female. Notice that swollen lymph nodes are characteristic of the disease in males while distortion of tissue and rectal narrowing are characteristic in females.

FOCUS

UREAPLASMA LINKED TO HUMAN INFERTILITY

It is estimated that 8 million or 15% of all couples in the United States are infertile, and that half of them are infertile because of infections. The mycoplasma *Ureaplasma urealyticum* is responsible for approximately 2.4 million or 60% of the infections that make couples infertile. Although *Ureaplasma* infects the genitals in both men and women, it is not clear how it causes infertility.

When both partners are treated with appropriate antibiotics, the bacteria are eliminated in over 80% of the couples. Sixty percent of the treated couples who become free of *Ureaplasma* are then able to have children, compared to only 5% of the untreated, *Ureaplasma* infected couples.

Chlamydia trachomatis (the agent of nonspecific urethritis and lymphogranuloma venereum) and *Neisseria gonorrhoeae* (the agent of gonorrhea) are believed to account for much of the remaining infertility (40%) due to infectious agents. Long-term *Chlamydia* and *Neisseria* infections destroy the fallopian tubes and consequently make women infertile.

Yearly, approximately 1-2.5 million people in the United States suffer from *Ureaplasma* infections, 1 million from *Neisseria* infections, and 3 million from *Chlamydia* infections.

SUMMARY

1. Diseases of the urogenital tract are common in human populations.
2. Urinary tract infections caused by *Escherichia coli* are the most common infections in humans.
3. Sexually transmitted diseases, in general, are currently on the rise.
4. Pathogens such as *Entamoeba histolytica* and *Giardia intestinalis*, not normally considered to be sexually-transmitted diseases, are now being encountered as the cause of some venereal diseases.

STRUCTURE AND FUNCTION OF THE UROGENITAL TRACT

1. The principal roles of the urinary tract are to maintain homeostasis in the blood and to void body wastes.
2. The flow of urine from the bladder through the urethra helps cleanse the urethra of adhering microorganisms.
3. The reproductive system serves to generate gametes and to nurture the embryo.
4. Secretions of the reproductive system, particularly that of the female, help maintain a population of microorganisms that may protect the host from infections.

NORMAL FLORA OF THE UROGENITAL SYSTEM

1. The urinary tract, except for the anterior urethra, normally is maintained free of microorganisms by body defenses. The anterior urethra contains an array of bacteria, which include gram positive cocci and diphtheroids.
2. The vagina and cervix have a complex flora composed primarily of lactobacilli.
3. The indigenous flora helps keep potential pathogens from causing disease.

DISEASES OF THE URINARY TRACT

1. Urinary tract infections (UTI) are very common in human populations. They are caused primarily by *Escherichia coli* and *Proteus mirabilis*.
2. UTIs are characterized by cystitis, urinary frequency, hematuria, low back pain, and fever.
3. Virulence factors, such as pili and capsular antigens, play a role in pathogenesis in that they help the pathogen adhere to epithelial surfaces or to be protected from host defense mechanisms.
4. UTIs can be controlled by preventing urethral contamination with feces. Large doses of amoxicillin, kanamycin, or trimethoprim-sulfamethoxathole are used to treat UTIs.

DISEASES OF THE GENITAL TRACT

1. *Candida albicans* is an endogenous pathogen that causes a form of vaginitis sometimes called thrush. The disease is characterized by a thick, cream-colored vaginal discharge. Predisposing factors such as

diabetes, hormonal imbalance, or depletion of the normal flora usually account for the abnormal proliferation of the opportunistic yeast.

2. *Trichomonas vaginalis*, a flagellate, causes a vaginitis that consists of a profuse, whitish exudate accompanied by burning and itching. The disease is transmitted sexually and is treated with Flagyl.
3. Nonspecific vaginitis is an inflammation of the vagina that generally is not caused by the usual pathogens. The exudate is homogeneous and malodorous and usually contains "clue cells." It is thought that the disease is caused by *Gardnerella vaginalis*, and it can be treated with broad-spectrum antibiotics or with Flagyl.
4. Toxic shock syndrome (TSS) is an intoxication caused by certain strains of *Staphylococcus aureus*. The disease usually ensues in menstruating individuals who use tampons and keep them in place for long periods of time. The bacteria multiply at the expense of nutrients that have been absorbed by the tampon and produce a toxin. The intoxication is characterized by hypotension, shock, fever, a rash, and several other constitutional symptoms. Treatment is by removing the tampon and immediately replacing electrolytes to reverse shock and hypotension.

SEXUALLY TRANSMITTED DISEASES (STD)

1. Gonorrhea, caused by *Neisseria gonorrhoeae*, is characterized by an acute inflammation of the genital mucosa with purulent discharge.
2. Virulence factors such as pili and IgA proteases are important in the pathogenicity of the bacterium.
3. The disease occurs in epidemic proportions in human populations, with more than 1 million cases reported annually in the U.S. alone.
4. Gonorrhea is treated successfully with penicillin, although some penicillinase-producing strains have been isolated. These are treated with tetracycline or spectinomycin.
5. Syphilis, caused by *Treponema pallidum*, is characterized by the formation of a firm, ulcerative lesion called a chancre.
6. The pathogenesis of the disease involves both the invasive properties of the pathogen and its ability to elicit allergic responses in the host.
7. Primary and secondary syphilis are due primarily to the invasion of host tissues by the parasite, while tertiary syphilis is due to delayed hypersensitivity responses to treponemal antigens.
8. There are an average of 35,000 new cases of syphilis reported in the U.S. each year. This figure, however, may represent only about 25% of the actual number of cases.
9. The treatment of choice for active cases of syphilis is penicillin.
10. Genital herpes is caused by herpes simplex type I or II, although the latter is much more frequent. The disease is characterized by a first episode of painful, prolonged ulceration of the genital area, with fever and a burning sensation with frequent recurrences of the lesions.
11. Genital herpes has reached epidemic proportions in the United States and has influenced sexual practices in this country.
12. The treatment of genital herpes is symptomatic, although a new drug called Zovirax has shown promising results when administered intravenously.
13. Nongonococcal urethritis afflicts more than 1 million persons in the United States each year. *Chlamydia trachomatis* and *Ureaplasma urealyticum* are the most common causative agents. *Chlamydia* is also responsible for lymphogranuloma venereum, while *Ureaplasma* is implicated in many cases of infertility.
14. Genital warts are caused by a number of viruses, in particular papilloma and polyoma viruses.

STUDY QUESTIONS

1. Explain the pathogenesis of urinary tract infections caused by *Escherichia coli.*
2. Describe the epidemiology of:
   a. UTI
   b. candidiasis
   c. TSS
   d. genital herpes
   e. syphilis
3. Describe the role(s) of the host's immune response in the following disease processes:
   a. syphilis
   b. gonorrhea
   c. genital herpes
   d. UTI
   e. candidiasis
   f. TSS
4. Outline a program of prevention for the following diseases:
   a. TSS
   b. candidiasis

c. syphilis
d. gonorrhea
e. genital herpes
f. STD

5. Describe the role (if any) of the carrier population in the following diseases:
   a. UTI
   b. gonorrhea (both males and females)
   c. syphilis (both males and females)
   d. candidiasis
   e. genital herpes (both males and females)
   f. trichomoniasis

6. Describe the typical signs and symptoms of the following diseases, and explain the mechanisms by which they occur:
   a. candidiasis
   b. gonorrhea
   c. tertiary syphilis
   d. TSS
   e. genital herpes

7. Why is genital herpes of such public health importance?

8. Construct a table of diseases of the urogenital tract containing the causative agent, epidemiology, and treatment of choice.

SUPPLEMENTAL READINGS

Karchmer, A. W. 1982. Sexually transmitted diseases. *Scientific American Medicine* Section 7-XXII.

Karchmer, A. W. 1983. Infections due to *Neisseria*. *Scientific American Medicine* Section 7-III.

Rubin, R. H. 1984. Infections of the urinary tract. *Scientific American Medicine* Section 7-XXIII.

National Institute of Allergies and Infectious Diseases. Study Group. 1981. *Sexually transmitted diseases: 1980 status reports.* Washington: U. S. Department of Health and Human Services, National Institutes of Health.

Persson, E. and Holmberg, K. 1984. Clinical evaluation of preciptin tests for genital actinomycosis. *Journal of Clinical Microbiology* 20(5):917–922.

Sewell, D. L. and Horn, S. A. 1985. Evaluation of a commercial enzyme-linked immunosorbent assay for the detection of Herpes Simplex virus. *Journal of Clinical Microbiology* 21(3): 457–458.

Turck, M. 1980. Urinary tract infections. *Hospital Practice* 15:49–58.

Wroblewsky, S. S. 1981. Toxic shock syndrome. *American Journal of Nursing* 81:82–85.

CHAPTER 27
DISEASES OF THE NERVOUS SYSTEM

CHAPTER PREVIEW
WHERE DO THE SLOW VIRUSES HIDE?

Certain viruses infect animals and humans for many years before the damage they cause becomes apparent. Since the incubation periods are sometimes many decades, these viruses are known as **slow viruses.** Scientists are interested in knowing how these viruses escape the host's immune system and where they reside during their long incubation periods.

A study of the slow virus that causes **visna** in sheep and goats has indicated that it hides in macrophages (cells that usually destroy invading microorganisms), where it replicates and accumulates in cytoplasmic vacuoles. Surprisingly, no cytopathic effects are visible in the macrophages.

Damage in the host occurs only after the infected macrophages fuse with cells in the brain, lungs, or joints. The visna viruses begin to replicate in the fusion-cells, lysing them. The virus replication and cell destruction draw cells of the immune system to the affected tissue. The attack on the visna virus by the cells of the immune system eventually causes most of the damage to the animal. As the disease develops, the animal becomes paralyzed and emaciated and dies.

Subacute sclerosing panencephalitis is a slowly-progressing, fatal human disease of the central nervous system caused by the long-term presence of measles virus antigens. Virus parts are seen in the nucleus and cytoplasm of infected brain cells.The infectious measles virus is not present, because one of the proteins required for its maturation is not synthesized, but all the other virus proteins are synthesized. The measles antigens associated with brain cells stimulate the immune system to attack the brain tissue.

The visna virus and the measles virus illustrate two ways in which viruses reside in host cells and cause slowly-progressing diseases of the central nervous system. The visna virus accumulates quietly in macrophages until the macrophages fuse with other cells, while the measles viruses fail to mature in brain cells. Since virus multiplication is delayed, the destruction of nervous tissue progresses slowly under the onslaught of the immune system.

In this chapter you will be introduced to other pathogens that cause neurological disorders. You will discover that microorganisms are responsible for a number of diseases of the nervous system that account for many thousands of deaths each year in the United States.

LEARNING OBJECTIVES

A STUDY OF THIS CHAPTER SHOULD ENABLE YOU TO:

EXPLAIN HOW INFECTIONS IN ANOTHER PART OF THE BODY MIGHT LEAD TO INFECTIONS OF THE NERVOUS SYSTEM

EXPLAIN HOW MICROORGANISMS CAN CAUSE NERVE DAMAGE WITHOUT INFECTING THE NERVOUS SYSTEM

LIST THE DISEASES OF THE NERVOUS SYSTEM THAT CAN BE PREVENTED BY VACCINATIONS

OUTLINE THE DEVELOPMENT OF EACH OF THE DISEASES OF THE NERVOUS SYSTEM

LIST THOSE DISEASES OF THE NERVOUS SYSTEM THAT ARE ZOONOSES

Diseases of the nervous system are caused by a variety of bacteria, viruses, protozoa, and fungi. These diseases often result when the infectious agent or some toxin spreads from the initial site of infection to nerve cells. Since infections and diseases of the nervous system are often fatal, they require prompt treatment. In this chapter we will discuss diseases in which the most obvious symptoms result from infections or disturbances of the nervous system.

ORGANIZATION OF THE NERVOUS SYSTEM

The nervous system is generally divided into two parts: the **central nervous system** and the **peripheral nervous system.** The central nervous system includes all of the nerve cells, supporting cells, cerebrospinal fluid, and membranes that make up and are associated with the brain and spinal cord (fig. 27-1a). The peripheral nervous system consists of all the nerve cells and supporting cells that innervate the body and that connect to the spinal cord or directly to the brain (fig. 27-1b). The **sympathetic nerves** are those in the autonomic nervous system that innervate smooth muscles, heart muscle, and glands.

The brain and spinal cord are protected by three membranes called **meninges** (singular, meninx) and by cerebrospinal fluid. The meninx that covers the surface of the brain and spinal cord is called the **pia mater.** This meninx is, in turn, covered by the **arachnoid.** Cerebrospinal fluid produced by brain cells flows between the pia mater and the arachnoid in what is called the **subarachnoid space.** The third meninx, called the **dura mater,** encloses another layer of cerebrospinal fluid around the brain, called the **superior sagittal sinus.** The brain and spinal cord are protected by the skull bones and the vertebrae, which make up the backbone. There are four interconnecting cavities within the brain called **ventricles,** which are filled with cerebrospinal fluid. Specialized cells and neurons lining the ventricles release hormones and neurotransmitters into the fluid. Cells in the ventricles are also believed to be involved in transporting hormones and neurotransmitters in the cerebrospinal fluid to various parts of the brain and to capillaries within the brain. It is thought that the brain, cerebrospinal fluid, and blood are integrated into one neuroendocrine unit by hormones and neurotransmitters. Even though small molecules such as hormones and neurotransmitters can pass back and forth between the blood and the central nervous system, large molecules (such as antibodies) and microorganisms (such as viruses and bacteria) are unable to penetrate the barriers between the brain capillaries and the blood system. Consequently, unlike most of the body, the central nervous system is immunologically isolated and there is a "blood-brain barrier" that inhibits the movement of some drugs and antibiotics into the brain. Thus, infections of the brain may be very severe because components of the immune system and many drugs and antibiotics are unable to fight infections in this organ.

BACTERIAL DISEASES OF THE NERVOUS SYSTEM

Acute Pyogenic Meningitis

Acute pyogenic meningitis is a severe inflammation of the meninges accompanied by the formation of pus, which can be caused by a number of

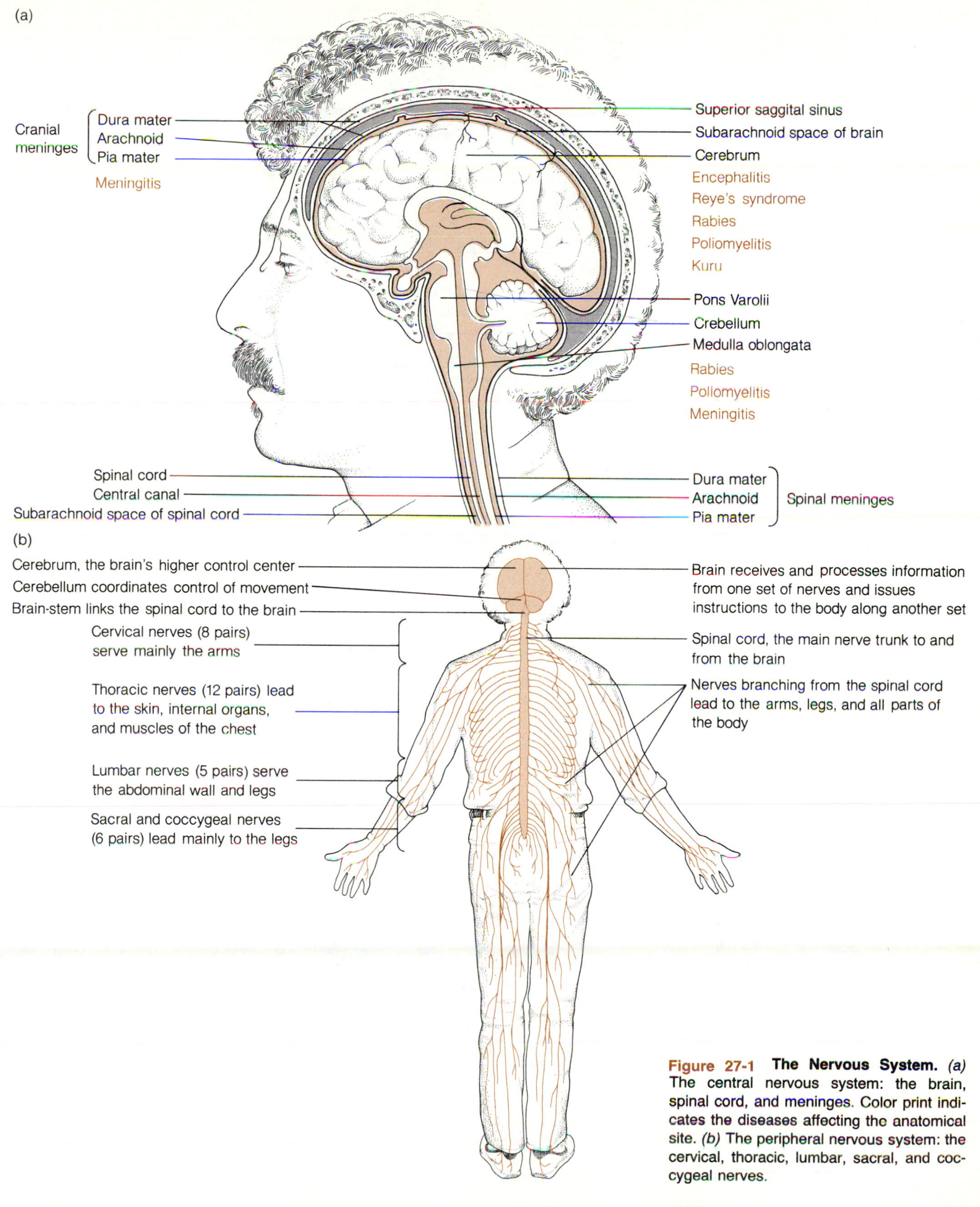

Figure 27-1 The Nervous System. *(a)* The central nervous system: the brain, spinal cord, and meninges. Color print indicates the diseases affecting the anatomical site. *(b)* The peripheral nervous system: the cervical, thoracic, lumbar, sacral, and coccygeal nerves.

bacteria (table 27-1). The bacteria causing meningitis in babies one month old or less are *Escherichia coli* (40%) and Group B *Streptococcus* (30%). In children 15 years old or younger, the major causes of meningitis are *Haemophilus influenzae* (50%), *Neisseria meningitidis* (30%), and *Streptococcus pneumoniae* (15%). In adults, meningitis is due mainly to *Streptococcus pneumoniae* (40%), *Neisseria meningitidis* (25%), and *Staphylococcus aureus* (10%).

Meningococcal meningitis (caused by *Neisseria meningitidis)* is the only kind of bacterial meningitis that causes epidemics □ 532 and small outbreaks (fig. 27-2). The outbreaks are common in crowded military camps and in crowded slum areas. In the U.S., the number of reported cases per year averages about 2,700 (1.5/100,000) (fig. 27-3). In 1943–1944 there was an epidemic of meningococcal meningitis in the U.S., related to the large number of recruits gathered together during the war. At least 35,000 persons suffered from meningococcal meningitis during those two years.

Pneumococcal meningitis (caused by *Streptococcus pneumoniae*) occurs most frequently in persons with ear infections (acute otitis media) and mastoid infections (mastoiditis) (30%); in persons with pneumonia (30%); in persons suffering from alcoholism and cirrhosis of the liver (20%); and in persons with severe nonpenetrating head injuries (10%).

The bacteria that cause meningitis are spread from one person to another in aerosols and on fomites. The bacteria generally enter the nasopharyngeal region and the respiratory tract, where they establish infections. Infections of the sinuses (sinusitis), mastoids (mastoiditis), and middle ear (otitis) can provide a direct access to the meninges. Infections of the lungs (pneumonia), bronchi (bronchitis), and pleura (pleuritis) can lead to bacteremia, which provides another route to the meninges. Frequently, the bacteria migrate from nasopharyngeal venules (blood vessels) to the meninges. The bacteria infecting the meninges rapidly spread throughout the subarachnoid space (fig. 27-1) and infect the ventricles (ventriculitis).

Meningitis caused by *N. meningitidis*, *H. influenzae*, and *S. pneumoniae* is usually preceded by an upper respiratory tract infection, an ear infection, or a lung infection. Headache, stiff neck, muscle pain (myalgia), back pain,

TABLE 27-1
FREQUENCY OF BACTERIAL CAUSES OF MENINGITIS BY AGE GROUP

	NEONATES (ONE MONTH OF AGE OR LESS)	CHILDREN (ONE MONTH TO 15 YEARS OF AGE)	ADULTS (OLDER THAN 15 YEARS OF AGE)
Streptococcus pneumoniae	0–5%	10–20%	30–50%
Neisseria meningitidis	0–1%	25–40%	10–35%
Haemophilus influenzae	0–3%	40–60%	1–3%
Streptococci*	20–40%	2–4%	1–5%
Staphylococci	1–5%	1–2%	5–15%
Listeria	2–10%	1–2%	5%
Gram negative rods**	50–60%	1–2%	1–10%

*Almost all streptococci isolated in neonatal meningitis are Group B.
***Escherichia coli* accounts for approximately 40% of all bacterial neonatal meningitis. *Enterobacter, Klebsiella, Proteus, Pseudomonas,* and *Serratia* are responsible for the remaining cases caused by gram negative bacteria.

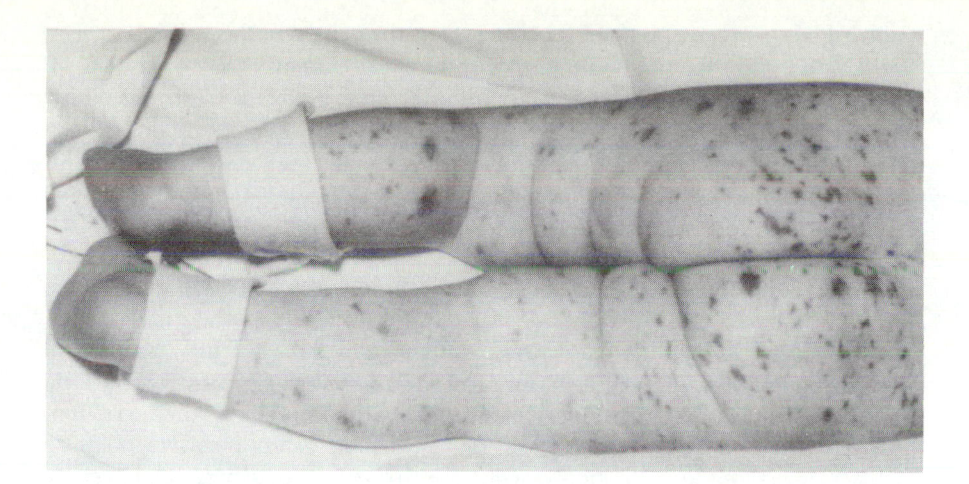

Figure 27-2 Meningococcal Disease. Acute meningococcemia in a young child with meningococcal meningitis.

and vomiting are common early symptoms before meningitis occurs. Meningitis begins when the bacteria spread to the meninges. Symptoms then include muscle ache, backache, stiff neck, fever, drowsiness, confusion, and unconsciousness. The pus and blood clots that accumulate in the brain block the flow of cerebrospinal fluid out of the brain, causing the subarachnoid space to swell. The increased pressure on the brain can cause seizures, paralysis, and coma. Meningococcal *(Neisseria meningitidis)* infections are generally accompanied by a purpuric (purplish) rash all over the body (fig. 27-2). The purple color is due to hemorrhaging capillaries just under the skin. This rash may be caused by a **bacterial protease** that is known to hydrolyze IgA, or it may be due to damage to blood vessels caused by endotoxin □ released by the gram negative *Neisseria meningitidis* (Shwartzmann phenomenon). A purpuric rash rarely occurs during a *Streptococcus pneumoniae* bacteremia and meningitis; however, the purpuric rash of echovirus **aseptic**

433

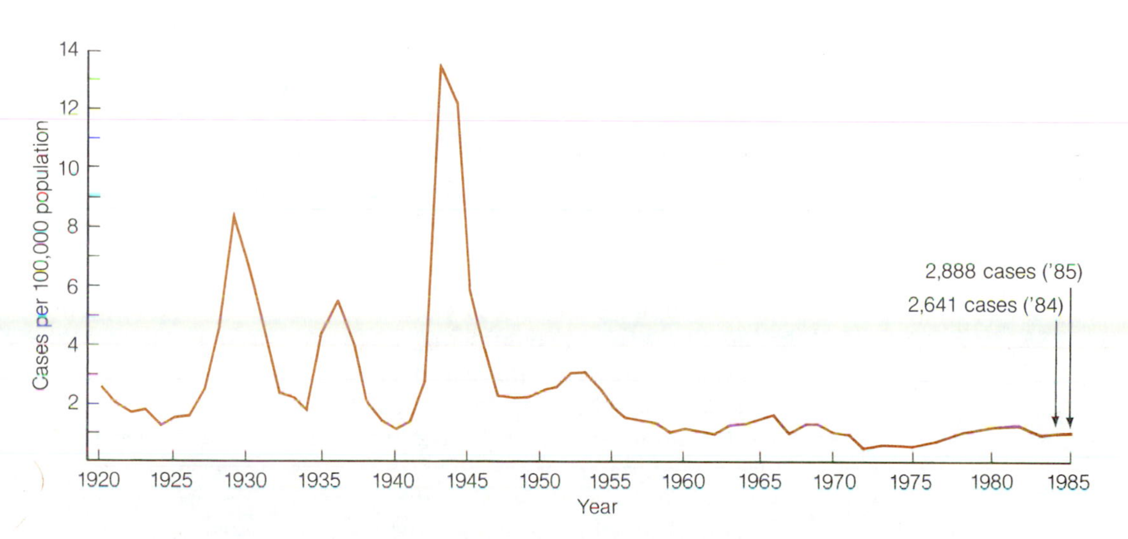

Figure 27-3 Incidence of Meningococcal Infections. During the last few years (1980–1985), approximately 2,700 cases of meningococcal cases have been reported per year. The last major epidemic of meningococcal disease in the U.S. occurred during the Second World War, when large numbers of young soldiers were brought together in army camps. Today most cases occur in infants less than one year old.

meningitis (an infection of the meninges that does not involve bacteria) looks very much like the meningococcal meningitis rash.

The risk of meningococcal meningitis can be reduced by avoiding crowded quarters with poor air circulation in which large numbers of persons gather, such as crowded infant day-care centers or military barracks. Meningitis caused by *S. pneumoniae*, *N. meningitidis*, and *H. influenzae* can be controlled by early diagnosis and treatment of ear infections, mastoid infections, and pneumonia. In the last few years, the incidence of meningococcal infections in the U.S. has been about 2,700 cases/year, while the number of deaths has ranged between 300 and 700 per year.

Treatment of bacterial meningitis involves the use of antibiotics, but shock can be brought on by antibiotic treatment during acute bacteremia (blood infection). The shock is due to an allergic reaction against the lipopolysaccharide (endotoxin) released by the lysing gram negative bacteria *(Neisseria, Hemophilus)*. Lipopolysaccharide from the membranous envelope that surrounds gram negative bacteria reacts with IgE, sometimes resulting in a severe anaphylactic □ reaction and shock known as the Shwartzmann phenomenon. The shock often is severe enough to kill the individual being treated.

482

Acute cerebral swelling generally is treated by intravenous infusion of mannitol and the use of corticosteroids (such as dexamethasone sodium phosphate). Cerebellar pressure is manifested by enlargement of the pupils and by rigid extension of the feet (bilateral extensor plantar responses). Seizures, which may cause fluids to go down the trachea and choke a patient, can be controlled by anticonvulsants.

Tetanus

Tetanus (lockjaw) is a disease characterized by painful cramps, convulsions, labored breathing, and spastic paralysis, caused by a neurotoxin produced by the bacterium *Clostridium tetani*. The neurotoxin binds to central nervous system nerve cells and inhibits nerve signals. *Clostridium tetani*, an anaerobic endospore former, is ubiquitous and lives in most soils in the warmer parts of the world.

When the endospores of *C. tetani* enter into a host tissue as a result of a deep puncture wound, they are able to germinate into vegetative cells because the damaged tissue provides an anaerobic environment. As the vegetative cells multiply, there is local tissue destruction and necrosis (tissue death) due to the exoenzymes released. The bacteria remain within the necrotic tissue, because they are unable to reproduce in normal tissue due to the high O_2 concentration. The bacteria multiply slowly for 7–14 days before any symptoms appear. As some of the bacteria die, they autolyse and release one of the most potent neurotoxins known to humans: less than 2.5×10^{-9} g/kg body weight of this neurotoxin is required to kill a human. Thus, a 150-pound (70-kg) person can be killed by only 1.75×10^{-7} g (0.175 mcg) of the neurotoxin, which diffuses into the lymph and then into the blood and spreads throughout the body. It may also travel inside peripheral nerves to the central nervous system; the neurotoxin has an affinity for the axons of neurons, to which it binds (fig. 27-4).

Nerve cells transmit signals by releasing the neurotransmitter acetylcholine from the ends of their axons. The neurotransmitter diffuses across the synaptic junction to the dendrite of another nerve cell, where it causes a

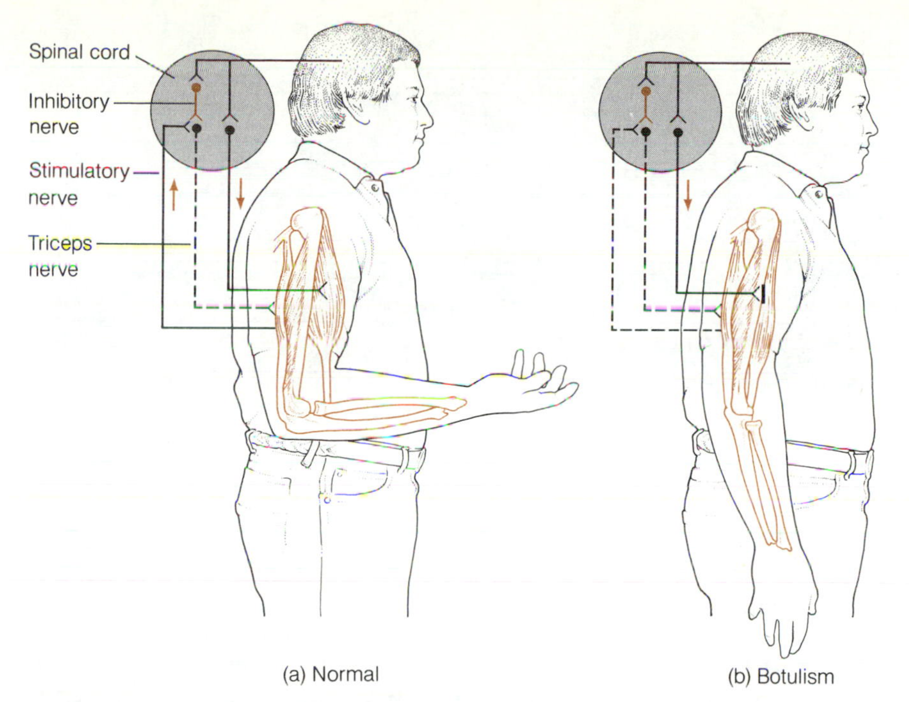

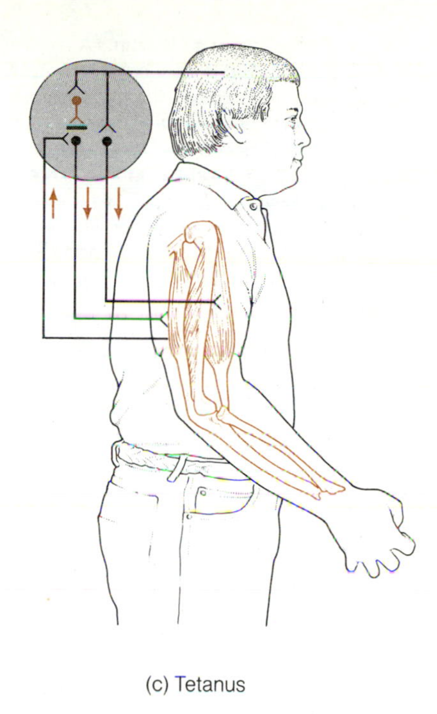

Figure 27-4 How Botulism and Tetanus Toxins Work. *(a)* Nervous control of the biceps and triceps are illustrated. Nerve impulses from the brain result in the contraction of the biceps. The stretching of the triceps sends a stimulatory signal to the triceps to prevent the stretching. This signal is inhibited by the inhibitory nerve, which prevents the triceps' contraction. If nerve impulses to the triceps were not inhibited, the triceps and biceps would work against each other. *(b)* The botulism toxins bind to nerves at the nerve-muscle junctions and block the release of acetylcholine. The biceps (and triceps) cannot be stimulated to contract, resulting in flaccid paralysis. *(c)* The tetanus toxins bind to the inhibitory nerve and block the release of inhibitory neurotransmitters. The triceps, which are normally inhibited, are now able to contract against the biceps, resulting in cramps known as tetanus.

transient collapse of the membrane potential. This effect initiates a nerve signal, which rapidly moves the length of the nerve. Then the neurotransmitter is rapidly destroyed by an enzyme called acetylcholinesterase, so that the nerve can reestablish its membrane potential and be ready to receive new signals.

It has been found that the tetanus neurotoxin binds mainly to the axons of inhibitory nerve cells and blocks the release of acetylcholine (fig. 27-4). Consequently, the nerves that go directly to the muscles constantly send signals to opposing muscles. **Tetanus** occurs when opposing muscles are constantly stimulated to contract. Muscles pulling against one another result in painful cramps all over the body (fig. 27-5). Overactivity of the sympathetic nerves produces drastic changes in blood pressure and heart rate which may result in death. The tremendous use of energy and the severe pain lead to exhaustion, shock, and death.

477

Tetanus is prevented by vaccination with **tetanus toxoid** □. Tetanus toxoid is a denatured form of the tetanus neurotoxin which produces no disease but is able to stimulate the immune response. Tetanus toxoid generally is given at the same time that the vaccines for diphtheria and whooping cough (pertussis) are administered. The combined vaccine is known as DPT (diphtheria-pertussis-tetanus vaccine). Immunization of children at 2, 4, 6, and 18 months of age is recommended. Booster shots □ at 5 years, and every 10 years thereafter, are supposed to supply ample protection against tetanus.

478

Figure 27-5 Tetanus Victim. A soldier wounded at the Battle of Corunna in 1809 is a victim of the effects of the tetanus toxin. The drawing is by a Scottish surgeon and anatomist, Sir Charles Bell, and was published in 1832. The soldier is in a state of spastic paralysis or tetanus. His muscles are working against one another, locking his jaw, bending his back, and curling his toes. The painful and exhausting cramps eventually lead to shock and death.

Persons who acquire deep puncture wounds, however, or who have serious compound fractures that break through the skin, should receive a booster shot of tetanus toxoid.

Persons who have never been vaccinated for tetanus, or who have not had a booster shot for more than 10 years, should be given **antiserum** as well as the **tetanus toxoid** if they get a deep puncture wound. Antiserum contains antibodies against the tetanus neurotoxin and is obtained from humans or animals (generally horses, sheep, or rabbits) that have been vaccinated with tetanus toxoid. The antiserum provides passive immunity against the tetanus neurotoxin until sufficient antibody titer can develop after vaccination.

The number of reported cases of tetanus per year in the U.S. has declined steadily from approximately 450 in 1955 to 100 in 1975. The number of cases per year since 1975 has ranged between 50 and 100. The decline in tetanus cases is due to the increased incidence of vaccination since the 1950s. The lack of adequate vaccination is blamed for the current incidence of tetanus. In the last few years, the mortality in the U.S. due to tetanus has ranged between 20 and 50 per year.

Once an individual shows signs of tetanus (spontaneous contractions of the muscles, convulsions, cramps, and labored breathing), there is little that can be done to alleviate the symptoms. Antiserum is not very effective, because it cannot remove neurotoxin already bound to the nerve cells. It can, however, inactivate newly synthesized neurotoxin. Cleaning of the wound and penicillin treatment can also block the production of new neurotoxin. Limiting tetanus neurotoxin production may prevent death in the less severe cases.

Botulism

Botulism is a foodborne disease that involves the progressive loss of muscle activity all over the body due to a number of neurotoxins produced by the anaerobic bacterium *Clostridium botulinum.* The bacterium and its endospores can be found in most soils, on plants, on animals, and in water. Consequently, these bacteria commonly contaminate foods. Under anaerobic conditions, the vegetative cells may produce a neurotoxin. If ingested, the neurotoxin causes disease. The neurotoxins bind to nerve cells, thus blocking the signals between nerve cells and between nerve and muscle cells (fig. 27-4).

Botulism in animals and humans is caused by the consumption of neurotoxins in food in which *Clostridium botulinum* was able to grow. The botulinal toxins may be produced in canned and bottled foods that have been improperly sterilized. The anaerobic environment and nutrients in canned and bottled foods promote the germination of *Clostridium* endospores that have escaped sterilization. Growth of *Clostridium* can also occur in sausages and hams that do not contain bacterial inhibitors.

Vegetative cells produce and release the botulism toxin into the food. When the food is consumed, the toxin passes from the intestines into the circulatory system and spreads throughout the body. The botulism toxin has strong affinity for nerve cells; it binds to the axons and prevents the release of acetylcholine, thus blocking nerve impulses to the muscle cells so that they are not stimulated to contract (fig. 27-4). Thus, botulism is a type of food poisoning that results in progressive flaccid paralysis. Twelve to 48 hours after ingestion of the botulism toxin, the first signs of the disease appear: blurred vision, progressive weakness throughout the body, and difficulty in chewing and swallowing. Death, generally due to respiratory or cardiac paralysis, occurs in 25% of those who show signs and symptoms of the disease. In the U.S., there is an average of 16 cases per year in humans, but the range is 5–80 cases annually (fig. 27-6). The number of deaths each year ranges between 1 and 5.

Many animals also suffer from botulism. For example, in parts of South Africa, where the soils are deficient in phosphorus, cattle frequently chew on animal bones and associated flesh to make up for their phosphorus deficiencies. Often the cattle die from consuming botulism toxin that has been produced in the carcasses. Any animal eating remains of this type ingests spores

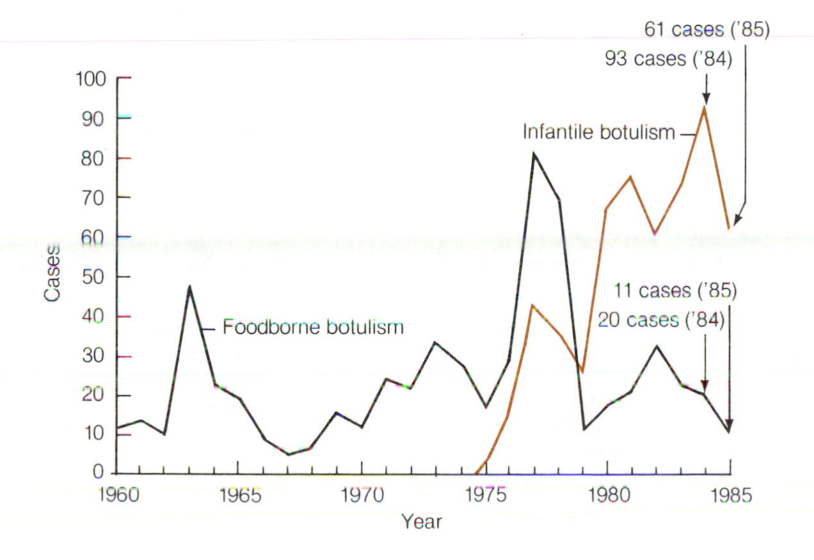

Figure 27-6 Morbidity Due to Botulism. The number of cases of foodborne botulism reported annually in the U.S. averages 16 per year. Every few years there is a drastic increase: 1951 (33), 1963 (45), 1973 (33), 1977 (80), and 1978 (65). During the last few years, the number of reported deaths has averaged about 5 per year. The number of reported cases of infant botulism in the last few years (1980–1985) in the U.S. has averaged about 70 per year. Infant botulism generally occurs in 3- to 40-week-old infants of both sexes. There is an average of one death per year, but some scientists believe that there are actually many more cases and deaths due to infant botulism than those reported. The actual number per year may be more than 400.

that germinate in its intestines after death. The vegetative cells then invade the rest of the carcass. Whole populations of wild waterfowl, mostly ducks, frequently are killed by botulism. Often, as many as 10,000–50,000 birds die per year. "Western duck sickness" occurs in the Western U.S. and Canada and is responsible, in some years, for the deaths of millions of wild waterfowl. It generally occurs when lakes and marshes are filled with decaying vegetation. Anaerobic conditions prevail at the bottom of these bodies of water, and this situation promotes the growth of *C. botulinum*. The waterfowl feeding in these environments pick up the bacteria and the toxin.

Botulism can be prevented by destroying spoiled canned foods, hams, and sausages. Spoiled food should not be given to animals, since they can also be affected by the botulism toxins. In order to prevent bovine botulism, carcasses should be disposed of immediately by burning or burial. To prevent fowl botulism, wild birds should be herded away from contaminated lakes and ponds. The contaminated areas should be drained, and feed should be distributed to keep wild birds from returning to these areas. Immunization of cattle with toxoids (Type D and Type C) is of value in South Africa and Australia. Immunization of mink with Type C and pheasants with Type A and Type C toxoids is also of use in controlling outbreaks in wild animal populations. Humans are not vaccinated against the botulinal toxins, but persons who have eaten tainted foods can be treated with a botulism antiserum. The antiserum contains antibodies that inactivate the circulating botulism toxin, but has little effect on botulism toxin attached to nerve cells. Consequently, once the disease symptoms appear, botulism antiserum is only partially effective.

Infant Botulism

Infants from birth to about 12 months of age are particularly susceptible to **infant botulism.** Apparently, *C. botulinum* endospores are able to germinate in babies' intestines, and the vegetative cells are able to produce neurotoxins there. The first nerves inhibited by the toxin are those that control the muscles in the neck and head. The infant suckles poorly, has difficulty in swallowing, and is unable to hold up its head. Thus, this syndrome is known as the "floppy baby syndrome." As the disease progresses, the child's arms and legs become paralyzed, and respiratory paralysis may lead to death. The number of reported cases each year in the U.S. (most are in California) is about 70 (fig. 27-6). Reported deaths in the U.S. due to infant botulism have been 1–2 per year; however, some studies have indicated that 5% of the infants reported to have died of respiratory failure, and classified as **sudden infant deaths,** were infected with *Clostridium* or contained botulism toxin. Since there are approximately 7,000 infants per year in the U.S. who die of sudden infant death, the actual number of infant botulism deaths may be more than 350 per year.

Infant botulism may be prevented by avoiding foods that contain large numbers of spores (such as honey) and those that promote germination of spores (such as spinach). Treatment of infant botulism includes administration of the appropriate antiserum and respiratory support units to help the infant continue breathing. *Clostridium botulinum* vegetative cells and toxins are found in the feces of infants who have recovered from the disease. Eventually, the clostridia and toxins disappear as a result of the competing microorganism in the feces and the children recover fully.

VIRAL DISEASES OF THE NERVOUS SYSTEM

Viral diseases of the nervous system are often due to viruses that initiated their attack elsewhere in the body. The initial infection generally occurs at the site of an animal bite or insect bite, or in the upper respiratory tract, which then spreads to the nervous system.

Rabies

Rabies is a disease of the central nervous system which results in the deterioration of the brain and eventually death. It is caused by the growth of a bullet-shaped **rhabdovirus** in the spinal cord and brain and is characterized by paralysis, acute renal failure, coma, and death. **Rabies viruses** are found all over the world, infecting wild animals such as skunks, raccoons, foxes, coyotes, bobcats, and some bats. Because the rabies viruses grow exceedingly well in the salivary glands, they are found in large numbers in the saliva of an infected animal. Rabies generally is spread from one animal to another, or from animals to humans, through bites that inject the virus into the tissue. Consequently, many domestic animals become infected when they are bitten by a rabid animal. Cattle, sheep, and oxen are frequently infected when they are bitten by a rabid bat, skunk, fox, coyote, or wolf. Dogs and cats become infected when they attack a rabid animal or disturb its carcass. Rabid domestic animals, in turn, can transmit the disease to humans through bites. It is believed that the rabies virus can also be transmitted by virus aerosols; in this case, the virus begins its infection in the epithelium lining the nasal passages and the mouth.

Once the rabies virus enters the body, it begins to replicate in host cells at the site of infection. It appears to proliferate most efficiently in peripheral nerve cells. Interferon or other components of the immune system generally do not inhibit the rabies virus infection, so the virus spreads slowly from the site of infection along nerve cells, eventually reaching the spinal cord or brain. If the bite is on the hand, it might take the virus 6–12 weeks to reach the brain (fig. 27-7). If the infection occurs on the head, however, it might take the virus only 2–4 weeks to reach the brain. In humans, death occurs about two weeks after symptoms appear. No symptoms are observed until the virus begins to grow in the spinal cord and brain. The clinical course of rabies in humans involves malaise, fever, weakness, paralysis, acute renal failure, coma, and death (fig. 27-7).

Dogs with **furious rabies** are initially restless and then become highly agitated. They bark, growl, and snap at anything that catches their attention. As nervous control is lost, swallowing becomes difficult, so saliva fills the animal's mouth and begins to foam if the animal is active. As the virus causes more damage to the brain, paralysis develops and death soon follows. In dogs, death generally occurs within 2–3 days after symptoms appear. Animals with **dumb rabies** or **paralytic rabies** do not show the excitability shown by animals with furious rabies, although they do bite when irritated. Paralysis of the throat and masseter (chewing) muscles, salivation, and the inability to swallow are common symptoms of this form of rabies.

The danger of rabies to humans and pets can be reduced by vaccination of domestic animals and of humans with high-risk occupations (skunk trappers, lab workers, veterinarians, dogcatchers) with a rabies virus vaccine. Vaccination of dogs and cats in rural areas is extremely important. The in-

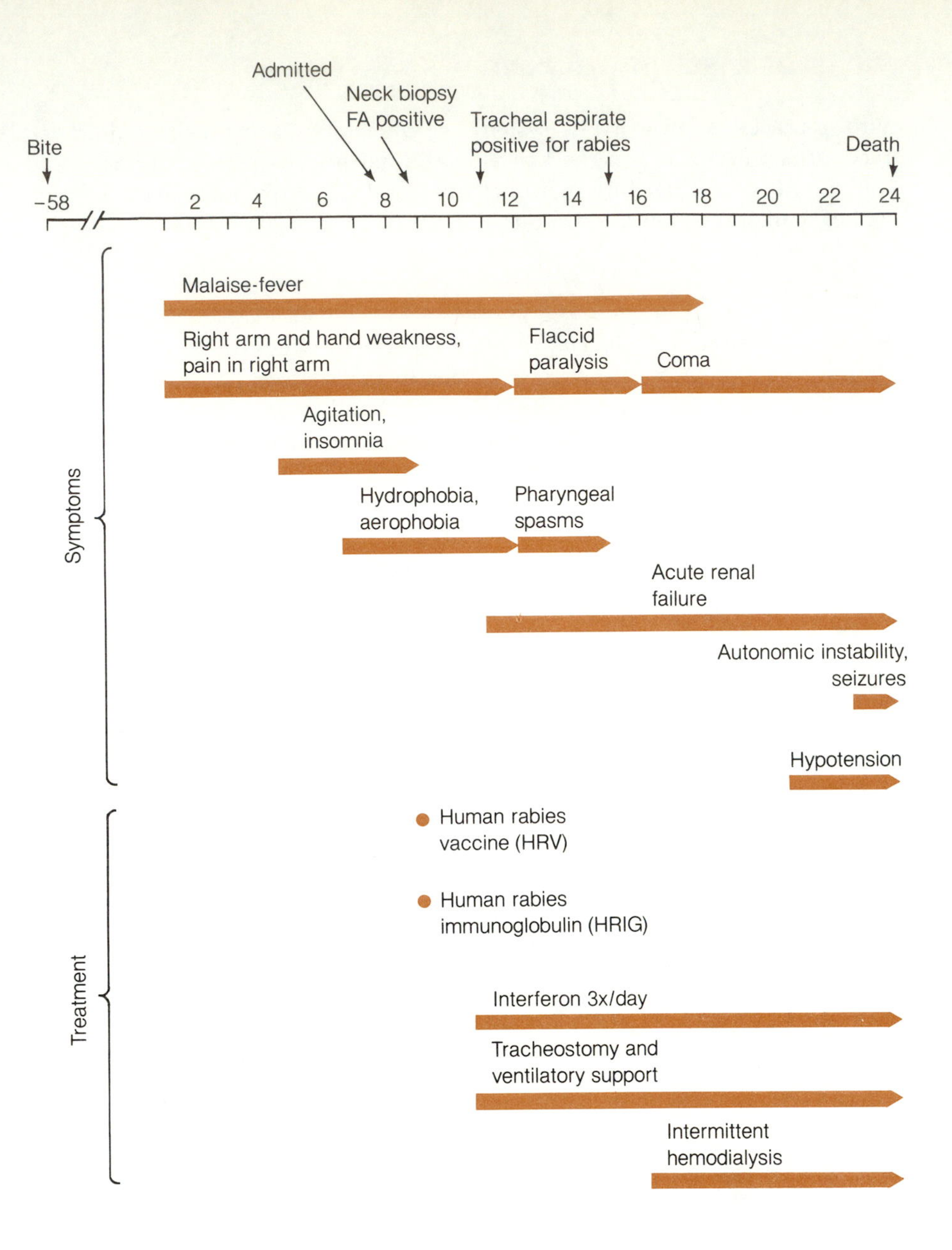

Figure 27-7 Clinical Course of a Human Rabies Case. This illustration schematizes the clinical course and management of a well-studied case of rabies. Bilateral bronchopneumonia, acute renal failure, hypotension, autonomic instability, and seizures caused sudden death due to respiratory and vascular collapse. Postmortem examination indicated fluorescent antibody positive rabies antigen in kidney parenchyma, bladder nerve tissue, skin of neck, skin from bite site, occipital nerve, pharynx, choroid plexus, hippocampus, pons, cervical spinal nerve, and dorsal root ganglion.

cidence of rabies can also be reduced by restricting dogs so that they do not run wild and attack rabid wild animals (fig. 27-8).

Louis Pasteur is credited with the development of the first vaccine against the rabies virus. The vaccine was prepared by inoculating rabbits with spinal cord extracts from animals that had died of rabies. The rabbits' spinal cords were then removed and dried for many days. The longer the

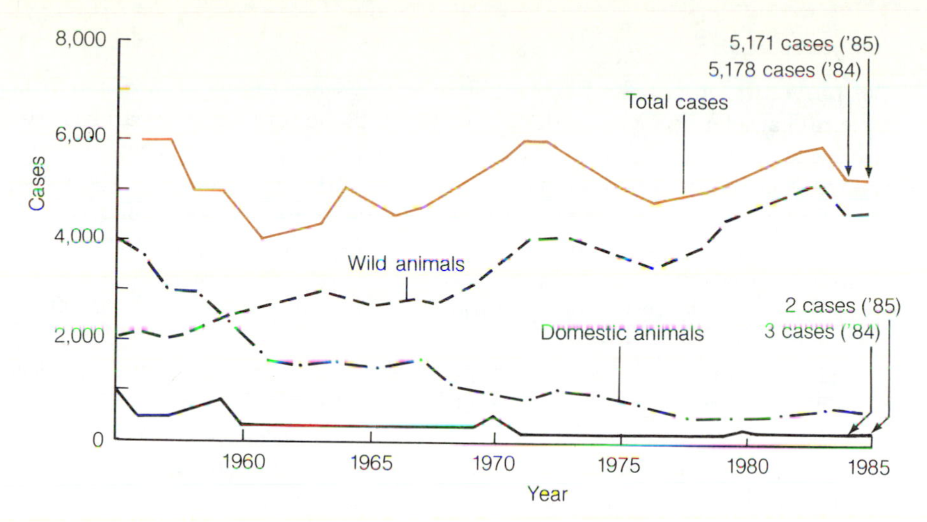

Figure 27-8 Incidence of Rabies. In the last few years the number of rabies cases in wild and domestic animals has averaged about 5,500 per year and the number of rabies cases in humans has averaged about 2 per year.

drying period, the fewer viruses survived. The vaccination procedure began with extracts from spinal cords that had few surviving viruses. Each day, for 14–21 days, extracts that had been dried for successively shorter periods of time were injected into the patient's abdomen. In 1884, Pasteur directed doctors in the vaccination of a young boy who had been bitten severely by a rabid dog and who was expected to contract rabies. The boy survived the painful vaccination procedure and demonstrated for the first time that vaccination against rabies was possible.

There are a number of rabies vaccines for animals that require only one vaccination per year. The viruses may be grown in tissue culture, extracted, and heat- and phenol-inactivated. Live virus vaccines are prepared in chicken or duck eggs. The growth of the viruses in chicken embryos leads to their 477 attenuation □. Caution must be used with live vaccines, however, because they can revert to virulent strains. Some are used only in dogs, because they become pathogenic in other animals.

Active immunization against rabies in humans still requires six inoculations of chemically inactivated viruses. Five intramuscular injections are given over a 30-day period, and one injection is given two months later. High-risk individuals, such as veterinarians and persons working with the virus, are immunized prophylactically. Persons in the general population who are exposed to the virus are passively immunized with **human rabies immune globulin** or with **horse antirabies serum,** and are then actively immunized with **human diploid cell vaccine** or with the older, infrequently used **duck embryo vaccine.** Human diploid cell vaccine is the vaccine of choice, since it requires only six intramuscular injections and it does not produce severe allergic reactions. Rabies is the only disease in which immunization started after exposure to the virus can prevent the disease. Protection against the rabies virus can develop before the disease symptoms appear because it takes such a long time for the viruses to reach the central nervous system. Once symptoms of the disease appear, however, rabies is nearly 100% fatal, so there is no treatment.

Poliomyelitis

Poliomyelitis, in its most severe form, is a disease of the central nervous system characterized by destruction of nerve tissue and muscle paralysis. It is caused by the growth of a small, nonenveloped, icosahedral virus **(picornavirus)** in the brain and spinal cord. The destruction caused by the virus in the central nervous system can lead to paralysis of various parts of the body and to death. In those who survive the disease, some part of the body is often permanently paralyzed. The paralyzed part often wastes away because the muscles are not being stimulated, and this in turn leads to deformity (fig. 27-9).

Humans appear to be the only host species for the polio virus. Because the virus often multiplies in the throat and intestines and causes mild symptoms such as sore throat, headache, fever, and nausea, the virus is readily spread from one person to another. Antibodies against the polio virus are found in approximately 99% of the human population, indicating that most people have come in contact with the virus at some time (either naturally or by vaccination). It is transmitted from one person to another in water or food contaminated with saliva or feces. Public swimming pools are a good reservoir of polio viruses, which are not readily destroyed by the chlorine concentrations used to control other microorganisms.

The virus initially infects the throat, tonsils, intestines, and lymph nodes of the ileum (opening into the large intestine). From these tissues it spreads into the blood, causing what is called **viremia.** Usually the infection does not proceed beyond viremia and the patient recovers fully. Occasionally, however, the viruses in the blood manage to infect nerve cells of the spinal cord or brain. The replication of the viruses in nerve cells causes their death, which in turn leads to paralysis. The virus does not infect or destroy peripheral nerves. Destruction of large numbers of nerve cells in the brain results in death.

Figure 27-9 Ancient Egyptian with Polio. This Egyptian hieroglyph illustrates a man with an atrophied leg, which closely resembles the results of a paralytic poliomyelitis infection.

The severe symptoms of "polio" are more prevalent in developed countries, where the level of sanitation is high, than in underdeveloped countries. One reason is that, in countries with poor sanitation, polio viruses generally infect infants soon after they are born when they are still protected by their mother's circulating antibodies. These infections stimulate the infant to develop an immune response against the polioviruses. It also appears that infants less than a year old are not very good hosts for the virus; thus, infections of infants lead to the mild form of the disease. In counties where hygiene is highly developed, immunization against polio is imperative. In the United States, vaccination against polio was initiated on a massive scale in 1955 (fig. 27-10). The incidence of paralytic polio in the U.S. in the 1970s and 1980s has ranged between 5 and 10 cases per year, and the number of deaths has averaged less than 1 per year. In 1980 there were 8 cases of paralytic poliomyelitis in the U.S.: 5 were due to the oral polio vaccine, 2 were brought in from Mexico, and one was recorded as endemic.

The average death rate for those who contact the severe form of polio is 10%, but it may be as high as 50% in some epidemics. Ten to 50% of those who survive have residual paralysis.

Once polio is contracted, treatment with vaccine may help to boost the immune system; however, there is no effective treatment for polio. Before the development of polio vaccines, the **Sister Kenney treatment** was used to inhibit virus replication. The treatment consisted of wrapping the polio

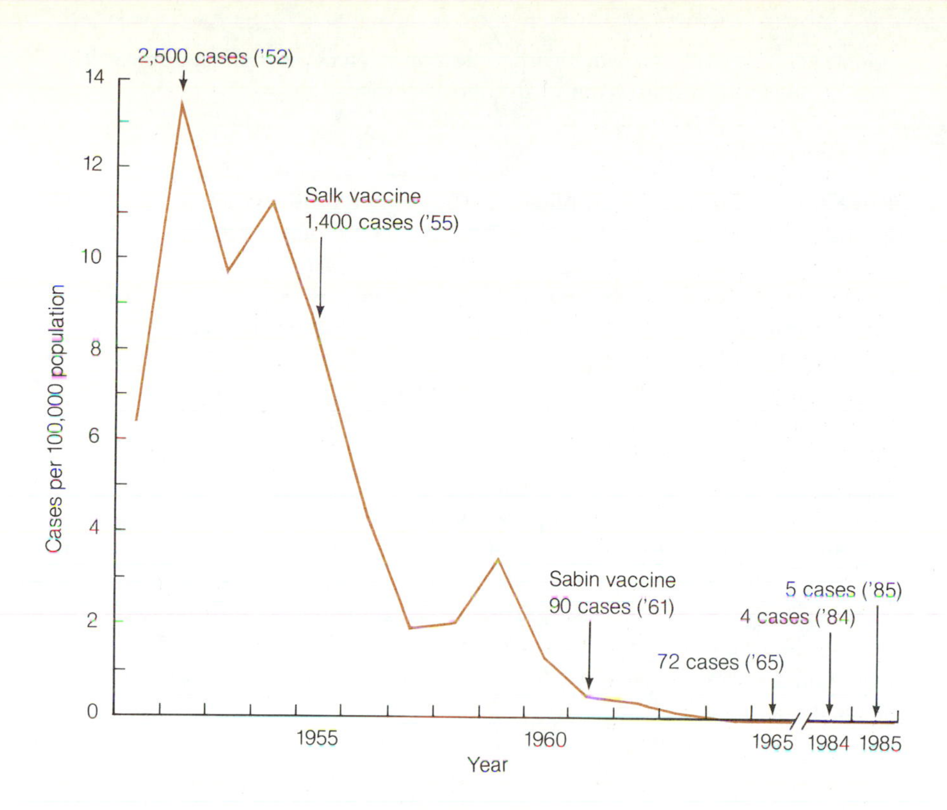

Figure 27-10 Incidence of Polio. The incidence of polio reached a peak in the early 1950s. The Salk inactivated virus vaccine was introduced in 1955. Within 5 years, it was responsible for decreasing the number of cases more than twentyfold. In 1961 the Sabin oral vaccine using attenuated polioviruses was introduced. The Sabin vaccine was responsible for reducing the number of cases in the U.S. to an average of 5 per year.

victim in very hot towels for 8–10 hours per day, in an attempt to raise the victim's temperature to 40°C (104°F). Polio viruses apparently grow poorly at temperatures above 37°C (normal human body temperature). This treatment caused extreme discomfort, but in a few instances it prevented the development of the severe symptoms of polio.

In 1955, formalin-inactivated polio viruses were used in a polio vaccine known as the **Salk vaccine.** Jonas Salk is given credit for the development of this vaccine, which is injected into the body. The dead viruses stimulate an immune reaction; however, a number of injections are required over a few months to develop a strong immune response to prevent viral infection, and booster shots are required every five years to maintain a high level of immunity.

In 1961, attenuated live polio viruses were used in a polio vaccine called the **Sabin oral polio vaccine.** Albert Sabin is given credit for the development of this vaccine, which is simply swallowed. The attenuated polio viruses invade the intestinal epithelium and multiply, but because of their attenuated form, they are unable to spread efficiently to other parts of the body. The growth of the viruses in the intestines is believed to stimulate the production of IgA in the intestines as well as a generalized immune response. IgA in the intestines inhibits the growth of the wild-type polio viruses, thus reducing their prevalence in the immunized population. The Sabin oral polio vaccine is very simple to administer, because it requires only the swallowing of a few drops of virus-laden syrup. Two separate inoculations are required over a period of a month to insure a strong immune response □. One disadvantage of the Sabin vaccine is that one of the three types of poliovirus (type 3) sometimes reverts back to its virulent form and causes polio. Polio is contracted from vaccines at a rate of about one in 5 million immuniza-

466

tions (1/5 × 10^6). In 1980, five people contracted paralytic poliomyelitis in the U.S. because of the Sabin oral polio vaccine.

Reye's Syndrome

Reye's syndrome is a complication that can follow influenza A, influenza B, and varicella (chickenpox) infections and is characterized by a swelling of the brain and a fatty degeneration of the liver. Approximately 95% of the cases occur in children 14 years old or younger. Reported cases in the U.S. range from 200 to 400 cases per year, or 1 to 2 cases per 100,000 children per year (fig. 27-11).

In a typical case, a child with influenza or chickenpox (varicella) who is beginning to recover suddenly begins to vomit repeatedly. Within a day, the child becomes lethargic, disoriented, and irritable. Soon afterward, the child may become unconscious, and death follows in about 25% of the cases (50 to 100 deaths/year in U.S.). Apparently, the influenza or varicella viruses infect the liver and brain. Damage to the liver causes an increase in blood ammonia and increases in the enzymes **serum glutamic oxaloacetic transaminase** (SGOT) and **serum glutamic pyruvic transaminase** (SGPT). Diagnosis of Reye's syndrome is made by measuring the levels of NH_3 and these liver enzymes in the serum. Growth of the viruses in the brain appears to cause an acute noninflammatory encephalopathy. The brain swelling is probably due to the accumulation of cerebrospinal fluid around the brain. As the volume of cerebrospinal fluid in the brain increases, there is an increase in pressure on the brain, which damages it. Some children who survive the disease have permanent brain damage. Death from Reye's syndrome can sometimes be avoided by controlling the swelling of the brain with drugs.

Guillain-Barré Syndrome

Guillain-Barré syndrome is a complication that can follow some virus infections. A reversible paralysis is associated with the Guillian-Barré syn-

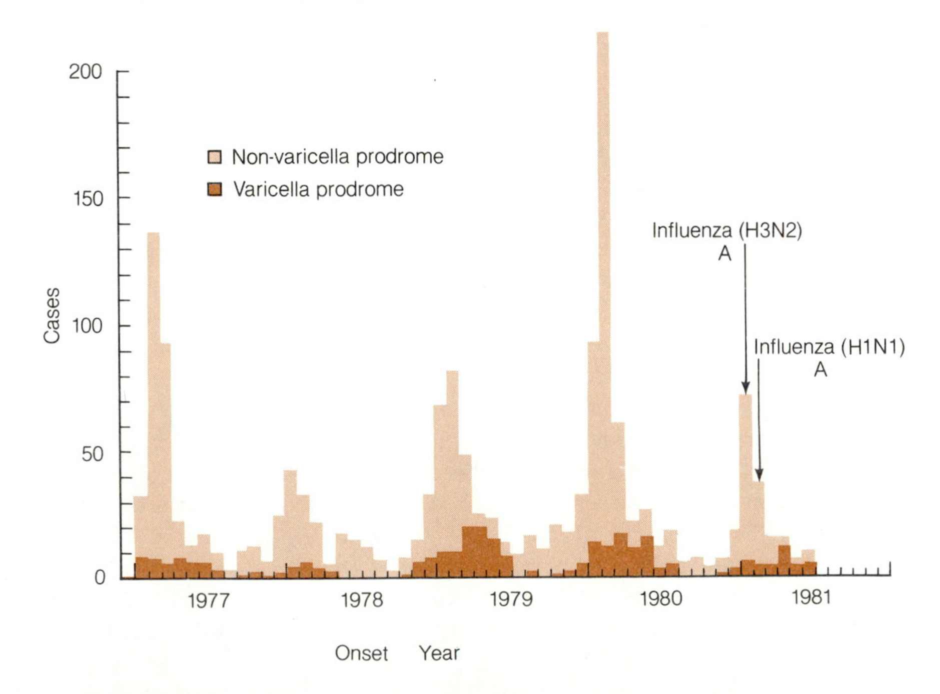

Figure 27-11 Incidence of Reye's Syndrome. Cases of Reye's syndrome and Influenza A and B isolates appear to peak in late January and early February, indicating a causal relationship between the two diseases. Varicella-related Reye's syndrome cases occur sporadically throughout the year; however, 63% of all varicella-associated cases occurred during the late winter (Jan.–Mar.) and early spring (Apr.–May) when chickenpox cases usually reach a peak.

drome, and death may result if the lung diaphragm is paralyzed. Approximately 500 cases are reported each year in the U.S. (fig. 27-12). Because Guillain-Barré syndrome occured in some persons after they received the swine flu vaccine in 1978, the vaccination program against swine flu was halted. A study of the situation eventually indicated, however, that Guillain-Barré syndrome was not associated with influenza vaccinations.

It is believed, however, that influenza viruses infect and kill Schwann cells, which myelinate peripheral nerves, thus causing a demyelination of these nerves. This effect results in the paralysis of some muscles, since many peripheral nerves do not transmit nerve impulses without a myelin sheath. Death can be prevented by artificially stimulating breathing. Since Schwann cells can proliferate and wrap around demyelinated nerves, nerves can be made to function once more, and recovery from the Guillain-Barré syndrome is generally complete.

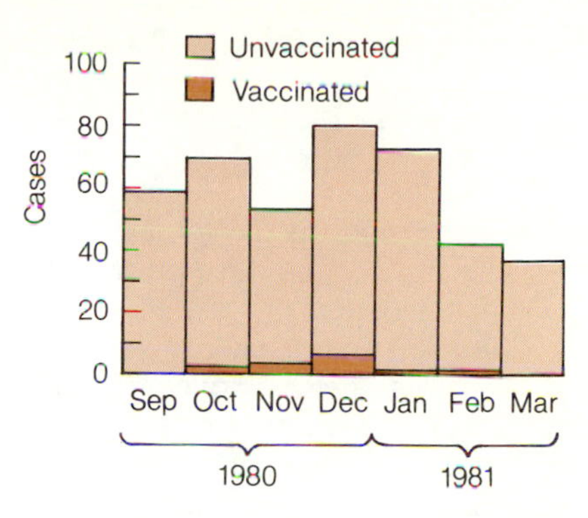

Figure 27-12 Incidence of Guillain-Barre Syndrome. Guillain-Barre Syndrome was believed to be related to the use of influenza vaccines. The data indicate, however, that most cases of Guillain-Barre Syndrome occur in persons not vaccinated with influenza vaccines.

Slow Viruses and Prion Diseases

There are several diseases of the brain and spinal cord which often take many years to develop after the infectious agent invades the nervous system. Some of the infectious agents that cause these "slow virus diseases" are conventional viruses, while others are nonconventional pathogens that appear to be fundamentally different from either viruses or cellular microorganisms (table 27-2). There is now evidence to indicate that some of the nonconventional pathogens are **prions** □, proteins that stimulate their own synthesis. Allergic reactions and inflammation stimulated by the conventional viruses are responsible for some viral diseases. The nonconventional pathogens, by contrast, do not seem to stimulate an immunologic response or inflammation.

420

Kuru

Kuru is a progressive degenerative disease of humans that is due to a nonconventional pathogen believed to be a prion. Kuru is distinguished by erratic

TABLE 27-2
DISEASES DUE TO SLOW VIRUSES AND PRIONS

DISEASE	HOST	AGENT	CHARACTERISTIC	INFLAMMATION
Scrapie	Sheep	Prion	Lysis of nerve cells	0
Mink encephalitis	Mink	Prion?		0
Visna	Sheep	Virus	Demyelination	+
Aleutian disease of mink	Mink	Virus	Demyelination	+
Kuru	Human	Prion?	Lysis of nerve cells	0
Creutzfeldt-Jakob disease	Human	Prion?	Lysis of nerve cells	0
Subacute sclerosing panencephalitis	Human	Rubeola virus	Demyelination	+
Progressive rubella Panencephalitis	Human	Rubella virus	Demyelination	+
Multiple sclerosis	Human	Rubeola virus	Demyelination	+
Amyotrophic lateral sclerosis	Human	Retrovirus	Lysis of motor neurons	+
Progressive multifocal leukoencephalopathy	Human	Picornavirus	Demyelination	+/0

FOCUS

THE INFLUENZA VIRUS MAY BE RESPONSIBLE FOR MAJOR NEUROLOGICAL DISEASES

Scientists at the Center for Disease Control (CDC) in Atlanta, Georgia have proposed that the influenza virus causes not only pneumonia but also encephalitis and chronic heart diseases that may affect people many decades after they contracted influenza.

The more than 20 million people who died worldwide in the 1918 influenza epidemic were only the first victims of the virus. During the years 1919–1928, approximately 500,000 people died worldwide of **encephalitis lethargica,** a severe neurological disease caused by influenza A virus. In addition, the CDC researchers propose that much of the heart disease in elderly people could be due to the 1918 influenza virus having damaged the nerves that regulate the heart. In the 1970s, there was a decline in the number of cases of chronic mental disorders and heart disease. The CDC researchers postulate that this was because many of the people alive during the 1918 influenza epidemic died because of old age and/or disease.

jerking, tremors, loss of muscular control, inability to walk or stand, tissue wasting, and death 9–12 months after the onset of the signs and symptoms.

Kuru is found infrequently today, but it did infect a large number of individuals in a primitive New Guinea tribe who practiced headhunting and cannibalism of brains. Kuru attacked primarily young boys and girls and adult women, but almost never adult men. The reason was that the women prepared infected brain tissue from persons who had died of kuru, and the children and women ate this tissue. Tribal laws prohibited the men from touching or consuming human tissues, so they infrequently developed the disease.

Creutzfeldt-Jakob Disease

Creutzfeldt-Jakob disease in humans is due to a nonconventional pathogen thought to be a prion. The disease usually affects persons between 50 and 60 years of age and is distinguished by a progressive mental deterioration, jerking, muscle tremors, and eventual death. The nonconventional pathogen causes lysis of nerve cells and the proliferation of astrocytes, so that a spongiform encephalitis develops.

A high incidence of this disease is found in Libyan Jews, who frequently consume ovine eyes and brains. Creutzfeldt-Jakob disease can infect recipients of corneal transplants when the donor is infected. Neurosurgeons can become infected while working on diseased patients. There is a high incidence of Creutzfeldt-Jakob disease in people who have had neurosurgery or eye surgery, indicating improperly sterilized equipment (brain electrodes) or surgical tools.

A reservoir for the nonconventional pathogen that causes Creutzfeldt-Jakob disease may be prion-infected sheep that have **scrapie.** Human infections could come from eating scrapie-infected sheep tissue. The scrapie agent is extremely resistant to inactivation by heat or chemicals, so it could be transmitted during preparation of infected meat or in undercooked lamb.

Alzheimer's Disease

Alzheimer's disease is distinguished by a progressive mental deterioration of persons in their 50s, 60s, and 70s characterized by loss of memory and apathy. It is an important disease because it affects over 3 million people in the U.S., and cases are occurring at a rate of approximately 100,000 per year. At this time it is not clear whether Alzheimer's disease is hereditary or due to a nonconventional agent. A number of researchers believe that it is caused by a prion. If this turns out to be correct, Alzheimer's disease may be considered an epidemic disease.

Alzheimer's disease destroys neurons and causes atrophy of the cerebral cortex and neurofibrillar tangles. A typical case of Alzheimer's disease might be diagnosed at 60, progress for 10 to 15 years, and result in a demented, mute, bedridden individual.

Subacute Sclerosing Panencephalitis and Progressive Rubella Panencephalitis

490

Subacute sclerosing panencephalitis □ is a neurologic disease of young children and adolescents, rarely affecting persons older than 14. It is initially characterized by mental deterioration, jerking, and the inability to control muscles. As the disease progresses, it leads to dementia, stupor, rigidity, and death. Serum and cerebrospinal fluid have high titers of antibodies against the measles virus. The virus causes the destruction of the myelin sheath around brain nerve cells, and it stimulates the proliferation of astrocytes and the invasion of the brain by mononuclear cells.

Subacute sclerosing panencephalitis is due to a mutant form of the rubeola (red measles) virus, while progressive rubella panencephalitis is due to a persistent infection by the rubella virus. In order for the viruses to cause these diseases, apparently they must infect the host early in life: before the age of two, in the case of the rubeola virus, and before birth, in the case of the rubella virus. High titers of antibodies that diminish host cell-mediated immune responses appear to select for mutant forms of the rubeola and rubella viruses which are able to persist in the brain.

Patients with congenital rubella infections may undergo progressive mental deterioration, muscle spasms, lack of muscle control, and seizures. The damage to the brain resembles that caused by subacute sclerosing panencephalitis.

Progressive Multifocal Leukoencephalopathy

Progressive multifocal leukoencephalopathy is caused by a virus that resembles the papovavirus SV40. The disease results in the destruction of the myelin sheath in persons in their 50s and 60s, with signs and symptoms that resemble those of subacute sclerosing panencephalitis.

Multiple Sclerosis

Multiple sclerosis is a general term describing a type of degeneration of the brain or spinal cord which involves the destruction of myelin sheaths around nerve cells. This results in muscular weakness, trembling, slurred speech, diminished vision and paralysis. Some types of multiple sclerosis are initiated by virus infections; the measles virus is a favorite candidate.

The normal brain or spinal cord appears white when sectioned, because of the large amount of myelin surrounding the nerve cells. If the **oligodendroglial cells** that form the myelin sheaths are destroyed, however, an abnormal growth of astrocytes surrounds the nerve cells and produces a discoloration and hardening (sclerosis) of the tissue.

It is hypothesized that a virus (measles, influenza, etc.) infects the brain during an illness. The virus infection is thought to damage the oligodendroglial cells and the "blood-brain barrier," so that myelin proteins, which normally are found only in the brain, pass into the body and initiate an autoimmune □ reaction. Apparently, stimulated macrophages and lymphocytes can penetrate the "blood-brain barrier" and attack the myelin sheath that covers the nerve cells in the brain and spinal cord. An experimentally induced multiple sclerosis has been produced in guinea pigs simply by injecting a single dose of white matter from other guinea pigs. Consequently, it is believed that multiple sclerosis is due to an autoimmune reaction that is initiated by a virus infection of the brain or spinal cord. This fatal disease in humans is often of very long duration, taking 15–20 years to kill the individual.

493

Viral Encephalitis

Viral encephalitis (swelling of the brain) is caused mainly by **enteroviruses** (mumps, polio-, coxsackie-, and echoviruses) and by **arboviruses** (eastern equine encephalitis (EEE), western equine encephalitis (WEE), Venezuelan equine encephalitis (VEE), dengue fever, Japanese B encephalitis, Saint Louis encephalitis, California encephalitis, and yellow fever viruses). These viruses infect the brain, meninges, and spinal cord. This leads to fever, headache, chills, vomiting, encephalitis, and sometimes death. All of these viruses not only infect humans but have a number of animal hosts as well as humans. The arboviruses (*ar*thropod *bo*rne *viruses*) occasionally infect horses and cause widespread illness and death among herds. The Saint Louis encephalitis, California encephalitis, and Japanese B encephalitis viruses, by contrast, usually infect humans. In the U.S., there are about 1,200 reported cases of human encephalitis per year (fig. 27-13), with a 10% mortality. In 1975 there was an epidemic of Saint Louis encephalitis in the U.S. which struck over 4,000 individuals and killed 387 of them.

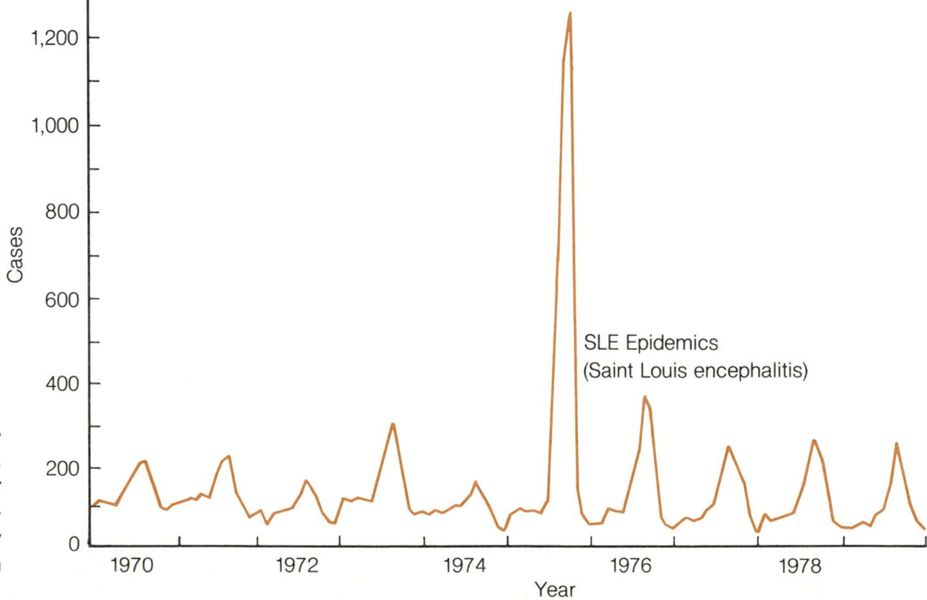

Figure 27-13 Incidence of Viral Encephalitis (Arboviruses and Enteroviruses). Viral encephalitis reaches a peak each year in the late summer (July–Sept.), when people are outdoors and likely to be infected by mosquito bites and in contact with each other.

386

The viral reservoirs are believed to be mainly wild and domestic fowl, rodents, horses, and the mosquito vector □. The bite of the infected mosquito introduces the viruses into the blood, where they can spread to all parts of the body and multiply in susceptible cells. The systemic replication and spread of the virus causes fever, headache, chills, and vomiting. If the virus spreads into the brain, it causes sufficient damage so that the brain swells. This encephalitis can result in loss of consciousness and in death.

Encephalitis caused by arboviruses can be controlled by eliminating the mosquito's breeding grounds and by using insecticides where breeding sites cannot be destroyed. There are vaccines that can be used to protect humans against the EEE, WEE, and VEE viruses. In addition, there are vaccines for horses against these viruses, and these are recommended during outbreaks of the diseases; mortality in horses is about 20–50% for WEE, as high as 75% for VEE, and over 90% for EEE. Two intradermal injections of vaccine 10 days apart give immunity that lasts approximately a year.

Aseptic Meningitis

Aseptic meningitis is an inflammation of the meninges which may be caused by the mumps, polio-, coxsackie-, or echoviruses. The term "aseptic" refers to the fact that bacteria cannot be cultured from the diseased meninges or spinal fluid. The echoviruses are responsible for the vast majority of aseptic meningitis cases. These viruses apparently spread from the throat and intestines into the blood, producing a viremia. From the blood, they penetrate to the meninges and infect these membranes. Outbreaks of aseptic meningitis occur in July, August, and September each year, the months during which large numbers of people are most likely to use pools, lakes, and streams that are contaminated by fecal matter.

FUNGAL AND PROTOZOAL DISEASES OF THE NERVOUS SYSTEM

Cryptococcosis

Cryptococcosis is a fungal disease that is usually accompanied by fever, cough, and pleural pain. In some cases the disease can become systemic and involve the brain (fig. 27-14).

Cryptococcus neoformans is a yeastlike fungus often found growing in soils and in pigeon droppings. The yeast has a thick polysaccharide capsule that is clearly visible in a capsule stain and aids in the yeast's identification. The yeast can be blown into the air on windy days and thus transmitted to humans. The inhaled *Cryptococcus* often is able to grow in the lungs; when it does so, it leads to a fever, a severe cough, and pleural pain. In relatively healthy humans, cryptococcosis is self-limiting. In older persons and immunosuppressed individuals, however, the yeast may manage to spread from the lungs into the blood and reach various organs. If the meninges are infected by the yeast, a chronic meningitis results that is fatal if untreated. Cryptococcosis can be treated effectively with amphotericin B.

African Sleeping Sickness

African sleeping sickness, or **trypanosomiasis** is a disease that is characterized by attacks of fever over a period of several months and then a

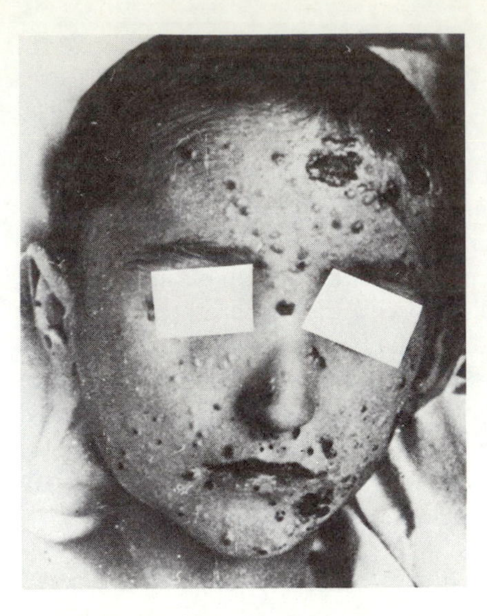

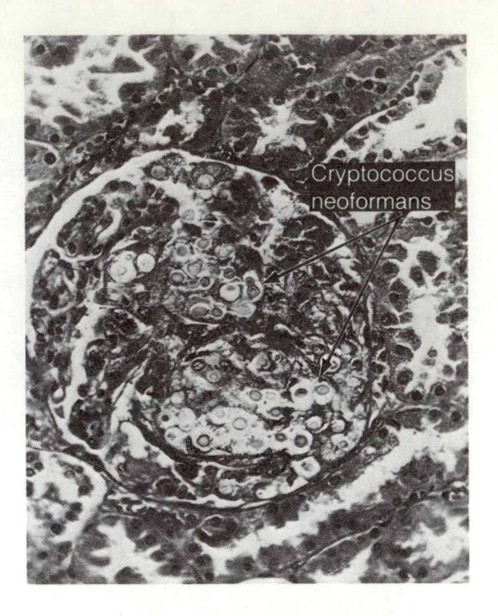

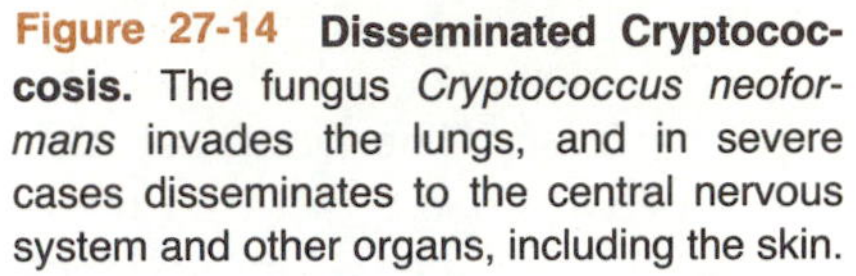
Figure 27-14 Disseminated Cryptococcosis. The fungus *Cryptococcus neoformans* invades the lungs, and in severe cases disseminates to the central nervous system and other organs, including the skin. *(a)* Child with disseminated cryptococcosis. Fungi have spread to the brain and skin from the lungs. *(b)* Large number of *Cryptococcus neoformans* cells in kidney glomerulus.

progressive disorientation, slurred speech, and difficulty in walking. The last stages of the disease include convulsions, paralysis, mental deterioration, extended periods of sleep, coma, and then death. These last stages of the disease generally occur over a number of months. The disease is caused by the protozoa *Trypanosoma gambiense* and *T. rhodesiense.*

Trypanosoma gambiense reproduces in humans and in the tsetse fly, while *T. rhodesiense* additionally reproduces in wild and domestic animals. Trypanosomes in the salivary gland of the tsetse fly are injected into animals by the bite of the fly. The trypanosomes develop in the blood and lymph and may become polymorphic. Multiplication of the organisms in the lymph cause periodic fevers □ 2–3 weeks after infection. Eventually, the trypanosomes invade the central nervous system, producing a mild meningitis and encephalitis that produces headaches, general weakness, loss of appetite, nausea, slurred speech, and mental deterioration. The infection may last for as long as a year before the individual becomes comatose and dies.

241

The trypanosomes are generally numerous enough in infected humans and animals that a biting tsetse fly will become infected. The trypanosomes reproduce in the fly's midgut and then migrate to its salivary glands. Tsetse flies with trypanosomes in their salivary glands act as the vectors for the disease.

The best ways to prevent the disease in Africa are to eradicate the tsetse fly's breeding sites wherever possible; wear clothing that protects against fly bites; and take prophylactic drugs during times of high risk. There are no effective vaccines against the trypanosome, because the protozoan changes its antigenic coat during the course of infection.

Trypanosomiasis can be diagnosed by observing the parasites in the blood, in the lymph nodes, or in the spinal fluid. Once African sleeping sickness is contracted, it can be treated with the drugs **pentamidine isothion-**

TABLE 27-3
DISEASES OF THE NERVOUS SYSTEM

DISEASE	CAUSATIVE AGENT	MICROBIAL GROUP
Meningitis	*Neisseria meningitidis*	Gram − cocci
	Streptococcus pneumoniae	Gram + cocci
	Haemophilus influenzae	Gram − rods
	Escherichia coli	Gram − rods
Aseptic meningitis	Viruses	Various groups
Tetanus	*Clostridium tetani*	Sporogenous rod
Botulism	*Clostridium botulinum*	Sporogenous rod
Rabies	Rabies virus	Rhabdoviruses
Poliomyelitis	Polioviruses	Picornaviruses
Reye's syndrome	Influenza viruses	Orthomyxoviruses
	Varicella-Zoster virus	Herpesvirus
Encephalitis	Arboviruses	Arboviruses
Cryptococcosis	*Cryptococcus neoformans*	Yeast
Sleeping sickness	*Trypanosoma gambiense*	Protozoa
	Trypanosoma rhodesiense	Protozoa

ate and **melarsoprol.** Treatment of *T. rhodesiense* infections also relies on the drug **suramin.**

Microorganisms cause many diseases that involve the human nervous system. The salient characteristics of some of these diseases are summarized in table 27-3.

SUMMARY

ORGANIZATION OF THE NERVOUS SYSTEM

1. The central nervous system includes the brain, spinal cord, cerebrospinal fluid, and meninges (the membranes that cover the brain and spinal cord).
2. The peripheral nervous system consists of all the nerve cells from the body that connect to the spinal cord or brain.
3. The "blood-brain barrier" inhibits the movement of antibodies and some drugs from the blood into the spinal cord and brain. Consequently, the brain is not directly protected by the immune system, and some drugs and antibiotics are useless in treating spinal cord or brain infections.

BACTERIAL DISEASES OF THE NERVOUS SYSTEM

1. Acute pyogenic meningitis is caused by *Escherichia coli* and Group B streptococci in babies younger than 1 month; by *Haemophilus influenzae*, *Neisseria meningitidis*, and *Streptococcus pneumoniae* in children less than 15 years old; and by *Streptococcus pneumoniae*, *Neisseria meningitidis*, and *Staphylococcus aureus* in adults. There are 2,700 cases/year and 300–700 deaths/year in the U.S. due to meningococcal meningitis.
2. *Neisseria meningitidis* is the only bacterium that causes meningitis epidemics.
3. Meningitis is preceded by upper respiratory tract infections, ear infections, or lung infections.
4. Meningococcal infections, in contrast to pneumococcal infections, are generally accompanied by a rash due to hemorrhaging capillaries.
5. Tetanus or lockjaw is due to a neurotoxin produced by the anaerobic bacterium *Clostridium tetani*.
6. The tetanus neurotoxin binds primarily to inhibitory nerve cells and blocks the release of acetylcholine. Thus, nerves send signals to opposing muscles and painful cramps develop all over the body.
7. The tetanus neurotoxin also blocks neuromuscular

transmissions, so that muscles become paralyzed.

8. Tetanus is prevented by vaccination with tetanus toxoid, a denatured form of the neurotoxin.
9. There are 50–100 human cases/year and 20–50 deaths/year in the U.S. due to tetanus.
10. Botulism is due to a neurotoxin produced by the anaerobic bacterium *Clostridium botulinum.*
11. The botulism neurotoxin blocks neuromuscular transmissions so that muscles become paralyzed.
12. Deaths due to botulism are generally due to respiratory or cardiac paralysis. There are 5–80 adult cases annually and 1–5 deaths each year.
13. Flocks of wild waterfowl, mostly ducks, are often wiped out by botulism.
14. Infant botulism may kill 70–350 infants/year in the U.S.

VIRAL DISEASES OF THE NERVOUS SYSTEM

1. Rabies is caused by a rhabdovirus that infects the central nervous system.
2. The rabies virus causes death because of its growth in the brain.
3. The original rabies vaccine was developed by Louis Pasteur in the 1880s. The vaccine was prepared from infected rabbit spinal cords.
4. The Pasteur treatment for rabies requires 23 subcutaneous injections, in contrast to the newer rabies vaccine that requires six intramuscular injections.
5. The morbidity and mortality of rabies in the U.S. is approximately five each year.
6. Rabies in wild animals has been increasing for a number of years, and has reached almost 5,900/year in the U.S.
7. Poliomyelitis is caused by three strains of a picornavirus that initially infect the throat and intestines and subsequently infect the central nervous system.
8. The Salk vaccine is made from formalin-inactivated polioviruses and was first used in 1955. The Sabin oral polio vaccine is made from attenuated live polioviruses and was introduced in 1961.
9. The polio vaccines have dramatically reduced the number of polio cases from approximately 2000 cases/year in the early 1950s to fewer than 10 cases/year in the 1980s.
10. Reversion of one of the polioviruses to a virulent form in the Sabin oral polio vaccine was responsible for 5 of the 8 cases of polio reported in 1980.
11. Reye's syndrome is characterized by a swelling of the brain and fatty degeneration of the tissue and is due to a complication following infections by influenza A, influenza B, and varicella (chickenpox). There are between 200 and 400 cases of Reye's syndrome/year in the U.S. Death occurs in 25% of the cases.
12. Guillain-Barré syndrome is characterized by a reversible paralysis that occurs after some virus infections. There are approximately 500 cases/year in the U.S.
13. Slow virus diseases include Aleutian disease of mink, subacute sclerosing panencephalitis, progressive rubella panencephalitis, and multiple sclerosis.
14. Sheep scrapie, mink encephalitis, Kuru, Creutzfeldt-Jakob disease, and possibly Alzheimer's disease appear to be caused by "infectious proteins" called prions.
15. Viral encephalitis is caused by enteroviruses and arboviruses that infect the meninges, spinal cord, and brain. The infection causes the brain to swell. Death occurs in 10% of cases.
16. There are approximately 1,200 cases of human encephalitis and 150 deaths/year in the U.S.
17. Arbovirus reservoirs are found mainly in wild and domestic fowl, rodents, horses, and the mosquito vectors.
18. Aseptic meningitis is an inflammation of the meninges, caused by the enteroviruses. There are approximately 1,200 cases/year in the U.S. The echoviruses are responsible for most of the cases of aseptic meningitis.

FUNGAL AND PROTOZOAL DISEASES OF THE NERVOUS SYSTEM

1. African sleeping sickness is caused by the protozoans *Trypanosoma gambiense* and *T. rhodesiense.*
2. The trypanosomes are spread by the bite of the tsetse fly, which functions both as a host and as a vector.
3. Cryptococcosis is due to the yeastlike fungus *Cryptococcus neoformans.* This fungus sometimes achieves a systemic infection that leads to meningitis.

STUDY QUESTIONS

1. Explain what the central nervous system, peripheral nervous system, and autonomic nervous system are.
2. How does the "blood-brain barrier" protect the spinal cord and brain? What is a disadvantage of this barrier?
3. Which bacteria are responsible for acute pyogenic men-

ingitis in babies less than a month old; in children less than 15 years old; and in adults?

4. Which bacterium, of the ones that cause meningitis, usually causes a rash?
5. Give the genus and species of the bacteria that cause tetanus and botulism.
6. How does the tetanus neurotoxin affect a human, and what is the cause of death?
7. How does a person acquire tetanus?
8. How does the botulism neurotoxin affect a human, and what is the cause of death?
9. How does a person acquire botulism?
10. What type of organism causes rabies, and how is the disease generally contracted?
11. Who developed the first rabies vaccine, and how was it prepared?
12. What is the difference between the Pasteur treatment and the modern treatment for rabies?
13. What type of organisms cause poliomyelitis, and how is the disease generally contracted?
14. What is the difference between the Salk and the Sabin polio vaccines?
15. How effective have the polio vaccines been in reducing the incidence of the disease?
16. What potential problem is there with the Sabin vaccine?
17. What kind of organisms cause Reye's syndrome, and how are they contracted?
18. What kind of organisms cause Guillain-Barré syndrome, and how are they contracted?
19. Scrapies, kuru, and Creutzfeldt-Jakob disease are due to what kind of infectious agent?
20. Compare and contrast subacute sclerosing panencephalitis, progressive rubella panencephalitis, multiple sclerosis, and viral encephalitis.
21. What kind of organisms are believed to be responsible for multiple sclerosis? How do they cause the disease?
22. What kind of organisms lead to viral encephalitis? How do they cause the disease?
23. Aseptic meningitis is caused by what type of viruses? How many cases are there each year in the U.S.?
24. What kind of an organism causes African sleeping sickness? How does the organism cause the disease?

SUPPLEMENTAL READINGS

Bennett, D. D. 1985. Like sheep virus, AIDS virus infects brain. *Science News* 127:22

Hirsch, M. S. 1984. Acute viral central nervous system diseases. *Scientific American Medicine* Section 7-XXVII. New York: Freeman.

Holland, J. S. 1974. Slow, inapparent, and recurrent viruses. *Scientific American 230*:32–40.

Joklik, W. K. and Willett, H. P. (eds.). 1984. *Zinsser Microbiology*. New York: Appleton Century Crofts.

McKinstry, D. W. 1983. The slow visna virus: where it hides. *Research Resources Reporter* 7:1–3

Merigan, T. C. 1984. Slow virus diseases. *Scientific American Medicine* Section 7-XXXII. New York: Freeman.

Shaw, G. M., *et al.* 1985. HTLV-III infection in brains of children and adults with AIDS encephalopathy. *Science* 227:177–182.

U.S. Department of Health and Human Services. 1980–1984. *Morbidity and Mortality Weekly Reports*. Washington D.C.: U.S. Department of Health and Human Services.

Van Heyningen, W. E. 1968. Tetanus. *Scientific American 218*:69–77.

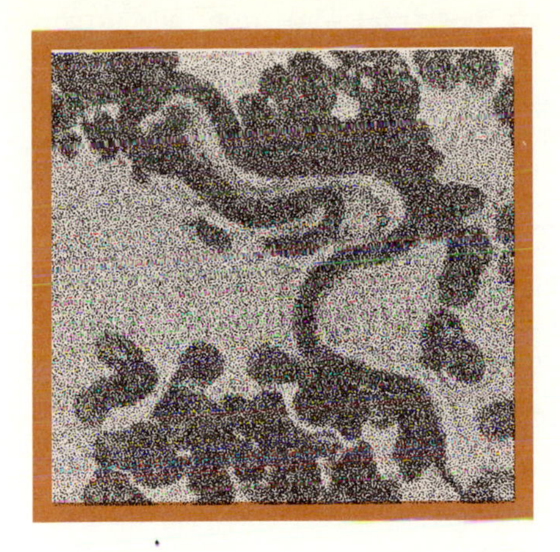

CHAPTER 28
DISEASES OF THE BLOOD, LYMPH, MUSCLE, AND INTERNAL ORGANS

CHAPTER PREVIEW
THE BLACK DEATH

The most severe plague epidemic, known as the Black Death, took place in Europe during the years 1347–1350. Millions of people died from a blood infection after suffering from a type of pneumonia (pneumonic plague) and/or swollen lymph nodes (bubonic plague). Eighty percent of those who came down with the plague died in great pain within two or three days. Studies indicate that the plague was introduced from Asia by rats or diseased seamen at one or more ports in southern France and Italy in 1347 and transmitted to humans by the bite of the rat flea. The plague spread rapidly throughout Europe.

The population of Europe in 1347 is estimated to have been 84 million, but by the end of 1350 only about 64 million people remained. Europe lost one-fourth of its population, while England is estimated to have lost one-half of its population. After the European pandemic of 1347–1350, there were recurrent outbreaks all over Europe every 10 years or so which reduced the population even further. After 1450, the epidemics killed fewer people and the population increased rapidly. Between 1347 and 1600, 25 epidemics of plague occurred in Venice and 20 in London. Between 1630 and 1670, serious epidemics in Europe broke out again and are estimated to have killed more than 10 million persons. In the late 1700s, the recurrent epidemics of plague in Europe came to an end, possibly due to the selection of more resistant hosts and vectors and/or to the selection of less virulent bacterial strains.

Plague is an example of a disease that involves the blood, lymph, and lungs. The bacterium that causes the disease is found in many wild rodents and is spread by fleas among animals and to humans. In this chapter you will learn more about plague and other diseases that affect the blood, lymph, muscle, and internal organs.

LEARNING OBJECTIVES

A STUDY OF THIS CHAPTER SHOULD ENABLE YOU TO:

EXPLAIN HOW MICROORGANISMS ENTER THE BLOOD AND LYMPHATIC SYSTEM

OUTLINE HOW MICROORGANISMS CAUSE DISEASE WHEN INFECTING BLOOD, LYMPH, MUSCLE, AND INTERNAL ORGANS

EXPLAIN HOW DISEASES OF THE BLOOD, LYMPH, MUSCLE, AND INTERNAL ORGANS CAN BE PREVENTED AND TREATED

DISCUSS THE RELATIONSHIP BETWEEN ZOONOSES AND HUMAN DISEASES

Superficial infections of the body, such as those that sometimes accompany cuts, insect bites, scratches, and tooth decay, can result in life-threatening diseases if microorganisms spread to internal organs such as the heart, brain, lungs, liver, spleen, kidneys, or bone. The purpose of this chapter is to discuss some of the mechanisms by which microorganisms infect blood, lymph, muscle, and internal organs and cause disease. As you shall see, many of these infections are initiated by vectors that introduce the microorganisms into the blood and lymph. Most of the diseases occur because toxins and microorganisms are disseminated throughout the body by the lymphatic and circulatory systems. Microorganisms that escape the host's defenses often establish multiple foci of infections.

STRUCTURE AND FUNCTION OF THE CIRCULATORY AND LYMPHATIC SYSTEMS

The circulatory system is a network of vessels that supply all cells of the body with nutrients and oxygen and remove wastes and toxic substances (fig. 28-1). Part of the blood plasma passes through blood capillaries and bathes the tissue cells in what is called **interstitial fluid** (fluid between cells).

The **lymphatic system** consists of **lymph vessels** that drain the tissues supplied by blood capillaries. The interstitial fluid, also called lymph, is picked up by the lymph vessels and carried to the **lymph nodes** (fig. 28-1), where the immune system reacts with toxins and microorganisms that may have been drawn into the lymph capillaries. After passing through the lymph nodes, the lymph flows into the blood. During infections, most of the toxins and microorganisms that escape from the initial site of infection enter the lymphatic system, where they are attacked by the immune system. Toxins and microorganisms that escape from the lymph nodes enter the venous blood and are pumped directly to the heart. From there, they are pumped to the internal tissues and organs. The immune system attacks toxins and microorganisms in the blood; for example, leukocytes and monocytes destroy microorganisms, while circulating antibodies and complement □ inactive toxins and also kill microorganisms. Microorganisms and toxins that escape destruction can initiate infections or cause damage in those tissues they reach through the blood system. Thus, some diseases of the heart (endocarditis), lungs (pleurisy), liver (hepatitis), kidney (glomerulonephritis), brain (meningitis, encephalitis), and bone (osteomyelitis) arise from superficial infections.

472

BACTERIAL DISEASES CAUSED BY INVASION OF THE LYMPH AND BLOOD

Bacteremia is the invasion of the blood by bacteria. The terms **bacteremia, septicemia,** and **bacterial sepsis** are often used interchangeably, but septicemia and bacterial sepsis are most commonly used to describe blood infections that produce signs and symptoms. Those of septicemia include fever, chills, hypotension (decreased blood pressure), and shock. Normal flora from some part of the body entering the lymph and blood are usually responsible for bacteremia. They often penetrate the normal barriers in persons suffering from leukemia, solid tumors, and multiple myelomas; the use of immuno-

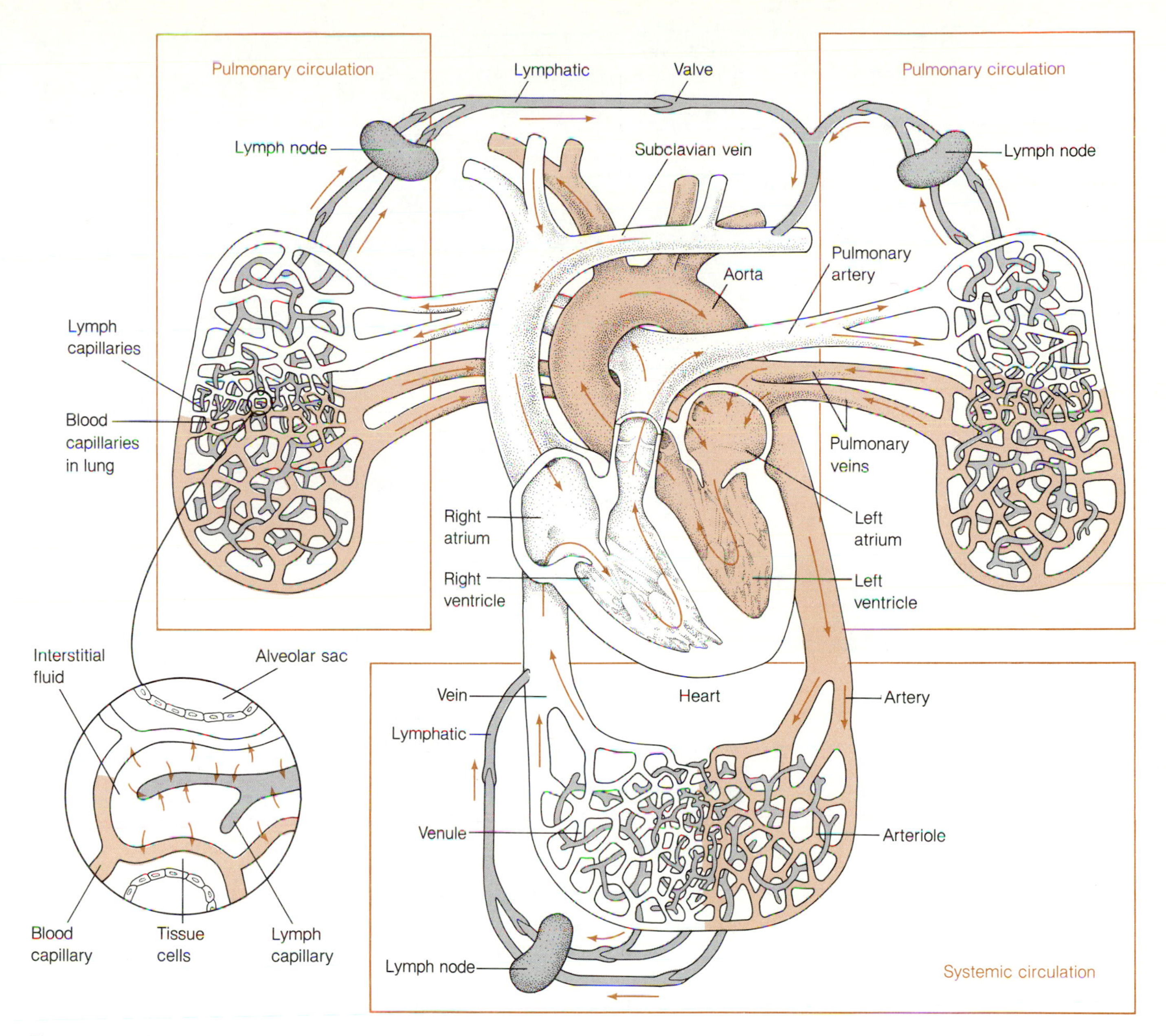

Figure 28-1 Relationship between the Blood and Lymphatic Circulation

suppressive drugs makes matters worse. Urinary catheterizations, intravenous tubes, and surgeries open the body and allow the entrance of bacteria into the lymph and blood. The extended use of broad-spectrum drugs such as tetracyclines, cephalosporins, and chloramphenicol often change the normal flora of the body, so that bacteria usually kept in check by the normal flora proliferate excessively and destroy the mucous membranes in the nasopharynx and gastrointestinal tract and then spread into the lymph and blood.

Since the introduction of antibiotics in the late 1940s and early 1950s, gram negative bacteria have become responsible for about 80% of the esti-

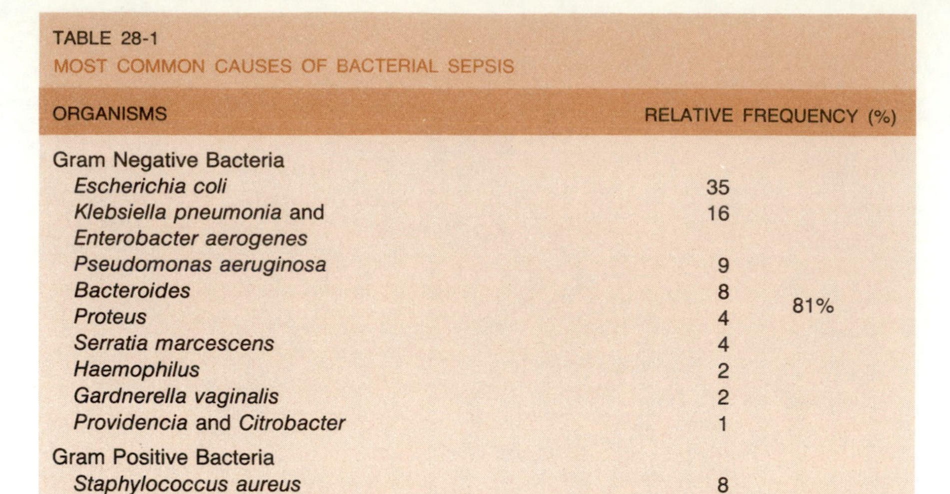

TABLE 28-1
MOST COMMON CAUSES OF BACTERIAL SEPSIS

ORGANISMS	RELATIVE FREQUENCY (%)	
Gram Negative Bacteria		
Escherichia coli	35	
Klebsiella pneumonia and *Enterobacter aerogenes*	16	
Pseudomonas aeruginosa	9	
Bacteroides	8	81%
Proteus	4	
Serratia marcescens	4	
Haemophilus	2	
Gardnerella vaginalis	2	
Providencia and *Citrobacter*	1	
Gram Positive Bacteria		
Staphylococcus aureus	8	
Streptococcus pneumoniae	5	
Group D streptococci	3	19%
Viridans group of streptococci (other than group D)	1	
Group A streptococci	1	

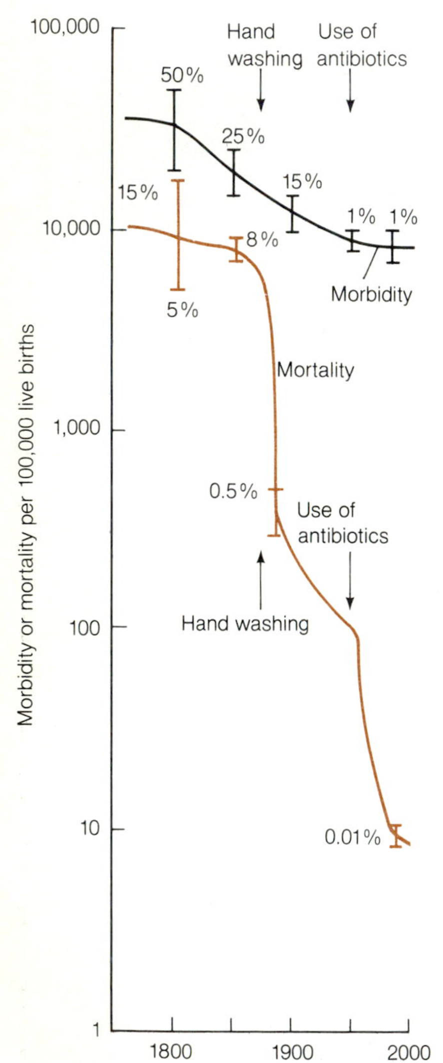

Figure 28-2 Morbidity and Mortality of Puerperal Sepsis in the United States and Europe

mated 300,000 septicemias/year in the U.S. Of these, approximately 100,000 result in death (table 28-1).

The lipopolysaccaride or endotoxin associated with the outer membrane of the gram negative bacteria is thought to be responsible for much of the pathology associated with gram negative sepsis. Studies indicate that endotoxin causes the release of kinins, activates complement by the alternate pathway □, and activates the fibrinolytic system and the coagulation systems. Kinins, complement, and histamines are released during the blood invasion, and intravascular coagulation throughout the body results in chills, fever, blockage in microvascular circulation, anaphylactic shock □, and death.

473

482

The antibiotic of choice for treatment of gram negative septicemia is the broad-spectrum antibiotic gentamicin. It is ineffective, however, against anaerobic bacteria such as *Bacteroides*, *Peptostreptococcus*, and *Peptococcus*. Clindamycin and carbenicillin are used to treat these anaerobes. The gram positive bacteria are best treated with a cephalosporin or a semisynthetic penicillin, in combination with an aminoglycoside (streptomycin) to inhibit the enterococci. Shock is treated by maintaining an adequate blood volume, by replacing lost plasma with plasma or whole blood, and by administering drugs to raise the blood pressure without producing vasoconstriction. Isoproterenol and dopamine are now being used for this purpose.

Childbed Fever or Puerperal Sepsis

Childbirth and abortion are processes that drastically disrupt the uterus and the birth canal, so that normal flora associated with these tissues, as well as those that might be introduced by mechanical means, may penetrate into the lymph and blood and cause disease. In the early 1800s, the incidence of **puerperal sepsis** (childbed fever) among new mothers often reached 50% and the death rate ranged between 5% and 15% (fig. 28-2). The morbidity

FOCUS

IGNATZ SEMMELWEISS (1816–1865) AND CHILDBED FEVER

Before Pasteur's and Koch's research showing that microorganisms were responsible for decomposition and disease, much of the medical profession was unaware of the importance of microorganisms in causing infection. Doctors frequently attended diseased patients and performed autopsies and then assisted in births, without washing their hands or changing their clothes. Because of ignorance about infection, many new mothers became infected following childbirth and a high percentage died of septicemia.

In the late 1840s, Ignatz Semmelweiss, working in the Vienna General Hospital, noticed that childbed fever was much more common in wards where physicians attended new mothers than in wards where midwives aided in births. Semmelweiss hypothesized that the high incidence of puerperal sepsis and death might be due to the doctors transferring decaying material from corpses to expectant mothers. The midwives did not perform autopsies, and they washed their hands occasionally during their shifts. After Semmelweiss insisted that midwives and other personnel in his obstetric ward wash their hands in chlorinated lime before they attended new mothers, the rate of puerperal sepsis and death in his ward dropped significantly.

Even after Semmelweiss's demonstration that hand washing could reduce the mortality and morbidity of puerperal sepsis, doctors at the Vienna General Hospital refused to institute aseptic procedures such as hand washing, and they continued to ignore the possibility that they were responsible for most of the cases of puerperal sepsis in their hospital. Semmelweiss's hypothesis and accusation stirred such hostility against him by doctors in the hospital that he was discharged. Nevertheless, within 25 years most doctors came to accept the validity of Semmelweiss's observation.

(incidence) and mortality (death) due to puerperal sepsis in the late 1800s were reduced to less than 15% and 0.5%, respectively, when the doctors and midwives began to wash their hands and sterilize surgical equipment used on women giving birth. Nevertheless, a 0.5% death rate amounts to 500 deaths/100,000 women giving birth. This death rate is 25–50 times higher than today. Before the use of antibiotics, puerperal sepsis was due mainly to the gram positive bacteria *Streptococcus* and *Staphylococcus*.

Presently, the incidence of puerperal sepsis in the U.S. is estimated to be about 30,000 cases/year and the death rate is approximately 300–600/year. These women die from infections by both gram positive and gram negative bacteria, which arise during and after delivery.

Endocarditis

Bacterial endocarditis is an infection of heart endothelium, usually in one of the valves. The infection results in the constant shedding of organisms that may spread to other parts of the body via the blood and infect internal organs such as the kidneys, spleen, and brain. Clinical symptoms include severe weakness and a persistent fever that lasts for months. The patient may have chills and sweats, difficulty in breathing, lesions on the lower legs, an enlarged spleen and heart, and a blowing sound associated with the aortic heart valve. If treatment is not effective or is not initiated, aortic insufficiency increases because of damage to the aortic valves, and the heart becomes unable to pump blood efficiently through the body. Consequently, the lymph does not pick up interstitial fluids and return them to the heart, so fluids accumulate in the tissue, particularly in the lungs. This situation is known as **congestive heart failure.**

TABLE 28-2
CHARACTERISTICS OF BACTERIAL ENDOCARDITIS

	ACUTE ENDOCARDITIS	SUBACUTE ENDOCARDITIS
Length of disease	<6 weeks	6 weeks or longer
Predisposing factors	Normal or prosthetic valve	Pheumatic or congenital heart disease; prosthetic valve
Microorganisms responsible	50% Viridans streptococci 35–40% *Streptococcus pyogenes* *Neisseria gonorrhoeae* *Pseudomonas aeruginosa* 10–15% *Staphylococcus aureus* *S. epidermidis*	80% Viridans streptococci 20% *Streptococcus faecalis* Anaerobic streptococci
Treatment	Penicillin and streptomycin	Penicillin and streptomycin

Subacute endocarditis generally progresses for two to three months before it flares up (Table 28-2). Eighty percent of cases of subacute endocarditis are due to alpha hemolytic streptococci (viridans streptococci), such as *Streptococcus mutans.* Infection of the heart tissue usually occurs in abnormal or damaged epithelium. The remainder of cases of subacute endocarditis appears to be caused by group D enterococci, such as *Streptococcus faecalis;* and anaerobic bacteria, such as *Propionibacterium acnes*, *Bacteroides*, *Fusobacterium*, and *Clostridium.*

Acute endocarditis usually reaches a fulminant stage within a month after the heart is infected (table 28-2). Fifty percent of cases are attributed to alpha hemolytic streptococci, and 10–15% are due to *Staphylococcus aureus* and *Staphylococcus epidermidis.* The Group A such as *Streptococcus pyogenes*, and the gram negative bacteria *Neisseria gonorrhoeae* and *Pseudomonas aeruginosa* account for the remaining cases.

Dental surgery and tooth decay can introduce in the bloodstream α-hemolytic streptococci, β-hemolytic streptococci, and some of the anaerobic species mentioned above. Inflammation, ulcers, or cancers of the intestinal tract generally introduce *S. faecalis* and anaerobic bacteria, while boils and wounds usually introduce *S. aureus.*

Gas Gangrene

Gas gangrene is a disease caused by the bacteria *Clostridium perfringens*, *C. septicum*, and *C. novyi.* The disease is characterized by an invasion of the lymph and blood, rotting away of tissue, the presence of ulcerating lesions, and gas bubbles often caught under the skin or within the muscle tissue being destroyed (fig. 28-3).

Clostridium perfringens, the organism responsible for most cases of gas gangrene, can be isolated from the skin of 45% of noninfected individuals and from the gastro-intestinal tract of 35% of normal persons. Invasion of the bloodstream by *C. perfringens* can result from serious wounds that are contaminated by soil bacteria, or from ulcerating lesions of the gastrointes-

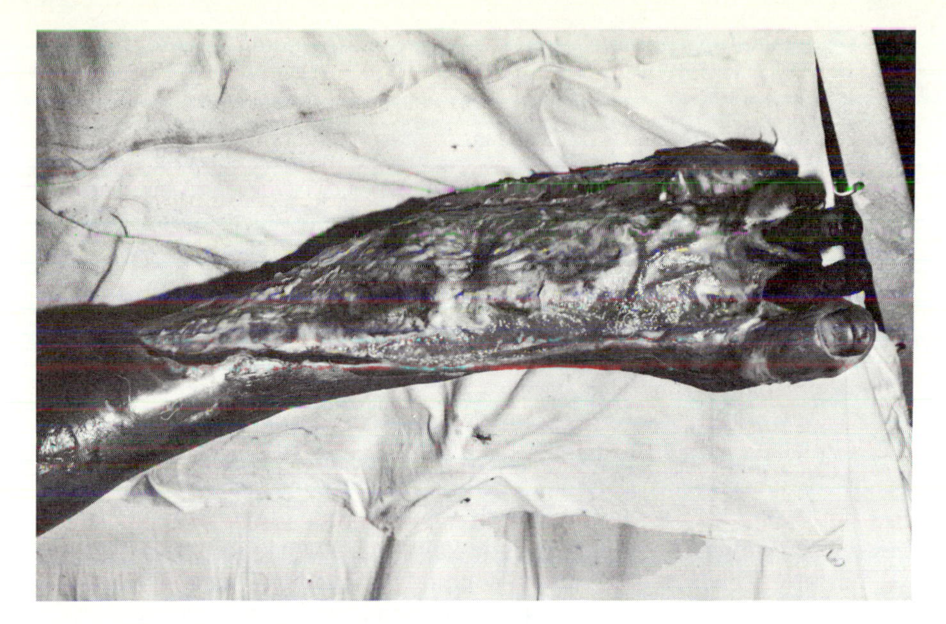

Figure 28-3 **Patient with Gas Gangrene**

tinal tract. Malignant tumors of the gastro-intestinal tract often lead to *C. perfringens* bacteremia. The bacteria do not reproduce in the blood because of the high concentration of O_2. If, however, they enter tissue that has a low oxygen concentration, they can cause gas gangrene. In addition, gas gangrene often occurs as a result of postoperative complications in patients with diabetes, with severe peripheral arteriosclerosis, or with amputations that result in decreased blood flow to the area.

Clostridium perfringens produces numerous exoenzymes □ that hydrolyze the matrix that holds tissues together (collagen, hyaluronic acid) and the lipids and proteins that make up cell membranes. The breakdown of tissue and the lack of blood flow into the area promotes the spread of the organism. The fermentation of sugars produced from the hydrolysis of cellular polysaccharide is the source of the gas caught in the decaying tissues.

Gas gangrene can be treated effectively with antibiotics and hyperbaric (high oxygen tension) therapy. Hyperbaric therapy increases oxygen pressure around infected tissues so as to increase the oxygen concentration in the damaged tissues. This extra oxygen concentration inhibits the growth and exotoxin release of the clostridia.

Pneumonic and Bubonic Plague (The Black Death)

The plague, a disease caused by the small gram negative bacterium *Yersinia pestis,* can involve the lymphatic system, the circulatory system, and the lower respiratory tract. When the infection develops in the lymph, the disease is known as **bubonic plague.** The lymph nodes in the groin and in the neck swell to about the size of golf balls; these swollen lymph nodes are known as **buboes,** hence the name bubonic plague. During the early stages of bubonic plague there is fever, delirium, and swelling of the lymph nodes. Eventually, a septicemia develops and hemorrhagic, blackened lesions appear. The hemorrhaging is the reason this disease is often called the "black death."

Yersinia pestis, sometimes spread from the blood into the lungs and initiate a **pneumonic plague,** which is generally fatal. In the early stages of pneumonic plague there is a high fever, a heavy cough that disperses the bacteria into the air, and thick mucus, often with blood. Bacteria in the air can be inhaled by others, so pneumonic plague is highly contagious. As the disease progresses, a septicemia or blood infection develops and hemorrhagic lesions may be visible in the mouth and on the lips.

Yersinia pestis is ubiquitous in rodent populations in the temperate parts of the world. From time to time, when rodent populations are stressed by overcrowding or disease, *Yersinia pestis* causes the plague in animals. The bacteria are spread from diseased and dead animals to other animals by fleas, principally *Xenopsylla cheopis,* which feed on the blood of diseased animals and become infected with the bacteria. *Yersinia pestis* is able to multiply in the flea, so if the flea bites another animal before it dies of the plague, it inoculates the animal with the plague bacterium. Diseased fleas can rapidly spread the plague through a rodent population and to other animals and humans. **Epidemic plague** is indicated when numerous rodents such as ground squirrels or rats are found dead due to *Y. pestis* infections.

When infected fleas bite animals or humans, the bacteria are injected subcutaneously. Granulocytes, monocytes, and macrophages attack the bacteria, but some escape destruction by the phagocytes and enter the blood and lymph. Because the blood and lymph are filtered by lymph nodes, *Y. pestis* ends up in the nodes, where it is attacked by macrophages. Macrophages often do not destroy the bacteria, however, so they are able to multiply in the nodes. *Yersinia pestis* also produces a powerful **necrotizing exotoxin** that kills many of the phagocytes. The toxic materials released during the multiplication of *Y. pestis* in the lymph nodes causes them to swell and also elicit a fever. If the immune system does not control the infection, *Yersinia* spills from the swollen lymph nodes into the lymph and then into the blood, causing a bacteremia, which causes fever, delirium, shock, and eventually death. The necrotizing exotoxin may cause hemorrhaging in various parts of the body.

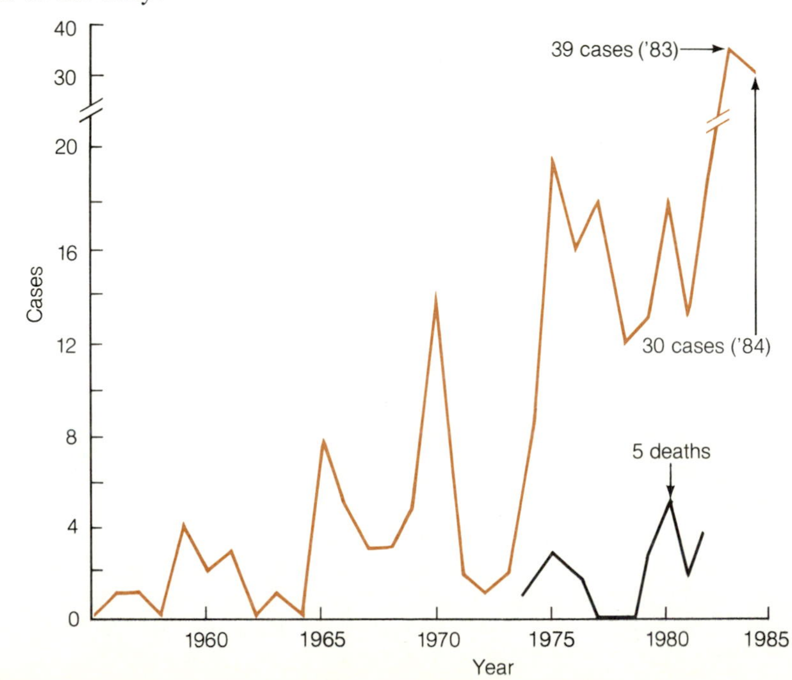

Figure 28-4 Occurrence of Plague in the United States. The incidence of plague (color) has been increasing since the early 1970s. Most of the cases are due to contact with wild animals. The number of deaths (black) fluctuates between 2 and 5 each year.

The plague can be prevented by eliminating slums and garbage heaps that support large populations of rats and by limiting the size of ground squirrel populations that come into contact with humans. Vaccination of persons at high risk, such as veterinarians and personnel monitoring ground squirrel and rodent populations, with a heat- or chemically inactivated strain of *Y. pestis* can prevent the spread of plague among humans. Treatment of plague includes the rapid diagnosis of the disease and the use of streptomycin or tetracycline.

Currently, there is an average of 15 reported cases of human plague in the U.S. each year, with about 1 death per year (fig. 28-4).

Relapsing Fever

Relapsing fever is a disease caused by the spirochete *Borrelia recurrentis* (fig. 28-5). The disease is characterized by recurrent periods of high fever that alternate with periods of normal body temperature. The infection causes headache, musclular pains, chills, nausea, and septicemia, and there may be a rash of rose-colored spots. The symptoms subside after a few days, but a day or so later they return in a milder form. Recoveries and relapses may occur three or four times, but eventually the symptoms disappear completely. Diagnosis of the disease is made by observing *Borrelia* in the blood.

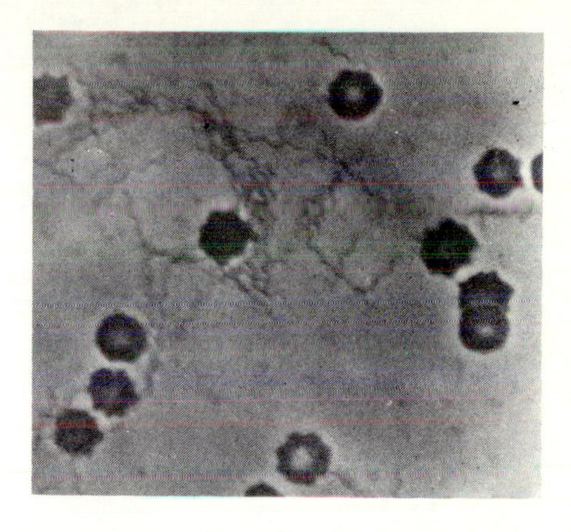

Figure 28-5 ***Borrelia*, the Cause of Relapsing Fever**

The disease is found primarily in the tropics. Reservoirs of *B. recurrentis* include ticks, rodents, and humans; ticks and lice function as vectors. In ticks, **transovarial** passage of *Borrelia* from the infected female to her eggs allows the bacteria to mulitply wherever the tick is found. Thus, ticks are an important reservoir. Most infections in the U.S. are caused by tick bites, but lice are responsible for spreading *Borrelia* from person to person during epidemics.

The disease can be controlled by removing ticks immediately and by maintaining a high level of hygiene to control lice. Once the disease is contracted, it can be treated with chloramphenicol or tetracycline.

Brucellosis

Brucellosis (undulant or Malta fever) is a disease often found in cattle, goats, swine, and sheep. The disease is caused by *Brucella*, a gram negative coccobacillus. In animals, brucellosis is characterized by abortion in the female, infertility in both sexes, a bacteremia (blood infection), and an undulating fever that drops during the day. In humans, the disease is characterized by recurring bouts of fever followed by periods of normal body temperature.

638

Many herds have carrier animals infected by *Brucella*. In bulls with infected seminal vesicles, testicles, and epididymidis □, semen is contaminated with *Brucella*. If contaminated semen is deposited in the uterus of a cow, she can become infected. Thus, it is important that semen used in artificial insemination come from disease-free bulls. In the cow, *Brucella* infects the uterus and developing fetus, causing abortion. The rate of abortion in an infected herd often reaches 50 percent. *Brucella* also infects the udders and is shed in the milk, sometimes for the life of the animal. The disease is frequently spread by ingestion of organisms that are present in milk, contaminated feed, and contaminated water. Feed and water are contaminated by uterine discharges, milk release, and feces. The licking of aborted fetuses and membranes is another important way that animals become infected.

It is believed that the bacteria multiply in the mucous membranes of the gastrointestinal tract (or the uterus) and then spread into the lymphatic system and blood, from which they infect many tissues and organs. Long-term infections are caused by the bacteria that establish a symbiotic relationship inside cells of the reticuloendothelial □ system. Carriers in a population generally produce 20% less milk than normal animals. 461

Humans can also become infected by *Brucella:* approximately 200 cases of brucellosis in humans **(Malta fever)** are reported each year in the U.S., and 60% of these cases occur in butchers and workers from meat processing plants. About 30% occur in ranchers and hunters, while fewer than 10% occur in persons who consume unpasteurized dairy products. Mortality in humans is low, with an average of 1 death/year in the U.S. Symptoms in humans generally include an undulating fever, enlarged lymph nodes and spleen, and muscle aches. The disease is often very severe in humans and may linger on for a year without antibiotic treatment.

In herds of cattle, the disease can be eliminated by vaccination of calves, annual booster injections of adults, and slaughtering and disposal of infected animals. The vaccines use killed or attenuated □ organisms. The most effective measures for controlling brucellosis in humans are inspection of herds, elimination of diseased animals, and pasteurization of dairy products. Chlortetracycline and streptomycin are effective for curing valuable animals, but antibiotic treatment is expensive in large animals and is rarely used to treat a herd. Tetracycline and ampicillin are used to treat human cases of brucellosis. 477

Tularemia (Rabbit Fever)

Tularemia is a disease characterized by fever, fatigue, coughing, diarrhea, and enlarged lymph nodes, spleen, and liver. Most often, however, an infection by the bacterial agent, *Francisella tularensis,* produces no clinical symptoms. The disease is endemic in the United States in wild animals such as rabbits, squirrels, deer, and birds. The gram negative coccobacilli are often transmitted to domestic animals such as cattle, sheep, goats, hogs, horses, dogs, and cats by ticks that feed on infected wild animals. Tularemia is transmitted to humans by bites of infected ticks and deer flies. Hunters who skin infected rabbits and deer can acquire the organism if it gets into cuts or onto mucous membranes of the mouth or nose.

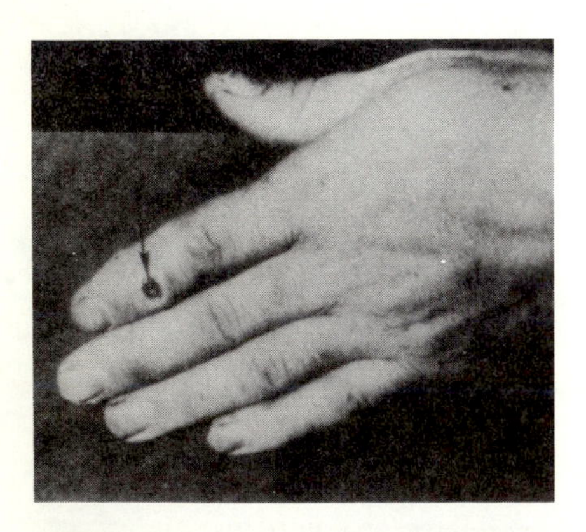

Figure 28-6 Characteristic Lesions of Tularemia Caused by *Francisella tularensis*

The bacteria usually reproduce at the site of entry and produce an ulcer (fig. 28-6). The bacteria are engulfed by phagocytes, but are not effectively eliminated; thus, *F. tularensis* often spreads to the lymph and lymph nodes in dead and dying phagocytic cells. From there the bacteria spread into the blood, causing a septicemia, and into various tissues and organs, such as the liver and spleen. About 200 human cases of tularemia are reported each year in the U.S.

Prevention of tularemia is through the control of ticks infesting herds or flocks. Domestic animals (cattle, sheep, goats, hogs, horses, dogs, and cats) can be dip-washed or sprayed with chemicals such as arsenic, rotenone, toxaphene, or malathion. The disease can be treated effectively with the antibiotic oxytetracycline.

Rocky Mountain Spotted Fever

Rocky Mountain spotted fever is characterized in humans by a fever and by a rash that starts at the extremities and then develops on the rest of the body. The rash is caused by damage to peripheral blood vessels, which leads to inflammation, hemorrhaging, and escape of blood into surrounding tissues. The disease is caused by the bacterium *Rickettsia rickettsii*, a gram negative coccobacillus.

Rocky Mountain spotted fever is prevalent in the South Atlantic states (60% of reported cases), as well as in the Rocky Mountain states (fig. 28-7). The disease can be confirmed by the **Weil-Felix agglutination** test. The Weil-Felix test is a bacterial agglutination test 516 in which serum from patients with typhus or spotted fever agglutinates OX-19, OX-2, and OX-K strains of *Proteus vulgaris*. This procedure is used commonly because the testing antigen *(P. vulgaris)* is more readily available than the rickettsial antigen. The disease can also be diagnosed using the complement fixation or indirect fluorescent antibody techniques. There are approximately 1,150 cases of Rocky Mountain spotted fever per year in the U.S. with a 5% fatality rate (about 55 deaths/year).

The reservoirs 534 for *Rickettsia rickettsii* are rodents, rabbits, and ticks. Infected ticks transmit the organism through the egg to offspring and also function as the vector. The bacteria are transmitted to animals and to humans almost exclusively by tick bites: When the ticks feed, the bacteria enter the lymph and blood and then reproduce in neutrophils and in the cells lining the peripheral blood vessels, causing **thrombosis** (coagulation of the blood) and **extravasation** (loss of blood into the tissues). The rash is due to the destruction of the endothelial cells that line the superficial blood vessels.

A vaccine consisting of formalin-killed bacteria has been developed for humans; it diminishes the severity of the disease but does not stop it. There

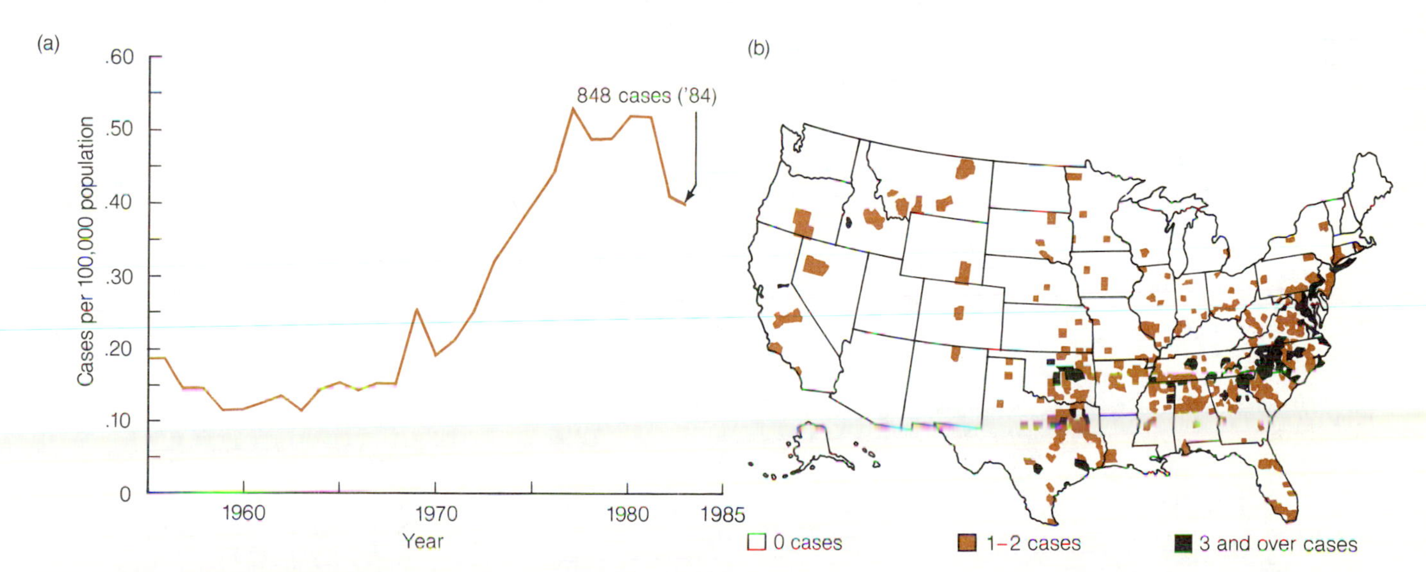

Figure 28-7 Incidence and Distribution of Rocky Mountain Spotted Fever. *(a)* Reported cases of Rocky Mountain spotted fever caused by *Rickettsia rickettsii. (b)* Distribution of cases of Rocky Mountain spotted fever in the United States. It is noteworthy that the vast majority of the cases occur in the eastern United States, not in the Rocky Mountains in the west. The name of the disease was coined because the spotted fever was first known to occur in the Rocky Mountains. The incidence of the disease parallels well the distribution of the tick vectors.

TABLE 28-3

DISEASES OF THE BLOOD, LYMPH, MUSCLE, AND INTERNAL ORGANS CAUSED BY RICKETTSIA-LIKE BACTERIA AND ARBOVIRUSES

DISEASE	NAME OF PATHOGEN	RESERVOIR	VECTOR	HOW ACQUIRED
Acquired immune deficiency syndrome (AIDS)	Human T-Cell leukemia virus (HTLV-III)	Humans	—	Homosexuality, transfusions, injections
African sleeping sickness	*Trypanosoma gambiense* *T. rhodesiense*		Tsetse fly	Fly bite
Brill's-Zinsser disease	*Rickettsia prowazekii*	Humans	—	Latent typhus
Brucellosis	*Brucella abortus*	Cattle, goats, sheep	—	Ingestion, tissue rupture
Cat scratch fever	*Chlamydia*	Cat	—	Cat scratch
Chagas' disease	*Trypanosoma cruzi*	Rodents, cats, raccoons, deer, dogs	Bedbugs, lice	Bite
Colorado tick fever	Arbovirus	Rodents	Tick	Tick bite
Dengue fever	Arbovirus	Mosquito	Mosquito	Mosquito bite
Elephantiasis (filariasis)	*Wuchereria bancrofti*	Humans	Mosquito	Bite
Endemic (murine) typhus	*Rickettsia mooseri*	Rodents	Flea	Flea bite
Endocarditis	*Streptococcus* *Staphylococcus*	Skin	—	Tissue rupture, infection
Epidemic typhus	*Rickettsia prowazekii*	Humans	Lice	Louse bite
Gas gangrene	*Clostridium perfingens*	Soil	—	Tissue rupture, infection
Loiasis	*Loa loa*	Humans	Horse fly	Fly bite
Malaria	*Plasmodium falciparum* *P. vivax* *P. malariae* *P. ovale*	Humans, monkeys	Mosquito	Mosquito bite
Mononucleosis	Herpes virus (Epstein-Barr)	Humans	—	Kissing, ingestion
Plague	*Yersinia pestis*	Rodents	Fleas	Flea bite
Puerperal sepsis	*Escherichia coli*	Intestines	—	Tissue rupture
Q fever	*Coxiella burnetii*	Cattle	—	Ingestion
Rat bite fever	*Spirillum minor* *Streptobacillus moniliformis*	Rats, turkeys, weasels	—	Rat bite
Relapsing fever	*Borrelia recurrentis*	Rodents, ticks, lice, opossums	Tick, louse	Tick or louse bite
Rickettsialpox	*Rickettsia akari*	Rodents	Mites	Mite bite
River blindness	*Onchocerca volvulus*	Humans	Black fly	Fly bite
Rocky Mountain spotted fever	*Rickettsia rickettsii*	Rodents, ticks	Tick	Tick bite
Schistosomiasis	*Schistosoma mansoni (japonicum) (haemotobium)*	Human, snails	—	Burrow through skin
Scrub typhus	*Rickettsia tsutsugamushi*	Rodents	Mites	Mite bite
Toxoplasmosis	*Toxoplasma gondii*	Cats, rodents, pigs, sheep	—	Ingestion, inhalation
Trench fever	*Rochalimaea quintana*	Humans	Lice	Louse bite
Trichinosis	*Trichinella spiralis*	Pigs, many carnivores	—	Ingestion
Tularemia	*Francisella tularensis*	Rabbits, deer	Tick, deer fly	Ingestion, bites, & tissue rupture
Weil's disease (leptospirosis)	*Leptospira interrogans*	Rodents, skunks, dogs, racoons, cattle, swine	—	Ingestion, through skin
Yellow fever	Arbovirus	Humans, monkeys	Mosquito	Mosquito bite

are no vaccines for domestic animals. Control is best achieved by prompt removal of ticks. Generally, ticks must feed for several hours before they transmit virulent rickettsia, and so removal twice a day can generally prevent the disease. The bite should be treated with tincture of iodine to prevent secondary bacterial infection. The use of tick repellents and tight, overlapping clothing can prevent tick bites. Treatment of the disease includes bed rest and the administration of antibiotics such as tetracycline.

Typhus Fevers

Typhus fevers are infectious diseases caused by various species of *Rickettsia*. These diseases spread by the bites of arthropods such as lice, fleas, ticks, and mites. Human typhus is characterized by septicemia, high fever, headache, haziness, confusion, and a rash, often associated with hemorrhaging beneath the skin. Table 28-3 lists the various typhus fevers and the bacteria and insect vectors that are responsible. The disease can be diagnosed using the Weil-Felix and complement-fixation tests.

Endemic typhus, also known as **murine typhus,** caused by *R. mooseri*, is widely distributed in warm climates and is transmitted to humans in the feces of infected fleas. The bacteria enter the body through a flea bite or other break in the skin. They invade neutrophils and endothelial cells of capillaries, where they reproduce, causing the symptoms and signs of typhus. There are approximately 60 cases of endemic typhus every year in the U.S. and a 5% mortality rate (fig. 28-8). The reservoir for murine typhus is rodents, such as rats and squirrels.

Epidemic typhus, caused by *R. prowazekii*, occurs in cool climates throughout the world and is transmitted to humans in the feces of infected body lice. The bacteria may enter through bites or other breaks in the skin. Epidemics of this disease generally occur in times of war and famine, when overcrowding and unsanitary conditions prevail. The mortality rate can reach 40 percent. Very few cases of epidemic typhus occur in the U.S. Humans are the reservoir for epidemic typhus, because the organisms can exist in a latent

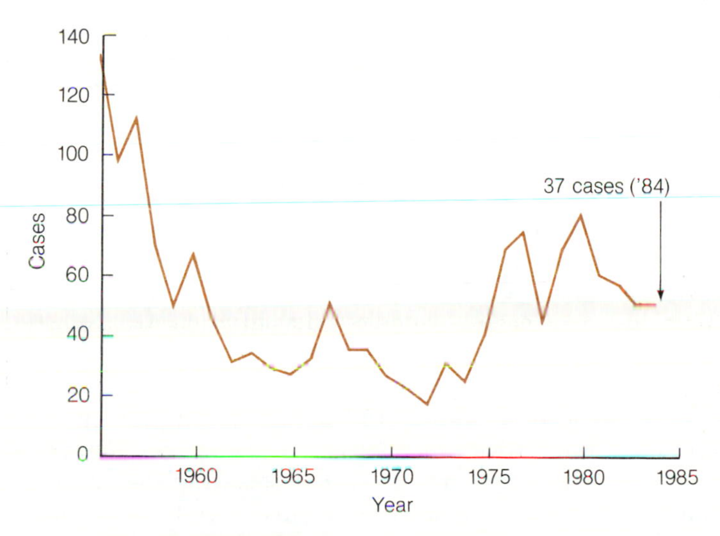

Figure 28-8 Cases of Endemic Typhus in the United States. Murine rodents are the reservoirs for the agent of murine typhus, *Rickettsia typhi.* A total of 58 cases were reported in 1982, of which 41 occurred in Texas and 9 in Hawaii.

form for many years. In some instances, infected persons who are clinically cured may exhibit a relapse of the disease many years later. This recurrent form of typhus is called the **Brill-Zinsser** disease.

Typhus fevers can be controlled by spraying with pesticides and by practicing personal hygiene and good sanitation. Pesticides and cleanliness eliminate body lice, mites, and fleas, the vectors for *Rickettsia*, while sanitation (garbage removal and deposition) can eliminate some of the reservoir animals, such as mice and rats. A vaccine of chemically-treated microorganisms is effective against both epidemic typhus and murine typhus. Treatment of the disease with tetracycline or chloramphenicol effectively destroys the organisms and generally eliminates the carrier state in epidemic typhus.

Rickettsias are obligate intracellular parasites, apparently because they cannot synthesize NAD and CoA. Therefore, chick embryos or mammalian cells must be used for their cultivation and isolation in the laboratory.

VIRAL DISEASES CAUSED BY INVASION OF THE LYMPH AND BLOOD

Colorado Tick Fever

Colorado tick fever in humans is an illness characterized by fever, chills, headache, and muscle aches, and it may be accompanied by a rash. The disease is caused by an arbovirus □ and is transmitted to humans from infected wild rodents by tick vectors (fig. 28-9). The disease occurs in western North America, and can be diagnosed by detecting antibodies against the virus or by isolating the virus in clinical specimens (blood).

676

Control of the disease is through the use of pesticides and tick repellents to eliminate the vector population. There is no treatment for this infection, which generally subsides after running its course. Since this is a non-notifiable disease in the United States, it is difficult to determine how many cases there are; in 1980, however, there were approximately 200 reported cases in the United States, none of which was fatal. A vaccine is prepared by growing the virus in chick embryos so that they are attenuated. Infection with either the virulent strain or the vaccine strain produces a long-lasting immunity.

Yellow Fever

Yellow fever is caused by an arbovirus and is spread among humans, or from monkeys to humans, by the mosquito *Aedes aegypti* (fig. 28-9). The disease is found in the tropical regions of South and Central America, India, Southeast Asia, and West, Central, and East Africa. It is characterized by fever, internal hemorrhaging, jaundice (due to extensive liver destruction), and vomiting. When mosquitoes bite, they inject the virus into the blood and lymph of the host. The viruses infect and multiply within macrophages, monocytes, and reticulocytes in lymph nodes and in the spleen, and then spread into the blood, where they infect the endothelial lining of the blood vessels and cause hemorrhaging. They also spread to the brain, causing an encephalitis. Infections of the internal organs cause hepatitis, nephritis, endocarditis, and osteomyelitis. Proliferation of the virus causes fever, chills, rash, headache, and muscle pains.

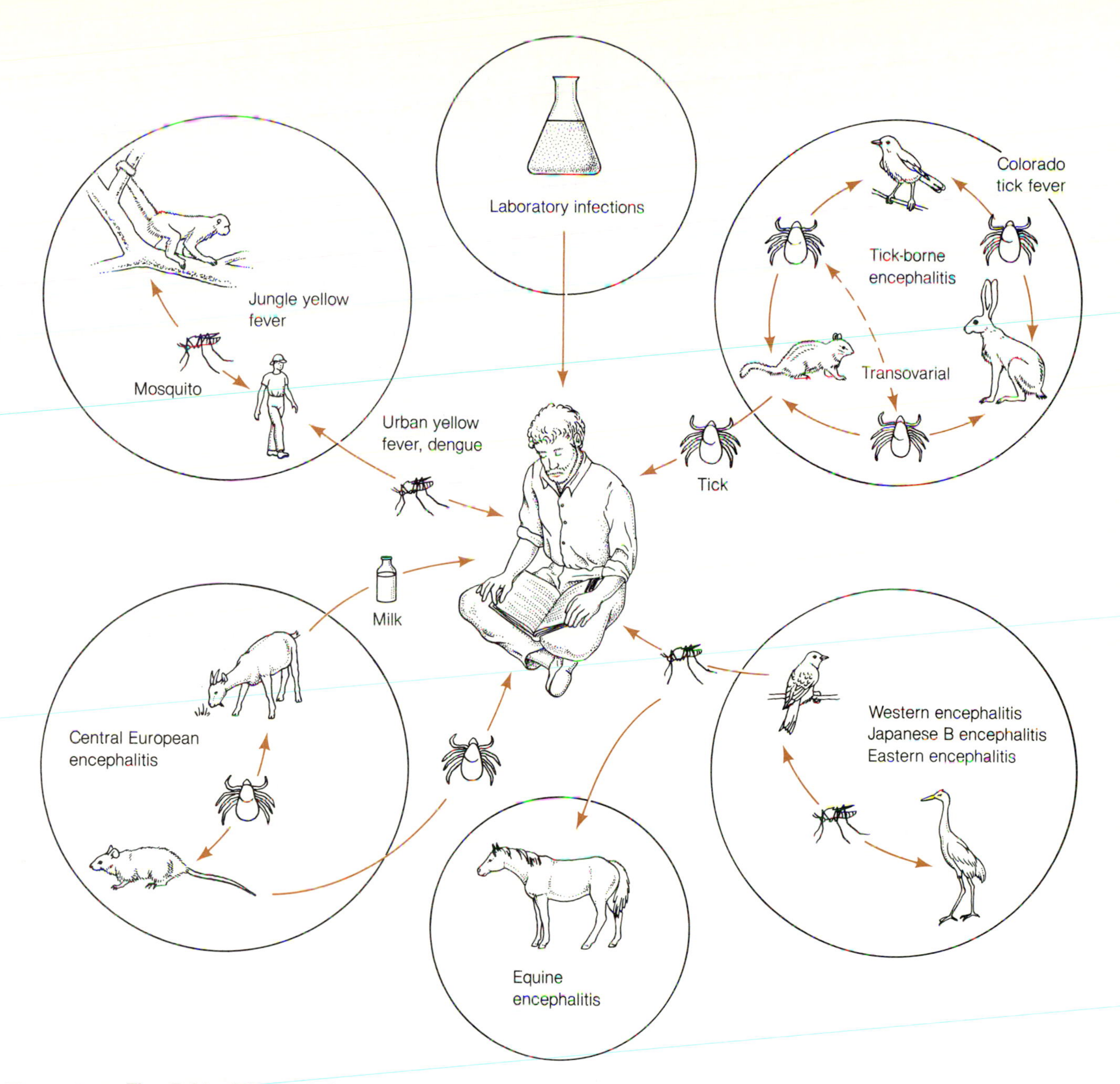

Figure 28-9 The Epidemiology of Arbovirus-Caused Encephalitis

In the Mississippi Valley in 1878, an epidemic of yellow fever killed approximately 13,000 persons. Although there has not been a reported case of yellow fever in the U.S. since 1924, suitable mosquito vectors can be found in the U.S., and therefore the possibility always remains that an infected immigrant will initiate an outbreak such as that of 1878. The yellow fever virus is of importance because it kills many people each year throughout the world. Those persons traveling to South America and Africa, where the disease is endemic □, should be vaccinated against the virus.

533

Dengue Fever

Dengue (hemorrhagic) fever is a disease defined by fever, joint pain, skin rash, and mental depression. These signs and symptoms may be accompanied by severe hemorrhaging. It is caused by an arbovirus found in tropical and subtropical areas in Southeast Asia and the Philippines. The dengue fever virus is transmitted by females of the mosquito *Aedes aegypti*, which also act as a reservoir for the virus (fig. 28-9). Occasionally, when a new serotype of the dengue virus infects an individual sensitized to another serotype, antibodies crossreact with the new virus but do not inactivate it. In fact, the antibodies appear to aid the adsorption of the virus to blood leukocytes and, consequently, the multiplication of the virus. Virus-antibody complexes sometimes activate 80% of the serum C3 and C5 □, causing intravascular coagulation and inflammatory reactions throughout the body. The disease caused by the immune complex is known as **dengue hemorrhagic shock syndrome.** Dengue is not common in the continental United States, but in Puerto Rico there has been an average of 11,000 cases reported/year in the last few years.

473

Infectious Mononucleosis

Infectious mononucleosis is caused by the Epstein-Barr virus (a herpes virus). The prodromal (early, nonspecific) symptoms include exhaustion and a general poor feeling. The disease itself is characterized by extreme exhaustion, fever, sore throat, pharyngitis, tonsillitis, swollen lymph nodes and spleen. The presence of more than 10% atypical T-lymphocytes (called **Downey cells**) in the blood, and a total WBC count of about 15,000–20,000 cells/mm^3 (5,000 is normal), are further signs of the disease. Recovery generally takes 4–8 weeks, during which extreme exhaustion after only a few hours of activity is one of the symptoms of the disease.

The human population serves as a reservoir for the Epstein-Barr virus (EB-virus). In areas where sanitation is poor, almost entire populations have been infected by the age of five without recognizable clinical disease. Eighty to 95% of Africans, for example, have antibodies against EB-virus by the age of two, but infectious mononucleosis is extremely rare. In contrast, in industrialized areas where sanitation practices are better, antibodies against EB-virus are found in only 25% of those under two and in less than 50% of those under 20 years of age. Thus, infection by the EB-virus is delayed until adolescence and early adulthood in high socioeconomic populations, and 50% of those infected develop the signs and symptoms of mononucleosis. It is believed that infected adolescents in Western countries spread the EB-virus within their age group when they begin to share drinks and cigarettes and when kissing first becomes common. EB-virus may be spread from infant to infant in saliva-contaminated breasts or communal food in those populations where individuals are are infected by the age of two.

The virus initially infects the throat and reproduces in epithelial cells of the throat and pharynx, then it spreads to the lymph and blood. Viruses are shed into the throat and pharynx beginning approximately two weeks after exposure and for up to two years. Excretion of the EB-virus can occur in the absence of symptoms of mononucleosis.

After infecting the oral pharynx, the EB-virus invades B-lymphocytes and circulates in a noninfectious form in these cells. In this form, the virus DNA is found within the B-lymphocytes, but progeny viruses are not produced.

468 Cell-mediated immunity □ is carried out by Downey Cells, which increase in the blood a week after infection. Downey cells account for 10–80% of the total white blood cells, and their numbers remain at a high level for 4–5 weeks. Humoral immunity against the viral capsid is mediated by IgM, which is found in 90% of adults at the beginning of an infection and lasts for a month or so.

Infectious mononucleosis can be diagnosed by the symptoms, the presence of Downey cells in the blood, and by serology □ 512. Since sore throat and tonsillitis are important symptoms, infections by group A streptococci (which cause strep throat and rheumatic fever) must be ruled out.

The disease may be very difficult to prevent in industrialized countries, since prevention would require adolescents and young adults to avoid kissing. A vaccine has not been developed so far, and there are no drugs that can be used as a prophylactic or treatment. Mononucleosis is generally self-limiting; complications such as ruptured spleen, Guillain Barré syndrome, and hemolytic anemia occur infrequently. Even though infectious mononucleosis is a nonreportable disease, approximately 20,000 cases of mononucleosis are reported per year in the U.S., along with an average of 20 deaths per year.

It is believed that the EB-virus may also cause two human cancers: **African Burkitt's lymphoma** and **nasopharyngeal carcinoma.** Burkitt's lymphoma is a malignant lymphoma of children which causes jaw tumors in more than 50% of cases. Its incidence is highest in Africa and in New Guinea, where 95% of the children have antibody titers against the EB-virus. Nasopharyngeal carcinoma occurs with an incidence of 24/100,000 per year in China, as compared with 1/100,000 per year in the U.S. Again, the incidence of nasopharyngeal carcinoma may be highest in persons subjected to an EB-virus in infancy. Thus, vaccines against this virus cannot be used until it has been learned if there is any danger from infant vaccination.

PROTOZOAL DISEASES CAUSED BY INVASION OF THE LYMPH AND BLOOD

Malaria

Malaria is due to an infection of the red blood cells (RBCs) by the protozoans □ 371 *Plasmodium falciparum* (50%), *P. vivax* (43%), *P. malariae* (6%), or *P. ovale* (1%) (fig. 28-10). One stage of the parasite infects RBCs, causing their lysis. This process leads to periodic chills and fever that leave the affected person drenched in sweat and exhausted. Hemolysis results in anemia and jaundice (yellowish pigmentation of the skin due to excessive bile). Infected RBCs, because they are no longer pliable, stick in small capillaries and cause blood clots. Anoxia (lack of oxygen) and electrolyte imbalance result and can lead to death. The liver and spleen enlarge to handle the debris from the RBCs; the spleen can become so large and fibrous that sometimes it ruptures spontaneously. In cases of acute hemolysis, large amounts of hemoglobin are released into the blood and plug up the kidneys, leading to renal failure. It is estimated that there are 100 million new cases of malaria/year and about 1 million deaths/year worldwide.

The major reservoir of the malarial parasites is humans, although some monkeys can also carry the protozoan. The female *Anopheles* mosquito is the vector for the disease, spreading it from human to human. When the female *Anopheles* bites, it releases crescent-shaped **sporozoites** (15 μm long) into

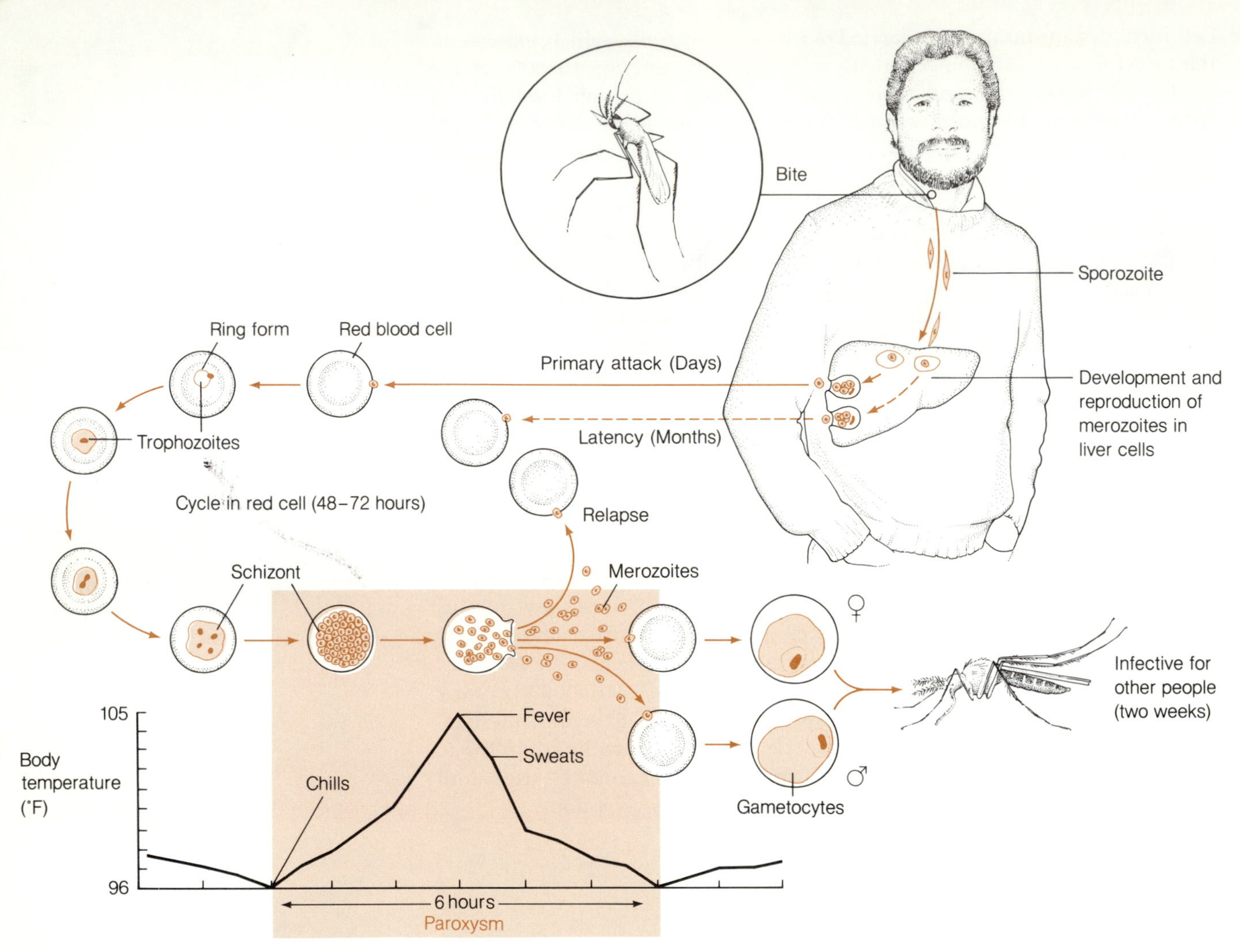

Figure 28-10 Life Cycle of *Plasmodium vivax* and Its Relationship to Human Malaria. Malaria is characterized by recurring bouts of fever and chills (paroxysms). The paroxysms begin with severe chills, followed by progressive increase in body temperature (104–105°F). The fever is followed by sweats, with a concurrent decline in body temperature. The paroxysms last about 6 hours (vivax malaria). The frequency of the paroxysms is related to the life cycle of the malaria parasite in the blood: the paroxysms coincide with the time during which the parasite is in schizogony.

the lymph and blood. (fig. 28-10). The sporozoites migrate directly to the liver, where they invade hepatic cells. Within the liver cells, the sporozoites multiply and differentiate into oval **merozoites,** which are released about 10 days after infection into blood capillaries. The merozoites spread throughout the blood and invade RBCs and develop into ring-shaped **trophozoites.** The trophozoites differentiate into **schinzonts** when they undergo multiple fission inside the red blood cells to form more merozoites. The RBCs lyse at this point, and the merozoites are released into the blood; they may invade other RBCs or return to the liver. The debris from the hemolysis of RBCs, and from the merozoites that are attacked and destroyed by white blood cells (WBCs), leads to the fever and chills of malaria. The reproduction of *Plasmodium vivax* in the blood takes approximately 48 hours and therefore causes spells of chills and fever every third day (tertian malaria), while *P. malariae* requires 72 hours to complete their cycle (quartan malaria).

FOCUS

THE DANGERS OF TRANSFUSION

Many accident victims, persons undergoing major surgeries, and hemophiliacs require blood transfusions. Unfortunately, each year hundreds of persons in the U.S. acquire serious diseases because of blood donated by individuals carrying infectious organisms.

One of the most serious diseases transmitted to patients during transfusion is hepatitis caused by a non-A non-B virus. Post-transfusion non-A and non-B hepatitis occurs in 40–50% of patients who receive more than six units of paid-donor blood, and in 6–10% of patients receiving volunteer blood. The problem with non-A non-B hepatitis is that 40–50% of the cases become chronic, and about half of these develop cirrhosis of the liver 5–10 years later. Death results from liver hemorrhaging, infection, and liver failure. It is estimated that there are over 100,000 cases of non-A non-B hepatitis per year and more than 4,000 annual deaths.

At least two distinct viral agents may be responsible for non-A non-B hepatitis. One type of non-A non-B hepatitis has an incubation period of 15–45 days, produces a mild disease, and can be transmitted by the fecal-oral route as well as by transfusions and injections. Ten percent of the cases become chronic. The second type has an incubation period of 90-180 days and is transmitted only by transfusions and injections. A majority of the cases become chronic.

At present, there is no way of screening blood donors to see if they are carriers for non-A non-B. In addition, there is no serological test that can identify non-A non-B hepatitis agents in blood. The only precaution used to avoid non-A non-B is to exclude donors with serological evidence of hepatitis A or B.

Another disease transmitted through blood transfusions is AIDS (acquired immune deficiency syndrome). The agent that causes this disease is believed to be a human T-cell leukemia virus. In the last few years, a number of persons who have had transfusions have acquired the disease. Most of the victims have been hemophiliacs and young children. In 1985 and ELISA screening test was licensed to detect the presence of antibodies against HTLV-III. This test detects 92–98% of the blood samples with anti HTLV-III antibodies. Because this test does not detect all possible blood donors that might have AIDS and because some of the positive results may be false positives, further tests are necessary to ascertain whether or not the donor indeed has AIDS.

The most severe malaria (pernicious) is caused by *P. falciparum*. In this type of malaria, the patient's fever does not subside completely between cycles of multiplication. Anoxia and kidney failure due to the breakdown of red blood cells frequently lead to shock and death. After a series of attacks, an individual sometimes gets well, apparently because the immune system has eliminated all the infected RBCs and merozoites. It is believed, however, that sporozoites lie dormant within hepatic cells. Thus, relapses can occur when dormant sporozoites become active and form merozoites. Ultimately, when all sporozoites introduced by the original mosquito bite have differentiated and all merozoites are destroyed, relapses no longer occur. Relapses lose their strength over a number of years.

Diagnosis of malaria may be confirmed by observing malarial parasites in Wright or Giemsa stained RBCs. The parasites show up as blue rings with attached red chromatin dots, or they may completely fill the RBC. Fluorescent antibodies that bind specifically to the malaria parasite are also used to find the protozoa.

Control of malaria can be achieved by screening doors and windows; by using mosquito netting over beds; and by using clothing that protects from mosquito bites. Malaria is also prevented by draining, spraying, or oiling ponds where mosquitos breed, or by planting fish in the ponds to eat the mosquito larvae. It is also important to detect and cure carriers of malaria, especially if there are mosquitoes that could spread the disease. During the

Figure 28-11 **Incidence of Malaria in the United States**

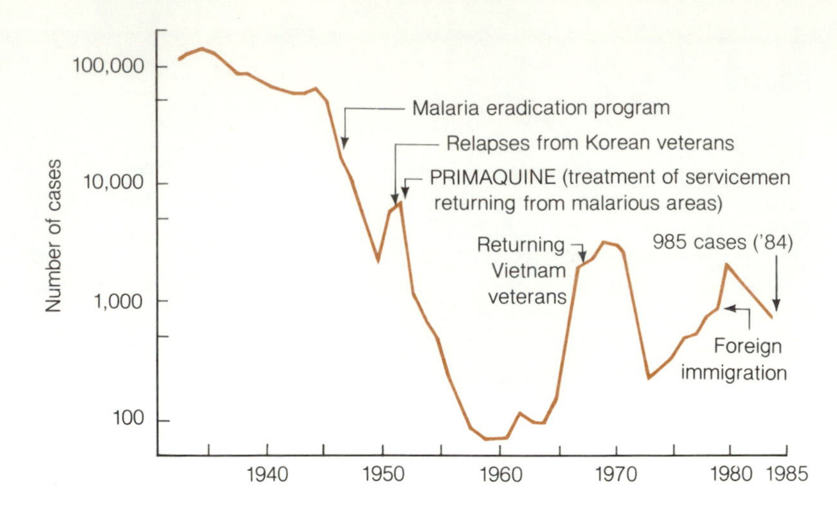

late 1970s in the United States, the number of cases of malaria has increased from about 375 in 1975 to approximately 2,000 in 1980 (fig. 28-11). About 80–85% of these cases were found in immigrants from Southeast Asia. Deaths in the United States average 5 per year. Malaria is treated with chemicals such as quinine and chloraquine, a derivative of quinine. Prophylactic treatment with such drugs is recommended for persons visiting endemic areas since at present there is no vaccine commercially available.

Toxoplasmosis

Toxoplasmosis is an infection by the protozoan *Toxoplasma gondii* □ characterized by mild fever, headache, muscle aches, and swollen lymph nodes and spleen. There may also be lesions of the gastrointestinal tract, myocarditis, necrosis of the liver, central nervous system disorders, or abortion. Cats are important reservoirs for *T. gondii*, although many wild and domestic animals and birds often become infected. For example, latent toxoplasmosis may occur in up to 50% of pigs, 50% of sheep, and 10% of cattle. This disease is an important cause of abortions and stillbirths in sheep.

371

Humans can become infected by accidentally ingesting dried particles of cat feces contaminated with oocysts (fig. 28-12). Subclinical infections in pregnant women can often result in the infection of the fetus, since the parasites can cross the placenta. Severe brain damage, blindness, and death of the fetus may result. Using serological procedures such as the complement fixation test and the fluorescent antibody test, it has been estimated that about 30% of the population has been infected with *T. gondii*, but most of these infections are asymptomatic.

Diagnosis of toxoplasmosis is confirmed by the demonstration of specific antibodies against the parasite, using serological procedures such as complement fixation, indirect fluorescent antibody, and hemagglutination tests. Toxoplasmosis in humans can be prevented by avoiding contact with cat feces, by covering children's sandboxes, and by carefully disposing of cat litter boxes. Thoroughly cooking lamb, pork, and beef (hamburger meat) also reduces the likelihood of picking up these parasites. Cats can be kept relatively free of the protozoan by feeding them only cooked meats and not

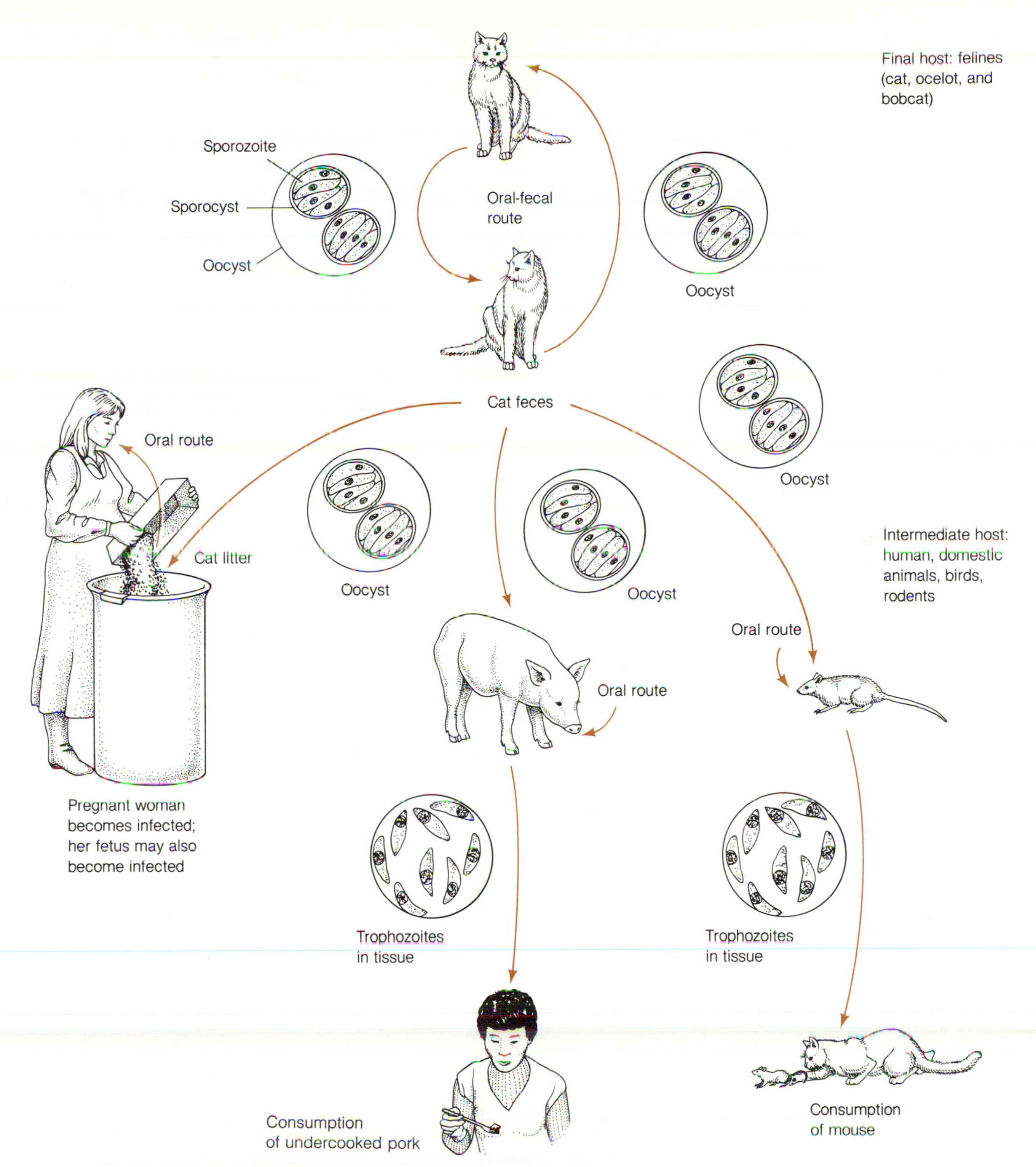

Figure 28-12 **Life Cycle of *Toxoplasma gondii.*** *Toxoplasma gondii* infects cats that consume animals infected with cysts of *Toxoplasma*. The parasite undergoes sexual reproduction in the cat's intestine, and the oocysts are passed in the feces. Children playing in sandboxes may become infected with the oocysts. Adults become infected when they consume the oocysts on their hands or the cysts in foods. *Toxoplasma* infections are dangerous when they occur in pregnant females, because the parasite can invade the fetus through the placenta and cause severe damage. It has also been noted that 52% of those AIDS patients that have serious neurological damage have *toxoplasma* cysts in the brain.

allowing them to eat mice and birds that they might catch. No vaccine is available against this organism. The drugs **sulfadiazine** and **pyrimethamine** (folinic acid antagonist), which affect only the trophozoites, are used to treat toxoplasmosis.

Trypanosomiasis (African Sleeping Sickness and Chagas' Disease)

African sleeping sickness is a trypanosomiasis transmitted to humans by the tsetse fly. The organisms causing the disease are *Trypanosoma gambiense* and *T. rhodesiense.* The symptoms, epidemiology, prevention, and treatment are described in Chapter 27.

Chagas' disease is found throughout Central and South America and is caused by *Trypanosoma cruzi.* The protozoan is spread among humans by insects commonly called "kissing bugs," (Reduviidae). The disease is characterized by fever, exhaustion, lack of apetite, muscle aches, a puffy face, heart irregularities, and swelling of the lymph nodes, liver, and spleen. The trypanosome often causes an infection of the eyes in children, resulting in swelling of the eyelids. (Romaña's sign). Organisms in the circulatory system infect the spleen, lymph nodes, liver, heart, and digestive tract. In chronic cases, cysts develop within heart tissue, leading to necrosis of the tissue, heart failure, and death. Infections of the esophagus interfere with swallowing, while those in the intestines can result in severe constipation. Esophagus and colon disease may require surgery to correct the damage.

There are a great number of reservoir hosts for *Trypanosoma cruzi,* including opossums, armadillos, rodents, racoons, monkeys, deer, cats, and dogs. Important vectors of *Trypanosoma cruzi* are hemipterans and other blood-sucking insects (bedbugs, kissing bugs, lice, etc.). These insects generally ingest the trypanosomes during a blood meal on an animal. The trypanosome multiplies in the insect's gut, attaches to the rectal region, and flows out in the feces. Since the insect vectors generally defecate while they are biting and feeding, the bite often becomes infected. The insects sometimes bite children on the eyelids and consequently cause an infection of the eyelids.

Diagnosis is by observing the trypanosome in the blood. When the trypanosome is not detectable in the blood, the infection can be verified by complement fixation tests □ using *T. cruzi* antigens. The drug nifurtimox is used to treat infections that have not progressed too far. 527

DISEASES CAUSED BY MULTICELLULAR PARASITES

River Blindness and Loaisis

Onchocerciasis or river blindness □, is caused by the parasitic nematode *Onchocerra volvulus* and is spread by black flies. The disease afflicts more than 150 million people worldwide. Most of the cases are in West Africa, but infections are also reported in east Africa, Yemen, Mexico, Central America, and South America. 375

An African disease called **loaisis** is caused by a nematode similar to *Onchocerra,* called *Loa loa,* and is spread by the bite of a horse fly or mango fly. The infection is similar to river blindness. The *Loa loa* microfilariae are easily seen crawling over the eye just under the corneal conjuctiva (fig. 28-13).

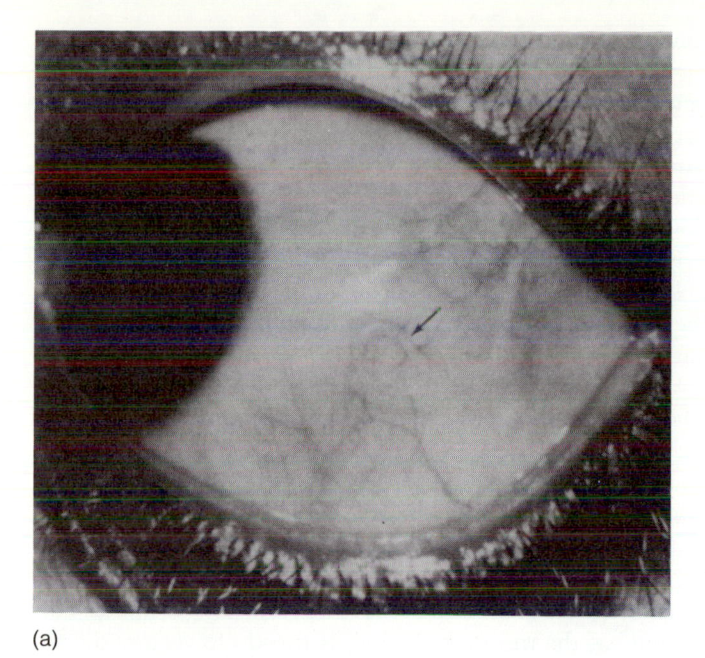

(a)

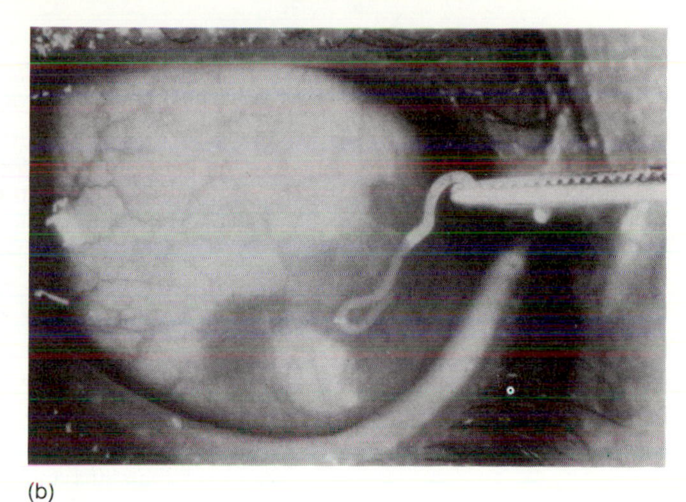

(b)

Figure 28-13 ***Loa loa* in the eye.** *(a)* Adult worm in the eye. *(b)* Surgical removal of worm.

Filariasis (Elephantiasis)

Filariasis is a disease caused by the nematode *Wuchereria bancrofti* (fig. 28-14). It is characterized by **microfilariae** infecting the lymph vessels and lymph nodes and obstructing the flow of lymph. The microfilariae also cause 443 an inflammatory □ reaction in the infected tissue. Connective tissue is stimulated to multiply, so that the skin of the lower extremities and external genitalia becomes enlarged, coarse, and thickened (fig. 28-14). Legs often become so swollen and altered that they resemble the legs of elephants, hence the name **elephantiasis.** Filariasis is very prevalent in tropical regions

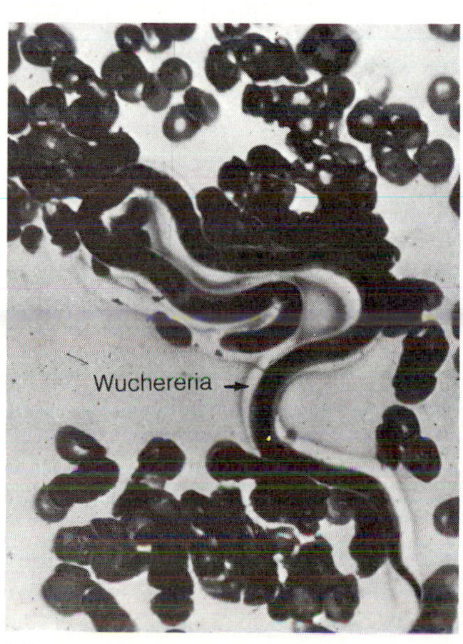

(a)

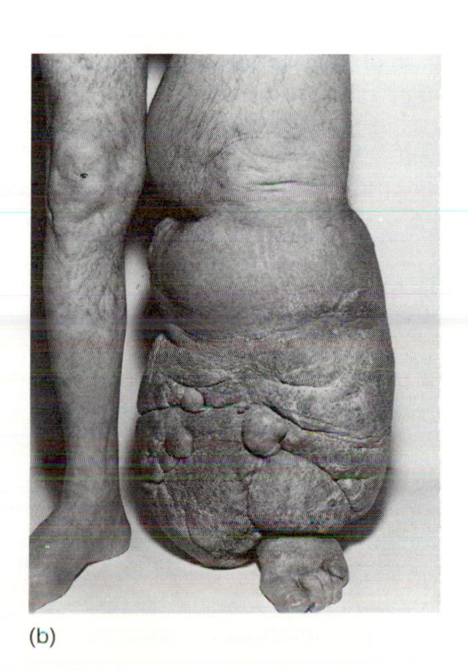

(b)

Figure 28-14 **Elephantiasis.** *(a)* Microfilaria of *Wuchereria bancroftii* in blood smear. *(b)* Patient with extensive elephantiasis of the leg.

all over the world and extends into some subtropical areas. Humans are the only known reservoir for this nematode, and the mosquito *Culex* is the vector for the filarial parasite.

Female (90 mm × 0.25 mm) and male (40 mm × 0.10 mm) *Wuchereria* reside in the lymph nodes. The female gives birth to threadlike embryos that develop into microfilariae (250 μm × 10 μm) in the lymph and blood. The microfilariae lodge in capillary walls and in subcutaneous tissue, but migrate into the peripheral blood at night, the time that the mosquito vector feeds. Thus, the mosquito sucks in microfilariae along with blood from an infected person. The larvae develop in the mosquito into a form that successfully infects humans when injected during a blood meal. Once in the human host, the larvae develop into adults that reside in lymph node sinuses, where they mate. The allergic reactions to the microfilariae and larvae growing in the subcutaneous tissue, lymph, and blood are responsible for many of the signs and symptoms of the disease.

Diagnosis of filariasis can sometimes be made by observing the microfilariae in wet mounts or in Wright stained smears □ of blood drawn in the middle of the night. Control of the disease can be achieved by spraying or draining pools of water where mosquitoes breed. Alternatively, the pools can be stocked with fish that eat the mosquito larvae. If mosquito vectors do not take in the microfilariae, the larvae die in the human host. Apparently, the development of the larvae in the mosquito is an essential part of the developmental process of these parasites.

504

Trichinosis

Trichinosis is a disease of connective tissue caused by the nematode (roundworm) *Trichinella spiralis*. About 24 hours after consuming *Trichinella*-infected meat, the person develops a nonspecific gastroenteritis that may include abdominal pain, nausea, vomiting, diarrhea, and fever. Beginning approximately a week after infection, and for several weeks thereafter, the infected person suffers from muscle pains, low grade fever, dyspnea (painful or difficult breathing), edema (swelling of tissues), skin eruptions, anorexia (loss of appetite), and emaciation. After the disease becomes chronic, muscular pains persist for months. Trichinosis is found worldwide in animals and in humans who consume raw or undercooked meat, especially pork and bear.

Infection generally occurs by consuming meat with *Trichinella* cysts embedded between muscle fibers. Stomach acids and intestinal enzymes digest the cyst wall and free the nematode larva, which takes up residence between the cells of the duodenal and jejunal mucosa. In two days or so the larvae develop into sexually mature adults, which then mate. The females (3–4 mm long) burrow further into the intestinal lining, where they give birth about a week later to 500–1,000 small larvae (0.1 mm long) over a 2- to 6-week period. Following reproduction, the adults die and are digested. The larvae migrate to all parts of the body through the connective tissue and through the lymphatic and circulatory systems. In most of the organs of the body, the larvae are destroyed by immune reactions; in the muscles, however, the larvae burrow between muscle cells, grow to 1 mm in length, coil up, and become encysted □. **Myositis** (inflammation of the muscle), rigidity, and loss of strength occurs. The nematode can remain viable within the cyst for a number of years.

383

Presumptive evidence for an infection is eosinophilia (over 500 eosinophils/cm^3 of blood). The most frequently used method of diagnosis, however, is by serological methods, such as the indirect fluorescent antibody test 522 □. Biopsy of human and animal muscle to observe the encysted nematode is generally not feasible. Sometimes adult nematodes can be observed in the feces during the first two or three days of the infection. Control can be achieved by cooking garbage and meat fed to pigs, cats, and dogs. Humans can avoid the disease by thoroughly cooking meats. Once the disease is contracted, treatment consists of alleviating the pain and discomfort with adrenocorticotrophic hormone (ACTH) and treating the infection with thiabendazole.

In the United States, about 150–200 cases of trichinosis are reported annually with a very low mortality. An average of less than 1 death/year has been reported in the last 10 years in the United States.

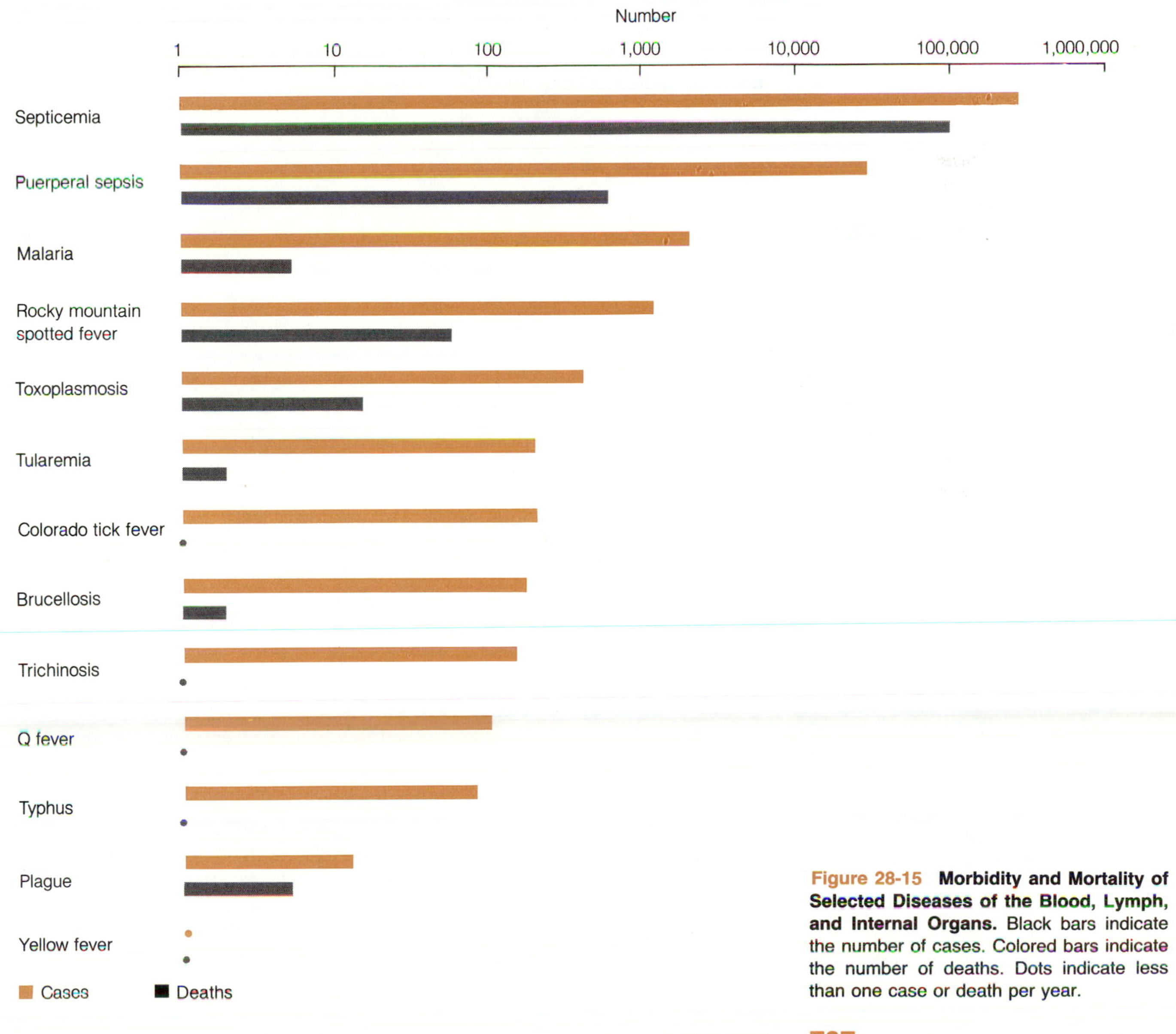

Figure 28-15 Morbidity and Mortality of Selected Diseases of the Blood, Lymph, and Internal Organs. Black bars indicate the number of cases. Colored bars indicate the number of deaths. Dots indicate less than one case or death per year.

CONCLUDING REMARKS

The great majority of infections that affect blood, lymph, muscle, and internal organs are initiated by insect vectors that inject the microorganisms into the lymph and blood (table 28-3). A few internal infections are due to microorganisms that penetrate intestinal or respiratory mucous membranes after being consumed or inhaled. A few organisms penetrate via animal bites (rabies) or scratches (cat scratch fever), open cuts (gangrene and sometimes tularemia), or by direct penetration (schistosomiasis).

Table 28-3 summarizes a number of characteristics of bacterial, viral, protozoal, and helminthic diseases that affect the lymph, blood, muscle, and internal organs. Notice that many of the bacteria are transmitted by fleas (plague), lice (typhus), and ticks (Rocky Mountain spotted fever). Viruses are transmitted by mosquitoes (yellow fever) and ticks (Colorado tick fever); protozoans are spread by flies (African sleeping sickness), bugs (Chagas' disease), and mosquitoes (malaria); and animals such as nematodes are disseminated by flies (river blindness) and mosquitoes (filariasis).

Fig. 28-15 illustrates the seriousness in the U.S. of the diseases of the muscles, lymph, blood, and internal organs.

SUMMARY

STRUCTURE AND FUNCTION OF THE CIRCULATORY AND LYMPHATIC SYSTEMS

1. A number of serious diseases can occur when bacteria penetrate the skin or mucous membranes and invade the lymph and blood. Access to the lymph and blood generally occurs because of wounds, microorganism-induced ulcers, or animal or insect bites.
2. The blood system, at times, provides a pathway for infecting microorganisms to spread throughout the body.
3. Infecting microorganisms that are not destroyed in the lymph nodes may escape from the lymph nodes and flow into the blood and then into the heart.
4. Infecting microorganisms that are not destroyed in the blood by the immune systems can spread to the internal organs and cause disease.

BACTERIAL DISEASES CAUSED BY INVASION OF THE LYMPH AND BLOOD

1. Septicemia, or bacterial sepsis, is a blood infection characterized by fever, chills, decreased blood pressure, shock, and death in 30–50% of cases. As many as 100,000 persons die each year in the U.S. from septicemia. Eighty percent of the infections are due to gram negative bacteria, such as *Escherichia*, *Klebsiella*, and *Enterobacter*.
2. Puerperal sepsis, or childbed fever, is a septicemia that results when a woman who has recently given birth becomes infected. Today, as many as 600 women die each year in the U.S. from puerperal sepsis. Before the extensive use of antibiotics in the late 1940s and early 1950s, puerperal sepsis was due mainly to the gram positive bacteria *Streptococcus pyogenes*, *Staphylococcus aureus*, and *Staphylococcus epidermidis*. Presently, most of the cases of puerperal sepsis are due to gram negative bacteria such as *Escherichia*, *Klebsiella* and *Enterobacter*.
3. Endocarditis is an infection of heart endothelium, generally in one of the valves that is damaged or abnormal. This infection leads to a speticemia and heart failure. Subacute endocarditis progresses for two to three months before it fulminates (produces severe symptoms). The gram positive alpha hemolytic streptococci, such as *Streptococcus mutans*, are responsible for 80% of the cases of subacute endocarditis. Acute endocarditis generally flares up within a month after the infection begins. The gram positive alpha hemolytic streptococci account for 50% of the cases.
4. Gas gangrene results from an infection of the muscles or internal organs by any one of a number of clostridia. The disease is characterized by the production of gas and necrotic tissue. The infection generally occurs in tissues that have poor blood circulation. Serious wounds, operations, arteriosclerosis, and diabetes predispose toward gas gangrene. *Clostridium perfringens* produces a number of ex-

oenzymes (exotoxins) that are responsible for the degradation of the matrix and tissue cells. The bacterium ferments the hydrolyzed polysaccharide and produces gas, which often gets caught in the necrotic muscle and under the skin.

5. The plague, or black death, is caused by the bacterium *Yersina pestis* and is spread from animal to animal and from animal to man by fleas. When the bacteria multiply in the lymph nodes and cause them to swell, the plague is known as bubonic plague. When the bacteria multiply in the lungs, the disease is called pneumonic plague. Pneumonic plague is extremely contagious because it can spread from person to person in respiratory secretions and aerosols.
6. Relapsing fever is caused by the spirochete *Borrelia recurrentis*. The bacterium is usually spread by ticks, which are also the reservoir for this pathogen.
7. Brucellosis, or undulant fever generally occurs in cattle, goats, swine, and sheep and is due to various species of the bacterium *Brucella*. In dairy herds, the disease frequently results in abortion. The bacterium can be spread to humans in unpasteurized milk from diseased animals, but butchers, workers in meat processing plants, ranchers, and hunters account for most of the brucellosis in humans (90%).
8. Tularemia, or rabbit fever, is caused by the bacterium *Francisella tularensis*. Tularemia is transmitted to man when infected ticks and deer flies bite. The bacteria sometimes enter small cuts or penetrate the mucous membranes when infected animals are skinned by hunters.
9. Rocky Mountain spotted fever is due to the obligate intracellular bacterium *Rickettsia rickettsi*. The reservoirs for this bacterium are rodents, rabbits, and ticks. The ticks also act as the vector.
10. Typhus fevers are caused by various *Rickettsia*. Epidemic typhus is caused by *R. prowazekii* and is transmitted to humans in the feces of infected body lice that bite. Endemic typhus is caused by *R. mooseri* in the feces of infected fleas when they bite. *Rickettsia* are obligate intracellular parasites.

VIRAL DISEASES CAUSED BY INVASION OF THE LYMPH AND BLOOD

1. Colorado tick fever is due to an arbovirus and is transmitted to humans from infected wild rodents by tick vectors.
2. Yellow fever is caused by an arbovirus and is spread among humans, or from monkeys to humans, by mosquito vectors. There have been no cases of yellow fever in the U.S. since 1924. The disease is endemic in Africa and the Far East.
3. Dengue fever is due to an arbovirus that is transmitted by a specific mosquito, which acts as a reservoir. There are few cases of dengue in the U.S., but recently in Puerto Rico an average of 11,000 cases has been reported/year.
4. Infectious mononucleosis is caused by a herpes virus called the Epstein-Barr virus. About 20,000 cases of mononucleosis, and 20 deaths, are reported per year in the U.S.

PROTOZOAL DISEASES CAUSED BY INVASION OF THE LYMPH AND BLOOD

1. Malaria is caused by the protozoans *Plasmodium falciparum* (50%), *P. vivax* (43%), *P. malariae* (6%), and *P. ovale* (1%). The major reservoir of *Plasmodium* is humans, although some monkeys can carry the protozoan. A specific mosquito is responsible for spreading the organisms from human to human when it bites.
2. Toxoplasmosis is due to the protozoan *Toxoplasma gondii*. Cats are an important reservoir for this protozoan, although wild and domestic animals (pigs, sheep, and cattle) and birds often become infected. Humans become infected when they ingest dried particles of cat feces contaminated with the *Toxoplasma* oocysts.
3. African sleeping sickness is caused by *Trypanosoma gambiense* and *T. rhodesiense*, which are spread from human to human by the bite of the tsetse fly. Humans, animals, and the tsetse fly act as reservoirs for the protozoan.
4. Chagas' disease is caused by *Trypanosoma cruzi*, which is spread among humans by biting insects such as kissing bugs, bedbugs, and lice.

DISEASES CAUSED BY MULTICELLULAR PARASITES

1. River blindness, or onchocerciasis, is caused by the parasitic nematode *Onchocerca volvulus*, which is spread by black flies. The disease afflicts more than 150 million people worldwide, mostly in West Africa. Humans appear to be the nematode's only reservoir.
2. Loiasis, a disease similar to onchocerciasis, is caused by nematode *Loa loa* that is spread by the bite of a horse fly or mango fly.
3. Filariasis, or elephantiasis, is caused by the nematode *Wuchereria bancrofti*. Humans are the only known reservoir for the nematode, which is spread from human to human by a specific mosquito.

STUDY QUESTIONS

1. For each of the following diseases, list a) the name of the pathogen, b) reservoirs, c) vectors, and d) how acquired.
   a. Puerperal sepsis
   b. Endocarditis
   c. Gas gangrene
   d. Plague
   e. Relapsing fever
   f. Brucellosis
   g. Tularemia
   h. Epidemic typhus
   i. Endemic typhus
   j. Tick fever
   k. Yellow fever
   l. Dengue fever
   m. Mononucleosis
   n. Malaria
   o. Toxoplasmosis
   p. Chagas' disease
   q. Sleeping sickness
   r. River blindness
   s. Loiasis
   t. Filariasis
2. From which activity could a person become infected by each disease listed?

Francisella tularensis	a. eating undercooked meat
Brucella melitensis	b. injecting drugs intraven-ouusly
Yersinia pestis	c. changing cat litter
Clostridium perfringens	d. drinking unpasteurized milk
Rickettsia rickettsi	e. butchering rabbits or deer
Rickettsia prowazekii	f. sleeping without mosquito netting
yellow fever virus	g. wading or swimming in waters that contain flukes and their snail hosts
dengue fever virus	h. collecting dead rats
Epstein-Barr virus	i. walking in heavy brush with bare legs and bare upper body
Plasmodium falciparum	j. infrequently washing clothes and body
Toxoplasma gondii	k. kissing or using contaminated cups or utensils
Trypanosoma gambiense	l. using bird or bat fertilizers
Trypanosoma cruzi	
Onchocerca volvulus	
Wuchereria bancrofti	
Trichinella spiralis	
Schistosoma mansoni	
Histoplasma capsulatum	

3. Most septicemias are due to what group of bacteria? How many persons in the U.S. die from septicemias each year?
4. Puerperal sepsis is caused mostly by which bacteria? How many women die each year in the U.S. because of puerperal sepsis?
5. Which group of bacteria is responsible for most cases of subacute endocarditis?
6. Which bacteria are responsible for most cases of acute endocarditis?
7. What is the difference between subacute and acute endocarditis?
8. Lung infections would be expected to occur most readily from endocarditis in which part of the heart?
9. Which of the following diseases do not occur naturally in the U.S.? a. malaria; b. plague; c. yellow fever; d. puerperal sepsis; e. relapsing fever; f. brucellosis; g. Q-fever; h. tularemia; i. Rocky Mountain spotted fever; j. epidemic typhus; k. endemic typhus; l. Colorado tick fever; m. dengue fever; n. mononucleosis; o. toxoplasmosis; p. onchocerciasis; q. elephantiasis (filariasis); r. trichinosis; s. schistosomiasis; t. histoplasmosis; u. blastomycosis.

SUPPLEMENTAL READINGS

Kolata, G. 1984. The search for a malaria vaccine. *Science* 226:679–682.

Langer, W. L. 1964. The black death. *Scientific American* 210:114–121.

Paterson, P. Y. 1982. Bacterial endocarditis. Chap. 50 in *Microbiology: Basic science and medical applications*, ed. A. Braude, C. Davis, and J. Fierer. Philadelphia: W. B. Saunders Co.

Rabinowitz, S. G. 1982. Bacterial sepsis and endotoxic shock. Chap. 36 in *Microbiology: Basic science and medical applications*, ed. A. Braude, C. Davis, and J. Fierer. Philadelphia: W. B. Saunders Co.

Rubin, R. H. 1985. Infection in the immunosuppressed host. *Scientific American Medicine* Section 7-X. New York: Freeman.

Sommers, H. M. 1982. Diseases due to anaerobic bacteria. Chap. 48 in *Microbiology: Basic science and medical applications*, ed. A. Braude, C. Davis, and J. Fierer. Philadelphia: W. B. Saunders Co.

Van Heynigen, W. E. 1968. Tetanus. *Scientific American* 218:69–77.

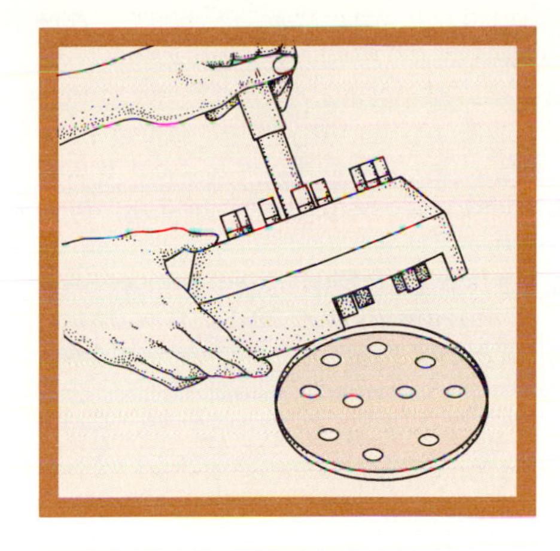

CHAPTER 29
CHEMOTHERAPY OF INFECTIOUS DISEASES

CHAPTER PREVIEW

CHEMOTHERAPEUTIC DRUGS FROM THE SEA

Humans have experimented with plants and animals for thousands of years to find drugs that might alleviate or cure their ills. The most notable discoveries have been the ipecac root from Brazil, which yields **emetine,** a chemical that kills *Entamoeba histolytica,* the cause of amebic dysentery and the cinchona bark from Peru, which produces **quinine,** an alkaloid that kills *Plasmodium*, the cause of malaria. Fungi and bacteria also produce chemicals which help alleviate or cure ills. For example, the fungus *Penicillium* produces the antibiotic **penicillin** and bacteria in the genus *Streptomyces* produce an array of antibiotics that includes **streptomycin** and **tetracycline.**

Now scientists are finding useful chemicals and drugs derived from sea life. Marine biologists have discovered a number of drugs that may be effective against various cancers. One species of animal that forms mats on rocks and on the hulls of ships, known as "false coral," produces a compound called **bryostatin-1** that doubles the lifespan of mice dying from leukemia. As little as 0.1 μg is effective in prolonging the life of these animals. Marine biologists have also discovered that soft corals and mollusks that lack an external shell produce toxic chemicals to avoid being eaten by predators. More than 30 chemicals that inhibit cell division have been isolated from soft sea corals and mollusks. Some of these chemicals may be useful in treating cancerous growths and parasitic infections.

A compound called **stypoldione,** which repels fish, has been extracted from a Caribbean seaweed. Because stypoldione also inhibits cell division, it is being investigated to see if it might be useful in treating some types of cancer or parasitic infections.

For more than 40 years, scientists have isolated many hundreds of drugs from bacteria and fungi. In fact, most useful drugs are produced by bacteria and fungi that live in the soil. The discovery of drugs produced by sea creatures will undoubtedly stimulate scientists to begin an extensive search for new chemicals. The main task ahead is to characterize these chemicals and to determine how they might be useful to humans in preventing and treating disease.

In this chapter you will be introduced to some chemicals that are used as chemotherapeutic agents. You will also learn what factors are important in a good drug and how various drugs function.

LEARNING OBJECTIVES

A STUDY OF THIS CHAPTER SHOULD ENABLE YOU TO:

- EXPLAIN WHAT AN ANTIBIOTIC IS
- LIST THE MAJOR GROUPS OF ANTIBIOTICS AND DRUGS
- OUTLINE HOW THE MAJOR ANTIBIOTICS AND DRUGS WORK
- DISCUSS THE MECHANISMS BY WHICH MICROORGANISMS BECOME RESISTANT TO ANTIBIOTICS AND DRUGS
- EXPLAIN WHY SOME ANTIBIOTICS AND DRUGS ARE EXTREMELY TOXIC TO THE PATIENT

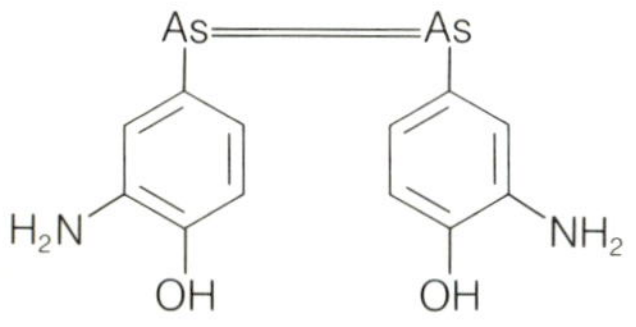

Salvarsan

Figure 29-1 **Salvarsan.** Compound 606, salvarsan, was Paul Ehrlich's "magic bullet." This arsenic-containing compound was effective in combating syphilis in experimental animals.

PERSPECTIVE

Since ancient times, humans have used medicines and potions to treat their diseases. For example, moldy leaves and bread were sometimes used by various cultures to treat festering wounds.

Modern chemotherapy began in 1910, when Paul Ehrlich announced his discovery of an arsenic compound he called **salvarsan** (fig. 29-1). This drug, also known as Ehrlich's "magic bullet," was useful in treating yaws, syphilis, and relapsing fever. It was Ehrlich's belief that chemicals could be synthesized that would specifically bind and kill microorganisms. Unfortunately, salvarsan was not specific, and it was extremely toxic because it had to be used in high concentrations. Long-term treatment at reduced dosages led to serious brain and nerve damage. At recommended doses, the patients became nauseated and weak and often developed unusual allergies. Salvarsan and neosalvarsan (which was less toxic) were nevertheless much better than the folk cures for syphilis being used at the time, and Ehrlich's arsenic compounds cured many people of this disease.

In 1935 Gerhard Domagk reported that **prontosil rubrum,** a red azo dye combined with a sulfonamide group (fig. 29-2), was effective against bacterial infections in experimental animals and in humans. He discovered its usefulness in humans when he used the chemical on his daughter, who was dying of septicemia: the drug effected a rapid and miraculous recovery. Although prontosil had no effect on bacteria growing in test tubes, it did have an effect *in vivo*, where it was cleaved into a sulfa drug called **sulfanilamide** (fig. 29-2). Sulfa drugs are effective only against organisms that synthesize the coenzyme □ folic acid from para-aminobenzoic acid (PABA). 54 Sulfanilamide competes with PABA for the enzyme that joins PABA to another compound in the synthesis of folic acid. Because the sulfa drug is unable to take the place of PABA, the formation of the coenzyme is blocked (fig. 29-3). Humans and animals do not synthesize their own folic acid but instead require it preformed, so they are not disturbed by the sulfa drug. Sulfonamides such as sulfanilamide, sulfapyridine, and sulfathiazole were found to be effective against a variety of bacterial infections. Sulfa drugs are effective against *Streptococcus pyogenes* (skin and throat infections), *Neisseria* (meningitis), *Escherichia coli* (urinary tract infections), *Chlamydia trachomatis* (trachoma), *Haemophilus ducreyi* (chancroid), *Nocardia asteroides* (nocardiosis), and *Actinomyces israelii* (actinomycosis). Domagk was awarded the 1939 Nobel Prize for medicine or physiology, but for political reasons he declined the honor and money.

In 1939, Rene Dubos discovered that the soil bacterium *Bacillus brevis* was producing a chemical he called **tyrothricin,** which killed *Streptococcus* in a liquid solution. Tyrothricin and another antibiotic, **gramicidin,** produced by another species of *Bacillus*, proved to be too toxic to be taken internally by humans and so they were used only for curing skin infections.

Selman Waksman, stimulated by Rene Dubos's successes, began a search for actinomycetes that might produce useful antimicrobial agents. Waksman coined the term **antibiotic** to describe the natural antimicrobial agents produced by microorganisms. Today, antimicrobial agents produced by microorganisms are called antibiotics to distinguish them from manmade chemicals, such as salvarsan and sulfanilamide. In 1940, Waksman and coworkers discovered an actinomycete that produced **actinomycin.** Although the antibiotic proved very useful in research, it was too toxic to treat animal infections. By 1943, another actinomycete called *Streptomyces griseus* was discovered by Albert Schatz, one of Waksman's assistants. It produced the antibiotic called **streptomycin** (fig. 29-4). Streptomycin was concentrated, purified, and crystallized, and was shown to be extremely effective in treating tuberculosis. Waksman and his associates developed strains of *S. griseus* that overproduced the antibiotic, making it economically feasible to produce the drug. Streptomycin turned out to be effective not only against *Mycobacterium tuberculosis* but also against *Francisella tularensis* (tularemia), *Neisseria meningitidis* (meningitis), *Streptococcus pneumoniae* (pneumonia), and *Escherichia coli.* Waksman and his colleagues are also credited with the discovery of **erythromycin, neomycin,** and **candicidin.** For his discovery, isolation, and testing of streptomycin, Waksman received the Nobel Prize for physiology or medicine in 1952.

One of the most significant developments in chemotherapy was the discovery of the antibiotic **penicillin G** (fig. 29-5), produced by the fungus *Penicillium notatum.* The discovery of penicillin occurred in 1928, but its purification and use did not take place until the early 1940s. Sir Alexander Fleming is credited with the discovery of penicillin, while Howard Florey and Ernst

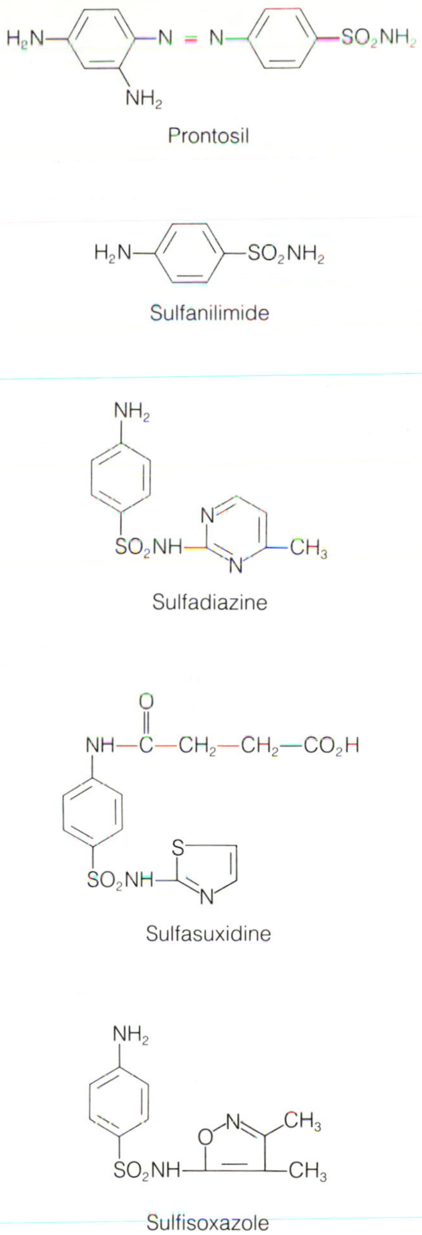

Prontosil

Sulfanilimide

Sulfadiazine

Sulfasuxidine

Sulfisoxazole

Figure 29-2 Prontosil and Other Sulfa Drugs

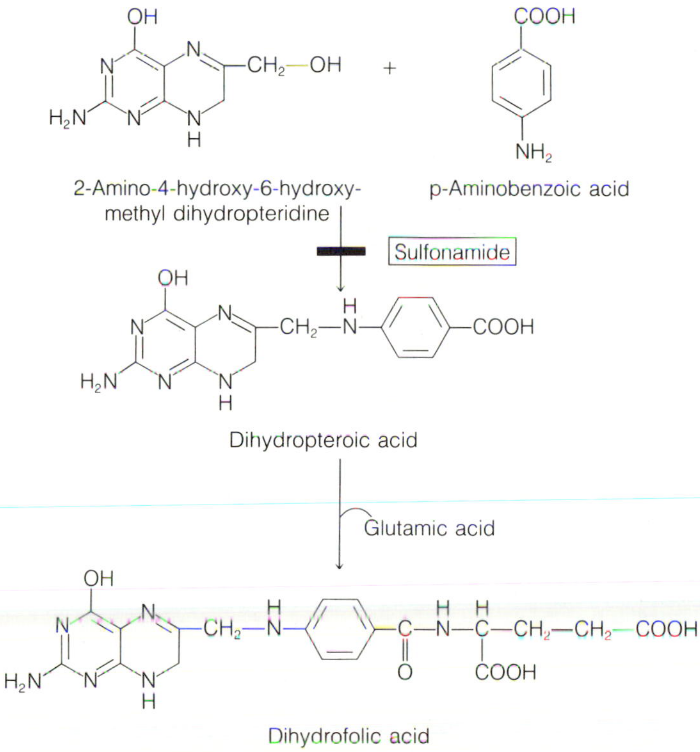

2-Amino-4-hydroxy-6-hydroxymethyl dihydropteridine

p-Aminobenzoic acid

Sulfonamide

Dihydropteroic acid

Glutamic acid

Dihydrofolic acid

Figure 29-3 Mode of Action of Sulfa Drugs. Sulfa drugs inhibit the synthesis of folic acid, a coenzyme needed for the synthesis of nucleic acids. The synthesis of folic acid requires that para-aminobenzoic acid (PABA) be present. Sulfa drugs like sulfonamide compete with PABA in the synthesis of folic acid. When sulfonamide is used instead of PABA, the resulting product is inactive, and hence the synthesis of nucleic acids is inhibited.

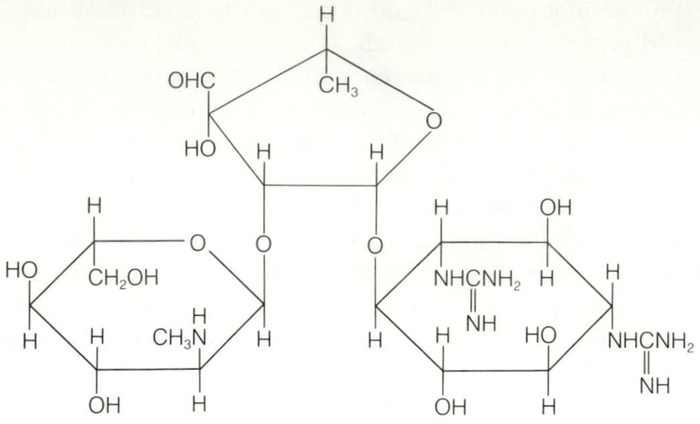

Streptomycin

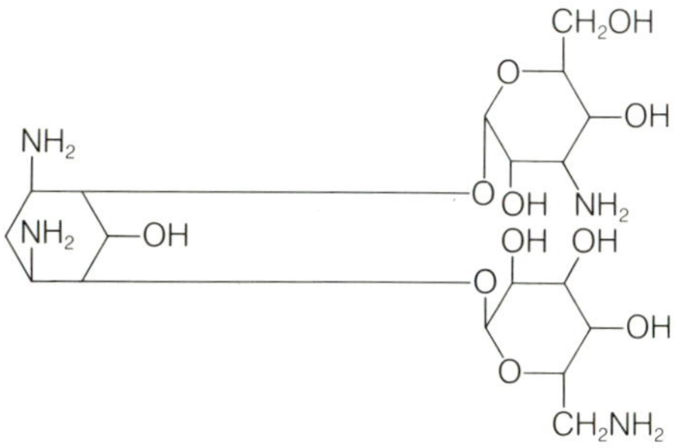

Kanamycin

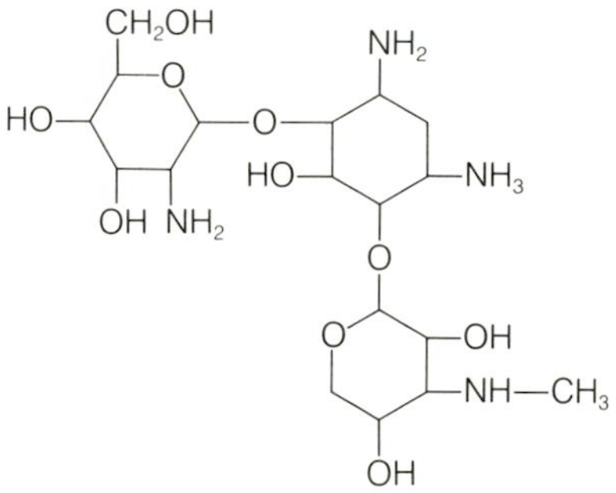

Gentamycin A

Figure 29-4 Streptomycin, Kanamycin and Gentamycin A

Chain are responsible for its isolation and purification. The discovery and successful use of penicillin during the last part of World War II was important, because it stimulated the search for other natural substances produced by fungi and soil bacteria which could inhibit infectious microorganisms. The search for new antibiotics was also spurred by the discovery that penicillin was not a panacea: the fungi, protozoans, and some bacteria are not affected by this antibiotic. Until the early 1940s, plant substances (quinine), poisonous arsenic compounds (salvarsan), and sulfonamides (sulfanilamide)

(a)

Penicillin G

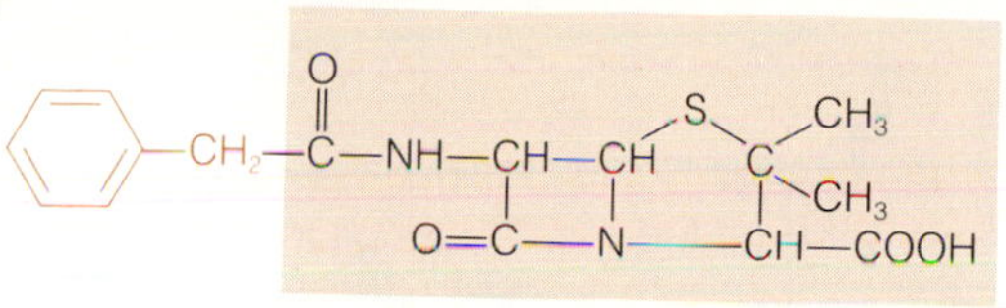

Penicillin V

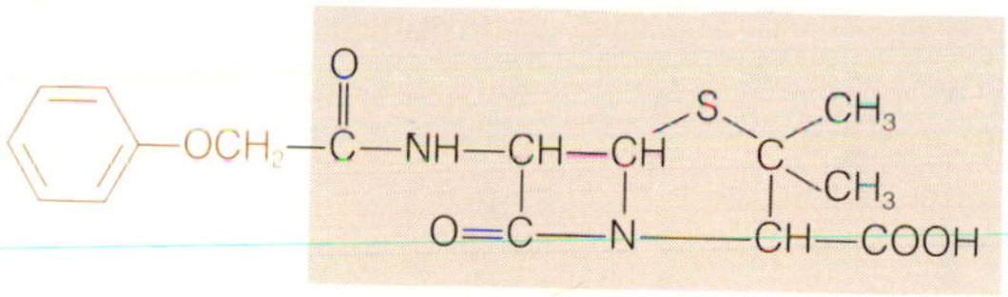

Penicillin F

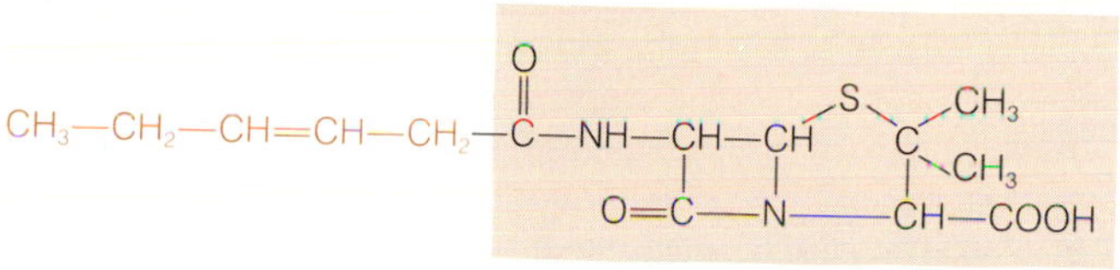

(b)

Ampicillin

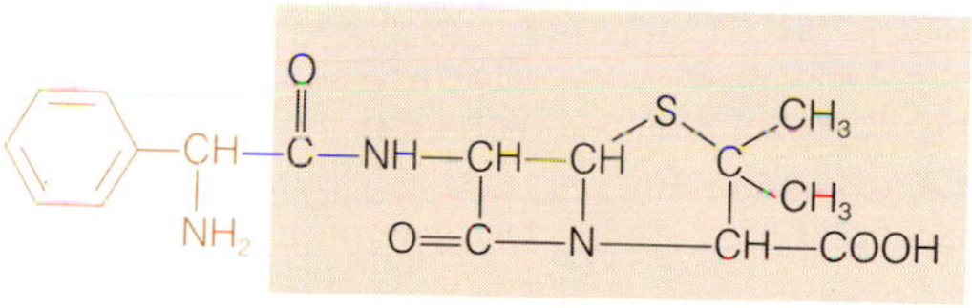

Methicillin

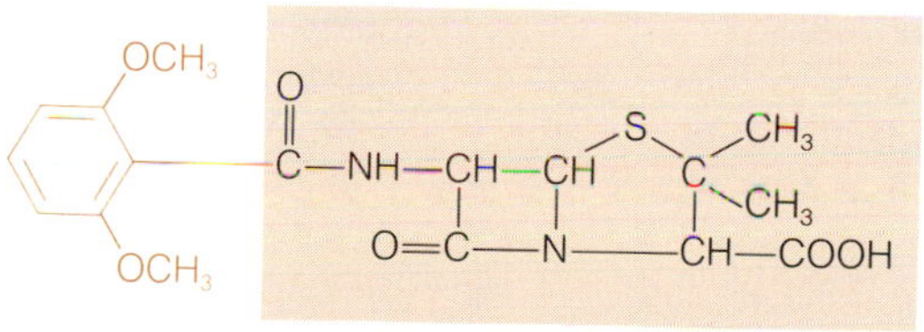

Carbenicillin

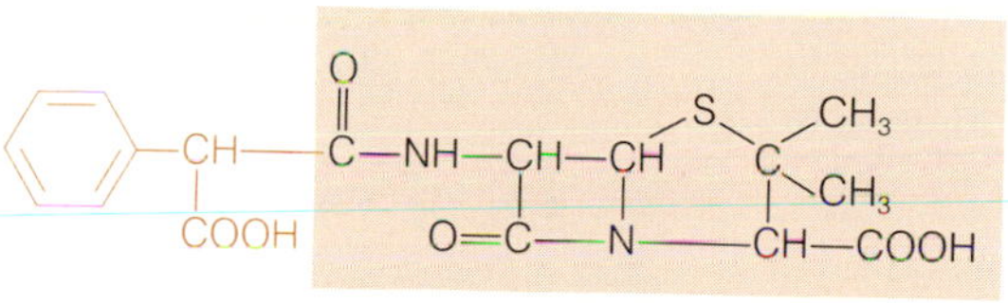

Oxacillin

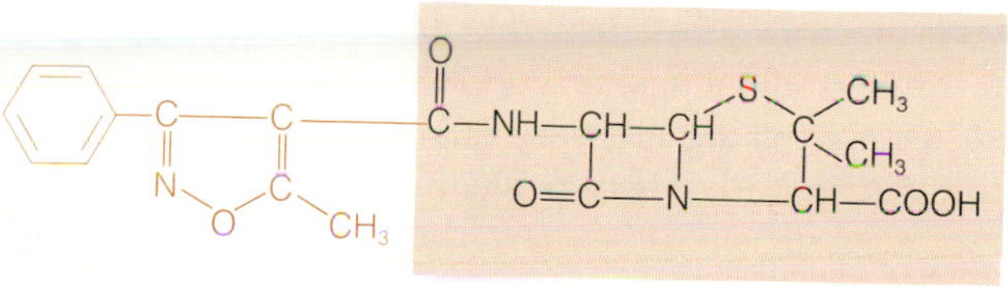

Figure 29-5 Chemical Structure of Penicillins. All penicillins, both naturally occurring *(a)* and synthetic *(b)*, include a β-lactam ring. The side chains (in color print), which make each penicillin type distinct, influence the activity and range of activity of the molecule.

FOCUS

THE DISCOVERY OF QUININE AND THE TREATMENT OF MALARIA

The discovery that extracts from certain barks were useful in the treatment of fevers is properly attributed to pre-Columbian Peruvian Indians who chewed these barks. The wife of the Spanish viceroy in Peru during the early 1600s, the Countess del Chinchon, supposedly used the Indian cure to recover from a severe fever. As the story goes, the tree and the curative substance were named "cinchona" after the Countess del Chinchon. This account of the discovery of quinine has been challenged by at least one historian; the real story behind the discovery of quinine may be less glamorous.

In the 1600s, Peruvian balsam became extremely popular in Spain as a fever remedy, ointment, and perfume. Balsam is an oily, gummy, aromatic, mintlike resin that can be extracted from a number of plants. Balsam was initially extracted from the bark of the quina-quina tree but because this tree was in short supply, merchants soon replaced it with barks from other trees. One of the substitute barks was from the cinchona tree. Eventually, it was discovered that the balsam from the bark of the cinchona tree (but not that from the quina-quina tree) was useful in the treatment of malaria. Thus, the discovery of the beneficial uses of extracts from the cinchona tree appears to have been accidental. Which bark the Indians and the Countess del Chinchon actually used will remain a mystery, because the (worthless?) quina-quina bark was used by the Indians before the cinchona bark, and the two barks were prescribed indiscriminately in Europe for many years.

In 1820 two French chemists, Pierre Pelletier and Joseph Caventou, extracted the antimalarial alkaloid from the cinchona bark. The antimalarial agent became known as **quinine,** and it turned out to be lethal to some of the malarial parasites. It could be used to kill *Plasmodium faciparium,* but not the other species of *Plasmodium.* It was also found, however, that quinine was effective in suppressing symptoms caused by all *Plasmodium* species. Like most alkaloids, quinine has detrimental side effects when taken in the large doses. Quinine causes vomiting, headaches, rashes, blurred vision, and ringing in the ears. Nevertheless, these symptoms are more acceptable than those of malaria.

Because of the British expansionism into India and Southeast Asia, where malaria was a severe problem, cinchona trees from Peru were planted and cultivated in these regions. India, Indonesia, and Java became the world's major source of quinine before the Second World War. In the 1930s and during the Second World War, when the Japanese occupied much of Southeast Asia, several drugs were developed to replace quinine. The synthetic quinacrine (Atabrine) was used extensively during the Second World War by soldiers from the U.S. and Europe. Quinacrine was effective in preventing and suppressing malaria, but it had major drawbacks: the drug turned the skin yellow and caused vomiting.

In Vietnam during the 1960s and 1970s, the U.S. used pyrimethamine (Daraprim) and sulphormethoxine to prevent and treat malaria. Presently, a number of antimalarial drugs are used. The most useful are primaquine, chloroquine, chloroguanide, cycloguanil, and pyrimethamine.

In the last 10 years there have been numerous reports of drug-resistant strains of *Plasmodium falciparum* in Southeast Asia. There are now strains that are resistant to chloroquine, primethamine, and sulfadoxine. These drug-resistant strains are a problem, because the usual treatments must be modified and effective drugs must be tested.

were the only substances that had been discovered after years of searching for antimicrobial agents. Suddenly, after the isolation of penicillin, researchers were checking all types of soil microorganisms to find other antibiotics that would work where penicillin did not.

FUNDAMENTALS OF CHEMOTHERAPY

A drug or antibiotic is useful only if it inhibits or kills a microorganism without doing harm to the host animal. The arsenic compounds atoxyl, salvarsan, and neosalvarsan, although useful against some spirochetes, were extremely toxic and made the patient severely ill. The problem with Ehrlich's antimicrobial agents was that they did not show a **selective toxicity** toward the infectious agents, but in fact inhibited eukaryotic cells almost as readily as prokaryotic cells. Thus, the dose required to kill the spirochetes was near the dose that would debilitate or kill the patient.

The most useful drugs and antibiotics show selective toxicity; that is, they specifically inhibit or kill microorganisms but do not harm the host. Some antimicrobial agents that show selective toxicity are penicillin, which inhibits bacterial cell wall synthesis; nalidixic acid, which inhibits only bacterial DNA replication □; and tetracycline, which inhibits only bacterial protein synthesis □. Even antimicrobial agents that show a selective toxicity can be toxic if taken in excessive amounts or for extended periods of time. Dosage, calculated according to body weight, is the maximum amount that does not cause harm and that eliminates the infectious agent. Every antimicrobial agent can be described by its **therapeutic ratio,** which is the highest dose that can be tolerated without harm to the patient divided by the minimum dose needed to eliminate the infectious agent. Ehrlich's arsenic compounds have a low therapeutic ratio, while many of the modern antibiotics have a high therapeutic ratio.

204
209

A number of factors besides selective toxicity affect an antimicrobial agent's therapeutic ratio. For example, an antimicrobial agent that is very reactive (such as penicillin), and that often stimulates a life-threatening allergic reaction, has no value in sensitized individuals. A drug that is insoluble or that is rapidly altered or excreted from the patient is generally of little use in treating internal infections. Antimicrobial agents used against resistant strains of microorganisms do not clear up the infection, even at very high doses, and may even promote other infections by killing the normal body flora that inhibits the growth of opportunistic microorganisms.

Antimicrobial agents that kill bacteria are said to be **bactericidal,** while those that simply inhibit growth are said to be **bacteriostatic** □. Bac-

152

TABLE 29-1

BREADTH OF ACTIVITY OF CERTAIN ANTIBIOTICS*

NAME OF ANTIBIOTIC	BROAD	NARROW	ORGANISMS AFFECTED
Tetracycline	X		Most bacteria
Chloroamphenicol (Chloromycetin)	X		Most bacteria
Democlocycline (Declomycin)	X		Most bacteria
Oxytetracycline (Terramycin)	X		Most bacteria
Kanamycin (Kantrex)	X		Most bacteria
Ampicillin	X		Most bacteria
Cephalothin (Keflin)	X		Most bacteria
Gentamycin	X		Most bacteria
Penicillin		X	Gram + bacteria, *Neisseria*
Streptomycin		X	*Streptococcus,* some gram − bacteria
Erythromycin (Ilotycin)		X	Gram + bacteria
Polymyxin B		X	Gram − bacteria
Nyastatin (Mycostatin)		X	Yeasts
Griseofulvin (Grisactin)		X	Dermatophyte fungi
Rifampin		X	*Mycobacterium*
Amphotericin B (Fungizone)		X	Systemic fungi

*Names in parentheses are trade names of antibiotics.

teriostatic substances generally must be taken for longer periods of time than bactericidal agents, because microorganisms may still be capable of multiplication if the bacteriostatic substances are removed.

Those antimicrobial agents that are effective against a small number of microorganisms are called **narrow spectrum** antimicrobials, while those drugs that are effective against a large number of very different organisms are called **broad spectrum** antimicrobials (table 29-1). For example, the antibiotic penicillin is most effective against gram positive organisms and a few special gram negative bacteria, so it is classified as a narrow spectrum antibiotic. On the other hand, tetracycline is active against most gram negative and gram positive organisms and so is classified as a broad spectrum antibiotic. If the presence of an infectious agent has not been determined before chemotherapy is initiated, it is recommended that a broad spectrum antimicrobial be used until the diagnosis is made and a more appropriate drug can be prescribed (if necessary).

DETERMINATION OF DRUG AND ANTIBIOTIC SENSITIVITIES

It is of great value in treating an infection to determine what antimicrobial agent is most effective. The sensitivity of an infectious agent can be determined in a number of ways. One method for determining sensitivities to drugs is the **tube dilution test** (fig. 29-6). In the tube dilution test, a number of tubes are inoculated with a dilute suspension of microorganisms and then varying concentrations of the antimicrobial agent are added to the tubes. In the tubes containing very low concentrations of the antimicrobial agent, the organisms grow and turn the medium turbid. In the tubes containing intermediate to high concentrations, however, no growth is visible. The **minimum inhibitory concentration (MIC)** of an antimicrobial agent can be determined from the tube with the lowest concentration that shows no turbidity (no growth). Similar procedures with other agents allow a comparison between antimicrobials. The antimicrobial agent with the lowest MIC is generally the most effective. The **minimum killing concentration** is established by transferring the organisms in the clear tubes into fresh, antimicrobial-free media. The tube with the lowest concentration of antimicrobial agent that shows no growth upon transfer to fresh media indicates the minimum killing concentration (fig. 29-7). Since these measurements generally are done on bacteria, the minimum killing concentration is often called the **minimum bactericidal concentration.**

By definition, if the concentration of antimicrobial agent needed to kill a population of organisms is $10\times$ higher than the amount necessary to inhibit the organisms, the agent is said to be **biostatic** (bacteriostatic, fungistatic, etc.). On the other hand, if the concentration needed to kill a population of organisms is less than $5\times$ the amount required to inhibit the organism, the agent is said to be **biocidal** (bactericidal, fungacidal, etc.).

In order to determine the dosage and how often the antimicrobial agent should be given, it is necessary to know not only the minimum killing concentration but also the concentration of the agent in the blood and tissues and how it changes with time. One procedure for determining *in vivo* concentrations of drugs involves sampling the blood several times before and after administration of the drug. The blood samples are placed in holes punched into an agar plate seeded with sensitive microorganisms (fig. 29-8).

Various known concentrations of the antibiotic are placed in other holes, so that a standard curve can be developed (fig. 29-8), using the diameter of inhibition around the holes as a measure of the concentration of the antibiotic in the blood. A graph relating the concentration of antibiotic in the blood to time can be constructed (fig. 29-8). This type of curve indicates what the dose should be and how often it should be given to maintain a minimum killing concentration.

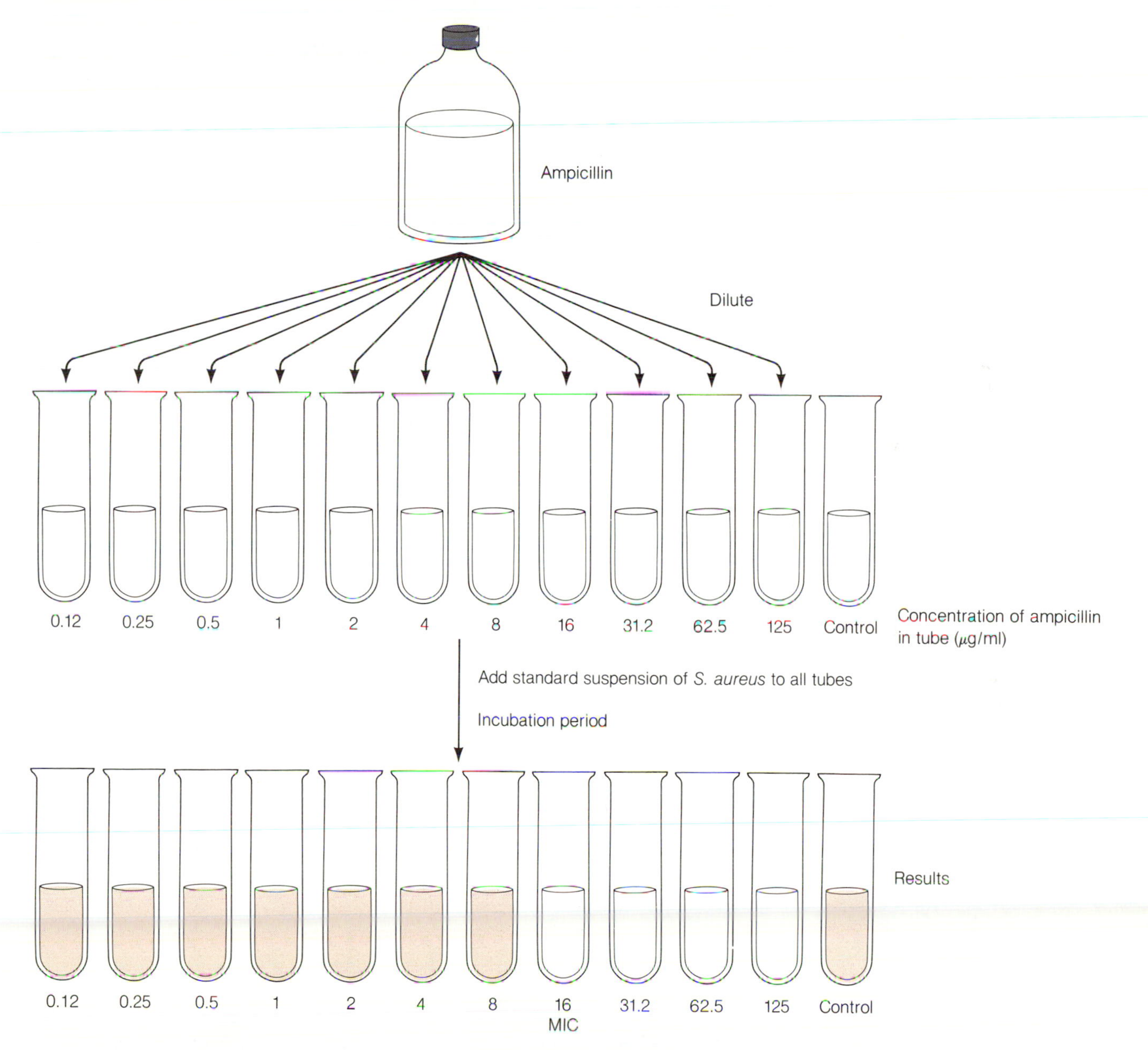

Figure 29-6 Determination of the Minimum Inhibitory Concentration of an Antimicrobial Agent. The minimum inhibitory concentration (MIC) of an antimicrobial consists of the lowest concentration that will inhibit the growth of a test organism. The diagram illustrates a common procedure for determining the MIC of ampicillin for *Staphylococcus aureus.* Various dilutions of ampicillin (ranging from 125 mcg/ml to 0.12 mcg/ml) in a liquid broth are inoculated with a standard suspension of *S. aureus* and incubated at 35°C. After the incubation period, the tubes are examined for growth. The highest dilution of ampicillin that inhibits bacterial growth (clear tube) is the MIC. In this example, the MIC is 16 mcg/ml.

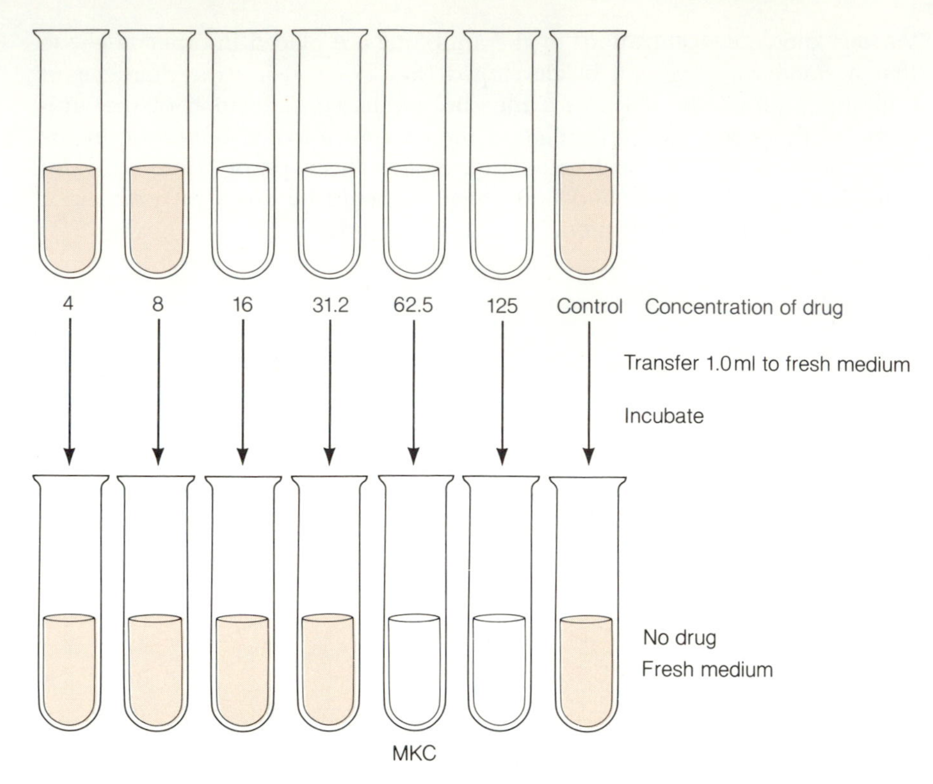

Figure 29-7 Determination of the Minimum Killing Concentration. The minimum killing concentration (MKC) is the lowest concentration that will kill a standard population of bacteria. In the example given, samples from tubes of the MIC test (fig. 29-6) are inoculated into antibiotic-free broths and incubated at 35°C. After an incubation period, the tubes are examined for growth. The MKC is the lowest concentration of antibiotic which will kill the bacterial population. In this example, the MKC is 62.5 mcg/ml. Notice that, although the MIC is 16 mcg/ml, the MKC is 62.5 mcg/ml, indicating that at concentrations of 16 mcg/ml and 31.2 mcg/ml the bacteria are merely inhibited from multiplying, but not killed.

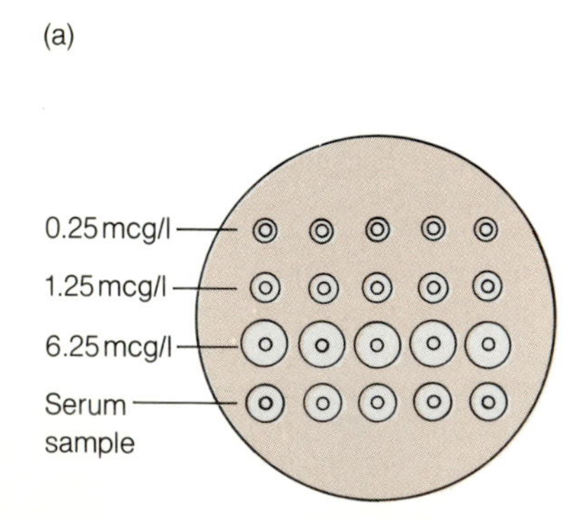

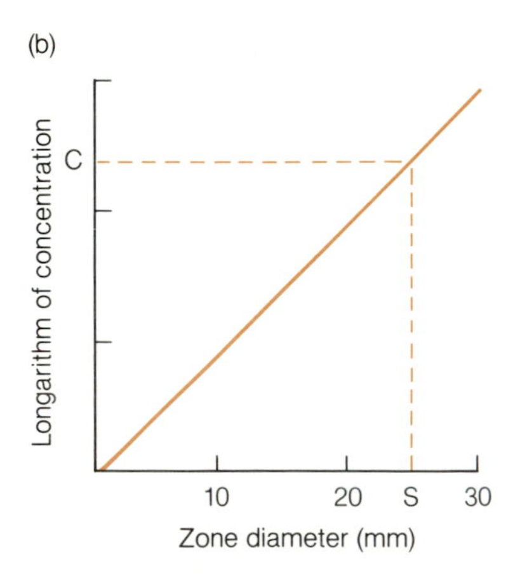

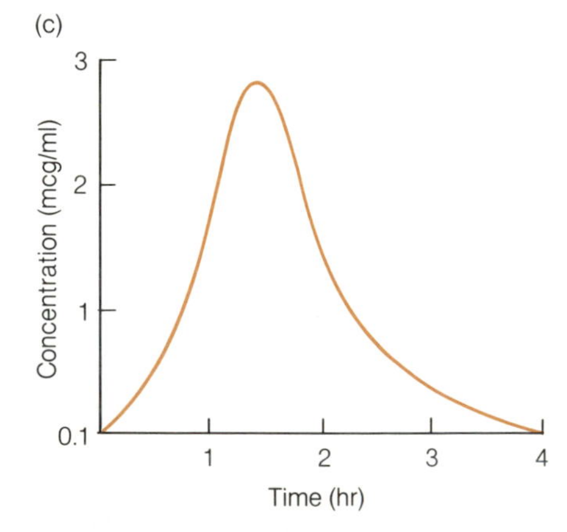

Figure 29-8 Determination of *in vivo* Concentrations of Antimicrobials. *(a)* Agar plates are seeded with test bacteria and wells are cut into the agar. The wells are filled with various concentrations of an antimicrobial or with serum samples. The plates are incubated overnight at 35°C. *(b)* The diameters of the zones of inhibition are plotted against the concentration of antimicrobials to obtain a standard curve. The antimicrobial concentration in serum can be determined by using the standard curve. *(c)* The concentration of an antimicrobial in serum varies with time.

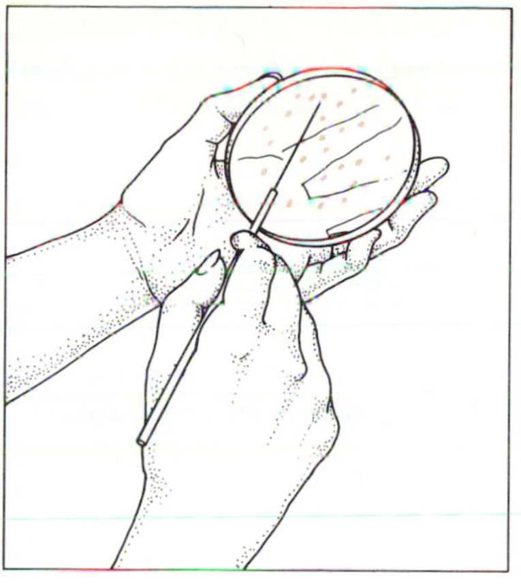

(a) Pick colony

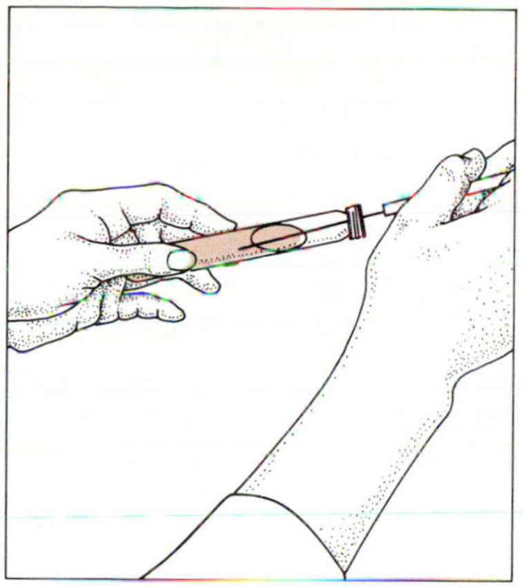

(b) Subculture in broth

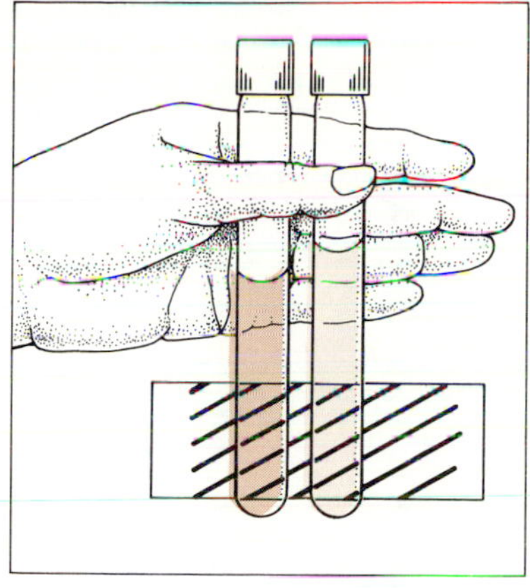

(c) Dilute culture to a standard turbidity

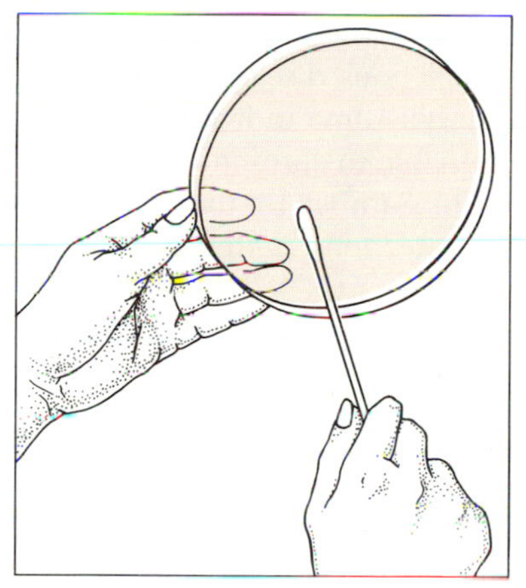

(d) Spread diluted culture over agar plate; allow to dry for a few minutes

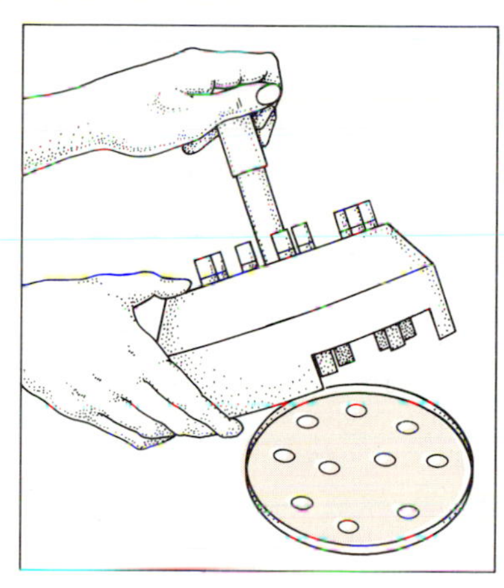

(e) Apply antibiotic impregnated disks to plate

(f) Incubate

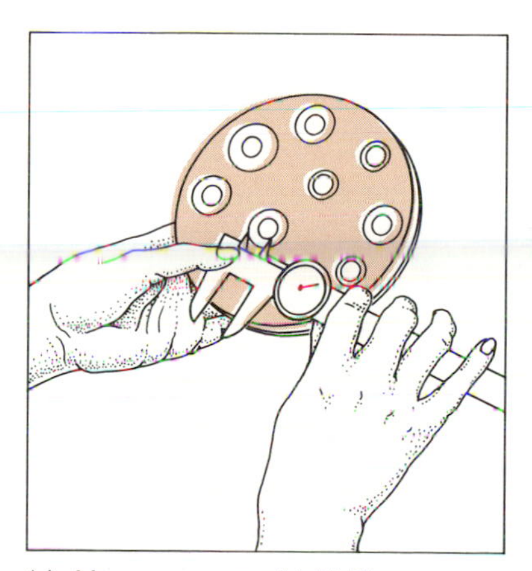

(g) Measure zones of inhibition

Figure 29-9 The Kirby-Bauer Method

A rapid procedure for determining the degree of sensitivity or resistance to a number of antibiotics is the **Kirby-Bauer test,** introduced by William Kirby and Alfred Bauer in 1966. In the Kirby-Bauer test, a lawn of the organism in question is seeded onto an agar petri dish. Antibiotic-impregnated filter disks are then placed onto the developing lawn and incubated for 16 hours (fig 29-9). During the incubation period, each of the antibiotics in the disks will diffuse in the agar, creating a concentration gradient, from high to low, as it diffuses radially from the disk. The bacteria on the plate will multiply throughout, except in those areas where the antibiotic is inhibiting concentration, producing a zone of no growth, called the **zone of inhibition.** The diameter of the zone of inhibition around the antibiotic disks is related to minimum inhibitory concentrations and minimum killing concentrations (fig. 29-10). Table 29-2 lists a number of antimicrobial agents and the data that indicate whether or not the agent would be effective against the organism tested. (The values in the table are for a specific strain of *Staphylococcus aureus* and give no information as to how another organism might respond.)

Although many hundreds of antimicrobial agents have been developed and extracted from microorganisms, and many of these are very effective against some organisms, there are still many infectious diseases that do not respond to treatment. One important reason is that some individuals have a depressed immune response and are therefore unable to eliminate the organisms, which may only be inhibited or reduced in number by the treatment. Another very commmon reason for persisitent infections is that a microorganism has become resistant to the antimicrobial agent.

One problem in the use of antimicrobial agents is that toxins may be released by the infecting agent when it is killed. In the case of septicemia by a gram negative bacterium such as *E. coli*, an antibiotic that kills all the organisms can lead to death because of an allergic reaction precipitated by the release of excessive endotoxin □. In some cases, the antimicrobial agent is unable to get into infected tissues at sufficiently high concentrations to kill

433

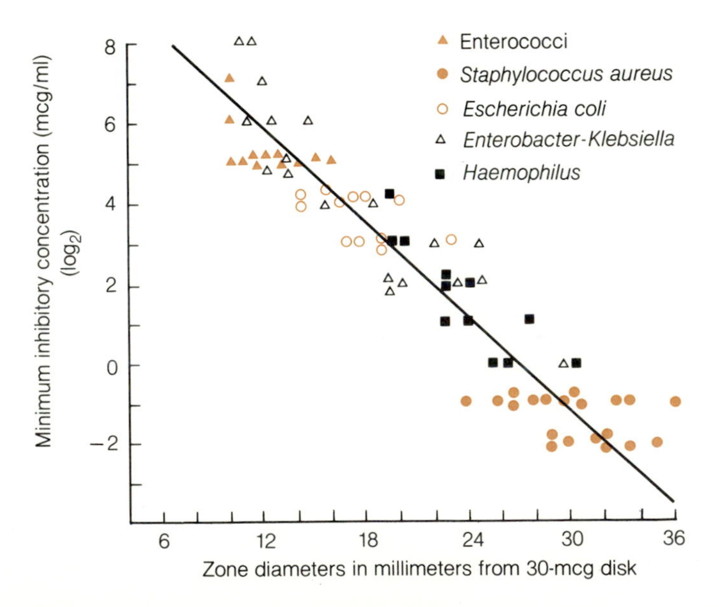

Figure 29-10 Relationship between the MIC and the Kirby-Bauer Test Results

TABLE 29-2

ZONES OF INHIBITION FOR CERTAIN ANTIBIOTICS USING *STAPHYLOCOCCUS AUREUS* AS THE TEST ORGANISM

ANTIMICROBIAL DRUG*	ZONE OF INHIBITION (mm) SENSITIVE	RESISTANT
Ampicillin (10 mcg disk)	>28	<21
Chloroamphenicol (30 mcg disk)	>17	<13
Colistin (10 mcg disk)	>10	< 9
Kanamycin (30 mcg disk)	>17	<14
Penicillin (10 unit disk)	>28	<21
Sulfadiazine (300 mcg disk)	>16	<13
Tetracycline (30 mcg disk)	>18	<15

*Values in parentheses indicate the concentration of the antibiotic in the paper disk.

the organisms. Consequently, these antibiotics are not very useful. Acid urine, nucleic acids in pus, and large abscesses are factors that interfere with the activity of antimicrobial agents.

DRUG RESISTANCE

Because of the extensive use of many antibiotics all over the world, numerous bacteria are becoming resistant to one or more antibiotics. In the 1940s, when penicillin was first introduced to treat gonorrhea, a single dose of 200,000 units of penicillin was sufficient to clear up all *Neisseria gonorrhoeae* infections. In the mid-1950s, partially resistant strains began to appear which required higher doses of penicillin. By 1975, 4.8 million units of pen-318icillin were required to eliminate most *Neisseria gonorrhoeae* □. A few outbreaks also occurred in which the bacteria were completely resistant to penicillin. The source of these penicillin-resistant strains is found in those countries were penicillin has been used as a prophylactic and as a treatment for many different infections. Penicillinase-producing *Neisseria gonorrhoeae* is endemic in many countries that have large camps of foreign military or businessmen and the associated prostitution. Penicillinase-producing *Neisseria gonorrhoeae* have been isolated in Korea (U.S. military), the Philippines (U.S. military), Vietnam (U.S. military), Japan (U.S. military), Hong Kong (British military and international businessmen), Singapore (British military and international businessmen), Australia, Great Britain, Norway, Canada, and the United States. Spectinomycin is often used in penicillin-resistant cases, but unfortunately some spectinomycin-resistant *Neisseria gonorrhoeae* have also been reported in Great Britain and in the U.S.

Similarly, many drug-resistant *Neisseria meningitidis* strains are appearing. Over 70% of the *N. meningitidis* now isolated are resistant to the sulfa drugs, which are the antimicrobial agents of choice because they work better than penicillin against meningitis. Since the 1960s, more and more cases have had to be treated with penicillin. It is feared that penicillin drug resistance will be transferred from *N. gonorrhoeae* to *N. meningitidis*. Since penicillin is one of the few drugs that works on *N. meningitidis*, an epidemic of untreatable meningitis is expected.

Since 1972, the bacterium *Haemophilus influenzae*, which causes infant meningitis at a rate of 40,000 cases/yr, has been found to be resistant to penicillin (ampicillin and methicillin) because of a penicillinase it produces. *Haemophilus influenzae* is also responsible for 30% of all middle ear infections (otitis media) in children. Since about 95% of all children have at least one ear infection before the age of 5, this bacterium accounts for a great deal of suffering. Ear infections can be very serious, because about one-third of all chronic cases lead to some permanent hearing loss. In the early 1980s, about 40% of *H. influenzae* strains checked turned out to be resistant to ampicillin.

Multiple drug resistance is being observed more frequently in bacteria, especially among the gram negative bacteria such as *Escherichia* and *Pseudomonas*. This multiple drug resistance is due to genes that are carried on plasmids □, often referred to as **R plasmids** because they confer drug resistance. The genes that make an organism resistant to a drug or antibiotic generally code for an enzyme that modifies the antimicrobial agent. Mutant **permeases** and **drug resistant enzymes** are common strategies for creating a resistant state. Many soil microorganisms normally carry drug-resistant genes to protect themselves from antibiotics in the soil and also from antibiotics they may produce. For example, *Bacillus circulans* produces the aminoglycoside antibiotic **butirosin,** as well as an enzyme that inactivates butirosin and several other aminoglycosides such as neomycin. Plasmids specifying penicillinase have been found in *Staphylococcus* strains isolated before the use of penicillin. Similarly, plasmids conferring resistance to tetracycline are found in several species of soil bacteria. When plasmids were first discovered, they specified resistance to one or two antibiotics. Now, however, many confer resistance to as many as 10 different antibiotics, drugs and heavy metals.

219

MECHANISMS OF DRUG ACTION AND RESISTANCE

Chemicals and antibiotics are characterized as to whether they affect viruses, bacteria, fungi, protozoans, or metazoans. In recent years, the modes of action of many chemicals and antibiotics have been worked out. Table 29-3 lists some of the ways in which chemicals and antibiotics affect microorganisms. As you can see, bacteria can be inhibited or killed by blocking DNA replication (naladixic acid, novobiocin), RNA synthesis (rifampin, actinomycin D), or protein synthesis (streptomycin, chloramphenicol, tetracycline); by disrupting membranes (valinomycin, gramicidin); by poisoning electron transport chains (rotenone, antimycin A); by poisoning ATPases (rutamycin); by blocking cell wall synthesis (cycloserine, bacitracin, vancomycin, penicillin); or by interfering with a specific step in metabolism (sulfanilamide). Although many of the drugs and antibiotics used to treat infections have been claimed to be "wonder drugs," there is potential danger in their improper use. Many of the "wonder drugs" can produce very serious side effects. For example, some enter mitochondria and inhibit the bacterial-like metabolism in these organelles.

Viral infections are generally untreatable with drugs, because most of the steps in a virus infection involve cellular components. If one attempts to inhibit some step in virus multiplication, the cell is disrupted also. The drugs that have been useful in treating some virus infections are analogues of nu-

TABLE 29-3
MODE OF ACTION OF SELECTED ANTIMICROBIAL DRUGS

NAME OF ANTIMICROBIAL	MODE OF ACTION: INHIBITS CELL WALL SYNTHESIS	INHIBITS PROTEIN SYNTHESIS	INHIBITS NUCLEIC ACID SYNTHESIS	INHIBITS ENZYME ACTIVITY	DAMAGES MEMBRANE	POISONS ELECTRON TRANSPORT SYSTEM
Penicillin	X					
Cephalosporin	X					
Cycloserine	X					
Ristocetin	X					
Bacitracin	X					
Vancomycin	X					
Chloroamphenicol		X				
Lincomycin		X				
Erythromycin		X				
Tetracycline		X				
Streptomycin		X				
Rifampin			X			
Rifamyoin B			X			
Nalidixic acid			X			
Trimethoprim			X			
Mitomycin			X			
Fluorocytocine			X			
Novobiocin			X			
Actinomycin D			X			
Sulfanilamide				X		
Isoniazid				X		
Nitrofuran				X		
Para-aminosalicylic acid				X		
Rutamycin				X		
Polymyxin B					X	
Valinomycin					X	
Nyastatin					X	
Amphotericin B					X	
Tryocidine					X	
Gramicidin					X	
Rotenone						X
Antimycin A						X

cleotides or chemicals that block DNA and RNA synthesis. These drugs inhibit virus DNA and RNA synthesis differentially because viruses replicate much more rapidly than eukaryotic cells. Drugs that inhibit virus uncoating and that block virus protein synthesis have also had limited usefulness.

Many fungal infections are particularly difficult to treat because of the toxicity of effective chemicals. Most chemicals that inhibit the fungi also inhibit the host's cells, since both are eukaryotic cells; however, a number of drugs have been found which inhibit and kill fungi differentially.

Antibacterial Drugs

Drugs That Inhibit Bacterial Cell Wall Synthesis

The penicillins kill growing bacteria by binding to and inhibiting enzymes that join the peptidoglycan polysaccharides into a rigid layer of the cell wall

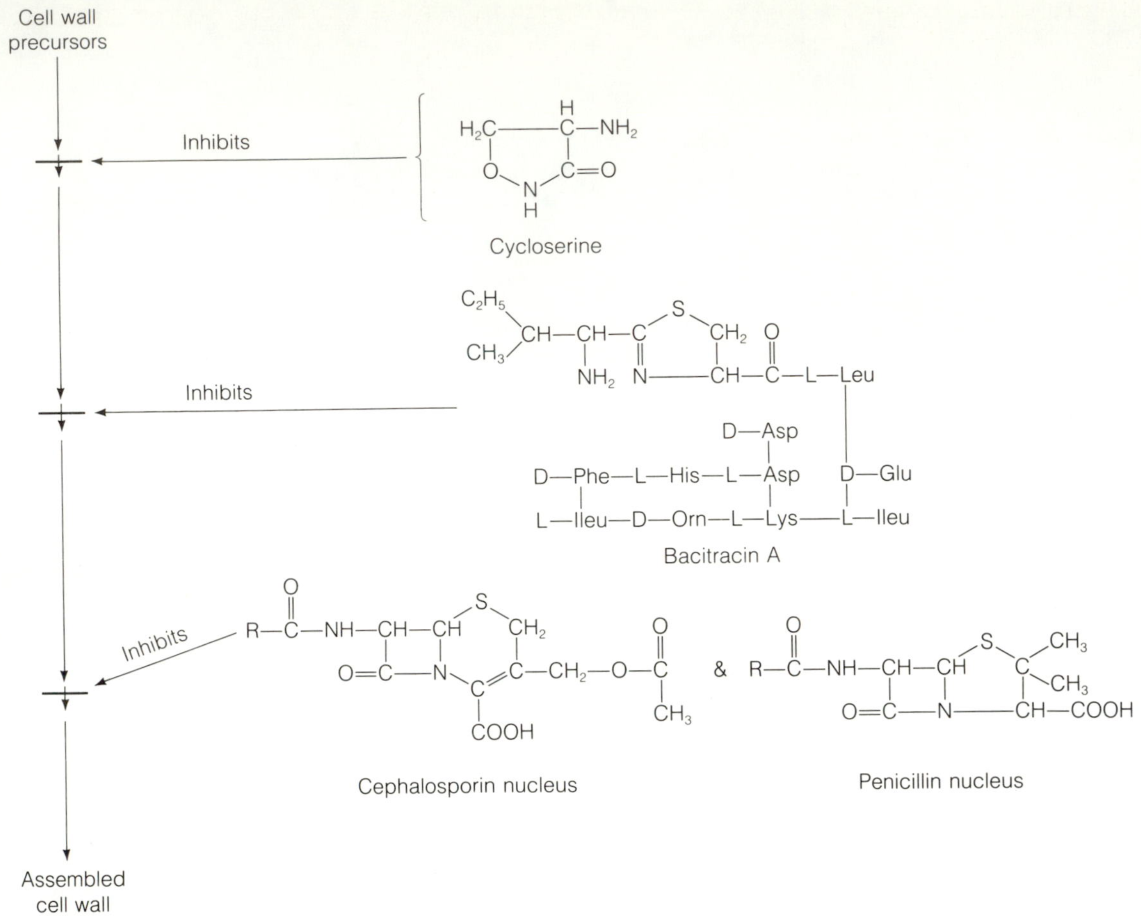

Figure 29-11 Effect of Antibiotics on Bacterial Cell Wall Synthesis

(fig. 29-11). Penicillin binds to approximately six different proteins in the bacterial membranes. Most of the penicillin binds to carboxypeptidase A and B, while the remainder binds to an endopeptidase, to the transpeptidase, and to other membrane proteins. Only the transpeptidase is involved in cross-linking peptidoglycan polysaccharides. Carboxypeptidases and endopeptidases are involved in modifying and breaking down the cross-linking peptides. If the peptidoglycan polysaccharides are not cross-linked in growing bacteria, autolysins (self-digesting enzymes) are activated, the cytoplasmic membrane balloons out through the weakened wall, and the membrane eventually gives way. Penicillin has no effect on stationary phase □ bacteria, since it does not cause the breakdown of the bacterial cell wall but only interferes with its synthesis. Penicillin also has no effect on plant or animal cells, because plant cells have very different enzymes operating in the synthesis of their cellulose walls, and animal cells lack cell walls.

One type of penicillin resistance occurs when the transpeptidase connecting peptidoglycan polysaccharides is replaced by a mutated enzyme that does not bind penicillin. The mutated transpeptidase continues to join the peptidoglycan polysaccharides together because it is not efficiently inhibited by penicillin. Mutated transpeptidases are believed to be responsible for most of the low-level resistance to penicillin. Tolerance to low levels of penicillin may also be due to missing autolysins or to overproduction of lipoteichoic acid polymers, which inhibit the autolysins.

Resistance to high levels of penicillin is due to the production of enzymes that modify or break down penicillin so that it is no longer capable of inhibiting transpeptidases. Enzymes that modify penicillin are called **penicillinases.** The most frequently encountered penicillinase is known as a **β-lactamase** because it cleaves the lactam ring of the penicillin molecule. The hydrolysis of penicillin G yields penicilloic acid, which has no antibacterial activity.

There are a number of antimicrobial agents that block cell wall synthesis in bacteria, in addition to penicillin (fig. 29-11 and table 29-3).

The antibiotic **cycloserine** (oxamycin) is bactericidal, like penicillin. Cycloserine blocks the synthesis of the pentapeptide necessary for peptidoglycan synthesis. Side effects of cycloserine include confusion and coma and occasionally liver damage, folate deficiency, peripheral nervous system disorders, and malabsorption syndrome.

Bacitracin is a polypeptide antibiotic that blocks the attachment of peptidoglycan polysaccharide to the membrane. It does this by inhibiting the dephosphorylation of a membrane phospholipid to which the peptidoglycan polysaccharide must attach (fig. 29-11). Bacitracin is most effective against gram positive bacteria, such as *Streptococcus* and *Staphylococcus,* but because it is so toxic to the kidneys it is used only topically, in creams and ointments to treat skin infections. Used internally, bacitracin may cause local pain at the injection site, gastrointestinal disturbances, and renal damage.

Vancomycin and **ristocetin** block the transfer of the peptidoglycan polysaccharide from the membrane phospholipid to the peptidoglycan layer of the wall (fig. 29-11). Vancomycin frequently causes thrombophlebitis, chills, and fever, and occasionally it causes kidney damage and damage to the eighth cranial nerve when large doses are given or treatment lasts for more than 10 days.

Cephalosporins (fig. 29-11) are antibiotics produced by the fungus *Cephalosporium acremonium.* The cephalosporins block the final step of peptidoglycan synthesis in growing bacteria, but turn out to be bacteriostatic. These antibiotics are broad spectrum, affecting many gram positive and gram negative bacteria. Some of the derivatives can be taken orally, but others must be injected, because they are inactivated by stomach acid as is penicillin G. The cephalosporins do not cause allergic reactions in persons sensitized to penicillin, and consequently are useful for treating patients who are hypersensitive to penicillin or who are infected by a penicillin-resistant bacterium. The cephalosporins can cause thrombophlebitis, a serum sickness, and occasional gastrointestinal disturbances.

Antibiotics That Block Bacterial Protein Synthesis

One of the best studied groups of antibiotics includes those that block protein synthesis in bacteria (table 29-3). Most of these antibiotics are synthesized by various species of soil bacteria in the genus *Streptomyces.* Most of these antibiotics bind to some component of the bacterial ribosome and thereby inhibit a specific step in protein synthesis (fig. 29-12). Resistance to the antibiotics can arise in a number of ways: 1) an alteration of the cell's membranes can keep the antibiotic out of the cell; 2) an enzyme can alter the antibiotic so that it is unable to bind to the ribosome; or 3) the antibiotic

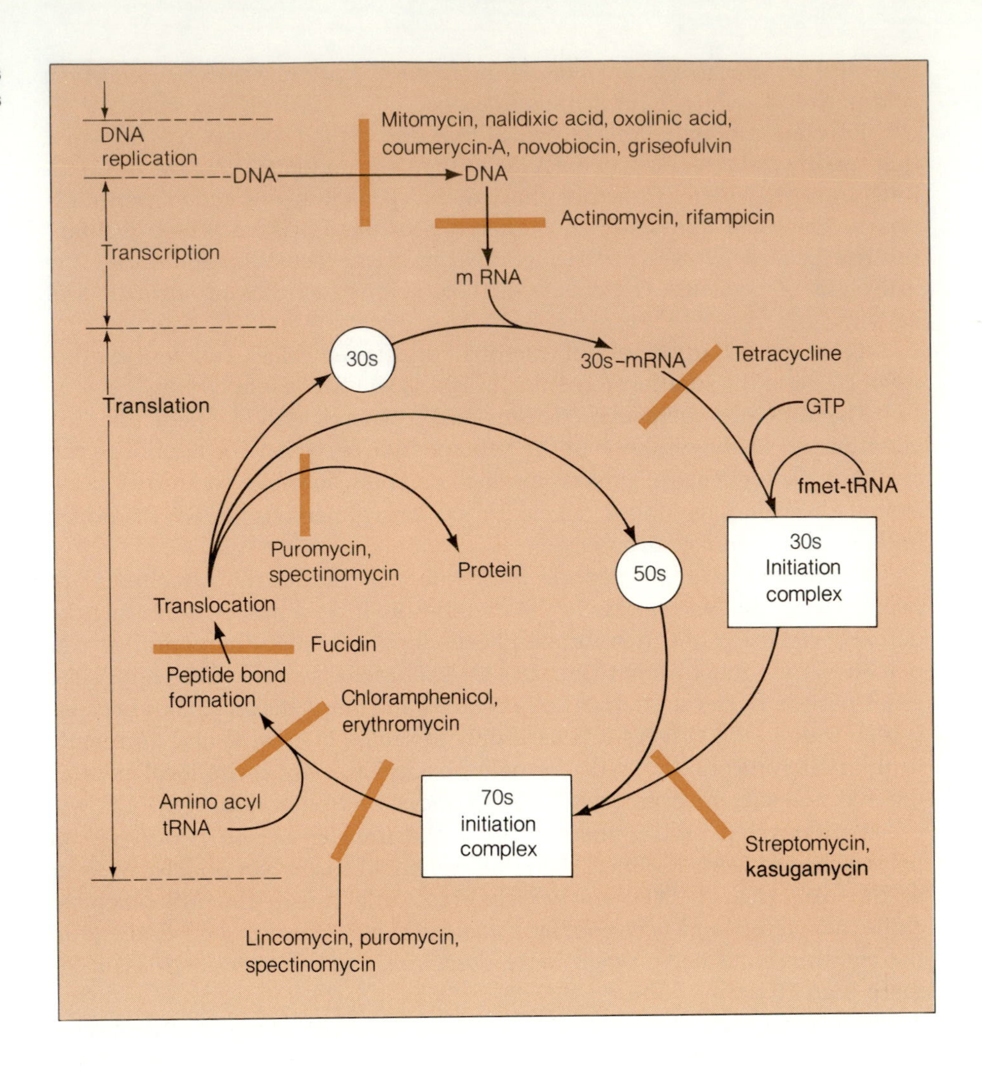

Figure 29-12 Mode of Action of Drugs That Affect Bacterial Protein Synthesis and Nucleic Acid Synthesis

binding site on the ribosome can be altered by mutations so that the antibiotic cannot bind.

Tetracycline, chlortetracycline (aureomycin), and **oxytetracycline** (terramycin) (see appendix) are believed to bind to the 30s subunit □ at the initiation of protein synthesis and block the attachment of the first aminoacyl-tRNA (N-formylmethionine-tRNA) to the 30s-mRNA complex (fig. 29-12). The tetracyclines are bacteriostatic. Resistance to these antibiotics occurs when membrane permeability is altered or when enzymes inactivate the antibiotics. The tetracyclines may cause gastrointestinal disturbances and bone lesions, as well as deformed and stained teeth in children up to 8 years of age. Newborn infants frequently have bone lesions and stained teeth when the antibiotics are given to the mother after the fourth month of pregnancy. 212

Spectinomycin binds to the 30s subunit of the ribosome before the 70s ribosome forms. It blocks aminoacyl-tRNA binding and consequently blocks the formation of a peptide. Protein synthesis initiated in the absence of spectinomycin is not inhibited, however the antibiotic is bacteriostatic rather than bactericidal because it dissociates, with a half-life of only 2 minutes. Spectinomycin resistance occurs when one of the proteins in the 30s subunit mutates.

212

Kasugamycin binds to the 16s rRNA in the 30s subunit and consequently prevents the formation of the 70s initiation complex □ (fig. 29-12). The antibiotic has no effect on translating ribosomes. Resistance to the antibiotic occurs when the 16s rRNA remains unmethylated. Thus, a defective rRNA methylase is responsible for kasugamycin resistance.

Streptomycin, neomycin, kanamycin, paromycin, and **gentamicin** are known as **aminoglycoside** antibiotics because they contain a number of glycosidic bonds. These antibiotics bind to a number of different proteins in the 30s ribosomal subunit. For example, streptomycin binds strongly to one part of the 30s subunit, causing extensive misreading by the 70s ribosome. It also causes false initiation of protein synthesis at internal sites in the mRNA. At high concentrations, streptomycin binds to a second site on the 30s subunit and blocks all initiation complexes on the mRNA. Resistance to the antibiotic is due to mutations in one of the 30s proteins and to enzymes that modify the antibiotic so that it is unable to bind to the 30s subunit. Enzymes such as acetyltransferases, phosphotransferases, and adenyltransferases, which add acetyl groups, phosphate, and nucleotides to these antibiotics, have been identified.

The aminoglycosides are bactericidal because they remain tightly attached to the bacterial ribosome long after levels of antibiotic drop. Even one ribosome blocking an mRNA can completely block translation. Streptomycin is bactericidal for many gram negative and gram positive bacteria; in particular, it kills *Mycobacterium tuberculosis*. The antibiotic is ineffective against anaerobic bacteria, however, and is very poorly absorbed when taken orally and consequently must be injected. One drawback to streptomycin is that it has a cumulative effect when given for long periods of time. It can destroy nerves in the inner ear (eighth cranial nerve), affecting balance and hearing. Occasionally there is renal damage.

Chloramphenicol binds to the 50s subunit of the 70s ribosome and inhibits the enzyme called **peptidyltransferase,** which catalyzes the formation of peptide bonds between amino acids. Resistance is due to a mutated 50s protein that blocks chloramphenicol binding; to enzymes that modify chloramphenicol; or to an altered membrane permeability. A bacterial acetyltransferase that adds acetyl groups to chloramphenicol makes an organism resistant to the antibiotic. Chloramphenicol is a bacteriostatic agent against a broad spectrum of bacteria. It is the antibiotic most frequently used to treat difficult cases of typhoid fever. Chloramphenicol has been reported to cause occasional gastrointestinal disturbances, blood clumping, and **gray syndrome** in infants.[1]

Erythromycin binds to the 50s ribosomal subunit, apparently to the peptidyltransferase, and blocks the formation of peptide bonds (fig. 29-12). The antibiotic is bacteriostatic and is most active against gram positive bacteria. It is frequently used as an alternative to penicillin, when penicillin-resistant organisms are involved or when the patient is allergic to penicillin. Erythromycin has been reported to cause mild damage to the liver, gastrointestinal disturbances, and transient hearing loss at high doses or when used for prolonged periods.

1. Vomiting, lack of sucking response, rapid and irregular respiration, cyanosis and abdominal distension in newborn infants treated with Chloroamphenicol. Infants appear ashen-gray in color. Death occurs in 40% of the patients.

Drugs and Antibiotics That Inhibit Bacterial DNA Replication

A number of drugs and antibiotics have been discovered which specifically block DNA replication (fig. 29-12, table 29-3).

Four drugs are known to affect bacterial DNA gyrase, the enzyme involved in unwinding DNA. They are **nalidixic acid, oxolinic acid, novobiocin,** and **coumerycin-A.** Nalidixic and oxolinic acids are synthetic chemicals that bind to the α-subunit of bacterial DNA gyrase. The binding of these chemicals inhibits the activity of the enzyme, which is required for bacterial DNA synthesis. Resistance to nalidixic and oxolinic acids occurs when the α-subunit undergoes a mutation, and also when the transport system that brings the drugs into the cell is altered. Naladixic acid is most effective against gram negative enteric bacteria and is used to treat urinary tract infections. Unfortunately, however, it sometimes causes gastrointestinal disturbances, visual problems, and a rash, and occasionally it induces convulsions, central nervous system disturbances, and hyperglycemia (high blood sugar). Oxolinic acid can cause central nervous system disturbances, nausea, vomiting, dizziness, insomnia, and blurred vision. Novobiocin and coumerycin-A bind to the β-subunit of DNA gyrase and inhibit the enzyme's activity. Novobiocin treatment results in side effects that include gastrointestinal problems, jaundice, neonatal hyperbilirubinemia, and allergic reactions. Occasionally, severe blood clumping is induced by novobiocin.

Drugs That Block Bacterial RNA Synthesis

Some drugs and antibiotics specifically block RNA synthesis □ in bacteria (table 29-3). **Rifampicin** (rifampin) is a semisynthetic antibiotic produced from the antibiotic rifamycin B. It binds to the β-subunit of bacterial RNA polymerases and inhibits the initiation of RNA synthesis (fig. 29-12). Resistant bacterial mutants arise when the β-subunit of the RNA polymerase undergoes an appropriate mutation. Rifampin is a broad spectrum antibiotic that is effective against *Mycobacterium* and that can be taken by mouth. Occasionally, it induces gastrointestinal disturbances, liver damage, and allergic reactions.

207

Drugs That Block Bacterial Metabolism

The **nitrofurans,** such as **nitrofurantoin,** are nitrated furfural derivatives that act as broad spectrum antimicrobial agents. They are effective against many different bacteria and some protozoans and fungi. Unfortunately, they are also toxic to animals and humans. The nitrofurans are thought to interfere with reactions involving acetyl coenzyme A □, and because of this effect they inhibit a number of steps in metabolism (table 29-3). Nitrofurantoin frequently causes gastrointestinal disturbances and allergic reactions, and occasionally hemolytic anemia, blood clumping, peripheral nervous system problems, and autoimmune reactions. Sometimes severe pulmonary fibrosis occurs.

193

Para-aminosalicylic acid (table 29-3) is a synthetic drug that inhibits the synthesis of folic acid by competing with para-aminobenzoic acid. Para-aminosalicylic acid and the antibiotic streptomycin are frequently used to treat tuberculosis. Para-aminosalicylic acid occasionally causes gastrointestinal disturbances, allergic reactions, liver damage, blood clumping, renal irritation, thyroid enlargement, and malabsorption syndrome.

Isonicotinic hydrazide (or isoniazid) is a synthetic drug that resembles pyridoxin (vitamin B6). Because it is an analogue of vitamin B6, isonicotinic hydrazide interferes with some of the reactions that require this vitamin (table 29-3). It is believed that it may also be incorporated into NAD or NADP. Isonicotinic hydrazide is bacteriostatic for *Mycobacterium tuberculosis*, but has little effect on other bacteria. This drug is generally used in conjunction with streptomycin and rifampin to treat tuberculosis in humans. Unfortunately, it occasionally causes a hepatitis that may be fatal. In addition, isonicotinic hydrazide is responsible for gastrointestinal disturbances, allergic reactions, and fever.

Drugs That Affect Bacterial Membranes

80
A number of drugs and antibiotics interact with bacterial membranes □ and thereby kill the bacteria (table 29-3). **Polymyxin** antibiotics, produced by *Bacillus polymyxa*, are used to treat topical infections by gram negative bacteria, in particular *Pseudomonas*. Polymyxin B destroys bacteria by interacting with the envelope and cytoplasmic membrane, disrupting transport and ionic balance. The antibiotic is bactericidal and very toxic to animals. It can cause kidney and brain damage when taken internally.

TREATMENT OF INFECTIONS CAUSED BY EUKARYOTES

The treatment of infections caused by eukaryotes requires a completely different group of drugs and antibiotics than those used to treat bacterial infections (table 29-4). The reason is that the most useful antibacterial agents are those that specifically interfere with prokaryotic metabolism and have little effect on eukaryotic organisms, except when used at toxic concentrations. The drugs and antibiotics that are useful against fungi and protozoans are those that affect these organisms more adversely than they do the patient's cells (table 29-5).

Nystatin is an antibiotic used to treat nonsystemic fungal infections. This fungicidal agent is a **polyene** antibiotic that interacts with sterols in the cytoplasmic membrane of many fungi, disrupting the ionic balance within the cells. The leakage of Ca^{++} into the cells and of K^{+} and Mg^{++} out of the cells drastically alters cell physiology so that RNA and protein synthesis is blocked. Some algae, protozoans, and mammalian cells are also sensitive to nystatin; however, there seems to be little toxicity associated with nystatin when it is used to treat intestinal, vaginal, or superficial candidiasis. The discovery of nystatin was reported in the 1950s by Elizabeth Hazen and Rachel Brown. Polyenes have no effect on bacteria, except for some of the mycoplasmas that contain sterols.

Another polyene antifungal agent is **amphotericin B.** The drug is believed to disrupt fungal cells in the same manner as nystatin. It is used to treat superficial and systemic candidiasis, as well as systemic aspergillosis, coccidioidomycosis, histoplasmosis, blastomycosis, and cryptococcosis. Unfortunately, it has a low theraputic ratio and causes damage to kidney tissue and to blood-forming □ tissues. Symptoms due to the treatment include nausea, vomiting, abdominal pain, convulsions, anemia, and cardiac arrest. Because of the toxicity of this drug, it is necessary that the patient be hospitalized for treatment.
459

Griseofulvin is a fungistatic antibiotic and is generally used to treat fungal skin infections, such as ringworm. It is taken orally, but must be taken for many weeks in order to eliminate the fungus. Griseofulvin is

TABLE 29-4
ANTIMICROBIALS EFFECTIVE AGAINST EUKARYOTES AND VIRUSES

ANTIMICROBIAL DRUG	MICROORGANISMS AFFECTED			
	VIRUSES	FUNGI	PROTOZOA	HELMINTHS
Amantidine	X			
Vidarabine	X			
Acyclovir	X			
Interferon	X			
Idoxuridine	X			
Methisazone	X			
Cytosine arabinoside	X			
2-deoxy-d-glucose	X			
Griseofulvin		X		
Mycostatin		X		
Ketoconazole		X		
Myconazole		X		
Natamycin		X		
Clotrimazole		X		
Tolnaftate		X		
Haloprogin		X		
Amphotericin B		X		
Quinine			X	
Chloroquine			X	
Amodiaquine			X	
Dihydroemetine			X	
Diiodohydroxyquin			X	
Melarsoprol			X	
Nifurtimox			X	
Primaquine			X	
Quinacrine			X	
Stibogluconate			X	
Suramin			X	
Emetine			X	
Metronidazole			X	
Niclosamine				X
Pyrivinium				X
Mebendazole				X
Pyrantel pamoate				X
Piperazine citrate				X
Avermectin				X

thought to accumulate within the fungal cell, where it disrupts components of the cytoskeleton, causing aberrant cell growth and mitosis. It is mildly toxic, occasionally causing fatigue, gastrointestinal disturbances (such as nausea, vomiting, and diarrhea), allergies, skin lesions, and photosensitive reactions.

Tolnaftate is a synthetic antifungal agent first introduced in 1965 for the treatment of athlete's foot (tinea pedis), jock itch (tinea cruris), and body ringworm (tinea corporis). Since this chemical is used topically, its toxic effects are very mild.

Bacteria and fungi are not the only living forms that can produce antimicrobials. The tropical plants **ipecacuanha** and **cinchona** produce emetine and quinine, respectively. Since very early times, American Indians

TABLE 29-5
ADVERSE REACTIONS TO ANTIMICROBIAL AGENTS

ADVERSE REACTION	DRUGS THAT MAY BE INVOLVED
Hypersensitivities (skin eruptions, fever, and anaphylaxis) and immunothrombocytopenias	Penicillins, streptomycin, isoniazid, sulfonamides, stibophen, chloroquine
Kidney dysfunction	Aminoglycosides, polymyxins, amphotericin B
Damage to bone marrow	Chloroamphenicol, amphotericin B.
Damage to nervous system	Streptomycin, gentamycin, neomycin, kanamycin, amikacin, aminoglycosides
Liver damage	Isoniazid, rifampin, PAS, pyrazinamide
Diarrheas and enterocolitis	Erythromycin, clindamycin, tetracycline, chloroamphenicol, neomycin

treated malaria simply by chewing the bark of the cinchona tree, and Europeans as early as the 1800s extracted quinine from cinchona bark to treat 699 malaria □.

Chloroquine, a synthetic derivative of quinine, was produced in 1934 for treatment of malaria. It kills the malaria parasite, apparently by blocking DNA and RNA synthesis and by creating frame shift mutations in the mRNA. Chloroquine is used as a prophylactic drug to avoid malaria when traveling through infested areas. A dose is taken weekly, beginning two weeks before departure and continuing for six weeks after leaving the infested area. *Plasmodium* residing in the liver must be treated with **primaquine phosphate** (table 29-4). Chloroquine is also effective against *Entamoeba histolytica,* which causes amebic abscesses and dysentery. Quinine causes headache, blurred vision, ringing in the ears, and blood disorders. Chloroquine causes vomiting, blurred vision, hair loss, and anemia, while primaquine phosphate may cause nausea and blood disorders. **Emetine** and **dihydroemetine** are used to treat *Entamoeba histolytica* infections, which cause amebic abscess and amebic dysentery.

TREATMENT OF VIRUS INFECTIONS

Most virus infections are difficult to treat with antimicrobial agents, because many of the enzymes used in their replication are host enzymes. Thus, a good viricidal or viristatic agent must inhibit a process or enzyme that is associated only with the virus. So far, most of the antiviral agents that have been developed (table 29-4) block DNA or RNA synthesis differentially, because the viruses replicate much more rapidly than the host cells. These drugs are generally very toxic to the patient, however.

Vidarabine is a nucleoside that resembles adenosine but has an attached sugar called arabinose (adenine arabinoside). The antibiotic is isolated from *Streptomyces antibioticus.* Vidarabine binds to the DNA polymerase used by the virus and inhibits its activity. It is most effective against herpes that causes encephalitis. In treating encephalitis, vidarabine is taken intravenously and is able to reduce the incidence of death by about 50 percent. Vidarabine is also used topically for treating herpes infections of the cornea. Vidarabine occasionally causes weakness, thrombophlebitis, and gastrointestinal disturbances that include nausea and vomiting.

Methisazone is a thiosemicarbazone that inhibits the assembly of pox viruses by interfering with protein synthesis. It frequently causes severe nausea and vomiting as well as weight loss. Diarrhea and allergic reactions occur occasionally.

Idoxuridine is a synthetic antiviral nucleoside, also called 2′-deoxy-5-iodouridine, which is incorporated into some viral DNAs instead of thymidine. This substitution inhibits the replication of viral DNA and results in noninfective DNA. The idoxuridine is used topically to treat herpes simplex I infections of the eye. It is ineffective on herpes simplex II, which is the cause of most genital infections. Idoxuridine also appears to have little effect on recurrent herpes infections. Some persons treated with idoxuridine develop edema of the eyes or eyelids, allergic reactions, photophobia, and occasional corneal clouding. Idoxuridine causes birth defects in animals when used on the pregnant female's eyes. Therefore, it should not be used on pregnant women.

Acyclovir (Zovirax) is a synthetic antiviral agent released in 1982 to treat genital herpes, which is estimated to afflict 20 million persons in the U.S. Acyclovir can also be used to treat herpes infections of the eye. Oral acyclovir and intravenous injection inhibit herpes simplex II replication, shorten the healing time and reduce pain. The drug attacks the virus selectively and so is not toxic to the patient. Unfortunately, the drug does not appear to eliminate the recurrent infections of the virus.

FOCUS

SYMMETREL FIGHTS INFLUENZA

Influenza, a disease that is responsible for the death of many thousands of elderly persons each year, can be prevented and treated with the synthetic drug Symmetrel (amantadine) (fig. 29-13). Studies indicate that Symmetrel is very effective in preventing influenza A infections. In fact, it appears that this drug may be more effective than the flu vaccines, because it affects all influenza A type viruses regardless of their surface antigens. Symmetrel prevents the influenza virus from penetrating cells, and so blocks the spread of the virus. Some researchers claim that vaccines protect only about 50% of those immunized, while Symmetrel protects about 90% of those taking the drug.

There are minor adverse reactions to Symmetrel. Seven to 10% of the persons taking the drug complain of insomnia, jitteriness, slurred speach, difficulty in concentrating, lethargy, and dizziness. These symptoms diminish, however, after the drug is used for a number of days and disappear after the drug is discontinued.

Doctors familiar with Symmetrel recommend that influenza vaccination be coupled with the use of 200 mg/day of the drug at the first signs of the flu, in order to prevent the thousands of deaths each year. Although Symmetrel has been produced since the early 1970s, it has not been very popular with doctors.

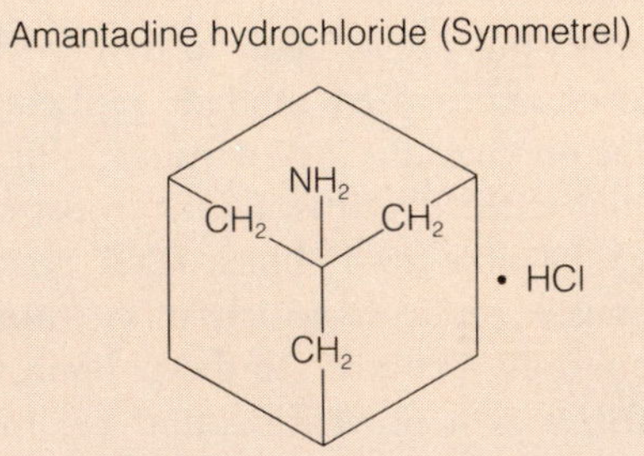

Amantadine hydrochloride (Symmetrel)

NH_2 CH_2 CH_2 CH_2 • HCl

Figure 29-13 The Structure of Amantidine Hydrochloride (Symmetrel)

SUMMARY

PERSPECTIVE

1. Paul Ehrlich discovered an arsenic compound called salvarsan that was effective against *Treponema pallidum*, the bacterium that causes syphilis.
2. Gerhard Domagk is credited with the discovery of prontosil, a dye combined with a sulfonamide. In animals, prontosil is cleaved into the dye and sulfanilamide.
3. Sulfonamides block the synthesis of the coenzyme folic acid and consequently inhibit those organisms that synthesize the coenzyme.
4. Sulfonamides are active against a number of bacteria.
5. Selman Waksman discovered that various species of the soil bacterium *Streptomyces* produced antibiotics, such as streptomycin, erythromycin, and neomycin. These antibiotics block protein synthesis in bacteria.
6. Alexander Fleming discovered that the fungus *Penicillium* produced an antibiotic he called penicillin. This antibiotic blocks cell wall synthesis in gram positive bacteria.

FUNDAMENTALS OF CHEMOTHERAPY

1. The best antimicrobial agents have a selective toxicity (they harm only the parasite) and a high therapeutic ratio (they can be used at high concentrations without harm to the patient, and low concentrations are effective against the pathogen).
2. Narrow spectrum antimicrobials affect only a small group of organisms, while broad spectrum antimicrobials affect a large number of different organisms.

DETERMINATION OF DRUG AND ANTIBIOTIC SENSITIVITIES

1. The minimum inhibitory concentration (MIC) of an antimicrobial agent can be determined from a tube dilution test. The MIC is the lowest concentration of an antimicrobial that does not allow growth.
2. The minimum killing concentration (MKC) is the lowest concentration that kills an organism.
3. In the Kirby-Bauer test, paper disks with various concentrations of antimicrobials are placed on a lawn of bacteria in order to determine the organism's sensitivity.
4. The zone of inhibition around antimicrobial disks is related to the *in vivo* minimum killing concentration.

DRUG RESISTANCE

1. Some strains of *Neisseria gonorrhoeae* are completely resistant to penicillin because they produce an exoenzyme called penicillinase. *Neisseria gonorrhoeae* are also becoming resistant to spectinomycin.
2. *Neisseria meningitidis* strains resistant to sulfa drugs are appearing. Sulfa drugs and penicillin are the most effective drugs against *Neisseria meningitidis.*
3. *Haemophilus influenzae* strains resistant to ampicillin are frequently being isolated.
4. Multiple drug resistance is due to genes carried on drug-resistant plasmids called R plasmids.
5. The extensive use of antibiotics all over the world has apparently selected for organisms that have R plasmids with multiple drug-resistant genes.

MECHANISMS OF DRUG ACTION AND RESISTANCE

1. DNA replication in bacteria is blocked by naladixic acid and novobiocin.
2. RNA replication in bacteria is blocked by rifampin and actinomycin D.
3. Protein synthesis in bacteria is inhibited by streptomycin, chloramphenicol, and tetracycline.
4. Membrane potentials are disrupted by valinomycin, gramicidin, and polymyxins.
5. The electron transport system is blocked by rotenone and antimycin.
6. ATPase activity is inhibited by rutamycin.
7. Cell wall synthesis in bacteria is inhibited by cycloserine, bacitracin, vancomycin, and penicillin.

TREATMENT OF INFECTIONS CAUSED BY EUKARYOTES

1. Nystatin and amphotericin B are polyene antibiotics used to kill fungi.
2. Griseofulvin is an antibiotic used extensively to treat fungal skin infections.
3. Tolnaftate is a synthetic antifungal agent used to treat athlete's foot, jock itch, and ringworm.
4. Quinine is one of the oldest antibiotics used by man to treat malaria.
5. Chloroquine is a synthetic derivative of quinine that blocks DNA and RNA synthesis in *Plasmodium* (malaria parasite) and in *Entamoeba* (amoebic dysentery parasite).

TREATMENT OF VIRUS INFECTIONS

1. Most drugs used to treat virus infections are analogues of nucleotides and consequently inhibit DNA and RNA replication.

2. Idoxuridine and acyclovir are used to treat herpes simplex I and II infections. Acyclovir appears to be the more effective of the two drugs.

STUDY QUESTIONS

1. What is salvarsan? Which diseases were treated with it? What problems were there with salvarsan? Who is credited with its discovery?
2. How do sulfonamides inhibit bacteria? Sulfonamides have no effect on what type of bacteria?
3. How does streptomycin inhibit bacteria? By what organism is it produced?
4. How does penicillin inhibit bacteria? What organism produces penicillin?
5. Explain the difference between an antibiotic and a man-made antimicrobial.
6. What was one of the first antibiotics used by humans?
7. What contributions did the following persons make to the science of chemotherapy? Paul Ehrlich, Gerhard Domagk, Rene Dubos, Selman Waksman, Alexander Fleming.
8. Is a drug that shows selective toxicity a good one or a poor one? Explain.
9. Is a drug that has a high therapeutic ratio a good one or a poor one? Explain.
10. What is the difference between a bacteriostatic and a bactericidal agent?
11. Explain what a broad spectrum antimicrobial is.
12. Explain how the minimum inhibitory concentration (MIC) of an antimicrobial is determined.
13. Explain how the minimum killing concentration (MKC) is determined.
14. How is it determined whether an antimicrobial is bacteriostatic or bactericidal?
15. Describe a typical Kirby-Bauer test.
16. What is one reason that penicillin-resistant *Neisseria* are increasing?
17. Which antibiotic is used to treat penicillin-resistant *Neisseria?*
18. Explain what multiple drug resistance is and what is causing it in bacteria.
19. How do the penicillins kill bacteria?
20. Describe the two ways that a bacterium can become resistant to penicillin.
21. Compare and contrast the action of cephalosporins and penicillin. Which type of organism produces cephalosporins?

SUPPLEMENTAL READINGS

American Medical Association. 1971. *AMA drug evaluation 1971.* 1st ed. Chicago: American Medical Association.

Berdy, J. (ed). 1982. *Handbook of Antibiotic Compounds.* Boca Raton, Florida: CRC Press Inc.

Bopp, C. A., Birkness, K. A., Wachsmuth, I. K., and Barrett, T. J. 1985. In vitro antimicrobial susceptibility, plasmid analysis, and serotyping of epidemic-associated *Campylobacter jejuni. Journal of Clinical Microbiology* 21:4–7.

Cartwright, F. and Biddiss, M. 1972. *Disease and history.* New York: Thomas Y. Crowell Co., 248 pp.

Collier, J. R. and Kaplan, D. A. 1984. Immunotoxins. *Scientific American* 251(1):56–64.

Culliton, B. J. 1976. Penicillin-resistant gonorrhea: New strain spreading worldwide. *Science* 194:1395–1397.

Davis, B., Tai, P., and Wallace, B. 1974. Complex interactions of antibiotics with the ribosome. In *Ribosomes*, ed. Nomura, Tissieres, and Lengyel, 771–789. New York: Cold Spring Harbor Laboratory.

Gorini, L. 1974. Streptomycin and misreading of the genetic code. In *Ribosomes*, ed. Nomura, Tissieres, and Lengyel, 791–803. New York: Cold Spring Harbor Laboratory.

Klein, J. O. 1975. Shifts in microbial sensitivity: Implications for pediatrics. *Hospital Practice* 10:81–88.

Langone, J. 1985. AIDS: The quest for a cure. *Discover* 6:75–77.

Maugh, T. H. 1981. A new wave of antibiotics. *Science* 214:1225–1228.

Merigan, T. C. 1985. Viral vaccines, immunoglobulins, and antiviral drugs. *Scientific American Medicine* Section 7-XXXIII.

Neu, H. C. 1984. Changing mechanisms of bacterial resistance. *The American Journal of Medicine* Section 77(1B):11–24.

Sobell, H. M. 1974. How actinomycin binds to DNA. *Scientific American* 231:82–91.

PART IV APPLIED AND ENVIRONMENTAL MICROBIOLOGY

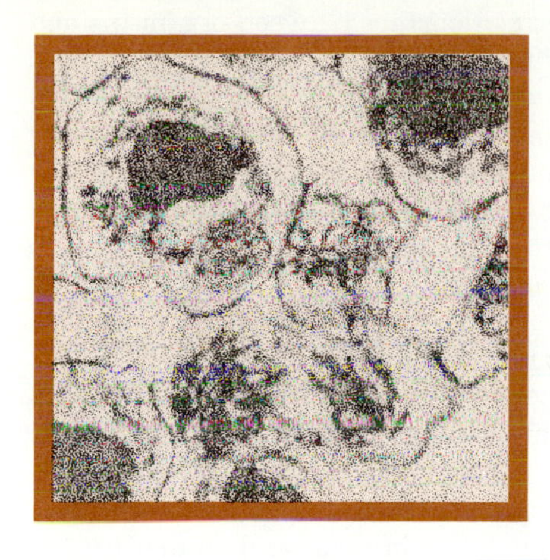

CHAPTER 30 INTRODUCTION TO ENVIRONMENTAL MICROBIOLOGY

CHAPTER PREVIEW

CAN TERRESTRIAL LIFE FORMS COLONIZE MARS?

In August 1975, the Viking 1 and 2 spacecrafts began transmitting data about Mars to Earth. Scientists at NASA-Ames Research Center, Moffett Field, California, examined the data to determine whether the Martian environment could support humans and other terrestrial life forms, and to develop a model Martian community that could make Mars habitable by humans.

Scientists found that the present Martian atmosphere has no oxygen and very little water vapor, but some carbon dioxide. The water on the planet is mostly ice in the polar ice caps and in the permafrost (frozen water below the surface of the soil). The average Martian temperature is well below freezing (0°C) and stays that way for much of the Martian year.

It is also known that the Martian surface is heavily bombarded with ultraviolet radiation and whipped by winds exceeding 200 miles per hour (100 meters/second). In essence, the present Martian environment is quite hostile to terrestrial life and in some ways resembles the conditions encountered in dry valleys in Antarctica on the earth.

Despite these harsh conditions, it was generally agreed by the researchers at NASA-Ames Research Center that certain terrestrial microorganisms could reproduce in the Martian environment. Furthermore, it was thought that these microorganisms could fill the atmosphere with oxygen, raise the ozone level high enough to filter out much of the ultraviolet irradiation, elevate the planet's temperature, and increase the availability of liquid water. It was their proposal that a **community** of terrestrial microorganisms could be seeded onto Martian soil which could change the Martian environment within 100,000 years so that humans could colonize the planet.

Which microorganisms would be suitable for life on Mars, and where on earth could they be found? Since the Martian environment resembles in some ways the dry valleys on Antarctica, the microbial communities reproducing in these terrestrial habitats could be likely candidates for the colonization of Mars.

Any pioneer Martian community must include photosynthetic organisms, such as the cyanobacteria or the lichens, because these organisms are not entirely dependent on oxygen for survival. They produce oxygen from their photosynthesis which would accumulate in the Martian atmosphere and create an aerobic environment. The accumulation of oxygen in the atmosphere would also result in the formation of an ozone layer that would shield the surface of the planet from irradiation. As the photosynthetic organisms multiplied, the amount of organic matter would accumulate on the planet's soils, providing nutrients for other pioneer populations of microorganisms. Once the organic matter had built up, other pioneer populations could recycle these nutrients and create a steady state equilibrium for biologically important chemical elements such as nitrogen, oxygen, carbon, and sulfur.

The studies conducted by these scientists, using computer-simulated models and laboratory experiments, were based on the fact that microorganisms can drastically alter the environment as a result of their metabolism. In this chapter, many of the concepts that served as foundations for this NASA project will be discussed. As you read this chapter, it will become obvious that without microorganisms life on the planet Earth would be impossible.

LEARNING OBJECTIVES

A STUDY OF THIS CHAPTER SHOULD ENABLE YOU TO:

DEFINE WHAT POPULATIONS AND COMMUNITIES ARE

DISCUSS THE CONCEPT OF AN ECOSYSTEM, AND CITE THE VARIOUS ELEMENTS THAT MAKE UP AN ECOSYSTEM

EXPLAIN THE ROLE THAT MICROORGANISMS HAVE IN THE FUNCTIONING OF ECOSYSTEMS

DESCRIBE SOME OF THE WAYS IN WHICH MICROORGANISMS INTERACT WITH HIGHER ORGANISMS IN A COMMUNITY

The science of **ecology** studies the interrelationships between organisms and their environment. The term "ecology," first coined by Ernst Haeckel in 1866, became popular during the 1960s, when people began to be aware of the effects that changes in the environment had on their lives. Biologists have been aware for many years of the central role microorganisms play in the environment. Microorganisms colonize and multiply on nearly every surface or material. As they multiply, they obtain nutrients from their immediate environment and in turn release metabolic wastes. These microbial activities are responsible for the many changes in the environment which ultimately affect the growth and health of other organisms.

The composition of the environment is not static, because some substances are used up by microorganisms during their multiplication while metabolic byproducts take their place. Some changes effected by microorganisms in animal and plant material are detrimental to the plant or animal. For example, when microorganisms grow on or in plant and animal tissues, they may alter the organism's metabolism and growth, causing disease. By and large, however, the growth of microorganisms in the environment is absolutely essential for the existence of all forms of life. It is this aspect of microbial activities that will be investigated in the ensuing chapters. In this chapter we will study the various facets of microbial ecology and the roles microorganisms play in our environment.

THE BIOSPHERE

The **biosphere** is that portion of the planet earth where living organisms can be found. It is a relatively thin layer around the earth that extends about 10 km into the atmosphere, to the depths of the ocean, and a few meters into the earth. The biosphere is subdivided into **biomes,** or areas of the biosphere such as the desert, the grasslands, and the oceans. Biomes, in turn, are composed of specific **communities** of organisms, and each community is made up of various **populations** (fig. 30-1).

Populations are aggregations of organisms of a single **species.** In essence, when you examine a bacterial colony growing on a plate of nutrient agar, you are actually examining a population of bacteria. This particular population, the colony □, arises from a single cell and is therefore made up of a single species (fig. 30-1).

143

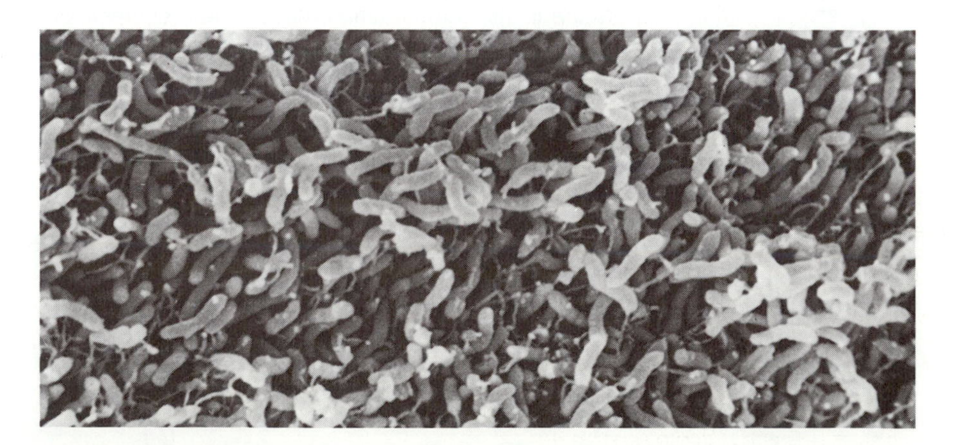

Figure 30-1 **A Bacterial Population**

The concept of a species is relatively simple to convey in multicellular organisms. It is generally agreed among biologists that a species is a group (or population) of organisms that closely resemble one another and are capable of interbreeding. Cats can breed with one another, giving rise to cat offspring, because all these animals are in the same species, *Felis catus*. In microbial populations, particularly bacteria, this species concept is difficult to apply because many bacteria that differ significantly from each other can nevertheless exchange genetic information (e.g., *Escherichia coli* and *Salmonella typhimurium*). In addition, this species criterion cannot be used to define a bacterial species because genetic transfer in many bacterial populations is a relatively rare event, and in some species is nonexistent. Thus, microbiologists generally have used characteristics such as cell morphology, gram staining properties, growth characteristics, and biochemical activities in order to define bacterial species. Since the phenotype (observed characteristics) of a given bacterium reflects the genetic makeup (the genotype) of that individual, the more phenotypic traits two bacteria have in common, the more closely related they are likely to be. Bacteria belonging to the same species have a large number of characteristics in common. This is not to say that two individuals of the same species must be identical; it is implied that there is some room for variation. In fact, variation within a species is expected.

In nature, populations generally are found interacting with other populations in some manner, forming a community and sharing a common environment. In a community, each population influences the size of other interacting populations by providing necessary nutrients, by competing for nutrients and space, or by releasing inhibitory substances. Communities often modify an environment so that new populations can become established. To get an idea of what is meant by a community, let us consider a small section of garden soil. In the soil, you might expect to find organic matter such as decomposing animals, insects, and vegetation. In addition, the soil contains many living animals, plants, and microorganisms. All these populations together make up the garden soil community. Each population takes nutrients from the soil and introduces new materials to it. For example, one population might obtain its energy and chemical building blocks from the decaying organic matter, and as a result of its growth the population might release waste products such as carbon dioxide or ammonia. The ammonia might be used as an energy source, and the CO_2 as a carbon source, by some populations of microorganisms. These microorganisms that use the ammonia may, in turn, release nitrates as byproducts of their metabolism. The nitrates do not accumulate in the soil, but are assimilated by growing vegetation and by microorganisms. When the vegetation dies, it provides organic nutrients to many microorganisms, plants, and animals. In this simplified description of interactions in a garden soil community, it is evident that the existence of each population is dependent upon that of some other population within the community.

In a community, each species resides in its own **habitat.** This is usually a physical space, such as a pine tree, garden soil, or salt marsh. In addition, each species has its own **ecological niche,** defined as the sum total of all the roles and interactions of the species in the community and the physical environment. Temperature, sunlight, moisture, food resources, predators, social organization of the species, and interspecies competition are all components of a species' ecological niche.

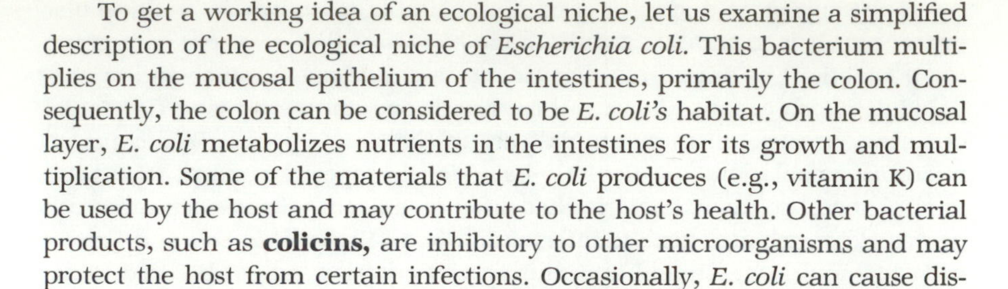

Figure 30-2 Competition between *Paramecium aurelia* and *P. caudatum*. *P. aurelia* and *P. caudatum* growing under the same conditions have different growth rates; *P. aurelia* has the faster rate. When both species are cultured in the same environment and competing for the same nutrients, *P. aurelia* eventually will replace *P. caudatum*.

To get a working idea of an ecological niche, let us examine a simplified description of the ecological niche of *Escherichia coli*. This bacterium multiplies on the mucosal epithelium of the intestines, primarily the colon. Consequently, the colon can be considered to be *E. coli's* habitat. On the mucosal layer, *E. coli* metabolizes nutrients in the intestines for its growth and multiplication. Some of the materials that *E. coli* produces (e.g., vitamin K) can be used by the host and may contribute to the host's health. Other bacterial products, such as **colicins,** are inhibitory to other microorganisms and may protect the host from certain infections. Occasionally, *E. coli* can cause disease by invading tissue or by releasing toxic substances. The sum total of all these activities can be considered to be the ecological niche of *E. coli*.

In a community, only one species occupies a given ecological niche. If two species attempt to occupy a single ecological niche, only the better-adapted species will survive. This concept is called **Gause's competitive exclusion principle.** Figure 30-2 illustrates this principle using two species of *Paramecium*, *P. aurelia* and *P. caudatum*. If these two species are placed in the same environment competing for the same resource, after 10–15 days *P. aurelia* will be the only survivor and *P. caudatum* will have become extinct.

Biomes are large patches of communities arranged in a characteristic manner due to the influences of temperature, seasonal changes, and annual rainfall. For example, a tropical rain forest (a type of biome) is an environment with lush vegetation, many exotic and colorful animals, lots of rain (200–400 cm/year), and a hot and humid climate. These biomes are found in Central America, South America, Asia, and Africa. Biomes may be terrestrial, freshwater, or marine.

THE CONCEPT OF AN ECOSYSTEM

The physical environment and the communities that occupy that environment make up what is called an **ecosystem** (fig. 30-3). The physical environment is considered to be the **abiotic** (nonliving) component of the ecosystem which ultimately provides the energy and nutrients that are needed for the **biotic** (living) component. While a limited amount of energy is recycled in an ecosystem, most of the energy comes from light that strikes the earth. Photosynthetic organisms convert light energy into chemical energy □, 194 which can then be used by other organisms. A knowledge of how energy and nutrients are used in an ecosystem is essential in order to understand what role(s) microorganisms play in nature. Both energy and nutrients must be perennial (always present), or the ecosystem eventually will cease to exist. Sunlight represents the energy source that powers almost all ecosystems. Light energy results from the fusion of hydrogen atoms in the sun. The amount of solar energy reaching the earth is about 10^{24} calories/year. This means that about every minute, each square centimeter of the earth's surface receives about 2 calories of energy. Most of this energy heats the earth's surface and radiates back out into space, but about 0.1% of the sunlight energy is captured by living systems. At its present rate of energy loss, it is estimated that the sun will burn itself out in about 3 billion years. Although energy available to ecosystems comes primarily from the sun, there are some ecosystems that exist in the dark and obtain their energy from chemical reactions. Energy, regardless of its source, must be in continuous supply. Living

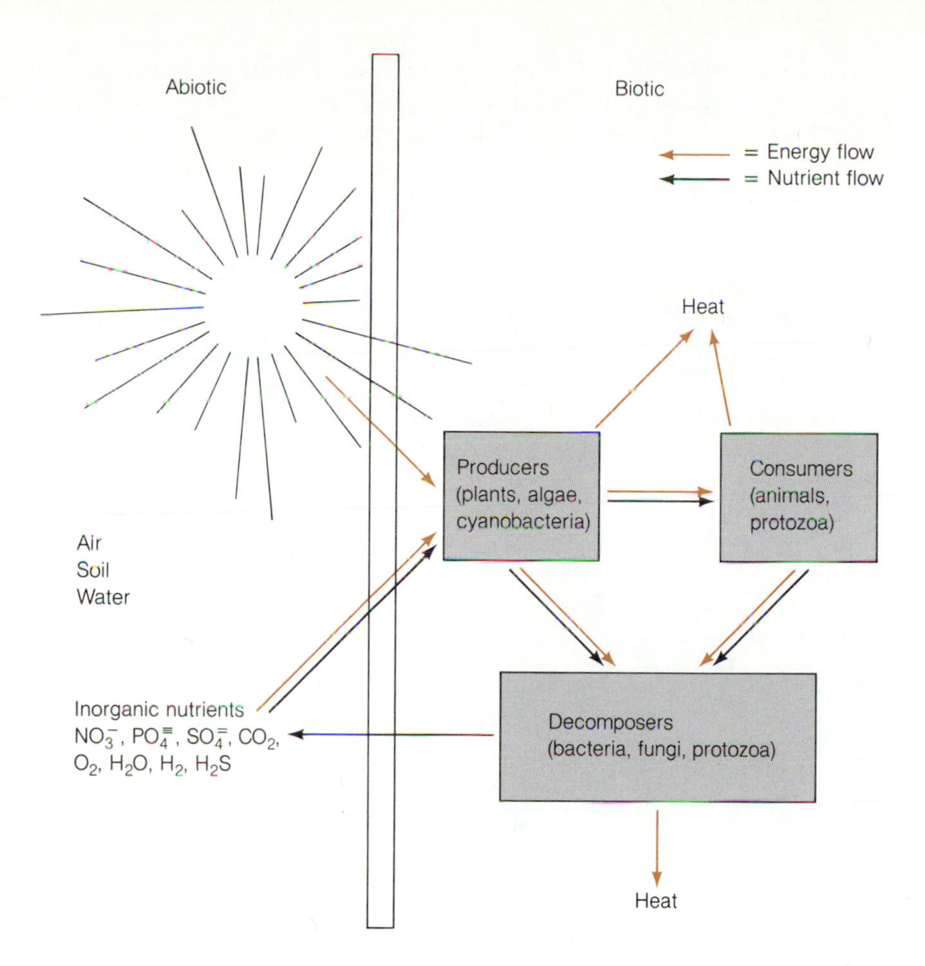

Figure 30-3 Diagrammatic Representation of an Ecosystem

organisms use up the available energy to power their activities and to provide heat for their metabolism.

Many of the abiotic (nonliving) components of the ecosystem, such as water, oxygen, nitrogen, carbon dioxide, nitrates, ammonia, sulfates, phosphates, and various other elements, serve as nutrients. These organic and inorganic compounds are found as part of air, water, and soils, which make up the physical environment of communities.

Community Structure

The biotic (living) component of the ecosystem can be divided into three major groups of organisms: the **producers,** the **consumers,** and the **decomposers.** The transfer of energy and nutrients within a community takes place from one population to another so that a food chain, or more accurately a **food web,** is established. The community is organized on the basis of "feeding" or **trophic levels** (fig. 30-4). The first trophic level serves as a source of food for the second level, which, in turn, serves as a source of food for the third level, and so on (fig. 30-4). Within the community, one can distinguish various trophic levels in the food web. The first trophic level in any community is the **producer** population. The producers provide much of the energy and organic nutrients for the rest of the community. The produc-

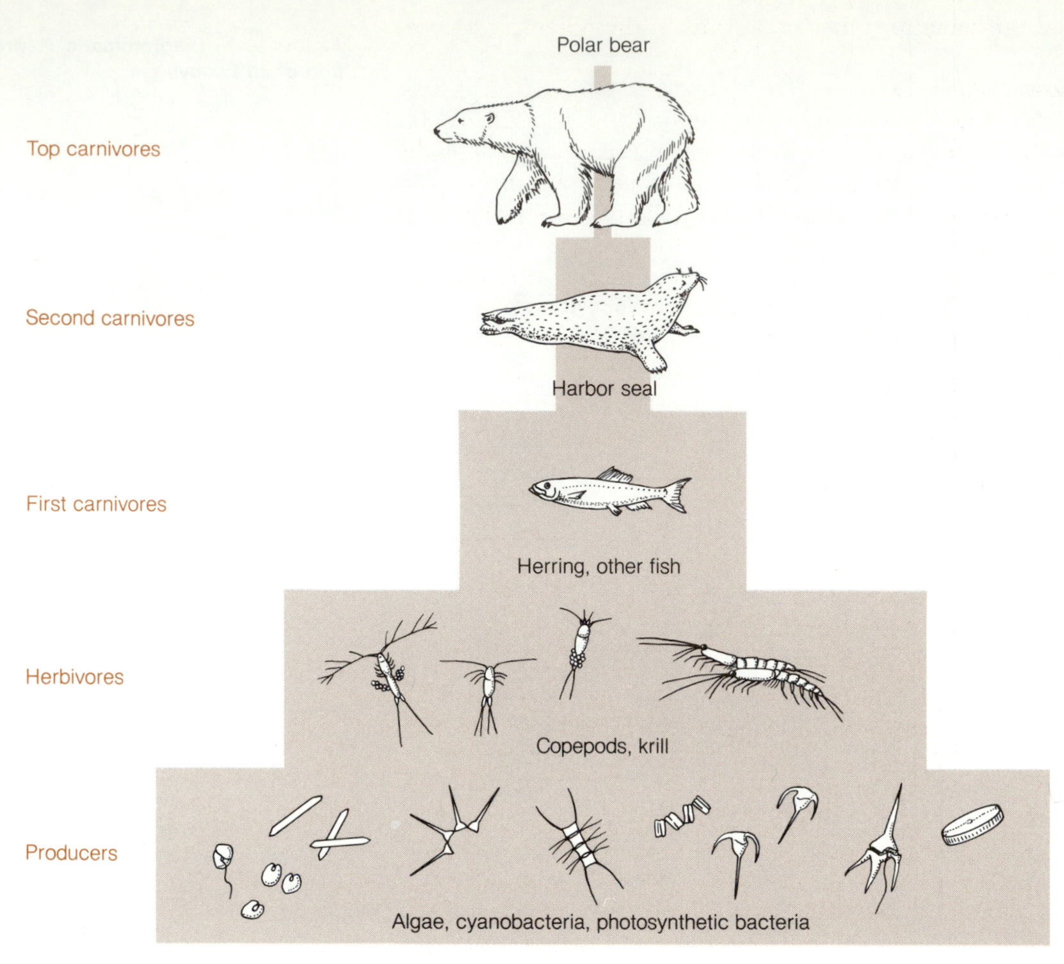

Figure 30-4 A Food Pyramid and the Various Trophic Levels. The complex feeding interactions that take place among organisms in a community constitute a food web. Food webs are arranged in tropic levels, with the producers at the first trophic level representing the bulk of the biomass available in the ecosystem. The food pyramid indicates the amount of biomass making up each trophic level in an ecosystem. The amount of biomass is usually visualized by the area of the rectangle representing the trophic level. As shown, the first trophic level has by far the largest amount of biomass (approximately 99%), while the fifth trophic level, corresponding to the top carnivore population, has the least.

ers consist of green plants, algae, and certain bacteria. The producer bacteria are autotrophic □ and obtain their energy either from the sun or from the oxidation of inorganic materials. Much of the energy that producers obtain from the sun is used to synthesize organic matter by photosynthesis: the producers convert solar energy into chemical energy to power the synthesis of organic molecules from CO_2. The organic matter can then be used by chemoheterotrophs □ in subsequent trophic levels. The producers are by far the largest group of organisms in a food chain (fig. 30-4), constituting almost 99% of all the available organic matter (biomass).

112
111

It was once thought that the producers in an ecosystem were exclusively photoautotrophic organisms. In 1977, however, a Woods Hole Oceanographic Institute research team discovered an ecosystem at the bottom of the Pacific Ocean in which the first trophic level (the producers) was made up not of photoautotrophs, but rather of chemolithotrophic □ bacteria. These bacteria use H_2S as a source of energy and fix CO_2. The chemolithotrophs serve as a

194

source of energy and nutrients for other organisms in the oceanic rift community.

Another chemosynthetic ecosystem has been discovered in Spirit Lake near Mount St. Helens. After the first volcanic eruption, the lake filled with mud and ash. The mud destroyed many of the original producer populations because it blocked the absorption of light by the aquatic organisms, but chemolithotrophic bacteria proliferated and became the primary producers.

Consumers constitute the second, third, fourth, and any subsequent trophic levels in a community. These organisms obtain their energy and nutrients from the producers. Consumers make a direct demand on the preceding trophic level. For example, grazing cattle feeding upon the grasses on a meadow make a direct demand on the grass because they consume it. The same can be said of a cheetah preying upon a gazelle herd. If the cheetah is successful in its hunt, the consumer makes a direct demand on that population of gazelles and reduces the number of animals in the population by one. There are also predatory microorganisms that feed upon other microorganisms; examples are nematode-trapping fungi and *Didinium*, a protozoan that feeds upon other protozoans such as *Paramecium*.

The consumers that feed upon producer populations are called herbivores or **primary consumers.** These consumers serve as food sources for the next trophic level, the carnivores or **secondary consumers.** The secondary consumers, in turn, serve as food sources for the **tertiary consumers.** Each higher trophic level has fewer organisms in it than the previous level, creating a **food pyramid** (fig. 30-4).

The **decomposers** consist mainly of bacteria, fungi, and protozoa. These microorganisms make no direct demand on any trophic level. Their sources of nutrients are decaying organic matter (dead organisms) and metabolic or other waste products resulting from the growth of organisms. Each population, as it multiplies, utilizes available nutrients and releases wastes. These waste products, in turn, are utilized by other microorganisms. In this way, the waste products from one population serve as sources of nutrients for other populations. Hence, the nutrients in an ecosystem are recycled and made available to all organisms in an ecosystem.

THE ROLE OF MICROORGANISMS AS DECOMPOSERS

There is a finite amount of nutrients available on the earth, and most of them are recycled over and over again by members of a community. Recycling of nutrients in an ecosystem is carried out principally by microorganisms. The cycling of nutrients throughout the ecosystem is often referred to as **biogeochemical cycles.** The next chapter will discuss some of the biogeochemical cycles in ecosystems. At this point in the discussion, however, an examination of a simplified version of the nitrogen cycle (fig. 30-5) might help clarify the function that certain microorganisms have in nutrient recycling.

Recycling of the Element Nitrogen in Ecosystems

A large quantity of nitrogen is tied up in proteins, nucleic acids, and other complex organic molecules that make up microorganisms, animals, and plants. Decomposers obtain nutrients from dead organic matter. As the de-

Figure 30-5 **Flow of Nitrogen Among Populations in a Terrestrial Ecosystem**

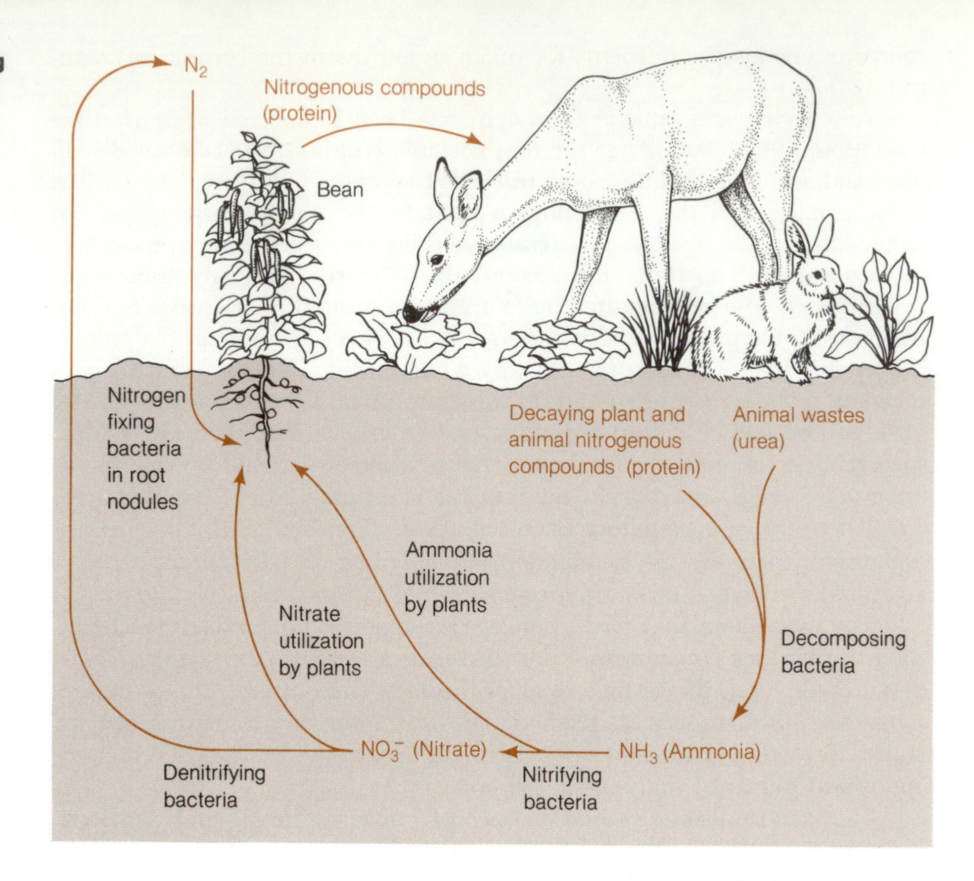

composers multiply at the expense of the organic matter, they excrete metabolic wastes such as ammonia and urea. Some of the ammonia released by decomposers is assimilated by other microorganisms and by certain green plants; a large portion of the ammonia, however, is used by **nitrifying bacteria** as their source of energy. Nitrifying bacteria commonly present in ecosystems use the excreted ammonia as an energy source, oxidizing it to nitrate, a waste product of their chemolithotrophic metabolism. The nitrate released by the nitrifying bacteria can then be assimilated by green plants and by microorganisms. In this way, a limited supply of nitrogen is passed around from one population to another, thus insuring that all populations have access to their needed nitrogen.

MICROBIAL INTERACTIONS

The existence of pure populations of microorganisms is more or less a laboratory phenomenon. In nature, microorganisms seldom exist alone and are usually part of complex communities. As we saw in the previous section, nutrient and energy transfer between populations is one way in which they interact. Interactions among populations in an ecosystem are complex phenomena and are not fully understood. In nature, microorganisms may interact with other populations (microbial, plant, or animal) in a variety of ways (table 30-1). Chapter 18 discussed three types of microbial interactions that were important in understanding the disease process: **mutualism, commen-**

TABLE 30-1
TYPES OF INTERACTIONS AMONG POPULATIONS IN NATURE

TYPE OF INTERACTION	EFFECT OF INTERACTION ON: POPULATION A	POPULATION B
Neutralism	No effect	No effect
Commensalism	Beneficial	No effect
Synergism	Beneficial	Beneficial
Mutualism	Beneficial	Beneficial
Competition	Detrimental	Detrimental
Amensalism	Beneficial or detrimental	No effect
Predation	Beneficial	Detrimental
Parasitism	Beneficial	Detrimental

salism, and **parasitism.** In this section, other microbial interactions that take place in nature will also be discussed.

When two microbial populations in a community are not interacting at all, they are demonstrating **neutralism.** The two populations must be physically apart so that neither one detects the presence (either physical or chemical) of the other. Even if two populations cannot directly detect each other, however, it does not mean that their relation is neutralistic. Consider, for example, two microbial populations inhabiting a plant, one in the root system and the other on young leaves. If one population infects the plant and kills it, this population destroys the other's habitat as well. This relationship is certainly not neutral. The preceding example points out the fact that it is often difficult to be sure that an interaction between two populations in an ecosystem is neutral. In aquatic habitats, where vast expanses of water support relatively little microbial growth, one might see neutralism, probably because the microorganisms are "diluted" in the environment.

Synergism and **mutualism** are relationships in which both populations derive some benefit from the association. In contrast to mutualism, however, synergism is not an obligatory association; each population can carry out a free-living existence without the other. The phenomenon of **cross-feeding** or syntrophism is an example of synergism. For example, *Lactobacillus arabinosus* and *Streptococcus faecalis* can grow independently of each other, but they grow much better together (fig. 30-6). *Streptococcus faecalis* provides *L. arabinosus* with high levels of the amino acid phenylalanine, and *L. arabinosus* provides *S. faecalis* with the vitamin folic acid, which they need. Certain algae and fungi develop mutualistic associations, called lichens, which allow these two organisms to live in environments that neither is capable of inhabiting on its own. In contrast to the free-living existence of *Lactobacillus* and *Streptococcus*, the lichen symbionts generally stay together.

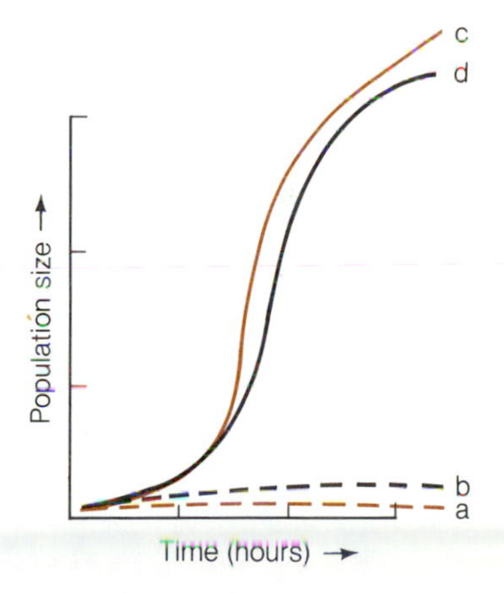

Figure 30-6 The Phenomenon of Cross-Feeding. In a medium devoid of folic acid and phenylalanine (an amino acid), neither *Streptococcus faecalis* (which requires folic acid for growth) nor *Lactobacillus arabinosus* (which requires phenylalanine for growth) grow (a and b). When both of these organisms are present in the same medium, however, they provide each other with the required growth factors (c and d).

Competition is a negative interaction between two populations. In this situation, one population usually induces a reduction in the other population's size by competing for available nutrients or space. Figure 30-2 shows the impact of competition between two microbial populations. Generally, two competing populations are affected by this interaction such that the population densities of both are reduced by the interaction. Closely related to competition is **amensalism.** In this case, one microbial population releases substances that are inhibitory to another population. One example is the fermentation of cabbage by the lactic acid bacteria to produce sauerkraut 814 □. The initial growth of the lactic acid bacteria on the cabbage results in the production of lactic acid, which lowers the pH to about 3.5–4.0 and inhibits the growth of other microorganisms. Many inhibitory chemical alterations of the environment by microorganisms, such as production of antibiotics, production of oxygen, consumption of oxygen, or release of toxic substances, are examples of amensalism.

In **predation,** one organism, the predator, engulfs and digests another, the prey (fig. 30-7). Generally, the predator is larger than the prey. In parasitism □, by contrast, the parasite is generally smaller than the host. Both predator and prey populations are dependent upon each other. As the prey population increases, there is more food available for the predator, whose population also increases. Conversely, as the prey population declines, it becomes increasingly difficult for the predator to find food, so its population becomes smaller. Over a long period of time, the result is a cyclic increase and decrease of both populations, the peaks and valleys of the predator population lagging behind those of the prey (fig. 30-7). Conditions that affect one population also affect the other. The predator, at least, is a consumer, since it has a direct impact on the previous trophic level.

427

Cyanobacteria, heterotrophic bacteria, yeasts, filamentous fungi, algae, and lichens often establish **commensalistic** relationships with plants. For instance, yeasts colonize the skin of grapes and benefit by obtaining nutrients, while the plant host does not appear to be affected. Lichens attached to tree trunks or hanging from branches benefit by having an attachment site, but do not appear to affect the host trees. We have already discussed the relationship between cyanobacteria and rice plants (see Chapter 13). These bacteria fix nitrogen, which can then be utilized by the plants. In essence, the cyanobacteria, particularly *Anabaena*, serve as a soil fertilizer. **Parasitism** □ is exemplified when microorganisms cause diseases in plants and animals. Some of these pathogens can have a significant impact on a nation's economy when they devastate crop and domestic animal populations. We will discuss some of the important plant and animal diseases in Chapter 34.

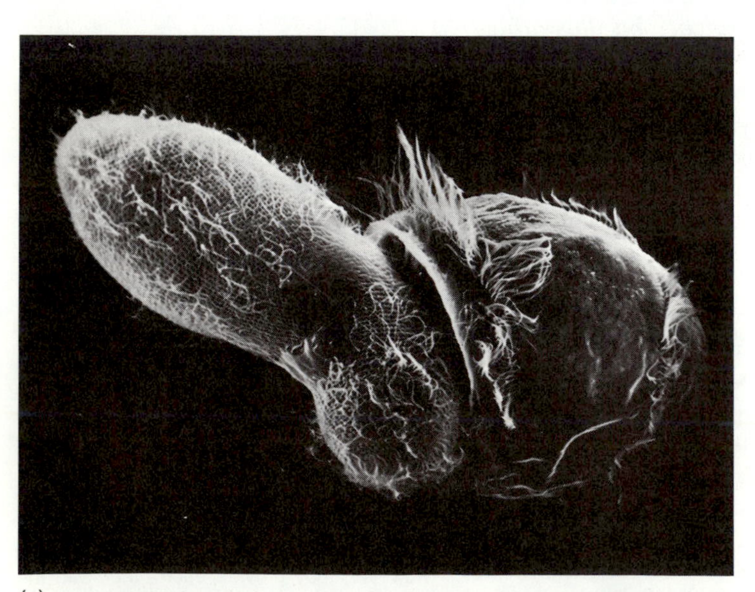

(a)

Predator
Prey
Population size
Time

(b)

Figure 30-7 Predator-Prey Relationships. *(a)* Predation of *Paramecium* by *Didynium.* *(b)* Oscillation of predator and pray populations.

INTERACTIONS OF MICROORGANISMS WITH PLANTS

As members of complex communities, microorganisms interact, one way or another, with plant populations in a community. Upon detailed examination of plants, one finds specific microbial populations associated with all parts of the plant: roots, stems, leaves, and flowers. Some of these interactions are mutualistic, synergistic, and commensalistic, while others are parasitic.

Mycorrhizae

339 **Mycorrhiza** □ (fungus root) results from a mutualistic relationship between certain fungi and plant roots. In this relationship, fungal and plant root material become integrated, forming the mycorrhizae (fig. 30-8). Both mutuals derive a benefit from the association. The fungal member has access to organic nutrients that are provided by the plant; in addition, the mycorrhizal fungus occupies a protected habitat, free from competition and away from stressful environmental conditions. The importance of mycorrhizal associations are best seen in plants colonizing barren soils. These plants are better able to take up nutrients, particularly phosphate, in uncultivated soils than are plants lacking mycorrhizae. The mycorrhizal association does not seem to provide a significant advantage to plants growing in cultivated lands, however. The importance of mycorrhizal associations may reside in the fact that crop plants, if they develop mycorrhizal associations, are able to grow as well in low-nutrient soils (fig. 30-8) as their counterparts in fertilized soils. This effect of mycorrhizae would obviously reduce the need for expensive fertilizers. In addition to nutrient uptake, the mycorrhizal fungus protects the plant symbiont from certain plant pathogens. The mycorrhyzal fungi have been found to produce allelopathic (toxic) substances that inhibit the growth of nearby plants and therefore reduce competition.

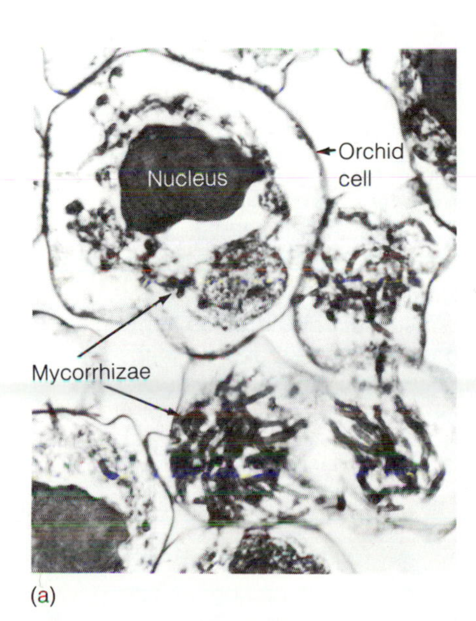

(a)

(b)

(c)

Figure 30-8 Mycorrhizal Interactions. *(a)* Mycorrhizal fungus in association with orchid root cells. *(b)* Influence of mycorrhizae on plant growth. The pine tree is growing in a nutrient-poor soil and has not developed a substantial mycorrhizal association. *(c)* The pine tree is growing in the same soil but has a well-developed mycorrhizal association.

Nitrogen Fixation

Some bacteria, for example those in the genera *Rhizobium* and *Clarkia*, are able to form symbiotic associations with green plants, in which the plant provides the bacterial symbiont with necessary nutrients while the bacterium provides the plant with organic nitrogen. The bacterium converts atmospheric nitrogen (N_2) into organic nitrogen. This process, called **nitrogen fixation** □, takes place only when both bacteria and green plant are forming a mutualistic association. Hence, this type of nitrogen fixation is called **symbiotic** nitrogen fixation.

761

INTERACTIONS OF MICROORGANISMS WITH ANIMALS

Microorganisms also develop symbiotic associations with animals. These associations involve a variety of functions, from nutrient digestion to protection. Although many of these associations are of benefit to both the animal and the microorganism, some are detrimental to at least one of the symbionts. In Part III of this textbook, we discussed in considerable detail symbiotic associations between microorganisms and the human animal. In this section we will discuss other types of associations between microorganisms and animals.

Food Digestion and Nutrition

Many animals develop long-lasting associations with microorganisms and use them as a source of digestive enzymes, while others nurture a resident flora that serves as a source of nutrients. The first association is well illustrated by the symbiotic association between termites and their symbionts.

FOCUS

MUTUALISM BETWEEN *LACTOBACILLUS* AND HUMANS

Bacteria of the genus *Lactobacillus* are gram positive rods that ferment carbohydrates, forming lactic acid. These bacteria are widely distributed in nature and are found in the adult human vagina. In fact, *Lactobacillus* is an important and essential component of the vaginal flora. Any changes that adversely affect the size of the lactobacilli population in the vagina can ultimately lead to a disease called **vaginitis** (inflammation of the vagina).

Lactobacillus utilizes some of the glycogen present on the vaginal mucosa as a source of energy and carbon. As a consequence of its metabolism, lactic acid is released, which in turn reduces the pH of the vagina. The lowered pH provides an environment that is hostile to many microorganisms, including those that can cause disease. Hence, humans are protected from many vaginal infections by their normal vaginal flora.

Conditions that tend to reduce the lactobacilli make humans more susceptible to disease. For example, prolonged administration of antibiotics such as penicillin can kill much of the resident lactobacilli flora, allowing resistant bacteria or yeasts to proliferate excessively and cause disease. This can also happen when the level of estrogen hormone drops, as is the case in postmenopausal women. These individuals become more susceptible to vaginal infections, partly because the lowered level of estrogen hormone cause a decrease in the amount of glycogen available in the vaginal mucosa. With less glycogen for metabolism, the lactobacilli population declines, allowing other microorganisms to become predominant. When the vaginal pH increases, acid-sensitive pathogens can colonize the vagina and cause disease.

The latter association is illustrated by the association between ruminants and microorganisms in the rumen.

Termites and Their Gut Flora

Termites are wood-eating insects that can cause severe damage to wooden structures. These insects chew up the wood and grind it into fine particles, which they subsequently swallow. They use the cellulose component of the wood for their energy and carbon sources and excrete the lignin. These insects are themselves unable to digest cellulose, however, and rely on a complex gut microflora to digest it for them. One organism in the gut microflora is the protozoan *Myxotricha paradoxa*. This protozoan itself harbors certain bacteria that appear to produce cellulose-decomposing enzymes (cellulases). Therefore, the cellulose digestion by termites is actually carried out by cellulolytic bacteria within the protozoan *M. paradoxa* which, in turn, is living inside the termite.

Insects that have rather restricted diets also carry with them a community of microorganisms that provide the insect with needed growth factors. Certain other organisms cultivate their own food (fig. 30-9). For example, some termites and certain ants have their own "fungus gardens" which they nurture for food.

Rumen Symbiosis

Animals such as deer, moose, caribou, cow, sheep, goats, and giraffes are called ruminants because they have a stomach consisting of four parts, one of which is the **rumen** (fig. 30-10). These animals are herbivores, and their primary source of food consists of grasses, leaves, and twigs, all of which are rich in cellulose. The ruminant, however, like the termite, is itself unable to digest this cellulose. The complex microbial community of the rumen is

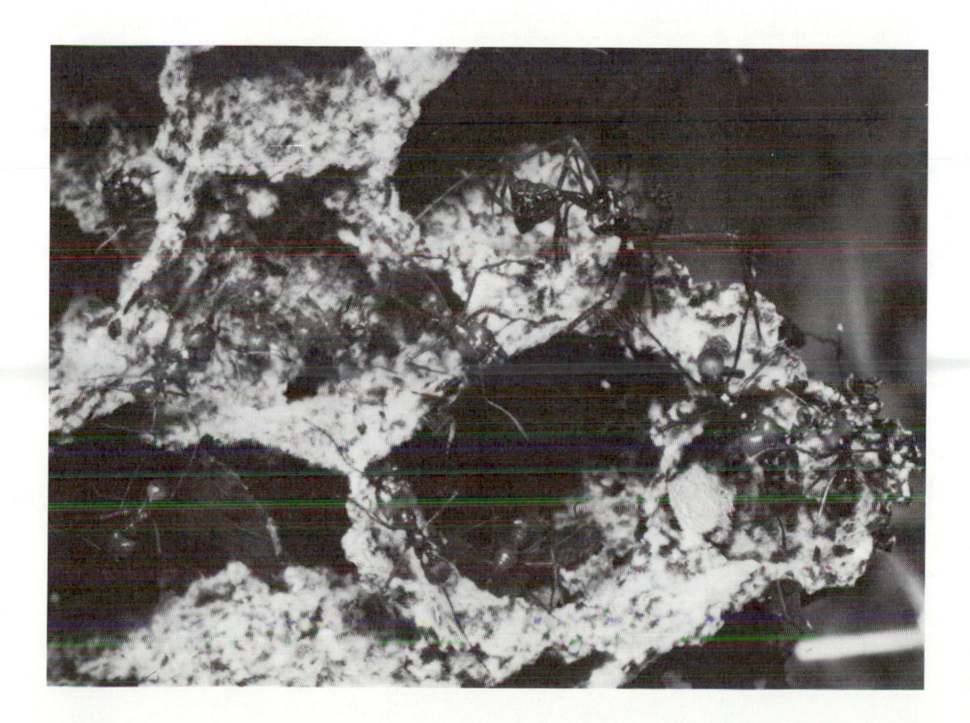

Figure 30-9 Ants Cultivating a Fungus Garden

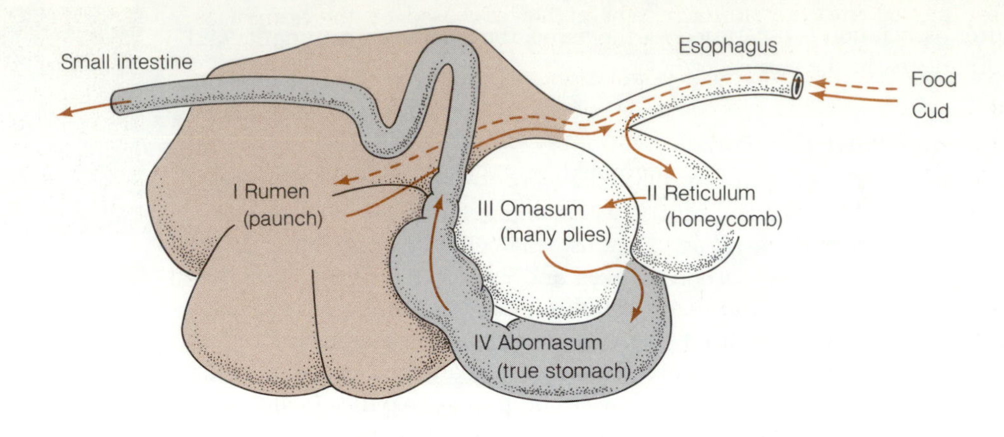

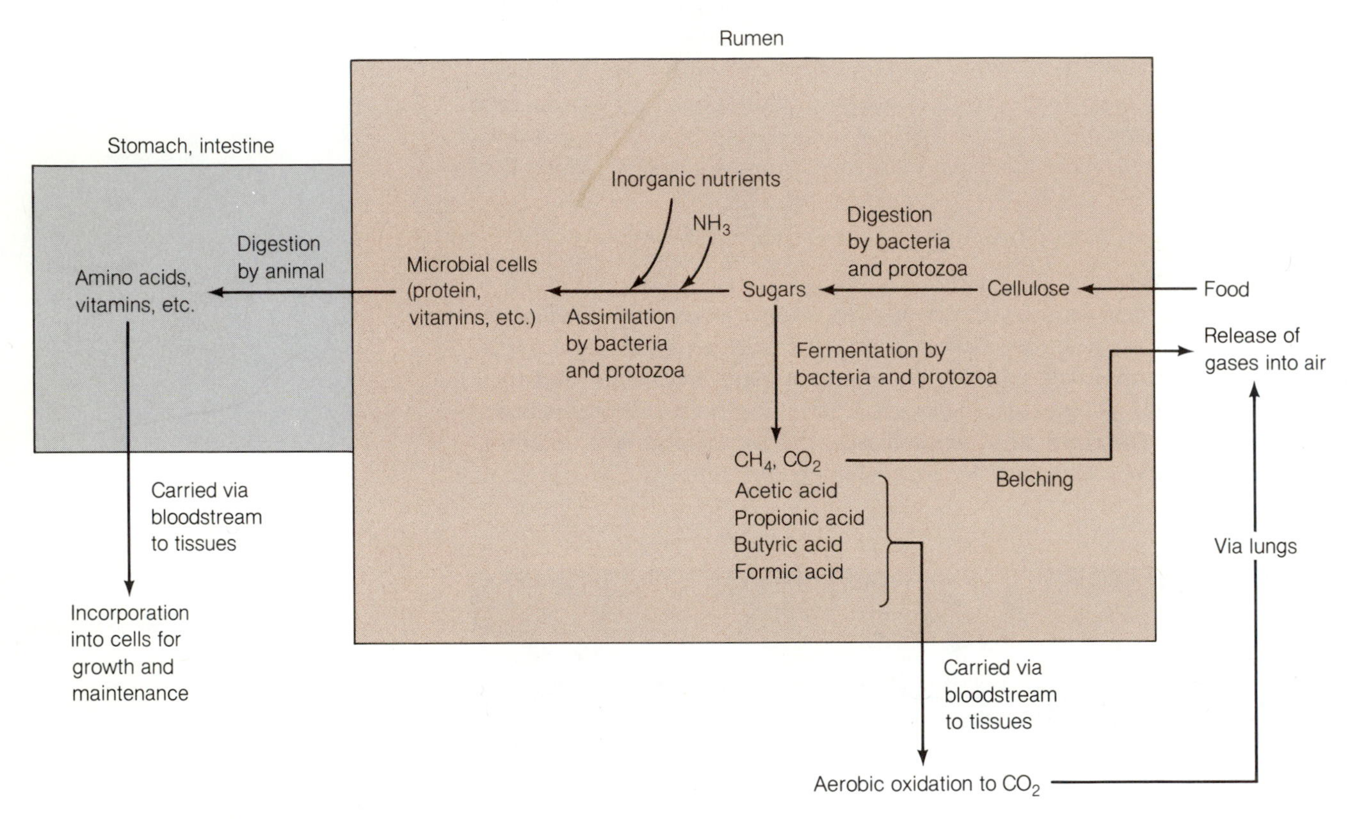

Figure 30-10 Microbial Interactions in the Rumen. Food enters the rumen, where it is digested by various microorganisms. The partially digested food or cud is regurgitated, chewed, and swallowed again. The chewed cud enters the reticulum, and then passes through the omassum, abomassum, and finally the intestines. In the rumen, the cellulose is digested to fermentable sugars by the rumen microbiota. The sugars are utilized by the microbiota as carbon and energy sources. The ruminant obtains its nutrients as fermentation byproducts of microbial metabolism or as excess microbial mass.

adapted to living in a highly anaerobic environment that is provided with a constant supply of cellulose and fermentable carbohydrates. The cellulose and carbohydrates, which certain microorganisms hydrolyze, are degraded further by other microorganisms to carbon dioxide, water, methane, a variety of organic acids (acetate, propionate, butyrate), and biomass (fig. 30-10). The animal uses the organic acids as energy and carbon sources by absorption directly from the rumen into the blood, while the microbial cells eventually are passed into the stomach, where they are used by the animal as sources of amino acids, nucleic acid bases, and vitamins. The ruminants do not consume the entire microbial population, but only the amount passing from the rumen to latter stomachs as new food is swallowed. Thus, the various populations in the rumen community maintain relatively constant sizes. The rumen community is maintained in a steady state, in much the same way that a microbial population is maintained in a steady state using a chemostat □.

143

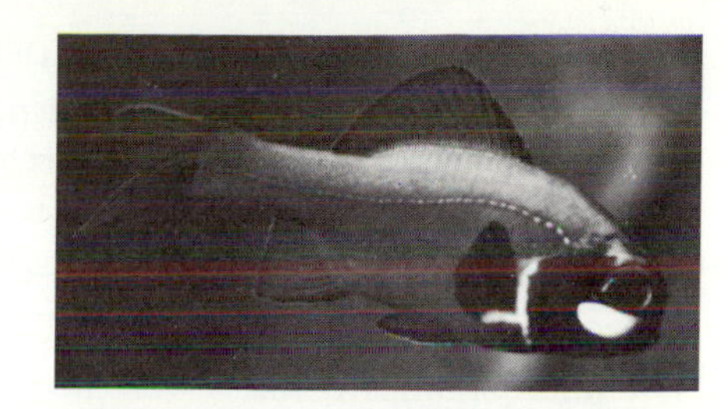

Figure 30-11 Light Organ in a Marine Fish. The marine fish *Photoblepharon* has a light organ which contains bioluminescent bacteria in a pouch beneath the eye.

Symbiotic Light Production

Many marine fishes nurture **bioluminescent** bacteria (fig. 30-11), which serve as a source of light for these animals that normally inhabit aquatic environments with little or no light. These bioluminescent microorganisms are generally bacteria of the genera *Photobacterium* or *Beneckea*. These bacteria are usually in "light organs" located near the eyes, mouth parts, abdomen, or rectal area of the fish. The fish are able to control the light from the bioluminescent bacteria by opening and closing epidermal folds. The light may be used to illuminate the environment, to warn antagonists, or to attract mates or prey.

Predatory Microorganisms

Certain microorganisms are able to feed directly on other organisms by killing them. In figure 30-7, the predatory ciliate *Didinium* is shown feeding upon *Paramecium*. A more interesting phenomenon is that of the nematode-trapping fungi □. These fungi, found in soils, produce loops of hyphae that serve to trap nematodes. The fungi lie in wait for a nematode to slither into one of the loops. Once the fungus "senses" the presence of the nematode inside the loop, it closes the loop and traps the worm. The nematode is then slowly digested by the fungus.

339

SUMMARY

1. Microorganisms play a central role in nutrient recycling in nature.
2. Ecology is the science that studies the interaction between organisms and their environment.

THE BIOSPHERE

1. The biosphere is that portion of the earth which supports life.
2. Populations are aggregations of organisms of a single species.
3. Two microorganisms are said to be in the same species if they have many genetic traits in common.
4. A community is a group of interacting populations within a given area.
5. A habitat is that area in an environment where a given population resides, while an ecological niche is the sum total of all the activities of a population in a community.
6. Biomes are large patches of communities arranged in a characteristic fashion.

THE CONCEPT OF AN ECOSYSTEM

1. An ecosystem is made up of the physical environment and a community of organisms.
2. In an ecosystem, chemical energy and nutrients are recycled, but much of the energy is lost as heat.
3. The biotic component of an ecosystem consists of producers, consumers, and decomposers.
4. The producers provide the nutrients and energy needed by all other organisms in the ecosystem community.
5. Consumers feed upon other members of the community, making a direct demand on the number of organisms in the community.
6. Microorganisms are the main decomposers in an ecosystem.

THE ROLE OF MICROORGANISMS AS DECOMPOSERS

Microorganisms participate in nutrient recycling by utilizing available nutrients and releasing metabolic wastes. The wastes can then be utilized as nutrient sources by other populations in the community.

MICROBIAL INTERACTIONS

1. Microorganisms rarely exist in pure cultures in nature. Rather, they exist as part of complex, interacting communities.
2. Interactions can benefit both populations, only one, or neither. The size of each interacting population is largely determined by other populations.

INTERACTIONS OF MICROORGANISMS WITH PLANTS

1. Microorganisms colonize various parts of plants. Some enhance the plant's chances for survival in the ecosystem, while others cause disease.
2. Mycorrhiza are mutualistic associations between plants and certain fungi. The plant's mycorrhiza increases the efficiency of nutrient absorption by the plant. The fungal symbiont is provided with a stable environment and a ready source of nutrients.

INTERACTIONS OF MICROORGANISMS WITH ANIMALS

1. Microorganisms colonize various parts of an animal and may cause changes in the animal's metabolism.
2. Many animals develop long-lasting associations with microorganisms that provide digestive enzymes or nutrients.
3. Termites have a gut flora that helps the termite digest cellulose.
4. Ruminants nurture a complex community of anaerobic microorganisms in their rumen which provides the animal with nutrients from the cellulose they digest.

STUDY QUESTIONS

1. Define the following terms:
   a. biosphere
   b. ecosystem
   c. biome
   d. community
   e. population
2. How are species of microorganisms defined?
3. Describe the ecological niche of *Escherichia coli.*
4. Diagram the major components of all ecosystems. Indicate the direction(s) of flow of energy and carbon.
5. What are trophic levels? What role do they play in the structure of a community?
6. What role do chemoheterotrophic microorganisms play in an ecosystem?
7. Give an example of the following:
   a. synergism
   b. competition
   c. predation
   d. parasitism
   e. mutualism
8. Why are mycorrhizae considered to be mutualistic relations?
9. How do the rumen bacteria benefit ruminants?
10. How is it that termites rely solely on cellulose as a source of food when they cannot digest this polymer?

SUPPLEMENTAL READINGS

Alexander, M. 1971. Microbial ecology. New York: John Wiley & Sons.

Atlas, R. M. and Bartha, R. 1981. Microbial Ecology. Menlo Park, CA: Addison-Wesley Publishing Co.

Burns, R. G. and Slater, J. H., eds. 1982. Experimental microbial ecology. Oxford: Blackwell Scientific Publications.

Campbell, R. E. 1977. Microbial ecology. Oxford: Blackwell Scientific Publications.

Delwiche, C. C. 1970. The nitrogen cycle. Scientific American 223:137–143.

Moore, J. 1984. Parasites that change the behavior of their host. *Scientific American* 250(5):108–115.

Ourisson, G., Albrecht, P., and Rohmer, M. 1984. The microbial origin of fossil fuels. *Scientific American* 251(2):44–51.

Stout, J. D. 1980. The role of protozoa in nutrient cycling and energy flow. Advances in Microbial Ecology 4:1–11.

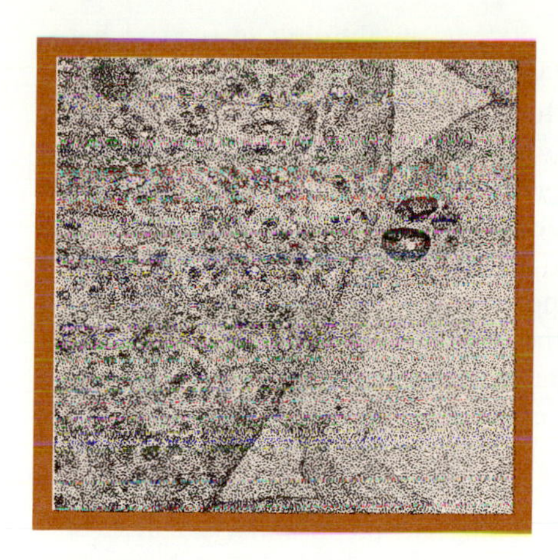

CHAPTER 31 MICROBIOLOGY OF TERRESTRIAL HABITATS

CHAPTER PREVIEW

MINER BACTERIA

Each gram of soil contains millions of microorganisms that contribute to the soil's fertility. Some of these soil microorganisms can be used to leach valuable metals such as copper and uranium from the soil. These microorganisms have been exploited since antiquity for the recovery of such valuable materials. For example, during the first century B.C., Mediterranean cultures collected acid mine drainages to recover the dissolved copper present in them. Similar mining operations were carried out by Spaniards during the eighteenth century A.D.

Although microbial mining has been conducted for many years, it was not until recently that the role of bacteria in the leaching process became widely known. Presently, thousands of tons of copper (and a lesser amount of uranium) are recovered from poor quality ores in mining operations in the United States. The copper yield from these operations accounts for approximately 10% of the total copper tonnage mined.

Because of the ever-increasing demand for metals such as copper and uranium, microorganisms have assumed a useful role in the mining industry. Their usefulness becomes apparent when one considers that much of the copper mined in this fashion comes from ores that would not have been mined otherwise.

The leaching process involves chemoautotrophic, aerobic bacteria, such as *Thiobacillus thiooxidans,* which oxidize ferrous ions into ferric ions, in acid environments, deriving energy from this oxidation. As a result of this metabolism, insoluble metal (e.g., copper) sulfides in the ore are converted to soluble metal sulfates. Sulfates are produced from metal sulfides because the ferric ion oxidizes the sulfides.

The soluble metal solutions are then collected in basins and the metals extracted. The mining industry is attempting to develop genetically engineered strains of bacteria that will leach out copper and uranium more efficiently from low grade ores, thus increasing the productivity of mining operations.

One type of bacteria-dependent mining operation, called **dump leaching,** is carried out by spreading tons of low-grade copper ore over a hillside. The low-grade ore soils, which contain abundant supplies of copper-leaching bacteria such as *Thiobacillus ferrooxidans* and *T. thiooxidans,* are sprayed with acidified water. The acid water and sufficient dissolved oxygen stimulate the chemical reactions that solubilize the metal. The metal solution is collected in basins and the metal is extracted by chemical means. The leaching solution, devoid of the metal ions, is recycled. In this manner, microorganisms native to certain types of soils are used to extract valuable products from the soil.

The ore-leaching bacteria represent an example of the many useful microorganisms present in terrestrial environments. Many microbial-directed processes have gone undetected and unexploited by humans. With a better understanding of the biology of microorganisms and of their remarkable ability to alter their environment, industries have made use of microorganisms to improve the extraction and manufacture of many products.

In this chapter we will study the role of microorganisms in the fertility and productivity of soils. This chapter will emphasize the fact that the metabolism of soil microorganisms serves to alter terrestrial habitats drastically. As a consequence of their activities in soils, microorganisms may produce valuable products.

LEARNING OBJECTIVES

A STUDY OF THIS CHAPTER SHOULD ENABLE YOU TO:

DISCUSS SOME OF THE VARIOUS ROLES MICROORGANISMS PLAY IN TERRESTRIAL ECOSYSTEMS

DESCRIBE WHAT SOIL IS, HOW IT DEVELOPS, AND THE MICROORGANISMS THAT COLONIZE IT

OUTLINE THE VARIOUS BIOGEOCHEMICAL CYCLES AND HOW THEY OCCUR

DISCUSS SOME OF THE USEFUL FUNCTIONS THAT MICROORGANISMS PERFORM IN TERRESTRIAL HABITATS

Microorganisms have played an important part in shaping terrestrial habitats. On the primitive earth, cyanobacteria, over a period of about 2 billion years (3.5 to 1.5 billion years before the present), fixed CO_2 into organic compounds and released O_2 into the atmosphere. The organic compounds and O_2 stimulated the evolution and growth of aerobic, terrestrial organisms. Part of the organic compounds that microorganisms made was sequestered under sediments and eventually became oil and natural gas. Another portion of the photosynthesized carbon remained tied up in the living mass **(biomass)** of organisms accumulating on the earth.

In the last 500 million years, plants and eukaryotic algae have replaced the prokaryotes as the major producers of organic materials and O_2. Some of the plant material has been sequestered under sediments, where part of it has become or will eventually become coal, oil, and natural gas. Although plant material is the major source of organic carbon in soils, soil microorganisms are largely responsible for converting the plant material into CO_2 and simple organic compounds that can be utilized by a variety of heterotrophs.

The fertility of soils, and the types of organisms that are able to colonize a particular habitat, are determined partly by the physical environment of the area and partly by the microbial composition of the soils. In this chapter we will study some of the biological processes carried out by microorganisms in terrestrial environments and how these processes affect soil fertility and human populations.

SOIL FORMATION

Soil is the surface layer of the earth which supports the growth of plants, animals, and microorganisms. It is composed of particles of rock, sand, clay, organic materials, and inorganic nutrients. The proportions of these components result in different types of soils.

The development of soils takes place by the combined action of weathering processes and the growth of successive populations of organisms (fig. 31-1). Microorganisms play an important role in this process. For example, in rocky areas, the process of soil development usually starts when photosynthetic microorganisms such as the cyanobacteria colonize rock crevices. The reproduction of cyanobacteria allows the colonization of the rock by certain other bacteria and fungi. The acids released from the chemical activities of these microorganisms help to dissolve the rock, creating a substrate suitable for the growth of higher organisms. Lichens and simple plants, such as mosses, grow on these rocks, decomposing them further. The dead and decaying lichens and mosses begin to create a thin soil in rocky areas, thus allowing larger plants to grow.

Areas where sediments accumulate are generally rich in nutrients that stimulate the growth of various populations of microorganisms and higher plants. If the plants grow well and their remains are not rapidly destroyed by the soil microorganisms, the soil becomes enriched with organic matter and **humus** from the partial breakdown of roots, stems, and leaves.

Humus consists of lignin, cellulose, hemicellulose, polypeptides, polysaccharides, nucleic acids, phospholipids, and other materials that have escaped microbial digestion. This material is important to plants because it acts as a reservoir of nutrients, soaks up water, and promotes soil aeration. As humus

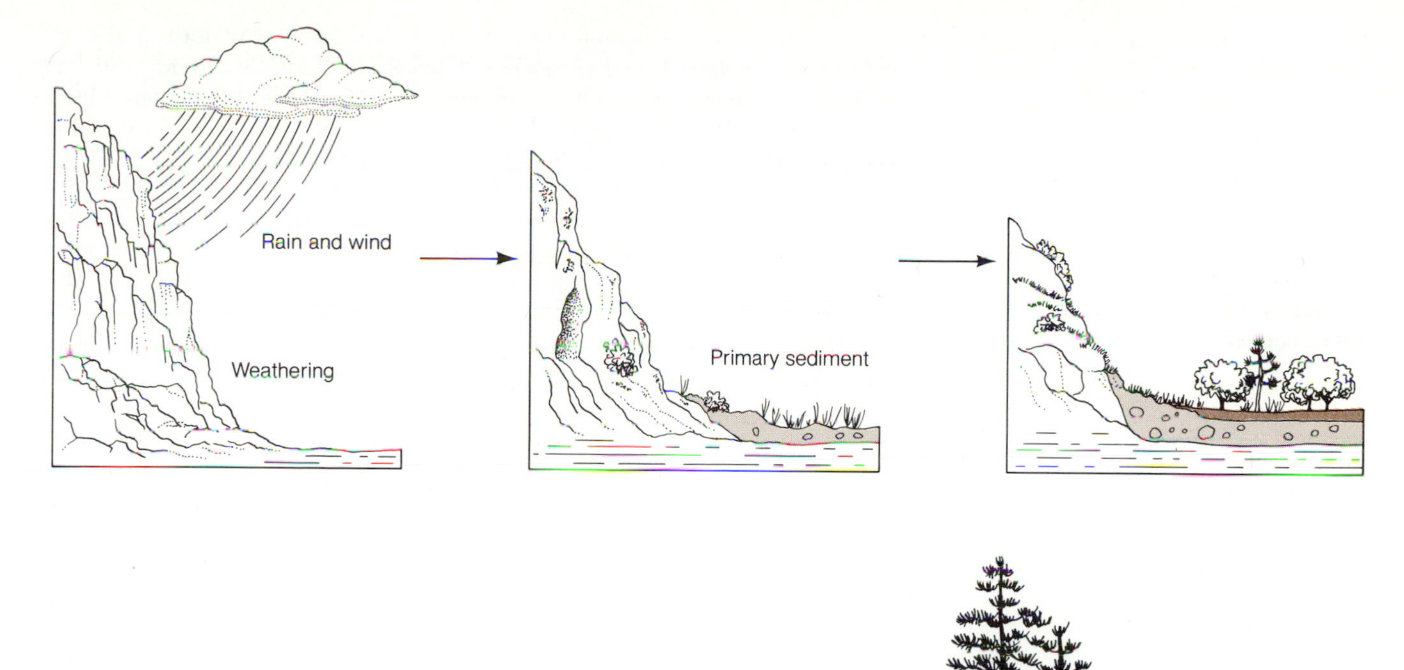

Figure 31-1 Process of Soil Formation from Bare Rock. Erosion, water, and wind weather the bare rock and break it into small particles. Microorganisms such as cyanobacteria, lichens, and mosses colonize these particles and continue to degrade them. Accumulation of organic matter and soil degradation by metabolic waste products continue, and eventually a primary sediment is formed. The primary sediment continues to build up and can support small vascular plants, in addition to cyanobacteria, lichens, and mosses. This leads to the formation of humus and the buildup of soil. Eventually, the soil has built up sufficiently to support the growth of large plants.

is slowly decomposed by microorganisms, it provides soil-dwelling organisms a steady supply of sugars, amino acids, nucleosides, phosphate, ammonia, and trace elements.

The type of soil that develops in a given area is dependent upon the type of sediment or bedrock that is weathered, the climate of the region, and the plant species that colonize the soil. For example, the soils of coniferous forests, which are common in the temperate areas of North America, are typically acid and low in nutrients. The soil is acid because the microorganisms that decompose the coniferous materials release considerable amounts of acid. This acid, together with percolating water from rains, leaches the nutrients out of the soil. Hence, the soils that develop in coniferous forests generally are unsuitable for agriculture because of their acidity and low mineral and nutrient content. If these soils are to be farmed, they must be treated with lime ($CaCO_3$), to raise the pH, and enriched with fertilizers.

In contrast, deciduous forests of oak, maple, beech, walnut, and chestnut, which are found in the eastern United States, usually develop rich soils because little acid is released at any one time. The forest litter, consisting of fallen leaves, shrubbery, and other plant material, creates a thick humus layer. The nutrients are released slowly from the humus so that a fertile soil is created. Many of the deciduous forests in North America were cleared in the 1800s and planted with crops. This rich land still provides nutrients for many of today's cereal crops and orchards.

Composing

When animal and plant materials are decomposed by microorganisms, minerals and organic compounds become part of the soil once more and contribute to its fertility. The release of minerals from the decomposition of animals and plants is called **mineralization.** When a thick layer of grass, leaves, twigs, and branches decomposes, a **compost** is formed. Composts are rich in minerals, cellulose, hemicellulose, lignin, partially degraded plant material, and microorganisms. In thick composts, the temperature often reaches 70°C because of the metabolic activities of the microorganisms carrying out the decomposition. Because of the high temperatures, thermophilic □ bacteria (mainly actinomycetes) and fungi carry out the decomposition of the heap. As the temperature of the compost declines, these microorganisms are replaced by mesophilic bacteria.

145

Bacteria in the genus *Streptomyces* are very important members of the compost heap. Sometimes these bacteria are so abundant that they form a white, powdery coating on the compost. In addition, these bacteria produce a chemical called **geosmin** that gives a musty, "earthy" odor to compost or soils rich in decomposing plant material.

BIOGEOCHEMICAL CYCLES

As we learned in Chapter 30, nutrients in an ecosystem are shared and recycled by members of the community. The recycling takes place as a consequence of the metabolism of plants, animals, and microorganisms. Most nutrients sequestered in the biomass of a community are recycled by microorganisms and thus made available to all organisms. In the following sections, we will discuss some of the ways in which microorganisms participate in nutrient recycling in ecosystems and the impact that they have on human populations.

Carbon Cycle

Biological molecules are made up mostly of carbon. Consequently, this element is of central importance to all living organisms. Most of the carbon in nature is found as CO_2 and in biological polymers. Organisms that convert carbon from one form to another provide various communities of organisms in an ecosystem with the form of carbon that they need to survive. When CO_2 in the atmosphere is fixed by photosynthetic organisms, O_2 is released into the atmosphere and carbon is incorporated into biological compounds. The organic compounds resulting from the CO_2 fixation are oxidized by chemoheterotrophic organisms to produce CO_2. As can be seen from figure 31-

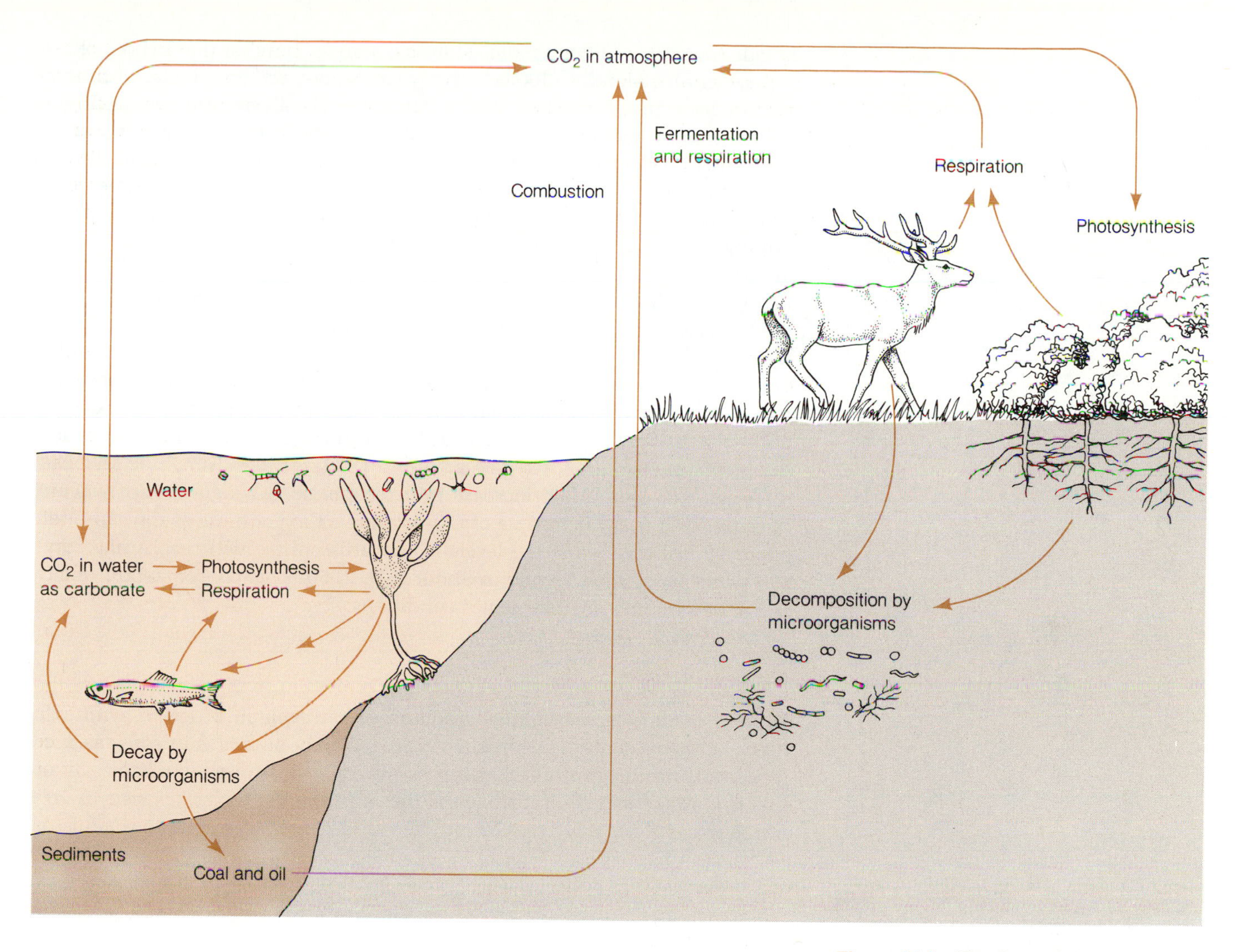

Figure 31-2 **The Carbon Cycle.** Much of the carbon available to living organisms is in the form of CO_2. The CO_2 is incorporated into organic matter through photosynthesis. The organic carbon, as part of living organisms, is oxidized to CO_2 during respiration and returned to the atmosphere. Some of the organic carbon may be in the form of fossil fuels, which are oxidized to CO_2 during combustion.

2, carbon is recycled among autotrophic and heterotrophic organisms; the autotrophs convert the gaseous CO_2 into biological molecules, while the heterotrophs return the CO_2 to the atmosphere.

Not all carbon is cycled immediately. Some of it, for example, accumulates as coal, oil, or natural gas. Based on pool sizes and utilization rates, it has been calculated that each molecule of CO_2 in the atmosphere is likely to be fixed by photosynthetic organisms every 300 years.

Methanogenic bacteria participate in the carbon cycle by producing methane from CO_2. Some of this methane is sequestered in the earth, but most of it is released into the atmosphere or respired by methane-oxidizing bacteria. Much of the atmospheric methane is oxidized by bacteria to CO_2 and water.

The movement of carbon from place to place on the earth and its conversion from one form to another are illustrated in the carbon cycle (fig. 31-2). Ocean and terrestrial photosynthesis is estimated to absorb about 200 billion tons of CO_2 from the atmosphere each year. This CO_2 becomes part of living organisms in the oceans and on the earth. Vast amounts of CO_2 (130 trillion tons) are dissolved in the oceans and form part of large deposits of calcium carbonate and magnesium carbonate. Some of the carbon tied up in biological molecules (100,000 tons/year) goes into fossil beds, to become part of the estimated 20,000 trillion tons of coal, oil, natural gas, and other buried organic matter on the earth. The unoxidized beds of organic carbon are very important to life on earth and to the environment. If significant proportions of it should be oxidized, the level of CO_2 in the atmosphere would increase dramatically and the level of O_2 would decrease. This shift in atmospheric gases would cause drastic changes in the earth's climate and in terrestrial life.

Carbon dioxide is returned to the atmosphere as a result of respiration in the oceans and on the earth (200 billion tons/year), the burning of fossil fuels (5 billion tons/year), the increased metabolism by microorganisms in cleared and tilled soils (2 billion tons/year), hot springs and volcanoes (1 billion tons/year), and the weathering of calcium carbonate in rocks (100 million tons/year). If the amount of CO_2 absorbed by living organisms and the oceans is compared with the amount of CO_2 released, it can be seen that there has been a gradual increase in atmospheric CO_2 as a consequence of burning fossil fuels. If the rise of CO_2 is allowed to continue at the same rate as in the last few years, the levels in the atmosphere will reach more than 600 parts per million (ppm) in about A.D. 2050. This concentration of CO_2 will seriously alter environmental conditions.

Nitrogen Cycle

Organisms require substantial amounts of nitrogen in order to synthesize many of their cellular constituents. Nitrogen may be found in soils, as inorganic salts of nitrogen (ammonium or nitrate), or as organic matter. In addition, a major source of nitrogen is the atmosphere, which consists of 78% nitrogen gas (N_2). This form of nitrogen is largely inaccessible to organisms,

FOCUS

METHANOGENIC TERMITES

At present, the earth's atmosphere contains less than 0.005% methane and approximately 0.034% CO_2. Scientists at the National Center for Atmospheric Research in Boulder, Colorado are noticing a gradual increase in atmospheric methane and CO_2. The increase in methane is approximately 2% per year, while that of CO_2 is about 0.5% per year.

The importance of methane and CO_2 in the atmosphere lies in the fact that slight increases of these gases could create what is called a "greenhouse effect" on the earth and lead to climate warming. If methane increased to 0.01% and CO_2 to 0.07%, the atmosphere would trap the solar rays reflected from the earth's surface and prevent the earth from cooling each day. These concentrations of methane and CO_2 would drastically affect life on earth. The amount of heat generated from such a greenhouse effect is directly proportional to the amount of these gases in the atmosphere.

What is the source of methane gas? Scientists have noted that as much as 50% of it comes from the activity of termites. These insects, whose diet consists entirely of cellulose (wood fiber), totally digest the cellulose and release waste products, primarily methane and CO_2. Termites produce approximately 10^{14} g of methane/year and 5×10^{16} g of CO_2/year. It is the methanogenic bacteria that convert carbon dioxide and hydrogen (resulting from the metabolism of cellulose) into methane. Since there are more than 2 trillion termites in the biosphere, and since they seem to be increasing in numbers, there is concern that they may significantly affect the concentration of methane in the atmosphere.

however, except for those that can incorporate this gas into organic molecules. The nitrogen-containing compounds not only serve as a source of nutrients for soil microorganisms but also be used as energy sources or electron acceptors. The conversion of nitrogenous compounds from one form to another by various communities results in what is called the **nitrogen cycle** (fig. 31-3).

Nitrogen Fixation

Nitrogen fixation is a process that involves the conversion of atmospheric nitrogen (N_2) into organic nitrogen (fig. 31-4). This process is carried out by many different types of bacteria and fungi. The bacterial genus *Rhizobium*, when it forms a symbiotic association with leguminous plants, is responsible for much of the nitrogen fixed on land. For example, species of *Rhizobium* associated with alfalfa plants may fix more than 100 kg (1 kg = 2.2 pounds) of nitrogen/acre/year. In contrast, *Azotobacter*, a free-living, nitrogen-fixing

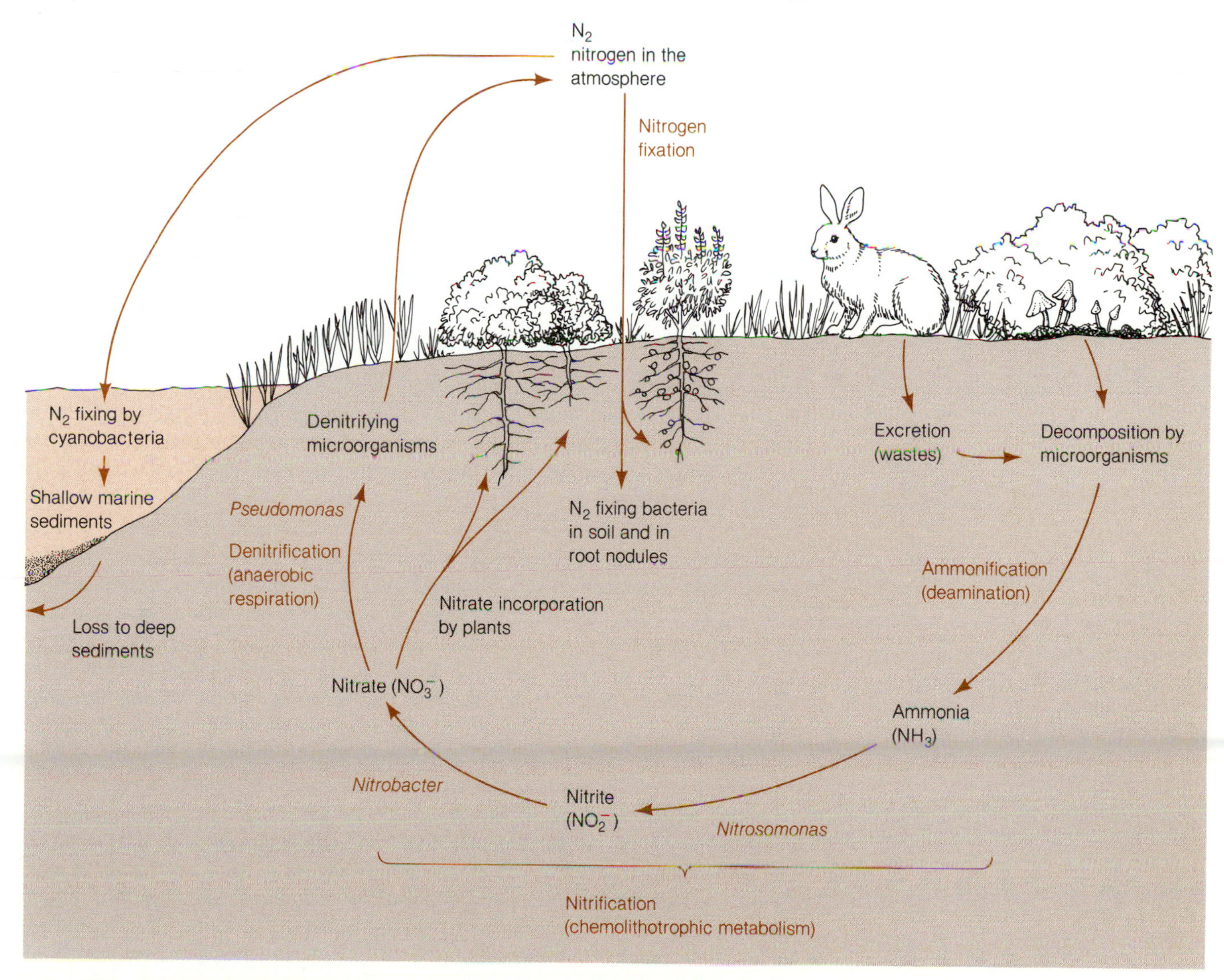

Figure 31-3 The Nitrogen Cycle. The nitrogen cycle is an important process that takes place in nature and insures an adequate supply of nitrogenous compounds to living organisms. The cycle involves a sequence of transformations involving various forms of organic and inorganic nitrogen compounds achieved by the metabolic activities of microorganisms.

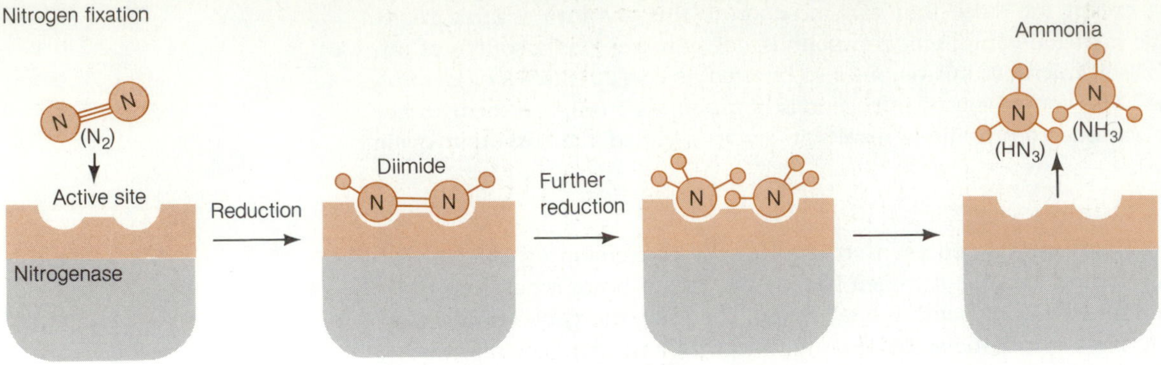

Figure 31-4 The Action of the Nitrogenase Enzyme. The nitrogenase enzyme reduces molecular nitrogen (N_2) to ammonia (NH_3). The ammonia can then be incorporated into organic acids to form amino acids.

bacterium, fixes less than 1 kg of nitrogen/acre/year in the same types of soils. Some actinomycetes fix nitrogen in symbiotic relationships with trees such as alder and bayberry. In aquatic environments, cyanobacteria such as *Anabaena*, *Rivularia*, and *Nostoc* fix as much as 10 kg of nitrogen/acre/year.

The soil bacterium *Rhizobium* can be isolated from tumorous growths (fig. 31-5) on the roots of leguminous plants such as alfalfa, clover, peas, lentils, and soybeans. These tumors, called **root nodules,** constitute the site of symbiotic nitrogen fixation. The bacteria induce the formation of root nodules when they invade plant root cells (fig. 31-5).

The development of the root nodule begins when *Rhizobium* growing around the plant roots metabolizes the amino acid tryptophan, released by the plant root, into indoleacetic acid (IAA). The IAA is a plant growth hormone that induces the root hairs to curl around the bacteria (fig. 31-6). The bacteria then penetrate the plant cell wall around the root hairs and reside between the cell wall and plasma membrane. The infection stimulates the root hair to develop a groove called the **infection thread.** The bacteria migrate down the infection thread to the plant cell body, where they invade the cytoplasm (probably by endocytosis). The bacteria multiply in the plant cell and spread as the plant cell divides. It is believed that the infecting bacteria stimulate DNA replication in the infected root cells, and as a consequence the infected root cells contain four copies of each chromosome □ and are said to be **tetraploid.** The tetraploid cells are also stimulated to multiply uncontrollably by the infecting bacteria. It is this uncontrolled growth that is responsible for the formation of the root nodules (fig. 31-6). Within the nodules, many of the bacteria become oddly shaped resembling X's, Y's, and T's. These oddly shaped cells are called **bacteroids.**

The bacteroids contain the enzymes necessary to reduce N_2 to NH_3, a first step in the fixation of nitrogen into organic molecules. The major enzyme involved in the fixation of nitrogen is **nitrogenase** (fig. 31-4). Nitrogenase is an oxygen-sensitive enzyme composed of a number of protein subunits. The subunits contain metal ions, such as iron and molybdenum, which are necessary for the reduction of N_2. The plant cell provides the bacteroids with an anaerobic environment and all the nutrients required for their metabolism and growth.

Nitrogen fixation is carried out in an anaerobic environment because the nitrogenase enzyme is very sensitive to oxygen. The oxygen concentration

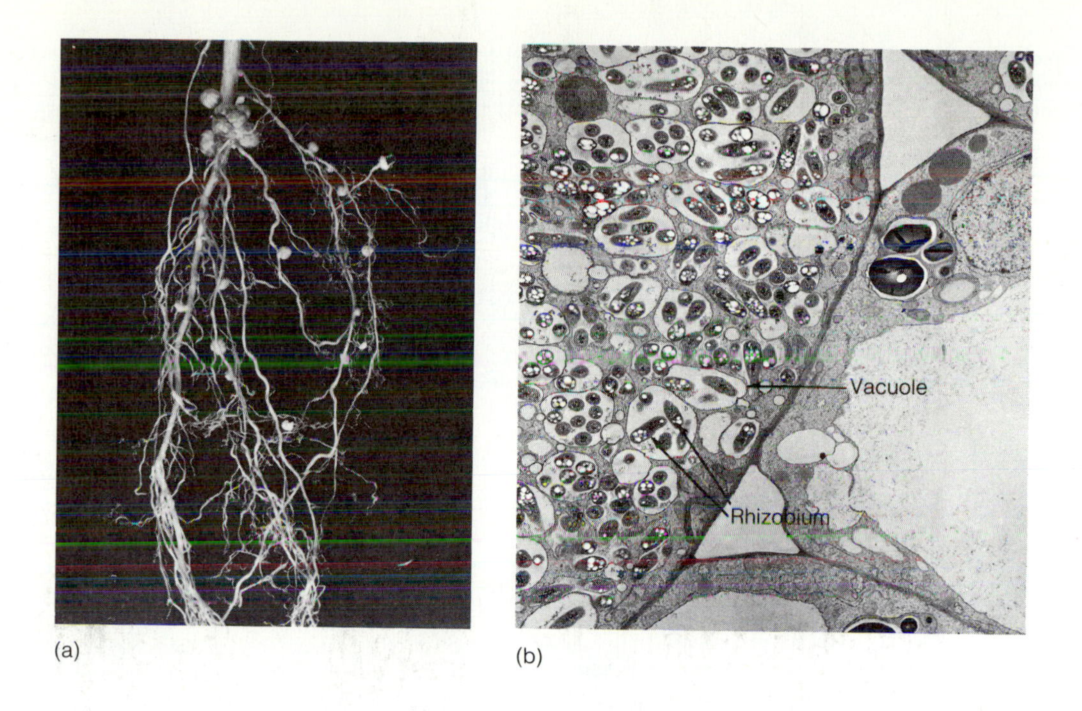

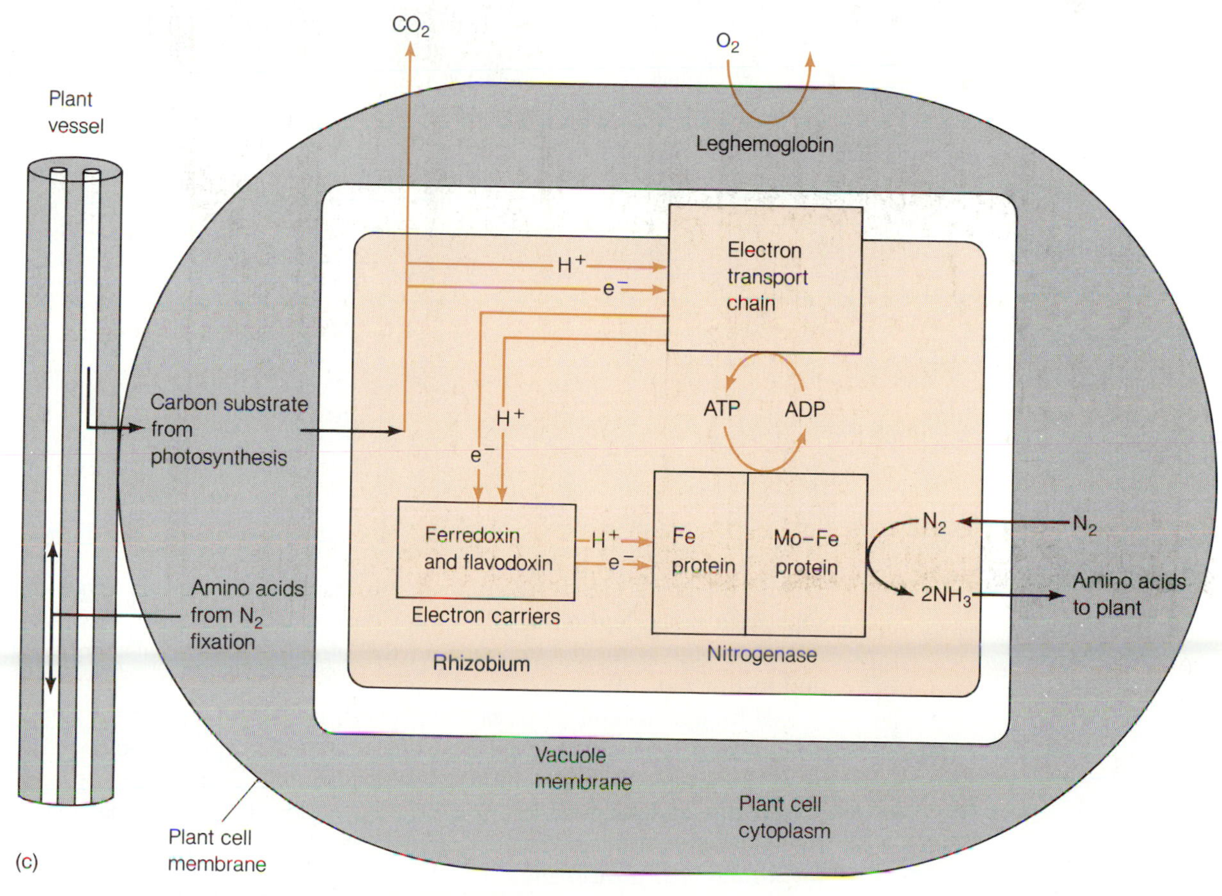

Figure 31-5 Root Nodules. *(a) Rhizobium* root nodules on a leguminous plant. The nodules are the site where nitrogen fixation takes place. *(b)* TEM of root nodule in cross section. Numerous vacuoles filled with *Rhizobium* can be seen in the plant cell. *(c)* Relationship between *Rhizobium's* electron transport system, nitrogenase activity, and plant symbiont nutrition.

Figure 31-6 **Sequence of Events Leading to the Formation of a *Rhizobium*-Induced Root Nodule**

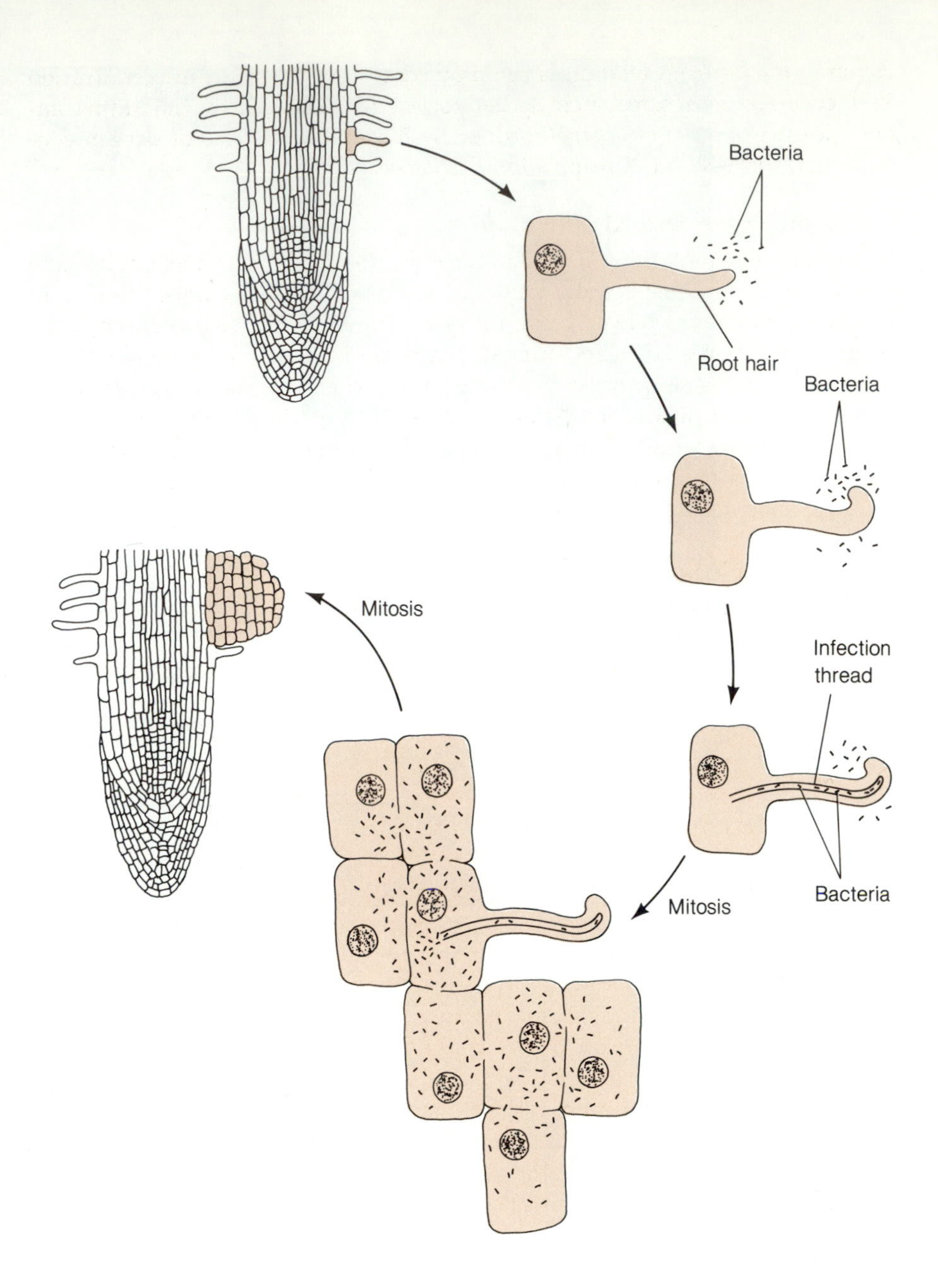

around the bacteroids is maintained at a very low level by an oxygen-binding substance called **leghemoglobin,** provided by the plant host. This substance resembles the hemoglobin found in the blood of mammals, both in form and in function.

Nitrogen fixation involves the expenditure of an enormous quantity of ATP, which is obtained from the bacterial chemotrophic metabolism. It has been estimated that as many as 24 molecules of ATP are used up each time a molecule of N_2 is reduced to 2 NH_3 molecules. The ammonia produced in the bacteroids diffuses into the plant's cytoplasm, where it becomes part of the amino acid **glutamine.**

Nitrogen fixation by other organisms also involves the presence of a nitrogenase enzyme like that found in *Rhizobium.* The anaerobic environment required for nitrogen fixation is maintained by carrying out nitrogen fixation in specialized structures such as heterocysts *(Anabaena);* by inhabiting anaerobic environments *(Clostridium);* or by having a high rate of oxygen consumption through rapid respiration *(Azotobacter).*

Ammonification and Nitrification

When nitrogen-containing organic molecules (amino acids, nucleotides, and urea) in excretions or cellular material are decomposed (catabolized), ammonium generally is released into the environment. This release of ammonia by microorganisms is called **ammonification** (fig. 31-3). Ammonification is a common biological process that takes place wherever microorganisms metabolize organic nitrogenous compounds. Many of the characteristic odors emanating from decaying organic matter or animal wastes result from ammonification. Much of the ammonia comes from the **deamination** of amino acid. Ammonia may be used by the microorganisms to satisfy their need for nitrogen.

194

A few bacteria also use ammonium as a source of energy for their metabolism. For example, the aerobic chemolithotrophic □ bacteria *Nitrosomonas* and *Nitrosococcus* oxidize (remove electrons from) ammonium ions, converting them to nitrite. These bacteria run the electrons obtained from the oxidation of ammonium through their electron transport system, thus making large amounts of ATP. Nitrite does not accumulate in the soil, however, because other aerobic chemolithotrophic soil bacteria, such as *Nitrobacter* and *Nitrococcus*, convert the nitrite to nitrate by removing more electrons from the nitrogen atom. These bacteria also pass the electrons through an electron transport system and make ATP. The conversion of ammonium into nitrate by microorganisms is called **nitrification.** The process of nitrification is quite common in well-aerated soils and provides a large quantity of nitrate for plant growth.

Both ammonium and nitrate are used by plants, although nitrate appears to be more readily available because it does not bind to clay particles in the soil as does ammonium. The addition of ammonium-containing fertilizers to grasses and crops is based on the expectation that nitrifying bacteria will oxidize the ammonium into nitrate, which can then be used as nutrients by grass plants. Soil microorganisms may also use nitrates and ammonium for their nitrogen metabolism and incorporate these inorganic ions into organic molecules.

Denitrification

Nitrate ions accumulate in aerobic soils because of nitrification and generally are assimilated by organisms in the terrestrial ecosystem. Under anaerobic conditions (as in waterlogged soils), most of the nitrate ions are used as electron acceptors in anaerobic respiration by certain soil bacteria. During this process, bacteria such as *Pseudomonas* reduce nitrate ions to nitrite, nitrous oxide, and eventually molecular nitrogen. In essence, nitrate reduction by anaerobic bacteria depletes the soil of its nitrates. The reduction of environmental nitrate and its subsequent loss to the atmosphere is called **denitrification.**

The nitrogen cycle (fig. 31-3) clearly shows that microorganisms are intimately involved in nitrogen fixation, ammonification, nitrification, and denitrification. These processes, which involve the reuse of a finite resource, demonstrate that the waste products of one population can serve as a nutrient or energy source for another population, which in turn provides nutrients for yet another population. Together, these organisms recycle nitrogen atoms among the various populations of the soil or aquatic community.

Farmers take advantage of the microbial activities that take place in soils. For example, plowing fields before fertilization and planting creates an aerobic environment that promotes the growth of ammonifiers and nitrifiers and discourages the growth of denitrifiers. Plowing helps to achieve a maximum amount of ammonium and nitrate in the soil. Inexpensive ammonium-containing fertilizers such as $(NH_4)_2SO_4$ can be applied to soils because the nitrifying bacteria in the soil will convert the ammonium to nitrate, thus promoting lush plant growth.

Sulfur Cycle

Sulfur is an essential element that is required for the synthesis of certain amino acids (methionine and cysteine), coenzymes (coenzyme A), and sulfate-containing polysaccharides. Sulfur is also used as a source of electrons (energy) and as an electron acceptor by various microorganisms. Sulfate is believed to be the preferred form of sulfur for assimilation, even though many microorganisms are able to utilize sulfur-containing organic molecules instead.

The conversion of sulfur from one form to another by microorganisms is called the **sulfur cycle** (fig. 31-7). The soil community that participates in the recycling of sulfur is made up of various populations.

Hydrogen sulfide, sodium thiosulfate, and hydrogen gas are used by photosynthetic bacteria, such as the green and purple sulfur bacteria, as a source of reducing electrons to fix CO_2. These bacteria oxidize hydrogen sulfide to elemental sulfur and eventually to sulfate. Nonphotosynthetic bacteria, such as *Thiothrix* and *Beggiatoa*, also can use hydrogen sulfide or thiosulfate as a source of electrons for their metabolism. Although the photosynthetic bacteria are anaerobic and the chemolithotrophic bacteria are aerobic, both groups of microorganisms release sulfate as a waste product of their metabolism. The sulfate does not accumulate in the soil, but instead is assimilated by a variety of soil inhabitants.

One of the most interesting soil bacteria, which is active in the sulfur cycle as well as in the nitrogen cycle, is *Thiobacillus.* Under aerobic conditions, this organism can remove electrons from sulfur (S) and use them to make ATP, producing sulfuric acid as a byproduct. Under anaerobic conditions, however these bacteria reduce nitrate to nitrite. Some species obtain their electrons from organic compounds, others from sulfur compounds.

Some chemoheterotrophic bacteria, such as *Proteus* and *Salmonella*, remove the sulfide from the amino acids methionine and cysteine and release H_2S. Another bacterium, called *Desulfovibrio*, in the absence of oxygen and in the presence of sulfate ions will carry out anaerobic respiration to produce hydrogen sulfide. These bacteria use sulfate ions as the terminal electron acceptor □ and consequently reduce the sulfate to hydrogen sulfide. The hydrogen sulfide may then react with iron in the soil to form ferrous sulfide, which makes anaerobic soils black. Since extensive ferrous sulfide forms in marshes and stagnant ponds, the soil in these areas is usually black. *Desulfovibrio* causes problems because it can corrode iron pipes and foundations in sulfate-containing soils. Under anaerobic conditions, this bacterium pro-

190

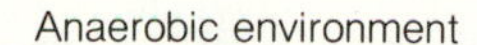

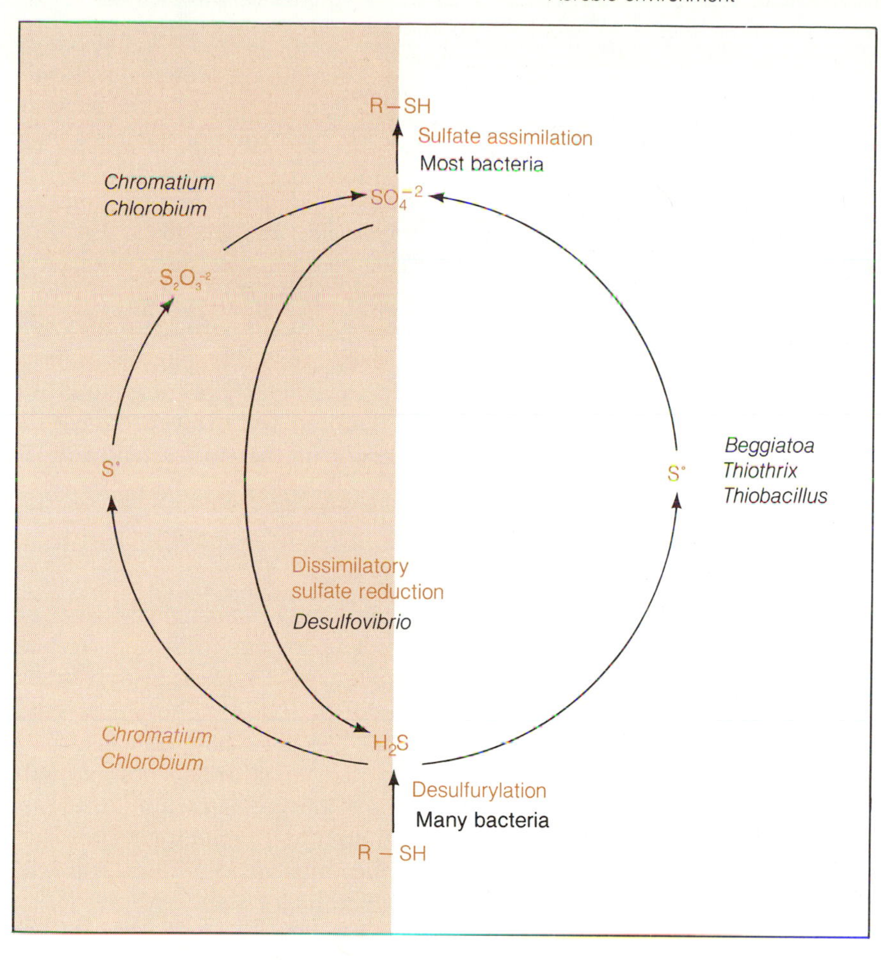

Figure 31-7 The Sulfur Cycle. The sulfur cycle is an important process that takes place in nature. It consists of a sequence of transformations of sulfur-containing compounds achieved by the metabolic activities of microorganisms.

duces H_2S, which then reacts with the $Fe(OH)_3$ layer on iron, converting it to FeS. The black FeS flakes off the iron, a process that leads to the disintegration of the pipe or foundation.

Sulfur oxidation frequently produces sulfuric acid, which solubilizes and mobilizes phosphates and other minerals beneficial to microorganisms and plants. The reduction of sulfur to H_2S, however, creates a poison that inhibits aerobic respiration in many organisms.

Fossil fuels such as coal and oil contain large amounts of sulfide and organic sulfur. The burning of coal and oil release most of the sulfur as sulfur dioxide. When the sulfur dioxide reacts with water in the atmosphere, sulfurous acid is formed. **Acid rains** in highly industrialized areas of the world cause considerable damage by corroding limestone and marble buildings, eroding plant leaves, and acidifying soils, streams, ponds, and lakes. A low pH changes the ecology of soils and waters by killing plants, animals, and many microbial populations. Presently, little research is devoted to "cleaning up" fossil fuels by having microorganisms remove the sulfur. It has been proposed, however, that coal might be ground up into a slurry and inoculated with *Thiobacillus*. This bacterium would convert the H_2S into sulfuric acid, which could then be removed from the coal.

Iron Cycle

Iron is an essential element required in trace amounts by all living organisms. Iron is part of many different Fe and S containing proteins and cytochromes, which are involved in oxidation-reduction reactions. In the environment, iron is generally found as insoluble ferric hydroxide or ferrous sulfide (fig. 31-8).

Some chemolithotrophic bacteria, such as *Thiobacillus ferrooxidans*, are able to obtain energy by oxidizing the ferrous ion into the ferric ion. Iron pyrite or "fool's gold" is also oxidized by *T. ferrooxidans* to ferrous ions and to sulfate. One of the iron bacteria, *Gallionella ferruginea*, oxidizes ferrous ions to the ferric ion and uses this ion to cover its sheath with insoluble ferric hydroxide. Iron-oxidizing bacteria are responsible for large deposits of ferric hydroxide. These deposits are often created in anaerobic swamps and are referred to as **bog-iron.** Many deposits of bog-iron were mined during the early Industrial Revolution because they were near the surface and easy to develop.

Phosphorous Cycle

Phosphate is required in large amounts by all living organisms (fig. 31-9). It is required for the synthesis of nucleic acids, phospholipids, and teichoic acids. **Apatite** is a large reservoir of phosphate. This material is used by the chemical and fertilizer industries as a source of sodium phosphoric acid, which is made by reacting apatite with acid. The phosphate that is derived from the chemical treatment is used as a fertilizer to stimulate crop growth. Microorganisms that release acids into the soil during fermentation □ also solubilize small amounts of phosphate from apatite. Chemolithotrophs such as *Nitrobacter* and *Thiobacillus*, which produce nitric acid and sulfuric acid, respectively, can also solubilize phosphate. The major source of phosphate, however, is that found dissolved in water and in living and dead organisms.

186

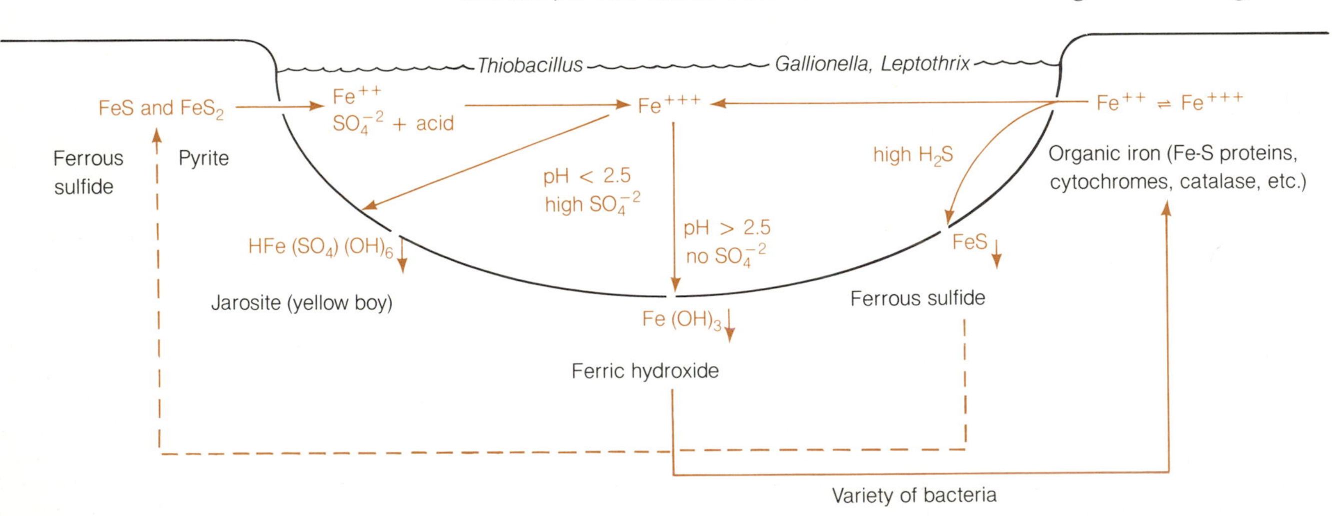

Figure 31-8 The Iron Cycle. In nature, iron is generally found as Ferrous sulfide, pyrite, and ferric hydroxide. The ferrous ion is converted to the ferric ion by bacteria such as *Thiobacillus.* In the presence of high concentrations of sulfate and under very acidic conditions (pH <2.5), the ferric ion precipitates as Jarosite (yellow boy). In the absence of sulfate, the ferric ion precipitates as ferric hydroxide. Bacteria such as *Gallionella* and *Leptothrix* oxide ferrous to ferric ions, which generally precipitate as ferric hydroxide. Insoluble ferric hydroxide is solubilized by various bacteria when they reduce the iron (during anaerobic respiration) to ferrous ions.

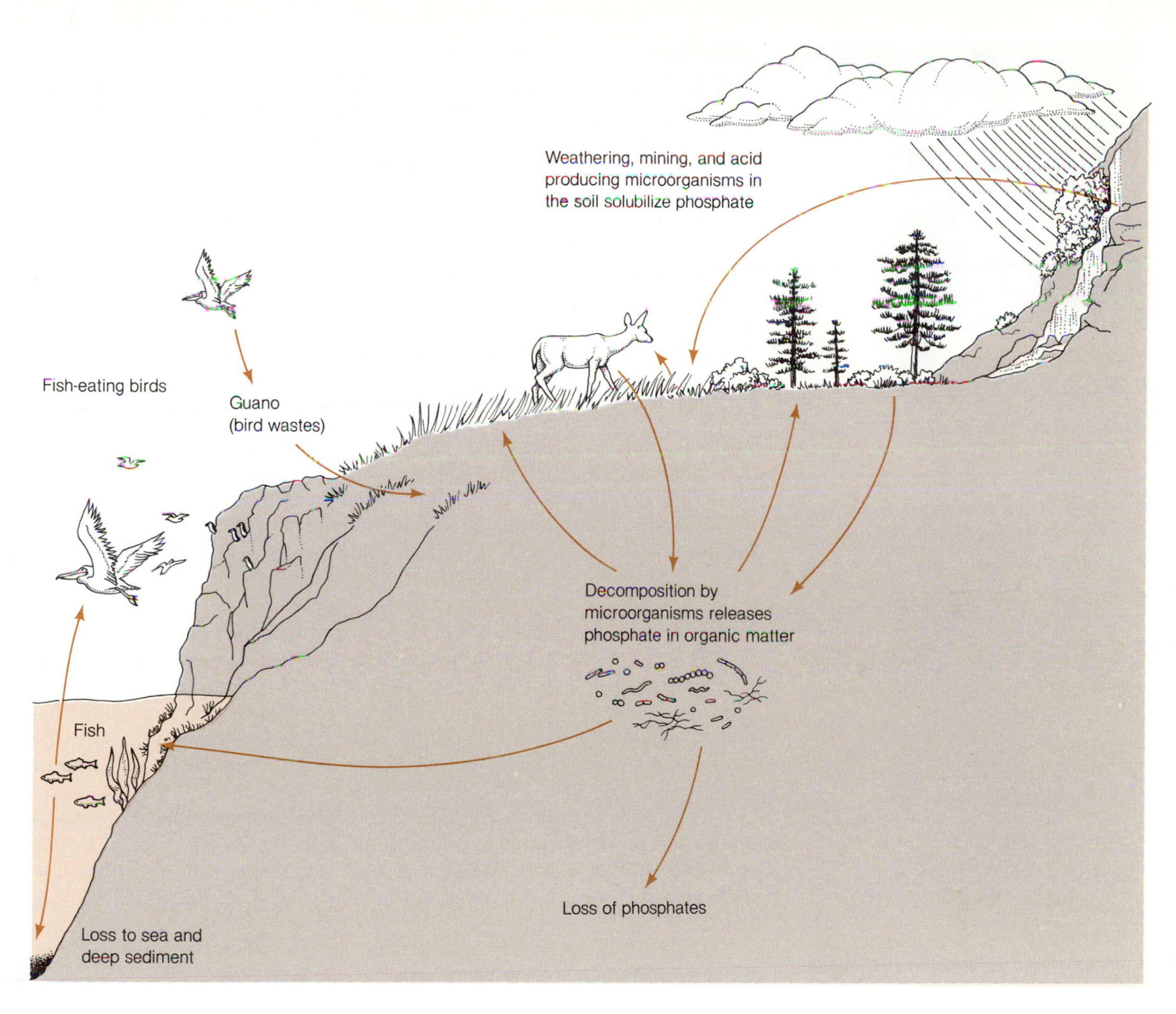

Figure 31-9 The Phosphorus Cycle. Phosphate in apatite, $Ca_5(PO_4)_3 \cdot (Cl \text{ or } F)$, is solubilized by acid-producing bacteria in the soil. Some of this phosphate is asimilated by plants and incorporated into their cells. The organic phosphate can then be used by other organisms when they consume plant products. Phosphates in body wastes, bones, spines, and so forth can be solubilized by acid-producing bacteria, but some of the phosphate is lost to the sea or leached out of the soil.

Most microorganisms do not alter the valence of the phosphorus atom. Some organisms, however, are able to use phosphate as an electron sink when oxygen, sulfate, and nitrate are not present. Phosphine (PH_3), a highly volatile and explosive compound, is produced when certain microorganisms reduce phosphate. This substance is sometimes produced in swamps and in mass graveyards, where extensive degradation of organic material is occurring in the absence of the usual electron acceptors (oxygen, nitrate, or sulfate). The oxidation of phosphine to phosphate produces a green glow and frequently ignities marsh gas (methane) so that flashes of light occur.

THE WINOGRADSKY COLUMN

The Winogradsky column (fig. 31-10) is a closed ecosystem in which nutrients are passed from one population of organisms to another. All that is required is a constant energy source. The column is a very useful device for demonstrating the relationships among soil bacteria. Light, the energy source for the photosynthetic microorganisms in the Winogradsky column, generally is provided by an incandescent bulb.

Figure 31-10 The Winogradsky Column. The various patches appearing in the Winogradsky column represent various populations of microorganisms that develop. The Winogradsky column is a good tool to visualize microbial activities in nature, especially the various transformations of the element sulfur.

A Winogradsky column can be made by packing a one-liter glass cylinder with mud mixed with plaster of Paris ($CaSO_4$), straw, and shredded filter paper (cellulose). The column is packed with about 800 ml of the enriched mud and filled with pond water. The cylinder can then be placed on a window ledge in direct sunlight or under an incandescent light. Distilled water can be added to replace any evaporated water.

Mud and straw are sources of many different types of bacteria, such as *Erwinia*, *Pseudomonas*, and *Cytophaga*, which break down cellulose, hemicellulose, pectin, and chitin. These decomposers create an anaerobic environment at the bottom of the column and also produce metabolic wastes such as lactic acid, acetic acid, butyric acid, ethanol, butanol, diacetyl, and 2,3-butanediol. These wastes are a source of electrons for bacteria such as *Pseudomonas* and *Desulfovibrio*. *Pseudomonas* carries out anaerobic respiration by reducing nitrate and nitrite, producing ammonium and nitrogen gas, while *Desulfovibrio* reduces the sulfate in the plaster of Paris, producing hydrogen sulfide. Some of the hydrogen sulfide reacts with iron, forming ferrous sulfide and blackening the mud.

The production of H_2S and CO_2, the absence of O_2, and the presence of light select for the obligate anaerobic, photosynthetic bacteria. These bacteria (purple and green sulfur bacteria) use hydrogen sulfide, thiosulfate, and hydrogen gas as sources of electrons. Some of the purple sulfur bacteria, such as *Chromatium*, actually fill with sulfur (S) granules as they oxidize the hydrogen sulfide. The green and purple sulfur bacteria form large colonies in the mud, and these colonies can be seen as green or purple patches against the glass.

The purple nonsulfur bacteria, such as *Rhodospirillum* and *Rhodopseudomonas*, also develop in the column, but they develop in areas where the H_2S concentration is low. These bacteria are sensitive to hydrogen sulfide, so they generally use organic molecules as their source of electrons for fixing CO_2.

The aerobic cyanobacteria that fix nitrogen are found in the upper part of the column, where there is atmospheric nitrogen. Bacteria such as *Beggiatoa* are also found in the upper layers of the column. These long, gliding bacteria are usually found associated with the decomposing paper and straw in the column, because the cellulose-decomposing bacteria provide them with acids and alcohols they require.

At the very top of the column, where there is plenty of oxygen and little or no hydrogen sulfide, one finds facultative anaerobic bacteria, eukaryotic algae, and protozoans.

USEFUL SOIL MICROORGANISMS

Much thought and research has focused on the use of soil microorganisms in economically or environmentally important tasks. For example, the mining industry is now investigating the possibility of genetically engineering bacteria to aid in the efficient mining of copper and uranium from low-grade ores (see Chapter Preview). The petroleum industry is also taking advantage of microorganisms to increase the efficiency with which they extract oils. Vast amounts of oil in shale **(shale oils)** are not exploited at present because it is rather expensive to extract the oil by chemical means. Instead, microorganisms are being tested that have been selected (or engineered) to solubilize the shale and hence extract the oil. The use of these bacteria is in the experimental stage; however, bacteria that produce xanthans, such as *Xanthomonas*, are being used to increase the viscosity of water pumped into depleted oil fields. This water is used to force any oil trapped in subterranean crevices into the well pump.

Certain other soil bacteria perform functions that are useful to the ecology of an area. For example, pesticides and herbicides that accumulate in the soil and affect the ecology of a region are often degraded by soil organisms. Soil bacteria such as *Achromobacter* and *Corynebacterium* inactivate the herbicide called 2,4-dichlorophenoxyacetic acid (2,4-D) in 3–7 weeks. Similarly, pseudomonads and *Agrobacterium* deactivate the herbicide Dalapon (2,2-dichloropropionic acid).

As can be seen from the previous discussion, soil microorganisms are actively involved in nutrient recycling and are essential to the well-being of terrestrial organisms. Soil microorganisms also perform other functions that are exploited by various industries to increase the yield of valuable resources. These microorganisms also cause diseases in plants and animals. In Chapter 34, we will see the impact that these pathogenic microorganisms have on domestic plants and animals.

SUMMARY

Microorganisms play an important role in shaping terrestrial environments. The fertility of soils, and the types of organisms that can grow in them, are dependent upon the microbial activities that take place in the soil.

SOIL FORMATION

1. Soil is the surface layer of the earth which supports the growth of plants, animals, and microorganisms.
2. There are many different types of soils. The differences are due to the type of rock or sediment that is decomposed, the climate, and the type of plants and microorganisms that live on and decompose the soil.

BIOGEOCHEMICAL CYCLES

1. Microorganisms and plants are important in establishing and changing the earth's environment because they are able to promote chemical conversions of large amounts of carbon, oxygen, nitrogen, phosphorus, sulfur, and metals.
2. The earth's organic matter, CO_2, H_2O, and O_2 are closely linked and interconnected by the processes we call photosynthesis and respiration.
3. O_2 has accumulated and CO_2 has decreased to present atmospheric levels because of the sequestering of organic compounds under sediment. If all the organic sediments should suddenly be oxidized, the oxygen would virtually disappear from the atmosphere and the carbon dioxide would increase drastically.
4. A sufficiently high concentration of CO_2 could increase the earth's temperature, because of the greenhouse effect, and could destroy life as we know it.
5. Methane is produced from CO_2 and H_2 by methanogenic bacteria or hydrogen bacteria, and is respired to CO_2 and H_2O by methane bacteria (methylotrophs).
6. All living organisms are dependent upon the organic nitrogen produced by a few groups of microorganisms and their plant partners.
7. Organic nitrogen is created when microorganisms fix atmospheric nitrogen gas into organic compounds. This process is called nitrogen fixation. Organic nitrogen can also be formed when microorganisms and plants assimilate ammonia or nitrates into their organic compounds.
8. Organic nitrogen can be converted to ammonia by numerous organisms, by a process called ammonification.

9. Ammonium ions can be converted to nitrite ions, and these in turn can be converted to nitrate ions, by a small group of bacteria. This process is called nitrification.
10. Nitrogen can temporarily be lost to organisms when it is converted into N_2 gas and escapes to the atmosphere. The process of converting nitrate to N_2 is called denitrification.
11. Sulfur is required to make two amino acids, a coenzyme, and sulfate-containing polysaccharides.
12. Various chemical forms of sulfur are interconvertible by microorganisms. Some organisms use H_2S as an electron source and convert it to S and this in turn to SO_4^{-2}. Other microorganisms use SO_4^{-2} as an electron acceptor to create H_2S.
13. Fossil fuels, such as coal and oil, contain large amounts of sulfide and organic sulfur. The sulfur is released as sulfur dioxide when the fossil fuels are burned. When the sulfur dioxide reacts with water, sulfurous acid is produced. Acid rains, resulting from this process, corrode limestone and marble buildings and decrease soil and water pH.
14. Iron is an essential element for a number of enzyme systems. In living organisms, it is generally converted back and forth between the ferrous ion (Fe^{+2}) and the ferric ion (Fe^{+3}).
15. Iron is often oxidized and reduced by microorganisms. Ferric hydroxide is a water-insoluble form that often occurs in large, red deposits. Ferrous sulfide is a black precipitate that is also quite common, especially in sulfide-rich soils.
16. Phosphate is required in large amounts for the synthesis of nucleic acids (DNA and RNA), energy storage nucleotides (ATP, GTP, UTP, etc.), and phospholipids.
17. Phosphate is the ionic form used by all organisms, and it is rarely oxidized or reduced by microorganisms.

THE WINOGRADSKY COLUMN

1. The Winogradsky column is a stable ecosystem that is useful in demonstrating the relationships that exist among many populations of soil microorganisms.
2. The process of nutrient recycling, particularly the sulfur cycle, is clearly seen in this closed ecosystem.

USEFUL SOIL MICROORGANISMS

Soil microorganisms are being studied extensively by various industries in the hope that some of them can provide useful materials or aid in the extraction of economically-important substances.

STUDY QUESTIONS

1. Describe three ways in which soil microorganisms affect humans.
2. Compare the microbial processes that take place in the nitrogen cycle with those in the sulfur cycle.
3. What happens to fertile, well-aerated soils when excessive rains make the soils waterlogged?
4. How might soil microorganisms be used in industrial processes?
5. Where in the nitrogen cycle do the following processes take place? Name one organism that carries out this process, and state if the process requires the input of energy or is energy producing.
   a. anaerobic respiration;
   b. aerobic respiration;
   c. chemolithotrophic metabolism;
   d. nitrate reduction;
   e. N_2 reduction.
6. What useful function(s) do soil microorganisms perform in nature?
7. What characteristics are shared by all the various biogeochemical cycles discussed in this chapter?

SUPPLEMENTAL READINGS

Atlas, R. M. and R. Bartha. 1981. *Microbial Ecology. Fundamentals and applications.* Menlo Park, Calif.: Addison-Wesley.

Brierly, C. L. 1982. Microbiological mining. *Scientific American* 247:44–53.

Brill, W. J. 1977. Biological nitrogen fixation. *Scientific American* 236:68–81.

Hardy, R. W. F. and U. D. Havelka. 1975. Nitrogen fixation research: A key to world food. *Science* 188:633–643.

Ourisson, G., Albrecht, P., and Rhomer, M. 1984. The microbial origin of fossil fuels. *Scientific American* 251(2):44–51.

Payne, W. J. 1983. Bacterial denitrification: Asset or defect. *BioScience* 33:319–325.

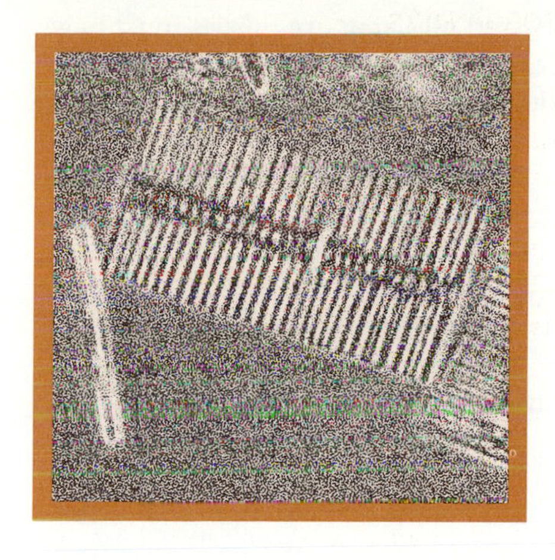

CHAPTER 32
MICROBIOLOGY OF AQUATIC ENVIRONMENTS

CHAPTER PREVIEW

THE ECOLOGY OF SPIRIT LAKE AFTER THE ERUPTION OF MT. ST. HELENS

The explosion of Mount St. Helens on May 18, 1980 had a significant impact on many communities in the state of Washington. The eruption sent millions of tons of volcanic ash into the atmosphere. This ash later fell to the ground, covering the state with 1–5 inches of ash. The eruption also sent massive quantities of ash, soil, organic matter, and wood debris into the waters of Spirit Lake and many other lakes neighboring Mount St. Helens. Spirit Lake, a pristine, subalpine lake, became almost instantly an anaerobic, dying lake, largely devoid of aquatic life.

Among the most dramatic changes in the lake's ecology was the sudden drop in the water's dissolved oxygen. The oxygen concentration of the water, which is essential for the survival of aerobic microorganisms, dropped to zero within three months after the eruption of the volcano. The drop in dissolved oxygen resulted from the sudden rise in water temperature caused by the blast and by the enrichment of the lake's water with organic matter. The increased water temperature reduced the solubility of oxygen. The influx of organic matter, from the tons of soil that contaminated the lake, promoted the growth of heterotrophic microorganisms that consumed the remaining oxygen. The total number of bacteria/ml of lake water rose to about 5×10^8, 2–3 orders of magnitude greater than before the eruption. By the end of 1980, a once highly aerobic young lake had become instantly "old."

The present microbial activities in Spirit Lake and in many other neighboring lakes are largely anaerobic. Denitrification and nitrogen fixation are common microbial activities found in these dark, anaerobic lakes. Aerobic processes, such as nitrification and biological oxidations, are as yet undetectable.

The increased lake turbidity due to volcanic ash created yet another pollution problem for the already burdened lake. Turbid water limited light penetration and hence limited the amount of photosynthesis. This change resulted in the development of chemolithotrophic bacteria just below the light penetration zone in which the primary energy source is inorganic material.

It is possible that, as the lake communities stabilize and humans intervene to remove the ash and organic nutrients, Spirit Lake may return to being a pristine lake with a water-purifying microbial community. In the meantime, Spirit Lake has become an "old," decaying lake as a consequence of a catastrophic event that caused a rapid aging process to occur, not over many thousands of years, but in a mere three months.

The natural aging process is a very gradual and insidious process that leads to the inevitable conclusion of a lake's life. As the lake changes and ages, however, the microbial populations change with it. Even when the lake conditions are no longer suitable for the life of most plant or animals, microbial populations continue to thrive and influence the ecology of the lake.

This chapter will discuss the microbial processes associated with aquatic habitats and how these processes participate in the ecology of aquatic environments. This chapter will also discuss in detail those microbial activities that participate in the purification of natural and manmade bodies of water.

LEARNING OBJECTIVES

A STUDY OF THIS CHAPTER SHOULD ENABLE YOU TO:

DESCRIBE SOME OF THE VARIOUS GROUPS OF MICROORGANISMS THAT COLONIZE AQUATIC HABITATS

DISCUSS THE ROLES THAT MICROORGANISMS PLAY IN THEIR AQUATIC ENVIRONMENT

OUTLINE THE STEPS INVOLVED IN THE PURIFICATION OF DRINKING WATER SUPPLIES

DESCRIBE HOW WATER IS TESTED FOR POLLUTION, AND INDICATE WHICH KINDS OF PATHOGENS ARE LIKELY TO BE FOUND IN WATER SUPPLIES

OUTLINE THE VARIOUS PROCEDURES AND STAGES IN SEWAGE TREATMENT

Water is essential for biological processes, and all living organisms must have continuous access to this substance in order to survive. The water used by organisms is recirculated in what is called the hydrologic cycle. The oceans, lakes, and streams serve as reservoirs for the water. Heat from the sun causes water to evaporate from these reservoirs. When the water vapor cools, clouds are formed, and the condensation of the water in the clouds can result in the formation of rain or snow. This precipitation refills the water reservoirs. Much of the water from the land manages to reach the oceans, and the hydrologic cycle begins again. It is this water that makes up the aquatic habitats in the biosphere.

Microorganisms in the water are important to humans, because some of them are pathogenic and others help maintain the purity of the water. Close examination of a drop of pond water reveals a myriad of microorganisms. Some of them come from the air and colonize the water, where they multiply and establish a more or less transient residence. Others come from the soil, and yet others from human and animal wastes. Most, however, are permanent residents of the pond environment.

Aquatic environments have unique communities of microorganisms, many of which help eliminate pollutants from the water. For example, such microorganisms often degrade or detoxify organic materials from industrial discharges and agriculture by using these organic materials as nutrients.

Aquatic communities live in a state of equilibrium within their environment. An increased concentration of nutrients in the environment, for instance, promotes the growth of aerobic microorganisms that increase in numbers and deplete the available oxygen. If this process continues unchecked, eventually it leads to an anaerobic environment. If the amount of nutrients introduced is not excessive, however, or if the source of pollution is removed, the aquatic habitat may recover. In this chapter, we will study some of the natural events in which aquatic microorganisms participate and the ways in which humans can affect these events and put them to good use.

SOME AQUATIC ENVIRONMENTS AND THEIR BIOLOGICAL COMPOSITION

The **hydrosphere** includes all the bodies of water on the earth, such as rivers, lakes, and oceans. These aqueous habitats contain an array of organic nutrients and salts from runoff, percolation, or dissolving rock. The distribution and quantities of these chemicals determine the nature of the habitat. The hydrosphere generally is divided into freshwater, marine, and estuarine habitats. Common freshwater habitats are lakes, ponds, rivers, streams, and swamps. Marine habitats are found within and along all the oceans, while estuarine habitats occur where freshwater habitats and oceans meet.

Freshwater Habitats

The science that studies freshwater habitats is called **limnology.** Limnological studies have shown that freshwater habitats have their characteristic resident microbiota. Because nutrients often are very dilute in fresh water, many of the microorganisms are adapted to feeding in nutrient-sparse **(oligotrophic)** habitats, and many are motile so that they can move toward nutrients. For example, *Caulobacter*, a common bacterial inhabitant of

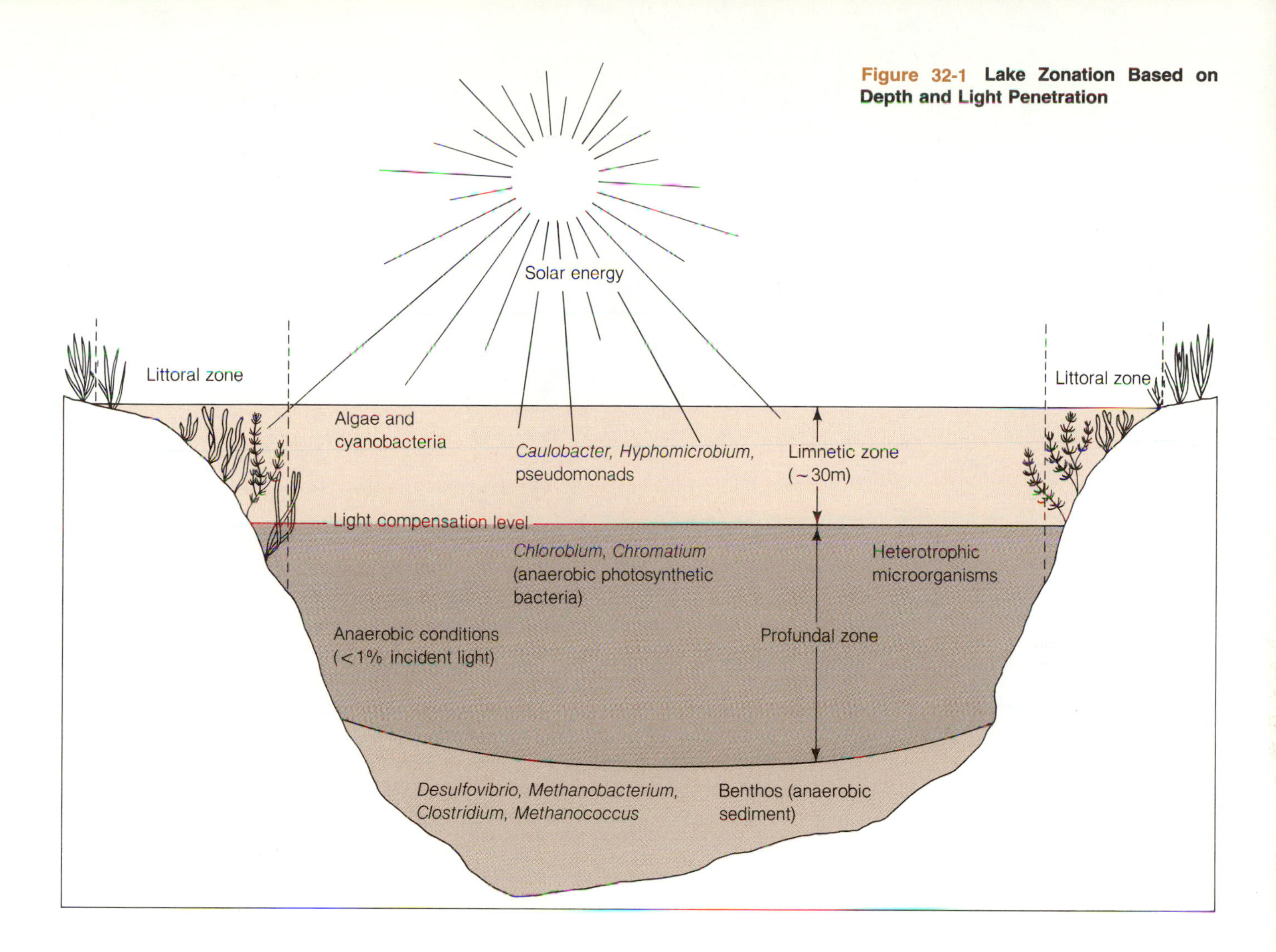

Figure 32-1 Lake Zonation Based on Depth and Light Penetration

aquatic environments, is motile and has an extension of the cytoplasm called a **prostheca** □ that increases their surface area so that dilute nutrients are absorbed efficiently.

89

Lakes

Lakes are inland bodies of water that are fed by rain, melting glaciers, snow packs, rivers, and streams. They range in size from a few square kilometers $(km)^2$ to tens of thousands of km^2. A lake can be divided into various parts, according to depth and amount of light available (fig. 32–1) or water temperature (fig. 32-2).

The **littoral zone** (shore) is an area of shallow water located at the edge of the lake (fig. 32-1). In the littoral zone, light penetrates all the way to the bottom, supplying phototrophs □ with energy for their metabolism. Since nutrients in the littoral zone are in high concentration, this part of the lake can support large numbers of microorganisms. The organisms found in the littoral zone include cyanobacteria, purple and green sulfur bacteria, eukaryotic algae, protozoa, and fungi. In this zone one also finds plants such as

110

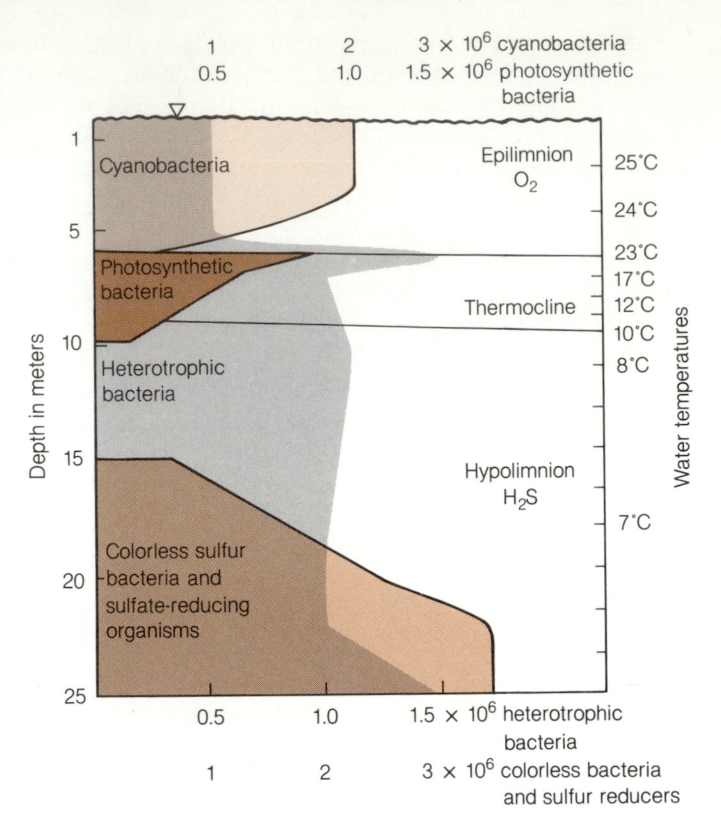

Figure 32-2 Lake Zonation Based on Temperature Changes. Temperature profile of a lake. During the summer months, the lake becomes stratified into three distinct layers. In the epilimnion the water is warm and its temperature declines gradually. In the thermocline, the water temperature drops rapidly. The hypolimnion is a zone below the thermocline where the water temperature remains relatively constant. The distribution of microorganisms is also illustrated. (After Rheinheimer, G. 1974. *Aquatic Microbiology.* New York: John Wiley & Sons.)

cattails, rushes, water lilies, and duckweed. Animals in the littoral zone include snails, insect larvae, clams, and worms.

The **limnetic zone** is an area of open water beyond the littoral zone, extending to a depth of a few meters below the surface, where less than 1% of the sunlight penetrates. This is called the **light compensation level.** The depth of the light compensation level depends upon the turbidity of the lake, but it rarely exceeds 25–30 meters even in the clearest of lakes. The majority of microorganisms in the limnetic zone are small, free-floating phytoplankton, such as cyanobacteria (*Microcystis*, *Anabaena*, and *Aphanizominon*) and eukaryotic algae. Fewer microoganisms are found in this zone than in the littoral zone.

The **profundal zone** extends from the light compensation level to the bottom of the lake. It has no plant life and is dominated by saprophytic fungi and bacteria. The bottom of the lake is a zone of decomposition, colonized primarily by anaerobes that decompose settling organic material.

Sometimes it is desirable to subdivide a lake into zones based on the water temperature at different depths (fig. 32-2). The water temperature of a lake during the summer declines gradually from the surface to a depth at which the temperature begins to decline rapidly. This area of rapid temperature decline is called the **thermocline.** The column of water above the thermocline is called the **epilimnion,** and that below the thermocline is called the **hypolimnion.** During the summer, the epilimnion is characterized by warm temperatures and a high concentration of oxygen. In this area, considerable photosynthesis is carried out by the phytoplankton. The cyanobacteria and photosynthetic bacteria are present in large numbers in the epilimnion (fig. 32-2). Autotrophic bacteria, such as *Nitrosomonas* and *Nitrobacter*, are present in aerobic portions of the epilimnion. These bacteria 745 are very important in nutrient recycling □ in lakes. They convert ammonium to nitrate, which is then made available to other organisms.

The hypolimnion is colonized primarily by photosynthetic heterotrophic, and sulfate-reducing bacteria. The distribution of the photosynthetic bacteria is determined by the availability of hydrogen sulfide and light and the lack of dissolved oxygen. Heterotrophic bacteria are distributed throughout the lake, although they predominate in the thermocline and hypolimnion. Anaerobic sulfur and sulfate-reducing bacteria generally are found in the lowest areas of the hypolimnion, where the concentration of organic matter is high. These bacteria release H_2S, which can then be used by phototrophic bacteria. The bottom of the lake **(benthos)** contains bacteria such as *Desulfovibrio*, which reduces sulfate to hydrogen sulfide. This chemical imparts the smell of "rotten eggs" to lakes and ocean muds. Methane-producing (methanogenic) bacteria and anaerobic spore formers may also be found in the benthos.

The number of resident microorganisms in a lake is generally determined by the amount of nutrients in the lake. Certain lakes are called oligotrophic because they have a low concentration of dissolved nutrients. Consequently, oligotrophic lakes have a relatively small microbial population and retain a high concentration of dissolved oxygen in the water. As the amount of dissolved nutrients in the lake increases, either by pollution or by **siltation** and runoff (a natural aging process of most lakes), the productivity of the lake increases until it supports a large microbial population. This large community of microorganisms is predominantly heterotrophic and consumes large amounts of dissolved oxygen. Because of the tremendous amount of microbial respiration taking place, the lake eventually becomes anaerobic and stagnant. Lakes of this sort are called **eutrophic** lakes.

Rivers and Streams

Rivers and streams are flowing bodies of fresh water that provide a number of very different habitats. The velocity of flow, the nutrient content, the oxygen concentration, and the temperature vary in different parts of the river and contribute to the diversity of habitats. The upper courses of rivers and streams are generally cool and rich in dissolved oxygen, but nutrient content is low. Most of the available nutrients are readily absorbed by microorganisms and other life forms attached to surfaces over which the water flows. The mineral and organic content is reduced primarily by aerobic, heterotrophic microorganisms. As the rivers slow down in the middle and lower courses, the oxygenation of the water diminishes, the water becomes warmer, and the nutrient content increases. These changes result in an upsurge in the number of microorganisms present in the water. In addition, rivers flowing through heavily populated areas accumulate nutrients and many microorganisms, including pathogens, from sewage outflows.

Swimming Pools, Spas, and Hot Tubs

Recreational waters such as pools, spas, and hot tubs provide a great deal of relaxation and fun. They can also present a health hazard, however, if they are not cared for properly. A constant surveillance of the numbers and types of microorganisms in the water must be maintained to insure that they are safe to use. Routine filtration and chlorination must be carried out to remove and kill possible pathogens, such as staphylococci, enterics, pseudomonads, herpes viruses, and polio viruses, which are constantly being introduced into the water by bathers. The high temperatures of hot tubs and spas (about 100–105° F) promote the luxuriant growth of numerous microorganisms, such as fungi, bacteria, protozoa, and algae. In the crevices, grouting, and porous surfaces of these containers, pathogenic microorganisms often breed and serve as a source of infection for bathers. Some cases of scabies have been reported in public hot tubs in which the parasite larvae were found in the wood pores above the water line. Outbreaks of amebic encephalitis have also been reported in bathers □ (see Chapter 15); the sources of infection eventually were traced to cracks in swimming pools where the protozoa multiplied unchecked. Prevention of these diseases requires occasional draining of the water and washing of the walls and bottom with disinfectants. This treatment reduces the microbial population adhering to the pool surfaces and therefore decreases the chances of bather infection. The waters should be filtered daily for at least two hours, and the filters should be cleaned weekly. The water should be chlorinated daily to insure a constant microbiocidal level of chlorine in the water. In addition, bathers should be encouraged to shower before entering pools.

353

The Oceans

The oceans play an important role in the hydrologic cycle. They are the major reservoir for water in the biosphere. The oceans are saline bodies of water, containing virtually every known element. The sodium chloride concentration of the oceans is generally around 2%, but ranges from 0.7% in the Baltic Sea to about 4% in the Red Sea. The oceans cover more than 70% of the surface of the earth and have an average depth of about 3 miles. Much of the ocean is sparsely inhabited by microorganisms. Oceans, like lakes, can be divided into zones based on water depth and light penetration. The **littoral** (shore line) zone represents an interphase between soil and water. It is rich in bacteria, algae, fungi, and protozoa. The **neritic** (shallow waters along the shore) and **pelagic** (open waters) zone contain a large number of photosynthetic microorganisms on their surfaces. These microorganisms are mostly **planktonic algae** (fig. 32-3) and are the primary producers of most marine ecosystems. These organisms also release large quantities of oxygen that contribute significantly to the oxygen in the atmosphere. Most of the resident bacteria in marine environments are gram negative aerobes or facultative anaerobes such as vibrios, pseudomonads, actinomycetes, *Hyphomicrobium*, *Cytophaga*, and *Flavobacterium*. Aquatic fungi with unique conidial structures are found in the littoral zone. The conidia are star-shaped because this form maximizes their contact with the substrate as they tumble in the surf.

The marine microorganisms participate actively in nutrient recycling and represent a complex community. Although most bacteria in marine ecosystems function as decomposers, some, such as cyanobacteria and desulfovibrios, are primary producers.

(a)

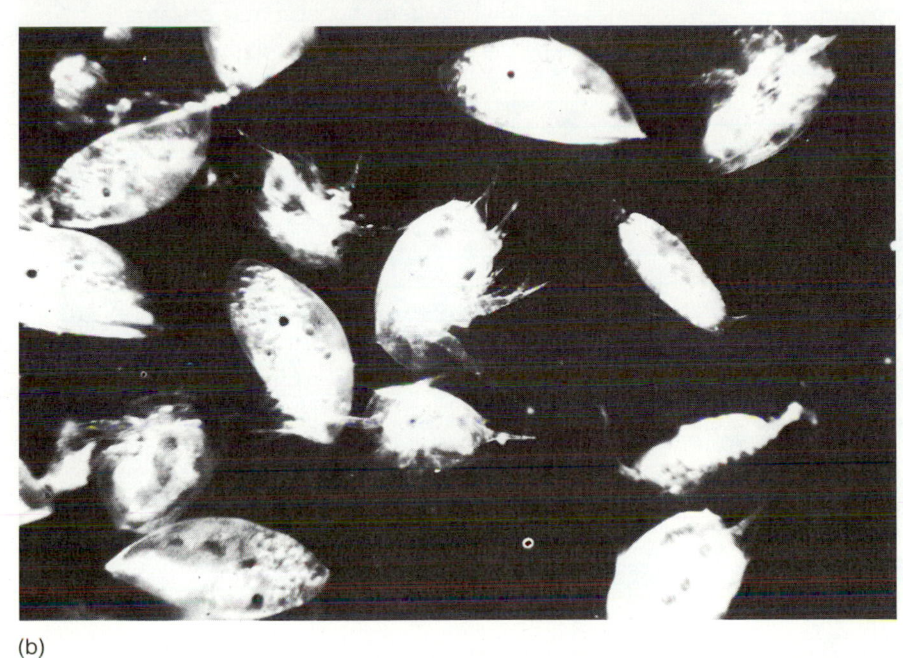

(b)

Figure 32-3 **Plankton.** *(a)* Phytoplankton. *(b)* Zooplankton.

MICROBIOLOGY OF DRINKING WATER

All communities must have a constant source of water from which to drink. Some people obtain their water from wells, others from rivers, lakes, or manmade reservoirs. Many of these water sources are unsuitable for human consumption and must be treated before they can be used. Sometimes simply adding chlorine to the water to kill pathogens is all that is needed to make the water safe to drink **(potable)** but in other cases further treatment is

FOCUS

UNDERWATER MARINE GARDENS

The bottom of the ocean normally is very cold (< 5°C) and dark where no light penetrates and hence no photosynthesis takes place. Most organisms that live at these depths must depend entirely on food from the ocean's surface. Thus, many sea bottom communities, like surface communities, are dependent on photosynthetic organisms as their source of food. In 1977, an oceanographic expedition, headed by Dr. John B. Corliss of Oregon State University, discovered a new type of submarine ecosystem based, not on light, but on chemical reactions carried out by chemolithotrophic bacteria.

The expedition—aboard the submarine *Alvin* from the Woods Hole Oceanographic Institute—discovered this new ecosystem at a place called the Galapagos Rift, located between two adjacent ocean bottom plates (fig. 32-4). At the rift, water seeps into fissures at the ocean's bottom and into the earth's mantle, where it becomes heated to temperatures above 350°C (662°F). The heated water also picks up inorganic materials like sulfur and hydrogen sulfide from the earth's crust. These inorganic materials are used by bacteria such as *Thermodiscus* (S) and *Thiobacillus* (H_2S) as a source of energy for chemolithotrophic metabolism. These chemolithotrophic bacteria serve as the primary producers for the ecosystem. The biomass created by these chemolithotrophs is the nutrient source for a complex food chain that includes gigantic tubeworms and many marine animals.

This type of chemosynthetic community was found not to be unique to the Galapagos Rift; apparently it is commonplace in mid-ocean ridges where molten rock from the earth's mantle comes in contact with ocean waters, resulting in underwater oases where the primary producers are chemolithotrophic bacteria. It is now thought that mid-oceanic rifts not only nourish a chemosynthetic ecosystem, but also may be largely responsible for the chemical composition of sea water.

(a)

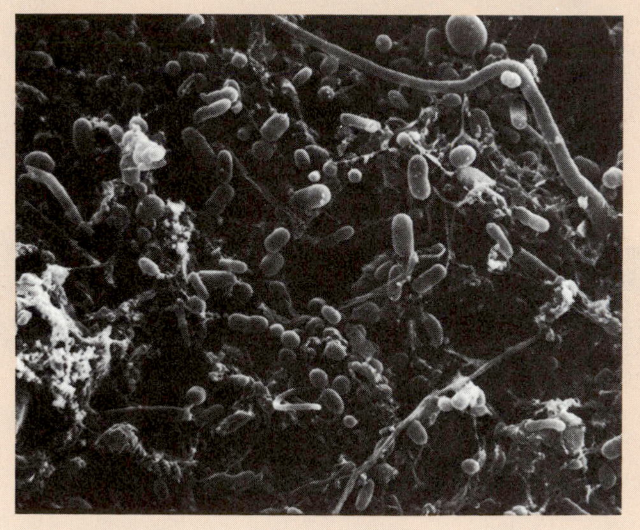

(b)

Figure 32-4 The Galapagos Rift Community. *(a)* Submarine community as photographed from *Alvin. (b)* Scanning electron photomicrographs of chemosynthetic bacteria which serve as the primary producers for the Galapagos Rift ecosystem.

needed. Biological contamination is an important consideration when determining the kind of water treatment necessary, because many water contaminants can cause severe diseases and death in human populations. Chemical contamination of water must also be dealt with, because many toxic chemicals and ions can pollute water and be as harmful as pathogenic microorganisms (table 32-1).

TABLE 32-1

SOME WATER POLLUTANTS AND THEIR POSSIBLE EFFECTS ON HUMANS

POLLUTANT	POSSIBLE SOURCE	ADVERSE EFFECTS
Physical		
Asbestos	Industrial waste	Cancer
Suspended clay	Runoff	Interferes with sanitation
Chemical		
Heavy metals	Pipes and industry	Various illnesses
Sulfate	Algicides and mines	Diarrhea
Nitrate	Fertilizers	Methemoglobinemia
Sodium	Water softeners	Fluid retention, congestive heart disease
Pesticides	Agriculture	Various illnesses
Chloroform	Industry	Cancer
Biological		
Bacteria	Feces and urine	Typhoid fever Shigellosis Salmonellosis Gastroenteritis Tularemia Leptospirosis
Viruses	Feces	Hepatitis Poliomyelitis Gastroenteritis
Protozoa	Feces	Amebic dysentery Giardiasis Balantidiasis

Water Purification

Water from lakes, rivers, reservoirs, and wells near highly polluted areas generally is not suitable for human consumption because of biological and/or chemical pollutants. For this reason, most communities purify their drinking water. As a result, the water obtained from the tap is generally safe because it has been subjected to some kind of water purification treatment so that it tastes and looks good as well as being safe to drink (fig. 32-5).

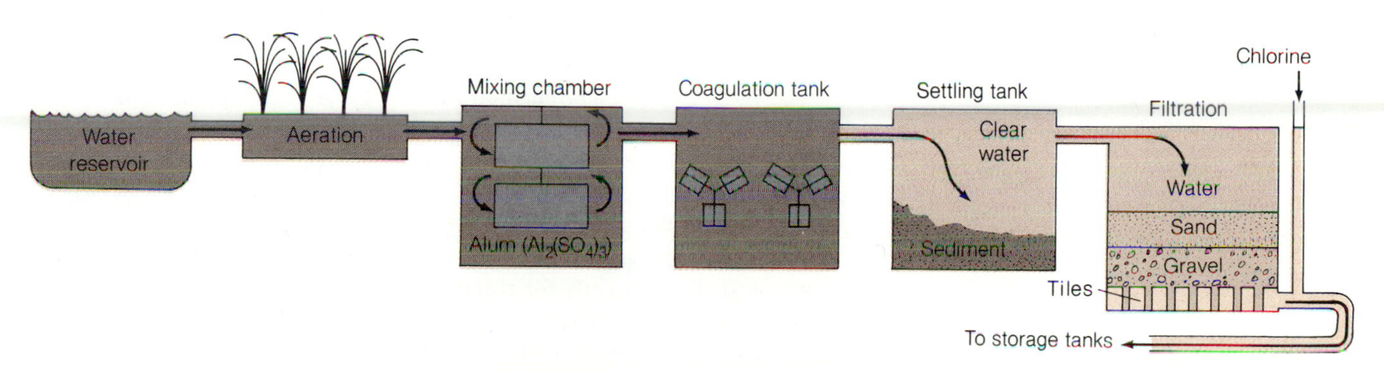

Figure 32-5 Typical Scheme for Drinking Water Purification. The various steps in water purification involve the removal of suspended particles and microorganisms from the raw water by coagulation or settling, filtration, and chlorination.

The first step in drinking water purification involves the removal of particles that make the water cloudy or turbid. This is accomplished by pumping the water into **sedimentation basins** where suspended particles can settle. With some types of water, this process may take quite some time and algal growth may give the water an unpleasant taste. To avoid this problem, **coagulation basins** are used instead. These are basins in which water is mixed with a **flocculant,** such as **alum** (aluminum sulfate), to promote sedimentation of suspended particles (solids). The sedimentation step removes not only inert particles but also many microorganisms.

From the sedimentation or the coagulation basins, the water is pumped into sand filters, where most microorganisms (more than 99%) are removed, as well as any finely suspended particles that do not settle out during the sedimentation procedure. The filters are essentially tanks containing large volumes of sand or **diatomaceous earth.** The water from the coagulation basins is pumped to the top of the sand bed and allowed to pass through the sand column, which traps particles and microorganisms. The rate of flow of the water through the filters can be regulated by controlling the height of water above the sand.

One last step in most water treatment plants involves **chlorination.** The filtered water is pumped into chlorination tanks or basins, where enough chlorine is added to the water so that it reaches a concentration of 0.5–1.0 parts per million (ppm), where 1 ppm = 1 mg/l. This step usually kills any remaining microorganisms in the water. One notable exception is the cyst of the flagellated protozoan called *Giardia intestinalis.* This protozoan has cysts that are resistant to the levels of chlorine usually maintained in drinking water supplies. It has been shown, however, that the normal filtration procedures usually remove these cysts.

After the filtration step, the water usually is pumped into **storage tanks** and from there to the community. Some communities opt to soften their water by removing calcium and magnesium ions in ion exchange columns **(water softeners),** replacing these ions with sodium. One interesting finding by the National Research Council indicates that calcium and magnesium ions present in hard water may help prevent heart disease. Soft water, however, which contains low levels of these ions but large quantities of sodium, may increase the chances of coronary heart disease. Other communities choose to **fluoridate** their water in order to reduce dental caries.

Monitoring Water Supplies

As you may already suspect, water serves as a vehicle for the transmission of many infectious agents (table 32-2). It is not unusual to detect epidemics of enteric fevers or gastroenterites that are traceable to a contaminated water supply. The source of the contamination can be a break in the water line which allows access of pathogens into the water supply; animals or humans defecating in or near drinking water sources; or pollution from industrial and agricultural sources. For this reason, most communities have a program to monitor their drinking water for the presence of pathogens.

Although it might seem that the easiest way to monitor water supplies would be to test them directly for the presence of pathogens generally this is not done. One important reason is that there are so many different pathogens that it would be impractical to test for each one individually. In addition, testing directly for pathogens frequently results in failure, because some of the pathogens are difficult to culture and others are present in very small

TABLE 32-2

MICROORGANISMS INVOLVED IN OUTBREAKS OF GASTROENTERITIS IN THE UNITED STATES

PATHOGEN	MICROBIAL GROUP	COMMON LOCATION OF OUTBREAKS
Giardia intestinalis	Protozoa	Camps, cities, towns
Entamoeba histolytica	Protozoa	Rural residences, camps
Shigella spp.	Gram – rods	Hotels, restaurants
Salmonella spp.	Gram – rods	Resorts, camps, restaurants
Escherichia coli	Gram – rod	Camps, cities, restaurants
Polioviruses	Picornaviruses	Rural residences, towns
Hepatitis-A	Picornaviruses	Camps, towns, cities
Campylobacter jejuni	Gram-curved rod	Camps, towns, restaurants

numbers or survive for very short periods of time. Thus, a pathogen such as *Salmonella* or *Shigella* may no longer be detectable 6–10 hours after its introduction because of dilution. Unless the sampling of water is done frequently or during the time when the pathogen is detectable, the presence of the pathogen will be missed. Even when the pathogen is undetectable in the drinking water, however, it may be consumed by individuals and cause disease. Furthermore, pathogens such as polioviruses and the hepatitis A virus, even if present in large numbers, cannot easily be cultured in the laboratory. Because of the many difficulties encountered when testing directly for pathogens, many communities test instead for **indicator** organisms that could herald the potential presence of human pathogens in the water.

Testing for Indicator Bacteria. An approach that is used routinely in most laboratories to monitor water supplies for the presence of fecal pollution is to test for the presence of microorganisms that indicate water pollution. Such indicator microorganisms are chosen because they are present in relatively high numbers in feces, persist in the water supply for extended periods, and are easily cultured in the laboratory. Their presence in the water indicates that there is fecal contamination of the water and therefore a potential for the presence of enteric pathogens.

Two groups of microorganisms normally are used as indicators of fecal contamination of water supplies: the **coliform** group and the **fecal streptococci** group. The coliform group is by far the commonest indicator used in the United States, while fecal streptococci are the most common indicator used in the United States, while fecal streptococci are the most common indicator microorganisms employed in European countries. The coliform group consists of all *aerobic and facultative anaerobic, gram negative, asporogenous bacilli that ferment lactose with the production of gas within 48 hours at* 35°C. In this group, *Escherichia coli*, *Enterobacter aerogenes*, and *Klebsiella pneumoniae* are commonly found coliforms. A distinction is sometimes made between total coliforms and fecal coliforms in order to pinpoint the source of contamination. Essentially, fecal coliforms are *E. coli* isolates which can grow at 44.5°C.

The fecal streptococci, such as *S. faecalis*, *S. faecium*, *S. bovis*, and *S. avium*, belong to groups D and Q. The habitats for these streptococci are human and animal intestines, so they can also serve as indicators of fecal water pollution. The fecal streptococci have not been used routinely in the United States as indicators of fecal contamination, but are now becoming increasingly important for monitoring fecal contamination in lakes, streams, reservoirs, and estuaries. They can be used together with coliforms to identify the source of pollution, thus providing more information than either indicator group alone.

Testing for indicator bacteria can be accomplished using any one of three different procedures: a) the multiple tube fermentation test (MTF); b) the membrane filter (MF) method; or c) the standard plate count (SPC).

MTF Test for Coliforms. Of the tests mentioned above, the multiple tube fermentation (MTF) test for coliforms is the most widely used in the U.S. The test is carried out in a series of three steps designed to ascertain the presence of coliforms in the water. Initially, various tubes of broth culture media containing lactose (lauryl sulfate broth is the most common) are in-

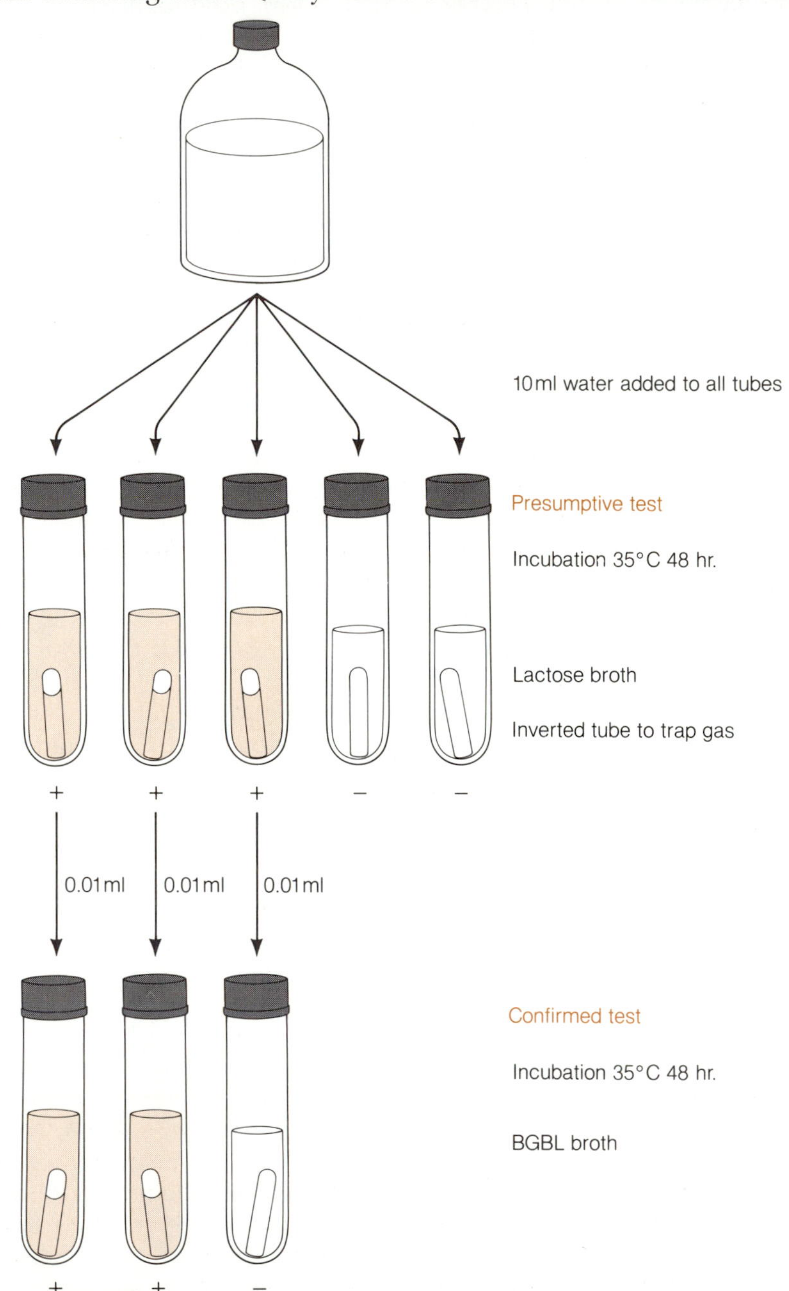

Figure 32-6 The Multiple Tube Fermentation Test for Coliforms in Potable Water. Presumptive test. Each of 5 tubes of lactose (or lauryl tryptose) broth is inoculated with 10 ml of the water sample and incubated at 35°C for 48 hours. Each tube is then examined for gas formation due to lactose fermentation. Confirmed test. A portion (1 loopful) of each positive presumptive tube (lactose broth) is inoculated into tubes of brilliant green bile lactose (BGBL) broth and incubated at 35°C for 48 hours. Each tube is then examined for gas formation, and the MPN/100 ml of coliforms is determined from MPN tables. In the example above there were three positive presumptive tubes, which then were used to inoculate three tubes of BGBL. Of these, only two were positive. The MPN per 100 ml is 5.1 (see Table 6-1). The completed test, when performed, consists of inoculating plates of EMB with portions of positive confirmed tubes. Typical coliform colonies are dark with a green sheen over their surface.

oculated with a standard volume of test water (usually 10 ml, 1 ml, or 0.1 ml) and incubated at 35°C. Each tube of broth contains a vial in an inverted position (fig. 32-6), called a **Durham tube,** which detects the production of gas from lactose fermentation. If potable water is being tested, each of 5 tubes of lauryl sulfate broth is inoculated with 10 ml of the water sample. After an incubation period of 48 hours, the tubes are examined for the presence of gas in the Durham tube. Tubes showing gas are considered positive. This portion of the MTF test is called the **presumptive** test for coliforms; i.e., it is presumed that the gas-producing microorganisms in the sample are coliforms. These results must be confirmed, however, and the **confirmed** test does just that.

In the confirmed test, portions of each tube of broth that contained gas after 48 hours are inoculated into fermentation tubes that contain brilliant green, bile, and lactose (BGBL). The ingredients in BGBL broth are selective for coliforms. After incubation at 35°C for 48 hours, the BGBL tubes are examined for evidence of gas, and the number of tubes with gas out of the total inoculated in the presumptive test (5 for potable water) is determined. From these data, one can estimate the number of coliforms in the sample

134

from MPN □ (most probable number) tables. The MPN values are statistical estimations and are based on the fact that the more microorganisms present in the sample of water, the more positive tubes there will be. From the MPN value given in figure 32-6, you can see that if a given water sample produces gas in 1 tube of BGBL out of a total of 5 inoculated, there will be, on the average, about 2.2 coliforms/100 ml of water.

The **completed** test is carried out on positive confirmed tubes to complete the identification of coliforms in the sample. The completed test involves streaking portions of a positive confirmed tube onto a plate of selective and differential medium for coliforms, called Eosin–Methylene Blue (EMB) agar. This procedure allows the isolation of colonies, which can then be tested to see if they are aerogenic (gas-producing) lactose fermenters and gram negative asporogenous rods.

Membrane Filter Method for Coliforms. A popular alternative method for the detection of coliforms in water is the membrane filter (MF) method (fig. 32-7). The procedure involves filtering a volume of water (usually 100 ml) with a membrane filter that has pores with a diameter of 0.45 μm. As the water is filtered, microorganisms are retained on the membrane filter. The membrane is then removed and placed in a petri dish (usually 60 × 15 mm) containing an agar medium or a filter pad soaked with a broth. The

121

media used, generally Endo or EMB, are selective and differential □ for coliforms. The plates are incubated for 24 hours at 35°C and then examined for the presence of coliform colonies. On Endo, typical coliform colonies are red and have a pink or red metallic sheen on their surfaces. On EMB, the colonies are dark purple and have a green metallic sheen. All coliform colonies are counted, and the results are expressed as total coliform colonies per 100 ml.

132

Standard Plate Count. The standard plate count □ is sometimes used to determine the total number of microorganisms present in a water sample. For instance, bottled water is routinely tested for the total number of bacteria using the standard plate count. For drinking water, the test is carried out by plating 1 ml of the water sample, using a culture medium such as plate count agar or standard methods agar. The plates are incubated at 35°C for 48 to 72 hours and then counted, using a colony counter.

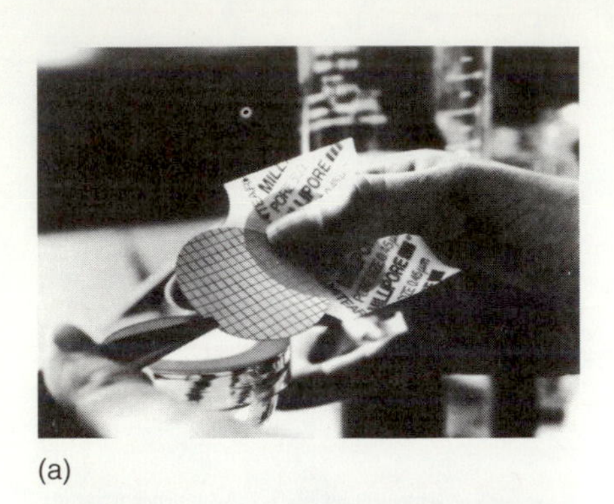

(a)

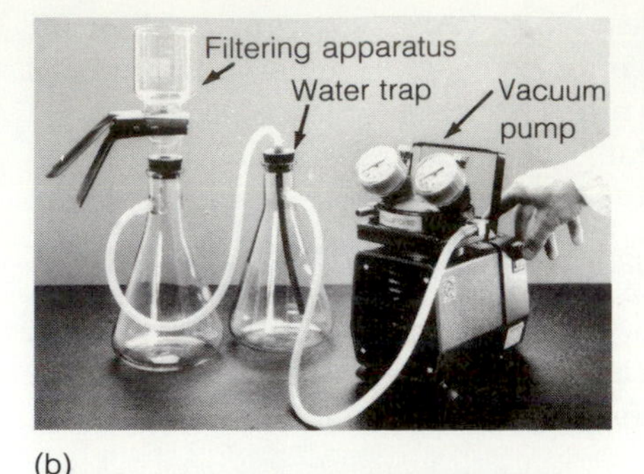

(b)

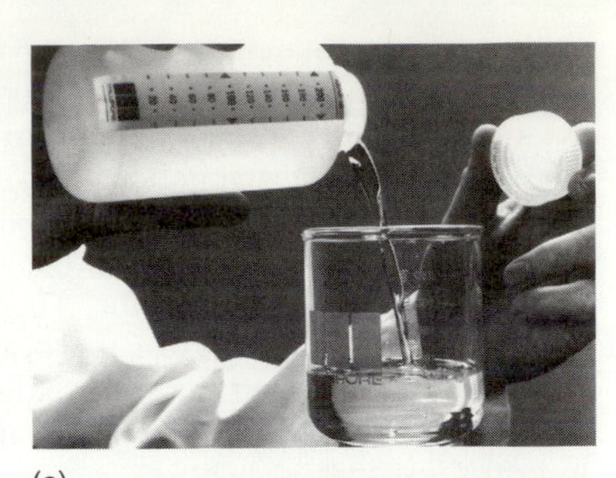

(c)

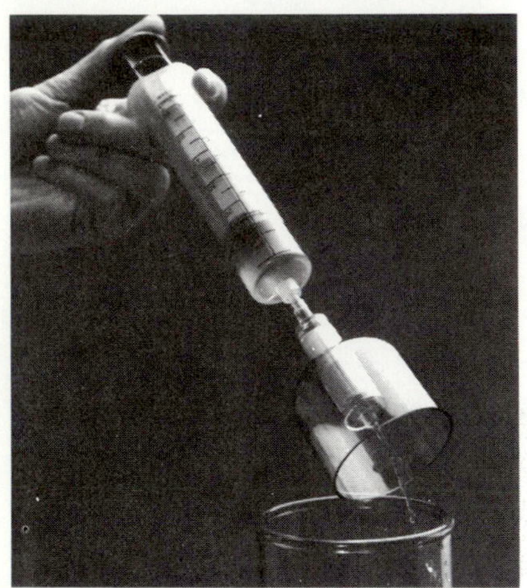

(d)

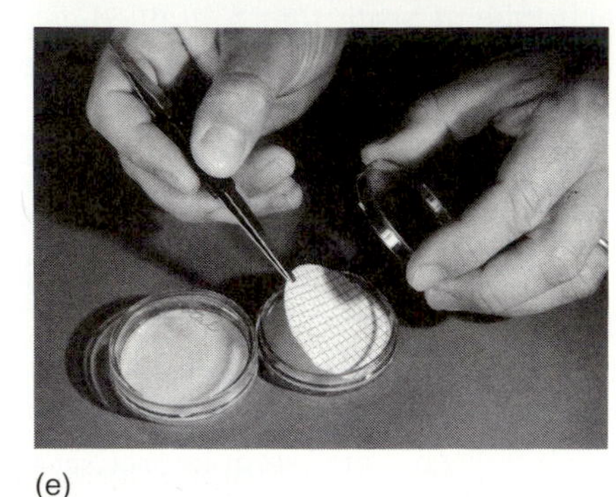

(e)

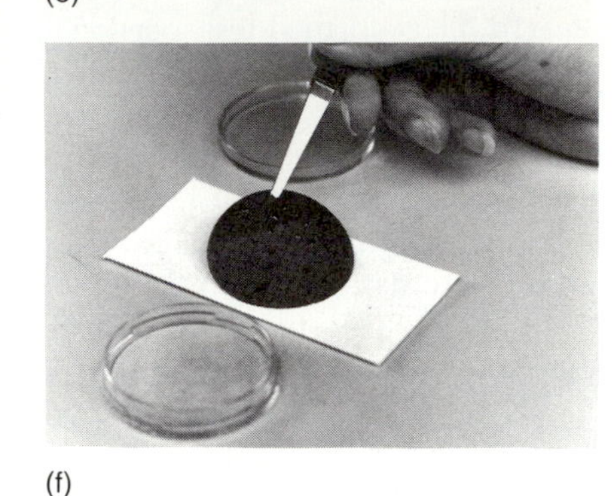

(f)

Figure 32-7 The Membrane Filter Test for Coliforms. *(a)* Remove a sterile membrane filter (0.45 μm pore size) from the package and place on the fritted glass surface of the funnel, using sterile forceps. *(b)* Assemble the filtering apparatus and connect it to a vacuum system. Apply suction. *(c)* Pour 100 ml of the water sample into the filtering apparatus container and allow the entire sample to be filtered through the membrane filter. *(d)* Rinse the walls of the container with sterile water so that any microorganisms adhering to the glass walls are washed into the fitler. *(e)* Remove the filter from the funnel and place it in a petri dish containing a culture medium such as m-ENDO or m-EMB. Incubate at 35°C for 24 hours. *(f)* Appearance of typical coliform colonies growing on the surface of the membrane filter. (Courtesy of Millipore Corporation.)

Drinking Water Quality

The U.S. Environmental Protection Agency (EPA) has set standards for drinking water quality which should be adhered to in order to assure safe drinking water. Using the multiple tube fermentation test for coliforms, an MPN of < 2.2 (no gas in any of the tubes) is the desired standard. If the MPN is ≥ 9.2—that is, 3 or more tubes of BGBL showing evidence of gas—the water sample is considered unsuitable for human consumption, and remedial action must be taken to make the water potable.

In the MF technique, the water is considered suitable for drinking if there is less than 1 colony/100 ml of sample. If there are ≥ 4 colonies/100 ml, the water is deemed unsatisfactory and remedial action must be taken. The EPA has also established limits of safety for chemical and physical constituents of water (table 32-3). In addition, the World Health Organization (WHO) has established international standards that apply worldwide.

SEWAGE TREATMENT

Sewage consists of human and domestic wastes as well as industrial and agricultural discharges. These liquid wastes, if not treated properly, find their way into natural bodies of water and cause very serious detrimental changes (e.g., eutrophication). Rivers, lakes, and artificial reservoirs are irreplaceable, valuable resources that must be saved at all costs and protected from pollution. They provide the water needed for drinking, bathing, laundering, irrigating, and manufacturing. Contaminated water can cause serious problems. For instance, water contaminated with pathogens spreads disease, and water contaminated with heavy metals and salts damages

TABLE 32-3
SAFE DRINKING WATER STANDARDS[1]

CONSTITUENT	EPA*	WHO**
Microbiological (Coliforms)		
Fermentation tube method	<2.2MPN/100 ml in 90% samples/mo	<2.2 MPN/100 ml in 95% samples/yr
Membrane filter technique	1 coliform/100 ml/mo	
Inorganic Chemicals		
Arsenic	0.05 mg/l	0.05 mg/l
Barium	1.0	
Cadmium	0.01	0.01
Chromium	0.05	
Cyanide		0.05
Lead	0.05	0.1
Mercury	0.002	0.001
Nitrate	10.0	
Selenium	0.01	0.01
Silver	0.05	
Organic Chemicals		
Chlorinated hydrocarbons		
Endrin	0.0002 mg/l	
Lindane	0.004	
Methoxychlor	0.1	
Toxophene	0.005	
Chlorophenoxys		
2,4-D	0.1	
2,4,5-TP Silvex	0.01	
Radiological		
Radium-226 + radium-228	5pCi/l	
Gross alpha particle activity	15pCi/l	3 pCi/l
Gross beta particle activity	4 mrem	30 pCi/l
Physical		
Color	15 color units	
Odor	3 threshold odor number	

*EPA = U.S. Environmental Protection Agency.
**WHO = World Health Organization.
1. Data selected from:
   (a) E.P.A. 1976. National Interim Primary Drinking Water Regulations U.S. Document #EPA-570/9-76-003.
   (b) W.H.O. 1971. International Standards for Drinking-Water. W.H.O., Geneva, Switzerland.

agricultural lands and fishing areas. For these reasons, many communities have sewage treatment facilities that are aimed at purifying liquid wastes before they are discharged into natural bodies of water.

The amount of organic material in sewage is important because it affects the rate of eutrophication of lakes and estuaries receiving the polluted waters. The **biochemical oxygen demand (BOD)** is a measure of the amount of organic nutrients available for the growth of microorganisms in sewage. It is measured by determining the amount of oxygen consumed by microorganisms oxidizing organic matter in a water sample.

A common procedure for the determination of the BOD is carried out by collecting the water sample in an airtight bottle and measuring the amount of **dissolved oxygen (DO)** in the sample. The container is then seeded with microorganisms (if necessary, since microorganisms may already be present in the sample) and allowed to incubate at a standard temperature (5–20°C) for 5 days. After that time, the amount of dissolved oxygen in the sample is again measured. The difference in DO between the first measurement and the second is used to determine the BOD. The greater the decline in the DO measurement, the higher the BOD of the sample.

The aim of modern sewage treatment is to reduce the BOD of liquid wastes as much as possible before the treated water is discharged into receiving waters. Figure 32-8 shows the steps that may be taken in the treatment of sewage.

Primary Treatment

Primary treatment of sewage involves the removal of solid wastes from raw sewage. As raw sewage (influent) enters the sewage treatment plant, it is passed through a series of screens that remove large objects such as marbles, paper, rocks, dentures, and glasses (fig. 32-8). The screened water is then diverted into settling tanks or basins, where the suspended solids are allowed to settle to the bottom. In many cases, skimmer arms cruise the surface of

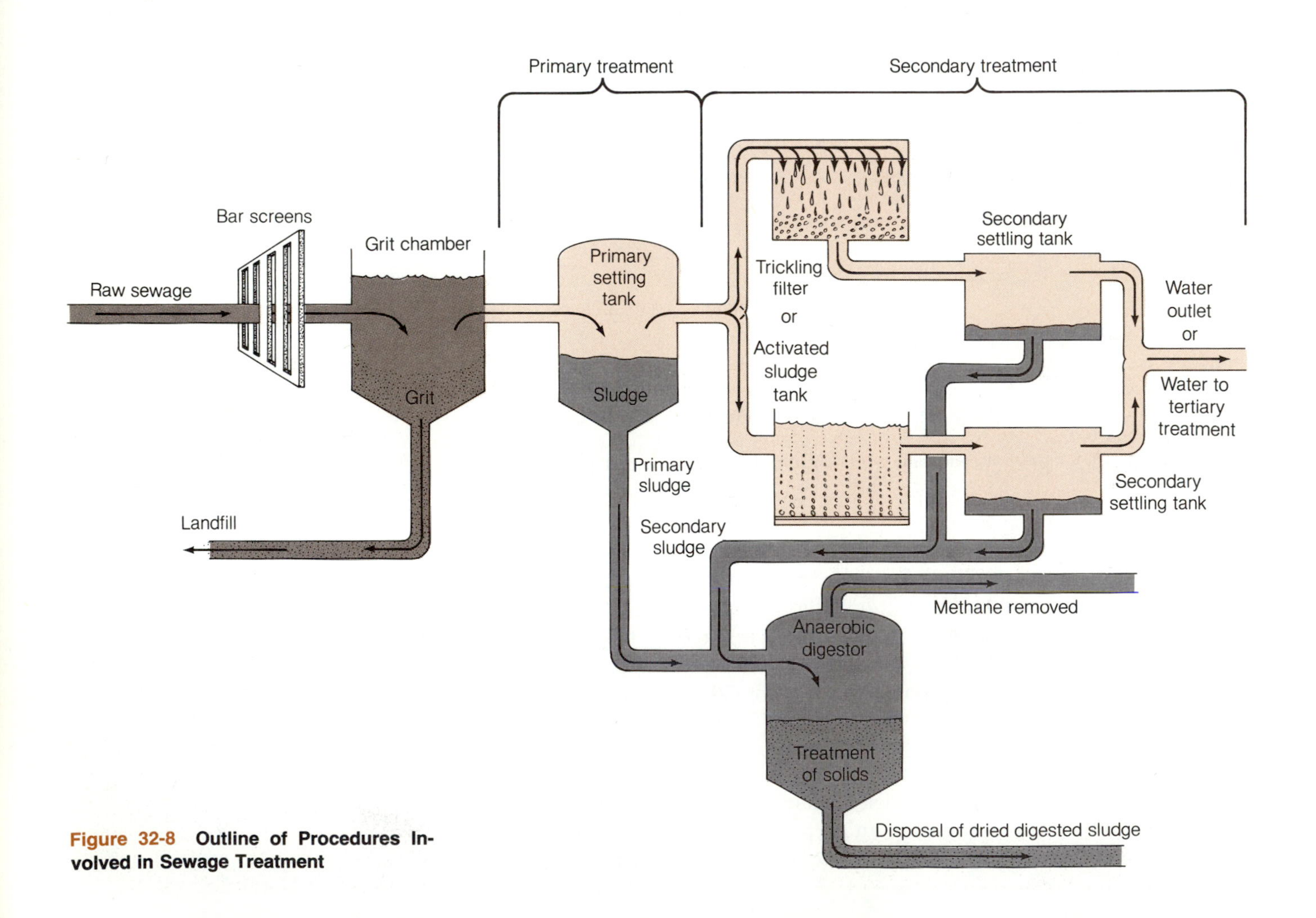

Figure 32-8 Outline of Procedures Involved in Sewage Treatment

the **settling tanks** to remove any floating objects such as feathers, grease, cellophane, or wood. The settled wastes are collected and digested anaerobically or composted. Primary treatment removes much of the organic material in the water, but a significant amount remains. In plants where only primary treatment of sewage occurs, the water is chlorinated to kill pathogens and then discharged into receiving waters. After primary treatment of sewage, about 30–40% of the BOD has been removed.

Natural waters have a resident microbiota of heterotrophic microorganisms that oxidize organic compounds and convert them into carbon dioxide, water, and inorganic salts. 759 Ammonium is oxidized to nitrate □ by various organisms, and hydrogen sulfide is converted to sulfate □. 767 Along with the removal of organic matter, many pathogens are outgrown by the resident microbiota and consequently are greatly reduced in numbers. Since wastewater from primary treatment is still heavily contaminated with organic matter (i.e., has a high BOD), the resident microbiota of the receiving waters cannot purify the water, generally because oxygen is limited in these waters.

Secondary Treatment

In **secondary treatment** of sewage, much of the remaining organic matter in the waste is oxidized by microorganisms before the wastewater is discharged from the plant. **Trickling filter** (fig. 32-9) and **activated sludge** (fig. 32-10) systems are common installations used to expedite the oxidation of organic wastes. If the secondary treatment is the final purification process, the water is drained off, chlorinated, and returned to the environment. After secondary treatment, 85–90% of the BOD has been removed from the sewage. Secondary treatment forces oxygen into wastewater to improve the oxidation of organic materials present.

Figure 32-9 A Trickling Filter System. (*a*) Rotating arms of the trickling filter system are spraying sewage water onto a rock bed that functions as the filter. As the sewage water trickles through the rock bed the dissolved organic nutrients in the water are oxidized by microbial communities attached to the rocks. The water collected at the bottom of the filter is free of much of the dissolved organic matter that was present in the sewage. (*b*) A close up view of water sprayed onto the trickling filter.

Figure 32-10 Activated Sludge System. The activated sludge system involves the use of flocs to oxidize the sewage water. Flocs are microbial populations linked together by mucilagenous sheaths and glycocalyces produced by bacteria. Sewage water containing the flocs is constantly aerated by blowing air through the water using aerators (indicated in photograph). The oxygen is necessary for the aerobic oxidation of dissolved organic matter.

The Trickling Filter

The **trickling filter** is a very successful method of purifying relatively small quantities of sewage, and it is used primarily in communities with fewer than 50,000 inhabitants (fig. 32-9). It consists of a large tank or basin filled with gravel or other porous material through which water from the primary treatment percolates. The water is sprayed over the surface of the porous bed and becomes aerated. As the water trickles down through the bed, the organic materials dissolved in it are oxidized by heterotrophic microorganisms adhering to the surfaces of the rocks. Each gravel piece is coated with microorganisms held together in a slime matrix produced by bacteria such as *Zooglea ramigera*. This bacterium produces large quantities of polysaccharides that hold together a complex community of bacteria, fungi, protozoa, rotifers, insect larvae, and nematodes. Biological oxidations convert organic molecules such as proteins, carbohydrates, and amino acids to carbon dioxide and mineral salts of nitrate, sulfate, and phosphate.

The Activated Sludge System

The **activated sludge system** is a very popular method for carrying out biological oxidations of liquid wastes in large quantities. It consists of a basin or tank with provisions for aeration (fig. 32-10). The tank is filled with wastewater from the primary treatment system and it is kept well aerated by blowing air through the water from the bottom of the tank. A large community of heterotrophic organisms develops in these waters and includes coliforms, pseudomonads, micrococci, *Zooglea*, *Thiothrix*, and *Sphaerotilus*. These microorganisms form **flocs** or aggregates held together by the extracellular polysaccharides produced by *Zooglea*. Ciliated protozoa, such as *Vor-*

ticella, are often found adhering to the flocs. During aeration, the flocs are suspended throughout the wastewater and the bacteria in the flocs oxidize the organic matter. The protozoa adhering to the film also feed on suspended bacteria in the water. Some of the organic nutrients are used by the floc microorganisms to increase their numbers. After treatment, the water is then allowed to stand so that the flocs settle to the bottom. A portion of the settled flocs is used to seed the next batch of wastewater.

Tertiary Treatment

Although effluent waters from secondary treatment have a low BOD, they contain mineral salts that are serious pollutants. Salts, when introduced into receiving waters, enhance productivity by promoting the growth of autotrophic microorganisms (e.g., cyanobacteria), leading to eutrophication. In order to reduce the rate of eutrophication in receiving waters, effluents from secondary treatment are subjected to a third stage of purification. This **tertiary treatment** removes much of the ammonia, nitrate, and phosphate from the water, as well as toxic organic substances, such as pesticides and herbicides, and any nonbiodegradable organic matter.

Many communities treat their secondary treatment waste in **oxidation ponds** or shallow lagoons (3–5 feet deep), in which microbial activities remove much of the organic and inorganic materials remaining after secondary treatment. The organisms present in these lagoons include algae, bacteria, protozoa, fungi, and macroscopic organisms. During the 30 days or so that the water is allowed to stand in the lagoon, microbial activities remove about 90–95% of the nutrients in the water. For example, the ammonium nitrogen in the pond may be oxidized to nitrate by nitrifying bacteria □, and the nitrate may then be removed from the water as molecular nitrogen by the process of denitrification □.

329

Other communities, fewer in number, opt to remove the toxic organic materials. The removal of organic materials usually involves passing the water through a bed of activated charcoal that absorbs these organic pollutants from the water. Phosphate, most of which comes from detergents, is removed by precipitation as salts of calcium or iron through the addition of lime (liming). Ammonium nitrogen, a major concern in lake eutrophication, is usually removed by volatilization of the ammonium ion as ammmonia at a high pH. This process is called **stripping.** After tertiary treatment the effluent, which is rarely chlorinated, is of very high quality and potable.

Effluents after secondary treatment might also be treated further by irrigating soils with them. Microorganisms present in the water are killed by heat and ultraviolet light from the sun. Organic substances in the irrigation water are oxidized by soil microorganisms, and the inorganic ions are either assimilated by plants or microorganisms or converted into other useful ions.

Digestion of Solid Wastes

Solid wastes (**sludge**) from primary and secondary treatment consists of undigested matter and large numbers of microorganisms (>10^{10} microorganisms/ml). Sludge can be digested further under anaerobic conditions in large fermentation tanks (fig. 32-11). Most of the anaerobic digestion of sewage is carried out by bacteria. Initially, the bulk of the solid waste, consisting of fibers and complex organic compounds, is digested by anaerobic bacteria to

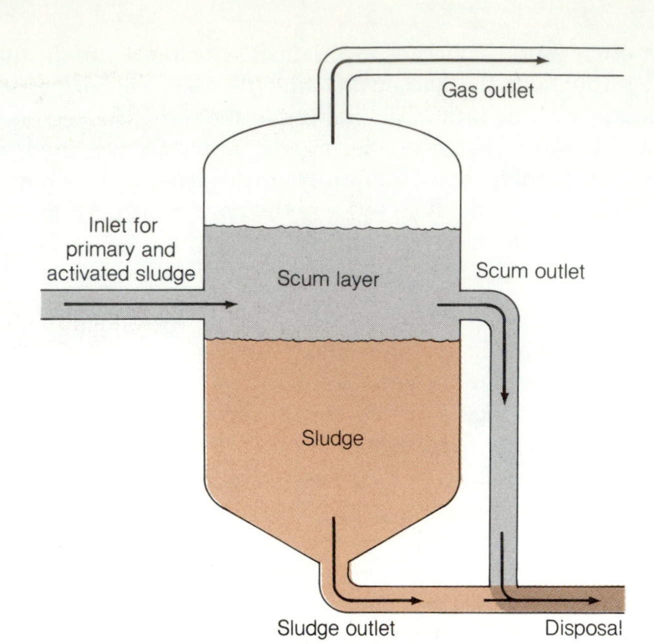

Figure 32-11 Anaerobic Sludge Digestion. This process involves the digestion of solids removed from sewage. It involves many different types of anaerobic microorganisms that effect the sequential breakdown of organic matter, ultimately producing methane gas as the principal waste product. The methane can then be used to heat the sewage treatment plant or to power some of its processes.

such molecules as simple sugars, amino acids, and nucleotides. After sludge digestion has taken place for a period of time, sufficient sugar and amino acids accumulate to allow the increase in the size of the population of fermentative bacteria. These ferment the sugars and amino acids to organic acids and alcohols, which, in turn, are converted by other microorganisms to carbon dioxide, hydrogen gas, and other small molecules. The carbon dioxide and hydrogen are then used by bacteria such as *Methanobacterium*, *Methanococcus*, and *Methanosarcina*, which convert them to methane. □ 329 The methane is vented and ignited to prevent explosions, or is used to heat or provide energy to the plant. Once the digestion has proceeded to completion and methane production drops off, the spent sludge is removed from the tanks, dried, and composted, after which it can be used as a soil conditioner unless it contains nonbiodegradable chemicals or radionuclides. Some plant operations allow the sludge to compost by piling the cakes of digested sludge onto wood chip beds. The composting process—which generates enough heat to kill many bacteria, even human pathogens—converts much of the digested sludge into useful soil amendment.

Septic Tanks

Septic tank systems (fig. 32-12) are commonly used to treat small amounts of sewage in rural areas or in suburban homes not connected to a sewage

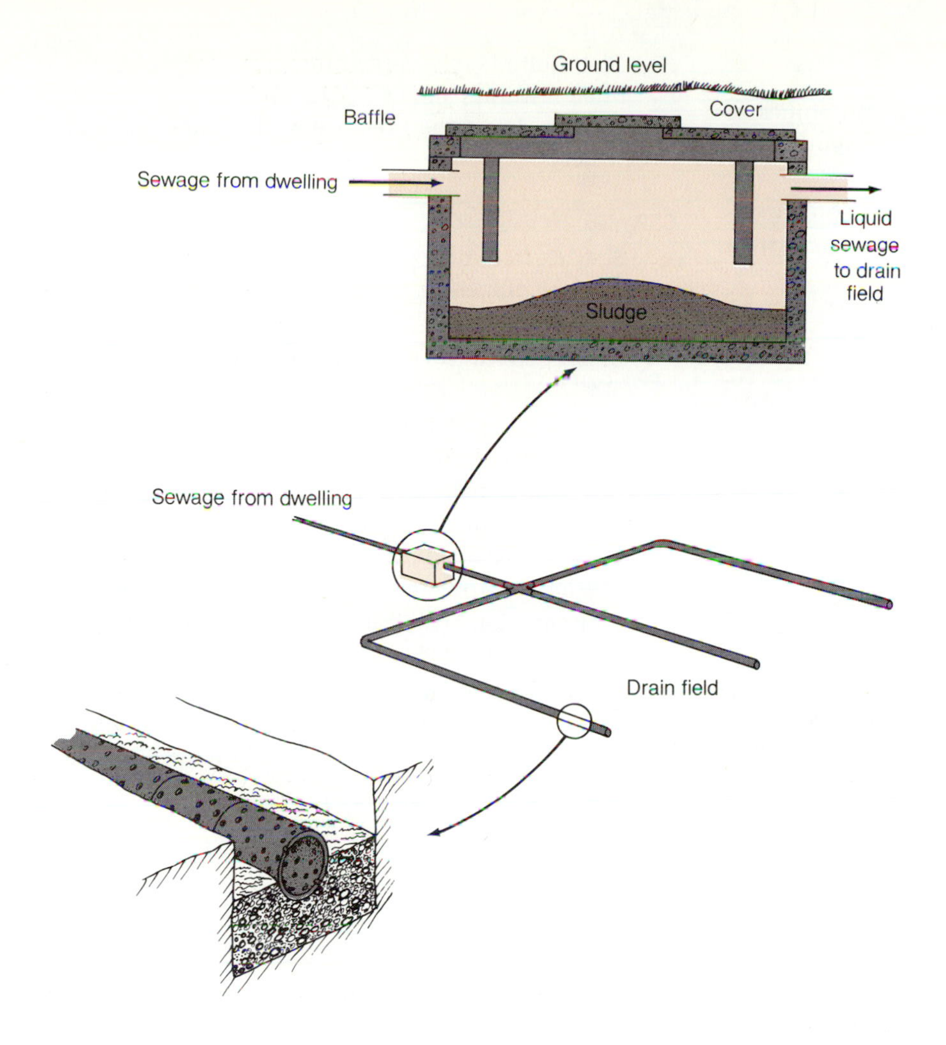

Figure 32-12 **A Septic Tank System**

treatment plant. A septic tank system consists of a holding tank with a methane vent and a wastewater outlet. The solids in the sewage settle to the bottom of the tank, where they are subjected to some anaerobic digestion. Carbon dioxide and hydrogen are used by the methanogenic bacteria, which release methane as a waste product. The undigested sludge and sedimenting material must periodically be pumped out when the tank fills. The liquid wastes from the tank flow out of the outlet into a leaching field or a seepage pit (a deep hole filled with gravel). As the wastewater flows through the leaching field or seepage pit, some of the organic wastes are oxidized by microoganisms. Less than 50% of the BOD is removed in septic tanks and enteric pathogens are not killed, so septic tank effluents that seep into nearby wells may contaminate the drinking water supply. In homes with septic tanks located near lakes, tank effluents may reach the lake. Thus, the nature of the drainage field is an important consideration in the design of a safe septic tank system. The effluent must pass though enough porous material so that biological oxidations can take place and microorganisms can be filtered out before they reach groundwater, streams, and lakes. If the drainage field becomes clogged, the sewage backs up and a new leaching field or seepage pit must be established.

SUMMARY

1. Aquatic environments have unique communities of microorganisms that help eliminate pollutants from the water.
2. Increased concentrations of nutrients in aquatic environments can lead to detrimental changes, such as eutrophication.

SOME AQUATIC ENVIRONMENTS AND THEIR BIOLOGICAL COMPOSITION

1. The hydrosphere is made up of the rivers, lakes, swamps, estuaries, oceans, and groundwater in the biosphere.
2. Freshwater habitats have their own resident microbiota.
3. The microbiota is made up of a variety of microorganisms, including phototrophs and chemotrophs.
4. Lakes are inland bodies of fresh water that are formed by ice melts, rivers, or precipitation.
5. Lakes are divided into various zones according to temperature, depth, and light penetration. Each zone has its own characteristic microbiota.
6. Rivers and streams are flowing bodies of fresh water with various velocities of flow, temperatures, and nutrient and oxygen concentrations, depending upon the geography of the river course.
7. The oceans are vast saline bodies of water that serve as reservoirs for the biosphere.
8. Marine microorganisms are found primarily in the littoral zone and are diluted in the open waters, where the predominant microbiota is the phytoplankton.

MICROBIOLOGY OF DRINKING WATER

1. Water must be purified and decontaminated before it is safe for drinking.
2. Water purification includes flocculation of suspended particles, filtration, and chlorination.
3. Drinking water supplies are constantly monitored for their biological and chemical quality.
4. Biological monitoring is carried out principally by looking for indicator bacteria that herald fecal contamination.
5. Indicator bacteria commonly used to monitor water supplies are coliforms and fecal streptococci.
6. Coliforms are gram negative, lactose fermenting bacteria that inhabit the intestines of humans and other animals.
7. The multiple tube fermentation test, the membrane filter test, and the standard plate count can all be used to detect the presence of indicator bacteria in water.

SEWAGE TREATMENT

1. Sewage treatment is aimed at reducing the BOD before the effluent is discharged into receiving waters.
2. Primary treatment of sewage involves the removal of suspended soils and lowers the BOD by 40%–50%.
3. Secondary treatment involves the biological oxidation of dissolved organic materials in the wastewater, which reduces the BOD by about 90%. This treatment can be accomplished by using trickling filters or activated sludge facilities.
4. Tertiary treatment is a process aimed at removing inorganic ions resulting from secondary treatment. It reduces the BOD by more than 98%.
5. Solid wastes are digested by fermenting microorganisms in anaerobic tanks.
6. Septic tanks are anaerobic sewage treatment installations best adapted for use in single-family dwellings. This treatment process reduces the BOD by about 40%.

STUDY QUESTIONS

1. Discuss the possible impact that aquatic microorganisms might have on humans.
2. Define the following terms:
   a. hydrosphere;
   b. limnology;
   c. potable;
   d. coliforms;
   e. eutrophication.
3. Diagram a typical lake, label the various zones, and indicate which microorganisms might be present in each zone.
4. Is the water found in lakes and reservoirs suitable for human consumption? Why?
5. Describe the process employed by many communities to make water potable.
6. What are indicator organisms?
7. Describe two procedures commonly used to monitor water supplies for fecal contamination.
8. What is sewage? How is it treated?
9. What is primary sewage treatment? Secondary sewage treatment? Tertiary sewage treatment?

SUPPLEMENTAL READINGS

American Public Health Association. 1985. *Standard methods for the examination of water and wastewater*, 16th ed. Washington: American Public Health Association.

Baross, J. A., Dahm, C. N., Ward, A. K., Lilley, M. D., and Sedell, J. R. 1982. Initial microbiological response to the Mt. St. Helens eruption. *Nature*, 296:49–52.

Edmond, J. M. and Damm, K. V. 1983. Hot springs on the ocean floor. *Scientific American*, 284:78–93.

Imhoff, K., Muller, N. J., and Thistlewayte, D. K. B. 1971. *Disposal of sewage and other water-borne wastes*, Ann Arbor, MI: Ann Arbor *Science.*

Taber, W. A. 1976. Waste water microbiology. *Annual Review of Microbiology*, 30:263–277.

U. S. Environmental Protection Agency. 1975. U.S. drinking water regulations. *Federal Register*, Vol. 40, No. 248.

CHAPTER 33
FOOD MICROBIOLOGY

CHAPTER PREVIEW

FOOD POISONING ON THE HIGH SEAS

Cruising is one of the newest American pastimes. Cruise liners, really luxurious floating restaurants, wine and dine more than 1,000 passengers with exquisite cuisine and plenty of it. Food preparation aboard cruise ships is an ongoing process. The foods generally are prepared fresh, but sometimes they are acquired from caterers at the various ports of call. It is remarkable that more food poisoning outbreaks do not occur aboard these ships—but a few do occur each year! Since most of the cruises originate in Miami, Florida, or Los Angeles, California, it is not surprising that most of the food poisoning reports are made in these two American ports.

These cruises usually include a midnight buffet and two or three buffet luncheons. It is usually during these meals that food poisoning outbreaks occur. The food poisonings are usually due to staphylococci, *Salmonella, Campylobacter,* or *Vibrio.* One such outbreak of food poisoning, encompassing both crew and passengers over the span of two consecutive cruises, was caused by *Salmonella heidelberg.*

The outbreak was noticed when 3% of the passengers on the cruise liner became ill with a diarrheal disease. By the end of the outbreak, 34% of the passengers and more than half of the crew had became ill with a watery diarrhea or with cramps and vomiting. Both of these signs and symptoms are consistent with *Salmonella* food poisoning. The patients whose stools were sampled yielded *Salmonella heidelberg.*

After the survey of the passengers was conducted, it was determined that a turkey and/or macaroni salads were the vehicles of transmission. It was also determined that the contaminated salads were probably consumed during the evening buffet. Although the original salads were no longer available for examination, several other foods remaining in the galley were cultured and found to contain *Salmonella heidelberg.* This finding indicated to the health authorities that the source of contamination was not an infected animal but contaminated food handlers. The health authorities cultured fecal specimens from more than 250 food handlers and found that approximately 22% of them yielded the same species of *Salmonella* responsible for the outbreak.

Once the contaminated foods were destroyed, the galley sanitized, and the food handlers screened for *Salmonella,* the outbreak was stopped.

It is the role of the food microbiologist to monitor and control microbial growth in foods, so that harmful microorganisms are removed or destroyed while beneficial ones are nurtured. In this chapter we will discuss the various ways in which microorganisms affect foods, spoiling them, making them more flavorful, or increasing their shelf life.

LEARNING OBJECTIVES

A STUDY OF THIS CHAPTER SHOULD ENABLE YOU TO:

NAME SOME OF THE MICROORGANISMS PRESENT IN FOODS

DISCUSS WAYS IN WHICH MICROORGANISMS SPOIL FOODS

DESCRIBE METHODS EMPLOYED IN FOOD PERSERVATION

OUTLINE SOME OF THE WAYS IN WHICH MICROORGANISMS PARTICIPATE IN FOOD PRODUCTION AND PRESERVATION

DISCUSS HOW MICROORGANISMS GET INTO FOODS AND CAUSE HUMAN DISEASE

Food plays a very important role in our lives. We reward ourselves with foods, and many of our social gatherings center around sharing them. Most of us make a conscious effort to know the nutritional composition (e.g. vitamins) and chemical makeup (e.g. preservatives) of the foods we eat, but we rarely are concerned with which microorganisms are present. We should be concerned with the microbiological composition and quality of foods, because microoganisms are an integral part of the foods we eat and contribute to their nutritional content. Microorangisms also play a central role in food spoilage, and some can even cause disease. In addition, foods such as sauerkraut, buttermilk, and cheeses are made by the intervention of microorganisms. The science of food microbiology deals with microorganisms that spoil foods, those that are used to make foods, and those that may cause disease. The control of harmful microbes and the growth and cultivation of beneficial ones are ongoing problems for the food microbiologist. This chapter discusses the role of microorganisms in food spoilage, food production, and foodborne disease.

MICROBIAL COMPOSITION OF FOODS

Foods generally are derived from plant and animal tissues that have a microbial flora or acquire one while handled by workers during processing. For example, the skin of grapes has a resident population of yeasts □ such as *Saccharomyces*, *Kloeckera*, and *Pichia*. Early winemakers, probably unaware of grape microorganisms, took advantage of the presence of yeasts on the skin of grapes to make wine. They knew that if the grape skins were mixed with the juice, for even a short time, the resident yeast population would ferment the juice and eventually, after aging, convert the grape juice into wine. We now know that this occurs because most fruits have a large yeast population that ferments the sugars to produce alcohol and CO_2. Other microorganisms often present on fruits are filamentous fungi, bacteria, and protozoa.

344

Meat is an animal product that also harbors microorganisms. Muscle tissue, which makes up most of the meat we eat, does not have its own resident microflora. These tissues are essentially sterile (unless the animal is infected with a parasite) until the animals are butchered. Contamination of meat occurs when it is processed by butchers and meat packers and by contamination from the environment. The bacteria and fungi that colonize meats are varied and include spore-formers, proteolytic bacteria, lipolytic microorganisms, and fungi. Occasionally, pathogens such as *Clostridium botulinum*, *C. perfringens*, *Staphylococus aureus*, and *Vibrio parahaemolyticus* contaminate meat, which then serves as a vehicle of infection for humans and animals.

Milk is an animal product that has its own characteristic flora, which is acquired from environmental contamination. Milk within the mammary glands (lacteals) of a healthy animal is sterile and becomes contaminated only during milking, storage, and processing. Microorganisms that are present in the milk include psychrophilic, gram negative bacteria (e.g., pseudomonads), lactic acid bacteria (e.g., lactobacilli and streptococci), micrococci, yeasts, coliforms, and spore formers. One important affliction of dairy cows is mastitis. This disease, characterized by an inflammation of the mammary glands, is most often caused by *Streptococcus agalactiae* or *Staphylococcus*

TABLE 33-1
PATHOGENS THAT MAY BE FOUND IN FOODS

PATHOGEN	DISEASE	SOURCE
Mycobacterium bovis	Tuberculosis	Milk
Brucella abortus	Brucellosis	Milk
Coxiella burnetti	Q-fever	Milk
Salmonella spp.	Gastroenteritis	Poultry, milk
Shigella spp.	Dysentery	Poultry, milk
Hepatitis-A virus	Hepatitis	Contaminated foods
Yersinia enterocolitica	Enterocolitis	Contaminated foods
Amanita spp.	Mycetismus	Mushrooms
Staphylococcus aureus	Food poisoning	Unrefrigerated, contaminated foods
Vibrio parahaemolyticus	Gastroenteritis	Shellfish
Trichinella spiralis	Trichinosis	Pork and bear meat
ECHO viruses	Gastroenteritis	Contaminated foods

aureus. These organisms cause severe financial losses to farmers in lowered milk production and the added expense of treating the sick animals. *Streptococcus agalactiae* also causes infections in humans. In addition, milk may contain other human and animal pathogens (table 33-1) such as *Mycobacterium*, *Brucella*, *Salmonella*, and *Coxiella.*

ROLE OF MICROORGANISMS IN FOOD SPOILAGE

Spoilage is any disagreeable change or departure from the normal state in the food that can be detected with the senses of smell, touch, taste, or vision (**organoleptic** testing). The changes that occur depend on the food composition and the microorganisms that are present (table 33-2). These changes occur as a result of chemical reactions in the food or metabolic activities □ of the microorganisms. Not all foods are equally susceptible to spoilage, and various physical, chemical, and biological factors play a role in spoilage (table 33-2). For example, lipolytic microorganisms, such as pseudomonads or certain fungi, growing in sweet butter (mostly fats) cause a form of spoilage called **rancidity.** This type of spoilage results from the microbial breakdown of butterfats to produce glycerol and fatty acids, which are responsible for the smell and taste of rancid butter. Similarly, proteolytic bacteria growing on hamburger meat (mostly protein) will break down the muscle protein and release products such as putrescine and cadaverine, which give the rotting meat its characteristic odor. This type of spoilage, called **putrescence** results from the incomplete microbial metabolism of amino acids. Another form of food spoilage is **sour spoilage.** For example, if raw milk is allowed to stand undisturbed, either in the refrigerator or outside, eventually it will sour. The sour spoilage results from the growth of lactic acid bacteria in the milk. These bacteria obtain their energy and carbon by fermenting the milk sugar (lactose) to lactic acid and other acids, which give milk an unpalatable, sour taste.

183

TABLE 33-2
MICROORGANISMS INVOLVED IN FOOD SPOILAGE

FOODS AFFECTED	TYPE OF SPOILAGE	MICROORGANISMS INVOLVED	MICROBIAL GROUP
Bacon, sausage	Greening	*Lactobacillus*	Gram + rod
		Leuconostoc	Gram + coccus
Bread	Ropiness (stringy)	*Bacillus subtilis*	Gram + rod
	Moldy	*Aspergillus, Rhizopus*	Fungi
Canned goods	Moldy	*Byssochlamys*	Yeast
Eggs	Rotting, greening	*Pseudomonas, Proteus*	Gram − rods
Fresh fish	Rotting (fishy)	*Pseudomonas, Serratia*	Gram − rods
		Flavobacterium	Gram − rod
		Micrococcus	Gram + coccus
	Discoloration	*Micrococcus*	Gram + coccus
		Pseudomonas, Serratia	Gram − rods
		Yeasts and molds	Fungi
Fresh meats	Slimy	*Flavobacterium, Pseudomonas*	Gram − rods
		Yeasts	Fungi
Milk	Souring	*Lactobacillus*	Gram + rod
		Streptococcus	Gram + coccus
	Ropiness	*Alcaligenes*	Gram − rod
	Gassy	*Bacillus, Clostridium*	Gram + rods
Preserves	Moldy	*Aspergillus*	Fungi
		Penicillium, yeasts	

Since much food spoilage is due to microorganisms, the quality of the food and the risk of spoilage can be estimated by knowing the number and types of microorganisms present. Traditionally, the microbiological quality of foods has been determined by three methods: a) dye reduction tests; b) direct microscopic counts; and c) viable counts.

The **dye reduction test** is a qualitative test that gives the microbiologist an idea of the number of metabolizing microorganisms present in the foods. The test is based on the observation that dyes such as methylene blue and resazurin change in color when they are reduced (accept electrons). When microorganisms are metabolizing in the presence of methylene blue, electrons □ donated by the microorganisms reduce the methylene blue, causing it to become colorless. This color change is slow and requires a large number of metabolizing microorganisms. The time required for the color change is dependent upon the number of microorganisms present in the sample. Dye reduction tests were once used extensively in the dairy industry to measure the quality of milk. For the most part, however, these methods have been replaced by more sensitive ones.

31

The **direct microscopic count** (DMC) is another method used to assess the microbiological quality of foods. It gives the microbiologist an estimate of the total number of microorganisms present in the food. The DMC is

TABLE 33-3
NEW TECHNIQUES EMPLOYED IN MEASURING THE QUALITY OF FOODS

PROCEDURE	PRINCIPLE INVOLVED
Bioluminescence	Measures ATP (a component of cells) by using it as a source of energy for bioluminescence.
Coulter counting	Measures number of microorganisms by counting changes in electrical conductivity.
DNA hybridization	Detects the presence of specific pathogens by determining the degree of DNA similarity between the sample and a standard.
Laser beam	Measures amount of light scattered by microorganisms.
Microcalorimetry	Measures heat released by microorganisms during reproduction.
Radiometry	Measures release of ^{14}C in CO_2 from microbial metabolism.

carried out by spreading a known volume (usually 0.01 ml) of food suspension or milk over a known area (1 cm^2). The smear is then fixed and stained, and the microorganisms present are counted with the aid of a microscope. The results of the DMC are expressed as microorganisms/g or microorganisms/ml. The DMC is still used to estimate the number of microorganisms present in milk and some foods. The DMC test is reliable when large numbers of microorganisms, more than 250,000/ml, are present in the food, but it is not useful when microorganisms are present in numbers much below 10,000/ml. Another problem with the DMC is that a certain amount of expertise is required to differentiate microoganisms from artifacts.

132

Viable counts □ remain the most popular procedures for determining the microbiological quality of foods. Several reasons account for this popularity. Viable counts are simple to perform and can easily be adapted to culture many of the microorganisms present in foods, including pathogens. In addition, viable counts are desirable because they detect the presence only of living microorganisms and hence offer a better indication of the potential for spoilage or disease. Viable counts are also very sensitive, since theoretically they can be used to detect the presence of even 1 bacterium/ml. One drawback is that a certain amount of skill is required in order for the test to yield accurate results. Proficiency in the performance of aseptic techniques is essential to obtain accurate and reliable results. In addition, results generally require a waiting period of at least one day before the quality of the food can be determined.

To obviate the need for long waiting periods before the results can be obtained, various tests such as fluorescent antibody techniques, microcalorimetry, DNA hybridization, laser beam spectrophotometry, and bioluminescence have been developed. The main advantage of these methods resides in the fact that they rapidly determine the presence of microorganisms in the food (table 33-3).

Prevention of Spoilage

Since microbial growth is the most common cause of food spoilage, any procedure that inhibits the growth of microorganisms without affecting the palatability of the food is desirable. In this section we will discuss some of the common ways in which foods are preserved, in industry and in the home, and the possible mechanisms of action of some preservatives.

Cold Storage

The advent of refrigeration revolutionized food preservation. Before iceboxes and refrigerators were developed, foods were destined to spoil if not eaten quickly. Cold storage is the most common form of food preservation in industrialized countries. This process capitalizes on the fact that microorganisms, like all organisms, require a certain amount of heat for their metabolism. At low temperatures, enzymatic reactions are very sluggish or do not occur to a significant extent and therefore microbial multiplication does not take place. Refrigeration does not prevent spoilage, however, but simply delays the inevitable, because many of the microorganisms responsible for spoilage of refrigerated foods are psychrophilic and therefore can grow at refrigerator temperatures (about 40°F or 5°C). These organisms cause many different types of spoilage.

Whereas refrigerator temperatures may permit the growth of psychrophiles □, deep freezing inhibits all microbial growth. The effectiveness of freezing as a means of food preservation is based on the fact that at low freezing temperatures (around −20°C), water is ice and is therefore not available for physiological reactions. Freezing is also a microbiocidal agent. Depending upon the species of microorganisms, over 50% of the microorganisms in food may be killed during freezing. The formation of ice crystals causes irreversible damage to the plasma membrane.

145

Frozen foods that have warmed enough to thaw partially may spoil. During the thawing period, hydrolytic enzymes released from damaged cells decompose the food, altering its flavor and texture. Also, upon thawing, psychrophilic microorganisms begin to grow on the foods and promote their spoilage. Pathogenic microorganisms generally do not grow at temperatures below 10°C, but food showing signs of spoilage should not be eaten, because the temperature may have risen high enough to permit the growth of pathogens. Freezing □, since it is not a 100% effective microbiocidal process, will not make previously frozen foods safe to eat.

161

Lowering the Water Activity

Microorganisms differ widely in the amount of water they need (fig. 33-1). In the absence of water, many microorganisms die. Those that are able to endure drying conditions are called **xeroduric. Water activity** □, a_w, is a measurement of the amount of water *available* to microorganisms for growth. Figure 33-1 illustrates the water activities of various foods and the microorganisms that are most likely to cause spoilage at various water activities. One can lower the water activity of foods by freezing, drying (dehydration), or increasing the osmotic pressure.

114

Drying. Drying is an old and popular method of food preservation which uses the sun's heat and moving air to extract the moisture from foods. It is believed that prehistoric humans used this method to preserve some foods. The foods, generally fruits, fish, vegetables, and meats, are cut into thin slices or chopped finely before drying in order to facilitate the process. Today, drying chambers are often used to dehydrate foods. Drying chambers are outfitted with a source of heat and a fan that circulates the air, both devices that expedite the exit of moisture from the foods. Meats are usually dried to make jerky. Fruits and vegetables are frequently dried and sold in that state.

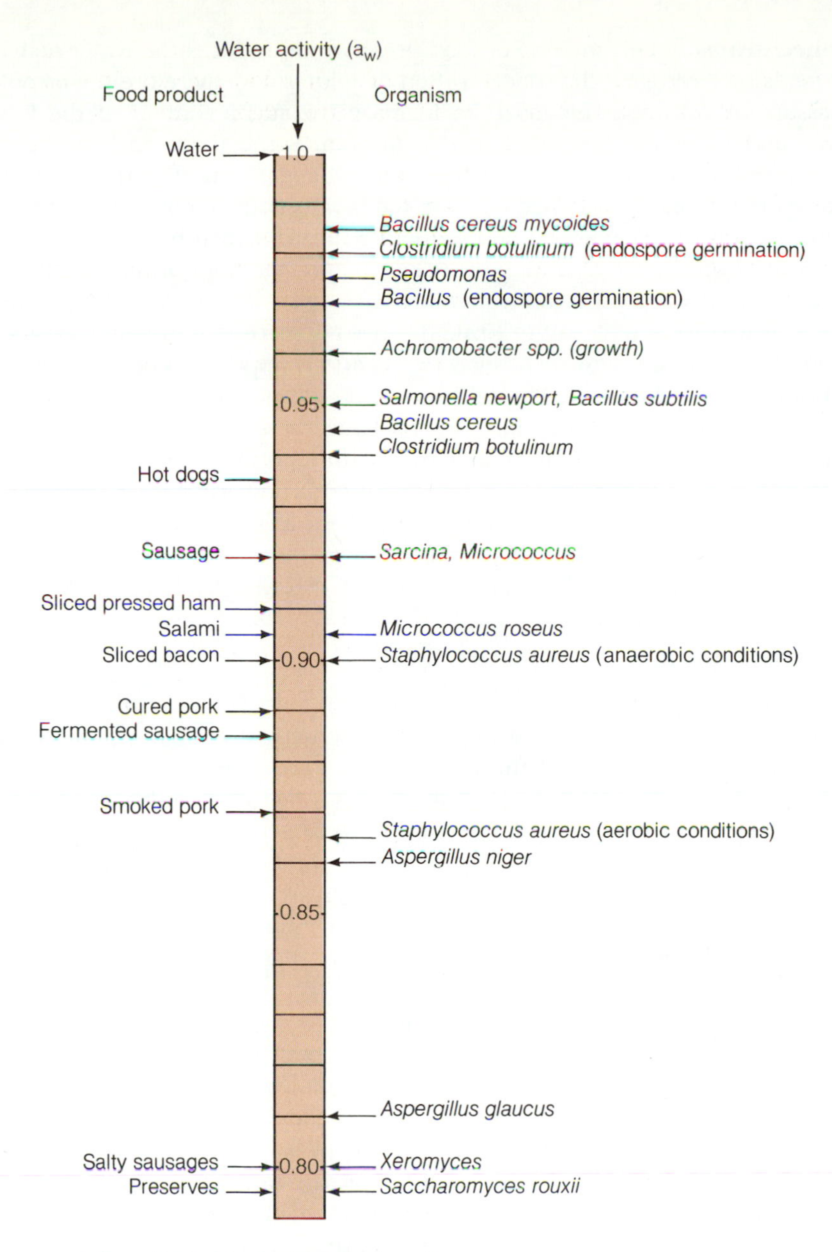

Figure 33-1 Water Activities of Some Common Foods and Microorganisms Involved in Their Spoilage

Freeze-drying or **lyophilization** is another method used to dehydrate foods. Freeze-drying is accomplished by sublimation of the water in the foods. The foods are frozen and then placed in a vacuum. Almost all of the water vapor from the frozen food escapes and is evacuated from the vacuum chamber. This is an expensive but popular way of processing coffee, milk, fruits, vegetables, and meats. Prepared meals are also freeze-dried and packaged. Because freeze-drying removes most of the water, it makes the food lighter and very stable at warm temperatures. Lyophilized foods are very popular with hikers because they can carry a supply that lasts for many days, is not heavy, occupies little space, and does not spoil.

Concentrates. This method of food preservation reduces the water activity of foods by increasing the concentration of solutes and therefore the osmotic pressure. A common method is to increase the sugar content of the food. Jams and jellies are preserved in this fashion. Some foods such as honey, dates, and raisins, which have a high natural sugar concentration, are very resistant to spoilage. Spoilage of these foods is usually carried out by osmophilic organisms. Condensed and evaporated milk, which have an increased amount of solids, are also more resistant to spoilage than ordinary milk because they have a high osmotic pressure.

Curing meats with salt or brine has been a successful preservation technique for centuries. Some fish, such as cod and snapper, can be preserved by rubbing salt on the meat. The salt draws the water from the muscle tissue, thus dehydrating it. Some countries package these salted fish and export them. Alternatively, vegetables and meats may be preserved by soaking in a salty brine. The dissolved salt enters the cells, lowering their water concentration. In addition, foods may be smoked. The smoking process coats the foods with aldehydes and phenolic compounds that are microbiocidal. The smoking process also removes some of the water from the food and flavors it.

Acidifying Foods

The addition of acids to foods lowers their pH. A low pH inhibits the growth of many microorganisms by affecting enzymatic activities □. Foods with high acid content, such as citrus fruits and tomatoes, are more resistant to spoilage than low acid foods, such as corn and beans. Low acid foods can be made more resistant to spoilage by immersing them in acids, commonly by adding acids like vinegar (acetic acid). This process is sometimes called pickling. Cucumbers, beets, and other vegetables can be preserved in this fashion. Sometimes garlic, dill, and spices are also added to enhance the flavor.

146

Some microorganisms, when growing in sugar-containing foods, can ferment the food's sugars and release acids. For example, the lactic acid bacteria ferment sugars to produce lactic acid. Allowing lactic acid bacteria to ferment milk sugar (lactose) is a simple way of preserving milk protein. The lactic acid produced by the bacteria coagulates the milk protein so that a curd is formed. The curd, mostly casein, is acidic and is therefore more resistant to microbial spoilage than normal milk proteins. Cheese, buttermilk, yogurt, koumiss, and kefir are a few of the milk products that can be prepared using lactic acid bacteria. Vegetables such as cabbage and cucumbers, when fermented, are converted into sauerkraut and pickles, respectively, which are more resistant to spoilage than the fresh vegetables.

Some acidophilic microorganisms, predominantly fungi, are able to grow on acid foods, using the acids as a carbon and energy source. This growth increases the pH, so the foods become more susceptible to spoilage.

Canning

Nicholas Appert, prompted by the prospect of a 12,000-franc reward for developing a form of food preservation that would be practical for the military, developed the process of canning in 1809. Canning is a form of food preservation in which high heat □ is used to kill microorganisms in the food and hermetically-sealed containers are used to protect the food from contamination. The amount of heat applied must be sufficient to kill the spores of spoilage and pathogenic microorganisms, but not so hot that the food loses its aesthetic qualities.

158

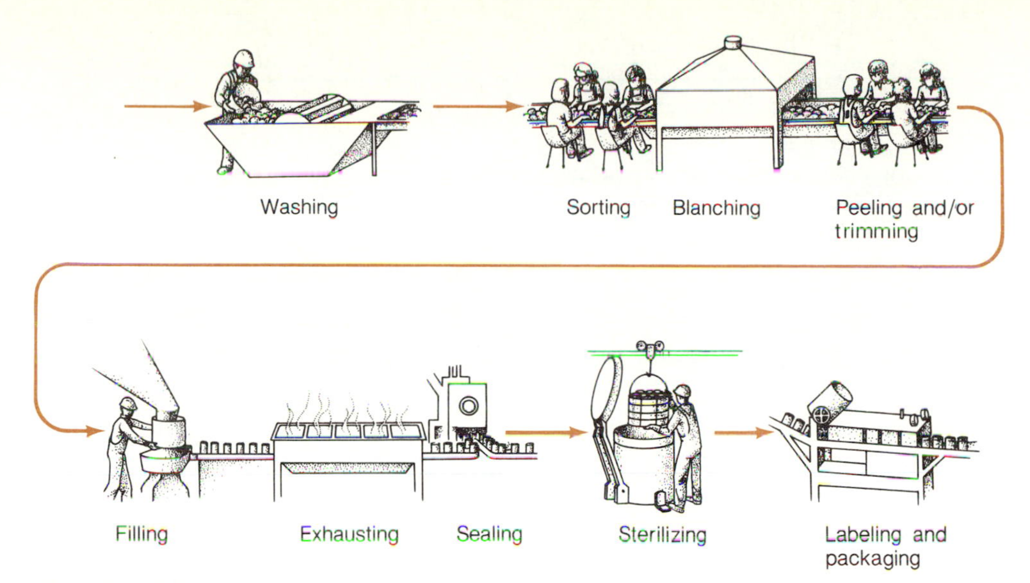

Figure 33-2 Steps Involved in the Large-Scale Preparation of Canned Foods

Soon after the development of canning, it became obvious that, even though an anaerobic environment within a sealed container inhibited the growth of aerobic spoilage microorganisms, it promoted the growth of anaerobes that might have survived the heating process. Of special concern was the multiplication of *Clostridium botulinum* in canned goods, because this organism produces the deadly botulin toxin when it grows anaerobically in canned goods.

The natural acidity of some foods expedites their sterilization. High acid foods are sterilized by heat more rapidly than low acid foods at the same temperature. The exact procedure that must be followed in canning depends upon several factors, such as the makeup of the product, its resident microbiota, its pH, and the rate at which heat penetrates it.

The process of canning (fig. 33-2) involves: a) preparing the food; b) filling the cans (or jars); c) exhausting the air from the food; d) sealing the cans; e) sterilizing the contents of the cans; and f) cooling the cans. Grading the quality of the product and separating and sizing the food constitute an integral part of food preparation. Once the food is graded and sorted, it is washed, then cut or sliced. The prepared raw food is briefly heated in boiling water or steam (blanched), in order to inactivate enzymes that may cause damage to the food while in the can; to fix the color of the food; to kill some spoilage microorganisms; and to prevent disagreeable changes in the food. After blanching, the cans are filled and the air is exhausted by placing the filled cans in a hot water bath. Next, the cans are sealed and sterilized. Sterilization of the sealed cans is the most crucial step in the canning process, because it is during this step that the remaining microorganisms are killed. The cans are exposed to high heat for a specified period of time to kill the microorganisms and to inactivate any enzymes that might cause disagreeable changes in the food. The amount of heat applied depends upon the type of food and volume of canned material, but it must be high enough and long enough to kill the endospores of *Clostridium botulinum*.

Acid foods with a pH of 3–4.5, such as tomatoes, apricots, pineapples, berries, and citrus fruits, need not be processed in a pressure cooker. Placing such canned goods in a boiling water bath for 15–30 minutes is sufficient to kill spoilage microorganisms, because acid foods promote a more rapid killing than low acid foods. Some recent varieties of tomatoes are low in acid and must be heated using a pressure cooker or supplemented with additional acid in order for them to be canned safely. In addition, fungi, asporogenous rods, and lactobacilli, which commonly colonize these foods, are readily killed at food temperatures around 80°C.

Foods such as meats, fish, vegetables, and eggs, with pH values above 4.5, are low acid foods and so require extensive heating for sterilization. These canned foods generally are placed in pressure cookers and heated to temperatures between 115° and 121°C. The appropriate sterilization times and temperatures are determined by whether or not *C. botulinum* or *Bacillus stearothermophilus* endospores are destroyed in the canned goods. Strict quality control must be exercised in order to insure that pathogens do not survive in the cans.

Modern canning procedures are very reliable, and little botulism results from commercial canning. Most cases of botulism are associated with home canning in which insufficient heat was applied to low acid foods contaminated with *C. botulinum* endospores.

Spoilage of Canned Goods. Most spoilage of canned goods is caused by anaerobic bacteria that have survived the canning process. Most of these canned goods have been understerilized or improperly sealed. One reason for understerlization is that occasionally a food is so contaminated with endospores or microorganisms that normal heating periods are not sufficient to kill them all.

There are basically two types of spoilage of canned products: a) **flat sour spoilage** and b) **gassy spoilage.** Bacteria such as *Bacillus stearothermophilus*, *B. coagulans*, and *Lactobacillus* sp., which are responsible for most flat sour spoilage, carry out a fermentation in which acid is produced but not gas. In this type of spoilage the can appears normal, but when the food is tasted, the contents are tart and have a disagreeable metallic flavor.

Microorganisms that cause gassy spoilage of canned goods do so because they produce gas, which causes the cans to swell. Acid foods, such as citrus foods and pickles, are spoiled primarily by yeasts. Lactobacilli often carry out gassy spoilage of tomatoes, while meat products are spoiled primarily by clostridia. Not all gassy fermentation is caused by **aerogenic** (gas-producing) **fermentors**. Some bacteria, such as *Bacillus polymyxa*, are capable of reducing nitrate to nitrogen gas □. The nitrogen gas produced during anaerobic respiration also causes the cans to bulge. The amount of gas released ultimately determines the extent of bulging.

765

The food industry has given names such as flippers, springers, soft swells, and hard swells to cans that have undergone various degrees of gassy spoilage (fig. 33-3). The term **flipper** is used to describe cans whose lids will flip in and out when they are compressed by a thumb. Flippers result when aerogenic spoilage microorganisms produce enough gas to balance the atmospheric pressure outside the can. If gassy spoilage proceeds beyond the flipper stage, the cans may bulge at one end and become **springers,** so called because, when the bulged lid is pressed, the bulge will spring to the

(a)

(b)

Figure 33-3 Gassy Spoilage of Canned Goods. *(a)* Can of tomato paste showing a slight swell (flipper). Only one end is swollen. (*b*) Can of tomato paste showing a severe swell (hard swell). Both ends are swollen. The swollen bottom causes the can to sit at an angle.

opposite end of the can. If sufficient gas is produced, both lids will bulge out. When the bulged lids can be pushed inward with only a slight thumb pressure, the can is said to have the **soft swells.** If the bulge cannot be moved easily, however, then the can is said to have the **hard swells.** When the amount of gas produced by microorganisms increases the internal pressure much above 5 atmospheres, the seals open and the can bursts, spewing the malodorous goods.

Radiation

High energy radiation is sometimes used to sterilize foods. The amount of ionizing radiation required to destroy microorganisms differs from food to food. For example, 3 mrads are required to inactivate 10,000 endospores of *C. botulinum* in peas, while in meats the dose must be raised to 3.5 mrads. Radiation sterilization or **radpasteurization** is sometimes employed on hams, bacon, and vegetables. One objectionable side effect of radiation is that it can change the flavor of radiated foods. It has been found that these flavor changes occur at radiation levels below microbiocidal levels.

Chemical Additives. During the preparation of foods, and before they are packaged, chemicals are sometimes added to increase shelf life. These chemicals must be chosen so that they are microbiocidal or microbiostatic 152 □ at levels that are not injurious to humans. Furthermore, the chemicals must not alter the appearance, smell, or taste of the food. These chemical additives can be either inorganic or organic compounds. Sulfur dioxide, nitrates, and nitrites are the most common inorganic additives. Sulfur dioxide is used to

FOCUS

ANTIFUNGAL ACTIVITIES OF GARLIC

Garlic *(Allium sativum)* has long been considered to be an important crop plant, not only because it is a tasty condiment, but also because of its medicinal powers.

Allium sativum has been credited with the ability to ward off insect pests, such as mosquitoes and flies. It has been said that individuals who eat large amounts of garlic are less likely to be bitten by mosquitoes than those who do not eat garlic. In addition, certain organic gardeners plant garlic between other crop plants, such as green peppers and tomatoes, in the belief that the garlic will prevent the other crops from becoming infested by insects and spiders.

The medicinal powers of garlic have also been documented since antiquity, although little scientific evidence has been presented. The curative powers of garlic, if real, are likely associated with its ability to inhibit or kill microorganisms.

Scientists at Loma Linda University in California have provided evidence that garlic extracts inhibit or kill human-pathogenic fungi. Their tests involved the use of commercially-prepared powdered garlic against an array of known fungal pathogens. These included the yeast *Candida albicans* and a pathogen of the central nervous system called *Cryptococcus neoformans.* The test was performed by mixing garlic powder with Sabouraud glucose broth (a medium that favors the growth of fungi) at a ratio of 1 part garlic to 40 parts broth. The garlic-broth mixture was filtered to remove undissolved particles and then filter-sterilized. The "garlic broth" was then inoculated with a standard amount of fungi and incubated for a period of time ranging from 48 hours to approximately 4 weeks. The viability of the fungi was tested by viable plate counts. The findings revealed that garlic was inhibitory to all fungi tested, *Cryptococcus neoformans* being the most sensitive to the garlic.

Whether or not garlic extracts can be used in the treatment of fungal infections or to preserve foods remains to be seen. The work of the scientists at Loma Linda University provides the cornerstone for further research into the preservative and medicinal powers of garlic and how this plant can be used to improve human health.

kill "wild yeasts" during the initial steps in winemaking □. Wine yeasts are more resistant to the microbiocidal action of sulfur dioxide than wild yeasts and therefore survive the treatment. In this way, winemakers can effectively control the fermentation process. 859

Salts of nitrate and nitrite are commonly used as chemical additives to preserve the fresh, red color of meats and to inhibit the growth of *C. botulinum* in sausages and other prepared meats. The usefulness of nitrite, especially as an antibotulinal agent, has been marred somewhat by reports that nitrite is converted into nitrosamines in the stomach, in the intestines, and during cooking (microwave cooking excepted) or prolonged storage. Nitrosamines have long been known to be carcinogenic agents. Nitrates are added to meats and are reduced by microorganisms into nitrites; thus, the nitrates serve as a continuous supply of nitrite.

The most common organic compounds used as chemical additives are monosodium glutamate (MSG), sorbic acid, sodium benzoate, and calcium propionate. Monosodium glutamate is used as an additive simply to enhance the flavors of some foods, and it has little or no preservative value. Calcium propionate is commonly employed as a preservative in breads. This compound inhibits the growth of surface molds and bacteria by interfering with their cellular structures (e.g., plasma membrane) and metabolism. Sodium benzoate and sorbic acid are used to inhibit microbial growth in dairy products, fruit juices, and soft drinks.

DAIRY MICROBIOLOGY

Milk is a secretion of the mammary glands of animals. Most milk in the United States is from cows, although a small but increasing amount of goat milk is also consumed. **Colostrum** is a milklike substance produced by the mammary glands during the first few days of lactation. It is rich in antibodies and leukocytes and confers passive immunity to certain infectious diseases.

Milk is sold in a variety of forms. Raw milk comes directly from the animal without any processing, while pasteurized milk has been heated to kill any pathogens that may be present. Some milks have their fat content modified and are sold as whole (3.5%), lowfat (1% to 2%), or skim (less than 0.25% fat) milk. There are many other dairy products, such as yogurt, buttermilk, sour cream, cheeses, kefir, and ice cream.

Microbiology of Raw Milk

Although lacteal secretions (milk) are initially sterile, they become contaminated rapidly during milking and processing. Health authorities, in cooperation with dairy farmers, have established guidelines of quality which assure the consumer a safe product with a reasonably long life and high quality (table 33-4). Automation and the use of easily maintained stainless steel storage tanks have also contributed to the production of good milk.

The dominant flora in refrigerated milk consists of gram negative, psychrophilic bacteria. Properly collected raw milk, from healthy cows should contain fewer than 15,000 (usually 5,000–10,000) microorganisms/ml. Most of these microorganisms are *Pseudomonas*, *Achromobacter*, *Flavobacterium*, lactobacilli, yeasts, and fungi. Coliforms may occcasionally be found at concentrations of less than 100–700 coliforms/ml. In improperly collected or stored milk, however, the coliform count may exceed 10,000 coliforms/ml. In addition to these microorganisms, pathogens may also be present in the milk (table 33-1). Before pasteurization of milk became widespread, tuberculosis, enteric fevers, gastroenteritis, brucellosis, Q fever, leptospirosis, streptococcal sore throat, scarlet fever, and diphtheria were sometimes acquired from drinking raw milk.

Pasteurization

Pasteurization (fig. 33-4) was introduced by Louis Pasteur to destroy wine-spoiling microorganisms. This process, first tried on milk in 1886, is now used to treat most milk sold in the United States. The pasteurization □ of milk involves heating it to temperatures below the boiling point to kill human pathogens.

160

The temperature and time of pasteurization can vary widely. For instance, a **flash pasteurization** can be carried out at 72°C for 15 seconds, or a batch pasteurization at 63°C for 30 minutes. The treatment must be able to kill not only *Mycobacterium tuberculosis* and *M. bovis* but also the more heat-resistant bacterium *Coxiella burnetii*. This heating process does not kill all microorganisms in milk, but is sufficient to kill all pathogens without altering the taste of the milk.

The **low temperature long time** (LTLT) process, also called the **kettle process,** is a batch pasteurization method in which the milk is heated to

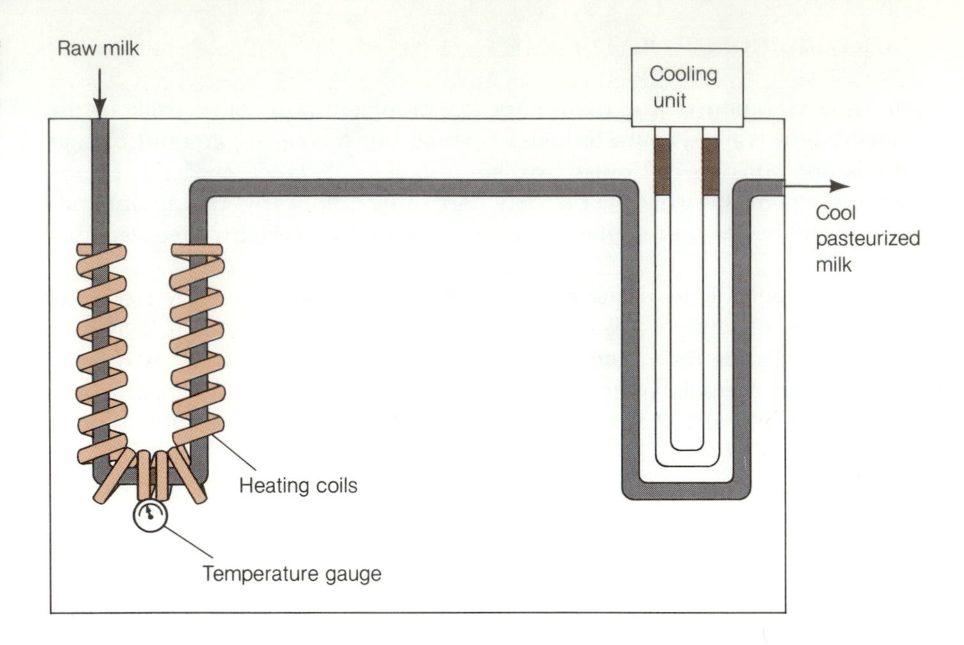

Figure 33-4 **Diagram of the Process of Bulk Pasteurization of Milk by the High Temperature, Short Time Method (Flash Pasteurization)**

about 63°C for 30 minutes. A continuous flow system called the **high temperature short time** (HTST) process carries out pasteurization at 72°C for 15 seconds, at 88°C for 1 second, at 90°C for 0.5 second, or at 96°C for 0.05 second. The HTST method is more desirable, because it is well suited for large volumes of milk, induces fewer changes in milk flavor, and kills more microorganisms. Thus, it not only assures the destruction of pathogens but

TABLE 33-4
MICROBIOLOGICAL STANDARDS FOR MILK AND DAIRY PRODUCTS

	MAXIMUM NUMBER OF MICROORGANISMS/ML ALLOWED	
DAIRY PRODUCT	TOTAL BACTERIA[1]	TOTAL COLIFORMS[2]
Grade A raw milk (before mixing with other batches)	100,000[3]	250–300[3]
Grade A raw milk (after mixing)	200–300,000	250–300[3]
Grade A pasteurized	20,000	10
Condensed milk	30,000	10
Ice cream	50,000	10
Frozen yogurt	50,000	10
Nonfat dry milk	30,000	10
Grade B raw milk	600,000	
Grade B Pasteurized	40,000	
Certified raw milk	10,000	10

[1]Total bacteria determined by the standard plate count.
[2]Total coliforms determined using Violet Red Bile Agar.
[3]Limits vary depending on the agency setting standard.

also increases the shelf life of the milk. Sterile milk has now been introduced commercially. This product, completely devoid of living microorganisms, is not likely to spoil and hence has a long shelf life.

Not all milk sold to the consumer is pasteurized. Raw milk is also sometimes sold, although its microbiological quality is well monitored. Health authorities have implemented a **Certified Raw Milk** (table 33-4) program in which the raw milk and the dairy operation are closely scrutinized for sanitary practices, herd vaccination against pathogens, and the microbiological quality of the milk. No raw milk, however, is completely safe. *Salmonella* and *Campylobacter* are very difficult to eliminate from a dairy herd and are likely to be present, even under carefully controlled conditions.

Dairy Products

Milk is a remarkably versatile food. It is nutritious in its unaltered state and can also be used as a starting point for a large variety of dairy products. Preparation of most of these foods involves allowing resident microorganisms or starter cultures to ferment the milk sugars (mostly lactose) and produce acid. The acid curdles the milk, forming a curd that can then be processed further, depending upon the product. In the next section of this chapter we will look at the role microorganisms play in making some dairy products.

ROLE OF MICROORGANISMS IN FOOD PRODUCTION

Microorganisms participate not only in the spoilage of foods, but also in the process of making certain foods (table 33-5). Their metabolism causes alterations in the texture, aroma, and flavor of food. The effects of microorganisms on foods are evaluated subjectively by humans; for example, the fermentation of cabbage by lactic acid bacteria to produce sauerkraut may be considered a positive step by some individuals while others may consider the product to be spoiled cabbage.

Buttermilk

Buttermilk, as it is now made, results from the souring of skim or lowfat milk by lactic acid bacteria. Originally, buttermilk was the liquid that remained at the bottom of a churn after the making of cultured butter. Buttermilk has a characteristic texture, acid taste, and aroma. The texture results from the broken-up curd. The aroma and the flavor are due to **diacetyl, acetaldehyde**, and other metabolic products released by the fermenting bacteria. Modern cultured buttermilk is made commercially utilizing a lactic acid culture consisting of *Streptococcus cremoris*, *S. diacetylactis*, and *Leuconostoc cremoris*. The nature of the starter culture varies among different manufacturers, and some use *Lactobacillus bulgaricus* to make Bulgarian buttermilk. The production of acid and the formation of the curd are due to *S. cremoris*, while the aroma and the flavor are due to the metabolism and multiplication of the other two bacteria in the curdled milk.

To make buttermilk, skim or lowfat milk that has been homogenized and pasteurized is inoculated with 1% starter culture and allowed to ferment at 18–22°C for about 14 hours. After the fermentation, the resulting product

TABLE 33-5

FOODS PREPARED USING MICROORGANISMS

FERMENTED FOOD	STARTING MATERIAL(S)	MICROORGANISMS INVOLVED INVOLVED IN FERMENTATION	MICROBIAL GROUP
Pickles	Cucumbers	*Lactobacillus* sp., *Pediococcus* sp.	Gram + rod, coccus
Acidophilus milk	Pasteurized milk	*Lactobacillus acidophilus*	Gram + rod
Bread	Wheat flour, sugars	*Saccharomyces cerevisiae*	Yeast
Bread, sourdough	Wheat flours	*Saccharomyces exiguus* *Lactobacillus sanfrancisco*	Yeast Gram + rod
Buttermilk	Pasteurized milk	*Lactobacillus bulgaricus*	Gram + rod
Koumiss	Mare's milk	*Lactobacillus bulgaricus* *Torula, Mycoderma*	Gram + rod Yeasts
Kefir	Fresh whole milk	*Streptococcus* sp., *Lactobacillus* sp., *Candida, Saccharomyces*	Gram + coccus, rod Yeasts
Yogurt	Pasteurized milk	*Lactobacillus bulgaricus,* *Streptococcus thermophilus*	Gram + rod Gram + coccus
Olives	Fresh olives	*Leuconostoc mesenteroides,* *Lactobacillus plantarum*	Gram + coccus Gram + rod
Poi	Taro roots	*Lactobacillus* spp.	Gram + rod
Shoyu (soy sauce)	Rice, soy beans, rice	*Lactobacillus delbrüeckii,* *Aspergillus oryzae,* *Saccharomyces rouxii*	Gram + rod Fungi Yeast
Cheeses	Curdled milk	Various bacteria & fungi	
Beer	Grains	*Saccharomyces carlsbergii*	Yeast
Wine	Grape juice	*Saccharomyces cerevisiae,* *S. champagnii*	Yeast Yeast
Cured hams and sausage			
Sausages	Pork and beef	*Pediococcus cerevisiae*	Gram + rods
Cured Hams	Pork	*Aspergillus, Penicillium*	Fungi

is shaken vigorously to break up the curd, cooled to 4°C, and packaged in milk containers. The final product is a homogeneous, thickened liquid that is slightly effervescent (due to carbon dioxide production), with an acid flavor and a buttery aroma.

Yogurt

Yogurt is a fermented milk with a puddinglike consistency that results from the action of *Streptococcus thermophilus* and *Lactobacillus bulgaricus.* Traditionally, the milk was heated for several hours to evaporate some of the water and increase the proportion of milk solids to liquid. After the evaporation step, the milk was cooled to about 108°F (40–42°C) and inoculated with a previous batch of yogurt. After an overnight incubation in a warm place, the product was cooled. The bacteria produce diacetyl, acetaldehyde, and a variety of other metabolic products that impart the characteristic flavor and aroma of yogurt. Today, the starting milk is thickened by adding powdered milk (milk solids) to pasteurized milk rather than by evaporating away the liquid. Thickeners such as gelatin or carrageenan □ may also be added. 360

Cheese

Cheese is made from milk by separating the **curd** from the **whey**. The milk protein is denatured either by lactic acid, produced by lactic acid bacteria added to the milk as a starter culture, or by added proteases such as **rennin.** The first method produces **acid curd**, while the second results in **sweet curd**. Cheesemaking is a controlled process (fig. 33-5) in which the curd is treated and aged. The manner in which the curd is processed determines the type of cheese that is made.

A simple cheese is made by briefly scalding milk so as to kill most milk spoilage microorganisms and then adding a starter culture to the cooled milk. The starter culture usually consists of lactic acid bacteria such as *Streptococcus lactis*, *S. cremoris*, *Leuconostoc citrovorum*, and *L. dextranicum*. The seeded milk is allowed to ferment at 18°C for approximately 24 hours so that a curd is produced. The liquid, or whey, can be removed by draining the curd in cheesecloth. The curd is then salted to inhibit further microbial growth. *Leuconostoc* releases diacetyl, a compound synthesized from citric acid, which is responsible for the aroma and flavor associated with cheeses. Other bacterial products also contribute to the flavors and aromas of various cheeses. Cheeses may be incubated for long periods to allow the maturation of the curd and the growth of microorganisms to add flavors.

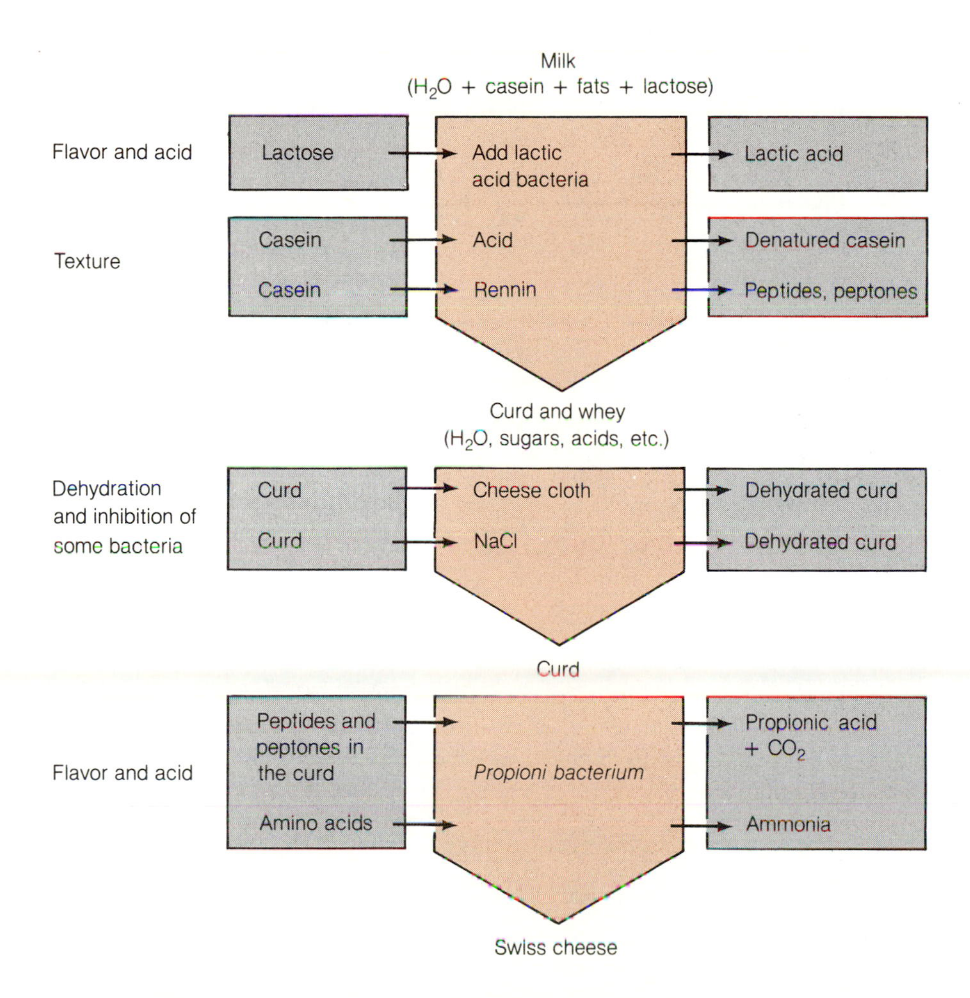

Figure 33-5 Steps Involved in the Preparation of Hard Cheese. Outline of the preparation of Swiss cheese.

Soft cheeses contain 50–80% water. They are consumed soon after they are made. Cottage cheese, an unripened soft cheese, is very moist, not heavily salted, and not very acidic since it is made using rennin. Because of its low acid content, it spoils relatively rapidly even when refrigerated. On the other hand, **Camembert** is a ripened soft cheese that does not spoil rapidly. The fungus *Penicillium camemberti* is grown on the surface of the curd to introduce the characteristic flavor and aroma of this cheese.

Hard cheeses have a water content of less than 40% and are generally ripened with bacteria or fungi. For example, Swiss cheese is made with *Streptococcus lactis, S. thermophilus, Propionibacterium shermanii, P. freudenreich, Lactobacillus helveticus,* and *L. bulgaricus.* The propionic acid bacteria give Swiss cheese its characteristic flavor and produce CO_2 pockets (the holes) in the curd.

Bread

Bread is one of the earliest processed foods made by humans. In the British Museum there are samples of bread made by Egyptians before 2000 B.C. Bread is made by mixing flour, water, salt (sometimes sugar), and yeast to make dough. The yeast most commonly used is *Saccharomyces cerevisiae.* It ferments the carbohydrates in the dough and produces carbon dioxide and ethanol. With refined flour, sugar must also be added, as the yeasts do not have the ability to synthesize amylases to break down the grain starch to fermentable carbohydrates. The carbon dioxide causes the bread to rise and gives the final product the light, porous texture characteristic of leavened bread. The bread is then baked and the ethanol evaporates. Although the yeasts are killed during the baking, metabolic products remain in the bread and add flavor to it.

Single-Cell Protein

Microbial cells can be grown on inexpensive substrates, such as waste products from other industries. One such substrate is whey, a byproduct of cheesemaking. Because of the increasing number of people suffering from starvation, considerable funds have been earmarked for research aimed at developing strains of microorganisms that are rich in proteins that could be suitable for human consumption. Yeasts such as *Saccharomyces cerevisiae, Candida utilis,* and *Kluyveromyces fragilis* have been grown in the United States as food yeasts. Also, the alga *Spirulina* has been used as a source of food protein.

Sauerkraut

The making of sauerkraut in northern Europe dates back to the 1500s. Sauerkraut results from the fermentation of salted cabbage by coliforms and lactic acid bacteria. The cabbage, which contains about 2% carbohydrate, is usually shredded and mixed with 2–3% salt. The salted cabbage is then packed firmly into vats to eliminate any air pockets and allowed to ferment for about three weeks. During the fermentation process, at least three successive species of halotolerant lactic acid bacteria, which are resident flora on the cabbage leaves, participate in the production of sauerkraut. The salt serves as a selective agent and promotes preferentially the growth of these bacteria. The fermentation is carried out at cool room temperatures.

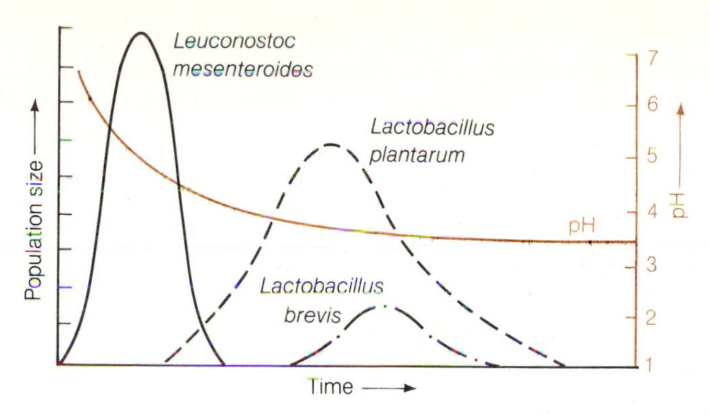

Figure 33-6 Microorganisms Involved in the Fermentation of Cabbage into Sauerkraut. *Leuconostoc mesenteroides* initiates the lactic acid fermentation of the cabbage carbohydrates, and the resulting acid lowers the pH of the brine. This change provides a stimulus for *Lactobacillus* to reproduce and proceed with the fermentation of the cabbage to produce sauerkraut.

Leuconostoc mesenteroides is the first population of microorganisms to appear in high numbers (fig. 33–6). It multiplies well on the salted cabbage, rapidly producing lactic acid and lowering the pH within three days to about 4.5. This acidity discourages the growth of contaminating microorganisms. Acetic acid, mannitol, and ethanol are produced during this stage in addition to lactic acid. The mannitol, which would give a bitter flavor to the sauerkraut, is fermented by other populations of lactic acid bacteria *(Pediococcus cerevisiae, Lactobacillus brevis,* and *Lactobacillus plantarum).* These bacteria continue to produce lactic acid, lowering the pH to about 3.5. After 21 days, the process is completed, possibly because the fermentable carbohydrates are depleted, and the sauerkraut has a pH of 3.5–3.6, 2% salt, and a little bit of ethanol.

ROLE OF MICROORGANISMS IN FOODBORNE DISEASES

Ingestion of contaminated foods and beverages can be the cause of a large number of human diseases (table 33-1). These diseases may arise either from infections or from intoxications. Foodborne diseases 632 □ are caused by many different microorganisms and can result in life-threatening situations. Foods such as poultry, fish, and pork may contain pathogenic organisms or may be contaminated during preparation by handlers and cooks (fig. 33-7).

Health authorities require that food be monitored for the presence of pathogens. For example, shellfish such as oysters and mussels are closely monitored for indicator 783 □ bacteria that could herald the presence of pathogenic microorganisms, such as *Campylobacter*, *Salmonella*, *Shigella*, *Vibrio*, hepatitis A virus, and polio virus. Because of this risk, shellfish are routinely tested for the presence of coliforms as indicators of fecal pollution of the shellfish beds. The procedure used to test for the microbiological quality of shellfish is similar to that used to test water, but also includes a standard plate count of the shellfish meat to determine the total number of microorganisms present (fig. 33-8). The procedure is carried out by homogenizing (blending) approximately 100 g of shellfish meat with an equal volume of sterile diluent. The homogenate is then diluted and inoculated into various

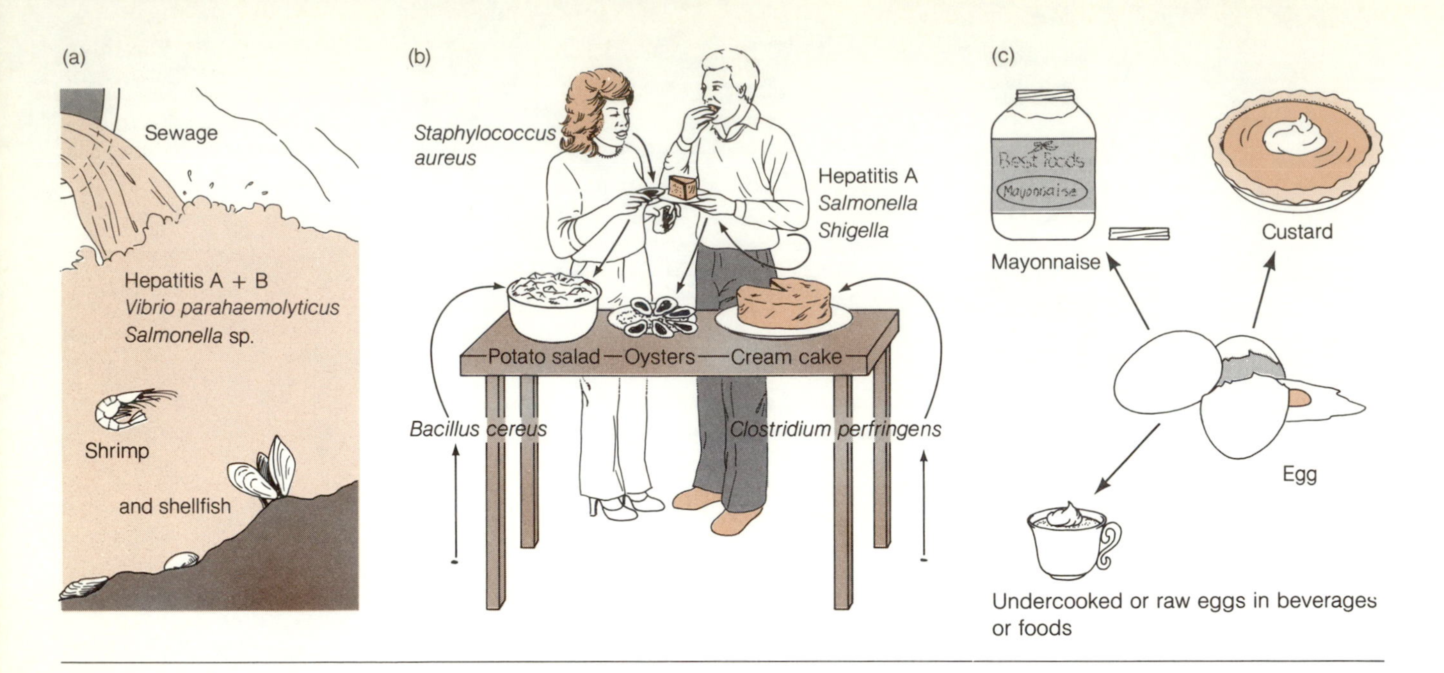

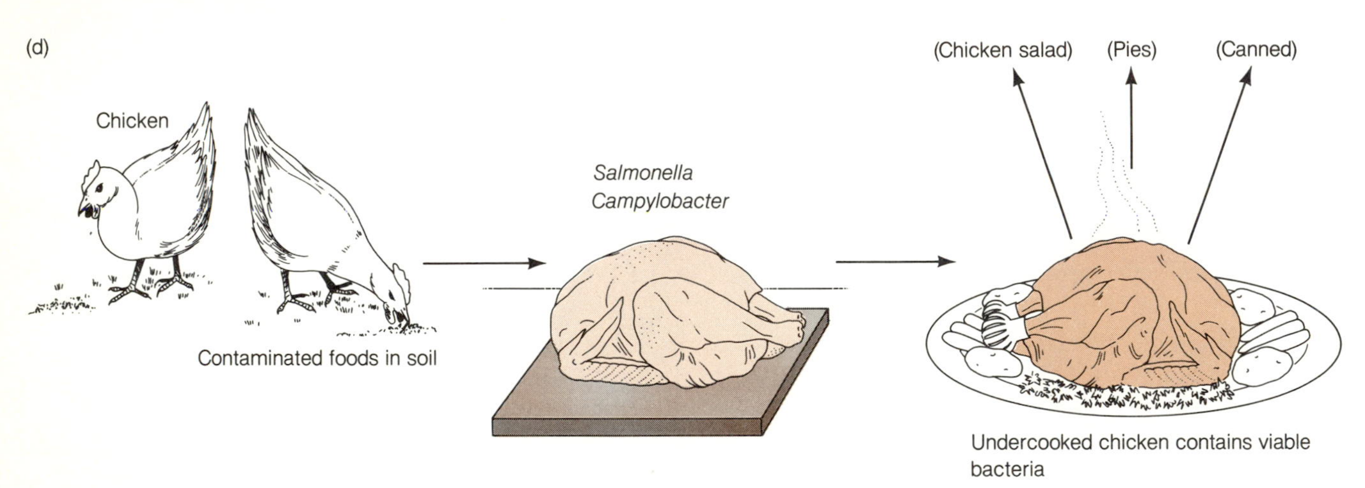

Figure 33-7 Possible Sources of Food Contamination

tubes of lauryl sulfate broth, as well as plated on a nutrient agar for a standard plate count. This procedure is outlined in figure 33-8. The results of the test are expressed as the coliform MPN/100 g shellfish meat and colony forming units/g of meat.

Foods implicated in outbreaks of food poisoning are tested for the presence of specific pathogens. For example, if a particular food is suspected of being the source of staphylococcal enterotoxin during an outbreak of staphylococcal food poisoning, it is highly desirable that live *Staphylococcus aureus* be isolated from the food in large numbers in order to support the suspicion. The suspected food is homogenized in a known amount of sterile diluent and the homogenate and subsequent dilutions are plated on *Staphylococcus* 110 Agar (table 33-6), a selective and differential medium □ for staphylococci Once typical staphylococcal colonies appear on the plate, they are iso- 121

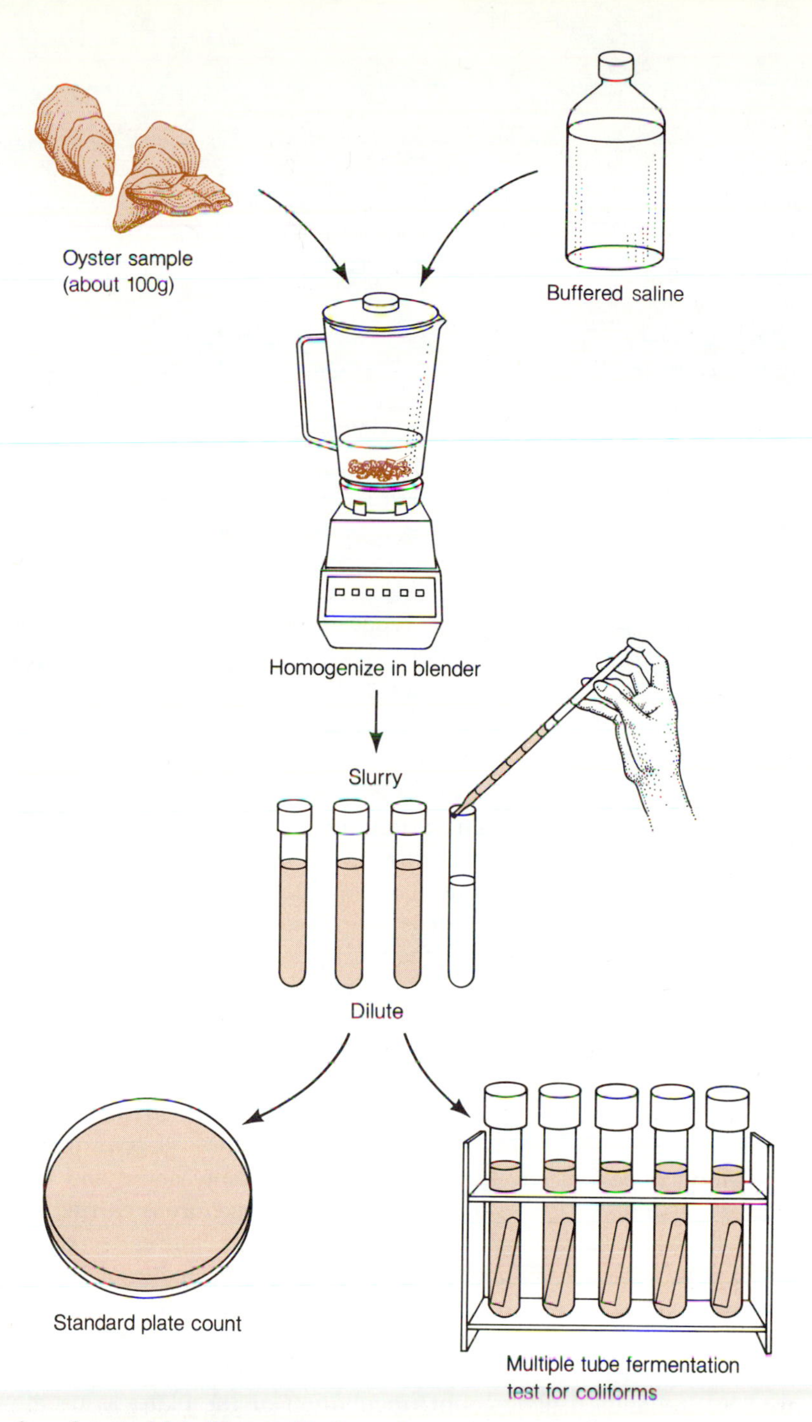

Figure 33-8 Outline of the Procedure for the Microbiological Examination of Oysters. Notice that the quality of the shellfish depends on both the total number of bacteria present in the meat and the total number of coliforms.

lated and tested biochemically in order to ascertain their identity as *S. aureus*. After the identity of the isolates has been verified, the staphylococci on the plates are counted and an estimate of the number of bacteria in the food is made. Another method of establishing the source of an outbreak of staphylococcal food poisoning is by testing for the presence of enterotoxins in the food.

Foods suspected of carrying salmonellae are homogenized and plated on Brilliant Green (BG) agar. This agar medium is the medium of choice for

TYPE 33-6
SELECTIVE MEDIA USED TO ISOLATE FOODBORNE PATHOGENS

PATHOGEN	SELECTIVE CULTURE MEDIA USED
Arizona	Deoxycholate agar (DCA), MacConkey
Bacillus cereus	KG Agar, MYP Agar
Brucella	Brucella Agar
Campylobacter	Campy-BAP Agar
Clostridium botulinum	Anaerobic Egg Yolk Agar
Clostridium perfringens	Sulfite Polymyxin Sulfadiazine Agar, TSN, SFP
Coliforms	DCA, MacConkey Agar, Violet Red Bile Agar
Enterococci	KF Streptococcal Agar
Salmonella	Bismuth Sulfite, Brilliant Green Agar, SS agar
Shigella	SS Agar, MacConkey, agar
Staphylococcus aureus	Tellurite-Glycine agar, Staphylococcus-110 agar
Vibrio (cholerae & parahaemolyticus)	Thiosulfate Citrate Bile salts Sucrose (TCBS)
Yersinia enterocolytica	SS, MacConkey agar (with cold enrichment at 4°C)

culturing salmonellae in foods. Hence, the same basic procedure can be employed to culture pathogens from various foods, and all that needs to be changed is the composition of the culture medium used (table 33-6).

When the suspected agent is not cultivatable, other methods for detecting it must be employed. In these cases, microbiologists must be familiar with the characteristic life stages of the pathogens and make decisions as to which procedure is best suited to determine the cause of the disease. For example, meats suspected of carrying *Trichinella spiralis* are examined with a microscope to establish the presence of the nematode's cysts □ in muscle tissue. 383

In addition to monitoring foods for their microbiological quality, health authorities also require public eating places to have their glassware and silverware sampled periodically to determine if the dishwashing and storage procedures are suitable (fig. 33-9). The sampling procedure is carried out by swabbing the surfaces of cups, glasses, dishes, forks, knives, spoons, and serving utensils with a moistened sterile swab, usually made of Dacron or nonabsorbent cotton. After sampling, the swab is placed into a tube containing a known volume of diluent and the tube is shaken to remove the microorganisms attached to the swab. A 1-ml portion of the diluent is then used to carry out a standard plate count. Incubation of the plates is usually at 35°C for 48 hours, after which time the number of colony-forming units (CFU) per utensil is estimated. The guidelines vary among health agencies, but usually counts of 50–150 CFU/utensil or surface area of equipment swabbed is considered an unsatisfactory number of microorganisms, and changes must be made in the dishwashing or storage procedures in order to reduce the microbial burden of the glass or silverware. Although this standard does not take into consideration the types of microorganisms present nor their public health significance, it offers a general index of utensil and equipment sanitation.

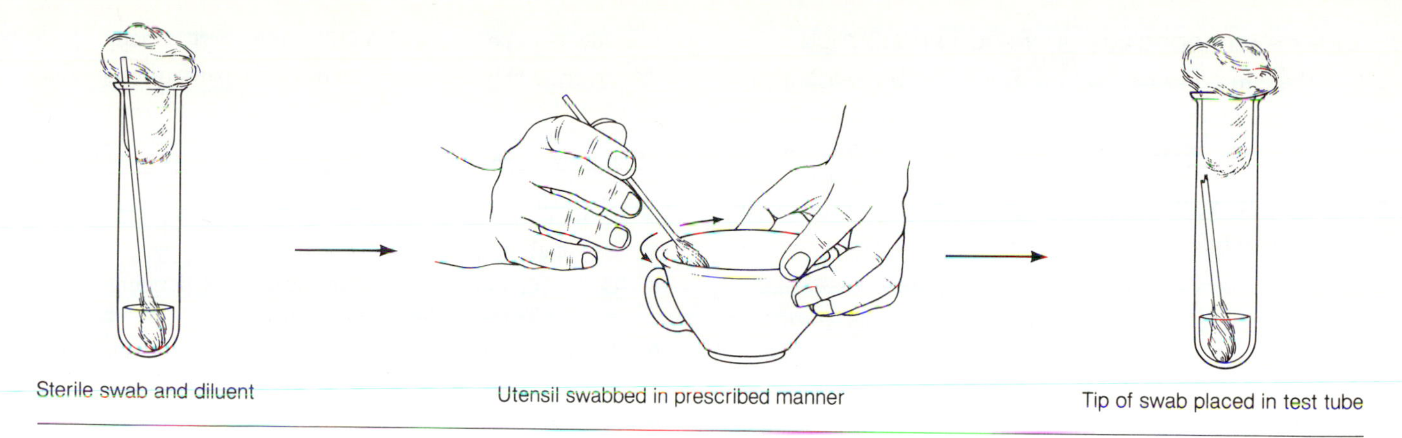

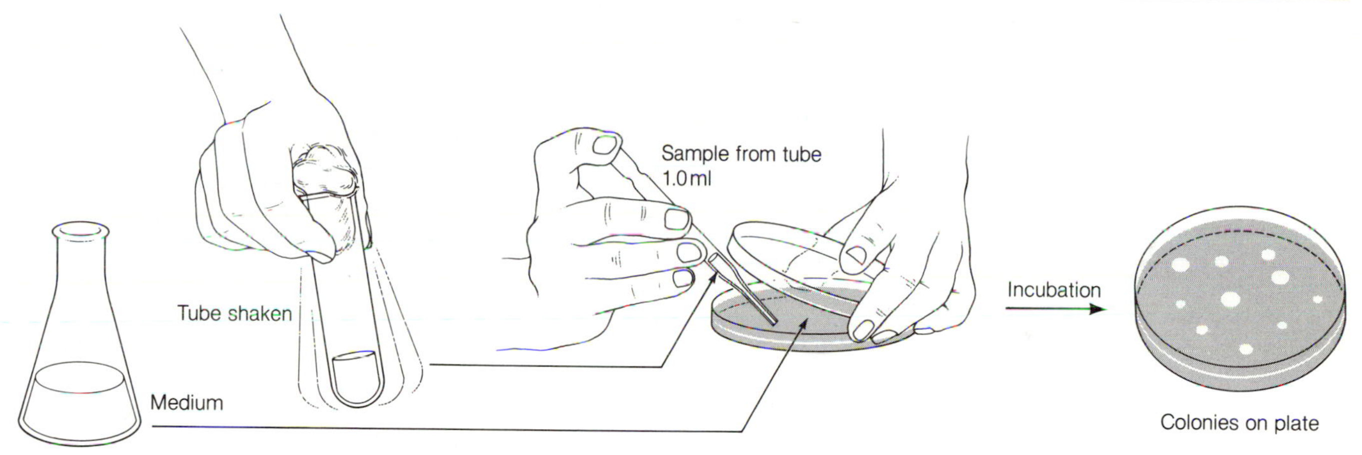

Figure 33-9 Procedure for Assaying the Presence of Microorganisms on Kitchen Utensils

SUMMARY

MICROBIAL COMPOSITION OF FOODS

Microorganisms are an integral part of the food we eat. They are responsible for the spoilage of most foods. Microorganisms or their products in foods can cause serious diseases. Not all microbial activities in foods are detrimental: some foods, such as fermented milk products and sauerkraut, are made through the intervention of microorganisms.

ROLE OF MICROORGANISMS IN FOOD SPOILAGE

1. Food spoilage is considered to be any undesirable organoleptic change that takes place in the food.
2. Microorganisms, while they multiply in food, can cause various types of organoleptic changes in the food, thus spoiling it.
3. The risk of spoilage for a particular food can be determined by counting the numbers and types of microorganisms present in the food.
4. Spoilage can be prevented by making the food uninhabitable for or free of microorganisms.
5. Preservation of foods can be accomplished by adding chemicals, lowering the pH, lowering the water activity, cold storage, or heating.

DAIRY MICROBIOLOGY

1. Milk, a secretion of the mammary glands of mammals, contains fats, sugars, proteins, and nonfat solids.
2. Raw milk contains a variety of microorganisms that contaminate it during milking, storage, or handling.
3. Pasteurization is a process in which milk is heated to destroy pathogens.
4. Using milk as a starting material, various dairy products can be made by allowing lactic acid bacteria to ferment the milk sugars and form a curd.

ROLE OF MICROORGANISMS IN FOOD PRODUCTION

1. Certain types of food are made by microorganisms.
2. Fermented foods are made when microorganisms multiply in the food, changing the flavor and texture of the original product. The changes are caused by metabolic products released by the microorganisms while fermenting a food substrate.

ROLE OF MICROORGANISMS IN FOODBORNE DISEASES

1. Foodborne diseases occur when pathogenic microorganisms or their toxic products in the food are ingested.
2. The presence of pathogens in foods can be determined using viable counts such as the standard plate count.
3. Public eating places are required to have their silverware, glassware, and utensils tested periodically to see if their dishwashing and storage procedures are safe.

STUDY QUESTIONS

1. Describe four ways in which microorganisms can contaminate foods.
2. How might microorganisms affect foods?
3. Describe three different ways in which microorganisms can spoil foods.
4. How can the microbiological quality of foods be assessed?
5. Describe five different processes commonly employed to preserve foods.
6. How does pasteurization differ from sterilization?
7. Describe how microorganisms participate in the production of the following foods:
   a. cheese;
   b. sauerkraut;
   c. yogurt;
   d. buttermilk;
   e. bread.
8. Construct a table, indicating various microorganisms and the foodborne diseases they cause.
9. In the previous table, indicate which foods are most likely to be contaminated.
10. Why are eating and cooking utensils routinely surveyed for their microbiological quality?

SUPPLEMENTAL READINGS

American Public Health Association. 1976. *Recommended methods for the microbiological examination of foods.* 2nd ed. New York: American Public Health Association.

Ayres, J. C., Mundt, J. O., and Sandine, W. E. 1980. *Microbiology of foods.* San Francisco: W. H. Freeman.

Block, E. 1985. The chemistry of garlic and onions. *Scientific American* 252(3):114–119.

Kharatyan, S. G. 1978. Microbes for humans. *Annual Review of Microbiology* 32:301–327.

Kolata, G. 1985. Testing for trichinosis. *Science* 227:621–624.

Rose, A. H. 1981. The microbiological production of food and drink. *Scientific American* 245:140–154.

CHAPTER 34
AGRICULTURAL MICROBIOLOGY

CHAPTER PREVIEW

WITCHCRAFT OR A CASE OF FOOD POISONING?

The Puritans living in New England during the 1600s and 1700s believed in witches and witchcraft. Almost every year, people were accused of witchcraft, and sometimes the accused witches were tried, found guilty, and hanged. In 1692 in Massachusetts around Salem, approximately 150 persons were accused of witchcraft, over 35 of them were brought to trial, and 12 were hanged. Some spent months in jail, and at least one died there. Most of the witchcraft accusations were made by teenage girls who were afflicted by symptoms that resemble those of persons taking hallucinogenic drugs. About 70% of the accusations were against women, and approximately 60% of those executed were women.

A fungus growing on rye is believed to have been the cause of the numerous strange deaths, bewitched persons, and accusations of witchcraft near Salem in 1692. The years 1690, 1691, and 1692 were cooler than average in eastern New England and the winters were extremely cold. The moist, cool spring along the coast is believed to have favored the growth of the fungus *Claviceps purpurea,* commonly known as **ergot,** on rye and other grasses. One of the fruiting structures of the fungus is called the **sclerotia** (fig. 34-1), and when consumed along with the rye it causes food poisoning with symptoms that could be interpreted as bewitchment by people who believe in witchcraft. The sclerotia contain alkaloids that cause poisoning when consumed. Ergot sclerotia is a source of lysergic acid diethylamide (LSD), a very potent hallucinogen. The symptoms of this food poisoning, called **convulsive ergotism,** range from mild to severe and can lead to death.

The victims of "bewitchment" in Massachusetts were mainly infants, young children, and teenagers. Seven infants and young children are known to have developed symptoms of convulsive ergotism and died. One boy suffered from a painful urinary difficulty caused by ergotism, and his brother died of a mysterious affliction. It is now known that nursing infants can develop ergotism from their mothers' milk. In addition, several cows developed symptoms of food poisoning and died.

The symptoms of "bewitchment" first began in Salem at the end of 1691 and were probably due to contaminated rye harvested in the summer of that year. Records indicate that rye crops in the Salem area frequently remained unthreshed in storage until November and December. This fact may explain why the symptoms did not appear soon after the crops were harvested. Twenty-four of the 30 persons who were bewitched in Massachusetts during 1692 had convulsions and feelings of being pinched, pricked, or bitten. Three girls felt that they were being torn to pieces and that their bones were being pulled out of their joints. Symptoms also included temporary blindness, deafness, pain, visions, and flying sensations. A number of girls suffered from hallucinations, fits, and laughing and crying spells.

Figure 34-1 Sclerotia on Rye

Since the Salem trials, a number of incidents of convulsive ergotism have occurred in the New England area. Most of the food poisoning epidemics were mild, but a Salem epidemic of convulsive ergotism called "nervous fever" killed more than 30 persons in 1795. Epidemics of ergotism occur throughout the world when cold, damp years weaken rye crops and stimulate the growth of the fungus.

These cases of convulsive ergotism should remind us of the importance of microorganisms that grow in association with animals and plants. In this chapter we will continue to stress the beneficial and detrimental aspects of these microorganisms.

LEARNING OBJECTIVES

A STUDY OF THIS CHAPTER SHOULD ENABLE YOU TO:

EXPLAIN HOW THE MYCORRHIZAL FUNGI STIMULATE PLANT GROWTH

DISCUSS HOW *RHIZOBIUM* PROMOTES PLANT GROWTH

LIST THE FACTORS THAT MAY MAKE A SOIL SUPPRESSIVE OR CONDUCIVE TO DISEASE-CAUSING MICROORGANISMS

DISCUSS THE SIGNS AND SYMPTOMS OF SOME PLANT DISEASES

EXPLAIN HOW INTESTINAL MICROORGANISMS BENEFIT ANIMALS

DISCUSS THE ECONOMIC EFFECTS OF ANIMAL DISEASES ON HUMANS

EXPLAIN WHICH ANIMAL DISEASES ARE CONTAGIOUS TO HUMANS

The healthy growth of plants is determined not only by the nutrients present in the soil and by the soil's physical characteristics but also by the microorganisms that associate with the roots, stems, and leaves. Some microorganisms are capable of invading plant tissues and causing disease. In the last few years, however, it has become clear that numerous soil microorganisms growing on and around plant roots inhibit the growth of pathogenic microorganisms. In addition, certain microorganisms can detoxify poisons in the soil, solubilize minerals, and absorb nutrients and water for the plant. Thus, the types of microorganisms in the soil can determine which crops can be grown most successfully, the type of crop rotation that is best, and even whether fertilizers and pesticides are necessary.

MICROORGANISMS AND PLANT NUTRITION

Nutrients

Plant roots secrete numerous organic compounds, such as amino acids, sugars, and vitamins, which stimulate the development of a community of microorganisms around the roots. The volume around the roots is known as the **rhizosphere,** while the root area is the **rhizoplane.** The spectrum of compounds released determines which microorganisms will predominate in the rhizosphere and on the rhizoplane. Some of these organisms are pathogens, while others protect the plant from invading microorganisms. Microorganisms often provide needed molecules to the plants from which they obtain nutrients. For example, *Azospirillum* and *Azotobacter* fix nitrogen □ 761 in the rhizosphere and release nitrogenous compounds to plant roots. Submerged shallow seawater plants, such as eelgrass and turtle grass, are supplied with fixed nitrogen by the anaerobic clostridia associated with their roots. Rice and other partially submerged plants, in some cases, are protected from poisonous H_2S by hydrogen sulfide–oxidizing bacteria such as *Beggiatoa*. Hydrogen sulfide can poison plants' electron transport systems, so the removal of this poison is essential to the plants' survival. Some of the rhizobacteria, such as *Agrobacterium*, *Arthrobacter*, and *Pseudomonas*, produce substances such as indoleacetic acid (IAA), auxins, and gibberellins, which stimulate plant growth. Other rhizosphere microorganisms increase the availability of phosphate by degrading phosphate-containing compounds and by dissolving apatite (a granular mineral of calcium fluorophosphate), which cannot be utilized by plants.

Just as some microorganisms donate nutrients to their host plants, some can create deficiencies in plants. Often mineral deficiencies arise when microorganisms bind or oxidize metals. For example, a disease of fruit trees known as "little leaf" is often due to microorganisms that sequester zinc, while "grey speck" of oats is caused by microorganisms that oxidize manganese. Some organisms inhibit plant growth by converting nitrate fertilizers to N_2.

Mycorrhizae

Mycorrhizae are symbiotic associations between certain species of fungi and the roots of many plants □. 339 The fungi found in the mycorrhizae are called **mycorrhizal fungi.** It appears that mycorrhizal fungi increase the rate of nutrient absorption from the soil; produce plant growth stimulators, such as auxins, gibberellins, cytokinins, vitamins, fatty acids, and antibiotics; and protect the roots from numerous plant pathogens. For example, **ectomycorrhizal fungi,** which form a sheath of fungal growth around roots and penetrate the roots intercellularly (between the cells), reduce the mortality of Douglas fir trees in nurseries. The fungi protect the roots from pathogens by producing antibiotics and organic acids and by constituting a physical barrier against invading microbes. The ectomycorrhizae also stimulate the roots to release compounds into the rhizosphere which inhibit other fungi. The coating of pine seeds with the fungus *Pisolithus tinctorius* reduces the mortality of many pine species used to replant unfavorable habitats, such as uncultivated soils low in phosphate (fig. 34-2). *Pisolithus tinctorius* increases the surface area of the root so that there is greater absorption of water and nutrients. Pine seedlings inoculated with mycorrhizal fungi grow more rapidly and are more likely to survive than uninoculated seedlings.

749

Vesicular-arbuscular (treelike) endomycorrhizal □ growth penetrates plant roots intercellularly and intracellularly. It is the most common of the mycorrhizal associations. These mycorrhizal fungi grow on most major agricultural crops, including wheat, corn, beans, tomatoes, potatoes, strawberries, oranges, apples, grapes, cotton, tobacco, tea, coffee, cacao, sugar cane, and rubber trees. The agricultural use of vesicular-arbuscular endomycorrhizae has been limited, because these mycorrhizal fungi cannot be grown in pure culture but must be grown on host plants.

Since the mycorrhizal fungi spread out into the rhizosphere, they increase the surface area for gathering water and dissolving nutrients. The mycorrhizal growth benefits plants in poor soils by solubilizing phosphate and increasing the uptake of nutrients. Thus, plants with a heavy mycorrhizal layer can withstand drought conditions and nutrient-poor soils better than plants that lack a mycorrhizal layer.

Root Nodules

Legumes such as sweetpea, peas, beans, peanuts, clover, and alfalfa have their roots infected by various species of the soil bacterium *Rhizobium*. These bacteria infect root hairs and become intracellular symbionts within tumorous growths known as root nodules. □ *Rhizobium* receives nutrients from the plant, and in turn fixes atmospheric nitrogen, providing the plant with nitrogenous compounds.

763

The bacterium *Frankia alni* also fixes nitrogen for alder trees and a few other nonleguminous plants. Present research is aimed at enabling the free-living, nitrogen-fixing bacterium *Azotobacter vinelandii*, which overproduces ammonia, to bind to roots of cereal crops (corn, wheat, barley) and become part of the rhizoplane microbiota. This association would allow the growth of these cereal crops without the addition of nitrogenous fertilizers. Researchers are also attempting to breed plants that promote a symbiotic relationship with *Azotobacter* in order to eliminate or reduce the need for expensive fertilizers.

SOIL MICROORGANISMS AND PLANT HEALTH

Soils that inhibit disease-causing microorganisms are called **suppressive soils,** while those that do not have this property known as **conducive soils.**

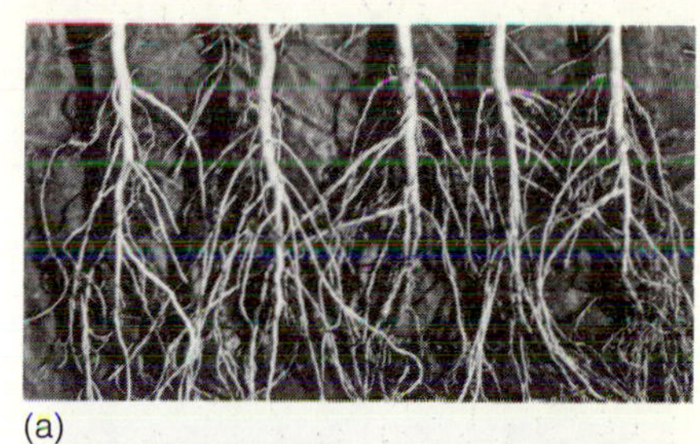
(a)

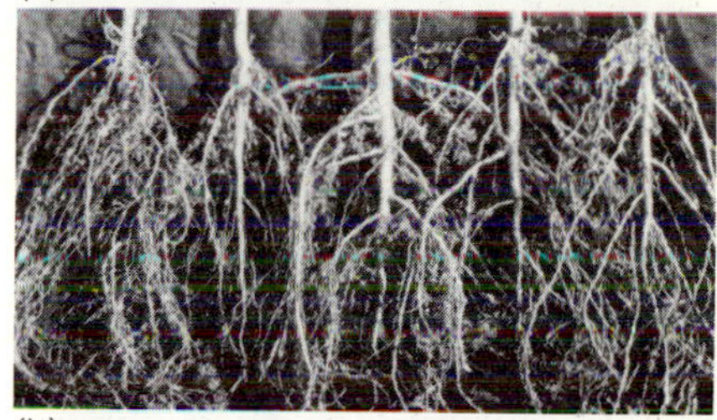
(b)

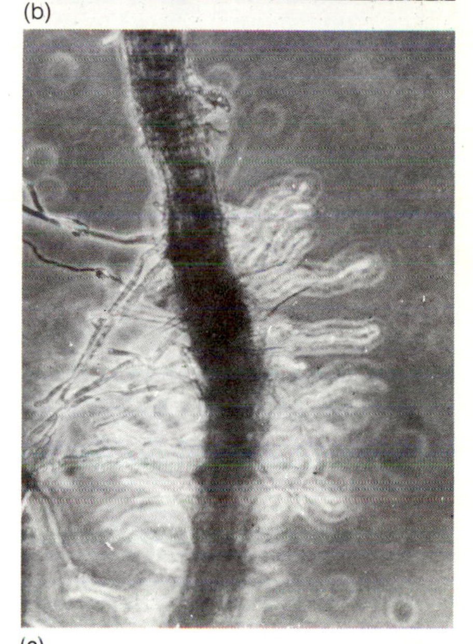
(c)

Figure 34-2 Effect of Mycorrhizae on Root Surface Area. *(a)* Pine seedlings not inoculated with mycorrhizal fungi. The root surface area is not extensive. *(b)* Pine seedlings that have been inoculated with the mycorrhizal fungus *Pisolithus tinctorius*. These pine seedlings have developed an extensive mycorrhizae, which has increased significantly the root surface area. The mycorrhizae increase the surface area of the root for the absorption of water and nutrients. The mycorrhizae are able to grow in many areas of the soil where the roots would have difficulty growing. The seedlings with mycorrhizae grow more rapidly and are more likely to survive in nutrient-poor environments than uninoculated seedlings. *(c)* Root hair of pine seedling exhibiting extensive ectomycorrhizae.

It is possible to test the disease-suppressiveness of soils by seeding them with varying amounts of plant pathogens. For example, figure 34-3 illustrates what happens to muskmelons when soils taken from various regions are inoculated with different concentrations of the fungus *Fusarium oxysporum melonis*, which causes *Fusarium* wilt. At very low concentrations of *Fusarium*, most of the plants are healthy in all the soils. As the population of *Fusarium* increases, however, most plants growing in some soils become diseased. Notice in figure 34-3 that the Chateaurenard soil is a very suppressive soil and that the other soils exhibit varying degrees of suppressiveness. In some of these soils, *Fusarium* wilt is sometimes so severe that entire crops are lost.

Many different microorganisms appear to be able to confer disease suppressiveness to soils. Bacteria such as *Bacillus*, *Agrobacterium*, and *Pseudomonas*, and fungi such as *Trichoderma*, *Pythium*, *Laetisaria*, and *Chaetomium*, have been used to inoculate seeds and roots to protect plants from various plant diseases. In field tests in which potato seeds were inoculated with *Pseudomonas* sp., potato yields were increased by more than 30%. Similarly, sugar beet seeds treated with *Pseudomonas fluorescens* and *P. putida* increased the yield of sugar beets by as much as 3 tons/acre and the carbohydrate content by 300–400 kilograms/acre. It is believed that the bacteria in the rhizosphere and on the rhizoplane stimulate plant health and growth by releasing antibiotics that inhibit pathogenic microorganisms. The antibiotic-producing microorganisms reduce fungi and bacteria on the root surfaces by 25–90%.

One of the inhibitory substances produced by the pseudomonads is a **siderophore,** an Fe^{+3} transport molecule that binds iron in the rhizoplane. The siderophore not only inhibits fungi and bacteria in the rhizosphere but also stimulates plant growth. The siderophores sequester Fe^{+3} in the rhizoplane and consequently deprive certain pathogens of their essential iron. Iron has been shown to be an important factor in making a soil suppressive. For example, if Fe^{+3} is added to suppressive soils, the survival of wheat seedlings challenged by *Gaeumannomyces graminis tritici* is reduced from 83% to 27%, and the survival of flax seedlings subjected to *Fusarium oxysporum lini* is decreased from 82% to 48%. Conducive soils can be made temporarily suppressive to both of these fungi by inoculating seeds with certain species of *Pseudomonas* or by adding the purified siderophore.

Whether there is a future in manipulating root microflora to benefit helpful microorganisms depends upon the development of microorganisms that can survive in the soils of large commercial farms. Much more research

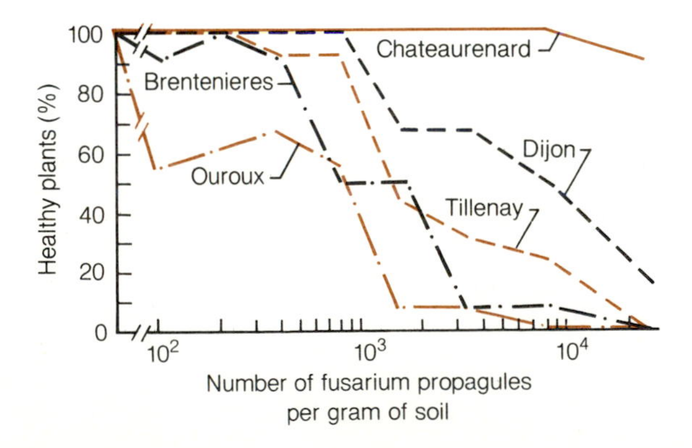

Figure 34-3 Comparison of Soil Suppressiveness. Some soils suppress the ability of the fungus *Fusarium oxysporum melonis* to cause muskmelon wilt. The most suppressive soil is Chateaurenard. At 10^4 infective units, nearly 100% of the muskmelons are healthy. A moderately suppressive soil is Dijon. At 10^4 infective units, about 50% of the muskmelons are healthy. The least suppressive soil is Ouroux. At 10^4 infective units, none of the muskmelons is healthy.

is still needed to discover what factors, in addition to iron-binding siderophores, can produce suppressive soils.

PLANT PATHOLOGY

A large number of viruses (table 34-1), bacteria (table 34-2), fungi (table 34-3) protozoans, and metazoans cause disease in plants and destroy billions of dollars worth of crops throughout the world each year. In the years 1975–1979, for example, insurance compaines paid between $60 million and $150 million each of those year to farmers in the United States. Insects accounted for about 4% of the loss ($2.4–$6.0 million), while disease accounted for about 3% ($1.8–$4.5 million). These numbers may not seem significant when one considers that the U.S. crops each year are worth more than $17 billion, but sometimes plant diseases can destroy crops upon which humans and animals depend for their livelihood, causing economic ruin, famine, pestilence, and death. For example, in 1845 the fungus *Phytophthora infestans*, which causes potato blight, began to destroy potato crops in Ireland (fig. 34-4). The total destruction of this mainstay crop over the next few years led to the complete breakdown of Irish society. It is estimated that more than 2 million Irish, out of a total population of 8 million, died of starvation and disease (typhus and relapsing fever) that became epidemic because of the weakened condition of the Irish and the breakdown of normal life. Of the more than 1 million Irish who tried to emigrate to the United States and Canada, more than 25% died during the trip because of starvation and disease.

Microorganisms and viruses may infect a plant from the roots or from the stem or leaves. Most viral diseases of plants are transmitted by contaminated or infected insect vectors that bite into the leaves or stem and introduce the virus into plant cells. Some viruses, however, infect roots when plant parasitic nematodes burrow into the roots, and other viruses are

TABLE 34-1
REPRESENTATIVE VIRAL DISEASES* OF PLANTS

DISEASE	CAUSE
Cauliflower mosaic disease	Caulimovirus
Tobacco rattle disease; pea early browning disease	Tobavirus
Tobacco mosaic disease	Tobamovirus
Potato virus X disease	Potexvirus
Carnation latent virus disease	Carlavirus
Potato virus disease	Potyvirus
Beet yellows	Beet yellows virus
Citrus tristerza	Citrus tristerza virus
Cucumber mosaic disease	Cucumovirus
Turnip yellow mosaic disease	Tymovirus
Tobacco ringspot disease	Nepovirus
Tomato bushy stunt	Tombushvirus
Alfalfa mosaic disease	Alfalfa mosaic virus
Wound tumors; rice dwarf disease	Wound tumor virus
Tomato spotted wilt	Tomato spotted wilt virus
Lettuce necrosis	Lettuce necrotic yellows virus

*See table 17-2 for a description of the viruses.

TABLE 34-2
REPRESENTATIVE BACTERIAL DISEASES OF PLANTS

DISEASE	CAUSE
Blight of beans	*Xanthomonas phaseoli*
Blight of peas	*Pseudomonas pisi*
Blight of rice	*Xanthomonas oryzae*
Blight of soybeans	*Pseudomonas glycinea*
Blight of walnut	*Xanthomonas juglandis*
Cane gall of raspberries	*Agrobacterium rubi*
Canker of citrus	*Xanthomonas citri*
Citrus stubborn disease	*Spiroplasma* sp.
Corn stunt	*Spiroplasma* sp.
Crown gall of various plants	*Agrobacterium tumefaciens*
Fire blight of pears and apples	*Erwinia amylovora*
Gumming of sugar cane	*Xanthomonas vascularum*
Hairy root of apple	*Agrobacterium rhizogenes*
Halo blight of beans	*Pseudomonas phaseolicola*
Leaf spot of fruits	*Xanthomonas pruni*
Leaf spot of tobacco	*Pseudomonas angulata*
Moko of banana	*Pseudomonas solanacearum*
Olive knot disease	*Pseudomonas savastanoi*
Peach X disease	*Mycoplasma* sp.
Peach yellows	*Mycoplasma* sp.
Pox of sweet potato	*Streptomyces ipomoeae*
Scab of potato	*Streptomyces scabies*
Slippery skin of onion	*Pseudomonas marginalis*
Soft rot of fruit, black leg of potato	*Erwinia carotovora*
Sour skin of onion	*Pseudomonas cepacia*
Wildfire of tobacco	*Pseudomonas tabaci*
Wilt of alfalfa	*Corynebacterium insidiosum*
Wilt of corn	*Erwinia stewartii*
Wilt of tomato	*Corynebacterium michiganese*

FOCUS

BACTERIAL DISEASE ENDANGERS CALIFORNIA CITRUS CROPS

Young citrus trees in southern California and the San Joaquin Valley are attacked by a mycoplasma-like bacterium called *Spiroplasma citri*. *Spiroplasma* causes **stubborn disease** of citrus and in 1979 made worthless approximately 2 million citrus trees. Stubborn disease is characterized by one or more of the following signs: stunted growth, abnormally shaped or small leaves, chlorosis (yellowing or bleaching) of the foliage, small fruits, fruit discoloration, premature mummification and fruit drop, insipid or bitter fruit flavor, and seed abortion. Stubborn disease generally attacks young orange, grapefruit, and tangelo trees. *Spiroplasma* is transmitted among citrus trees, many vegetables, weeds, and ornamental plants by insect leafhoppers. Growers are advised that the only "cure" for stubborn disease is to pull out promptly all trees that develop symptoms of the disease. This procedure is expected to slow down the spread of the disease.

Scientists and growers fear that this bacterial disease might make the growth of citrus crops in some parts of California economically unprofitable. There is a precedent for this idea. Temple oranges, once a very important commercial fruit in the U.S., were virtually eliminated by an infectious viroid. In recent years it has been possible to grow meristem cells from Temple orange trees in tissue culture and obtain plants free of the viroid. Thus, in the next few years this orange may make a comeback in Florida and California.

Being able to identify plant diseases and the organisms that cause them is important to growers so that they can make appropriate decisions to save their crops and avoid financial ruin. In the following pages, various plant diseases are discussed.

Extensive algal bloom on pond.

Thermal pond in Yellowstone National Park. Colored patches adjacent to the pond are blooms of thermophilic cyanobacteria.

Algal bloom on agricultural drainage pond. Half Moon Bay, California.

Cyanobacterial mats in Laguna Figueroa, Baja California, Mexico.

Section through a cyanobacterial mat from Laguna Figueroa, Baja California, Mexico.

Romal stromatolites approximately $1{,}800 \times 10^6$ years old in the East Arm of Great Slave Lake, Canada.

Harvest of the aquacultured kelp *Laminaria groenlandica*.

Bloom of the cyanobacaterium *Anabaena* on Drum's pond, California Polytechnic State University, San Luis Obispo.

Bloom of the cyanobacterium *Azolla* on an agricultural drainage pond.

Close-up of *Azolla* on Cal Poly pond.

Extensive growth of purple sulfur bacteria on an agricultural drainage pond.

The purple color of the water is due to the extensive growth of purple sulfur bacteria.

Dinner roll showing extensive growth of mold *Penicillium*.

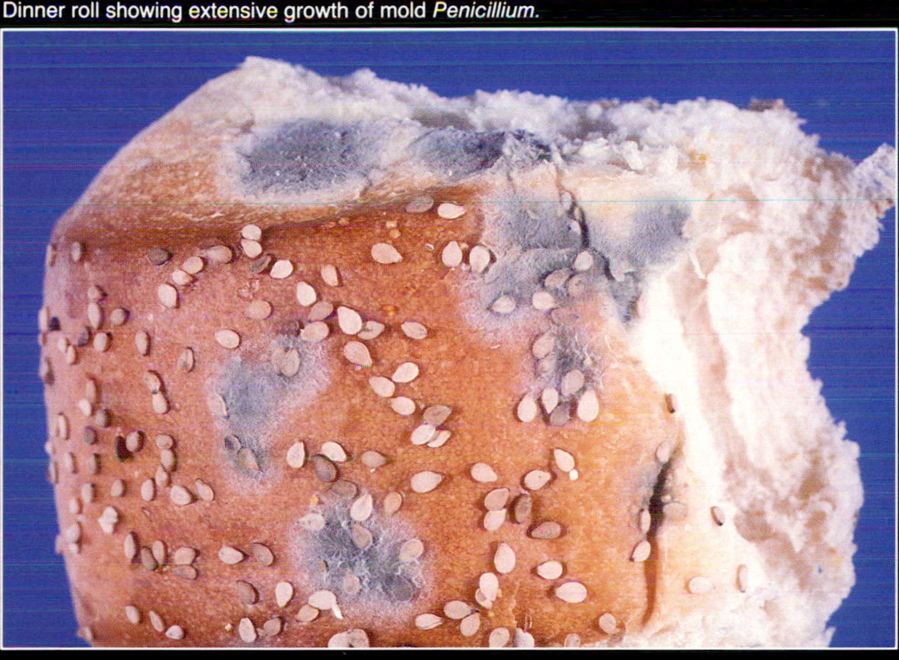

Various mold colonies growing on leftover lamb stew.

Powdery mildew fungus on squash leaf.

Lower surface of Lombardy poplar leaf showing the release of spores of the fungus *Melampsora,* the poplar rust agent.

Copper vats used in the production of beer. Carlsberg Brewery, Copenhagen, Denmark.

Shiitake mushroom culture. Edible mushroom is fruiting on oak log.

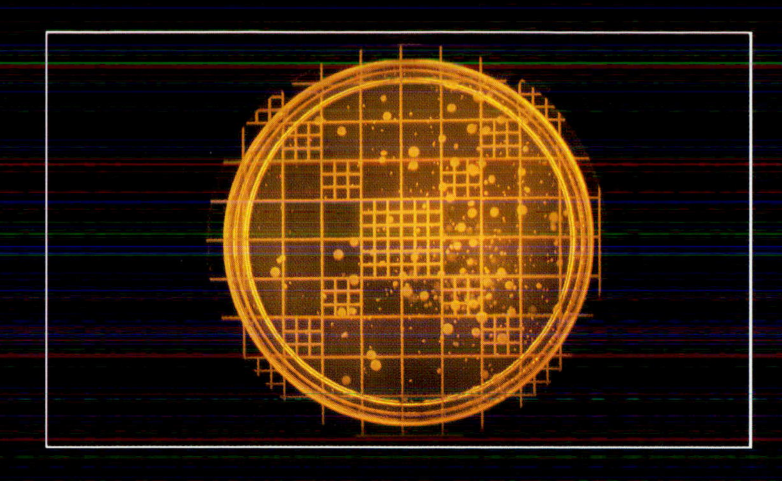

Nutrient agar plate with colonies as seen using a colony counter grid.

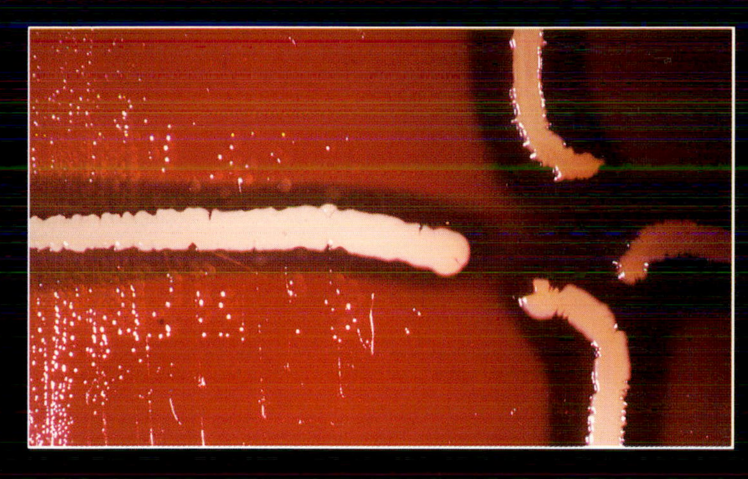

Satellite phenomenon. *Haemophilus influenzae* (small colonies) grows only adjacent to areas where the necessary growth factors, produced by the other bacterium, are available.

Separation of *Escherichia coli* plasmic from chromosomal DNA by $CsCl_2$ density gradient centrifugation. The lower yellow band consists of plasmid DNA. The upper yellow band consists of chromosomal DNA. The rec band is probably the bacterial cell wall.

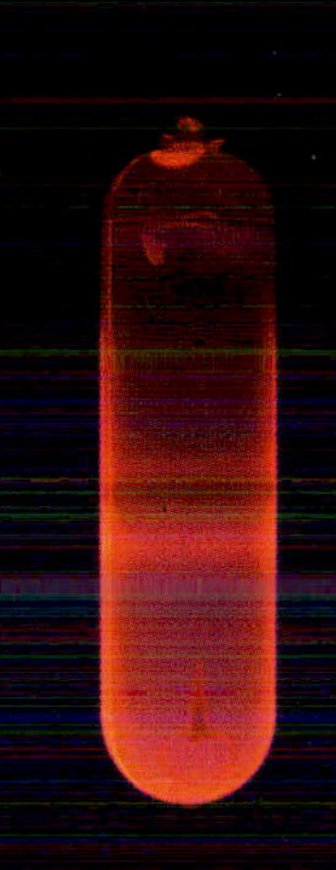

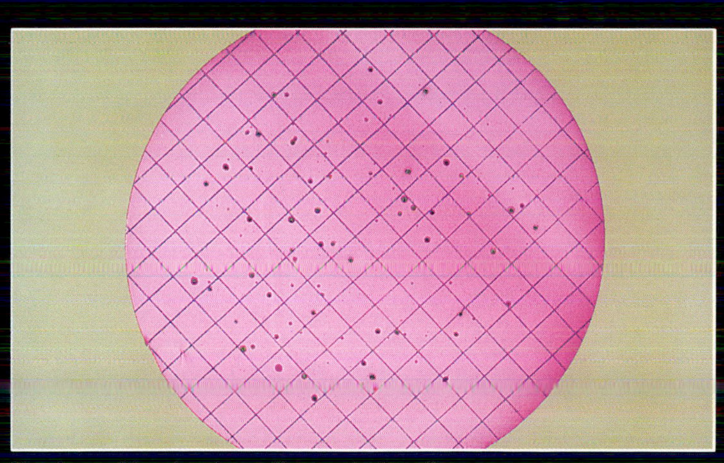

Membrane filter showing coliform colonies after 24 hours of incubation on MF-Endo medium.

Lichens growing on the fruiting bodies of a basidiomycete.

The lichen *Letharia vulpina*.

TABLE 34-3
REPRESENTATIVE FUNGAL DISEASES OF PLANTS

DISEASE	AGENT	FUNGUS GROUP
Root disease of cereals	*Polymyxa*	Slime molds
Powdery scab of potato	*Spongospora*	
Downy mildew of grapes	*Plasmopara*	
Late blight of potato	*Phytophthora*	Oomycetes
Seed decay, root rots	*Pythium*	
Brown spot of corn	*Physodermai*	Chitridiomycetes
Black wart of potato	*Synchytrium*	
Soft rot of fruits	*Rhizopus*	Zygomycetes
Dutch elm disease	*Ceratocystis*	Ascomycetes
Ergot of rye	*Claviceps*	
Bean pod blight	*Diaporthe*	
Black knot of cherries	*Dibotryon*	
Chestnut blight	*Endothia*	
Powdery mildew of wheat	*Erysiphe*	
Stone fruit brown rot	*Monilinia*	
Take all of wheat	*Ophiobolus*	
Powdery mildew of apple	*Podosphaera*	
Soft rot of vegetables	*Sclerotinia*	
Peach leaf curl	*Taphrina*	
Apple scab	*Venturia*	
Root rots of trees	*Armillaria*	Basidiomycetes
Heart rot of trees	*Fomes*	
Wheat rust	*Puccinia*	
Loose smut of sorghum	*Sphacelotheca*	
Stinking smut of wheat	*Tilletia*	
Smut of onion	*Urocystis*	
Rust of beans	*Uromyces*	
Smut of corn, wheat, and barley	*Ustilago*	
Leaf spots and blight	*Alternaria*	Deuteromycetes
Rots of seeds	*Aspergillus*	
Blights	*Botrytis*	
Leaf mold of tomato	*Cladosporium*	
Leaf spots of various plants	*Cylindrosporium*	
Root rot of many plants	*Fusarium*	
Blight of cereals	*Helminthosporium*	
Blue mold rot of fruits	*Penicillium*	
Root rot of various plants	*Rhizoctonia*	
Black root rot of tobacco	*Thielaviopsis*	
Wilt of various plants	*Verticillium*	

splashed into plant wounds by rain from the soil or from other infected plants. Certain viruses are associated with pollen or seeds and so do not rely on vectors for their transmission; this form of transmission from parent to offspring is called **vertical transmission.** Bacterial diseases can occur when hydrolytic enzymes or mechanical breaks provide openings in the roots, stems, or leaves. Many fungal plant pathogens are spread through the air by spores □ and penetrate the plant leaves through the leaf openings called **stomata** (singular **stoma**). A few fungi penetrate plant leaves and stems through the action of hydrolytic enzymes. For example, powdery mildews caused by *Erysiphe* and soft rot due to *Botrytis* are associated with enzymatic degradation of the leaf cuticle, while *Phytophthora* potato blight and *Fusar-*

332

ium root rot occur when the fungi bore their way into the root system and invade plant tissue. Nematodes are responsible for a large number of plant diseases when they feed on or penetrate the root system.

Plant pathogens produce a number of signs and symptoms that are helpful in diagnosing the cause of the disease and ultimately allowing the plant pathologists to take remedial actions. In the following sections we will explain briefly how some of these signs and symptoms develop.

Soft Rot

Soft Rot is generally due to bacteria and fungi that are able to hydrolyze plant cell walls. The hydrolysis of the wall polysaccharides results in a hypertonic environment □, so that water flows out of the cells and causes them to shrink. The loss of water causes a loss of turgidity and a slimy area where the plant tissue is being degraded. 118

Wilts

Wilts are generally caused by bacteria and fungi that block the flow of water within the plant. When parts of a plant do not get enough water, the cells shrink and the tissue loses its rigidity. Wilts occur when organisms grow in the vessels (xylem) that conduct water and either plug them up or damage them. Some bacteria produce a heavy polysaccharide slime that plugs up the xylem vessels.

Spots

Spots (fig. 34-4) can be due to viruses, bacteria, and fungi that cause limited destruction of cells in the leaves, stems, and seed pods. Badly spotted leaves can turn yellow and drop from the plant. Viruses cause spots by invading cells and lysing them. Bacteria may cause spots by producing toxins that kill the cells at the site of infection. Fungi may grow at the site of infection, destroying cells and creating a visible mass of growth in the tissue. Infections occur through stomata, wounds, bites, or enzymatic action.

Blight

Blights (fig. 34-4) are caused by bacteria and fungi that destroy large portions of a plant so that much or all of the plant dies. The pathogens infect the plant through wounds, stomata, or destruction of plant tissue. Some bacterial infections are due to organisms associated with the seeds. The tissue destruction that follows the initial infection is due to hydrolytic exoenzymes, toxins that inhibit metabolism, or slime that blocks the movement of water and nutrients within the plant. Blights generally occur within a few days after infection. A seemingly healthy plant suddenly turns brown and dies.

Galls

Galls generally are caused by bacteria and nematodes that induce a rapid and uncontrolled proliferation of plant tissue. Root-knot nematodes cause galling of roots (fig. 34-4) by growing within them and stimulating rapid

Figure 34-4 Representative Plant Diseases Caused by Microorganisms. *(a)* Western gall rust on a pine tree due to an infection by the fungus *Peridermium harknessii* (basidiomycete). *(b)* Gummosa of apricot due to the bacterium *Pseudomonas syringiae.* *(c)* Apple scab on apple leaves caused by the fungus (ascomycete) *Venturia inaequalis.* *(d)* Fire blight of *Pyracantha* caused by the bacterium *Erwinia amylovora.* *(e)* Root knot of cellery caused by the nematode *Meloidogyne* sp. *(f)* Tobacco mosaic disease caused by the tobacco mosaic virus. *(g)* Peach leaf curl caused by the fungus (ascomycete) *Taphrina deformans.* *(h)* Canker on a branch of an English apple tree. The canker results from the fire blight caused by *Erwinia amylovora.* *(i)* Powdery mildew on the leaves of a pea plant caused by the fungus (ascomycete) *Erysiphe polygoni.* *(j)* Corn smut showing swollen kernels and smut galls caused by the fungus (basidiomycete) *Ustilago maydis.* *(k)* Brown rot of peach due to the fungus (ascomycete) *Monilinia fructicola.* *(l)* Late blight of potato due to the fungus (oomycete) *Phythophthora infestans* infecting the leaves, stems, roots, and tubers of a sensitive variety (right). A resistant variety (left) survives the infection.

multiplication of the plant tissue. *Agrobacterium tumefaciens* is the bacterium that causes crown gall in plants such as tomatoes, tobacco, sugar beets, and certain fruit trees. This bacterium causes the loss of millions of dollars worth of crops in the United States each year because of a decrease in productivity. Galls are plant tumors that develop on the roots or on the stem where *Agrobacterium* has penetrated the outer layer of cells. The growth of galls frequently results in distorted roots and stems, as well as stunted or sickly plants. The loss of productivity occurs because the galls take up many of the nutrients and thus interfere with the uptake of nutrients by the rest of the plant.

Ti-plasmids □ (tumor-inducing plasmids) from *Agrobacterium*, which penetrate the plant cells and integrate into the plant DNA, are responsible for blocking cell differentiation and stimulating cell multiplication. The Ti-plasmid contains genes for the synthesis of indoleacetic acid and cytokinins. These hormones are believed to be responsible for the multiplication of the tumor cells and possibly for the transformation of the plant cells. One of the genes on the Ti-plasmid codes for the enzymes **octopine synthase** and **nopaline synthase,** which catalyze the synthesis of the unusual amino acids octopine and nopaline, respectively. The tumor cells that contain Ti-plasmids secrete large amounts of one of these two amino acids, which are then used by *Agrobacterium* as a source of carbon and energy. Thus, *A. tumefaciens* causes plant cells to produce a special nutrient that most other bacteria are unable to utilize. The Ti-plasmid also codes for permeases and catabolic enzymes that allow *Agrobacterium* to use the octopine and nopaline as a source of carbon and energy.

261

Chlorosis

Chlorosis, or loss of green color, may be due to viruses, bacteria, fungi, or nematodes and is caused by the loss or destruction of chloroplasts. Viruses often cause extensive mottling or yellowing of the leaves and stems. For example, the tobacco mosaic virus causes necrosis and yellowing of the tobacco leaf (fig. 34-4). Flower coloring is sometimes altered by virus infections, as in the case of the yellow streaks in tulip flowers called tulip break.

Curly Leaf and Misshapen Fruits

Curly leaves and distorted fruits (fig. 34-4) are caused by viruses, bacteria, and fungi. The organisms infecting the leaves or fruits may kill cells and induce shrinkage or expansion of the cells so that the tissue becomes deformed.

Cankers

Cankers (fig. 34-4) are necrotic or dead areas on the stem caused by exoenzymes and toxins produced by pathogenic bacteria and fungi. The bacterium *Erwinia amylovora*, which causes fire blight of pears and apples, produces cankers on branches and stems. The associated gummy exudate blocks water flow in the plants and causes a wilt.

Mildew

Mildews (fig. 34-4) are caused by fungi that grow on and into the surfaces of leaves, stems, and flowers. The extensive growth of the fungus leads to the loss of the fruits or vegetables, wilting, and eventually death of the plant.

Ground Rots

Ground rots are caused by bacteria and fungi that grow on and into the surfaces of seeds, roots, stems, and vegetables. The extensive growth of the organisms that cause ground rots causes the death of the tissue. An important cause of ground rots is the fungus *Rhizoctonia solani*, which is responsible for a number of plant diseases: seed rot, damping off, wire stem, stem canker, crater rot, cabbage bottom rot, potato stem rot, tuber black scurf, and soil rot of tomato.

Smuts

Smuts are due to fungi that grow into various plant tissues and stimulate the tissues to grow in an uncontrolled manner. Malformed flowers, seeds, and stems generally result (fig. 34-4).

Rusts

Rusts are due to fungi that grow into leaf or stem tissues and cause the destruction of the tissue. One of the best known rusts is wheat rust, caused by the fungus *Puccinia gramanis*. This fungus is of interest because its life cycle involves the infection of two different hosts: wheat and barberry.

Brown Rot of Stone Fruits

342
Brown rot of stone fruits is induced by fungi belonging to the genus *Monilinia* (fig. 34-4). The ascospores □ of the fungus germinate into a hypha, which grows on the surfaces and into the tissues of developing fruit. The hyphal growth causes mummification (drying up) of the fruit.

MICROORGANISMS AND ANIMAL NUTRITION

426
Microorganisms have established relationships not only with plants but also with animals. For example, in mutualistic □ associations, both organisms evolved into new life forms that exist in harmony with one another. Examples of animals that harbor essential symbionts are the termites and the ruminants. Termites are unable to exist without their gut bacteria, and the gut bacteria are unable to live without the termites. Similarly, ruminants cannot exist without the microorganisms that are found in their rumen □, and many of these microorganisms cannot live outside the rumen.
751

In synergistic relationships both of the partners benefit but generally are able to exist on their own. For instance, intestinal bacteria in animals and humans provide nutrients that promote good health, but both the bacteria and the animals can exist without each other. In this section we will consider a few examples of how microorganisms foster good health or induce disease in animals.

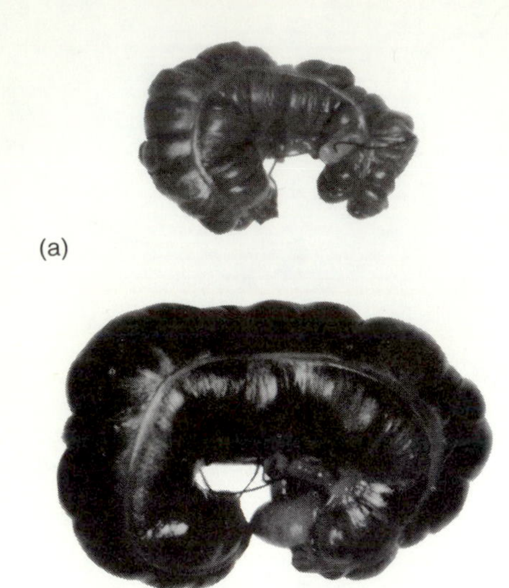

(b)

Figure 34-5 Enlarged Cecum of Germ Free Animal. *(a)* Cecum from a normal rodent. *(b)* Cecum from a germ-free rodent.

Intestinal Bacteria

The microbial flora in the gastrointestinal tract of many animals provides the animal with acids, polysaccharides, proteins, and vitamins that it requires. Such microorganisms also aid in digestion and protect the intestines from many potentially pathogenic microorganisms. Animals raised in the absence of microorganisms (germ-free or **gnotobiotic** animals) develop abnormally. For instance, the reticuloendothelial □ organs that provide cells for the immune system are not developed, probably because they have not been stimulated by the great number of microbial antigens. In gnotobiotic rabbits, mice, and guinea pigs, the cecum is greatly enlarged (fig. 34-5). This enlargement may be a response to the lack of bacterial proliferation and lack of nutrients produced by the bacteria; by increasing the organ's size, the absorption of nutrients may be improved. In addition, the entire intestine has a thin wall and does not respond normally to mechanical stimuli. Vitamin K must be supplied to gnotobiotic animals, because microorganisms such as *Escherichia coli*, which synthesize much of the vitamin K that animals require, are not present.

Gnotobiotic animals are abnormally prone to diseases caused by common pathogens, and they are also extremely sensitive to usually nonpathogenic organisms. For example, gnotobiotic animals become ill when exposed to common organisms such as *Bacillus subtilis* and *Micrococcus luteus*, which are harmless to normal animals. Germ-free animals are free of tooth decay, however, when fed large amounts of sucrose.

Ruminants

The ruminant intestine includes a rumen in which large numbers of microorganisms □ break down cellulose and hemicellulose. These organisms convert plant material into acids that can be absorbed by the animal:

752

$$C_6H_{12}O_6 \longrightarrow \text{acetate} + \text{propionate} + \text{butyrate} + CO_2 + H_2 + CH_4$$

The microorganisms in the rumen also produce vitamins, proteins, and sugars that are utilized by the ruminant when the organisms are digested in the stomach. The partially-digested organisms are moved to the small intestine and large intestine, where they are further digested and sugars, amino acids, short peptides, fats, vitamins, and other nutrients are absorbed into the blood.

Bacterial genera found in the rumen include *Bacteroides, Butyrivibrio, Ruminococcus, Clostridium, Selenomonas, Succinimonas, Streptococcus, Peptidococcus, Lachnospira, Methanobacterium, Succinivibrio.* The Enterobacteriaceae (*Escherichia, Enterobacter, Klebsiella*, etc.) are almost nonexistent in the rumen (<0.00001%). Most of the protozoans in the rumen are ciliates that can grow anaerobically by fermenting digested cellulose or starch or by preying on bacteria. Even though the protozoans are less numerous than the bacteria, they represent 40% of the cell mass and protein. Anaerobic fungi are prevalent in the rumen of animals fed high-roughage diets and are believed to help in the digestion of the lignocellulose.

Microbial metabolism in the rumen can sometimes be modified through the use of antibiotics in order to benefit breeders. For example, antibiotics such as monensin are used in feeds for beef cattle to increase the production of propionate and decrease the production of CH_4 in the rumen. The production of methane is wasteful, because this material escapes into the atmosphere. Antibiotic-fed animals eat less food than control animals but gain weight at the same rate. It is believed that monensin inhibits those bacteria that produce H_2 and CO_2. Without H_2 and CO_2, the methanogenic bacteria are unable to produce methane. Since more propionate is made and is not wasted as carbon dioxide and hydrogen, more nutrients are available to the animal. Much of the propionate is converted to glucose in the liver, and this effect in turn causes an increase in animal protein but a decrease in milk fat. The use of either the antibiotic monensin or starchy foods decreases milk fat. This effect is of no benefit to dairy farmers, however, because the highest prices are paid for milk with high fat content.

Carnivores and Omnivores

744
In carnivores □ and omnivores, the glycogen, fat, and protein found in ingested meat is degraded by enzymes in the oral cavity, stomach, and small intestine. Plant proteins and carbohydrates are also degraded in these organs. Cellulose, hemicelluloses, and vegetable and fruit pectins are not hydrolyzed by host enzymes; the undegraded macromolecules are moved into the large intestine, where they are acted upon by microorganisms. Urea is released from the blood into the intestines and serves as the major source of nitrogen for the microorganisms. The fermentation wastes produced in the large intestines are similar to those produced in the rumen: acetate, propionate, butyrate, H_2, CO_2, and CH_4. There are approximately 10^{11} bacteria/g of feces, and since about 120 g of feces are released per day by a 70-kg animal, more than 10^{13} bacteria are also excreted. Most of the acid the bacteria produce is absorbed into the bloodstream. The acid that remains in the feces inhibits several bacteria that cause intestinal infections. Much of the H_2 and CH_4 produced in the large intestine is absorbed by the blood, carried to the lungs, and exhaled. Some of the gas is also released as flatus. Various species of *Bacteroides* (45%), *Eubacterium* (25%), *Peptococcus* (15%), and *Fusobacterium* (10%) together account for 95% of the bacteria in the human large intestine. Less than 0.5% of the bacteria belong to the family Enterobacteriaceae (*Escherichia*, *Klebsiella*, *Enterobacter*, etc.). *Streptococcus* species also account for less than 0.5% of the bacteria. Notice that the vast majority of the bacteria ($>95\%$) are obligate anaerobes. Unlike the rumen, the large intestines of omnivores and carnivores contain few methanogenic bacteria □.
329

VETERINARY MICROBIOLOGY

Animals suffer from many of the same diseases that afflict humans. A number of animal diseases, however, are specific for certain groups of animals that share similar physiologies. Animal morbidity and mortality are very important to humans, whose livelihood often is completely dependent upon animals that they use for food, labor, and transportation. In addition, several hundred pathogenic agents of animals can cause diseases in humans. Sick animals or carrier animals serve as reservoirs □ for microorganisms that can, from time to time, cause outbreaks in human populations. Animal diseases that can be transmitted to humans and other animals are known as **zoonoses** (singular, **zoonosis**).
534

FOCUS

EASTERN U.S. MOSQUITOES CARRY A KILLER VIRUS

Outbreaks of Eastern equine encephalitis, St. Louis encephalitis, La Crosse encephalitis, and California encephalitis have been reported in the eastern and southeastern United States. These types of encephalitis are due to a number of viruses that cause an inflammation of the brain and spinal cord. The viruses are carried by birds and transmitted to horses and humans by mosquitoes.

In 1982, large swarms of mosquitoes were spawned by the heavy summer rains in many parts of the United States. These mosquitoes were responsible for spreading the diseases. The outbreaks of encephalitis led to the death of three people in New York, one in Georgia, and one in Florida. Eastern equine encephalitis resulted in the deaths of more than 70 horses in Georgia and more than 150 in Florida. Many of the infected horses staggered, went blind, wandered around, and died, since there is no treatment for the disease.

Caged quail and pheasants on six Georgia farms and young wild birds in Kentucky, Mississippi, Ohio, and Tennessee died by the thousands. St. Louis encephalitis is reported to have been responsible for the deaths of the wild birds, while Eastern equine encephalitis is believed to have been responsible for the deaths of the quail and pheasants.

How important is animal morbidity and mortality to humans? We have already mentioned (Chapter 28) that in the years 1347–1350 more than 10 million persons died in Europe from the bubonic plague carried by rats and transmitted by fleas. In the 1600s, Europe was hit by cattle plague **(rinderpest),** caused by a virus that is estimated to have killed about 200 million cattle. This loss of livestock devastated European life, since a healthy animal population was needed to supply food and power to plow, harvest, and transport food. In the early 1900s, bovine tuberculosis was a very serious disease in both animals and humans in the United States. About 20% of tuberculosis in humans was due to bovine tuberculosis; approximately 190,000 persons died from tuberculosis each year in the 1900s, so bovine tuberculosis was responsible for as many as 38,000 human deaths each year in the United States. In the years since 1900, the spread of bovine tuberculosis to humans has been reduced drastically in countries that pasteurize their milk.

Pasteurization, good hygiene, and eradication programs eventually reduced the incidence of tuberculin-positive □ cattle in the U.S. to less than 0.2% in the 1960s. Childhood tuberculosis of animal origin has been virtually eliminated in countries that pasteurize their milk. In countries where animal tuberculosis is still prevalent, however, humans contract the disease, there is a shortage of meat and milk products, and there is an economic loss to the individuals and communities involved.

529

In the United States, many millions of dollars are lost each year because of animals that die of infectious diseases and because of medical treatments and culling. Inspection of butchered animals reveals many thousands of diseased carcasses each year that must be condemned (table 34-4), thus resulting in the yearly loss of millions of dollars in the United States.

A number of animal diseases are host specific and so are not a direct threat to the health of humans. Some of these strictly animal diseases, however, have at times killed so many animals that human societies have been drastically affected because a basic food source was eliminated and economies were disrupted. Some of the more important diseases of wild and domestic animals, and their impact on human populations, are considered in the following sections.

Mastitis

Mastitis is an inflammation of the mammary gland of cows, goats, camels, and other mammals, caused by bacteria or fungi infecting the tissue. Mastitis 564 frequently results in the release of pus □, microorganisms, and blood into

TABLE 34-4
INFECTIOUS AGENTS FOUND IN CARCASSES

	NUMBER OF CARCASSES CONDEMNED			
CAUSE OF CONDEMNATION	CATTLE	CALVES	SHEEP & LAMBS	SWINE
Infectious diseases:				
Actinomycosis, Actinobacillosis	833	2		10
Anaplasmosis	308			
Caseous lymphadenitis			11,637	
Coccidioidal granuloma	23		1	1
Swine erysipelas				3,165
Tuberculosis, nonreactor	86	1		4,996
Tuberculosis Reactor with lesions	60			
Miscellaneous	184	79	8	54
Inflammatory diseases:				
Enteritis, gastritis, peritonitis	5,087	1,617	289	9,129
Eosinophilic myositis	3,878	9	28	28
Mastitis, mammitis	1,012	2	4	43
Metritis	1,932	5	161	904
Nephritis, pyelitis	3,825	257	639	2,661
Pericarditis	4,585	76	223	1,350
Pneumonia	13,398	4,368	6,912	12,553
Miscellaneous	462	249	404	424
Parasitic conditions:				
Cysticercosis	135		189	4
Sarcosporidiosis			3,695	
Stephanuriasis				2,322
Miscellaneous	88	3	133	74
Septic conditions:				
Abscess, pyemia	11,582	534	2,116	28,136
Septicemia	7,749	2,318	828	3,878
Toxemia	3,535	384	627	2,738
Miscellaneous	169	154	32	53
Other:				
Arthritis, polyarthritis	1,468	2,786	2,102	19,332
Asphyxia	56	272	41	2,727
Bone conditions	21	1	2	38
Contamination	1,135	36	316	5,111
Icterus	477	2,133	1,688	9,884
Uremia	1,364	45	1,265	874
Miscellaneous general	4,445	1,564	374	5,308

SOURCE: Food Safety and Quality Service, U.S. Dept. of Agriculture, 1977.

the milk. This disease is estimated to cause a 1-billion-dollar loss each year in the United States due to decreased milk production and destruction of contaminated milk (65%), treatment (25%), culling and death (10%). The bacteria commonly responsible for bovine mastitis are *Staphylococcus aureus* and *Streptococcus agalactiae*. These organisms enter the mammary glands through the teat openings and reproduce in the milk and on the surface of the milk channels (lacteals). Toxins produced by the bacteria irritate the tissue, and the irritation leads to inflammation, inhibition of milk production, and the proliferation of connective tissue. The bacteria generally are spread from cow to cow during milking.

Staphylococcus mastitis is the most severe form of mastitis in dairies. In most herds, as many as 40% of the cows may have subclinical infections. Chronic infections often are difficult to treat because tissue sometimes develops around the bacteria and prevents antibiotics from reaching the infected site. Cloxacillin, a penicillinase-resistant drug, is used to treat staphylococcal and streptococcal mastitis. Coliform mastitis due to *Escherichia coli* and *Enterobacter aerogenes* is treated with dihydrostreptomycin sulfate or ampicillin, while *Pseudomonas aeruginosa* mastitis is treated with carbenicillin.

Mastitis can be controlled by good hygienic practices. For example, washing teats with clean running water and antiseptic soaps; disinfecting milking equipment between cows; disinfecting hands between cows; detecting and treating infected animals; milking infected animals last; and culling cows with five or more relapses of mastitis are practices that help to reduce and control mastitis in a herd.

Brucellosis

Brucellosis, or undulant fever, is still a problem in the world. Most cattle herds and swine have infected individuals, and it is estimated that about 75% (300 million) of the 400 million goats in the world are infected by *Brucella*. In the U.S., this disease is a major problem in swine because about 5% of the approximately 70 million swine are infected and pose a threat to ranchers, veterinarians, butchers, and consumers. In the last few years there have been about 200–300 human cases and one to three deaths per year in the U.S.

Up to 75% of the cattle herds from some southern states, in particular Texas, have *Brucella*-infected animals that are a threat to herds in other states that have eliminated the disease or reduced it to a very low level by slaughter and calf vaccination. Importation of diseased dairy cows can be a costly venture, as indicated by what happened in California in the 1970s. In 1969, California dairy herds were certified free of brucellosis. In the early 1970s, however, between 25,000 and 55,000 cattle were imported each year from southern states to meet the increased demand for milk. Beginning in 1973, 20–25 California dairy herds each year were discovered to contain infected animals. The disease can be controlled by vaccination of calves and by slaughtering and disposing of infected animals.

Foot and Mouth Disease

Foot and mouth disease is a viral disease that causes severe debilitation, weight loss, cessation of lactation, and abortion in cattle, sheep, water buffaloes, and other ruminants. Pigs can also catch the disease, but horses, dogs, cats are not susceptible. Humans can contract foot and mouth disease (especially laboratory workers handling the virus) and can serve as important carriers of the virus. Death in cattle occurs when lesions of the alimentary tract become extensive. The most frequent signs of the disease are lesions in and around the mouth which prevent eating and cause the accumulation of saliva that drips from the animal's mouth in long, tenuous strands. The virus also infects the feet, producing lesions that make it difficult for the animal to walk (fig. 34-6).

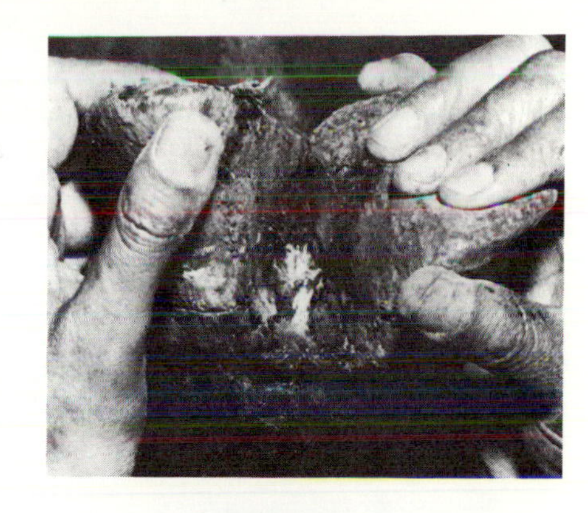

Figure 34-6 Hoof and Mouth Disease in a Cow

North America and Australia are the only areas presently free of the disease. In the past, slaughter and burial of all infected and exposed animals was the only effective way of controlling the disease. Recently, however, a vaccine against one type of foot and mouth virus has been made, using the virus protein produced in genetically engineered 261 □ bacteria. This vaccine promises to be the first of many vaccines that will not revert to the virulent form and that do not have to be refrigerated.

Leptospirosis

Leptospirosis is a systemic disease caused by spirochetes in the genus *Leptospira*. These bacteria cause fever, hemoglobinuria (hemoglobin in the urine), jaundice, abortion, and sometimes death in infected animals. Cattle, sheep, swine, horses, dogs, and humans are all susceptible to infection. Infection generally occurs when animals consume urine-contaminated feed or water, but the bacteria may enter the human body through skin abrasions, mucous membranes (genitals and mouth), and the conjunctiva (a membrane that covers the eye). The bacteria produce hemolysins, which rupture red blood cells and may account for the hemoglobin in the urine. Infection of the fetus causes abortion, while localization in the kidneys results in the bacteria being shed in the urine in very large numbers, sometimes for months to years.

Leptospira can often be observed in urine by dark field microscopy. Diagnosis of leptospirosis is usually made by detecting antibodies 463 □ against the bacterium and a rise in the antibody titer in consecutive serum samples. Vaccination of dairy herds every six months prevents the disease, while injections of streptomycin are used to treat animal cases.

Morbidity (sickness) in cattle may exceed 75% in old stock but often approaches 100% in calves. Death occurs in 5–15% of the animals. Swine are a problem reservoir for *Leptospira* because infected animals, seemingly healthy, often excrete large numbers of the bacteria in their urine (fig. 34-7). These carrier pigs are a threat to ranchers, to other pigs, and to cattle herds. Transmission of the bacteria from pig to pig occurs in contaminated urine, through sexual contact, and by artificial insemination with contaminated semen. Annual vaccinations of breeding animals with *L. pomona* prevents abortion, but some vaccinated pigs may still become carriers when exposed to the bacteria in nature.

Newcastle Disease

Newcastle disease is a viral disease of fowl which is characterized by coughing, sneezing, lesions on various mucous membranes, drooping wings, leg dragging, circling, walking backward, and complete paralysis. Mortality in the flock depends upon the virulence of the strain of Newcastle disease virus and the health of the flock, but can approach 100%. The virus is spread

Figure 34-7 **Mode of transmission and Life History of the Spirochete *Leptospira***

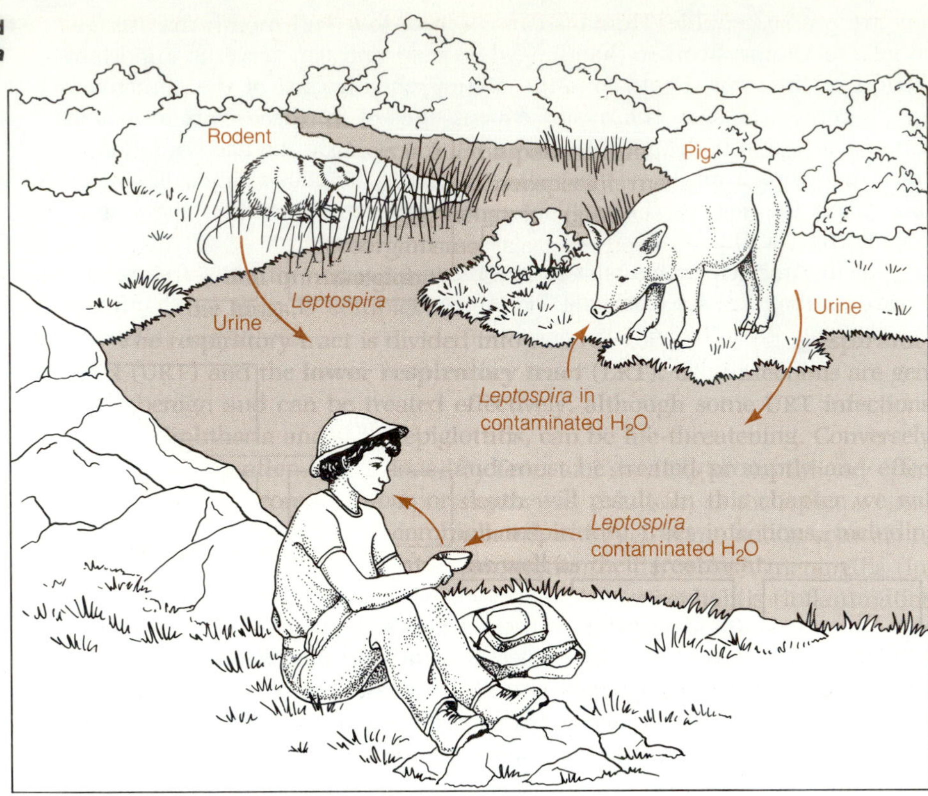

among animals in respiratory secretions and excretions. The disease is diagnosed from the signs of the rapidly spreading respiratory and nervous system disease. Isolation of a hemagglutinating paramyxovirus and a rise in hemagglutination-inhibiting □ antibodies in the infected fowl confirm the diagnosis. A number of vaccines, all quite effective in controlling this disease, are used to protect poultry.

519

Fowl Plague or Avian Influenza

Fowl plague is caused by orthomyxoviruses in the influenza A group. The symptoms of the disease include labored breathing, bloodstained oral and nasal discharges, and lesions on the mucous membrane. The disease has a high mortality rate, which in certain epidemics approaches 100%. The virus spreads in secretions from bird to bird and possibly between mammals and birds. The disease is controlled by quarantine □ of unaffected flocks and destruction of all exposed birds. Vaccines have not always been successful, since a number of influenza A viruses can cause the disease.

548

Psittacosis-Ornithosis

Psittacosis and ornithosis are similar diseases in psittacine birds (parrots) and ornithocine birds (other birds), both caused by the bacterium *Chlamydia psittaci*, which can also infect humans. The signs and symptoms of the disease in birds include weakness, loss of appetite, weight loss, nasal discharge, diarrhea, lesions, and inflamed respiratory passages. The bacteria are transmitted from bird to bird and from birds to mammals (including hu-

FOCUS

A HERPES VIRUS STRIKES LIPPIZANER HORSES

In 1983 Austrian veterinarians did not respond rapidly enough to isolated outbreaks of a deadly herpes virus that attacks horses, and the result was the death of thirty-four Lippizaner mares and foals. Twenty-three other horses were infected and in danger of dying or losing their foals. The Lippizaner stallions are famous throughout the world because of their beauty and ability to perform precision ballet routines and aerial feats. They are one of Vienna's prime tourist attractions and have toured many countries, including the United States.

The herpes virus that infected the Lippizaner mares and foals causes the disease equine rhinopneumonitis, which is characterized by coughing and peneumonia. The foals killed accounted for about half of the year's births. Once horses are infected, it is very difficult to obtain a cure, but a number of infected mares received interferon treatments daily and recovered. One of the infected mares gave birth to a foal that was healthy and disease-free.

The loss of the valuable and beautiful Lippizaner mares and foals illustrates once more the importance of veterinarians' being able to identify outbreaks of a disease rapidly and to respond rapidly, either with a vaccination of the herd or with isolation and treatment of the diseased animals.

mans) in aerosols and respiratory discharges. Domestic poultry (chickens, turkeys, pigeons, ducks) are often infected by wild birds. *Chlamydia psittaci* infects most of the internal organs and can be found in stained smears sampled from liver, spleen, kidney, and lungs. There are no effective vaccines against this disease, but it can be treated successfully with tetracycline.

Fowl Cholera

Fowl cholera, caused by the bacterium *Pasteurella multocida*, may also infect mammals, including humans. The signs and symptoms of the disease vary greatly but may include difficulty in breathing, diarrhea, lameness and swollen joints, lesions on the mucous membranes, and sudden death. The bacterium comes from contaminated soil, feed, water, and decaying bird carcasses. A polyvalent vaccine is used to prevent the disease in chickens, turkeys, and rabbits. Sulfa drugs and tetracyclines are used in water and feed to reduce mortality.

Feline Distemper

Feline distemper, which affects cats and also raccoons, is caused by a DNA virus and is characterized by fever, weakness, vomiting, diarrhea, emaciation, dehydration, and destruction of white blood cells. The disease runs for 5–7 days and death occurs in 75% of the cases. Lymph nodes undergo swelling and necrosis, while the bone marrow becomes semifluid and appears fatty. The intestines, liver, and kidneys show degeneration. Annual vaccination of cats protects them from this deadly virus. The disease can be treated by providing nutrients, electrolytes, and water intravenously and by whole blood transfusions.

Feline Conjunctivitis and Rhinotracheitis and Feline Pneumonitis

Feline conjunctivitis and rhinotracheitis are most frequently caused by the herpes-like virus called feline rhinotracheitis virus. It is a common cause of upper respiratory infections in cats. Picornaviruses and reoviruses also have been isolated from cats with rhinitis and conjunctivitis. The signs and symptoms of the disease caused by these pathogens are very similar and often include fever, sneezing, conjunctivitis, rhinitis, nasal and ocular discharge, and weight loss. Lesions occur in the respiratory tract and conjunctiva and on the surface of the tongue. There are no vaccines that can be used against the virus infections.

The signs and symptoms of feline pneumonitis can be mistaken for feline rhinotracheitis. Feline pneumonitis is caused by the bacterium *Chlamydia psittaci*, however, and can be treated with antibiotics. There is also a live attenuated vaccine for feline pneumonitis.

Canine Rickettsial Disease

Canine rickettsial disease is caused by the bacterium *Ehrlichia canis*, which is transmitted by ticks. The disease is characterized by a fever, discharges from the nose and eyes, vomiting, emaciation, and spleen enlargement. The death rate is sometimes high when there is concurrent babesiasis or trypanosomiasis. Diagnosis is by observing rickettsial bodies within monocytes and neutrophils in stained blood smears. The spread of the disease can be stopped by destroying ticks. Treatment is with tetracyclines and drugs that are effective against concurrent protozoan infections.

Canine Distemper

Canine distemper is caused by a highly contagious virus. The signs of the disease vary widely but can include fever, mucopurulent discharge from the eyes and nose, convulsive seizures, and the inability to stand. Dogs can be protected from this virus by immunization. The disease can be treated with anticanine distemper serum if diagnosed early enough.

Infectious Canine Hepatitis

Infectious canine hepatitis is caused by an adenovirus that produces varying signs, which may include fever, destruction of white blood cells, thirst, conjunctivitis, discharges from the eyes and nose, weight loss, bleeding lesions in the mouth and around the teeth, and a prolonged clotting time. The virus is transmitted by the fecal-oral route and is present in all secretions and excretions; it may be found in the urine for months. Virus-containing urine often spreads the disease. A viral vaccine is used to protect dogs.

Rabies

Rabies, caused by a rhabdovirus afflicts approximately 4,000 wild animals and about 600 domestic animals each year in the United States □. Skunks (59%), bats (14%), raccoons (10%), cattle (5%), dogs (4%), cats (3%), foxes (2%), and horses (1%) account for most of the animal rabies cases. In Europe, fox rabies recurs in three-year cycles and kills about 50% of the fox population.

667

It is estimated that each year more than a million persons worldwide are treated for possible rabies by the painful and dangerous series of fourteen to twenty-one inoculations of rabies vaccine. The rabies vaccine made from viruses grown in duck embryos causes allergic encephalitis in about 1 in 25,000 persons vaccinated.

The duck embryo vaccine is no longer used in the United States, where more than 30,000 people per year undergo rabies vaccinations with a vaccine made in human tissue cultures. This vaccine requires only five injections, so there is a far lower risk of allergic reactions to the vaccine. Humans can be protected from rabies simply by vaccinating any pets and domestic animals that might be bitten by rabid skunks, bats, raccoons, or foxes. None of the rabies vaccines are effective for animals other than livestock, dogs, and cats. Thus, pet skunks and raccons that are allowed to roam free in a rural environment are a danger.

SUMMARY

MICROORGANISMS AND PLANT NUTRITION

1. Rhizosphere microorganisms can promote plant growth in a number of ways. Bacteria such as *Azotobacter*, *Klebsiella*, and *Clostridium* can fix nitrogen, thus enriching the soil with nitrogenous compounds. Bacteria such as *Agrobacterium*, *Arthrobacter*, and *Pseudomonas* produce plant growth stimulators such as indoleacetic acid, auxins, and gibberellins. Some bacteria destroy toxic compounds such as H_2S.
2. Mycorrhizal fungi increase the rates of water and nutrient absorption, produce plant growth stimulators, release vitamins and antibiotics, and solubilize phosphate.

SOIL MICROORGANISMS AND PLANT HEALTH

1. Soils that inhibit plant pathogens are called suppressive soils, while those that do not are known as conducive soils.
2. A number of bacteria and fungi are responsible for making a soil suppressive and protecting a plant from pathogens. *Pseudomonas* species have been shown, for example, to produce an antibiotic called a siderophore that binds the Fe^{+3} ion. The siderophore sequesters Fe^{+3} in the rhizosphere and so deprives certain pathogens of essential iron.

PLANT PATHOLOGY

1. Bacteria are responsible for a few of the many diseases of plants. Bacterial soft rot is generally caused by *Erwinia* species, while bacterial wilts can be due to *Xanthomonas*, *Pseudomonas*, or *Erwinia*. Bacterial blights that destory large portions of the plant or much of the fruit are also caused by *Xanthomonas*, *Pseudomonas*, or *Erwinia* species. Crown gall is induced by *Agrobacterium* when part of its Ti-plasmid becomes part of the cells' host genetic information.
2. Fungi are responsible for most of the major crop diseases. Late blight is caused by *Phytophthora*. Downy mildew is due to a number of fungi, and soft rot is associated with *Rhizopus*, *Penicillium*, *Aspergillus*, and *Monilia* species. Vascular wilts are induced by six genera of fungi. The most important of these, *Erysiphe*, attacks apples, squashes, melons, wheat, oats, barley, and rye. Two common diseases are peach leaf curl, due to *Taphrina*, and brown rot of stone fruits, which is caused by *Monilinia*. *Ustilago* is responsible for smuts in corn, wheat, and barley. *Puccinia* is responsible for many of the rusts, such as wheat rust.
3. The bacterial and fungal diseases can generally be treated and controlled by various bactericides and fungicides or by the use of resistant varieties of plants.
4. Nematodes can be controlled by potent poisons called nematocides and by the use of resistant varieties of plants.
5. Viruses are responsible for a number of plant diseases. The most studied plant virus is tobacco mosaic virus, which causes (depending upon the plant stock) extensive chlorosis or individual plaques on tobacco leaves.

MICROORGANISMS AND ANIMAL NUTRITION

1. Animals raised so that they are not colonized by microorganisms are called germ-free or gnotobiotic.
2. Intestinal microorganisms provide the host with nutrients and protect it from many pathogens.
3. Many microorganisms ferment plant material in the rumen of ruminants to acetate, propionate, butyr-

ate, CO_2, and H_2. Methanogenic bacteria may produce CO_2 and H_2 from CH_4 which is released by the ruminant as flatus.

4. Ruminants absorb the acids released by fermenting bacteria. In their stomachs and large intestines, many of the microorganisms are degraded and the nutrients are absorbed by the ruminant.
5. Some antibiotics are used in animal feed to inhibit those bacteria that waste energy by producing CO_2 and H_2. This treatment increases the beef yield from a given amount of feed.

VETERINARY MICROBIOLOGY

1. Animal morbidity and mortality affect humans because many animal diseases can be transmitted to humans and because many human populations are dependent upon their animals for milk, meat, hides, and work.
2. Animal diseases transmitted to humans and to other animals are called zoonoses.
3. Tuberculosis, brucellosis, and Q fever are diseases of domestic animals that frequently are transmitted to humans in milk.
4. Equine encephalitis is a disease found in wild animals which can be spread to humans by insect vectors.
5. Mastitis is an inflammation of the mammary glands generally associated with large milk producers such as cows, goats, and camels. The infection is usually caused by *Streptococcus agalactiae* and *Staphylococcus aureus*. The inflammation results in the milk being contaminated with pus, microorganisms, and blood.
6. Brucellosis is a worldwide problem. Most cattle and swine herds have infected individuals, and approximately 75% of the world's goats are infected. The disease in dairy cows is controlled by calf vaccination and destruction of infected individuals.
7. Foot and mouth disease, caused by a virus, has from time to time taken a heavy toll of cattle all over the world.
8. Leptospirosis is caused by the spirochete *Leptospira* and is characterized by fever, jaundice, and sometimes abortion and death of infected animals. Swine, cattle, sheep, horses, and humans are all susceptible. Swine generally serve as a reservoir for the bacteria.
9. Newcastle disease and fowl plague are economically important diseases of wild and domestic birds which are caused by viruses. Wild birds serve as the reservoir for these viruses. Vaccination is effective against Newcastle disease but not against fowl plague.
10. Psittacosis is due to the bacterium *Chlamydia*. Wild birds are the reservoir for this bacterium. There is no effective vaccine, but tetracycline treatment is useful.
11. Feline and canine distemper are caused by different viruses. These viruses cause the death of many young cats and dogs, even though the diseases can be prevented by vaccination.
12. Rabies, due to a virus, causes thousands of wild and domestic animal deaths worldwide each year. Over a million humans each year undergo the series of painful rabies vaccinations to protect them from bites inflicted by possibly rabid animals.

STUDY QUESTIONS

1. Outline three ways in which rhizosphere bacteria may promote plant growth.
2. Explain how mycorrhizal fungi promote plant growth.
3. Define a suppressive soil and a conducive soil.
4. Explain what a siderophore is and how it can make a soil suppressive.
5. Which organism is responsible for crown gall?
6. Compare and contrast bacterial soft rot with bacterial wilts and bacterial blights.
7. What kind of organisms are responsible for a) late blight of potatoes; b) peach leaf curl; c) brown rot of stone fruits; d) corn, wheat, and barley smuts; e) wheat rust?
8. Explain what stubborn disease of citrus is and how it can be controlled.
9. What causes convulsive ergotism?
10. How do gnotobiotic animals differ from regular animals?
11. How do intestinal microorganisms contribute to an animal's nutrition?
12. How do intestinal microorganisms protect an animal from pathogenic microorganisms?
13. What waste products are produced by microorganisms in the rumen?
14. What do the methanogenic bacteria produce in the rumen?
15. Explain how the ruminant obtains its vitamins and amino acids.
16. Explain how an antibiotic in animal feed could increase beef yield.
17. What are zoonoses?
18. Name three animal infectious diseases that occasionally are transmitted to humans in unpasteurized cow's milk.

19. What is mastitis? What effect does it have on milk production?

20. What kind of organisms cause rabies, foot and mouth disease, psittacosis, leptospirosis, and bovine tuberculosis? Which of these diseases can be transmitted to humans?

SUPPLEMENTAL READINGS

Drummond, M. Crown gall disease. 1979. *Nature* 281:343–347.

Kaplan, M. M. and Koprowski, H. 1980. Rabies. *Scientific American* 242(1):120–134.

Matossian, M. K. 1982. Ergot and the Salem witchcraft affair. *American Scientist* 70:355–357.

Mlot, C. 1984. For the sake of citrus. *Science News* 126:380–381.

Schroth, M. and Hancock, J. G. 1982. Disease-suppressive soil and root-colonizing bacteria. *Science* 216: 1376–1381.

Shepard, J.F. 1982. The regeneration of potato plants from leaf-cell protoplasts. *Scientific American* 246(5):154–166.

Sun, M. 1984. New study adds to antibiotic debate. *Science* 226 (4676):818.

Wolin, M. J. 1981. Fermentation in the rumen and human large intestine. *Science* 213:1463–1468.

CHAPTER 35 INDUSTRIAL MICROBIOLOGY

CHAPTER PREVIEW

BEER- AND BREAD-MAKING ARE ANCIENT INDUSTRIES

Humans have been using microorganisms since the dawn of history to make useful products. For example, the Sumerians and Babylonians used "natural" yeast to make beer as far back as 6000 B.C., while the Egyptians, as long ago as 4000 B.C., used brewer's yeast to leaven their bread. The Babylonians also knew how to convert the ethanol in their beers into acetic acid (vinegar).

Egyptian tomb reliefs dated at about 2400 B.C. indicate that the preparation of leavened bread and beer was not just a family project, but probably a major industry employing many men and women (fig. 35-1). The top panel of the relief illustrates how the Egyptians manually separated the stalks from the heads of grain and how they pounded the heads of grain to separate the chaff. Flour was produced by grinding the grain between two rocks. An Egyptian depicted in the center relief is believed to be soaking the pounded, winnowed grain in a basket, so that it will sprout (malt) and yield starch-degrading enzymes and subsequently sugar for natural or added yeasts to ferment.

The central relief in the second panel indicates that some of the malted grain was added to flour to produce leavened dough. The malted grain contained sugar that yeast could ferment to ethanol and CO_2.

Figure 35-1 Egyptian Tomb Relief

The CO_2 caused the dough to rise. Two Egyptians are shown kneading the dough to stimulate the growth of yeast and the production of CO_2. Above the Egyptians are spherical loaves being allowed to rise before baking. Another individual is shown tending the oven where the bread was baked.

In the bottom panel, two individuals are preparing beer in the leftover malted grain in the basket. The Egyptian may be pouring warm water over the malted grain to wash sugar and natural yeasts through the basket into the clay fermenting vat below. The grain in the basket may have been used to feed cattle. The sugar extract (wort) was converted to ethanol and CO_2 by the yeast in the clay vat over a period of a few days. When the alcoholic beverage was ready, it was poured into numerous deep pottery jars that were sealed with clay.

Many of the steps in beer production today are similar to those used by the Egyptians over 4,000 years ago. In this chapter you will learn how beers are currently produced and how their flavor can be influenced by the use of various grains and by the addition of plant materials such as hops.

LEARNING OBJECTIVES

A STUDY OF THIS CHAPTER SHOULD ENABLE YOU TO:

DISCUSS THE ROLE MICROORGANISMS PLAY IN THE PRODUCTION OF ANTIBIOTICS

EXPLAIN HOW MICROORGANISMS ARE USED TO SYNTHESIZE STEROIDS

LIST SOME OF THE CHEMICALS THAT ARE PRODUCED BY MICROORGANISMS

OUTLINE HOW MICROORGANISMS ARE INVOLVED IN THE PRODUCTION OF ALCOHOLIC BEVERAGES

EXPLAIN HOW VACCINES ARE MADE FROM GENETICALLY ENGINEERED MICROORGANISMS AND TOXIN-PRODUCING MICROORGANISMS

Through the years, microorganisms have become important in many industries. The products and chemical changes produced by microorganisms are worth many billions of dollars each year, and many millions are being spent to find and develop microorganisms that may become the cornerstones of new industries or make old industries more profitable.

From antiquity, milk has been made into dairy products such as yogurt, buttermilk, and cheese so as to reduce the rate of spoilage. Often, cakes of dried yeast and bacteria were traded among people and used to make these dairy products. By the 1300s, ethanol and alcoholic beverages were being made by distilling the alcohol from fermented grains. During World War I, the Germans discovered that they could make glycerol, needed for the manufacture of explosives, by adding small amounts of sodium bisulfite to an alcoholic fermentation. Sodium bisulfite causes yeast to produce mostly glycerol rather than ethanol during fermentation. The Germans at one time produced as much as 1,000 tons of glycerol per month. The British developed an acetone-butanol fermentation that was carried out by the bacterium *Clostridium acetobutylicum*. The acetone was used as a solvent in the manufacture of explosive cordite, a stringy, smokeless explosive that contains nitroglycerine, guncotton, petroleum jelly, and acetone.

Folklore often mentions the use of moldy breads, meats, and cheeses to help heal superficial infections and wounds. Possibly, some of the moldy foods were contaminated with the antibiotic-producing *Penicillium* and contained enough penicillin to do some good. There are also reports of primitive peoples using organically rich soil, moldy leaves, or moldy bark to treat wounds. If these treatments worked at all, perhaps *Bacillus* and *Streptomyces* produced sufficient quantities of antibiotics to inhibit the growth of microorganisms. It was not until 1929, however, that Alexander Fleming reported that colonies of the fungus *Penicillium notatum* and fungus-free growth medium in which the fungus had formally been grown inhibited and killed bacteria. The first attempts to isolate penicillin were unsuccessful because this drug is unstable. In the early 1940s, however, Florey and Chain at the University of Oxford succeeded in isolating penicillin. This achievement initiated the search for, and the isolation of, a great many antibiotics that were effective against bacteria and some eukaryotes. For example, Selman Waksman in the early 1940s discovered that the soil bacteria *Bacillus* and *Streptomyces* produced antibiotics. Since the 1940s, a multi–billion dollar industry has developed around the isolation of antibiotics and the synthesis of drugs from the antibiotics.

THE IMPORTANCE OF MICROORGANISMS IN INDUSTRY

Natural products synthesized by microorganisms are becoming increasingly important in world markets. Anticoagulants, antidepressants, vasodilators, herbicides, insecticides, plant hormones, enzymes, and enzyme inhibitors have all been isolated from microorganisms (table 35-1). Microorganisms are being used more and more frequently in the production of enzymes, such as **amylases,** used in brewing, baking, and textile production, and **proteases,** used for tenderizing meats, preparing leathers, and making detergents and cheeses.

The food, petroleum, cosmetic, and pharmaceutical industries are also using microorganisms to make polysaccharides. For example, the bacterium

TABLE 35-1
COMMERCIALLY IMPORTANT MICROBIAL PRODUCTS

PRODUCT	USE OF PRODUCT	SOURCE
Phialocin	Anticoagulant	*Phialocephala repens*
1,3,Diphenethylurea	Antidepressant	*Streptomyces* sp.
Avermectin	Antihelminthic	*Streptomyces avermitilus*
Vitamin B_{12}	Antipernicious anemia	*Streptomyces griseus*
Naematolin	Coronary vasodialator	*Naematoloma fasciculare*
Monascin	Food pigment	*Monascus* sp.
Herbicidin	Herbicide	*Streptomyces saganonensis*
Fusaric acid	Lowers blood pressure	*Fusarium* sp.
N-acetylmuramyl-tripeptide	Immune enhancer	*Bacillus cereus*
Pericidin	Insecticide	*Streptomyces mobaraensis*
Tetranactin	Miticide	*Streptomyces aureus*
Gibberillic acid	Plant hormone	*Gibberella fujikuroi*
Lactic acid	Chemical reagents	*Lactobacillus delbruekii*
Acetic acid	Vinegar, solvent	*Acetobacter* sp.
Citric acid	Food preservative	*Aspergillus niger*
Glutamic acid	Growth factor, preservative	*Micrococcus glutamicus*
Lysine	Growth factor	*Micrococcus glutamicus*
Valine	Growth factor	*Escherichia coli*
Amylase	Hydrolysis of starch (brewing)	*Aspergillus* sp.
Pectinases	Degrading pectin	*Aspergillus* sp. & *Rhizopus* sp.
Streptokinase	Degradation of blood clots	*Streptococcus pyogenes*
Proteases	Proteins to amino acids (tenderizing meats)	*Bacillus subtilis, B. licheniformis*
Invertase	Sucrose degraded to glucose & fructose	*Saccharomyces cerevisiae*
Penicillinase	Degradation of penicillin	*Bacillus subtilis*
Glycerol	Preserving foods	*Saccharomyces cerevisiae*
2,3-Butanediol	Chemical laboratories	*Bacillus polymyxa*
Butanol, Acetone	Chemical laboratories, manufacturing	*Clostridium acetobutylicum*
Ethanol	Chemical laboratories	*Saccharomyces* sp.

Xanthomonas campestris produces the polysaccharide called **xanthan,** which is used to stabilize and thicken foods, as part of drilling muds, as a base for cosmetics, as a binding agent in many pharmaceuticals, and for textile printing and dyeing. *Leuconostoc mesenteroides*, when grown on sucrose, produces dextran, a polysaccharide that is used to extend blood plasma. Cross-linked chains of dextran can be used as molecular sieves for separating molecules in column chromatography.

The chemical industry produces tons of amino acids, nucleotides, vitamins, and organic acids, which are sold to many different types of laboratories and to health food stores. L-lysine, which is prescribed by some doctors to treat herpes infections, is synthesized by the bacterium *Corynebacterium glutamicum*. Vitamins such as B_{12} (cyanocobalamin) and B_2 (riboflavin) are synthesized by the bacterium *Pseudomonas denitrificans* and the yeast *Ashbya gossypii*, respectively.

The pharmaceutical companies are now manipulating bacteria genetically so that they produce proteins such as insulin and interferon that are identical to the human proteins. B-cells and cancer cells are also being fused to produce **hybridomas** ☐ 274 that make single specificity antibodies known as a **monoclonal antibodies.** These specific antibodies are being used to diagnose infectious diseases, to measure exceedingly small amounts of biological materials, and to purify minute quantities of these valuable substances.

TABLE 35-2
DRUGS PRODUCED BY MICROORGANISMS

CATEGORY OF DRUG	MAJOR U.S. PRODUCERS	MARKET VALUE
Penicillins	Ayerst Laboratories Lederle Laboratories Eli Lilly and Company Smith, Kline & French Laboratories E. R. Squibb & Sons, Inc. Warner-Lambert Company Wyeth Laboratories	$220,943,000
Other broad- and medium-spectrum antibiotics	Abbott Laboratories Bristol Laboratories Lederle Laboratories Eli Lilly and Company Merck Sharp & Dohme Schering-Plough Corporation E. R. Squibb & Sons, Inc. The Upjohn Company Warner-Lambert Company Wyeth Laboratories	$638,297,000
Antibiotics in combination with sulfonamides	Bristol Laboratories Burroughs Wellcome Co. Ross Laboratories	$ 16,921,000
Topical antibiotics	Lederle Laboratories Eli Lilly and Company Marion Laboratories, Inc. Schering-Plough Corporation The Upjohn Company Warner-Lambert Company	$ 17,064,000
Vaccines	Lederle Laboratories Merck Sharp & Dohme Warner-Lambert Company Wyeth Laboratories	$ 90,000,000
Sulfonamides	Alcon Laboratories, Inc. Lederle Laboratories Hoffmann-La Roche, Inc. Smith, Kline & French Laboratories E. R. Squibb & Sons, Inc. Warner-Lambert Company	$ 47,562,000
Antifungal drugs	Ayerst Laboratories Barnes-Hind Pharmaceuticals, Inc. Ciba-Geigy Corporation Lederle Laboratories Ortho Pharmaceutical Corporation Hoffmann–La Roche, Inc. Schering-Plough Corporation E. R. Squibb & Sons, Inc.	$103,911,000
Antiseptic preparations	Burroughs Wellcome Co. Norwich-Eaton Pharmaceuticals Ortho Pharmaceutical Corporation Sterling Drug Inc. E. R. Squibb & Sons, Inc.	$ 15,000,000
Tuberculostatic agents	Ciba-Geigy Corporation Dow Chemical U.S.A. Lederle Laboratories E. R. Squibb & Sons, Inc. Warner-Lambert Company	$ 12,835,000

Digestive Enzymes	Armour and Company B. F. Ascher & Company, Inc. Hoechst-Roussel Pharmaceuticals, Inc. Organon Inc. Reed & Carnrick Warner-Lambert Company	$ 16,999,000
Vitamins (prescription only)	Abbott Laboratories The Central Pharmacal Company Lederle Laboratories Mead Johnson & Company Hoffmann–La Roche, Inc. Ross Laboratories E. R. Squibb & Sons, Inc. Warner-Lambert Company	$133,891,000

From "The Microbiological Production of Pharmaceuticals," by Aharonowitz and Cohen.

In the future, monoclonal antibodies are expected to be useful in treating cancers, since they can be made to bind specifically to cancerous cells.

Microorganisms themselves are valuable because of the jobs they can carry out for humans. For example, microorganisms such as the bacterium *Pseudomonas putida* are being used to degrade various components of oil in order to clean up oil spills.

872

The bacterium *Bacillus thuringiensis* is being used successfully as a pesticide and as a pesticide producer □. *Bacillus thuringiensis*, when consumed by many insect larvae, produces an exoenzyme that degrades hyaluronic acid, the polysaccharide that holds cells together. This exoenzyme severely damages the digestive tracts of the larvae and kills them. In addition, a toxin has been isolated from *Bacillus thuringiensis* that can be sprayed on plants to kill insect larvae in the absence of the bacteria.

755

Microorganisms are being used to leach low-grade ores so as to extract valuable metals □. For example, copper and uranium can be leached from ores by bacteria such as *Thiobacillus*. First, *Thiobacillus* frees ferrous ions (Fe^{+2}) in FeS ores by oxidizing the ferrous ion to the ferric ion (Fe^{+3}). Then, the ferric ion oxidizes the sulfur in pyrite (FeS_2), the yellow mineral often called fool's gold, so that sulfuric acid is produced. In addition, the ferric ion oxidizes insoluble CuS to the soluble $CuSO_4$, which can easily be collected.

THE PRODUCTION OF PHARMACEUTICALS

Microorganisms are involved in the synthesis of more than $1.5 billion worth of drugs in the United States each year (table 35-2). When you compare this to the $7.5 billion of prescription drugs sold each year in the U.S., it is clear that industrial microbiology is big business. Worldwide sales of just four groups of antibiotics—the tetracyclines, the penicillins, the cephalosporins, and erythromycin—total over $4 billion per year. The industrial production of antibiotics relies upon a small group of spore-forming microorganisms. Various species of the fungus *Cephalosporium* synthesize the cephalosporins; the fungus *Penicillium* produces some of the penicillins; and species of the bacterium *Streptomyces* synthesize numerous antibiotics, such as streptomycin, tetracycline, erythromycin, and chloroamphenicol.

214

In order to produce sufficient and economical antibiotics, it has been necessary to obtain mutants □ that produce 100 to 10,000 times the amount of antibiotics that are produced by the wild-type (naturally-occurring) strains. In the 1940s, wild-type strains of *Penicillium* produced only a few mg of penicillin per liter of media. By the early 1950s, mutant strains of *Penicillium chrysogenum* were made to produce 60 mg of penicillin per liter. After a series of mutations and selections for enhanced penicillin production, however, mutants were isolated that produced 20,000 mg of penicillin per liter. In the last few years, the natural penicillins and cephalosporins have been modified chemically to produce semisynthetic antibiotics that are effective against a broader spectrum of bacteria, and that are not inactivated as easily by bacterial enzymes.

Figure 35-2 Fermentation Tanks. These fermentation tanks are used to grow microorganisms that produce antibiotics. The ports and gauges are used to monitor and maintain optimal growth conditions for the production of antibiotics.

The manufacture of antibiotics takes place in large growth tanks (fig. 35-2). For example, *Penicillium chrysogenum* may be grown in 100,000-liter (26,000-gallon) fermenters for approximately 200 hours (8 days). The removal of the fungus from such large amounts of growth media takes approximately 15 hours. The fungus grows on sugar and phenylacetic acid, which are added continuously. The phenylacetic acid provides the benzyl side chain of penicillin G. Penicillin G is extracted from the filtrate and crystallized. In order to make the semisynthetic penicillins, penicillin G is mixed with a bacterium that secretes enzymes called **acylases.** These enzymes remove the benzyl group from penicillin G and so convert it to **aminopenicillanic acid** (6-APA). Aminopenicillanic acid is the core molecule that is used to make other penicillins. Various chemical groups are added to aminopenicillanic acid (fig. 35-3). Similarly, cephalosporin C has an amide side chain removed, and then new groups are added to the core 7-α-aminocephalosporanic acid (fig. 35-3).

It has been discovered that the bacterium *Streptomyces clavaligerus* produces antibiotics that resemble cephalosporins such as cefoxitin (fig. 35-3). These new β-lactam antibiotics are called **cephamycins.** The cephamycins are active against many gram positive and gram negative bacteria by blocking the bacterial transpeptidases. *Streptomyces cattleya* produces antibiotics that inhibit the β-lactamases, enzymes that inactivate the penicillins and cephalosporins. These compounds, called **thienamycins,** include clavulanic acid, olivalic acid, and thienamycin (fig. 35-3). Clavulanic acid and a penicillin called **amoxicillin** are used together to create an antibiotic hybrid that inhibits β-lactamases as well as transpeptidases.

Microorganisms that produce antibiotics can be discovered by spotting them on lawns of sensitive bacteria and observing the inhibition of growth they create. In this way, over 5,000 antibiotics have been discovered that are effective against many different types of bacteria. The search for antibiotics that can be used against viruses, fungi, protozoans, helminths, and cancerous cells has not been as successful, however, because these antibiotics are almost as toxic to the patient as to the disease-causing organisms.

THE MANUFACTURE OF STEROIDS

Steroid hormones have become very important in medicine. For example, cortisone and similar steroids are known to relieve pain and reduce swelling □. Consequently, they are useful in treating such diverse conditions as poison oak, asthma, and rheumatoid arthritis. Steroid hormones are also used in contraceptive pills and for treating hormone imbalances.

443

a) Penicillins

Penicillin G

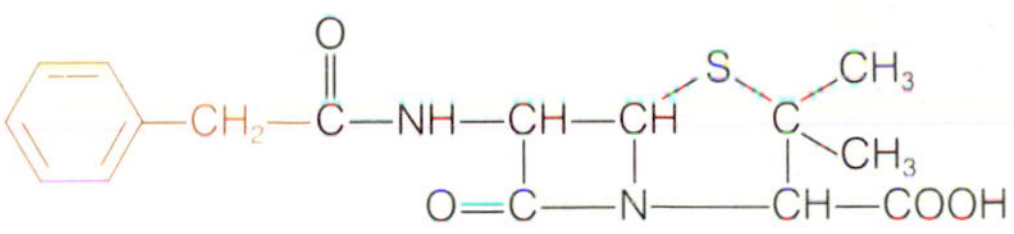

Penicillin V

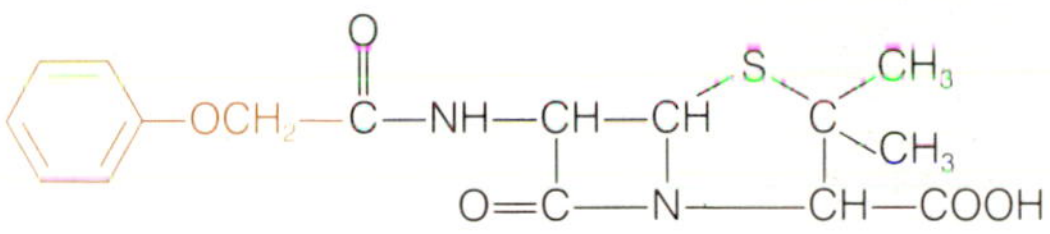

Penicillin F

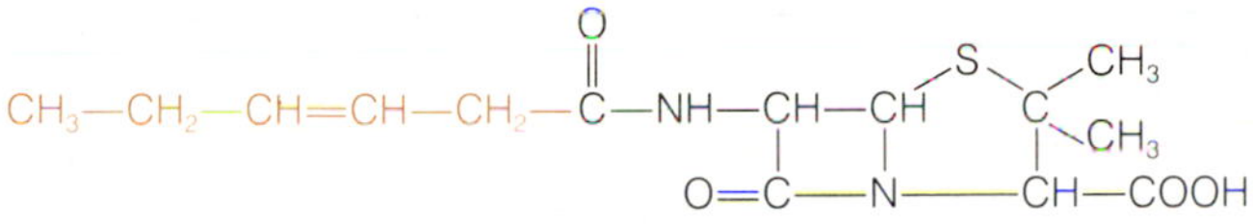

Ampicillin

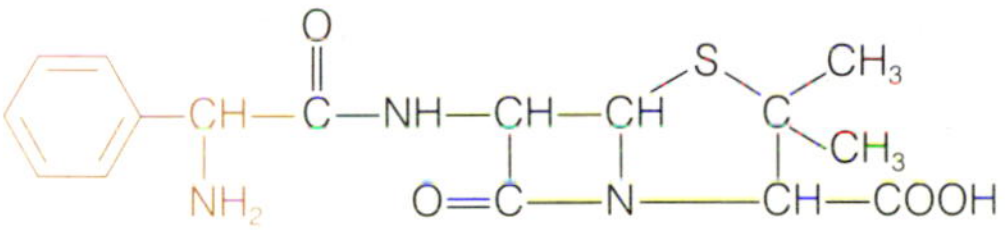

Methicillin

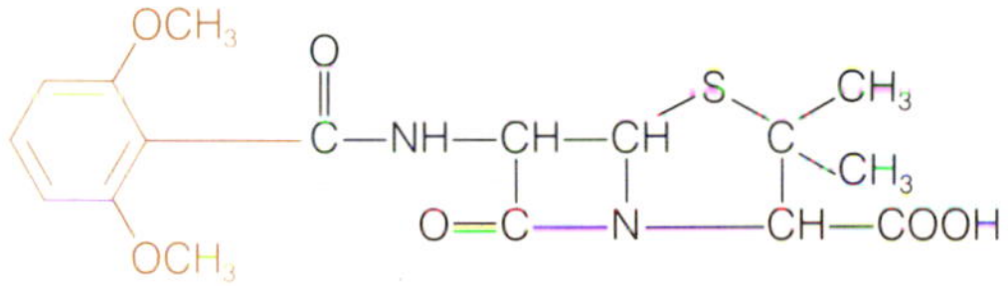

Carbenicillin

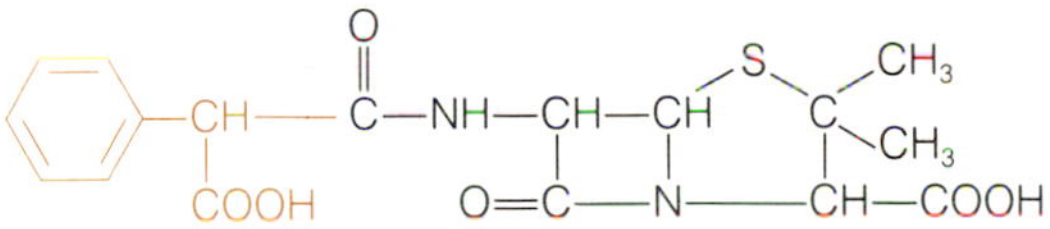

Oxacillin

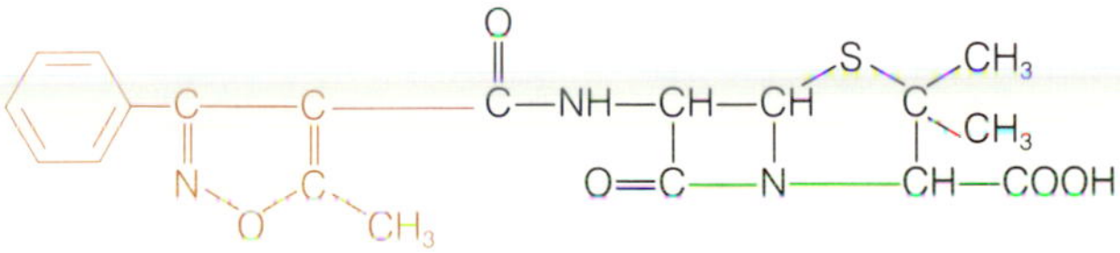

Figure 35-3 Penicillins, Cephalosporins, Cephamycins, and Thienamycins. *(a)* The chemical structure of penicillins is based upon the β-lactam ring and a 5-member ring containing sulfur. Penicillin G serves as the core structure upon which new side chains (color print) are attached after the removal of the benzyl group. Many organisms chemically alter penicillin G, making it ineffective as an antibiotic. In addition, penicillin G is not very effective against most gram negative bacteria, because it does not effectively penetrate the outer membrane. Methicillin, because of its side group, is not readily inactivated by β lactamases. Ampicillin is effective against gram negative bacteria. *(b)* The chemical structure of cephalosporins is based upon the β-lactam ring and a 6-member ring containing sulfur. Cephalosporin C serves as the core structure upon which new side chains are attached. *(c)* Cephamycins are cephalosporin antibiotics produced by the bacterium *Streptomyces*. *(d)* Thienamycins are β-lactam antibiotics produced by the bacterium *Streptomyces*. These β-lactam antibiotics differ from penicillin G, cephalosporin C, and cefoxitin in that the rings attached to the β-lactam group lack sulfur. The thienamycins bind to β-lactamases and inhibit them.

b) Cephalosporins

Cephalosporin *C*

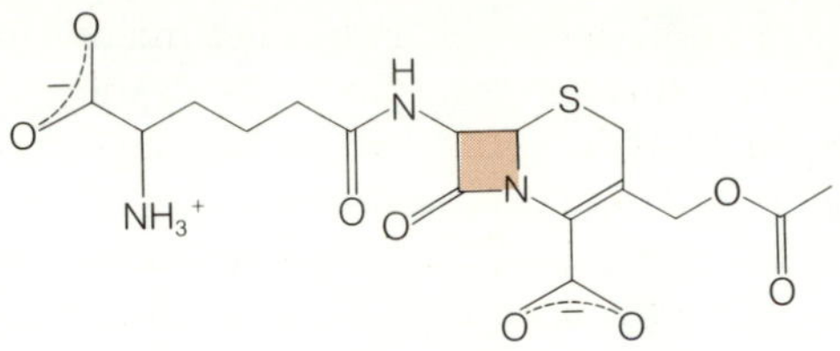

Cephalothin

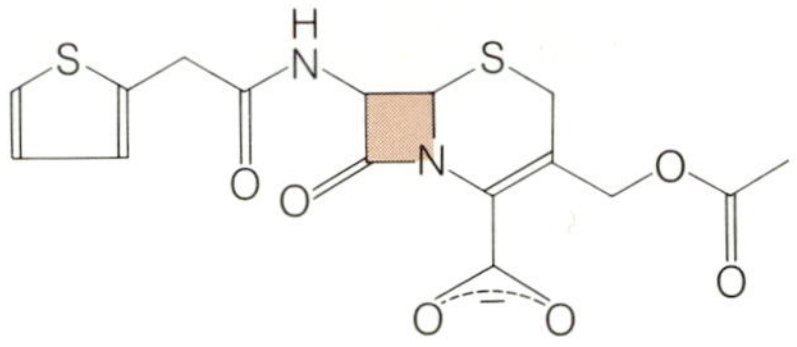

Cephalexin

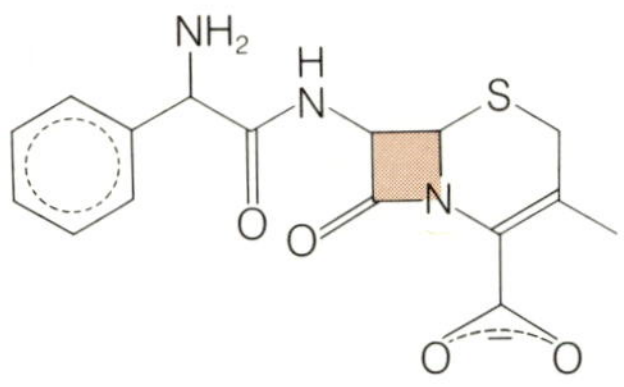

c) Cephamycin

Cefoxitin

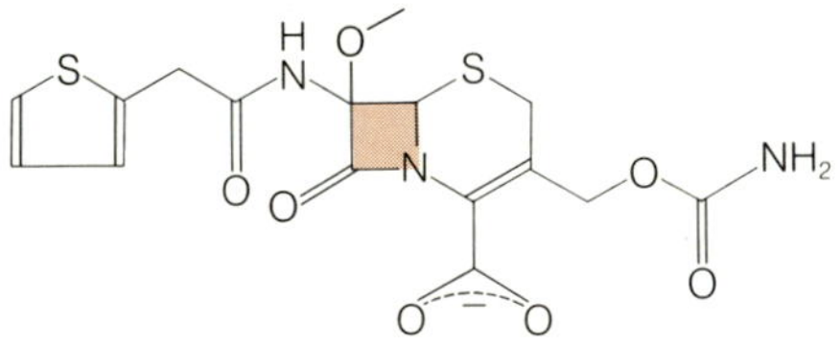

d) Thienamycins

Clavulanic acid

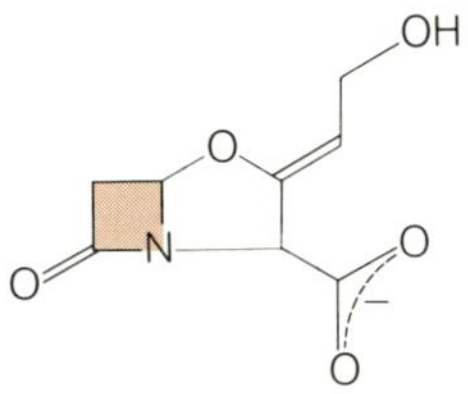

Thienamycin

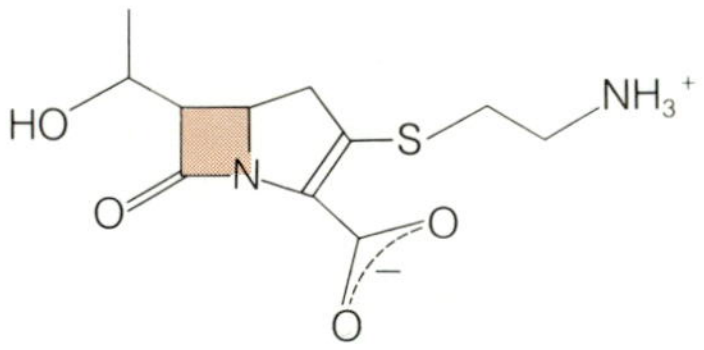

Because the chemical synthesis of steroids such as cortisone requires more than 35 steps, steroids are very expensive to produce chemically. Microorganisms can be used to carry out some of the chemical steps, thus reducing the number of steps to 11 and hence the expense of making steroids (fig. 35-4). The use of microorganisms to alter chemicals is known as **bioconversion,** and it has helped to reduce the cost of making steroids from over $200/gram to around $0.50/gram. One example of a bioconversion is that carried out by the bread mold *Rhizopus arrhizus.* This fungus is able to hydroxylate progesterone and reduce the synthesis of cortisone to 11 steps. Because this reaction can be carried out at 37°C, in water, and at atmospheric pressures, it is much cheaper and produces less pollution than the chemical method. The raw material for making steroid hormones is a waste product, stigmasterol, which accumulates during the production of soybean oil, or diosgenin, found in the roots of Mexican barbasco plants. Over $300 million worth of the steroid hormones cortisone, aldosterone, prednisone, and prednisolone are sold worldwide per year and are made by using bioconversion methods.

PROTEINS MADE IN GENETICALLY ENGINEERED BACTERIA

262 In the past few years, recombinant DNA □ technology has allowed pharmaceutical companies to use genetically engineered bacteria to synthesize viral, bacterial, and human proteins. This technology has resulted in the manufacture of cheaper and better vaccines and hormones.

Insulin, used to treat diabetes, has been extracted from the pancreas of cattle and swine. Although these animal insulins alleviate many of the symptoms of diabetes, they differ slightly from the human insulin and sometimes 482 cause allergic reactions □. It is hoped that human insulin synthesized by bacteria will be cheaper and safer than animal insulins. Over $200 million worth of insulin is sold each year worldwide.

Only one of the interferons that block virus replication in mammalian cells has been extracted from white blood cells. Because of the large amount of blood needed to obtain usable amounts of the interferon, and because of the difficulty in extracting it, the cost of this antiviral agent has been prohibitive. Two liters of human blood yields only 1 microgram (mcg) of leukocyte interferon. A number of different human interferon genes have been inserted into bacteria, however, and the bacteria make large quantities of the interferons economically. As much as 1 mg (1,000 mcg) of interferon can be synthesized from a 1-liter culture of genetically engineered bacteria. 100,000-liter fermenters are expected to yield 100 grams (10^8 mcg) of interferon per batch. As soon as the various types of interferons are synthesized and produced in sufficient quantities, research into their usefulness in treating virus infections and cancers will progress much more rapidly, and the experiments will give more definitive results.

Vaccines against infectious diseases are also being produced in bacteria. For example, a vaccine against foot and mouth disease virus (FMDV), which affects great numbers of cattle, swine, and sheep worldwide, has been prepared from one of the viral proteins. In the past, foot and mouth disease vaccines have been prepared by growing the virus in baby hamster kidney cells or in bovine tongue epithelium and then inactivating the virus. Attenuated viruses have also been used. Outbreaks of foot and mouth disease oc-

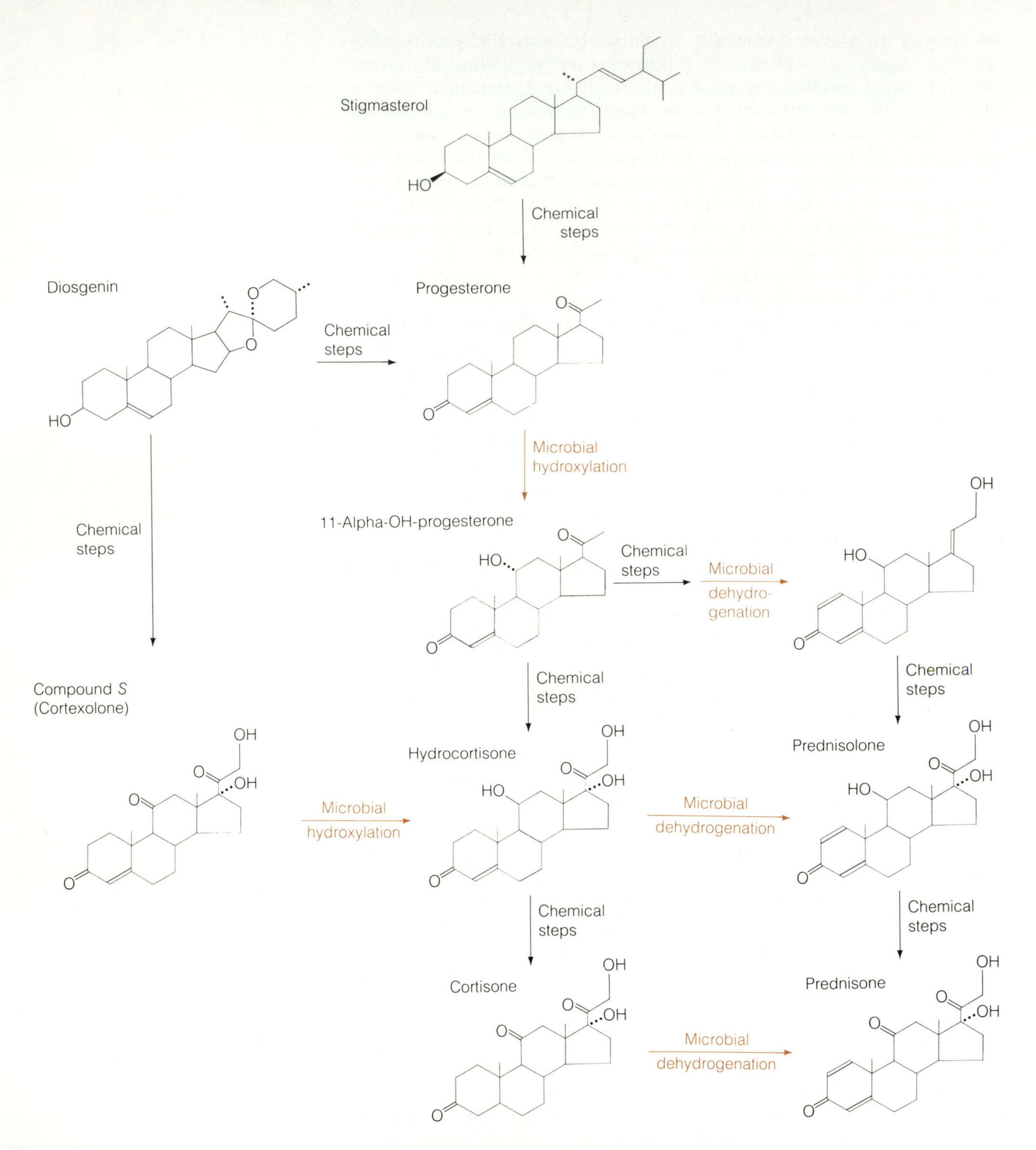

Figure 35-4 Bioconversions in the Making of Steroids. Cortisone, prednisolone, and prednisone are valuable steroid hormones. To make these hormones from stigmasterol using purely chemical methods requires 37 steps. By using microorganisms, the synthesis is reduced to 11 steps, thus greatly reducing the cost of the product. Progesterone is converted to 11-α-hydroxyprogesterone by the fungus *Rhizopus.* Compound S is converted to hydrocortisone by the fungus *Curvularia lunata. Hydrocortisone is converted to prednisolone by the fungus Rhizopus.* Cortisone is converted to prednisone by the bacterium *Corynebacterium simplex.*

cur frequently, however, because of incompletely inactivated vaccine or reversion of attenuated strains to virulent types. Another problem with whole virus vaccines is that they are unstable and require refrigeration. This means that they are not very useful in underdeveloped countries, where refrigeration is not often available and the disease is a major problem. Genetic engineering 390 has made it possible to synthesize a FMDV capsid □ protein in the bacterium *E. coli*. This capsid protein is used to make a safe, stable, and effective vaccine for foot and mouth disease. This is the most widely used antiviral vaccine in the world, as indicated by the estimated 500 million doses used each year. The new vaccine protects against only one of the seven types of foot and disease viruses, however, so research is in progress to create vaccines with a broader spectrum of activity.

Recently, genetically engineered yeast cells have been used to produce a hepatitis viral coat protein that forms hepatitis B surface antigen (HBsAg particles) within the yeast. The yeast glycosylates the protein, making it immunogenic. It is hoped that this antigen will be useful as a vaccine against hepatitis B, a virus that is responsible for 80,000–100,000 new cases of hepatitis and 800–1,000 deaths each year in the U.S. Presently, HBsAg particles are being isolated from the blood of hepatitis carriers and used as a vaccine. This is a very expensive and involved procedure, and the vaccine may contain other dangerous viruses or antigens that can elicit serious allergic reactions in the liver and kidneys of patients.

Another genetically engineered bacterium produces a protein that can be used as a vaccine to prevent colibacillosis, a widespread disease that kills many newborn calves, foals, lambs, and piglets. Colibacillosis is due mainly to *Escherichia coli*, but *Salmonella*, and *Chlamydia* may also be involved.

INDUSTRIAL CHEMICALS AND ENZYMES

Enzymes

Enzymes isolated from microorganisms have a large market worldwide (table 35-1). For example, proteases isolated from *Bacillus licheniformis* are used in many detergents as cleaning aids. The proteases break down and solubilize proteins that soil clothing and other fabrics. In addition, proteases from fungi and bacteria are used to a limited extent as digestive aids in animal feed and as meat tenderizers. Three enzymes made by industrial fermentations—α-amylase, glucamylase, and glucose isomerase—are used to convert starches to fructose corn syrup, the sweetener that is replacing sucrose in soft drinks. Two million tons of high-fructose corn syrup are being synthesized per year. The enzyme **α-amylase,** extracted from the bacterium *Bacillus subtilis*, cleaves starch molecules to short glucose polymers; β-amylase converts the short glucose polymers to maltose, a disaccharide of glucose. **Amyloglucosidase** splits off single glucose sugars, while **glucose isomerase** converts glucose to fructose. Worldwide sales of α-amylase, glucamylase, and glucose isomerase amount to over $100 million per year.

Genetic engineering requires the use of numerous enzymes to cut DNA 264 molecules at precise locations (restriction endonucleases) □, to digest the DNA (exonucleases), and to repair the DNA once it has been spliced (ligases). Most of these enzymes are isolated from various bacteria. Multiple copies of the gene for DNA ligase have been introduced into *E. coli*, thus increasing the production of this enzyme more than 500-fold. The enzymes used in genetic engineering are prepared in huge quantities, using fermenters.

Alcohols

Ethanol is a solvent and extractant used in many chemical processes. It is also used to make antifreeze, and serves as a substrate for the synthesis of other compounds such as vinegar. About 100 million gallons of alcohol are currently being used as fuel, such as gasohol, which is a 9:1 blend of gasoline and ethanol. Over 600,000 tons (about 300 million gallons) of industrial ethanol are made per year, and almost $300 million worth of it was sold in the United States in 1980. The synthesis and sale of alcoholic beverages are not included in these figures.

In the microbiological synthesis of industrial alcohol, the yeast *Saccharomyces cerevisiae* ferments □ starches and sugars. At the present time, only 30% of industrial alcohol is made by fermentation, because the starches and sugars cost more than petroleum and conversion to biological alcohol production is expensive. Researchers are attempting to reduce the cost of making fermented alcohol by using garbage and waste materials, such as used paper and wood, as substrates for the yeast. The expense comes from the fact that the cellulose and hemicellulose in paper and wood must be separated and then converted to glucose and xylulose, respectively, before the material can be fermented.

189

FOCUS

GASOHOL AND ANIMAL FEED FROM WOOD

The scientific and business communities are working together to produce ethanol cheaply so that it can function as an alternative fuel source. Numerous filling stations in the United States are already selling gasohol (a blend of 90% unleaded gasoline and 10% ethanol), but the price is still higher than that of high-octane unleaded gasoline. Gasohol can readily be substituted for unleaded gasoline in vehicles with no engine adjustments and with little or no change in the performance of the automobile or truck.

Brazil is experimenting with ethanol as a fuel source. Most automobiles in Brazil burn gasohol that contains 20% ethanol, and the country is building 250,000 automobiles and trucks that will use pure ethanol. Presently, the sugar for ethanolic fermentation comes from sugar cane, but the Brazilians are also studying the use of cassava roots and eucalyptus trees in making ethanol.

In order to make cheap ethanol, old newspapers, farm crop residue, wood, and garbage are being used as the starting materials in ethanolic fermentations. Scientists are constantly searching for fungi and yeasts that will efficiently break down the various polysaccharides that make up the starting materials and that will efficiently ferment all the sugars to ethanol. A mutant strain of the fungus *Trichoderma viride,* which produces excess amounts of a number of different cellulases and hemicellulases, is now being tested to determine whether the fungus will efficiently degrade the cellulose, hemicellulose, and pectin in wood to hexoses and pentoses. Forty to 50% of paper, wood, and agricultural residue consists of glucose in the form of cellulose, while 25–35% is made up of xylose and arabinose in the form of hemicellulose. Presently, the yeast *Saccharomyces* is being used to ferment the glucose to ethanol and CO_2, but it is unable to ferment the pentoses xylose and arabinose. Since there is so much xylose and arabinose in paper, wood, and agricultural residue, it is important to isolate organisms that can break down the hemicelluloses and ferment the pentoses. In an attempt to make the fermentation process economical, the yeast *Pachysolen tannophilus* is being tested to see if it ferments the pentoses efficiently.

The leftover stillage that results from the growth of the fungi and yeast is very high in protein and vitamins and consequently can be used for animal feed. Pigs, goats, and geese do well on the stillage, but the magnesium content must be reduced before it can be given to cattle.

The organic solvents n-butanol and acetone can be made by the fermentation of starch by *Clostridium acetobutylicum*, or by the metabolism of sugar by *C. saccharoacetobutylicum*. Very little use is made of these processes, however, because the cost of petroleum is still below that of starch and sugar sources, such as corn and molasses.

Glycerol is used in preserving many soft foods, such as breakfast snacks and pet foods. It is also used in making candy, toothpaste, adhesives, cellophane, and some paper. As mentioned previously, glycerol was made in a biological process by the Germans during World War I and was used to make explosives. Today, however, the United States makes glycerol from petroleum derivatives, because they are cheaper than sugars. The United States produces approximately 75,000 tons of glycerol, and sales amount to about $70 million per year.

Acids

More than 1.4 million tons of acetic acid, with a market value of $500 million, are synthesized each year in the United States. This acid is used in the

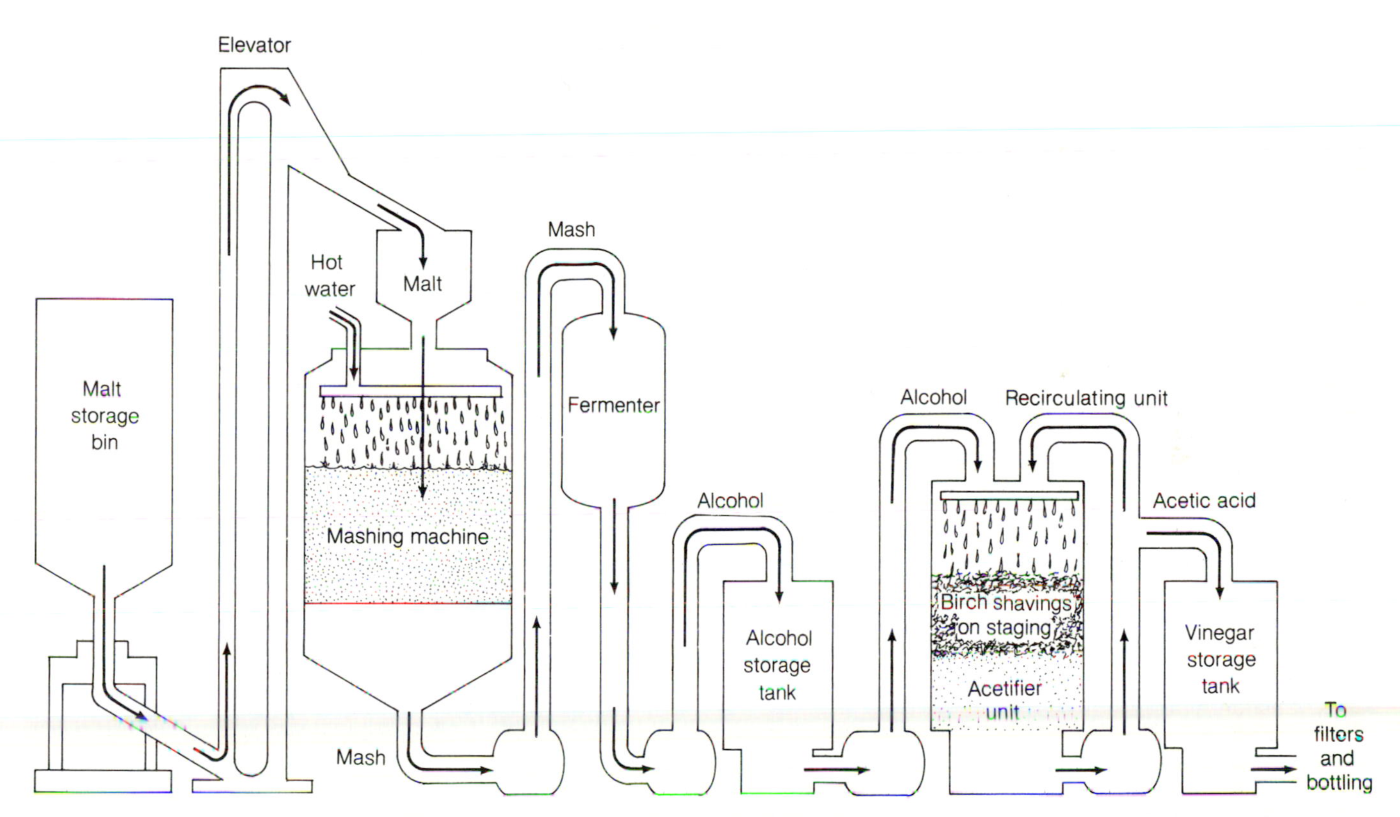

Figure 35-5 Manufacture of Vinegar. Barley malt (sprouted barley seeds) can be the starting material used to make acetic acid (vinegar). The sprouted barley seeds produce enzymes that can break down the starches in the seeds to glucose and maltose. The malted barley is crushed in the mashing machine to release the enzymes that break down the starch. After the enzymes have broken down the starch, hot water is added to denature the enzymes and to create a sugar medium that will support the growth of fermenting yeasts. The mash is pumped to a fermenter, where yeasts *(Saccharomyces)* ferment the sugar to ethanol and CO_2. The alcoholic product may be stored for a number of months before being used to make vinegar. The ethanol is pumped to a tower and allowed to trickle down through shelves of beech- or birchwood shavings upon which *Acetobacter* is growing. *Acetobacter* oxidizes the alcohol to acetic acid. The ethanol and acetic acid are recycled until there is approximately 5% vinegar. The vinegar is stored in large tanks before it is filtered and bottled.

manufacture of rubber, plastics, dyes, and insecticides, and as a substrate for the production of amino acids. Presently, very little of the industrial acetic acid is made by microorganisms converting ethanol to acetic acid. Much of the acetic acid used as vinegar, however, is made from ethyl alcohol under aerobic conditions by the bacteria *Acetobacter aceti*, *A. orleanense*, and *A. schutzenbachii*, which are widely distributed in the soil and on plant material.

Commercial vinegar is made in large tanks that contain beechwood shavings (fig. 35-5). Alcohol is trickled from the top of the tank through wood shavings that have *Acetobacter* growing on them. The alcohol that accumulates at the bottom of the tank is recirculated and allowed to trickle through the shavings a number of times. Air enters from the bottom of the tank to provide oxygen for the bacteria. The tanks are kept at a temperature 25–30°C for 8–10 days. Oxidation of the alcohol produces a 4–5% solution of vinegar.

Lysine and glutamic acid are two commercially valuable amino acids that are made by biological processes (table 35-1). L-lysine is a valuable nutrient that is used as an additive to animal feed, while glutamic acid, used in the salt form (monosodium glutamate), is a widely used flavor enhancer. Over 40,000 tons of L-lysine and 300,000 tons of monosodium glutamate are produced each year worldwide.

Most monosodium glutamate is produced commercially in Japan by cultures of *Corynebacterium glutamicum* and *Brevibacterium flavum*. The substrates used are glucose, acetic acid, or n-paraffins obtained from petroleum. Figure 35-6 illustrates how glucose is converted through the Embden-Meyerhof pathway □ to pyruvic acid, and how pyruvic acid subsequently enters the citric acid cycle □. Glutamic acid is made from α-ketoglutaric acid, one of the intermediates of the citric acid cycle. Because the synthesis of glutamic acid is feedback-repressed □, it must leak out of the cells if large amounts are to be synthesized. The cell membranes can be made leaky by growing the bacteria in a medium with limited biotin, thus inducing phospholipid deficiency. Membranes can also be made leaky by growing the bacteria in the presence of saturated fatty acids, detergents, or penicillin.

186
191
254

A mutant of *Corynebacterium glutamicum* that is unable to synthesize methionine, threonine, or isoleucine is used to produce lysine. The mutant is necessary if large amounts of lysine are to be made, because threonine and lysine together block the beginning of the biosynthetic pathways that lead to the synthesis of these amino acids (fig. 35-6). If threonine is not present in high concentrations, the first enzyme that reacts with aspartate is not feedback inhibited; thus, pools of lysine are able to accumulate in the mutant. Lysine cannot accumulate in the properly regulated **prototroph** □, because as soon as lysine and threonine begin to be produced in excess, their synthesis is feedback inhibited.

222

In the future, it is expected that many amino acids will be produced commercially by microorganisms. There is a demand for methionine as a feed supplement, for glycine and alanine as flavor enhancers, and for cysteine to give texture to bread. The synthesis of amino acids is a big business, as indicated by the fact that approximately 105,000 tons of methionine are produced per year in the U.S. by chemical means. Much of this methionine is used as a feed supplement and in vitamin pills. So far, it has not been possible to engineer bacteria genetically to produce methionine in large quanti-

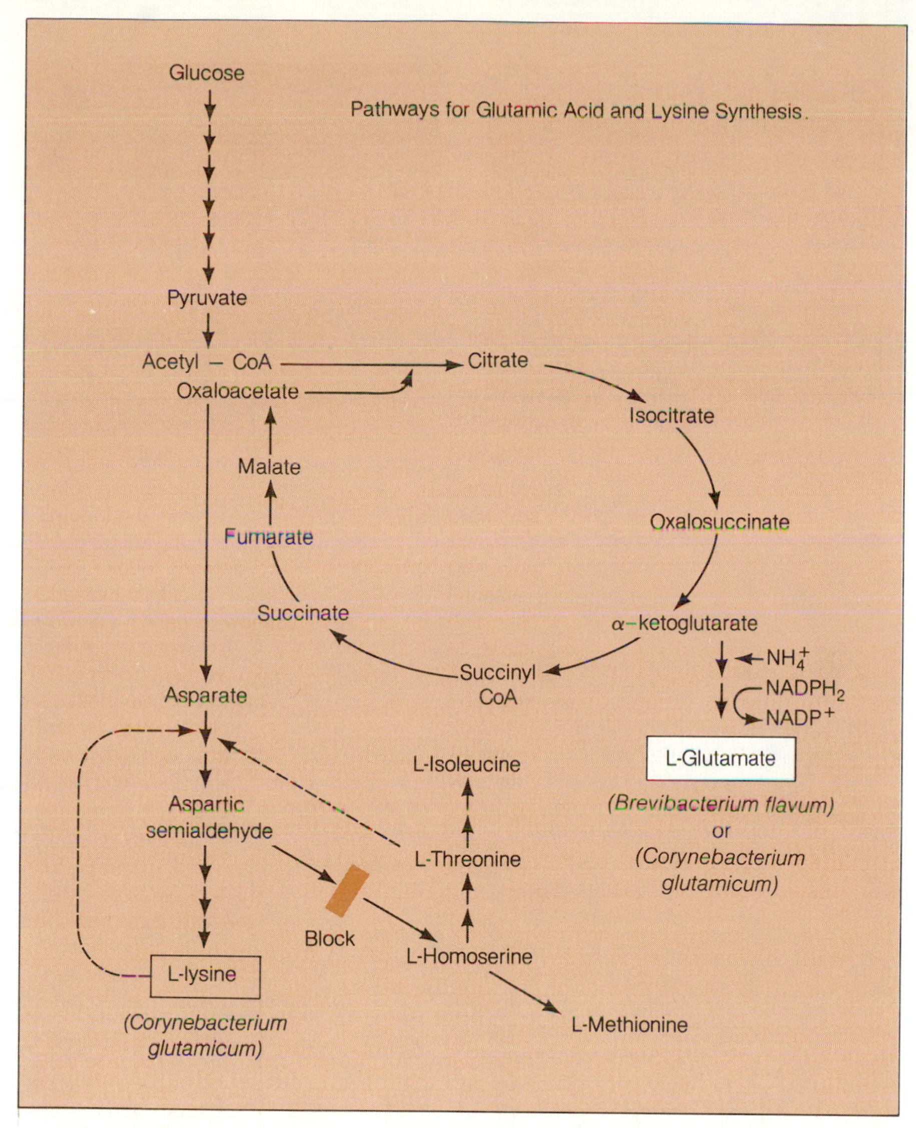

Figure 35-6 Chemical Pathways in the Production of L-Glutamate and L-Lysine. L-glutamate is made commercially using *Brevibacterium flavum* or *Corynebacterium glutamicum*. L-glutamate is made from α-ketoglutarate, one of the intermediates in the Krebs cycle, by reducing α-ketoglutarate and adding an amino group. Monosodium glutamate (MSG) is L-glutamate with a single sodium associated with one of the carboxyl groups. L-lysine is made commercially using a mutated *Corynebacterium glutamicum* that is unable to synthesize L-methionine, L-threonine, or L-isoleucine. Consequently, if this organism is to be grown, it must be supplied with these amino acids or with their precursor, L-homoserine. The mutation prevents L-threonine from accumulating to levels that feedback-repress the first enzyme in the pathway to L-lysine. Thus, L-lysine can accumulate to higher levels in the mutant than in the prototroph. L-methionine is made commercially by chemical synthesis rather than by biological methods.

ties. No doubt, the synthesis of proteins high in methionine will be achieved in the future by genetic engineering of suitable microorganisms.

BEVERAGE AND FOOD PRODUCTION

Winemaking

Wine is the aged product of the alcoholic fermentation of grape juice by yeasts (table 35-3). The term "wine" has been applied loosely to any alcoholic beverage that results from the fermentation of fruit juices by yeasts. Wines have gained tremendous popularity in the United States. It was once believed that "good" wines had to be imported from France, Italy, or Spain; today, however, the states of California, New York, and Washington produce wines of extremely high quality which are comparable to the imported varieties.

Winemaking is a well-controlled industrial process involving several steps (fig. 35-7). Winemaking begins with the crushing and destemming of the grapes. In the United States, this step is usually carried out in September or October, right after the grape harvest. The crushed grapes, called the

TABLE 35-3
PRINCIPAL COMPONENTS OF WINES

INGREDIENTS	MICROBIAL METABOLISM	GRAPE	ADDED
Alcohols	Ethanol 12%. Glycerol <0.5%, acetoin 0.5%, <2,3 butanediol 0.4–1.5%, methanol 0.02% (hydrolysis of pectins)*, isoamyl, isobutyl, n-propyl, n-butyl, n-hexyl, heptyl, nonyl, decyl, etc. 0.03% (300ppm**)		Ethanol 8% (fortified wines)
Carbon dioxide	CO_2 1.5% (champagne)		
Aldehydes	Acetaldehyde 45–90ppm		
Esters	Ethyl laurate, ethyl acetate		
Hydrogen ions	pH = 2.8–4.1 (3.5 average)		
Acids	Acetic, lactic, succinic, formic, propionic, carbonic, valeric, capric, vanillic 0.03–0.5% (300–500ppm)	Tartaric 0.4–0.6% (5000ppm), malic, citric, tannic, phosphoric	
Vitamins	Nicotinic acid 1ppm, pyridoxin 0.8ppm, pantothenic acid 0.45ppm, thiamin 0.25ppm, riboflavin 0.20ppm, folic acid 20ppb**, biotin 5ppb, cobalamin 25ppt.**		
Amino acids	Amino acids 0.01–0.22% (naturally found in grapes & from hydrolysis of proteins)*		
Minerals		K^+ 900ppm, Na^+ 110ppm, Ca^{+2} 60ppm, Fe^{+3} 5ppm, Cu^{+2} 0.3ppm.	
Polyphenols pigments		Tannins 0.03–0.2% (dry white & sherry 0.03%, rose 0.05%, port 0.1%, dry red 0.2%), phenolic acids	
Carbohydrates		Glucose & fructose 0.1–14% (dry reds 0.1–1.5%, champagne 1.5%, sweet white 1–3%, desert wines 2.5–14%)	

Sulfur dioxide	100ppm–150ppm SO_2

*Nonfermentative metabolism
**100mg/L = 100ppm = 0.01% (ppm = parts per million, ppb = parts per billion, ppt = parts per trillion).

must, are treated with sodium metabisulfite (or another form of sulfur dioxide) to kill wild yeasts, bacteria, and fungi. The must consists of grape juice (90–95%), grape skins, and seeds. It is placed in fermentation tanks (usually made out of stainless steel), inoculated with wine yeasts (*Saccharomyces cerevisiae* or *S. ellipsoideus*), and allowed to ferment (anaerobically) under carefully controlled conditions of temperature (27–30°C), and humidity. It is dur-

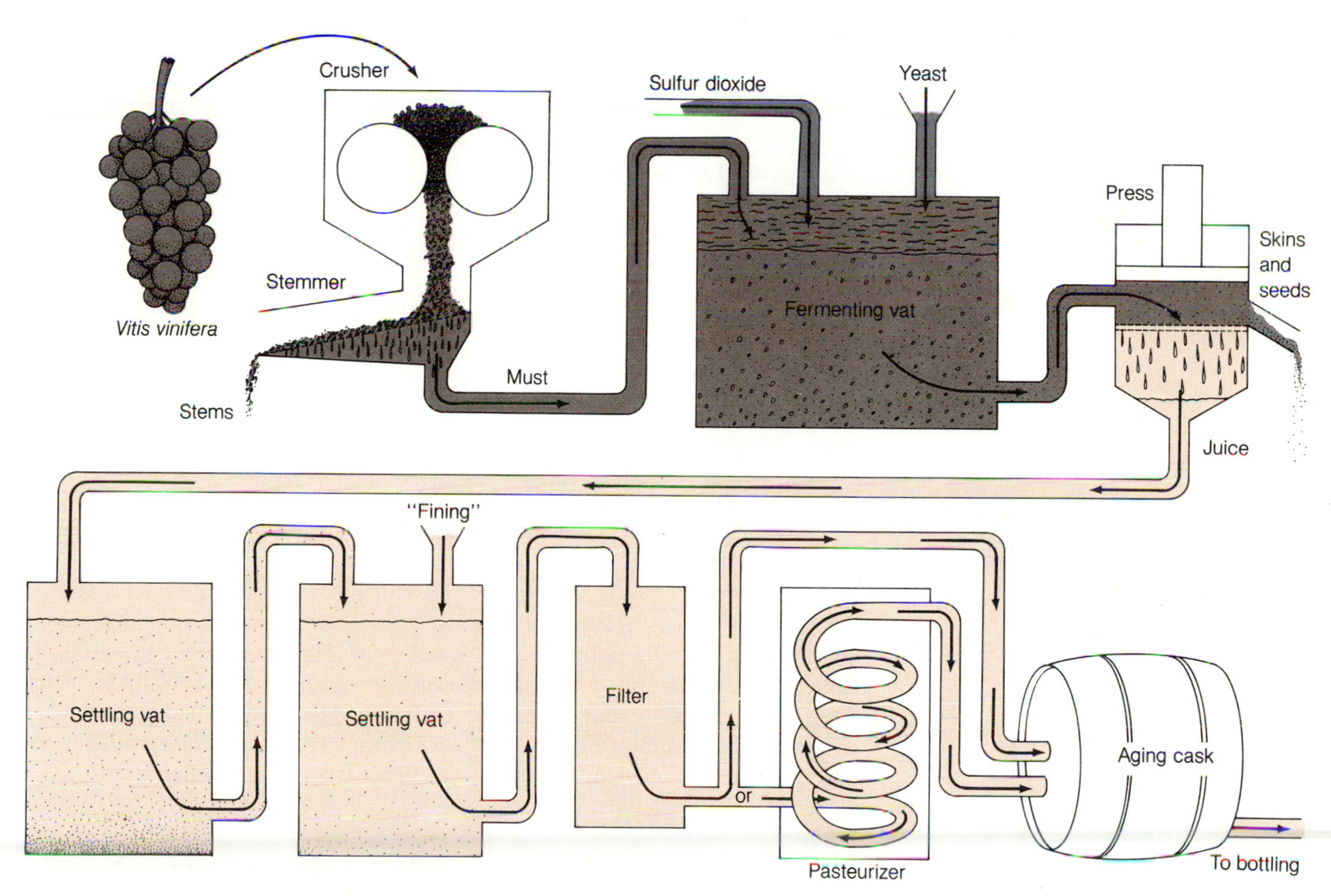

Figure 35-7 Winemaking. White wines are usually made from white grapes, while red wines are made from red grapes. Grapes are crushed to release the juice and the stems are removed. The juice, skins, and seeds (called the must) are treated with sulfur dioxide (SO_2) at about 100 ppm, to inhibit undesirable yeasts and filamentous fungi. The must is transferred to a fermenting vat, where yeasts such as *Saccharomyces cerevisiae* are added. The juice is separated from the grape skins and seeds after a few days of fermentation in the press. The time at which skins and seeds are removed determines the color of the wine and the amount of tannins: the red wines are fermented at about 22°C for 3–5 days with the skins; the light red wines have the skins removed earlier; the white wines are fermented at about 15°C for 7–14 days without the skins. Dry wines contain little or no sugar, while sweet wines contain sugar. Settling vats are used to allow the sedimentation of yeast, proteins, and tannins. The process of adding chemicals to promote settling of particles is called fining. The wine is then filtered and pasteurized. The aging of wine occurs as chemical reactions change some of the alcohol and remaining organic acids.

ing the fermentation step that red, rosé, or white wines are made. The skins and seeds of red grapes contain chemicals called **tannins** (table 35-3), which, if allowed to stay in the fermentation tanks for 5–10 days, will make the wine red and flavor it. If the grape skins are removed shortly after the fermentation begins (1–2 days), the wine is pink and is called rosé wine. If the skins are excluded from the must altogether, white wines will result.

Sometimes, if the sugar content of the grapes is low or if a sweeter wine is desired, sugar is added to the must at the time the yeasts are added. The fermentation is allowed to take place until the desired alcohol concentration is achieved, usually ranging from about 8–9% in sweet wines to about 12–14% in dry wines. Dry wines are those in which all of the available sugar has been fermented to ethanol and carbon dioxide. During the fermentation of most wines, the carbon dioxide is allowed to escape; however, sparkling wines are prepared so that they retain some of the carbon dioxide. The carbon dioxide is responsible for the bubbles in sparkling wines, such as champagne, spumante, and cold duck.

After the fermentation has reached the desired point, the fermented grape juice is clarified (fining), filtered, pasteurized, and placed into casks (stainless steel or oak) for aging. The fermented grape juice is allowed to age for months to years before the resulting wine is bottled. The aging process is essential to winemaking, because it is during this step that the subtle changes or **secondary fermentations** take place, making the fermented grape juice into wine. As a general rule, red wines are aged for much longer periods of time than white wines (2–10 years vs. 1–2 years) before they are bottled. After aging and bottling, the wines are distributed to the consumer or aged further in the bottle. Sparkling wines, for example, are allowed to ferment further in the bottle (secondary fermentation of sugars) so that a small amount of carbon dioxide is produced while in the bottle, giving the wine its sparkle.

Wine varies considerably in color, flavor, and bouquet, depending upon the type of grapes used, the geographic area where the grapes were grown, the strain of wine yeast(s) used, and the care the ferment received during winemaking.

A number of things can go wrong during winemaking which result in distasteful wines. If the final product is not aesthetically pleasing and of desirable taste or bouquet, no one will buy it, and therefore it must be discarded. This outcome can result in severe economic loss, especially in wineries where tens of thousands of gallons of grape juice are fermented each year. Some of the common defects of wine include off flavors, ropiness, sour flavors, turning to vinegar, and the formation of flocculent deposits that make the wine cloudy.

Beer Production

Approximately 18.5 billion gallons of beer are produced each year worldwide. Most beers today are made from barley seeds and a mixture of other grains. Beer-making begins when the barley seeds are mixed with water in large mixing tanks and spread out to make them **malt** (sprout) (fig. 35-8). The malting of the seeds stimulates the production of enzymes such as α-amylase, β-amylase, and amyloglucosidases, which break down starches to maltose and glucose (fig. 35-9). The malted barley is crushed and ground to release the starch-degrading enzymes. This mashed material, called the **mash,** is then transferred to the **mash tun** (fig. 35-8), where the barley enzymes are allowed to degrade the barley starch and proteins. The nondegradable parts of the grain are separated from the aqueous extract, called **malt wort.** The spent grain is made into cattle cakes, while the wort is transferred to a brew kettle, where it is mixed with **hops** and sugar. Hops are dried flower petals from a climbing vine. The mixture is simmered or boiled to extract flavors from the hops and to denature the enzymes in the wort. The **hopped wort,** which is very sugary and full of flavor, is inoculated with a strain of *Saccharomyces cerevisiae* that begins to ferment the sugar into ethanol.

Lager beers are fermented for 8–10 days at 6–8°C, while **ale** beers are fermented for 5–7 days at 14–23°C. As the yeast grows in the hopped wort,

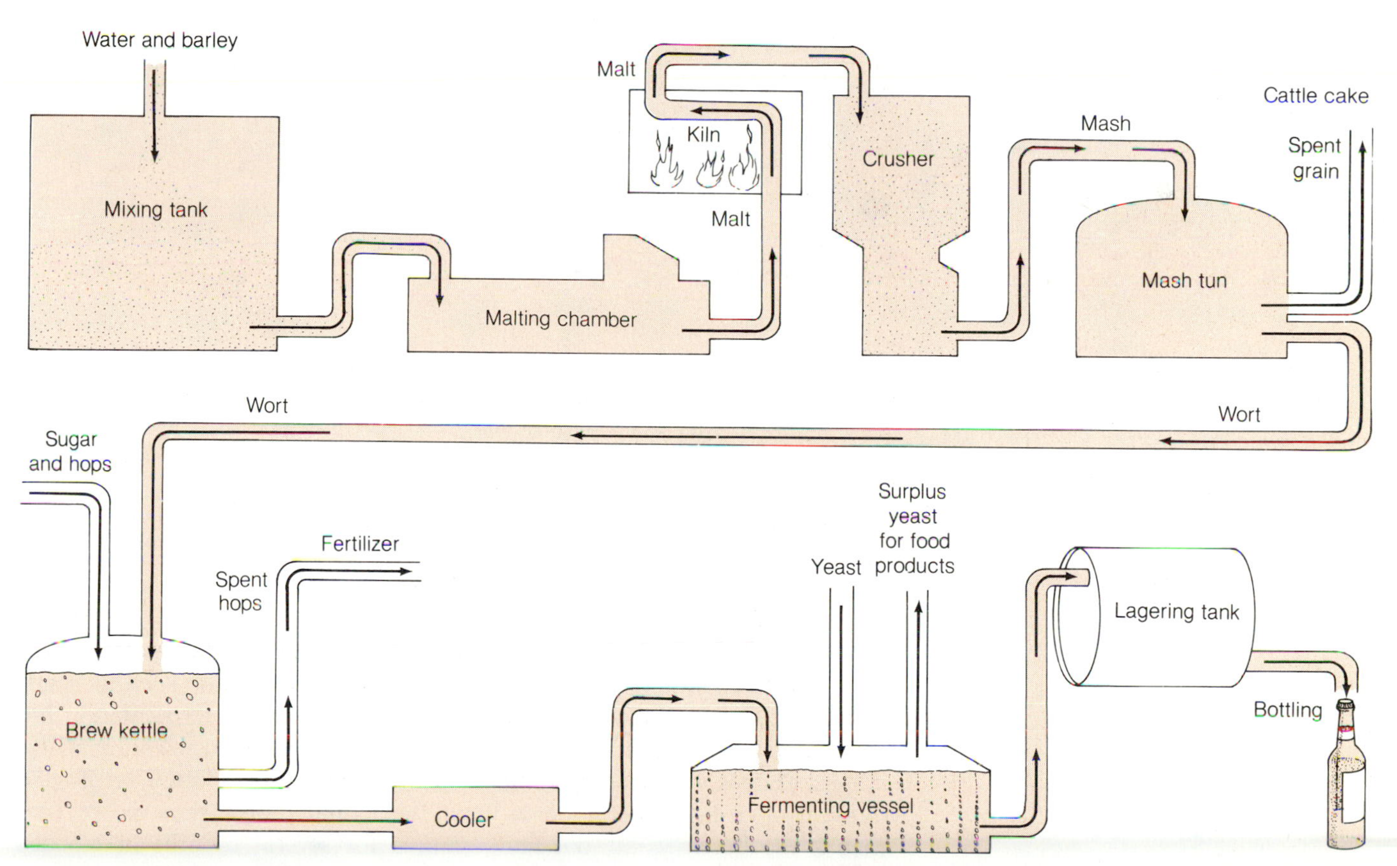

Figure 35-8 Beer Production. Barley grains are soaked in water for 2 days in the mixing tank, then transferred to the malting chamber for 5–7 days, where the grains germinate. During the germination (malting process), the starch in the seeds is partially degraded to maltose and dextrins by the amylases produced during germination. The germinating barley is next transferred to a kiln, where the germinated seeds are dried. Often, the malt is mixed. The malt is subsequently crushed to release the amylases and starch. The crushed malt, called the mash, is transferred to the mash tun, where the starch is converted to maltose. The sugar and dextrin solution, called the wort, is separated from the crushed grain, which is made into cattle cakes. The wort is mixed with sugar and hops to add flavor. The acids, resins, and tannins in hops give beer its bitter taste. After the spent hops are separated from the sugar solution, the sugar and dextrin solution is transferred to a fermenting vessel, and yeasts are added to convert the sugar to ethanol. Lagering tanks store the beer at 6–8°C so that it can become mellow. Mellowing occurs as yeasts, proteins, and other materials sediment and chemical changes take place in the suspended material.

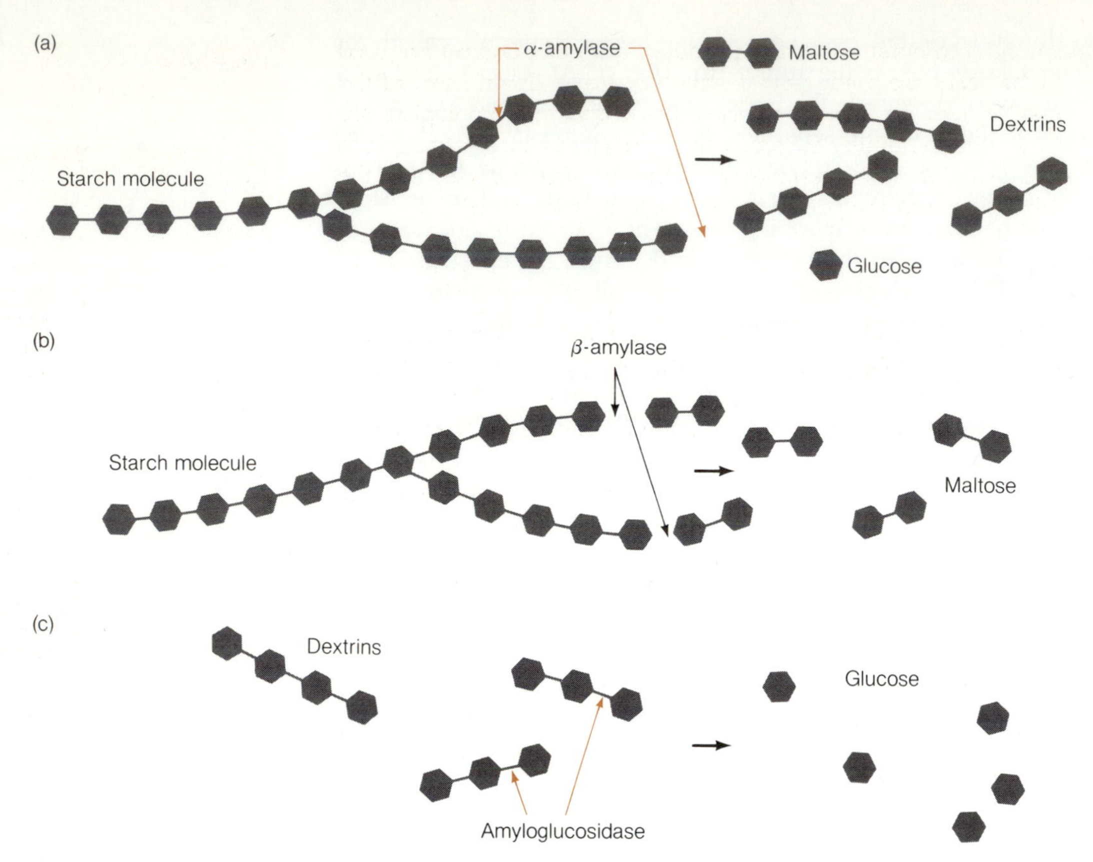

Figure 35-9 Starch Hydrolysis. *(a)* α-amylase splits the starch into dextrins. *(b)* β-amylase splits the starch and dextrins into maltose. *(c)* Amyloglucosidase splits dextrins into glucose. This enzyme is used to break down the dextrins in a mash so that a light beer, low in calories, can be produced.

it releases numerous chemicals that flavor the mixture. Higher alcohols such as amyl, isoamyl, and phenylethyl alcohols, as well as acids such as acetic and butyric acids, add to the taste of the beer. After the fermentation has ended, the beer is separated from the yeast and transferred to tanks, where it ages. The alcoholic content of most lager beers is less than 6%, while that of ale is often higher. Beer has a harsh flavor that can be removed by **maturing** the beverage at cool temperatures for several months in wooden casks. Some of the harshness is due to the higher alcohols, which are oxidized or esterified during the aging. Other unpalatable substances such as peptides, resins, and yeast precipitates are produced during the maturing period and can easily be separated from the beer. The beer is **finished** by filtering, pasteurizing, and carbonating the beverage.

Lager beers are made with yeasts that ferment on the bottom of the fermentation vats. These **bottom yeasts** are called *Saccharomyces carlsbergensis*, after the Carlsberg Institute in Copenhagen, where they were first isolated. Top-fermenting yeasts, such as *Saccharomyces cerevisiae*, carry out

[189] a vigorous alcoholic fermentation □ that produces a lot of CO_2 and are used to make ale. Light beers are made with yeasts that break down most of the starch and sugar and so are low in calories. Presently, microbiologists are attempting to transfer the ability to break down dextrans in the beer from *Saccharomyces diastaticus* to *S. cerevisiae*, without also transferring the closely linked ability to ruin the beer with foul-tasting substances.

Sake

The making of sake (rice wine) resembles beer-making. Steamed rice and *Aspergillus oryzae* are incubated together at 35°C for approximately six days. *Aspergillus* growing in the rice breaks down the rice starch to sugars that can be fermented by yeasts, which are then added. The *Aspergillus oryzae–Saccharomyces cerevisiae*–rice mixture is a starter culture, used to begin the fermentation of large batches of steamed rice. The fermentation proceeds for approximately three weeks, during which time the alcohol content often reaches 20%.

Distilled Liquors

The ethanol in distilled liquors such as whiskey, rum, and vodka is made from a number of starting materials, such as barley, sugar cane, and potatoes (table 35-4). What makes distilled liquors different from beer and wine is that the ethanol solution is distilled in order to increase its concentration to 30–45%. Beer contains 3–7% ethanol, while wine usually contains about 12% ethanol.

The initial steps in the production of distilled liquors resemble those followed in making beer or wine (fig. 35-10). If a grain is used, it is allowed to **malt** (sprout) so that amylases are produced. Then the grain is mashed to break the starches down to glucose and maltose. The wort from the **mash** is subsequently transferred to a fermenting vat, where yeasts are added. If potatoes are used, a small amount of malted grain is added and the material is mashed. The mashed grain supplies the amylases necessary to break down the potato starch efficiently.

After the yeasts have completed their fermentation, the ethanolic solution (beer or wine) is heated in a still to eliminate some of the water. Because alcohol has a lower boiling temperature than water, more alcohol vaporizes than water; therefore, the vapor condenses to a liquid that contains an increased concentration of ethanol, as high as 95%. Most liquors, however, have an alcoholic concentration of 30–45%.

A number of volatile compounds in the ethanolic solution distill along with the ethanol. These compounds are partly responsible for the characteristic tastes of the various liquors. The manner in which the malt is prepared also contributes to a liquor's unique flavor. For example, the barley malt that is used to make "true" Scotch whiskey is kiln-dried at a high temperature over a peat moss fire. The aging process adds the final flavors to a liquor.

Tempeh and Sufu

Soybean, peanut, or coconut fermentations improve the taste and flavor of these seeds and also increase their protein content because of the growth of

TABLE 35-4
LIQUORS AND MATERIALS USED TO MAKE THEM

LIQUOR	MATERIAL FERMENTED
Applejack	Apple juice
Aquavit	Grain or potatoes, flavored with caraway seeds
Bourbon	Corn
Brandy	Various fruits
Canadian whiskey	Rye
Cognac	White grapes produced in the Cognac region of France
Gin	Grain mash, flavored with juniper berries, anise, etc.
Irish whiskey	Malt, unmalted barley, wheat, rye, and oats
Kirschwasser	Cherry juice
Rum	Sugar cane
Scotch whiskey	Barley
Tequila (Mescal)	Juice from the hearts of the cactus *Agave tequilana*
Vodka	Grain or potatoes, not flavored

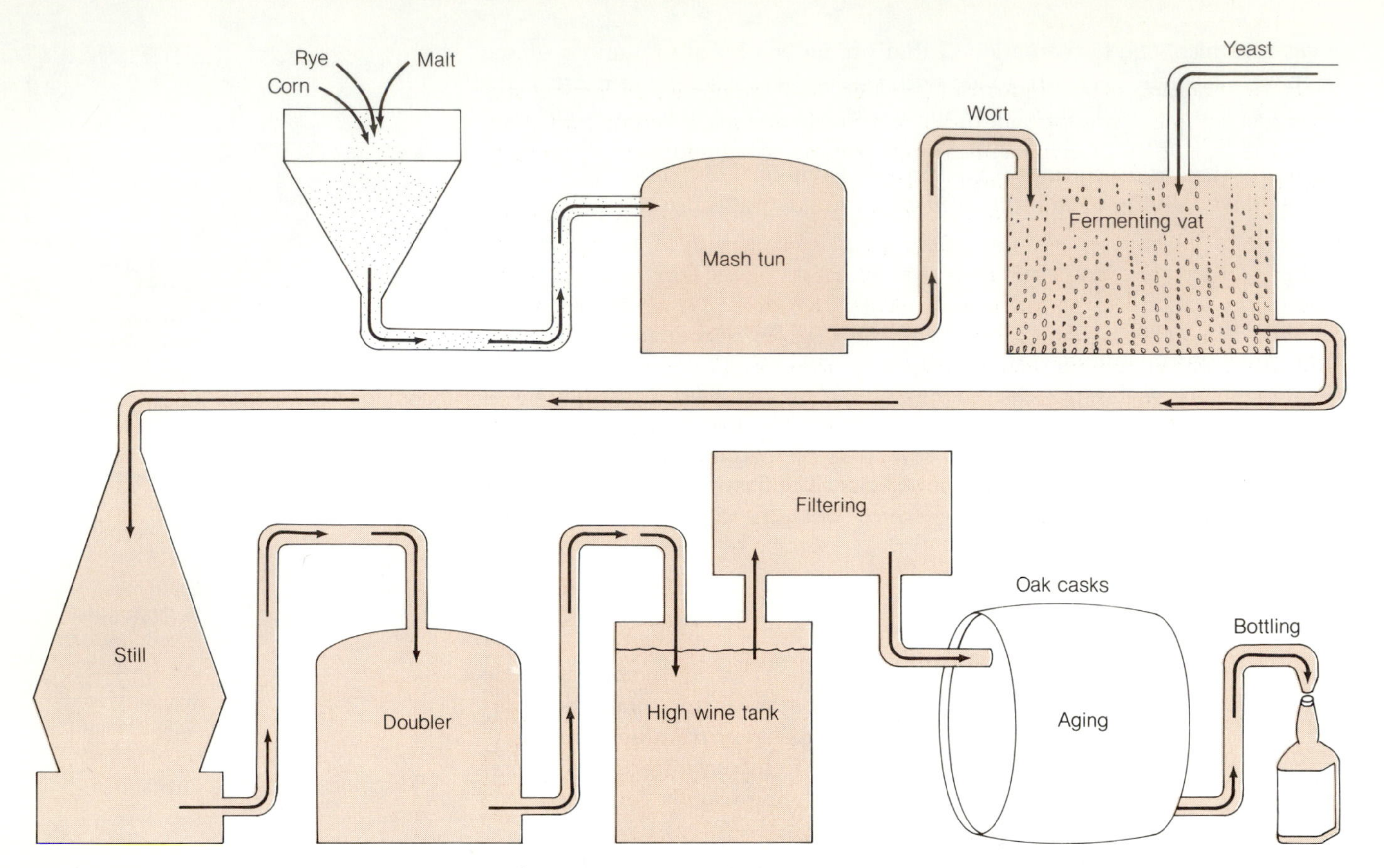

Figure 35-10 Distilled Liquors. Distilled liquors are made from fruit juices, grains, vegetables, sugar cane syrup, or molasses. The production of alcohol from these starting materials proceeds much like the making of wine or beer. After a fermentation in the vat, the alcohol is distilled to increase the alcohol content to as much as 95%. Distilled liquors are generally categorized as brandies, whiskeys, neutral spirits, and rums. Brandies are made from various fruit juices; whiskeys are made from malted cereal grains; neutral spirits are made from vegetable or grain starches; and rums are made from sugar cane syrup or molasses.

microorganisms. Tempeh is a cake of ground seed and mycelium produced by the fungus *Rhizopus.* Tempeh is prepared by soaking the seed in water, removing the seed coverings, boiling, draining, mashing, and then adding a starter culture from a previous batch of tempeh. The mash is incubated at 20°C in trays until the fungus has grown throughout the material. Tempeh usually is not served "raw," but is deep-fried in oil. Sufu is made by fermenting soybean curd with fungi such as *Mucor.*

Soy Sauce

Soy sauce, used as a flavoring in a variety of oriental dishes, is made by allowing the fungus *Aspergillus oryzae* to begin the fermentation of a salted, mashed mixture of soybeans and wheat. This material is then fermented for 8–12 months by the bacterium *Pediococcus soyae* and the yeasts *Saccharomyces rouxii* and *Torulopsis* sp., at low temperatures and with agitation. When the fermentation is complete, the resulting mass is pressed. The liquid or soy sauce that comes from the press contains numerous acids, the most prominent of which is lactic acid, and alcohols such as ethanol.

TOXINS FOR VACCINES

477 Vaccines □ are important industrial products that are needed in large quantities and in constant supply in order to provide adequate health care. Vaccines are made by pharmaceutical companies, using industrial processes similar to those used to make antibiotics.

Toxins for making the tetanus and the diphtheria toxoids are obtained by growing *Clostridium tetani* and *Corynebacterium diphtheriae*, respectively, in large fermenters. The organisms are cultivated under conditions that enable them to produce the toxins and release them into the growth medium. The bacterial cells are then removed from the growth medium by centrifugation, and the remaining toxin is converted into toxoid by treating the material with 0.5% formalin until the toxin shows no activity when injected into test animals. The toxoid is purified by precipitating it with ammonium sulfate and by chromatography. The purified toxoid is often adsorbed to aluminum hydroxide or aluminum phosphate, which makes it a more potent 462 antigen □. The vaccines are then tested for their activity, bottled, and distributed.

TEXTILE PRODUCTION

Microorganisms are used in the preparation of plant fibers from flax and hemp, which are used to make textiles and rope. Fiber bundles in the stem cortex of flax and hemp can be freed from the plant stem by **retting,** which involves the digestion of the polysaccharide pectin that holds the cortex together. Many bacteria and fungi produce the enzymes that break down pectin. **Anaerobic retting** is carried out by bacteria such as *Clostridium felsineum* and *C. pectinovorum* in tanks of media that support growth of these organisms. **Aerobic retting** is carried out by fungi and bacteria in well-aerated tanks.

FERMENTERS

All industrial processes involving microorganisms are called "fermentations," even though they may involve respiring microorganisms. The commercial production of alcoholic beverages, antibiotics, and amino acids are examples of industrial processes that are carried out in large fermentation tanks (fermenters), containing as many as 20,000 to 60,000 gallons of media. Each time one of these tanks is filled with media and inoculated, there is a considerable investment in time and money. It is the microbiologist's job to insure that the tank and the media are sterile before the fermentation begins and that the inoculum consists of only the microorganisms that are necessary to carry out the desired fermentation. If contaminating organisms get into the tanks, they could ruin a very valuable product by producing toxic or unpalatable contaminants. Tanks 100 feet high and all the connecting pipes must be made sterile and kept free of contaminating microorganisms (fig. 35-11). Also, thousands of gallons of growth media and large volumes of air must be made sterile. Finally, the inoculum must be kept free of contaminating microorganisms.

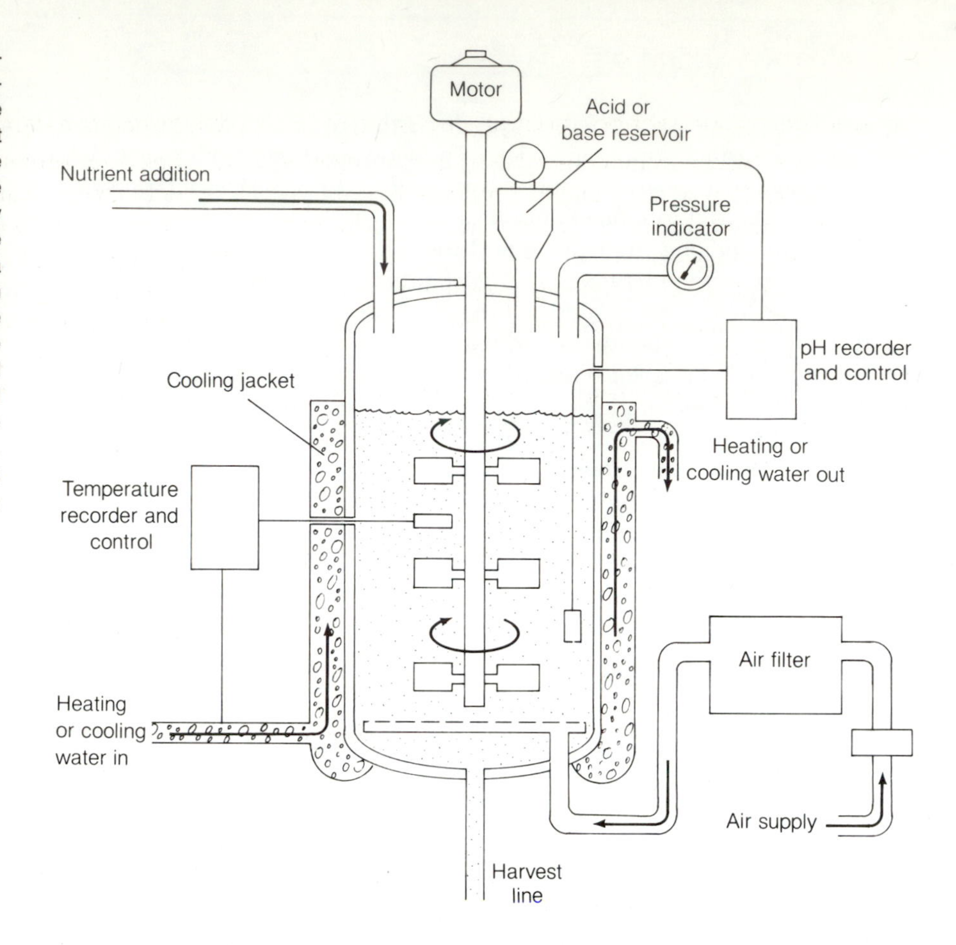

Figure 35-11 Fermenter. A batch fermenter is used to grow many industrially important microorganisms. The chamber of the fermenter and all lines leading into and out of it are sterilized with steam. The sterile medium enters the fermenter through the nutrient addition line. Microorganisms may be added to the nutrient before it enters the chamber, or they may be added through a port. Filtered air, if required, enters through the air supply line. Paddle wheels mix the nutrients and microorganisms to insure rapid growth. The temperature is controlled by the flow of warm or cold water around the fermenter. The pH is controlled by the release of acid or base from a reservoir. The microorganisms and any substances they have released are harvested through the harvest line.

Figure 35-11 illustrates a typical fermentation tank. Steam passing into the tank under pressure is used to sterilize it. Autoclaved medium enters the tank and material is harvested from the bottom. Air, if necessary for an aerobic process, passes through air filters and then into the bottom of the tank. Propellers within the tank help to aerate the organisms or move them around so that they do not settle to the bottom, where they would use the nutrients inefficiently. Temperature sensors automatically record and control the temperature of the tank by controlling the circulation of cooling or heating water. Sensors automatically record and control the pH by adding base of acid to the medium. The microbiologist is responsible for maintaining pure cultures of the microorganisms to be used and for inoculating the tanks without contaminating the media.

There are two basic types of industrial fermentations that use these tanks: **batch fermentations** and **continuous process fermentations.** In batch fermentations, such as winemaking, the fermentation tanks are used for the process and then the tanks are drained, cleaned, sterilized, and refilled for the next batch. In continuous process fermentations, such as antibiotic synthesis, the fermenters are used for extended periods of time, during which the product and microorganisms are continuously harvested and replaced with fresh, sterile growth medium. Occasionally, the fermenters are drained, cleaned, and sterilized before they are used again.

SUMMARY

THE IMPORTANCE OF MICROORGANISMS IN INDUSTRY

1. Microorganisms are now being used in a number of industries to produce valuable products such as enzymes, polysaccharides, amino acids, hormones, and monoclonal antibodies.
2. Microorganisms are also being used to degrade toxic chemicals, disintegrate oil spills, serve as pesticides, and aid in mining.
3. Enzymes are used to tenderize meat, prepare leathers, produce detergents, and make cheeses.
4. Polysaccharides are used to stabilize and thicken foods, as part of drilling muds, as cosmetic bases, as binding agents in many pharmaceuticals, as extenders of blood plasma, and as molecular sieves.
5. Hormones such as insulin and growth hormone are used to treat persons who have genetic defects.

THE PRODUCTION OF PHARMACEUTICALS

1. The fungi *Penicillium* and *Cephalosporium* produce the β-lactam antibiotics penicillin G and cephalosporin C, respectively. Bacteria in the genus *Streptomyces* also produce cephalosporins.
2. Penicillins and cephalosporins inhibit the synthesis and assembly of peptidoglycan in growing bacteria, by attaching to at least three enzymes that catalyze steps in the synthesis of the cell wall. The enzymes inhibited are transpeptidase, carboxypeptidase, and endopeptidase.
3. Thienamycins are antibiotics produced by various genera of *Streptomyces* that inhibit bacterial enzymes (β-lactamases) that inactivate the β-lactam antibiotics.

THE MANUFACTURE OF STEROIDS

1. Steroids are used in medicine to relieve pain, to reduce swelling, in treatment of hormone imbalances, and as contraceptives.
2. Microorganisms are used to make steroid hormones from plant steroids, because bioconversions are less costly than chemical conversions.

PROTEINS MADE IN GENETICALLY ENGINEERED BACTERIA

1. Genes for human insulin and interferons have been introduced into bacteria. The insulin and interferon produced in bacteria are cheaper than those produced from other sources.
2. The genes for a number of viral proteins have been introduced into bacteria. The viral proteins produced are useful as vaccines against the viruses.

INDUSTRIAL CHEMICALS AND ENZYMES

1. Enzymes produced by microorganisms are used as detergents, cleaning aids, meat tenderizers, and nucleases, and to make corn syrup.
2. Alcohols produced by microorganisms are used as solvents, antifreeze, fuel (gasohol), and substrates for vinegar production.
3. *Acetobacter* species are used to make vinegar from ethanol under aerobic conditions.
4. *Corynebacterium* and *Brevibacterium* are used to make monosodium glutamate from glucose, acetic acid, or paraffins. A mutant form of *Corynebacterium* is used to make the animal feed additive L-lysine.

BEVERAGE AND FOOD PRODUCTION

1. Wines are generally made from the juice of grapes.
2. Yeasts of the genus *Saccharomyces* ferment the grape sugar, turning it into ethanol.
3. The skins and seeds from red grapes contain tannins and pigments that turn the ferment red.
4. If the skins and seeds are totally excluded from the fermentation, white wine results.
5. Beers are generally made from sprouted barley seeds (malt). The malt provides starch and enzymes that turn the starch into sugar. An aqueous extract (wort) of glucose and maltose is separated from the spent grain and mixed with hops (dried flowers). Boiling of the mixture extracts flavors from the hops and denatures the enzymes in the wort. Eventually, the hops are removed from the wort, which is inoculated with yeasts belonging to the genus *Saccharomyces*. The yeasts ferment the sugar to ethanol and CO_2.
6. The higher alcohols, such as amyl, isoamyl, and phenylethyl, and acids such as acetic and butyric give fresh beer a harsh taste. This taste is removed by aging, which oxidizes and esterifies the compounds. Harsh-tasting particulate matter precipitates during aging and can be removed.
7. Distilled liquors are made from various types of grain, fruit juices, or sugar extracts. After a beer or wine is produced from the sugar by yeast, the mixture is distilled to increase the alcoholic content. While most beers contain less than 6% ethanol and wines contain 9–12% ethanol, most distilled liquors contain 30–45% ethanol.

8. Tempeh is made from a ground soybean mash upon which the fungus *Rhizopus* is grown. Sufu is made by using the fungus *Mucor* to ferment a soybean curd.
9. Soy sauce is made by the fermentation of a soybean and wheat mixture by *Aspergillus oryzae.*

TOXINS FOR VACCINES

1. A number of bacteria, such as *Corynebacterium diphtheriae* and *Clostridium tetani*, are used to obtain toxins.
2. These toxins are inactivated and then used as vaccines to protect against the diseases caused by the toxins.

TEXTILE PRODUCTION

1. Plant fibers from flax and hemp are used to make textiles and rope.
2. Species of *Clostridium* and certain fungi are used to separate the fibers from the stem cortex.
3. The bacteria and fungi digest the pectin that holds the cortex together. The separation process is known as retting.

FERMENTERS

Industrial processes involving the growth of microorganisms generally takes place in fermenters. An understanding of microbiology is important in the operation of a fermenter. The tank and media must remain sterile until the fermentation begins, and the inoculum must be a pure culture. Any air (oxygen) used in the fermentation must be sterile.

STUDY QUESTIONS

1. What is the difference between penicillin G and semisynthetic penicillins?
2. How do the penicillins and cephalosporins differ chemically and in the manner in which they affect bacteria?
3. How do the thienamycins affect bacteria, and what value might they have in treating penicillin- or cephalosporin-resistant bacteria?
4. Explain how microorganisms that produce antibiotics can be discovered.
5. For what purposes are steroid hormones used?
6. Why is the synthesis of steroid hormones cheaper when bioconversions are used rather than chemical conversions?
7. Which human proteins are being produced in genetically engineered bacteria?
8. Name 5 of the useful bacterial proteins being produced in large amounts?
9. Which microorganisms are used to make ethanol, vinegar, monosodium glutamate, and L-lysine?
10. What materials and microorganisms are generally used to make wines?
11. What procedure accounts for red, rosé, and white wines?
12. What materials and microorganisms are generally used to make beer?
13. Explain what barley malt is, and what part it plays in beer production.
14. Explain why hops and sugar are added to malt during the beer-making process.
15. What is the difference between lager beers and ales?
16. Explain the difference between tempeh and sufu.
17. Outline the steps in soy sauce production.
18. Explain how toxins can be of value to humans.
19. Explain how thousands of gallons of media and a large fermenter are sterilized.
20. Explain how air that enters the fermenter is sterilized.
21. Explain the difference between a batch fermentation and a continuous process fermentation, and in which situations they are used.

SUPPLEMENTAL READINGS

Abraham, E. P. 1981. The beta-lactam antibiotics. *Scientific American* 244(6):76–86.

Aharonowitz, Y. and Cohn, G. 1981. The microbiological production of pharmaceuticals. *Scientific American* 245(3):140–152.

Demain, A. L. and Solomon, N. A. 1981. Industrial microbiology. *Scientific American* 245(3):66–75.

Eveleigh, D. E. 1981. The microbiological production of industrial chemicals. *Scientific American* 245(3):154–178.

Gaden, E. L. 1981. Production methods in industrial microbiology. *Scientific American* 245(3):180–196.

Hopwood, D. A. 1981. The genetic programming of industrial microorganisms. *Scientific American* 245(3):90–102.

Phaff, H. 1981. Industrial microorganisms. *Scientific American* 245(3):76–89.

Rose, M. J. 1981. The microbiological production of food and drink. *Scientific American* 245(3):126–138.

Webb, A. D. 1984. The science of making wine. *American Scientist* 72:360–367.

Woodruff, H. B. 1980. Natural products from microorganisms. *Science* 208:1225–1229.

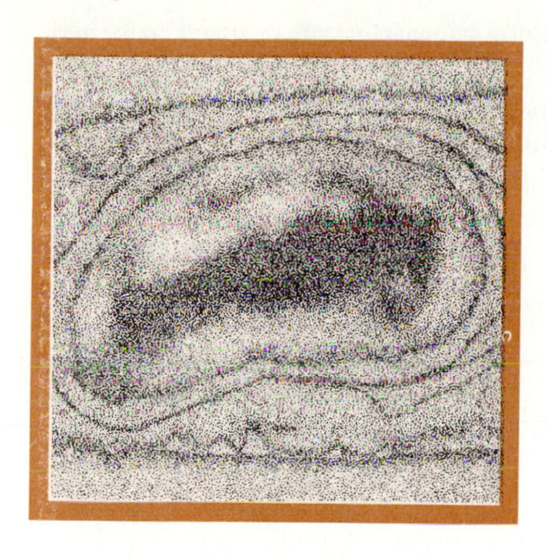

CHAPTER 36
NOVEL USES OF MICROORGANISMS

CHAPTER PREVIEW

DANGERS ASSOCIATED WITH THE USE OF MICROORGANISMS?

Before microorganisms are used in medicine or in nature, they should be carefully tested so as to avoid the outbreak of diseases or adverse environmental changes. Often there is a lack of knowledge among scientists concerning the safe use of microorganisms. A case in point is the development of a herpes vaccine from live attenuated herpes viruses. In the development of a vaccine, its safety and its effectiveness are of paramount importance.

Past experience has shown that immunity generated by live attenuated vaccines generally lasts longer and is stronger than that induced by killed vaccines or even by purified protein vaccines. Often, vaccines made from heat-killed organisms or from purified proteins result in little or no protection. Given only this information, it would seem that a herpes vaccine should be developed from attenuated viruses. Since there is strong evidence indicating that herpes viruses cause cervical carcinoma in some women, however, it would be very dangerous to administer attenuated herpes viruses or even heat-killed viruses to women, because this treatment could cause a rise in the number of cases of cervical carcinoma. At this time, no one knows the risk that a person runs when viral DNA is introduced into his or her body.

A herpes vaccine developed from recombinant DNA expressed in a bacterium would not cause cervical carcinoma, because there would be no herpes DNA in the vaccine. Herpes DNA introduced into a bacterium such as *E. coli* would be used to make one of the virus's surface antigens. The surface antigen would be separated from the *E. coli* and herpes DNA and used to make a vaccine. The cost of preparing the recombinant DNA vaccine would be much less than the cost of growing and preparing herpes viruses in tissue culture. The recombinant DNA vaccine could be made in fermentation tanks, and yields would amount to hundreds of milligrams per liter. Scientists preparing to make the herpes recombinant DNA vaccine claim that they will be able to make it into an effective vaccine by developing an appropriate adjuvant.

Time will tell whether an effective vaccine can be made by recombinant DNA methods. In any case, the herpes vaccine problem demonstrates one of the dangers that might exist when microorganisms are used in a novel way. When reading about some of the novel uses of microorganisms, in this chapter, try to envision problems that might arise from their use.

LEARNING OBJECTIVES

A STUDY OF THIS CHAPTER SHOULD ENABLE YOU TO:

EXPLAIN HOW MICROORGANISMS ARE USED AS INSECTICIDES

EXPLAIN HOW BACTERIA MAY BE USED TO CLEAN UP OIL SPILLS

UNDERSTAND HOW BACTERIA MAY BE USED TO ENHANCE OIL DRILLING AND OIL PRODUCTION

DESCRIBE HOW BACTERIA MAY BE USED TO ELIMINATE PESTICIDES AND OTHER CHEMICAL WASTES

EXPLAIN HOW MICROORGANISMS MIGHT BE USED IN WARFARE

OUTLINE HOW THE METHANE BACTERIA MIGHT BE USED TO PRODUCE ENERGY

During the last decade, the fact that microorganisms can be used to benefit humans, animals, and plants has been amply illustrated. For example, researchers in all areas of biology are attempting to find microorganisms whose association with plant roots will protect the plants from viral, bacterial, and fungal pathogens. Many geneticists are attempting to mutate nitrogen-fixing bacteria so that they will associate with the cereals (wheat, barley, rice, corn, etc.), thus allowing these crops to grow in nitrogen-deficient soil in underdeveloped countries that cannot afford fertilizers. Agricultural engineers are working on projects that will allow the economic production of methane gas from animal wastes, thus providing an inexpensive source of energy. Cheap methane could be used to heat animal barns and run machinery and consequently help reduce the cost of animal products. Biologists and agriculturalists are looking for biological insecticides that will kill the pests that destroy billions of dollars worth of food each year. We have discussed in previous chapters how microorganisms are now being used to mine copper and uranium from low-grade ores and how some organisms may be used to remove ions and organic matter from polluted well water, streams, and lakes. In this chapter, a number of novel uses for microorganisms will be examined and their usefulness to humans, animals, and plants evaluated.

BIOLOGICAL PESTICIDES

Bacillus Thuringiensis

The bacterium *Bacillus thuringiensis* has been used successfully to kill many different insect larvae (moths, butterflies, and mosquitoes) that harm trees and crops, destroy packaged seeds, grains, and flour in warehouses, food stores, and seed storage houses, and infest animals and humans. The bacterium appears to kill insect larvae in two ways. Vegetative cells of *B. thuringiensis* growing in the insect's gut penetrate the gut and grow in other parts of the body. Bacterial proteases and a chitinase contribute to the destruction of the larvae; gut destruction and bacteremia usually kill the insect within 3–7 days. *Bacillus thuringiensis* strains produce at least two endotoxins that affect insect larvae. One of the toxins is associated with the endospore coat, while the other is found in **parasporal bodies** (fig. 36-1). The toxic material in parasporal bodies is a crystalline protein. In some insects, the parasporal bodies cause a paralysis that blocks feeding and a disintegration of midgut epithelium within a day of being contaminated. These effects lead to the death of the larvae, sometimes within a few hours after gut disintegration.

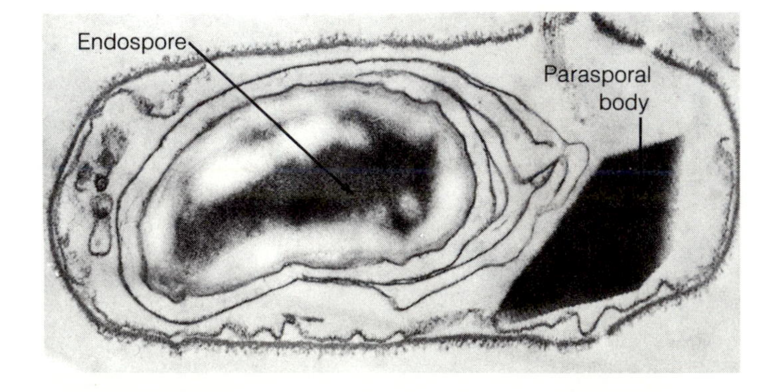

Figure 36-1 Parasporal Body and Endospore. A vegetative cell of *Bascillus thuringiensis* may contain an endospore and a parasporal body. The parasporal bodies, also called crystals, contain a protein endotoxin that is lethal to many insect larvae.

Moth larvae destroy tons of stored grains and seeds each year worldwide. It has been found that as few as 10^4 *B. thuringiensis* endospores concentrated in one location are sufficient to inhibit insect destruction of 200 grams of packaged food. More than 10^9 endospores/200 g of food are needed, however, if the endospores are spread evenly throughout the food. It is believed that, when a larva eats endospore-contaminated materials, the endospores enter the insect's gut and germinate there. The bacteria replicate in the gut and are released in the insect's excrement, thus further contaminating the insects' environment. Other larvae, eating food contaminated by the diseased larvae, pick up great numbers of endospores, parasporal bodies, and bacteria. *Bacillus thuringiensis kenyae* has been found to kill large numbers of moth larvae within a week in warehouses where corn is stored.

"Purified" parasporal bodies (99.5% pure with <0.1% spores) from *B. thuringiensis alesti,* used at a concentration of 40 mcg/770 mm^2 of diet surface, induce 99% mortality in cabbage looper larvae *(Trichoplusia ni)* (fig. 36-2). The addition of antibiotics to the "purified" parasporal bodies, to block germination of endospores, demonstrates that bacterial growth in the midgut and bacteremia are not the cause of death. The crystalline protein of the parasporal bodies, rather than an endotoxin associated with the small number of contaminating endospores, appears to be responsible for the death of the larvae.

Studies indicate that commercial preparations of *B. thuringiensis* endospores and crystals used on loopers infecting cotton give better and more economical control over the loopers than the insecticides phosdrin, sevin, DDT, or toxaphene-DDT (2:1). A little more than 2.5 quarts/acre of *B. thuringiensis* preparation results in about 30,000 crystals/mm^2 and 20,000 endospores/mm^2 of leaf surface. This dosage kills more than 90% of the larvae within 5 days. Those larvae that survive are sickly and grow very poorly (fig. 36-3).

In addition to bacteria that destroy insect larvae, many viruses have been discovered that are useful as pesticides. The virus known as *Baculovirus heliothis* has been found to be very destructive to the cotton bollworm *(Heliothis zea).*

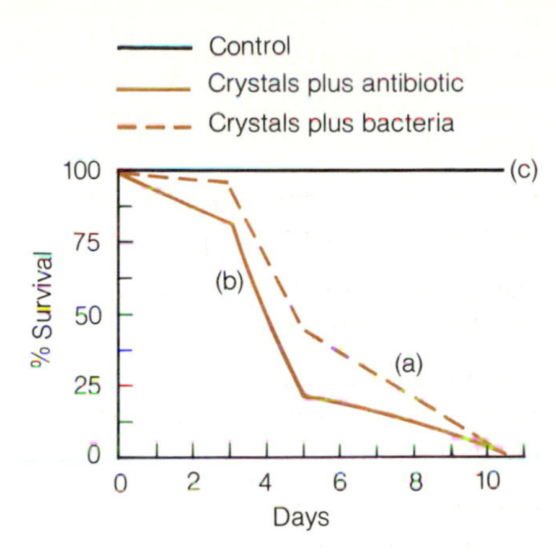

Figure 36-2 Effect of Parasporal Bodies (Crystals) on Cabbage Looper Larvae. *(a)* "Purified" parasporal bodies and vegetative cells of *B. thuringiensis* were spread on the surface of cabbage leaves. Ninety-nine percent of the cabbage looper larvae that fed on the inoculated diet (0.1 ml parasporal preparation/770 mm^2 diet surface) died within 10 days. *(b)* "Purified" parasporal bodies, and an antibiotic to inhibit the germination of contaminating spores, were spread on the surface of the diet. Ninety-nine percent of the cabbage looper larvae that fed on the inoculated diet died within 10 days. *(c)* In a control experiment, cabbage looper larvae were fed uninoculated diet. No larvae died during the course of the experiment.

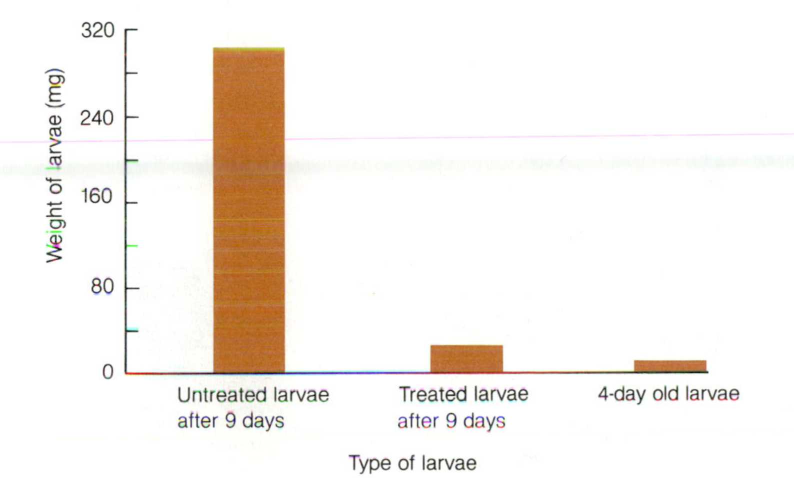

Figure 36-3 Comparison of Cabbage Looper Control Larvae and Larvae Fed on Diet Inoculated with Endospores and Parasporal Bodies. Control 4-day-old larvae, after feeding on uncontaminated diet for 5 days, had an average weight of 286 mg. Those 4-day-old larvae that survived for 5 days after feeding on diet contaminated with endospores and parasporal bodies (33,600 crystals/mm^2) had an average weight of 25 mg. This difference in weight indicates that the larvae feeding on the treated leaves were not able to feed normally. Normal 4-day-old larvae have an average body weight of 10.4 mg, while normal one-day-old larvae have an average body weight of 0.45 mg.

Sunlight (visible and ultraviolet), leaf temperature, and moisture all affect the viability or field persistence of endospores and viruses. Field persistence for *B. thuringiensis* is greater in hot, dry climates than in cool, humid environments. The type of plant and differences in commercial preparations also produce a wide variation in persistence. These insecticides are valuable control agents because they can kill more than 90% of the larvae, they are specific, and they are not toxic or pathogenic to animals or humans. Table 36-1 lists some of the plant pathogens that have been effectively treated with biological pesticides.

Mosquito and Black Fly Killer

Mosquitoes and black flies are insect vectors □ for diseases that cause a great deal of human suffering and death. Malaria is estimated to affect 200 million persons and to kill 1 million each year worldwide; river blindness afflicts more than 50 million, mostly in Africa; and trypanosomiasis affects more than 10 million in Africa and the Americas. In light of these representative morbidity and mortality figures, it is clear that insecticides and ecological

536

TABLE 36-1
BIOLOGICAL PESTICIDES

PEST ORGANISM/DISEASE	PESTICIDE
Pineapple root knot nematode (Hawaii)	Nematode-trapping fungi *(Arthrobotrys dactyloides)*
Silkworm/Pebrine	*Microsporidia* (protozoan)
Colorado beetle (Russia)	*Beauveria bassiana* (fungus)
Codling moth	*Beauveria bassiana*
Corn earworm	*Beauveria bassiana*
Citrus rust mite (Russia)	*Hirsutella* (fungus)
Leaf hoppers (Brazil)	*Metarrhizium* (fungus)
Honeybee/America & Europe foul broods	*Bacillus larvae* (bacterium) *Bacillus alvei*
Locusts & grasshoppers	*Enterobacter aerogenes*
Japanese beetle/milky white disease	*Bacillus popilliae* *Bacillus lentimorbus*
Mediterranean flour moth	*Bacillus thuringiensis*
Alfalfa caterpillar	*B. thuringiensis*
Cotton boll worms	*B. thuringiensis*
Corn earworms	*B. thuringiensis*
Cabbage worms	*B. thuringiensis*
Tobacco budworms	*B. thuringiensis*
Orange fruit tree	*B. thuringiensis*
Leaf rollers	
California oakworm	*B. thuringiensis*
Gypsy moth	*B. thuringiensis*
Mosquito (vector for malaria)	*B. thuringiensis*
Black fly (vector for river blindness)	*B. thuringiensis*
Corn earworm	Baculovirus
Soybean pod-feeding caterpillars	Baculovirus
Cotton boll worm	Baculovirus
Spruce and pine sawflies	Baculovirus
European rabbit	Myxoma virus

changes, such as draining breeding grounds, have not solved the problem. Researchers continue to look for other means to destroy the insect vectors.

Once again a bacterium, *Bacillus thuringiensis israelensis*, has been found to kill mosquito and black fly larvae selectively. When bacterial endospores are spread over the surfaces of ponds or lakes where the insects breed, most of the insects are destroyed. Although the endospores germinate into vegetative cells in the insects' guts, it is believed that the toxic protein crystals (parasporal bodies) are responsible for the deaths of the larvae, because they begin to be paralyzed within 30 minutes of consuming the bacterial material. *Bacillus thuringiensis israelensis* is believed to be safe for the environment, as fish and other aquatic life do not appear to be affected in any way.

Dutch Elm Disease

Dutch elm disease is caused by the fungus *Ceratocystis ulni*, which grows in the tree's vascular system and secretes a group of glycoprotein toxins that block the movement of water from the xylem to the leaves of the affected tree. This causes the uneven wilting of leaves and eventually the death of the leaves and stems. Yellowing or wilting in the upper branches of the tree are the first signs of the disease.

342

Ceratocystis ulmi, is an ascomycete □ that is spread from tree to tree by bark beetles that bore into the bark and breed in the phloem or inner bark. In the absence of the insect vector, the fungus is unable to cause Dutch elm disease; similarly, in the absence of the fungus, the bark beetles do not cause the disease. The disease was first noticed in the Netherlands in 1919 and soon became epidemic in Europe. Commerce in elm logs, used in making furniture veneers, is reputed to be responsible for spreading the disease to England (1927) and to the United States (1930). The disease spread rapidly through the northeastern states and southern Quebec. By the mid 1970s, the disease had reached the West Coast of the United States. Dutch elm disease has killed millions of elm trees in Europe and North America and has economically hurt businesses that rely on elm wood. Millions of dollars have been spent by communities to stop this disease. The use of antifungal agents such as benzimidazole has not been very effective, because the fungus often is not killed by them but only inhibited from reproducing while the fungistatic substance is present. In addition, fungicides have been ineffective because some fungal strains are totally resistant to the fungicides, treatment must be repeated as the tree grows, and repeated treatment weakens the tree so that other pathogens are able to cause disease.

Figure 36-4 Inhibition of *Ceratocystis ulmi*, which Causes Dutch Elm Disease, by *Pseudomonas syrigae*. Four colonies of *P. syringae* were established on the medium in the petri dish. A suspension of spores and hyphal fragments of *C. ulmi* was sprayed on the dish. After several days of incubation, the fungus became visible as a white layer covering all of the dish except the circular zone around each bacterial colony. An antimycotic agent synthesized by the bacterium effectively inhibits the growth of the fungus.

Presently, researchers are investigating bacteria that produce antimycotic agents to stop Dutch elm disease. The bacterium *Pseudomonas syrigae*, when injected into elms, grows in the trees and produces an antimycotic agent that inhibits *Ceratocystis ulmi* (fig. 36-4). Neither the pseudomonad □ nor its antimycotic agent appears to be pathogenic for the elm tree. Bacteria injected into elm seedlings two weeks before they are infected with the fungus pathogen protect all the trees from the fungus (fig. 36-5). The effectiveness of the bacteria in diseased mature trees, however, is not quite as impressive. Seven out of 8 lightly infected trees that were treated with *Pseudomonas* showed little or no sign of decline over two growing seasons. All of 14 heavily infected trees that were treated with the bacterium showed a decline in the same time period.

314

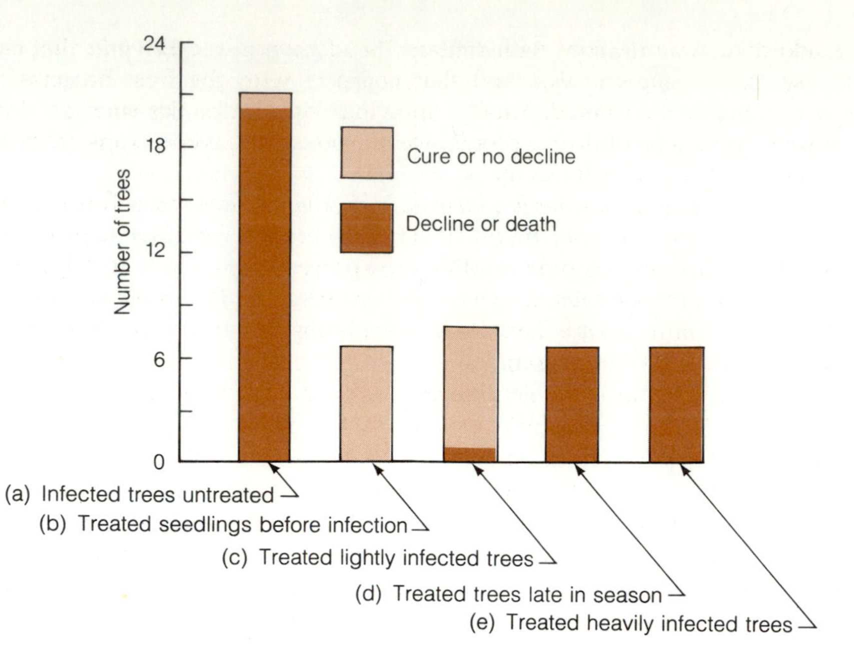

Figure 36-5 Effectiveness of *Pseudomonas syrigae* in Treating Dutch Elm Disease in Trees. *(a)* Most infected trees (20/21) that were not treated with *Pseudomonas syrigae* showed a decline or died over a two-year period. *(b)* All seedlings (7/7) treated wtih *P. syrigae* showed no decline and no death over a two-year period. *(c)* Most lightly infected trees (7/8) which were treated with *P. syrigae* showed no decline or death over a two-year period. *(d)* and *(e)* All trees (14/14) treated with *P. syrigae* late in the growing season, or which were heavily infected, showed a decline or death over a two-year period.

FROST BUGS

Almost every winter, freezing temperatures in California and Florida cause hundreds of thousands of dollars to be spent on the operation of oil-powered heaters and giant propellers to circulate the air. Nevertheless, millions of dollars worth of avocados, oranges, corn, almonds, beans, squash, and tomatoes are lost each year due to frost damage.

As the temperature drops below 32°F (0°C), moisture on plants becomes organized around particles on the leaves and turns to ice. Frost and ice do not form unless particles are present, as shown by the fact that very pure water can be supercooled to almost −36°F (−38°C). It has been found that, at temperatures above 23°F (−5°C), the majority of the particles on plants that promote freezing are actually bacteria, such as *Pseudomonas syringae* and *Erwinia herbicola.* These gram negative bacteria have a glycocalyx □ on their surfaces that arrange water molecules into the configuration that stimulates the formation of ice crystals. In the absence of these bacteria, plants can supercool to 23°F without damage, because ice crystals do not form. Below 23°F, however, other particles stimulate freezing and the formation of ice crystals that damage the leaves. The bacteria can act as ice formers even if they are dead.

84

Research is being carried out on sprays that will alter the glycocalyces of these bacteria so that they do not promote the formation of ice crystals.

In addition, dusting crops with mutant *Pseudomonas* and *Erwinia* that cannot synthesize a glycocalyx, and that compete with the frost bacteria for space, is being considered. Finally, spraying with antibiotics such as streptomycin may be useful and economical in protecting some crops from the bacteria that promote freezing.

At least one atmospheric physicist is studying these frost-forming bacteria in the hope that they might be useful in seeding clouds and producing rain. Presently, silver iodide crystals are sprayed over clouds to trigger ice formation. As the ice falls, it melts and becomes rain. Bacteria may be more efficient than silver iodide crystals in stimulating ice formation, especially if they do so at higher temperatures.

A genetic engineering firm located in California is planning to market killed cells, *Pseudomonas syringae*, as a "snow maker". The product, called Snomax, has been tested as an artificial snow maker at the Copper Mountain Resort in Colorado. It is hoped that through the use of bacteria for making artificial snow more efficiently, the amount of energy required to make the snow is reduced, and hence, the cost of snow making. Apparently, a compound called phosphatady linositol, found on the surfaces of the bacteria, promotes the formation of ice on the bacterial cell.

GLUTTONOUS MICROORGANISMS

Industrial nations produce millions of tons of toxic materials and wastes each year which are buried or enter the environment by being discharged into sewage systems, streams, lakes, and oceans. Many bacteria have been adapted to destroy some of the human wastes polluting the world. For example, aerobic and facultative anaerobic microorganisms are being used to degrade synthetic detergents, benzenes, phenols, cresols, naphthalenes, gasoline, kerosene, and oil spilled in soil and water. Saline-tolerant bacterial strains have been developed that break down crude oils to CO_2 and water. Aromatic compounds, asphaltics, paraffins, and naphthalines are all attacked by these organisms. Such bacteria are being used to restore soils, beaches, and saltwater estuaries polluted by oil spills and refinery wastes.

Phosphate and Oil-Eating Bacteria

Processed sewage, nitrogen- and phosphorus-containing fertilizers, and phosphate-containing detergents that are released into lakes or estuaries supply algae with vast amounts of nitrogenous compounds and phosphates. The resulting large algal blooms deplete the waters of O_2, so that animals die and organic matter and silt accumulate at an abnormal rate. The natural process of eutrophication □ by which lakes and estuaries are converted into swamps and eventually into dry land, can take place in a few hundred years, rather than the normal 100,000 years. 777

In order to slow down the destruction of lakes and estuaries, microorganisms are being isolated, which can reduce the phosphorus content of lakes and estuaries to low levels. This would reduce the algal growth. A bacterium known as *Acinetobacter phosphadevorus* has been isolated that accumulates phosphate in the form of volutin granules. The addition of this bacterium to activated sludges results in the removal of 16 times more phosphorus from sewage than is normally removed. *Acinetobacter* can also use a wide variety of hydrocarbons, with chain lengths running from C-1 to C-40, although it grows most rapidly on C-12 hydrocarbons. The hydrocarbons are stored in droplets within the cells. This bacterium may be useful in revitalizing sewage sludges that have been inactivated by motor oil and kerosene dumped into sewage systems. *Acinetobacter* reduces the amount of oil and phosphorus in oil-contaminated sludges to undetectable levels in a few days.

Acinetobacter's usefulness in treating oil spills in both fresh waters and oceans is being investigated. Nitrogen and phosphorus must be provided in these environments, because the concentrations of these essential nutrients are too low to stimulate the growth of large populations of *A. phosphadevorus.* Isolation of *Acinetobacter* mutants that can attack oil at low temperatures is in progress. Psychrophilic □ or cold-tolerant organisms would be most useful, because many spills occur in cold oceanic water off Alaska and Nova Scotia.

145

A genetically engineered *Pseudomonas* is being tested to see if it can be used to enhance oil recovery. It has a number of useful properties: (a) it digests solid waxes and paraffins found in heavy crude oil, but does not significantly metabolize the lighter liquid hydrocarbons; (b) it produces surfactants that reduce the petroleum's surface tension, so that it easily flows from porous rock where it is bound; and (c) it synthesizes an external polysaccharide that increases the viscosity of water used to drive the oil from wells.

The engineered *Pseudomonas* may not be very useful in anaerobic oil wells, however, because it is an aerobe. An additional problem may be that it is not very salt- or heat-tolerant. Many oil fields contain 2–14% sodium, and high temperatures are often reached in wells, depending upon their depth. In spite of these difficulties, the surfactants and polysaccharide could be produced by fermentation processes and used without the bacterium itself.

Sulfur-Digesting Bacteria

Another genetically engineered bacterium is being tested for its ability to digest organic sulfur compounds from high-sulfur oil and coal. It is desirable to find an economic method of eliminating sulfur from oil and coal, because this sulfur, when burned, yields a serious pollutant, sulfur dioxide. Since there are vast reserves of high-sulfur coal in the U.S., a bacterium that could economically remove this pollutant would open up this source of fuel to the nation.

Herbicide Eaters

Agent Orange, which was used in Vietman to defoliate forests, contains dioxin, 2,4,5-T, and 2,4-D (fig. 36–6). It has been implicated in causing crippling diseases in hundreds of Vietnam veterans and birth defects in some of their children. Among the Vietnamese, it is believed to be responsible for similar diseases and for a great number of miscarrages, aborted, deformed fetuses, and birth defects. Vietnam veterans exposed to high doses of Agent Orange have suffered from virulent skin eruption, skin rashes, temporary blindness, palsy, loss of sensation in the extremities (hands, arms, legs, and head), rectal bleeding, ocular bleeding, stomach cancer, liver failure, nausea,

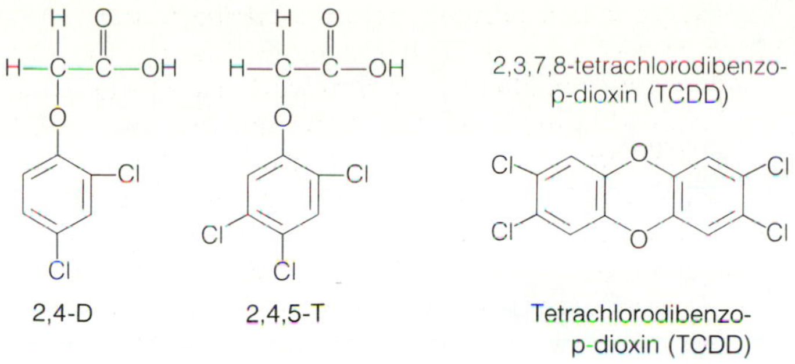

2,4-D

2,4,5-T

2,3,7,8-tetrachlorodibenzo-p-dioxin (TCDD)

Tetrachlorodibenzo-p-dioxin (TCDD)

Figure 36-6 Structure of Some Environmental Contaminants

headaches, permanent loss of hair, fatigue, respiratory problems, anorexia, deterioration of bones and muscle, premature aging, and infertility. Birth defects in the children of Vietnam veterans exposed to high doses of Agent Orange include deformed hands, partial deafness, face tumors, general malaise, lack of bones in the wrists and arms, and lack of fingers or arms.

It is unclear at this time which chemicals (dioxin, 2,4,5-T, 2,4-D) or combinations of chemicals are responsible for the veterans' disease. Dioxin is a likely candidate, because it is known to be a carcinogen as well as being a deadly poison. Ingestion of 0.1 mg of dioxin will kill an adult. It is probably significant that more than 1×10^8 mg of dioxin are reported to have been dropped on Vietnam during the period 1962–1971. This is more than enough dioxin, if consumed, to kill 1 billion people!

Dioxin may still be affecting people in Vietnam, because it has a half-life of approximately 18 months in some soils, much longer than originally estimated. The half-life of a chemical is the time that it takes for 50% of it to become altered chemically. Chemicals with a half-life of 18 months can have an effect 10–20 years after they enter the ecosystem, if the half-life remains constant. Factors that affect the half-life of a chemical include its exposure to light, heat, and degrading soil microorganisms. If a soil sample contains 10,000 parts per million (ppm) of dioxin (and the half-life of 18 months remains constant), it will take approximately 15 years for the concentration to drop to 10 ppm and about 20 years for the concentration to drop to 1 ppm. Thus, even after 20 years, a soil initially contaminated with 10,000 ppm might still have 1,000 parts per billion. Cancer-causing chemicals are considered dangerous when present in a few parts per billion.

During the Vietnam War, more than 12 million gallons of Agent Orange were sprayed by the U.S. military on South Vietnam. 2,4-D (2,4-dichlorophenoxyacetic acid), a widely used herbicide, was a major component (about 50%) of Agent Orange. 2,4-D causes long-lasting nerve damage, nausea, flu-like symptoms, numbness, and weakness or complete paralysis of arms and legs. About 1,000 tons/year (total U.S. production is 30,000 tons/year) are used in California to control underbrush in forests, highway weeds, waterways, gardens, and agricultural lands. 2,4-D has a half-life of about four days.

2,4,5-T (2,4,5-trichlorophenoxyacetic acid) is a hazardous plant-killing chemical that produces many of the same symptoms as 2,4-D. Because 2,4,5-T has a half-life of 2–5 weeks, depending upon the soil, it remains in some soils at dangerous concentrations for as long as a year (table 36-2, fig. 36-7).

A bacterium has recently been developed that breaks down 2,4,5-T. This bacterium belongs to the genus *Pseudomonas* and contains a plasmid with a group of genes that allows it to consume the herbicide. It may be useful in

TABLE 36-2

HALF-LIFE OF SELECTED HAZARDOUS CHEMICALS

COMMON NAME	CHEMICAL DESCRIPTION	ESTIMATED HALF-LIFE*	DEGRADING ORGANISM
	Chlorinated Insecticides		
Aldrin	hexachloro-dimethano-naphthalene	3–18 months	
Chlordane	Octochloro-hexahydro-methanoindene	5–18 months	
DDT	Trichloro-chlorophenyl-ethane	5–18 months	
Endrin	Hexachloro-epoxy-octahydro-dimethano-naphthalene	18 months	
Heptachlor	Heptachloro-tetrahydro-endomethanoindene	2–18 months	
Lindane	Hexachloro-cyclohexane	3–18 months	
	Organophosphate Insecticides		
Malathion		1–15 days	
Parathion	Diethyl-nitrophenyl-phosphoro-thioate	1–15 days	
	Chlorinated Herbicides		
2,4-D	Dichloro-phenoxy-acetic acid	4 days	*Achromobacter* *Corynebacterium*
2,4,5-T	Trichloro-phenoxy-acetic acid	5 weeks	*Pseudomonas*
2,3,5-TBA	Trichloro-benzoic acid	10 weeks	
Dalapon	Dichloro-propionic acid	3–6 days	*Agrobacterium* *Pseudomonas*
Monuron	Chlorophenyl-dimethyl-urea	1–4 months	*Pseudomonas*
Simazine	Chloro-ethylamino-triazine	5–10 weeks	
Propazine	Chloro-isopropylamino-triazine	8–14 weeks	
	Contaminants in 2,4,5-T		
TCDD	Tetrachloro-dibenzo-dioxin	18 months	

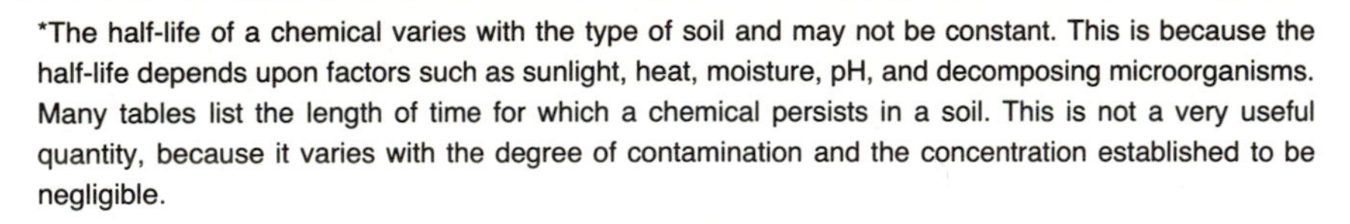

*The half-life of a chemical varies with the type of soil and may not be constant. This is because the half-life depends upon factors such as sunlight, heat, moisture, pH, and decomposing microorganisms. Many tables list the length of time for which a chemical persists in a soil. This is not a very useful quantity, because it varies with the degree of contamination and the concentration established to be negligible.

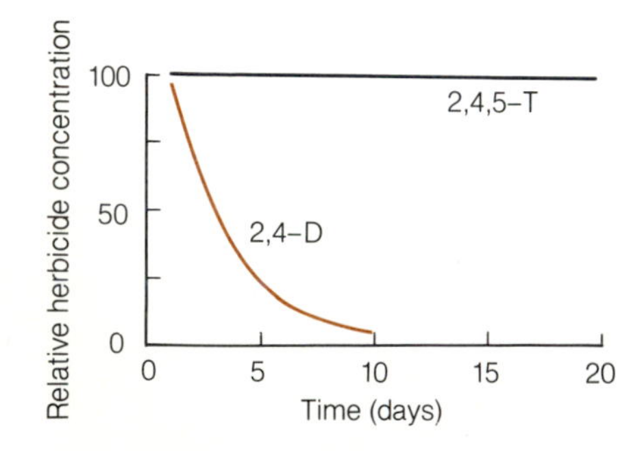

Figure 36-7 Decomposition of 2,4,5-T and 2,4-D. (From Whiteside, J. S. and Alexander, M. 1960. *Weeds* 8:204.)

reclaiming areas where Agent Orange has been used. Soils at target sites used by the U.S. Air Force have been reported to contain as much as 20,000 ppm of 2,4,5-T (2% of the soil). In soils with concentrations of 1,000 ppm of 2,4,5-T, bacteria are able to destroy more than 98% of the chemical in about six weeks of weekly applications. The residue is about 20 ppm. A mixed culture of *Pseudomonas* and *Achromobacter* is able to degrade 2,4-D, the most widely used herbicide in the world. At the present time, only a few bacterial genera—*Pseudomonas*, *Arthrobacter*, and *Acinetobacter*—have complete enzyme systems on plasmids for degrading some of the unusual manmade pollutants.

OIL-PRODUCING MICROBES

"Hibernating" bacteria isolated from shale have been found to produce hydrocarbons similar to those found in fossil fuels. Some of the bacteria studied are able to produce their weight in oily hydrocarbons in about two weeks. Research is being carried out to determine which hydrocarbons are being synthesized and the biochemical pathways involved in producing them. Future experimentation will focus on the possibility of increasing the rate of hydrocarbon production so that it might be economically feasible.

GERM WARFARE

The world's human population and domestic livestock depend upon grains and other crops for their livelihood. The military has long considered the usefulness of destroying a country's crops or its livestock during times of war. Crops and animals could be destroyed by dropping virulent microorganisms on agricultural lands. The prime targets in any of the large industrialized countries would be wheat, rice, potato, corn, poultry, cattle, sheep, and goats. Humans could also be attacked directly be virulent pathogens or toxic compounds made from microorganisms. Table 36-3 summarizes some of the organisms considered for germ warfare.

Anthrax

One organism that has been considered useful as a biological weapon is *Bacillus anthracis*, the bacterium that causes anthrax in animals and in hu-

TABLE 36-3
MICROORGANISMS THAT MIGHT BE USED IN GERM WARFARE

DISEASE	INCUBATION PERIOD (DAYS)	MORTALITY	CHEMOTHERAPY	VACCINATION
Viral				
Dengue fever	5–8	Low (<1%)	–	–
Eastern equine encephalitis	5–15	High (>60%)	–	+
Influenza	1–3	Usually low (2%)	+	+
Smallpox	7–16	Variable (up to 30%)	–	+
Tick-borne encephalitis	7–14	Variable (up to 30%)	–	+
Venezuelan equine encephalitis	2–5	Low (<1%)	–	+
Yellow fever	3–6	High (up to 40%)	–	+
Bacterial				
Anthrax (pulmonary)	1–5	Almost 100% fatal	+	+
Brucellosis	7–21	Low (<25%)	+	+
Cholera	1–5	Usually high (up to 80%)	+	+
Dysentery	1–3	Variable (up to 10%)	+	–
Epidemic typhus	6–15	Variable (up to 70%)	+	+
Plague (pneumonic)	2–5	Almost 100% fatal	+	+
Psittacosis	4–15	Moderately high (up to 10%)	+	+
Q-fever	10–21	Low (usually <1%)	+	+
Rocky Mountain spotted fever	3–10	Usually high (up to 80%)	+	+
Tularemia	1–10	Variable (up to 60%)	+	+
Typhoid fever	7–21	Moderately high (up to 10%)	+	+
Fungal				
Coccidioidomycosis	7–21	Low (<1%)	+	+

FOCUS

SERRATIA MARCESCENS AND GERM WARFARE

In the late 1940s, the U.S. Army was involved in research to determine the feasibility of using micoorganisms in war and the impact these microorganisms might have when used on U.S. cities. In 1950, the U.S. Army released aerosols of *Serratia marcescens* near the Golden Gate Bridge in order to determine how the bacteria would spread over San Francisco. *Serratia* was used in the experiment because it was considered nonpathogenic and because it produces a bright red pigment when grown at 18°C. The pigment allowed investigators to distinguish easily between *Serratia* and other bacteria in the air that might fall onto solid nutrient media placed around the city.

As it turns out, *Serratia marcescens* is not as innocuous as was once thought. *Serratia* is a common cause of diarrhea, pulmonary infection, and urinary tract infections. Records gathered at the time of the experiment indicate that 11 patients at a San Francisco hospital developed urinary tract infections and one patient developed a pneumonia due to *Serratia* during the five months after the release of the bacteria. Apparently, the bacteria did a good job of contaminating the hospital. How many other infections in San Francisco were caused by the Army experiment will probably remain unknown forever. This experiment points out, however, that aerosols of pathogenic microoganisms can be very dangerous and cause disease over a long period of time.

mans. Endospores of this virulent bacterium could be dropped over military installations, cities, stockyards, and major grazing lands. The endospores □, which can be inhaled, would cause epidemics of pulmonary anthrax that would devastate cities and leave grazing lands permanently unusable.

94

The virulence of disseminated anthrax was demonstrated recently in Sverdlovsk, Russia. It is believed that an accident in 1979 at a research station that produces anthrax vaccine (and possible biological weapons) resulted in the escape of anthrax. The bacterium and its spores subsequently spread through Sverdlovsk. It is reported that as many as 1,000 persons died of pulmonary anthrax over a period of one month. Those who contracted the disease suffered severe flulike symptoms that led to their death within a few days. The Russians claim that the outbreak was not an epidemic of pulmonary anthrax, but instead an outbreak of enteric anthrax from contaminated meat. Anthrax is normally a problem for the Russians, and more than 1,000,000 persons are vaccinated against the organism each year.

Pneumonia and Influenza

Pneumonia and influenza normally account for more than 1 million cases and 50,000 deaths each year in the United States. Virulent strains of the viruses and bacteria that cause pneumonia could be dropped over military installations and cities to paralyze a country.

Fungi as Plant and Human Pathogens

Plant pathogens that attack wheat, rice, potatoes, and corn could lead to starvation and a crippled economy. Virulent strains of *Phytophthora* could be used to cause potato blight; *Erysiphe* might be used to destroy oats, barley rye, and wheat, by causing powdery mildew; *Ustilago* could cause loose smut of wheat, barley, and corn; and *Puccinia* might cause wheat stem rust.

Some species of the fungus *Fusarium* produce a powerful toxin when they grow on cereals. It is thought that this toxin may have poisoned as many as 1 million persons who ate moldy foods at the end of World War II. More recently (1980–1983), *Fusarium* toxins have been blamed for unidentified ailments among villagers in Laos, Cambodia, and Thailand. Some scientists have hypothesized that *Fusarium* growing on moldy food may have been the cause of the ailments, while others have proposed that the fungus growing on bee excrement produced the toxins, which were subsequently washed and blown into the villagers' water and food. Still other scientists concluded, however, that the toxins found in Southeast Asia during 1980–1983 were examples of toxins used in chemical warfare. The U.S. State Department accused the Russians of producing the *Fusarium* toxins, called **trichothecenes**, and accused the Vietnamese and Russians of using these toxins in Cambodia, Laos, Thailand, and Afghanistan. One of the toxins in question is called **T2** and is made by *Fusarium*, which commonly grows on wheat and corn. It has been found in samples of water, vegetation, blood, and urine collected in battle areas in Cambodia, Thailand, and Laos. The half-life of T2 in cattle and swine is about 10 minutes, while that of **ochratoxin**, another *Fusarium* toxin, is about three days. Persons whose water, food, skin, and hair are contaminated may absorb the toxins over long periods of time and so be continuously affected. T2 causes vomiting, diarrhea, and convulsions in humans and animals. Russian and Israeli scientists produce the T2 mycotoxin for use in controlling forest and agricultural pests.

Foot and Mouth Disease

836

The foot and mouth disease □ (FMD) virus might be a devastating weapon against cattle, buffalo, sheep, and goats. Foot and mouth disease has not existed in the U.S. since the 1940s, when it was briefly introduced in contaminated garbage. Although Australia and New Zealand are also free of FMD, the disease is found throughout the rest of the world and is one of the most serious animal diseases. Recently, recombinant DNA techniques have resulted in the development of a safe, heat-stable vaccine against one of the seven types of viruses. This vaccine does not have to be refrigerated and so is of value in poorly developed areas that lack means of refrigeration. The value of this vaccine can be understood in light of the fact that more than 500 million vaccinations are carried out annually worldwide.

METHANE-PRODUCING MICROORGANISMS

329

Methane is produced under anaerobic conditions by a small group of bacteria, commonly called the methanogenic bacteria □. The methanogenic bacteria generally use H_2 as their source of energy (electrons) and CO_2 as their ultimate electron acceptor. During their anaerobic metabolism, these bacteria carry out the following reaction:

$$4H_2 + CO_2 \longrightarrow CH_4 + 2H_2O$$

Methane is a valuable fuel that can be used for heating buildings and for powering all sorts of machinery. Approximately 8 trillion cubic feet of CH_4/year are extracted from natural gas wells in the United States. Presently, a number of large cities are tapping the CH_4 being produced in their garbage dumps and using it as a source of natural gas.

Methane production occurs after a variety of compounds such as cellulose, starch, proteins, and lipids are decomposed anaerobically by various fermenting bacteria. The waste H_2 gas, acetic acid, and CO_2 resulting from fermentation is used by the methane bacteria to produce methane. Thus, methane production is dependent upon a mixture of bacteria, including those that ferment organic compounds to hydrogen gas and carbon dioxide and those that use these products to make methane.

A large amount of methane is produced by microorganisms, some of which becomes part of the atmosphere. It has been estimated that microorganisms in the gut of termites □ are responsible for producing nearly half of all methane in the atmosphere—4 to 8 trillion cubic feet/year. Most of the remaining methane comes from microorganisms in swamps, estuaries, rice paddies, and deep sea ocean vents.

751

A number of anaerobic wastewater treatment facilities are being built to clean up wastewater from sugar and petroleum refineries, food processing plants, breweries and distilleries, pulp and paper mills, and chemical plants. Anaerobic systems of water purification require 10–25% of the electricity needed for aerobic processes and only about 10% of the nitrogen and phosphorus. The electricity is used to circulate the water through anaerobic activated sludges. Almost six cubic feet of methane are produced per pound of chemical oxygen demand (COD). Anaerobic activated sludge plants now operating produce about 170 million ft^3 of methane per year. The methane is then used to run the plant's utility boilers and saves approximately $500,000 annually. In addition, some plants are set up to extract fermentation wastes such as methanol, formaldehyde, butyraldehyde, butanol, trimethylolpropane, methylethyl ketone, ethyl acetate, and methyl formate.

ORGANISMS THAT PRODUCE HEAT-STABLE ENZYMES

A number of assumptions scientists have held regarding life have been challanged by experiments indicating that certain bacteria can proliferate at temperatures and pressures exceeding 250°C and 265 atmospheres. At a temperature of 250°C, however, the lipid membranes of typical bacteria run like oil, the proteins denature, the DNA melts into separate strands and most of the molecules of the cell decompose within minutes.

Some scientists, however, have been unable to repeat the experiments that showed bacteria growing at 250°C. They have also shown that artifacts could account for the original results. Hence, whether or not super thermophiles really do exist is still uncertain. Microorganisms that are able to grow at 120°C have been isolated and cultivated, however.

Heat-stable enzymes from the thermophiles that grow above 100°C may be of value in those industries in which high temperatures are a problem. For instance, heat-stable proteases could be of use in hot water laundry detergents to remove tough stains. Possibly, the thermophiles could be seeded into garbage dumps to speed up the production of methane and make its production from garbage more economical. The thermophiles would not be inhibited by the high temperatures that develop under garbage dumps, and consequently would be able to produce methane when the normal methanogens are inhibited.

FOCUS

THE USE OF CLONING AND GENETIC ENGINEERING TO RECREATE EXTINCT ANIMALS

In parts of northern Europe and Russia, numerous extinct animals have been discovered frozen. The best-known finds are those of complete Siberian mammoths in 1938 and in 1977. In recent years, Russian cytologists and molecular biologists have been attempting to revive some of the frozen cells from mammoth carcasses, so that nuclei from the revived cells might be exchanged with nuclei in fertilized elephant eggs. The altered elephant egg might then be implanted in an elephant, in the hope that it would develop into a baby mammoth. The Russian scientists have so far been unsuccessful in reviving any of the mammoth cells in tissue culture. Most likely, the mammoths were not frozen rapidly enough or maintained at low enough temperatures to avoid cellular damage.

Another approach to recreating a mammoth might be to isolate nuclei from the frozen mammoth cells and introduce them into elephant cells in tissue culture. If this could be achieved, then the next step would be to introduce the nuclei into fertilized elephant eggs and implant these in elephants. If this second procedure fails, genetic engineering might be tried in order to create a mammoth out of an elephant cell.

Most of the cells from the frozen mammoths probably contain undamaged DNA from which a mammoth DNA library could be created by cloning the DNA in *E. coli*. The mammoth DNA could be compared to elephant DNA and the differences determined. Then elephant cells in tissue culture could be transformed with the mammoth DNA that differed from the elephant DNA, and the transformed elephant nuclei could be transferred to anucleated, fertilized elephant eggs.

Even if it turns out to be impossible to recreate a mammoth by cloning and genetic engineering, these studies should reveal much about the evolutionary relationship between mammoths and modern-day elephants and the relationship between DNA and an animal's final form.

AMES TEST

Each year in the United States, more than 700,000 people are diagnosed to have cancer and approximately 400,000 people die of the disease. Available evidence indicates that 90% of the cases of cancer are due to chemicals and radiation that humans could avoid. The major cancer-causing chemicals are those found in tobacco smoke. Chemicals such as benzopyrene, found in tobacco smoke, chewing tobacco, and snuff, are responsible for more than 100,000 deaths per year due to lung cancer (100,000), pancreas cancer (20,000), bladder cancer (10,000), oral cancer (8,000), esophagus cancer (8,000), and larynx cancer (4,000). It is estimated that nearly one-third of all cancer deaths in men in the United States could be prevented by eliminating only one environmental carcinogen: tobacco smoke.

Thousands of chemical carcinogens are found in our water, food, and air. These chemicals often act synergistically (work together) to produce more cancers than each would produce alone. Consequently, it is imperative to detect which chemicals are capable of causing mutations that often lead to cancer and to limit the release to these chemicals into our environment. The testing of a chemical to determine whether it causes cancer generally costs hundreds of thousands of dollars and many years of work. In order to test the thousands of chemicals (already more than 50,000) that are entering our water, food, and air, many billions of dollars and many trained scientists would be required. So far, the money to carry out an extensive investigation has not been forthcoming.

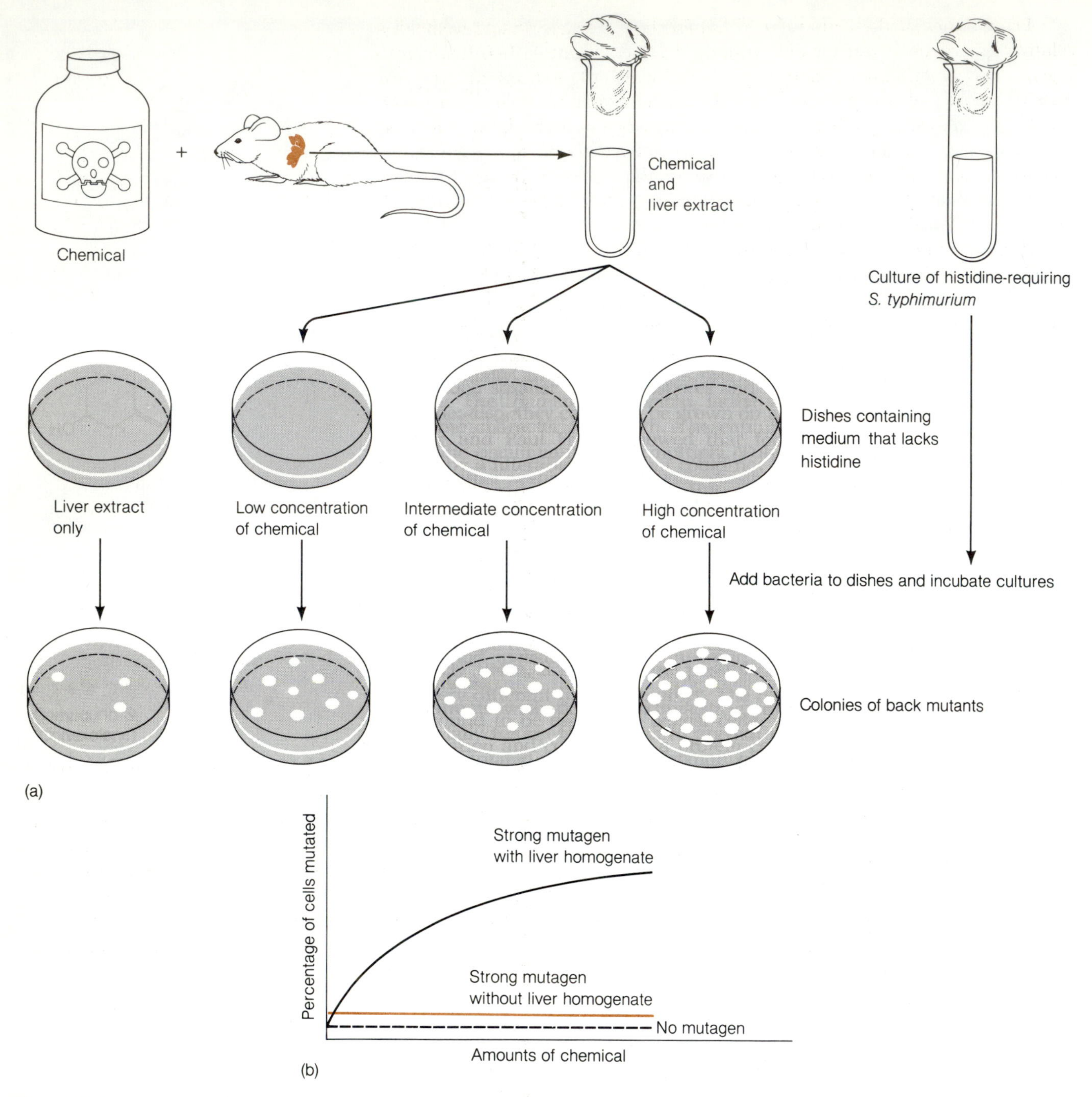

Figure 36-8 **Ames Test.** The Ames test is a simple method to determine whether or not a chemical is a mutagen. Since mutagens are potential carcinogens, the Ames test indicates which chemicals may be carcinogens. *(a)* In order to test a chemical, a sample is mixed with mouse liver homogenate and various concentrations of the mixture are poured into petri dishes containing a culture medium lacking histidine. An auxotrophic mutant of *Salmonella* requiring histidine is then spread over the surface of the culture medium and incubated at 35°C for 24–48 hours. Since the bacteria are unable to synthesize the amino acid histidine, they will not grow on a minimal medium that contains only glucose and salts. *Salmonella* that undergo a mutation hat reverts them to prototrophy will begin to grow and form a colony. The potency of a mutagen is indicated by the number of revertants to histidine prototrophy; the more colonies, the more efficient is the mutagen. Some chemicals must be incubated with homogenized liver before they are tested, because it is not the original chemical that is mutagenic and carcinogenic but rather a breakdown product. Since bacteria lack liver enzymes that alter compounds, the effect of these compounds in animals would be missed if the compounds were not mixed with liver extract.

Bruce Ames of the University of California at Berkeley has developed a relatively inexpensive test for determining whether or not a chemical is mutagenic and potentially carcinogenic (cancer-causing). Instead of using expensive animals to test potential cancer-causing agents, this test uses bacteria. The Ames test (fig. 36–8) indicates whether or not a chemical can cause mutations in a histidine auxotroph of *Salmonella typhimurium*. A histidine auxotroph is unable to grow on a medium that lacks histidine; however, those individuals that acquire a mutation that allows them to synthesize histidine will be able to grow and multiply.

Because many chemicals are not mutagenic or carcinogenic until they have been biochemically altered by animal cells, chemicals are first metabolized to other forms by extracts from mammalian liver cells before they are tested on bacteria. Metabolic conversion of chemicals into mutagens and carcinogens is due to various enzymes that normally destroy toxic materials that enter cells. Usually beneficial enzymes convert certain chemicals into mutagens and carcinogens.

The Ames test for mutagens and carcinogens is based upon the observation that most carcinogens are mutagens. Consequently, if a chemical is found to be mutagenic in bacteria, there is a good chance that it is carcinogenic in humans and animals. These chemicals can then be tested more carefully on laboratory animals.

SUMMARY

BIOLOGICAL PESTICIDES

1. *Bacillus thuringiensis* has been used successfully as an insecticide. The growth of this bacterium and the endotoxin found in parasporal bodies are responsible for the death of certain insects.
2. Certain insect viruses are effective as insecticides.
3. A *Pseudomonas* has been tested and found to be effective against the fungus that causes Dutch elm disease.
4. A *Bacillus thuringiensis* strain has been found that kills mosquito and black fly larvae.

FROST BUGS

1. Some *Pseudomonas* and some *Erwinia* on the surfaces of plants promote the freezing of water at temperatures below 0°C, because of a glycocalyx the bacteria produce. In order to reduce frost formation, antibiotics or other bacteria might be sprayed onto plants so as to reduce the amount of frost and protect crops from freezing.
2. The frost-precipitating bacteria are being tested to see if they might be used to seed clouds and cause rain to form.

GLUTTONOUS MICROORGANISMS

1. *Acinetobacter* species have been isolated which accumulate phosphate in the form of volutin granules and hydrocarbons as intracellular droplets. This bacterium may be useful in removing phosphate and oil from sewage systems.
2. *Pseudomonas* species have been isolated that are able to decompose, 2,4,5-T, a long-lasting herbicide that is toxic to humans and animals.
3. A mixed culture of *Pseudomonas* and *Achromobacter* is able to decompose 2,4-D, another widely used herbicide that is toxic to humans and animals.

OIL-PRODUCING MICROBES

1. Bacteria isolated from shale have been found to produce hydrocarbons.
2. These bacteria might be used in the future to produce some of the oil needed.

GERM WARFARE

1. *Bacillus anthracis* vegetative cells or endospores could be used against animals and humans, since their inhalation can result in deadly enteric and pulmonary anthrax.
2. Various fungi could be exploited as plant pathogens to destroy cereal crops and vegetables.
3. Foot and mouth disease virus might be used to attack cattle, buffalo, sheep, and goats.
4. Instead of using living microorganisms, biological

toxins might be used, such as the trichothecenes produced by the plant fungus *Fusarium*.

METHANE-PRODUCING MICROORGANISMS

1. Methane bacteria (methanogens) associated with termites, swamps, estuaries, rice paddies, and hot vents in the sea floor are responsible for much of the methane that is released into the atmosphere each year.
2. There are many projects attempting to use waste materials and methane bacteria to produce methane to run machinery and heat buildings.

ORGANISMS THAT PRODUCE HEAT-STABLE ENZYMES

1. Thermophiles, associated with 300°C water vents in the sea floor, are methane bacteria that use reduced sulfur as their source of energy and reducing electrons.
2. Heat-stable enzymes are of value in many industries.

AMES TEST

1. The Ames test is an economical procedure for determining whether or not chemicals are mutagenic and potentially carcinogenic in animals.
2. A histidine auxotroph of *Salmonella* is used to test for a chemical's capacity to induce mutations. If a chemical reverts the histidine auxotroph to prototrophy, the chemical is a mutagen.
3. Most mutagens are also carcinogens.

STUDY QUESTIONS

1. Explain how *Bacillus thuringiensis* is used as a pesticide.
2. How does *Bacillus thuringiensis* kill insects?
3. Of what value is *Pseudomonas syrigae* in preventing Dutch elm disease?
4. How are bacteria sometimes responsible for frost damage to crops?
5. Which bacteria might be useful in seeding clouds to create rain in areas affected by drought?
6. Which bacterium may be useful in removing phosphate and oil from sewage?
7. Which bacteria may be useful in degrasdng the herbicides 2,4-D and 2,4,5-T?
8. What evidence is there that *Bacillus anthracis* might be very effective as a weapon in germ warfare?
9. What would be the long-term effect of using *Bacillus anthracis* as a weapon?
10. What kind of organisms might be used to destroy a nation's crops?
11. What virus might be used to destroy a nation's cattle, sheep, and goat herds?
12. What kind of biological toxins might be used as weapons?
13. What do most methanogens use as their source of energy and reducing electrons?
14. Outline a couple of ways in which methane bacteria are being used to benefit humans.

SUPPLEMENTAL READINGS

Armstrong, J. L., Rohrmann, G. F., and Beaudreau, G. S. 1985. Delta endotoxin of *Bacillus thuringiensis* subsp. *israelensis. Journal of Bacteriology* 161(1):39–46.

Budiansky, S. 1983. Is yellow rain simply bees' natural excreta? *Nature* 303:3–4.

Herrman, L. 1981. Alien worlds on the ocean floor. *Science Digest* 89:52–122.

Kellogg, S. T., Chatterjee, D. K., and Chakrabarty A. M., A. M. 1981. Plasmid-assisted molecular breeding: New technique for enhanced biodegradation of persistent toxic chemicals. *Science* 214:1133–1135.

Marshall, E. 1982. Yellow rain: Filling in the gaps. *Science* 217:31–34.

Olson, S. 1981. Fighting frost bugs. *Science81* 2:94–95.

Ourisson, G., Albrecht, P., and Rohmer, M. 1984. The microbial origin of fossil fuels. *Scientific American* 251(2):44–51.

Rasmussen, R. A. and Khalil, M. A. 1983. Global production of methane by termites. *Nature* 301:700–702.

Strobel, G. and Lanier, G. 1981. Dutch elm diseaqrse. *Scientific American* 245:56–66.

Van Brunt, J. 1985. Biochips: the ultimate computer. *Biotechnology* 3:209–215.

Walton, S. 1980. Germ warfare violation hard to pin down. *BioScience* 30:485–487.

Weisburd, S. 1985. *Cereus* bacteria go for the gold. *Science News* 127:102–103.

Zimmerman, P., Greenberg, J., Wandiga, S., and Crutzen, P. 1982. Termites: A potentially large source of atmospheric methane, carbon dioxide, and molecular hydrogen. *Science* 218:563–565.

APPENDIX A
TEMPERATURE CONVERSION CHART

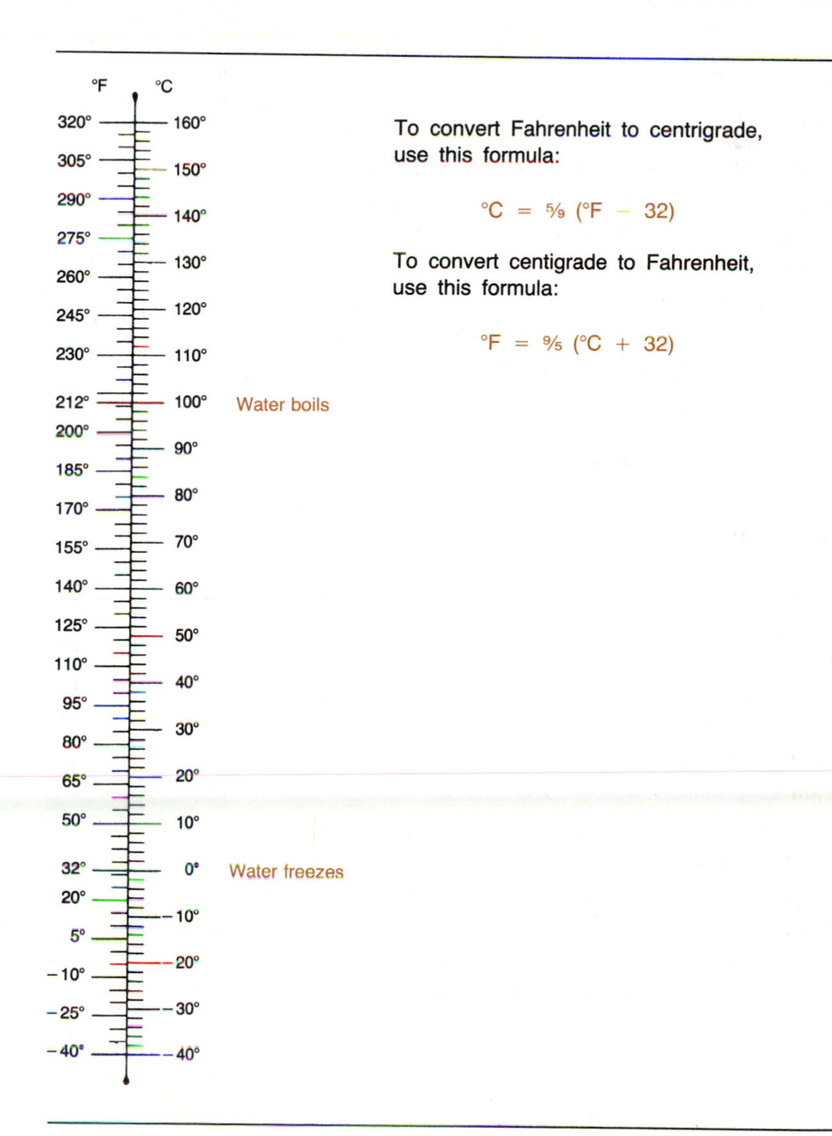

To convert Fahrenheit to centrigrade, use this formula:

$$°C = \frac{5}{9}(°F - 32)$$

To convert centigrade to Fahrenheit, use this formula:

$$°F = \frac{9}{5}(°C + 32)$$

APPENDIX B
CLASSIFICATION OF BACTERIA

KINGDOM *PROCARYOTAE* (MURRAY, 1968)

The kingdom is comprised of organisms with cells whose genome (genophore) is not separated from the cytoplasm by a unit membrane. Organisms in this kingdom also lack unit membrane-bounded intracellular organelles (e.g., mitochondria and chloroplasts) and usually measure less than 10 μm in diameter. Prokaryotes possess 70S ribosomes and might be motile by rotating flagella or by gliding. See page 76 in the text and Table 4–1 for further details of the kingdom *Procaryotae.*

Organisms in kingdom *Procaryotae* are grouped into four divisions, based primarily on cell wall characteristics.

Division I. Gracilicutes

Prokaryotes in this division have a Gram-negative type cell wall with an outer membrane containing lipopolysaccharides and a thin inner layer of peptidoglycan. Asexual reproduction is usually by binary fission but some species reproduce by budding.

Division II. Firmicutes

Prokaryotes in this division have a Gram-positive type cell wall composed mainly of peptidoglycan. Most stain positive when Gram-stained (but not always). Some species form endospores. Asexual reproduction is usually by binary fission.

Division III. Tenericutes

Prokaryotes in this division lack a cell wall. They do not synthesize the chemical procursors for peptidoglycan synthesis. The cells are highly pleomorphic and are surrounded only by the plasma membrane. These cells stain Gram negative. Asexual reproduction is by budding, fragmentation, or binary fission. The group is usually referred to as the mycoplasmas.

Division IV. Mendosicutes

Prokaryotes in this division are sometimes called the archaeobacteria. They have cell walls that lack muramic acid and hence, no conventional peptidoglycan. Some species have cell walls that are made out only of protein, others of heteropolysaccharides, and still others lack a cell wall altogether. They may stain Gram positive or Gram negative. These prokaryotes are phylogenetically distinct from other bacteria and represent a very ancient line of microorganisms.

SECTION 1
THE SPIROCHETES

Order I. Spirochaetales
- Family I. Spirochaetaceae
  - Genus I. *Spirochaeta*
  - Genus II. *Cristispira*
  - Genus III. *Treponema*
  - Genus IV. *Borrelia*
- Family II. Leptospiraceae
  - Genus I. *Leptospira*

Other organisms. Hindgut spirochetes of termites and *Cryptocercus*

SECTION 2
AEROBIC/MICROAEROPHILIC, MOTILE, HELICAL/VIBRIOID GRAM-NEGATIVE BACTERIA

- Genus *Aquaspirillum*
- Genus *Spirillum*
- Genus *Azospirillum*
- Genus *Oceanospirillum*
- Genus *Campylobacter*
- Genus *Bdellovibrio*
- Genus *Vampirovibrio*

SECTION 3
NONMOTILE (OR RARELY MOTILE), GRAM-NEGATIVE CURVED BACTERIA

- Family I. Spirosomaceae
  - Genus I. *Spirosoma*
  - Genus II. *Runella*
  - Genus III. *Flectobacillus*
- Other Genera
  - Genus *Microcyclus*
  - Genus *Meniscus*
  - Genus *Brachyarcus*
  - Genus *Pelosigma*

SECTION 4
GRAM-NEGATIVE AEROBIC RODS AND COCCI

- Family I. Pseudomonadaceae
  - Genus I. *Pseudomonas*
  - Genus II. *Xanthomonas*
  - Genus III. *Frateuria*
  - Genus IV. *Zoogloea*
- Family II. Azotobacteraceae
  - Genus I. *Azotobacter*
  - Genus II. *Azomonas*
- Family III. Rhizobiaceae
  - Genus I. *Rhizobium*
  - Genus II. *Bradyrhizobium*
  - Genus III. *Agrobacterium*
  - Genus IV. *Phyllobacterium*
- Family IV. Methylococcaceae
  - Genus I. *Methylococcus*
  - Genus II. *Methylomonas*
- Family V. Halobacteriaceae
  - Genus I. *Halobacterium*
  - Genus II. *Halococcus*
- Family VI. Acetobacteraceae
  - Genus I. *Acetobacter*
  - Genus II. *Gluconobacter*
- Family VII. Legionellaceae
  - Genus I. *Legionella*
- Family VIII. Neisseriaceae
  - Genus I. *Neisseria*
  - Genus II. *Moraxella*
  - Genus III. *Acinetobacter*
  - Genus IV. *Kingella*
- Other Genera
  - Genus *Beijerinckia*
  - Genus *Derxia*
  - Genus *Xanthobacter*
  - Genus *Thermus*
  - Genus *Thermomicrobium*
  - Genus *Halomonas*
  - Genus *Alteromonas*
  - Genus *Flavobacterium*
  - Genus *Alcaligenes*
  - Genus *Serpens*
  - Genus *Janthinobacterium*
  - Genus *Brucella*
  - Genus *Bordetella*
  - Genus *Francisella*
  - Genus *Paracoccus*
  - Genus *Lampropedia*

SECTION 5
FACULTATIVELY ANAEROBIC GRAM-NEGATIVE RODS

- Family I. Enterobacteriaceae
  - Genus I. *Escherichia*

Genus II. *Shigella*
Genus III. *Salmonella*
Genus IV. *Citrobacter*
Genus V. *Klebsiella*
Genus VI. *Enterobacter*
Genus VII. *Erwinia*
Genus VIII. *Serratia*
Genus XI. *Hafnia*
Genus X. *Edwarsiella*
Genus XI. *Proteus*
Genus XII. *Providencia*
Genus XIII. *Morganella*
Genus XIV. *Yersinia*

Other Genera of the Family Enterobacteriaceae
Genus *Obesumbacterium*
Genus *Xenorhabdus*
Genus *Kluyvera*
Genus *Rahnella*
Genus *Cedecea*
Genus *Tatumella*

Family II. Vibrinoaceae
Genus I. *Vibrio*
Genus II. *Photobacterium*
Genus III. *Aeromonas*
Genus IV. *Plesiomonas*

Other Genera
Genus *Zymomonas*
Genus *Chromobacterium*
Genus *Cardiobacterium*
Genus *Calymmatobacterium*
Genus *Gardnerella*
Genus *Eikenella*
Genus *Streptobacillus*

SECTION 6
ANAEROBIC GRAM-NEGATIVE STRAIGHT, CURVED, AND HELICAL RODS

Family I. Bacteroidaceae
Genus I. *Bacteoides*
Genus II. *Fusobacterium*
Genus III. *Leptotrichia*
Genus IV. *Butyrivibrio*
Genus V. *Succinomonas*
Genus VI. *Succinivibrio*
Genus VII. *Anaerobiospirillum*
Genus VIII. *Wolinella*
Genus XI. *Selenomonas*
Genus X. *Anaerovibrio*
Genus XI. *Pectinatus*
Genus XII. *Acetovibrio*
Genus XIII. *Lachnospira*

SECTION 7
DISSIMILATORY SULFATE- OR SULFUR-REDUCING BACTERIA

Genus *Desulfuromonas*
Genus *Desulfovibrio*
Genus *Desulfomonas*
Genus *Desulfococcus*
Genus *Desulfobacter*
Genus *Desulfobulbus*
Genus *Desulfosarcina*

SECTION 8
ANAEROBIC GRAM-NEGATIVE COCCI

Family I. Veillonellaceae
Genus I. *Veillonella*
Genus II. *Acidaminococcus*
Genus III. *Megasphaera*

SECTION 9
THE RICKETTSIAS AND CHLAMYDIAS

Order I. Rickettsiales
Family I. Rickettsiaceae
Genus I. *Rickettsia*
Genus II. *Rochalimaea*
Genus III. *Coxiella*
Genus IV. *Ehrlichia*
Genus V. *Cowdria*
Genus VI. *Neorickettsia*
Genus VII. *Wolbachia*
Genus VIII. *Rickettsiella*

Family II. Bartonellaceae
Genus I. *Bartonella*
Genus II. *Grahamella*

Family III. Anaplasmataceae
Genus I. *Anaplasma*
Genus II. *Aegyptianella*
Genus III. *Haemobartonella*
Genus IV. *Eperythrozoon*

Order II. Chlamydiales
Family I. Chlamydiaceae
Genus I. *Chlamydia*

SECTION 10
THE MYCOPLASMAS

Order I. Mycoplasmatales
Family I. Mycoplasmataceae
Genus I. *Mycoplasma*
Genus II. *Ureaplasma*

Family II. Acholeplasmataceae
Genus I. *Acholeplasma*

Family III. Spiraplasmataceae
Genus I. *Spiroplasma*

Other Genera
Genus *Anaeroplasma*
Genus *Thermoplasma*

SECTION 11
ENDOSYMBIONTS

A. Endosymbionts of Protozoa
Genus I. *Holospora*
Genus II. *Caedibacter*
Genus III. *Pseudocaedibacter*
Genus IV. *Lyticum*
Genus V. *Tectibacter*

B. Endosymbionts of Insects
Genus *Blattabacterium*

C. Endosymbionts of Invertebrates Other than Arthropods

SECTION 12
GRAM-POSITIVE COCCI

Family I. Micrococcaceae
Genus I. *Micrococcus*
Genus II. *Stomatococcus*
Genus III. *Planococcus*
Genus IV. *Staphylococcus*

Family II. Deinococcaceae
Genus I. *Deinococcus*

Other Organisms
Pyogenic streptococci
Oral streptococci
Lactic streptococci and enterococci
Genus *Leuconostoc*
Genus *Pediococcus*
Genus *Aerococcus*
Genus *Gemella*
Genus *Peptococcus*
Genus *Peptostreptococcus*
Genus *Ruminococcus*
Genus *Coprococcus*
Genus *Sarcina*

SECTION 13
ENDOSPORE-FORMING GRAM-POSITIVE RODS AND COCCI

Genus *Bacillus*
Genus *Sporolactobacillus*
Genus *Clostridium*
Genus *Desulfotomaculum*
Genus *Sporosarcina*
Genus *Oscillospora*

SECTION 14
REGULAR, NON-SPORING, GRAM-POSITIVE RODS

Genus *Lactobacillus*
Genus *Listeria*
Genus *Erysipelothrix*
Genus *Brochothrix*
Genus *Renibacterium*
Genus *Kurthia*
Genus *Caryophanon*

SECTION 15
IRREGULAR, NON-SPORING, GRAM-POSITIVE RODS

Animal and Saprophytic Corynebacteria (*Corynebacterium*)
Plant Corynebacteria (*Corynebacterium*)
Other Genera
Genus *Gardnerella*
Genus *Arcanobacterium*
Genus *Arthrobacter*
Genus *Brevibacterium*
Genus *Curtobacterium*
Genus *Caseobacter*
Genus *Microbacterium*
Genus *Microbacterium*
Genus *Aureobacterium*
Genus *Cellulomonas*
Genus *Agromyces*
Genus *Arachnia*
Genus *Rothia*
Genus *Propionibacterium*
Genus *Eubacterium*
Genus *Acetobacterium*
Genus *Lachnospira*
Genus *Butyrivibrio*
Genus *Thermoanaerobacter*
Genus *Actinomyces*
Genus *Bifidobacterium*

SECTION 16
MYCOBACTERIA

Family I. Mycobacteriaceae
Genus I. *Mycobacterium*

SECTION 17
NOCARDIFORMS

Genus *Nocardia*
Genus *Rhodococcus*
Genus *Nocardioides*
Genus *Pseudonocardia*
Genus *Oerskovia*
Genus *Saccharopolyspora*
Genus *Micropolyspora*

Genus *Promicromonospora*
Genus *Intrasporangium*

SECTION 18
GLIDING, NON-FRUITING BACTERIA
Order I. Cytophagales
Family I. Cytophagaceae
Genus I. *Cytophaga*
Genus II. *Sporocytophaga*
Genus III. *Capnocytophaga*
Genus IV. *Flexithrix*
Genus V. *Flexibacter*
Genus VI. *Microscilla*
Genus VII. *Saprospira*
Genus VIII. *Herpetosiphon*
Order II. Lysobacterales
Family I. Lysobacteraceae
Genus I. *Lysobacter*
Order III. Beggiatoales
Family I. Beggiatoaceae
Genus I. *Beggiatoa*
Genus II. *Thioploca*
Genus III. *Thiospirillopsis*
Genus IV. *Thithrix*
Genus V. *Achromatium*
Family II. Simonsiellaceae
Genus I. *Simonsiella*
Genus II. *Alysiella*
Family III. Leucotrichaceae
Genus I. *Leucothrix*
Family IV. Pelonemataceae
Genus I. *Pelonema*
Genus II. *Achroonema*
Genus III. *Peloploca*
Genus IV. *Desmanthus*
Families and genera *incertae sedis*
Genus *Toxothrix*
Genus *Vitreoscilla*
Genus *Chitinophagen*
Genus *Desulfonema*

SECTION 19
ANOXYGENIC PHOTOTROPHIC BACTERIA
PURPLE BACTERIA
Family I. Chromatiaceae
Genus I. *Chromatium*
Genus II. *Thicystis*
Genus III. *Thispirillum*
Genus IV. *Thiocapsa*
Genus V. *Amoebobacter*
Genus VI. *Lamprobacter*
Genus VII. *Lamprocystis*
Genus VIII. *Thiodictyon*
Genus IX. *Thiopedia*
Family II. Ectothiorhodospirqceae
Genus I. *Ectothiorhodospira*
Purple nonsulfur bacteria
Genus *Rhodospirillum*
Genus *Rhodopseudomonas*
Genus *Rhodobacter*
Genus *Rhodomicrobium*
Genus *Rhodopila*
Genus *Rhodocyclus*
GREEN BACTERIA
Green sulfur bacteria
Genus *Chlorobium*
Genus *Prosthecochloris*
Genus *Ancalochloris*
Genus *Pelodyction*
Genus *Chloroherpeton*
Genus Symbiotic consortia
Multicellular filamentous green bacteria
Genus *Chloroflexus*
Genus *Heliothrix*
Genus *Oscillochloris*
Genus *Chloronema*
GENERA *INCERTAE SEDIS*
Genus *Heliobacterium*
Genus *Erythrobacter*

SECTION 20
BUDDING AND/OR APPENDAGED BACTERIA
PROSTHECATE BACTERIA
Budding bacteria
Genus *Hyphomicrobium*
Genus *Hyphomonas*
Genus *Pedomicrobium*
Genus *"Filomicrobium"*
Genus *"Dicotomicrobium"*
Genus *"Tetramicrobium"*
Genus *Stella*
Genus *Ancalomicrobium*
Genus *Prosthecomicrobium*
Non-budding bacteria
Genus *Caulobacter*
Genus *Asticcacaulis*
Genus *Prosthecobacter*
Genus *Thiodendron*
NON-PROSTHECATE BACTERIA
Budding bacteria
Genus *Planctomyces*
Genus *Pasteuria*
Genus *Blastobacter*
Genus *Angulomicrobium*
Genus *Gemmiger*
Genus *Ensifer*
Genus *Isosphaera*
Non-budding stalked bacteria
Genus *Gallionella*
Genus *Nevskia*
Morphologically unusual budding bacteria involved in iron and manganese deposition
Genus *Seliberia*
Genus *Metallogenium*
Genus *Caulococcus*
Genus *Kuznezovia*
Others
Spinate bacteria

SECTION 21
ARCHAEOBACTERIA
METHANOGENIC BACTERIA
Genus *Methanococcus*
Genus *Methanobrevibacter*
Genus *Methanococcus*
Genus *Methanomicrobium*
Genus *Methanospirillum*
Genus *Methanosarcina*
Genus *Methanococcoides*
Genus *Methanothermus*
Genus *Methanolobus*
Genus *Methanoplanus*
Genus *Methanogenium*
Genus *Methanothrix*
EXTREME HALOPHILIC BACTERIA
Genus *Halobacterium*
Genus *Halococcus*
EXTREME THERMOPHILIC BACTERIA
Genus *Thermoplasma*
Genus *Sulfolobus*
Genus *Thermoproteus*
Genus *Thermofilum*
Genus *Thermococcus*
Genus *Desulfurococcus*
Genus *Thermodiscus*
Genus *Pyrodictium*

SECTION 22
SHEATHED BACTERIA
Genus *Sphaerotilus*
Genus *Leptothrix*
Genus *Haliscominobacter*
Genus *Lieskeella*
Genus *Phargmidiothrix*
Genus *Crenothrix*
Genus *Clonothrix*

SECTION 23
GLIDING, FRUITING BACTERIA
Order I. Myxobacteriales
Family I. Myxococcaceae
Genus I. *Myxococcus*
Family II. Archangiaceae
Genus I. *Archangium*
Family III. Cystobacteraceae
Genus I. *Cystobacter*
Genus II. *Melittangium*
Genus III. *Stigmatella*
Family IV. Polyangiaceae
Genus I. *Polyangium*
Genus II. *Nannocystis*
Genus III. *Chondromyces*
Genus *incerta sedis*
Genus *Angiococcus*

SECTION 24
CHEMOLITHOTROPHIC BACTERIA
NITRIFIERS
Family I. Nitrobacteraceae
Genus I. *Nitrobacter*
Genus II. *Nitrospina*
Genus III. *Nitrococcus*
Genus IV. *Nitrosomonas*
Genus V. *Nitrosopira*
Genus VI. *Nitrosococcus*
Genus VII. *Nitrosolobus*
SULFUR OXIDIZERS
Genus *Thibacillus*
Genus *Thiomicrospira*
Genus *Thiobacterium*
Genus *Thiospira*
Genus *Micromonas*
OBLIGATE HYDROGEN OXIDIZERS
Genus *Hydrogenbacter*
METAL OXIDIZERS AND DEPOSITERS
Family I. Siderocapsaceae
Genus I. *Siderocapsa*
Genus II. *Naumaniella*
Genus III. *Ochrobium*
Genus IV. *Siderococcus*
OTHER MAGNETOTACTIC BACTERIA

SECTION 25
CYANOBACTERIA

SECTION 26
OTHERS
Order I. Prochlorales
Family I. Prochloraceae
Genus I. *Prochloron*

SECTION 27
ACTINOMYCETES THAT DIVIDE IN MORE THAN ONE PLANE

Genus *Geodermatophilus*
Genus *Dermatophilus*
Genus *Frankia*
Genus *Tonsilophilus*

SECTION 28
SPORANGIATE ACTINOMYCETES

Genus *Actinoplanes* (including *Amorphosporangium*)
Genus *Streptosporangium*
Genus *Ampullariella*
Genus *Spirillospora*
Genus *Pilimelia*
Genus *Dactylosporangium*
Genus *Planomonospora*
Genus *Planobispora*

SECTION 29
STREPTOMYCETES AND THEIR ALLIES

Genus *Streptomyces*
Genus *Streptoverticillium*
Genus *Actinopycnidium*
Genus *Actinosporangium*
Genus *Chainia*
Genus *Elytrosporangium*
Genus *Microellobosporia*

SECTION 30
OTHER CONIDIATE GENERA

Genus *Actinopolyspora*
Genus *Actinosynnema*
Genus *Kineospora*
Genus *Kitasatosporia*
Genus *Microbisporia*
Genus *Micromonospora*
Genus *Microtetrospora*
Genus *Saccharomonospora*
Genus *Sporichthya*
Genus *Streptoalloteichus*
Genus *Thermomonospora*
Genus *Actinomadura*
Genus *Nocardiopsis*
Genus *Exellospora*
Genus *Thermoactinomyces*

APPENDIX C
DISEASE REFERENCE GUIDE

DISEASE Acquired immunodeficiency syndrome

CAUSE(S) Human T-cell leukemia virus (HTLV-III)

SIGNS AND SYMPTOMS Inability to develop a strong immune response to infectious agents. Susceptibility to many infectious agents

MODE OF TRANSMISSION Personal contact. Blood transfusions

PREVENTIVE MEASURES Avoid contact with infected individuals. Receive blood that has been screened for HTLV-III

TREATMENT OF CHOICE None known

DISEASE Actinomycosis

CAUSE(S) *Actinomyces israelii*

SIGNS AND SYMPTOMS Formation of sinuses, slow-forming granulomas, discharge of pus. Affects face, neck, lungs, or intestines

MODE OF TRANSMISSION Traumatic implantation

PREVENTIVE MEASURES Avoid contact with contaminated material; no vaccine available

TREATMENT OF CHOICE Penicillin or tetracyclines

DISEASE African sleeping sickness

CAUSE(S) *Trypanosoma gambiense* or *T. rhodesiense*

SIGNS AND SYMPTOMS Fever, chancre (at site of bite), rash, splenomegaly, headache, confusion, somnolence, and coma

MODE OF TRANSMISSION Bite of tsetse fly

PREVENTIVE MEASURES Sleep with mosquito netting, and wear clothing that covers ankles and wrists

TREATMENT OF CHOICE Melarsoprol and suramin

DISEASE Amebiasis

CAUSE(S) *Entamoeba histolytica*

SIGNS AND SYMPTOMS Dysentery-like symptoms. Fever, malaise, diarrhea with blood and mucus, painful intestinal contractions

MODE OF TRANSMISSION Fecal-oral route

PREVENTIVE MEASURES Sanitize foods and water before consumption

TREATMENT OF CHOICE Quinacrine hydrochloride, metrodinazole

DISEASE Anthrax

CAUSE(S) *Bacillus anthracis*

SIGNS AND SYMPTOMS Development of a malignant pustule at site of infection, necrosis of involved lymph nodes, respiratory distress, shock, coma, death

MODE OF TRANSMISSION Traumatic implantation or inhalation

PREVENTIVE MEASURES Vaccination

TREATMENT OF CHOICE Penicillin or tetracycline

DISEASE Babesiosis

CAUSE(S) *Babesia* sp. (a protozoan)

SIGNS AND SYMPTOMS Fever, sweats, chills, malaise, pain in muscles and joints, and changes in mood

MODE OF TRANSMISSION Bite of an *Ixodes* tick associated with rodents and pets

PREVENTIVE MEASURES Avoid contact with tick-infested animals

TREATMENT OF CHOICE Cloroquine phosphate

DISEASE Bacillary dysentery

CAUSE(S) *Shigella sp., Campylobacter jejuni, Escherichia coli*

SIGNS AND SYMPTOMS Abdominal pain and cramps, vomiting, diarrhea with blood and mucus, fever, and malaise

MODE OF TRANSMISSION Fecal-oral route

PREVENTIVE MEASURES Cook foods well; sanitize water

TREATMENT OF CHOICE Ampicillin, trimethoprim-sulfamethoxazole, gentamycin

DISEASE Botulism

CAUSE(S) *Clostridium botulinum*

SIGNS AND SYMPTOMS Fatigue, weakness, dizziness, nausea, vomiting, diarrhea, and cardiac and respiratory paralysis

MODE OF TRANSMISSION Ingestion of foods containing botulinal toxin

PREVENTIVE MEASURES Cook foods well, especially canned goods

TREATMENT OF CHOICE Administration of polyvalent antiserum

DISEASE Brucellosis

CAUSE(S) *Brucella abortus*, *B. melitensis*, and *B. suis*

SIGNS AND SYMPTOMS Fever, often fluctuating; body aches and anorexia (loss of appetite)

MODE OF TRANSMISSION Ingestion of contaminated dairy products or other foods; skin abrasions or inhalation

PREVENTIVE MEASURES Sanitize foods

TREATMENT OF CHOICE Tetracycline and streptomycin administered together

DISEASE Candidiasis

CAUSE(S) *Candida albicans*

SIGNS AND SYMPTOMS Development of a pseudomembrane and a yellowish exudate

MODE OF TRANSMISSION Disease usually endogenous. Some forms (vaginitis and balanitis) are transmitted sexually

PREVENTIVE MEASURES Maintain overall good health

TREATMENT OF CHOICE Mycostatin, amphotericin B, miconazole

DISEASE Chagas' disease

CAUSE(S) *Trypanosoma cruzi*

SIGNS AND SYMPTOMS Fever, enlargement of lymph nodes, enlargement of the liver, and facial edema

MODE OF TRANSMISSION Bite of reduviid bug

PREVENTIVE MEASURES Sanitation of dwelling; vaccination (experimental)

TREATMENT OF CHOICE Nifurtimox

DISEASE Chancroid

CAUSE(S) *Haemophilus ducreyi*

SIGNS AND SYMPTOMS Pustule or ulcer at site of infection. Scar remains after healing. Progresses rapidly

MODE OF TRANSMISSION Sexually transmitted

PREVENTIVE MEASURES Avoid sexual contact with infected individuals

TREATMENT OF CHOICE Sulfonamides, chloroamphenicol, tetracyclines

DISEASE Cholera

CAUSE(S) *Vibrio cholerae*

SIGNS AND SYMPTOMS Violent vomiting, copious rice water stool (diarrhea). Hypovolemic shock and death in severe cases

MODE OF TRANSMISSION Fecal-oral route

PREVENTIVE MEASURES Vaccination; sanitize water before drinking

TREATMENT OF CHOICE Tetracyclines, and replacement of fluids and electrolytes

DISEASE Coccidioidomycosis

CAUSE(S) *Coccidioides immitis*

SIGNS AND SYMPTOMS Chronic cough, fever, chest pains, malaise, weight loss

MODE OF TRANSMISSION Inhalation of conidia

PREVENTIVE MEASURES Vaccination (experimental); wear face masks during dust storms in endemic areas

TREATMENT OF CHOICE Amphotericin B, ketoconazole

DISEASE Common cold

CAUSE(S) Rhinoviruses, coronaviruses, and adenoviruses

SIGNS AND SYMPTOMS Cough, runny nose, fever, malaise

MODE OF TRANSMISSION Hand-to-nose, mouth or eye. Some via aerosols

PREVENTIVE MEASURES Wash hands; avoid contact with infected individuals

TREATMENT OF CHOICE None

DISEASE Conjunctivitis

CAUSE(S) *Haemophilus influenzae*, *Streptococcus pneumoniae*, *Moraxella lacunata*, or *Chalamydia trachomatis*

SIGNS AND SYMPTOMS Tearing, itching, and inflammation of the conjunctive. Sometimes blindness

MODE OF TRANSMISSION Hand-to-eye, aerosols

PREVENTIVE MEASURES Wash hands; avoid contact with infected individuals

TREATMENT OF CHOICE Chloroamphenicol, penicillin, tetracyclines

DISEASE Croup

CAUSE(S) Respiratory syncytial virus, parainfluenza virus, and Influenza A and B viruses

SIGNS AND SYMPTOMS Laryngitis, loss of voice, deep, noisy cough, difficult breathing, rapid pulse, and fever

MODE OF TRANSMISSION Aerosols

PREVENTIVE MEASURES Avoid contact with infected individuals

TREATMENT OF CHOICE Hot compresses on throat, humidify air; establishment of airway if necessary

DISEASE Cystitis

CAUSE(S) *Escherichia coli*, *Proteus mirabilis*, *Enterobacter aerogenes*, *Klebsiella pneumoniae*

SIGNS AND SYMPTOMS Painful and frequent urination, hematuria

MODE OF TRANSMISSION Endogenous infection

PREVENTIVE MEASURES Practive good hygiene

TREATMENT OF CHOICE Gentamycin or tobramycin

DISEASE Dengue

CAUSE(S) Group B arbovirus

SIGNS AND SYMPTOMS High fever of sudden onset, headache, prostration, joint and muscleache, enlargement of lymphnodes, and a rash

MODE OF TRANSMISSION Bite of mosquito

PREVENTIVE MEASURES Control mosquito population; sleep within mosquito net

TREATMENT OF CHOICE None proven useful

DISEASE Dermatophytosis

CAUSE(S) Fungi in the genera *Microsporum*, *Trichophyton*, and *Epidermophyton*

SIGNS AND SYMPTOMS Athlete's foot, jock itch, ringlike scalp and skin lesions

MODE OF TRANSMISSION Contact with humans, animals, or fomites

PREVENTIVE MEASURES Avoid contact with animals or objects carrying infectious agents

TREATMENT OF CHOICE Oral griseofulvin

DISEASE Diphtheria

CAUSE(S) *Corynebacterium diphtheriae*

SIGNS AND SYMPTOMS Fever, headache, malaise. Sore throat, with yellowish membrane in throat and tonsils. Adenitis. Death due to cardiac failure

MODE OF TRANSMISSION Direct contact with human carrier, or via fomites

PREVENTIVE MEASURES Vaccination

TREATMENT OF CHOICE Erythromycin and administration of diphtheria antitoxin

DISEASE Epiglottitis

CAUSE(S) *Haemophilus influenzae*

SIGNS AND SYMPTOMS Sore throat, fever, cough, drooling, cyanosis, coma, and (if airway is not reestablished) death

MODE OF TRANSMISSION Endogenous or by aerosols

PREVENTIVE MEASURES Avoid contact with infected individuals

TREATMENT OF CHOICE Establish airway and then treat with chloroamphenicol

DISEASE Erysipelas

CAUSE(S) *Streptococcus pyogenes*

SIGNS AND SYMPTOMS Fever, chills, nausea, painful and warm skin. Blisters on skin may develop

MODE OF TRANSMISSION Aerosols, person-to-person contact

PREVENTIVE MEASURES Avoid contact with infected individuals

TREATMENT OF CHOICE Penicillin or erythromycin

DISEASE Erysipeloid

CAUSE(S) *Erysipelothrix rhusiopathiae*

SIGNS AND SYMPTOMS Painful, reddish skin eruptions

MODE OF TRANSMISSION Scratches and cuts after handling infected animals (e.g., fish, shellfish, and poultry)

PREVENTIVE MEASURES Wear gloves when handling shellfish, etc.

TREATMENT OF CHOICE Penicillin or gentamycin

DISEASE Gastroenteritis

CAUSE(S) *Salmonella* sp., *Campylobacter jejuni*, *Shigella* sp., *Escherichia coli*

SIGNS AND SYMPTOMS Diarrhea, vomiting, fever, and malaise

MODE OF TRANSMISSION Fecal-oral route

PREVENTIVE MEASURES Sanitize or cook foods and beverages before consuming

TREATMENT OF CHOICE Usually none necessary; ampicillin, chloroamphenicol, gentamycin, or tobramycin for severe cases

DISEASE Genital herpes

CAUSE(S) *Herpes simplex* virus

SIGNS AND SYMPTOMS Painful, recurring blisters on genitals

MODE OF TRANSMISSION Sexually transmitted

PREVENTIVE MEASURES Avoid sexual contact with actively infected individuals

TREATMENT OF CHOICE Zovirax (topical, oral, or parenteral)

DISEASE Glanders

CAUSE(S) *Pseudomonas mallei*

SIGNS AND SYMPTOMS Fever, inflammation of skin and mucous membranes, resulting in abscesses and ulcers. Discharge is foul-smelling

MODE OF TRANSMISSION Contact with infected horses, donkeys, or mules

PREVENTIVE MEASURES Avoid contact with infected animals

TREATMENT OF CHOICE Sulfadiazine

DISEASE Gonorrhea

CAUSE(S) *Neisseria gonorrhoeae*

SIGNS AND SYMPTOMS In males: Yellow, mucopurulent discharge from urethra with painful urination. In females: urethral or vaginal discharge; painful, frequent urination

MODE OF TRANSMISSION Sexually transmitted

PREVENTIVE MEASURES Avoid sexual contact with infected individuals

TREATMENT OF CHOICE Penicillin or ampicillin

DISEASE Granuloma inguinale

CAUSE(S) *Calymmatobacterium inguinale*

SIGNS AND SYMPTOMS Ulcer that appears in genital region. Initial lesion is painless nodule

MODE OF TRANSMISSION Sexually transmitted

PREVENTIVE MEASURES Avoid sexual contact with infected individuals

TREATMENT OF CHOICE Tetracyclines or streptomycin

DISEASE Giardiasis

CAUSE(S) *Giardia intestinalis*

SIGNS AND SYMPTOMS Malodorous flatus, greasy stools, diarrhea

MODE OF TRANSMISSION Fecal-oral route

PREVENTIVE MEASURES Sanitize water

TREATMENT OF CHOICE Metrodinazole, quinacrine

DISEASE Hepatitis, viral

CAUSE(S) Hepatitis A, B, or non-A non-B viruses

SIGNS AND SYMPTOMS Loss of appetite, jaundice, fever, weakness

MODE OF TRANSMISSION Fecal-oral route or parenteral

PREVENTIVE MEASURES Sanitize foods or beverages; practice good hygiene

TREATMENT OF CHOICE Treat symptoms to relieve discomfort

DISEASE Histoplasmosis

CAUSE(S) *Histoplasma capsulatum*

SIGNS AND SYMPTOMS Fever, malaise, cough, chest pain

MODE OF TRANSMISSION Inhalation of conidia or hyphal fragments

PREVENTIVE MEASURES Wear masks during dust storms and demolition of old buildings

TREATMENT OF CHOICE Amphotericin B or miconazole

DISEASE Impetigo

CAUSE(S) *Staphylococcus aureus* or *Streptococcus* group A.

SIGNS AND SYMPTOMS Development of pustules on the skin, which become encrusted and rupture

MODE OF TRANSMISSION Direct contact with nasal secretions of carriers or exudates of infected individuals.

PREVENTIVE MEASURES Avoid contact with infected individuals

TREATMENT OF CHOICE Penicillin or erythromycin

DISEASE Infectious mononucleosis

CAUSE(S) Epstein-Barr virus

SIGNS AND SYMPTOMS Increased number of mononuclear white blood cells; weakness, fever, sore throat, and lymphoadenopathy

MODE OF TRANSMISSION Kissing infected individuals, or contact with contaminated utensils or glassware

PREVENTIVE MEASURES Sanitize glassware, avoid contact with infected individuals

TREATMENT OF CHOICE Treat symptoms, bed rest

DISEASE Influenza

CAUSE(S) Myxoviruses

SIGNS AND SYMPTOMS Fever, chills, headache, cough, runny nose, muscle aches, and sore throat

MODE OF TRANSMISSION Aerosols

PREVENTIVE MEASURES Vaccination; avoid contact with infected individuals

TREATMENT OF CHOICE Amantadine, bed rest, and symptomatic treatment

DISEASE Legionellosis

CAUSE(S) *Legionella pneumophila* and other legionellae

SIGNS AND SYMPTOMS Fever, chills, pneumonia, cough, and chest pain. Onset very rapid

MODE OF TRANSMISSION Aerosols created by air-conditioning systems

PREVENTIVE MEASURES Sanitize water in air-cooling systems

TREATMENT OF CHOICE Erythromycin or rifampin

DISEASE Leishmaniasis (kala-azar)

CAUSE(S) *Leishmania donovani* (a protozoan)

SIGNS AND SYMPTOMS Weakness, weight loss, diarrhea, fever (spikes twice daily), and enlargement of liver and spleen

MODE OF TRANSMISSION Bite of *Phlebotomus* fly

PREVENTIVE MEASURES Wear protective clothing to avoid bite of fly

TREATMENT OF CHOICE Sodium stibogluconate or pentamidine

DISEASE Leprosy

CAUSE(S) *Mycobacterium leprae*

SIGNS AND SYMPTOMS Anesthesia of affected parts of the skin, paralysis, ulceration, and deformity of affected areas

MODE OF TRANSMISSION Direct contact with infected individuals or formites

PREVENTIVE MEASURES Avoid close personal contact with infected individuals

TREATMENT OF CHOICE Dapsone

DISEASE Leptospirosis

CAUSE(S) *Leptospira* sp.

SIGNS AND SYMPTOMS Influenza-like symptoms. Complications may result in liver or kidney disease

MODE OF TRANSMISSION Through abrasions of skin, after contact with contaminated water or animal urine

PREVENTIVE MEASURES Sanitize implements, wear gloves, avoid contact with infected materials

TREATMENT OF CHOICE Penicillin

DISEASE Lyme disease

CAUSE(S) *Borrelia* sp.

SIGNS AND SYMPTOMS Fever, joint pain, and a ringlike rash at site of infection

MODE OF TRANSMISSION Bites of ticks

PREVENTIVE MEASURES Wear long sleeves and head coverings when venturing into tick-infected areas

TREATMENT OF CHOICE Tetracyclines

DISEASE Malaria

CAUSE(S) *Plasmodium vivax, P. falciparum, P. malariae, P. ovale*

SIGNS AND SYMPTOMS Bouts of alternating fever and chills every 2-3 days; sweating and prostration

MODE OF TRANSMISSION Bite of female *Anopheles* mosquito

PREVENTIVE MEASURES Prophylaxis of chloroquine phosphate once a week

TREATMENT OF CHOICE Chloroquine phosphate and quinine sulfate

DISEASE Melioidosis

CAUSE(S) *Pseudomonas pseudomallei*

SIGNS AND SYMPTOMS Pneumonia, multiple abscesses, septicemia, and sometimes death

MODE OF TRANSMISSION Inhalation

PREVENTIVE MEASURES Avoid contact with infected individuals

TREATMENT OF CHOICE Carbenicillin

DISEASE Meningitis

CAUSE(S) *Neisseria meningitidis, Haemophilus influenzae, streptococcus pneumoniae.* Viruses, protozoa, and fungi also can cause meningitis.

SIGNS AND SYMPTOMS Fever, loss of appetite, hyperesthesia, delirium, convulsions, and coma

MODE OF TRANSMISSION Direct contact, aerosols

PREVENTIVE MEASURES Prophylaxis with antibiotics; isolating active cases

TREATMENT OF CHOICE Ampicillin, penicillin, kanamycin, chloroamphenicol, and gentamycin

DISEASE Mumps

CAUSE(S) Mumps virus

SIGNS AND SYMPTOMS Malaise, chills, headache, swelling of parotid glands. Orchitis is a complication

MODE OF TRANSMISSION Aerosols or contact with secretions contaminated with the mumps virus

PREVENTIVE MEASURES Vaccination

TREATMENT OF CHOICE Rest and analgesics

DISEASE Mycetoma

CAUSE(S) *Pesudoallescheria boydii, Madurella*, and some actinomycetes

SIGNS AND SYMPTOMS Swelling of limb, draining sinuses, and formation of grains

MODE OF TRANSMISSION Traumatic implantation

PREVENTIVE MEASURES Wear shoes and gloves when gardening or handling vegetable materials

TREATMENT OF CHOICE Amphotericin B or surgical excision of infected areas. Penicillin for bacterial disease

DISEASE Nocardiosis

CAUSE(S) *Nocardia asteroides* or *N. brasiliensis*

SIGNS AND SYMPTOMS Productive cough, chest pain, weakness, malaise

MODE OF TRANSMISSION Aerosols

PREVENTIVE MEASURES Avoid contact with infected individuals

TREATMENT OF CHOICE Sulfisoxazole

DISEASE Nongonococcal urethritis

CAUSE(S) *Chlamydia trachomatis, Ureaplasma urealyticum*

SIGNS AND SYMPTOMS Yellowish urethral discharge, with painful urination

MODE OF TRANSMISSION Sexually transmitted

PREVENTIVE MEASURES Avoid sexual contact with infected individuals

TREATMENT OF CHOICE Tetracyclines, erythromycin

DISEASE Nonspecific vaginitis

CAUSE(S) *Gardnerella vaginalis* and certain anaerobes

SIGNS AND SYMPTOMS Profuse, malodorous vaginal discharge; clue cells (Epithelical cells with gram-variable rods attached) present.

MODE OF TRANSMISSION Sexually transmitted

PREVENTIVE MEASURES Avoid sexual contact with infected individuals

TREATMENT OF CHOICE Metrodinazole or ampicillin

DISEASE Pharyngotonsillitis

CAUSE(S) *Streptococcus pyogenes*

SIGNS AND SYMPTOMS Sore throat, fever, and swollen lymph nodes of neck region

MODE OF TRANSMISSION Contact with carriers, or ingestion of contaminated foods.

PREVENTIVE MEASURES Avoid contact with infected individuals or eating foods handled by them

TREATMENT OF CHOICE Penicillin or erythromycin

DISEASE Pinta

CAUSE(S) *Treponema carateum*

SIGNS AND SYMPTOMS Depigmented skin spots on hands and feet

MODE OF TRANSMISSION Direct contact with infected individuals

PREVENTIVE MEASURES Avoid contact with infected individuals

TREATMENT OF CHOICE Penicillin

DISEASE Plague

CAUSE(S) *Yersinia pestis*

SIGNS AND SYMPTOMS High fever, restlessness, confusion, prostration, inflammation of lymph nodes (buboes), delirium, shock, and death

MODE OF TRANSMISSION Bite of infected fleas

PREVENTIVE MEASURES Vaccination; removal of fleas and rodents

TREATMENT OF CHOICE Streptomycin

DISEASE Pneumonia

CAUSE(S) *Streptococcus pneumoniae, Klebsiella pneumoniae, Mycoplasma pneumoniae,* Influenza viruses

SIGNS AND SYMPTOMS Chills, high fever of rapid onset, chest pain, purulent exudate from cough, blood in sputum

MODE OF TRANSMISSION Endogenous or from aerosols

PREVENTIVE MEASURES Vaccination, or avoid contact with infected individuals

TREATMENT OF CHOICE Penicillin, gentamycin, erythromycin

DISEASE Poliomyelitis

CAUSE(S) Polio viruses

SIGNS AND SYMPTOMS Coldlike symptoms or diarrhea, vomiting; low-grade fever; paralysis may develop

MODE OF TRANSMISSION Fecal-oral route

PREVENTIVE MEASURES Vaccination

TREATMENT OF CHOICE None specific. Alleviate symptoms

DISEASE Pseudomembranous enterocolitis

CAUSE(S) *Clostridium difficile*

SIGNS AND SYMPTOMS Diarrhea, vomiting, abdominal pain, and plaque-formation in colon mucosa

MODE OF TRANSMISSION Complication following treatment with certain antibiotics

PREVENTIVE MEASURES None that are effective

TREATMENT OF CHOICE Penicillin

DISEASE Psittacosis

CAUSE(S) *Chlamydia psittaci*

SIGNS AND SYMPTOMS Headache, nausea, chills and fever, cough, and chest pains

MODE OF TRANSMISSION Contact with infected birds

PREVENTIVE MEASURES Avoid contact with infected birds (parrots and parakeets)

TREATMENT OF CHOICE Tetracyclines

DISEASE Puerperal sepsis

CAUSE(S) *Streptococcus pyogenes* or *Clostridium* sp.

SIGNS AND SYMPTOMS Fever above 38°C on two consecutive days after parturition or abortion

MODE OF TRANSMISSION Contamination from gloves or surgical instruments during delivery or abortion

PREVENTIVE MEASURES Practice of aseptic techniques during delivery or abortion

TREATMENT OF CHOICE Penicillin, erythromycin

DISEASE Q-fever

CAUSE(S) *Coxiella burnetii*

SIGNS AND SYMPTOMS Headache, fever, myalgia, weight loss, and malaise

MODE OF TRANSMISSION Drinking contaminated milk; inhalation of cells; handling infected animals

PREVENTIVE MEASURES Pasteurize milk; avoid handling infected animals

TREATMENT OF CHOICE Tetracyclines

DISEASE Rabies

CAUSE(S) Rabies virus (a rhabdovirus)

SIGNS AND SYMPTOMS Malaise, depression, hydrophobia, hyperesthesia, aggressiveness, and death

MODE OF TRANSMISSION Bite of rabid dog, skunk, bat, or other animals

PREVENTIVE MEASURES Vaccination

TREATMENT OF CHOICE Vaccination (Pasteur treatment) immediately after bite

DISEASE Relapsing fever

CAUSE(S) *Borrelia recurrentis*

SIGNS AND SYMPTOMS Fever of sudden onset, lasting about 3 days. Bouts of fever recur every 10 days or so

MODE OF TRANSMISSION Bite of body louse (*Pediculus humanus*) or ticks (*Ornithodorus*)

PREVENTIVE MEASURES Practice good hygiene; avoid contact with louse-infested individuals; wear protective clothing when visiting tick-infested areas

TREATMENT OF CHOICE Tetracyclines

DISEASE Rheumatic fever

CAUSE(S) *Streptococcus pyogenes*

SIGNS AND SYMPTOMS Fever, arthralgia, myocarditis, following streptococcal infection

MODE OF TRANSMISSION Allergic response to previous streptococcal infection

PREVENTIVE MEASURES Prophylaxis with penicillin; treat streptococcal infections promptly

TREATMENT OF CHOICE Penicillin, with bed rest

DISEASE Rocky Mountain Spotted Fever

CAUSE(S) *Rickettsia rickettsii*

SIGNS AND SYMPTOMS Rash, fever, headache, myalgia

MODE OF TRANSMISSION Bite of infected ticks

PREVENTIVE MEASURES Wear protective clothing when visiting tick-infested areas

TREATMENT OF CHOICE Tetracyclines

DISEASE Rubella

CAUSE(S) Rubella virus

SIGNS AND SYMPTOMS Upper respiratory tract disease, accompanied by a rash. Fetal defects or miscarriages occur if mother is infected during first trimester of pregnancy

MODE OF TRANSMISSION Aerosols

PREVENTIVE MEASURES Vaccination

TREATMENT OF CHOICE Treat symptoms

DISEASE Rubeola

CAUSE(S) Measles virus

SIGNS AND SYMPTOMS Upper respiratory tract disease, accompanied by mouth lesions called "Koplik spots"

MODE OF TRANSMISSION Aerosols and person-to-person contact

PREVENTIVE MEASURES Vaccination

TREATMENT OF CHOICE Treat symptoms

DISEASE Scalded skin syndrome

CAUSE(S) *Staphylococcus aureus*

SIGNS AND SYMPTOMS Blistering and shedding of outermost layer of skin

MODE OF TRANSMISSION Occurs mostly in infants. Contact with persons or objects carrying exfoliatin-producing pathogens

PREVENTIVE MEASURES None that are effective

TREATMENT OF CHOICE Penicillin

DISEASE Scarlet fever

CAUSE(S) *Streptococcus pyogenes*

SIGNS AND SYMPTOMS Complication of "strep throat." Development of rash, roughening of the skin, and a "strawberry" tongue. Vomiting may also occur

MODE OF TRANSMISSION Direct contact with carriers; ingestion of contaminated foods

PREVENTIVE MEASURES Avoid contact with carriers or foods handled by infected individuals

TREATMENT OF CHOICE Penicillin or erythromycin

DISEASE Schistosomiasis

CAUSE(S) *Schistosoma mansoni*, *s. haematobium*, or *s. japonicum* (trematodes)

SIGNS AND SYMPTOMS Dermatitis (at site of infection), followed by fever, cough, spleen and liver-enlargement, malaise, and eosinophilia. Hematuria, dysuria, and frequency occur in chronic cases.

MODE OF TRANSMISSION Cercaria penetrate through the skin (usually feet) of wading individuals

PREVENTIVE MEASURES Avoid wading in host snail–infested waters

TREATMENT OF CHOICE Praziquantel or oxamniquine

DISEASE Syphilis

CAUSE(S) *Treponema pallidum*

SIGNS AND SYMPTOMS Development of a chancre (at site of treponemal entrance). Healing of chancre followed by a rash, runny nose, and pronounced tearing

MODE OF TRANSMISSION Sexually transmitted

PREVENTIVE MEASURES Avoid sexual contact with infected individuals

TREATMENT OF CHOICE Penicillin

DISEASE Taeniasis

CAUSE(S) *Taeniarhynchus saginatus* (beef tapeworm), *Taenia solium* (pork tapeworm)

SIGNS AND SYMPTOMS Abdominal pain, weight loss, and weakness. Cysticercosis may develop after *Taenia solium* infections

MODE OF TRANSMISSION Eating uncooked meats with cysticerci of tapeworms

PREVENTIVE MEASURES Cook meats well

TREATMENT OF CHOICE Niclosamide

DISEASE Tetanus

CAUSE(S) *Clostridium tetani*

SIGNS AND SYMPTOMS Stiffness of jaw and neck muscles. Rigidity of jaw, alteration of voice, tetanic contractions of muscles of back and extremities, arching the body

MODE OF TRANSMISSION Traumatic implantation of endospores

PREVENTIVE MEASURES Vaccination

TREATMENT OF CHOICE Booster shot, injection of antitoxin, and treatment with penicillin

DISEASE Toxic shock syndrome

CAUSE(S) *Staphylococcus aureus*

SIGNS AND SYMPTOMS Fever, diarrhea, low blood pressure, myalgia, and a rash, which may be followed by peeling of skin; shock

MODE OF TRANSMISSION Endogenous, or in contaminated tampons

PREVENTIVE MEASURES Change tampons frequently; practice good hygiene

TREATMENT OF CHOICE Penicillin, and treatment to reverse shock

DISEASE Toxoplasmosis

CAUSE(S) *Toxoplasma gondii*

SIGNS AND SYMPTOMS Swelling of lymph nodes, inflammation of the eye, chorioretinitis, and sometimes neurological symptoms

MODE OF TRANSMISSION Ingestion of oocysts of *Toxoplasma gondii*, in foods or from fingers.

PREVENTIVE MEASURES Wear gloves when changing "kitty litter"; wash hands

TREATMENT OF CHOICE Pyrimethamine and sulfonamides

DISEASE Trichinosis

CAUSE(S) *Trichinella spiralis* (a nematode)

SIGNS AND SYMPTOMS Nausea, vomiting, constipation, or abdominal pain, followed by fever, myalgia, facial edema, and hemorrhaging in eye or tongue

MODE OF TRANSMISSION Eating poorly cooked pork or bear meat (or other meats) with encysted larvae

PREVENTIVE MEASURES Cook meats well (especially pork and bear)

TREATMENT OF CHOICE Serious cases are treated with prednisone and thiabendazole

DISEASE Tuberculosis

CAUSE(S) *Mycobacterium tuberculosis* or *M. bovis*

SIGNS AND SYMPTOMS Persistent cough, sputum production, weakness, prostration, weight loss. X-rays reveal many lung lesions

MODE OF TRANSMISSION Inhalation of aerosol particles, or ingestion of contaminated milk.

PREVENTIVE MEASURES Avoid contact with infected individuals; pasteurize milk

TREATMENT OF CHOICE Streptomycin, Para-amino salicylic acid, rifampin

DISEASE Tularemia

CAUSE(S) *Franciscella tularensis*

SIGNS AND SYMPTOMS Ulcerating lesion, enlargement of lymph nodes, fever, malaise. Various organs may become involved

MODE OF TRANSMISSION Aerosols, bites of ticks, ingestion of contaminated foods, or direct contact

PREVENTIVE MEASURES Avoid contact with infected animals

TREATMENT OF CHOICE Streptomycin

DISEASE Typhoid fever

CAUSE(S) *Salmonella typhi*

SIGNS AND SYMPTOMS High fever, abdominal pain, headache. Accompanying lesions on Peyer's patches and spleen

MODE OF TRANSMISSION Ingestion of contaminated water or foods

PREVENTIVE MEASURES Vaccination; sanitize foods and beverages

TREATMENT OF CHOICE Chloroamphenicol

DISEASE Typhus

CAUSE(S) *Rickettsia prowazekii*

SIGNS AND SYMPTOMS Severe headache, backache, prostration, high fever, and formation of a rash. Bronchopneumonia is a complication

MODE OF TRANSMISSION Bite of lice

PREVENTIVE MEASURES Vaccination; avoid contact with louse-infested individuals

TREATMENT OF CHOICE Tetracyclines

DISEASE Trichomoniasis

CAUSE(S) *Trichomonas vaginalis*

SIGNS AND SYMPTOMS Profuse, whitish, frothy vaginal discharge.

MODE OF TRANSMISSION Sexually transmitted

PREVENTIVE MEASURES Avoid sexual contact with infected individuals

TREATMENT OF CHOICE Metrodinazole

DISEASE Varicella (chickenpox)

CAUSE(S) Varicella-zoster virus

SIGNS AND SYMPTOMS Fever, headache, malaise, and a vesicular rash that later encrusts

MODE OF TRANSMISSION Aerosols, or contact with exudates from rash

PREVENTIVE MEASURES Avoid contact with infected individuals

TREATMENT OF CHOICE Bed rest and treatment of symptoms

DISEASE Whooping cough

CAUSES(S) *Bordetella pertussis*

SIGNS AND SYMPTOMS Bouts of violent coughing that last for almost a minute, causing the patient to suck air forcibly producing a "whooping" sound. Vomiting and convulsions

MODE OF TRANSMISSION Aerosols

PREVENTIVE MEASURES Vaccination

TREATMENT OF CHOICE Administration of pertussis immune human globulin

DISEASE Yaws

CAUSE(S) *Treponema pertenue*

SIGNS AND SYMPTOMS Painless, red papule surrounded by a red ring. It ulcerates, encrusts, and ultimately heals. Lesions may recur

MODE OF TRANSMISSION Direct contact with infected individuals

PREVENTIVE MEASURES Avoid contact with ulcerating lesions on infected individuals

TREATMENT OF CHOICE Penicillin

DISEASE Yellow fever

CAUSE(S) Yellow fever virus (on arbovirus)

SIGNS AND SYMPTOMS Fever and chills, head- and backache; skin becomes yellow; vomiting, rapid pulse, uremia

MODE OF TRANSMISSION Bite of mosquito *Aedes aegypti*

PREVENTIVE MEASURES Vaccination and mosquito control

TREATMENT OF CHOICE Treat symptoms

GLOSSARY

Acetic acid bacteria Exclusively fermentative, catalase negative, gram positive, rod-shaped and spherical bacteria that release acetic acid. Examples are *Acetobacter* and *Acetobacterium.*

Acetobacter pasteurianum (ah-SEE-toe-BACK-ter pass-TUR-ee-ah-num) Gram negative, aerobic, ellipsoid bacterium that can oxidize ethanol to acetic acid (vinegar).

Acetobacterium (ah-SEE-toe-back-TER-ee-um) Gram negative, anaerobic, ellipsoid bacterium that may oxidize hydrogen gas and reduce carbon dioxide or ferment carbohydrates to acetic acid (vinegar).

Acetoin (as-SEE-toe-in) The same as acetomethylcarbinol. An intermediate in the 2, 3-butanediol fermentation which is used to detect the pathway. See Voges-Proskauer test.

Acetomethylcarbinol The same as acetoin.

Acetone-butanol fermentation A fermentation pathway in which the major waste products are carbon dioxide, acetone, and butanol. Species of *Clostridium* carry out this type of fermentation.

Acid dyes Dyes in which the colored portion is negatively charged. Generally do not stain cells and are used to stain the background.

Acid-fast characteristic The property, shared by most mycobacteria and some nocardiae, of retaining heated carbofuchsin, even after treating with an acidified decolorizing (leaching) agent.

Acidophile (as-SID-doe-file) An organism that grows at (or requires) low pH values.

Acinetobacter calcoaceticus (ah-sin-ET-toe-BAK-ter kal-koh-a-SIT-tee-kus) Gram negative, aerobic, short bacilli in pairs and short chains, which oxidize carbohydrates.

Acrasiomycetes (ah-KRAY-see-oh-MY-seats) A class of fungi known as cellular slime molds. During part of their life cycle, they resemble amebae. Upon nutrient deprivation, the amebae congregate and form a multicellular structure called a pseudoplasmodium. The pseudoplasmodium forms a fruiting body that contains spores. The spores germinate into amebae. The cells may develop cellulose cell walls—unusual for fungi, which generally have chitin cell walls.

Actinomycetes (ACK-tin-no-MY-seats) Gram positive, irregularly-staining, rod-shaped, diphtheroid or branched, nonmotile, aerotolerant or anaerobic bacteria. There are eight families in the order Actinomycetales: Actinomycetaceae, Mycobacteriaceae, Frankiaceae, Actinoplanaceae, Dermatophilaceae, Nocardiaceae, Streptomycetaceae, and Micromonosproaceae.

Actinomyces antibioticus (ack-tin-no-MY-sees an-tie-by-OT-tee-kus) Gram positive, irregularly staining, rod-shaped, diphtheroid or branched, microaerophilic bacterium.

Actinomyces israelii (is-rye-ELL-ee-eye) Causes actinomycosis.

Adenosine triphosphate ATP.

Adjuvant (ADD-jew-vant) Compound that increases the efficiency of antigens to induce an immune response.

***Aedes* sp.** (EH-des) Mosquito vectors for encephalitis (EEE, WEE, yellow fever).

Aedes aegypti (eh-GIP-tee) Mosquito vector for the yellow fever virus and the Dengue fever virus.

Aerobe (EH-robe) An organism that requires molecular oxygen for life. Examples are *Micrococcus* and *Pseudomonas.*

Aerotolerant anaerobe (eh-row-TOL-er-ant AN-eh-robe) Usually, a fermentative organism that is able to grow in the presence of molecular oxygen. Examples are *Streptococcus* and *Lactobacillus.*

Aerococcus viridans (eh-row-KOK-kus veer-EE-dans) Gram positive, nonmotile, microaerophilic, fermentative, catalase negative (or weakly catalase positive) cocci.

Aeromonas hydrophila (eh-row-MO-nas hi-dro-FILL-la) Gram negative, polarly flagellated, facultatively anaerobic, oxidase-positive, rod-shaped bacterium.

Aeromonas shigelloides (she-gehl-LOYD-dees)

Agglutinins Antibodies that bind cells together, causing them to clump.

Aflatoxins (AFF-la-tox-ins) Toxins harmful to animals and humans, produced by fungi growing on wheat, rye corn, and other grains. Some of the toxins cause cancer in animals and humans.

Agrobacterium tumefaciens (ag-grow-TEAR-ee-um too-ma-FAY-see-ens) Gram negative, peritrichously flagellated, aerobic, oxidase positive, plant parasitic, rod-shaped bacterium.

AIDS Acquired immunodeficiency syndrome. A virus (HTLV-III) caused disease which results from the inability of helper T-lymphocytes to initiate an immune response.

Akinetes (A-kin-neats) Thick-walled, elongated spores formed by cyanobacteria and some green algae.

Alcaligenes denitrificans (al-ka-LI-jen-eez dee-NI-tree-fee-cans) Gram negative, peritrichously flagellated, aerobic, oxidase positive coccobacilli.

Alcaligenes faecalis (fee-KAL-is)

Alga [pl. algae] (AL-gah [pl. should be AL-geh but always pronounced AL-gee]) Includes both microscopic, single-celled, prokaryotic and eukaryotic photosynthetic organisms and macroscopic, multicelled, eukaryotic photosynthetic organisms.

Alkalinophile An organism that grows under basic conditions or at a high pH.

Alcoholic (ethanolic) fermentation A fermentation in which the major waste products are carbon dioxide and alcohol (ethanol).

Allele (ah-LEEL) An alternate form of a gene.

Allergens Antigens that initiate a hypersensitive (allergic) reaction.

Allergy An overactive response by the immune system to an antigen which results in a diseased state (e.g. asthma, rash, welts). An allergy may be mediated by humoral antibodies or T-lymphocytes.

Allosteric proteins Proteins that bind effector molecules, which alter the structure of the protein and render it active or inactive. Examples are CRP (catabolite repressor protein), which becomes active when it binds cAMP, and the lactose repressor, which becomes inactive when it binds allolactose or another effector. The effector-binding sities and the substrate-binding sites are distinct, often on separate protein subunits.

Alpha helix Helical portion of a protein, stabilized by hydrogen bonding.

Alpha hemolysis A greening and partial lysis of red blood cells around certain colonies growing on blood agar plates. See Beta hemolysis and Gamma reaction.

Aminoacyl-tRNA (ah-ME-no-A-seal-tee-R-N-A) An amino acid–tRNA complex.

Aminoacyl-tRNA synthetase (SIN-tha-tayz) An enzyme that hooks together a specific tRNA and a specific amino acid.

Ammonification (ah-MOAN-ni-fi-KAY-shun) The release of ammonium from organic molecules.

Amoeba proteus (ah-MEE-ba PRO-tee-us) A protozoan that moves by extending pseudopodia and retracting terminal portions of the cell.

Amoeboid motion (ah-MEE-boyd) Movement due to the extension of one edge of a cell and the retraction of another edge.

Anabaena (an-na-BEH-na)

Anabolism (an-NEH-bow-liz-um) The synthesis of compounds within a cell. See also Catabolism.

Anaerobe (an-EH-robe) An organism that is inhibited or killed by the presence of molecular oxygen and obtains its energy by fermentative processes or anaerobic respiration.

Anaerobic respiration Respiration in which the final electron acceptor is an inorganic molecule (sulfate or nitrate) other than molecular oxygen.

Anamnestic response (ah-nam-NESS-tik) Rapid immune response against a specific antigen by B-cells or T-cells because they have previously been sensitized to the antigen.

Anaphylactic shock (AN-eh-feh-LAK-tik) Severe contraction of smooth muscle and excessive loss of fluid from the blood into the tissues due to an immediate allergic reaction (atopic hypersensitivity) mediated by IgE.

Anaphylatoxin (AF-feh-lah-TOX-in) The C3 and C5 complement proteins that cause basophils and mast cells to release various molecules, such as histamine, which cause inflammation.

Annealing of DNA The joining together, through hydrogen bonding, of complementary strands of DNA.

***Anopheles* sp.** (ah-NOF-fell-lees) Mosquito vectors for the malaria parasite and the western equine encephalitis (WEE) virus.

Anopheles gambiae (ah-NOF-fell-lees GAM-bee-eh) A mosquito vector for *Plasmodium*, the protozoan that causes malaria.

Antibiotic Organic compounds produced by microorganisms which inhibit or kill other organisms.

Antibodies Immunoglobulins. Serum proteins that are synthesized by plasma cells in response to stimulation by antigens. They react specifically with the inducing antigen.

Antigens Compounds that are able to elicit an immune response.

Antimicrobial agent Any physical or chemical agent that inhibits or kills microorganisms.

Antiseptic A chemical agent that inhibits or kills microorganisms and that can be used safely on the surface of an animal.

Antitoxins Antibodies that inactivate toxins.

Appertization (ah-pear-tie-ZEY-shun) The heating of foods in closed containers so as to preserve them. The process of canning was developed by Nicholas Appert in 1809. See Sterilization and Pasteurization.

Arboviruses Viruses carried by arthropod vectors which multiply in the arthropod and in the vertebrate host. The arboviruses cause various types of encephalitis (EEE, VEE, WEE, St. Louis) and yellow fever.

Archaebacteria (are-cheh-back-TEAR-ee-ah) Prokaryotes that lack peptidoglycan in their cell walls, and that have phospholipids with ether bonds rather than the usual ester bonds. Many of the archaebacteria are chemoautotrophic methanogens, reducing carbon dioxide with electrons from hydrogen gas to make energy and also using electrons from hydrogen gas to fix carbon dioxide. The archaebacteria are generally found in extreme environments where high temperatures, high salinity, and high acid content are the rule. Examples of archaebacteria are *Sulfolobus*, *Thermoplasma*, *Halobacterium*, and *Methanobacterium*.

Arthrobacter (are-throw-BACK-ter) Gram positive, aerobic coccobacilli, showing great variation in size; may be flagellated.

Arthropods Invertebrate animals with jointed legs, such as the insects and arachnids.

Ascomycetes A class of fungi that produces sexual spores called ascospores inside a sack known as the ascus. Ascomycetes produce asexual spores called conidia. Examples include the mold *Neurospora* and the yeast *Saccharomyces*.

Asepsis (a-SEP-sis) The absence of contaminating or infecting microorganisms. Asepsis does not imply sterility. See Sepsis.

Aseptic technique Procedures that reduce the risk of contaminating materials or infecting patients.

Aspergillus flavus (as-per-JILL-us FLAY-vus) A fungus belonging to the class Deuteromycetes.

Aspergillus niger (NYE-jer)

Aspergillus oryzae (or-RYE-zeh)

Assimilation The uptake of nutrients and their conversion to cellular material.

Atopic hypersensitvity IgE-mediated immediate allergic reaction.

Autoimmunity A situation in which an animal's immune system destroys its own tissues.

Autotroph An organism that uses carbon dioxide as its source of carbon.

Auxotroph A mutant that cannot synthesize simple compounds, such as amino acids, nucleosides, and vitamins. A medium must be supplemented with these growth factors if the organism is to proliferate.

Axenic culture (a-ZEE-nik) Pure culture of microorganisms.

Azotobacter vinelandi (ah-zo-toe-BACK-ter vin-LAN-dee) Gram negative, aerobic, oval cocci; may be peritrichously flagellated; molybdenum or vanadium required for nitrogen fixation.

B-cells B-lymphocytes that circulate in the lymph and blood and mature in the bone marrow and spleen. B-cells, when stimulated by foreign materials, differentiate into plasma cells and release antibodies into the lymph and blood.

Bacillus (ba-SILL-lus) A rod-shaped bacterium. When capitalized and italicized (or underscored), the term refers to a bacterial genus.

Bacillus (ba-SILL-lus) Gram positive, rod-shaped, flagellated or nonmotile, aerobic or facultatively anaerobic, endospore-forming bacteria.

Bacillus anthracis (ann-THRAY-sis) Facultatively anaerobic, nonmotile; causes anthrax in sheep, cattle, and humans.

Bacillus cereus (SEER-ree-us)

Bacillus fastidiosus (fas-tid-dee-OH-sus)

Bacillus licheniformis (lie-kin-nee-FORM-iss)

Bacillus megaterium (meg-ga-TER-ee-um)

Bacillus polymyxa (poly-MIX-ah)

Bacillus stearothermophilus (stair-oh-ther-MA-fill-us)

Bacillus thuringiensis (ther-in-gee-EN-sis)

Bacteriocins (back-TER-ee-oh-sins) Proteins produced by bacteria which kill sensitive members of related bacterial species. A special class of antibiotics.

Bacteriophage A bacterial virus, usually referred to as a phage.

Bacteroids Irregularly shaped bacteria, found within the plant cells that make up root nodules.

Bacteroides fragilis (back-ter-OYE-dees frah-JILL-iss) Gram negative, rod-shaped, nonmotile, anaerobic, catalase negative (or weakly catalase positive) bacterium.

Bacteroides succinogenes (suck-SIN-no-jen-is)

Balantidium coli (bal-lan-TID-dee-um CO-lee) Ciliated protozoan that may cause diarrhea and dysentery.

Basic dye A dye in which the colored portion is positively charged. A dye that stains cells.

Basidiomycetes (ba-sid-dee-oh-MY-seats) A class of fungi that produces sexual spores, called basidospores, from a structure called the basidium. The meadow mushroom, *Agaricus compestris*, is an example of a Basidiomycete.

Basophil White blood cell that releases histamines and other compounds that act on smooth muscle and other tissues.

Beta hemolysis Extensive lysis around certain colonies of bacteria growing on blood agar plates. See Alpha hemolysis and Gamma reaction.

Bdellovibrio bacteriovorus (del-low-VIB-bree-oh back-ter-ee-oh-VOR-us) Small curved or spiral, gram negative, polarly flagellated, aerobic bacteria.

Beggiatoa (bej-gee-ah-TOE-ah) Gram negative, filamentous, gliding, aerobic to microaerophilic, hydrogen sulfide–oxidizing bacteria with numerous sulfur granules within their cytoplasms. These bacteria require organic compounds as their carbon source.

Behring, Emil von (BEH-rink, EH-meel von)

Beijerinck, Martinus (BY-jer-rink, mar-TEE-nus)

Beijerinckia (by-jer-RINK-kia) Gram negative, oval or rod-shaped, peritrichously flagellated or nonmotile, aerobic bacteria. Molybdenum required for nitrogen fixation.

Bergey's Manual A book in four volumes that characterizes and classifies bacteria.

Bifidobacterium (by-fid-dough-back-TER-ee-um) Gram positive, highly variable, rod-shaped, nonmotile, anaerobic, catalase negative bacteria.

Bioluminescence The production and release of light by some microorganisms and insects.

Blue-green algae Photosynthetic bacteria that release oxygen as a consequence of their photosynthesis. Blue-green algae are also known as cyanobacteria.

BOD Biochemical oxygen demand. The amount of oxygen required to metabolize dissolved organic material to carbon dioxide and water is referred to as the biological oxygen demand.

Bordetella pertussis (bore-de-TELL-la pair-TUSS-sis) Gram negative, small coccobacilli; nonmotile, aerobic bacterium. Causes whooping cough.

Borrelia recurrentis (bore-REL-lee-ah ree-cur-REN-tis) Gram negative spirochete with axial filaments; anaerobic bacterium. Causes relapsing fever.

Branhamella catarrhalis (bran-ha-MEL-la ka-ta-RAH-lis) Gram negative, flattened, nonmotile, aerobic cocci.

Brucella abortus (bru-SELL-la ah-BORE-tus) Gram negative coccobacilli or short rod-shaped bacteria; nonmotile, aerobic, oxidase positive. Facultatively intracellular. Cause of brucellosis (undulant fever, Bang's disease).

Brucella melitensis (mel-ee-TEN-sis) Cause of brucellosis.

Brucella suis (SUE-is) Cause of brucellosis.

Bubonic plague A bacterial infection of the lymphatic system which causes swelling of the lymph nodes. A swollen lymph node is known as a bubo.

2,3-Butanediol fermentation A fermentation in which the major waste products are 2,3-butanediol and carbon dioxide. Bacteria such as *Enterobacter aerogenes* carry out a 2,3-butanediol fermentation.

Butyric acid bacteria Bacteria that release butyric acid when they ferment. An example is *Clostridium.*

Butyric acid–butanol fermentation A fermentation in which the major waste products are butyric acid, or butanol, and carbon dioxide. Bacteria such as *Clostridium* carry out a butyric acid–butanol fermentation.

Calvin cycle A cyclic series of chemical reactions in which large amounts of carbon dioxide are fixed into organic molecules. The dark reactions of photosynthesis.

Campylobacter jejuni (kam-pill-low-BACK-ter je-JEW-nee) Gram negative, curved, rod-shaped, flagellated, microaerophilic, oxidase positive bacterium. Causes dysentery and enterocolitis.

Candida albicans (KAN de dah al bee kanz) A yeast belonging to the class Deuteromycetes that causes thrush of the mouth and yeast infections of the vagina.

CAP Catabolite activator protein or cyclic AMP binding protein, which binds cAMP and regulates the expression of numerous operons involved in catabolic reactions. Identical to CRP (catabolite repressor protein).

Capsid The protein coat of viruses, made up of capsomeres.

Capsule A thick layer of polysaccharide (infrequently protein) that surrounds some bacteria. See also Glycocalyx.

Carcinogen A physical or chemical agent that causes cancer.

Catabolism (kah-TAB-bow-liz-im) The breakdown of nutrients or cellular material. See also Anabolism.

Catabolite repression The deactivation of catabolic operons due to the rapid entrance of metabolites and their breakdown. The deactivation is mediated by a cAMP-binding protein (CRP) that must bind to promoter sites for the initiation of transcription. The rapid entrance of metabolites is balanced by a rapid exit of cAMP, which is required for CRP binding. In the absence of CRP binding to promoters, the initiation of transcription ceases.

Caulobacter crescentus (COW-low-back-ter kre-SENT-tus) Gram negative, aerobic, flagellated, rod-shaped or nonmotile, prosthecate bacterium.

Cellular immunity Immunity mediated by thymus-derived lymphocytes (T-cells) and macrophages.

Cell wall A rigid layer of material outside the cytoplasmic membrane which protects the cell. The cell wall of most bacteria contains a layer of peptidoglycan. The gram positive bacteria have a wall that consists of peptidoglycan and teichoic acids, but most gram negative bacteria have a wall that consists of a peptidoglycan layer and an outer membrane containing proteins and lipopolysaccharide.

Gene The segment of DNA or RNA that codes for a protein, rRNA, or tRNA. See also Cistron.

Cephalosporium acremonium (sef-fa-low-SPORE-ee-um ah-kreh-MO-nee-um) A fungus that produces the antibiotic cephalosporin C.

Chemiosmotic hypothesis The theory that proton gradients are created by the movement of electrons and protons through an electron transport system, and that this proton gradient provides energy for doing work (concentrating nutrients and moving flagella) or synthesizing ATP from ADP and inorganic phosphate.

Chemoautotroph An organism that derives its energy from various chemicals (vs. sunlight) and its carbon from carbon dioxide.

Chemoheterotroph An organism that derives its energy from various chemicals and its carbon from organic molecules other than carbon dioxide. Usually, the same chemicals supply both the energy and the carbon.

Chemolithotroph An organism that derives its energy from inorganic chemicals, such as ferrous iron, hydrogen sulfide, or ammonium.

Chemotaxis Movement toward or away from a chemical.

Chitin (KAI-tin) A homopolymer of N-acetyl-D-glucosamine, which is the major structural molecule in the cell walls of most classes of fungi and in the exoskeletons of insects and crustaceans.

Chlamydia psittaci (klah-MID-dee-ah SIT-tah-see) Gram negative, spherical, nonmotile, ATP-requiring, intracellular bacterium. Causes lung infections in birds (ornithosis or psittacosis) and humans.

Chlamydia trachomatis (trah-ko-MA-tiss) Causes trachoma, inclusion conjunctivitis, infant pneumonia, nonspecific urethritis, and lymphogranuloma venereum.

Chlamydomonas rhinehardii (klah-mid-doe-MOW-nas rine-HARD-dee) A flagellated green alga.

Chlorobium limicola (klor-OH-bee-um lim-me-KO-la). Gram negative, rod-shaped, nonmotile, anaerobic, hydrogen sulfide–oxidizing, photosynthetic bacterium that generates sulfur or sulfate rather than molecular oxygen. Sulfur granules are deposited outside the cell. Also known as green sulfur bacteria. Growth is yellow-green to brown.

Chloroflexus aurantiacus (klor-oh-FLEX-us are-rant-TEE-ah-kus) Filamentous, gliding, anaerobic, nonsulfur, photosynthetic bacterium. Also known as green nonsulfur bacteria. Growth is yellow-green to orange.

Chromatophore Vesicular membranes found in the cytoplasm of the anaerobic photosynthetic bacteria.

Chromatium okenii (crow-MAY-she-um oh-KEN-nee-eye)

Gram negative, rod-shaped, flagellated or nonmotile, purple, anerobic, hydrogen sulfide–oxidizing, photosynthetic bacterium that generates sulfur or sulfate rather than molecular oxygen. Sulfur granules are stored within the cell. Also known as purple sulfur bacteria. Growth is red-violet, blue-violet, or brownish orange.

Chromophore The colored portion of a dye.

Chromoplast A colored organelle (plastid), such as a chromosome or a chromoplastid.

Chromosome Chromosomes are DNA molecules that are coiled and condensed into organized structures, such as the metaphase chromosomes in mammals and the polytene chromosomes in certain insects. In eukaryotes, protein also makes up a major portion of the chromosomes. The bulk of the hereditary material in bacteria is found in a single circular molecule of DNA, sometimes referred to as the chromosome but more frequently called the genome. Small, autonomous pieces of hereditary material are called plasmids. See also genome.

Chromatin Fibers of DNA.

Cistron A gene or region of DNA that codes for a single polypeptide.

Citric acid cycle Also known as the Krebs cycle and the tricarboxylic acid cycle. A cyclic sequence of chemical reactions that produces carbon dioxide, reducing electrons, and ATP. The citric acid cycle also provides building blocks for the synthesis of cellular material.

Citrobacter freundii (sit-trow-BACK-ter FRO-een-dee) Gram negative, rod-shaped, flagellated, facultative anaerobic bacterium.

Claviceps purpurea (KLAH-vee-seps poor-POOR-ee-ah) A fungus that belongs to the class Ascomycetes. Causes ergot of rye.

Clostridia (kloss-TRID-dee-ah) Gram positive, rod-shaped, flagellated or nonmotile, endospore forming, anaerobic or microaerotolerant, generally catalase negative bacteria.

Clostridium acetobutylicum (kloss-TRID-dee-um ah-see-toe-bue-TILL-lee-kum)

Clostridum botulinum (bot-chew-LIE-num) Causes botulism.

Clostridium histolyticus (hiss-toe-LIT-tee-cus)

Clostridium kluyveri (kluy-VE-ree)

Clostridium perringens (per-FRIN-jens) Causes intestinal infections, gas gangrene, and food poisoning.

Clostridium sporogenes (spore-AH-jen-ease)

Clostridium tetani (TET-ah-neye) Causes tetanus.

Coenocytic (seen-no-SIT-tic) The condition in which a cell contains many nuclei.

Coenzyme A small organic molecule that works with enzymes, carrying chemical groups to or away from chemical reactions. Coenzymes are required for enzyme activity. Many vitamins are used to make coenzymes.

Cofactor Generally, a metal ion that is required for enzyme activity.

Coccidioidomycosis (cock-sid-dee-OYE-doe-my-co-sis) Valley fever.

Coccidioides immitis (cock-sid-dee-OYE-dees IM-me-tis) The fungus that causes coccidioidomycosis.

Colicin A bacteriocin produced by coliform bacteria (and some enterics).

Complement A group of serum proteins involved in immune reactions which have cytolytic and chemotactic properties.

Complementation In genetics, a situation in which two defective DNA molecules, together, supply a missing function.

Complex medium A medium that contains an unknown mixture of many different nutrients. Generally, a rich medium that is used to grow fastidious (demanding) microorganisms. See Minimal medium.

Conjugation In bacteria, the unidirectional transfer of hereditary material from one bacterium (donor) to another (recipient). Transfer requires cell-to-cell contact.

Constitutive operons (enzymes) Operons or enzymes that are always expressed.

Contact inhibition Cells of higher animals generally cease growing and proliferating when they touch one another.

Continuous cultures Cultures that are maintained in the exponential phase of growth for extended periods of time by continuously supplying the culture with fresh medium.

Corepressor A compound that binds to a protein and converts it into a repressor, which turns off genes. An example of a corepressor is the amino acid tryptophan, which binds to an inactive regulatory protein and converts it into a repressor of the tryptophan operon.

Corynebacteria Gram positive, rod-shaped, nonmotile, aerobic or facultative anaerobic bacteria.

Corynebacterium diphtheriae (core-in-nee-back-TER-ee-um dip-THER-ee-eh) Causes diphtheria.

Corynebacterium hofmanni (hoff-MAN-neye)

Corynebacterium xerosis (zer-ROW-sis)

Coxiella burnetii (kocks-ee-ELL-ah ber-NET-tee-eye) Gram negative, rod-shaped, nonmotile, spore forming, obligate intracellular bacterium. Causes Q-fever.

CRP Catabolite repressor protein. Same as CAP.

Cryptococcus neoformans (krip-toe-COCK-kus nee-oh-FOR-mans) A fungus in the class Deuteromycetes or Fungi Imperfecti. Causes cryptococcosis, a mild pneumonia, meningitis, or systemic infection. *Fillobasidiella neoformans* is the perfect state (a Basidiomycete) of *C. neoformans*.

***Culex* sp.** (KU-lex) Mosquito vectors for encephalitis (EEE, WEE, St. Louis).

Culex pipiens (pip-PEE-ens)

Culture Population of microorganisms growing in an artificial medium.

Cyanobacteria (sigh-ann-no-back-TER-ee-ah) Photosynthetic bacteria that release molecular oxygen as a consequence of their photosynthesis. Also known as blue-green bacteria, blue-green algae.

Cyclic AMP (cAMP) 3′,5′-cyclic adenosine monophosphate. The phosphate group is connected to the 3′ and 5′ carbons of the ribose sugar, thus creating a cyclic structure.

Cytochrome Complex metal-containing porphyrin-protein molecules that are readily reduced and oxidized. They are important components of electron transport systems.

Cytopathic effect A visible change in cell cultures due to infection by viruses, bacteria, fungi, or protozoans.

Cytophaga (sigh-TOE-fay-ga) Gram negative, rod-shaped, gliding, facultative anaerobic or aerobic bacteria.

Cytoplasmic membrane See Plasma membrane.

D-value (decimal reduction time) The time that it takes for an antimicrobial agent to reduce a population to 10 percent of its original size at a specified temperature.

Denature To alter the normal shape or activity of a molecule.

Denitrifying bacteria Bacteria that convert nitrate to nitrite or to nitrogen gas by anaerobic respiration.

Dental caries Areas of tooth decay.

Dermatomycosis A fungal skin infection.

Dermatophyte A fungus that grows in the skin and causes infections. Dermatophytes digest keratin.

Desulfotomaculum (dee-SULL-fo-toe-MA-cue-lum) Gram positive, rod-shaped, flagellated, endospore forming, anaerobic, sulfate- and sulfite-reducing bacteria that produce hydrogen sulfide.

Desulfovibrio (dee-SULL-fo-VIB-bree-oh) Gram negative, curved, flagellated, anaerobic, sulfate- and sulfite-reducing bacterium that produces hydrogen sulfide.

Dextran Polysaccharides in which most of the glucose is held together by alpha (1,6) linkages, whereas alpha (1,2), alpha (1,3), or alpha (1,4) linkages are responsible for the branching. The dextrans form viscous, slimy solutions.

Deuteromycetes (DUE-ter-oh-MY-seats) A class of fungi also known as the Fungi Imperfecti. These fungi have no known sexual stage; they produce asexual spores called conidia. Examples are *Penicillium* and *Aspergillus*.

Dextrin A branched polysaccharide of glucose which remains after limited amylase hydrolysis of starch or glycogen. Most of the glucose is held together by alpha (1,4) linkages, whereas alpha (1,6) linkages are responsible for the branching.

Dextrose Glucose.

Diacetyl A compound released by certain bacteria and yeast which gives an off-flavor to beer and a buttery aroma and taste to butter.

Dialysis The removal of low molecular weight ions and molecules by diffusion across a membrane.

Diatomaceous earth Earth consisting of diatom cell walls.

Diauxic (di-OX-ick) The two-step growth curve that results when a population of organisms metabolizes one carbon source completely before beginning to metabolize another.

Dictyostelium (dick-tee-oh-STILL-ee-um) A slime mold that produces a pseudoplasmodium.

Didinium (die-DIN-ee-um) A large ciliated protozoan.

Diffusion The net movement of a solute from a region of high concentration to a region of low concentration. In passive diffusion across a membrane, the rate of diffusion is directly proportional to the concentration of the solute; in facilitated diffusion across a membrane, the rate of diffusion at low solute concentrations is dependent upon the concentration of the solute, but at high solute concentrations the rate of diffusion is dependent upon the mechanisms that are aiding the diffusion (e.g., concentration of carrier molecules on membrane).

Dikaryon A cell that contains two different types of nuclei.

Dimorphic fungus A fungus that has two different morphologies that change, depending upon the environmental conditions. For example, the formation of a mycelium at room temperature and yeastlike structures at 35°C.

Dissimilatory sulfate reduction The conversion of cellular sulfate to environmental hydrogen sulfide.

Domagk, Gerhard (DOE-mak, GER-hart)

Double helix The double-stranded DNA molecule.

Doubling time The time that it takes a population to double its size. Same as generation time.

Dysentery Clinical syndrome resulting from an inflamation of the large intestine. It is characterized by diarrhea with blood and mucus, tenesmus, fever, and malaise.

Eclipse period The time from the initiation of a virus infection until new infectious virus particles appear within the host cell. See Latent period.

Ectothiorhodospira (eck-toe-thy-oh-row-doe-SPY-rah) Gram negative, rod-shaped, flagellated, anaerobic, hydrogen sulfide–oxidizing, photosynthetic bacteria that produce sulfur and sulfate rather than molecular oxygen. Sulfur granules are deposited outside the cell. Also known as purple sulfur bacteria.

Ehrlich, Paul (ER-lick)

Effector molecule A compound that binds to a protein and alters the protein's activity. See Allosteric proteins.

Elementary bodies Intracellular chlamydia cells with an atypical structure which are the infectious form of the bacterium. Elementary bodies are sporelike structures that do not multiply, do not transport ATP, and do not carry out protein synthesis. The number of ribosomes in the elementary bodies is greatly reduced as compared with the number in the reticular bodies (initial bodies). The outer membrane proteins of elementary bodies are linked by disulfide bonds, making the cells rigid. See Reticular bodies (initial bodies).

Embden-Meyerhof pathway A specific sequence of chemical reactions that converts glucose to pyruvic acid. Same as glycolysis.

Endergonic reaction A reaction that absorbs heat or energy from its environment. See Exergonic reaction.

Endonuclease An enzyme that cuts within a DNA or RNA molecule. See Exonuclease.

Endosymbiont An organism that lives within another in a symbiotic relationship.

Endotoxin The lipopolysaccharide fraction from the outer membrane in gram negative bacteria which is toxic to many animals.

End product inhibition A situation in which the product of a biosynthetic pathway inhibits one of the first enzymes of the biosynthetic pathway.

End product repression A situation in which the product of a biosynthetic pathway is involved in the repression of the genes that are coding for the enzymes in the biosynthetic pathway.

Enrichment Cultural selection for an organism, so that its relative abundance increases.

Entamoeba coli (ent-ah-ME-ba KO-lee) A nonpathogenic ameba that colonizes the intestines of about 30% of the human population.

Entamoeba histolytica (his-toe-LIT-tee-kah) An ameba that causes dysentery.

Enteric bacteria Bacteria that normally live in the intestines.

Enterobacter aerogenes (en-ter-oh-BACK-ter air-RAH-jen-eez) Gram negative, rod-shaped, flagellated, facultatively anaerobic bacterium. Sometimes involved in urinary tract infections.

Enterococcus Spherical bacteria that live in the intestines, such as *Streptococcus faecalis*.

Enterotoxin A toxin that causes gastrointestinal problems when ingested or produced in the intestines.

Enterovirus Viruses that live or are commonly found in the intestines, such as the polioviruses and hepatitis A virus.

Epidermiology The study of disease incidence, distribution, and modes of transmission.

Epidermophyton floccosum (eh-pee-der-MO-fy-ton flock-KO-sum) A fungus that causes skin infections, such as jock itch (ringworm of the groin) and athlete's foot (ringworm of the foot).

Episome A plasmid that is able to integrate into the host's chromosomes.

Epizootic (eh-pee-zow-OT-tick) Disease that affects a large number of animals.

Ergot A fungus that grows on wheat and rye. See *Claviceps purpurea*.

Erysipelas A serious infection of the skin caused by Group A *Streptococcus*.

Erwinia amylovora (err-WIN-nee-ah ah-me-LOV-or-ah) Gram negative, rod-shaped, flagellated, facultatively anaerobic, oxidase negative bacterium.

Erwinia carotovora (ka-row-TOV-or-ah)

Etiology (et-tee-OL-lo-gee) The cause of a disease.

Escherichia coli (eh-share-REE-kee-ah KO-lee) Gram negative, rod-shaped, flagellated, facultatively anaerobic bacterium commonly found in the intestines.

Ethanol fermentation A fermentation in which the major waste products are carbon dioxide and ethanol.

Eubacteria Classical bacteria that have peptidoglycan in their cell walls, lipids with unbranched fatty acids connected by ester bonds in their cell membranes, and distinctive rRNAs, tRNAs, and RNA polymerases. The mycoplasmas, which lack a cell wall, and the chlamydia, which lack peptidoglycan in their cell walls, are considered eubacteria but the cyanobacteria are not. The archaebacteria lack peptidoglycan in their cell walls, have lipids with branched hydrocarbon chains connected by ether bonds, and have distinctive rRNAs, tRNAs, and RNA polymerases are not considered to be eubacteria either.

Euglena gracilis (you-GLEE-nah gra-SILL-lis) Single-celled alga.

Eukaryote A cell that has one or more nuclei and generally contains organelles and structures such as mitochondria, chloroplasts, Golgi bodies, endoplasmic reticulum, and vacuoles.

Eutrophication (you-trow-fi-KAY-shun) A massive growth of algae in lakes and other bodies of water, which leads to a decrease in dissolved oxygen and a decrease in animal life in the water. Eutrophication is due to an excess of nutrients from erosion or sewage.

Exergonic reaction A reaction that gives off heat or energy. See Endergonic reaction.

Exoenzyme Enzyme that functions outside the cell.

Exons The regions of mRNA or DNA that code for a polypeptide. See Introns.

Exonuclease An enzyme that removes nucleotides from the ends of DNA or RNA molecules. See Endonuclease.

Exotoxin Large molecules, usually proteins, that are toxic to animals and plants.

Exponential phase A period of growth during which a population is growing at a constant rate. Same as log phase. See also Lag, Stationary, and Death phases.

F-factor (fertility factor) A plasmid that makes a bacterium able to transfer hereditary material to another bacterium during conjugation. In a conjugation, F^- cells receive hereditary information; F^+ cells donate plasmid genes; F' cells donate plasmid genes and some genes also found on the main chromosome; and Hfr cells donate genes on the main chromosome and some genes on the integrated plasmid.

Facilitated diffusion Diffusion of molecules through a membrane by means of special pores or proteins that penetrate the membrane. See Diffusion.

Facultative anaerobe Organism that usually respires aerobically, but occasionally proliferates under anaerobic conditions by fermenting or anaerobically respiring.

Feedback inhibition See End product inhibition.

Feedback repression See End product repression.

Fermentation The production of energy exclusively by substrate level phosphorylation and the use of organic molecules as electron acceptors. Fermentations generally result in the release of organic and inorganic molecules, such as carbon dioxide and molecular hydrogen (gases); lactic, formic, acetic, succinic, butyric, and propionic acids; and ethanol, butanol, and propanol (alcohols).

Fimbria [pl. fimbriae] (FIM-bree-ah, FIM-bree-eh) Various types of pili that extend from the surface of bacteria.

Fix (1) To attach bacteria to a slide. (2) To alter material such as carbon dioxide and molecular nitrogen, and incorporate the carbon and nitrogen into cellular material. See Nitrogen fixation and Photosynthesis.

Flavobacterium (flay-vo-back-TER-ee-um) Gram negative, sperical or rod-shaped, flagellated or nonmotile, aerobic bacteria.

Flavoprotein A protein that contains a covalently bonded flavin group. A component of electron transport systems.

Flexibacter columnaris (flex-ee-BACK-ter koll-lum-NAR-iss) Gram negative, rod-shaped or filamentous, gliding, aerobic bacteria.

Fomite (FOE-might) An inanimate object or sputum that may be the source of microorganisms.

Francicella tularensis (fran-sis-SELL-lah too-lar-REN-sis) Gram negative, ellipsoid, rod–shaped, nonmotile, aerobic bacteria. Causes tularemia.

Free energy Energy that can do useful work.

Fungi Imperfecti (FUN-jeye im-per-FEK-teye) See Deuteromycetes.

Fungus [pl. fungi] (FUN-gus; FUN-jee or FUN-jeye) Eukaryotic, nonphotosynthetic, micro- and macroscopic organisms that form tubular cells with a rigid cell wall. The cells are interconnected, forming hyphae. Fungi grow typically by extention of hyphal tips.

Furuncle (fir-UNK-kul) An abscess in the skin, similar to a boil, which sometimes results from an infection of a hair follicle.

Fusobacteria Gram negative, cigar-shaped, nonmotile or flagellated, anaerobic, catalase negative bacteria.

Fusobacterium polymorphus (few-so-back-TER-ee-um pollee-MORE-fus)

Gaffkya (GAFF-key-ah) Gram positive, spherical, nonmotile, facultative anaerobic bacterium, usually found in groups of four.

Galactoside Molecule that has the sugar galactose attached to it. Examples are lactose, isopropyl thiogalactoside (IPTG), and thiomethyl galactoside (TMG).

Gallionella (gal-yawn-NELL-ah) Gram negative, kidney-shaped, flagellated or nonmotile, microaerophilic, ferrous oxidizing bacteria that form a fibrous (noncellular) stalk.

Gardnerella vaginalis (gard-ner-RELL-ah vah-jin-NAL-iss) Gram negative, rod-shaped, nonmotile, facultatively anaerobic, catalase negative bacteria with many growth requirements. Causes nonspecific vaginitis.

Gamma globulins A group of serum proteins, some of which function as antibodies.

Gamma hemolysis The absence of hemolysis around bacterial colonies growing on blood agar plates.

Generation time The time it takes for a population to double its size. Same as doubling time.

Genetic code The representation of amino acids by specific sequences of three nucleotides in DNA or RNA. The specified amino acids are incorporated during protein synthesis.

Genetic map An ordered sequence of mutations or genes in a chromosome.

Genome A complete set of an organism's hereditary material. In polyploid organisms, a monoploid set of the chromosomes. In bacteria, the circular chromosome and any plasmids. In viruses, the DNA or RNA molecules.

Geotrichum candidum (gee-oh-TRICK-um KAN-dee-dum) A filamentous fungus in the class Deuteromycetes which forms arthrospores.

Giardia intestinalis (gee-ARE-dee-ah in-tess-tea-NAL-liss) A flagellated protozoan. The cause of beaver fever and traveler's diarrhea. Also called *G. lambia.*

Gliding bacteria A large group of diverse bacteria that move slowly and smoothly over moist surfaces. They lack axial filaments, flagella, or other protrusions that could be involved in their movement.

Glycocalyx (gly-ko-KAY-licks) Stringy polysaccharides that emanate from the surfaces of cells and help the cells attach to one another and to inanimate objects. In bacteria, the glycocalyx can become so thick that it is indistinguishable from a capsule.

Glycolysis Same as Embden-Meyerhof pathway.

Granulocytes White blood cells that contain numerous large granules in their cytoplasm: neutrophils, basophils, and acidophils.

Growth curve A curve tracing the growth of a population in a broth medium. It is divided into phases: lag, exponential (log), stationary, and death.

Growth factors Various organic molecules such as amino acids, nucleosides, and vitamins which an organism is unable to synthesize and with which it must be supplied.

Growth rate The increase in the number of individuals in a population as a function of time.

Growth yield The mass of cells that can be harvested.

Haemophilus (he-MAW-fill-us) Gram negative, spherical to rod-shaped, nonmotile, aerobic or facultatively anaerobic, intracellular bacteria that require growth factors.

Haemophilus ducreyi (due-KRAY-ee) Requires hemin (X-factor). Causes soft chancre known as chancroid, a venereal disease.

Haemophilus influenzae (in-flew-ENZ-eh) Requires hemin (X-factor) and NAD^+ (V-factor). Causes meningitis, sinusitis, and pneumonia.

Hafnia (HAF-nee-ah) Gram negative, rod-shaped, peritrichously flagellated, facultatively anaerobic bacteria that resemble *Enterobacter*.

Halobacterium (hal-low-back-TER-ee-um) Gram negative, rod-shaped, polarly flagellated or nonmotile, sodium chloride–requiring (more than 2 M), aerobic, oxidase positive bacteria.

Halococcus (hal-low-KOCK-kus) Gram negative, spherical, nonmotile, sodium chloride–requiring (more than 2.5 M), aerobic bacteria.

Halophiles Organisms that grow best at (or require) high concentrations of salt.

Hapten A molecule that is specifically bound by antibodies, but is unable to induce the production and release of antibodies by plasma cells. A hapten may induce a specific immune response if it is attached to a large protein (carrier molecule).

Hemicelluloses Polymers of D-xylose held together by beta (1,4) linkages with side chains of arabinose and other sugars.

Heterokaryosis A condition in which a cell contains more than one type of nucleus.

Heterofermentation A fermentation in which there are a number of waste products. Used in connection with lactic acid bacteria that have major waste products other than lactic acid.

Histoplasma capsulatum (hiss-toe-PLAZ-ma cap-sue-LA-tum) A fungus in the class Deuteromycetes which causes histoplasmosis, usually a mild coldlike disease.

Homofermentation A fermentation in which there is only one waste product. Used in connection with lactic acid bacteria that produce lactic acid and very minor amounts of other acids.

Humoral immunity Immunity mediated by antibodies that circulate through the lymphatic system and the blood system. "Humor" refers to the body fluids.
Hydrogen bacteria Bacteria that derive their energy by oxidizing molecular hydrogen. Examples include some species of *Alcaligenes*, *Xanthobacter*, *Nocardia*, and *Paracoccus*.
Hydrophilic forces Forces that arise from chemical groups or molecules that arrange themselves in an aqueous environment in such a way as to be surrounded by water.
Hydrophobic forces Forces that arise from chemical groups or molecules that arrange themselves in an aqueous environment in such a way as to exclude water.
Hypertonic A solution with a relatively high concentration of solutes or dissolved materials. See Hypotonic and Isotonic.
Hypha [pl. hyphae] (HI-fah; plural should be HI-feh, but pronounced HI-fee) A filament of fungal growth.
Hyphomicrobium (HI-fo-my-CROW-bee-um) Rod-shaped, hyphae forming, nonmotile or motile, aerobic bacteria that reproduce by a budding process.
Hypotonic A solution with a relatively low concentration of solutes or dissolved materials as compared to the cellular environment. See Hypertonic and Isotonic.
Icosahedron A closed three-dimensional structure with 20 triangular faces (surfaces) and 12 corners.
IgA Antibodies of the A class, found in secretions.
IgD Antibodies of the D class, found attached to the surfaces of B-lymphocytes.
IgE Antibodies of the E class, found in the lymph and blood, which are involved in allergic reactions.
IgG Antibodies of the G class, found in the lymph and blood, which bind complement. IgG are the most abundant antibodies.
IgM Antibodies of the M class, found attached to the surfaces of B-lymphocytes and in the lymph and blood, which bind complement. They are the first antibodies to be produced by B-lymphocytes.
Immediate-type hypersensitivity An antibody-mediated allergic reaction that occurs within minutes after presentation of an antigen to which the individual is sensitized.
Immunodeficiency The lack of part or all of the immune system.
Immunofluorescence Fluorescence from a dye that is attached (conjugated) to an antibody. This type of fluorescing antibody is used to visualize the presence of specific antigens.
Immunoglobulin Same as antibody.
IMViC A group of four biochemical tests (indole, methyl red, Voges-Proskauer, and citrate) used to differentiate enteric bacteria.
Inducer A physical or chemical agent that causes the development of lysogenic viruses or turns on gene systems (operons). For example, ultraviolet light is an inducer of the lambda prophage, whereas lactose if the inducer of the lactose operon. See Effector.
Induction (enzymes, phage) The process of turning on gene systems or causing the development of phage in a lysogen.
Initial bodies Same as reticular bodies: intracellular chlamydia cells that grow and divide. They transport ATP and carry out protein synthesis, but they are not infectious because they are extremely sensitive to mechanical and osmotic shock. In electron micrographs, they have the appearance of typical prokaryotes except that they appear to lack a peptidoglycan layer. Inital bodies are also known as reticulate bodies (RBs). See Elementary bodies.
Insertion sequence Regions of DNA that are able to recombine readily with different regions within and between chromosomes. Insertion sequences in plasmids are believed to be involved in the plasmid's ability to insert into the cell's chromosomes.
Interferon A class of proteins, released by virus-infected cells, which stimulate uninfected cells to produce proteins that inhibit the proliferation of viruses.
Intracytoplasmic membranes Membranes found in the cytoplasm of some bacteria. Many of these membranes are invaginations of the cytoplasmic membrane. See Chromatophore, Photosynthetic membrane, Thylakoid, and Chlorobium vesicle.
Intron Regions of mRNA or DNA within a gene which do not code for a polypeptide. Generally, introns are found only in the eukaryotes.
Isotonic A solution with the same osmotic pressure (usually the same concentration of solutes) as another solution. See Hypertonic and Hypotonic.
Iron bacteria Bacteria that derive energy by oxidizing ferrous iron (Fe^{+2}) to ferric iron (Fe^{+3}). An example is *Thiobacillus ferrooxidans*.
In vitro Within an artificial container.
In vivo Within a living organism.
Isoantibodies Antibodies against foreign isoantigens. For instance, antibodies against A-type blood in persons with B- or O-type blood.
Isoantigens Antigens that are present in some individuals of a population but not in other individuals. For example, people with A-type blood have antigens that people with B- and O-type blood do not have.
Isotonic solution A solution with the same concentration of solute (atoms, small molecules, and ions) as another solution.
Iwanowski, Dmitri (i-van-NOW-ski, deh-MEE-tree)
Ixodes persulcatus (icks-OH-deeze pair-sull-CAT-us) A tick that ransmits the virus responsible for Russian spring-summer encephalitis (RSSE).
Jacob, Francois (jah-KOBE, FRAHN-swah)
K-antigen The capsule material from enteric bacteria.
Khorana, Har Gobind (ko-RAH-nah, har GO-bind)
Killer T-Cells One class of T-lymphocytes involved in cell-mediated immunity.
Kinins Small peptides, released into the blood and lymph upon trauma, which are involved in blood clotting and inflammatory reactions.
Klebsiella pneumoniae (kleb-see-EL-lah new-MO-nee-eh) Gram negative, rod-shaped, nonmotile, encapsuled, facultatively anaerobic bacterium. A cause of pneumonia.
Koch, Robert (KAWK or KAWH; breathe out on the H)

Koch's postulates A set of procedures for proving that an organism is responsible for a disease.
Krebs cycle Same as the citric acid cycle and the tricarboxylic acid cycle.
Kuru A progressive degenerative brain disease caused by a prion.
L-forms Common bacteria that do not produce a normal peptidoglycan layer or have lost the ability to manufacture the peptidoglycan layer of their cell walls. See Protoplasts and Spheroplasts.
Lactic acid bacteria A group of gram positive, catalase negative, exclusively fermentative, spherical and rod-shaped bacteria that carry out a lactic acid fermentation. Examples are *Streptococcus* and *Lactobacillus.*
Lactic acid fermentation A fermentation in which the major waste product is lactic acid.
Lactobacilli (should be lack-toe-ba-SILL-ee, but always pronounced lack-toe-ba-SILL-eye) Lactic acid bacteria. Gram positive, rod-shaped, nonmotile, aerotolerant, anaerobic, catalase negative bacteria.
Lactobacillus acidophilus (lack-toe-ba-SILL-us ah-sid-DA-fill-us)
Lactobacillus brevis (BRE-vis)
Lactobacillus bulgaricus (bull-GAR-ree-kus)
Lactobacillus casei (KAY-seh-ee)
Lactobacillus delbruckii (dell-BROOK-ee-eye)
Lactobacillus lactis (LAK-tis)
Lag phase The period of time during which a population does not increase in number. The period of time during which a population is adapting to new environmental conditions before it reproduces exponentially. See Exponential, Stationary, and Death phases.
Latent period The time from the initiation of a virus infection until virus progeny are released.
Leeuwenhoek, Antony van (leh-OYE-ven-hook, an-TOE-nee van)
Leghemoglobin A hemoglobin-like molecule coded for by both the endosymbiotic bacteria in root nodules of leguminous plants and the host plant cells. Leghemoglobin binds molecular oxygen and protects the bacterial nitrogenase, which is oxygen sensitive.
Legionella pneumophila (lee-jon-NELL-ah new-mow-FILL-ah) Gram negative, rod-shaped, nommotile, aerobic, weakly oxidase positive bacteria with numerous growth requirements. Causes Legionnaires' disease, a pneumonia.
Legionnaires' disease A pneumonia. See *Legionella.*
Leishmania donovani (liesh-main-NEE-ah don-oh-VAN-ee) A flagellated protozoan that reproduces in phagocytes and causes a serious leishmaniasis, called kala-azar.
Leptospira interrogans (lep-toe-SPY-rah in-TER-row-gans) Gram negative, helical, motile by axial filament, microaerophilic, aerobic bacterium. Causes leptospirosis.
Leptospira interrogans canicola (kan-nee-KO-la) A serotype of *L. interrogans.* Causes leptospirosis.
Leptothrix (lep-tow-THRIX) Gram negative, rod-shaped, nonmotile, anaerobic, catalase negative, carbon dioxide–requiring bacteria.
Loeffler, Friedrich (LOF-fler, FRIED-rehh)
Leucocytes White blood cells, such as neutrophils, basophils, and eosinophils.
Leuconostoc mesenteroides (lou-ko-NOS-stock mez-zen-ter-OYE-deez) A lactic acid bacterium that is gram positive, spherical, nonmotile, aerotolerant, anaerobic, and catalase negative.
Leucothrix (lou-kow-THRIX) Gram negative, oval, filament-forming, nonmotile, aerobic, sulfide-oxidizing bacteria that do not form sulfur granules.
Levans A homopolymer of D-fructose connected by beta (1,2) linkages.
Lichen (LIE-ken) A symbiotic association between eukaryotic algae or cyanobacteria and fungi which creates an organism distinct from either of the mutuals.
Lignin A polymer of aromatic alcohols that makes up 25 percent of wood.
Limit of resolution The minimum distance that must separate two points so that their image can be resolved. See Resolving limit.
Lipopolysaccharide A molecule found in the outer membrane of gram negative bacteria, responsible for allergic reactions in some animals. An endotoxin from gram negative bacteria.
Lipoteichoic acid Teichoic acids attach to lipids in the cytoplasmic membrane rather than to peptidoglycan. Lipoteichoic acids help some bacteria to attach to inanimate objects and to other organisms.
Log phase Same as exponential phase.
LPS Lipopolysaccharides found in the outer membrane of gram negative bacteria.
Lymphocytes Cells derived from the lymphatic system which are involved in the immune response, specifically B-cells and T-cells.
Lyngbya (LING-by-ah) A filamentous, nonmotile cyanobacterium.
Lysogen A cell that carries a provirus.
M-protein A protein on the surface of the bacterium *Streptococcus pyogenes* which helps it attach to host tissue.
Macrophage A large ameboid-like cell generally found in the reticulo-endothelial system. Involved in phagocytosis and in the immune response.
Malassezia furfur (mal-es-SEZ-zia FER-fer) A fungus belonging to the class Deuteromycetes which is responsible for tinea versicolor, a pigmented fungal infection of the skin on the torso and upper legs.
Matthaei, Heinrich (mat-THEH-ee, HINE-rehh)
Memory cell B-lymphocytes or T-lymphocytes that have been sensitized to a specific antigen and which can mount an immune response more rapidly than unsensitized lymphocytes involved in a primary immune response. See Anamnestic response.
Meninges [sing. meninx] (meh-NIN-gees, MEN-inks) The three membranes that cover the brain and spinal cord.
Meningitis An inflammation of the membranes that surround the brain and spinal cord.
Metchnikoff, Elie (METCH-nee-kof, EH-lee)

Methane-oxidizing bacteria Bacteria that utilize methane as a source of carbon and energy. Examples are *Methylobacter* and *Methylococcus*. See Methylotrophs.

Methane-producing bacteria Bacteria that release methane. These bacteria belong to the family Methanobacteriaceae. Examples are *Methanobacterium*, *Methanosarcna*, and *Methanococcus*. See Methanogenic bacteria.

Methanobacterium (meh-thane-no-bak-TER-ee-um) Archaebacterium. Gram negative, rod-shaped, anaerobic, hydrogen- and carbon dioxide–requiring bacterium that releases methane.

Methanococcus (meh-thane-no-KOCK-kus) Archaebacterium.

Methanogenic bacteria Bacteria that generate methane during their metabolism. Same as methane-producing bacteria.

Methanosarcina (meh-thane-no-sar-SIN-ah) Archaebacterium.

Methanospirillum (meh-thane-no-spear-RILL-um) An archaebacterium.

Methylococcus (meh-THILL-low-KOCK-kus) Gram negative, spherical, nonmotile, aerobic, methane- and methanol-utilizing, oxidase positive bacteria.

Methylomonas (meh-thill-low-MOW-nas) Gram negative, rod-shaped, polarly flagellated, methane-utilizing, oxidase positive bacteria.

Methylotrophs Bacteria that obtain their energy by metabolizing methane or methanol. These bacteria belong to the family Methylomonadaceae.

Microaerophiles Organisms that require reduced concentrations of oxygen, and elevated concentrations of carbon dioxide, in order to grow.

Micrococcus luteus (my-crow-KOCK-kus LU-teh-us) Gram positive, spherical, nonmotile, aerobic, oxidase positive bacterium.

Micrococcus roseus (ROW-seh-us)

Microorganism Any microscopic single-celled organism or microscopic multicellular animal or mold: bacteria, protozoans, algae, fungi, and metazoans. May also be used when referring to viruses. "Infectious agent" is used when referring to viruses, viroids, and prions.

Microsporum canis (my-krow-SPORE-um KAN-nis) A fungus belonging to the Deuteromycetes which is responsible for ringworm of the face, arms, and body.

MIC Minimum inhibitory concentration of a chemical that inhibits the growth of an organism.

Minimal medium A medium in which all the constituents and their concentrations are known. Also known as a defined medium. A minimal medium generally consists of salts that supply phosphate, nitrate (or ammonium), sulfate, magnesium, and trace elements, and a carbon source such as glucose. See Complex medium.

Mixed acid fermentation A fermentation in which a mixture of acids, alcohols, and gases is released. Carried out by bacteria such as *Escherichia coli*. Tested for by the methyl red test.

Monocyte Mononuclear phagocytic leukocyte.

Monod, Jacques (mah-NO, JAH-kez)

Morbidity The number of diseased individuals existing during a given period or during a particular incident.

Mortality The number of deaths occurring during a given period or during a particular incident.

Mucopeptide See Peptidoglycan.

Murein See Peptidoglycan.

Mutagen Physical or chemical agents that induce mutations. See also Carcinogen.

Mutualism A symbiotic association between two organisms in which both benefit.

Myco- Having to do with fungi. Many bacteria showing filamentous and/or branching growth originally were thought to be related to fungi, or were mistaken for fungi, and consequently contain the prefix referring to fungi in their genus designation.

Mycelium [pl. mycelia] (my-SEE-lee-um, my-SEE-lee-ah) The mass of intertwined, branched, filamentous growth associated with many fungi.

Mycobacteria One group of gram positive, rod-shaped, nonmotile, aerobic, acid-fast, slow-growing bacteria.

Mycobacterium leprae (my-ko-back-TER-ee-um LEP-preh) Causes leprosy.

Mycobacterium phlei (FLAY-ee)

Mycobacterium smegmatis (smeg-MA-tiss)

Mycobacterium tuberculosis (too-ber-cue-LOW-sis) Causes tuberculosis.

Mycology The study of fungi.

Mycoplasmas (my-ko-PLAZ-mahz) A group of bacteria that lack cell walls.

Mycoplasma pneumonia (my-ko-PLAZ-ma new-MO-nee-eh) Gram negative, pleiomorphic (spherical, filamentous, and branched), nonmotile, anaerobic, sterol-requiring, bacterium that lacks a cell wall. It causes primary atypical pneumonia.

Mycorrhiza (my-ko-RYE-zah) A mutualistic association between a fungus and plant roots.

Mycosis A fungal infection.

Mycotoxin A fungal toxin.

Myxobacteria (mix-oh-back-TER-ee-ah) Gliding bacteria that congregate upon nutrient deprivation and form fruiting bodies. The fruiting bodies contain microcysts, which are more tolerant of high heat, desiccation, and radiation than the vegetative cells. An example of a myxobacterium is *Chondromyces crocatus*.

Myxomycetes (mix-oh-my-SEATS) A class of fungi known as the plasmodial slime molds. During part of their life cycle, the Myxomycetes exist as flat, macroscopic cells, as much as 5 cm in diameter, with numerous diploid nuclei. This slimelike cell is called a plasmodium. Upon nutrient deprivation and/or exposure to light, the plasmodium develops into numerous fruiting bodies, which contain spores with a single nucleus. The spores germinate into a vegetative cell that develops into a plasmodium. The plasmodium is amebalike in that it is able to phagocytize bacteria and other small single-celled organisms.

Necrotic tissue Dead tissue.

Nematodes Wormlike animals, often known as roundworms.

Neisseria (nye-SEAR-ee-ah) Gram negative, spherical, nonmotile, aerobic, oxidase positive bacteria.

Neisseria catarrhalis (ka-ta-RAH-liss)

Neisseria flava (FLAH-vah)

Neisseria gonorrhoeae (gon-nah-REE-eh) Causes the venereal disease gonorrhea.

Neisseria meningitidis (meh-nin-JEYE-tid-dis) Causes an inflammation of the meninges, known as meningitis.

Neisseria sicca (SEEK-kah)

Neutrophil White blood cells involved in phagocytosis, in which the granules stain poorly with basic and acidic dyes.

Nitrification The conversion of ammonium to nitrate by soil microorganisms.

Nitrate-reducing bacteria Bacteria the convert nitrate to nitrite and nitrogen gas. Examples are *Pseudomonas* and *Thiobacillus.* See Denitrifying bacteria.

Nitrobacter winogradski (neye-trow-BACK-ter vin-no-GRAD-ski) Gram negative, rod-shaped, nonmotile, aerobic, ammonium-oxidizing, carbon dioxide requiring bacterium. Chemolithotroph. Nitrifying bacterium.

Nitrococcus mobilis (neye-trow-KOCK-kus moh-BILL-iss) Gram negative, spherical, nonmotile, aerobic, nitrate-oxidizing, carbon dioxide–requiring bacterium. Chemolithotroph. Nitrifying bacterium.

Nitrogen cycle A cyclic series of chemical reactions carried out by various microorganisms in which molecular nitrogen is reduced to ammonium, the ammonium is converted to nitrite, the nitrite is further oxidized to nitrate, and the nitrate is reduced back to molecular nitrogen.

Nitrogen fixation The reduction of atmospheric molecular nitrogen to ammonium by enzyme nitrogenase, and the incorporation of the ammonium nitrogen into organic compounds.

Nitrogen-fixing bacteria Bacteria that are able to reduce molecular nitrogen to ammonium and then assimilate (fix) the nitrogen into organic compounds. Examples include *Azotobacter, Klebsiella, Rhizobium, Anabaena,* and *Nostoc.*

Nitrosococcus (neye-trow-so-KOCK-kus) Gram negative, spherical, nonmotile, aerobic, nitrite-oxidizing, carbon dioxide–requiring bacteria. Chemolithotroph. Nitrifying bacteria.

Nitrosomonas europaea (neye-trow-so-MO-nas your-OH-pah-eh) Gram negative, rod-shaped, nonmotile, aerobic, nitrite-oxidizing, carbon dioxide–requiring bacterium. Chemolithotroph. Nitrifying bacterium.

Nocardia asteroides (no-CAR-dee-ah ass-ter-OYE-dees) Gram positive, acid-fast, rod-shaped, mycelium-forming, nonmotile, aerobic bacterium. Causes pulmonary, cutaneous, and subcutaneous infections. May become systemic and infect various organs.

Nodule A small mass of cells found on the roots of legumes in which nitrogen-fixing bacteria reside.

Nosocomial infection An infection obtained while a patient or worker in a hospital.

Nostoc A cyanobacterium.

Nucleus (pl. nuclei) (NEW-klee-us; pl. should be NEW-kleh-ee, but always pronounced NEW-klee-eye). Membrane-enclosed organelle of eukaryotic cells containing the cell's genetic information.

O-antigen The outer portion of the polysaccharide that extends from the lipopolysaccharide found in the outer membrane of gram negative bacteria.

Oligotroph (OH-lee-go-trowf) An organism that can grow at very low concentrations of nutrients.

Oligotrophic waters (oh-lee-go-TROW-fik) Bodies of water that are low in nutrients.

Oomycetes (oh-oh-my-SEAT) A class of unicellular or filamentous fungi that produce flagellated gametes. The walls contain cellulose rather than chitin. Filamentous forms show nonseptated hyphae.

Oscillatoria (ah-sill-la-TOR-ee-ah) A cyanobacterium.

Onchocerca volvulus (on-ko-SIR-kah VOL-view-lus) A parasitic roundworm (nematode) that causes onchocerciasis and is responsible for the disease known as river blindness.

Oncogenic viruses Viruses that can cause cancer.

One-gene-one-enzyme hypothesis The idea that a region of DNA, called a gene, contains the information for making a functional enzyme. It was originally proposed by Beadle and Tatum.

One-step growth curve The growth curve shown by viruses that lyse their host cells.

Operon One or more structural genes controlled by one or more controlling sites. The smallest operon consists of a gene that codes for a polypeptide and a promoter (RNA polymerase binding site is a controlling site).

Osmosis The diffusion of solvent (water) across a membrane from a region of high concentration to a region of low concentration.

Osmotic pressure The pressure required to prevent the influx of water into a volume that is hypertonic. The internal water pressure on the cell membrane and wall due to the influx of water into the cell because of a hypotonic environment.

Oxidation The removal of electrons.

Oxidative phosphorylation Occurs when electrons and protons from oxidized substrates are used to create a proton gradient across a membrane, and this proton gradient is used to power the synthesis of ATP from ADP and inorganic phosphate. See Substrate phosphorylation and Photophosphorylation.

Paramecium aurelia (par-ah-MEE-see-um ah-RELL-lee-ah) A ciliated protozoan.

Passive diffusion See Diffusion.

Pasteur, Louis (pass-TUR, lu-EES)

Pasteurization (pass-tur-rye-ZAY-shun) The heating of material such as wine and milk to temperatures below the boiling point, in order to kill spoilage organisms of disease-causing organisms. Thermoduric organisms survive pasteurization. See Sterilization and Appertization.

Pectin A polymer of D-galacturonate.

Pediculus humanus (peh-DICK-ku-lus hue-MAN-us) Body louse. Vector of the bacterium that causes relapsing fever, *Borrelia recurrentis.*

Pediococcus (peh-dee-oh-KOCK-kus) Gram positive, spherical, nonmotile, microaerotolerant, anaerobic bacterium with complex growth requirements, often arranged in tetrads.

Pellicle (PEL-lah-kul) (1) A film of microorganisms on the top of a broth. (2) An envelope that surrounds a microorganism. (3) A convoluted and cytoskeleton-strengthened plasma membrane found in some protozoans.
Penicillium (pen-neye-SILL-lee-um) A fungus belonging to the class Deuteromycetes which forms conidia from annellides.
Penicillium chrysogenum (kry-so-JEN-um)
Penicillium notatum (no-TAH-tum)
Penicillium roquefortii (rowk-FOUR-tee-eye)
Peptidoglycan A polysaccharide layer in the walls of most eubacteria, constructed from alternating units of N-acetyl-glucosamine and N-acetyl-muramic acid. Peptidoglycan is also known as murein or mucopeptide.
Periplasmic space In gram negative bacteria, the region between the cytoplasmic membrane and the outer membrane of the wall. In gram positive bacteria, the region between the cytoplasmic membrane and the peptidoglycan layer.
Permease A protein involved in the transport of materials into or out of a cell.
Phage (FAGE) A bacterial virus.
Photophosphorylation The synthesis of ATP using energy derived from light. See Oxidative phosphorylation.
Photosynthesis The process that uses light to generate energy (ATP) and reducing power (NADPH), and that uses ATP and NADPH to fix carbon dioxide.
Phytophthora infestans (fie-TOFF-thor-ah in-FESS-tans) A fungus belonging to the class Oomycetes which causes late blight of potato.
Plaque (1) A clearing in a bacterial lawn due to a virus infection. (2) A virus colony. (3) The bacteria, glycocalyx, food particles, and minerals that form a mat on teeth.
Plasma cells Cells that release antibodies and that develop when a B-lymphocyte is stimulated by antigen.
Plasma membrane The membrane that contains the cytoplasm and defines a cell.
Plasmid Small, autonomous piece of hereditary material distinct from the main genome. In bacteria, small circular chromosomes 5,000–100,000 base pairs long (5–100 genes).
Plasmodium See Myxomycetes and Acrasiomycetes.
Plasmodium (plaz-MO-dee-um) Flagellated protozoans.
Plasmodium falciparum (fall-SIP-are-um) Causes malaria.
Plasmodium malariae (mah-LAIR-ee-eh) Causes malaria.
Plasmodium vivax (VEE-vahz) Causes malaria.
Plastids Organelles found in eukanyotic cells, such as chromosomes and chromoplasts.
Primary structure In proteins and nucleic acids, the linear sequence of amino acids and nucleotides, respectively.
Prion (PREE-on) An infectious agent that lacks nucleic acids and appears to consist entirely of protein. See Virus and Viroid.
Prophage (PRO-fage) The phage DNA that is integrated into the host's hereditary material.
Propionic acid bacteria Bacteria that carry out a propionic acid fermentation.
Propionic acid fermentation A fermentation in which the major waste products are propionic acid and carbon dioxide. Bacteria such as *Propionibacterium* carry out this type of fermentation.
Propionibacterium (pro-pee-on-nee-back-TER-ee-um) Gram positive, rod- to club-shaped diphtheroid, nonmotile, aerotolerant and anaerobic bacteria.
Prostheca [pl. prosthecae] (praws-THEE-ka, praws-THEE-keh) Cellular extensions of the cell, such as the stalk in *Caulobacter* and the hypha in *Rhodomicrobium.*
Proteus mirabilis (PRO-tee-us mere-AH-bill-iss) Gram negative, rod-shaped, peritrichously flagellated, facultatively anaerobic bacterium. Causes urinary tract infections.
Proteus vulgaris (vul-GAR-is) Causes urinary tract infections.
Proton-motive force (p) An energized state of a membrane due to electrical potentials and proton gradients (pH differences). The potential created across a membrane when protons (hydrogen ions) are concentrated on one side of the membrane is: p = U − 2.3RT pH/F, where U = membrane potential, R = gas constant, T = absolute temperature, pH = pH difference across the membrane, and F = faraday.
Protoplast Gram positive cell that lacks its peptidoglycan cell wall because of mutation or treatment with chemicals. In an isotonic solution, protoplasts become spherical. Lysozyme treatment of gram positive cells results in protoplasts. See Spheroplast.
Prototroph Naturally occurring strain of microorganisms (wild type).
Providencia (pro-vee-DEN-see-ah) Gram negative, rod-shaped, peritrichously flagellated or nonmotile, facultatively anaerobic bacterium.
Provirus The virus DNA that is integrated into the host's hereditary material.
Pseudomonads (sue-doe-MO-nads) Gram negative, rod-shaped, polarly flagellated, aerobic bacteria.
Pseudomonas aeruginosa (sue-doe-MO-nas ah-rue-jin-NO-sah) Gram negative, rod-shaped, polarly flagellated, aerobic bacterium. Produces a green pigment. An opportunistic pathogen.
Pseudomonas fluorescens (floor-ES-sense)
Pseudoplasmodium See Acrasiomycetes; Slime mold.
Psychrophile (1) An organism that has its maximum growth rate between 0°C and 20°C. (2) An organism that grows at temperatures below 5°C.
Puerperal sepsis (PWEAR-per-ul SEP-sis) A bacterial invasion of the blood, acquired during childbirth. Also known as childbed fever.
Purple nonsulfur bacteria Generally anaerobic, photosynthetic bacteria that use light to generate their energy (ATP) by cyclic photophosphorylation only. They may use simple organic acids such as acetate to generate their reducing electrons to fix carbon dioxide.
Purple sulfur bacteria Anaerobic, photosynthetic bacteria that use light to generate their energy (ATP) by cyclic photophosphorylation and hydrogen sulfide as a source of reducing electrons to fix carbon dioxide. They release sulfate rather than molecular oxygen as a result of their metabolism. The purple sulfur bacteria may use organic compounds as a source

of reducing electrons and carbon, as do the purple nonsulfur bacteria.

Putrefaction The decomposition of proteins by microorganisms, with resulting production of foul-smelling compounds from the amino acids.

Q-fever A disease caused by the bacterium *Coxiella burnetii* sometimes transmitted to humans through milk from diseased cattle. The bacterium is highly resistant to heat because of the sporelike structures it forms.

Quaternary structure The structure that results when two or more folded proteins form a complex molecule, such as an antibody molecule or the hemoglobin molecule.

R-factor A plasmid that confers fertility on the host cell and that carries genes that make the host cell resistant to antibiotics and drugs.

Reduction The addition of electrons to a molecule. See Oxidation.

Refractive index The ratio of the speed of light in a vacuum to the speed of light in a material. The index of refraction of a lens is approximately 1.4.

Resolving limit The resolving limit (R) is the minimum distance between two points such that the images of the two points can be distinguished from each other. The resolving limit (R) of a lens system depends upon the wavelength of light (λ), the index of refraction (n) of the material between the object and the objective lens, and the diameter and working distance of the lens, given by sin θ: $R = \lambda/n \sin \theta$. Same as the limit of resolution.

Resolving power The ability to distinguish between two closely spaced points. See Resolving limit.

Respiration Involves the oxidation of inorganic or organic molecules, the generation of energy (ATP) by running electrons (and hydrogen ions) through an electron transport system, and the donation of electrons to an inorganic electron acceptor. Aerobic respiration occurs when the electron acceptor is molecular oxygen, and anaerobic respiration occurs when the electron acceptor is an inorganic molecule other than oxygen, such as sulfate or nitrate.

Restriction endonucleases Bacterial enzymes that cut DNA within the DNA molecule. The most useful endonucleases are those that cut at specific sites. These enzymes are used in genetic engineering to splice genes. See Endonuclease and Exonuclease.

Reticular bodies The dividing, noninfectious chlamydia found in infected cells. Same as initial bodies. See Elementary bodies.

Rhizobium leguminosarum (rye-ZOH-bee-um leh-goo-min-OH-sar-um) Gram negative, rod-shaped, flagellated, aerobic bacteria able to grow within certain plant cells. When growing within root nodules, the cells are pleomorphic. Carries out symbiotic nitrogen fixation.

Rhizopus nigricans (RYE-zoh-puss NEYE-gree-kans) A fungus in the class Zygomycetes which forms asexual and sexual sporangiospores and sexual zygospores. The mycelium is coenocytic and nonseptated.

Rhizopus stolonifer (stow-law-NIF-fer)

Rhodomicrobium vannielii (row-doe-my-CROW-bee-um van-NEEL-lee-eye) Gram negative, rod-shaped, hyphae-forming, nonmotile or peritrichously flagellated, anaerobic, photosynthetic bacterium that oxidizes simple organic compounds to generate reducing power for fixing carbon dioxide. Molecular oxygen is not evolved during photosynthesis. Also known as purple nonsulfur bacteria.

Rhodopseudomonas sphaeroides (row-doe-sue-doe-MO-nass sfair-OYE-dees) Gram negative, oval or rod-shaped, polarly flagellated, anaerobic, photosynthetic bacterium that oxidizes simple organic compounds to generate reducing power for fixing carbon dioxide. Molecular oxygen is not evolved during photosynthesis. Also known as purple nonsulfur bacteria.

Rhodospirillum rubrum (row-doe-speer-RIL-lum RUE-brum) Gram negative, spiral-shaped, polarly flagellated, anaerobic, photosynthetic bacterium that oxidizes simple organic compounds to generate reducing power for fixing carbon dioxide. Molecular oxygen is not evolved during photosynthesis. Also known as purple nonsulfur bacteria.

Richettsia (rick-KET-see-ah) Gram negative, oval to rod-shaped, nonmotile, obligately intracellular bacteria. See also *Coxiella.*

Rickettsia prowazekii (pro-wah-ZE-key-eye) Causes epidemic typhus.

Rickettsia typhi (TIE-fee) Causes rat and mouse (murine) typhus.

Rickettsia rickettsii (rick-KET-see-eye) Causes Rocky Mountain spotted fever.

Sabouraud, Raymond (SAH-ba-rod, RAY-mond)

Saccharomyces cerevisiae (sak-kah-row-MY-sees ser-reh-VIS-ee-ee) A yeast belonging to the class Ascomycetes which forms asci and sexual ascospores and carries out alcoholic fermentation. It is used in making bread, wine, and beer.

Salmonella typhi (sal-mo-NEL-lah TIE-fee) Gram negative, rod-shaped, peritrichously flagellated, facultatively anaerobic bacteria. Causes typhoid fever.

Salmonella typhimurium (tie-fee-MUR-ee-um) Causes paratyphoid fever: gastroenteritis and bacteremia.

San Joaquin Valley Fever (SAN hwah-KEEN VA-lee FEE-ver) A pulmonary infection caused by the fungus *Coccidioides.* See Valley fever and Coccidioidomycosis.

Sarcina ventriculi (sar-SIN-na ven-tree-KUL-lee) Gram positive, spherical, nonmotile, anaerobic, catalase negative bacteria that are found in packets of four, eight, or more cells. They have complex growth factors. The cell wall of this species has a thick outer layer of cellulose. This is the only bacterium known to have a wall with cellulose. Only one other bacterium, *Acetobacter xylinum*, is known to synthesize cellulose.

Sarcoptes scabei (sar-COP-tees SKAY-bee-ee) The mite (arachnid) that burrows under the outer layer of skin and causes scabies.

Schistosoma haematobium (shis-toe-SOW-mah he-mah-TOE-bee-um) A flatworm (fluke) that is responsible for schistosomiasis, an infection of the intestinal veins and liver which leads to liver destruction.

Schistosoma japonicum (jah-PON-nee-kum) Causes schistosomiasis.

Schistosoma mansoni (man-SOW-nee) Causes schistosomiasis.

Schwann, Theodor (SHOE-ann, TEE-oh-dor)

Secondary structure In proteins, the structure formed from the folding of the primary structure due to disulfide bonds, and the alpha helix or pleated sheets formed from the hydrogen bonding. In nucleic acids, the structure formed from the folding of the primary structure due to the hydrogen bonding between complementary bases. See Tertiary structure.

Semmelweis, Ignaz (seh-mel-*VISE*, EEG-nahs)

Sepsis Generally, infected tissue. See Asepsis.

Septa [pl. septae] (SEP-tah, SEP-teh) Generally used to describe the walls that segment fungal hyphae and filamentous and branching bacteria.

Septicemia A bacterial infection of the blood. See Viricemia.

Serratia marcescens (ser-RAY-she-ah mar-SES-sens) Gram negative, rod-shaped, peritrichously flagellated, facultatively anaerobic bacterium. Colonies form a red pigment at 18°C but not at 37°C.

Sex pilus Pilus that is necessary if a bacterium is to be able to conjugate and donate genetic information to a recipient cell.

Shigella dysenteriae (she-GELL-lah dis-sen-TER-ee-eh) Gram negative, rod-shaped, nonmotile, facultatively anaerobic bacterium. Causes dysentery.

Shigella flexneri (flex-NER-ee) Causes dysentery.

Simonsiella (see-mon-SEE-el-lah) Gram negative, filament-forming, gliding, aerobic bacteria with complex growth requirements.

Slime molds Two classes of fungi: Acrasiomycetes (which form pseudoplasmodia) and Myxomycetes (which form plasmodia). The Acrasiomycetes exist as amebae or as pseudoplasmodia (multicellular structures) during one part of their life cycle and form fruiting bodies, similar to molds, during another part. The Myxomycetes exist as plasmodia (single, macroscopic cells with multiple nuclei) during one part of their life cycle and form fruiting bodies when deprived of nutrients.

Solutes The dissolved ions and small molecules in a solution.

Solution A liquid mixture of compounds in which one material, the solvent, uniformly distributes and dissolves other compounds (solutes). Examples are salt in water and ethanol in water.

Solvent A material, usually water, that uniformly distributes and dissolves another material.

Sphaerotilus natans (sfer-AH-til-us NAY-tans) Gram negative, rod-shaped, chain-forming, sheathed, aerobic bacterium. Individual cells may have subpolar flagella.

Spheroplast Gram negative cell that has lost its peptidoglycan layer. In an isotonic solution, spheroplasts become spherical. Spheroplasts can be formed by treating gram negative cells with lysozyme.

Spirillum volutans (spear-RILL-um vol-you-TANS) Gram negative, spiral, polarly flagellated, aerobic, oxidase positive bacterium.

Spirochete (SPY-row-keet) A group of helical bacteria that have axial filaments rather than flagella. Examples are *Treponema* and *Leptospira*.

Spirogyra (spy-row-JI-rah) A filamentous eukaryotic green alga with spiral-shaped chloroplasts.

Spirulina (spee-rue-LIE-na) A photosynthetic prokaryotic cyanobacterium that forms trichomes. Molecular oxygen is evolved from photosynthesis.

Sporolactobacillus (spore-row-LAK-toe-bah-SILL-us) Gram positive, rod-shaped, peritrichously flagellated, endospore-forming, microaerophilic, catalase negative bacteria that lack cytochromes and ferments by the lactic acid route.

Sporosarcina ureae (spore-row-sar-SIN-ah your-EE-eh) Gram positive, spherical, tetrad-forming, endospore forming, nonmotile, aerobic bacterium.

Staphylococcus (staff-fee-low-KOCK-kus) Gram positive, spherical, cluster-forming, nonmotile, facultatively anaerobic bacteria.

Staphylococcus albus (AHL-bus)

Staphylococcus aureus (AH-ree-us) Commonly found in the nose. Causes skin infections and food poisoning.

Staphylococcus epidermidis (eh-pee-der-MID-dis) Commonly found on the skin. A cause of endocarditis.

Sterilization The killing of all microorganisms and infectious agents, such as prions and viroids.

Streptococcus (strep-toe-KOCK-cus) Gram positive, spherical, chain-forming, nonmotile, aerotolerant anaerobic, catalase negative bacteria that lack cytochromes.

Streptococci, Group A: *S. pyogenes.*

Streptococci, Group B: *S. agalactiae.*

Streptococci, Group C: *S. equi* and *S. dysgalactiae.*

Streptococci, Group D: *S. faecalis*, *S. faecium*, *S. durans.* Cause bacterial endocarditis and urinary tract infections.

Streptococci, Group F: *S. anginosus.*

Streptococci, Viridans group: *S. salivarius*, *S. sanguis*, and *S. mutans.* Alpha-hemolytic on blood agar plates. Normally found in the oral cavity. May cause subacute bacterial endocarditis.

Streptococcus agalactiae (a-gah-LACK-tee-eh) Group B. Causes bovine mastitis and infections in newborn infants.

Streptococcus anginosus (an-jee-NO-sus) Group F.

Streptococcus bovis (BOW-vis) Group D nonenterococcus. Causes subacute bacterial endorcarditis.

Streptococcus cremoris (kre-MOR-is) Used in the fermentation of dairy products.

Streptococcus durans (dur-RANS) Group D.

Streptococcus dysgalactiae (des-gah-LAK-tee-eh) Group C. Causes bovine mastitis.

Streptococcus equi (EH-qui) Group C.

Streptococcus equinus (eh-QUI-nus) Group D nonenterococcus. Causes subacute bacterial endocarditis.

Streptococcus faecalis (fee-KAH-lis) Group D enterococcus.

Streptococcus faecium (fay-SEE-um) Group D enterococcus.

Streptococcus lactis (LACK-tis) Group N.

Streptococcus mitis (MY-tis) Viridans group.

Streptococcus mutans (MU-tans) Viridans group. Responsible for tooth decay.

Streptococcus pneumoniae (new-MOAN-nee-eh) Viridans group. A major cause of pneumonia.

Streptococcus pyogenes (pie-AH-jen-nees) Group A. Causes strep throat, scarlet fever, rheumatic fever, impetigo, skin infections, and puerperal sepsis.

Streptococcus salivarius (sal-lee-VAR-ee-us) Viridans group.

Streptococcus sanguis (SANG-gwis) Viridans group.

Streptococcus thermophilus (ther-MAH-fill-us) Participates in the fermentation of milk to produce yogurt.

Streptococcus viridans (vir-ree-DANS)

Streptomyces (strep-toe-MY-seez) Gram positive, rod-shaped, mycelia-forming, aerobic bacteria. Many species produce antibiotics.

Streptomyces griseus (GRIS-say-us)

Streptomyces venezuelae (veh-nez-WHALE-eh)

Substrate level phosphorylation Occurs when a phosphorylated substrate donates the phosphate in the synthesis of ATP from ADP. See Oxidative phosphorylation and fermentation.

Sulfate-reducing bacteria Bacteria that use sulfate as an electron acceptor during anaerobic respiration. Examples include *Desulfovibrio* and *Desulfotomaculum.*

Sulfur-oxidizing bacteria Bacteria that use reduced forms of sulfur, such as hydrogen sulfide, elemental sulfur, and thiosulfate, as a source of reducing electrons or energy. Examples include the photosynthetic bacteria, such as *Chromatium* and *Chlorobium*, and the aerobic *Beggiatoa* and *Thiothrix*.

Systemic infection An infection that has spread throughout the body.

T-cells (T-lymphocytes) Thymus-derived lymphocytes protect animals from intracellular infections agents, such as viruses and some bacteria and protozoans. T-lymphocytes also modulate the antibody response to infectious agents.

Taenia (TEH-nee-ah) Tapeworms. Cause taeniasis, an infection of the intestines, heart, spinal cord, and brain. The symptoms are diarrhea, abdominal pain and weight loss.

Taenia solium (SO-lee-um) Pig tapeworm.

Taeniarrhyncus saginatus (teh-nee-ah-RINK-us sah-jin-NAH-tus) Beef tapeworm.

Tautomeric shift In chemistry, the transient shift of electrons and protons in a molecule which creates a transient form of the molecule with different chemical properties. For example, when the base adenine undergoes a tautomeric shift, the amino group become an imino group, and the transient form of adenine no longer pairs with thymine but with cytosine.

Teichoic acids Large polymers that are attached to the peptidoglycan layer in gram positive bacteria and may contribute as much as 50 percent to the weight of the cell wall. Teichoic acids are polymers of phosphate and molecules such as glycerol or ribitol. When teichoic acids are attached to membrane lipids, they are referred to as lipoteichoic acids.

Temperate bacteriophage A bacterial virus that generally becomes lysogenic rather than lytic.

Tertiary structure In a protein, the final three-dimensional structure that forms from the folding of the polypeptide. The final folding is determined by the disulfide bonds, the hydrogen bonds, and the hydrophobic and hydrophilic interactions of the amino acids with water and with themselves.

Tetrahymena pyriformis (teh-tra-HI-men-ah pie-rye-FORM-is) A single-celled green eukaryotic protozoon.

Thermoactinomyces sacchari (ther-mo-ak-tin-no-MY-seez SACK-kar-ee) Gram variable, rod-shaped, branching, endospore-forming, aerobic bacterium that grows best at about 55°C.

Thermoduric Highly resistant to heat. Vegetative cells and many types of spores are thermoduric.

Thiobacillus (thi-oh-ba-SILL-us) Gram negative, rod-shaped, polarly flagellated or nonmotile, aerobic, carbon dioxide–fixing bacteria that oxidize hydrogen sulfide, sulfur, or thiosulfate to sulfate for energy and for reducing electrons.

Thiobacillus ferrooxidans (fair-ro-OX-ee-dans)

Thiobacillus thiooxidans (thi-oh-OX-ee-dans)

Thiospirillium jenese (thi-oh-speer-RIL-ee-um je-NE-see) Gram negative, spiral-shaped, polarly flagellated, anaerobic, photosynthetic bacterium that oxidizes hydrogen sulfide for reducing electrons.

Thiothrix Gram negative, oval to rod-shaped, filament-forming, aerobic, carbon dioxide–fixing, hydrogen sulfide–and sulfur-oxidizing bacteria that accumulate sulfur granules within the cell.

Tinea Fungal infections of the skin.

Toxoplasma gondii (talks-oh-PLAZ-mah GONE-dee-eye) The protozoan responsible for the disease toxoplasmosis. The protozoan is an intracellular parasite, often infecting various organs of the body. It is often spread to humans from cats.

Trace element Minerals required in very small amounts by living organisms. Examples are zinc, copper, cobalt, and molybdenum.

Transduction The transfer of hereditary material from one cellular organism to another by a virus, with subsequent recombination of the hereditary material with the recipient's genome and the transformation of the recipient.

Transformation The alteration of an organism's hereditary material. In microbiology, the uptake of naked DNA by an organism and the subsequent recombination of the hereditary material with the organism's genome and alteration of the organism's genetics and physiology.

Treponema carateum (treh-po-NEE-mah kah-rah-TEH-um) Gram negative, long helical, axially filamented (motile), anaerobic, catalase negative bacterium. Causes pinta, a skin disease that resembles syphilis. Generally found in tropical America.

Treponema dentium (DEN-tee-um)

Treponema pallidium (PAL-lee-dum) The bacterium responsible for the venereal disease syphilis.

Treponema pertenue (per-TEH-new) Causes yaws, a skin disease that resembles syphilis. Generally found in tropical Africa.

Tricarboxylic acid cycle See Citric acid cycle and Krebs cycle.

Trichome A filament of cells.

Trichomonas vaginalis (trick-ka-MO-nas vah-jin-NAH-lis) A flagellated protozoan that infects the urinary and genital tract and is responsible for the venereal disease trichomoniasis.

Trichophyton canis (trick-KO-fee-ton KAH-nees) A fungus that causes ringworm (tinea).

Trichophyton mentagrophytes (men-ta-GROW-fi-teez) Causes ringworm.

Trophic level Feeding level. Relative position of a population in a food chain.

Trophozoites An infectious vegetative (growing and reproducing) form of some protozoans.

Trypanosoma cruzi (try-pan-oh-SO-mah CREW-zee) The flagellated protozoan responsible for Chagas' disease.

Trypanosoma brucei gambiense (bru-SAY-ee gam-bee-EN-seh) A flagellated protozoan responsible for African sleeping sickness.

Trypanosoma brucei rhodesiense (row-DEE-zee-EN-seh) A flagellated protozoan responsible for African sleeping sickness.

Tubercle A mass of cells (granuloma) that develops in the lungs. The mass may contain the tubercle bacillus *(Mycobacterium tuberculosis).*

Twort, William (tah-WORT)

Ureaplasma urealyticum (you-ree-ah-PLAS-mah you-ree-ah-LEET-eit-cum) A mycoplasma. Gram negative, spherical, nonmotile, wall-less, anaerobic, catalase negative bacterium. Causes urethritis.

Vaccination The inoculation of an animal or human with microorganisms or material from microorganisms in order to induce an immune response that protects the animal or human from infections by a specific microorganism.

Variolation Vaccination against smallpox, using the smallpox virus.

Vector (1) The organism that carries and transmits a disease-causing organism. (2) The DNA (plasmid or virus) that is used to carry and clone specific pieces of DNA.

Vegetative cell A cell that grows and divides.

Veillonella (veye-lon-NEL-ah) Gram negative, spherical, nonmotile, anaerobic, carbon dioxide–requiring bacteria, unable to ferment carbohydrates or polymeric alcohols, with complex growth requirements.

Vibrio cholerae (VEE-bree-oh KAW-ler-eh) Gram negative, curved or rod-shaped, polarly flagellated, facultatively anaerobic, oxidase positive bacterium. Responsible for cholera.

Vibrio parahaemolyticus (pah-rah-heh-mo-LIT-tee-kus) A cause of gastroenteritis usually acquired from seafoods.

Virchow, Rudolf (vir-CHOW, RUE-dolf)

Viremia (vy-REE-me-ah) A virus infection of the blood. See Septicemia.

Virion A virus.

Viroids Naked, double-stranded RNA molecules that infect many different types of plant cells.

Virus Protein-covered nucleic acids that infect all types of cells.

Voges-Proskauer test (VOGE-PROS-cow-er) A chemical test for acetoin (acetylmethylcarbinol), which indicates whether or not a bacterium carries out the 2,3-butanediol fermentation. One of the IMViC tests for differentiating among enterics.

Vorticella (vor-tee-SELL-ah) A ciliated protozoan common in pond water.

Waksman, Selman (VAHKS-mahn, SELL-mahn)

Wild-type The type found in nature. The organism from which mutants (which have growth requirements, or cannot utilize a carbohydrate) have been derived in the laboratory.

Winogradsky, Sergei (vee-no-GRAD-skee, ser-GAY)

Wuchereria bancrofti (wu-cher-AIR-ee-ah, ban-KROF-tee) A nematode (roundworm), transmitted by mosquito, that infects the lymphatics and connective tissue. In chronic cases, it may cause elephantiasis.

Xanthomonas pharmicola (zan-tho-MO-nas farm-ME-ko-lah) Gram negative, rod-shaped, polarly flagellated, aerobic, oxidase negative bacteria with complex growth requirements.

Yeast Fungi that usually exist as single-celled organisms. They may form short hyphae, but a mycelium never develops.

Yersinia pestis (i-er-SIN-nee-ah PES-tis) Gram negative, rod-shaped, nonmotile, facultatively anaerobic bacterium. Causes the plague.

Zoogloea ramigera (zo-O-glee-a rah-ME-ger-ah) Gram negative, rod-shaped, polarly flagellated, aerobic bacterium. Produces a slime that causes microorganisms to cling to rocks in trickling filters or to each other forming flocs (activated sludge).

Zygomycetes (zy-go-my-SEE-tees) A class of fungi that produces sexual spores called zygospores and asexual spores called basidiospores. An example is *Rhizopus*.

INDEX

FIGURE CREDITS

Chapter 1. Opener and Figures 1-3, 1-4a, 1-7, 1-12a: Culver Pictures, Inc. Figures 1-1a left and center: Z. Skobe/BPS. Figures 1-1a right and 1-1b left center: P. W. Johnson and J. M. Sieburth, Univ. of Rhode Island/BPS. Figure 1-1b left: Harvey J. Marchant. Figure 1-1b right center: V. Cassie. Figure 1-1b right: J. N. A. Lott, McMaster Univ./BPS. Figure 1-1c: S. G. Baum/BPS. Figure 1-1d left: G. T. Cole, Univ. of Texas/BPS. Figure 1-1d right: P. Dayanandan. Figure 1-1e: A. N. Broers, B. J. Panessa, and J. F. Gennaro, Jr. Figures 1-4b and c: BPS. Figures 1-9, 1-12b, 1-13a and b: National Library of Medicine. Figures 1-11a and b: H. A. Lechevalier, Waksman Institute of Microbiology.

Chapter 2. Figures 2-13 and 2-16: C. Starr and R. Taggart, *Biology: The Unity and Diversity of Life,* 3rd ed., Wadsworth Pub. Co., 1984.

Chapter 3. Opener and Figure 3-9: P. W. Johnson and J. M. Sieburth, Univ. of Rhode Island/BPS. Figure 3-1: Dept. of Molecular, Cellular, and Developmental Biology, Univ. of Colorado, Boulder. Figures 3-6a, b, c, and d: J. R. Waaland, Univ. of Washington/BPS. Figures 3-8, 3-14: Centers for Disease Control, Atlanta, GA. Figures 3-11a, 3-13: S. C. Holt, Univ. of Texas Health Science Center, San Antonio/BPS. Figure 3-11b: Z. Skobe/BPS. Figures 3-15a and b: J. J. Cardamone, Jr., and B. A. Phillips, Univ. of Pittsburgh/BPS. Figure 3-16: G. T. Cole, Univ. of Texas/BPS.

Chapter 4. Opener and Figures 4-12b, 4-19: H. S. Pankratz, Michigan State Univ./BPS. Figure 4-1a: M. E. Bayer, Inst. for Cancer Research, Philadelphia. Figure 4-1b: M. W. Steer, Univ. of Wisconsin/BPS. Figure 4-2: C. L. Sanders, Battelle Pacific Northwest Labs/BPS. Figures 4-6, 4-7, 4-17, 4-22a: T. J. Beveridge, Univ. of Guelph/BPS. Figure 4-8: J. F. M. Hoeniger, Univ. of Toronto. Figures 4-9b and c: American Society for Microbiology. Figures 4-12a, 4-28b: P. W. Johnson and J. M. Sieburth, Univ. of Rhode Island/BPS. Figure 4-13: S. Abraham and E. H. Beachey, V. A. Medical Center, Memphis, TN. Figure 4-14: R. L. Moore, BioTechniques Labs/BPS. Figure 4-15: R. Kavenoff, Designergenes Posters, Ltd./BPS. Figure 4-16: R. Welch, Univ. of Wisconsin Medical School/BPS. Figures 4-18a and c, 4-22b: S. C. Holt, Univ. of Texas Health Science Center, San Antonio/BPS. Figure 4-18b: S. W. Watson, Woods Hole Oceanographic Inst. Figure 4-20: L. E. Roth and G. Stacey, The Univ. of Tennessee, Knoxville/BPS. Figure 4-24: R. Blakemore and N. Blakemore. Figures 4-25a and b: O. Wyss, M. G. Neumann, and M. S. Socolofsky, *J. Biophys. Biochem. Cytol.,* 10–555 (1961), Elsevier Biomedical Press. Figures 4-26a, b, and c: W. L. Dentler, The Univ. of Kansas/BPS. Figure 4-27: S. L. Wolfe, *Biology of the Cell,* 2nd ed., Wadsworth Pub. Co., 1983. Figure 4-29: E. H. Newcomb, Univ. of Wisconsin—Madison/BPS. Figure 4-31: Osborn, Webster, and Weber, *J. Cell Biol.,* 77:R27–R34 (1978).

Chapter 5. Opener and Figure 5-10 upper left, upper right, and lower right: Centers for Disease Control, Atlanta, GA. Figure 5-6: GCA/Instruments & Equipment Group. Figures 5-10 lower left and center right: E. H. Runyon, V. A. Hospital, Salt Lake City. Figure 5-10 center left: R. J. Hawley, Holy Cross Hospital of Silver Spring, Maryland.

Chapter 6. Opener and Figure 6-12: J. J. Cardamone, Jr., and B. K. Pugashetti: Univ. of Pittsburgh/BPS. Figure 6-4: New Brunswick Scientific Co., Inc.

Chapter 7. Opener and Figure 7-7: H. W. Jannasch, Woods Hole Oceanographic Inst. Figure 7-10: R. Humbert/BPS. Figure 7-11: American Sterilizer Co., Erie, PA.

Chapter 8. Figure 8-15: P. W. Johnson and J. M. Sieburth, Univ. of Rhode Island/BPS.

Chapter 9. Opener and Figure 9-14: J. Carnahan and C. Brinton, Univ. of Pittsburgh Figure 9-1: The Bettmann Archive.

Chapter 10. Opener and Figure 10-13: Stanley N. Cohen, Stanford Univ. School of Medicine.

Chapter 11. Opener and Figure 11-1: Raúl J. Cano.

Chapter 12. Opener and Figure 12-10: W. T. Hall, National Institutes of Health. Figure 12-1a: S. M. Awramik, Univ. of California/BPS. Figures 12-1b, 12-2a, b, c, and d: J. William Schopf, ed., *Earth's Earliest Biosphere,* Princeton Univ. Press, 1983. Figures 12-5a and b: S. W. Fox, Inst. for Molecular and

Cellular Evolution, Univ. of Miami. Figures 12-16a–e: Vitek Systems, McDonnell Douglas Health Systems Co.

Chapter 13. Opener and Figures 13-3a and c, 13-17a: P. W. Johnson and J. M. Sieburth, Univ. of Rhode Island/BPS. Figures 13-1a, b and c: J. R. Waaland, Univ. of Washington/BPS. Figure 13-2: N. Lang, Univ. of California, Davis/BPS. Figures 13-3b, 13-16a, b and c: S. C. Holt, Univ. of Texas Health Science Center, San Antonio/BPS. Figures 13-4a, 13-14: Raúl J. Cano. Figure 13-4b: L. Thomashow, Washington State Univ. Figures 13-4c, 13-19a: T. J. Beveridge, Univ. of Guelph/BPS. Figure 13-5: J. J. Cardamone, Jr., and B. K. Pugashetti, Univ. of Pittsburgh/BPS. Figure 13-6: J. G. Hancock and M. N. Schroth, *Science*, 216:1378; copyright 1982 by the AAAS. Figures 13-7a and b, 13-12a, b, and c, 13-13, 13-15a and b: Centers for Disease Control, Atlanta, GA. Figure 13-8: W. Burgdorfer, Rocky Mountain Lab., Hamilton, MT. Figure 13-9: R. C. Cutlip, National Animal Disease Center. Figure 13-10a: M. G. Gabridge, W. Alton Jones Cell Sci. Ctr., Lake Placid, NY. Figure 13-10b: Karl Maramorosch, Rutgers Univ. Figure 13-11a: Z. Skobe/BPS. Figure 13-11b: D. Selinger and W. P. Reed, V. A. Medical Center, Albuquerque, NM. Figure 13-15c: J. J. Duda and J. M. Slack, *J. Gen. Microbiol.*, 71:63–68(1972). Figure 13-16d: Peter Hirsch, Michigan State Univ. Figure 13-17b: K. Stephens, Stanford Univ./BPS. Figure 13-18: R. G. Mulder and M. H. Deinema, in *The Prokaryotes* (Starr, Stolp, Truper, Balows, and Schegel, eds.), Springer-Verlag, 1981. Figure 13-19b: R. L. Moore, BioTechniques Labs./BPS.

Chapter 14. Opener and Figure 14-3 upper left: Centers for Disease Control, Atlanta, GA. Figures 14-1, 14-2, 14-8a, 14-13c: G. T. Cole, Univ. of Texas/BPS. Figures 14-3 upper right and both lower, 14-8: Raúl J. Cano. Figure 14-9: D. Pramer, Rutgers Univ. Figure 14-13a: J. R. Waaland, Univ. of Washington/BPS. Figure 14-13b: C. Rovinow, Univ. of Western Ontario. Figure 14-15: Reprinted by permission from Mycologià 55:35–38, copyright 1963, C. E. Bracker and E. E. Butler and The New York Botanical Garden.

Chapter 15. Opener and Figure 15-8: P. Dayanandan. Figure 15-3: J. N. A. Lott, McMaster Univ./BPS. Figures 15-5a and b: J. R. Waaland, Univ. of Washington/BPS. Figure 15-6: L. E. Roth, Univ. of Tennessee, Knoxville. Figures 15-10b and c: D. J. Cross/BPS. Figure 15-10d: R. K. Burnard, Ohio State Univ./BPS. Figure 15-13: R. N. Band and H. S. Pankratz, Michigan State Univ./BPS. Figures 15-15a and 15-17a: Centers for Disease Control, Atlanta, GA. Figures 15-15b, 15-17b and c: Lawrence Ash. Figure 15-15c: I. Armstrong, Washington Hospital Center, Washington, DC.

Chapter 16. Opener and Figures 16-1, 16-5b, 16-10, 16-14c: Photographs by Carolina Biological Supply Company. Figure 16-5a: J. H. Applegarth. Figures 16-5c, 16-8a, 16-9a, and b: Wards Natural Science Establishment, Inc. Figures 16-8b and 16-11: L. R. Ash and T. C. Orihel, *Atlas of Human Parasitology*, 2nd ed., The American Soc. of Clinical Pathologists, © 1984, used by permission. Figure 16-12: Office of the Surgeon General/Army. Figures 16-13a, and b, 16-14a: Centers for Disease Control, Atlanta, GA. Figure 16-14b: Raúl J. Cano.

Chapter 17. Opener and Figures 17-3c, 17-11a, 17-15: J. D. Griffith, Cancer Research Center, Univ. of North Carolina. Figure 17-1: National Cancer Institute, NIH. Figure 17-2a: R. Humbert, Stanford Univ./BPS. Figure 17-2b: Raúl J. Cano. Figure 17-2c: Centers for Disease Control, Atlanta, GA. Figure 17-3a: L. Caro and R. Curtiss. Figures 17-3b and 17-4: L. Simon, Rutgers Univ. Figure 17-7: M. E. Bayer, Inst. for Cancer Research, Philadelphia. Figure 17-16b: G. Wertz, Univ. of North Carolina School of Medicine. Figure 17-17: Courtesy T. O. Diener, U.S. Dept. of Agriculture. Figure 17-18: S. B. Prusiner et al., *Cell*, 35:353, copyright by M.I.T. Press, 1983.

Chapter 18. Opener and Figure 18-12: C. L. Sanders, Battelle Pacific Northwest Labs/BPS. Figure 18-3: J. W. Costerton, Univ. of Calgary. Figure 18-4a: Runk/Schoenberger, Grant Heilman Photography. Figure 18-4b: Luvenia Miller, Camera M. D. Studios. Figure 18-5: G. T. Cole, Univ. of Texas/BPS. Figure 18-6: J. W. Rippon, *Medical Mycology*, 2nd ed., W. B. Saunders Co., 1982. Figure 18-7: T. J. Beveridge, Univ. of Guelph/BPS. Figure 18-9: L. J. LeBeau, Univ. of Illinois Hospital/BPS. Figures 18-10a and b: R. Rodewald, Univ. of Virginia/BPS. Figure 18-10c: J. J. Cardamone, Jr., and S. Salvin, Univ. of Pittsburgh/BPS. Figure 18-11b: Raúl J. Cano.

Chapter 19. Opener and Figures 19-2a, b and c: R. Rodewald, Univ. of Virginia/BPS. Figure 19-6: Wallenius, *J. Biol. Chem.*, 225:253–267, by the American Society of Biological Chemists, © 1957. Figure 19-9: Davis, Kim, and Hood, *Cell*, 22:1–2, by M.I.T. Press, © 1980. Figures 19-14, 19-15: Mayer, Scientific American Medicine, 229:58–59, by Scientific American, Inc., © 1973.

Chapter 20. Opener and Figures 20-4a and b: Cedric S. Raine, Albert Einstein College of Medicine, Bronx, NY. Figure 20-1: Baylor College of Medicine, Houston, TX. Figures 20-5a and b: Armed Forces Inst. of Pathology. Figure 20-6a: L. Winograd, Stanford Univ./BPS.

Chapter 21. Opener and Figure 21-3: T. Huber. Figures 21-1a, c, and d: Wards Natural Science Establishment, Inc. Figure 21-1b: Raúl J. Cano. Figure 21-2: Centers for Disease Control, Atlanta, GA. Figure 21-8a: L. J. LeBeau, Univ. of Illinois Hospital/BPS. Figure 21-8b: J. R. Barrett, *Textbook of Immunology*, 4th ed., C. V. Mosby, 1983. Figure 21-10: BPS.

Chapter 22. Opener and Figure 22-3a and b: © 1985 Chris Grajczyk.

Chapter 23. Opener and Figure 23-8: G. T. Cole, Univ. of Texas/BPS. Figures 23-2a, b and c, 23-3a and b, 23-7a and b, 23-10: Centers for Disease Control, Atlanta, GA. Figures 23-4a, 23-13a: L. Winograd, Stanford Univ./BPS. Figure 23-4b: L. J. LeBeau, Univ. of Illinois Hospital/BPS. Figure 23-4c: Carroll H. Weiss, Camera M.D. Studios. Figures 23-4d, 23-6, 23-11, 23-12, 23-13b: Armed Forces Inst. of Pathology. Figure 23-4e: Charles Stoer, Camera M.D. Studios. Figure 23-9a: N. Goodman, Univ. of Kentucky Medical College. Figure 23-9b: J. W. Rippon, *Medical Mycology*, 2nd ed., W. B. Saunders Co., 1982.

Chapter 24. Opener and Figure 24-6: K. E. Muse/BPS. Figure 24-3: J. W. Rippon, *Medical Mycology*, 2nd ed., W. B. Saunders Co., 1982. Figure 24-8a: R. B. Morrison, M.D., Austin, TX. Figures 24-8b, 24-12: Raúl J. Cano. Figure 24-9: G. G Douglas in *Antiviral Agents and Viral Diseases in Man* (Galasso, Buchanan, and Merigan, eds.), Raven Press, 1979. Figure 24-11a: Raúl J. Cano. Figure 24-11b: L. Winogrand, Stanford Univ. School of Medicine/BPS.

Chapter 25. Opener and Figure 25-12: R. L. Owen, *Gastroenterology*, 76:757–759 (1979). Figure 25-2b: D. C. Savage, Univ. of Illinois, Urbana. Figure 25-3e: M. Listgarten, Univ. of Pennsylvania. Figures 25-3f and g: Z. Skobe/BPS. Figure 25-4: Camera M. D. Studios. Figures 25-7a and b: G. T. Cole, Univ. of Texas/BPS. Figure 25-11: Armed Forces Inst. of Pathology.

Chapter 26. Opener and Figures 26-5a and b, 26-9a, b and c, 26-12: Centers for Disease Control, Atlanta, GA. Figures 26-10a, 26-13c: Armed Forces Inst. of Pathology. Figure 26-10b: Centers for Disease Control, Atlanta, GA. Figure 26-11: L. Winograd, Stanford Univ./BPS. Figure 26-13a: Carroll H. Weiss, Camera M.D. Studios. Figure 26-13b: Harvey Blank, Camera M.D. Studios.

Chapter 27. Opener and Figures 27-5, 27-9: The Bettmann Archive. Figure 27-2: W. E. Bell and W. F. McCormick, *Neurologic Infections in Children*, W. B. Saunders Co., 1975. Figure 27-14: Armed Forces Inst. of Pathology.

Chapter 28. Opener and Figure 28-14a: Stephen Lerner, Univ. of Chicago School of Medicine. Figure 28-3: L. J. LeBeau, Univ. of Illinois Hospital/BPS. Figure 28-5: Raúl J. Cano. Figure 28-6: Armed Forces Inst. of Pathology. Figures 28-13a and b: P. J. Fenton, *Arch. Ophthal.*, 76:867; © 1966, American Medical Assn. Figure 28-14b: World Health Organization.

Chapter 29. Figure 29-10: K. J. Ryan, F. D. Schoenknecht, and W. M. M. Kirby, "Disc Sensitivity Testing," *Hospital Practice* 5:2. 1970. Adapted from a drawing by Albert Miller.

Chapter 30. Opener and Figure 30-8: Shirley R. Sparling, California Polytechnic State Univ. Figure 30-1: Z. Skobe/BPS. Figure 30-7a: H. S. Wessenberg and G. A. Antipa, *J. Protozool.*, 17:250–270 (1970). Figure 30-9: B. L. Thorne, Mus. of Comparative Zoology, Harvard Univ. Figure 30-11: D. Powell, Monterey Bay Aquarium.

Chapter 31. Opener and Figure 31-5b: E. H. Newcomb, Univ. of Wisconsin—Madison/BPS. Figure 31-5a: H. E. Evans, Oregon State Univ. Figure 31-10: Raúl J. Cano.

Chapter 32. Opener and Figure 32-3a: J. R. Waaland, Univ. of Washington/BPS. Figure 32-3b: H. W. Pratt/BPS. Figure 32-4a: Woods Hole Oceanographic Inst. Figure 32-4b: H. W. Jannasch, Woods Hole Oceanographic Inst. Figure 32-7: Courtesy of Millipore Corporation. Figure 32-9: Raúl J. Cano. Figure 32-10: V. Paulson/BPS.

Chapter 33. Opener and Figure 33-3: C. W. May/BPS.

Chapter 34. Opener and Figures 34-1, 34-2c, 34-4: Raúl J. Cano. Figure 34-2a and b: Inst. for Mycorrhizal Research and Development, USDA Forest Service. Figure 34-3: J. G. Hancock and M. N. Schroth, *Science*, 216:1378; copyright 1982 by the AAAS. Figure 34-5: Walter Reed Army Inst. of Research, Washington, DC. Figure 34-6: United States Dept. of Agriculture.

Chapter 35. Opener and Figure 35-2: Pfizer Inc. Figure 35-1: M. J. Vinkesteyn, National Mus. of Antiquities, Leyden, Holland. Figure 35-11: E. L. Gaden, "Production Methods in Industrial Microbiology," *Scientific American* 245(3), by Scientific American, Inc. © 1981.

Chapter 36. Opener and Figure 36-1: P. C. Fitz-James, Univ. of Western Ontario. Figure 36-4: G. Strobel, Montana State Univ.

Part-opening photos. Part I: H. S. Pankratz, Michigan State Univ. Part II: Centers for Disease Control, Atlanta, GA. Part III: C. L. Sanders, Battelle Pacific Northwest Labs. Part IV: E. H. Newcomb, Univ. of Wisconsin–Madison/BPS.

COLOR PLATES

Contents. Chapter 1: J. R. Waaland, Univ. of Washington/BPS. Chapters 6 and 22: Centers for Disease Control, Atlanta, GA. Chapters 7, 12, 19: L. J. LeBeau, Univ. of Illinois Hospital/BPS. Chapters 8, 17, 29, 33: Raúl J. Cano. Chapter 9: © Phillip Harrington. Chapter 30: R. K. Burnard, Ohio State Univ./BPS.

Page II-1. Upper left: J. R. Waaland, Univ. of Washington/BPS. Upper right, lower left, and lower right: P. W. Johnson & J. McN. Sieburth, Univ. of Rhode Island/BPS.

Page II-2. Upper right: Centers for Disease Control, Atlanta, GA. Middle: L. J. LeBeau, Univ. of Illinois Hospital/BPS. Lower left: Raúl J. Cano. Lower right: J. R. Waaland, Univ. of Washington/BPS.

Page II-3. Upper left, lower left, lower right: J. R. Waaland, Univ. of Washington/BPS. Middle: P. W. Johnson & J. McN. Sieburth, Univ. of Rhode Island/BPS.

Page II-4. Upper left: B. J. Miller, Fairfax, VA/BPS. Upper right: J. N. A. Lott, McMaster Univ./BPS. Middle: R. K. Burnard, Ohio State Univ./BPS. Lower right: J. R. Waaland, Univ. of Washington/BPS.

Page II-5. Upper left, middle: R. Humbert/BPS. Upper right: Charles E. Schmidt/Taurus Photos. Lower left: J. R. Waaland, Univ. of Washington/BPS.

Page II-6. Upper left, upper right, lower right, lower left: Centers for Disease Control, Atlanta, GA.

Page II-7. Upper right, upper left: Centers for Disease Control, Atlanta, GA. Lower left, lower right: J. R. Waaland, Univ. of Washington/BPS.

Page II-8. Upper left, upper right, middle left: J. R. Waaland, Univ. of Washington/BPS. Lower right: M. Murayama, Murayama Research Lab./BPS. Lower left: P. W. Johnson & J. McN. Sieburth, Univ. of Rhode Island/BPS.

Page III-1. Upper, middle, lower: Martin M. Rotker/Taurus Photos.

Page III-2. Upper right, lower right, upper left, lower left: Centers for Disease Control, Atlanta, GA.

Page III-3. Upper left, lower left: Centers for Disease Control, Atlanta, GA. Upper right: L. J. LeBeau, Univ. of Illinois Hospital/BPS. Lower right: L. M. Pope & D. R. Grote, Univ. of Texas, Austin/BPS.

Page III-4. Upper left: Centers for Disease Control, Atlanta, GA. Upper right, middle, lower: L. J. LeBeau, Univ. of Illinois Hospital/BPS.

Page III-5. Upper: Raúl J. Cano. Middle: I. Armstrong, Washington Hospital Center, Washington, DC. Lower: Centers for Disease Control, Atlanta, GA.

Page III-6. Upper left, middle, lower left, lower right: Centers for Disease Control, Atlanta, GA. Upper right: I. Armstrong, Washington Hospital Center, Washington, DC.

Page III-7. Upper, middle left: R. K. Burnard, Ohio State Univ./BPS. Middle right: Alfred Pasieka/Taurus Photos. Lower: Raúl J. Cano.

Page III-8. Upper: Raúl J. Cano. Middle: L. J. LeBeau, Univ. of Illinois Hospital/BPS. Lower left: R. A. Finklestein, Univ. of Missouri, School of Medicine. Lower middle, lower right: Centers for Disease Control, Atlanta, GA.

Page IV-1. Upper: Patrick Grace/Taurus Photos. Middle: L. L. T. Rhodes/Taurus Photos. Lower: C. W. May/BPS.

Page IV-2. Upper, middle, lower: S. M. Awramik, Univ. of California/BPS.

Page IV-3. Upper left: J. R. Waaland, Univ. of Washington/BPS. Middle right, lower right: Shirley R. Sparling.

Page IV-4. Upper, middle, lower: Shirley R. Sparling.

Page IV-5. Upper: Martin M. Rotker/Taurus Photos. Lower: C. W. May/BPS.

Page IV-6. Upper right, middle right: J. N. A. Lott, McMaster Univ./BPS. Lower right: J. R. Waaland, Univ. of Washington/BPS. Lower left: Vance Henry/Taurus Photos.

Page IV-7. Upper left: Martin M. Rotker/Taurus Photos. Middle left, lower left: L. J. LeBeau, Univ. of Illinois Hospital/BPS. Lower right: B. J. Miller, Fairfax, VA/BPS.

Page IV-8. Upper: B. Miller. Lower: J. N. A. Lott, McMaster Univ./BPS.